# Fundamentals of Industrial Hygiene

## OCCUPATIONAL SAFETY AND HEALTH SERIES

The National Safety Council's OCCUPATIONAL SAFETY AND HEALTH SERIES helps readers establish and maintain safety health programs. The latest information on establishing priorities, collecting and analyzing data to help identify problems, and developing methods and procedures to reduce or eliminate illness and accidents, thus mitigating injury and minimizing economic loss resulting from accidents, is contained in all volumes in this series:

*ACCIDENT PREVENTION MANUAL FOR BUSINESS & INDUSTRY*
    (3-volume set)
    *Administration & Programs*
    *Engineering & Technology*
    *Environmental Management*
*STUDY GUIDE: ACCIDENT PREVENTION MANUAL FOR BUSINESS & INDUSTRY*
*OCCUPATIONAL HEALTH & SAFETY*

Other safety and health references published by the Council include:

*ACCIDENT FACTS* (published annually)
*LOCKOUT/TAGOUT: THE PROCESS OF CONTROLLING HAZARDOUS ENERGY*
*SUPERVISORS' SAFETY MANUAL*
*OUT IN FRONT: EFFECTIVE SUPERVISION IN THE WORKPLACE*
*PRODUCT SAFETY: MANAGEMENT GUIDELINES*
*OSHA BLOODBORNE PATHOGENS EXPOSURE CONTROL PLAN*
    (National Safety Council/CRC-Lewis Publication)
*COMPLETE CONFINED SPACES HANDBOOK* (National Safety Council/CRC-Lewis
    Publication)

FOURTH
EDITION

# Fundamentals of Industrial Hygiene

Barbara A. Plog, MPH, CIH, CSP

*Editor-in-Chief*

Jill Niland, MPH, CIH, CSP
*Associate Editor*

Patricia J. Quinlan, MPH, CIH
*Associate Editor*

Itasca, Illinois

*Editor-in-Chief:* Barbara A. Plog, MPH, CIH, CSP
*Associate Editor:* Jill Niland, MPH, CIH, CSP
*Associate Editor:* Patricia J. Quinlan, MPH, CIH
*Project Editor:* Jodey B. Schonfeld
*Interior Design and Composition:* Publication Services, Inc.
*Cover Design:* Anderson Creative Services

**Library of Congress Cataloging-in-Publication Data**

Fundamentals of industrial hygiene. —4th ed. / Barbara A. Plog,
   editor ; Jill Niland, asst. editor, Patricia Quinlan, asst. editor.
      p.     cm. — (Occupational safety and health series)
   Includes bibliographical references and index.
   ISBN 0-87912-171-8 (he)
   1. Industrial hygiene.   I. Plog, Barbara A.   II. Niland, Jill,
1951– .   III. Quinlan, Patricia, 1951– .   IV. Series;
Occupational safety and health series (Chicago, Ill.)
RC967.F85   1995                                          95–30403
613.6'2–dc20                                               CIP

2M301                                      Product Number: 15134-0000

# Contents

**Part 1    History and Development    1**

   1    Overview of Industrial Hygiene    3

**Part 2    Anatomy, Physiology, and Pathology    33**

   2    The Lungs    35

   3    The Skin and Occupational Dermatoses    53

   4    The Ears    83

   5    The Eyes    103

**Part 3    Recognition of Hazards    121**

   6    Industrial Toxicology    123

   7    Gases, Vapors, and Solvents    153

   8    Particulates    175

   9    Industrial Noise    197

  10    Ionizing Radiation    247

  11    Nonionizing Radiation    273

  12    Thermal Stress    319

  13    Ergonomics    347

  14    Biological Hazards    403

**Part 4    Evaluation of Hazards    451**

  15    Evaluation    453

  16    Air Sampling    485

  17    Direct-Reading Instruments for Gases, Vapors, and Particulates    509

**Part 5    Control of Hazards    529**

  18    Methods of Control    531

  19    Local Exhaust Ventilation of Industrial Occupancies    553

20 General Ventilation of Industrial Occupancies    581

21 General Ventilation of Nonindustrial Occupancies    595

22 Respiratory Protection    619

**Part 6   Occupational Health and Safety Programs    657**

23 The Industrial Hygienist    659

24 The Safety Professional    675

25 The Occupational Physician    701

26 Occupational Health Nursing    729

27 The Industrial Hygiene Program    749

28 Computerizing an Industrial Hygiene Program    759

**Part 7   Government Regulations and Their Impact    767**

29 Government Regulations    769

30 History of the Federal Occupational Safety and
Health Administration    787

**Appendixes**

A Additional Resources    843

B Threshold Limit Values and Biological
Exposure Indices (ACGIH)    865

C Conversion of Units    905

D Review of Mathematics    911

E European Union Initiatives in Occupational
Health and Safety    917

F Glossary    929

**Index   987**

# Foreword

No matter what the current government regulatory climate is, it is critical to the work of the industrial hygienist to implement key elements or fundamentals of industrial hygiene. The fourth edition of the National Safety Council's *FUNDAMENTALS OF INDUSTRIAL HYGIENE* can make that responsibility easier and more effective.

*FUNDAMENTALS OF INDUSTRIAL HYGIENE* uses a clear and concise approach to assist the reader, regardless of his or her knowledge of industrial hygiene, in the anticipation, recognition, evaluation, and control of occupational health hazards.

I personally have used previous editions of *FUNDAMENTALS OF INDUSTRIAL HYGIENE,* which provided me with a road map of how I could apply the knowledge gained to the companies and plants for which I had responsibility. *FUNDAMENTALS OF INDUSTRIAL HYGIENE* also made clear the risk assessment process, which made it easier for me to provide guidance for operations and manufacturing people.

Past experience has proven that *FUNDAMENTALS OF INDUSTRIAL HYGIENE* can reduce employees' exposure to health hazards while having a positive influence on the cost effectiveness of company programs. Therefore, I feel a responsibility to make sure this new fourth edition enjoys wide distribution to professionals and nonprofessionals alike.

I highly endorse and recommend *FUNDAMENTALS OF INDUSTRIAL HYGIENE* (fourth edition) to all health and safety professionals, whether you are a seasoned veteran or a professional who is beginning a career in the industrial hygiene and safety field.

GERARD F. SCANNELL
PRESIDENT, NATIONAL SAFETY COUNCIL

# Preface

The fourth edition of *FUNDAMENTALS OF INDUSTRIAL HYGIENE* comes at a time of intense activity that may profoundly change the structure and function of the federal Occupational Safety and Health Administration (OSHA) and the National Institute for Occupational Safety and Health (NIOSH). These actions will have far reaching effects on the scope and practice of industrial hygiene in the United States.

The future of OSHA and NIOSH may be unclear as of this writing, but what is clear—in the words of OSHA head Joseph Dear to the American Industrial Hygiene Conference in Kansas City, Missouri, on May 24, 1995—is that OSHA strives to "guarantee that each worker who leaves for work in the morning arrives home safely each night." (*The Synergist,* American Industrial Hygiene Association, Volume 6, Number 6–7, p. 10, June/July, 1995.)

Also clear is that in times of such uncertainty about the federal health and safety apparatus, the fundamental principles of industrial hygiene deserve more emphasis than at any time before.

This edition of *FUNDAMENTALS OF INDUSTRIAL HYGIENE* presents original new chapters on The Skin and Occupational Dermatoses (Chapter 3), Nonionizing Radiation (Chapter 11), Thermal Stress (Chapter 12), Biological Hazards (Chapter 14), General Ventilation of Nonindustrial Occupancies (Chapter 21), Occupational Health Nursing (Chapter 26), and a new appendix, European Union Initiatives in Occupational Health and Safety, (Appendix E). All other chapters have been extensively updated and revised.

The primary purpose of this book is to provide a reference for those who have either an interest in or a direct responsibility for the recognition, evaluation, and control of occupational health hazards. Thus, it is intended to be of use to industrial hygienists, industrial hygiene students, physicians, nurses, safety personnel from labor and industry, labor organizations, public service groups, government agencies, and manufacturers. Others who may find this reference helpful include consultants, architects, lawyers, and allied professional personnel who work with those engaged in business, industry, and agriculture. It is hoped that this book will be of use to those responsible for planning and carrying out programs to minimize occupational health hazards.

An understanding of the fundamentals of industrial hygiene is very important to anyone involved in environmental, community, or occupational health. This manual should be of help in defining the magnitude and extent of an industrial hygiene problem; it should help the reader decide when expert help is needed.

*FUNDAMENTALS OF INDUSTRIAL HYGIENE* is also intended to be used either as a self-instructional text or as a text for an industrial hygiene fundamentals course, such as the ones offered by the National Safety Council, various colleges and universities, and professional organizations.

The increase in the number and complexity of substances found in the workplace—substances that may spill over into the community environment—makes imperative the dissemination, as efficiently and conveniently as possible, of certain basic information relating to occupational health hazards and resultant occupational diseases.

The book is organized into seven parts; each can stand alone as a reference source. For that reason, we have permitted a certain amount of redundancy.

*Part One* introduces the subject areas to be covered, in an overview of the fundamentals of industrial hygiene.

*Part Two* includes chapters on the fundamental aspects of the anatomy, physiology, hazards, and pathology of the lungs, skin, ears, and eyes. This background lays the groundwork for understanding how these organ systems interrelate and function.

*Part Three* is concerned with the recognition of specific environmental factors or stresses. The chemical substances, physical agents, and biological and ergonomic hazards present in the workplace are covered. The basic concepts of industrial toxicology are also presented in this section. Anticipation of these hazards is the desired result.

*Part Four* describes methods and techniques of evaluating the hazard. Included is one of the more important aspects of an industrial hygiene program: the methods used to evaluate the extent of exposure to harmful chemical and physical agents. Basic information is given on the various types of instruments available to measure these stresses and on how to use the instruments properly to obtain valid measurements.

*Part Five* deals with the control of the environmental hazards. Although industrial hygiene problems vary, the basic principles of health hazard control, problem-solving techniques, and the examples of engineering control measures given here are general enough to have wide application. To augment the basics, specific information is covered in the chapters on industrial ventilation.

*Part Six* is directed specifically to people responsible for conducting and organizing occupational health and safety programs. The fundamental concepts of the roles of the industrial hygienist, the occupational health nurse, the safety professional, and the occupational physician in implementing a successful program are discussed in detail. Particular attention is paid to a discussion of the practice of industrial hygiene in the public and private sectors and to a description of the professional certification of industrial hygienists. Computerizing an industrial hygiene program is also discussed.

*Part Seven* contains up-to-date information on government regulations and their impact on the practice of industrial hygiene.

*Appendix A* provides additional resources. One of the most difficult parts of getting any project started is finding sources of help and information. For this reason, we have included a completely updated and comprehensive annotated bibliography and a listing of professional and service organizations, government agencies, and other resources.

A new addition to the book, *Appendix E,* details health and safety initiatives in the European Union. Other appendixes include Threshold Limit Values and Biological Exposure Indices, a review of mathematics, instructions on conversion of units, and a glossary of terms used in industrial hygiene, occupational health, and pollution control. An extensive index is included to assist the reader in locating information in this text.

We would like to gratefully acknowledge the work of the contributors to previous editions of the *FUNDAMENTALS OF INDUSTRIAL HYGIENE.*

**First Edition.** 1—Fundamental Concepts of Industrial Hygiene—Julian B. Olishifski, PE   2—Solvents and Health in the Occupational Environment—Donald R. McFee, ScD   3—Pneumoconiosis-Producing Dusts—Fred Cook, MS   4—Industrial Dermatitis—Charles W. Wyman, BS   5—Industrial Noise—Herbert T. Walworth, MS   6—Basic Concepts of Ionizing Radiation Safety—E. L. Alpaugh, PE   7—Nonionizing Radiation: Lasers, Microwaves, Light—Julian B. Olishifski, PE   8—Effects of Temperature Extremes—E. L. Alpaugh, PE   9—Ergonomics Stresses: Physical and Mental—Julian B. Olishifski, PE   10—Evaluating the Hazard—J. B. Olishifski, PE   11—Toxicology—J. B. Olishifski, PE   12—General Methods of Control—J. B. Olishifski, PE   13—Respiratory Protective Equipment—A. M. Lundin, BS   14—Industrial Ventilation—W. G. Hazard, AM   15—General Ventilation and Special Operations—W. G. Hazard, AM   20—Setting Up an Industrial Hygiene Program—Julian B. Olishifski, PE   21—Sources of Information on Industrial Hygiene—Julian B. Olishifski, PE

**Second Edition.** 1—Fundamental Concepts—Julian B. Olishifski, PE   2—The Lungs—Julian B. Olishifski, PE   3—The Skin—Julian B. Olishifski, PE   4—The Ear—Julian B. Olishifski, PE   5—The Eyes—Julian B. Olishifski, PE   6—Solvents—Donald R. McFee, ScD   7—Particulates—Edwin L. Alpaugh, PE   8—Industrial Dermatoses—Larry L. Hipp   9—Industrial Noise—Julian B. Olishifski, PE   10—Ionizing Radiation—C. Lyle Cheever, MS, MBA   11—Nonionizing Radiation—Edward J. Largent and Julian B. Olishifski, PE   12—Temperature Extremes—Edwin L. Alpaugh, PE   13—Ergonomics—Bruce A. Hertig   14—Biological Hazards—Alvin L. Miller, PhD, CIH and Anne C. Leopold   15—Industrial Toxicology—Ralph G. Smith and Julian B. Olishifski, PE   16—Evaluation—Edward R. Hermann, CE, PhD, PE, CIH and Jack E. Peterson, PhD, PE, CIH   17—Methods of Evaluation—Julian B. Olishifski, PE   18—Air-Sampling Instruments—Julian B. Olishifski, PE   19—Direct-Reading Gas and Vapor Monitors—Joseph E. Zatek, CSP, CIH, CHCM, CHM   20—Methods of Control—Julian B. Olishifski, PE   21—Industrial Ventilation—Willis G. Hazard, AM   22—General Ventilation—Willis G. Hazard, AM   23—Respiratory Protective Equipment—Allen M. Lundin, BS   24—Governmental Regulations—M. Chain Robbins, BS, MPH, CSP, PE   25—The Industrial Hygienist—Clyde M. Berry   26—The Safety Professional—Willis T. McLean   27—The Occupational Physician—Carl Zenz, MD, ScD   28—The Occupational Health Nurse—Jeanette M. Cornyn   29—Industrial Hygiene Program—Edward J. Largent and Julian B. Olishifski, PE   30—Sources of Information—Julian B. Olishifski, PE and Robert Pedroza

**Third Edition.** 1—Overview of Industrial Hygiene—Barbara Plog, MPH, CIH, CSP   2—The Lungs—George S. Benjamin, MD, FACS   3—The Skin—James S. Taylor, MD   4—The Ears—George S. Benjamin, MD, FACS   5—The Eyes—George S. Benjamin, MD, FACS   6—Solvents—Donald R. McFee, ScD, CIH, PE, CSP; Peter Zavon, CIH   7—Particulates—Theodore J. Hogan, PhD, CIH   8—Industrial Dermatoses—James S. Taylor, MD   9—Industrial Noise—John J. Standard, MS, MPH, CIH, CSP   10—Ionizing Radiation—C. Lyle Cheever, MS, MBA   11—Nonionizing Radiation—Larry E. Anderson, PhD   12—Temperature Extremes—Theodore J. Hogan, PhD, CIH   13—Ergonomics—Karl H. E. Kroemer, PhD   14—Biological Hazards—Alvin L. Miller, PhD, CIH; Cynthia S. Volk   15—Industrial Toxicology—Carl Zenz, MD, ScD   16—Evaluation—Edward R. Hermann, CE, PhD, PE, CIH; Jack E. Peterson, PhD, PE, CIH   17—Methods of Evaluation—Julian B. Olishifski, MS, PE, CSP   18—Air-Sampling Instruments—Maureen A. Kerwin (now Maureen A. Huey), MPH   19—Direct-Reading Gas and Vapor Monitors—Joseph E. Zatek, CSP, CIH, CHCM, CHM   20—Methods of Control—Julian B. Olishifski, MS, PE, CSP   21—Industrial Ventilation—D. Jeff Burton, PE, CIH   22—General Ventilation —D. Jeff Burton, PE, CIH   23—Respiratory Protective Equipment—Craig E. Colton, CIH   24—The Industrial Hygienist—Barbara A. Plog, MPH, CIH, CSP   25—The Safety Professional—Fred A. Manuele, PE, CSP   26—The Occupational Physician—Carl Zenz, MD   27—The Occupational Health Nurse—Larry Hannigan, RN   28—The Industrial Hygiene Program—Maureen A. Kerwin (now Maureen A. Huey), MPH   29—Computerizing an Industrial Hygiene Program—Adrienne Whyte, PhD   30—Governmental Regulations—M. Chain Robbins, BS, MPH, CSP, PE   31—Occupational Safety and Health: The Federal Regulatory Program—A History—Benjamin W. Mintz

We would also like to thank the following individuals, who reviewed material for the fourth edition: Bassam H. Atieh, ScD, Hal Barrett, John F. Beltz, CIH, George Benjamin, MD, FACS, Eva Bernard, Steven R. Brehio, Beverly Brennan-Roy, Joan Leslie Davis, Kenneth C. Eck, Sandra Lee Filippi, A. Keith Furr, Gary Gilbertson, Thomas L. Grieco, Charles Hart, MA, CIH, CSP, RS, John W. Herrington, Michael J. Horowitz, MS, CIH, Hubert L. Ivie, Michele J. Johnson, Patricia A. Kandziora, Ausrine A. E. Karaitis, FAIC, Scott D. Keimig, PhD, CIH, Sandra F. Kulik, Frank J. Labato, CIH, CSP, Bruce L. Macdonald, Chris A. McGuffin, Gary L. Monroe, John Morrell, Norbert Norman, Dale O. Ritzel, PhD, Margaret G. Schemm, Carol Scott, CIH, John E. Shanks, Robert D. Soule, CIH, CSP, Gary T. Staffo, CSP, PE, Ruth Helen Sullivan, Eric L. Van Fleet, Stefan Wawzynieck, Jr., CIH, and Del Weed.

A special thanks to George Benjamin, Jill Niland, Patty Quinlan, and Jodey Schonfeld, whose excellent work, tireless attention to technical detail, and professionalism helped to make this edition the best yet.

And finally, this book is dedicated to my family, Michael and Max, who supported having "the book" in our lives for the past three years; and my mother and father, Doris and Henry Plog; and to the working women and men who are, after all, the point of it all.

Because this manual will be revised periodically, contributions and comments from readers are welcome.

BARBARA A. PLOG, MPH, CIH, CSP
EDITOR-IN-CHIEF
AUGUST 1995

## PART 1

# History and Development

# CHAPTER 1

# Overview of Industrial Hygiene

by Barbara A. Plog, MPH, CIH, CSP

**Professional Ethics**
*The Occupational Health and Safety Team*

**Federal Regulations**

**Environmental Factors or Stresses**
*Chemical Hazards* ▪ *Physical Hazards* ▪ *Ergonomic Hazards* ▪ *Biological Hazards*

**Harmful Agents–Route of Entry**
*Inhalation* ▪ *Absorption* ▪ *Ingestion*

**Types of Airborne Contaminants**
*States of Matter* ▪ *Respiratory Hazards*

**Threshold Limit Values**
*Skin Notation* ▪ *Mixtures* ▪ *Federal Occupational Safety and Health Standards*

**Evaluation**
*Basic Hazard-Recognition Procedures* ▪ *Information Required* ▪ *Degree of Hazard* ▪ *Air Sampling*

**Occupational Skin Diseases**
*Types* ▪ *Causes* ▪ *Physical Examinations* ▪ *Preventive Measures*

**Control Methods**
*Engineering Controls* ▪ *Ventilation* ▪ *Personal Protective Equipment* ▪ *Administrative Controls*

**Sources of Help**

**Summary**

**Bibliography**

INDUSTRIAL HYGIENE IS THAT SCIENCE and art devoted to the anticipation, recognition, evaluation, and control of those environmental factors or stresses arising in or from the workplace that may cause sickness, impaired health and well-being, or significant discomfort among workers or among the citizens of the community (AIHA *1994–1995 Membership Directory*). Industrial hygienists are occupational health professionals who are concerned primarily with the control of environmental stresses or occupational health hazards that arise as a result of or during the course of work. The industrial hygienist recognizes that environmental stresses may endanger life and health, accelerate the aging process, or cause significant discomfort.

The industrial hygienist, although trained in engineering, physics, chemistry, environmental sciences, safety, or biology, has acquired through postgraduate study or experience a knowledge of the health effects of chemical, physical, biological, and ergonomic agents. The industrial hygienist is involved in the monitoring and analysis required to detect the extent of exposure, and the engineering and other methods used for hazard control.

Evaluation of the magnitude of work-related environmental hazards and stresses is done by the industrial hygienist, aided by training, experience, and quantitative measurement of the chemical, physical, ergonomic, or biological stresses. The industrial hygienist can thus give an expert opinion as to the degree of risk the environmental stresses pose.

Industrial hygiene includes the development of corrective measures in order to control health hazards by either reducing or eliminating the exposure. These control procedures may include the substitution of harmful or toxic materials with less dangerous ones, changing of work processes to eliminate or minimize work exposure, installation of exhaust ventilation systems, good housekeeping (including appropriate waste disposal methods), and the provision of proper personal protective equipment.

An effective industrial hygiene program involves the anticipation and recognition of health hazards arising from work operations and processes, evaluation and measurement of the magnitude of the hazard (based on past experience and study), and control of the hazard.

*Occupational health hazards* may mean conditions that cause legally compensable illnesses, or it may mean any conditions in the workplace that impair the health of employees enough to make them lose time from work or to cause significant discomfort. Both are undesirable. Both are preventable. Their correction is properly a responsibility of management.

# Professional Ethics

In late 1994, the four major U.S. industrial hygiene organizations gave final endorsements to a revised Code of Ethics for the Practice of Industrial Hygiene. These organizations are the American Conference of Governmental Industrial Hygienists (ACGIH), the American Academy of Industrial Hygiene (AAIH), the American Board of Industrial Hygiene (ABIH), and the American Industrial Hygiene Association (AIHA).

The new code defines practice standards (Canons of Ethical Conduct) and applications (interpretive guidelines). The Canons of Ethical Conduct are as follows:

Industrial Hygienists shall practice their profession following recognized scientific principles with the realization that the lives, health, and well-being of people may depend upon their professional judgment and that they are obligated to protect the health and well-being of people.

Industrial Hygienists shall counsel affected parties regarding potential health risks and precautions necessary to avoid adverse health effects.

Industrial Hygienists shall keep confidential personal and business information obtained during the exercise of industrial hygiene activities, except when required by law or overriding health and safety considerations.

Industrial Hygienists shall avoid circumstances where a compromise of professional judgment or conflict of interest may arise.

Industrial Hygienists shall perform services only in the areas of their competence.

Industrial Hygienists shall act responsibly to uphold the integrity of the profession.

The interpretive guidelines to the Canons of Ethical Conduct are a series of statements that amplify the code. (Figure 1–1). These guidelines may be supplemented when necessary, as ethical issues and claims arise.

## The Occupational Health and Safety Team

The chief goal of an occupational health and safety program in a facility is to prevent occupational injury and illness by anticipating, recognizing, evaluating, and controlling occupational health and safety hazards. The medical, industrial hygiene, and safety programs may have distinct, additional program goals but all programs interact and are often considered different components of the overall health and safety program. The occupational health and safety team consists, then, of the industrial hygienist, the safety professional, the occupational health nurse, the occupational physician, the employees, senior and line management, and others depending on the size and character of the particular facility. All team members must act in concert to provide information and activities, supporting the other parts to achieve the overall goal of a healthy and safe work environment. Therefore, the separate functions must be administratively linked in order to effect a successful and smoothly run program.

The first vital component to an effective health and safety program is the commitment of *senior management* and *line management*. Serious commitment is demonstrated when management is visibly involved in the program both by management support and personal compliance with all health and safety practices. Equally critical is the assignment of the authority, as well as the responsibility, to carry out the health and safety program. The health and safety function must be given the same level of importance and accountability as the production function.

The function of the *industrial hygienist* has been defined above. (Also see Chapter 23, The Industrial Hygienist.) The industrial hygiene program must be made up of several key components: a written program/policy statement, hazard recognition procedures, hazard evaluation and exposure assessment, hazard control, employee training, employee involvement, program evaluation and audit, and recordkeeping. (See Chapter 27, The Industrial Hygiene Program, for further discussion.)

The *safety professional* must draw upon specialized knowledge in the physical and social sciences. Knowledge of engineering, physics, chemistry, statistics, mathematics, and principles of measurement and analysis is integrated in the evaluation of safety performance. The safety professional must thoroughly understand the factors contributing

# Code of Ethics for the Practice of Industrial Hygiene

## Objective

These canons provide standards of ethical conduct for Industrial Hygienists as they practice their profession and exercise their primary mission, to protect the health and well-being of working people and the public from chemical, microbiological, and physical health hazards present at, or emanating from, the workplace.

## Canons of ethical conduct

### CANON 1

Industrial Hygienists shall practice their profession following recognized scientific principles with the realization that the lives, health, and well-being of people may depend upon their professional judgment and that they are obligated to protect the health and well-being of people.

INTERPRETIVE GUIDELINES

- Industrial Hygienists should base their professional opinions, judgments, interpretations of findings, and recommendations upon recognized scientific principles and practices which preserve and protect the health and well-being of people.
- Industrial Hygienists shall not distort, alter, or hide facts in rendering professional opinions or recommendations.
- Industrial Hygienists shall not knowingly make statements that misrepresent or omit facts.

### CANON 2

Industrial Hygienists shall counsel affected parties factually regarding potential health risks and precautions necessary to avoid adverse health effects.

INTERPRETIVE GUIDELINES

- Industrial Hygienists should obtain information regarding potential health risks from reliable sources.
- Industrial Hygienists should review the pertinent, readily available information to factually inform the affected parties.
- Industrial Hygienists should initiate appropriate measures to see that the health risks are effectively communicated to the affected parties.
- Parties may include management, clients, employees, contractor employees, or others, dependent on circumstances at the time.

### CANON 3

Industrial Hygienists shall keep confidential personal and business information obtained during the exercise of industrial hygiene activities, except when required by law or overriding health and safety considerations.

INTERPRETIVE GUIDELINES

- Industrial Hygienists should report and communicate information which is necessary to protect the health and safety of workers and the community.
- If their professional judgment is overruled under circumstances where the health and lives of people are endangered, Industrial Hygienists shall notify their employer, client, or other such authority, as may be appropriate.
- Industrial Hygienists should release confidential personal or business information only with the information owners express authorization, except when there is a duty to disclose information as required by law or regulation.

### CANON 4

Industrial Hygienists shall avoid circumstances where a compromise of professional judgment or conflict of interest may arise.

INTERPRETIVE GUIDELINES

- Industrial Hygienists should promptly disclose known or potential conflicts of interest to parties that may be affected.
- Industrial Hygienists shall not solicit or accept financial or other valuable consideration from any party, directly or indirectly, which is intended to influence professional judgment.
- Industrial Hygienists shall not offer any substantial gift, or other valuable consideration, in order to secure work.
- Industrial Hygienists should advise their clients or employer when they initially believe a project to improve industrial hygiene conditions will not be successful.
- Industrial Hygienists should not accept work that negatively impacts the ability to fulfill existing commitments.
- In the event that this Code of Ethics appears to conflict with another professional code to which Industrial Hygienists are bound, they will resolve the conflict in the manner that protects the health of affected parties.

### CANON 5

Industrial Hygienists shall perform services only in the areas of their competence.

INTERPRETIVE GUIDELINES

- Industrial Hygienists should undertake to perform services only when qualified by education, training, or experience in the specific technical fields involved, unless sufficient assistance is provided by qualified associates, consultants, or employees.
- Industrial Hygienists shall obtain appropriate certifications, registrations, and/or licenses as required by federal, state, and/or local regulatory agencies prior to providing industrial hygiene services, where such credentials are required.
- Industrial Hygienists shall affix or authorize the use of their seal, stamp, or signature only when the document is prepared by the Industrial Hygienist or someone under their direction and control.

### CANON 6

Industrial Hygienists shall act responsibly to uphold the integrity of the profession.

INTERPRETIVE GUIDELINES

- Industrial Hygienists shall avoid conduct or practice which is likely to discredit the profession or deceive the public.
- Industrial Hygienists shall not permit use of their name or firm name by any person or firm which they have reason to believe is engaging in fraudulent or dishonest industrial hygiene practices.
- Industrial Hygienists shall not use statements in advertising their expertise or services containing a material misrepresentation of fact or omitting a material fact necessary to keep statements from being misleading.
- Industrial Hygienists shall not knowingly permit their employees, employers, or others to misrepresent the individuals' professional background, expertise, or services which are misrepresentations of fact.
- Industrial Hygienists shall not misrepresent their professional education, experience, or credentials.

**Figure 1–1.** The joint Code of Ethics for the Practice of Industrial Hygiene endorsed by the AIHA, the ABIH, the AAIH, and the ACGIH. (From *ACGIH Today!* 3(1), January 1995.)

to accident occurrence and combine this with knowledge of motivation, behavior, and communication in order to devise methods and procedures to control safety hazards. Because the practice of the safety professional and the industrial hygienist are so closely related, it is rare to find a safety professional who does not practice some traditional industrial hygiene and vice versa. At times, the safety and industrial hygiene responsibilities may be vested in the same individual or position. (See Chapter 24, The Safety Professional.)

The *occupational health nurse* (OHN) is the key to the delivery of comprehensive health care services to workers. Occupational health nursing is focused on the promotion, protection, and restoration of workers' health within the context of a safe and healthy work environment. The OHN provides the critical link between the employee's health status, the work process, and the determination of employee ability to do the job. Knowledge of health and safety regulations, workplace hazards, direct care skills, counseling, teaching, and program management are but a few of the key knowledge areas for the OHN, with strong communication skills of the utmost importance. OHNs deliver high-quality care at worksites and support the primary prevention dictum that most workplace injuries and illnesses are preventable. If injuries occur, OHNs use a case-management approach to return injured employees to appropriate work on a timely basis. The OHN often functions in multiple roles within one job position, including clinician, educator, manager, and consultant. (See Chapter 26, The Occupational Health Nurse.)

The *occupational physician* has acquired, through graduate training or experience, extensive knowledge of cause and effect relationships of chemical, physical, biological, and ergonomic hazards, the signs and symptoms of chronic and acute exposures, and the treatment of adverse effects. The primary goal of the occupational physician is to prevent occupational illness and, when illness occurs, to restore employee health within the context of a healthy and safe workplace. Many regulations provide for a minimum medical surveillance program and specify mandatory certain tests and procedures.

The occupational physician and the occupational health nurse should be familiar with all jobs, materials, and processes used. An occasional workplace inspection by the medical team enables them to suggest protective measures and aids them in recommending placement of employees in jobs best suited to their physical capabilities. (See discussion of the Americans with Disabilities Act in Chapter 25, The Occupational Physician.)

Determining the work-relatedness of disease is another task for the occupational physician. The industrial hygienist provides information about the manufacturing operations and work environment of a company to the medical department as well. In many cases it is extremely difficult to differentiate between the symptoms of occupational and nonoccupational disease. The industrial hygienist supplies information on the work operations and their associated hazards and enables the medical department to correlate the employee's condition and symptoms with potential workplace health hazards.

The *employee* plays a major role in the occupational health and safety program. Employees are excellent sources of information on work processes and procedures and the hazards of their daily operations. Industrial hygienists benefit from this source of information and often obtain innovative suggestions for controlling hazards.

The *safety and health committee* provides a forum for securing the cooperation, coordination, and exchange of ideas among those involved in the health and safety program. It provides a means of involving employees in the program. The typical functions of the safety and health committee include, among others, to examine company safety and health issues and recommend policies to management, conduct periodic workplace inspections, and evaluate and promote interest in the health and safety program. Joint labor–management safety and health committees are often used where employees are represented by a union. The committee meetings also present an opportunity to discuss key industrial hygiene program concerns and to formulate appropriate policies.

# ■ Federal Regulations

Before 1970, government regulation of health and safety matters was largely the concern of state agencies. There was little uniformity in codes and standards or in the application of these standards. Almost no enforcement procedures existed.

On December 29, 1970, the Occupational Safety and Health Act, known as the OSHAct, was enacted by Congress. Its purpose was to "assure so far as possible every working man and woman in the nation safe and healthful working conditions and to preserve our human resources." The OSHAct sets out two duties for employers:

■ Each employer shall furnish to each employee a place of employment, which is free from recognized hazards that are causing or are likely to cause death or serious harm to their employees.

■ Each employer shall comply with occupational safety and health standards under the Act.

For employees, the OSHAct states that "Each employee shall comply with occupational safety and health standards and all rules, regulations, and orders issued pursuant to the Act which are applicable to his own actions and conduct."

The Occupational Safety and Health Administration (OSHA) came into official existence on April 28, 1971, the date the OSHAct became effective. It is housed within the

U.S. Department of Labor. The OSHAct also established the National Institute for Occupational Safety and Health (NIOSH), which is housed within the Centers for Disease Control (CDC). The CDC is part of the U.S. Public Health Service.

OSHA was empowered to promulgate safety and health standards with technical advice from NIOSH. OSHA is empowered to enter workplaces to investigate alleged violations of these standards and to perform routine inspections. Formal complaints of standards violation may be made by employees or their representatives. The OSHAct also gives OSHA the right to issue citations and penalties, provide for employee walk-around or interview of employees during the inspection, require employers to maintain accurate records of exposures to potentially hazardous materials, and to inform employees of the monitoring results. OSHA is also empowered to provide up to 50/50 funding with states that wish to establish state OSHA programs that are at least as effective as the federal program. As of this date, there are 23 approved state plans and approved plans from Puerto Rico and the Virgin Islands.

NIOSH is the principal federal agency engaged in occupational health and safety research. The agency is responsible for identifying hazards and making recommendations for regulations. These recommendations are called Recommended Exposure Limits (RELs). NIOSH also issues criteria documents and health hazard alerts on various hazards and is responsible for testing and certifying respiratory protective equipment.

Part of NIOSH research takes place during activities called Health Hazard Evaluations. These are on-the-job investigations of reported worker exposures that are carried out in response to a request by either the employer or the employee or employee representative. In addition to its own research program, NIOSH also funds supportive research activities at a number of universities, colleges, and private facilities.

NIOSH has training grant programs in colleges and universities across the nation. These are located at designated Educational Resource Centers (ERCs). ERCs train occupational physicians, occupational health nurses, industrial hygienists, safety professionals, ergonomists, and others in the safety and health field. They also provide continuing professional education for practicing occupational health and safety professionals. ***Editor's note:*** **This information is current as of June 1995. However, as we go to press, congressional legislation has been introduced that may substantially change both the mandate and function of the federal safety and health agencies.** (See Chapter 29, Government Regulations, and Chapter 30, History of the Federal Occupational Safety and Health Administration, for a full discussion of federal agencies and regulations.)

# Environmental Factors or Stresses

The various environmental factors or stresses that can cause sickness, impaired health, or significant discomfort in workers can be classified as chemical, physical, biological, or ergonomic.

***Chemical hazards.*** These arise from excessive airborne concentrations of mists, vapors, gases, or solids in the form of dusts or fumes. In addition to the hazard of inhalation, some of these materials may act as skin irritants or may be toxic by absorption through the skin.

***Physical hazards.*** These include excessive levels of nonionizing and ionizing radiations, noise, vibration, and extremes of temperature and pressure.

***Ergonomic hazards.*** These include improperly designed tools, work areas, or work procedures. Improper lifting or reaching, poor visual conditions, or repeated motions in an awkward position can result in accidents or illnesses in the occupational environment. Designing the tools and the job to fit the worker is of prime importance. Engineering and biomechanical principles must be applied to eliminate hazards of this kind.

***Biological hazards.*** These are any living organism or its properties that can cause an adverse response in humans. They can be part of the total environment or associated with a particular occupation. Work-related illnesses due to biological agents have been widely reported, but in many workplaces their presence and resultant illness is not well-recognized. It is estimated that the population at risk for occupational biohazards may be several hundred million workers worldwide.

Exposure to many of the harmful stresses or hazards listed can produce an immediate response due to the intensity of the hazard, or the response can result from longer exposure at a lower intensity.

In certain occupations, depending on the duration and severity of exposure, the work environment can produce significant subjective responses or strain. The energies and agents responsible for these effects are called environmental stresses. An employee is most often exposed to an intricate interplay of many stresses, not to a single environmental stress.

## Chemical Hazards

The majority of occupational health hazards arise from inhaling chemical agents in the form of vapors, gases, dusts, fumes, and mists, or by skin contact with these materials. The degree of risk of handling a given substance depends on the magnitude and duration of exposure. (See Chapter 15, Evaluation, for more details.)

To recognize occupational factors or stresses, a health and safety professional must first know about the chemicals used as raw materials and the nature of the products and by-products manufactured. This sometimes requires great effort. The required information can be obtained from the Material Safety Data Sheet (MSDS) (Figure 1–2) that must be supplied by the chemical manufacturer or importer for all hazardous materials under the OSHA hazard communication standard. The MSDS is a summary of the important health, safety, and toxicological information on the chemical or the mixture ingredients. Other stipulations of the hazard communication standard require that all containers of hazardous substances in the workplace be labeled with appropriate warning and identification labels. See Chapter 29, Government Regulations, and Chapter 30, History of the Federal Occupational Safety and Health Administration, for further discussion of the hazard communication standard.

If the MSDS or the label does not give complete information but only trade names, it may be necessary to contact the manufacturer to obtain this information.

Many industrial materials such as resins and polymers are relatively inert and nontoxic under normal conditions of use, but when heated or machined, they may decompose to form highly toxic by-products. Information about these hazardous products and by-products must also be included in the company's hazard communication program.

Breathing of some materials can irritate the upper respiratory tract or the terminal passages of the lungs and the air sacs, depending upon the solubility of the material. Contact of irritants with the skin surface can produce various kinds of dermatitis.

The presence of excessive amounts of biologically inert gases can dilute the atmospheric oxygen below the level required to maintain the normal blood saturation value for oxygen and disturb cellular processes. Other gases and vapors can prevent the blood from carrying oxygen to the tissues or interfere with its transfer from the blood to the tissue, thus producing chemical asphyxia or suffocation. Carbon monoxide and hydrogen cyanide are examples of chemical asphyxiants.

Some substances may affect the central nervous system and brain to produce narcosis or anaesthesia. In varying degrees, many solvents have these effects. Substances are often classified, according to the major reaction they produce, as asphyxiants, systemic toxins, pneumoconiosis-producing agents, carcinogens, irritant gases, and so on.

**Solvents.** This section discusses some general hazards arising from the use of solvents; a more detailed description is given in Chapter 7, Solvents, Gases, and Vapors.

Solvent vapors enter the body mainly by inhalation, although some skin absorption can occur. The vapors are absorbed from the lungs into the blood and are distributed mainly to tissues with a high content of fat and lipids, such as the central nervous system, liver, and bone marrow. Solvents include aliphatic and aromatic hydrocarbons, alcohols, aldehydes, ketones, chlorinated hydrocarbons, and carbon disulfide.

Occupational exposure can occur in many different processes, such as the degreasing of metals in the machine industry and the extraction of fats or oils in the chemical or food industry, dry cleaning, painting, and the plastics industry.

The widespread industrial use of solvents presents a major problem to the industrial hygienist, the safety professional, and others responsible for maintaining a safe, healthful working environment. Getting the job done using solvents without hazard to employees or property depends on the proper selection, application, handling, and control of solvents and an understanding of their properties.

A working knowledge of the physical properties, nomenclature, and effects of exposure is absolutely necessary in making a proper assessment of a solvent exposure. Nomenclature can be misleading. For example, benzine is sometimes mistakenly called benzene, a completely different solvent. Some commercial grades of benzine may contain benzene as a contaminant.

Use the information on the MSDS (Figure 1–2) or the manufacturer's label for the specific name and composition of the solvents involved.

The severity of a hazard in the use of organic solvents and other chemicals depends on the following factors:

- How the chemical is used
- Type of job operation, which determines how the workers are exposed
- Work pattern
- Duration of exposure
- Operating temperature
- Exposed liquid surface
- Ventilation rates
- Evaporation rate of solvent
- Pattern of airflow
- Concentration of vapor in workroom air
- Housekeeping

The hazard is determined not only by the toxicity of the solvent or chemical itself but by the conditions of its use (who, what, how, where, and how long).

The health and safety professional can obtain much valuable information by observing the manner in which health hazards are generated, the number of people involved, and the control measures in use.

After the list of chemicals and physical conditions to which employees are exposed has been prepared, determine which of the chemicals or agents may result in hazardous exposures and need further study.

Dangerous materials are chemicals that may, under specific circumstances, cause injury to persons or damage to property because of reactivity, instability, spontaneous decomposition, flammability, or volatility. Under this definition, we will consider substances, mixtures, or compounds that are explosive, corrosive, flammable, or toxic.

**Material Safety Data Sheet**
May be used to comply with
OSHA's Hazard Communication Standard,
29 CFR 1910.1200. Standard must be
consulted for specific requirements.

**U.S. Department of Labor**
Occupational Safety and Health Administration
(Non-Mandatory Form)
Form Approved
OMB No. 1218-0072

| IDENTITY (As Used on Label and List) | Note: Blank spaces are not permitted. If any item is not applicable, or no information is available, the space must be marked to indicate that. |
|---|---|

**Section I**

| Manufacturer's Name | Emergency Telephone Number |
|---|---|
| Address (Number, Street, City, State, and ZIP Code) | Telephone Number for Information |
| | Date Prepared |
| | Signature of Preparer (optional) |

**Section II — Hazardous Ingredients/Identity Information**

| Hazardous Components (Specific Chemical Identity; Common Name(s)) | OSHA PEL | ACGIH TLV | Other Limits Recommended | % (optional) |
|---|---|---|---|---|
| | | | | |
| | | | | |
| | | | | |
| | | | | |
| | | | | |
| | | | | |
| | | | | |
| | | | | |
| | | | | |
| | | | | |

**Section III — Physical/Chemical Characteristics**

| Boiling Point | | Specific Gravity ($H_2O$ = 1) | |
|---|---|---|---|
| Vapor Pressure (mm Hg.) | | Melting Point | |
| Vapor Density (AIR = 1) | | Evaporation Rate (Butyl Acetate = 1) | |
| Solubility in Water | | | |
| Appearance and Odor | | | |

**Section IV — Fire and Explosion Hazard Data**

| Flash Point (Method Used) | Flammable Limits | LEL | UEL |
|---|---|---|---|
| Extinguishing Media | | | |
| Special Fire Fighting Procedures | | | |
| Unusual Fire and Explosion Hazards | | | |

| (Reproduce locally) | OSHA 174, Sept. 1985 |
|---|---|

**Figure 1–2.** Material Safety Data Sheet. Its format meets the requirements of the federal hazard communication standard. (*Continues*)

## Section V — Reactivity Data

| Stability | Unstable | | Conditions to Avoid |
|---|---|---|---|
| | Stable | | |

Incompatibility (*Materials to Avoid*)

Hazardous Decomposition or Byproducts

| Hazardous Polymerization | May Occur | | Conditions to Avoid |
|---|---|---|---|
| | Will Not Occur | | |

## Section VI — Health Hazard Data

Route(s) of Entry:                Inhalation?                          Skin?                          Ingestion?

Health Hazards (*Acute and Chronic*)

Carcinogenicity:                NTP?                          IARC Monographs?                          OSHA Regulated?

Signs and Symptoms of Exposure

Medical Conditions
Generally Aggravated by Exposure

Emergency and First Aid Procedures

## Section VII — Precautions for Safe Handling and Use

Steps to Be Taken in Case Material Is Released or Spilled

Waste Disposal Method

Precautions to Be Taken in Handling and Storing

Other Precautions

## Section VIII — Control Measures

Respiratory Protection (*Specify Type*)

| Ventilation | Local Exhaust | Special |
|---|---|---|
| | Mechanical (*General*) | Other |

| Protective Gloves | Eye Protection |
|---|---|

Other Protective Clothing or Equipment

Work/Hygienic Practices

☆ U S G P O 1986–491–529/45775

**Figure 1–2.** (*Continued*)

Explosives are substances, mixtures, or compounds capable of entering into a combustion reaction so rapidly and violently as to cause an explosion.

Corrosives are capable of destroying living tissue and have a destructive effect on other substances, particularly on combustible materials; this effect can result in a fire or explosion.

Flammable liquids are liquids with a flash point of 100 F (38 C) or less, although those with higher flash points can be both combustible and dangerous.

Toxic chemicals are gases, liquids, or solids that, through their chemical properties, can produce injurious or lethal effects on contact with body cells.

Oxidizing materials are chemicals that decompose readily under certain conditions to yield oxygen. They may cause a fire in contact with combustible materials, can react violently with water, and when involved in a fire can react violently.

Dangerous gases are those that can cause lethal or injurious effects and damage to property by their toxic, corrosive, flammable, or explosive physical and chemical properties.

Storage of dangerous chemicals should be limited to one day's supply, consistent with the safe and efficient operation of the process. The storage should comply with applicable local laws and ordinances. An approved storehouse should be provided for the main supply of hazardous materials.

For hazardous materials, MSDSs can be consulted for toxicological information. The information is useful to the medical, purchasing, managerial, engineering, and health and safety departments in setting guidelines for safe use of these materials. This information is also very helpful in an emergency. The information should cover materials actually in use and those that may be contemplated for early future use. Possibly the best and earliest source of information concerning such materials is the purchasing agent. Thus, a close liaison should be set up between the purchasing agent and health and safety personnel so that early information is available concerning materials in use and those to be ordered, and to ensure that MSDSs are received and reviewed for all hazardous substances.

**Toxicity versus hazard.** The toxicity of a material is not synonymous with its hazard. Toxicity is the capacity of a material to produce injury or harm when the chemical has reached a sufficient concentration at a certain site in the body. Hazard is the probability that this concentration in the body will occur. This degree of hazard is determined by many factors or elements.

The key elements to be considered when evaluating a health hazard are as follows:

- What is the route of entry of the chemical into the body?
- How much of the material must be in contact with a body cell and for how long to produce injury?

- What is the probability that the material will be absorbed or come in contact with body cells?
- What is the rate of generation of airborne contaminants?
- What control measures are in place?

The effects of exposure to a substance depend on dose, rate, physical state of the substance, temperature, site of absorption, diet, and general state of a person's health.

## Physical Hazards

Problems caused by such things as noise, temperature extremes, ionizing radiation, nonionizing radiation, and pressure extremes are physical stresses. It is important that the employer, supervisor, and those responsible for safety and health be alert to these hazards because of the possible immediate or cumulative effects on the health of employees.

**Noise.** Noise (unwanted sound) is a form of vibration conducted through solids, liquids, or gases. The effects of noise on humans include the following:

- Psychological effects (noise can startle, annoy, and disrupt concentration, sleep, or relaxation)
- Interference with speech communication and, as a consequence, interference with job performance and safety
- Physiological effects (noise-induced hearing loss, or aural pain when the exposure is severe)

*Damage risk criteria.* If the ear is subjected to high levels of noise for a sufficient period of time, some loss of hearing may occur. A number of factors can influence the effect of the noise exposure:

- Variation in individual susceptibility
- Total energy of the sound
- Frequency distribution of the sound
- Other characteristics of the noise exposure, such as whether it is continuous, intermittent, or made up of a series of impacts
- Total daily duration of exposure
- Length of employment in the noise environment

Because of the complex relationships of noise and exposure time to threshold shift (reduction in hearing level) and the many contributory causes, establishing criteria for protecting workers against hearing loss is difficult. However, criteria have been developed to protect against hearing loss in the speech-frequency range. These criteria are known as the Threshold Limit Values for Noise. (See Chapter 9, Industrial Noise, and Appendix B, Threshold Limit Values, for more details.)

There are three nontechnical guidelines to determine whether the work area has excessive noise levels:

- If it is necessary to speak very loudly or shout directly into the ear of a person in order to be understood, it is possible that the exposure limit for noise is being exceeded. Conversation becomes difficult when the noise level exceeds 70 decibels (dBA).

- If employees say that they have heard ringing noises in their ears at the end of the workday, they may be exposed to too much noise.
- If employees complain that the sounds of speech or music seem muffled after leaving work, but that their hearing is fairly clear in the morning when they return to work, they may be exposed to noise levels that cause a partial temporary loss of hearing, which can become permanent with repeated exposure.

***Permissible levels.*** The criteria for hearing conservation, required by OSHAct in 29 *CFR* 1910.95, establishes the permissible levels of harmful noise to which an employee may be subjected. The permissible decibel levels and hours (duration per day) are specified. For example, a noise level of 90 dBA is permissible for 8 hours, 95 dBA for 4 hours, etc. (See Chapter 9, Industrial Noise, for more details.)

The regulations stipulate that when employees are subjected to sound that exceeds the permissible limits, feasible administrative or engineering controls shall be used. If such controls fail to reduce sound exposure within permissible levels, personal protective equipment must be provided and used to reduce sound levels to within permissible levels. Exposure to impulsive or impact noise should not exceed 140 dBA peak sound pressure level.

According to the Hearing Conservation Amendment to 29 *CFR* 1910.95, in all cases when the sound levels exceed 85 dBA on an 8-hour time-weighed average (TWA), a continuing, effective hearing conservation program shall be administered. The Hearing Conservation Amendment specifies the essential elements of a hearing conservation program. (See Chapter 9, Industrial Noise, for a discussion of noise and OSHA noise regulations.)

Administering a hearing conservation program goes beyond the wearing of earplugs or earmuffs. Such programs can be complex, and professional guidance is essential for establishing programs that are responsive to the need. Valid noise exposure information correlated with audiometric tests results is needed to help health and safety and medical personnel to make informed decisions about hearing conservation programs.

The effectiveness of a hearing conservation program depends on the cooperation of employers, employees, and others concerned. Management's responsibility in such a program includes noise measurements, initiation of noise control measures, provision of hearing protection equipment, audiometric testing of employees to measure their hearing levels (thresholds), and information and training programs for employees.

The employee's responsibility is to properly use the protective equipment provided by management, and to observe any rules or regulations on the use of equipment in order to minimize noise exposure.

**Extremes of temperature.** Probably the most elementary factor of environmental control is control of the thermal environment in which people work. Extremes of temperature, or thermal stress, affect the amount of work people can do and the manner in which they do it. In industry, the problem is more often high temperatures rather than low temperatures. (More details on this subject are given in Chapter 12, Thermal Stress.)

The body continuously produces heat through its metabolic processes. Because the body processes are designed to operate only within a very narrow range of temperature, the body must dissipate this heat as rapidly as it is produced if it is to function efficiently. A sensitive and rapidly acting set of temperature-sensing devices in the body must also control the rates of its temperature-regulating processes. (This mechanism is described in Chapter 3, The Skin and Occupational Dermatoses.)

Heat stress is a common problem, as are the problems presented by a very cold environment. Evaluation of heat stress in a work environment is not simple. Considerably more is involved than simply taking a number of air-temperature measurements and making decisions on the basis of this information.

One question that must be asked is whether the temperature is merely causing discomfort or whether continued exposure will cause the body temperature to fall below or rise above safe limits. It is difficult for a person with only a clipboard full of data to interpret how another person actually feels or is adversely affected.

People function efficiently only in a very narrow body temperature range, a core temperature measured deep inside the body, not on the skin or at body extremities. Fluctuations in core temperatures exceeding 2 F below or 3 F above the normal core temperature of 99.6 F (37.6 C), which is 98.6 F (37 C) mouth temperature, impair performance markedly. If this 5-degree range is exceeded, a health hazard exists.

The body attempts to counteract the effects of high temperature by increasing the heart rate. The capillaries in the skin also dilate to bring more blood to the surface so that the rate of cooling is increased. Sweating is an important factor in cooling the body.

Heatstroke is caused by exposure to an environment in which the body is unable to cool itself sufficiently. Heatstroke is a much more serious condition than heat cramps or heat exhaustion. An important predisposing factor is excessive physical exertion or moderate exertion in extreme heat conditions. The method of control is to reduce the temperature of the surroundings or to increase the ability of the body to cool itself, so that body temperature does not rise. In heatstroke, sweating may cease and the body temperature can quickly rise to fatal levels. It is critical to undertake emergency cooling of the body even while medical help is on the way. Studies show that the higher the body temperature on admission to emergency rooms, the higher the fatality rate. Heatstroke is a life-threatening medical emergency.

Heat cramps can result from exposure to high temperature for a relatively long time, particularly if accompanied by heavy exertion, with excessive loss of salt and moisture from the body. Even if the moisture is replaced by drinking plenty of water, an excessive loss of salt can cause heat cramps or heat exhaustion.

Heat exhaustion can also result from physical exertion in a hot environment. Its signs are a mildly elevated temperature, pallor, weak pulse, dizziness, profuse sweating, and cool, moist skin.

**Environmental measurements.** In many heat stress studies, the variables commonly measured are work energy metabolism (often estimated rather than measured), air movement, air temperature, humidity, and radiant heat. See Chapter 12, Thermal Stress, for illustrations and more details.

Air movement is measured with some type of anemometer and the air temperature with a thermometer, often called a *dry bulb thermometer.*

Humidity, or the moisture content of the air, is generally measured with a psychrometer, which gives both dry bulb and wet bulb temperatures. Using these temperatures and referring to a psychrometric chart, the relative humidity can be established.

The term *wet bulb* is commonly used to describe the temperature obtained by having a wet wick over the mercury-well bulb of an ordinary thermometer. Evaporation of moisture in the wick, to the extent that the moisture content of the surrounding air permits, cools the thermometer to a temperature below that registered by the dry bulb. The combined readings of the dry bulb and wet bulb thermometers are then used to calculate percent relative humidity, absolute moisture content of the air, and water vapor pressure.

Radiant heat is a form of electromagnetic energy similar to light but of longer wavelength. Radiant heat (from such sources as red-hot metal, open flames, and the sun) has no appreciable heating effect on the air it passes through, but its energy is absorbed by any object it strikes, thus heating the person, wall, machine, or whatever object it falls on. Protection requires placing opaque shields or screens between the person and the radiating surface.

An ordinary dry bulb thermometer alone will not measure radiant heat. However, if the thermometer bulb is fixed in the center of a metal toilet float that has been painted dull black, and the top of the thermometer stem protrudes outside through a one-hole cork or rubber stopper, radiant heat can be measured by the heat absorbed in this sphere. This device is known as a globe thermometer.

***Heat loss.*** Conduction is an important means of heat loss when the body is in contact with a good cooling agent, such as water. For this reason, when people are immersed in cold water, they become chilled much more rapidly and effectively than when exposed to air of the same temperature.

Air movement cools the body by convection: The moving air removes the air film or the saturated air (which is formed very rapidly by evaporation of sweat) and replaces it with a fresh air layer capable of accepting more moisture from the skin.

***Heat stress indices.*** The methods commonly used to estimate heat stress relate various physiological and environmental variables and end up with one number that then serves as a guide for evaluating stress. For example, the effective temperature index combines air temperature (dry bulb), humidity (wet bulb), and air movement to produce a single index called an effective temperature.

Another index is the wet bulb globe temperature (WBGT). The numerical value of the WBGT index is calculated by the following equations.

Indoors or outdoors with no solar loads:

$$WBGT = 0.7\ WB + 0.3\ GT$$

Outdoors with solar load:

$$WBGT = 0.7\ WB + 0.2\ GT + 0.1\ DB$$

where   WB = natural wet bulb temperature
GT = globe temperature
DB = dry bulb temperature

In its *Criteria Document on Hot Environments* (see Bibliography), NIOSH states that when impermeable clothing is worn, the WBGT should not be used because evaporative cooling would be limited. The WBGT combines the effects of humidity and air movement, air temperature and radiation, and air temperature. It has been successfully used for environmental heat stress monitoring at military camps to control heat stress casualties. The measurements are few and easy to make; the instrumentation is simple, inexpensive, and rugged, and the calculations are straightforward. It is also the index used in the *ACGIH Threshold Limit Value and Biological Exposure Indices* booklet (see Appendix B). The ACGIH recommends TLVs for continuous work in hot environments as well as when 25, 50, or 75 percent of each working hour is at rest. Regulating allowable exposure time in the heat is a viable technique for permitting necessary work to continue under heat-stress conditions that would be intolerable for continuous exposure. The NIOSH criteria document also contains a complete recommended heat stress control program including work practices.

Work practices include acclimation periods, work and rest regimens, distribution of work load with time, regular breaks of a minimum of one per hour, provision for water intake, protective clothing, and application of engineering controls. Experience has shown that workers do not stand a hot job very well at first, but develop tolerance rapidly through acclimation and acquire full endurance in a week to a month. (For more details, see Chapter 12, Thermal Stress, and the NIOSH criteria document.)

**Cold stress.** Generally, the answer to a cold work area is to supply heat where possible, except for areas that must be cold, such as food storage areas.

General hypothermia is an acute problem resulting from prolonged cold exposure and heat loss. If an individual becomes fatigued during physical activity, he or she will be more prone to heat loss, and as exhaustion approaches, sudden vasodilation (blood vessel dilation) occurs with resultant rapid loss of heat.

Cold stress is proportional to the total thermal gradient between the skin and the environment because this gradient determines the rate of heat loss from the body by radiation and convection. When vasoconstriction (blood vessel constriction) is no longer adequate to maintain body heat balance, shivering becomes an important mechanism for increasing body temperature by causing metabolic heat production to increase to several times the resting rate.

General physical activity increases metabolic heat. With clothing providing the proper insulation to minimize heat loss, a satisfactory microclimate can be maintained. Only exposed body surfaces are likely to be excessively chilled and frostbitten. If clothing becomes wet either from contact with water or due to sweating during intensive physical work, its cold-insulating property is greatly diminished.

Frostbite occurs when the skin tissues freeze. Theoretically, the freezing point of the skin is about 30 F (1 C); however, with increasing wind velocity, heat loss is greater and frostbite occurs more rapidly. Once started, freezing progresses rapidly. For example, if the wind velocity reaches 20 mph, exposed flesh can freeze within about 1 minute at 14 F (10 C). Furthermore, if the skin comes in direct contact with objects whose surface temperature is below the freezing point, frostbite can develop at the point of contact despite warm environmental temperatures. Air movement is more important in cold environments than in hot because the combined effect of wind and temperature can produce a condition called *windchill*. The windchill index should be consulted by everyone facing exposure to low temperature and strong winds. (See Chapter 12, Thermal Stress.)

**Ionizing radiation.** A brief description of ionizing radiation hazards is given in this section; for a complete description, see Chapter 10, Ionizing Radiation.

To understand a little about ionization, recall that the human body is made up of various chemical compounds that are in turn composed of molecules and atoms. Each atom has a nucleus with its own outer system of electrons.

When ionization of body tissues occurs, some of the electrons surrounding the atoms are forcibly ejected from their orbits. The greater the intensity of the ionizing radiation, the more ions are created and the more physical damage is done to the cells.

Light consisting of electromagnetic radiation from the sun that strikes the surface of the earth is very similar to x-rays and gamma-radiation; it differs only in wavelength and energy content. (See description in Chapter 11, Nonionizing Radiation.) However, the energy level of sunlight at the earth's surface is too low to disturb orbital electrons, so sunlight is not considered ionizing even though it has enough energy to cause severe skin burns over a period of time.

The exact mechanism of the manner in which ionization affects body cells and tissue is complex. At the risk of oversimplifying some basic physical principles and ignoring others, the purpose of this section is to present enough information so the health and safety professional will recognize the problems involved and know when to call on health physicists or radiation safety experts for help.

At least three basic factors must be considered in such an approach to radiation safety:

- Radioactive materials emit energy that can damage living tissue.

- Different kinds of radioactivity present different kinds of radiation safety problems. The types of ionizing radiation we will consider are alpha-, beta-, x-ray, and gamma-radiation, and neutrons.

- Radioactive materials can be hazardous in two different ways. Certain materials can be hazardous even when located some distance away from the body; these are external hazards. Other types are hazardous only when they get inside the body through breathing, eating, or broken skin. These are called internal radiation hazards.

Instruments are available for evaluating possible radiation hazards. Meters or other devices are used for measuring radiation levels and doses.

***Kinds of radioactivity.*** The five kinds of radioactivity that are of concern are alpha, beta, x-ray, gamma, and neutron. The first four are the most important because neutron sources usually are not used in ordinary manufacturing operations.

Of the five types of radiation mentioned, alpha-particles are the least penetrating. They do not penetrate thin barriers. For example, paper, cellophane, and skin stop alpha-particles.

Beta-radiation has considerably more penetrating power than alpha radiation. A quarter of an inch of aluminum can stop the more energetic betas. Virtually everyone is familiar with the penetrating ability of x rays and the fact that a barrier such as concrete or lead is required to stop them.

Gamma-rays are, for all practical purposes, the same as x rays and require the same kinds of heavy shielding materials.

Neutrons are very penetrating and have characteristics that make it necessary to use shielding materials of high hydrogen atom content rather than high mass alone.

Although the type of radiation from one radioactive material may be the same as that emitted by several other different radioactive materials, there may be a wide variation in energies.

The amount of energy a particular kind of radioactive material possesses is defined in terms of MeV (million electron volts); the greater the number of MeV, the greater the energy. Each radioactive material emits its own particular kinds of radiation, with energy measured in terms of MeV.

***External versus internal hazards.*** Radioactive materials that emit x rays, gamma-rays, or neutrons are external hazards. In other words, such materials can be located some distance from the body and emit radiation that produces ionization (and thus damage) as it passes through the body. Control by limiting exposure time, working at a safe distance, use of barriers or shielding, or a combination of all three is required for adequate protection against external radiation hazards.

As long as a radioactive material that emits only alpha-particles remains outside the body, it will not cause trouble. Internally, it is a hazard because the ionizing ability of alpha particles at very short distances in soft tissue makes them a veritable bulldozer. Once inside the body—in the lungs, stomach, as an open wound, for example—there is no thick layer of skin to serve as a barrier and damage results. Alpha-emitting radioactive materials that concentrate as persisting deposits in specific parts of the body are considered very hazardous.

Beta-emitters are generally considered an internal hazard although they also can be classed as an external hazard because they can produce burns when in contact with the skin. They require the same precautions as do alpha-emitters if there is a chance they can become airborne. In addition, some shielding may be required.

***Measuring ionizing radiation.*** Many types of meters are used to measure various kinds of ionizing radiation. These meters must be accurately calibrated for the type of radiation they are designed to measure.

Meters with very thin windows in the probes can be used to check for alpha-radiation. Geiger–Müller and ionization chamber-type instruments are used for measuring beta-, gamma-, and x-radiation. Special types of meters are available for measuring neutrons.

Devices are available that measure accumulated amounts (doses) of radiation. Film badges are used as dosimeters to record the amount of radiation received from beta-, x-ray, or gamma-radiation and special badges are available to record neutron radiation.

Film badges are worn by a worker continuously during each monitoring period. Depending on how they are worn, they allow an estimate of an accumulated dose of radiation to the whole body or to just a part of the body, such as a hand or arm.

Alpha-radiation cannot be measured with film badges because alpha-particles do not penetrate the paper that must be used over the film emulsion to exclude light. (For more details on measurement and government regulations for ionizing radiation, see Chapter 10, Ionizing Radiation.)

**Nonionizing radiation.** This is a form of electromagnetic radiation with varying effects on the body, depending largely on the wavelength of the radiation involved. In the following paragraphs, in approximate order of decreasing wavelength and increasing frequency, are some hazards associated with different regions of the nonionizing electromagnetic radiation spectrum. Nonionizing radiation is covered in detail by OSHAct regulations 29 *CFR* 1910.97, and in Chapter 11, Nonionizing Radiation.

*Low frequency.* Longer wavelengths, including powerline transmission frequencies, broadcast radio, and shortwave radio, can produce general heating of the body. The health hazard from these radiations is very small, however, because it is unlikely that they would be found in intensities great enough to cause significant effect. An exception can be found very close to powerful radio transmitter aerials.

Microwaves are found in radar, communications, some types of cooking, and diathermy applications. Microwave intensities may be sufficient to cause significant heating of tissues.

The effect is related to wavelength, power intensity, and time of exposure. Generally, longer wavelengths produce a greater penetration and temperature rise in deeper tissues than shorter wavelengths. However, for a given power intensity, there is less subjective awareness to the heat from longer wavelengths than there is to the heat from shorter wavelengths, because of the absorption of the longer wavelength radiation beneath the body's surface.

An intolerable rise in body temperature, as well as localized damage to specific organs, can result from an exposure of sufficient intensity and time. In addition, flammable gases and vapors can ignite when they are inside metallic objects located in a microwave beam.

Infrared radiation does not penetrate below the superficial layer of the skin, so its only effect is to heat the skin and the tissues immediately below it. Except for thermal burns, the health hazard of exposure to low-level conventional infrared radiation sources is negligible. (For information on possible damage to the eye, consult Chapter 11, Nonionizing Radiation.)

Visible radiation, which is about midway in the electromagnetic spectrum, is important because it can affect both the quality and accuracy of work. Good lighting conditions generally result in increased product quality with less spoilage and increased production.

Lighting should be bright enough for easy and efficient sight, and directed so that it does not create glare. Illumination levels and brightness ratios recommended for manufacturing and service industries are published by the Illuminating Engineering Society. (See Chapter 11, Nonionizing Radiation, for further information.)

One of the most objectionable features of lighting is glare (brightness in the field of vision that causes discomfort or interferes with seeing). The brightness can be caused by either direct or reflected light. To prevent glare, the source of light should be kept well above the line of vision or shielded with opaque or translucent material.

Almost as problematic is an area of excessively high brightness in the visual field. A highly reflective white paper in the center of a dark, nonreflecting surface or a brightly illuminated control handle on a dark or dirty machine are two examples.

To prevent such conditions, keep surfaces uniformly light or dark with little difference in surface reflectivity. Color contrasts are acceptable, however.

Although it is generally best to provide even, shadow-free light, some jobs require contrast lighting. In these cases, keep the general (or background) light well-diffused and glareless and add a supplementary source of light that casts shadows where needed.

Ultraviolet radiation in industry can be found around electrical arcs, and such arcs should be shielded by materials opaque to ultraviolet. The fact that a material can be opaque to ultraviolet has no relation to its opacity to other parts of the spectrum. Ordinary window glass, for instance, is almost completely opaque to the ultraviolet in sunlight although transparent to the visible wavelengths. A piece of plastic dyed a deep red-violet may be almost entirely opaque in the visible part of the spectrum and transparent in the near-ultraviolet.

Electric welding arcs and germicidal lamps are the most common strong producers of ultraviolet radiation in industry. The ordinary fluorescent lamp generates a good deal of ultraviolet inside the bulb, but it is essentially all absorbed by the bulb and its coating.

The most common exposure to ultraviolet radiation is from direct sunlight, and a familiar result of overexposure—one that is known to all sunbathers—is sunburn. Most people are familiar with certain compounds and lotions that reduce the effects of the sun's rays, but many are unaware that some industrial materials, such as cresols, make the skin especially sensitive to ultraviolet rays. After exposure to cresols, even a short exposure in the sun usually results in a severe sunburn.

Lasers emit beams of coherent radiation of a single color or wavelength and frequency, in contrast to conventional light sources, which produce random, disordered light wave mixtures of various frequencies. The laser (an acronym for light amplification by stimulated emission of radiation) is made up of light waves that are nearly parallel to each other, all traveling in the same direction. Atoms are "pumped" full of energy, and when they are stimulated to fall to a lower energy level, they give off radiation that is directed to produce the coherent laser beam. (See Chapter 11, Nonionizing Radiation, for more details.)

The maser, the laser's predecessor, emits microwaves instead of light. Some companies call their lasers "optical masers." Because the laser is highly collimated (has a small divergence angle), it can have a large energy density in a narrow beam. Direct viewing of the laser source or its reflections should be avoided. The work area should contain no reflective surface (such as mirrors or highly polished furniture) because even a reflected laser beam can be hazardous. Suitable shielding to contain the laser beam should be provided. The OSHAct covers protection against laser hazards in its construction regulations.

***Biological effects.*** The eye is the organ that is most vulnerable to injury by laser energy because the cornea and lens focus the parallel laser beam on a small spot on the retina. The fact that infrared radiation of certain lasers may not be visible to the naked eye contributes to the potential hazard.

Lasers generating in the ultraviolet range of the electromagnetic spectrum can produce corneal burns rather than retinal damage, because of the way the eye handles ultraviolet light. (See Chapter 11, Nonionizing Radiation.)

Other factors that affect the degree of eye injury induced by laser light are as follows:

- Pupil size (the smaller the pupil diameter, the less laser energy reaches the retina)
- The ability of the cornea and lens to focus the incident light on the retina
- The distance from the source of energy to the retina
- The energy and wavelength of the laser
- The pigmentation of the eye of the subject
- The location on the retina where the light is focused
- The divergence of the laser light
- The presence of scattering media in the light path

A discussion of laser beam characteristics and protective eyewear can be found in Chapter 11.

**Extremes of pressure.** It has been recognized from the beginning of caisson work (work performed in a watertight structure) that people working under pressures greater than normal atmospheric pressure are subject to various health effects. Hyperbaric (greater than normal pressure) environments are also encountered by divers who work under water, whether by holding the breath while diving, breathing from a self-contained underwater breathing apparatus (SCUBA), or by breathing gas mixtures supplied by compression from the surface.

Occupational exposures occur in caisson or tunneling operations, where a compressed gas environment is used to exclude water or mud and to provide support for structures. Humans can withstand large pressures if air has free access to lungs, sinuses, and the middle ear. Unequal distribution of pressure can result in barotrauma, a kind of tissue damage resulting from expansion or contraction of gas spaces within or adjacent to the body, which can occur either during compression (descent) or during decompression (ascent).

The teeth, sinuses, and ears are often affected by pressure differentials. For example, gas spaces adjacent to tooth roots or fillings may be compressed during descent. Fluid or tissue forced into these spaces can cause pain during descent or ascent. Sinus blockage caused by occlusion of the sinus aperture by inflamed nasal mucosa prevents equalization of pressures.

Under some conditions of work at high pressure, the concentration of carbon dioxide in the atmosphere can be considerably increased so that the carbon dioxide acts as a narcotic. Keeping the oxygen concentration high minimizes this condition, but does not prevent it. The procedure is useful where the carbon dioxide concentration cannot be kept at a proper level.

Decompression sickness, commonly called the bends, results from the release of nitrogen bubbles into the circulation and tissues during decompression. If the bubbles lodge at the joints and under muscles, they cause severe cramps. To prevent this, decompression is carried out slowly and by stages so that the nitrogen can be eliminated slowly, without forming bubbles.

Deep-sea divers are supplied with a mixture of helium and oxygen for breathing, and because helium is an inert diluent and less soluble in blood and tissue than is nitrogen, it presents a less formidable decompression problem.

One of the most common troubles encountered by workers under compressed air is pain and congestion in the ears from inability to ventilate the middle ear properly during compression and decompression. As a result, many workers subjected to increased air pressures suffer from temporary hearing loss; some have permanent hearing loss. This damage is believed to be caused by obstruction of the eustachian tubes, which prevents proper equalization of pressure from the throat to the middle ear.

The effects of reduced pressure on the worker are much the same as the effects of decompression from a high pressure. If pressure is reduced too rapidly, decompression sickness and ear disturbances similar to the diver's conditions can result.

## Ergonomic Hazards

*Ergonomics* literally means the customs, habits, and laws of work. According to the International Labor Office, it is "the application of human biological science in conjunction with the engineering sciences to achieve the optimum mutual adjustment of man [sic] and his [sic] work, the benefits being measured in terms of human efficiency and well-being." The topic of ergonomics is covered briefly here. (For more details, see Chapter 13, Ergonomics.)

The ergonomics approach goes beyond productivity, health, and safety. It includes consideration of the total physiological and psychological demands of the job on the worker.

In the broad sense, the benefits that can be expected from designing work systems to minimize physical stress on workers are as follows:

- Reduced incidence of repetitive motion disorders
- Reduced injury rate
- More efficient operation
- Fewer accidents
- Lower cost of operation
- Reduced training time
- More effective use of personnel

The human body can endure considerable discomfort and stress and can perform many awkward and unnatural movements for a limited period of time. However, when awkward conditions or motions are continued for prolonged periods, they can exceed the worker's physiological limitations. To ensure a continued high level of performance, work systems must be tailored to human capacities and limitations.

Ergonomics considers the physiological and psychological stresses of the task. The task should not require excessive muscular effort, considering the worker's age, sex, and state of health. The job should not be so easy that boredom and inattention lead to unnecessary errors, material waste, and accidents. Ergonomic stresses can impair the health and efficiency of the worker just as significantly as the more commonly recognized environmental stresses.

The task of the design engineer and health and safety professional is to find the happy medium between "easy" and "difficult" jobs. In any human–machine system, there are tasks that are better performed by people than by machines and, conversely, tasks that are better handled by machines.

Ergonomics deals with the interactions between humans and such traditional environmental elements as atmospheric contaminants, heat, light, sound, and tools and equipment. People are the monitoring link of a human–machine environment system.

In any activity, a person receives and processes information, and then acts on it. The receptor function occurs largely through the sense organs of the eyes and the ear, but information can also be conveyed through the senses of smell, touch, or sensations of heat or cold. This information is conveyed to the central mechanism of the brain and spinal cord, where the information is processed to arrive at a decision. This can involve the integration of the information, which has already been stored in the brain, and decisions can vary from automatic responses to those involving a high degree of reasoning and logic.

Having received the information and processed it, the individual then takes action (control) as a result of the decision, usually through muscular activity based on the skeletal framework of the body. When an individual's activity involves the operation of a piece of equipment, the person often forms part of a "closed-loop servosystem," displaying many of the feedback characteristics of such a system. The person usually forms the part of the system that makes decisions, and thus has a fundamental part to play in the efficiency of the system.

**Biomechanics–physical demands.** Biomechanics can be a very effective tool in preventing excessive work stress. *Biomechanics* means the mechanics of biological organisms. It deals with the functioning of the structural elements of the body and the effects of external and internal forces on the various parts of the body.

Cumulative effects of excessive ergonomic stress on the worker can, in an insidious and subtle manner, result in physical illnesses and injuries such as "trigger finger," tenosynovitis, bursitis, carpal tunnel syndrome, and other cumulative trauma disorders.

Cases of excessive fatigue and discomfort are, in many cases, forerunners of soreness and pain. By exerting a strong distracting influence on a worker, these stresses can render the worker more prone to major accidents. Discomfort and fatigue tend to make the worker less capable of maintaining the proper vigilance for the safe performance of the task.

Some of the principles of biomechanics can be illustrated by considering different parts of the human anatomy, such as the hand.

***Hand anatomy.*** The flexing action in the fingers is controlled by tendons attached to muscles in the forearm. The tendons, which run in lubricated sheaths, enter the hand through a tunnel in the wrist formed by bones and ligaments (the carpal tunnel) and continue on to point of attachment to the different segments, or phalanges, of the fingers (Figure 1–3).

When the wrist is bent toward the little finger side, the tendons tend to bunch up on one side of the tunnel through which they enter the hand. If an excessive amount of force is continuously applied with the fingers while the wrist is flexed, or if the flexing motion is repeated rapidly over a long period of time, the resulting friction can produce inflammation of the tendon sheaths, or tenosynovitis. This can lead to a disabling condition called carpal tunnel syndrome. (See Chapter 13, Ergonomics.)

The palm of the hand, which contains a network of nerves and blood vessels, should never be used as a hammer or subjected to continued firm pressure. Repetitive or prolonged pressure on the nerves and blood vessels in this area can result in pain either in the palm itself or at any point along the nerve pathways up through the arm and shoulder. Other parts of the body, such as the elbow joints and shoulders, can become painful for similar reasons.

***Mechanical vibration.*** A condition known to stonecutters as "dead fingers" or "white fingers" (Raynaud's phenomenon) occurs mainly in the fingers of the hand used to guide the cutting tool. The circulation in this hand becomes impaired, and when exposed to cold the fingers become

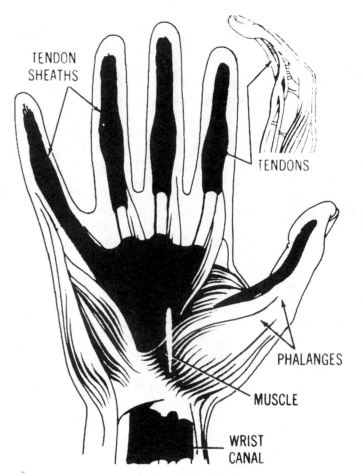

**Figure 1–3.** Diagram of hand anatomy.

white and without sensation, as though mildly frostbitten. The white appearance usually disappears when the fingers are warmed for some time, but a few cases are sufficiently disabling that the victims are forced to seek other types of work. In many instances both hands are affected.

The condition has been observed in a number of other occupations involving the use of vibrating tools, such as the air hammers used for scarfing metal surfaces, the air chisels for chipping castings in the metal trades, and the chain saws used in forestry. The injury is caused by vibration of the fingers as they grip the tools to guide them in performing their tasks. The related damage to blood vessels can progress to nearly complete obstruction of the vessels.

Prevention should be directed at reducing the vibrational energy transferred to the fingers (perhaps by the use of padding) and by changing energy and the frequency of the vibration. Low frequencies, 25–75 hertz, are more damaging than higher frequencies.

***Lifting.*** The injuries resulting from manual handling of objects and materials make up a large proportion of all compensable injuries. This problem is of considerable concern to the health and safety professional and represents an area where the biomechanical data relating to lifting and carrying can be applied in the work layout and design of jobs that require handling of materials. (For more details, see Chapter 13, Ergonomics, and the NIOSH *Guide to Manual Lifting*.)

The relevant data concerning lifting can be classified into task, human, and environmental variables.

- *Task variables*
  Location of object to be lifted
  Size of the object to be lifted
  Height from which and to which the object is lifted
  Frequency of lift
  Weight of object
  Working position

- *Human variables*
  Sex of worker
  Age of worker
  Training of worker
  Physical fitness or conditioning of worker
  Body dimensions, such as height of the worker

- *Environmental variables*
  Extremes of temperature
  Humidity
  Air contaminants

***Static work.*** Another very fatiguing situation encountered in industry, which unfortunately is often overlooked, is static, or isometric, work. Because very little outward movement occurs, it seems that no muscular effort is involved. Often, however, such work generates more muscular fatigue than work involving some outward movement. A cramped working posture, for example, is a substantial source of static muscular loading.

In general, maintaining any set of muscles in a rigid, unsupported position for long periods of time results in muscular strain. The blood supply to the contracted muscle is diminished, a local deficiency of oxygen can occur, and waste products accumulate. Alternating static and dynamic work, or providing support for partial relaxation of the member involved, alleviates this problem.

Armrests are usually needed in two types of situations. One is the case just mentioned—to relieve the isometric muscular work involved in holding the arm in a fixed, unsupported position for long periods of time. The second case is where the arm is pressed against a hard surface such as the edge of a bench or machine. The pressure on the soft tissues overlaying the bones can cause bruises and pain. Padded armrests have solved numerous problems of both types (see Figure 1–4).

**Workplace design.** Relating the physical characteristics and capabilities of the worker to the design of equipment and to the layout of the workplace is another key ergonomic concept. When this is done, the result is an increase in efficiency, a decrease in human error, and a consequent reduction in accident frequency. However, several different types of information are needed: a description of the job,

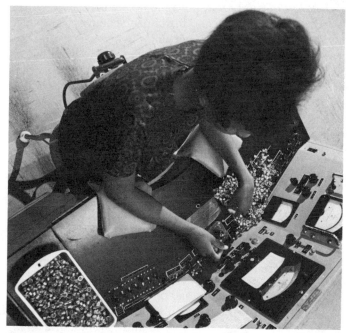

**Figure 1–4.** Worker uses pads to keep her forearm off the sharp table edge.

an understanding of the kinds of equipment to be used, a description of the kinds of people who will use the equipment, and the biological characteristics of these people.

In general, the first three items—job, equipment, and users—can be defined easily. The biological characteristics of the users, however, can often be determined satisfactorily only from special surveys that yield descriptive data on human body size and biomechanical abilities and limitations.

***Anthropometric data.*** Anthropometric data consist of various heights, lengths, and breadths used to establish the minimum clearances and spatial accommodations, and the functional arm, leg, and body movements that are made by the worker during the performance of the task.

**Behavioral aspects—mental demands.** One important aspect of industrial machine design directly related to the safety and productivity of the worker is the design of displays and controls. Displays are one of the most common types of operator input; the others include direct sensing and verbal or visual commands. Displays tell the operator what the machine is doing and how it is performing. Problems of display design are primarily related to the human senses.

A machine operator can successfully control equipment only to the extent that the operator receives clear, unambiguous information, when needed on all pertinent aspects of the task. Accidents, or operational errors, often occur because a worker has misinterpreted or was unable to obtain information from displays. Displays are usually visual, although they also can be auditory (for example, a warning bell rather than a warning light), especially when there is danger of overloading the visual sensory channels.

***Design of controls.*** An operator must decide on the proper course of action and manipulate controls to produce any desired change in the machine's performance. The efficiency and effectiveness—that is, the safety with which controls can be operated—depend on the extent to which information on the dynamics of human movement (or biomechanics) has been incorporated in their design. This is particularly true whenever controls must be operated at high speed, against large resistances, with great precision, or over long periods of time.

Controls should be designed so that rapid, accurate settings easily can be made without undue fatigue, thereby avoiding many accidents and operational errors. Because there is a wide variety of machine controls, ranging from the simple on–off action of pushbuttons to very complex mechanisms, advance analysis of the task requirements must be made. On the basis of considerable experimental evidence, it is possible to recommend the most appropriate control and its desirable range of operation.

In general, the mechanical design of equipment must be compatible with the biological and psychological characteristics of the operator. The effectiveness of the human–machine combination can be greatly enhanced by treating the operator and the equipment as a unified system. Thus, the instruments should be considered as extensions of the operator's nervous and perceptual systems, the controls as extensions of the hands, and the feet as simple tools. Any control that is difficult to reach or operate, any instrument dial that has poor legibility, any seat that induces poor posture or discomfort, or any obstruction of vision can contribute directly to an accident or illness.

## Biological Hazards

Approximately 200 biological agents, such as infectious microorganisms, biological allergens, and toxins, are known to produce infections and allergenic, toxic, and carcinogenic reactions in workers. Most of the identified biohazardous agents belong to these groups:

- Microorganisms and their toxins (viruses, bacteria, fungi, and their products) resulting in infection, exposure, or allergy
- Arthropod (crustaceans, arachnids, insects) associated with bites or stings resulting in skin inflammation, systemic intoxication and transmission of infectious agents, or allergic response
- Allergens and toxins from higher plants, producing dermatitis, rhinitis, or asthma
- Protein allergens from vertebrate animals (such as urine, feces, hair, saliva, and dander)

Other groups with the potential to expose workers to biohazards include lower plants other than fungi (lichen, liverworts, ferns) and invertebrate animals other than arthropods (parasites such as protozoa, Schistoma) and roundworms (Ascaris).

Workers engaging in agricultural, medical, and laboratory work have been identified as most at risk to occupational biohazards but many varied workplaces present the potential for such exposure. For example, at least 24 of the 150 zoonotic diseases known worldwide are considered to be a hazard for agricultural workers in North America. Risk of infection varies with the type and species of animal and geographic location. Disease may be contracted directly from animals, but more often it is acquired in the workplace environment. Controls include awareness of specific hazards, use of personal protective equipment, preventive veterinary care, worker education, and medical monitoring or prophylactic therapy, where appropriate.

The potential for exposure to occupational biohazards exists in most work environments. The following are but a few examples in very diverse workplaces:

- Workers maintaining water systems can be exposed to *Legionella pneumophila* and *Naegleria spp.*
- Workers associated with birds (parrots, parakeets, pigeons) in pet shops, aviaries, or on construction and public works jobs near perching or nesting sites can be exposed to *Chlamydia psittaci.*
- Workers in wood processing facilities can be exposed to endotoxins, allergenic fungi growing on timber, and fungi causing deep mycoses.
- Sewage and compost workers can be exposed to enteric bacteria, hepatitis A virus, infectious or endotoxin-producing bacteria, parasitic protozoa, and allergenic fungi.

**Building-related illnesses due to biological hazards.** The sources of biological hazards may be fairly obvious in occupations associated with the handling of microorganisms, plants, and animals and in occupations involving contact with potentially infected people. However, recognizing and identifying biological hazards may not be as simple in other situations such as office buildings and nonindustrial workplaces. Building-related illness (BRI) is a clinically diagnosed disease in one or more building occupants, as distinguished from sick-building syndrome (SBS), in which building occupants' nonspecific symptoms cannot be associated with an identifiable cause. Certain BRI such as infectious and hypersensitivity diseases are clearly associated with biological hazards, but the role of biological materials in SBS is not as well understood.

The conditions and events necessary to result in human exposure to bioaerosols are presence of a reservoir that can support the growth of microorganisms or allow accumulation of biological material, multiplication of contaminating organisms or biological materials in the reservoir, generation of aerosols containing biological material, and exposure of susceptible workers. (See Chapter 14, Biological Hazards, for a full discussion.)

**Industrial sanitation.** The requirements for sanitation and personal facilities are covered in the OSHAct safety and health regulations 29 *CFR* 1910, Subpart J—General Environmental Controls. The OSHAct regulations for carcinogens require special personal health and sanitary facilities for employees working with potentially carcinogenic materials.

***Water supply.*** Potable water should be provided in workplaces when needed for drinking and personal washing, cooking, washing of foods or utensils, washing of food preparation premises, and personal service rooms.

Drinking fountain surfaces must be constructed of materials impervious to water and not subject to oxidation. The nozzle of the fountain must be located to prevent the return of water in the jet or bowl to the nozzle orifice. A guard over the nozzle prevents contact with the nozzle by the mouth or nose of people using the drinking fountain.

Potable drinking water dispensers must be designed and constructed so that sanitary conditions are maintained; they must be capable of being closed and equipped with a tap. Ice that comes in contact with drinking water must be made of potable water and maintained in a sanitary condition. Standing water in cooling towers and other air-moving systems should be monitored for Legionella bacteria (see Chapter 14, Biological Hazards, for details).

Outlets for nonpotable water, such as water for industrial or firefighting purposes, must be marked in a manner that indicates clearly that the water is unsafe and is not to be used as drinking water. Nonpotable water systems or systems carrying any other nonpotable substance should be constructed so as to prevent backflow or backsiphonage.

## Harmful Agents–Route of Entry

In order to exert its toxic effect, a harmful agent must come into contact with a body cell and must enter the body via inhalation, skin absorption, or ingestion.

Chemical compounds in the form of liquids, gases, mists, dusts, fumes, and vapors can cause problems by inhalation (breathing), absorption (through direct contact with the skin), or ingestion (eating or drinking).

### Inhalation

Inhalation involves airborne contaminants that can be inhaled directly into the lungs and can be physically classified as gases, vapors, and particulate matter including dusts, fumes, smokes, aerosols, and mists.

Inhalation, as a route of entry, is particularly important because of the rapidity with which a toxic material can be absorbed in the lungs, pass into the bloodstream, and reach the brain. Inhalation is the major route of entry for hazardous chemicals in the work environment.

### Absorption

Absorption through the skin can occur quite rapidly if the skin is cut or abraded. Intact skin, however, offers a reason-ably good barrier to chemicals. Unfortunately, there are many compounds that can be absorbed through intact skin.

Some substances are absorbed by way of the openings for hair follicles and others dissolve in the fats and oils of the skin, such as organic lead compounds, many nitro compounds, and organic phosphate pesticides. Compounds that are good solvents for fats (such as toluene and xylene) also can be absorbed through the skin.

Many organic compounds, such as TNT, cyanides, and most aromatic amines, amides, and phenols, can produce systemic poisoning by direct contact with the skin.

### Ingestion

In the workplace, people can unknowingly eat or drink harmful chemicals. Toxic compounds can be absorbed from the gastrointestinal tract into the blood. Lead oxide can cause serious problems if people working with this material are allowed to eat or smoke in work areas. Thorough washing is required both before eating and at the end of every shift.

Inhaled toxic dusts can also be ingested in hazardous amounts. If the toxic dust swallowed with food or saliva is not soluble in digestive fluids, it is eliminated directly through the intestinal tract. Toxic materials that are readily soluble in digestive fluids can be absorbed into the blood from the digestive system.

It is important to study all routes of entry when evaluating the work environment—candy bars or lunches in the work area, solvents being used to clean work clothing and hands, in addition to airborne contaminants in working areas. (For more details, see Chapter 6, Industrial Toxicology.)

## Types of Airborne Contaminants

There are precise meanings of certain words commonly used in industrial hygiene. These must be used correctly in order to understand the requirements of OSHAct regulations, effectively communicate with other occupational health professionals, recommend or design and test appropriate engineering controls, and correctly prescribe personal protective equipment. For example, a fume respirator is worthless as protection against gases or vapors. Too often, terms (such as gases, vapors, fumes, and mists) are used interchangeably. Each term has a definite meaning and describes a certain state of matter.

### States of Matter

**Dusts.** These are solid particles generated by handling, crushing, grinding, rapid impact, detonation, and decrepitation (breaking apart by heating) of organic or inorganic materials, such as rock, ore, metal, coal, wood, and grain.

*Dust* is a term used in industry to describe airborne solid particles that range in size from 0.1–25 µm in diameter

(1 μm = 0.0001 cm or 1/25,400 in.). Dusts more than 5 μm in size usually do not remain airborne long enough to present an inhalation problem (see Chapter 8, Particulates).

Dust can enter the air from various sources, such as when a dusty material is handled (as when lead oxide is dumped into a mixer or talc is dusted on a product). When solid materials are reduced to small sizes in processes such as grinding, crushing, blasting, shaking, and drilling, the mechanical action of the grinding or shaking device supplies energy to disperse the dust.

Evaluating dust exposures properly requires knowledge of the chemical composition, particle size, dust concentration in air, how it is dispersed, and many other factors described here. Although in the case of gases, the concentration that reaches the alveolar sacs is nearly like the concentration in the air breathed, this is not the case for aerosols or dust particles. Large particles, more than 10 μm aerodynamic diameter, can be deposited through gravity and impaction in large ducts before they reach the very small sacs (alveoli). Only the smaller particles reach the alveoli. (See Chapter 2, The Lungs, for more details.)

Except for some fibrous materials, dust particles must usually be smaller than 5 μm in order to penetrate to the alveoli or inner recess of the lungs.

A person with normal eyesight can detect dust particles as small as 50 μm in diameter. Smaller airborne particles can be detected individually by the naked eye only when strong light is reflected from them. Particles of dust of respirable size (less than 10 μm) cannot be seen without the aid of a microscope, but they may be perceived as a haze.

Most industrial dusts consist of particles that vary widely in size, with the small particles greatly outnumbering the large ones. Consequently (with few exceptions), when dust is noticeable in the air near a dusty operation, probably more invisible dust particles than visible ones are present. A process that produces dust fine enough to remain suspended in the air long enough to be breathed should be regarded as hazardous until it can be proved safe.

There is no simple one-to-one relationship between the concentration of an atmospheric contaminant and duration of exposure and the rate of dosage by the hazardous agent to the critical site in the body. For a given magnitude of atmospheric exposure to a potentially toxic particulate contaminant, the resulting hazard can range from an insignificant level to one of great danger, depending on the toxicity of the material, the size of the inhaled particles, and other factors that determine their fate in the respiratory system.

**Fumes.** These are formed when the material from a volatilized solid condenses in cool air. The solid particles that are formed make up a fume that is extremely fine, usually less than 1.0 μm in diameter. In most cases, the hot vapor reacts with the air to form an oxide. Gases and vapors are not fumes, although the terms are often mistakenly used interchangeably.

Welding, metalizing, and other operations involving vapors from molten metals may produce fumes; these may be harmful under certain conditions. Arc welding volatilizes metal vapor that condenses as the metal or its oxide in the air around the arc. In addition, the rod coating is partially volatilized. These fumes, because they are extremely fine, are readily inhaled.

Other toxic fumes, such as those formed when welding structures that have been painted with lead-based paints or when welding galvanized metal, can produce severe symptoms of toxicity rather rapidly unless fumes are controlled with effective local exhaust ventilation or the welder is protected by respiratory protective equipment.

Fortunately, most soldering operations do not require temperatures high enough to volatilize an appreciable amount of lead. However, the lead in molten solder pots is oxidized by contact with air at the surface. If this oxide, often called dross, is mechanically dispersed into the air, it can produce a severe lead-poisoning hazard.

In operations when lead dust may be present in air, such as soldering or lead battery-making, preventing occupational poisoning is largely a matter of scrupulously clean housekeeping to prevent the lead oxide from becoming dispersed into the air. It is customary to enclose melting pots, dross boxes, and similar operations, and to ventilate them adequately to control the hazard. Other controls may be necessary as well.

**Smoke.** This consists of carbon or soot particles less than 0.1 μm in size, and results from the incomplete combustion of carbonaceous materials such as coal or oil. Smoke generally contains droplets as well as dry particles. Tobacco, for instance, produces a wet smoke composed of minute tarry droplets.

**Aerosols.** These are liquid droplets or solid particles of fine enough particle size to remain dispersed in air for a prolonged period of time.

**Mists.** These are suspended liquid droplets generated by condensation of liquids from the vapor back to the liquid state or by breaking up a liquid into a dispersed state, such as by splashing, foaming, or atomizing. The term *mist* is applied to a finely divided liquid suspended in the atmosphere. Examples are the oil mist produced during cutting and grinding operations, acid mists from electroplating, acid or alkali mists from pickling operations, paint spray mist in painting operations, and the condensation of water vapor to form a fog or rain.

**Gases.** These are formless fluids that expand to occupy the space or enclosure in which they are confined. Gases are a state of matter in which the molecules are unrestricted by cohesive forces. Examples are arc-welding gases, internal combustion engine exhaust gases, and air.

**Vapors.** These are the volatile form of substances that are normally in the solid or liquid state at room temperature and pressure. Evaporation is the process by which a liquid is changed into the vapor state and mixed with the surrounding atmosphere. Solvents with low boiling points volatilize readily at room temperature.

In addition to the definitions concerning states of matter that are used daily by industrial hygienists, terms used to describe degree of exposure include the following:

- ppm: parts of vapor or gases per million parts of air by volume at room temperature and pressure
- mppcf: millions of particles of a particulate per cubic foot of air
- mg/m³: milligrams of a substance per cubic meter of air

The health and safety professional recognizes that air contaminants exist as a gas, dust, fume, mist, or vapor in the workroom air. In evaluating the degree of exposure, the measured concentration of the air contaminant is compared to limits or exposure guidelines that appear in the published standards on levels of exposure (see Appendix B).

## Respiratory Hazards

Airborne chemical agents that enter the lungs can pass directly into the bloodstream and be carried to other parts of the body. The respiratory system consists of organs contributing to normal respiration or breathing. Strictly speaking, it includes the nose, mouth, upper throat, larynx, trachea, and bronchi (which are all air passages or airways) and the lungs, where oxygen is passed into the blood and carbon dioxide is given off. Finally, it includes the diaphragm and the muscles of the chest, which perform the normal respiratory movements of inspiration and expiration (see Chapter 2, The Lungs).

All living cells of the body are engaged in a series of chemical processes; the sum total of these processes is called metabolism. In the course of its metabolism, each cell consumes oxygen and produces carbon dioxide as a waste product.

Respiratory hazards can be broken down into two main groups:

- Oxygen deficiency, in which the oxygen concentration (or partial pressure of oxygen) is below the level considered safe for human exposure
- Air that contains harmful or toxic contaminants

**Oxygen-deficient atmospheres.** Each living cell in the body requires a constant supply of oxygen. Some cells are more dependent on a continuing oxygen supply than others. Some cells in the brain and nervous system can be injured or die after 4–6 min without oxygen. These cells, if destroyed, cannot be regenerated or replaced, and permanent changes and impaired functioning of the brain can

result from such damage. Other cells in the body are not as critically dependent on an oxygen supply because they can be replaced.

Normal air at sea level contains approximately 21 percent oxygen and 79 percent nitrogen and other inert gases. At sea level and normal barometric pressure (760 mm Hg or 101.3 kPa), the partial pressure of oxygen would be 21 percent of 760 mm, or 160 mm. The partial pressure of nitrogen and inert gases would be 600 mm (79 percent of 760 mm).

At higher altitudes or under conditions of reduced barometric pressure, the relative proportions of oxygen and nitrogen remain the same, but the partial pressure of each gas is decreased. The partial pressure of oxygen at the alveolar surface of the lung is critical because it determines the rate of oxygen diffusion through the moist lung tissue membranes.

Oxygen-deficient atmospheres may exist in confined spaces as oxygen is *consumed* by chemical reactions such as oxidation (rust, fermentation), *replaced* by inert gases such as argon, nitrogen, and carbon dioxide, or *absorbed* by porous surfaces such as activated charcoal.

Deficiency of oxygen in the atmosphere of confined spaces can be a problem in industry. For this reason, the oxygen content of any tank or other confined space (as well as the levels of any toxic contaminants) should be measured before entry is made. Instruments are commercially available for this purpose. See Chapter 16, Air Sampling, Chapter 17, Direct-Reading Instruments for Gases, Vapors, and Particulates, and Chapter 22, Respiratory Protection, for more details.

The first physiological signs of an oxygen deficiency (anoxia) are an increased rate and depth of breathing. A worker should never enter or remain in areas where tests have indicated oxygen deficiency without a supplied-air or self-contained respirator that is specifically approved by NIOSH for those conditions. (See Chapter 22, Respiratory Protection, for more details.)

Oxygen-deficient atmospheres can cause an inability to move and a semiconscious lack of concern about the imminence of death. In cases of abrupt entry into areas containing little or no oxygen, the person usually has no warning symptoms, immediately loses consciousness, and has no recollection of the incident if rescued in time to be revived. The senses cannot be relied on to alert or warn a person of atmospheres deficient in oxygen.

Oxygen-deficient atmospheres can occur in tanks, vats, holds of ships, silos, mines, or in areas where the air may be diluted or displaced by asphyxiating levels of gases or vapors, or where the oxygen may have been consumed by chemical or biological reactions.

Ordinary jobs involving maintenance and repair of systems for storing and transporting fluids or entering tanks or tunnels for cleaning and repairs are controlled almost entirely by the immediate supervisor, so that person should be particularly knowledgeable of all rules and precautions to ensure the safety of those who work in such atmospheres. Safeguards should be meticulously observed.

For example, there should be a standard operating procedure for entering tanks. Such procedures should be consistent with OSHAct regulations and augmented by in-house procedures, which may enhance the basic OSHAct rules. The American National Standards Institute (ANSI) lists confined space procedures in its respiratory protection standard and NIOSH has also issued guidelines for work in confined spaces including a criteria document for working in confined spaces (see Bibliography). Even if a tank is empty, it may have been closed for some time and developed an oxygen deficiency through chemical reactions of residues left in the tank. It may be unsafe to enter without proper respiratory protection.

**The hazard of airborne contaminants.** Inhaling harmful materials can irritate the upper respiratory tract and lung tissue, or the terminal passages of the lungs and the air sacs, depending on the solubility of the material.

Inhalation of biologically inert gases can dilute the atmospheric oxygen below the normal blood saturation value and disturb cellular processes. Other gases and vapors may prevent the blood from carrying oxygen to the tissues or interfere with its transfer from the blood to the tissue, producing chemical asphyxia.

Inhaled contaminants that adversely affect the lungs fall into three general categories:

- Aerosols (particulates), which, when deposited in the lungs, can produce either rapid local tissue damage, some slower tissue reactions, eventual disease, or physical plugging
- Toxic vapors and gases that produce adverse reaction in the tissue of the lungs
- Some toxic aerosols or gases that do not affect the lung tissue locally but pass from the lungs into the bloodstream, where they are carried to other body organs or have adverse effects on the oxygen-carrying capacity of the blood cells

An example of an aerosol is silica dust, which causes fibrotic growth (scar tissue) in the lungs. Other harmful aerosols are fungi found in sugar cane residues, producing bagassosis.

An example of the second type of inhaled contaminant is hydrogen fluoride, a gas that directly affects lung tissue. It is a primary irritant of mucous membranes, even causing chemical burns. Inhalation of this gas causes pulmonary edema and direct interference with the gas transfer function of the alveolar lining.

An example of the third type of inhaled contaminant is carbon monoxide, a toxic gas passed into the bloodstream without harming the lung. The carbon monoxide passes through the alveolar walls into the blood, where it ties up the hemoglobin so that it cannot accept oxygen, thus causing oxygen starvation. Cyanide gas has another effect—it prevents enzymatic utilization of molecular oxygen by cells.

Sometimes several types of lung hazards occur simultaneously. In mining operations, for example, explosives release oxides of nitrogen. These impair the bronchial clearance mechanism so that coal dust (of the particle sizes associated with the explosions) is not efficiently cleansed from the lungs.

If a compound is very soluble—such as ammonia, sulfuric acid, or hydrochloric acid—it is rapidly absorbed in the upper respiratory tract and during the initial phases of exposure does not penetrate deeply into the lungs. Consequently, the nose and throat become very irritated.

Compounds that are insoluble in body fluids cause considerably less throat irritation than the soluble ones, but can penetrate deeply into the lungs. Thus, a very serious hazard can be present and not be recognized immediately because of a lack of warning that the local irritation would otherwise provide. Examples of such compounds (gases) are nitrogen dioxide and ozone. The immediate danger from these compounds in high concentrations is acute lung irritation or, possibly, chemical pneumonia.

There are numerous chemical compounds that do not follow the general solubility rule. Such compounds are not very soluble in water and yet are very irritating to the eyes and respiratory tract. They also can cause lung damage and even death under certain conditions.

# ■ Threshold Limit Values

Threshold Limit Values (TLVs) are exposure guidelines established for airborne concentrations of many chemical compounds. The health and safety professional or other responsible person should understand something about TLVs and the terminology in which their concentrations are expressed. (See Chapter 15, Evaluation, Chapter 6, Industrial Toxicology, and Appendix B for more details.)

TLVs are airborne concentrations of substances that are believed to represent conditions under which nearly all workers may be repeatedly exposed, day after day, without adverse effect. Control of the work environment is based on the assumption that for each substance there is some safe or tolerable level of exposure below which no significant adverse effect occurs. These tolerable levels are called Threshold Limit Values. The introduction to the TLV booklet states that because individual susceptibility varies widely, a small percentage of workers may experience discomfort from some substances at concentrations at or below the threshold limit and a smaller percentage may be affected more seriously by aggravation of a preexisting condition or by development of an occupational illness. Smoking may enhance the biological effects of chemicals encountered in the workplace and may reduce the body's defense mechanisms against toxic substances.

Hypersusceptible individuals or those otherwise unusually responsive to some industrial chemicals because of genetic factors, age, personal habits (smoking and use of alcohol or other drugs), medication, or previous exposures may not be adequately protected from adverse health effects of chemicals at concentrations at or below the threshold limits.

These limits are not fine lines between safe and dangerous concentration, nor are they a relative index of toxicity. They should not be used by anyone untrained in the discipline of industrial hygiene.

The copyrighted trademark *Threshold Limit Value* refers to limits published by ACGIH. The TLVs are reviewed and updated annually to reflect the most current information on the effects of each substance assigned a TLV. (See Appendix B and the Bibliography of this chapter.)

The data for establishing TLVs come from animal studies, human studies, and industrial experience, and the limit may be selected for several reasons. As mentioned earlier in this chapter, the TLV can be based on the fact that a substance is very irritating to the majority of people exposed, or the fact that a substance is an asphyxiant. Still other reasons for establishing a TLV for a given substance include the fact that certain chemical compounds are anesthetic or fibrogenic or can cause allergic reactions or malignancies. Some additional TLVs have been established because exposure above a certain airborne concentration is a nuisance.

The amount and nature of the information available for establishing a TLV varies from substance to substance; consequently, the precision of the estimated TLV continues to be subject to revision and debate. The latest documentation for that substance should be consulted to assess the present data available for a given substance.

In addition to the TLVs set for chemical compounds, there are limits for physical agents such as noise, radiofrequency/microwave radiation, segmental vibration, lasers, ionizing radiation, static magnetic fields, light, near-infrared radiation, subradiofrequency ($\leq$ 30 kHz) magnetic fields, subradiofrequency and static electric fields, ultraviolet radiation, cold stress, and heat stress; there are also biological exposure indices (BEIs) (see Chapter 9, Industrial Noise, Chapter 11, Nonionizing Radiation, and Appendix B).

The ACGIH periodically publishes a documentation of TLVs in which it gives the data and information on which the TLV for each substance is based. This documentation can be used to provide health and safety professionals with insight to aid professional judgment when applying the TLVs.

The most current edition of the ACGIH *Threshold Limit Values and Biological Exposure Indices* should be used. When referring to an ACGIH TLV, the year of publication should always preface the value, as in "the 1995 TLV for nitric oxide was 25 ppm." Note that the TLVs are not mandatory federal or state employee exposure standards, and the term *TLV* should not be used for standards published by OSHA or any agency except the ACGIH.

Three categories of Threshold Limit Values are specified as follows:

**Time-weighted average (TLV–TWA).** This is the time-weighted average concentration for a normal 8-hour workday or 40-hour workweek, to which nearly all workers may be repeatedly exposed, day after day, without adverse effect.

**Short-term exposure limit (TLV–STEL).** This is the maximal concentration to which workers can be exposed for a period of up to 15 min continuously without suffering from any of the following:

■ Irritation

■ Chronic or irreversible tissue damage

■ Narcosis of sufficient degree to increase the likelihood of accidental injury, impair self-rescue, or materially reduce work efficiency and provided that the daily TLV–TWA is not exceeded

A STEL is a 15-min TWA exposure that should not be exceeded at any time during a workday, even if the 8-hour TWA is within the TLV. Exposures at the STEL should not be longer than 15 min and should not be repeated more than four times daily. There should be at least 60 min between successive exposures at the STEL.

The TLV–STEL is not a separate, independent exposure limit; it supplements the TWA limit when there are recognized acute effects from a substance that has primarily chronic effects. The STELs are recommended only when toxic effects in humans or animals have been reported from high short-term exposures.

*Note:* None of the limits mentioned here, especially the TWA–STEL, should be used as engineering design criteria.

**Ceiling (TLV–C).** This is the concentration that should not be exceeded during any part of the working exposure. To assess a TLV–C if instantaneous monitoring is not feasible, the conventional industrial hygiene practice is to sample during a 15-min period, except for substances that can cause immediate irritation with exceedingly short exposures.

For some substances (such as irritant gases), only one category, the TLV–C, may be relevant. For other substances, two or three categories may be relevant, depending on their physiological action. If any one of these three TLVs is exceeded, a potential hazard from that substance is presumed to exist.

Limits based on physical irritation should be considered no less binding than those based on physical impairment. Increasing evidence shows that physical irritation can initiate, promote, or accelerate physical impairment via interaction with other chemical or biological agents.

The amount by which threshold limits can be exceeded for short periods without injury to health depends on many factors: the nature of the contaminant; whether very high concentrations, even for a short period, produce acute poisoning;

whether the effects are cumulative; the frequency with which high concentrations occur; and the duration of such periods. All factors must be considered when deciding whether a hazardous condition exists.

## Skin Notation

Nearly 25 percent of the substances in the TLV list are followed by the designation *Skin*. This refers to potential exposure through the cutaneous route, usually by direct contact with the substance. Vehicles such as certain solvents can alter skin absorption. This designation is intended to suggest appropriate measures for the prevention of cutaneous absorption.

## Mixtures

Special consideration should be given in assessing the health hazards that can be associated with exposure to mixtures of two or more substances.

## Federal Occupational Safety and Health Standards

The first compilation of the health and safety standards promulgated by OSHA in 1970 was derived from the then-existing federal standards and national consensus standards. Thus, many of the 1968 TLVs established by the ACGIH became federal standards or permissible exposure limits (PELs). Also, certain workplace quality standards known as ANSI maximal acceptable concentrations were incorporated as federal health standards in 29 *CFR* 1910.1000 (Table Z–2) as national consensus standards.

In adopting the ACGIH TLVs, OSHA also adopted the concept of the TWA for a workday. In general:

$$\text{WA} = \frac{C_a T_a + C_b T_b + \ldots + C_n T}{8}$$

where
$T_a$ = the time of the first exposure period during the shift
$C_a$ = the concentration of contaminant in period $a$
$T_b$ = another time period during the shift
$C_b$ = the concentration during period $b$
$T_n$ = the $n$th or final time period in the shift
$C_n$ = the concentration during period $n$

This simply provides a summation throughout the workday of the product of the concentrations and the time periods for the concentrations encountered in each time interval and averaged over an 8-hour standard workday.

## ■ Evaluation

Evaluation can be defined as the decision-making process resulting in an opinion on the degree of health hazard posed by chemical, physical, biological, or ergonomic stresses in industrial operations. The basic approach to controlling occupational disease consists of evaluating the potential hazard and controlling the specific hazard by suitable industrial hygiene techniques. (See Chapter 15, Evaluation, for more details.)

Evaluation involves judging the magnitude of the chemical, physical, biological, or ergonomic stresses. Determining whether a health hazard exists is based on a combination of observation, interviews, and measurement of the levels of energy or air contaminants arising from the work process as well as an evaluation of the effectiveness of control measures in the workplace. The industrial hygienist then compares environmental measurements with hygienic guides, TLVs, OSHA PELs, NIOSH RELs, or reports in the literature.

Evaluation, in the broad sense, also includes determining the levels of physical and chemical agents arising out of a process to study the related work procedures and to determine the effectiveness of a given piece of equipment used to control the hazards from that process.

Anticipating and recognizing industrial health hazards involve knowledge and understanding of the several types of workplace environmental stresses and the effects of these stresses on the health of the worker. Control involves the reduction of environmental stresses to values that the worker can tolerate without impairment of health or productivity. Measuring and quantitating environmental stress are the essential ingredients for modern industrial hygiene, and are instrumental in conserving the health and well-being of workers.

## Basic Hazard-Recognition Procedures

There is a basic, systematic procedure for recognizing and evaluating environmental health hazards, which includes the following questions:

- What is produced?
- What raw material is used?
- What materials are added in the process?
- What equipment is involved?
- What is the cycle of operations?
- What operational procedures are used?
- Is there a written procedure for the safe handling and storage of materials?
- What about dust control, cleanup after spills, and waste disposal?
- Are the ventilating and exhaust systems adequate?
- Does the facility layout minimize exposure?
- Is the facility well-equipped with safety appliances such as showers, masks, respirators, and emergency eyewash fountains?
- Are safe operating procedures outlined and enforced?
- Is a complete hazard communication program that meets state or federal OSHA requirements in effect?

Understand the industrial process well enough to see where contaminants are released. For each process, perform the following:

- For each contaminant, find the OSHA PEL or other safe exposure guidelines based on the toxicological effect of the material.

- Determine the actual level of exposure to harmful physical agents.

- Determine the number of employees exposed and length of exposure.

- Identify the chemicals and contaminants in the process.

- Determine the level of airborne contaminants using air-sampling techniques.

- Calculate the resulting daily average and peak exposures from the air-sampling results and employee exposure times.

- Compare the calculated exposures with OSHA standards, the TLV listing published by the ACGIH, the NIOSH RELs, the hygienic guides, or other toxicological recommendations.

All of the above are discussed in detail in the following chapters.

## Information Required

Detailed information should be obtained regarding types of hazardous materials used in a facility, the type of job operation, how the workers are exposed, work patterns, levels of air contamination, duration of exposure, control measures used, and other pertinent information. The hazard potential of the material is determined not only by its inherent toxicity, but also by the conditions of use (who uses what, where, and how long?).

To recognize hazardous environmental factors or stresses, a health and safety professional must first know the raw materials used and the nature of the products and by-products manufactured. Consult MSDSs for the substances.

Any person responsible for maintaining a safe, healthful work environment should be thoroughly acquainted with the concentrations of harmful materials or energies that may be encountered in the industrial environment for which they are responsible.

If a facility is going to handle a hazardous material, the health and safety professional must consider all the unexpected events that can occur and determine what precautions are required in case of an accident to prevent or control atmospheric release of a toxic material.

After these considerations have been studied and proper countermeasures installed, operating and maintenance personnel must be taught the proper operation of the health and safety control measures. Only in this way can personnel be made aware of the possible hazards and the need for certain built-in safety features.

The operating and maintenance people should set up a routine procedure (at frequent, stated intervals) for testing the emergency industrial hygiene and safety provisions that are not used in normal, ordinary facility or process operations.

## Degree of Hazard

The degree of hazard from exposure to harmful environmental factors or stresses depends on the following:

- Nature of the material or energy involved

- Intensity of the exposure

- Duration of the exposure

The key elements to be considered when evaluating a health hazard are how much of the material in contact with body cells is required to produce injury, the probability of the material being absorbed by the body to result in an injury, the rate at which the airborne contaminant is generated, the total time of contact, and the control measures in use.

## Air Sampling

The importance of the sampling location, the proper time to sample, and the number of samples to be taken during the course of an investigation of the work environment cannot be overstressed.

Although this procedure might appear to be a routine, mechanical job, actually it is an art requiring detailed knowledge of the sampling equipment and its shortcomings; where and when to sample; and how to weigh the many factors that can influence the sample results, such as ambient temperature, season of the year, unusual problems in work operations, and interference from other contaminants. The sample must usually be taken in the breathing zone of an employee (see Figure 1–5).

The air volume sampled must be sufficient to permit a representative determination of the contaminant to properly compare the result with the TLV or PEL. The sampling period must usually be sufficient to give a direct measure of the average full-shift exposure of the employees concerned. The sample must be sealed and identified if it is to be shipped to a laboratory so that it is possible to identify positively the time and place of sampling and the individual who took the sample.

Area samples, taken by setting the sampling equipment in a fixed position in the work area, are useful as an index of general contamination. However, the actual exposure of the employee at the point of generation of the contaminant can be greater than is indicated by an area sample.

**Figure 1–5.** Portable pump with intake positioned to collect continuous samples from the breathing zone of an employee. (Courtesy MSA)

To meet the requirement of establishing the TWA concentrations, the sampling method and time periods should be chosen to average out fluctuations that commonly occur in a day's work. If there are wide fluctuations in concentration, the long-term samples should be supplemented by samples designed to catch the peaks separately.

If the exposure being measured is from a continuous operation, it is necessary to follow the particular operator through two cycles of operation, or through the full shift if operations follow a random pattern during the day. For operations of this sort, it is particularly important to find out what the workers do when the equipment is down for maintenance or process change. Such periods are often also periods of maximum exposure. (See Chapter 16, Air Sampling.)

As an example of the very small concentrations involved, the industrial hygienist commonly samples and measures substances in the air of the working environment in concentrations ranging from 1 to 100 ppm. Some idea of the magnitude of these concentrations can be appreciated when one realizes that 1 in. in 16 mi is 1 part per million; 1 cent in $10,000, 1 ounce of salt in 62,500 pounds of sugar, 1 ounce of oil in 7,812.5 gallons of water all represent 1 part per million.

# Occupational Skin Diseases

Some general observations on dermatitis are given in this chapter, but more detailed information is given in Chapter 3, The Skin and Occupational Dermatoses. Occupational dermatoses can be caused by organic substances, such as formaldehyde, solvents or inorganic materials, such as acids and alkalis, and chromium and nickel compounds. Skin irritants are usually either liquids or dusts.

## Types

There are two general types of dermatitis: primary irritation and sensitization.

**Primary irritation dermatitis.** Nearly all people suffer primary irritation dermatitis from mechanical agents such as friction, from physical agents such as heat or cold, and from chemical agents such as acids, alkalis, irritant gases, and vapors. Brief contact with a high concentration of a primary irritant or prolonged exposure to a low concentration causes inflammation. Allergy is not a factor in these conditions.

**Sensitization dermatitis.** This type results from an allergic reaction to a given substance. The sensitivity becomes established during the induction period, which may be a few days to a few months. After the sensitivity is established, exposure to even a small amount of the sensitizing material is likely to produce a severe reaction.

Some substances can produce both primary irritation dermatitis and sensitization dermatitis. Among them are organic solvents, chromic acid, and epoxy resin systems.

## Causes

Occupational dermatitis can be caused by chemical, mechanical, physical, and biological agents and plant poisons.

Chemical agents are the predominant causes of dermatitis in manufacturing industries. Cutting oils and similar substances are significant because the oil dermatitis they cause is probably of greater interest to industrial concerns than is any other type of dermatitis.

*Detergents* and solvents remove the natural oils from the skin or react with the oils of the skin to increase susceptibility to reactions from chemicals that ordinarily do not affect the skin. Materials that remove the natural oils include alkalis, soap, and turpentine.

*Dessicators,* hygroscopic agents, and anhydrides take water out of the skin and generate heat. Examples are sulfur dioxide and trioxide, phosphorus pentoxide, strong acids such as sulfuric acid, and strong alkalis such as potash.

*Protein precipitants* tend to coagulate the outer layers of the skin. They include all the heavy metallic salts and those that form alkaline albuminates on combining with the skin, such as mercuric and ferric chloride. Alcohol, tannic acid, formaldehyde, picric acid, phenol, and intense ultraviolet rays are other examples of protein-precipitating agents.

*Oxidizers* unite with hydrogen and liberate nascent oxygen on the skin. Such materials include nitrates, chlorine, iodine, bromine, hypochlorites, ferric chloride, hydrogen peroxide, chromic acid, permanganates, and ozone.

*Solvents* extract essential skin constituents. Examples are ketones, aliphatic and aromatic hydrocarbons, halogenated hydrocarbons, ethers, esters, and certain nitro compounds.

*Allergic or anaphylactic proteins* stimulate the production of antibodies that cause skin reactions in sensitive people. The sources of these antigens are usually cereals, flour, and pollens, but can include feathers, scales, flesh, fur, and other emanations.

Mechanical causes of skin irritation include friction, pressure, and trauma, which may become infected with either bacteria or fungi.

Physical agents leading to occupational dermatitis include heat, cold, sunlight, x rays, ionizing radiation, and electricity. The x rays and other ionizing radiation can cause dermatitis, severe burns, and even cancer. Prolonged exposure to sunlight produces skin changes and may cause skin cancer.

Biological agents causing dermatitis can be bacterial, fungal, or parasitic. Boils and folliculitis caused by staphylococci and streptococci, and general infection from occupational wounds, are probably the best known among the bacterial skin infections. These can be occupationally induced infections.

Fungi cause athlete's foot and other types of dermatitis among kitchen workers, bakers, and fruit handlers; fur, hide, and wool handlers or sorters; barbers; and horticulturists. Parasites cause grain itch and often occur among handlers of grains and straws, and particularly among farmers, laborers, miners, fruit handlers, and horticulturists.

Plant poisons causing dermatitis are produced by several hundred species of plants. The best known are poison ivy, poison oak, and poison sumac. Dermatitis from these three sources can result from bodily contact with any part of the plant, exposure of any part of the body to smoke from the burning plant, or contact with clothing or other objects previously exposed to the plant.

## Physical Examinations

Preplacement examinations help identify those especially susceptible to skin irritations. The examining physician should be given detailed information on the type of work for which the applicant is being considered. If the work involves exposure to skin irritants, the physician should determine whether the prospective employee has deficiencies or characteristics likely to predispose him or her to dermatitis (see Chapter 25, The Occupational Physician, for more details).

## Preventive Measures

Before new or different chemicals are introduced in an established process, possible dermatitis hazards should be carefully considered. Once these hazards are anticipated, suitable engineering controls should be devised and built into the processes to avoid them.

The type, number, and amounts of skin irritants used in various industrial processes affect the degree of control that can be readily obtained, but the primary objective in every case should be to eliminate skin contact as completely as possible. The preventive measures discussed in Chapter 18, Methods of Control, can be adapted to control industrial dermatitis.

# ■ Control Methods

With employment in the United States shifting from manufacturing to the service sector, many workplaces today present nontraditional occupational health hazards. Industrial hygienists need to possess the skills to implement control methodology in both industrial settings and in workplaces such as laboratories, offices, health care facilities, and environmental remediation projects. Hazards can change with time as well, so that hazard control systems require continual review and updating.

Control methods for health hazards in the work environment are divided into three basic categories:

1. Engineering controls that engineer out the hazard, either by initial design specifications or by applying methods of substitution, isolation, enclosure, or ventilation. In the hierarchy of control methods, the use of engineering controls should be considered first.

2. Administrative controls that reduce employee exposures by scheduling reduced work times in contaminant areas (or during cooler times of the day for heat stress exposure, for example). Also included here is employee training that includes hazard recognition and specific work practices that help reduce exposure. (This type of training is required by law for all employees exposed to hazardous materials in the course of their work.)

3. Personal protective equipment the employees wear to protect them from their environment. Personal protective equipment includes anything from gloves to full body suits with self-contained breathing apparatus, and can be used in conjunction with engineering and administrative controls.

Engineering controls should be used as the first line of defense against workplace hazards wherever feasible. Such built-in protection, inherent in the design of a process, is preferable to a method that depends on continual human implementation or intervention. The federal regulations, and their interpretation by the Occupational Safety and Health Review commission, mandate the use of engineering controls to the extent feasible; if they are not sufficient to achieve acceptable limits of exposure, the use of personal protective equipment and other corrective measures may be considered.

Engineering controls include ventilation to minimize dispersion of airborne contaminants, isolation of a hazardous operation or substance by means of barriers or enclosures, and substitution of a material, equipment, or process to provide hazard control. Although administrative control measures can limit the duration of individual exposures, they are not generally favored by employers because they are difficult to implement and maintain. For similar reasons, control of health hazards by using respirators and protective clothing is usually considered secondary to the use of engineering control methods.

## Engineering Controls

Substituting or replacing a toxic material with a harmless one is a very practical method of eliminating an industrial health hazard. In many cases, a solvent with a lower order of toxicity or flammability can be substituted for a more hazardous one. In a solvent substitution, it is always advisable to experiment on a small scale before making the new solvent part of the operation or process.

A change in process often offers an ideal chance to improve working conditions as well as quality and production. In some cases, a process can be modified to reduce the hazard. Brush painting or dipping instead of spray painting minimizes the concentration of airborne contaminants from toxic pigments. Structural bolts in place of riveting, steam-cleaning instead of vapor degreasing of parts, and airless spraying techniques and electrostatic devices to replace hand-spraying are examples of process change. In buying individual machines, the need for accessory ventilation, noise and vibration suppression, and heat control should be considered before the purchase.

Noisy operations can be isolated from the people nearby by a physical barrier (such as an acoustic box to contain noise from a whining blower or a rip saw). Isolation is particularly useful for limited operations requiring relatively few workers or where control by any other method is not feasible.

Enclosing the process or equipment is a desirable method of control because it can minimize escape of the contaminant into the workroom atmosphere. Examples of this type of control are glove box enclosures and abrasive shot blast machines for cleaning castings.

In the chemical industry, isolating hazardous processes in closed systems is a widespread practice. The use of a closed system is one reason why the manufacture of toxic substances can be less hazardous than their use.

Dust hazards often can be minimized or greatly reduced by spraying water at the source of dust dispersion. "Wetting down" is one of the simplest methods for dust control. However, its effectiveness depends on proper wetting of the dust and keeping it moist. To be effective, the addition of a wetting agent to the water and proper and timely disposal of the wetted dust before it dries out and is redispersed may be necessary.

## Ventilation

The major use of exhaust ventilation for contaminant control is to prevent health hazards from airborne materials. OSHA has ventilation standards for abrasive blasting, grinding, polishing and buffing operations, spray finishing operations, and open-surface tanks. For more details, see Chapter 19, Local Exhaust Ventilation of Industrial Occupancies, and Chapter 20, General Ventilation of Industrial Occupancies.

A local exhaust system traps and removes the air contaminant near the generating source, which usually makes this method much more effective than general ventilation. Therefore, local exhaust ventilation should be used when exposures to the contaminant cannot be controlled by substitution, changing the process, isolation, or enclosure. Even though a process has been isolated, it still may require a local exhaust system.

General or dilution ventilation—removing and adding air to dilute the concentration of a contaminant to below hazardous levels—uses natural or forced air movement through open doors, windows, roof ventilators, and chimneys. General exhaust fans can be mounted in roofs, walls, or windows (see Chapters 19 and 20 for more details).

Consideration must be given to providing replacement air, especially during winter. Dilution ventilation is feasible only if the quantity of air contaminant is not excessive, and is particularly effective if the contaminant is released at a substantial distance from the worker's breathing zone. General ventilation should not be used where there is a major, localized source of contamination (especially highly toxic dusts and fumes). A local exhaust system is more effective in such cases.

Air conditioning does not substitute for air cleaning. Air conditioning is mainly concerned with control of air temperature and humidity and can be accomplished by systems that accomplish little or no air cleaning. An air-conditioning system usually uses an air washer to accomplish temperature and humidity control, but these air washers are not designed as efficient air cleaners and should not be used as such. (See Chapter 21, General Ventilation of Nonindustrial Occupancies.)

Processes in which materials are crushed, ground, or transported are potential sources of dust dispersion, and should be controlled either by wet methods or enclosed and ventilated by local exhaust ventilation. Points where conveyors are loaded or discharged, transfer points along the conveying system, and heads or boots of elevators should be enclosed as well as ventilated. (For more details, see Chapter 19, Local Exhaust Ventilation of Industrial Occupancies.)

## Personal Protective Equipment

When it is not feasible to render the working environment completely safe, it may be necessary to protect the worker from that environment by using personal protective equipment. This is considered a secondary control method to engineering and administrative controls and should be used as a last resort.

Where it is not possible to enclose or isolate the process or equipment, ventilation or other control measures should be provided. Where there are short exposures to hazardous concentrations of contaminants and where unavoidable spills may occur, personal protective equipment must be provided and used.

Personal protective devices have one serious drawback: They do nothing to reduce or eliminate the hazard. They interpose a barrier between worker and hazard; if the barrier fails, immediate exposure is the result. The supervisor must be constantly alert to make sure that required protective equipment is worn by workers who need supplementary protection, as may be required by OSHA standards.

## Administrative Controls

When exposure cannot be reduced to permissible levels through engineering controls, as in the case of air contaminants or noise, an effort should be made to limit the employee's exposure through administrative controls.

Examples of some administrative controls are as follows:

■ Arranging work schedules and the related duration of exposures so that employees are minimally exposed to health hazards

■ Transferring employees who have reached their upper permissible limits of exposure to an environment where no further additional exposure will be experienced

Where exposure levels exceed the PEL for one worker in one day, the job can be assigned to two, three, or as many workers as needed to keep each one's duration of exposure within the PEL. In the case of noise, other possibilities may involve intermittent use of noisy equipment.

Administrative controls must be designed only by knowledgeable health and safety professionals, and used cautiously and judiciously. They are not as satisfactory as engineering controls and have been criticized by some as a means of spreading exposures instead of reducing or eliminating the exposure.

Good housekeeping plays a key role in occupational health protection. Basically, it is a key tool for preventing dispersion of dangerous contaminants and for maintaining safe and healthful working conditions. Immediate cleanup of any spills or toxic material, by workers wearing proper protective equipment, is a very important control measure.

Good housekeeping is also essential where solvents are stored, handled, and used. Leaking containers or spigots should be fixed immediately, and spills cleaned promptly. All solvent-soaked rags or absorbents should be placed in airtight metal receptacles and removed daily.

It is impossible to have an effective occupational health program without good maintenance and housekeeping. Workers should be informed about the need for these controls. Proper training and education are vital elements for successful implementation of any control effort, and are required by law as part of a complete federal or state OSHA hazard communication program.

## ■ Sources of Help

Specialized help is available from a number of sources. Every supplier of products or services is likely to have competent professional staff who can provide technical assistance or guidance. Many insurance companies that carry workers' compensation insurance provide industrial hygiene consultation services, just as they provide periodic safety inspections.

Professional consultants and privately owned laboratories are available on a fee basis for concentrated studies of a specific problem or for a facilitywide or companywide survey, which can be undertaken to identify and catalog individual environmental exposures. Lists of certified analytical laboratories and industrial hygiene consultants are available from the AIHA.

Many states have excellent industrial hygiene departments that can provide consultation on a specific problem. Appendix A, Additional Resources, contains names and addresses of state and provincial health and hygiene agencies. NIOSH has a Technical Information Center that can provide information on specific problems. Scientific and technical societies that can help with problems are listed in Appendix A. Some provide consultation services to nonmembers; they all have much accessible technical information. A list of organizations concerned with industrial hygiene is included in Appendix A.

## ■ Summary

No matter what health hazards are encountered, the approach of the industrial hygienist is essentially the same. Using methods relevant to the problem, he or she secures qualitative and quantitative estimates of the extent of hazard. These data are then compared with the recommended exposure guidelines. If a situation hazardous to life or health is shown, recommendations for correction are made. The industrial hygienist's recommendations place particular emphasis on effectiveness of control, cost, and ease of maintenance of the control measures.

Anticipation, recognition, evaluation, and control are the fundamental concepts of providing all workers with a healthy working environment.

# ■ Bibliography

American Conference of Governmental Industrial Hygienists. *Threshold Limit Values and Biological Exposure Indices.* Cincinnati: ACGIH, published annually.

American Conference of Governmental Industrial Hygienists. *Air Sampling Instruments,* 8th ed. Cincinnati: ACGIH, 1995.

American Conference of Governmental Industrial Hygienists and Committee on Industrial Ventilation. *Industrial Ventilation: A Manual of Recommended Practice,* 22nd ed. Lansing, MI: ACGIH, 1995.

American Conference of Governmental Industrial Hygienists. *Documentation of Threshold Limit Values,* 6th ed. Cincinnati: ACGIH, 1991.

American Industrial Hygiene Association. *Biosafety Reference Manual,* 2nd ed. Fairfax, VA: AIHA, 1995.

American Industrial Hygiene Association. *Respiratory Protection: A Manual and Guideline,* 2nd ed. Fairfax, VA: AIHA, 1991.

American Industrial Hygiene Association. *Chemical Protective Clothing, Vol. 1.* Fairfax, VA: AIHA, 1990.

American Industrial Hygiene Association. *Chemical Protective Clothing, Vol. 2: Product and Performance Information.* Fairfax, VA: AIHA, 1990.

American Industrial Hygiene Association. *Industrial Noise Manual,* 4th ed. Fairfax, VA: AIHA, 1986.

American Industrial Hygiene Association. *Engineering Field Reference Manual.* Fairfax, VA: AIHA, 1984.

American Industrial Hygiene Association. *Ergonomics Guide Series.* Fairfax, VA: AIHA, published periodically.

American Industrial Hygiene Association. *Hygienic Guide Series.* Fairfax, VA: AIHA, published periodically.

American Industrial Hygiene Association. *1994–1995 Membership Directory.* Fairfax, VA: AIHA, 1995.

American National Standards Institute, 1430 Broadway, New York, NY 10017.
*Respiratory Protection* Standard Z88.2-1992.
*Fire Department Self-Contained Breathing Apparatus Program,* ANSI/NFPA Standard 1404-1989.
*Physical Qualifications for Respirator Use,* Standard Z88.6-1984.
*Practices for Respiratory Equipment During Fumigation,* Standard Z88.3-1983.

Burgess WA. *Recognition of Health Hazards in Industry: A Review of Materials and Processes,* 2nd ed. New York: Wiley, 1995.

Clayton GD, Clayton FE, Cralley LJ, et al., eds. *Patty's Industrial Hygiene and Toxicology,* 4th ed. Vols. 1A–B, 2A–F, 3A–B. New York: Wiley, 1991–1995.

Cralley LJ, Cralley LV, series eds. *Industrial Hygiene Aspects of Plant Operations:* Vol. 1: *Process Flows;* Mutchler JF, ed. Vol. 2: *Unit Operations and Product Fabrication;* Caplan KJ, ed.

Vol. 3: *Selection, Layout, and Building Design.* New York: Macmillan, 1986.

Gosselin RE, et al. *Clinical Toxicology of Commercial Products: Acute Poisoning,* 5th ed. Baltimore: Williams & Wilkins, 1984.

Grandjean E. *Fitting the Task to the Man: A Textbook of Occupational Ergonomics,* 4th ed. London, New York: Taylor & Francis, 1988.

Key MM et al., eds. *Occupational Diseases—A Guide to Their Recognition,* rev. ed. Washington, DC: U.S. Government Printing Office, 1977.

LaDou J, ed. *Occupational Health & Safety,* 2nd ed. Itasca, IL: National Safety Council, 1993.

Levy BS, Wegman DH. *Occupational Health: Recognizing and Preventing Work-Related Disease,* 3rd ed. Boston: Little, Brown, 1995.

McDermott HJ. *Handbook of Ventilation for Contaminant Control,* 2nd ed. Stoneham, MA: Butterworth, 1985.

Merchant JA, Bochlecke BA, Taylor G, eds. *Occupational Respiratory Diseases.* Cincinnati: NIOSH Publications Dissemination, 1986.

National Institute for Occupational Safety and Health, USDHHS Division of Safety Research. *A Guide to Safety in Confined Spaces.* Morgantown, WV: NIOSH Pub. no. 87–113, 1987.

National Institute for Occupational Safety and Health, USDHHS Division of Safety Research. *Criteria for a Recommended Standard, Occupational Exposure to Hot Environments,* revised criteria, NIOSH Pub. no. 86–113. Cincinnati: NIOSH Publications Dissemination, 1986.

National Institute for Occupational Safety and Health, USDHHS Division of Safety Research. *Work Practices Guide for Manual Lifting,* NIOSH Pub. no. 81–122. Cincinnati: NIOSH Publications Dissemination, March 1981.

National Institute for Occupational Safety and Health, USDHHS Division of Safety Research. *Criteria for a Recommended Standard: Working in Confined Spaces.* NIOSH Pub. no. 80–106. Cincinnati: NIOSH Publications Dissemination, 1979.

National Safety Council. *Accident Prevention Manual for Business & Industry,* 10th ed. Vol. 1: *Administration & Programs* (1992); Vol. 2: *Engineering & Technology* (1992); Vol. 3: *Environmental Management* (1995). Itasca, IL: National Safety Council, 1992, 1995.

National Safety Council. *Protecting Workers' Lives: A Safety and Health Guide for Unions.* Itasca, IL: National Safety Council, 1992.

Procter NH, Hughes JP. *Chemical Hazards of the Workplace,* 3rd ed. Philadelphia: J.B. Lippincott, 1991.

Rekus JF. *Complete Confined Spaces Handbook.* National Safety Council. Boca Raton, FL: Lewis Publishers, 1994.

Schilling RS, ed. *Occupational Health Practice,* 2nd ed. Stoneham, MA: Butterworth, 1981.

Woodson WE. *Human Factors Reference Guide for Process Plants.* New York: McGraw-Hill, 1987.

Zenz C, Dickerson OB, Horvath EP, eds. *Occupational Medicine,* 3rd ed. St. Louis, MO: Mosby-Year Book Medical Publishers, 1994.

P A R T  2

# Anatomy, Physiology, and Pathology

# CHAPTER 2

# The Lungs

by George S. Benjamin, MD, FACS

THE MATERIAL IN THIS CHAPTER on human respiration is intended primarily for engineers and health and safety professionals who must evaluate and control industrial health hazards.

Establishing an effective industrial hygiene program calls for an understanding of the anatomy and physiology of the human respiratory system. The respiratory system is a quick and direct avenue of entry for toxic materials into the body because of its intimate association with the circulatory system and the constant need to oxygenate human tissue cells. Anything affecting the respiratory system, whether it is insufficient oxygen or contaminated air, affects the entire human organism.

Humans can survive for weeks without food and for days without water, but for only a few minutes without air. Air must reach the lungs almost constantly so oxygen can be extracted and distributed via the blood to every body cell. The life-giving component of air is oxygen, which constitutes a little less than one-fifth of its volume.

All living cells of the body are engaged in a series of chemical processes. The total of these processes is metabolism. In the course of the body's metabolism, each cell consumes oxygen and produces carbon dioxide as a waste substance.

Each living cell in the body requires a constant supply of oxygen. Some cells, however, are more vulnerable than others; cells in the brain and heart may die after 4–6 minutes without oxygen. These cells can never be replaced, and permanent changes result from such damage. Other cells in the body are not so critically

**Anatomy**
Nose ▪ Pharynx ▪ Larynx ▪ Trachea ▪ Bronchi ▪ Lungs

**Respiration**
Gas Exchange ▪ Mechanics of Breathing ▪ Pressure Changes ▪ Control of Breathing ▪ Lung Volumes and Capacities

**Hazards**

**Natural Defenses**

**AMA Guides for Evaluating Impairment**
Rating of Impairment ▪ Tests of Pulmonary Function

**Summary**

**Bibliography**

dependent on an oxygen supply because they are replaceable.

Thus, the respiratory system by which oxygen is delivered to the body and carbon dioxide removed is a very important part of the body. The respiratory system consists of all the organs of the body that contribute to normal respiration or breathing. Strictly speaking, it includes the nose, mouth, upper throat, larynx, trachea, and bronchi, which are all air passages or airways. It includes the lungs, where oxygen is passed into the blood and carbon dioxide is given off. Finally, it includes the diaphragm and the muscles of the chest, which permit normal respiratory movements (Figure 2–1).

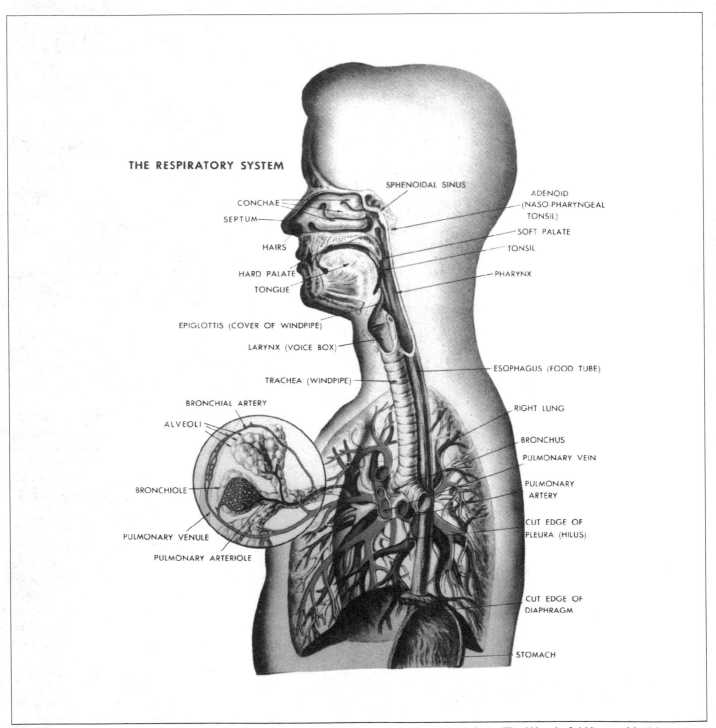

**Figure 2–1.** Schematic drawing of the respiratory system. (Reprinted with permission from *The Wonderful Human Machine.* Chicago: American Medical Association, 1971.)

# ■ Anatomy

## Nose

The nose consists of an external and an internal portion. The external portion of the nose protrudes from the face and is highly variable in shape. The upper part of this triangular structure is held in a fixed position by the supporting nasal bones that form the bridge of the nose. The lower portion is movable because of its pliable framework of fibrous tissue, cartilage, and skin.

The internal portion of the nose lies within the skull between the base of the cranium and the roof of the mouth, and is in front of the nasopharynx (the upper extension of the throat). The skull bones that enter into the formation of the nose include the frontal, the sphenoid, the ethmoid, the nasal, the maxillary, the lacrimal, the vomer, and the palatine and inferior conchae.

The nasal septum is a narrow partition that divides the nose into right and left nasal cavities. In some people the nasal septum is markedly deflected to one side, causing the affected nasal cavity to be almost completely obstructed; this condition is called a deviated nasal septum.

The nasal cavities are open to the outside through the anterior nares (or nostrils); toward the rear, they open into the nasopharynx by means of the posterior nares, or conchae. The vestibule of each cavity is the dilated portion just inside the nostril. Toward the front, the vestibule is lined with skin and presents a ring of coarse hairs that serve to trap dust particles. Toward the rear, the lining of the vestibule changes from skin to a highly vascular ciliated mucous membrane, called the nasal mucosa, which lines the rest of the nasal cavity.

Extending into the nasal cavity from the base of the skull are large nerve filaments, which are part of the sense organ for smell. From these filaments, information on odors is relayed to the olfactory nerve, which goes to the brain.

**Turbinates.** Near the middle of the nasal cavity and on both sides of the septum are a series of scroll-like bones called the conchae, or turbinates. The purpose of the turbinates is to increase the amount of tissue surface within the nose so that incoming air has a greater opportunity to be conditioned before it continues to the lungs.

Respiration begins with the nose, which is specially designed for the purpose, although there are times when you breathe through the mouth as well. When you perform any vigorous activity and begin to puff and pant, you are breathing rapidly through the mouth to provide the blood with the extra oxygen needed.

However, the mouth is not designed for breathing. You may have noticed this on cold days when you make a deliberate effort to keep your mouth tightly closed, because if you take air in through the mouth you can feel its coldness. Cold air passing through the mouth has no chance to become properly warmed. But cold as the air may be, you can breathe comfortably through the nose.

Air enters through the nares or nostrils, passes through a web of nasal hairs, and flows posteriorly toward the nasopharynx. The air is warmed and moistened in its passage and partially depleted of particles. Some particles are removed by impaction on the nasal hairs and at bends in the air path, and others by sedimentation.

In mouth breathing, some particles are deposited, primarily by impaction, in the oral cavity and at the back of the throat. These particles are rapidly passed to the esophagus by swallowing.

**Mucus.** The surfaces of the turbinates, like the rest of the interior walls of the nose, are covered with mucous membranes. These membranes secrete a fluid called mucus. The film of mucus is produced continuously and drains slowly into the throat. The mucus gives up heat and moisture to incoming air and serves as a trap for bacteria and dust in the air. It also helps dilute any irritating substances in the air.

The common cold involves an inflammation of the mucous membrane of the nose. It is characterized by an acute congestion of the mucous membrane and increased secretion of mucus. It is difficult to breathe through the nose because of the swelling of the mucous membrane and the accumulated secretions clogging the air passageway.

In cold weather, the membranes can increase the flow of mucus. If the atmosphere is unusually dry, as in an improperly heated building, the mucus may lose its moisture too rapidly and the membrane may become dry and irritated.

**Cilia.** In addition to the mucus, the membrane is coated with cilia, or hairlike filaments, that move in coordinated waves to propel mucus and trapped particles toward the nostrils. The millions of cilia lining the nasal cavity help the mucus clean the incoming air. When you breathe through the mouth, the protective benefits of the cilia and mucus are lost.

In summary, the nose serves not only as a passageway for air going to and from the lungs but also as an air conditioner and as the sense organ for smell. The importance of breathing through the nose is obvious as it moistens, filters, and warms or cools the air that is on its way to the lungs (Figure 2–2).

## Pharynx

From the nasal cavity, air moves into the pharynx, or throat. Seven tubes enter the pharynx: the two from the nasal cavity, the eustachian tubes (which lead to the ears), the mouth cavity, the opening of the esophagus, and the opening of the windpipe.

The pharynx, or throat, is a tubular passageway attached to the base of the skull and extending downward behind the nasal cavity, the mouth, and the larynx to

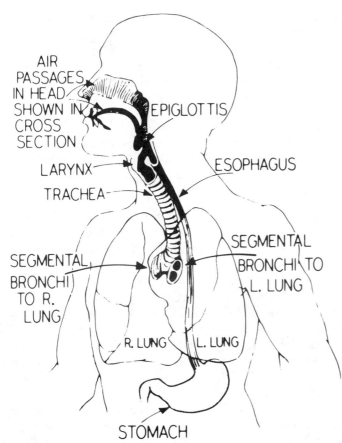

**Figure 2–2.** Parts of the human respiratory system. Air enters through the mouth and nose, passes down the trachea and into the lungs. (Reproduced with permission from *Emergency Care and Transportation of the Sick and Injured,* ed. 1. Chicago, IL, American Academy of Orthopaedic Surgeons, 1971.)

continue as the esophagus (the food tube). Its walls are composed of skeletal muscle and the lining consists of mucous membrane. The nasal passage joins the food canal just behind the mouth. The union of the two passageways at this point makes it possible to breathe with reasonable comfort through the mouth when the nasal passages are blocked because of a cold or allergy.

The nasopharynx is the superior portion of the pharyngeal cavity; it lies behind the nasal cavities and above the level of the soft palate. This portion of the upper respiratory tract serves as a major defense against infectious organisms. Its ciliated mucosal lining is continuous with that of the nasal cavities. Immediately beneath the mucosa are collections of lymphoid tissue, the adenoids. The adenoids and tonsils, lower down in the throat, are part of the immune system and serve as a first defense against infectious organisms.

At the bottom of the throat are two passageways: the esophagus behind and the trachea in front. Food and liquids entering the pharynx pass into the esophagus, which carries them to the stomach. Air and other gases enter the trachea to go to the lungs.

**Epiglottis.** Guarding the opening of the trachea is a thin, leaf-shaped structure called the epiglottis (Figure 2–3). This structure helps food glide from the mouth to the esophagus.

Everyone is aware that swallowing food and breathing cannot take place at the same time without danger of choking. But nature has devised a way for food and air to use the same general opening, the pharynx, with only an occasional mix-up.

The incoming air travels through the nasal cavity and through the larynx by crossing over the path used by food on its way to the stomach. Similarly, food crosses over the route of air. When food is swallowed, the larynx rises against the base of the tongue to help seal the opening.

If food accidentally starts down the wrong way, into the lungs rather than the stomach, there are explosive protests from the lungs. Any contact of a sizable liquid or solid particle with the trachea sets off a cough, an explosive expulsion of air that blows it out again. A cough can be a very powerful force. A slight breathing in, closing of the glottis, buildup of pressure, and a sudden release of the trapped air are involved. Also, stimulation of the larynx can cause spasm of the vocal cords, with total obstruction of breathing.

Normally, swallowing blocks off the glottis, halts breathing briefly, and ensures correct division of air and food. However, an unconscious person may lack this automatic response, and if a drink is given, it may proceed straight into the lungs.

The diaphragm is sometimes subject to periodic spasms of contraction that enlarge the lung cavities and lead to a quick inrush of air. The vocal cords come together to stop the flow, and the air, so suddenly set into motion and so suddenly stopped, makes the sharp noise called the hiccup.

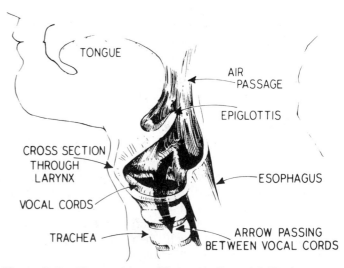

**Figure 2–3.** The anatomy of the neck: the epiglottis, larynx, vocal cords, trachea, and esophagus. (Reproduced with permission from *Emergency Care and Transportation of the Sick and Injured,* ed. 1. Chicago, IL, American Academy of Orthopaedic Surgeons, 1971.)

Hiccups may be due to indigestion, overloaded stomach, irritation under the surface of the diaphragm, too much alcohol, or many other possible causes, including heart attacks.

## Larynx

The larynx, or voice box, serves as a passageway for air between the pharynx and the trachea. It lies in the midline of the neck, below the hyoid bone and in front of the laryngopharynx. The unique structure of the larynx enables it to function somewhat like a valve on guard duty at the entrance to the windpipe, controlling air flow and preventing anything but air from entering the lower air passages. Exhalation of air through the larynx is controlled by voluntary muscles; thus, the larynx is the organ of voice.

The larynx is a triangular box composed of nine cartilages joined together by ligaments and controlled by skeletal muscles. The larynx is lined with ciliated mucous membrane (except the vocal folds), and the cilia move particles upward to the pharynx.

**Vocal cords.** The larynx, or voice box, is at the top of the windpipe, or trachea, which takes air to the lungs. Although incoming air passes through the boxlike larynx, it is actually air expelled from the lungs that makes voice sounds. In the front of the larynx, two folds of membranes, the vocal cords, are attached and held by tiny cartilages. Muscles attached to the cartilages move the vocal cords, which are made to vibrate by air exhaled from the lungs.

During ordinary breathing, the vocal cords are held toward the walls of the larynx so that air can pass without being obstructed. During speech, the vocal cords swing over the center of the tube and muscles contract to tense the vocal cords.

**Speech.** Sounds are created as air is forced past the vocal cords, making them vibrate. These vibrations make the sound. You can feel these vibrations by placing your fingers lightly on your larynx (Adam's apple) while speaking.

The vibrations are carried through the air upward into the pharynx, mouth, nasal cavities, and sinuses, which act as resonating chambers. The greater the force and amount of air from the lungs, the louder the voice. Pitch differences result from variations in the tension of the cords. The larger the larynx and the longer the cords, the deeper the voice. The average man's vocal cords are about 0.75 in. (1.9 cm) long. Shorter vocal cords give women higher-pitched voices. Words and other understandable sounds are formed by the tongue and muscles of the mouth.

Infections of the throat and nasal passages alter the shape of the resonating chambers and change the voice, roughening it so that it sounds hoarse. When the membranes of the larynx themselves are affected (laryngitis), speech may be reduced to a whisper. In whispering, the vocal cords are not involved; sound is produced by tissue folds, sometimes called false vocal cords, that lie just above the vocal cords themselves.

## Trachea

The windpipe, or trachea, is a tube about 4.5 in. (11.5 cm) long and 1 in. (2.5 cm) in diameter, extending from the bottom of the larynx through the neck and into the chest cavity. At its lower end it divides into two tubes, the right and left bronchi. The esophagus, which carries food to the stomach, is immediately behind the trachea.

Rings of cartilage hold the trachea and bronchi open. If the head is tilted back, the tube can be felt as the fingers run down the front of the neck. The ridges produced by the alternation of cartilage and fibrous tissue are also felt, giving the tube a feeling of roughness.

The windpipe wall is lined with mucous membrane and with many hairlike cilia fanning upward toward the throat, moving dust particles that have been caught in the sticky membrane away from the lungs.

The path of the esophagus, which carries food to the stomach, runs immediately behind the trachea. At the point behind the middle of the breastbone, where the aorta arches away from the heart, the trachea divides into two branches: the right and the left bronchi.

Respiratory infections such as colds and sore throats may sometimes extend down into the trachea; they are then called tracheitis. Inflammation of the walls of these passages causes harsh breathing and deep cough.

## Bronchi

The trachea divides into the right and left main stem bronchi under the sternum (breastbone), approximately where the second and third ribs connect to the sternum. Each bronchus enters the lung of its own side through the hilus (an opening through which vessels or nerves enter or leave an organ).

The right main stem bronchus is wider and shorter than the left. Its direction is almost identical to that of the trachea. That is why most aspirated material enters the right lung.

Each bronchus leads to a separate lung, and in doing so divides and subdivides into increasingly smaller, finer, and more numerous tubes, something like the branches of a tree; the whole structure is sometimes called the bronchial tree. In the larger branches there also is stiffening by rings of cartilage, but as the branches get smaller the cartilage diminishes to small plates and finally disappears.

The smaller branches of the bronchial tree, bronchioles, are another possible source of discomfort. The fine subdivisions of the air passages are lined by circular muscles, which through contraction or relaxation can alter the diameter, thus helping to control the flow of air through the lungs. Sometimes, as a result of infection or an allergic reaction to some foreign substance, there is a spasmodic contraction of the small muscles and a swelling of the mucous membrane of the bronchioles. The air passages narrow and airflow is reduced.

## Lungs

There are two lungs, one on each side of the thoracic cage (Figure 2–4). The lungs are suspended within the thoracic cage by the trachea, by the arteries and veins running to and from the heart, and by pulmonary ligaments.

The lungs extend from the collarbone to the diaphragm, one on the right side of the body and one on the left. Taken together, they fill almost all of the thoracic cavity. The two lungs are not quite mirror images of each other. The right lung, slightly the larger of the two, is partially divided into three lobes; the left lung is divided into only two.

The mediastinum is the compartment between the left and right lung. It contains the heart, great vessels (aorta, vena cava, pulmonary veins, and arteries), nerves, trachea, main stem bronchi, and esophagus.

**Pleura.** The lungs are covered by a double membrane. One, the pleural membrane, lies over the lungs; the other lines the chest cavity (Figure 2–5). They are separated by a thin layer of fluid that, during breathing, prevents the two membranes from rubbing against each other. Inflammation of the pleura can cause roughness and irritation, the condition called pleurisy.

The potential intrapleural space (between the two pleural layers) has a negative atmospheric pressure. An introduction of air between the pleural layers (pneumothorax) would decrease or disrupt this negative pressure and the lung would partially or totally collapse.

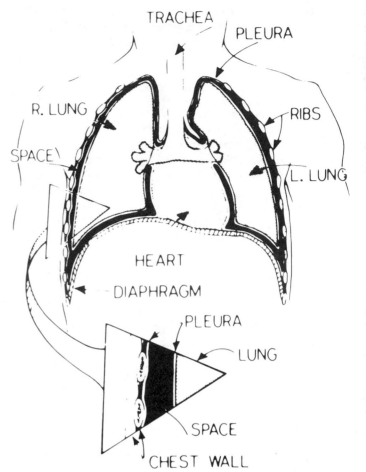

**Figure 2–5.** The lungs, pleura, and pleural space. Inset shows the chest wall relationships. (Reproduced with permission from *Emergency Care and Transportation of the Sick and Injured,* ed. 1, Chicago, IL, American Academy of Orthopaedic Surgeons, 1971.)

The tendency to collapse is counteracted by a pull in the opposite direction. The lung surface is held tenaciously to the chest wall not by physical bonds, but by the negative pressure of the intrapleural space. Normally, this negative pressure acts somewhat like a suction cup to pull the lung against the chest wall and keep it expanded.

**Alveoli.** Within a lung, the bronchi divide and subdivide, becoming smaller and smaller, until the branches reach a very fine size and are called bronchioles. The respiratory bronchioles lead into several ducts; each duct ends in a cluster of air sacs, which resemble a tiny bunch of grapes called alveoli (Figure 2–6).

The walls of the alveoli are two cells thick and oxygen can pass freely across those thin membranes. It can pass freely in both directions, of course, but the blood coming to the lungs has a lower partial pressure of oxygen than inspired air, so the net exchange is from the lungs to the bloodstream.

The human respiratory tract branches successively from the trachea to 25–100 million branches. These

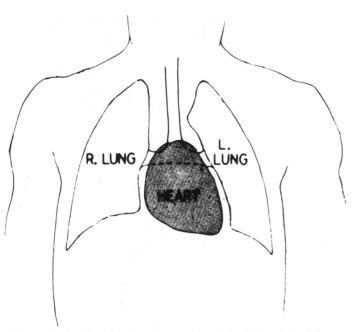

**Figure 2–4.** The relative size and spatial relationship of the human heart and lungs. (Reproduced with permission from *Emergency Care and Transportation of the Sick and Injured,* ed. 1. Chicago, IL, American Academy of Orthopaedic Surgeons, 1971.)

The respiratory tract, with its successive branches and tortuous passageways, is a highly efficient dust collector. Essentially all particles entering the respiratory system larger than 4 or 5 micrometers (μm) are deposited in it. About half of those of 1-μm size appear to be deposited and the other half exhaled. The sites of deposition in the system are different for various sizes. Discussion of dust deposition in the respiratory system is simplified by the concept of equivalent size of particles. The equivalent size of a particle is the diameter of a unit density sphere, which has the same terminal settling velocity in still air as does the particle.

Particles greater than 2.5 or 3 μm equivalent size are deposited, for the most part, in the upper respiratory system—that is, the nasal cavity, the trachea, the bronchial tubes, and other air passages—whereas particles 2 μm in equivalent size are deposited about equally in the upper respiratory system and in the alveolar or pulmonary air spaces. Particles about 1 μm in equivalent size are deposited more efficiently in the alveolar spaces than elsewhere; essentially none are collected in the upper respiratory system. For more details, see Chapter 8, Particulates.

## Respiration

The process through which the body combines oxygen with food substances, and thus produces energy, is called metabolism (Figure 2–7). The term *respiration* refers to the tissue enzyme oxidation processes that use oxygen and produce carbon dioxide. More generally, this term designates the phases of oxygen supply and carbon dioxide removal. The following are the general subdivisions of the overall process:

- Breathing—movement of chest/lung complex to ventilate the alveoli
- External respiration—exchange of gas (oxygen and carbon dioxide) between lung (alveolar) air and blood
- Internal respiration—exchange of gas between tissue blood and the tissue cells
- Intracellular respiration—ultimate utilization of oxygen by the cells with the coincident release of carbon dioxide.

To a biochemist, *respiration* refers to the enzymatic processes in the tissues that use oxygen and produce carbon dioxide. The blood contains a chemical that is part protein and part iron pigment, called hemoglobin. The hemoglobin binds oxygen when the blood flows through regions where oxygen is plentiful—as in the alveoli—and releases it to tissues that are consuming oxygen. Similarly, the carbon dioxide produced when the body cells burn fuel is dissolved in the bloodstream as it flows through tissues where carbon dioxide is plentiful, and is released in the lungs, where carbon dioxide is comparatively scarce.

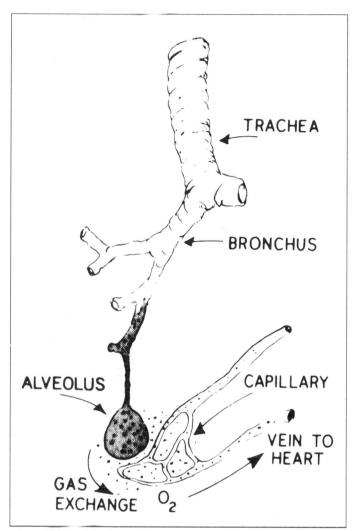

**Figure 2–6.** The branching characteristic of the trachea into smaller airways ending in an alveolus is shown. (Reproduced with permission from *Emergency Care and Transportation of the Sick and Injured,* ed. 1, Chicago, IL, American Academy of Orthopaedic Surgeons, 1971.)

branches terminate in some 300 million air sacs, or alveoli. The cross section of the trachea is about 31 sq in. (2 cm²) and the combined cross sections of the alveolar ducts, which handle about the same quantity of air, are about 8 ft² (8,000 cm²).

The respiratory surface in the lungs ranges from about 300 ft² (28 m²) at rest to about 1,000 ft² (93 m²) at deepest inspiration. The membrane separating the alveolar air space from circulating blood may be only one or two cells thick. In the course of an eight-hour day of moderate work, a human breathes about 300 ft³ (8.5 m³) of air. Contrast the forced ventilation exposure of the large delicate lung surface with the ambient air exposure of the skin, which has some 20 ft² (1.9 m²) of surface and a thickness measured in millimeters. It is evident that the lungs represent by far the most extensive and intimate contact of the body with the ambient atmosphere.

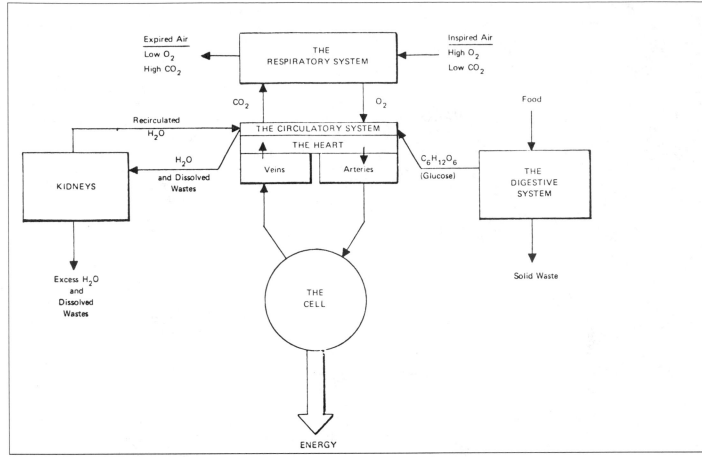

**Figure 2–7.** The conversion of food into energy (the metabolic process) is illustrated. (Reprinted from *A Guide to Industrial Respiratory Protection*, NIOSH Publication No. 76–189, 1976.)

Carbon dioxide is always present in the atmosphere, but the proportion of carbon dioxide in air exhaled from the lungs is 100 times greater. The proportion of water vapor in air exhaled from the lungs is about 10 times greater than that of the normal atmosphere. Everyone has no doubt noticed the moisture that accumulates on a glass window when the nose and mouth are close to it. Breath appears as a white cloud on cold days because the low temperature of the air causes the exhaled water vapor to condense.

## Gas Exchange

Gases diffuse rapidly from areas of higher to lower concentrations. The concentration of oxygen is higher in alveolar air than it is in the blood coming to the lungs from the right ventricle; therefore, oxygen diffuses into the blood from the alveolar air. On the other hand, the concentration of oxygen is low in the cells of the body tissues and in tissue fluid; therefore, oxygen diffuses from the blood in the capillaries into the tissue fluid and into cells.

If there is a pressure difference across a permeable membrane such as that separating the alveoli from the pulmonary capillaries, gas molecules pass from the high- to the low-pressure region until the pressures are equalized (Figure 2–8).

The concentration of carbon dioxide in the tissue cells and the tissue fluid is higher than in the blood in the capillaries. Therefore, carbon dioxide diffuses from tissue cells and tissue fluid into the blood. The concentration of carbon dioxide is higher in blood coming to the lungs from the right ventricle than it is in alveolar air; therefore, it diffuses from blood in pulmonary capillaries into the alveolar air.

On entering the bloodstream, both oxygen and carbon dioxide immediately go into simple physical solution in the plasma. However, because the plasma can hold only a small amount of gas in solution, most of the oxygen and carbon dioxide quickly enter into chemical combinations with other blood constituents.

**Oxygen tension.** Only a small amount of oxygen is carried in solution in the plasma. However, it is this oxygen that exerts tension or pressure and is available for immediate diffusion when blood reaches the systemic capillaries (Figure 2–9). The remaining oxygen in the blood is combined with hemoglobin in the red blood cells to form oxyhemoglobin ($HHbO_2$). This oxygen is given up readily by the hemoglobin whenever the oxygen tension of the plasma decreases, so that as oxygen diffuses from

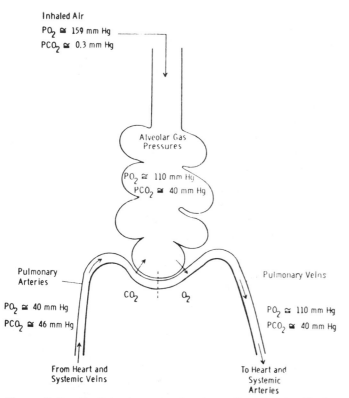

**Figure 2–8.** Partial pressures of various gases involved in the gas exchange in the lungs are shown. (Reprinted from *A Guide to Industrial Respiratory Protection,* NIOSH Publication No. 76–189, 1976.)

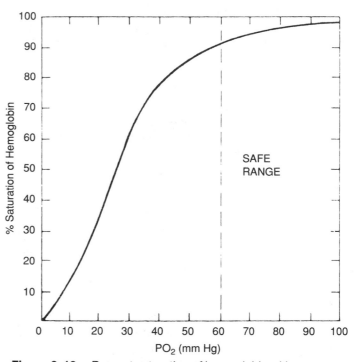

**Figure 2–10.** Percent saturation of hemoglobin with oxygen at various partial pressures is shown in the hemoglobin saturation curve. (Reprinted from *A Guide to Industrial Respiratory Protection,* NIOSH Publication No. 76–189, 1976.)

the plasma in tissue capillaries it is replenished by more from the oxyhemoglobin. Hemoglobin that has given up its load of oxygen is called *reduced* hemoglobin (HHb) (Figure 2–10). See Chapter 6, Industrial Toxicology, for more details.

In most people during routine activities, the depth and rate of breathing movements are regulated for the maintenance of carbon dioxide in the arterial blood. Oxygen want can be regulating, but only when the oxygen content

of the inspired gases is reduced to nearly half that in air at sea level. Oxygen partial pressure, except in some unusual circumstances, should always be high enough that the breathing is regulated by the body requirements for carbon dioxide.

The oxygen content of lung air is determined by the oxygen content of the inspired gases, the flushing of the lungs required for carbon dioxide regulation, and the rate of oxygen uptake by the blood as it passes through the lungs.

## Mechanics of Breathing

Breathing is the act of taking fresh air into and expelling stale air from the lungs. Breathing is accomplished by changes in the size of the chest cavity. Twelve pairs of ribs surround and guard the lungs. They are joined to the spine at the back and curve around the chest to form a cage. In front, the top seven pairs are connected to the breastbone. The next three pairs are connected to the rib above. The last two pairs, unconnected in front, are called floating ribs. The entire cage is flexible and can be expanded readily by special muscles. The rib cage forms the wall of the chest; the dome-shaped diaphragm forms the floor of the chest cavity. The diaphragm is attached to the breastbone in front, the spinal column in back, and the lower ribs on the sides.

## Pressure Changes

The basic principle underlying the movement of any gas is that it travels from an area of higher pressure to an area of

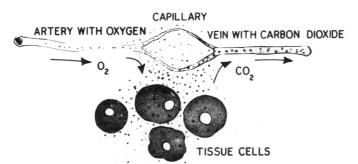

**Figure 2–9.** Exchange of oxygen and carbon dioxide between blood vessels, capillaries, and tissue cells. Oxygen passes from the blood to the capillaries to the tissue cells. Carbon dioxide passes from the tissue cells to the capillaries and into the blood. (Reproduced with permission from *Emergency Care and Transportation of the Sick and Injured,* ed. 2. Chicago, IL, American Academy of Orthopaedic Surgeons, 1977.)

lower pressure, or from a point of greater concentration of molecules to a point of lower concentration. This principle applies not only to the flow of air into and out of the lungs but also to the diffusion of oxygen and carbon dioxide through alveolar and capillary membranes. The respiratory muscles and the elasticity of the lungs make the necessary changes in the pressure gradient possible, so that air first flows into the air passages and then is expelled.

Atmospheric pressure is the pressure exerted against all parts of the body by the surrounding air. It averages 760 mm of mercury (760 mmHg) at sea level. Any pressure that falls below atmospheric pressure is called a negative pressure and represents a partial vacuum.

Intrapulmonic pressure is the pressure of air within the bronchial tree and the alveoli. During each respiratory cycle this pressure fluctuates below and above atmospheric pressure as air moves into and out of the lungs. Intrapulmonic pressure is below atmospheric pressure during inspiration, equal to atmospheric pressure at the end of inspiration, above atmospheric pressure during expiration, and again equal to atmospheric pressure at the end of expiration.

This series of changes in intrapulmonic pressure is repeated with each respiratory cycle. Whenever the size of the thoracic cavity remains constant for a few seconds, or in a position of rest, the intrapulmonic pressure is equal to atmospheric pressure.

Lungs have one way of filling themselves. Movement of the thoracic cage and the diaphragm permits air to enter the lungs. The thoracic cage is a semirigid bony case enclosed by muscle and skin. The diaphragm is a muscular partition separating the chest and abdominal cavities.

The chest cage can be compared to a bellows. The ribs maintain the shape of the chest bellows. The opening of the chest bellows is through the trachea. Air moves through the trachea to and from the lungs to fill and empty the air sacs (Figure 2–11). When a bellows is opened, the volume it can hold increases, causing a slight vacuum. This lowers the air pressure inside the bellows and causes the higher pressure outside the bellows to drive air through the opening, thereby filling the bellows.

When the air pressure inside equals the pressure outside, air stops moving into the bellows. Air will move from a high-pressure area to a low-pressure area until the pressure in both areas is equal. Therefore, as the bellows is closed, the pressure inside becomes higher than outside and air is expelled (Figure 2–11).

During inspiration (inhaling), the diaphragm and rib muscles contract. When the diaphragm contracts, it moves downward and enlarges the thoracic cavity from top to bottom. When the rib muscles contract, they raise the ribs. This enlarges the chest cavity (bellows) in all dimensions. This enlargement of the thoracic cavity reduces the pressure within the chest. The action is identical to that of

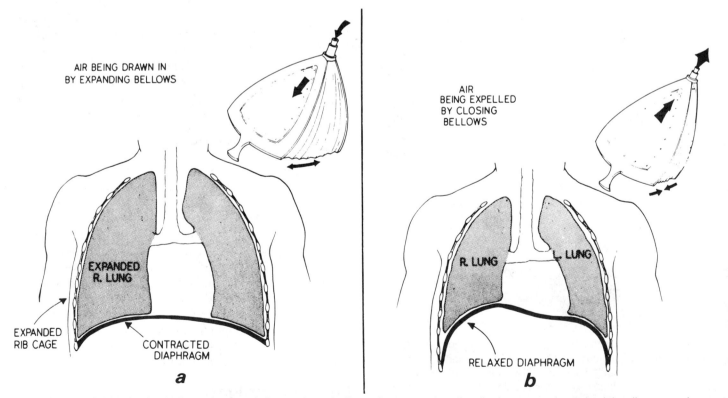

**Figure 2–11.** Inhalation is similar to the act of air entering a bellows. It occurs when the diaphragm contracts and the ribs expand. Exhalation is similar to the act of air leaving a bellows. It occurs when the diaphragm and ribs relax. (Reproduced with permission from *Emergency Care and Transportation of the Sick and Injured,* ed. 1. Chicago, IL, American Academy of Orthopaedic Surgeons, 1971.)

opening a bellows. Air rushes into the lungs. Take a deep breath to see how the chest increases in size. This is the active muscular part of breathing.

During expiration (exhaling), the diaphragm and the rib muscles relax. As these muscles relax, the chest cavity decreases in size in all dimensions. As it does so, the air in the lungs is pressed into a smaller space, the pressure increases, and air is pushed out through the trachea. Decrease in size of the chest cavity after relaxation is accomplished largely by action of elastic tissue in the lung, which stretches for inhalation and recoils after muscular relaxation.

## Control of Breathing

Breathing is controlled by a series of respiratory centers in the nervous system. One center is the medulla, the part of the brain at the top of the spinal cord (Figure 2–12).

**Respiratory center.** Nervous impulses originating in the motor areas of the cerebral cortex and traveling to the respiratory center enable us to consciously alter the rate and the depth of breathing. For example, during speaking or singing, breath control is very important.

You can hold your breath voluntarily for a short period of time. However, voluntary control is limited, and the respiratory center will ignore messages from the cortex when breathing is necessary to meet the body's basic needs.

**Carbon dioxide.** Breathing action can be triggered by the respiratory centers when the amount of carbon dioxide in the blood increases or when the oxygen level of the blood decreases.

If you hold your breath, carbon dioxide accumulates in the blood until, finally, it so strongly stimulates the respiratory control center of the brain that you are forced to breathe again. The length of time the breath can be held varies from 25 to 75 seconds; some people can hold their breath even longer.

**Rate.** Even in a relaxed state, you breathe in and out 10–14 times a minute, with each breath lasting 4–6 seconds. In a minute, 9–12 pt (4.3–5.7 l) of air are taken in. The fact is that the body has small reserves of oxygen; all of it is consumed within less than half a minute after the start of vigorous exertion. With such exertion, the need for air increases many times so that the breathing rate may speed up to one breath per second and a total intake of 31 gal (120 l) of air per minute.

In a normal day, you breathe some 3,300 gal (12,491 l) of air—enough to occupy a space of about 441 ft$^3$; in a lifetime, you will consume enough to occupy 13 million ft$^3$ (368,120 m$^3$) of space.

## Lung Volumes and Capacities

For descriptive convenience the total capacity of the lung at full inspiration is divided into several functional subdivisions. These are illustrated in Figure 2–13.

The four primary lung volumes that do not overlap are as follows:

- Tidal volume (TV)—the volume of gas inspired or expired during each respiratory cycle

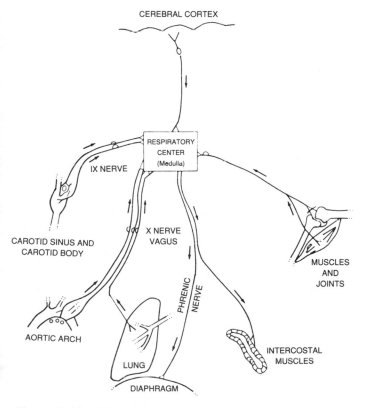

**Figure 2–12.** Normal rhythmic breathing is controlled by the requirement to ventilate the lungs to remove carbon dioxide as fast as it is produced by metabolic activity. The factors effective in controlling breathing are illustrated schematically. (Reprinted from Parker JF. *Bioastronautics,* 2nd ed. Washington, DC: National Aeronautics and Space Administration, 1973.)

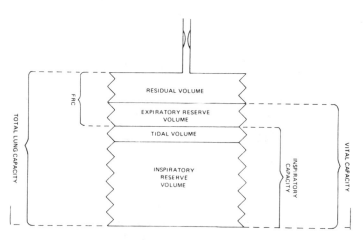

**Figure 2–13.** Inspiratory capacity and tidal capacity. (Reprinted from Parker JF. *Bioastronautics,* 2nd ed. Washington, DC: National Aeronautics and Space Administration, 1973.)

- Inspiratory reserve volume (IRV)—the maximal volume that can be forcibly inspired following a normal inspiration (from the end-inspiratory position)

- Expiratory reserve volume (ERV)—the maximum amount of air that can be forcibly expired following a normal expiration

- Residual volume (RV)—the amount of air remaining in the lungs following a maximum expiratory effort.

Each of the four following capacities includes two or more of the primary volumes:

- Total lung capacity (TLC)—the sum of all four of the primary lung volumes

- Inspiratory capacity (IC)—the maximum volume by which the lung can be increased by a maximum inspiratory effort from midposition

- Vital capacity (VC)—the maximum amount of air that can be exhaled from the lungs after a maximum inspiration (the sum of the inspiratory reserve volume, tidal volume, and expiratory reserve volume)

- Functional residual capacity (FRC)—the normal volume at the end of passive exhalation; that is, the gas volume that normally remains in the lung and functions as the residual capacity.

In an ordinary inhalation the first air to enter the lungs is the air that was in the bronchi, throat, and nose—air that had left the lungs in the previous expiration but had not been pushed out as far as the outside world. Then, after an inspiration is complete, some of the fresh air that entered through the nostrils remains in the air passages; here it is useless, and is expired again before it can get to the lungs. The dead space represents the air passages between the nostrils and lungs. Fresh air actually entering the lungs with each breath may amount to no more than 21.4 in.$^3$ (350 cm$^3$). This represents only 1/18 of the lungs' total capacity and is called the tidal volume.

The partial replacement of the air in the lungs (alveolar air) by the shallow breathing we normally engage in is sufficient for ordinary purposes. We are quite capable of taking a deep drag of air as well, forcing far more into the lungs than would ordinarily enter. After about 30.5 in.$^3$ (500 cm$^3$) of air have been inhaled in a normal quiet breath, an additional 153 in.$^3$ (2,500 cm$^3$) can be sucked in. On the other hand, you can force 42.7 in.$^3$ (7,000 cm$^3$) of additional air out of the lungs after an ordinary quiet expiration is completed. By forcing all possible air out of the lungs and then drawing in the deepest possible breath, you can bring well over 1.7 ft$^3$ (4,000 cm$^3$) of new air into the lungs in one breath. This is the vital capacity.

Even with the utmost straining, the lungs cannot be completely emptied of air. After the last bubble has been forced out, about 73 in.$^3$ (1,200 cm$^3$) remain. This is the residual volume and is a measure of the necessary inefficiency of the lungs.

Vital lung capacity is measured by inhaling as deeply as possible and blowing as much as possible into a spirometer. The quantity of expelled air varies with body size and age. A medium-sized man may have a vital capacity of 1.7–1.9 ft$^3$ (4,000–4,500 cm$^3$) between the ages of 20 and 40 years. However, as the elasticity of tissues decreases with age, the vital capacity diminishes and may be as much as 20 percent less at age 60 and 40 percent less at age 75.

*Spirometry* means measurement of air—the ventilatory capacity of the lungs. The spirometer achieves this by measuring volumes of air and relating them to time.

Change in the ability to move air into and out of the lungs in a normal manner results in what is called either obstructive or restrictive ventilatory defect, or a combination of the two. In obstructive bronchopulmonary disease, such as reduced forced expiratory flow (FEF) and forced expiratory volume in 1 s (FEV$_1$), there is reduction of airflow rates and prolongation of expiration.

Forced vital capacity (FVC) is the maximal volume of air that can be exhaled forcefully after a maximal inspiration. For all practical purposes, the VC without forced effort and the FVC are identical in most people.

In the early detection of pneumoconiosis, the FVC is of variable use. In asbestosis, the FVC is regarded as the most sensitive indicator of early disease and is often impaired before there are radiographic abnormalities. Conversely, x-ray changes may be evident in the silicotic while the FVC is still normal.

Forced expiratory volume in 1 s (FEV$_1$) is the volume of air that can be forcibly expelled during the first second of expiration.

Forced expiratory flow during the middle half of the FVC (FEF$_{25-75\%}$) can be defined as the average rate of flow during the middle two quarters of the forced expiratory effort. Compared with FEV$_1$, it may be more sensitive in detecting early airway obstruction and tends to reflect changes in airways less than 2 mm in diameter. Airborne substances are thought to exert their initial deleterious effects in these smaller bronchi and bronchioles. Smoking one cigarette can lower the FEF for several hours.

Peak expiratory flow (PEF) is another measure of expiratory flow. PEF is the peak of the expiration flow volume curve. The peak expiratory flow rate is the rate of maximal expiratory flow. The development of portable peak flow meters that directly measure flow rates has sparked interest in the PEF as a useful parameter for epidemiological studies in the workplace. It has the advantages of sensitivity to early obstructive changes and ease of measurement. A significant drawback is inability to derive acceptable standards due to wide variation in healthy individuals.

# ■ Hazards

Let's look at some of the unhealthy conditions to which the lung is subject, the associated terminology, and some typical hazardous substances.

The membrane lining of the nasal passages can be affected by a number of causes. The resultant condition is called rhinitis. Inflammation in the larynx is called laryngitis; that of the bronchial tubes is called bronchitis. Constriction of the tube muscles in response to irritation, allergy, or other stimulus is called asthma.

In the lung sacs, a number of conditions can develop:

- *Atelectasis* means incomplete expansion. It refers to imperfect expansion of the lungs at birth or partial collapse of the lungs after birth. The latter is caused by occlusion of a bronchus, perhaps by a plug of heavy mucus, with subsequent absorption of the air, or by external compression, as from a pleural effusion or a tumor. The atelectatic portion of the lung will pass blood through without adding oxygen or removing carbon dioxide.

- The term *emphysema* derives from Greek words meaning *overinflated*. The overinflated structures are microscopic air sacs of the lungs (alveoli). Tiny bronchioles through which air flows to and from the air sacs have muscle fibers in their walls. These structures can become hypertrophied and lose elasticity. Then air flows into the air sacs easily but cannot flow out easily because of the narrowed diameter of the bronchioles. The patient can breathe in, but cannot breathe out efficiently; this leaves too much stale air in the lungs. As pressure builds up in the air cells, their thin walls are stretched to the point of rupture, so several air spaces communicate and the area of surfaces where gas exchange takes place is decreased.

- Pleurisy is caused when the outer lung lining (the visceral pleura) and the chest cavity's inner lining (the parietal pleura) lose their lubricating properties. The resultant friction causes irritation and pain. The thin, glistening layer of pleura that is inseparably bound to the lung has no pain fibers, but the opposing pleura is richly supplied. Normally the pleural layers glide over each other on a thin film of lubricating fluid. Disease may cause the pleura to become inflamed and adherent, or fluids may accumulate in the interspace, separating the layers.

- Pneumonitis is any inflammation of the lung. It is essentially equivalent to the term *pneumonia,* which is reserved for certain types of pneumonitis, usually infectious.

- Bronchitis is inflammation of the lining of the bronchial tubes.

- Pneumoconiosis (dusty lung) is a general word for various pulmonary manifestations of dust inhalation, whether the dust is harmful or not. Two common forms of pneumoconiosis are silicosis and asbestosis. The typical pathological condition in harmful pneumoconiosis is the existence of fibrotic (stringy) tissue in the alveolar sacs or at lymph nodes in the lungs. This fibrotic tissue, caused by some dust particles, reduces the efficiency of the lungs by making them less resilient and by reducing the effective working surface for gaseous exchange. A simple benign pneumoconiosis with little disturbance of pulmonary function can also predispose to such diseases as pneumonia and tuberculosis.

The fate of the air contaminant once it reaches the deep portion of the lung (alveoli) depends on its solubility and reactivity. The more soluble, reactive substances may evoke acute inflammatory reactions and pulmonary edema. Most of the particles that reach the deep lung (alveoli) are engulfed by macrophages that migrate proximally to the airways and are either expectorated or swallowed or may enter the interstitial tissues. Once in the deep lung, however, chemical components in particles or in the vapor state can be absorbed into the bloodstream.

Inhaled contaminants that adversely affect the lungs fall into three general categories:

- Aerosols and dusts, which, when deposited in the lungs, may produce either tissue damage, tissue reaction, disease, or physical obstruction

- Toxic gases that produce adverse reaction in the tissue of the lungs themselves

- Toxic aerosols or gases that do not affect the lung tissue, but are passed from the lung into the bloodstream, where they are carried to other organs or have adverse effects on the oxygen-carrying capacity of the bloodstream (see Chapter 6, Industrial Toxicology).

An example of the first type is asbestos fiber, which causes fibrotic growth in the alveolar tissue, narrowing the ducts or limiting the effective area of the alveolar lining. Other harmful aerosols are certain fungi found in sugar cane residues, which produce bagassosis.

Potential health hazards from dust occur on three levels. The inhalation of sufficient quantities of dust, regardless of its chemical composition, can cause a person to choke or cough; it can also accumulate in the lungs and the pores of the skin. Depending on its chemical composition, dust can cause allergic or sensitization reaction in the respiratory tract or on the skin. Depending on both its size and chemical composition, dust can, by physical irritation or chemical action, damage the vital internal tissues.

Fibrosis can be produced by certain insoluble and relatively inert fibrous and nonfibrous solid particulates found in industry. It is now thought that one of the prerequisites for particulate-induced bronchogenic carcinoma may be the insolubility of the particulate in the fluids and tissues of the respiratory tract. More insoluble particulates reside in the lung long enough to induce tumors.

Some of the highly reactive industrial gases and vapors of low solubility can produce an immediate irritation and inflammation of the respiratory tract and pulmonary edema. Prolonged or continued exposure to these gases and vapors can lead to chronic inflammatory or neoplastic changes or to fibrosis of the lung.

Hydrogen fluoride is a gas that directly affects lung tissue. It is a primary irritant of mucous membranes, causing chemical burns. Inhalation of this gas causes pulmonary edema and direct interference with the gas transfer function of the alveolar lining.

Very soluble gases, such as sulfur dioxide and ammonia, seldom proceed much further down the respiratory tract than the bronchi. Less soluble gases, such as nitrogen dioxide, phosgene, and ozone, reach the deeper recesses of the respiratory tract, affecting mainly the bronchioles and the adjacent alveolar spaces, where they may produce pulmonary edema within a few hours.

Carbon monoxide (CO) is a toxic gas that is passed via the lungs into the bloodstream but does not damage the lungs. Carbon monoxide passes through the alveolar walls into the blood, where it ties up red corpuscles so they cannot accept oxygen, thus causing oxygen starvation.

Many metal oxides of submicron particle size (called fume) produce both immediate and long-term effects; the latter can occur in organs and tissues remote from the site of entry. For example, cadmium oxide fume inhaled at concentrations well above the threshold limit value (TLV) may produce immediate pulmonary edema that can be fatal; in addition, inhalation for many years of the fume at concentrations of a few multiples of the TLV can result in eventual renal injury and pulmonary emphysema.

Individual susceptibility to respiratory toxins is difficult to assess. In the occupational setting, workers exposed to the same environment for equal periods of time may develop different degrees of pulmonary disease. This can be due to the variation of the rate of clearance from the lung, the effect of cigarette smoking, coexistent pulmonary disease, and genetic factors.

## ■ Natural Defenses

The respiratory system has a rather complete set of mechanisms for shrugging off insults: the warming and humidifying effects of the nasal and throat passages (as defenses against very cold or overly dry air), the mucous lining, and the mechanical valves in the throat.

Because the mucous lining plays an important role in the cleansing of aerosols from the lungs, it deserves closer inspection. Cells in the windpipe and bronchi produce mucus that is constantly being carried toward the mouth by tiny hairlike projections, called cilia, waving in synchrony. This moving blanket acts as a vehicle to carry foreign substances up and out of the system to the throat, where they can be expectorated or swallowed.

In a healthy lung, aerosols that get into a bronchiole can be carried back out of the system in a matter of hours. Given adequate recovery time (about 16 hours) after an 8-hour exposure to dust, the healthy lung can thus cleanse itself.

Other defense mechanisms include muscular contraction of the bronchial tubes upon irritation—this reaction restricts the airflow and thus minimizes intake of the irritating substance—and the cough and sneeze, which tend to rid the upper respiratory tract of irritants.

Thus far we have discussed only the defenses of the airways leading to the alveoli. In general, only very fine particles and gases reach the alveolar sacs. The larger the particle, the sooner it will be deposited through impaction or gravity on the lining of the airway tubes leading to the sacs.

In the case of gases, the concentration that reaches the alveolar sacs will be nearly the same as the concentration in the air breathed. With aerosols, this is not the case. Large particles, more than 10 μm, will be deposited long before they reach the sacs, through gravity and impaction. Only the smaller particles will reach the alveoli. In the sacs, Brownian movement of the particles results in deposition by diffusion.

Because the very small aerosol particles are the only ones likely to reach the alveoli in great quantities, and because the alveoli are the most important area in the lungs, it is clear that the minute aerosols are potentially more harmful than larger aerosols. What happens to the very small suspended particles and gases that do reach the alveolar sacs?

First, because the air in the sacs is nearly quiescent (because the sacs are dead ends), the majority of the aerosols will be deposited. The deposition may be through diffusion (these aerosols act almost like gases) or gravitational settling.

Particles may fall prey to the mobile phagocyte cells, which are white blood cells capable of ingesting particles. Once laden with foreign matter, these cells can:

■ migrate to the bronchioles, where the mucous lining carries them out of the system

■ pass through the alveolar membrane into the lymph vessels associated with the blood capillaries

■ be destroyed (if the contaminant is cytotoxic) and break up, releasing the particles into the alveolar sac.

If the aerosol is not removed by these means, it can form a deposit in the sac. Such deposits may or may not acutely affect the health of the lungs.

All of the defense mechanisms are subject to some deterioration and slowing down with age or ill health. Thus, an older worker's lungs will not cleanse themselves as quickly or efficiently as those of a younger person.

Also, some contaminants may impede the defense mechanisms themselves, increasing the rate of retention of the contaminant in the lungs.

# AMA Guides for Evaluating Impairment

This section on determining the percent impairment is adapted from the American Medical Association's (AMA) *Guides to the Evaluation of Permanent Impairment* and is included to assist health and safety professionals in interpreting and understanding medical reports of workers' compensation cases.

The AMA publication assists physicians in evaluating permanent impairment of the respiratory system and the effect such impairment has on a person's ability to perform the activities of daily life. Permanent impairment of the respiratory system is not necessarily a static condition. A changing process can be present, so that it may be desirable to reevaluate the patient's impairment at appropriate intervals.

The measurable degree of dysfunction of the respiratory system does not necessarily parallel either the extent and severity of the anatomic changes of the lungs or the patient's own account of difficulties in carrying out the activities of daily life. Among the reasons for this phenomenon are the large pulmonary reserves normally present, the existence of disease in other systems (particularly the cardiovascular system), the wide variation in certain physiological measurements in normal individuals, and the patient's emotional response to respiratory disease or injury.

Many tests of pulmonary function have value and interest as guides to therapy and prognosis. For most patients, however, most of these are neither practical nor necessary for assignment to a particular class of impairment. Judicious interpretation of the results of ventilatory function tests and diffusion studies, combined with the clinical impression gained from weighing all the information gathered, should permit a physician to place the patient in the proper class of impairment.

## Rating of Impairment

The classification of respiratory impairment is based primarily on spirometric tests of pulmonary function and gas diffusion studies. Significant symptoms to be considered are dyspnea, cough with sputum or blood, and wheezing.

Procedures useful in evaluating impairment of the respiratory system include but are not limited to complete history and physical examination with special reference to cardiopulmonary symptoms and signs; chest roentgenography (posteroanterior, PA) in full inspiration, lateral, and other procedures as indicated; hematocrit or hemoglobin determination; electrocardiogram; tests of one-second forced expiratory volume and forced vital capacity (both defined in Table 2-A; and other pulmonary function tests, such as blood gas studies and diffusion studies, as indicated.

## Tests of Pulmonary Function

**Ventilation.** The tests of ventilatory function have certain limitations:

■ They require maximal voluntary effort by the patient, who may be unable or reluctant to perform the tests as well as ventilatory capacity permits. For example, the performance may be affected by the patient's lack of understanding of the test; state of physical training; fear of cough, chest pain, hemoptysis, or worsening of dyspnea; motivation and cooperation; the effects of

**Table 2–A.** Terminology of Certain Ventilatory Measurements

| Terms Used | Symbol | Description | Remarks |
|---|---|---|---|
| Vital capacity | VC | The largest volume of air measured on complete expiration after the deepest inspiration without forced or rapid effort. | Test not recommended for rating purposes, because result may be normal in those with severe respiratory impairment. |
| Forced vital capacity | FVC | The vital capacity performed with expiration as forceful and rapid as possible. | Formerly called timed vital capacity. |
| One-second forced expiratory volume | $FEV_1$ | Volume of air exhaled during the performance of a forced expiratory maneuver in the first second. | |
| One-second forced expiratory volume expressed as a percentage of FVC | $\frac{FEV_1}{FVC} \times 100$ | The observed $FEV_1$ expressed as a percentage of the observed FVC. | This value normally should exceed 70 percent. A lower value suggests the presence of some degree of obstructive airway disease. |
| Peak expiratory flow rate | PEFR | The rate of maximal expiratory flow. | Small, portable devices can be used by patients. Useful for epidemiological studies. No acceptable standards available. |

(Adapted with permission from the American Medical Association. *Guides to the Evaluation of Permanent Impairment*, 1st ed. Chicago: AMA, 1971)

other illness, particularly heart disease; and the effects of certain temporary factors on the day of the test, such as the presence of a respiratory infection or bronchospasm.

■ The results of these tests vary considerably among normal people of the same sex, age, and height.

■ Infrequently, significant impairment of respiratory function can exist even though the patient can perform the tests of ventilatory function normally; that is, the bellows action of the lungs and thorax is normal, but there are abnormalities of pulmonary circulation or gas exchange that give rise to the impairment and necessitate other evaluation procedures.

Various types of apparatus are available that give a permanent record and that readily permit measurement of the $FEV_1$ and the FVC. These tests can be understood by patients after a short explanation and instruction period, but most patients must be encouraged to put forth their best effort. The $FEV_1$ and FVC should each be administered three times, with the best test result considered most representative of the patient's ability. The test should not be considered valid unless the best two curves agree within 5 percent.

If the forced expiratory volume test is interpreted as showing airflow obstruction, the test might be repeated 5–10 min after the patient has inhaled a nebulized bronchodilator. If there is at least 15 percent improvement in the performance of the test, the possible reversibility of the airway obstruction and, incidentally, the presumed efficiency of bronchodilator therapy are established. However, the best results of tests before bronchodilation should be used in determining the degree of impairment.)

Results of tests of ventilatory function should be expressed both in liters or liters per minute and as a percentage of the predicted normal. The FVC as a percentage of the predicted normal is taken as a measure of restrictive impairment. The ratio of actual $FEV_1$ to actual FVC is a criterion for diagnosing obstructive impairment, but the value of measured $FEV_1$ either by itself or as a percentage of predicted $FEV_1$ is considered the best measure of severity. Tables for "normal" values of FVC, $FEV_1$, and $FVC/FEV_1$ are given in the AMA *Guides to the Evaluation of Permanent Impairment.*

Diffusion studies, determination of exercise capacity, and arterial blood-gas determinations are useful when a patient's symptoms do not correlate well with spirometric studies. The test for diffusing capacity of carbon monoxide (single-breath $D_{CO}$) is available in most pulmonary function laboratories. It detects interference with passage of gases across the alveolar membrane, as may occur in interstitial fibrosis. Tables of normal values are given in the AMA Guides, but many factors not related to pulmonary impairment can affect this measurement.

Quantitative exercise capacity measurements can be done using a treadmill or stationary bicycle, but such testing can be hazardous to people in poor health.

Determinations of partial pressures of oxygen and carbon dioxide in arterial blood, particularly before and after exercise, can be useful in certain cases. These measurements require arterial puncture, so they are not suitable for routine evaluation.

Other measurements of pulmonary function are available in specialized laboratories, but they are not sufficiently standardized for evaluation of impairment.

The AMA Guides classify respiratory impairment into four classes—none, mild, moderate, and severe—based on exercise capacity or FVC, $FEV_1$, $FVC/FEV_1$, and $D_{CO}$. Special criteria independent of pulmonary function are assigned to asthma, pulmonary hypersensitivity, pneumoconiosis, or lung cancer.

## ■ Summary

**The nose.** This external organ is lined by an extensive mucous membrane that warms, moistens, and filters the air passing through. It is the organ of smell and gives resonance to the voice.

**The pharynx.** Located at the back of the nose, mouth, and larynx, this cylindrical tube allows passage of food and air.

**The larynx.** This anterior structure in the neck is the voice box. Its cartilaginous walls hold it open during inspiration and expiration.

**The trachea and bronchi.** These airways are lined with ciliated mucous membrane and have rings of cartilage to maintain patency. At midsternal level the trachea divides into two bronchi, one going to each lung. The left bronchus is longer and more horizontal than the right to accommodate the heart; consequently, inhaled foreign bodies find their way more easily into the right bronchus. These structures are the main sensory area for the initiation of the cough reflex. Their ciliated linings sweep mucus upward to the throat.

**The lungs.** These two spongy cone-shaped organs occupy the major portion of the thoracic cavity. The space between them is called the mediastinum, and contains the heart, blood vessels, and all tubes passing to and from the abdomen. The lungs are made up of ever-branching bronchioles; at the end of each there is an alveolar duct, from which clusters of alveoli open like balloons.

**The alveoli.** Alveoli are clustered at the ends of the bronchioles like bunches of grapes. They have a rich blood supply from the pulmonary arteries, because the blood and the air are in close contact, thus permitting the interchange of oxygen into the blood and carbon dioxide into the air. The bronchioles, alveoli, and blood vessels are supported by elastic connective tissue, which, with lymphatic vessels, glands, and nerves, form the substance of the lungs.

The vital capacity of the lungs is 3–4 l of air, but only half a liter is changed with each quiet respiration. As well as the gaseous exchange in the lungs, heat and moisture are lost from the body.

Spontaneous inspiratory nervous impulses arise from a center in the brain stem. This center is influenced by stimuli from many chemical and mechanical receptors. The stimulus for expiration is of a nervous origin and arises from stretching of the nerve endings in the alveolar wall. This stimulus cuts out the impulses that produced inspiration; by elastic recoil and relaxation of muscle, expiration is produced.

Human lungs are size-selective dust collectors. Only relatively small particles, generally those less than 5 μm in diameter, reach the alveolar spaces. Such small particles move with the air currents; they settle very slowly in still air, and even when thrown into the air with high velocity they travel only a short distance.

The lungs have a very large surface, 300–1,000 ft$^2$ (28–92 m$^2$) of very delicate tissue. This surface is exposed to contaminants in the air breathed. The lungs have good defenses against particulates; when unimpaired, these clearance mechanisms remove about 99 percent of the insoluble dust deposited in the lungs.

## ■ Bibliography

American Medical Association. *Guides to the Evaluation of Permanent Impairment,* 4th ed., Rev. Chicago: AMA, 1993.

American Thoracic Society. Lung function testing: Selection of reference values and interpretive strategies. *Ann Rev Resp Dis* 144:1202–1218, 1991.

American Thoracic Society. Standardization of spirometry (update). *Ann Rev Resp Dis* 136:1285–1298, 1987.

Bateman HE, Mason RM. *Applied Anatomy and Physiology and the Speech and Hearing Mechanism.* Springfield, IL: Charles C. Thomas, 1984.

Eisen EA, Wegman DH, Kriebl D. Application of peak expiratory flow in epidemiologic studies of occupation. *Occupational Medicine: State of the Art Reviews* 8:265–277, 1993.

Harber P, Schenker M, Balmes JR. *Occupational & Environmental Respiratory Diseases.* St. Louis: Mosby–Yearbook, 1995.

Merchant JA, ed. *Occupational Respiratory Diseases.* U.S. DHHS #86–102, 1986.

Murray JF. *The Normal Lung,* 2nd ed. Philadelphia: W.B. Saunders, 1986.

Rom WN, ed. *Environmental and Occupational Medicine,* 2nd ed. Boston: Little, Brown, 1992.

**Anatomy**
*Epidermis ▪ Dermis ▪ Subcutaneous Layer ▪ Glands in the
Skin ▪ Sweat Glands ▪ Sebaceous Glands ▪ Blood Vessels ▪
Hair ▪ Nails*

**Physiology and Functions**
*Temperature Regulation ▪ Sweat ▪ Ultraviolet Light ▪
Skin Absorption*

**Defense Mechanisms**

**Definitions and Incidence of Occupational
Skin Disorders**

**Direct Causes of Occupational Skin Disease**
*Chemical ▪ Mechanical ▪ Physical ▪ Biological ▪ Botanical*

**Predisposing Factors**
*Age and Experience ▪ Skin Type ▪ Sweating ▪ Gender ▪ Seasons
and Humidity ▪ Hereditary Allergy (Atopy) ▪ Personal Hygiene ▪
Preexisting Skin Disease*

**Classification of Occupational Skin Disease**
*Contact Dermatitis ▪ Contact Urticaria/Latex Allergy ▪
Photosensitivity ▪ Occupational Acne ▪ Pigmentary
Abnormalities ▪ Sweat-Induced Reactions, Including Miliaria and
Intertrigo ▪ Cutaneous Tumors ▪ Ulcerations ▪ Granulomas ▪
Alopecia ▪ Nail Disease ▪ Systemic Intoxication*

**Burns**
*Nature of Chemical Burns ▪ Classification of Burns ▪
Complications of Burns*

**Diagnosis**
*Appearance of the Lesion ▪ Sites of Involvement ▪ History and
Course of the Disease ▪ Ancillary Diagnostic Tests ▪ Treatment*

**Workers' Compensation**

**Evaluation of Occupational Dermatoses for Workers'
Compensation**
*Diagnosis ▪ Causation ▪ Impairment Evaluation ▪ Conclusions and
Recommendations ▪ Physical Examinations*

**Prevention and Control**
*Environment ▪ Monitoring and Control Technology ▪ Personal
Cleanliness ▪ Personal Protective Equipment ▪ Responsibility for
Control ▪ Case Examples of Control*

**Summary**

**Bibliography**

# CHAPTER 3

# The Skin and Occupational Dermatoses

by James S. Taylor, MD

*Portions of this chapter in the first and second editions of this
book were written by Julian B. Olishifsky and Larry L. Hipp.*

THE SKIN IS THE LARGEST ORGAN of the body. Its surface area
is about 2 m² and in most places it is no more than 2 mm
thick, yet its mass exceeds that of all other organs. Skin is a
tough, flexible cover and is the first body barrier to make
contact with a wide variety of industrial hazards. The skin is
subject to attack from heat, cold, moisture, radiation, bacteria, fungi, and penetrating objects. Health and safety professionals should have a basic understanding of the anatomy,
physiology, and defense mechanisms of the skin before recommending proper control measures.

The skin performs a number of important functions.
Among these are protecting the body from invasion by
microorganisms (fungi, bacteria, etc.), injury to vital internal organs, the rays of the sun, and the loss of moisture.
The skin is also an organ of sensory perception; the sensations of pain, touch, itch, pressure, heat, cold, and warmth
may be elicited in human skin.

Temperature regulation is yet another job performed
by the skin. Blood vessels dilate (widen) when the body
needs to lose heat or constrict (narrow) when the body
must reduce the amount of heat loss through the skin.

When the surrounding air is comparatively warm, the skin is cooled by evaporation of moisture excreted by the sweat glands. There are between 2 and 3 million sweat glands over the surface of the body, excluding mucous membranes. The greatest concentration of sweat glands is on the palms of the hands and the soles of the feet. Their function depends on an intact nerve supply. Thermoregulatory sweating is controlled by a heat regulator in the brain. Emotions stimulate sweating primarily on the palms and soles.

The surface of the skin may look smooth, but if it is examined under a magnifying glass, countless ridges and valleys can be seen in which the many small openings of pores, hair follicles, and sweat glands are found (Figure 3–1). There are also different patterns of skin texture; compare the palm of the hand with the back of the hand, for example. The skin generally is soft, flexible, and elastic, particularly in young people.

A number of predisposing factors interact to determine the degree to which a person's skin responds to chemical, physical, and biological insults. These include type of skin (pigmentation, dryness, amount of hair), age, sex, season, previous skin diseases, allergies, and personal hygiene.

A worker's skin is very vulnerable to occupational hazards. Surveys indicate that dermatological conditions other than injuries are the second most common cause of all occupational diseases, accounting for 14 percent of all cases reported to the Bureau of Labor Statistics in 1992. Occupational skin disease is underreported and results in considerable lost time from work.

Although most occupational skin disorders are treated by primary care and occupational physicians, dermatologists are often consulted. Dermatology is the branch of medicine concerned with the diagnosis, treatment including surgery, and prevention of diseases of the skin, hair, and nails. Some dermatologists have had special training in occupational skin disorders.

Some disorders that are visible in the skin do not arise primarily in the skin but in other organs. Thus, the skin is an early warning system and its examination is very important in physical diagnosis, occasionally furnishing the first clue to identification of systemic diseases.

## ■ Anatomy

Three distinct layers of tissue make up the skin; from the surface downward, they are the epidermis, the dermis, and the subcutaneous layer.

The thickness of the skin varies from 0.5 mm on the eyelid (the dermis is thinnest here) to 3–4 mm on the palms of the hands and soles of the feet (the epidermis is thickest here). Skin is also relatively thin in the skin folds: the axillae (armpits), under the breasts, the groin, and between the fingers and toes.

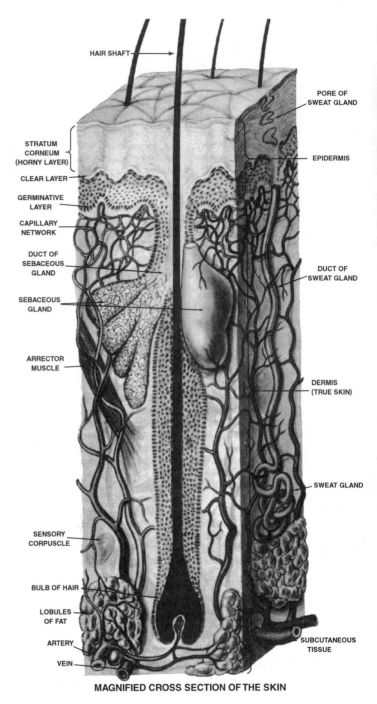

**Figure 3–1.** Magnified cross section of the skin. (Reprinted with permission of the AMA. *Today's Health Guide. Chicago: AMA, 1965.*)

## Epidermis

The top layer of the epidermis is composed of dead cells called the horny or keratin layer or the stratum corneum. This layer resists chemical attack fairly well, with the notable exception of alkali. It serves as the chief rate-limiting barrier against absorption of water and aqueous solutions, but offers little protection against lipid-soluble materials (such as organic solvents) or gases.

The horny layer gradually flakes off, or soaks off when wet. It is constantly being replaced by cells pushed toward the surface as new cells are formed in the deeper, germinative layer of the epidermis. This regenerative and sloughing characteristic serves to some extent as a protection against chemicals and microorganisms.

This constant shedding of flaky material goes mostly unnoticed unless a person has dandruff or must peel off insensate skin patches after a sunburn. Brisk rubbing with a towel peels off little rolls of material composed of dead outer skin cells that are never missed.

There are three cell types in the epidermis:

- Keratinocytes, which make up the bulk of the epidermis, form from below and move up to become dead horny cells.

- Melanocytes, or pigment-forming cells, synthesize melanin (pigment) granules, which are then transferred to keratinocytes. The amount of melanin in keratinocytes determines the degree of pigmentation of skin and hair. The absolute number of melanocytes in human skin is the same for all races. Differences in coloration among races result from differences in the number, size, degree of pigment formation, distribution, and rate of degradation of pigment granules within keratinocytes. Melanin proliferates under stimulus of certain wavelengths of sunlight and becomes visible as suntan or freckles. Moles are growths that contain melanin. Some people with little or no pigment in their skin have albinism, an inherited abnormality in which melanin (pigment) production is decreased. Vitiligo is a more common disorder in which loss of melanocytes also results in areas of cutaneous pigment loss. Some chemicals, such as phenolic germicides, can destroy pigment after occupational or environmental exposure.

- Langerhans' cells, located in the mid-epidermis, account for a relatively small percentage of all epidermal cells and play an important role in various immune processes, especially allergic contact dermatitis. Although the epidermis is an active tissue, it is not richly supplied with blood because blood vessels are absent. The blood supply to the epidermis is through the candelabra pattern of blood vessels in the upper dermal papillae (Figure 3–1).

The epidermis is thin enough that the nerve endings in the dermis are sufficiently close to the surface to supply the fine sense of touch (Merkel cell). Some of this sensation is lost where areas of the skin are chronically subjected to friction, resulting in subsequent thickening of the epidermis that provides protection in the form of a callus.

Thus, the soles of the feet are commonly callused among those who habitually walk barefoot, as are the palms of the hands of those who do heavy work.

## Dermis

Beneath the epidermis is the dermis, which is much thicker than the overlying epidermis, in most locations. It contains connective tissue composed of collagen elastic fibers and ground substance, is strong and elastic, and is the part of animal skins that makes leather when tanned. It is laced with blood vessels, nerve fibers, and receptor organs (for sensations of touch, pain, heat, and cold), and contains muscular elements, hair follicles, and oil and sweat glands (see Figure 3–1).

The dermis is tough and resilient, and is the main natural protection against trauma. When injured, it can form new tissue—a scar—to repair itself.

The top of the dermis is made up of a layer of tiny cone-shaped objects, called papillae. Thousands of papillae are scattered over the body. They are more numerous in areas such as the fingertips, where the skin appears to be more sensitive. Nerve fibers and special nerve endings are found in many of the papillae. As a result, the sense of touch is best developed in areas where papillae with nerve endings are most abundant.

The papillary layer fits snugly against the outer layer of skin, the epidermis, which has ridges corresponding to those of the papillae. The ridges prevent the skin layers from slipping against one another.

The ridges on the surfaces of the fingertips form the whorls, loops, and arches that make up fingerprints; dermatoglyphics is the study of the patterns of the ridges of the skin. Similar ridges appear on the soles of the feet. Because it is unlikely that two people will have the same pattern of ridges, fingertip patterns are used by the police to identify individuals.

The larger component of the dermis (reticular dermis) extends from the base of the papillary dermis to the subcutaneous fat. Muscle fibers are commonly seen in the reticular dermis on the face and neck.

## Subcutaneous Layer

Beneath the dermis is a layer of subcutaneous tissue with fatty and resilient elements that cushions and insulates the skin above it. The distinguishing feature of the subcutaneous layer is the presence of fat. Also present are the lower parts of some eccrine and apocrine sweat glands and hairs, as well as hairs, nerves, blood and lymphatic vessels and cells, and fibrous partitions composed of collagen, elastic tissue, and reticulum. This layer links the dermis with tissue covering the muscles and bones.

Loss of subcutaneous fat and softer parts of the skin removes bouncy supporting material, and because the external skin does not shrink at the same rate, it tends to collapse and become enfolded in wrinkles.

## Glands in the Skin

Two main types of glands are located in the dermis. One, already mentioned, is the sweat gland. Under the microscope it appears as a tightly coiled tube deep in the dermis with a corkscrew-like tubule that rises through the epidermis to the surface of the skin.

The second type is the sebaceous or oil gland, which is usually located in or near a hair follicle. Sebaceous glands are located in all parts of the skin except on the palms and soles. They are particularly numerous on the face and scalp.

## Sweat Glands

Sweat glands excrete a fluid known as sweat, or perspiration. The working or secreting parts of the glands are intricately coiled tubules in the dermis. There are two kinds of sweat glands, which produce different kinds of sweat.

**Apocrine.** One kind, called apocrine sweat, is not very important physiologically but has some social significance. Apocrine sweat glands open into hair follicles and are limited to a few regions of the body, particularly the underarm and genital areas. Apocrine sweat is sterile when excreted but decomposes when contaminated by bacteria from the skin surface, resulting in a strong and characteristic odor. The purpose of the many cosmetic underarm preparations is to remove these bacteria or block gland excretion.

**Eccrine.** The other kind of sweat, called eccrine, is of great importance to our comfort and, in some cases, our lives. Multitudes of eccrine sweat glands are present everywhere in the skin except the lips and a few other areas. They are crowded in largest numbers into skin of the palms, soles, and forehead.

Eccrine sweat is little more than extremely dilute salt water. Its function is to help the body to dissipate excessive internal heat by evaporation from the surface of the skin.

## Sebaceous Glands

There are many sebaceous (oil-secreting) glands in the skin. They are distributed over almost the entire body, and are most common in regions of the forehead, face, neck, and chest—the areas typically involved in common acne, a condition associated with cell-clogged sebaceous glands. The primary function of this oil substance, or sebum, is lubrication of the hair shaft and the horny surface layers of the skin. A certain amount of natural skin oil is necessary to keep skin and hair soft and pliable.

A strap of internal, plain, involuntary muscle tissue, the arrectores pilorum ("raiser of hair"), is located in the lower portion of the hair follicle below the sebaceous glands and originates in the connective tissue of the upper dermis. Goose bumps appear on the skin when these muscles attempt to produce heat.

## Blood Vessels

The skin is richly supplied with small blood vessels. The blood supply in the skin accounts for the reddening of sunburn and the coloration of the fingers beneath the nails. Engorgement of the blood vessels accounts for the reddening of the skin when we blush.

Vascular birthmarks, such as hemangiomas, strawberry marks, and port wine stains, derive their coloration from unusually large numbers of tiny blood vessels concentrated in a small area of the skin.

## Hair

Hair and nails are modified forms of skin cells containing keratin as their major structural material. Keratin is produced by the same processes that change living epidermal cells into dead, horny cells. However, hair and nails are made up almost entirely of keratin.

With the exception of the palms and soles, hair follicles populate the entire cutaneous surface, although in many areas they are so inconspicuous or vestigial that they are never noticed. Hair ranges in texture from the soft, almost invisible hair on the forehead to the long hair of the scalp and the short, stiff hair of the eyelashes.

Hair follicles develop as downgrowths of the epidermis. The hair then grows outward from the bottom of the follicle. Each hair has a root, which is anchored at the bottom of the follicle, and a shaft, which extends past the top of the follicle. The hair follicle enters the epidermis and passes deep into the dermis at an angle. The follicles of long hairs can extend into the subcutaneous layer. Sebaceous glands empty into the follicle. At the root of the hair is a cone-shaped papilla that is similar to the peg-like papillae that underlie the ridges of the fingers, palms, and soles.

The hair shaft is covered with tiny, overlapping scales. An inner layer of cells contains pigment that gives the hair its color. Most hair tips project from the skin at a slant. Minute arrectores pilorum muscles attached to the follicle have the fascinating ability to make the hair stand on end, as in goose bumps.

Hair follicles and the sweat glands also serve as routes for percutaneous absorption of chemicals. Physicians sometimes use this absorptive ability of the skin in administering certain drugs, such as nitroglycerine, scopolamine, estrogen, and nicotine. Some chemicals placed on the skin can be detected in the saliva a few minutes later. In the workplace the skin is a potential route of entry for a number of hazardous chemicals.

## Nails

The fingernails and toenails, like hair, are specialized forms of the skin. The fully developed nail overlays a modified part of the dermis called the nail bed.

Nails are essentially the same in structure as hair. Like hair, nails also contain keratin, but nails are flat, hard plates. The living part of a nail lies in the matrix in back of the half-moon, or lunula. If the dead nail plate, which constitutes most of the visible part of the nail, is destroyed by injury, a new nail will grow if the matrix is intact.

The growth rate of nails varies and depends on such factors as the person's age and health. Nails grow faster in young people and grow slower with serious illnesses.

# Physiology and Functions

The skin performs a number of important functions. It protects the body from invasion of bacteria, injury to vital internal organs, the rays of the sun, and the loss of moisture.

## Temperature Regulation

For a discussion of the role of the skin heat regulation of the body, see Chapter 12, Thermal Stress.

## Sweat

Sweat is produced constantly, usually in proportion to the temperature of the environment. In hot environments, the body must lose heat by evaporation, which is more effective than simple radiation. In cool, dry weather the amount of sweat produced is relatively small and the skin remains dry to the touch. We are not aware of sweating, so the small amount of sweat produced is called insensible perspiration.

When heat production of the body is increased or when the ambient temperature is unusually high, the sweat glands produce more perspiration. The rate of production outstrips the rate of evaporation, particularly if humidity is high, because the rate of evaporation declines with the rise in humidity. Perspiration then collects on the body in visible drops and we are conscious of sweating.

However, heat is lost only when the sweat evaporates. All sweat glands are innervated by fibers of the sympathetic nervous system, ultimately controlled by the hypothalamus. Emotional stimulation from anxiety or fright may stimulate sweating in the palms and soles.

One way of increasing the rate at which water is evaporated from the body is to breathe rapidly, thus moving larger quantities of air from the moist surfaces of the mouth, throat, and lungs. Humans cannot do this in comfort, but it is the chief method of cooling available to dogs; in warm weather, dogs sit with mouth open, tongue extended, and pant.

## Ultraviolet Light

Skin protects not only against mechanical shocks, but also against various forms of ultraviolet (UV) light. (See Chapter 11, Nonionizing Radiation, for a discussion of the forms.) Most animals are protected from sunlight by scales, hair, and feathers, which absorb the sun's rays without harm to themselves.

Humans have only the skin as protection from the sun's UV rays. Ultraviolet light energy produces chemical changes within the skin's cells; the effects vary with the time of the year, the geographic area, and the hour of the day.

Generally, after initial exposure to summer sun at midday, skin shows reddening or erythema, which may not appear for several hours. If the dose of sunlight is intense, the erythema may be followed by blistering and peeling of the outer layer of epidermal cells.

If the erythema is not severe, it fades in a few days and the skin gradually acquires a tan coloration (suntan). The tan color is produced by darkening of existing pigment (immediate pigment darkening) and by increase in pigment formation. When skin is exposed to the sun, it is believed that melanin pigment moves toward the surface of the skin and is replaced by new melanin in the lower cell layer. Along with pigmentation increase, the stratum corneum thickens to furnish additional protection against solar radiation injury. One or two weeks may be required to develop a suntan by moderate daily doses of sunlight; the tan fades if occasional exposure to sunlight is not continued.

As some protection against repeated UV light exposure, human skin is equipped with the capacity to form pigment (melanin), which absorbs UV light and thus acts as a protective umbrella over the regions beneath (delayed pigment darkening).

Solar UV radiation can induce actinic (solar) degeneration and skin cancer and is a major hazard of chronic sun exposure. Chronic exposure to artificial UV light in tanning salons may induce similar changes. Additionally, sunlight and artificial UV light may induce a number of abnormal cutaneous reactions in patients with certain hereditary or acquired diseases or in those taking certain medication. Photoaging and natural chronologic aging are different entities.

There is evidence now that the immune system of humans is affected by UV radiation and that environmental sources of radiation can have similar effects, such as contact photoallergy.

One beneficial normal effect of UV radiation on skin is the photochemistry that leads to the production of vitamin $D_3$. In most industrial countries, sufficient vitamin D is added to food to meet normal daily requirements.

## Skin Absorption

A waxy type of mixture composed of sebum, breakdown products of keratin, and sweat, called the surface lipid film, coats the outer surface of the keratin layer, but there is no evidence that this normal coating has any barrier function.

The epidermis, especially the stratum corneum, acts as the major permeability barrier to the entry of foreign chemicals into the body. Overall, the skin is selectively permeable—more impermeable than permeable—and shows regional variation in absorptive capacity. Absorption of materials through the skin markedly increases when the continuity of the skin is disrupted by dermatitis, lacerations, or punctures. The hair follicles and sweat ducts may play only a minor role in skin absorption. However, they act as diffusion shunts—that is, relatively easy pathways through the skin for certain substances such as polar compounds, very large molecules that move across the stratum corneum very slowly, and pharmacologically active substances, especially in very hairy areas. After this initial phase, however, most of the percutaneous absorption of all substances takes place across the stratum corneum, which has a much greater surface area than that of the hair follicles and sweat ducts. Absorption of fat-soluble chemicals and oils can also occur via the hair follicle.

## Defense Mechanisms

Anyone who works is a candidate for occupational skin disease, yet most workers are not affected by such disorders because the skin is a primary organ of defense. The skin is able to perform its many defense functions because of its location, structure, and physiological activity.

These are the specific defenses the skin has in terms of its protection against typical industrial hazards.

- *Bacteria:* The skin is a naturally dry terrain (except in places such as armpits and the groin, and during abnormal sweating) and has a normal contingent of bacteria that tends to destroy pathogenic bacteria. Free fatty acids in the surface oil also can have some antibacterial value. The immune defenses of the skin also defend against infections.

- *Sunlight:* The skin has two defenses: an increase in pigmentation, and thickening of the stratum corneum.

- *Primary irritants:* The skin resists acids but offers much less protection against organic and inorganic alkalis. Sweat can act as a diluent to decrease the effect of water-soluble toxins. Conversely, it enhances hydration and maceration of the barrier, thereby promoting percutaneous absorption.

- *Injury:* The skin's resilience, especially of the dermis, provides a measure of resistance to forceful impact. The cutaneous nerves also provide information about the state of the external environment through sensations of touch and temperature.

- *Excessive increase or decrease in body heat:* The body's thermoregulatory mechanisms include the activity of sweat glands and blood vessels.

- *The absorption of chemicals through the skin:* This is where the most important function is performed. The skin is a flexible body envelope and the epidermal barrier, especially the stratum corneum, provides a significant blockade against water loss from the body and penetration of the skin by chemical agents.

## Definitions and Incidence of Occupational Skin Disorders

A dermatosis is any abnormal condition of the skin, ranging from the mildest redness, itching, or scaling to an eczematous (superficial inflammation), ulcerative (ulcer-forming), acneiform (resembling acne), pigmentary (abnormal skin color), granulomatous (tumor-like mass, nodule), or neoplastic (new, abnormal tissue growth) disorder. Occupational dermatoses include any skin abnormalities resulting directly from or aggravated by the work environment. *Dermatitis* is a more limited term referring to any inflammation of the skin, such as contact dermatitis or cement dermatitis.

Occupational skin diseases can occur in workers of all ages and in any work setting, and cause a great deal of illness, personal misery, and reduced productivity. Although the frequency of occupational skin disease often parallels the level of hygiene practiced by employers, occupational skin diseases are largely preventable. Many consider this type of disease trivial and insignificant, but occupational skin disorders can result in complex impairment. Data compiled by the U.S. Bureau of Labor Statistics (BLS) for 1991 indicate that approximately 94 percent (5,977,400) of all occupational disorders are injuries and almost 6 percent (368,300) are diseases. Because large surface areas of skin are often directly exposed to the environment, the skin is particularly vulnerable to occupational insults. Although complete data on the extent and cost of dermatological injuries are not available, the National Institute for Occupational Safety and Health (NIOSH) estimated in 1986 that skin injuries may account for 23–35 percent of all injuries. An estimated 1–1.65 million dermatological injuries may occur annually, with an estimated annual rate of skin injury of 1.4–2.2 per 100 full-time workers. The highest percentage is due to lacerations and punctures (82 percent) followed by burns (chemical and other, 14 percent) (Table 3–A).

In the mid-1950s, skin disorders other than injuries accounted for 50–70 percent of all occupational diseases. This figure has been gradually decreasing and was 14

**Table 3–A.** Occupational Dermatological Injuries in the United States, 1983

| Type of Injury | No. | (%) |
|---|---|---|
| Lacerations and punctures | 253,141 | ( 82.3 ) |
| Burns (nonchemical) | 36,477 | ( 11.9 ) |
| Abrasions | 10,576 | ( 3.4 ) |
| Burns (chemical) | 6,828 | ( 2.2 ) |
| Cold injuries | 566 | ( 0.2 ) |
| Radiation injuries | 135 | ( 0.04) |
| Total | 307,723 | (100.0 ) |

(Reported by the Supplementary Data System of the Bureau of Labor Statistics from 29 participating states.)

percent, or 62,900 cases, in 1992. NIOSH attributes the decline of skin diseases during the past 30 years to a continuing trend toward automation, enclosure of industrial process, and educational efforts. Despite these figures, dermatitis is the second most common cause of reported occupational disease in the United States. National data indicate that as many as 20–25 percent of all occupational skin diseases involve lost time from work, with an average of 11 workdays lost per case. California and South Carolina have reported similar data based on workers' compensation claims. The results of two studies show a serious underreporting of occupational disease of all types, which may mean that the true incidence is 10 to 50 times greater than that reported by the BLS.

NIOSH has included work-related dermatological conditions on its list of 10 leading work-related diseases and injuries in the United States (Table 3–B). Reasons include the fact that 10–15 percent of requests NIOSH receives for health hazard evaluation involve skin complaints and the fact that the economic impact of dermatological conditions is substantial. The annual cost resulting from lost worker productivity, medical care, and disability payments has been estimated to range between $222 million and $1 billion.

Table 3–C gives incidence of occupational dermatoses (disease) by industry group for the United States in 1984. The highest incidence rate was in agriculture (28.5 cases

**Table 3–C.** Cases and Incidence Rate of Occupational Dermatological Conditions in a Segment of Workers, by Major Industrial Divisions–United States, 1984

| Industrial Division | No. | Incidence Rate |
|---|---|---|
| Agriculture/forestry/fishing | 2,233 | 28.5 |
| Manufacturing | 23,017 | 12.3 |
| Construction | 2,456 | 6.6 |
| Services | 7,973 | 5.0 |
| Transportation/utilities | 2,114 | 4.3 |
| Mining | 393 | 4.0 |
| Wholesale/retail trade | 3,770 | 2.1 |
| Finance/insurance/real estate | 563 | 1.1 |

Bureau of Labor Statistics Annual Survey.
*Per 10,000 full-time workers (2,000 employment hours/full-time worker/year).

per 10,000 full-time workers). The most hazardous industrial processes for skin disorders are as follows:

- Use of cutting oils and coolants in machine tool operation
- Plastics manufacturing
- Rubber manufacturing
- Food processing
- Leather tanning and finishing
- Agriculture
- Metal plating and cleaning
- Construction
- Printing
- Forest products manufacturing

## ■ Direct Causes of Occupational Skin Disease

There are unlimited substances and conditions capable of inducing a skin disorder in the workplace. Each year, new causes are reported and most can be classified under one of the following five broad headings:

- Chemical
- Mechanical
- Physical
- Biological
- Botanical

### Chemical

Organic and inorganic chemicals are the predominant causes of dermatoses in the work environment (Table 3–D). The list of such chemicals is endless because each year additional agents capable of injuring the skin are added. Chemical agents may be divided into two groups: primary irritants and sensitizers.

**Table 3–B.** The 10 Leading Work-Related Diseases and Injuries in the United States, 1982

**Occupational lung diseases:** asbestos, byssinosis, silicosis, coal worker's pneomoconiosis, lung cancer, occupational asthma

**Musculoskeletal injuries:** disorders of the back, trunk, upper extremity, neck, lower extremity, traumatically induced Raynaud's phenomenon

**Occupational cancers (other than lung):** leukemia, mesothelioma, cancers of the bladder, nose, and liver

**Severe occupational traumatic injuries:** amputations, fractures, eye loss, lacerations, and traumatic deaths

**Cardiovascular diseases:** hypertension, coronary artery disease, acute myocardial infarction

**Disorders of reproduction:** infertility, spontaneous abortion, teratogenesis

**Neurotoxic disorders:** peripheral neuropathy, toxic encephalitis, psychoses, extreme personality changes (exposure-related)

**Noise-induced loss of hearing**

**Dermatological conditions:** dermatoses, burns (scalding), chemical burns, contusions (abrasions)

**Psychological disorders:** neuroses, personality disorders, alcoholism, drug dependency

The conditions listed under each category are to be viewed as selected examples, not comprehensive definitions of the category, and are *not* in order of incidence or importance.

NIOSH has recently developed a suggested list of the 10 leading work-related diseases and injuries (Table 8-B). Three criteria were used to develop the list: (1) the frequency of occurrence of the disease or injury; (2) its severity in the individual case; and (3) its amenability to prevention.

**Table 3–D.** Selected Chemical Causes of Skin Disorders

| Chemical | Primary Irritants | Sensitizers | Selected Skin Manifestations (Some also have important systemic effects on other organs.) | Selected Occupations, Trades, or Processes Where Exposure Can Occur |
|---|---|---|---|---|
| **ACIDS** | | | | |
| Acetic | X | ? | Dermatitis and ulceration | Manufacturing acetate rayon, textile printing and dyeing, vinyl plastic makers |
| Carbolic (phenol) | X | | Corrosive action on skin, local anesthetic effect | Carbolic acid makers, disinfectant manufacturing, dye makers, pharmaceutical workers, plastic manufacturing |
| Chromic | X | X | Ulcers ("chrome holes") on skin, inflammation and perforation of nasal septum | Platers, manufacturing organic chemicals and dyestuffs |
| Cresylic | X | | Corrosive to skin, local anesthetic effect | Manufacturing disinfectants, coal tar pitch workers, foundry workers |
| Formic | X | | Severe irritation with blisters and ulcerations | Rubber and laundry workers, mordanters, cellulose formate workers, airplane dope makers |
| Hydrochloric | X | | Irritation and ulceration of skin | Bleachers, picklers (metals), refiners (metals), tinners, chemical manufacturing, masons (clean cement) |
| Hydrofluoric | X | | Severe chemical burn with blisters, erosion, or ulceration | Enamel manufacturing, etchers, hydrofluoric acid makers, flurochemical workers |
| Lactic | X | | Ulceration (if strong solutions are used) | Adhesives, plastics, textiles |
| Nitric | X | | Severe skin burns and ulcers | Nitric acid workers, electroplaters metal cleaners, acid dippers, nitrators, dye makers |
| Oxalic | X | | Severe corrosive action on skin, cyanosis (bluish discoloration), and brittleness of nails | Tannery workers, blueprint paper makers, oxalic acid makers |
| Picric | X | X | Erythema, dermatitis, scaling, yellow discoloration of skin and hair | Explosives workers, picric acid makers, dyers and dye makers, tannery workers |
| Sulfuric | X | | Corrosive action on skin, severe inflammation of mucous membranes | Nitrators, picklers (metals), dippers, chemical manufacturing |
| **ALKALIS** | | | | |
| Ammonia | X | | Irritation including airborne dermatitis of face from vapors | Ammonia production, fertilizers, photocopying (blueprint, diazo); gas and liquid forms |
| Calcium cyanamide | X | | Irritation and ulceration | Fertilizer makers, agricultural workers, nitrogen compound makers |
| Calcium oxide | X | | Dermatitis, burns, or ulceration | Lime workers, manufacturing of calcium, salts, glass, and fertilizer |
| Potassium hydroxide | X | | Severe corrosion of skin, deep-seated persistent ulcers, loss of fingernails | Potassium hydroxide makers, electroplaters, paper, soap, and printing ink makers |
| Sodium hydroxide | X | | Severe corrosion of skin, deep-seated persistent ulcers, loss of fingernails | Sodium hydroxide makers, bleachers, soap and dye makers, petroleum refiners, mercerizers, plastic manufacturing |
| Sodium or potassium cyanide | X | | Blisters, ulcers | Electroplaters, case hardening, extraction of gold |
| Trisodium phosphate | X | | Blisters, ulcers | Photographic developers, leather tanning, industrial cleaning detergents |
| **SALTS OR ELEMENTS** | | | | |
| Antimony and its compounds | X | ? | Irritation and lichenoid eruptions of skin | Antimony extractors, glass and rubber mixers, manufacturing of various alloys, fireworks, and aniline colors |
| Arsenic and its compounds | X | X | Spotty pigmentation of skin, perforation of nasal septum, skin cancer, keratoses especially on palms and soles, dermatitis, pustules | Leather workers, manufacturing insecticides, glass industry, agriculture, pesticides, tanning, taxidermy, alloy, lubricating oils |
| Barium and its compounds | X | | Irritation of skin | Barium carbonate, fireworks, textile dyes, and paint makers |
| Bromine and its compounds | X | | Irritation, vesicles, and ulceration; acne | Bromine extractors, bromine salts makers, dye and drug makers, photographic trades |

*(Continues)*

**Table 3–D.** Selected Chemical Causes of Skin Disorders (*Continued*)

| Chemical | Primary Irritants | Sensitizers | Selected Skin Manifestations (Some also have important systemic effects on other organs.) | Selected Occupations, Trades, or Processes Where Exposure Can Occur |
|---|---|---|---|---|
| **SALTS OR ELEMENTS (continued)** | | | | |
| Chromium and its compounds | X | X | Pitlike ulcers (chrome holes) on skin, perforation of nasal septum, dermatitis | Chromium platers, dye industry workers, chrome manufacturing, leather tanners |
| Mercury and its compounds | X | X | Corrosion and irritation of skin, dermatitis | Explosives manufacturing, silver and gold extractors, manufacturing electrical appliances and scientific equipment |
| Nickel salts | X | X | Folliculitis, dermatitis | Nickel platers, alloy makers |
| Sodium and certain of its compounds | X | | Burns and ulceration | Bleaching: detergent, paper, glass, tetraethyl lead manufacturing |
| Zinc chloride | X | ? | Ulcers of skin and nasal septum | Manufacturing chemicals, dyestuffs, paper, disinfectants |
| **SOLVENTS** | | | | |
| Acetone | X | | Dry (defatted) skin | Spray painters, celluloid industry, artificial silk and leather workers, acetylene workers, lacquer and varnish makers, garage mechanics |
| Benzene and its homologues (toluene and xylene) | X | | Dry (defatted) skin | Chemical and rubber manufacturing |
| Carbon disulfide | X | X | Dry (defatted) irritated skin | Extraction of oils, fats, and a wide range of other materials, manufacture of rayon, rubber, rubber cements, germicides, and other chemicals |
| Trichloroethylene | X | ? | Dermatitis | Degreasers, chemical intermediates |
| Turpentine | X | X | Dermatitis | Painters, furniture polishers, lacquerers, artists |
| Alcohols (such as ethanol) | X | X | Dermatitis | Chemical manufacture; painters |
| **SOME DYE INTERMEDIATES** | | | | |
| Dinitrobenzene | X | | Yellow discoloration of skin, hair, and eyes | Dye manufacturing |
| Nitro and nitroso compounds | X | X | Dermatitis | Dye manufacturing |
| Phenyl hydrazine | X | X | Severe chemical burns, dermatitis | Dye and pharmaceutical manufacturing |
| **PETROLEUM AND COAL-TAR DERIVATIVES** | | | | |
| Petroleum oils | X | | Dermatitis, folliculitis | Petroleum workers, machinists, mechanics |
| Pitch and asphalt | X | | Dermatitis, folliculitis, keratoses, skin cancer | Manufacturing pitch and asphalt, roofers |
| Tar (coal) | X | X | Dermatitis, folliculitis, skin cancer, eye inflammation (keratitis) | Tar manufacturing: manufacturing roofing paper and pitch; road building and repairing |
| **DYES (such as paraphenylenediamine)** | X | | Contact dermatitis (erythema, blisters, edema) | Dye workers, cosmetologists |
| **RUBBER ACCELERATORS AND ANTIOXIDANTS** | | | | |
| Mercaptobenzothiazole, tetramethylthiuram disulfide, diethylthiourea, and paraphenylenediamine | X | | Contact dermatitis (erythema, blisters, edema) | Rubber workers, such as compound mixers and calendar and mill operators; fabricators of rubber products |
| **SOAPS AND SOAP POWDERS** | X | X | Dermatitis, dry skin, paronychia (inflammation around finger-nails); allergy from fragrance, germicides, or dyes | Soap manufacturing, dishwashers, soda fountain clerks, maintenance workers—all associated with wet work |
| **INSECTICIDES** | | | | |
| Arsenic | X | | See above under salts or elements | Manufacturing and applying insecticides |
| Pentachlorophenols | X | ? | Dermatitis, chloracne | Pesticide and wood preservative |
| Creosote | X | X | Dermatitis, folliculitis, keratoses, hyperpigmentation, skin cancer | Manufacturing wood preservatives, railroad ties, coal tar lamp black and pitch workers |
| Fluorides | X | | Severe burns, dermatitis | Manufacturing insecticides, enamel manufacturing |
| Phenylmercury compounds | X | X | Dermatitis | Manufacturing and applying fungicides and disinfectants |
| Pyrethrum | | X | Dermatitis | Manufacturing and applying insecticides |
| Rotenone | X | | Dermatitis | Manufacturing and applying insecticides |

(*Continues*)

**Table 3–D.** Selected Chemical Causes of Skin Disorders (*Continued*)

| Chemical | Primary Irritants | Sensitizers | Selected Skin Manifestations (Some also have important systemic effects on other organs.) | Selected Occupations, Trades, or Processes Where Exposure Can Occur |
|---|---|---|---|---|
| **RESINS (Natural)*** | | | | |
| Cashew nut oils | | X | Severe poison ivylike dermatitis | Handlers of unprocessed cashew nuts, varnish |
| Rosin | | X | Dermatitis | Adhesive and paper mill workers, dentists, rubber industry |
| Shellac | | X | Dermatitis | Coating, cosmetics |
| Synthetic resins such as phenolformaldehyde, urea-formaldehyde, epoxy, vinyl, polyurethane, polyester, acrylate, cellulose esters | X | X | Dermatitis | Plastic workers, varnish makers, adhesives, coatings, rubber, cosmetology |

\* The skin reactions from this group of chemicals in some instances are due to the essential composition of the synthetic resin, but in other cases are due to the presence of added compounds such as plasticizers and other modifying agents.

| Chemical | Primary Irritants | Sensitizers | Selected Skin Manifestations | Selected Occupations, Trades, or Processes |
|---|---|---|---|---|
| **EXPLOSIVES** | | | | |
| Nitrates, mercury fulminate, tetryl, lead azide, TNT, nitroglycerin | X | X | Severe irritation, dermatitis, skin discoloration | Explosives manufacturing, shell loading |
| **METAL WORKING FLUIDS** | | | | |
| Cutting oils | X | | Oil acne (folliculitis), rare dermatitis | Machinists |
| Coolants—synthetic and semisynthetic | X | X | Dermatitis | Machinists |
| **OXIDIZING AGENTS** | | | | |
| Hydrogen peroxide | X | ? | Dermatitis | Chemical industry; medical disinfectant; cosmetology |
| Benzoyl peroxide | X | X | Dermatitis | Chemical industry; polyester manufacture |
| **OTHER** | | | | |
| Isocyanates such as TDI, MDI, HDI | X | X | Dermatitis | Polyurethane makers, adhesive workers, organic chemical synthesizers |
| Vinyl chloride | X | | Dermatitis, acro-osteolysis | Polyvinyl resin, rubber and organic chemical makers |
| Formaldehyde | X | X | Dermatitis | Undertakers, biologists, textile workers |
| Plants, weeds (such as poison oak, ivy, sumac) | X | X | Dermatitis | Outdoor workers (such as fire fighters, utility workers) |

**Primary irritants.** These are likely to affect most people; some actually affect everyone. These agents react on contact. The reaction alters the chemistry of the skin by dissolving a portion of it by precipitating the protein of the cells, or by some other chemical reaction. The result can range from tissue destruction (chemical burn) to inflammation (dermatitis) depending on the strength of the agent and the duration of the exposure.

Primary irritants damage skin because they have an innate chemical capacity to do so. Many irritants are water-soluble and thus react with certain components of the skin. The water-insoluble compounds, including many solvents, react with the lipid (fatty) elements within skin. The precise mechanism of primary irritation on the skin is not known, but some useful generalizations explain the activity of groups of materials in the irritant category. About 80 percent of all occupational dermatoses are caused by primary irritants. Dermatitis caused by a primary irritant is referred to as irritant contact dermatitis because the skin irritation is normally confined to the area of direct contact.

Most inorganic and organic acids act as primary irritants. Certain inorganic alkalis, such as ammonium hydroxide, calcium chloride, sodium carbonate, and sodium hydroxide, are skin irritants. Organic alkalis, particularly amines, also are active irritants. Metallic salts, especially arsenicals, chromates, mercurials, nickel sulphate, and zinc chloride, severely irritate the skin. Organic solvents include many substances, such as chlorinated hydrocarbons, petroleum-based compounds, ketones, alcohols, and terpenes, that irritate the skin because of their solvent qualities (Figure 3–2).

***Keratin solvents.*** All of the alkalis, organic and inorganic, injure the keratin layer with sufficient concentration and exposure time. These agents soften, dehydrate, and destroy the keratin cells, resulting in dry, cracked skin. This prepares the way for secondary infection and, at times, for the development of allergic contact dermatitis.

***Keratin stimulants.*** Several chemicals stimulate the skin so that it undertakes growth patterns that can lead to tumor or cancer formation. Certain petroleum products, a number of coal tar-based materials, arsenic, and some poly-

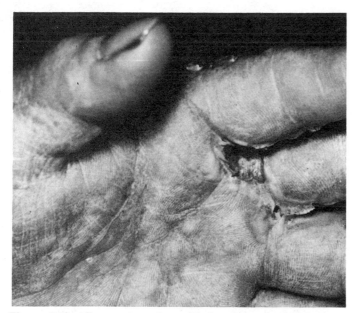

**Figure 3–2.** Eczematous dermatitis is a form of contact dermatitis caused by contact with organic solvents. It is one of the most prevalent types of dermatitis.

cyclic aromatic hydrocarbons can stimulate the epidermal cells to produce these effects (Figure 3–3).

***Fats and oil solvents.*** Just as organic solvents dissolve oily and greasy industrial soils, they remove the skin's surface lipids and disturb the keratin layer of cells so that they lose their water-holding capacity. Workers exposed each day to organic solvents develop exceedingly dry and cracked skin.

***Protein precipitants.*** Several of the heavy metal salts precipitate protein and denature it. The salts of arsenic, chromium, mercury, and zinc are best known for this action.

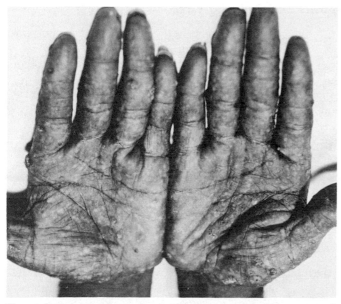

**Figure 3–3.** Nodules in the keratin layer of the skin may result from repeated exposure to certain tars or coal tar derivatives.

***Reducers.*** In sufficient concentration, salicylic acid, oxalic acid, urea, and other substances can actually reduce the keratin layer so that it is no longer protective, and an occupational dermatosis results.

**Sensitizers.** Some primary skin irritants also sensitize. Certain irritants sensitize a person so that a dermatitis develops from a very low, nonirritating concentration of a compound that previously could have been handled without any problem.

Some chemical and many plant substances and biological agents are classified as sensitizers. Initial skin contact with them may not produce dermatitis, but after repeated or extended exposure some people develop an allergic reaction called allergic contact dermatitis. Clinically, allergic contact dermatitis is often indistinguishable from irritant contact dermatitis (see pp. 66–67 for further discussion of allergic contact dermatitis and patch testing).

Substances that are both irritants and allergens include turpentine, formaldehyde, chromic acid, and epoxy resin components. Common sensitizers are plant oleoresins such as poison ivy, epoxy resins, azo dyes, certain spices, certain metals such as nickel and chromium, and topical medicaments such as neomycin.

Other chemicals can sensitize the skin to light. Known as photosensitizers, these chemicals include coal tar and pitch derivatives, fluorescent dyes, salicylanilides, musk ambrette, sunscreens containing p-aminobenzoic acid (PABA) and benzophenone, and some plants.

## Mechanical

Trauma at work can be mild, moderate, or severe and occur as a single or repeated event. Friction results in the formation of a blister or callus, pressure in thickening and color change, sharp objects in laceration, and external force in bruising, punctures, or tears. A commonly cited example is fibrous glass, which can cause irritation, itching, and scratching. Secondary infection may complicate blisters, calluses, or breaks in the skin.

## Physical

Physical agents such as heat, cold, and radiation can cause occupational dermatoses. For example, high temperatures cause perspiration and softening of the outer horny layer of the skin. This can lead to miliaria, or heat rash, common among workers exposed to hot humid weather, electric furnaces, hot metals, and other sources of heat.

High temperatures can also cause systemic symptoms and signs such as heat cramps, heat exhaustion, and even heat stroke. Burns can result from electric shock, sources of ionizing radiation, molten metals and glass, and solvents or detergents used at elevated temperatures.

Exposure to low temperatures can cause frostbite and result in permanent damage to blood vessels. The ears, nose, fingers, and toes are the most often frostbitten.

Electric utility and telephone line workers, highway maintenance workers, agricultural workers, fishermen, police officers, letter carriers, and other outdoor workers are most often affected.

Sunlight is the greatest source of skin-damaging radiation and is a source of danger to construction workers, fishermen, agricultural workers, foresters, and all others who work outdoors for extended periods of time. The most serious effect on the skin is skin cancer.

Increasing numbers of people come into casual or prolonged contact with artificial UV light sources such as molten metals and glass, welding operations, and plasma torches. A wide variety of newer lasers are being used in medicine and other scientific disciplines. Because lasers can injure the skin, eye, and other biological tissue, it is important to use appropriate protective devices.

Ionizing radiation sources include the following:

- Alpha-radiation is completely stopped by the skin and thus does not injure skin. However, alpha-radiation-emitting radioactive substances, such as plutonium, are harmful when ingested or inhaled.

- Beta-radiation can injure the skin by contact and substances such as phosphorus-32 are dangerous when inhaled or ingested. Beta-particles are usually localized at the surface or within the outer layers of skin, with the depth of penetration depending on the energy of the beta particle.

- Gamma-radiation and x-rays are well-known skin (radiodermatitis and skin cancer) and systemic (internal) hazards when sufficient exposure occurs. Radiodermatitis is characterized by dry skin, hair loss, telangiectasia, spider-like angiomas, and hyperkeratosis. Skin cancer may ultimately develop (see Chapter 10, Ionizing Radiation, for more information).

## Biological

Bacteria, viruses, fungi, and parasites can produce cutaneous or systemic disease of occupational origin. Animal breeders, agricultural workers, bakers, culinary employees, florists, horticulturists, laboratory technicians, and tannery workers are among those at greater risk of developing infections. Examples include anthrax in hide processors, yeast infections of the nail in dishwashers, bartenders, and others engaged in wet work, and animal ringworm in agricultural workers and veterinarians. Parasitic mites are common inhabitants of grain and other foodstuffs and attack those handling such materials, such as grocers, truckers, longshoremen, and agricultural workers. Outdoor workers such as bricklayers and plumbers in southeastern states risk contracting animal hookworm via larvae deposited by infected animals in sandy soil. Health care workers, medical laboratory workers, and emergency medical technicians are exposed to a number of microorganisms, especially hepatitis B, HIV, herpes simplex (herpetic whitlow of the fingers from direct viral exposure and inoculation), fungi (candida and superficial and deep fungi), and bacteria (staphylococci and tuberculosis), which may be acquired from patients or from biological specimens.

## Botanical

Many plants and woods, of which poison ivy and poison oak are the most common, can cause contact dermatitis. Irritant contact dermatitis can also be caused by some plants, and although the chemical identity of many of the toxins is not known, the allergen or irritant occurs in the leaves, stems, flowers, bark, or other part of the plant. Other plants, such as wild parsnip and fresh or diseased celery (pink-rot) are photosensitizers. Several outbreaks of photodermatitis have been reported in grocery produce workers, especially those visiting tanning parlors. With woods, dermatitis occurs especially when they are being sandpapered, polished, and cut. Fomites can carry and transmit these allergens, which can also be dispersed by the smoke from burning.

# ■ Predisposing Factors

In classifying and determining the severity of occupational dermatoses, a number of factors should be considered: the nature, duration, and extent of exposure to an environmental agent, the potential toxic effects of the agent, its chemical stability, and its potential for being absorbed through the skin. Other variables include preexisting skin disease or exposure to more than one agent. Indirect or predisposing factors leading to the development of occupational dermatoses are generally associated with age, sex, skin type, perspiration, season of the year, personal hygiene, and allergy.

## Age and Experience

Younger, inexperienced, and inadequately trained workers have a higher prevalence of occupational dermatoses than older workers. However, older workers may be prone to chronic skin irritation because their skin is generally drier.

## Skin Type

Workers with naturally dry skin cannot tolerate the action of solvents and detergents as well as persons with oily skin (Figure 3–4). Hairy arms and legs are common sites of folliculitis and acne induced by cutting oils (Figure 3–5).

## Sweating

Hyperhidrosis, or increased sweating, can produce maceration with softening and resultant separation of skin already irritated by rubbing in adjacent body areas, as occurs in the armpit and the groin. This predisposes the skin to second-

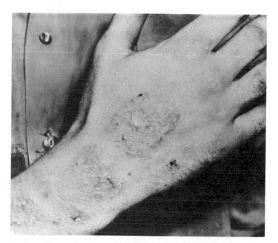

**Figure 3–4.** Cleaning hands with a strong petroleum solvent instead of a good industrial cleanser caused this case of dermatitis.

ary fungal and bacterial infection. Some materials, such as caustics, soda ash, and slaked lime, become irritants in solution. However, sweating can also serve a protective function by diluting the toxic substances.

## Gender

Because the incidence of nickel allergy is much greater in women (due to ear-piercing), they are more susceptible to developing dermatitis when handling coins or when in contact with nickel salts and metal alloys. The incidence of nickel allergy in men, even those with earrings, is lower, for reasons that are not known. A recent study suggested that women are more easily sensitized than men.

## Seasons and Humidity

Occupational dermatoses are more common in warm weather, when workers wear less clothing and are more

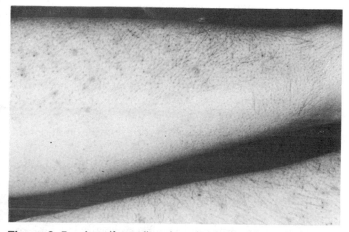

**Figure 3–5.** Acneiform disorder, shown on this worker's forearm, is often caused by exposure to cutting oils. Lack of splash guards and poor personal hygiene can be factors.

likely to come in contact with external irritants. Excessive perspiration, with resulting skin damage, is also more common in warm weather. When a work area is hot, workers may not use protective clothing. Warm weather also means that many workers have greater exposure to sunlight, poisonous plants, and insects, the effects of which may or may not be related to the job.

Winter brings chapping from exposure to cold and wind. Heated rooms usually are low in relative humidity, so skin loses moisture. Large-scale outbreaks of dermatitis in some factories has been traced to nothing more than low humidity. Clothing can keep dust particles and mechanical irritants in close contact with the skin. Infrequent bathing and changing of clothing can increase the incidence of skin irritation. (See Chapter 12, Thermal Stress, for more information.)

## Hereditary Allergy (Atopy)

Atopy (the name means *uncommon* or *out of place*) is a relatively common genetic tendency toward the development of atopic dermatitis, asthma, and hay fever. Atopic people are predisposed to developing dermatitis because of their reduced skin resistance to chemical irritants, inherent dry skin, dysfunctional sweating, and a high skin colonization rate of the bacterium *Staphylococcus aureus*. In an atopic adult, the hands are the main location for dermatitis. Atopic people, especially those with eczema, are more prone to irritant dermatitis than those who are not atopic. Atopy is common among hairdressers, health care workers, and others performing wet work. Contact allergy and contact urticaria—especially to latex—also occur in atopic workers.

## Personal Hygiene

Poor personal hygiene is believed to be a major factor causing occupational skin disorders. Unwashed skin covered with unwashed and unchanged clothes may be in prolonged contact with chemicals. Responsibility for maintaining clean skin is shared by employer and employee. Thus, adequate facilities for maintaining personal cleanliness should be provided in every place of employment. Educating workers in the preventive aspects of personal hygiene is imperative. On the other hand, excessive skin cleansing with harsh agents can produce an irritant contact dermatitis or aggravate preexisting dermatitis.

## Preexisting Skin Disease

Other forms of skin irritation (eczema), such as atopic eczema, nonoccupational contact dermatitis, palmar psoriasis, and lichen planus, can be aggravated by chemicals in the work environment. Ultraviolet light-sensitive disease, such as lupus erythematosus, and cold-induced disease, such as Raynaud's phenomenon, can be aggravated and precipitated by sunlight and cold exposure, respectively.

# ■ Classification of Occupational Skin Disease

Skin disorders are relatively easy to recognize because they are visible. However, accurate diagnosis and classification of disease type and its relationship to employment usually requires a high level of clinical skill and expertise. The varied nature of skin responses causing occupational skin disorders takes several forms. The appearance and pattern of the dermatosis seldom indicates the provoking substance definitively, but can provide clues to the class of materials involved. Diagnosis depends on appearance, location, and (most importantly) on the history. Preexisting skin disorders, adverse effects of treatment, and secondary infections add to the difficulty in diagnosis. The following grouping includes most occupational dermatoses.

## Contact Dermatitis

Contact dermatitis is the most frequent cause of occupational skin disease, accounting for most reported cases. Two types are generally recognized: irritant and allergic. Approximately 80 percent of all cases of occupational contact dermatitis result from irritation and 20 percent from allergy. Both are difficult to differentiate clinically because each can appear as an acute or chronic eczematous dermatitis. The acute form is erythematous (increased redness), vesicular (small blisters) to bullous (large vesicles), edematous (swollen), and oozing, and of short duration, lasting days or weeks. The chronic form is lichenified (thickened skin), scaly, and fissured, and may last for weeks, months, or years. Itching is usually a major symptom.

Contact dermatitis most often occurs on the hands, wrists, and forearms, although any area can be affected. Dusts, vapors, and mists can affect the exposed areas, including forehead, eyelids, face, ears, and neck, and often collect in areas where the body bends, such as under the collar and at the tops of shoes. The palms and soles are partially protected by a thick stratum corneum. The scalp tends to be protected by the hair, but the male genitalia are commonly affected, as irritants are often transferred by the hands. Contact dermatitis also localizes under rings and between fingers, toes, and other cutaneous areas that rub together.

**Irritant contact dermatitis.** A primary skin irritant is a substance that causes damage at the site of contact because of its direct chemical or physical action on the skin. Irritants are generally divided into strong (absolute) and weak (marginal) types. Strong (absolute) irritants include strong acids, alkalis, aromatic amines, phosphorus, ethylene oxide, riot-control agents, and metallic salts, and produce an observable effect within minutes. In contrast, marginal irritants such as soap and water, detergents, solvents, and oils can require days before clinical changes appear. Cumulative exposure to marginal irritants causes most cases of occupational irritant dermatitis and is a major skin problem in the workplace. (See the following pages for further discussion and lists of irritants and sensitizers).

Important factors to consider in irritant dermatitis are the nature of the substance (pH, solubility, physical state, concentration, duration of contact, and host and environmental factors). Despite the prevalence of irritant dermatitis, much is unknown about the precise mechanisms of how irritants disturb the skin. Several points merit emphasis:

■ Contact dermatitis can occur from contact with several marginal irritants, the effects of which are cumulative.

■ Cumulative irritant contact dermatitis can lead to skin fatigue, a condition in which even mild substances can irritate the skin, or to "hardening," in which the skin eventually accommodates repeated exposure to an offending agent.

■ The clinical and histological differentiation of irritant and allergic contact dermatitis is often difficult or impossible.

■ Constant exposure to irritants impairs the barrier function of the skin and allows penetration of potential allergens.

■ Irritant and allergic contact dermatitis often coexist in the same patient.

**Allergic contact dermatitis.** A variety of industrial chemicals are potential contact allergens. The incidence of allergic contact dermatitis varies depending on the nature of the materials handled, predisposing factors, and the ability of the physician to accurately use and interpret patch tests. Allergic contact dermatitis, in contrast with primary irritation, is a form of cell-mediated, antigen–antibody immune reaction. Sensitizing agents differ from primary irritants in their mechanism of action and their effect on the skin. Unless they are concomitant irritants, most sensitizers do not produce a skin reaction on first contact. Following this sensitization phase of one week or longer, further contact with the same or a cross-reacting substance on the same or other parts of the body results in an acute dermatitis (elicitation phase).

Other essential points about allergic contact dermatitis include the following:

■ As a general rule, a key difference between irritation and allergic contact dermatitis is that an irritant usually affects many workers, whereas a sensitizer generally affects few. Exceptions exist with potent sensitizers, such as poison oak oleoresin, or epoxy resin and components.

■ Differentiation of marginal irritants from skin allergens also can be difficult. Marginal irritants may require repeated or prolonged exposure before a dermatitis appears; allergic contact dermatitis also may not develop for months or years after exposure to an agent.

■ Many skin sensitizers, such as chromates, nickel salts, and epoxy resin hardeners, are also primary irritants.

- However, sensitization (allergy) can be produced or maintained by allergens such as nickel, chromates, formaldehyde, and turpentine in minute amounts and in concentrations insufficient to irritate the nonallergic skin.

- Cross-sensitivity is an important phenomenon in which a worker sensitized to one chemical also reacts to one or more closely related chemicals. A number of examples exist: rhus antigens such as poison oak, ivy, sumac, Japanese lacquer, mango, and cashew nutshell oil; aromatic amines such as p-phenylenediamine, procaine, benzocaine, and p-aminobenzoic acid (sunscreens); and perfume or flavoring agents such as balsam of Peru, benzoin, cinnamates, and vanilla.

- Systemic contact dermatitis is a widespread, eczematous contact-like dermatitis that can result from oral or parenteral (intravenous or intramuscular) administration of an allergen to which a worker is sensitized topically (such as oral administration of sulfonamides and thiazide diuretics in patients with contact allergy to p-phenylenediamine and benzocaine-containing topical anesthetics).

- Patch testing is used to differentiate allergic contact dermatitis from irritant dermatitis. The sine qua non for the diagnosis of allergic contact dermatitis is a properly performed and interpreted positive patch test.

The most common contact sensitizers in the general population have been determined from clinical experience and from published studies on the prevalence of positive patch test reactions in dermatology departments. Major sensitizers include the following:

- Rhus (poison oak, ivy, and sumac)
- P-phenylenediamine
- Nickel
- Rubber chemicals
- Quaternium-15 (a formaldehyde-releasing preservative)
- Topical medicaments containing benzocaine, antihistamines such as diphenhydramine, and antibiotics such as neomycin and bacitracin

Additional industrial allergens include the following:

- Chromates
- Plastics and adhesives (especially epoxy and acrylic resins)
- Formaldehyde and other preservatives
- Mercury
- Cobalt

## Contact Urticaria/Latex Allergy

Contact urticaria is characterized by the appearance of urticaria, or hives, usually within several minutes at the site of contact with a wide variety of substances. There are three types: immunologic, nonimmunologic, and contact urticaria of uncertain mechanism. The nonimmunologic type is most common; causes include substances that release histamine or other vasoactive substances, such as plants (nettles), insects (caterpillars and moths), cobalt chloride, cinnamic aldehyde, nicotinic acid esters (trafuril), and dimethyl sulfoxide. Causes of contact urticaria of uncertain type include ammonium persulfate, certain types of solar urticaria (caused by sun exposure) and aquagenic urticaria (caused by water exposure). Examples of agents that may produce immunologic contact urticaria include penicillin, nitrogen mustard, neomycin, and the insect repellant diethyl toluamide (DEET).

The paradigm for the immunologic type of contact urticaria is natural rubber latex allergy, which has become a significant medical and occupational health problem. Most affected patients have contact urticaria, but others have experienced generalized urticaria, angioedema, asthma, and anaphylaxis including vascular collapse and death. Latex allergy is an immunoglobulin (Ig) E-mediated hypersensitivity to one or more of a number of proteins present in raw or cured natural rubber latex. The highest prevalence, 34–50 percent, is found in children with congenital defects of the spine, such as spina bifida; such children are exposed to latex during multiple operations and procedures. Two to ten percent of medical and dental personnel may be affected, probably from repeated wearing of natural rubber latex gloves. Surgical glove and rubber band factory workers are also at risk, along with homemakers who regularly wear natural rubber latex gloves. Diagnosis is made by a compatible history of urticaria, angioedema (especially angioedema or swelling of the lips from inflating balloons or of the genitalia following condom use), asthma or anaphylaxis from latex exposure and a positive wear test, skin test, or blood test (RAST) for latex-specific antibody. Because latex hyposensitization is not yet possible, latex-sensitive people must avoid latex exposure and substitute gloves and other medical devices made of other materials (vinyl, synthetic latex, and other polymers, Tactyl 1 and 2, or Elastyrn). (See the section on Gloves, pp. 78–80.)

## Photosensitivity

Photosensitivity is the capacity of an organ or organism or certain chemicals and plants to be stimulated to activity by light or to react to light. Two types are generally recognized: phototoxicity and photoallergy. Phototoxicity, like primary irritation, can affect anyone, although darkly pigmented people are more resistant. Photoallergens, like contact allergens, involve immune mechanisms and affect fewer people.

Industrial sources of photosensitivity can be obscure, requiring careful epidemiological and clinical investigation including photopatch testing. An example is phototoxicity from p-aminobenzoic acid used in the manufacture of ultraviolet-cured inks. Medical personnel may be occupationally exposed to photosensitizing drugs. Other workers who can have contact with topical photosensitizers include outdoor and field workers (photosensitizers in plants and chemicals), machinists (antimicrobials in metalworking fluids), pharmaceutical workers (drugs, dyes, and fragrances), and oil field, road construction, and coal tar workers (tars, pitch, and other hydrocarbons).

## Occupational Acne

Occupational acne results from contact with petroleum and its derivatives, coal tar products, or certain halogenated aromatic hydrocarbons (Table 3–E). The eruption can be mild, involving localized, exposed, or covered areas of the body, or severe and generalized, with acne involving almost every follicular orifice. Chloracne, in addition to being a difficult cosmetic and therapeutic problem, is of considerable concern because it is caused by highly toxic chemicals.

Occupational acne is seen most commonly in workers exposed to cutting oils in the machine tool trades. The insoluble (straight) oils are the most common cause (see Figure 3–5). Oil acne typically starts as comedones and an inflammatory folliculitis affecting the tops of the hands and extensor surfaces of the forearms. However, covered areas of the body (thighs, lower abdomen, and buttocks) can be affected by contact with oil-saturated clothing. Although the lesions are commonly called oil boils, they almost never develop from bacteria present in the oils.

Any form of occupational acne or preexisting or coexisting acne vulgaris (nonoccupational) can be aggravated by heat (acne tropicalis and aestivalis); constant friction (acne mechanica), with acne localized to the forehead (hard hat), waist (belt), or other area; excessive scrubbing with harsh soaps (acne detergicans); cosmetics (acne cosmetica); pomade and vaseline (pomade acne); and topical corticosteroids (steroid rosacea). Acneiform eruptions from systemic medication containing bromides, iodides, and corticosteroids and the syndrome of senile or solar comedones on the face are also to be considered in the differential diagnoses.

Coal tar oils, creosote, and pitch can produce extensive acne in coal tar facility workers, roofers, and road mainte-nance and construction workers. Comedones are typical of this form of acne. Phototoxic reactions involving both the skin and eye (keratoconjunctivitis) can complicate the picture and produce coal tar melanosis and exacerbations of the acne. Pitch keratoses and acanthomas (precancerous and cancerous skin lesions) can develop later.

Certain halogenated aromatic chemicals (some chloronaphthalenes, PCBs and dibenzofurans, dibenzo-p-dioxins, and chlorobenzenes) are the most potent acnegens and are among the most toxic environmental chemicals. These chemicals can produce chloracne, a type of acne that is often resistant to therapy, and can be accompanied by systemic toxicity. Chloracne is one of the most sensitive indicators of biological response to these chemicals, and acts as a marker of the medical and environmental impact of contamination of technical-grade chemicals with potentially highly toxic intermediates.

## Pigmentary Abnormalities

Pigmentary abnormalities can result from exposure to certain chemical, physical, and biological agents. They not only represent difficult cosmetic problems, but can indicate exposure to potential systemic toxins. Differentiation from various nonoccupational, genetic, metabolic, endocrine, inflammatory, and neoplastic pigmentary conditions is necessary.

**Hyperpigmentation.** Hyperpigmentation (skin darkening) can follow almost any dermatitis as a postinflammatory event. Chemical photosensitizers (tar, pitch, plant, and drug photosensitizers), physical agents (ultraviolet light and thermal and ionizing radiation), and trauma (chronic itching) are common causes. Exposure to certain chemicals (arsenic and acnegenic aromatic hydrocarbons) can also cause hyperpigmentation.

**Hypopigmentation.** Pigment loss can also follow inflammation (Figure 3–6). Physical or chemical damage to the

---

**Table 3–E.** Some Causes of Occupational Acne

**Petroleum and its derivatives** (crude oil and fractious cutting oils)
**Coal tar products** (coal tar oils, pitch, creosote)
**Halogenated Aromatic Compounds** (chloracnegens)
Polyhalogenated naphthalenes
Polyhalogenated biphenyls (PCBs, PBBs)
Polyhalogenated dibenzofurans
Contaminants of polychlorophenol compounds especially herbicides (2,4,5-T and pentachlorophenol) and herbicide intermediates (trichlorophenols), e.g., dioxin
Contaminants of 3,4-dichloroaniline and related herbicides (Propanil and Methazole) azo- and azoxybenzenes

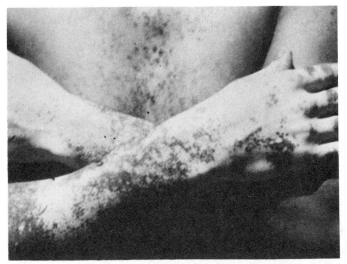

**Figure 3–6.** Pigment loss in the skin (hypopigmentation) was caused by exposure to a known depigmenting chemical.

skin from thermal, ultraviolet, radiation, or chemical burns may cause not only loss of pigment, but also scarring. These changes usually pose no diagnostic problem.

However, pigment loss from certain chemical exposures can be difficult to differentiate from idiopathic vitiligo (a patchy loss of pigment from otherwise healthy skin). Occupational leukoderma (white skin) of this type was first reported from exposure to monobenzyl ether of hydroquinone (agerite alba), once used as an antioxidant in industrial gloves. During the past 20 years, a number of phenolic compounds have caused leukoderma among exposed workers. Sources include hospital and industrial germicidal cleaners, metalworking fluids, oils, latex glues, inks, paints, and plastic resins. Table 3–F lists some of these compounds. These chemicals interfere with melanin pigment biosynthesis, destruction, or both. Hands and forearms are usually affected, although covered parts can also be affected, possibly from ingestion or inhalation of the chemicals.

## Sweat-Induced Reactions, Including Miliaria and Intertrigo

Miliaria (prickly heat or heat rash) results from obstruction of sweat ducts and is an inflammatory reaction to retained extravasated sweat. It is a common reaction of people who sweat profusely while exposed to heat. The lesions consist of pinpoint to pinhead-sized papules and vesicles (blisters) on the chest, back, submammary, inguinal, and axillary folds.

Intertrigo represents maceration that occurs on apposing skin surfaces and is a scaling, erythematous eruption. Superimposed yeast or superficial fungal infection can also be present. Obesity and heat exposure are aggravating factors.

## Cutaneous Tumors

Neoplastic growths of the skin are classified as benign lesions, precancers, or cancers. Benign viral warts (*verrucae vulgaris*) are more common among workers in certain occupations associated with wet work (such as butchers). Keratoacanthomas can be occupationally associated with exposure to sunlight or contact with various tars, pitch, and oils (pitch warts and acanthomas). Although classed as benign lesions, keratoacanthomas can be extremely difficult

**Table 3–F.** Some Chemicals Producing Occupational Leukoderma

Monobenzyl ether of hydroquinone
Monomethyl ether of hydroquinone
Hydroquinone
*P*-Tertiary amyl phenol
*P*-Tertiary butyl phenol
*P*-Tertiary butyl catchol
Alkyl phenols
Selected other phenolic compounds

to differentiate clinically and pathologically from squamous cell carcinoma. Pitch and tar warts (keratoses) and acanthomas can be premalignant lesions.

Excessive exposure to sunlight is the most common cause of precancers and cancers in human skin. Additionally, inorganic arsenic compounds, polycyclic aromatic hydrocarbon compounds associated with asphalt, paraffins, coal tars, oils (creosote, shale, hydrogenated, petroleum, insoluble cutting, and mineral), and ionizing radiation can cause cancer of the skin and other organs. Precancerous actinic keratoses (caused by rays of light that produce chemical effects) appear in sun-exposed areas, can be extensive in workers with outdoor jobs (such as utility line workers, agricultural workers, construction workers, ranchers, fishermen, and sailors), and can progress to squamous cell and basal cell carcinomas. Such workers often have other signs of sun exposure from solar degeneration of collagen, including hyperpigmentation, thin and wrinkled skin, and telangiectasia (a spider-like growth composed of blood or lymph vessels). Epidemiological studies show that sunlight can also be a factor in the increased incidence of malignant melanoma. Skin biopsy is absolutely essential for the diagnosis of all types of skin cancer.

## Ulcerations

Tissue injury on a skin or mucous membrane surface can result in erythema, blisters, or pustules, which may result in necrosis and ulceration. This can be caused by trauma, thermal or chemical burns, cutaneous infection, and a number of chemicals, including certain chromium, beryllium, nickel, and platinum salts, calcium oxide, calcium arsenate, calcium nitrate, and strong acids. Cutaneous tumors can also ulcerate. Self-inflicted or unintentionally produced skin disorders commonly appear as ulcerations.

## Granulomas

These represent chronic, indolent areas of inflammation and can be localized or generalized. Scar formation often results. Causes include a variety of bacterial (anthrax), mycologic (sporotrichosis), viral (herpes simplex), parasitic (protothecosis), and botanical (thorns) sources. Other causes include minerals (silica, beryllium, zirconium), bone, chitin, and grease.

## Alopecia

Alopecia (absence of hair from the skin areas where it is normally present) has many causes: trauma, cutaneous and systemic disease, drugs, chemicals, and other physical factors, including ionizing radiation. Industrially caused hair loss is rare and the differential diagnosis is long. Chemicals or medications can cause extensive hair shedding by precipitating telogen (resting hair) development, directly poisoning the anagen (growing) hair, or acting in other unknown ways. Other alopecia-producing chemicals include thallium (rodent poison) and boric acid. Medications, primarily cancer chemotherapeutic agents, can precipitate anagen hair loss (immediate loss). Drugs capable of causing telogen hair loss (delayed loss) include oral contraceptives, anticoagulants, propranolol, and thallium.

## Nail Disease

Chronic inflammation of the folds of tissue surrounding the fingernail (paronychia), with associated nail dystrophy, is a common occupational disorder associated with wet work (bartenders, maintenance workers, and kitchen workers). This disorder is commonly associated with *Candida* species, Pseudomonas species, other bacteria, or dermatophyte fungi.

Nail discoloration can result from exposure to chemicals such as bichromates (accompanied by nail dystrophy), formaldehyde, certain amines, picric acid, nicotine, mercury, resorcinol, or iodochlorhydroxyquin (Vioform).

Nail dystrophy can also accompany exposure to a number of chemicals, especially solvents; it is also caused by trauma or occupational marks in certain occupations such as weaving and the fur industry. Nail dystrophy can also be secondary to Raynaud's phenomenon, vibratory trauma, and acro-osteolysis.

## Systemic Intoxication

A number of chemicals with or without direct toxic effect on the skin itself can be absorbed through it and cause (or contribute to, when a substance is also inhaled) systemic intoxication; the severity depends on the amount absorbed. A partial list of substances and their systemic effects includes the following:

- Aniline (red blood cells and methemoglobinemia)
- Benzidine (carcinoma of urinary bladder)
- Carbon disulfide (nervous system and psychological disturbances)
- Carbon tetrachloride (central nervous system, or CNS, depression, hepatotoxicity, and nephrotoxicity)
- Dioxane (CNS depression)
- Ethylene glycol ethers (CNS depression, pulmonary edema, hepatotoxicity, and nephrotoxicity)
- Halogenated naphthalenes, diphenyls, and dioxins (neurotoxicity and hepatotoxicity, altered metabolism)
- Methyl butyl ketone (CNS depression and peripheral neuritis)
- Organophosphate pesticides (inhibition of enzyme cholinesterase with cardiovascular, gastrointestinal, neuromuscular, and pulmonary toxicity)
- Tetrachloroethylene (CNS depression, suspected carcinogen)
- Toluene (CNS depression).

Chemicals whose absorption may contribute to the total exposure are designated with a "Skin" notation in the American Conference of Governmental Industrial Hygienists (ACGIH) Threshold Limit Values for Chemical Substances and Physical Agents and Biological Exposure Indices (see Appendix B). More information can also be found in the ACGIH publication *Documentation of the Threshold Limit Values and Biological Exposure Indices*, 6th ed. Contact the ACGIH for details. Cutaneous absorption here refers to absorption by skin, mucous membranes, and eyes through either airborne or direct contact.

## Burns

Because all burns have essentially the same features, they are usually classified by degree according to depth of injury as first-, second-, or third-degree.

The main types of burns are as follows:

- Explosion burns, usually affecting exposed areas (hands, face)
- Steam burns, often superficial on exposed areas (more serious if with eye or respiratory contact)
- Hot-water burns, often leading to blistering depending on water temperature, more severe if victim is wearing heavy, permeable clothing
- Molten-metal burns, often affecting lower limbs, often extremely deep with metal encrusted in skin
- Hot-solid burns, normally not extensive, can be very deep
- Flame burns, almost always deep, often extensive, with the type of clothing being a major factor in severity.
- Electricity and radiant energy burns, almost always severe, often with complications; ordinary clothing offers little protection

### Nature of Chemical Burns

Burns caused by chemicals are similar to those caused by heat. In fact, some chemicals, such as sodium hydroxide, cause not only chemical burns because of their caustic action, but also thermal burns because of the heat they can generate when they react with moisture in the skin. After patients with chemical burns have been given emergency first aid, their treatment is the same as for patients with thermal burns.

Both thermal burns and chemical burns destroy body tissue. Some chemicals continue to cause damage until reaction with body tissue is complete or until the chemical is washed away by prolonged flushing with water. Strong alkalis penetrate tissue deeply, and strong acids corrode tissue with a characteristic stain.

Many concentrated chemical solutions have an affinity for water. When they come in contact with body tissue, they withdraw water from it so rapidly that the original chemical composition of the tissue (and hence the tissue itself) is destroyed. In fact, a strong caustic may dissolve even dehydrated animal tissue. The more concentrated the solution, the more rapid is the destruction.

Sulfuric, nitric, and hydrofluoric acids are the most corrosive of the inorganic acids, even more corrosive than hydrochloric acid. Some chemicals, such as phenol, are doubly hazardous. In addition to being highly corrosive, they are poisonous when absorbed through the skin. The severity of chemical burns depends on the following factors:

- Corrosiveness of the chemical
- Concentration of the chemical
- Temperature of the chemical or its solution
- Duration of the contact

The first three factors are determined by the very nature of the chemical and the requirements of the process in which it is used. The fourth factor, duration of the contact, can be controlled by the proper first-aid treatment administered without delay.

## Classification of Burns

Burns are commonly classified as first-, second-, or third-degree. Second-degree burns may be further classified as superficial or deep dermal. However, for purposes of this chapter, the common classifications of first-, second-, and third-degree are adequate and are described here.

**First-degree burns.** These are characterized by redness and heat accompanied by itching, burning, and considerable pain. Only the outer layer of the epidermis is involved.

**Second-degree burns.** These are highly painful and involve deeper portions of the epidermis and the upper layer of dermis. Generally, the skin is mottled red with a moist surface and blisters form. Such burns are easily infected.

**Third-degree burns.** These are very severe forms of injury, involving loss of skin and deeper subcutaneous tissue. They are pearly-white or charred in appearance and the surface is dry. They are not exceedingly painful at first because nerve endings are usually impaired or destroyed.

**Special types of burns.** Cement and hydrofluoric acid deserve special mention. Recently, there have been many reports of severe burns from kneeling in wet cement or from wet cement becoming trapped inside boots. Pressure and occlusion are important factors, as well as the need to work fast with premixed cement, which can encourage prolonged contact. Symptoms may be delayed, so workers

must be alert to the danger. Adherent cement should be removed by copious and gentle irrigation with water.

Hydrofluoric acid is one of the strongest acids known and is widely used in industry. Hydrofluoric acid burns are characterized by intense pain, often delayed, and progressive deep tissue destruction (necrosis). Immediate treatment with topical magnesium sulfate or benzalkonium chloride and calcium gluconate gels and injections is recommended.

## Complications of Burns

The dangers to life that result from extensive burns are infection (which causes most burn complications), loss of body fluid (plasma or lymph from the blood), and subsequent shock. Finally, the functional, cosmetic, and psychological sequelae may require the full attention of a rehabilitation team.

# Diagnosis

Anyone who works can develop a skin disorder, but not all skin disorders occurring in the workplace are occupational. Arriving at the correct diagnosis is not generally difficult, but it is more than a routine exercise. The following criteria are generally used.

## Appearance of the Lesion

The dermatosis should fall into one of the accepted clinical types with respect to its morphological appearance.

## Sites of Involvement

Common sites are the hands, wrists, and forearms, but other areas can be affected. Widespread dermatitis can indicate heavy exposure to dust because of inadequate protective clothing or poor hygiene habits.

## History and Course of the Disease

A thorough and pertinent clinical history is the most important aspect in diagnosis of occupational dermatoses. This includes a description of the eruption, response to therapy, medical history and review of systems, a detailed work history (description of present and past jobs, moonlighting, preventive measures, cleansers, and barrier creams), and a detailed description of nonoccupational exposures. The behavior of the eruption on weekends, vacation, and sick leave can be very helpful in assessing the occupational component.

## Ancillary Diagnostic Tests

When indicated, selected laboratory tests and office procedures are used for detecting skin disorders. These may include direct microscopic examination and bacterial and fungal cultures of the skin, skin biopsy for histopathological diagnosis, and patch tests and photopatch tests to detect any occupational or nonoccupational allergens and photosensitizers. Patch tests should be performed by physicians experienced with the procedure.

## Treatment

Therapy of occupational skin disorders is essentially no different from that of the same nonoccupational disorder. However, two key factors are often overlooked when a worker's skin clears and he or she returns to work: identifying the cause of the disease and preventing a recurrence. Patients with skin disorders not responding to initial treatment should be referred for specialty evaluation.

## ■ Workers' Compensation

State workers' compensation laws are no-fault statutes that hold employers responsible for the cost of occupational injury and disease claims while guaranteeing benefits to covered workers who meet the laws' requirements. All states now recognize responsibility for occupational diseases, and health professionals should become familiar with their own state workers' compensation laws and regulations. The American Medical Association's *Guides to the Evaluation of Permanent Impairment* is mandated, recommended, or often used by authorities in 40 of 53 jurisdictions (38 states and 2 territories). In general, three types of payment may be made when a claim is approved: temporary total disability payments to compensate for lost wages, payment of medical bills, and payment for permanent partial or permanent total disability.

Most claims based on occupational skin disease involve temporary total or permanent partial disability. Temporary total disability usually ceases when the patient has reached maximum medical improvement, has a valid job offer within his or her physical capabilities, or actually returns to work. Some states allow third-party liability suits arising out of workers' compensation cases.

## ■ Evaluation of Occupational Dermatoses for Workers' Compensation

In the evaluation of workers' compensation cases, the key elements are diagnosis, causation, impairment, and conclusions and recommendations.

### Diagnosis

An accurate diagnosis of the claimant's condition is imperative. This often involves obtaining a detailed history of the present illness, reviewing of work exposures including material safety data sheets, and, whenever possible, visiting the workplace. Evaluation of the cutaneous findings should include an examination of the skin and skin biopsies, cultures, patch testing, or other ancillary tests if warranted. If diagnosis is in doubt, specialized consultation is warranted.

### Causation

Determination of the cause-and-effect relationship between a skin disorder and an occupation is not always clear-cut. Questions to be considered include the following:

- Is the clinical appearance compatible with an occupational dermatosis?
- Are there workplace exposures to chemical, physical, mechanical, or biologic agents that may affect the skin?
- Is the anatomic distribution of the eruption compatible with job exposure? Many occupational dermatoses involve the hands.
- Is the temporal relationship between exposure and onset consistent with an occupational skin disease?
- Have nonoccupational exposures been excluded as causes?
- Does the dermatitis improve away from work exposure to the suspected agent(s)?
- Do patch or provocation tests identify a probable cause?

A "yes" answer to at least four of these questions is probably adequate to establish probable cause.

### Impairment Evaluation

As a prelude to impairment evaluation, the physician must determine the impact of the medical condition on life activities and the stability of the condition. If the worker has a new or recent onset condition that significantly precludes working on the current job, then temporary total impairment may exist and an appropriate amount of time away from work may be warranted under most workers' compensation laws. Unduly restrictive limitations, such as avoiding all contact with a particular substance, may jeopardize a worker's job; before writing such recommendations, the physician should generally discuss them with the worker and employer.

The AMA's *Guides to the Evaluation of Permanent Impairment* are used to evaluate *permanent* impairment of any body system(s), from both occupational and nonoccupational causes; they are not designed for use in evaluating *temporary* impairment. They are guidelines, are not absolute recommendations, and are designed to bring objectivity to an area of great subjectivity. The *Guides* include clinically sound and reproducible criteria useful to physicians, attorneys, and adjudicators. They espouse the philosophy that all physical and mental impairments affect the whole person. A 95–100 percent whole person impairment is considered to represent almost total impairment, a state that is approaching death. Before using the *Guides* for evaluating cutaneous impairment, the health professional should read the two introductory chapters, the glossary and then Chapter 13, The Skin. Chapter 2, Records and Reports, lists a suggested outline for a medical evaluation report.

Permanent impairment of the skin is "any anatomic or functional abnormality or loss that persists after medical treatment and rehabilitation and after a length of time sufficient to permit regeneration and other physiologic adjustments." (pp. 277–78 of the 4th ed.) The *Guides* popularized the concept that impairment is a medical issue assessed by medical means, in contrast to disability, which is a nonmedical assessment generally determined by adjudicating authorities.

Disability is defined as an "alteration... of the capacity of an individual to meet personal, social, or occupational demands or to meet statutory or regulatory requirements..., refers to an activity an individual cannot accomplish..., and may be thought of as the gap between what a person can do and what the person needs or wants to do." (p. 2 of 4th ed.) It is important to remember that an impaired person is not necessarily disabled. The classic example often cited is that of two people who lose the distal portion of the phalanx of the same finger. Although impairment for both people is the same, disability is likely to be greater for a concert pianist than for a bank president.

The AMA chapter on the skin lists five classes of impairment, ranging from 0 percent to 95 percent (Table 3-G.)

The impact of the disorder on the activities of daily living should be the major consideration in determining the class of impairment. The frequency and intensity of signs and symptoms and the frequency and complexity of medical treatment should guide the selection of an appropriate impairment percentage and estimate within any class.

The activities of daily living (ADL) include self-care and personal hygiene, communication, physical activity, sensory function, hand functions (grasping, holding, pinching, percussive movements, and sensory discrimination), travel, sexual function, sleep, and social and recreational activities. Other examples of specific ADLs are listed in the glossary of the *Guides*.

The examples within each class are very important guides for the first time user. It is critically important to remember that impairment is not determined by diagnosis alone, but also by the effect of the disease on ADLs along with the frequency and intensity of disease and the frequency and complexity of therapy. Most cutaneous impairment falls within the first three classes ranging from 0 percent to 54 percent.

Unique to the *Guides'* skin chapter are the discussions of pruritus (itching), disfigurement, scars and skin grafts, and patch testing. Itching is evaluated by determining its interference with the ADLs and the extent to which the description of pruritus is supported by objective findings such as lichenification, excoriation, or hyperpigmentation. Disfigurement usually involves no loss of body function and little or no effect on the ADLs. Disfigurement may well impair self-image, cause life-style alteration, and result in social rejection. These changes are best evaluated

**Table 3-G.** Impairment Classes and Percents for Skin Disorders

| Class 1: 0%–9% impairment | Class 2: 10%–24% impairment | Class 3: 25%–54% impairment | Class 4: 55%–84% impairment | Class 5: 85%–95% impairment |
|---|---|---|---|---|
| Signs and symptoms of skin disorder are present or only intermittently present; | Signs and symptoms of skin disorder are present or intermittently present; | Signs and symptoms of skin disorder are present or intermittently present; | Signs and symptoms of skin disorder are *constantly* present; | Signs and symptoms of skin disorder are *constantly* present; |
| **and** | **and** | **and** | **and** | **and** |
| There is no limitation or minimal limitation in the performance of *few* activities of daily living, although exposure to certain chemical or physical agents might increase limitation temporarily; | There is limitation in the performance of *some* of the activities of daily living; | There is limitation in the performance of *many* of the activities of daily living; | There is limitation in the performance of *many* of the activities of daily living that may include intermittent confinement at home or other domicile; | There is limitation in the performance of *most* of the activities of daily living, including occasional to constant confinement at home or other domicile; |
| **and** | **and** | **and** | **and** | **and** |
| *No* treatment or intermittent treatment is required. | Intermittent to constant treatment may be required. | Intermittent to constant treatment may be required. | Intermittent to constant treatment may be required. | Intermittent to constant treatment may be required. |

The signs and symptoms of disorders in classes 1 and 2 may be intermittent and not present at the time of examination. The impact of the skin disorder on daily activities should be the primary consideration in determining the class of impairment. The frequency and intensity of signs and symptoms and the frequency and complexity of medical treatment should guide the selection of an appropriate impairment percentage and estimate within any class (see chapter introduction). Permission to reprint from American Medical Association.

in accordance with the criteria in the chapter on mental and behavioral conditions. Evaluation of scars and skin grafts is made according to the impact on ADLs. When impairment is based on peripheral nerve dysfunction or loss of range of motion, it may be evaluated according to the criteria in the chapters on the nervous system and musculoskeletal system. When properly performed and interpreted, patch tests can make a significant contribution to the diagnosis of allergic contact dermatitis

## Conclusions and Recommendations

All diagnoses should be listed and summarized. A summary statement regarding causation is then made that states whether, within a reasonable degree of medical certainty, the disease is related to work. The diagnosis should include a description of specific clinical findings related to the impairment and how they relate to and compare with the criteria in the *Guides*. The impairment value also should be explained. Specific recommendations for therapy should be included along with a brief explanation of the treatment. Recommendations for prevention, including work restrictions, are next; possible suggestions are environmental modification (exhaust ventilation, splash guards) and personal protective equipment. The effect of future exposures to chemical, physical, and biologic agents should be addressed, along with any need for rehabilitation. In conclusion, impairment evaluation is an important and sometimes daunting task for physicians evaluating workers with putative occupational diseases.

## Physical Examinations

Preplacement examinations will help identify those who may be especially susceptible to skin irritations. The physician in charge of the examination should be provided with detailed information regarding the type of work for which a person is being considered.

Routine use of preplacement patch tests to determine sensitivity to various materials is not recommended. Patch tests cannot predict whether new workers will become sensitized to certain materials and develop dermatitis, but only tell whether people who have previously worked on similar jobs are sensitized to the chemicals with which they have worked.

The industrial physician has the primary responsibility for determining whether an applicant may be predisposed to skin irritations and for recommending suitable placement on the basis of these findings. Nevertheless, considerable responsibility also may fall to the safety and personnel departments, supervisors, industrial hygienists, and other people functionally responsible for accident prevention work and control of industrial diseases.

Care should be taken in restricting people who are not specifically sensitive to the agents involved in the job just because of a history of skin trouble unless there is active skin disease at the time of placement. In many cases, the physician is limited to simply counseling the person about risk.

## ■ Prevention and Control

Dermatoses caused by substances or conditions present in the work environment are largely preventable, but only through the combined effort of management and workers. This type of combined effort is best demonstrated in large industrial firms.

There are two major approaches to the prevention and control of occupational diseases in general and dermatoses in particular: environmental control measures and personal hygiene methods. In both cases, the key is cleanliness, both environmental and personal.

### Environment

Environmental cleanliness includes good housekeeping (discussed later in this section). Its primary function in preventing industrial dermatitis (and other industrial diseases) is to reduce the possibility of contact with the offending agent.

**Planning.** Proper design of equipment during construction is of great importance in the reduction of dermatitis and other industrial health problems. Ventilation must meet the industry requirements.

Provisions must be made for the safe handling of irritant chemicals. Pumps, valves, pipes, fittings, and the like must be maintained to eliminate (as much as possible) the contact of workers with irritants. Empty drums or bags used to transport incoming materials of a hazardous nature should be properly disposed of to prevent accidental exposure. Containers being readied for shipment should be filled in a manner that prevents contact with workers and left clean so that truckers, warehouse workers, and others cannot accidentally contact a harmful material on contaminated surfaces. Containers with harmful materials should be labeled with proper precautions.

**Process control.** Before any new process or work procedure is introduced and before new substances are adopted in an established process, an industrial hygienist or chemist should carefully consider every aspect of the operation for possible or known dermatitis hazards, including those that can be caused by trace impurities. Analyzing work procedures and processes often requires specialized equipment and techniques.

Once the potential dermatitis-causing factors have been determined, suitable engineering controls can be instituted and built into the work processes or operations.

The best way of controlling dermatitis is to prevent skin contact with offending substances; if there is no exposure, there will be no dermatitis. Unfortunately, this is more easily said than done.

Operations should be planned and engineered to ensure minimal worker contact with irritants and sensitizing chemicals. When possible, chemicals of low toxicity and low irritant potential should be substituted. Enclosure guards and mechanical handling facilities may be necessary when an operation involves highly corrosive materials. Operations that give off dust, fumes, or vapors need suitable exhaust ventilation to minimize exposure. Low ambient relative humidity in the workplace may cause or contribute to some occupational skin diseases.

**Selection of materials.** Much can be done to minimize hazardous conditions through careful selection of materials. Dry sodium and potassium hydroxide, for example, are now available in virtually dust-free forms; for many uses, they can be purchased in solution and handled with pumps. Other products are available as prills (beads), pellets, granules, or solutions that do an adequate job and reduce the dust hazard. Concentrated solutions are also finding favor not only for safety, but for economy in handling.

Some compounds can be successfully used when the percentage of the irritant in the compound is reduced. In other cases, a less irritating or nonirritating material, or a less sensitizing material, can be substituted. The supplier should be asked to provide a closely related and generally satisfactory substitute for the irritating or sensitizing material. Examples include substituting one germicide for another in metalworking fluids when an allergy to the first germicide is found and using vinyl gloves in place of rubber gloves if a rubber allergy is found.

## Monitoring and Control Technology

In order to correctly measure and sample skin exposure to chemicals, it is important to understand the methods of such exposure. These include contact with chemicals in a container (spill), contact with contaminated surfaces (tools or rags), exposure to aerosols (fallout of mist or soluble powder), exposure to sprays (ballistic droplets versus mist), and permeation through clothing (protective and personal). Other factors to be considered in the evaluation of skin exposure to chemicals include variable deposition rates onto the body, the effect of clothing, duration of skin contact with the chemical, and time of skin retention and permeation through the skin.

Current sampling procedures are often difficult to apply to prevention of occupational skin disease. Some exceptions include use of wipe samples for chemical analysis, hand or skin rinses, dermal dosimeters or patches, sampling cotton socks when shoe contamination is suspected, and sampling air inside a suit when the air is contaminated. Air levels of dusts and chemicals may have some limited application. The use of fluorescent tracers, fiber-optic luminescence skin monitoring, and charcoal cloth absorptive padosimeters are currently being evaluated. Color indicator soaps have been used to detect exposure to tetryl (used in munitions facilities) and mercury. A scientific study established methods to determine relative benefits of equipment such as gloves and clothing to protect skin against styrene in a reinforced plastics facility. Another study dealt with the biological surveillance of workers exposed to dimethylformamide and the influence of skin protection on its percutaneous absorption. Standardized techniques have improved measurement of the effectiveness of protective material such as gloves and clothing against carcinogens and polychlorinated biphenyls.

**Good housekeeping.** Environmental cleanliness is nothing more than good housekeeping and it is maintained by frequently cleaning floors, walls, ceilings, windows, and machinery. Good housekeeping work is usually performed by a special maintenance group that is given direct responsibility for maintenance cleaning. In order to be effective, cleaning should be part of a plan and should be performed on schedule. The necessary equipment and materials to do the most effective job possible in a reasonable amount of time should be assigned, and housekeeping workers should be trained so that they perform their operations efficiently and safely.

Environmental cleanliness is important to maintain good morale, reduce contact dermatitis, and set an example for workers. Floors, walls, ceilings, and light fixtures should be cleaned regularly in order to maintain the best possible conditions in the facility. (The requirements of Part 1910 of the Occupational Safety and Health Standards, Section 1910.14, "Sanitation," contain details on housekeeping, waste disposal, vermin control, water supply, toilet facilities, washing facilities, change rooms, consumption of food and beverages on premises, and food handling.) As pointed out in the OSHA standards, washrooms, showers, toilets, and locker rooms should be kept clean and sanitary.

Many types of cleaners are available, from simple cleaning agents to complex formulations. These come in solid, liquid, or paste, and contain cleaners and sanitizing agents using synthetic detergents, soaps, and alkaline salts in combinations. Some mixtures include sanitizing agents to help prevent the spread of bacteria, fungi, and other biological agents. Environmental cleanliness and good housekeeping are also beneficial because they set an example for the workers and encourage personal cleanliness.

## Personal Cleanliness

The importance of personal cleanliness in the prevention and control of occupational dermatoses cannot be overemphasized. When investigating contact dermatitis, one should also consider the possibility of irritants contacted at home or with a hobby.

**Prevention of contact.** When facility and process design cannot eliminate all contact with irritants, personal protective equipment must be used. Included are gloves, gauntlets, aprons, and boots made of a material that is impervious to the particular substance. These, along with goggles, afford sufficient protection in most cases. Disposable gauntlets, aprons, boots, and gloves are available, but they are more subject to tears than heavier safety gear. Other gear may provide insulation against heat or light. All personal protective equipment required for a job should be carefully maintained and replaced when it becomes worn.

In order to minimize contact with harmful agents, workers must have access to facilities for washing hands and be furnished with other means of keeping clean. It is up to the employer to provide adequate washing facilities, good cleansing materials, and education on the need for good hygiene practices. Washbasins must be well-designed, conveniently located, and kept clean; otherwise they will be used infrequently, if at all. The farther workers must walk to clean up, the less likely they are to do so. Inconveniently located washbasins invite such undesirable practices as washing with more easily available solvents, mineral oils, or industrial detergents, none of which is intended for skin cleansing. For workers to keep their skin reasonably free of injurious agents, they must use washing facilities at least four times a day: during work (before eating, drinking, smoking, or using the restroom), before lunch, after lunch, and before leaving the facility (see Figure 3–7).

Those who work with toxic chemicals and radioactive substances must receive specific safe handling instructions and should take a shower after their work shift and change their clothing. Workers should be instructed in specific procedures for cleanliness. They should be told where, how, and when to wash and should be given sufficient time to wash, advised that they will be rated on this part of their job performance, and informed of the possible health hazards involved.

**Figure 3–7.** Wash facilities should be conveniently located and should be adequate for all needs.

For many exposures, frequent washing alone is a successful preventive, particularly when the dermatitis is caused by plugging of the pores, as from dust. In all cases, however, the use of large quantities of water on the skin following exposure to irritants is necessary. Safety showers and eyewash fountains should be available, and flushing should continue for at least 15 minutes.

It may be advisable in some instances to use neutralizing solutions after a thorough flushing with water. However, because some neutralizing solutions are themselves irritants, they should be used only on the advice of a physician.

The type of soap used is important. Even a generally good soap can cause irritation on certain types of skin. Harsh mineral abrasives can cause dermatitis in many people.

The choice of a good soap may in some cases involve technical considerations, which are better left to the medical department or other qualified department than to laypeople. The basic requirements of industrial skin cleansers are as follows:

- They should remove industrial soil quickly and efficiently.
- They should not harmfully dehydrate, abrade, or irritate the skin by normal application.
- They should flow easily through dispensers.
- They should be adequately preserved against microbial contamination.

Additional desirable qualities include the following:

- They should have aesthetic appeal (color and odor).
- They should have good foaming qualities.

A number of cases of industrial dermatitis are reported to be caused not by substances used in the workplace, but by cleansing materials used to remove those substances. A worker may be inclined to wash the hands with the cleaning agents that are most available and work the fastest, but these are often dermatitis-producing solvents. Overuse of waterless cleansers can irritate and dry the skin. Generally, workers should apply a good hand lotion after applying waterless cleansers.

The installation in work-area washing places of soap-dispensing units containing properly selected cleansing agents has proved to be a valuable measure. Such units should be placed in convenient locations, and enough of them should be provided to accommodate all employees who are exposed to skin irritants. Where soap-dispensing units are furnished, workers should be required to use them.

**Barrier creams.** A barrier cream is the least effective way of protecting skin (Figure 3–8). However, there are instances when a protective cream may be used for preventing contact with harmful agents when the face cannot be covered by a shield or gloves cannot be worn (Figure 3–9). Several manufacturers compound a variety of products, each designed for a certain type of protective purpose.

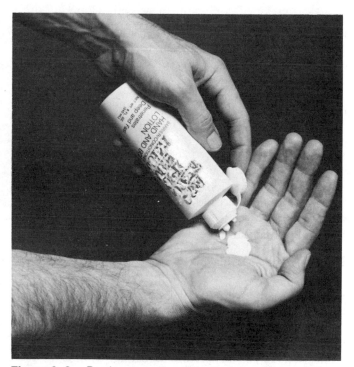

**Figure 3–8.** Barrier cream is applied by an employee before starting work and is washed off before lunch; it is reapplied after lunch and washed off before the worker leaves. Barrier cream is the least effective way of protecting the skin.

**Figure 3–9.** Most glove manufacturers provide a hand protection counseling service. They assess hazards and match them with gloves or other devices. It is up to the employer to educate employees to use the proper protection for the job performed.

Thus, there are barrier creams that protect against dry substances and those that protect against wet materials. Using a barrier cream to protect against a solvent is not as effective as using an impervious glove; however, there are compounds that offer some protection against solvents when applied with sufficient frequency.

Barrier creams and lotions should be used to supplement, but not to replace, personal protective equipment. Protective barrier agents should be applied to clean skin. When skin becomes soiled, both the barrier and any soil should be washed off and the cream reapplied.

Three main types of barrier creams and lotions are available.

*Vanishing cream* usually contains soap and emollients that coat the skin and cover the pores to make subsequent cleanup easier.

*Water-repellent* cream leaves a thin film of water-repellent substance such as lanolin, beeswax, petrolatum, or silicone on the skin, and helps to prevent ready contact with water-soluble irritants such as acids, alkalies, and certain metallic acids. Remember, however, that the protection may not be complete, especially when the barrier has been on the skin for some time. Alkaline cleaning solutions tend to emulsify and remove the barrier rapidly, thus leaving the skin unprotected.

*Solvent-repellent* creams contain ingredients that repel oil and solvent. Lanolin has some oil-repellent and water-repellent properties and can be used as an ingredient. There are two types of solvent-repellent barrier preparations; one leaves ointment film and the other leaves a dry, oil-repellent film. Sodium alginates, methyl cellulose, sodium silicate, and tragacanth are commonly used. Lanolin offers some protection against oils as well as water.

*Special types.* In addition to these three main types, a number of specialized barriers have been developed. Creams and lotions containing ultraviolet screening and blocking agents are used to help prevent overexposure to sun or other ultraviolet sources. Others have been developed to afford protection from such diverse irritants as insects, gunfire backflash, and poison oak dermatitis.

## Personal Protective Equipment

**Clothing.** Whenever irritating chemicals are likely to contaminate clothing, care must be taken to provide clean clothing at least daily. Because workers' families have developed contact dermatitis or chloracne from contact with clothing worn home from the job, clothing worn on the job should not be worn at home. Clothing contaminated with chemicals should always be thoroughly laundered before it is worn again. Clothing contaminated on the job should be changed at once.

**Protective clothing.** Sometimes handling irritant, allergenic, or toxic materials cannot be avoided; in this situation, protective clothing may provide a good barrier against exposure. OSHA requirements for protective clothing are described in Subpart I, Section 1910.132, General Requirements for Personal Protective Equipment. Other protective clothing requirements appear in standards covering specific hazards.

High-quality clothing should be obtained. Manufacturers provide a large selection of protective garments made of rubber, plastic film, leather, cotton, or synthetic fiber that are designed for specific purposes. For example, there is clothing that protects against acids, alkalis, extreme exposures of heat, cold, moisture, and oils. When such garments must be worn, management should purchase them and enforce their use. Management should make sure that the clothing is mended and laundered often enough to keep it protective. If work clothes are laundered at home, the worker's other clothes may be contaminated with chemicals, glass fiber, or other dusts.

Closely woven fabrics also protect against irritating dust. Gloves and aprons of impervious materials (such as rubber or plastic) protect against liquids, vapors, and fumes (Table 3–H). Natural rubber gloves, aprons, boots, and sleeves are impervious to water-soluble irritants, but soon deteriorate if exposed to strong alkali and certain solvents.

Synthetic rubbers, such as neoprene and many of the newer plastics, are more resistant to alkalis and solvents than is natural rubber; however, some materials are adversely affected by chlorinated hydrocarbon solvents. The protection used should be based on the particular solvents that are used. For workers who wear rubber or plastic, the irritant will eventually penetrate and be trapped next to the skin, causing repeated exposure every time the garment is worn. So gloves should be used to protect against splashes and wet items, but will not protect against immersions. Instruments or containers should be used for items that must be immersed. Reusable gloves should be washed according to the manufacturer's instructions.

Disposable paper and plastic garments can also be used for some tasks. Garments are also necessary in sterile areas to keep products from being contaminated (Figure 3–10).

**Fabrics.** Fabrics without coatings are generally unsuitable as protective clothing for toxic chemical exposures because they are all permeable and have other weaknesses. Cotton and rayon, for instance, are degraded by acids; wool is degraded by alkalis.

**Gloves.** Because a high percentage of occupational contact dermatitis cases involve the hands, it is imperative for health and safety professionals to be knowledgeable about the types of glove materials available, their selection and use, and the types of adverse dermatological reactions to gloves.

Glove materials include the following:

- Natural rubber latex
- Synthetic rubber latex and synthetic rubber: butyl, chloroprene (Neoprene®), fluon (Viton®), nitrile, styrene-butadiene block polymer (Elastyren®), and styrene-ethylene-butadiene (Tactylon®, Tactyl 1®, and Tactyl 2®)
- Plastic polymers: ethylene-methyl methacrylate, polyethylene, polyvinyl alcohol, and polyvinyl chloride
- Laminated plastic polymers: a laminate of polyethylene-ethylenevinylalcohol copolymer-polyethylene (Silver Shield® and 4 H® gloves)

**Table 3-H.** Physical Performance Chart of Selected Glove Materials

| Coating | Abrasion Resistance | Cut Resistance | Puncture Resistance | Heat Resistance | Flexibility | Ozone Resistance | Tear Resistance | Relative Cost |
|---|---|---|---|---|---|---|---|---|
| Natural rubber | E | E | E | F | E | P | E | Medium |
| Neoprene | E | E | G | G | G | E | G | Medium |
| Chlorinated polyethylene (CPE) | E | G | G | G | G | E | G | Low |
| Butyl rubber | F | G | G | E | G | E | G | High |
| Polyvinyl chloride | G | P | G | P | F | E | G | Low |
| Polyvinyl alcohol | F | F | F | G | P | E | G | Very high |
| Polyethylene | F | F | P | F | G | F | F | Low |
| Nitrile rubber | E | E | E | G | E | F | G | Medium |
| Nitrile rubber/Polyvinyl chloride (Nitrite PVC) | G | G | G | F | G | E | G | Medium |
| Polyurethane | E | G | G | G | E | G | G | High |
| Styrene-butadiene rubber (SBR) | E | G | F | G | G | F | F | Low |
| Viton | G | G | G | G | G | E | G | Very high |

Grip/slip is related to glove surface and is enhanced when the glove is rough.

Dexterity is related to glove thickness and decreases as the glove thickness increases.

KEY TO CHART: E=excellent; G=good; F=fair; P=poor (Ratings are subject to variation, depending on formulation, thickness, and whether the material is supported by fabric).

The listings were taken from various glove manufacturers and NIOSH, and are ONLY A GENERAL GUIDE. When selecting gloves for any application, contact the manufacturer, giving as much information as possible.

This table shows the *physical* properties of selected glove materials. For *chemical resistance* properties of glove materials see section *Gloves* and published reference guides in that section.

A number of standards, guides, and rules have recently been published for gloves in both the United States and Europe. Categories of glove standards depend on the type of glove material used, type of work being done, and type of hazard encountered. Parameters of glove performance evaluated include physical strength, dexterity, abrasion, and heat and cold resistance. Other physical resistance factors include cut and puncture resistance, tear and tensile strength, and flammability (see Table 3-H). Resistance to swelling, degradation, permeation, and penetration are some of the more important chemical parameters evaluated; biological resistance to liquids and microorganisms also has been evaluated. Consult the references by Henry (1994) and Mellstrom & Carlsson (1994) for a list of current standards.

Forsberg & Mansdorf (1993) outline a number of factors to consider in the selection of gloves and other protective clothing:

- Chemicals ultimately penetrate protective barriers and can do so without evidence of damage to the barrier.

- Protective equipment cannot be used if torn or damaged.

- A barrier may protect against a single chemical but not necessarily against a mixture of chemicals.

- Temperatures higher than room temperature decrease breakthrough time of chemicals. The authors' *Quick Selection Guide* (1993) is based on room temperature data.

- Generally, thicker materials or layers of material are better for chemical resistance.

- Beware of look-alikes with chemical protective clothing.

- If a chemical has penetrated a material, the garment must be decontaminated before reuse. Gloves used in health care must be disposed of properly after each patient contact and cannot be decontaminated.

- Users of chemical protective clothing should check with their suppliers to confirm the performance standards of their products, to verify information on proper storage and care, and to select materials.

Another critically important factor is safety. Gloves should not be used where they may be caught in machinery, with potential for serious injury or loss of fingers or hands. Safety and health professionals should review any recommendation for glove use by health care providers who may not be familiar with a specific job.

Reference guidelines for the selection of gloves include the following:

- Schwope AD, et al. *Guidelines for the Selection of Chemical Protective Clothing,* 3rd ed., Vols. I and II. Cincinnati: American Conference of Governmental Industrial Hygienists, 1987.

- Forsberg K, Mansdorf SZ. *Quick Selection Guide to Chemical Protective Clothing,* 2nd ed. New York: Van Nostrand, Reinhold, 1993.

**Figure 3–10.**   Protective clothing is worn over work clothes to prevent contamination of both the worker and, in sterile rooms, the product.

- Leather (chrome or vegetable tanned)
- Textiles: natural or synthetic; woven from fabrics, knit or terry cloth; also may be coated with rubber or plastic; fabric such as cotton or nylon may also be used as glove liners
- Other materials: wire cloth made of stainless steel or nickel as used by autopsy prosectors

■ Computer data bases: Consult Mellstrom & Carlsson (1994) for a list of data bases.

■ Other sources: Consult Mellstrom et al. (1994) for lists of other general texts on chemical protective clothing and the reference by Hamann & Kick (1994) for the selection of gloves for health care workers. Also see the discussion of contact urticaria/latex allergy above and of dermatologic reactions to gloves below.

Dermatologic reactions to gloves are classified as irritation from occlusion, friction, and maceration; allergy to glove materials and their chemical additives causing allergic contact dermatitis or allergy to certain natural rubber latex proteins causing contact urticaria; aggravation of preexisting skin diseases; and penetration of chemicals through gloves. Infrequent reactions to endotoxins and ethylene oxide, and to potentially depigmenting chemical constituents of gloves have been reported. Accurate diagnosis of any suspected adverse reactions to gloves is imperative before recommending alternatives to gloves. This is especially true of workers with contact urticaria to latex (latex allergy) who may still develop a severe reaction from gloves labeled as hypoallergenic. Because so-called hypoallergenic gloves may still contain allergenic latex proteins, synthetic gloves are necessary for these workers. See Table 4 in the Hamann & Kick (1994) and other chapters in the book *Protective Gloves of Occupational Use* (Mellstrom et al., 1994) for help in the selection of medical examination and sterile surgical gloves. *As this information may change, it is important to consult the glove manufacturer for the latest data.*

**Safety.** It is imperative that all safety gear be worn only when safely possible. Protective clothing, especially gloves, can be caught in moving machinery, resulting in serious injury.

## Responsibility for Control

Top management, the safety department, the purchasing department, the medical department or company physician, the supervisors, and the workers all have specific responsibilities for the prevention of industrial skin diseases and the control of exposure to skin irritants.

To control or eliminate dermatitis in the workplace, management should first recognize the scope of the problem and then delegate authority for action to the proper employees. When it is necessary to have more than one department work on phases of dermatitis control or elimination, the activities of those departments should be coordinated. Periodic reports on the status of the dermatitis problem within the organization should be made to management by its delegated representatives.

The industrial hygienist (or people doing this type of work, such as the safety professional, safety committee members, the nurse, or the industrial hygienist) should gather information on dermatitis hazards of materials used

in the plant and should disseminate this information among supervisors and other operating personnel. The industrial hygienist should make periodic surveys to check for exposure to skin irritants and should suggest means to correct any hazards found.

## Case Examples of Control

The following examples of the use of controls to reduce or eliminate occupational dermatoses were taken from the 1978 *Report of the OSHA Advisory Committee on Cutaneous Hazards* and are still applicable today.

**Powered epoxy spraying operation.** A manufacturer of household washing machines began using an epoxy material as a finished surface on its products. The epoxy material came in powdered form and was sprayed on the parts to be assembled, which were then baked in an oven to form an extremely hard surface. The spraying was done automatically, inside a booth. The parts passed through the booth hanging from an overhead conveyor. Overspray was exhausted out the bottom of the booth and into a barrel; some overspray remained on the inside walls of the booth. The only worker in the area during the spraying was an operator who sat inside an enclosed control booth and thus was not exposed to the epoxy powder.

On the midnight shift, however, when production was stopped, a cleanup crew entered the area to perform a number of duties:

■ They used air hoses to blow out the overspray that had accumulated on the inside walls of the spray booth.

■ They dumped barrels of exhausted overspray back into the supply system for reuse.

■ They swept floors and other surfaces outside the booth to clean up some spray that had escaped the booth.

The powder was very fine and the slightest turbulence caused it to become airborne; consequently, a great concentration of epoxy dust was in the air. The cleanup crew was equipped with disposable respirators, hair covers, boots, and complete coveralls. Despite the personal protection, several members of the cleanup crew broke out in rashes after spraying had been performed for a few weeks.

The problem was solved, after an investigation, by changing the overspray exhaust system to return the overspray directly into the supply system, thus eliminating one major source of dust. Using a vacuum system rather than sweeping or air hoses eliminated the other sources of dust. No cases of dermatitis recurred.

**Machining operations.** Exposure to cutting fluids in machining operations constitutes one of the major causes of industrial dermatitis. Controls that have virtually eliminated dermatitis have been instituted in many machining operations. For example, in one well-controlled plant that produces diesel engines, more than 2,000 workers on two shifts operating approximately 1,000 machines had not a single case of recordable occupational dermatitis in 1977, in contrast with some poorly controlled operations in which roughly 30 percent of the work force have skin problems. Control programs put into effect included the following:

- Careful identification, by generic name, of all ingredients in the cutting fluids used
- Programs to keep the coolant free of tramp oil, foreign particles, and dirt through the use of effective filters and redesign of the coolant flow system to eliminate eddies and backwash of coolant
- Daily programs to monitor coolant characteristics such as pH and bacteria count
- Daily programs such as hosing down to keep machinery clean
- Redesign of spray application to minimize coolant splash and spray
- Use of splash goggles and curtains
- Use of local exhaust systems and oil collectors to reduce airborne oil mist
- Use of abundant quantities of shop rags
- Provision of paid wash time to allow operators to keep clean

Experience shows that when the coolant is well-controlled and measures are taken to reduce the amount of coolant splashed on the worker, the rate of dermatitis is reduced.

**Rubber manufacturing.** Improvements in rubber manufacturing operations have problems with skin disease in facilities where the improvements were made. These improvements have taken many forms:

- Methods of material handling have been improved to reduce the amount of skin contact with rubber and related chemicals.
- Known skin sensitizers have been replaced by less hazardous chemicals, such as the replacement of isopropyl-phenylparaphenyl-diamine (IPPD), used as an antioxidant in tires, with less toxic derivatives of paraphenylene diamine.
- Methods of mixing rubber chemicals have been improved to reduce exposure to a wide variety of known skin irritants and sensitizers; such improvements have included preblending chemicals, using exhaust-ventilated mixing booths, and automating the mixing process.
- In one plant, an air-conditioned isolation booth was installed for a worker who was strongly sensitized to an antiozonent.

**Chemical manufacturing.** A major producer of industrial chemicals has instituted a wide variety of controls that have resulted in a reduced rate of dermatitis. Their program includes the following controls:

- Extensive use of self-contained systems to handle chemicals to eliminate worker exposure to dermatitis-producing substances, designed with a goal of zero emissions
- Mechanization of material-handling systems to eliminate worker exposure to chemicals
- Emphasis on good housekeeping
- Use of wipe testing to check equipment surfaces for films of toxic materials
- Adoption of extensive employee education programs to inform employees of the risks of chemicals
- Implementation of programs of personal hygiene that, in the case of one particularly hazardous material, included three daily showers for the exposed employee
- For handling liquid chemicals, use of sealless pumps and, where leaks cannot be permanently sealed, use of local exhaust systems; grouping of all pumps in one central area for better control; and scaling of floors around the pumps
- Preparation of educational materials to be supplied to purchasers of chemicals, including proper controls for the materials

## Summary

Occupational skin disorders are a significant cause of impairment and disability that in many cases are entirely preventable. Accurate diagnosis and a complete knowledge of the workplace are the keys to appropriate treatment and prevention. The industrial hygienist is a major player in this cooperative effort, along with the occupational physician, dermatologist, safety professional, and occupational health nurse.

## Bibliography

Adams RM, ed. *Occupational Skin Diseases,* 2nd ed. Philadelphia: W.B. Saunders, 1990.

American Medical Association. *Guides to the Evaluation of Permanent Impairment,* 4th ed. Chicago: American Medical Association, 1993.

Berardinelli S. Chemical protective gloves. *Dermatol Clin* 6: 11–120, 1988.

Birmingham DJ. Occupational dermatoses. In *Patty's Industrial Hygiene and Toxicology,* vol. 1, Part A, *General Principles,* 4th ed., edited by Clayton GD, Clayton FE. New York: Wiley Interscience, 1991.

Birmingham DJ, ed. *The Prevention of Occupational Skin Diseases.* New York: Soap and Detergent Association, 1981.

Burman LG, Fryklund B. The selection and use of gloves by health care professionals. In *Protective Gloves for Occupational Use,* edited by Mellstrom GA, Wahlberg JE, Maibach HI. Boca Raton, FL: CRC Press, 1994.

Centers for Disease Control. *Morbidity and Mortality Weekly Reports.* Atlanta: Centers for Disease Control, January 21, 1983, and September 5, 1986.

Cohen BSM, Popendorf WA. A method for monitoring dermal exposure to volatile chemicals. *AIHAJ* 50:216–223, 1989.

Cohen SR. Risk factors in occupational skin disease. In *Occupational and Industrial Dermatology,* 2nd ed., edited by Maibach HI. Chicago: Year Book Medical Publishers, 1987.

Cooley DG, ed. *Family Medical Guide.* New York: Better Homes and Gardens Books, 1978.

Estlander T, Jolanki R. How to protect the hands. *Dermatol Clin* 6:105–113, 1988.

Fisher AA. *Contact Dermatitis.* Philadelphia: Lea & Febiger, 1986.

Fitzpatrick TB, et al. *Dermatology in General Medicine,* 3rd ed. New York: McGraw-Hill, 1987.

Forsberg K, Mansdorf SZ. *Quick Selection Guide to Chemical Protective Clothing,* 2nd ed. New York: Van Nostrand Reinhold, 1993.

Goldsmith LA, ed. *Biochemistry and Physiology of the Skin.* New York: Oxford, 1983.

Hamann CP, Kick SA. Diagnosis-driven management of natural rubber latex glove sensitivity. In *Protective Gloves for Occupational Use,* edited by Mellstrom GA, Wahlberg JE, Maibach HI. Boca Raton, FL: CRC Press, 1994.

Henry N. Protective gloves for occupational use—U.S. rules, regulations, and standards. In *Protective Gloves for Occupational Use,* edited by Mellstrom GA, Wahlberg JE, Maibach HI. Boca Raton, FL: CRC Press, 1994.

Hurley HJ. Permeability of the skin. In *Dermatology,* edited by Moschella SL, Hurley HJ. Philadelphia: Saunders, 1985.

ILO. *Encyclopedia of Occupational Health and Safety,* 3rd ed. Geneva, Switzerland: International Labor Organization, 1983.

Industrial dermatitis: Part 1: The skin. *National Safety News* 112:59–62, 1975a.

Industrial dermatitis: Part 2: Primary irritation. *National Safety News* 112:107–112, 1975b.

Industrial dermatitis: Part 3: Sensitization dermatitis. *National Safety News* 112:65–68, 1975c.

Jackson EM, Goldner R. *Irritant Contact Dermatitis.* New York: Marcel Dekker, 1990.

Jakubovic HR, Ackerman AB. Structure and function of skin. In *Dermatology,* 2nd ed., edited by Moschella SL, Hurley HJ. Philadelphia: Saunders, 1985.

Key MM et al., eds. *Occupational Diseases—A Guide to Their Recognition.* DHEW NIOSH Publication No. 77–181, 1977.

Leinster P. The selection and use of gloves against chemicals. In *Protective Gloves for Occupational Use,* edited by Mellstrom GA, Wahlberg JE, Maibach HI. Boca Raton, FL: CRC Press, 1994.

Litt JZ. *Your Skin and How to Live in It.* Cleveland: Corinthian Press, 1980.

Maibach H. *Occupational and Industrial Dermatology,* 2nd ed. Chicago: Year Book Medical Publishers, 1987.

Mathias CGT. Contact dermatitis and workers' compensation: Criteria for establishing occupational causation and aggravation. *J Am Acad Dermatol* 20:842–848, 1989.

Mellstrom GA, Carlsson B. European standards on protective gloves. In *Protective Gloves for Occupational Use,* edited by Mellstrom GA, Wahlberg JE, Maibach HI. Boca Raton, FL: CRC Press, 1994.

Mellstrom GA, Wahlberg JE, Maibach HE, eds. *Protective Gloves for Occupational Use.* Boca Raton, FL: CRC Press, 1994.

Mitchell JA, Rook A. *Botanical Dermatology.* Vancouver: J.A. Mitchell, 1979.

National Safety Council, Occupational Safety and Health Data Sheet: *Poison Ivy, Poison Oak, and Poison Sumac, 12304-0304,* Itasca, IL: National Safety Council, 1983.

Nethercott JR. The Americans with Disabilities Act. *Amer J Contact Dermatitis* 4:185–186, 1993.

Nethercott JR. Disability due to occupational contact dermatitis. *Occupational Medicine; State of the Art Reviews* 1:199–203, 1986.

The number one occupational illness: Dermatitis. *National Safety News* 105:38–43, 1972.

*Report of the Advisory Committee on Cutaneous Hazards to the Assistant Secretary of Labor, OSHA.* Washington, DC: Occupational Safety and Health Administration, U.S. Department of Labor, December 19, 1978.

Rook A et al. *Textbook of Dermatology,* 4th ed. Oxford: Blackwell, 1986.

Rycroft RJG. Low-humidity occupational dermatoses. *Dermatol Clin* 2:553–559, 1984.

Samitz MH. Assessment of cutaneous impairment and disability. In *Occupational and Industrial Dermatology,* 2nd ed., edited by Maibach HI. Chicago: Year Book Medical Publishers, 1987.

Schwope AD et al. *Guidelines for the Selection of Chemical Protective Clothing,* 3rd ed., Vols I and II. Cincinnati: American Conference of Governmental Industrial Hygienists, 1987.

Shmunes E. Predisposing factors in occupational skin diseases. *Dermatol Clin* 6:7–14, 1988.

Taylor JS. Evaluation of impairment due to work-related skin disease. *Occupational Medicine; State of the Art Reviews* 9:1–10, 1994.

Taylor JS. Other reactions from gloves. In *Protective Gloves for Occupational Use,* edited by Mellstrom GA, Wahlberg JE, Maibach HI. Boca Raton, FL: CRC Press, 1994.

Taylor JS, ed. Occupational dermatoses. *Dermatol Clin* 6:1–129, 1988.

Taylor JS. The pilosebaceous unit. In *Occupational and Industrial Dermatology,* 2nd ed., edited by Maibach HI. Chicago: Year Book Medical Publishers, 1987.

Taylor JS. Occupational dermatoses. In *Clinical Medicine for the Occupational Physician,* edited by Alderman MH, Hanley JJ. New York: Dekker, 1982.

Taylor JS, Parrish JA, Blank IH. Environmental reactions to chemical, physical, and biological events. *J Am Acad Dermatol* 11:1007–1021, 1984.

Toxic plants. *National Safety News* 105:45–59, 1972.

Tucker SB. Prevention of occupational skin disease. *Dermatol Clin* 6:87–96, 1988.

Wilkening GM. Ionizing radiation. In *Patty's Industrial Hygiene and Toxicology,* Vol. I, *General Principles,* 3rd ed., edited by Clayton GD, Clayton FE. New York: Wiley Interscience, 1978.

CHAPTER 4

# The Ears

by George S. Benjamin, MD, FACS
and Barry J. Benjamin, MD, FACS

**Anatomy**
*External Ear* ▪ *Middle Ear* ▪ *Inner Ear*

**Physiology**
*External Ear* ▪ *Middle Ear* ▪ *Inner Ear*

**Pathology**
*External Ear* ▪ *Ear Canal* ▪ *Eardrum* ▪ *Eustachian Tube* ▪ *Middle Ear* ▪ *Inner Ear* ▪ *Tinnitus (Head Noises)* ▪ *Nonauditory Effects*

**The Hearing Process**
*Audiometer* ▪ *Audiogram*

**Hearing Loss**
*Bone-Conduction Tests* ▪ *Types of Hearing Loss*

**Effects of Noise Exposure**
*Temporary Threshold Shift* ▪ *Permanent Threshold Shift* ▪ *Noise-Induced Hearing Loss*

**Communication Problems**
*Hearing Versus Understanding* ▪ *Loudness* ▪ *Clarity* ▪ *Speech Sounds* ▪ *Treatment*

**AMA Guides for Evaluating Impairment**

**OSHA Hearing Conservation Program (29 *CFR* 1910.95)**

**Summary**

**Bibliography**

THE ANATOMY OF THE HUMAN EAR is very complex, and its minute size and protective encasement in hard, dense bone further complicate scientific study of this delicate organ. Many aspects of ear function are unknown, particularly those involving the inner ear and the pathways leading to the brain. For safety and health professionals, this chapter presents an overview of the structure of the auditory mechanism, how it seems to work, and the common causes of impairment of this sensitive organ.

The auditory mechanism enables us to hear sound. In air, sound is defined as variations of pressure above and below the ambient atmospheric pressure. These air pressure fluctuations, or sound waves, vary in intensity, harmonic content, frequency, and direction. The word *sound* also indicates the sensation experienced when pressure fluctuations strike the ear. (Chapter 9, Industrial Noise, covers this subject in detail.)

Through the ear, sound waves are detected within a range of 20–20,000 Hz and then are converted into electrical impulses that are transmitted to the brain for interpretation. The brain performs this function by receiving the sensory information as a series of electrical impulses and sending or shunting these impulses from one brain cell to another.

The organ of hearing is divided into three parts—the external, the middle, and the inner ear (Figure 4–1).

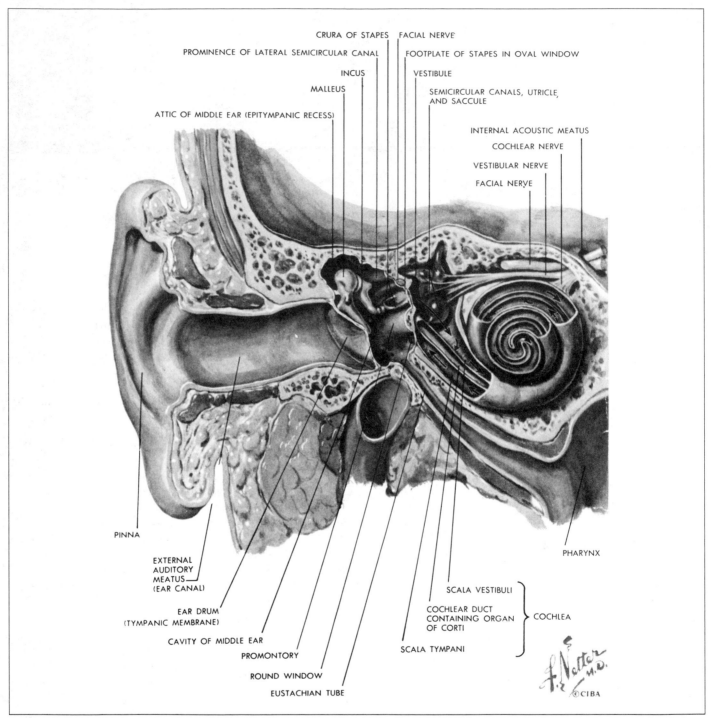

**Figure 4–1.** Illustration of the outer, middle, and inner ear. (Reprinted with permission from Netter FH, *Clinical Symposia.* CIBA Pharmaceutical Co.)

## Anatomy

### External Ear

The external ear is divided into two sections: the portion attached to the outer surface of the head, called the auricle or pinna, and the external auditory canal.

**The pinna.** This is the most visible part of the ear, a delicately folded cartilaginous structure with a few small muscles, covered by subcutaneous tissue and skin. The pinna functions as a funnel to collect sound waves. Many animals can voluntarily control the ear muscles and aim the pinnae to enhance the collection of sound waves.

**The external auditory canal.** This canal, or meatus, is a skin-lined pouch about 1.5 in. (3.8 cm) long, supported in its outer third by the cartilage of the pinna and in its inner two thirds by bone of the skull. At its innermost end lies the tympanic membrane, or eardrum, which separates the external from the middle ear.

The small hairs, or vibrissae, and ceruminal glands, which secrete a waxy substance called cerumen, are located in the skin of the outer third of the ear canal. The hairs serve a protective function by filtering out particulate matter and other large pieces of debris. Cerumen, both sticky and bactericidal, prevents smaller particles from entering the ear canal and keeps the canal healthy and free of infection.

## Middle Ear

The middle ear is the space or cavity, about 1–2 ml in volume, between the eardrum and the bony wall of the inner ear (Figure 4–2). The middle ear is lined with mucous membrane essentially the same as that lining the mouth. The ossicles, which are the smallest bones in the body, are located within the middle ear cavity (Figure 4–3). The ossicles connect the eardrum to an opening in the wall of the inner ear called the oval window.

Picture the middle ear space as a cube:

- The outer wall is formed by the eardrum.
- The inner wall is the bony partition separating the inner ear from the middle ear. The round and oval windows fit into this wall and include the only two movable barriers between the middle and inner ear.
- The front wall opens into the eustachian tube.
- The back wall opens into the mastoid air cells.
- The roof separates the middle ear from the temporal lobe of the brain.
- The floor separates the middle ear from the jugular vein and the internal carotid artery, which lies high in the neck.

The sound-conducting mechanism in the middle ear includes the eardrum and three ossicles (bones) that are supported by ligaments and two muscles.

**The eustachian tube.** This tube equalizes the pressure in the middle ear with the external atmospheric pressure. It opens during swallowing and yawning. It tends to remain closed when the pressure is increasing, as during rapid descent in an airplane. In some people, after frequent

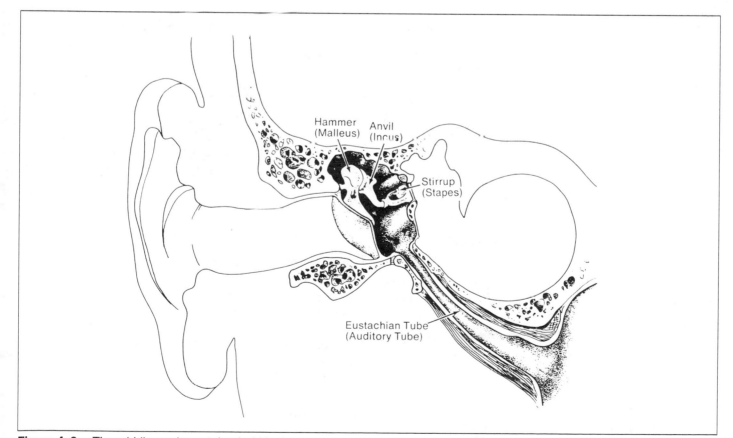

**Figure 4–2.**   The middle ear is contained within the temporal bone and is made up of the eustachian tube, the middle ear space, and the mastoid air cell system. (Adapted from Figure 4–1.)

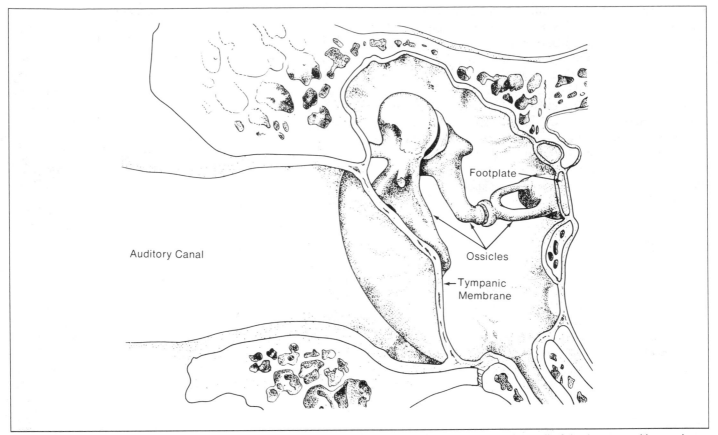

**Figure 4–3.** The ossicles, located in the middle ear cavity, link the eardrum to an opening in the wall of the inner ear (the oval window). (Adapted from Figure 4–1.)

infection, the tube is closed permanently by adhesions and their hearing is impaired greatly. When the pressure is unequal on the two sides of the eardrum, it cannot vibrate freely in response to sound waves.

**The eardrum.** This membrane separates the external ear canal from the middle ear. It consists of an inner layer of mucous membrane, a middle layer of fibrous tissue, and an outer layer of squamous epithelium. It is shaped like a spiderweb, with radial and circular fibers for structural support.

**Ossicular chain.** The ossicles, which together are called the ossicular chain, are the malleus, incus, and the stapes.

- The malleus, or hammer, is fastened to the eardrum by the handle. The head lies in the upper area of the middle ear cavity and is connected to the incus.

- The incus, also called the anvil, is the second ossicle and has a long projection that runs downward and joins the stapes.

- The stapes, also called the stirrup, lies almost perpendicular to the long axis of the incus. The two branches of the stapes, anterior and posterior, end in the footplate that fits into the oval window.

When the handle of the hammer is set into motion by movement of the eardrum, the action is transferred mechanically through the ossicular chain to the oval window.

**The oval window and the round window.** These are located on the inner wall of the middle ear. The round window is covered by a very thin membrane that moves out as the footplate in the oval window moves in. As the action is reversed and the footplate in the oval window is pulled out, the round window membrane moves inward.

**Mastoid air cell system.** On the back wall of the middle ear space is an opening that extends into the mastoid. This opening resembles a honeycomb of spaces filled with air. The mucous membrane lining the middle ear is continuous with that of the pharynx and the mastoid air cells; thus, it is possible for infection to travel along the mucous membrane from the nose or the throat to the middle ear and to the mastoid air cells.

## Inner Ear

The inner ear contains the receptors for hearing and position sense. It consists of a bony labyrinth that contains a membranous labyrinth.

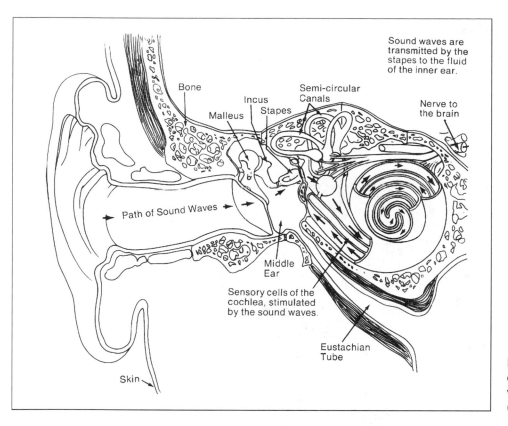

**Figure 4–4.** The major components of the inner ear, including the vestibular system and the cochlea. (Adapted from Figure 4–1.)

The bony labyrinth consists of a series of tiny canals and cavities hollowed out of the petrous portion of the temporal bone. They contain a watery fluid called perilymph. There are three bony divisions: the cochlea, the vestibule, and the semicircular canals (Figure 4–4).

**The cochlea.** This is shaped like a snail shell (see Figure 4–5). It is a bony tube winding around a central pillar of bone called the modiolus. The membranous cochlear duct resembles a lopsided triangle as it lies within the bony cochlea. It extends like a shelf across the bony canal and is attached to the sides. The portion of the bony canal above the cochlear duct is called the scala (stairway) vestibuli, and the portion below is called the scala tympani. They are continuous with each other through a tiny opening (helicotrema) in the apical end of the cochlea.

The scala vestibuli begins beneath the footplate of the stapes and is separated from the scala media below it by Reissner's membrane. The scala vestibuli continues to the helicotrema, where it joins the scala tympani.

The scala tympani lies below the scala media, separated from it by the basilar membrane, and ends at the round window. The oval and round windows lie at opposite ends of the perilymphatic space of the cochlea (Figure 4–6).

**Organ of Corti.** The cochlear duct is connected with the inner wall of the bony canal by the osseous spiral lamina and with the outer wall by the spiral ligament. The roof of the cochlear duct is thin and is called Reissner's membrane.

The floor is composed of the basilar membrane. Resting on this basilar membrane is the spiral organ of Corti, the essential receptor end organ for hearing (Figure 4–5). It is a very complicated structure, consisting of a supporting framework on which the hair cells rest.

## ■ Physiology

### External Ear

The function of the outer ear in the hearing process is relatively simple: The external portion of the ear collects sound waves from the air and funnels them into the ear canal, where they are transported to the eardrum. The collected sound waves cause the eardrum to move back and forth in a vibrating mechanical motion that is passed on to the bones of the middle ear (Figure 4–6).

### Middle Ear

The primary function of the middle ear in the hearing process is to transfer sound energy from the outer to the inner ear. As the eardrum vibrates, it transfers its motion to the attached hammer (malleus). Because the bones of the ossicular chain are connected to one another, the movements of the hammer are passed on to the anvil, and finally to the stirrup embedded in the oval window.

As the stirrup moves back and forth in a rocking motion, it passes the vibrations on to the inner ear through the

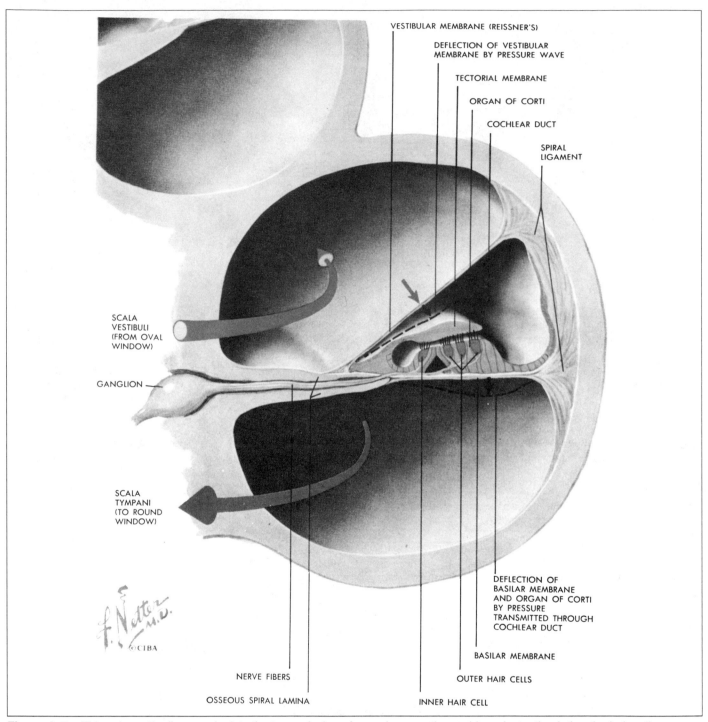

VESTIBULAR MEMBRANE (REISSNER'S)

DEFLECTION OF VESTIBULAR
MEMBRANE BY PRESSURE WAVE

TECTORIAL MEMBRANE

ORGAN OF CORTI

COCHLEAR DUCT

SPIRAL
LIGAMENT

SCALA
VESTIBULI
(FROM OVAL
WINDOW)

GANGLION

SCALA
TYMPANI
(TO ROUND
WINDOW)

DEFLECTION OF
BASILAR MEMBRANE
AND ORGAN OF CORTI
BY PRESSURE
TRANSMITTED THROUGH
COCHLEAR DUCT

BASILAR MEMBRANE

OUTER HAIR CELLS

NERVE FIBERS

OSSEOUS SPIRAL LAMINA

INNER HAIR CELL

**Figure 4–5.** This schematic diagram depicts the transmission of sound across the cochlear duct, stimulating the hair cells. (Reprinted with permission from Netter FH, *Clinical Symposia.* CIBA Pharmaceutical Co.)

oval window. Thus, the mechanical motion of the eardrum is effectively transmitted through the middle ear and into the fluid of the inner ear.

**Amplification.** The sound-conducting mechanism also amplifies sound by two main mechanisms. First, the large surface area of the base of the stapes (footplate) creates a hydraulic effect. The eardrum has about 25 times as much

surface area as the oval window. All of the sound pressure collected on the eardrum is transmitted through the ossicular chain and is concentrated on the much smaller area of the oval window. This produces a significant increase in pressure (Figure 4–7).

The bones of the ossicular chain are arranged in such a way that they act as a series of levers. The long arms are nearest the eardrum and the shorter arms are near the oval

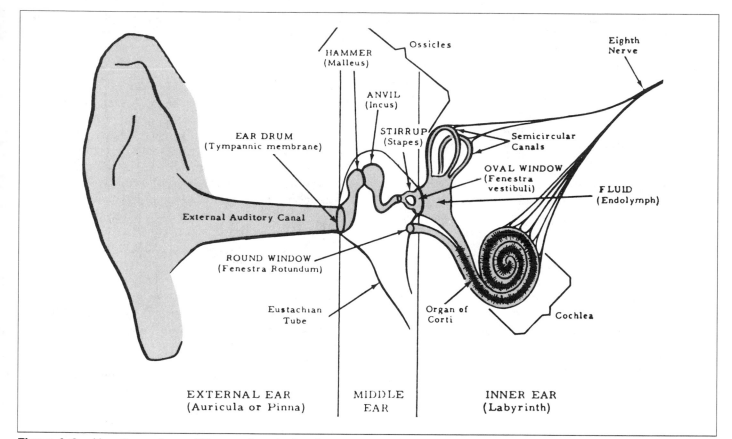

**Figure 4–6.** How the ear hears: Wave motions in the air set up sympathetic vibrations that are transmitted by the eardrum and the three bones in the middle ear to the fluid-filled chamber of the inner ear. In the process, the relatively large but feeble air-induced vibrations of the eardrum are converted to much smaller but more powerful mechanical vibrations by the three ossicles, and finally into fluid vibrations. The wave motion in the fluid is sensed by the nerves in the cochlea, which transmit neural messages to the brain. (Reprinted with permission from the American Foundrymen's Society.)

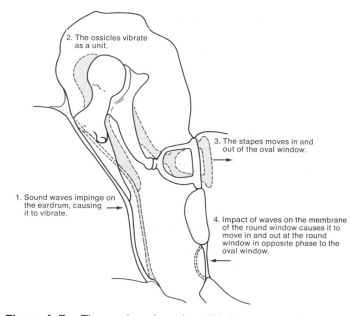

**Figure 4–7.** The eardrum has about 25 times as much surface area as the oval window. All of the sound energy collected on the eardrum is transmitted through the ossicular chain to the smaller area of the oval window.

window. The fulcrums are located where the individual bones meet. A small pressure on the long arm of the lever produces a much stronger pressure on the shorter arm. Because the longer arm is attached to the eardrum and the shorter arm is attached to the oval window, the ossicular chain acts as an amplifier of sound pressure. The magnification effect of the entire sound conducting mechanism is about 22 to 1.

Two tiny muscles attach to the ossicular chain: the stapedius to the neck of the stapes bone and the tensor tympani to the malleus. Loud sounds cause these muscles to contract, which stiffens and diminishes the movement of the ossicular chain.

## Inner Ear

The major components of the inner ear include the vestibular receptive system and the cochlea, housed within the compact temporal bone called the osseous labyrinth. It is filled with fluid (perilymph) in which a tubular membrane (membranous labyrinth) floats. The membranous labyrinth is filled with a fluid of a slightly different chemical composition called endolymph. Endolymph bathes the balance receptors and the hearing organ (organ of Corti) located within the membranous labyrinth.

The back portion of the inner ear consists of the three semicircular canals, perpendicular to one another, each containing a single balance receptor. The front section (cochlea) is shaped like a snail shell, coiled two and one-half times around its own axis; it houses the organ of Corti. These two regions are separated by the vestibule, which contains two additional receptors for balance.

**Vestibular system.** Our sense of balance is dependent not on hearing, but on organs of equilibrium. Near the cochlea are three semicircular canals lying in planes perpendicular to each other. The canals contain fluid that responds to movements and, over intricate nerve pathways to the brain, gives information about positions of the body (Figure 4–8).

The vestibular branch of the acoustic nerve transmits impulses to the cerebral cortex and we recognize the position of our head in space as it relates to the pull of gravity.

If you are rotated in a chair, the endolymph in the semicircular canals is set in motion and stimulates the hair cells. A sensation of vertigo or dizziness occurs, together with a peculiar movement of the eyes called nystagmus. Nystagmus consists of a rapid movement of the eyes in one direction and a slow movement in the opposite direction; they appear to oscillate. Nausea may occur. We call these rotation sensations motion sickness.

**The cochlea.** This is a tubular bony structure lined with a membrane containing thousands of feathery hair cells tuned to vibrate to different sound frequencies. Nerve endings are contained in a complex, slightly elevated structure over the floor of the tube forming the cochlea. This structure (organ of Corti) is the center of the sense of hearing.

Vibrations of the stapedial footplate set into motion the fluids of the inner ear. As the basilar membrane is displaced, a shearing movement occurs on the tectorial surface that drags the hair cells attached to the nerve endings. This sets up electrical impulses that are appropriately coded and transmitted to the brain via the auditory (cochlear) nerve (Figure 4–9).

The nerve endings in the cochlea are sensitive to different frequencies. Those sensitive to high frequencies are located at the large, base end of the cochlea near the oval and round windows. The nerve endings that respond to low frequencies are located at the small end of the cochlea.

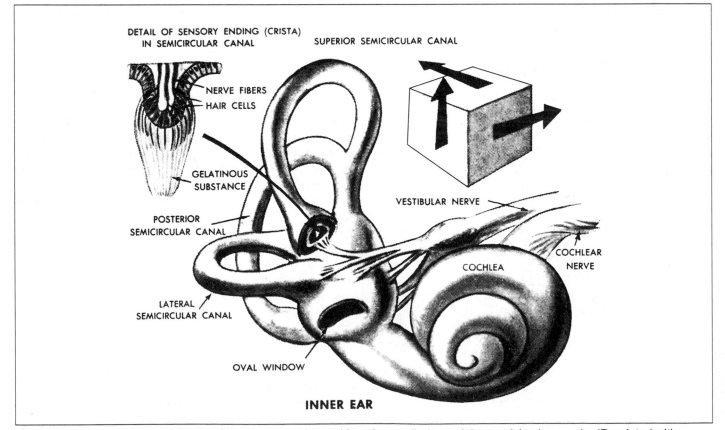

**Figure 4–8.** The three subdivisions of the inner ear: the cochlea, the vestibule, and the semicircular canals. (Reprinted with permission from the American Medical Association, *The Wonderful Human Machine.* Chicago: AMA, 1971.)

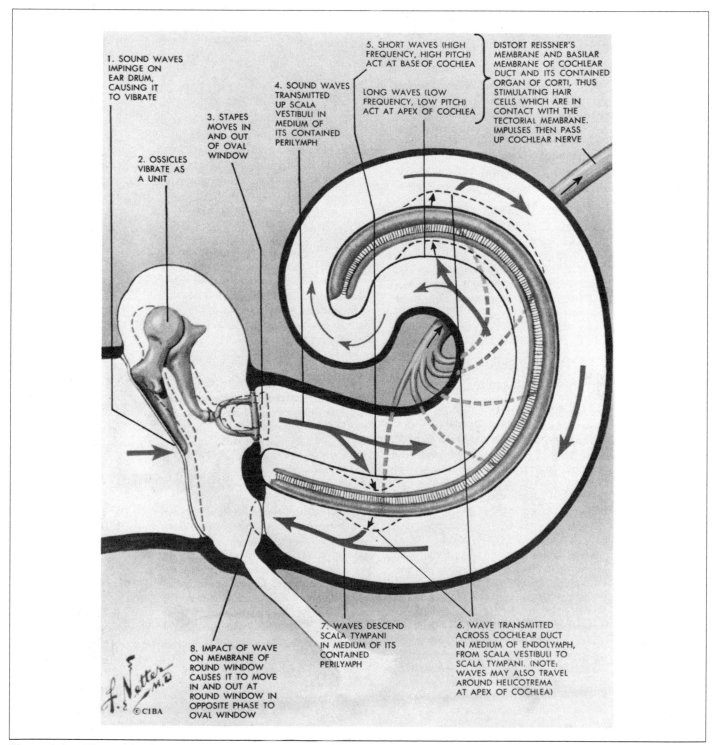

**Figure 4–9.**   The mechanism for transmission of sound vibrations from the eardrum through the cochlea. (Reprinted with permission from Netter FH, *Clinical Symposia.* CIBA Pharmaceutical Co.)

# ■ Pathology

Although the human ear is subject to a number of disorders that can cause hearing loss, the major occupational hazard is excessive unwanted sound (noise).

However, there are many nonjob-related causes of hearing loss. It is estimated that at least 25 percent of all adults have some sort of hearing abnormality.

Hearing impairments not induced by noise can arise from the following causes:

- Physical blockage of the auditory canals, as with excessive wax or foreign bodies
- Traumatic damage such as punctured eardrums or displacement of the ossicles
- Disease damage: childhood diseases such as mumps, congenital rubella, infections of the inner ear; degenerative diseases; tumors
- Hereditary or prenatal damages
- Drug-induced damages, such as from use of streptomycin or quinine
- Presbycusis: natural reduced hearing sensitivity due to aging

Off-the-job noise exposure to loud music, motorcycles, snowmobiles, private airplanes, and other sources can be a significant cause of hearing loss. All health and safety professionals responsible for hearing tests should know about the effects of these nonoccupational noise exposures in order to avoid confusing nonoccupational with occupational causes. Under the workers' compensation laws of many states, a hearing loss may be compensable if it is due to any occupational exposure.

## External Ear

Because of its prominence and its thin, tight skin, the external ear is especially subject to sunburn and frostbite. Thus, it must be protected from the elements. Injured cartilage is replaced by fibrous tissue and repeated injuries result in the cauliflower-shaped ear seen on many boxers. Disorders of the auricle include congenital malformations (in which the cartilage is misshapen) and protruding or lop ears, both of which may be surgically corrected. The auricle is the most common site in the ear for malignancies. Dermatitis and infection are common in this area.

## Ear Canal

The ear canal leading to the middle ear is normally kept healthy by wax (cerumen), which protects it from drying and scaling or from damage by water while swimming or bathing.

Foreign objects accidentally lodged in the ear can be dangerous and should be removed only by a physician. A live insect in the ear canal can be especially annoying or painful. If this happens, drop light mineral oil into the canal to suffocate and quiet the insect until it can be removed.

The external ear canal is prone to infection because of its high skin temperature and humidity. In this area, bacterial infections are most common; fungus (otomycosis) and viral causes are less common. Skin disorders (dermatitis) are also common ear canal problems. Generalized skin disorders, usually of the scalp, may extend to the outer ear.

An abnormal narrowing of the ear canal is called stenosis and may be caused by congenital malformation or infection. Tumors are rare in this area.

Normally, ear canals are self-cleaning but occasionally this mechanism fails, resulting in wax impaction. The use of cotton-tipped swabs for cleaning tends to pack wax into the ear canal. Also, swabbing stimulates excess production of wax. However, the external ear canal must be almost totally occluded (blocked) before attenuation of sound occurs; it takes a considerable amount of wax to cause hearing loss. Wax impaction should be removed by a properly trained health care professional because of the risks of injury and infection inherent in the cleaning process.

## Eardrum

Infections localized to the eardrum are rare; when they do occur, they are caused by viruses such as the varicella zoster virus (shingles). However, the eardrum is often included in infections of the external auditory canal or the middle ear.

Perforations are most often caused by infections or injuries. A blow to the ear, compressing air in the external auditory canal, can rupture the eardrum. Sudden pressure changes, such as occur in scuba diving or explosions, cause perforations. Perforations of the eardrum can be repaired by surgical grafting procedures. Most perforations heal spontaneously except for those caused by hot substances such as splatter from welding.

## Eustachian Tube

Retractions of the eardrum are caused by poor middle ear ventilation due to eustachian tube dysfunction.

Failure of the eustachian tube to ventilate creates a vacuum in the middle ear space, which in turn causes one of two pathological events to occur: It pulls fluid into the middle ear, resulting in a condition called nonsuppurative otitis media (noninfectious inflammation of the middle ear), or it pulls the eardrum inward, interferes with mobility, and causes hearing loss. Eventually the fluid thickens and hearing loss persists.

The opposite condition, which is uncommon, is a patent eustachian tube in which the tube constantly remains open. This condition results in the annoying symptom of hearing one's own voice and breath sounds (autophonia) in the involved ear.

## Middle Ear

The middle ear space is prone to infectious diseases, especially in childhood. These are predominantly bacterial in origin and called suppurative otitis media. Because the middle ear space connects with the mastoid air cell system, infection can easily spread to this area (mastoiditis). Before the days of antibiotics, these were serious, often life-threatening problems because the infection could spread to the brain or major vessels surrounding the ear. Though less likely to occur today, these dangers still exist.

Congenital deformities of the middle ear are not very common and are usually associated with structural abnormalities of the sound conducting system. Tumors in the middle ear are rare.

**Ossicles.** Disease can impair hearing by affecting the ossicular chain in two ways: fixation (the chain cannot vibrate or vibrates inefficiently) and interruption (a gap in the chain). Fixation can result from developmental errors, adhesions, or scars from old middle ear infections or bone diseases that affect this area.

Otosclerosis is a prime example of fixation. It usually begins in early adult life. Interruptions are usually caused by middle ear infections, cholesteatomas, or head injuries.

It is important to realize that when disease is confined to the outer ear and middle ear, the resulting hearing loss is conductive. Conductive losses are losses of loudness, not clarity. Most disorders that cause conductive losses are medically or surgically correctable. Some are dangerous and some are progressive. Therefore, all affected employees should have the benefit of otologic evaluation and care.

## Inner Ear

Disorders of the inner ear result in a sensory hearing loss. Inner ear hearing losses may or may not have associated losses of clarity. There are many types of congenital sensory hearing losses (inner ear hearing losses present at birth). Some of these are inherited and are called familial sensory hearing losses. Inflammatory disorders include suppurative labyrinthitis, a very rare condition; viral labyrinthitis caused by such organisms as mumps or measles viruses; and toxic labyrinthitis, which relates to inner ear dysfunction caused by hormonal (hypothyroidism) and allergic factors and to drugs that have an adverse effect on the hearing mechanism.

Concussions of the inner ear may also occur and usually cause an incomplete sensory loss with partial, if not full, recovery of hearing. Hearing acuity, like visual acuity, tends to decrease with age after the fourth decade. This loss, called presbycusis, is of the same sensorineural type and may be caused by the noise trauma of everyday life. Tumors usually do not originate within the cochlea but affect hearing by extension into the inner ear or by compression of the auditory nerve.

Ménière's disease affects both parts of the inner ear (hearing and balance) and its cause is unknown. It is characterized by episodic dizziness (often severe) and associated with nausea and vomiting, fluctuant sensory hearing loss that is generally progressive, noise in the ear, or tinnitus, and a peculiar sensation of fullness in the involved ear.

## Tinnitus (Head Noises)

Tinnitus is a symptom, not a disease. It is a perception of sound arising in the head. It may be heard only by the affected person (subjective tinnitus) or it may be audible to the examiner also (objective tinnitus). Objective tinnitus is usually a symptom not of disease of the ear but of a tumor or vascular malformation. All cases should be evaluated by a qualified physician.

Subjective tinnitus is usually perceived as a ringing or hiss. Occasionally, no explanation can be found, but most cases are secondary to high-frequency hearing loss. Some cases can be caused by wax, perforation of the drum, or fluid in the middle ear. Drugs or stimulants such as caffeine, aspirin, or alcohol can also cause the symptom.

Tinnitus can lead to psychological stress and be disabling. Maskers that match the frequency of the tinnitus, combined with a hearing aid, can be helpful. Patient education is essential to successful treatment.

## Nonauditory Effects

Research on other effects of noise has addressed interference with communication, altered performance, annoyance, and physiologic responses such as elevated blood pressure and sleep disturbances. Definitive studies have yet to be done on most of these issues. Certainly, levels of background noise above 80 dBA reduce the intelligibility of speech to workers with normal hearing. (See Chapter 9, Industrial Noise, for more information on noise measurement and the distinction between dB and dBA.) Furthermore, repeated shouting to overcome noise has been observed to lead to chronic laryngitis and even traumatic vocal cord polyps. On the other hand, stress effects (poor performance, annoyance, elevated blood pressure, and disturbed sleep) are inconsistent and not well-correlated with noise level. In field studies it has not been possible to separate noise from other stresses and job dissatisfactions. Whether the source is considered a nuisance or whether the worker has some control over the exposure influences the effects. It seems that a person's feelings about the noise are as significant as the intensity.

## ■ The Hearing Process

The outer ear funnels and conducts sound vibrations to the eardrum through the ear canal. The eardrum vibrates in response to the sound waves that strike it. This vibratory movement, in turn, is transmitted to the chain of three tiny bones in the middle ear. These small bones, the ossicles, conduct the sound vibration across the air-filled middle ear cavity to a fluid in the delicate inner ear. The vibration of the ossicle creates waves in the inner ear fluid that stimulate microscopic hair cells. The stimulation of these hair cells generates nerve impulses, which pass along the auditory nerve to the brain for interpretation.

The outer and middle sections of the ear conduct sound energy to the deeper structures. Therefore, the outer and middle ear together act as the conductive hearing mechanism. In contrast, the deeper structures, including the inner ear and the auditory nerve, are referred to as the sensorineural mechanism. The words *conductive* and *sensorineural* describe two major types of hearing impairment.

As long as the hearing mechanism functions normally, the ear can detect sounds of minute intensity while tolerating sounds of great intensity. The loudest sound the normal ear can tolerate is more than 100 million ($10^8$) times more powerful than the faintest sound the ear can detect (see Figure 4–10). Furthermore, a young listener with normal hearing can detect sounds across a very wide frequency range—from very low-pitched sounds of 20 Hz to very high-pitched sounds of 20,000 Hz.

Although nature has surrounded the delicate ear mechanism with hard, protective bone, any portion of the ear can become impaired. The part of the hearing mechanism affected and the extent of damage has a direct bearing on the type of hearing loss that results.

## Audiometer

An audiometer is a frequency-controlled audio-signal generator. It produces pure tones (the simplest form of a sound) at various frequencies and intensities for use in measuring hearing sensitivity. When hearing thresholds are measured, essentially it is a person's ability to hear pure tones that is being measured.

The audiometer was developed to provide an electronic pure-tone sound similar to that of the tuning fork. In one respect, the audiometer is superior to the tuning fork; intensities can be controlled much more accurately and, therefore, the results can be more carefully quantified.

## Audiogram

The audiometer is used to test hearing by finding the minimum intensity levels at which a person is able to distinguish various sounds. The results are recorded on a standard chart, called the audiogram.

**Frequency.** Shown across the top of the audiogram in Figure 4–11 are several numbers (125–8,000 Hz). These numbers represent the frequency or pitch of sounds, expressed in hertz (Hz). The lower numbers (125, 250) to the left represent low-pitched sounds. For example, a 250-Hz tone sounds like middle C on a piano. The tones become higher in pitch as one moves to the higher frequencies. A 4,000-Hz tone sounds much like a piccolo hitting a high note.

**Intensity.** The numbers on the left side of the audiogram in Figure 4–11 indicate the intensity or loudness of the sound, which is measured in decibels (dB). The smaller the number, the fainter the sound. When measuring a person's hearing, the level is established at each test frequency where the sound can just barely be heard. This level is called the threshold of hearing.

In audiometry, the further a person's threshold is below the zero line of the audiogram, the greater is the loss of hearing. Common practice is to record the pure-tone thresholds by air conduction for the right ear on the

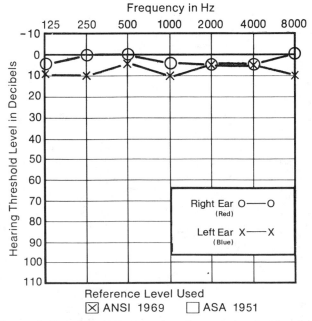

Figure 4–11. A typical manual audiogram showing hearing thresholds within the normal range.

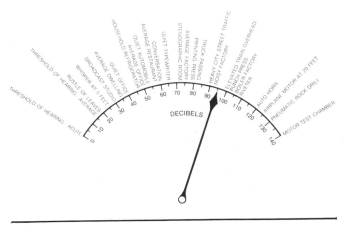

**Figure 4–10.** Typical sound levels associated with various activities.

audiogram as red circles. In contrast, a blue X is recorded to reflect each threshold for the left ear.

If a sound must be made louder than 25 dB in the speech-important frequencies for a person to detect its presence, the thresholds begin to fall into the range of hearing impairment. Thus, the more intense (louder) the sound from the audiometer must be for a person to hear it, the greater is that individual's hearing loss. However, as long as hearing is normal or nearly normal across the speech frequency range (500–2,000 Hz), the person should have little difficulty hearing speech in ordinary listening situations.

The American Academy of Ophthalmology and Otolaryngology has recommended that audiograms be drawn to a scale like the illustration shown in Figure 4–11. For every 20-dB interval measured along one side and for one octave measured across the top (250–500 Hz, for example), there is a perfect square. The reason for a standardized scale is that the apparent hearing loss can be altered a good deal by changing the dimensions of an audiogram. If the proportions of the audiogram are different from standard dimensions, a person's hearing loss may look quite different than if it were plotted on the standard audiogram format. Customarily, audiograms are scaled in 10-dB steps. Of course, if a person has a threshold of 55 dB, it is plotted on the appropriate frequency line at the halfway point between 50 and 60 dB.

Hearing losses plotted on a chart produce a profile of a person's hearing. A trained person can review an audiogram to determine the type and degree of hearing loss and can estimate the difficulty in communication this loss will cause.

**Calibration.** Hearing level for a pure tone is the listener's threshold of hearing (in decibels) over the standard audiometric zero for that frequency. It is the reading on the hearing threshold level (hearing loss) dial of an audiometer that is calibrated according to American National Standard Institute (ANSI) standard *Specifications for Audiometers* (S 3.6–1989).

# Hearing Loss

A steady loss of hearing acuity occurs as we grow older. The normal young ear can hear tones within a range of 20 Hz—the lowest bass note of a piano—up to high-pitched sounds of 20,000 Hz. People in their sixties are lucky to hear normal level sounds at 12,000 Hz. This hearing loss is greater for high-frequency sounds and is considered normal because it happens to practically everybody as the years roll on.

A slight loss of perception of high-pitched sounds muffles some of the shrillness of the world. However, hearing impairment severe enough to make ordinary conversation difficult or impossible to understand is quite another matter.

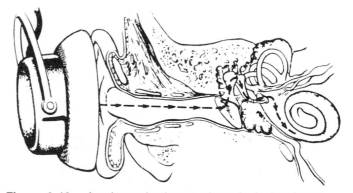

**Figure 4–12.** An air-conduction earphone is depicted; note that the earphone is placed directly over the external ear canal and the sound waves are conducted (by air) to the eardrum and through the middle ear to the inner ear.

The pitch of human speech ranges between 300 and 4,000 Hz. These are the frequencies most vital for communication. Inability to hear well within this range is a serious personal and social handicap.

## Bone-Conduction Tests

Until now, only one of the two ways that sound reaches the inner ear has been discussed: air conduction, in which sound travels from the outer ear through the bones of the middle ear into the fluid of the inner ear (Figure 4–12). However, there is another way to introduce sound and to measure hearing. This is by bone conduction, where sound travels directly to the inner ear via the bones of the skull, bypassing the outer and middle ear. Bone-conduction audiometry is rarely performed by the industrial audiometric technician; however, a basic understanding of bone-conduction tests can help when interpreting audiograms.

During such a test, a bone vibrator is placed on the mastoid bone behind the auricle (outer ear), held in place by a headband. This unit vibrates the skull (Figure 4–13).

Obviously, bone has more resistance to vibration than the air column in the outer ear canal. Thus, it takes a good deal more intensity for a listener to detect sound from a bone vibrator than from an earphone. This increase in sound output is built into the audiometer when it is manufactured and calibrated at the factory.

## Types of Hearing Loss

**Conductive.** When test results show that a person has depressed hearing by air conduction but normal hearing by bone conduction, the presence of a conductive hearing loss is indicated (Figure 4–14). In other words, the conductive mechanism is impaired in some way, but the deeper structures of the ear are intact.

**Sensorineural.** If the person hears just as poorly by bone conduction as by air conduction, then the hearing

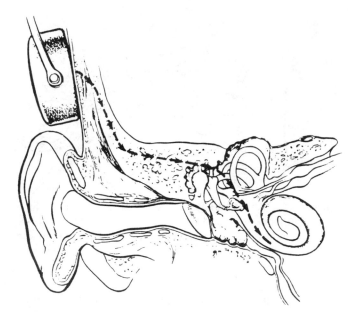

**Figure 4–13.** Sound can be transmitted directly to the inner ear through the bones of the skull using a bone-conduction vibrator placed on the mastoid bone behind the outer ear. The broken line (with arrows) shows the path taken by the sound waves through the bony areas of the head to the inner ear.

loss can be due to damage in the deep structures of the ear. (No matter how sound is presented to the sensorineural mechanism, it is met by an insufficient receiver in the inner ear.) This then would indicate a sensorineural loss (Figure 4–15).

**Mixed hearing loss.** A third type of hearing loss is a combination of conductive and sensorineural. This is referred to as a mixed hearing loss. Mixed hearing loss shows on both types of tests.

A conductive loss is due simply to some impairment of sound transmission before it reaches the inner ear. A conductive impairment, then, is one that results from some interference with the function of the outer or the middle ear.

Any blockage—usually ear wax or infection—of the outer ear that results in a loss of sound energy conducted to the middle ear can cause a conductive hearing loss. Similarly, any impairment in the sound transmission system of the middle ear can cause a conductive hearing loss. Of course, such a loss could also be due to malfunction in both the outer and the middle ear.

In contrast, hearing impairment that involves only the inner ear or the auditory nerve is classified as sensori-

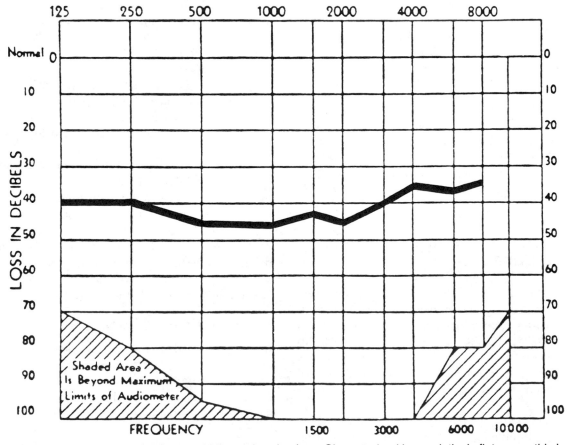

**Figure 4–14.** Audiogram shows conductive or middle ear hearing loss. Characterized by a relatively flat curve, this loss is not caused by noise exposure.

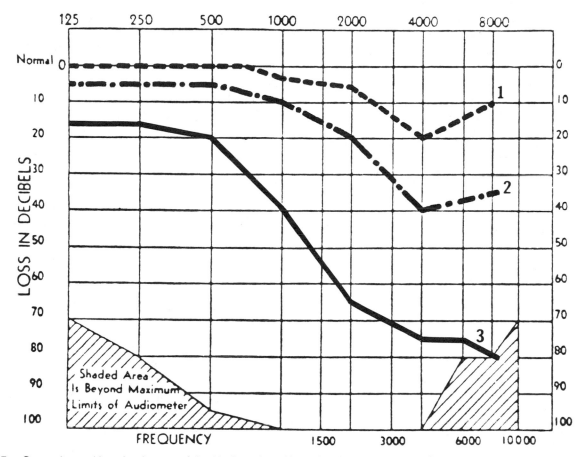

**Figure 4–15.** Sensorineural hearing losses of the kind produced by noise or other causes. Curve 1 = early; curve 2 = intermediate; curve 3 = advanced. (As shown, curve 3 might include some involvement of presbycusis.)

neural impairment. (*Sensori-* refers to the sense organ in the inner ear; *-neural* refers to the nerve fibers.) Sensorineural loss can involve impairment of the cochlea, the auditory nerve, or both.

It is virtually impossible to tell from an audiogram whether the damage is in the inner ear or in the auditory nerve, which is the transmission line to the brain. The loss is labeled sensorineural because the specific area of damage cannot be determined from audiometric findings.

Of course, it is possible for a person to have damage in both the conductive hearing mechanism and the sensorineural hearing mechanism. In fact, the damage does not have to occur at the same time. For example, a person can have a sensorineural loss and then later develop a middle ear infection, which would then produce a conductive hearing loss. If the person had both types of hearing loss, it would be called mixed hearing loss.

Certain antibiotics (aminoglycosides) can cause permanent sensorineural hearing loss. Sensorineural hearing loss is a rare but permanent complication of mumps. Blows and skull fractures can damage ear structures. Advanced infections of the middle ear are less common since the advent of antibiotics, but are still a significant cause of hearing impairment.

## ■ Effects of Noise Exposure

Effects of noise on humans can be classified in various ways. For example, they can be treated as health or medical problems because of their underlying biological basis. Noise-induced hearing loss involves damage to the cochlea and medical remedies may be possible. On the other hand, when noise interferes with oral communication or causes stressful annoyance, engineering or administrative changes should be considered.

The ear is especially adapted and very responsive to the pressure changes caused by airborne sounds or noise. The outer and middle ear structures are rarely damaged by exposure to intense sound energy, although explosive sounds or blasts can rupture the eardrum. More commonly, excessive exposure produces hearing loss by injuring the hair cells of the inner ear.

### Temporary Threshold Shift

Temporary threshold shift (TTS) of the hearing level can be produced by a brief exposure to high-level sound. Temporary threshold shift is greatest immediately after exposure to excessive noise and progressively diminishes with increasing rest time as the ear recovers from the apparent noise overstimulation. A noise capable of causing significant TTS with brief exposures is probably capable of causing a significant permanent threshold shift (PTS) with prolonged or recurrent exposure.

## Permanent Threshold Shift

Permanent threshold shift resembles TTS except that the recovery of hearing is incomplete. Important variables in the development of temporary and permanent hearing threshold changes include the following:

- Sound level: Sound levels must exceed 60–80 dB before the typical person will experience TTS.
- Frequency distribution of sound: Sounds having most of their energy in the speech frequencies are more potent in causing a threshold shift than are sounds having most of their energy below the speech frequencies.
- Duration of sound: The longer the sound lasts, the greater the amount of threshold shift.
- Temporal distribution of sound exposure: The longer and more numerous the quiet periods between periods of sound, the lower the potential for threshold shift.
- Individual differences in tolerance of sound: These vary greatly among individuals.
- Type of sound (steady-state, intermittent, impulse, or impact): The tolerance to peak sound pressure is greatly reduced by increasing the rise time or burst duration of the sound.

## Noise-Induced Hearing Loss

When a person is first exposed to hazardous levels of noise, the initial change usually observed is a loss of hearing in the higher frequency range, usually a dip or a notch at about 4,000 Hz. After a rest period away from the noise, the hearing usually returns to its former level. For practical purposes, a rest period of 14 hours or so away from the noise is adequate to return the threshold to previous levels.

Permanent damage from noise is generally classified as noise-induced hearing loss or acoustic trauma, depending on the nature of exposure. The long-term cumulative effects of repeated and prolonged hazardous noise exposure result in permanent pathological changes in the cochlea and irreversible threshold shifts in hearing acuity. This is called noise-induced hearing loss. It is usually represented audiometrically by a notch at 4,000 Hz. Because the hearing loss does not necessarily stop here, further exposure may result in a deepening and widening of the notch. When the hearing loss involves the speech frequency range, considerable difficulty in hearing conversational speech results.

The effect of noise on hearing depends on the amount and characteristics of the noise as well as the duration of exposure. A single loud sound such as an explosion may produce a permanent hearing loss. This is called acoustic trauma and typically affects one ear much more than the other. In most cases, however, deterioration of hearing occurs during the initial 5–10 years of employment in a noise-risk environment.

Noise is a pervasive, insidious cause of hearing loss. It causes no particular pain unless it is as loud as a rifle blast. The ears have considerable comeback power from temporary brief exposure to noise and ordinarily recover overnight. However, prolonged exposure to intense noise gradually damages the inner ear.

Susceptibility to noise-induced hearing loss varies greatly from one individual to another. Beyond certain levels of extremely high intensity, it is generally agreed that all individuals are susceptible when the exposure is long enough. Generally speaking, tests of temporary threshold shift throw no specific light on the susceptibility of individuals to permanent threshold damage.

## ■ Communication Problems

The communicative problem of people with noise-induced hearing loss is very frustrating to them and is easily misunderstood by family and friends. This problem causes a great deal of inconsistent auditory behavior—the person appears to hear very well at some times and very poorly at others, and is thus often accused of not paying attention. It is helpful to understand the kind of communicative problem imposed by a hearing loss caused by substantial exposure to noise.

### Hearing Versus Understanding

There are two important characteristics of normal hearing: the ability to hear sounds as loud as they truly are and the ability to hear sounds with complete clarity. It is important that the distinction between these two characteristics of hearing be understood.

#### Loudness

If Figure 4–16 is held at arm's length, the printing on the left side is obvious but almost impossible to read. A close look confirms that it is indeed a word; an even closer look reveals that the word is *Loudness*. This example illustrates a very common problem associated with hearing loss: the inability to hear soft sounds. If the sounds are made louder, a person with only a loss of hearing sensitivity will have much less difficulty.

#### Clarity

If Figure 4–16 is held at arm's length, there is no difficulty in seeing everything on the right side. But can the word be read? If not, the reason is that too much of the word is miss-

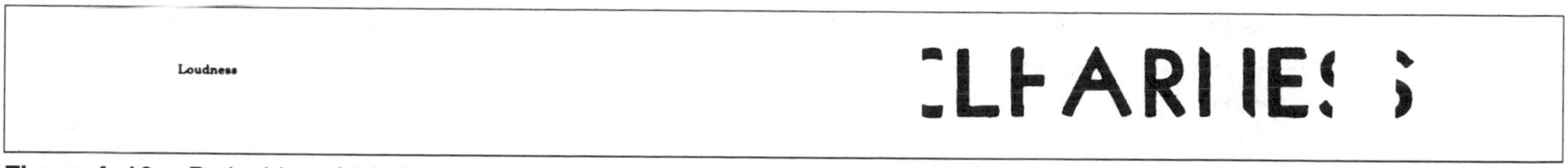

Loudness

CLFARIIES

**Figure 4–16.** Both sides of this illustration visually represent two important dimensions of normal hearing: the ability to hear sounds as loudly as they really are and the ability to hear sounds with complete clarity. *Left:* It is almost impossible to read the word *Loudness;* this is comparable to not being able to hear faint sounds. Moving the figure closer makes it easier to read, just as increasing volume makes sounds easier to hear. *Right:* No matter how closely the illustration is held, it is difficult to interpret. The word is not clear because some important parts of the letters are missing. This shows, by analogy, hearing difficulty caused by a loss in the ability to distinguish between various sounds (the word is *CLEARNESS*).

ing, even though it is large enough. No matter how closely you look, it is difficult to read because some of the important parts are missing. This illustrates the problem of loss of clarity, or an inability to distinguish between the various sounds in spoken language. (The word is *CLEARNESS.*)

If any portion of the outer or middle ear is damaged, the primary result is an inability to hear soft speech. Clarity, however, is preserved. The key to clarity in hearing is held by the inner ear mechanism and the nerve fibers that carry the message to the brain.

If the inner ear or auditory nerve is damaged, a loss not only in the loudness of sounds but also (in many cases) in the clarity of sounds results. In such dual problems associated with sensorineural hearing loss, speech seems muffled or fuzzy no matter how loud it is. In these cases, the major problem is trouble understanding what is being said.

Hearing loss from noise exposure often results in this kind of problem. The tiny hair cells that respond to specific speech sounds may be so severely damaged that they cannot react when the vibrations from the outside strike them. At the same time, hair cells for other speech sounds may be functioning normally. People with this kind of hearing impairment miss parts of words and parts of sentences and often misunderstand what is being said. This can be a very subtle problem. In fact, people with this loss often do not know that they do not hear everything. To summarize, people with high-frequency, noise-induced loss can hear speech but may not understand what is being said.

## Speech Sounds

Characteristic speech sounds can be classified as vowels or consonants. The vowel sounds—located in the lower frequencies—are the more powerful speech sounds. Therefore, vowels carry the energy for speech. In contrast, the consonant sounds—located in the higher frequencies—are the keys to distinguishing one word from another, especially if the words sound alike. This is the heart of the communicative problem of people with noise-induced (high-frequency) hearing loss. They cannot distinguish between similar words such as *stop* and *shop.*

It is quite easy to miss a key sound in a word. In turn, this could change the meaning of a key word in a sen-

tence. As a result, the entire sentence might be misunderstood.

People with high-frequency hearing loss often get along fairly well in quiet listening situations. But as soon as they are in a place where there is a lot of background noise, such as in traffic or on the job, it becomes difficult to communicate through hearing alone.

If a speaker were talking in the presence of typical background sound, a listener with no hearing impairment could hear quite well. However, if the listener developed a hearing loss for all speech sounds above 1,000 Hz, he or she would notice a marked loss in hearing. As long as it is quiet, a person can use good hearing below the midpitch range to an advantage. Unfortunately, most noises around us interfere with the low-pitched speech sounds and cause us to miss many of these sounds as they occur in words and sentences. Thus, the hearing loss caused by the noise and the hearing loss from damage to the ear result in a greater hearing problem than occurs with just one of these conditions.

## Treatment

Although there is no medical or surgical cure for sensorineural deafness, auditory rehabilitation methods can be used to help people compensate for hearing disabilities and lead normal lives, with minimal effect on social and economic status.

Fitting of a proper type of hearing aid, when indicated, is a vital part of an auditory rehabilitation program. However, a patient must be psychologically prepared to accept a hearing aid. Many are reluctant to use hearing aids and never use them, or use them ineffectively. Before recommending a hearing aid, determine whether the patient will be helped by it enough to justify purchasing one. This is particularly important in cases of sensorineural hearing loss, in which the problem is one of discrimination rather than amplification.

In people whose hearing losses resulted from noise exposure, one of the most important benefits of a hearing aid is that it enables the individual to hear what is already heard, but with greater clarity; it minimizes the stress of listening. In sensorineural hearing loss, broad amplification further muddies the lost frequencies with noise. Hearing aids are now available with digital processors that can be programmed to selectively amplify the hearing deficiencies of the individual patient. These devices can be fitted in the auricle or the canal. Cochlear implants do not yet correct discrimination problems, but in cases of severe deafness acquired after learning to recognize speech, cochlear implants give some degree of hearing acuity.

Patients should seek early medical attention for hearing loss. If the condition can be helped by medical or surgical means, early treatment improves the prognosis. If a hearing aid is necessary, the sooner it is acquired, the less disturbing environmental noises will be when audible again.

Once purchased, thousands of hearing aids, given too brief and halfhearted a trial, are relegated to a desk drawer. Overlong postponement in acquiring the aid is sometimes a factor. In other cases, the patient expects to hear normally with a hearing aid when the condition of his or her hearing organs makes such a result impossible. The dispenser should make it clear to the patient that a hearing aid is not a substitute for a normal ear, especially in patients with sensorineural deafness. One common cause of disappointment with a hearing aid is patients' desire for invisible or inconspicuous aid, when what they should be looking for is the aid that will best enable them to understand conversation.

## AMA Guides for Evaluating Impairment

Chapter 9.1 of the American Medical Association's *Guides to the Evaluation of Permanent Impairment* includes criteria for evaluating permanent impairment resulting from the principal dysfunctions of the ear. Permanent impairment is expressed in terms of the whole person.

Although the ear and related structures have multiple functions, some of which are interdependent, a rating of permanent impairment depends on clinically established deviation from normal in hearing or equilibrium.

As in other types of impairments, competent evaluation requires use of the proper equipment. It should be emphasized, however, that the actual level of function should be determined without the aid of a prosthetic device.

Furthermore, the physician should be able to explain and substantiate the conclusions regarding the patient's condition. When reports are required, the physician should clearly indicate the date of the findings and conclusions. The report should include a history, findings by physical and special examinations, consultations, diagnoses, classifications (when used), and the resulting impairment rating. Clinical judgment has a role but objective observations should be emphasized.

Equilibrium and hearing are considered separately. Only general criteria are provided for disturbances of equilibrium, but the criteria for evaluating hearing impairment are relatively specific.

Such disturbances of the ear as chronic otorrhea, otalgia, and tinnitus are not quantifiable, so the physician should assign a value based on severity, importance, and consistency with established values. Up to 5 percent unilateral hearing loss may be assigned for impaired speech discrimination due to tinnitus. Deformities of the auricle or other cosmetic defects that do not alter function are not considered.

Impairment of equilibrium due to vestibular dysfunction is rated as Class 1 through Class 5, ranging from an impairment of the whole person of 0–95 percent. Classification depends on objective findings and the degree of impairment of daily activities.

Hearing loss for each ear is calculated at thresholds of 500, 1,000, 2,000, and 3,000 Hz. No allowance is made for impairment of speech discrimination due to deficits at higher frequencies. Binaural hearing loss must be used in determining impairment of the whole person. The *Guides to the Evaluation of Permanent Impairment* include a formula for binaural hearing loss that is based on the percentage impairment of each ear, tested separately. The percentage of impairment of the poorer ear is added to five times the percentage of impairment of the better ear and the total divided by six. Thus, a person totally deaf in one ear but with normal hearing in the other would have a binaural hearing impairment of 17 percent $((100 + 0) \div 6)$. A table is provided for converting hearing impairment into impairment of the whole person. Because many state agencies have adopted the use of the AMA formula for determining hearing impairment, it is important that the industrial hygienist or occupational physician understand it.

## OSHA Hearing Conservation Program (29 *CFR* 1910.95)

On April 7, 1983, the Hearing Conservation Amendment (HCA), included in the *Federal Register 29 CFR* 1910.95, became effective. The Occupational Health and Safety Administration (OSHA) Noise Regulation provides specifics on the content of required hearing conservation programs. The regulation covers the following:

- Noise monitoring
- Audiometric testing program
- Definition of standard (permanent) threshold shift
- Employee follow-up and referral
- Hearing protection
- Employee training
- Recordkeeping

This amendment requires that all OSHA-covered workers be included in hearing conservation programs if they are exposed at or above an eight-hour time-weighted average (TWA) of 85 dBA or more. Chapter 9 contains a complete discussion of the requirements of the HCA. Unfortunately, a number of people develop threshold shifts on prolonged exposure to levels as low as 80 dBA. For a discussion of compensation laws as they relate to hearing, see Chapter 9, Industrial Noise.

## ■ Summary

*External ear.* This consists of two sections: the auricle, or pinna, and the external auditory canal (Figure 4–1).

*Middle ear.* This is the cavity between the eardrum and the bony wall of the inner ear (Figure 4–2), which consists of the eustachian tube, the eardrum, the ossicular chain, the oval and round windows, and the mastoid air cell system.

*Inner ear.* This contains the receptors for hearing and position sense, and contains the cochlea, the vestibule, and the semicircular canals.

This chapter covered the anatomy, physiology, and pathology of the ear and the hearing process. The effects of noise exposure (see Chapter 9, Industrial Noise) and communications problems were also mentioned, as well as the *AMA Guides for Evaluating Impairment* and the OSHA Hearing Conservation Program (29 *CFR* 1910.95).

## ■ Bibliography

Alberti PW, ed. *Personal Hearing Protection in Industry.* New York: Raven Press, 1982.

American Medical Association. *Guides to the Evaluation of Permanent Impairment,* 4th ed. Chicago: AMA, 1993.

Berger EH, Ward WD, Morrill JC, et al., eds. *Noise and Hearing Conservation Manual,* 4th ed. Akron, OH: American Industrial Hygiene Association, 1986.

Council for Accreditation in Occupational Hearing Conservation. *Hearing Conservation Manual,* 3rd ed. Milwaukee: Council for Accreditation in Occupational Hearing Conservation, 1993.

Gray RF, Hawthorne M. *Synopsis of Otolaryngology: Part I,* 5th ed. Boston: Butterworth-Heinemann, 1992.

Katz J. *Handbook of Clinical Audiology,* 3rd ed. Baltimore: Williams & Wilkins, 1985.

Sataloff RT, Sataloff J. *Hearing Loss,* 3rd ed. New York: Marcel Dekker, 1993.

Schuknecht HF. *Pathology of the Ear,* 2nd ed. Philadelphia: Lea & Febiger, 1993.

Smith A. A review of the non-auditory effects of noise on health. *Work & Stress* 5:49–62, 1991.

CHAPTER 5

# The Eyes

by George S. Benjamin, MD, FACS

**Anatomy**
*Eyeball* ▪ *Retina* ▪ *Binocular Vision*

**Eye Problems**
*Specialists* ▪ *Examining Instruments* ▪ *Snellen Chart* ▪ *Eye Defects* ▪ *Eyeglasses*

**Visual Performance**
*Visual Acuity* ▪ *Dark Adaptation* ▪ *Depth Perception*

**Eye Disorders**
*Conjunctivitis* ▪ *Glaucoma* ▪ *Cataracts* ▪ *Excessive Brightness* ▪ *Night Blindness* ▪ *Eyestrain* ▪ *Nystagmus*

**Physical Hazards**
*Blows from Objects* ▪ *Corneal Lacerations and Abrasions* ▪ *Blink Reflex* ▪ *Foreign Bodies* ▪ *Thermal Burns* ▪ *Irradiation Burns* ▪ *Chemical Hazards* ▪ *Chemical Burns* ▪ *Evaluating Eye Hazards* ▪ *First Aid* ▪ *ANSI Z358.1–1990*

**Protective Equipment**
*ANSI Z87.1–1989* ▪ *Impact Protection* ▪ *Eye Protection for Welding* ▪ *Laser Beam Protection* ▪ *Video Display Terminals (VDTs)* ▪ *Plastic Versus Glass Lenses* ▪ *Sunglasses* ▪ *Contact Lenses* ▪ *Comfort and Fit*

**Vision Conservation Program**
*Environmental Survey* ▪ *Vision Screening Program* ▪ *Remedial Program* ▪ *Professional Fitting* ▪ *Guidelines*

**AMA Guides for Evaluating Impairment**
*Criteria and Methods for Evaluation*

**Summary**

**Bibliography**

THE EYE MAY BE THE ORGAN MOST VULNERABLE to occupational injuries. Although the eye has some natural defenses, they do not compare with the healing properties of the skin, the automatic cleansing abilities of the lungs, or the recuperative powers of the ear. Consequently, the eye is at greater risk and eye and face protection is a major occupational health issue.

The demands placed on the eye in modern workplaces and practices, such as prolonged viewing at close distances or at distances neither near nor far, are great. Both of these conditions can cause acute and chronic eye fatigue and visual discomfort. Hazardous substances can be absorbed into the eye system. Machinery, if guarding mechanisms fail, can propel objects capable of causing traumatic injury to the eye.

The eye, with its remarkable ability to translate radiant light energy into neural impulses, which are transmitted to the visual cortex of the brain, is certainly one of the most valued organs. Protection of the sensory organs of this complex system should be a high priority in every occupational health and safety program.

## Anatomy

A look at the structure of the human eye and how it can be affected by industrial hazards clarifies the need for eye protection programs. The eyeball is housed in a cone of cushioning fatty tissue that insulates it from the skull's bony eye socket. The skull has brow and cheek ridges projecting in front of the eyeball, which is composed of specialized tissue that does not react to injury like other body tissue (Figure 5–1).

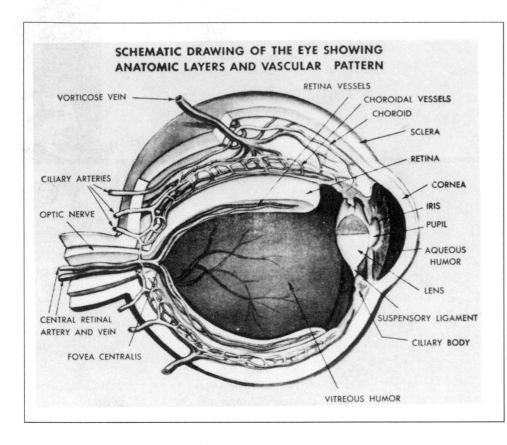

Figure 5–1. Schematic drawing of the eye showing anatomic layers and vascular pattern. (Reprinted with permission from the American Medical Association, *The Wonderful Human Machine.* Chicago: AMA, 1971.)

## Eyeball

The eyeball consists of three coats, or layers, of tissue surrounding the transparent internal structures. There is an external fibrous layer, a middle vascular layer, and an inner layer of nerve tissue.

The outermost fibrous layer of the eyeball consists of the sclera and the cornea. The sclera, also called the white of the eye, is composed of dense fibrous tissue and is the protective and supporting outer layer of the eyeball. In front of the lens this layer is modified from a white, opaque structure to the transparent cornea. The cornea is composed of a dense fibrous connective tissue and has no blood vessels. The cornea must be transparent to let light through to the receptors in the eyeball.

The middle vascular layer of the eyeball is heavily pigmented and contains many blood vessels that help nourish other tissues.

The nerve layer, or retina, is the third and innermost layer of the eyeball. Toward the rear, the retina is continuous with the optic nerve; toward the front, it ends a short distance behind the ciliary body in a wavy border called the ora serrata. The retina is composed of two parts: the outer part is pigmented and attached to the choroid layer and the inner part consists of nerve tissue.

The front of the eyeball is protected by a smooth, transparent layer of tissue called the conjunctiva. A similar membrane covers the inner surfaces of the eyelids. The eyelids also contain dozens of tiny tarsal glands that secrete an oil to lubricate the surfaces of the eyeball and eyelids. Still further protection is provided by the lacrimal gland, located at the outer edge of the eye socket. It secretes tears to clean the protective membrane and keep it moist (Figure 5–2).

The region between the cornea and the lens is filled with a salty, clear fluid known as the aqueous humor. The

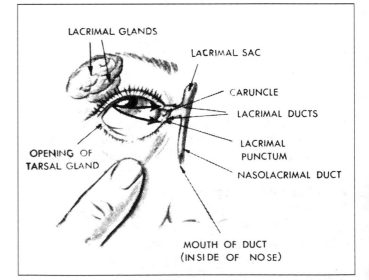

Figure 5–2. Illustration of the eye and tear ducts. (Reprinted with permission from the AMA, *The Wonderful Human Machine.* Chicago: AMA, 1971.)

eyeball behind the lens is filled with a jelly-like substance called the vitreous humor.

Light rays enter the transparent cornea and are refracted at the curved interface between air and the fluid bathing the cornea. After passing through the cornea and the clear liquid, (the aqueous humor, contained in the anterior chamber), the bundle of rays is restricted by a circular variable aperture, the pupil. Its size is changed by action of the iris muscles.

The light rays are further refracted by passage through the lens, traversing the clear, jelly-like vitreous humor of the posterior chamber so that, in a properly focused eye, a sharp image is formed on the retina. Scattering of light within the eye is minimized by a darkly pigmented layer of tissue underlying the retina, called the choroid. The choroid contains an extremely rich blood supply that is believed to dissipate the heat resulting from absorbed light energy. The shape of the eyeball is maintained by its enclosure in an elastic capsule, the sclera, and by the fluids within that are maintained at positive pressure.

The lens is attached by suspensory ligaments to the ciliary body, a muscular organ attached to the sclera. The ciliary body muscles alter the lens shape to fine-focus the incoming light beam. Ordinarily, these muscles are active only when looking at objects closer than 20 ft (6.1 m). Consequently, when doing close work, it is restful to pause occasionally and look out a window into the distance. Many complaints of eye fatigue are really complaints about tired ciliary muscles. With age, the lens gradually loses some of its accommodative power and no amount of ciliary effort can replace holding a book at arm's length to read it.

The pigmented iris, overlying the lens, is a muscular structure designed to expand or contract and thus regulate the amount of light entering the eye. The circular aperture formed by the iris is called the pupil.

The aqueous and vitreous humor and other eye tissues are composed primarily of water, so their absorption characteristics are similar to those of water.

## Retina

The retina, a thin membrane lining the rear of the eye, contains the light-sensitive cells. These cells are of two functionally discrete types: rods and cones. They get their names from the rod and cone shapes seen when the layer is viewed under a microscope. The rods are more sensitive to light than the cones; the cones are sensitive to colors.

There are more rod cells than cones; each eye has about 120 million rods and only 6 million cones. The rods are incapable of color discrimination because they contain a single photosensitive pigment. There are fewer cone cells, and they are less sensitive to low levels of luminance. There are three types of cones in the human eye; each contains a different photopigment, with peak response to a particular part of the visible spectrum. Thus,

by differential transmission of nerve impulses on stimulation, the cones encode information about the spectral content of the image so that the observer experiences the sensation of color.

## Binocular Vision

Binocular vision refers to vision with two eyes. The advantages of binocular vision are a larger visual field and a perception of depth, or stereoscopic vision. There is a slight difference in the images on the two retinas; there is a right-eyed picture on the right retina and a left-eyed picture on the left retina. It is as if the same landscape were photographed twice, with the camera in two positions a slight distance apart. The two images blend in consciousness and give us an impression of depth or solidity. Binocular vision is not identical to depth perception but is an important clue to that visual function. Depth perception is further discussed in the section on visual performance of this chapter.

## ■ Eye Problems

A few definitions of the specialists involved in this field are necessary.

## Specialists

**Ophthalmologist.** An ophthalmologist (oculist) is a doctor of medicine who is licensed to practice all branches of medicine and specializes in the examination of the eye and its related structures and in the prevention, diagnosis, and medical and surgical treatment of eye defects and diseases. An ophthalmologist also prescribes whatever medication and correction is required, including eyeglasses and contact lenses.

Education and training qualify an ophthalmologist to relate findings observed in an examination of the eye to diseases in other parts and systems of the body, which may have an effect on the eye. Almost every medical doctor of this type has passed American Board of Ophthalmology examinations and is qualified as an expert in this field.

**Optometrist.** An optometrist has the education, training, and licensure necessary to examine eyes for abnormal visual problems not due to disease. Optometrists can prescribe, fit, and supply eyeglasses and contact lenses; the extent to which an optometrist can prescribe drugs is limited and varies from state to state.

**Optician.** An optician is a person who manufactures eyeglasses at the request of an ophthalmologist or an optometrist. Most states license opticians only after a period of training and require them to follow standards for measuring frames and grinding lenses for eyeglasses. In addition, some opticians fit contact lenses, following the prescription of an ophthalmologist or optometrist.

## Examining Instruments

The vision tests used in industrial preplacement examinations are screening tests. They detect possible problems in visual performance and are not by themselves diagnostic. They should not be the basis of any job restrictions without further evaluation by a qualified physician or optometrist and consideration of reasonable accommodations.

## Snellen Chart

The most common industrial test for distance acuity is the Snellen wall chart in its several variations. The Snellen chart consists of block letters in diminishing sizes so that at various distances, the appropriate letter subtends a visual angle of 5 minutes at the nodal point of the eye. Thus, the top large letter appears to be the same size when it is 200 ft (61 m) away as the standard appears at 20 ft (6.1 m).

The distance of 20 ft (6.1 m) is considered to be infinity. This means that the rays of light coming from an illuminated object are parallel; they neither diverge nor converge. If the object is closer than 20 ft (6.1 m), the light rays diverge and must be made parallel by action of the lens within the eye or by the addition of a supplementary lens held in front of the eye; otherwise, they do not come to a sharp focus on the area of central visual acuity of the retina.

The cornea and lens of the eye bend the parallel rays to converge to a focus on the retina. Looking at an object at 20 ft (6.1 m) or more, the normal lens is relaxed into its usual biconvex shape. The parallel rays of light that are bent (refracted) by the cornea and lens cross at the nodal point of the eye (about 7 mm behind the cornea) and, continuing their straight course, fall on the retina, forming an inverted image.

It is important to check several factors. The distance from the chart to the person being tested must be 20 ft (6.1 m), or 10 ft (3 m) if a mirror and reversed chart are used. The lighting should be uniform, its source not visible to the person being tested. The chart should be clean. Finally, the tester must be trained to hold the cover correctly over the eye not being tested, to vary the order of lines and letters, and to be alert for any unusual factors. It is important to separately test and record the vision of each eye and the vision when reading the chart with both eyes.

Satisfactory vision at a distance does not ensure adequate near-point vision, so it is important to recognize near-point abnormalities. Many industrial work situations involve near-point seeing even though they are not confined to near-point work.

Industrial vision testing should be done with standardized and foolproof tests to detect and identify substandard visual functions. Accuracy is vital because, in some jobs, workers with visual defects can put themselves and others at risk. A competent examination of the eyes requires the use of a number of special examining instruments.

- An ophthalmoscope permits study of the interior structures and the slit-lamp microscope allows study of the structures in the anterior through high magnification.
- A tonometer is used to measure the pressure in the eyeball.
- A perimeter is used to map the limits of the fields of vision.
- A gonioscope views the angle of the chamber where the outflow drainage apparatus of the eye is found.
- A refractor can detect a refractive aberration such as myopia and astigmatism or modifications of the same. The instrument can be fitted with test lenses to determine which lenses will improve the vision (Figure 5–3).
- A vision screening device, in which the test slides are typically set in two drums rotated by a crank on the outside of the instrument and are not visible except through the viewing box. Tests in the distance drum are at the optical equivalent of 8 m. Tests in the near drum are at a downward angle and at the optical equivalent of 13 in. (0.32 m) from the instrument lenses. Adapters are available for midrange (30 in., or 0.75 m) screening. This distance is critical for testing those involved in prolonged use of video display terminals (VDTs) (Figure 5–4). VDTs are discussed later in this chapter.

## Eye Defects

At the time of this writing, according to the National Society to Prevent Blindness, 890,000 Americans are legally blind and 11.4 million are visually impaired (having vision not fully correctable with lenses). Eye diseases associated with aging, diabetes, and glaucoma are the leading causes, followed by injuries. The society states that nearly one half of all cases of blindness could be prevented by taking advantage of current medical technology and proper safety measures.

It has also been estimated that 40 percent of the population wear glasses, indicating that nearly one in two people have some visual defect. Three common eye defects—farsightedness, nearsightedness, and astigmatism—are the results of simple optical aberrations in the eye.

**Farsightedness.** When the eyeball is too short from front to back, the light rays come to a focus behind the retina. Light rays coming from a distant object may reach their focus at the retina, so that distant vision is good, but near vision is blurred. The treatment is to wear a convex lens that converges the light rays from near objects so that they are brought to a focus on the retina (Figure 5–5).

The closer an object is brought to the eye, the more convex the human eye lens must become in order to focus on it. Through aging, the human lens loses its elasticity and its power of thickening, so that by the time we are in our forties it may seem as if our arms are too short to read a book. This condition is called presbyopia, and is

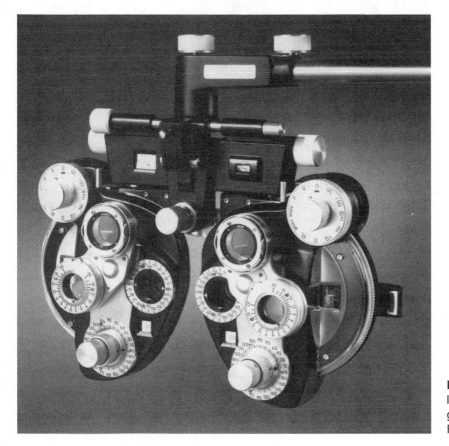

**Figure 5–3.** A refractor is used with a set of test lenses to find which ones aid vision. (This photograph has been reprinted with permission of Reichert Ophthalmic Instruments.)

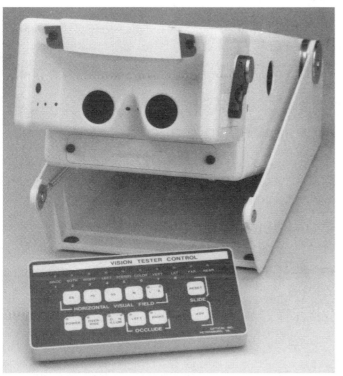

**Figure 5–4.** Vision screening device tests acuity, depth, color, and vertical and lateral balance between the two eyes at various distances. (Courtesy of Titmus Optical, Inc.)

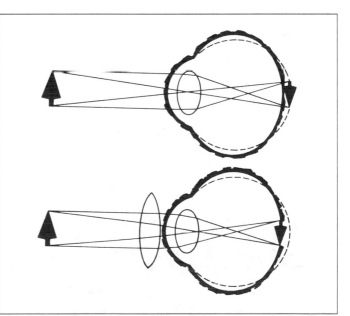

**Figure 5–5.** In farsightedness, or hyperopia, the eyeball is too short, so the image of an object is focused behind the retina. A convex lens brings the light rays into focus on the retina. Although hyperopic people may be able to see things sharply by thickening the lens of the eye (accommodation), this involves effort of inner muscles of the eye and may cause eye fatigue.

overcome by wearing convex lenses, often in a bifocal. People often need progressively stronger lenses as they age.

**Nearsightedness.** If the eyeball is too long from front to back, as it is in nearsightedness, or myopia, the image of an object 20 ft (6.1 m) or more away falls somewhere in front of the retina. The eye can focus on it sharply only by looking through a concave lens, which diverges the rays coming from the object. By bringing the object near enough to the eyes, a myopic person can get a good focus (Figure 5–6).

**Astigmatism.** If the curvature of the cornea is irregular so that some rays of light are bent more in one direction than in another, the resulting image is blurred because if one part of the ray is focused, the other part is not. This is something like the distortion produced by a wavy pane of glass, and is called astigmatism. It is corrected by using a lens that bends the rays of light in only one diameter (axis). This lens is called a cylindrical lens and it can be turned in the trial frame to its proper axis to even up the focusing of the light rays in all parts (Figure 5–7).

## Eyeglasses

The purpose of wearing eyeglasses is to help focus the rays of light on the retina. Glasses cannot change the eye or produce any disease even if they are badly fitted.

A prescription for glasses may look something like this:

$$+ 2.0 + 0.50 \text{ cyl ax90} \tag{1}$$

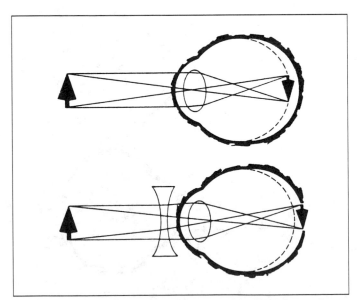

**Figure 5–6.** In nearsightedness, or myopia, the image of an object (unless it is held close to the eyes) falls in front of the retina instead of on it, and the object is seen indistinctly. The condition is corrected by using a concave lens of proper curvature to bring the image into focus on the retina. (Reprinted with permission from Cooley DG, ed. *Family Medical Guide.* New York: Better Homes and Gardens Books, 1973.)

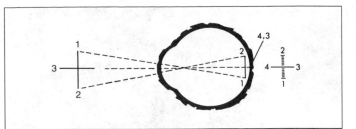

**Figure 5–7.** Astigmatism resulting from irregular curvature of the cornea is something like distortion produced by a wavy pane of glass. Drawing shows light rays (3,4) in sharp focus on retina, with light rays (1,2) focused in front of the retina, resulting in a blurred image. The small diagram on the right shows horizontal image in focus, vertical image out of focus. A cylindrical lens placed in the proper axis to bring light rays to even focus corrects astigmatism.

The + sign indicates a convex lens suitable for a far-sighted person. A − sign would indicate a concave lens for a myopic person. The *2.0* indicates the *diopters,* which indicate the strength or power of the lens. A diopter is a unit of measurement of the refractive or light-bending power of a lens. The normal human lens in its relaxed biconvex shape has a power of about 10 D.

Thus, the prescription in the example means that the optician grinds a convex spherical lens of 2.0 diopters combined with a convex cylindrical lens of half a diopter (0.50) situated vertically (axis 90 degrees). Lens prescriptions may look strange, but opticians everywhere know what they mean.

Incidentally, it is a good idea to carry your lens prescription when traveling in the event your glasses are lost or smashed. A spare set of glasses is good insurance, particularly for people who would be seriously handicapped without them.

## ■ Visual Performance

Normal visual performance involves a number of interdependent discriminations made in response to the visual environment and mediated by the visual system.

### Visual Acuity

There are many definitions of the term *visual acuity;* all, however, incorporate the concept of detail resolution. Many test patterns have been used to measure acuity, from single dots to twin stars, gratings, broken rings, checkerboards, and letters. There is no general agreement on the choice of a test and results from the different patterns are often at odds. The Snellen Letter Test is probably the most familiar and is widely used despite the fact that it tests letter recognition rather than retinal resolution. The most satisfactory expression for visual acuity is the amount of critical detail that can just be discriminated.

Some important variables affecting visual acuity are as follows:

**Luminance.** The level of adaptation of the eye has a profound effect on visual acuity.

**Position in the field.** At phototropic levels, acuity is best at the fovea and drops off as the retinal periphery is approached because there are fewer cones at the periphery. Nocturnal acuity is quite poor, with essential blindness at the fovea and best resolution in the periphery, where rods are more plentiful.

**Duration.** When the pattern is viewed for only a short time, measured acuity diminishes.

**Contrast.** Visual acuity decreases as the contrast between pattern and background diminishes. The strength of this correlation depends on the adapting luminance.

## Dark Adaptation

Optimal visual discrimination under conditions of very low light can be made only if the eyes are adapted to the level of the prevailing light or lower. If a fully light-adapted eye is suddenly plunged into darkness, its initial sensitivity is poor. With time, however, sensitivity increases as a result of photochemical regeneration, certain functional neural changes, and, to a much smaller degree, enlargement of the pupil. After the eye remains in total darkness for 30–60 minutes, the adaptation process is nearly complete; the sensitivity of the eye in parts of the retina where both rods and cones are present is increased by a factor of 10,000 for white light.

There are several important operational consequences of the dark adaptation process:

■ Best performance on a task in low light requires that the eye be preadapted to an appropriately low level long enough to attain maximum sensitivity.

■ Because the rods are more sensitive than the cones at low light levels, detection capability is highest on the parts of the retina where rods abound (10–30 degrees from the fovea) and averted vision is required for optimal performance.

■ Because the rods are relatively insensitive to extreme red wavelengths, dark adaptation is facilitated if the observer wears red goggles or if the illumination provided is a very deep red, as in a spacecraft or photographic darkroom. By this means, the observer can continue to use the high-acuity capability of the central fovea at elevated luminance levels for reading instruments and such while the adaptation process goes on, although vision naturally is monochromatic in this case.

■ Because the two eyes are essentially independent in adaptation, it is possible to maintain dark adaptation in one (using an eye patch, for example) while the other is used at high light levels.

## Depth Perception

Depth can be estimated by an experienced observer through use of various cues. Some of these cues are provided by the nature of the scene of interest; others are inherent in the observer.

In cases where only internal cues to distance are available—that is, when objects of interest are of unknown size and shape—the observer must depend on his or her stereoscopic acuity, accommodation and convergence, and, where possible, movement parallax. Accommodation is the only effective cue to distances at ranges of a meter or less, and even here it is inaccurate. Convergence alone is a somewhat more useful cue, but only within about 20 m of the observer. Stereoscopic acuity provides a powerful clue to distance, but it should not be assumed that people with monocular vision cannot do tasks normally assumed to require stereoscopic vision; actual performance should be tested.

# ■ Eye Disorders

## Conjunctivitis

Various types of conjunctivitis, or inflammation of the mucous membrane, can develop beneath the eyelids. The eye becomes scratchy and red and has a discharge. Most often, the cause is bacterial or viral.

A fairly common type of infection, especially in adults, is caused by the herpes simplex virus on the anterior surface of the eyeball. This can lead to blurred vision, scarring, and permanent damage to vision. It may affect one or both eyes and quite often recurs. Topical steroids often used for other types of conjunctivitis greatly aggravate herpes infections. Therefore, any signs suggesting inflammation near the cornea should be immediately referred to a physician experienced in treating eye disorders.

Inflammation of the interior eye is common in adults. One of the most common areas of infection is the uveal tract, which is the middle coat of the eye. Inflammation of this type damages the retina, the lens, and the cornea. Uveal inflammations are quite often associated with diseases of the joint, lung, or intestinal tract. A search must be made for disease elsewhere in the body that might be the cause of the eye problem. When found, the primary cause should be treated. The eye should also be treated to prevent damage to vision.

## Glaucoma

Glaucoma is a leading cause of blindness. The most common form develops when the fluid that normally fills the eyeball, the aqueous humor, fails to drain properly. Ordinarily, the fluid is continuously produced in the eye and excess drains off through a small duct near the iris. Aging, infection, injuries, congenital defects, and other causes can constrict or block the duct. Fluid pressure then builds up and the pressure, if great and of long duration, can damage the optic nerve.

In acute glaucoma, vision dims suddenly, the eyeball becomes painful, and the victim feels quite ill. Insidious glaucoma causes no pain, injuring vision very slowly. Sometimes symptoms include the perception of colored rings and halos about bright objects or dimming of side vision.

Much can be done to preserve vision in most cases when glaucoma is diagnosed in time. Medication is often effective in controlling the pressure, or adequate drainage of the aqueous humor can be established by surgery.

## Cataracts

Cataracts are opaque spots that form on the lens and impair the vision of many elderly and some younger people. Many cases are associated with metabolic disease or aging, but there are also traumatic cases associated with industrial exposures to ionizing radiation, ultraviolet radiation, infrared radiation, foreign bodies, and certain chemicals. If the vision impairment is severe, the diseased lens can be removed and often replaced with a plastic implant.

## Excessive Brightness

Good sunglasses can protect the eyes in bright sunlight. Poor ones compound or create problems. Do not wear glasses with scratches and irregularities. Some glasses are too lightly tinted to do much good; good glasses reduce the invisible as well as the visible light. (Injury from light will be discussed later in this chapter under irradiation burns.)

If a person uses regular glasses, it is worthwhile to have a pair of sunglasses made to the prescription rather than to use possibly inferior sunglasses over a regular correction.

## Night Blindness

Inability to see well or at all in dim light can mean something is wrong not only with the eye but with the entire visual system. Night blindness, as it is called, is a threat to safety, particularly on the highway, because a driver may have 20/20 vision and not realize that his or her vision is somewhat impaired at night. The condition produces no discernible change in eye tissues, so it cannot be diagnosed unless a patient tells the physician of difficulty in reading road signs or picking out objects at night. It is not normal to have trouble seeing in dim light because sufficient accommodation occurs in two or three minutes.

## Eyestrain

Eyestrain can lead to severe signs of local irritation, headaches, fatigue, vertigo, and digestive and psychological reactions. This condition can result from a need for eyeglasses or from using glasses with the wrong correction.

Eye muscle strain may also result from unfavorable conditions such as improper lighting while reading or doing close work. To avoid strain when reading, do not face the light; it should come from behind and to the side. Be sure light bulbs are strong enough (75–100 watts). Hold the book or paper about 16–18 in. (0.4 m) away and slightly below eye level. Avoid glare and occasionally rest the eyes by shifting focus and looking off into the distance. ANSI/IES-RP-7 provides detailed recommendations for the design of industrial lighting.

## Nystagmus

Nystagmus, involuntary movement of the eyeballs, may occur among workers who, for extended periods, subject their eyes to abnormal and unaccustomed movements. Complaints of objects dancing before the eyes, headaches, dizziness, and general fatigue are associated symptoms; all can clear up quickly if a change of work is made. The involuntary movements of the eyeball characteristic of nystagmus can sometimes be induced by occupational causes affecting the eyes through the central nervous system or by some extraneous cause. The most prevalent form of occupational nystagmus is seen in miners.

# ■ Physical Hazards

The eye is subject to many kinds of physical injury—blows from blunt objects, cuts from sharp objects, and damage from foreign bodies.

## Blows from Objects

A blow from a blunt object can produce direct pressure on the eyeball or, if the object delivering the blow seals the rim of the bony orbit on impact, it can exert hydraulic pressure. Such blows may cause contusion of the iris, lens, retina, or even the optic nerve. Violent blows might rupture the entire globe or fracture the thin lower plate of the bony orbit, entrapping the eye muscles.

Contusions may result in serious, irreversible injury if not treated promptly and adequately. Hemorrhaging releases blood, which can be toxic to eye tissues, and physical dislocations of lens, retina, and other parts are unlikely to repair themselves. Lacerations of the cornea, lid, or conjunctiva can be caused by any sharp object, from a knife to the corner of a piece of typing paper.

## Corneal Lacerations and Abrasions

Corneal lacerations, if full-thickness, may allow the aqueous solution behind the cornea to gush out until the iris, which has the consistency of wet tissue paper, is pulled toward the laceration and plugs the wound. The iris can be put back in place, the laceration sutured, and the eye made nearly as good as new. More common corneal injuries are scrapes or abrasions that do not penetrate to the chamber behind the cornea. Such abrasions are very painful, but heal within several days if treated properly. If they are too deep or allowed to become infected, scars that interfere with vision result.

Lacerations of the lid heal, but the scar tissue can pull the lid into an unnatural position. In addition to cosmetic deformity, the lids might not close completely or lashes might turn in against the eyeball. Vertical lacerations are more serious in this respect.

Because it is composed of highly differentiated tissues, the eye is more likely to suffer permanent damage from injury than a finger.

This does not mean the eye has no natural defenses. The bony ridges of the skull protect the eyeball from traumatic injury caused by massive impact. A baseball, for instance, is too big to crush the eyeball—it is stopped by the bony orbit.

The cushioning layers of conjunctiva and muscle around the eyeball absorb impact. The fact that the eyeball can be displaced in its socket is also a defense against injury. In addition, the optic nerve is long enough to allow some displacement of the eyeball without rupture of the nerve.

## Blink Reflex

The eye is most vulnerable to attack at the corneal surface. Here, the eye is equipped with an automatic wiper and washer combination. The washers are the lacrimal glands; the wiper is the blinking action.

The teary blink washes foreign bodies from the corneal or conjunctival surfaces before they can become embedded. The triggering mechanism is irritation.

The reflex blink can also act like a door to shut out a foreign object heading for the eye if the eye can see it coming and it isn't coming too fast. Protective equipment for the eye is used in industry to improve or extend these natural defenses. These defenses might be adequate protection against light and small foreign objects and small quantities of mildly toxic liquids, but they are no match for industrial eye hazards such as small high-speed particles or caustic powders and liquids.

## Foreign Bodies

Invasion by a foreign body is the most common type of physical injury to the eye. Not all foreign bodies, however, affect the eye in the same way.

Foreign bodies affecting the conjunctiva are not usually very serious. They may result in redness and discomfort, but not vision damage. Bodies on the conjunctiva, however, can be transferred to the cornea and become embedded if a person rubs the eye. Even with minor irritations of the conjunctiva, a trip to the nurse is advisable. If there is obvious irritation and no object can be found, it is advisable to see a physician immediately.

Some industrial eye injuries may appear trivial, but can become serious due to complications. The most common complication is infection, which can cause delayed healing and corneal scarring. The infection can be carried into the intraocular tissues by a foreign substance and the bacteria can originate either from sources outside of the eye or from pathogenic organisms already present on the lids, conjunctiva, or in the lacrimal apparatus.

Foreign bodies in the cornea can cause the following problems:

- *Pain.* Because the cornea is heavily endowed with nerves, an object sitting on the surface of the cornea constantly stimulates the nerves.

- *Infection.* Bacteria or fungi can be carried by the foreign particle or by fingers used to rub the eye. Such infections used to be much more common, but antibiotics have greatly reduced the problem.

- *Scarring.* Corneal tissue will heal, but the scars are optically imperfect and may obscure vision.

Intraocular foreign bodies can cause the following problems:

- *Infection.* Infection is much less of a problem with low-speed, low-mass particles, but in some cases, the speed of small metallic particles often creates enough heat to sterilize them. Wood particles, however, do not heat up; if they penetrate the eye, they can cause dangerous infection, which usually causes a marked reduction in vision.

- *Damage.* Depending on its angle, point of entry, and speed, an intraocular particle may cause traumatic damage to the cornea, iris, lens, or retina. Damage to the lens is especially serious because it is not supplied with blood and is slow to heal. Also, any damage to the lens can act as a catalyst for protein coagulation, resulting in opacity and loss of vision.

Pure copper particles can cause serious damage to the eye because the toxic copper molecules become deposited in the lens, cornea, and iris (chalcosis). Copper alloys do not seem to have any toxic effects.

Pain cannot be relied on to alert the worker that there is a foreign body in his or her eye. The cornea is very sensitive, but if the object has penetrated into the eyeball, there may be no acute pain.

## Thermal Burns

Heat can destroy eye and eyelid tissue just as it does any other body tissue, but eye tissues do not recover as well as skin and muscle from such trauma. The lids are more likely to be involved in burns than the eye itself because involuntary closing of the eye is an automatic response to excessive heat.

## Irradiation Burns

**Damage mechanisms.** Light in sufficient amounts may damage eye tissue, ranging from barely detectable impairment to gross lesions. The degree of damage depends on the tissue involved and the energy of the incident light photon. Far-infrared light usually effects damage through a general increase in tissue temperature, whereas far-ultraviolet light generally causes specific photochemical reactions.

Damage to lens cells may not be apparent for some time after insult because of the low level of metabolic activity. Low-degree damage is evidenced by vision clouding or cataract and usually is not reversible. When recovery does occur, it is a slow process. Lens damage may be cumulative because dead cells cannot be eliminated from the lens capsule, causing a progressive loss of visual acuity.

Retinal damage can take a number of forms. Generally, the neural components of the retina, such as the photoreceptors, may regenerate when slightly injured, but usually degenerate when extensively injured.

**Ultraviolet radiation.** Harmful exposures to ultraviolet (UV) light usually occur in welding operations, particularly in electric arc welding. The effects include acute keratoconjunctivitis (welder's flash), an acute inflammation of the cornea and conjunctiva which develops in about 6 hours after even a momentary exposure to the arc light. The welder rarely is involved, being too close to the arc to look at it without an eyeshield, but welders' helpers and other bystanders often suffer from exposure.

**Infrared radiation.** Unlike UV, infrared (IR) radiations pass easily through the cornea and their energy is absorbed by the lens and retina. With automation of metals operations, eye damage from IR radiation is not as common as it once was.

**Visible light.** Various combinations of light sources, exposure durations, and experimental animals have been used to determine the threshold level of light capable of producing a visible retinal lesion. Unfortunately, the experiments recorded were not systematically designed or standardized, leaving numerous gaps and inconsistencies in the reports. Such parameters as pulse duration and irradiated spot area or diameter on the retina are notably lacking. Methods of measurement are often unstated, making comparisons and appraisals of accuracy difficult.

Many models have been proposed to explain the production of visible lesions on the retina from exposure to laser light. Most of the models consider thermal injury to be the only cause of damage. (See Chapter 11, Nonionizing Radiation, for more details.)

## Chemical Hazards

The effects of accidental contamination of the eye with chemicals varies from minor irritation to complete loss of vision. In addition to accidental splashing, some mists, vapors, and gases produce eye irritation, either acute or chronic. In some instances, a chemical that does no damage to the eye can be sufficiently absorbed to cause systemic poisoning.

Exposure to irritant chemicals provokes acute inflammation of the cornea (acute keratitis), with pinpoint vacuoles (holes) of the cornea, which rapidly break down into erosions. Some industrial chemicals irritate the mucous membrane, stimulating lacrimation (excessive watering of the eyes). Other results can include discoloration of the conjunctiva, disturbances of vision, double vision from paralysis of the eye muscles, optic atrophy, and temporary or permanent blindness.

## Chemical Burns

Because caustics are much more injurious to the eyes than acids, the medical prognosis of caustic burns is always guarded. An eye might not look too bad on the first day after exposure to a caustic, but later it may deteriorate markedly. This is in contrast to acid burns, in which the initial appearance is a good indication of the ultimate damage.

This is because strong acids tend to precipitate a protein barrier that prevents further penetration into the tissue. The alkalis do not do this; they continue to soak into the tissue as long as they are allowed to remain in the eye.

The ultimate result of a chemical burn may be a scar on the cornea. If this is not in front of the window in the iris, vision may not be greatly hampered. If the scar is superficial, a corneal transplant can alleviate burn damage. Densely scarred corneal tissue cannot be repaired by transplants, but plastic implants are now available.

When the chemical penetrates the anterior chamber of the eye, the condition is called iritis (irritation caused by bathing the iris with the chemical agent). Glaucoma may be a complication of chemical iritis.

## Evaluating Eye Hazards

It does not take special training or engineering skills to identify most eye hazards. When people handle acids or caustics, when airborne particles of dust, wood, metal, or stone are present, or when blows from blunt objects are likely, eye protection is necessary.

Workers directly involved with operations producing these hazards are usually included in protective equipment programs, but often workers on the perimeter of eye-hazardous operations are left unprotected, with costly results. Who has not heard of pieces of broken metal tools propelled from the drill press into a worker's eye or flash burns from reflected radiation?

The danger from agents with delayed or cumulative effects is even less likely to be recognized. A host of new technologies carry risk of exposure to a portion of the ultraviolet spectrum: industrial photo processes, sterilization and disinfection, UV therapy and diagnosis in ambulatory medicine and dentistry, polymerization of dental and orthopedic resins, research labs, and insect traps. Nonoccupational exposures from outdoor activities or tanning parlors may enhance borderline exposures in the workplace.

Although work may only occasionally bring an employee near eye hazards, the safest policy is to encourage all-day eye protection. The outdated technique of hanging a pair of community goggles near the grinding wheel is an example of the *eye-hazard job* approach to eye protection, in which jobs that involved eye hazards were identified and eye protection was required only for the worker actually doing the job.

The *eye-hazard area concept* is a better approach. An eye-hazard area is an area where the continuous or intermittent work being performed can cause an eye injury to anyone in the area. The concept emphasizes the need for process and environmental controls such as enclosures and radiation-absorbing surfaces and provides eye protection equipment for workers, neighboring workers, supervisors, and visitors. With proper enforcement and designation of areas, this approach is the most effective way to prevent eye injuries.

### First Aid

Propelled object injuries require immediate medical attention. Even for foreign bodies on the corneal surface, self-help should be discouraged; removal of such particles is a job for a trained medical staff member.

Chemical splash injuries require a different approach. Here the extent of permanent damage depends almost entirely on how the victim reacts. If the victim of a concentrated caustic splash gets quickly to an eyewash fountain, properly irrigates the eye for at least 15 minutes, and promptly receives expert medical attention, the chances are good for a clear cornea or, at most, minimal damage.

Such irrigation should be with plain water from standard eyewash fountains, emergency showers, hoses, or any other available sources. Water for eye irrigation should be clean and within certain temperature limits for comfort. Tests show that 112 F (33 C) is about the upper threshold limit for comfort, but colder water, even ice water, apparently causes no harm and is not uncomfortable enough to discourage irrigation.

### ANSI Z358.1–1990

The American National Standard Institute (ANSI) standard Z358.1–1990 covers the design and function of eyewash fountains; the water should meet potable standards. It has been noted that acanthamoebae capable of infecting traumatized eyes can be present in potable water. No cases have been directly attributed to the presence of these organisms in eyewash stations, but it seems prudent to follow the ANSI recommendation of a weekly systemic flushing. At least three minutes of flushing significantly reduces the number of organisms.

Portable units are intended for brief irrigation of an injury. A full 15-minute flushing of the injured eye at a stationary station should follow. It has been suggested that water for a portable station be treated with calcium hypochlorite up to 25 ppm free chlorine to eliminate acanthamoebae.

Some industrial medical units use sterile water for irrigation. Use of water substitutes such as neutralizing solutions, boric acid solutions, and mineral oil is discouraged by nearly all industrial ophthalmologists because in many instances, such preparations can cause eye damage greater than if no irrigation were used at all.

## Protective Equipment

All eye-protection equipment is designed to enhance one or more of the eye's natural defenses. Chipper's cup goggles extend the bony ridge protecting the eye socket and provide an auxiliary, more penetration-resistant cornea. Chemical splash goggles are better than a blinking eyelid.

There is a tremendous variety of eye protection available, from throwaway visitor's eye shields to trifocal prescription safety spectacles and from welder's helmets to clip-on, antiglare lenses. But the classic safety glasses, with or without sideshields, are probably adequate for 90 percent of general industrial work.

The requirement for proper eye protection should be vigorously enforced to ensure maximum protection for the degree of hazard involved. On certain jobs, 100 percent eye protection must be insisted on.

Protection of the eyes and face from injury by physical and chemical agents or by radiation is vital in any occupational safety program. Eye-protective devices must be considered optical instruments and should be carefully selected, fitted, and used.

Unfortunately, the very term *safety glasses* can be confusing. A Food and Drug Administration (FDA) ruling, effective January 1, 1972, requires that all prescription eyeglass and sunglass lenses be impact-resistant. However, such lenses are not the equivalent of industrial-quality safety lenses and they should not be used in an industrial environment where protection is mandatory.

### ANSI Z87.1–1989

Only safety eyewear that meets or exceeds the requirements of ANSI standard Z87.1–1989, *Practice for Occupational and Educational Eye and Face Protection* (referenced in OSHAct Regulations), is approved for full-time use by industrial workers.

The Z87 standard specifies that industrial safety lenses must be at least 3 mm thick and capable of withstanding impact from a 1-in. diameter steel ball dropped 50 in. (1.3 m). The FDA ruling does not mention lens thickness, and requires that a 5/8-in. (16-mm) diameter steel ball be used to verify impact resistance. To pass this test, the lens cannot become chipped or displaced from the frame. For more information, refer to OSHA regulations on eye protection in the general industry codes, in particular, 29 *CFR* 1910.133.

## Impact Protection

Three types of equipment are used to protect eyes from flying particles: spectacles with impact-resistant lenses, flexible or cushion-fitting goggles, and chipping goggles.

**Spectacles.** Spectacles without sideshields should be used for limited hazards requiring only frontal protection. Where side as well as frontal protection is required, the spectacles must have sideshields. Full-cup sideshields are designed to restrict side entry of flying particles. Semifold or flatfold sideshields can be used where only lateral protection is required. Snap-on and clip-on sideshield types are not acceptable unless they are secured (Figure 5–8).

**Flexible-fitting goggles.** These should have a wholly flexible frame forming the lens holder. Cushion-fitting goggles should have a rigid plastic frame with a separate, cushioned surface on the facial contact area (Figure 5–9). Both flexible and cushioned goggles usually have a single plastic lens. These goggles are designed to give the eyes frontal and side protection from flying particles. Most models fit over ordinary ophthalmic spectacles.

**Chipping goggles.** These have contour-shaped rigid plastic eyecups and come in two styles: one for people who do not wear eyeglasses, and one to fit over corrective glasses. Chipping goggles should be used where maximum protection from flying particles is needed.

**Figure 5–8.** Full-cup sideshields are designed to restrict the entry of flying objects from the side of the wearer.

**Figure 5–9.** Flexible-fitting goggles should have a flexible frame forming the lens holder.

If lenses will be exposed to pitting from grinding wheel sparks, a transparent and durable coating can be applied to them.

## Eye Protection for Welding

In addition to damage from physical and chemical agents, the eyes are subject to the effects of radiant energy. Ultraviolet, visible, and infrared bands of the spectrum all produce harmful effects on the eyes, and therefore require special attention.

Welding processes emit radiation in three spectral bands. Depending on the flux used and the size and temperature of the pool of melted metal, welding processes emit UV, visible, and IR radiation; the proportion of the energy emitted in the visible range increases as the temperature rises.

All welding presents problems, mostly in the control of IR and visible radiation. Heavy gas welding and cutting operations as well as arc cutting and welding exceeding 30 amperes also emit UV radiation.

Welders can choose the shade of lenses they prefer within one or two shade numbers:

- Shades numbered 1.5–3.0 are intended to protect against glare from snow, ice, and reflecting surfaces and against stray flashes and reflected radiation from nearby cutting and welding operations. These shades also are recommended for use as goggles or spectacles with sideshields worn under helmets in arc-welding operations, particularly gas-shielded arc-welding operations.

- Shade number 4 is the same as shades 1.5–3.0, but for greater radiation intensity.

For welding, cutting, brazing, or soldering operations, the guide for the selection of proper shade numbers of filter lenses or windows is given in ANSI Z87.1–1979, *Eye and Face Protection*. (For more details, see Chapter 11, Nonionizing Radiation.)

## Laser Beam Protection

No one type of glass offers protection from all laser wavelengths. Consequently, most laser-using firms do not depend on safety glasses to protect employees' eyes from laser burns. Some point out that laser goggles or glasses might give a false sense of security, tempting the wearer to expose himself or herself to unnecessary hazards (Figure 5–10).

Nevertheless, researchers and laser technicians often need eye protection. Both spectacles and goggles are available and glass for protection against nearly all known lasers can be ordered from eyewear manufacturers. Typically, the eyewear has maximum attenuation at a specific

**Figure 5–10.** One type of safety goggles suitable for protection against laser beams.

laser wavelength; protection falls off rapidly at other wavelengths. (For more details, see Chapter 11, Nonionizing Radiation.)

## Video Display Terminals (VDTs)

Much concern has focused on health problems associated with the use of VDTs. There is no doubt that use of these devices can cause increased eye fatigue and visual discomfort. Factors that lead to visual discomfort are poor contrast between the characters and background, high contrast between the screen and other surfaces (such as the documents), and glare from and flicker on the screen. Long periods of eye fixation and refractive errors are also significant contributors.

Often, the displays are at a distance of about 30 in. (76 cm) from the operator's eye. Special corrective lenses may be required because the distance is neither near nor far vision.

Regulations for the design and operation of VDT workstations have been issued by some countries, but most authorities oppose rigid rules. Because of the variety of visual problems and work practices of operators, proper design should allow for flexibility in the placement of the screen, keyboard, source documents, and work surfaces. Ambient lighting must be adjustable, bright backgrounds, such as windows, should be eliminated, and appropriate rest periods are indicated. ANSI/HSF 100–1988 gives detailed recommendations for the design of a visual display terminal workstation. (See Chapter 13, Ergonomics, for additional information.)

## Plastic Versus Glass Lenses

When making a decision between plastic and glass lenses, there are several issues to consider:

- Both can pass impact tests when of certain formulation and thickness.

- Glass has a slightly lower resistance than plastic to breakage by sharp objects.

- Tests show that plastic lenses have more favorable resistance to small objects moving at high rates of speed than glass lenses.

- Plastics have low abrasion resistance, which can be increased with the use of coatings.

- Plastics are resistant to hot materials. Hot metal invariably shatters glass but not plastic. Hot metal also tends to adhere to glass.

- Plastics generally show surface reaction to some chemicals, but satisfactorily stop splashes and protect the eyes.

- Whereas fogging occurs on both glass and plastic, it usually takes longer for plastic to fog. Plastic goggles are available with a hydrophilic coating that tends to prevent fogging. There are also double-lens plastic goggles that operate on the thermopane principle and suppress fogging to a large extent.

## Sunglasses

Use of safety sunglasses is a common practice for people who work outdoors, but they are not appropriate for indoor work. No tinted lenses of any kind should be worn indoors unless specifically required because of excessive glare or eye-hazardous radiation. There may be some employees whose eyesight would benefit from lightly tinted lenses because of a particular eye condition, but not very many.

**Phototropic lenses.** Phototropic or photochromic lenses automatically change tint from light to dark and back again, depending on their exposure to UV light. The convenience of sunglasses with variable-tint lenses is obvious, even though such lenses do not react indoors, in a car, or anywhere else that UV light cannot reach.

The ANSI Z87 standard recommends a variety of fixed-density tinted lenses for specific job situations involving radiation harmful to vision. Each tint is assigned an individual shade number, which is inscribed on the front surface of each such lens. The current Z87 standard makes no mention of phototropic lenses; this fact alone should give pause to health and safety specialists who might be pressured to shift a work force immediately into phototropic lenses.

Such lenses may have a future for outdoor use by telephone line and brush crews and gas-line transmission workers. However, until safety eyewear manufacturers are willing—or able—to certify phototropic lenses as being fully in compliance with the Z87 standard, industry would be well-advised to follow ANSI guidelines.

## Contact Lenses

Contact lenses have certain advantages in many applications outside the industrial environment. However, with rare exceptions, contact lenses have few practical applications for people employed in manufacturing and other industries.

Contact lenses do not provide eye protection in the industrial sense; their use without appropriate eye or face protective devices of industrial quality should not be permitted in a hazardous environment, according to the National Society for the Prevention of Blindness and ANSI standard Z87.1–1979.

Many contact lens wearers are employed in industry; a great many others will accept such work in the future. Thus, the possible effects of an industrial environment on the eyesight of workers who wear contact lenses are of particular concern to health and safety professionals.

Workers who use contact lenses should also have a pair of prescription safety glasses available. Accidental displacement or loss of a contact lens may occur without warning, thereby causing immediate incapacitation by sudden change of vision, excessive tearing, light sensitivity, and involuntary squeezing together of the eyelids.

Another important factor often overlooked when contact lenses are worn by an industrial worker is spectacle blur. The vision of either eye may be blurred when contact lenses are removed. Although the change in visual acuity is not extreme, the blurred vision may persist for as long as an hour.

## Comfort and Fit

To be comfortable, eye-protective equipment must be properly fitted. Corrective spectacles should be fitted only by a qualified, licensed practitioner. However, a technician can be trained to fit, adjust, and maintain eye-protective equipment. Of course, each worker should be taught the proper care of the device being used.

To give the widest possible field of vision, goggles should be close to the eyes but the eyelashes should not touch the lenses (Figure 5–11).

Various defogging materials are available. Before a selection is made, test to determine the most effective type for a specific application.

In areas where goggles or other types of eye protection are extensively used, goggle-cleaning stations should be conveniently located, along with defogging materials, wiping tissues, and a waste receptacle.

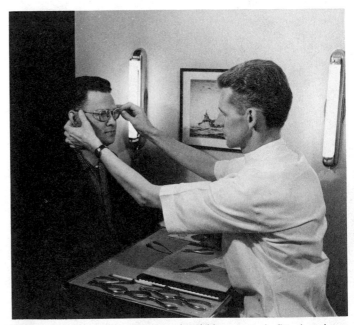

**Figure 5–11.** Safety glasses should be properly fitted and adjusted.

# ■ Vision Conservation Program

There are four steps in a vision conservation program:

1. The environmental survey
2. Vision screening program
3. Remedial program
4. Professional fitting and follow-up procedures

## Environmental Survey

The environment should be surveyed by people qualified in industrial vision. The survey should assess the likelihood of injury and potential severity of injury from the worker's job operation, the potential for injury from adjacent operations, and the optimum visual acuity requirements for fast, safe, efficient operations.

The environmental survey includes illumination measurements and recommendations for improvements to make the workplace safer. Often, simply cleaning existing lighting can increase illumination 100 percent.

Job working distances and viewing angles should be measured so eye doctors have the necessary information to prescribe lens strengths affording optimum comfort, efficiency, safety, and ergonomic advantage (Figure 5–12).

Each workstation should be free of toxic or corrosive materials and employees should be instructed in the correct use of eyewash facilities. Eyewash fountains should be examined to make certain they work properly and provide an even flow of water.

All environmental factors influencing an employee's visual performance and safety should be written in a visual job description, in terms understandable to the eye doctors who will care for employees with deficient vision.

**Figure 5–12.** Measuring the distance from the eye to the work plane is essential for proper safety vision prescriptions.

## Vision Screening Program

The next step, after working conditions and visual requirements are known, is to determine the visual status of the work force. Reliable vision screening instruments are available for use by nurses or trained technicians.

These instruments test the visual acuity in each eye separately, both eyes together, both at near point (usually working distance), both at far point (distant vision), and binocular coordination (the ability to make the two eyes work together). Additional specific tests—such as color vision, field of vision, glaucoma testing, and depth perception—can be added when a need is indicated.

Often, an employee actually performs many jobs. In these instances, recommendations are made after the job most frequently performed is compared with the one most visually demanding.

## Remedial Program

Each employee is told the result of his or her vision screening; people showing deficiencies are referred to the eye doctor of their choice. The employee goes to the doctor with the prescription form for safety glasses and a written description of the visual aspect of the job or duties at work. The job description includes recommendations concerning the type of prescription that will make the worker more comfortable and more efficient.

## Professional Fitting

The final step in a vision conservation program is professionally fitting the protective and corrective safety eyewear to the employee. Proper fitting and sizing are essential and cannot be correctly done by the tool crib attendant or purchasing agent. Workers come in various sizes and shapes, as do safety glasses. Proper measurement and fitting can be the difference between a successful and an unsuccessful program. A perfect prescription is useless if the frame hurts so much that it cannot be worn.

This fitting procedure is equally important for employees required to wear nonprescription safety glasses. These employees usually are the most difficult to fit. They are not accustomed to having anything in front of their eyes or feeling weight on their nose or ears. Considerable care must be exercised in the fitting for such employees. Adjustment and alignment should be readily available.

Human vision is not static; it changes constantly. Employees, especially those over age 40, should be advised that gradual, sometimes unnoticed, changes in their vision may affect their safety and efficiency.

Continued testing must be instituted as a health and safety department policy. If no health specialist is employed at the facility, make arrangements for this service with your local health care community. The local academy of ophthalmology or optometry can usually provide trained personnel.

## Guidelines

The following guidelines should be a part of every vision conservation program:

- Make it a 100 percent program; include everyone. Employees will accept it more readily and it will be easier to administer. Promote it well in advance; get union cooperation.

- Make certain that safety eyewear is properly fitted. A few jobs (welding, labs, lasers) require special types. Optical companies will assist in fitting eyewear and explaining maintenance.

- Include eye care stations for first aid and for cleaning lenses.

- Control eye hazards at the source; install safety glass guards on machines to prevent flying chips or splashing liquids and install enclosures to control fine dusts, mists, or vapors.

- Make sure all areas have adequate lighting, are free from glare, and are painted in colors that emphasize depth perception and highlight potential hazards.

- Post signs such as ALL PERSONNEL AND VISITORS MUST WEAR PROTECTIVE EYEWEAR in all hazardous areas.

- All employees should be given preplacement eye examinations. Periodic follow-up examinations should be scheduled, especially for employees over 40.

## ■ AMA Guides for Evaluating Impairment

This section is adapted from the AMA's *Guides to Evaluation of Permanent Impairment*. It is included in this chapter to assist health and safety professionals in interpreting and understanding medical reports of workers' compensation cases.

This section includes criteria for use in evaluating permanent impairment of the visual system.

This guide provides a simplified method for determining permanent visual impairment and its effect on a person's ability to perform the activities of daily living. Diminished visual ability is expressed as a percentage of impairment of the visual system. Diminished ability of the individual is expressed as a percentage of impairment of the whole person.

Visual impairment results from a deviation from normal in one or more of three eye functions: corrected visual acuity for distance and near, visual fields, and ocular motility with diplopia. Evaluation of visual impairment is based on these three functions. Although they are not equally important, vision is imperfect without their coordinated function.

Other ocular functions and disturbances are considered to the extent they are reflected in one or more of the three coordinated functions. Such ocular disturbances include a slight paralysis of movement (paresis of accommodation), a paralysis of the sphincter of the iris (iridoplegia), distorted image (metamorphopsia), inversion of the eyelid causing the lashes to rub against the eyeball (entropion), eversion of the eyelid (ectropion), a persistent flow of tears from excessive secretion or impeded outflow (epiphora), and a condition where the eyelids cannot be entirely closed (lagophthalmia).

To the extent that any ocular disturbance causes impairment not reflected in visual acuity, visual fields, or ocular motility without double vision (diplopia), it must be evaluated by the physician and combined with the impairment of the visual system, as determined in the calculations that follow. In such circumstances, a physician will combine the value of the measurable functions on which evaluation is based in this guide with the value assigned to other ocular impairments in the individual case. Deformities of the orbit and cosmetic defects that do not alter ocular function are not considered.

The following equipment is necessary to test the functions of the eyes:

- Visual acuity test charts for distance (such as the Snellen test chart with block letters or numbers, the illiterate E chart, or Landolt's broken-ring chart) and near vision (using a similar Snellen notation for inches or Jaeger print or point type notation)

- A standard perimeter with radius of 330 mm and a tangent screen

- Refraction equipment

### Criteria and Methods for Evaluation

**Central visual acuity.** Illumination of the test chart should be at least 5 footcandles (50 lux) and the chart or reflecting surface should not be dirty or discolored. The test distances should be 20 ft (6 m) for distance and 14 in. (36 cm) for near vision.

Central vision should be measured and recorded for distance and near vision with and without correction. The use of contact lenses might further improve vision impaired by irregular astigmatism from corneal injury or disease. However, the practical difficulties of fitting and the expense and tolerance of such lenses, in addition to the fact that they are sometimes medically contraindicated, are sufficiently important to recommend regular ophthalmic lenses to obtain the best corrected vision. In the absence of contraindications, contact lenses are acceptable.

Visual acuity for distance should be recorded in the notation of a fraction in which the numerator is the test distance in feet or meters and the denominator is the distance at which the smallest letter discriminated by the patient subtends 5 minutes (the distance at which an eye with 20/20 vision would see that letter).

This fractional designation is purely a convenient form of notation and does not imply percentage of visual acuity. A similar Snellen notation with use of inches or a comparable Jaeger or point-type notation can be used in designating near vision acuity.

The visual acuity notations for distance and near vision with corresponding percentages of loss of central vision shown in Table 5–B are included solely to indicate the basic values used in developing Table 5–A.

Simple addition of two percentages of loss corresponding to appropriate notations for distance and near vision does not provide the true percentage loss of central vision.

In accordance with accepted principles, true loss of central vision is the mean of the two percentages (Table 5–B).

Aphakia, absence of the lens, is considered an additional visual handicap and, if present, is assigned a value of 50 percent decrease in the remaining corrected central vision (Table 5–B).

To determine loss of central vision in one eye, measure and record central visual acuity for distance and near vision with and without corrective lenses, either conventional or contacts. Then consult Table 5–B for corresponding loss of central vision depending on the presence of monocular aphakia (absence of the crystalline lens).

For example, without allowance for monocular aphakia, 14/56 for near vision and 20/200 for distance equals 80

**Table 5–A.**  Loss of Central Vision

| Snellen Rating for Distance in Feet | Approximate Snellen Rating for Near Vision in Inches | | | | | | | | | | | | | |
| --- | --- | --- | --- | --- | --- | --- | --- | --- | --- | --- | --- | --- | --- | --- |
| | 14/14 | 14/18 | 14/21 | 14/24 | 14/28 | 14/35 | 14/40 | 14/45 | 14/60 | 14/70 | 14/80 | 14/88 | 14/112 | 14/140 |
| 20/15 | 0% / 50 | 0% / 50 | 3% / 52 | 4% / 52 | 5% / 53 | 25% / 63 | 27% / 64 | 30% / 65 | 40% / 70 | 43% / 72 | 44% / 72 | 45% / 73 | 48% / 74 | 49% / 75 |
| 20/15 | 0 / 50 | 0 / 50 | 3 / 52 | 4 / 52 | 5 / 53 | 25 / 63 | 27 / 64 | 30 / 65 | 40 / 70 | 43 / 72 | 44 / 72 | 46 / 73 | 48 / 74 | 49 / 75 |
| 20/25 | 3 / 52 | 3 / 52 | 5 / 53 | 6 / 53 | 8 / 54 | 28 / 64 | 30 / 65 | 33 / 67 | 43 / 72 | 45 / 73 | 46 / 73 | 48 / 74 | 50 / 75 | 52 / 76 |
| 20/30 | 5 / 53 | 5 / 53 | 8 / 54 | 9 / 54 | 10 / 55 | 30 / 65 | 32 / 66 | 35 / 68 | 45 / 73 | 48 / 74 | 49 / 74 | 50 / 75 | 53 / 76 | 54 / 77 |
| 20/40 | 8 / 54 | 8 / 54 | 10 / 55 | 11 / 56 | 13 / 57 | 33 / 67 | 35 / 68 | 38 / 69 | 48 / 74 | 50 / 75 | 51 / 76 | 53 / 77 | 55 / 78 | 57 / 79 |
| 20/50 | 13 / 57 | 13 / 57 | 15 / 58 | 16 / 58 | 18 / 59 | 38 / 69 | 40 / 70 | 43 / 72 | 53 / 77 | 55 / 78 | 56 / 78 | 58 / 79 | 60 / 80 | 62 / 81 |
| 20/60 | 16 / 58 | 16 / 58 | 18 / 59 | 20 / 60 | 22 / 61 | 41 / 70 | 44 / 72 | 46 / 73 | 56 / 78 | 59 / 79 | 60 / 80 | 61 / 81 | 64 / 82 | 65 / 83 |
| 20/80 | 20 / 60 | 20 / 60 | 23 / 62 | 24 / 62 | 25 / 63 | 45 / 73 | 47 / 74 | 50 / 75 | 60 / 80 | 63 / 82 | 64 / 82 | 65 / 83 | 68 / 84 | 69 / 85 |
| 20/100 | 25 / 63 | 25 / 63 | 28 / 64 | 29 / 64 | 30 / 65 | 50 / 75 | 52 / 76 | 55 / 78 | 65 / 83 | 68 / 84 | 69 / 84 | 70 / 85 | 73 / 87 | 74 / 87 |
| 20/125 | 30 / 65 | 30 / 65 | 33 / 67 | 34 / 67 | 35 / 68 | 55 / 78 | 57 / 79 | 60 / 80 | 70 / 85 | 73 / 87 | 74 / 87 | 75 / 88 | 78 / 89 | 79 / 90 |
| 20/150 | 34 / 67 | 34 / 67 | 37 / 68 | 38 / 69 | 39 / 70 | 59 / 80 | 61 / 81 | 64 / 82 | 74 / 87 | 77 / 88 | 78 / 89 | 79 / 90 | 82 / 91 | 83 / 92 |
| 20/200 | 40 / 70 | 40 / 70 | 43 / 72 | 44 / 72 | 45 / 73 | 65 / 83 | 67 / 84 | 70 / 85 | 80 / 90 | 83 / 91 | 84 / 92 | 85 / 93 | 88 / 94 | 89 / 95 |
| 20/300 | 43 / 72 | 43 / 72 | 45 / 73 | 46 / 73 | 48 / 74 | 68 / 84 | 70 / 85 | 73 / 87 | 83 / 91 | 85 / 93 | 86 / 93 | 88 / 94 | 90 / 95 | 92 / 96 |
| 20/400 | 45 / 73 | 45 / 73 | 48 / 74 | 49 / 74 | 50 / 76 | 70 / 85 | 72 / 86 | 75 / 88 | 85 / 93 | 88 / 94 | 89 / 94 | 90 / 95 | 93 / 97 | 94 / 97 |
| 20/800 | 48 / 74 | 48 / 74 | 50 / 75 | 51 / 76 | 53 / 77 | 73 / 87 | 75 / 88 | 78 / 89 | 88 / 94 | 90 / 95 | 91 / 96 | 93 / 97 | 95 / 98 | 97 / 99 |

*Note:* The upper figure = percent loss of central vision without allowance for monocular aphakia; the lower figure = loss of central vision with allowance for monocular aphakia. (Reprinted with permission from Guides to the Evaluation of Permanent Impairment, 4th ed. Revised. American Medical Association, copyright 1993.)

**Table 5–B.** Visual Acuity Notations with Corresponding Percentages of Loss of Central Vision for Distance

### Snellen Notations

| English | Metric 6 | Metric 4 | % Loss |
|---------|----------|----------|--------|
| 20/15 | 6/5 | 4/3 | 0 |
| 20/20 | 6/5 | 4/4 | 0 |
| 20/25 | 6/7.5 | 4/5 | 5 |
| 20/30 | 6/10 | 4/6 | 10 |
| 20/40 | 6/12 | 4/8 | 15 |
| 20/50 | 6/15 | 4/10 | 25 |
| 20/60 | 6/20 | 4/12 | 35 |
| 20/70 | 6/22 | 4/14 | 40 |
| 20/80 | 6/24 | 4/16 | 45 |
| 20/100 | 6/30 | 4/20 | 50 |
| 20/125 | 6/38 | 4/25 | 60 |
| 20/150 | 6/50 | 4/30 | 70 |
| 20/200 | 6/60 | 4/40 | 80 |
| 20/300 | 6/90 | 4/60 | 85 |
| 20/400 | 6/120 | 4/80 | 90 |
| 20/800 | 6/240 | 4/160 | 95 |

### For Near

| Near Snellen | | Revised Jager Standard | American Point-Type | % Loss |
|---|---|---|---|---|
| Inches | Centimeters | | | |
| 14/14 | 35/35 | 1 | 3 | 0 |
| 14/18 | 35/45 | 2 | 4 | 0 |
| 14/21 | 35/53 | 3 | 5 | 5 |
| 14/24 | 35/60 | 4 | 6 | 7 |
| 14/28 | 35/70 | 5 | 7 | 10 |
| 14/35 | 35/88 | 6 | 8 | 50 |
| 14/40 | 35/100 | 7 | 9 | 55 |
| 14/45 | 35/113 | 8 | 10 | 60 |
| 14/60 | 35/150 | 9 | 11 | 80 |
| 14/70 | 35/175 | 10 | 12 | 85 |
| 14/80 | 35/200 | 11 | 13 | 87 |
| 14/88 | 35/220 | 12 | 14 | 90 |
| 14/112 | 35/280 | 13 | 21 | 95 |
| 14/140 | 35/350 | 14 | 23 | 98 |

(Reprinted with permission from *Guides to the Evaluation of Permanent Impairment,* 4th ed. Revised. American Medical Association copyright 1993.) Chicago: AMA, 1993.

percent loss of central vision. With allowance for monocular aphakia (applicable to corrected vision only), 14/56 for near vision and 20/200 for distance equals 90 percent loss of central vision.

**Visual fields.** The extent of the visual field is determined by the usual perimetric methods with a white target that subtends a 0.5-degree angle (a 3-mm white disk at a distance of 330 mm) under illumination of at least 7 footcandles. A 6/330 white disk (6-mm disk at 330 mm) should be used for aphakia. The test object is brought from the periphery to the seeing area. At least two peripheral fields should be obtained that agree within 15 degrees in each meridian. The reliability of the patient's responses should be noted. The result is plotted on an ordinary visual field chart on each of the eight 45-degree principal meridians. For more complete information including perimetric and other methods of calculating percent loss of visual field, consult Chapter 8 in the AMA's *Guides to the Evaluation of Permanent Impairment.*

**Ocular motility.** Unless diplopia (double vision) is present within 30 degrees of the center of fixation, it rarely causes significant visual loss except on looking downward. The extent of diplopia in the various directions of gaze is determined on a perimeter of 330 mm or on a bowl perimeter 30 cm from the patient's eyes in each of the 45-degree meridians, with use of a small test light and without colored lenses or correcting prisms.

## ■ Summary

Ocular anatomy, visual performance, disorders, and problems were discussed. Also covered were potential hazards and how to protect and conserve vision, as well as how to evaluate visual impairment.

## ■ Bibliography

American Medical Association (AMA). *Guides to the Evaluation of Permanent Impairment,* 4th ed. Chicago: AMA, 1993.

American National Standards Institute. *Practice for Industrial Lighting.* New York: ANSI/IES RP-7, 1990.

American National Standards Institute. *Practice for Occupational and Educational Eye and Face Protection.* New York: ANSI Z87.1, 1989.

Davson H, ed. *The Eye,* 3rd ed. Orlando, FL: Academic Press, 1984.

Fraunfelder FT, Roy FH, Hampton F, et al. *Current Ocular Therapy.* Philadelphia: W.B. Saunders, 1990.

Grant WM. *Toxicology of the Eye,* 3rd ed. Springfield, IL: Charles C. Thomas, 1986.

Hart WM, ed. *Adler's Physiology of the Eye,* 9th ed. St. Louis: Mosby Year Book, 1992.

Rom WN, ed. *Environmental and Occupational Medicine,* 2nd ed. Boston: Little, Brown, 1992.

Tyndall RL, Ironside KS, Lyle MM. The presence of acanthamoebae in portable and stationary eyewash stations. *AIHAJ* 48:933–934, 1987.

PART 3

# Recognition
# of Hazards

CHAPTER 6

# Industrial Toxicology

by Richard Cohen, MD, MPH and
Kameron Balzer, CIH

**Definition**
*Toxicity Versus Hazard*

**Entry Into the Body**
*Inhalation ▪ Skin Absorption ▪ Ingestion ▪ Injection*

**Dose–Response Relationship**
*Threshold Concept ▪ Lethal Dose ▪ Lethal Concentration ▪ Responses*

**Action of Toxic Substances**
*Acute Effects ▪ Chronic Effects ▪ Exposures*

**Effects of Exposure to Air Contaminants**
*Irritation ▪ Asphyxiants ▪ Central Nervous System Depressants ▪ Other Effects*

**Neoplasms and Reproductive Toxicity**
*Carcinogenesis ▪ Mutagenesis ▪ Reproductive Toxicity*

**Basis for Workplace Standards**
*Chemical Analogy ▪ Animal Experimentation ▪ Human Epidemiological Data*

**Federal Regulations**
*Occupational Safety and Health Act (OSHAct) ▪ TSCA ▪ Toxic Substances List ▪ NIOSH/OSHA Standards*

**Discussion of ACGIH Threshold Limit Values**
*Guides ▪ Time-weighted Average ▪ Ceiling Values ▪ Mixtures ▪ Carcinogens ▪ Physical Factors ▪ Unlisted Substances ▪ Basic Data Used for TLVs ▪ Documentation*

**Biological Standards**
*Urine Tests ▪ Blood Analysis ▪ Breath Analysis ▪ Biological Limits*

**Sources of Toxicological Information**
*Material Safety Data Sheet*

**Summary**

**Bibliography**

**Other Resources**
*Computer Accessible Data Bases ▪ Organizations*

TOXICOLOGY IS THE SCIENCE THAT STUDIES the poisonous, or toxic, properties of substances. Everyone is exposed on and off the job to a variety of chemical substances; most are not hazardous under ordinary circumstances, but they all have the potential to cause injury at some concentration. How a material is used is the major determinant of its hazard potential. Any substance contacting or entering the body is injurious at an excessive level of exposure and theoretically can be tolerated without effect at some lower exposure.

## Definition

A toxic effect is any reversible or irreversible noxious effect on the body—any chemically induced tumor or any mutagenic or teratogenic effect or death—as a result of contact with a substance via the respiratory tract, skin, eye, mouth, or any other route. Toxic effects are undesirable disturbances of physiological function caused by an overexposure to chemical or physical agents. Toxic effects can also arise as side effects in response to medication and vaccines. Toxicity is the capacity of a chemical to harm or injure a living organism by other than mechanical means. Toxicity entails a definite dimension (quantity or amount); the toxicity of a chemical depends on the degree of exposure.

Many chemicals essential for health in small quantities are highly toxic in larger quantities. Small amounts of zinc, manganese, copper, molybdenum, selenium, chromium, nickel, tin, potassium, and many other chemicals are essential for life. However, severe acute and chronic toxicity results from an uptake of large amounts of these materials. For example, nickel and chromium in some of their forms are considered carcinogens.

The responsibility of the industrial toxicologist is to define how much is too much and to prescribe precautionary measures and limitations so that normal, recommended use does not result in the absorption of too much of a particular material. From a toxicological viewpoint, the industrial hygienist must consider all types of exposure and the subsequent effects on the living organism.

## Toxicity Versus Hazard

A distinction must be made between toxicity and hazard. Toxicologists generally consider toxicity as the ability of a substance to produce an unwanted effect when the chemical has reached a sufficient concentration at a certain site in the body; hazard is regarded as the probability that this concentration in the body will occur. Many factors contribute to determining the degree of hazard—route of entry, dosage, physiological state, environmental variables, and other factors. Assessing a hazard involves estimating the probability that a substance will cause harm. Toxicity, along with the chemical and physical properties of a substance, determines the level of hazard. Two liquids can possess the same degree of toxicity but present different degrees of hazard. One may be odorless and not irritating to the eyes and nose whereas the other may produce a pungent or disagreeable odor at a harmless concentration, or be an eye or respiratory irritant. The material with the warning properties at low or harmless concentrations presents a lesser degree of hazard. Its presence can be detected in time to avert injury.

Many chemical agents are not selective in their action on tissues or cells; they can exert a harmful effect on all living matter. Other chemical agents act only on specific cells. Some agents are harmful only to certain species; other species have built-in protective mechanisms.

The term *toxicity* is commonly used in comparing one chemical agent with another but is meaningless without data designating the biological species used and the conditions under which the harmful effects were induced.

A chemical stimulus can be considered to have produced a toxic effect when it satisfies the following criteria:

- An observable or measurable physiological deviation has been produced in any organ or organ system. The change can be anatomic in character and may accelerate or inhibit a normal physiological process, or the deviation can be a specific biochemical change.

- The observed change can be duplicated from animal to animal even though the dose–effect relationships vary.

- The stimulus has changed normal physiological processes in such a way that a protective mechanism is impaired in its defense against other adverse stimuli.

- The effect is either reversible or at least attenuated when the stimulus is removed.

- The effect does not occur without a stimulus or occurs so infrequently that it indicates generalized or nonspecific response. When high degrees of susceptibility are noted, equally significant degrees of resistance should be apparent.

- The observation must be noted and must be reproducible by other investigators.

- The physiological change reduces the efficiency of an organ or function and impairs physiological reserve in such a way as to interfere with the ability to resist or adapt to other normal stimuli, either permanently or temporarily.

Although the toxic effects of many chemical agents used in industry are well-known, many other commonly used chemicals are not as well-defined. The toxicity of a material is not a physical constant—such as boiling point, melting point, vapor pressure, or temperature—and often only a general statement can be made concerning the harmful nature of some chemical agents.

If the toxicity of a chemical could be predicted from its chemical constitution or structural formula, it would certainly make the job easier. Although certain important analogies are apparent between structure and toxicity, important differences exist that require individual study of each compound.

In addition to establishing toxicity, evaluation of a chemical hazard involves establishing the amount and duration of exposure, the physical characteristics of the substance, the conditions under which exposure occurs, and the determination of the effects of other substances in a combined exposure. All of these may significantly influence the toxic potency of a substance.

The chemical properties of a compound are often one of the main factors in its hazard potential. Vapor pressure (an indicator of how quickly a liquid or solid evaporates) determines whether a substance has the potential to pose a hazard from inhalation. Many solvents are troublesome because they are quite volatile and vaporize readily into the air to produce high concentrations of vapor. Hence, a solvent with a low boiling point would be a greater hazard than an equally toxic solvent with a high boiling point simply because it is more volatile and it evaporates faster.

Chemical injury can be local or systemic. Local injury results from direct contact of the irritant with tissue. The skin can be severely burned or the surface of the eye can be injured to the extent that vision is impaired. In the respiratory tract, the lining of the trachea and the lungs can be injured as a result of inhaling toxic amounts of vapors, fumes, dusts, or mists. These are all examples of direct chemical contact with tissues and the toxicological reactions can be slight or severe.

# Entry Into the Body

In discussing toxicity, it is necessary to describe how a material gains entrance into the body and then into the bloodstream. For an adverse effect to occur, the toxic substance must first reach the organ or bodily site where it can cause damage. Common routes of entry are inhalation, skin absorption, ingestion, and injection. Depending on the substance and its specific properties, however, entry and absorption can occur by more than one route, such as inhaling a solvent that can also penetrate the skin. Where absorption into the bloodstream occurs, a substance may elicit general effects or, most likely, the critical injury will be localized in specific tissues or organs.

## Inhalation

For industrial exposures to chemicals, the most important route of entry is usually inhalation. Nearly all materials that are airborne can be inhaled.

The respiratory system is composed of two main areas: the upper respiratory tract airways (the nose, throat, trachea, and major bronchial tubes leading to the lobes of the lungs) and the alveoli, where the actual transfer of gases across thin cell walls takes place. Only particles smaller than about 5 μm in diameter are likely to enter the alveolar sac.

The total amount of a toxic compound absorbed via the respiratory pathways depends on its concentration in the air, the duration of exposure, and the pulmonary ventilation volumes, which increase with higher work loads. If the toxic substance is present in the form of an aerosol, deposition and absorption occur in the respiratory tract. For more details, see Chapter 2, The Lungs.

Gases and vapors of low water solubility but high fat solubility pass through the alveolar lining into the bloodstream and are distributed to organ sites for which they have special affinity. During inhalation exposure at a uniform level, the absorption of the compound into the blood reaches an equilibrium with metabolism and elimination.

## Skin Absorption

An important route of entry is absorption through either intact or abraded skin. Contact of a substance with skin results in four possible actions: The skin can act as an effective barrier, the substance can react with the skin and cause local irritation or tissue destruction, the substance can produce skin sensitization, or the substance can penetrate to the blood vessels under the skin and enter the bloodstream.

For some substances (such as parathion), the skin is the main portal of entry in typical occupational exposure. For other substances (such as aniline, nitrobenzene, and phenol), the amounts absorbed through the skin are roughly equivalent to the amounts absorbed through inhalation. For the majority of other organic compounds, the contribution from skin (cutaneous) absorption to the total exposure is significant. Results from animal studies using phospho-organic compounds show that toxic effects can occur because of cutaneous penetration.

The cutaneous absorption rate of some organic compounds rises when temperature or perspiration increases. Therefore, absorption can be higher in warm climates or seasons. The absorption of liquid organic compounds may follow surface contamination of the skin or clothes; for other compounds, it may directly follow the vapor phase, in which case the rate of absorption is roughly proportional to the air concentration of the vapors. The process may be a combination of deposition of the substances on the skin surface followed by absorption through the skin.

The physiochemical properties of a material mainly determine whether a material is absorbed through the skin. Among the important factors are pH of the skin and the chemical's extent of ionization, aqueous and lipid solubilities, and molecular size.

Human skin shows great differences in absorption at different anatomic regions. The skin on the palm of the hand shows approximately the same penetration as that of the forearm for certain organic phosphates. The skin on the back of the hand and the skin of the abdomen have twice the penetration potential of the forearm, whereas follicle-rich sites such as the scalp, forehead, angle of the jaw, postauricular area (behind the ear), and the scrotum show a much greater penetration potential. High temperatures generally increase skin absorption by increasing vasodilation. If the skin is damaged by scratching or other abrasion, the normal protective barrier to absorption of chemicals is lessened and penetration occurs more easily. (See Chapter 3, The Skin, for more information.)

## Ingestion

Anything swallowed moves into the intestine and can be absorbed into the bloodstream and thereafter prove toxic. The problem of ingesting chemicals is not widespread in industry; most workers do not deliberately swallow materials they handle.

Workers can ingest toxic materials as a result of eating in contaminated work areas; contaminated fingers and hands can lead to accidental oral intake when a worker eats or smokes on the job. They can also inhale materials when contaminants deposited in the respiratory tract are carried out to the throat by the action of the ciliated lining of the respiratory tract. These contaminants are then swallowed and significant absorption of the material may occur through the gastrointestinal tract. Approximately one quart of mucus is produced daily in an adult's lungs. This constant flow of mucus can carry contaminants out of the lungs into the throat to be swallowed with the saliva or coughed up and expectorated.

Oral toxicity is generally lower than the inhalation toxicity for the same material because absorption of many materials from the intestines into the bloodstream is relatively poor. Food and liquid mixed with a toxic substance not only provide dilution, but can combine with it to form less soluble substances. Also, action of digestive acids and enzymes can chemically alter the substance and reduce its toxicity. After absorption from the gastrointestinal system into the bloodstream, the toxic material goes to the liver, which metabolically alters, degrades, or detoxifies many substances. This detoxification process is an important body defense mechanism. Basically, detoxification involves a sequence of reactions: deposition in the liver, conversion to a nontoxic water-soluble compound, transportation to the kidneys via the bloodstream, and excretion through the kidney and urinary tract.

## Injection

A material can be injected into some part of the body. This can be done directly into the bloodstream, the peritoneal cavity, or the pleural cavity. The material can also be injected into the skin, muscle, or any other place a needle can be inserted. The effects produced vary with the location of administration. In industrial settings, injection is an infrequent route of worker chemical exposure.

There is increasing attention to prevention of skin puncture and injection injuries associated with bloodborne pathogens (hepatitis B, HIV, and hepatitis C). Risk of infection is significant following accidental skin puncture by a needle or instrument contaminated with infected blood or tissue.

In the laboratory, toxic substances are injected into animals because it is far more convenient and less costly than establishing blood levels by inhalation exposures. Injection is a rather imperfect way of getting toxic materials into the body if the end result is to gain knowledge about industrial hazards. Intravenous injection sidesteps protective mechanisms in the body that prevent substances from entering the blood.

## ▪ Dose–Response Relationship

All toxicological considerations are based on the dose–response relationship. A dose is administered to test animals and, depending on the outcome, is increased or decreased until a range is found where at the upper end all animals die and, at the lower end, all animals survive. The data collected are used to prepare a dose–response curve relating percent mortality to dose administered (Figure 6–1).

The doses given are expressed as the quantity administered per unit body weight, quantity per skin surface area, or quantity per unit volume of respired air. In addition, the length of time during which the dose was administered should be listed.

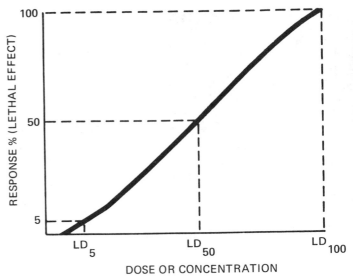

**Figure 6–1.** Dose–response curves for a chemical agent administered to a uniform population of test animals.

The dose–response relationship can also be expressed as the product of a concentration ($C$) multiplied by the time duration ($T$) of exposure. This product is proportional more or less to a constant ($K$); or mathematically, $C \times T \approx K$. The dose involves two variables—concentration and duration of exposure. For certain chemicals, a high concentration breathed for a short time produces the same effect as a lower concentration breathed for a longer time. The $CT$ value provides a rough approximation of other combinations of concentration of a chemical and time that would produce similar effects. Although this concept must be used very cautiously and cannot be applied at extreme conditions of concentration or time, it can be useful in predicting safe limits for airborne contaminants in the workplace. Regulatory exposure limits are set so that the combination of concentrations and time durations are theoretically below the levels that produce injury to exposed individuals.

## Threshold Concept

A small amount of most chemicals is not harmful; there is a threshold of effect or a no-effect level. The most toxic chemical known, if present in small enough amounts (a few molecules) produces no measurable effect. It can damage one cell or several cells, but no measurable effect, such as kidney dysfunction, will result. As the dose is increased, there is a point at which the first measurable effect is noted. The toxic potency of a chemical is defined by the relationship between the dose (the amount) of the chemical and the response produced in a biological system. Thus, a high concentration of toxic substance in the target organ causes a severe reaction and a low concentration causes a less severe or no reaction.

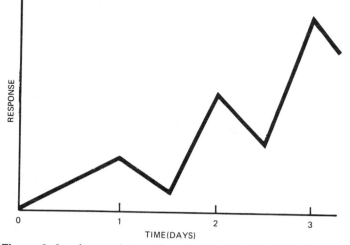

**Figure 6–2.** Accumulation of a substance in a body is shown in this curve; the level of the substance increases with the duration of exposure.

Although many exposures to substances in industry occur by way of the respiratory tract or skin, most published reports of exposures concern studies of experimental animals in which the test substances were introduced primarily through the mouth (in food, in drinking water, or by intubation [tube] directly into the stomach).

The harmfulness of a material depends on its chemical composition, the type and rate or degree of exposure, and the fate of the material in the body. A single large dose of a toxic substance usually produces a greater response than the same total dose administered in small amounts over a long period of time. Each of the small amounts can be detoxified quickly but a large dose produces its detrimental action before appreciable detoxification occurs. A toxic substance that is detoxified or excreted at a rate slower than the rate of intake becomes a cumulative poison.

Accumulation of a substance in the body is understood as a process in which the level of the substance increases with the duration of exposure and can apply to both continuous and to repeated exposure. Biological tests of exposure show that an accumulation is taking place when rising levels of the substance are seen in the urine, blood, or expired air (Figure 6–2).

Exposure thresholds are most easily determined (and more available) for effects occurring soon after exposure. Other effects such as birth defects and cancer occur months or years after exposure began. Dose-related data are often imprecise in human epidemiologic studies. For these and other reasons, thresholds for some carcinogens (such as asbestos) have not been identified and are considered to be zero.

Because different biologic mechanisms are involved in reproductive toxicity, attempts are being made to identify exposure levels (mostly from animal studies) below which

no evidence of injury or impairment can be found; these are called the No Observable Effect Level (NOEL).

## Lethal Dose

If a number of animals are exposed to a toxic substance, when the concentration reaches a certain level, some but not all of those animals will die. Results of such studies are used to calculate the lethal dose (LD) of toxic substances.

If the only variable being studied is the number of deaths, it is possible to use the concept of the LD. The $LD_{50}$ is the calculated dose of a substance that is expected to kill 50 percent of a defined experimental animal population, as determined from the exposure to the substance by any route other than inhalation.

Several designations can be used, such as $LD_{50}$, $LD_0$, $LD_{100}$, and so on. The designation $LD_0$, which is rarely used, is the concentration that produces no deaths in an experimental group and is the highest concentration tolerated in animals; $LD_{100}$ is the lowest concentration that kills 100 percent of the exposed animals.

Although $LD_{50}$ is the concentration that kills half of the exposed animals, it does not mean that the other half are in good health.

Normally, $LD_{50}$ units are weight of substance per kilogram of animal body weight, usually milligrams per kilogram or micrograms per kilogram. The $LD_{50}$ value should be accompanied by an indication of the species of experimental animal used, the route of administration of the compound, the vehicle used to dissolve or suspend the material, if applicable, and the time period during which the animals were observed.

The slope of the dose–response curve provides useful information. It suggests an index of the margin of safety, or the magnitude of the range of doses between a noneffective dose and a lethal dose. If the dose–response curve is very steep, this margin of safety is slight. One compound could be rated as more toxic than a second compound because of the shape and slope of the dose–response curve (Figure 6–3).

## Lethal Concentration

When considering inhalation exposures, the $LD_{50}$ is not very useful; the dose by inhalation is needed. A similar designation, lethal concentration (LC), is used for airborne materials. An $LC_{50}$ might be 500 parts of the substance per million parts of air (ppm), which means that when a defined experimental animal population is exposed to a calculated concentration of a substance, that concentration of the substance is expected to kill 50 percent of the animals in a stated length of time.

The duration of exposure is very important because a half-hour exposure might produce an effect that is significantly different from that of a 24-hour exposure.

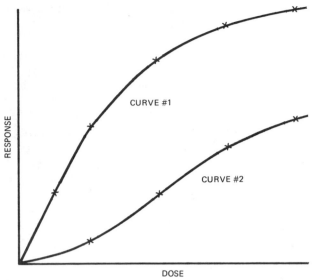

**Figure 6–3.**   The chemical represented by curve #1 has a lower margin of safety and greater toxicity than the curve #2 chemical.

Any publication dealing with LCs should state the species of animal studied, the length of time the exposure was maintained, and the length of time observation was carried out after exposure.

In one study, animals were exposed for a short time to nitrogen dioxide ($NO_2$). At first there was no observable response, but 36 hours after the exposure, the animals developed a chemical pneumonia, became very sick, and died. If the animals had been observed for only the 24-hour period after the exposure, the significant effects that occurred in the next 24 hours would have been missed. Thus, there must be an adequate postexposure time period. The dose should be delivered in a specified length of time with the animals under observation for another specified period of time—this may be 24 hours or 30 days, or even several years when testing for carcinogenesis.

## Responses

After the toxic material has been administered, there are various criteria the toxicologist can use to evaluate the response.

Examining the organs removed from exposed animals reveals the site of action of the toxic agent, the mode of action, and the cause of death. Important pathological changes in tissues can be observed following dose levels below those needed to produce the death of animals. The liver and the kidney are particularly sensitive to the action of many toxic agents.

The effect of the toxic agent on the growth rate of the animals is another criterion of adverse response. Relatively low levels of compounds that do not produce death or signs of serious illness can result in a diminished rate of growth. The food intake must also be measured to learn whether loss of appetite was a cause of diminished growth.

Changes in the ratio of organ weight to body weight can be used as a criterion of adverse response. In some instances, such alterations are specific to the chemical being tested; for example, an increase of lung weight to body weight ratio can result from the pulmonary edema produced by irritants such as ozone or oxides of nitrogen.

Physiological function tests also provide useful criteria of response, both in experimental studies and in assessing the response of exposed workers. They can be especially useful in studies of populations with chronic conditions.

Substances can then be rated according to their relative toxicity, as shown in animal experiments, and the probable LD for humans can be estimated. These examples are based on the results of short-term (acute) exposures only. It is quite possible that long-term (chronic) exposures to a substance could produce serious tissue damage even though short-term exposure tests indicated a low order of toxicity.

Animal experimental data are sometimes difficult to interpret and apply to human exposure. Such data are valuable as guides for an industrial toxicologist in estimating the likely range of toxicity of a substance as well as in guiding further investigation.

## ■ Action of Toxic Substances

The toxic action of a substance can be arbitrarily divided into acute and chronic effects. In addition to acute and chronic toxicity, we can distinguish acute and chronic exposures (Table 6–A). Factors other than immediate effects often determine the type and severity of a chemical's adverse effects. For example, acute benzene toxicity has a

**Table 6–A.**   Brief Comparison of Acute, Prolonged, and Chronic Toxicity Tests

|  | *Acute* | *Prolonged* | *Chronic* |
| --- | --- | --- | --- |
| Exposure lasts . . . | ≤ 24 hours, usually single dose | Typically 2, 4, or 6 weeks | ≥ 3 months |
| Typically yields . . . | Single lethal dose, clinical signs of toxicity | Cumulative dose (if any), major metabolic routes, detoxification or excretion | Potential for carcinogenic effect or other delayed effects |
| Exemplified by . . . | Potassium cyanide rapidly depriving tissues of oxygen | Carbon tetrachloride causing destruction of liver cells after repeated exposure over several weeks | Carbon tetrachloride causing liver cancer following prolonged exposure (and observation) |

different clinical picture from that of chronic toxicity. Although ethyl alcohol has a somewhat greater systemic toxicity than methyl alcohol, the latter is much more dangerous when ingested because methyl alcohol can cause serious damage to the optic nerve.

## Acute Effects

Acute exposures and acute effects generally involve short-term high concentrations and immediate results of some kind (illness, irritation, or death). Acute occupational exposures are often related to an accident.

Acute exposures typically are sudden and severe and are characterized by rapid absorption of the offending material. For example, inhaling high levels of carbon monoxide or swallowing a large quantity of cyanide compound produces acute toxicity very rapidly. The critical period, during which survival of the victim is uncertain, occurs suddenly. Such incidents generally involve a single exposure in which the chemical is rapidly absorbed and damages one or more of the vital organs. The effect of a chemical hazard is considered acute when it appears within a short time, such as within minutes or hours, and is relatively short-lived.

## Chronic Effects

In contrast to acute effects, chronic effect or illness is characterized by symptoms or disease of long duration or frequent recurrence. Chronic effects often develop slowly. The meaning of each term conforms to its derivation—the Latin word *acutus* (sharpened) for *acute* and the Greek word *chronikos* (time) for chronic.

The term *chronic exposure* refers to exposure continued for a prolonged period, usually years. Standard-setting organizations now try to establish limits that control chronic as well as acute exposures.

Chronic poisoning means that some level of material is continuously present in the tissues. Chronic poisoning can also be produced by exposure to a harmful material that produces irreversible damage so that the injury, rather than the poison, accumulates or progresses. The symptoms of chronic poisoning are usually different from those seen in acute poisoning by the same toxic agents and because the level of contaminant is relatively low, the worker is often unaware of the exposures as they occur.

Sensitizers are another example of agents that cause recurrent effects after the worker becomes sensitized (allergic) to the toxin. The first few exposures may cause no reaction, but once a person becomes sensitized, reactions can occur from later contact with very small quantities for very short periods of time.

## Exposures

Levels of exposure to air contaminants can also be referred to in terms of acute and chronic exposure. Acute exposure generally refers to exposure to very high concentrations during very short time periods; chronic exposure involves repetitive or continuous exposure during long time periods.

Whether such chronic exposure is a potential health hazard generally depends on the material and the worker's cumulative exposure.

Chronic effects of air contaminants are not necessarily less serious than acute effects simply because they result from exposure to lower concentrations. In fact, the opposite is often true: Although the onset of damage to health can be slow, the ultimate effect can be quite serious and irreversible.

An effect is local if it harms only the part of the body it comes in contact with, as with an acid burn of the skin. A systemic effect is generalized and changes the normal functioning of related organs operating as a system. Examples include the action of carbon monoxide on the blood or hydrogen cyanide on tissue oxidation, and the ultimate effects of both on the central nervous system.

## ■ Effects of Exposure to Air Contaminants

Air contaminants can be classified on the basis of physiological action into irritants, asphyxiants, central nervous system (CNS) depressants, and others not fitting into these three groups.

Physiological responses to toxic materials depend on the concentration and duration of exposure. For example, a vapor or gas at one concentration can exert its principal action on the body as an anesthetic, whereas at a lower concentration for a longer exposure time, the same gas or vapor can injure some internal organ or the blood system without causing CNS depression or anesthesia.

## Irritation

Irritation is an inflammation or aggravation of the tissue the material contacts. Contact of some materials with the face and upper respiratory system affects the eyes, the cells lining the nose, and the mouth.

There are many industrial chemicals that at fairly low concentrations irritate tissues with which they come in contact.

Many irritants are liquids; for many of these, the degree of local irritation is unrelated to their systemic toxicities. Sometimes differences in viscosity are the determining factors in the type of injury. This applies especially in the lungs, where the inhalation hazard from a substance of low viscosity, such as kerosene, is quite different and more severe than the hazard from a higher viscosity substance such as mineral oil.

To a large extent, the solubility of an irritant gas influences the part of the respiratory tract that is affected (Table 6–B). Ammonia, which is very soluble in water, irritates

**Table 6–B.** Comparison of Several Irritants Affecting the Respiratory Tract

| Substance | Description | TLV | Concentrations Exceeding TLV |
|---|---|---|---|
| *A. Irritants Affecting Upper Respiratory Tract* | | | |
| Formaldehyde (HCHO) | Aldehyde, colorless gas at ordinary temperatures. Soluble in water up to 55%. (Formalin is aqueous solution). | TLV = 0.3 ppm ceiling based on complaints of irritation <1 ppm, constant prickling irritation, disturbed sleep. Suspected carcinogen. | 10–20 ppm causes severe difficulty in breathing, intense lacrimation, severe cough. |
| Acrolein (CH2=CHCHO) | Aldehyde, colorless or yellowish liquid. Water soluble. | TLV = 0.1 ppm low enough to minimize, but not entirely prevent, irritation in exposed individuals. | 1 ppm may be strongly irritating to eyes and nose within five minutes or less. 8–10 ppm lethal within four hours or less; 100 ppm and above may be lethal within a short time. |
| Ammonia (NH3) | Alkali, colorless gas. Soluble in water; pungent odor detected as low as 1 ppm. | TLV = 25 ppm should protect against irritation to eyes and respiratory tract, minimize complaints of discomfort among unacclimated individuals. | Irritation of respiratory tract and conjuctiva in workers inhaling 100 ppm. Severe eye damage, lung and airway dysfunction at higher concentrations. |
| Sulfur dioxide (SO2) | A colorless nonflammable gas with acid odor, pungent taste, one of the most common community air pollutants. | TLV $SO_2$ = 2 ppm expected to prevent irritation and accelerated loss of pulmonary function in most workers. | High acute exposure causes intense irritation, death may follow from suffocation due to respiratory paralysis or pulmonary edema. Industrial poisoning usually chronic—may develop as pulmonary dysfunction progressing to emphysema. |
| *B. Irritants Affecting Both Upper Respiratory Tract and Lung Tissues* | | | |
| Chlorine ($Cl_2$) | Halogen, greenish-yellow gas with suffocating odor, which may be noticeable 1–4 ppm. Soluble in water up to 0.8% by weight. | TLV = 0.5 ppm to minimize chronic lung changes, accelerated aging, teeth erosion. | 30 ppm produces intense coughing. |
| Ozone ($O_3$) | Bluish or colorless explosive gas or blue liquid. Pleasant characteristic odor in concentrations of less than 2 ppm. Slightly water-soluble, used as disinfectant. | TLV = 0.1 ppm ceiling, which causes no symptoms but may result in slightly reduced lung function. | Daily intermittent exposure above 5 ppm may cause incapacitating pulmonary congestion. Lung function changes are dose-dependent and reversible at lower doses. |
| *C. Irritants Affecting Primarily Terminal Respiratory Passages and Air Sacs* | | | |
| Nitrogen dioxide ($NO_2$) | Reddish-brown gas with irritating odor. Decomposes in water, nitric acid ($HNO_3$), and nitric oxide (NO). | TLV = 3 ppm ceiling, considered sufficiently low to ensure against reduced respiratory function. | 10–20 ppm may cause mucosal irritation or chronic disease. 100–500 ppm may lead to sudden death, insidious, delayed, and potentially lethal pulmonary edema (most characteristic), delayed inflammatory changes leading to death several weeks after exposure. |
| Phosgene (carbonyl chloride) ($COCl_2$) | Colorless, nonflammable gas. Suffocating odor when concentrated, otherwise odor suggestive of decaying fruit or moldy hay. Slightly soluble in water and hydrolyzed by it. | TLV = 0.1 ppm because of its irritating effects on the respiratory tract at levels slightly above 0.1 ppm. | 3 ppm causes immediate throat irritation, 50 ppm rapidly lethal. |

the nose and throat, primarily because the moisture on the surface absorbs and reacts with ammonia. Nitrogen dioxide, which is much less water soluble, acts mainly on the tissues in the lungs by traveling deep into the lungs before any significant absorption on moist surfaces occurs.

Some irritants produce acute pulmonary edema (fluid in lungs), which usually begins as an immediate or intense inflammation that is later manifested by coughing, difficulty breathing, shortness of breath, cyanosis, or coughing up large amounts of mucus. With other chemicals, irritation can be delayed or an immediate reaction can be followed by a period of remission, typically a few hours for phosgene or 24–48 hours for nitrogen oxides. Sensitizing irritants, such as toluene diisocyanate, induce asthmatic bronchitis.

Respiratory irritants can be inhaled in gaseous form, as a mist, or as particles with a coating of absorbed liquid. Irritants are often grouped according to their site of action (Table 6–B).

Irritants can be subdivided into primary and secondary irritants. A primary irritant is a material that exerts little systemic toxic action, either because the products formed on the tissues of the respiratory tract are nontoxic or because the irritant action at the contact site is far greater than any systemic toxic action.

A secondary irritant produces irritant action on mucous membranes, but this effect is overshadowed by systemic effects resulting from absorption. Examples of materials in this category are many of the aromatic hydrocarbons and other organic compounds. The direct contact of liquid hydrocarbons with the lung can cause chemical pneumonitis. Thus, in the case of accidental ingestion of these materials, inducing vomiting is not recommended because some of the vomited hydrocarbon could be breathed (aspirated) into the lungs.

Normally, irritation is completely reversible. Reversible means that if the victim is taken out of the exposure quickly enough, the irritation will go away with no residual damage. If a person breathes carbon tetrachloride in a sufficient quantity to produce liver damage and is then taken out of the exposure, any existing pulmonary irritation may quickly disappear. However, if the liver damage persists even after sufficient time out of the exposure and subsequent treatment, then it is said to be irreversible.

Irritation is generally reversible after short-term exposures. If a worker goes into a cloud of ammonia, immediate irritation is experienced and unless the worker is greatly overexposed, the sensation of pain and irritation will be largely gone very shortly after removal from exposure. However, temporary damage to the respiratory epithelium can make the worker susceptible to other irritants that would otherwise be tolerated. Corrosive substances, such as strong acids and caustics, by contrast, can cause irreversible tissue damage or destruction.

## Asphyxiants

Asphyxiants interfere with oxygenation of the tissues and the affected individual may suffocate. This class is generally divided into simple asphyxiants and chemical asphyxiants

Simple asphyxiants are physiologically inert gases that dilute or displace atmospheric oxygen below that required to maintain blood levels sufficient for normal tissue respiration. Common examples are carbon dioxide, ethane, helium, hydrogen, methane, and nitrogen.

Asphyxiants deprive the body of the needed oxygen that must be transported from the lungs via the bloodstream to the cells. With complete deprivation of oxygen, brain cells perish in 3–5 min. Total asphyxiation leads to complete absence of oxygen in the blood (anoxia). Partial asphyxiation leads to low levels of oxygen in the blood (hypoxia). If allowed to continue too long, hypoxia also can result in brain damage or death.

Normal air contains approximately 21 percent oxygen. As the oxygen level goes lower, it interferes with the life process and, if it gets too low, it can produce death. At higher altitudes, higher percentages of oxygen are required; 100 percent is needed at 33,000 feet.

**Chemical asphyxiants.** Through their direct chemical action, chemical asphyxiants prevent the uptake of oxygen by the blood, interfere with the transportation of oxygen from the lungs to the tissues, or prevent normal oxygenation of tissues even when the blood is well-oxygenated. Carbon monoxide prevents oxygen transport by preferentially combining with hemoglobin. Hydrogen cyanide inhibits enzyme systems, particularly the cytochrome oxidase system necessary for cellular oxygen use. Hydrogen sulfide paralyzes the respiratory center of the brain and the olfactory nerve. At sufficiently high levels, all three of these chemical asphyxiants can cause almost instantaneous collapse and unconsciousness (Table 6–C).

The principal action of carbon monoxide is its interference with the delivery of oxygen to the tissues. The concentration of carbon monoxide required to cause death is small compared with the lethal amount for simple asphyxiants. Carbon monoxide combines with hemoglobin to form carboxyhemoglobin. Carbon monoxide occupies oxygen's usual binding site on hemoglobin, thus preventing the transport of oxygen through the bloodstream to all cells. Hemoglobin combines with carbon monoxide much more readily than it does with oxygen by a ratio of approximately 300 to 1.

Beyond the familiar acute effects of carbon monoxide, there is concern about how low-level exposures affect performance of such tasks as automobile driving. The blood has a certain oxygen-carrying capacity, called percent oxygen saturation. The actual amount of oxygen that can be transported varies with amount of hemoglobin in a person's blood.

In nonsmokers, a small amount of the hemoglobin is tied up with carbon monoxide at any given time. This occurs because in the normal life process, some carbon monoxide is always being formed in the body (it is a normal process of metabolism). This carbon monoxide as it evolves forms a little bit of carboxyhemoglobin in all people.

Pack-a-day smokers often have 5–10 percent carboxyhemoglobin in the hemoglobin. For this reason, studies involving carbon monoxide exposure in industrial operations must take into account the difference between smokers and nonsmokers. The effects of smoking usually overshadow and outweigh the environmental effects expected from carbon monoxide as an air pollutant. The smoker's carboxyhemoglobin is comparable to that produced by exposure to carbon monoxide at the Threshold Limit Value (TLV) of 50 ppm in air. Additionally, methylene chloride can contribute to carboxyhemoglobin levels because the body metabolizes this solvent to carbon monoxide.

Another example of a chemical asphyxiant is hydrogen cyanide. It is transported by the bloodstream to the individual cells of the body, where it blocks oxygen uptake

**Table 6–C.** Comparison of Some Chemical Asphyxiants

| Substance | Description | TLV | Concentrations Exceeding TLV |
|---|---|---|---|
| Hydrogen cyanide (HCN) | Colorless liquid or gas, flammable, inhibits cellular respiration, almond-like odor. | TLV = 4.7 ppm (5 mg/m³) ceiling STEL, which may give a seven or eightfold margin against lethal effects. | 18–36 ppm causes slight symptoms after several hours, 90 ppm fatal after 30 to 60 minutes, 270 ppm immediately fatal. |
| Carbon monoxide (CO) | Colorless, odorless gas, sparingly soluble in water, that combines with hemoglobin to form carboxyhemoglobin (COHb), which interferes with oxygen transport to tissues and removal of $CO_2$ from tissues. | TLV = 25 ppm (29 mg/m³), based on an air concentration that should not generally result in COHb levels above 10%. Heavy labor, high temperatures, or altitudes 5,000–8,000 feet above sea level may require 25 ppm TLV. | Fatal in 1 minute at 1% concentration (= 10,000 ppm), which causes approximately 20% COHb. Severe poisoning from short exposure often followed by complete recovery but neurological, cardiovascular, pulmonary, other complications may occur. |
| Hydrogen sulfide ($H_2S$) | Colorless flammable gas, burns to sulfur dioxide. Soluble in water but solutions unstable heavier than air. Characteristic odor of rotten eggs detectable at concentrations of 0.02 ppm or appreciable less. Higher toxic concentrations can rapidly deaden sense of smell. Inhibits cellular respiration. | TLV = 10 ppm (14 mg/m³), based primarily on eye effects sometimes reported from slightly lower concentrations. | Concentrations of 300–1000 ppm cause rapid unconsciousness and death through respiratory paralysis. Associated with an unusual diversity of symptoms including chronic keratoconjunctivitis, nausea, insomnia, pulmonary edema, balance disorders, polyneuritis, and gray-green discoloration of the teeth. |

at the cellular level by combining with the enzymes that control cellular oxidation. Oxygen uptake at the cellular level is blocked only as long as the cyanide is present. Normal cellular oxygen uptake resumes if the cells have not died.

Cyanide is toxic to body tissues. It poisons any enzyme system it comes in contact with. The whole life process is a very delicate balance of thousands of systems working together, catalyzed by thousands of organic molecules called enzymes. Many things can interfere with this catalytic action or enzymatic action and many toxic effects can be traced to a particular enzyme system being compromised by a given substance; the result is the same as when that particular enzyme is inhibited.

## Central Nervous System Depressants

Central nervous system depressants (CNSDs) can produce unconsciousness and many of the same symptoms that asphyxiants cause. They prevent the central nervous system (brain and spinal cord) from doing its normal job.

CNSDs exert their principal action by causing simple anesthesia without serious systemic effects, unless the dose is massive. Depending on the concentration, the depth of anesthesia ranges from mild symptoms to complete loss of consciousness and death. In accidents involving very high concentrations, death may be due to simple asphyxiation.

CNSDs include aliphatic alcohols (such as ethyl and propyl), aliphatic ketones (such as acetone and methyl-ethyl-ketone), aromatic hydrocarbons, ethers (such as ethyl and isopropyl), and short-chain halogenated hydrocarbons.

Substances such as ether, chloroform, and other anesthetics that are very effective in producing anesthesia are selected when the intent is to make someone uncon-

scious. A successful anesthetic is one that produces narcosis but does so with a lot of room for error. Some narcotics are quite good but do not have a sufficient margin of safety so they are not intentionally used as anesthetics. In industry, there are many substances that have narcotic properties, such as nearly all organic solvents.

## Other Effects

There are many other substances with a variety of toxicological actions that do not fit into any of the three groups described previously.

**Cardiac sensitization.** Cardiac sensitization from inhalation of certain volatile hydrocarbons can make the heart abnormally sensitive to epinephrine (adrenalin). Some people exposed to these materials can develop abnormal, dangerous cardiac rhythms, usually ventricular in origin. Cardiac sensitization has been observed with the anesthetic use of chloroform and cyclopropane, solvent misuse such as sniffing aerosols or glue, and exposure to some industrial solvents and chlorofluorocarbon (CFC) refrigerants at levels grossly exceeding the recommended TLV.

**Neurotoxic effects.** These are caused by agents that produce significant toxic effects on the nervous system. Metals such as manganese, lead, and mercury are examples. The central nervous system seems particularly sensitive to organometallic compounds.

A different neurotoxic effect involves acetylcholine, a neurotransmitter. In the transfer of energy from one part of the nerve to the next part, the chemical acetylcholine is essential, and the enzyme cholinesterase sees to it that the level of acetylcholine is maintained at the proper levels. As soon as that enzyme is inhibited, the acetylcholine level starts increasing and reaches the level where it is incompatible with the transfer of nervous energy, and the nervous system undergoes a collapse and fails to work. The triggers do not fire the way they should. Some cholinesterase inhibitors include organic phosphate and organophosphate pesticides such as parathion or carbamate pesticides such as carbofuran.

Other neurotoxic effects include neurasthenia and peripheral neuropathy. Neurasthenia involves emotional irritability and loss of intellectual function; it is associated with prolonged (years) exposure to some hydrocarbon solvents (such as styrene and toluene). Peripheral neuropathy involves loss of limb strength or sensation; it is associated with exposure to agents such as lead, hexane, and acrylamide.

In considering health effects from inhaled dust, primary concern is given to solid material that is small enough to enter the alveoli. A certain amount of filtration by the upper respiratory system prevents the large particles from getting into the lung. In the workplace, particles are dispersed in a nonuniform way and in a full spectrum of sizes; only a portion of them are small enough to get into the lung. There is a need, therefore, for standards for the respirable fraction of dusts.

Dusts less than 1 or 2 $\mu$m in diameter reach the deep lung readily and can be expected to exert some effect. The simplest effect is deposition in the lung without tissue damage. Inert dusts such as calcium carbonate (the principal ingredient in limestone, marble, and chalk) and calcium sulfate (the principal ingredient in gypsum) are considered relatively harmless unless the exposure is severe.

Inert dusts are sometimes called nuisance dusts. The definition of inert is relative because all particulates evoke a tissue response when inhaled in sufficient amount. The TLV for nuisance dusts or nuisance aerosols has been set at 10 mg/m$^3$ (total particulates) or 5 mg/m$^3$ when the dust sample is made up of particulates of respirable diameter, provided that their inhalation does not alter the structure of lung air spaces, does not cause collagen or scar formation to any significant extent, and results in a potentially reversible tissue reaction.

Some inert dusts may cause radiopaque deposits in the pulmonary system that are visible on x-ray films, but they produce little or no tissue reactions unless the exposure is overwhelming (as with iron).

Acute reactions to inhaled dust can be described as irritant, toxic, or allergenic. Chronic exposure to dust is associated with various types of pneumoconiosis (see Chapter 8, Particulates).

Far more serious is the effect of insoluble materials that damage the lung. When these get into the lung, they have no place to go. They are either removed by the cleansing mechanisms of the lung or they are transported from the alveoli to a lymph node.

Pneumoconioses, or "dusty lungs," are differentiated into the simple or nonfibrotic type and the complicated, fibrotic, or fibrogenic type. Pneumoconioses associated with inert dusts are sometimes called benign pneumoconioses.

Fibrotic changes are produced by materials such as free silica, which produces the typical silicotic nodule or small area of scar-like tissue. Asbestos also produces typical fibrotic damage to lung tissue (asbestosis) as well as cancerous lung changes. There is concern about the possible effects of nonoccupational low-level exposure to asbestos.

In combination with smoking, exposure to airborne asbestos particles leads to an excessive incidence of lung cancer. The incidence of lung cancer to nonsmokers exposed to asbestos, though lower, is still abnormally high. Asbestos exposure has also been shown to cause mesothelioma, which in nonasbestos-exposed people is a very rare cancer of the lining of the abdomen or the lung. (See Chapter 8, Particulates, for more information on asbestos-related diseases and other pneumoconioses.)

Some inhaled particles that gain entrance to the body through the lung are soluble in body fluids. More specifically, they are soluble in the fluid of the tissues that line the lung. Although they may not damage the lung tissues, they are absorbed into the blood and distributed throughout the body and can damage the nervous system, the kidney, the liver, or other organs.

The effects that result from heavy metals being absorbed through the respiratory tract vary appreciably from substance to substance. Often there is slow, cumulative absorption and retention of metal in the body; however, the insidious toxic symptoms develop so slowly that the source or cause of the symptoms is often not initially recognized.

Inorganic lead poisoning differs physiologically from organic lead poisoning. Excessive exposure to the inorganic form usually results from ingestion or inhalation of dust or fume, causing physical abnormalities that are characterized by anemia, headache, anorexia, weakness, and weight loss. With chronic, high exposures, more serious reactions include bone marrow changes and peripheral neuropathy colic, which can simulate acute appendicitis.

Organic lead, unlike inorganic lead, tends to concentrate in the brain. Some organic lead materials can easily be absorbed through the skin and add to the hazard from ingestion or inhalation. A single exposure to tetraethyl lead can cause symptoms in a few hours and absorption of a relatively small amount can be fatal.

Although mercury poisoning was described by Paracelsus among miners several centuries ago, it has received much more attention recently. Inorganic compounds of mercury are readily absorbed from the intestine and tend to concentrate primarily in the kidneys, but can also damage the brain. Organic mercury tends to be especially concentrated in blood and the brain.

Industrial manganese poisoning, except for its action related to metal fume fever, is primarily a chronic disease resulting from inhalation of fume or dust in the mining or refining of manganese ores. It also can be caused by cutting and welding metals containing high manganese content while in a confined space without respiratory protection. Manganese poisoning is noted for its peculiar neurological effects, especially psychomotor instability.

# Neoplasms and Reproductive Toxicity

Although any new and abnormal tissue growth can be classed as a neoplasm, this term is most often used to describe cancerous or potentially cancerous tissue. The cells of a neoplasm are to some extent out of control. If neoplastic cells invade tissues or spread to new locations in the body (metastasize) the neoplasm has become cancerous or malignant.

## Carcinogenesis

It is well-established that exposure to some chemicals can produce cancer in laboratory animals and humans. In common usage, *carcinogen* refers to any agent that can produce or accelerate the development of malignant or potentially malignant tumors or malignant neoplastic proliferation of cells. *Carcinogen* refers specifically to agents that cause carcinoma, but the current trend is to broaden its usage to indicate an agent that possesses carcinogenic potential. The terms *tumorigen, oncogen,* and *blastomogen* are all used synonymously with *carcinogen.*

There are a number of factors that have been related to the incidence of cancer—the genetic makeup of the host, viruses, radiation including sunshine, hormone imbalance, and exposure to certain chemicals. Other factors such as cocarcinogens and tumor accelerators can be involved. It is also possible that some combination of factors must be present to induce cancers. There is good clinical evidence that some cancers are virus-related.

**Definitions.** *Carcinogenesis* has several possible definitions. A *carcinogen* can be defined as a substance that will induce a malignant tumor in humans following a reasonable exposure.

A *carcinogen* has also been defined as a substance that will induce any neoplastic growth in any tissue of any animal at any dose by any method of application applied for as long as the lifetime of the animal.

Problems arise when all substances that would fulfill the second definition in the experimental laboratory are classified as carcinogens and it is implied that they will cause malignancies in humans in accordance with the first definition. According to NIOSH, a substance is considered a suspected carcinogen to humans if it produces cancers in two or more animal species.

Even if we could extrapolate from a specific strain of a laboratory animal to all species including humans, taking into consideration such factors as weight, surface area, metabolic profiles, and drug-induced changes in metabo-

lism, carcinogenic potential is often difficult to estimate because of interactions with other agents or biologic susceptibilities. One of the best examples is the effect of cigarette smoking and asbestos on the lung: Their synergistic relationship results in a multiplicative increase in lung cancer risk.

Chemicals that induce cancer do so by mechanisms that are not completely defined. It is generally believed that the transformation of a cell from a normal to a carcinogenic state is multistaged and influenced by both internal and external factors. For example, some materials can produce cancer in the lungs after inhalation whereas others pass through the lung as a route of entry and produce the cancer elsewhere.

A toxicologist can look at the structure of an organic chemical and speculate that because of certain functional groups the chemical is carcinogenic but another chemical having similar functional groups is not carcinogenic. The theories that explain why a chemical is carcinogenic are very involved and, as yet, imperfect.

**Typical carcinogens.** Coal tar and various petroleum products have been identified as skin and subcutaneous carcinogens. Pitch, creosote oil, anthracene oil, soot, lamp black, lignite, asphalt, bitumen, certain cutting oils, waxes, and paraffin oils have also been implicated as potential carcinogens.

Workers can be exposed to arsenic, also recognized as a carcinogen, in the manufacture or use of roasting metallic sulfide ores as well as certain paints or enamels, dyes or tints, pesticides, and miscellaneous chemicals.

Inorganic salts of metals such as chromium, and to a lesser extent nickel compounds, are associated with cancer of the respiratory tract, usually the lungs. Other metals such as beryllium and cobalt are suspect, but their direct toxic effects in humans can obscure carcinogenic potential.

Leukemia is a group of diseases characterized by widespread, uncontrolled proliferation and abnormal accumulation of white blood cells and the failure of many of these cells to reach maturity. Exposures to ionizing radiation and benzene are principal occupational causes. Benzene exposure is also associated with blood dyscrasias (diseased state of the blood, generally involving abnormal or deficiently formed cellular elements), which may progress to leukemia or aplastic anemia.

Osteogenic sarcomas (bone tumors) have been detected in workers who applied radioactive luminous paint to instrument and watch dials. Angiosarcomas (a relatively rare malignant growth) of the liver have been found to be associated with human exposure to vinyl chloride monomer. Oat-cell carcinomas of the lung have been found in workers exposed to bis (chloromethyl) ether (BCME). BCME exposure can occur as an unsought intermediate in certain reactions involving formaldehyde and hydrochloric acid. BCME is carcinogenic by inhalation, skin, and subcutaneous routes in animals.

The result of exposure to chemical carcinogens is such that people who absorb a chemical carcinogen such as benzidine are at increased risk of getting bladder cancer. With many carcinogens (Table 6–D). it is clear that there is a higher incidence of cancer in certain groups of people who are exposed to carcinogenic materials. For a more complete listing and discussion of carcinogens, consult the references by the IARC and the National Toxicology Program (1991).

**Environmental factors.** Cancer is considered so insidious and has such severe end results that carcinogenic chemicals are isolated and looked at more carefully than all others. The statement that 80–90 percent of all cancers are environmentally caused does not mean that 80–90 percent of cancers are caused by industry. The environment includes not only the air we breathe and water we drink but our diet and all elements of our lifestyle, on and off the job. The predominant causes of environmental cancer are tobacco smoke and diet.

## Mutagenesis

A mutagen is an agent that affects the genetic material of the exposed organism. It may cause cancer, birth defects, or undesirable effects in later generations. People who work with a certain chemical may not be harmed, but their offspring can be.

The problem of time lag between exposure and effect is particularly severe for mutagenic agents. Mutations will not show up until the next generation at the earliest, and may not appear for several generations. The long latency

**Table 6–D.** Carcinogens with Possible Occupational Relevance

*Substances Known to Be Carcinogenic*

| | | |
|---|---|---|
| 4-aminobiphenyl | Bis (chloromethyl) ether and technical-grade | Erionite |
| Arsenic and certain arsenic compounds | chloromethyl methyl ether | 2-naphthylamine |
| Asbestos | 1-(2-chloroethyl)-3-(4-methylcyclohexyl)-1- | Thorium dioxide |
| Benzene | nitrosourea (MeCCNU) | Vinyl chloride (monomer) |
| Benzidine | Chromium and certain chromium compounds | |

*Substances That May Reasonably Be Assumed to Be Carcinogenic*

| | | |
|---|---|---|
| Acetaldehyde | Dichloromethane (methylene chloride) | 4,4'-methylendianiline and its dihydrochloride |
| 2-acetylaminofluorene | 1,3-dichloropropene (technical grade) | Methyl methanesulfonate |
| Acrylamide | Diepoxybutane | Lindane and other hexachlorocyclohexane |
| Acrylonitrile | Di(2-ethylhexyl)phthalate | isomers |
| 2-aminoanthraquinone | Diethyl sulfate | Michler's ketone |
| O-aminoazotoluene | Diglycidyl resorcinol ether | Nickel and certain nickel compounds |
| 1-amino-2-methylanthraquinone | 3,3'-dimethoxybenzidine and 3,3'-dimethoxy- | 4,4'-methylenebis (N,N-dimethyl)benzenamine |
| Amitrole | benzidine | 2-nitropropane |
| O-anisidine hydrochloride | 3,3'-dimethylbenzidine | 4,4'-oxydianiline |
| Benzotrichloride | Dimethylcarbamoyl chloride | N-methyl-n'-nitro-n-nitrosoguanidine |
| Beryllium and certain beryllium compounds | 4-dimethylaminoazobenzene | Polychlorinated biphenyls |
| Bromodichloromethane | Dimethyl sulfate | Polycyclic aromatic hydrocarbons (such as |
| 1,3-butadiene | Dimethylvinyl chloride | benz(a)anthracene and benzo(a)pyrene) |
| Butylated hydroxyanisole | 1,1-dimethylhydrazine | Nitrilotriacetic acid |
| Cadmium and certain cadmium compounds | Epichlorohydrin | B-propiolactone |
| Carbon tetrachloride | Ethyl acrylate | Propylene oxide |
| Chlorendic acid | 1,4-dioxane | Polybrominated biphenyls |
| Chlorinated paraffins (C₁₂, 60% chlorine) | Ethyl methanesulfonate | Silica, crystalline (respirable) in the form of |
| Chloroform | Formaldehyde (gas) | quartz, cristobalite, and tridymite |
| 3-chloro-2-methylpropene | Ethylene oxide | Sulfallate |
| 4-chloro-o-phenylenediamine | Hexamethylphosphoramide | 1,3-propane sultone |
| P-cresidine | Hydrazine and hydrazine sulfate | Tetrachloroethylene (perchloroethylene) |
| DDT | Ethylene thiourea | Thiourea |
| 2,4-diaminotoluene | Kepone (chlordecone) | Selenium sulfide |
| 1,2-dibromo-3-chloropropane | Lead acetate and lead phosphate | O-toluidine and o-toluidine hydrochloride |
| 1,2-dibromoethane (EDB) | Hexachlorbenzene | Toxaphene |
| 1,4-dichlorobenzene | 2-methylaziridine (propyleneimine) | 2,3,7,8-tetrachlorodibenzo-p-dioxin (TCDD) |
| 1,4-dichlorobenzidine and 3,3'-dichloro- | 4,4'-methylenebis (2-chloroaniline) (MBOCA) | Toluene diisocyanate |
| benzidine dihydrochloride | Hydrazobenzene | 2,4,6-trichlorophenol |
| 1,2-dichloroethane | | |

*Occupational Exposures Associated with a Technological Process That Are Known to Be Carcinogenic*

| | |
|---|---|
| Coke oven emissions | Soots |
| Mineral oils | Tars |

Known carcinogens are substances for which evidence from human studies indicates that there is a causal relationship between exposure to the substance and human cancer. Substances that may reasonably be expected to be carcinogens are those for which there is limited evidence of carcinogenicity in humans or sufficient evidence of carcinogenicity in experimental animals.

(Adapted from *Sixth Annual Report on Carcinogens, 1991 National Toxicology Program.*)

makes it difficult to discover the connection between the exposure and the manifestation of genetic damage.

Mutagens are chemical or physical agents that cause inheritable changes in the chromosomes. A mutagen might have an effect on somatic cells but not on germ cells. In this case, its effects are not passed on to offspring, but depend on the kind of cell affected. For example, the cells of the bone marrow go on multiplying through life and shed the products of division into the blood, where they function for a time as red and white blood cells before they are removed and replaced. Gross interference with the genetic material of such cells may make cell division ineffective

Another type of interference with the genetic materials of these cells by a virus or chemical can make them capable of more rapid growth and multiplication, so that they are formed far more rapidly than they can be removed from the blood, where they interfere with normal body functions. If the white cells are affected in this way, the outcome is a leukemia.

Similar interference with the genetic material could theoretically start up division in cells that do not normally divide during adult life. If the products of such division displace or invade normal tissues, the result is a cancer. In both these instances, the mutagen responsible would have manifested activity as a carcinogen.

## Reproductive Toxicity

Reproduction results from a complex series of events involving both parents. It begins with each parent's genetic contribution (chromosome) and ends with expression of the genes acquired by the offspring. Every step in the reproductive process is vulnerable to effects from external physical and chemical agents. Chromosomal replication, sexual function, ovulation, conception/fertilization, embryo implantation, placental function, fetal development, labor, delivery, and even child development are components of the reproductive process. Table 6–E lists known or suspected human reproductive toxins. Reproductive abnormalities include changes in sperm count, sperm motility, libido, menstruation and cycle length, and fertility rate; these and other changes can result in miscarriage, embryo toxicity, developmental defects, and stillbirth (Tables 6–F, 6–G).

Teratogenesis (congenital malformation) results from interference with normal embryonic development by a biological, chemical, or physical agent. Exposure of a pregnant female may, under certain conditions, produce malformations of the fetus without inducing damage to the mother or killing the fetus. Such malformations are not hereditary. In contrast, congenital malformations resulting from changes in the genetic material are mutations and are hereditary.

**Typical teratogens.** Agents currently identified as human teratogens include infections such as rubella, metals such as

**Table 6–E.** Occupational Agents Known or Strongly Suspected to Cause Human Reproductive Toxicity

| Developmental Effects | Female Reproductive Toxicity |
|---|---|
| Carbon disulfide | Alkylating/antineoplastic agents |
| Carbon monoxide | Arsenic |
| Ethylene glycol monoethyl ether | Carbon disulfide |
| Ethylene glycol monoethyl ether acetate | Ethylene oxide |
| Ethylene glycol monomethyl ether | Ionizing radiation |
| Ethylene glycol monomethyl ether acetate | Mercury |
| Ionizing radiation | |
| Lead | *Male Reproductive Toxicity* |
| Mercury (compounds) | |
| Methyl bromide | Carbon disulfide |
| Polychlorinated biphenyl (PCB) | 1,2-dibromo-3-chloropropane (DBCP) |
| Ribavirin | Dinitrobenzene |
| 2,3,7,8-tetrachlorodibenzo para dioxin (TCDD) | Ethylene glycol monoethyl ether |
| | Ethylene glycol monoethyl ether acetate |
| | Ethylene glycol monomethyl ether |
| | Ethylene gylcol monomethyl ether acetate |
| | Lead |

lead and mercury, chemicals including PCBs, and ionizing radiation.

**Pregnant women in the workplace.** A teratogen, by definition, is different from a mutagen in that it must affect a developing fetus. This is extremely important today because of the very considerable pressure to address the topic of pregnant women in the workplace.

The fetus is protected from some toxic chemicals because the placenta prevents them from entering the fetal bloodstream; however, many toxic chemicals, such as lead, easily cross the placenta. Damage to the fetus (embryo) is most likely to occur in early pregnancy, particularly during the first 8–10 weeks. During much of this critical period, many women are not even aware that they are pregnant.

It can be extremely difficult to establish specific cause-and-effect relationships between a teratogen and the birth defect it can produce. Animal studies must be supplemented with epidemiological data and it may be decades before researchers know with certainty what substances hold how much risk for which unborn infants

The fact that there are pregnant women in the workplace and that they can be exposed to teratogens leads to a problem in setting occupational health standards. An embryo of a few weeks or a fetus of a few months should be given consideration and should not be exposed to a toxic environment. Although one way to solve this problem is to restrict the activities of fertile women in the workplace, this practice is no longer legally acceptable (*Auto Workers vs. Johnson Controls*). Also, the potential for adverse effects on the male reproductive system cannot be overlooked. The workplace should be such that fertile men and women are able to work there without likelihood of harm.

**Table 6–F.** Measures of Reproductive Function Readily Obtainable Before Fertilization

| | | Affected Individual | |
| --- | --- | --- | --- |
| Endpoint | Male | Both | Female |
| Sexual function | Erection<br>Ejaculaton | Libido<br>Behavior | |
| Endocrine system | | Luteinizing hormone<br>Follicle-stimulating hormone<br>Steroid hormones (androgens, estrogens, and progestins) | Cervical mucus quality |
| Germ cells | Sperm number<br>Sperm motility<br>Sperm shape<br>Chromosomal integrity<br>Fertilizing ability | | |
| Fecundity (ability to conceive) | Testicular integrity<br>Semen quality | Integrity of external genitalia | Ovarian integrity<br>Blockage of oviduct<br>Menstrual regularity<br>Amenorrhea<br>Anovulatory cycles |
| Secondary sexual characteristics | | Breast development<br>Facial and axillary hair growth<br>Sebaceous glands | |
| Reproductive lifespan | | Age at puberty | Age at menopause |

(From U.S. Congress, Office of Technology Assessment. *Reproductive Health Hazards in the Workplace,* Chapter 3, Table 3–1. Washington, DC: U.S. Government Printing Office, OTA- BA-266, Dec. 1985.)

**Table 6–G.** Measures of Reproductive Function Readily Available After Fertilization

| | | Affected Individual | |
| --- | --- | --- | --- |
| Endpoint | Female | Both | Offspring |
| Endocrine system | Human chorionic gonadotropin<br>Steroid hormones, especially progesterone | | |
| Health during pregnancy | Hemorrhage<br>Toxemia | Fetal death<br>Spontaneous abortion | Morphology<br>Chromosomal aberrations |
| Perinatal period | | Premature birth<br>Postmature birth | Death<br>Chromosomal aberrations<br>Birth defects<br>Birth weight<br>Apgar score |
| Postnatal period | Lactation | | Infant death<br>Childhood morbidity<br>Childhood malignancies<br>Development<br>Behavior |
| Reproductive lifespan | Age at menopause | | |

(From U.S. Congress, Office of Technology Assessment.*Reproductive Health Hazards in the Workplace,* Chapter 3, Table 3–2. Washington, DC: U.S. Government Printing Office, OTA-BA-266, Dec. 1985.)

# Basis for Workplace Standards

## Chemical Analogy

When dealing with a new chemical, animal or human toxicity data are usually unavailable. Therefore, the nature of response to a chemical can be assumed to be analogous to that produced by contact with a substance with a similar chemical and biological structure. Chemicals that are similar have been assumed initially to produce similar biological responses.

## Animal Experimentation

Before introducing chemical agents into the workplace, it is advisable to know their toxic effects. Then preventive measures can be designed to protect workers and emergency procedures can be put in place to minimize accidental exposure. Because there is often little or no information available about new chemicals, an important method of developing such new information quickly is animal experimentation.

**Exposure standards.** The toxicological effects of vapors, gases, fumes, and dusts are initially determined in the laboratory by actually exposing animals to known concentrations. Range-finding studies are usually conducted on animals by exposing them to known concentrations for controlled periods of time.

Groups of animals can be exposed to controlled concentrations for 8 hours a day, 5 days a week, for weeks, months, or years. In such cases, extreme care must be taken not only in the selection but also in the subsequent care of the animals throughout the experimental period.

Animals must be observed daily to ascertain any untoward physiological responses during exposure and postexposure periods. On terminating chronic experiments, all animals are sacrificed and the internal organs are weighed and examined histopathologically. In this manner, the toxicologist gains information regarding no-effect levels as well as levels that produce systemic injury.

**Screening procedures.** Toxicological screening should include both acute toxicity studies and studies of repeated administration at short intervals. Long-term studies performed during the lifespan of the animal, and in some instances over several generations, are part of the complete test program.

Because biological variations influence the reaction to a foreign chemical in different species, it is difficult to duplicate in animal experiments the precise situation to which humans can be exposed. In the development of a specific test program, preliminary studies are necessary to select species that absorb and metabolize related classes of chemicals in ways similar to humans.

A route of administration different from that usually occurring in humans, such as parenteral (intra-abdominal but outside the intestine) administration instead of inhalation or ingestion, can give misleading results.

Chronic toxicity studies involve repeated administration of test substances. However, chronic effects can also be expected from a single exposure to a substance if the body stores the material so that it remains in the organism for long periods of time. Repeated administration of the test substance is useful in the investigation of such problems as cumulative toxicity, tolerance, and enzyme-induction phenomena.

**Problem areas.** Animal testing provides only an estimate of the toxicity of a chemical for humans. It is very difficult to extrapolate an $LD_{50}$ or an $LC_{50}$ to an acceptable Threshold Limit Value.

Because the primary concern is the prevention of harm to humans, the limitations inherent in animal-derived data should be recognized. Whether human response will resemble that of the most reactive or the least reactive species tested is often not known. Finally, whether the animal response is an exact parallel to human response cannot always be predicted.

## Human Epidemiological Data

Records of human experience for exposures to many substances are available. This is particularly true for older chemicals such as carbon monoxide and lead. Epidemiological data can be descriptive, retrospective, or prospective.

*Descriptive studies* identify a change or difference in prevalence of a disease in a subgroup of the population.

*Retrospective studies* reveal a relationship between a chemical and a certain effect caused by exposure that occurred months or years before the initiation of data collection.

*Prospective studies* can define more precisely the time relationship and the magnitude of risk. Prospective studies are present and future continuing studies that measure health effects as the exposures occur in work areas.

Epidemiological analysis reveals the relationship between time of occurrence of an adverse effect and age at the time of the first exposure, which helps to establish or clarify the influence of variables other than the agent under study. For example, cigarette smoking in the study of lung cancer among asbestos workers is such a variable.

Finally, if a specific chemical is removed from the environment, it should be followed by epidemiological evidence of a decline in the frequency of the effect.

# ■ Federal Regulations

## Occupational Safety and Health Act (OSHAct)

The Occupational Safety and Health Act (OSHAct, enacted in 1970) is administered by the Occupational Safety and Health Administration (OSHA), which has the regulatory authority to protect workers from physical hazards and hazardous substances and forms of energy (such as noise and radiation) in the work environment. OSHA monitors health and safety in the workplace, setting standards for worker exposure to specific chemicals, for permissible exposure levels (PELs), and for monitoring procedures. It also provides research, information, education, and training in occupational safety and health. By establishing the National Institute for Occupational Safety and Health (NIOSH), the act provided for studies to be conducted so that regulatory decisions can be based on the best available information.

In 1983, OSHA enacted the Hazard Communication Standard, 29 *CFR* 1910.1200, which sets standards for worker notification and training for chemicals in the workplace. This also is known as right-to-know legislation. For more information on right-to-know legislation, see Chapter 29, Government Regulations, and Chapter 30, History of the Federal Occupational Safety and Health Administration.

OSHA adopted as federal standards (PELs) the 1968 Threshold Limit Values (TLVs) suggested for industrial chemical exposures by the American Conference of Governmental Industrial Hygienists (ACGIH).

The ACGIH TLVs were directed primarily at substances that caused physiological reactions such as poisoning, irritation of eyes and respiratory tract, and skin rashes. The 1968 TLVs were not established on the basis of carcinogenic, teratogenic, or mutagenic properties, and synergistic effects of chemical mixtures were not included in tests used to determine the TLVs. Approximately 400 TLVs were adopted as OSHA standards. Since that time, a few new standards for carcinogenic and other chemical agents have been promulgated by OSHA. ACGIH has also been updating and adding to its original list of TLVs—the ACGIH TLVs may represent more current research than do the OSHA PELs. See Appendix B for a complete listing of the ACGIH TLVs.

The American National Standards Institute (ANSI), through its former Committee on Acceptable Concentrations of Toxic Dusts and Gases, also published standards on levels of materials in the air in work areas. Wherever possible, OSHA adopted ANSI standards because ANSI was considered to be a consensus standard-setting group.

## TSCA

The Toxic Substances Control Act of 1976 (TSCA) is administered by the Environmental Protection Agency (EPA) and covers almost all chemicals manufactured in the United States, excluding certain compounds covered under other regulations such as the Federal Insecticide, Fungicide and Rodenticide Act (FIFRA). The act requires that chemical manufacturers and processors develop adequate data on the health and environmental effects of the chemicals they produce. The EPA is required to establish standards for the testing of chemicals.

Companies are required to notify the EPA 90 days before manufacturing any new chemical and to provide test data and other information about the safety of the product. The EPA has the authority to ban or regulate such chemicals if test information is insufficient and if the chemical is to be produced in substantial quantities with wide distribution. The EPA is required to ban or restrict the use of any chemical presenting an unreasonable risk of injury to health or the environment.

**Unreasonable risk.** The term *unreasonable risk* is vague and subject to various interpretations. The definition of *unreasonable* emerges from regulatory and administrative decisions of EPA and there are debates around the issue. Consumer advocates and environmentalists prefer a definition leading to stringent and rigidly enforced control over hazardous chemicals; industry representatives prefer a definition that allows for individual discretion and voluntary compliance.

## Toxic Substances List

The Occupational Safety and Health Act (OSHAct) of 1970, Section 20(a)(6), requires that "the Secretary of Health, Education, and Welfare shall publish within six months of enactment of this act and thereafter as needed, but at least annually, a list of all known toxic substances by generic, family or other useful grouping, and the concentrations at which such toxicity is known to occur." The first such list was prepared in 1971. Under the OSHAct, the Secretary of Labor must issue regulations requiring employers to monitor employee exposure to toxic materials and to keep records of such exposure. This requirement is set forth in Section 8(c)(3) of the act (see Chapter 30, A History of the Federal Occupational Safety and Health Administration, for more details).

The Registry of Toxic Effects of Chemical Substances (RTECS) is a compendium of toxicity data extracted from the scientific literature by the U.S. Department of Health and Human Services (formerly the Department of Health, Education and Welfare).

**Purpose.** The purpose of the Toxic Substances List is to identify all known toxic substances in accordance with standardized definitions that can be used to describe toxicity. The entry of a substance on the list does not automatically mean that it is to be avoided but that the listed substance has been found to be hazardous if misused.

The absence of a substance from the list does not necessarily indicate that a substance is not toxic. Some hazardous substances may not qualify for the list because the dose that causes the toxic effect is not known.

Other chemicals associated with skin sensitization and carcinogenicity do not appear on the list because the effects have not been reproduced in experimental animals or because the human data are not definitive. Thus, the published comments and evaluations of the scientific community are relied on; also, there has been no attempt at an evaluation of the degree of hazard that might be expected from substances on the list; that is a goal of the hazard-evaluation studies.

**Hazard evaluation.** Hazard evaluation involves far more than the recognition of a toxic substance and a knowledge of its relative toxic potency. It involves a measurement of the quantity available for absorption by the user, the amount of time available for absorption, the frequency with which the exposure occurs, the physical form of the substances, and the presence of other substances (toxic or nontoxic), additives, or contaminants.

Ventilation, appropriate hygienic practices, housekeeping, protective clothing, and pertinent training for safe handling may eliminate or diminish hazards that might exist.

Hazard evaluation is performed by engineers, chemists, industrial hygienists, toxicologists, and physicians trained in toxicology, industrial hygiene, and occupational medicine who strive to recognize, measure, and control these hazards.

## NIOSH/OSHA Standards

Since the passage of the OSHAct, both NIOSH and OSHA have been committed to establishing permissible standards for the workplace that are far more complete than the TLVs issued by the ACGIH.

A complete standard should include the exposure limit of the substance that has been determined to provide a safe, healthful work environment; the methods for collecting, sampling, and analyzing the substance; the engineering controls necessary for maintaining a safe environment; appropriate equipment and clothing for safe handling of the substance; emergency procedures in the event of an accident; medical surveillance procedures necessary for the prevention of illness or injury from inadvertent overexposure; and the use of signs and labels to identify hazardous substances.

**NIOSH criteria documents.** Except in the case of emergency standards, the normal first step in the standard-setting process is the creation of a criteria document for the substance by NIOSH. Such documents are forwarded to OSHA for consideration as permanent OSHA standards.

NIOSH also develops recommended exposure limits (RELs) for hazardous substances in the workplace. Unless noted otherwise, RELs are time-weighted average (TWA) concentrations for up to a 10- hour workday during a 40-hour workweek. RELs are published in NIOSH criteria documents along with appropriate measures to reduce adverse health effects.

Since 1976, NIOSH criteria documents have incorporated animal and human data, when available, on carcinogenicity, mutagenicity, teratogenicity, and effects on reproduction. When possible, attempts are made to correlate these adverse reactions with exposures and effects.

Current Intelligence Bulletins (CIBs) have been issued by NIOSH since 1975 for more rapid dissemination of new scientific information about occupational hazards. A CIB may draw attention to a hazard previously unrecognized or may report new data suggesting that a known hazard is either more or less dangerous than was previously thought (see Bibliography).

**OSHA standards.** The U.S. secretary of labor is responsible for promulgating standards. In some cases, a recommended standard is referred to an advisory committee for study and review in accordance with provisions of the act. OSHA standards are arrived at after extensive review including public hearings. Regardless of the status of the

proposed standards in the criteria documents, these documents constitute valuable and readily available sources of information and should be consulted whenever there is interest in a substance for which a criteria document has been written.

Although the standard-setting process just described is an extremely thorough one that allows all interested groups an opportunity to express opinions, it is also a lengthy and very costly activity that has resulted in the promulgation of very few permanent standards.

## ■ Discussion of ACGIH Threshold Limit Values

Many guides for exposure to airborne contaminants have been proposed and some of them have been used throughout the years. The most widely accepted are those issued annually by the American Conference of Governmental Industrial Hygienists, and are termed Threshold Limit Values (TLVs).

Contrary to popular opinion, these TLVs are not the work of a government agency, but are instead the product of a committee whose members are associated with a government or educational industrial hygiene activity. Only OSHA PELs are legally binding.

Appendix B presents the Threshold Limit Values (TLVs) for Chemical Substances and Physical Agents and the Biological Exposure Indices (BEIs) that were adopted by the ACGIH for 1994–1995. The Notice of Intended Changes for TLVs and the Notice of Intent to Establish BEIs are also included.

The *Threshold Limit Values and Biological Exposure Indices* booklet is published annually by ACGIH and is available for a nominal cost. It is copyrighted by the ACGIH and is reproduced here with permission. For information about this publication, contact ACGIH at Kemper Meadow Center, 1330 Kemper Meadow Drive, Cincinnati, OH 45240.

### Guides

The TLV committee intended the TLVs they issued to be used as *guides* in the control of health hazards, *not as fine lines between safe and dangerous concentrations.* Many reasons for the inadequacy of these numbers for such purposes were noted by the committee. Despite this admonition, however, the TLV list gradually became incorporated into state and federal regulations. For this reason, it is worth examining the nature of the TLVs in some detail.

### Time-weighted Average

It is implicit in all TLVs that measurements are made in the breathing zone of a worker and are obtained in such a way that a TWA can be calculated. In general, for an 8-hour

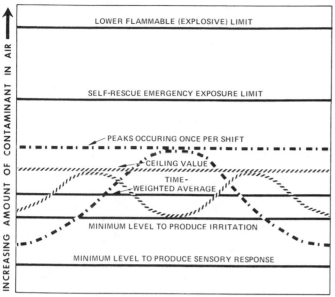

**Figure 6–4.** Knowledge of the type of injury that would result from exposure to various contaminant levels is important to health and safety professionals.

workday,

$$\text{TWA} = \frac{C_a T_a + C_b T_b + \ldots C_n T_n}{8}$$

where $T$ is the time of exposure period and $C$ is the concentration of contaminant during that period. This concept has proven to be a useful means of estimating the long-term chronic effects of exposure to most substances in the workplace. Although the TWA does not necessarily predict the amount of a substance that will be absorbed, it does measure the amount that can be inhaled during a workday; considerations of the extent of absorption aid in the selection of a value that affords the desired degree of protection.

It is inherent in the definition of a TWA that concentrations higher than the recommended value can be permitted for some periods of time as long as these levels are offset by periods of lesser concentration. The degree of permissible excursion is related to the TLV of the substance. The relationship between threshold limit and permissible excursion is a general rule that may not apply in certain cases (Figure 6–4).

## Ceiling Values

For some substances, it is not advisable to permit concentrations substantially above the recommended TWA; the TLV committee designates these substances with the letter *C*, which stands for *ceiling value*. Most substances designated with a *C* tend to be irritants for which a TLV has been set only slightly below the level where irritation will be noticed by the most sensitive individuals.

The manner of sampling used to determine whether the exposures are within the limits for each group usually differ. A single brief sample that is applicable to a C limit is not appropriate for calculating a TWA; here, a sufficient number of samples are needed to permit calculation of the TWA concentration throughout the workshift. The ceiling limit places a definite boundary that should not be exceeded.

The TLV list also contains another listing of values for many substances that are called short-term exposure limits (STELs).

One of the most fundamental tasks confronting the industrial hygienist, therefore, is assessing the possible degrees of exposure to a bewildering array of chemicals in the work environment. It is almost universally accepted that there is a threshold level for any of these substances below which no impairment of health or other undesirable effects occur. Because the most common route of entry for a chemical in the workplace is inhalation, the practice for many years has been to sample the air being breathed by the workers and compare the result with a suitable standard.

Although air standards and guides such as those developed by ACGIH, NIOSH, and OSHA are most widely used in industrial hygiene practice, certain shortcomings are inherent in any air standard; this limits their applicability to all situations. Some of the more common recognized problems include the following:

1. Difficulty in acquiring a truly representative breathing zone sample

2. Uncertainties about the extent of absorption of the amount inhaled

3. Nonroutine or nonrepetitive work; air samples can characterize work operations only on the day the sample is taken

4. Misleading information resulting from variations in particle size and particle solubility

5. Accidental or deliberate contamination of sample

The TLV list is basically an alphabetical listing of substances with the recommended limits expressed either in parts per million by volume or milligrams per cubic meter; see Appendix B. It is the practice to express the TLV for all substances expected to be present in the air as particulate suspensions in milligrams per cubic meter. For substances expected to be present as gases or vapors, the TLV is expressed in parts per million, and for convenience the equivalence in milligrams per cubic meter is also presented.

Formulas are available for determining TLVs for the inhalable-thoracic and respirable-particulate fractions, obtained by means of a suitable particle-size-discriminating device. The asbestos TLV is unique, and is expressed in fibers per cubic centimeter of air.

## Mixtures

When two or more hazardous substances that act on the same body organ system are present, their combined effect, rather than that of either component, should be given primary consideration. In the absence of information to the contrary, the effects of the different hazards should be considered additive. Exceptions can be made when there is a good reason to believe that the chief effects of the different harmful substances are independent, as when purely local effects on different organs of the body are produced by the various components of the mixture.

The formula for additive effects is as follows:

$$\frac{C_1}{T_1} + \frac{C_2}{T_2} + \frac{C_3}{T_3} + \dots = 1$$

where  $C$ = the observed atmospheric concentration
$T$ = the corresponding Threshold Limit Value

If the sum of the fractions is greater than 1, then the Threshold Limit Value has been exceeded.

Antagonistic action or potentiation may occur with some combinations of contaminants. At present, such cases must be determined individually. Potentiating or antagonistic agents may not necessarily be harmful by themselves. Potentiating effects of exposure by routes other than that of inhalation are also possible, such as the effect of imbibed alcohol on an inhaled narcotic (trichloroethylene). Potentiation is characteristically exhibited at high concentrations and is probably less at low concentrations.

When a given operation or process emits a number of harmful dusts, fumes, vapors, or gases, it is often feasible to attempt to evaluate the hazard by measuring a single substance. In such cases, the threshold limit used for this substance should be reduced by a suitable factor, the magnitude of which depends on the number, toxicity, and relative quantity of the other contaminants ordinarily present.

Examples of processes that are typically associated with two or more harmful atmospheric contaminants are welding, automobile repair, blasting, painting, lacquering, certain foundry operations, reinforced plastic fabrication, and shipbuilding.

## Carcinogens

The ACGIH categories for carcinogens are as follows:

■ Confirmed human carcinogen: The agent is carcinogenic to humans based on the weight of evidence from epidemiologic studies of, or convincing clinical evidence in, exposed humans.
■ Suspected human carcinogen: The agent is carcinogenic in experimental animals at dose levels, by routes of administration, at sites, of histologic types, or by mechanisms considered relevant to worker exposure.

Available epidemiologic studies are conflicting or insufficient to confirm an increased risk of cancer in exposed humans.

■ Animal carcinogen: The agent is carcinogenic in experimental animals at a relatively high dose, by routes of administration, at sites, of histologic types, or by mechanisms that are not considered relevant to worker exposure. Available epidemiologic studies do not confirm an increased risk of cancer in exposed humans. Available evidence suggests that the agent is not likely to cause cancer in humans except under uncommon or unlikely routes or levels of exposure.
■ Not classifiable as a human carcinogen: There are inadequate data on which to classify the agent in terms of its carcinogenicity in humans and animals.
■ Not suspected as a human carcinogen: The agent is not suspected to be a human carcinogen on the basis of properly conducted epidemiologic studies in humans. These studies have sufficiently long follow-up, reliable exposure histories, sufficiently high doses, and adequate statistical power to conclude that exposure to the agent does not convey a significant risk of cancer to humans. Evidence suggesting a lack of carcinogenicity in experimental animals is considered if it is supported by other relevant data.

Substances for which no human or experimental animal carcinogenic data have been reported are assigned no carcinogenic designation.

## Physical Factors

It is recognized that such physical factors as heat, ultraviolet and ionizing radiation, and work under high atmospheric pressure or at a high altitude may place added stress on the body so that the effects of exposure to chemical agents may be altered. Certain physical stresses may increase the response to a toxic substance. Although most threshold limits have built-in safety factors to guard against moderate deviations from normal environments, the safety factors of most substances are not large enough to account for gross deviations. For example, continuous work at temperatures above 32 C (90 F) or overtime extending the workweek by more than 25 percent might be considered gross deviations. In such instances, judgment must be exercised in the proper downward adjustments of the TLVs.

## Unlisted Substances

Many substances present or produced as by-products in industrial processes do not appear on the TLV list. In a number of instances, the material is rarely present as a particulate, vapor, or other airborne contaminant and a TLV is not necessary. More often, the committee does not have sufficient information to warrant development of a TLV, even on a tentative basis. Some substances of considerable toxicity have been omitted primarily because only a limited number of workers, such as employees of a single plant, have potential exposure to possibly harmful concentrations.

## Basic Data Used for TLVs

Whenever such data appear in the published literature, the ACGIH TLV committee selects a value based on human experience. Good epidemiological studies, complete with environmental data as well as morbidity and mortality data, are perhaps the best possible basis for a TLV, but in most cases, such studies do not exist. In the absence of epidemiological studies, individual cases involving human exposures are considered, but the majority of the literature available to justify any given value is based on animal toxicological studies. The preferred studies for determining acceptable exposure limits are those based on long-term inhalation tests involving several animal species at concentrations both above and below the recommended level.

However, scientists often must rely on short-term inhalation data or, in many cases, toxicity studies in which the substance was introduced into the experimental animals by routes other than inhalation. The least useful toxicological data are those based on short-term oral intake, intended to measure the acute toxicity, or the ability of the substance to kill the exposed animals. For some chemicals listed, there are no human exposure data whatsoever; only meager data based on animal feeding studies are available. It is not surprising, therefore, that the publication of new information often results in dramatic changes in some TLVs.

## Documentation

The policy of the TLV committee is to prepare a justification for each proposed TLV. From time to time, these are published in a document titled *Documentation of Threshold Limit Values,* available from the ACGIH (see Bibliography). In these rather brief discussions, the principal data that the committee considered significant are reviewed and references are cited. This document should be consulted whenever a particular TLV is to be applied, for it is important to be aware of the basis for each standard. In most cases, a particular value is selected on the basis of one or more of the consequences of overexposure listed in Table 6–H.

One of the advantages of the TLV list is its timeliness. The list contains a section titled "Notice of Intended Changes"; as the name suggests, all changes, including additions, are listed for a period of at least two years. During this period, the TLV committee solicits comments from interested parties concerning the suggested changes.

Threshold Limit Values are intended only for use in the practice of industrial hygiene as guidelines or recommendations in the control of potential health hazards. The TLVs should be interpreted and applied only by a person trained in industrial hygiene. They are not intended for use, or for modification for use, as a relative index of hazard or toxicity, in the evaluation or control of community air pollution nuisances, in estimating the toxic potential of continuous, uninterrupted exposures or other extended work periods, as proof or disproof of an existing disease or physical condition, or for adoption by countries in which working conditions differ from those in the United States and where substances and processes differ. These limits are *not* fine lines between safe and dangerous concentrations. See Chapter 15, Evaluation, for further discussion of the use of TLVs.

## ■ Biological Standards

A useful means of assessing occupational exposure to a harmful material is the analysis of biological samples

---

**Table 6–H.** Classification of Criteria for ACGIH TLVs Applicable to Humans and Animals

| | Applied Criteria | | |
|---|---|---|---|
| *Morphologic* | *Functional* | *Biochemical* | *Miscellaneous* |
| Systems or organs affected: lung, liver, kidney, blood, skin, eye, bone, CNS, endocrine, exocrine, reproductive<br>Carcinogenesis<br>Roentgenographic changes | Changes in organ function: lung, liver, kidney<br>Irritation<br>Mucus membranes (epithelial linings, eye, and skin)<br>Narcosis<br>Odor | Changes in amounts of biochemical constituents, including hematologic<br>Changes in enzyme activity<br>Immunochemical allergic sensitization | Nuisance:<br>Visibility<br>Cosmetic<br>Comfort<br>Aesthetic<br>(Analogy) |
| | Potentially Useful Criteria | | |
| Altered reproduction<br>Body-weight changes<br>Organ/body weight changes<br>Food consumption | Behavioral changes:<br>Cerebral functions<br>Conditioned and unconditioned reflexes—learning<br>Audible and visual responses<br>Endocrine glands<br>Exocrine glands | Radiomimetic effects<br>Teratogenesis<br>Mutagenesis | |

(Adapted from Stokinger, H.E. Criteria and procedures for assessing the toxic responses to industrial chemicals. *Permissible Levels of Toxic Substances in the Working Environment, Occupational Safety and Health,* ser. 20. Geneva, Switzerland: International Labor Office, 1970.)

obtained from exposed workers. Biological sampling, however, should not be considered a substitute for air sampling. Ethical considerations prohibit what some have called the use of the worker as an "integrated air-sampling device." Biological analysis may provide an indication of the body burden of the substance, the amount circulating in the blood, or the amount being excreted. Virtually every tissue and fluid in the body can be analyzed, but for practical reasons, most bioassays are confined to specimens of urine or blood. For substances such as carbon monoxide and many solvents, the analysis of exhaled breath samples indicates the level of previous exposure. Occasionally, analysis of samples of hair, nails, feces, or other tissues may be useful.

Whereas air monitoring measures the composition of the external environment surrounding the worker, biological monitoring measures the amount of chemical absorbed by the body. Substances absorbed through the skin, pulmonary system, and gastrointestinal tract are accounted for. In addition, the effects of added stress (such as increased work load resulting in a higher respiration rate with increased intake of the air contaminant) are reflected in the results. The total exposure (both on and off the job) to harmful materials is accounted for. In the case of sub-

stances with a long biological half-life, the concentration in tissues or fluids is more independent from variations in concentrations in workroom air. For some chemicals, biological assays, in addition to air measurements, can be much more reliable indicators of health risks than measurements of air contaminants alone (Table 6–I).

Analyses that can be performed on biological samples include the following:

- Analysis for the unchanged substance (such as lead, arsenic, mercury) in body fluids and tissues
- Analysis for a metabolite of the substance in body fluids or tissues, such as phenol in urine resulting from exposure to benzene Analysis to determine the variations in the level of a naturally occurring enzyme or other biochemical substance normally present in body fluids or tissues, such as depression of cholinesterase activity as a result of exposure to organic phosphate compounds

The rates of absorption, metabolism, and excretion for a particular substance determine when it is most appropriate to analyze samples in relation to duration and time of exposure. For rapidly excreted or exhaled substances,

---

**Table 6–I.** Body Tissues and Fluids Suitable for Biological Analysis

*Analysis of Urine Samples May Be Useful for the Following Compounds:*

| | | |
|---|---|---|
| Acetone | Ethyl benzene (mandelic acid) | Parathion (nitrophenol) |
| Acrylonitrile | Fluoride | Pentachlorophenol |
| Aniline | Furfural (furoic acid) | Phenol |
| Antimony | N-hexane (2-5-hexanedione) | Selenium |
| Arsenic | Hydrogen bromide | Styrene (mandelic acid) |
| Arsine | Hydrogen cyanide | Tellurium |
| Benzene (phenol) | Hydrogen fluoride | Tetrachloroethylene (trichloroacetic acid) |
| Cadmium | Hydrogen selenide | Thallium |
| Carbon disulfide (2-thiothiazolidine-4-carboxylic acid) | Isopropyl alcohol (acetone) | 1,1,1-trichloroethane (trichloroacetic acid, trichloroethanol) |
| Chlorinated benzene (4-chlorocatectol) | Lead | Trichloroethylene (trichloroacetic acid, trichloroethanol) |
| Chromium $H_2O$-soluble compounds | Manganese | |
| Cobalt | Mercury | Triethylamine (triethylamic and TEA n-oxide) |
| Cyanide (thiocyanate) | Methanol (or formic acid) | Uranium |
| Cyclohexane (cyclohexanol) | Methyl ethyl ketone | Vanadium |
| Dimethylacetamide (methylacetamide) | Methyl isobutyl ketone | Xylene (methyl hippuric acid) |
| Dimethyl formamide (N-methylformamide) | Nickel | Zinc |
| 2-ethoxyethanol (2-ethoxyacetic acid) | Nickel carbonyl (nickel) | |
| 2-ethoxyethanol acetate (2-ethoxyacetic acid) | Nitrobenzene (methemoglobin or p-nitrophenol) | |

*Analysis of Blood Samples May Be Useful for the Following Compounds:*

| | | |
|---|---|---|
| Acetone | Cyclohexane | Pentachlorophenol |
| Aluminum | Dichloromethane (carboxyhemoglobin) | Styrene |
| Aniline (methemoglobin) | Dimethylformamide | Tetrachloroethylene |
| Cadmium dust, fume | Ethylene oxide | Toluene |
| Carbon monoxide (carboxyhemoglobin) | Lead | Trichloroethylene (or trichloroethanol) |
| Cholinesterase inhibitors (RBC cholinesterase) | Manganese | Xylene |
| Cobalt | Mercury | Zinc |

*Breath Analysis May Be Useful for the Following:*

| | | |
|---|---|---|
| Benzene | N-hexane | 1,1,1-trichloroethane |
| Dichloromethane | Tetrachloroethylene | Trichloroethylene |
| Ethyl benzene | | |

Note: Metabolites are given in parentheses where they are the best indicators of exposure and absorption of a compound.

peak concentrations are found during or immediately after exposure. Peak excretion rates for metabolites of some organic solvents and some inorganic substances may occur 1 to 3 days after exposure. Biological levels of metals with cumulative properties (such as lead or mercury) may reflect the response to several weeks' prior exposure.

People with virtually identical exposure histories can show a wide variation in response due to subtle differences in their rates of absorption, tissue storage, or metabolism. Greater significance should be given to the variations in an individual's level from period to period than to the variations between different individuals within a group.

Many harmful substances can be stored for long periods of time in various parts of the body. The concentrations are unlikely to be evenly distributed throughout the body. In many cases, the organ with the highest concentration of the material is the liver or kidney; for a number of other substances, it is both the liver and bones.

Many organic solvents are stored in body fat in the lipid-containing sheaths of nervous tissues, including the brain. Certain substances have been studied in humans, demonstrating storage and excretion days to weeks after exposure has ceased. The "half-lives" of various chemicals in the body have been established and are an important consideration in the total assessment of exposure.

Many materials, including organic compounds, undergo detoxification in the body. The body converts the material to something else that usually reduces its ability to cause injury. Occasionally, the conversion enhances the toxicity, but in any event, the process helps the body to dispose of the material. The conversion products may appear in the urine or blood as metabolites (see Table 6–I). When the body metabolizes benzene, increased levels of phenols are found in the urine; more specifically, the metabolites are conjugated sulfates (the phenols are linked to sulphur molecules). Either the sulfate or the phenols can be determined as an index of exposure to benzene. For many years, this has been a very useful test to determine the extent of benzene exposure.

The enzyme cholinesterase is inhibited by organic phosphates. Cholinesterase activity in the red blood cells can be measured and a certain reduction of activity below normal is significant.

## Urine Tests

Tests for the level of metabolites of toxic agents in the urine have found wide use in industrial toxicology as a means of evaluating exposure of workers. The concentration of the metabolic product is related to the exposure level of the toxic agent. Because normal values of such metabolites have been established, an increase above normal levels indicates that an overexposure has occurred. This provides a valuable screening mechanism for estimating the hazard from continued or excessive exposure. Because lead, for example, interferes in the porphyrin metabolism, erythro-

cyte protoporphyrin can be a useful measure and the results are useful as an indicator of lead absorption.)

## Blood Analysis

One of the best documented examples of the effectiveness of biological sampling is that of analysis for exposure to lead and its compounds. In most work operations, there is rather poor correlation between breathing zone levels of lead and blood-lead levels. It is almost universally agreed that the level of lead in the blood is the best index of the probability of damage resulting from lead exposure.

The purpose of a workplace air sample is to determine whether the airborne lead concentration exceeds an acceptable level. If it does, then steps can be taken to reduce the lead concentration in the air to a safe value and prevent lead intoxication in workers.

It is possible to have air concentration of lead above the TLV but still not have a hazardous exposure; the reason is that some lead particles can be very large and do not get into the alveoli of the lung, where they would contribute to significant intake. Although large particles may get as far as the nasal passages, they are eventually excreted without coming in contact with lung tissue and thus are not absorbed. Conversely, air concentrations of lead can be lower than the TLV, yet result in high blood-lead levels. This may be due to poor hygiene practices that allow lead to be ingested and inhaled. Personal cleanliness, good housekeeping, and prohibitions against eating, drinking, and smoking in work areas are very important.

An air-lead program is necessary to define the airborne exposure potential and a blood-lead program is needed to keep track of each employee's actual absorption. For example, in the case of a battery plant using lead, a blood-lead program is essential.

For many other substances there is no known biological test as useful as the measurement of lead in blood, whereas for other substances the correlation between bioassay tests and symptoms is so poor as to render a biological analysis of little value. The aim of an industrial hygienist or safety professional is to control exposure to harmful materials and it is likely that both air sampling and biological monitoring programs sometimes will be needed.

The industrial hygienist or safety professional is charged with maintaining a safe, healthful environment and should probably do air sampling and, where appropriate, biological sampling. In addition, air sampling at preselected locations should be performed to monitor the controls. Air sampling can be used to find sources of dispersion.

## Breath Analysis

If inhaled gases and vapors are fat-soluble and are not metabolized, they are cleared from the body primarily through the respiratory system. Examples of these are the volatile halogenated hydrocarbons; the volatile aliphatic, olefinic, and aromatic hydrocarbons; some volatile aliphatic saturated ketones and ethers; esters of low molecular weight; and certain other organic solvents such as carbon disulfide.

For industrial solvents that continue clearing from the body in exhaled breath for several hours after exposure, analysis of progressive decrease in the rate of excretion in the breath can be very helpful in showing not only the nature of the substances to which the worker was exposed, but also the magnitude of the exposure and probable blood levels. By the use of gas chromatography or infrared analysis of the breath samples, the identification of the substance is established, permitting comparison of the exposed workers' breath decay rate with published excretion curves. There is, however, considerable individual variation and it is not easy to set standard values.

## Biological Limits

There now exists a considerable body of knowledge about a large number of substances used in the workplace; for many of these substances, biological analysis, in addition to air-sampling, is more useful for evaluating exposure to a toxic substance. Many organic chemicals of high molecular weight and low vapor pressure are not found in workroom air at elevated concentrations under normal conditions of work, but the same substances can be absorbed through the intact skin, giving rise to excessive absorption that cannot be measured by air sampling. In such cases, a suitable biological analysis can be an excellent means of detecting the failure of skin-protection measures.

The ACGIH has adopted a set of advisory biological limit values called the Biological Exposure Indices (BEIs) for a limited number of substances. These indices use urine, blood, or expired air sampled under strictly defined conditions. The user should become familiar with the extensive documentation that accompanies these indices in the ACGIH *Documentation of Threshold Limit Values* (1992) and *Threshold Limit Values for Chemical Substances and Agents and Biological Exposure Indices* (1993–1994). (See Chapter 15, Evaluation, for a more detailed discussion of Biological Exposure Indices.)

Because the collection of blood samples requires the use of medical personnel, most programs of biological monitoring become a cooperative effort between safety, industrial hygiene, and medical departments.

The analysis and interpretation of biological samples is obviously of great importance and because the quantities involved are almost always very slight, great care must be taken in performing such analyses. Ordinarily, existing plant laboratories are not equipped or trained to perform these analyses in a satisfactory manner, and it is advisable to use a laboratory that has proven capability in this area.

## ■ Sources of Toxicological Information

The health and safety professional can turn to several sources for information when a question arises about the toxicity and hazard of a material.

### Material Safety Data Sheet

Material Safety Data Sheets (MSDSs) are a prime source of information on the hazardous properties of chemical products, although the quality of such information is highly variable. The OSHA Hazard Communication Standard requires that all chemical manufacturers and importers supply an appropriate MSDS to their customers. The MSDS is usually developed by the chemical manufacturer. Additionally, all users of the product (employers) must have an MSDS for every hazardous chemical used in the workplace.

Although OSHA does not specify the format of the MSDS, it does require certain specific information. A sample form approved by OSHA for compliance with the Hazard Communication Standard is shown in Figure 6–5.

There are eight categories of information on the MSDS:

**Section I**

- Name and address of the manufacturer (the originator of the MSDS)
- Emergency telephone number, which can be used to contact a "responsible party" for information about the product
- Information telephone number, to be used in nonemergency cases to contact the manufacturer
- Signature of the person responsible for the MSDS and the date it was developed or revised

**Section II: Hazardous Ingredients**

- Common name used as identification on the label (the code name or number, trade, brand, or generic name)
- Chemical name: the scientific designation of a chemical in accordance with the nomenclature systems of the International Union of Pure and Applied Chemistry (IUPAC) or the Chemical Abstracts Service (CAS)
- CAS number: the identification number that is unique to a particular chemical and is assigned by the Chemical Abstracts Service

**Section III: Physical and Chemical Characteristics**

- Physical and chemical data such as boiling point, vapor pressure, vapor density, solubility appearance, melting point, and evaporation rate

**Section IV: Fire and Explosion Hazard Data**

- Information needed for planning fire and explosion prevention, including flash point, flammable limits, and special fire-fighting procedures

## Material Safety Data Sheet

May be used to comply with
OSHA's Hazard Communication Standard,
29 CFR 1910.1200. Standard must be
consulted for specific requirements.

## U.S. Department of Labor

Occupational Safety and Health Administration
(Non-Mandatory Form)
Form Approved
OMB No. 1218-0072

IDENTITY (As Used on Label and List)

Note: Blank spaces are not permitted. If any item is not applicable, or no information is available, the space must be marked to indicate that.

### Section I

| Manufacturer's Name | Emergency Telephone Number |
|---|---|
| Address (Number, Street, City, State, and ZIP Code) | Telephone Number for Information |
| | Date Prepared |
| | Signature of Preparer (optional) |

### Section II — Hazardous Ingredients/Identity Information

| Hazardous Components (Specific Chemical Identity; Common Name(s)) | OSHA PEL | ACGIH TLV | Other Limits Recommended | % (optional) |
|---|---|---|---|---|
| | | | | |
| | | | | |
| | | | | |
| | | | | |
| | | | | |
| | | | | |
| | | | | |
| | | | | |
| | | | | |
| | | | | |
| | | | | |

### Section III — Physical/Chemical Characteristics

| Boiling Point | | Specific Gravity ($H_2O$ = 1) | |
|---|---|---|---|
| Vapor Pressure (mm Hg.) | | Melting Point | |
| Vapor Density (AIR = 1) | | Evaporation Rate (Butyl Acetate = 1) | |

Solubility in Water

Appearance and Odor

### Section IV — Fire and Explosion Hazard Data

| Flash Point (Method Used) | Flammable Limits | LEL | UEL |
|---|---|---|---|

Extinguishing Media

Special Fire Fighting Procedures

Unusual Fire and Explosion Hazards

(Reproduce locally)                                                                 OSHA 174, Sept. 1985

**Figure 6–5.**   Material Safety Data Sheet (*Continues*)

## Section V — Reactivity Data

| Stability | Unstable | | Conditions to Avoid |
|---|---|---|---|
| | Stable | | |

Incompatibility (*Materials to Avoid*)

Hazardous Decomposition or Byproducts

| Hazardous Polymerization | May Occur | | Conditions to Avoid |
|---|---|---|---|
| | Will Not Occur | | |

## Section VI — Health Hazard Data

Route(s) of Entry:     Inhalation?          Skin?          Ingestion?

Health Hazards (*Acute and Chronic*)

Carcinogenicity:     NTP?          IARC Monographs?          OSHA Regulated?

Signs and Symptoms of Exposure

Medical Conditions
Generally Aggravated by Exposure

Emergency and First Aid Procedures

## Section VII — Precautions for Safe Handling and Use

Steps to Be Taken in Case Material Is Released or Spilled

Waste Disposal Method

Precautions to Be Taken in Handling and Storing

Other Precautions

## Section VIII — Control Measures

Respiratory Protection (*Specify Type*)

| Ventilation | Local Exhaust | Special |
|---|---|---|
| | Mechanical (*General*) | Other |

| Protective Gloves | Eye Protection |
|---|---|

Other Protective Clothing or Equipment

Work/Hygienic Practices

Figure 6–5.  (*Continued*)

### Section V: Reactivity Data

- Outline of the stability of the product and the potential for hazardous polymerization and decomposition and a list of materials and conditions to avoid during use

### Section VI: Health Hazards

- Explanation of the most common sensations or symptoms a person might experience from acute and chronic *overexposure* to the material or its components, emergency and first aid procedures, any TLVs or PELs are listed, and, if the chemical is a carcinogen, the source of this designation

### Section VII: Safe Handling and Use

- Designated special handling and disposal methods and storage and spill precautions
- Section VIII: Control Measures
- Manufacturer recommendations for the use of ventilation, personal protective equipment, and hygienic practices

All required sections must be completed. If the required information is not available or not applicable, this must be shown on the form. Additionally, if the ingredients of a chemical mixture are trade secrets, their identity can be withheld, but their hazardous properties must be given.

These forms are required to be readily available to employees. Training in their use should be included in employee training required under the Hazard Communication legislation (see Chapter 29, Government Regulations).

## ■ Summary

The word *toxicity* is used to describe the ability of a substance to have an adverse effect on the health or well-being of a human. Whether any ill effects occur depends on the properties of the chemical, the dose (the amount of the chemical acting on the body or system), the route by which the substance enters the body, and the susceptibility or resistance of the exposed individual.

There are four routes of entry or means by which a substance may enter or act on the body: inhalation, skin absorption or contact, ingestion, and injection. Of these, inhalation is the most important occupational exposure route.

When a toxic chemical acts on the human body, the nature and extent of the injurious response depends on the dose received—that is, the amount of the chemical that actually enters the body or system and the time interval during which this dose is administered. Response can vary widely and might be as slight as a cough or mild respiratory irritation or as serious as unconsciousness and death.

The practice of industrial hygiene is based on the concept that for each substance there is a level of exposure below which significant injury, illness, or discomfort rarely or never occurs. The industrial hygienist protects the health of workers by assessing potential chemical and physical agent exposures and controlling the environmental conditions so that the risk of exposure is minimized.

## ■ Bibliography

Adams RM, *Occupational Skin Disease,* 2nd ed. Philadelphia: D.B. Saunders, 1990.

Amdur MO, Doull J, Klassen CD, eds. *Casarett and Doull's Toxicology—The Basic Science of Poisons,* 4th ed. New York: Macmillan, 1991.

American Conference of Governmental Industrial Hygienists. *Documentation of Threshold Limit Values,* 6th ed. Cincinnati: American Conference of Governmental Industrial Hygienists, 1992.

American Conference of Governmental Industrial Hygienists. *Threshold Limit Values for Chemical Substances and Agents and Biological Exposure Indices.* Cincinnati: American Conference of Governmental Industrial Hygienists, 1994–1995.

American Industrial Hygiene Association. *Biohazards Reference Manual.* Akron, OH: American Industrial Hygiene Association, 1985.

American Industrial Hygiene Association. *Hygienic Guide Series.* Fairfax, VA: American Industrial Hygiene Association, 1988–1991.

Calabrese EJ. *Multiple Chemical Interactions.* Chelsea, MI: Lewis Publishers, 1991.

Carson BL, et al. *Toxicology and Biological Monitoring of Metals in Humans Including Feasibility & Need.* Chelsea, MI: Lewis Publishers, 1986.

Clayton GD, Clayton FE, Cralley LJ, et al., eds. *Patty's Industrial Hygiene and Toxicology,* 4th ed., Vols. 1A, 1B, 2A, 2B, 2C, 3. New York: Wiley, 1979, 1981, 1982, 1991, 1993.

Council on Scientific Affairs. *Effects of toxic chemicals on the reproductive system. JAMA* 253:23, 1985.

Dreisbach RH. *Handbook of Poisoning: Prevention, Diagnosis and Treatment,* 12th ed. Los Altos, CA: Lange Medical Publication, 1987.

Ellenhorn MJ, Barceloux DG. *Medical Toxicology.* New York: Elsevier, 1988.

Gosselin RE. *Clinical Toxicology of Commercial Products,* 5th ed. Baltimore, MD: Williams & Wilkins, 1984.

Grant WM, Schuman JS. *Toxicology of the Eye: Effects on the Eyes and Visual System from Chemicals, Drugs, Metals and Materials, Plants, Toxins and Venoms: Also Systemic Side Effects from Eye Medications.* 4th ed. Springfield, IL: Charles C. Thomas, 1993.

Guyton AC. *Textbook of Medical Physiology,* 8th ed. Philadelphia: W.J. Saunders, 1990.

Hathaway GJ, Procter NH, et al. *Chemical Hazards of the Workplace,* 3rd ed. Philadelphia: J.P. Lippincott, 1991.

Hayes WJ, Laws ER, eds. *Handbook of Pesticide Toxicology.* San Diego, CA: Academic Press, 1990.

*The Industrial Environment, Its Evaluation and Control,* 3rd ed. DHHS/NIOSH Pub. No. 74–117. Rockville, MD: NIOSH, 1973. (Order from GPO No. 017–0011–00396–4.)

Key MM, et al. *Occupational Diseases: A Guide to Their Recognition,* rev. ed., DHHS (NIOSH) Pub. No. 77–181., Washington, DC: U.S. Department of Health, Education and Welfare, June 1977. (Order from NTIS No. PB-83–129–528/A99.)

LaDou J, ed. *Occupational Medicine.* San Mateo, CA: Appleton & Lange, 1990.

Lauwerys RR, Hoet P. *Industrial Chemical Exposure: Guidelines for Biological Monitoring,* 2nd ed. Ann Arbor, MI: Books on Demand, 1993.

Loomis TA. *Essentials of Toxicology.* Philadelphia: Lea & Febiger, 1978.

National Institute for Occupational Safety and Health, U.S. Department of Health and Human Services. *NIOSH Pocket Guide to Chemical Hazards.* Washington, DC: GPO, June 1990.

National Toxicology Program, U.S. Department of Health and Human Services. *6th Annual Report on Carcinogens Summary 1991.* Research Triangle Park, NC: National Institute of Environmental Health Sciences, 1991.

U.S. Congress, Office of Technology Assessment. *Reproductive Health Hazards in the Workplace.* Washington DC: U.S. Government Printing Office, OTA-BA-266, Dec. 1985.

Parmeggiani L, ed. *The Encyclopedia of Occupational Health and Safety,* 3rd ed., vol. 2. Geneva, Switzerland: International Labor Office, 1983.

Piotrowski JK. *Exposure Tests for Organic Compounds in Industrial Toxicology.* Cincinnati: NIOSH, 1977.

Rom WN, ed. *Environmental and Occupational Medicine,* 2nd ed. Boston: Little, Brown, 1992.

Sax IN, Lewis RJ Sr. *Dangerous Properties of Industrial Materials,* 8th ed. New York: Van Nostrand Reinhold, 1992.

Schmahl D. *Combination Effects in Chemical Carcinogenesis.* New York: VCH Publishers, 1988.

Scialli AR. *A Clinical Guide to Reproductive and Developmental Toxicology.* Boca Raton, FL: CRC Press, 1991.

Shepard TH. *Catalog of Teratogenic Agents,* 6th ed. Baltimore: The Johns Hopkins University Press, 1989.

Wexler P. *Information Resources in Toxicology,* 2nd ed. Bethesda, MD: Toxicology Information Program, National Library of Medicine; New York: Elsevier, 1988.

Zenz C, Dickerson OB, Horvath EP, eds. *Occupational Medicine,* 3rd ed. St. Louis: Mosby-Year Book, 1994.

## ■ Other Resources

### Computer Accessible Data Bases

**MEDLARS.** MEDLARS (Medical Literature Analysis and Retrieval System) is a computerized database maintained by the National Library of Medicine. Through a personal computer and modem, one can gain access through MEDLARS to more than 30 data bases including MEDLINE, TOXLINE, RTECS, and others (Table 6–J). The National Library of Medicine has published a software program titled *Grateful Med* that assists with formatting searches. For further infor-

**Table 6–J.** MEDLARS (National Library of Medicine) Data Bases

| | |
|---|---|
| AIDSDRUGS | Descriptions of substances used in AIDS-related trials |
| AIDSLINE | References to the recent AIDS literature |
| AIDSTRIALS | AIDS-related clinical trials |
| ALERT | Clinical alerts |
| AVLINE | Audiovisual materials for health professionals |
| CANCERLIT | References to the journal literature in cancer |
| CATLINE | Books |
| CCRIS | Chemical Carcinogenesis Research Information System |
| CHEMID | Dictionary of chemical compounds (non-royalty) |
| CHEMLINE | Dictionary of chemical compounds |
| DIRLINE | Directory of organizations |
| HEALTH | Administration and planning information |
| HSDB | Hazardous Substances Data Bank |
| MEDLINE | Biomedical journals (current and old) |
| RTECS | Registry of Toxid Effects of Chemicals |
| SDILINE | The most recent month of MEDLINE |
| SERLINE | Journals currently indexed |
| TOXLINE | References to toxicology information |
| TOXLINE65 | References to toxicology from 1965 to 1980 |
| TOXLIT | Toxicology references from royalty files |
| TOXLIT65 | Toxicology royalty references from 1965 to 1980 |
| TOXNET | Access to TOXNET databases (HSDB, CCRIS, RTECS, and TRI) |
| TRI | Toxic Chemical Release Inventory (1987, 1988, 1989) |

mation, contact the MEDLARS Service Desk at the National Library of Medicine at 800–638–8480.

**Other data bases.** Dialog (800–334–2564) is a service that makes selected databases available for on-line searching. Dialog allows access to Biosis, Embase, Sci Search, Toxline, Medline, NIOSH publications, TSCA Initial Inventory, and the Federal Register.

STN International (800–848–6533) provides on-line access to chemical abstracts.

Micromedex (800–525–9083) distributes a CD-ROM quarterly to its subscribers. It is called TOMES and contains medical and hazardous substance information related to toxicologic and management of health, spill, and related chemical emergencies.

### Organizations

Recommended standards or guidelines specific to many individual physical, chemical, and biologic agents are developed, periodically updated, and available from the following organizations:

World Health Organization
WHO Publications Center
49 Sheridan Avenue
Albany, NY 12210
Phone: (518) 436–9686
Fax: (518) 436–7433

American National Standards Institute
11 West 42nd Street
New York, NY 10036
Phone: (212) 642–4900
Fax: (212) 302–1286

American Conference of Governmental Industrial Hygienists
Kemper Woods Center
1330 Kemper Meadow Drive
Cincinnati, OH 45240
Phone: (513) 742–2020
Fax: (513) 742–3355

American Industrial Hygiene Association
2700 Prosperity Avenue, Suite 250
Fairfax, VA 22031
Phone: (703) 849–8888
Fax: (703) 207–3561

U.S. Government Agency catalogues and publications (OSHA, NIOSH, EPA, and NTP) can be obtained through the U.S. Government Printing Office. Also, the respective agencies and state government offices can be contacted directly for publication catalogues.

U.S. Government Printing Office
Superintendent of Documents
Washington, DC 20402
Phone: (202) 512–1800
Fax: (202) 512–2250

The following organizations also publish reports and guidelines in the field of occupational health and safety:

American Chemical Society
1155 Sixteenth Street N.W.
Washington, DC 20036

American Society of Safety Engineers
1800 E. Oakton Street
Des Plaines, IL 60018

National Safety Council
1121 Spring Lake Drive
Itasca, IL 60143

# CHAPTER 7

# Gases, Vapors, and Solvents

by George P. Fulton, MS, CIH

**Critical Exposure Factors**
*Mode of Use and Potential for Exposure ▪ Temperature and Volatility ▪ Concentration ▪ Reactivity ▪ Exposure Guidelines*

**Solvents**
*Aqueous Systems ▪ Organic Systems*

**Gases**
*Cryogenics ▪ Simple Asphyxiants ▪ Chemical Asphyxiants*

**Flammable and Combustible Liquids**
*Flammable Liquid ▪ Combustible Liquid ▪ Flash Points ▪ Fire Point ▪ Flammable Range ▪ Requirements and Guidelines*

**Effects**
*Physiological Effects ▪ Hazard Potential ▪ Air Pollution*

**Evaluation of Hazards**

**Control of Hazards**
*Responsibility of Health and Safety Personnel ▪ Process Controls ▪ Engineering Controls ▪ Personal Protective Equipment*

**Summary**

**Bibliography**

A POTENTIAL THREAT TO THE HEALTH, productivity, and efficiency of workers in most occupations and industries is their exposure to gases and vapors from solvents, chemical products, by-products of chemical use, and chemical processes. No one fully comprehends the total effect, yet all of us are exposed and we all are affected.

Exposures to volatile chemicals occur throughout life, from conception to death. For example, organic solvent vapors inhaled by a mother can reach the fetus. Exposures also occur in the course of daily living, ranging from the inhalation of vapors from a newspaper freshly off the press, to exposure to cleaning solvents by all routes of entry, to a worker manufacturing computer chips, to a researcher in a laboratory, to a farm worker hoeing weeds. It may occur at home or at work. Effects from the exposure may range from a simple objection to an odor to death at high concentrations. In between, there is a whole spectrum of effects.

Solvents convert substances into a form suitable for a particular use. Solvents are significant because many substances are most useful when in solution.

Organic and inorganic compounds are used in the home as cleaning agents, paint thinners, coatings, and spot removers; in the office as typewriter key cleaners, desktop cleaners, and wax removers; in commercial laundries as dry cleaning liquids; on the farm as pesticides; in laboratories as chemical reagents and drying, cleaning, and liquid extraction agents; in shops as cleaners, solvents, by-products from processes such as welding, and paints. Many consumer products packaged in cans and drums contain mixtures of organic chemicals.

Because of the nearly infinite number of combinations possible for the variables involved—hundreds of different compounds, degree of concentration, duration of exposure, combined effects with other solvents, gases and vapors, and the health and age of an exposed person—generalizations about effects of exposure on a particular person are difficult to make. The problem lies not so much in the effect itself, but rather in determining which effects are harmful and at what level.

## Critical Exposure Factors

### Mode of Use and Potential for Exposure

The most important factors in exposure potential are how the material is used and what controls (engineering or personal protective equipment), if any, are in place. Processes that use gaseous reactants (as in a semiconductor chip manufacturing process) are completely enclosed; the gases are often too reactive to coexist with air and the purity of the product demands it. In the event of a system failure, a compressed gas at pressures of several hundred pounds per square inch (psi) poses greater risk than a liquefied gas with a pressure of a few psi. Painting operations pose risk of exposure to solvents, reactive chemicals such as the isocyanates, or suspected carcinogens such as hexavalent chromium (used in primers). Spray painting poses a greater exposure risk for the worker than brush or roller application. When inhalation exposures are controlled, dermal exposure may be the major route of entry. Many organics readily permeate the skin or glove materials; some, such as dimethyl sulfoxide, are of limited toxicity but may be a vehicle for transporting other toxic materials into the body.

### Temperature and Volatility

The vapor pressure of any chemical compound is directly related to temperature. For organics that are liquid at room temperature, this can make a significant difference in exposure risk. For example, methyl ethyl ketone (MEK) has a vapor pressure of 100 mm Hg at 25 C and 400 mm Hg at 60 C. At a compound's normal boiling point, the vapor pressure is 760 mm Hg. Boiling points and vapor pressure are related to molecular weight; in a homologous series (such as acetone, methyl ethyl ketone, and methyl isobutyl ketone), the lighter compound has the lower boiling point (56.5, 79.6, and 127.5 C, respectively) and vapor pressure. Thus, for two similar processes, the one taking place at lower temperature has less potential for exposure.

### Concentration

The effect of concentration may be manifested in several ways. From the point of view of chemical kinetics, reaction rates depend on some factor of concentration; for a given concentration $x$, the rate is proportional to $x$, $x^2$, or some other factor depending on the reaction mechanism. Nitric oxide (NO, TLV 25 ppm) at higher concentrations reacts rapidly with oxygen to produce nitrogen dioxide ($NO_2$, TLV 3 ppm), but at very low concentrations the rate is very slow. The vapor pressure of a solute over a solution varies directly as the mole fraction (concentration) of the solute according to Raoult's law. In both cases, reducing concentration reduces the potential for exposure by limiting the amount of toxic product formed or volatilized. Some chemicals, such as sulfuric acid, are available in a variety of concentrations; the standard grade is approximately 98 percent acid. The vapor pressure is so low at room temperature that it poses minimal risk. The grade marketed as oleum, or fuming sulfuric acid, is much more reactive and off-gases sulfur dioxide; it poses significant risk to the worker. Chemical protective clothing for those working with highly reactive materials must be selected carefully; butyl rubber is satisfactory for sulfuric acid, but has significantly diminished breakthrough times for oleum.

### Reactivity

Chemical reactivity can enhance or reduce health hazard potential. Other physical hazards, such as fire, become significant for pyrophoric materials such as silane ($SiH_4$) and yellow phosphorus. Acids or bases can react with volatile compounds to stabilize them. In a strongly alkaline solution (high pH), cyanide salts cannot form volatile hydrogen cyanide (HCN); in an acid solution (low pH), volatile amines (ammonia or organic amines) are converted to nonvolatile ammonium salts in solution. Liquid metal halides (tungsten or rhenium hexafluoride) react rapidly with moisture in the air to form gaseous hydrogen fluoride (HF).

### Exposure Guidelines

The Threshold Limit Value (TLV) is the concentration in air under which it is believed that nearly all workers can be repeatedly exposed day after day without adverse effects. A list of TLVs is published yearly by the American Conference of Governmental Industrial Hygienists (see Appendix B). For further discussion, see Chapter 6, Industrial Toxicology.

The TLV should not be confused with the permissible exposure limit (PEL), set forth in the regulations promulgated by the Occupational Safety and Health Administration (OSHA) (see Appendix B). Most of the OSHA levels were adopted initially from the 1968 TLV list. The OSHA standards are a part of the law and are not updated annually. The TLV list is updated annually, so you should consult the latest listing for guidance. However, some OSHA standards contain more information on chemicals and their control than the single TLV, and this information should be used.

# Solvents

Solvents are materials used to dissolve other materials; they include aqueous and nonaqueous systems. A solution is a mixture of two or more substances. A solution has uniform chemical and physical properties. There are two components to every solution: the solvent and the solute. As a matter of convenience, we designate the major component of a solution as the solvent; the solute is the lesser component. Thus, we have a gaseous solution when a substance is dissolved in a gas, a liquid solution when a substance is dissolved in a liquid, and a solid solution when a substance is dissolved in a solid.

However, for simplicity, in this chapter the word *solvent* includes only liquids commonly used to dissolve other materials. Organic solvents include naphtha, mineral spirits, turpentine, benzene, alcohol, perchloroethylene, and trichloroethane.

## Aqueous Systems

These solvent systems are based on water. Examples are solutions of acids, alkalis, and detergents dissolved in water. In general, aqueous systems have low vapor pressures at ambient temperatures; thus, the potential hazard by inhalation and subsequent systemic toxicity is not great. The volatility of salts from solution is so low that they can be discounted unless the process, such as aerating a tank, produces a mist. Due to the trend to reduce volatile organic compounds (VOC) in paints, coatings, and other systems, aqueous systems are seeing an increase in use.

## Organic Systems

Many organic compounds are very widely used, with or without a solvent. Polymer precursors may be any of a large variety of compounds, including pesticides, coatings, and adhesives. The organic chemistry of solvents applies to these species as well.

Organic chemistry is the chemistry of the compounds of carbon. The carbon atom can form single, double, and triple bonds to other carbon atoms and to atoms of other elements. The bonds are covalent and have definite directions in space. A molecular chain (or skeleton) consists of a line of carbon atoms that can have branches of carbon atoms, or functional groups. These functional groups can contain oxygen (O), nitrogen (N), phosphorus (P), and sulfur (S), among others. A functional group in an organic molecule is a region where reactions can take place. Double and triple bonds and the presence of atoms other than carbon make up typical functional groups.

Organic compounds are named according to the number of carbon atoms in the basic skeletal chain. The location of the functional groups is designated by the number of the carbon atom to which it is attached. Carbon skeletons with straight chains of carbon with all carbon–hydrogen bonds filled with a hydrogen atom are called alkanes. These are further identified as aliphatic or paraffin hydrocarbons. Those forming rings are identified by the prefix *cyclic-* or *cyclo-*. One specific stable ring structure, that of benzene, contains six carbon atoms with carbon–carbon bonds of intermediate length and bond energy between carbon–carbon single and double bonds. Skeletons containing this ring are described as aromatic.

Isomers are molecules that have the same basic skeletal structure with the same number and kinds of atoms, but the atoms are arranged differently, and thus have different physical and chemical properties.

The common organic solvents can be classified as aliphatic, cyclic, aromatic, halogenated hydrocarbons, ketones, esters, alcohols, and ethers. Each class has a characteristic molecular structure, as shown in Table 7–A.

A good working knowledge of the nomenclature, the characteristic molecular structure, and the different toxicities is helpful in making a proper assessment of an exposure.

Nomenclature itself can often be misleading to health and safety professionals. For example, trichloroethane and trichloroethylene are chlorinated hydrocarbons differing in the degree of saturation arrangement of chlorine atoms. Trichloroethane is saturated whereas trichloroethylene has a carbon–carbon double bond. Trichloroethane is less toxic (TLV = 350 ppm) than trichloroethylene (TLV = 50 ppm). These two chemicals are easily confused due to slight difference in names, which might not be apparent to a worker with inadequate training. Common names do not impart any information as to the structure of a molecule (muriatic acid as compared to hydrochloric acid) and can even be misleading (ethylene dichloride is actually the completely saturated 1,2-dichloroethane); systematic names, using the International Union of Pure and Applied Chemistry rules, unambiguously tell the structure of a molecule. The *CRC Handbook of Chemistry and Physics* describes this system in detail.

Even a scientifically trained user often has only a vague and sometimes completely erroneous knowledge of the chemical preparation in use. It is a good practice to verify the specific name and composition of the solvents involved with direct evidence from the label, from the manufacturer's Material Safety Data Sheet, or from the laboratory. Only after verification of name and composition should one attempt to evaluate the potential effect or hazard of a solvent.

Manufacturers now are required by government regulations to provide information on the composition of their trade name materials. The minimum information is that contained in the Material Safety Data Sheet (MSDS) (Figure 7–1). Manufacturers can withhold proprietary information from an MSDS for general use, but are required to furnish it in cases of medical necessity.

Hawley's *Condensed Chemical Dictionary,* Windholz's *The Merck Index,* Gleason's *Clinical Toxicology of Com-*

**Table 7–A.** Major Classes of Organic Compounds

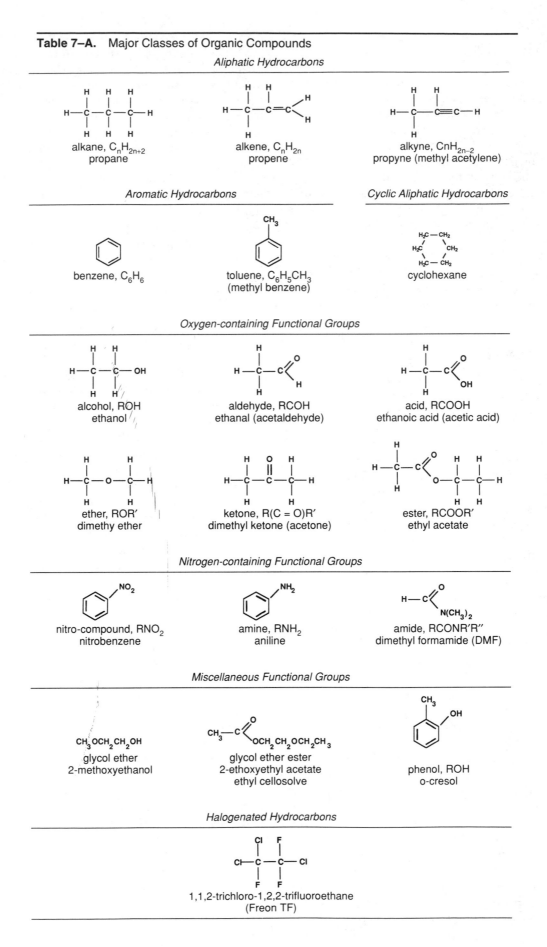

*Aliphatic Hydrocarbons*

alkane, $C_nH_{2n+2}$
propane

alkene, $C_nH_{2n}$
propene

alkyne, $C_nH_{2n-2}$
propyne (methyl acetylene)

*Aromatic Hydrocarbons*

benzene, $C_6H_6$

toluene, $C_6H_5CH_3$
(methyl benzene)

*Cyclic Aliphatic Hydrocarbons*

cyclohexane

*Oxygen-containing Functional Groups*

alcohol, ROH
ethanol

aldehyde, RCOH
ethanal (acetaldehyde)

acid, RCOOH
ethanoic acid (acetic acid)

ether, ROR′
dimethy ether

ketone, R(C = O)R′
dimethyl ketone (acetone)

ester, RCOOR′
ethyl acetate

*Nitrogen-containing Functional Groups*

nitro-compound, $RNO_2$
nitrobenzene

amine, $RNH_2$
aniline

amide, RCONR′R″
dimethyl formamide (DMF)

*Miscellaneous Functional Groups*

$CH_3OCH_2CH_2OH$
glycol ether
2-methoxyethanol

glycol ether ester
2-ethoxyethyl acetate
ethyl cellosolve

phenol, ROH
o-cresol

*Halogenated Hydrocarbons*

1,1,2-trichloro-1,2,2-trifluoroethane
(Freon TF)

## Material Safety Data Sheet

May be used to comply with
OSHA's Hazard Communication Standard,
29 CFR 1910.1200. Standard must be
consulted for specific requirements.

## U.S. Department of Labor

Occupational Safety and Health Administration
(Non-Mandatory Form)
 Form Approved
OMB No. 1218-0072

| IDENTITY (As Used on Label and List) | Note: Blank spaces are not permitted. If any item is not applicable, or no information is available, the space must be marked to indicate that. |
|---|---|

### Section I

| Manufacturer's Name | Emergency Telephone Number |
|---|---|
| Address (Number, Street, City, State, and ZIP Code) | Telephone Number for Information |
| | Date Prepared |
| | Signature of Preparer (optional) |

### Section II — Hazardous Ingredients/Identity Information

| Hazardous Components (Specific Chemical Identity; Common Name(s)) | OSHA PEL | ACGIH TLV | Other Limits Recommended | % (optional) |
|---|---|---|---|---|
| | | | | |

### Section III — Physical/Chemical Characteristics

| Boiling Point | | Specific Gravity (H₂O = 1) | |
|---|---|---|---|
| Vapor Pressure (mm Hg.) | | Melting Point | |
| Vapor Density (AIR = 1) | | Evaporation Rate (Butyl Acetate = 1) | |

Solubility in Water

Appearance and Odor

### Section IV — Fire and Explosion Hazard Data

| Flash Point (Method Used) | Flammable Limits | LEL | UEL |
|---|---|---|---|

Extinguishing Media

Special Fire Fighting Procedures

Unusual Fire and Explosion Hazards

(Reproduce locally)    OSHA 174, Sept. 1985

**Figure 7–1.** Material Safety Data Sheet. This form or an equivalent is used by manufacturers and distributors to provide health and safety information on solvents. (*Continues*)

## Section V — Reactivity Data

| Stability | Unstable | | Conditions to Avoid |
|---|---|---|---|
| | Stable | | |

Incompatibility (*Materials to Avoid*)

Hazardous Decomposition or Byproducts

| Hazardous Polymerization | May Occur | | Conditions to Avoid |
|---|---|---|---|
| | Will Not Occur | | |

## Section VI — Health Hazard Data

Route(s) of Entry:                    Inhalation?                    Skin?                    Ingestion?

Health Hazards (*Acute and Chronic*)

Carcinogenicity:                    NTP?                    IARC Monographs?                    OSHA Regulated?

Signs and Symptoms of Exposure

Medical Conditions
Generally Aggravated by Exposure

Emergency and First Aid Procedures

## Section VII — Precautions for Safe Handling and Use

Steps to Be Taken in Case Material Is Released or Spilled

Waste Disposal Method

Precautions to Be Taken in Handling and Storing

Other Precautions

## Section VIII — Control Measures

Respiratory Protection (*Specify Type*)

| Ventilation | Local Exhaust | | Special |
|---|---|---|---|
| | Mechanical (*General*) | | Other |

| Protective Gloves | | Eye Protection |
|---|---|---|

Other Protective Clothing or Equipment

Work/Hygienic Practices

**Figure 7–1.**  (*Continued*)

mercial *Products*, and the NFPA *Fire Hazard Properties of Flammable Liquids, Gases, and Vapors* provide general information and descriptions of many solvents, including trade name materials. These are helpful references for classifying and understanding the composition of a solvent.

# Gases

A gas is a formless fluid that completely fills its container and whose exerted pressure is the same in all directions. Gases are materials (elements or chemical compounds) whose physical state (or phase) is a gas at normal temperature and pressure (25 C and 1 atm); all materials exist in the gas phase if the temperature is high enough. A vapor is a gas formed when a volatile chemical (liquid or solid) vaporizes. For practical purposes, there is no difference between a gas and a vapor. In a gas mixture, the total pressure is equal to the sum of the partial pressures of the component gases. Gases and vapors are both described by the same equation of state, known as the Ideal Gas Law:

$$PV = nRT$$

where $P$ = pressure
$V$ = volume
$n$ = number of moles of gas
$R$ = ideal gas constant
$T$ = absolute temperature

Commercially, gases are available as compressed or liquefied gases. Certain gases, such as nitrogen, helium, and some others that are available as liquefied gases, are used as a source of high-purity gas from boil-off or as a cryogenic fluid at the gas's boiling point (77 K or –196 C for liquid nitrogen). Table 7–B gives common industrial gases, cylinder pressures, and physical state. All of these readily volatilize, although some organometallic compounds typically available as liquefied gases must be heated to give an adequate working pressure.

Gases can be divided broadly into two groups:

- Gases whose predominant health effect is asphyxia, such as the simple asphyxiants He, Ne, $N_2$, $H_2$, $CH_4$, or the chemical asphyxiants CO, HCN, $H_2S$
- All other gases whose health effects and toxicity are determined by their chemical functional group and reactivity

## Cryogenics

Cryogenic liquids pose several safety concerns in addition to frostbite from extreme cold. Spills of cryogenics rapidly vaporize, producing a gas that is initially significantly more dense than air, resulting in potential oxygen deficiency hazards in pits, vaults, and enclosed spaces. Given sufficient time, the gas reaches thermal equilibrium with its surroundings and disperses throughout the available space. Liquid nitrogen, the most common cryogenic fluid, boils at a lower temperature than liquid oxygen, providing a location for oxygen to condense out of the atmosphere into dewars (double-walled flasks with a vacuum between the walls) with the nitrogen. This creates a potential explosion hazard if the oxygen comes in contact with an oxidizable material. Cryogen dewars and cryogenic systems must have proper pressure relief to prevent pressure buildup and possible rupture as the liquid vaporizes. Full containment of a liquefied gas is usually not possible; helium requires about 18,000 pounds per square inch (psi) and nitrogen requires 43,000 psi. Table 7–C lists common cryogenic liquids.

## Simple Asphyxiants

Inert gases, or simple asphyxiants, exert their hazardous effects by diluting or displacing oxygen. Other hazards related to the partial pressure or concentration of nitrogen are nitrogen narcosis and the bends.

At pressures greater than about 1.5 atmospheres, nitrogen dissolves in fat-containing brain cells, which may result in nitrogen narcosis, a condition similar to inhalation anesthesia. Other gases, such as nitrous oxide and

**Table 7–B.**  Selected Compressed and Liquefied Gases

| Gas | Formula | Form | Cylinder Pressure (psig) |
|-----|---------|------|--------------------------|
| Ammonia | $NH_3$ | Liquid | 114 |
| Argon | Ar | Gas | 225 – 6,000 |
| Arsine | $AsH_3$ | Liquid | 190 – 205 |
| Carbon Dioxide | $CO_2$ | Liquid | 830 |
| Helium | He | Gas | 225 – 6,000 |
| Hydrogen | $H_2$ | Gas | 225 – 3,500 |
| Hydrogen chloride | HCl | Liquid | 613 |
| Hydrogen flouride | HF | Liquid | 0.6 |
| Methane | $CH_4$ | Gas | 1,500 – 2,300 |
| Neon | Ne | Gas | 225 – 1,900 |
| Nitrogen | $N_2$ | Gas | 225 – 6,000 |
| Nitrous oxide | $N_2O$ | Liquid | 745 |
| Oxygen | $O_2$ | Gas | 225 – 2,200 |
| Phosphine | $PH_3$ | Liquid | 400 – 590 |
| Silane | $SiH_4$ | Gas | 150 – 1,200 |

Reprinted with permission from Matheson Gas Products, *Matheson Gases & Equipment*, 1993.

**Table 7–C.**  Common Cryogenic Liquids

| Cryogen | Boiling Point (C) |
|---------|-------------------|
| Argon | –186 |
| Helium | –287 |
| Hydrogen | –252 |
| Neon | –245 |
| Nirtogen | –196 |
| Oxygen | –183 |

(Reprinted with permission from Lide DR, Frederickse HPR, eds. *CRC Handbook of Chemistry and Physics*, 75th ed. Boca Raton, FL: CRC Press, 1994.)

xenon, which are more fat soluble, cause the same effect but at lower pressures. Divers prevent nitrogen narcosis by using breathing air with a higher concentration of oxygen for shallow diving or using a mixture of nitrogen, oxygen, and helium for deep diving. The condition is readily reversible.

The bends, or decompression sickness, is caused by sudden decompression after adjusting to elevated pressure, as in deep diving work. This condition is due to formation of bubbles of gas, mainly nitrogen, as the gas rapidly comes out of solution in body tissues or fluids with the reduction in pressure. This can be prevented by breathing enriched oxygen to "wash" the nitrogen out of tissue, using breathing air of oxygen with less soluble diluent gases, or very slow decompression. Proper breathing air mixtures are preferred, as decompression can take several days.

## Chemical Asphyxiants

Several gases, including CO, HCN, and $H_2S$, are known as chemical asphyxiants. These gases readily pass through the alveolar membranes into blood cells and tissue. These compounds readily combine with iron proteins. Carboxyhemoglobin, the product of carbon monoxide (CO) reacting with hemoglobin, is a more stable complex than oxyferrohemoglobin and thus prevents blood from carrying oxygen. The reaction is reversible; if the exposure to CO is eliminated, the equilibrium favors uncomplexed hemoglobin and the oxygen-carrying capability is restored. These gases also complex with iron in the cytochromes, stopping respiration at the cellular level. For details of oxygen metabolism and respiration, see Lehninger (1975).

# ■ Flammable and Combustible Liquids

Liquids are divided into two general categories by the National Fire Protection Association (see NFPA 30, *Flammable and Combustible Liquids Code* and NFPA 325, *Fire Hazard Properties of Flammable Liquids, Gases and Volatile Solids)* and are defined as follows:

## Flammable Liquid

A liquid with a closed-cup flash point below 100 F (37.8 C) and a vapor pressure not exceeding 40 psi absolute (psia) at 37.8 C is a Class I liquid. Class I liquids are subdivided as follows:

- Class IA liquids include those with a flash point below 73 F (22.8 C) and a boiling point below 100 F (37.8 C).
- Class IB liquids include those with a flash point below 73 F (22.8 C) and a boiling point at or above 100 F (37.8 C).

- Class IC liquids include those with a flash point at or above 73 F (22.8 C) and below 100 F (37.8 C).

## Combustible Liquid

Liquids with a closed-cup flash point at or above 100 F (37.8 C) are called combustible liquids, which are subdivided as follows:

- Class II liquids include those with a flash point at or above 100 F (37.8) and below 140 F (60 C).
- Class IIIA liquids include those with a flash point at or above 140 F (60 C) and below 200 F (93.4 C).
- Class IIIB liquids include those with a flash point at or above 200 F (93.4 C).

## Flash Points

The flash point of a liquid is the lowest temperature at which it gives off enough vapor to form an ignitable mixture with the air near the surface of the liquid or in a vessel capable of flame propagation away from the source of ignition. Some evaporation takes place below the flash point, but not in sufficient quantities to cause an ignitable mixture.

The most commonly used device for determining flash points is the Tag (Tagliabue) Closed Tester for testing liquids with flash points below 200 F (93.4 C) and viscosities below 45 Saybolt Universal Seconds (SUS, the efflux time for viscosity measurements) at 100 F (37.8 C) (ASTM D56–93, *Standard Test Method for Flash Point by Tag Closed Tester).*

The Pensky–Martens Closed Tester is considered the most accurate for testing liquids with flash points about 200 F (93.4 C) and viscosities equal to or greater than 45 SUS (see ASTM D93–90, *Standard Test Methods for Flash Point by Pensky–Martens Closed Tester).*

As an alternative to closed-cup test methods, the Setaflash Closed Tester can be used for aviation turbine fuels; see ASTM D3828–93, *Standard Test Methods for Flash Point by Small Scale Closed Tester.* For paints, enamels, lacquers, varnishes, and related products or components with flash points between 32 F (0 C) and 230 F (145 C), see ASTM D3278–89, *Standard Test Method for Flash Point of Liquids by Setaflash Closed-Cup Apparatus.*

The Cleveland Open-Cup Tester is commonly used for petroleum products, except fuel oils with an open-cup flash below 175 F (79 C) (see ASTM D92–90, *Standard Test Method for Flash and Fire Points by Cleveland Open Cup).*

Another type is the Tag Open-Cup Apparatus, which is sometimes used for low-flash liquids to make tests representative of conditions in open tanks and for labeling and transportation purposes (see ASTM D1310–86, *Standard Test Method for Flash Point of Liquids by Tag Open-Cup Apparatus).*

Open-cup flash points are determined with the liquid in the open air and are generally 10 to 20 percent higher than closed-cup flash point figures for the same substance. When open-cup flash point figures are given, they are usually identified by the initials *OC*.

## Fire Point

The fire point of a liquid is the lowest temperature at which vapors evolve fast enough to support continuous combustion. The fire point temperature is usually about 5 F above the flash point temperature.

## Flammable Range

A prominent factor in rating the fire hazard of a flammable liquid or gas is its flammable range, sometimes called the explosive range. For each flammable liquid or gas, there is a minimum concentration of its vapor, in air, below which propagation of flame does not occur on contact with a source of ignition because the mixture is too lean. Propagation of flame is the self-sustaining spread of flame through the body of the flammable vapor–air mixture after introduction of the source of ignition; a vapor–air mixture at or below its lower explosive limit can burn at the point of ignition without propagating. There is also a maximum concentration of vapor, in air, above which propagation of flame does not occur because the mixture is too rich. The mixtures of vapor with air that, if ignited, just propagate flame are known as the lower and upper flammable (or explosive) limits, and are usually expressed in terms of percentage by volume of vapor in air.

The flammable (or explosive) range includes all the concentrations of a vapor in air between the lower explosive limit (LEL) or lower flammable limit (LFL) and the upper explosive limit (UEL) or upper flammable limit (UFL). The lower flammable limit is important because if this percentage is small, it takes only a small amount of the liquid vaporized in air to form an ignitable mixture.

It also should be noted that if the concentration of vapor in the vapor–air mixture is above the upper flammable limit, introduction of air (by ventilation or other means) produces a mixture within the flammable range before a safe concentration of vapor below the lower flammable limit can be reached.

For a large number of common liquids or gases, the LEL is a few percent and the UEL is 6–12 percent, although there are notable exceptions. The LEL and UEL for hydrogen are 4 and 75 percent and for anhydrous hydrazine they are 4.7 and 100 percent, respectively. For specific materials, consult the material safety data sheet or NFPA 325. For hazardous vapors, if the airborne concentration is kept below the PEL or TLV, the concentration is less than the LEL (note that a concentration of 1 percent by volume is 10,000 ppm).

## Requirements and Guidelines

The occupational safety requirements for the handling and use of flammable and combustible liquids and gases are given in Subpart H of 29 *CFR* 1910. These rules are based on the 1965–1970 editions of the NFPA guidelines current at the time the OSHA standards were first written. Best management practice suggests that current NFPA guidelines should be followed to the extent feasible. Compliance with Subpart H is a minimum.

# ■ Effects

## Physiological Effects

This information indicates only the general toxicological effects to be used in determining hazard potential and establishing a frame of reference. Actual effects of a specific solvent or from a given mixture of solvents can vary considerably. In each classification, there are usually solvents that are more hazardous than other homologues.

The physiological effects of different solvents are far too complex and variable to be discussed in detail here; however, certain generalizations can be made.

**Aqueous systems.** These are known for their irritant effects after prolonged exposure. Contact dermatitis from aqueous solutions is quite common, usually appearing as "dishpan hands." Excessive levels of mists in the air (resulting from heating, agitation, and spraying) can cause throat irritation and bronchitis. Many other effects and hazards are possible if chemicals react with their containers. Bretherick (1985) and Lewis (1993) cite a number of examples. As a rule, aqueous systems, because of their low vapor pressure and ease of control, are a lesser problem, but they cannot be dismissed as potential hazards.

**Organic compounds.** These present a different type of problem. Vapor pressures are usually higher and the potential for inhalation of toxic quantities is much greater. Some of the effects are outlined in the following paragraphs. For detailed information on specific organic solvents, consult Clayton and Clayton's *Patty's Industrial Hygiene and Toxicology,* Gerarde's *Toxicology and Biochemistry of Aromatic Hydrocarbons,* Browning's *Toxicity and Metabolism of Industrial Solvents,* the American Industrial Hygiene Association (AIHA) *Hygienic Guide Series* and the NIOSH *Registry of Toxic Effects of Chemical Substances* or the NIOSH criteria document on the subject solvent (see Bibliography).

All organic solvents affect the central nervous system (CNS) to some extent because they act as depressants and anesthetics. They also cause other effects. Depending on the degree of exposure and the solvent involved, these effects may range from mild narcosis to death from respiratory arrest.

**Aliphatic hydrocarbons.** The aliphatic compounds take their name from the Greek word *aliphe,* meaning fat, because fats are derivatives of this class of hydrocarbons (see Table 7–A).

Aliphatic hydrocarbons are further classified as alkanes, alkenes, cycloalkanes, cycloalkenes, alkynes (acetylenes), and arenes (an aliphatic group is bonded to an aromatic ring). Petroleum and natural gas are the most important sources of alkanes, alkenes, and cycloalkanes. Coal tar is an important source of arenes. High-molecular-weight alkanes are broken down (cracked) catalytically to increase the yield of gasoline from petroleum. Ethylene ($H_2CCH_2$) is an important by-product of cracking and is used to make plastics and ethanol ($CH_3CH_2OH$).

The saturated aliphatic hydrocarbons, $C_nH_{2n+2}$, known as alkanes, are those with all bond positions saturated by bonding with hydrogen. Compounds in this series have the characteristic *-ane* suffix, as in isobutane, 2-methylpentane and 2,2-dimethylpentane. They are as inert biochemically as they are chemically. Even as air pollutants, they are among the least reactive and do not pose a significant problem. The alkanes are good solvents for natural rubber. They act primarily as CNS depressants.

A relatively high level is required for toxic effects. The TLVs generally range from 100 ppm and higher. The exception is n-hexane (or normal hexane, the straight-chained isomer), with a TLV of 50 ppm. Repeated exposures to excessive concentrations may cause peripheral neuropathy, although other isomers have not been found to have this health effect.

The unsaturated aliphatic hydrocarbons, the alkenes ($C_nH_{2n}$) and the alkynes ($C_nH_{2n-2}$), with double and triple bonds respectively, are similarly inert in the body. However, they are more chemically reactive than the saturated hydrocarbons. As air pollutants, they are reactive and create a control problem. The primary health problem associated with the aliphatics is dermatitis.

Crude oil (petroleum) is mainly a very complex mixture of aliphatic compounds. It contains alkanes, alkenes, cycloalkenes, and arenes, as well as small amounts of nitrogen and sulfur compounds, which vary depending on the source.

Petroleum is separated into mixtures of hydrocarbons by fractional distillation. Gasoline is the fraction of petroleum boiling between room temperature and 200 C and is mainly made up of $C_5$ to $C_{11}$ hydrocarbons, with $C_8$ predominating. It has been estimated that there are as many as 500 different hydrocarbons in gasoline alone. About 150 of them have been separated and identified.

**Cyclic hydrocarbons.** The cyclic hydrocarbons act much in the same manner as the aliphatics, but they are not quite as inert. A significant percentage of cyclic hydrocarbons are metabolized to compounds with a low level of toxicity.

The lower molecular weight cycloalkanes (cyclopropane and cyclobutane) have been used as anesthetics. The cycloalkanes typically are CNS depressants, but as molecular weight increases, the margin of safety between anesthesia and adverse health effects decreases. The unsaturated cyclic hydrocarbons generally are more irritating than the saturated forms. This may be due in part to the reactivity of the carbon–carbon double bond. For example, cyclopentane causes slight reddening and drying of the skin and cyclopentene causes moderate to severe irritation of the skin and eyes.

**Aromatic hydrocarbons.** These get their names from *aroma,* meaning pleasant odor. The molecules are usually characterized by one or more six-carbon rings (benzene) or fused rings (naphthalene or larger rings). This classification once served to distinguish petroleum and coal-tar hydrocarbon solvents. Now, however, aromatics are derived from both sources.

Benzene and other aromatics do not undergo the addition reactions shown by alkenes and alkynes, but do undergo aromatic substitution reactions in which an atom or group of atoms replaces one of the hydrogen atoms on the ring. Aromatic substitution reactions of benzene produce a wide variety of useful products (see Table 7–A). The aromatic hydrocarbon benzene is notorious for its effect on the blood-forming tissues of the bone marrow. Gerarde (see Bibliography) has shown that, in animals, injury may result from a single exposure. Benzene is now indicated as a leukemogenic agent. This has greatly reduced the extent of its use as a solvent. Toxic levels of benzene are easily absorbed through the skin and inhaled. Benzene should not be used for cleaning processes or for any process requiring skin contact or where the concentration in the air is not controlled by proper ventilation. The 1994–95 TLV (Notice of Intended Changes) for benzene is 0.3 ppm, with an A1 classification. However, NIOSH recommended that the PEL be reduced to 0.1 ppm, averaged over an 8-hour period, and that benzene be regulated as an occupational carcinogen.

The aromatic hydrocarbons in general are local irritants and vasodilators that cause severe pulmonary and vascular injury when absorbed in sufficient concentrations. They also are potent narcotics. The primary problems with common aromatic solvents other than benzene are dermatitis and effects on the central nervous system (CNS).

**Halogenated hydrocarbons.** The halogens are a group of five elements: fluorine, chlorine, bromine, iodine, and astatine. The halogens are a remarkable family of elements, marked by their great chemical activity and unique properties. Stability, nonflammability, and a wide range of solvency are but a few of the characteristics imparted by their application (see Table 7–A). Halogenated hydrocarbons are organic compounds in which one or more hydrogens have been replaced by fluorine, chlorine, or bromine, or rarely, iodine. The effects of the halogenated hydrocarbons vary considerably with the number and type of halogen atoms present in the molecule. Carbon tetrachloride at one end of the scale is highly toxic, causing acute injury to the kidneys, liver, CNS, and gastrointestinal tract. The 1994–95 TLV for carbon tetrachloride is 5 ppm (A3 classification—animal carcinogen); however, NIOSH recommends that the PEL be reduced to a short-term exposure limit (STEL) of 2 ppm averaged over a 1-hour period and that the chemical be regulated as an occupational carcinogen.

Chronic exposure to carbon tetrachloride also damages the liver and kidneys and is suspected of causing liver cancer. Carbon tetrachloride has become the classic liver toxicant for use in studies on the effects of damage to the liver. As with benzene, this solvent should not be used for open cleaning processes where there is skin contact or where the concentration in the breathing zone may exceed recommended levels. Its use should be avoided altogether.

Replacing some of the chlorine atoms with fluorine as in 1,1,2-trichloro-1,2,2-trifluoroethane (Freon TF) produces a compound with a low level of toxicity. Its present TLV is 1,000 ppm as a ceiling limit. The depressant effect on the CNS and cardiac arrhythmias occur at concentrations much greater than the TLV. Because it is nonflammable and of low toxicity, it may be a suitable substitute for the more hazardous chlorinated solvents; however, environmental considerations such as ozone depletion preclude its application in many cases.

The chlorinated hydrocarbons, in general, are more toxic than the common fluorinated hydrocarbon solvents, although there are significant exceptions such as perfluoroisobutylene (TLV = 0.01 ppm, ceiling). Specific effects and toxicities vary widely, but the most common effects from the chlorinated hydrocarbons of intermediate toxicity (trichloroethylene, for example) are CNS depression, dermatitis, and injury to the liver. There is disagreement as to the carcinogenicity of 1,1,1-trichloroethane; the ACGIH TLV carcinogen classification is A5, not suspected as a human carcinogen, but other organizations, such as NIOSH, consider 1,1,1-trichloroethane to be a carcinogen.

In addition, the chlorinated hydrocarbons, especially trichloroethylene, are noted for their synergistic effects with alcohol consumption. These include flushed, red face and personality changes. These effects must be taken into account when evaluating industrial exposure.

Perchloroethylene (tetrachloroethylene), commonly used in dry cleaning, textile processing, and other industrial processes, has been suggested as a potential human carcinogen. Recent animal lifetime exposure studies (National Toxicology Program, 1986) did show an increased tumor incidence. Earlier animal data, in some cases, is confounded by simultaneous exposure to epichlorohydrin (a known mutagen); human epidemiology is confounded by concomitant exposures to other chlorinated solvents, smoking, and other factors. In its *Seventh Annual Report on Carcinogens, 1994* the National

Toxicology Program stated that there is sufficient evidence for the carcinogenicity of perchloroethylene in experimental animals and cited an International Agency for Research on Cancer (IARC) working group conclusion that there are insufficient data to evaluate its carcinogenicity in humans (National Toxicology Program, 1994).

Refrigerants (Freon, chlorofluorocarbons or CFCs, and hydrochlorofluorocarbons or HCFCs) are a subclass of the halogenated hydrocarbons. The majority of these are methane or ethane derivatives, with a few higher carbon analogues and some other inorganic and organic gaseous compounds. The materials in current use are generally of low toxicity and exposure limits and guidelines tend to be high; the TLVs for trichlorofluoromethane (R-11) and chlorodifluoromethane (R-22) are 1,000 ppm. New refrigerants have come on the market as a result of the EPA incentives to produce nonozone-depleting refrigerants. There are no exposure standards or guidelines for these compounds other than those suggested by the manufacturer. Safety of some of the new refrigerants has come under question as benign tumors have been seen in long-term animal studies. Refrigerant safety has been reviewed by Calm (1994) and engineering controls have been addressed by ASHRAE (see Bibliography).

**Nitrohydrocarbons.** These vary in their toxicological effects, depending on whether the hydrocarbon is an alkane or an aromatic hydrocarbon. Nitroalkanes are known more for their irritant effects accompanied by nausea; effects on the CNS and liver become significant during acute exposures. 2-nitropropane is listed as a suspect human carcinogen, but with an assigned TLV and low carcinogenic potency. The nitroaromatics (such as nitrobenzene) are much more acutely hazardous. They cause the formation of methemoglobin and act on the CNS, the liver, and other organs.

**Oxygen-containing functional groups.** These are found in the alcohols, aldehydes, ketones, carboxylic acids and their esters, anhydrides, and the ethers.

*Alcohols.* One of the most important classes of industrial solvents is characterized by the presence of a hydroxyl group (-OH). Saturated alcohols are widely used as solvents. All alcohols are formed by the replacement of one or more hydrogen atoms by one or more hydroxyl groups.

These polar compounds are classified on the basis of both the number of hydroxyl groups and the nature of the radicals attached to the hydroxyl groups. The monohydric alcohols, which contain one hydroxyl group, are known simply as alcohols; dihydric alcohols have two hydroxyl groups and are known as glycols; trihydric alcohols have three hydroxyl groups and are called glycerols or polyols.

The alcohols are noted for their effect on the CNS and the liver but they vary widely in their degree of toxicity. Methanol ($CH_3OH$) and ethanol ($CH_3CH_2OH$) are the two most important industrial alcohols. Methanol is made by catalytic hydrogenation of carbon monoxide and may one day replace gasoline and natural gas as a fuel because it can be made from coal. Ethanol is made by fermentation of starch (or other carbohydrates) and by hydration of ethene.

Methanol causes several types of injuries, notably impairment of vision and injury of the optic nerve. Methanol slowly produces toxic metabolites. For this reason, its chronic toxicity is greater than that of ethanol.

Ethanol is used industrially in a denatured form. It is quickly metabolized in the body and largely converted to carbon dioxide, and is the least toxic of the alcohols. Any toxicity it causes can be more related to the denaturants (such as benzene or methanol). The undesirable effects of ethanol primarily are related to its recreational use, which affects the drinker's physical safety and can compound the effects of other solvents or medications.

Alcohol is a depressant, not a stimulant. Medically, alcohol depresses the CNS, slowing down the activity of the brain and spinal cord. A large enough dose of alcohol can sedate the brain to a point where involuntary functions such as breathing are lost, causing death.

Propanol is metabolized to toxic by-products and is more toxic than ethanol when taken internally but less toxic than the higher homologues.

**Aldehydes.** These are well-known causes of skin and mucosal irritation and CNS effects. Dermatitis from the aldehydes is common. The aldehydes also are characterized by their sensitizing properties. Allergic responses are common.

**Ketones.** These have become increasingly important solvents for acetate rayon and vinyl resin coatings. Ketones are stable solvents with high dilution ratios for hydrocarbon diluents. They are freely miscible with most lacquer solvents (low-molecular-weight alcohols, aromatics, and esters) and diluents, and their compatibility with lacquer ingredients gives an acceptable finish to many products. Generally, ketones are good solvents for cellulose esters and ethers and many natural and synthetic resins.

The common ketones (such as acetone or methyl ethyl ketone, MEK) generally exert a narcotic-type action. All are irritating to the eyes, nose, and throat, so high concentrations are not usually tolerated. Methyl ethyl ketone in conjunction with toluene or xylene has been reported to cause vertigo and nausea. Lower tolerable concentrations may impair judgment and thereby create secondary hazards. The lower saturated aliphatic ketones are rapidly excreted and for this reason cause only minor systemic effects. Methyl n-butyl ketone received widespread attention during the 1970s, when it was pinpointed as the etiological agent producing a high incidence of peripheral neuropathy in one working population.

**Esters.** These are produced by the reaction of an organic acid with an alcohol. The particular properties of the esters are, therefore, partly determined by the parent alcohol. Esters often have pleasant odors. Esters in low concentrations are used as artificial fruit essences, flavorings, or components of perfumes. The esters are good solvents for surface coatings. Esters in high concentrations are noted for their irritating effects to exposed skin surfaces and to the respiratory tract. They also are potent anesthetics. Cumulative effects of the common esters used as solvents are not significant except for conditions resulting from irritation. Esters of some mineral acids, such as dimethyl sulfate, are highly toxic.

**Ethers.** These are made up of two hydrocarbon groups held together by an oxygen atom. They are made by combining two molecules of the corresponding alcohol. Compared with alcohols, ethers are characterized by their greater volatility, lower solubility in water, and higher solvent power for oils, fats, and greases. Because of their nonreactivity with solutes and ease of recovery, ethers are widely used for extraction. Mixtures of the lower alkyl ethers with alcohols make efficient solvents for cellulose esters. The epoxides (cyclic ethers) differ from other ethers, which are chemically inert, because their unstable three-membered rings make them highly active chemically. Because the epoxides react with the unstable hydrogen atom from water, alcohols, amines, and similar substances, they form a wide range of industrially important compounds.

The primary reactions to the saturated and unsaturated alkyl ethers, such as ethyl and divinyl ether, are anesthesia and irritation of the mucous membranes. However, the greatest safety hazard of these ethers is their tendency to form explosive peroxides. Once opened, ethers should be used within a short period of time, stored so that peroxides are destroyed as they are formed, or checked for the presence of peroxides. The ethers vary in their rate of peroxide formation; diisopropyl ether is one of the most rapid. Halogenated ethers (such as bis-chloromethyl ether) generally are highly toxic and the reader should refer to the more comprehensive references for information on these materials (see Bibliography, especially Clayton and Clayton, Browning, the AIHA *Hygienic Guide Series,* and the OSHA standards).

**Glycols, glycol ethers, and their esters.** The glycols, like the cellosolves and the carbitols, are colorless liquids of mild odor. They are miscible with most liquids (organic and aqueous) and owe this wide solubility to the presence of the hydroxyl, the ether, and alkyl groups in the molecule. The glycol dialkyl ethers are pure ethers with a mild and pleasant odor. They are better solvents for resins and oils than are the monoethers. As a rule, these compounds are more volatile than the monoethers with the same boiling point.

The glycol ethers exert their effects on the brain, the blood, the reproductive system, and the kidneys. Of these, 2-methoxyethanol (ethylene glycol monomethyl ether), 2-ethoxyethanol (ethylene glycol monoethyl ether), and their acetates are the most toxic. They are rapidly absorbed through the skin and elicit neurological symptoms, including changes in personality. They also affect the reproductive system in both men and women. The higher-molecular-weight glycol ethers (such as propylene glycol and ethylene glycol monobutyl ether), on the other hand, are much less toxic. Work with 2-methoxyethanol, 2- ethoxyethanol, or their acetates should be conducted so as to preclude *any* skin contact. OSHA is proposing to reduce the PELs for 2-methoxyethanol and its acetate from 25 ppm to 0.1 ppm and for 2-ethoxyethanol and its acetate from 200 and 100 ppm, respectively, to 0.5 ppm.

**Inorganic acids.** The common inorganic acids include the hydrogen halides (HF, HCl, HI, HBr), the oxygen acids (nitric [$HNO_3$], phosphoric [$H_3PO_4$], and sulfuric [$H_2SO_4$]), and others such as hydrogen sulfide ($H_2S$) and hydrogen cyanide (HCN). These are commercially available as compressed gases, liquids, aqueous solutions of various concentrations, or, in some cases, as solids. These acids may be strong (completely ionized in aqueous solution) or weak (parent acid in equilibrium with its acid anion; the parent acid predominates).

All of the acids off-gas from solution due to the equilibrium between the gas phase and the solution. The more volatile, such as HCl, do so readily; the less volatile, such as $H_2SO_4$ and $H_3PO_4$, do so only at elevated temperatures. As temperatures are increased, equilibrium is driven to the gas phase; the simple acids are driven off as the gas and the oxygen acids can decompose to produce oxides such as $NO/NO_2$ and $SO_2/SO_3$. As the pH is lowered (more acidic), volatility from solution is increased; as the pH is raised (more alkaline), the acid anion is stabilized in solution and it is essentially not volatile.

Health effects are variable and concentration-dependent, most often on the site of contact with tissue; they include irritation of mucous membranes or respiratory tract by HCl, chemical burns by the concentrated solutions, oxidation by $HNO_3$ (an oxidizing acid), and dehydration by $H_2SO_4$. The highly toxic acids $H_2S$ and HCN act differently from other acids; they complex with metal-containing enzymes (cytochromes), preventing cellular oxygen metabolism.

Concentrated HF is particularly corrosive to tissue and bone. Pain from HF solutions stronger than 50 percent is felt within a few minutes; lower concentrations may not produce pain for several hours. Serious tissue damage may result without the person being aware of it. HF burns require immediate action: Irrigate the exposed area to flush away as much HF as possible and seek medical attention immediately. Treatment is dependent on the severity of the burn; mild cases can be managed with magnesium oxide but more severe burns may require infiltration of the affected tissue with calcium gluconate.

Engineering controls and protective equipment should be used to limit exposure. Use of personal protective equipment is essential for working with concentrated acids. Chemical compatibility of the protective clothing with the acid must be considered to ensure protection of the worker (see Johnson, Swope, Goydan, et al., 1991).

**Organic and inorganic gases.** Simple gases (low-molecular-weight hydrocarbons such as methane, ethane, and propane), nitrogen, hydrogen, the inert gases (helium, neon, and argon), and some compounds ($CO_2$) have no significant toxicity of their own. They are simple asphyxiants; they dilute oxygen in the atmosphere. These elements and compounds have either no or very minimal odor and thus have poor warning properties.

The oxides of carbon, nitrogen, and sulfur can be produced by use of oxygen acids, combustion, welding, chemical cleaning or electroplating, or a variety of other processes. Decomposition of organic material may produce toxic (hydrogen sulfide) or flammable (methane) atmospheres. Others, such as ammonia, boron halides, phosphine, arsine, and silane, are used as reactants in industrial and manufacturing processes. Others, such as the reactive, volatile metal halides, are used in research.

In all of these cases, reactivity of the gas is important. The oxides may react with moisture in the mucous membranes to form acids; ammonia, an alkaline gas, acts as a primary irritant. Boron or volatile metal halides react spontaneously in moist air to form gaseous hydrogen halides. Phosphine and silane are spontaneously flammable in air. Arsine and phosphine are also highly toxic.

Exposure control relies primarily on engineering controls. Dilution ventilation *may be* sufficient for the simple asphyxiants. Local exhaust ventilation is suitable for many processes. The more reactive or highly toxic compounds require complete control (exhaust ventilation, gas sensors, all-welded construction for gas lines, excess flow-controllers, automatic shut-down systems, etc.) to ensure protection of the workers.

## Hazard Potential

The toxicological effects alone are not adequate to assess the hazard potential of a solvent. The vapor pressure, ventilation, and manner of use determine the concentration in air, and thus the amount of material available to produce an effect.

**The vapor/hazard ratio number.** This is one approach toward a numerical comparison of potential hazard under a given set of conditions. This number is the ratio of the equilibrium vapor concentration at 77 F (25 C) to the TLV (ppm/ppm)—the lower the ratio, the lower the potential hazard. For example, 4-methyl-2-pentanone (methyl isobutyl ketone or MIBK), with a TLV of 50 ppm, might be judged potentially more hazardous than 2-butanone (methyl ethyl ketone or MEK), which has a TLV of 200 ppm, if judged on the basis of the TLV only. When the potential for vaporization is taken into account, it becomes apparent that MIBK should have a lower hazard potential. The vapor hazard ratio for MIBK is only 186, whereas the vapor hazard ratio for MEK is 625 (Table 7–D).

**Other factors.** Other factors must be taken into consideration. For example, handling procedures and type of clothing determine the degree of skin contact and absorption.

**Table 7–D.** Organic Liquids Arranged in Order of Vapor Hazard

| Substance | Vapor Hazard* | Threshold Limit** (ppm by volume) | B.P. (C)+ | Substance | Vapor Hazard* | Threshold Limit** (ppm by volume) | B.P. (C)+ |
|---|---|---|---|---|---|---|---|
| Tetranitromethane | 3,160,000 | 0.005 | 126.0 | Toluene | 736 | 50 | 110.8 |
| Ethyl bromide++ | 121,200 | 5 | 38.4 | P-tert-Butyl toluene | 720 | 1 | 200 |
| Acrylonitrile++ | 112,000 | 2 | 78.9 | 2-Butanone (MEK) | 625 | 200 | 79.6 |
| Diethylamine | 63,000 | 5 | 55.5 | Phenylhydrazine++ | 550 | 0.1 | 243.5 |
| Carbon disulfide | 46,000 | 10 | 46.3 | Methylal (dimethoxymethane) | 526 | 1,000 | 42 – 43 |
| Butylamine | 34,000 | 5 | 77.8 | 1,1,1-trichloroethane | 489 | 350 | 74.1 |
| Carbon tetrachloride++ | 28,340 | 5 | 76.8 | (methyl chloroform) | | | |
| Chloroform++ | 24,850 | 10 | 61.2 | Nitrobenzene | 474 | 1 | 210.9 |
| Allyl alcohol | 16,450 | 2 | 96.6 | Cyclohexane | 427 | 300 | 80 – 81 |
| 1,1-dichloroethane | 14,500 | 10 | 57.3 | Propul acetate | 390 | 200 | 88.4 |
| Ethylene dichloride | 11,600 | 10 | 83.7 | Acetone | 387 | 750 | 56.5 |
| (1,2-dichloroethane) | | | | Aniline | 330 | 2 | 184.4 |
| Methylene chloride++ | 8,640 | 50 | 40.1 | Ethyl acetate | 303 | 400 | 77.1 |
| 1,1,2,2-tetrachloroethane | 8,420 | 1 | 146.3 | Dichloroethyl ether | 288 | 5 | 178.5 |
| Hexane (n-hexane) | 4,100 | 50 | 69.0 | Propyl alcohol | 280 | 200 | 82.5 |
| Ethyl formate | 3,160 | 100 | 54.0 | Ethyl acetate | 303 | 400 | 77.1 |
| Methyl cellosolve | 3,150 | 5 | 124 – 125 | Dichloroethyl ether | 288 | 5 | 178.5 |
| (2-methoxyethanol) | | | | Propyl alcohol | 280 | 200 | 82.5 |
| 2-nitropropane++ | 2,240 | 10 | 120.3 | Nitroethane | 263 | 100 | 114.8 |
| Nitromethane | 2,170 | 20 | 101.0 | Methyl isobutyl carbinol | 263 | 25 | 131.8 |
| Trichloroethylene | 2,000 | 50 | 87.2 | (methylamyl alcohol) | | | |
| Acetic acis | 1,970 | 10 | 118.1 | Cyclohexanone | 226 | 25 | 155 – 156 |
| Dioxane (diethylene dioxide) | 1,960 | 25 | 101.0 | Isoamyl alcohol | 220 | 100 | 132.0 |
| Ethylene diamine | 1,710 | 10 | 55.5 | Styrene monomer | 194 | 50 | 145 – 146 |
| Cellosolve (2-ethoxyethanol) | 1,640 | 5 | 135.1 | (phenyl ethylene) | | | |
| Benzyl chloride | 1,580 | 1 | 179.4 | Cresol (all isomers) | 184 | 5 | 191 – 203 |
| Chlorobenzene | 1,575 | 10 | 132.1 | Butyl alcohol (n-butanol) | 184 | 50 | 117.0 |
| (monochlorobenzene) | | | | O-toluidine++ | 165 | 2 | 199.7 |
| Methyl acetate | 1,380 | 200 | 57.1 | Methylcyclohexane | 151 | 400 | 101.0 |
| Ethyl ether | 1,380 | 400 | 34.6 | Heptane (n-heptane) | 151 | 400 | 98.4 |
| Acetic anhydride | 1,340 | 5 | 139.6 | Diisobutyl ketone | 142 | 25 | 168.1 |
| Hexanone (methyl butel ketone) | 1,000 | 5 | 128.0 | Nitrotoluene | 138 | 2 | 222 – 238 |
| | | | | Isophorone | 130 | 5 | 215.0 |
| Perchlorethylene | 948 | 25 | 120.8 | Ethyl benzene | 126 | 100 | 136.2 |
| (tetrachloroethylene)++ | | | | Pentanone | 112 | 200 | 95 |
| Propylene dichloride | 910 | 75 | 96.8 | (methylpropyl ketone) | | | |
| (1,2-dichloropropane) | | | | Xylenes | 100 | 100 | 139 – 44 |
| Mesityl oxide | 893 | 15 | 130.0 | Methylcyclohexanone | 94 | 50 | 162 – 70 |
| Dimethylaniline | 870 | 5 | 193.0 | Ethyl alcohol (ethanol) | 76 | 1,000 | 78.4 |
| (N,N-dimethylaniline) | | | | Turpentine | 66 | 100 | 120 – 80 |
| Isopropyl ether | 840 | 250 | 69.0 | Octane | 57 | 300 | 99 – 125 |
| Cellosolve acetate | 840 | 5 | 156.3 | Amyl acetate | 53 | 100 | 142.0 |
| (ethoxyethyl acetate) | | | | Cyclohexanol | 47 | 50 | 160 – 61 |
| Methyl alcohol | 820 | 200 | 64.7 | Stoddard solvent | 35 | 100 | 150 – 90 |
| Methyl cellosolve acetate | 815 | 5 | 144.5 | O-dichlorobenzene | 18 | 25 | 179 |
| (ethylene glycol monome-thyl ether acetate) | | | | Methylcyclohexanol | 14 | 50 | 165 – 75 |
| Butyl acetate (n-butyl acetate) | 788 | 20 | 127 | Diacetone alcohol (4-hydroxy-4-methyl-2-pentanone) | 8 | 50 | 167.9 |
| Pentane | 750 | 600 | 36.3 | | | | |

* Ratio (ppm/ppm) of equilibrium vapor concentration at 25 C to the TLV, computed from vapor pressure data.
** From ACGIH Threshold Limit Values for 1994–1995.
+ Boiling point at 760 mm Hg. Observed boiling points in mixtures may be lower due to formation of azeotropes.
++ Suspected carcinogen; TLV may be lowered. Consult the latest TLV list.
Data adapted from Dean (1992) and Lide and Frederikse (1994).

Even the degree of a user's respect for the hazard potential can be a decisive factor.

Ignition temperature, flash point, and other factors determining the potential for fire and explosion also must be considered. Although concentrations that are safe from a toxicological viewpoint are much lower than the lower flammable limits of flammable solvents, concentrations at potential points of ignition may be far higher than concentrations in the user's breathing zone.

Evaluation of hazard potential requires assessment of the consequences of exposure, the degree of exposure, and all factors contributing to the exposure.

## Air Pollution

Solvents and other chemicals may become hazardous to the public in the form of air pollutants when released outdoors. Hydrocarbons are a major factor in the formation of photochemical smog. In the presence of sunlight, they react with atomic oxygen and ozone to produce aldehydes, acids, oxides of nitrogen and sulfur, and a series of other irritant and noxious compounds.

The greatest portion of hydrocarbons contributing to air pollution originates from automobiles, but a significant amount also comes from the tons of solvents exhausted daily from industrial cleaning and surface-coating processes. In a global sense, natural sources of pollution (volcanoes, biological decay, biological activity in soils and the oceans, forest or grassland fires, and other sources) outweigh artificial sources; however, in a restricted area such as the Los Angeles Basin or most major metropolitan areas, the effect of artificial sources predominates.

Nitric oxide (NO) is produced by the reaction of nitrogen with oxygen in high-temperature combustion, as in automobiles and fuel-burning power plants. Nitric oxide is photochemically oxidized to nitrogen dioxide ($NO_2$), a corrosive and an irritant. Nitrogen dioxide is an energy trap, reacting with sunlight to form nitric oxide and atomic oxygen:

$$NO_2 + h\upsilon \rightarrow NO + O$$

Atomic oxygen is highly reactive, forming ozone and initiating a host of secondary photochemical reactions. The nitric oxide produced can again react to produce more nitrogen dioxide, propagating the process. The yellow-brown haze seen over many cities is made up of nitrogen dioxide and its reaction products. Ozone in the troposphere (the atmosphere less than 10 km altitude) detracts from air quality.

Some compounds are more reactive to sunlight and contribute heavily to the smog problem. The use of such solvents is being curtailed in more and more areas, especially large cities. Other solvents are less reactive and are exempt from stringent control. Here, they are listed in decreasing order of photochemical reactivity as a general guide.

- Alkenes (unsaturated open-chain hydrocarbons containing one or more double bonds)
- Aromatics (except benzene)
- Branched ketones, including methyl isobutyl ketone
- Chlorinated ethylenes, including trichloroethylene (except perchloroethylene)
- Normal ketones (for example methyl ethyl ketone)
- Alcohols and aldehydes
- Branched alkanes
- Cyclic alkanes
- Normal alkanes
- Benzene, acetone, perchloroethylene, and the saturated halogenated hydrocarbons

Opinion is divided as to the exact order of reactivity and many solvents have yet to be tested. The trend is toward the development and use of nonreactive solvent blends.

**Upper atmosphere effects.** In addition to the smog-related materials discussed previously, fluorocarbons such as trichlorotrifluoroethane and related materials catalyze the destruction of ozone in the upper atmosphere. Although the extent of this reaction is not well-established, their production and use has been reduced. Ozone in the stratosphere (the atmosphere 10–50 km in altitude) absorbs solar ultraviolet radiation at the 290-nanometer (nm) wave length. Should the destruction of ozone by fluorocarbons and other materials prove to be significant, the amount of solar ultraviolet radiation reaching the earth's surface will increase. This would impair agricultural production and increase the incidence of skin cancer. In 1987, the industrialized nations met and signed the Montreal Protocol on Substances that Deplete the Ozone Layer. The Montreal Protocol calls for the reduction of use and elimination of the major ozone-depleting chemicals.

**Global warming.** Carbon dioxide, the product of combustion of carbon-based fuels, contributes to the greenhouse effect. Put simply, solar radiation penetrates the atmosphere and is absorbed by the earth; a portion is radiated back into space, and a portion is consumed in life processes and atmospheric chemical reactions, thus setting up a thermal equilibrium. Carbon dioxide absorbs the shorter wavelength energy re-radiated into space; this energy is then manifested as heat. It has been estimated that an increase of $CO_2$ concentration to 370 ppm from the present value of about 320 ppm would increase the temperature 0.5 C. In reality, other factors, such as cloud cover, atmospheric water vapor and particulates, and weather patterns all affect the process and may offset the warming trend of carbon dioxide.

## ■ Evaluation of Hazards

A prime question regarding any process using a volatile chemical is whether the concentration of the solvent in the air exceeds acceptable limits. Getting the answer to this is not as difficult as it may seem.

A knowledge of the chemical, its properties, and the process in which it is used should give the investigator some idea of the potential hazards.

If a chemical of high-hazard potential is being used, if the equipment and ventilation system is poorly designed, or if performance of the system is questioned, then there is a greater probability of physiologic injury and immediate action should be taken to evaluate and reduce the hazard before it becomes a problem.

The evaluation procedure, where the industrial hygienist assesses the degree of risk in the workplace, is based on the following factors:

- The toxicity of the substance
- The concentration in the breathing zone
- The manner of use
- The length of time of the exposure
- The controls already in place and their effectiveness
- Any special susceptibilities on the part of the employees

Samples can be collected in the field and returned to the laboratory for analysis or (and this is the trend) they may be collected and analyzed on the spot with direct-reading instrumentation (Figure 7–2). In this way, a much greater number of samples and much more information can be obtained and evaluated immediately. Recorders can be connected to direct-reading instruments to obtain a continuous record of the concentration. Peak concentrations become apparent. Such peaks are likely to be missed with grab samples and integrated samples that are returned to the laboratory. Peak concentrations are espe-

cially important when the vapor is an irritant or is highly odorous, or if a subjective complaint is involved. The concentration above the norm must be reduced to achieve satisfaction as quickly as possible.

Because direct-reading field instruments often require considerable laboratory backup for maintenance, testing, and calibration, they are not as expedient as they seem. The cost for a field evaluation may be much more than the cost of a laboratory analysis, but this may be offset because much more information is obtained.

See Chapter 15, Evaluation, Chapter 16, Air Sampling, and Chapter 17, Direct-Reading Instruments for Gases, Vapors, and Particulates for more information.

## ■ Control of Hazards

### Responsibility of Health and Safety Personnel

Personnel concerned with health and safety should recognize that the use of organic and other chemicals can be a major threat to health and that hazard assessment and control are necessary to prevent detrimental physiological effects.

Exposure evaluation and workplace inspection should be a routine part of any health and safety program. Exposure evaluations should be performed for new processes to ensure that controls are adequate to protect workers. Surveys or searches should also be made for evidence of disease. Dermatitis, unusual behavior, coughing, or complaints of irritation, headache, and ill feeling are all outward signs of potential disease that warrants further investigation. Positive findings justify the effort and provide convincing evidence for educating personnel to the need for corrective actions.

Note conditions and practices that contribute to excessive exposure and call them to the attention of responsible personnel. Train users to handle chemicals properly to prevent injurious exposures. Set guidelines to direct operating personnel in the selection, use, and handling of chemicals. Prohibit general use of highly toxic chemicals, highly flammable solvents, or solvents that are extremely hazardous, unless special evaluation or authorization is obtained.

Finally, provide technical assistance to help the user select the least hazardous chemicals, design and obtain proper ventilation, eliminate the risk of fire, eliminate skin contact, and evaluate situations when workers might be exposed to excessive levels.

### Process Controls

**Selection of chemicals.** One of the most effective means of controlling chemical exposure is to use the least hazardous material. By simply substituting a less toxic or less vola-

**Figure 7–2.** The industrial hygienist assesses the degree of risk in the workplace based on the toxicity of the substance, the concentration in the breathing zone, the manner of use, the duration of the exposure, the controls already in place and their effectiveness, and any special susceptibilities of the employees in the workplace. (Courtesy Fermilab Visual Media Services.)

tile solvent, for example, one can minimize or eliminate a hazard. The fact that a certain chemical has been specified does not mean that it is the only one or even the best one for a particular use. At times, the one specified is the most familiar one.

This fact is more apparent if one compares the TLVs and the vapor pressures or distillation ranges of different solvents in each class. Toluene and xylene are solvents that can usually be substituted for benzene, for example. If an aromatic hydrocarbon is not required, then it can be replaced with less toxic aliphatic mineral spirits. Low-molecular-weight glycol ethers are used in semiconductor manufacturing processes as a solvent for photoresists, but higher-molecular-weight analogues that have little, if any, reproductive toxicity can be suitable in some situations. The potential for fire can be minimized by the introduction of nonflammable 1,1,2-trifluoro-2,2,1-trichloroethane or 1,1,1- trichloroethane.

The best all-around solvent is water. It is nontoxic and nonflammable and (with the proper additives) it forms an aqueous solvent system that is a good solvent for many organic materials. For the cleanup of inorganic soils, aqueous solvent systems are still the best. The disadvantages are corrosivity of many aqueous solutions and the slow evaporation rate of water. Also, additives may leave a residue on a manufactured item, necessitating further cleaning.

Aliphatic hydrocarbons are good for dissolving nonpolar organic materials such as oils and lubricants. The aliphatics, however, are not effective cleaners for dissolving or removing many tenacious inorganic materials.

Aromatic hydrocarbons are especially effective on resins and polymeric materials. Between the aromatic and aliphatic hydrocarbons in solvent power are the cyclic hydrocarbons. Halogenated hydrocarbons are effective solvents for a wide range of nonpolar and semipolar compounds.

The nitrohydrocarbons have not been used to a large extent as cleaning agents. Their greatest use has been as solvents for esters, resins, waxes, paints, and the like. Because the ketones, alcohols, esters, ethers, aldehydes, and glycols are more water soluble than the other classes, they are good solvents for the more polar compounds. These solvents are often used as cleaning agents alone or combined with other solvents, especially water. They are useful as solvents for paints, varnishes, and plastics. The publication *Proceedings of the 4th International Symposium on Contamination Control* (International Committee of Contamination Control Societies, 1978) is a good source of information for the effectiveness of the different solvents against various materials.

Remember that for nearly every process there is an effective solvent or solvent blend that has low toxicity and low flammability. For example, several companies have switched to water containing an alkaline cleaner as a replacement for naphtha and other such organic solvents

for cleaning hydraulic tubing, tanks, and other containers. Inhibited 1,1,1-trichloroethane has replaced carbon tetrachloride as a household spot remover. As a guide, the following suggestions might be used:

- Use an aqueous (water) solution if possible.
- If water is not suitable, use a so-called safety solvent, but make sure there is adequate ventilation. Safety solvents include inhibited 1,1,1-trichloroethane, the aliphatic hydrocarbons with high flash points, and the fluorinated hydrocarbons.
- When possible, consider a different process altogether, one that does not involve chemicals.
- Solvents that are more toxic than the safety solvents are to be used only with properly engineered local exhaust systems. Solvents such as trichloroethylene, toluene, and ethylene dichloride are in this category.
- Highly toxic or highly flammable solvents, such as benzene, carbon tetrachloride, and gasoline, should be prohibited as general cleaning solvents.

Definite dividends will result from this policy. The number of employees who might have exposures exceeding the TLV can be reduced significantly. The number of small fires resulting from the use of flammable bench solvents also can be reduced.

## Engineering Controls

**Enclosure and ventilation.** The major route of entry for chemicals into the body is the lungs (see Chapter 2, The Lungs). The lungs have a surface area of about 85,000–115,000 sq in. (55–75 $m^2$); much of this area is permeated with thin-walled capillaries. Chemicals in the breathing zone are drawn into the lungs during breathing, quickly absorbed into the bloodstream, and distributed to other parts of the body. The most effective way to prevent inhalation of gases and vapors is to keep them out of the breathing zone. This is done by using closed systems and local exhaust ventilation. All open vessels should be kept covered except when in use. Systems should be designed to prevent leakage and spillage and to collect and contain the solvent in the event of a leak or spill. For open and closed system design parameters, see Cralley and Cralley (1986). Proper ventilation must be installed for any process using solvents. Even storage areas require adequate general ventilation to prevent accumulation and buildup of flammable or toxic concentrations (see Chapters 19 through 21 on industrial and general ventilation).

If subambient temperature storage of solvents is recommended, they should be stored only in refrigerators constructed and designated for that use (explosion-proof or explosion-safe). Such refrigerators have had their ignition sources removed. Refrigerators used for storage of food and beverages should not be used for any other purpose.

Local exhaust ventilation is necessary to capture the vapors at their point of origin and thus prevent excessive concentrations in the breathing zone. If general ventilation is good, a simple pedestal fan blowing the vapors away from the user's breathing zone (dilution ventilation) is often sufficient for solvents of low toxicity. If a highly toxic solvent or gas is being used, or if general ventilation is poor, a local exhaust system, or completely enclosing the process, is necessary to remove the vapors. All control measures should maintain concentrations of hazardous chemicals in the breathing zone well below the OSHA-specified levels. Present trends in worker's compensation insurance and federal regulations justify designs that are well on the safe side. Ventilation systems are a topic in themselves and the reader should refer to Chapters 19 through 21 and to the ACGIH Industrial Ventilation Manual or American National Standards Institute (ANSI) series Z9 standards on industrial ventilation (see Bibliography). Remember that the local exhaust ventilation system of removing vapors at their point of origin is usually the most satisfactory means of control.

## Personal Protective Equipment

**Respirators.** Do not use respirators as the primary or only means of protection against hazardous chemical vapors because too many factors limit their use. They can be used as emergency or backup protection. Respiratory protective equipment, especially the air-purifying type, is limited by leakage around the mask edges, surface contamination, impaired efficiency with use, and need for adequate oxygen. Unless it is correctly used and properly cared for, a respirator may present a greater danger to an employee than no protection at all. Too often, such equipment gives a false sense of security and the wearer becomes careless and may be exposed to highly hazardous levels. Respirators should be controlled through a program that provides for proper selection, fitting, testing, education, and maintenance under the surveillance of competent personnel. Such a program is mandatory under present federal occupational safety and health standards. (See Chapter 22, Respiratory Protection.) Make sure that the level of gases or vapors in the air does not exceed the protective factor of the respirator. Air-purifying respirators should not be used for operations where the solvent is air-sprayed unless there is supplementary mechanical ventilation. The AIHA's *Respiratory Protection—A Manual and Guideline,* 2nd edition, is an excellent guide for such equipment. See Chapter 22 for more details.

**Protective clothing and gloves.** Another major route of entry for hazardous chemicals is through the skin. Dermatitis is the leading industrial disease, and solvents are second only to cutting oils and lubricants in causing this disease. (See Chapter 3, The Skin and Occupational Dermatoses.) Skin contact occurs through direct immersion, splashing, spilling, contact with chemical-soaked clothing, improper gloves, and contact with solvent-wet objects. Some solvents such as benzene, carbon tetrachloride, and methyl alcohol can be absorbed in amounts great enough to cause physiological injury to organs other than the skin. The most effective way and often the only way to prevent harm is to keep the solvent from the skin. This can be done by using mechanical handling devices, such as tongs and baskets, and by using impermeable protective clothing, such as aprons, face shields, and gloves.

The use of gloves requires caution. A common mistake is to recommend "rubber" or neoprene gloves for use as hand protection against a solvent, regardless of the kind of solvent in use. Many solvents can quickly penetrate latex rubber or neoprene gloves and come in contact with the skin.

The permeability of gloves to certain solvents and chemicals is the most important characteristic to consider when selecting gloves for protection. Chemical manufacturers include permeability information with their product, often in the form of permeability tables or computer software for glove selection. They suggest appropriate glove materials for particular chemicals. The abrasion resistance of glove materials is also given in tables and this information is often more widely available. Note both permeability and abrasion resistance when considering the type of gloves to use with certain solvents.

Permeability measurements should be made on the complete glove if the effect of weak or thin spots is to be detected. A rough comparison of the permeability of gloves plus an indication of some of the other characteristics can easily be made by turning the gloves inside out, filling them three-fourths full of solvent, sealing the cuff, and measuring the loss of weight, the stretch, and other parameters.

More precise methods for measuring glove permeability are available, but require the use of an analytical laboratory. A standard method has been published by the American Society for Testing and Materials, ASTM F739–91, *Standard Test Method for Resistance of Protective Clothing Materials to Permeation by Liquids or Gases Under Conditions of Continuous Contact* (see also references cited therein).

The time required for a chemical to penetrate a glove is affected by the glove's thickness and its composition. In some cases, the time to break through can be as brief as 5 minutes. For example, it has been shown that benzene breaks through a 0.03-mm polyethylene glove in five minutes. Conversely, the same glove material had a 2-hour breakthrough time when tested against butyl acetate.

Note that gloves made of the same material and nominal thickness, but from different manufacturers, may have significantly different breakthrough times. This difference may result from differences in formulation of glove materials or manufacturing procedures used.

Examples of the glove performance are given in Tables 7–E and 7–F. The data in Table 7–E give ranges of break-

**Table 7–E.**  Solvent Breakthrough Time in Hours for Various Glove Materials

| Solvent | Glove Material | | | | | |
|---|---|---|---|---|---|---|
| | Natural Rubber | Neoprene | Neoprene + Natural Rubber | Nitrile | PVC | PVA |
| Carbon tetrachloride | .03 – .5 | .08 – > 1.0 | .07 | 1 – > 8 | .01 – .66 | > 3 |
| Chloroform | .01 – .05 | .01 – .36 | .05 – .11 | .07 – .30 | .01 | > 6 |
| Methylene chloride | .01 – .10 | .01 – .22 | .03 – .07 | .03 – .15 | .01 – .20 | .28 – > 8 |
| Methyl iodide | .03 – .05 | .01 – .28 | .03 – .09 | .01 – .13 | .02 | > 8 |
| 1,1,2,2-tetrachloroethane | .03 – .35 | .09 – > 1 | .15 | .22 – 1.2 | .01 – .10 | > 8 |
| 1,1,2-trichloroethane | .02 | .12 | — | .03 | .03 | .25 – > 8 |
| Perchloroethylene | < .02 – .10 | .10 – .80 | .05 – .30 | .22 – 7.3 | < .01 – .75 | .35 – > 16 |
| Methanol | .02 – 6 | .25 – > 8 | .10 – 1.1 | .18 – 3.2 | .02 – 6 | .02 – .04 |
| Ethanol | .02 – > 8 | 1 – 12.6 | .17 – > 1 | > 1 – > 8 | .03 – .33 | 1.67 |
| 2-propanol | .02 – > 8 | > 1 – > 6 | .10 – .83 | > 6 | .5 – > 3 | — |
| N-butanol | .02 – 1.2 | 4 – > 8 | .25 – 1.2 | > 6 | .42 – 3 | .5 – > 8 |
| Benzene | .01 – .18 | .02 – 3.1 | .05 | .07 – 1.0 | < .01 – .50 | .05 – > 33 |
| Toluene | .01 – .68 | .02 – .52 | .07 – .20 | .13 – 1.2 | < .01 – .47 | .02 – > 25 |
| Aniline | .25 – > 8 | .25 – > 8 | .09 – > 6 | .30 – 5.4 | .05 – > 16 | > 1 – > 16 |
| Phenol | .27 – > 8 | .35 – > 8 | > 6 | .53 – > 8 | .05 – 1.3 | > 6 |
| Acetone | .02 – > 3 | .04 – > 1 | .05 – .43 | .07 – .30 | < .02 – .30 | .07 – > 4 |
| Methyl ethyl ketone | .02 – .17 | .04 – 2.8 | .08 – .10 | .06 – .33 | .02 – .27 | .10 – 7 |
| Tetrahydrofuran | .02 – .11 | .02 – 2.5 | .02 – .06 | < .01 – .28 | .01 – .10 | 4.7 |
| Dimethyl sulfoxide | .02 – > 6 | < .01 – > 8 | .7 – > 6 | < .01 – > 4 | .05 – 1.0 | — |
| Dimethyl formamide | .25 – .11 | .02 – > 6 | .25 – .77 | .02 – > 5 | .23 – 1.0 | .08 – .37 |
| Pyridine | .03 – .43 | .03 – .63 | .14 – .23 | .09 – .25 | .02 | .52 |
| Dioxane | .08 – .45 | .09 – 1.8 | .10 – .30 | .28 – 1.1 | .01 – .11 | > 16 |
| N-hexane | .08 | .06 – 3.3 | > .08 | > 1 – > 8 | .20 – .42 | > 6 |
| Water | — | — | — | — | .25 | — |

Data adapted from Johnson, Swope, Goydan, et al., 1991.

**Table 7–F.**  Solvent Permeation Rates in µg/sq cm/min Through Glove Materials

| Solvent | Glove Material | | | | | |
|---|---|---|---|---|---|---|
| | Natural Rubber | Neoprene | Neoprene + Natural Rubber | Nitrile | PVC | PVA |
| Carbon tetrachloride | 2,664 | 1,036 | 3,858 | 22 | 967 | 3 |
| Chloroform | 7,856 | 3,227 | 5,711 | 6,079 | 12,606 | — |
| Methylene chloride | 4,529 | 2,360 | 1,348 | 4,269 | 8,312 | 1 |
| Methyl iodide | 10,794 | 4,896 | 8,917 | 7,976 | — | — |
| 1,1,2,2-tetrachloroethane | 2,765 | 1,173 | 3,206 | 2,204 | 3,466 | — |
| 1,1,2-trichloroethane | — | — | — | — | 1,236 | — |
| Perchloroethylene | 2,222 | 770 | 995 | 72 | 516 | 5 |
| Methanol | 32 | 15 | 11 | 77 | 30 | 545 |
| Ethanol | 12 | 3 | 4 | 5 | 24 | 30 |
| 2-propanol | 4 | 5 | 4 | 5 | 24 | 30 |
| N-butanol | 15 | 5 | 1 | — | 18 | 49 |
| Benzene | 2,185 | 545 | 1,739 | 549 | 2,037 | 12 |
| Toluene | 1,579 | 1,439 | 1,803 | 478 | 2,030 | 85 |
| Aniline | 13 | 7 | 15 | 215 | 75 | — |
| Phenol | 17 | 25 | — | 283 | 104 | — |
| Acetone | 330 | 243 | 96 | 2,066 | 340 | 37 |
| Methyl ethyl ketone | 746 | 641 | 432 | 1,546 | 721 | 3 |
| Tetrahydrofuran | 11,202 | 6,456 | 16,699 | 2,650 | — | 3 |
| Dimethyl sulfoxide | 78 | 30 | 4 | 78 | 102 | — |
| Dimethyl formamide | 298 | 97 | 126 | 120 | 138 | 716 |
| Pyridine | 621 | 732 | 350 | 3,266 | — | 402 |
| Dioxane | 300 | 345 | 224 | 770 | 2,793 | — |
| N-hexane | 751 | 130 | — | 284 | 180 | — |
| Water | — | 1 | — | — | — | — |

Data adapted from Johnson, Swope, Goydan, et al., 1991.

through times for common glove materials; Table 7–F gives average permeation rates for the same solvents and glove materials. It must be noted that these data have been extracted from a wide variety of sources, for many materials, and using varied test protocols.

Neoprene is good for protection against most common oils, aliphatic hydrocarbons, and certain other solvents, but is not satisfactory for use against the aromatic hydrocarbons, halogenated hydrocarbons, ketones, and many other solvents. Natural rubber is not effective against these solvents.

Polyvinyl alcohol (PVA) gloves provide adequate protection against the aromatic and chlorinated hydrocarbons, but they must be kept away from water, acetone, and other solvents miscible in water to prevent deterioration. Butyl rubber gloves can be a suitable compromise when polyvinyl alcohol cannot be used.

Regular periodic cleaning and drying of gloves is as important as using the proper type. Keep an extra pair of gloves available for use while the cleaned pair is being aired and dried. When the gloves become soiled with hard-to-remove hazardous materials such as insecticides and epoxy resins, it is often better to discard the glove than to try to clean it. In some situations, gloves must be replaced after only a few minutes' work. If the outside of any glove becomes thoroughly wetted, remove it promptly.

Disposable gloves are useful for light laboratory or assembly work, but are too easily torn or punctured for heavier work. Latex medical gloves provide good manual dexterity, but they tear easily and are permeable to many solvents. Their use is not recommended in an industrial setting. If they are used, an effective practice is to change immediately after a splash or contamination or change gloves frequently even if direct contact has not occurred. There is no set recommendation for the use of gloves; what works well for one group of workers may not work for another. In many cases, a certain amount of trial and error is required.

Barrier creams are the least effective way of protecting skin. Barrier creams are not a substitute for gloves, except when there is only occasional and minor contact with a solvent, or around rotating machinery when gloves cannot be worn because of the catching hazard. Barrier creams are *not* as effective as an impervious glove. (See Chapter 3 for more information on barrier creams.)

Good personal hygiene is important whenever chemicals are used. Remove spills and splashes immediately with soap and water. This includes showering and replacing solvent-soaked or splattered clothing with clean clothing immediately and as often as necessary.

**Protective eyewear.** Workers at risk for a splash of chemicals in the eyes must wear appropriate protective eyewear. It must be noted that protective eyewear should not be used as the sole protection, but in conjunction with engineering control, guards, and good manufacturing practice. OSHA, in 29 *CFR* 1910.133, requires eye and face protection when injury can be prevented by its use. The standard practice is given by ANSI Z87.1, *Practice for Occupational and Educational Eye and Face Protection.*

For chemical splash or irritating mists, eye protection should be selected from unvented chemical goggles, indirect-vented chemical goggles, or indirect-vented eyecup goggles. Direct vented goggles and spectacle-type eye protection do not provide protection against liquid exposures and should not be used. For severe exposures, a face shield should be used in conjunction with goggles; a face shield by itself does not provide adequate protection against liquid splashes. Where both an inhalation and splash hazard exists, full-face respiratory protection is preferable over a half-mask and goggles.

Contact lenses should not be worn in eye hazard areas. Contact lenses can trap harmful substances and decrease the effectiveness of eyewashes. Soft contact (gas permeable) lenses may absorb gases and be contaminated by chemicals, leading to further injury.

## ■ Summary

Critical exposure factors include how the material is used and what controls (engineering or personal protective equipment) are in place, temperature and volatility, concentration, and reactivity. Guidelines for exposure are discussed. In this chapter, solvents are classified as aqueous or organic systems. Gases as cryogenic liquid and simple and chemical asphyxiants and their characteristics are discussed. Flammable and combustible liquids, flash points, flammable range, and requirements and guidelines are given. The physiological effects of aqueous systems, organic compounds, aliphatic hydrocarbons, cyclic hydrocarbons, aromatic hydrocarbons, halogenated hydrocarbons, nitrohydrocarbons, oxygen-containing functional groups, inorganic acids, and organic and inorganic gases as well as hazard potential, evaluation, and control are covered.

## ■ Bibliography

American Conference of Governmental Industrial Hygienists, Committee on Industrial Ventilation. *Industrial Ventilation— A Manual of Recommended Practice,* 22nd ed. Lansing, MI: ACGIH, 1995.

American Conference of Governmental Industrial Hygienists. *Threshold Limit Values and Biological Exposure Indices for 1994–95.* Cincinnati: ACGIH, 1994.

American Conference of Governmental Industrial Hygienists. *Documentation of the Threshold Limit Values and Biological Exposure Indices,* 6th ed. Cincinnati: ACGIH, 1991.

American Industrial Hygiene Association. *Hygienic Guide Series on specific materials.* Fairfax, VA: AIHA.

American Industrial Hygiene Association. *American National Standard for Laboratory Ventilation,* ANSI/AIHA Z9.5–1992. Fairfax, VA; AIHA, 1993.

American Industrial Hygiene Association. *Respiratory Protection—A Manual and Guideline,* 2nd ed. Fairfax, VA: AIHA, 1991.

American National Standards Institute. *ANSI Z87.1–1989, Practice for Occupational and Educational Eye and Face Protection* (supplement). New York: ANSI, 1991.

American National Standards Institute. *ANSI Z87.1–1989, Practice for Occupational and Educational Eye and Face Protection,* corrections. New York: ANSI, 1990.

American National Standards Institute. *ANSI Z87.1–1989, Practice for Occupational and Educational Eye and Face Protection.* New York: ANSI, 1989.

American National Standards Institute. *ANSI Z9.2–1979, Fundamentals Governing the Design and Operation of Local Exhaust Systems.* New York: ANSI, 1979.

American Society of Heating, Refrigeration and Air Conditioning Engineers. *ASHRAE Standard 15–1992, Safety Code for Mechanical Refrigeration.* Atlanta: ASHRAE, 1992.

American Society of Heating, Refrigeration and Air Conditioning Engineers. *1991 ASHRAE Handbook: Heating, Ventilating, and Air-Conditioning Applications.* Atlanta: ASHRAE, 1991.

American Society of Heating, Refrigeration and Air Conditioning Engineers. *1989 ASHRAE Handbook: Fundamentals.* Atlanta: ASHRAE, 1989.

American Society for Testing and Materials. *1994 Annual Book of ASTM Standards.* Philadelphia: ASTM, 1994.

American Society for Testing and Materials. *ASTM D3828–93, Standard Test Methods for Flash Point by Small Scale Closed Tester.* Philadelphia: ASTM, 1993.

American Society for Testing and Materials. *ASTM D56–93, Standard Test Method for Flash Point by Tag Closed Tester.* Philadelphia: ASTM, 1993.

American Society for Testing and Materials. *ASTM F739–91, Standard Test Method for Resistance of Protective Clothing Materials to Permeation by Liquids or Gases Under Conditions of Continuous Contact.* Philadelphia: ASTM, 1991.

American Society for Testing and Materials. *ASTM D92–90, Standard Test Method for Flash and Fire Points by Cleveland Open Cup.* Philadelphia: ASTM, 1990.

American Society for Testing and Materials. *ASTM D93–90, Standard Test Method for Flash Point by Pensky–Martens Closed Tester.* Philadelphia: ASTM, 1990.

American Society for Testing and Materials. *ASTM D3278–89, Standard Test Method for Flash Point of Liquids by Setaflash Closed-Cup Apparatus.* Philadelphia: ASTM, 1989.

American Society for Testing and Materials. *ASTM D1310–86, Standard Test Method for Flash Point and Fire Point of Liquids by Tag Open-Cup Apparatus.* Philadelphia: ASTM, 1986 (reapproved 1990).

Braker W, Mossman AL. *Matheson Gas Data Book,* 6th ed. Secaucus, NJ: Matheson Gas Products. 1980.

Bretherick L. *Handbook of Reactive Chemical Hazards,* 3rd ed. London: Butterworths, 1985.

Browning E. *Toxicity and Metabolism of Industrial Solvents.* New York: Elsevier.

Calm JM. Refrigerant safety: The alternative refrigerants are as safe or safer than those they replace, but more care is needed with all refrigerants. *ASHRAE J,* 1994. Reprint: TECH-R-135.

Clayton GD, Clayton FE, eds. *Patty's Industrial Hygiene and Toxicology,* 4th ed., vols. 2A–2F. New York: Wiley, 1993, 1994.

Cralley LV, Cralley LJ, eds. *Industrial Hygiene Aspects of Plant Operations,* vol. 3. New York: Macmillan, 1986.

Dean JA. *Lange's Handbook of Chemistry,* 14th ed. New York: McGraw-Hill, 1992.

Gerarde HW. *Toxicology and Biochemistry of Aromatic Hydrocarbons.* New York: Elsevier, 1960.

Gerarde HW. Toxicological studies on hydrocarbons: 111. The biochemorphology of phenylalkanes and phenylalkenes. *Arch Ind Health* 19:403, 1959.

Gleason MN, Gosslin RE, Hodge HC. *Clinical Toxicology of Commercial Products,* 5th ed. Baltimore: Williams & Wilkins, 1981.

Hamming WJ. *Photochemical Reactivity of Solvents.* Paper No. 670809 presented at the October 2–6, 1969, Aeronautic and Space Engineering and Manufacturing meeting sponsored by the Society of Automotive Engineers.

International Committee of Contamination Control Societies. *Proceedings of the 4th International Symposium on Contamination Control,* September 1978, Washington, DC. Mt. Prospect, IL: Institute of Environmental Sciences (formerly American Association for Contamination Control).

Johnson JS, Swope AD, Goydan R, et al. *Guidelines for the Selection of Chemical Protective Clothing.* Washington, DC: U.S. Department of Energy, Office of Environment, Safety, and Health, 1991.

Lehninger AL. *Biochemistry,* 2nd ed. New York: Worth Publishers, 1975.

Lewis RJ. *Hawley's Condensed Chemical Dictionary,* 12th ed. New York: Van Nostrand Reinhold, 1993.

Lide DR, Frederikse HPR, eds. *CRC Handbook of Chemistry and Physics,* 75th ed. Boca Raton, FL: CRC Press, 1994.

Lunche RG et al. L.A.'s rule 66 nips air pollution due to solvents. *SAE J* 76:25, 1968.

Matheson Gas Products. *Matheson Gases & Equipment,* 1993.

McFee D, Garrison RP. Process characteristics—Open systems. In Cralley LV, Cralley LJ, eds., *Industrial Hygiene Aspects of Plant Operations,* vol. 3. New York: Macmillan, 1986.

McFee DR. How well do gloves protect hands against solvents? *J Am Soc Safety Eng,* May 1964.

National Draeger, Inc. *Dräger Detector Tube Handbook,* 8th ed. Pittsburgh: National Draeger, 1992.

National Fire Protection Association (Batterymarch Park, Quincy, MA 02269). *Fire Protection Handbook,* latest edition.

National Fire Protection Association. *Fire Hazard Properties of Flammable Liquids, Gases, and Volatile Solids,* NFPA no. 325.

National Fire Protection Association. *National Electrical Code,* NFPA no. 70.

National Fire Protection Association. *Flammable and Combustible Liquids Code,* NFPA no. 30.

National Institute for Occupational Safety and Health (Cincinnati, OH 45226). *Certified Equipment List.* Pub. no. DHHS (NIOSH) 91–105.

National Institute for Occupational Safety and Health. *Criteria Documents* (listing available).

National Institute for Occupational Safety and Health. *Registry of Toxic Effects of Chemical Substances,* 1985–86 ed., DHHS (NIOSH) Pub. no. 87–114.

National Institute for Occupational Safety and Health. *Occupational Diseases: A Guide to Their Recognition,* Revised ed. Pub. no. DHEW (NIOSH) 77–181, 1977.

National Institute for Occupational Safety and Health. *The Industrial Environment—Its Evaluation and Control,* 1973.

National Research Council. *Rethinking the Ozone Problem in Urban and Regional Air Pollution.* Washington, DC: National Academy Press, 1991.

National Toxicology Program. *Seventh Annual Report on Carcinogens, 1994 Summary.* Research Triangle Park, NC: NTP, 1994.

National Toxicology Program. *Toxicology and Carcinogenesis of Tetrachloroethylene (Perchloroethylene) in F344/N Rats and B6C3F1 Mice (Inhalation Studies).* NTP TR 3111 DHHS (NIH) Pub no. 86–2567.

Research Triangle Park, NC: NTP, 1986.

Nelson GO et al. Glove permeation by organic solvents. *AIHAJ* 42:217–225, 1981.

Sansone EB, Tewari YB. Differences in the extent of solvent penetration through natural rubber and nitrile gloves from various manufacturers. *AIHAJ* 41:527, 1980.

Sansone EB, Tewari YB. The permeability of laboratory gloves to selected solvents. *AIHAJ* 39:169–174, 1978.

Sax NI. *Dangerous Properties of Industrial Chemicals,* 8th ed. New York: Van Nostrand Reinhold, 1992.

Urone P. The primary air pollutants—Gaseous: Their occurrence, sources, and effects. In Stern A, ed. *Air Pollution,* 3rd ed. New York: Academic Press, 1976.

Van Dolah RW et al. Flame propagation, extinguishment, and environmental effects on combustion. *Fire Technol* 1(2):138–145, 1965.

Williams JR. Permeation of glove materials by physiologically harmful chemicals. *AIHAJ* 40:877–882, 1979.

Windholz M, ed. *The Merck Index,* 10th ed. Rahway, NJ: Merck & Co., 1983.

Zabetakis MG. *Safety with Cryogenic Fluids.* New York: Plenum, 1967.

## CHAPTER 8

# Particulates

by Theodore J. Hogan, PhD, CIH

**Critical Exposure Factors**
*Type of Particulate Involved* ▪ *Length of Exposure* ▪ *Particulate Concentration* ▪ *Particle Size*

**Biological Reaction**

**Selected Particulates**
*Silica* ▪ *Asbestos* ▪ *Lead* ▪ *Beryllium* ▪ *Coal Dusts* ▪ *Miscellaneous Dusts* ▪ *Toxic Dusts and Fumes* ▪ *Welding Fumes* ▪ *Radioactive Dusts* ▪ *Bacteria and Fungi* ▪ *Allergens*

**Measurement of Particulate Exposures**
*Microscopic Count Procedures* ▪ *"Total" Mass Concentration Methods* ▪ *Respirable Mass Size-Selection Measurement (Personal Sampling)*

**Control of Particulates**
*Engineering Controls* ▪ *Administrative Controls* ▪ *Personal Protective Equipment*

**Summary**

**Bibliography**

THE AIR WE BREATHE contains particulates in the form of dust, and a portion of that dust is retained in the lungs. Inhaling too much dust can cause pneumoconiosis. *Pneumoconiosis*—a tongue-twisting term of Greek derivation—means "dusty lung."

Fume, another form of a particulate, can also cause pneumoconiosis. Fume differs from dust only in the way it is generated and in its particle size. Dust normally involves a wide range of particle sizes that are the result of some mechanical action, such as crushing or grinding. A fume consists of extremely small particles, less than a micron ($\mu$m) in diameter, and is generated by processes such as combustion, condensation, and sublimation. Mists are finely divided, suspended liquids formed via condensation or a dispersion process such as atomizing. Mists, such as the oil mist produced in cutting operations or mists from spray painting, are so light they float in the air. Fibers connote thinness and elongation, as in the case of asbestos. Thread-like fibers are separated from the base rock during crushing, cutting, and mining.

Dusts, fumes, fibers, and mists are all particulates. (For more information on the effects of inhaled particulates see Chapter 2, The Lungs.)

Some fibrosis or scarring is present in almost every lung. Because of their greater total exposure, the lungs of older people generally exhibit a greater degree of fibrosis than those of young people. The term *pneumoconiosis* is restricted to cases in which fibrosis results from an exposure to inorganic particulates. One of the first symptoms of pneumoconiosis is shortness of breath, which can have many causes. Consequently, a diagnosis of pneumoconiosis should not be made solely on the basis of shortness of breath or any other single symptom.

Except for skin diseases, most occupational diseases are caused by inhalation of materials used in a work area (Table 8–A). This is because lung tissue is by far the most efficient medium the body possesses for capturing and absorbing airborne contaminants. One reason is that the surface area of lung tissue averages 590–920 ft$^2$ (55–75 m$^2$), whereas surface area of the skin averages about 22 ft$^2$ (2 m$^2$). Consequently, particulates that are soluble in body fluids and that reach the lungs can be absorbed and eventually pass directly into the bloodstream. Other particulates that are not soluble may stay in the lungs and cause local or irritant damaging action.

Toxic and irritant particulates can also be ingested in amounts that affect health. If the toxic particulate is swallowed with food or saliva and is not soluble in body fluids, it is eliminated directly through the intestinal tract.

**Table 8–A.** Selected Toxic Dusts and Fumes

| Substance | Description and Effects | Threshold Limit in Milligrams per Cubic Meter of Air* |
|---|---|---|
| Antimony | Often associated with lead and arsenic. Hazardous from inhalation and ingestion. Soluble salts may cause dermatitis. Antimony trioxide may cause lung cancer. | 0.5 |
| Arsenic | Silvery brittle crystalline metal. Hazardous from inhalation and ingestion. Usually encountered as arsenic trioxide. Arsenic trioxide production is associated with increased cancer rates in humans. | 0.01, A1 |
| Asbestos | Fibrous mineral. Inhalation may cause lung disease including cancers. May cause mesothelioma, a rare cancer of the lining of the lung and intestines. | (Intended change: 0.2 fibers/cc, A1) (OSHA PEL: 0.1fibers/cc) |
| Barium (soluble compounds) | Soluble barium chloride and sulfide are toxic when taken orally. | 0.5 |
| Beryllium | Light weight gray metal. The metal, low-fired oxides, soluble salts, and some alloys are toxic by inhalation. Suspected carcinogen. | 0.002, A2 |
| Cadmium oxide fume | Used in some silver solders and as a metal coating. Very high acute fume exposures can be fatal. | 0.01, A2 (total dust) 0.002, A2 (respirable dust) |
| Chromic acid | Red, brown, or black crystals. Caustic action on mucous membranes or skin. Carcinogen. | 0.05, A1 |
| Cyanide (as CN) | Nonvolatile cyanides are ingestion hazards. Cyanides inhibit tissue oxidation when inhaled and cause death. | 5.0 (skin**), ceiling |
| Dinitrobenzene | Yellowish crystal. Hazardous as a result of skin absorption, inhalation, and ingestion. | 1.0 (skin**) |
| Fluorides | Inorganic fluorides are highly irritant and toxic. | 2.5 |
| Hydroquinone | Colorless hexagonal crystals. Contact with the skin may cause sensitization and irritation. Excessive exposure to dust may cause corneal injury. | 2.0 |
| Iron oxide fume | Major sources are cutting and welding. | 5.0 |
| Lead | Lead fumes and lead compounds cause poisoning after prolonged exposure. Most important means of entry into body is by inhalation. Skin absorption is of significance only from such organic compounds as lead tetraethyl. | 0.15 (Intended: 0.05, A3) |
| Lead arsenate | White crystals. Highly toxic. | 0.15 |
| Magnesium oxide fume | White powder. Inhalation of freshly generated fume may cause metal fume fever. | 10.0 |
| Manganese | Silver gray metal. Hazardous if fumes or dust are inhaled. | 5.0 (dust) (Intended change: 0.2) |
| Pentachlorophenol | Dark-colored flakes. Harmful dust. Emits toxic fumes when heated. | 0.5 (skin**) |
| Phosphorus (yellow) | Poisonous mainly on inhalation. Severe burn hazard from skin contact. | 0.1 |
| Picric acid | Yellow crystals or liquid. Explosive-particularly metallic salts. Toxic fumes on decomposition. | 0.1 |
| Selenium compounds | Toxicity varies somewhat according to the solubility of the specific compound. Often causes contact dermatitis. | 0.2 |
| Sodium Hydroxide | White, deliquescent pieces or lumps. Has severe action on all body tissue. | 2.0 ceiling |
| Tellurium | Similar to selenium chemically and in physiological effects. | 0.1 |
| Titanium dioxide | White to black powder. Considered in the nuisance category. | 10.0 |
| Trinitrotoluene | Colorless to yellow monoclinic crystals. Emits toxic fumes of oxides or nitrogen when heated to decomposition. Highly poisonous explosive. | 0.5 (skin**) |
| Uranium | Highly toxic-particularly to kidneys-and a radiation hazard that requires special consideration. Soluble and insoluble compounds as U (natural). | 0.2, 0.6 ceiling |
| Vanadium pentoxide (respirable dust and fume) | Yellow to red crystals. Acts chiefly as an irritant to the conjunctiva and respiratory tract. | 0.05 |
| Zinc oxide fume | Amorphous white or yellow powder. The powder is of low toxicity but freshly generated fume may cause metal fume fever. | 5.0, 10.0 ceiling |
| Zirconium compounds | Most compounds are insoluble and have low toxicity. | 5.0, 10.0 ceiling |

* These Threshold Limit Values were adopted by the American Conference of Governmental Industrial Hygienists in 1994–1995.
** The word *skin* in this table indicates that the substance can penetrate the skin and contribute to the exposure.
A1 = Confirmed human carcinogen as designated by the ACGIH.
A2 = Suspected human carcinogen as designated by the ACGIH.
A3 = Animal carcinogen as designated by the ACGIH.

Toxic materials that are readily soluble in body fluids, such as lead oxide particulate or lead fume, can, if swallowed, be absorbed by the digestive system and absorbed and transported by the blood. However, both the ingestion of and skin contact with particulates are of relatively minor importance in occupational disease. Inhalation is the primary route of exposure.

# Critical Exposure Factors

Health problems associated with the various kinds of particulate exposures are influenced by four critical factors:

- The type of particulate involved
- The length of exposure time (possibly in years)
- The concentration of airborne particulates in the breathing zone of those exposed
- The size of the particles present in the breathing zone

Each of these factors can be critical. However, the factors are so interrelated that each must be considered in evaluating particulate exposures. For example, an airborne particulate of a potentially toxic material will not cause pulmonary illness if its particle size is too large to gain access to the lungs, if it is present in low concentrations, or if exposure time is very short. Thus, the importance of any one factor, as it affects health, must be evaluated in terms of the other three.

When particulate exposure is measured, these four factors must be taken into account. A "total dust" measurement provides very little information about the health consequences of the exposure. However, exposure measurement techniques can be specific enough to identify the type and size of the particulate matter as well as the overall particulate concentration.

Because an understanding of the four critical factors is important to anyone concerned with the recognition, evaluation, and control of industrial particulate problems, each will be discussed in detail.

## Type of Particulate Involved

Industrial dusts can be classified in two very broad categories: organic and inorganic. Organic particulates originate from plant and animal material or from synthetic material. In general, naturally occurring organic particulates tend to produce allergic responses after acute and chronic exposures. Synthetic organic particulates can produce irritation and allergic responses as well as a whole range of local effects (such as dermatitis) or systemic toxic effects (such as liver damage), depending on the chemical involved. Inorganic particulates can be classified as metallic or nonmetallic. Nonmetallic particulates that contain silica are further classified as crystalline or amorphous. Figure 8–1 shows the types of particulates associated with each category.

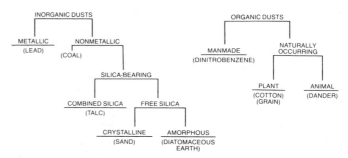

**Figure 8–1.** Classification for sampling and evaluating respirable dusts. (Adapted from Bureau of Mines Circular 8503, February 1971.)

Inorganic metallic particulates can produce local dermatitis and sensitization (such as nickel itch) and systemic toxicity, particularly to the kidneys, blood, and central nervous system (CNS) (Table 8–A). Inorganic silica-bearing particulates are normally not an acute dermatologic or systemic hazard. However, particulates containing crystalline, or "free," silica can cause pneumoconiosis as a result of chronic exposure. The silicate fibers of asbestos can cause lung scarring and cancers.

## Length of Exposure

Pneumoconioses, such as silicosis, asbestosis, and coal workers pneumoconiosis, normally become disabling only after several years of particulate exposure. Toxic metal particulates, such as lead and manganese, can cause problems after a shorter exposure time (from several days to several months) depending on how much of the toxic metal particulate is absorbed in a specified period of time. A few hours of overexposure to some metal fumes can cause metal fume fever, a transient illness that is similar to the flu. Particulates that cause allergic reactions or severe irritation can cause serious problems with only a brief exposure time at relatively low concentrations. Air sampling is needed to evaluate actual exposure times; sampling can be a simple process or it can be difficult, depending on the mobility of the person exposed and the fluctuations in exposure patterns of his or her job.

## Particulate Concentration

Another critical factor in evaluating an exposure to particulates is the actual concentration of particulate in the breathing zone. For many years, industrial hygienists have been guided by the Threshold Limit Values (TLVs) established by the American Conference of Governmental Industrial Hygienists (ACGIH) and by the permissible exposure limits (PELs) for particulates in the workplace established by the Occupational Safety and Health Act (OSHA). (Information on the collection and measurement of particles is presented later in this chapter.)

TLVs are exposure levels that should not be exceeded. The values are guidelines to help control exposures; they are not absolute measures of safety. TLVs are reviewed annually and updated when necessary. (For more information on TLVs, see Chapter 6, Industrial Toxicology, Chapter 15, Evaluation, and Appendix B.)

## Particle Size

The last factor in the evaluation of particulate exposure is the actual size of airborne particles (Figure 8–2). Only very small particles are respirable (capable of being breathed into the lungs). The industrial hygienist measures the size of airborne dust particles in microns (μm), a unit that is one thousandth of a millimeter or approximately one twenty-five

thousandth of an inch. By using a size-selective device (such as a cyclone, described later in this chapter) in front of a filter at a specific airflow sampling rate, it is possible to collect respirable-sized particles (less than 5 μm) on the filter. See Appendix D of the TLV booklet reprinted in Appendix B for more detailed discussion of particle size and respirability.

When a solid is broken into fine particles, its surface area increases many times. For example, when one solid cubic centimeter of quartz (with a surface area of 6 cm$^2$) is crushed into 1-μm cubes, one trillion particles, with a total surface area of 6 m$^2$, are yielded.

When a solid is broken into fine particles, the volume occupied by the mass is also increased because of the

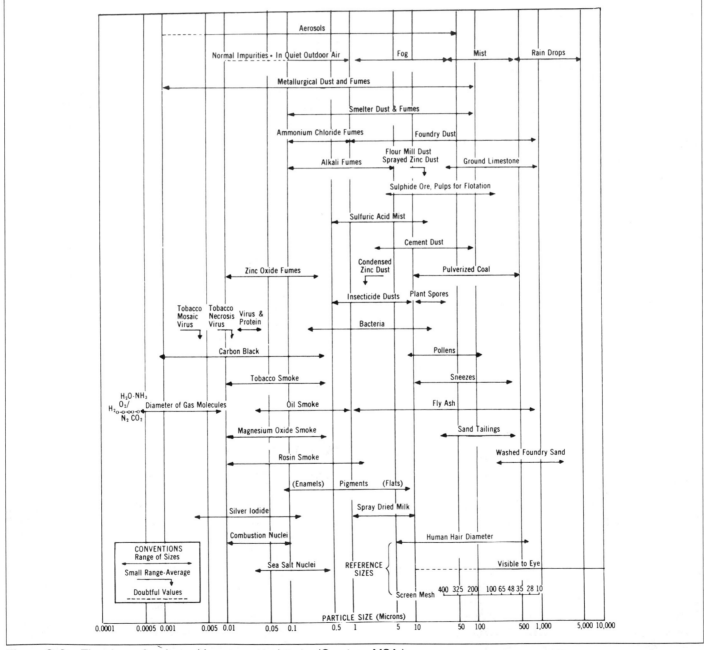

**Figure 8–2.** The sizes of various airborne contaminants. (Courtesy MSA.)

voids between the particles. A particulate concentration of 50 million particles per cubic foot (50 mppcf) of air, which results when 1 cm³ of material is reduced to particles 1 µm³ in size, occupies an air space of 560 m³.

A person with normal eyesight can detect individual dust particles as small as 50 µm in diameter. Smaller airborne particles can be detected individually by the naked eye only when strong light is reflected from them. Particulates of respirable size (usually considered to be below 5 µm) cannot be seen as individual particles without the aid of a microscope. However, high concentrations of suspended small particles may be perceived as a haze or have the appearance of smoke.

Most industrial particulates consist of particles that vary widely in size; the small particles greatly outnumber the larger ones. Consequently, when dust is noticeable in the air around an operation, more invisible dust particles than visible ones are probably present.

Particulates in the air may not have the same composition as the parent material. The factors that determine composition are particle size and density of each component in the original mixture and the hardness of the materials.

Although dust particles are, of course, subject to gravity, their settling rate through still air varies according to their size, density, and shape. Microscopically small particles settle more slowly than do larger particles because they are less dense and are influenced by Brownian movement. Mineral particles larger than 10 µm settle relatively quickly. As Table 8–B shows, respirable particles released into still air can remain airborne for many hours. Particles fall faster in the workplace due to air movement. Still, if the rate of particulate generation is constant, the particulate level (and, therefore, exposure) increases throughout the workday unless ventilation or other methods of control are used.

With the exception of such fibrous materials as asbestos, dust particles usually must be smaller than 5 µm in order to enter the alveoli or inner recesses of the lungs. Although a few particles up to 10 µm in size occasionally enter the lungs, nearly all larger particles become trapped in the nasal passages, throat, larynx, trachea, and bronchi, from which they are expectorated or swallowed into the digestive tract.

Ragweed pollen, which varies from 18 to 25 µm in diameter, can cause hay fever through its action in the upper respiratory system. This type of allergenic particulate, as well as bacterial and irritant particulates, can cause difficulty even when the airborne particles are large.

When particulate-laden air is inhaled, some of the larger particles are trapped by the hairs in the nose. Other dust particles are removed from the air as it passes over the mucous membranes of the nose, throat, and other portions of the upper respiratory system.

The bronchi and other respiratory passages are covered with a large number of tiny, hairlike cilia, which aid in the removal of particulate trapped on these surfaces. The cilia, all bending in one direction, make a fast stroke toward the mouth and a slower stroke away from the mouth. This action pushes mucus and deposited particulates upward to the mouth so that the particles can be expectorated or swallowed.

## ■ Biological Reaction

Because there are many different types of dust, fumes, and mists, the biological reaction caused by exposure to any one of them depends on the type. A reaction may include any of the following:

- Lung diseases are caused by the body's reaction to an accumulation of particulates in the lungs. These diseases include fibrosis (scar tissue formation), bronchitis (the overproduction of mucus), asthma (the constriction of the bronchial tubes), and cancer. Restriction of lung function places an additional burden on the right side of the heart, which tries to pump more blood to the lungs to maintain an adequate oxygen supply. This additional strain can cause permanent heart damage.

- Systemic reactions are caused when the blood absorbs inorganic toxic particulates of such elements as lead, manganese, cadmium, and mercury, and certain organic compounds.

- Metal fume fever results from the inhalation of finely divided and freshly generated fumes of zinc, magnesium, copper, or their oxides. The inhalation of aluminum, antimony, cadmium, copper, iron, manganese, nickel, selenium, silver, and tin have also been reported to cause metal fume fever.

- Allergic and sensitization reactions are caused by inhalation of or skin contact with such materials as organic particulates from flour and grains and some woods and particulates of a few organic and inorganic chemicals.

- Bacterial and fungus infections result from inhalation of particulates containing active organisms, such as wool or fur particulates containing anthrax spores or wood bark, or grain particulates containing parasitic fungi.

- Irritation of the nose and throat is caused by acid, alkali, or other irritating dusts or mists. Some particu-

**Table 8–B.** Estimated Settling Rates for Silica Dust in Still Air

| Size in Micrometers | Time to Fall 1 Foot (0.30 m) (minutes) |
| --- | --- |
| 0.25 | 590.0 |
| 0.50 | 187.0 |
| 1.00 | 54.0 |
| 2.00 | 14.5 |
| 5.00 | 2.5 |

lates such as soluble chromate dusts may cause ulceration of nasal passages or even lung cancer.

■ Damage to internal tissues can result from the inhalation of radioactive materials such as radium and its products and from the inhalation of other particulate radioisotopes that emit highly ionizing radiation.

# ■ Selected Particulates

## Silica

The term *silicon dioxide* usually refers to amorphous silica (noncrystalline), crystallized silica such as sand (quartz), and silicates such as clay (aluminum silicate). Only the crystalline (free silica) material found in quartz, tridymite, cristobalite, and a few other nonsilicate materials causes silicosis. The crystalline structure in tridymite and cristobalite differs from that in quartz. For that reason, tridymite and cristobalite are more potent than quartz in causing silicosis. Uncombined or free silica (quartz) is the most significant factor in industrial dust exposure.

Silicosis is a lung disease caused by the inhalation of free silica particulates. The risk of silicosis is present in industries and occupations where the crystalline form of free silica particulates is found, as in foundries, glass manufacturing facilities, granite-cutting operations, and mining and tunneling sites in quartz rock. Because of the presence of free silica in many materials, silicosis is found throughout the world, and in the past it has had many names such as miner's asthma, grinder's consumption, miner's phthisis, potter's rot, and stonemason's disease. All these names, however, describe the same disease, caused by particulates from the crystalline form of free silica, usually quartz.

Crystallized silicon dioxide ($SiO_2$) most commonly occurs as sand, but it is also widely distributed in hard rocks and minerals. The percentage of crystalline $SiO_2$ in a dust mixture is the usual basis for evaluating the hazards associated with breathing in the mixture (Table 8–C).

**Action of silica on the lungs.** Over the years, many theories have been advanced to explain why the crystalline form of free silica acts as it does in the lungs. These theories have been based on the hardness of the material and the effect of sharp edges, solubility phenomena, electrochemical action of the crystals, and immunological reactions. Free silica particles have a toxic effect on macrophage cells, which try to engulf and remove foreign matter from the lungs. Free silica causes macrophage cells to release free radicals and enzymes into the lung tissue, which can induce fibrosis.

Some particulates move out of the air spaces into other portions of the lung and, at the several points in the lung where silica particulates are deposited and accumulate, a fibrous tissue develops and grows around the particle,

**Table 8–C.** Crystalline Silicon Dioxide in Various Materials

| Material | Normal Range Percent $SiO_2$ |
|---|---|
| Foundry molding sand | 50–90 |
| Pottery ware body | 15–25 |
| Brick and tile compositions | 10–35 |
| Buffing wheel dressings | 0–60 |
| Road rock | 0–80 |
| Limestone (agricultural) | 0–3 |
| Feldspar | 12–25 |
| Clay | 0–40 |
| Mica | 0–10 |
| Talc | 0–5 |
| Slate and shale | 5–15 |

forming a silicotic nodule. This fibrous tissue is not as elastic as normal lung tissue and does not permit the ready passage of oxygen and carbon dioxide. As it proliferates, the fibrous tissue reduces the amount of normal lung tissue. As a result, the available functional volume of the lung is reduced.

In some advanced cases, the fibrous tissue slows down or even prevents the diffusion of oxygen from the lung to the blood in the capillaries, and the blood in the area is not completely oxygenated. The fibrous tissue can also obliterate the blood vessels or reduce the flow of blood so that ultimately the lungs do not readily oxygenate sufficient blood for the body's needs. When the body's oxygen demand is increased by exertion, the individual experiences shortness of breath.

In very severe cases, fibrous tissue can so hinder the flow of blood in vessels of the lung that the heart enlarges in an effort to pump more blood. Serious enlargement of the heart is called cor pulmonale. Death can result from the cardiopulmonary effects of chronic silicosis.

**Diagnosis of disease.** This subsection is taken from the National Institute for Occupational Safety and Health (NIOSH) book *Occupational Respiratory Diseases* (Publication no. DHHS (NIOSH) 86–102).

In the presence of an adequate occupational history revealing work exposure to silica, the diagnosis of simple silicosis is usually straightforward. Most clinicians rely on roentgenographic changes for the diagnosis. Although it is often argued that changes on chest x-rays precede other clinical findings, there is now some reason to doubt this argument. In fact, mild restrictive disease detected by pulmonary function may occur in workers with early x-ray manifestations. The clinician must remember that a forced vital capacity (FVC) of $\geq 80$ percent of the predicted value is considered normal. On the other hand, if there are 100 workers who have early x-ray evidence of silicosis whose average FVC is 90 percent of predicted, this is an abnormal population. Clinicians are usually at a disadvantage when they see individuals one at a time, and they often have no baseline chest x-ray or pulmonary function test results for comparing current clinical findings.

In a person with exposure to silica, a mild restrictive defect suggests early silicosis; likewise, small round opacities on the chest x-ray support the diagnosis. With minimal loss of pulmonary function or minimal chest x-ray abnormality, symptoms such as shortness of breath are unlikely to be a clinical feature unless associated with underlying chronic airways disease.

For complicated silicosis, progressive massive fibrosis (PMF), or Caplan's syndrome, the diagnosis is more difficult. The possibility of a lung tumor or tuberculosis must be considered. Bacteriological testing of sputum usually reveals mycobacterium tuberculosis. A lung biopsy may be necessary to diagnose carcinoma.

Abnormalities of diffusing capacity are not common or profound in early silicosis. Similarly, clubbing and physical signs in the chest do not rule out silicosis, but they suggest other diseases. Hyperinflation, reduced breath sounds, prolonged expiratory phase, and reduced expansion of the chest are among the most common physical findings in advanced silicosis.

Acute silicosis should be suspected in a worker with massive exposure to silica (such as an unprotected sandblaster). Mycobacterial infection occurs in about one-quarter of these cases. Progressive shortness of breath is a common symptom. Weakness, weight loss, diffuse rales, and even cyanosis can be seen. Usually there is evidence of massive disease on chest x-ray, with the diaphragm often being high. Pulmonary function is severely compromised.

### Action of silicates.

Silicates contain silicon and oxygen combined with other elements to form a complex molecule. Analyses of minerals are sometimes reported as percentages of oxides, which may include silicon dioxide ($SiO_2$), aluminum oxide ($Al_2O_3$), potassium oxide ($K_2O$), and ferric oxide ($Fe_2O_3$). The silicon dioxide reported in such chemical analyses is the total of both the free silica (if present) and the silica present in the mineral. Such analyses are not reliable indications of the silicosis potential of the mineral.

To properly evaluate an exposure, the percentage of uncombined silica must be determined using x-ray diffraction analyses or other special analytical chemical procedures.

With the exception of asbestos and some talcs, the silicate dusts do not ordinarily cause serious disabling lung conditions such as those produced by free silica. Much higher levels of silicate dusts can be tolerated. In many industries, people have worked with silicate dusts that contained no free silica and did not develop a disability or nodulation in the lungs. An x-ray may show shadows indicating dust deposits in the lungs.

Disabling pneumoconioses due to exposure to abnormally high concentrations of mica, tremolite talc, and kaolin dusts have been described in the literature. The clinical signs for these silicate dusts are not the same as they are for free silica. The body does not have an adequate defense against large quantities of dust; consequently, although specific symptoms have not been described for many mineral dusts, the general experience indicates that dust levels should be kept below the most current TLVs.

### PELs and TLVs.

The TLVs for chemical substances and physical agents in the workroom are guides that have been adopted by the ACGIH for use in the control of occupational hazards. (The 1994–1995 TLVs are given in Appendix B, Tables of Threshold Limit Values.)

OSHA establishes permissible exposure limits (PELs) for materials. The OSHA PEL for dusts containing crystalline silica (quartz) is based on a formula that takes into account the percentage of silica present in the sample. The formula found in Table Z–3 of 29 *CFR* 1910.1000 is as follows:

$$\text{PEL (mg/m}^3), \text{respirable dust} = \frac{10 \text{ mg/m}^3}{(\%SiO_2 + 2)}$$

For example, to calculate the PEL for a sample of respirable dust containing 5.5 percent $SiO_2$ (quartz), substitute 5.5 percent in the formula.

$$\text{PEL} = \frac{10 \text{ mg/m}^3}{5.5 + 2}$$

$$\text{PEL} = 1.3 \text{ mg/m}^3$$

Therefore, measured levels of respirable particulate in the air that are above 1.3 mg/m$^3$ exceed the PEL for this particular sample containing 5.5 percent respirable silica.

The formula is based on the collection of dust by size-selective sampling devices. These instruments collect a fraction of the dust that is capable of penetrating to the gas exchange portion of the lung, where long-term retention of dust occurs. The concentration of airborne free silica in this fraction corresponds closely to the degree of health hazard. A constant is added to the denominator to prevent excessively high respirable dust concentrations when the fraction of free silica in the dust is low. The constant 2 limits the concentration of respirable dust with less than 1 percent free silica to 5 mg/m$^3$.

The TLV time-weighted average (TWA) for crystalline silica is 0.1 mg/m$^3$ of the respirable quartz particulate in the air. The formula for the silica TLV used to be the same as the standard OSHA formula. The common practice of analyzing particulate samples for the mass and percentage of free silica has eliminated the need for the TLV formula. The PEL established by OSHA and the TLV established by the ACGIH are virtually the same. This can be seen by substituting 100 percent in the PEL formula. This yields a PEL of 0.1 mg/m$^3$ for respirable particulate composed of 100 percent quartz. Because industrial hygiene laboratories are equipped to perform free silica analyses and these tests are commonly performed on each sample collected, the TLV for pure respirable free silica has been established at 0.1 mg/m$^3$, regardless of the total dust concentration.

In addition to quartz, other forms of free silica have been assigned a specific TLV. These values are based on experimental data or on industrial experience that indicated a need for individual identification.

- Cristobalite (above 5 percent): This free silica was originally listed in 1960 with a TLV of 5 mppcf, based on studies in the diatomite industry, analogy with the TLV for silica, and experimental studies using animals. In 1968, the TLV was reduced to one half the value obtained from either the count or mass formula for quartz, following a review of existing documentation and information produced by the TLV committee. This information suggested that the limit of 5 mppcf for cristobalite did not allow a sufficient safety factor for the prevention of pneumoconiosis. The current TLV for respirable cristobalite is 0.05 mg/m$^3$.

- Tridymite: This was also assigned half the quartz value based on animal toxicity data. When tridymite dust was administered by intratracheal injection into the lungs of rats, evidence indicated that it was a more active form of free silica than quartz. Analogy was also made with cristobalite. The current TLV for respirable particles of tridymite is 0.05 mg/m$^3$.

- Fused silica dust: Although insufficient industrial experience was available to assess the degree of hazard presented by fused silica dust, the same limit as that required by the quartz formulae was adopted in 1969. Intratracheal injection studies with rats indicated that fused silica was considerably less active than quartz. Respirable fused silica has a TLV of 0.1 mg/m$^3$.

- Tripoli and silica flour: These were added to the TLV list in 1972, with the recommendation that the standard for these materials be derived using the respirable mass formula for quartz. Documentation for inclusion of tripoli on the list came from study by McCord et al. (1943), in which tissue proliferation was induced by direct intraperitoneal implantation of tripoli dust in rats and guinea pigs. The tissue proliferation was similar to that produced by quartz. Silica flour was included on the list as a result of a study that showed that silica flour has a significant fibrogenic potential because of its fine particle size. The TLV here is 0.1 mg/m$^3$ for respirable dust.

## Asbestos

Another kind of pneumoconiosis, which involves specific lung changes, is asbestosis, caused by the inhalation of asbestos fibers.

*Asbestos* is a generic term used to describe a number of naturally occurring, fibrous, hydrated mineral silicates that differ in chemical composition. These may be divided into two mineral groups: pyroxenes and amphiboles. Pyroxenes, which include chrysotile (3MgO · SiO$_2$ · 2H$_2$O), are the most widely used in U.S. industry. Amphiboles include amosite ([FeMg]SiO$_3$), crocidolite (NaFe[SiO$_3$]2H$_2$O), tremolite (Ca$_2$Mg$_5$Si$_8$O$_{22}$[OH]$_2$), anthophyllite ([MgFe]$_7$Si$_8$O$_{22}$), and actinolite (CaO · 3[MgFe]O · SiO$_2$).

Asbestos fibers generally have high tensile strength, flexibility, heat and chemical resistance, and favorable frictional properties. Certain grades of asbestos can be carded, spun, and woven; others can be pressed to form paper or used for structural reinforcement of materials such as cement, plastic, and asphalt.

- Chrysolite (white asbestos) is the fibrous form of the mineral serpentine. It is the most common variety of asbestos and is widely distributed geographically. The largest deposits are in Canada, Russia, and Rhodesia.

- Crocidolite (blue asbestos) is another important, although more specialized, form of asbestos. It is the fibrous form of riebeckite, and has fine, resilient fibers of a characteristic blue color.

- Amosite is the fibrous variety of the mineral grunerite, a ferrous magnesium silicate mined only in South Africa. Amosite can be readily broken down into long, somewhat harsh fibers that range in color from a brownish yellow to almost white, depending on the quality.

- Anthophyllite is magnesium silicate of somewhat variable composition that has rather fragile brownish or off-white fibers.

- Tremolite, a calcium magnesium silicate, is sometimes used in the production of industrial and commercial talc.

- Actinolite, a calcium magnesium iron silicate, is rarely used in industry.

**Effects of asbestos exposure.** Asbestos in its several commercial forms is associated with the development of several disease entities. Asbestosis is a diffuse, interstitial, nonmalignant scarring of the lungs. Bronchogenic carcinoma is a malignancy of the lining of the lung's air passages. Mesothelioma is a diffuse malignancy of the lining of the chest cavity (pleural mesothelioma) or the lining of the abdomen (peritoneal mesothelioma).

The link between asbestos and cancer of the stomach, colon, and rectum has not been well-established. In its advanced stages, asbestos exposure is manifested by its characteristic appearance on x-ray films, by restrictive pulmonary function, or by clinical signs, such as finger clubbing and rales (dry, cracking sounds within the lung). Its most important symptom is dyspnea, or undue shortness of breath. The disease resulting from asbestos exposure can be progressive, even in the absence of further exposure; the inhaled fibers trapped in the lung continue their biological action. In severe cases, death results from the inability of the body to obtain enough oxygen or from the heart's failure to pump blood through the scarred lungs.

Mesotheliomas are diffuse and spread rapidly through the cavity of origin. They cannot be cured by any known treatment, including chemotherapy, radiation, or surgery. Death usually results within a year of diagnosis. In the general population, mesothelioma is very rare. It may account for one death in several hundred thousand in the absence of an environmental or occupational asbestos exposure.

Once established, the other asbestos-associated cancers differ little from those that occur in the general population, although there may be variations in the location of the primary site. Appropriate treatment and prognosis follow the usual pattern of the particular tumor.

Lung cancer, pleural mesothelioma, and peritoneal mesothelioma usually do not become clinically evident until more than 20 years after onset of exposure. This time lag is now widely recognized. Although some of these cancers can appear during the second decade following onset of occupational exposure, peak incidence is often not noted until 30 or more years later. This is true for regular, long-term, brief, or intermittent exposures. Variations in the time of occurrence depend on the intensity and duration of exposure (with heavier exposure often associated with shorter latency periods), but variations among individual cases make it impossible to predict the latency period for any particular worker.

**OSHA asbestos regulations.** Because of its severe health consequences, exposure to asbestos has been regulated by OSHA since 1971. A major revision to the asbestos regulations for general industry (29 *CFR* 1910.1001) and the construction industry (29 *CFR* 1910.1101) was published August 10, 1994, along with a new standard for the shipyard industry (29 *CFR* 1915.1001). The regulations have become more comprehensive and more stringent as data on occupational exposure have been gathered and evaluated. Interpretation and enforcement of the regulations vary. This overview of the general industry and construction standards was prepared by Sandra L. Mattson, licensed architect and asbestos project designer and president of Mattson Associates, Ltd., Downers Grove, IL.

The basis for much of the current asbestos awareness is the Asbestos-Containing Materials in Schools Rule (40 *CFR* 763) published by the EPA on October 30, 1987, in response to the Asbestos Hazard Emergency Response Act (AHERA). This document established standards for training and accreditation of asbestos professionals, protocols for building inspections for asbestos-containing materials, requirements for response actions to asbestos hazards, and clearance tests to determine completions of asbestos abatement activities. The provisions of this rule have become the industry standards. The subsequent Asbestos Schools Hazard Abatement Reauthorization Act (ASHARA) (1990) law

and Model Accreditation Plan (1994) increased training requirements and extended many of the provisions to public and commercial building. Portions of these rules have been incorporated into the OSHA regulations by reference.

**Applicability of regulations.** The OSHA general industry standard applies to all occupational exposures to asbestos except for construction and shipbuilding. Specifically included are workers involved in the primary or secondary manufacture of asbestos products, those engaged in brake and clutch repair, and those involved in care of asbestos-containing flooring. In addition, employers and building owners are made responsible for exposures to installed asbestos-containing materials (ACM). Thermal system insulation (TSI) and sprayed-on or troweled-on surfacing materials are presumed asbestos containing materials (PACM). Resilient flooring materials installed no later than 1980 must also be treated as ACM. The identification of ACM and PACM is made by an industrial hygienist or other qualified person and the PACM designation can be rebutted only by having an inspection conducted by an EPA-accredited inspector or Certified Industrial Hygienist (CIH) according to the requirements of AHERA, which include sampling and testing of the materials. The owner/employer is responsible for informing workers who perform housekeeping activities in areas where ACM/PACM is located and to provide asbestos awareness training annually. Specific requirements are given for the maintenance of ACM flooring. Low-abrasion pads, speeds lower than 300 rpm, and wet methods must be used for floor stripping, and floors cannot be sanded. For brake and clutch activities, the employer must institute engineering controls and work practices using either a negative-pressure/high-efficiency particulate aerosol (HEPA) vacuum method, a low-pressure/wet cleaning method, or the equivalent.

The OSHA construction industry standard regulates the following types of work in structures where asbestos is present: demolition or salvage; removal or encapsulation of ACM; construction, alteration, repair, maintenance, or renovation; installation of products containing asbestos; asbestos spill/emergency cleanup; and transportation, disposal, storage, containment, and housekeeping at construction sites. Before work subject to the standard is begun, building owners must identify the presence, location, and quantity of ACM/PACM at the worksite. Asbestos work is divided into four categories, each with specific requirements. Class I work is the removal of TSI and surfacing ACM/PACM. Class II includes the removal of other ACM such as wallboard, resilient flooring, roofing, siding, and construction mastics. Class III is repair and maintenance work where TSI and surfacing material are likely to be disturbed. Class IV means maintenance and custodial activities in contact with ACM.

On multiemployer worksites, the employer performing the asbestos abatement work is responsible for informing the other employers about such work and the protective measures required and for abating any asbestos contamination hazards. All employers are responsible for protecting their own employees by daily ascertaining the integrity of work area enclosures and effectiveness of control methods. In addition, all general contractors are deemed to exercise supervisory control over the abatement work and must determine whether the asbestos contractor is in compliance with the standard.

**Exposure monitoring.** The essential requirement of these regulations is to ensure that no employee is exposed to airborne asbestos fibers in excess of the PEL, as follows:

- 0.1 fiber/cc of air, 8-hour TWA
- 1.0 fiber/cc of air, 30-minute short-term excursion limit (STEL)

Monitoring is performed and evaluated according to the procedures specified in Appendix A of the standard, with all samples taken from the worker's breathing zone.

The general industry employer must determine when workers may reasonably be expected to be exposed to concentrations at or above the PEL. If there has been no previous monitoring of workers performing the activity in question and if there are no objective data indicating that the asbestos is not capable of being released in concentrations exceeding the PEL, then initial monitoring must be performed. Additional monitoring must be performed when changes in the work practices, equipment, or personnel could result in exposures above the PEL and monitoring must be performed periodically in order to accurately represent the levels of exposure.

The OSHA construction standard establishes criteria for a negative exposure assessment to demonstrate that employee exposure is expected to be consistently below the PEL during an operation. This includes the use of monitoring data obtained during work operations closely resembling the process in question within the past 12 months. Daily monitoring is required for Class I and Class II operations unless a negative exposure assessment has been made for the entire operation. Monitoring is not required when all employees are equipped with supplied air respirators operated in the positive pressure mode, except when employees are performing Class I work using control methods other than those listed in the regulation. Periodic monitoring must be conducted for Class III and Class IV work. Monitoring must always be sufficient to demonstrate that the PEL has not been exceeded and employees must be given the results of the monitoring and an opportunity to observe the process.

**Training.** *Competent person* means one who is capable of identifying asbestos hazards in the workplace and selecting the appropriate control strategy for asbestos exposure, and who has the authority to take prompt corrective measures. For Class I and II work, the competent person must have completed a training course that meets the criteria of the EPA Model Accreditation Plan for Project Designer or Supervisor and, for Class III and Class IV work, training in a 16-hour operations and maintenance course developed by the EPA.

Workers performing Class I and Class II work must have completed the equivalent of an EPA-approved asbestos worker training course. Those who perform Class II work involving only the removal of one category of ACM, such as roofing, must be trained in the specific work practices and engineering controls related to their work as well as in basic information about recognizing asbestos, health effects, the relationship between asbestos and smoking in producing lung cancer, and pertinent information about controls, personal protective equipment, hygiene facilities, decontamination, emergency procedures, and waste disposal. Training for employees performing Class III operations must be equivalent to the EPA 16-hour operations and maintenance course. Class IV operations require only the equivalent of an EPA 2-hour awareness training course for custodial workers.

**Medical surveillance.** The employer must provide medical monitoring for all employees engaged in Class I, II, and III work for 30 or more days per year, those who are exposed at or above the PEL or STEL, and those who wear negative pressure respirators. Examination must be performed annually and must include work history, initial and periodic chest x-ray, pulmonary function test, and physician's examination directed to the pulmonary and gastrointestinal systems.

**Personal protection.** The employer must provide respirators and ensure that they are used in the following circumstances:

- During all Class I asbestos jobs
- During all Class II work where the ACM is not removed in a substantially intact state
- During all Class II and III work that is not performed using wet methods
- During all Class II and III asbestos jobs where the employer does not have a negative exposure assessment
- During all Class II jobs where TSI or surfacing ACM or PACM is being disturbed
- During all Class IV work performed in regulated areas where other employees are required to wear respirators
- During all work where employees are exposed above the PEL or STEL
- In emergencies

Selection of appropriate respirators is based on airborne concentration of asbestos fibers or on conditions of use.

- A half-mask air-purifying respirator with HEPA filters shall be provided for Class II and III jobs where there is no negative exposure assessment and Class III jobs where TSI or surfacing ACM/PACM is being disturbed.

- A powered air-purifying respirator must be provided in place of a negative pressure respirator where an employee chooses to use it.

- A full-facepiece supplied-air respirator operated in pressure demand mode and equipped with auxiliary positive-pressure self-contained breathing apparatus is required for all employees in the regulated area where Class I work is being performed for which a negative exposure assessment has not been produced.

The employer must provide and require the use of coveralls, head coverings, gloves, and foot coverings for employees exposed to airborne concentrations that exceed the PEL or STEL or for which a required negative exposure assessment has not been produced and for any employee performing Class I operations that involve the removal of over 25 linear ft or 10 ft$^2$ of TSI or surfacing ACM. The competent person must examine the worksuits at least once per shift for rips or tears, which must immediately be mended. Contaminated clothing must be treated as ACM.

A three-chambered decontamination facility connected to the regulated area and consisting of an equipment room, shower, and clean room must be provided for employees performing Class I asbestos jobs involving over 25 linear ft or 10 ft$^2$ of TSI or surfacing ACM. For smaller Class I jobs and for Class II, III, and IV work where exposures exceed the PEL or where there is no negative exposure assessment, the employer must establish an equipment room adjacent to the regulated area for the decontamination of employees and their equipment. This room must consist of an area covered by an impermeable dropcloth on the floor.

**Engineering controls.** Regulated areas must be established for all Class I, II, and III work and for all other work in which the PEL is exceeded. Class I and II work requires critical barriers and all classes require signage and barrier tape. Heating, ventilation, and air conditioning (HVAC) systems must be isolated in the regulated area by sealing with a double layer of 8-mil-thick polyethylene; impermeable dropcloths must be placed on surfaces beneath all removal activity. Access must be limited to authorized personnel and the work in regulated areas must be supervised by a competent person. Smoking, eating, drinking, chewing gum, and applying cosmetics are prohibited in the area and workers must decontaminate themselves before leaving.

Negative pressure enclosures (NPEs) are required for all Class I work where the layout does not make it infeasible. The NPE must have a minimum of four air changes per hour and a negative pressure of at least 0.02 in. of water measured with a manometer. The NPE must be smoke-tested at the beginning of each shift and all electrical circuits must be deactivated or equipped with ground fault circuit interrupters. Negative pressure glovebag and glovebox systems are considered NPEs. HEPA-filtered exhaust is used to move air away from the breathing zone of employees for Class I jobs where the PEL is exceeded.

**Work practices.** Prohibited activities include the following:

- Sanding
- High-speed abrasive disk saws that are not equipped with point-of-cut HEPA-filtered exhaust
- Compressed air used to remove asbestos unless in conjunction with an enclosed ventilation system
- Dry sweeping, shoveling, or other dry cleanup of dust and debris
- Employee rotation as a means of reducing exposure

For all classes of work, wet methods are required except where they can be proven to be infeasible or a safety hazard. All waste and debris must be cleaned up promptly and placed in leakproof containers. HEPA-filtered vacuums are used to remove all loose dust in the area.

Glovebags may be used for the removal of TSI or other material that can be totally enclosed within the bag. They must be smoke-tested for leaks and any leaks must be sealed before use. Glovebags may be used only once and they may not be moved (Figure 8–3).

Resilient flooring must be removed using wet methods, unless heat is being used. Sheet flooring must be wet during delamination and not ripped up. All scraping of adhesive and residual backing must be done using wet methods. Tiles must be removed intact unless it is demonstrated that this is not possible. Mechanical chipping may be performed only in an NPE.

Roofing material must be removed intact using wet methods to the extent feasible. Cutting machines must be continuously misted and any dust must be immediately cleaned with a HEPA vacuum. Removed roofing shall be bagged, wrapped in plastic sheeting, or immediately lowered to the ground via a dust-tight chute, and placed in a closed receptacle. Roof-level HVAC air intakes must be isolated while the work is being performed.

Asbestos cement siding, shingles, and panels must not be cut, abraded, or broken unless the employer can demonstrate that methods less likely to result in fiber release cannot be used. The material must be sprayed with amended water before removal and must be bagged, wrapped in plastic sheeting, or immediately lowered to the ground via a dust-tight chute, and placed in a closed receptacle.

When planning to do work covered under this regulation, please refer to the full text of the standard, including appendices, for more detail on the requirements.

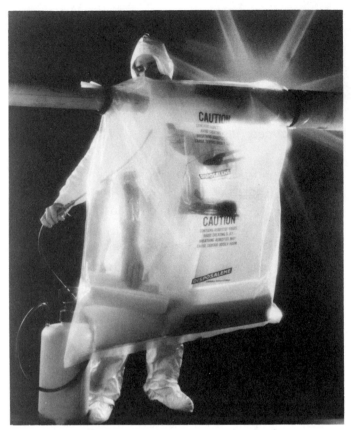

**Figure 8–3.** Disposable glove bags can be used to remove thermal system insulation or other asbestos-containing materials that can be totally enclosed in the bag. (Courtesy of Omni Sales and Manufacturing.)

## Lead

Exposure to lead dust and fume can present a severe hazard. Unlike many other metals that are necessary nutrients at low concentrations in food, lead serves no biological function. Inhaled or, to a lesser extent, ingested, lead is readily absorbed and distributed throughout the body. Repeated exposures can cause a gradual accumulation of lead, particularly in the bones. Symptoms of chronic overexposure include anxiety, weakness, headaches, tremors, excessive tiredness, and other indicators of nervous system damage. Anemia, kidney damage, and reproductive defects in both men and women (such as sterility, miscarriages, and birth defects) can also be caused by lead.

OSHA has developed an exposure standard for lead (29 *CFR* 1910.1025). As one of OSHA's completed standards, it not only specifies an 8-hour PEL for lead (50 μg/cu m), but also establishes requirements for air sampling, medical monitoring, respiratory protection, protective clothing, engineering controls, employee training, and hygiene facilities and practices. If there is a potential for exposure, the OSHA lead standard should be consulted. It provides an excellent summary of the potential hazards and preventive measures. OSHA also has adopted a lead in construction standard, *CFR* 1926.62.

**Effects of exposure to lead.** Both short- and long-term exposure to lead can have effects. Acute large doses lead to systemic poisoning or seizures. Structural brain changes and disease can develop from both kinds of exposure.

Chronic overexposure damages the blood-forming bone marrow and the urinary, reproductive, and nervous systems. Severe damage to the CNS and brain are likely. Structural damage to the brain, or encephalopathy, can lead to memory loss, irritability, convulsions, and coma.

Kidney and urinary system damage can progress without symptoms until damage is severe. Loss of kidney function can result. Lead affects the reproductive system and results in impotence, sterility, and decreased fertility. Birth defects and mental retardation in offspring can occur. Anemia can occur as lead builds up in and impedes the blood-forming system in bone marrow. Also, lead is a suspected human carcinogen.

**Biological evaluation.** A blood lead level measures the amount of lead absorbed into a person's bloodstream. There is no apparent safe blood level of exposure. Blood testing gives only an indication, however, because lead is stored in bones and tissue. At 40 μg of lead per dl blood, the risk of damage to peripheral nerves, brain, kidneys, reproductive organs, and bone marrow becomes more likely.

**Lead regulations and removal.** OSHA's general industry exposure standard (29 *CFR* 1910.1025) applies to inorganic lead and organic lead soaps. Construction work (including installation, maintenance, and demolition) involving lead is covered by OSHA Standard 1926.62. The standards differ in their requirements. Both should be consulted to determine whether both are applicable in the same workplace. The general industry standard requires initial monitoring of worksites by collecting personal samples for each shift and job. Based on this monitoring, the employer continues monitoring if exposure is above the PEL or if a change takes place that could increase exposure.

Engineering and administrative controls and a written compliance plan are required to help reduce exposures above the PEL to as low a level as is feasible. Where controls cannot sufficiently cut down exposures, the employer must provide respiratory protection at the appropriate level.

At exposures above the PEL, full-body suits, gloves, hats, shoes, and appropriate face shields must be provided. Procedures for removing and cleaning contaminated clothing and for the setting up of change rooms, showers, and lunchrooms are included.

Requirements for lead and zinc blood-level testing—every six months for employees exposed above the action level for more than 30 days each year, and more frequently as exposures increase—are listed. Medical exams must be conducted annually. A schedule for temporarily removing an employee who has elevated lead blood levels, and the return schedule, are set out.

Employees must be trained on the OSHA standard, operations that lead to exposure, the use of respirators, the medical surveillance program, controls used, and informational materials once a year.

Warning signs that read "Warning/Lead Work Area/ No Smoking or Eating" must be posted in areas where exposure is above the PEL. Records of monitoring, medical surveillance, and medical removal must be kept and be available on request for 40 years or the length of employment plus 20 years, whichever is longer. Under the standard, employees have the right to observe monitoring.

## Beryllium

Beryllium intoxication is a severe systemic disease that results from the inhalation of dust or fumes from metallic beryllium, beryllium oxide, or soluble beryllium compounds.

There are two forms of the disease. One is an acute form of chemical pneumonitis with cough, pain, difficulty in breathing, cyanosis, and weight loss. The other is the chronic type, known as berylliosis, in which there may be loss of appetite and weight, weakness, cough, extreme difficulty in breathing, cyanosis, and cardiac failure. Formerly, mortality was high in cases of chronic beryllium intoxication, and many of those who survived suffered from pulmonary distress.

Beryllium intoxication has never been found in people who mine or handle ore only; there is also no evidence of intoxication as a result of the ingestion of beryllium oxide, beryllium metal, or any of the beryllium alloys. Only the inhalation of beryllium-bearing dusts or fumes produces the systemic disease. Accordingly, control of such dusts and fumes at or below the TLVs should be recognized as a basic protective measure.

When the soluble salts of beryllium, especially beryllium fluoride, come in contact with cuts or abrasions on the skin, deep ulcers can form that heal very slowly. Complete surgical excision of the ulcer is sometimes required before the wound can heal. Hypersensitivity (allergic-like) reactions of the skin can also occur.

### Chronic beryllium disease (berylliosis).

The clinical nature of chronic beryllium disease differs from acute pneumoconiosis in that the former often develops several years after exposure. A number of case histories have revealed a delay of 5 to 10 years between the last beryllium exposure and the appearance of detectable evidence of disease. In some cases, a delay of 20 years or more has occurred. Furthermore, chronic illness has been characterized as a systemic disease that is prolonged in duration and commonly progressive in severity despite cessation of exposure.

Chronic beryllium disease results from inhalation of beryllium particulates. This disease is characterized by granulomas in the lungs, skin, and other organs. Symptoms include cough, chest pain, and general weakness.

Pulmonary dysfunction and systemic effects such as heart enlargement (cor pulmonale, leading to cardiac failure), enlargements of the liver and spleen, cyanosis, and the appearance of kidney stones also characterize the chronic illness. Beryllium has also been shown to cause lung cancer in rats and rhesus monkeys.

The present OSHA standard prescribes an 8-hour TWA of 2.0 $\mu g/m^3$ with a ceiling concentration of 5.0 $\mu g/m^3$. In addition, the present standard allows a peak concentration above the acceptable ceiling concentration for an 8-hour shift of 25 $\mu g/m^3$, for a maximum duration of 30 minutes. The 1994–1995 TLV for beryllium is an 8-hour TWA of 2.0 $\mu g/m^3$. Beryllium has been classified as an industrial substance suspected of carcinogenic potential in humans.

## Coal Dusts

*Black lung* is the name given to all lung diseases associated with chronic overexposure to coal dust. These diseases include chronic bronchitis, silicosis, and coal workers' pneumoconiosis. A coal miner may have all three diseases at the same time. The exact cause of chronic bronchitis among coal workers is unknown, although cigarette smoking may be an aggravating factor. Coal dust (particularly from anthracite coal) can contain free silica, and exposure to such coal dust can cause silicosis.

Exposure to coal dust that does not contain free silica was once thought to be harmless. Since the 1940s, however, it has been shown that even coal dust with minimal free silica content can cause the fibrotic lung disease known as coal workers' pneumoconiosis (CWP). The disease mechanism of CWP is not well-understood and the symptoms are hard to distinguish from other lung diseases. People with early stages of CWP may have no symptoms that can be directly related to fibrotic changes. Instead, any respiratory deficiencies are often due to the chronic bronchitis associated with coal dust exposure.

Varying degrees of fibrotic changes and symptoms can occur with CWP. In a small number of those exposed, CWP can develop from simple fibrosis to progressive massive fibrosis, a condition in which small, discrete fibrotic nodules conglomerate, resulting in a severe restriction of lung capacity.

Exposure to coal dust in mines is regulated by the Mine Safety and Health Administration (MSHA). The MSHA standard for respirable coal dust is 2 $mg/m^3$ over an 8-hour work shift. Black lung in miners is a federally compensable occupational disease under the 1969 Federal Coal Mine Safety and Health Act.

Coal dust exposure away from mining operations is regulated by OSHA. The OSHA PEL for coal dust that contains less than 5 percent free silica is 2.4 mg of respirable dust per cubic meter of air. If the free silica content of coal amounts to more than 5 percent, the coal dust limit becomes the same as for crystalline silica.

## Miscellaneous Dusts

Limestone, marble, lime, gypsum, and portland cement dusts apparently have minimal effects even after long exposures. These materials have little free silica. Also, many silicates and other minerals have not caused impairment in people who have inhaled the dusts, and the resulting pneumoconioses have generally been classed as benign. These and many other dusts used to be classified as nuisance particulates, but it has been recognized that even if the dusts do not cause fibrosis, they can still affect the lungs. A TLV of 3 mg/m$^3$ for a respirable particulate not otherwise classified (PNOC) (listed as an Intended Change in the 1994–1995 TLV booklet) acknowledges the fact that excessive amounts of any dust can be hazardous.

Some cements can contain diatomaceous earth, which can be converted to cristobalite silica with high heat. Other cements contain asbestos. It is important to consult the Material Safety Data Sheet of any dusty material to determine the potential health hazards.

Even when a dust is considered generally innocuous and not recognized as the direct cause of a serious pathological condition, its level should be kept as low as possible. Examples of particulate material generally considered to be nuisance dusts are listed in the ACGIH TLV booklet as PNOCs. A concentration of 10 mg/m$^3$ of total dust (or 3 mg/m$^3$ respirable dust) that contains less than 1 percent silicon dioxide is suggested as the threshold limit for a number of nuisance dusts. Good engineering practices ensure that this level is not exceeded. Any reduction below this level increases the comfort of employees and improves facility housekeeping.

Mica dust and kaolin dust are good examples of dusts that ordinarily are considered benign but, in excessive amounts, can cause a troublesome pneumoconiosis.

**Mica.** Mica pneumoconiosis has been observed in grinding operations where mica dust was present but free silica was not. Marked changes appeared in x-ray films of workers' lungs and some disability following exposure occurred. These cases resulted from massive dust exposures over many years.

**Kaolin.** Kaolinosis is a condition induced by inhalation of dust released in the grinding and handling of kaolin (china clay). In the facility where the cases occurred, dust levels of several hundred million particles per cubic foot of air were common.

**Bauxite.** Also known as Shaver's disease, bauxite pneumoconiosis is found only in workers exposed to fumes containing aluminum oxide and minute or ultramicroscopic silica particles arising from smelting bauxite in the manufacture of corundum, an impure form of aluminum oxide that may contain small amounts of aluminum silicate. The disease does not result from the use of corundum grinding wheels or from exposure to other forms of aluminum oxide.

Some pneumoconioses show marked shadows on an x-ray film that, without the necessary information on the exposure of the worker, can be alarming in general x-ray screening program. Clinical examination, however, often discloses no disability or symptoms. These shadows are often encountered when the dusts contain atoms of a relatively high molecular weight because the heavier atoms are fairly opaque to x-rays. Insoluble barium dusts and tin oxide dusts, for example, can show very marked shadows on x-ray films without producing signs of significant pathology. (Barium dust that is soluble in body fluids, however, can produce a toxic reaction.)

**Iron oxide.** Iron oxide, particularly excessive fumes from welding operations, can produce siderosis with a pigmentation of the lungs (black welders' lungs and red in the lungs of iron ore miners), usually without causing disability. Some people with siderosis may have symptoms of chronic bronchitis and shortness of breath. The shadows produced by iron oxide in x-ray films of the lungs are somewhat similar to the shadows produced by silicosis. Because of this similarity, differential diagnosis is often difficult, and heavy exposures to iron oxide dust and fumes may lead to medicolegal problems.

## Toxic Dusts and Fumes

Systemic reactions are caused by the inhalation of toxic dusts and fumes from various elements and their compounds. All metallic fumes are irritating, especially when freshly generated. Industrially important metals and their compounds that can have a toxic effect when the dust or fume is inhaled include arsenic, antimony, cadmium, chromium, lead, manganese, mercury, selenium, tellurium, thallium, and uranium.

The effect of some metals, such as magnesium and zinc, appears to be transient. Although the dusts and fumes from metals with low toxicity do not require as much caution as the dusts and fumes from highly toxic metals, the former should not be disregarded. Metals with low toxicity also should be kept at levels that are as low as possible because excessive amounts of any of them can be harmful (Table 8–A).

## Welding Fumes

The American Welding Society has published an extensive study called *Fumes and Gases in the Welding Environment* (see Bibliography) that lists the possible constituents of welding fumes.

Welding fumes cannot be classified simply. The composition and quantity of welding fumes depend on the alloy being welded and the process and electrodes being used. Reliable analysis of fumes cannot be made without considering the nature of the welding process and the system being examined. Reactive metals and alloys such as aluminum and titanium are arc-welded in a protective, inert atmosphere such as argon. Although these arcs create relatively little fume, they do produce an intense radiation that can produce ozone.

Similar processes are used to arc weld steels, which also produce a relatively low level of fumes. Ferrous alloys are also arc-welded in oxidizing environments that generate considerable fume and that can produce carbon monoxide. Such fumes generally are composed of discrete particles of amorphous slags containing iron, manganese, silicon, and other metallic constituents.

Chromium and nickel compounds are found in fumes when stainless steels are arc-welded. Some coated and flux-cored electrodes are formulated with fluorides, and the fumes associated with them can contain significantly more fluorides than oxides.

Because of these factors, arc-welding fumes often must be tested for the presence of likely constituents to determine whether specific TLVs have been exceeded. For example, stainless steel metal fumes can contain a particularly hazardous form of chromium compound called hexavalent chromium (sometimes called $Cr^{VI}$). Some hexavalent chromium compounds (particularly the water-insoluble ones) are carcinogenic. The TLVs for chromium and chromium compounds vary according to the type of compound. When arranging for an analysis of a chromium-containing fume, it is important to obtain a test for both hexavalent and total chromium.

Conclusions based on total fume concentration are generally adequate if no toxic elements are present in welding rods, metal, or metal coatings and conditions are not conducive to the formation of toxic gases.

**Metal fume fever.** Metal fume fever is an acute condition of short duration caused by a brief high exposure to the freshly generated fumes of metals such as zinc or magnesium or their oxides. Symptoms appear 4–12 hours after exposure and include fever and shaking chills. Recovery is usually complete within one day, and ordinarily the employee can return to the same job without suffering a recurrence of the symptoms. Oddly enough, daily exposure confers immunity, but not permanently. If the daily exposure is interrupted over a long 3-day weekend, for example, subsequent exposure will result in a recurrence of the symptoms. The severity of the attack depends on the length of the interruption.

Heavy concentrations of fumes cause metal fume fever. Zinc oxide fume is the most common source, but cases caused by the inhalation of fumes from magnesium oxide, copper oxide, and other metallic oxides have also been reported. The condition does not result from the handling of these oxides in powder form. Apparently, the condition results only from the inhalation of extremely fine particles that have been freshly formed as fume (nascent fume). Nickel, mercury, and other metals can also produce a fever.

## Radioactive Dusts

A radioactive contaminant in the form of a gas, dust, fume, or mist can pose a chemical toxicity hazard in addition to an ionizing radiation exposure. For example, uranium can be a radioactive hazard and can cause kidney damage similar to that produced by other heavy metals.

Radioactive contaminants taken into the body can be deposited in various organs, where they constitute sources of internal radiation. The chemical characteristics of the radioactive contaminant or isotope determine the organ in which it will be deposited. If a radioisotope has been deposited in the body, the internal exposure is regarded as continuous until the isotope is lost through radiological or biological decay.

Because radioisotopes are selectively deposited in individual organs, they may cause only local irradiation. Solubility and particle size determine how much of the active material will gain access to and remain in the bloodstream and organs.

If radioactive airborne contamination is present, control measures are essential to prevent inhalation. Access to such areas must be restricted by law to admit only trained personnel equipped with radiation-monitoring dosimeters. As a minimal precaution, the use of respirators with HEPA filters approved for radioactive particulate matter is required.

If the presence of contamination is unknown but suspected, determine whether the airborne concentrations of the radioisotope are below acceptable legal limits (see Chapter 10, Ionizing Radiation).

Good housekeeping, personal hygiene, and good operating techniques are much more important in the safe handling of radioactive materials than in the handling of most other materials used in the industry.

Engineering controls for radioactive dusts are similar to those for other dusts and depend primarily on capture at the point of generation. The difference lies in the fact that controls for radioactive dusts must be extremely efficient. Exposure limits for radioactive particulates are very low, and in some cases 100 percent efficiency in capture and retention is required. (Additional details are given in Chapter 10, Ionizing Radiation.)

## Bacteria and Fungi

The risk of lung infections from the inhalation of bacteria and fungi exists in several industries. Pulmonary anthrax as a result of the inhalation of dust that contains anthrax spores has occurred among workers who handle wool and crush bones of infected animals.

In addition to contracting infections, workers can develop a specific sensitization to mold spores. This sensitization is similar in nature to an allergic reaction, except that the alveoli (air sacs) of the lungs instead of the bronchial tubes are constricted. A few hours after a sensitized worker is exposed, he or she experiences flulike symptoms and shortness of breath. The total lung capacity is reduced. Although these symptoms are initially reversible, repeated exposures can result in irreversible symptoms and eventually respiratory or cardiac failure.

Fungi (molds) growing on grain have been found in the sputum of workers who shovel grain and are believed to be the cause of respiratory disorders. Likewise, fungi found in sugar cane residues (bagasse) are believed to be part of the cause of bagassosis. Fungal spores found under the bark of some trees have been blamed for respiratory difficulties among workers who strip bark from dry logs.

Although the incidence of occupationally related bacterial and fungal infections and sensitizations is relatively low, the respiratory effects can be troublesome, and in the case of pulmonary anthrax, even fatal. Some indoor air quality complaints are related to excessive levels of fungi in the air. The basic methods of control are the same as those for controlling the pneumoconiosis-producing dusts, but to these methods, source reduction, moisture control, sterilization, and disinfection must be added.

### Allergens

When they are present as dust, many materials cause allergic reactions in susceptible people. Such agents include certain animal products, foods, drugs, and chemicals. Western red cedar dust, proteolytic enzyme detergents, and ethanolamines (found in some aluminum solder fluxes) can cause allergic reactions. The allergic reactions, which can be quite severe, usually involve the skin, respiratory system, and gastrointestinal system. Occasionally, two or more systems are involved. Some of the allergic reactions are dermatitis, hay fever, asthma, and hives.

Usually, the worker is subjected to a series of exposures without any reactions; during this time, sensitization is built up. These exposures can occur continuously for years. Then, at the end of the "incubation period," which varies with each individual, a reaction occurs.

Two factors are required for a true allergic reaction:

- Prior exposure to the material involved (sometimes this is not known even by the affected employees)
- A challenge dose of the material, which provokes the allergic reaction

Continuous exposure may act as desensitizing doses and, under these conditions, an allergic person can work without incident for long periods of time. However, reexposure following removal from the sensitizing material (such as after a vacation) causes an allergic response. Unfortunately, a reaction known as allergic alveolitis develops in some people subjected to continuous exposure. This condition can be progressive and can develop into a severely disabling obstructive disease.

Medical and engineering recommendations intended to prevent allergic reactions are based on the prevention of exposure by means of personal protective equipment, ventilation methods, or the removal of sensitized workers from the exposure.

## ■ Measurement of Particulate Exposures

To evaluate dust exposure, it is first necessary to determine the composition of the dust that remains suspended in the air that workers breathe.

Operations that involve the crushing, grinding, or polishing of minerals or mineral mixtures often do not produce airborne dusts that have the same composition as the material being worked on. Of primary concern is the percentage of crystalline silicon dioxide that is suspended with other dusts in the air.

In many mineral mixtures, crystalline silicon dioxide is harder to crush than the remainder of the mixture. Thus, in most crushing and grinding operations, airborne dust is produced with a lower percentage of crystalline silicon dioxide than is found in the original mineral mixture. Occasionally, such as when the grinding operation involves a great deal of impact force, the situation is reversed. For example, measurements obtained in coal breaker houses (where coal is prepared for combustion in industrial boilers) often show that silica levels are higher in the respirable coal dust than in the original coal.

In pulverized materials, crystalline silicon dioxide is often found in larger particle sizes than are the softer components; when this is the case, subsequent handling may result in a higher percentage of the softer components in the airborne dust.

Therefore, it is necessary to obtain a sample of airborne dust and to analyze it if the composition of the dust is to be used in evaluating the hazard associated with breathing the dust.

The extremely small size of the dust—0–5 μm in diameter—makes chemical methods of analysis (which depend on the differential solubility of the various components) too unreliable to be used alone. The x-ray diffraction and infrared analysis techniques are usually preferred. Table 8–C indicates the normal range of crystalline silicon dioxide determined by analysis of many minerals and industrial compositions.

As mentioned earlier in this chapter, when the standard technique is used to evaluate dust concentrations, particles larger than 10 μm are not considered because particles this size or larger do not remain suspended in still air for appreciable lengths of time, nor do they penetrate deeply into the lung if they are inhaled. Larger particles seldom reach the lungs because they are filtered out in the nose and throat. The physiological effects of extremely small dust particles, however, are uncertain.

When air samples are collected in the immediate vicinity of a dust-producing operation, larger particles that have not yet settled from the air may be collected. If a large number of these particles appear in the dust sample, the effect of their presence may have to be evaluated separately.

Particle sampling is often done using a cyclone and filter cassette collection device and a filter. Air to be sampled is drawn past a filter medium with a battery- operated pump. The particles collect on the fiber or membrane.

Filter selection depends on the material, filter properties, efficiency, availabilities, and analytical procedure used. Membrane filters, made of cellulose esters or polyethylene, are useful because of their narrow range of pore diameter. Glass fiber filters are used in collecting organic chemical dusts.

Some filter applications can be seen in the table below:

| Application | Filter Type | Filter Size |
|---|---|---|
| Metal Fume | Mixed Cellulose ester (MCE) | 0.8 µm, 37 mm |
| Asbestos | MCE | 0.1–1.2 µm, 25 mm filter with 50 mm cowl |
| Silicia | Polyvinylchloride (PVC) | 0.8 µm, 37 mm |
| PCB Dust | Glass Filter (binderless) | 13 mm |
| Nuisance Dust | PVC | often 5.0 micron, 37 mm |

Using a gravimetric procedure (precise before and after weighing of the filter) is one way to determine the concentration of particulates collected. This is usually followed up by chemical analyses to determine the types of particles collected.

Atomic absorption spectroscopy (AA) can identify metal contaminants. Particulates on a filter are dissolved, burned, and analyzed for identifiable metallic color. Quantity is determined by the color intensity. X-ray diffraction is used for measurement of inorganic compounds, whose crystalline forms give distinct diffraction patterns as the electrons scatter the rays.

Where particles settle in the lungs is a function of their size. Large particles are removed before reaching the lungs. Small particles (less than 5 µm) reach the alveoli. Therefore, it is important to separate nonrespirable particles from the smaller ones to be analyzed. Cyclones use a process in which the air is drawn into an orifice and whirled against the side so the heavier particles drop down and can be removed while respirable particles are drawn up onto a filter. Control of the air flow rate is important for accuracy. A 10-mm diameter cyclone is used to collect dust and silica particles. A 2-in. cyclone is also available.

Horizontal elutriators are placed in front of a sampler and used to remove coarse particles by means of gravity. They are commonly used in cotton dust sampling. Cascade impactors use a series of impactor plates to capture different-sized particles in a sample. In impactors, air flowing at a specific rate strikes the plates, which are set at a 90-degree angle to it and separate out the particles. These are used for area samples (large units) or for personal breathing zone samples (mini-cascade impactors).

## Microscopic Count Procedures

The concern of industrial hygienists over the years has been to determine the fraction of a dust that causes pneumoconiosis. Because only dust particles smaller than approximately 5 µm in aerodynamic diameter are deposited and retained in the lung, methods have been sought to measure the concentration of these tiny dust particles. Microscopic counting of dust collected by impingement used to be the common method for this type of analysis. It has been replaced in most cases by gravimetric and chemical analyses. It is included here for historical reasons.

Dust counting as an index of dust concentration, and consequently of workers' exposure, was performed in South Africa using the konimeter and in Australia using a jet dust sampler. In the United Kingdom, thermal precipitation was often used for dust collection. In the United States, Greenburg–Smith and midget impingers were commonly used.

In these investigations, the lower limit of dust size included in counts was determined by the counting procedure used. In impinger counts, where a 10 x 16- mm objective lens was used with light-field counting, the usual lower limit of particles seen was approximately 1.0 µm in diameter. Others studies used dark field illumination techniques that enable investigators to see particles as small as 0.1 µm in diameter.

## "Total" Mass Concentration Methods

The simplest method of measuring dust concentration is to determine the total weight of dust collected in a given volume of air. The "total" mass, however, is determined to a considerable extent by the large dust particles, which cannot penetrate to the pulmonary spaces and cause disease. The proportion of dust that is small enough to penetrate into the pulmonary spaces (respirable dust) is extremely variable; it ranges in industrial dust clouds from as little as 5 percent to more than 50 percent (by weight). Thus, the "total" dust concentration by weight is not a reliable index of respirable dust concentrations, nor is it an index of respiratory hazard.

## Respirable Mass Size-Selection Measurement (Personal Sampling)

For evaluating a silicosis hazard, the method generally preferred is personal (breathing zone) respirable mass sampling. Dust collection devices now available for this method of sampling also provide a means for a size-frequency analysis of the collected dust.

The dust sample is size-separated by the design and flow characteristics of the sampling device. Such equipment includes impactors, centrifugal and gravitational separators, and a range of miniature cyclones. In addition to particle-size separation, these instruments are also capable of collecting a quantity of dust sufficient for an analysis of the free silica content of the dust.

Respirable mass samples are preferably taken over a full 8-hour shift. However, multiple shorter-period samples (over a 2–4 hour period) can be collected during a worker's full shift. The samples are then pooled for analytical purposes and the average respirable mass concentration for free silica is calculated for a full shift. The recommended equipment and the method for collecting respirable dust are presented in Figure 8–4.

# ■ Control of Particulates

The methods used to control occupational particulate exposures are numerous and varied, and their application involves extensive technical knowledge.

The handling of dust-laden air in ventilating processes involves principles that must take into account the weight of the dust in the air and the separation of the dust from the air before it is exhausted outside the building. This can require the use of expensive and complicated equipment.

Exhausting the air without dust removal can create a dust load in the outside area that can contaminate the facility on reentry or create neighborhood problems. Applicable EPA regulations should be consulted. (See Chapter 19, Local Exhaust Ventilation of Industrial Occupancies, and Chapter 20, General Ventilation of Industrial Occupancies.)

## Engineering Controls

**Closed processes.** The surest and most positive dust-control method is a total enclosure of the dust-producing process that includes an exhaust of the enclosure to maintain a negative pressure. This is often impractical, although certain pieces of equipment can be enclosed. In other instances, a total enclosure with only a feed or hopper opening can be used with a sufficient exhaust system to make sure air moves into these openings at all times.

Large equipment that does not require constant attention can be enclosed in buildings with separate exhausts; workers can wear personal protective equipment when they must enter these buildings. Consideration should be given to preventing dust explosions in such operations.

**Local exhaust ventilation.** Local exhaust ventilation is often used at points of high dust production; when it can

be combined with a hooded enclosure, such ventilation can be quite effective. In these instances, the enclosure should be designed with as small an opening as practical so that workers can stand outside of the hood. Generally, hooded enclosures that do not extend to the floor or the work bench, or those with more than one open side, require large amounts of air to be effective.

The ventilated enclosure of buffing and grinding wheels illustrates an application of hood design that must take into account the mechanical generation of dust (Figure 8–5). The ventilation of the wheel enclosure would be totally ineffective in collecting the high-speed particles that emerge from the point of grinding contact if they were not directed into the exhaust system (Figure 8–6). From a practical standpoint, no amount of air movement supplied by the suction applied to the hood could deflect the high-speed particles. The dust generated by the wheel is, in effect, collected by a hood designed so the mechanically generated particles are projected into it.

Exhaust slots must sometimes be used because of operational requirements. In these applications, the slots are not completely effective if the dust is generated more than 1 ft from the slot or if the method of generation involves a mechanical agitation of the dust. Slots are somewhat more effective if they are set in baffle plates or other applications that direct air collection. Generally, exhaust slots require high suction pressures, which are expensive to maintain. In some applications of exhaust slots, the push–pull principle of supplying air from one side and directing it toward the slot on the opposite side can be helpful. However, in this application, the balance of air volumes and the dispersing effects of supplied air must be carefully considered. (See Chapter 19 for further discussion of local exhaust ventilation).

**General dilution ventilation.** In instances in which the sources of dust generation are numerous, widely distributed general or dilution ventilation may be the best solution. Such a method can be both expensive and ineffective if exhaust locations are not properly placed and if the exhaust is not adequately balanced with well-placed, heated supplied air.

All air exhausted from a facility must be supplied from some source. Because the incoming air usually reaches plant air temperatures before it is exhausted, it is normally more economical—and almost always more effective—to supply the working area with heated makeup air by mechanical means than it is to rely on facility leakage to make up the supply of exhausted air.

For some reason, this principle is the least understood of all factors involved in industrial ventilation. Heat and fuel are not conserved when the system exhausts the warmest air from the roof of the building and makeup air enters the plant through windows and doors at "no cost." Even in warm weather, when it is desirable to remove warm air from the upper strata for comfort control, a

### Figure 7-4. Respirable Dust Sampling

**A. Prepare cyclone assembly and mount sampling apparatus on employee.**

1. Loosen set screw on back of cyclone and remove rubber stopper (B).
2. Identify preweighed polyvinylchloride (PVC) filter cassette by number or letter.
3. Remove end plugs from filter cassette (save) and insert cassette (filter on top side) into cyclone (A).
4. Tighten set screw (make sure that cassette inlet and outlet are seated into each holder of cyclone—upper and lower [A]).
5. Attach pump to worker (B,C).
6. Connect sample hose to cyclone outlet. *NOTE: Be careful to keep cyclone in an upright position. This will prevent dust in the bottom of the cyclone from falling into filter cassette.*
7. Clip cyclone assembly to worker's collar, allowing cyclone to hang free (A,B).

**B. Calibrate sampling train**

1. Connect free end of sample hose to lower port of rotameter (C).
2. Connect upper port of rotameter to pump inlet.
3. Determine rotameter float elevation from calibration chart of rotameter (C). *NOTE: Flow rate should be 1.8 L/min.*
4. Hold rotameter vertically, start pump, and adjust slotted flow valve on pump with screwdriver until float elevation (reading center of ball) determined in step B-3 is established (C).
5. *Remove* rotameter and connect pump directly to cyclone (B).

**C. Collect sample**

1. Record start time.
2. Collect sample as detailed in attached sampling procedure, while *periodically checking flow rate* by repeating steps B-1 through B-5. (Not required with constant-flow pumps.)
3. Stop pump and record time.
4. Remove filter cassette from cycle and insert plugs. *(Caution: Do not invert cyclone while cassette is in place.)*

(Adapted with permission from NATLSCO Environmental Sciences)

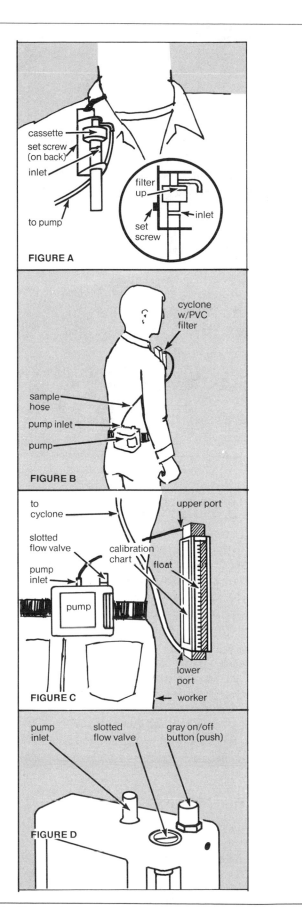

**Figure 8-4.** Collecting dust containing respirable free silica. (Adapted with permission from NATLSCO Environmental Sciences.)

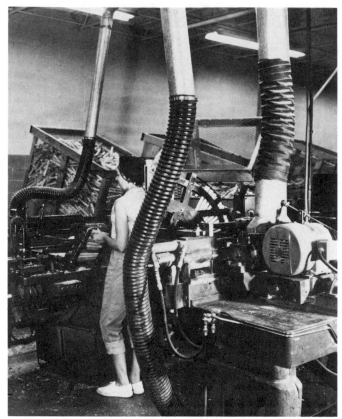

**Figure 8–5.** Typical in-plant installations show flexible ducts exhausting dusts and gases from machinery sections.

greater degree of comfort is attained if outside air is supplied at the level (position) of the worker. (See Chapter 20 for more information on general ventilation.)

**Moisture control.** When pulverized materials are handled, the amount of dust dispersed by any given mechanical operation varies with the material's moisture content. Wet drilling and grinding are typical examples of this application that involve excessive moisture. Foundry molding in which moist sand is used is an example of moisture control required by the process.

***Substituting wet process for dry process.*** The advantage of wet processing over dry processing is that dust hazards are reduced by as much as 75 percent. Water is often used in drilling operations where dust is a hazard. Water under high pressure can clean effectively and eliminate the need for sandblasting. Sand-like materials, kept moist before use and in the process of mixing, are relatively dustless.

## Administrative Controls

Wet methods have their place in housekeeping. Wet sweeping involves a thorough, even wetting and possibly an added wetting agent. The floor must be kept moist until swept and the dust properly collected to avoid redispersing it. Additionally, HEPA vacuums are recommended for specific dust hazards such as asbestos and lead.

Areas where eating, drinking, and smoking are allowed must be kept dust free and these activities should be prohibited in areas where particulates are present. Clearly demarcated areas for removing and disposing of contaminated clothing are essential. Clean and dirty areas in work changing rooms are often used. Procedures for taking breaks should be planned with contamination avoidance in mind.

## Personal Protective Equipment

**Respirators.** Because most dusts are hazardous to the lungs, respirators are a common method of primary or secondary protection. Respirators are appropriate as a primary control during intermittent maintenance or cleaning activities when fixed engineering controls are not feasible. Respirators can also be used as a supplement to good engineering and work practice controls for dusts to increase employee protection and comfort.

To be effective, respirators must be carefully matched to the type of particulate hazard present. The critical exposure factors that determine the degree of hazard (type of dust, length of exposure, dust concentration, and particle size) are also used to determine the type of respirator to be used. For example, metal fumes are extremely small particles. A respirator approved by NIOSH for dusts and mists cannot protect against these fumes. Only respirators approved by NIOSH for fumes can filter out these small particles (see Chapter 22, Respiratory Protection).

Very high concentrations of dust can present both a respiratory and eye irritation hazard. Full-face respirators, which cover the eyes, nose, mouth, and chin, protect the eyes and provide a higher degree of respiratory protection than a half-mask because a full-face respirator maintains a better fit.

It is important to consider how long a worker is required to wear a respirator. Wearing a respirator for prolonged periods can be uncomfortable and physically stressful. If exposure time is prolonged and the respirator provided is uncomfortable, the worker may decide not to wear the respirator or remove it for brief periods. Repeated brief breakings of the respirator-to-face seal can greatly increase the worker's overall exposure. This can be a critical problem when exposure to highly toxic dusts is involved. Reducing respirator wearing time through job redesign, allowing the worker to select a comfortable respirator, instructing the worker on the importance of continuously maintaining an adequate respirator-to-face seal, and monitoring of respirator use by supervisors are ways to increase worker compliance.

The OSHA regulations for some particulates (such as lead, asbestos, and cadmium) have specific respirator requirements. These regulations describe what types of respirators are appropriate for various levels of exposure, as well as how and when respirators can be used as part of a comprehensive approach to controlling exposures. These regulations must be consulted before respirators are used. Respirator use is also covered under the general

**Figure 8–6.** Because the swing-frame grinder is used in many positions, the local exhaust system must be adjustable. The flexible duct (A) permits movement of the exhaust hood (B). (Courtesy of the American Foundrymen's Society.)

OSHA respirator provision in 29 *CFR* 1910.134. These regulations apply to the use of all respirators and specify mandatory requirements in a minimally acceptable respirator program.

**Protective clothing.** Protective clothing is selected based on the severity of the dust hazard. For hazardous waste site work, OSHA has established four levels of hazard and their appropriate protective clothing. The level for highest hazards requires a fully encapsulating, vapor-protective suit, two layers of chemical-resistant gloves, a hard hat, steel-toed and -shanked boots under the suit, and an appropriate respirator. The next level, where expected skin hazard is lower, requires a hooded, chemical-resistant suit, two layers of gloves, a hard hat, boots, and a respirator. The third level requires a chemical-resistant suit, two layers of gloves, a hard hat, boots, and a respirator. The least hazardous level provides minimal skin protection with work clothes, gloves, boots, and eye protection, with no respirator required.

The general requirements for personal protective equipment are in 29 *CFR* 1910.132. This standard requires that a hazard assessment be performed. The assessment identifies hazards that require the use of personal protective equipment (PPE). The hazard assessment must be documented in writing and include the following information:

- The work sites evaluated
- The name of the person verifying that the industrial hygienist performed the hazard assessment
- The date performed the hazard assessment was performed
- Indication that the purpose of the hazard assessment was to determine PPE requirements

Once the hazard assessment is performed and a need for PPE is determined, the following steps must be taken:

- Select the PPE that provides protection for the hazards identified in the assessment
- Select PPE that properly fits each employee
- Inform the affected employees of the PPE required for each hazard

## Summary

The critical exposure factors (type, size, and concentration of particulate involved and length of exposure) are discussed. Biological reactions to particulates (dusts, fumes, and mists) and selected particulates (silica, asbestos, lead, beryllium, coal dusts, miscellaneous dusts, toxic dusts and fumes, welding fumes, radioactive dusts, bacteria and fungi, and allergens) are outlined. Measurement of particulate exposures and control of particulates are presented.

# Bibliography

Amdur MO et al. *Casarett and Doull's Toxicology: The Basic Science of Poisons*, 4th ed. New York: McGraw-Hill, 1991.

American Conference of Governmental Industrial Hygienists. *1994–1995 Threshold Limit Values for Chemical Substances and Physical Agents and Biological Exposure Indices*. Cincinnati: American Conference of Governmental Industrial Hygienists, 1994.

American Conference of Governmental Industrial Hygienists. *Documentation of the Threshold Limit Values and Biological Exposure Indices,* 6th ed. Cincinnati: American Conference of Governmental Industrial Hygienists, 1993.

American Conference of Governmental Industrial Hygienists. *Industrial Ventilation: A Manual of Recommended Practice,* 21st ed. Cincinnati: American Conference of Governmental Industrial Hygienists, 1992.

American Conference of Governmental Industrial Hygienists. *Air-Sampling Instruments for Evaluation of Atmospheric Contaminants,* 7th ed. Cincinnati: American Conference of Governmental Industrial Hygienists, 1989.

American Industrial Hygiene Association. *Arc Welding and Your Health: A Handbook of Health Information for Welding.* Akron, OH: American Industrial Hygiene Association, 1984a.

American Industrial Hygiene Association. *Welding Health and Safety Resource Manual.* Akron, OH: American Industrial Hygiene Association, 1984b.

American Welding Society, Inc. *Fumes and Gases in the Welding Environment.* Miami, FL: American Welding Society, 1979.

Clayton GD, Clayton FE. *Patty's Industrial Hygiene and Toxicology,* Vol. 2, 4th Ed. New York: Wiley-Interscience, 1993.

McCord CP, Meeke SF, Harrold GC. Tripoli and Silicosis. *Industrial Medicine* 12: 373–378, 1943.

National Institute for Occupational Safety and Health. Cincinnati: NIOSH, *Occupational Respiratory Diseases*. Pub. NO. PHHS (NIOSH) 86–102.

CHAPTER

# 9

# Industrial Noise

Revised by John J. Standard, MS, MPH, CIH, CSP

**Compensation Aspects**

**Properties of Sound**
*Noise ▪ Sound Waves ▪ Frequency ▪ Wavelength ▪ Velocity ▪ Sound Pressure ▪ Decibels and Levels ▪ Loudness*

**Occupational Damage-Risk Criteria**
*Hearing Ability ▪ Risk Factors ▪ Analysis of Noise Exposure*

**Sound-Measuring Instruments**
*Sound Level Meters ▪ Octave-Band Analyzers ▪ Noise Dosimeters*

**Sound Surveys**
*Source Measurements ▪ Preliminary Noise Survey ▪ Detailed Noise Survey ▪ General Classes of Noise Exposure*

**Noise-Control Programs**
*Source ▪ Noise Path ▪ Enclosures ▪ Control Measures*

**Industrial Audiometry**
*Threshold Audiometry ▪ Who Should Be Examined ▪ Effective Programs*

**Noise-Exposure Regulations**
*Background ▪ The Occupational Noise Standard ▪ Hearing Conservation Programs*

**Summary**

**Bibliography**

**Addendum:**
*29 CFR 1910.95 Occupational Noise Exposure*

THE SOUNDS OF INDUSTRY, growing in volume over the years, have heralded not only technical and economic progress, but also an ever-increasing incidence of hearing loss and other noise-related hazards to exposed employees. Noise is not a new hazard. Indeed, noise-induced hearing loss was observed centuries ago. In 1700, Ramazzini in "De Morbis Artificium Diatriba" described how workers who hammer copper "have their ears so injured by that perpetual din . . . that workers of this class become hard of hearing, and if they grow old at this work, completely deaf." Before the Industrial Revolution, however, comparatively few people were exposed to high levels of noise in the workplace. The advent of steam power during the Industrial Revolution first brought general attention to noise as an occupational hazard. Workers who fabricated steam boilers were found to develop hearing loss in such numbers that the malady was dubbed boilermakers' disease. The increasing mechanization that has occurred in all industries and in most trades has since aggravated the noise problem. Noise levels in the workplace, particularly those maintained in mechanized industries, are likely to be more intense and sustained than any noise levels experienced outside the workplace. The recognition, evaluation, and control of industrial noise hazards are introduced in this chapter. Basically, this involves assessing the extent of the noise problem, setting objectives for a noise abatement program, controlling exposure to excessive noise, and monitoring the hearing of exposed employees.

# Compensation Aspects

The trend toward covering hearing losses under state workers' compensation laws has stimulated interest on the part of employers in controlling industrial noise exposures. Compensation laws that cover loss of hearing due to noise exposure have been enacted in many states; compensation is being awarded in other states even though hearing loss is not specifically defined in many compensation laws.

Occupational hearing loss can be defined as a hearing impairment of one or both ears, partial or complete, that results from one's employment. It includes acoustic trauma as well as noise-induced hearing loss.

*Acoustic trauma* denotes injury to the sensorineural elements of the inner ear. Acoustic trauma is produced by one or a few exposures to sudden intense acoustic forms of energy resulting from blasts and explosions or by direct trauma to the head or ear. The worker should be able to relate the onset of hearing loss to one single incident. For details on ear anatomy, see Chapter 4, The Ears.

Noise-induced hearing loss, on the other hand, describes the cumulative permanent loss of hearing—always of the sensorineural type—that develops over months or years of hazardous noise exposure.

Noise-induced hearing loss usually affects both ears equally in the extent and degree of loss. It should also be kept in mind that the onset of hearing loss, its progression, its permanency, and the characteristics of the audio-grams obtained, vary depending on whether the injury is a noise-induced hearing loss or acoustic trauma.

To establish a diagnosis of noise-induced hearing loss and a causal relationship to employment, the physician considers the following factors:

- The employee's history of hearing loss—onset and progress
- The employee's occupational history, type of work, and years of employment
- The results of the employee's otological examination
- The results of audiological and hearing studies performed (preplacement, periodic, and termination)
- The ruling out of nonindustrial causes of hearing loss

It has been estimated that 1.7 million workers in the United States between 50 and 59 years of age have compensable noise-induced hearing loss. Assuming that only 10 percent of these workers file for compensation and that the average claim amounts to $3,000, the potential cost to industry could exceed $500 million.

Estimates show that 14 percent of the working population are employed in jobs where the noise level exceeds 90 dBA (Table 9–A). At present, no test can predict which individuals will incur a hearing loss. If enough people are placed in an environment where the predominant noise level exceeds 90 dBA for a sufficient period of time, some individuals incur a hearing impairment greater than that due to presbycusis (loss of hearing due to aging). The

**Table 9–A.** Noise Exposures Above 90 dBA in Manufacturing

| Code | Number of Plants in Sample | Total Number of Employees in Sample | Number Located in Areas 90 dBA and Above | Percent of Work Force Exposed | Total Work Force | Number Projected to Be Located in Areas 90 dBA and Over |
|---|---|---|---|---|---|---|
| Textile mill products | 23 | 12,764 | 5,634 | 44.1 | 963,300 | 424,815 |
| Petroleum and coal products | 16 | 20,493 | 5,875 | 28.6 | 192,800 | 55,140 |
| Lumber and wood products | 14 | 5,654 | 1,460 | 25.8 | 601,000 | 155,058 |
| Food and kindred products | 17 | 23,690 | 5,959 | 25.1 | 1,898,600 | 476,549 |
| Furniture and fixtures | 11 | 10,374 | 1,849 | 17.8 | 465,400 | 82,841 |
| Fabricated metal products | 56 | 41,371 | 7,079 | 17.1 | 1,335,000 | 228,285 |
| Stone, clay, and glass products | 5 | 2,502 | 416 | 16.6 | 643,800 | 106,870 |
| Primary metal industries | 51 | 71,208 | 11,001 | 15.4 | 1,190,000 | 183,260 |
| Rubber and plastic products | 4 | 7,671 | 1,105 | 14.4 | 589,500 | 84,888 |
| Transportation equipment | 46 | 199,212 | 23,445 | 11.7 | 1,705,500 | 199,543 |
| Electrical equipment and supplies | 7 | 8,790 | 973 | 11.0 | 1,778,100 | 195,591 |
| Chemicals and allied products | 8 | 3,081 | 324 | 10.5 | 1,014,400 | 106,512 |
| Apparel and other textile products | 1 | 50 | 5 | 10.0 | 1,353,100 | * |
| Paper and allied products | 21 | 14,997 | 1,385 | 9.2 | 687,400 | 63,420 |
| Ordnance and accessories | 12 | 39,403 | 3,480 | 8.8 | 93,900 | 17,063 |
| Instruments and related products | 6 | 3,254 | 193 | 5.9 | 433,800 | 25,594 |
| Machinery except electrical | 38 | 25,016 | 1,144 | 4.5 | 1,768,000 | 79,560 |
| Printing and publishing | 5 | 5,597 | 237 | 4.2 | 1,085,900 | 45,607 |
| TOTAL | 341† | 495,127 | 71,564 | 14.5 | 17,799,500 | 2,530,596 |

* Insufficient data for projection.

† 2,709 questionnaires were sent to the manufacturing industries listed: 1,550 were returned, and 341 of these respondents answered this question.

(Reprinted from NIOSH *Criteria Document, Occupational Exposure to Noise*, 1972.)

number of workers subjected to noise hazards exceeds that of those exposed to any other significant occupational hazard.

The audiometer provides an easily reproducible means of measuring the status of an individual's hearing with appreciable accuracy. Partial hearing losses are easily measurable by commercially available audiometers.

# Properties of Sound

Sound can be defined as any pressure variation (in air, water, or some other medium) that the human ear can detect. The number of pressure variations over time is called the frequency of sound. Frequency is measured in cycles per second, or hertz (Hz).

Sound can be defined as a stimulus that produces a sensory response in the brain. The perception of sound resulting in the sensation called hearing is the principal sensory response; however, under certain conditions, additional subjective sensations ranging from pressure in the chest cavity to actual pain in the ears can be produced (see Chapter 4, The Ears). There are certain effects produced by sounds that appear to be universally undesirable for all people. These effects include the following:

- The masking of wanted sounds, particularly speech
- Auditory fatigue
- Damage to hearing
- Annoyance

### Noise

What we call noise is usually sound that bears no information and whose intensity usually varies randomly in time. The word *noise* is often used to mean unpleasant sound that the listener does not want to hear. Noise interferes with the perception of wanted sound and is likely to be physiologically harmful.

Noise does not always have particular physical characteristics that distinguish it from wanted sound. No instrument can distinguish between a sound and a noise—only human reaction can.

A variety of methods have been devised to relate objective physical measurements of sound to subjective human perception. The purpose of this section is to outline both the objective physical properties of sound and its important subjective aspects.

The term *sound* usually refers to the form of energy that produces a sensation perceived by the sense of hearing in humans, whereas *vibration* usually refers to nonaudible acoustic phenomena that are recognized by the tactile experience of touch, or feeling. However, there is no essential physical difference between the sonic and vibratory forms of sound energy.

The generation and propagation of sound are easily visualized by means of a simple model. Consider a plate suspended in midair (Figure 9–1). When struck, the plate vibrates rapidly back and forth. As the plate travels in either direction, it compresses the air, causing a slight increase in its pressure. When the plate reverses direction, it leaves a partial vacuum, or rarefaction, of the air. These alternate compressions and rarefactions cause small but repeated fluctuations in the atmospheric pressure that extend outward from the plate. When these pressure variations strike an eardrum, they cause it to vibrate in response to the slight changes in atmospheric pressure. The disturbance of the eardrum is translated into a neural sensation in the inner ear and is carried to the brain, where it is interpreted as sound (Figure 9–2).

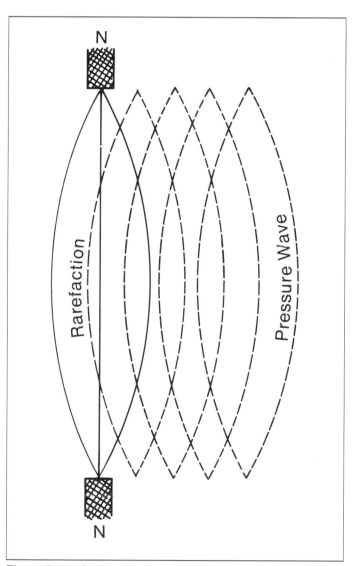

**Figure 9–1.** As the vibrating plate moves back and forth, it compresses the air in the direction of its motions. When it reverses direction, it produces a partial vacuum, or rarefaction, imparting energy to the air, which radiates away from the plate as sound.

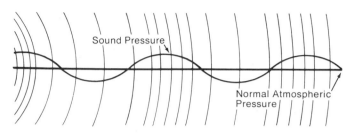

**Figure 9–2.** Air is an elastic medium and behaves as if it were a succession of adjoining particles. The resulting motion of the medium is known as wave motion, and the instantaneous form of the disturbance is called a sound wave.

Sound is invariably produced by vibratory motion of some sort. The sounding body must act on some medium to produce vibrations that are characteristic of sound. Any type of vibration can be a source of sound, but by definition, only longitudinal vibration of the conducting medium is a sound wave.

## Sound Waves

Sound waves are a particular form of a general class of waves known as elastic waves. Sound waves can occur only in media that have the properties of mass (such as inertia) and elasticity. Because air possesses both inertia and elasticity, a sound wave can be propagated in air. One sound wave may have three times the frequency and one-third the amplitude of another sound wave. However, if both the waves cross their respective zero positions in the same direction at the same time, they are said to be in phase (Figure 9–3).

## Frequency

Frequency is the number of times per second that an air molecule at the sound source is displaced from its position of equilibrium, rebounds through the equilibrium position to a maximum displacement opposite in direction to the initial displacement, and then returns to its equilibrium position. In other words, frequency is the number of times per second a vibrating body traces one complete cycle of motion (Figure 9–4). The time required for each cycle is known as the period of the wave and is simply the reciprocal of the frequency. The phrase formerly used to describe frequency, *cycles per second,* has now been replaced by *Hertz,* abbreviated *Hz.*

Frequency is perceived as pitch. The audible range of frequencies for humans with good hearing is between 20 Hz and 20,000 Hz. Most everyday sounds contain a mixture of frequencies generated by a variety of sources. A sound's frequency composition is called its spectrum. The frequency spectrum can be a determining factor in the level of annoyance caused by noise; high-frequency noise generally is more annoying than low-frequency noise. Also, narrow frequency bands or pure tones (single

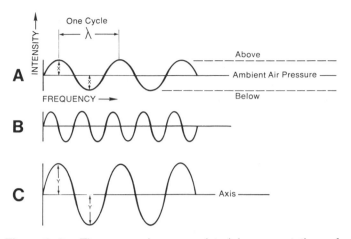

**Figure 9–3.** The curves shown are pictorial representations of sound waves. Pitch is related to frequency and loudness is related to the intensity of a sound. Curve B represents a sound that has a higher frequency—a higher perceived pitch—than the sound represented by curve A because the variations in air pressure, as represented by a point on the curve, cross the axis more often. The intensity of a sound can be shown by the height of the curve. Curve C represents a sound that has a greater intensity—a greater perceived loudness—than the sound represented by curve A (distance Y is greater than distance X).

frequencies) can be somewhat more harmful to hearing than broadband noise.

## Wavelength

Wavelength is the distance measured between two analogous points on two successive parts of a wave. In other words, wavelength is the distance that a sound wave travels in one cycle. The Greek letter lambda ($\lambda$) is used to express wavelength, and it is measured in feet or meters (Figure 9–3).

Wavelength is an important property of sound. For example, sound waves that have a wavelength that is much larger than an obstacle are little affected by the presence of that obstacle; the sound waves bend around it. This bending of the sound around obstacles is called diffraction.

If the wavelength of the sound is small in comparison with the size of an obstacle (small wavelengths are characteristic of high-frequency sounds), the sound is reflected or scattered in many directions, and the obstacle casts a "shadow." Actually, some sound is diffracted into the

**Figure 9–4.** Relative positions of an air molecule during one complete cycle of motion.

shadow, and there is significant reflection of the sound. As a consequence of diffraction, a wall is of little use as a shield against low-frequency sound (long wavelength), but it can be an effective barrier against high-frequency sound (short wavelength) (Figure 9–5).

## Velocity

The velocity at which the analogous pressure points on successive parts of a sound wave pass a given point is called the speed of sound. The speed of sound is always equal to the product of the wavelength and the frequency:

$$c = f\lambda \qquad (1)$$

where  $c$ = speed of sound (feet or meters per second),
  $f$ = frequency of sound (Hz), and
  $\lambda$ = wavelength (feet or meters).

The speed with which the sound disturbance spreads depends on the mass and elastic properties of the medium. In air at 72 F, the speed of sound is about 1,130 ft/sec (344 m/sec). Its effects are commonly observed in echoes and in the apparent delay between a flash of lightning and the accompanying thunder.

In a homogeneous medium, the speed of sound is independent of frequency; that is, in such a medium, sounds of all frequencies travel at the same speed. However, the speed of sound varies with the density and compressibility of the medium through which it is traveling. Speed increases as medium density increases and medium compressibility decreases. For example, the speed of sound is approximately 1,433 m/sec in water, 3,962 m/sec in wood, and 5,029 m/sec in steel. Sound, therefore, can be transmitted through many media before it is eventually transmitted through air to the ear of the receiver.

## Sound Pressure

Sound is a slight, rapid variation in atmospheric pressure, caused by some disturbance or agitation of the air. The sounds of normal conversation amount to sound pressure of only a few millionths of a pound per square inch, yet they can be easily heard because of the remarkable sensitivity of the human ear. The sounds that can damage our hearing have sound pressures of only a few thousandths of a pound per square inch.

Most common sounds consist of a rapid, irregular series of positive pressure disturbances (compressions)

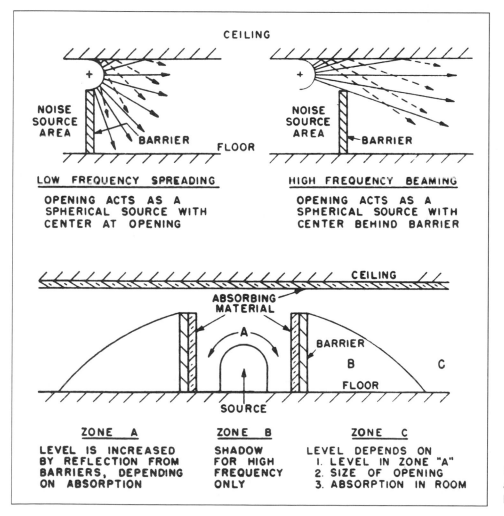

**Figure 9–5.** The effects of a barrier as a shield to contain noise of low or high frequency.

and negative pressure disturbances (rarefactions) measured against the equilibrium pressure value. If we were to measure the mean value of a sound pressure disturbance, we would find it to be zero because there are as many positive compressions as negative rarefactions. Thus, the mean value of sound pressure is not a useful measurement. We must look for a measurement that permits the effects of rarefactions to be added to (rather than subtracted from) the effects of compressions.

The root-mean-square (rms) sound pressure is one such measurement. The rms sound pressure is obtained by squaring the value of the sound pressure disturbance at each instant of time. The squared values are then added and averaged over the given time. The rms sound pressure is the square root of this time average. Because the squaring operation converts all the negative sound pressures to positive squared values, the rms sound pressure is a useful, nonzero measurement of the magnitude of the sound wave. The units used to measure sound pressure are micropascals (μPa), newtons per square meter (N/sq m), microbars (μbar), and dynes per square centimeter (d/sq cm). Relations among these units are as follows: 1 μbar = 1 d/sq cm = 0.1 N/sq m = 0.1 Pa.

## Decibels and Levels

Even though the weakest sound pressure perceived as sound is a small quantity, the *range* of sound pressure perceived as sound is extremely large. The weakest sound that can be heard by a person with very good hearing in an extremely quiet location is known as the *threshold of hearing*. At a reference tone of 1,000 Hz, the threshold of hearing for an average person is taken to be a sound pressure of 20 μPa. The *threshold of pain,* or the greatest sound pressure that can be perceived without pain, is approximately 10 million times greater. It is therefore more convenient to use a *relative scale* of sound pressure rather than an absolute scale.

For this purpose, the bel, a unit of measure in electrical-communications engineering, is used. The decibel, abbreviated dB, is the preferred unit for measuring sound. One decibel is one-tenth of a bel and is the minimum difference in loudness that is usually perceptible. By definition, the decibel is a dimensionless unit used to express the logarithm of the ratio of a measured quantity to a reference quantity. In acoustics, the decibel is used to describe the level of quantities that are proportional to sound power.

*Sound power* (*W*) is the amount of energy per unit time that radiates from a source in the form of an acoustic wave. *Sound power level* ($L_W$), which is expressed in decibels relative to the reference power of $10^{-12}$ watt ($W_0$), expresses the total amount of sound power radiated by a sound source, regardless of the space into which the source is placed. The relationship is shown below.

$$L_W = 10 \log \frac{W}{W_0} \qquad (2)$$

where  *W* = sound power (watts)
  $W_0$ = reference power ($10^{-12}$ watts)
  log = a logarithm to the base 10

Consider for example, a large chipping hammer having a sound power of 1 watt. Expressing this sound power in decibels,

$$L_W = 10 \log \frac{W}{W_0} = 10 \log \frac{1}{10^{-12}}$$
$$= 10 \log 10^{12} = 120 \text{ db} \qquad (3)$$

As sound power is radiated from a point source in free space, the power is distributed over a spherical surface, so that at any given point there exists a certain sound power per unit area. This is designated as *intensity* and is measured in units of watts per square meter. Although intensity diminishes as distance from the source increases, the power that is radiated, being the product of the intensity and the area over which it is spread, remains constant.

Sound power cannot be measured directly. It is possible to measure intensity, but the instruments are expensive and must be used carefully. Under most conditions of sound radiation, sound intensity is proportional to the square of sound pressure. Sound pressure can be measured more easily, so sound level meters are built to measure *sound pressure level* ($L_p$) in decibels.

The sound level meter directly indicates sound pressure level referenced to a sound pressure of 20 μPa, the approximate threshold of hearing. The equation for sound pressure level is:

$$L_p = 10 \log \frac{p^2}{p_0^2} = 20 \log \frac{p}{p_0} \qquad (4)$$

where  *p* = measured root-mean-square (rms) sound pressure
  $p_0$ = reference rms sound pressure (20 μPa)

Note that the multiplier is 20 and not 10 as in the case of the sound power level equation. This is because sound power is proportional to the square of sound pressure and because $10 \log p^2 = 20 \log p$.

Table 9–B shows the relationship between sound pressure in micropascals and sound pressure level in decibels for some common sounds. The table also illustrates the advantage of using decibel notation rather than the wide range of pressure (or power). Note that a change of sound pressure by a factor of 10 corresponds to a change in sound pressure level of 20 dB. Also note that any range over which the sound pressure is doubled is equivalent to 6 dB whether at low or high levels. For example, sound pressures of 20 μPa and 40 μPa are equivalent to the following sound pressure levels:

*p* = 20μPa:

$$L_p = 20 \log \frac{p}{p_0} = 20 \log \frac{20}{20} = 20 \log 1 = 0 \text{ dB} \qquad (5)$$

$p = 40\mu\text{Pa}$:

$$L_p = 20 \log \frac{p}{P_0} = 20 \log \frac{40}{20} = 20 \log 2 = 6 \text{ dB} \qquad (6)$$

Although a doubling of sound pressure represents an increase of 6 dB in the sound pressure level, doubling the *sound power* results in an increase of 3 dB in the sound power level:

$$W = 1 \text{ watt: } L_W = 10 \log \frac{1}{10^{-12}} = 120 \text{ dB} \qquad (7)$$

$$W = 2 \text{ watt: } L_W = 10 \log \frac{2}{10^{-12}} = 123 \text{ dB} \qquad (8)$$

Again, as seen in the above examples, doubling the sound power increases the sound power level 3 dB, whereas doubling the sound pressure increases the sound pressure level 6 dB. These results are not contradictory because doubling the sound power is equivalent to doubling the *square* of the sound pressure. Remember, sound power is proportional to the *square* of sound pressure.

There is a common tendency to confuse sound power with sound pressure. Sound power and sound pressure can be illustrated simply with an analogy between light and sound.

Light bulbs are rated in terms of their power consumption (60-W bulbs, 25-W bulbs, etc.). From experience, we know that, in a given location, the intensity or illumination of a 60-W bulb is greater than that of a 25-W bulb at a given distance. Analogously, a sound source of 60 W produces a greater sound pressure level than a 25-W source at a given distance.

**Table 9–B.** Sound Pressure and Sound Pressure Level Values for Some Typical Sounds

| Sound Pressure (µPa) | Overall Sound Pressure Level (dB, re: 20 µPa) | Example |
|---|---|---|
| 20 | 0 | Threshold of Hearing |
| 63 | 10 | |
| 200 | 20 | Studio for sound pictures |
| 630 | 30 | Soft whisper (5 feet) |
| 2,000 | 40 | Quiet office; Audiometric testing booth |
| 6,300 | 50 | Average residence; Large office |
| 20,000 | 60 | Conversational speech (3 ft) |
| 63,000 | 70 | Freight train (100 ft) |
| 200,000 | 80 | Very noisy restaurant |
| 630,000 | 90 | Subway; Printing press plant |
| 2,000,000 | 100 | Looms in textile mill; Electric furnace area |
| 6,300,000 | 110 | Woodworking; Casting shakeout area |
| 20,000,000 | 120 | Hydraulic press; 50-HP siren (100 ft) |
| 200,000,000 | 140 | Threshold of pain; Jet plane |
| 20,000,000,000 | 180 | Rocket-launching pad |

A change of sound pressure by a factor of 10 corresponds to a change in sound pressure level of 20 dB.

Sound power is somewhat analogous to the power rating of the light bulb. A "weak" sound source would produce low sound levels, whereas a "stronger" sound source would produce higher sound levels. Sound power level is independent of the environmental surroundings. Sound pressure, on the other hand, is related to intensity and is analogous to the illumination produced by the light bulb. The magnitude of the sound pressure from a given sound source depends on the distance from the source. As discussed earlier, sound pressure is readily measured by a sound level meter, but sound power cannot be measured directly.

It is important to note that the decibel scale of measurement is not used only in the description of sound pressure level and sound power level. By definition, the decibel is a dimensionless unit related to the logarithm of the ratio of a measured quantity to a reference quantity. The decibel has no meaning unless a reference quantity is specified. Because of the mathematical properties of the logarithmic function, the decibel scale can compress data involving entities of large and small magnitude into a relative scale involving a small range of numbers. The decibel is commonly used to describe levels of such things as acoustic intensity, hearing thresholds, electrical voltage, electrical current, and electrical power, as well as sound pressure and sound power.

Because decibels are logarithmic values, it is not proper to add them by normal algebraic addition. For example, 60 dB plus 60 dB *does not* equal 120 dB but only 63 dB.

In order to show how to combine decibel levels of sound sources when given the power pressure, we present some examples.

### Example 1

Two sources are radiating noise in a free field. One source has a sound power level of 123 dB and the other source has a sound power level of 117 dB (re: $10^{-12}$ W). What is the combined sound power level of the two sources?

**Solution:**

$$L_W = 10 \log \frac{W}{W_0} \qquad (9)$$

or

$$\frac{W}{W_0} = 10^{L_W/10} \qquad (10)$$

$$\frac{W_1}{W_0} = 10^{L_{W_1}/10} = 10^{123/10} = 10^{12.3} = 1.995 \times 10^{12} \qquad (11)$$

$$\frac{W_2}{W_0} = 10^{L_{W_2}/10} = 10^{117/10} = 10^{11.7} = 5.012 \times 10^{11} \qquad (12)$$

$$\frac{W_1}{W_0} + \frac{W_2}{W_0} = 2.496 \times 10^{12} \qquad (13)$$

The combined sound power level of the two sources, $L_W$ (total), can then be calculated as

$$L_W \text{ (total)} = 10 \log\left(\frac{W_1}{W_0} + \frac{W_2}{W_0}\right)$$

$$= 10 \log (2.496 \times 10^{12}) \qquad (14)$$

$$= 10 (12.40)$$

$$L_W \text{ (total)} = 124 \text{ dB}$$

The same process can be used for sound pressure levels.

## Example 2

Suppose the sound pressure level of each of three individual noise sources is measured at a point such that with only the first source running, the $L_p = 86$ dB, with only the second source running it is 84 dB, and with only the third source running it is 89 dB (re: 20 µPa). What will the sound pressure level at the point be with all three sources running concurrently?

$$L_p = 10 \log\left(\frac{p}{p_0}\right)^2 \qquad (15)$$

or

$$\left(\frac{p}{p_0}\right)^2 = 10^{L_p/10} \qquad (16)$$

$$\left(\frac{p_{\text{total}}}{p_0}\right)^2 = 10^{L_{p_1}/10} + 10^{L_{p_2}/10} + 10^{L_{p_3}/10}$$

$$= 10^{8.6} + 10^{8.4} + 10^{8.9}$$

$$= (3.982 + 2.512 + 7.944)\, 10^8 \qquad (17)$$

$$\left(\frac{p_{\text{total}}}{p_0}\right)^2 = 14.438 \times 10^8$$

The sound pressure level at the point with all three sources running is then equal to

$$L_p \text{ (total)} = 10 \log\left(\frac{p_{\text{total}}}{p_0}\right)^2$$

$$= 10 \log (14.438 \times 10^8) \qquad (18)$$

$$= 10 (9.16)$$

$$L_p \text{ (total)} = 91.6 \text{ dB}$$

In general then, the procedure for adding decibels can be summarized as follows:

$$L_{\text{total}} = 10 \log\left(\sum_{i=1}^{N} 10^{L_i/10}\right) \qquad (19)$$

where $L$ can be sound power level or sound pressure level.

It is often adequate to use the simplified schedule shown below for adding decibels.

| Difference in Decibel Values | Add to Higher Value |
|---|---|
| 0 or 1 dB | 3 db |
| 2 or 3 dB | 2 dB |
| 4 to 9 dB | 1 dB |
| 10 dB or more | 0 dB |

Examples: 83 dB + 82 dB = 86 dB
83 dB + 80 dB = 85 dB
83 dB + 78 dB = 84 dB
83 dB + 73 dB = 83 dB

More than two levels can be combined using the above simplified schedule by taking the combinations in pairs.

## Example 3

When measured at the same location, four noise sources have sound pressure levels of 89, 87, 78, and 81 dB, respectively. What would the sound pressure level at this location be if all four sources were running concurrently? Using the simplified method,

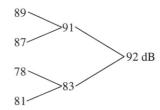

Using the calculated method,

$$L_p = 10 \log (10^{8.9} + 10^{8.7} + 10^{7.8} + 10^{8.1}) = 91.7 \text{ dB} \qquad (20)$$

Although the simplified method is less accurate than the calculated method, the difference may not be significant. Even so, when accurate results are required, the calculated method should be used.

## Loudness

Although loudness depends primarily on sound pressure, it is also affected by frequency. (Pitch is closely related to frequency.) The reason for this is that the human ear is more sensitive to high-frequency sounds than it is to low-frequency sounds.

The upper limit of frequency at which airborne sounds can be heard depends primarily on the condition of a person's hearing and on the intensity of the sound. For young adults, this upper limit is usually somewhere between 16,000 and 20,000 Hz. For most practical purposes, the actual figure is not important. It is important, however, to realize that most people lose sensitivity for the higher-frequency sounds as they grow older (presbycusis).

The complete hearing process seems to consist of a number of separate processes that, in themselves, are fairly complicated. No simple relationship exists between the physical measurement of a sound pressure level and the human perception of the sound. One pure tone may sound louder than another pure tone, even though the measured sound pressure level is the same in both cases.

Sound pressure levels, therefore, are only a part of the story and can be deceiving. The fundamental problem is that the quantities to be measured must include a person's reaction to the sound—a reaction that can be determined by such varied factors as the state of the person's health, characteristics of the sound, and the person's attitude toward the device or the person that generates the sound. In the course of time, various loudness level–rating methods have been suggested, and a number of different criteria for tolerable noise levels have been proposed.

A complete physical description of sound must include its frequency spectrum, its overall sound pressure level, and the variation of both of these quantities over time. Loudness is the subjective human response to sound pressure and intensity. At any given frequency, loudness varies directly as sound pressure and intensity vary, but not in a simple, straight-line manner.

The physical characteristics of a sound as measured by an instrument and the "noisiness" of a sound as a subjective characteristic may bear little relationship to one another. A sound level meter cannot distinguish between a pleasant sound and an unpleasant one. A human reaction is required to differentiate between a pleasant sound and a noise. Loudness is not merely a question of sound pressure level. A sound that has a constant sound pressure can be made to appear quieter or louder by changing its frequency.

**Equal-loudness contours.** Results of experiments designed to determine the response of the human ear to sound were reported by Fletcher and Munson in 1933. A reference tone and a test tone were presented alternately to the test subjects (young men), who were asked to adjust the level of the test tone until it sounded as loud to them as the reference tone (1,000 Hz). The results of these experiments yielded the familiar Fletcher–Munson, or equal-loudness, contours (Figure 9–6). The contours represent the sound pressure level necessary at each frequency to produce the same loudness response in the average listener. The nonlinearity of the ear's response is represented by the changing contour shapes as the sound pressure level is increased, a phenomenon that is particularly noticeable at low frequencies. The lower, dashed, curve indicates the threshold of hearing, which represents the sound pressure level necessary to trigger the sensation of hearing in the average listener. The actual threshold varies as much as ±10 dB among healthy individuals.

**Sound pressure weighting.** It would seem relatively simple to build an electronic circuit whose sensitivity varied

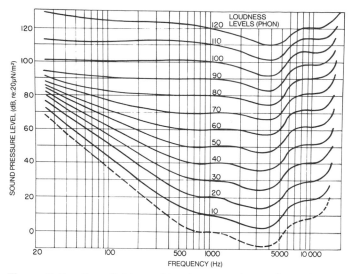

**Figure 9–6.** Free-field equal-loudness contours of pure tones. Because the human ear is more sensitive to the higher frequencies of sound, changing the frequency of a sound changes its relative loudness. These are also called Fletcher–Munson contours. (Adapted from the *Handbook of Noise Measurement,* 9th ed. GenRad, Inc., 1980.)

with frequency in the same way as the human ear. This has in fact been done and has resulted in three different internationally standardized characteristics called weighting networks A, B, and C. The A-network was designed to approximate the equal-loudness curves at low sound pressure levels, the B-network was designed for medium sound pressure levels, and the C-network was designed for high levels.

The weighting networks are the sound level meter's means of responding to some frequencies more than to others. The very low frequencies are discriminated against (attenuated) quite severely by the A-network, moderately attenuated by the B-network, and hardly attenuated at all by the C-network (Figure 9–7). Therefore, if the measured sound level of a noise is much higher on C-weighting than

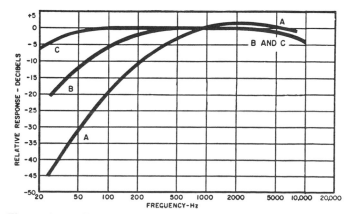

**Figure 9–7.** Frequency-response attenuation characteristics for the A-, B-, and C-weighting networks.

on A-weighting, much of the noise energy is probably of low frequency.

By definition, a weighted-frequency scale is simply a series of correction factors that are applied to sound pressure levels on an energy basis as a function of frequency. Shown in Table 9–C are the corrections for the A-weighting network at each of the octave-band center frequencies commonly used in noise measurements.

The A-weighted sound level measurement has become popular in the assessment of overall noise hazard because it is thought to provide a rating of industrial broadband noises that indicates the injurious effects such noise has on the human ear.

As a result of its simplicity in rating the hazard to hearing, the A-weighted sound level has been adopted as the measurement for assessing noise exposure by the American Conference of Governmental Industrial Hygienists (ACGIH). The A-weighted sound level as the preferred unit of measurement was also adopted by the U.S. Department of Labor as part of its Occupational Safety and Health Standards. The A-weighted sound levels have also been shown to provide reasonably good assessments of speech interference and community disturbance conditions and have been adopted by the EPA for these purposes (Figure 9–8).

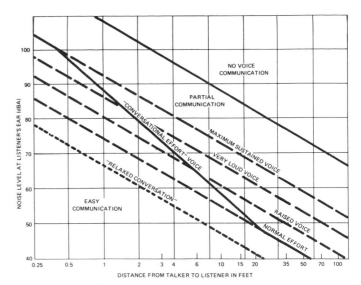

**Figure 9–8.** Distance at which ordinary speech can be understood (as a function of the A-weighted sound levels of the masking noise in an outdoor environment). (Reprinted from *Public Health and Welfare Criteria of Noise*, July 27, 1973, U.S. Environmental Protection Agency.)

## ■ Occupational Damage-Risk Criteria

The purpose of damage-risk criteria is to define maximum permissible noise levels during given periods that, if not exceeded, would result in acceptable small changes in the hearing levels of exposed employees over a working lifetime. The acceptability of a particular noise level is a function of many variables.

Increasing attention is being given by regulatory agencies and industrial and labor groups to the effects of noise exposures on employees; therefore, equitable, reliable, and practical damage-risk noise criteria are needed.

A criterion is a standard, rule, or test by which a judgment can be formed. A criterion for establishing levels for

damage risk noise requires one or more standards for judgment. Damage-risk criteria can be developed once standards are selected by which the effects of occupational noise exposure on employees can be judged.

### Hearing Ability

Tests for evaluating the ability to hear speech have been developed. These tests generally fall into two classes: those that measure the hearing threshold or the ability to *hear* very faint speech sounds and those that measure discrimination, or the ability to *understand* speech (see Chapter 4, The Ears).

Ideally, hearing impairment should be evaluated in terms of an individual's ability (or inability) to hear normal speech under everyday conditions. The ability of an individual to hear sentences and to repeat them correctly in a quiet environment is considered to be satisfactory evidence of adequate hearing ability. Hearing tests using pure tones are extensively employed to monitor the status of a person's hearing and the possible progression of a hearing loss. A person's ability to hear pure tones is related to the hearing of speech.

People who work in noisy environments should have their hearing checked periodically to determine whether the noise exposure is producing a detrimental effect on hearing. The noise-induced hearing losses that can be measured by pure-tone audiometry are the threshold shifts that constitute a departure from a specified baseline. This baseline, or normal hearing level, can be defined as the average hearing threshold of a group of young people who have no history of previous exposure to intense noise and no otological malfunction.

**Table 9–C.** Octave-Band Correction Factors of the A-Weighted Network

| Octave-Band Center Frequency (Hz) | A-Network Correction Factor (dB) |
| --- | --- |
| 31.5 | −39.4 |
| 63 | −26.2 |
| 125 | −16.1 |
| 250 | −8.6 |
| 500 | −3.2 |
| 1,000 | 0 |
| 2,000 | +1.2 |
| 4,000 | +1.0 |
| 8,000 | −1.1 |

**AAOO–AMA guide.** After years of studying the various methods and procedures used to determine the presence and extent of hearing damage, the Committee on Conservation of Hearing of the American Academy of Ophthalmology and Otolaryngology (AAOO) issued a report that was published by the American Medical Association (AMA): the *Guide for the Evaluation of Hearing Impairment* (see Chapter 4). The report recommended that the average hearing level for pure tones at 500 Hz, 1,000 Hz, and 2,000 Hz be used as an indirect measure of the probable ability to hear everyday speech. If the average monaural hearing level at 500, 1,000, and 2,000 Hz is 25 dB or less, the AMA report stated that no impairment in the ability to hear everyday speech usually occurs under everyday conditions (see Chapter 4, The Ears).

## Risk Factors

If the ear is subjected to high levels of noise for a sufficient period of time, some loss of hearing will occur. There are many factors that affect the degree and extent of hearing loss, including the following:

- The intensity of the noise (sound pressure level)
- The type of noise (frequency spectrum)
- The period of exposure each day (worker's schedule per day)
- The total work duration (years of employment)
- Individual susceptibility
- The age of the worker
- Coexisting hearing loss and ear disease
- The character of the surroundings in which the noise is produced
- The distance from the source
- The position of the ear with respect to sound waves

The first four factors are the most important, and they are called *noise exposure* factors. Thus, it is necessary to know not only how much noise is present, but also what kind of noise it is and its duration.

Because of the complex relationship of noise and exposure time to threshold shift (reduction in hearing level) and its many possible contributory causes, the criteria designed to protect workers from hearing loss took many years to develop and establish.

A relatively recent effort to establish a basis for reliable noise criteria was that of the Intersociety Committee on Guidelines for Noise Exposure Control. A significant part of their report is shown graphically in Figure 9–9. The curves in the figure relate the incidence of significant hearing loss to age and the magnitude of noise exposure over a working lifetime.

Without attempting to explain the full significance of the graph, it can be stated that 20 percent of the general population between the ages of 50 and 59 experience

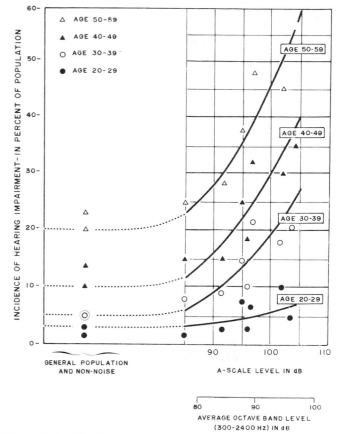

**Figure 9–9.** The incidence of hearing impairment in the general population and in selected populations by age group and by occupational noise exposure. (Reprinted with permission from the *American Industrial Hygiene Association Journal.*)

hearing losses without having had any exposure to industrial noise, but groups of workers exposed to steady-state industrial noise over a working lifetime show a greater increase in the incidence of hearing loss.

For example, exposure to steady-state noise at 90 dB on the A-scale of the sound level meter (90 dBA) results in significant hearing losses in 27 percent of the exposed group. If the working lifetime exposure is 95 dBA, 36 percent of the group shows significant hearing loss.

Essentially, the graph in Figure 9–9 supplies industry and other interested groups with information from which the risk of developing compensable hearing loss among groups of workers exposed to noises of different magnitudes can be predicted.

## Analysis of Noise Exposure

The critical factors in the analysis of noise exposure are the A-weighted sound level; the frequency composition, or spectrum, of the noise; and the duration and distribution of noise exposure during a typical workday.

It is currently believed that any exposure of the unprotected ear to sound levels above 115 dBA is hazardous and should be avoided. Exposure to sound levels below 70 dBA can be assumed to be safe and do not produce any permanent hearing loss. The majority of industrial noise exposures fall within this 45-dBA range; thus, additional information is required for evaluation of damage risk, such as the type of noise and duration of exposure.

It would be very helpful to know the predominant frequencies present and the contributions from each of the frequency bands that make up the overall level. It is currently believed that noise energy with predominant frequencies above 500 Hz has a greater potential for causing hearing loss than noise energy concentrated in the low-frequency regions. It is also believed that noises that have a sharp peak in a narrow-frequency band (such as a pure tone) present a greater hazard to hearing than noises of equal energy levels that have a continuous distribution of energy across a broad frequency range.

The incidence of noise-induced hearing loss is directly related to total exposure time. In addition, it is believed that intermittent exposures are far less damaging to the ear than are continuous exposures, even if the sound pressure levels for the intermittent exposures are considerably higher than those during continuous exposures. The rest periods between noise exposures allow the ear to recuperate.

At present, the deleterious effects of noise exposure and the energy content of the noise cannot be directly equated. For example, doubling the energy content does not produce twice the hearing loss. In general, however, the greater the total energy content of the noise, the shorter the time of exposure required to produce the same amount of hearing loss. However, the exact relation between time and energy is not known.

Another factor that should be considered in the analysis of noise exposures is the type of noise. For instance, impact noise is generated by drop hammers and punch presses, whereas steady-state noise is generated by turbines and fans. Impact noise is a sharp burst of sound; therefore, sophisticated instrumentation is necessary to determine the peak levels for this type of noise. Additional research must be done to fully define the effects of impact noise on the ear (Table 9–D).

The total noise exposure during a person's normal working lifetime must be known to arrive at a valid judgment of how noise will affect that person's hearing. Instruments such as noise dosimeters can be used to determine the exposure pattern of a particular individual. Instruments such as sound level meters can be used to determine the noise exposure at a given instant in time (that is, during the time the test is being taken). An exposure pattern can be established using a series of such tests and the work history of the individual.

# ■ Sound-Measuring Instruments

A wide assortment of equipment is available for noise measurements, including sound survey meters, sound level meters, octave-band analyzers, narrowband analyzers, noise dosimeters, tape and graphic level recorders, impact sound level meters, and equipment for calibrating these instruments.

For most noise problems encountered in industry, the sound level meter and octave-band analyzers provide ample information (Figure 9–10).

## Sound Level Meters

The basic instrument used to measure sound pressure variations in air is the sound level meter. This instrument contains a microphone, an amplifier with a calibrated attenuator, a set of frequency-response networks (weighting networks), and an indicating meter (Figure 9–11). The sound level meter is a sensitive electronic voltmeter that measures the electrical signal emitted from a microphone, which is ordinarily attached to the instrument. The alternating electrical signal emitted from the microphone is amplified sufficiently so that, after conversion to direct current by

**Table 9–D.** Acceptable Exposure to Noise (in dBA) as a Function of the Number of Occurrences per Day

| Daily Duration | | Number of Times the Noise Occurs per Day | | | | | | |
|---|---|---|---|---|---|---|---|---|
| Hours | Minutes | 1 | 3 | 7 | 15 | 35 | 75 | 160 + |
| 8 | | 90 | 90 | 90 | 90 | 90 | 90 | 90 |
| 6 | | 91 | 93 | 96 | 98 | 97 | 95 | 94 |
| 4 | | 92 | 95 | 99 | 102 | 104 | 102 | 100 |
| 2 | | 95 | 99 | 102 | 106 | 109 | 114 | |
| 1 | | 98 | 103 | 107 | 110 | 115 | | |
| | 30 | 101 | 106 | 110 | 115 | | | |
| | 15 | 105 | 110 | 115 | | | | |
| | 8 | 109 | 115 | | | | | |
| | 4 | 113 | | | | | | |

This table summarizes the results of TTS studies, which can be used to estimate the effect of intermittency of noise exposures on the risk of hearing impairment. The information in the table can be approximated by the simple rule that for each halving of daily exposure time, the noise level can be increased by 5 dB without increasing the hazard of hearing impairment. To use the table, select the column headed by the number of times the noise occurs per day, read down to the average sound level of the noise, and locate directly to the left, in the first column, the total duration of noise permitted for any 24-hour period. It is permissible to interpolate if necessary. Noise levels are in dBA. (Adapted from Intersociety Committee Report [1970] "Guidelines for Noise Exposure Control," *Journal of Occupational Medicine,* July 1970, Vol. 12, No. 7.)

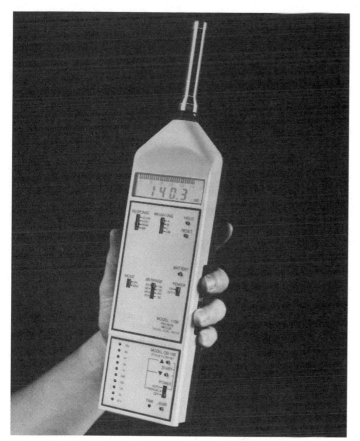

**Figure 9–10.** The multipurpose instrument shown here can be used as a sound level meter, octave-band analyzer, and impact/impulse noise meter. (Courtesy Quest Technologies.)

means of a rectifier, the signal can deflect a needle on an indicating meter. An attenuator controls the overall amplification of the instrument. The response-versus-frequency characteristics of the amplified signal are controlled by the weighting networks.

Some sound level meters have a measurement range of about 40–140 dB (re: 20 μPa) without the aid of special accessory equipment. Special microphones permit measurement of lower or of considerably higher sound levels. An amplifier that can register the electrical output signal of the microphone is usually provided with the sound level

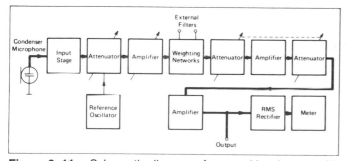

**Figure 9–11.** Schematic diagram of a sound level meter with auxiliary output.

meter so that it can be hooked up to other instruments for recording and analysis. The sound level meter is designed to be a device for field use and as such should be reliable, rugged, reasonably stable under battery operation, and lightweight.

**Microphone.** The microphone responds to sound pressure variations and produces an electrical signal that is processed by the sound level meter.

**Amplifier.** The amplifier in a sound level meter must have a high available gain so that it can measure the low-voltage signal from a microphone in a quiet location. It should have a wide frequency range, usually on the order of 20–20,000 Hz. The range of greatest interest in noise measurements is 50–6,000 Hz. The inherent electronic noise floor and hum level of the amplifier must be low.

**Attenuators.** Sound level meters are used for measuring sounds that differ greatly in level. A small portion of this range is covered by the relative deflection of the needle on the indicating meter. The rest of the range is covered by an adjustable attenuator, which is an electrical resistance network inserted into the amplifier to produce known ranges of signal level. To simplify use, it is customary to have the attenuator adjustable in steps of 10 dB.

In some instances, the attenuator can be split into sections among various amplifying stages to improve the signal-to-noise ratio of the instrument and to limit the dynamic range.

**Weighting networks.** The sound level meter response at various frequencies can be controlled by electrical weighting networks. The response curves for these particular networks have been established in the American National Standards Institute's publication ANSI S1.41–1983. C-weighting approximates a uniform response over the frequency range from 25–8,000 Hz. Changes in the electronic circuit are sometimes made to compensate for the response of particular microphones, so that the net response is uniform (flat) within the tolerance allowed by the standards. The C-weighting network is generally used when the sound level meter supplies a signal to an auxiliary instrument for a more detailed analysis. (The weighting networks are shown in Figure 9–7.) The A-weighting network is used to determine compliance with the OSHA standard.

**Metering system.** After the electrical signal from the microphone is amplified and sent through the attenuators and weighting networks, the signal is used to drive a metering circuit. This metering circuit displays a value that is proportional to the electrical signal applied to it. The ANSI S1.4–1983 standard for sound level meters specifies that the rms value of the signal should be indicated. This requirement corresponds to adding up the different components of the sound wave on an energy basis. When measuring sound, the rms value is a useful indication of the general energy content.

Because the indicating needle on the dial of a sound level meter cannot follow rapidly changing variations in sound pressure, a running average of the rectified output of the metering circuit is shown. The average time (or response speed) is determined by the meter ballistics and the response circuit (which is chosen with a switch).

Two meter-needle ballistic modes of operation—fast and slow—are provided on every sound level meter. In the fast mode, the needle responds relatively quickly to rapidly changing noise levels, whereas in the slow mode, the needle responds rather slowly. The use of each mode is best illustrated by example. If one were to measure the noise level of a passing vehicle, the fast mode would be used to obtain the maximum level. In a factory, where an average noise level is often more useful, the slow mode would be selected to reduce rapid, hard-to-read needle excursions. OSHA requires the use of slow response for measurements to check for compliance with its regulations.

## Octave-Band Analyzers

For many industrial noise problems, it is necessary to use some type of analyzer to determine where the noise energy lies in the frequency spectrum. This is especially true if engineering control of noise problems is planned, because industrial noise is made up of various sound intensities at various frequencies (Figure 9–12).

In order to properly represent the total noise of a noise source, it is usually necessary to break the total noise down into its various frequency components—low frequency, high frequency, or middle frequency. This is necessary for two reasons: People react differently to low-frequency and high-frequency noises (for the same sound pressure level, high frequency noise is much more disturbing and is more capable of producing hearing loss than is low-frequency noise); and the engineering solutions for reducing or controlling noise are different for low-frequency and high-frequency noise (low-frequency noise is more difficult to control, in general).

It is conventional practice in acoustics to determine the frequency distribution of a noise by passing that noise successively through several different filters that separate the noise into 8 or 9 octaves on a frequency scale. Just as with an octave on a piano keyboard, an octave in sound analysis represents the frequency interval between a given frequency (such as 250 Hz) and twice that frequency (500 Hz).

A young, healthy ear is sensitive to sound frequency in the range from about 20 to 20,000 Hz. Most octave-band analyzing filters now cover the audio range from about 22 Hz to about 11,300 Hz in nine octave-frequency bands. These filters are identified by their geometric mean frequencies as shown in Table 9–E.

Notice that these filters are constant-percentage filters. The width of the band being utilized (bandwidth) is a fixed percentage of the frequency at which the instrument is operating. Octave-band filters have bandwidths that are 70.7 percent of the mean frequency (this is easily seen in the 1-kHz band).

For a more detailed analysis of the distribution of sound energy as a function of frequency, still narrower bands are used. The next commonly used division is a split of the octave into three parts.

Some of the mean frequencies for such a series would be, for example, 100, 125, 160, 200, 250, 315, 400, 500, 630, and 800 Hz. One-third–octave filters have bandwidths that are 23.2 percent of the mean frequency.

Still narrower band filters are available, such as one-tenth octave-band. The narrower the band for analysis, the more sharply defined the data.

The identification of pure tone components, when present, is an extremely useful diagnostic tool for locating and quieting the noise source.

Some noise sources have a well-defined frequency content. For example, the hum generated by a fan or blower is usually centered at the blade-passage frequency, which is the product of the number of blades of the fan multiplied by the speed (revolutions per second) of the fan.

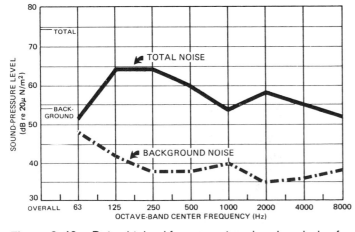

**Figure 9–12.** Data obtained from an octave-band analysis of a noise source, showing the total noise and the background noise levels when the noise source is not in operation.

**Table 9–E.** Octave-Band Mean Frequencies and Corresponding Band Limits ANSI S1.11–1986 (R1993)

| Lower Band Limit (Hz) | Geometric Mean Frequency of Band (Hz) | Upper Band Limit (Hz) |
|---|---|---|
| 22 | 31.5 | 44 |
| 44 | 63 | 88 |
| 88 | 125 | 177 |
| 177 | 250 | 354 |
| 354 | 500 | 707 |
| 707 | 1,000 | 1,414 |
| 1,414 | 2,000 | 2,828 |
| 2,828 | 4,000 | 5,656 |
| 5,656 | 8,000 | 11,312 |

The fundamental blade-passage frequency ($f_B$) of fans is given by the following equation:

$$f_B = \frac{(\text{rpm})\,(N)}{60} \qquad (21)$$

where  rpm = shaft rotational speed (revolutions per minute)
  $N$ = number of blades
  60 = a constant to convert rpm to revolutions per second

Higher harmonics are usually present with diminished sound pressure amplitude at integral multiples of the fundamental (that is, $2f_B$, $3f_B$, . . .).

The above relation can also be used to predict the fundamental tones from blowers, gears, and so on by letting $N$ represent the number of impeller lobes or gear teeth. For example, consider the following.

### Example

What is the frequency of the predominant tone that would be emitted from an axial fan with four blades rotating at 6,000 rpm?

$$f_B = \frac{(\text{rpm})\,(N)}{60} = \frac{(6,000)\,(4)}{60} = 400 \text{ Hz} \qquad (22)$$

The fundamental blade-passage frequency is then 400 Hz. One would also expect additional tones at integral multiples of the predominant tone at

$$f_2 = (2)(400) = 800 \text{ Hz}$$
$$f_3 = (3)(400) = 1200 \text{ Hz}$$
$$\vdots$$
$$f_N = (N)(400) \text{ Hz}$$

The higher-frequency tones would have progressively diminished sound pressure amplitude.

Electrical transformers usually hum in frequencies that are multiples of 60 Hz. Positive-displacement pumps have a sound pressure distribution that is directly related to the pressure pulses on either the inlet or the outlet of the pump.

Noise resulting from the discharge of steam- or air-pressure relief valves has a frequency peak that is related to the pressure in the system and the diameter of the restriction preceding the discharge to the atmosphere. A peak energy content in any single octave band would provide information as to the predominant frequency of a particular noise source.

Sound level meters have evolved from relatively simple devices capable of measuring weighted sound levels and performing octave-band frequency analysis to highly sophisticated instruments that serve as the front end of a data acquisition system.

The instrument shown in Figure 9–13, for example, not only functions as a sound level meter and octave- or one-third octave-band analyzer but is also capable of measuring all the relevant parameters defined by OSHA and ANSI standards for industrial and environmental noise. The output of this instrument can be recorded, printed, or transferred to a personal computer for additional analysis or graphical presentation.

## Noise Dosimeters

In many work environments, it may not be adequate to measure noise exposure at a fixed location for the duration of a workshift. Some workers move about to several locations in the course of their duties or perform a variety of operations during the day and are therefore subjected to different noise levels. The practical way to measure the noise exposure in these circumstances is with a noise-exposure monitor, or dosimeter, that can be worn by the worker and that moves with the worker during the day. The noise dosimeter records the noise energy to which the worker is exposed during the workshift.

A dosimeter (Figure 9–14) includes a microphone placed in the person's hearing zone and the remainder of the instrument, which automatically computes the desired noise measures—most commonly the daily noise dose. This measure is used to check for compliance with the OSHA noise standard.

Dosimeters have evolved from simple devices that compute single-number exposure measures to highly sophisticated monitors that compute and store comprehensive data on the sound field encountered by the subject. The instrument shown in Figure 9–15, for example, not only functions as a noise dosimeter, but also as a multipurpose integrating sound level meter. It is also capable of producing several types of printed reports, including statistical analyses of data and graphical time history reports.

**Figure 9–13.** Precision integrating sound level meter. (Courtesy Bruel & Kjaer Instruments, Inc.)

**Figure 9–14.** Dosimeter being worn to monitor noise exposure. (Courtesy Quest Technologies.)

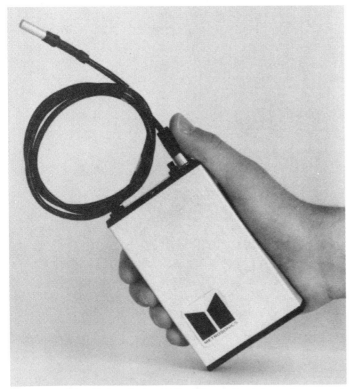

**Figure 9–15.** Noise dosimeter/multipurpose integrating sound level meter. (Courtesy Metrosonics, Inc.)

It is important to recognize the fact that dosimeters were derived directly from sound level meters. Dosimeters were developed to simplify measurement and computational procedures. In order to obtain comparable results, dosimeters must correctly duplicate the dynamic characteristics of sound level meters. Specific requirements prescribing these characteristics are set forth in the American National Standards Institute (ANSI) Standard S1.25–1991, *Specification for Personal Noise Dosimeters.*

## Sound Surveys

Sound measurement falls into two broad categories: source measurement and ambient-noise measurement. Source measurements involve the collection of acoustic data for the purpose of determining the characteristics of noise radiated by a source. The source might be a single piece of equipment or a combination of equipment or systems. For example, a single electrical motor or an entire facility can be considered a noise source.

Ambient-noise measurements can be used to study a single sound level or to make a detailed analysis showing hundreds of components of a complex vibration. The number of measurements taken and the type of instruments needed depend on the information that is required. If compliance with a certain noise specification must be checked, the particular measurement required is reasonably clear. Only some guidance as to the selection of instruments and their use is needed. But if the goal is to reduce the noise produced by industrial operations in general, the situation is more complex, and careful attention to the acoustic environment is essential.

Measurement of the noise field may require using different types of sound level–measuring instruments. These measurements must be repeated as changes in noise-producing equipment or operating procedures occur.

The use of the dBA scale for preliminary noise measurement greatly simplifies the collection of sound level survey data. Detailed sound level survey and octave-band analysis data are necessary to provide sufficient information so that the proper remedial measures for noise-control procedures can be determined. Calibration checks of the instruments should be made before, during, and after the sound level survey.

### Source Measurements

Source measurements frequently are made in the presence of noise created by other sources that form the background- or ambient-noise level. Although it is not always possible to make a clear distinction between source- and ambient-noise measurement, it is important to understand that source measurements describe the characteristics of a particular sound source, while ambient-noise measurements describe the characteristics of a sound field of largely unspecified or unknown sources.

A uniform, standard reporting procedure should be established to ensure that sufficient data are collected in a form suitable for subsequent analysis. To be effective, this standard reporting procedure should include detailed descriptions of the techniques of measurement position, operating conditions, instrument calibration, exposure time, amplitude patterns, and other important variables.

Several forms have been devised to record data obtained during a screening survey. Use of these forms facilitates the recording of pertinent information that will be extremely useful if more detailed studies are conducted later. An employee noise-exposure survey is conducted by measuring noise levels at each workstation that an employee occupies throughout the day or by acquiring a sufficient sampling of data at each workstation so that the exposure of an employee while at that workstation can be evaluated. Workstations that pose particular noise-exposure hazards can be readily identified by using measured sound level contours if these are obtained in adequate detail.

In many industrial situations, however, it is extremely difficult to accurately evaluate the noise exposure to which a particular worker is subjected. This is due in part to the fact that the noise level to which the stationary worker is exposed throughout the workday fluctuates, making it difficult to evaluate compliance or noncompliance with the OSHA regulations. Another problem arises when a worker's job requires that he or she spend time in areas where the noise levels vary from very low to very high.

Because of the fluctuating nature of many industrial noise levels, it would not be accurate or meaningful to use a single sound level meter reading to estimate the daily time-weighted average (TWA) noise level.

## Preliminary Noise Survey

A hearing conservation program should start with a preliminary facilitywide noise level survey using appropriate sound level–measuring equipment to locate operations or areas where workers may be exposed to hazardous noise levels.

Those conducting the survey have to decide whether to purchase sound level–measuring equipment and train personnel to use it or to contract the work to an outside firm. The extent of the noise problem, the size of the facility, and the nature of the work affect this decision. In most facilities, noise surveys are conducted by a qualified engineer, an audiologist, an industrial hygienist, or a safety and health professional.

A noise survey should be carried out at work areas where it is difficult to communicate in normal tones. A common rule of thumb is that if you have to shout to communicate at a distance of three feet, noise levels may be excessive. A noise survey should also be performed if,

after being exposed to high noise levels during their workshift, workers notice that speech and other sounds are muffled for several hours or they develop ringing in the ears.

As a general guideline for conducting a noise survey, the information recorded should be sufficient to allow another individual to take the report, use the same equipment, find the various measurement locations, and, finally, reproduce the measured and/or recorded data.

The preliminary noise survey normally does not define the noise environment in depth and therefore should not be used to determine employee exposure time and other details. The preliminary noise survey simply supplies sufficient data to determine whether a potential noise problem exists and, if so, to indicate how serious it is.

## Detailed Noise Survey

From the preliminary noise survey, it is relatively easy to determine specific locations that require more detailed study and attention. A detailed noise study should then be made at each of these locations to determine employees' TWA exposures.

The purposes of a detailed noise survey are the following:

■ To obtain specific information on the noise levels existing at each employee's work station

■ To develop guidelines for establishing engineering and/or administrative controls

■ To define areas where hearing protection will be required

■ To identify those work areas where audiometric testing of employees is desirable and/or required

In addition, detailed noise survey data can be used to develop engineering control policies and procedures and to determine whether specific company, state, or federal requirements have been complied with.

An effective hearing conservation program always starts with the question, "Does a noise problem exist?" The answer must not be based simply on the subjective feeling that the problem exists but on the results of a careful technical definition of the problem. Answers to the following questions must be obtained:

■ How noisy is each work area?

■ What equipment or process is generating the noise?

■ Which employees are exposed to the noise?

■ How long are they exposed?

Line supervisors can provide basic job function information concerning the duration of operation, the types of noise-producing equipment in work areas, and the percentage of time a worker spends in each of the areas.

**Figure 9–16.** Take sound level measurements near an employee's workstation.

Production records can be examined, and on-site evaluation can provide information as to the extent of the noise problems.

The noise survey should be made using a general-purpose sound level meter that meets standards set by ANSI S1.4 1983. The sound level meter should be set for A-scale slow response.

Measurements of noise exposure should be taken at approximate ear level (Figure 9–16). No worker should be exposed to steady-state or interrupted steady-state sound levels that exceed the maximum listed in the current noise standard. Other information should include the name of the individual making the noise survey as well as the date, location of measurement, and time the measurement was made. The serial numbers of the sound level meters and the date of calibration are also essential for compliance records.

The noise survey procedure is a three-step process.

**Step 1: Area measurements.** Using a sound level meter set for A-scale slow response, the regularly occurring maximum noise level and the regularly occurring minimum noise level are recorded at the center of each work area. (For measurement purposes, the size of the work area should be limited to 1,000 sq ft [93 sq m] or smaller.) If the maximum sound level in a work area does not exceed 80 dBA, it can be assumed that all employees in that area are working in an environment with an acceptable noise level. If the noise levels measured at the center of the work area fall between 80 and 92 dBA, then more information is needed.

Sound level contours (Figure 9–17) can be used during this screening survey to identify workstations that may pose particular noise-exposure hazards. Sound level contours provide a visual depiction of the degree of the noise hazard in a work area, at a particular time.

To construct a sound level contour, the work area is divided into a grid whose lines are evenly spaced at an approximate distance of 10 ft (3.05 m). A-weighted sound level measurements are recorded at each measurement position. When the observed sound levels vary significantly, then the grid spacing should be decreased. It is usually necessary to decrease the grid spacing in the proximity of dominant noise sources due to rapidly changing sound levels. The contour lines drawn in Figure 9–17 are based on 2-dBA changes in the measured sound level.

**Step 2: Workstation measurements.** To evaluate the noise exposure for people working in locations where measurements at the center of the work area range from 80–92 dBA, measurements should be made at each employee's normal workstation. If the level varies on a regular basis, both the maximum and minimum levels should be recorded. If the noise level never goes below 90 dBA, an unsatisfactory noise exposure is indicated. If the measured level is never greater than 85 dBA, the noise exposure to which the employee is subjected can be regarded as satisfactory.

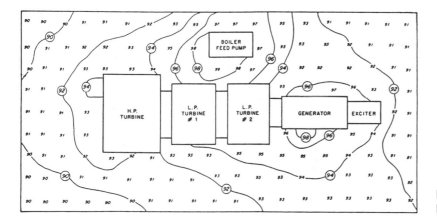

**Figure 9–17.** Sound level contours: operating-level turbine building. (From Di Blasi et al., 1983.)

**Step 3: Exposure duration.** At workstations where the regularly occurring noise varies above and below the 85-dBA level, further analysis is needed.

If an employee has varying work patterns in different work areas, it is necessary to ascertain the sound level and duration of noise exposure within each work area. A breakdown of hours worked in each area can be obtained by consulting the employee or the employee's supervisor, or by visual monitoring. A briefing/debriefing approach for a particular day's activities can also be used. This approach consists of requesting each employee to keep a general work area/time log of his or her daily activities. The employee is then debriefed at the end of the work period to ensure that sufficient information was logged. In many cases, it may be desirable for an employee to wear a noise dosimeter that records daily exposure in terms of the current OSHA requirements.

The procedure for determining an employee's daily noise-exposure rating is discussed in the section that follows.

## General Classes of Noise Exposure

There are three general classes into which occupational noise exposures can be grouped: continuous noise, intermittent noise, and impact-type noise.

**Continuous noise.** Continuous noise is normally defined as broadband noise of approximately constant level and spectrum to which an employee is exposed for a period of 8 hours per day, 40 hours per week. A large number of industrial operations fit into this class of noise exposure. Most damage-risk criteria are written for this type of noise exposure because it is the easiest to define in terms of amplitude, frequency content, and duration.

The OSHA Noise Standard, 29 CFR 1910.95(a) and (b), established permissible employee noise exposures in terms of duration in hours per day at various sound levels. The standard requires that the employer reduce employee exposures to the allowable level by use of feasible engineering or administrative controls. The standard defines the permissible exposure level (PEL) as that noise dose that would result from a continuous 8-hour exposure to a sound level of 90 dBA. This is a dose of 100 percent. Doses for other exposures that are either continuous or fluctuating in level are computed relative to the PEL based on a 5-dBA trading relationship between noise level and exposure time (Table 9–F). Every 5-dBA increase in noise level cuts the allowable exposure time in half. This is known as a 5-dB exchange rate.

When employees are exposed to different noise levels during the day, the mixed exposure, $E_{\mathrm{m}}$, must be calculated by using the following formula:

$$E_m = \frac{C_1}{T_1} + \frac{C_2}{T_2} + \frac{C_3}{T_3} + \cdots + \frac{C_n}{T_n} \quad (23)$$

where $C_n$ = the amount of time an employee was exposed to noise at a specific level

$T_n$ = amount of time the employee can be permitted to be exposed to that level

If the sum of the fractions equals or exceeds 1, the mixed exposure is considered to exceed the allowable limit value, according to the OSHA standard. Daily noise dose ($D$) is an expression of $E_m$ in percentage terms. For example, $E_m = 1$ is equivalent to a noise dose of 100 percent. (Note that OSHA does not consider noise levels below 90 dBA in determining the need for engineering controls.)

■

### Example 1

An employee is exposed to the following noise levels during the workday:

- 85 dBA for 3.75 hr
- 90 dBA for 2 hr
- 95 dBA for 2 hr
- 110 dBA for 0.25 hr

Thus, the daily noise dose is as follows:

$$D = 100\left(\frac{3.75}{\text{no limit}} \quad \text{or} \quad 0 + \frac{2}{8} + \frac{2}{4} + \frac{0.25}{0.50}\right) = 125\% \quad (24)$$

Because the dose exceeds 100 percent, the employee received an excessive exposure during the workday.

■

The permissible exposures given in Table 9–F are based on the presence of continuous noise rather than intermittent or impact-type noise. By OSHA definition, "if the variations in noise level involve maxima at intervals of one second or less, it is considered to be continuous."

**Table 9–F.** Permissible Noise Exposures

| Duration Per Day (hours) | Sound Level, Slow Response (dBA) |
|---|---|
| 8 | 90 |
| 6 | 92 |
| 4 | 95 |
| 3 | 97 |
| 2 | 100 |
| 1½ | 102 |
| 1 | 105 |
| ½ | 110 |
| ¼ or less | 115 |

**Note:** When the daily noise exposure is composed of two or more periods of noise exposure of different levels, their combined effect should be considered, rather than the individual effect of each. If the sum of the fractions $C_1/T_1 + C_2/T_2 + \cdots + C_n/T_n$ exceeds 100 percent, the mixed exposure should be considered to exceed the limit value. $C_n$ indicates the total time of exposure at a specified noise level, and $T_n$ indicates the total time of exposure permitted at that level.

## Example 2

A drill runs for 15 seconds and is off 0.5 second between operations. This noise is rated at its "on" level for an entire 8-hour day. The noise generated by the drill would be "safe" only if the level were 90 dBA or less.

Further interpretation of the OSHA standard indicates that exposure above 115 dBA is not permissible for any length of time.

As discussed earlier, daily noise dose can be measured using a noise dosimeter. For OSHA use, the dosimeter must have a 5-dB exchange rate, 90-dBA criterion level, slow response, and either an 80-dBA or 90-dBA threshold gate for the appropriate standard to be evaluated. OSHA prescribes a 90-dBA threshold level for compliance with 29 *CFR* 1910.95 (a) and (b), which require implementation of engineering or administrative controls. An 80-dBA threshold is prescribed in 29 *CFR* 1910.95 (d) for monitoring situations for hearing conservation.

### Intermittent noise.

Exposure to intermittent noise can be defined as exposure to a given broadband sound-pressure level several times during a normal working day. The inspector or facility supervisor who periodically makes trips from a relatively quiet office into noisy production areas may be subject to this type of noise. Criteria established for this type of noise exposure are shown in Table 9–D.

With steady noises, it is sufficient to record the A-weighted sound level attained by the noise. With noises that are not steady, such as impulsive noises, impact noises, and the like, the temporal character of the noise requires additional specification. Both the short-term and long-term variations of the noise must be described. Non-steady noise-exposure measurements are most easily made using dosimeters.

### Impact-type noise.

Impact-type noise is a sharp burst of sound, and sophisticated instrumentation is necessary to determine the peak levels for this type of noise. Noise types other than steady ones are commonly encountered. In general, sounds repeated more than once per second can be considered as steady. Impulsive or impact noise, such as that made by hammer blows or explosions, is generally less than one-half second in duration and does not repeat more often than once per second. Employees should not be exposed to impulsive or impact noise that exceeds a peak sound pressure level of 140 dB.

Compliance with this provision requires the use of an instrument capable of measuring unweighted peak sound pressure levels. These analyzers are either integral with or used in conjunction with a sound level meter.

The American Conference of Governmental Industrial Hygienists (ACGIH) recommends that exposure to impulsive or impact noise not exceed the limits listed in Table 9–G.

**Table 9–G.** Threshold Limit Values for Impulsive or Impact Noise

| Sound Level dB* | Permitted Number of Impulses or Impacts per Day |
|---|---|
| 140 | 100 |
| 130 | 1,000 |
| 120 | 10,000 |

* Decibels peak sound pressure level (re: 20 μPa).

Brief, intermittent sounds that occur at intervals greater than 1 second should be evaluated in terms of level and total duration during an 8-hour day. If the noise level exhibits peaks at intervals of 1 second or less, the noise should be considered continuous, and the highest value should be used in determining exposure in accordance with Table 9–F.

Individual impulse and impact sounds can be characterized in terms of their rise time, peak sound level, and pulse duration. The rate and number of the impact sounds that occur during an exposure period are factors in judging the possible hazards of these types of sound.

It is not easy to accurately measure the sound levels of rapidly varying staccato noises. Sound level meters do not follow sudden peaks in sound pressure, and they can systematically distort and misrepresent the true sound pressure levels reached by that noise. This is particularly true when measuring impact noise—for example, the noise made by a drop forge.

## Noise-Control Programs

The degree of noise reduction required is determined by comparing the measured levels with acceptable noise levels. The next step is to consider various noise-control measures such as making alterations in engineering design, limiting the time of exposure, or using personal protective devices to achieve the desired level of reduction.

Every noise problem can be broken down into three parts: a source that radiates sound energy, a path along which the sound energy travels, and a receiver such as the human ear (Figure 9–18). The "system" approach to noise-problem analysis and control assists in understanding both the problem and the changes that are necessary for noise reduction. If each part of the system—source, path, and receiver—is examined in detail, the overall problem is greatly simplified. To help translate these principles into practical terms, specific examples of controlling industrial noise exposure are outlined in this section.

### Source

The most desirable method of controlling a noise problem is to minimize the noise at the source. This generally means modifying existing equipment and structures or possibly

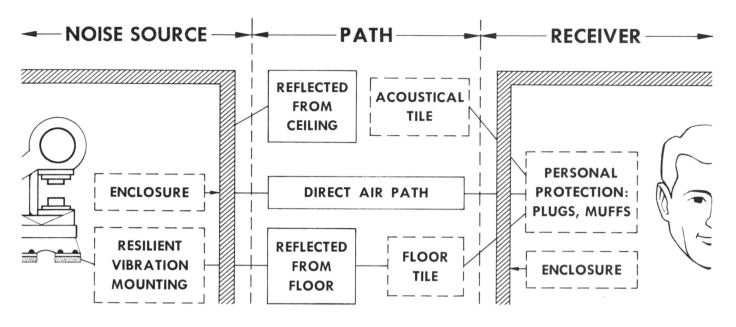

**REDUCTION OF NOISE
AT SOURCE BY:**

1. Acoustical design
   a. Decrease energy for driving vibrating system.
   b. Change coupling between this energy and acoustical radiating system.
   c. Change structure so less sound is radiated.
2. Substitution with less noisy equipment.
3. Change in method of processing.

**REDUCTION OF NOISE
BY CHANGES IN PATH:**

1. Increase distance between source and receiver.
2. Acoustical treatment of ceiling, walls and floor to absorb sound and reduce reverberation.
3. Enclosure of noise source.

**REDUCTION OF NOISE
AT RECEIVER BY:**

1. Personal protection.
2. Enclosures — isolating the worker.
3. Rotation of personnel to reduce exposure time.
4. Changing job schedules.

**Figure 9–18.** Every noise problem can be broken down into three component parts: a source that radiates sound energy, a path along which the sound energy travels, and a receiver such as the human ear.

introducing noise-reduction measures at the design stage of new machinery and equipment.

## Noise Path

Because the desired amount of noise reduction cannot always be achieved by control at the source, modification along the noise path and at the receiver must also be considered.

Noise reduction along the path can be accomplished in many ways: by shielding or enclosing the source, by increasing the distance between the source and the receiver, or by placing a shield between the source and the receiver. Noise can be reduced along the path by means of baffles and enclosures placed over noise-producing equipment to minimize the transmission of noise to areas occupied by employees. Use of acoustical material on walls, ceilings, and floors to absorb sound waves and to reduce reverberations can result in significant noise reduction.

Noise produced by a source travels outward in all directions. If all of the walls, the floor, and the ceiling are hard, reflecting surfaces, all the sound is reflected again and again. The sound level measured at any point in the room is the sum of the sound radiated directly by the source plus all the reflected sounds. Practically all industrial machine installations are located in such environments. These locations are known as semireverberant. Noise measured around a machine in a semireverberant location is the sum of two components: the noise radiated directly by the machine and the noise reflected from the walls, floor, and ceiling.

Close to the machine, most of the noise is radiated directly by the machine. Close to the walls, the reflected component may be predominant. Sound-absorption materials applied to the walls and ceiling can reduce the reflected noise but has no effect on the noise directly radiated by the source.

## Enclosures

In many cases, the purpose of an acoustic enclosure is to prevent noise from getting inside. Soundproof booths for machine operators and audiometric testing booths for testing the hearing of employees are examples of such enclosures. More often, however, an enclosure is placed around a noise source to prevent noise from getting outside. Enclosures are normally lined with sound-absorption material to decrease internal sound pressure buildup.

Noise can best be prevented from entering or leaving an enclosure by sealing all outlets. In extreme cases, double structures can be used. Special treatment, including the use of steel and lead panels, is available to prevent noise leakage in certain cases. Gaskets around doors can also reduce noise transmission from one space to another.

## Control Measures

Noise control can often be designed into equipment so that little or no compromise in the design goals is required. Noise control measures undertaken on existing equipment are usually more difficult. Engineering control of industrial noise problems requires the skill of individuals who are highly proficient in this field.

Noise-control strategies require careful objective analysis on both a practical and economic basis. Complete redesign requires that product and equipment designers consider noise level a primary product or equipment specification in the design of all new products. Full replacement of all products or equipment would eventually take place, the schedule depending on the service life of each. Many designers feel this approach minimizes the cost increases associated with noise-control measures. Existing products or equipment modifications would require manufacturers to modify or replace existing products and equipment to lower the noise levels of noisy equipment.

The existing equipment within any facility was probably selected because it was economical and efficient. However, careful acoustic design can result in quieter equipment that would even be more economical to operate than noisier equipment. Examples of noise-control measures applied at the source include the substitution of quieter machines, the use of vibration-isolation mountings, and the maximum possible reduction of the external surface areas of vibrating parts. Machines mounted directly on floors and walls can cause them to vibrate, resulting in sound radiation. Proper machine mounting can isolate the machines and reduce the transmission of vibrations to the floors and walls.

Although substitution of less noisy machines may have limited application, there are certain areas in which substitution has a potentially wider application. Examples include using "squeeze"-type equipment instead of drop hammers, welding instead of riveting, and instituting chemical cleaning of metal rather than high-speed polishing and grinding.

**Engineering.** When starting a noise-reduction program, it is most desirable to apply engineering principles that are designed to reduce noise levels. The application of known noise-control principles can usually reduce any noise to any desired degree. However, economical considerations or operational necessities can make some applications impractical.

Engineering controls are procedures other than administrative or personal protection procedures that reduce the sound level either at the source or within the hearing zone of workers. The following are examples of engineering principles that can be applied to reduce noise levels.

1. Maintenance:
   a. replacement or adjustment of worn, loose, or unbalanced parts of machines
   b. lubrication of machine parts and use of cutting oils
   c. use of properly shaped and sharpened cutting tools

2. Substitution of machines:
   a. larger, slower machines for smaller, faster ones
   b. step dies for single-operation dies
   c. presses for hammers
   d. rotating shears for square shears
   e. hydraulic presses for mechanical presses
   f. belt drives for gears

3. Substitution of processes:
   a. compression riveting for impact riveting
   b. welding for riveting
   c. hot working for cold working
   d. pressing for rolling or forging

4. Reduction of the driving force of vibrating surfaces:
   a. reduction of the forces
   b. minimization of rotational speed
   c. isolation

5. Reduction of the response of vibrating surfaces:
   a. damping
   b. additional support
   c. increased stiffness of the material
   d. increased mass of vibrating members
   e. change in the size to change resonance frequency

6. Reduction of the sound radiation from vibrating surfaces:
   a. reduction of the radiating area
   b. reduction of the overall size
   c. perforation of the surfaces

7. Reduction of the sound transmission through solids:
   a. use of flexible mountings
   b. use of flexible sections in pipe runs
   c. use of flexible-shaft couplings
   d. use of fabric sections in ducts
   e. use of resilient flooring

8. Reduction of the sound produced by gas flow:
   a. use of intake and exhaust mufflers
   b. use of fan blades designed to reduce turbulence

c. use of large, low-speed fans instead of smaller, high-speed fans

d. reduction of the velocity of fluid flow (air)

e. increase in the cross section of streams

f. reduction of the pressure

g. reduction of air turbulence

9. Reduction of noise by reducing its transmission through air:

a. use of sound-absorptive material on walls and ceiling in work areas

b. use of sound barriers and sound absorption along the transmission path

c. complete enclosure of individual machines

d. use of baffles

e. confinement of high-noise machines to insulated rooms

10. Isolation of the operator by means of a relatively soundproof booth

Some of the noise-control measures described here can be executed quite inexpensively by facility personnel. Other controls require considerable expense and highly specialized technical knowledge to obtain the required results. The services of competent acoustical engineers should be contracted when planning and carrying out engineering noise-control programs.

The possibility that excessive facility noise levels exist should be considered at the planning stage. Vendors supplying machinery and equipment should be advised that specified low noise levels will be considered in the selection process. Suppliers should be asked to provide information on the noise levels of currently available equipment. The inclusion of noise specifications in purchase orders has been used successfully to obtain quiet equipment. If purchasers of industrial equipment demand quieter machines, designers will give more consideration to the problem of noise control.

It is not enough to specify that the sound pressure level of a single machine shall be 90 dBA or less at the operator's station; if another identical machine is placed nearby, the sound level produced by the two machines could be 93 dBA at the operator's station. (As mentioned earlier in this chapter, an increase of 3 dB represents a doubling of sound energy.)

To estimate the effect of a given machine on the total work environment, it is necessary to know the sound power that the machine produces. If there is no operator's work station in the machine's immediate vicinity, the sound power specifications may be sufficient; if, however, there is an operator in the near-sound field, more information is generally needed.

Objectionable noise levels that are by-products of manufacturing operations are found in almost every industry. Practical noise-control measures are not easy to develop, and few ready-made solutions are available. Unfortunately, a standard technique or procedure that can be applied to all or even most situations cannot be presented here. The same machine, process, or noise source in two different locations can present two entirely different problems that must be solved in two entirely different ways.

Noise-control techniques are now being incorporated into products during the design stage. Machine tool buyers are one group who currently specify maximum noise levels in their purchase orders. Equipment can be designed with lower noise levels, but performance tradeoffs involving weight, size, power consumption, and perhaps increased maintenance costs may be necessary. These new, quieter products will probably weigh more, be bigger and bulkier, cost more, and be more difficult and expensive to service and maintain.

To attain quieter products, the engineer must be prepared to trade off, to some degree, many of the design goals that have been achieved in response to market demands. However, lightweight, low-cost, portable machines that are easily operated and simply maintained should not be cast aside lightly, even though price increases may be inevitable. The cost/benefit relationship should be examined in each case. In addition to paying higher prices for original equipment, the user (both as a consumer and as a taxpayer) pays for increased indirect costs.

The success of a noise-reduction project usually depends on the ingenuity with which basic noise-control measures can be applied without decreasing the maximum use and accessibility of the machine or other noise source that is being quieted.

**Administrative controls.** There are many operations in which the exposure of employees to noise can be controlled administratively; for instance, production schedules can simply be changed or jobs rotated so that individual workers' exposure times are reduced. Employees can be transferred from job locations with high noise levels to job locations with lower ones if this procedure would make their daily noise exposure acceptable.

Administrative controls also include scheduling machine operating times so as to reduce the number of workers exposed to noise. For example, if an operation is performed during only one 8-hour day per week, and the operator is overexposed on that one day, it might be possible to perform the operation in two half-days of 4 hours each. The employee might then not be overexposed.

Employees who are particularly susceptible to noise can be transferred and allowed to work in a less noisy area. The benefits from transferring employees can be limited, however, because personnel problems can be caused due to loss of seniority and prestige and lower productivity and pay.

Administrative controls include any administrative decision that results in lower noise exposure, such as complying with purchase agreements that specify maximum noise levels at the operator's position.

The sound level specification that is made part of a purchasing agreement must be more than just a general compliance statement such as "Must meet the requirements of the OSHA." It is important to realize that OSHA sets allowable noise limits relative to the exposure of the people involved. OSHA does not set specific standards for noise-generating equipment. The OSHA noise standard is not intended to be used as an equipment design specification and thus cannot be used as such.

**Personal hearing protection.** Pending the application of engineering control measures, employee exposure to noise can be reduced by the mandatory use of hearing-protective devices. Occupational noise regulations require that whenever employees are exposed to excessive noise levels, feasible administrative or engineering controls should be used to reduce those levels. When these control measures cannot be completely accomplished, or while such controls are being initiated, personnel should be protected from the effects of excessive noise levels. Such protection can, in most cases, be provided by wearing suitable hearing-protection devices. Once management has decided that hearing protectors should be worn, the success of such a program depends largely on the method of initiation used and on the proper indoctrination of supervisory personnel and workers. Supervisors should set an example by wearing their hearing protectors when they go into noisy areas.

Some companies have found it very helpful to meet with employees or their representatives to thoroughly review the contemplated protection program and reach an understanding of the various problems involved. This process includes reviewing work areas where hearing protection will be provided or required and complying with state and federal regulations that require the use of hearing-protective devices.

*Hearing-protective devices.* Hearing-protective devices such as earplugs and earmuffs have one serious drawback—they do nothing to reduce or eliminate the hazard. Their failure means immediate exposure to the hazard. The fact that a hearing protector can become ineffective without the knowledge of the wearer is particularly serious. Training on the purpose, benefits, proper fitting, use, and care of hearing protectors is essential to the success of the program and is required by OSHA. Distributing a flyer (Figure 9–19) highlighting the care and use of the hearing protector is also helpful.

Personal hearing-protective devices are acoustic barriers that reduce the amount of sound energy transmitted through the ear canal to receptors in the inner ear.

The sound attenuation (reduction) capability of a hearing-protective device (in decibels) is the difference in the measured hearing threshold of an observer wearing hearing protectors (test threshold) and the measured hearing threshold when the observer's ears are uncovered (reference threshold).

Inserts or muffs are hearing-protective devices that are in common use today. The insert-type protector attenuates noise by plugging the external ear canal, whereas the muff-type protector encloses the auricle of the ear to provide an acoustic seal. The effectiveness of hearing-protective devices depends on several factors that are related to the manner in which the sound energy is transmitted through or around the device. Figure 9–20 shows four pathways by which sound can reach the inner ear when hearing-protective devices are worn: seal leaks, material leaks, hearing-protective device vibration, and conduction through bone and tissue.

---

**LET'S REVIEW THE FACTS**

1. It is necessary for employees in certain noisy areas to wear ear protectors.
2. Prolonged exposure to excessive noise can harm the delicate hearing mechanism.
3. Ear protectors such as ear plugs or ear muffs will reduce the noise before it reaches the ear drum.
4. Your job assignment will determine whether you should wear ear plugs (inserts) or muffs (covers).
5. Speech and warning signals can be fully heard with ear protectors in noisy shop areas.

**WEAR YOUR EAR PROTECTORS**

1. The nurse will fit them and instruct you how to wear them.
2. Wear them for short periods to start and gradually increase the wearing time. After a few days you will be able to wear them all day with minimum discomfort.

Suggested Wearing Time Schedule

|  | A.M. | P.M. |
|---|---|---|
| 1st day | = 30 minutes | — 1 hour |
| 2nd day | = 1 hour | — 1 hour |
| 3rd day | = 2 hours | — 2 hours |
| 4th day | = 3 hours | — 3 hours |
| 5th day | = all day | — all day thereafter |

3. If after five days the ear protectors feel uncomfortable, come in and see the nurse in the Company hospital.
4. Ear protectors should be replaced when they become worn, stiff or lose their shape.
5. If ear protectors are misplaced, a new pair should be obtained without delay.
6. Never put soiled ear plugs into your ears. Wash the ear plugs at least once a day with soap and water.
7. With proper care, ear plugs should last for several months and ear muffs should last for several years.

**OTHER POINTS TO REMEMBER**

1. The best ear protector is the one that is properly fitted and worn.
2. Good protection depends on a snug fit. A small leak can destroy the effectiveness of the protection.
3. Ear plugs tend to work loose as a result of talking or chewing, and they must be re-seated from time to time during the working day.
4. If ear plugs are kept clean, skin irritations and other reactions should not occur.

**YOUR HEARING IS PRICELESS**

**PROTECT IT**

**Figure 9–19.** An example of a flyer distributed to all company employees who are required to wear some form of hearing protection. The flyer highlights care and use of the device.

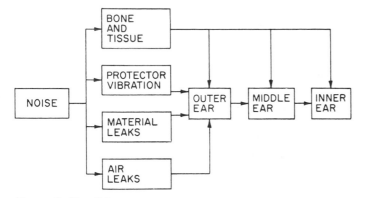

**Figure 9–20.** When a person is wearing a hearing protector, sound reaches the inner ear by different pathways.

***Seal leaks.*** For maximum protection, the device must form a virtually airtight seal against the ear canal or the side of the head. Inserts must accurately fit the contours of the ear canal, and muffs must accurately fit the area surrounding the external ear. Small air leaks in the seal between the hearing protector and the skin can significantly reduce the low-frequency sound attenuation or permit a greater proportion of the low-frequency sounds to pass through. As the air leak becomes larger, attenuation lessens at all frequencies.

***Material leaks.*** Another possible transmission pathway for sound is directly through the material of the hearing-protective device. Although the hearing-protective device can attenuate or prevent the passage of most of the sound energy, some sound is still allowed to pass through.

***Vibration of the hearing-protective device.*** Sound can also be transmitted to the inner ear when the hearing-protective device itself is set into vibration in response to exposure to external sound energy.

Because of the flexibility of the flesh in the ear canal, earplugs can vibrate in a piston-like manner within the ear canal. This limits their low-frequency attenuation. Likewise, an earmuff cannot be attached to the head in a totally rigid manner. Its cup vibrates against the head like a mass/spring system. The muff's effectiveness is governed by the flexibility of the muff cushion and the flesh surrounding the ear, as well as by the air volume entrapped under the cup.

***Bone conduction.*** If the ear canal were completely closed so that no sound entered the ear by that path, some sound energy could still reach the inner ear by means of bone conduction. However, the sound reaching the inner ear by such means would be about 50 dB below the level of air-conducted sound received through the open ear canal. It is therefore obvious that no matter how the ear canal is blocked, the hearing-protective device will be bypassed by the bone-conduction pathway through the skull. A perfect hearing-protective device cannot provide more than about 50 dB of effective sound attenuation.

When a hearing-protective device is properly sized and carefully fitted and adjusted for optimum performance on a laboratory subject, air leaks are minimized, and material leaks, hearing-protective device vibration, and bone conduction are the primary sound transmission paths. In the workplace, however, this is usually not the case; sound transmission through air leaks is often the primary pathway.

All hearing protectors must be properly fitted when they are initially dispensed. Comfort, motivation, and training are also very important factors to consider if hearing protectors are to be successfully used.

**Classes of hearing protection.** Personal hearing-protective equipment can be divided into four classifications:

- Enclosures (entire head)
- Aural inserts, or earplugs
- Superaural protectors, or canal caps
- Circumaural protectors, or earmuffs

***Enclosures.*** The enclosure-type hearing-protective device entirely envelops the head. A typical example is the helmet worn by an astronaut. In this case, attenuation at the ear is achieved through the acoustic properties of the helmet.

The maximum amount that a hearing protector can reduce the sound reaching the ear is from about 35 dB at 250 Hz to about 50 dB at the higher frequencies. By wearing hearing protectors and then adding a helmet that encloses the head, an additional 10-dB reduction of sound transmitted to the ears can be achieved.

Helmets can be used to support earmuffs or earphones and cover the bony portion of the head in an attempt to reduce bone-conducted sound. Helmets are particularly well-suited for use in extremely high–noise level areas and where workers need to protect their heads from bumps or missiles. With good design and careful fitting of the seal between the edges of the helmet and the skin of the face and neck, 5–10 dB of sound attenuation can be obtained beyond that already provided by the earmuffs or earphones worn inside the helmet. This approach to protection against excessive noise is practical only in very special applications. Cost, as well as bulk, normally precludes the use of helmet-type hearing protectors in a general industrial hearing conservation program.

***Aural insert protectors.*** Aural insert hearing-protective devices are normally called inserts or earplugs. This type of protector is generally inexpensive, but the service life is limited, ranging from single-time use to several months. Insert-type protectors or plugs are supplied in many different configurations and are made from such materials as rubber, plastics, fine glass down, foam, and wax-impregnated cotton. The pliable materials used in these aural inserts are quite soft, and there is little danger of injury resulting from accidentally forcing the plug against the tender lining of the ear canal.

It is desirable to have the employee's ears examined by qualified medical personnel before earplugs are fitted. Occasionally, the physical shape of the ear canal precludes the use of insert-type protectors. There is also the possibility that the ear canal is filled with hardened wax. If wax (cerumen) is a problem, it should be removed by qualified personnel. In some cases, the skin of the ear may be sensitive to a particular earplug material, and earplugs that do not cause an allergic response should be recommended.

Earplugs fall into three broad categories of general classification: formable, custom-molded, and premolded.

Formable earplugs (Figure 9–21) can provide good attenuation and fit any ear. Many of the formable types are designed for one-time use only, after which they are thrown away. Materials from which these disposable plugs are made include very fine glass fiber (often called Swedish wool), wax-impregnated cotton, expandable plastic, and foam.

These materials are generally rolled into a conical shape before being inserted into the ear. However, while adequate instruction must be given to emphasize the importance of a snug fit, the user must be careful not to push the material so far into the ear canal that it has to be removed by medical personnel.

One type of formable earplug is made from a plastic-like substance similar in consistency to putty. The preparation of this material requires that the individual take a quantity of it and mold or form it so that it can be inserted into the ear canal. The user should be shown the correct method of forming the material. In addition, the user must be cautioned to have clean hands when forming the material and placing it in the ear. If the hands are dirty, foreign material can get into the ear.

Custom-molded earplugs are, as the name implies, custom fit for the individual user. Generally, two or more materials (packaged separately) are mixed together to form a compound that resembles soft rubber when set. For use as a hearing-protective device, the mixture is care-

fully placed into the outer ear with some portion of it in the ear canal, in the manner prescribed by the manufacturer. As the material sets, it molds itself to the shape of the individual ear and external ear canal. In some cases, the materials are premixed and come in a tube from which they can be injected into the ear.

Premolded earplugs are often referred to as prefabricated, because they are usually made in large quantities in a multiple-cavity mold. The materials of construction range from soft silicone or rubber to other plastics.

There are two versions of premolded insert protector. One is known as the universal-fit type. In this type, the plug is designed to fit a wide variety of ear canal shapes and sizes. The other type of premolded protector is supplied in several different sizes to ensure a good fit (Figure 9–22). The design of the plug is important. For example, the smooth bullet-shaped plug is very comfortable and provides adequate attenuation in straight ear canals; however, its performance falls off sharply in many irregularly shaped canals.

The use of premolded earplugs requires proper fitting by trained personnel. In many individuals, the right and left ear canals are not the same size. For this reason, properly trained personnel must prescribe the correct protector size for each ear canal. Sizing devices are available to aid in proper fitting.

The premolded type of earplug has a number of disadvantages that limit its practical acceptability. To be effective, it has to fit snugly and, for some users, this is uncomfortable. Because the plug must fit tightly and because many people have irregularly shaped ear canals, an incorrect size of plug can be selected, or the plug may not be inserted far enough, and a good fit is not obtained.

Some premolded earplugs can shrink and become hard. This is caused primarily by ear wax (present in all ear canals). The wax extracts the plasticizer from some plug materials, causing hardening and possible shrinkage

**Figure 9–21.** Formable earplugs. (Courtesy Cabot Safety Corp.)

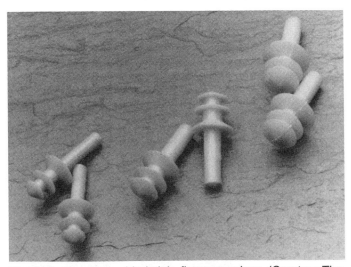

**Figure 9–22.** Premolded triple-flange earplugs. (Courtesy The Bilsom Group.)

of the plug. The degree of hardening and shrinkage varies from one individual user to another, depending on such factors as temperature, duration of use, and the personal hygiene of the user. Regular cleaning of the protectors with mild soap and water prolongs their useful life. To keep the plugs clean and free from contamination, most manufacturers provide a carrying case for storing the plugs when they are not in use.

***Superaural protectors.*** Hearing-protective devices in this category (commonly known as canal caps) seal the external opening of the ear canal to achieve sound attenuation (Figure 9–23). A soft, rubberlike material is held in place by a lightweight headband. The tension of the band holds the superaural device against the external opening of the ear canal.

***Circumaural protectors.*** Circumaural hearing-protective devices, or earmuffs, consist essentially of two cup- or dome-shaped devices that fit over the entire external ear, including the lobe, and a cushion or pad that seals against the side of the head. The ear cups are generally made of a molded rigid plastic and are lined with a cell-type foam material. The size and shape of the ear cup vary from one manufacturer to another (Figure 9–24). The cups are usually held in place by a spring-loaded suspension assembly or headband. The force applied against the head is directly related to the degree of attenuation desired. The width, circumference, and material of the earmuff cushion must be considered to maintain a proper balance of performance and comfort. To provide a good acoustic seal, the required width of the contact surface depends to a large degree on the material used in the cushion. The cup with the smallest possible circumference that can accommodate the largest ear lobes should be chosen. A slight pressure on the lobe can become painful in time, so it is very important to select a muff dome that is large enough.

The earmuffs currently on the market come with replaceable ear seals or cushions that are filled with either

**Figure 9–24.** Earmuffs. (Courtesy The Bilsom Group.)

foam, liquid, or air—the foam-filled type is the most common. The outer covering of these seals is vinyl or a similar thermoplastic material. Human perspiration tends to extract the plasticizer from the seal material, which results in an eventual stiffening of the seals. For this reason, the seals require periodic replacement; the frequency of replacement depends on the conditions of exposure.

**Selection of protector.** The attenuation characteristics of a particular hearing protector must be considered before it is used for a specific application (Figure 9–25). As part of a well-planned hearing conservation program, characteristics of the noise levels in various areas should be known. From these data and from the attenuation information available from manufacturers, it can be determined whether a given device is suitable for the intended application. One must also consider the work area where the individual will use the hearing-protective device. For example, a large-volume earmuff would not be practical for an individual who must work in confined areas where there is very little head clearance. In such instances, a very small or flat ear cup or insert-type protector would be more practical.

When using muff-type protectors in special hazard areas (such as power-generating stations where there are electrical hazards), it may be desirable to use nonconductive suspension systems with muff-type protectors. Also, if other personal protective equipment such as safety hats or safety spectacles must be worn, the degree of hearing protection required must not be compro-

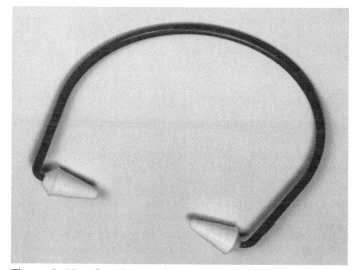

**Figure 9–23.** Canal caps. (Courtesy Cabot Safety Corp.)

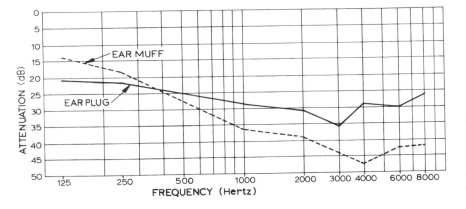

**Figure 9–25.** Comparison of the attenuation properties of a molded-type earplug and an earmuff protector. Note that the earplug offers greater attenuation of the lower frequencies, while the earmuff is better at the higher frequencies.

mised. The efficiency of muff type protectors is reduced when they are worn over the frames of eye-protective devices. In these cases, the amount of reduction in noise attenuation depends on the type of glasses being worn as well as the size and shape of the individual wearer's head. When eye-protective devices are required, it is recommended that ones with cable-type temples be used, because they create the smallest possible opening between the seal and the head.

When selecting a hearing-protective device, one should also consider how often a worker is exposed to excessive noise. If exposure is relatively rare (once a day or once a week), an insert or plug device will probably satisfy the requirement. On the other hand, if the noise exposure is relatively frequent and the employee must wear the protective device for an extended period of time, the muff-type protector might be preferable. If the noise exposure is intermittent, the muff-type protector is probably more desirable, because it is somewhat more difficult to remove and reinsert earplugs.

When determining the suitability of a hearing-protective device for a given application, the manufacturer's reported test data must be examined carefully. It is necessary to correlate that information with the specific noise exposure the device is intended to control. The manufacturer should provide attenuation characteristics of the individual hearing-protective devices over a range of frequencies.

The most convenient method by which to gauge the adequacy of a hearing protector's attenuation capacity is to check its Noise Reduction Rating (NRR), a rating that was developed by the EPA (U.S. Environmental Protection Agency). According to the EPA regulation, the NRR must be printed on the hearing protector's package. The NRR can be correlated with an individual worker's noise environment to assess the adequacy of the attenuation characteristics of the particular hearing-protective device. Appendix B of 29 *CFR* 1910.95 describes methods of using the NRR to determine whether a particular hearing-protective device provides adequate protection within a given exposure environment.

## Industrial Audiometry

Audiometry, or the measurement of hearing, is central to industrial hearing conservation programs because all follow-up activities and program evaluations are based on such test results. Briefly, the objectives in industrial audiometry are as follows:

- Obtain a baseline audiogram that indicates an individual's hearing ability at the time of the preplacement examination.
- Provide a record of an employee's hearing acuity.
- Check the effectiveness of noise-control measures by measuring the hearing thresholds of exposed employees.
- Record significant hearing threshold shifts in exposed employees during the course of their employment.
- Comply with government regulations.

An audiometer is required to help assess an individual's hearing ability. An audiometer is an electronic instrument that converts electrical energy into sound energy in precisely variable amounts. It should meet the standards set forth in ANSI S3.6–1989, *Specifications for Audiometers.*

An audiometer consists of an oscillator, which produces pure tones at predetermined frequencies; an attenuator, which controls the intensity of the sound or tone produced; a presenter switch; and earphones, through which the person whose hearing is being tested hears the tone.

### Threshold Audiometry

Threshold audiometry is used to determine an employee's auditory threshold for a given stimulus. Measurements of hearing are made to determine hearing acuity and to detect abnormal function in the ear. Before hearing can be described as abnormal, a reference point, or normal value, must be designated.

The quantity that is of interest, however, is not the sound pressure level of the normal hearing threshold, but rather the magnitude of departure from a standard reference threshold. Levels that depart from the norm can be easily detected by their divergence from the reference threshold, which shows up directly on the audiometer attenuator dial.

Hearing threshold levels are those intensities at specific frequencies at which a sound or a tone can just barely be heard. The term *air conduction* refers to the air path by which the test sounds generated at the earphones are conducted through air to stimulate the eardrum.

The record of measured hearing thresholds is called a threshold audiogram. Audiometric tests can also be recorded in the form of audiograms, on which are plotted both sound intensity (in dB) and frequency (in Hz). A sample audiogram is shown in Figure 9–26.

## Who Should Be Examined

Preplacement hearing-threshold tests should be taken by all job applicants, not just those who are to work in noisy

areas. This establishes a baseline hearing threshold for each employee for future comparison. Preplacement hearing tests are essential if a company is to protect itself from liability for preexisting hearing loss incurred elsewhere. If an employee is hired with hearing damage and he or she is subsequently exposed to high noise levels, the company may be liable for all the employee's hearing loss—unless it can be proved that the employee had a preexisting hearing loss when hired. In some states, the most recent employer is liable for all compensable hearing loss, regardless of past exposures.

Periodic follow-up hearing tests should be administered to persons stationed in areas where noise exposures exceed permissible levels.

The schedule for periodic follow-up hearing tests depends largely on an employee's noise exposure. Assuming that a record of the worker's hearing status was established at the time of employment or placement,

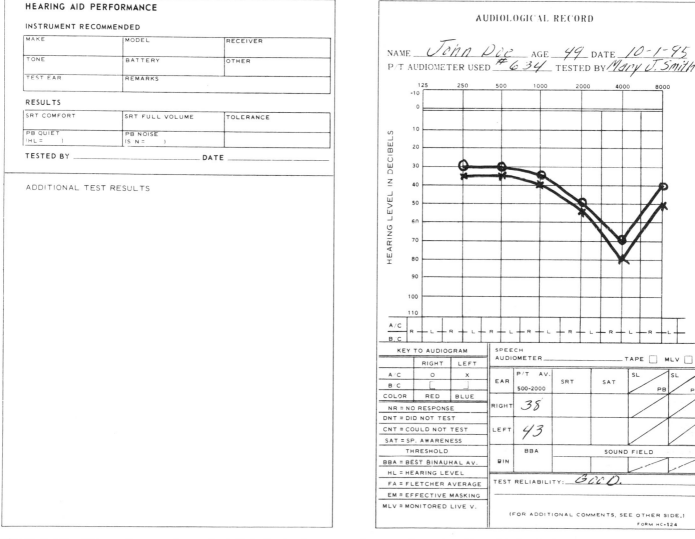

**Figure 9–26.** This audiogram shows the initial effects of exposure to excessive noise. Note the decided notch at the 4,000-Hz frequency.

the first reexamination should be made 9–12 months after placement. If no significant threshold shifts relative to the preplacement audiogram are noted, subsequent follow-up tests can be administered at yearly intervals. If the noise exposure is relatively low, the interval between follow-up tests can be increased. This decision should be based on the combination of the conditions of the exposure and the results of previous audiograms and clinical examinations.

Noise exposure is by no means the only reason for a change in an individual's audiogram. When a change in hearing status is confirmed, its cause must be determined. Improper placement of earphones and excessive ambient noise in the test room can affect audiogram results. Physiological changes as a result of the employee's age and state of health can also affect audiogram results. The individual's motivation and attitude toward the test can affect performance.

An industrial audiometric program can identify people who are experiencing hearing threshold changes that are not related to noise exposure on the job. These workers should be referred to their family physician for diagnosis and treatment. However, when threshold shifts related to noise exposure are identified, this procedure should be followed:

1. Check the fit of the hearing-protective device, if one is worn by the worker.

2. Repeat or initiate educational sessions to encourage the employee to wear a hearing-protective device, if it has not been worn.

3. Investigate the noise levels in the work area, particularly if a previous sound level survey failed to reveal noise hazards.

Noise exposure information, correlated with audiometric test results, is necessary to make intelligent decisions about a firm's hearing conservation program. If all hearing tests and medical opinion point to a progressive deterioration of an individual's hearing, the safety and health professional should provide and enforce the use of hearing-protection equipment and/or recommend that the individual's exposure to excessive noise be controlled.

Conclusions about the general noise environment should not be based on changes in the hearing of a single individual, because the variation in individual susceptibility to noise is broad. Conclusions can, however, be drawn from the average changes—or lack of them—in a group of employees exposed to the same noise environment.

## Effective Programs

An effective industrial audiometric program should include consideration of the following components:

- Medical surveillance
- Qualified personnel
- Suitable test environment
- Calibrated equipment
- Adequate record keeping

**Medical surveillance.** Medical surveillance is essential in a hearing-testing program so that the program can fulfill its dual purpose of detecting hearing loss and providing valid records for compensation claims. Although many smaller companies do not have a medical department, they can satisfy the general medical surveillance requirement by using part-time medical consultants.

Noise-susceptible workers are employees who suffer handicapping hearing losses more quickly than do their colleagues under equivalent noise exposures. These workers constitute the group from whom compensation claims are most likely to arise and for whom the risk of hearing damage is likely to be greatest.

During the preplacement examination, the applicant should provide a detailed history covering his or her prior occupational experience and a personal record of illnesses and injuries. For applicants who will work in noisy environments, the history should include noise exposures in previous jobs, including any in the military services. The medical phase of the history should detail frequency of earache, ear discharge, ear injury, surgery (ear or mastoid), head injury with unconsciousness, ringing in the ears, hearing loss in the immediate family, the use of drugs, and history of allergy and toxic exposures. A standard form can be created for this purpose.

**Qualified personnel.** Audiometric tests should be administered by a qualified individual such as a specially trained nurse, an audiologist, or an occupational hearing conservationist. An occupational hearing conservationist is an individual who has satisfactorily completed a course of training that meets, as a minimum, the guidelines established by the Council for Accreditation in Occupational Hearing Conservation.

The duties of the Occupational hearing conservationist (OHC) are to perform baseline and periodic pure-tone air-conduction threshold tests. Systematic supervision and encouragement of the OHC by the physician, audiologist, or other qualified person in charge of the audiometric program is recommended to maintain the high motivation required for good audiometric testing. The supervision should include periodic review of the testing procedures used by the audiometric technician to make sure that they conform to established procedures.

**Suitable test environment.** Hearing measurements must be made in a test room or booth that conforms to the requirements established by ANSI S3.1–1991, *Maximum Permissible Ambient Noise Levels for Audiometric Test Rooms*. It must be sufficiently quiet within the enclosure so that external noises do not interfere with the employee's perception of the test sounds. This usually requires a special sound-treated enclosure (Figure 9–27).

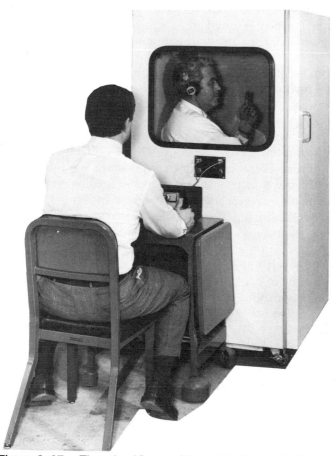

**Figure 9–27.** The raised finger of the subject's hand indicates that he is, in all probability, hearing the test tone. The technician must decide whether this is an accurate or inaccurate response. (Courtesy Eckel Industries, Inc.)

Hearing-testing rooms should be located away from outside walls, elevators, and locations with heating and plumbing noises. If the background noise levels in the test area do not exceed the sound levels allowed by the standard, the background noise does not affect the hearing test results. The hearing test booth or room can be either a prefabricated unit or one that is built on the premises. Doors, gaskets, and other parts of the room or booth that can deteriorate, warp, or crack should be carefully inspected periodically.

In addition to proper acoustical standards, the booth or room should allow for ease of access and egress and be provided with good, comfortable ventilation and lighting. The audiometric technician should be able to sit outside the room or booth but be able to see the interior of the room through a window.

To select the proper room, it is necessary to conduct a noise survey at the proposed test location. Noise levels at each test frequency should be measured and recorded using an octave-band sound level meter. The audiometric booth selected must have sufficient noise attenuation so that the background noise levels present at each test fre-

quency are reduced and do not exceed the maximum permissible background levels listed in the current ANSI standard.

**Calibrated equipment.** Limited-range, pure-tone audiometers must conform to the current ANSI standard listed in *Specifications for Audiometers*. Two basic types of audiometers are available: automatic recording audiometers and manually operated audiometers.

The audiometer should be subjected to a biological check each day before the instrument is used. The biological check is done by testing the hearing of a person whose hearing threshold is known and stable. The check should include the movement and bending of cords and wires, knob turning, switch actuating, and button pushing to make sure that no sounds are produced in the earphones other than the test tones.

An exhaustive electronic calibration of the audiometric test instrument should be made annually by a repair and calibration facility that has the specialized equipment and skilled technical personnel necessary for this work. A certificate of calibration should be kept with the audiometer at all times.

**Adequate records.** The medical form used in audiometric testing programs should include all basic data related to the hearing evaluation. Hearing threshold values, noise-exposure history, and pertinent medical history should be accurately recorded each time an employee's hearing is tested. The employee should be identified by name, social security number, sex, and age. Additional information such as the date and time of the test (day of the week, time of day), conditions under which the test was performed, and the name of the examiner should also be included.

Audiometric test records for an employee should be kept for at least the duration of employment. The records could become the basis for a settlement of a hearing loss claim.

Periodic audiograms are a profile of the employee's hearing acuity. Any change from the results of previous audiograms should be investigated. One possible reason for a hearing loss is that the employee's hearing protectors are inadequate or improperly worn.

The audiometric testing program should be both practical and feasible. In small companies, where the total number of employees to be tested is small, it would be impractical to purchase a booth and audiometer. It would be more economical to consider a mobile audiometer testing service or to refer the employees to a local, properly equipped and staffed hearing center or to a qualified physician or audiologist for an audiometric examination.

Audiometric testing is an integral part of a comprehensive hearing conservation program. The OSHA Hearing Conservation Standard discussed in the following section details specific requirements for audiometric testing.

# Noise-Exposure Regulations

## Background

The federal regulation of occupational noise exposure started with the rules issued by the Bureau of Labor Standards under the authority of the Walsh–Healey Public Contracts Act. These rules required that occupational noise exposure be reasonably controlled to minimize fatigue and the probability of accidents. The federal occupational noise-exposure regulations were originally written to apply only to contractors under the Walsh–Healey Public Contracts Act and the McNamara–O'Hara Service Contracts Act. Under the Williams–Steiger Occupational Safety and Health Act of 1970, the Bureau of Labor Standards was replaced by the Occupational Safety and Health Administration (OSHA).

The National Institute for Occupational Safety and Health (NIOSH) was established within the Department of Health and Human Services (formerly the Department of Health, Education and Welfare) by the Occupational Safety and Health Act of 1970 to conduct research and to recommend new occupational safety and health standards. The recommendations are transmitted to the Department of Labor, which is responsible for the final setting, promulgation, and enforcement of the standards.

In 1972, NIOSH provided the Department of Labor with a document called *Criteria for a Recommended Standard: Occupational Exposure to Noise.* Subsequently, the assistant secretary of labor determined that a standard advisory committee on noise should be formed. The purpose of this OSHA Advisory Committee was to obtain and evaluate additional recommendations from labor, management, government, and independent experts. The committee considered written and oral comments directed to it by interested parties. The committee then transmitted its recommendations for a revised standard to OSHA on December 20, 1973.

In 1974, OSHA published a proposed standard in the *Federal Register* that limited an employee's exposure level to 90 dBA, calculated as an 8-hour, time-weighted average (TWA). NIOSH commented on OSHA's proposed standard, stating that there was a need for reducing the 8-hour exposure level to 85 dBA. However, NIOSH was unable to recommend a specific future date after which the 85-dBA noise level should become mandatory for all industries. Sufficient data were not available to demonstrate the technological feasibility of this level.

The EPA reviewed the OSHA proposal and recommended that the limit not exceed 85 dBA. They reviewed the proposed noise standard and recommended that additional studies be undertaken to explore the efficacy of reducing the permissible level still further at some future date. The proposed revisions to the OSHA rules for occupational noise exposure were published in the *Federal Register* on October 24, 1974.

After years of collecting oral and written testimony, which resulted in an unwieldy public record of almost 40,000 pages, OSHA promulgated revisions for the noise standard (46 *FR* 4078) in January 1981. These revisions were followed by deferrals, stays (46 *FR* 42622), further revisions, further public hearings, and a multiplicity of lawsuits, all of which culminated in the promulgation of a hearing conservation amendment (48 *FR* 9738) on March 8, 1983, with an effective date of April 7, 1983.

It is estimated by OSHA (46 *FR* 4078) that there are 2.9 million workers in American production industries who experience 8-hour noise exposures exceeding 90 dBA. An additional 2.3 million experience exposure levels in excess of 85 dBA. The Hearing Conservation Amendment (HCA) applies to all those 5.2 million employees except for those in oil and gas well drilling and servicing industries, which are specifically exempted. Additionally, the Amendment does not apply to those engaged in construction or agriculture, although a Construction Industry Noise Standard exists (29 *CFR* 1926.52 and 1926.101). This standard is essentially identical to paragraphs (a) and (b) of the General Industry Noise Standard.

## The Occupational Noise Standard

Prior to promulgation of the HCA the existing Noise Standard (29 *CFR* 1910.95[a] and [b]) established a permissible noise-exposure level of 90 dBA for 8 hours and required the employer to reduce exposure to that level by use of feasible engineering and administrative controls. In all cases in which sound levels exceeded the permissible exposure, regardless of the use of hearing-protective devices, "a continuing, effective hearing conservation program" was required. However, the details of such a program were never mandated. Paragraphs (c) through (p) of the HCA replaced paragraph (b)(3) of 29 *CFR* 1910.95 and supplemented OSHA's definition of an "effective hearing conservation program."

## Hearing Conservation Programs

An effective hearing conservation program prevents hearing impairment as a result of noise exposure on the job. In terms of existing workers' compensation laws, an effective hearing conservation program is one that limits the amount of compensable hearing loss in the frequency range over which normal hearing is necessary for communication. It should be noted that "compensable" loss at present does not include frequencies over 4,000 Hz, although such loss impairs enjoyment of sound and may interfere with speech discrimination. In compliance with the OSHA requirements, an effective hearing conservation program must be instituted if any employee's noise exposure exceeds current limits as defined in the OSHA Noise Exposure Standard 29 *CFR* 1910.95.

All employees whose noise exposures equal or exceed an 8-hour TWA of 85 dBA must be included in a hearing conservation program comprised of five basic components: exposure monitoring, audiometric testing, hearing protection, employee training, and record keeping. Note that although the 8-hour TWA permissible exposure remains 90 dBA, a hearing conservation program becomes mandatory at an 8-hour TWA exposure of 85 dBA.

The following summary briefly discusses the required components of the hearing conservation program.

**Monitoring.** The HCA requires employers to monitor employee noise-exposure levels in a manner that can accurately identify employees who are exposed at or above an 8-hour TWA exposure of 85 dBA. The exposure measurement must include all noise within an 80–130-dBA range. The requirement is performance oriented and allows employers to choose the monitoring method that best suits each situation.

Employees are entitled to observe monitoring procedures, and, in addition, they must be notified of the results of exposure monitoring. However, the method used to notify employees is left to the discretion of the employer.

Employers must remonitor workers' exposures whenever changes in exposures are sufficient to require new hearing protectors or whenever employees not previously included, because they were not exposed to an 8-hour TWA of 85 dBA, are included in the program.

Instruments used for monitoring employee exposures must be calibrated to ensure that the measurements are accurate. Because calibration procedures are unique to each instrument, employers should follow the manufacturer's instructions to determine when and how extensively to calibrate.

**Audiometric testing.** Audiometric testing not only monitors employee hearing acuity over time but also provides an opportunity for employers to educate employees about their hearing and the need to protect it. The audiometric testing program includes obtaining baseline and annual audiograms and initiating training and follow-up procedures. The audiometric testing program should indicate whether hearing loss is being prevented by the employer's hearing conservation program. Audiometric testing must be made available to all employees who have time-weighted average–exposure levels of 85 dBA. A professional (audiologist, otolaryngologist, or physician) must be responsible for the program, but need not be present when a qualified occupational hearing conservationist is actually conducting the testing. Professional responsibilities include overseeing the program and the work of the OHCs, reviewing problem audiograms, and determining whether referral is necessary. Either a professional or an OHC can conduct audiometric testing. In addition to administering audiometric tests, the tester (or the supervis-

ing professional) is also responsible for ensuring that the tests are conducted in an appropriate test environment, for seeing that the audiometer works properly, for reviewing audiograms for standard threshold shifts (as defined in the HCA), and for identifying audiograms that require further evaluation by a professional.

***Audiograms.*** There are two types of audiograms required in the hearing conservation program: baseline and annual audiograms. The baseline audiogram is the reference audiogram against which subsequent audiograms are compared. Baseline audiograms must be provided within six months of an employee's first exposure at or above a TWA of 85 dBA. However, when employers use mobile test vans to do audiograms, they have up to one year after an employee's first exposure to workplace noise at or above a TWA of 85 dBA to obtain the baseline audiogram. Additionally, when mobile vans are used and employers are allowed to delay baseline testing for up to a year, those employees exposed to time-weighted average levels of 85 dBA or more must be issued and fitted with hearing protectors six months after their first exposure. The hearing protectors are to be worn until the baseline audiogram is obtained. Baseline audiograms taken before the effective date of the amendment are acceptable as baselines in the program if the professional supervisor determines that the audiogram is valid. The annual audiogram must be conducted within one year of the baseline. It is important to test hearing on an annual basis to identify changes in hearing acuity so that protective follow-up measures can be initiated before hearing loss progresses.

***Audiogram evaluation.*** Annual audiograms must be routinely compared to baseline audiograms to determine whether the audiogram is accurate and whether the employee has lost hearing ability; that is, to determine whether a standard threshold shift, or STS, has occurred. An effective program depends on a uniform definition of an STS. An STS is defined in the amendment as an average shift (or loss) in either ear of 10 dB or more at the 2,000-, 3,000-, and 4,000-Hz frequencies. A method of determining an STS by computing an average was chosen because it diminishes the number of persons identified as having an STS who are later shown not to have had a significant change in hearing ability.

---

## Example

An example of computing the STS is shown in Table 9–II. Considering the values for 2,000, 3,000, and 4,000 Hz, there are changes in hearing threshold of 10, 15, and 25 dB, respectively. Thus,

$$\text{STS} = \frac{(10 + 15 + 25)}{3} = \frac{50}{3} = 16.7 \text{ dB} \qquad (25)$$

**Table 9–H.** Computing the Standard Threshold Shift (STS)

| Frequency (Hz) | Baseline Audiogram Threshold (dB) | Annual Audiogram Threshold (dB) | Change |
|---|---|---|---|
| 500 | 5 | 5 | 0 |
| 1,000 | 5 | 5 | 0 |
| 2,000 | 0 | 10 | +10 |
| 3,000 | 5 | 20 | +15 |
| 4,000 | 10 | 35 | +25 |
| 6,000 | 10 | 15 | +5 |

■
**Conclusion**

The STS is +16.7 dB; hearing has deteriorated; the employee must be notified in writing within 21 days; and, depending on professional discretion, the employer can elect to revise the baseline.

■

If an STS is identified, the employee must be fitted or refitted with adequate hearing protectors, shown how to use them, and required to wear them. In addition, employees must be notified within 21 days from the time the determination is made that their audiometric test results indicate an STS. Some employees with an STS should be referred for further testing if the professional determines that their test results are questionable or if they have an ear problem of a medical nature caused or aggravated by wearing hearing protectors. If the suspected medical problem is not thought to be related to wearing protectors, employees must merely be informed that they should see a physician. If subsequent audiometric tests show that the STS identified on a previous audiogram is not persistent, employees exposed to a TWA of less than 90 dBA can discontinue wearing hearing protectors.

A subsequent audiogram can be substituted for the original baseline audiogram if the professional supervising the program determines that the employee has experienced a persistent STS. The substituted audiogram becomes known as the revised baseline audiogram. This substitution ensures that the same shift is not repeatedly identified. The professional may also decide to revise the baseline audiogram after an improvement in hearing has occurred, which ensures that the baseline reflects actual thresholds as much as is possible. When a baseline audiogram is revised, the employer must, of course, also retain the original audiogram. To obtain valid audiograms, audiometers must be used, maintained, and calibrated according to specifications detailed in appendices C and E of the standard.

**Hearing protectors.** Hearing protectors must be made available to all workers exposed at or above a TWA of 85 dBA. This requirement ensures that employees have access

to protectors before they experience a loss in hearing. When baseline audiograms are delayed because it is inconvenient for mobile test vans to visit the workplace more than once a year, protectors must be worn by employees for any period exceeding six months from the time they are first exposed to 8-hour average noise levels of 85 dBA or above until their baseline audiograms are obtained. The use of hearing protectors is also mandatory for employees who have experienced threshold shifts, because these workers are particularly susceptible to noise.

With the help of a person who is trained in fitting hearing protectors, employees should decide which size and type protector is most suitable for their working environment. The protector selected should be comfortable to wear and offer sufficient attenuation to prevent hearing loss. Employees must be shown how to use and care for their protectors, and they must be supervised on the job to ensure that they continue to wear them correctly.

Hearing protectors must provide adequate attenuation in each employee's work environment. The employer must reevaluate the suitability of an employee's present protector whenever there is a change in working conditions that might render the hearing protector inadequate. If workplace noise levels increase, employees must be given more effective protectors. The protector must reduce the level of exposure to at least as low as 90 dBA, or to 85 dBA or below when an STS has occurred.

**Training.** Employee training is important because when workers understand the hearing conservation program's requirements and why it is necessary to protect their hearing, they are better motivated to actively participate in the program. They are more willing to cooperate by wearing their protectors and by undergoing audiometric tests. Employees exposed to TWAs of 85 dBA and above must be trained at least annually in the following: the effects of noise; the purpose, advantages, disadvantages, and attenuation characteristics of various types of hearing protectors; the selection, fitting, and care of protectors; and the purpose and procedures of audiometric testing. Training does not have to be accomplished in one session. The program can be structured in any format, and different individuals can conduct different parts as long as the required topics are covered. For example, audiometric procedures could be discussed immediately before audiometric testing. The training requirements are such that employees must be reminded on a yearly basis that noise is hazardous to hearing, and that they can prevent damage by wearing a hearing protector, where appropriate, and by participating in audiometric testing.

**Record keeping.** Records of noise-exposure measurement must be kept for two years. It may be prudent, however, to keep these records for a longer time in accordance with other medical records requirements under OSHA. Records of audiometric test results must be maintained for the duration of the affected employee's employment. Audiometric test records must include the name and job classification of the employee, the date the test was performed, the examiner's name, the date of acoustic or exhaustive calibration, measurements of the background sound pressure levels in audiometric test rooms, and the employee's most recent noise-exposure measurement.

## Summary

Because industrial noise problems are extremely complex, there is no one "standard" program that is applicable to all situations. In order to protect the hearing of employees and to avoid compensation costs, it behooves industry to consider and evaluate its noise problems and to take steps toward the establishment of effective hearing conservation procedures. The OSHA regulations require the control of noise exposures, employee protection against the effects of noise exposures, and the initiation of comprehensive and effective hearing conservation programs.

As outlined in this chapter, an effective hearing conservation program consists of the following:

- Noise measurement and analysis
- Engineering control of noise exceeding permissible levels
- Hearing protection for those employees working in areas where noise cannot be feasibly controlled
- Audiometric examinations for all employees
- Employee training
- Record keeping

The effectiveness of a hearing conservation program depends on the cooperation of employers, supervisors, employees, and others concerned. Management's responsibility in this type of program includes taking noise measurements, initiating noise-control measures, undertaking the audiometric testing of employees, providing hearing-protective equipment where it is required, enforcing the use of such protective equipment with sound policies and by example, and informing employees of the benefits to be derived from a hearing conservation program.

It is the employee's responsibility to make proper use of the protective equipment provided by management. It is also the employee's responsibility to observe any rules or regulations in the use of equipment designed to minimize noise exposure.

Detailed references to noise and its management, effects, and control can be found in a great many books and periodicals. For those companies needing assistance in establishing hearing conservation programs, consultation services are available in a number of professional areas through private consultation, insurance, and governmental groups.

## Bibliography

*Acoustics Handbook.* Palo Alto, CA: Hewlett-Packard, 1968.

Alberti PW, ed. *Personal Hearing Protection in Industry.* New York: Raven Press, 1982.

American Academy of Ophthalmology and Otolaryngology. *Guide for Conservation of Hearing.* Rochester, MN: AAOO, 1970.

American Industrial Hygiene Association (AIHA). *Noise and Hearing Conservation Manual,* 4th ed. Fairfax, VA: AIHA, 1986. American National Standards Institute, New York.

*Maximum Permissible Ambient Noise Levels for Audiometric Test Rooms,* S3. 1–1991.

*Method for the Measurement of Real-Ear Protection of Hearing Protectors and Physical Attenuation of Earmuffs,* S3.19–1974 (R1979).

Bell LH. *Industrial Noise Control.* New York: Marcel Dekker, 1982.

Harris CM, ed. *Handbook of Acoustical Measurements and Noise Control,* 3rd ed. New York: McGraw-Hill, 1991.

Jones RS. *Noise and Vibration Control in Buildings.* New York: McGraw-Hill, 1984.

Kryter KD. *The Effects of Noise on Man,* 2nd ed. New York: Academic Press, 1985.

Michael PL. Ear protectors—their usefulness and limitations. *Archives of Environmental Health* 10:612–618, 1965.

National Institute for Occupational Safety and Health. *Criteria for a Recommended Standard—Occupational Exposure to Noise.* Washington, DC: GPO, 1972, HSM 73–1101.

Newby HA, Popelka GR. *Audiology,* 5th ed. Englewood Cliffs, NJ: Prentice-Hall, 1985.

Occupational Safety and Health Administration. *Occupational Noise Exposure and Hearing Conservation Amendment.*

*Federal Register* 46(11) (1981):4,078–4,181.

*Federal Register* 46(162) (1981):42,622–42,639.

*Federal Register* 48(46) (1983):9,738–9,783.

Olishifski JB, Harford ER. *Industrial Noise and Hearing Conservation.* Chicago: National Safety Council, 1975.

Petersen APG. *Handbook of Noise Measurement,* 9th ed. Concord, MA: GenRad, 1980.

Sataloff J, Michael P. *Hearing Conservation.* Springfield, IL: Charles C. Thomas, 1973.

U.S. Department of Health and Human Services, CDC, National Institute for Occupational Safety and Health, Division of Technical Services, Cincinnati, Ohio.

*The NIOSH Compendium of Hearing Protection Devices, 1994,* Pub. No. 95–105.

*Occupational Noise and Hearing: 1968–1972,* Pub. No. 74–116.

Ward WD, Fricke FE, eds. *Noise as a Public Hazard.* Washington, DC: American Speech and Hearing Association, 1969.

# 1910.95—OCCUPATIONAL NOISE EXPOSURE

**(a)** Protection against the effects of noise exposure shall be provided when the sound levels exceed those shown in Table G–16 when measured on the A scale of a standard sound level meter at slow response. When noise levels are determined by octave band analysis, the equivalent A-weighted sound level may be determined as follows:

TABLE G–16—PERMISSIBLE NOISE EXPOSURES [1]

| Duration per day, hours | Sound level dBA slow response |
|---|---|
| 8 | 90 |
| 6 | 92 |
| 4 | 95 |
| 3 | 97 |
| 2 | 100 |
| 1½ | 102 |
| 1 | 105 |
| ½ | 110 |
| ¼ or less | 115 |

[1]When the daily noise exposure is composed of two or more periods of noise exposure of different levels, their combined effect should be considered, rather than the individual effect of each. If the sum of the following fractions: $C_1/T_1 + C_2/T_2 ... C_n/T_n$ exceeds unity, then, the mixed exposure should be considered to exceed the limit value. $C_n$ indicates the total time of exposure at a specified noise level, and $T_n$ indicates the total time of exposure permitted at that level.

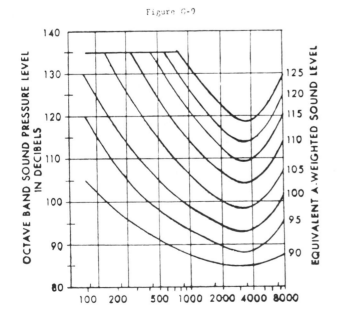

Figure G–9

Equivalent sound level contours. Octave band sound pressure levels may be converted to the equivalent A-weighted sound level by plotting them on this graph and noting the A-weighted sound level corresponding to the point of highest penetration into the sound level contours. This equivalent A-weighted sound level, which may differ from the actual A-weighted sound level of the noise, is used to determine exposure limits from Table G–16.

**(b)**

**(1)** When employees are subjected to sound exceeding those listed in Table G–16, feasible administrative or engineering controls shall be utilized. If such controls fail to reduce sound levels within the levels of Table G–16, personal protective equipment shall be provided and used to reduce sound levels within the levels of the table.

Exposure to impulsive or impact noise should not exceed 140 dB peak sound pressure level.

**(2)** If the variations in noise level involve maxima at intervals of 1 second or less, it is to be considered continuous.

## (c) Hearing conservation program.

**(1)** The employer shall administer a continuing, effective hearing conservation program, as described in paragraphs (c) through (o) of this section, whenever employee noise exposures equal or exceed an 8-hour time-weighted average sound level (TWA) of 85 decibels measured on the A scale (slow response) or, equivalently, a dose of fifty percent. For purposes of the hearing conservation program, employee noise exposures shall be computed in accordance with Appendix A and Table G–16a, and without regard to any attenuation provided by the use of personal protective equipment.

**(2)** For purposes of paragraphs (c) through (n) of this section, an 8-hour time-weighted average of 85 decibels or a dose of fifty percent shall also be referred to as the action level.

## (d) Monitoring.

**(1)** When information indicates that any employee's exposure may equal or exceed an 8-

hour time-weighted average of 85 decibels, the employer shall develop and implement a monitoring program.

(I) The sampling strategy shall be designed to identify employees for inclusion in the hearing conservation program and to enable the proper selection of hearing protectors.

(II) Where circumstances such as high worker mobility, significant variations in sound level, or a significant component of impulse noise make area monitoring generally inappropriate, the employer shall use representative personal sampling to comply with the monitoring requirements of this paragraph unless the employer can show that area sampling produces equivalent results.

(2)

(I) All continuous, intermittent and impulsive sound levels from 80 decibels to 130 decibels shall be integrated into the noise measurements.

(II) Instruments used to measure employee noise exposure shall be calibrated to ensure measurement accuracy.

(3) Monitoring shall be repeated whenever a change in production, process, equipment or controls increases noise exposures to the extent that:

(I) Additional employees may be exposed at or above the action level; or

(II) The attenuation provided by hearing protectors being used by employees may be rendered inadequate to meet the requirements of paragraph (j) of this section.

(e) **Employee notification.** The employer shall notify each employee exposed at or above an 8-hour time-weighted average of 85 decibels of the results of the monitoring.

(f) **Observation of monitoring.** The employer shall provide affected employees or their representatives with an opportunity to observe any noise measurements conducted pursuant to this section.

(g) **Audiometric testing program.**

(1) The employer shall establish and maintain an audiometric testing program as provided in this paragraph by making audiometric testing available to all employees whose exposures equal or exceed an 8-hour time-weighted average of 85 decibels.

(2) The program shall be provided at no cost to employees.

(3) Audiometric tests shall be performed by a licensed or certified audiologist, otolaryngologist, or other physician, or by a technician who is certified by the Council of Accreditation in Occupational Hearing Conservation, or who has satisfactorily demonstrated competence in administering audiometric examinations, obtaining valid audiograms, and properly using, maintaining and checking calibration and proper functioning of the audiometers being used. A technician who operates microprocessor audiometers does not need to be certified. A technician who performs audiometric tests must be responsible to an audiologist, otolaryngologist or physician.

(4) All audiograms obtained pursuant to this section shall meet the requirements of Appendix C: *Audiometric Measuring Instruments.*

(5) **Baseline audiogram.**

(I) Within 6 months of an employee's first exposure at or above the action level, the employer shall establish a valid baseline audiogram against which subsequent audiograms can be compared.

(II) **Mobile test van exception.** Where mobile test vans are used to meet the audiometric testing obligations, the employer shall obtain a valid baseline audiogram within 1 year of an employee's first exposure at or above the action level. Where baseline audiograms are obtained more than 6 months after the employee's first exposure at or above the action level, employees shall wear hearing protectors for any period exceeding six months after first exposure until the baseline audiogram is obtained.

(III) Testing to establish a baseline audiogram shall be preceded by at least 14 hours without exposure to workplace noise. Hearing protectors may be used as a substitute for the requirement that baseline audiograms be preceded by 14 hours without exposure to workplace noise.

(IV) The employer shall notify employees of the need to avoid high levels of non-occupational noise exposure during the 14-hour period immediately preceding the audiometric examination.

**(6) Annual audiogram.** At least annually after obtaining the baseline audiogram, the employer shall obtain a new audiogram for each employee exposed at or above an 8-hour time-weighted average of 85 decibels.

**(7) Evaluation of audiogram.**

**(I)** Each employee's annual audiogram shall be compared to that employee's baseline audiogram to determine if the audiogram is valid and if a standard threshold shift as defined in paragraph (g)(10) of this section has occurred. This comparison may be done by a technician.

**(II)** If the annual audiogram shows that an employee has suffered a standard threshold shift, the employer may obtain a retest within 30 days and consider the results of the retest as the annual audiogram.

**(III)** The audiologist, otolaryngologist, or physician shall review problem audiograms and shall determine whether there is a need for further evaluation. The employer shall provide to the person performing this evaluation the following information:

(a) A copy of the requirements for hearing conservation as set forth in paragraphs (c) through (n) of this section;

(b) The baseline audiogram and most recent audiogram of the employee to be evaluated;

(c) Measurements of background sound pressure levels in the audiometric test room as required in Appendix D: Audiometric Test Rooms.

(d) Records of audiometer calibrations required by paragraph (h)(5) of this section.

**(8) Follow-up procedures.**

**(I)** If a comparison of the annual audiogram to the baseline audiogram indicates a standard threshold shift as defined in paragraph (g)(10) of this section has occurred, the employee shall be informed of this fact in writing, within 21 days of the determination.

**(II)** Unless a physician determines that the standard threshold shift is not work related or aggravated by occupational noise exposure, the employer shall ensure that the following steps are taken when a standard threshold shift occurs:

(a) Employees not using hearing protectors shall be fitted with hearing protectors, trained in their use and care, and required to use them.

(b) Employees already using hearing protectors shall be refitted and retained in the use of hearing protectors and provided with hearing protectors offering greater attenuation if necessary.

(c) The employee shall be referred for a clinical audiological evaluation or an otological examination, as appropriate, if additional testing is necessary or if the employer suspects that a medical pathology of the ear is caused or aggravated by the wearing of hearing protectors.

(d) The employee is informed of the need for an otological examination if a medical pathology of the ear that is unrelated to the use of hearing protectors is suspected.

**(III)** If subsequent audiometric testing of an employee whose exposure to noise is less than an 8-hour TWA of 90 decibels indicates that a standard threshold shift is not persistent, the employer:

(a) Shall inform the employee of the new audiometric interpretation; and

(b) May discontinue the required use of hearing protectors for that employee.

**(9) Revised baseline.** An annual audiogram may be substituted for the baseline audiogram when, in the judgment of the audiologist, otolaryngologist or physician who is evaluating the audiogram:

**(I)** The standard threshold shift revealed by the audiogram is persistent; or

**(II)** The hearing threshold shown in the annual audiogram indicates significant improvement over the baseline audiogram.

**(10) Standard threshold shift.**

**(i)** As used in this section, a standard threshold shift is a change in hearing threshold relative to the baseline audiogram of an average of 10 dB or more at 2000, 3000, and 4000 Hz in either ear.

**(ii)** In determining whether a standard threshold shift has occurred, allowance may be made for the contribution of aging (presbycusis) to the change in hearing level by correcting the annual audiogram according to the procedure described in Appendix F: *Calculation and Application of Age Correction to Audiograms.*

## (h) Audiometric test requirements.

**(1)** Audiometric tests shall be pure tone, air conduction, hearing threshold examinations, with test frequencies including as a minimum 500, 1000, 2000, 3000, 4000, and 6000 Hz. Tests at each frequency shall be taken separately for each ear.

**(2)** Audiometric tests shall be conducted with audiometers (including microprocessor audiometers) that meet the specifications of, and are maintained and used in accordance with, American National Standard Specification for Audiometers, S3.6–1969.

**(3)** Pulsed-tone and self-recording audiometers, if used, shall meet the requirements specified in Appendix C: *Audiometric Measuring Instruments.*

**(4)** Audiometric examinations shall be administered in a room meeting the requirements listed in Appendix D: *Audiometric Test Rooms.*

**(5) Audiometer calibration.**

**(i)** The functional operation of the audiometer shall be checked before each day's use by testing a person with known, stable hearing thresholds, and by listening to the audiometer's output to make sure that the output is free from distorted or unwanted sounds. Deviations of 10 decibels or greater require an acoustic calibration.

**(ii)** Audiometer calibration shall be checked acoustically at least annually in accordance with Appendix E: *Acoustic Calibration of Audiometers.* Test frequencies below 500 Hz and above 6000 Hz may be omitted from this check. Deviations of 15 decibels or greater require an exhaustive calibration.

**(iii)** An exhaustive calibration shall be performed at least every two years in accordance with sections 4.1.2; 4.1.3.; 4.1.4.3; 4.2; 4.4.1; 4.4.2; 4.4.3; and 4.5 of the American National Standard Specification for Audiometers, S3.6–1969. Test frequencies below 500 Hz and above 6000 Hz may be omitted from this calibration.

## (i) Hearing protectors.

**(1)** Employers shall make hearing protectors available to all employees exposed to an 8-hour time-weighted average of 85 decibels or greater at no cost to the employees. Hearing protectors shall be replaced as necessary.

**(2)** Employers shall ensure that hearing protectors are worn:

**(i)** By an employee who is required by paragraph (b)(1) of this section to wear personal protective equipment; and

**(ii)** By any employee who is exposed to an 8-hour time-weighted average of 85 decibels or greater, and who:

**(a)** Has not yet had a baseline audiogram established pursuant to paragraph (g)(5)(ii); or

**(b)** Has experienced a standard threshold shift.

**(3)** Employees shall be given the opportunity to select their hearing protectors from a variety of suitable hearing protectors provided by the employer.

**(4)** The employer shall provide training in the use and care of all hearing protectors provided to employees.

**(5)** The employer shall ensure proper initial fitting and supervise the correct use of all hearing protectors.

## (j) Hearing protector attenuation.

**(1)** The employer shall evaluate hearing protector attenuation for the specific noise environments in which the protector will be used. The employer shall use one of the evaluation methods described in Appendix B: *Methods for Estimating the Adequacy of Hearing Protection Attenuation.*

**(2)** Hearing protectors must attenuate employee exposure at least to an 8-hour time-weighted average of 90 decibels as required by paragraph (b) of this section.

**(3)** For employees who have experienced a standard threshold shift, hearing protectors must attenuate employee exposure to an 8-hour time-weighted average of 85 decibels or below.

**(4)** The adequacy of hearing protector attenuation shall be re-evaluated whenever employee noise exposures increase to the extent that the hearing protectors provided may no longer provide adequate attenuation. The employee shall provide more effective hearing protectors where necessary.

## (k) Training program.

**(1)** The employer shall institute a training program for all employees who are exposed to noise at or above an 8-hour time-weighted average of 85 decibels, and shall ensure employee participation in such program.

**(2)** The training program shall be repeated annually for each employee included in the hearing conservation program. Information provided in the training program shall be updated to be consistent with changes in protective equipment and work processes.

**(3)** The employer shall ensure that each employee is informed of the following:

**(i)** The effects of noise on hearing;

**(ii)** The purpose of hearing protectors, the advantages, disadvantages, and attenuation of various types, and instructions on selection, fitting, use, and care; and

**(iii)** The purpose of audiometric testing, and an explanation of the test procedures.

## (l) Access to information and training materials.

**(1)** The employer shall make available to affected employees or their representatives copies of this standard and shall also post a copy in the workplace.

**(2)** The employer shall provide to affected employees any informational materials pertaining to the standard that are supplied to the employer by the Assistant Secretary.

**(3)** The employer shall provide, upon request, all materials related to the employer's training and education program pertaining to this standard to the Assistant Secretary and the Director.

## (m) Recordkeeping.

**(1) Exposure measurements.** The employer shall maintain an accurate record of all employee exposure measurements required by paragraph (d) of this section.

**(2) Audiometric tests.**

**(i)** The employer shall retain all employee audiometric test records obtained pursuant to paragraph (g) of this section:

**(ii)** This record shall include:

**(a)** Name and job classification of the employee;

**(b)** Date of the audiogram;

**(c)** The examiner's name;

**(d)** Date of the last acoustic or exhaustive calibration of the audiometer; and

**(e)** Employee's most recent noise exposure assessment.

**(f)** The employer shall maintain accurate records of the measurements of the background sound pressure levels in audiometric test rooms.

**(3) Record retention.** The employer shall retain records required in this paragraph (m) for at least the following periods.

**(i)** Noise exposure measurement records shall be retained for two years.

**(ii)** Audiometric test records shall be retained for the duration of the affected employee's employment.

**(4) Access to records.** All records required by this section shall be provided upon request to employees, former employees, representatives designated by the individual employee, and the Assistant Secretary. The provisions of 29 CFR 1910.20(a)–(e) and (g)–(i) apply to access to records under this section.

**(5) Transfer of records.** If the employer ceases to do business, the employer shall transfer to the successor employer all records required to be maintained by this section, and the successor employer shall retain them for the remainder of the period prescribed in paragraph (m)(3) of this section.

## (n) Appendices.

**(1)** Appendices A, B, C, D, and E to this section are incorporated as part of this section and the contents of these Appendices are mandatory.

**(2)** Appendices F and G to this section are informational and are not intended to create any additional obligations not otherwise imposed or to detract from any existing obligations.

**(o) Exemptions.** Paragraphs (c) through (n) of this section shall not apply to employers engaged in oil and gas well drilling and servicing operations.

**(p) Startup date.** Baseline audiograms required by paragraph (g) of this section shall be completed by March 1, 1984.

---

## APPENDIX A: NOISE EXPOSURE COMPUTATION

**This Appendix is Mandatory**

**I.** Computation of Employee Noise Exposure

**(1)** Noise dose is computed using Table G–16a as follows:

**(I)** When the sound level, L, is constant over the entire work shift, the noise dose, D, in percent, is given by: D=100 C/T where C is the total length of the work day, in hours, and T is the reference duration corresponding to the measured sound level, L, as given in Table G–16a or by the formula shown as a footnote to that table.

**(II)** When the workshift noise exposure is composed of two or more periods of noise at different levels, the total noise dose over the work day is given by:
D=100 ($C_1/T_1 + C_2/T_2 + \ldots + C_n/T_n$),
where $C_n$ indicates the total time of exposure at a specific noise level, and $T_n$ indicates the reference duration for that level as given by Table G–16a.

**(2)** The eight-hour time-weighted average sound level (TWA), in decibels, may be computed from the dose, in percent, by means of the formula: TWA=16.61 $\log_{10}$ (D/100)+90. For an eight-hour workshift with the noise level constant over the entire shift, the TWA is equal to the measured sound level.

**(3)** A table relating dose and TWA is given in Section II.

**TABLE G–16A**

| A-weighted sound level, L (decibel) | Reference duration, T (hour) |
|---|---|
| 80 | 32 |
| 81 | 27.9 |
| 82 | 24.3 |
| 83 | 21.1 |
| 84 | 16.4 |
| 85 | 16 |
| 86 | 13.9 |
| 87 | 12.1 |
| 88 | 10.6 |
| 89 | 9.2 |
| 90 | 8 |
| 91 | 7.0 |
| 92 | 6.1 |
| 93 | 5.3 |
| 94 | 4.6 |
| 95 | 4 |
| 96 | 3.5 |
| 97 | 3.0 |
| 98 | 2.6 |
| 99 | 2.3 |
| 100 | 2 |
| 101 | 1.7 |
| 102 | 1.5 |
| 103 | 1.3 |
| 104 | 1.1 |
| 105 | 1 |
| 106 | 0.87 |
| 107 | 0.76 |
| 108 | 0.66 |
| 109 | 0.57 |
| 110 | 0.5 |
| 111 | 0.44 |
| 112 | 0.38 |
| 113 | 0.33 |
| 114 | 0.29 |
| 115 | 0.25 |
| 116 | 0.22 |
| 117 | 0.19 |
| 118 | 0.16 |
| 119 | 0.14 |
| 120 | 0.125 |
| 121 | 0.11 |
| 122 | 0.095 |
| 123 | 0.082 |
| 124 | 0.072 |
| 125 | 0.063 |
| 126 | 0.054 |
| 127 | 0.047 |
| 128 | 0.041 |
| 129 | 0.036 |
| 130 | 0.031 |

In the above table the reference duration, T, is computed by

$$T = \frac{8}{2^{(L-90)/5}}$$

where L is the measured A-weighted sound level.

**II.** Conversion Between "Dose" and "8-Hour Time-Weighted Average" Sound Level

Compliance with paragraphs (c)–(r) of this regulation is determined by the amount of exposure to noise in the workplace. The amount of such exposure is usually measured with an audiodosimeter which gives a readout in terms of "dose." In order to better understand the requirements of the amendment, dosimeter readings can be converted to an "8-hour time-weighted average sound level." (TWA).

In order to convert the reading of a dosimeter into TWA, see Table A–1, below. This table applies to dosimeters that are set by the manufacturer to calculate dose or percent exposure according to the relationships in Table G–16a. So, for example, a dose of 91 percent over an eight hour day results in a TWA of 89.3 dB, and, a dose of 50 percent corresponds to a TWA of 85 dB.

If the dose as read on the dosimeter is less than or greater than the values found in Table A–1, the TWA may be calculated by using the formula: $TWA = 16.61 \log_{10}(D/100) + 90$ where TWA=8-hour time-weighted average sound level and D=accumulated dose in percent exposure.

**Table A–1.—Conversion From "Percent Noise Exposure" or "Dose" to "8-Hour Time-Weighted Average Sound Level" (TWA)**

| Dose or percent noise exposure | TWA |
|---|---|
| 10 | 73.4 |
| 15 | 76.3 |
| 20 | 78.4 |
| 25 | 80.0 |
| 30 | 81.3 |
| 35 | 82.4 |
| 40 | 83.4 |
| 45 | 84.2 |
| 50 | 85.0 |
| 55 | 85.7 |
| 60 | 86.3 |
| 65 | 86.9 |
| 70 | 87.4 |
| 75 | 87.9 |
| 80 | 88.4 |
| 81 | 88.5 |
| 82 | 88.6 |
| 83 | 88.7 |
| 84 | 88.7 |
| 85 | 88.8 |
| 86 | 88.9 |
| 87 | 89.0 |
| 88 | 89.1 |
| 89 | 89.2 |
| 90 | 89.2 |
| 91 | 89.3 |
| 92 | 89.4 |
| 93 | 89.5 |
| 94 | 89.6 |
| 95 | 89.6 |
| 96 | 89.7 |
| 97 | 89.8 |
| 98 | 89.9 |
| 99 | 89.9 |
| 100 | 90.0 |
| 101 | 90.1 |
| 102 | 90.1 |
| 103 | 90.2 |
| 104 | 90.3 |
| 105 | 90.4 |
| 106 | 90.4 |
| 107 | 90.5 |
| 108 | 90.6 |
| 109 | 90.6 |

**Table A–1.—Conversions From "Percent Noise Exposure" or "Dose" to "8-Hour Time-Weighted Average Sound level" (TWA)—Continued**

| Dose or percent noise exposure | TWA |
|---|---|
| 110 | 90.7 |
| 111 | 90.8 |
| 112 | 90.8 |
| 113 | 90.9 |
| 114 | 90.9 |
| 115 | 91.1 |
| 116 | 91.1 |
| 117 | 91.1 |
| 118 | 91.2 |
| 119 | 91.3 |
| 120 | 91.3 |
| 125 | 91.6 |
| 130 | 91.9 |
| 135 | 92.2 |
| 140 | 92.4 |
| 145 | 92.7 |
| 150 | 92.9 |
| 155 | 93.2 |
| 160 | 93.4 |
| 165 | 93.6 |
| 170 | 93.8 |
| 175 | 94.0 |
| 180 | 94.2 |
| 185 | 94.4 |
| 190 | 94.6 |
| 195 | 94.8 |
| 200 | 95.0 |
| 210 | 95.4 |
| 220 | 95.7 |
| 230 | 96.0 |
| 240 | 96.3 |
| 250 | 96.6 |
| 260 | 96.9 |
| 270 | 97.2 |
| 280 | 97.4 |
| 290 | 97.7 |
| 300 | 97.9 |
| 310 | 98.2 |
| 320 | 98.4 |
| 330 | 98.6 |
| 340 | 98.8 |
| 350 | 99.0 |
| 360 | 99.2 |
| 370 | 99.4 |
| 380 | 99.6 |
| 390 | 99.8 |
| 400 | 100.0 |
| 410 | 100.2 |
| 420 | 100.4 |
| 430 | 100.5 |
| 440 | 100.7 |
| 450 | 100.8 |
| 460 | 101.0 |
| 470 | 101.2 |
| 480 | 101.3 |
| 490 | 101.5 |
| 500 | 101.6 |
| 510 | 101.8 |
| 520 | 101.9 |
| 530 | 102.0 |
| 540 | 102.2 |
| 550 | 102.3 |
| 560 | 102.4 |
| 570 | 102.6 |
| 580 | 102.7 |

**Table A-1.—Conversions From "Percent Noise Exposure" or "Dose" to "8-Hour Time-Weighted Average Sound level" (TWA)—Continued**

| Dose or percent noise exposure | TWA |
|---|---|
| 590 | 102.8 |
| 600 | 102.9 |
| 610 | 103.0 |
| 620 | 103.2 |
| 630 | 103.3 |
| 640 | 103.4 |
| 650 | 103.5 |
| 660 | 103.6 |
| 670 | 103.7 |
| 680 | 103.8 |
| 690 | 103.9 |
| 700 | 104.0 |
| 710 | 104.1 |
| 720 | 104.2 |
| 730 | 104.3 |
| 740 | 104.4 |
| 750 | 104.5 |
| 760 | 104.6 |
| 770 | 104.7 |
| 780 | 104.8 |
| 790 | 104.9 |
| 800 | 105.0 |
| 810 | 105.1 |
| 820 | 105.2 |
| 830 | 105.3 |
| 840 | 105.4 |
| 850 | 105.4 |
| 860 | 105.5 |
| 870 | 105.6 |
| 880 | 105.7 |
| 890 | 105.8 |
| 900 | 105.8 |
| 910 | 105.9 |
| 920 | 106.0 |
| 930 | 106.1 |
| 940 | 106.2 |
| 950 | 106.2 |
| 960 | 106.3 |
| 970 | 106.4 |
| 980 | 106.5 |
| 990 | 106.5 |
| 999 | 106.6 |

## APPENDIX B: METHODS FOR ESTIMATING THE ADEQUACY OF HEARING PROTECTOR ATTENUATION

### This Appendix is Mandatory

For employees who have experienced a significant threshold shift, hearing protector attenuation must be sufficient to reduce employee exposure to a TWA of 85 dB. Employers must select one of the following methods by which to estimate the adequacy of hearing protector attenuation.

The most convenient method is the Noise Reduction Rating (NRR) developed by the Environmental Protection Agency (EPA). According to EPA regulation, the NRR must be shown on the hearing protector package. The NRR is then related to an individual worker's noise environment in order to assess the adequacy of the attenuation of a given hearing protector. This Appendix describes four methods of using the

NRR to determine whether a particular hearing protector provides adequate protection within a given exposure environment. Selection among the four procedures is dependent upon the employer's noise measuring instruments.

Instead of using the NRR, employers may evaluate the adequacy of hearing protector attenuation by using one of the three methods developed by the National Institute for Occupational Safety and Health (NIOSH), which are described in the "List of Personal Hearing Protectors and Attenuation Data," HEW Publication No. 76-120, 1975, pages 21-37. These methods are known as NIOSH methods #1, #2 and #3. The NRR described below is a simplification of NIOSH method #2. The most complex method is NIOSH method #1, which is probably the most accurate method since it uses the largest amount of spectral information from the individual employee's noise environment. As in the case of the NRR method described below, if one of the NIOSH methods is used, the selected method must be applied to an individual's noise environment to assess the adequacy of the attenuation. Employers should be careful to take a sufficient number of measurements in order to achieve a representative sample for each time segment.

Note.—The employer must remember that calculated attenuation values reflect realistic values only to the extent that the protectors are properly fitted and worn.

When using the NRR to assess hearing protector adequacy, one of the following methods must be used:

**(I)** When using a dosimeter that is capable of C-weighted measurements:

**(A)** Obtain the employee's C-weighted dose for the entire workshift, and convert to TWA (see Appendix A, II).

**(B)** Subtract the NRR from the C-weighted TWA to obtain the estimated A-weighted TWA under the ear protector.

**(II)** When using a dosimeter that is not capable of C-weighted measurements, the following method may be used:

**(A)** Convert the A-weighted dose to TWA (see Appendix A).

**(B)** Subtract 7 dB from the NRR.

**(C)** Subtract the remainder from the A-weighted TWA to obtain the estimated A-weighted TWA under the ear protector.

**(III)** When using a sound level meter set to the A-weighting network:

**(A)** Obtain the employee's A-weighted TWA.

**(B)** Subtract 7 dB from the NRR, and subtract the remainder from the A-weighted TWA to obtain the estimated A-weighted TWA under the ear protector.

**(Iv)** When using a sound level meter set on the C-weighting network:

**(A)** Obtain a representative sample of the C-weighted sound levels in the employee's environment.

**(B)** Subtract the NRR from the C-weighted average sound level to obtain the estimated A-weighted TWA under the ear protector.

**(v)** When using area monitoring procedures and a sound level meter set to the A-weighing network.

**(A)** Obtain a representative sound level for the area in question.

**(B)** Subtract 7 dB from the NRR and subtract the remainder from the A-weighted sound level for that area.

**(vi)** When using area monitoring procedures and a sound level meter set to the C-weighting network:

**(A)** Obtain a representative sound level for the area in question.

**(B)** Subtract the NRR from the C-weighted sound level for that area.

---

## APPENDIX C: AUDIOMETRIC MEASURING INSTRUMENTS

**This Appendix is Mandatory**

**1.** In the event that pulsed-tone audiometers are used, they shall have a tone on-time of at least 200 milliseconds.

**2.** Self-recording audiometers shall comply with the following requirements:

**(A)** The chart upon which the audiogram is traced shall have lines at positions corresponding to all multiples of 10 dB hearing level within the intensity range spanned by the audiometer. The lines shall be equally spaced and shall be separated by at least ¼ inch. Additional increments are optional. The audiogram pen tracings shall not exceed 2 dB in width.

**(B)** It shall be possible to set the stylus manually at the 10-dB increment lines for calibration purposes.

**(C)** The slewing rate for the audiometer attenuator shall not be more than 6 dB/sec except that an initial slewing rate greater than 6 dB/sec is permitted at the beginning of each new test frequency, but only until the second subject response.

**(D)** The audiometer shall remain at each required test frequency for 30 seconds (±3 seconds). The audiogram shall be clearly marked at each change of frequency and the actual frequency change of the audiometer shall not deviate from the frequency boundaries marked on the audiogram by more than ±3 seconds.

**(E)** It must be possible at each test frequency to place a horizontal line segment parallel to the time axis on the audiogram, such that the audiometric tracing crosses the line segment at least six times at that test frequency. At each test frequency the threshold shall be the average of the midpoints of the tracing excursions.

---

## APPENDIX D: AUDIOMETRIC TEST ROOMS

**This Appendix is Mandatory**

Rooms used for audiometric testing shall not have background sound pressure levels exceeding those in Table D-1

when measured by equipment conforming at least to the Type 2 requirements of American National Standard Specification for Sound Level Meters, S1.4–1971 (R1976), and to the Class II requirements of American National Standard Specification for Octave, Half-Octave, and Third-Octave Band Filter Sets, S1.11–1971 (R1976).

**Table D–1.—Maximum Allowable Octave-Band Sound Pressure Levels for Audiometric Test Rooms**

| Octave-band center frequency (Hz) | 500 | 1000 | 2000 | 4000 | 8000 |
|---|---|---|---|---|---|
| Sound pressure level (dB) | 40 | 40 | 47 | 57 | 62 |

---

## APPENDIX E: ACOUSTIC CALIBRATION OF AUDIOMETERS

**This Appendix is Mandatory**

Audiometer calibration shall be checked acoustically, at least annually, according to the procedures described in this Appendix. The equipment necessary to perform these measurements is a sound level meter, octave-band filter set, and a National Bureau of Standards 9A coupler. In making these measurements, the accuracy of the calibrating equipment shall be sufficient to determine that the audiometer is within the tolerances permitted by American Standard Specification for Audiometers, S3.6–1969.

**(1) Sound Pressure Output Check**

**A.** Place the earphone coupler over the microphone of the sound level meter and place the earphone on the coupler.

**B.** Set the audiometer's hearing threshold level (HTL) dial to 70 dB.

**C.** Measure the sound pressure level of the tones that each test frequency from 500 Hz through 6000 Hz for each earphone.

**D.** At each frequency the readout on the sound level meter should correspond to the levels in Table E–1 or Table E–2, as appropriate, for the type of earphone, in the column entitled "sound level meter reading."

**(2) Linearity Check**

**A.** With the earphone in place, set the frequency to 1000 Hz and the HTL dial on the audiometer to 70 dB.

**B.** Measure the sound levels in the coupler at each 10-dB decrement from 70 dB to 10 dB, noting the sound level meter reading at each setting.

**C.** For each 10-dB decrement on the audiometer the sound level meter should indicate a corresponding 10 dB decrease.

**D.** This measurement may be made electrically with a voltmeter connected to the earphone terminals.

## (3) Tolerances

When any of the measured sound levels deviate from the levels in Table E-1 or Table E-2 by ±3 dB at any test frequency between 500 and 3000 Hz, 4 dB at 4000 Hz, or 5 dB at 6000 Hz, an exhaustive calibration is advised. An exhaustive calibration is required if the deviations are greater than 10 dB at any test frequency.

### Table E-1.—Reference Threshold Levels for Telephonics—TDH-39 Earphones

| Frequency, Hz | Reference threshold level for TDH-39 earphones, dB | Sound level meter reading, dB |
|---|---|---|
| 500............. | 11.5 | 81.5 |
| 1000............. | 7 | 77 |
| 2000............. | 9 | 79 |
| 3000............. | 10 | 80 |
| 4000............. | 9.5 | 79.5 |
| 6000............. | 15.5 | 85.5 |

### Table E-2.—Reference Threshold Levels for Telephonics—TDH-49 Earphones

| Frequency, Hz | Reference threshold level for TDH-49 earphones, dB | Sound level meter reading, dB |
|---|---|---|
| 500............. | 13.5 | 83.5 |
| 1000............. | 7.5 | 77.5 |
| 2000............. | 11 | 81.0 |
| 3000............. | 9.5 | 79.5 |
| 4000............. | 10.5 | 80.5 |
| 6000............. | 13.5 | 83.5 |

## APPENDIX F: CALCULATIONS AND APPLICATION OF AGE CORRECTIONS TO AUDIOGRAMS

**This Appendix is Non-Mandatory**

In determining whether a standard threshold shift has occurred, allowance may be made for the contribution of aging to the change in hearing level by adjusting the most recent audiogram. If the employer chooses to adjust the audiogram, the employer shall follow the procedure described below. This procedure and the age correction tables were developed by the National Institute for Occupational Safety and Health in the criteria document entitled "Criteria for a Recommended Standard ... Occpational Exposure to Noise," ((HSM)-11001).

For each audiometric test frequency;
**(I)** Determine from Tables F-1 or F-2 the age correction values for the employee by:

**(A)** Finding the age at which the most recent audiogram was taken and recording the corresponding values of age corrections at 1000 Hz through 6000 Hz;

**(B)** Finding the age at which the baseline audiogram was taken and recording the corresponding values of age corrections at 1000 Hz through 6000 Hz.

**(II)** Subtract the values found in step (i)(A) from the value found in step (i)(B).

**(III)** The differences calculated in step (ii) represented that portion of the change in hearing that may be due to aging.

Example: Employee is a 32-year-old male. The audiometric history for his right ear is shown in decibels below.

| Employee's age | 1000 | 2000 | 3000 | 4000 | 6000 |
|---|---|---|---|---|---|
| 26............. | 10 | 5 | 5 | 10 | 5 |
| *27............. | 0 | 0 | 0 | 5 | 5 |
| 28............. | 0 | 0 | 0 | 10 | 5 |
| 29............. | 5 | 0 | 5 | 15 | 5 |
| 30............. | 0 | 5 | 10 | 20 | 10 |
| 31............. | 5 | 10 | 20 | 15 | 15 |
| *32............. | 5 | 10 | 10 | 25 | 20 |

The audiogram at age 27 is considered the baseline since it shows the best hearing threshold levels. Asterisks have been used to identify the baseline and most recent audiogram. A threshold shift of 20 dB exists at 4000 Hz between the audiograms taken at ages 27 and 32.

(The threshold shift is computed by subtracting the hearing threshold at age 27, which was 5, from the hearing threshold at age 32, which is 25). A retest audiogram has confirmed this shift. The contribution of aging to this change in hearing may be estimated in the following manner:

Go to Table F-1 and find the age correction values (in dB) for 4000 Hz at age 27 and age 32.

| | 1000 | 2000 | 3000 | 4000 | 6000 |
|---|---|---|---|---|---|
| Age 32......... | 6 | 5 | 7 | 10 | 14 |
| Age 27......... | 5 | 4 | 6 | 7 | 11 |
| Difference ..... | 1 | 1 | 1 | 3 | 3 |

The difference represents the amount of hearing loss that may be attributed to aging in the time period between the baseline audiogram and the most recent audiogram. In this example, the difference at 4000 Hz is 3 dB. This value is subtracted from the hearing level at 4000 Hz, which in the most recent audiogram is 25, yielding 22 after adjustment. Then the hearing threshold in the baseline audiogram at 4000 Hz (5) is subtracted from the adjusted annual audiogram hearing threshold at 4000 Hz (22). Thus the age-corrected threshold shift would be 17 dB (as opposed to a threshold shift of 20 dB without age correction).

#### Table F-1.—Age Correction Values In Decibels For Males

| Years | Audiometric Test Frequencies (Hz) | | | | |
|---|---|---|---|---|---|
| | 1000 | 2000 | 3000 | 4000 | 6000 |
| 20 or younger | 5 | 3 | 4 | 5 | 8 |
| 21 | 5 | 3 | 4 | 5 | 8 |
| 22 | 5 | 3 | 4 | 5 | 8 |
| 23 | 5 | 3 | 4 | 6 | 9 |
| 24 | 5 | 3 | 5 | 6 | 9 |
| 25 | 5 | 3 | 5 | 7 | 10 |
| 26 | 5 | 4 | 5 | 7 | 10 |
| 27 | 5 | 4 | 6 | 7 | 11 |
| 28 | 6 | 4 | 6 | 8 | 11 |
| 29 | 6 | 4 | 6 | 8 | 12 |
| 30 | 6 | 4 | 6 | 9 | 12 |
| 31 | 6 | 4 | 7 | 9 | 13 |
| 32 | 6 | 5 | 7 | 10 | 14 |
| 33 | 6 | 5 | 7 | 10 | 14 |
| 34 | 6 | 5 | 8 | 11 | 15 |
| 35 | 7 | 5 | 8 | 11 | 15 |
| 36 | 7 | 5 | 9 | 12 | 16 |
| 37 | 7 | 6 | 9 | 12 | 17 |
| 38 | 7 | 6 | 9 | 13 | 17 |
| 39 | 7 | 6 | 10 | 14 | 18 |
| 40 | 7 | 6 | 10 | 14 | 19 |
| 41 | 7 | 6 | 10 | 14 | 20 |
| 42 | 8 | 7 | 11 | 16 | 20 |
| 43 | 8 | 7 | 12 | 16 | 21 |
| 44 | 8 | 7 | 12 | 17 | 22 |
| 45 | 8 | 7 | 13 | 18 | 23 |
| 46 | 8 | 8 | 13 | 19 | 24 |
| 47 | 8 | 8 | 14 | 19 | 24 |
| 48 | 9 | 8 | 14 | 20 | 25 |
| 49 | 9 | 9 | 15 | 21 | 26 |
| 50 | 9 | 9 | 16 | 22 | 27 |
| 51 | 9 | 9 | 16 | 23 | 28 |
| 52 | 9 | 10 | 17 | 24 | 29 |
| 53 | 9 | 10 | 18 | 25 | 30 |
| 54 | 10 | 10 | 18 | 26 | 31 |

#### Table F-1.—Age Correction Values In Decibels For Males—Continued

| Years | Audiometric Test Frequencies (Hz) | | | | |
|---|---|---|---|---|---|
| | 1000 | 2000 | 3000 | 4000 | 6000 |
| 55 | 10 | 11 | 19 | 27 | 32 |
| 56 | 10 | 11 | 20 | 28 | 34 |
| 57 | 10 | 11 | 21 | 29 | 35 |
| 58 | 10 | 12 | 22 | 31 | 36 |
| 59 | 11 | 12 | 22 | 32 | 37 |
| 60 or older | 11 | 13 | 23 | 33 | 38 |

#### Table F-2.—Age Correction Values In Decibels For Females

| Years | Audiometric Test Frequencies (Hz) | | | | |
|---|---|---|---|---|---|
| | 1000 | 2000 | 3000 | 4000 | 6000 |
| 20 or younger | 7 | 4 | 3 | 3 | 6 |
| 21 | 7 | 4 | 4 | 3 | 6 |
| 22 | 7 | 4 | 4 | 4 | 6 |
| 23 | 7 | 5 | 4 | 4 | 7 |
| 24 | 7 | 5 | 4 | 4 | 7 |
| 25 | 8 | 5 | 4 | 4 | 7 |
| 26 | 8 | 5 | 5 | 4 | 8 |
| 27 | 8 | 5 | 5 | 5 | 8 |

#### Table F-2.—Age Correction Values In Decibels For Females —Continued

| Years | Audiometric Test Frequencies (Hz) | | | | |
|---|---|---|---|---|---|
| | 1000 | 2000 | 3000 | 4000 | 6000 |
| 28 | 8 | 5 | 5 | 5 | 8 |
| 29 | 8 | 5 | 5 | 5 | 9 |
| 30 | 8 | 6 | 5 | 5 | 9 |
| 31 | 8 | 6 | 6 | 5 | 9 |
| 32 | 9 | 6 | 6 | 6 | 10 |
| 33 | 9 | 6 | 6 | 6 | 10 |
| 34 | 9 | 6 | 6 | 6 | 10 |
| 35 | 9 | 6 | 7 | 7 | 11 |
| 36 | 9 | 7 | 7 | 7 | 11 |
| 37 | 9 | 7 | 7 | 7 | 12 |
| 38 | 10 | 7 | 7 | 7 | 12 |
| 39 | 10 | 7 | 8 | 8 | 12 |
| 40 | 10 | 7 | 8 | 8 | 13 |
| 41 | 10 | 8 | 8 | 8 | 13 |
| 42 | 10 | 8 | 9 | 9 | 13 |
| 43 | 11 | 8 | 9 | 9 | 14 |
| 44 | 11 | 8 | 9 | 9 | 14 |
| 45 | 11 | 8 | 10 | 10 | 15 |
| 46 | 11 | 9 | 10 | 10 | 15 |
| 47 | 11 | 9 | 10 | 11 | 16 |
| 48 | 12 | 9 | 11 | 11 | 16 |
| 49 | 12 | 9 | 11 | 11 | 16 |
| 50 | 12 | 10 | 11 | 12 | 17 |
| 51 | 12 | 10 | 12 | 12 | 17 |
| 52 | 12 | 10 | 12 | 13 | 18 |
| 53 | 13 | 10 | 13 | 13 | 18 |
| 54 | 13 | 11 | 13 | 14 | 19 |
| 55 | 13 | 11 | 14 | 14 | 19 |
| 56 | 13 | 11 | 14 | 15 | 20 |
| 57 | 13 | 11 | 15 | 15 | 20 |
| 58 | 14 | 12 | 15 | 16 | 21 |
| 59 | 14 | 12 | 16 | 16 | 21 |
| 60 or older | 14 | 12 | 16 | 17 | 22 |

## APPENDIX G: MONITORING NOISE LEVELS NON-MANDATORY INFORMATIONAL APPENDIX

This appendix provides information to help employers comply with the noise monitoring obligations that are part of the hearing conservation amendment.

### What is the purpose of noise monitoring?

This revised amendment requires that employees be placed in a hearing conservation program if they are exposed to average noise levels of 85 dB or greater during an 8 hour workday. In order to determine if exposures are at or above this level, it may be necessary to measure or monitor the actual noise levels in the workplace and to estimate the noise exposure or "dose" received by employees during the workday.

### When is it necessary to implement a noise monitoring program?

It is not necessary for every employer to measure workplace noise. Noise monitoring or measuring must be conducted only when exposures are at or above 85 dB. Factors which suggest that noise exposures in the workplace may be at this level include employee complaints about the loudness of noise, indications that employees are losing their hearing, or noisy conditions which make normal conversation difficult. The employer should also consider any information available regarding noise emitted from specific machines. In addition, actual workplace noise measurements can suggest whether or not a monitoring program should be initiated.

### How is noise measured?

Basically, there are two different instruments to measure noise exposures: the sound level meter and the dosimeter. A sound level meter is a device that measures the intensity of sound at a given moment. Since sound level meters provide a measure of sound intensity at only one point in time, it is generally necessary to take a number of measurements at different times during the day to estimate noise exposure over a workday. If noise levels fluctuate, the amount of time noise remains at each of the various measured levels must be determined.

To estimate employee noise exposures with a sound level meter it is also generally necessary to take several measurements at different locations within the workplace. After appropriate sound level meter readings are obtained, people sometimes draw "maps" of the sound levels within different areas of the workplace. By using a sound level "map" and information on employee locations throughout the day, estimates of individual exposure levels can be developed. This measurement method is generally referred to as *area* noise monitoring.

A dosimeter is like a sound level meter except that it stores sound level measurements and integrates these measurements over time, providing an average noise exposure reading for a given period of time, such as an 8-hour workday. With a dosimeter, a microphone is attached to the employee's clothing and the exposure measurement is simply read at the end of the desired time period. A reader may be used to read-out the dosimeter's measurements. Since the dosimeter is worn by the employee, it measures noise levels in those locations in which the employee travels. A sound level meter can also be positioned within the immediate vicinity of the exposed worker to obtain an individual exposure estimate. Such procedures are generally referred to as *personal* noise monitoring.

Area monitoring can be used to estimate noise exposure when the noise levels are relatively constant and employees are not mobile. In workplaces where employees move about in different areas or where the noise intensity tends to fluctuate over time, noise exposure is generally more accurately estimated by the personal monitoring approach.

In situations where personal monitoring is appropriate, proper positioning of the microphone is necessary to obtain accurate measurements. With a dosimeter, the microphone is generally located on the shoulder and remains in that position for the entire workday. With a sound level meter, the microphone is stationed near the employee's head, and the instrument is usually held by an individual who follows the employee as he or she moves about.

Manufacturer's instructions, contained in dosimeter and sound level meter operating manuals, should be followed for calibration and maintenance. To ensure accurate results, it is considered good professional practice to calibrate instruments before and after each use.

How often is it necessary to monitor noise levels?

The amendment requires that when there are significant changes in machinery or production processes that may result in increased noise levels, remonitoring must be conducted to determine whether additional employees need to be included in the hearing conservation program. Many companies choose to remonitor periodically (once every year or two) to ensure that all exposed employees are included in their hearing conservation programs.

Where can equipment and technical advice be obtained?

Noise monitoring equipment may be either purchased or rented. Sound level meters cost about $500 to $1,000, while dosimeters range in price from about $750 to $1,500. Smaller companies may find it more economical to rent equipment rather than to purchase it. Names of equipment suppliers may be found in the telephone book (Yellow Pages) under headings such as: "Safety Equipment," "Industrial Hygiene," or "Engineers-Acoustical." In addition to providing information on obtaining noise monitoring equipment, many companies and individuals included under such listings can provide professional advice on how to conduct a valid noise monitoring program. Some audiological testing firms and industrial hygiene firms also provide noise monitoring services. Universities with audiology, industrial hygiene, or acoustical engineering departments may also provide information or may be able to help employers meet their obligations under this amendment.

Free, on-site assistance may be obtained from OSHA-supported state and private consultation organizations. These safety and health consultative entities generally give priority to the needs of small businesses. See the attached directory for a listing of organizations to contact for aid.

#### OSHA Onsite Consultation Project Directory

| State | Office and address | Contact |
|---|---|---|
| Alabama | Alabama Consultation Program, P.O. Box 6005, University, Alabama 35486 | (205) 348-7136, Mr. William Weems, Director. |
| Alaska | State of Alaska, Department of Labor, Occupational Safety & Health, 3301 Eagle St., Pouch 7-022, Anchorage, Alaska 99510. | (907) 276-5013, Mr. Stan Godhos, Project Manager (By Mail). |
| American Samoa | Service not yet available. | |
| Arizona | Consultation and Training, Arizona Division of Occupational Safety and Health, P.O. Box 19970, 1624 W. Adams, Phoenix, Ariz. 85005. | (602) 255-5795, Mr. Thomas Ramsley, Manager. |
| Arkansas | OSHA Consultation, Arkansas Department of Labor, 1022 High St., Little Rock, Ark. 72202 | (501) 371-2992, Mr. George Smith, Project Director. |
| California | CAL/OSHA Consultation Service, 2nd Floor, 525 Golden Gate Avenue, San Francisco, Calif. 94102. | (415) 557-2870, Mr. Emmett Jones, Chief. |
| Colorado | Occupational Safety & Health Section, Colorado State University, Institute of Rural Environmental Health, 110 Veterinary Science Building, Fort Collins, Colo. 80523. | (303) 491-6151, Dr. Roy M. Buchan, Project Director. |
| Connecticut | Division of Occupational Safety & Health, Connecticut Department of Labor, 200 Folly Brook Boulevard, Wethersfield, Conn. 06109. | (203) 566-4550, Mr. Leo Alix, Director. |
| Delaware | Delaware Department of Labor, Division of Industrial Affairs, 820 North French Street, 6th Floor, Wilmington, Del. 19801. | (302) 571-3908, Mr. Bruno Salvadori, Director. |
| District of Columbia | Occupational Safety & Health Division, District of Columbia, Department Employment Services, Office of Labor Standards, 2900 Newton Street NE., Washington, D.C. 20018. | (202) 832-1230, Mr. Lorenzo M. White, Acting Associate Director. |
| Florida | Department of Labor & Employment Security, Bureau of Industrial Safety and Health, LaFayette Building, Room 204, 2551 Executive Center Circle West, Tallahassee, Fla. 32301. | (904) 488-3044, Mr. John C. Glenn, Administrator. |
| Georgia | Economic Development Division, Technology and Development Laboratory, Engineering Experiment Station, Georgia Institute of Technology, Atlanta, Ga. 30332. | (404) 894-3806, Mr. William C. Howard, Assistant to Director. Mr. James Burson, Project Manager. |

### OSHA ONSITE CONSULTATION PROJECT DIRECTORY—Continued

| State | Office and address | Contact |
|---|---|---|
| Guam | Department of Labor, Government of Guam, 23548 Guam Main Facility, Agana, Guam 96921 | (671) 772-8291, Joe R. San Agustin, Director. |
| Hawaii | Education and Information Branch, Division of Occupational Safety and Health, Suite 910, 677 Ala Moana, Honolulu, Hawaii 96813. | (808) 548-2511, Mr. Don Alper, Manager (Air Mail). |
| Idaho | OSHA Onsite Consultation Program, Boise State University, Community and Environmental Health, 1910 University Drive, Boise, Idaho 83725. | (208) 385-3929, Dr. Eldon Edmundson, Director. |
| Illinois | Division of Industrial Services, Dept. of Commerce and Community Affairs, 310 S. Michigan Avenue, 10 Floor, Chicago, Ill. 60601. | (800) 972-4140/4216 (Toll-free in State), (312) 793-3270, Mr. Stan Czwinski, Assistant Director. |
| Iowa | Bureau of Labor, 307 E. Seventh Street, Des Moines, Iowa 50319 | (515) 281-3666, Mr. Allen J. Meier, Commissioner. |
| Indiana | Bureau of Safety, Education and Training, Indiana Division of Labor, 1013 State Office Building, Indianapolis, Indiana 46204. | (317) 633-5845, Mr. Harold Mills, Director. |
| Kansas | Kansas Dept. of Human Resources, 401 Topeka Ave., Topeka, Kans. 66603 | (913) 296-4066, Ms. Jerry Abbott, Secretary. |
| Kentucky | Education and Training, Occupational Safety and Health, Kentucky Department of Labor, 127 Building, 127 South, Frankfort, Ky. 40601. | (502) 564-6895, Mr. Larry Potter, Director. |
| Louisiana | No services available as yet (Pending FY 83). | |
| Maine | Division of Industrial Safety, Maine Dept. of Labor, Labor Station 45, State Office Building, Augusta, Maine 04333. | (207) 289-3331, Mr. Lester Wood, Director. |
| Maryland | Consultation Services, Division of Labor & Industry, 501 St. Paul Place, Baltimore, Maryland 21202 | (301) 659-4210, Ms. Ileana O'Brien, Project Manager, 7(c)(1) Agreement. |
| Massachusetts | Division of Industrial Safety, Massachusetts Department of Labor and Industries, 100 Cambridge Street, Boston, Massachusetts 02202. | (617) 727-3587, Mr. Edward Noseworthy, Project Director |
| Michigan (Health) | Special Programs Section, Division of Occupational Health, Michigan Dept. of Public Health, 3500 N. Logan, Lansing, Mich. | (517) 373-1410, Mr. Irving Davis, Chief |
| Michigan (Safety) | Safety Education & Training Division Bureau of Safety and Regulation, Michigan Department of Labor, 7150 Harris Drive, Box 30015, Lansing, Michigan 48909. | (517) 322-1809, Mr. Alan Harvie, Chief. |
| Minnesota | Training and Education Unit, Department of Labor and Industry, 5th Floor, 444 Lafayette Road, St. Paul, Minn. 55101. | (612) 296-2973, Mr. Timothy Tierney, Project Manager |
| Mississippi | Division of Occupational Safety and Health, Mississippi State Board of Health, P.O. Box 1700, Jackson, Mississippi 39205. | (601) 982-6315, Mr. Henry L. Laird, Director. |
| Missouri | Missouri Department of Labor and Industrial Relations, 722 Jefferson Street, Jefferson City, Missouri 65101 | 1-(800) 392-0208, (314) 751-3403, Ms. Paula Smith, Mr. Jim Brake. |
| Montana | Montana Bureau of Safety & Health, Division of Workers Compensation, 815 Front Street, Helena, Montana 59601. | (406) 449-3402, Mr. Ed Gatzemeier, Chief. |
| Nebraska | Nebraska Department of Labor, State House Station, State Capitol, P.O. Box 94600, Lincoln, Nebraska 68509 | 475-8451 Ext. 258, Mr. Joseph Carroll, Commissioner |
| Nevada | Department of Occupational Safety and Health, Nevada Industrial Commission, 515 E. Musser Street, Carson City, Nev. 89714. | (702) 885-5240, Mr. Allen Traenkner, Director. |
| New Hampshire | For information contact | Office of Consultation Programs, Room N3472 200 Constitution Avenue, N.W. Washington, D.C. 20210, Phone: (202) 523-8985. |
| New Jersey | New Jersey Department of Labor and Industry Division of Work Place Standards, CN-054, Trenton, New Jersey 08625. | (609) 292-2313, FTS-8-477-2313, Mr. William Clark, Assistant Commissioner. |
| New Mexico | OSHA Consultation, Health and Environment Department, Environmental Improvement Division, Occupational Health & Safety Section, 4215 Montgomery Boulevard, NE., Albuquerque, New Mexico 87109. | (505) 842-3387, Mr. Albert M. Stevens, Project Manager. |
| New York | Division of Safety and Health, New York State Department of Labor, 2 World Trade Center, Room 6995, New York, New York 10047. | (212) 488-7746/7, Mr. Joseph Alleva, Project Manager, DOSH. |
| North Carolina | Consultation Services, North Carolina Department of Labor, 4 West Edenton Street, Raleigh, N.C. 27601. | (919) 733-4885, Mr. David Pierce, Director. |
| North Dakota | Division of Environmental Research, Department of Health, Missouri Office Building, 1200 Missouri Avenue, Bismarck, N. Dak. 58506. | (701) 224-2348, Mr. Jay Crawford, Director. |
| Ohio | Department of Industrial Relations, Division of Onsite Consultation, P.O. Box 825, 2323 5th Avenue, Columbus, Ohio 43216. | (800) 282-1425 (Toll-free in State), (614) 466-7485, Mr. Andrew Doehrel, Project Manager. |
| Oklahoma | OSHA Division, Oklahoma Department of Labor, State Capitol, Suite 118, Oklahoma City, Okla. 73105. | (405) 521-2461, Mr. Charles W. McGilton, Director. |
| Oregon | Consultative Section, Department of Workers' Compensation, Accident Prevention Division, Room 102, Building 1, 2110 Front Street NE., Salem, Oregon 97310. | (503) 378-2890, Mr. Jack Buckland, Supervisor. |
| Pennsylvania | For information contact | Office of Consultation Programs, Room N3472, 200 Constitution Avenue NW., Washington, D.C. 20210, Phone: (202) 523-8985. |
| Puerto Rico | Occupational Safety & Health, Puerto Rico Department of Labor and Human Resources, 505 Munoz Rivera Ave., 21st Floor, Hato Rey, Puerto Rico 00918. | (809) 754-2134, Mr. John Cinque, Assistant Secretary, (Air Mail). |
| Rhode Island | Division of Occupational Health, Rhode Island Department of Health, The Cannon Building, 206 Health Department Building, Providence, R.I. 02903. | (401) 277-2438, Mr. James E. Hickey, Chief. |
| South Carolina | Consultation and Monitoring, South Carolina Department of Labor, P.O. Box 11329, Columbia, S.C. 29211. | (803) 758-8921, Mr. Robert Peck, Director, 7(c)(1), Project. |
| South Dakota | South Dakota Consultation Program, South Dakota State University, S.T.A.T.E.-Engineering Extension, 201 Pugsley Center-SDSU, Brookings, S. Dak. 57007. | (605) 688-4101, Mr. James Ceglian, Director. |
| Tennessee | OSHA Consultative Services, Tennessee Department of Labor, 2nd Floor, 501 Union Building, Nashville, Tennessee 37219. | (615) 741-2793, Mr. L. H. Craig Director. |
| Texas | Division of Occupational Safety and State Safety Engineer, Texas Department of Health and Resources, 1100 West 49th Street, Austin, Texas 78756. | (512) 458-7287, Mr. Walter G. Martin, P.E. Director. |
| Trust Territories | Service not yet available. | |
| Utah | Utah Job Safety and Health Consultation Service, Suite 4004, Crane Building, 307 West 200 South, Salt Lake City, Utah 84101. | (801) 533-7927/8/9, Mr. H. M. Bergeson, Project Director. |
| Vermont | Division of Occupational Safety and Health, Vermont Department of Labor and Industry, 118 State Street, Montpelier, Vt. 05602. | (802) 828-2765, Mr. Robert Mcleod, Project Director. |
| Virginia | Department of Labor and Industry, P.O. Box 12064, 205 N. 4th Street, Richmond, Va. 23241 | (804) 786-5875, Mr. Robert Beard, Commissioner. |
| Virgin Islands | Division of Occupational Safety and Health, Virgin Islands Department of Labor, Lagoon Street, Room 207, Frederiksted, Virgin Islands 00840. | (809) 772-1315, Mr. Louis Llanos, Deputy Director-DOSH. |
| Washington | Department of Labor and Industry, P.O. Box 207, Olympia, Wash. 98504 | (206) 753-6500, Mr. James Sullivan, Assistant Director. |
| West Virginia | West Virginia Department of Labor, Room 451B, State Capitol, 1900 Washington Street, Charleston, W. Va. 25305. | FTS 8-885-7890, Mr. Lawrence Barker, Commissioner. |
| Wisconsin (Health) | Section of Occupational Health, Department of Health and Social Services, P.O. Box 309, Madison, Wisconsin 53701. | (608) 266-0417, Ms. Patricia Netzke, Acting Chief. |
| Wisconsin (Safety) | Division of Safety and Buildings, Department of Industry, Labor and Human Relations, 1570 E. Moreland Blvd., Waukesha, Wis. 53186. | (414) 544-8886, Mr. Richard Michalski, Supervisor. |
| Wyoming | Wyoming Occupational Health and Safety Department, 200 East 8th Avenue, Cheyenne, Wyo. 82002. | (307) 777-7786, Mr. Donald Owsley, Health and Safety Administrator. |

## APPENDIX H: AVAILABILITY OF REFERENCED DOCUMENTS

Paragraphs (c) through (o) of 29 CFR 1910.95 and the accompanying appendices contain provisions which incorporate publications by reference. Generally, the publications provide criteria for instruments to be used in monitoring and audiometric testing. These criteria are intended to be mandatory when so indicated in the applicable paragraphs of Section 1910.95 and appendices.

It should be noted that OSHA does not require that employers purchase a copy of the referenced publications. Employers, however, may desire to obtain a copy of the referenced publications for their own information.

The designation of the paragraph of the standard in which the referenced publications appear, the titles of the publications, and the availability of the publication are as follows:

| Paragraph designation | Referenced publication | Available from— |
|---|---|---|
| Appendix B | "List of Personal Hearing Protectors and Attenuation Data," HEW Pub. No. 76-120, 1975, NTIS-PB267461. | National Technical Information Service, Port Royal Road, Springfield, VA 22161. |
| Appendix D | "Specification for Sound Level Meters," S1.4-1971 (R1976). | American National Standards Institute, Inc., 1430 Broadway, New York, NY 10018. |
| §1910.95(k)(2), appendix E | "Specifications for Audiometers," S3.6-1969. | American National Standards Institute, Inc., 1430 Broadway, New York, NY 10018. |
| Appendix D | "Specification for Octave, Half-Octave and Third-Octave Band Filter Sets," S1.11-1971 (R1976). | Back Numbers Department, Dept. STD, American Institute of Physics, 333 E. 45 St., New York, NY 10017; American National Standards Institute, Inc., 1430 Broadway, New York, NY 10018. |

The referenced publications (or a microfiche of the publications) are available for review at many universities and public libraries throughout the country. These publications may also be examined at the OSHA Technical Data Center, Room N2439, United States Department of Labor, 200 Constitution Avenue, NW., Washington, D.C. 20210, (202) 523-9700 or at any OSHA Regional Office (see telephone directories under United States Government—Labor Department).

## APPENDIX I: DEFINITIONS

These definitions apply to the following terms as used in paragraphs (c) through (n) of 29 CFR 1910.95.

Action level—An 8-hour time-weighted average of 85 decibels measured on the A-scale, slow response, or equivalently, a dose of fifty percent.

Audiogram—A chart, graph, or table resulting from an audiometric test showing an individual's hearing threshold levels as a function of frequency.

Audiologist—A professional, specializing in the study and rehabilitation of hearing, who is certified by the American Speech-Language-Hearing Association or licensed by a state board of examiners.

Baseline audiogram—The audiogram against which future audiograms are compared.

Criterion sound level—A sound level of 90 decibels.

Decibel (dB)—Unit of measurement of sound level.

Hertz (Hz)—Unit of measurement of frequency, numerically equal to cycles per second.

Medical pathology—A disorder or disease. For purposes of this regulation, a condition or disease affecting the ear, which should be treated by a physician specialist.

Noise dose—The ratio, expressed as a percentage, of (1) the time integral, over a stated time or event, of the 0.6 power of the measured SLOW exponential time-averaged, squared A-weighted sound pressure and (2) the product of the criterion duration (8 hours) and the 0.6 power of the squared sound pressure corresponding to the criterion sound level (90 dB).

Noise dosimeter—An instrument that integrates a function of sound pressure over a period of time in such a manner that it directly indicates a noise dose.

Otolaryngologist—A physician specializing in diagnosis and treatment of disorders of the ear, nose and throat.

Representative exposure—Measurements of an employee's noise dose or 8-hour time-weighted average sound level that the employers deem to be representative of the exposures of other employees in the workplace.

Sound level—Ten times the common logarithm of the ratio of the square of the measured A-weighted sound pressure to the square of the standard reference pressure of 20 micropascals. Unit: decibels (dB). For use with this regulation, SLOW time response, in accordance with ANSI S1.4-1971 (R1976), is required.

South level meter—An instrument for the measurement of sound level.

Time-weighted average sound level—That sound level, which if constant over an 8-hour exposure, would result in the same noise dose as is measured.

# CHAPTER 10

# Ionizing Radiation

by C. Lyle Cheever, MS, MBA

**Ionizing Radiation Terms**

**Types of Ionizing Radiation**
*Alpha-Particles ▪ Beta-Particles ▪ Neutrons ▪ X-Radiation ▪ Gamma-Radiation ▪ Radioactive Decay Calculations*

**Biological Effects of Radiation**
*Types of Injuries ▪ Relating Dosage to Damage*

**Standards and Guides**

**Monitoring Instruments**
*Film Badge ▪ Thermoluminescence Detectors ▪ Pocket Dosimeter ▪ Other Dosimeters ▪ Ionization Chambers ▪ Geiger–Mueller Counters ▪ Other Monitoring Instruments ▪ Calibration*

**Basic Safety Factors**
*Time ▪ Distance ▪ Shielding*

**Control Programs**
*Sources of Radiation ▪ Operational Factors ▪ Employee Exposure Potential ▪ Records*

**Summary**

**Bibliography**

BASIC CONCEPTS OF IONIZING RADIATION and safe handling of radioactive materials are presented in this chapter. The aim is to provide the framework within which the safety of ionizing radiation conditions can be evaluated. The chapter is directed to health and safety professionals who need a basic understanding of radiation safety. Health and safety professionals should also know where to find consultation and technical help on specific radiation problems. The control of ionizing radiation exposures requires special training in health physics, and people qualified in health physics should be obtained for this work.

Radiation safety should be part of an organization's total health and safety program. The introduction of radiation devices or radioactive materials entails radiation safety reviews, engineering studies, and facility modifications. It may be necessary to alter traffic patterns of personnel and mobile equipment to minimize the spread of radioactive materials in the event of an accident.

The health and safety professional should have general knowledge of the nature of radiation, the detection of radiation, permissible exposure limits (PELs), biological effects of radiation, monitoring techniques, procedures, and control measures. Facility personnel need to be properly advised of radiation hazards and safe procedures. The health and safety professional should ensure that health physicists review or oversee radiation installations to ensure compliance with federal, state, and local regulations and company policies. Medical and emergency plans must be in place.

Effective accident prevention techniques are required for the control of radiation exposures to personnel. Radiation hazards can to some extent be prevented with common safety measures. In other cases, health physicists must provide expert guidance. Some radiation control operations entail extraordinarily expensive facilities and equipment. The organization and the health and safety professional must see that the proper actions are taken.

Basic information on the characteristics of radiation, standards for exposure limitation, safety factors, and control measures are presented here. As an aid to further study, consult the Bibliography at the end of this chapter.

## ■ Ionizing Radiation Terms

In order to discuss intelligently the health and safety aspects of ionizing radiation, an understanding of some basic terminology is necessary. Brief definitions of some important terms are given here. In the text, English units of radiation exposure are listed first, followed by the SI (metric) units in parentheses. Figure 10–1 illustrates the process of radioactive disintegration.

***Activity.*** The number of nuclear disintegrations occurring in a given quantity of material per unit of time.

***Alpha-particle (alpha-radiation, α).*** An alpha-particle is made up of two neutrons and two protons that give it a unit charge of +2. It is emitted from the nucleus of a radioactive atom and causes high-density ionization. Alpha-particles transfer their energy in a very short distance and are readily deflected by a piece of paper or the top, dead layer of the skin. Alpha-radioactivity is therefore primarily an internal radiation hazard.

***Annihilation.*** The process by which a negative electron and a positive electron, or positron, combine and dis-

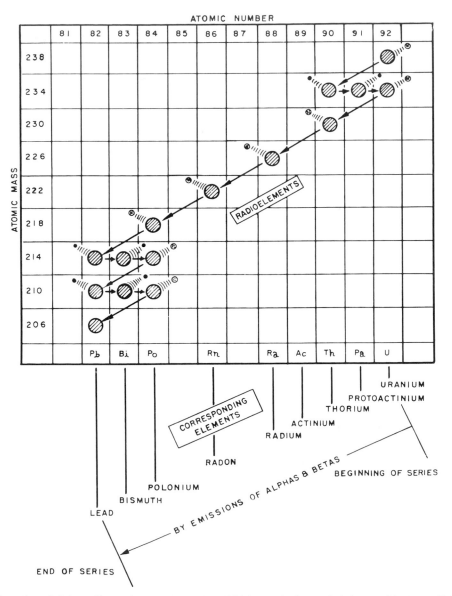

**Figure 10–1.** The radioactive disintegration scheme of uranium-238 by emissions of alpha- and beta-particles. (Reprinted with permission from *Atomic Radiation,* RCA Service Co., Inc., 1957.)

appear. The process results in the emission of electro-magnetic radiation.

**Annual limit on intake (ALI).**   The activity of a radionuclide that, taken into the body over the course of 1 year, constitutes a committed effective dose equivalent equal to the annual occupational effective dose equivalent limit. This is based on Reference Man.

**Atomic number.**   The atomic number is the number of protons (positively charged particles) in the nucleus of an atom. Each element has a different atomic number. The atomic number of hydrogen is 1, that of oxygen 8, iron 26, lead 82, and uranium 92. The atomic number is also called the charge number.

**Atomic weight.**   The atomic weight is approximately the sum of the number of protons and neutrons in the nucleus of an atom. This sum is also called the mass number. The atomic weight of oxygen, for example, is approximately 16; most oxygen atoms contain 8 protons and 8 neutrons.

**Background radiation.**   The radiation coming from sources other than the radioactive material to be measured. Background radiation is primarily a result of cosmic rays, which constantly bombard the earth from outer space. It also comes from such sources as soil and building materials.

**Becquerel (Bq).**   One disintegration per second (dps). This unit is used in measuring the rate of radioactive disintegration. There are $3.7 \times 10^{10}$ becquerels per curie of radioactivity.

**Beta-particle (beta-radiation, β).**   Beta-particles are small, electrically charged particles emitted from the nucleus of radioactive atoms. They are identical to electrons and have a negative electrical charge of 1. Beta-particles are emitted with various kinetic energies. They pose an internal exposure hazard and are often penetrating enough to cause skin burns.

**Bremsstrahlung.**   The electromagnetic radiation associated with the deceleration of charged particles. The term can also be applied to electromagnetic radiation produced by acceleration of charged particles.

**Compton effect.**   The glancing collision of a gamma-photon with an orbital electron. The gamma-photon gives up part of its energy to the electron, ejecting the electron from its orbit.

**Controlled area.**   A specified area in which exposure of personnel to radiation or radioactive material is controlled. Controlled areas should be under the supervision of a person who has knowledge of and responsibility for applying the appropriate radiation protection practices.

**Counter.**   A device for counting nuclear disintegrations, thereby measuring the amount of radioactivity. The electronic signal announcing disintegration is called a count.

**Curie (Ci).**   A measure of the rate at which a radioactive material emits particles. One curie corresponds to $3.7 \times 10^{10}$ becquerels.

**Derived air concentration (DAC).**   The ALI of a radionuclide divided by the volume of air inhaled by Reference Man in a working year ($Bq/m^3$).

**Disintegration.**   When a radioactive atom disintegrates, it emits a particle from its nucleus. What remains is a different element. For example, when an atom of polonium disintegrates, it ejects an alpha-particle and changes to a lead atom by this process (see Figure 10–1).

**Dose.**   A general term denoting the quantity of radiation or energy absorbed by a specified mass. For special purposes, its meaning should be specified, as in *absorbed dose*.

**Dosimeter (dose meter).**   An instrument used to determine the radiation dose a person has received.

**Electron.**   A minute atomic particle possessing the smallest possible amount of negative electric charge (–1). Orbital electrons rotate around the nucleus of an atom. Electrons have only about 1/1,820 the mass of protons or neutrons.

**Electron volt (eV).**   A small unit of energy—the amount of energy that an electron gains when it is acted upon by one volt. Radioactive materials emit radiation in energies of up to several million electron volts, or MeV. Gamma-ray energies from radioisotopes range up to 4 MeV or higher. Some are emitted at relatively low energies and are correspondingly less hazardous.

**Element.**   All atoms of a given element contain the same number of protons and therefore have the same atomic number. Various isotopes of an element result from a change in the number of neutrons in the nucleus. However, the electrical charge and chemical properties of the various isotopes of an element are identical.

**Film badge.**   A piece of masked photographic film worn as a badge for personal monitoring of radiation exposure. It is darkened by penetrating radiation, and radiation exposure can be checked by developing and interpreting the film. The type of masking depends on the type of radiation to be measured.

**Gamma-rays (Gamma-radiation, γ).**   A class of electromagnetic photons emitted from the nuclei of radioactive atoms. Gamma-rays are highly penetrating and present an external radiation exposure hazard.

**Gray (Gy).**   Unit of absorbed radiation dose equal to one joule of absorbed energy per kilogram of matter. 1 Gray = 100 rad.

**ICRP.**   International Commission on Radiological Protection and Measurements.

**Half-life.**   A means of classifying the rate of decay of radioisotopes according to the time it takes them to lose half their strength (intensity). Half-lives range from fractions of a second to billions of years. Cobalt-60, for example, has a half-life of 5.3 years.

**Half-value layer.**   The thickness of a specified substance that, when introduced into the path of a given beam of radiation, reduces the value of the radiation quantity by one-half. It is sometimes expressed in terms of mass per unit area.

**Ion.** An atom or molecule that carries either a positive or negative electrical charge.

**Ionizing radiation.** Electromagnetic or particulate radiation capable of producing ions, directly or indirectly, by interaction with matter. In biological systems, such radiation must have a photon energy greater than 10 eV. This excludes most of the ultraviolet bands and all longer wavelengths.

**Ionization chamber.** A basic counting device to measure radioactivity.

**Isotope.** These are nuclei that have the same atomic number. Isotopes of a given element contain the same number of protons but a different number of neutrons. For example, the isotope uranium-238 contains 92 protons and 146 neutrons whereas U-235 contains 92 protons and 143 neutrons. Thus the atomic weight, or mass, of U-238 is 3 higher than that of U-235.

**Moderator.** A material used to slow neutrons; it is used, for example, in reactors. Slow neutrons are particularly effective in causing fission. Neutrons are slowed down when they collide with atoms of light elements such as hydrogen, deuterium, and carbon—three common moderators.

**Molecule.** The smallest unit of a compound or element as it exists in nature. A water molecule consists of two hydrogen atoms combined with one oxygen atom, hence the well-known formula $H_2O$. The element oxygen exists in the form of diatomic molecules, $O_2$.

**NCRP.** National Council on Radiation Protection and Measurements.

**Neutron.** An atomic particle. The neutron weighs about the same as the proton. As its name implies, the neutron has no electrical charge. Neutrons make effective atomic projectiles for the bombardment of nuclei. Neutrons can also present unique external exposure hazards to personnel.

**NRC.** Nuclear Regulatory Commission.

**Nucleus.** The inner core of an atom. The nucleus consists of neutrons and protons tightly bound together.

**Pair production.** The conversion of a gamma-ray into a pair of particles—an electron and a positron. This is an example of direct conversion of energy into matter, and is quantified by Einstein's famous formula $E = Mc^2$; Energy = Mass × Velocity of Light (squared).

**Photoelectric effect.** Occurs when an electron is ejected from the orbit of an atom by a photon that imparts all of its energy to the electron.

**Photon.** A class, or quantum, of electromagnetic radiation, such as x-rays, gamma-rays, visible light, and radio waves.

**Plutonium.** A man-made heavy element that undergoes fission under the impact of neutrons. It is a useful fuel in nuclear reactors. Plutonium can be produced by the capture of slow neutrons in uranium. It is a highly hazardous alpha-emitter.

**Proton.** An elementary particle found in an atom's nucleus. Its positive charge of 1 is opposite that of the electron.

**Quality factor (Q).** A function of the linear collision stopping power (L ∞) in water at the point of interest and with a specified energy dependence. It weights the absorbed dose for the biological effectiveness of the charged particles producing the absorbed dose.

**Radioactivity.** The emission of very fast atomic particles or rays by nuclei. Some elements are naturally radioactive; others become radioactive after bombardment with neutrons or other particles. The three major forms of radioactivity are alpha (α), beta (β), and gamma (γ), named for the first three letters of the Greek alphabet.

**Radioisotope.** A radioactive isotope of an element. A radioisotope can be produced by placing material in a nuclear reactor and bombarding it with neutrons. Many fission products are radioisotopes. Radioisotopes are sometimes used as tracers or as energy sources for chemical processing, food pasteurization, and nuclear batteries.

**Radium.** One of the earliest-known naturally radioactive elements. Radium is far more radioactive than uranium and is found in the same ores. It is a highly hazardous alpha-emitter.

**Roentgen (R).** The amount of x- or gamma-radiation that produces ionization resulting in one electrostatic unit of charge in one cubic centimeter of dry air at standard conditions.

**Roentgen absorbed dose (rad).** The mean energy per unit of mass imparted by ionizing radiation in a mass. One rad is 100 ergs absorbed per gram. 1 rad = 0.01 Gy.

**Roentgen equivalent man (rem).** A unit of absorbed dose (in rad) times a quality factor that is used to express the relative biological effect of the particular radiation as compared to gamma-radiation. Personnel exposure limits are often expressed in rem.

**Scintillation counter.** A radiation-counting device that registers the tiny flashes of light (scintillations) that particles produce when they strike certain crystals or liquids.

**Shielding.** A barrier that protects workers from harmful radiations released by radioactive materials. Lead bricks, dense concrete, water, and earth are examples of materials used for shielding.

**Sievert (Sv).** Unit of absorbed radiation dose (in Gy) times the quality factor of the radiation as compared to gamma-radiation. It is equal to the Gray times the quality factor and is equivalent to 100 rem.

**Strontium-90.** An isotope of strontium having a mass number of 90. Strontium-90 is an important fission product. It has a half-life of 25 years and is a highly hazardous beta-emitter.

**Tracer.** A radioisotope that is mixed with a stable material. Radioisotopes enable scientists to trace chemical and physical changes in materials. Tracers are widely used in science, industry, and agriculture. For example, when radioactive phosphorus is mixed with a chemical fertilizer, the uptake of radioactive phosphorus from fertilized soil can be measured in the plants as they grow, thereby also indicating the rate of uptake of the fertilizer.

***Tritium.*** Often called hydrogen-3. Tritium is an extra-heavy hydrogen whose nucleus contains two neutrons and one proton. It is radioactive as a beta-emitter.

***Uranium.*** A heavy metal, the two principal natural isotopes of which are U-238 and U-235. U-235 has the only readily fissionable nucleus that occurs in appreciable quantities in nature, hence its importance as a nuclear reactor fuel. Only one part in 140 of natural uranium is U-235.

***X ray.*** Highly penetrating electromagnetic radiation similar to the gamma-ray. X rays are produced by electron bombardment of target materials. They are commonly used to produce shadow pictures (roentgenograms) of dense portions of objects.

# Types of Ionizing Radiation

Radiation is a form of energy. Familiar forms of radiation energy include light (a form of radiation we can see) and infrared (a form of radiation we can feel as heat). Radio and television waves are forms of radiation that we can neither see nor feel. The relationship between the various categories of electromagnetic radiation is shown in Figure 10–2.

Gamma-rays and x rays overlap and occupy a common range in the electromagnetic spectrum. X-radiation is produced in the orbiting electron portion of the atom or from free electrons, whereas gamma-radiation is produced in the nucleus. X rays generally are machine produced and gamma-rays are emitted spontaneously from radioactive materials.

All matter is composed of atoms, each of which has two basic parts: a heavy core, or nucleus, containing positively charged particles called protons and neutral particles called neutrons; and relatively lightweight, negatively charged particles called electrons, which spin around the nuclear core (Figure 10–3). Neutrons and protons were once considered basic particles. However, they have been found to be composed of even smaller particles.

Ionization is an energy transfer process that changes the normal electrical balance in an atom. If a normal atom (electrically neutral) were to lose one of its orbiting electrons (one negative charge), the atom would no longer be neutral. It would have more positive charges than negative charges, making it a positive ion. Electrons thus removed are called free electrons. If the free electron then attached to another atom, that atom would become a negative ion. The positive and negative ions thus produced are known as an ion pair.

The term *nuclear radiation* describes all forms of radiation energy that originate in the nucleus of a radioactive atom. In addition to gamma-rays, fast-moving particles are sometimes emitted from radioactive atoms.

Some materials are naturally radioactive; others can be made radioactive in a nuclear reactor or accelerator. Some nonradioactive atoms can be converted to radioactive atoms when an extra neutron is captured by a nucleus. The resulting radioactive atom is unstable because of the extra energy that the neutron added to the nucleus. The excited, or radioactive, atoms get rid of their excess energy and return to a stable state by emitting subatomic particles and gamma-rays from the nucleus. The most important of these particles are alpha-particles, beta-particles, and neutrons.

The hazardous properties of radioactive materials are usually thought of in terms of nuclear radiation. All types of radiation share the common properties of being absorbed and transferring energy to the absorbing body.

The most commonly encountered types of ionizing radiation are alpha-, beta-, and neutron particles and x- or gamma-electromagnetic radiation. Other types of ionizing radiation are encountered in specialized facilities.

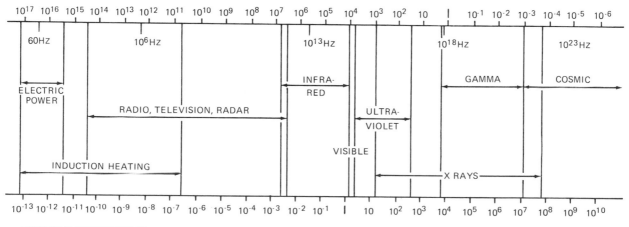

**Figure 10–2.** Electromagnetic spectrum illustrates energy and wavelength of the various categories of electromagnetic radiation.

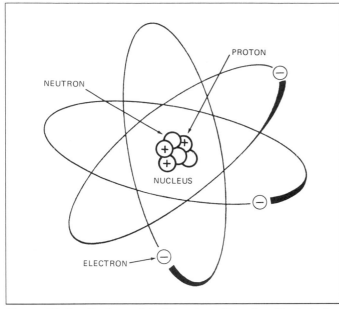

**Figure 10–3.** Basic model of the atom. The atom illustrated here is lithium-6.

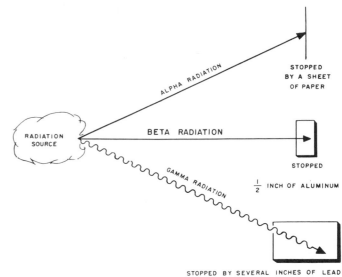

**Figure 10–4.** Relative penetrating power of alpha-, beta-, and gamma-radiation. (Reprinted with permission from *Atomic Radiation,* RCA Service Co., Inc., 1957.)

## Alpha-Particles

Alpha-particles originate in the nuclei of radioactive atoms during the process of disintegration. These particles consist of a cluster of two protons and two neutrons (giving a mass number of 4, which is structurally the same as the nucleus of a helium atom). Alpha-particles, on slowing down, combine with electrons from the material through which they are passing, and thus become helium atoms.

The mass and electrical charge characterize the hazardous properties of alpha-particles. They have a positive charge of 2 units and interact electrically with human tissues and other matter. Alpha-particles range in energy to over 7 MeV. Because of their large mass and the dense ionization along their path through a material, they travel only a short distance. Their range is at most about 4 in. (10 cm) in air. They are stopped by the dead, outer layer of the skin; a film of water; a sheet of paper; or other paper-thin material (Figure 10–4).

To detect alpha-particles with a radiation survey meter, the instrument probe must be held close to the source, and the window on the probe must be very thin and designed specifically for alpha-detection.

Alpha-particles are produced by elements with high atomic numbers (Figure 10–1). Alpha-emitters are hazardous when taken into the body. Because they are chemically similar to calcium in their action within the body, some alpha-emitters are absorbed into the bones, where they remain for long periods of time. As they disintegrate, they emit alpha-particles, which can damage tissue. Other alpha-emitters are not bone seekers but concentrate in body organs such as the kidney, liver, lungs, and spleen.

If the alpha-emitting material is kept outside the body, little damage results because generally alpha-particles cannot penetrate the outermost, dead layer of skin. Alpha-emitters are considered to be only internal radiation hazards.

## Beta-Particles

Beta-particles are electrically charged particles ejected from the nuclei of radioactive atoms during disintegration. They generally have a negative electrical charge of 1 unit and the same mass as an electron. Negative beta-particle emission causes the disintegrating atom to change into an element of a higher atomic number. Thus, strontium-90 changes to yttrium-90 on disintegration with ejection of a beta-particle (electron) from a neutron in the nucleus. The atomic number of the atom (the number of protons in the nucleus) is changed from 38 to 39 by the beta-particle emission from a neutron, causing it to gain a proton. (A neutron, by emitting a beta-particle, becomes a proton.)

Particles similar to beta-particles, with the same mass as electrons but with a positive unit charge, are called positrons, which can also be ejected from nuclei by disintegration. A proton from which a positron is emitted subsequently becomes a neutron, and the atomic number of the atom is reduced by 1. However, positrons are readily annihilated by combination with electrons, yielding gamma-radiation.

Beta-particles do not penetrate to the depth that x rays or gamma-radiation of similar energy do (Figure 10–4). Their maximum range in wood is about 1.5 in. (4 cm), and they can penetrate into the human body from 0.1–0.5 in. (0.2–1.3 cm).

Skin burns result from extremely high doses of beta-radiation, because it requires only about 70 keV

(1 keV = 0.001 MeV) of energy for a beta-particle to penetrate the dead, outer layer of skin. Beta-emitters are internal radiation hazards when taken into the body.

Beta-particles have a broad distribution of energies ranging from near zero up to the maximum value specific for the particular radionuclide. The maximum range in air of the beta-particles emitted from one radionuclide may be 6 in. (15 cm), and the maximum range of emissions from another radionuclide may be 60 ft (18 m). The higher-energy beta-particles penetrate farther, transfer more energy, and cause more damage.

Beta-particles, or high-energy electrons, are emitted from a wide range of light and heavy radioactive elements. Beta-particles are much smaller than alpha-particles, and some have a velocity approaching the velocity of light. The energy level of beta-particles can be 4 MeV or higher. The energy range of beta-particles is similar to that of gamma-rays.

Beta-particles have an electrical charge of 1 unit, which tends to limit their penetration range somewhat, but penetration is greater than that of the doubly charged alpha-particles. Beta-particles are relatively more hazardous externally than alpha-particles because their penetration power is greater.

When a beta-particle is slowed down or stopped, secondary x-radiation, known as bremsstrahlung, may be produced.

Light metals such as aluminum are preferred for shielding from beta-particles because light metals produce less bremsstrahlung radiation. Plexiglas™ is another shielding material that is effective against such radionuclides as P-32. Common beta-radiations have ranges of less than 30 ft (9 m) in air. Depending on their energy, they can be stopped by the walls of a room or by a sheet of aluminum 0.5 in. (1.3 cm) thick.

## Neutrons

Neutrons are not encountered as commonly as alpha- and beta-particles. The neutron particle, as its name implies, has no electrical charge. Neutrons exist within the nuclei of all atoms except those of the lightest isotope of hydrogen.

Neutrons are released on disintegration of certain radioactive materials (the fissionable isotopes). They have long or short ranges in air depending on their kinetic energy, which in turn depends on the method by which a particular neutron was produced. Furthermore, their range depends on the characteristics of the material through which they pass, the way the atoms of that material interact with the neutrons, and the kind of collisions that occur. In human tissue, the average depth of penetration of neutrons varies from about 0.25 in. (0.6 cm) to several inches (centimeters), depending on the neutrons' energy.

Very high-energy neutrons collide with atoms of the material through which they are passing, often breaking these nuclei into high-energy fragments. The neutron itself is unstable and emits a beta-particle as it decays to a proton. Interaction with the atoms of the material through which neutrons pass is the main way neutrons are removed from a beam.

Absorption of neutrons results when neutrons are deflected by, or collide with, nuclei. They collide repeatedly, which slows them down. The loss of energy by neutrons that suffer such collisions leads to an increased probability of absorption by a nucleus. This absorption process is called neutron capture.

In the human body, most of the captures that occur take place in hydrogen or nitrogen atoms. The result is that the nucleus of the atom is in an excited state. That is, an excess of energy is available. An atom can exist in this state only for a short period of time and returns to the ground (unexcited) state by releasing the excess energy. In the process, a proton, gamma-ray, beta-particle, or alpha-particle is emitted, depending on what type of atom captures the neutron. Because it is these secondary emissions that produce damage in tissue, the task of determining the neutron dose is difficult. The health hazard that neutrons present arises from the fact that they cause the release of secondary radiation.

Human exposure to neutrons occurs around reactors, accelerators, and sources designed to produce neutrons. Determination of the extent of damage from neutron exposures must be made by highly skilled people using specialized equipment. The amount of harm caused by a dose is dependent not only on the number of neutrons absorbed but also on their energy distribution.

## X-Radiation

X-radiation is commonly thought of as electromagnetic radiation produced by an x-ray machine. When high-speed electrons are suddenly slowed down when they strike a target, they lose energy in the form of x-radiation. In an x-ray machine, the voltage across the electrodes of the vacuum tube determines the energy of the electrons, which principally determines the wavelength and penetrating quality of the resulting x rays.

The character of x-radiation is also affected by the target material inside the x-ray tube. That is, the wavelength of a portion of the x-radiation is affected by the kind of material composing the target. Because the electrons strike and interact at various speeds, the x-ray beam has a variety of wavelengths and energies. The energy of an x ray is inversely proportional to its wavelength; the more energy an x ray possesses, the shorter its wavelength.

The extent of penetration of x rays depends on wavelength and the material being irradiated. The x rays of short wavelength are called hard, and they penetrate several centimeters of steel. The x rays of long, or soft, wavelengths are less penetrating. The power to penetrate through matter is called the *quality*. *Intensity* is the energy flux density. The physical properties of a beam of x or gamma-rays are often summarized by the two concepts of intensity and quality.

The range of penetration can be expressed in terms of *half-value layer*. This is the thickness of material necessary to reduce the incident radiation by one-half. The half-value layers for x-radiation range up to several centimeters of concrete (Table 10–A).

## Gamma-Radiation

Gamma-radiation is similar to x-radiation in that it is also electromagnetic and ionizing. In fact, it is identical to x-radiation except for its source being the nucleus of an atom. X-radiation is electromagnetic radiation that originates outside the nucleus.

Radioactive materials, by definition, spontaneously emit one or more characteristic radiations, possibly includ-

**Table 10–A.** Half-Value Layer for Five Shielding Materials

| Material | Cobalt-60 | Cesium-137 |
|---|---|---|
| Lead | 1.24 cm | 0.64 cm |
| Copper | 2.10 cm | 1.65 cm |
| Iron | 2.21 cm | 1.73 cm |
| Zinc | 2.67 cm | 2.06 cm |
| Concrete | 6.60 cm | 5.33 cm |

Note: 1 cm = 0.394 in.

ing gamma-radiation. The radiation comes from an excited or unstable nucleus of an atom.

A gamma-ray emitted by a given radionuclide has a fixed energy specific to that radionuclide (Table 10–B). Gamma-rays from various radionuclides cover a wide range of wavelengths, or energies. They present an exter-

**Table 10–B.** Isotopes Commonly Available, Listed by Increasing Half-Life

| Half-Life | Element and Symbol | Atomic Number | Mass Number | Gamma-Radiation Energy (MeV) |
|---|---|---|---|---|
| 88 days | Sulfur (S) | 16 | 35 | none |
| 115 days | Tantalum (Ta) | 73 | 182 | 0.068, .10, .15, .22. 1.12, 1.19, 1.22 |
| 120 days | Selenium (Se) | 34 | 75 | 0.12, .14, .26, .28, .40 |
| 130 days | Thulium (Tm) | 69 | 170 | 0.084 |
| 138 days | Polonium (Po) | 84 | 210 | 0.80 |
| 165 days | Calcium (Ca) | 20 | 45 | none |
| 245 days | Zinc (Zn) | 30 | 65 | 1.12 |
| 270 days | Cobalt (Co) | 27 | 57 | 0.12, .13 |
| 253 days | Silver (Ag) | 47 | 110 | 0.66, .68, .71, .76, .81, .89, .94, 1.39 |
| 284 days | Cerium (Ce) | 58 | 144 | 0.08, .134 |
| 303 days | Manganese (Mn) | 25 | 54 | 0.84 |
| 367 days | Ruthenium (Ru) | 44 | 106 | none |
| 1.81 years | Europium (Eu) | 63 | 155 | 0.09, .11 |
| 2.05 years | Cesium (Cs) | 55 | 134 | 0.57, .60, .79 |
| 2.6 years | Promethium (Pm) | 61 | 147 | none |
| 2.6 years | Sodium (Na) | 11 | 22 | 1.277 |
| 2.7 years | Antimony (Sb) | 51 | 125 | 0.18, .43, .46, .60, .64 |
| 2.6 years | Iron (Fe) | 26 | 55 | none |
| 3.8 years | Thallium (Tl) | 81 | 204 | none |
| 5.27 years | Cobalt (Co) | 27 | 60 | 1.3, 1.12 |
| 12.46 years | Hydrogen (H) | 1 | 3 | none |
| 12 years | Europium (Eu) | 63 | 152 | 0.12, .24, .34, .78, .96, 1.09, 1.11, 1.41 |
| 16 years | Europium (Eu) | 63 | 154 | 0.123, .23, .59, .72, .87, 1.00, 1.28 |
| 28.1 years | Strontium (Sr) | 38 | 90 | none |
| 21 years | Lead (Pb) | 82 | 210 | 0.047 |
| 30 years | Cesium (Cs) | 55 | 137 | 0.661 |
| 92 years | Nickel (Ni) | 28 | 63 | none |
| 1602 years | Radium (Ra) | 88 | 226 | 0.186 |
| 5730 years | Carbon (C) | 6 | 14 | none |
| $2.12 \times 10^5$ years | Technetium (Tc) | 43 | 99 | none |
| $3.1 \times 10^5$ years | Chlorine (Cl) | 17 | 36 | none |

(Reprinted from *Radiological Health Handbook*, revised ed., January 1970. U.S. Dept. of Health & Human Services, Rockville, MD.)

nal exposure problem because of their deep penetration. The half-value layer of shielding for 1.0 MeV gamma- or x-radiation is slightly more than 0.5 in. (1.3 cm) of steel.

## Radioactive Decay Calculations

Radioactive materials decay—that is, they give off alpha-particles, beta-particles, and photon energy and lose a portion of their radioactivity with a characteristic half-life. The half-life can be brief, for example, 55 seconds for R-220; a matter of days, such as 8 days for I-131; or many years, such as 4.51 billion years for U-238. The amount of radioactivity remaining after a given period of time is calculated as follows:

$$A = A_0 \, e\left(-0.693\frac{t}{T_{1/2}}\right) = \frac{A_0}{e \, 0.693\frac{t}{T_{1/2}}} \qquad (1)$$

where
$A$ = radioactivity remaining after time $t$
$A_0$ = radioactivity at a given original time
$e$ = base of natural logarithms (2.718)
$t$ = elapsed time
$T_{1/2}$ = half-life of the radionuclide

Note that radioactivity and time units must be consistent: If $A_0$ is given in curies (Ci), then $A$ should also be in curies. If $T_{1/2}$ is given in years, then $t$ must also be in years.

■

### Example

What radioactivity would remain from 1 Ci (curie) of Co-60 (5.24 years half-life) after a 20-year period?

### Solution

$$A = \frac{1 \text{ Ci}}{e^{\frac{0.693\,(20\,y)}{5.24\,y}}} = \frac{1 \text{ Ci}}{e \, 2.645} = \frac{1 \text{ Ci}}{14.08} = 0.071 \text{ Ci} \quad (2)$$

The amount of radioactivity remaining from 1 Ci of cobalt-60 after 20 years is 71 mCi (0.071 Ci).

■

Note that in some instances radioactive decay results in another radioisotope, and there is a decay chain of radioisotopes before a nonradioactive (stable) isotope is formed (see Figure 10–1). The published tables of radioisotopes should be checked to determine the decay chains.

## ■ Biological Effects of Radiation

The human body can apparently tolerate a certain amount of exposure to ionizing radiation without its overall functions being impaired as a result. We are continuously exposed to ionizing radiation from natural sources such as cosmic radiation from outer space and from radioactive materials in the earth and materials around us and in us. This "background radiation" is part of our normal environment; we have evolved under its effects and are continuously exposed to it. The average annual dose from background radiation varies across the country, but it is commonly around 300 mR/yr.

One fundamental property of ionizing radiation is that when it passes into or through a material, it transfers energy by the ionization process. The intensity of radiation to which a material is subjected is known as the radiation field. The term *dose* is generally used to express a measure of radiation that a body or other material absorbs when exposed to a radiation field.

If the incoming and outgoing energies are almost identical in nature and amount, then little energy is transferred and the dose is small.

External radiation sources, which are located outside of the body, present an entirely different set of conditions than radionuclides that have entered the body. Once inside the body, radionuclides are absorbed, metabolized, and distributed throughout the tissues and organs according to their chemical properties. Their effects on organs or tissues depend on the type and energy of the radiation and their residence time within the body.

The effects of irradiation on living systems are studied by looking for effects on the living cells, for changes in biochemical reactions, for evidence of production of disease, and for changes in life or normal growth patterns (Figures 10–5 and 10–6). Interpretation of such findings is not simple. Extensive research in this field continues to be conducted in many laboratories.

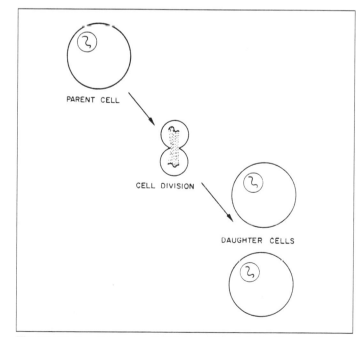

**Figure 10–5.** Normal cell division. (Reprinted with permission from *Atomic Radiation*, RCA Service Co., Inc., 1957.)

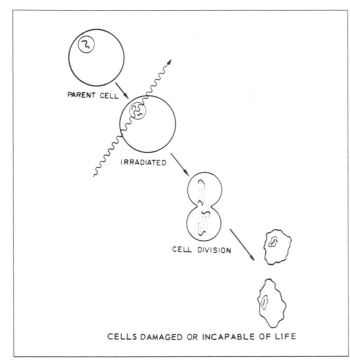

**Figure 10–6.** Damage or death of cells after division of irradiated parent cell. (Reprinted with permission from *Atomic Radiation,* RCA Service Co., Inc., 1957.)

The effect of ionizing radiation on living tissue is generally assumed to be almost entirely due to the ionization process, which destroys the capacity of reproduction or division in some cells and causes mutation in others (Figure 10–6). The human body is a complex chemical machine that constantly produces new cells to replace those that have died or been damaged. The body has a tremendous capability for repairing cell damage. Therefore, our survival depends on our ability to keep cell damage within the body's repair capabilities.

Ionization strips electrons from atoms and breaks their chemical bonds with other atoms. A simple molecule such as that of water recombines after ionization. This, however, is not the case in a complicated living cell. Here, ionization can result in many possible atomic recombinations. The rupture of a few bonds in the elaborate structure of the molecules of the living cell can have profound effects.

## Types of Injuries

We started to learn about the effects of radiation on humans in 1895, when Wilhelm Roentgen discovered the x ray. Radiation from uranium was discovered by Antoine-Henri Becquerel in 1896. Alpha- and beta-particles and gamma-rays were identified shortly thereafter, and the neutron was discovered in 1930. Injurious effects of radiation were experienced by these early workers.

There are two points of view for consideration of the injurious effects of ionizing radiation: the somatic effects (injury to individuals) and the genetic effects, which are passed on to future generations.

The degree of injury inflicted on an individual by radiation exposure depends on such factors as the total dose, the rate at which the dose is received, the kind of radiation, and the body part receiving it.

Tissues such as the bone marrow, where blood cells are produced; the lining of the digestive tract; and some cells of the skin are more sensitive to radiation than those of bone, muscle, and nerve.

In general, if the total amount of radiation is received slowly over a long period of time, a larger dose is required to produce the same degree of damage than if the same total is received over a short period. If the rate at which the dose is received is low enough, the body recovery processes are able to keep abreast of the slight damage.

Some relatively small doses have no effect if given only once but will shorten life span and produce abnormalities if continued long enough. The time between exposure and the first signs of radiation damage is called the latent period. The larger the dose, the shorter the latent period.

The various tissues and organs of the body are not affected equally by equal irradiation. Their responses vary considerably. For radiation protection purposes, it is essential that the dose to the most sensitive organs be given primary consideration.

Over the years, allowable radiation levels have been consistently reduced, as researchers have obtained more information on the effects of radiation exposure and as judgments involving acceptable risks have become more conservative. As radiation was applied to humans for healing purpose, effects such as skin redness, dermatitis, and skin cancer were noted, as were hair loss and eye inflammation. It was found that the incidence of cancer and of certain blood diseases was higher among radiologists than would be expected for a chance distribution in the population studied.

The pool of health experience data was obtained from the following sources:

■ Early radiation workers

■ Medical personnel who routinely administered radiation for diagnostic and therapeutic purposes

■ Patients who were treated with radiation

■ A group of workers who painted dials with luminous paints containing radium

■ Studies of Japanese atomic bomb survivors

With data collected from these sources, it became apparent that exposure to ionizing radiation was associated with a higher than normal incidence of certain diseases such as skin, lung, and other cancers; of bone damage; and of cataracts. There was also some evidence of life shortening.

Geneticists consider the population as a whole rather than particular individuals. Their concern is the effect of radiation on future generations. Radiation damage to human reproductive materials can be transmitted to succeeding generations.

Genetic damage is sometimes caused by disease or toxic chemicals as well as by ionizing radiation. Birth defects are usually the result of defective genes, but it is not possible at this time to determine what caused the damage to a given defective gene.

## Relating Dosage to Damage

Extensive work has been done in an attempt to relate radiation dose to resulting damage. Researchers have done a great deal of laboratory work with various biological systems (both plants and animals) in order to learn more about the conditions of irradiation, the effects it produces, and the relative effectiveness of the kinds and energies of radiation.

These studies provide the basis for determining maximum permissible levels of exposure. The maximum permissible levels denote the maximum radiation dose that can be tolerated with little chance of later development of adverse effects. The BEIR V report listed in the Bibliogra-

phy at the end of the chapter gives information on biological effects of ionizing radiation.

## ■ Standards and Guides

Maximum permissible levels of external and internal radiation have been published by the National Council on Radiation Protection and Measurements (NCRP), with exposure limits as shown in Table 10–C. The Nuclear Regulatory Commission establishes "Permissible Doses, Levels, and Concentrations" in *10 CFR 20*. (See Bibliography at the end of the chapter.)

Guides for maximum permissible external exposure to ionizing radiation have also been published by the International Commission on Radiological Protection and Measurements (ICRP).

Levels far below the maximum exposure limits recommended by the Federal Radiation Council can be achieved with no substantial inconvenience. The accepted practice is to keep radiation exposure as low as reasonably achievable (ALARA). Present limiting values are given separately for protection against different types of effects on health

**Table 10–C.** NCRP Recommendations on Limits for Exposure to Ionizing Radiation (Excluding medical exposures)

| | | |
|---|---|---|
| A. Occupational exposure | | |
|   1. Effective dose limits | | |
|     a. Annual | 5 rem | 50 mSv |
|     b. Cumulative | 1 rem × age | 10 mSv × age |
|   2. Equivalent dose annual limits for tissues and organs | | |
|     a. Lens of eye | 15 rem | 150 mSv |
|     b. Skin, hands, and feet | 50 rem | 500 mSv |
| B. Guidance for emergency occupational exposure* (see Section 14 of NCRP report 116) | | |
| C. Public exposure (annual) | | |
|   1. Effective dose limit, continuous or frequent exposure* | 100 mrem | 1 mSv |
|   2. Effective dose limit, infrequent exposure* | 500 mrem | 5 mSv |
|   3. Equivalent dose limits for tissues and organs | | |
|     a. Lens of eye | 1.5 rem | 15 mSv |
|     b. Skin, hands, and feet | 5 rem | 50 mSv |
|   4. Remedial action for natural sources: | | |
|     a. Effective dose (excluding radon) | > 500 mrem | > 5 mSv |
|     b. Exposure to radon decay products | | > 7 x 10$^{-3}$ Jh m$^{-3}$ |
| D. Education and training exposures (annual)* | | |
|   1. Effective dose limit | 100 mrem | 1 mSv |
|   2. Equivalent dose limit for tissues and organs | | |
|     a. Lens of eye | 1.5 rem | 15 mSv |
|     b. Skin, hands, and feet | 5 rem | 50 mSv |
| E. Embryo/fetus exposures (monthly)* | | |
|   1. Equivalent dose limit | 50 mrem | 0.5 mSv |
| F. Negligible individual dose (annual)* | 1 mrem | 0.01 mSv |

**Note:** Recommendations on $W_R$ and $W_T$ are in tables 4.2 and 5.1 of NCRP report 116.
*Sum of external and internal exposures but excluding doses from natural sources.

(From NCRP Report No. 116, *Limitation of Exposure to Ionizing Radiation,* 1993, with English units added for comparison to metric units.)

and apply to the sum of doses from external and internal sources of radiation. For cancer and genetic effects, the limiting value is specified in terms of a derived quantity called the effective dose equivalent. The effective dose equivalent received in any year by an adult worker should not exceed 5 rem (0.05 Sv). For other health effects, the limiting values are specified in terms of the dose equivalent to specific organs and tissues. (See *Recommendations for the Safe Use and Regulation of Radiation Sources in Industry, Medicine, Research and Teaching,* ICRP Report 102, 1990.)

Occupational dose equivalents to individuals under the age of 18 should be limited to educational and training purposes and should be less than 0.1 rem per year (1 mSv per year) (NCRP).

There is disagreement on the magnitude, extent, and cause of effects, if any, at exposures below the guide levels. Much of this discussion centers on theory, rather than on measured cause and demonstrated effect in human beings.

Because of the sensitivity of the fetus to radiation damage, the NCRP has recommended an occupational monthly equivalent dose limit of 50 mrem (0.5 mSv) to fetuses (excluding medical and natural background radiation) once pregnancy is known. The reason for the lower limit of exposure is that rapidly dividing cells are more susceptible to radiation damage than are mature cells. Women of childbearing age who may be exposed to radiation should be informed of the need to protect the fetus from excessive or unnecessary radiation exposure. In the event that occupational exposure exceeds PELs, action should be taken to reduce the exposures to within the guidelines. If this is not practicable, it is necessary to transfer the female worker out of the radiation exposure area for the duration of the pregnancy.

In summary, radiation dose limits established by any official organization or government body that has authority with respect to the user should be considered upper limits; the objective should be to keep exposure as low as reasonably achievable.

The NCRP recommends a whole-body dose limit of 5 rem per year (50 mSv) for occupational exposure. Cumulative exposure is not to exceed 1 rem (0.01 Sv) multiplied by the individual's age in years.

For further explanation of radiation protection standards and guides, refer to the Bibliography at the end of this chapter.

# Monitoring Instruments

There are a variety of detectors and readout devices used for monitoring and measuring radiation. None is universally applicable, and selection of the most appropriate detector or detectors for each radiation measurement (or type of measurement) is a matter of great importance.

Radiation monitoring involves the routine or special measurement of radiation fields in the vicinity of a radiation source, measurement of surface contamination, and measurement of airborne radioactivity. Such monitoring procedures are sometimes called radiation surveys.

## Film Badge

The film badge, worn on the outer clothing, is an example of a personal radiation monitor for gamma-, x-, and high-energy beta-radiation. It consists of a small piece of photographic film wrapped in an opaque cover and supported with a metal backing. It can be pinned to the clothing or worn as a ring. Radiation interacts with the silver atoms in the photographic film to affect (expose) the film the same way that light rays do. The badge is removed at regular intervals and the film developed. The amount of darkening of a film is then compared to a control film, that was not exposed to radiation during the same time period, to determine the amount of radiation exposure. Figure 10–7 shows a typical film-badge arrangement for monitoring personnel; the film badge provides a permanent record of dosage.

## Thermoluminescence Detectors

Thermoluminescence detectors (TLD) have come into widespread use for radiation exposure monitoring for gamma-, x-, and beta-radiation. These dosimeters can be worn by the person as body badges or finger rings. Most commonly they are small chips of lithium fluoride. A major advantage of the TLDs is that for x- and gamma-radiation, they essentially require no energy source to operate in exposure from 20 keV up. The ionizing radiation energy absorbed by the TLD displaces electrons on it from their ground state (valence state). The electrons are trapped in a metastable state but can be returned to the ground state by heating. When electrons return to the ground state, light is emitted. A TLD readout instrument is used for precise control of heating the chip and of measuring the light that is then emitted from it. The amount of light released is related to the absorbed radiation dose and, in turn, to the radiation exposure of the individual. Because the stored energy is released on readout, the readings cannot be repeated. It is therefore common practice to include two or more chips in a dosimeter.

## Pocket Dosimeter

The pocket dosimeter is a direct-reading portable unit shaped like a pen with a pocket clip. It is generally used to measure x- and gamma-radiation (although it may respond to beta-radiation).

A dosimeter consists of a quartz fiber, a scale, a lens to observe the movement of the fiber across the scale, and an ionization chamber. The fiber is charged electrostatically until it reaches zero on the scale. When the dosimeter is exposed to radiation, some of the air atoms in the

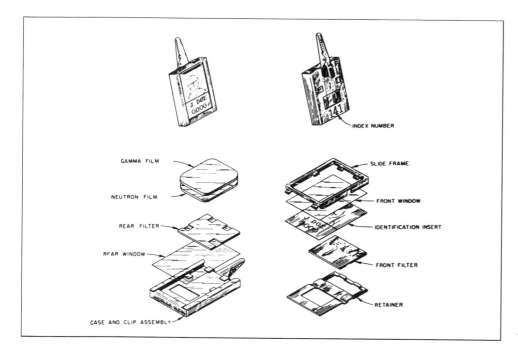

**Figure 10–7.** Exploded view of radiation film badge. Upper left is front view; upper right is back view.

chamber become ionized. This allows the static electricity charge to leak from the quartz fiber in direct relationship to the amount of radiation present. As the charge leaks away, the fiber deflects to some new position on the scale that indicates the amount of radiation exposure.

The main advantage of the pocket dosimeter is that it allows the individual to determine the radiation dose while working with radiation, rather than waiting until after periodic processing of a film badge or thermoluminescent dosimeter.

## Other Dosimeters

Personal electronic alarm dosimeters are now available to monitor the presence of x- and gamma-radiation. These units, usually Geiger–Mueller tubes, have an automatic audible alarm if significant exposure rates are encountered. These units can also indicate total integrated exposure rates at any time or provide a digital readout.

## Ionization Chambers

Radiation can be measured very conveniently and accurately by measuring the ionization in a small volume of air. If two plates or electrodes with an electrical potential between them are placed in a container filled with air, an ionization chamber is formed. If the ionization chamber is exposed to a beam of radiation, a current flows in the circuit because the electrons that are knocked out of the air atoms by the radiation are attracted by the positive electrode. In other words, ionized air becomes conductive (Figure 10–8).

The ionization chamber measures ionization directly and is energy independent. These units can measure gamma-, x-, beta-, and, if the window is thin enough, alpha-radiation. It is a very useful and popular tool for

radiation safety work. Use of an ionization chamber instrument for gamma-radiation measurements is illustrated in Figures 10–9 and 10–10.

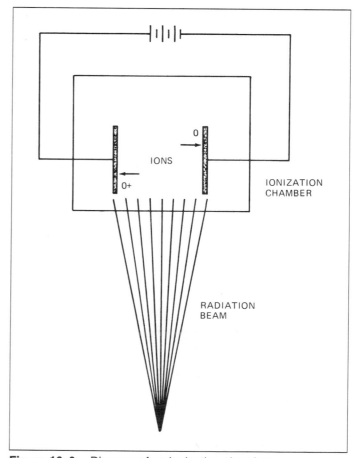

**Figure 10–8.** Diagram of an ionization chamber.

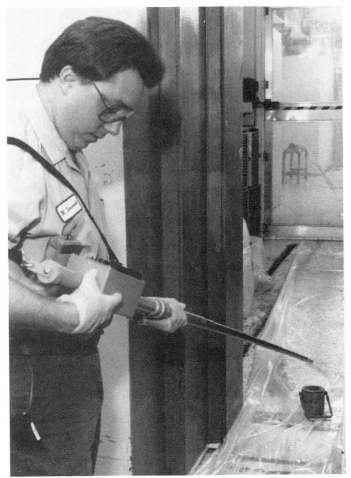

**Figure 10–9.** Health Physics technician using telescopic high-range ion chamber survey instrument. (Note the thermal luminescent dosimetry badge on shirt pocket and electronic alarming dosimeter on trousers pocket.)

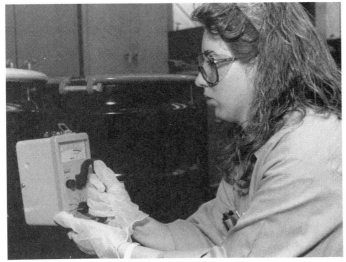

**Figure 10–10.** Health physics technician taking exposure rate readings with an ion chamber survey instrument.

## Geiger–Mueller Counters

A Geiger–Mueller counter is used for beta-, gamma-, and x-radiation survey measurements because it is capable of detecting very small amounts of radiation. It is especially sensitive to beta-radiation. It uses an ionization chamber but it is filled with a special gas and has a greater voltage supplied between its electrodes. It only takes a very small number of ions to put it into discharge. Electrons are freed by the initial ionization process, and they acquire enough extra energy from the applied voltage to create more ions. This instrument does not give a uniform response for different radiation energy levels and is accurate only for the type of radiation energy for which it is calibrated. A radiation survey with a Geiger–Mueller counter is shown in Figure 10–11.

## Other Monitoring Instruments

Scintillator types of radiation-monitoring instruments are designed to measure light flashes created by the interaction of ionizing radiation and scintillator materials. This type of survey instrument, which is useful for sensitive measure-

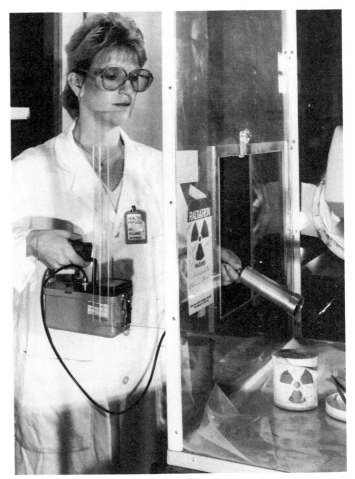

**Figure 10–11.** Use of a Geiger–Mueller counter. (Courtesy Argonne National Laboratory.)

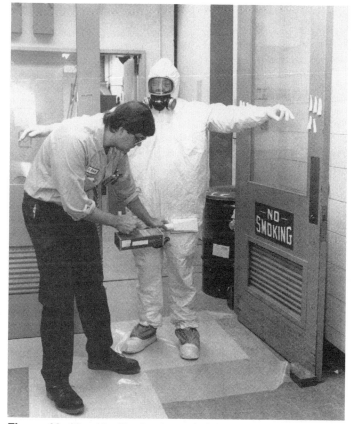

**Figure 10–12.** Health physics technician monitoring protective clothing after removal of outer protective suit. Survey meter has a dual plastic scintillator to detect alpha-, beta-, and gamma-radioactivity.

ments of alpha- and beta-gamma–radiation, is shown in Figure 10–12.

## Calibration

Usually, the calibration of radiation meters is a laboratory procedure that should be carried out by qualified experts. Under certain circumstances, however, it is possible and permissible to calibrate meters by comparing a radiation-measuring instrument with a standard radiation source of known output.

Consult up-to-date manuals and become familiar with regulations concerning the use and calibration of radiation-monitoring devices.

## ■ Basic Safety Factors

For external radiation exposure hazards, the basic protection measures are associated with time, distance, and shielding.

### Time

The longer the exposure, the greater the chance for radiation injury. Because there is a direct relationship between exposure dose and duration of exposure, reducing the exposure time by one-half reduces the dose received by one-half.

A common practice is to reduce the exposure time and thus the exposure. From the knowledge of the dose rate in a given location and the maximum dose that is acceptable for the time period under consideration, the maximum acceptable exposure time can be calculated. If the exposure rate is 2.5 mrem/hr (0.025 mSv/hr), 40 hours results in 100 mrem (1 mSv) exposure. If the exposure rate is 10 times higher, then the time must be reduced to one-tenth (to 4 hr) for the same dose (Figure 10–13).

In practice, the dose received in accomplishing a task can be spread over several employees so that no one person's exposure exceeds the guide levels. Only the minimum necessary exposure should be planned for a work task. Exposure rate is measured with one of several types of instruments. Direct-reading dosimeters are available, as are those that emit alarm sounds at a preset exposure rate and at an accumulated exposure setting.

### Distance

The inverse square law can be applied to determine the change in external penetrating radiation exposure with change in distance from a radiation source. Figure 10–14 indicates that by doubling the distance from the source of radiation the exposure would be decreased to $(1/2)^2$, or one-fourth, of the original amount. By increasing the distance from 2 to 20 m, the exposure would be decreased to $(2/20)^2$, or 1 percent, of the original amount. While the inverse square law can be used as an approximation, it should be recognized that it applies only to a point source

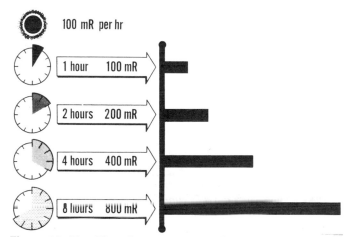

100 mR per hr

1 hour    100 mR

2 hours   200 mR

4 hours   400 mR

8 hours   800 mR

**Figure 10–13.** The effect of time on radiation exposure is easy to understand. An individual in an area where the radiation level from penetrating x- or gamma-radiations is 100 mrem/hr would get 100 mrem in an hour. After 2 hours, the exposure would be 200 mrem, and so on. (Reprinted from U.S. Nuclear Regulatory Commission, *Living with Radiation—Fundamentals*.)

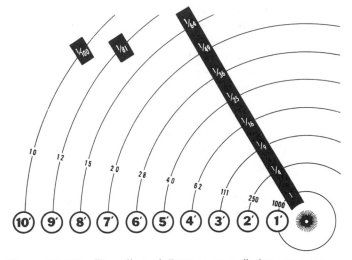

**Figure 10–14.** The effect of distance on radiation exposure follows the inverse square law—the intensity of radiation falls off by the square of the distance from the source. If we had a point source of radiation giving off 1,000 units of penetrating external radiation at 1 ft (or any other unit of distance), we would receive only one-fourth as much, or $(1/2)^2$, if we double our distance. If we triple our distance, we reduce the dose to one-ninth, or $(1/3)^2$. (Reprinted with permission from *Safe Handling of Radioisotopes in Industrial Radiography,* Picker X-Ray Corp., 1962.)

in free space where there is no scattering of radiation. In practical applications, the radiation source may not be equivalent to a point source, and surrounding surfaces may reflect some radiation so that free space does not apply. However, in most instances, the approximation is adequate.

Maintaining a safe distance is especially critical for employees who must work near inadequately shielded sources of radiation. This applies to both portable and nonportable source types (Table 10–D). In such cases, a radiation survey should be made with an appropriate survey meter by a qualified health and safety professional to establish minimum safe distances that workers must abide by.

Work involving penetrating types of radiation should be performed where permanent barriers and shielding protect workers from harmful exposure. However, it is possible that certain operations cannot be performed without some exposure of employees. In these cases, all

unnecessary exposure to radiation should be avoided, for example, by barring workers, including technicians, from active areas during exposure time such as during use of x-ray equipment.

A safe distance is the distance nonoperating workers must maintain from the radiation source in order to receive no more exposure than that specified in the NCRP *Radiation Protection Guides,* even if personnel were to remain at that distance continually. The distances specified for a given job do not necessarily remain constant if the radiation source in use is modified.

Hazardous areas indicated by the protection survey must be barricaded or roped off to form a restricted section that cannot be entered by nonoperating workers or bystanders. Large signs bearing the standard radiation symbol with proper wording should be posted (Figure 10–15).

## Shielding

Shielding is commonly used to protect against radiation from radioactive sources. The more mass that is placed between a source and a person, the less radiation the person will receive (Figure 10–16). For a high-density material such as lead, the barrier thickness required for a given attenuation of x- or gamma-radiation is less than it is for a less dense material such as concrete (Table 10–E).

Shielding from neutrons requires a different approach than does shielding from x- or gamma-radiation. Neutrons transfer energy to, and are stopped most effectively

**Table 10–D.** Gamma-Emitters and Radiation Levels at Various Distances from the Source

| Isotope | 0.3 m | 0.6 m | 1.2 m | 2.4 m | 3.0 m |
|---------|-------|-------|-------|-------|-------|
| Cobalt-60 | 14.5 | 3.6 | 0.9 | 0.23 | 0.145 |
| Radium-226 | 9.0 | 2.3 | 0.6 | 0.14 | 0.09 |
| Cesium-137 | 4.2 | 1.1 | 0.26 | 0.07 | 0.042 |
| Iridium-192 | 5.9 | 1.5 | 0.4 | 0.09 | 0.059 |
| Thulium-170 | 0.027 | 0.007 | 0.002 | 0.0004 | 0.00027 |

**Note:** Distances given in meters; 3.28 ft = 1 m.
(Reprinted from *Safe Handling of Radioisotopes in Industrial Radiography,* Picker X-Ray Corp., 1962.)

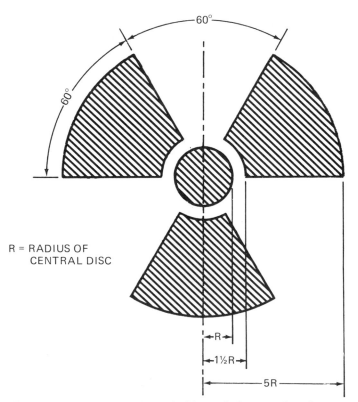

R = RADIUS OF CENTRAL DISC

**Figure 10–15.** Standard symbol for radiation warning signs. (Courtesy American National Standards Institute.)

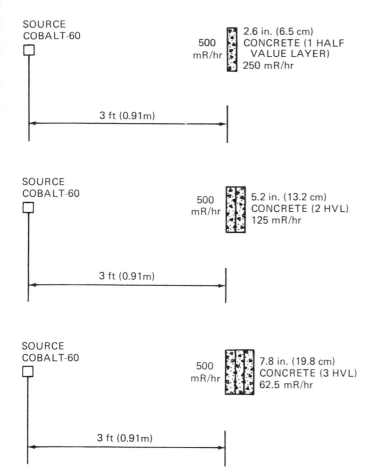

**Figure 10–16.** Illustration shows attenuation of cobalt-60 gamma-radiation by half-value layers of concrete.

by, light nuclei (hydrogen atoms are the most effective). Therefore, water or other materials rich in hydrogen content are used as shields for neutrons. Carbon atoms such as in graphite are also often applied for neutron shielding.

Shielding can take many forms. These include cladding on radioactive material, containers with heavy walls and covers for radioactive sources, cells with thick, high-density–concrete walls having viewing windows filled with high-density transparent liquid for remote handling of high-level gamma-emitters, and a deep layer of water for shielding against gamma-radiation from spent nuclear reactor fuel. Shielding calculations are often highly technical and require the services of an expert in this area.

Tables of half-value layers are given in radiation handbooks, and typical values for two radioactive materials, cobalt-60 and cesium-137, are given in Table 10–A.

Table 10–A states that 2.6 in. (6.6 cm) thickness of concrete can reduce the gamma-radiation coming from any cobalt-60 source to a factor of 1/2. If the gamma-emitter is cesium-137, then the half-value layer for concrete becomes 2.1 in. (5.3 cm). The gamma-radiation from cesium-137 is lower in energy (not as many MeVs) than the gamma-radiation from cobalt-60. Figure 10–16 illustrates how additional mass reduces radiation levels. The example used is a cobalt-60 source that gives a meter reading of 0.5 roentgen per hour (5 mGy/hr) at a distance of 3 ft (0.91 m).

A formula used for gamma-emitters is

$$\text{R/hr at 1 ft} = (6)(C)(E)(f) \qquad (3)$$

This relationship states that the number of roentgens (R) per hour measured at a distance of 1 ft (0.3 m) from a source is approximately equal to six times the curie strength ($C$) of the source times the energy ($E$) of the gamma-radiation in MeV times the fractional yield ($f$) of the gamma-radiation per disintegration. It cannot be used for estimating beta-radiation levels.

Given the name of the radioactive source and its quantity (or activity) in curies, the value for its energy in MeVs can be obtained from handbooks. For example, Table 10–B gives the energy of radiation from several radioactive isotopes. The fractional yield ($f$) of the gamma-radiation per disintegration can be found in the *Health Physics and Radiological Handbooks, 1992.*

**Table 10–E.** Approximate Tenth- and Half-Value Thicknesses for Shielding Radiographic Sources

| Radiographic Source | Tenth- and Half-Value Thicknesses (cm) | | | | | |
|---|---|---|---|---|---|---|
| | Lead | | Iron | | Concrete | |
| | 1/10 | 1/2 | 1/10 | 1/2 | 1/10 | 1/2 |
| Co-60 | 4.11 | 1.24 | 7.36 | 2.21 | 22.9 | 6.90 |
| Ra-226 | 4.70 | 1.42 | 7.70 | 2.31 | 24.4 | 7.37 |
| Cs-137 | 2.13 | 0.64 | 5.72 | 1.73 | 18.0 | 5.33 |
| Ir-192 | 1.63 | 0.48 | * | * | 15.7 | 4.83 |

\* = No data.

**Notes:** The thicknesses for tenth- and half-value layers provide shielding protection from the scattered radiation resulting from deflection of the primary gamma-rays within the shield as well as protection from primary radiation from the source. The tenth-value layers were determined from the reduction factor *v* shield thickness curves. The tenth-value thicknesses were taken as one-third the thickness of shielding material necessary to give a reduction factor of 1,000. The half-value thickness is equal to the tenth-value thickness divided by 3.32.
1 cm = 0.394 in.
Density of concrete assumed to be 2.35 g/cm² (147 lb ft³).
(Reprinted from the Nuclear Regulatory Commission Publication U-2967.)

Two cautions are required for proper application of this formula:

- Some radioactive materials such as cobalt-60 emit more than one gamma, each with different energies. The sum of the energy of the total emissions must be used. This information can be found in handbooks.

- Terms must be consistent. If the source activity is given in millicuries or microcuries, it should be converted to curies. If millicuries are used in the source term, then the answer is in milliroentgens.

### Example

What radiation reading would be expected at a distance of 1 ft from an unshielded 100-millicurie cobalt-60 source?

$$R/hr \text{ at } 1 \text{ ft} = (6)(C)(E)(f) \qquad (4)$$

where $C = 100$ millicuries (0.1 curie)
$E = 1.2$ MeV + 1.3 MeV (from handbook tables)
$F = 1$ (from handbook tables)

### Solution

$$R/hr \text{ at } 1 \text{ ft} = 6(0.1)(1.2 + 1.3)(1) = 1.5 \text{ R/hr} \qquad (5)$$

### Example

What radiation reading would be expected at a distance of 1 ft from an unshielded 1-millicurie cesium- 137 source?

### Solution

$$R/hr \text{ at } 1 \text{ ft} = 6(0.001)(0.66)(0.9) = 0.0036 \qquad (6)$$

which is 3.6 milliroentgens per hour (36 μGy/hr) at 1 ft (0.3 m).

The preceding examples illustrate the use of the formula in calculating dose rates and show how the application of the formula can assist in the interpretation of fractional amounts of curies.

The fact that a 1-millicurie source of cesium-137 is used doesn't quantify the exposure problem. However, by calculating the exposure rate of 3.6 milliroentgens per hour at 1 foot, and then applying the distance rule of decreasing radiation (1 divided by the distance squared— the inverse square rule) it becomes apparent that 4 feet (1.2 m) away from the source the radiation level would be $3.6/(4)^2$ 0.22 mR/hr (2.2 μGy/hr), which is very low.

Table 10–D lists a number of gamma-emitters and their radiation levels measured at various distances from a 1-curie source. The numbers were obtained by using the previous formula and then applying the inverse square law for distances of other than one foot.

Knowledge of existing radiation levels at various distances from various radioactive sources facilitates use of control procedures to ensure that cumulative doses do not exceed the maximum permissible dose (MPD).

### Example

As a final example of how control calculations can be used, assume that a 0.5-curie cobalt-60 source is involved in a fire. The radiation meters have been damaged and there is reason to believe that the source container is not functioning as an effective shield. Until someone with a good meter can conduct a survey and determine how severe the problem is, at what distance should the area be fenced off?

### Solution

First, calculate R/hr at 1 ft for a 0.5-curie cobalt-60 source.

$$R/hr \text{ at } 1 \text{ ft} = 6(0.5)(2.5)(1) = 7.5 \text{ or } 7,500 \text{ milliroentgen per hour at one foot} \qquad (7)$$

Second, at a control level of 40 milliroentgens per week (0.4 mSv per week) for a 40-hour week, the dose rate per hour should not exceed 40 mR/40 hr or 1.0 mR/hr (10 μGy/hr).

Third, use the inverse square law as follows:

$$x \text{ ft} = \left( \frac{\text{mrem/hr at 1 ft}}{1.0 \text{ mrem/hr}} \right)^{0.5}$$
$$= \sqrt{\frac{7,500 \text{ mrem/hr at 1 ft}}{1 \text{ mrem/hr at Control Ft.}}} \qquad (8)$$

The distance in feet to place the barricade equals the square root of the quotient (milliroentgen per hour at one foot divided by 1.0 milliroentgen per hour). Thus,

$$x \text{ ft} = \left( \frac{7,500}{1.0} \right)^{0.5} = 86.6 \text{ ft} \qquad (9)$$

Workers should not be allowed within 86.6 ft (26.2 m) of the source location. (All examples assume that exposures are occupational to workers with accurately maintained radiation exposure records.)

## Control Programs

A radiation health and safety program should establish safe working procedures, detect and measure radiation, make surveys, be concerned with decontamination and disposal, laboratory and other special services, and record keeping.

Ionizing radiation cannot be felt, seen, heard, tasted, or smelled. However, radioactive isotopes can be measured and identified with instruments, and therefore can be adequately controlled.

A basic concept in radiation protection practice is the establishment of a controlled area. Areas where radiation dose rates are excessive can be guarded during exposure times by suitable methods such as the erection of barriers and warning signs (Figure 10–15), stationing of attendants to keep personnel out of restricted localities, and, in extreme cases, complete closing off of the areas. Safe exposure times and safe practices can be established through measurement and control and by learning from past experience.

Work with radiation can and should be so planned and managed that radiation exposures of employees and of the general public are kept to a minimum. Controls are needed to prevent the release of radioactive materials that would result in exposures above guide levels.

Potential avenues of exposure to the public are contaminated air or water, waste materials, or employees unknowingly leaving the place of work with contamination on their persons, clothing, or shoes. Use of a hand-and-shoe monitor is shown in Figure 10–17.

## Sources of Radiation

One must always consider the amount and kind of radiation sources used in a contemplated operation. Suitable radiation safety operating conditions can then be designed. Special circumstances can greatly affect the safety requirements.

Table 10–F lists a number of radioactive materials according to their toxicity. Table 10–G classifies the laboratory or work area required for radioactive materials of differing toxicity.

Radionuclides are often used because measurements of their radiation can disclose useful information. One example is the employment of radioactive sources to effect quantitative or dimensional control of industrial materials. For example, radionuclides can measure fluid-flow rates or thickness of semi-finished or finished products.

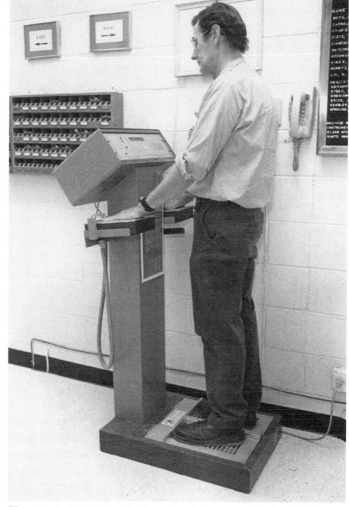

**Figure 10–17.** Use of a hand-and-shoe monitor for final radiological checkout.

---

**Table 10–F.** Classification of Isotopes According to Relative Radiotoxicity per Unit Activity

The isotopes in each class are listed in order of increasing atomic number

| | |
|---|---|
| CLASS 1 (very high toxicity) | Sr-90 + Y-90, *Pb-210 + Bi-210 (Ra D + E), Po-210, At-211, Ra-226 + percent *daughter products, Ac-227, *U-233, Pu-239, *Am-241, Cm-242. |
| CLASS 2 (high toxicity) | Ca-45, *Fe-59, Sr-89, Y-91, Ru-106 + *Rh-106, *I-131, *Ba-140 + La-140, Ce-144 • *Pr-144, Sm-151, *Eu-154; *Tm-170, *Th-234 + *Pa-234, *natural uranium. |
| CLASS 3 (moderate toxicity) | *Na-22, *Na-24, P-32, S-35, Cl-36, *K-42, *Sc-46, Sc-47, *Sc-48, *V-48, *Mn-52, *Mn-54, *Mn-56, Fe-55, *Co-58, *Co-60, Ni-59, *Cu-64, *Zn-65, *Ga-72, *As-74, *As-76, *Br-82, *Rb-86, *Zr-95 - *Nb-95, *Nb-95, *Mo-99, Tc-98, *Rh-105, Pd-103 - Rh-103, *Ag-105, Ag-11, Cd-109 - *Ag-109, *Sn-113, *Te-127, *Te-129, *I-132, Cs-137 - *Ba-137, *La-140, Pr-143, Pm-147, *Ho-166, *Lu-177, *Ta-182, *W-181, *Re-183, *Ir-190, *Ir-192, Pt-191, *Pt-193, *Au-198, *Au-199, Tl-200, Tl-202, Tl-204, *Pb-203. |
| CLASS 4 (Slight toxicity) | H-3, *Be-7, C-14, F-18, *Cr-51, Ge-71, *Tl-201. |

**Note:** The isotopes in each class are listed in order of increasing atomic number.
*Gamma- emitters.

(Reprinted from *Safe Handling of Radionuclides*, Safety Series No. 1, International Atomic Energy Agency, Vienna.)

**Table 10–G.** Laboratory or Work Area Required for Isotopes of Increasing Radiotoxicity

| Radiotoxicity of Isotopes | Minimum significant quantity | Type of Laboratory or Work Area Required | | |
| --- | --- | --- | --- | --- |
| | | Type C Good Chemical Laboratory | Type B Radioisotope Laboratory | Type A High-Level Laboratory |
| Very high | 0.1 μc | 10 μc or less | 10 μc–10 mc | 10 mc or more |
| High | 1.0 μc | 100 μc or less | 100 μc–100 mc | 100 mc or more |
| Moderate | 10 μc | 1 mc or less | 1 mc–1 c | 1 c or more |
| Slight | 100 μc | 10 mc or less | 10 mc–10 c | 10 c or more |

| Procedure | | | | Modifying factor |
| --- | --- | --- | --- | --- |
| Storage (stock solutions) | | | X | 100 |
| Very simple wet operations | | | X | 10 |
| Normal chemical operations | | | X | 1 |
| Complex wet operations with risk of spills | | | X | 0.1 |
| Simple dry operations | | | X | |
| Dry and dusty operations | | | X | 0.01 |

**Note:** Modifying factors should be applied to the quantities shown in the last three columns, according to the complexity of the procedures to be followed. Factors are suggested only because due regard must be paid to the circumstances affecting individual cases.

(Reprinted from *Safe Handling of Radionuclides*, Safety Series No. 1, Atomic Energy Agency, Vienna.)

**Sealed sources.** Sealed sources with widely varying amounts of radioactivity are available. A high-intensity source may require so much shielding that it is not considered portable. Shielding, control devices, and procedures should be designed by someone experienced in such work. Keeping external exposures to personnel under control can be accomplished with the aid of alarms, interlocks, strict control of access, and thorough monitoring (Bibliography, NCRP Report 30, *Safe Handling of Radioactive Materials*).

Sources of low intensity should only require keeping track of their presence.

Sources of intermediate intensity are more troublesome. They are portable and in some cases require so little shielding that they can easily be moved by hand. Persons operating a portable source must consistently follow the necessary precautions, including the use of dosimeters to measure their own exposure. They must operate the source so that other persons are not accidently irradiated (Bibliography, NSC Industrial Data Sheet 461, *Beta-Particle Sealed Sources*, rev. 1986).

Some sealed sources are subject to breakage and spillage, and even welded seals have been known to fail. The hazard can be severe, and great care should be taken in handling sealed sources. Even if no known accident has occurred, such sources should be tested for leaks by a standardized procedure at scheduled intervals by an experienced technician.

**Radiation-producing machines.** For many years, radiation-producing machines have been in use, and proper operation of x-ray machines has been described in many publications. Remote actuation and shielding are common ways to control personnel exposures. If an accelerator is used, radiation safety personnel with special training or experience are required to ensure proper installation and operation. (Bibliography, NCRP Reports 51, 88 and 102.)

Portable x-ray units present many of the same kinds of radiation safety problems as nonportable sources of corresponding intensity. Special problems related to the portability and lack of stationary shields must also be considered.

**Radioisotopes.** Use of radioisotopes in the laboratory encompasses a wide range of hazards. The degree of hazards depends on the quantities and types of radioisotopes used as well as the kinds of operations to be performed.

Radioisotopes must be considered individually when hazards are being analyzed because their effect on human health varies with the type of radiation emitted, the process or work being conducted, and the safety practices being employed by those conducting the work. No general statements can be made until specific information is gathered and analyzed. Controls and procedures should be established by a health physicist.

Table 10–F groups various radioisotopes according to their relative radiotoxicity per unit activity in accordance with the International Atomic Energy Agency's (IAEA) publication, *Safe Handling of Radionuclides*. The IAEA explains the hazard classification for unsealed radioactive sources as follows:

> Hazards arising out of the handling of unsealed sources depend on factors such as the types of compounds in which these isotopes appear, the specific activity, the volatility, the complexity of the procedures involved, and of the relative doses of radiation to the critical organs and tissues, if an accident should occur that gives rise to skin penetration, inhalation, or ingestion.

Broad classifications for the radiotoxicity of various isotopes and their significant amounts are given in Tables 10–F and 10–G

Spent fuel elements from a nuclear power reactor can contain on the order of a million curies, depending on the length of time they have been irradiated in the reactor and the interval of time between their removal from the reactor and measurement of their radioactivity. Shielding is required for handling, storage, and shipment of spent fuel materials and other high-level gamma-radiation sources. The use of manipulators, shielding cells, and shielding windows for remote work is shown in Figures 10–18 and 10–19. This illustrates confinement and shielding for high-level gamma-radioactivity. The photographs show the use of an in-cell remote camera and external video screen in Figure 10–18 and the remote use of strippable decontamination paint in Figure 10–19.

**Radioactive metals.** Radioactive metals vary greatly in degree of hazard. Some have such low radiation rates that they can safely be held in the hand (if they are solids that are not flaking or dusting). A piece of normal uranium or an alloy of this material, for example, can be safely handled without personal protection for a few hours per week. It is good practice, nevertheless, to wear gloves when handling such materials, because metals (like uranium) can oxidize and develop a flaky surface. Use of metal cladding or a paint-type surface coating may be necessary to avoid loose contamination. Respiratory protection may also be needed.

Certain radioactive metals have such high surface dose rates that they must be handled only with remote control devices. If a solid radioactive metal has a loose contaminating layer of radioactivity on its surface, it must be handled in a ventilated enclosure equipped with a high-efficiency exhaust filter. In each case, the radiation level of the particular material and its composition must be known before safe handling practices can be determined.

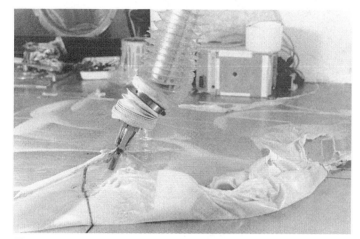

**Figure 10–19.** Remote removal of strippable decontamination paint in a hot cell.

Instrument measurement of the dose rate from the material help determine the control measures needed. Exposure time, distance, and shielding are the basic control measures. Confinement and other means to avoid intake of radioactive material constitute other important control measures. Review of processes and handling history, visual inspections, and smear and air sampling is useful when determining the probability of contamination incidents and the necessary control measures.

Uranium metal will burn under certain conditions and, along with some alloys used in chemical operations, poses a chemical explosion hazard. Casting of uranium metal must be done in a vacuum. It is necessary to use an inert gas, rather than air, to break the vacuum. Wastes and scraps must be carefully handled from the time they are generated until they are ultimately disposed of, not only for reasons of safety, but also because of their monetary value.

Cutting of uranium with abrasive cutoff wheels requires special local exhaust ventilation. Because chips, turnings, and finely divided uranium burn spontaneously under some conditions, it is advisable to convert uranium into oxide daily, as the scrap is accumulated. Uranium-contaminated waste requires packaging that meets U.S. Department of Transportation and disposal site criteria.

**Criticality.** Criticality is the fissioning, or breaking apart, of nuclei, with emission of neutrons at a rate faster than neutrons are absorbed or lost from the system. It results in a highly hazardous instantaneous burst of neutrons with associated high-level gamma-radiation. Whenever the presence of fissionable radioactive materials such as uranium-235 and plutonium-239 necessitates criticality controls, an expert in this area of radiation safety must be involved.

A number of fatalities have resulted from criticality incidents, and even personnel at considerable distances

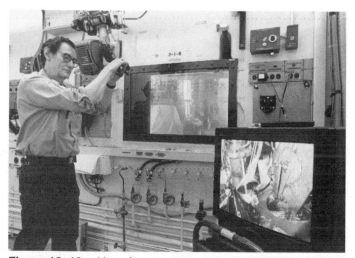

**Figure 10–18.** Use of a remote camera and external video screen for viewing areas in a hot cell.

from the source can receive severe radiation exposure. Stringent controls over amounts and movement of fissionable materials are necessary. Immediate local sensing of a critical incident and audible alarms to trigger quick evacuation are other precautions employed to protect against this hazard.

**Plutonium.** Plutonium is a highly hazardous alpha-emitter that must be handled under rigidly controlled conditions. Enclosures, some of which are called glove boxes or dry boxes (Figure 10–20), must be carefully designed and are relatively expensive. Glove boxes must be maintained at a negative pressure, and may necessitate an inert atmosphere such as argon or nitrogen to avoid fires. Plutonium is pyrophoric under certain conditions. Plutonium-239 is fissionable and presents a criticality control problem when sufficient amounts are present (>300 grams).

Glove-box exhaust gas is usually filtered by two or more high-efficiency particulate air (HEPA) filters in series. Emergency power for exhaust ventilation and alarms for loss of negative pressure in the boxes are system safety features.

Because glove-box gloves can develop pinholes, cracks, tears, or punctures, workers wear another pair of protective gloves when using them. Workers should monitor their gloved hands periodically and monitor and remove protective clothing when leaving the work area.

## Operational Factors

Ascertain the required level of radiation protection and evaluate the problems that might arise by analyzing the radiation work in terms of the following factors:

- Area involved (in square feet or meters), and number of rooms and buildings.
- Number of employees exposed to radiation and in what locations.
- Chemical and physical states of the radioactive material and the nature of its use.
- Incidents that might occur and their possible locations.
- Nonradiation hazards involved.
- Nature of the probable radiation exposure or release of radioactive material:

    Controlled, or supervised, release (as in a disposal operation)

    Accidental release or exposure that cannot be sensed by radiation-detection instruments

**Figure 10–20.**   A glove box type of hood permits rigid control of conditions when radioactive metals are worked. (Courtesy Argonne National Laboratory.)

Violent release of dust, droplets, gases, or vapors through fire, increase in temperature, or explosion

Spread of contamination as the result of its adherence to other material or through spillage

- Inherent danger of the material because of its internal or external effect on humans. This danger depends on the isotopic and chemical forms of the material (Bibliography, NCRP Reports 22).
- Probability of detection of hazardous situation by routine surveys or monitoring.
- Knowledge of current conditions is essential to determine the acceptability of the risks involved.
- Possible effects of accidents on operations, such as interruption of production, loss of occupancy of space, loss of equipment, and cost of cleanup.

## Employee Exposure Potential

The industrial accidents that can produce radioactive contamination are usually no more difficult to prevent than the more common types of industrial accidents, but there is an added, compelling reason to control them—the danger of intake of radioactive materials by personnel.

Monitors for detecting radioactive contamination of personnel are shown in Figures 10–12 and 10–17. It is important that the appropriate monitoring instruments be available and properly maintained and calibrated.

**External hazards.** Under ordinary operating conditions, and barring accidents, the level of exposure to external radiation can be determined through use of continuously monitoring instruments or dosimeters, or from previous measurements of radiation fields for identical operations. The problem is then reduced to limiting the rate of exposure or the length of exposure time. Instrument measurement of the exposure indicates what advantage can be obtained by installing additional shielding or by having personnel work farther from the source of radiation.

Under emergency conditions with high-exposure levels, it is advisable to rotate personnel so as to prevent exposure of any one individual above guide levels. Emergency exposure levels for extreme conditions have been published in NCRP Report No. 116, *Limitation of Exposure to Ionizing Radiation.* These are essentially once-in-a-lifetime allowances for handling an extremely serious situation.

Occasionally, an unexpectedly high reading appears on a personal dosimeter. While the reading may be invalid for any one of a number of reasons, an investigation should be conducted to determine whether or not the reading was valid and what should be done to prevent recurrences.

**Internal hazards.** Inhalation is the most frequent route of entry of radioactive material into the body. Routine air sampling can detect the concentration, and a constant air-monitoring system can be designed to sound an alarm if hazardous levels are detected. Also, employees are likely to recognize conditions that in the past have resulted in exposure, and can act promptly to minimize exposure.

In the event that a spill of radioactive material is detected by immediate recognition, hand and foot monitoring, or through air or room surface monitoring, prompt action is essential. If there is a potential inhalation hazard, personnel should evacuate to the nearest safe location. Further movement of workers should be restricted until a monitoring survey can be done to determine the extent of radioactive contamination. Quick arrangements should be made to restrict exit from and entry into the building, wing, or laboratory until there is a determination that radioactivity has not and will not be tracked or carried out to other areas. Corridors should be roped off or posted to control movement until monitoring of personnel, surface, and air shows that contamination is controlled.

It is important to isolate and clean up a spill as soon as possible. Personnel who conduct the monitoring and cleanup should wear appropriate protective equipment and clothing. The incident can be resolved as soon as there is assurance that the potential hazards have been controlled.

A less common route by which radioactive material enters the body is through broken skin, and in rare cases this can be more serious than entry through the lungs. When radioactive materials gain entry through a minute cut or abrasion, they enter the bloodstream and are then dispersed throughout the body. The result can be serious if there is a high concentration of the radioisotope at the skin break. If the workers have been appropriately educated, the injured person will be aware of the possible danger and can request a radiation survey of the wound area.

By and large, absorption through intact skin is not a significant route of entry. Tritiated water vapor is one of the exceptions; it is absorbed rapidly through unbroken skin. It is important to scrub up immediately when significant skin contamination occurs. This minimizes the possibility of entry through inhalation, ingestion, or a subsequent skin break and prevents the spread of the radioactive material.

Analysis of the radionuclide content of body fluids and excreta and whole-body counting are important for after-the-fact detection of intake of radioactivity. When the presence of radionuclides within the body is detected or an accidental intake of radionuclides becomes known, steps should be taken to determine the cause and prevent recurrence. This serves as a backup to environmental monitoring.

## Records

Some diseases that are caused by exposure to radiation or radioactive materials occur after many years of exposure. Retention of suitable records relating to exposures and working conditions is desirable and is usually required.

# Summary

It is impossible to discuss all of the important aspects of radiation safety in a fundamentals manual such as this. Nevertheless, health and safety professionals will have a good background for further study and assistance in radiation safety if they do the following:

- Recognize that ionizing radiation control requires the expertise of qualified health physics personnel.

- Treat radiation and radioactive materials with respect because of the recognized potential hazards.

- Recognize the two distinct types of exposure hazards involved: external and internal.

- Learn about the various types of ionizing radiation

- Know that calibrated monitoring instruments appropriate for the specific type of ionizing radiation must be used for measurements.

- Recognize the importance of basic exposure control measures including time, distance, and shielding.

- Know that control guidelines and exposure limits must be followed.

# Bibliography

American Industrial Hygiene Association, 2700 Prosperity Avenue, Suite 250, Fairfax, VA 22031. *Respiratory Protection—A Manual and Guideline,* 2nd ed. Fairfax, VA: AIHA, 1991.

American National Standards Institute, 11 West 42nd St., New York, NY 10036.
*Administrative Practices in Radiation Monitoring, Guide to* (R1988), ANSI N13.2–69.
*Inspection and Test Specifications for Direct Reading and Indirect Reading Quartz Fiber Pocket Dosimeters* (R1991), ANSI N322–77.
*Occupational Radiation Exposure Records Systems, Practice for* (R1989), ANSI N13.6–66.
*Personnel Neutron Dosimeters (Neutron Energies Less Than 20 MeV)* (R1984) ANSI N319–76.
*Pocket-Sized Alarm Dosimeters and Alarm Ratemeters, Performance Requirements for* (R1992) ANSI N13.27–81.
*Radiation Detectors—Personnel Thermoluminescent Dosimetry Systems—Performance* ANSI N1315–85.
*Radioactive Materials Leakage Tests on Packages for Shipments* ANSI N14.5–87.
*Radiation Protection—Photographic Film Dosimeters—Criteria for Performance* ANSI N13.7–83.
*Recommended Fire Protection for Facilities Handling Radioactive Materials* ANSI/NFPA No. 801–91.
*Sampling Airborne Radioactive Materials in Nuclear Facilities, Guide for* (R1993) ANSI N13.1–69.
*Sealed Radioactive Sources, Classification (NBS126)* ANSI N542–77.
*Specification and Performance of On-site Instrumentation for Continuously Monitoring Radioactivity in Effluents,* (R1991) ANSI N42.18–80.
*Warning Symbols—Radiation Symbol,* ANSI N2.1–89.

Cameron JR, Suntharalingam, Kenney. *Thermoluminescent Dosimetry.* Madison, WI: University of Wisconsin Press, 1968.

Cember H. *Introduction to Health Physics,* 2nd ed. New York: McGraw-Hill, 1992.

*Code of Federal Regulations (CFR).* Title 10, Chapter 1—Energy, Part 20. Standards for Protection Against Radiation. Washington, DC: Nuclear Regulatory Commission (NRC), Superintendent of Documents, 1993.

*CFR 49.* Transportation: Part 100–177, Subchapter C—Hazardous Materials Regulations. Revised Oct. 1, 1992. Washington, DC: U.S. Government Printing Office, 1992.

Cralley LJ, Cralley LV. "Evaluation of Exposure to Ionizing Radiation." In *Patty's Industrial Hygiene and Toxicology.* New York: John Wiley & Sons, 1991.

*Hazardous Materials Regulations of the DOT.* Tariff No. BOE-6000-L. Effective Apr. 23, 1992. Available through Keller CL, Agent, 1920 L St., N.W., Washington, DC.

*Health Physics,* official journal of the Health Physics Society. Baltimore: Williams and Wilkins.

International Atomic Energy Agency and the World Health Organization, Unipub, P.O. Box 443, New York, 10016. Safety Series:
*Basic Factors for the Treatment and Disposal of Radioactive Wastes,* No. 24, 1967.
*Basic Requirements for Personnel Monitoring,* No. 14, 1980.
*Basic Safety Standards for Radiation Protection,* No. 9, 1982.
*Manual on Safety Aspects of the Design and Equipment of Hot Laboratories,* No. 30, 1969.
*Medical Supervision of Radiation Workers,* No. 25, 1968.
*Planning for the Handling of Radiation Accidents,* No. 32, 1969.
*Radiation Protection Procedures,* No. 38, 1973.
*Recommendations for the Safe Use and Regulation of Radiation Sources in Industry, Medicine, Research and Teaching,* No. 102, 1990.
*Regulations for the Safe Transport of Radioactive Materials,* No. 6, 1985.
*Respirators and Protective Clothing,* No. 22, 1967.
*Safe Handling of Radionuclides,* No. 1, 1973.
*Safe Use of Radioactive Tracers in Industrial Processes,* No. 40, 1974.

Kocher DC. "Radioactive Decay Data Tables," DOE/TIC-11026. Technical Information Center, U.S. Department of Energy, 1981.

Martin A, Harbison SA. *An Introduction to Radiation Protection,* 3rd ed. New York: Chapman & Hall, 1986.

Morgan KZ, Turner JE. *Principles of Radiation Protection.* Kriezer, 1973.

National Academy of Sciences. BEIR V, 1990: Health Effects of Exposure to Low Levels of Ionizing Radiation, Committee on the Biological Effects of Ionizing Radiations. Washington, DC: the Academy, 1990.

National Council on Radiation Protection and Measurement:
*Alpha-Emitting Particles in Lungs,* No. 46, 1975.
*Cesium-137 From the Environment to Man: Metabolism and Dose,* No. 52, 1977.
*Control and Removal of Radioactive Contamination in Laboratories,* No. 8, 1951.
*Dental X-Ray Protection,* No. 35, 1970.
*Environmental Radiation Measurements,* No. 50, 1976.

*General Concepts for the Dosimetry of Internally Deposited Radionuclides,* No. 84, 1985.

*A Handbook of Radioactivity Measurements Procedures,* No. 58, 2nd ed., 1985.

*Instrumentation and Monitoring Methods for Radiation Protection,* No. 57, 1978.

*Krypton-85 in the Atmosphere—Accumulation, Biological Significance, and Control Technology,* No. 44, 1975.

*Limitation of Exposure to Ionizing Radiation,* No. 116, 1993.

*Maintaining Radiation Protection Records,* No. 114, 1992.

*Management of Persons Accidentally Contaminated with Radionuclides,* No. 65, 1980.

*Maximum Permissible Body Burdens and Maximum Permissible Concentrations of Radionuclides in Air and in Water for Occupational Exposure,* No. 22, 1959. (Includes Addendum I, issued in Aug. 1963.)

*Measurement of Absorbed Dose of Neutrons and of Mixtures of Neutrons and Gamma Rays,* No. 25, 1961.

*Measurement of Neutron Flux and Spectra for Physical and Biological Applications,* No. 23, 1960.

*Medical Radiation Exposure of Pregnant and Potentially Pregnant Women,* No. 54, 1977.

*Medical X-Ray, Electron Beam and Gamma-Ray Protection for Energies up to 50 MeV (Equipment Design, Performance, and Use),* No. 102, 1989.

*Operational Radiation Safety Program,* No. 59, 1978.

*Operational Radiation Safety—Training,* No. 71, 1983.

*Precautions in the Management of Patients Who Have Received Therapeutic Amounts of Radionuclides,* No. 37, 1970.

*Protection Against Neutron Radiation,* No. 38, 1971.

*Protection Against Radiation from Brachytherapy Sources,* No. 40, 1972.

*Protection of the Thyroid Gland in the Event of Releases of Radioiodine,* No. 55, 1977.

*Radiation Protection Design Guidelines for 0.1–100 MeV Particle Accelerator Facilities,* No. 51, 1977.

*Radiation Protection in Educational Institutions,* No. 32, 1966.

*Radiation Protection in Veterinary Medicine,* No. 36, 1970.

*Radiological Factors Affecting Decision-Making in a Nuclear Attack,* No. 42, 1974.

*Review of NCRP Radiation Dose Limit for Embryo and Fetus in Occupationally Exposed Women,* No. 53, 1977.

*Specification of Gamma-Ray Brachytherapy Sources,* No. 41, 1974.

*Stopping Powers for Use with Cavity Chambers,* No. 27, 1961.

*Structural Shielding Design and Evaluation for Medical Use X-Rays and Gamma Rays of Energies up to 10 MeV,* No. 49, 1976.

*Tritium Measurement Techniques,* No. 47, 1976.

National Safety Council:

*Beta-Particle Sealed Sources,* Data Sheet 12304-461, 1986.

*Sources of Information on Nuclear Energy and Energy Related Activities,* Data Sheet 12304-685, 1985.

Purrington RG, Patterson HW. *Handling Radiation Emergencies.* Boston: National Fire Protection Agency, 1977.

Shapiro J. *Radiation Protection, A Guide for Scientists and Physicians,* 3rd ed. Cambridge, MA: Harvard University Press, 1990.

Stewart DC. *Data for Radioactive Waste Management and Nuclear Applications.* New York: John Wiley & Sons, 1985.

Wade JE, Cunningham GE. *Radiation Monitoring, A Programmed Instruction Book.* Oak Ridge, TN: Division of Technical Information Extension, 1967.

**Electric and Magnetic Fields**
*Electric Fields* ▪ *Magnetic Fields*

**Electromagnetic Radiation**
*Electromagnetic Spectrum*

**When Are Fields Important and When Is Radiation Important?**
*Parts of an Electromagnetic Device*

**Subradiofrequency Fields: 0–3,000 Hz**
*Field Strengths* ▪ *How Do Fields Interact with the Body?*

**Biological Effects and Standards for Steady (DC) Electric Fields**

**Biological Effects and Exposure Standards for Static Magnetic Fields**
*Other Safety Concerns of Static Magnetic Fields*

**Biological Effects and Exposure Standards for Time-Varying Subradiofrequency Fields**

**Measuring Subradiofrequency Fields**

**Controls and Shielding**

**Radiofrequency/Microwave Radiation and Fields (RF/MW)**
*Industrial, Scientific, and Medical Bands and Frequency Nomenclature* ▪ *Interactions of Radiation and Matter*

**Biological Effects and Exposure Standards for Radiofrequency Fields**
*Dosimetry* ▪ *Target Organs* ▪ *Standard-Setting Rationale* ▪ *Averaging Time and Pulsed Fields* ▪ *Regulatory Considerations*

**Measuring Radiofrequency Radiation and Fields**

**Video Display Terminals and Microwave Ovens**
*Video Display Terminals* ▪ *Microwave Ovens*

**RF/MW Controls**

**Optical Radiation and Lasers**
*CIE Bands*

**Biological Effects and Exposure Standards for Optical Radiation**
*The Eye* ▪ *The Skin* ▪ *Standards* ▪ *Controls for Nonlaser Sources*

**Laser**
*Biological Damage Mechanisms of Lasers* ▪ *Standards* ▪ *Controls* ▪ *Laser Pointers* ▪ *Nonbeam Hazards of Lasers* ▪ *Other Regulatory Concerns with Lasers*

**Measuring Optical Radiation**

**Lighting**

**Summary**

**Bibliography**
*Periodicals*

**Internet Worldwide Web and Turbogopher Services**

CHAPTER 11

# Nonionizing Radiation

by Gordon Miller, CIH

*The author thanks Scott Hildum, former laser safety officer at Lawrence Livermore National Laboratory, and Robert Curtis, OSHA's leading expert on nonionizing radiation, for reviewing this chapter.*

Rᴀᴅᴀʀ ᴡᴀs ᴏɴᴇ ᴏғ ᴛʜᴇ sᴄɪᴇɴᴛɪғɪᴄ ᴡᴏɴᴅᴇʀs to emerge from World War II, but there is controversy over whether it is safe. Microwaves are finding more and more uses as portable communications change from luxury to necessity. There is a radar in a box in most kitchens and no food products supplier can ignore the need to package food so it is microwavable. Lasers were a scientific wonder of the 1960s, but now laser CDs have displaced phonograph disks and most offices have at least one laser printer. One hallmark of technological change is the increasing use of the electromagnetic spectrum and it is certain that new uses of electromagnetic energy will be found after this edition is published. Perhaps the scientific controversy regarding the safety of power line fields will be at least partially resolved before this chapter is revised for the next edition. This chapter presents the basic principles of electromagnetic fields and radiation and how to protect people from the hazards associated with this energy.

## ■ Electric and Magnetic Fields

### Electric Fields

Atoms are divided into a central part, the nucleus, and electrons that orbit around the nucleus. The protons in the nucleus carry what is arbitrarily called a positive charge and the electrons orbiting about the nucleus carry what is called a negative charge. Negative charges attract positive charges, negative charges repel other negative charges, and positive charges repel other positive charges. These parts are able to exert a force on other distant objects; they create a force field. This field is an electric field, often visualized as lines of force that originate at a positively charged object and end at a negatively charged object, as shown in Figure 11–1. The charge state (positive or negative) is the polarity. The force exerted by a charged object on another charged object at a distance depends on the amount of charge in the objects, the polarity of the charges in the objects, and the distance between them. Force is directly proportional to the amount of charge and decreases as the distance between charged objects increases according to an inverse square relationship. If some level of electric force exists at a given distance, the electric force is one quarter as great at twice the distance, one ninth as much at three times the distance, and so on. Any charged object creates an electric field whether it is stationary or in motion.

The electric force is very powerful. Opposite charges attract and meet to cancel each other out. For example, the protons in the nuclei of atoms attract an identical number of electrons so the atom is neutral. Electrical imbalances are rare, so a much weaker force created by all matter—gravity—dominates. An object carrying an electric charge is an ion. An ion can be an atom carrying too few or too many electrons or a bigger object, such as a dust particle, with a deficit or surplus of electrons.

## Magnetic Fields

A *moving* electric charge creates yet another field: magnetism. Imagine electric charges are moving in some direction, such as through a wire. The amount of charge flowing past a given point is the current. Magnetic fields exist in a direction perpendicular to the direction of the current flow. The orientation of the field is such that the north-seeking end of a compass needle points as shown in Figure 11–2a. Magnetic field polarities are given as north and south and magnetic fields are usually visualized as lines of force that start at the north pole of a magnet and come back around, land on the south pole of the magnet, and complete a circuit by going through the body of the magnet until they reemerge at the north pole. Thus, magnetic fields don't radiate out into space as electric field lines do; they return to the other pole, as shown in Figure 11–2b. The north pole of a magnet points to the north pole of the Earth.

A magnetic field exerts a force on moving electrically charged particles in a direction perpendicular to the field. The force is proportional to the amount of moving charge (the current flow) and the distance between the system carrying that current and a charged object.

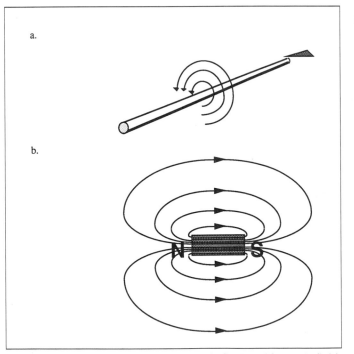

**Figure 11–2.** Depictions of a magnetic field. a. Magnetic field created by current flow. In this figure, current flows from negative to positive, as shown by the arrow pointing right at the upper end of the wire. The north-pointing needle of a compass will point in the direction of the smaller arrows around the wire. b. Fields around a common bar magnet. Fields form loops that are assumed to start at the north pole, come around to land at the south pole, and move through the magnet. The current flow that creates the field of a permanent magnet is the movement of electrons around atoms.

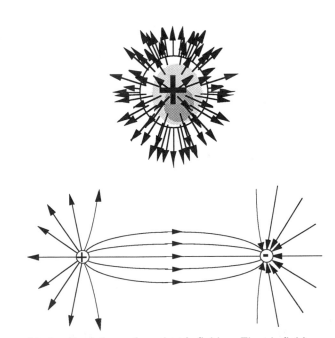

**Figure 11–1.** Depictions of an electric field. a. Electric field lines originating at a positively charged object and extending out equally in all directions. b. Electric field lines leaving a positively charged object and landing on a negatively charged object. If the objects had the same charge, the field lines would push away from each other.

When a magnetic field changes, it causes nonmoving charges to move or induces a current flow in objects that conduct electricity, including the human body. The human body is mostly made of salt water, a good conductor filled with ions. Thus, a person in a changing magnetic field experiences current flow in loops oriented perpendicular to the direction of the magnetic field, as shown in Figure 11–3. People are unaware of this current flow, but it is always happening because they are immersed in magnetic fields that change with time (time-varying fields) arising from the transmission and use of alternating current (AC) electricity. The polarity and strength of the current keeps alternating between positive and negative, so it is called alternating current. In the United States and Canada, polarity changes occur 60 times a second (50 times a second elsewhere).

The history of AC power is worth describing because it underlies concerns about power line fields and introduces the basic law of electrical engineering: Ohm's law. Resistance to the flow of current is proportional to the current—more current means more resistance. Electric current flows from one pole of a source to the other pole, through the source, and back out again for further rounds in the pathway or circuit. The current is measured in units of amperes (A, amps). A useful mechanical analogy is that of the closed-circuit water pump in an aquarium. The pump requires a source of energy to give the water pressure to keep moving because there is resistance to flow in the system. Electrical circuits also have resistance to flow. The unit of electrical resistance (R) is the ohm $\Omega$ and the unit of pressure or potential that keeps the current flowing is the volt (V). According to Ohm's law, volts equals amps times ohms (V = I × R).

The power loss dissipated in an electrical circuit due to resistance is proportional to the square of the current. Electrical engineers in the early twentieth century faced a problem figuring out how to transmit electric current to distant points. The answer is to keep changing the polarity of the current and use the resulting changes in magnetic fields to induce another current to flow. The device that does this is the transformer, which swaps current for voltage and vice versa. A utility generates current at a voltage of 20,000 V at a power plant and sends the current to a transformer, which steps up the voltage to create a much smaller current that is transmitted over large distances with low resistance losses. The transmitted current then passes through a sequence of other transformers that step down the voltage in exchange for higher currents and is finally distributed to users at a modest 115 volt potential, as shown in Figure 11–4. Transmission line voltages range from 69,000 V up to 765,000 V and the intermediate distribution voltages created by the step-down transformers range from 4,000 to 35,000 V. AC current makes long-distance electrical power systems practical; otherwise, resistance losses would become intolerable.

## Electromagnetic Radiation

So far, we have discussed electric and magnetic fields as single entities. There is a common phenomenon where the two exist together: electromagnetic radiation.

Electric and magnetic fields change polarities as the fields pass by any given location in space, as shown in Figure 11–5. The number of times the fields change polarity and return to the beginning polarity in a given time is the frequency. The frequency ($\nu$) is usually specified in full cycles of polarity reversals and returns that occur in one second. If the electric field is zero at a given moment and starts climbing to a peak positive value, it will peak, fall back through zero, reach a maximum negative value, and climb back to zero. This is a full cycle. If this happened once in a second, then the frequency would be one cycle per second. If it happens one thousand times a second, then the frequency is one thousand cycles per second. One cycle per second is also called 1 hertz (Hz), after Heinrich Hertz, who discovered radio waves. One cycle per second equals 1 Hz, 1,000 cycles per second equals 1,000 Hz or, using proper metric prefixes, 1 kilohertz (kHz). See Table 11–A for a list of metric prefixes.

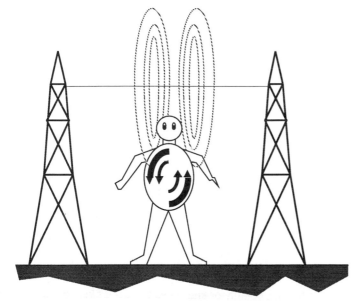

**Figure 11–3.** Interaction of an AC magnetic field with a person. The field will induce a flow of current at a right angle to the direction of the field. The field radiates out from the line overhead, as shown by dotted lines, but changes direction as the line current changes polarity. The current induced in the body also changes direction. The magnitude of the current is proportional to the radius of the loop it is traveling in so magnetic fields would be of greatest concern in big organs that are electrically active, such as the brain and heart. (Adapted from Nair, Morgan, and Florig, 1989.)

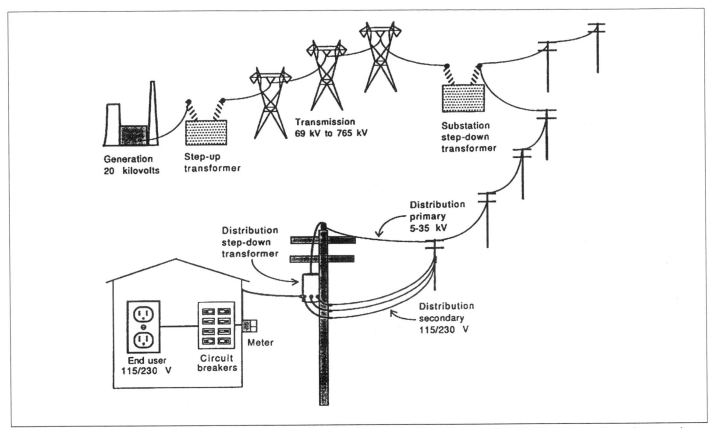

**Figure 11–4.** Electrical transmission and distribution. Transformers are used at several points, first to step up the voltage and reduce the current so the electricity can be transmitted over large distances with minimal resistance losses, then stepped down to lower voltages at higher current levels before final distribution. Transmission lines are designed to operate with little variation in current, but distribution systems are designed to take broad ranges of current flows. The voltages are fixed at each step and currents vary. (From Nair, Morgan, and Florig, 1989.)

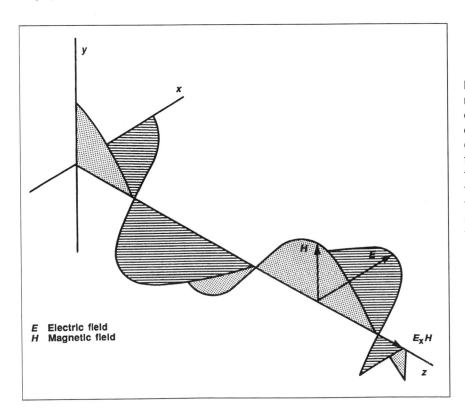

E Electric field
H Magnetic field

**Figure 11–5.** The components of electromagnetic radiation. Note that the electric field is oriented in one direction (shown by line-filled curves) whereas the magnetic field (shown by dot-filled curves) is perpendicular to the electric field and the direction of travel is perpendicular to the other two. The number of times that the fields change from a given strength and polarity to the opposite polarity and back to the starting strength and polarity in a second is the frequency. Polarization (not to be confused with polarity or charge state) is the direction of the electric field in relation to the surface of the earth. If the earth were below this page, then this wave would be horizontally polarized. If the electric field were turned 90 degrees (vibrating perpendicularly to the earth's surface), then it would be vertically polarized. The electric field can be caused to change direction constantly; it is circularly polarized if it points in all directions equally, or elliptically polarized if it points more in one direction than another.

**Table 11–A.** Metric Prefixes

| Prefix | Abbreviation | Definition | Scientific Notation | Common Uses in Nonionizing Radiation |
|--------|--------------|------------|---------------------|--------------------------------------|
| Atto | a | pentillionth | $10^{-18}$ | — |
| Femto | f | quadrillionth | $10^{-15}$ | — |
| Pico | p | trillionth | $10^{-12}$ | Picometer (ionizing radiation wavelengths) |
| Nano | n | billionth | $10^{-9}$ | Nanometer (UV, visible, and IR-A wavelengths) |
| Micro | μ | millionth | $10^{-6}$ | Micrometer (IR-B, IR-C wavelengths), microjoule (energy level), microwatt (power level) |
| Milli | m | thousandth | $10^{-3}$ | Millimeter (microwave or IR-C wavelength), milliwatt (power level) |
| Centi | c | hundredth | $10^{-2}$ | Centimeter (microwave wavelength) |
| Kilo | k | thousand | $10^{3}$ | Kilohertz (radiofrequencies), kilometers (wavelengths of radiofrequency and extremely low-frequency radiation), kilowatt (power level) |
| Mega | M | million | $10^{6}$ | Megahertz (higher microwave frequencies), megawatt (power level), megajoule (energy level) |
| Giga | G | billion | $10^{9}$ | Gigahertz (higher microwave frequencies), gigawatt (power level) |
| Tera | T | trillion | $10^{12}$ | Terahertz (IR and visible frequencies) |
| Peta | P | quadrillion | $10^{15}$ | Petahertz (UV frequencies) |
| Exa | E | pentillion | $10^{18}$ | Exahertz (ionizing radiation frequency) |

Not only do the fields change polarity and strength, but they are rigidly bound together. If the electric field is pointing in some direction, then the magnetic field must move in a direction perpendicular to the electric field and the wave must travel in yet another direction perpendicular to the other two, as shown in Figure 11–5. Note how the shape of the fields is similar to that of waves on the ocean. They can be drawn by plotting the trigonometric sine or cosine functions and are often called sine waves.

In air and a vacuum, the speed of electromagnetic radiation is 299,792,458 m/s, or about 186,000 mi/s. The speed of light is often rounded off to 300,000,000 m/s. An important way of expressing frequency is the distance the wave travels through one cycle, or the wavelength (l). The amount of time elapsed during one cycle is called the period, which is calculated by dividing 1 by the frequency, in Hz. The frequency times the wavelength must equal the speed of light, so as frequency increases, wavelength decreases. Wavelength can be calculated by dividing the speed of light, 300,000,000 m/s, by the frequency. Thus, radiation with a frequency of 300,000,000 Hz has a wavelength of 1 m; 60 Hz radiation, if encountered, would have a wavelength of 5,000,000 m (about 3,100 mi).

## Electromagnetic Spectrum.

The electromagnetic spectrum is summarized in Table 11–B. Note the immense variation in frequencies. A gamma-ray might have a frequency of about 3,000,000,000,000,000,000,000 Hz (300 EHz, wavelength about 0.1 pm), a helium–neon laser pointer has a frequency of about 474,400,000,000,000 Hz (about 474 THz, wavelength 0.632 μm), a microwave oven has a frequency of 2,450,000,000 Hz (2.45 GHz, wavelength 12.2 cm), a well-known New York AM radio station broadcasts at 710,000 Hz (wave length 422 m), and the U.S. Navy's submarine communications system operates at 76 Hz (wavelength about 3,945 km). This is an immense span of frequencies. Frequency is extremely important because the energy of a parcel of electromagnetic radiation, or photon, is directly proportional to its frequency. The electromagnetic spectrum is divided into four parts in this chapter: Subradiofrequency ranges from 0 to 3 kHz, radiofrequency/microwave (RF/MW) ranges from 3 kHz to 300 GHz, optical radiation ranges from 300 GHz to $3 \times 10^{15}$ Hz, and ionizing radiation exists at higher frequencies. This chapter covers all of the electromagnetic spectrum except ionizing radiation, which is covered in Chapter 10.

The strength of the magnetic field and the strength of the electric field are related to one another. The ratio of the two is set, in accordance with Ohm's law, by the resistance of the medium in which the radiation is traveling. In air and a vacuum, the resistance is 377.

The two fields together transmit power through an area of space that can be expressed in watts of power passing through an area of space or striking a given surface area. It is customary to adjust the units to milliwatts of power per $cm^2$ of area. Radiofrequency and microwave specialists call this power *density;* laser and optical specialists call it *irradiance*. These different terms have the same units and same meaning.

Because power losses are proportional to current, today we use AC electricity. This results from the second basic relationship of electrical engineering: Power (P), in watts, equals current (I), in amps, times resistance (R), in ohms. Recall Ohm's law, which states that volts equals amps times ohms. Combining these two relationships gives the relationship watts equals amps times amps times ohms, or $P = I^2 \times R$. Given that the resistance of air and space is 377 Ω, the relationships given in Table 11–C are widely used by people who do field surveys of radiofrequency and microwave radiation.

**Table 11–B.** Two Views of the Electromagnetic Spectrum

| Frequency (Hz) | | Name | Wavelength (m) | 1 (other units) | Photon Energy (eV) | Type |
|---|---|---|---|---|---|---|
| $3 \times 10^{21}$ | | Gamma rays | $10^{-13}$ | | $10^7$ | |
| $3 \times 10^{20}$ | | "Hard" x rays | $10^{-12}$ | | $10^6$ | |
| $3 \times 10^{19}$ | | | $10^{-11}$ | | $10^5$ | |
| $3 \times 10^{18}$ | x rays | | $10^{-10}$ | 1 Ångstrom | $10^4$ | Ionizing |
| $3 \times 10^{17}$ | | | $10^{-9}$ | 1 nanometer | 1,000 | |
| $3 \times 10^{16}$ | | "Soft" x rays | $10^{-8}$ | | 100 | |
| $3 \times 10^{15}$ | | Ultraviolet | $10^{-7}$ | | 10 | |
| $3 \times 10^{14}$ | Optical | Visible (400–760 nm)* | $10^{-6}$ | 1 µm | 1 | |
| $3 \times 10^{13}$ | | Infrared | $10^{-5}$ | | $10^{-1}$ | |
| $3 \times 10^{12}$ | | | $10^{-4}$ | | $10^{-2}$ | |
| $3 \times 10^{11}$ | | Millimetric microwaves | $10^{-3}$ | 1 mm | $10^{-3}$ | |
| $3 \times 10^{10}$ | | | $10^{-2}$ | | $10^{-4}$ | |
| $3 \times 10^9$ | | Microwaves | $10^{-1}$ | | $10^{-5}$ | |
| $3 \times 10^8$ | | | 1 | 1 m | $10^{-6}$ | Nonionizing |
| $3 \times 10^7$ | Radiofrequency | | 10 | | $10^{-7}$ | |
| $3 \times 10^6$ | | | $10^2$ | | $10^{-8}$ | |
| $3 \times 10^5$ | | | $10^3$ | | $10^{-9}$ | |
| $3 \times 10^4$ | | | $10^4$ | | $10^{-10}$ | |
| 3,000 | | | $10^5$ | | $10^{-11}$ | |
| 300 | | Subradiofrequency ELF | $10^6$ | | $10^{-12}$ | |
| 30 | | | $10^7$ | | $10^{-13}$ | |
| 0 | | DC | — | | 0 | DC |

* LIA/ANSI Z136.1-1993 lists the range of visible radiation as 400–700 nm. The eye can perceive longer-wave radiation, but the visual response to it is poor.

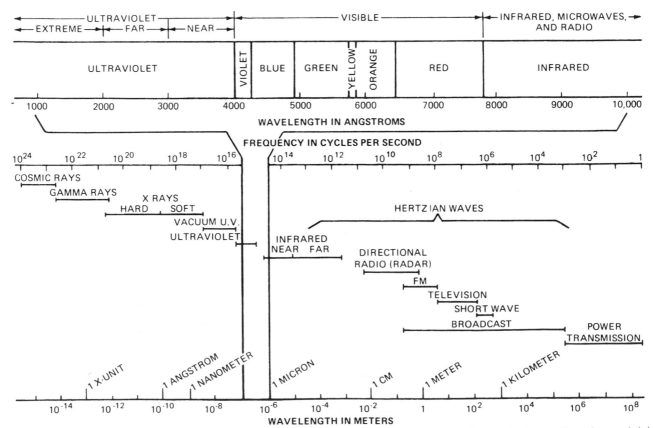

The electromagnetic spectrum, encompassing the ionizing radiations and the nonionizing radiations (expanded portion and right). Top portion expands spectrum between $10^{-7}$ and $10^{-8}$ m. Note: cycles per second (cps) = hertz (Hz).

**Table 11–C.** Relationships Between Electric Field Strength, Magnetic Field Strength, and Power Density (Irradiance) for Electromagnetic Radiation

$$S_m = E^2/3{,}770$$
$$S_m = 37.7\,H^2$$
$$E^2 = 3{,}770\,S_m$$
$$E = (3{,}770\,S_m)^{1/2}$$
$$H^2 = S_m/37.7$$
$$H = (S_m/37.7)^{1/2}$$

$S_m$ = Power density in units of mW/cm
$E^2$ = Electric field strength$^2$ (V$^2$/m$^2$)
$E$ = Electric field strength (V/m)
$H^2$ = Magnetic field strength$^2$ (A$^2$/m$^2$)
$H$ = Magnetic field strength (A/m)

# When Are Fields Important and When Is Radiation Important?

This question is important in the radiofrequency and microwave portion of the spectrum for two reasons:

- The strengths of the fields in radiation are rigidly related to each other so only one field must be measured (usually the electric field). Both the electric and magnetic field must be measured where radiation does not exist.

- Radiation obeys the inverse square law unless the source happens to be a laser, so a field strength measurement at one location can be used to calculate field strengths at other distances in the same direction from the source.

Separate fields are commonly found at lower frequencies (longer wavelengths). Radiation is commonly found at higher frequencies (shorter wavelengths). As a rule, separate fields are found within one to a few wavelengths of a source. Light has wavelengths ranging between 400 and 760 nm, so light is always found as radiation. A 60-Hz wave has a wavelength of 3,100 miles, so it follows that fields, rather than radiation, are found around power lines. Electrical engineers refer to separate fields as near fields (because they are found close to sources) and to places where radiation is found as far fields. Near fields are divided into *reactive* near fields, which are electric fields created by the voltages and magnetic fields created by current flows in the source, and *radiating* near fields, where electric and magnetic fields combine to form radiation that travels away from the source. The relationship between electric and magnetic field strengths in both near fields is not rigid, so both fields must be measured separately. Electrical engineers have formulas and rules for calculating the distance from specific antennas where far fields

exist. It is customarily assumed that separate fields exist when the wavelength is more than about 1 m (frequency of 300 MHz) and that radiation exists at shorter wavelengths or higher frequencies. As a result, *separate electric and magnetic field surveys are needed for frequencies of 300 MHz and less.*

## Parts of an Electromagnetic Device

Any electromagnetic device, whether it is a microwave antenna, junkyard magnet, or laser, can be visualized as having three parts: a source, a transmission path, and a receiver. A laser scanner in a CD player has a laser embedded in the device, optics to transmit the energy to the disk, and a sensor to receive the reflected energy. A 27.12 MHz plasma etcher has an energy source, a cable to transfer the energy, and a chamber where the energy is deposited to do work; these are often all located in one cabinet. A surveyor always needs to be aware of these parts and their locations. Particular attention should be paid to the transmission path, which could pass through open air, an enclosed passage such as a microwave waveguide that could leak at joints or connections, or a fiber-optic cable that could be cut or broken.

# Subradiofrequency Fields: 0–3,000 Hz

Because the distance where radiation is dominant is about 1 wavelength, fields, rather than radiation, are considered in this section. The wavelength of the highest frequency in this section, 3 kHz, is 100,000 m, or about 61 mi. Static fields do not produce radiation. As shown in Table 11–D, the subradiofrequency portion of the spectrum includes the extremely low frequency (ELF) band, which includes AC fields and radiation up to 300 Hz, and the voice frequency band, which includes frequencies from 300 to 3,000 Hz. Power lines use 50 or 60 Hz currents, which create fields at these frequencies; these frequencies are often called power frequencies.

## Field Strengths

Electric fields can be measured by inserting a displacement sensor, a pair of flat conductive plates, into the field and measuring the electric potential between the plates. The electric field lines land on one plate and create voltage that drives a current through the meter to the other plate, where the field lines continue as shown in Figure 11–6. The electric field can be calculated by dividing the voltage between the two plates by the distance between the plates. If a 1-kV potential were found to exist and the plates were 1 m apart, then the electric field strength would be 1 kV/m; if the plates were 0.5 m apart and the potential were 1 kV, then the electric field strength would be 2 kV/m. Instruments

**Table 11–D.** Uses of Radiofrequency and Subradiofrequency Fields

| Frequency, ν | Wavelength, l | Name | Uses | ISM Bands Center Frequency and ± Range | Radar Band and Frequencies | WWII Radar Bands and Frequencies |
|---|---|---|---|---|---|---|
| *>300 GHz* | *<1mm* | *Infrared* | | | | |
| 300 GHz | 1mm | EHF | Satellite communications, radio relay, navigation aids | None | M 60–100 GHz<br>L 40–60 GHz | H 44–56 GHz<br>Q 36–46 GHz |
| 30 GHz | 1 cm | SHF | Satellite communications, radar, fire, police speed guns (24.15 GHz) | 22.125 GHz<br>± 0.125 GHZ<br>5.86 GHz<br>± 0.075 | K 20–40 GHz<br>J 10–20 GHz<br>I 8–10 GHz<br>H 6–8 GHz<br>G4–6 GHz | Ka 22–36 GHz<br>X 5,2–10.9 GHz<br>C 5.9–6.2 GHz |
| 3 GHz | 10 cm | UHF | TV 14–82, CB, taxi dispatch, radar, ovens, cellular phones | 2.45 GHz<br>± 0.05<br>915 MHz<br>± 0.025 | F 3–4 GHz<br>E 2–3 GHz<br>D 1–2 GHz | S 1.55–5.2 GHz<br>L 0.36–1.55 GHz |
| *>300 MHz* | *<1m* | *Microwaves* | | | | |
| 300 MHz | 1 m | VHF | TV 2–13, FM radio, fire, police | 40.68 MHz<br>± 0.02 | B 250–500 MHz<br>A 0–250 MHz | P 220–390 MHz |
| 30 MHz | 10 m | HF | CB radio, diathermy (VDT flyback fields) | 27.12 MHz<br>± 0.16<br>13.56 MHz<br>± 0.00678<br>(both used for plasma etch) | A | None |
| 3 MHz | 100 m | MF | AM radio, amateur radio, navigation aids | None | A | None |
| 300 kHz | 1 km | LF | Navigation aids, marine and long-range communications | None | A | None |
| 30 kHz | 10 km | VLF | Communications, long-range navigation, induction heating (flyback rate of VDTs) | None | A | None |
| *<3 kHz* | *>100 km* | *Subradiofrequency* | | | | |
| 3 kHz | 100 km | Voice | Modulation, induction heating | None | A | None |
| 300 Hz | 1,000 km | ELF | Submarine communications, induction heating (electric power, refresh fields of VDTs) | None | A | None |
| 0 Hz | DC | | | | | |

like this are widely used for measuring electric fields at frequencies ranging up to 100 kHz.

Magnetic fields are often measured with loops of conducting wire, as shown in Figure 11–7. The lines of magnetic field passing through the loop induce current flow. The field can be calculated by measuring the amperes of induced current and dividing that by the circumference of the loop. The common unit of magnetic field strength in the United States was the gauss (G), but it is being replaced by an SI unit, the tesla (T): 1 T = 10,000 G, 1 mT = 10 G, and 1 μT = 10 mG. One G ≅ 80 A of induced current per meter of circumference of conducting loop (A/m).

## How Do Fields Interact with the Body?

The body is a conductor of electricity, but it does not have well-known magnetic structures. (At the time this chapter was prepared, research suggested that microscopic magnetic structures exist in the human brain.) We have seen that electric charge imbalances are balanced as quickly as possible. Imagine you were below a high-voltage electric transmission line, as shown in Figure 11–8, and the line's polarity was positive. Your body would develop a corresponding negative charge because the positive charge above would attract any moveable negative charge toward it. These charges would come from your body and from the infinite pool of charges you were standing on, the ground. The electric field lines would originate at the power line and land on your body. As the polarity reversed, the charge in your body would subside and reverse; the rush of negative charges into your body would be replaced by an exodus of negative charges, so you would take on a positive charge as the line overhead became negatively charged. The electric field lines are perpendicular to the surface they

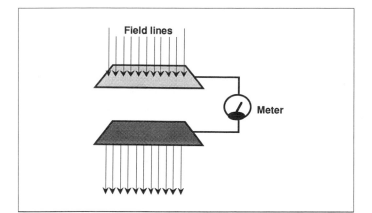

**Figure 11–6.** Displacement (split-plate) electric sensor used for AC field instruments in extremely low-frequency through medium-frequency regions (>0 Hz–3 MHz). Note that field lines impinge on the upper plate, drive a current through the meter, and reemerge at the other plate. The field strength is the voltage difference between the two plates divided by the distance (V/m).

end at or leave. The result of this perpendicularity phenomenon is that a person in an electric field created by a source above the person, such as a power line, draws field lines in from nearby space so they land perpendicular to the head. As a result, the electric field around the head becomes highly concentrated and intense. If this person was holding

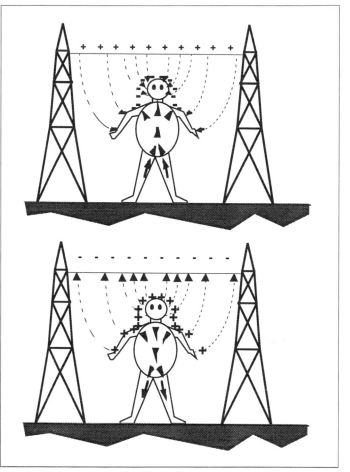

**Figure 11–8.** Induction of current by an external magnetic field. a. The overhead transmission line has a positive voltage. The electric force (dashed lines) induces negative charges to accumulate in the person's body. The negative charges enter the body as a flow of electrical current from the ground. This happens 60 times a second. b. 1/120 of a second later, the line's charge is negative, inducing a positive charge in the body created by a flow of electrical current into the ground. Radiofrequency fields can create similar current flows, but at higher frequencies. (Adapted from Nair, Morgan, and Florig, 1989.)

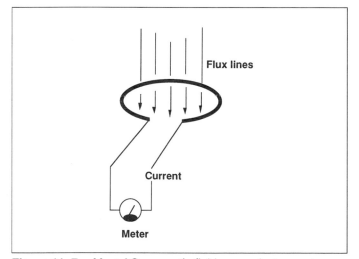

**Figure 11–7.** Most AC magnetic field survey instruments use looped conductors. The changing magnetic field induces a flow of current in the loop. The unit of field measurement is the amperes of current induced to flow through a specified distance of loop (A/m). The induced current is proportional to the sine of the angle between the loop and field. If the loop is parallel to the field, the sine of 0° is 0 and no current is induced to flow. If the loop is perpendicular to the field, as shown, the sine of 90° is 1, so the loop will give a maximum response. Thus, measurements made with one loop are highly directional. Rotating a loop like this in a magnetic field produces a sinusoidally changing current.

an electric field meter at waist height, the surveyor's body would largely shield it and the reading would be falsely low. The reading could also be falsely increased if the surveyor was the tallest object in the area and the field source was above the surveyor.

Magnetic fields are easier to measure because the human body does not perturb the magnetic field as it perturbs the electric field. We are filled with conductive brine, so a magnetic field that changes with time induces loop-shaped current flows in a direction perpendicular to the field, as shown in Figure 11–3. The strength of the induced current is proportional to the strength of the magnetic field and the radius of the loop in which the current is flowing. The current at the center of the loop is zero and reaches a maximum level at the rim of the loop.

# Biological Effects and Standards for Steady (DC) Electric Fields

Steady fields are created with charges and currents that do not change polarity or strength with time. Steady electric fields are not a significant area of concern today; in fact, no instrument is marketed that is intended for use in safety surveys for DC fields (although there are instruments that can be adapted for exposure surveys). The adverse effects identified by the Non-Ionizing Radiation Committee of the International Radiation Protection Association (now the International Commission for Non-Ionizing Radiation Protection, or ICNIRP) are as follows:

- Irritating sparks could occur at electric field strengths of 5 kV/m or more.
- Painful sparks could occur at electric field strengths of 15 kV/m.

The 1994–1995 Threshold Limit Value (TLV) for DC electric fields from 0 Hz to 100 Hz is 25 kV/m as a ceiling limit (a limit that should not be exceeded for any length of time). The field strengths in the TLV are the field levels present in air away from the surfaces of conductors (where spark discharges and contact currents pose significant hazards). It applies to both partial-body and whole-body exposures.

# Biological Effects and Exposure Standards for Static Magnetic Fields

Magnetic effects are caused by charges in motion or changes in magnetic fields. The static fields from permanent magnets or superconducting magnets do not change with time, so any interaction would need to occur where charges are in motion. Blood is briny material, so magnetic field effects can be expected in the circulatory system where charges are in motion. The effect can be seen on an electrocardiogram (ECG) made while an animal is in a static magnetic field. The entire output of the heart is pumped into the aorta at high speed. This flow induces a magnetohydrodynamic (MHD) voltage that appears on the ECG at the same time as the T wave of the heart. The added MHD voltage makes the T wave look bigger. The bigger the magnetic field, the bigger the T wave; when the magnetic field is shut off, the T wave reverts to normal.

Based on this effect, the current TLV is 60 millitesla (mT), or 600 G, which is based on limiting the average MHD voltage to 1 mV. It is an 8-hour time-weighted average (TWA) criterion. The peak exposure is set at 2 T, based on medical imaging exposure limits. The International Commission for Non-Ionizing Radiation Protection

announced new guidelines for static magnetic fields in the January 1994 issue of *Health Physics* that retain the 2 T peak exposure limit but raise the TWA exposure limit to 200 mT (2,000 G). This change accounts for MHD voltage and is based on human experience and on the results of lifetime exposure studies involving mice exposed to 1 T, which demonstrated no effects. Humans exposed to 4 T fields are reported to experience symptoms such as nausea, metallic taste in the mouth, dizziness, and, when the head is moved, magnetic phosphenes.

Another effect is worth noting. Rotating the earth's magnetic field, which is usually about 0.5 G, disrupts the circadian rhythms of test animals and fluctuations in body function that are associated with time of day, such as hormone levels and body temperature. This is related to our sense of time, which is affected by jet lag. This effect occurs if the animal is unaware of the day–night cycles around it, but even 9 T fields didn't break the sense of time of animals aware of light–dark cycles. People who work in strong magnetic fields should avoid unusual shift work that causes them to lose track of day–night cycles.

## Other Safety Concerns of Static Magnetic Fields

Before leaving static magnetic fields, it is necessary to consider two related issues: the effects on medical electronic devices and classic safety concerns. Artificial cardiac pacemakers work by amplifying the electrical activity of the natural pacemaker tissues of the heart. The electrical output of the pacemaker changes as the electrical activity of the heart's pacemaker tissue changes. Artificial cardiac pacemakers can be fooled by ambient electric and magnetic fields, including very strong AC fields. When this happens, they could amplify the AC fields instead of boosting the activity of the natural pacemaker and the heart, now trying to work at 60 Hz, would not circulate blood. Pacemakers have built-in protective circuits that sense malfunctions and cause the pacemaker to send impulses to the heart at a fixed rate. The wearer tests these circuits in a cardiologist's office (or at home while in telephone/telemetry contact with a cardiologist) by changing the setting of a reed switch in the artificial pacemaker so it fires at the fixed rate. A permanent magnet held over the pacemaker is used to reset the reed switch. Thus, strong static fields could cause this reed switch setting to inadvertently change in places where no medical supervision is available. Magnetic fields stronger than 0.31 mT (3.1 G) could reset some very susceptible pacemakers, so a pacemaker safety criterion of 0.5 mT (5 G) has been set; this criterion may be revised to 0.3 mT in the near future.

Magnetic fields exert forces on objects that can be magnetized. The degree to which material can be magnetized depends on its permeability. Aluminum, stainless steel, plastics, and organisms are not permeable. Soft iron, steel, and various transition metal alloys (such as those of nickel and cobalt) are permeable. Tools and some medical implants are made of permeable alloys so they can move in a strong magnetic field. The force of such a field is proportional to the strength of and gradient of the field (the rate at which that field changes intensity through some part of space). A steel wrench in a perfectly uniform super-strong magnetic field would not move because there is no gradient, but it would move as it was taken from the strong field through a weak field. The force is also proportional to the object's permeability and volume. A nonpermeable stainless steel wrench would not move even in a strong and rapidly changing field. Thus, tool and compressed gas cylinder controls, and limits on workers with metal prosthetic implants, are needed. The simplest test devised to locate hazardous locations is to tie a washer or other small permeable object to a string and tie off the other end to a belt loop and watch for places where the washer is pulled out by magnetic fields. Stainless steel, widely used in prosthetic implants, is normally nonpermeable but can be made permeable where it is machined. ICNIRP advises that mechanical hazards due to flying tools and movement of metallic medical implants become a potential hazard when fields are as low as 3 mT (30 G). ICNIRP also advises that magnetic media, such as the magnetic stripes on the backs of credit cards and diskettes, can be erased by fields above 1 mT (10 G).

## ■ Biological Effects and Exposure Standards for Time-Varying Subradiofrequency Fields

Time-varying fields have emerged as an area of concern because of widespread fear that they can cause cancer. The initial concern arose as a result of epidemiological studies. These studies are not discussed further here because they either found no effect or effects in such small numbers of cases as to leave reasonable doubt. A large body of research results from in vivo and in vitro experiments has accumulated about the possible biological effects of such fields, but much of it is contradictory. The Bioelectromagnetics Society issued a primer on biophysics and exposure systems to assist researchers in 1993.

It is now generally, but not universally, accepted that power line fields influence cell membranes. This effect is observable as an increase in the rate at which calcium ions are moved from the inside of a cell through the cell membrane to the outside. The possible effects of these interactions are potentially far-reaching and include promotion of tumor growth, but adverse effects have not been demonstrated in replicated experiments at the time this chapter was written.

It is also generally accepted that power line fields affect circadian rhythms much as static fields do. Recent research suggests that the prolonged exposures do not exert this effect; rather, the rapid changes that occur when the field is turned on and off exerts the effect. Circadian rhythm is controlled by the pineal gland, located in the center of the brain. The pineal receives electrical impulses from the eyes via the superchiasmatic nucleus, located where the optic nerves cross. The pineal secretes melatonin at night. It has been speculated that altered pineal function could reduce the ability of the immune system to eliminate infections or tumors, but this has not been proven at the time this chapter was written.

Before setting science-based exposure standards, it is necessary to demonstrate a mechanism by which the fields exert their effects on living tissue. (At the time this chapter was written, it was becoming evident that exposure standards could be set by litigation.) Many possible mechanisms have been proposed, but none have been proven to work at the time this chapter was written. This uncertainty about mechanisms is important because some of the proposed mechanisms do not involve either of the dose–response models mentioned in Chapters 6 and 10. It is possible that moderate field intensities are hazardous whereas stronger or weaker fields are not hazardous. Regulations have assumed that less is safer, so the possibility of windowed effects is something regulators have not seen before.

The standards that do exist are based on the currents induced in the body. According to WHO Environmental Health Criterion 69, Magnetic Fields, the following associations can be drawn between induced currents and reported biological effects:

| Induced Current Density (mA/m²) | Effect |
| --- | --- |
| <1 | None established |
| 1–10 | Minor biological effects |
| 1–10 | Magnetophosphenes, possible nervous system effects, enhancement of bone fracture healing |
| 1–10 | Changes in CNS excitability, stimulation thresholds, possible health hazards |
| >1,000 | Extrasystoles, possible ventricular fibrillation, definite health hazards |

Health regulations fall into two broad categories: exposure criteria, which specify how much a person can be exposed to, and emission criteria, which describe how much can be released into or be present in the environment near a source. Both types of regulations exist for nonionizing radiation. ICNIRP has issued general public and occupational exposure criteria based on induced current flow considerations, as shown in Table 11–E. The American Conference of Governmental Industrial Hygienists (ACGIH) also issued TLV exposure criteria for workers, based on avoiding induced currents that are stronger than those that already exist in the body due to the normal functioning of nerves and muscles. A number of states have established emission criteria for power transmission lines by specifying maximum fields that can exist along the edges of the right of way occupied by a power

**Table 11–E.** Exposure Criteria for DC and Subradiofrequency Fields

| Frequency | Exposure Group | Exposure Duration | Exposed Part of Body | Exposure Criterion Electric (kV/m) | Magnetic (mT) |
|---|---|---|---|---|---|
| Static | Occupational | 8-hr shift | Trunk | | 200* |
| Static | Occupational | <8-hr shift | Limbs | | 5,000 |
| Static | Occupational | Ceiling | All | | 2,000 |
| 60 Hz | Public | 24-hr day | All | 5 | 0.1 (80 A/m)[†] |
| 60 Hz | Public | <24-hr** | All | 10 | 1 (800 A/m)[†] |
| 1 Hz–300 Hz[††] | Occupational | Ceiling[‡] | All | | 60/f (600/f G)[‡] |
| 1 Hz–300 Hz[††] | Occupational | Ceiling[‡] | Extremities | | 300/f (3,000/f G)[††] |
| DC to 100 Hz | Occupational | Ceiling[‡] | All | 25 | |
| 100 Hz–4,000 Hz[§] | Occupational | Ceiling[‡] | All | 2,500/f[‡‡] | |

* Time-weighted average.
** Peak allowed for times below 24 hrs as long as the average exposure during an 8-hr shift does not exceed 5 kV/m or 0.1 mT (1 G).
[†] 1 mT = 10 G $\cong$ 800 A/m.
[††] ACGIH considers frequencies <30 kHz to be subradiofrequencies, but their standard matches the 163 A/m standard of ANSI/IEEE C95.1-1991 at a frequency of 294 Hz, which ACGIH rounds up to 300 Hz. ACGIH advises that pacemakers can malfunction at field strengths of 0.1 mT (1G), but only at 50 and 60 Hz.
[‡] Maximum exposure allowed for any time period.
[‡‡] Frequency in Hz.
[§] ACGIH considers frequencies <30 kHz to be subradiofrequencies, but their standard matches the 614 V/m standard of ANSI/IEEE C95.1-1991 at 4,071 Hz, which ACGIH rounds down to 4,000 Hz. ACGIH advises that pacemakers can malfunction at field strengths of 1 kV/m, but only at 50 and 60 Hz.

transmission line. These tend to be around 20 µT (200 mG) for magnetic fields and 1 kV/m for electric fields.

The clamor for action has prompted researchers and regulators to develop a number of ideas. Two are worth noting. The first, prudent avoidance, relies on reducing magnetic fields when possible by means that do not involve great expense, similar to the As Low As Reasonably Achievable (ALARA) concept applied to reducing ionizing radiation exposures. The second, 2 mG, is an averaged exposure level used to define high average exposure scenarios in epidemiological studies. It may not define the field strengths that might actually cause harmful effects; such levels could be much higher. Thus, caution is advisable when deciding whether to apply the prudent avoidance or 2 mG approaches in the absence of definitive scientific data.

Before leaving biological effects, it is necessary to return to the subject of pacemakers. ICNIRP advises that power frequency AC fields could interfere with normal pacemaker function at field strengths of 2.5 kV/m and 1 G. This is because such fields create potentials that the pacemaker could confuse with the heart's natural electrical activity, which has a similar ELF frequency. ACGIH advises that pacemakers could be influenced by 1 kV/m electric fields and 1 G magnetic fields. There is no generally accepted guidance about how to address this potential hazard.

## ■ Measuring Subradiofrequency Fields

Electric fields are measured using variations of the displacement sensor described earlier. The surveyor must stand away from the detector when measuring electric fields because the surveyor's body will shield the detector and create a falsely low measurement. Thus, these instruments come with long nonconductive handles so they can be held as far from the operator's body as possible. One vendor supplies a long pole, one supplies a long handle and instructs the surveyor to hold the detector as far away as possible and to use a fiber-optic readout, and a third uses an electrician's hot stick. The sensors of these instruments resemble paddles or clamshells, with a nonconducting seam separating the shell halves. These instruments are directional; that is, they do not respond equally to fields coming from all directions. These instruments must be held so that the electric field is perpendicular to the paddle or to the seam between the clam shells. The reading reaches a peak level when the sensor is aligned properly.

Magnetic fields are measured with loops. The induced current can be boosted by increasing the number of wire turns in the loop or by putting a core of permeable material in the loop. Loops are highly directional. The current induced in a loop reaches a peak value when the loop opening is perpendicular to the flux lines of a magnetic field and does not respond when the opening is parallel to the lines of flux. The response is proportional to the sine of the angle between the loop opening and the flux lines. (The sine of 0° is 0, the sine of 90° is 1, and sine values cannot be higher than 1.)

When using a single-loop detector, it is necessary to either know how the field is oriented (from a knowledgeable person or other reliable source) or to measure the field with the detector pointed in one direction, then in another perpendicular direction, and finally in a third direction perpendicular to the other two. The center of the detector must be the same for these three measurements. The field is estimated by taking the squares of each measurement, adding the squares, and by taking the square root of the sum of the squares. This calculation is easily done using a spreadsheet.

A word of caution about loops: An instrument using a loop or coil sensor will display the average magnetic field in the area surrounded by the loop; magnetic fields at specific points in the loop may be different.

Another method of magnetic field measurement relies on the Hall effect. An object in a magnetic field will develop a voltage in a direction perpendicular to the magnetic field, which can be measured. Hall effect sensors are less sensitive than loops, but are used for DC field surveys where concern begins at a few G, whereas AC field concerns begin at 1 mG or less. Hall effect instruments also read super-strong tesla strength fields. Hall effect probes come in axial response and transverse response types. Transverse response probes are handled just like the single-axis loops just mentioned; peak response occurs when the field is perpendicular to the flat side of the tip of the sensor blade. Axial probes are aligned to give peak response when the field is parallel to the long axis of the probe, so the measurement protocols mentioned earlier must be modified to account for the fact that the peak response occurs when the field is parallel with, rather than perpendicular to, the probe. A personal dosimeter is now available that has three mutually orthogonal Hall effect sensors for isotropic response.

Loops are often ganged together in mutually orthogonal arrays of three loops so that the detector does not operate in a directional manner; such an instrument is shown in Figure 11–9. Magnetic field survey instruments using orthogonal triple-loop detector arrays are now available for all frequencies of interest. Orthogonal loops (or Hall effect sensors) confer isotropic response; that is, the response is about equal for all probe orientations in the field and the sensor is largely nondirectional.

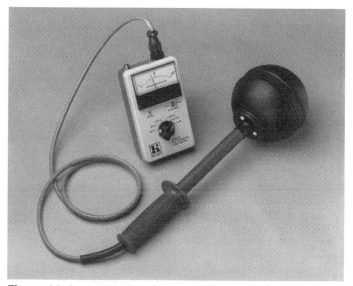

**Figure 11–9.** Loops are often ganged together in mutually orthogonal arrays of three loops. (Courtesy Holaday Industries, Inc.)

Field loggers, often called dosimeters, are available for ELF magnetic field measurements. The EMDEX loggers for ELF fields were developed for the Electric Power Research Institute and closely resemble modern audio dosimeters in the kind of data they provide. They use mutually orthogonal loop sensors (for time-varying fields). The EMDEX loggers read out through a computer that uses or simulates MS/DOS to provide a minute-by-minute summary of exposures in all three axes plus the overall field. Several vendors offer wheeled harnesses for EMDEX loggers, which can be used to precisely measure off locations when the logger is used to log field strengths at various points in a measurement area.

## Controls and Shielding

Stopping electric fields is easy. Imagine that a person is standing below a power transmission line. If a sheet of conductor were placed between the person and the source and that conductor were grounded, induced charges would flow between the conductor and the earth. As a result, the field lines would begin at the conductor above the person, induce current flow in the sheet through the ground, and not reach down to the person. The material can be solid, but mesh will do as long as the opening of the mesh is smaller than about 1/4 wavelength. This means that chicken wire will block a 60 Hz electric field. Operational shields must be grounded in accordance with electrical safety codes. A practical result of the ease of blocking electric fields for surveyors is that the electric fields in most structures are created by appliances inside the structure, even if a high-voltage transmission line passes overhead.

Shields used to block magnetic fields are different from other forms of shielding, which work by stopping an agent with a barrier. Magnetic fields are controlled using permeable alloy that confines the magnetic flux lines and diverts them. Recall that magnetic fields exist as circuits; they do not reach out into space as electric field lines and radiation do. Magnetic shielding can be made using high nickel alloys called mu metal or soft iron. Forming mu metal into complex shapes is expensive and mu metal is easily damaged. Magnetic field shielding alloys are less permeable at low field strengths than at high field strengths, so they work best at high field strengths. Such shielding is best applied near the field source, whenever practical. Materials that work at lower magnetic field strengths could become commercially available in the future. Another approach is to use nonpermeable metals such as copper or aluminum to produce eddy currents that cancel out the original magnetic field.

Exposures of people to magnetic fields is routinely but unintentionally reduced by canceling fields, as in appliance cords with closely spaced conductors. If a current flows in one direction through a conductor, say from the power outlet in a wall to an appliance, and back in the opposite direction through the conductor next to the first one, each conductor creates a magnetic field, but the orientations of the fields are opposite and the fields cancel each other out. The closer the two conductors are, the more cancellation occurs. Overhead power lines use widely spaced conductors to avoid arcing, so relatively little cancellation occurs compared with underground power lines, where the insulated conductors are close together. This is why magnetic fields around underground lines are lower than those of overhead lines. Soil has no shielding value for magnetic fields although, as a conductor, it does block electric fields.

Field cancellation technology is being increasingly used for AC fields. Some low-field video display terminals (VDTs) use it in the form of additional field coils next to the coils that steer the electron beam to create a canceling field. Utilities can rewire transmission lines to obtain more cancellation. One utility company has developed a technique for creating canceling fields around transmission lines. Field cancellation may also prove useful in households where poorly wired appliances are leaking currents to the ground. The current leaving a house through the ground often does not enter where the current entered the house, so it cannot cancel the magnetic fields created by the incoming current. A house with leaking appliances can have two magnetic hot spots: one where the service enters the house and the other by the ground carrying the leakage current, as shown in Figure 11–10. Any changes made to household wiring to reduce magnetic field exposure must comply with electrical safety codes. It may be necessary to change electrical safety codes if 60 Hz magnetic fields prove to be hazardous.

# ■ Radiofrequency/Microwave Radiation and Fields (RF/MW)

This portion of the electromagnetic spectrum covers a huge range of frequencies ranging from 3 kHz to 300 GHz or wavelengths ranging from 100 km to just 1 mm. Recall the earlier section about when fields are important and when radiation is important. Some RF/MW scenarios involve fields and others involve radiation. It is generally agreed that radiation is likely to be found at frequencies above 300 MHz (wavelength = 1 m) whereas fields are likely to be found at lower frequencies. The practical consequence of this is that two surveys, one for electric fields and the other for magnetic fields, are required at lower frequencies but only one survey is needed at higher frequencies.

See Table 11–D for a summary of divisions and uses of the RF/MW portion of the spectrum, which includes very low frequency, or VLF (3–30 kHz), low frequency, or LF (30–300 kHz), medium frequency, or MF (300 kHz–3 MHz), high frequency, or HF (3–30 MHz), very high frequency, or VHF (30–300 MHz), ultrahigh frequency, or UHF (300 MHz–3 GHz), super high frequency, or SHF (3–30 GHz), and extremely high frequency, or EHF (30–300 GHz). Microwaves are the portion of the radiofrequency spectrum ranging from 300 MHz to 300 GHz; strictly speaking, the radiofrequencies extend from 3 kHz to 300 GHz.

## Industrial, Scientific, and Medical Bands and Frequency Nomenclature

A band is a part of the entire electromagnetic spectrum. Safety and health specialists often find equipment working in the industrial, scientific, and medical (ISM) bands, which

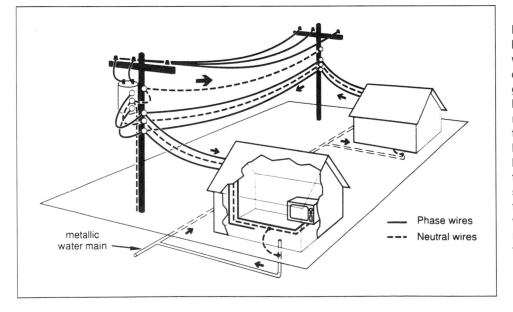

**Figure 11–10.** Current can leave a house next to where it came in, in which case there is good field cancellation. If an appliance is poorly grounded, however, some current leaves through the ground and returns to the utility by flowing through the earth. The current flow into the house is not balanced and canceled by the return current, so the magnetic fields are higher at the unbalanced supply wires and where the current flows into the ground. The phase wires bring the current in and the neutral wires return it. (Reprinted from Electric Power Research Institute, *Electric and Magnetic Field Fundamentals—An EMF Effects Resource Paper.*)

metallic water main ⟶

—— Phase wires
- - - Neutral wires

anyone can freely use because no licensing is required. The most popular ISM band is the 2.45 GHz band used by microwave ovens. Electrical engineers often use the traditional band designations originally used for radars and military electronics in World War II. Thus, you may hear or read about a Ka band police radar speed gun that works at 24.15 or 35 GHz. These designations were not user-friendly; the modern designation system uses letters of the alphabet, in ascending order, to describe increasing frequencies. For example, the MiG-25 radar operates in the J band, which is somewhere between 10 and 20 GHz. Table 11–D lists the ISM bands and the traditional and new band designations.

## Interactions of Radiation and Matter

The interactions of radiation and matter can be described in terms of how much energy in the radiation is lost to the matter it strikes, as shown in Figure 11–11. If all of the energy in the radiation is lost to the matter, it is absorbed. If some energy, but not all, passes from a chunk of matter, then the radiation is scattered because the remaining, less energetic radiation often leaves in a different direction. If none of the energy is lost, then the radiation was transmitted. Electrical engineers call objects such as radio broadcast towers and radar sets emitters rather than transmitters. Finally, radiation does not pass from one medium to another when the electrical properties are too dissimilar. When this occurs, the radiation is reflected back into the medium it came from. Reflection is the basis of radar; controlling reflections from mirrors and other shiny objects is a major concern in laser safety.

RF/MW that is absorbed or scattered can impart energy to living matter by induction of current flow, which in turn encounters resistive parts of tissues such as cell membranes, or by interactions between the electric field and charged portions of water or organic molecules such as proteins. Electrically charged portions of molecules can be

caused to vibrate and ions of metals dissolved in water can be caused to move in response to the electric fields. In both cases, the energy finally appears as heat.

## ■ Biological Effects and Exposure Standards for Radiofrequency Fields

The status of biophysics research in the radiofrequencies is somewhat more certain that it is for the subradiofrequencies. More research is needed to answer concerns about the safety of radiofrequency energy used by devices such as VDTs (which use ELF fields in the 60 Hz range and VLF fields in the 20–30 kHz range), cellular phones (which work around 840 MHz), wireless office technology (which uses the 915 MHz ISM band), and police radars (which have used 10, 24.15, and now 35 GHz). A great deal of research has been conducted and more is under way. The research has been sufficient to develop generally accepted safety standards. Radiofrequency and microwave energy cause a wide variety of biological effects (summarized in Table 11–F), especially if

**Table 11–F.** Biological Effects Reported for Radiofrequency and Microwave Radiation

| Target Organ/ Overall Effect | Effect | Exposure |
|---|---|---|
| Eyes (animals and humans) | Cataracts | Hours at 120 mW/cm$^2$ |
| | Keratitis | 40 mW/cm$^2$ |
| Behavioral (animals only) | Various test changes | ≥1.1 W/kg |
| | Behavioral thermoregulation | ≥1.1 W/kg |
| Endocrine (animals only) | Corticosteroid and thyroid increases | >8.3 W/kg |
| Immune (animals only) | B and T cell activity changes | ≥1.4 W/kg |
| Neurological (animals only) | Tests of blood/brain barrier contradictory | |
| Mutations | Not found in replicated studies to date | |
| Cancer | Not found in humans or animals to date | |
| Reproduction | Temporary male sterility | 5.6 W/kg |
| | Testicular changes | ≥15 W/kg |
| | Leutenizing hormone changes | >2 W/kg |
| Teratology (animals only) | Malformed offspring | ≥31 W/kg |
| Thermoacoustic/inner ear (pulsed only) | Observed in radar operators in WWII as perceived clicking sound while in beam | |
| | Possible cause of neurological effects observed in test animals | 0.6 W/kg |

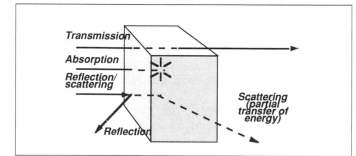

**Figure 11–11.** Electromagnetic energy can be completely absorbed by matter (absorption), pass through without any interaction (transmission), lose some of its energy (scattering), or not enter at all (reflection). Reflection occurs when the electrical properties of the material the radiation comes from and those of the material it strikes are too different. Radar works when electromagnetic radiation in air, a fair insulator, strikes metal, a superb conductor.

exposures are intense enough to cause significant heating. The health significance of doses that are not sufficient to cause measurable heating (athermal effects) is uncertain.

## Dosimetry

It is not enough in radiofrequency biophysics research to state that rats were exposed to 10 mW/cm² of 2.45 GHz continuous wave radiation and certain effects were observed. Researchers must address physics problems that determine how the power density in the rats' ambient environment turns into power deposited in the rats' tissues. The issues to be addressed are the size of the organism compared with the wavelength of the radiation, the polarization of the radiation (how the electric field is aligned to the earth's surface) as compared with the orientation of the exposed organism, and the interaction of exposed tissues with the radiation or fields. The response of tissue to radiation and fields is determined by the electrical properties of the tissue. The discipline that addresses this concern is RF/MW dosimetry.

The dose rate (rate at which energy is transferred to tissue) is called the specific absorption rate (SAR), expressed in watts of power deposited per kilogram of tissue (W/kg). The term for the quantity of energy transferred to tissue is specific absorption (SA), expressed in joules of energy per kilogram (J/kg) of tissue. Note that energy is equal to power times time. Power is expressed in watts and 1 watt times 1 second equals 1 joule of energy. One kilowatt hour of energy use equals 3,600 s × 1,000 W, or 3.6 million J. Joules are used in exposure standards for pulsed laser energy.

An organism acts as an antenna. An object absorbs the most radiation energy if it is about 40 percent of the wavelength and not well-grounded or when it is about 20 percent of the wavelength and well-grounded. Thus, the resonant frequency for a rat is much higher than for a human. Present regulations assume that peak absorption, or resonance, occurs at frequencies of 30–300 MHz. At lower frequencies (<30 MHz), absorption diminishes in proportion to the square of the frequency so the object absorbs 25 percent as much radiation at 50 percent of the frequency and 1 percent as much radiation at 10 percent of the frequency. At higher frequencies (>300 MHz), absorption also falls off for a while and then flattens out at about 8 percent of that at the resonant peak. This is illustrated in Figure 11–12.

Another significant factor is polarization. Radiofrequency radiation is absorbed most when the electric field is parallel to the long axis of the organism and is absorbed least when the magnetic field is parallel to the long axis of the organism, as shown in Figure 11–13. Rats typically best absorb horizontally polarized radiation whereas upright humans best absorb vertically polarized radiation. Some absorption occurs within "H" (magnetic field).

Dosimetry can be done by measurement or mathematically. Measurement dosimetry was hampered by using

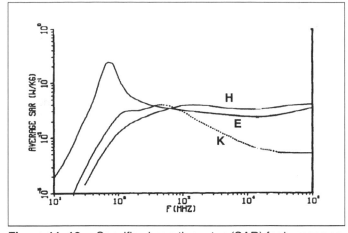

**Figure 11–12.** Specific absorption rates (SAR) for humans as a function of frequency. The letters *E, H,* and *K* show that the radiation is aligned so the electric field axis points along the long axis of the organism (vertically polarized), the magnetic field points along the long axis of the organism, and the radiation travels through the long axis of the organism. *E* = electronic field, *H* = magnetic field, and *K* = direction of travel. Note that there are three major regions. In the *subresonant region,* the body does not function well as an antenna. Absorption drops in an inverse square relationship to wavelength. In the *resonant region,* the body is a good antenna that maximizes absorption. In the *superresonant region* (6–300 GHz), the body no longer acts as a good antenna and absorption levels off at about 10 percent of peak absorption. Quasi-optical focusing occurs from 6 to 15 GHz and skin absorption predominates at >15 GHz (penetration <1 cm).

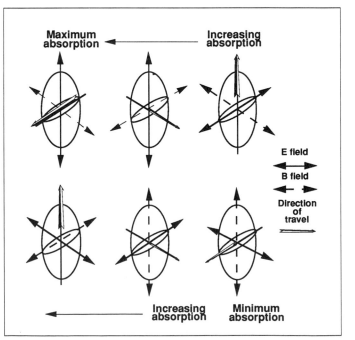

**Figure 11–13.** An elongated object (prolate spheroid) absorbs the most energy when the electric field is parallel to its long axis and the least energy when the magnetic field is parallel to its long axis.

common thermocouples, which include a pair of conducting wires trailing from the object being tested. The wires interact with the electric fields so only one temperature measurement could be made even in a human-sized object. Fiber-optic devices are now available so several concurrent temperature measurements can be made in one animal. Another way of making dosimetric measurements is to sacrifice animals and cut the remains in two longitudinal halves. The halves are joined, irradiated at intense levels, separated, and the temperature change is measured using an infrared device. The irradiation takes place in the same place as the irradiation of live animals, but at higher levels to produce more easily measured heating.

Mathematical dosimetry is also progressing. Early efforts were limited by computer capacity and by the need to use immense amounts of computer memory. Newer mathematical models make more efficient use of computer memory and supercomputers offer adequate capacity to do complex, tedious calculations. The original models used to develop the relationships just described between wavelength and polarization and the size and orientation of the organism simply estimated the average whole body SAR. Newer models estimate the SAR for specific organs and parts of the body. This will make it easier to interpret the results of future animal experiments. The University of Utah pioneered the development of mathematical dosimetry with funding from the U.S. Air Force School of Aerospace Medicine and published a series of dosimetry handbooks.

The leading dosimetric standard in the United States today is 0.4 W/kg as a whole body average. This objective is based on heat avoidance behavior of animals at an SAR of 4 W/kg, divided by a safety factor of 10. Higher local SARs are permitted (8 W/kg for specific parts of the trunk and nonextremity parts of the body and up to 20 W/kg at the extremities).

## Target Organs

A variety of effects are known to occur at SARs above 4 W/kg, but the target organs are the eyes and testes, based on limits of circulation and heat dissipation at these two organ systems. It is assumed that the thermal equilibrium time for these target organs could be as brief as 6 min, so the standards specify exposure limits that apply for 6-min time intervals (until the frequency rises above 15 GHz, where the time interval gradually drops to 10 seconds). The only proven adverse effects for humans, regardless of SAR, are eye cataracts, facial burns, and electric shocks and burns. Cataracts occur in animals at power densities above 120 mW/cm$^2$ if the exposures last for hours, but only minutes of exposure are needed at 350 mW/cm$^2$. Electric shocks and strong current flows through the ankles have occurred in humans and are important at frequencies below 100 MHz. Other effects have been demonstrated in animals, such as teratogenic effects in animals subjected to intense SARs and neurochemical and eye effects in animals exposed to pulsed radiofrequency and microwave radiation, but these effects have not been duplicated in humans.

## Standard-Setting Rationale

We can now see that the radiofrequency portion of the electromagnetic spectrum can be divided into three major parts, as shown in Figure 11–14. At lower frequencies, current flow considerations dominate. The goal of the standard

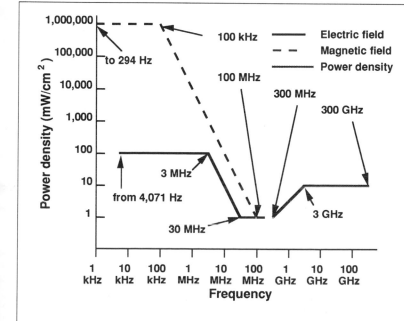

**Figure 11–14.** The inverse of the absorption curves shown in Figure 11–12. These are the C95.1 standards, which are most stringent in the frequencies where absorption is at its peak and level off at higher frequencies. The leveling off at lower frequencies reflects the risk of RF electric burns caused by contact with conductors immersed in strong fields. The *subresonant region* (<3 MHz) is the region in which current flow considerations are most important. In the *resonant region* (3 MHz–6 GHz), current flow and SAR are both important considerations where current is ≤100 MHz. Absorption falls off in proportion to the square of the frequency between 3 MHz and 30 MHz. Standards are most stringent between 30 MHz and 300 MHz. Between 300 MHz and 3 GHz, absorption falls off linearly as frequency increases. The *superresonant region* ranges from 3 GHz to 300 GHz. (From ANSI/IEEE C95.1-1991, 1992.)

is to prevent burns caused by radiofrequency electric current (at frequencies of a few MHz and below) and excessive ankle heating caused by the surge of electricity to and from the ground through the feet (from a few MHz through the resonant frequencies). There was no convenient way to measure contact current until about 1991. Regulators approached the problem by limiting the ambient field strengths allowed to strengths that were equivalent to 100 mW/cm² if the fields existed as electromagnetic radiation. The risk of electrical burns caused by touching an object energized by radiofrequency fields and the risk of significant ankle heating are covered by current flow criteria.

At the resonant frequencies, where our bodies act as good antennas, controlling SAR becomes the main concern and remains a serious concern at higher frequencies until skin absorption becomes dominant. For regulatory purposes, the resonant portion of the spectrum ranges from 3 MHz (for electric fields) or 100 kHz (for magnetic fields) to 3 GHz and is most restrictive at frequencies ranging from 30 to 300 MHz, wavelengths where an adult or child functions as a good antenna.

At higher frequencies (above 3 GHz) skin absorption predominates and mathematical modeling reveals that absorption is about 8 percent of absorption at the resonant peak. Above about 6 GHz, most of the radiation is absorbed in the skin and outer tissue layers; at frequencies above 15 GHz, most radiation is absorbed in the first centimeter of tissue. The properties of microwaves become similar to those of infrared (IR) radiation and thermal damage to the skin becomes a major concern.

The regulatory community had also examined the question of how magnetic fields should be regulated in near-field scenarios. Magnetic fields passing through conductors induce currents in loop shapes, as discussed in the ELF section of this chapter. Based on this and calculations of the current density induced by magnetic fields, it was agreed that magnetic fields would be subject to relatively lenient regulations.

Separate criteria were set for the general public and for occupational exposures. All earlier versions of the Institute of Electrical and Electronic Engineers' C95.1 standard, *American National Standard Safety Levels with Respect to Human Exposure to Radio Frequency Electromagnetic Fields,* did not differentiate between occupational and nonoccupational exposures. For frequencies from 1.34 MHz to 15 GHz, nonoccupational standards are generally five times more restrictive than occupational standards. The Institute of Electrical and Electronic Engineers (IEEE) replaced the general public and occupational standards approach used for chemicals with an uncontrolled access versus controlled access approach, which allowed higher exposures for those who knew they were being exposed and could be provided with hazard awareness information, warnings, and training.

The standards are presented in Tables 11–G, 11–H., and 11–I. These are whole-body average exposures that

---

**Table 11–G.** Exposure Standards for Radiofrequency and Microwave Radiation and Fields Where Access is Controlled

*Part A: Electromagnetic Fields\**

| Frequency Range (MHz) | Electric Field E (V/m)** | Magnetic Field E (A/m)** | Power Density, S (mW/cm²) E | Power Density, S (mW/cm²) H | Averaging Time E², H², or S (min.) |
|---|---|---|---|---|---|
| 0.000294†–0.1 | — | 163 | 100 | 1,000,000†† | 6 |
| 0.004071†–0.1 | 614 | — | 100 | 1,000,000†† | 6 |
| 0.1–3.0 | 614 | $16.3/f_m$ | 100 | $10,000/f_m$†† | 6 |
| 3–30 | $1,842/f_m$ | $16.3/f_m$ | $900/f_m^2$ | $10,000/f_m^2$†† | 6 |
| 30–100 | 61.4 | $16.3/f_m$ | 1.0 | $10,000/f_m^2$†† | 6 |
| 100–300 | 61.4 | 0.163 | 1.0 | 1.0 | 6 |
| 300–3,000 | — | — | $f_m/300$ | $f_m/300$ | 6 |
| 3,000–15,000 | — | — | 10 | 106 | 6 |
| 5,000–300,000 | — | — | 10 | 10 | $616,000/f_m^{1.2}$ |

*Part B: Induced and Contact Radiofrequency Currents*

| Frequency range (MHz) | Maximum Current (mA) Through Both Feet | Maximum Current (mA) Through Each Foot | Maximum Current (mA) Contact |
|---|---|---|---|
| 0.003–1 | $2,000f_m$ | $1,000f_m$ | $1,000f_m$ |
| 0.1–100 | 200 | 100 | 100 |

\* The exposure values in terms of electric and magnetic field strength are obtained by spatially averaging values over an area equivalent to the vertical cross-section of the human body (projected area).
\*\* $f_m$ = frequency in MHz.
† ANSI/IEEE C95.1-1991 covers frequencies between 3 kHz and 300 GHz; at 294 Hz and 4,071 Hz, C95.1 magnetic and electric field standards match the subradiofrequency TLVs.
†† These plane-wave equivalent power density values, though not appropriate for near-field conditions, are commonly used as a convenient comparison with maximum permissible exposures at higher frequencies and are displayed on some instruments.
(From ANSI/IEEE C95.1-1991, 1992.)

**Table 11–H.** Exposure Standards for Radiofrequency and Microwave Radiation and Fields Where Access Cannot Be Controlled

*Part A: Electromagnetic Fields**

| Frequency Range (MHz) | Electric Field E (V/m)** | Magnetic Field E (A/m)** | Power Density, S (mW/cm²) | Averaging Time (min) $E^2$, $H^2$ | Averaging Time (min) S |
|---|---|---|---|---|---|
| 0.003–0.1 | 614 | 163 | † | 6 | 6 |
| 0.1–1.34 | 614 | $16.3/f_m$ | † | 6 | 6 |
| 1.34–3.0 | $823.8/f_m$ | $16.3/f_m$ | † | $f_m^2/0.3$ | 6 |
| 3.0–30 | $823.8/f_m$ | $16.3/f_m$ | † | 30 | 6 |
| 30–100 | 27.5 | $158.3/f_m^{1.668}$ | † | 30 | $0.0636/f_m^{1.337}$ |
| 100–300 | 27.5 | 0.0729 | 0.2 | 30 | — |
| 300–3,000 | — | — | $f_m/1,500$ | 30 | — |
| 3,000–15,000 | — | — | $f_m/1,500$ | $90,000/f_m$ | — |
| 15,000–300,000 | — | — | 10 | $616,000/f_m^{1.2}$ | — |

*Part B: Plane-Wave Equivalent Power Density Values†*

| Frequency (MHz) | E | H |
|---|---|---|
| 0.003–0.1 | 100 | 1,000,000 |
| 0.1–1.34 | 100 | $10,000/f_m^2$ |
| 1.34–30 | $180/f_m^2$ | $10,000/f_m^2$ |
| 3.0–30 | $180/f_m^2$ | $10,000/f_m^2$ |
| 30–100 | 0.2 | $940,000/f_m^{3.336}$ |

*Part C: Induced and Contact Radiofrequency Currents*

| Frequency Range (MHz) | Maximum Current (mA) Through Both Feet | Through Each Foot | Contact |
|---|---|---|---|
| 0.003–0.1 | $900f_m$ | $450f_m$ | $450f_m$ |
| 0.1–100 | 90 | 45 | 45 |

* The exposure values in terms of electric and magnetic field strength are obtained by spatially averaging values over an area equivalent to the vertical cross-section of the human body (projected area).
** $f_m$ = frequency in MHz.
† These plane-wave equivalent power density values, though not appropriate for near-field conditions, are commonly used as a convenient comparison with maximum permissible exposures at higher frequencies and are displayed on some instruments.
(From ANSI/IEEE C95.1-1991, 1992.)

are calculated based on measurements made at 20-cm vertical intervals beginning at the floor and reaching to the head at distances of 20 cm or more from radiating or energized objects.

## Averaging Time and Pulsed Fields

Industrial hygienists are used to thinking in terms of 8-hr time-weighted averages or 15-min intervals for short-term exposure limits. Radiofrequency standards rely on another interval, the thermal equilibrium of the target organs (the eyes and testes), which is assumed to be 6 min. Thus, time-averaging calculations can be made much as they are for chemicals, but the exposures must not exceed the C95.1 limit over a 6-min interval rather than an 8-hr shift. Put another way, a person can be exposed to twice the limit for 3 min if no other radiation exposures occur for the other 3 min. The averaging interval for uncontrolled exposure situations is 30 min for frequencies ranging between 3 MHz and 3 GHz. IR exposures at the end facing microwave radiation are set at 10 mW/cm² for 10 s, so the averaging time changes from 6 min to 10 s for frequencies between 15 and 300 GHz.

**Table 11–I.** Relaxations for Partial Body Exposures for Table 7 and 8* of the ANSI/ IEEE Standard

| | Frequency (GHz) | Peak Value of Mean Squared Field | Equivalent Power Density (mW/cm²) |
|---|---|---|---|
| Controlled Access Exposures | $0.001 \leq f_g < 0.3$ | $<20E^2$ or 20 $H^2$** | |
| | $0.3 < f_g \leq 6$ | | <20 |
| | $6 < f_g \leq 96$ | | $<20 (f_g/6)^{1/4}$†† |
| | $96 < f_g \leq 300$ | | 40 |
| Uncontrolled Access Exposures | $0.0001 \leq f_g \leq 0.3$ | $<20 E^2$ or 20 $H^2$† | |
| | $0.3 < f_g \leq 6$ | | 4 |
| | $6 < f_g \leq 30$ | | $f_g/1.5$†† |
| | $30 < f_g \leq 300$ | | 20 |

* These relaxations do not apply to the eyes and testes.
** E and H are spatially averaged values from Table 2 of the standard.
† E and H are spatially averaged values from Table 3 of the standard.
†† $f_g$ in GHz.
(From ANSI/IEEE C95.1-1991, 1992.)

The idea of limiting above-average exposures so the 6-min time-weighted average is below the standard has limits. Much as the chemical TLV committee has attempted to address the topic of brief but intense exposures above a TLV, the C95.1 committee considered pulsed electromagnetic radiation and fields. Pulsed fields are common and recent military devices that simulate the pulses from nuclear weapons stimulated public concern about pulsed electromagnetic fields. As with chemical exposures, a simple TWA allows for extremely intense exposures if they are brief enough. Thus, an exposure to a 360-ns pulse could be a billion times more than the C95.1 limit if there were no other exposures for 6 min. The resolution has been to limit electric fields to 100 kV/m and to ensure that the power density does not exceed that given by the following formula:

$$\frac{(\text{cw limit of table 7 or 8} \times \text{average time of Table 11–F or 11–G in secs})}{(5 \times \text{pulse width in secs})} \quad (1)$$

The Europeans, led by Finland, are in the process of adopting pulsed field standards. The proposed standard sets limits on electric fields induced in the body by external pulses. Application of this standard entails resolving how pulses behave when they enter tissue and calculating the induced fields that result. Further developments in regulations for exposures to pulsed fields and radiation can be expected.

## Regulatory Considerations

The standard paperback volume of Title 29 *Code of Federal Regulations (CFR)*, Part 1910, the OSHA general industry regulations, includes the original 1970 radiofrequency/ microwave exposure standard, 29 *CFR* 1910.97. This standard, based on ANSI C95.1-1966, is obsolete. The original regulation was struck down by the Occupational Safety and Health Review Commission in 1981, but the *UAW, Brock v. General Dynamics Land Systems Division* decision allows OSHA to apply state-of-the-art standards developed by others when OSHA has no standard of its own. IEEE adopted a revision of C95.1 in 1992 as IEEE C95.1-1991 and the American National Standards Institute (ANSI) adopted the 1991 revision in 1992. Hence, OSHA can enforce ANSI/IEEE C95.1-1991, so the reader should become acquainted with this standard (Figure 11–15).

## Measuring Radiofrequency Radiation and Fields

Recall that fields rather than radiation are presumed to exist at frequencies below 300 MHz so two surveys, one for electric fields and the other for magnetic fields, are needed. Only one survey is needed for frequencies above 300 MHz.

**Figure 11–15.**   The internationally recognized radiofrequency hazard icon, "the radiator." OSHA has not updated its published radiofrequency standard, but has informed IEEE that it will not cite an organization using this symbol instead of the obsolete symbol found in the OSHA regulations.

Electric fields at frequencies above 100 kHz are measured using small dipole antennas. The electric field induces a surge of current in the dipole, which is connected to a diode. A diode is the electrical equivalent of a one-way valve so the induced current is allowed to leave the detector and goes to an amplifier. The amplified current drives a display. Alternatively, the electric field induces current flow and heating in an array of thermocouples and the change of resistance caused by heating is measured to drive the display. Dipoles and diodes can be used at lower frequencies (below about 100 kHz), but displacement sensors are also used.

Most dipole/diode field survey instruments gang three sensors together so they are mutually orthogonal (perpendicular to each other) to provide an isotropic response (a response that is about equal regardless of how the sensor is aligned in the field), as shown in Figure 11–16. Microwave oven survey instruments have two dipole/diode sensors that are perpendicular to each other. The lack of the third sensor makes them directional and they must be held so the probe handle is perpendicular to the surface being measured. When measuring electric fields using a single detector, it is often apparent where the field source is located so alignment of the detector is simple. A displacement sensor is oriented so that its flat surface faces the source; such an instrument is shown in Figure 11–17. Most instruments used for radiofrequency measurements below 300 MHz have displays marked to read in V/m. Instruments operating in the microwave region can be marked to read out in $V^2/m^2$ or $mW/cm^2$; microwave power density meters usually measure the electric field intensity and convert it to power density at the display.

Magnetic fields below 300 MHz are measured with single loops or triple mutually orthogonal loops. Single loops

a.

b.

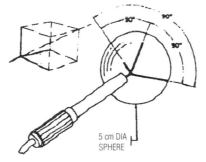

5 cm DIA
SPHERE

c.

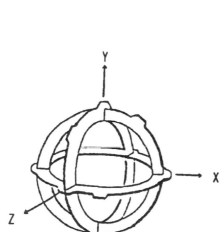

**Figure 11–16.** Isotropic (nondirectional) antenna arrays. a. Mutually orthogonal dipoles respond to electric fields (Holaday Industries HI 3102 probe). b. Mutually perpendicular array of thermocouples (Loral/Narda 8621 series probes). c. Mutually perpendicular magnetic field loops.

**Figure 11–17.** A displacement sensor is oriented so that its flat surface faces the source. (Courtesy Holaday Industries, Inc.)

are highly directional, so the surveyor must follow the protocol on pages 284–285. Triple orthogonal loops are isotropic, so probe orientation is not critical.

Radiofrequency field and radiation measurements should be made where the worker would normally be but while he or she is away so reflections and induced fields do not create falsely high results. (This is similar to noise measurements, which are best done where the ears would normally be, but while the operator is away.)

Either the sensor is held just above floor level and raised up through the body position at 8-in. (20-cm) vertical intervals and the result recorded for each point, or the probe is steadily moved through the body positions using an instrument with an add-on module that averages the readings. The sensor must be kept 20 cm or one sensor dimension (whichever is greater) from energized objects. At this time, measurements must be made at the locations of the eyes and testes. Exposures above the standard and their locations must also be recorded. Excursions up to 20 times the standard are allowed at the extremities (hands and feet) and excursions up to 8 times the standards are allowed elsewhere (except at the locations of the eyes and testes).

Contact and induced current meters are now available. The contact current meter has electrodes that are pressed against an object that may be electrically charged due to ambient RF fields, as well as a cable and built-in ammeter. Induced current meters look like bathroom scales and are meant for people to stand on. The induced current flows through the instrument and drives an AC ammeter before going to ground. One manufacturer sells a cylindrical antenna that electrically simulates an upright human and can be placed on an induced current meter. Soon, an instrument that measures current flow through the ankle with a cuff-shaped coil will be available. At this time, ANSI/IEEE C95.1-1991 specifies that current flows are to be averaged over a 1-s time interval.

# Video Display Terminals and Microwave Ovens

## Video Display Terminals

These are given their own section because they operate in both the ELF and VLF portions of the spectrum. A VDT includes a cathode ray tube (CRT) very similar to the picture tube in a TV set. The CRT has an electron source or electron gun at the back and uses a high-voltage electric field to accelerate a beam of electrons from the gun to the screen and impart enough energy to the beam to cause phosphors on the inside of the screen to glow. The beam is steered to specific parts of the screen by a pair of fields created by magnets in a yoke near the back of the CRT. One magnetic field is aligned vertically the other horizontally. Remember that magnetic fields act at right angles. The vertical magnetic field moves the beam left and right while the horizontal field moves it up and down. The image on the screen contains about 500 lines (this varies with manufacturer) and it is refreshed or replaced with a brand-new image about 60 times a second (which also varies with manufacturer) to avoid flicker. Thus, the magnetic field that deflects the beam vertically has a frequency of about 60 Hz. The field that deflects the beam horizontally has a frequency of 60 images per second times the number of lines in the image (about 500), for about 30,000 Hz. The horizontal scanning field is generated by the "flyback" transformer. Most VDTs work at frequencies below 30,000 Hz, whereas a commercial TV image has 525 lines refreshed only 30 times a second, for a flyback frequency of 15.75 kHz. See Figure 11–18 for a depiction of the fields around a VDT or TV.

If you watched how the field strength of most VDTs changed with time, you would not see the familiar sine (ocean) wave. You would see the field slowly change as the beam was dragged from the left of the screen to the right, then abruptly change to the starting value so the beam could be swept across the next line (the beam itself is actually shut off during this abrupt reversal). The result is called a sawtooth waveform. The refresh field also has a sawtooth shape, but at a lower frequency.

Thus, a survey of VDTs is more complex than many other surveys. The electric and magnetic fields at two frequencies can be assessed. One manufacturer has developed a family of meters using paddle-shaped displacement electric field and single-loop magnetic field sensors for this application. A worker's electric field exposure is assessed by placing the sensor squarely over his or her chest so the electric field lines land on the sensor. Of course, the surveyor must stand away during electric field measurements (a fiber-optic remote readout is provided), but the surveyor can be near the VDT during magnetic field surveys.

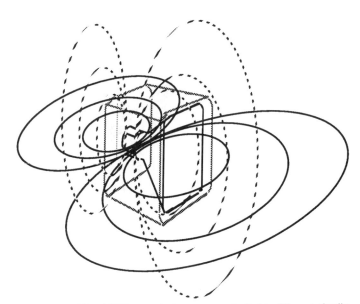

**Figure 11–18.** VDTs create two magnetic fields. The vertically aligned flyback field moves the beam horizontally at a frequency of 20,000–30,000 Hz. The horizontally aligned refresh field moves the beam from top to bottom at a frequency of about 60 Hz.

The Swedish MPR or Swedish Board for Technical Accreditation (SWEDAC) standards are emission standards rather than exposure standards, and are summarized below:

| | |
|---|---|
| ELF electric field | 25 V/m |
| ELF magnetic field | 2.5 mG |
| Electric field | 2.5 V/m |
| Magnetic field | 0.25 mG |

The Swedish standard is given in two Swedish Board for Technical Accreditation publications cited in the Bibliography. It is a convenience standard; that is, it is based on what is attainable rather than on an estimate of what constitutes a safe exposure. SWEDAC wrote:

Knowledge of the biological effects of weak low-frequency magnetic fields is limited. The debate on this that occurred in Sweden during 1986 and 1987 resulted in the publishing of guidelines of acceptable values of fields from VDUs. These guidelines were not based on biological effects, but rather on what was technically possible. The results obtained from the first three years of voluntary testing show that it is possible to obtain VDUs that give rise to only low magnetic field intensities without significantly degrading other performance requirements. It has also been found that, for good-quality VDUs, there does not appear to be any correlation between the magnetic field and the important visual ergonomic characteristics.

SWEDAC specifies conducting a series of measurements at 50 cm in a laboratory with a controlled electromagnetic environment. Any field measurement made to assess approximate compliance with SWEDAC must be made at 50 cm rather than at the 30 cm customary in the United States. TCO, the Swedish Confederation of Professional Employees, has developed guidelines for VDT electromagnetic emissions and ergonomics. The TCO emission criteria are somewhat stricter than MPR's. TCO has established an office in Chicago to support distribution of its guidelines, which have been translated into English.

## Microwave Ovens

Microwave ovens use the 2.45 GHz ISM band. Power is supplied by a magnetron tube and routed to the cooking chamber via a waveguide. The radiation leaving the waveguide passes over a rotor (or rotating waveguide) set in motion by moving air. The rotor is shaped to reflect the radiation around the cooking chamber to obtain a somewhat even distribution of energy. This caused concern when microwave ovens first appeared because the radiation leaking from an oven was amplitude-modulated by the rotor and could be picked up by cardiac pacemakers of that era. The pacemakers could amplify the leaking radiation rather than the heart's own electrical activity, with serious consequences for the user. Since that time, pacemaker manufacturers have added capacitors and protective circuits to block out extraneous fields and most pacemakers are now tested against electric fields equivalent to 10 mW/cm$^2$, the ANSI C95.1-1966 exposure standard in effect when microwave ovens first appeared. The present emission standard for microwave ovens, 21 *CFR* 1930.10, from the Food and Drug Administration, allows new ovens to leak no more than 1 mW/cm$^2$ when measured at 5 cm and old ovens to leak no more than 5 mW/cm$^2$ when measured at 5 cm (microwave oven survey meters have Styrofoam spacer cones to guarantee this 5-cm separation).

Excessive leakage from a microwave oven could occur if the oven was mechanically damaged, so oven users should check the condition of the door, door jamb, vision screen, and interlock when they use an oven and have damaged ovens repaired or replaced. Annual surveys are not warranted.

## ■ RF/MW Controls

RF/MW shielding is relatively simple and can be installed using relatively inexpensive materials. Lead is not needed. The most effective form of shielding is applied at the source as an enclosure. Absorbing foams, porous polyurethane foam filled with carbon in a manner that causes the arriving radiation to pass through gradually lowering impedance, can be used. Electromagnetic radiation does not pass from one medium to another if the difference in electrical conductivities is large; it is simply reflected from the object. Air is a fairly good insulator, whereas metals are

good conductors, so tools such as radar work by causing radiation arriving from air to strike and be reflected by metal. The foam manufacturer makes the impedance change gradually so the radiation is not reflected from the absorber. This can be done by forming the foam into pyramid shapes so the arriving radiation passes the tips and travels towards the bases of the pyramids. The impedance gradually drops as the radiation encounters less air and more carbon. The arriving radiation eases into the foam and induces a flow of electric current in the foam. Graphite is a poor conductor, so the current induced by the radiation is dissipated as heat. The foam manufacturer can also make the carbon content of the foam increase from nothing at the outside of the foam to a maximum level at the base of the foam. Then the foam can be sold as flat sheet because pyramids are not needed.

Metal screens or sheets can also be used for shielding or enclosures. The key is that the mesh openings be no more than 1/4 wavelength in dimension. Metal screens and sheets must be electrically bonded to each other where they join and the whole assembly grounded; otherwise, fields can pass through the gaps. A Faraday cage is a grounded enclosure made of continuously bonded conductors. An object inside is protected from electric fields and radiation on the outside. The doors and door jambs of Faraday cages are lined with resilient alloy strips that make contact when the door is shut so the door is bonded to the jamb and the rest of the cage.

Waveguides and coaxial cables are used to transfer power and act as enclosures. Waveguides are open metallic conduits with flanged ends. The flanged ends allow waveguides to be fastened to other waveguides, sources, or receivers. Waveguides must be snugly attached or leakage will occur. Similarly, coaxial cables can be bent and fail. Leakage can occur where they fail.

Copper tape is most commonly used in electrical engineering. It is used to patch leak points in enclosures or in runs of waveguide. Very-high-power devices have been successfully enclosed in wooden boxes lined on the inside with well-joined copper sheets and absorbing foam and finished outside with copper tape at the seams.

Distance is also very effective in far fields, due to the inverse square law, and can also be used in near fields. Examples of using distance as a control include enclosing a source in a barricade to keep people away from the source or using long-handled tools to manipulate objects immersed in strong radiation fields.

Time limitations are possible, based on a 6-min averaging time, but often are not practical. Another control, the sign found in the OSHA nonionizing radiation standard, became obsolete in 1992; the symbol shown in Figure 11–15 should be used instead of the aluminum, black, and red diamond found in the old OSHA standard. The magenta and yellow trefoil associated with ionizing radiation should never be used!

# ■ Optical Radiation and Lasers

## CIE Bands

Optical radiation includes infrared/visible and ultraviolet radiation. Optical radiation is manipulated with nonconductive optics (with the exception of mirrors, which use metal to reflect the radiation). It is customary to describe optical radiation by wavelength rather than frequency. Nomenclature has been developed by the Commission Internationale d'Eclairage (CIE), or International Lighting Commission. The band designations for infrared and ultraviolet radiation begin at the border of each band starting with visible radiation (wavelength of 400–750 nm) and extend each way from the visible band:

| CIE Band | Wavelength | Non-CIE Nomenclature |
|---|---|---|
| Microwaves | >1 mm | |
| IR-C | 3 μm–1 mm | Far IR: 25μm–1 mm |
| IR-B | 1.4 μm–3 μm | Intermediate IR: 2500 nm–25 μm |
| IR-A | 760–1400 nm | Near IR: 760–2500 nm |
| visible | 400–760 nm | |
| UV-A | 315–400 nm | Near UV, black light |
| UV-B | 280–315 nm | Middle UV, actinic UV |
| UV-C | 100–280 nm | Far UV, actinic UV, Vacuum UV: <200 nm |
| x rays | <4 nm | |

Note that the ANSI standard for laser safety, Z136.1-1993, uses the non-CIE terms *near, intermediate,* and *far* to define IR bands.

# ■ Biological Effects and Exposure Standards for Optical Radiation

Biological effects of optical radiation result from thermal and photochemical mechanisms. (The scientific controversy and uncertainty that plague efforts to protect people from microwave, RF, and ELF do not apply to optical radiation.) Thermal effects are dominant in the IR portion of the spectrum, photochemical effects dominate in the UV, thermal effects are more important at the red end of the visible spectrum, and photochemical effects are dominant at the blue end of the visible spectrum. The target organs are the eyes and skin. Visible and IR-A radiation, 400–1400 nm, is particularly hazardous to the eye because it is transmitted through the cornea, is focused by the lens, and strikes the retina. The potential for retinal damage is great because the radiation can be highly concentrated by focusing.

## The Eye

The exterior of the eye is protected by a transparent layer that contains living tissue, the cornea. The cornea is wetted by continual blinking of the eyelids when they are open. The vitreous humor, a sack of watery material, lies inside the cornea. The cornea and vitreous humor form the ante-rior portion of the eye, which is divided from the posterior region by the iris, which is opaque and pigmented. The iris can absorb and be heated by radiation. The iris contains muscles that adjust the size of the central opening, the pupil, to control the amount of light passing into the rest of the eye. Behind the iris lies the lens, a flexible mass that is made thicker or thinner by the ciliary muscles to provide the sharpest possible image on the retina. It transmits visible and IR-A radiation, but other bands of IR and UV are absorbed here or at the cornea. Light passes through the aqueous humor and strikes the retina, where photosensitive cells convert the energy of light to electrical signals that are fed to the vision centers of the brain and the pineal. There are two types of photoreceptor cells in the retina: rods, which work in a broad range of lighting conditions but cannot differentiate between wavelengths (see colors), and cones, which perceive colors in brightly lit conditions. Rods cover the entire retina, but cones cover only the macula, a small area in the back of the retina. The fovea, a small pit rich in cones and essentially rod-free, is located in the center of the macula directly opposite the center of the iris, which makes color vision particularly vulnerable to overexposures to visible and IR-A radiation. When the lens is removed surgically, creating an aphakic condition, then UV-A that normally is blocked by the lens also passes through and evokes a retinal response. The rods lie on a bed of highly pigmented tissue, the retinal pigmented epithelium (RPE); whereas the rods are essentially transparent and do not absorb light, the underlying RPE can absorb light, become hot, and damage the rods. (See Chapter 5, The Eyes, for anatomical illustrations of the human eye.)

The target portions of the eye are summarized below:

| CIE Band | Wavelengths | Primary Visual Hazard | Other Visual Hazards |
|---|---|---|---|
| IR-C | 1 mm–3.0 μm | Corneal burns | |
| IR-B | 3 μm–1.4 μm | Corneal burns | |
| IR-A | 760–1400 nm | Retinal burns | Cataracts of lens (glassblowers cataracts) |
| Visible | 400–760 nm | Retinal burns | Night and color vision impairment (chronic exposures to intense sunlight) |
| UV-A | 315–400 nm | Cataracts of lens | |
| UV-B | 280–315 nm | Corneal injuries (welder's flash) | Cataracts of lens (at longer wavelengths of UV-B band) |
| UV-C | 100–280 nm | Corneal injuries (welder's flash) | |

See Figure 11–19 for a summary of the CIE bands and target organs of optical radiation.

The blood circulating through the choroid below the retina is an important defense mechanism. The capillaries in the choroid have larger diameters than typical capillaries elsewhere, but the oxygen demand of the living retinal and other eye tissues is modest; this capillary net is a vast liquid cooling system that extracts heat from the retina.

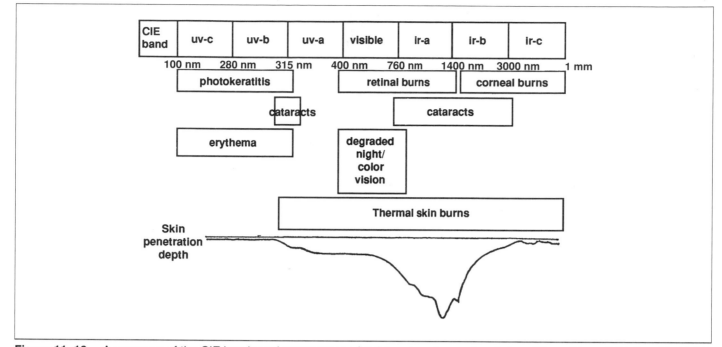

**Figure 11–19.** A summary of the CIE bands and target organs for the various types of radiation. (From Sliney & Wolbarsht, 1985.)

This defense can be overwhelmed by an extremely brief but intense pulse of light into the retina at a rate above the rate at which the heat can be extracted by blood flow. Pulsed lasers are particularly hazardous; they can deposit energy so rapidly that the water in the retinal tissue flashes to a boil and "explodes," causing local tissue damage. If the same amount of energy is deposited in a shorter time period in briefer pulses, more damage is created by the shorter pulses because less heat dissipation can occur. This thermoacoustic mechanism of damage was unknown until the laser was invented. Alternatively, large quantities of light can be hazardous when deposited over larger areas of the retina; this is why it is unsafe to look at the sun through a telescope. The choroid circulation can remove the heat of glimpsing at the sun for a brief period because the image of the sun on the retina is so small that only a small part of the retina is strongly illuminated. A telescope makes the retinal image much larger and the brightness that was limited to just a small retinal area now strikes a much larger area. This can overwhelm the heat-removing capacity of choroid circulation. Visible radiation can cause retinal injuries or burns because it is transmitted and focused. If the resulting lesion, or scotoma, damages some portion of most of the retina, very little visual impairment will result, but if the scotoma occurs in the macula, serious visual impairment results.

Another defense is afforded by aversion reflex actions such as blinking or looking away from bright light. Excessively bright light prompts these responses, which end exposure in about 1/4 s; this is enough to provide protection from injuries due to sunlight and most artificial light sources other than lasers. One reason why lasers are haz-ardous is because they can deposit damaging amounts of energy into the eye well before the aversion reflex ends the exposure.

## The Skin

See Chapter 3, The Skin and Occupational Dermatoses, for anatomical illustrations of the skin. The outermost layer of skin, the epidermis, contains a single sheet of cells at its base, consisting of keratinocytes (cells that divide move outward) and melanocytes (cells that form dark granules of melanin pigment that are transferred to the keratinocytes). The keratinocytes divide and are pushed outward by newer cells being created by the basal cells. As they move outward they flatten, develop pigment granules, and finally die. Thus, the dividing and moving keratinocytes carry pigment with them. This pigment absorbs UV and prevents the generation of excessive levels of vitamin D and UV injuries to the dermal and subcutaneous tissues below. The inner layer of living cells is called the stratum malphigii; the outermost layer of dead cells the stratum corneum.

IR and visible light skin injuries are confined to thermal burns. About 2/3 mW/cm$^2$ irradiance produces a feeling of warmth. UV also causes skin effects through photochemical mechanisms. Sunbathers now use sunscreen lotions to allow UV-A to reach the skin and stimulate melanin production by the melanocytes, but absorb UV-B before it can reach the skin. UV-B and UV-C produce two undesirable effects: the skin toughening evident among desert dwellers and skin cancer. (Outdoor exposures do not include UV-C because it is absorbed by oxygen in the atmosphere to produce ozone). Research now shows that UV-A can also cause skin cancer. Thermal and photochemical damage mechanisms again dominate. The most common skin effect is erythema, or reddening, which is commonly called sunburn. UV penetrates to the living cells of the dermis and causes damage, which is repaired. Blood supply to the skin is increased as part of the repair process, causing the reddening. Repeated exposure to UV, particularly UV-B, causes the skin to thicken and harden. This is why people who live in sunny areas can develop leathery skin.

The exposure standards that address UV and far IR exposure to the skin are about the same as the standards addressing eye exposures, but the standards for visible and near IR skin exposure are much more lenient than those for eye exposure because of the possibility of retinal damage caused by focusing.

## Standards

It is necessary to digress into mathematics here. Optical radiation and laser safety standards use radians rather than degrees as a measure of angle. The length of an arc (segment of a circle) created (subtended) by a given angle is equal to the radius of the circle times the size of the angle in radians. Put another way, the size of a source that is emitting radiation or a surface that is reflecting radiation is equal to the distance between the source or surface and the viewer times the angle in radians. The customary symbol for the angle is alpha ($\alpha$). The use of radian units is shown in Figure 11–20. For comparison purposes, half a circle, 180°, is $\pi$ (about 3.1416) radians and a full circle, 360°, contains $2\pi$ (about 6.2832) radians. Another angular measure is also used to describe portions of the surface of a sphere: the steradian. Radians and degrees cover familiar two

dimensional circles and arcs, but the steradian unit applies to three-dimensional situations such as areas that are reflecting or generating radiation. The size of an area on the surface of a sphere is equal to the solid angle subtended by that area times the radius of the sphere. A sphere subtends a solid angle of $4\pi$ steradians. Put another way, the size of an illuminated or radiating area is equal to the square of the distance between the area and the observer times the solid angle, in steradians.

The principal standards for nonlaser optical radiation are found in the back of the TLV booklets issued annually by the ACGIH. ACGIH groups visible radiation with IR radiation and addresses UV separately. These TLVs are fundamentally identical in concept to the more familiar noise standard in that they use spectral effectiveness factors to account for the difference in damage caused by energy of different wavelengths (much as the A-weighting curve does for noise exposures) and they trade exposure intensity off against permissible exposure time. Instruments are now on the market that have wavelength response characteristics that approximate the spectral effectiveness curves of the TLVs; such an instrument is shown in Figure 11–21. See Appendix B for the 1994–1995 TLVs.

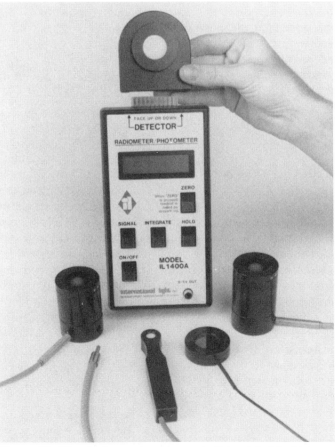

**Figure 11–21.** Instruments are now on the market that have wavelength response characteristics that approximate the spectral effectiveness curves of the TLVs. (Courtesy International Light, Inc.)

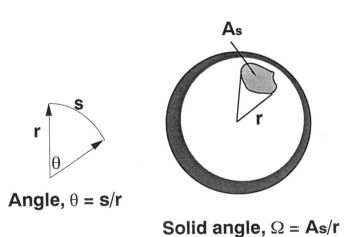

**Angle, $\theta = s/r$**

**Solid angle, $\Omega = As/r$**

**Figure 11–20.** An angle in radians can be calculated by dividing its arc by the radius. A solid angle can be calculated in steradians by dividing the area subtended by the solid angle by the radius squared. In optical safety applications, the arc is the long dimension of an illuminated or luminous area. The area for a solid angle calculation is the illuminated or luminous area. In both cases, the radius is the distance between the illuminated or luminous area and an observer.

The TLV for visible/IR radiation has two basic elements: one for situations where visible radiation is present and a much simpler standard that applies when only IR is present. The visible standards must account for both photochemical injuries at the blue end and thermal injuries at the red end and IR. The thermal injury standard covers wavelengths from 400 to 1,400 nm and balances ambient irradiances multiplied by the appropriate spectral effectiveness factors (listed in a table that shows that the maximum thermal hazard exists almost at the UV end, wavelength = 435 and 440 nm) against the angle being viewed and the exposure time. Thus, stronger exposures to more hazardous wavelengths from bigger sources warrant shorter permissible exposures. The TLV also specifies making a second comparison to account for blue light photochemical hazards of visible radiation, wavelengths from 400 to 700 nm. A table of spectral effectiveness factors is given for red light and blue light injury. Stronger exposures of more hazardous wavelengths from bigger sources (sources covering larger solid angles) are balanced against permissible exposure duration. Special spectral effectiveness factors are required for those who have had their lenses removed (aphakes); these require assessments down to 305 nm (through all of UV-A and just into the high wavelength end of UV-B).

The IR TLV does not extend to wavelengths >1.4 μm; one must consult LIA/ANSI laser safety standard Z136.1 for guidance about wavelengths between 1.4 μm and 1 mm. The Z136 standard can be applied to nonlaser sources at wavelengths above 1.4 μm because the concern about retinal focusing does not exist above 1.4 μm so eye and skin hazards are the same for laser and nonlaser sources.

Two formulas are given for situations such as heat lamps where IR is not accompanied by visible radiation. One formula addresses corneal hazards from 770 nm to 3 μm by limiting exposure times if irradiances exceed those allowed for viewing intervals greater than 1,000 seconds (10 mW/cm$^2$). The other addresses retinal hazards from 770 to 1400 nm.

The TLV for UV uses similar logic and relies on a combined eye and skin hazard weighting curve so only one set of measurements is needed. The combined curve shows maximum potency at 270 nm. Neither the skin nor the eye hazard curve peaks at 270 nm, but the combined curve does. Instruments have been developed in which the sensor *response* approximates the combined UV spectral effectiveness curve.

## Controls for Nonlaser Sources

It is necessary to discuss optical density before proceeding to controls. Optical density (OD) is the base-10 logarithm of the ratio of the intensity of the radiation leaving the filter divided by the intensity of the radiation entering the filter (the attenuation provided by the filter). If a filter absorbs all but 1 percent, or 1/100, of the radiation entering it, then the OD is the logarithm of 100, or 2. If a (1-kW) laser beam

strikes a filter and only 1 mW is transmitted through the filter, then the filter reduced the beam intensity by a factor of 1,000,000 and its OD is 6.

A variation of optical density, *shade number,* has been used in ANSI Z49.1 to describe welders' eye protection for several decades. The shade number is 1 plus the product of 7/3 multiplied by the optical density of the filter.

Baffles and sight barriers are common engineering controls for optical radiation, particularly for welding areas. A number of vendors sell absorbing plastic panels that allow observers to see welders without being exposed to hazards from arcs.

A wide variety of personal protective equipment is now available for nonlaser optical radiation hazards. Some points of interest follow:

- The shade numbers for welding eyewear are summarized in Table 11–J. They are excerpted from ANSI Z49.1-1988, which OSHA incorporated in the 1994 revisions of 29 *CFR* Subpart I personal protective equipment standards.

- Common glass does not offer complete protection against UV-A, although it is very effective against UV-B and UV-C. Tinted eyewear is often effective against UV-A. Eyewear vendors will send transmission curves for their products that can be used to select UV-A protective eyewear for applications that do not involve intense exposures to UV-A.

- The use of photochemically darkened lenses should be avoided because the lenses can darken fairly rapidly in response to sunlight, but become paler slowly. One can enter a building or drive into a tunnel and experience impairment of vision by such glasses, particularly immediately after entry, until the tinting adjusts to the reduced light levels.

- Special lenses are available for nonlaser light and IR sources such as glass blowing and steel making.

- OSHA standard 29 *CFR* 1910.133 requires eye and face protectors to be distinctly marked to facilitate identification of the manufacturer and to meet ANSI Z87.1–1989 impact criteria. OSHA standard 29 *CFR* 1910.133 also requires that protective eyewear for users with prescriptions either have the prescription built in or be worn over prescription lenses without disturbing the prescription lenses. Several eyewear vendors now offer clip-on adapters so prescription lenses can be worn behind eyewear that filters optical radiation.

## ■ Laser

*Laser* is an acronym for *light amplification by stimulated emission of radiation*. Its original derivation has been obscured by years of common use. The original lasers followed microwave devices called masers (the *m* stands for

**Table 11–J.** Guide for Welding Shade Numbers

| Process | Electrode Size in 1/32 in. (mm) or Process Description | Arc Current (A) | Minimum Protective Shade** | Suggested Shade (Comfort)* |
|---|---|---|---|---|
| Shielded metal arc | <3 (2.5) | <60 | 7 | — |
| | 3–5 (2.5–4) | 60–160 | 8 | 10 |
| | 5–8 (4–6.4) | 160–250 | 10 | 12 |
| | >8 (6.4) | 250–550 | 11 | 14 |
| Gas metal arc or flux covered arc | | <60 | 7 | — |
| | | 60–160 | 10 | 11 |
| | | 160–250 | 10 | 12 |
| | | 250–550 | 10 | 14 |
| Gas tungsten arc | | <50 | 8 | 10 |
| | | 50–150 | 8 | 12 |
| | | 150–500 | 10 | 14 |
| Air carbon | Light | <500 | 10 | 12 |
| Arc cutting | Heavy | 500–1,000 | 11 | 14 |
| Plasma arc welding | | <20 | 6 | 6–8 |
| | | 20–100 | 8 | 10 |
| | | 100–400 | 10 | 12 |
| | | 400–800 | 11 | 14 |
| Plasma arc cutting | Light† | <300 | 8 | 9 |
| | Medium† | 300–400 | 9 | 12 |
| | Heavy† | 400–800 | 10 | 14 |
| Torch brazing | | | | 3** or 4 |
| Torch soldering | | | | 2 |
| Carbon arc welding | | | | 14 |
| | | Plate thickness in. (mm) | | |
| Gas welding | Light | <1/8 (<3.2) | | 4** or 5 |
| | Medium | 1/8–1/2 (3.2–12.7) | | 5** or 6 |
| | Heavy | >1/2 (>12.7) | | 6** or 8 |
| Oxygen cutting | Light | <1 (<25) | | 3** or 4 |
| | Medium | 1–6 (25–150) | | 4** or 5 |
| | Heavy | >6 (>150) | | 5** or 6 |

* Start with a shade too dark to see the weld zone. Then go to a lighter shade that gives a sufficient view of the weld zone without going below the minimum. In oxyfuel gas welding or cutting, where the torch produces a harsh yellow light, use a filter lens that absorbs the yellow or sodium line in the visible radiation.
** This is listed as the minimum protective shade in the OSHA General Industry Standard for Eye and Face Protection, 29 *CFR* 1910.133 of April 6, 1994.
† These values apply where the arc can be seen clearly. Experience has shown that lighter filters can be used when the arc is hidden by the workpiece.
(From AWS/ANSI Z49.1-1988.)

*microwave*) and were called optical masers. Masers and lasers work by pumping the electrons in a suitable material, the lasing medium, with strong energy and directing some of that energy out in the form of a beam of radiation. To do this, most lasers include the parts shown in Figure 11–22: a source of pumping energy such as a flash lamp or electrodes embedded in a gas-filled tube, a lasing medium that emits radiation when pumped, and a resonant cavity that is a multiple of the product radiation's wavelength and is formed by placing a mirror at each end of the lasing material. The pumping energy strikes electrons of atoms in the lasing medium, which quickly lose the pumping energy as radiation. Some of the radiation produced hits electrons in other pumped atoms in the lasing medium and makes them yield their stored energy as more radiation. The radiation

reaches one end of the cavity and is reflected off of a mirror back into the lasing medium, where it strikes more electrons in other pumped atoms and causes them to yield their energy until it strikes a mirror at the opposite end of the cavity, where the process begins again. The radiation surges back and forth along the cavity, gaining strength until it emerges from one end (with a partially reflective mirror) as a beam with parallel sides.

Laser radiation has unique properties. It is monochromatic, which means that it is made of one or a few wavelengths of radiation, determined by the lasing material. It is coherent; that is, the waves are matched to one another as soldiers in a column are matched to one another. It is also bright: It can have very high irradiance and be focused to deposit intense energy on small surfaces.

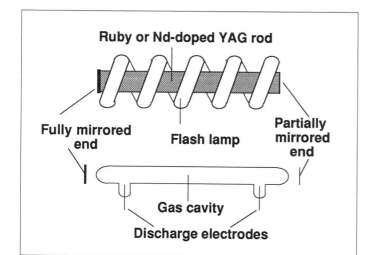

**Figure 11–22.** Parts of two types of lasers: solid state and gas. The top drawing is a solid-state laser, the bottom a gas laser. Both lasers fit in protective housings that include cooling equipment for high-power lasers and a beam-blocking device for Class 4 lasers. Power supplies and control panels are usually located elsewhere; the power supply and cooling cables leading to the laser enclosure give a rough indication of the laser's power.

Coherent radiation can be very sharply focused whereas common noncoherent radiation cannot. The parallel-sided and coherent nature of the product radiation means that the beam does not spread as rapidly as radiation from other sources. Thus, a laser beam keeps its strength over long distances. Visible and IR-A laser beams can be focused to create extremely intense exposures on the retina that deliver up to 300,000 times as much power or energy per unit area on the retina as on the outside of the eye. This is why laser safety standards are much stricter in this band than at other bands and why laser standards are stricter than standards for noncoherent sources in this band.

Lasers can be operated in two major modes, much like other radiation sources: pulsed and continuous wave (CW). Exposure standards and laser outputs for pulsed and relatively brief exposures are expressed in $J/cm^2$ of lit area; exposure standards and outputs for continuous wave lasers are expressed in $W/cm^2$ of lit area.

All forms of matter are used as lasing media. The first lasers used ruby rods; the pumping energy came from high-intensity flash lamps. Yttrium aluminum garnets treated with neodymium (Nd-doped YAG) followed for IR-A lasers. Helium-neon (He–Ne) lasers, essentially neon lamps filled with helium–neon mixtures surrounded by mirrors to create an appropriate resonant cavity, are very common. Similar lamps filled with krypton, argon, or carbon dioxide gas are also in common use. Carbon dioxide lasers produce 10.6 μm beams in the IR-C band; the other lasing media just mentioned produce mostly visible wavelengths, as shown in Tables 11–K and 11–L. Organic dyes

are used because the output wavelength can be precisely adjusted at the discretion of the user, unlike other media, which have fixed product wavelengths. Solid-state diodes made from selectively blended compounds of gallium arsenide can be used to produce radiation in the IR-A band and diode lasers are now used in supermarket bar code scanners and lecture hall pointers. Diode lasers producing dangerous outputs can be smaller then a fingertip. Excimer lasers use ionized halogen atoms in inert gases to emit UV radiation.

Energy sources usually use electricity to fire flash lamps or energize electrodes in a gas cavity, but other forms of pumping energy can be used. The energy released by the combustion of deuterium and fluorine is harnessed in gas dynamic lasers, rocket engine-like devices that exhaust in one direction and produce an IR beam in a perpendicular direction. Dye lasers use light from a flash lamp or from another laser for pumping.

## Biological Damage Mechanisms of Lasers

Laser beams produce biological damage by the two mechanisms mentioned earlier (thermal burns and photochemical injuries) and visible and IR-A lasers produce retinal damage by a third mechanism unique to lasers. The highly focused beam generates a steam bubble near the retina that pops, sending shock waves into the retinal tissue that produce thermoacoustic tissue damage. Briefer pulses are more hazardous than longer pulses of equal energy content because the heat does not dissipate as much during a shorter pulse. Q-switching is a method of reducing the pulse duration of a laser. Q-switched pulses can last tenths of microseconds whereas the original unswitched beams can last milliseconds. Thus, joule for joule, Q-switched pulses are more hazardous than non-Q-switched pulses. The bursting bubble can damage blood vessels and blood is toxic to nerve tissue, so the degree of lasting impairment caused by thermoacoustic injury is a matter of luck determined by what part of the retina was struck and whether blood reached nerve tissue.

It is important to note that coherent visible and IR-A radiation with wavelengths of 400–1,400 nm can be focused into an ultra-small (diffraction-limited) spot at the back of the retina. The brightness of the radiation striking the retina can be as much as 300,000 times stronger than that entering the eye. Thus, visible and IR-A lasers are particularly hazardous to the eye. This explains why laser exposure standards are so strict.

## Standards

The dominant standard for laser safety in the United States is the Laser Institute of America's LIA/ANSI Z136.1 standard, *American National Standard for Safe Use of Lasers,* last revised in 1993. Regulations for commercial laser products promulgated by the Center for Devices and Radiological Health (CDRH) of the Food and Drug Administration, 21 *CFR* Part 1040, are important for commercial laser devices.

**Table 11–K.** Common Continuous Wave Laser and Division into Classes by Accessible Emission Limits *

| Lasing Medium | Output Wavelength (nm) | Class 1 (W) | Class 2 (W) | Class 3b (W) | Class 4 (W) |
|---|---|---|---|---|---|
| **UV-B and C** | | | | | |
| Argon | 275 | $\leq 9.6 \times 10^{-9}$ for 8 hr | — | >Class I and <0.5 | >0.5 |
| | | **UV-A** | | | |
| Helium–cadmium | 325 | | | | |
| Argon | 351 and 363 | $\leq 3.2 \times 10^{-6}$ | — | >Class I and <0.5 | >0.5 |
| Krypton | 350.7 and 356.4 | | | | |
| **Visible** | | | | | |
| Helium–cadmium | 441.6 | | | | |
| Argon | 457, 476, 488, 514, etc. | | | | |
| Krypton | 530 | | | | |
| Neodymium: YAG first harmonic | 532 | $\leq 0.4 \times 10^{-6}$ | | | |
| Helium–neon | 543 | | | | |
| Dyes | 400–550 | | >Class I, but $\leq 1 \times 10^{-3}$ | >Class I and < 0.5 | >0.5 |
| Helium–selenium | 460–550 | | | | |
| Helium–neon | 632 | $\leq 7 \times 10^{-6}$ | | | |
| Dyes | 550–700 | $< 0.4 \times 10^{-6}$ to $7 \times 10^{-6}$ | | | |
| Indium gallium aluminum phosphide | 670 | $\leq 2.4 \times 10^{-5}$ | | | |
| Titanium–sapphire | 670 | $\leq 2.4 \times 10^{-5}$ | | | |
| Krypton | 647, 676.4 | $1.1 \times 10^{-5}, 3 \times 10^{-5}$ | | | |
| **Near IR** | | | | | |
| Gallium aluminum arsenide | 780 | $\leq 0.18 \times 10^{-3}$ | | | |
| As above | 850 | $\leq 0.25 \times 10^{-3}$ | | | |
| Gallium arsenide | 905 | $\leq 0.32 \times 10^{-3}$ | — | >Class I and <0.5 | >0.5 |
| Neodymium–YAG | 1.064 μm | $\leq 0.64 \times 10^{-3}$ | | | |
| Helium–neon | 1.080 and 1.152 μm only | $\leq 0.64 \times 10^{-3}$ | | | |
| Indium gallium arsenic phosphide | 1.310 μm | $\leq 4.40 \times 10^{-3}$ | | | |
| **Far IR** | | | | | |
| Indium gallium arsenic phosphide | 1.550 μm | | | | |
| Holmium | 2.10 μm | | | | |
| Erbium | 2.94 μm | | | | |
| Hydrogen fluoride | 2.6–3 μm | $\leq 9.6 \times 10^{-3}$ | | | |
| Helium–neon | 3.39 μm only | | — | >Class I and <0.5 | >0.5 |
| Carbon monoxide | 5.00–5.50 μm | | | | |
| Carbon dioxide | 10.6 μm | | | | |
| Water vapor | 118 | $\leq 9.5 \times 10^{-2}$ | | | |
| Hydrogen cyanide | 337 | | | | |

*Assumes no mechanical or electrical design incorporated into laser system to present exposures from lasting 8 hours; otherwise, the Class 1 AEL could be larger than tabulated.

(From LIA/ANSI Z136.1-1993.)

The ANSI standard addresses facility and program elements as well as laser safety features. 21 *CFR* 1040 addresses product safety features. Both standards were early uses of the hazard control class type of regulation. Control class regulations require assigning a classification that reflects the severity of the hazard. For lasers, the principal parameters that describe the anticipated hazard severity are wavelength and output power or energy. Other measures of anticipated severity are pulse duration and the size of extended sources such as groups of diode lasers or the reflection of a spread laser beam from a surface. The LIA/ANSI and CDRH standards specify precautions based on the anticipated hazard of the accessible beam by setting accessible emission limits (AELs) for the amount of laser energy people could encounter from each class of laser. Precautions can be relaxed when the beam is thoroughly enclosed or reduced in power before it can enter a place where people could be exposed to it; this gives laser equipment suppliers a strong incentive to apply effective engineering controls. Precautions become more stringent in a series of defined steps (Classes) as the output of the laser increases from Class 1 (so low-powered as to be intrinsically safe) through Class 4 (very dangerous).

Classifications for several lasers are given in Tables 11–K and 11–L. Extremely weak beams are deemed nonhazardous (Class 1) and no precautions are needed. Low-power visible beams that could be hazardous if viewed for prolonged periods (Class 2 and 2a) warrant limited precautions. Class 2a visible lasers are those that could cause excessive exposures if the beam was viewed for 1,000 s or more; Class 2 lasers are visible light lasers that could produce excessive exposures if viewed for more than the 0.25-s

**Table 11–L.** Common Single-Pulsed Lasers and Division into Classes by Accessible Emission Limits

| Lasing Medium | Output Wavelength (nm) | Pulse Duration | Class 1 (J) | Class 3b (J) | Class 4 (W) |
|---|---|---|---|---|---|
| **UV** | | | | | |
| Argon fluoride excimer | 193 | 20 ns | $\leq 1.9 \times 10^{-6}$ | | |
| Krypton fluoride excimer | 248 | 20 ns | $\leq 1.9 \times 10^{-6}$ | | |
| Xenon chloride excimer | 308 | 20 ns | $\leq 4.3 \times 10^{-6}$ | >Class 1, but <0.125 | >0.125 |
| Nitrogen | 337 | 10 ns | $\leq 3.6 \times 10^{-6}$ | | |
| Xenon fluoride excimer | 351 | 20 ns | $\leq 4.3 \times 10^{-6}$ | | |
| **Visible** | | | | | |
| Rhodamine 6G dye | 450–650 | 1 µs | $\leq 0.2 \times 10^{-6}$ | | |
| Copper vapor | 510 and 578 | 25 ns | $\leq 2 \times 10^{-7}$ | >Class 1, but $\leq 0.03\ C_A$* | |
| Neodymium–YAG, 1st harmonic | 532 | 20 ns** | $\leq 2 \times 10^{-7}$ | | >0.03 |
| Ruby (Q-switched) | 694.3 | 20 ns | $\leq 0.2 \times 10^{-6}$ | >Class 1,* but $\leq 0.03\ C_A$ | |
| Ruby | 694.3 | 1 ms | $\leq 4 \times 10^{-6}$ | | |
| **Near IR** | | | | | |
| Titanium sapphire | 700–1,000 | 6 µs | $\leq 1.9 – 7.7 \times 10^{-7}$ | | |
| Alexandrite | 720–800 | 100 µs | $\leq 0.76 – 1.1 \times 10^{-6}$ | Class 1, but $\leq 0.03$ | >0.03 |
| Neodymium–YAG (Q-switched) | 1.064 µm | 20 ns** | $\leq 2 \times 10^{-6}$ | >Class 1, but $\leq 0.15$ | >0.15 |
| **Far IR** | | | | | |
| Erbium–glass | 1.54 µm | 10 ns* | $\leq 9.7 \times 10^{-2}$ | | |
| Cobalt–magnesium fluoride | 1.75–2.50 µm | 80 µs | $\leq 97 – 9.7 \times 10^{-3}$ | | |
| Holmium | 2.10 µm | 0.25 ms | $\leq 9.7 \times 10^{-3}$ | | |
| Hydrogen fluoride | 2.60–3.00 µm | 400 ns | $\leq 1.1 – 0.11 \times 10^{-3}$ | >Class 1, but $\leq 0.125$ | >0.125 |
| Erbium | 2.94 µm | 0.25 ms | $\leq 6.8 \times 10^{-3}$ | | |
| Carbon dioxide | 10.6 µm | 100 ns** | $\leq 9.7 \times 10^{-4}$ | | |
| Carbon dioxide | 10.6 µm | 1 ms | $\leq 9.6 \times 10^{-3}$ | | |

* See Table 11–P for the definition of $C_A$.
** Q-switched.
(From LIA/ANSI Z136.1-1993.)

response time of the aversion reflexes. Class 3 moderate-power lasers warrant more precautions and include moderately high-powered visible lasers and moderately powered invisible UV and IR lasers. Class 3 lasers are subdivided into Class 3a and 3b; Class 3a includes lasers that could be hazardous only if stared at or viewed through an optical device such as a telescope. All other Class 3 lasers are Class 3b and are presumed to have beams powerful enough to harm the eye during incidental exposure without an optical instrument. Class 4 high-power lasers warrant rigorous precautions because the beam can harm the skin and eyes and even diffuse reflections could be harmful. The precautions specified by LIA/ANSI Z136.1-1993 for each class are summarized in Table 11–M.

The use of hazard control classes reduces the need for measurements to determine personnel exposures and puts reliance on being able to estimate exposures by calculations.

LIA/ANSI Z136.1-1993 also sets maximum permissible exposures (MPEs) for lasers, as listed in Tables 11–N, 11–O, and 11–P. The MPEs are directly equivalent to OSHA PELs for chemicals or the MPEs for radiofrequency/microwave radiation set by ANSI/IEEE C95.1-1991. Adjustment factors are applied to calculate MPEs for extended sources, based on the size of the angle subtended by that source in radian units. The AELs used to define laser classes are equal to the MPEs for the maximum allowable viewing time listed in the standard for visible and IR-A lasers, multiplied by the limiting aperture (typically the size of the pupil opening, which can range from 1 to 7 mm, as defined in the standard [see Table 11–P]). The MPEs for wavelengths above 1,400 nm are the only available guidance for nonlaser IR sources emitted at frequencies above 1,400 nm. Limiting aperture dimensions depend on wavelength and the expected duration of an exposure.

LIA/ANSI Z136.1-1993 uses different averaging times depending on the wavelength of the radiation involved. The averaging time for UV exposures is the duration of the exposure throughout a shift, similar to the averaging time for chemicals. The standard actually is based on accumulating a dose of UV energy rather than measuring or calculating the average exposure. The averaging time for visible exposures is 0.25 s, the operating time of aversion reflexes. The averaging time for IR exposures is 10 s. This standard also gives protocols for evaluating repetitively pulsed exposures that are used when the pulse repetition rate exceeds one pulse per second. Pulses that occur less frequently than once per second are addressed by the standards shown in Table 11–Q. Computer applications are available to do routine laser safety calculations, but the user should know how to do the calculations on paper before beginning to routinely use the application

**Table 11–M.**  Summary of Principal Laser Hazard Control Measures

| Control Measure | Laser Class | | | | | |
| --- | --- | --- | --- | --- | --- | --- |
| | 1 | 2a | 2 | 3a | 3b | 4 |
| *Engineering Controls* | | | | | | |
| Protective housing | Shall | Shall | Shall | Shall | Shall | Shall |
| Without protective housing | | | LSO shall establish alternate controls | | | |
| Interlocks on protective housing | Δ | Δ | Δ | Δ | Shall | Shall |
| Service access panels interlocked or require special tools for removal | Δ | Δ | Δ | Δ | Shall | Shall |
| Key control | >MPE | >MPE | >MPE | >MPE | Should | Shall |
| Viewing portals and display screens have means to reduce levels to <MPE | — | — | >MPE | >MPE | >MPE | >MPE |
| Collecting optics have means to reduce levels to <MPE | >MPE | >MPE | >MPE | >MPE | >MPE | >MPE |
| Totally open beam path | — | — | — | — | NHZ | NHZ |
| Limited open beam path | — | — | — | — | NHZ | NHZ |
| Enclosed beam path | | *Not required if engineering or LSO-approved hazards analysis performed and administrative controls are in place* | | | | |
| Remote interlock connector to allow for multiple shutdown devices such as "panic buttons" | — | — | — | — | Should | Shall |
| Attached beam stop or attenuator to reduce output <MPE if full output of laser isn't needed | — | — | — | — | Should | Shall |
| Activation warning systems such as lights or audible devices or a countdown | — | — | — | — | Should | Shall |
| Emission delay between annunciation of warning and laser firing | — | — | — | — | — | Shall |
| Indoor laser controlled area determination made by hazard analysis of LSO | — | — | — | — | NHZ | NHZ |
| Temporary laser controlled area during maintenance, etc. when protective housing and other engineering controls must be bypassed | Δ >MPE | Δ >MPE | Δ >MPE | Δ >MPE | — | — |
| Remote firing and monitoring | — | — | — | — | — | Should |
| Labels posted on laser equipment | Shall | Shall | Shall | Shall | Shall | Shall |
| Area posting of laser hazard warning signs/devices | — | — | — | Should | NHZ | NHZ |
| *Administrative and Other Controls* | | | | | | |
| Standard/safe operating procedures | — | — | — | — | Should | Shall |
| Limiting output if laser is overpowered for job | — | — | — | | Shall if LSO determined need | |
| Training operators and maintenance/service personnel | — | — | Should | Should | Shall | Shall |
| Operation/servicing by authorized personnel only | — | — | — | — | Shall | Shall |
| Standard/safe operating procedures governing alignment | — | — | Shall | Shall | Shall | Shall |
| Laser protective eyewear required if other measures are insufficient | — | — | — | — | Should if >MPE | MPE |
| Skin protection | — | — | — | — | MPE | MPE |
| Protective windows installed | — | — | — | — | NHZ | NHZ |
| Warning signs and labels of standard design | — | — | Should | Should | NHZ | NHZ |
| Fiber optic system controls | MPE | MPE | MPE | MPE | Shall | Shall |
| Public demonstration controls | MPE† | — | Shall | Shall | Shall | Shall |

KEY:
- Shall — Must be done
- Δ — Shall if enclosed Class 3b or 4
- MPE — Shall if MPE is exceeded
- — — No requirement
- Should — Should be done
- NHZ — Analysis to define nominal hazard zone required
- † — Applicable to UV and IR lasers

(From LIA/ANSI Z136.1–1993.)

and verify that the application works for the pulsing and wavelength scenarios of interest by comparing computer results to the results of paper calculations.

Pulsed laser exposure criteria are complicated and hard to apply. LIA/ANSI Z136.1-1993 specifies that the exposure interval for visible lasers is 0.25 s (it defines visible as ranging between 400 and 700 nm rather than 760 nm) whereas the exposure interval for IR lasers is 10 s. These exposure durations are applied concurrently in two procedures for determining MPEs and AELs for pulsed

lasers, per paragraph 8.2.2 of the standard. People working with pulsed lasers need to understand this part of the standard completely, but a full discussion of paragraph 8.2.2 is beyond the scope of this chapter.

## Controls

Engineering and administrative controls and personal protective equipment are used for lasers and optical radiation. The main engineering control for lasers is enclo-

**Table 11–N.**  Maximum Permissible Exposures for Laser Radiation for Intrabeam Viewing

| Wavelength (μm) | Exposure Duration | MPE (J/cm²) | MPE (W/cm²) | Notes |
|---|---|---|---|---|
| **UV** | | | | |
| 0.180–0.302 | 1 ns–30,000 s | $3 \times 10^{-3}$ | — | |
| 0.303 | " | $4 \times 10^{-3}$ | — | |
| 0.304 | " | $6 \times 10^{-3}$ | — | |
| 0.305 | " | $10 \times 10^{-3}$ | — | |
| 0.306 | " | $16 \times 10^{-3}$ | — | |
| 0.307 | " | $25 \times 10^{-3}$ | — | |
| 0.308 | " | $40 \times 10^{-3}$ | — | or $0.56t^{1/4}$, whichever is less |
| 0.309 | " | $63 \times 10^{-3}$ | — | See Table 11–Q for definitions of limiting apertures. |
| 0.310 | " | 0.1 | — | |
| 0.311 | " | 0.16 | — | |
| 0.312 | " | 0.25 | — | |
| 0.313 | " | 0.40 | — | |
| 0.314 | " | 0.63 | — | |
| 0.315–0.400 | 1 ns–10 s | $0.56\ t^{1/4}$ | — | |
| | 10–30,000 s | 1.0 | — | |
| **Visible and Near IR** | | | | |
| 0.400–0.700 | 1 ns–18 μs | $0.5 \times 10^{-6}$ | — | |
| | 18 μs–10 s | $1.8\ t^{3/4} \times 10^{-3}$ | — | |
| 0.400–0.550 | 10–10,000 s | $10 \times 10^{-3}$ | — | |
| 0.550–0.700 | 10–$T_1$ | $1.8\ t^{3/4} \times 10^{-3}$ | — | |
| | $T_1$–10,000 s | $10\ C_B \times 10^{-3}$ * | — | |
| 0.400–0.700 | 10,000–30,000 s | | $C_B \times 10^{-6}$ * | See Table 11–Q for definitions of limiting apertures. |
| 0.700–1.050 | 1 ns–18 μs | $0.5\ C_A \times 10^{-6}$ * | — | |
| | 18 μs–1,000 s | $1.8\ C_A\ t^{3/4} \times 10^{-3}$ * | — | |
| | 1,000–30,000 s | | $320\ C_A \times 10^{-6}$ * | |
| 1.050–1.400 | 1 ns–50 μs | $5\ C_C \times 10^{-6}$ * | — | |
| | 50 μs–1,000 s | $9\ C_C\ t^{3/4} \times 10^{-3}$ * | — | |
| | 1,000–30,000 s | | $1.6\ C_C \times 10^{-3}$ * | |
| **Far IR** | | | | |
| 1.4–1.5 | 1ns–1 ms | 0.1 | — | |
| | 1 ms–10 s | $0.56\ t^{1/4}$ | — | |
| | 10–30,000 s | | 0.1 | |
| 1.5–1.8 | 1 ns–10 s | 1 | — | |
| | 10–30,000 s | | 0.1 | |
| 1.8–2.6 | 1 ns–1 ms | 0.1 | — | Only exposure standard >1.4 μm |
| | 1 ms–10 s | $0.56\ t^{1/4}$ | — | See Table 11–Q for definitions of limiting apertures. |
| | 10–30,000 s | | 0.1 | |
| 2.6 μm–1 mm | 1–100 ns | $10 \times 10^{-3}$ | — | |
| | 100 ns–10 s | $0.56\ t^{1/4}$ | — | |
| | 10–30,000 s | | 0.1 | |

* See Table 11–P for definitions of $C_A$, $C_B$, $C_C$, $C_E$, and $T_1$.
(From LIA/ANSI Z136.1-1993.)

sure, often in the form of interlocked rooms and protective housings. Interlocked laser beam enclosures can range from plastic panels on a framework for research setups to sturdy metal boxes with person- or vehicle-sized access doors for IR laser cutting tools. Laser beams can be routed over or below walkways using mirrors and, possibly, elevated enclosures or tunnels. LIA/ANSI Z136.1-1993 calls for fail-safe interlocks, such as double interlocks connected in series, on access panels of enclosures of Class 3b or 4 laser beams. The interlocks either shut off the electrical power to the laser or drop a shutter into the beam at or within the housing of the laser itself. An important extension of interlocks is the remote interlock connection, which allows additional interlocks, including those located remotely, to trigger a shutdown of electric power to a laser or drop a shutter or filter into the beam. Remote interlocks can make up safety chains for laser systems covering large areas and include emergency panic button shutoffs.

Viewing portals, viewing screens, and optical instruments must be connected to interlocks or reduce beam intensity to acceptable levels for Class 4 lasers. LIA/ANSI Z136.1-1993 recognizes the use of fasteners requiring special tools as providing equivalent safety to interlocks for access panels. It is convenient to add room status lights (green for safe to enter, yellow for possibly unsafe to enter, and red for unsafe to enter) to the laser interlock system and to install a loudspeaker and buzzer near the room status lights so visitors can talk to the room occupants during laser operations. These boxes were

**Table 11–O.** Maximum Permissible Exposures for Laser Radiation Skin Exposures

| Wavelength (μm) | Exposure Duration | MPE (J/cm²) | MPE (W/cm²) | Notes |
|---|---|---|---|---|
| *UV* | | | | |
| 0.180–0.302 | 1 ns–30,000 s | $3 \times 10^{-3}$ | — | |
| 0.303 | " | $4 \times 10^{-3}$ | — | |
| 0.304 | " | $6 \times 10^{-3}$ | — | |
| 0.305 | " | $10 \times 10^{-3}$ | — | |
| 0.306 | " | $16 \times 10^{-3}$ | — | |
| 0.307 | " | $25 \times 10^{-3}$ | — | |
| 0.308 | " | $40 \times 10^{-3}$ | — | |
| 0.309 | " | $63 \times 10^{-3}$ | — | or $0.56t^{1/4}$, whichever is less; 3.5 mm limiting aperture |
| 0.310 | " | 0.1 | — | |
| 0.311 | " | 0.16 | — | |
| 0.312 | " | 0.25 | — | |
| 0.313 | " | 0.40 | — | |
| 0.314 | " | 0.63 | — | |
| 0.315–0.400 | 1 ns–10 s | $0.56\ t^{1/4}$ | — | |
| | 10–1,000 s | 1.0 | — | |
| | 1,000–30,000 s | | $1 \times 10^{-3}$ | |
| *Visible and Near IR* | | | | |
| 0.4–1.4 | 1–100 ns | $2\ C_A \times 10^{-2}$ * | — | |
| | 100 ns–10 s | $1.1\ C_A t^{1/4}$ * | — | 3.5 mm limiting aperture |
| | 10–30,000 s | | $0.2\ C_A$ * | |
| *Far IR* | | | | |
| 1.4–1 mm | 1–100 ns | $10^{-2}$ | — | Only exposure standard >1.4 μm |
| | 100 ns–10 s | $0.56\ t^{1/4}$ | — | See Table 11–Q for definitions of limiting apertures. |
| | 10–30,000 s | | 0.1 | |

* See Table 11–P for definition of $C_A$, $C_B$, $C_C$, $C_E$, and $T_1$.
(From LIA/ANSI Z136.1-1993.)

developed and are used extensively at Lawrence Livermore National Laboratory, and are also commercially available.

A simple control, especially useful for lasers located in places where unauthorized people could find them, is a key-in-lock control, which makes it very unlikely that the laser could be inadvertently operated by unauthorized people. Of course, the operator must remove the key and take it when the laser is turned off.

Engineering controls are effective when a laser system is set up, but not while it is being set up or during maintenance. Thus, special caution, including heavy reliance on administrative controls and special warning signs, is needed during setup and maintenance. A recent review of laser accidents shows that 37.2 percent of laser accidents occurred during alignment. Techniques for reducing hazards during alignment include using low-powered alignment lasers. The alignment laser beam is brought into the path of the main laser beam and system adjustments are made using the less dangerous alignment beam; the accessible main laser beam power can also be reduced for setup or alignment, if practical. Alternatively, special laser alignment eyewear that blocks most of the light can be used during alignment after steps have been taken to ensure that the main beam cannot enter the eyes. ANSI Z136.1 specifies special warning signs to be used during setup. A variety of aids are available to make alignment easier even with protective eyewear:

■ An IR disc that is mounted in an optical-component holder.

■ IR or UV cards that glow with visible light when struck by an IR or UV laser beam.

■ IR viewers that make the place where the beam strikes a surface show as a visible dot.

**Table 11–P.** Correction Factors for Intrabeam Viewing MPEs of Table 11–N and Table 11–O

| Correction Factor | | Definition | Wavelength (μm) |
|---|---|---|---|
| $C_A$ | = | 1 | 0.400–0.700 |
| $C_A$ | = | $10^{2(\lambda-0.700)}$ | 0.700–1.050 |
| $C_A$ | = | 5 | 1.050–1.400 |
| $C_B$ | = | 1 | 0.400–0.550 |
| $C_B$ | = | $10^{15(\lambda-0.550)}$ | 0.550–0.700 |
| $C_C$ | = | 1 | 1.050–1.150 |
| $C_C$ | = | $10^{18(\lambda-1.150)}$ | 1.150–1.200 |
| $C_C$ | = | 8 | 1.200–1.400 |
| $C_E$ | = | 1 (if the subtended angle, $\alpha$, is $<\alpha_{min}$) | 0.4–1.4 |
| $C_E$ | = | $\alpha/\alpha_{min}$ if the subtended angle is between $\alpha_{min}$ and 100 mradians | 0.4–1.4 |
| $C_E$ | = | $\alpha/(100\ \alpha_{min})$ if the subtended angle is > 100 mradians | 0.4–1.4 |
| $T_1$ | = | $10 \cdot 10^{20(\lambda-0.550)}$ | 0.550–0.700 |

$\alpha_{min} = 1.5$ milliradians when $t \le 0.7$ s; $\alpha_{min} = 2\ t^{3/4}$ milliradians when $0.7$ s $< t < 10$ s; $\alpha_{min} = 11$ milliradians when $t \ge 10$ s.
(From LIA/ANSI Z136.1-1993.)

**Table 11–Q.** Limiting Apertures Used to Develop AELs

| Spectral Region (µm) | Duration (s) | Eye Aperture | | Skin Aperture | |
|---|---|---|---|---|---|
| | | Diameter (mm) | Area (cm²)** | Diameter (mm) | Area (cm²)** |
| ≥0.180 to <0.400 | $10^{-9}$ to 0.25 | 1 | 0.0078 | 3.5 | 0.0962 |
| | 0.25 to $3 \times 10^4$ | 3.5 | 0.0962 | 3.5 | 0.0962 |
| ≥0.400 to <1.400 | $10^{-9}$ to $3 \times 10^4$ | 7 | 0.3848 | 3.5 | 0.0962 |
| ≥1.400 to <100 | $10^{-9}$ to 0.3 | 1 | 0.0078 | 3.5 | 0.0962 |
| | 0.3 to 10* | $1.5\, t^{3/8}$ | | 3.5 | 0.0962 |
| | 10 to $3 \times 10^4$ | 3.5 | 0.0962 | 3.5 | 0.0962 |
| ≥100 to <1 mm | $10^{-9}$ to $3 \times 10^4$ | 11.0 | 0.9503 | 11.0 | 0.9503 |

\* This exposure duration normally would not be used for hazard evaluation.
\*\* Area in cm² is used to calculate AELs.
(From LIA/ANSI Z136.1-1993.)

- Strong white paper or colored paper that fluoresces in different wavelengths (colors) when struck by a visible laser beam. The fluorescent radiation can pass through the lenses of laser safety eyewear intended for the laser wavelength. Some experimentation is needed to find which papers work.

Specular (mirror-like) reflections are more hazardous than diffuse reflections because virtually all of the beam's power is retained, as shown in Figure 11–23, and safety precautions always include eliminating all unnecessary specular reflectors. Most optical instrument manufacturers mount their products on shiny posts, but the posts are cylindrical, so a reflected beam will still be spread out. LIA/ANSI Z136.1-1993 includes formulae for calculating reflections from rounded mirrors. Painted walls and stipple-finished tools can easily be specular reflectors for $CO_2$ laser beams because the surface roughness that causes the diffuse reflections we see is smaller than 10.6 µm so the surface acts as if it is slick and causes specular reflections at 10.6 µm.

Administrative controls are used during setup and maintenance. Z136.1-1993 calls for calculating the nominal hazard zone (NHZ), where people could be exposed to laser beam levels above the MPEs, and excluding people from the NHZ by barriers, signs, flashing beacons, and the diligence of personnel authorized to be inside the NHZ. Warning signs should follow international conventions, applied in the United States through the ANSI Z535 series of standards; these specify colors, warning words, standard symbols, and warning sign layout. The laser warning symbol is the familiar sunburst. The word *CAUTION* is used on signs and labels for Class 2 and 3a lasers and laser systems that do not exceed the appropriate MPE. The word *DANGER* must be used for other Class 3a lasers and all Class 3b and 4 lasers. Existing laser-safety signs and labels are grandfathered in by Z136.1-1993. The signs must specify essential precautions above the tail of the sunburst (shown in Figure 11–24) and describe the properties of the laser and the laser class below the center of the tail and the lower-right-hand corner of the sign, respectively.

**Figure 11–24.** The internationally recognized sunburst laser warning icon. ANSI Z535 standards eliminate the tradition of using special colors for specific hazard warnings, but ANSI Z136.1-1993 makes limited use of a red symbol for *DANGER* signs. Symbols and narrative text now appear in black and colors are used to indicate degree of hazard (yellow and black for *CAUTION,* orange and black for *WARNING,* and red, black, and white for *DANGER*). The long tail symbolizes how laser beams retain their power over long distances. The same symbol, but without the tail, was proposed for intense nonlaser sources of optical radiation, but it is not widely used.

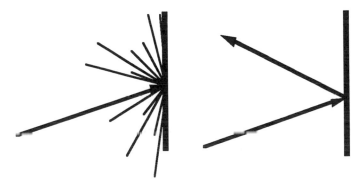

**Figure 11–23.** Diffuse reflections (left) dissipate the energy of incident radiation, so the reflection is not as bright as the incident radiation because it is scattered in many directions. Specular reflections (right) are much more dangerous because the beam bounces in only one direction, retaining most of its original intensity and hazard.

Medical surveillance and training programs are required for users of Class 3b and 4 lasers. The laser safety program must include the following:

- A laser safety officer and, if the number and diversity of laser operations warrants, a laser safety committee
- Education of authorized personnel
- Application of the controls specified in the standard
- Accident reporting and investigation and action plans to prevent recurrence of incidents
- Medical surveillance as specified in the standard

Laser protective goggles must show the manufacturer, wavelengths they work against and optical density, and must meet the provisions of the latest edition of ANSI Z87.1 for safety eyewear. As with all personal protective equipment, it is necessary to select laser eyewear of appropriate optical density for the type (wavelengths) of radiation encountered and the severity of exposure. Color coding is recommended for multilaser environments and eyewear manufacturers offer color-coded frames. Laser protective eyewear usually includes glass or plastic lenses. In general, glass lenses are heavier, but offer more resistance to direct strikes by laser beams and often let through more light. As a rule, glass is often used when the average power of a laser exceeds about 100 mW.

## Laser Pointers

Laser pointers are now available at retail office supply stores. The first laser pointers were helium–neon (He–Ne) lasers emitting at 632 nm (red). But they were delicate because of the glass envelope of the He–Ne laser. Diode lasers rapidly replaced He–Ne lasers because they are more rugged and allow for compact, pen-sized pointers. Unfortunately, the early diodes emitted at 670 nm, where the visual response of the eye is much poorer than it is at 632 nm. This difficulty was overcome by raising the laser output from somewhere below one mW to a few mW. Although the visual response of the eye is exquisitely wavelength-sensitive, the vulnerability of the eye to thermal injury is not. The old He–Ne pointers were Class 2 lasers; the original diode pointers were Class 3a. The original He–Ne pointers bore a caution label; the diode laser pointers bear danger labels. Stories of cavalier use of laser pointers at lectures are common and these lasers are accessible to children. Fortunately, diode pointers emitting at 635 nm are now on the market and even shorter wavelength pointers can be expected as solid state laser technology advances. The new Class 2 laser diode pointers are both more brilliant and safer.

Safety officers should:

- Warn employees about the hazard of misusing diode laser pointers—they could cause injury if someone stares into the beam for a prolonged period.
- Replace Class 3a 670 nm pointers with Class 2 635 nm pointers.

Safety officers working for retailers should advise management to market only Cass 2 laser pointers. The possibility that a child could misuse a Class 3a pointer cannot be overlooked.

## Nonbeam Hazards of Lasers

The two greatest hazards of lasers, other than laser weapons, are electricity and fire. Electrocution from high-voltage power supplies and capacitor banks is a real hazard and can kill while most laser beam exposures cause some loss of vision at worst. A draft NFPA standard for laser fire protection advises that a continuous wave laser radiation creating an irradiance above beam irradiances >2 W/cm$^2$ is an ignition hazard. The beam could ignite flammable substances such as paper and solvents, so flammable materials must be kept out of Class 4 and some focused Class 3 laser areas. One partial exception is plastic walls of enclosures that glow where they are struck by visible or strong invisible laser beams while the area is occupied; fire-resistant plastics should be used for these applications. Objects that could be struck by laser beams must be selected to avoid toxic pyrolysis products; for example, some polyurethanes and epoxies produce hydrogen cyanide when they burn. Dye lasers are particular fire hazards because most use solutions of dye in an alcohol, dimethyl sulfoxide (DMSO), or some other flammable or combustible solvent. The dye solution may be located close to pumping energy sources such as flash lamps, which can add the energy needed to ignite the solvent.

Some lasers use flash lamps to supply pumping energy. Flash lamps can explode if dropped, struck, or improperly handled, posing a laceration hazard.

Laser beams can pyrolyze organic materials and investigators who have looked at the pyrolysis products report finding polycyclic aromatic hydrocarbons (PAHs), the carcinogenic substances found in chimney soot. Thus, pyrolysis products must be controlled both in the air and as deposits on surfaces. For example, Kwan (1990) found that epoxy-reinforced graphite composite materials emit a variety of PAHs when cut by CO$_2$ lasers. It is also now known that viable organisms can be found in the plumes created when lasers are used to cut tissue.

Many laser dyes were brought to the market with essentially no toxicology screening and often come from chemical families that include mutagens or highly toxic materials. Mosovsky and Miller, 1986, working with Avila, Felton, and Lewis, found that a number of laser dyes are mutagenic; the chemistries of these dyes are such that a positive mutagenicity finding is cause for concern that the material may be a carcinogen. The original dye recipes used DMSO solvent, which carries other materials through the skin. However, most butyl and some neoprene and nitrile gloves are effective for blocking DMSO. DMSO can increase absorption through the skin of the dye, which may be toxic or mutagenic; it also contains sulfur, which makes disposal by incineration difficult. Dioxane, a potential carcinogen that also forms explosive peroxides if left in contact with air for prolonged time periods, should be avoided as a solvent. Chlorinated solvents are costlier to dispose of than glycols or alcohols because it is difficult to send chlorinated solvents to incinerators. Precautions for laser dyes developed at Lawrence Livermore National Laboratory are summarized in Tables 11–R and 11–S. A variation on this theme is found in the recipe for Q-switching dyes. Dye-filled cells can be used for Q-switching. Some Q-switching dyes use ethylene dichloride, a potential carcinogen, as the solvent.

Excimer lasers use halogen mixtures such as 5 percent or less of fluorine or hydrogen chloride in inert gas. These mixtures are toxic, but fortunately the corrosiveness of the halogens is moderated by dilution in inert gases. Care is still needed in selecting corrosion-resistant materials and consideration must be given to venting these gases during cavity refilling and emergencies. Passivation procedures, which use small quantities of the reactive gas to lay down a nonreactive deposit in surfaces that will later be exposed to larger flows of that gas, must be used to prepare piping for halogen gas service and avoid serious piping failures. Gas-handling equipment that is corrosion-resistant, highly automated, and designed to maximize safety and convenience is now very common in the semiconductor industry. Excimer laser users can obtain information about this technology from gas suppliers affiliated with gashandling equipment manufacturers.

Other forms of radiation, such as flash lamp radiation, electromagnetic fields from power supplies, and x rays from high-voltage devices, can also be hazardous. It may be necessary to provide additional shielding or implement additional controls for these fields and other forms of radiation. Radiation from flash lamps, which can contain 100–1,000 times the energy of the laser beam, could be a hazard during setup and maintenance.

Hazardous materials can be found in optical components (such as heavy metals in detectors), components of lasers (such as beryllia heat sinks of electronic parts), or miscellaneous applications (such as coolants). Thus, an industrial hygiene review of laser maintenance procedures is warranted.

## Other Regulatory Concerns with Lasers

OSHA has promulgated construction safety standard 29 *CFR* 1926.54 for the use of lasers in tasks such as alignment, although general industry uses of lasers are covered by ANSI Z136.1-1993 through the *UAW, Brock v. General Dynamics Land Systems Division* decision because OSHA general industry standards do not address lasers. The use of laser protective eyewear is covered by the eye and face protection standard, 29 *CFR* 1910.132. The OSHA construction standard specifies that those who could be exposed to direct or reflected light above 5 mW must be provided with

**Table 11–R.** Summary of Laser Dye Handling Precautions Developed at Lawrence Livermore National Laboratory

| Action | Dye hazard class | | |
|---|---|---|---|
| | L | M | S |
| *Work Practices* | | | |
| Avoid eating, drinking, smoking, or storing food, drinks, materials, or cosmetics in dye work areas | X | X | X |
| Post signs to this effect | — | — | X |
| Keep dye areas clean; clean up after work | X | X | X |
| Cap off dye lines not in use | X | X | X |
| Minimize quantities of dyes and solutions >1% in use or storage | — | — | X |
| Limit access to dye and work areas with signs | — | — | X |
| Avoid having janitors do dye cleanup | — | X | X |
| Store dyes and solutions >1% in double, labeled containers | — | — | X |
| Use toxic-dust vacuum to clean up dye spills | — | — | X |
| *Personal Protective Equipment* | | | |
| Use safety eyewear | X | X | X |
| Use lab coat | — | X | X |
| Use nitrile or neoprene gloves to handle solutions, impervious gloves to handle powders | X | X | X |
| *Fire Safety* | | | |
| Keep heat, flames, ignition sources away | X | X | X |
| Keep solvents and dye solutions in colored, labeled containers | X | X | X |
| Keep alcohol wastes in labeled safety cans | X | X | X |
| Avoid oxidizers | X | X | X |
| *Equipment* | | | |
| Pressure test systems | X | X | X |
| Install drip pans under systems, enclose systems if possible | X | X | X |
| Provide safety shower/eyewash | — | X | X |
| Mix dyes in hood or glove box | — | X | X |
| Mix dyes in glove box | — | — | X |
| Avoid cracks, crevices, matte-textured surfaces, and dark colors | — | — | X |
| Avoid false floors, if possible | — | — | X |
| Provide a designated dye storage area | — | — | X |
| Provide separate storage for dye-soiled equipment | — | — | X |
| Use mechanical pipetting aids | — | — | X |
| *Spills* | | | |
| Clean up spills | X | X | X |
| Call for help if more than 100 ml of solution is spilled, or if exposure to dye powder is possible | X | X | X |
| Notify safety and health department of all spills | — | — | X |

X = listed precaution;
— = not a listed precaution.
L = limited precaution class (good chemical lab practice, assigned to dyes known to be neither highly toxic nor mutagenic/carcinogenic).
M = moderate precaution (dyes known to be highly toxic or to have unknown toxic, mutagenic, or carcinogenic properties).
S = strict precaution (mutagenic materials, treated with same precautions as carcinogens).

laser eye protection and that only mechanical or electronic devices may be used to guide the alignment of a laser. The standard also sets exposure criteria summarized below:

| | |
|---|---|
| Direct staring limit | $1 \ \mu W/cm^2$ |
| Incidental observing limit | $1 \ mW/cm^2$ |
| Diffused reflected light | $2.5 \ W/cm^2$ |

**Table 11–S.** Laser Dye and Solvent Control Classes in Use at Lawrence Livermore National Laboratory

| Materials/Synonyms | Control Class | Comments |
|---|---|---|
| BBQ | M | Nonmutagenic, unknown toxicity |
| Benzyl alcohol | L | Moderate toxicity, low vapor pressure |
| Carbazine 720 | M | Nonmutagenic, unknown toxicity |
| Coumarin 1/460 | M | Nonmutagenic, moderately toxic |
| Coumarin 2/45 | M | Nonmutagenic, unknown toxicity |
| Coumarin 30/515 | S | Mutagenic, unknown toxicity |
| Coumarin 102/480 | S | Strong mutagen, unknown toxicity |
| Coumarin 120/440 | M | Nonmutagenic, unknown toxicity |
| Coumarin 314/504 | M | Nonmutagenic, unknown toxicity |
| Coumarin 420 | M | Nonmutagenic, unknown toxicity |
| Coumarin 481 | M | Nonmutagenic, unknown toxicity |
| Coumarin 498 | M | Unknown mutagenicity, unknown toxicity |
| Coumarin 500 | S | Mutagenic, unknown toxicity |
| Coumarin 535 | S | Mutagenic, unknown toxicity |
| Coumarin 540A | M | Nonmutagenic, unknown toxicity |
| Cresyl violet 670 | S | Very strong mutagen, unknown toxicity |
| 1,3,5,7-Cyclooctatetrene (COT) | M | Unknown mutagenicity, unknown toxicity |
| DCM | S | Very strong mutagen, unknown toxicity |
| p,p′-diaminoquaterphenyl | M | Nonmutagenic, unknown toxicity |
| p,p′-diaminoterphenyl | S | Mutagenic, unknown toxicity |
| Dioxane | M | Moderate toxicity |
| DMSO | M | Moderate toxicity |
| DODCI | M | Unknown mutagenicity, unknown toxicity |
| DQOCI | M | Unknown mutagenicity, unknown toxicity |
| DPS | M | Doubtful bacterial mutagen, unknown toxicity |
| Ethylene dichloride (1,2 dichloroethane) | M | Suspected carcinogen; avoid inhalation of vapors |
| Ethyl alcohol | L | Low toxicity |
| Ethylene glycol | M | Moderate toxicity, low vapor pressure |
| Fluorescein 548 | M | Unknown mutagenicity, unknown toxicity |
| IR-26 | M | Unknown mutagenicity, unknown toxicity |
| IR-125 | M | Unknown mutagenicity, unknown toxicity |
| IR-132 | M | Unknown mutagenicity, unknown toxicity |
| IR-140 | M | Unknown mutagenicity, unknown toxicity |
| IR-144 | M | Unknown mutagenicity, unknown toxicity |
| Kiton Red 620 | L | Nonmutagenic, practically nontoxic |
| Kodak Q-Switch #2 | M | Unknown mutagenicity, unknown toxicity |
| Kodak Q-Switch #5 | M | Unknown mutagenicity, unknown toxicity |
| LD-390 | M | Unknown mutagenicity, unknown toxicity |
| LD-490 | S | Mutagenic, unknown toxicity |
| LD-688 | S | Mutagenic, unknown toxicity |
| LD-700 | M | Nonmutagenic, unknown toxicity |
| LDS-698 | S | Mutagenic, unknown toxicity |
| LDS-722 | S | Strong mutagen, unknown toxicity |
| LDS-750 | M | Unknown mutagenicity, unknown toxicity |
| LDS-751 | M | Unknown mutagenicity, unknown toxicity |
| LDS-820 | M | Unknown mutagenicity, unknown toxicity |
| LDS-867 | M | Unknown mutagenicity, unknown toxicity |
| 9-Methylanthracene | S | Mutagenic, unknown toxicity |
| Methyl alcohol | L | Moderate toxicity |
| Bis-MSB | M | Nonmutagenic, unknown toxicity |
| Nile Blue 690 | S | Commercial grade is strongly mutagenic; purified dye is not. Unknown toxicity. |
| Oxazine 720 | M | Nonmutagenic, unknown toxicity |
| Rhodamine 6G/590 | M | Moderately toxic. National Toxicology Program tests did not demonstrate strong carcinogenicity. Commercial-grade dye has been found to induce injection-site tumors. |
| Rhodamine 110/560 | S | Weak mutagen, unknown toxicity |
| Rhodamine 610/B | M | Nonmutagenic, moderately toxic |
| Rhodamine 640 | M | Nonmutagenic, unknown toxicity |
| Sulforhodamine | M | Unknown mutagenicity, unknown toxicity |
| Stilbene 420/3 | L | Nonmutagenic, practically nontoxic |
| N,N,N′,N′-tetraethyldiaminoquaterphenyl | M | Nonmutagenic, unknown toxicity |
| N,N,N′,N′-tetraethyldiaminoterphenyl | S | Strong mutagen, unknown toxicity |

L = limited control class; M = moderate control class; S = strict control class. Precautions to be followed for the L class are given in Table 11–R. Dye/solvent mixtures with less than 1 percent dye must be handled as appropriate to the solvent, with the exception that strict-class requirements for container and equipment labeling and spill cleanup must be followed when strict-class dyes are used. When using dye/solvent mixtures with more than 1 percent dye, follow the precautions for the component in the strictest control class.

The appropriate mutagen potency of the dye in the standard Ames *Salmonella* assay is as follows: weak mutagen = <100 revertants/mg; mutagenic = 100–1,000 revertants/mg; strong mutagen = 1,000–10,000 revertants/mg; very strong mutagen = ≥10,000 mutagens/mg. The Ames test is a reliable predictor of whether a compound is a carcinogen in mammals, but it does not predict the potency of the carcinogen. Thus, a weak Ames mutagen could be a strong carcinogen. Ames test data are used because animal testing is more costly and has not been done on most dyes.

The LIA/ANSI Z136 committee has also issued standards Z136.2 and Z136.3 for use of lasers in fiber-optic communications systems and medical applications, respectively. The Z136.3 standard addresses the same concerns as the Z136.1 standard, but allows for greater flexibility and use of administrative controls.

The Center for Devices and Radiological Health (CDRH) of the Food and Drug Administration sets standards for commercial laser products that specify the types of safety devices that must be installed for commercial equipment using lasers. The CDRH also uses control classes based on AELs that are defined in their standard for laser devices (21 *CFR*, part 1040). CDRH specifies using interlocks, beam stops/shutters, and key-in-lock controls. CDRH also has very explicit requirements for audible and visible alarms to announce when a laser is functioning and, for Class 4, when a laser is about to function. The CDRH regulations also include an incident reporting procedure.

## ■ Measuring Optical Radiation

Two types of detector are widely used: thermal and quantum. Thermal detectors are fundamentally no different from globe thermometers used in heat stress studies. They consist of a sensor embedded in an object that is dark-colored to absorb IR radiation, warm up, and produce a measurable response in the detector. Relatively small, lightweight objects are desirable for thermal sensors because they change temperature rapidly, leading to faster response times (but still not fast enough to measure pulsed lasers). A variant of the thermal detector is the pyroelectric detector, which measures the rate of temperature change in crystals. This is much faster than a conventional thermal detector and, with some caution, these can be used for repetitively pulsed lasers. Thermal detectors are best for IR measurements. The heat-absorbing coatings vary in how well they absorb IR, so one should obtain information about the absorption properties of the coating before buying a thermal sensor. Another note of caution is that these detectors respond to temperature changes, so changing the room thermostat can produce a response in the instrument.

Quantum detectors emit electrons in response to being struck by radiation and are best suited for use in the UV, visible, and IR-A bands (up to 1,100 nm). These can be very fast. The detectors are often made of alloys such as cesium telluride, lead telluride, or lead selenide and the responses of the detectors to radiation of differing wavelengths are different. A buyer should check with the instrument vendor about the suitability of a detector for the radiation of interest. Note that no procedure exists for field verification of the ODs of laser eyewear.

## ■ Lighting

Insufficient light causes accidents and reduces work performance. One needs adequate lighting to see hazards in the workplace and to read information such as text and dials. Most lighting concerns are quantitative, but some qualitative concerns may also arise such as contrast, reflections, and color.

The Illumination Engineering Society (IES) advises that 20 footcandles of illuminance are needed for tasks requiring sustained seeing. This is also one aspect of nonionizing radiation covered by various OSHA regulations, as summarized in Table 11-T. IES/ANSI RP-7-1991 specifies the following illuminance levels for safety in normal conditions (where light will not ruin a process or pose a safety hazard):

| | Illuminance Level (footcandles) | |
| --- | --- | --- |
| Degree of Hazard | Low Activity Level | High Activity Level |
| Slight Hazard | 0.5 | 2 |
| High Hazard (obstacle in path) | 1 | 5 |

Illuminance is similar to irradiance and power density, but the levels of light of various wavelengths are weighted in terms of their impact on the functioning of the cone cells of the retina, which are involved in color and detailed vision. The units of illuminance are the lux and the footcandle; 1 footcandle = 10.8 lux. The dominant quantitative standards for industrial lighting in the United States are IES/ANSI RP-1-1982 and RP-7-1991, which address office and industrial lighting, respectively. These standards replace the historical telephone-directory-style list of tasks and listed lighting levels with a procedure in which workplaces, work force, and tasks are analyzed for type of task, age, reflectance of room surfaces, and (in some cases) whether the seeing task is unimportant, important, or critical. Important aspects of any seeing task are object size (the bigger it is, the easier it is to see), contrast, time available to do the seeing job, and luminance. RP-7 notes that luminance is often the only factor that can be controlled. Illuminances are specified in ranges for nine categories (A through I); ranges of three levels of illuminance are specified for each category. The analyst reviews the other factors to determine whether to select from the low, middle, or high end of the illuminance range for each category. The categories and specified illuminances are listed in Table 11-U; the work force and environmental/task factors are summarized in Table 11-V.

Common qualitative concerns include glare (particularly off of VDT screens), contrast, and color. Glare, either reflected or direct, is still a major concern. Reflected glare is usually a specular reflection of a sunlit window or lamp off of a screen or other shiny surface that partially obscures or veils the scene at the reflection. Direct glare is

**Table 11–T.** Summary of OSHA Regulations Concerning General Lighting

| Regulation | Summary | |
|---|---|---|
| 1910.179(c)(4): Cranes | Cab lighting shall be adequate. | |
| 1910.303(g)(1)(v): Electrical work areas | Illumination shall be provided. | |
| 1910.303(h)(3)(ii): Lighting maintenance | Adequate illumination shall be provided. | |
| 1910.333(c)(4): Electrical work practices | Spaces containing energized parts shall be illuminated; work shall not be performed in and people shall not reach into unlit spaces. | |
| 1910.38(q): Exits | Exit signs shall be lit by a reliable light source >5 footcandles. | |
| 1910.120: Hazardous waste site operations | Quantitative specifications: Area | Illumination (footcandles) |
| | General work areas | 5 |
| | Excavation and waste areas, loading platforms, refueling, and field maintenance areas. | 3 |
| | Indoors: warehouses, corridors, hallways, and exit ways. | 5 |
| | Tunnels, shafts, and general underground work areas. (Exception: ≥10 footcandles is required at tunnel and shaft heading during drilling, mucking, and scaling. MSHA-approved cap lights acceptable for use in the tunnel heading). | 5 |
| | General shops (mechanical and electrical equipment rooms, active storerooms, barracks or living quarters, locker or dressing rooms, dining areas, indoor toilets, and workrooms). | 10 |
| | First-aid stations, infirmaries, and offices. | 30 |
| 1910.142(g): Labor camps | Each habitable room shall have at least one ceiling light fixture and a floor or wall outlet. Laundry rooms, toilets, and rooms where people congregate shall contain a ceiling or wall fixture. Toilets and storage rooms shall be lit at ≥20 footcandles 30 in. above the floor. Other rooms, including kitchens and living quarters, shall be lit ≥30 footcandles 30 in. above the floor. | |
| 1910.219(c)(5)(iii): Power transmission equipment in basements | Lighting shall conform to ANSI A11.1-1970. | |
| 1910.68(b)(6)(iii): Manlifts | Lighting ≥5 footcandles shall be provided at each landing when lift is operating. | |
| 1910.68(b)(14): Manlifts | Lighting ≥1 footcandles shall be provided in runs when lift is operating. | |
| 1910.178(h): Industrial trucks | Supplemental lighting shall be provided where lighting is ≤2 lumens/ft². | |
| 1910.261(b)(7): Pulp and paper mills | Emergency lighting shall be provided where operators must stay during emergencies, in passageways, stairways, and aisles used for emergency exit, and at first-aid and medical facilities. | |
| 1910.261(c)(10): Pulp and paper mills | Loading/unloading areas shall be lit in accordance with ANSI A11.1-1970. | |
| 1910.266(e)(15): Pulpwood logging | Lighting shall be provided for night work if needed. | |
| 1910.265(c)(5)(iii): Sawmills | Stairway shall be adequately illuminated. | |
| 1910.265(c)(9): Sawmills | Work areas shall be provided with adequate illumination. | |
| 1910.265(c)(23)(iii): Sawmills | Fuel houses and bins shall have adequate exits and lighting. | |
| 1910.268: Telecommunications | Adequate lighting shall be provided. | |
| 1926.56: Construction | Areas not covered by the following table shall be illuminated in accordance with ANSI A11.1-1970. | Illumination (footcandles) |
| | Quantitative specifications: Area | |
| | General construction area lighting | 5 |
| | General construction areas, concrete placement, excavation and waste areas, access ways, active storage areas, loading platforms, refueling, and field maintenance areas. | 3 |
| | Indoors: warehouses, corridors, hallways, and exit ways. | 5 |
| | Tunnels, shafts, and general underground work areas. (Exception: ≥10 footcandles is required at tunnel and shaft heading during drilling, mucking, and scaling. MSHA-approved cap lights acceptable for use in the tunnel heading). | 5 |
| | General construction plant and shops (batch plants, screening plants, mechanical and electrical equipment rooms, carpenter shops, rigging lofts and active storerooms, barracks or living quarters, locker or dressing rooms, dining areas, and indoor toilets and workrooms). | 10 |
| | First-aid stations, infirmaries, and offices. | 30 |

a relatively bright object, such as an unshaded window, in an otherwise dark area that prevents the eyes from adapting to the dark area. Reflected glare can be controlled by locating the screen or other surface of interest so it does not reflect the images of windows or lamps.

A screen can be angled so it does not reflect the images of lamps or windows into the user's eyes. In some cases, visors or partitions can be used to block light from lamps or windows. Another option is to place a textured surface above the object that breaks up specular reflec-

tions while allowing the light from the object below it to pass through. Dimpled plastic is often used. RP-7 devotes an entire Annex to the subject of glare.

RP-7 also gives guidance for lighting contrast, listed in Table 11–W. A gradation of contrasts is sought between the task and its immediate and more remote visual surroundings. In essence, strong lighting can exist in an area of moderate lighting, which can be surrounded by a dimly lit or unlit expanse, but darkness should not immediately surround brightness. An example of harsh contrast is a lit desk

**Table 11–U.** Illuminance Categories of IES/ANSI RP-7-1991

| Type of Activity | Examples | Illuminance Category | Range of Illuminance (footcandles) | Reference Workplane |
|---|---|---|---|---|
| Public spaces with dark surroundings | Aircraft ramp area | A | 2–3–5 | |
| Simple orientation for short visits | Active storage area of a farm, VDT screens | B | 5–7.5–10 | General lighting throughout spaces |
| Working spaces where visual tasks are only occasionally performed | Active traffic area of a garage, elevators/escalators | C | 10–15–20 | |
| Performance of visual tasks of high contrast or large size | Simple assembly or inspection | D | 20–30–50 | |
| Performance of visual tasks of medium contrast or small size | Hand decorating in a bakery, mail sorting | E | 50–75–100 | Illuminance on task |
| Performance of visual tasks of low contrast or very small size | Finished lumber grading, model making | F | 100–150–200 | |
| Performance of visual tasks of low contrast and very small size over a prolonged period | Sewing clothes | G | 200–300–500 | Illuminance on task obtained by a combination of general and local (supplementary) lighting |
| Performance of very prolonged and exacting visual tasks | Exacting inspection | H | 500–750–1,000 | |
| Performance of very special visual tasks of extremely low contrast and small size | Cloth inspection and examining (perching) of sewn products | I | 1,000–1,500–2,000 | |

in a poorly lit warehouse. The lamp at the desk should be supplemented by area lighting to avoid contrast problems. People look away from their visual tasks from time to time, so the person at this desk would probably not wish to stay there because the visual contrast between the desk and its visual surroundings is too great for comfort.

Nonlaser lighting adheres to the inverse square law. This means that lights must be placed close to areas being lit if the lighting is needed to perform a task. Sometimes, this cannot be done by area lighting alone. A warehouse may need area lights placed above the heights of shelves and forklift trucks, so supplemental lighting, provided by floor or desk lamps, may be needed at desks located in the warehouse. The limits of supplemental lighting include harsh contrast, already mentioned, and also the possibility that the supplemental lighting could cause direct glare for people in the surrounding area. Energy conservation and safety needs can be reconciled by using

motion detectors to activate area lighting when a person enters an area. Building and lighting cleaning and painting can make a tremendous difference in lighting by causing more of the light emitted from lamps to be reflected to places where people are working. Light, matte-textured surfaces are preferred. Matte textures avoid specular reflections and reflected glare. RP-1 and RP-7 give guidance about how reflective surfaces should be.

Color can be a problem if unusual fluorescent tubes or colored incandescent bulbs are installed. White light contains radiation associated with every color we can see; colored lights radiate selected wavelengths more intensely. Colored lighting is useful for some jobs, such as blue-enhanced fluorescent tubes for greenhouse lighting or yellow-orange low-pressure sodium lamps for abundant yet cheap safety lighting at night. Colored lighting without some benefit can create difficulties. Colors may be harder to perceive when nonwhite lighting is used. Yellow and

**Table 11–V.** Factors Influencing Assigning Illumination Levels Within Illuminance Categories of IES/ANSI RP-1-1982 and RP-7-1991

*For Illuminance Categories A Through C in Industrial and Office Settings*

| Room and Occupant Characteristics | Weighting | | |
|---|---|---|---|
| | Low End of Range | Mid-Range | High End of Range |
| Occupant ages | <40 | 40–55 | >55 |
| Room surface reflectances | >70% | 30–70% | <30% |

*For Illuminance Categories D Through I in Industrial and Office Settings*

| Task and Worker Characteristics | Weighting | | |
|---|---|---|---|
| | Low End of Range | Mid-Range | High End of Range |
| Occupant ages | <40 | 40–55 | >55 |
| Speed or accuracy of seeing | Not important | Important | Critical |
| Reflectance of task background | >70% | 30–70% | <30% |

**Table 11–W.** IES/ANSI RP-7-1991 Recommended Maximum Luminance Ratios

| Situation | Environmental Group | | |
| | A | B | C |
| --- | --- | --- | --- |
| Between tasks and adjacent darker surroundings | 3 to 1 | 3 to 1 | 5 to 1 |
| Between tasks and adjacent lighter surroundings | 1 to 3 | 1 to 3 | 1 to 5 |
| Between tasks and adjacent more remote darker surfaces | 10 to 1 | 20 to 1 | Control not practical |
| Between tasks and adjacent more remote darker surfaces | 1 to 10 | 1 to 20 | Control not practical |
| Between luminaires (or windows, skylights, etc.) and surfaces adjacent to them | 20 to 1 | Control not practical | Control not practical |
| Anywhere within normal field of view | 40 to 1 | Control not practical | Control not practical |

Environmental groups: A-interior areas where reflectances of entire space can be controlled in line with recommendations for optimum seeing conditions; B-areas where reflectances of immediate work area can be controlled, but control of remote surroundings is limited; C-areas (indoor or outdoor) where it is impractical to control reflectances and difficult to alter environmental conditions.

white objects could, for example, both appear the same in yellow or red lighting, so yellow signs and warning devices could become unreliable and blue surfaces would appear to be black.

Concern has been expressed about the safety of fluorescent tubes. Fluorescent tubes contain a minute amount of mercury that conducts electric current and glows in the ultraviolet portion of the spectrum. The UV is absorbed by phosphors that line the tubes and is reradiated as visible light. ICNIRP issued a position statement that advises that UV emissions from fluorescent tubes are not a problem. Mercury vapor lamps are also used for lighting and they generate UV that is absorbed by an outer glass sheath. Mercury vapor lamps are safe unless the outer sheath is broken. A mercury vapor lamp with a broken sheath should be turned off and replaced immediately. It should be noted that disposing of fluorescent tubes is associated with other industrial hygiene concerns. Tube-breaking equipment can be noisy and heavily contaminated with toxic mercury metal, which can accumulate in tube breaking equipment. Fluorescent tubes contained highly toxic beryllium salts in the 1940s. Cadmium compounds were used, but their use was discontinued in the late 1980s, so cadmium exposure remains a potential hazard for tube breakers.

Lighting measurements are usually made 30 in. above the floor to measure the illumination striking surfaces that are to be seen. Special measurements can be made on surfaces of interest, such as desktops and working surfaces. The instruments used are typically inexpensive photoelectric devices. It is noted that IES developed guidance for lighting VDT workplaces, RP-24-1989, which is listed in the Bibliography.

# ■ Summary

- Nonionizing "radiation" is often not radiation, but rather discrete electric and magnetic fields that exist independently of one another, whereas the fields in radiation are rigidly interrelated. The frequency below which it is assumed one is dealing with fields rather than radiation is 300 MHz.

- Electric fields are caused by nonmoving or moving electric charges and the electric field increases as the quantity of charge increases. Magnetic fields are caused by the flow of electric charges, or electric current, and increase as more current flows. In other words, electric charges create electric fields and moving them creates magnetic fields.

- The frequency of electromagnetic radiation times its wavelength equals its speed of travel, very close to 300,000,000 m/s in air or a vacuum. The frequency of electromagnetic radiation is the number of times the fields go through a complete cycle of polarity and strength change. Frequencies are expressed in hertz (Hz), the number of polarity and strength changes that occur in a second. The wavelength is the distance traveled as the radiation goes through that cycle of polarity and strength change. Higher frequencies mean shorter wavelengths and lower frequencies mean longer wavelengths.

- The various frequencies of the electric and magnetic field are divided into the electromagnetic spectrum. This is divided into subradiofrequency fields, which have frequencies below 3,000 Hz, radiofrequencies from 3,000 to 300,000,000,000 Hz including microwaves, which span 300,000,000 to 300,000,000,000 Hz, and optical radiation with higher frequencies and energies.

- Electromagnetic devices can be thought of as containing a source of energy, a transmission path, and a receiver of the energy.

- The exposure guideline for static magnetic fields was set by ICNIRP at 200 mT (2,000 [G]). The exposure guideline for static electric fields was set by ACGIH at 25 kV/m.

- ACGIH and ICNIRP issued ELF exposure criteria for workers based on avoiding induced currents that are stronger than those already created in the body by the normal functioning of nerves and muscles.

- Radiofrequency exposure standards are based on avoiding hazardous electric current flows at frequencies below a few MHz and on avoiding excessive rates of energy deposition at higher frequencies. Energy deposition is most significant at frequencies where a person's height is 20–40 percent of the wavelength of the radiation in air (30–300 MHz). At frequencies above a few GHz, energy deposition occurs mainly in the skin. The frequency of 300 GHz has a wavelength of 1 mm and higher frequencies are classified as IR radiation.

- Optical radiation is described by its wavelength and is divided into IR (760 nm–1 mm), visible (400–700 nm), and UV (variously 4 or 100 nm–400 nm). There are TLVs for all nonlaser optical radiation from 180–1,400 nm and LIA/ANSI Z136.1-1993 addresses wavelengths from 180 nm to 1 mm. Shorter wavelengths are absorbed by oxygen to make ozone.

- Laser radiation is monochromatic (literally one color) or has just a few wavelengths, coherent (well-organized), and bright. Coherent radiation is very directional, so lasers can project intense or hazardous energies over longer distances than nonlaser sources of optical radiation.

- Laser radiation with wavelengths between 400 and 1,400 nm, the ocular hazard region, can be intensely focused on the retina. This makes lasers operating in this wavelength range particularly hazardous.

- Lasers are controlled by dividing them into hazard classes ranging from 1 (essentially harmless) to 4 (very hazardous). Precautions become more stringent as the hazard class increases.

- OSHA has limited standards for nonionizing radiation, but can enforce consensus standards according to the *UAW, Brock v. General Dynamics Land Systems Division* decision. These standards come from the ACGIH, the ICNIRP, the IEEE, and the LIA and are referenced in the Bibliography. European international organizations, particularly Swedish and Finnish organizations, are also active in this area.

- Industrial lighting standards relate to safety and productivity. They are promulgated by the IES.

Nonionizing radiation is becoming more and more a part of our lives on and off the job. Dealing with it will be one of the larger challenges facing the industrial hygiene profession in the future.

# Bibliography

American Conference of Governmental Industrial Hygienists. *Threshold Limit Values* (latest ed.). Cincinnati, OH: American Conference of Governmental Industrial Hygienists.

American Conference of Governmental Industrial Hygienists. *Documentation of the Threshold Limit Values and Biological Exposure Indices*, 6th Ed., Vol. III, PA-45–PA-70. Cincinnati, OH: American Conference of Governmental Industrial Hygienists, 1993.

Anderson LE. ELF: Exposure levels, bioeffects, and epidemiology. *Health Phys* 61:41–46, 1991.

Anderson LE. Biological effects of extremely low-frequency electromagnetic fields: In vivo studies. *AIHAJ* 54:186–196, 1993.

Bates MN. Extremely low frequency electromagnetic fields and cancer: The epidemiologic evidence. *Environ Health Perspect* 95:147–156, 1991.

Bonneville Power Administration. *Electrical and Biological Effects of Transmission Lines—A Review*, DOE/BP-945. Portland, OR: Bonneville Power Administration, 1993.

Bracken TD. Exposure assessment for power frequency electric and magnetic fields. *AIHAJ* 54:165–177, 1993.

Breysse PN, Gray R. Video display terminal exposure assessment. *Appl Occup Environ Hygiene* 9:671–677, 1994.

Bridges JE, Frazier MJ. *The Effects of 60-Hz Electric and Magnetic Fields on Implanted Cardiac Pacemakers*, EPRI EA-1174. Palo Alto, CA: Electric Power Research Institute, 1979.

Cahill DF, Elder JA, eds. *Biological Effects of Radiofrequency Radiation*, EPA-600/8-83-026F. Washington, DC: Environmental Protection Agency, 1984.

Cleary SF. A review of in vitro studies: Low-frequency electromagnetic fields. *AIHAJ* 54:178–185, 1993.

Confederation of Professional Employees. *Screen Facts*. Chicago: TCO, 1993.

David A, Savitz DA. Overview of epidemiologic research on electric and magnetic fields and cancer. *AIHAJ* 54:197–204, 1993.

Davis JG et al. Health effects of low-frequency electric and magnetic fields. *Environ Sci Technol* 27:42–51, 1993.

Department of Energy and Public Policy, Carnegie Mellon University. *Part 1: Measuring Power Frequency Magnetic Fields*. Pittsburgh: Carnegie Mellon University, 1993.

Department of Energy and Public Policy, Carnegie Mellon University. *Part 2: What Can We Conclude from Measurements of Power Frequency Magnetic Fields?* Pittsburgh: Carnegie Mellon University, 1993.

Duchene AS, Lakey JRA, Repacholi MH, eds. *IRPA Guidelines on Protection Against Non-Ionizing Radiation*. New York: McGraw-Hill, 1991.

Environmental Protection Agency. *Laboratory Testing of Commercially Available Power Frequency Magnetic Field Meters*, Final Report, 400R-92-010. Washington, DC: Environmental Protection Agency, 1991.

Feero WE. Electric and magnetic field management. *AIHAJ* 54:205–210, 1993.

Floderus B et al. *Occupational Exposures to Electromagnetic Fields in Relation to Leukemia and Brain Tumors, a Case Control Study*. Solna, Sweden: Department of Neuromedicine, National Institute of Occupational Health, 1992.

Franeschetti G, Gandhi OP, Grandolfo M, eds. *Electromagnetic Biointeraction—Mechanisms, Safety Standards, Protection Guides*. New York: Plenum, 1989.

Gandhi OP, ed. *Biological Effects and Medical Applications of Electromagnetic Energy*. Englewood Cliffs, NJ: Prentice Hall, 1990.

Goldhaber MK, Polen MR, Hiatt RA. The risk of miscarriages and birth defects among women who use visual display terminals during pregnancy. *Am J Ind Med* 13:695–706, 1988.

Grandolfo M, Michaelson SM, Rindi A, eds. *Biological Effects and Dosimetry of Static and ELF Electromagnetic Fields*. New York: Plenum, 1985.

Hankin NN. *The Radiofrequency Radiation Environment: Environmental Exposure Levels and RF Emitting Sources*, EPA

520/1-85-014. Washington, DC: Environmental Protection Agency, 1986.

Hitchcock RT, McMahan S, Miller GC. *Nonionizing Radiation Guide for Extremely Low Frequency (ELF) Electric and Magnetic Fields.* Fairfax, VA: American Industrial Hygiene Association,1995.

Hitchcock RT, *Nonionizing Guide for Radio-Frequency and Microwave Radiation.* Fairfax, VA: American Industrial Hygiene Association, 1994.

Illumination Engineering Society. *American National Standard Practice for Industrial Lighting,* RP-7. New York: Illumination Engineering Society, 1990.

Illumination Engineering Society. *VDT Lighitng—IES Recommended Practices for Lighting Offices Containing Computer Visual Display Terminals,* IES RP-24. New York: Illumination Engineering Society, 1989.

Illumination Engineering Society. *Office Lighting—American National Standard Practice,* ANSI/IES RP-1-1982. New York: Illumination Engineering Society, 1982.

Institute of Electrical and Electronic Engineers. *American National Standard Safety Levels with Respect to Human Exposure to Radio Frequency Electromagnetic Fields, 3 kHz to 300 GHz,* C95.1-1991. Piscatatway, NJ: Institute of Electrical and Electronic Engineers, 1992.

International Commission for Non-Ionizing Radiation Protection. Guidelines on limits of exposure to static magnetic fields. *Health Phys* 66:100–106, 1994.

Juutilainen J. Effects of low-frequency magnetic fields on embryonic development and pregnancy. *Scand J Work Environ Health* 17:149–158, 1991.

Kaufman JE, ed. *IES Lighting Handbook—1981 Application Volume.* New York: Illumination Engineering Society, 1981.

Kaufman JE, ed. *IES Lighting Handbook—1981 Reference Volume.* New York: Illumination Engineering Society, 1981.

Kavet R, Tell R. VDTs: Field levels, epidemiology, and laboratory studies. *Health Phys* 61:47–57, 1991.

Kwan JK, *Health Hazard Evaluation of the Postcuring Phase of Graphite Composite Operations of the Lawrence Livermore National Laboratory, Livermore, CA* (PhD thesis) UCR2-LR-104684. Livermore, CA: Lawrence Livermore National Laboratory, 1990.

Laser Institute of America. *American National Standard for Safe Use of Lasers,* Z136.1-1993. Orlando, FL: Laser Institute of America, 1993.

Laser Institute of America. *Safe Use of Optical Fiber Communication Systems Using Laser Diode and LED Sources,* Z136.2-1988. Orlando, FL: Laser Institute of America, 1988.

Laser Institute of America. *Safe Use of Lasers in Health Care Facilities,* Z136.3-1988. Orlando, FL: Laser Institute of America, 1988.

Lindbohm M-L et al. Magnetic fields of video display terminals and spontaneous abortions. *Am J Epidemiol* 136:1041–1051, 1992.

Luben RA. Effects of low-energy electromagnetic fields (pulsed and DC) on membrane signal transduction processes in biological systems. *Health Phys* 61:15–28, 1991.

Marriott IA, Stuchly MA. Health aspects of work with visual display terminals. *J Occup Med* 28:833–848, 1986.

McMahan S, Ericson J, Meyer J. Depressive symptomatology in women and residential proximity to high-voltage transmission lines. *Am J Epidemiol* 139:58–63, 1994.

Michaelson SM. Biological effects of radiofrequency radiation: Concepts and criteria. *Health Phys* 61:3–14, 1991.

Miller GC. Exposure guidelines for magnetic fields. *AIHAJ* 48:957–968, 1987a.

Miller GC Precautions for Handling Laser Dyes. In *Hazards Control Annual Technology Review 1986,* UCRL-50007-86. Livermore, CA: Lawrence Livermore National Laboratory, 1987.

Mosovsky JA. *Laser Dye Toxicity, Hazards and Recommended Controls,* UCRL-89148. Livermore, CA: Lawrence Livermore National Laboratory, 1983.

Murray WE. Video display terminals: Radiation issues. *Library High Tech* 12(3):43–47, 1985.

Nair I, Morgan MG, Florig HK. *Biological Effects of Power Frequency Electric and Magnetic Fields.* Background paper prepared for U.S. Congress Office of Technology Assessment. Technical Information Service accession no. PB89-209985. Springfield, VA: National Technical Information Service, 1989.

National Council for Radiological Protection and Measurements. *A Practical Guide to the Determination of Human Exposure to Radiofrequency Fields,* NCRP Report No. 119. Washington, DC: National Council for Radiological Protection and Measurements, 1993.

National Council for Radiological Protection and Measurements. *Biological Effects and Exposure Criteria for Radiofrequency Electromagnetic Fields,* NCRP Report No. 86. Washington, DC: National Council for Radiological Protection and Measurements, 1986.

Occupational Safety and Health Administration. 21 *CFR* Subchapter J, "Radiological Health."

Olsen JH, Nielsen A, Schulgen G. Residence near high voltage facilities and risk of cancer in children. *Br Med J* 307:891–895, 1993.

Petersen RC. Radiofrequency/microwave protection guides. *Health Phys* 61:59–67, 1991.

Pool R. Is there an EMF–cancer connection? *Science* 249:1096–1098, 1990.

Pool R. Electromagnetic fields: The biological evidence. *Science* 249:1378–1381, 1990.

Pool R. Flying blind: The making of EMF policy. *Science* 250:23–25, 1990.

Reilly JP. *Peripheral Nerve and Cardiac Excitation by Time-Varying Magnetic Fields: A Comparison of Thresholds.* Silver Spring, MD Metatec Associates, 1990.

Rockwell RJ. Laser accidents: Reviewing 30 years of incidents—What are the concerns old and new. *Laser Appl* 6:203–211, 1994.

Savitz DA. Health effects of low-frequency electric and magnetic fields. *Environ Sci Technol* 27:52–54, 1993.

Savitz DA et al. Case-control study of childhood cancer and exposure to 60 Hz magnetic fields. *Am J Epidemiol* 128:21–38, 1988.

Savitz DA, Pearce NA, Poole C. Methodological issues in the epidemiology of electromagnetic fields and cancer. *Epidemiol Rev* 11:59–78, 1989.

Schnorr TM et al. Video display terminals and the risk of spontaneous abortions. *N Engl J Med* 324:727–733, 1991.

Sliney D, Wolbarsht M. *Safety with Lasers and Other Optical Sources—A Comprehensive Handbook.* New York: Plenum, 1985.

Steneck NH et al. The origins of U.S. safety standards for microwave radiation. *Science* 208:1230–1237, 1980.

Suess MJ, ed. *Nonionizing Radiation Protection,* WHO Regional Publications European Series No. 10. Copenhagen, Denmark: WHO Regional Office for Europe, 1982.

Swedish Board for Technical Accreditation. *Test Methods for Visual Display Units,* MPR 1990:8 1990-12-01. Northridge, CA: Standards Sales Group, 1990.

Swedish Board for Technical Accreditation. *User's Handbook for Evaluating Visual Display Units,* MPR 1990:10 1990-12-31. Northridge, CA: Standards Sales Group, 1990.

Tenforde TS. Health effects of low-frequency electric and magnetic fields. *Environ Sci Technol* 27:56–58, 1993.

Theriault G et al. Cancer risks associated with occupational exposure to magnetic fields among utility workers in Ontario and Quebec, Canada, and France: 1970–1989. *Am J Epidemiol* 139:550–572, 1994.

Varanelli AG, Electrical hazards associated with lasers. *Laser Appl* 7:62–64, 1995.

Verkasalao PK et al. Risk of cancer in Finnish children living close to power lines. *Br Med J* 307:895–899, 1993.

Wertheimer N, Leeper E. Electrical wiring configurations and childhood cancer. *Am J Epidemiol* 109:273–284, 1979.

Wiley MJ et al. The effects of continuous exposure to 20-kHz sawtooth magnetic fields on the litters of CD-1 mice. *Teratology* 46:391–398, 1992.

World Health Organization. *Environmental Health Criteria 69—Magnetic Fields.* Albany, NY: WHO Publications Centre, 1987.

World Health Organization. *Environmental Health Criteria 35—Extremely Low Frequency (ELF) Fields.* Albany, NY: WHO Publications Centre, 1984.

Yost M et al. California protocol for measuring 60 Hz magnetic fields in residences. *Appl Occupation Environ Hygiene* 7:772–777, 1992.

## Periodicals

*Bioelectromagnetics,* Subscription Department, John Wiley and Sons, Inc., 9th Floor, 605 Third Avenue, New York, NY 10158-0012, phone (212) 850-6543. *Journal of the Bioelectromagnetics Society and the Society for Physical Regulation in Biology and Medicine.* The Bioelectromagnetics Society administrative office is managed by W/L Associates, Ltd., at 7519 Ridge Road, Frederick, MD 21702-3519, phone (301) 663-4252.

*Biological Effects of Nonionizing Electromagnetic Radiation (BENER Digest Update—A Digest of Current Literature),* Information Ventures, Inc., 1500 Locust St., Suite 3216, Philadelphia, PA 19102, phone (215) 732-9003, fax (215) 732-3754. Quarterly listing of literature abstracts worldwide.

*EMF Health & Safety Digest* (formerly *Transmission/Distribution Health & Safety Report*), EMF Information Project, 2701 University Avenue Southeast, Minneapolis, MN 55414-0501, phone (612) 623-4600, fax (612) 623-3645. Project's support includes utility organizations worldwide. Reviews of conferences, literature developments, and lists of papers are very useful.

*EMF Health Report,* Information Ventures, Inc., 1500 Locust St., Suite 3216, Philadelphia, PA 19102, phone (215) 732-9083, fax (215) 732-3754. Aimed at lay audiences.

*Journal of Laser Applications,* Laser Institute of America, 12424 Research Parkway, Suite 125, Orlando, FL 32826, phone (407) 380-1553, fax (407) 380-5588. Journal of the Laser Institute of America.

*Laser Focus World,* 10 Tara Boulevard, Fifth Floor, Nashua, NH 03062, phone (603) 891-0123, fax (603) 891-0574. Free to those who make or influence procurement decisions.

*Microwave News,* PO Box 1799, Grand Central Station, New York, NY 10163, phone (212) 517-2800, fax (516) 734-0316. Editorial viewpoint contrasts with that of EMF Health & Safety Digest.

*Photonics Spectra,* Berkshire Common, P. O. Box 4949, Pittsfield, MA 01202, phone (413) 499-0514, fax (413) 442-3180. Free to those who make or influence procurement decisions.

*VDT News,* PO Box 1799, Grand Central Station, New York, NY 10163, phone (212) 517-2800, fax (516) 734-0316. This is published by the Microwave News organization and also covers ergonomic issues.

## ■ Internet Worldwide Web and Turbogopher Services

Worldwide Web home page for EMF issues: http://archive.xrt.upenn.edu:1000/0h/emf/top/emf-link.html. Provided by Information Ventures.

Worldwide Web home page for laser safety: http://iac.net:80/~rli/index.html. Provided by Rockwell Laser Industries.

The University of Minnesota's Turbogopher service has an EMF-BIO bulletin board. The address is: emf-bio@net.bio.net.

# CHAPTER 12

# Thermal Stress

by Thomas E. Bernard, PhD, CIH

*Acknowledgment: The chapter on Temperature Extremes was first prepared by Mr. Edwin L. Alpaugh and was revised in the 3rd edition by Dr. Theodore J. Hogan. Dr. Francis N. Dukes-Dobos provided valuable review of the material in this revision of the chapter. Although direct reference is not made to specific sources, many ideas and concepts in this chapter have been adapted from sources in the bibliography.*

**Degrees of Thermal Stress**

**Thermal Balance**
*Model of Thermal Balance* ▪ *Factors Affecting Thermal Balance*

**Heat Stress**
*Recognition of Heat Stress* ▪ *Evaluation of Heat Stress* ▪ *Computation of Heat Exchange* ▪ *Control of Heat Stress* ▪ *Hot Surfaces, Breathing Hot Air, and Respirators*

**Cold Stress**
*Model of Thermal Balance* ▪ *Measurement of Cold Stress* ▪ *Recognition of Cold Stress* ▪ *Evaluation of Cold Stress* ▪ *Control of Cold Stress*

**Thermal Comfort**

**Summary**

**Bibliography**

THERMAL STRESS IS A SIGNIFICANT physical agent in many working environments. For routine work outdoors, air temperatures between –20 and 110 F are routinely expected over different regions of the United States. Other countries can reasonably expect temperatures beyond that range. Artificial environments from freezers to ovens extend the range of thermal environments in which work is expected to be performed. Because tasks must be performed under adverse thermal conditions, this chapter provides guidance for recognition, evaluation, and control of work in thermal extremes.

## Degrees of Thermal Stress

Work can occur in one of five zones along the continuum of thermal stress. In the middle is the comfort zone. Here, most people report thermal sensations as being acceptable (neither hot nor cold). In the comfort zone, the demands for physiological adaptation are modest, and productivity should be the greatest. The comfort zone is described at the end of this chapter in order to guide health and safety professionals who have been asked to evaluate thermal conditions with comfort as a goal.

On either side of the comfort zone are the discomfort zones for heat and cold stress. Under these conditions, most people are able to safely work without experiencing a disorder related to the stress (that is, heat- or cold-related disorders). They do, however, report sensations of cold or heat, and productivity and quality of work may decrease, and the risk of accidents increase. The goal of most evaluation schemes for occupational heat and cold stress is to limit exposures to the discomfort zone.

The health risk zones for heat and cold stress are the outer zones of the thermal stress continuum. Physiological adaptations have reached their limits, and work capacity is severely limited. In the health risk zone, the likelihood of heat and cold stress–related disorders increases markedly. Health and safety professionals should manage exposures in the health risk zone. Management of exposures in the health risk zone is the principal theme of this chapter.

Of course, there are no firm boundaries to these zones, because the boundaries depend on the environment, individuals, and season, as well as many unknown variables. But we should try to control the thermal stress factors for the less tolerant workers, in order to minimize the risk of injury and illness to the lowest reasonable level. The major emphasis of evaluation and control is on the transition from the discomfort zone to the health risk zone, for both heat and cold stress.

## ■ Thermal Balance

### Model of Thermal Balance

Three factors influence the degree of thermal stress. The most obvious is the climatic conditions of the environment. The other two factors are work demands and clothing. The tradition for more than 40 years has been to describe thermal balance with an equation in which the major avenues of heat exchange between the body and the environment are represented by variables. (There is no uniformly accepted version, but it is easy to reconcile different versions as they are found.)

$$S = (M + W) + R + C + K + (C_{resp} + E_{resp}) + E \qquad (1)$$

where $S$ = heat storage rate
$M$ = metabolic rate
$W$ = external work rate
$R$ = radiant heat exchange rate
$C$ = convective heat exchange rate
$K$ = conductive heat exchange rate
$C_{resp}$ = rate of convective heat exchange by respiration
$E_{resp}$ = rate of evaporative heat loss by respiration
$E$ = rate of evaporative heat loss

Most versions of the heat balance equation use ± instead of +, especially in front of $R$ and $C$. The purpose is to emphasize that heat exchange represented by $R$, $C$, $K$, and $C_{resp}$ can be in either direction. A more rigorous sign

convention is used in this chapter. A positive value for any of these terms (as opposed to the sign in front of the term) means that the heat is gained by the body and a negative value means that heat is lost from the body. The values for $M$ and $(M + W)$ can only be positive. The values for $W$, $E_{resp}$, and $E$ are always negative, meaning that only heat loss is associated with these terms. Each term has the unit of energy per unit of time; that is, the terms represent rates of energy transfer. The international units (SI units) are Watts, and other units that are reported include kcal/hr, kcal/min and Btu/hr. Sometimes the rates are reported as normalized values per body surface area.

***S—Heat storage rate.*** If the value for $S$ is zero, the body is in thermal equilibrium, and any heat gain is balanced by loss from the body. If $S$ is positive, the body is gaining heat at the rate indicated by the value of $S$. If $S$ is negative, the body is losing heat, and body temperature is decreasing.

***M—Metabolic rate.*** Chemical reactions occur continuously inside the body. These serve to sustain life (as in the basal metabolism) and meet the demands of work (as in muscle metabolism). As muscle metabolism increases to meet work demands, the rate of energy conversion from chemical energy to kinetic energy rises. Because energy conversion from chemical energy to kinetic energy is inefficient, increased metabolism results in increased rates of heat gain to the person. The rate of metabolism depends directly on the rate and type of external work demanded of the body.

***W—External work rate.*** The external work rate is the amount of energy that is successfully converted from chemical energy to mechanical work outside of the body. This route of energy transfer is called external work, and it does not contribute to body heat. The rate of external work depends directly on external forces applied against the body and distance moved. $W$ is usually about 10 percent of $M$.

***R—Radiant heat exchange rate (radiation).*** Solid bodies of different temperatures have a net heat flow from the hotter to the cooler surface by electromagnetic radiation (primarily infrared radiation). The rate of heat transfer by radiation depends on the average temperature of the surrounding solid surfaces, on skin temperature, and on clothing.

***C—Convective heat exchange rate (convection).*** The exchange of heat between the skin and the surrounding air is known as convection. The direction of heat flow depends on the temperature difference between the skin and air. If air temperature is greater than that of the skin, $C$ is positive, and heat flows from the air to the skin. If the air is cooler than the skin, $C$ is negative, and heat flows from the body. The rate of convective heat exchange depends on the magnitude of the temperature difference, the amount of air motion, and clothing.

***K—Conductive heat exchange rate (conduction).*** When two solid bodies are in contact, heat flows from the warmer body to the cooler body. The rate of heat transfer depends on the difference in the temperatures of the skin and the solid surface, the thermal conductivity of the solid body that the person contacts, and clothing that separates the person from the solid surface.

**$C_{resp}$—Rate of convective heat exchange by respiration.** The movement of air in and out of the lungs, which have a large surface area, results in an opportunity to gain or lose heat. The rate of heat exchange depends on the air temperature and volume of air inhaled.

**$E_{resp}$—Rate of evaporative heat loss by respiration.** The large surface area of the lungs provides an opportunity to lose body heat through evaporation. The rate of heat exchange depends on air humidity and volume of air inhaled.

**$E$—Rate of evaporative heat loss.** Sweat on the skin surface absorbs heat from the skin when evaporating into the air. The process of evaporation cools the skin and, in turn, the whole body. The rate of evaporative heat loss depends on the amount of sweating, air movement, ambient humidity, and clothing.

Because $W$, $K$, $C_{resp}$, and $E_{resp}$ are small relative to the other routes of heat exchange in industrial applications, they are usually ignored. Equation 1 thus becomes Equation 2 as a general statement of heat balance:

$$S = M + R + C + E \qquad (2)$$

Excessive heating or cooling of a small portion of the skin can occur when it comes in contact with a hot or cold surface. The contact can be either intentional or incidental. Injury occurs when there is sufficient heat gain to cause a burn or heat loss to cause the tissue to freeze (or at least become very cold for a period of time). In these cases, the local storage rate ($S_{local}$) becomes important.

$$S_{local} = K + D \qquad (3)$$

where $K$ is conductive heat transfer between the skin and an object and $D$ is the rate of heat transfer to or from the local area by conduction through the local tissue and by the heat supplied or removed via local blood flow.

## Factors Affecting Thermal Balance

As mentioned at the beginning of the discussion on thermal balance, three factors play an important role: climatic conditions of the environment, work demands, and clothing. Climatic conditions are widely used to describe the degree of stress, as is seen in casual descriptions that include air temperature, relative humidity, and wind chill. They are not the only determinants.

The role of metabolic rate in heat balance is very important because it is a substantial contributor to heat gain. In heat stress, metabolic rate can add 10 to 100 times more heat to the body than radiation and convection combined. In the case of cold stress, metabolic rate affects heat balance to the same extent as radiation and convection loss.

Clothing is also a major contributor to thermal balance. Clothing has three relevant characteristics: insulation, permeability, and ventilation. Insulation is a measure of the resistance to heat flow by radiation, convection, and conduction. The greater the amount of insulation, the less the rate of heat flow from the warmer mass to the cooler one. During heat stress exposures, insulation reduces heat flow by radiation and convection. It also reduces heat flow by conduction if a substantial portion of a person's body is in contact with a warm surface. Insulation plays a very important role both in preventing burns by contact with a hot surface and in cold stress. In cold stress, insulation reduces heat loss by convection and radiation as well as conduction, and insulation prevents cold injury to local tissues by preventing direct contact with cold surfaces.

Permeability is a measure of the resistance to water vapor movement through the clothing. It is a factor in thermal stress because it influences the amount of evaporative cooling that takes place. Permeability is related to both insulation characteristics and the clothing fabrics. Generally, as insulation increases, permeability decreases. In addition, some clothing fabrics designed as contamination barriers reduce the magnitude of permeability. This means that there is sometimes a trade-off between the risks of heat stress and the risks from skin contact with harmful chemicals. New protective clothing fabrics entering the market provide protection against some chemical hazards while permitting water vapor transmission. These new fabrics provide a better opportunity to achieve an acceptable balance between prevention of chemical exposure and prevention of heat stress.

Ventilation is the third factor in clothing's role in thermal balance. Depending on the nature of the fabric, garment construction, and work demands, ambient air can move through the fabric itself or through garment openings. Clothing ensembles that enable the movement of air enhance evaporative and convective cooling, whereas garments that are designed to limit air movement limit evaporative and convective cooling. A good example of using ventilation characteristics to regulate heat balance is that of arctic parkas with drawstrings around the waist, cuffs, and hood. As metabolism heats the wearer, cooling can be achieved by loosening some of the closures to increase the amount of air flow (ventilation) under the clothing.

## Heat Stress

Heat stress is caused by a combination of factors (affected by environmental, work, and clothing factors) and tends to increase body temperature, heart rate, and sweating. These physiological adaptations are collectively known as heat strain. Figure 12–1 is a schematic representation of the physiological responses to heat stress.

We will first look at metabolism. The heat generated by muscular work heats the deep body tissues, which means

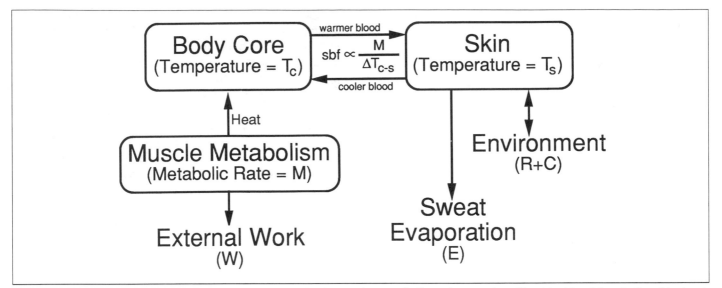

**Figure 12–1.** Heat flow through the body: heating the body core by metabolism, the transfer of heat by blood flow to the skin, heat gain or loss to the skin from the environment by radiation and convection, heat loss by sweat evaporation, and cooler blood returning to the core. Skin blood flow (sbf), which promotes heat transfer, is proportional to metabolic rate ($M$) divided by the difference between core and skin temperatures ($\Delta T_{c-s}$).

that work tends to increase body core temperature. Blood circulating through the core picks up heat energy, and the warmer blood is directed to the skin, where the blood is cooled. The cooler blood then returns to the core to pick up more heat energy. The skin is the site of heat exchange with the environment. Convection and radiation depend on the temperature difference between the skin and the environment, and the net heat exchange by $R + C$ (radiation plus convection) can be either positive (heat gain) or negative (heat loss). In addition, the skin secretes sweat onto its surface. As the water evaporates, it removes more heat energy from the skin, cooling the skin surface. Under ideal conditions, the body balances heat gains with losses so that the storage rate, $S$, is zero. This is accomplished by increasing the sweating rate until evaporative cooling is sufficient to remove the heat generated by metabolism plus any heat gained from (or lost to) the environment through $R + C$. The required evaporative cooling is denoted by $E_{req}$. (Remember that the value of $E_{req}$ is negative because heat flow is away from the body.) Then Equation 2 becomes

$$E_{req} = -(M + R + C) \qquad (4)$$

Thus, $E_{req}$ is indicative of physiological adjustment required to establish a thermal equilibrium between the body and the environment so that the body does not store heat. In many heat stress exposures, $M$ is the dominant term, and $E_{req}$ increases to meet additional cooling requirements of the work demands.

Heart rate is another important physiological parameter of heat strain, because it reflects the demands on the cardiovascular system to move blood (and heat) from the body core to the skin. The total blood flow through the heart is proportional to the metabolic rate and inversely proportional to the temperature difference between the body core and the skin. As work demands and metabolic rate increase, cardiac output increases, as seen in the heart rate. Sometimes skin temperature increases because evaporative cooling is limited or the net heat gain from $R + C$ is high. As skin temperature increases, approaching core temperature, more blood must be delivered to the skin to achieve the same rate of cooling.

Finally, sweat rate (and total sweat volume) is another important measure of physiological strain. The greater the level of heat stress, the greater is the sweat rate. The body has a natural ability to increase its tolerance of heat stress exposures through acclimation (often called acclimatization). As people become acclimated, they are able to sweat more and therefore increase their cooling capability. With increased cooling, heart rate and body core temperature decrease under the same work conditions.

The following material on heat stress describes recognition, evaluation, and control of heat stress as it affects the whole body. These sections are followed by one on special topics including contact with hot surfaces and breathing of hot air.

## Recognition of Heat Stress

Heat stress in the workplace can be recognized by noting workplace risk factors and by the effects it has on workers. The workplace risk factors, broadly stated, are hot environments, high work demands, and protective clothing requirements. These factors are the traditional considerations in the evaluation of heat stress and more details concerning them are in the section on evaluation of heat stress. In essence, if a workplace is generally considered hot in the subjective judgment of workers and supervisors, then heat stress may be present. If the demands for external work are high (resulting in a high metabolic rate), heat stress may be a factor in environments that are considered comfortable by casual observers (those not exerting themselves in that environment). Clothing is the third workplace risk factor. Although lightweight, loose-fitting cotton clothing is the best choice during exposures to heat stress, many workplaces require protective clothing that decreases permeability and ventilation and increases insulation. The added weight of personal protection increases the metabolic heat load and therefore the level of heat stress.

The responses of workers provide a good cue for the recognition of heat stress in the workplace. At the extreme

is a pattern of heat-related disorders. Intermediate markers are physiological adjustments and worker behaviors.

**Heat-related disorders.** Heat-related disorders are manifestations of overexposure to heat stress. Table 12–A is a list of common or important heat-related disorders. The table includes the signs a trained observer might observe, the symptoms the worker might report, the likely cause of the disorder, first aid, and steps for prevention. Figure 12–2 is a simple illustration of normal responses to heat stress and how these responses may lead to a heat-related disorder.

Heatstroke is the most serious heat-related disorder. Although it is relatively rare, it must be immediately recognized and treated to minimize permanent damage. The risk of death is high with heatstroke. Heat exhaustion is the

**Table 12–A.** Heat-Related Disorders Including the Symptoms, Signs, Causes, and Steps for First Aid and Prevention

| Disorder | Symptoms | Signs | Cause | First Aid | Prevention |
|---|---|---|---|---|---|
| **Heat stroke** | Chills<br>Restlessness<br>Irritability | Euphoria<br>Red face<br>Disorientation<br>Hot, dry skin (usually, but not always)<br>Erratic behavior<br>Collapse<br>Shivering<br>Unconsciousness<br>Convulsions<br>Body temperature ≥104 F (40 C) | Excessive exposure<br>Subnormal tolerance (genetic or acquired)<br>Drug / alcohol abuse | Immediate, aggressive, effective cooling.<br>Transport to hospital.<br>Take body temperature. | Self-determination of heat stress exposure.<br>Maintain a healthy life-style.<br>Acclimation. |
| **Heat exhaustion** | Fatigue<br>Weakness<br>Blurred vision<br>Dizziness, headache | High pulse rate<br>Profuse sweating<br>Low blood pressure<br>Insecure gait<br>Pale face<br>Collapse<br>Body temperature: Normal to slightly increased | Dehydration (caused by sweating, diarrhea, vomiting)<br>Distribution of blood to the periphery<br>Low level of acclimation<br>Low level of fitness | Lie down flat on back in cool environment.<br>Drink water.<br>Loosen clothing. | Drink water or other fluids frequently.<br>Add salt to food.<br>Acclimation. |
| **Dehydration** | No early symptoms<br>Fatigue / weakness<br>Dry mouth | Loss of work capacity<br>Increased response time | Excessive fluid loss caused by sweating, illness (vomiting or diarrhea), alcohol consumption | Fluid and salt replacement. | Drink water or other fluids frequently.<br>Add salt to food. |
| **Heat syncope** | Blurred vision (grey-out)<br>Fainting (brief black-out)<br>Normal temperature | Brief fainting or near-fainting behavior | Pooling of blood in the legs and skin from prolonged static posture and heat exposure | Lie on back in cool environment.<br>Drink water. | Flex leg muscles several times before moving.<br>Stand or sit up slowly. |
| **Heat cramps** | Painful muscle cramps, especially in abdominal or fatigued muscles | Incapacitating pain in muscle | Electrolyte imbalance caused by prolonged sweating without adequate fluid and salt intake | Rest in cool environment.<br>Drink salted water (0.5% salt solution).<br>Massage muscles | If hard physical work is part of the job, workers should add extra salt to their food. |
| **Heat rash** (prickly heat) | Itching skin<br>Reduced sweating | Skin eruptions | Prolonged, uninterrupted sweating<br>Inadequate hygiene practices | Keep skin clean and dry.<br>Reduce heat exposure. | Keep skin clean and periodically allow the skin to dry. |

**Note:** Salting foods is encouraged as both treatment and prevention of some heat-related disorders. Workers on salt-restricted diets must consult their personal physicians.

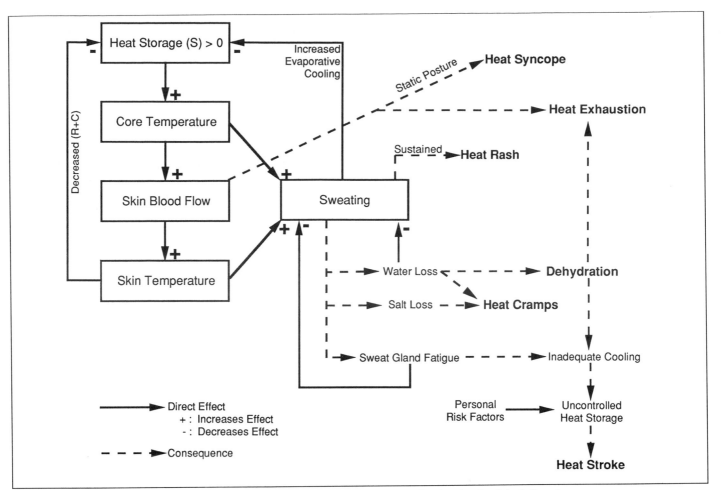

**Figure 12–2.** Normal responses to heat stress exposure and how they can lead to heat-related disorders.

most commonly seen heat-related disorder for which treatment is sought. Dehydration is a precursor to heat exhaustion, but it is usually not noticed or reported by workers.

As part of the recognition process, the health and safety professional should examine records at the medical or first-aid facility that workers and supervisors report to. If no specific heat-related disorders are listed, this does not mean heat stress is not present. It is worthwhile to examine the records for reports of faintness, weakness, nausea, cramps, headaches, and skin rashes. If temperatures were taken, some may have been elevated. If urine samples were taken, some may have had high specific gravity because of dehydration. The records may also show an increase in the number of accidental injuries that are related to heat stress conditions.

**Physiological markers.** Physiological responses to heat exposure can be used as a recognition tool. The most easily measurable ones are oral temperature, heart rate, and water loss. When noting one or more physiological responses of members of the work force, the health and safety professional can begin to determine whether a more detailed evaluation is necessary. When selecting workers to sample,

it is important not to be biased toward those that appear to be the most tolerant. The sampling process should either be random or favor those who appear to be having more problems than others (see behaviors below).

Although rectal, esophageal, and tympanic temperatures are the preferred measures of body core temperature, oral temperature has long been used as an indicator of core temperature in industrial environments. Oral temperature can be measured with a traditional mercury-in-glass thermometer or an electronic one. (Precautions should be taken with oral temperatures: The person should not eat or drink for 15 min before taking the temperature, and the person must keep the mouth closed.) Core temperature is estimated by adding 1 F (0.5 C) to the measured value of oral temperature. There is also a thermometer on the market that measures body temperature from the ear canal with an infrared sensor. Care should be taken when using ear temperatures, however, because they can be influenced by the environment. If core temperature is above 100.4 F (38 C), then heat stress is high enough to warrant further evaluation.

Heart rate can be estimated by palpation of an artery near enough the surface of the skin that the pulse can be felt by the fingertips. There are also devices that can measure heart rate by electronic means. Among heart rate measurements, recovery heart rate is most useful as a tool for recognizing stress. Recovery heart rate measurements require that the worker stop in or at the end of a work cycle, sit down, and determine the pulse rate after a given amount of time. One guideline for recovery heart rate stipulates that the heart rate at 1 min after sitting down should be at or below 110 beats per minute (bpm). Another guideline is that the heart rate at 3 min should be below 90 bpm. If a worker's heart rate is higher at either of these two points, then the work that includes the heat stress may be excessive, and evaluation is appropriate.

Because devices for measuring and logging heart rate are readily available, it is a reasonable task to determine the average heart rate of a worker over an 8-hour day. If the average heart rate over a day is greater than 110 bpm, the work and heat stress may be excessive. Examination of the log for peak heart rates is also informative. If heart rates are above a nominal threshold of 160 bpm, then the demands of the work should be evaluated. (For information on individually set thresholds, see the section on evaluation of physiological strain later in the chapter.)

Monitoring dehydration is a third means of recognizing potential heat stress conditions. Dehydration can be detected by noting the change in body weight from the beginning to the end of a workshift. If there is more than a 1.5 percent loss of body weight, then excessive dehydration is likely and evaluation is advisable.

**Worker behaviors.** Heat stress not only induces physiological changes but also affects behavior. Behaviors commonly associated with heat stress are actions that reduce exposure, such as adjusting the clothing to increase evaporative loss, slowing the work rate or taking small breaks to lower the metabolic rate, and using shortcuts in work methods. Effects on attitude show up as irritability, low morale, and absenteeism. There are also an increased number of errors and machine breakdowns, and the frequency of unsafe behavior increases.

**Summary of recognition.** Basically, there are four kinds of questions you should ask to determine whether work conditions should be evaluated for heat stress:

1. Is the environment subjectively judged as being hot? Are the work demands high? Is protective clothing required?

2. Are worker behaviors indicative of attempts to reduce heat stress? Is morale low or absenteeism high? Are people making mistakes or getting hurt?

3. Do medical records show a pattern of fatigue, weakness, headache, rashes, or high body temperature?

4. Is body temperature, heart rate, or sweat loss high among a sample of workers?

When the answers to any of these questions is yes, an evaluation is probably in order.

## Evaluation of Heat Stress

In 1969, the World Health Organization (WHO) set the tone for worker protection against heat stress. One of their recommendations focuses on body core temperature, which can be estimated from oral temperature as described in the preceding discussion of physiological markers. Core temperature should not exceed 100.4 F (38 C) during prolonged daily exposure to heat stress. The WHO panel did recognize that 102.2 F (39 C) is acceptable as an upper limit for short periods followed by adequate recovery, and that the average heart rate over a day should not exceed 110 bpm. The foundation of evaluation schemes proposed by the National Institute for Occupational Safety and Health (NIOSH) in a 1986 Criteria Document and the American Conference of Governmental Industrial Hygienists (ACGIH) in a current booklet on Threshold Limit Values is the WHO acceptable core temperature limit for prolonged exposures (100.4 F [38 C]).

When prolonged daily exposure to heat stress is evaluated, it is assumed that the given work conditions are prevalent for a full 8-hour workshift with nominal breaks. Time-limited exposures to heat stress are evaluated in terms of safe exposure times for a given level of heat stress. Safe exposure times are maintained by prescribing work–rest cycles based on prolonged daily exposure criteria or determined through heat balance analyses that compute the amount of time to reach an upper limit of body core temperature. Methods for evaluating heat stress require at least an assessment of metabolic rate and some measures of the thermal environment. Some methods assume one particular type of clothing is worn by the workers, and others have provisions for different types.

Evaluation of workplace heat stress can also be accomplished by determining whether there is excessive physiological strain on the workers.

**Assessment of metabolic rate.** Metabolic rate is the rate of internal heat generation. Heat must be dissipated from the body to maintain thermal equilibrium. (In situations of cold stress, metabolic rate is important to maintain deep body temperature.) A base level of metabolism is necessary to support life. Beyond basal metabolism, there is a work-driven metabolism that is largely the result of muscular effort. The greatest amount of muscle metabolism occurs when the muscles exert a force with motion (dynamic work). Much less metabolism is required to exert a force with no motion (isometric contraction, or static work). Therefore, the greatest metabolic rates occur when the body moves over a distance, especially upwards, and when objects are moved in space by the body. Lower metabolic rates occur during sedentary activities or when one is simply holding objects without any motion.

One very simple method of assessing the metabolic rate that is acceptable for a given task is to classify the work demands into three to five categories of metabolic rate (for example, light, moderate, and heavy). Another simple method is to look for similar activities in published tables of metabolic rates for specific activities (or assume the demands are equivalent by subjective matching). Both of these methods are expedient, but neither is very accurate and they are prone to overestimation.

Some discipline is required in the assessment of metabolic rate. The first step is to divide the job being assessed into discrete, homogeneous tasks, and then determine the frequency and duration of each. Then a metabolic rate can be assigned to each task. Finally, the time-weighted average for metabolic rate can be determined (TWA-$M$). Using categories or tables to assign metabolic-rate values to a task reduces the possibility of some errors, but the method described in the ACGIH TLV documentation is recommended (it is outlined in the next paragraph). In addition, there are other published methods that provide good results.

According to the NIOSH/ACGIH method, the metabolic rate for a given task can be estimated by summing four components: basal metabolism ($B$); posture ($P$); activity, based on degree of body involvement ($A$); and vertical motion ($V$). That is,

$$M_{task} = B + P + A + V \qquad (5)$$

Table 12–B describes values of metabolic rate associated with each of the components. The SI unit for metabolic rate is the watt (W), and formulas for conversion from watts to other units (such as kcal/min, kcal/hr, Btu/hr, l/min of oxygen consumption) and other units to watts are provided at the bottom of the table. (The ACGIH TLV booklet uses kcal/min and it provides an example of how to use this method.)

**Assessment of environmental conditions.** The environmental factors that are central to the assessment of heat stress are air temperature, humidity, air speed, and average temperature of the solid surroundings. How these factors are incorporated into the evaluation of heat stress depend on the evaluation tool. The SI unit for temperature is the Centigrade degree.

***Dry bulb temperature ($T_{db}$).*** This is a direct measure of air temperature. The temperature sensor is surrounded by air, which is allowed to flow freely around the sensor. The sensor, however, can be influenced by radiant heat sources and should therefore be shielded from them.

***Psychrometric wet bulb temperature ($T_{pwb}$).*** This measure is based on the degree of evaporative cooling that can occur. A wetted wick is wrapped around a temperature sensor, and enough air (> 3 m/s) is forced over the wick to maximize the rate of evaporative cooling. The amount of temperature reduction that can be achieved depends directly on the amount of water vapor in the air. When humidity is high, the reduction in temperature is low. As humidity decreases, temperature reduction increases.

***Ambient water vapor pressure ($P_v$).*** This is commonly known as humidity. There are two ways humidity is expressed: relative and absolute. At any given temperature, the pressure of water vapor that can be in the air has a maximum value, which is called the saturation pressure. At low temperatures, the saturation pressure is low, and it increases exponentially with temperature. Relative humidity is the ratio of the water vapor pressure in the air to the saturation pressure at that temperature. So 50 percent relative humidity means that the water vapor pressure is 50 percent of the saturation pressure. Unfortunately, relative humidity is not very useful as a tool in assessing heat stress, because the water vapor pressure represented by a relative humidity value varies widely, depending on the air temperature. Absolute humidity is expressed as the amount of water vapor in the air in terms of either partial pressure or weight-per-unit volume of air. The usual practice for heat stress evaluation is to use the partial pressure, and the SI unit for this is the kilopascals (kPa). (To convert from kPa to mmHg, the value in kPa is multiplied by 7.5.) Usually, a psychrometric chart is used to determine humidity from $T_{db}$ and $T_{pwb}$. (Figure 12–7 provides an equation for estimating water vapor pressure from these temperatures.)

***Natural wet bulb temperature ($T_{nwb}$).*** The instrument used for this measurement is similar to the psychrometric wet bulb, except that air is allowed to flow over the sensor naturally rather than being forced. When air flow is less than 3 m/s, the temperature reduction is less than that achieved with a psychrometric wet bulb at the same abso-

**Table 12–B.** Values Needed to Estimate the Metabolic Rate for a Task (in watts)

| | | Average (watts) | Range (watts) |
|---|---|---|---|
| Basal metabolism ($B$) | | 70 | |
| Posture metabolism ($P$) | | | |
| Sitting | | 20 | |
| Standing | | 40 | |
| Walking | | 170 | 140–210 |
| Activity ($A$) | | | |
| Hand | light | 30 | 15–85 |
| | heavy | 65 | |
| One arm | light | 70 | 50–175 |
| | heavy | 120 | |
| Both arms | light | 105 | 70–245 |
| | heavy | 175 | |
| Whole body | light | 245 | 175–1,050 |
| | moderate | 350 | |
| | heavy | 490 | |
| | very heavy | 630 | |

Climbing ($V$)
$V_{vert}$ = rate of vertical ascent (meters/min)
$V = 56 V_{vert}$

| Unit Conversions for Metabolic Rate | |
|---|---|
| *From Watts to Other Units* | *From Other Units to Watts* |
| kcal/min = 0.014 × W | W = 70 × kcal/min |
| kcal/hr = 0.86 × W | W = 1.2 × kcal/hr |
| Btu/hr = 3.4 × W | W = 0.29 × Btu/hr |
| l O$_2$/min = 0.0029 × W | W = 350 × l O$_2$/min |

**Note:** Values are based on a worker weighing 70 kg.

lute humidity. That is, natural wet bulb temperature is sensitive to both humidity and air movement.

***Air speed (V*ₐᵢᵣ*).*** This is measured using an appropriate anemometer. The anemometer should not be unidirectional. A unidirectional anemometer is sensitive to air motion in one direction, so it may not measure air motion in other directions. Because air speed varies with space and time, an average measured value should be used for the evaluation.

***Globe temperature (T*g*).*** This is a measure of radiant heat from the solid surroundings and convective heat exchange with the ambient air. The globe temperature is classically measured using a 6-inch, thin-walled copper sphere painted matte black on the outside. The temperature sensor is placed at the center of the globe. The globe is then suspended in the air in a location near the workspace. When all the surrounding surfaces are the same temperature as the air, the globe temperature is equal to air temperature. If one or more of the surfaces are different, then the globe temperature increases or decreases, depending on the average temperature of the solid surroundings. Finally, for a given level of radiant heat exchange with the globe, the globe temperature differs more from air temperature when there is little air movement, and it differs less when there is significant air motion, because the globe thermometer is also sensitive to convective heat exchange with the air. Globe temperature is used to estimate the average wall temperature of the surroundings.

***Effective temperature (ET) and corrected effective temperature (CET).*** These indices of the thermal environment were first developed to quantify subjective thermal sensation and were later used to describe thermal stress. The ET is determined from a nomogram whose use requires knowledge of the values for $T_{db}$, $T_{pwb}$, and $V_{air}$. Equal values of ET would predict similar sensations of warmth. Because ET was widely accepted, it was used in the early studies of heat stress as an index of the environment.

Because ET does not account for radiant heat, CET, or corrected effective temperature, was proposed. In this method, $T_{db}$ is replaced with $T_g$ on the nomogram. ET and CET are no longer used to evaluate heat stress. A new index was needed, one that could be more easily determined and was indicative of thermal stress from the environment. The wet bulb globe temperature was the next evolutionary step from ET and CET.

***Wet bulb globe temperature (WBGT).*** This index of environmental heat is widely used to evaluate industrial heat stress. In environments that are either indoors or outdoors in the shade or on a cloudy day, it is computed as

$$\text{WBGT}_{in} = 0.7\,T_{nwb} + 0.3\,T_g \qquad (6)$$

Under conditions of direct sunlight (outdoors and no cloud cover), it is computed as

$$\text{WBGT}_{out} = 0.7\,T_{nwb} + 0.2\,T_g + 0.1\,T_{db} \qquad (7)$$

The original instrumentation for assessing WBGT consisted of mercury-in-glass thermometers and a large cop-

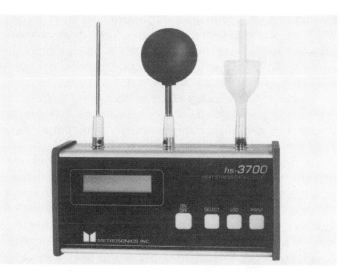

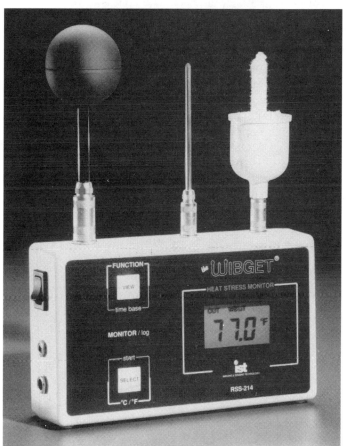

**Figure 12–3.** Examples of electronic devices used to measure WBGT. (Courtesy of IST and Metrosonics, Inc.)

per globe. There are several manufacturers of these instruments who now use smaller globes. In addition to computing the WBGT directly, some wet bulb globe temperature instruments can perform data-logging functions as well as real-time analysis of the environment (such as determining safe work times). Electronic instrumentation has virtually replaced the conventional WBGT cluster of individual sensors on a tripod.

## Evaluation of prolonged exposures to heat stress.

When the goal is to limit the heat stress dose (body core temperature not to exceed 100.4 F [38 C]), the problem that arises is that of relating exposure (combinations of environment, metabolism, and clothing) to dose. In 1963 Alexander Lind proposed the concept of the upper limit of the prescriptive zone. In a set of classic experiments, he demonstrated several important relationships between work, environment, and body core temperature. In essence, he found that for a given metabolic rate, core temperature remains relatively constant for increasing levels of environmental heat until a critical level is reached. Then, the core temperature steeply rises with increasing levels of environmental heat, creating an increased risk for heat disorders. This critical level of heat stress is called the upper limit of the prescriptive zone, and a person can work 8 hours at or below this level without significant risk of a heat disorder. The limit is lower for people with higher metabolic rates, and vice versa.

By investigating the upper limit of the prescriptive zone for different worker populations, Dukes-Dobos developed protective limits for the 95th percentile of the general worker population. At these protective limits, body core temperature should not exceed 100.4 F (38 C), and first-minute recovery heart rates should not exceed 110 bpm. The upper limit of the prescriptive zone was first adapted by NIOSH in 1972 for exposure limits, which were revised in 1986. The NIOSH limits were adopted by the ACGIH in 1973 (and revised in 1990) for their TLV for Heat Stress. The International Organization for Standardization (ISO) also adapted the NIOSH thresholds in its 1983 standard. The limits are expressed in hourly time-weighted averages for both WBGT and metabolic rate. The thresholds are illustrated in Figure 12–4 using both the NIOSH and ACGIH nomenclature.

The middle curve is called the recommended exposure limit (REL) by NIOSH. For workers wearing ordinary cloth summer-weight work clothes and who are acclimated to heat, there should be practically no risk of heat-related disorders when working for 8 hours with nominal breaks every 2 hours. Notice that as the metabolic rate increases, the threshold for WBGT decreases. This means that to maintain body core temperature below 100.4 F (38 C), a cooler environment is necessary for those workers with high internal heat generation. The ACGIH calls this threshold the TLV for acclimated workers.

The lower curve demonstrates what is called the recommended alert limit by NIOSH and the TLV for unacclimated workers by the ACGIH. This curve reflects the fact that unacclimated workers are less able to tolerate heat stress exposure, and shows that lower tolerance can be accommodated by a lower level of environmental heat but the same level of work demands.

The upper curve is the ceiling limit that NIOSH has proposed (but was not included by the ACGIH in their TLV documentation). It is a level that NIOSH recommends should not be exceeded for more than 15 minutes without personal protection.

Because clothing is a factor in determining the level of heat stress, the ACGIH has proposed a table of adjustment factors for different types of clothing that can be subtracted from the threshold values (or, equivalently, added to the measured values of WBGT) in the environments being considered. These are given in Table 12–C. An important note about using the adjustment factors is that they represent the current best guess about the effects of clothing on heat stress, and some caution is necessary in using them. Using physiological strain indicators (that is, personal monitoring) to confirm the adjustments may be appropriate.

To evaluate a job for heat stress, a 1- or 2-hour interval for time-weighted averaging should be selected. If the work is repeated in an hourly pattern, a 1-hour TWA can be used. If the work is intermittent, a 2-hour TWA may be more representative of the demands. For the selected interval, TWA-$M$ and TWA-WBGT must be calculated. Adjust the WBGT for each location by adding the clothing adjustment factor to the measured value as appropriate. With these TWA values, the location on the graph of Figure 12–4

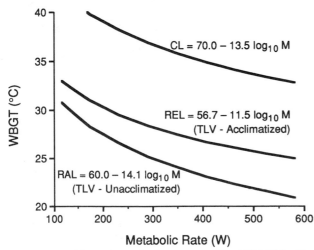

**Figure 12–4.** Heat stress limits proposed by NIOSH (ceiling limit/CL, recommended exposure limit/REL, and recommended alert limit/RAL) and ACGIH (TLV for acclimated and unacclimated workers).

**Table 12–C.** WBGT Adjustment Factors (in degrees Centigrade) for Different Clothing Ensembles

| Clothing Type | ACGIH | Other Sources |
| --- | --- | --- |
| Work clothes | 0 | 0 |
| Coveralls | 2 | |
| Winter uniform | 4 | |
| Particle barrier | | 4–7 |
| Vapor transmitting | 6 | |
| Vapor barrier | | 8 |
| Encapsulating suit | | 11 |

can be determined for the combination of TWA-WBGT and TWA-*M*. If it is below the RAL/TLV curve for unacclimated workers, there is no practical risk of heat-related disorders developing, even in the least heat tolerant but otherwise healthy workers. If it is between the RAL/TLV for unacclimated workers and the REL/TLV for acclimated workers, then a program of heat stress management should be put in place that includes at least the general controls described in the section on controls. If it is above the REL/TLV for acclimated workers, then heat stress is a hazard in the work environment and control actions should be taken. If it is above the ceiling limit, then special precautions should be taken.

### Evaluation of time-limited exposures to heat stress.

For heat stress exposures above the REL/TLV, the question becomes, How long can someone safely work? This question can be answered either by WBGT-based methods or by heat balance analysis.

*WBGT techniques.* WBGT techniques fall into two categories. One method uses TWAs and the TLV. With this method, the problem is posed in terms of how much time in an hour can someone work above the TLV, as opposed to how much time should work or recovery be below the TLV. This question is generally concerned with work–rest cycles. For the simple case of one work condition and one recovery condition, refer to Figure 12–5. At work, Point A is (TWA-*M* and TWA-WBGT)$_{work}$, and recovery (rest or light-duty work) (TWA-*M* and TWA-WBGT)$_{recovery}$ is at Point C. Point B is at the intersection of Line A–C with the threshold curve. The ratio of recovery time to heat stress exposure time is equal to the ratio of the length of A–B to the length of B–C.

The second method that uses WBGT and TLV is to test various combinations of exposure time and recovery time

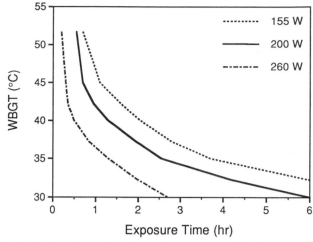

**Figure 12–6.** Example of U.S. Navy permissible heat exposure limit (PHEL) chart. The example assumes ordinary summer-weight work clothes.

that equal 1 hour, until a combination of TWA-*M* and TWA-WBGT is found that lands on the TLV curve. (A spreadsheet solution can be developed by using the equation for the REL/TLV line provided in Figure 12–4.)

The U.S. Navy and the Electric Power Research Institute have proposed WBGT methods to determine safe exposures times from charts. An example of the Navy's permissible heat exposure limit (PHEL) chart is illustrated in Figure 12–6.

*Heat balance analysis.* Heat balance analysis uses a model of heat exchange between a hypothetical person and the environment. Whereas the WBGT methods previously described are empirical methods of heat stress evaluation, heat balance analysis is considered a rational method. In a hypothetical work situation, if thermal equilibrium can be established, there is no risk of an excessive level of heat stress. If thermal equilibrium cannot be established, then the amount of time to reach an upper limit of heat storage (nominally to a core temperature of 101.3 F [38.5 C]) should be determined. Heat balance analysis methods provide a way to determine this amount of time. A further advantage of heat balance analysis is that desktop evaluations of potential countermeasures can be performed.

One method proposed by Belding and Hatch and used by many industrial hygienists is called the Heat Stress Index (HSI). It starts from the premise that Equation 2 describes heat balance and Equation 4 describes evaporative cooling requirements. Equation 4 is repeated here:

$$E_{req} = -(M + R + C) \qquad (4)$$

The HSI is based on simple equations, which have been updated over the years for computing *R* and *C*. The equations that follow are for workers wearing ordinary cloth work clothes and the units are Watts, degrees C,

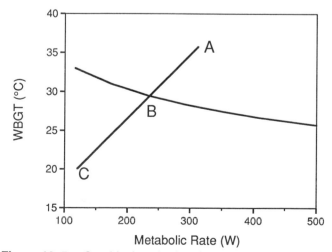

**Figure 12–5.** Graphical method to determine the recommended ratio of recovery to exposure time. Point A represents work conditions and Point C, recovery conditions. Ratio of rest time to work time is equal to the length of A–B divided by the length of B–C.

meters per second (m/s) for air speed ($V_{air}$) and kiloPascals (kPa) for water vapor pressure.

Basically, $R$ is equal to a clothing-related constant times the difference between the mean temperature of the surroundings ($T_r$) and a mean skin temperature of 95 F (35 C). Obviously, if the average surrounding temperature is greater than skin temperature, there is a gain of heat by radiation.

$$R = 7.7 (T_r - 35) \qquad (8)$$

where

$$T_r = T_g + 1.8 V_{air}^{0.5} (T_g - T_{db}) \qquad (9)$$

$C$ is equal to a clothing-related constant times a power function of air speed times the difference between air and skin temperatures, as follows:

$$C = 8.1 V_{air}^{0.6} (T_{db} - 35) \qquad (10)$$

If air temperature is greater than skin temperature, there is a heat gain to the worker by convection; if skin temperature is less than air temperature, there is a heat loss.

The method also provides for the determination of the maximum rate of evaporative cooling ($E_{max}$), which has either an environmental or a physiological limit. The environmental limit on $E_{max}$ is determined as a clothing-related constant times a power function of air speed times the difference between skin and air water vapor pressure ($P_v$).

$$E_{max} = 122 V_{air}^{0.6} (P_v - 5.6) \qquad (11)$$

Because $P_v$ is less than 5.6, $E_{max}$ has a negative value, representing a heat loss. The physiological limit is based on a limiting sweat rate of 1 l/hr, which is equivalent to a heat flow of –675 W. Therefore, $E_{max}$ is the greater negative number of that computed from Equation 6 or –675. Then,

$$HSI = 100 \frac{E_{req}}{E_{max}} \qquad (12)$$

If the HSI is less than 40, heat stress is low and no further action is required. If the HSI is between 40 and 70, heat stress is a significant workplace hazard. If HSI is between 70 and 100, heat stress is high and workers are at risk for heat-related disorders. If HSI is greater than 100, there is significant heat storage and exposure should be time limited.

A number of variations to the HSI method, as well as alternatives to it, have been proposed. The International Organization for Standardization has published the most recent and comprehensive rational method for heat balance analysis. It is based on the required sweat rate ($SR_{req}$) and its steps are described in Figure 12–7. Terms used in the figure are defined in Table 12–D.

**Table 12–D.** Definition of Terms Used in Figure 12–7

| Term | Description | Units |
|------|-------------|-------|
| $M$ | Metabolic rate—normalized | $W/m^2$ |
| SA | Body surface area | $m^2$ |
| $P_a$ | Ambient water vapor pressure | kPa |
| $T_{pwb}$ | Psychrometric wet bulb temperature | °C |
| $T_{db}$ | Dry bulb (air) temperature | °C |
| $T_r$ | Mean radiant temperature | °C |
| $T_g$ | Globe temperature | °C |
| $V_{air}$ | Air speed | m/s |
| $T_{sk}$ | Average skin temperature | °C |
| $I_{cl}$ | Clothing insulation | $(m^2 \cdot °C)/W$ |
| $V$ | Adjusted air speed | m/s |
| $h_{cn}$ | Coefficient of convection—natural | $W/(m^2°C)$ |
| $h_{cf}$ | Coefficient of convection—forced | $W/(m^2°C)$ |
| $h_c$ | Coefficient of convection | $W/(m^2°C)$ |
| $h_r$ | Coefficient of radiation | $W/(m^2°C)$ |
| $F_{cl}$ | Clothing reduction factor | — |
| $C$ | Convective heat exchange | $W/m^2$ |
| $R$ | Radiant heat exchange | $W/m^2$ |
| $E_{req}$ | Required evaporative cooling | $W/m^2$ |
| $F_{pcl}$ | Clothing reduction factor for water vapor permeation | — |
| $P_{sk}$ | Water vapor pressure on skin | kPa |
| $E_{max}$ | Maximum rate of evaporative cooling | $W/m^2$ |
| $w_{req}$ | Required skin wetness | — |
| $\eta$ | Sweating efficiency | — |
| $SW_{req}$ | Required sweat rate | $W/m^2$ |
| $SW_{max}$ | Maximum sweat rate | $W/m^2$ |
| $w_{max}$ | Maximum skin wetness | — |
| $Q_{max}$ | Maximum heat storage | $(W \cdot hr)/m^2$ |
| $SV_{max}$ | Maximum sweat loss | $(W \cdot hr)/m^2$ |
| $w_p$ | Predicted skin wetness | — |
| $E_p$ | Predicted rate of evaporative cooling | $W/m^2$ |
| $SW_p$ | Predicted sweat rate | $W/m^2$ |
| $SR$ | Sweat rate | l/hr |
| $t_w$ | Time limit based on skin wetness | min |
| $t_d$ | Time limit based on dehydration | min |
| $t$ | Time limit | min |

The principal purposes of the ISO method are to determine the amount of evaporative cooling that is required for thermal equilibrium ($E_{req}$), whether the required cooling can be achieved by sweating and evaporation, and what time limits apply if sweating or evaporation is insufficient. The limits are based on inadequate cooling and potential dehydration. (The standard was developed with cotton clothing in mind, but experimentally derived values for insulation [$I_{cl}$] and permeability [$F_{pcl}$] of other types of clothing are becoming available for use in these computations.)

If the heat balance analysis indicates that thermal equilibrium can be achieved, then heat stress does not play a limiting role in the work. If the time limits are less than 6 hours, then serious consideration must be given to heat stress controls.

The following description of the ISO analysis of required sweat rate is divided into three sections. The first describes the steps in computing the individual terms of the heat balance equation (see Equation 1). The second section presents the criteria for unacclimated and acclimated workers at two levels of heat stress risk. The final section describes the computations for the time limit.

## Computation of Heat Exchange

**Adjust metabolic rate for body surface area.** The following values of heat exchange are normalized to body surface area.

$$M = \frac{\text{Metabolic rate}}{1.8} \qquad (1)$$

**Adjust air velocity for work activities if $M > 58$.** The method recognizes that air motion around the body results from both environmental conditions and body motion during work.

$$V = V_{air} + 0.0052 (M - 58) \qquad (2)$$

Limit the increase due to metabolism to 0.7 m/s (at M > 193 W/m²) and V to 3.0 m/s.

**Compute coefficient for convection.** The convective heat exchange can be caused by either natural or forced convection. The greater value for natural or forced convection is used.

$$h_{cn} = 2.38 \left( |T_{db} - 36| \right)^{0.25} \qquad (3)$$

If $V \leq 1$ m/s, $h_{cf} = 3.5 + 5.2V$; if $V > 1$ m/s, $h_{cf} = 8.7V^{0.6}$

$$h_c = \text{Greater value of } (h_{cn}, h_{cf}) \qquad (4)$$

**Compute mean radiant temperature.** The following equation computes the mean radiant temperature from globe temperature, recognizing that natural convection and air movement influence globe temperature.

$$\zeta = \text{Greater value of } \{0.4 (|T_g - T_{db}|)^{0.25}, 2.5 V_{air}^{0.6}\} \qquad (5)$$

$$T_r = 100 \left[ \left( \frac{T_g + 273}{100} \right)^4 + \zeta (T_g - T_{db}) \right]^{0.25} - 273 \qquad (6)$$

**Compute coefficient of radiation for a standing person.** The rate of radiant heat exchange is lower for other postures.

$$h_r = 4.1 \times 10^{-8} \frac{(T_{sk} + 273)^4 - (T_r + 273)^4}{T_{sk} - T_r} \qquad (7)$$

**Compute clothing reduction factor for insulation.** Because clothing insulation influences (reduces) the rate of heat transfer by radiation and convection with reference to a nude person, the reduction factor must be computed.

$$F_{cl} = \frac{1}{(h_c + h_r) I_{cl} + \frac{1}{(1 + 1.97 I_{cl})}} \qquad (8)$$

where $I_{cl}$ is the clothing insulation ("clo" is a unit of clothing insulation: clo × 0.156 = m² · °C/W ) For example,

$$\text{light work clothes (reference: 0.6 clo) 0.093 m}^2 \text{ °C/W}$$
$$\text{coveralls (reference: 0.8 clo) 0.125 m}^2 \text{ °C/W} \qquad (9)$$

Average skin temperature is often assumed to be between 35 and 36 C. It can also be estimated as

$$T_{sk} = 30.0 + 0.093 T_{db} + 0.045 T_r - 0.571 V_{air} + 0.254 P_a$$
$$+ 0.00128 M - 3.57 I_{cl} \qquad (10)$$

**Compute convective heat exchange.**

$$C = h_c F_{cl} (T_{db} - T_{sk}) \qquad (11)$$

**Compute radiant heat exchange.**

$$R = h_r F_{cl} (T_r - T_{sk}) \qquad (12)$$

**Compute required evaporative cooling.**

$$E_{req} = -(M + C + R) \qquad (13)$$

**Compute clothing reduction factor for water vapor permeation for cotton clothing.** Because clothing insulation influences (reduces) the rate of heat transfer by evaporation with reference to a nude person, the reduction factor must be computed.

$$F_{pcl} = 1 + 2.22 h_c \left( \frac{I_{cl} - \left( 1 - \frac{1}{1 + 1.97 I_{cl}} \right)}{h_c + h_r} \right) \qquad (14)$$

**Compute ambient water vapor pressure ($P_a$).** (An alternative is to determine this from a psychrometric chart.)

$$P_a = 0.6105 \exp \frac{17.27 T_{pwb}}{T_{pwb} + 237.3}$$
$$-0.0669 (T_{db} - T_{pwb})(1 + 0.00115 T_{pwb}) \qquad (15)$$

To estimate $T_{pwb}$ from $T_{nwb}$,

$$T_{pwb} = T_{db} - \frac{T_{db} - T_{nwb}}{d} \qquad (16)$$

where  $d = 0.85$ if $V_{air} \leq 0.03$ m/s
$d = 1.00$ if $V_{air} \geq 3.0$ m/s
$d = 0.69 \log_{10} V_{air} + 0.96$ if $0.03 < V_{air} < 3.0$ m/s

**Compute water vapor pressure on the skin.** (An alternative is to get this from a psychrometric chart or tables.)

$$P_{sk} = 0.6105 \exp \frac{17.27 T_{sk}}{T_{sk} + 237.3} \qquad (17)$$

**Compute maximal evaporative cooling to the environment.**

$$E_{max} = -16.7 h_c F_{pcl} (P_{sk} - P_a) \qquad (18)$$

**Figure 12–7.** Description of computational steps for the heat balance analysis proposed by the International Organization for Standardization (ISO) for the evaluation of time-limited heat stress. (*Continues*)

**Compute required skin wetness.** Skin wetness is the degree to which the skin is covered by water.

$$w = \frac{E_{req}}{E_{max}} \tag{19}$$

$$w_{req} = \text{Lesser value of } (w, 1.0) \tag{20}$$

**Compute sweating efficiency.** Because some water is lost by dripping, sweating efficiency depends on skin wetness.

$$\eta = 1.0 - \frac{(w_{req})^2}{2} \tag{21}$$

**Compute required sweat rate.**

$$SW_{req} = \frac{-E_{req}}{\eta} \tag{22}$$

**Criteria.** Select criteria by acclimation state and by heat strain level (warning or danger).

| Criteria | Units | Unacclimated Warning | Danger | Acclimated Warning | Danger |
|---|---|---|---|---|---|
| $SW_{max}$ | $W/m^2$ | 200 | 250 | 300 | 400 |
| — | l/hr | 0.52 | 0.65 | 0.78 | 1.04 |
| $w_{max}$ | — | 0.85 | 0.85 | 1.0 | 1.0 |
| $Q_{max}$ | $W \cdot hr/m^2$ | 50 | 60 | 50 | 60 |
| $SV_{max}$ | $W \cdot hr/m^2$ | 1,000 | 1,250 | 1,500 | 2,000 |
| — | l | 2.5 | 3.25 | 3.9 | 5.2 |

Unacclimated: Workers have not had at least 5 days of 2-hour heat stress exposures over the preceding 7 days.

Acclimated: Workers have had at least 5 days of 2-hour heat stress exposures over the past 7 days.

Warning: If work time exceeds recommended time for Warning, controls should be considered. Adequate recovery should be allowed.

Danger: If work time reaches Danger time, controls are necessary.

*If $M < 65$, the $SW_{max} = 0.6 \, SW_{max}$.

Once the criteria are selected, predicted values for skin wetness, evaporative cooling, and sweat rate can be determined.

If $w_{req} \leq w_{max}$ and $8SW_{req} \leq SV_{max}$, then the heat stress is not time limited.

Compute Condition A. The predicted values equal required values as follows:

$$w_p = w_{req} \tag{23}$$

$$E_p = E_{req} \tag{24}$$

$$SW_p = SW_{req} \tag{25}$$

If $w_{req} > w_{max}$, compute Condition B (time is limited by skin wetness).

$$w_p = w_{max} \tag{26}$$

$$E_p = w_p E_{max} \tag{27}$$

$$\eta_p = 1.0 - \frac{(w_p)^2}{2} \tag{28}$$

$$SW_p = \frac{-E_p}{\eta_p} \tag{29}$$

If $SW_{req} > SW_{max}$ from Condition A or $SW_p > SW_{max}$ from Condition B, compute Condition C, which is limited by sweat rate.

$$\phi = \frac{E_{max}}{SW_{max}} \tag{30}$$

$$w^*_p = (\phi + 2)^{0.5} - \phi \tag{31}$$

$$w_p = \text{Lesser value of } (w^*_p, w_{max}) \tag{32}$$

$$\eta_p = 1.0 - \frac{(w_p)^2}{2} \tag{33}$$

$$E_p = w_p E_{max} \tag{34}$$

$$SW_p = \frac{-E_p}{\eta_p} \text{ (which is usually } SW_{max}) \tag{35}$$

**Compute time limit.** The computed time limits are based on the ability to support a sufficient rate of evaporative cooling and on the risk of dehydration.

If Condition C was computed, use Condition C values to compute the time limit. If Condition C was not computed and Condition B was computed, use Condition B values to compute the time limit. If neither Condition B nor C was computed, use Condition A values to compute the time limit.

**Evaporative cooling time limit.**

$$t_e = \frac{60 Q_{max}}{E_p - E_{req}} \tag{36}$$

**Sweat volume time limit.**

$$t_d = \frac{60 SV_{max}}{SW_p} \tag{37}$$

**Sweat rate for standard man (SR).**

$$SR = 0.0026 SW_p \tag{38}$$

The ISO method is best computed on a personal computer using a spreadsheet or computational language such as BASIC or FORTRAN. A commercial software package developed by J. B. Malchaire, Heat Stress Evaluation, Version 2, is available from Lewis Publishers and other packages developed by individuals are also available.

**Figure 12–7.** *(Continued)*

The accuracy of heat balance analysis is limited because certain variables are not easily measured, but fair assumptions can be made. The important advantage of heat balance analysis is the ability to compare the relative advantages of proposed changes in the environment, work demands, and clothing requirements.

**Evaluation of physiological strain.** As mentioned earlier, physiological strain caused by heat stress exposures results in elevations in body core temperature, heart rate, and sweating. These factors are therefore good evaluation tools for heat stress exposure. Physiological evaluation is a valid approach in that it uses direct assessment of the effects of heat stress (dose) rather than an index of exposure that is then related to dose through empirical evidence and models. If no excessive physiological strain is demonstrated in the working population at a workplace, then it can be inferred that heat stress is controlled by the work practices in place.

Physiological evaluation as an alternative or confirming evaluation of heat stress may be worthwhile when protective clothing is required. It can also be used to demonstrate compliance with the spirit of the NIOSH and ACGIH thresholds when work occurs above the thresholds.

In selecting workers to sample for the physiological evaluation, the choice should be random and sufficient in size to ensure statistical reliability.

*Core temperature.* Core temperature is a physiological construct used to describe internal body temperature. There are several laboratory methods used to assess core temperature that are not acceptable in a workplace. Acceptable alternative methods are available. The alternative with the longest history is oral temperature. To register an accurate oral temperature, the individual must not eat or drink for 15 minutes prior to the sample, and the mouth must remain closed. Core temperature is approximately equal to oral temperature plus 1 F (0.5 C).

Electronic thermometers are now available. One has an infrared sensor that estimates core temperature from the ear canal. There is also a personal monitor with datalogging capabilities that monitors ear canal temperature with a thermistor held in place by a disposable ear plug. Another commercial device estimates core temperature from a surface-mounted sensor placed on the chest (and is part of one of the heart rate–monitoring devices described in the following section on heart rate). Finally, there are swallowable "pills" that can transmit a value for temperature to a receiver outside of the body.

As a criterion for core temperature, 100.4 F (38 C) is the limit if the temperature is sustained over the course of the workday. If the work is intermittent, then transient increases up to 102.2 F (39 C) should be acceptable as long as there is sufficient recovery time to allow core temperature to return toward 98.6–99.5 F (37–37.5 C). That is, the time-weighted average should not exceed 100.4 F (38 C). As a matter of practice, core temperature should not exceed 102.2 F (39 C) for industrial exposures to heat stress.

*Heart rate.* Another indicator of heat strain is heart rate. Four methods proposed for assessing heart rate are in use. Three methods mentioned above are recovery heart rate, peak heart rate and average heart rates over 8 hours. Another method evaluates a set of averaged heart rates over a typical exposure period.

Recovery heart rate was mentioned above as a tool for recognition of heat stress. It has also been used to evaluate workplaces. To demonstrate effective control of heat stress, the recovery heart rate at 1 minute ($HRR_1$) should be less than 110 bpm. Alternatively, the heart rate at 3 minutes ($HRR_3$) should be less than 90 bpm, or the value of $HRR_1 - HRR_3$ should be at least 10 bpm.

If the daily average heart rate exceeds 110 bpm, heat stress and/or very strenuous work may be the cause. This limiting average has been recommended by the WHO experts and confirmed in laboratory and field studies.

Setting a threshold for heart rate is a third approach. A recording of heart rate during a heat stress exposure is shown in Figure 12–8. As a rule of thumb, peak heart rates should not exceed 90 percent of a person's maximal heart rate ($HR_{max}$). Sometimes this value is known from a stress test; other times it must be estimated from age (the equation is $HR_{max} = 195 - 0.67[Age - 25]$). Brief periods (of less than 1 minute) above this threshold are not significant, but more than 1 minute may be excessive. When looking over a history of heart rate during a day, a sustained trend toward higher heart rates is also indicative of heat stress above the thresholds, because the body is having trouble maintaining thermal equilibrium. The momentary peaks in Figure 12–8 are not significant, but the trend in the second half is a classic representation of the response to moderately high heat stress.

There is a personal monitor commercially available that sets multiple thresholds for heart rate based on averages in seven different time windows. The purpose of this approach is to examine the peak demands as well as the overall trends for a broader evaluation. This device is the one mentioned

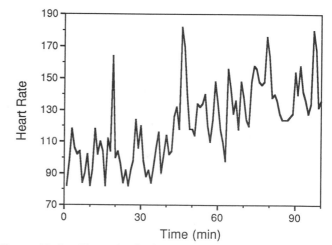

**Figure 12–8.** Example of a heart rate response to heat stress exposure.

earlier that also has a surface sensor for estimating core temperature. If the workers in the sample stay below the thresholds set on this device, it can be inferred that heat stress is not significant.

*Sweat rate and volume.* In theory, sweat rate and volume can serve as measures of physiological strain, but they are less practical than core temperature and heart rate. Sweat volume over a given period of time is equal to an initial body weight plus the weight of food and drink consumed minus the weight of anything excreted minus a final body weight. The overall weight change in kilograms is equal to the sweat volume in liters. If there is more than a 5-liter sweat volume, heat stress is sufficient to cause dehydration and therefore must be controlled. Over a 2– to 4-hour interval, the sweat rate should be less than 1 liter per hour.

## Control of Heat Stress

Control of heat stress and heat strain centers around the causes of heat stress and the resulting physiological strain. It takes the form of general controls that are applicable to all heat-related jobs and specific controls that must be evaluated and selected based on the constraints of the working conditions.

**General controls.** General controls are those actions that are universally applicable to heat stress work. The general controls are training, heat stress hygiene practices, medical surveillance, and a heat-alert program. Any time a group of workers might be exposed to heat stress that is above the RAL (or equivalently the TLV for unacclimated workers), the general controls should be implemented.

*Training.* Training is an essential feature of heat stress management. It should be conducted for both employees working on heat-related jobs and for their supervisors. The information gained from the training enables them to recognize heat stress and to control the risks associated with it. Training is divided into two types: preplacement and periodic.

Preplacement training is directed toward employees who are reporting to heat-related jobs for the first time. The preplacement training can be combined with other job training including safety or skill training. It is not necessary to repeat preplacement training for an employee who has had it once. The formal content of the preplacement training is the same as that for annual training (see below). A possible complement to preplacement training is counseling by medical personnel that relates to an employee's physical condition.

Annual heat stress training should be given to employees working on heat-related jobs to refresh their knowledge of heat stress and controls. The following topics should be covered during training:

- Description of heat stress
    Environment, work demands, and clothing
    Physiological responses including acclimation

- Recognition of and first aid for heat-related disorders
    Description of heat-related disorders, including their symptoms and causes
    Description of first-aid measures for each disorder

- Heat stress hygiene practices (see the next section for details)
    Description of heat stress hygiene practices
    Emphasis on individual responsibility

- Overview of heat stress policy and guidelines
    Company policy
    Management responsibilities
    Employee responsibilities

The format of the training can be similar to other health and safety training. Using commercially available videotapes and written materials is effective and efficient. It is important to point out any issues that are unique to the work site.

*Heat stress hygiene practices.* Heat stress hygiene practices are the actions taken by individuals to reduce the risks of a heat disorder. The individuals are responsible for practicing good heat stress hygiene. Site management informs the workers of good practices and helps the workers implement them.

*Fluid replacement.* A great deal of water is lost from the body in the form of sweat for the purpose of evaporative cooling. Losses can range up to 6 quarts or liters of water in one day, equivalent to about 13 pounds. This water should be replaced by drinking cool water or flavored drinks such as dilute iced tea, artificially sweetened lemonade, or commercial fluid-replacement drinks. Because thirst is not sufficient driver for adequate water replacement, workers should drink small quantities as frequently as possible. This helps instill drinking as a habit, and the volumes do not cause discomfort. If work is to be performed in a drinking-restricted area, drinking about one pint per hour of work before the work begins helps meet the demands for water during the work.

Unacclimated workers in job situations near the TLV can benefit from beverages containing about 0.1 percent salt. Workers on physician-advised salt restricted diets should consult their physician. (Note: Salt tablets should never be taken.)

*Self-determination.* One aspect of self-determination is limiting exposure to heat stress. It is the responsibility of the worker and the supervisor. In self-determination, the person terminates an exposure to heat stress at the first symptom of a heat-related disorder, or with extreme discomfort. Serious injury can occur if the onset of symptoms is ignored.

Another aspect of self-determination is reducing the effects of heat stress by lowering peak work demands and making the work demands lighter. For instance, when a fixed amount of work is assigned for a portion of the shift, peak demands can be reduced by leveling the work effort out over the allocated time or by taking more frequent breaks. For those working in crews, those setting the pace should consider the least heat-tolerant worker.

*Diet.* A well-balanced diet is important to maintain the good health needed to work under heat stress. Large meals should not be eaten during work breaks, because they increase circulatory load and metabolic rate.

Weight-loss diets should be directed by a physician who understands that the patient is working under conditions of potential heat stress. Weight control for overweight workers is recommended because obesity increases the risk of heat-related disorders.

Salt intake as part of a normal diet is usually sufficient to meet the salt demands during heat stress work. Added salt may be desirable when repeated heat stress exposures are first experienced, that is, during acclimation. If salt is restricted by a physician's order, the physician should be consulted.

*Life-style.* A healthy life-style is important to lowering the risk of a heat-related disorder. A worker should have adequate sleep and a good diet. Exercise helps. A healthy life-style also means no abuse of alcohol or drugs, which have been implicated in heatstroke. In addition, exposures to heat stress immediately prior to work increase the risk of a heat disorder at work.

*Health status.* All workers should recognize that chronic illnesses such as heart, lung, kidney, or liver disease indicate a potential for lower heat tolerance, and therefore an increased risk of experiencing a heat-related disorder during heat stress exposures. As a matter of principle, workers suffering from any chronic disorder should inform their physician of occupational exposures to heat stress and seek advice about the potential effects of the disorder or drugs used to treat it.

If a worker is experiencing the symptoms of any acute illness and still reports to work, that worker should inform the immediate supervisor of his or her condition.

*Acclimation.* Acclimation is the adaptation of the body to prolonged daily heat stress exposures. The ability to work increases and the risk of heat disorders decreases with acclimation. Acclimation is lost when there are no more heat exposures. The loss is accelerated when an illness occurs. This process should be recognized and expectations adjusted. Workers will be able to work better after several days of heat exposures, and they should expect less of themselves in the early days.

Table 12–E provides a framework for how acclimation can be induced and reinduced after an absence from heat exposures. Recommendations for new worker acclimation usually start at lower levels to further account for the lack of familiarity with the job and the resulting greater risk of accidents. This is reflected in the 5-day schedule that begins at 20 percent of daily exposure. For experienced workers, recent OSHA recommendations prescribe 3 days of increasing exposures followed by full exposures. Recognizing that full acclimation is normally lost over three weeks, the re-acclimation schedule in Table 12–E is a recommendation by the author.

**Table 12–E.** Basic Acclimation Schedule and a Schedule for Reacclimation after Periods Away from Heat Stress Exposures Due to Routine Absence or Illness

*Basic Acclimation Schedule*

| Day | Activity (% of full work assignment) | |
|---|---|---|
| | Experienced | New |
| Day 1 | 50% | 20% |
| Day 2 | 60 | 40 |
| Day 3 | 80 | 60 |
| Day 4 | 100 | 80 |
| Day 5 | | 100 |

*Re-acclimation Schedule*

| Days Away from Heat-Related Schedule | | Exposure Sequence (% of full work assignment) | | | |
|---|---|---|---|---|---|
| Routine Absence | Illness | Day 1 | Day 2 | Day 3 | Day 4 |
| <4 | — | 100% | | | |
| 4–5 | 1–3 | R/E* | 100% | | |
| 6–12 | 4–5 | 80 | 100% | | |
| 12–20 | 6–8 | 60 | 80 | 100% | |
| >20 | >8 | 50 | 60 | 80 | 100% |

\* Reduce expectations, some diminished capacity

***Medical surveillance.*** Medical surveillance includes the evaluation of individual risk for adverse effects from heat stress exposures, treatment for heat-related disorders, and assessment of the information collected from heat-related disorder incidents. Medical surveillance should be under the direction of a licensed physician.

*Evaluation of risk.* Medical surveillance includes identifying workers who are at extraordinary risk for heat-related disorders. Preplacement and routine physicals under the direction of a physician are used to identify these people. The physician should consult the NIOSH Criteria Document for more information, but ultimately the physician must set the criteria.

Before an employee is placed in a heat-related job, the employee should receive a preplacement physical examination that covers the following items:

■ Comprehensive work and medical history with an emphasis on past intolerance to heat stress and relevant information on the cardiovascular, respiratory, and nervous systems; skin; liver; and kidneys

■ Comprehensive physical examination that gives special attention to the cardiovascular, respiratory, and nervous systems; skin; liver; kidneys; and obesity

■ Assessment of the use of prescription and over-the-counter drugs as well as the abuse of alcohol or other drugs that may increase the risk of heat intolerance

■ Assessment of ability to wear and use the necessary personal protection

■ Assessment of other factors that may affect heat tolerance as deemed important by the physician in charge

The physician should provide a written opinion of the results, and it should be placed in the employee's medical file. A copy should be provided to the employee. The written opinion should contain the following:

- Results of the examination and tests
- Physician's opinion on potential risk to the employee
- Physician's opinion on the employee's capability to work on heat-related jobs
- Any recommended limitations or restrictions
- A statement that the employee has been informed of the results

Because an employee's health status can change over time, periodic reevaluations are appropriate. These periodic physicals should be scheduled approximately yearly. When the physician or management believes that the ability of an individual to tolerate heat stress has diminished, the physician may perform a timely physical outside the schedule for periodic physicals.

*Response to heat-related disorders.* The medical department is responsible for providing emergency response to reported heat-related disorders directly through medical department facilities, by medical department personnel at the job site, or by providing first-aid training to selected personnel.

In addition to providing for emergency response, the physician or designee (for example, an employee in the safety department) should periodically review heat stress incidents to update the program.

***Heat-alert programs.*** Heat-alert programs are a collection of activities undertaken in anticipation of heat stress conditions or an unusually high level of heat stress. These conditions might be the approach of summer, of a maintenance outage, of special operating conditions, or of a heat wave. The first step in a heat-alert program is the appointment of a committee that is responsible for the annual review of heat stress management activities and for making adjustments as necessary. This committee should be comprised of management representatives from such departments as operations, maintenance, engineering, medical, industrial hygiene, safety, and human resources, as well as representatives of labor from different departments that are affected. It should meet well before the anticipated presence of heat stress in the workplace. The committee should complete at least the following activities:

- Review training materials and set training schedule for the current year
- Oversee the preparation of the facility for heat stress conditions (such as reversing winterization) and check the operability of heat stress controls (such as fans, air conditioners, drinking stations, and personal protection)
- Oversee the preparations for changes in staffing and work practices if appropriate

- Review policies and procedures regarding heat-related disorders
- Prepare for extraordinary heat stress conditions by doing the following:

  Set criteria for a Heat-Alert State (such as a sudden increase in ambient temperatures from a heat wave) and how it will be announced

  Prepare special administrative controls (see following section) such as rescheduling work, increasing the number of workers, further restricting overtime, and personal monitoring for excessive heat strain.

  Closely monitor workers for heat-related disorders

**Specific controls.** The two major factors in heat stress are work demands and environmental conditions (air temperature, humidity, air movement, and hot surfaces). Clothing requirements are a third factor when multiple layers, nonwoven clothing, or vapor-barrier fabrics are worn. For specific jobs, the control of heat stress and its physiological strain on workers is accomplished through engineering controls, administrative controls, and personal protection.

Table 12–F is a list of controls suggested by NIOSH in the revised criteria document. Studying the table is one way to begin to understand what might be done to manage the level of heat stress. Although each job must be examined in light of the work to be accomplished and the constraints of the workplace, the following discussion highlights the principles of the control measures and can be used to focus discussion of them.

To select controls for specific jobs, the first step is to discuss aspects of production, engineering, and health and safety functions using the following discussion of control measures. A "long list" of ideas should be generated that places emphasis on engineering controls, followed by administrative controls, and finally personal protection. Imagination is essential during the development of ideas for the long list, and no proposed control should be rejected out of hand unless it quite obviously will not work. Controls should then be judged as to their effectiveness and technical and economical feasibility. The result is a "short list" of controls that can be prioritized and implemented over a reasonable time frame. It is advisable to have short-term solutions in place while long-term solutions are being planned and executed.

***Engineering controls.*** Engineering controls are the kind of controls that reduce or contain a hazard. For heat stress, engineering controls are directed toward reducing physical work demands, reducing external heat gain from the air and hot surfaces, and enhancing external heat loss by increasing sweat evaporation and decreasing air temperature.

*Reduce physical work demands.* The metabolic cost of doing work is the greatest contributor to heat gain by a worker. Reducing the physical work demand can greatly reduce the level of heat stress.

**Table 12–F.** Overview of Specific Controls for Heat Stress Provided by NIOSH in the Criteria Document for Heat Stress

| Item | Actions for Consideration |
|---|---|
| **Controls** | |
| M, Body heat production of task | Reduce physical demands of the work; powered assistance for heavy tasks. |
| R, Radiative load | Interpose line-of-sight barrier; furnace wall insulation, metallic reflecting screen, heat reflective clothing, cover exposed parts of body. |
| C, Convective load | If air temperature is above 35 C (95 F); reduce air temperature, reduce air speed across skin, wear clothing. |
| | If air temperature is below 35 C (95 F); increase air speed across skin and reduce clothing. |
| $E_{Max}$, Maximum evaporative cooling by sweating | Increase by: decreasing humidity, increasing air speed |
| | Decrease clothing |
| **Work practices** | Shorten duration of each exposure; more frequent short exposures better than fewer long exposures. |
| | Schedule very hot jobs in cooler part of day when possible. |
| Exposure limit | Self-limiting, based on formal indoctrination of workers and supervisors on signs and symptoms of overstrain. |
| Recovery | Air-conditioned space nearby. |
| **Personal protection** R, C, and $E_{max}$ | Cooled air, cooled fluid, or ice cooled conditioned clothing |
| | Reflective clothing or aprons |
| **Other considerations** | Determine by medical evaluation, primarily of cardiovascular status |
| | Careful break-in of unacclimatized workers |
| | Water intake at frequent intervals to prevent hypohydration |
| | Fatigue or mild illness not related to the job may temporarily contraindicate exposure (e.g., low-grade infection, diarrhea, sleepless night, alcohol ingestion) |
| **Heat wave** | Introduce heat alert program |

(Reprinted from DS *NIOSH Criteria for a Recommended Standard . . . Occupational Exposure to Hot Environments—Revised Criteria 1986.* Washington, DC: U.S. Government Printing Office, 1986.)

Ways in which the physical work demand can be reduced usually include use of powered tools or new processes to reduce manual effort.

*Reduce air temperature.* When air temperature is above 104 F (40 C), workers gain a significant amount of heat from the air. If the air temperature is below 90 F (32 C), there is a significant loss of body heat. Lowering air temperature serves to either reduce heat gain or enhance the loss of heat. It is a significant factor in the control of heat stress.

Air temperature can be reduced by dilution ventilation and active cooling. Dilution ventilation brings in a supply of cooler air from another area and reduces the temperature in the work area by diluting the hot air with cooler air. This can be accomplished with general area ventilation or local (spot) ventilation. Active cooling means that mechanical refrigeration, evaporative cooling, or a water chiller is employed to reduce the temperature of supplied air for dilution ventilation. Cool rooms are an example of providing a local area for cooling near work areas. By spending some time of the work cycle in the cooler area, the effective exposure to heat stress is reduced.

*Reduce air humidity.* The rate of evaporative cooling of sweat is affected by the air humidity. The rate of cooling is often sufficiently restricted that excessive heat strain occurs. The rate of evaporative cooling can be enhanced by lowering the water content of the air. Water is best removed from air by cooling the air by means of water chillers or mechanical refrigeration. Thus heat stress is reduced by both removing water vapor and lowering air temperature. Again, the use of cool rooms reduces heat stress because of the lower air temperature and lower humidity, and thus an increased rate of evaporative cooling, in them.

*Change clothing.* Clothing is an important contributor to heat stress if it is not a lightweight cotton work uniform. Frequently, when clothing is chosen for good barrier properties against contaminants, not enough thought is given to the effects on heat stress. For instance, when the WBGT is 90 F (32 C) ($T_{db}$ = 100 F, or 38 C) at a moderate rate of work (about 260 W), a person in work clothes can work for about 2 hours, whereas a person in vapor-barrier clothing can work for only about 30 minutes. Changing from vapor-barrier to water-barrier (vapor-transmitting) clothing can increase the tolerance time to 70 minutes.

*Reduce radiant heat.* When the globe temperature is greater than 109 F (43 C), radiant heat is a significant source of heat stress. Radiant heat can come from well-defined or diffuse sources with high surface temperatures.

If a source of radiant heat is well defined and localized, it can be effectively controlled by shielding. Diffuse sources of radiant heat are more difficult to control. For diffuse sources, control can also be provided by shielding, but two other means are also available. One is to insulate surfaces to reduce surface temperature, and the other is to decrease emissivity of the surface. Increasing insulation also reduces air temperature and decreases energy costs.

*Increase air movement.* The advantage of increasing air movement is that it enhances evaporative cooling and convective cooling if the air temperature is less than 95 F (35 C). Between 95 and 104 F (35 and 40 C), heat gain by convection increases with increases in air movement, but it is more than offset by increases in evaporative cooling. Above 104 F (40 C), increases in air movement actually increase the overall heat stress. The greatest reduction in heat stress occurs when air motion is increased from less than 1 meter per second (m/s) to 2 m/s. There is no further improvement in evaporative cooling for air speeds above 3 m/s. When clothing is fairly heavy, higher air speeds can increase clothing ventilation, that is, air better penetrates the clothing and moves near the skin.

The chief mechanism for increasing air movement around a worker is the use of a fan in the work space. Another means of increasing air motion is local ventilation. Increasing air movement, however, often results in an undesirable increase in the level of airborne particles.

***Administrative controls.*** Administrative controls change the way work is performed in order to limit exposures or risks. For heat stress, administrative controls are directed toward limiting exposures so that increases in heart rate and core temperature do not exceed accepted limits.

*Acclimation.* Acclimation is the process that allows a worker to become accustomed to heat stress; after acclimating, the worker is better able to work in the heat. Acclimation is a powerful adaptation that comes naturally to more than 95 percent of the work force. Acclimation is usually achieved through a schedule of increasing exposures. A schedule for acclimation and reacclimation is provided in Table 12–E.

An alternative method is to control the exposures to a level equivalent to the NIOSH-recommended alert limit. This alternative approach requires more attention to the work methods and environment and is not routinely used.

*Pacing of work.* Because work metabolism is an important contributor to heat stress, methods to reduce the rate of metabolism can go a long way toward reducing heat stress. The rate is reduced when the same amount of work is performed over a longer period of time. Any idle time inherent in the work process should be spent in cooler areas to realize the full benefit.

For instance, many jobs have a fixed amount of work to be accomplished, and the workers are given an allotted amount of time in which to do it. The tendency in cool conditions is to work very fast and have the remaining time idle. The same pace in hot environments can cause excessive heat strain. So when the environment is hot, the work should be leveled out to reduce the rate of metabolism and the potential for excessive heat stress.

*Sharing work.* Another way to reduce metabolism is to share or distribute the work among other workers. This may require that some work be postponed to another time. In scheduling the work, thought should be given to how to use the staff most efficiently and effectively. For instance, it may be possible for an extra worker to move between two crews during the same work period and to reduce heat stress. Furthermore, if extra workers can straddle work shifts, they can work the second half of the day shift and the first half of the afternoon shift, when heat stress is most likely to be a problem.

*Scheduling of work.* An administrative control to reduce environmental heat's contribution to heat stress is to schedule nonessential work at cooler times of day or during cooler periods.

*Work times, self-determination, and personal monitoring.* Preplanned work times, self-determination, and personal monitoring are ways to control a heat stress exposure that is known to be high. Predetermined work times are assigned to a worker or crew before a job begins. They may be allowed to extend the work time with the knowledge that heat stress will eventually affect their ability to work, and there is a risk of heat-related disorder. The extension should be under the controls of self-determination aided by personal monitoring.

The purpose of self-determination with personal monitoring is to allow more heat-tolerant workers to work longer than less tolerant ones by letting workers stop exposures when they feel it necessary. These kinds of administrative controls apply better to self-paced and nonroutine work, and may be more difficult to manage during externally paced work. Self-determination is best instituted by periodic queries to individual workers about their subjective judgment of heat strain and their ability to continue. Because subjective decisions are unreliable, objective data on heat strain should be obtained from personal monitoring of body temperature and/or heart rate. In 1995, there were at least two electronic personal monitors designed for industrial use that were commercially available. One device measures body temperature through ear canal temperature, and the second examines both heart rate and body temperature (through a surface sensor). These are shown in Figure 12–9. There are also other heart rate monitors and methods to measure temperature that were not specifically designed for industrial heat exposures.

***Personal protection.*** Personal protection is a control that provides protection for individual workers. For heat stress, personal protection is primarily some form of personal cooling, but can include reflective clothing for high–radiant heat conditions. Personal cooling systems, if chosen to match the job situation, can significantly increase the safe exposure time. When personal protection is used, conventional evaluation methods do not apply, and work practices must be developed for the successful use of personal protection for heat stress.

*Circulating air systems.* Circulating air as a personal cooling method consists of circulating air under the clothing and around the torso. It requires the delivery of air to the individual either through a high-pressure air line and a pressure reducer or by a portable (self-contained) blower. Circulating air under the clothing effectively increases the amount of convective and evaporative cooling of the body. (Note: The circulating air must be breathing-grade air.)

Air-line systems can be used continuously for work that is relatively stationary, and they can also be used profitably by workers as temporary relief during pauses in the work. If there is a sufficient supply of compressed air, vortex tubes can be used to significantly reduce (by approximately 18 F [10 C]) the temperature of the air going into the clothing. The vortex tube allows compressed air to expand and separate into two streams of air (one cool and the other warm). The cool air is delivered

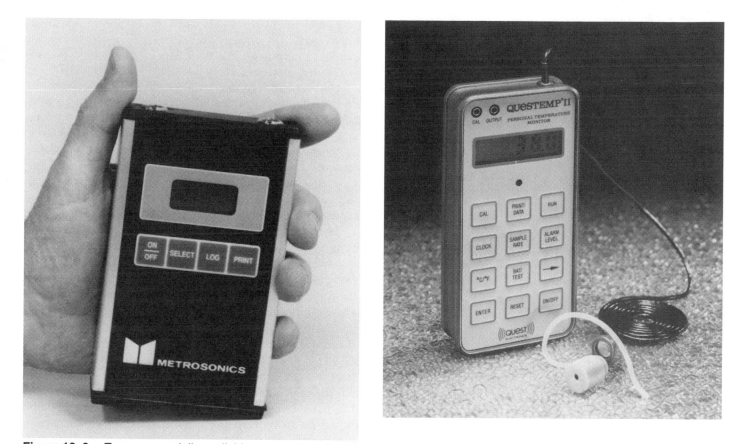

**Figure 12–9.** Two commercially available personal monitors developed for use with heat stress exposure: Metrosonics hs3800 and Quest Electronics QuesTemp II. (Photos provided courtesy of Metrosonics, Inc., and Quest Technologies.)

to the user and the warm stream is exhausted to the environment.

Portable blower systems are starting to be used in the workplace. Because they use air in the work locale, they often do not provide as much cooling capacity as air-line systems delivering the same volume flow rate.

*Circulating water systems.* A second type of personal cooling is a system that circulates cool water through tubes and channels around the body. There are a variety of systems available. One kind of system covers virtually the whole body; others cover only portions of the back and chest. There are also portable versions as well as versions with a heat sink that the worker is connected to by a tether. The degree of cooling that can be achieved depends on surface area covered, rate of water circulation, and the capacity of the heat sink.

The selection of circulating water systems should be made in consultation with the vendors and someone familiar with personal cooling in order to obtain a good match for your needs.

*Ice garments.* Several frozen-water garments, often called ice vests, are commercially available. They control heat strain by removing body heat via conduction from the skin to packets of ice. The typical vest weighs about 5 kg. The vests provide good mobility, with some bulk around the torso. The ability to cool and the length of

time an ice vest is effective depend on the rate of work, the amount of ice, and the design of the particular garment. For a given amount of work, the time is limited by the ice. Service times up to 2 hours are reasonable from some models.

*Reflective clothing.* Whereas personal cooling is designed to absorb body heat, reflective clothing is designed to reduce the amount of heat reaching the individual. Reflective clothing is best suited for sources of high radiant heat. There is a trade-off with reflective clothing in that it reduces sweat evaporation. That means that the level of heat stress can actually increase if the reflective clothing is not selected to best match the source of radiant heat and the job.

## Hot Surfaces, Breathing Hot Air, and Respirators

Hot work environments usually include hot surfaces with the potential to cause burns. The potential for a burn (or to elicit pain at a somewhat lower level of heat transfer) depends on the thermal conductivity of the solid, the temperature of the solid, and the contact time with the skin. Table 12–G gives approximate surface temperatures for common surfaces that elicit pain or a burn with brief (1 s) contact time and burns with longer

**Table 12–G.** Limits on Surface Temperature (in degrees Centigrade) to Avoid Pain and Burns with Different Contact Periods Against Skin

| Material | Pain (1 s) | Burn (1 s) | Burn (4 s) | Burn (10 min) | Burn (prolonged) |
|---|---|---|---|---|---|
| Metals | 45 | 65 | 60 | | |
| Glass | 55 | 85 | 75 | | |
| Wood | 75 | 115–140 | 95–120 | | |
| | | (depends on dryness of wood) | | | |
| Any material | | | | 48 | 43 |

(Adapted from Eastman Kodak, 1983; Siekmann, 1990)

contact times. For prolonged contact, surface temperature is the dominant characteristic. To avoid tissue injury, the surface temperature should be less than 118 F (48 C) for up to 10 minutes of contact time and 109 F (43 C) for prolonged contact. There will be reports of extreme discomfort with surface temperatures greater than 100 F (38 C).

Sometimes there is a concern about the temperature of the air that is being breathed. If the wet bulb temperature of the air is less than 113 F (45 C), breathing the air is not likely to cause extreme discomfort or ill effects. There are laboratory observations of sustained breathing at wet bulb temperatures of 122 F (50 C) without complaint. Above these temperatures, the probability of individual discomfort increases.

There are also concerns about the effects of tight-fitting, full-face respirator facepieces on heat stress; usually that negative-pressure respirators increase the level of heat stress, and that air-supplied respirators can give a false sense of cooling. There are small changes in the level of physiological response, but not enough that the heat stress guidelines for evaluation described above should be adjusted to account for respirator use. There is no doubt, though, that subjectively measured discomfort increases while wearing a respirator facepiece in hot environments, and this may affect performance.

## Cold Stress

Cold stress is a fundamentally different kind of problem than heat stress. Whereas adaptive mechanisms such as sweating and acclimation are crucial during heat stress exposures, the physiological adaptations to cold stress have less dramatic effects. The first physiological response to cold stress is to conserve body heat by reducing blood circulation through the skin. This effectively makes the skin an insulating layer. A second physiological response is shivering, which increases the rate of metabolism. Shivering is a good sign that cold stress is significant and that hypothermia may be present. These responses, how-

ever, are relatively weak as a protection mechanism. Behavior is the primary human response to preventing excessive exposure to cold stress. Behaviors include increasing clothing insulation, increasing activity, and seeking warm locations.

Insulation is a critical characteristic of clothing worn during cold stress exposures. Clothing materials used for their insulating characteristics include cotton, wool, silk, nylon, down, and polyester. Generally, better insulation is achieved by layering clothes rather than having one garment. The further advantage of layers is that a person can add or remove layers to adjust for differing insulation needs during the work period.

The insulating value of clothing is greatly diminished by moisture. Sources of water are the work environment and sweat. Water vapor permeability is also important. If sweat is allowed to evaporate through the clothing, it does not accumulate in the clothing. Once clothing becomes wet, it is important to replace it immediately.

Like layering, clothing ventilation is a valuable means to adjust the heat transfer properties of the ensemble. During low levels of work, insulation demands are high; as the work rate increases, insulation must decrease to maintain thermal equilibrium. In addition to removing layers, the effective insulation of an ensemble can be reduced by increasing air movement under the clothing by increasing clothing ventilation.

Hazards associated with cold stress are manifested in two distinct fashions: systemic (hypothermia) and local (localized tissue damage). Also, manual dexterity decreases. The disorders related to cold stress exposures are described in Table 12–H.

As hypothermia progresses, depression of the central nervous system becomes more severe. This accounts for the progression of signs and symptoms from sluggishness through slurred speech and unsafe behaviors to disorientation and unconsciousness. The ability to sustain metabolic rate and reduced skin blood flow is diminished by fatigue. Thus fatigue increases the risk of severe hypothermia through decreasing metabolic heat and increased heat loss from the skin. Because blood flow through the skin is reduced to conserve heat, the skin and underlying tissues are more susceptible to local cold injury.

### Model of Thermal Balance

Systemic cold stress can be quantified in terms of heat exchange and how much heat the body can store ($S$):

$$S = M + (R + C) + K + E \qquad (13)$$

$M$ is metabolic rate and represents a source of internal heat gain. $(R + C)$ is the combination of heat loss due to cooler air and surroundings. $K$ is conduction to a solid surface in contact with the body. $E$ is evaporative cooling by

**Table 12–H.** Cold-Related Disorders Including the Symptoms, Signs, Causes, and Steps for First Aid

| Disorder | Symptoms | Signs | Causes | First Aid |
|---|---|---|---|---|
| Hypothermia | Chills<br>Pain in extremities<br>Fatigue or drowsiness | Euphoria<br>Slow, weak pulse<br>Slurred speech<br>Collapse<br>Shivering<br>Unconsciousness<br>Body temperature <95 F (35 C) | Excessive exposure<br>Exhaustion or dehydration<br>Subnormal tolerance (genetic or acquired)<br>Drug/alcohol abuse | Move to warm area and remove wet clothing<br>Modest external warming (external heat packs, blankets, etc.)<br>Drink warm, sweet fluids if conscious<br>Transport to hospital |
| Frostbite | Burning sensation at first<br>Coldness, numbness, tingling | Skin color white or grayish yellow to reddish violet to black<br>Blisters<br>Response to touch depends on depth of freezing | Exposure to cold<br>Vascular disease | Move to warm area and remove wet clothing<br>External warming (e.g., warm water)<br>Drink warm, sweet fluids if conscious<br>Treat as a burn, do not rub affected area<br>Transport to hospital |
| Frostnip | Possible itching or pain | Skin turns white | Exposure to cold (above freezing) | Similar to frostbite |
| Trench Foot | Severe pain<br>Tingling, itching | Edema<br>Blisters<br>Response to touch depends on depth of freezing | Exposure to cold (above freezing) and dampness | Similar to frostbite |
| Chilblain | Recurrent, localized itching<br>Painful inflammation | Swelling<br>Severe spasms | Inadequate clothing<br>Exposure to cold and dampness<br>Vascular disease | Remove to warm area<br>Consult physician |
| Raynaud's disorder | Fingers tingle<br>Intermittent blanching and reddening | Fingers blanch with cold exposure | Exposure to cold and vibration<br>Vascular disease | Remove to warm area<br>Consult physician |

Note: Hypothermia is related to systemic cold stress, and the other disorders are related to local tissue cooling.

sweat evaporation; it has a negative value because the flow of heat is away from the body. Thermal equilibrium is established when $S = 0$.

$M$ can be increased as a behavioral response to cold stress, and significant contributions to thermal balance can be made by reducing $(R + C)$ and $K$ with behavioral adaptations like clothing and avoidance of cold environments. For a given level of clothing, the greater the work demands (greater metabolic rate), the greater the level of cold stress that can be tolerated.

The goal of systemic cold stress control is to avoid hypothermia by limiting the minimum core temperature to 96.8 F (36 C) for prolonged exposures and to 95 F (35 C) for occasional exposures of short duration.

## Measurement of Cold Stress

Two climatic factors in the environment influence the rate of heat exchange between a person and the environment. These factors are air temperature and air speed. As the difference between skin and ambient temperatures increases and/or the air speed increases, the rate of heat loss from exposed skin increases. The rate of heat loss is approximated by Equation 14.

$$H = 1.16(10 V_{air}^{0.5} + 10.45 - V_{air})(33 - T_{db})(W/m^2) \quad (14)$$

The equivalent chill temperature (ECT) (also known as the windchill index) was developed by the US Army to account for both factors. Table 12–I gives the ECT for different combinations of air temperature and speed. The table was generated by determining the heat loss under conditions of each combination of air temperature and speed, and then computing an equivalent air temperature with no air motion with which the same rate of heat loss occurs.

## Recognition of Cold Stress

Subjective responses of workers provide a good tool for recognition of cold stress in the workplace. If the workplace is generally described as cold, then cold stress may be present. Worker behaviors to cold stress exposures are generally those of seeking warm locations, adding layers of clothing, or increasing the work rate. Other behaviors are loss of manual dexterity, shivering, accidents, and unsafe behaviors.

Using the first-aid logs and other records, determine whether there is a pattern of signs and symptoms that

**Table 12–I.** Equivalent Chill Temperature (ECT) in Degrees C for Different Combinations of Air Temperature and Air Speed (also known as the windchill index)

| Air Speed | | Air Temperature (C) | | | | | | | | | | | |
|---|---|---|---|---|---|---|---|---|---|---|---|---|---|
| | | 10 | 4 | −1 | −7 | −12 | −18 | −23 | −29 | −34 | −40 | −46 | −51 |
| | | Air Temperature (F) | | | | | | | | | | | |
| | | 50 | 40 | 30 | 20 | 10 | 0 | −10 | −20 | −30 | −40 | −50 | −60 |
| m/s | mph | Equivalent Chill Temperature (ECT) (C) | | | | | | | | | | | |
| 0.0 | 0 | 10 | 4 | −1 | −7 | −12 | −18 | −23 | −29 | −34 | −40 | −46 | −51 |
| 2.2 | 5 | 9 | 3 | −3 | −9 | −14 | −21 | −26 | −32 | −38 | −44 | −49 | −56 |
| 4.5 | 10 | 4 | −2 | −9 | −16 | −23 | −31 | −36 | −43 | −50 | −57 | −64 | −71 |
| 6.7 | 15 | 2 | −6 | −13 | −21 | −28 | −36 | −43 | −50 | −58 | −65 | −73 | −80 |
| 8.9 | 20 | 0 | −8 | −16 | −23 | −32 | −39 | −47 | −55 | −63 | −71 | −79 | −85 |
| 11 | 25 | −1 | −9 | −18 | −26 | −34 | −42 | −51 | −59 | −67 | −76 | −83 | −92 |
| 13 | 30 | −2 | −11 | −19 | −28 | −36 | −44 | −53 | −62 | −70 | −78 | −87 | −96 |
| 16 | 35 | −3 | −12 | −20 | −29 | −37 | −46 | −55 | −63 | −72 | −81 | −89 | −98 |
| >16 | >35 | −3 | −12 | −21 | −29 | −38 | −47 | −56 | −65 | −73 | −82 | −91 | −100 |

**Little Danger**
If exposures with dry skin and less than 60 min. Caution: Avoid false sense of security.

**Increasing Danger**
Exposed flesh may freeze within 1 min.

**Great Danger**
Flesh may freeze within 30 s.

**Caution:** Trenchfoot and immersion foot may occur anywhere on this chart. (Developed by the U.S. Army Research Institute of Environmental Medicine, Natick, MA.)

might be attributed to hypothermia. A physiological marker is reduced core temperature below 96.8 F (36 C).

If there is a noticeable drop in manual dexterity reported by workers or supervision, local cold stress is possible. In addition, if there is a pattern of cold-related disorders reported in the first-aid logs, injury and illness logs, and workers compensation records, the work conditions should be evaluated.

## Evaluation of Cold Stress

**Workplace monitoring.** When temperatures fall below 61 F (16 C), workplace monitoring should be instituted. Below 30 F (–1 C), the dry bulb temperature and air speed should be measured and recorded at least every 4 hours. When air speed is greater than 5 miles per hour (2 m/s), the ECT should be determined from Table 12–I. When the ECT falls below 19 F (–7 C), it should also be recorded.

**Systemic cold stress.** Hypothermia can occur with air temperatures up to 50 F (10 C). The ACGIH recommends that employers require protective measures when air temperature is less than 41 F (5 C). Equation 15 can be used to estimate the amount of clothing insulation ($I_{clo}$ in clo units, where 1 clo = 0.156 m² · C/W) required for a specific task in a given air temperature ($T_{db}$ in degrees C) and metabolic rate ($M$ in watts). Figure 12–10 illustrates the relationships among temperature, work rate, and clothing necessary to maintain thermal equilibrium based on Equation 15.

$$I_{clo} = 11.5 \, (33 - T_{db}) \, / \, M \qquad (15)$$

Remember that the clothing must be kept dry and that $I_{clo}$ will vary with different tasks and environments.

As the environmental conditions become very cold, rewarming periods should be provided. Specific rewarming schedules are provided in the TLV for Cold Stress (see Appendix B, ACGIH Threshold Limit Values and Biological Exposure Indices, Cold Stress, Table 3). The maximum exposure time depends on air temperature and ambient air movement, and is followed by a 10-min warmup break. Reductions in the working time are recommended if the work rate is low to moderate because internal heat generation will be lower. There is a point at which non-emergency work should not be performed.

**Local cold stress.** Skin cannot freeze until the air temperature is less than 30 F (–1 C), and there is little risk of local cold injury associated with ECTs greater than –22 F

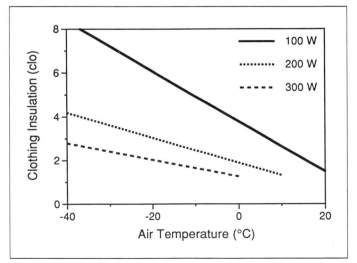

**Figure 12–10.** Estimation of clothing insulation (clo) required as a function of air temperature for different levels of work.

(–30 C) (or heat loss rates less than 1,750 W/m²). The temperature limit for surfaces with which exposed skin makes incidental contact is 19 F (–7 C). If the contact is prolonged (for example, for tools), the limit is 30 F (–1 C).

Manual dexterity drops when there is uninterrupted work for 10–20 minutes at temperatures below 61 F (16 C).

## Control of Cold Stress

As with heat stress, general controls are actions that should be taken when workers are exposed to cold stress. The general controls include training, hygiene practices, and medical surveillance.

**Training.** When the air temperature is below 41 F (5 C), workers should be informed that cold stress is a hazard and told what the proper clothing is and that self-determination and cold stress hygiene should be practiced.

When work is performed at or below 10 F (–12 C) ECT, additional training topics include safe work practices and the recognition and first-aid treatment of hypothermia and other cold-related disorders.

**Hygiene practices.** Cold stress hygiene practices center around fluid replacement with warm, sweet, non–caffeine-containing drinks and self-determination. Warm drinks are palatable in cold stress, and the fluid replacement is important because significant dehydration can occur. Sweetened drinks provide a readily usable energy source to reduce the risk of fatigue. In addition, employees should be encouraged to eat a normal, balanced diet. Any worker who experiences extreme discomfort or any of the symptoms of hypothermia (or any other cold-related disorder) should stop work and seek a place to rewarm.

Safe work practices include at least the following. In air temperatures below 36 F (2 C), anytime clothing becomes wet, it must be replaced immediately, and the workers should be treated as if they are experiencing hypothermia. When handling liquids with boiling points below 39 F (4 C), special precautions should be taken to ensure that clothes do not become soaked in the liquid.

**Medical surveillance.** Medical certification is suggested for those who are routinely exposed below –11 F (–24 C) ECT. Certification should be based on a physical that considers fitness, weight, cardiovascular health, and other conditions that can affect cold stress tolerance. Furthermore, any employee who is under care for chronic disease should consult their personal physician.

If there is reason to suspect that a worker cannot properly thermoregulate, a medical restriction is appropriate in air temperatures below 30 F (–1 C).

### Specific controls

*Engineering controls.* Engineering controls attempt to reduce heat loss from the person as a whole and from exposed skin. Engineering controls include the following:

- Providing for general or spot heating including hand warming

  Hand warming for fine hand work below 61 F (16 C)

  Minimizing air movement in such ways as providing shielding and adjusting ventilation

- Reducing conductive heat transfer (for example, not allowing the use of metal chairs or uninsulated tools)

- Redesigning equipment, processes, and so on to control systemic and local cold stress

- Providing warming shelters if exposures occur below 19 F (–7 C) ECT

*Administrative controls.* Administrative controls attempt to reduce the exposure time, allow individual control over the work, and provide for mutual observation. Recommended administrative controls include

- Set up work–rest cycles

- Schedule work at warmest times

- Move work to warmer areas

- Assign additional workers

- Encourage self-pacing and extra breaks if required

- Establish a buddy system, emphasizing mutual observation

- Avoid long periods of sedentary effort

- Allow for productivity reductions and extra effort required when wearing protective clothing

- Provide an adjustment or conditioning period for new employees

- Monitor weight changes for dehydration

*Personal protection.* Because clothing is so important, personal protection is fundamental in managing cold stress. Clothing should be carefully selected in consultation with knowledgeable vendors and with the workers themselves. The workers must be educated as to the roles of clothing items and what compromises effectiveness. Personal protection includes the following:

- Properly selected insulated clothing

  Wind barriers

  Special attention to feet, fingers, ears, nose, and face

  Gloves when air temperature is less than 61 F (16 C) for light work, 39 F (4 C) for moderate work and 19 F ( 7 C) for heavy work

  Mittens when air temperature is less than 1 F (–17 C)

- Water barriers to external liquids

- Appropriate active warming systems such as circulating air or liquids, or electric heaters

- Appropriate eye protection for snow- or ice-covered terrain

## Thermal Comfort

Thermal comfort is a condition of mind in which a person will express satisfaction with the thermal environment. Factors that affect thermal comfort include air temperature, humidity, air motion, surface temperatures, metabolic rate, and clothing. Age, sex, season, cultural background, and intraindividual variation play minor roles once the previously mentioned factors are accounted for. For instance, what appear to be differences based on sex and seasonal changes in a comfortable environment can often be explained by differences in clothing. Factors that can disrupt a theoretically comfortable environment are asymmetric thermal radiation, drafts, vertical temperature gradients, and floor temperatures.

The ASHRAE Standard 55–1981 describes comfort zones based on operative temperature and humidity. These are illustrated in Figure 12–11. One zone is for the winter season, and another, overlapping zone is for the summer season. The seasonal zones are specified to account for changes in clothing habits in the winter and summer. The operative temperature is approximately the globe temperature, and with little radiant heat, it corresponds to the air (dry bulb) temperature.

Under ideal conditions, no more than 95 percent of the working occupants are satisfied with the thermal environment. As the climatic conditions deviate from the ideal, more people are dissatisfied with the conditions. For those who wish to gain a better understanding of how groups of people respond to different environments, the ISO has published a standard on thermal comfort. The standard considers such factors as climate, clothing, and metabolic rate.

## Summary

This chapter was divided into three sections: Heat Stress, Cold Stress, and Comfort. The Heat Stress and Cold Stress sections described how to recognize whether stress is present and significant in the workplace. Recognition comes through knowledge of the risks entailed by types of work that are usually associated with the stress, and through manifestations of disorders and worker behaviors resulting from the stress. Methods to evaluate the stress were then described, with an emphasis on those recommended by the American Conference of Governmental Industrial Hygienists (ACGIH) in their documentation of the Threshold Limit Values (TLVs) for Heat and Cold Stress. Controls for heat and cold stress were then described in passages on general controls and specific controls.

General controls are broadly applicable when there is a potential for excessive exposure. General controls include training, hygiene practices, and medical surveillance. With regard to training, annual sessions within the format for other health and safety training are encouraged. Training sessions should cover the nature of the hazard, when and how it can occur, physiological responses to the stress, disorders related to the stress, hygiene practices, and proper use of personal protection and other controls.

Specific controls are controls that are appropriate for a specific job and are selected based on job and site constraints. They include the traditional hierarchy of controls: engineering controls, administrative controls, and personal protection.

Finally, means to evaluate thermal comfort were described. The factors examined in the evaluation can be manipulated in order to improve the degree of comfort in the work environment.

## Bibliography

American Conference of Governmental Industrial Hygienists. *Threshold Limit Values and Biological Exposure Indices for 1994–1995.* Cincinnati, OH: American Conference of Governmental Industrial Hygienists, 1994 (updated annually).

Eastman Kodak. *Ergonomic Design for People at Work,* Vol. 1. New York: Van Nostrand Reinhold, 1983.

Holmér I. Assessment of cold stress in terms of required clothing insulation—IREQ. *Int J Industrial Ergonomics* 3:159–166, 1988.

Holmér I. Cold Stress. Part I—Guidelines for the practitioner, and Part II—The scientific basis (knowledge base) for the guide. *Int J Industrial Ergonomics* 14:139–159, 1994.

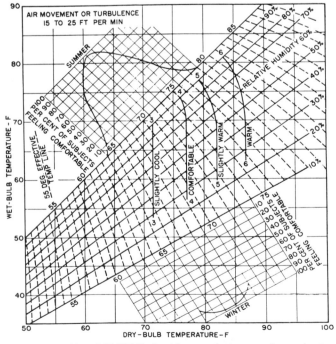

**Figure 12–11.** ASHRAE chart for comfort zones for sedentary activities. (Reprinted with permission from ASHRAE *Handbook.*)

International Organization for Standardization. *Hot Environments—Analytical Determination and Interpretation of Thermal Stress Using Calculation of Required Sweat Rate*. Geneva: International Organization for Standardization 7933, 1989.

International Organization for Standardization. *Thermal Environments—Instruments and Methods for Measuring Physical Quantities*. Geneva: International Organization for Standardization 7726, 1985.

International Organization for Standardization. *Moderate Thermal Environments—Determination of the PMV and PPD Indices and Specification of the Conditions for Thermal Comfort*. Geneva: International Organization for Standardization 7730, 1984.

National Institute for Occupational Safety and Health. *Criteria for a Recommended Standard—Occupational Exposure to Hot Environments. Revised Criteria 1986*. Washington, DC: U.S. Department of Health and Human Services (NIOSH) 86–113, 1986.

Ramsey JD, Dukes-Dobos FN, Bernard TE. Evaluation and control of hot working environments. Part I—Guidelines for the practitioner, and Part II—The scientific bases (knowledge base) for the guide. *Int J Industrial Ergonomics* 14:119–138, 1994.

Siekmann H. Recommended maximum temperatures for touchable surfaces. *Appl Ergonomics* 21: 69–73, 1990.

U.S. Army. Tri-Services Document. *Prevention, Treatment and Control of Heat Injury*. U.S. Army TB Med 507, 1980.

World Health Organization. *Health Factors Involved in Working Under Conditions of Heat Stress*. Technical Report Series 42. Geneva: World Health Organization, 1969.

CHAPTER 13

# Ergonomics

by Karl H. E. Kroemer, PhD

**Matching Person and Task**

**The Human as Information Processor**
*The Nervous System ▪ Sensors ▪ The Signal Loop ▪ Responding to Stimuli*

**Mental Workload**

**Human Capacity for Physical Work**
*Energy Cost of Work ▪ Work Classification ▪ Work/Rest Cycles ▪ Fatigue*

**Engineering Anthropometry**
*Civilian Body Dimensions ▪ Designing for Percentiles ▪ Population Changes*

**Biomechanics**
*Muscle Strength*

**Handling Loads**
*The Body as Energy Source ▪ Training for Safe Lifting Practices ▪ Personnel Selection for Material Handling ▪ Assessment Methods ▪ Screening Techniques ▪ Permissible Load Handling ▪ Limits for Lifting, Lowering, Pushing, Pulling, and Carrying ▪ Comparing NIOSH with the Snook–Ciriello Recommendations.*

**Hand Tools**

**Workstation Design**
*General Principles ▪ Standing or Sitting*

**Workplace Design**
*Work Space Dimensions ▪ Office (Computer) Workstations ▪ Work Task ▪ Positioning the Body in Relation to the Computer ▪ Healthy Work Postures ▪ Biomechanical Actions on the Spine ▪ Experimental Studies*

**Ergonomic Design of Office Workstations**

**Controls and Displays**

**Light Signals**

**Labels**

**Avoiding Cumulative Trauma Disorders**
*History ▪ CTDs in Industry ▪ CTDs of Keyboard Users ▪ CTDs in Medical Terms ▪ Points of View ▪ Cumulative Injuries to the Body ▪ Which Body Components are at CTD Risk? ▪ Causes of CTDs ▪ Carpal Tunnel Syndrome ▪ Occupational Activities and Related Disorders ▪ Ergonomic Countermeasures ▪ Research Challenges ▪ Safe Thresholds and Doses*

**Summary**

**Bibliography**

THE TERM *ERGONOMICS* WAS COINED in 1950 by a group of physical, biological, and psychological scientists in the United Kingdom to describe their interdisciplinary efforts to design equipment and work tasks to fit the operator. The term is derived from the Greek roots *ergon,* (human) work and strength, and *nomos,* indicating law or rule. In 1957, American behavioral scientists, anthropometrists, and engineers working in this emerging scientific discipline decided to call their new professional association the Human Factors Society, which became the Human Factors and Ergonomics Society in 1992.

Ergonomics is the study of human characteristics for the appropriate design of the living and work environment. Ergonomic researchers strive to learn about human characteristics (capabilities, limitations, motivations, and desires) in order to adapt a human-made environment to the people involved. This knowledge may affect complex technical systems or work tasks, equipment, and workstations, or the tools used at work and at home. Hence, ergonomics is human-centered, transdisciplinary, and application-oriented.

The goals of ergonomics and human factors range from making work safe and humane to increasing human efficiency to promoting human well-being. The Committee on Human Factors of the National Research Council (1983) stated:

Human factors engineering can be defined as the application of scientific principles, methods, and data drawn from a variety of disciplines to the development of engineering systems in which people play a significant role. Successful application is measured by improved productivity, efficiency, safety, and acceptance of the resultant system design. The disciplines that may be applied to a particular problem include psychology, cognitive science, physiology, biomechanics, applied physical anthropology, and industrial and systems engineering. The systems range from the use of a simple tool by a consumer to multiperson sociotechnical systems. They typically include both technological and human components.

Human factors specialists . . . are united by a singular perspective on the system design process: that design begins with an understanding of the user's role in overall system performance and that systems exist to serve their users, whether they are consumers, system operators, production workers, or maintenance crews. This user-oriented design philosophy acknowledges human variability as a design parameter. The resultant designs incorporate features that take advantage of unique human capabilities as well as build in safeguards to avoid or reduce the impact of unpredictable human error.

Figure 13–1 shows how ergonomics interacts with related applied disciplines. Among the primary foundations of ergonomics are the biological sciences, particularly anatomy and physiology. Leonardo da Vinci in the sixteenth century, Giovanni Alfonso Borelli in the seventeenth century, and Lavoisier, Amar, Rubner, Johannson, and many others in the nineteenth and early twentieth centuries contributed ideas, theories, and practical data to forward the understanding of the role of the human body in a work environment. Among the social and behavioral sciences, anthropologists, psychologists, and sociologists have contributed toward modeling and understanding the human role in societal and technological systems, including management theories. Among the engineering disciplines, industrial engineers, mechanical and computer engineers, and designers are the major users of ergonomic knowledge. A typical application is computer-aided design (CAD),

which incorporates the systematic consideration of human attributes, especially anthropometric and biomechanical information. Ergonomists have developed their own theories, methods, techniques, and tools to perform scientific research. Formal college degrees are offered by a number of universities, usually in departments of industrial engineering or psychology. Productivity, health and safety at work, the quality of work life, and participatory management are some well-known programmatic aspects of ergonomics.

There are three levels at which ergonomic knowledge can be used. Tolerable conditions do not pose known dangers to human life or health. Acceptable conditions are those on which (according to the current scientific knowledge and under given sociological, technological, and organizational circumstances) the people involved can voluntarily agree. Optimal conditions are so well-adapted to human characteristics, capabilities, and desires that physical, mental, and social well-being are achieved. The aim of ergonomics is to achieve ease and efficiency at work.

## ■ Matching Person and Task

People perform widely differing tasks in daily work situations. These tasks must be matched with human capabilities to avoid underloading, in which human capabilities are not used fully, as well as overloading, which may cause the employee to break down and suffer reduced performance capability or even permanent damage. Engineering psychologists, work physiologists, and occupational biomechanists evaluate the capacities and limitations of the worker

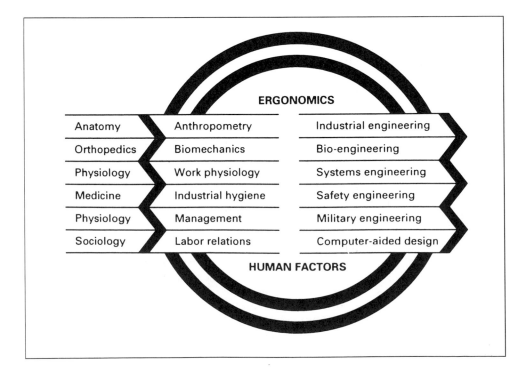

**Figure 13–1.** Origins and applications of ergonomics and human factors.

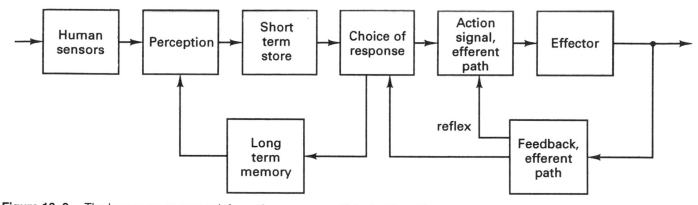

**Figure 13–2.** The human as energy or information processor. (Adapted from Kroemer, Kroemer, Kroemer-Elbert, *Ergonomics: How to Design for Ease and Efficiency.* Englewood Cliffs, NJ: Prentice Hall, 1994. Used with permission by the publisher. All rights reserved.)

for performing work; they also determine human tolerance to stresses produced by the environment.

Note: The following text contains material from Kroemer, Kroemer, and Kroemer-Elbert, *Ergonomics: How to Design for Ease and Efficiency,* Englewood Cliffs, NJ: Prentice Hall, 1994. Used with permission by the publisher. All rights reserved.

## The Human as Information Processor

In the traditional system concept of engineering psychology, the human is considered a receptor and processor of information or energy who outputs information or energy. Input, processing, and output follow each other in sequence. The output can be used to run a machine, which may be a simple hand tool or a spacecraft. This basic model is depicted in Figure 13–2.

The actual performance of this human–technology system (in the past often called a man–machine system) is monitored and compared with the desired performance. Feedback loops connect the output side with the input side. The difference between output and input is registered in a comparator and corrective actions are taken to minimize any output/input difference. The human in this system controls, compares, makes decisions, and corrects.

*Affordance* is the property of an environment that has value to the human. An example is a stairway, which affords passage for a person who can walk but not for a person using a wheelchair. Thus, passage is a property of the stairway, but its affordance value is specific to the user. Accordingly, ergonomics or human engineering provides affordances.

Traditional engineering psychologists describe our activities as a linear sequence of stages from perception to decision to response. Research is done separately on each of these stages, on their substages, and on their connections. Such independent, stage-related information is then combined into a linear model. This behavioristic model is thought invalid by ecological psychologists, who see the events from human perception to action as simultaneous rather than sequential. According to ecological theory, information about affordances can be perceived directly and simultaneously by various human senses as part of the intimate coupling of perception and action. Thus, the closed-loop coupling of perception and action of the human in the environment is *not* modeled as the traditional linear sequence of perception–encoding–decision–response. However, about all of our current data developed by engineering psychologists fit the sequential model.

### The Nervous System

Anatomically, the nervous system has three major subdivisions. The *central* nervous system (CNS) includes the brain and spinal cord; it has primarily control functions. The *peripheral* nervous system (PNS) includes the cranial and spinal nerves; it transmits signals, but usually does not control. The *autonomic* nervous system consists of the sympathetic and the parasympathetic subsystems, which regulate involuntary functions such as of the smooth and cardiac muscles, blood vessels, digestion, and glucose release in the liver. The autonomic system is not under conscious control. Functionally, there are two subdivisions of the nervous system: the autonomic system (just discussed) and the *somatic* nervous system (from Greek *soma,* meaning body), which controls mental activities, conscious actions, and skeletal muscle.

The brain is usually divided into *forebrain, midbrain,* and *hindbrain.* Of particular interest for the neuromuscular control system is the forebrain, with the cerebrum, which consists of the two (left and right) cerebral hemispheres, each divided into four lobes. Control of voluntary movements, sensory experience, abstract thought, memory, learning, and consciousness are located in the cerebrum. The multifolded *cortex* covers most of the cerebrum. The cortex controls voluntary movements of the skeletal muscle and interprets sensory inputs. The *basal ganglia* of the midbrain are composed of large pools of neurons, which control semivoluntary complex activities such as walking. Part of the hindbrain is the *cerebellum,* which integrates and distributes impulses from the cerebral association centers to the motor neurons in the spinal cord.

The *spinal cord* is an extension of the brain. The uppermost section of the spinal cord contains the twelve pairs of *cranial nerves,* which serve structures in the head and neck, as well as the lungs, heart, pharynx and larynx, and many abdominal organs. They control eye, tongue, facial movements, and the secretion of tears and saliva. Their main inputs are from the eyes, the tastebuds in the mouth, the nasal olfactory receptors, and touch, pain, heat, and cold receptors of the head. Thirty-one pairs of *spinal nerves* pass out between the appropriate vertebrae and serve defined sectors of the rest of the body. Nerves are mixed sensory and motor pathways, carrying both somatic and autonomic signals between the spinal cord and the muscles, articulations, skin, and visceral organs. Figure 13–3 shows how the spinal nerves emanating from sections of the spinal column (known as the spinal nerve roots) innervate defined areas of the skin (dermatomes).

## Sensors

The central nervous system receives information from internal receptors, or *interoceptors,* which report on changes in the body, the functions of digestion, circulation, and excretion, and feelings of hunger, thirst, and sexual arousal, and wellness or sickness. *Exteroceptors* respond to light, sound, touch, temperature, electricity, and chemicals. Because all of these sensations come from various parts of the body, external and internal receptors together are called somesthetic sensors.

Internal receptors include the *proprioceptors.* Among these are the muscle spindles (nerve filaments wrapped around small muscle fibers that detect the amount of stretch of the muscle). Golgi organs are associated with muscle tendons and detect their tension and report to the central nervous system information about the strength of contraction of the muscle. Ruffini organs are kinesthetic receptors located in the capsules of articulations, or joints. They respond to the degree of angulation of the joints (joint position), to change in general, and also to the rate of change.

The vestibular sensors are also proprioceptors. They detect and report the position of the head in space and respond to sudden changes in its attitude. To relate the position of the body to that of the head, proprioceptors in the neck are triggered by displacements between trunk and head.

Another set of interoceptors, called *visceroceptors,* reports on the events in the visceral (internal) structures of the body, such as organs of the abdomen and chest, as well as on events in the head and other deep structures. The usual modalities of visceral sensations are pain, burning, and pressure.

External receptors provide information about the interaction between the body and the outside: sight (vision), sound (audition), taste (gustation), smell (olfaction), temperature, electricity, and touch (taction). The sensations of touch, sight, and sound can be used as feedback to the body regarding the direction and intensity of muscular activities transmitted to an outside object. Free nerve endings, Meissner's and Pacinian corpuscles, and other receptors are located throughout the skin, but in different densities. They transmit the sensations of touch, pressure, and pain. Because the nerve pathways from the free endings interconnect extensively, the sensations reported are not always specific for a modality; for example, very hot or cold sensations can be associated with pain, which may also be caused by hard pressure on the skin.

Because the pathways of interoceptors and exteroceptors are closely related and anatomically intertwined, information about the body is often integrated with information about the outside.

## The Signal Loop

Following the traditional concept, one can model the human as processor of signals in some detail, as shown in Figure 13–4. Information (energy) is received by a sensor and impulses are sent along the afferent (sensory) pathways of the PNS to the CNS. Here, the signals are perceived and compared with information stored in the short- or long-term memories. The signal is processed in the CNS and an action (including "no action") is chosen. Appropriate action (feedforward) impulses are generated and transmitted along the efferent (motor) pathways of the PNS to the *effectors* (such as the voice and hands). Of course, many feedback loops exist.

Both sides of the processor model can be analyzed further. Figure 13–5 shows how *distal stimuli* provide information, which can be visual, auditory, or tactile. To be sensed, the stimulus must appear in a form to which human sensors can respond; it must. have suitable qualities and quantities of electromagnetic, mechanical, electrical, or chemical energy. If the distal events do not generate *proximal stimuli* that can be sensed directly, the distal stimuli must be transformed into energies that can trigger human sensations. For this, *transducers* are designed by the ergonomist: For example, a display of some kind, such as a computer screen, dial, or a light can act as a transducer.

On the output side, the actions of the human effector (such as hand or foot) may directly control the machine, or one may need another transducer; for example, movement of a steering wheel by the human hand can be amplified by power steering. Figure 13–6 portrays the model. Of course, recognizing the need for a transducer

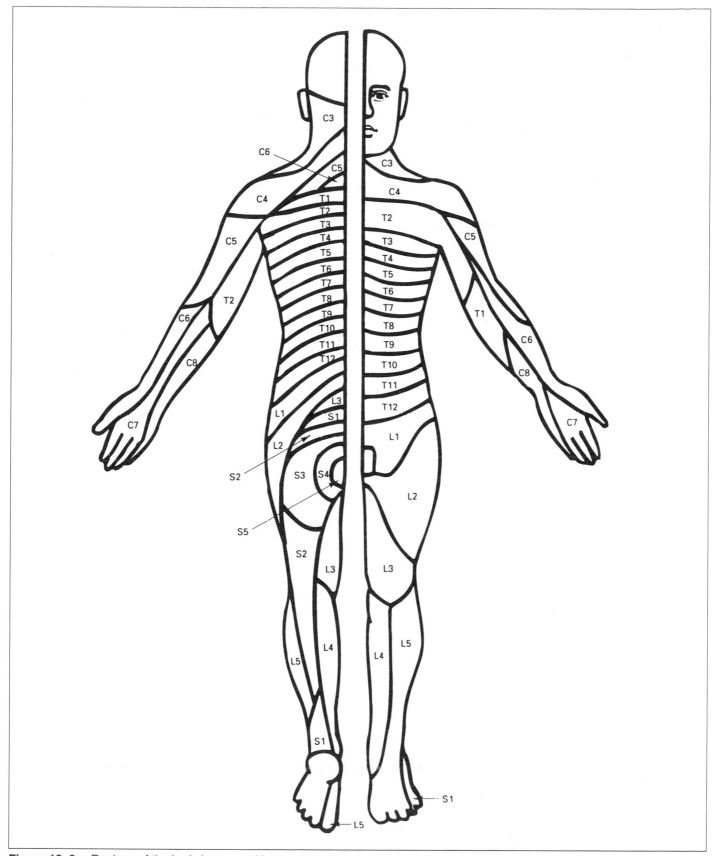

**Figure 13–3.** Regions of the body innervated by nerves emanating from sections of the spinal column. (Adapted from Kroemer, Kroemer, Kroemer-Elbert, *Ergonomics: How to Design for Ease and Efficiency.* Englewood Cliffs, NJ: Prentice Hall, 1994. Used with permission by the publisher. All rights reserved.)

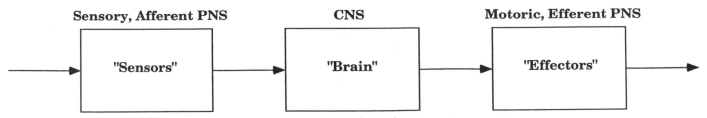

**Figure 13–4.** The human as receptor, processor, and generator of signals.

and providing information for its design are primary tasks of the ergonomist or human factors engineer.

## Responding to Stimuli

The time between the appearance of a proximal stimulus (such as a traffic light) and an effector action (such as foot movement) is called *reaction time*. The additional time to perform an appropriate movement (e.g., stepping on a brake pedal) is called *motion* or *movement time*. Motion time added to reaction time results in the *response time*. (Note that in everyday use, these terms are often not clearly distinguished.)

Experimental analysis of reaction time goes back to the very roots of experimental psychology; many of the basic results were obtained in the 1930s, with additional experimental work done in the 1950s and 1960s. Innumerable experiments have been performed and many tables of such times have been published. In engineering handbooks, some tables apparently were consolidated from various sources; however, the origin of the data, the experimental conditions (e.g., the intensity of the stimulus) under which they were measured, the measuring accuracy, and the subjects participating are no longer known.

There appears to be little practical time difference in reactions to electrical, tactile, and sound stimuli. The slightly longer times for sight and temperature sensations are well within the range measuring accuracy or within the variability among individuals. However, the time following a smell stimulus is distinctly longer, and the time for taste is considerably longer, and the time for pain response is longest.

If a person knows that a particular stimulus will occur, is prepared for it, and knows how to react to it, the resulting reaction time (RT) is called simple reaction time. Its duration depends on the stimulus modality and the intensity of the stimulus. If one of several possible stimuli occurs, or if the person has to choose between several possible reactions, the time is called choice reaction time.

Under optimal conditions, simple auditory, visual, and tactile reaction times are nearly 0.2 seconds. If conditions deteriorate, as when the subject is uncertain about the appearance of the signal, reaction slows. For example, simple reaction time to tones near the lower auditory threshold may increase to 0.4 seconds. Similarly, visual reaction time depends on intensity, size, and flash duration of the stimulus. Reactions to visual signals in the periphery of the visual field (such as 45 degrees from the fovea) are about 15 to 30 ms slower than to centrally located stimuli (Boff & Lincoln, 1988).

Reaction times of different body parts to tactile stimuli vary only slightly (about 10 percent) for finger, forearm, and upper arm. Reaction time slows if it is difficult to distinguish between several stimuli that are quite similar, but only one of which should trigger the response. Also, reaction time increases if one has to choose between several possible responses to one signal. Simple reaction time changes little between 15 and 60 years of age, but is substantially slower at younger ages and slows moderately as one ages.

Motion time follows reaction time. Movements may be simple, such as lifting a finger in response to a stimulus, or quite complex, such as swinging a tennis racket. Swinging the racket contains not only more complex movement elements, but also larger body and object masses that must be moved, which takes time. Movement time also depends on the distance covered and the precision required. Related data are contained in many systems of time and motion analyses, often used in industrial engineering.

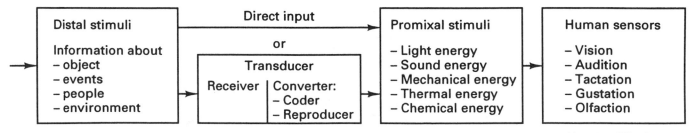

**Figure 13–5.**   Input from distal and proximal stimuli to human sensors. (Adapted from Kroemer, Kroemer, Kroemer-Elbert, *Ergonomics: How to Design for Ease and Efficiency*. Englewood Cliffs, NJ: Prentice Hall, 1994. Used with permission by the publisher. All rights reserved.)

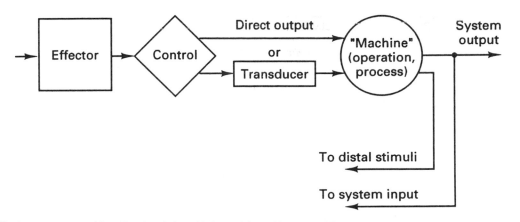

**Figure 13–6.** Effector outputs and feedback origins. (Adapted from Kroemer, Kroemer, Kroemer-Elbert, *Ergonomics: How to Design for Ease and Efficiency.* Englewood Cliffs, NJ: Prentice Hall, 1994. Used with permission by the publisher. All rights reserved.)

## Mental Workload

The assessment of workload, whether psychological or physical, commonly relies on the resource construct of a given quantity of available capability, of which a certain percentage is demanded. If less is required than available, a reserve exists (see Figure 13–7). Accordingly, workload often is defined as the portion of a resource (i.e., of the maximal performance capacity) expended in performing a given task.

Workload is empirically assessed using four different approaches: objective measures of primary task performance, of secondary task performance, and of physiological events, and subjective assessment. Subjective assessment and measures of task performance require knowledge of both zero capacity and full capacity. Performance measurement on a secondary concurrent task is intended to assess the capacity that remains after capacity resources are allocated to the primary task. If the subject allocates some of the resources needed for the primary task to the secondary task, the secondary task is *intrusive* (or *invasive*) on the primary task.

Physiological measures such as heart rate, eye movements, pupil diameter, and muscle tension can often be done without intruding on the primary task (Kramer, 1991). However, these measures may be insensitive to the task requirements, or may be difficult to interpret (Wierwille & Eggemeyer, 1993).

## Human Capacity for Physical Work

An individual's physical tolerance of physical work is usually determined by the capacities of his or her respiratory and cardiovascular systems to deliver oxygen to the working muscles and by the capacity of the metabolic system to use chemically stored energy to do muscular work. Maximal oxygen uptake is often used to describe the upper limit of the aggregate capacity. If a person is pushed beyond this limit, the additional energy required is provided by anaerobic processes. The oxygen deficit must be replenished following cessation of the effort. Tolerance times for maximal efforts are measured in minutes, even in seconds for a sprint runner. In a modern industrial setting, maximal effort may be required for brief periods, such as when an employee must lift heavy loads onto a handtruck. During an 8-hour shift, however, the average energy required usually falls well below human peak capacity.

The biochemical steps to transform food into energy available for work are quite complex; some are anaerobic but, altogether, the process is aerobic. Consequently, measurement of the volume of oxygen consumed provides an overall index of energy consumption and hence of the energy demands of work. Use of 1 l of oxygen yields approximately 5 kilocalories (kcal), equivalent to

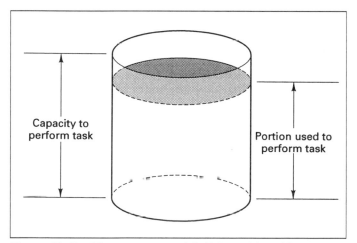

**Figure 13–7.** The resource model. (Adapted from Kroemer, Kroemer, Kroemer-Elbert, *Ergonomics: How to Design for Ease and Efficiency.* Englewood Cliffs, NJ: Prentice Hall, 1994. Used with permission by the publisher. All rights reserved.)

**Table 13–A.** Metabolic Energy Costs of Several Activities

| Activity | kcal/hr | Btu/hr |
|---|---|---|
| Resting, prone | 80–90 | 320–360 |
| Resting, seated | 95–100 | 375–397 |
| Standing, at ease | 100–110 | 397–440 |
| Drafting | 105 | 415 |
| Light assembly (bench work) | 105 | 415 |
| Medium assembly | 160 | 635 |
| Driving automobile | 170 | 675 |
| Walking, casual | 175–225 | 695–900 |
| Sheet metal work | 180 | 715 |
| Machining | 185 | 730 |
| Rock drilling | 225–550 | 900–2,170 |
| Mixing cement | 275 | 1070 |
| Walking on job | 290–400 | 1,150–1,570 |
| Pushing wheelbarrow | 300–400 | 1,170–1,570 |
| Shoveling | 235–525 | 930–2,070 |
| Chopping with axe | 400–1,400 | 1,570–5,550 |
| Climbing stairs | 450–775 | 1,770–3,070 |
| Slag removal | 630–750 | 2,500–2,970 |

Values are for a male worker of 70 kg (154 lb). (Reprinted with permission from "Ergonomics guide to assessment of metabolic and cardiac costs of physical work." *American Industrial Hygiene Association Journal*, 32 [1971]:560-564)

about 21 kilojoules (kJ). To put oxygen consumption and energy demands into proper perspective, consider that a trained athlete may reach an oxygen uptake of 6 l/min. Aside from a person's physique and training, age and gender influence the oxygen intake capacity substantially. Men in their twenties have an average maximal capacity of 3–3.5 l/min; women of the same age have an average capacity of 2.3–2.8 l/min. At age 60, the capacities diminish to about 2.2–2.5 l/min for men and 1.8–2.0 l/min for women. As with most physiological characteristics, there is considerable individual variability.

## Energy Cost of Work

Translation of maximal oxygen consumption figures into maximal energy production yields about 1,000 kcal/hr for the industrial population. Roughly half of this is needed to sustain the body functions and the other half is available for the work effort. Typically, the heaviest work a young, fit man can sustain is about 500 kcal/hr. Among the general male population, this figure is about 400–425 kcal/hr. Women, on average, can sustain about 80 percent of these values. These figures are equivalent to about 40 percent of the maximal uptake capability.

Industrial jobs seldom demand such a high energy expenditure continuously over the course of a workday. Rest pauses, fetching tools, mopping the brow, and receiving instruction reduce the average energy expenditure considerably. Table 13–A lists several typical activities and their average metabolic costs; resting values are included for reference. The given values must be adjusted according to one's body weight; the table applies to a man of 155 lb (70 kg). Other tables published in the physiological

literature may be in different units (calories, joules, or Btu, per hour or minute), and may or may not include basal rates. Be careful to use these tables correctly.

When intermittent tasks are performed, the total expenditure can be calculated using the following formula:

$$M = M_1 t_1 + M_2 t_2 + \cdots + M_n t_n \qquad (1)$$

In this formula, $M$ is the total metabolic energy cost, $M_1 \ldots M_n$ are the metabolic costs of individual tasks, and $t_1 \ldots t_n$ indicate the durations of the individual task. Using the values listed in Table 13–A, an example of estimating energy expenditures during the day is shown in Table 13–B.

There is close interaction between the human circulatory and metabolic systems. Nutrients and oxygen must be brought to the working muscles and metabolic by-products removed from them to ensure proper functioning. Therefore, heart rate (an important indicator of circulatory functions) and oxygen consumption (representing the metabolic processes taking place in the body) have a linear relationship in the range between light and heavy work. (When very light work or extremely heavy work is done, that relationship may not be reliable. It is also not reliable under severe environmental conditions or when workers are under mental stress.) Given such a linear relationship, one can often simply substitute heart rate recording for measurement of oxygen consumption. This is a very attractive shortcut because the heart rate can be recorded rather easily. The heart rate reacts faster to work demands and therefore more easily indicates quick changes in body functions due to changes in work requirements.

The simplest technique for heart rate assessment is to palpate an artery, often in the wrist. The measurer counts the number of heartbeats over a given period of time, such as 15 seconds, and then calculates the average heart rate per minute. More refined methods use plethysmo-

**Table 13–B.** Example of Energy Expenditure During Leisure and at Work

| Activity | Duration (hr) | Energy Cost kcal/hr | Energy Cost kcal/Duration |
|---|---|---|---|
| Sleeping | 8 | 85 | 680 |
| Sitting, resting | 4.5 | 100 | 450 |
| Walking | 1.5 | 170 | 255 |
| Driving automobile | 1 | 200 | 200 |
| Subtotal for leisure | 15 | | 1,585 |
| Work, light assembly | 7 | 105 | 735 |
| Work, walking | 1 | 200 | 200 |
| Work breaks, sitting | 1 | 100 | 100 |
| Subtotal for work | 9 | | 1,035 |
| Total per day | 24 | | 2,620 |

graphic techniques, which rely on the deformation of tissue due to changes that result when the imbedded blood vessels fill with blood with each pulse. Such techniques range from mechanically measuring the change in the volume of tissues to using photoelectric techniques that react to changes in the transmissibility of light caused by the blood filling, such as in the earlobe. More expensive techniques rely on electric signals generated by the nervous control systems that control heart rate. In this technique, electrodes are usually placed on the person's chest.

Obviously, it is important to match human capabilities with the related requirements of a given job. If the job demands equal or exceed the worker's capabilities, the person will be under strain and may not be able to perform the task. Hence, various stress tests have been developed to assess an individual's capability to perform physically demanding work. Bicycle ergometers, treadmills, or steps are used to simulate stressful demands. The reactions of the individual in terms of oxygen consumption, heart rate, and blood pressure are used to assess that person's ability to meet or exceed such demands. However, the examining physician needs to know what the actual work demands are. The industrial hygienist may be called on to help in the assessment of the existing work demands.

For more detailed information on the assessment of physical work capacity and of energy cost of work, consult Astrand & Randahl (1986), Eastman Kodak (1983, 1986), Grandjean (1988), Kroemer et al. (1990, 1994), and *Ergonomics Guides* published by the American Industrial Hygiene Association (AIHA).

## Work Classification

The work demands listed in Table 13–C are rated from light to extremely heavy in terms of energy expenditure per minute; the relative heart rate in beats per minute is also given. Light work is associated with small energy expenditures and is accompanied by a heart rate of approximately 90 beats/min. At this level of work, the energy needs of the working muscles are supplied by oxygen available in the blood and by glycogen in the muscle. There is no buildup of lactic acid or other metabolic by-products that would limit a person's ability to continue such work.

**Table 13–C.** Classification of Light to Heavy Work According to Energy Expenditure and Heart Rate

| Classification | Total Energy Expenditure (kcal/min) | Heart Rate (beats/min) |
|---|---|---|
| Light work | 2.5 | 90 or less |
| Medium work | 5 | 100 |
| Heavy work | 7.5 | 120 |
| Very heavy work | 10 | 140 |
| Extremely heavy work | 15 | 160 or more |

At medium work, which is associated with about 100 heart beats/min, the oxygen required by the working muscles is still covered and lactic acid developed initially at the beginning of the work period is resynthesized to glycogen during the activity.

In heavy work, during which the heart rate is about 120 beats/min, the oxygen required is still supplied if the person is physically capable of such work and specifically trained in this job. However, the lactic acid concentration produced during the initial phase of the work remains high until the end of the work period, when it returns to normal levels.

In the course of light, medium, and even heavy work (if the person is capable and trained), the metabolic and other physiological functions can attain a steady-state condition. This indicates that all physiological functions can meet the demands and remain essentially constant for the duration of the effort.

However, no steady state exists in the course of very heavy work, during which the heart rate reaches or exceeds 140 beats/min. In this case, the oxygen deficit incurred during the early phase of the work increases throughout the work period and metabolic by-products accumulate, making intermittent rest periods necessary or even forcing the person to stop completely. At even higher energy expenditures, which are associated with heart rates of 160 beats/min or more, the lactic acid concentration in the blood and the oxygen deficit achieve such magnitudes that frequent rest periods are needed. Even highly trained and capable persons are usually unable to perform such a demanding job throughout a full work day.

Hence, energy requirements or heart rate allow one to judge whether a job is energetically easy or hard. Of course, the labels light, medium, and heavy reflect judgments of physiological events (and of their underlying job demands) that rely very much on the current socioeconomic concept of what is comfortable, acceptable, permissible, difficult, or excessive.

*Rating the perceived effort* is another way to classify work demands. We are all able to perceive the strain generated in our body by a given task and we can make absolute and relative judgments about this perceived effort. Around the middle of the nineteenth century, Weber and Fechner described models of the relationship between physical stimulus and the perceptual sensation of that stimulus, known as the psychophysical correlate. Weber suggested that the just-noticeable difference (expressed as $\Delta I$) that can be perceived increases with the absolute magnitude of the physical stimulus $I$:

$$\Delta I = \alpha \cdot I \qquad (2)$$

with $\alpha$ a constant.

In the 1950s, Stevens at Harvard and Ekman in Sweden introduced ratio scales, which assume a zero point and

**Table 13–D.** Borg's Ratings of Perceived Exertion (RPE) Scales

| The 1960 Borg RPE | The Borg General Scale (1980) |
|---|---|
| 6 | 0—nothing at all |
| 7—very, very light | 0.5—extremely weak (just noticeable) |
| 8 | 1—very weak |
| 9—very light | 2—weak |
| 10 | 3—moderate |
| 11—fairly light | 4—somewhat strong |
| 12 | 5—strong |
| 13—somewhat hard | 6 |
| 14 | 7—very strong |
| 15—hard | 8 |
| 16 | 9 |
| 17—very hard | 10—extremely strong (almost maximal) |
| 18 | |
| 19—very, very hard | |
| 20 | |

equidistant scale values. These scales have since been used to describe the relationships between the perceived intensity ($p$) and the physically measured intensity ($I$) of a stimulus in a variety of sensory modalities (related to sound, lighting, and climate) as follows:

$$p = \beta \cdot I^n \qquad (3)$$

where $\beta$ is a constant and $n$ ranges from 0.5 to 4.

Since the 1960s, Borg and his co-workers have modified these relationships to take deviations from previous assumptions (such as zero point and equidistance) into account and to describe the perception of different kinds of physical efforts (Borg, 1982). Borg's general function is as follows:

$$P = a + c(I + b)^n \qquad (4)$$

In the formula, the constant $a$ represents the basic conceptual noise (normally less than 10 percent of $I$), the constant $b$ indicates the starting point of the curve, and $c$ is a conversion factor that depends on the type of effort.

Ratio scales indicate proportions between percentages but do not indicate absolute intensity levels; they allow neither intermodal comparisons nor comparisons between intensities perceived by different individuals. Borg has tried to overcome this problem by assuming that the subjective range and intensity level are about the same for each subject at the level of maximum intensity. In 1960, this led to the development of a category scale for rating the perceived exertion (RPE). The scale ranges from 6 to 20 and matches heart rates from 60–200 beats/min. A new category begins at every second number. In 1982, Borg proposed a category scale with ratio properties that could yield ratios and levels and allow comparisons but still retain the same high correlation with heart rate as the 1960 scale. Table 13–D shows two of Borg's best-known RPE scales.

The following exemplifies how the Borg scales are used. While the subject looks at the rating scale, the person administering the test says, "I will not ask you to specify the feeling but do select a number that most accurately corresponds to your perception of [experimenter specifies symptom]. If you don't feel anything, for example, if there is no [symptom], you answer zero—nothing at all. If you start feeling something that is only just about noticeable, you answer 0.5—extremely weak, just noticeable. If you have an extremely strong feeling of [symptom], you answer 10—extremely strong, almost maximal. This is the strongest you have ever experienced. The more you feel, the stronger the feeling, the higher the number you choose. Keep in mind that there are no wrong numbers. Be honest. Do not overestimate or underestimate your ratings. Do not think of any sensation other than the one I ask you about. Do you have any questions?"

Let the subject get well-acquainted with the rating scale before the test. During the test, let the subject do the ratings toward the end of every work period—about 30 seconds before stopping or changing the work load. If the test must be stopped before the scheduled end of the work period, let the subject rate the feeling at the moment of stoppage.

## Work/Rest Cycles

If a task demands more of the worker than can be sustained, rest pauses must be taken. A general principle governing the schedule of work/rest cycles is to break up excessively heavy work into bouts of work that are as short as practical for the task at hand. Frequent short rest periods reduce cumulative fatigue better than a few long breaks. The worst procedure is to let the worker go home early, exhausted.

A formula has been used to estimate the percentage of time that should be allotted to rest:

$$T_{rest}(\%) = \frac{M_{max} - M}{M_{rest} - M} 100 \qquad (5)$$

In the formula, $T_{rest}$ is the percentage of rest time, $M_{max}$ is the upper limit of the metabolic cost for sustained work, $M$ is the metabolic cost of the task, and $M_{rest}$ represents the resting (sitting) metabolism.

For example, suppose that $M_{max}$ = 350 kcal/hr and that an average value for $M_{rest}$ is 100 kcal/hr. Then assume that the task requires 525 kcal/hr, which is obviously too high. Apply these values to the formula as follows:

$$T_{rest}(\%) = \frac{350 - 525}{100 - 525} 100 = \frac{-175}{-425} 100 = 41\% \qquad (6)$$

Thus, rest pauses should be scheduled to last a total of 41 percent (24 min) of the hour.

As an alternative to the idle rest pause, one may consider intermingling a light task with the heavy task. To calculate the proportion of time that should be allocated to

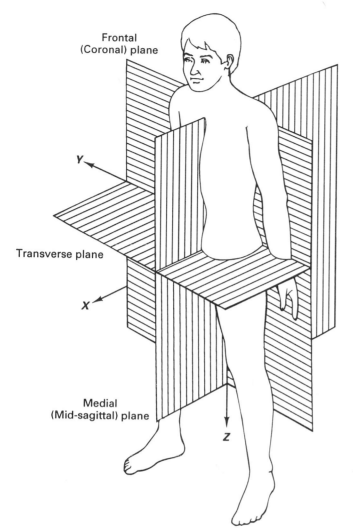

Frontal
(Coronal) plane

Y

Transverse plane

X

Medial
(Mid-sagittal) plane

Z

**Figure 13–8.** Measuring planes and terms used in anthropometry. (Adapted from Kroemer, Kroemer, Kroemer-Elbert, *Ergonomics: How to Design for Ease and Efficiency.* Englewood Cliffs, NJ: Prentice Hall, 1994. Used with permission by the publisher. All rights reserved.)

the two tasks, consider a heavy task that requires 500 kcal/hr, interrupted by a light task that requires 250 kcal/hr. Again, assume that $M_{max}$ = 350 kcal/hr.

Accordingly, the hard work could consume 40 percent of the time, the light task 60 percent. The light, secondary work task actually constitutes rest time from the heavy, primary task. Sharpening tools or walking to get material or cleaning the work place can provide productive respites from heavy work.

## Fatigue

Fatigue is an overexertion phenomenon that leads to a temporary decrease in physical performance. It is often associated with a buildup of lactic acid in the body, which can be metabolized to carbon dioxide and water during reduced activity or rest. The subjective sensation of fatigue is feeling tired. When tired, a person has reduced capability and desire for either physical or mental work, and feels heavy and sluggish. The sensation of fatigue has a protective function similar to hunger and thirst. Feeling fatigued forces one to avoid further stress and allows recovery to take place.

Subjective feelings of lowered motivation and deteriorated mental and physical activities can result from fatigue. Fatigue can occur together with monotony, a sensation associated with the lack of stimuli. Fatigue-induced low performance can be completely restored to its full level by rest.

The most important factors that produce fatigue are physical work intensity (static and dynamic work); illness, pain, lack of rest (sleep), and poor eating habits; and psychological factors such as worry, conflict, and possibly monotony. Hence, many different sources can be responsible for the sensation and the state of fatigue, which may be the result of an accumulation of effects from various sources.

## Engineering Anthropometry

*Anthropometry* literally means *measuring the human,* traditionally in metric units of heights, breadths, depths, and distances—all straight-line, point-to-point measurements between landmarks on the body or reference surfaces. Also, curvatures and circumferences following contours are measured, as is weight.

For the measurement, usually the body is placed in one of two defined postures, with body segments at 180° or 90° with respect to each other. In one standard posture, the subject is required to stand erect with heels together, buttocks, shoulder blades, and back of head touching the wall, arms vertical (or extended straight forward), and fingers straight. This is similar to the so-called anatomical position. The second standard posture is for measurements taken on a seated subject: the flat and horizontal surfaces of seat and foot support are so arranged that the thighs are horizontal, the lower legs are vertical, and the feet are flat on their support. The subject is nude, or nearly so, and unshod. Figure 13–8 shows reference planes and descriptive terms often used in anthropometry. Figures 13–9 and 13–10 show important anatomical landmarks of the human body. The NASA–Webb *Anthropometric Sourcebook* (1978) contains much information on measurement techniques in general and on military anthropometric data in particular. More recent books by Kroemer et al. (1994) and Roebuck (1995) condense and update this information.

The existing anthropometric information has been gathered while the subjects assume highly stylized and standardized postures that are quite different from the body positions assumed while working, particularly when the worker is moving around. Hence, current data must be

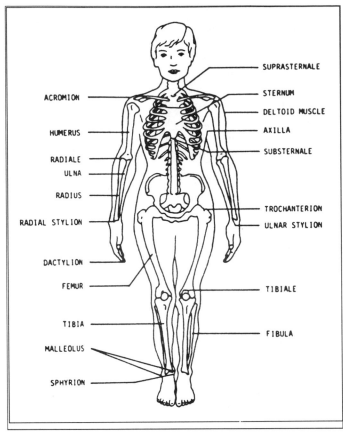

**Figure 13–9.** Landmarks on the human body in the frontal view. (Reprinted with permission from NASA–Webb, 1978.)

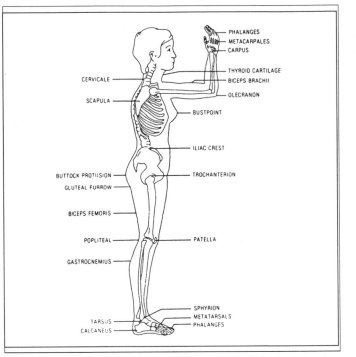

**Figure 13–10.** Landmarks on the human body in the lateral view. (Reprinted from NASA–Webb, 1978.)

interpreted by the ergonomic designer for practical applications.

## Civilian Body Dimensions

The anthropometric literature abounds with data on military personnel. No large, complete, and reliable surveys of the U.S. civilian population have been performed during the last decades. Hence, many of the anthropometric data applied to industrial populations were obtained from soldiers who are generally young and healthy. Head, hand, and foot dimensions are essentially the same for soldiers and civilians; the other body sizes measured in the 1988 survey of the U.S. Army seem to be rather normal as well, as a comparison with the few data available on the working population shows (Marras & Kim, 1993). However, body weight is distinctly different between soldiers and civilians, with civilians having more extreme values. An up-to-date compilation of estimated civilian body dimensions is presented in Table 13–E. The table shows the 5th, 50th, and 95th percentile dimensions as well as the standard deviations. Assuming a normal distribution of the data, the 50th percentile coincides with the average (mean) value.

It is not suitable to design for the "average person" because no one is average in many or all body dimensions. It is good ergonomic practice to design equipment,

work spaces, and tools so that they fit the range from the small body (usually that of women in the 5th percentile) to the large body (usually that of men in the 95th percentile). The 5th and 95th percentile values correspond to points 1.65 standard deviations below and above the mean. Use Table 13–F to calculate other percentile points.

## Designing for Percentiles

Percentiles are very convenient to determine exactly which percentage of a known population is fitted by a design range. For example, the adjustment range of a work seat should accommodate the popliteal height (a measurement of lower leg length) from a woman in the 5th percentile to a man in the 95th percentile. Hence, seat height is between about 14 in. (35 cm) and 19 in. (48 cm) for bare feet, with heel height to be added to this range.

The correct design approach is to select the specific anthropometric dimensions to be fitted and to establish (for each dimension) the design range (such as the 5th to the 95th percentile) so that proper fit is achieved. Reference books by Eastman Kodak Company (1983, 1986), Kroemer et al. (1990, 1994), and Roebuck (1995) provide guidance for ergonomic design to fit the body.

## Population Changes

Since World War I, increases in certain body dimensions have been observed. Many children grow to be taller than their parents. This stature increase has been about 1 cm per decade, but seems to have leveled off now. Another increase

**Table 13–E.**   Body Dimensions of U.S. Civilian Adults (Female/Male, in cm)

| Dimensions | | 5th | 50th | 95th | Standard Deviation |
|---|---|---|---|---|---|
| | | *Percentile* | | | |
| | | | | | |

*Heights, Standing*

| | | | | | |
|---|---|---|---|---|---|
| 99 | Stature ("height") | 152.8/164.7 | 162.94/175.58 | 173.7/186.6 | 6.36/6.68 |
| D19 | Eye | 141.5/152.8 | 151.61/163.39 | 162.1/174.3 | 6.25/6.57 |
| 2 | Shoulder (acromion) | 124.1/134.2 | 133.36/144.25 | 143.2/154.6 | 5.79/6.20 |
| D16 | Elbow | 92.6/99.5 | 99.79/107.25 | 107.4/115.3 | 4.48/4.81 |
| 127 | Wrist | 72.8/77.8 | 79.03/84.65 | 85.5/91.5 | 3.86/4.15 |
| 38 | Crotch | 70.0/76.4 | 77.14/83.72 | 84.6/91.6 | 4.41/4.62 |
| 84 | Overhead fingertip reach (on toes) | 200.6/216.7 | 215.34/32.80 | 231.3/249.4 | 9.50/9.99 |

*Heights, Sitting*

| | | | | | |
|---|---|---|---|---|---|
| 93 | Sitting height | 79.5/85.5 | 85.20/91.39 | 91.0/97.2 | 3.49/3.56 |
| 49 | Eye | 68.5/73.5 | 73.87/79.02 | 79.4/84.8 | 3.32/3.42 |
| 3 | Shoulder (acromion) | 50.9/54.9 | 55.55/59.78 | 60.4/64.6 | 2.86/2.96 |
| 48 | Elbow rest | 17.6/18.4 | 22.05/23.06 | 27.1/27.4 | 2.68/2.72 |
| 73 | Knee | 47.4/51.4 | 51.54/55.88 | 56.0/60.6 | 2.63/2.79 |
| 86 | Popliteal | 35.1/39.5 | 38.94/43.41 | 42.9/47.6 | 2.37/2.49 |
| 104 | Thigh clearance | 14.0/14.9 | 15.89/16.82 | 18.0/19.0 | 1.21/1.26 |

*Depths*

| | | | | | |
|---|---|---|---|---|---|
| 36 | Chest | 20.9/21.0 | 23.94/24.32 | 27.8/28.0 | 2.11/2.15 |
| 187 | Elbow–fingertip | 40.6/44.8 | 44.35/48.40 | 48.3/52.5 | 2.36/2.33 |
| 26 | Buttock–knee sitting | 54.2/56.9 | 58.89/61.64 | 64.0/66.7 | 2.96/2.99 |
| 27 | Buttock–popliteal sitting | 44.0/45.8 | 48.17/50.04 | 52.8/54.6 | 2.66/2.66 |
| 106 | Thumbtip reach | 67.7/73.9 | 73.46/80.08 | 79.7/86.7 | 3.64/3.92 |

*Breadths*

| | | | | | |
|---|---|---|---|---|---|
| 53 | Forearm–forearm | 41.5/47.7 | 46.85/54.61 | 52.8/62.1 | 3.47/4.36 |
| 66 | Hip, sitting | 34.3/32.9 | 38.45/36.68 | 43.2/41.2 | 2.72/2.52 |

*Head Dimensions*

| | | | | | |
|---|---|---|---|---|---|
| 62 | Length | 17.6/18.5 | 18.72/19.71 | 19.8/20.9 | 0.64/0.71 |
| 60 | Breadth | 13.7/14.3 | 14.44/15.17 | 15.3/16.1 | 0.49/0.54 |
| 61 | Circumference | 52.3/54.3 | 54.62/56.77 | 57.1/59.4 | 1.46/1.54 |
| 68 | Interpupillary breadth | 5.7/5.9 | 6.23/6.47 | 6.9/7.1 | 0.36/0.37 |

*Hand Dimensions*

| | | | | | |
|---|---|---|---|---|---|
| 65 | Wrist circumference | 14.1/16.2 | 15.14/17.43 | 16.3/18.8 | 0.69/0.82 |
| 59 | Length, stylion to tip 3 | 16.5/17.09 | 18.07/19.11 | 19.8/21.1 | 0.98/0.99 |
| 63 | Breadth, metacarpal | 7.4/8.4 | 7.95/9.04 | 8.6/9.8 | 0.38/0.42 |
| 60 | Circumference, metacarpal | 17.3/19.8 | 18.65/21.39 | 20.1/23.1 | 0.86/0.98 |
| 4 | Digit 1: breadth, distal joint | 1.9/2.2 | 2.06/2.40 | 2.3/2.6 | 0.13/0.13 |
| 1 | Length | 5.6/6.2 | 6.35/6.97 | 7.2/7.8 | 0.48/0.48 |
| 15 | Digit 2: breadth, distal joint | 1.5/1.8 | 1.73/2.01 | 1.9/2.3 | 0.12/0.15 |
| 10 | Length | 6.2/6.7 | 6.96/7.53 | 7.7/8.4 | 0.46/0.49 |
| 27 | Digit 3: breadth, distal joint | 1.5/1.7 | 1.71/1.98 | 1.9/2.2 | 0.11/0.14 |
| 22 | Length | 6.9/7.5 | 7.72/8.38 | 8.6/9.3 | 0.51/0.54 |
| 39 | Digit 4: breadth, distal joint | 1.4/1.6 | 1.58/1.85 | 1.8/2.1 | 0.11/0.14 |
| 34 | Length | 6.4/7.1 | 7.22/7.92 | 8.1/8.8 | 0.50/0.52 |
| 51 | Digit 5: breadth, distal joint | 1.3/1.5 | 1.47/1.74 | 1.7/2.0 | 0.11/0.13 |
| 46 | Length | 5.1/5.7 | 5.83/6.47 | 6.6/7.3 | 0.46/0.49 |

*Foot Dimensions*

| | | | | | |
|---|---|---|---|---|---|
| 51 | Length | 22.4/24.9 | 24.44/26.97 | 26.5/29.2 | 1.22/1.31 |
| 50 | Breadth | 8.2/9.2 | 8.97/10.06 | 9.8/11.0 | 0.49/0.53 |
| 75 | Lateral malleolus height | 5.2/5.8 | 6.06/6.71 | 7.0/7.6 | 0.53/0.55 |
| 124 | Weight (kg) U.S. Army | 49.6/61.6 | 62.01/78.49 | 77.0/98.1 | 8.35/11.10 |
| | Weight (kg) civilians* | 39/58* | 62.0/78.5* | 85/99* | 13.8/12.6* |

* Estimated (from Kroemer, 1981).
Note that all values (except for civilians' weight) are based on measured, not estimated, data that may be slightly different from values calculated from average plus or minus 1.65 standard deviation.

(Excerpted from Gordon, Churchill, Clauser, et al., 1989; Greiner, 1991.)

**Table 13–F.** Calculation of Percentiles

| Percentile p | | Central Percentage Included in the Range $x_{pi}$ to $x_{pj}$ | k |
|---|---|---|---|
| $x_i = \bar{x} - kS$ (below mean) | $x_j = \bar{x} + kS$ (above mean) | | |
| 0.5 | 99.5 | 99 | 2.576 |
| 1 | 99 | 98 | 2.326 |
| 2 | 98 | 96 | 2.06 |
| 2.5 | 97.5 | 95 | 1.96 |
| 3 | 97 | 94 | 1.88 |
| 5 | 95 | 90 | 1.65 |
| 10 | 90 | 80 | 1.28 |
| 15 | 85 | 70 | 1.04 |
| 16.5 | 83.5 | 67 | 1.00 |
| 20 | 80 | 60 | 0.84 |
| 25 | 75 | 50 | 0.67 |
| 37.5 | 62.5 | 25 | 0.32 |
| 50 | 50 | 0 | 0 |

has occurred in body weight, with increments of 2 to 3 kg per decade. Whether this increase in average mass is affected by nutrition and exercise trends remains to be seen.

Altogether, such developments of body dimensions are rather small and slow. Hence, for most engineers, the changes in body data should have little practical consequences for the design of tools, equipment, workstations, or work clothes because virtually none are designed to be used over many decades. However, one may have to consider increased ranges of variability in certain dimensions, and this is best expressed by percentile values.

The working population has been changing in gender, occupation, age, and composition. For example, the U.S. work force today has many more women in occupations that were dominated by men just a few decades ago. Occupations have changed drastically; computers and service industries are pulling people from blue-collar work. Average life expectancy has increased in the United States since 1900 by 27 years to nearly 75 years. Approximately 20 percent of all Americans are 65 years and older; thus, the number of elderly workers in the U.S. work force is expected to increase in the very near future. With the current low birth rate, a reduction in the U.S. population within a few decades could take place, but immigration is likely to keep the number of U.S. citizens from decreasing. Immigration (of relatively short Central and South Americans and Asians, for example) can have a pronounced effect on the anthropometry of the working population in certain geographical regions. Hence, the anthropometric data that describe local workers on the shop floor and the office and users and operators of equipment can be quite different from national statistics.

# ■ Biomechanics

Biomechanics explains characteristics of the human body in mechanical terms. More than 300 years ago, Giovanni

Alfonso Borelli described a model of the human body that consisted of links (bones) joined in their articulations and powered by muscles bridging the joints. This "stick person," refined and embellished with mass properties and material characteristics, still underlies most current biomechanical models of the human body. More than 100 years ago, Harless determined the masses of body segments; Braune & Fischer investigated the interactions between mass distribution, body posture, and external forces applied to the body; and von Meyer discussed the body's statics and mechanics. Biomechanical research has addressed responses of the body to vibrations and impacts, functions and strain properties of the spinal column, and human motion and strength characteristics.

When treating the human body as a mechanical system, many gross simplifications are necessary, and many functions, such as mental processes, are disregarded. However, within its limitations and simplifications, a large body of useful biomechanical information is already available, and this scientific and engineering field is developing rapidly. (See Chaffin & Andersson, 1991; Kroemer et al., 1990, 1994; Pope et al., 1984; Nordin et al., 1995; Oezkaya & Nordin, 1991.)

## Muscle Strength

Assessment of human muscle strength is a biomechanical procedure. This assessment uses Newton's second and third laws, which state that force is proportional to mass times acceleration and that each action is opposed by an equivalent reaction. Because human muscle strength is not measurable at the muscle in vivo and in situ, human strength is measured as the amount of force or torque applied to an external measuring instrument. Inside the body, muscular force vectors develop torque (also called moment) around the body joint bridged by the muscle. Torque is the product of force and its lever arm to the body joint; the direction of the force is measured at a right angle to its lever arm. In kinesiology, the lever arm is often called the mechanical advantage.

Figure 13–11 uses the example of elbow flexion to illustrate these relationships. The primary flexing muscle (biceps brachii) exerts a force ($M$) at the forearm at its lever arm ($m$) about the elbow joint. This generates a torque: $T = mM$. By torque definition, a right angle must exist between lever arm and force vector, so the useful lever arm is smaller when the elbow angle is wide open, or acute, than when the elbow angle is a right angle.

The correct unit to express force is the newton (N); torque (moment) is measured in newton-meters (Nm). A 1-lb force is approximately 4.45 N, and 1-kg force ($kg_f$) equals 9.81 N. The pound, ounce, and gram are not force but mass units. According to Newton's second law, force equals mass times acceleration; hence, weight (of a mass) generates a force proportional to gravitational acceleration. Force (as well as torque) has vector qualities, which means

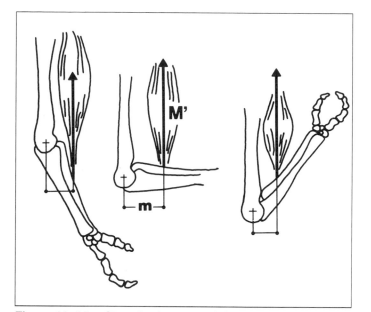

**Figure 13–11.** Changing lever arm (*m*) of the muscle force (*M*) with varying elbow angle.

that it must be described not only in terms of magnitude, but also by direction, the line of force application, and the point of its application.

Figure 13–12 shows a more realistic model of the muscle forces that flex or extend the forearm about the elbow joint. It indicates an external force (*E*) and hand/forearm weight (*W*) at their respective lever arms (*e* and *w*). It also shows the force vectors of the two major flexor muscles: the biceps (*B*) and the radiobrachialis (*R*) acting at their specific lever arms (*b* and *r*) about the elbow. Also shown is the force vector of the triceps (*T*) with its lever arm (*t*).

With a change in elbow angle (*a*), all force vectors change their angles of application ($\tau$, $\rho$, $\beta$, and $\varepsilon$). (Note that the brachialis muscle is not shown; its contribution would be similar to *B* and *R*.)

For static equilibrium, all forces and all torques must sum to zero. This provides three equilibrium equations, one each for the horizontal forces, the vertical forces, and the moments about the elbow joint. Including the joint reactions, indicated by *H* for horizontal and *V* for vertical force, and *M* for moment, these equations are as follows:

$$\sum (\text{Horizontal forces}) = 0$$
$$= T \sin \tau + R \sin \rho + B \sin \beta - E \sin \varepsilon - H \qquad (7)$$

$$\sum (\text{Vertical forces}) = 0$$
$$= T \cos \tau + R \cos \rho + B \cos \beta - E \cos \varepsilon - W - V \qquad (8)$$

$$\sum (\text{Moments about joint}) = 0$$
$$= tT \cos \tau - rR \cos \rho - bB \cos \beta + eE \cos \varepsilon + wW - M \ (9)$$

These equations can be solved only if there are no more unknowns than equations. One can determine the length of the lever arms from anatomy (either measure the angles or take them from anatomy and geometry); the weight of the forearm can be calculated from geometry or taken from biometric tables; the external force can be measured (as the load to be lifted, for example). Still, this leaves the muscular force vectors (*T*, *R*, and *B*) unknown, as well as the joint reaction forces (*H* and *V*), and the joint reaction moment (*M*). Various possibilities exist to reduce that number of unknowns to three (to equal the number of equations). These approaches include assumption of no coactivation or of certain coactivation ratios (for example, according to the cross-sections of the muscles) or

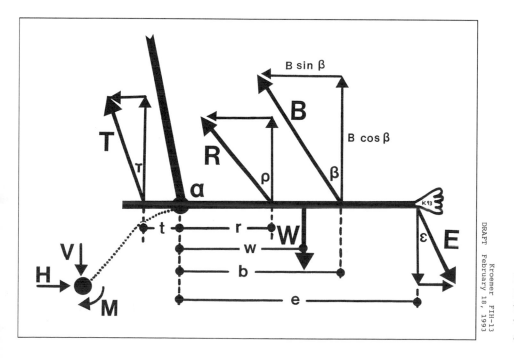

Kroemer FTH-13
DRAFT February 18, 1993

**Figure 13–12.** A more sophisticated biomechanical model of elbow flexion with indications of force vectors, vector directions, and lever arms.

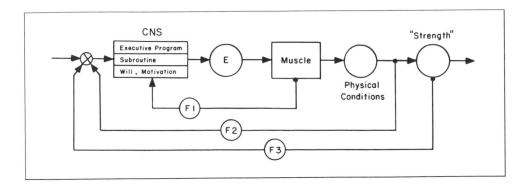

**Figure 13–13.** Model of generation and control of muscle strength exertion. (Reprinted with permission from Kroemer, Kroemer, and Kroemer-Elbert, 1990.)

according to electromyogram (EMG) signals; other techniques rely on statistical optimization procedures.

The biomechanical conditions become more complex than just described for the static case when movement occurs. In dynamics, additional forces must be considered as body segments rotate about their proximal joints. This introduces tangential, centrifugal, and Coriolis forces that may also generate new torques, as does the inertia of the segment and possibly of an external load.

The principles of vector algebra can be applied to a chain of body segments; for example, pushing or lifting with the hands transmits torques (forces) from the wrist to the elbow joint, to the shoulder, down the spinal column, and across hip, knee, and ankle joints to the ground where the torques (forces) are generated that counteract the hand effort. Such kinematic chain models have been used to assess the weak links of the human body; these are often in the spinal column, where especially low-back injuries are so common. (For more information, see Chaffin & Andersson, 1991; Kroemer et al., 1990, 1994; Oezkaya & Nordin, 1991.)

**Measurement of muscle strength.** Voluntary muscle strength is of much practical interest because it moves the segments of the body and generates energy that is exerted on outside objects when one performs work. There are over 200 skeletal muscles in the body. They consist of bundles of muscle fibers, each of which is wrapped—as is the total muscle—in connective tissue in which nerves and blood vessels are embedded. At the ends of the muscle, the tissues combine to form tendons that connect the muscle to bones.

The only active action a muscle can take is to contract; elongation is brought about by force external to the muscle. Contraction actually occurs in fine structures of the muscle, called filaments, that slide along each other. The nervous control for muscular contraction is provided by the neuromuscular system, which carries signals from the brain to the muscle and provides feedback. Electrical events associated with contraction of motor units of the muscle are observable in an EMG.

Figure 13–13 shows a simple model of the generation of muscular strength. Activation signals ($E$) travel from the central nervous system (CNS) along the efferent pathways

to the muscle. Here, they generate contraction, which, modified by the existing physical conditions, generates the strength that is applied to a measuring instrument (or a workpiece or hand tool). Three feedback loops are shown: The first one indicates reflexes; the second indicates transmitted sensations of touch, pressure, and kinesthetic signals; the third indicates sound and vision events related to the strength exerted.

Consider the feed-forward section of the model. With current technology, no suitable means exist to measure the executive program or the subroutines in the CNS or the effects of will or motivation on the motor signals generated. The efferent excitation impulses travel along the efferent nerves to the muscles, where they generate contraction signals that are observable in an EMG but are difficult to interpret, especially under dynamic conditions. The resulting contraction activities at the muscles can be qualitatively observed but not quantitatively measured in humans; no instruments are available that can directly measure the muscle tension in situ. Although it is difficult to record and control the mechanical conditions in the body (the lever arms of the tendon attachments or the pull angles of muscles with respect to the bones to which they are attached), the mechanical conditions outside the body can be observed and controlled; this concerns the kind and location of the coupling between the body and the measuring device, the direction of exerted force or torque, the time history of this exertion, the position and support of the body, and environmental conditions such as temperature and humidity.

Hence, with current means, only the output of this complex system, called muscle strength, can be defined and measured. Because strength has vector qualities and is time-dependent, it must be recorded not only in magnitude, but also in direction, point of application, and time history.

**Measuring techniques.** According to Newton's second law (force equals mass times acceleration, or torque equals moment of inertia times rotational acceleration), the measurer first has to decide whether acceleration is present.

If there is no acceleration, there is no change in speed. If speed is set to zero, then adjacent body segments do not move with respect to each other. Hence, the length of the muscles spanning the joint remains constant. Physically, this means that the measurement of muscle strength is performed in the static condition. Biologically, this condition is described as isometric (meaning constant muscle length). In this case, the words *static* and *isometric* are factually synonymous. Measurement of static strength is straightforward and involves only simple instrumentation. For this reason, almost all information on muscle strength currently available reflects isometric exertions.

If velocity is not zero but is constant, the condition is described as isokinematic (meaning constant motion). Measurement devices that establish a constant angular velocity around a given body articulation are commercially available. During their constant speed phase (but neither at the beginning nor the end of the motion), these devices provide a defined condition for which the exerted strength can be recorded. Only the angular velocity is controlled; the amount of strength actually exerted at any moment remains under control of the subject. (Hence, the devices are not isokinetic.)

If acceleration is present, meaning that the velocity is variable, strength is exerted under dynamic conditions that must be defined and controlled depending on the circumstances. Such experimental control is possible in the laboratory, but is very difficult and often impractical. Extreme cases of dynamic conditions are feats of strength at sports events, which cannot be easily measured or controlled because they are highly specific to the situation and the person. Table 13–G presents an overview of classifications of strength exertions and of their experimental control and measurement.

One technique commonly used to control dynamic conditions is to have the subject work isoinertially, that is, move constant masses (weights). To assess lifting capability, the weight is usually increased from test to test until one can determine the largest mass the subject can lift (see the section on material handling later in this chapter).

**The strength test protocol.** For selecting the type of strength test to be administered and determining the measurement techniques and measurement devices to be used, an experimental protocol must be devised. (Of course, established guidelines and legislation covering experiments with human subjects should be followed.) This concerns selecting subjects and guaranteeing their protection, planning and controlling the experimental conditions, calibrating and applying the measurement devices, and recording, analyzing, and statistically describing the results. Regarding the selection of subjects, care must be taken that the people who participate in the tests are a representative sample of the population for which data are to be gathered. Control over motivational aspects is particularly difficult. It is widely accepted (outside sports) that the experimenter should not exhort or encourage the subject to perform (Caldwell et al., 1974). The revised Caldwell regimen for static strength assessment reads as follows:

1. Strength Exertion and Duration
   a. Require a maximal steady exertion sustained for four seconds.
   b. Disregard the transient periods before and after the steady exertion.

**Table 13–G.** Dependent and Independent Variables in the Measurement of Muscle Strength

| Variables | Isometric (Static) Indep. | Dep. | Isovelocity (Dynamic) Indep. | Dep. | Isoacceleration (Dynamic) Indep. | Dep. | Isojerk (Dynamic) Indep. | Dep. | Isoforce (Static or Dynamic) Indep. | Dep. | Isoinertia (Static or Dynamic) Indep. | Dep. | Free Dynamic Indep. | Dep. |
|---|---|---|---|---|---|---|---|---|---|---|---|---|---|---|
| Displacement, linear/angular | constant* (zero) | | C or X | | C or X | | C or X | | C or X | | C or X | | | X |
| Velocity, linear/angular | 0 | | constant | | C or X | | C or X | | C or X | | C or X | | | X |
| Acceleration, linear/angular | 0 | | 0 | | constant | | C or X | | C or X | | C or X | | | X |
| Jerk, linear/angular | 0 | | 0 | | 0 | | constant | | C or X | | C or X | | | X |
| Force, torque | C or X | | C or X | | C or X | | C or X | | constant | | C or X | | | X |
| Mass, moment of inertia | C | | C | | C | | C | | C | | constant | | C or X | |
| Repetition | C or X | | C or X | | C or X | | C or X | | C or X | | C or X | | C or X | |

*Legend*
Indep = independent
Dep = dependent
C = variable can be controlled
* = set to zero
0 = variable is not present (zero)
X = can be dependent variable
The boxed constant variable provides the descriptive name.
(Adapted from Kroemer, Marras, McGlothlin, et al., 1990.)

c. Record the strength datum as the mean score during the first three seconds of the steady exertion.

2. Subject Instructions
   a. Inform the subject about the test purpose and procedures.
   b. Keep instructions factual; do not include emotional appeals.
   c. Instruct the subject to increase to maximal exertion (without jerking) in about 1 s and to maintain this effort during a 4-s count (see #1).
   d. Inform the subject during the test session about his or her general performance in qualitative, noncomparative, positive terms. Do not give instantaneous feedback during the exertion.
   e. Avoid rewards, goal setting, competition, spectators, fear, noise, and other distractions because they can affect the subject's motivation and performance.

3. Rest Periods
   Provide a rest period between related efforts of at least two minutes—more if symptoms of fatigue are apparent.

4. Test Conditions
   Describe and control the conditions existing during the strength testing in terms of the following:
   a. Body parts and muscles chiefly used
   b. Body position
   c. Body support and reaction force available
   d. Coupling of the subject to the measuring device (to describe location of the strength vector)
   e. Strength measuring and recording device

5. Subject Description
   a. Population and sample selection
   b. Current health status: a medical examination and questionnaire are recommended
   c. Gender
   d. Age
   e. Anthropometry (at least height and weight)
   f. Training/skill related to the strength testing

6. Data Reporting
   a. Mean (median, mode)
   b. Standard deviation
   c. Skewness
   d. Minimum and maximum values
   e. Sample size
   f. Repetitions
   g. Results of statistical post-hoc analysis, including significant differences

While this regimen applies to static (isometric) strength testing, a similar procedure should be applied for dynamic measurement.

## ■ Handling Loads

We all handle material daily. We lift, hold, carry, push, pull, and lower while moving, packing, and storing objects. The objects may be soft or solid, bulky or small, smooth or with corners and edges; the objects may be bags, boxes, or containers that come with or without handles. We may handle material occasionally or repeatedly. We may handle material during leisure activities or as part of paid work. Manual handling involves lifting light or heavy objects. Heavy loads pose additional strain on the body because of their weight, bulk, or lack of handles. Even lightweight and small objects can strain us because we have to stretch, move, bend, or straighten out body parts, using fingers, arms, trunk, and legs.

Material handling is one of the most common causes of injury—and of the most severe injuries—in U.S. industry. The direct and indirect costs are enormous.

Four keys of material-handling ergonomics are described by Kroemer (1984) as follows

- Key 1: Initial layout or improvement of *facilities* contributes to safe and efficient material transfer. What process is selected and how the flow of material is organized and designed in detail determines the involvement of people and how they must handle material.

- Key 2: *Job design* determines the stress imposed on the worker by the work. Initially, the engineer must decide whether to assign certain tasks to a person or a machine. Furthermore, the layout of the task, the kind of material-handling motions to be performed, the organization of work and rest periods, and many other engineering and managerial techniques determine whether a job is well-designed (whether it is safe, efficient, and agreeable for the operator).

- Key 3: Selection, use, and improvement of *equipment,* machines, and tools strongly affect material-handling requirements. Ergonomic principles concerning operator space requirements, control design, visibility, and color and sign coding must be considered.

- Key 4: This concerns *people* as material handlers, particularly with regard to body size, strength, and energy capabilities. People are critical in manual material activities because they supervise, control, operate, drive, and actually handle material. If people are not needed in the system, then it should be automated. If they are needed, the system must be designed for them.

These four keys guide a systematic analysis of material-handling problems. Optimizing each of these keys (with the aid of seminars or workshops) can lead to dramatic improvements. Many of the strains and risks caused by manual material operation in the workplace can be reduced or avoided by proper design of facilities, appropriate selection and use of equipment, and intelligent job design.

A systematic way to evaluate material handling is as follows:

- Describe the material movement process, from receiving to distribution.

- Break down the process into its separate functions.

- Within these functions, chart and tabulate the activities to determine manual materials-handling details.
- Allocate tasks between humans and machines and determine job requirements if people must handle the material.

In this procedure, several principles of material movement are important. The unit size principle is of particular interest. According to it, one can either increase the quantity (size, weight) of the unit load so that equipment use becomes feasible and appropriate (this is called the big unit outcome) or reduce the size and weight of the load so that one operator can safely handle the material (the small unit outcome).

If all opportunities to automate or mechanize the movement of material have been exhausted, some material handling may have to be assigned to people. In this case, job requirements must be established that do not overload the person or pose hazards. One must organize the task, establish job procedures, and determine details to enable the operator to perform the work safely and efficiently. Here are some guidelines:

- If people must move material, make sure the movement is predominantly horizontal. Push and pull, rather than lift or lower, and avoid severe bending of the body.
- If people must lift or lower material, let them do so between knuckle height and shoulder height. Lifting and lowering below knuckle height and above shoulder height are most likely to result in overexertion injuries.
- If people must lift and lower, make sure these activities occur close to and in front of the body. Bending forward or, worse, twisting the body sideways, causes most overexertion injuries.
- If people must move material, make sure the material is light, compact, and safe to grasp. Light objects strain the spinal column and body tissues less than heavy objects. Compact material can be held more closely to the body than a bulky object. A solid object with good handles is more safely held and more easily moved than pliable material.
- If people must handle material, make sure it does not have sharp edges, corners, or pinch points.
- If material is delivered in bins or containers, make sure it can be easily removed from them; the operator should not have to "dive" into the container to reach the material.

## The Body as Energy Source

The human body must maintain an energy balance between the external demands produced by work and the work environment and the capacity of the internal body functions to produce energy. The body is an energy factory that converts chemical energy derived from food into externally useful energy; the final stages of this metabolic process take place at skeletal muscles. Oxygen is transported from the lungs to the energy-conversion sites (mitochondria) at the muscles by the blood, which also removes by-products of the energy conversion such as carbon dioxide, water, and heat. Carbon dioxide dissipates in the lungs. Although some heat and water are dispelled in the lungs, most is released through the skin via sweat. The blood circulation system is powered by the heart. Thus, the pulmonary system, the circulatory system, and the metabolic system establish *central* limitations of a person's ability to perform strenuous work.

A person's capability for labor is limited also by muscular strength, the ability for movement in body joints, and the strength and stability of the spinal column. These are *local* limitations of the force or work that a person can exert. As discussed earlier, today's work demands do not usually tax one's central capabilities, but local limitations often establish the upper limits for performance capability.

While a person is handling material, the force exerted with the hands must be transmitted through the whole body (across wrists, elbows, shoulders, trunk, hips, knees, ankles, and feet to the floor). In this chain of force factors, the weakest link determines the capability of the whole body to do the job. A particularly weak link in this chain is the spinal column, particularly at the low back. Muscular strains, ligament sprains, displacements in the facet joints, or deformations of the intervertebral disks often curtail a person's ability to handle material.

To ensure that physical work, particularly load handling, is performed as safely and efficiently as possible, one should match people with their tasks, following three basic guidelines:

- Design tasks and equipment to ensure ease and efficiency of manual material handling.
- Train people in safe and efficient procedures.
- Select people who are capable of performing the labor.

Obviously, these approaches are closely related. Only people who possess the basic physical capabilities to do the labor can be trained for the job. Selection and training can be carried out only if the specific job demands are known. Hence, knowledge of job requirements is a basic ergonomic requirement. On the other hand, job design can be executed properly only if human capabilities and limitations are known. Information on body dimensions (Table 13–C) and on work capabilities (see the sections on physiology and biomechanics) is the other basic ergonomic requirement for matching people and tasks.

## Training for Safe Lifting Practices

Numerous attempts have been made to train material handlers to do their work, particularly lifting, in a safe manner. (See *Ergonomics,* Numbers 7, 8, and 9 of Vol. 35, 1992.) Unfortunately, hopes for significant and lasting reductions of overexertion injuries through the use of training generally have not been fulfilled. There are several reasons for the disappointing results:

- If the job requirements are stressful, "doctoring the system" through behavioral modification does not eliminate the inherent risk. Designing a safe job is better than training people to behave safely in an unsafe job.

- People tend to revert to previous habits and customs if new ones are not reinforced and refreshed periodically.

- Emergency situations, the unusual case, the sudden quick movement, increased body weight, or impaired physical well-being may overly strain the body because training usually does not include these conditions.

Thus, training for safe material handling cannot be expected to solve the problem, but if it is properly applied and periodically reinforced, training should help to alleviate some problems.

The idea of training workers in safe and proper techniques of manual material movement has been propagated for decades. Originally, workers were told to lift with a straight back by unbending the knees. However, the frequency and intensity of back injuries was not reduced by this lifting method. Biomechanical and physiological research has shown that leg muscles used in this lifting technique do not always have the needed strength. Also, awkward and stressful postures may be assumed if this technique is applied when the object is bulky. Hence, the straight back–bent knees action evolved into the "kinetic" lift, in which the back is kept mostly straight and the knees are bent, but the positions of the feet, chin, arms, hands, and torso are prescribed.

Another variant is the "free style" lift. New techniques, often with catchy names, come and go, but it appears that no single lifting method is best for all situations. Simply providing back braces (lift belts) as quick fixes is not a general solution either (Enos & Mitchell, 1993; McGill, 1993).

A thorough review of the existing literature indicates that the issue of training for prevention of back injuries is confused, at best. Training may not be effective in injury prevention, or its effect may be so uncertain and inconsistent that money and effort paid for training programs might be better spent on research and implementation of techniques for worker selection and ergonomic job design. According to the National Institute for Occupational Safety and Health (NIOSH, 1981, p. 99),

> The importance of training in manual materials handling in reducing hazard is generally accepted. The lacking ingredient is largely a definition of what the training should be and how this early experience can be given to a new worker without harm. The value of any training program is open to question as there appear to have been no controlled studies showing a consequent drop in the manual material handling

(MMH) accident rate or the back injury rate. Yet so long as it is a legal duty for employers to provide such training or for as long as the employer is liable to a claim of negligence for failing to train workers in safe methods of MMH, the practice is likely to continue despite the lack of evidence to support it. Meanwhile, it may be worth considering what improvements can be made to existing training techniques.

Two principal training approaches appear most likely to be successful. One involves awareness and attitude training on the physics involved in manual materials handling and on the related biomechanical and physiological events going on in one's body. The other approach is the improvement of individual physical fitness through exercise and warm-ups (which indirectly influences awareness and attitude).

Training in proper lifting techniques is an unsettled issue. It is unclear what exactly should be taught, who should be taught, how and how often a technique should be taught. This uncertainty concerns both the objectives and methods, as well as the expected results. Claims about the effectiveness of one technique or another are frequent but are usually not supported by convincing evidence.

There are no comprehensive, sure-fire rules for safe lifting. Manual load handling is a very complex combination of moving body segments, changing joint angles, tightening muscles, and loading the spinal column. The following Dos and Don'ts apply, however:

- DO design manual lifting and lowering out of the task and workplace. If it nevertheless must be done by a worker, perform it between knuckle and shoulder height.

- DO be in good physical shape. If you are not used to lifting and vigorous exercise, do not attempt to do difficult lifting or lowering tasks.

- DO think before acting. Place material conveniently within reach. Have handling aids available. Make sure sufficient space is cleared.

- DO get a good grip on the load. Test the weight before trying to move it. If it is too bulky or heavy, get a mechanical lifting aid or somebody else to help, or both.

- DO get the load close to the body. Place the feet close to the load. Stand in a stable position with the feet pointing in the direction of movement. Lift mostly by straightening the legs

- DO NOT twist the back or bend sideways.

- DO NOT lift or lower awkwardly.

- DO NOT hesitate to get mechanical help or help from another person.

- DO NOT lift or lower with the arms extended.

- DO NOT continue heaving when the load is too heavy.

## Personnel Selection for Material Handling

Selecting people who are unlikely to suffer an overexertion injury is one method to attempt to reduce the risk of musculoskeletal disorders in manual materials activities. The purpose of screening is to assign only the people who can do the job safely. The basic premise is that the risk of overexertion injury in manual load movement decreases as the handler's capability to perform such activity increases. This means that the test should be designed to match a person's capabilities with the actual demands of the job. This matching process requires that the test administrator know quantitatively both the job requirements and the related capabilities to be tested. Care must be taken to observe equal employment opportunity regulations and legislation such as the Americans with Disabilities Act (ADA). See Chapter 25, The Occupational Physician, for discussion of the ADA.

Scientists usually rely on the development and use of *models*. A model is an abstract (mathematical–physical) system that obeys specific rules and conditions and is used to understand the real system (in this case, the worker-task) to which it is analogous (in physiological, biomechanical, psychophysical, or other traits). A model usually represents a theory. Without proper models, reliable and suitable methods cannot be developed. A *method* is a systematic, orderly way of arranging thoughts and executing actions. A *technique* is the specific, practical manner in which actions are done; it implements the methods that are derived from models. Test techniques involve specific procedures and instruments used to measure the subject's ability to perform manual handling of materials.

Many models have been developed to describe the central and local limitations just discussed. In the following section, these models are simplified and categorized by major disciplines for convenience.

**Anatomical/anthropometric models.** Since Alfonso Giovanni Borelli's *De Motu Animalium* (about 1680), the human body has been described as links (representing the long bones and vertebrae), connected at articulations (joints), and powered by muscles (engines) that span one or two articulations and move the links.

**Physiological models.** Physiological models primarily provide information on oxygen consumption (in l/min or l/kg/min) as a measure of energy conversion and expenditure (in kcal/min or kJ/min) and on the loading of the circulatory system (in beats/min). These primarily reflect central functions.

**Orthopedic (musculoskeletal) models.** Orthopedic models concentrate on musculoskeletal functions, often related to deformities, diseases, and loss of limbs. Of particular interest are the body joints, including the spinal column, often in connection with surgical or rehabilitory treatments or in connection with prostheses. Thus, these models commonly address local functions.

**Biomechanical models.** During the last 100 years, many models have been developed (usually following the Borelli concept) in an effort to explain the behavior of the body in mechanical terms. Hence, these models rely primarily on anatomical, anthropometric, and orthopedic paradigms and inputs, with much use of computer technology in recent years. Such models usually reflect local functions.

**Statistical models.** In contrast to the more normative models just discussed, many statistical models have been developed on the basis of empirical data. These models try to describe the combined effects of parametric inputs into a model that is essentially defined by the same parameters. Typical examples are multiple-regression models, which predict lift capacity from observed anthropometric, biomechanical, and muscle-strength parameters. The predictive power of these models is usually expressed by the square of the (combined) correlation coefficient, $R^2$, among the involved variables.

**Psychophysical models.** The psychophysical concept suggests that human capabilities are synergistically determined by bodily and mental functions and the person's ability to rate the perceived strain and to control his or her actions accordingly. The psychophysical approach to the understanding of central and local functions has been successfully used to describe the effects of sound, climate, illumination, vibration, and physical exertion.

It is interesting to note that the assessment of maximal voluntary muscle strength is implicit to nearly all models and is essential for anatomical, orthopedic, and biomechanical models. However, its measurement depends on the voluntary cooperation (motivation) of the subject. Because there is currently no feasible technique to assess true (structural) muscle strength, all models rely (by concept or input) in this respect on the psychophysical approach.

## Assessment Methods

Based on models such as those just described, various methods have been developed to assess an individual's capabilities for performing specified handling tasks. Such assessments rest on a foundation of general epidemiological information or on more specific etiological information.

The *medical examination* primarily evaluates the individual physiologically and orthopedically in relation to specific job tasks. The *physiological examination* is often combined with medical testing. It identifies individual limitations in central capabilities (such as pulmonary, circulatory, or metabolic functions). The *biomechanical examination* addresses mechanical functions of the body, primarily of the musculoskeletal type (such as load-bearing capacities of the spine or muscle strength exertable in certain postures or motions). *The psychophysical examination* addresses all (local or central) functions strained in the test; it "filters" the strain experienced through the sensation of the subject. Thus, this examination may include all or many of the systems checked via medical, physiological, or biomechanical methods.

## Screening Techniques

Several screening techniques exist that should select people who are able to perform defined materials-handling activities with no or acceptably small risks of overexertion injuries; that is, they should select people whose capabilities match the job demands with a safety margin. Primarily, these tests differ in the techniques used to generate the external stresses that strain the (local and central) function capabilities to be measured.

**Static techniques.** Static techniques require that the subject exert isometric (static) muscle strength against an external measuring instrument. Because muscle length does not change, there is no displacement of body segments involved; hence, no time derivatives of displacement exist. This establishes a mechanically and physiologically simple case that allows the straightforward measurement of isometric muscle strength. Unfortunately, static strength testing has been shown to have rather low predictive power for dynamic tasks ($R^2 < 0.5$). A further consideration with *any* screening technique is whether it complies with laws concerning equal employment opportunity and disabilities, (See Chapter 25, The Occupational Physician, for a discussion of ADA requirements.) Under some laws, job requirements must be task-specific.

**Dynamic techniques.** Dynamic techniques appear more relevant to actual material handling activities, but they are also more complex because many possible displacements and time derivatives (velocities, accelerations) exist. Hence, current dynamic testing techniques employ one of two ways to generate dynamic tests stresses:

- In the *isokinematic* technique (often mislabeled as isokinetic), the subject moves limb or trunk at constant angular velocity about a specific joint while exerting maximal voluntary torque. The test equipment is designed so that the torque actually exerted can be monitored continuously throughout the angular displacement.

- In the *isoinertial* technique, the subject moves a constant mass (weight) between two defined points. The maximal load that can be lifted (or lowered, pushed, pulled, carried, or held) by the subject is the measurement of that person's capability, whereas the forces or torques exerted depend (according to Newton's second law) on mass and acceleration.

The advantage of suitable dynamic measuring is its similarity to actual material handling, performed in motion. The difficulty of dynamic measuring lies in the fact that the movement of body parts can generate a variety of dynamic conditions. Practical dynamic tests have been developed by McDaniel et al. (1983) and Kroemer

(1985). Their technique has been used to test millions of military recruits, but has been ignored by industry.

As in all areas of developing scientific and applied knowledge, the body of knowledge changes, often rapidly. With respect to testing for lift strength, static testing of the 1980s is being replaced by new dynamic testing techniques. *A major problem in all techniques of strength testing is that of validity (the predictive power of the tests for true working conditions).* One should be cautious about applying any one technique exclusively and should also use other physiological, biomechanical, psychological, and medical assessments as deemed appropriate.

## Permissible Load Handling

Tables of lift weights for men, women, and children were used in the United States until the NIOSH *Work Practices Guide for Manual Lifting* appeared in 1981. Based on epidemiological, medical, physiological, biomechanical, and psychological approaches, new knowledge about human material handling capabilities has been gained since then. However, even new guidelines are still based on assumptions and approaches that need refinement and further evaluation.

**Limits for lifting and lowering.** The NIOSH *Work Practices Guide for Manual Lifting* contained distinct recommendations for acceptable masses to be lifted that differed from the previous assumptions that one could establish *one* given weight each for men, women, or children. In this 1981 guide, two different threshold curves were established: the lower, called action limit (AL) was thought to be safe for 99 percent of working men and 75 percent of women in the United States. The AL values depended on the starting height of the load, the length of its upward path, its distance in front of the body, and the frequency of lifting. If the existing weight was above the AL value, engineering or managerial controls had to be applied to bring the load value down to the acceptable limit. However, under no circumstances was lifting allowed if the load was three times larger than the AL value. This threshold was called the maximum permissible load (MPL).

A decade later, NIOSH revised the technique for assessing overexertion hazards of manual activity (Putz-Anderson & Waters, 1991; Waters et al., 1993). The new NIOSH guideline no longer contains two separate weight limits, but has only one recommended weight limit (RWL). It represents the maximal weight of a load that may be lifted or lowered by about 90 percent of American industrial workers, male or female, physically fit and accustomed to physical labor.

The 1991 equation used to calculate the RWL resembles the 1981 formula for AL, but includes new multipliers to reflect asymmetry and the quality of hand-load coupling. However, the 1991 equation allows as maximum a load constant (LC), permissible only under the most favorable circumstances, with a value of 51 lbs (23 kg). This is quite a reduction from the maximal 40 kg in the 1981 NIOSH guidelines.

The following assumptions and limitations apply:

- The equation does not include safety factors for such conditions as unexpectedly heavy loads, slips, or falls, or for temperatures outside the range of 66 F (19 C) to 79 F (26 C) and for humidity between 35 and 65 percent.

- The equation does not apply to one-handed tasks while seated or kneeling, or to tasks in a confined work space.

- The equation assumes that other manual handling activities and body motions requiring high energy expenditure, such as pushing, pulling, carrying, walking, climbing, or static efforts as in holding, are less than 20 percent of the total work activity for the work shift (Waters, 1991).

- The equation assumes that the worker/floor surface coupling provides a coefficient of static friction of at least 0.4 between the shoe sole and the standing surface.

- The equation can be applied under the following circumstances:

   Objects are lifted or lowered (the worker manually grasps and moves an object of definable size without mechanical aids to a different height level.

   The time duration of such an act is normally between two and four seconds. The load is grasped with both hands.

   The motion is smooth and continuous.

   The posture is unrestricted (see above).

   The foot traction is adequate (see above).

   The temperature and humidity are moderate (see above).

   The horizontal distance between the two hands is no more than 25 in. (65 cm).

For these conditions, NIOSH provides an equation for calculating the RWL:

$$RWL = LC \cdot HM \cdot VM \cdot DM \cdot AM \cdot FM \cdot CM \quad (10)$$

Where LC = load constant of 51 lb (23 kg)
   Each multiplier can assume values between 0 and 1

   HM = horizontal multiplier
   ($H$ is the horizontal distance of the hands from the midpoint between the ankles at the start and the end points of the lift)

   VM = vertical multiplier
   ($V$ is the height of the hands above the floor at the start and end points of the lift)

DM = Distance multiplier
   ($D$ is the vertical travel distance from the start to the end point of the lift)

AM = asymmetry multiplier
   ($A$ is the angular displacement of the load from the mid-sagittal plane that forces the operator to twist the body; it is measured at the start and end points of the lift, projected onto the floor)

FM = FM = frequency multiplier
   ($F$ is the frequency rate of lifting, expressed in lifts/min; it depends on the duration of the lifting task)

CM = coupling multiplier
   ($C$ indicates the quality of coupling between hand and load)

The following values are entered in the equation for RW:

|  | Metric | U.S. Customary |
|---|---|---|
| LC = Load constant = | 23 kg | 51 lb |
| HM = Horizontal multiplier = | 25/H | 10/H |
| VM = Vertical multiplier = | $1 - (0.003\|V-75\|)$ | $1 - (0.0075\|V-30\|)$ |
| DM = Distance multiplier = | $0.82 + (4.5/D)$ | $0.82 + (1.8/D)$ |
| AM = Asymmetry multiplier = | $1 - (0.0032A)$ | $1 - (0.0032A)$ |
| FM = Frequency multiplier | | |
| CM = Coupling multiplier | | |

These variables can have the following values:

- $H$ is between 10 in. (25 cm) and 25 in. (63 cm). Although objects can be carried or held closer than 25 cm in front of the ankles, most objects that are closer cannot be lifted or lowered without encountering interference from the abdomen. Objects farther away than 25 in. (63 cm) cannot be lifted or lowered without loss of balance, particularly when the lift is asymmetrical and the operator is small.

- $V$ is between 10 in. (25 cm) and $70 - V$ in. $(175 - V$ cm). For a lifting task, $D = V_{end} - V_{start}$; for a lowering task, $D = V_{start} - V_{end}$.

- $A$ is between 0° and 135°.

- $F$ is between 1 lift or lower every 5 min (over an 8-hr work day) to 15 lifts or lowers every minute (over 1 hr or less), depending on the vertical location $V$ of the object. Table 13–H lists the frequency multipliers (FM).

- $C$ is between 1.00 ("good") and 0.90 ("poor"). The effectiveness of the coupling may vary as the object is being lifted or lowered: a "good" coupling can quickly become "poor." Three categories are defined in detail

**Table 13–H.** Frequency Multipliers for the 1991 NIOSH Equation

| Frequency, lifts/min | Work Duration (Continuous) | | | | | |
| | ≤8 hr | | ≤2 hr | | ≤1 hr | |
| | V < 75 * | V ≥ 75 | V < 75 | V ≥ 75 | V < 75 | V ≥ 75 |
|---|---|---|---|---|---|---|
| 0.2 | 0.85 | 0.85 | 0.95 | 0.95 | 1.00 | 1.00 |
| 0.5 | 0.81 | 0.81 | 0.92 | 0.92 | 0.97 | 0.97 |
| 1 | 0.75 | 0.75 | 0.88 | 0.88 | 0.94 | 0.94 |
| 2 | 0.65 | 0.65 | 0.84 | 0.84 | 0.91 | 0.91 |
| 3 | 0.55 | 0.55 | 0.79 | 0.79 | 0.88 | 0.88 |
| 4 | 0.45 | 0.45 | 0.72 | 0.72 | 0.84 | 0.84 |
| 5 | 0.35 | 0.35 | 0.60 | 0.60 | 0.80 | 0.80 |
| 6 | 0.27 | 0.27 | 0.50 | 0.50 | 0.75 | 0.75 |
| 7 | 0.22 | 0.22 | 0.42 | 0.42 | 0.70 | 0.70 |
| 8 | 0.18 | 0.18 | 0.35 | 0.35 | 0.60 | 0.60 |
| 9 | 0 | 0.15 | 0.30 | 0.30 | 0.52 | 0.52 |
| 10 | 0 | 0.13 | 0.26 | 0.26 | 0.45 | 0.45 |
| 11 | 0 | 0 | 0 | 0.23 | 0.41 | 0.41 |
| 12 | 0 | 0 | 0 | 0.21 | 0.37 | 0.37 |
| 13 | 0 | 0 | 0 | 0 | 0 | 0.34 |
| 14 | 0 | 0 | 0 | 0 | 0 | 0.31 |
| 15 | 0 | 0 | 0 | 0 | 0 | 0.28 |
| >15 | 0 | 0 | 0 | 0 | 0 | 0 |

* $V$ is expressed in centimeters.
(From Putz-Andersson & Waters, 1991.)

in the NIOSH publication and result in the following listing of values for the coupling multiplier (CM):

| Couplings | V < 30 in. (75 cm) | V ≥ 30 in. (75 cm) |
|---|---|---|
| Good | 1.00 | 1.00 |
| Fair | 0.95 | 1.00 |
| Poor | 0.90 | 0.90 |

To help apply the 1991 NIOSH recommended weight limit, a lifting index (LI) is calculated: $LI = L/RWL$, with $L$ the actual load. If LI is at or below 1, no action must be taken. If LI exceeds 1, the job must be ergonomically redesigned.

## Limits for Lifting, Lowering, Pushing, Pulling, and Carrying

In 1978, Snook published extensive tables of loads and forces found acceptable by male and female workers for continuous manual material handling jobs. These data were first updated in 1983 by Ciriello & Snook and revised in Snook & Ciriello (1991). For application of their 1991 data, the following prerequisites apply:

- Two-handed symmetrical material handling in the medial (mid-sagittal) plane (directly in front of the body), with a slight body twist allowed during lifting or lowering
- Moderate width of the load (75 cm or less)
- Good couplings of hands with handles and of shoes with floor
- Unrestricted working postures

- Favorable physical environment, such as about 21 C at a relative humidity of 45 percent
- Only minimal other physical work activities
- Material handlers who are physically fit and accustomed to labor

There are several differences in format between Snook & Ciriello's recommendations compared to the NIOSH guidelines. The NIOSH values are unisex, whereas the Snook & Ciriello data are separated for female and males. The Snook & Ciriello (1991) data are also grouped with respect to the percentage of the worker population to whom the values are acceptable. The data do not indicate individual capacity limits; rather, they represent the opinions of more than 100 experienced material handlers as to what they would do willingly and without overexertion.

Tables 13–I through 13–M show, in abbreviated form, Snook & Ciriello's 1991 recommendations for suitable loads and forces in lifting, lowering, pushing, pulling, and carrying. The tables are shown here only as examples; consult Snook & Cieriello's original tables for complete information.

It is interesting to note that, like NIOSH recommendations, the data in Snook & Ciriello's (1991) study indicate that lack of handles reduces the loads people are willing to lift and lower by an average of about 15 percent. If the objects become so wide or so deep as to be difficult to grasp, the lifting and lowering values are again considerably reduced. If several material-handling activities occur together, the most strenuous task establishes the handling limit.

If actual loads or forces exceed table values, engineering or administrative controls should be applied. Snook

**Table 13–I.** Maximal Acceptable Lift Weights (kg)

| | | | Floor Level to Knuckle Height — One Lift Every | | | | | | | | Knuckle Height to Shoulder Height — One Lift Every | | | | | | | | Shoulder Height to Overhead Reach — One Lift Every | | | | | | | |
| | | | s | | | min | | | | hr | s | | | min | | | | hr | s | | | min | | | | hr |
| Width* | Distance** | Percent† | 5 | 9 | 14 | 1 | 2 | 5 | 30 | 8 | 5 | 9 | 14 | 1 | 2 | 5 | 30 | 8 | 5 | 9 | 14 | 1 | 2 | 5 | 30 | 8 |
|---|---|---|---|---|---|---|---|---|---|---|---|---|---|---|---|---|---|---|---|---|---|---|---|---|---|---|
| *Males* | | | | | | | | | | | | | | | | | | | | | | | | | | |
| 34 | 51 | 90 | 9 | 10 | 12 | 16 | 18 | 20 | 20 | 24 | 9 | 12 | 14 | 17 | 17 | 18 | 20 | 22 | 8 | 11 | 13 | 16 | 16 | 17 | 18 | 20 |
| | | 75 | 12 | 58 | 18 | 23 | 26 | 28 | 29 | 34 | 12 | 16 | 18 | 22 | 23 | 23 | 26 | 29 | 11 | 14 | 17 | 21 | 21 | 22 | 24 | 26 |
| | | 50 | 17 | 20 | 24 | 31 | 35 | 38 | 39 | 46 | 15 | 20 | 23 | 28 | 29 | 30 | 33 | 36 | 14 | 18 | 21 | 26 | 27 | 28 | 31 | 34 |
| *Females* | | | | | | | | | | | | | | | | | | | | | | | | | | |
| 34 | 51 | 90 | 7 | 9 | 9 | 11 | 12 | 12 | 13 | 18 | 8 | 8 | 9 | 10 | 11 | 11 | 12 | 14 | 7 | 7 | 8 | 9 | 10 | 10 | 11 | 12 |
| | | 75 | 9 | 11 | 12 | 14 | 15 | 15 | 16 | 22 | 9 | 10 | 11 | 12 | 13 | 13 | 14 | 17 | 8 | 8 | 9 | 11 | 11 | 11 | 12 | 14 |
| | | 50 | 11 | 13 | 14 | 16 | 18 | 18 | 20 | 27 | 10 | 11 | 13 | 14 | 15 | 15 | 17 | 19 | 9 | 10 | 11 | 12 | 13 | 13 | 14 | 17 |

* Handles in front of the operator (cm).
** Vertical distance of lifting (cm).
† Acceptable to 50, 75, or 90 percent of industrial workers.
(Adapted from Snook & Ciriello, 1991.)

believes that industrial back injuries could be reduced by about 33 percent if the loads could be eliminated that lie above the values acceptable to 75 percent of the material handlers.

## Comparing NIOSH with the Snook–Ciriello Recommendations

The guidelines by NIOSH are based mostly on biomechanical considerations, a threshold compression of force in the lower spine of 3,400 N, with some consideration of physiological strains. The calculation of the RWL is done twice, once for the beginning point and once for the ending point of the lifting or lowering.

The guidelines by Snook & Ciriello, in contrast, rely on the psychophysical assessments of experienced industrial

material handlers performing controlled material-handing activities in a laboratory. These activities go beyond lifting and lowering; they include pushing/pulling, carrying, and holding. Thus, the Snook–Ciriello tables have wider applicability than what can be calculated from the NIOSH formula.

Direct comparisons are possible only between the lifting and lowering recommendations. In general, the results are quite similar, but with some larger deviations in the extremes of frequencies. To be prudent, one should take the lower values of either set of recommendations.

In general, one should prefer pushing and pulling to carrying, and to lifting and lowering. Figure 13–14 schematically shows engineering interventions applied to solve material handling problems. The main intent is to eliminate, or at least to reduce, the overexertion injury risks to

**Table 13–J.** Maximal Acceptable Lower Weights (kg)

| | | | Knuckle Height to Floor Level — One Lower Every | | | | | | | | Shoulder Height to Knuckle Height — One Lower Every | | | | | | | | Overhead Reach to Shoulder Height — One Lower Every | | | | | | | |
| | | | s | | | min | | | | hr | s | | | min | | | | hr | s | | | min | | | | hr |
| Width* | Distance** | Percent† | 5 | 9 | 14 | 1 | 2 | 5 | 30 | 8 | 5 | 9 | 14 | 1 | 2 | 5 | 30 | 8 | 5 | 9 | 14 | 1 | 2 | 5 | 30 | 8 |
|---|---|---|---|---|---|---|---|---|---|---|---|---|---|---|---|---|---|---|---|---|---|---|---|---|---|---|
| *Males* | | | | | | | | | | | | | | | | | | | | | | | | | | |
| 34 | 51 | 90 | 10 | 13 | 14 | 17 | 20 | 22 | 22 | 29 | 11 | 13 | 15 | 17 | 20 | 20 | 20 | 24 | 9 | 10 | 12 | 14 | 16 | 16 | 16 | 20 |
| | | 75 | 14 | 18 | 20 | 25 | 28 | 30 | 32 | 40 | 15 | 18 | 21 | 23 | 27 | 27 | 27 | 33 | 12 | 14 | 17 | 19 | 22 | 22 | 22 | 27 |
| | | 50 | 19 | 24 | 26 | 33 | 37 | 40 | 42 | 53 | 20 | 23 | 27 | 30 | 35 | 35 | 35 | 43 | 16 | 19 | 22 | 24 | 28 | 28 | 28 | 35 |
| *Females* | | | | | | | | | | | | | | | | | | | | | | | | | | |
| 34 | 51 | 90 | 7 | 9 | 9 | 11 | 12 | 13 | 14 | 18 | 8 | 9 | 9 | 10 | 11 | 12 | 12 | 15 | 7 | 8 | 8 | 8 | 10 | 11 | 11 | 13 |
| | | 75 | 9 | 11 | 11 | 13 | 15 | 16 | 17 | 22 | 9 | 11 | 11 | 12 | 14 | 15 | 15 | 19 | 8 | 9 | 10 | 10 | 12 | 13 | 13 | 16 |
| | | 50 | 10 | 13 | 14 | 16 | 18 | 19 | 20 | 27 | 11 | 13 | 13 | 14 | 16 | 18 | 18 | 22 | 10 | 11 | 11 | 12 | 14 | 15 | 15 | 19 |

* Handles in front of the operator (cm).
** Vertical distance of lowering (cm).
† Acceptable to 50, 75, or 90 percent of industrial workers.
(Adapted from Snook & Ciriello, 1991.)

**Table 13–K.** Maximal Acceptable Push Forces (kg$_f$)

| | | One 2.1-m Push Every | | | | | | | One 30.5-m Push Every | | | | |
| | | s | | min | | | | hr | min | | | | hr |
| Height* | Percent** | 6 | 12 | 1 | 2 | 5 | 30 | 8 | 1 | 2 | 5 | 30 | 8 |
|---|---|---|---|---|---|---|---|---|---|---|---|---|---|
| | | | | | | Initial Forces | | | | | | | |
| *Males* | | | | | | | | | | | | | |
| 95 | 90 | 21 | 24 | 26 | 26 | 28 | 28 | 34 | 17 | 19 | 22 | 22 | 27 |
| | 75 | 28 | 31 | 34 | 34 | 36 | 36 | 44 | 21 | 24 | 28 | 28 | 35 |
| | 50 | 34 | 38 | 43 | 43 | 45 | 45 | 54 | 27 | 30 | 35 | 35 | 44 |
| *Females* | | | | | | | | | | | | | |
| 89 | 90 | 14 | 15 | 17 | 18 | 20 | 21 | 22 | 12 | 14 | 15 | 16 | 18 |
| | 75 | 17 | 18 | 21 | 22 | 24 | 25 | 27 | 15 | 16 | 18 | 19 | 21 |
| | 50 | 20 | 22 | 25 | 26 | 29 | 30 | 32 | 18 | 20 | 21 | 23 | 26 |
| | | | | | | Sustained Forces | | | | | | | |
| *Males* | | | | | | | | | | | | | |
| 95 | 90 | 10 | 13 | 16 | 17 | 19 | 19 | 23 | 8 | 10 | 12 | 13 | 16 |
| | 75 | 14 | 18 | 22 | 22 | 25 | 26 | 31 | 11 | 13 | 16 | 18 | 21 |
| | 50 | 18 | 23 | 28 | 29 | 33 | 34 | 40 | 15 | 17 | 20 | 23 | 27 |
| *Females* | | | | | | | | | | | | | |
| 89 | 90 | 6 | 7 | 9 | 9 | 10 | 11 | 13 | 5 | 6 | 6 | 7 | 9 |
| | 75 | 8 | 11 | 13 | 13 | 15 | 16 | 19 | 8 | 9 | 9 | 10 | 13 |
| | 50 | 11 | 15 | 18 | 18 | 20 | 21 | 26 | 10 | 12 | 12 | 13 | 17 |

* Vertical distance from floor to hands (cm).
** Acceptable to 50, 75, or 90 percent of industrial workers.
(Adapted from Snook & Ciriello, 1991.)

material handlers. Kroemer et al. (1994) provide detailed recommendations for activities other than industrial material handling, including outdoor load transport and moving of patients in hospitals and nursing homes. They also discuss ergonomic selection of material transport and handling equipment and the design and use of trays and containers.

# ■ Hand Tools

Many hand tools are really extensions of the hand. Pliers, for example, amplify the hand's strength, extend its reach, and protect sensitive tissues. Other tools are able to do things that the hand cannot—such as soldering—but they are held and directed by the hand. Too often, design efforts are focused on the working end of the tool rather than how it interacts with the hand. Some hand tools are difficult to use because of inappropriate design (such as tools that force the wrist to bend) or pressure points between the hand and the handle. Many of our everyday hand tools are acceptable if we use them only occasionally, but must be redesigned for frequent use over long periods of time, as in occupational tasks where hand tools are used as part of the job. Figure 13–15 shows an example of straight-nose pliers that require an acute bend in the wrist because, as shown, the line of thrust of a hand tool is at approximately 30 degrees from the line of thrust of the forearm. The handle

should be bent so that the tool can be manipulated with a straight wrist. Furthermore, a handle surface fitted to the enclosing surfaces of the hand helps the user to hold the tool securely and use it efficiently.

There are many ways to manipulate an object. Figure 13–16 organizes these from a simple touch applied by finger, thumb, or palm (couplings 1–3), to tip and pinch grips (couplings 4–7) including the so-called precision grip, to powerful grasps in which thumb, fingers, and palmar surfaces are involved (couplings 8–10). Any of these couplings evolves easily from those shown adjacently.

For touch-type couplings, relatively little attention must be paid to fitting the grip surface of the handle to the touching surface of the hand. However, one may want to put a slight cavity into the top of a push button so that the fingertip does not slide off, to hollow out the handle of a scalpel slightly so that the tips of the digits can hold on securely, or to roughen the grip surface of a dentist's tool and, instead of making it round in cross-section, flatten it out or otherwise contour it for a secure hand-hold.

Such considerations of secure tool handling are most important for the enclosure couplings, which are used to transmit large energies between hand and tool. The tool designer's task is to make sure that the handle can be held securely (without fatiguing muscles unnecessarily and avoiding pressure points) while one exerts linear force or rotating torque to it. In most tools, force must be applied by the hand in two directions: one perpendicular to the

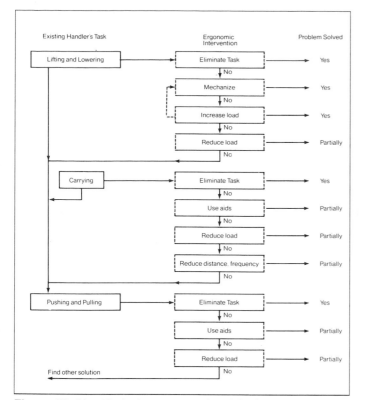

**Figure 13–14.** Reducing overexertion risks in material handling.

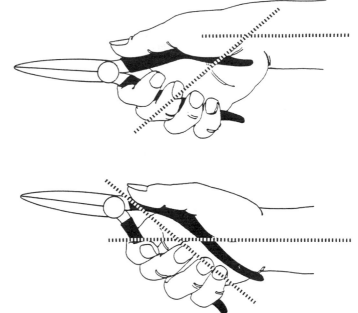

**Figure 13–15.** Common and ergonomic pliers. (Modified from Kroemer, Kroemer, Kroemer-Elbert, *Ergonomics: How to Design for Ease and Efficiency.* Englewood Cliffs, NJ: Prentice Hall, 1994. Used with permission by the publisher. All rights reserved.)

handle surface (by the palm closing the handles of pliers, for example), the other perpendicular to that direction (in pulling an object with the pliers. Hence, both the cross-sectional and longitudinal shape of the tool must be considered. Furthermore, the presence of grease or dirt between the hand and the handle or the wearing of gloves can have profound effects on the coupling.

Several publications address the problem of handle design: Chaffin & Anderson (1990), Cochran & Riley (1986) Drury (1980), Greenberg & Chaffin (1979), Fraser (1980), Konz (1990), Mital (1991), and Woodson et al. (1991). Kroemer et al. (1994) have supplied information about forces that can be generated by males in various hand–handle couplings; for female operators, about 2/3 of these strengths can be expected (see Tables 13–N, 13–O, and 13–P).

**Table 13–L.** Maximal Acceptable Pull Forces (kg$_f$)

| | | | One 2.1-m Pull Every | | | | | | |
|---|---|---|---|---|---|---|---|---|---|
| | | | *s* | | *min* | | | | *hr* |
| | | | 6 | 12 | 1 | 2 | 5 | 30 | 8 |
| | Height* | Percent** | | | *Initial Pull* | | | | |
| | | 90 | 19 | 22 | 25 | 25 | 27 | 27 | 32 |
| Males | 95 | 75 | 23 | 27 | 31 | 31 | 32 | 33 | 39 |
| | | 50 | 28 | 32 | 36 | 36 | 39 | 39 | 47 |
| | | 90 | 14 | 16 | 18 | 19 | 21 | 22 | 23 |
| Females | 89 | 75 | 16 | 19 | 21 | 22 | 25 | 26 | 27 |
| | | 50 | 19 | 23 | 25 | 26 | 29 | 30 | 32 |
| | | | | | *Sustained Pull* | | | | |
| | | 90 | 10 | 13 | 16 | 17 | 19 | 20 | 24 |
| Males | 95 | 75 | 13 | 17 | 21 | 22 | 25 | 26 | 30 |
| | | 50 | 16 | 21 | 26 | 27 | 31 | 32 | 37 |
| | | 90 | 6 | 9 | 10 | 10 | 11 | 12 | 14 |
| Females | 89 | 75 | 8 | 12 | 13 | 13 | 15 | 16 | 19 |
| | | 50 | 10 | 15 | 16 | 17 | 19 | 20 | 25 |

* Vertical distance from floor to hands (cm).
** Acceptable to 50, 75, or 90 percent of industrial workers.
(Adapted from Snook & Ciriello, 1991.)

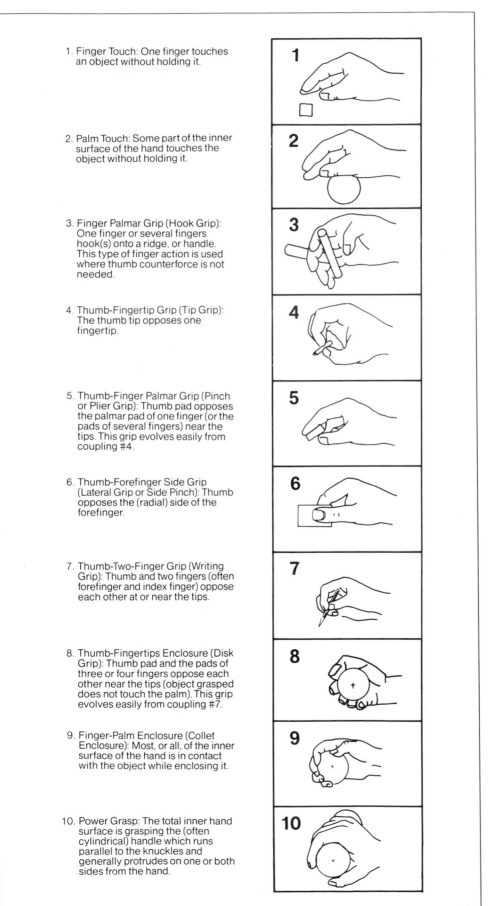

1. Finger Touch: One finger touches an object without holding it.

2. Palm Touch: Some part of the inner surface of the hand touches the object without holding it.

3. Finger Palmar Grip (Hook Grip): One finger or several fingers hook(s) onto a ridge, or handle. This type of finger action is used where thumb counterforce is not needed.

4. Thumb-Fingertip Grip (Tip Grip): The thumb tip opposes one fingertip.

5. Thumb-Finger Palmar Grip (Pinch or Plier Grip): Thumb pad opposes the palmar pad of one finger (or the pads of several fingers) near the tips. This grip evolves easily from coupling #4.

6. Thumb-Forefinger Side Grip (Lateral Grip or Side Pinch): Thumb opposes the (radial) side of the forefinger.

7. Thumb-Two-Finger Grip (Writing Grip): Thumb and two fingers (often forefinger and index finger) oppose each other at or near the tips.

8. Thumb-Fingertips Enclosure (Disk Grip): Thumb pad and the pads of three or four fingers oppose each other near the tips (object grasped does not touch the palm). This grip evolves easily from coupling #7.

9. Finger-Palm Enclosure (Collet Enclosure): Most, or all, of the inner surface of the hand is in contact with the object while enclosing it.

10. Power Grasp: The total inner hand surface is grasping the (often cylindrical) handle which runs parallel to the knuckles and generally protrudes on one or both sides from the hand.

**Figure 13–16.** Couplings between hand and handle. (Adapted from Kroemer. 1986.)

**Table 13–M.**  Maximal Acceptable Carrying Weights (kg)

| | Height* | Percent** | One 2.1-m Carry Every | | | | | | |
|---|---|---|---|---|---|---|---|---|---|
| | | | s | | min | | | | hr |
| | | | 6 | 12 | 1 | 2 | 5 | 30 | 8 |
| Males | 79 | 90 | 13 | 17 | 21 | 21 | 23 | 26 | 31 |
| | | 75 | 18 | 23 | 28 | 29 | 32 | 36 | 42 |
| | | 50 | 23 | 30 | 37 | 37 | 41 | 46 | 54 |
| Females | 72 | 90 | 13 | 14 | 16 | 16 | 16 | 16 | 22 |
| | | 75 | 15 | 17 | 18 | 18 | 19 | 19 | 25 |
| | | 50 | 17 | 19 | 21 | 21 | 22 | 22 | 29 |

\* Vertical distance from floor to hands (cm).
\*\* Acceptable to 50, 75, or 90 percent of industrial workers.

(Adapted from Snook & Ciriello, 1991.)

**Table 13–N.**  Average Digit Poke Forces* Exerted by 30 Subjects in Direction of the Straight Digits

| Digit | 10 Skilled Males | 10 Unskilled Males | 10 Unskilled Females |
|---|---|---|---|
| 1, Thumb | 83.8 (25.19) A | 46.7 (29.19) C | 32.4 (15.36) D |
| 2, Index finger | 60.4 (25.81) B | 45.0 (29.99) C | 25.4 (9.55) DE |
| 3, Middle finger | 55.9 (31.85) B | 41.3 (21.55) C | 21.5 (6.46) E |

Entries with different letters are significantly different from each other ($p \leq 0.05$).
\* Means and standard deviations, in N.

(Adapted from Kroemer, Kroemer, Kroemer-Elbert, *Ergonomics: How to Design for Ease and Efficiency*. Englewood Cliffs, NJ: Prentice Hall, 1994. Used with permission by the publisher. All rights reserved.)

Given the various tasks and uses of hand tools, only a few general guidelines can be provided; details must be decided according to the given conditions:

- In manipulating the hand tool, the wrist should stay straight with respect to the forearm; that is, it should be neither rotated (pro- or supinated) nor bent (flexed-extended or laterally deviated). This often requires that the working side of the hand tool be at an oblique angle with the handle. (However, this may make the tool useful only for certain tasks.)

- The handle should be of such cross-sectional size that the hand nearly encircles the handle, with no more space than about 0.5 in. (1.3 cm) between the fingertips and the thumb side. This means that the diameter, if the handle is circular, or the largest distance between

**Table 13–O.**  Grip and Grasp Forces* Exerted by 21 Male Students and 12 Machinists

| Couplings | Digit 1 (Thumb) | Digit 2 (Index) | Digit 3 (Middle) | Digit 4 (Ring) | Digit 5 (Little) | All Digits Combined: |
|---|---|---|---|---|---|---|
| Push with digit tip in direction of the extended digit ("poke") | 91 (39)** | 52 (16)** | 51 (20)** | 35 (12)** | 30 (10)** | |
| | 138 (41) | 84 (35) | 86 (28) | 66 (22) | 52 (14) | |
| Digit touch (Coupling #1) perpendicular to extended digit | 84 (33)** | 43 (14)** | 36 (13)** | 30 (13)** | 25 (10)** | — |
| | 131 (42) | 70 (17) | 76 (20) | 57 (17) | 55 (16) | |
| Same, but all fingers press on one bar | — | digits 2, 3, 4, 5 combined: 162 (33) | | | | |
| Tip force (as in typing; angle between distal and proximal phalanges about 135 degrees) | — | 30 (12)** | 29 (11)** | 23 (9)** | 19 (7)** | — |
| | 65 (12) | 69 (22) | 50 (11) | 46 (14) | | |
| Palm touch (Coupling #2) perpendicular to palm (arm, hand, digits extended and horizontal) | — | — | — | — | — | 233 (65) |
| Hook force exerted with digit tip pad (Coupling #3, "scratch") | 61 (21) | 49 (17) | 48 (19) | 38 (13) | 34 (10) | 108 (39)** |
| | 118 (24) | 89 (29) | 104 (26) | 77 (21) | 66 (17) | 252 (63) |
| Thumb–fingertip grip (Coupling #4, "tip pinch") | | all digits combined: | | | | |
| | — | 1 on 2 | 1 on 3 | 1 on 4 | 1 on 5 | — |
| | | 50 (14)** | 53 (14)** | 38 (7)** | 28 (7)** | |
| | | 59 (15) | 63 (16) | 44 (12) | 30 (6) | |
| Thumb–finger palmar grip (Coupling #5, "pad pinch") | 1 on 2 and 3 | 1 on 2 | 1 on 3 | 1 on 4 | 1 on 5 | — |
| | 85 (16)** | 63 (12)** | 61 (16)** | 41 (12)** | 31 (9)** | |
| | 95 (19) | 34 (7) | 70 (15) | 54 (15) | 34 (7) | |
| Thumb–forefinger side grip (Coupling #6, "side pinch") | — | 1 on 2 | — | — | — | |
| | | 98 (13)** | | | | |
| | | 112 (16) | | | | |
| Power grasp (Coupling #10, "grip strength") | — | — | — | — | — | 318 (61)** |
| | | | | | | 366 (53) |

\* Means and standard deviations in N.
\*\* Students' results; all others are machinists' results.

(Adapted from Kroemer, Kroemer, Kroemer-Elbert, *Ergonomics: How to Design for Ease and Efficiency*. Englewood Cliffs, NJ: Prentice Hall, 1994. Used with permission by the publisher. All rights reserved.)

**Table 13–P.** Average Forces* Exerted with the Fingertip by Angle of the Proximal Intraphalangeal (PIP) Joint

| | PIP Joint at 30 Degrees | | | PIP Joint at 60 Degrees | | |
|---|---|---|---|---|---|---|
| | *Fore* | *Aft* | *Down* | *Fore* | *Aft* | *Down* |
| DIGIT | | | | | | |
| 2 Index | 5.4 (2.0) | 5.5 (2.2) | 27.4 (13.0) | 5.2 (2.4) | 6.8 (2.8) | 24.4 (13.6) |
| 2 Nonpreferred hand | 4.8 (2.2) | 6.1 (2.2) | 21.7 (11.7) | 5.6 (2.9) | 5.3 (2.1) | 25.1 (13.7) |
| 3 Middle | 4.8 (2.5) | 5.4 (2.4) | 24.0 (12.6) | 4.2 (1.9) | 6.5 (2.2) | 21.3 (10.9) |
| 4 Ring | 4.3 (2.4) | 5.2 (2.0) | 19.1 (10.4) | 3.7 (1.7) | 5.2 (1.9) | 19.5 (10.9) |
| 5 Little | 4.8 (1.9) | 4.1 (1.6) | 15.1 (8.0) | 3.5 (1.6) | 3.5 (2.2) | 15.5 (8.5) |

* Average forces and standard deviations, in N.
Adapted from Kroemer, Kroemer, Kroemer-Elbert, *Ergonomics: How to Design for Ease and Efficiency.* Englewood Cliffs, NJ: Prentice Hall, 1994. Used with permission by the publisher. All rights reserved.

two opposing sides of the handle should be between 1 and 2.5 in. (2.5 and 6.5 cm).

- The shape of the handle, in cross section, depends on the task to be performed, that is, on the motions involved in opening and closing the handle and on the magnitude of force or torque (moment) to be developed for use of the tool. In many cases, elliptical shapes (or rectangular ones with well-rounded edges) are advantageous if twisting (torquing or turning such as with screwdrivers) must be performed. However, more circular cross sections are preferred if the tool must be grasped in many different manners.

- The handle should easily accommodate the length of the hand in contact with it; for example, a knife handle should be at least as long as the hand enclosing it is broad. The contour of the handle can follow the contour of the inside of the hand enfolding the handle. However, strongly form-fitting the handle shape might prevent people with different hand sizes or people who grasp the handle in a different way from using the tool comfortably.

- Pressure points should be avoided. These are often present if the form of the handle has pronounced shapes such as deep indentations for the fingers or sharp edges or contours.

- Rough surfaces of the handle might be uncomfortable for sensitive hands but can counteract the effects of grease that make the handle slippery.

- Flanges at the end of the handle can guide the hand to the correct position and prevent the hand from sliding off the handle.

Improperly designed hand tools, particularly when combined with ill-conceived workplaces and job procedures, can cause acute injuries to surface tissues and the musculoskeletal system of the hand–arm complex. Often, the repetition of biomechanical insults, each in itself insignificant, can lead to cumulative trauma disorders in the upper extremities, a major cause of lost work in many hand-intensive industries. Many of these disorders can be averted easily by paying attention to the following recommendations:

- Avoid repetitive or sustained exertions, particularly if they are accompanied by deviations from a straight wrist and/or by forceful exertions.

- Keep the shoulder relaxed, the elbow at the side of the body, the forearm semipronated, and the wrist straight.

- Use tools with handles of appropriate size and shape.

- Round off all edges and sharp corners on the hand tool or at the workstation with which the worker might come in contact.

- Avoid excessive cooling of the hand, either by a cold environment or by strong air movement (particularly important with air-powered tools) or by contact with metal handles that easily conduct heat away from the hand.

- Ensure that gloves worn actually help the activity but do not hinder motion or enforce awkward wrist positions.

- Minimize vibrations of hand tools.

# ■ Workstation Design

The goal in designing a workstation is to promote ease and efficiency for the worker. Productivity suffers in quantity and quality if the operator is uncomfortable or if the layout of the workstation or the job procedures are awkward. Conversely, productivity is enhanced if the operator is comfortable physiologically and psychologically and if the layout of the workstation is conducive to performing the task well. Keeping this in mind, it is advisable to try to establish an ideal workstation, task, and work environment first and to make concessions to practical limitations only if absolutely necessary.

## General Principles

Six general rules govern the design of workplaces:

- Plan the ideal, then the practical.
- Plan the whole, then the detail.
- Plan the work process and equipment to fit the worker.

- Plan the workplace layout around the process and equipment.
- Plan the final enclosure around the workplace layout.
- Use mockups to evaluate alternative solutions and to check the final design.

For this, the following design aspects must be considered primarily:

- *Space:* clearance for the operator's body entrance and egress (including emergency exit), suitable body posture at work (including the space needed for changing positions), operation of controls and equipment (without bumping elbows, knees, or head), and the avoidance of excessive forces or inadvertent operation of controls
- *Manipulation:* operation of tools, controls, workpieces (including seat adjustment), and emergency items (stop button, flashlight, survival equipment) by hand or foot
- *Seeing:* visual field and information both inside (displays and control settings) and outside (road, machine being controlled), visual contact with co-workers, and lighting (illumination, luminance, shadows, avoidance of glare)
- *Hearing:* auditory information such as oral communication with other workers, signals (including warning signals), and sounds from equipment (engine, cutting tool)

More detailed design guidelines depend on the special workstation, the specific work task, and the environment. Such guidelines can be found in books by Cushman & Rosenberg (1991), Eastman Kodak Company (1983, 1986), Fraser (1989), Grandjean (1988), Konz (1985, 1990), Kroemer et al. (1994), Sanders & McCormick (1993), Salvendy (1987), and Woodson et al. (1991). *Military Standards 759* and *1472* also provide a wealth of human engineering information.

## Standing or Sitting

Whether the operator should stand or sit at the workstation depends on several factors: the mobility required, the forces needed, the size of the workpiece, and the required precision. The advantages of standing over sitting include more mobility, more body strength available, less front-to-rear room required, no seat needed, and greater latitude in workstation design. The advantages of sitting are that pedals can be operated with the foot more effectively (with more force and precision), it is less fatiguing to maintain the sitting posture (if a good seat is available), and manipulation and vision may be more precise. Unless the specific work task, environment, or conditions strictly demand either sitting or standing, provisions should be made to allow the operator to sit or stand at will. Figure 13–17 shows stand-seats, which can offer a useful compromise.

## Workplace Design

The most basic requirement for a workplace is that it must accommodate the person working in it. Specifically, this means that the work space for the hands should be between hip and chest height in front of the body. In this region, the lower locations are preferred for heavy manual work and the higher locations are suitable for tasks that require close visual observation. Contours of reach envelopes indicate the maximum distances at which objects can be manipulated or placed. Figure 13–18 shows an example of such reach capabilities.

Work objects should be located close to the front edge of the work surface so that the worker does not have to bend over and lean across the surface to grasp items. To allow the person to be close to the front edge of the work surface, sufficient room must be provided so that thighs, knees, and toes can be placed somewhat under the work surface if the work is performed while standing. For sitting operators, deeper and wider leg room must be provided under the bench, table, or desk. If foot controls are used, additional room for foot and leg motions may be needed. Pedals that must be operated continuously or frequently normally require a seated operator because if a person operated them while standing, the body weight would have to be supported on one foot.

Visual displays including instruments, counters, dials, and signal lights should be placed in front of the body and below eye level so that the line of sight (which aligns the eyes with the visual target) is declined 10°–40° below the horizontal level. Table 13–Q lists general principles for workstation design, but specifics for computer workstations are detailed later in this chapter.

### Work Space Dimensions

Work space dimensions can be grouped into three basic categories: minimal, maximal, and adjustable dimensions. *Minimal* work space dimensions provide clearance for the worker. Many minimal clearance dimensions, such as the open leg space under a work bench, can be determined using 95th percentiles values from anthropometric tables (such as those provided in Table 13–E). Larger values must be considered for other, more critical clearances: if the height of the door frame were at the 95th percentile value, at least 5 percent of all users would bump their heads. *Maximal* dimensions, in contrast, permit smaller workers to use the equipment by selecting work space dimensions over which a small person can reach or establishing control forces small enough that even a weak person can operate the equipment, for example. Usually, the 5th percentile value of the relevant body attribute is used for this design.

*Adjustability* permits the operator to modify the work environment and equipment so that it conforms to that individual's physical (anthropometric, biomechanical) characteristics as well as to subjective preferences. A six-

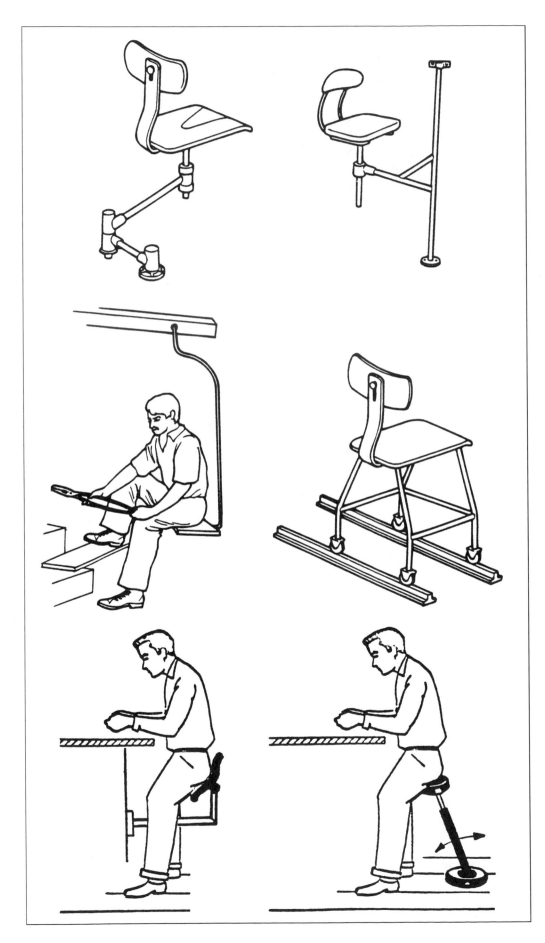

**Figure 13–17.** Examples of stand-seats. (Adapted from Kroemer, Kroemer, Kroemer-Elbert, *Ergonomics: How to Design for Ease and Efficiency*. Englewood Cliffs, NJ: Prentice Hall, 1994. Used with permission by the publisher. All rights reserved.)

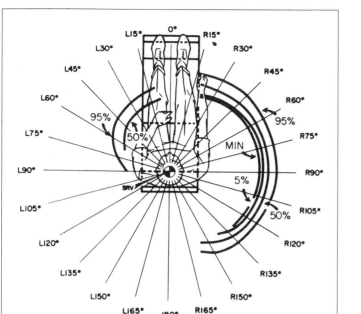

**Figure 13–18.** Grasping reach contours of the right hand in a horizontal plane 25 cm above the seat reference point. (Reprinted with permission from Damon, Stoudt, and McFarland, *Human Body in Equipment Design,* 1966.)

way adjustable seat in a truck is an example of the proper adjustment capabilities available to individual operators. Adjustable dimensions are particularly important when optimal performance with minimum effort is necessary to accomplish the work task.

Approximate dimensions for industrial workplaces are presented in Table 13–R and in Figures 13–19 and 13–20. Of course, it is most desirable to allow the worker as much free space as desired to position the body and the workpiece; however, space limitations or structural requirements might limit the dispensable space. Still, severely limiting the leg room is not suitable because it forces the operator to assume a fixed posture.

The requirements for an industrial work seat are essentially the same as those for an office chair. However, the industrial work seat is probably somewhat more rugged and has soil-resistant upholstery. As shown in Figure 13–21, the seat pan height of an industrial work seat should be adjustable between about 15 and 20 in. (38 and 51 cm). Its front edge should be well-rounded to avoid pressure to the underside of the thighs. A backrest should be provided, if the work activities allow it. The backrest should be adjustable in height and in distance from the front edge of the seat pan. To allow free mobility of the arms and shoulder blades, the backrest probably should not extend up to the neck (as the office chair may); however, a larger seat can be used for relaxation during a break from the work activities. The backrest should have a protrusion or pad at lumbar height, just like an office chair.

Objects that must be seen and observed (displays, signal lights, controls, dials, keyboards, and written documents) should be placed within the worker's visual field. The more important ones and those that must be read exactly should be placed within the preferred viewing area. Figure 13–22 and Table 13–S identify these areas.

## Office (Computer) Workstations

Complaints related to posture and vision are by far the most frequent health problems voiced by computer operators (see also Chapter 5, The Eyes, and Chapter 11, Nonionizing Radiation). Musculoskeletal pain and discomfort, eye strain,

---

**Table 13–Q.** Ergonomic Guidelines for Workplace Design

1. In the design of the facility, assure a proper match between the facility and the operator to avoid static efforts, such as holding a work piece or hand tool. Static (isometric) muscle tension is inefficient and leads to rapid fatigue.

2. The design of the task and the design of the workplace are interrelated. The work system should be designed to prevent overloading the muscular system. Forces necessary for dynamic activities should be kept to less than 30% of the maximal forces the muscles are capable of generating. Occasionally, forces of up to 50% are acceptable when maintained for only short durations (approximately 5 minutes or less). If static effort is unavoidable, the muscular load should be kept quite low—less than 15% of the maximal muscle force.

3. Aim for the best mechanical advantage in the design of the task. Use postures for the limbs and body that provide the best lever arms for the muscles used. This avoids muscle overload.

4. Foot controls can be used by the seated operator. They are not recommended for continuous use by a standing operator because of the imbalanced posture imposed on the operator. If a pedal must be used by the standing operator, it should be operable with either foot. Avoid hard floors for the standing operator; a soft floor mat is recommended, if feasible.

5. Maintain a proper sitting height, which is usually achieved when the thighs are about horizontal, the lower legs vertical, and the feet flat on the floor. Use adjustable chairs and, if needed, footrests. When adjusting the chair, make sure that:
   a. elbows are at proper height in relation to work surface height;
   b. the footrest is adjusted to prevent pressure at undersides of the thighs;
   c. the backrest is large enough to be leaned against, at least for a break; and
   d. special seating devices are used if the task warrants them.

6. Permit change of posture—static posture causes problems in tissue compression, nerve irritation, and circulation. The operator should be able to change his or her posture frequently to avoid fatigue. Ideally, the operator should be able to alternate between sitting and standing; therefore, a workplace that can be used by either a sitting or standing operator is recommended.

7. In designing the facility, accommodate the large operator first and give that operator enough space. Then provide adjustments and support so that the smaller operator fits into the work space. For standing work, the work surface should be designed to accommodate the taller operator; use platforms to elevate shorter operators. (But watch out for stumbles and falls!) For reach, design to accommodate the shorter operator.

8. Instruct and train the operator to use good working postures whether sitting or standing, working with machines and tools, lifting or loading, or pushing or pulling loads.

**Table 13–R.** Approximate Dimensions (in cm) for Industrial Workplaces for Sitting or Standing Operators

| | Sitting | | Standing |
|---|---|---|---|
| | *Sitting* | | *Standing* |
| | *Outside Dimensions* | | |
| a | 15 to 25 | a | 15 to 38 |
| b | 50 to 71 (if adjustable) | n | 89 |
| | 71 (if fixed) | | |
| c | 63 to 102 | o | 96 to 122 |
| | *Leg Room* | | |
| d | 48 minimum | p | 81 |
| e | 63 or more | q | 10 |
| f | 30 minimum if "e" is impossible | — | — |
| g | 63 or more | r | 63 or more |
| h* | 30 (20 minimum) | — | — |
| i* | 2.5 at front edge | — | — |
| | 7 at 12 from front edge | — | — |
| | 15 at 25 from front edge | — | — |
| | *Foot Room* | | |
| k | 25 or more | s | 20 |
| m | 63 or more | t | 12 |
| g | 63 or more | r | 63 or more |
| h* | 30 (20 minimum) | | |

*Note:* Letters correspond to measurements illustrated in Figures 13–19 and 13–20. For alternate sitting and standing, combine all appropriate dimensions.

* Avoid divider rib or pedestal.

and fatigue constitute at least half of the health problems; in some surveys, they constitute up to 80 percent of all subjective and objective symptoms in offices in North America and Europe. Apparently, some of these complaints are related. Difficulties in viewing (focusing distance, angle of the line of sight, glare), together with straining curvatures of the spinal column, particularly in the neck and lumbar regions, joined by fatiguing postures of shoulders and arms result in a stress–strain combination in which causes and effects intermix, alternate, and build on each other, especially if sociopsychological conditions are faulty (Carayon, 1993; Smith, 1987).

Postural problems appear to be largely or partly caused by improperly designed and poorly arranged workstation furniture. The questionable convention of sitting upright at the desk has been carried over to computer workplaces as a design idol. Even the 1988 ANSI 100 standard on visual display terminal workstations used an upright or nearly straight posture for deriving furniture dimensions. In reality, many other working postures exist that are suitable for the work and are subjectively comfortable and hence may be individually preferred. The 1994 draft version of the ANSI 100 revision acknowledges the diversity of working postures.

Successful ergonomic design of the office workstation depends on proper consideration of several interrelated aspects, sketched in Figure 13–23. The postures a person assumes and the ways he or she performs activities are strongly influenced by workstation conditions including furniture, equipment, and environment. All must fit the person and the task.

The actual work postures and motions affect physical and psychological well-being, whereas the work activities determine the output of the person working with the equipment. Feeling well, both physically and mentally, affects health, work attitudes, and work output. Of course, these interactive relationships are not static but vary with time.

This brief discussion of the many and multilevel relationships among work variables is meant to emphasize the need for carefully designing the workstation—particularly furniture, equipment, and lighting—so that the desired results of well-being and high performance are achieved. Of course, the interpersonal and organizational climate is a major factor as well.

## Work Task

The now-proverbial office revolution has been largely brought about by the use of computers. Hence, the tasks associated with certain office job categories have changed, dramatically in some cases (Helander, 1991). In many offices, typing is now performed on a word processor rather than on a traditional typewriter. This allows the operator to control the layout of texts and graphics and eliminates the frustrating and time-consuming job of retyping large chunks of material to incorporate minor changes. However, information processing requires different and more complex skills and more control functions and responsibilities.

The interactions between the human and the computer impose special requirements on the operator's visual and motor capabilities. The eyes scan source documents (for input into the computer) and the display screen of the monitor (either to obtain new information or to receive feedback about the material already transmitted to the computer system). The fingers input information to the system via keyboard, mouse, trackball, light pen, or touch panel. Voice communication with the computer, both as input and output, is being developed.

The intensity with which the visual and motor communication links are used identifies groups of office tasks. Table 13–T lists visual and motor links between the user and the computer. Such a list can be used to identify the priority needs for ergonomic design of these links, as indicated in Table 13–U.

## Positioning the Body in Relation to the Computer

While the body is supported by the chair, the operator interfaces with the computer mostly through eyes and fingertips. The majority of sensory reception occurs through the eyes as they fix their sight on the monitor, on the source document, and even on the keys. (Many computer keyboards have an exorbitant number of keys that require

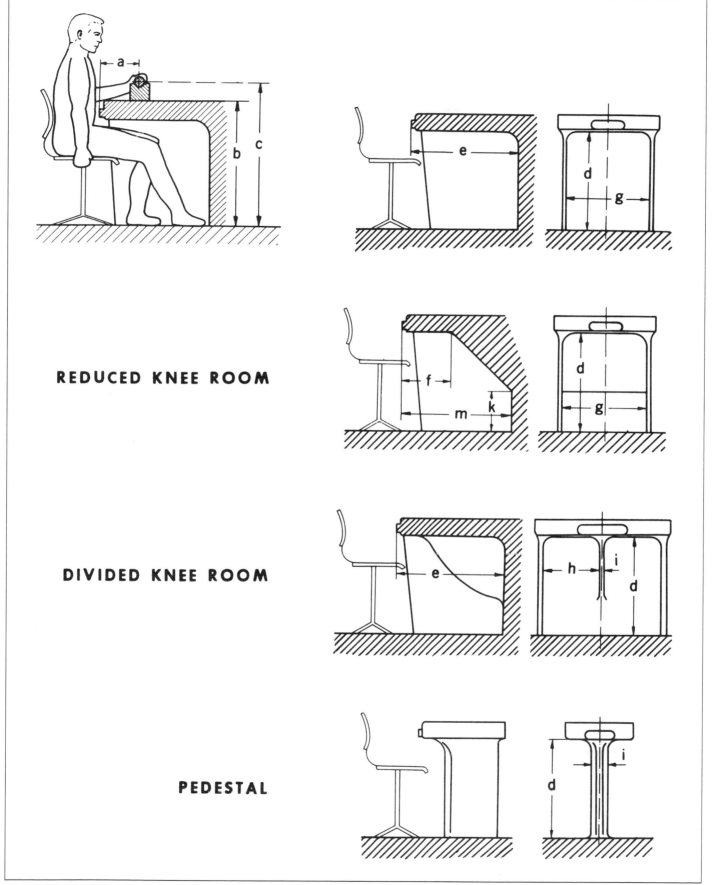

**REDUCED KNEE ROOM**

**DIVIDED KNEE ROOM**

**PEDESTAL**

**Figure 13–19.** Workplace dimensions for sitting operation.

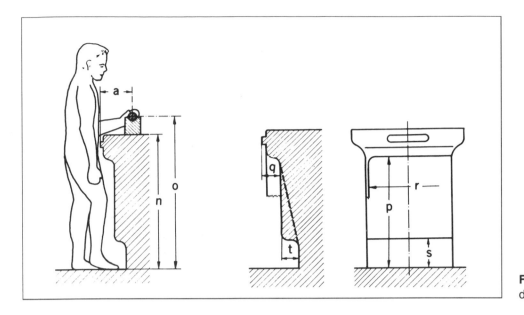

**Figure 13–20.** Workplace dimensions for standing operation.

visual search and identification; touchtyping is often not feasible.) Thus, the position of the person's eyes is relatively fixed with respect to the visual target. This eye fixation has rather stringent consequences for neck and trunk postures.

The ears are input channels for the operator, and the mouth is a natural output device. However, sound or speech signals are not often used with current computer systems, even though they do not restrict the position of the head to any specified location because acoustic signals travel through the air or can be transmitted through speakers or headphones. With present technology, the digits of the hands are the user's major output links to the computer. If keys, mousepad, trackball, or touch panel are fixed within the workstation, the operator has no choice but to keep the hands on them. This often determines the person's body posture, as does the location of foot controls.

## Healthy Work Postures

For about a hundred years it was generally believed that an upright trunk, as when standing, was part of a healthy posture. This idea had been used—usually together with presuming right angles at hips, knees, and ankles—to design office chairs and other furniture. However, are there physiological or orthopedic reasons compelling enough to make people sit straight even if, left alone, hardly anyone chooses this posture?

Sitting upright means that, in the lateral view, the spinal column forms an S-curve with slight forward bends (lordosis) in the neck and in the low back regions and a slight rearward bulge (kyphosis) in the chest region. Although there appears to be no reason to doubt that this is, in the current evolutionary condition of humans, a normal and therefore desirable posture, it is essential to discuss how such a posture can be achieved, supported, or even enforced.

Staffel (1984) proposed a forward-declining seat surface. In the 1960s, a seat pan design with a Schneider wedge on its rear edge was popular for about a decade in Europe. More recently, again seat surfaces were promoted that slope downward and forward. The underlying idea is that the desired lumbar lordosis be achieved by rotating the pelvis forward, opening the hip angle to more than 90 degrees. To prevent the buttocks from sliding off the forward-declined seat, the seat surface may be shaped to fit the human underside, or the downward-forward thrust can be counteracted by either bearing down on the feet or by propping the upper shins on special pads.

Another way to bring about lordosis of the lumbar region is to push that section of the back and spinal column forward with a specially designed backrest. The Akerblom pad of the seatback upholstery or the inflatable lumbar cushion incorporated in some car and airplane seats are examples of this design feature. Of course, one can shape the total backrest. Around 1960, Ridder in the United States and Grandjean in Switzerland independently found rather similar backrest shapes to be acceptable by experimental subjects. In essence, these shapes followed the curvature of the rear side of the human body: at the bottom, concave (or open) to accept the buttocks; above the seat, slightly convex to fill in the lumbar lordosis and then nearly straight but reclined backward to support the thoracic area; at the top, again convex to follow the neck lordosis. This shape has been used successfully in automobiles, aircraft, passenger trains, cars, and easy chairs. In the traditional office, these "first-class" chairs were available to managers; other employees had to make do with simpler designs that ranged down to the miserable small board attached to the so-called secretarial chair. (Extensive bibliographies and reviews of recommendations for seat designs encompassing the last three decades can be found in Grandjean, 1986;

Kroemer et al., 1994; Lueder, 1983; Lueder & Noro, 1994; Wilson et al., 1986; and Zacharkow, 1988.)

## Biomechanical Actions on the Spine

Because of their mechanical connections via bones, ligaments, and muscles, the lower spine, the pelvis, and the thighs do not assume independent positions, but rather influence each other. The postures of the upper trunk, the head, and arms affect the spine posture. When a person stands, the upper surface of the lowest part of the spine, the sacrum, is severely inclined forward, thus providing a downward-slanted support basis (at the L5–S1 interface) for the lumbar section of the spine. This helps to achieve lumbar lordosis. When the pelvis, to which the sacrum is attached, rolls backward (as when one sits upright on a flat surface), the S1–L5 interface becomes horizontal or even slightly slanted backward. This brings about flattening or kyphosis of the lumbar spine (Figure 13–24). An important muscular connection is established by the hamstring muscles, which run from the pelvis posteriorly along the thighs to their attachment at the back of the calves. Thus, they cross both the knee and hip joints and, particularly if the thigh is elevated (small hip angle) and the lower legs are brought forward (large knee angle), enforce a rear rotation of the pelvis that is usually accompanied by flattening or even kyphotic bending of the lumbar spine.

Physiological and mechanical events associated with these postures have been recorded under a variety of conditions and by many researchers. Radiographic studies have established that the rotation of the pelvis and the curvature of the spinal column are closely associated. When a person sits on a flat surface with the hip angle at approximately 90°, the pelvis naturally rotates backward and the lumbar spine flattens. If desired, such flattening can be avoided by muscle tension, by the design of the backrest, or by providing a seat surface that mechanically tilts the pelvis forward. Lumbar lordosis can also be achieved by opening the hip angle (for example, by thrusting the feet forward and the knees down). This is helped by elevating the seat pan and declining it, that is, making the front section lower than the rear part.

Disk compression and trunk muscle activities are related because the stability of the vertebrae stacked on top of each other is secured by ligaments and by the contraction of muscles that generate essentially vertical forces in the trunk (primarily the latissimus dorsi, the erector spinae, the oblique internus and externus abdominis, and the rectus abdominis). Directly or indirectly, these muscles pull down on the spine and keep the vertebrae aligned on top of each other. Each vertebra rests on the lower one, cushioned by the spinal disk between their main bodies and supported lateroposteriorly in the two facet joints of the articulation processes. Because the downward pull of the muscles generates disk and facet–joint compression forces (in response to upper body weight and external forces), there must be a

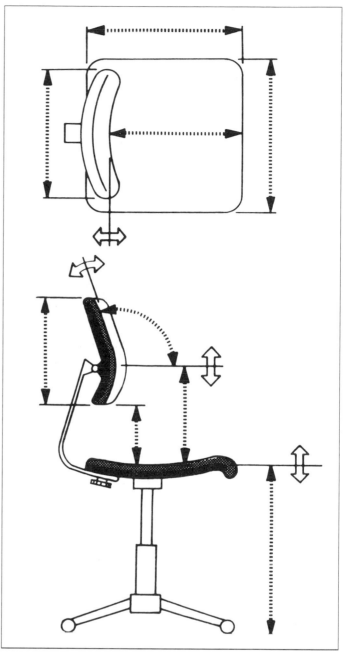

**Figure 13–21.** Main design features of an industrial work seat. (Reprinted with permission of the American Industrial Hygiene Association.)

relationship between trunk muscle activities and disk pressures. These strains are reduced when the trunk, neck, head, and arm weight are at least partially supported by a suitably high backrest (Chaffin & Andersson, 1984; Grandjean, 1986; and Kroemer et al., 1990, 1994).

## Experimental Studies

Since the empirical studies of Ridder and Grandjean, many analytical experiments have been performed to measure the physiological responses of the human body to certain postures. Lundervold (1951) was among the first to record and

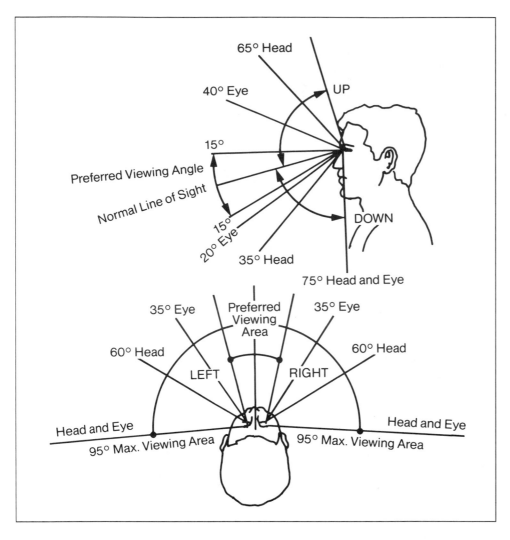

**Figure 13–22.** Visual field. (Reprinted from *Military Standards 759A,* 1981.)

interpret electromyographic activities of the upper body muscles associated with defined seated positions. Based on these early studies, many EMG recordings have been obtained (summarized by Chaffin & Andersson, 1984; Winkel & Bendix, 1986; and Soderberg et al., 1986). As may be expected, varying activities were found in the muscles that stabilize the body, particularly in the trunk. However, involvement of muscles in the hip and lower trunk seems to be rather unimportant for regular seated postures; the observed EMG activities indicate very low demands on muscular capabilities, typically well below 10 percent of the maximal contraction capabilities. (However, muscular strains in the low back area could be important for unusual conditions and postures, such as leaning over the desk at which one is seated to lift an object such as a computer monitor with extended arms.)

Furthermore, the interpretation of these weak EMG signals is controversial because one cannot necessarily assume that little muscle use (flat EMG signals) should be preferred over more extensive muscle use. Dynamic sitting is desired by some physiologists and biomechanists to obtain suitable muscle tone and training; to obtain orthostatic tolerance and electrolyte and fluid balance (Grieco,

1986; Kilbom, 1986); to benefit intervertebral disk metabolism (Hansson and Attebrant, 1986); and to benefit macro- and microcirculatory aspects (Winkel, 1986), including blood pooling (Thompson et al., 1986). In light of these considerations, bursts of muscular activities while sitting should be encouraged instead of maintaining the same posture over long periods of time. Continued muscle tension and prolonged spinal compression (even at the low levels just mentioned) become uncomfortable and should

**Table 13–S.** Preferred Line of Sight and Viewing Area

| Direction of the Normal Line of Sight | Preferred | *Maximal Deviations from the Preferred Angle* | | |
|---|---|---|---|---|
| | | Only Eye Rotation | Only Head Rotation | Combined Eye and Head Rotation |
| Right or Left, each | 15° | 35° | 60° | 95° |
| Up | 15° | 40° | 65° | 90° |
| Down to distant target | 15° | 20° | 35° | 85° |
| to close target, (e.g., computer monitor) | 10–30° | 25° | 35° | 85° |

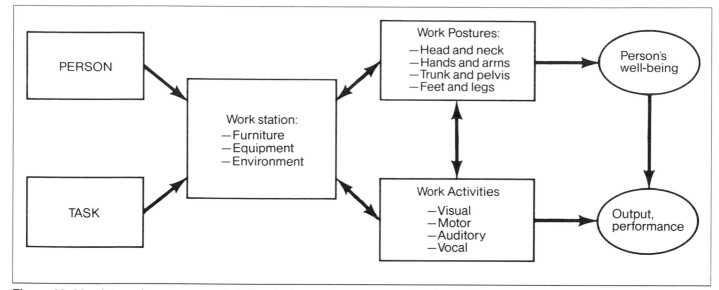

**Figure 13–23.** Interactions among workstation design, work postures, and work activities and their effect on the computer operator's well-being and performance.

be avoided by interspacing physical activities and exercises, followed by rest periods.

Tension and pain in the neck area are among the most frequently mentioned health complaints of computer operators. In contrast to the events in the lumbar region, EMG activities in the neck and shoulders are often considerably higher than the 10 percent level reported for lower trunk muscles and often must be maintained over considerable periods of time while the head is kept in a fixed position relative to the viewed object. Intensity, frequency, and the length of time that such muscle contractions are maintained can generate intense discomfort, pain, and related musculoskeletal health complaints that may persist over long periods of time.

Other analytical studies have addressed the pressure in the intervertebral disks, dependent on trunk posture. The most famous experiments are those performed in the 1970s in Scandinavia, during which pressure transducers were pushed into spinal disks. (A thorough compilation

and review was provided by Chaffin & Andersson, 1984). These experiments showed that the amount of intradisk force in the lumbar region depends on trunk posture and support. As Figures 13–25 and 13–26 show, the forces in the lumbar spine are in the neighborhood of 330 N when one is standing at ease. This force increases by about 100 N when one sits on a stool without a backrest where it makes little difference if one sits erect with the arms hanging or relaxed with the lower arms resting on the thighs. Sitting relaxed but letting the arms hang down increases the internal force to nearly 500 N. Thus, there is a significant increase in the spinal compression force in the lumbar region when one changes from a standing position to a sitting one, but the differences among several sitting positions are not very pronounced.

About the same force values are obtained when one sits on an office chair with a small lumbar support. Sitting with the arms hanging, writing with the arms resting on a table, and activating a pedal all result in forces of around 500 N. The spinal forces are increased by typing, when the forearms and hands must be lifted to keyboard height. A further increase is seen when a weight is lifted in the hands with the arms extended forward (Eklund et al., 1983). None of these postures makes use of the backrest.

**Table 13–T.** Linking the Human with the Computer in Different Work Tasks

| Task | Motor Input Requirements | Visual Requirements | Work Interruptions |
|---|---|---|---|
| Data entry | High (keyboard) | High (source & screen) | Few |
| Data acquisition | Medium (keyboard) | High (screen) | Varies |
| Word processing | High (keyboard) | High to medium (source & screen) | Few |
| Interactive communication | Medium | Medium (screen) | Varies |
| CAD | Low | High (screen & source) | Frequent |

(Adapted from National Research Council, 1983.)

**Table 13–U.** Links Between Human and Computer that Influence Work Posture

| Input to the Operator | Output from the Operator | Requirement of Locating the Operator Relative to the Computer |
|---|---|---|
| Eyes | — | High |
| — | Hands, feet | Medium to high |
| Ears | — | Low |
| — | Mouth | Low |

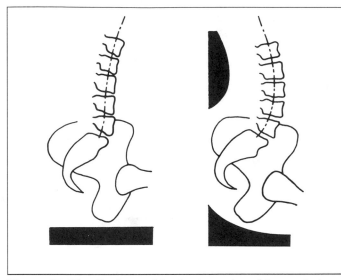

**Figure 13–24.** Postures of the lumbar spine of the seated operator (kyphosis, left, and lordosis, right). (Courtesy of the *American Industrial Hygiene Journal*.)

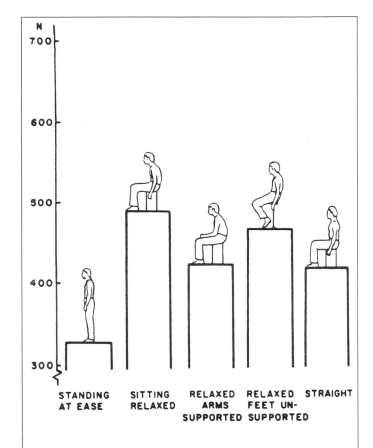

**Figure 13–25.** Forces in the third lumbar disc when standing or sitting on a stool without backrest. (Reprinted with permission from Chaffin & Andersson, 1984.)

However, if one leans back over the small backrest and lets the arms hang down, the internal compression forces are reduced to approximately 400 N, again close to the values measured when standing.

Figure 13–27 shows the effect of backrest use even more dramatically. When the backrest is upright, it cannot support the body and rather high disk forces may occur. When the straight backrest is declined behind the vertical position, the internal compression force is reduced because part of the upper body weight rests on the backrest and thus must not be transmitted through the spinal column. An even more pronounced effect can be brought about by making the backrest protrude toward the lumbar lordosis. A protrusion of 5 cm nearly cuts the disk compression force associated with a flat backrest in half; protrusions of 1–4 cm into the lumbar region produce intermediate effects.

These experimental results yield three important findings. The first is that sitting down from a standing position can increase disk pressure by one-third to one-half. The second is that there are no dramatic disk pressure differences among sitting straight, sitting relaxed, or sitting with supported arms if there is no backrest or only a small lumbar board. The third finding is that the use of a suitably designed backrest brings about disk pressures that are as low as those measured in a standing person. Certainly, these findings do not at all support the theory that sitting upright, as opposed to sitting relaxed or leaning back, reduces disk pressure.

If the backrest consists of only a small lumbar board, it is nearly worthless unless the person is draped by leaning backward over it. A large backrest is also nearly useless when it is upright but highly beneficial when it is reclined

backward so that it can support a large portion of the weight of the upper trunk, head, and arms. Its positive effects are dramatically enhanced if it is shaped to bring about the S-curve of the spinal column, particularly lumbar lordosis. Relaxed leaning against a reclined backrest is the least stressful sitting posture. This is a condition that is often freely chosen by people working in an office: "an impression which many observers have already perceived when visiting offices or workshops with VDT workstations: Most of the operators do not maintain an upright trunk posture. . . . In fact, the great majority of the operators lean backwards even if the chairs are not suitable for such a posture" (Grandjean et al., 1984, pp. 100–101.)

This observation indicates the problems associated with trying to find objective measures that reflect the complex, holistic, subjective, and variable feeling of comfort and well-being. To arrive at comprehensive criteria for judging the adequacy, acceptance, and comfort of design measures, subjective evaluations appear most sensitive (Life & Pheasant, 1984; Bhatnager et al., 1985; Drury & Francher, 1985; Kroemer et al., 1994).

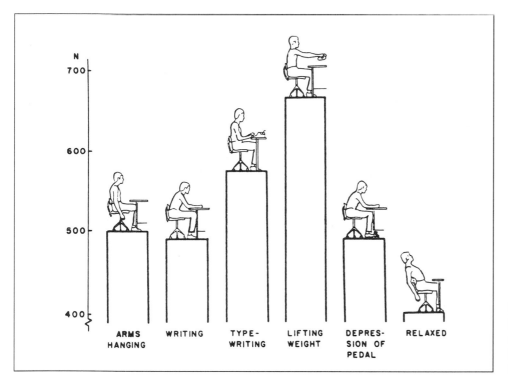

**Figure 13–26.** Forces in the third lumbar disc when sitting at a desk on an office chair with a small lumbar backrest. (Reprinted with permission from Chaffin & Andersson, 1984.)

## Ergonomic Design of Office Workstations

Several variables combine to determine the ergonomics of office workstations. These include psychological and attitudinal as well as organizational conditions. Another important variable is the physical environment, which includes climate, illumination, and general facility and work space design (ANSI, 1988; Grandjean, 1987, 1988; Kroemer et al., 1994; Smith et al., 1991).

Today's workstations consist of a display unit, a data entry unit, supports on which these rest, a chair, and the operator. Together with the task, these are the main system components that must be considered for ergonomic design recommendations.

The operator is the most important component in the system because she or he drives the output. The design of the workplace components should fit all operators and must facilitate variations in working postures that may be individually preferred. The fiction of one healthy upright posture, good for everybody, must be abolished.

Among the first steps in designing or selecting office furniture is establishment of the main clearance and external dimensions, which can be derived from body measurements. Table 13–E presents the best currently available body size information for U.S. female and male civilians. Body dimensions of the operator determine related dimensions of the workstation as follows:

- Eye height primarily determines the location (height and viewing distance from the eye) of the visual target, especially the monitor, and of the source document, notepad, and keyboard.
- Elbow height and forearm length are related to the location of such motor activities as keying and writing and to the operation of hand controls such as a keyboard, mouse, or trackball.
- Knee height and thigh thickness determine the needed clearance height of the leg room underneath tables or support surfaces.
- The forward protrusions of the knees (buttock–knee depth) and of the foot determine the needed clearance depth of the leg room under the equipment.
- Thigh breadth determines the minimal clearance width for the open leg room and for the seat.
- Buttock–popliteal depth determines the depth of the seat pan.
- Lower leg length (popliteal height) determines the height of the seat.

In some cases, the application of the anthropometric information to the design task is straightforward; for example, the seat pan depth must be shorter than the buttock–popliteal length to avoid uncomfortable pressure of the seat front on the soft tissues behind the knee. Other dimensions, such as functional reach, also have significant effects on workstation design dimensions, but their consideration may be more complex because of the variety of motions and tasks that are executed.

Three main strategies can be pursued to determine major dimensions of office furniture. The first is to make the seat height, the support heights for the equipment (primarily the keyboard and other working surfaces), and the support height for the display adjustable. The second strategy assumes that the support height must be fixed (as table heights in traditional offices usually are) but that seat height and display height are adjustable. The third strategy assumes that the seat height is fixed but that the support and display heights are adjustable.

In the commonly used first strategy for workstation design, seat height is determined from the popliteal height, taken from Table 13–E, with shoe heel height added. Thigh clearance height is then added to provide necessary clearance height underneath the support structure. Allowing some thickness for the support structure adds up to the total support surface height. The next step is to determine eye height above the seat pan, considering various possible postures. Then the center height of the display is determined for the preferred viewing distance and the preferred angle of sight (Kroemer, 1993).

Well-designed and carefully selected furniture allows the user to sit any way he or she likes, from bending forward to leaning back and holding the legs in any posture within available leg room. Using either conventional chairs or a seat of personal choice, some people prefer to semi-sit on a forward-declining surface; some may even want to stand at work, at least for a while.

Ergonomic seat design is of primary importance for reducing physiological and biomechanical stresses on the body when seated. Yet, the seat must be treated as just one of the system components, along with the workstation with display, input devices, and supports. The seat should provide a wide range of adjustments and must allow a range of postures to suit the individual, thus promoting well-being and performance. Given these far-ranging goals and wide personal variability, it is clearly impossible to recommend one particular seat design; in fact, various designs with varying features are needed for individual selection. Hence, only some basic features and dimensions can be mentioned here.

To fit most Americans (described in Table 13–E) (from the 5th to the 95th percentile), the height of the seat surface should be adjustable in the range of about 15 in. (38 cm) to 20 in. (50 cm). The seat surface should be approximately 16 in. (40 cm) to 20 in. (50 cm) deep and at least 18 in. (45 cm) wide. (These recommendations apply to a conventional seat whose pan is basically horizontal, or inclined or declined by only a few degrees. Less conventional designs, such as those used for semi-sitting, will probably deviate from these measurements.) The seat surface should be comfortably but firmly upholstered to distribute pressure and to allow various sitting postures. Particular attention should be paid to the front of the seat pan, which must not generate undue pressure to the area behind the knees.

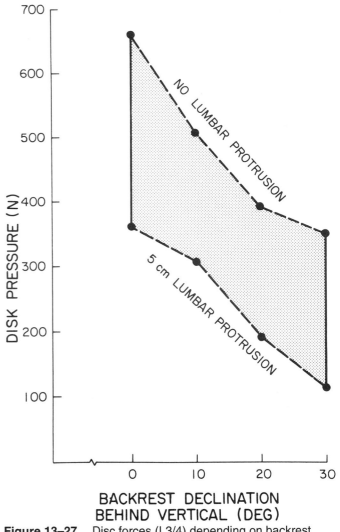

**Figure 13–27.** Disc forces (L3/4) depending on backrest angle and lumbar pad size. (Adapted from Chaffin & Andersson, 1984.)

The backrest should provide a large and well-formed surface to support the back and neck. At its lowest part, the backrest must provide room for the buttocks. Above, it should have a slight protrusion (of not more than 5 cm, preferably adjustable) to fit the lumbar concavity of the body. The height of this lumbar pad should be adjustable between 6 in. (15 cm) and 9 in. (23 cm) above the seat pan. Above the lumbar pad, the backrest can be nearly flat but must be able to be declined from a nearly upright position to 20 or possibly 30 degrees behind vertical. At the backrest's upper part, the surface should follow the concave form of the neck. This cervical pad should be adjustable to heights between 20 in. (50 cm) and 28 in. (70 cm) above the seat surface and also be adjustable in its protrusion to allow individual variation. The width of the backrest is not critical if it is at least 12 in. (30 cm). Figure 13–28 illustrates the major features of a suitable backrest.

Obviously, many seat dimensions must be adjustable to suit body form and the preferences of the individual. It

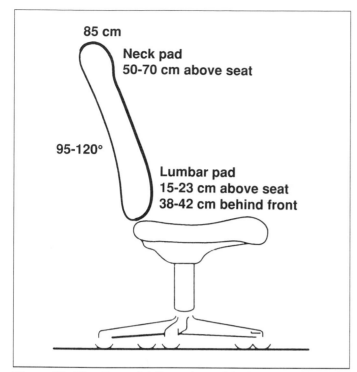

**Figure 13–28.** Essential dimensions of the backrest. (Adapted from Kroemer, Kroemer, Kroemer-Elbert, *Ergonomics: How to Design for Ease and Efficiency*. Englewood Cliffs, NJ: Prentice Hall, 1994. Used with permission by the publisher. All rights reserved.)

has been shown in the laboratory and through practical use that adjustment features will be used if they are easy to understand and operate; otherwise, they will be disregarded. Whether adjustment features should be coupled (for example, backrest tilt linked with seat-pan tilt) is a matter of preference and convenience. The same holds true for armrests, which may or may not be deemed desirable; however, the data shown in Figures 13–25 and 13–26 indicate that propping the arm on a support can reduce the compression load on the spinal column.

With regard to semi-sitting seats, diverse opinions have been voiced. Few scientific studies have been published on the subject; Drury & Francher (1985) found no general advantages of semi-sitting over conventional sitting. It appears that individual preferences in working postures vary widely.

If vision must be focused on the screen as well as on source document and on keyboard, all visual targets should be located close to each other: at the same distance from the eyes and in about the same direction of gaze. If the visual targets are spaced apart in direction or distance, the eye must be redirected and refocused continuously while sweeping from one target to another. This is particularly critical if reading glasses or other eye-correction lenses are worn because these are usually shaped for a special focusing distance and an assumed direction of sight. Often, the computer screen is arranged too high, forcing the operator (particularly when wearing reading glasses) to tilt his or her

neck severely backward. This position causes muscle tension and generates strain on the cervical part of the spinal column, which in turn often results in complaints about headaches and pains in the neck and shoulder region. Similar postural complaints are often voiced by people who have to hold their arms and hands in stressful positions when the keyboard or other tool is improperly placed. Proper ergonomic design, selection, adjustment, and use of the work equipment can prevent most postural complaints.

A major ergonomic concern is to provide the opportunity and means to change body posture frequently during the work period. Maintaining a particular posture, even if it is comfortable in the beginning, becomes stressful as time passes. Changes in posture are necessary and are best facilitated by brief periods of physical activity. To permit position changes for the hands, arms, and eyes, the input device (such as the keyboard) should be movable. Also, one should be able to adjust the display screen to various heights (and angles), which requires an easily adjustable possibly motor or spring-driven suspension system.

Another way to change working postures is to allow the computer operator, at his or her own choosing, to stand up for a period of time. A stand-up workstation should be adjustable so that the input device is approximately at elbow height when the worker is standing, that is, between 35 in. (90 cm) and 48 in. (120 cm). As in the sit-down workplace, the display unit should be located close to the other visual targets. In workplaces designed for standing operation, a footrest positioned at about half of knee height (approximately 12 in., or 25 cm) is often provided so that the operator can prop one foot up on it temporarily. This causes changes in pelvis rotation and spine curvature.

All components of the workstation must fit each other, and each must suit the operator. This requires easy adjustability. Figure 13–29 demonstrates various adjustment features that allow the user to match seat height with the height of the table or other support of the input devices—possibly while using a footrest—and to match eye position with the monitor display resting on its support.

These recommendations for the design and use of furniture assume flexibility in work organization and management attitudes. Ergonomic design of the computer workstation and its proper selection and use can further promote workers' well-being and performance.

## ■ Controls and Displays

Much research has been performed on controls and displays; in fact, in human factors engineering, the period after World War II is often called the knobs and dials era. Summaries of the findings have been published in Military Standards 759 and 1472, Van Cott & Kinkade (1972), Woodson (1981), Cushman & Rosenberg (1991), Woodson et al. (1991), Kroemer et al. (1994), and the Ergonomics chapter

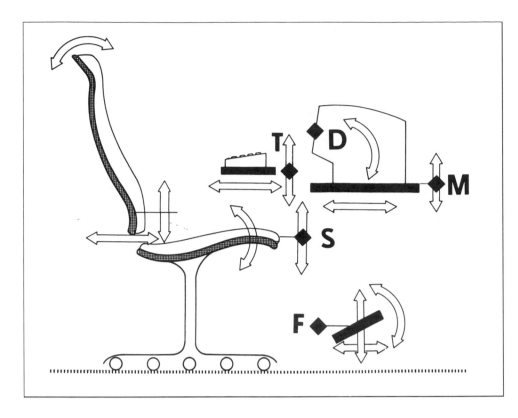

**Figure 13–29.** Adjustment features of a computer workstation. Key: S = seat height, T = table height, F = footrest height, D = monitor height, M = support height. (Reprinted with permission from Kroemer, Kroemer, & Kroemer-Elbert. *Ergonomics: How to Design for Ease and* Efficiency. Englewood Cliffs, NJ Prentice Hall, 1994.)

in the 1988 edition of this handbook. The revision of ANSI/HFS 100 expected in 1995 will provide up-to-date information on computer displays.

## Light Signals

A *red* signal light shall be used to alert an operator that the system or any portion of the system is inoperative or that a successful mission is not possible until appropriate corrective or override action is taken. Examples of indicators that should be red are those that display no-go, error, failure, and malfunction messages.

A *flashing red* signal light shall be used only to denote emergency conditions that require immediate operator action or to avert impending personnel injury, equipment damage, or both.

A *yellow* signal light shall be used to advise an operator that a marginal condition exists. Yellow shall also be used to alert the operator to situations for which caution, rechecking, or unexpected delay is necessary.

A *green* signal light shall be used to indicate that the monitored equipment is in satisfactory condition and that it is all right to proceed; green signifies "go ahead," "in tolerance," "ready," "function activated," "power on," and the like.

A *white* signal light shall be used to indicate system conditions that do not have right or wrong implications, such as alternative functions (such as "rear steering on") or transitory conditions (such as "fan on"), provided such

indication does not imply the success or failure of the operation.

A *blue* signal light may be used as an advisory signal, but common use of blue should be avoided.

Table 13–V lists recommended dimensions and suitable functions of indicator lights.

## Labels

Controls, displays, and any other items of equipment that must be located, identified, manipulated, or read shall be appropriately and clearly labeled to permit rapid and accurate performance. No label is required on equipment or controls whose use is obvious to the user.

Labeling characteristics are determined by such factors as the accuracy of identification required, the time available for recognition or other responses, the distance at which the labels must be read, the illumination level and color characteristics of the illuminant, the critical nature of the function labeled, and the consistency of label design within and between systems.

*Orientation.* Labels and the information printed on them should be oriented horizontally so that the labels can be read quickly and easily from left to right.

*Location.* Labels shall be placed on or very near the items they identify so as to eliminate confusion with other items and labels.

*Standardization.* Placement of labels shall be consistent throughout the equipment and system.

**Table 13–V.** Coding of Simple Indicator Lights

| Size/Type | Color | | | |
|---|---|---|---|---|
| | Red | Yellow | Green | White |
| 13 mm diameter or smaller/steady | Malfunction; action stopped; failure; stop action | Delay; check; recheck | Go ahead; in tolerance; acceptable; ready | Functional or physical position; action in progress |
| 25 mm diameter or larger/steady | Master summation (system or subsystem) | Extreme caution (impending danger) | Master summation (system or subsystem) | |
| 25 mm diameter or larger/flashing | Emergency condition (impending personnel or equipment disaster) | | | |

*Equipment functions.* Labels should primarily describe the functions of equipment items. The engineering characteristics or nomenclature may also be described.

*Abbreviations.* Standard abbreviations shall be used. If a new abbreviation is required, its meaning shall be obvious to the intended reader. Capital letters shall be used. Periods shall be omitted except when needed to preclude misinterpretation. The same abbreviation shall be used for all tenses and for both singular and plural forms of a word.

*Brevity.* Labels shall be as concise as possible without distorting the intended meaning or information and shall be unambiguous. Redundancy shall be minimized. If the general function is obvious, only the specific function shall be identified (for example, use *frequency* rather than *frequency factor*).

*Familiarity.* Words shall be chosen on the basis of operator familiarity whenever possible, provided the words express exactly what is intended. Brevity shall not be stressed if the results will be unfamiliar to operating personnel. Common, meaningful symbols (such as % and +) may be used as necessary.

*Visibility and legibility.* Labels and placards must be designed to be read easily and accurately at the anticipated operational reading distances, within the anticipated vibration/motion environment, and at minimally expected illumination levels. The following factors must be taken into consideration: contrast between the lettering and its immediate background; the height, width, stroke width, spacing, and style of letters; and the specular reflection of the background, cover, or other components.

# ■ Avoiding Cumulative Trauma Disorders

Single-event injuries, called acute or traumatic, differ from disorders that stem from repeated actions whose cumulative effects result in an injury. In the United States, OSHA is making their reduction a major goal of its investigation and enforcement program, which is of great concern to industry and the legal profession.

Cumulative strain injuries are the result of a series of *microtraumata,* each of which can "insult" the body, but not lead to a discernible damage. In their accumulation over time, however, the microstresses can cause discernible health complaints and result in clinically manifest disorders or injuries that reduce the ability to perform related work. Different terms are used to describe these conditions, such as *cumulative trauma disorder* (or *injury* or *syndrome*), *overuse disorder, repetitive motion injury, repetitive strain injury, occupational motion-related injury, regional musculoskeletal disorder, work-related disorder, osteoarthrosis,* and *rheumatic disease.* In this text, the term *cumulative trauma disorder* (CTD) will be used.

Putz-Anderson (1988) defined CTD as a "disorder of the muscular and/or tendinous and/or osseous and/or nervous system(s); caused, precipitated, or aggravated by repeated exertions or movements of the body."

## History

Nearly 300 years ago, Bernadino Ramazzini described health problems that we would call CTDs as appearing in workers who do violent and irregular motions and assume unnatural postures. Ramazzini also reported CTDs to occur among office clerks, believing that these events were caused by repetitive movements of the hands, constrained body postures, and excessive mental stress. Such activity-related disorders have been known for a long time as washer woman's sprain, gamekeeper's thumb, telegraphist's cramp, writer's cramp, trigger finger, tennis elbow, or golfer's elbow.

## CTDs in Industry

Early in the twentieth century, many CTDs were reported from various kinds of industrial and agricultural work. Peres wrote in his treatise "Process Work Without Strain" (1961, p. 6),

It has been fairly well established, by experimental research overseas and our own experience in local industry, that the continuous use of the same body movement and sets of muscles responsible for that movement during the normal working shift (notwithstanding the presence of rest breaks), can lead to the onset initially of fatigue, and ultimately of immediate or cumulative muscular strain in the local body area.

Peres noted that the muscles most often affected are those that control the movements of the fingers, of the hand, the forearm, and the upper arm, in that order. His special concerns were movements of the hand and fingers and injuries to tendons and tendon sheaths due to excessively repetitive motions of fingers. He said that fairly few cases were the result of a direct trauma, whereas many were related to repetitive use.

Thus, around 1960, it was known that cumulative muscle strain in industry was responsible for more injuries than has been generally thought. Peres said, "It is sometimes difficult to see why experienced people, after working satisfactorily for, say, 15 years at a given job, suddenly develop pains and strains. In some cases these are due to degenerative arthritic changes and/or traumatic injury of the bones of the wrist or other joints involved. In other cases, the cause seems to be compression of a nerve in the particular vicinity, as for example, compression of the median nerve in carpal tunnel syndrome. However, it may well be that many more are due to cumulative muscle strain arising from wrong methods of working" (p. 11).

## CTDs of Keyboard Users

In 1964, Kroemer reported,

> Steno-typists and other persons working with keyboards are often afflicted with disorders of tendons, tendon sheaths, and synovial tissues of tendons, and of the tendon and muscle attachments. . . . At this moment it is unknown what causes or aggravates these disorders. Possible sources may be, for example, the force needed to operate the keys, the displacement of the keys, or the frequency of operation. Hettinger (1957) attributes particular importance to the frequency. . . . Practical experiences give reason to assume that "electrical" typewriters are advantageous over "mechanical" ones. With electrical machines, the typing frequency is certainly not lower, but the key displacement and the operational force are smaller. . . . The body posture is indicted in several publications. Inappropriate posture and extensive muscle tension of the arms are mentioned. . . . , and the working position and direction of arms and hands are indicated. . . . The disadvantageous posture of arms and hands appears to be an important but generally little considered attribute of the work with typewriters and other keyboard machines.

## CTDs in Medical Terms

In 1883, Gray described inflammations of the extensor tendons of the thumb in their sheaths after excessive exercise (Armstrong, 1991); Hammer (1934) described diseases of the tendons in their sheaths. After Tanzer's classic report on carpal tunnel syndrome in 1959, Phalen discussed it in 1966 and 1972, as did Posch & Marcotte in 1976 and Birkbeck & Beer in 1975. Recent reviews were published by Armstrong (1991), Ayoub & Wittels (1989), Chatterjee (1987), Putz-Anderson (1988), and Rempel et al. (1992).

## Points of View

Although the occurrence of CTDs, their diagnoses, and their medical treatments were fairly well-known and established in the mid-twentieth century, their relationships to occupational activities have been hotly debated. Some believe that body usage within reason should not lead to repetitive injuries. Individuals may be predisposed to injury, particularly those with arthritis, diabetes, and endocrinological disorders. Such events as pregnancy, use of oral contraceptives, and gynecological surgery seem to be related to statistical occurrence. The fear of contracting CTDs or other psychosocial circumstances may lead to a sudden lowering of thresholds for discomfort that are normally acceptable, as in the so-called repetition strain injury (RSI) epidemic in Australia between 1983 and 1988. Some people who claim repetitive injuries are suspected to suffer from normal fatigue, to be malingerers, or to have compensation neurosis. However, the prevalent position taken in the current literature is that repetitive activities are causative, precipitating, or aggravating.

## Cumulative Injuries to the Body

Biomechanically, one can model the human body as consisting of a bony skeleton whose segments are connected in joints and are powered by muscles that bridge the joints. The muscle actions are controlled by the nervous and hormonal systems and maintained through a network of blood vessels.

Cumulative injuries occur often in connective soft tissues, particularly tendons and their sheaths. They may irritate or damage nerves and impede blood flow. They are common in the hand, wrist, and forearm area (as in carpal tunnel syndrome, discussed in more detail below), in the shoulder and neck, and in the back (often associated with lifting). Repetitive loadings may even damage bone, such as the vertebrae of the spinal column. A brief discussion of common cumulative injuries helps to understand not only their effects, but also their causes and prevention.

**Bones.** The skeletal system of the human body is made up of more than 200 bones, their articulations, and connective tissue. *Bones* provide the stable internal framework for the body. One distinguishes between flat axial bones (such as in the skull, sternum, ribs, and pelvis) and long cylindrical bones (such as in the arms and legs). The long bones act, in mechanical terms, as the lever arms at which muscles pull about the articulations that join the bones.

Although bone is firm and hard, and thus can resist high strain, it still has certain elastic properties. In childhood, when mineralization is low, bones are highly flexible. In contrast, bones of old people are highly mineralized and changed in geometry, and therefore more brittle; these effects are often connected with osteoporosis. Bones can be shattered or broken through sudden impact or torque, and they can be damaged through continual stresses such as vibration.

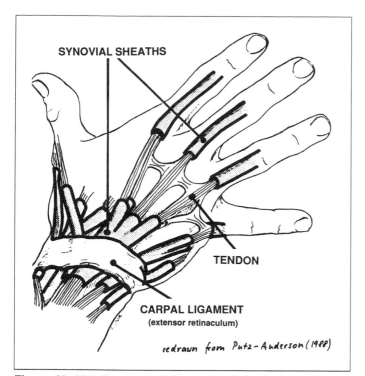

**Figure 13–30.** Tendons and their sheaths in the back of the hand. (Adapted from Putz-Anderson, *Cumulative Trauma Disorders: A Manual for Musculoskeletal Diseases of the Upper Limbs*. London: Taylor and Frances, 1988.)

**Connective tissues.** Bones are connected to each other and with other elements of the body through connective tissues. *Cartilage* is a translucent elastic material found at the ends of the ribs, in joint surfaces of the articulations, and as discs between the vertebrae of the spinal column. Other connective tissues that are less elastic are called *ligaments* when they connect bones and *tendons* when they connect muscle with bone. *Fascia* wraps organs or muscles. Fascial tissue condenses at the origin and insertion ends of muscle to tendons, which are usually encapsuled by a fibrous *sheath*. They allow a gliding motion of the tendon against surrounding materials that is facilitated by a viscous fluid, synovia, that reduces friction with the inner lining of the sheath. Particularly in the wrist and digits, sheaths keep tendons close to bones, acting as guides or pulleys for the pulling actions of the muscles. Figure 13–30 shows tendons and their sheaths in the back of the hand.

**Nervous control.** The peripheral nervous system transmits information about events outside and inside the body from various sensors along its feedback pathways to its central part, the spinal cord and brain. Here, decisions about appropriate reactions and actions are made. Accordingly, action signals are generated and sent along the feed-forward pathways to the muscles. Thus, proper functioning of the sensors, such as in the hands, and of the nervous pathways to and from the brain is essential for the flow of information and control signals in the nervous system.

**Blood network.** A network of arterial blood vessels provides oxygen and nutrients contained in working muscles and other organs. While passing through a muscle, metabolic waste products (such as lactic acid, carbon dioxide, heat, and water) are removed into the venous network. In addition, arterial blood transports hormones and other products of the internal glands that, parallel to the nervous system, regulate actions of body organs. Thus, blood flow is essential to working muscles for their proper functioning.

## Which Body Components Are at CTD Risk?

Although bones (except vertebrae) are not usually injured in the context of CTD, joints, muscles, and tendons and their related structures are at risk, as well as nerves and blood vessels.

**Soft Tissues.** A *strain* is an injury to muscle or tendon. *Muscles* can be stretched, which is associated with aching and swelling. A group of fibers torn apart is a more serious injury. If blood or nerve supply is interrupted for an extended time, the muscle atrophies.

*Tendons* contain collagen fibers that neither stretch nor contract. Tendon surfaces can become rough, impeding their motion along other tissues. If overly strained, tendon fibers can be torn and scar tissue may form that creates chronic tension and is easily reinjured.

Gliding movement of a tendon in its sheath, caused by muscle contraction and relaxation, can be quite large (5 cm in the hand when a finger is moved from fully extended to completely flexed). Synovial fluid in the tendon sheath, which acts as a lubricant, may be diminished, causing friction between tendon and its sheath. First signs of inflammation are feelings of tenderness and warmth, followed by discomfort and pain.

*Inflammation* of a tendon sheath is a protective response of the body. The feeling of warmth and swelling stems from the influx of blood. Compression of tissue from either swelling or mechanical pressure produces pain. Movement of the tendon within its compressive surroundings is hindered. Forced movement, particularly when often repeated, may cause the inflammation of additional fiber tissue that, in turn, can establish a permanent (chronic) condition of a swollen tendon or sheath that impedes tendon movement.

A *bursa* is a small, flat, synovia-filled sac lined with a slippery cushion that prevents rubbing of a muscle or tendon against bone. An often used muscle or tendon, particularly if it has become roughened, may irritate its adjacent bursa, setting up an inflammatory reaction that inhibits free movement.

When a joint is displaced beyond its regular range, fibers of a *ligament* can be stretched, torn apart, or pulled from the bone. This is called a sprain, often resulting from a single trauma but also possibly caused by repetitive actions. Injured ligaments can take weeks or even months to heal because their blood supply is poor. A ligament sprain can bring about a lasting joint instability and hence increases the risk of further injury.

**Nerves.** *Nerves* can also be affected by repeated or sustained pressure. Such pressure may come from bones, ligaments, tendons, tendon sheaths, and muscles in the body, or from hard surfaces and sharp edges of workplaces, tools, and equipment. Pressure within the body can occur if the position of a body segment reduces the passage opening through which a nerve runs. Another source of compression may be swelling due to inflammation of other structures in this opening, often of tendons and tendon sheaths. Carpal tunnel syndrome (discussed below) is a typical case of nerve compression.

There are three different systems of nerve fibers, which serve different functions: motor, sensory, and autonomic. Impairment of a *motor nerve* reduces the ability to transmit signals to the innervated motor units in muscle. Thus, motor nerve impairment impedes the controlled activity of muscles, and hence reduces the ability to generate force or torque to tools, equipment, and other external objects.

*Sensory nerve* impairment reduces the information that can be transmitted from sensors to the central nervous system. Sensory feedback is very important for hand activities because it contains information about force applied, position assumed, and motion experienced. Sensory nerve impairment usually brings about sensations of numbness, tingling, or even pain. The ability to distinguish hot from cold may be reduced.

Impairment of an *autonomic nerve* reduces the ability to control such functions as temperature by sweat production in the skin. A common sign of autonomic nerve impairment is dryness and shininess of skin areas controlled by that nerve.

**Blood vessels.** Like nerves, *blood vessels* may be compressed. Compressing an artery results in reduced blood flow to the affected area; compression of a vein hinders the return of blood. Blood vessel compression thus means reduced supply of oxygen and nutrients to muscles, and impaired temperature control of tissues near tendons and ligaments. Such ischemia limits the possible duration of muscular actions and impairs muscle recovery from fatigue after activity. Vascular compression, together with pressure on nerves, is often found in the neck, shoulder, and upper arm. Names such as *thoracic outlet* (or *hyperabduction*) *syndrome, cervicobrachial disorder, brachial plexus neuritis,* and *costoclavicular syndrome* describe the location of the condition.

*Vibrations* of body members, particularly of the hand, can trigger vasospasms, which reduce the diameter of arteries down to their complete closure. Of course, this impedes blood flow to the body areas supplied by the vessels, which becomes visible by a blanching of the area known as white finger or Raynaud's phenomenon. Exposure to cold can aggravate the problem because it also can trigger vasospasms. Associated symptoms include intermittent or continued numbness and tingling in the fingers, with the skin turning pale and cold, and eventually loss of sensation and control in the fingers. The symptoms are often caused by vibrating tools such as pneumatic hammers, chainsaws, power grinders, and power polishers. Frequent operation of keyboards may also constitute a source of vibration strain to the hand and wrist area.

## Causes of CTDs

*Cumulative trauma disorder* (CTD) is used here as a collective term for syndromes with or without physical manifestations. It is commonly characterized by discomfort, persistent pain, impairment, or disability in joints, muscles, tendons, and other soft tissues. It is often caused, precipitated, or aggravated by repetitive or forceful motions. It can occur in diverse occupational activities such as assembly, manufacturing, meat processing, agricultural work, packing, sewing, and keying. Some individuals may be predisposed. Also, CTDs are quite often associated with leisure and sports activities; "tennis elbow" is one of the better-known examples.

## Carpal Tunnel Syndrome

Among the best known cumulative trauma injuries is carpal tunnel syndrome (CTS), first described 125 years ago (Armstrong, 1991). In 1959, Tanzer discussed 22 cases. Two of his patients had been working in large kitchens with much stirring and ladled soup twice daily for about 600 students. Two patients had recently started to milk cows on a dairy farm, two had done gardening with considerable hand weeding, one had been using a spray gun with a finger trigger, and three worked in a shop in which objects were handled on a conveyor belt.

In 1966 and 1972, Phalen published reviews of 1,252 cases of CTS. These publications have become classics in the field; one of the most often-used tests for signs of CTS is called Phalen's test. In 1975, Birkbeck & Beer described the results of their survey of work and hobby activities of 658 patients who suffered from CTS. Seventy-nine percent of these patients were employed in work requiring light, highly repetitive movements of the wrists and fingers. The authors indicated that this type of manual activity can be a causal factor in the development of CTS. They referred to a report published in 1947 by Brain et al. that described six cases of compression of the median nerve in the carpal tunnel. According to Birkbeck & Beer, even then it had been concluded that occupation is a causal factor. In 1976, Posch & Marcotte analyzed another 1,201 cases of CTS.

In 1964, Kroemer described the occurrence of syndromes in typists. In 1980, Huenting et al. stated that impairments in hands and arms were commonly found in operators of accounting machines and that the disorders were related to working posture and operation of the keys. In 1981, Cannon et al. described a case-control study on the factors associated with the onset of CTS. They linked the occurrence of CTS to diverse etiological factors that included injury and illness, use of drugs, and hormonal changes in women. However, they stated that ergonomic theory relates the occurrence of CTS to repetitive movements of the wrists, performance of tasks with the wrists in ulnar deviation, and chronic exposure to low-frequency vibrations. To support their findings, they listed twelve publications from the 1970s.

In 1983, the American Industrial Hygiene Association acknowledged the prevalence and importance of CTS by publishing Armstrong's *An Ergonomics Guide to Carpal Tunnel Syndrome*. Armstrong describes CTS as an occupational illness of the hand and arm system among other repetitive trauma disorders such as tendinitis, synovitis, tenosynovitis, bursitis, trigger finger, and epicondylitis.

Thus, in the 1960s, 1970s, and early 1980s, CTS was recognized as an often occurring, disabling condition of the hand that can be caused, precipitated, or aggravated by certain work activities in the office and on the shop floor. Of course, leisure activities may be involved as well.

Silverstein (1985) and Silverstein et al. (1987) listed conditions associated with CTS and other repetitive strain injuries, with further details supplied by Putz-Anderson (1988) and Kroemer (1989, 1992). In 1990, Coe & Fisher stated that of 656 cases seen in their clinical practice in Chicago, repetitive motion at work was related to CTS in 71 percent of their 334 female patients and 49 percent of their 332 male patients. Also in 1990, Draganova reported from Bulgaria that 69 of 89 VDU data entry operators mentioned musculoskeletal complaints. However, Konz and Mital (1990) stated that the majority of CTS cases are not work-related.

On the palmar side of the wrist, near the base of the thumb, the carpal bones form a concave "floor" and the "walls" of the carpal canal. It has a "roof" that consists of three ligaments (the radial carpal, intercarpal, and carpometacarpal) covered by the transverse carpal ligament, which is firmly fused to the carpal bones. Thus, these bones and ligaments form a covered canal or tunnel, called the carpal tunnel. It is shown in Figure 13–31. In cross-section, it is roughly oval in shape. Through it pass the flexor tendons of the digits, as well as the median nerve. The radial nerve and artery, as well as the ulnar nerve and artery, are also confined in similar fashion. This crowded space is reduced if the wrist is bent up or down (flexed or extended) or pivoted to either side (ulnarly or radially.) Any swelling of the tendons and of their sheaths reduces the space available for the tendons, blood vessels, and nerves.

The median nerve innervates the thumb, much of the palm, and the index and middle fingers, as well as most of the ring finger, but not its ulnar side. Pressure on the nerve reduces its ability to transmit signals (feedforward to control proper functioning and feedback to report on existing conditions to the brain).

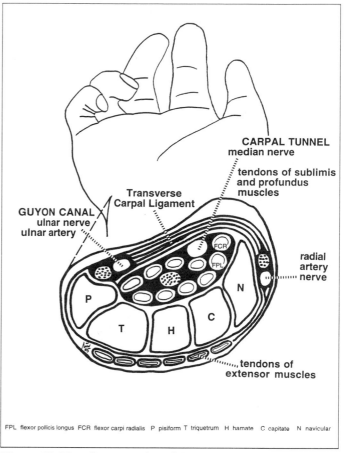

FPL flexor pollicis longus  FCR flexor carpi radialis  P pisiform  T triquetrum  H hamate  C capitate  N navicular

**Figure 13–31.** Cross-section of the carpal tunnel. The carpal bones (P, T, H, C, N) form the carpal "canal," which ligaments cover. Also shown are the tendons of the superficial and profound finger flexor muscles, flexors of the thumb (FCR, FPL), nerves, and arteries. (Adapted from Kroemer, 1989.)

## Occupational Activities and Related Disorders

Table 13–W lists conditions that are often associated with cumulative trauma (Kroemer, 1992). Of course, this list is neither complete nor exclusive. New occupational activities occur and several activities may be part of the same job.

Repetitive and forceful exertions, particularly if combined, are generally thought to be responsible for a large portion of the CTD. Silverstein (1985) proposed that *high repetitiveness* may be defined as a cycle time of less than 30 s, or as more than 50 percent of the cycle time spent performing the same fundamental motion. Silverstein also suggested that *high force* by itself (for hand force, more than 45 N) may be a causative factor.

If muscles must remain contracted at more than about 15–20 percent of their maximal capability, circulation is impaired. This can result in tissue ischemia and delayed dissipation of metabolites, which constitute conditions of general physiological strain. *Posture* may be highly important. For example, dorsiflexion of the wrist ("dropped wrist") generates a condition likely to cause CTS. Maintained isometric

**Table 13–W.** Common Cumulative Trauma Disorders

| Disorder Name | Description | Typical Job Activities |
|---|---|---|
| Carpal tunnel syndrome (writer's cramp, neuritis, median neuritis) (N) | The result of compression of the median nerve in the carpal tunnel of the wrist. This tunnel is an opening under the carpal ligament on the palmar side of the carpal bones. Through this tunnel pass the median nerve, the finger flexor tendons, and blood vessels. Swelling of the tendon sheaths reduces the size of the opening of the tunnel and pinches the median nerve and possibly blood vessels. The tunnel opening is also reduced if the wrist is flexed or extended or ulnarly or radially pivoted. | Buffing, grinding, polishing, sanding, assembly work, typing, keying, cashiering, playing musical instruments, surgery, packing, housekeeping, cooking, butchering, hand washing, scrubbing, hammering. |
| Cubital tunnel syndrome (N) | Compression of the ulnar nerve below the notch of the elbow. Tingling, numbness, or pain radiating into ring or little fingers. | Resting forearm near elbow on a hard surface or sharp edge or reaching over obstruction. |
| DeQuervain's syndrome (or disease) (T) | A special case of tendosynovitis that occurs in the abductor and extensor tendons of the thumb where they share a common sheath. This condition often results from combined forceful gripping and hand twisting, as in wringing cloths. | Buffing, grinding, polishing, sanding, pushing, pressing, sawing, cutting, surgery, butchering, use of pliers, 'turning' control such as on motorcycle, inserting screws in holes, forceful hand wringing. |
| Epicondylitis ("tennis elbow") (T) | Tendons attaching to the epicondyle (the lateral protrusion at the distal end of the humerus bone) become irritated. This condition is often the result of impacting or jerky throwing motions, repeated supination and pronation of the forearm, and forceful wrist extension movements. The condition is well-known among tennis players, pitchers, bowlers, and people hammering. A similar irritation of the tendon attachments on the inside of the elbow is called medical epicondylitis, also known as "golfer's elbow." | Turning screws, small parts assembly, hammering, meat cutting, playing musical instruments, playing tennis, pitching, bowling. |
| Ganglion (T) | A tendon sheath swelling that is filled with synovial fluid, or a cystic tumor at the tendon sheath or a joint membrane. The affected area swells up and causes a bump under the skin, often on the dorsal or radial side of the wrist. (Because it was in the past occasionally smashed by striking with a bible or heavy book, it was also called a "bible bump.") | Buffing, grinding, polishing, sanding, pushing, pressing, sawing, cutting, surgery, butchering, use of pliers, 'turning' control such as on motorcycle, inserting screws in holes, forceful hand wringing. |
| Neck tension syndrome (M) | An irritation of the levator scapulae and trapezius group of muscles of the neck, commonly occurring after repeated or sustained overhead work. | Belt conveyor assembly, typing, keying, small parts assembly, packing, load carrying in hand or on shoulder. |
| Pronator (teres) syndrome (N) | Result of compression of the median nerve in the distal third of the forearm, often where it passes through the two heads of the pronator teres muscle in the forearm; common with strenuous flexion of elbow and wrist. | Soldering, buffing, grinding, polishing, sanding. |
| Shoulder tendinitis (rotator cuff syndrome or tendinitis, supraspinatus tendinitis, subacromial bursitis, subdeltoid bursitis, partial tear of the rotator cuff) (T) | This is a shoulder disorder located at the rotator cuff. The cuff consists of four tendons that fuse over the shoulder joint, where they pronate and supinate the arm and help to abduct it. The rotator cuff tendons must pass through a small bony passage between the humerus and the acromion, with a bursa as cushion. Irritation and swelling of the tendon or of the bursa are often caused by continuous muscle and tendon effort to keep the arm elevated. | Punch press operations, overhead assembly, overhead welding, overhead painting, overhead auto repair, belt conveyor assembly work, packing, storing, construction work, postal letter carrying, reaching, lifting, carrying load on shoulder. |
| Tendinitis (tendinitis) (T) | An inflammation of a tendon. Often associated with repeated tension, motion, bending, being in contact with a hard surface, vibration. The tendon becomes thickened, bumpy, and irregular in its surface. Tendon fibers may be frayed or torn apart. In tendons without sheaths, such as within elbow and shoulder, the injured area may calcify. | Punch press operation, assembly work, wiring, packaging, core making, use of pliers. |

*(Continues)*

**Table 13–W.**   (*Continued*)

| Disorder Name | Description | Typical Job Activities |
|---|---|---|
| Tendosynovitis (tenosynovitis, tendovaginitis) (T) | This disorder occurs to tendons inside synovial sheaths. The sheath swells. Consequently, movement of the tendon with the sheath is impeded and painful. The tendon surfaces can become irritated, rough, and bumpy. If the inflamed sheath presses progressively onto the tendon, the condition is called stenosing tendosynovitis. DeQuervain's syndrome is a special case occurring in the thumb, while the trigger finger condition occurs in flexors of the fingers. | Buffing, grinding, polishing, sanding, punch press operation, sawing, cutting, surgery, butchering, use of pliers, 'turning' control such as on motorcycle, inserting screws in holes, forceful hand wringing. |
| Thoracic outlet syndrome (neurovascular compression syndrome, cervicobrachial disorder, brachial plexus neuritis, costoclavicular syndrome, hyperabduction syndrome) (V, N) | A disorder resulting from compression of nerves and blood vessels between clavicle and first and second ribs at the brachial plexus. If this neurovascular bundle is compressed by the pectoralis minor muscle, blood flow to and from the arm is reduced. This ischemic condition makes the arm numb and limits muscular activities. | Buffing, grinding, polishing, sanding, overhead assembly, overhead welding, overhead painting, overhead auto repair, typing, keying, cashiering, wiring, playing musical instruments, surgery, truck driving, stacking, material handling, postal letter carrying, carrying heavy loads with extended arms. |
| Trigger finger (or thumb) (T) | A special case of tendosynovitis where the tendon becomes nearly locked so that its forced movement is not smooth but snaps or jerks. This is a special case of stenosing tendosynovitis crepitans, a condition usually found with digit flexors at the A1 ligament. | Operating finger trigger, using hand tools that have sharp edges pressing into the tissue or whose handles are too far apart for the user's hand so that the end segments of the fingers are flexed while the middle segments are straight. |
| Ulnar artery aneurysm (V, N) | Weakening of a section of the wall of the ulnar artery as it passes through the Guyon tunnel in the wrist; often from pounding or pushing with heel of the hand. The resulting "bubble" presses on the ulnar nerve in the Guyon tunnel. | Assembly work. |
| Ulnar nerve entrapment (Guyon tunnel syndrome) (N) | Results from the entrapment of the ulnar nerve as it passes through the Guyon tunnel in the wrist. It can occur from prolonged flexion and extension of the wrist and repeated pressure on the hypothenar eminence of the palm. | Playing musical instruments, carpentering, brick laying, use of pliers, soldering, hammering. |
| White finger ("dead finger," Raynaud's syndrome, vibration syndrome) (V) | Stems from insufficient blood supply bringing about noticeable blanching. Finger turns cold, numb, tingles, and sensation and control of finger movement may be lost. The condition is due to closure of the digit's arteries caused by vasospasms triggered by vibrations. A common cause is continued forceful gripping of vibrating tools, particularly in a cold environment. | Chain sawing, jack hammering, use of vibrating tool, sanding, paint scraping, using vibrating tool too small for the hand, often in a cold environment. |

N = nerve disorder; T = tendon disorder; M = muscle disorder; V = vessel disorder. (Adapted from Kroemer, 1992. Reprinted with permission of the American Industrial Hygiene Association.)

contraction of muscles needed to keep a body part in an extreme position is often associated with a CTD condition. Inward or outward rotation of the forearm, especially with a bent wrist, any severe deviation of the wrist from its neutral position, and the pinch grip can be stressful.

## Ergonomic Countermeasures

Exercise as a prophylactic measure is of questionable value. Proponents hope to strengthen tissues, but inappropriate exercise may worsen conditions. Lee et al. (1992) evaluated the large number of proposed exercises and selected those that have been shown to be useful.

Ergonomic intervention depends on the stage. Early symptoms of CTD are reversible through work modification, rest breaks, and possibly exercises. In the later stages, the patient must rest and abstain from the work that caused the condition. Major changes in life-style and working capacity may result. Further medical treatments are physiotherapy, drug administration, and in some cases surgery, such as carpal tunnel release.

Of course, one should avoid the conditions that may lead to CTD. This is best done in the planning stage of a new job. For existing jobs, it is important to identify problems and symptoms early, to prevent injury through ergonomic work reorganization and work redesign.

At work, various ergonomic interventions can be taken. They start with suitable design of the work object and the equipment and tools used. They include training in proper habits. Managerial interventions include work diversification (the opposite of job simplification and specialization, which lead to repetition), relief workers, and rest pauses.

Suspicious jobs should be analyzed for their movement and force requirements, using variants of the well-established industrial engineering procedure of motion and time study. Each element of the work should be screened for factors that can contribute to CTD (Hahn et al., 1991; Jegerlehner, 1991). After the job analysis has been completed, workstations, equipment, and work procedures can be ergonomically reengineered and reorganized to reduce the stress on the operator's body. Table 13-X provides an overview of generic ergonomic countermeasures.

## Research Challenges

Our current knowledge about the relationships between activities (work or leisure) and CTDs is mostly limited to exertions of fairly large forces associated with high-exertion frequencies and unbecoming body postures. However, even for those gross muscular activities some uncertainty exists about the causal relationships between them and CTDs: The exact job factors are not well-defined, nor are the critical threshold values established. For example, the *forcefulness* of a job exertion may be measured statically (isometrically) or in dynamic terms. Silverstein's definitions apparently apply to static exertions, but most industrial activities are dynamic. Exertion of energy in either static or in dynamic conditions strains muscles in different ways; some CTDs are explicitly related to motion, as indicated by the term *repetitive motion injury*. However, our current knowledge base seems to be largely dependent on the assumption of isometric muscle efforts, that is, on a static condition.

Another unsolved quantitative problem, both in concept and in application, is the *repetitiveness* of activities thought to be related to CTDs. Repetitiveness or frequency of activities presumes, implicitly or explicitly, that the activities occur at regular intervals so that their occurrence can be expressed appropriately by an average number of exertions per time unit. However, the activity may not be regular, particularly if it occurs over many hours or a full work shift. Considerable variations in the frequency may occur during that time. It is unknown how an uneven distribution of activities over the working time may be related to the occurrence of CTDs.

Obviously, the *duration* of an activity is related to the strain it places on the active person. Rest pauses (such as 10 min every hour or 15 min every 2 hours) are thought to be beneficial for people doing keying tasks. Accordingly, an uneven distribution of keying movements (generated by interspaced rest pauses) might counter the occurrence of CTD. However, is a burst of 10,000 keystrokes done over a period of 3 hours followed by 5 hours of not keying less or more conducive to CTD than an even distribution of the same 10,000 keystrokes over 8 hours? Obviously, the concept of an average daily activity frequency is questionable. The durations of activity periods and their sequences must be considered. But how?

The case of keyboarding brings up another research challenge. Compared to many shop activities (which so far have been mostly linked with CTDs), the operation of keys (on typewriters, word processors, adding machines, telephone key sets) requires smaller energies per keystroke to be exerted by the fingers of the operator, but the number of such activations per time unit can be very high. This brings up the unresearched problem of *interrelations* between force, displacement, repetition, and posture with respect to CTDs. It is very likely that these interrelationships are rather complex, and that they include factors beyond (static) force, displacement, frequency, and posture. Marras & Schoenmarklin (1991) have developed instruments to measure such relationships. Their preliminary findings support that indeed certain dynamic wrist motion parameters (specifically, velocity and acceleration) appear to be closely related to the development of repetitive trauma syndromes.

## Safe Thresholds and Doses

What are the effects of activities (expressed in physical terms such as force, displacement, direction, repetition, body movement, and posture) on body tissues? For instance, why does rapid and often repeated depressing of the fingertip (to press a key) cause strain in the carpal tunnel? What kind of strain is it: the rubbing of tendons against their sheaths, the magnitude of frictional force, the pressure exchanged between the surfaces of the tendon and its sheath, the tension within the tendon, or the heat energy generated? How do theses stresses relate to the posture of the wrist, for example, and how is undue pressure generated within the carpal tunnel? Armstrong et al. (1991) have developed an experimental procedure to measure the pressure in the carpal tunnel.

The American National Standards Institute (ANSI) has accredited a standards committee, ASC Z365, on control of cumulative trauma disorders. The committee is trying to determine what is known about CTDs and what standard procedures could be established to avoid them. The deliberations of the ASC Z365 in the early 1990s indicated that much research must be conducted to single out and describe the components of activities that may lead to CTDs and to understand how physical events overload body structures and tissues, and how, if at all, psychosocial factors contribute. When these relationships are understood, it should be possible to establish threshold values or exposure doses for activity job factors such as force, displacement, repetition, duration, and posture, which separate suitable from unacceptable conditions. Such thresholds or doses could replace generic guidelines by specific ergonomic recommendations.

## ■ Summary

The industrial hygienist is largely responsible for the health and well-being of employees. Of course, this concern must be seen in light of the organization's productivity goals. Fortunately, ergonomic and human factors recommendations

**Table 13–X.**  Ergonomic Measures to Avoid Cumulative Trauma Disorders

| CTD | Avoid in General | Avoid in Particular | Do | Design |
|---|---|---|---|---|
| Carpal tunnel syndrome | Rapid, often-repeated finger movements, wrist deviation | Dorsal and palmar flexion, pinch grip, vibrations between 10 and 60 Hz | | |
| Cubital tunnel syndrome | Resting forearm on sharp edge or hard surface | | | |
| DeQuervain's syndrome | Combined forceful gripping and hard twisting | | | |
| Epicondylitis | "Bad tennis backhand" | Dorsiflexion, pronation | *Do* | *Design* |
| Pronator syndrome | Forearm pronation | Rapid and forceful pronation, strong elbow and wrist flexion | Use large muscles, but infrequently and for short durations | The work object properly |
| Shoulder tendinitis, rotator cuff syndrome | Arm elevation | Arm abduction, elbow elevation | Let wrists be in line with the forearm | The job task properly |
| Tendinitis | Often-repeated movements, particularly with force exertion; hard surface in contact with skin; vibrations | Frequent motions of digits, wrists, forearm, shoulder | Let shoulder and upper arm be relaxed | Hand tools properly ("bend tool, not the wrist") |
| Tendosynovitis, DeQuervain's syndrome, ganglion | Finger flexion, wrist deviation | Ulnar deviation, dorsal and palmar flexion, radial deviation with firm grip | Let forearms be horizontal or more declined | Round corners, pad |
| Thoracic outlet syndrome | Arm elevation, carrying | Shoulder flexion, arm hyperextension | | Place work object properly |
| Trigger finger or thumb | Digit flexion | Flexion of distal phalanx alone | | |
| Ulnar artery aneurism | Pounding and pushing with the heel of the hand | | | |
| Ulnar nerve entrapment | Wrist flexion and extension | Wrist flexion and extension, pressure on hypothenar eminence | | |
| White finger, vibration syndrome | Vibrations, tight grip, cold | Vibrations between 40 and 125 Hz | | |
| Neck tension syndrome | Static head posture | Prolonged static head/neck posture | Alternate head/neck postures | |

(Adapted from Kroemer, 1992. Reprinted with permission of the American Industrial Hygiene Association.)

usually bring about, directly or indirectly, improved job performance together with increased safety, health, and well-being. In recent years, both management and employee representatives, including unions, have cooperated in using ergonomics to increase ease and efficiency at work.

# ■ Bibliography

American National Standards Institute, 11 West 42nd Street, New York, NY 10036. *Human Factors Engineering of Visual Display Terminal Workstations*. ANSI/HFS 100-1988.

Armstrong TJ. Work related to cumulative trauma disorders. In *Proceedings, Occupational Ergonomics*. San Diego, CA: American Industrial Hygiene Association, San Diego Local Section, 1991.

Armstrong TJ *An Ergonomics Guide to Carpal Tunnel Syndrome*. Fairfax, VA: American Industrial Hygiene Association, 1983.

Armstrong TJ, Werner RA, Waring WP, et al. Intra-carpal canal pressure in selected hand tasks. In *Proceedings of the 11th Congress of the International Ergonomics Association*. London: Taylor and Francis, 1991.

Astrand PO, Randahl L. *Textbook of Work Physiology*, 3rd ed. New York: McGraw Hill, 1986.

Ayoub MA, Wittels NE. Cumulative trauma disorders. *Int Rev Ergonomics* 2:217–272, 1989.

Bhatnager V, Drury CG, Schiro SG. Posture, postural discomfort, and performance. *Hum Factors* 27:189–199, 1985.

Birkbeck MW, Beer TC. Occupation in relation to the carpal tunnel syndrome. *Rheumatol Rehab* 14:218–221, 1975.

Boff KR, Lincoln JE, eds. *Engineering Data Compendium: Human Perception and Performance*. Wright-Patterson AFB, OH: Armstrong Aerospace Medical Research Laboratory, 1988.

Borg GAV. Psychophysical bases of perceived exertion. *Med Sci Sports Exerc* 14:377–301, 1982.

Brain W, Wright A, Wilkinson M. Spontaneous compression of the median nerve in the carpal tunnel. *Lancet* 1:277–282, 1947.

Caldwell LS, Chaffin DB, Dukes-Dobos FN, et al. A proposed standard procedure for static muscle strength testing. *AIHAJ* 35:201–206, 1974.

Cannon LJ, Bernacki EJ, Walter SD. Personnel and occupational factors associated with carpal tunnel syndrome. *J Occup Med* 23(4):225–258, 1981.

Carayon P. Job design and job stress in office workers. *Ergonomics* 36:463–477, 1993.

Chaffin DB, Andersson GBJ. *Occupational Biomechanics,* 2nd ed. New York: Wiley, 1991.

Chaffin DB, Andersson GBJ. *Occupational Biomechanics.* New York: Wiley, 1984.

Chatterjee DS. Repetition strain injury—A recent review. *J Soc Occup Med* 37:100–105, 1987.

Ciriello VM, Snook SH. A study of size, distance, height, and frequency effects on manual handling tasks. *Hum Factors* 25:473–483, 1983.

Cochran DJ, Riley MW. The effects of handle shape and size on exerted forces. *Hum Factors* 28:253–265, 1986.

Coe JE, Fisher L. Occupational carpal tunnel syndrome: Clinical characteristics of an affected population and utility of screening techniques. In *Abstracts, International Conference on Occupational Musculo-Skeletal Disorders and Prevention of Low Back Pain.* Milan, Italy: University of Milan, Institute of Occupational Medicine, 1990.

Cushman WH, Rosenberg DJ. *Human Factors in Product Design.* Amsterdam: Elsevier, 1991.

Draganova N. Musculo-skeletal complaints in some professions. In *Abstracts, International Conference on Occupational Musculo-skeletal Disorders and Prevention of Low Back Pain.* Milan, Italy: University of Occupational Medicine, 1990.

Drury CG. Handles for manual material handling. *Appl Ergonomics* 11:35–42, 1980.

Drury CG, Francher M. Evaluation of a forward sloping chair. *Appl Ergonomics* 16:41–47, 1985.

Eastman Kodak Company. *Ergonomic Design for People at Work,* vol. 2. New York: Van Nostrand Reinhold, 1986.

Eastman Kodak Company. *Ergonomic Design for People at Work,* vol. 1. Belmont, CA: Lifetime Learning Publications, 1983.

Eklund JAE, Corlett EN, Johnsson FA. Method for measuring the load imposed on the back of a sitting person. *Ergonomics* 26:1063–1076, 1983.

Enos L, Mitchell DR. Back braces: The quick fix, Part 2. *Bull Hum Factors Ergonomics Soc* Aug.1993, p. 6.

Fraser TM. *The Worker at Work.* London: Taylor and Francis, 1989.

Fraser TM. *Ergonomic Principles in the Design of Hand Tools* (Occupational Safety and Health Series No. 44). Geneva, Switzerland: International Labour Office, 1980.

Gordon CC, Churchill T, Clauser CE et al. *1988 Anthropometric Survey of U.S. Army Personnel* (TR-89-027). Natick, MA: U.S. Army Natick Research, Development, and Engineering Center, 1989.

Grandjean E. *Fitting the Task to the Man,* 4th ed. London: Taylor and Francis, 1988.

Grandjean E. *Ergonomics in Computerized Offices.* Philadelphia: Taylor and Francis, 1987.

Grandjean E, Huenting W, Nishiyama K. Preferred VDT workstation settings, body posture and physical impairment. *Appl Ergonomics* 15:99–104, 1984.

Greenberg L, Chaffin DB. *Workers and Their Tools.* Midland, MI: Pendall, 1979.

Greiner TM. *Hand Anthropometry of U.S. Army Personnel* (TR-92-011). Natick, MA: U.S. Army Natick Research, Development, and Engineering Center, 1991.

Grieco A. Sitting posture: An old problem and a new one. *Ergonomics* 29:345–362, 1986.

Hahn K, Chin D, Ma P, et al. Cumulative trauma disorders in the industrial workplace—A systems approach. In Karwowski W, Yates JW, eds. *Advances in Industrial Ergonomics and Safety III.* Philadelphia: Taylor and Francis, 1991.

Hammer AW. Tenosynovitis. *Med Record* Oct. 3:353–355, 1934.

Hansson JE, Attebrant M. The effect of table height and table top angle on head position and reading distance. In *Proceedings of the Conference "Work With Display Units."* Stockholm: Swedish National Board of Occupational Safety and Health, 1986.

Helander M, ed. *Handbook of Human–Computer Interaction.* Amsterdam: North-Holland, 1991.

Huenting H, Grandjean E, Maeda K. Constrained postures in accounting machine operators. *Appl Ergonomics* 11(3):145–149, 1980.

Jegerlehner JL. Ergonomic analysis of problem jobs using computer spreadsheets. In Karwowski W, Yates JW, eds. *Advances in Industrial Safety III.* Philadelphia: Taylor and Francis, 1991.

Kilbom A. Physiological effects of extreme physical inactivity. In *Proceedings of the Conference Work on Display Units.* Stockholm: Swedish National Board of Occupational Safety and Health, May 12–15, 1986.

Konz S. *Work Design: Industrial Ergonomics.* Worthington, OH: Publishing Horizon, 1990.

Konz S. *Facility Design.* New York: Wiley, 1985.

Konz SA, Mital A. Carpal Tunnel Syndrome. *Int J Ind Ergonomics* 5:175–180, 1990.

Kramer AF. Mental workloads: A review of recent papers. In Damos DL, ed. *Multiple Task Performance.* London: Taylor and Francis, 1991.

Kroemer KHE. Locating the computer screen: How high, how far? *Ergonomics in Design* Oct. 1993, pp. 7–8.

Kroemer KHE. Avoiding cumulative trauma disorders in shop and office. *AIHAJ* 53(9):596–604, 1992.

Kroemer KHE. Cumulative trauma disorders. *Appl Ergonomics* 20:274–280, 1989.

Kroemer KHE. Testing individual capability to lift material: Repeatability of a dynamic test compared with static testing. *J Safety Res* 16:1–7, 1985.

Kroemer KHE. *Ergonomics Manual for Manual Material Handling,* 2nd ed. Radford, VA: Ergonomics Research Institute, 1984.

Kroemer KHE. Engineering anthropometry: Designing the work place to fit the human. In *Proceedings of the Annual Conference,* American Institute of Industrial Engineers, Detroit, May 17–20, 1981. Norcross, GA: AIIE, 1981, pp. 119–126.

Kroemer KHE. Ueber den Einfluss der raeumlichen Lage von Tastenfeldern auf die Leistung an Schreibmaschinen (On the effect of the spatial arrangement of keyboards on typing output). *Int Zeitschrift Angewandte Physiol einschl Arbeitsphysiol* 20:240–251, 1964.

Kroemer KHE, Kroemer HB, Kroemer-Elbert KE. *Ergonomics: Designing for Ease and Efficiency.* Englewood Cliffs, NJ: Prentice Hall, 1994.

Kroemer KHE, Kroemer HJ, Kroemer-Elbert KE. *Engineering Physiology: Bases of Human Factors/Ergonomics,* 2nd ed. New York: Van Nostrand Reinhold, 1990.

Lee K, Swanson N, Sauter S, et al. A review of physical exercises for VDT operators. *Appl Ergonomics* 23:387–408, 1992.

Life MA, Pheasant ST. An integrated approach to the study of posture in keyboard operation. *Appl Ergonomics* 15:83–90, 1984.

Lueder RK. Seat comfort: A review of the construct in the office environment. *Hum Factors* 26:339–345, 1983.

Lueder RK, Noro K, eds. *Hard Facts About Soft Machines: The Ergonomics of Seating*. London: Taylor and Francis, 1994.

Lundervold A. Electromyographic investigations of position and manner of working in typewriting. *Acta Physiol Scand* 24:84, 1951.

Marras WS, Kim JY. Anthropometry of industrial populations. *Ergonomics* 36:371–378, 1993.

Marras WS, Schoenmarklin RW. Wrist motions and CTD risk in industrial and service environments. In *Proceedings of the 11th Congress of the International Ergonomics Association*. London: Taylor and Francis, 1991.

McDaniel JW, Scandis RJ, Madole SW. *Weight Lifting Capabilities of Air Force Basic Trainees* (AFAMRL-TR-83-0001). Wright-Patterson AFB, OH: Aerospace Medical Research Laboratory, 1983.

McGill SM. Abdominal belts in industry: A position paper on their assets, liabilities, and use. *AIHAJ* 54:752–754, 1993.

Military Standards 759 and 1472. National Technical Information Service, 5285 Port Royal Road, Springfield, VA 22161.

Mital A. Hand tools: Injuries, illnesses, design, and usage. In Mital A, Karwowski W, eds. *Workplace, Equipment, and Tool Design*. Amsterdam: Elsevier, 1991.

National Aeronautics and Space Administration/Webb. *Anthropometric Sourcebook*, 3 vols. (NASA Ref. Pub. No. 1024). Houston, TX: NASA, 1978.

National Institute for Occupational Safety and Health. *Work Practices Guide for Manual Lifting*. DHHS (NIOSH) Pub. No. 81-122. Washington, DC: U.S. Government Printing Office, 1981.

National Research Council. *Video Displays, Work, and Vision*. Washington, DC: National Academy Press, 1983.

Nordin M, Andersson GBJ, Pope M, eds. *Occupational Musculoskeletal Disorders: Assessment, Treatment, and Prevention*, in press.

Oezkaya N, Nordin M. *Fundamentals of Biomechanics*. New York: Van Nostrand Reinhold, 1991.

Peres NJV. Process work without strain. *Aust Factory* 1, 1961.

Phalen GS. The carpal-tunnel syndrome—Clinical evaluation of 598 hands. *Clin Orthop* 83:29–40, 1972.

Phalen GS. The carpal tunnel syndrome—Seventeen years' experience in diagnosis and treatment of 644 hands. *J Bone Joint Surg* 48A(2):211–228, 1966.

Pope MH, Frymoyer JW, Andersson G. *Occupational Low Back Pain*. Philadelphia: Praeger, 1984.

Posch JL, Marcotte DR. Carpal tunnel syndrome—An analysis of 1,201 cases. *Orthop Rev* 5(5):25–35, 1976.

Putz-Anderson V. *Cumulative Trauma Disorders: A Manual for Musculoskeletal Diseases of the Upper Limbs*. London: Taylor and Francis, 1988.

Putz-Anderson V, Waters T. Revisions in *NIOSH Guide to Manual Lifting*. Paper presented at the conference "A National Strategy for Occupational Musculoskeletal Injury Prevention," Ann Arbor, MI, April 1991.

Rempel DM, Harrison RJ, Barnhart S. Work-related cumulative trauma disorders of the upper extremity. *JAMA* 267(6):838–842, 1992.

Roebuck JA. *Anthropometric Methods—Designing to Fit the Human Body*. Santa Monica, CA: Human Factors and Ergonomics Society, in press.

Salvendy G. *Handbook of Human Factors*. New York: Wiley, 1987.

Sanders MS, McCormick EJ. *Human Factors in Engineering and Design*, 7th ed. New York: McGraw-Hill, 1993.

Silverstein BA. *The Prevalence of Upper Extremity Cumulative Trauma Disorders in Industry*. Unpublished PhD dissertation, University of Michigan, 1985.

Silverstein BA, Fine LJ, Armstrong TJ. Occupational factors and carpal tunnel syndrome. *Am J Ind Med* 11:343–358, 1987.

Smith MJ. Occupational stress. In Salvendy G, ed. *Handbook of Human Factors*. New York: Wiley, 1987.

Smith MJ, Carayon-Saintfort P, Yang CL. Stability of the relationships between job characteristics and computer user well-being. In *Proceedings of the 11th Congress of the International Ergonomics Association*. London: Taylor and Francis, 1991.

Snook SH. The design of manual handling tasks. *Ergonomics* 21:963–985, 1978.

Snook SH, Ciriello VM. The design of manual handling tasks: Revised tables of maximum acceptable weights and forces. *Ergonomics* 34:1197–1213, 1991.

Soderberg GI, Blanco MK, Cosentino KA, et al. An EMG analysis of posterior trunk musculature during flat and anteriorly inclined sitting. *Hum Factors* 28:483–491, 1986.

Staffel F. On the hygiene of sitting (in German). *Zbl Allgemeine Gesundheitspflege* 3:403–421, 1985.

Tanzer RC. The carpal-tunnel syndrome—A clinical and anatomical study. *Am J Ind Med* 11:343–358, 1959.

Thompson FJ, Yates BJ, Frazen OG. Blood pooling in key skeletal muscles prevented by a "new" venopressor reflex mechanism. In *Proceedings of the Conference "Work with Display Units."* Stockholm: Swedish National Board of Occupational Safety and Health, May 12–15, 1986.

Van Cott HP, Kinkade RG, eds. *Human Engineering Guide to Equipment Design*, rev. ed. Washington, DC: U.S. Government Printing Office, 1972.

Waters TR. Strategies for assessing multi-task manual lifting jobs. In *Proceedings of the Human Factors Society 35th Annual Meeting*. Santa Monica, CA: Human Factors Society, 1991.

Waters TR, Putz-Andersson B, Garg A, et al. Revised NIOSH equation for the design and evaluation of manual lifting tasks. *Ergonomics* 36:749–776, 1993.

Wierwille WW, Eggemeyer TF. Recommendations for mental workload measurement in a test and evaluation environment. *Hum Factors* 35:263–281, 1993.

Wilson J, Corlett N, Manenica I, eds. *Ergonomics of Working Postures*. Philadelphia: Taylor and Francis, 1986.

Winkel J. Macro- and micro-circulatory changes during prolonged sedentary work and the need for lower limit values for leg activity. In *Proceedings of the Conference "Work with Display Units."* Stockholm: Swedish National Board of Occupational Safety and Health, May 12–15, 1986.

Winkel J, Bendix T. Muscular performance during seated work evaluated by two different EMG methods. *Eur J Appl Physiol* 55:167–173, 1986.

Woodson WE. *Human Factors Design Handbook*. New York: McGraw-Hill, 1981.

Woodson WE, Tillman B, Tillman P. *Human Factors Design Handbook*, 3rd ed. New York: McGraw-Hill, 1991.

Zacharkow D. *Posture: Sitting, Standing, Chair Design and Exercise*. Springfield, IL: Thomas, 1988.

# CHAPTER 14

# Biological Hazards

by A. Lynn Harding, MPH
Diane O. Fleming, PhD
Janet M. Macher, ScD, MPH

**Biological Safety**

**Hazard Identification**
*Microorganisms ▪ Infection ▪ Epidemiology of Work-Associated Infections ▪ Potentially Hazardous Workplaces*

**Risk Assessment**
*Factors Affecting Infection and Exposure ▪ Other Factors*

**Hazard Classification**
*Hazard Categories*

**Hazard Control**
*Containment ▪ Biosafety Program Management*

**Assessing Compliance**
*Use of Audits to Identify Problem Areas ▪ Annual Biosafety Review ▪ Incident/Accident Statistics*

**Current Topics in Biosafety**
*Bloodborne Pathogens ▪ Tuberculosis ▪ Legionnaires' Disease ▪ Investigating Work- or Building-Related Illnesses Caused by Biological Hazards*

**Regulations and Guidelines**

**Role of the Industrial Hygienist in Biosafety**

**Summary**

**Bibliography**

**Addendum: Large-Scale Biosafety Guidelines**
*Large-Scale Biosafety Containment*

EXPOSURE TO BIOLOGICAL HAZARDS in the workplace results in a significant amount of occupationally associated disease. Work-related illnesses due to biological agents such as infectious microorganisms, biological allergens, and toxins have been widely reported. However, in many workplaces their presence and resultant illnesses are not recognized. It has been estimated that the population at risk from occupational biohazards may be several hundred million workers worldwide (Dutkiewicz et al., 1988).

Dutkiewicz et al. (1988), in a review of occupational biohazards, noted that some 193 biological agents are known to produce infectious, allergenic, toxic, and carcinogenic reactions in workers. Most of the identified biohazardous agents belong to the following groups:

- Microorganisms and their toxins (viruses, bacteria, fungi, and their products): infection, exposure, or allergic reaction

- Arthropods (crustaceans, arachnids, and insects): bites or stings resulting in skin inflammation, systemic intoxication, transmission of infectious agents, or allergic reaction

- Allergens and toxins from higher plants: dermatitis from skin contact or rhinitis or asthma as a result of inhalation

- Protein allergens from vertebrate animals (urine, feces, hair, saliva, and dander): allergic reaction

Other groups that pose a potential biohazard include lower plants other than fungi (lichens, liverworts, and ferns) and invertebrate animals other than arthropods (parasites such as protozoa, flatworms such as *Schistosoma,* and roundworms such as *Ascaris*).

Workers engaging in agricultural, medical, and laboratory activities have been identified as being most at risk to occupational biohazards, but many varied workplaces have the potential for such exposure. A number of potentially hazardous workplaces are described in this chapter. Although biological hazards encompass a wide variety of biological agents, in large part this chapter concentrates on exposures to microorganisms and the substances associated with them. Although the chapter deals extensively with laboratory and medical environments, the concepts, principles, and exposure controls discussed here can be extrapolated to suit most occupations.

# Biological Safety

Biological safety, or biosafety, as a discipline grew out of research involving biological warfare agents at Fort Detrick in Frederick, MD. The Chemical Warfare Service (for chemical and biological warfare) was established by the U.S. Army in 1941, and in 1942 the National Academy of Sciences formed a biological warfare committee. Microbiologists from the American Society of Microbiology (ASM) served as advisors for the biological warfare activities instituted by the Chemical Warfare Service.

From 1955 to 1968, the ASM maintained an advisory committee for the Fort Detrick Biological Defense Research Program (BDRP) to provide scientific advice on the research program, peer review for publications, and assistance with staff recruitment. Members of the safety staff at Fort Detrick were well-trained microbiologists capable of advising on safety matters related to work with virulent microorganisms. Because of the considerable concern for worker safety, as well as the need to protect the surrounding community, great care was taken to prevent accidental exposure and release of infectious agents. The containment principles developed at Fort Detrick by Arnold G. Wedum and his colleagues form the framework for the discipline of biosafety today.

Following the discovery of recombinant DNA technology in the 1970s, the era of biotechnology began. The ability to insert into host cells specific pieces of donor DNA that replicate to produce a desired product has revolutionized the field of biology. Recombinant DNA technology, hybridoma technology, and protein and enzyme engineering are all components of the new biotechnology. The impact of biotechnology on research in health care, diagnostics, and agriculture, as well as many other industries such as food, chemical, mining, and petroleum, has been significant.

Whether real, potential, or imaginary, genetic manipulation of microorganisms brought a renewed concern for biological hazards. Employers and regulatory agencies took note of these concerns and began implementing or strengthening workplace biosafety programs. The appearance of a virus capable of destroying the human immune system (human immunodeficiency virus, or HIV) in the 1980s, coupled with the high incidence of occupationally acquired hepatitis B virus infection among health care workers, prompted the Occupational Safety and Health Administration (OSHA) to establish a standard that mandates protection of workers from occupational exposure to bloodborne pathogens (OSHA, 1991). This standard has heightened awareness and improved biosafety controls in the workplace considerably.

The principles and control methods used by biosafety specialists and industrial hygienists are similar: anticipation, recognition, evaluation, and control. This chapter covers the topics of hazard identification, risk assessment, hazard classification, and hazard control. Current issues in biosafety, such as bloodborne pathogens, tuberculosis, work- or building-related illness, and legionnaires' disease, are also addressed. Finally, topics such as assessing compliance, current regulations that apply to microbial activities, and the role of the industrial hygienist in biosafety are reviewed.

# Hazard Identification

Microorganisms are a diverse group of microscopic organisms that includes bacteria, fungi, algae, protozoa, and viruses. Although pathogenic or disease-producing microorganisms represent only a small portion of the total microbial population, attention is often focused on them because of their negative impact on humans, plants, and animals. In addition to their ability to produce infectious diseases, microorganisms such as fungi produce spores capable of causing allergic reactions among workers. Toxins such as endotoxin, a component in the cell walls of gram-negative bacteria, and mycotoxin, a natural product produced by fungi, have also been identified as occupational biohazards. Other biological agents such as pollen, mites, urine proteins, animal dander, and snake venoms, to list only a few, also fit within the broad scope of biological hazards.

## Microorganisms

Microorganisms are divided into two categories: prokaryotes (organisms in which DNA is not physically separated from the cytoplasm) and eukaryotes (organisms containing a membrane-bound nucleus). Prokaryotes and eukaryotes are organisms because they contain all of the enzymes required for their own replication, as well as the biological equipment necessary to produce metabolic energy. This distinguishes them from viruses, which depend on host cells for these necessities.

Prokaryotes, characterized by their relatively small size (around 1 μm in diameter) and the absence of a nuclear membrane, are divided further into two major groups: bacteria and archaebacter (Jawetz et al., 1991). The bacterial genera, or subgroups, of medical interest include spirochetes such as *Leptospira* and *Borrelia* (Lyme disease); *Mycoplasma,* which are pleomorphic bacteria that lack a cell wall; and a large group of rigid cells in which most bacterial pathogens are found. Rigid-cell bacteria include stalked, budding, and mycelial organisms (*Mycobacterium, Actinomyces, Nocardia,* and *Streptomyces*); simple unicellular, obligate, intracellular parasites (*Rickettsia, Coxiella,* and *Chlamydia*); and the largest group, which are free-living, gram-negative and -positive, aerobic and anaerobic bacteria such as *Staphylococcus, Bacillus, Neisseria, Brucella, Legionella, Salmonella,* and *Pseudomonas.*

In contrast to prokaryotes, eukaryotes are larger and contain a membrane-bound nucleus and organelles such as mitochondria. The microbial eukaryotes, or protists, fall into four major groups: algae, protozoa, fungi, and slime molds, with protozoa such as *Giardia, Trypanosoma, Toxoplasma,* and *Plasmodium* and fungi such as *Histoplasma, Aspergillus, Cryptococcus,* and *Coccidioides* being organisms of medical importance.

Viruses, whose unique properties distinguish them from other microorganisms, are totally dependent on their hosts for replication. They are inert outside of a host cell, and host–virus interactions are highly specific. Viruses, the smallest infectious agents, are 20–300 nm in diameter. Viral particles consist of nucleic acid molecules, either DNA or RNA, enclosed in a protein coat, or capsid. The capsid protects the nucleic acid and facilitates attachment to and penetration into the host cell by the virus. Once inside the cell, viral nucleic acid uses the host's enzymatic machinery for functions associated with viral replication. The host range of a given virus may be broad or extremely narrow, and viruses can infect unicellular organisms such as mycoplasmas, bacteria, algae, and all higher plants and animals. Classification of viruses is based on a number of properties such as nucleic acid type, size and morphology, susceptibility to physical and chemical agents, and presence of enzymes.

Agents smaller by an order of magnitude than viruses have been reported that have properties similar to viruses and cause degenerative disease in humans and animals. Scrapie, a disease of the nervous system of sheep, is caused by such an agent. Because of the novel properties of this agent, a new term, *prion,* has been designated to denote these small proteinaceous infectious particles, which are resistant to inactivation by most procedures that modify nucleic acids (Prusiner, 1982). The tropical disease kuru and Creutzfeldt–Jakob (a form of human dementia) are caused by similar agents.

Although the amount of information provided here is necessarily limited, a wealth of accessible material is readily available. Readers are encouraged to consult microbiology references such as *Medical Microbiology* (Jawetz et al., 1991), *The Microbial World* (Stanier et al., 1986), *Environmental Microbiology* (Mitchell, 1992), *Microbial Ecology* (Atlas & Bartha, 1993), *Manual of Clinical Microbiology* (Balows et al., 1991), *Bioaerosols Handbook* (Cox & Wathes, 1995), and *Bioaerosols* (Burge HA, 1995).

## Infection

*Infection* is a general term applied to invasion of the body by microorganisms such as bacteria, protozoa, and the larval forms of multicellular organisms such as the helminths (intestinal parasites including roundworms and tapeworms). It is further defined as an invasion of the body by pathogenic microorganisms and the reaction of the tissues to their presence and to the toxins generated by them (Arey et al., 1957).

Although human beings have microorganisms on every surface and in every external orifice of their bodies, only a small proportion of those agents are capable of producing an infection that could lead to disease in that person or, if communicable, in others. If the disease-causing agent arises from the microbial flora normally present in or on the body of a person (indigenous flora), its resulting infection is called endogenous. For example, most urinary tract infections are caused by agents such as *E. coli* or *Pseudomonas* spp., which are normally found in the feces of the patient. This example demonstrates the disease-causing potential of normal flora when they are able to reach a different site in the body. Normal flora can also take advantage of a lowering of host immunity to produce an infectious disease. Such infections occur in those who are immunocompromised by underlying disease processes or certain medications such as steroids and chemotherapy. Further information on normal flora can be found in Isenberg and D'Amato, 1991.

Individuals harboring communicable infectious agents without exhibiting signs of disease are called carriers. They can be a source of infection in co-workers, especially if the agent is transmitted by the aerosol route, as with measles or tuberculosis. Certain of these "wild-type" agents—strains found in nature, and therefore in the community—can sometimes also contaminate a sterile product in a laboratory environment. A vaccine strain being grown in cell culture can be contaminated by a worker who carries a different, potentially more virulent strain obtained in the community. The restriction of visitors and the use of approved vaccines is thus recommended when appropriate to protect the work being done.

Infections from microorganisms not normally found in or on the human body, but which gain entrance from the environment, are called exogenous infections. These agents gain entry into the host by inhalation, indirect or direct contact, penetration, or ingestion. Some agents routinely cause disease in healthy adult humans, whereas others, known as opportunists, require special circumstances of lowered host defense or overwhelming dose of exposure. Thus, infectious disease is not always the end result of the exposure to and colonization by an infectious agent. The end result depends on the virulence of the agent, the route of infection, and the relative immunity and health of the host.

Workers are expected to be healthy human adults, but should be medically assessed during a preplacement examination to determine fitness for specific work. Because exposures can occur both to those involved with the work and to those who merely enter the work area, all such individuals should be identified and assessed in order to prevent exposure and infection. The spread of infectious agents used in research or production to the outside environment, including the neighboring community, is rare.

## Epidemiology of Work-Associated Infections

An unfortunate consequence of working with infectious microorganisms or materials contaminated with them is the potential for acquiring a work-associated infection. The literature includes numerous descriptions of occupational exposures indicating that persons who handle infectious materials are clearly at higher risk for infection than the general population.

It is generally accepted that work-associated infections are underreported in the scientific literature. This may be the result of employees' unwillingness to report such incidents for fear of loss of employment, issues of liability, or an employer's refusal to publish such material. Literature on work-associated infections, reported as case studies, usually focuses on diagnosis and treatment of the patient and frequently fails to assess the circumstances related to the occupational exposure.

In the absence of a detailed database on work-associated infections, epidemiological methods provide the tools to evaluate the extent and nature of worker exposure. Defining the event or illness/infection, determining the population at risk, establishing the factors affecting exposure, and developing intervention controls are all part of the process to prevent occurrence or recurrence of infections.

A comprehensive survey of work-associated infections gathered by Sulkin and Pike (Sulkin & Pike, 1951; Pike, 1978; Pike, 1979) focuses specifically on laboratory-acquired infections. Data compiled through 1978 revealed 4,079 cases of clinically apparent infection classified by agent, source of infection (when known), and type of work involved. Of the total, 168 cases resulted in the worker's death. These numbers are considered low because the reporting of work-related infections is not required, and data on seroconversion (the production of antibodies in response to an infectious agent) or asymptomatic response

to occupationally acquired microorganisms are rarely reported.

The Sulkin and Pike surveys revealed the most common routes of exposure to be percutaneous inoculation (needles/syringes, cuts or abrasions from contaminated items, and animal bites), inhalation of aerosols generated by accidents or by work practices and procedures, contact between mucous membranes and contaminated material (hands and surfaces), and ingestion (Pike, 1979). Eighteen percent of the infections were attributable to known accidents caused by either carelessness or other human error. Twenty-five percent of these acknowledged accidents involved needles or syringes. Most of the remaining accidents involved spills and sprays, injury with broken glass or other sharp objects, accidental aspiration using a pipet, and animal bites, scratches, or contact with ectoparasites. Unfortunately, the sources of exposure for the remaining 82 percent of infections were not easily identifiable. Although some could be attributed to handling infectious animals, clinical specimens, and discarded glassware and to aerosols, all that was known about most exposures was that the person had worked with or was in the vicinity of work with the agent.

From these laboratory data, it is apparent that people engaged in research activities acquired the greatest number of infections. Trained investigators, technical assistants, animal caretakers, and graduate students experienced over three-quarters of the research-associated illnesses. The remainder occurred among clerical staff, dishwashers, janitors, and maintenance personnel. With rare exceptions, laboratory-acquired infections were not spread to the outside community.

The most frequently reported laboratory-acquired infections described by Pike include brucellosis, Q fever, hepatitis, typhoid fever, tularemia, and tuberculosis. Today, a number of these agents are rarely associated with workplace infections. Although there has been a resurgence of tuberculosis globally, including in the United States, there are no reports of a proportional rise in laboratory-associated infections. Ninety-seven percent of the cases of brucellosis and typhoid fever were reported before 1955. Forty percent of hepatitis cases attributed to handling human blood and blood products and most of the lymphocytic choriomeningitis infections have occurred since 1955. More recently, workplace infections have been associated with new or emerging viruses such as human immunodeficiency virus (HIV), the etiologic agent associated with AIDS (acquired immune deficiency syndrome), and the hantavirus that causes Korean hemorrhagic fever. Herpes B virus continues to infect workers who handle certain nonhuman primates and their tissues, and Ebola-related filovirus was associated with workplace asymptomatic seroconversions following an exposure to nonhuman primates in 1989 in Virginia (CDC, 1989).

Attempts to determine incidence rates of occupationally acquired infections among laboratory personnel must be interpreted cautiously, because estimates of the number of infections and the population at risk are imprecise. In 1969, it was estimated (using U.S. and European data) that the frequency rates for laboratory-acquired infections were on the order of 1 to 5 infections per million working hours, or $10^{-2}$ to $10^{-3}$ per year, for laboratory workers (Phillips, 1969). In 1988, in a survey of laboratory-associated infections and injuries among public health and hospital clinical laboratory employees, it was determined that the annual incidence rate for full-time–equivalent (FTE) employees was 1.4 infections per thousand for public health and 3.5 infections per thousand for hospital laboratories, whereas rates for those working directly with infectious agents were 2.7 and 4.0 per thousand for public health and hospital laboratories, respectively (Vesley & Hartmann, 1988). Another study showed an estimated annual incidence rate for clinical laboratories of 3 infections per thousand employed and 9.4 infections per thousand when only microbiologists were considered (Jacobson et al., 1985).

Despite the admitted flaws in the existing data and the fact that much of the information on work-associated infections focuses specifically on laboratory infections, it seems reasonable to accept the fact that some workers handling infectious materials may become infected by them. Epidemiological data provide the information necessary to make decisions regarding prevention or minimization of work-related infections. Accurate estimates of and surveillance for occupationally associated infections are lacking because reports of infections usually do not include occupational data. A recent report of proportionate mortality from pulmonary tuberculosis as associated with occupation reflects a new effort to identify the potential for exposure (CDC, 1995). Ideally, though, infections rather than death should have been used to identify the risk of exposure in certain workplaces.

## Potentially Hazardous Workplaces

Although most pathogenic microorganisms have the potential to cause occupationally acquired infections, knowledge of the hazard, containment practices, and preventive therapeutic measures such as use of vaccines greatly reduce their incidence. In workplaces where awareness of the hazard is high and the potential risk is understood, compliance with control practices minimizes exposure. However, there are some workplaces where controls are difficult to implement or are not readily available and where hazard recognition is low. For example, the focus in agricultural activity and its associated subsequent processing facilities is not on human pathogenic microorganisms. However, there are organisms that are intrinsically associated with some of the animals or plants concerned and these represent a hazard to the worker. Worker protection becomes more challenging in such an environment than it is in a research laboratory or hospital.

Because workplaces are varied and microbial habitats diverse, it can be difficult to find concise, detailed information

on microbial agents. The American Public Health Association publication *Control of Communicable Diseases in Man* is an excellent resource for information on a disease, the infectious agent, its occurrence, reservoir, mode of transmission, incubation period, and methods of control (Benenson, 1990).

**Microbiology, public health, and molecular biology laboratories.** The potential threat of occupational infection has long been recognized by microbiologists. However, new potential for exposure exists with the increasing number of nonmicrobiologists engaged in biotechnological activities. The earlier review of laboratory-related infections summarized experiences to date. Exposures tend to be directly related to the hazard classification of the organisms being manipulated, the potential for release of the organism during required manipulations, and the level of competency of personnel.

Staff in research laboratories, the type of workplace where the majority of laboratory-acquired infections have occurred, tend to work with more hazardous agents, including those of emerging diseases. They often handle concentrated preparations of infectious microorganisms, and some test procedures require complex manipulations. Microbial exposures (infections) mimic the focus of research at the time. As laboratories focus on the disease problems of the 1990s, laboratory-acquired infections involving the hepatitis viruses, human immunodeficiency virus, *Mycobacterium tuberculosis,* other *Mycobacterium* spp., and those of emerging infectious diseases such as coccidioidomycosis, cryptosporidiosis, hantavirus pulmonary syndrome, and antibiotic resistant bacterial disease can be expected (CDC, 1994). Because of inherent containment difficulties, the use of infected laboratory animals, including those taken from the wild, also increases the rate of worker exposures to infectious agents. More information on this subject is covered later in this section.

**Hospitals and health care establishments.** In addition to infectious agents, health care facilities (which include physicians' and dentists' offices, blood banks, and outpatient clinics) may expose their personnel to multiple hazards including cytotoxic drugs, anesthetic gases, ethylene oxide, radiation sources, steam, lifting heavy objects, and electrical shock. Infections in hospitals can be categorized as community acquired (transmitted to either patients or workers); occupationally acquired (resulting from worker exposure); and nosocomial (hospital-acquired infections of patients).

The Joint Commission on Accreditation of Healthcare Organizations (JCAHO), the CDC guidelines for hospital infection control, the OSHA bloodborne pathogens standard, and the CDC/NIH guidelines (U.S. Public Health Service [USPHS], 1993) all provide guidance concerning the control of nosocomial infections in patients and the protection of health care personnel. Because of the nature of hospital activities, nosocomial infections have become a complication of hospitalization. To prevent or reduce the incidence of such complications, infection control programs were developed and implemented in U.S. hospitals during the 1950s and 1960s. The Centers for Disease Control and Prevention (CDC) coordinate surveillance of hospital infections and recommend infection control practices and procedures.

In most instances a hospital epidemiologist—usually an MD who specializes in infectious diseases—and an infection control practitioner—often a nurse or, occasionally, a microbiologist—manage and oversee infection control activities. The prevalence of hospital infections has created a need for infection control procedures (barriers), rigorous disinfection and sterilization techniques, meticulous cleaning and waste-handling procedures, and, in some cases, special design criteria. The role of the industrial hygienist in hospital infection control may include assisting in the selection and testing of personal protective equipment, environmental testing in outbreak situations such as nosocomial fungal infections in oncology patients, and design and engineering controls such as ventilation and containment systems.

Because it is not within the scope of this chapter to provide detailed material on the many topics covered here, the reader is referred to the Hospital Infection Program at the CDC and the JCAHO in Oakbrook, IL, for guidelines and standards covering infection control programs and practices in the United States. For more information, two journals in the United States provide current information on hospital infection control: the *American Journal of Infection Control,* from the Association of Professionals in Infection Control (APIC) in Washington, DC; and *Infection Control and Hospital Epidemiology,* the journal of the Society for Healthcare Epidemiology of America, published by Slack, Inc., in Thorofare, NJ.

**Biotechnology facilities.** With the discovery of recombinant DNA technology and the resulting advances in the field of molecular biology, many opportunities for the development of products in medicine, industry, agriculture, and environmental management are now possible. Industrial microbiology, long associated with the chemical and pharmaceutical industries, has attained a position of prominence with the advent of "the age of biotechnology."

From the early stages of discovery to the ultimate marketing of a pharmaceutical, large volumes of product, whether it be a metabolite or an organism, are required. Depending on the hazard level (pathogenicity or biological activity), increase in production or concentration brings with it the need for adequate barriers to protect personnel, the product, and the community.

With some exceptions, the microorganisms most often used in manufacturing operations are those requiring minimum containment, such as genetically engineered bacteria (*E. coli,* K12), molds, yeasts, plant and animal cells.

Production operations usually involve the use of closed systems (either a primary container or a combination of primary and secondary containers) and validated inactivation of waste materials and contaminated by-products.

In addition to the possibility of experiencing the direct effects of the biological activity of an agent, workers may develop allergies to proteins (biological products derived from raw materials, fermentation products, or enzymes), to other chemicals, or to animal dander or (aerosolized) urine proteins from animals. Allergic responses following exposure to proteins in the work environment can and do produce significant health effects, but they are not addressed here. For additional information on allergic response and safety issues in biotechnology, consult Van Houten (1989), Ducatman and Liberman (1991), and American Industrial Hygiene Association (1995).

**Animal facilities and veterinary practices.** Although generally only work activities in research, medical, and industrial facilities involve handling laboratory animals, there are a wide range of occupations in which workers are exposed to animal-related allergens and infectious agents or their toxins. Agricultural workers, veterinarians, workers in zoos and museums, taxidermists, and workers in animal product–processing facilities are all at risk for occupational exposure to animal-related biological hazards. A number of these workers may be exposed to wild (captured) or exotic animal populations. Factors to be considered when handling animals include the nature of the animal (its aggressiveness and tendency to bite or scratch), the normal flora and natural ecto- and endoparasites of the animal, the zoonotic diseases to which they are susceptible, and the possible dissemination of allergens.

The development of laboratory animal allergy (LAA) is a significant and common problem for laboratory personnel, veterinarians, and others who work with animals (Bland et al., 1987). The manifestations of LAA include cough, wheezing (asthma), watery and itchy eyes, itchy skin, sneezing, and skin rash. Following contact with animals, symptoms can develop in less than one year, but may take up to several years to develop. Often, workers react to the proteins in shed animal dander and hair or to those in animal urine, serum, saliva, or tissues. Aerosolized mold spores and proteins from animal food and bedding can also act as allergens. The prevalence in Europe and the United States of LAA among lab workers and animal handlers has been reported to range between 11 and 30 percent (Bland et al., 1987). Specific equipment is available to reduce exposures to animal fur and dander, such as electric shavers with built-in vacuum attachments.

During the past 50 years, diseases that affect both humans and animals (zoonotic diseases) have been among the most commonly reported occupational illnesses of laboratory workers. Most of these have been caused by viral and bacterial (including rickettsial) agents.

Infection is most often the result of one of the following types of exposure:

- Animal bites or scratches
- Contaminated needlesticks or scalpels
- Infectious aerosols resulting from respiration, excretion, or infectious bedding
- Contact with infected tissue and cells during histological procedures, homogenization, or manipulation of cells in culture.

Literature documenting zoonoses among veterinarians is common (Constable & Harrington, 1982; Schnurrenberger et al., 1978). Several excellent references listing zoonotic diseases of laboratory animals and zoonotic pathogens causing diseases in man have been published (Fox & Lipman, 1991; National Research Council, 1989, pp. 175–186).

Animal-related infections can be expected at certain kinds of worksites. The infections frequently observed among personnel involve microorganisms with a low infectious dose (ID); where exposure results from aerosolized infectious materials. Numerous accounts of *Coxiella burnetii,* the rickettsial agent that causes Q fever, have been reported. *C. burnetii* has an estimated ID$^{25-50}$ of 10 organisms by inhalation (Wedum et al., 1972). This means that 25–50 percent of a population becomes infected after the introduction of only 10 *C. burnetii* bacteria. Many hospital and laboratory personnel have been exposed to *C. burnetii* as the result of research involving naturally infected asymptomatic sheep. In addition to being extremely infectious, the organism is very resistant to drying and remains viable for long periods of time. Q fever control measures for research facilities using sheep were published in 1982 (Bernard et al., 1982).

Hantavirus, the etiologic agent of Korean hemorrhagic fever, produces an asymptomatic infection in wild rodents. During the 1980s, over 70 work-related hantavirus infections, apparently resulting from inhalation of aerosols produced by chronically infected laboratory animals, were reported in the scientific literature (Desmyter et al., 1983; Lee & Johnson, 1982; Lloyd & Jones, 1986; Wong et al., 1988). Recent reports of a pulmonary illness caused by a unique hantavirus, the sin nombre virus, which was initially detected in the southwestern United States, underscore the infectious potential of this zoonotic agent (CDC, 1993). Cases continue to be reported from throughout the United States now that the virus has been identified.

Work-acquired infections contracted while handling nonhuman primates have been a concern for many years. Serious health consequences, including hemorrhagic disease and death, have resulted from Marburg virus (a human filovirus) infections in Europe (Martini & Siegert, 1971) and Ebola and Marburg viruses in Africa (WHO, 1978; Baron et al., 1983; Gear et al., 1975). Recently,

Ebola-related filovirus seroconversion was documented among several animal handlers in U.S. primate facilities (CDC, 1990b, p. 221). No evidence of clinical disease was detected, but these events warrant close scrutiny of nonhuman primate colonies. Guidelines for handling nonhuman primates during transit and quarantine have been published (CDC, 1990b, pp. 22–30).

Twenty-five cases of *Herpesvirus simiae* (B virus) infections have been documented, most (16 cases) with lethal or serious outcomes. In all instances exposure was related to activities involving macaques. Six infections occurring between 1987 and 1989 (CDC, 1987b, p. 289; CDC, 1989b, p. 453) prompted the CDC to publish guidelines for prevention of *Herpesvirus simiae* (B virus) infection in monkey handlers (CDC, 1987a, p. 680).

Significant similarities between the simian and human immunodeficiency viruses have led to the development of guidelines to prevent simian immunodeficiency virus infection in laboratory workers and animal handlers (CDC, 1988). No illness has been noted in workers to date, but seroconversions have occurred (CDC, 1992).

Certainly not all agents associated with zoonotic disease carry the same potential for occupational exposure as some of those described above. Nevertheless, it is critical to evaluate the risk and to determine the control measures necessary to contain the hazard before initiating work with potentially infectious or experimentally infected animals.

**Agriculture.** Agriculture, mining, and construction are considered to be among the most hazardous occupations of this century. Agricultural workers and those who process agricultural products are exposed to numerous safety and physical hazards as well as chemical and biological agents. Workers are readily exposed to infectious microorganisms and their spores and toxins through inhalation of bioaerosols, ingestion resulting from contact with contaminated materials, direct exposure of nonintact skin and mucous membranes, and inoculation resulting from traumatic injury. Factors such as host susceptibility, virulence, and dose all influence the development of disease.

Biological hazards associated with fungal disease (such as coccidioidomycosis, histoplasmosis, and blastomycosis) are found on plants and animals and in soils, and cause disease among farmers and horticultural workers. Food and grain handlers and farmers and laborers are exposed to parasitic diseases such as echinococcosis and toxoplasmosis. Processors who handle animal products may acquire bacterial skin diseases such as anthrax from working with contaminated hides, tularemia from skinning and dressing infected animals, and erysipelas from skin abrasions infected during contact with contaminated fish, shellfish, meat, or poultry. Infected turkeys, geese, squab, and ducks or the aerosolized feces from these birds expose poultry processing workers and farmers to psittacosis, a bacterial infection caused by *Chlamydia psittaci.*

Agricultural workers frequently experience allergic and/or respiratory symptoms in response to mycotoxins associated with fungi such as *Penicillium, Aspergillus,* and *Fusarium* spp. and to endotoxins produced by gram-negative bacteria in stored grains.

At least 24 out of the 150 zoonotic diseases known worldwide are considered to be a hazard for agricultural workers in North America (Donham, 1985; Acha & Szyfres, 1980). These diseases can be contracted directly from animals, but more often they are acquired in the work environment. Risk of infection varies with the type and species of animal and with geographic location. Controls include awareness of specific hazards, use of personal protective equipment (PPE), preventive veterinary care, worker education, and medical monitoring or prophylactic therapy where appropriate.

There is an extensive literature on occupational exposure of agricultural workers. For additional information, the reader is referred to Popendorf & Donham (1991), AIHA (1995), and the Agricultural Respiratory Hazards Education Series published by the American Lung Association (1986).

**Miscellaneous worksites.** The potential for exposure to occupational biohazards exists in most work environments. The following list, though incomplete, cites many of the diverse workplaces where the potential for exposure to biohazardous agents exists, along with the diseases or agents to which they may be exposed. A number of additional occupations and associated biohazardous agents are cited in the Dutkiewicz review of occupational biohazards (Dutkiewicz et al., 1988, pp. 612–615).

- Workers maintaining water systems: exposure to *Legionella pneumophila* and *Naegleria* spp.
- Workers associated with birds (such as parrots, parakeets, and pigeons) in pet shops, aviaries, zoos with avian exhibits, or on construction and public works jobs near perching or nesting sites: *Chlamydia psittaci.*
- Workers in wood-processing facilities: endotoxins, allergenic fungi growing on timber, and fungi that cause deep mycoses.
- Miners: zoonotic bacteria, mycobacteria, dermatophytic fungi, fungi causing deep mycoses, mycotoxin-producing fungi. (Miners may be immunocompromised because of exposure to coal dust, which results in black lung disease.)
- Sewage and compost workers: enteric bacteria, hepatitis A virus, infectious and/or endotoxin-producing bacteria, parasitic protozoa such as *Giardia* spp., allergenic fungi.
- Renovators of items such as books, buildings, and paintings; librarians: endotoxin-producing gram-negative bacteria, allergenic and toxigenic fungi growing on surfaces.

- Workers in textile manufacturing who process plant fibers (such as cotton, flax, hemp): gram-negative bacteria that produce endotoxins.
- Workers in the fishing industry: zoonotic bacteria (*such as Leptospira interrogans, Erysipelothrix rhusiopathiae,* and *Mycobacterium marinum*) and parasitic flukes (*such as Schistosoma*).
- Forestry workers: zoonotic diseases or agents (such as rabies virus, Russian spring summer fever virus, rocky mountain spotted fever, lyme disease, and tularemia), viruses and bacteria transmitted by ixodid ticks, and fungi that cause deep mycoses.
- Workers who handle animal hair and rough leather: zoonotic diseases (such as Q fever, anthrax, and tularemia) and dermatophytic fungi.
- Workers who handle products of plant origin: gram-negative bacteria with allergenic and endotoxic properties, actinomycete with allergenic properties, storage fungi with allergenic and mycotoxic properties, allergenic and toxic substances of plant origin, and allergenic storage mites.

# ■ Risk Assessment

In the late nineteenth century, as scientists became aware of the presence of microorganisms and their potential to cause illness, many microbiologists suffered significant health consequences as the result of their work. Diseases such as cholera, typhus, yellow fever, Rocky Mountain spotted fever, and tuberculosis claimed the lives of those dedicated to studying and/or eradicating them. Health care and agricultural workers suffered ongoing exposure to biological hazards before advances in medical science and animal husbandry could reduce the consequences of infectious diseases.

It is possible to work with infectious agents (or people, animals, or substances contaminated by them) and still avoid exposure and subsequent infection or illness. Infections do not necessarily occur simply because the exposed person works with a disease-producing agent or substance. A series of circumstances are necessary for exposure that leads to infection or illness. By performing a risk assessment on an operation or event, it is possible to make a systematic evaluation of the exposure potential, and then to make decisions as to how the exposure can best be avoided, reduced, or otherwise managed.

## Factors Affecting Infection and Exposure

**Modes of transmission.** The principal modes of transmission for infectious microorganisms and other biological materials include contact transmission (direct or indirect), vector-borne transmission, and airborne transmission. Direct contact of an infected person with another person is rare in the laboratory environment, but such transmission occurs commonly in the community and in medical settings where patients are treated. Animal-to-human (zoonotic) transmission through bites and scratches can occur when animals are associated with work activities. Spills or splashes of infectious materials (gross contamination) onto a receptive site such as an open wound, cut, eczematous skin, or mucous membranes are an effective means of transmitting microorganisms. Indirect transmission occurs when common environmental surfaces (such as equipment, work benches, laboratory accessories) become contaminated, and the infectious material is transferred to a host.

Vector-borne infection results when a causative agent is mechanically or biologically transmitted by a living vector such as a mosquito or tick, biologically, through a bite, directly through the skin in rare cases, or by mechanical means, to the host. Biological transmission involves propagation, multiplication, cyclic development, or a combination of these in the host before the arthropod can transmit the infective form of the agent. Infected ticks and mosquitoes have transmitted Rocky Mountain spotted fever, malaria, and yellow fever to investigators in the laboratory and in the field and are a potential hazard for other outdoor workers.

The inhalation of airborne infectious particles into the respiratory system constitutes airborne transmission. This mode is important in the transmission of certain pathogens such as *Mycobacterium tuberculosis*. The contaminated air in a room may escape to the outside and act as a conduit for contamination of the environment. Additional information on airborne transmission can be found later in this section of the chapter, under the heading "Aerosols."

**Routes of entry.** The routes of entry for microorganisms associated with occupationally acquired infection include inhalation, ingestion, penetration through skin (intact or nonintact), and contact with mucous membranes of the eyes, nose, and mouth. Many technical procedures (such as pressurizing liquids, sonicating, and grinding or sawing tissue) and equipment and spills in the workplace release microbes into the air, where they can be inhaled by workers.

Ingestion of infectious materials can occur when workers mouth-pipet or suction infectious materials or by hand-to-mouth contamination as the result of eating, drinking, smoking, or applying cosmetics in a contaminated work area. Hand-washing minimizes the opportunity for oral and ocular exposure.

Infectious agents are introduced into the body when contaminated objects (such as hypodermic needles, broken glassware, and scalpels) or animals puncture, cut, or scratch the skin (percutaneous exposure). This type of exposure also occurs through skin surfaces that are not intact, that is, when open wounds, cuts, hangnails, dermatitis, or eczema are present. Unbroken skin is a barrier to infectious agents. Exceptions occur only in instances

**Table 14–A.** Infectious Doses for 25–50 Percent of Volunteers

| Disease or Agent | Inoculation Route | Dose* |
|---|---|---|
| Scrub typhus | Intradermal | 3 |
| Q fever | Inhalation | 10 |
| Tularemia | Inhalation | 10 |
| Malaria | Intravenous | 10 |
| Syphilis | Intradermal | 57 |
| *Shigella flexneri* | Ingestion | 180 |
| Anthrax | Inhalation | ≥1,300 |
| Typhoid fever | Ingestion | $10^5$ |
| Cholera | Ingestion | $10^8$ |
| *Escherichia coli* | Ingestion | $10^8$ |
| Shigellosis | Ingestion | $10^9$ |

*Dose is in number of organisms.
(Adapted from Wedum et al., 1972, p. 1558.)

where skin penetration is the normal route of entry for an agent, such as with the infective cercariae stage of the parasitic agents *Schistosoma* spp.

Mucous membranes of the eyes, nose, and mouth are readily exposed to agents when rubbed with contaminated fingers or gloved hands and when splashes or sprays of infectious material occur. There have been reports of HIV infections related to splashes of the eyes and mucous membranes, and also of such infections with *T. cruzi* in a recent review by Herwaldt and Juranek (1995).

**Infectious dose.** The infectious, or infective, dose is the number of microorganisms required to initiate an infection. Although there are data available from animal studies on $ID^{50}$ (the number of organisms needed to infect 50 percent of a test population), only a modest amount of information exists for humans. Data accumulated by the National Institutes of Health on route of entry and infectious dose are shown in Tables 14–A and 14–B (Wedum et al., 1972).

**Viability and virulence of the agent.** Viability and virulence of the agent are also important in determining whether a person becomes infected. If a microorganism is not viable and able to replicate, the opportunity for infection does not exist. The external environment is critical in the replication of microorganisms. Factors such as temperature, humidity, and the presence or absence of growth factors or other chemicals all play an important role in viability. For example, some bacterial agents, such as *Bacillus anthracis,* are capable of producing spores that survive under adverse conditions, and agents such as *Mycobacterium tuberculosis* or *Staphylococcus aureus* are unaffected by drying and remain viable on environmental surfaces, whereas the herpesviruses are very susceptible to drying.

The virulence, or relative pathogenicity, of microorganisms varies greatly among types and strains. Some microbes are highly pathogenic, even in healthy adults, whereas others are opportunistic pathogens, able to infect only hosts with lowered immunity or sites other than their normal habitat. Some microbial strains are attenuated, or

**Table 14–B.** Minimal Human Infective Dose in Volunteers

| Viral Agent | Inoculation Route | Dose* |
|---|---|---|
| Measles virus | Intranasal spray | 0.2** |
| Rhinovirus | Nasal drops | ≤1 |
| Venezuelan encephalitis virus | Subcutaneous | 1† |
| West Nile fever virus | Intramuscular | 1†† |
| Parainfluenza 1virus | Nasal drops | ≤1.5 |
| Poliovirus 1 | Ingestion | 2**§ |
| Rubella virus | Pharyngeal spray | ≤10** |
| Coxsackie A21virus | Inhalation | ≤18 |
| Rubella virus | Subcutaneous | 30** |
| Adenovirus | Conjunctival swab | ≤32 |
| Rubella virus | Nasal drops | 60** |
| Adenovirus 7 | Nasal drops | ≤150 |
| Respiratory syncytial virus | Intranasal spray | ≤160–640 |
| Influenza A2 virus | Nasopharyngeal | ≤790 |
| SV-40 virus | Nasopharyngeal | 10,000 |

**Note:** There was illness after all inoculations except poliovirus, rubella virus (nasal drops), adenovirus (nasal drops), and SV-40 virus; in these four there was serologic conversion.

*Median infectious tissue culture dose
**Children
†Guinea pig infective unit
††Mouse infective unit
§Plaque-forming unit
(Adapted from Wedum et al., 1972, p. 1558.)

weakened, after reproducing through numerous generations in the laboratory. Certain vaccine strains, selected because they are immunogenic and do not produce significant disease, are also examples of attenuated organisms. Even though the vaccine strain of an organism may be attenuated, it is best to limit exposure to planned circumstances such as vaccination rather than by a work-related exposure.

**Host susceptibility.** Host susceptibility is often underestimated because the majority of persons working with potentially infectious material are healthy. The risk assessments and biosafety levels recommended by the Centers for Disease Control and Prevention (CDC) and National Institutes of Health (NIH) presume a population of immunocompetent individuals (USPHS, 1993). Employees working with infectious agents can be put at increased risk of infection because of a variety of medical conditions such as diseases, allergies, inability to receive particular vaccines, and pregnancy or by taking drugs that alter host defenses.

Conditions that alter host defenses at body surfaces or impair the functioning of the immune system may put a worker at risk for certain infections. Skin disorders such as chronic dermatitis, eczema, and psoriasis leave a worker without an intact skin barrier against infection. The gastrointestinal mucosa, colonized by a resident population of normal bacterial flora, offers protection against infection by pathogenic microorganisms. However, this protection is usually disrupted when antibiotic therapy is administered. The body's immune system, consisting mainly of antibody-mediated B-cells, cell-mediated T-cells, and phagocytic cells, offers a significant line of defense against invading microorganisms (Ammann, 1987).

A common concern for pregnant employees is the potential for congenital infection of the fetus, which can be caused by cytomegalovirus (CMV), rubella, hepatitis B virus (HBV), and toxoplasmosis (Sheretz & Hampton, 1986).

The development of allergies to proteins such as biological products from raw plant and animal materials, fermentation products, or enzymes, to chemicals or animal dander, or to aerosolized animal urine proteins also presents a risk to employees. If, because of an allergy to a constituent of a vaccine, it is not possible to immunize an employee, then safe working conditions for that person may be compromised. A higher level of work practices and personal protective equipment may provide the required level of protection for such a worker. All of these factors must be recognized and evaluated in relation to an employee's potential exposure. Decisions should be made on a case-by-case basis, with input from the employee, the employee's physician, institutional management, and an occupational health service professional (Goldman, 1989).

## Other Factors

Additional factors associated with risk assessment of microbial work include the ability to tolerate prophylactic or therapeutic measures (vaccines and effective interventions), knowledge of the host range of the microorganism, and an understanding of the organism's potential for escape to the community. The importance of assessing the work activity (including the facility, contamination potential, volume of material, and the agent concentration) in relation to the host and agent cannot be overemphasized.

**Aerosols—a special factor.** As noted earlier, Pike's survey of laboratory-associated infections indicated that only 18 percent of the documented infections had resulted from known accidents (Pike, 1976). For the remaining 82 percent, a connection between the infected person and the causative agent was difficult to determine. In many instances, all that is known is that the person worked with or was in the vicinity of work being done with the causative agent. The fact that no specific event could be associated with so many infections led earlier reviewers to implicate many routine laboratory procedures as the source of airborne contamination. Workers are exposed to airborne microorganisms through direct contact with or inhalation of minute airborne particles or by contact following the deposition of droplets, through splashing or spilling, onto surfaces, equipment, and personnel. (Contamination through splashing and spilling can also occur through transmission routes other than inhalation.)

Liquid, when under pressure and passed through a small opening or when dropped onto a solid surface, is aerosolized into a cloud of very small droplets. The droplets vary in size; the larger ones settle quickly onto surfaces, inanimate objects, clothing, and skin, whereas the smaller ones evaporate rapidly. Bacteria and other material in the droplets remain in a dried state as droplet nuclei. These particles, or droplet nuclei, can remain suspended in air for some time and be moved to remote areas by air currents or ventilation systems.

Laboratory procedures involving the manipulation of infectious materials generate infectious particles of various sizes. For example, particles released by opening or dropping lyophilized cultures have a 10-μm particle size, and particles generated by mixing, sonicating, or blending cultures range from 1.9–4.8 μm, depending on the operation (Kenny & Sabel, 1968; Reitman & Wedum, 1956). It has been shown that particles <5 μm in diameter are most effective in producing respiratory infection in animals (Hatch, 1961). See Chapter 2, The Lungs, and Chapter 8, Particulates, for additional aerosol information.

Infectious airborne particles can be generated not only from aerosolized liquids but also from lyophilized cultures, dried bacterial colonies, dried material on stoppers and caps of culture tubes and bottles, dried exudates, fungal and actinomycete spores released when cultures are opened, and dusts from animal cages. Tuberculosis, psittacosis, Q fever, pulmonary mycoses, and, in special circumstances, brucellosis are diseases documented to have been associated with airborne infection in the laboratory and other workplaces as well as in the community.

The epidemiological review of workplace infections verifies the observation that people whose work brings them into contact with pathogenic microorganisms are at greater risk of infection than the general population. However, the mere presence of an agent does not necessarily lead to occupational exposure and infection. Certain conditions must be satisfied before an infection occurs. They involve the multiple interrelated factors of route of entry, dose, viability, virulence, mode of transmission, and host susceptibility. For an infection to occur, the agent must be pathogenic and viable, present in sufficient numbers to produce infection, and be transmitted successfully and delivered to a susceptible host at a suitable entry site. It may be possible to reduce or eliminate susceptible hosts through immunization.

## Hazard Classification

Biological agents are not all equally dangerous to workers. An understanding of the potential for the hazardous agent to cause human disease, known as its pathogenicity, and through what routes of infection the agent is efficiently delivered to a worker permits one to develop a classification based on risk.

### Hazard Categories

**Biosafety containment levels—background.** In the 1970s, etiologic agents were classified by the CDC on the basis of hazard (CDC, 1974). The list compiled for that

purpose still exists in slightly modified form in many government documents, including a document describing interstate shipment of etiologic agents (CDC, 1980). For appropriate identification of the categories to be used in packaging, labeling, and shipping etiologic agents, lists were provided by the CDC in 1974 and in 42 *CFR* Part 72 (CDC, 1980). The 1974 list came to be used as a classification scheme in which an organism, genus, or group of microorganisms can be categorized into a specific hazard group by users and other federal agencies. The 1980 list identified organisms that were required to be sent by registered mail. At the present time, etiologic agents and other infectious materials are considered hazardous materials under the U.S. Department of Transportation (USDOT) 49 *CFR* Parts 171–180, although they are not included as hazardous agents under the OSHA hazard communication standard. The USDOT inclusion and the OSHA hazard communication MSDS (material safety data sheet) exemptions cause some confusion. Obviously, information similar to that found in certain sections of an MSDS is needed to allow appropriate spill cleanup and to alleviate inappropriate public perception of risk for any shipment of microorganisms. On the other hand, if the organism being shipped is not pathogenic, there should be a mechanism for declaring an exemption from the USDOT list so as not to restrict packaging to the small size limitations required for etiologic agents.

Lists currently in use should be reviewed at least annually and updated accordingly because of numerous taxonomic changes and the recognition that any species of organism may have avirulent and virulent strains. Information on the more commonly recognized human pathogens can be found in peer-reviewed scientific literature and in numerous microbiology textbooks. Unfortunately, it is rare to find an assessment of the level of risk or any directives regarding the containment to be used in working with pathogens in such references. However, if the agent is listed in one of the guidelines, the first estimation of its biosafety containment level can be ascertained (OSHA, 1991; NIH, 1994; USPHS,1993; USPHS,1974).

Opportunistic microbes and normal flora that can cause disease are often inferred to be at the same risk level as a frank pathogen, because the case reports in the literature do not routinely account for host factors by differentiating between an immunocompromised host and a healthy worker. Organisms that cause infections in immunocompromised adults should be evaluated to determine whether they pose a risk to healthy adults as well. Organisms that have been attenuated to make live vaccines may no longer require the same containment level as their wild-type, parent organism (*Official Journal of the European Communities, 1990,* Annex VI). For example, BCG (an attenuated strain of *Mycobacterium bovis* that may confer immunity against TB) is handled at biosafety level 2 (BSL-2), which is less restrictive than BSL-3, the level at which its parent organism, *Mycobacterium bovis,* is classified

(CDC, personal communication, 1994). Poliovirus vaccine that is given orally to children and adults should be considered to be BSL-1, although wild-type strains of polio are handled at BSL-2. Influenza vaccine strains are handled at BSL-1, whereas the parent virus is BSL-2 (CDC, 1974). The CDC Center for Infectious Diseases, Bacterial and Mycotic Diseases Branch (or other branches as appropriate) should be able to provide information on human pathogenicity of microbial agents.

After an estimated biosafety containment level for a microbial agent is provided to the appropriate supervisor, the remainder of the risk assessment involving host–environment interactions should be done by the institution (the biosafety committee and/or a biosafety officer).

Although not microorganisms, cell lines and primary cells in culture must be mentioned because of their potential to be contaminated with infectious agents. When assessing the potential hazard associated with cells in culture, consider the source of the cells (human, rodent, etc.), the potential for the cells to harbor viruses or mycoplasma, whether they are tumor cells or have been transformed with virus, or whether they are established lines or only recently isolated from a host (primary cells).

If allergens or chemical agents are used, appropriate precautions to prevent sensitization should be based on exposure control limits that have been established by a reputable association, as was done by the working party of the European Federation of Biotechnology (Kuenzi et al., 1985). Biosafety levels were developed for use in protection against living microorganisms that have the potential to cause infectious disease in healthy human adults; they are not usually appropriate for the control of other hazards.

**Published guidelines of containment levels.** The biosafety guidelines most commonly used in the United States for containment of biohazardous agents in the workplace are those recommended by the CDC, The National Institutes of Health, and the National Research Council (NIH, 1994; USPHS, 1993; CDC, 1974; National Research Council [NRC], 1989). Other guidelines found in the literature are based on interpretations of the recommendations of these agencies (OSHA, 1991; Kent and Kubica, 1985; Kruse et al., 1991; Fleming et al., 1995; National Committee for Clinical Laboratory Standards [NCCLS], 1991).

*The CDC/NIH Guidelines for Microbiological and Biomedical Laboratories (USPHS, 1993).* The CDC/NIH guidelines (*Biosafety in Microbiological and Biomedical Laboratories [BMBL]*) recommend that laboratory directors establish work practices involving containment equipment and facilities in the workplace. In doing this, they must take into account interactions of the virulence of the agent, immune status of potential hosts, and the hazards of the procedure. In addition, laboratory directors are responsible for making appropriate risk assessment of agents not included among the *BMBL* agent summary statements. The director

must be familiar with the subject of risk assessment regarding biological agents or seek the advice of one who has such expertise.

Although the guidelines assign the responsibility for risk evaluation to the laboratory director, no method is suggested for determining the virulence of agents. Except for agents that have caused laboratory-acquired infections or those that pose a serious hazard to healthy adults, the CDC/NIH guidelines do not account for the use of most agents. Although a mechanism for updating the agent summary statements in *BMBL* was proposed through publication in the *Morbidity and Mortality Weekly Report (MMWR)*, this has not been carried out in a timely manner. The only agent summary statement added between 1986 and 1993 was one for the human immunodeficiency virus. Information on the epidemiology and safety recommendations for the new hantavirus sin nombre was published in a special report in 1994 (CDC, 1994).

Risk management is achieved through use of practices, facilities, and equipment specified in defined biosafety containment levels. Biosafety practices are an important part of a program to manage the risk of exposure to potentially infectious agents. The general agent descriptions and applicable work environments described by the CDC and NIH in each of the biosafety levels (both small- and large-scale) are listed below as compiled directly from the NIH guidelines (NIH, 1991, 1994), the CDC (CDC, 1974), and the CDC/NIH (USPHS, 1993). An addendum at the end of this chapter outlines large-scale guidelines applicable to nonrecombinant organisms.

*Biosafety level 1 (BSL-1)* is used for work involving defined and well-characterized strains of viable microorganisms of no known or of minimal potential hazard to laboratory personnel or the environment. This level is appropriate for high school and undergraduate college teaching and training laboratories. No special competence is required, although training in the specific procedures should be provided, and there should be supervision by a scientist with general training in microbiology or a related science. The laboratory is not separated from general building traffic, and work is conducted on the open bench. Examples of organisms used under these conditions are *Bacillus subtilis, Naegleria gruberi,* and canine hepatitis virus. Much of the recombinant DNA work with *E. coli,* K12, and *Saccharomyces cerevisiae* has been approved at BSL-1.

*BSL-1LS* is for large-scale work (greater than 10 liters or production volumes), BSL-1LS is used for the agents that can be handled at BSL-1 on a small scale. Microbial agents that have been safely used for large-scale industrial production for many years may qualify for good large-scale practices (GLSP) status. Examples include *Lactobacillus casei, Penicillium camembertii, Saccharomyces cerevisiae, Cephalosporium acremonium, Bacillus thuringiensis,* and *Rhizobium mellioti.* The criteria for GLSP were originally developed for the European Economic Community (now known as the European Union) by the Organization for Economic Co-operation and Development (OECD) as "good industrial large-scale practices" (GILSP) (Frommer et al., 1989; OECD, 1986) and were slightly revised by the NIH for acceptance and use in the United States (NIH, 1991).

*Biosafety level 2 (BSL-2)* is used for work with many moderate-risk agents present in the community (indigenous) and associated with human disease of varying degrees of severity. Agents are usually of moderate potential hazard to personnel and the environment. This level is appropriate for clinical, diagnostic, teaching, and other research facilities in which work is done by individuals with a level of competency equal to or greater than one would expect in a college department of microbiology. Workers must be trained in good microbiological techniques in order for the handling of these agents on the open bench to be allowed when the potential for aerosol production is low. Laboratory personnel must have specific training in handling pathogenic agents and must be directed by competent scientists. They must be trained in proper use of biological safety cabinets or other appropriate primary containment equipment when the risk of aerosol production is high, as in such tasks as centrifuging, grinding, homogenizing, blending, vigorously shaking or mixing, performing sonic disruption, opening containers with increased internal pressure, inoculating animals intranasally, harvesting infected tissues from animals or eggs, and harvesting human cells from tissues using a cell separator. Access to the laboratory should be limited when work is in progress. Primary hazards to workers include accidental inoculation, exposure of nonintact skin or mucous membranes, and ingestion. Examples of organisms used under BSL-2 conditions are hepatitis B virus, *Salmonella* spp., and *Toxoplasma* spp.

*BSL-2LS* is for large-scale, in vitro work with agents that require BSL-2 containment when work is at small scale. A detailed description of the containment requirements for BSL-2LS can be found in Appendix K-III of the NIH's recombinant DNA guidelines (NIH, 1991, 1994).

*Biosafety level 3 (BSL-3)* is used for work with indigenous or exotic agents where the potential for infection by aerosols is real and the disease may have serious or lethal consequences. Indigenous and exotic agents vary by country, and within regions of some countries, so there must be some flexibility for assignment of containment levels. Biosafety level 3 is appropriate for clinical diagnostic microbiology work when tuberculosis or brucellosis is suspected and also for special teaching and research situations that require the handling of such agents. Partial containment equipment such as Class I or Class II biological safety cabinets is used for all manipulations of infectious material at BSL-3. There are special engineering design criteria and work practices associated with BSL-3 containment. Worker competency must equal or exceed that of college-level microbiologists and workers must have special training in handling these potentially lethal human

pathogens and infectious materials. Supervisors must be competent scientists who are experienced in working with these agents.

Primary routes of exposure of workers to BSL-3 hazards include inhalation, with relatively few infections reported as a result of accidental autoinoculation or ingestion. The extra personal protective clothing probably serves as a reminder of the hazard level and promotes an awareness that reduces such incidents. Examples of organisms used under BSL-3 conditions are *Mycobacterium tuberculosis, Brucella* spp., St. Louis encephalitis virus, Borna virus (an exotic agent when used in the United States), and *Coxiella burnetii.*

*BSL-3LS:* The detailed requirements for BSL-3LS are to be found in Appendix K-IV of the NIH recombinant DNA guidelines (NIH, 1991, 1994).

*Biosafety level 4 (BSL-4)* is used for work with dangerous and exotic agents that pose a high individual risk of life-threatening disease. Such an agent has a low infectious dose and poses a danger for the community from person-to-person spread. BSL-4 containment is appropriate for all manipulations of potentially infectious diagnostic materials, isolates, and naturally or experimentally infected animals. Maximum containment equipment, such as a Class III biological safety cabinet, or partial containment equipment in combination with a full-body, air-supplied, positive-pressure personnel suit is used for all procedures and activities. Because of the stringent requirements associated with BSL-4 containment, only a few facilities that meet this standard have been built and are operational.

The main hazard to laboratory or animal care personnel working with agents requiring such extreme caution and containment is respiratory exposure to infectious aerosols. Mucous membrane exposure to infectious droplets and accidental parenteral inoculation also play a role in transmission of infections. Worker competency must equal or exceed that of college-level microbiologists, and workers must receive specific, thorough training in handling extremely hazardous infectious agents. They must understand the function of the primary and secondary containment equipment and the facility design. Supervisors must be competent scientists trained and experienced in such work.

Laboratory access is strictly controlled. The facility is either separated from other buildings or completely isolated from other areas of the building. A separate facility operations manual is required. The maximum containment facility has special design and engineering features that prevent dissemination of microbes to the environment. Some examples of organisms used under these conditions are the agents of viral hemorrhagic fevers (Lassa, Machupo, Marburg, and Ebola), filoviruses, and certain arboviruses.

The requirements for laboratory-scale BSL-4 are described in *BMBL* (USPHS, 1993). Appendix K of the NIH recombinant guidelines (Physical containment for large scale uses of organisms) does not include a description of requirements for BSL-4LS because the requirements should be determined on a case-by-case basis if they are requested.

*Animal biosafety levels:* Other special biosafety precautions described in these guidelines apply to the use of naturally or experimentally infected animals. Animals are restricted from the laboratory unless they are part of the experiment at BSL-2 and higher, and decorative plants are restricted from use at BSL-3 and higher. Animals and plants harbor their own microbial flora, which could infect the worker or contaminate the work, so it is prudent to prohibit their use in microbiology laboratories.

There are intrinsic hazards associated with the use of certain animals (for example, from herpes B virus in macaques), which must be taken into account in assessing the risk to the worker. There are also extrinsic hazards when infectious agents are purposely used to infect animals and vector–host interactions are being studied (in vivo) as opposed to the "controlled work" in culture media (in vitro). These hazards must be addressed.

**NIH Guidelines for research involving recombinant DNA molecules (NIH, 1994 and revisions).** The CDC's *Classification of Etiologic Agents on the Basis of Hazard* (1974) was slightly modified and incorporated into Appendix B of the NIH guidelines. The original CDC classification system provided "points to consider" in estimating the degree of hazard, noting that it depended on the etiologic agent and its nature and use. These notes were not included in the NIH document, implying to the user that all members of the groups, species, and strains on that list were pathogenic. Because the NIH guidelines could not take into account all existing and anticipated information on special procedures, users were encouraged to recommend changes to the guidelines.

The NIH guidelines provide the same message as the CDC/NIH guidelines (*BMBL*) that the agent/product rather than the process of recombinant DNA work should be evaluated for worker safety (NRC, 1987). The basic biosafety requirements for work with all microorganisms are the same. However, nonpathogenic, genetically modified organisms are currently registered and regulated at most institutions because of the public's perception of risk, whereas work with true pathogens is often done without internal oversight and review because the public has not been made aware of the risk. For large-scale guidelines, see appendix K (NIH, 1994).

**National Research Council Guidelines (NRC, 1987, 1989).** *Biosafety in the Laboratory: Prudent Practices for the Handling and Disposal of Infectious Materials* (NRC, 1989): To avoid exposure to infectious agents, the NRC's Committee on Hazardous Biological Substances in the Laboratory recommended seven basic prudent biosafety practices, which are listed later in the chapter. These prudent practices provide barriers against the known routes of exposure for most diseases and are the basic recommendations for working with biohazardous agents. The recommendations are supplemented with additional practices, equipment, and facility design as the severity of the hazard increases. The practices recommended by the NRC, when accompanied by recommendations for facility design and containment equipment, are compatible with the CDC and NIH biosafety levels.

*Introduction of Recombinant DNA-Engineered Organisms into the Environment: Key Issues* (NRC, 1987): The NRC has concluded that there is no evidence of any unique hazards posed by recombinant DNA techniques. The risks associated with recombinant DNA are the same in kind as those associated with unmodified organisms or organisms modified by other means. The NRC recommended that risk assessments be based on the nature of the organism and the environment into which it is introduced, and not on the method by which it was produced. This recommendation was also accepted by the Office of Science and Technology Policy (OSTP, 1986). Given this conclusion, our two sets of guidelines, one for recombinant work and the other for work with human pathogens, are redundant. Guidelines for a single code of practice for protection of workers from exposure to biohazardous agents are appropriate and have already been accepted in Europe (*Official Journal of the European Communities*, 1990; OECD, 1986).

**Guideline interpretation.** Using the CDC/NIH guidelines previously described, decisions on containment levels for work at BSL-1 through BSL-3 at small and large scale can be made at the institutional level. BSL-4, because it is limited to so very few facilities, is not considered here. The expertise of an institution's biosafety committee and/or biosafety officer is needed for risk assessment of pathogenic agents and infectious materials. A professional biosafety consultant may be needed for facilities that do not have such in-house expertise. (The American Biological Safety Association in Mundelein, IL, can be contacted for a list of registered biosafety professionals.)

Risk assessment for agents to be used at large scale, especially for industrial production, should begin early as an integral part of the research and development process. The level of pathogenicity should be determined before production of the organism is scaled up, because special large-scale containment facilities are too costly to be built and maintained if they are not needed.

The Coordinated Framework for Biotechnology (Office of Science and Technology Policy, 1986) has added EPA, the U.S. Department of Agriculture, the Federal Drug Administration, and OSHA to the list of federal agencies such as the CDC and NIH who can provide oversight and information on risk assessment. Work with certain agents associated with terrorist activities or biological warfare may require oversight from federal defense agencies as well (Nettleman, 1991). The Department of Commerce also restricts the export of such agents/materials because of their potential use for biological warfare.

**Table 14–C.**  Summary of Recommended Biosafety Levels for Infectious Agents

| Biosafety Level | Agents | Practices | Safety Equipment (Primary Barriers) | Facilities (Secondary Barriers) |
|---|---|---|---|---|
| 1 | Not known to cause disease in healthy adults | Standard microbiological practices | None required | Open bench—top sink required |
| 2 | Associated with human disease; hazard is from auto-inoculation, ingestion, mucous membrane exposure | BSL-1 practice plus<br>• limited access<br>• biohazard warning signs<br>• "sharps" precautions<br>• biosafety manual defining any needed waste decontamination or medical surveillance policies | Primary barriers: Class I or II BSCs or other physical containment devices used for all manipulations of agents that cause splashes or aerosols of infectious materials; PPE; laboratory coats; gloves; face protection as needed | BSL-1 plus<br>  autoclave available |
| 3 | Indigenous or exotic agents with potential for aerosol transmission; disease may have serious or lethal consequences | BSL-2 practice plus<br>• controlled access<br>• decontamination of all waste<br>• decontamination of lab clothing before laundering<br>• baseline serum | Primary barriers: Class I or II BCSs or other physical containment devices used for all manipulations of agents; PPE; protective lab clothing; gloves; respiratory protection as needed | BSL-2 plus<br>• physical separation from access corridors<br>• self-closing, double-door access<br>• exhausted air not recirculated<br>• negative airflow into laboratory |
| 4 | Dangerous/exotic agents that pose high risk of life-threatening disease; aerosol-transmitted lab infections; or related agents with unknown risk of transmission | BSL-3 practices plus<br>• clothing change before entering<br>• shower on exit<br>• all material decontaminated on exit from facility | Primary barriers: All procedures conducted in Class III BCSs or Class I or II BSCs *in combination with* full-body, air-supplied, positive-pressure personnel suit | BSL-3 plus<br>• separate building or isolated zone<br>• dedicated supply/exhaust, vacuum, and decon systems<br>• other requirements outlined in the text |

(Adapted from CDC, *BMBL,* 1993.)

# ▪ Hazard Control

The process of developing controls to prevent or minimize occupational exposure to infectious agents or other biological agents becomes straightforward once the actual risk of work with the organism or agent is known and the risk category established. Prevention of exposure to potentially infectious agents can be achieved by source control, minimization of accidental release, and protection of the worker. Containment or barriers, used along with the other components of a comprehensive biosafety program, provide the means to work with biological agents without adverse effect.

## Containment

Workplace activities involving infectious or biological agents require containment so that workers, the immediate work environment, and the community including those outside the immediate workplace are protected or shielded from exposure. Facility design, safety equipment, and work practices are the building blocks of containment. Varying configurations of these components are used depending on the hazard category of the work.

Protection of workers and the immediate work environment, or primary containment, is achieved through the use of good work practices and appropriate safety equipment. Effective vaccines also decrease worker risk. Protection of the environment external to the workplace (the community),

or secondary containment, is attained by using adequately designed facilities and operational practices.

The CDC and NIH have designated the four biosafety levels (BSLs) previously outlined for work involving infectious agents or activities in which experimentally or naturally infected vertebrate animals are manipulated. Each biosafety level consists of a combination of laboratory practices and techniques, safety equipment, and facility design. The combination must be specifically appropriate for the operations performed, the documented or suspected routes of transmission of the agent, and the laboratory function or activity. The use of increasingly stringent procedures and more complex laboratory facilities permits microorganisms in higher-risk categories to be handled safely. Tables 14–C and 14–D summarize the recommended biosafety levels for handling different categories of infectious agents and experimentally or naturally infected animals (USPHS, 1993).

**Facility design.** The laboratory facility provides the shell, or barrier, necessary to protect the community and those outside the immediate work area from exposure to hazardous materials. When agents of increasing hazard are manipulated, facility design plays a more important role in reducing the potential for dissemination of the agent, particularly when an accidental release within the laboratory occurs.

Biosafety level–1 and –2 laboratories have no special design features beyond an ordinary laboratory, except that

**Table 14–D.** Summary of Recommended ABSLs—Biosafety Levels for Activities Using Experimentally or Naturally Infected Vertebrate Animals

| Biosafety Level | Agents | Practices | Safety Equipment (Primary Barriers) | Facilities (Secondary Barriers) |
|---|---|---|---|---|
| 1 | Not known to cause disease in healthy adults. | Standard animal care and management practices, including appropriate medical surveillance programs | As required for normal care of each species | Standard animal facility: • no recirculation of exhaust air • directional air flow recommended |
| 2 | Associated with human disease. Hazard from percutaneous exposure, ingestion, mucous membrane exposure. | ABSL-1 practice plus • limited access • biohazard warning signs • "sharps" precautions • biosafety manual • decontamination of all infectious wastes and of animal cages prior to washing | ABSL-1 equipment plus primary barriers; containment equipment appropriate for animal species; PPE; laboratory coats; gloves; face and respiratory protection as needed | ABSL-1 facility plus • autoclave available • handwashing sink available in the animal room |
| 3 | Indigenous or exotic agents with potential for aerosol transmission; disease may have serious health effects. | ABSL-2 practice plus • controlled access • decontamination of clothing before laundering • cages decontaminated before bedding removed • disinfectant foot bath as needed | ABSL-2 equipment plus containment equipment for housing animals and cage-dumping activities; Class I or II BSCs available for manipulative procedures (inoculation, necropsy) that may create infectious aerosols; PPE; appropriate respiratory protection | ABSL-2 facility plus • physical separation from access corridors • self-closing, double-door access • sealed penetrations • sealed windows • autoclave available in facility |
| 4 | Dangerous/exotic agents that pose high risk of life-threatening disease; aerosol transmission; or related agents with unknown risk of transmission | ABSL-3 practices plus • entrance through change room where personal clothing is removed and laboratory clothing is put on; shower on exiting. • decontamination of all wastes before removal from the facility | ABSL-3 equipment plus maximum containment equipment (i.e., Class III BSC or partial containment equipment in combination with full-body, air-supplied, positive-pressure personnel suit) used for all procedures and activities | ABSL-3 facility plus • separate building or isolated zone • dedicated supply/exhaust, vacuum, and decontamination systems • other requirements outlined in the text |

(Adapted from CDC, *BMBL*, 1993.)

a hand-washing sink is required and an autoclave must be available for BSL-2 activities.

The World Health Organization (WHO, 1993) characterizes agents handled in BSL-1 and BSL-2 laboratories as those having no or very low individual or community risk (BSL-1) or moderate individual risk and low community risk (BSL-2); therefore, the need for special design features to protect the community does not arise in these labs.

The biosafety level–3 laboratory includes design features of BSL-1 and BSL-2 laboratories plus controlled access (double-door entry), a specialized ventilation system that creates a directional one-pass airflow into the laboratory from surrounding "clean" areas, and special hand-washing controls (elbow, foot, or knee operated). Microorganisms classified at risk level 3 present a high risk to individuals, usually by respiratory exposure, although risk to the community is low.

BSL-3 facility requirements impose additional expense for containment features. In some circumstances where existing facilities do not meet BSL-3 requirements or cannot be made to do so, some accommodation may be made. Work involving routine or repetitive operations (such as diagnostic procedures involving propagation of agents for identification, typing, and susceptibility testing) can be carried out in a BSL-2 facility as long as the work practices and safety equipment required at BSL-3 are used. The decision to alter containment conditions should only be made by the laboratory director. The publication *Biosafety in Microbiological and Biomedical Laboratories* (USPHS, 1993) contains agent summary statements for most microorganisms handled at BSL-3 and, where appropriate, includes the option to modify containment conditions.

The biosafety level–4 facility, though rare, draws a great deal of attention, perhaps because it conjures visions of an "Andromeda strain." Agents handled in such a facility pose a high risk to both workers and the community, so design criteria must prevent both worker and community exposure. In addition to having the design components of level-3 facilities, a BSL-4 facility is housed in a separate building or in an isolated zone that has dedicated ventilation, vacuum, and decontamination systems. For more detailed information, see USPHS (1993) and Kuehne (1973).

**Safety equipment.** Because most experimental procedures are recognized as having the potential to generate aerosols, safety equipment designed to reduce the likelihood

**Table 14–E.**    Selection of a Biosafety Cabinet Through Risk Assessment

| Risk | | | Select | | |
|---|---|---|---|---|---|
| | | | % of Airflow | | |
| Agent | BSC Class | Face Velocity | Recirculated | Exhausted | Exhaust Airflow |
| BSL-1–3 agents | II/A | 75–100 | 70 | 30 | Exhaust into room; exhaust to outside with a thimble unit |
| BSL-1–3 agents; minute amounts of radionuclides and volatile toxic chemicals | II/B1* | 100 | 30 | 70 | Exhaust vented to outside with a hard duct |
| BSL-1–3 agents; radionuclides and volatile toxic chemicals | II/B2* | 100 | 0 | 100 | Exhaust vented to outside with a hard duct |
| BSL-1–3 agents; minute amounts of radionuclides and volatile toxic chemicals | II/B3* | 100 | 70 | 30 | Exhaust vented to outside with a hard duct |
| BSL-1–4 agents; radionuclides and chemicals | III* | NA | 0 | 100 | Exhaust vented to outside with a hard duct |

*All biologically contaminated ducts and plenums are under negative pressure or are surrounded by negative-pressure ducts and plenums.
(Adapted from CDC, Office of Health and Safety, 1994.)

of worker and environmental exposure has become standard equipment in biological laboratories during the last decade.

***Biological safety cabinets (BSCs).*** The most frequently used and effective example of laboratory containment equipment is the biological safety cabinet, which provides a primary barrier to prevent escape of infectious aerosols into the work environment. When used and maintained properly they provide a combination of worker, product, and environmental protection that varies according to the class and type of cabinet selected.

All three classes of biological safety cabinets (Class I, II, and III) have high-efficiency particulate air, or HEPA, filters for exhaust air. Of these, the Class II cabinet is most widely used. Selection of the class and type of cabinet must be based on the hazard level of the microorganism to be manipulated, the nature of the work activity (the potential of a technique to produce aerosols), and the need to protect the worker and/or the work environment from airborne contamination (see Table 14–E). Class I and II cabinets, when used in conjunction with good microbiological practices, provide an effective means to safely manipulate moderate- and high-risk microorganisms (BSL-2 and BSL-3).

Class I ventilated cabinets provide personnel and environmental protection by means of a nonrecirculated inward airflow away from the operator. The minimum face velocity at the work opening is at least 75 linear feet per minute (lfpm). The cabinet exhaust air is HEPA filtered to protect the environment before it is discharged either to the laboratory or through duct work to the outside atmosphere. In practice this cabinet functions in a manner similar to a chemical fume hood, except for the additional HEPA filtration of exhaust air. The use of Class I cabinets is relatively rare, although they continue to be adapted to provide containment for aerosol-producing equipment (or procedures) such as centrifuges, pressurized apparatus, and

necropsy of infected animals. Figure 14–1 shows the design and airflow patterns of this cabinet.

Because Class II biosafety cabinets provide protection to workers, experimental materials, and the environment and are easily accessed through a front work opening, they are often used in biological laboratories for manipulation of microorganisms and tissue cultures. Class II cabinets have a face velocity of 75–100 lfpm and are divided into types A and B, with type B cabinets designated as B1, B2, or B3 (see Figures 14–2a through d). Class II, type A cabinets may be used for microbiological activities when volatile or toxic substances and radionuclides are *not* used. The exhaust air from this cabinet is usually discharged to the work environment, although it may be connected to exhaust duct-work by a thimble connection. The design criteria for this cabinet permit contaminated ducts and plenums under positive pressure.

All Class II, type B cabinets are hard-ducted to the outside atmosphere and have 100 lfpm face velocity, and when contaminated plenums exist they are under negative pressure or are surrounded by negative-pressure ducts or plenums. Therefore, work associated with varying amounts of volatile and toxic chemicals and radionuclides may be handled in this type of cabinet. Whereas the Class II, type B1 cabinet is the basic B design, B2 and B3 have been developed to provide some useful alternatives. Because the B2 cabinet is a "total exhaust" cabinet with no air recirculation, it can be used for cell work involving hazardous or toxic chemicals, such as carcinogens and radionuclides. The B3 cabinet, known as the "convertible cabinet," is basically a type A design that can be converted to meet the type B criteria.

The Class III cabinet is a totally enclosed, ventilated, negative-pressure cabinet of gas-tight construction that is used for work requiring the highest level of containment (see Figure 14–3). It offers maximum protection for personnel, the environment, and work materials. Personnel protection

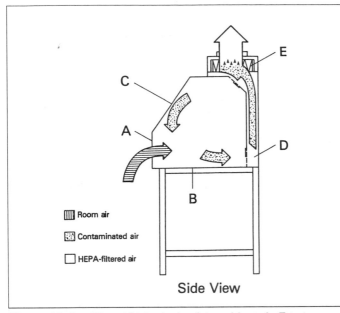

**Figure 14–1.** Class I biological safety cabinet. A: Front opening; B: Work surface; C: Window; D: Exhaust plenum; E: HEPA filter. (Reprinted from *Biosafety in Microbiological and Biomedical Laboratories* [*BMBL*]. CDC, 1993, p. 142.)

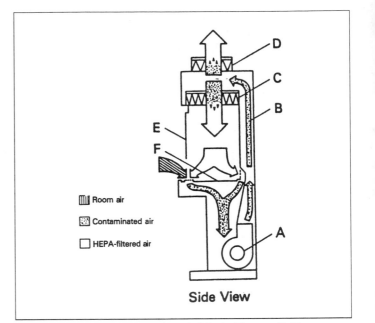

**Figure 14–2a.** Class II, type A BSC. A: Blower; B: Rear plenum; C: Supply HEPA filter; D: Exhaust HEPA filter; E: Sash; F: Work surface. (Reprinted from *BMBL*. CDC, 1993, p. 143.)

equivalent to that provided by the Class III cabinet can also be attained by using a positive-pressure ventilated suit in a maximum containment facility in conjunction with a Class I or II cabinet (USPHS, 1993, p. 141).

***Certification of biosafety cabinets.*** The effectiveness of biological safety cabinets depends on a combination of airflow velocity, filter integrity, and location in the laboratory, because ventilation currents and even workers' movements can disrupt cabinet air patterns. It should not be assumed that equipment (or a facility) is providing worker protection merely because it is designed to do so. To be assured that a biological safety cabinet is functioning as designed, it must be certified on a regular basis (prior to use, when moved, and after a filter change, or at least annually).

The National Sanitation Foundation Standard #49 (NSF, 1992) for Class II (laminar-flow) biohazard cabinetry specifies materials, design and construction, and performance criteria for manufacturers. It also outlines recommended field tests for certifiers. Manufacturers must submit new models for NSF testing. Approved models carry the NSF seal. Only approved cabinets should be purchased and used for work activities involving potentially hazardous biological agents

Correct design, materials, and construction; rigorous testing; and proper placement are not enough to prevent workplace and environmental exposures. The missing ingredient is a well-trained, knowledgeable, and conscientious worker. Poor work practices can easily cancel containment features designed into the cabinet, permitting the release of infectious particles into the environment. It is for this reason that worker training is critical. The reader

can find additional information on the design and use of these cabinets in a variety of excellent references (NRC, 1989; Fleming et al., 1995; Lupo, 1989; USPHS, 1993; WHO, 1993).

Horizontal- or vertical-flow clean benches, which force air out of the front opening into the room, should not be confused with biosafety cabinets. They *do not* protect workers from exposure; they protect the work product. Therefore, clean benches *must not* be used for work with materials that are potentially infectious, toxic, allergenic, or irritating.

***Centrifugation.*** Centrifugation can present two serious hazards: mechanical failure and dispersion of aerosols. A mechanical failure such as a broken drive shaft, a faulty bearing, or a damaged rotor can produce not only aerosols but also fast-moving fragments. Even when functioning correctly, a centrifuge is capable of producing hazardous aerosols if improperly operated or when poor laboratory practices are used. Mechanical failure can be minimized by routine maintenance and meticulous observance of the manufacturer's instructions. Generation of aerosols is avoided by using good work practices such as balancing and not overfilling tubes, checking containers for cracks and signs of stress, and checking and greasing O-rings where applicable. Aerosolization is minimized by placing primary containers into centrifuge safety cups and opening rotors and centrifuge containers in biosafety cabinets.

Laboratory procedures where force or energy act on cell walls or tissue to disrupt them result in the dispersion of aerosols and splatter. Particular attention should be paid to containment when procedures such as homogenization

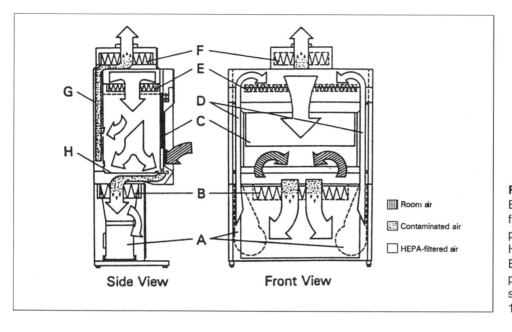

**Figure 14–2b.** Class II, type B1 BSC. A: Blowers; B: Supply HEPA filters; C: Sliding sash; D: Positive-pressure plenums; E: Additional supply HEPA filter or back-pressure plate; F: Exhaust HEPA filter; G: Negative-pressure exhaust plenum; H: Work surface. (Reprinted from *BMBL*. CDC, 1993, p. 143.)

and sonication are planned. In the case of homogenizers, some manufacturers design models that prevent the release of infectious material, and specially gasketed blenders are also available. The use of properly designed and maintained containment equipment is essential in minimizing release of infectious aerosols to the environment.

**Work practices.** Work practices—how one actually does the work—are the most important component in preventing occupational exposure. Understanding the concepts of transmission, infectious dose, and route of entry, as well as the potential for various procedures to release infectious material, is critical to the implementation of appropriate containment practices. When a risk assessment is performed on

work activities in advance, it becomes straightforward to identify the potential hazard and implement the safeguards necessary to protect workers.

The safeguards known as the seven basic rules of biosafety (NRC, 1989) are summarized here:

- Do not mouth pipette.
- Manipulate infectious fluids carefully to avoid spills and the production of aerosols and droplets.
- Restrict the use of needles and syringes to procedures for which there are no alternatives; use needles, syringes, and other sharps carefully to avoid self-inoculation; and dispose of sharps in leak- and puncture-resistant containers.

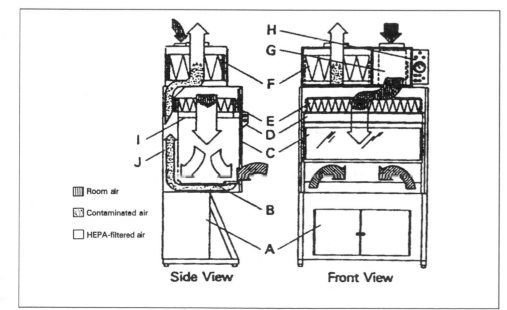

**Figure 14–2c.** Class II, type B2 BSC. A: Storage cabinet; B: Work surface; C: Sliding sash; D: Lights; E: Supply HEPA filter; F: Exhaust HEPA filter; G: Supply blower; H: Control panel; I: Filter screen; J: Negative-pressure plenum. (Reprinted from *BMBL*. CDC, 1993, p. 144.)

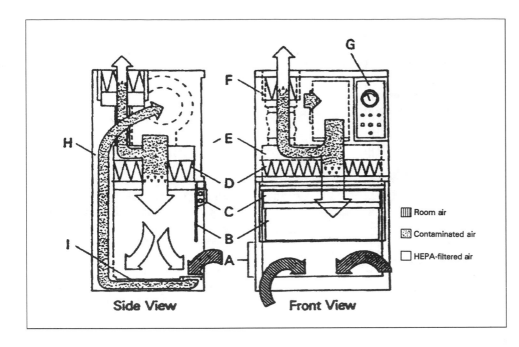

**Figure 14–2d.** Tabletop model of a Class II, type B3 BSC. A: Front opening; B: Sliding sash; C: Light; D: Supply HEPA filter; E: Positive-pressure plenum; F: Exhaust HEPA filter; G: Control panel; H: Negative-pressure plenum; I: Work surface. (Reprinted from *BMBL.* CDC, 1993, p. 144.)

- Use protective laboratory coats and gloves.

- Wash hands after all laboratory activities, after removing gloves, and immediately following contact with infectious materials.

- Decontaminate work surfaces before and after use, and immediately after spills.

- Do not eat, drink, store food, apply cosmetics, or smoke in the laboratory.

Although not a comprehensive listing, these rules represent baseline or minimum practices to be followed. They can be amplified with additional protective clothing such as goggles, full-face shields, or masks when face protection from splatter is needed, and with differing types of clothing such as back-fastening gowns, jumpsuits, impervious aprons, sleeve covers, and head and foot covers as suitable. Personal habits such as nail biting and eye and nose rubbing must be avoided because they offer an excellent means for ingesting pathogens and contaminating mucous membranes. The use of good microbiological practices is critical for a worker's own protection and the protection of adjacent colleagues.

**Decontamination.** The protection of personnel and the environment from exposure to infectious agents and the prevention of contamination of experimental materials by a variable, persistent, and unwanted background of

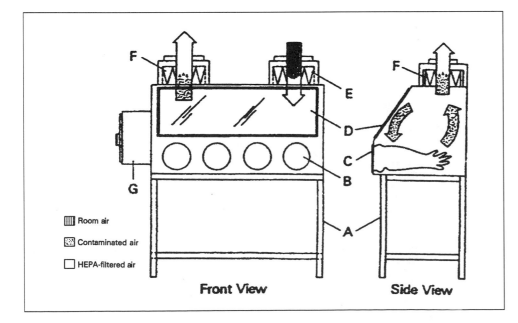

**Figure 14–3.** Class III BSC. A: Stand; B: Glove ports; C: O-ring for attaching arm-length gloves to cabinet; D: Sloped glass viewing window; E: Supply HEPA filter; F: Exhaust HEPA filter (Note that the second exhaust HEPA filter required for Class III cabinets is not depicted in this diagram.); G: Double-ended autoclave. (Reprinted from *BMBL.* CDC, 1993, p. 145.)

microorganisms is an integral part of good microbiological procedure. Decontamination, the use of physical or chemical means to render materials safe for further handling by reducing the number of organisms present, must be differentiated from disinfection, a process that kills infectious agents outside the body. Neither of these terms should be confused with sterilization, which implies complete elimination or destruction of all forms of microbial life. In the laboratory setting, the application of heat, either moist or dry, is the most effective method of sterilization. Steam at 250 F (121 C) under pressure in an autoclave is the most widely used and convenient method of rapidly achieving sterilization. However, many variables such as time, temperature, configuration and size of load, and permeability and dimensions of containers must be taken into account in order to successfully sterilize materials.

Chemical disinfectants inactivate microorganisms by one or more of a number of chemical reactions, primarily coagulation and denaturation of protein, lysis, or inactivation of an essential enzyme by either oxidation, binding, or destruction of the enzyme substrate. The level of effectiveness of chemical disinfectants is altered by changes in the concentration of active ingredients, contact duration, temperature, and humidity and the concentration of organic matter and the pH of the material being disinfected. Chemical disinfectants, classified by their active ingredients, include halogens, acids and alkalis, alcohols, heavy-metal salts, quaternary ammonium compounds, phenolics, aldehydes, ketones, and amines.

Specific terminology and classification schemes for chemical disinfectants used by the medical community and licensed by the EPA can be found in several excellent references and texts (Rutala, 1990; Garner & Favero, 1986; Spaulding, 1972; Block, 1991; Favero & Bond, 1991; Klein & Deforest, 1963).

The most frequently used disinfectants in the workplace include sodium hypochlorite (household bleach), isopropyl or ethyl alcohol, iodophors (Wescodyne), and phenolics (Lysol and amphyl). It is essential when choosing a disinfectant to review the manufacturer's literature to determine the disinfectant's efficacy (what microorganisms the disinfectant inactivates) and the recommended application (as an inanimate surface disinfectant, topical disinfectant, surgical scrub, liquid sterilant, or sanitizer). Table 14–F highlights various chemical disinfectants and their uses. By definition, chemical disinfectants are toxic to viable cells, so it is critical that users be familiar with the hazard potential of compounds they use and take necessary precautions to prevent workplace exposure. Compounds such as ethylene oxide, formaldehyde, glutaraldehydes, and concentrated acids and bases require special handling procedures. Consult OSHA regulations 29 *CFR* 1910.1000 (Table Z–1–A, Limits for Air Contaminants), 29 *CFR* 1910.1047 (Ethylene Oxide), 29 *CFR* 1910.1048 (Formaldehyde), 29 *CFR* 1910.1000 (Table Z–2), 29 *CFR* 1910.1450 (Occupational Exposure to Hazardous Chemicals in Laboratories), or 29 *CFR* 1910.1200

(Hazard Communication Standard) for regulations and information on chemical hazards.

**Infectious waste.** During the past decade the management of infectious waste has come under scrutiny from regulatory agencies. Public fear of exposure to AIDS and hepatitis has prompted many to demand the implementation of rigorous controls for infectious hospital and medical wastes. Most states have promulgated what can only be described as a patchwork of infectious waste regulations. Such waste is usually treated by autoclaving before disposal into a sanitary landfill, by incineration, or, in the case of some liquid wastes, by chemical disinfection. Local regulations should be consulted for individual state requirements. Federal regulations such as the Department of Transportation Hazardous Materials Regulations, 49 *CFR* 170 series, establish some packaging and volume limitations for the shipment of regulated medical wastes.

Because of the many misconceptions regarding hospital waste, the CDC published the following statement (CDC, 1987,1987c; Garner & Favero, 1986):

> There is no epidemiologic evidence that most hospital waste is any more infective than residential waste. Moreover, there is no epidemiological evidence that hospital waste has caused disease in the community as the result of improper disposal. Therefore, identifying wastes for which special precautions are indicated is largely a matter of judgment about the relative risk of disease transmission. The most practical approach to the management of infectious waste is to identify those wastes with a reasonable potential to cause infection during handling and disposal, and for which some special precautions appear prudent. Hospital wastes for which special precautions appear prudent include microbiology laboratory waste, pathology waste, and blood specimens and blood products. Although any item that has had contact with blood, exudates, or secretions may be potentially infective, it is not usually considered practical or necessary to treat all such wastes as infective.

***Spill management.*** The management of spills in the laboratory usually consists of flooding the contaminated area with liquid disinfectant, being careful not to generate aerosols, allowing adequate contact time for disinfection, cleaning up the spill, reapplying fresh disinfectant, and final cleanup. Protective clothing is always worn. When spills involve large volumes of infectious agents, personnel should leave the area until any aerosols have settled before cleanup is begun. Variations of this procedure are used for hospital spills, where emphasis is placed on first absorbing and cleaning the spill and then disinfecting the area. The reason for the difference is the concentration of the microorganisms present in the spill and the likelihood that the general public may be exposed and inadvertently spread contamination elsewhere.

**Summary.** Containment of microorganisms in laboratories (or other workplaces) is critical to the health of workers and to the community. Engineering controls such as safety

**Table 14–F.** Summary of Practical Disinfectants

| | Ethylene Oxide | Paraform-aldehyde (gas) | Vaporized Hydrogen Peroxide | Quaternary Ammonium Compounds | Phenolic Compounds | Chlorine Compounds | Iodophor Compounds | Alcohol (ethyl or isopropyl) | Formaldehyde (liquid) | Glutaral-dehyde | Hydrogen Peroxide (liquid) |
|---|---|---|---|---|---|---|---|---|---|---|---|
| **Use parameters** | | | | | | | | | | | |
| Concentrations of active ingredients | 400–800 mg/l | 0.3 g/ft³ | 2.4 mg/l | 0.1–2% | 0.2–3% | 0.01–5% | 0.47% | 70–85% | 4–8% | 2% | 6% |
| Temperature, °C | 35–60 | >23 | 4–50 | | | | | | | | |
| Relative humidity, % | 30–60 | >60 | <30 | | | | | | | | |
| Contact time, min. | 105–240 | 60–180 | 8–60 | 10–30 | 10–30 | 10–30 | 10–30 | 10–30 | 10–30 | 10–600 | 10–600 |
| **Effective against*** | | | | | | | | | | | |
| Vegetative bacteria | + | + | + | + | + | + | + | + | + | + | + |
| Bacterial spores | + | + | + | | | ± | | | ± | + | + |
| Lipo viruses | + | + | + | + | + | + | + | + | + | + | + |
| Hydrophilic viruses | + | + | + | | ± | + | ± | ± | + | + | + |
| Tubercle bacilli | + | + | | | + | + | + | + | + | + | + |
| HIV | + | + | | + | + | + | + | ± | + | + | + |
| HBV | + | + | | | ± | + | ± | ± | + | + | |
| **Applications*** | | | | | | | | | | | |
| Contaminated liquid discard | | | | + | + | + | | | ± | | |
| Contaminated glassware | ± | | | | | + | | + | ± | + | + |
| Contaminated instruments | ± | | | | + | | | | ± | + | + |
| Equipment total decontamination | ± | + | + | | | | | | | | |

*A + denotes a very positive response; a ±, a less positive response; and a blank, a negative response or not applicable.

(National Research Council. *Biosafety in the Laboratory: Prudent Practices for the Handling and Disposal of Infectious Materials.* Washington, DC: The Council, 1989, p. 40; and Miller et al., *Laboratory Safety: Principles and Practices.* Washington, DC: American Society for Microbiology, 1986, pp. 188–189 and 1995, pp. 226–227.)

equipment and facility design are important because, except for monitoring and appropriate maintenance, they do not require worker input to be effective. Despite this, experience indicates that the use of worker-initiated workplace controls in the form of good work practices and carefully executed techniques is critically important in minimizing biohazardous exposures in the work environment.

## Biosafety Program Management

The primary focus of an institutional biosafety program is to ensure that workers, their colleagues, and the community (which includes the general population and the environment) are not adversely affected by potentially hazardous microorganisms or their toxins. Biosafety program components usually include program support, a biosafety officer or specialist, an institutional biosafety committee, a biosafety manual of written policies and procedures, an occupational health program for relevant employees, and employee training or information communication. Biosafety programs vary markedly depending on the size of the institution and its activities (such as education, industrial research and development, manufacturing, medical patient care, or food service).

**Program support.** Without strong program support, even the best biosafety program has little chance of succeeding. It is critical to have administrative and financial support. Without support from upper management it is impossible to implement committee decisions and biosafety policies. Inadequate financial support of biosafety activities is equally problematic. Program financing usually comes from one of two general sources: fee for service or institutional overhead. Each method of support has its pros and cons, but it is important to prevent a situation where health and safety services are not used because of their cost.

**Biosafety specialist.** As institutional activities vary, so do the responsibilities and duties of the biosafety officer/specialist. A strong background in microbiology is critical so that this person may interact successfully with the technical community. In addition to microbiology, knowledge in the disciplines of molecular biology, infectious diseases, public health, sanitation, environmental microbiology, and epidemiology is extremely helpful. In addition to these academic credentials, it is equally important to have worked with microorganisms. An understanding of workplace procedures and equipment becomes invaluable when performing a risk assessment on a work activity or designing workable containment for experiments. A health and safety professional who is not trained in biosafety may misinterpret a perceived hazard because of lack of experience with the work activities or agents involved.

Because it is often necessary to receive input not only from the scientific staff but also from the biosafety committee and the administration, the biosafety professional must be able to lead diverse groups to a consensus. Biosafety officers work with architects, contractors, facility engineers, medical staff, animal facility personnel, and custodial services, to mention only a few groups.

**Institutional biosafety committee (IBC).** The original impetus to form these committees is found in the first publication (1976) of the "NIH guidelines for research involving recombinant DNA molecules." The NIH mandated that institutions receiving NIH funding conduct recombinant DNA research in compliance with their guidelines. Formation of an IBC was required. Committees were tasked to review research activities involving recombinant technology for compliance with the guidelines, and to oversee the safe conduct of work. Initially, many institutions chose to have their committees oversee only work activities involving the use of recombinant technology; consequently, work with infectious agents was not scrutinized. Today, most biosafety committees set policy and procedures for all activities involving infectious agents, materials, and animals, in addition to experiments using recombinant technology.

As defined by the NIH, IBCs overseeing recombinant activity must have no fewer than five members selected for their experience and expertise in the technology and their capability to assess the safety of the work and any potential risk to public health and the environment. At least two members must be chosen from outside the institution to represent the interests of the community. As committee responsibilities broaden, in addition to a strong core of microbiologists, experts in other disciplines (such as animal resources and plant pathology) are needed.

Institutional biosafety committees must interact with other review bodies within the institution. The use of infected animals alters the safety considerations to be addressed and the questions that must be answered, so institutional animal care and use committees (IACUCs) also become involved. As recombinant manipulations lead to the treatment of disease and human subjects become involved, human subject review boards (IRBs), radiation safety committees (RSCs) and radiation safety officers, chemical safety committees (CSCs) and chemical hygiene officers, and IBCs may need to interact to approve research proposals. In the interest of facilitating the timely review of work, it is imperative that the committees work together effectively.

**Biosafety manual.** One of the more challenging tasks of a health and safety professional is getting health and safety policy and other information into the workplace. A biosafety manual is one means of handling the communication of information and policy.

At the CDC Third National Symposium on Biosafety in 1994, a workshop was held covering the topic of biosafety manuals and their use and maintenance. Most of the health and safety professionals attending the workshop agreed that it is useful to have some type of biosafety manual. Many participants indicated that their institution had manuals and that these manuals are not used for a variety of reasons. Reasons for this included the fact that the manuals were difficult to use or out of date; employees did not know of their existence; or policies outlined in the manual did not have administrative and supervisory support. To help avoid some of these problems, including spending a great deal of time writing a manual that is destined to be underused, some preparation and planning is helpful. The following series of questions will help define the type of document appropriate for an institution:

- Why have a manual?
- What is the manual supposed to accomplish? (What should its scope be?)
- What support is needed for acceptance?
- Who should write it?
- What should be included?
- Should the manual be printed or available electronically?
- How should the manual be distributed to users?
- How can the document be kept current?

No single type of biosafety document is suitable for all work settings, so choose the one that best fits the institution or workplace. At least five manual formats were identified at the biosafety manual workshop. They included a formal, lengthy, administrative manual that covers all safety topics; a user manual or handbook that has safety information specific to a worksite such as a laboratory; a complete reference document on one subject, such as a bloodborne pathogen exposure control plan; a worksite procedure–specific manual for a defined group of people (waste handlers, glass washers); and a booklet or binder of work practices with all SOPs. Manuals should be updated frequently, and one way to make this easy to do is to place material that may change over time in three-ring binders.

**Occupational health program.** When setting up an occupational health program, decisions must be made regarding the focus of the program and whether it will include acute exposure-related problems and/or disease prevention and wellness programs. The purpose of an occupational health program is fourfold:

- To provide a mechanism to detect job-related illnesses
- To determine the adequacy of protective equipment and procedures and verify that hazardous agents are not being released into the general environment
- To establish baseline preexposure status
- To assess the presence of preexisting conditions that would put an employee at increased risk.

Health surveillance programs vary greatly, depending on the microorganisms being handled, the nature of the technical activities being conducted, the volume or concentration of material, and available medical facilities. Exposures to toxic chemicals, radionuclides, physiologically reactive biological and pharmaceutical products, animal allergens, and physical stresses also require consideration.

Some possible components of an occupational health surveillance program include taking medical and occupational histories, conducting physical examinations, laboratory testing, immunization, and serum storage, where indicated. If possible, it is advantageous to include the services of specialists in the fields of occupational health medicine and infectious diseases.

Work with infectious agents requires the evaluation of specific immunizations. Although there is no question that persons working with human blood and related products should receive hepatitis B immunization, there may be reasons to waive immunization with some less effective vaccines. An infectious disease physician should be consulted on such matters. Recommendations for the use of vaccines are included in the agent summary statements in Section VII of the CDC/NIH biosafety document (USPHS, 1993) and a comprehensive listing of immunoprophylaxis for personnel at risk is found in the National Research Council resource, *Biosafety in the Laboratory* (NRC, 1989, pp. 60–62). See also CDC, Recommendations of ACIP: for use of vaccines and immune globulins in persons with altered immunocompetence (1991a) and their update on adult immunizations (1993c).

Another aspect of the medical surveillance program is the controversial issue of employee serum banking. Although storage of serum can provide invaluable information regarding work-related exposures, this benefit must be weighed against such basic considerations as whether adequate facilities and technical support for long-term storage exist. Decisions regarding serum storage and testing should be based on agents handled, availability of reliable tests, and likelihood that infection will produce a serological change in exposed persons. Results of employee medical evaluations must remain confidential, with information being released only with the employee's consent.

Considerations involving host susceptibility must be taken into account, and some of these have been mentioned earlier in the risk assessment section. When decisions regarding employee health issues are made, the decision-making process should include the employee, the physician, and the employer. For additional information on medical surveillance programs for research and biotechnology settings, refer to Goldman (1989, 1991).

**Information and education.** All people whose work involves the handling of infectious organisms or materials must receive adequate information and education to enable them to work safely. Biosafety programs should include a mechanism to provide safety information to employees at all stages of their employment (new, altered work tasks, and long-term) as well as periodic safety updates.

In instances where regulations mandate training, such as the OSHA bloodborne pathogen standard, training content may be specified and written control plans required. Many regulations address frequency of training, record-keeping requirements, and qualifications of trainers.

Training materials developed for work with infectious materials must include a description of the biology of the agent(s), including symptoms of the disease; a review of the operations and procedures, with emphasis on potential sources of exposures and means of control; the correct use of containment equipment when applicable; review of acceptable work practices; decontamination methods and waste disposal; and emergency procedures. Equally importantly, personnel need to be made aware of the human factors (such as fatigue, inattentiveness, and haste) that predispose workers to accidents. Employees should understand that although employers must provide appropriate facilities and equipment to conduct work safely, they are responsible for following safety practices and procedures in order to protect themselves, their colleagues, and the community.

One challenge of health and safety training is to provide appropriate and factual information that is geared to the language and educational level of the employees. An obvious point is that information is more readily received when it is presented in an interesting or creative format. Although trainers now have access to an excellent collection of tools such as professionally produced videotapes and interactive computer educational programs, it is critical that training not be a solitary event.

Computers are now making new sources of health and safety information available to the health and safety community. In 1994 the biosafety group at the Massachusetts Institute of Technology, in conjunction with the American Biological Safety Association (ABSA), initiated an electronic discussion group on biosafety issues on the Internet. For additional information, contact R. Fink at rfink@mitvma.mit.cdu. A second electronic link exists at the University of Vermont; contact R. Stuart at rstuart-@moose.uvm.edu. Although the focus of this latter service is general safety and industrial hygiene, occasional biosafety points of interest are noted.

**Ideal biosafety programs.** The ideal biosafety program has strong administrative support from upper management and is funded adequately so that worker safety is not compromised. Depending on the volume of microbiological activity and the hazard associated with the microorganisms handled, a biosafety specialist should become involved in health and safety considerations. In some cases, it is not feasible to have a dedicated biosafety staff person, because work activities require only the lowest containment levels. Under these conditions it may be advisable to seek the assistance of a consultant who can work with the existing health and safety staff. When clinical activities are involved, infection control practitioners are an excellent resource for safety departments. In hospital settings, infection control practitioners may oversee clinical laboratory biosafety matters using policies approved by an infection control committee.

Administrative support and mechanisms must be in place in order to implement institutional biosafety committee (IBC) policies and decisions. Committee members must be knowledgeable and have credibility among their peers in order for committee decisions to be implemented effectively. Committee business and, ultimately, research activity are facilitated by timely attention to pertinent matters and, where needed, interaction with other institutional bodies such as animal resources, IACUCs, RSCs, CSCs, and IRBs.

No one biosafety manual or training program is appropriate for all work settings. Rather, institutions must strive to prepare material and training that provides their employees with pertinent material in a format that will be utilized by them. It is the challenge of every safety professional to develop these educational components so that they are not only relevant but interesting.

For the occupational health program to be effective and used, providers must offer services as required by law, be familiar with workplace activities, be responsive to the needs of employees, and interact with the IBC and biosafety specialist. Employees who are indifferent to occupational health programs such as immunization and serum storage are more likely to use them if, where possible and practical, services are performed at the worksite.

## ■ Assessing Compliance

Although there is often urgency to develop and implement health and safety programs and controls, the need to assess their efficacy following implementation is sometimes overlooked. Such a review is imperative in order to be assured that the safety practices have been incorporated into work activities and that they are minimizing potential hazards.

### Use of Audits to Identify Problem Areas

Self-audits of required safety practices provide a measure of compliance achievement. In work environments where hazardous aerosols are generated, such as agricultural processing facilities, one would begin by monitoring for compliance with personal protective equipment and clothing requirements. In laboratory settings the criteria for the designated biosafety level(s) can be used for the critical elements of the audit.

Routine operating procedures should include a safety check, for example, to determine that equipment (such as a biosafety cabinet or hematocrit centrifuge) is functioning properly or that work surface disinfectants and spill kits are on hand. When agents requiring higher containment are handled, ventilation system function must be checked and actual work practices and techniques reviewed to ensure containment.

Regular safety audits should be carried out quarterly or semiannually by designated safety specialists accompanied by the laboratory supervisor. Deficiencies can be pointed out and abated during the inspection. A written report, suggesting corrective actions, can be sent to the laboratory supervisor, who should report progress on remediation within a designated period of time. Notification of biohazards in use and a list of associated personnel should be obtained from the laboratory supervisor. The inspection program can be used to review information on the facility, work, and workers and serve as a reminder to update the biohazard database.

## Annual Biosafety Review

An annual renewal of biohazard registration also helps to remind each responsible supervisor to review the work in progress and keep the information updated. Pathogen or biohazard registration programs used by many institutions provide supervisors with a form to expedite such an update.

## Incident/Accident Statistics

Although small statistical changes in incident/accident figures do not usually indicate real deficiencies in a biosafety program, some institutions only judge the status of their safety program by a statistical review of changes in OSHA-recordable incidents.

The positive changes brought about through education and training in preventive methods can be measured with specific outcome audits. For example, in determining the effect of training and safety equipment on the number of needlestick incidents, trends in reports on such injuries show the cost–benefit of the changes. Observations of increases in such injuries, or the reporting of sentinel events (events whose single occurrence is of sufficient concern to trigger systematic response), highlight the need for intervention efforts. Such events can be used as tools for continuous quality improvement if action limits (the criteria for intervention) are defined (Birnbaum, 1993).

Work practices must be assessed for efficacy so that protective practices can be reinforced and unsafe practices altered.

# ■ Current Topics in Biosafety

## Bloodborne Pathogens

Hepatitis B virus has been and continues to be the most significant occupational infector of health care and laboratory personnel during the past 50 years. This fact, coupled with the identification of the human immunodeficiency virus (HIV), which is the causative agent of acquired immune deficiency syndrome (AIDS), prompted the development and implementation of measures that would promote worker protection. OSHA's publication of the standard for occupational exposure to bloodborne pathogens (OSHA, 1991) was the most significant regulation of work environments involved with potentially infectious materials to date. In addition to the standard, OSHA published compliance assistance instructions for their enforcement officers (DOL, 1992) that provides useful information for those covered by the regulation (OSHA, 1992).

The standard applies not only to the health care community but to all occupations (such as emergency responders and morticians) in which there is a potential for exposure to human blood. Bloodborne pathogens are microorganisms that may be present in human blood and body fluids and are capable of causing disease in human beings. Although HIV and hepatitis B and C viruses are the bloodborne pathogens most frequently associated with occupational infections, other microorganisms, such as syphilis and malaria, that may be present in blood have also caused work-related infections.

### Epidemiology

***Hepatitis B virus (HBV).*** Hepatitis B virus infects 200,000 persons among the general population annually, with 6,800 of these infections involving health care workers. High-risk health care workers (those with frequent contact with blood at work) account for 5,100 of the infections, and 5 percent, or 255, of these require hospitalization. It is estimated that 10 percent, or 510, of the high-risk health care workers are chronic HBV carriers and 21 percent (107) die of cirrhosis or hepatocellular carcinoma (Short & Bell, 1993). Hepatitis B virus infects liver cells and causes liver damage. The resulting disease can range from inapparent to severe to fatal. Symptoms of apparent HBV disease include flu-like illness, jaundice, extreme fatigue, anorexia, fever, nausea, and joint pain.

***Hepatitis C virus (HCV).*** Although there are no reliable estimates for the number of cases of occupational exposure to hepatitis C virus (formerly designated non-A, non-B), the virus, because of its similarities to HBV, also presents an occupational risk to persons whose work activities involve handling human blood and body fluids.

***Human immunodeficiency virus (HIV).*** Through June 1994, the CDC had received reports of 130 cases of documented or "possible" occupationally acquired HIV infection. Documented occupational exposure has been identified in 42 of these cases: Thirty-six of these resulted from percutaneous exposures, 4 from mucocutaneous exposures, 1 from a dual percutaneous and mucocutaneous exposure, and 1 of unknown source. Thirty-eight of the 42 workers were exposed to blood from HIV patients, 2 to an unspecified fluid, and 2 to concentrated virus. Fifteen of the 42 documented exposures have developed AIDS. The remaining 88 workers were classified as "possible" occupational HIV exposures because neither the date of infection nor its source has been documented. By occupation, clinical laboratory technicians (14 out of 88) and nurses (19 of 88) sustained the greatest number of seroconversions. Persons infected with HIV may remain healthy for years before the virus, which infects the cells of the immune system, ultimately destroys that system. Early infection may be asymptomatic or appear as a flu-like illness. Because symptoms disappear and individuals often remain healthy for a number of years, many infected individuals are unaware of their HIV status.

These viruses are transmitted through exposure to human blood and certain body fluids. In the work setting, infection can result from parenteral, nonintact skin, or mucous membrane (eyes, nose, and mouth) exposure to contaminated materials. Data from needlestick studies indicate that persons exposed to HIV-contaminated needles have a 0.3 percent chance of seroconversion to HIV (Short & Bell, 1993), whereas similar reviews involving HBV indicate that 6–30 percent of exposed persons become infected. Epidemiological data on the consequence of parenteral exposure to HIV and HBV indicate that one of the most critical control components must be the reduction of sharps-related incidents. This is particularly important with HIV exposure, because currently no effective vaccine or long-term curative therapy is available.

**OSHA standard.** The focus of the OSHA bloodborne pathogens standard is to prevent or minimize parenteral, nonintact skin, and mucous membrane exposure to human blood and body fluids in the workplace. To achieve this goal the standard requires employers whose employees are potentially exposed as the result of their work activities to implement an administrative mechanism for compliance and a series of workplace controls to prevent or minimize exposure (OSHA, 1991). Employers must develop a method for determining who among their employees is at risk from occupational exposure (perform an exposure determination) and an exposure control plan that outlines how their institution will meet all the requirements of the standard. Compliance methods specific to an institution are included in an exposure control plan.

**Controls.** The concept of universal precautions—that all human blood and certain human body fluids are to be treated as if known to be infectious for HIV, HBV, and other bloodborne pathogens—is a key component in prevention of work-related exposure. The OSHA standard outlines the potentially infectious materials (including HIV- and HBV-infected cells and animals) and specifies control measures.

The standard mandates engineering controls (needle disposal containers and equipment for reducing aerosols and splatter); work practice controls with special emphasis on personal protective clothing and equipment (such as gloves and face- and eyewear) and hand-washing; all aspects of sharps management (needle-handling procedures such as not resheathing needles by hand and the use of rigid sharps-disposal containers); labeling and transport requirements; housekeeping (including disinfection, disposal of infectious waste, and spill management);

handling of contaminated laundry; communication of hazard to employees (training); and general workplace practices and procedures. Biosafety level 2 (BSL-2) practices are required for laboratory activities involving clinical materials. Work may be performed on the open bench as long as procedures do not generate significant aerosols. Aerosol-generating procedures must be performed in a BSC or otherwise contained. The CDC/NIH document *Biosafety in the Microbiological and Biomedical Laboratory* provides guidance (agent summary statements) for handling hepatitis viruses and HIV (retroviruses) (USPHS, 1993). Additional information regarding practical disinfectants for HIV and HBV is available in Rutala (1990).

Immunization is a critical component of bloodborne pathogen hazard control. There is currently no means of immunizing people against HIV, but an effective vaccine for HBV is available. The standard requires employers to offer hepatitis B vaccination to employees whose work activities bring them into contact with bloodborne pathogens. Should an employee decline to be immunized, this must be documented. Some latitude on HBV immunization is given to employers whose employees, in addition to their regular nonhazardous work responsibility, also provide first aid at the worksite (Clark, 1992). The standard requires employers to provide for immediate and subsequent medical and counseling needs of employees who sustain an occupational exposure.

Because of the scope of the OSHA standard and the widespread agreement among the medical and microbiological research community that workers are indeed at risk from bloodborne pathogens, significant progress is being made toward reducing the hazard of bloodborne pathogens in the workplace.

## Tuberculosis

Tuberculosis (TB) is a bacterial disease, caused by *Mycobacterium tuberculosis* (M. tb.), that is responsible for morbidity and mortality worldwide. Man is the primary source of infection, although in some cases the source is infected cattle; it occurs only rarely in primates, badgers, and other animals (Benenson, 1990).

**Epidemiology.** TB, common outside the United States, is estimated to affect one-third of the world's population (CDC, 1993a). If prevention and control methods are not improved, during the present decade approximately 90 million new cases of tuberculosis can be expected worldwide. The emergence of drug-resistant TB is also being reported worldwide and is a serious problem in the United States (Stratton, 1993). In the United States, only 10 percent of M. tb. strains were resistant to one or more drugs in the years prior to 1984. Since 1988, there have been at least six outbreaks caused by multiple drug-resistant strains (MDR-TB). The mortality rate is approximately the same (40–60 percent) in those with MDR-TB, despite treatment, as in TB cases that go untreated (CDC, 1991b).

The number of tuberculosis cases had declined by 74 percent between 1953 and 1984, but decline slowed in 1985, and new cases of TB in the United States began to increase significantly. There was an increase in the cases reported in every racial or ethnic group except non-Hispanic whites and American Indians/Alaskan natives from 1985 to 1992. There was a 20 percent increase in TB cases in 1992, when a total of 26,673 new cases were reported. The CDC has estimated that there are between 10 and 15 million asymptomatic infected people in the United States. These facts indicate that the routine processing of sputum in high-risk urban areas is a potential source of laboratory-acquired infection.

Risk assessment and epidemiological studies (CDC, 1994a) show that the prevalence of TB is not distributed evenly throughout a population. Some groups are at higher risk of TB because of increased risk of exposure; others have a higher risk of progressing to active TB following infection. Those with increased risk of exposure include the foreign born from areas with high prevalence of TB (Asia, Africa, the Caribbean, and Latin America); the medically underserved, such as African Americans, Hispanics, Asians and Pacific Islanders, and American Indians and Alaska Natives; homeless persons; current or former correctional-facility inmates; alcoholics; intravenous drug users; and the elderly. Those who are at higher risk of progressing to active TB from latent infection include those recently infected (within the previous two years); young children (under 4 years); persons with fibrotic lesions that show up on chest radiographs; and persons with certain underlying medical conditions such as HIV infection, silicosis, gastrectomy or jejunoileal bypass, being 10 percent below ideal body weight, chronic renal failure with renal dialysis, diabetes mellitus, immunosuppression from receipt of high-dose corticosteroid or other immunosuppressive therapy, and some malignancies.

Much of the current increase in cases of tuberculosis has been attributed to HIV-infected people, particularly in Africa and Southeast Asia. Among persons coinfected with HIV and *M. tuberculosis,* the risk for developing active TB is increased because of the concurrent immunosuppression induced by HIV. The annual risk of progression into active TB among individuals infected with both HIV and TB is 5–15 percent, depending on the degree of immunosuppression (Raviglione et al., 1995). The CDC Advisory Committee on the Elimination of Tuberculosis (1989) recommended that HIV-infected individuals be screened for active TB as well as latent TB and be offered appropriate curative or preventive therapy (CDC, 1993a).

**Transmission.** Tuberculosis is usually transmitted by the inhalation of infectious droplet nuclei suspended in the air from coughing, sneezing, singing, or talking by an individual who has a pulmonary or laryngeal TB infection. Prolonged close contact with an infectious person may expose individuals such as family members or co-workers and lead to their infection. Although direct exposure to mucous membranes or invasion through breaks in the skin can result in infection, it is extremely rare. With the exception of laryngeal infections, extrapulmonary TB infection, even with a draining sinus, is usually not communicable. Bovine tuberculosis, caused by *Mycobacterium bovis,* results from drinking unpasteurized milk or dairy products from infected cattle, although there have been cases in which farmers or animal handlers have been exposed to infectious aerosols.

**Disease symptoms and progress.** Symptoms of tuberculosis include fatigue, fever, and weight loss early in the disease, and hoarseness, cough, and hemoptysis (blood-tinged sputum) appear later as the disease is localized in the respiratory tract.

It can take from one to four months from the time of infection to a demonstrable pulmonary lesion or a positive tuberculin reaction. The initial infection with the tubercle bacillus is usually asymptomatic, but in a few weeks sensitivity to the tuberculin (a purified protein derivative of *M. tuberculosis* used for skin testing) usually develops, as manifested in a positive skin test.

Progression to active disease is most likely in the first two years after infection, but can occur any time throughout life. Those who are actively shedding viable tubercle bacilli in sputum, including those who are inadequately treated, are a risk to others, but children with primary TB do not usually infect others. When effective treatment is given, communicability can be eliminated in several days or a few weeks, when tubercle bacilli are no longer visible in an acid-fast smear of patient sputum.

The internal lesions that develop in the respiratory tract usually heal with minor or no change, except for occasional calcifications in pulmonary or tracheobronchial lymph nodes. There is a lifelong risk of reactivation in 95 percent of those infected who enter this latent stage. In about 5 percent of those infected, the initial infection progresses to pulmonary TB or bacteremia with dissemination to other organs. Infants, adolescents, and young adults have a more serious outcome from the initial infection in tuberculosis. Tuberculosis is not very infectious in terms of unit of time exposed to the bacillus; that is, brief exposure rarely results in infection. However, long terms of exposure to chronic, asymptomatic cases, as with household contacts, lead to an overall 30 percent risk of infection and a 1–5 percent risk of active disease within a year. Reinfection or reactivation of the latent disease leads to progressive pulmonary tuberculosis, which can lead to death within two years if untreated (Benenson, 1990). The lifetime risk of developing active disease for those infected as infants is estimated at 10 percent.

**Risk of occupational exposure.** The key element in protecting workers from the risk of occupational exposure is risk assessment. The current CDC guidelines (CDC, 1994a) address the health care industry. A risk assessment can be done at any worksite, after which the appropriate administrative and engineering controls and personal respiratory protection can be implemented.

In the workplace, health care workers, including nursing home and emergency personnel, who provide patient care are at risk of aerosol-borne infectious droplet nuclei from patients. Those who work in clinical, research, or production situations with *Mycobacterium tuberculosis* or *M. bovis* are at risk from contact and percutaneous routes as well as from inhalation of droplet nuclei in aerosols produced during common work procedures (Benenson, 1990). Others who provide service to high-risk individuals such as those in shelters and prisons are also at increased risk.

A report on the risk for occupational exposure to TB was recently published (CDC, 1995). Although there are certain recognized limitations to this study, it is encouraging to find published data on the relationship between occupational exposure and illness. Of the 2,206 deaths from TB from 1979 to 1990, 1,024 (46.4 percent) were in workers in 21 groups that met the criteria for occupational risk. These groups were further categorized into four risk groups: high potential for exposure to TB, potential for exposure to silica, low socioeconomic status (SES) occupation without other recognized risk factors, and unknown risk factors. It should be noted that the high-risk groups include funeral directors, and health care service workers such as nursing aides, orderlies, and attendants. A list of the occupations in the four risk groups and the proportionate TB mortality rates is included in the CDC (1995) reference.

## Employee protection from exposure.

***Personal protective equipment—respirator selection and use.*** According to the CDC Guidelines (1994a), facilities that do not have isolation rooms and do not perform cough inducing procedures on patients who might have TB may not need to have a respiratory protection program for TB. Such facilities should have written protocols for the early identification of patients with signs and symptoms of TB and procedures for referring such patients to a facility for proper evaluation and management.

The CDC (1994) recommends the use of respirators by health care workers (HCWs) when entering the room of a patient in isolation for TB or suspected TB, when present during cough-inducing or aerosol-generating procedures on such patients, and in other settings where administrative and engineering controls cannot be ensured. These would include emergency transport, urgent surgical care, or urgent dental care of such patients.

According to the CDC (1994), the standards to be met by respiratory protective equipment used in health care settings to protect against TB include the following:

■ The ability to filter particles as small as 1 μm in size in the unloaded state with a filter of ≥ 95 percent (filter leak of ≤ 5 percent), given flow rates of up to 50 l/min.

■ The ability to quantitatively or qualitatively test fit in a reliable way to obtain a face-seal leakage of ≤ 10 percent.

■ The ability to fit different facial sizes and characteristics of HCWs; that is, to be available in at least three sizes.

■ The ability to be checked for facepiece fit, according to OSHA and good industrial hygiene practices, by the HCW each time the respirator is put on.

At the time of this writing, NIOSH has proposed respirator certification criteria that will probably be approved in the near future. The proposed criteria, described in Supplement 4 of the CDC TB guidelines (1994a), would establish three classes of NIOSH-certified respirators that would be acceptable for use around infectious aerosols. Respirator filter material would be tested with particles of a median aerodynamic diameter of 0.3 μm at a flow rate of 85 l/min. Filters would be certified into one of three acceptable classes. Acceptable filtration criteria for the Class A respirator would be ≥ 99.97 percent retention at 0.3μm particle size, which would include the standard, NIOSH-approved HEPA respirator. Criteria for Class B would be retention at ≥ 99 percent and that for Class C ≥ 95 percent efficiency. NIOSH is not only establishing the specific particle size challenge but also maximum face seal leakage requirements.

Risk assessment procedures can identify settings (bronchoscopy performed on suspected TB patients, autopsies on deceased patients who had active TB) in which a higher level of protection is needed, and if so, the situation should be documented and the protection implemented. Although certain regional OSHA requirements may still include mandatory positive-pressure air-purifying particulate respirator (PAPR) or full-face, HEPA-filtered respirators, the most recent recommendations from the CDC (1994a) outline the performance-based criteria listed above for the selection of respiratory protection (see CDC, 1994, Supplement 4).

## Workplace containment.

***Laboratory.*** Exposure to laboratory-generated aerosols is the most insidious hazard. Sputa and other clinical specimens from suspected or known cases should be handled with appropriate precautions to preclude the release of infectious droplets and spatter. Organisms can survive in heat-fixed smears and can be aerosolized during the preparation of frozen sections and the manipulation of liquid cultures.

According to the CDC/NIH guidelines (USPHS, 1993; CDC, 1994), BSL-2 practices, containment equipment, and facilities are required for activities at American Thoracic Society (ATS) laboratory level 1. This implies that a biosafety cabinet (BSC) is required for aerosol-producing activities with these specimens. Activities that can be safely contained using BSL-2 practices include the preparation of acid-fast (AF) smears and the culturing of sputa or other clinical specimens. If sputa are treated for 15 minutes in a

BSC with an equal amount of undiluted household bleach (5 percent sodium hypochlorite) prior to centrifugation, then liquification and concentration of sputa for acid-fast staining can be carried out on the open bench.

Biosafety level 3 practices, containment equipment, and facilities are required for activities at American Thoracic Society levels II and III. These include the propagation and manipulation of cultures of *M. tuberculosis* or *M. bovis* and animal studies, especially those utilizing nonhuman primates, that have been experimentally or naturally infected with these agents. According to the ATS, there are three levels of complexity of laboratories that correspond with three phases in the detection and isolation of mycobacteria. Level I and II laboratories must refer specimens or cultures to the Level III specialized laboratories, which have a full spectrum of bacteriologic support, professional expertise, and complete and safe facilities for further processing.

In a more practical approach to control of aerosol hazards in laboratories, Gilchrist et al. (1994) provide guidance on the personal protective equipment and the safe procedures necessary for handling liquid-amplified cultures, as opposed to those needed for the less hazardous work of planting primary clinical specimens on solid media.

*Patient care.* Infection control guidelines for the care of patients with tuberculosis have been provided and updated periodically since the mid-1970s by the CDC Hospital Infections Branch (Garner & Simmons, 1983). Concern for employee health in the hospital environment was the subject of a separate set of CDC guidelines (Garner & Favero, 1986). The CDC has recently published guidelines for preventing the transmission of TB in health care facilities, extending them to include protection of both patients and personnel (CDC, 1993a; CDC, 1994a). The CDC made it clear that the purpose of the guidelines was to make recommendations to reduce the risk of transmitting TB to HCWs, patients, volunteers, visitors, and other persons in health care settings. The recommendations were written to apply to inpatient facilities where health care is provided, such as hospitals, prison medical wards, nursing homes, and hospices.

In patient care settings, it is important to control TB at the source by identifying TB patients or those at high risk and taking the time to train such patients to control the formation and release of infectious droplets, for example, by using tissues to cover sneezes and coughs.

The CDC guidelines were adopted by OSHA in an advance notice of proposed rulemaking as requirements to protect employees from exposure and disease (Decker, 1993). Requirements include a written TB infection control plan, assignment of responsibility for the program, exposure determinations, evaluation of risk to employees, development of a written exposure control plan based on the risk assessment, and periodic reassessment of risk to evaluate the effectiveness of the TB infection control program (Clark, 1993).

*Ambulatory care facilities.* Ambulatory care facilities are of special importance because of the increase in patient users and the front-line health care worker status in the United States. Health care employers in outpatient settings should be aware of the risk of TB among their patient population, especially those who have both HIV and TB infections. Infection control policies should be developed accordingly.

Those who are HIV-positive or are otherwise at risk for contracting TB should receive a tuberculin skin test, and the results should be noted in the patient's medical record. Tuberculosis diagnostic procedures should be initiated if signs and symptoms of tuberculosis develop.

Ambulatory patients who have pulmonary symptoms of uncertain etiology should be instructed to cover their mouths and noses when coughing or sneezing; they should spend a minimum time in common waiting areas. Personnel who are the first point of contact in facilities serving patients at risk for tuberculosis should be trained to recognize, and bring to the attention of the appropriate person, any patients with symptoms suggestive of tuberculosis, such as a productive cough of greater than three weeks duration, especially when accompanied by other tuberculosis symptoms such as weight loss, fever, fatigue, or anorexia.

Ventilation systems in clinics serving patients who are at high risk for tuberculosis should be designed and maintained to reduce tuberculosis transmission. This is particularly important if immunosuppressed patients are treated in the same or a nearby area. In some settings, enhanced ventilation or air disinfection techniques (HEPA filters or indirect or contained UV germicidal irradiation [UVGI] [CDC, RR-13, p.88, 1994a]) may be appropriate for common areas such as waiting rooms. Air from clinics serving patients at high risk for tuberculosis should not be recirculated unless it is first passed through an effective decontamination system such as a HEPA filtration system.

In outpatient settings where cough-inducing procedures are carried out, infection control precautions for TB (respiratory precautions) should be implemented. A special concern is the drug treatment facility. TB patients who have substance abuse problems are likely to be noncompliant with TB therapy, and may develop drug-resistant disease as a result (Raviglione et al., 1995).

*Emergency medical services.* Emergency medical services (EMS) personnel should be included in a respiratory program and in a comprehensive tuberculin skin-testing program with a baseline test and follow-up testing according to risk assessment (CDC, 1994a).

EMS personnel and others who provide patient services should ensure that a surgical mask is placed over the patient's mouth and nose (if possible) when a patient who has confirmed or suspected TB is being transported. Because of the lack of engineering controls in the transport vehicle, and because administrative controls cannot be ensured under such circumstances, EMS personnel are advised to wear respiratory protection as well.

# Legionnaires' Disease

Legionnaires' disease (a type of pneumonia) is an infection of the lungs caused by inhaling bacteria in the genus *Legionella,* most often *L. pneumophila* serogroup 1 (APHA, 1989; Benenson, 1990; Kreiss, 1989). The term *Legionella* (plural *Legionellae*) refers to any of the bacteria in this genus. Symptoms of legionnaires' disease begin within two to ten days after exposure and typically include fever, cough, headache, muscle aches, and abdominal pain. People usually recover from legionnaires' disease in a few weeks and suffer no long-term consequences. Antibiotic treatment of legionnaires' disease reduces mortality. Legionnaires' disease principally affects people with other underlying illness or increased susceptibility.

Inhalation of legionellae can also cause a less serious, flu-like illness called Pontiac fever. Pontiac fever affects a wider range of people than legionnaires' disease, but it is not fatal. *Legionellosis* is a term for any *Legionella* infection. Both legionnaires' disease and Pontiac fever are acute respiratory illnesses, unlike sick building syndrome (SBS) (see the section on work- and building-related illness that follows) and are not chronic or recurrent conditions.

**Sources of legionellae.** Legionellae are common in nature, and can be found in lakes and rivers. These bacteria are often present in low numbers in drinking water supplies, but apparently cause no problems. Surveys have found legionellae in large percentages of tested hospitals, large buildings, and residences, often in hot-water supplies and cooling waters for heat-transfer systems (ASM, 1993). Factors known to enhance legionellae colonization of artificial water environments include warm temperature (77–108 F, or 25–45 C), suitable pH (2.5 to 9.5), water stagnation followed by agitation, and the presence of other organisms, sediment, and scale. Some algae and bacteria encourage legionella multiplication in the environment. In humans, legionellae infect a type of white blood cell in the lungs, whereas in the environment, the bacteria infect some free-living aquatic amoebae and other protozoa. Legionellae inside protozoa can be protected from biocides, desiccation, and other environmental stresses.

**Transmission of legionellosis.** Natural transmission of legionnaires' disease or Pontiac fever occurs by inhalation of legionellae in airborne droplets. Typical sources of exposure are airborne bacteria in sprays from cooling towers or evaporative condensers, and fine mists from showers and some types of humidifiers. Neither legionnaires' disease nor Pontiac fever is transmitted from person to person.

**Preventing legionellosis.** The risk of legionellosis can be reduced by proper design and operation of ventilation, humidification, and water-cooled heat-transfer equipment and of other water systems and equipment. Good system maintenance is also required, including regular cleaning and, where applicable, disinfection of possible sources.

Precautions specific to preventing legionellae multiplication in water systems include the following:

- Keeping hot water above 120 F (60 C) and cold water below 70 F (20 C).
- Separating or insulating water lines to prevent heat transfer.
- Avoiding tepid water systems (for example, deliver hot and cold water in separate lines and mix them at the point of use rather than in a warm-water tank).
- Flushing faucets and showers briefly before use, and flushing infrequently used water supply lines on a regular basis (weekly or monthly).
- Removing "deadlegs" in water systems (by disconnecting and draining unused plumbing and equipment).

Additional recommendations for reducing the risks of legionellosis that also help to prevent other problems associated with biological contamination include the following:

- Choosing heating, ventilating, and air-conditioning (HVAC) systems, water systems, and other equipment of the best design and capacity for what is needed in a particular facility
- Labeling equipment for easy identification
- Keeping up-to-date blueprints or schematic drawings available that identify control equipment and access points
- Operating and maintaining (inspecting, cleaning, and repairing) equipment according to the manufacturer's recommendations
- Outlining responsibilities in writing and seeing that staff understand and are trained for their assignments
- Resolving identified problems promptly
- Outlining emergency responses in writing and having important names and phone numbers readily available
- Keeping good records and seeing that reports are dated and signed
- Seeking expert advice when needed

# Investigating Work- or Building-Related Illnesses Caused by Biological Hazards

The sources of biological hazards are fairly obvious in occupations associated with the handling of microorganisms, plants, and animals (such as laboratory work, agriculture, animal handling, and food processing) and in occupations that involve contact with potentially infected people (such as health care and emergency-response settings). However, recognizing and identifying biological hazards is not as simple in other settings such as office buildings and nonindustrial workplaces (WHO, 1990). Building-related illness (BRI) is a clinically diagnosed disease in one or more building occupants (Hodgson, 1989; Kreiss, 1989; Morey & Singh, 1991). It is distinguished from sick building syndrome (SBS), in which building occupants' nonspecific symptoms cannot be associated with an identifiable cause (AIHA, 1993). Certain BRIs such as infectious and hypersensitivity diseases are associated clearly with biological hazards, but the role of biological materials in SBS is not as well understood (ASTM, 1990; Burge et al., 1987; Hodgson, 1989; Kreiss, 1989; Pope et al., 1993). For more information on this topic, see Chapter 21, General Ventilation of Nonindustrial Occupancies.

This section focuses on inhaled biological hazards rather than those that gain entry by ingestion or skin contact or through the mucous membranes. Currently there are no occupational exposure limits for airborne biological material (other than cotton, grain, and wood dusts) in any workplace, and there are no universally accepted sampling methods, as is explained in the following excerpt from the 1994–1995 American Conference of Governmental Industrial Hygienists *Threshold Limit Values and Biological Exposure Indices* booklet (ACGIH, 1994, pp. 9–11).

> Biologically derived airborne contaminants include bioaerosols (airborne particulates composed of or derived from living organisms) and volatile organic compounds released from living organisms. Bioaerosols include microorganisms (culturable, nonculturable, and dead microorganisms) and fragments, toxins, and particulate waste products from all varieties of living things. Biologically derived airborne contaminant mixtures are ubiquitous in nature and may be modified by human activity. All persons are repeatedly exposed, day after day, to a wide variety of such contaminants. At present, gravimetric Threshold Limit Values (TLVs) exist for some wood dusts, which are primarily of biological origin, and for cotton dust, which is at least in part biological. There are no TLVs for concentrations of total culturable or countable organisms and particles (e.g., "bacteria" or "fungi"); specific culturable or countable organisms and particles (e.g., *Aspergillus fumigatus*); infectious agents (e.g., *Legionella pneumophila*); or assayable biological-source contaminants (e.g., endotoxin or volatile organic compounds).
>
> A. A general TLV for a concentration of culturable (e.g., total bacteria and/or fungi) or countable bioaerosols (e.g., total pollen, fungal spores, and bacteria) is not scientifically supportable because:
>
>   1. Culturable organisms or countable spores do not comprise a single entity, i.e., bioaerosols are complex mixtures of different kinds of particles.
>
>   2. Human responses to bioaerosols range from innocuous effects to serious disease and depend on the specific agent and susceptibility factors within the person.
>
>   3. Measured concentrations of culturable and countable bioaerosols are dependent on the method of sample collection and analysis. It is not possible to collect and evaluate all of these bioaerosol components using a single sampling method.
>
> B. Specific TLVs for individual culturable or countable bioaerosols, established to prevent irritant, toxic, or allergic responses have not been established. At present, information relating culturable or countable bioaerosol concentrations to irritant, toxic, or allergic responses consists of case reports containing only qualitative exposure data. The epidemiologic data that exist are insufficient to describe exposure–response relationships. Reasons for the absence of good epidemiologic data on exposure–response relationships include:
>
>   1. Most data on concentrations of specific bioaerosols are derived from indicator measurements rather than from measurement of actual effector agents. For example, culturable fungi are used to represent exposure to allergens. In addition, most measurements are either from reservoir or from ambient air samples. These approaches are unlikely to accurately represent human exposure to actual effector agents.
>
>   2. The components and concentrations of bioaerosols vary widely. The most commonly used air sampling devices collect only "grab" samples over short periods of time and these single samples may not represent human exposure. Short-term grab samples may contain an amount of a particular bioaerosol that is orders of magnitude higher or lower than the average environmental concentration. Some organisms release aerosols as "concentration bursts" and can be detected only rarely using grab samples. Yet, such episodic bioaerosols may produce significant health effects.
>
> C. Dose–response data are available for some infectious bioaerosols. At present, air sampling protocols for infectious agents are limited and suitable only for research endeavors. Traditional public health methods, including immunization, active case finding, and medical treatment, remain the primary defenses against infectious bioaerosols. Certain public and medical facilities with high risk for transmission of infection (e.g., tuberculosis) should employ exposure controls to reduce possible airborne concentrations of virulent and opportunistic pathogens.
>
> D. Assayable, biologically derived contaminants are substances produced by living things that can be detected using either chemical, immunological, or biological assay and include endotoxin, mycotoxins, allergens, and volatile organic compounds. Evidence does not yet support TLVs for any of the assayable substances. Assay methods for certain common aeroallergens and endotoxin are steadily improving. Also, innovative molecular techniques are rendering assayable the concentration of specific organisms currently detected only by culture or counting. Dose–response relationships for some assayable bioaerosols have been observed in experimental studies and occasionally in epidemiologic studies. Validation of these assays in the field is progressing.

The ACGIH Bioaerosols Committee actively solicits information, comments, and especially data that will assist it in evaluating the role of bioaerosols in the environment.

The conditions and events necessary to result in human exposure to bioaerosols can be summarized as follows (Burge et al., 1989):

1. Presence of a reservoir that can support the growth of microorganisms or allow accumulation of biological material

2. Multiplication of contaminating organisms or accumulation of biological material in the reservoir

3. Generation of aerosols containing biological material

4. Exposure of susceptible workers to the biological material

It is important for industrial hygienists to realize that, although sources of biological material are abundant, there may be little hazard if microorganisms do not multiply or if materials do not accumulate to harmful levels, if there is no means for material to become airborne, or if aerosolized material does not reach susceptible people.

The steps involved in investigating work- or building-related biological hazards are identical to those for other industrial hygiene investigations:

1. Identify the types of biological agents or materials that could cause the symptoms that affected people report.

2. Conduct a walk-through inspection of the facility to find sources of the suspected materials or to observe mechanisms by which people could be exposed.

3. Interpret initial observations.

4. Collect samples from recognized sources to identify and quantify the material, if appropriate.

5. Formulate recommendations to control identified problems, and outline a follow-up program to confirm problem resolution and to prevent a recurrence.

**Medical evaluation of symptoms, illnesses, and complaints.** The first step of a work- or building-related biohazard investigation is a medical evaluation to classify reported symptoms into one of three possible categories: an infectious disease, a hypersensitivity disease, or other type of response, such as irritant or toxic reaction to biological material (Burge et al., 1989; Kreiss, 1989). Associated with recognition of a disease process is identification of potential sources of the causative agents or materials.

Reservoirs for infectious disease agents can be people or infected animals (as in the case of the measles virus or the Q-fever agent) or the environment (for example, cooling water contaminated with the legionnaires' disease bacterium or bird droppings supporting the growth of the histoplasmosis fungus). Common respiratory infections are a leading cause of worker absence, and the incidence of colds and flus rises during the winter months when people spend more time indoors in tightly sealed buildings (WHO, 1990).

A wide range of plant and animal material, as well as many microorganisms, can trigger hypersensitivity diseases, e.g., allergic asthma or rhinitis, hypersensitivity pneumonitis, and skin rashes. Sources for biological materials responsible for hypersensitivity diseases typically come from the environment, such as animals (e.g., dander from laboratory animals and livestock), plants (e.g., wood and grain dusts, and poison ivy), and microbiological contamination of plant materials and building furnishings (e.g., bacterial or fungal growth on wood chips or damp building materials).

Workers may be exposed to airborne irritants and potentially toxic biological materials from the environment (e.g., odorous compounds, and endo- and exotoxins from bacteria, and mycotoxins from fungi. Exposure to these materials by the respiratory route may be lower than by ingestion or skin contact. (See Chapter 3, The Skin and Occupational Dermatoses.) Malodors sometimes arise when microorganisms multiply, and exposures to biological toxins may occur in agricultural operations and during the large-scale production or processing of bacterial, fungal, plant, or animal materials.

**Walk-through inspection to identify routes of disease transmission and sources of exposure.** Inspectors should concentrate on understanding air movement and work practices in the areas of a facility linked epidemiologically with infectious diseases transmitted from person to person. Infectious diseases often are diagnosed, and active and recovered cases often are identified, through the use of clinical tests such as specimen culture, detection of serum antibodies, and TB skin test conversion. Environmental sampling generally is not necessary in these investigations.

Inspectors should concentrate on identifying potential environmental reservoirs and aerosol-generating work practices when investigating cases of infectious or hypersensitivity diseases or other reactions associated with environmental sources. Environmental sampling may be indicated in these investigations when sources of biological materials are not obvious. Some of the signs of biological contamination in buildings are the following:

- Water stains and evidence of standing water
- Slick or sticky biofilms on wet surfaces
- Algae or moss on surfaces that receive sunlight
- Cottony, wooly, or sooty fungal growth with or without colored spores on damp materials
- Odors such as moldy, mildewy, musty, yeasty, sour, foul, spoiled, swampy, or earthy odors
- Accumulated animal or plant debris, and signs of rodent, bird, or arthropod pests

**Interpreting observations.** Inspectors who observe improper handling of biological material, obvious problems with ventilation systems, or definite reservoirs or accumu-

**Table 14–G.** Field Sampling Kit

| | |
|---|---|
| Camera and film | Flashlight |
| Disposable respirator | Disposable gloves |
| Coveralls | Magnifying glass |
| Bleach solution and dropper | Cotton balls/absorbent paper |
| Alcohol wipes | Scissors |
| Scalpels | Forceps |
| Sterile swabs | Spatulas/wooden blades |
| Air and water thermometers | Sterile sample containers |
| New paper and plastic bags | Contact plates |
| Clear adhesive tape | Microscope slides |
| Data forms | Pens and pencils |

**Table 14–H.** Information on Samples Collected

| General | Air Samples |
|---|---|
| Collector's name | Air sampler and sample pump |
| Date | identification numbers |
| Study site (street address or | Sampling air flow rate, |
| building name) | sample start and finish times, |
| Sample identification number | volume of air collected |
| Sample type: air, surface, | Indoor and outdoor air |
| liquid, material, etc. | temperatures, relative |
| Sample collection site: mark | humidities, $CO_2$ |
| location on map or drawing of | concentrations, etc. |
| test area, or photograph site | Weather conditions: |
| with sampling equipment in | barometric pressure, wind |
| place | direction and velocity |
| Number of workers and | |
| activity at sampling site | *Material/Water Samples* |
| Sample transportation method | |
| and conditions; | Sample temperature, pH, |
| sample storage conditions | turbidity, etc. |
| Type of analysis requested | Sample amount/volume |
| Date and time samples | Total amount/volume/area of |
| delivered to laboratory | contaminated material |

lations of biological materials generally can recommend that the deficiencies be addressed without further testing or collection of samples for confirmation of contamination. However, identification of material as biological is not always possible in the field; for example, not all slick coatings on wet surfaces are biofilms, microbiological growth on some surfaces can look much like scale and corrosion, and one typically finds many potential sources for the legionnaires' disease bacteria in a facility, of which none may actually present a hazard. It may be appropriate to examine environmental samples in these cases before drawing conclusions on the hazards involved and the type of remediation needed.

**Environmental sampling to identify and quantify biological hazards.** Table 14–G provides a general list of field sampling equipment for inspections of possible biological hazards, and Table 14–H lists the types of information that inspectors may wish to collect along with environmental samples. Using standardized data collection forms (USEPA, 1991) can reduce the chance of overlooking important details during an investigation and helps ensure that everyone on an inspection team collects comparable information. Inspectors should follow written protocols for collecting air and material samples. Sampling programs should include control (apparently uncontaminated) and blank samples to provide comparison measurements and to detect improper sample handling. Possible biological contamination can be examined in the five ways outlined as follows.

*Visual inspection in the field.* Inspectors can examine suspect materials with a magnifying glass to distinguish, for example, algal or fungal growth from soil or dirt particles (ASTM, 1993a).

*Ability of sodium chloride to bleach color.* Inspectors can add a drop of 5 percent aqueous sodium hypochlorite to suspected fungal or algal growth, when possible, to see if color bleaches away (ASTM, 1993b). This test is not specific for biological contamination, but materials and discoloration that do not bleach probably are not biological, for example, mineral scale, metal corrosion, or soot.

*Bulk samples of contaminated materials.* Inspectors can excise sections of apparently contaminated materials

with a knife or scissors cleaned with alcohol and allowed to air dry. Inspectors can collect surface samples of loose materials with wetted sterile swabs or other tools, and can collect water samples in sterile containers (APHA, 1989). Field personnel should submit a sufficient volume of water, preferably ≥100 ml, or amount of sediment or bulk material to allow the laboratory to run several analytical tests. Samples generally can be shipped at ambient indoor temperatures if delivered to a laboratory within 24 hours, but samples should be refrigerated (not frozen) if transport will take longer or if samples might deteriorate.

*Adhesive tape specimens.* Inspectors can collect samples of possible biological growth from dry surfaces using clear adhesive tape when a bulk sample cannot be collected, for example, if one is not allowed to remove a piece of carpet, wallboard, or section of acoustical insulation. This is done by attaching the tape to a clean glass slide after touching the sticky side to the sample surface. Laboratory personnel examining a tape sample through a microscope can recognize such materials as fungal spores, pollen grains, or insect fragments, and thereby identify the contamination as biological material.

*Subculture using contact plates.* Inspectors can press special plates of raised culture medium to the surface of apparent biological growth and return the plates to a laboratory for incubation and examination (APHA, 1992; McGowan, 1985).

**Air sampling for biological hazards.** Investigators setting up an air-sampling program must decide what sampling equipment and procedures they will use, where they will monitor, how frequently and at what times they will collect samples, and how many replicate samples they will collect at each sampling site. Investigators are cautioned against undertaking air sampling for biological material without a clear purpose, a well-defined method, and predetermined criteria on how to interpret the findings. There is a discussion later in this section on the purpose of environmental sampling for biological hazards.

*Air-sampling instruments.* Clearly, no single sampling method allows recovery of all bioaerosols or is ideal in every test situation (ACGIH, 1994; Burge et al., 1989; Macher et al., 1995; Stetzenbach et al., 1992). Investigators choosing an air sampling method should consider air sampler collection efficiency, anticipated particle concentration and aerodynamic diameter, available analytical methods and their detection limits, and circumstances at the test site, e.g., availability of electrical power and need to keep sample collection unobtrusive. Air samplers commonly used indoors to collect culturable bioaerosols include slit, multiple-hole (sieve) and centrifugal impactors, impingers, cyclones, and cassette and high-volume filters (ASTM, 1990; Burge et al., 1989; Macher et al., 1995; Stetzenbach et al., 1992). (See also Chapter 16, Air Sampling.) Impaction directly onto an agar-based culture medium is the most commonly chosen method to collect culturable bacteria and fungi. Hand-held, battery-operated samplers have the advantages of portability and independence from a power supply, and although they sample at fairly high flow rates (40–180 l/min), these devices are quiet and fairly inconspicuous.

*Where to collect air samples.* Suggestions on where investigators should collect samples include outdoors near supply air intakes, indoors in the environments of people experiencing symptoms or where contamination is suspected, and indoors in comparison or control locations such as in noncomplaint areas or sites where no contamination was observed (Ayer, 1989; Woods et al., 1989). Sampling with and without people present and before and after HVAC and other equipment starts up is helpful in assessing the bioaerosol contributions from these sources. Data from questionnaires and observation of work procedures may identify specific areas warranting thorough inspection in buildings served by more than one air-handling unit, and for operations involving several steps.

*When to collect air samples.* The time at which samples are collected should reflect, as much as possible, the conditions the investigators wish to study. Experienced investigators recommend sampling at several different times of day and on multiple days. Sampling frequency at a test site depends on the purpose of the study and on how rapidly the parameters that influence bioaerosol release change. In other words, investigators should collect multiple samples at close intervals when they expect conditions to vary quickly, and the goal is to identify peak exposures rather than long-term averages. Less frequent sampling would suffice for routine surveillance after one has established a concentration range and variability.

*Number of air samples.* The number of replicate samples one should collect depends on the sensitivity and reliability of the analytical method and on the required certainty of detection (Ayer, 1989). It may be more informative, in some cases, to collect one or two samples at each of many sites or at a few sites on several occasions than it would be to determine very precisely the bioaerosol concentration at fewer locations or times. Investigators should collect at least two samples at each sampling site for each assay they run. They should decide before beginning a sampling program how they will summarize the resulting data and what criteria they will use to interpret the results.

*Expected air concentration and air sample volume.* Investigators should adjust sampling time to collect a detectable concentration of the material under study. For example, an investigator could collect a total volume of $\geq 0.1$ m$^3$ when there is no reason to expect unusually high concentrations of culturable microorganisms (Burge et al., 1989). One can detect concentrations on the order of 100 colony-forming units (cfu)/m$^3$ with this sample volume. Concentrations in occupied buildings are seldom lower than 10 cfu/m$^3$ except in ultraclean areas; concentrations above $10^3$ cfu/m$^3$ may not be unusual outdoors and in some indoor workplaces (Morey et al., 1990; Burge et al., 1989).

*Detection and assay methods.* A variety of methods are available to detect, identify, and quantify biological materials. The choice of method depends on the material under study and the type of information needed from its examination, for example, results of culture, direct examination by light or electron microscopy, or immunochemical or other assays (APHA, 1989; APHA, 1992; Burge et al., 1989; Nash & Krenz, 1991). Investigators might consider testing air, surface, bulk material, and liquid samples for mesophilic and thermophilic bacteria (which prefer moderate and elevated temperatures) and saprophytic fungi (which grow on dead organic material) when no information is available to suggest that certain other microorganisms are present. Sampling for human-commensal bacteria (which people shed normally from the skin and scalp) is useful in hospitals or laboratories, but in other occupied areas the enumeration of these bacteria may be less informative.

A single laboratory should process all of the samples for a study, and it would be appropriate to choose a licensed clinical laboratory if both human and environmental specimens will be tested. Otherwise, investigators may find that laboratories that routinely culture food, soil, or water samples have more experience with the saprophytic microorganisms found in environmental samples than do laboratories specializing in the identification of human pathogens (APHA, 1989, 1992; Morey et al., 1990; Burge et al., 1989).

**Purpose of environmental sampling.** Investigators conducting environmental monitoring for biological hazards have the greatest chances for success if they can state their purpose clearly, plan the study well, use methods that are scientifically sound, analyze and interpret the results correctly, and communicate the findings effectively to all involved parties (Kundsin, 1977). Inspectors can use air and source sampling to confirm suspected sources of contamination (not all apparently contaminated material actually is contaminated), to identify contaminants (control measures depend on the organisms present), and to evaluate the effectiveness of control measures. Investigators also should consider reasons to *not* monitor: Sampling can be expensive and time consuming, investigators may overlook the cause of a problem if they do not comprehensively consider all types of biological hazards, findings from improperly designed or executed monitoring programs may be difficult to interpret, and information needed to solve a problem often can be obtained by other means.

***Interpreting environmental sampling results.*** Investigators should seek explanations when biological materials are found in material or air samples from test sites but not in those from control sites, or when test site concentrations are many times higher than elsewhere. The presence indoors of biological material not found in outdoor or control-site air samples suggests a building- or work-related source. However, concentrations of human-commensal bacteria (from building occupants), and saprophytic microorganisms (from outdoor air or from building- and work-related sources) must be fairly high to be of health significance, except in hospitals (for patient safety) and in some laboratories and cleanrooms (for process and product purity).

The concentration of fungal spores in air usually exceeds that of other bioaerosols in outdoor samples, although pollen, bacteria, algae, and insect fragments are also present (Morey et al., 1990; Burge et al., 1989). Investigators routinely recover molds from the genera *Alternaria, Aspergillus, Aureobasidium, Cladosporium, Epicoccum, Fusarium, Mucor, Penicillium,* and *Rhizopus,* and yeasts such as *Rhodotorula* and other yeasts from indoor and outdoor air samples.

The concentration of bacteria in occupied buildings routinely exceeds outdoor concentrations (Morey et al., 1990; Burge et al., 1989). Human-commensal bacteria such as *Micrococcus* and *Staphylococcus* spp., which people shed from their skin and hair, and *Streptococcus* spp., which people expel in respiratory secretions while sneezing, coughing and talking, predominate indoors in the absence of building- and work-related sources of other bacteria. Higher indoor than outdoor air concentrations of environmental bacteria (*Bacillus* spp., for example) should prompt a search for an indoor source such as damp carpets, wall boards, or stored papers. Actinomycetes are unusual in nonagricultural indoor environments, and their presence indicates an indoor source when these bacteria are not found at similar concentrations in outdoor air samples.

An investigation team might find it revealing to compare test and control areas for other building-related conditions that affect occupant comfort and well-being, whether or not potential biological hazards are found. Examples of such factors are air temperature and relative humidity, outdoor air supply rates, occupancy levels, types of occupant activities, amount of time spent in the building, and job satisfaction (Quinlan et al., 1989; USEPA, 1991; AIHA, 1993).

**Recommendations and control measures for work- and building-related biological hazards.** Bioaerosol-related problems clearly are preventable; for example, laboratory personnel can work safely with biological agents, health care workers can deliver good care to infected patients without putting themselves at unnecessary risk, proper equipment maintenance and operation can control microbiological contamination, and accumulation and aerosolization of biological material can be avoided.

Infectious diseases should be controlled, whenever possible, by seeing that people are properly immunized to prevent infection and that infected people receive prompt medical treatment (Benenson, 1990; WHO, 1990). People with respiratory infections that can be transmitted from person to person should be isolated until they no longer present a hazard to others; they should remain at home or be hospitalized in properly ventilated rooms. Crowding plays a role in transmission of airborne infections, and an inadequate supply of outdoor air increases the chances that uninfected people will inhale microorganisms released by infected people. Measures to clean or disinfect indoor air (air filtration or germicidal UV irradiation) may be appropriate in some high-risk settings such as health care facilities and high-containment laboratories. Methods of prevention of hypersensitivity diseases focus on minimizing exposure, which includes preventing growth and accumulation of contaminants in buildings and ventilation systems, and removing hypersensitive people from situations where they might be exposed. When workers cannot be protected from infectious or allergenic bioaerosols by any other means, they should be provided with appropriate personal protection. (See also Chapter 22, Respiratory Protection.)

Investigators asked to control building-related biological hazards can recommend that architects and contractors design and construct buildings and ventilation systems so that they are easy to clean and to maintain; that engineers design HVAC systems to control moisture and so that biological contaminants do not multiply or accumulate and are not disseminated; that facility managers maintain clean and safe buildings; and that building owners deal with problems promptly when they learn of them.

An investigator asked to control work-related biological hazards can recommend that employers and work supervisors develop work practices that minimize direct skin contact and respiratory exposure to biological hazards; that they supply appropriate source controls where needed, such as biological safety cabinets and local exhaust ventilation; that they provide appropriate personal protection when needed, such as skin, eye, and respiratory protection; and that they see that workers adhere to safety precautions to protect themselves and others. Finally, administrators should relocate sensitive individuals when other control measures cannot ensure worker safety. (See also Part 5 of this text, Control of Hazards.)

**Personal protection for potential exposure to biological hazards.** An earlier section of this chapter described personal protection for laboratory work with different classes of biological agents. Industrial hygienists and safety personnel who must inspect potentially contaminated equipment (for example, HVAC systems and water-cooled, heat-exchange equipment) should request that the system be turned off while they examine it, if possible. Inspectors should wear disposable garments, slip-proof footwear, gloves, and eye protection while examining areas that are wet, potentially contaminated, or recently treated with biocides, disinfectants, or other chemicals. Anyone near operating equipment suspected of contamination and which might generate aerosols should wear a respirator approved for the situation. (See Chapter 22, Respiratory Protection.) The use of personal protection and of air monitoring may be advisable during the sampling and removal of extensively contaminated materials (such as building furnishings showing visible mold growth, and bird, insect, or rodent nests and droppings) and during the sampling and cleaning of contaminated HVAC and water systems and equipment.

# Regulations and Guidelines

There are few specific regulations that target work environments where employees might be exposed to infectious microorganisms. The OSHA general duty clause (employers shall provide a workplace free of recognized hazards that cause or are likely to cause death or serious physical harm) is an example of an early, nonspecific regulation (OSHA, 1970). Various government agencies (such as NIH, The National Cancer Institute [NCI], and the CDC) published guidelines in the 1970s that addressed issues relevant to microbiological safety. Although guidelines do not have the same impact as regulations, they are considered the accepted standard of practice at the time of their publication. Activities conducted in a manner contrary to published guidelines are generally considered unacceptable. Unlike guidelines, regulations define detailed requirements for specific activities, with penalties for noncompliance.

The following lists highlight relevant agencies, regulations, guidelines, and standards applicable to manipulation, transport, and disposal of infectious microorganisms or materials and professional organizations where biosafety and infection control assistance can be obtained. See also Appendix A, Additional Resources, in this text.

## Regulations That Affect Laboratory Biosafety Practices

- OSHA: 29 *CFR* 1910.1030: Occupational exposure to bloodborne pathogens, 1991.
- OSHA: Occupational Health and Safety Act; general duty clause, Section 5(a)(1), 1970.
- OSHA: 29 *CFR* 1910.132–133: Personal protective equipment including eye and face protection, 1970.
- OSHA: 29 *CFR* 1910.134: Respiratory protection, 1970.
- EPA: Genetically engineered organisms in industry and registration of disinfectants (TOSCA). (Infectious wastes are not currently regulated at the federal level.)
- US DOT: 49 *CFR* Parts 106–107 and 171–180: Packaging and transport of hazardous substances, including infectious substances. (See US DOT in Bibliography for modifications and extensions.) (Available from the Office of Hazardous Materials Standards, (202) 366–4488.)
- USPHS: 42 *CFR* Part 72: Interstate shipment of etiologic agent, July 1980; update pending (proposed March 1990).
- USPHS: 42 *CFR* Part 71.54: Foreign quarantine, etiologic agents, hosts, and vectors. (Contact CDC Biosafety Branch for assistance.)
- U.S. Department of Commerce: Export of infectious materials; revised regulations March 1994. (For information, call (202) 462–1309.)
- U.S. Postal Service: 39 *CFR* Part 111: Mailability of etiologic agents, 1989.
- U.S. Department of Agriculture: Plant and animal pathogens including genetically engineered pathogens (Plant Pest Act). (Contact the Animal and Plant Health Inspection Service.)
- Clinical Laboratory Improvement Act (CLIA) PL 100–578, 1988. CDC published "Regulations for implementing the clinical laboratory improvement amendments of 1988: a summary" in *MMWR* 41:#RR-2, pp. 1–17, 1992.
- Other: The FDA regulates antiseptics.

## Guidelines for the Safe Use of Pathogenic or Oncogenic Microorganisms

- NIH: Guidelines for research involving recombinant DNA molecules. *Federal Register* 59, #127:34498–34547, July 5, 1994. (Updates available from the Office of Recombinant DNA Activities at NIH.)
- National Cancer Institute: Safety standard for research involving oncogenic viruses. October 1974 (out of print).
- CDC/NIH: Biosafety in the microbiological and biomedical laboratory; U.S. safety standard for work involving etiologic agents.
- National Research Council of the National Academy of Science: *Biosafety in the Laboratory—Prudent Practices for Handling and Disposal of Infectious Materials.*
- CDC: Infection control guidelines. (Hospital Infections Branch.)

### Standard-Setting or Credentialing Groups

- Joint Commission on Accreditation of Healthcare Organizations (certification of laboratories in health care organizations)
- National Sanitation Foundation (Standard #49—Class II laminar flow) (Biohazard cabinetry; Ann Arbor, MI)
- National Committee for Clinical Laboratory Standards, Villanova, PA:

  Draft guidelines for equipment associated with biohazards

  Standard for handling clinical specimens

  Standard for transport of clinical specimens

  Guideline for the protection of laboratory workers from infectious diseases transmitted by blood, body fluids, and tissue
- American Society for Testing and Materials, Philadelphia, PA

### Professional Associations

- American Biological Safety Association, Mundelein, IL
- American Society for Microbiology, Public and Scientific Affairs Board, Laboratory Practices Committee, Laboratory Safety Subcommittee, ASM, Washington, DC
- American Industrial Hygiene Association, Biosafety Committee, Fairfax, VA
- American Conference of Governmental Industrial Hygienists, Committees on Infectious Agents; Agricultural Health and Safety; Bioaerosols; and Air Sampling, Cincinnati, OH
- Association of Professionals in Infection Control, Washington, DC
- American Society of Heating, Refrigerating and Air Conditioning Engineers, Atlanta, GA
- Campus Safety Association (associated with the National Safety Council)
- National Safety Council, Itasca, IL
- Society for Healthcare Epidemiologists of America, Woodbury, NJ

## ■ Role of the Industrial Hygienist in Biosafety

The industrial hygienist is trained to identify workplace hazards, evaluate their significance, and recommend programs and controls to eliminate or minimize occupational exposures. In addition to focusing on chemical and physical hazards, industrial hygienists (IHs) are often required to evaluate work-related illness caused by biological agents. Evaluation of illness or exposure in environments associated with agricultural work, mining, textile manufacturing,

water systems, and sewage treatment, to list a few, have routinely been undertaken by industrial hygienists.

Depending on technical background and interest, some IHs cover the broad range of biohazardous materials described in this chapter. However, the majority, because they lack training in pathogenic microbiology, infection control, and medical epidemiology, restrict their activities to biohazard control, defining biohazards in a more limited fashion. This definition of biohazard may only include agents such as *Legionella,* molds, and other biological agents involved in indoor air quality, including potential allergens or toxins from bacteria (endotoxins), fungi (mycotoxins), and plant pollens.

IHs can play an important role in the development and implementation of a biosafety program. Their training and experience enable them to understand and monitor biological activities in relatively low-risk biohazard situations (BSL-1 or -2). When work involves more hazardous agents such as *Mycobacterium tuberculosis,* requiring a higher level of laboratory containment such as BSL-3, assistance should be obtained from a professional in biological safety. This is similar to the role of the IH in radiation safety, who requires the assistance of a professional in health physics in certain high-risk situations. The names of registered biosafety professionals can be obtained from the American Biological Safety Association (ABSA) in Mundelein, Illinois. Presently, the pool of biosafety specialists is relatively small. Fortunately, both professions have successfully used each other's expertise to resolve issues related to occupational exposure to biological agents.

## ■ Summary

Agricultural, medical, and laboratory workers are most at risk to occupational biohazards, but many workplaces have the potential for such exposure, for example, microbiology, public health, and molecular biology laboratories; hospitals and other health care institutions; biotechnology facilities; veterinary practices; farms; wood-processing facilities; mines; textile manufacturing; and fishing and forestry industries. Therefore, containment of microorganisms in laboratories (and other workplaces) is critical to the health of workers and to the community. Engineering controls—safety equipment and facility design—and worker-initiated workplace controls—good work practices and carefully executed techniques—can minimize occupational biohazardous exposures.

## ■ Bibliography

Acha PN, Szyfres B. *Zoonoses and Communicable Diseases Common to Man and Animals.* Scientific Publication #354. Washington, DC: Pan American Health Organization, 1980.

American Conference of Governmental Industrial Hygienists. *1994–1995 Threshold Limit Values for Chemical Substances and Physical Agents and Biological Exposure Indices.* Cincinnati, OH: American Conference of Governmental Industrial Hygienists, 1994.

American Industrial Hygiene Association. *Biosafety Reference Manual.* Fairfax, VA: American Industrial Hygiene Association, 1995.

Technical Committee on Indoor Environmental Quality. *The Industrial Hygienist's Guide to Indoor Air Quality Investigations.* Fairfax, VA: American Industrial Hygiene Association, 1993.

American Lung Association. Agricultural Respiratory Hazards Education Series (9 parts). Ames American Lung Association of Iowa, 1986.

American Public Health Association. *Compendium of Methods for the Microbiological Examination of Foods,* 3rd ed. Washington, DC: American Public Health Association, 1992, pp. 51–134.

American Public Health Association/American Water Works Association & Water Pollution Control Federation. *Standard Methods for the Examination of Water and Wastewater,* 17th ed. Washington, DC: American Public Health Association, 1989, pp. 9.4–9.8, 9.52–9.66.

American Society for Testing and Materials. Test method of evaluating degree of surface disfigurement of paint films by microbial (fungal or algal) growth or soil and dirt accumulation. Method D 3274-82(1988). In *Standard Methods on Materials and Environmental Microbiology.* Philadelphia: American Society for Testing and Materials, 1993a, pp. 132–133.

American Society for Testing and Materials. Guide for determining the presence of and removing microbial (fungal or algal) growth on paint and related coatings. Method D 4610–86. In *Standard Methods on Materials and Environmental Microbiology.* Philadelphia: American Society for Testing and Materials, 1993b, pp. 151–152.

American Society for Testing Materials. *Emergency Test Method for Resistance of Protective Clothing Materials to Synthetic Blood* (ASTM Designation ES21–92). Philadelphia: American Society for Testing and Materials, 1992a.

American Society for Testing and Materials. *Emergency Test Method for Resistance of Protective Clothing Materials to Penetration by Blood-Borne Pathogens Using Viral Penetration as a Test System.* (ASTM Designation ES22–92). Philadelphia: American Society for Testing and Materials, 1992b.

American Thoracic Society. Levels of laboratory services for mycobacterial diseases. Reprinted by *Am Thorac Soc News* from *Am Rev Respir Dis* 128:213, 1983.

Ammann AJ. Immunodeficiency diseases. In Stites DP, Stobo JD, Wells JU, eds. *Basic and Clinical Immunology.* Norwalk, CT: Appleton & Lange, 1987.

Arey LB, Burrows W, Greenhill JP, et al., eds. *Dorland's Illustrated Medical Dictionary,* 23rd ed. Philadelphia: W.B. Saunders, 1957.

Atlas RM, Bartha R. *Microbial Ecology: Fundamentals and Applications,* 3rd ed. Redwood City, CA: Benjamin Cummings, 1993.

Ayer HE. Occupational Air Sampling Strategies. In Hering SV, ed. *Air Sampling Instruments for Evaluation of Atmospheric Contaminants.* Cincinnati, OH: American Conference of Governmental Industrial Hygienists, 1989, pp. 21–31.

Balows A, Hausler WJ, Herrmann KL, et al., eds. *Manual of Clinical Microbiology,* 5th ed. Washington, DC: American Society for Microbiology, 1991.

Barbaree JM, Breiman RF, Dufour AP, eds. *Legionella. Current Status and Emerging Perspectives.* Washington, DC: American Society for Microbiology, 1993.

Baron RC, McCormick JB, Zubeir OA. Ebola virus disease in southern Sudan: Hospital dissemination and intrafamilial spread. *Bull World Health Organ* 61:997–1003, 1983.

Benenson AS, ed. *Control of Communicable Diseases in Man,* 15th ed. Washington, DC: American Public Health Association, 1990.

Bernard KW, Parham GL, Winkler WG, et al. Q fever control measures: Recommendations for research facilities using sheep. *Infect Control* 3:461–465, 1982.

Birnbaum D. Statistics for hospital epidemiology: CQI tools, sentinel events, warning and action limits. *Infect Control Hosp Epidemiol* 14: 537–539, 1993.

Bland SM, Evans R, Rivera JD. Allergy to laboratory animals in health care personnel. *Occup Med* 2:525–546, 1987.

Block SS. *Disinfection, Sterilization and Preservation.* Philadelphia: Lea & Febiger, 1991.

Burge HA. *Bioaerosols.* Boca Raton, FL: Lewis Publishers, 1995.

Burge HA. The fungi. In Morey PR, Feeley JC, Otten JA, eds. *Biological Contaminants in Indoor Environments.* Philadelphia: American Society for Testing and Materials, 1990, pp. 136–162.

Burge HA, Feeley JC, Kreiss K, et al. *Guidelines for the Assessment of Bioaerosols in the Indoor Environment.* Cincinnati, OH: American Conference of Governmental Industrial Hygienists, 1989.

Burge S, Hedge A, Wilson S, et al. Sick building syndrome: A study of 4373 office workers. *Ann Occup Hyg* 31(4A):493–500, 1987.

Centers for Disease Control and Prevention. Proportionate mortality from pulmonary tuberculosis associated with occupations—28 states, 1979–1990. *MMWR* 44:14–19, 1995.

Centers for Disease Control and Prevention. Guidelines for preventing the transmission of *Mycobacterium tuberculosis* in health-care facilities. *MMWR* 43:RR-13, 1994a.

Centers for Disease Control and Prevention. Laboratory management of agents associated with hantavirus pulmonary syndrome: Interim biosafety guidelines. *MMWR* 43:RR-7, 1994b.

Centers for Disease Control and Prevention. Addressing emerging infectious disease threats: A prevention strategy for the United States. *MMWR* 43:RR-5, 1994c.

Centers for Disease Control and Prevention. Draft guidelines for preventing the transmission of tuberculosis in health-care facilities. (2nd ed. Notice of Comment Period). *Federal Register* 58:52810–52854, Oct. 12, 1993a.

Centers for Disease Control and Prevention. Hantavirus infection—Southwestern United States: interim recommendations for risk reduction. *MMWR* 42:RR-11, 1993b.

Centers for Disease Control and Prevention. Recommendations of the advisory committee on immunization practices (ACIP): Use of vaccines and immune globulins in persons with altered immunocompetence. *MMWR* 42:RR-4, 1993c.

Centers for Disease Control and Prevention. Update: Hantavirus pulmonary syndrome—U.S. *MMWR* 42:816–820, 1993d.

Centers for Disease Control and Prevention. Tuberculosis morbidity—United States, 1992. *MMWR* 42:696–704, 1993e.

Centers for Disease Control. Seroconversion to simian immunodeficiency virus in two laboratory workers. *MMWR* 41:678–681, 1992.

Centers for Disease Control. Update on adult immunization: recommendations of the immunization practices advisory committee (ACIP). *MMWR* 40:RR-12, 1991a.

Centers for Disease Control. Nosocomial transmission of multidrug-resistant tuberculosis among HIV-infected persons—Florida and New York, 1988–1991. *MMWR* 40:585–591, 1991b.

Centers for Disease Control. Update: Ebola-related filovirus infection in nonhuman primates and interim guidelines for handling nonhuman primates during transit and quarantine. *MMWR* 39:22–30, 1990a.

Centers for Disease Control. Update: Filovirus infection in animal handlers. *MMWR* 39:221, 1990b.

Centers for Disease Control. Ebola virus infection in imported primates—Virginia, 1989. *MMWR* 38:831–832, 837–838, 1989a.

Centers for Disease Control. B virus in humans—Michigan. *MMWR* 38:453–454, 1989b.

Centers for Disease Control. Guidelines to prevent simian immunodeficiency virus infection in laboratory workers and animal handlers. *MMWR* 37:693–694, 699–704, 1988.

Centers for Disease Control. Guidelines for prevention of *Herpesvirus simiae* (B virus) infection in monkey handlers. *MMWR* 36:680–689, 1987a.

Centers for Disease Control. B virus infections in humans—Pensacola, FL. *MMWR* 36:289–290, 295–296, 1987b.

Centers for Disease Control. Recommendations for prevention of HIV transmission in health care settings. *MMWR* 36:2 Supplement, 1987c.

Centers for Disease Control. Interstate shipment of etiologic agents (42 *CFR* Part 72). Part 72.2—Transportation of diagnostic specimens, biological products and other material; Part 72.3—Transportation of materials containing certain etiologic agents: minimum packaging requirements. *Federal Register* 45(141), July 21, 1980.

Center for Disease Control. *Classification of Etiologic Agents on the Basis of Hazard,* 4th ed. Washington, DC: U.S. Dept. of Health, Education and Welfare, Public Health Service, Office of Biosafety, 1974.

Clark PK. OSHA Memorandum: Changes to OSHA Instruction CPL 2–2.44C regarding first responders, July 1, 1992.

Clark RA. OSHA directorate of compliance programs, OSHA enforcement policy and procedures for occupational exposure to tuberculosis. *Infect Control Hosp Epidemiol* 14:694–699, 1993.

Constable PJ, Harrington JM. Risks of zoonoses in a veterinary service. *Br Med J* 284:246–248, 1982.

Cox CS, Wathes CM. *Bioaerosols Handbook.* Boca Raton, FL: CRC/Lewis Publishers, 1995.

Decker MD. OSHA enforcement policy for occupational exposure to tuberculosis. *Infect Control Hosp Epidemiol* 14:689–693, 1993.

Desmyter J, Johnson KM, Deckers C, et al. Laboratory rat associated outbreak of haemorrhagic fever with renal syndrome due to hantaan-like virus in Belgium. *Lancet* ii:1445–1448, 1983.

Donham KJ. Zoonotic diseases of occupational significance in agriculture: A review. *Int J Zoonoses* 12:163–191, 1985.

Ducatman AM, Liberman DF. The biotechnology industry. *Occup Med* 6(2), 1991.

Dutkiewicz J, Jablonski L, Olenchock SA. Occupational biohazards: a review. *Am J Ind Med* 14:605–623, 1988.

European Economic Community. Document No 4645/1/91 EN Draft proposal for a council directive amending directive 90/679/EEC on the protection of workers from risks related to exposure to biological agents at work. Brussels, Belgium: EEC, 1991.

Favero MS, Bond WW. Sterilization, disinfection and antisepsis in the hospital. In Balows A, ed. *Manual of Clinical Microbiology.* Washington, DC: American Society for Microbiology, 1991, pp. 183–200.

Fleming DO, Richardson JH, Tulis JJ, et al., eds. *Laboratory Safety: Principles and Practices,* 2nd ed. Washington, DC: American Society for Microbiology, 1995.

Fox JG, Lipman NS. Infections transmitted by large and small laboratory animals. *Infect Dis Clin North Am* 5: 131–163, 1990.

Frommer W, Ager B, Archer L, et al. Safe biotechnology. III. Safety precautions for handling microorganisms of different risk classes. *Appl Microbiol Biotechnol* 30:541–552, 1989.

Garner JS, Favero MS. Guideline for handwashing and hospital environmental control. *Am J Infect Control* 14:110–126, 1986.

Garner JS, Simmons BP. CDC guideline for isolation precautions in hospitals. *Am J Infect Control* 4:245–325, 1983.

Gear JJS, Cassel GA, Gear AJ, et al. Outbreak of Marburg virus disease in Johannesburg. *Br Med J* 4:489–493, 1975.

Gilchrist M, Fleming D, Hindler J. 1994. Laboratory safety management update: Aerosol borne microorganisms. In Eisenberg HD, ed. *Clinical Microbiology Procedures Handbook,* Supplement #1. Washington, DC: American Society for Microbiology, 1994, p. xxix.

Goldman RH. Medical surveillance in the biotechnology industry. *Occup Med* 6(2):209–225, 1991.

Goldman RH. Medical surveillance program. In Liberman DF, Gordon JG, eds. *Biohazard Management Handbook.* New York: Marcel Dekker, 1989.

Hatch TF. Distribution and deposition of inhaled particles in the respiratory tract. *Bacteriol Rev* 25:237–240, 1961.

Herwaldt BL, Juranek DD. Protozoa and helminths. In Fleming DO et al., eds. *Laboratory Safety: Principles and Practices,* 2nd ed. Washington, DC: American Society for Microbiology, 1995.

Hodgson MJ. Clinical diagnosis and management of building-related illness and the sick building syndrome. *Occup Med* 4(4):593–606, 1989.

Isenberg HD, d'Amato RF. Indigenous and pathogenic microorganisms of humans. In Balows A, Hausler WJ, Herrmann KL, et al., eds. *Manual of Clinical Microbiology.* Washington, DC: American Society for Microbiology, 1991, pp. 2–14.

Jacobson JT, Orlob RB, Clayton JL. Infections acquired in clinical laboratories in Utah. *J Clin Microbiol* 21:486–489, 1985.

Jawetz E, Melnick JL, Adelberg EA. *Medical Microbiology,* 19th ed. Norwalk, CT: Appleton & Lange, 1991.

Kenny MT, Sabel FL. Particle size distributions of *Serratia marcescens* aerosols created during common laboratory procedures and simulated laboratory accidents. *Appl Microbiol* 16:1146–1150, 1968.

Kent PT, Kubica GP. Safety in the laboratory. In *Public Health Mycobacteriology. A Guide for the Level III Laboratory.* Atlanta: Centers for Disease Control (DHHS, PHS), 1985, pp. 5–10.

Klein M, Deforest A. The inactivation of viruses by germicides. *Chem Specialists Manuf Assoc Proc* 49:116–118, 1963.

Kreiss K. The epidemiology of building-related complaints and illness. *Occup Med* 4(4):575–592, 1989.

Kruse RH, Puckett WH, Richardson JH. Biological safety cabinetry. *Clin Microbiol Rev* 4:207–241 (Tables 5–10), 1991.

Kuehne RW. Biological containment facility for studying infectious diseases. *Appl Microbiol* 26:239–243, 1973.

Kuenzi M, et al. Safe biotechnology—general considerations. A report prepared by the Safety in Biotechnology Working Party of the European Federation of Biotechnology. *Appl Microbiol Biotechnol* 21:1–6, 1985.

Kundsin RB. Microbiological monitoring of the hospital environment. In Cundy KR, Ball W, eds. *Infection Control in Health Care Facilities: Microbiological Surveillance.* Baltimore, MD: University Park Press, 1977.

Lee HW, Johnson KM. Laboratory-acquired infections with hantaan virus, the etiologic agent of Korean hemorrhagic fever. *J Infect Dis* 146:645–651, 1982.

Lloyd G, Jones N. Infection of laboratory workers with hantavirus acquired from immunocytomas propagated in laboratory rats. *J Infect Dis* 12:117–125, 1986.

Lupo D. Certification of biosafety cabinets. In Liberman DF, Gordon JG, eds. *Biohazards Management Handbook.* New York: Marcel Dekker, 1989.

Macher JM, Chatigny MA, Burge HA. Sampling airborne microorganisms and aeroallergens. In Cohen BS, ed. *Air Sampling Instruments for Evaluation of Atmospheric Contaminants,* 8th ed. Cincinnati, OH: American Conference for Governmental Industrial Hygienists (in press).

Martini GA, Siegert R, eds. *Marburg Virus Disease.* Berlin: Springer-Verlag, 1971.

McGowan JE. Role of the microbiology laboratory in prevention and control of nosocomial infections. In Lennette EH, ed. *Manual of Clinical Microbiology,* 4th ed. Washington, DC: American Society for Microbiology, 1985, pp. 110–122.

Miller BM, Groschel DHM, Richardson JH, et al., eds. *Laboratory Safety: Principles and Practices.* Washington, DC: American Society for Microbiology, 1986.

Mitchell R, ed. *Environmental Microbiology.* New York: Wiley-Liss, 1992.

Morey PR, Feeley JC, Otten JA, eds. *Biological Contaminants in Indoor Environments.* Philadelphia: American Society for Testing and Materials, 1990.

Morey PR, Singh J. Indoor air quality in nonindustrial occupational environments. In Clayton GD, Clayton FE, eds. *Patty's Industrial Hygiene and Toxicology,* 4th ed., Vol. 1, Part A. New York: Wiley, 1991. pp. 531–594.

Nash P, Krenz MM. Culture media. In Balows A, ed. *Manual of Clinical Microbiology,* 5th ed. Washington, DC: American Society for Microbiology, 1991, pp. 1226–1288.

National Committee for Clinical Laboratory Standards. *Protection of Laboratory Workers from Infectious Disease Transmitted by Blood, Body Fluids and Tissue* (Tentative Guideline M29-T2). Villanova, PA: National Committee for Clinical Laboratory Standards, 1991.

National Institutes of Health. Guidelines for research involving recombinant DNA molecules. *FR* 59:34496–34547, July 5, 1994.

National Institutes of Health. Actions under the guidelines, guidelines for research involving recombinant DNA molecules. *FR* 56:33174–33183, July 18, 1991.

National Research Council—Committee on Hazardous Biological Substances in the Laboratory. *Biosafety in the Laboratory: Prudent Practices for the Handling of Infectious Materials.* Washington, DC: National Academy Press, 1989.

National Research Council—Committee on the Introduction of Genetically Engineered Organisms into the Environment. *Introduction of Recombinant DNA–Engineered Organisms into the Environment: Key Issues.* Washington, DC: National Academy Press, 1987.

National Sanitation Foundation. *Standard 49—Class II (Laminar Flow) Biohazard Cabinetry.* Ann Arbor, MI: The Foundation, 1992.

Nettleman MD. Biological warfare and infection control. *Infect Control Hosp Epidemiol* 12(6):368–372, 1991.

Occupational Safety and Health Administration. Enforcement procedures for the occupational exposure to bloodborne pathogens standard. OSHA Instruction CPL 2–2.44C, 1992.

Occupational Safety and Health Adminstration. Protection from bloodborne pathogens (29 *CFR* Part 1910.1030). *FR* 56: #235, 64175–64182, Dec. 6, 1991.

Occupational Safety and Health Administration. Occupational Health and Safety Act of 1970. General Duty Clause, Section 5 (a)(1). Public Law 91–596. December 29, 1970.

Office of Science and Technology Policy. Coordinated framework for the regulation of biotechnology. *FR* 51:23302–23393, 1986.

Official Journal of the European Communities. Directive 90/679/EEC on the protection of workers from risks related to exposure to biological agents at work. *Official J Eur Communities* 374:1–12, 31.12.1990, 1990.

Organization for Economic Co-operation and Development. *Recombinant DNA Safety Considerations.* Paris: Organization For Economic Co-operation and Development, 1986.

Phillips GB. Control of microbiological hazards in the laboratory. *Am Ind Hyg Assoc J* 30:170–176, 1969.

Pike RM. Past and present hazards of working with infectious agents. *Arch Pathol Lab Med* 102:333–336, 1978.

Pike RM. Laboratory-associated infections: Summary and analysis of 3921 cases. *Health Lab Sci* 13:105–114, 1976.

Pope AM, Patterson R, Burge H. *Indoor Allergens—Assessing and Controlling Adverse Health Effects.* Washington, DC: National Academy Press, 1993.

Popendorf W, Donham KJ. Agricultural hygiene. In Clayton GD, Clayton FE, eds. *Patty's Industrial Hygiene and Toxicology 1:* Part A—General Principles. New York: Wiley, pp. 721–759, 1991.

Postgate J. *Microbes and Man.* Cambridge, UK: Cambridge University Press, 1992.

Prusiner SB. Novel proteaceous infectious particles cause scrapie. *Science* 216:136–144, 1982.

Quinlan P, Macher JM, Alevantis LE, et al. Protocol for the comprehensive evaluation of building-associated illness. *Occup Med* 4(4):771–797, 1989.

Raviglione MG, Snider DE, Kochi A. Global epidemiology of tuberculosis. *JAMA* 273:220–226, 1995.

Reitman M, Wedum AG. Microbiological safety. *Public Health Rep* 71:659–665, 1956.

Rutala WA. APIC guidelines for selection and use of disinfectants. *Am J Infect Control* 18:99–117, 1990.

Schnurrenberger PR, Grigor JK, Walker JF, et al. The zoonosis-prone veterinarian. *J Am Vet Med Assoc* 173:373–376, 1978.

Sheretz RJ, Hampton AL. Infection control aspects of hospital employee health. In Werzel RP, ed. *Preventional Control of Nosocomial Infections*. Baltimore, MD: Williams & Wilkins, 1986.

Short LJ, Bell DM. Risk of occupational infection with blood-borne pathogens in operating and delivery room settings. *Am J Infect Control* 21:343–350, 1993.

Spaulding EH. Chemical disinfection and antisepsis in the hospital. *J Hosp Res* 9:5–31, 1972.

Stanier RY, Adelberg EA, Ingraham J. *The Microbial World*. Englewood Cliffs, NJ: Prentice-Hall, 1986.

Stetzenbach LD, Hern SC, Seidler RJ. Field sampling design and experimental methods for the detection of airborne microorganisms. In Levin MA, Seidler RJ, Rogul M, eds. *Microbial Ecology: Principles, Methods, and Applications*. New York: McGraw Hill, 1992, pp. 543–555.

Stratton CW. Topics in clinical microbiology: tuberculosis, infection control, and the microbiology laboratory. *Infect Control Hosp Epidemiol* 14(6):481–487, 1993.

Sulkin SE, Pike RM. Survey of laboratory-acquired infections. *Am J Public Health* 41:769–780, 1951.

U.S. Department of Transportation, Research and Special Programs Administration. Amendment Docket HM-181G No 171–178, 59 *FR* 48762, Sept. 22, 1994.

U.S. Department of Transportation. ANPRM Hazardous materials regulations. Docket HM-181G. 58 *FR* 12207, March 3, 1993.

U.S. Department of Transportation. 49 *CFR* Parts 171–80. Hazardous materials regulations. Docket HM-181. Final Rule, 55 *FR* 52402, Dec. 21, 1990. [Revisions 56 *FR* 66124, December 20, 1991; 57 *FR* 45442, October 1, 1992; 58 *FR* 12182, March 3, 1993; and 58 *FR* 66302, December 20, 1993.]

U.S. Environmental Protection Agency/U.S. Department of Health and Human Services. *Building Air Quality. A Guide for Building Owners and Facility Managers*. S/N 055–000–00390–4. Washington, DC: USEPA/USDHHS, 1991.

U.S. Public Health Service. *Biosafety in Microbiological and Biomedical Laboratories*, 3rd ed. HHS Publication no. (CDC)93–8395. Washington, DC: USGPO (Stock# 017–040–00523–7), 1993.

U.S. Public Health Service. *National Cancer Institute Safety Standards for Research Involving Oncogenic Viruses—Manual*. Publication No. (NIH) 75–790. Washington, DC: USGPO, 1974.

Van Houten J. New frontiers in biosafety: the industrial prospective. In Liberman DJ, Gordon JG, eds. *Biohazards Management Handbook*. New York: Marcel Dekker, 1989.

Vesley D, Hartmann HM. Laboratory-acquired infections and injuries in clinical laboratories: a 1986 survey. *Am J Public Health* 78:1213–1215, 1988.

Wedum AG, Barkley WE, Hellman A. Handling of infectious agents. *J Am Vet Med Assoc* 161:1557–1567, 1972.

Wong TM, Chan YC, Yap Etl, et al. Serological evidence of hantavirus infection in laboratory rats and personnel. *Int J Epidemiol* 17:887–890, 1988.

Woods JE, Morey PR, Rask DR. Indoor air quality diagnostics: qualitative and quantitative procedures to improve environmental conditions. In Nagda NL, Harper JP, eds. *Design and Protocol for Monitoring Indoor Air Quality*. Philadelphia: American Society for Testing and Materials, 1989, pp. 80–98.

World Health Organization. *Laboratory Biosafety Manual*, 2nd ed. Geneva: World Health Organization, 1993.

World Health Organization. *Indoor Air Quality: Biological Contaminants*. Report on WHO Meeting, Rautavaara, 29 August–2 September, 1988. WHO Regional Publications, European Ser. No. 31, 1990.

World Health Organization. Ebola haemorrhagic fever in Sudan and Zaire, 1976: report of a WHO/international study team. *Bull World Health Organ* 56:247–293, 1978.

# ■ Addendum: Large-Scale Biosafety Guidelines

## Large-Scale Biosafety Containment

*Modification by D. Fleming of Appendix K of the NIH Guidelines for research involving recombinant DNA molecules, FR 56:138, pp. 33174–33183, July 18, 1991, to apply to nonrecombinant pathogens.*

For purposes of large-scale research or production, four large-scale physical containment levels, established by the NIH for research with recombinant DNA molecules, set containment conditions at those appropriate for the degree of hazard to health or the environment posed by the organism, and judged by experience with similar organisms and consistent with good large-scale practices. These are known as good large-scale practices (GLSP), Biosafety Level 1 (BSL-1LS), Biosafety Level 2 (BSL-2LS), and Biosafety Level 3 (BSL-3LS). The containment conditions for BSL-4LS are not defined, but are to be determined on a case-by-case basis.

These guidelines, in addition to the CDC/NIH guidelines (Biosafety in Microbiological and Biomedical Laboratories, *BMBL* [USPHS, 1993]) for large-scale work, shall apply to large-scale research or production activities. Only the biological hazard associated with the viable organisms is addressed here. Other hazards accompanying the large-scale cultivation of such organisms (such as toxic properties of products and physical, mechanical, and chemical aspects of downstream processing) are not addressed and shall be considered separately.

All provisions shall apply to large-scale research or production activities with the following modifications:

- These guidelines shall supersede the relevant sections of the CDC/NIH guidelines (*BMBL*) when quantities exceeding 10 liters of culture are involved in research or production. Sections of these guidelines apply to good large-scale practices.

- The institution shall appoint a biological safety officer if it engages in large-scale research or production activities involving viable organisms. The duties of the biological safety officer shall include those specified in these guidelines.

- The institution shall establish and maintain a health surveillance program for personnel engaged in large-scale research or production activities involving viable organisms that require Biosafety Level (BSL) 3 containment at the laboratory scale. The program shall include preassignment and periodic physical and medical examinations; collection, maintenance, and analysis of serum specimens (where appropriate) for monitoring serologic changes that may result from the employee's work experience; and provisions for the investigation of any serious, unusual, or extended illnesses of employees to determine possible occupational origin.

A. Good Large-Scale Practices (GLSP). The GLSP level of physical containment is recommended for large-scale research or production involving viable, nonpathogenic, nontoxigenic strains derived from host organisms that have an extended history of safe large-scale use. Likewise, the GLSP level of physical containment is recommended for organisms such as those in Appendix C of the Recombinant DNA Guidelines (NIH, 1994), which have built-in environmental limitations that permit optimal growth in the large-scale settings but limited survival without adverse consequences in the environment.

1. Institutional codes of practice shall be formulated and implemented to ensure adequate control of health and safety matters.

2. Written instructions and training of personnel shall be provided to ensure that cultures of viable organisms are handled prudently and that the workplace is kept clean and orderly.

3. In the interest of good personal hygiene, facilities (such as hand-washing sink, shower, and changing room) and protective clothing (such as uniforms and laboratory coats) shall be provided that are appropriate for the risk of exposures to viable organisms. In addition, eating, drinking, smoking, applying cosmetics, and mouth pipetting shall be prohibited in the work area.

4. Cultures of viable organisms shall be handled in facilities intended to safeguard health during work with organisms that do not require containment.

5. Discharges containing viable organisms shall be handled in accordance with applicable government environmental regulations.

6. Additions of materials to a system, sample collection, transfer of culture fluids within or between systems, and processing of culture fluids shall be conducted in a manner that maintains employee exposure to viable organisms at a level that does not adversely affect the health and safety of employees.

7. The facility's emergency response plan shall include provisions for handling spills. Spills and accidents that result in overt exposures to organisms are immediately reported to the laboratory

director, who is to notify the biological safety officer, the biosafety committee, and other appropriate authorities (if applicable).

B. Large-Scale Biosafety Level 1 (BSL-1LS). BSL-1LS is recommended for large-scale research or production of viable organisms that require BSL-1 containment at the laboratory scale and that do not qualify for GLSP.

1. Institutional codes of practice shall be formulated and implemented to ensure adequate control of health and safety matters.

2. Written instructions and training of personnel shall be provided to ensure that cultures of viable organisms are handled prudently and that the workplace is kept clean and orderly.

3. In the interest of good personal hygiene, facilities (such as hand-washing sink, shower, and changing room) and protective clothing (such as uniforms and laboratory coats) shall be provided that are appropriate for the risk of exposures to viable organisms. Eating, drinking, smoking, applying cosmetics, and mouth pipetting shall be prohibited in the work area.

4. Spills and accidents that result in overt exposures to organisms are immediately reported to the laboratory director, who is to notify the biological safety officer, the biosafety committee, and other appropriate authorities (if applicable). Medical evaluation, surveillance, and treatment are provided as appropriate, and written records are maintained.

5. Cultures of viable organisms shall be handled in a closed system (such as a closed vessel used for the propagation and growth of cultures) or other primary containment equipment (for example, a biological safety cabinet containing a centrifuge used to process culture fluids) that is designed to *reduce* the potential for escape of viable organisms. Volumes less than 10 liters may be handled outside of a closed system or other primary containment equipment, provided all physical containment requirements specified under BSL-1 laboratory scale are met.

6. Culture fluids (except as allowed in number 7 below) shall not be removed from a closed system or other primary containment equipment unless the viable etiologic agent has been inactivated by a validated inactivation procedure. A validated inactivation procedure is one that has been demonstrated to be effective when working with the organism that will be propagated at large scale.

7. Sample collection from a closed system, the addition of materials to a closed system, and the transfer of culture fluids from one closed system to another shall be done in a manner that *minimizes* the release of aerosols or the contamination of exposed surfaces.

8. Exhaust gases removed from a closed system or other primary containment equipment shall be treated by filters that have efficiencies equivalent to high-efficiency particulate air (HEPA) filters or by other equivalent procedures (such as incineration) to minimize the release of viable organisms to the environment.

9. A closed system or other primary containment equipment that has contained viable organisms shall not be opened for maintenance or other purposes unless it has been sterilized by a validated sterilization procedure. A validated sterilization procedure is one that has been demonstrated to be effective when working with the organism that will be propagated at large scale.

10. Emergency plans for handling accidental spills and personnel contamination shall include methods and procedures for handling large losses of culture on an emergency basis.

C. Large-Scale Biosafety Level 2 (BSL-2LS). BSL-2LS is recommended for large-scale research or production of viable organisms that require BSL-2 containment at the laboratory scale.

1. Institutional codes of practice shall be formulated and implemented to ensure adequate control of health and safety matters.

2. Written instructions and training of personnel shall be provided to ensure that cultures of viable organisms are handled prudently and that the workplace is kept clean and orderly.

3. In the interest of good personal hygiene, facilities (such as hand-washing sink, shower, and changing room) and protective clothing (such as uniforms and laboratory coats) shall be provided that are appropriate for the risk of exposures to viable organisms. Eating, drinking, smoking, applying cosmetics, and mouth pipetting shall be prohibited in the work area.

4. Spills and accidents that result in overt exposures to organisms are immediately reported to the biological safety officer, the biosafety committee, and other appropriate authorities (if applicable). Medical evaluation, surveillance, and treatment are provided as appropriate, and written records are maintained.

5. Cultures of viable organisms shall be handled in a closed system (such as a closed vessel used for the propagation and growth of cultures) or other primary

containment equipment (for example, a Class III biological safety cabinet containing a centrifuge used to process culture fluids) that is designed to *prevent* the escape of viable organisms. Volumes less than 10 liters may be handled outside of a closed system or other primary containment equipment, provided all physical containment requirements specified under BSL-2 are met.

6. Culture fluids (except as allowed in number 7 below) shall not be removed from a closed system or other primary containment equipment unless all viable organisms have been inactivated by a validated inactivation procedure. A validated inactivation procedure is one that has been demonstrated to be effective when working with the organism that will be propagated at large scale.

7. Sample collection from a closed system, the addition of materials to a closed system, and the transfer of culture fluids from one closed system to another shall be done in a manner that *prevents* the release of aerosols or contamination of exposed surfaces.

8. Exhaust gases removed from a closed system or other primary containment equipment shall be treated by filters that have efficiencies equivalent to high-efficiency particulate air (HEPA) filters or by other equivalent procedures (such as incineration) to *prevent* the release of viable organisms to the environment.

9. A closed system or other primary containment equipment that has contained viable organisms shall not be opened for maintenance or other purposes unless it has been sterilized by a validated sterilization procedure. A validated sterilization procedure is one that has been demonstrated to be effective when working with the organism that will be propagated at large scale.

10. Rotating seals and other mechanical devices directly associated with a closed system used for the propagation and growth of viable organisms shall be designed to *prevent* leakage or shall be fully enclosed in ventilated housings that are exhausted through filters that have efficiencies equivalent to high-efficiency particulate air (HEPA) filters or through other equivalent treatment devices.

11. A closed system used for the propagation and growth of viable organisms, and other primary containment equipment used to contain operations involving viable human etiologic agents, shall contain sensing devices that monitor the integrity of containment during operations.

12. A closed system used for the propagation and growth of viable organisms shall be tested for integrity of the containment features using the organism that will be propagated at large scale. Testing shall be accomplished prior to the introduction of viable organisms and following modification or replacement of essential containment features. Procedures and methods used in the testing shall be appropriate for the equipment design and for recovery and demonstration of the test organism. Records of tests and results shall be maintained on file.

13. A closed system used for the propagation and growth of viable organisms shall be permanently identified. This identification shall be used in all records reflecting testing, operation, and maintenance and in all documentation relating to use of this equipment for research or production activities involving viable organisms.

14. The universal biohazard sign shall be posted on each closed system and all primary containment equipment when used to contain viable organisms.

15. Emergency plans required for handling accidental spills and personnel contamination shall include methods and procedures for handling large losses of culture on an emergency basis.

D. Large-Scale Biosafety Level 3 (BSL-3LS). BSL-3LS is recommended for large-scale research or production of viable organisms that require BSL-3 containment at the laboratory scale.

1. Institutional codes of practice shall be formulated and implemented to ensure adequate control of health and safety matters.

2. Written instructions and training of personnel shall be provided to ensure that cultures of viable organisms are handled prudently and that the workplace is kept clean and orderly.

3. In the interest of good personal hygiene, facilities (such as hand-washing sink, shower, and changing room) and protective clothing (such as uniforms and laboratory coats) shall be provided that are appropriate for the risk of exposures to viable organisms. Eating, drinking, smoking, applying cosmetics, and mouth pipetting shall be prohibited in the work area.

4. Spills and accidents that result in overt exposures to organisms are immediately reported to the laboratory director, who is to inform the biological safety officer, the biosafety committee, and other appropriate authorities (if applicable). Medical evaluation, surveillance, and treatment are provided as appropriate, and written records are maintained.

5. Cultures of viable organisms shall be handled in a closed system (such as a closed vessel used for the propagation and growth of cultures) or other primary

containment equipment (for example, a Class III biological safety cabinet containing a centrifuge used to process culture fluids) that is designed to *prevent* the escape of viable organisms. Volumes less than 10 liters may be handled outside of a closed system or other primary containment equipment, provided all physical containment requirements specified under BSL-3 laboratory scale are met.

6. Culture fluids (except as allowed in number 7 below) shall not be removed from a closed system or other primary containment equipment unless the viable organisms have been inactivated by a validated inactivation procedure. A validated inactivation procedure is one which has been demonstrated to be effective when working with the organism that will be propagated at large scale.

7. Sample collection from a closed system, the addition of materials to a closed system, and the transfer of culture fluids from one closed system to another shall be done in a manner that *prevents* the release of aerosols or contamination of exposed surfaces.

8. Exhaust gases removed from a closed system or other primary containment equipment shall be treated by filters that have efficiencies equivalent to high-efficiency particulate air (HEPA) filters or by other equivalent procedures (such as incineration) to *prevent* the release of viable organisms to the environment.

9. A closed system or other primary containment equipment that has contained viable organisms shall not be opened for maintenance or other purposes unless it has been sterilized by a validated sterilization procedure. A validated sterilization procedure is one that has been demonstrated to be effective when working with the organism that will be propagated at large scale.

10. A closed system or other primary containment equipment containing viable organisms shall be operated so that the space above the culture level will be maintained at a pressure as low as possible, consistent with equipment design, in order to maintain the integrity of containment features.

11. Rotating seals and other mechanical devices directly associated with a closed system used to contain viable etiologic organisms shall be designed to *prevent* leakage or shall be fully enclosed in ventilated housings that are exhausted through filters that have efficiencies equivalent to high-efficiency particulate air (HEPA) filters or through other equivalent treatment devices.

12. A closed system used for the propagation and growth of viable organisms, and other primary containment equipment used to contain operations involving viable organisms, shall include monitoring and sensing devices that monitor the integrity of containment during operations.

13. A closed system used for the propagation and growth of viable organisms shall be tested for integrity of the containment features using the organism that will be propagated at large scale. Testing shall be accomplished prior to the introduction of viable organisms and following modification or replacement of essential containment features. Procedures and methods used in the testing shall be appropriate for the equipment design and for recovery and demonstration of the test organism. Records of tests and results shall be maintained on file.

14. A closed system used for the propagation and growth of viable organisms shall be permanently identified. This identification shall be used in all records reflecting testing, operation, and maintenance and in all documentation relating to use of this equipment for research or production activities involving viable organisms.

15. The universal biohazard sign shall be posted on each closed system and primary containment equipment when used to contain viable organisms.

16. Emergency plans required for BSL-3LS shall include those for BSL-3 laboratory scale and methods and procedures for handling large losses of culture on an emergency basis.

17. Closed systems and other primary containment equipment used in handling cultures of viable organisms shall be located within a controlled area that meets the following requirements:
    a. The controlled area shall have a separate entry area. The entry area shall be a double-doored space such as an air lock, anteroom, or change room that separates the controlled area from the rest of the facility.
    b. The surfaces of walls, ceilings, and floors in the controlled area shall be such as to permit ready cleaning and decontamination.
    c. Penetrations into the controlled area shall be sealed to permit liquid or vapor decontamination.
    d. All utilities and service or process piping and wiring entering the controlled area shall be protected against contamination.
    e. Hand-washing facilities equipped with foot, elbow, or automatically operated valves shall be located at each major work area and near each primary exit.
    f. A shower facility shall be provided. This facility shall be located in close proximity to the controlled area.

g. The controlled area shall be designed to preclude release of culture fluids outside the controlled areas in the event of an accidental spill or release from the closed systems or other primary containment equipment.

h. The controlled area shall have a ventilation system that is capable of controlling air movement. The movement of air shall be from areas of lower contamination potential to areas of higher contamination potential. If the ventilation system provides positive-pressure supply air, the system shall operate in a manner that prevents the reversal of the direction of air movement or shall be equipped with an alarm that would be actuated in the event that a reversal in the direction of air movement were to occur. The exhaust air from the controlled area should not be discharged to the outdoors without being HEPA filtered, subjected to thermal oxidation, or otherwise treated to prevent the release of viable organisms.

18. The following personnel and operational practices shall be required:

a. Personnel entry into the controlled areas shall be through an entry area as specified above (17.a).

b. Persons entering the controlled area shall exchange or cover their personal clothing with work garments such as jumpsuits, laboratory coats, pants and shirts, head cover, and shoes or shoe covers. On exit from the controlled area, the work clothing may be stored in a locker separate from that used for personal clothing, discarded into biohazard waste, or decontaminated before laundering.

c. Entry into the controlled area during periods when work is in progress shall be restricted to those persons required to meet program or support needs. Prior to entry, all personnel shall be informed of the operating practices and emergency procedures, and the nature of the work being conducted.

d. Persons under 18 years of age shall not be permitted to enter the controlled areas.

e. The universal biohazard sign shall be posted on entry doors to the controlled area and all internal doors when any work involving the organism is in progress. This includes periods when decontamination procedures are in progress. The sign posted on the entry doors to the controlled areas shall include a statement of agents in use and personnel authorized to enter the controlled area.

f. The controlled area shall be kept neat and clean.

g. Eating, drinking, smoking, and storage of food are prohibited in the controlled area.

h. Animals and plants shall be excluded from the controlled area.

i. An effective insect and rodent control program shall be maintained.

j. Access doors to the controlled area shall be kept closed, except as necessary for access, while work is in progress. Service doors leading directly outdoors shall be sealed and locked while work is in progress.

k. Persons shall wash their hands when leaving the controlled area.

l. Persons working in the controlled areas shall be trained in emergency procedures.

m. Equipment and materials required for the management of accidents involving the viable organisms in use shall be available in the controlled area.

n. The controlled area shall be decontaminated in accordance with established procedures following spills or other accidental release of viable organisms.

E. Large-Scale Biosafety Level 4 (BSL-4LS). No provisions are made for large-scale research or production of viable organisms that require BSL-4 containment at the laboratory scale. If necessary, these requirements will be established by appropriate consultation with experts at the CDC on an individual, case-by-case basis.

F. Biological Safety Officer (BSO)

1. Institutions that engage in large-scale research or production activities involving viable organisms shall appoint a biological safety officer.

2. Institutions that engage in research at BSL-3 or BSL-4 shall appoint a biological safety officer. The BSO shall be a member of the biosafety committee.

3. The biological safety officer's duties include, but are not limited to, the following:

a. Periodic inspections to ensure that standards are rigorously followed

b. Reporting to the biosafety committee and the institution any significant problems and any significant research-related accidents or illnesses of which the biological safety officer becomes aware, unless the BSO determines that a report has already been filed by the principal investigator or responsible scientist

c. Developing emergency plans for handling accidental spills and personnel contamination, and investigating laboratory accidents involving research with organisms

d. Providing advice on laboratory security

e. Providing technical advice to principal investigators or scientists and the institutional biosafety committee on research safety procedures.

PART 4

# Evaluation of Hazards

CHAPTER

# 15

# Evaluation

Revised by Elizabeth R. Gross, CIH
Elise Pechter Morse, CIH

IN INDUSTRIAL HYGIENE, evaluation is the decision-making
process that assesses the level of risk to workers from exposure to chemical, physical, and biological agents. The final
judgment is based on a combination of observation, interviews, and measurement of both the levels of energy or air
contaminants arising from a process or work operation and
the effectiveness of control measures used.

## ◼ General Principles

The need to evaluate hazards is driven by the acknowl-
edgement that chemical, biological, and physical agents
cause injury, disease, and premature death among exposed
workers. The U.S. Department of Labor's Bureau of Statistics reported 6.3 million occupational work-related injuries
and illnesses for 1991. The actual number is much larger,
because many occupational illnesses go unrecognized. The
task of evaluating the nature and severity of hazards relies
on judgment based on many factors:

- *Toxicity:* the inherent capacity of an agent to cause
  harm, the nature of that harm, and target organs
  affected.
- *Exposure levels, or dose:* the amount that workers
  absorb through all routes of entry during work.
- *Process or operation analysis:* the awareness of operations, from raw materials through their transformation
  to products and by-products, that may result in the
  release of chemicals or energy that could cause harm.

**General Principles**
*Basic Approach to Hazard Recognition* ▪ *Review of Literature* ▪
*Inventory*

**Description of Process or Operation**
*Process Flow Sheet* ▪ *Checklists* ▪ *Cleaning Methods* ▪ *Process
Safety Management*

**Field Survey**
*Sensory Perception* ▪ *Control Measures in Use* ▪ *Observation and
Interview*

**Monitoring and Sampling**
*Rationale* ▪ *Monitoring* ▪ *Sampling*

**Industrial Hygiene Calculations**
*Gases and Vapors* ▪ *Vapor Equivalents* ▪ *Weight-per-Unit
Volume* ▪ *Time-Weighted Average (TWA) Exposure* ▪ *Excursions*

**Interpretation of Results**
*Comparison with Standards and Guidelines* ▪ *Limitations of
Standards* ▪ *Comparison of Results with Other Data*

**Summary**

**Bibliography**

- *Maintenance activities, spills, and accidents:* the knowledge of acute incidents, infrequent events, leaks, and releases that are missed in routine evaluations.

- *Epidemiology and risk assessment:* a literature review of population-based research that may provide information about adverse health effects not yet noticed in a small work force.

- *Interview:* the information provided by workers, regarding health symptoms, tasks, and changes in conditions, that can provide essential details regarding process analysis, health impact, and other stressors on the job that may be chemical, physical, ergonomic, or biological.

- *Variability of response:* the way individuals vary in their susceptibility because of factors such as age, size, respiratory rate, and general health status. Recognition of the nonuniversality of response helps maintain awareness of possible unrecognized hazards.

The purpose of evaluation is the prevention of hazardous exposures and resulting adverse health effects. Unlike health care providers, whose job is to treat existing conditions, industrial hygienists can prevent illness through recognition and correction of hazards before they cause harm. The industrial hygienist uses many sources of information and methods in evaluating the workplace. The next section describes some of them.

## Basic Approach to Hazard Recognition

Almost any work environment has either potential or actual environmental hazards that the health and safety professional must recognize, measure, and monitor. The first step toward recognition of these potential hazards is the consideration of the raw materials being used, including any known impurities, and the potential of those materials to do harm. The next consideration is how these raw materials are modified through intermediate steps. Finally, an evaluation of the finished product or by-products must also be done, both under normal conditions and under anticipated emergency conditions, to determine whether any hazards might exist at this point.

A basic, systematic procedure can be followed in the recognition of occupational health hazards. Hazard recognition methods are similar whether a chemical, physical, or biological agent is involved. Questions should be formulated to organize information:

- What are the raw materials?
- What is produced?
- What intermediate products are formed in the process?
- What by-products may be released?
- What are the usual cleaning or maintenance procedures at the end of the day, end of a run, or changeover to another product?

There is a wealth of health and safety information that should be researched to help anticipate potential hazards in any work setting. Included in this search should be a review of the known hazards associated with industrial processes or job classifications. An inventory can then be made of previously identified hazards by category, and any relevant standards or guidelines can be referenced. Armed with this information, the next step is to study the specific operation or process and consider where air contaminants are released, as well as where and when employees are exposed.

Any job can include physical hazards as well as chemical hazards. Energy uses, electromagnetic fields, noise sources, fire hazards, physically demanding tasks, and material-handling jobs must all be noted. Hazards that could result in acute traumatic injuries also include vehicles and sources of energy. Vehicles include automobiles, forklift trucks, and overhead cranes. Energy sources could be mechanical (pneumatic or hydraulic), electrical, thermal, or chemical. The pattern of work may include sitting, standing, or lifting. Physical hazards could also include vibration, radiation, barometric pressure alterations, and hazardous motions or postures that could cause cumulative trauma disorders. Temperature extremes, lighting levels, and machine pacing are additional factors to consider in the initial hazard surveillance approach.

Biological hazards include infectious agents (bacteria, viruses, parasites, and fungi), toxins associated with plants or animals, and pharmacoactive substances such as enzymes, hormones, or other biological materials. Infectious agents include tuberculosis in shelters, clinics, hospitals, or offices; bloodborne pathogens for first aid providers; and mold or mildew in a basement office after a flood. In evaluations of agricultural work areas, one may need to consider other agents. In biotechnology or pharmaceutical companies, exotic endotoxins or biological materials may be employed.

Psychological hazards such as high job demands and low control can cause stress and should also be considered. Some factors that can result in emotional strain include machine pacing; boring, repetitive tasks; complex, highly demanding requirements; shift work; fear of layoffs or physical violence; computer monitoring of performance; and the absence of social or co-worker support.

## Review of Literature

Prior to evaluating any workplace, it is useful to know what hazards are anticipated. A literature review of the industry in question or the type of operation being done can facilitate this analysis. Recommended general resources include Burgess (1995); Cralley & Cralley (1986, 1986a, & 1986b); the International Labour Organisation (ILO) (1983); Weeks et al. (1991); and U.S. Dept. of Health, Education and Welfare (USDHEW), NIOSH (1977).

Other resources include current publications; review articles in journals; and computer sources such as compact disc, or CD-ROM, services. Sources to be consulted include technical journals in the fields of industrial hygiene, occupational medicine, environmental analysis, and epidemiology. Some trade associations incorporate health and safety articles into their regular communications with members. The National Institute for Occupational Safety and Health (NIOSH) and the Occupational Safety and Health Administration (OSHA) have published very useful pamphlets, criteria documents, reports, and technical bulletins on many subjects, which can be consulted prior to a workplace evaluation. The Office of Technology Assessment (OTA) also has several publications. NIOSH maintains a mailing list to inform interested persons about new publications and courses offered for further training.

## Inventory

A list should be prepared of all chemicals present in the facility. The list should include all raw materials and final products. This chemical inventory is required by OSHA's Hazard Communication Standard, 29 *CFR* 1910.1200, for the purpose of anticipating possible hazards and ensuring that these risks are communicated to employers and employees before they are encountered in the workplace. The manufacturer or supplier of each chemical must send a material safety data sheet (MSDS) for every product.

For every chemical, the relevant standards should be looked up; they cover many of the most hazardous materials currently in use and often highlight possible chemical exposures that must be controlled. The legal standards are OSHA's permissible exposure limits (PELs), which set the maximum boundaries for allowable worker exposures. There are different types of limits. Time-weighted average limits (TWAs) are used to evaluate average sampling results covering a whole shift. Short-term exposure limits (STELs) or ceiling levels are used to evaluate brief exposure times or peak releases.

Although not enforceable by law, other guidelines are often more current and therefore more protective of workers. These include recommended exposure limits (RELs), developed by NIOSH to guide OSHA in promulgating its legal standards, and Threshold Limit Values (TLVs), offered by the American Conference of Governmental Industrial Hygienists (ACGIH) annually in a pocket-sized booklet as a guideline for good workplace control. A convenient reference for OSHA and NIOSH limits is the NIOSH *Pocket Guide to Chemical Hazards* (1994).

Unfortunately, there are no standards or guidelines for most chemicals that reflect experience or research with them. It has been estimated that new chemical products are introduced into the workplace at a rate of 1,000 to 3,000 every year, yielding a total of 60,000 chemicals in widespread commercial use in the Western nations. Of

**Table 15–A.** Categories of Potential Hazards Found in Hospitals

| Hazard Categories | Definition | Examples Found in the Hospital Setting |
|---|---|---|
| Biological | Infectious/biological agents, such as bacteria, viruses, fungi, or parasites, that can be transmitted by contact with infected patients or with contaminated body secretions/fluids | Human Immuno-deficiency Virus (HIV) Hepatitis B virus Tuberculosis |
| Ergonomic | Ergonomics attempts to fit the job to the worker instead of the traditional method of fitting the worker to the job. It is the study of human characteristics, both behavioral and biological, for the appropriate design of the living and working environment. | Lifting Standing for long periods of time Poor lighting |
| Chemical | Chemicals that are potentially toxic or irritating to the body, including medications, gases, laboratory reagents. | Ethylene oxide Formaldehyde Glutaraldehyde Waste anesthetic gases Cytotoxic agents Pentamidine Ribavirin |
| Psychological | Factors/situations encountered in the workplace that create or potentiate stress, emotional strain, or interpersonal problems. | Stress Shiftwork |
| Physical | Physical agents that can cause tissue trauma | Radiation Lasers Noise Electricity Extreme temperatures |

(Adapted from OSHA, 1993.)

these, OTA estimates that only 5,000 chemicals have ever been tested for toxicity. Only 454 chemicals have specific limits promulgated by OSHA. Maintenance of a complete inventory helps to provide oversight and to keep track of any previously unrecognized problems and health effects.

The inventory can be extended to include physical, ergonomic, biological, and psychological hazards as well as chemical ones. For example, see Table 15–A for an inventory of potential hazards in a hospital setting. Any inventory should be maintained, updated, and used to develop, manage, and evaluate the appropriate programs and to ensure awareness of the broad range of hazards that may be present in the workplace.

## ■ Description of Process or Operation

The inventory provides information about the identity of the hazards present, but it cannot indicate the degree of risk from exposure to those materials. It does not quantify the amounts employed in the process; indicate how or where they are used or produced; or detail at what point, via what route, or for how long employees are potentially exposed. The severity of the hazards present depends on the potential for worker contact as well as the duration and concentration of exposure to the hazardous materials; therefore, information about the industrial processes and operations is needed to link the hazardous materials to their use in production and to personnel contact. Facility engineering and manufacturing personnel should be consulted regarding usual operations, abnormal operating conditions, and other factors that can affect exposures.

There are numerous industrial operations that should immediately alert the health and safety professional to a potential health hazard. Lists of industrial operations such as the one shown in Table 15–B are helpful in reviewing processes that might create special risks such as the aerosolization of a hazardous material. After a list of process operations that possibly produce harmful air contaminants has been prepared, certain operations should be selected for closer scrutiny.

## Process Flow Sheet

A simple process flow sheet should be drawn that shows in a stepwise fashion how and where each material is introduced and at what point products and by-products are made (Figure 15–1). Process flow sheets and the standard operating procedures (SOPs) that describe the particular operations involved should be obtained and studied. They not only provide a good description of the general operations involved, but also serve as an excellent source for the terminology used in that particular industry. In many industrial operations, many different hazards exist simultaneously. Therefore, it is necessary to carefully examine the overall process so that potentially hazardous conditions are not overlooked.

It is important to identify the air contaminants produced and to pinpoint the location and tasks of personnel that might be exposed to them. Repetitive operations, wherein a worker remains in one location and repeats the same task, can be relatively straightforward to analyze. In operations during which several contaminants are generated, the evaluation process involves identifying the points at which each material is released and the duration of each release, and factoring in the maximum number of times per workshift these exposures occur. This allows prediction of the amount of contaminant potentially released into the environment, and can help target areas for personal or area air sampling. In workplaces where tasks vary from day to day, depending on the products being made, work assignment, and other factors, a process flow sheet may be less useful, and individual assessments are necessary.

Chemical process companies involved with the manufacture of large volumes of chemicals use closed systems.

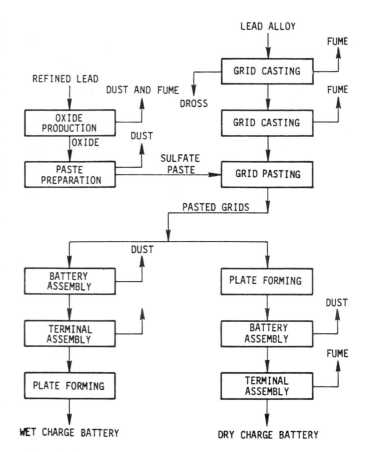

**Figure 15–1.** Flow diagram of a lead-acid battery plant. (Reprinted from Environmental Protection Agency. *Control Techniques for Lead Air Emissions,* Vol. 2, Chapter 4, Appendix B. Research Triangle Park, NC: U.S. EPA, EPA-450/ 2-77-012, 1977.)

Although chemicals are not routinely released to the atmosphere, exposure to air contaminants in work areas arises from the following:

- Leaks from joints, fittings, closures, and other components that allow release from the otherwise closed system
- The process of charging the system or preparing and loading the raw materials
- Intentional releases of contaminants from vents, process sampling points, or quality control checkpoints
- Stack gases from combustion processes
- Accidental or unintentional releases resulting from equipment malfunction or failure
- Maintenance or repair activities or infrequently performed functions without standard operating procedures

Many valves leak even when they are supposed to be shut, and such leaks can release significant concentrations of chemical air contaminants. Purges, minor overpressures, and system breathing into the atmosphere should be contained

## Table 15–B. Typical Industrial Operations and Their Associated Health Hazards

**Abrasive blasting.** Abrasive blasting equipment may be automatic, or it may be manually operated. Either type may use sand, steel, shot, or artificial abrasives. The dust levels of workroom air should be examined to make sure that the operators are not overexposed.

**Abrasive machining.** An abrasive machining operation is characterized by the removal of material from a workpiece by the cutting action of abrasive particles contained in or on a machine tool. The workpiece material is removed in the form of small particles and, whenever the operation is performed dry, these particles are projected into the air in the vicinity of the operation.

**Assembly operations.** Improper positioning of equipment and handling of work parts may present ergonomic hazards due to repeated awkward motion and resulting in excessive stresses.

**Bagging and handling of dry materials.** The bagging of powdered materials (such as plastic resins, paint pigments, pesticides, cement, and the like) is generally accompanied by the generation of airborne dusts. This occurs as a result of the displacement of air from the bag, spillage, and motions of the bagging machine and the worker. The conveying, sifting, sieving, screening, packaging, or bagging of any dry material may present a dust hazard. The transfer of dry, finely divided powder may result in the formation of considerable quantities of airborne dust. Inhalation and skin contact hazards may be present.

**Ceramic coating.** Ceramic coating may present the hazard of airborne dispersion of toxic pigments plus hazards of heat stress from the furnaces and hot ware.

**Coating operations.** Whenever a substance containing volatile constituents is applied to a surface in an industrial environment, there is obviously potential for any vapors evolved to enter the breathing zones of workers. If the volatiles evaporate at a sufficient rate and/or the particular operation is such that workers must remain in the immediate vicinity of the wet coating, these vapors may result in excessive exposures.

**Crushing and grinding.** Size reduction is the mechanical reduction in size of solid particulate material. Two of the principal methods of achieving size reduction are crushing and grinding, but the terms are not synonymous. Crushing generally refers to a relatively slow compressive action on individual pieces of coarse material ranging in size from several feet to under one inch. Grinding is performed on finer pieces and involves an attrition or rubbing action as well as interaction between individual pieces of material. Pulverizing and disintegrating are terms related to grinding. The former applies to an operation that produces a fine powder; the latter indicates the breakdown of relatively weak interparticulate bonds such as those present in caked powders. Dry grinding operations should be examined for airborne dust, noise, and ergonomic hazards.

**Dry mixing.** Mixing of dry material may present a dust hazard and should take place in completely enclosed mixers whenever air sampling indicates excessive amounts of airborne dust are present.

**Drying ovens.** Much of the equipment used for drying purposes is also used for curing, i.e., the application of heat to bring about a physical or chemical change in a substance. The first major category includes direct dryers in which hot gases are in direct contact with the material and carry away any vaporized substances to be exhausted. Limited-use dryers include radiant-heat and dielectric-heat dryers. The operation of the former is based on the generation, transmission, and absorption of infrared rays. The latter rely on heat generation within the solid when it is placed in a high-frequency electric field. Oven vapors (sometimes including carbon monoxide) are often released into the workroom, as are a variety of solvents and other substances found in the drying or curing products.

**Electron-beam welding.** Any process involving an electric discharge in a vacuum may be a source of ionizing radiation. Such processes involve the use of electron-beam equipment and similar devices.

**Fabric and paper coating.** The coating and impregnating of fabric and paper with plastic or rubber solutions may involve evaporation into the workroom air of large quantities of solvents.

**Forming and forging.** Hot bending, forming, or cutting of metals or nonmetals may involve hazards of lubricant mist, decomposition products of the lubricant, skin contact with the lubricant, heat stress (including radiant heat), noise, and dust.

**Gas furnace or oven heating operations (annealing, baking, drying, and so on).** Any gas- or oil-fired combustion process should be examined to determine the level of by-products of combustion that are released into the workroom atmosphere. Noise measurements should also be made to determine the level of burner noise.

**Grinding operations.** Grinding, crushing, or comminuting of any material may contaminate workroom air as a result of the dust produced from the material being processed or from the grinding wheel.

**High temperatures from hot castings, unlagged steam pipes, process equipment, and so on.** Any process or operation involving high ambient temperatures (dry-bulb temperature), radiant heat load (globe temperature), or excessive humidity (wet-bulb temperature) should be examined to determine the magnitude of the physical stresses that are present.

**Materials handling, warehousing.** Work areas should be checked for levels of carbon monoxide and oxides of nitrogen arising from internal combustion engine fork-lift operations. Operations should also be evaluated for ergonomic hazards.

**Metalizing.** Uncontrolled coating of parts with molten metals presents hazards of dust and fumes of metals and fluxes in addition to heat and nonionizing radiation.

**Microwave and radio-frequency heating operations.** Any process or operation involving microwaves or induction heating should be examined to determine the magnitude of heating effects and, in some cases, noise exposure to the employees.

**Molten metals.** Any process involving the melting and pouring of molten metals should be examined to determine the level of air contaminants of any toxic gas, metal fume, or dust produced in the operation.

**Open-surface tanks.** Open-surface tanks are used by industry for numerous purposes. Among their applications are included degreasing, electroplating, metal stripping, fur and leather finishing, dyeing, and pickling. An open-surface tank operation is defined as "any operation involving the immersion of materials in liquids that are contained in pots, tanks, vats, or similar containers." Excluded from consideration in this definition, however, are certain similar operations, such as surface-coating operations and operations involving molten metals, for which different engineering control requirements exist.

**Paint spraying.** Spray-painting operations should be examined for the possibility of hazards from inhalation and skin contact with toxic and irritating solvents and inhalation of toxic pigments. The solvent vapor evaporating from the sprayed surface may also be a source of hazard because ventilation may be provided only for the paint spray booth.

**Plating.** Electroplating processes involve risk of skin contact with strong chemicals, and in addition may present a respiratory hazard if mist or gases from the plating solutions are dispersed into the workroom air.

**Pouring stations for liquids.** Wherever volatile substances are poured from a spout into a container, some release of contaminants can be expected. In the paint and other coatings industries, solvents may be released when the contents of mills or mixers are poured into portable change cans. Although ladles and change cans can be covered while being moved through a workplace, they must of necessity be open at the points they are filled or discharged.

**Punch press, press brake, drawing operations, and so on.** Cold bending, forming, or cutting of metals or nonmetals should be examined for hazards of contact with lubricant, inhalation of lubricant mist, and excessive noise.

**Vapor degreasing.** The removal of oil and grease from metal products may present hazards. This operation should be examined to determine that excessive amounts of vapor are not being released into the workroom atmosphere.

**Welding—gas or electric arc.** Welding operations generally involve melting of a metal in the presence of a flux or a shielding gas by means of a flame or an electric arc. The operation may produce gases or fumes from the metal, the flux, metal surface coatings, or surface contaminants. Certain toxic gases such as ozone or nitrogen dioxide may also be formed by the flame or arc. If there is an arc or spark discharge, the effects of nonionizing radiation and the products of destruction of the electrodes should be investigated. These operations also commonly involve hazards of high-potential electric circuits of low internal resistance.

**Wet grinding.** Wet grinding of any material may produce possible hazards of mist, dust, and noise.

**Wet mixing.** Mixing of wet materials may present possible hazards of solvent vapors, mists, and possibly dust. The noise levels produced by the associated equipment should be checked.

(Adapted with permission from the OSHA *Compliance Operations Manual*, 1972.)

by collection, scrubbing, reaction, incineration, or other measures that safely dispose of the products or eliminate their release altogether. Environmental Protection Agency (EPA) emission requirements may require further control measures. Efforts by the environmental movement have influenced state and local legislation to discourage the use of hazardous materials completely. These toxic use reduction (TUR) efforts have emphasized review of chemical operations to eliminate or reduce the use of hazardous chemicals and substitute safer materials wherever possible.

## Checklists

A checklist for evaluating environmental hazards that can arise from industrial operations is presented here. It should be modified to fit each particular situation.

**Overall process or operation.** List all hazardous chemical or physical agents used or formed in the process. Carry out the following tasks, answering all of the appropriate questions:

- List the conditions necessary for the agent to be released into the workroom atmosphere. Does it usually occur in the process as a dust, mist, gas, fume, vapor, a low-volatile liquid, or a solid (Table 15–C)? What process conditions could cause material to be sprayed or discharged into the air as a liquid aerosol or dust cloud? Have the consequences of the exposure of raw materials or intermediates on people or operations been considered? Are incompatible materials, such as acids and cyanides in plating operations, kept separate from one another?

- Review storage of raw materials and finished product. Are unstable materials such as methyl ethyl ketone peroxide properly stored? Have chemical incompatibilities been considered? Are containers appropriate? Has flammability been considered?

- Consider transport and disposal. Have provisions been made for the safe disposal of toxic materials in compliance with all relevant regulations? Can reactants be removed and disposed of promptly in an emergency? Can spills be quickly and effectively contained?

- List the background airborne concentration levels in the workroom that would usually be present as a function of time. List the peak airborne concentrations as a function of task duration. List the appropriate PELs, RELs, TLVs, and STELs.

- Review fire safety. Are fire extinguishers the correct type and size for the materials present? Are they inspected and recharged on a scheduled basis? Are the extinguishing agents compatible with process materials? Is an evacuation plan prepared and disseminated? Are alarms visible, audible, and understandable? Are employees trained in fire safety? Do they know where emergency equipment is located?

**Table 15–C.** Potentially Hazardous Operations and Air Contaminants

| Process Types | Contaminant Type | Contaminant Examples |
|---|---|---|
| **Hot operations** | | |
| Welding | Gases (g) | Chromates (p) |
| Chemical reactions | Particulates (p) | Zinc and compounds (p) |
| Soldering | (Dusts, fumes, mists) | Manganese and compounds (p) |
| Melting | | Metal oxides (p) |
| Molding | | Carbon monoxide (g) |
| Burning | | Ozone (g) |
| | | Cadmium oxide (p) |
| | | Fluorides (p) |
| | | Lead (p) |
| | | Vinyl chloride (g) |
| **Liquid operations** | | |
| Painting | Vapors (v) | Benzene (v) |
| Degreasing | Gases (g) | Trichloroethylene (v) |
| Dipping | Mists (m) | Methylene chloride (v) |
| Spraying | | 1,1,1-trichloroethylene (v) |
| Brushing | | Hydrochloric acid (m) |
| Coating | | Sulfuric acid (m) |
| Etching | | Hydrogen chloride (g) |
| Cleaning | | Cyanide salts (m) |
| Dry cleaning | | Chromic acid (m) |
| Pickling | | Hydrogen cyanide (g) |
| Plating | | TDI, MDI (v) |
| Mixing | | Hydrogen sulfide (g) |
| Galvanizing | | Sulfur dioxide (g) |
| Chemical reactions | | Carbon tetrachloride (v) |
| **Solid operations** | | |
| Pouring | Dusts (d) | Cement |
| Mixing | | Quartz (free silica) |
| Separations | | Fibrous glass |
| Extraction | | |
| Crushing | | |
| Conveying | | |
| Loading | | |
| Bagging | | |
| **Pressurized spraying** | | |
| Cleaning parts | Vapors (v) | Organic solvents (v) |
| Applying pesticides | Dusts (d) | Chlordane (m) |
| Degreasing | Mists (m) | Parathion (m) |
| Sand blasting | | Trichloroethylene (v) |
| Painting | | 1,1,1-trichloroethane (v) |
| | | Methylene chloride (v) |
| | | Quartz (free silica, d) |
| **Shaping operations** | | |
| Cutting | Dusts (d) | Asbestos |
| Grinding | | Beryllium |
| Filing | | Uranium |
| Milling | | Zinc |
| Molding | | Lead |
| Sawing | | |
| Drilling | | |

Note: d = dusts, g = gases, m = mists, p = particulates, v = vapors.
(Reprinted from *Occupational Exposure Sampling Strategy Manual,* NIOSH Pub. No. 77-173.)

- List the levels of those physical agents that are normally present (such as ionizing and nonionizing radiation, temperature extremes, vibration, and noise). List any relevant standards or guidelines.

**Equipment.** Conduct all of the following procedures, listing those pieces of equipment that contain sufficient hazardous material or energy that a hazard would be produced if their contents were suddenly released to the environment:

- List the equipment that could release hazardous levels of physical agents during normal operations or abnormal situations such as power outages.

- List the equipment that can produce hazardous concentrations of airborne contaminants. For each item, indicate the control measures installed to minimize the hazard. Is the health and safety control measure adequate, fail-safe, and reliable? Is it checked on a routine basis?

- List process equipment with components that are likely to fail due to corrosion or to leak hazardous materials, such as valves, pump packing, and tank vents. What safeguards have been taken to prevent expected leakage? Is each safeguard adequate, fail-safe, and reliable? Is there a preventive maintenance program in place to ensure routine examination and replacement of these components?

- Label all chemical containers, transport vessels, and piping systems in accordance with the OSHA hazard communication standard. Are labels appropriate for the literacy level and language of the work force?

- Ensure that all equipment can be correctly locked out and tagged out during necessary procedures. Are emergency disconnect switches properly marked?

## Cleaning Methods

Cleaning operations should be noted to identify hazardous materials and processes. The primary cleaning methods used in industry include the following:

- Manual wiping of parts or equipment with a solvent-soaked rag
- Chemical stripping, degreasing, or removal by dissolving
- Use of hand-held or mechanical brushes
- Scraping or sanding
- Dry sweeping or wet mopping
- Wet sponging
- Abrasive blasting
- Steam cleaning
- Using compressed air to blow off dust
- Using vacuum-cleaning devices

The common feature of all these operations is that by some physical and/or chemical action, a contaminant is dislodged from the surface to which it was adhering and could be released into the work environment. In addition, the cleaning agent used to remove a hazardous material might introduce another, equally hazardous, chemical into the workplace.

Rag cleaning and stripping with organic solvents often expose the worker performing the task to vapors. They can also result in skin contact with the solvent, which might present a problem. A brush used to sweep up dusty substances or a scraper used to dislodge built-up cakes of dry substances can disperse dust into the air. The use of compressed air to blow dust from surfaces reentrains dust that had settled out and will probably produce the greatest concentration of air contaminants. OSHA regulations prohibit the use of compressed air to clean except when the pressure is reduced to 30 psi, and then only if an effective chip guard and personal protective equipment are used. The use of compressed air with asbestos dust is forbidden.

High-pressure water blasting and steam cleaning are essentially wet methods that might initially appear to be designed to suppress the generation of air contaminants. However, hydroblasting equipment to remove a solid can produce substantial concentrations of air contaminants, as can steam cleaning, because of the temperatures and forces involved.

The use of vacuum cleaners, which collect and contain material for removal, appears to be the most satisfactory method of cleaning dry, dusty materials without producing excessive amounts of contaminants. However, a special vacuum with a high-efficiency particulate filter (HEPA) is needed when dealing with certain highly toxic dusts such as asbestos, lead, arsenic, or cadmium. HEPA vacuuming is often accompanied by low pressure, wet methods such as misting or airless spraying, in order to keep levels of airborne toxic dusts to a minimum.

A preferable step is to reduce the dispersal of airborne dusts in the first place. Some power sanders and soldering guns can be specially fitted for dust or fume removal at the point of generation. The need for cleaning can sometimes be further reduced by analysis of the source of the material that must be removed. For example, degreasing is not necessary if oil or coolant is not applied to a part in the first place. A toxics use-reduction approach would seek the source of the soil and use the safest method to remove it; for example, with a water-based cleaner rather than a volatile organic solvent.

## Process Safety Management

Inherent in the use or storage of large quantities of highly hazardous or flammable chemicals is the risk of catastrophic releases that would prove injurious or fatal both to employees and to those living in the immediate vicinity of the facility. The experience at Bhopal's methyl isocyanate facility taught the world a tragic lesson about worker and community consequences of an unexpected chemical release.

Process safety management is a systematic approach to evaluating an entire process for the purpose of preventing such unwanted releases of hazardous chemicals into locations that could expose employees and others to serious hazards. In 1992, OSHA promulgated a standard for general industry, 29 *CFR* 1910.119, which mandates process safety management of highly hazardous chemicals for companies that use or store large quantities of flammable or highly hazardous chemicals in one location. It requires them to implement a program that incorporates analyses, written operating procedures, training, inspection and testing, and safety reviews for their own employees and for contractors. OSHA defines process safety management as the proactive identification, evaluation, and mitigation or prevention of chemical releases that could occur as a result of failures in process, procedures, or equipment.

OSHA acknowledges several methodologies to evaluate hazards of the process being analyzed: What-if scenarios, checklists, what-ifs in a checklist format, hazard and operability studies (HAZOPs), failure mode and effects analysis (FMEA), fault tree analysis, or other equivalent methodologies are all acceptable. Employers are required to determine and document the priority order for conducting process hazard analyses based on a rationale that considers the extent of the process hazards, the number of potentially affected employees, the age of the processes, and their operating history.

This approach requires the development of expertise, experience, and proactive initiative by a team of concerned individuals. These are attained primarily by conducting process hazard analyses, directed toward evaluating potential causes and consequences of fires, explosions, releases of toxic or flammable chemicals, and major spills of hazardous chemicals. The health and safety professional, with the assistance of process managers, employees, and others, must determine the potential failure points or modes in a process. The focus is on equipment, instrumentation, utilities, human actions (routine and nonroutine), and external factors that might impact the process.

For the health and safety professional, a process hazard analysis can provide a starting point in an overall hazard evaluation. Such analyses should be performed whenever possible. For those evaluating businesses where quantities of hazardous materials in use fall below the OSHA requirements for a written plan, the guidelines contained in the Process Safety Standard may still serve as a useful tool in the evaluation process.

# Field Survey

Thus far in the evaluation process, most of the research recommended has probably been conducted outside of the workplace. Process diagrams, literature searches, and inventories can be reviewed in the office, but evaluation requires on-site, direct observation, measurement, and assessment. It is at this point that the anticipation of hazards must be integrated with actual conditions. Whether the motivation for the evaluation is compliance, insurance, expert testimony, complaint investigation, or development of a comprehensive health and safety plan, direct, on-site involvement on the part of the industrial hygienist is required. This usually begins with a walkthrough of the workplace.

The walkthrough, or initial field survey, follows the flow of materials into the facility, through all the various processes involved in the operation, to the shipping of finished product, as well as tracking unwanted by-products. It should also include nonproduct areas such as maintenance and other service operations. It should be conducted with the facility or process manager, someone familiar with both the process design and usual operations. The walkthrough introduces the industrial hygienist to the language of facility operations, establishes a baseline of current conditions, and allows an initial assessment of hazards or areas that may require further evaluation.

The industrial hygienist can use this opportunity to meet operators of key processes, area supervisors, and other health and safety personnel. Communication with these individuals is essential to understanding the sources and effects of hazards on the job, and for planning future sampling and analyses. A checklist, a sketch of the facility layout, preliminary notes, or a tape recording is useful for documenting initial impressions and serves as a reminder to return to areas that require more intensive inspection.

As our society becomes more and more service oriented, the traditional facility tour may be conducted less frequently. The field survey will, however, remain a vital tool of the industrial hygienist, with the focus of the initial walkthrough being individual operations within a building. Conducting the initial assessment of the operation with an area supervisor who can help obtain a sketch of the physical operation of interest and familiarize the industrial hygienist with the current problems will remain invaluable, as will developing a checklist to use in the walkthrough itself.

## Sensory Perception

Field surveys also allow industrial hygienists to make use of their sensory perceptions (vision, hearing, and sense of smell) to note unwanted or unforeseen conditions. Observing dusty operations, patterns of shavings or powder on the floor, overspray on walls, puddles underneath valves, or wetness around an area not currently in use alerts the industrial hygienist to problems not considered before. The exact location of processes of concern, such as welding stations, degreasers, flammable storage, exits, and break areas, can be precisely located and added to the facility layout for later consideration.

The absence or presence of visible dust should not sway initial judgment excessively. Because dust particles of respirable size are not visible to the unaided eye, lack of a visible dust cloud does not guarantee an atmosphere free of respiratory hazards. Timing of dry sweeping and shaking out of dust collection devices should be noted. The need for air sampling for dusts should be determined by the source, identity, toxicity, health complaints, and processes of concern.

Whenever the tour guide must move closer to the industrial hygienist in order to be understood, it is likely that noise levels are excessive, and this fact should be noted. Patterns of hearing protector use should also be recorded during the walkthrough.

The presence of many vapors and gases is detectable by smell. The odor thresholds for some chemicals are in the parts per billion range, which helps serve as an early warning of exposure. This is especially true for someone entering an area from elsewhere and for certain aromatic or strong-smelling chemicals such as ethyl acrylate or hydrogen sulfide. The sense of smell fatigues with time and is variable from person to person. Odor thresholds listed in resource tables can vary by a factor of 100 from one person to another. Detecting an odor or experiencing eye or throat irritation should indicate to the occupational health professional that a chemical is present to some degree in the air, and an attempt should be made to identify the process or chemical. These sensory impressions do not necessarily reveal an overexposure, but they can provide important clues to a potentially hazardous source. Also, it is important to note that absence of an odor or irritation does not necessarily mean the absence of a chemical exposure.

## Control Measures in Use

During the walkthrough, or initial field survey, the types, locations, and effectiveness of control measures should be appraised. Controls include local exhaust and general dilution ventilation, shielding, and personal protective equipment such as gloves, respirators, hearing protectors, and safety glasses. Storage of respirators and availability of replacement cartridges are good indicators of a company's oversight of their respirator program.

Ventilation design should be appropriate for the hazard and the process. Homemade ventilation may be inadequate. For example, it might consist of a canopy hood for unheated processes or of hoods distant from the source, both of which would provide less than adequate capture velocity. Ventilation and airflow patterns can be visualized with an air current tube or estimated by taking air velocity measurements. The distance between the air collection device and source should be observed, and any turbulence created by portable fans or locations on aisles should be recorded.

## Observation and Interview

Observation and interviews with workers can reveal the best information regarding hazard evaluation and adequacy of controls. During the walkthrough, or while conducting sampling, the occupational health professional must carefully observe workers performing their jobs and note all opportunities for exposure by each route of entry. Without jeopardizing worker confidentiality, employees should be interviewed regarding the content of their jobs, how they spend their time, exposures of concern, and any health symptoms, especially as they relate to contact with various chemical products or processes. Identifying variations in production levels, assignments, shifts, seasonal work, and ventilation patterns helps in determining when peak exposures might occur and where sampling would be most useful (Table 15–D).

## ■ Monitoring and Sampling

### Rationale

There are a number of reasons why environmental measurements should be taken in the workplace. Of primary concern to the health and safety professional is evaluating the degree of employee exposure to hazardous materials on the job. Other important reasons include identification of the tasks or processes that could be sources of peak exposures, evaluation of the impact of process changes and control measures, and compliance with occupational and environmental regulations. Environmental sampling can be used to clear an area for reoccupancy, decide if a confined space is safe for entry, establish background or usual concentrations, or warn of a peak release of a hazardous product.

### Monitoring

Monitoring is a continuous program of observation, measurement, and judgment. It requires an awareness of the presence of potential health hazards as processes undergo change, and constant assessment of the adequacy of the control measures in place.

Monitoring is more than simply sampling the air to which an employee is exposed or examining the medical status of that employee. It is a combination of observation, interview, and measurement that permits a judgment to be made relative to the potential hazards and the adequacy of protection afforded employees. Included in the process are both personal and environmental monitoring, performed during a given operation where hazardous materials may be released, and follow-up biological and medical monitoring of the employees involved in that process.

**Personal monitoring.** Personal monitoring is the measurement of a particular employee's exposure to airborne contaminants and, in theory, reflects actual exposure to the employee. It is usually done during a specific time period, often an 8-hour shift or a 15-minute period, to ensure compliance with OSHA PELs or STELs; it can therefore include times when the employee is at break or involved in activities where the contaminant of interest is not in use. It is because of this variability that it is extremely important to observe individuals being monitored and to interview them about their work, both before and after the monitoring is done.

In personal monitoring, the measurement device, or dosimeter, is placed as close as possible to the contaminant's

---

**Table 15–D.** Checklist for Evaluating Chemical Exposure

---

A. Evaluate the Potential for Airborne Exposure.
   1. Exposure Sources (rank high/medium/low):
     a. Types and amounts of chemicals in use or created by combustion or decomposition.
     b. Visible leaks, spills, or emissions from process equipment, vents, stacks, or from containers.
     c. Settled dust, which may be resuspended into the air.
     d. Open containers from which liquids may evaporate.
     e. Heating or drying, which may make a chemical more volatile or dusty.
     f. Odors. Consult an odor threshold table to get an estimate of concentration.
     g. Do air monitoring where the presence of a contaminant is suspected but cannot be verified by sight or smell.
     h. Visualize exposure by taking photographs or videotape.
   2. Job Functions (estimate hours/day):
     a. Manual handling in general.
     b. Active verb job tasks such as grinding, scraping, sawing, cutting, sanding, drilling, spraying, measuring, mixing, blending, dumping, sweeping, wiping, pouring, crushing, filtering, extracting, packaging.
   3. Control Failures:
     a. Visible leaks from ventilation hoods, ductwork, collectors.
     b. Hoods that are located too far from the source or are missing or broken.
     c. Ductwork that is clogged, dented or has holes.
     d. Insufficient make-up air to replace exhausted air.
     e. Contamination inside respirators.
     f. Improperly selected, maintained, or used respirator.
     g. Lack of or inadequate housekeeping equipment.
     h. Lack of or inadequate doffing and laundering procedures for clothing contaminated by dust.
B. Evaluate the Potential for Accidental Ingestion.
   1. Exposure Sources (rank high/medium/low):
     a. Types and amounts of chemicals in use or created by combustion or decomposition. Solids are of primary concern.
     b. Contamination of work surfaces, which may spread to food, beverage, gum, cigarettes, hands or face.
     c. Contamination of hands or face, which may enter mouth.
     d. Do wipe sampling to verify the presence of a contaminant on work surfaces, hands, face, and so forth.
   2. Control Failures:
     a. Contamination of inside of respirator, which may enter mouth.
     b. Contamination of lunchroom surfaces, which may spread to food, beverage, gum, cigarettes, hands or face.
C. Evaluate the Potential for Skin Contact and Absorption.
   1. Exposure Sources:
     a. Types and amounts of chemicals in use or created by combustion or decomposition. Check dermal absorption potential. Do not rely on OSHA skin notations. Assume most liquids will penetrate skin.
     b. Consider whether one chemical can act as a carrier for other chemicals.
     c. Visualize dermal exposure by taking photographs or videotape.
   2. Job Functions:
     a. Dipping hands into material.
     b. Handling of wet objects or rags.
   3. Control Failures:
     a. Contamination of inside of gloves.
     b. Improperly selected, maintained, or used gloves.
     c. Improperly selected, maintained, or used chemical protective clothing.
     d. Lack of or inadequate facilities for washing of hands and face close to work areas.
     e. Lack of or inadequate shower facilities.

---

(Reprinted from *New Solutions,* Spring 1991, p. 77, P.O. Box 281200, Lakewood, CO 80228–8200.)

route of entry into the body. For example, when monitoring an air contaminant that is toxic if inhaled, the measurement device is placed on the employee's lapel or as close to the breathing zone as possible. When monitoring noise, the dosimeter should be placed close to the ear.

Even with the proper placement of the dosimeter, there is no guarantee that results of personal sampling will reflect actual exposure levels. Some materials are absorbed through the skin or mucous membranes in addition to being inhaled. The release of contaminants is often not uniform, and the side of the employee where the monitor is placed may not be the side closest to the point

of release of the contaminant. The results would therefore underestimate the exposure. On the other hand, if the sampling device is placed outside a respirator or face shield, the result might overestimate the true exposure to the worker.

Personal sampling relies on portable, battery-operated sampling pumps that the employee wears throughout the sampling. This offers freedom of movement because there is no need to maintain proximity to electrical outlets. The pumps, however, can be noisy and heavy, and employees are sometimes not willing to wear them on a continuous basis. In addition, because the pumps are battery oper-

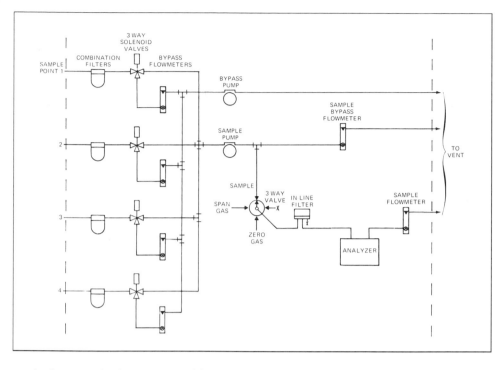

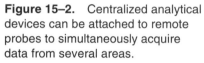

Figure 15–2. Centralized analytical devices can be attached to remote probes to simultaneously acquire data from several areas.

ated, they might have a variable output throughout the day, or might actually stop operating in the middle of sampling. The effective use of personal sampling pumps relies on proper calibration and maintenance and consistent supervision by well-trained professionals during the monitoring process.

**Area monitoring.** Area sampling is another method used by industrial hygienists to evaluate exposure. Here, however, exposure is measured not in terms of a particular employee, but rather in terms of the ambient air concentration of a particular substance in a given area at a given period of time. The measurement device, which does not have to be battery operated and can be larger and more rugged than those used in personal sampling, is placed adjacent to a worker's normal workstation. Centralized analytical devices can be attached to remote probes so that data can be acquired from several areas simultaneously and monitored from a central location (Figure 15–2). An alarm can be sounded if a preset limit is exceeded (as shown in Figure 15–3). Area sampling is an important technique to determine the need to develop, implement, or improve control measures.

Ideally, area sampling would be so thorough and the pattern of potential exposure to workers so well defined that, in any given work space, knowledge of a worker's activity would be sufficient to estimate that person's exposure, and personal monitoring would not be necessary. If, for instance, vapor concentrations and their duration around equipment were known and could be superimposed on a floor plan, then a worker's exposure could be determined from observing that worker's movements and plotting the frequency and duration spent in each area. The employee's daily exposure could then be found by adding short-term

exposures to compute the time-weighted average (TWA). This in-depth area exposure analysis is not routinely done, however. It requires a tremendous amount of time and monitoring equipment and may still miss crucial contributions to a worker's exposure on any given day.

In most processes, airborne concentrations of materials usually vary over time. The fluctuations may be large, and continue for hours, or they may be brief, sometimes lasting only seconds or minutes. Only extensive, continuous sampling can provide information about such fluctuations

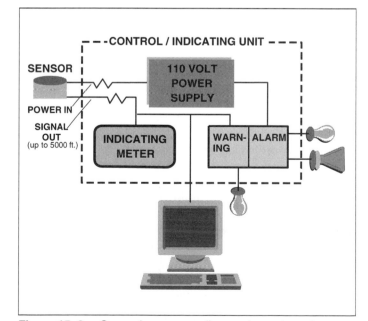

Figure 15–3. General area–sampling equipment can be connected to a computer to obtain a permanent record of contamination levels. (Courtesy MSA.)

in any given location. The data, if collected with a real-time monitor or printed on a strip-chart recorder, provide valuable clues about the main sources and timing of exposure, and thus a means to design controls that should be used in a process. The computer printout or strip chart can be used to estimate an individual's exposure and can also serve as a historical record.

Area sampling is also used to establish usual background concentrations for chemicals that are ubiquitous in our environment. An incident that occurred in Boston, Massachusetts, illustrates this rationale. A transformer fire released polychlorinated biphenyls (PCBs) into an office basement and ventilation system in October 1981. In June 1985 a new tenant, prior to occupying the building, performed testing that revealed contamination, including dioxins and dibenzofurans, which required extensive cleaning. Sampling methods were so sensitive, and the chemicals so persistent, that they were detected, even after thorough cleaning, four years after the incident. Questions about the adequacy of the cleanup and the attendant risk led researchers to consider what normal background concentrations in similar settings might be. In order to establish normal background concentration levels, area air sampling was conducted in similar buildings with no history of PCB release. These results were subsequently used to develop a criterion for reentry into a previously contaminated area, with assurance that the exposure and risk would be no greater than usual (U.S. Dept. of Health & Human Services [USDHHS], NIOSH, 1987 & 1988). Area sampling for the same purpose is used after asbestos or lead abatement in commercial or residential settings.

Area sampling has its disadvantages, though. Sampling equipment can be made rugged and reliable, but often it is not, and leaving it unattended for hours or days at a time without the supervision of a trained technician could result in no reliable data collection during a crucial period in the process. Area sampling may underestimate exposure if the worker works close to a process but the measurement probe or collection device is at a farther distance from the exposure point.

**Biological monitoring and screening.** Biological monitoring extends the concept of sampling to the affected individuals. Analysis of material such as blood, urine, or exhaled air can provide information about the impact of absorption of hazardous material. Alterations in blood or urine concentrations reflect absorption by all routes of entry, reflect the physiological response, and offer information beyond that provided by air monitoring alone. The values measured can be compared to background population values or to guidelines or standards in the same way air-sampling data are evaluated.

In general, there are three categories of biological monitoring: measurement of the contaminant itself (for example, lead, cadmium, or mercury), measurement of a metabolite of the chemical, and measurement of enzymes or functions that reflect harm caused by a hazardous exposure. Biological monitoring is appropriate in conjunction with industrial hygiene sampling, so that workplace sampling results can be correlated with the measured biological effects.

Often, the hazardous chemical cannot be measured directly, but a metabolite can be. For example, methylene chloride is metabolized to carboxyhemoglobin in the human body, and testing a venous blood sample for carboxyhemoglobin could reveal overexposure to methylene chloride. Another example is styrene, for which exposure and absorption can be evaluated by the concentration of mandelic acid in urine at the end of a workshift. A third example is the measurement of trichloroacetic acid in urine; when the concentration is greater than 10 mg per liter of urine at the end of the workweek, it indicates overexposure to methyl chloroform. It should be noted that many of these indirect biological markers are subject to variability in laboratory handling and analysis, in worker physiological and health status, and in reflecting multiple chemicals and exposure sources. For example, although an elevated level of carboxyhemoglobin may indicate overexposure to methylene chloride, it might also reveal an unwanted exposure to carbon monoxide at the same job site. This limits the usefulness of biological sampling as an individual diagnostic tool or as a definitive evaluation of a workplace exposure.

Sometimes the adverse effect of a workplace exposure is only revealed when medical evaluation reveals an unusual laboratory result or abnormal function test. For example, a pulmonary function test on an autobody spray painter can reveal deficits in forced expiratory volume. Such a reduced ability to exhale may be a marker of occupational asthma caused by contact with the isocyanates in polyurethane paint.

Another abnormal medical test result that may indicate occupational injury or disease is a slowed nerve conduction velocity, which may indicate trauma, repetitive strain, or peripheral neuropathy. Abnormal liver function enzymes may reflect hepatitis or liver injury from chronic solvent exposure. Such results reveal a disease process or harm but do not necessarily indicate the cause of the harm, which may be attributable to factors outside the workplace or to preexisting medical conditions previously unrecognized. Therefore, the results are nonspecific and require interpretation by a trained occupational physician or occupational health nurse. Unfortunately, many physicians have not been trained to identify the occupational or environmental causes of diseases, and some fail to ask their patients about their work and exposure hazards. For example, lead as a causative agent for abdominal pain, insomnia, infertility, and high blood pressure frequently goes undiagnosed and untested.

The significance of results from biological monitoring is open to speculation; alterations in function or unusual laboratory findings can be viewed as evidence of harm, or they can be viewed as only a marker that exposure has occurred. For example, the indication on chest x rays of pleural plaques, which can exist in the absence of disease, is a marker of past asbestos exposure. Interpretation of radiological findings in the lungs is known to be inconsistent; abnormalities are difficult to detect and even more difficult to interpret, so x rays should not be the sole determinant in diagnosis of occupational disease (Figure 15–4).

Interpretation of biological monitoring is complicated by the fact that the concentration of material measured is not exactly equivalent to the exposure dose. When a chemical is absorbed into the body, excretion of earlier ingestion of that substance may be occurring at the same time. How the body absorbs, integrates, stores, detoxifies, or modifies a contaminant during and after an exposure is reflected in the biological result and is important in the final analysis. Guidelines for biological monitoring must reflect an understanding of the biochemical dynamics of the contaminant in relation to physiological processes. Measurements may represent peak exposures and absorption prior to any significant clearance, or they may reflect equilibrium levels attained only after steady state has been reached. Obtaining information about the relationship between the timing of exposure and biological testing is very important.

For example, lead concentration in blood is used as an index of lead exposure by inhalation and ingestion in the previous days or weeks, whereas zinc protoporphyrin (ZPP) is used as a measure of lead exposure during the previous three or four months. Research shows that bone x-ray fluorescence (XRF) (Gerhardsson et al., 1993; Hu et al., 1989) may reveal the total body burden of lead, including that portion stored in the skeleton. These measurements can be used to identify hazardous exposures, dangerous work practices, or inadequacies in ventilation and personal protective equipment. Differences in blood lead and ZPP concentrations provide information about the timing of exposure and which tasks pose the greatest risk. Blood lead values can be used to identify individuals at risk, who should be removed from any further exposure.

***Medical surveillance.***    Medical surveillance can extend beyond biological monitoring of individuals to incorporate screening of exposed populations for the adverse effects of those exposures. For example, audiometric testing can be used to determine the extent of temporary or permanent shifts in thresholds of hearing acuity caused by noise exposure. Liver enzymes can be measured to assess the effect of solvents suspected of causing hepatitis or other liver injury. The appearance of the lungs on x-ray films can reveal pneumoconiosis, hypersensitivity pneumonitis, or other respiratory diseases. Baseline skin testing, followed by further skin testing after a potential exposure, is essential for those working in health care or other occupations at high risk for exposure to tuberculosis. Positive changes from baseline may indicate further evaluation or medical treatment.

One purpose of medical surveillance is the early detection of disease or conditions for which treatment can prevent further illness. The affected individual should be removed from the hazardous exposure and receive needed medical treatment and supervision. OSHA has incorporated this concept of medical removal protection (MRP) in its standards to protect workers overexposed to lead and cadmium. OSHA requires that if a worker's medical evaluation indicates overexposure or the adverse health effects associated with these substances, the employer must either provide alternative work in an area where there is no risk of exposure or allow the employee to stay home with full compensation during the period of treatment.

Medical screening can also be an invaluable preventive tool in hazard control. For example, regular testing of urine for mercury in an exposed worker population allows identification of individual work stations or work practice sources. Routine analysis of the group may allow early detection of subtle increases in mercury absorption that might reflect a breakdown in controls or a weakness in the training program.

OSHA has proposed that medical screening and evaluation be used to measure the effectiveness of its PELs. If workers exhibit adverse health effects, while at the same time air-sampling results show compliance with OSHA's standards, then OSHA will use the results to reexamine the adequacy of the PEL.

Twenty-one OSHA standards now have requirements for medical examinations or tests, focusing on either medical screening of individuals or surveillance of an entire exposed group. The medical evaluation required may involve screening of an exposed employee group for an individual agent, or it may include a more comprehensive examination of employee health in the workplace. The more hazardous the exposure, the more in-depth the

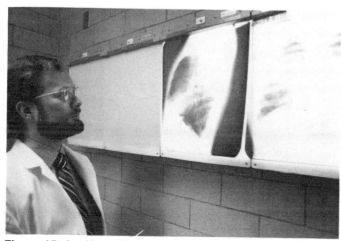

**Figure 15–4.**    X-ray films are used to diagnose pneumoconiosis.

**Table 15–E.** Recommended Medical Program

| Component | Recommended | Optional |
|---|---|---|
| Preplacement screening | Medical history.<br>Occupational history.<br>Physical examination.<br>Determination of fitness to work while wearing protective equipment.<br>Baseline monitoring for specific exposures. | Freezing preplacement serum specimen for later testing (limited to specific situations). |
| Periodic medical examinations | Yearly update of medical and occupational history; yearly physical examination; testing based on examination results, exposures, and job class and task.<br>More frequent testing based on specific exposures. | Yearly testing with routine medical tests. |
| Emergency treatment | Provide emergency first aid on site.<br>Develop liaison with local hospital and medical specialists.<br>Arrange for decontamination of victims.<br>Arrange in advance for transport of victims.<br>Transfer medical records; give details of incident and medical history to next care provider. | |
| Nonemergency treatment | Develop mechanism for nonemergency health care. | |
| Record keeping and review | Maintain and provide access to medical records in accordance with OSHA and state regulations.<br>Report and record occupational injuries and illnesses.<br>Review site safety plan regularly to determine whether additional testing is needed.<br>Review program periodically. Focus on current site hazards, exposures, and industrial hygiene standards. | |

health evaluation should be. For example, hazardous waste workers should receive preplacement screening and periodic medical examinations, with testing for specific exposures as necessary. Table 15–E describes a recommended medical program. In work environments where respirator use is necessary, workers should be evaluated medically for fitness to wear a respirator. In most cases, medical surveillance should include a medical and occupational history and a physical examination, with attention paid to the target organs and functions potentially affected. Medical records should be maintained to allow for review of deviations from the baseline of preplacement health status.

***Biological exposure indices.*** The concept of biological monitoring has led the ACGIH to develop a list of biological exposure indices (BEIs), published annually in their TLV booklet. Similarly, OSHA has incorporated required biological monitoring into several standards: the Lead Standards in General Industry (29 *CFR* 1910.1025) and in Construction (29 *CFR* 1926.62), the Cadmium Standard (29 *CFR* 1910.1027) and the Formaldehyde Standard (29 *CFR* 1910.1048). Several of OSHA's standards (such as those for benzene and ethylene oxide) only require medical surveillance when air sampling has revealed a pattern of exposure above either the action level (AL) or PEL during a specific number of days per year. In 1988, OSHA published an advance notice of proposed rulemaking on medical surveillance programs. To date, no standard has been adopted.

NIOSH, in conjunction with state departments of health, has promoted the use of biological sampling results to investigate occupational exposure and illnesses.

Many states now have occupational lead registries that facilitate investigation of the number of workers poisoned by lead on the job and promote analyses of those industries most responsible. These results can be used to develop educational materials for small businesses needing assistance in controlling hazards, and in selecting sites for government intervention for public health purposes. For example, the Massachusetts Occupational Lead Registry found that 60 percent of their registrants with blood lead concentrations greater than 40 micrograms of lead per deciliter of blood worked in painting, deleading, and other construction jobs (Rabin et al., 1994). This discovery led to efforts to work with the state's highway department to more closely supervise bridge-painting contracts.

***Combined effects.*** At present, very little is known about how the body integrates two different types of stress and the resultant strain, even if both stressors are chemical. The usual assumption is that chemicals affecting different organs or tissues should be considered independently, whereas those that affect the same organ or tissue should be considered jointly because they may produce additive or synergistic effects.

Synergism is known to occur with certain exposures. The best-known synergistic effect is that of smoking combined with asbestos exposure. The risk of lung cancer increases greatly, beyond that expected from adding the risks together. Similarly, in vitro studies of organophosphorus pesticides have shown that a combined exposure to malathion and diazinon results in cholinesterase inhibition significantly greater than a mere summation of the effects would predict (Iyaniwura, 1990).

Other research has focused on less obvious combined effects. One study looked at the effects of different chemicals on hearing and found that trichloroethylene, arsenic, heavy metals, organotin compounds, and manganese all caused some degree of hearing loss or audiometric abnormalities in occupationally exposed workers. Carbon disulfide interacted with noise to cause sensorineural hearing loss; toluene and noise acted synergistically to increase the incidence of hearing loss (Ryback, 1992). Another study, looking at the combined effects of chemicals commonly found at hazardous waste sites, saw both synergistic and antagonistic interactions. Whereas lead tetraacetate and arsenic trioxide produced antagonistic effects in one assay, tetrachloroethylene and dieldrin produced synergistic effects. The authors of this genotoxicity study cautioned that compounds may behave differently in a mixture than when alone (Ma et al., 1992).

The OSHA airborne exposure limits, as well as the RELs and TLVs, have been developed under the assumption that workers are exposed to chemicals one at a time. In fact, exposure to just a single chemical rarely occurs. One method to calculate the alteration in guidelines necessary to evaluate combined exposure is to add concentrations as a fraction of their respective TLVs. If the total equals or exceeds one, then an overexposure has been detected. This is not a conservative approach, because it assumes additive effects and allows excessive exposures if the effects are synergistic or if other stressors are present.

In most workplace exposure assessments, chemical, physical, biological, and psychological hazards are present at the same time. For example, the process of tunneling can involve simultaneous exposures to high atmospheric pressure, dust, noise, heat, high humidity, carbon monoxide, and physical safety hazards. An assessment of strain produced by any one of these stressors would be complicated by the presence of any or all of the others.

***Limitations of biological monitoring.*** Biological monitoring is one way to compare exposure to dose. However, it must be remembered that it measures exposure only after it occurs, and after the contaminant has affected bodily functions in some way. It must be used properly and in conjunction with other environmental controls and not as the sole control measure, as is sometimes the case when employers want to spare the cost of a more comprehensive, and therefore more expensive, monitoring program. Biological monitoring and medical surveillance are not replacements for environmental or personal sampling but should be used to complement them.

When biological monitoring is required by an OSHA comprehensive standard, the health care provider conducting the monitoring must be given a copy of the requirements. In some cases, the medical personnel will want to tour the workplace to enhance their awareness of potential hazards. The California Department of Health has developed model language to ensure that employers and health care providers develop a contract that accurately reflects the expectations and needs of both parties. (California Occupational Health Program, 1990). It is available in a workbook developed for the radiator repair industry, and available from the California Occupational Health Program (COHP) Dept. of Health Services, 2151 Berkeley Way, Annex Eleven in Berkeley, California 94704.

## Sampling

**Strategy.** The preliminary research and initial field survey help identify potential hazards to which workers may be exposed. The next task is to devise a sampling strategy to determine the intensity of exposure, the source of the hazards, and the adequacy of controls in place. Included in the plan must be a consideration of the sources of error, the desired precision and accuracy of measurements, and the degree of confidence needed for interpretation of the results.

If the industrial hygiene sampling is conducted to evaluate a problem, the sampling strategy can be designed to measure the "worst case." An example that occurred in central Massachusetts in 1990 illustrates this approach. Periodic use of a degreaser had resulted in dizziness and headaches in its two operators, as well as complaints from a neighboring department. Because the use of the degreaser was limited to three hours in the morning, it seemed unlikely that the 8-hour time-weighted average exposure exceeded the relevant PEL. However, the health symptoms and complaints indicated a problem with the operation of the degreaser, its cooling coils, or the local exhaust ventilation. Air sampling was planned to capture the particular solvent used during the worst-case exposure, when the smallest parts were being cleaned. Before sampling proceeded for this suspect carcinogen, the industrial hygienist made sure that the work practices and ventilation were exactly the same as they had been the day before, when the complaints had occurred.

Evaluating the worst case first, during the time of greatest exposure, at a location known to have caused problems, offers three advantages. First, this sampling is designed to solve a problem. Measuring the concentration of the chemical believed to have caused health symptoms and concerns helps identify the source, improve the controls, and correct the problem. Second, such results teach employees valuable lessons about indicators of equipment malfunction, the warning signs of overexposure, and the impact of work practices on airborne solvent levels, such as reducing drag-out of solvent. Finally, the process of evaluating the worst case during the longest exposure time to the highest expected concentration—if lower than the referenced PELs, STELs, TLVs, and RELs—allows assumptions and assurances to be made regarding shorter-term, lower-level exposures.

Another approach to air sampling is to capture "typical" conditions. This is not always as easy as it sounds. Day-to-day variations may make a typical exposure difficult to define and measure. In addition, managers and employees being monitored may take extra precautions when they know they are being observed by health and safety professionals. Concerns about being "sampled" or evaluated may serve to encourage companies to present their best face by adding ventilation or opening doors and windows that are usually closed. Preliminary air sampling is usually done on the day shift, when supervision is better, shipping doors are more likely to be open, and the timer for the ventilation is on the occupied setting.

A good sampling strategy makes use of both worst-case and typical sampling methods, each selected to answer the questions what, where, when, how, and whom to sample.

**What and how to sample.** The first and key principle to keep in mind is that samples should represent workers' exposures. Decisions about which chemicals to evaluate should be based on such factors as quantities and methods of use; worker reports of adverse experiences; concerns regarding high toxicity, volatility, carcinogenicity, or teratogenicity; and percent representation in mixtures.

For time-weighted average sampling, the *NIOSH Manual of Analytical Methods* (1994) can be referred to for the correct sampling technique. There is a wide choice of collection media, from charcoal or silica gel tubes for organic nonpolar vapors or polar vapors, respectively, to cellulose ester or fiberglass filters for fumes and particulate materials. The appropriate medium for a specific reagent is stated in the NIOSH manual if there is an approved method. These are methods that have proved to be reproducible, given certain flow rates and sampling and analytical conditions.

Also available to the industrial hygienist are grab sampling methods, in which a specific release point is monitored at a specific time in the process. (See Figure 15–5.) This is done with colorimetric tubes (such as Draeger or Sensidyne tubes) or by "grabbing" a volume of air in a sampling bag, canister, or other container, which is then analyzed in an accredited laboratory. The colorimetric tube method has limitations in its accuracy and should be used only as an initial, rough exposure estimate. The sampling bag method has limitations in collection efficiency and should be performed under the guidance of a laboratory accredited by the American Industrial Hygiene Association (AIHA).

Direct-reading instruments are available for a number of different chemical, physical, and radiation hazards (Figure 15–6). They provide immediate information about current conditions or concentrations, and can therefore be used to locate a source or detect a leak. Given this instant feedback, changes can be made in operating conditions, and the worksite can be evaluated for improvement. For example, it might be possible to alter ventilation settings and observe the impact on airborne hazard levels, or turn off a compressor and note a drop in noise levels.

Materials that are not listed in the NIOSH manual are more difficult to evaluate, and an accredited analytical laboratory should be consulted. All samples taken should be sent to an accredited analytical laboratory, and results will be reported back at a later date.

**Where to sample.** Personal monitoring is used to evaluate actual exposures to an individual by sampling for specific agents in the worker's immediate vicinity for durations corresponding to the process of concern or the appropriate occupational exposure limit (such as TWA, STEL, or ceiling). The sampling device is attached directly to the employee and is worn throughout the sampling period, reflecting worker movements in relation to the source of contamination, during both work and rest periods. The results from personal monitoring should be used to determine the effectiveness of control measures (engineering, work practices, and administrative) implemented to prevent overexposure.

Area monitoring is used to measure the contaminants found in the work area that is generally occupied by employees. Also called environmental monitoring, it provides information about the amount and type of exposures found in a fixed area of interest. It reflects the effectiveness of engineering controls put into place to control the release of hazardous materials. It only reflects actual employee exposure to the extent that the time period monitored represents the time most employees spend in a given area.

Monitoring conducted for the purpose of measuring employee exposure is normally done with personal sampling. The recommended sampling method or equipment may, however, be inconvenient to use. If the industrial hygienist wishes to determine what air concentrations are in an area where the highest levels of contaminant release are anticipated, or where continuous exposure close to a particular point source may occur, then area monitoring is often useful. If results from these types of "worst-case" exposures are less than the upper regulatory or recommended limits, and if the contaminant in question is released only in that area, then an assumption can usually be made that workers spending their day in this area have exposures below the acceptable upper limit. This type of assumption is frequently made.

**Whom to sample.** If the initial determination indicates the possibility of excessive exposure to airborne concentrations of a toxic substance, measurements of the most highly exposed employee should be made. This can be determined by observing the point of release and selecting the employee who is closest to the source of the contaminant in question.

Air movement patterns within a workroom must be considered when evaluating potential exposures to workers. Especially in operations or processes involving heating or combustion, the natural air circulation could be such that the maximum-risk employee might be located at consider-

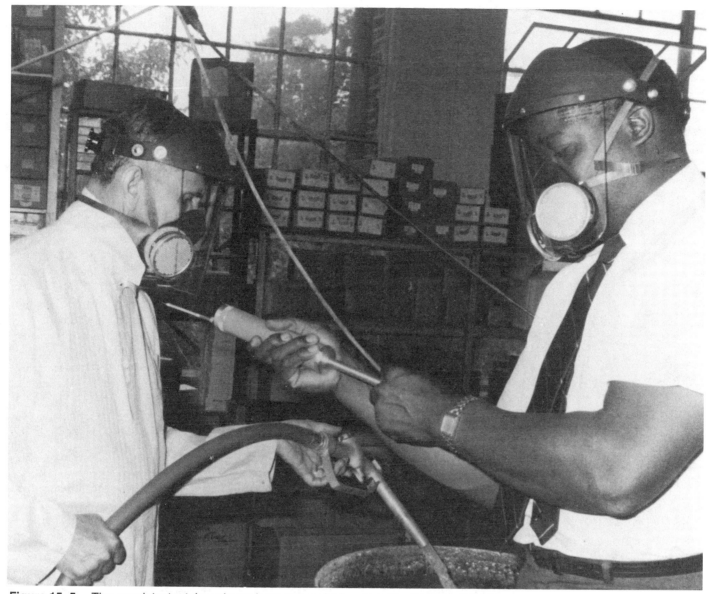

**Figure 15–5.** The gas detector tubes shown here are useful for obtaining direct readings of gas or vapor contamination in the workplace air.

able distance from the source. The location of ventilation booths, air supply inlets, and open doors and windows and the size and shape of the work area are all factors that affect workroom airflow patterns and can produce elevated concentrations at locations far removed from the source.

Differences in work habits of individual workers can significantly affect levels of exposure. Even though several workers are performing essentially the same tasks with the same materials, their individual methods of performing their work could affect the contaminant concentration to which each is exposed. Initial monitoring is often limited to a representative sample of the exposed population, usually those considered at greatest risk. Exposure results over the action level or PEL indicate that more extensive sampling is needed.

**When to sample.** Another factor that must be considered is when to sample. If temperature varies greatly from season to season, with windows kept open during one season and not another, then sampling should be done during both periods. Or, in this case, because more dilution of the contaminant occurs with windows open, worst-case exposure monitoring should be done with the windows closed. If air conditioning is used, levels of contaminant may be fairly constant throughout the year. However, this is not necessarily the case with variable air volume (VAV) systems that restrict fresh airflow during the coldest and hottest periods of the year. (See Chapter 21, General Ventilation of Nonindustrial Occupancies, for further discussion.) If the facility has more than one workshift, samples should be collected during each shift. Concentrations can vary considerably

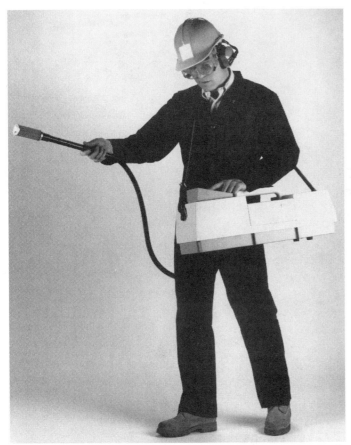

**Figure 15–6.** This portable ambient air analyzer can be used to measure concentrations at the operator's workstation. (Courtesy Foxboro Co.)

from time to time during the day, because of such factors as differences in production rate, degree of supervision, and ventilation provided during off-peak shifts.

**How long to sample.** The volume of air sampled and the duration of sampling is based on the sensitivity of the analytical procedure or direct-reading instrument, the estimated air concentration, and the OSHA standard or the TLV for that particular agent. Again, the *NIOSH Manual of Analytical Methods* or an accredited analytical laboratory should be consulted.

The duration of the sampling period should represent some identifiable period of time; for example, a complete cycle of an operation or a full shift. Often, the appropriate time period is specified in the regulatory upper limits; when looking at a PEL, a full 8-hour shift of monitoring is called for. For comparison to an OSHA short-term exposure limit (STEL), 15-minute samples during a worst-case exposure scenario are required. Longer workshifts require recalculation of the relevant standard, because the total time exposed is increased. For example, a 10-hour workshift requires that the PEL or TLV be modified to reflect the extra exposure time and be reduced to four-fifths of the original 8-hour standard. To illustrate this point, OSHA's

lead standard, in which the 8-hour TWA is 50 $\mu g/m^3$, requires employers to calculate the permissible exposure limit for workers exposed to lead for more than 8 hours in any workday, using the following formula: Maximum permissible limit (in $\mu g/m^3$) = 400/hours worked in a day.

The concentration of contaminant in the workplace is sometimes low. Direct-reading instruments and other devices used to collect samples for subsequent analysis must collect a sufficient quantity of the sample so that the chemist doing the analysis can accurately determine the presence of minute amounts (parts per million or sometimes parts per billion) of the contaminant.

**What to note during sampling.** Accurate record keeping is essential for the correct interpretation of air-sampling results. The fundamental records include total time sampled; pump flow rate, both at the beginning and end of the sampling period; location of the area or identification of the person being monitored; and a description of the process being evaluated. In addition, sampling notes should include the engineering controls present and the location of any local or general exhaust ventilation, as well as any measurements of these taken at the time of sampling. If other processes are located close enough to affect the sampling results, they should be described.

Use of personal protective equipment should be documented. Observations of work practices can help explain differences between results for workers performing the same task. An air-sampling worksheet can be developed to help prompt such notes (see Figure 15–7).

**How many samples to take.** There is no predetermined number of samples that must be taken in order to adequately evaluate a worker's exposure. The number of samples to be taken depends on the purpose of the sampling, the number of different tasks a worker performs in a given day, and the variability inherent in the contaminant generation process.

There are guidelines in *The Occupational Exposure Sampling Strategy Manual* (USDHEW, NIOSH, 1977) that can help in this decision-making process. They direct the industrial hygienist to ask pertinent questions such as whether the PEL or STEL is to be considered a ceiling value, not to be exceeded in any workday. Then, if this is the case, three nonrandom, worst-case exposure periods should be sampled. And if none of the results exceeds the PEL or STEL, one can be personally confident that they are not normally exceeded. However, one cannot be statistically confident of these results because the sampling periods were not randomly chosen. Without this random selection, a confidence limit for a worst-case exposure estimate cannot be computed because the results are not statistically representative of the entire exposed group.

**When to stop monitoring.** For the chemicals it regulates, OSHA requires that monitoring be conducted on a routine basis; the frequency depends on the substance and

## Air Sampling Worksheet

| Sample ID | Employee/Job Description | Flow Rate(Start/Stop) | Time (Start/Stop) | Time |
|-----------|------------------------|----------------------|-------------------|------|
|  |  |  |  |  |
|  |  |  |  |  |
|  |  |  |  |  |
|  |  |  |  |  |
|  |  |  |  |  |
|  |  |  |  |  |

**Process Description:**

**Engineering Controls:**

**Work Practice Controls:**

**Ventilation Measurements:**

**Personal Protective Equipment Used:**

**Figure 15–7.** Example of an air-sampling worksheet. (Printed with permission from Nancy Comeau, Massachusetts Division of Occupational Hygiene.)

the results from the initial or most recent monitoring. For example, monitoring for formaldehyde can be terminated if results from two consecutive sampling periods, taken at least 7 days apart, show that employee exposure is below both the action level and the STEL. Any change in process or engineering controls requires additional sampling to assess the effects of the change.

If initial sampling results are low, it is not necessary to repeat routine monitoring of employee exposure, as long as monitoring of other factors crucial to the overall health and safety program continues. Areas of interest should include the adequacy of engineering controls, work practices, the use of personal protective equipment, and training in all of these aspects. Documentation of this oversight should be part of any effective health and safety management program. This continued monitoring also serves to meet the requirements of many OSHA standards, including the Hazard Communication Standard (29 *CFR* 1910.1200), Occupational Exposure to Hazardous Chemicals in Laboratories (29 *CFR* 1910.1450), the Respirator Standard (29 *CFR* 1910.134), and other, more specific ones that may apply to a given workplace.

**Who should conduct sampling.** Although the concept of air sampling and the use of air-monitoring devices may at first appear to be simple, there are many considerations that must be balanced when devising a sampling strategy and interpreting the results, and it is often previous experiences that allow a final judgment to be made. It is therefore crucial that those conducting the sampling be adequately trained and supervised by a professional industrial hygienist. They must be cognizant of the potential for error and ensure proper calibration, maintenance, and use of sampling equipment. They must be familiar with potential problems and be available to resolve them if they occur. They must be aware of the limitations of sampling alone, know how to integrate observation and interviews with quantitative measurements, and know when it is not necessary to sample. The initial sampling strategy may lead to further questions or contradictions and significantly alter the overall plan. A comprehensive evaluation of the workplace depends on the judgment of the industrial hygienist.

The title *Certified Industrial Hygienist* (CIH) indicates that the professional has at least five years experience in the field of industrial hygiene, is currently in active practice, has met certain educational requirements, and has passed the series of professional exams required by the American Board of Industrial Hygiene (ABIH). Certification rosters are maintained by the ABIH. Membership in other professional organizations, such as the AIHA and ACGIH, indicates active participation in the current field of industrial hygiene but does not guarantee the CIH title. Both the AIHA and ACGIH maintain rosters of their members.

**Required accuracy and precision.** Although the word *sampling* is commonly used, its full implications are not always realized. To sample means to measure only part of

the environment, and, from the measurements taken, infer conclusions about the whole. In all sampling methodologies, there are both systematic and random errors to consider that can affect the interpretation of results and, therefore, final judgment about the work environment as a whole. Any exposure average calculated from air-sampling measurements is only an estimate of the true exposure. It is important to recognize, preferably in advance, where possible sources of error lie; to eliminate or control them to the degree possible; and to account for them in the interpretation of results.

*Accuracy.* Accuracy concerns the relationship between a measured value and the true value. For a measurement to be accurate, it must be close to the true value.

*Precision.* Precision is the degree of agreement among results obtained by repeated measurements under the same conditions and under a given set of parameters. It is possible for a measurement to be precise but not accurate, and vice versa (Figure 15–8).

Accuracy is affected by controllable sources of error. These are called determinate or systematic errors and include method error, personal error, and instrument error. Incorrect calculations, personal carelessness, poorly calibrated equipment, and use of contaminated reagents are examples of systematic error. They contribute a consistent bias to the results that renders them inaccurate. Where possible, these must be identified before sampling is performed, and eliminated or controlled.

Precision is affected by indeterminate or random errors, which cannot be controlled. These include intra- or inter-day concentration fluctuations, sampling equipment variations such as random pump flow fluctuations, and analytical method fluctuations such as variation in reagent addition or instrument response. These factors cause variability among the sample results. Statistical techniques are used to account for random error. For example, increasing the number of samples taken minimizes the effect of random error.

In several of the OSHA substance-specific standards, accuracy ranges for the sampling methods are specified for both the PEL and the STEL. For example, the Ethylene Oxide Standard (29 *CFR* 1910.1047) requires that a sampling method with accuracy to a confidence level of 95 percent (within 25 percent) be obtained for airborne concentrations of ethylene oxide at the 1.0 ppm PEL and within 35 percent at the action level of 0.5 ppm.

In order to ensure accuracy and precision, the following guidelines should be used:

■ Manufacturers' data for direct-reading instruments should be obtained whenever possible, stating the accuracy and precision of their method.

■ A calibration schedule should be established and documented for all sampling equipment.

■ The *NIOSH Manual of Analytical Methods* should be consulted for accuracy and precision of the methods

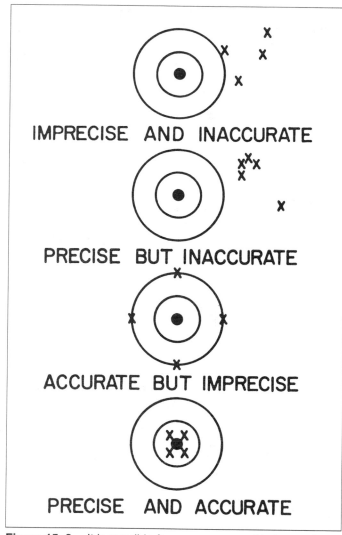

**IMPRECISE AND INACCURATE**

**PRECISE BUT INACCURATE**

**ACCURATE BUT IMPRECISE**

**PRECISE AND ACCURATE**

**Figure 15–8.** It is possible for a measurement to be precise but not accurate, and vice versa. (Reprinted with permission from Powell CH & Hosey AD, eds. *The Industrial Environment—Its Evaluation and Control,* 2nd ed., U.S. Public Health Services Pub. no. 614, 1965.)

chosen. When reporting the results of the sampling, the NIOSH sampling method followed should be cited.

■ Only laboratories that participate in industrial hygiene quality control programs, such as the one conducted by the AIHA, should be used.

In addition, to ensure compliance or violation, OSHA compliance officers use one-sided confidence limits (upper and lower confidence limits, UCL and LCL) whenever sampling is performed (OSHA Technical Manual (OSHA), 1991, Appendix 1-F). This practice recognizes that the sample measured on the employee is rarely the same as the "true" exposure, because of sampling and analytical errors (SAEs). The UCL and LCL incorporate these error factors statistically in order to obtain the lowest (LCL) and the highest (UCL) value that the true exposure could be, within a 95-percent confidence interval. The

UCL and LCL are called one-sided limits because they are used by both OSHA and employers to ensure that the true exposure lies on one side of the OSHA permissible exposure limit (PEL), either above or below it.

For example, if neither the measured results nor its UCL exceed the PEL, then one can be 95 percent confident that the exposure does not exceed the PEL. On the other hand, if both the measured exposure and its LCL exceed the PEL, then one can be 95 percent confident that the exposure exceeds the PEL, and a violation is established. Also listed in Appendix 1-F are grayer areas of evaluation; for example, when the UCL of an exposure exceeds the PEL but the measured exposure does not. OSHA offers guidance in these instances, including suggesting that further monitoring be conducted.

In order to compute the UCL and the LCL, the coefficient of variation (CV) for each analytical method must be computed. These can also be found in the *NIOSH Manual of Analytical Methods.*

$$CV = 100\frac{sd}{m} \tag{1}$$

where   sd = standard deviation of the method
   $m$ = mean (or analytical result)
   100 = factor to convert from fraction to percent

SAEs are often listed in OSHA report forms, but can be derived from the $CV_{total}$:

$$SAE = 1.645(CV_{total}) \tag{2}$$

where $CV_{total}$ = the coefficient of variation of the sampling method plus the coefficient of variation of the analytical method.

In general, the formula for the LCL at the 95th percentile level is

$$LCL\,(95\%) = \frac{x_{mean}}{PEL} - 1.645\frac{CV_{total}}{\sqrt{n}} \tag{3}$$

where   LCL (95%) = lower confidence limit at the 95th percentile
   $x_{mean}$ = average airborne concentration
   $CV_{total}$ = coefficient of variation including sampling error
   $n$ = number of data points to determine $x_{mean}$
   1.645 = appropriate factor from large sample statistics

In a similar fashion, the general formula for the UCL at the 95th percentile level is

$$UCL\,(95\%) = \frac{x_{mean}}{PEL} + 1.645\frac{CV_{total}}{\sqrt{n}} \tag{4}$$

OSHA uses simplified versions of the above formulas and distinguishes between three types of samples: full-period, continuous single samples; full-period consecutive samples; and grab samples. For a complete discussion of the calculations, refer to Appendix 1-F of the *OSHA Technical Manual*.

## Example

A charcoal tube and personal sampling pump were used to sample for xylene for an 8-hour period. The laboratory reported results of 105 ppm of xylene. The PEL for xylene is 100 ppm. The SAE for the sampling and analytical method is 0.10.

## Solution

The steps required to calculate the UCL and the LCL for this full-period, single sample are as follows:

1. Determine the standardized concentration, $Y$:

$$Y = \frac{X}{PEL} \tag{5}$$

where $X$ is the full-period sampling result. Therefore, for our example,

$$Y = \frac{105}{100} = 1.05 \tag{6}$$

2. Compute the UCL (95%) and the LCL (95%):

$$UCL\ (95\%) = Y + SAE \tag{7}$$

$$LCL\ (95\%) = Y - SAE \tag{8}$$

Therefore, for our example,

$$UCL = 1.05 + 0.10 = 1.15 \tag{9}$$

$$LCL = 1.05 - 0.10 = 0.95 \tag{10}$$

3. When the UCL ≤ 1 a violation does not exist, according to OSHA. When the LCL > 1 a violation exists, according to OSHA. If the LCL ≤ 1 and UCL > 1, the result is classified as a possible overexposure. In our example, because the LCL ≤ 1 and the UCL > 1, a possible overexposure exists.

## ■ Industrial Hygiene Calculations

### Gases and Vapors

Calculations for gas and vapor concentrations are based on the gas laws. Briefly, these are as follows

- The volume of gas under constant temperature is inversely proportional to the pressure: $P_1 V_1 = P_2 V_2$.
- The volume of gas under constant pressure is directly proportional to the Kelvin temperature, which is based

on absolute zero. ( 0 C = 273 K). The Rankine temperature scale is also used, where 0 C = 492 R, or degrees R = degrees F + 460 .

$$\frac{V_1}{T_1} = \frac{V_2}{T_2} \tag{11}$$

- The pressure of a gas of a constant volume is directly proportional to the Kelvin (or Rankine) temperature:

$$\frac{P_1}{T_1} = \frac{P_2}{T_2} \tag{12}$$

and $PV = nRT$.

Thus, when measuring contaminant concentrations, it is necessary to know the atmospheric temperature and pressure under which the samples were taken. At standard temperature (0 C) and pressure (760 mm Hg) (STP), 1 g-mol of an ideal gas occupies 22.4 liters. If the temperature is increased to 25 C (normal room temperature) and the pressure is the same, then 1 g-mol occupies 24.45 liters.

$$(22.4\ \text{liters})\frac{273 + 25}{273} = 24.45 \tag{13}$$

The concentration of gases and vapors is usually expressed in parts of contaminant per million parts of air, or parts per million (ppm).

$$ppm = \frac{\text{Parts of contaminant}}{\text{Million parts of air}} \tag{14}$$

This is a volume-to-volume relationship. Equivalent parts per million expressions include

$$\frac{\text{liters}}{10^6\ \text{liters}} = \frac{\text{centimeter}^3}{10^6\ \text{centimeter}^3} = \frac{10^{-3}\ \text{l}}{10^3\ \text{l}} =$$

$$\frac{\text{milliliters}}{\text{meter}^3} = \frac{\text{feet}^3}{10^6\ \text{feet}^3} \tag{15}$$

Sometimes it is necessary to convert milligrams per cubic meter (mg/m$^3$), a weight-per-unit volume ratio, into a volume-per-volume ratio. To begin, milligrams per cubic meter must be converted to millimoles per cubic meter and to milliliters per cubic meter, or parts per million. It is helpful in making this conversion to use dimensional analysis.

$$\left(\frac{\text{mg}_x}{\text{m}^3\ \text{air}}\right)\left(\frac{\text{mmol}_x}{\text{mg}_x}\right)\left(\frac{24.45\ \text{ml}_x}{\text{mmol}_x}\right) = \frac{\text{ml}_x}{\text{m}^3\ \text{air}} = ppm \tag{16}$$

At room temperature, to convert from ppm to mg/m$^3$, a similar conversion can be performed:

$$\frac{\text{mg}}{\text{m}^3} = \frac{\text{molecular weight}}{24.45\ \text{at 25 C}} ppm \tag{17}$$

Another method to predict gas or vapor concentration in parts per million is the partial pressure method. By dividing the vapor pressure of the material in question by the barometric pressure, the resultant percent fraction can then be multiplied by one million ($10^6$) in order to give a volume percent in ppm.

$$\frac{\text{vapor pressure of one constituent}}{\text{total barometric pressure}} 10^6 \qquad (18)$$

$$= \text{ppm of constituent}$$

## Example

Given the concentration of a vapor at STP in grams per liter, convert this to parts per million (ppm).

## Solution

Given that the gram-molecular volume at STP (0 C and 760 mm Hg) is 22.4 l, and that molecular weight is g/mol, the concentration of vapor at STP is

$$\frac{\text{grams of vapor}}{\text{liters of vapor}} = \left(\frac{g}{\text{mole}}\right)\left(\frac{\text{mole}}{1}\right)$$

$$= \frac{\text{molecular wt (g)}}{22.4 \text{ (l)}} \qquad (19)$$

Rearranging terms,

$$\text{liters of vapor} = \frac{(\text{grams of vapor})(22.4)}{\text{molecular wt}} \qquad (20)$$

$$\text{ppm} = \frac{\text{parts of vapor}}{1,000,000 \text{ parts of air}} = \frac{\text{liters of vapor}}{10^6 \text{ l of air}} \qquad (21)$$

Substituting liters of vapor from equation (21) into equation (22),

$$\text{ppm} = \frac{\dfrac{(\text{grams of vapor})(22.4)}{\text{molecular wt (g)}}}{10^6 \text{ l of air}}$$

$$= \frac{\dfrac{(10^3 \text{ mg of vapor})(22.4)}{\text{molecular wt (g)}}}{10^6 \text{ l}} \qquad (22)$$

Given that $10^6 \text{ l} = 10^3 \text{ m}^3$,

$$\text{ppm} = \frac{(10^3 \text{ mg})(22.4)}{(10^3 \text{ m}^3)(\text{molecular wt of vapor})}$$

$$= \left(\frac{\text{mg}}{\text{m}^3}\right)\frac{22.4}{\text{molecular wt}} \qquad (23)$$

For some chemicals, the analytical method requires the collection of material into a fixed volume of absorbing or reacting solution. The laboratory to which the sample is sent first analyzes the concentration of contaminant in the collection medium, then multiplies the volume of solution by the contaminant concentration and reports the total amount of contaminant collected during the sampling period. This can be converted to air concentration by dividing the total amount of contaminant sampled by the total amount of air collected.

## Example

At 25 C and 755 mm Hg, 15 l of air is bubbled through 30 ml of a solution that has 100 percent collection efficiency for HCl ( molecular weight = 36.5 ). The analytical laboratory reports the solution concentration as 15 μg/ml. What is the air concentration of HCl in ppm?

## Solution

First, the total amount of HCl is

$$\frac{15 \text{ μg}}{\text{ml}}(30 \text{ ml}) = 450 \text{ μg} \qquad (24)$$

Correcting for temperature and pressure in micromoles (μmol), the volume of 1 μmol of HCl is as follows:

$$1 \text{ μmol} \times 22.4 \times \frac{298}{273} \times \frac{760}{755} = 24.6 \text{ μl} \qquad (25)$$

Finally, the air concentration sampled in ppm is

$$\frac{450 \text{ μg HCl}}{15 \text{ l of air}} \times \frac{\text{μmol HCl}}{36.5 \text{ μg HCl}} \times \frac{24.6 \text{ μl HCl}}{\text{μmol HCl}}$$

$$= \frac{11,070 \text{ μl HCl}}{547.5 \text{ l of air}} = 20.22 \text{ ppm} \qquad (26)$$

Another useful equation to derive is the vapor concentration of a given amount of material in a chamber or a room, given the following:

$V_T$ = chamber volume in liters

MW = molecular weight of a substance, in g/mol

$T$ = absolute temperature in degrees Kelvin (K − C + 273)

$P$ = pressure in mm Hg

ρ = density, in g/ml

$V_x$ = volume of material in chamber or room, in ml

$C$ = concentration, in ppm

To find liters of pure vapor

$$\frac{(V_x \text{ ml})(\rho)(22.4 \text{ l/mol})}{\text{molecular wt of material}}\left(\frac{T}{273}\right)\left(\frac{760}{P}\right) \qquad (27)$$

$$C = \left(\frac{\text{liters of pure vapor}}{V_T}\right)10^6 \text{ parts of air}$$

$$= \frac{(V_x)(\rho)\left(\dfrac{22.4 \text{ l}}{\text{g-mol}}\right)\left(\dfrac{\text{g-mol}}{\text{MW}}\right)\left(\dfrac{T}{273}\right)\left(\dfrac{760}{P}\right)}{V_T \text{ l}} \times 10^6$$

$$= \frac{(V_x)(\rho)\left(\dfrac{22.4}{\text{MW}}\right)\left(\dfrac{T}{273}\right)\left(\dfrac{760}{P}\right)}{V_T} \times 10^6 \qquad (28)$$

One can also calculate the volume of liquid necessary to produce a desired concentration in a given volume at room temperature and standard pressure:

$$V_x = \frac{C \times \text{MW} \times 273 \times P \times V_T}{\rho \times 22.4 \times T \times 760 \times 10^6} \quad (29)$$

## Example

How much acetone (MW = 58.08 g/mol; density = 0.7899 g/ml) is needed to generate a concentration of 200 ppm in a 20-l container at 25 C and 740 mm Hg?

## Solution

$$V_x = \frac{(200)(58.08)(273)(740)(20)}{(0.7899)(22.4)(298)(760)} \times \frac{1}{10^6}$$

$$= 0.012 \text{ ml} \quad (30)$$

## Vapor Equivalents

When a liquid is released into a space of known dimensions, it is useful to determine the volume it will occupy when evaluating potential exposures from this release. The following formula is often helpful, because it establishes the amount of pure vapor formed at sea level by the complete evaporation of a know volume or weight of a liquid into an area, based on the following assumptions:

liters/mole of vapor at STP = 22.4

grams/pound = 453.6

liters/cubic foot = 28.32

grams/gram-mole = MW

$$\frac{\text{cubic feet of vapor}}{\text{pound of liquid}} = \left(\frac{\text{ft}^3}{\text{l}}\right)\left(\frac{\text{l}}{\text{mol}}\right)\left(\frac{\text{mol}}{\text{g}}\right)\left(\frac{\text{g}}{\text{lb}}\right)$$

$$= \left(\frac{1 \text{ ft}^3}{28.3 \text{ l}}\right)\left(\frac{22.4 \text{ l}}{\text{mol}}\right)\left(\frac{\text{mol}}{\text{g-MW}}\right)\left(\frac{453.6 \text{ g}}{\text{lb}}\right) = \frac{359}{\text{MW}} \quad (31)$$

This can be calculated for different temperatures and pressures.

## Example

At 70 F, what volume would one pound of toluene (MW = 92) occupy?

## Solution

$$\frac{\text{cubic feet}}{\text{pound}} \text{ at 70 F} = \frac{(530 \text{ R})(359)}{(492 \text{ R})(\text{mol wt})}$$

$$= \frac{387}{92} = 4.163 \text{ ft}^3 \quad (32)$$

Note that the Rankine scale was used here, and that °R = °F + 460. Therefore, 70° = 530°R, and 0°C = 32°F = 492°R. However, it must be noted that quantities of liquids are often stated as volumes, for example in pints or liters, and that, in order to use equation (31), liters or pints must first be converted into pounds.

## Example

A 1-pint container of toluene breaks in a room 50 feet by 100 feet by 15 feet. Assuming complete evaporation and no ventilation, what would you expect the concentration of toluene to be in the room, assuming the following:

$T = 70°F$

mass of water = 1.041 pounds/pint

specific gravity (sp gr) of toluene = 0.866 (the ratio of the mass of toluene to the mass of water at that temperature)

$$\left(\frac{\text{cubic feet}}{\text{pound}}\right)\left(\frac{\text{pound}}{\text{pint}}\right) = \frac{\text{cubic feet}}{\text{pint}} \quad (33)$$

$$= \frac{(387)(1.041)(\text{sp gr})}{\text{molecular wt}} = \frac{(403)(0.866)}{92} = 3.79 \text{ ft}^3$$

and the room volume is (50 ft)(100 ft)(15 ft) = 75,000 ft³. The concentration is then

$$\frac{3.79 \text{ cubic feet of toluene}}{75,000 \text{ cubic feet air}} \times 10^6 = 50.53 \text{ ppm} \quad (34)$$

## Example

A half-pound cylinder of chlorine fell and broke in a closed room 60 feet by 45 feet by 15 feet. What is the concentration of chlorine in ppm?

## Solution

The room volume is

$$(60 \text{ ft})(45 \text{ ft})(15 \text{ ft})\left(\frac{1 \text{ m}^3}{35.31 \text{ ft}^3}\right) = 1,147 \text{ m}^3 \quad (35)$$

Therefore, the concentration of chlorine in the room is

$$(0.5 \text{ lb Cl}_2)\left(453.6\frac{\text{g}}{\text{lb}}\right)\left(\frac{\text{mol}}{71 \text{ g}}\right)\left(\frac{24.45 \text{ l}}{\text{mol}}\right)$$

$$\left(\frac{\text{m}^3}{10^3 \text{ l}}\right)\left(\frac{1}{1,147 \text{ m}^3}\right) \times 10^6 = 68.2 \text{ ppm} \quad (36)$$

## Weight-per-Unit Volume

When a contaminant is released into the atmosphere as a solid or liquid and not as a vapor—for example as a dust, mist, or fume—its concentration is usually expressed as a weight per volume. Outdoor air pollutants and stack effluents are usually expressed in grams, milligrams, or micrograms per cubic meter of air (g, mg, or $\mu g/m^3$), ounces per thousand cubic feet ($oz/1,000\ ft^3$), pounds per thousand pounds of air (lb/1,000 lb), or as grains per cubic foot (gcf).

## Time-Weighted Average (TWA) Exposure

The time-weighted average exposure evolved as a method to calculate daily or full-shift average exposures, given that employees' job tasks may vary during a day and that facility operating conditions may also vary. In typical work environments, workers may experience several different, short-term exposures to the same material. By taking a time-weighted average of these exposures, the industrial hygienist can estimate or integrate the short-term measurements into an 8-hour exposure estimate and compare this to the relevant health and safety regulations or information. The TWA is determined by the following formula, where

$C$ = concentration of the contaminant

$T$ = time period during which this concentration was measured

$$TWA = \frac{C_1 T_1 + C_2 T_2 + C_n T_n}{8\ hrs} \qquad (37)$$

The TWA is usually expressed in ppm or in $mg/m^3$. Because OSHA's PELs and the ACGIH's TLVs are both based on an 8-hour workday, the denominator in this formula is usually 8 hours. However, any TWA can be determined, using the following formula:

$$\frac{\sum\limits_{i=1}^{n}(T_i)(C_i)}{T_{total}\ work\ time} = TWA \qquad (38)$$

where $i$ is an increment of time and $C$ is the concentration measured during that time. In this way, sequential incremental measurements can be made, allowing analysis of short-term exposures at the same time as a longer TWA is being computed. The total time covered by the samples should be as close to the total exposure time as possible.

### Example

A TWA of a foundry worker's exposure to particulates can be evaluated by the following series of short-term samples:

| Sample Number | Time |
|---|---|
| 1 | 7:00 A.M. to 8:00 A.M. |
| 2 | 8:00 A.M. to 9:30 A.M. |
| 3 | 9:30 A.M. to 11:00 A.M. |
| 4 | 11:00 A.M. to 1 P.M. |
| | (turned off and covered during 30-min lunch) |
| 5 | 1:00 P.M. to 3:30 P.M. |

The measurement obtained is a full period consecutive-sample measurement because it covers the entire time period applicable to the PEL or TLV.

In some cases, because of limitations in measurement methodology—for example, direct-reading instruments or charcoal tubes—it is impossible to collect consecutive samples whose total sampling duration equals that of the required time period stated in the relevant standard. In these cases, the grab methods are used for time periods that are felt to be representative of the entire workshift.

### Example

It is necessary to use charcoal tubes to estimate an employee's exposure to chloroform. Each charcoal tube is limited to 60 minutes' collection time. Out of the possible eight samples that could have been taken, only six were collected. The following results were obtained:

| Sample Number | Results (ppm) |
|---|---|
| 1 | 55 |
| 2 | 65 |
| 3 | 55 |
| 4 | 60 |
| 5 | 45 |
| 6 | 60 |

### Solution

The 6-hour TWA for these exposures is

$$\frac{1}{360\ min}[\,(60\ min)(55\ ppm) + (60)(65)$$

$$+ (60)(55) + (60)(60) + (60)(45) + (60)(60)\,]$$

$$= 57\ ppm \qquad (39)$$

If there is not much variation in the levels of air contamination measured, and it is certain that the employee's entire workday is spent in one area, then it is probably acceptable to assume that this represents an 8-hour TWA. If there is significant variation, however, resampling should be done for the entire 8-hour day.

### Example

An employee spends 4 hours of an 8-hour shift in an area where measured CO air concentrations remain fairly constant at 50 ppm. For the remaining 4 hours, the employee works in an area where there is no measurable CO in the air. What is the employee's 8-hour TWA?

### Solution

$$\frac{(4\ hr)(50\ ppm) + (4\ hr)(0\ ppm)}{8\ hr} = \frac{200\ ppm \cdot hr}{8\ hr}$$

$$= 25\ ppm \qquad (40)$$

## Example

A machinist works from 7:00 A.M. to 4:00 P.M. tending an automatic screw machine. The following levels of oil mist were measured:

| Time | Average Level of Oil Mist (mg/m³) | Time | Average Level of Oil Mist (mg/m³) |
|---|---|---|---|
| 7:00–8:00 | 0 | 11:00–12:00 | 2.0 |
| 8:00–9:00 | 1.0 | 12:00–1:00 | 0.0* |
| 9:00–10:00 | 1.5 | 1:00–3:00 | 4.0 |
| 10:00–11:00 | 1.5 | 3:00–4:00 | 5.0 |

*lunch period, no exposure

## Solution

The TWA of the machinist's exposure to oil mist is calculated as follows:

$$\frac{\overset{i=8}{\underset{i=1}{\sum}} (T_i)(C_i)}{T_{total} = 8\ hr} = TWA \tag{41}$$

**Time (hrs) × Concentration (mg/m³)**

| | | |
|---|---|---|
| (1) | (0) | = 0 |
| (1) | (1) | = 1 |
| (1) | (1.5) | = 1.5 |
| (1) | (1.5) | = 1.5 |
| (1) | (2.0) | = 2.0 |
| (1)* | (0.0) | = 0.0 |
| (2) | (4.0) | = 8.0 |
| (1) | (5.0) | = 5.0 |
| | | 19.0 (hr)(mg/m³) |

$$\frac{\overset{i=8}{\underset{i=1}{\sum}} (T_i)(C_i) = (19.0\ hr)(mg/m^3)}{8\ hr} = 2.38\ mg/m^3 \tag{42}$$

## Example

An employee is exposed to an average level of 100 ppm of xylene for 10 minutes out of every hour; during the remaining 50 minutes of each hour, there is no exposure to xylene. What is the TWA for xylene for this employee?

## Solution

Because there are 8 hours in a workday, each of which includes 10 minutes' exposure to 100 ppm and 50 minutes' exposure to 0 ppm, an 8-hr TWA can be calculated as follows:

$$\frac{(8)(10\ min)(100\ ppm) + (8)(50\ min)(0\ ppm)}{480\ min}$$

$$= \frac{8,000\ min \cdot ppm}{480\ min} = 16.7 = 17\ ppm \tag{43}$$

## Example

An employee is exposed to the same material at two work locations during an 8-hour shift. Monitoring of this worker's exposure was conducted by taking grab samples at each of the locations. The following results were obtained:

| Operation | Duration | Sample | Results (ppm) (5-min sample) |
|---|---|---|---|
| Cleaning room | 8:00–11:30 A.M. | A | 150 |
| | | B | 120 |
| | | C | 190 |
| | | D | 170 |
| | | E | 210 |
| Print shop | 12:30–4:30 | F | 90 |
| | | G | 70 |
| | | H | 120 |
| | | I | 110 |

## Solution

The average exposure ($C_2$) in the cleaning room:

$$C_i = \frac{120 + 150 + 170 + 190 + 210}{5} = 168\ ppm \tag{44}$$

The average exposure in the print shop:

$$C_2 = \frac{70 + 90 + 110 + 120}{4} = 98\ ppm \tag{45}$$

Thus the TWA exposure for the 8-hour shift, excluding 60 minutes for lunch, is as follows:

$$TWA = \frac{(168\ ppm)(3.5\ hr) + (98\ ppm)(4.0\ hr)}{8}$$

$$= 122.5\ ppm \tag{46}$$

## Example

As part of her job, a hospital central supply worker unloads sterilized materials from an ethylene oxide (EtO) sterilizer. She does this 4 times per 8-hour shift, it takes 15 minutes each time, and she has no other exposure to EtO during the shift. The 8-hour PEL for EtO is 1 ppm; the 15-minute excursion limit is 5 ppm. The following 15-minute sampling results were obtained: 4.8, 3.5, 4.9, and 3.4. None of the results exceeded the 5 ppm excursion limit. What is the 8-hr PEL for this worker?

## Solution

$$\frac{1}{480\ min}[(15\ min)(4.8\ min) + (15)(3.5)$$

$$+ (15)(4.9) + (15)(3.4) + (420\ min)(0\ ppm)]$$

$$= 0.52\ ppm \tag{47}$$

## Excursions

TWA concentrations imply fluctuations in the level of airborne contaminant. Excursions above the TLV are permissible if equivalent excursions below the TLV occur. The TLV booklet stipulates that short-term exposures may exceed three times the TLV for no more than a total of 30 minutes during the workday; under no circumstances should exposures exceed five times the TLV. This stipulation is valid if TLV-TWA is not exceeded. In some cases, a specific short-term exposure limit (STEL) has been established, for example, for formaldehyde and ethylene oxide.

## ■ Interpretation of Results

Interpretation of the results obtained from sampling is the final step in evaluating the environment to which a worker is exposed. The chemicals monitored, the sites chosen for the sampling, and the timing of the monitoring all reflect the industrial hygienist's best judgment about which exposures might be significant. Potential sources of error and the limitations of the sampling and analytical methods have been taken into consideration. How the results are interpreted depends on the original purpose of the sampling: whether it was for compliance with regulations, problem investigation, determination of the adequacy of control measures, or development of baseline assessment and/or comprehensive health and safety plan.

### Comparison with Standards and Guidelines

The first step in evaluating sampling results is to compare them with the relevant standards and guidelines. The legally enforceable maximum allowed exposures in general industry are the OSHA permissible exposure limits, which have been determined for over 400 air contaminants and are listed in three tables in the *Code of Federal Regulations* (29 *CFR* 1910.1000). Additional comprehensive standards have been promulgated for other chemicals. Sampling results greater than the PEL and its lower confidence limit can result in citations and fines.

Because of the role of sampling results in legal proceedings, they must be analyzed in a cookbook fashion. For example, unless documentation exists that exposure levels are constant, any work time for which no sampling was conducted must be considered as unexposed time, and a zero is factored in any calculation of the time-weighted average. Consider, for example, the sampling results for chloroform presented in equation (39). The time-weighted average calculated for the six hours sampled was 57 ppm. The PEL for chloroform is 2 ppm. The ACGIH TLV-TWA is 10 ppm. Clearly, the sampling results exceed both limits. NIOSH identifies chloroform as an occupational carcinogen, and therefore recommends that exposure be kept to the lowest feasible limit. NIOSH has not identified thresholds for carcinogens that will protect 100 percent of the population. A situation like the one in the example, that would result in such a high concentration of an occupational carcinogen, would require immediate action to prevent continued overexposure.

If the solvent measured had been methyl isobutyl ketone (hexone), which has a PEL of 50 ppm and a STEL of 75 ppm, and the 6-hour TWA had also been 57 ppm, then the interpretation of the results would be different. At first, this might also appear to exceed the 8-hour PEL. However, there is an additional consideration in this case. Two hours of the employee's workday had not been sampled. If there is no documentation to prove that he was similarly exposed during the remaining time, a 0 ppm concentration could be factored into the 8-hour PEL calculation:

$$\frac{(57\ ppm)(360\ min) + (0\ ppm)(120\ min)}{480\ min}$$

$$= 43\ ppm \tag{48}$$

The sampling results remain the same, but the interpretation has changed. This result, 43 ppm, is in compliance with the PEL. Such a result would not lead to a citation for violation of 29 *CFR* 1910.1000, but it can be interpreted as a significantly high exposure to a volatile solvent, which should be controlled. Such an exposure has the potential to harm the respiratory system, eyes, skin, and central nervous system. If it is possible that the worker is exposed to methyl isobutyl ketone at some concentration during the time not sampled, it is possible that the sampling omission will allow workers to remain overexposed indefinitely.

The results calculated in the example described in equation (40) can be analyzed in a similar manner. The concentration of carbon monoxide (CO) is in compliance with the OSHA PEL and the NIOSH REL for an 8-hour TWA, despite the fact that during four hours of the day the worker is exposed to 50 ppm. The ACGIH TLV for CO is 25 ppm (this standard has been reduced over the years as the adverse effects of carbon monoxide exposure have been demonstrated at lower levels). At first glance, this result might be considered satisfactory because it does not exceed the OSHA PEL. However, because there is evidence that a lower level is recommended by the ACGIH, other questions might be triggered by these results: Are there excursions during the four hours over the STEL? Do results vary from day to day? Would the results be viewed as acceptable if the worker in this example were pregnant? Is it acceptable to leave the hazard in place and simply rotate different employees into the area, so that no single individual is overexposed, but all of them are exposed for part of the day? The best actions in response to these sampling data would be to identify the source of the CO for the four hours of exposure measured, to evaluate others who may be at risk of exposure, and to attempt to reduce exposure to the lowest feasible level.

Exposure limits can be compared to speed limits. Traveling one mile per hour less than the posted limit does not guarantee safety. In addition, chemical exposure has a cumulative effect if the time between exposures has not been sufficient to allow clearance of the chemical and its metabolites from the body and recovery from the adverse physiological effects.

## Limitations of Standards

Any sampling result that is less than the PEL is considered to be in compliance with the law. This evaluation is often misinterpreted as meaning a clean bill of health. A review of OSHA's sampling results shows that 92 percent of them were in compliance (Senn, 1992), but OSHA estimates that hundreds of thousands of new cases of occupational illness occur annually. There have been many criticisms of the OSHA standards (Castleman & Ziem, 1988; Roach & Rappaport, 1990; Robinson et al., 1991; Tarlau, 1991), including the following:

- They evaluate only inhalation exposures.

- They are often out of date because updates take years and proceed very slowly; for example, by 1987, the ACGIH's TLV list, which in 1968 formed the basis for OSHA's original PELs, contained 168 substances not regulated by OSHA, and had reduced guidelines for an additional 234 substances (Robinson et al., 1991).

- They have been based on inadequate research that fails to consider chronic toxicity data, including immune or endocrine system function, reproductive toxicity, and neurological changes.

- Standards are inadequate to protect employees who become sensitized to chemicals that may cause asthma, dermatitis, and other immunologically mediated effects.

- Standards were often adopted based on epidemiological data on workers who were mainly white and male, excluding analysis of nonwhite and female employees.

- They allow a level of risk not tolerated for general environmental exposure, such as a risk for cancer of one in a thousand compared to one in a million for environmental exposures.

- They fail to account for multiple exposures that are additive or synergistic.

- They offer limits for less than 10 percent of the chemicals in widespread commercial use.

- The sampling results reflect conditions on one day and may miss excursions that occur irregularly or peak exposures occurring only during maintenance, leaks, and emergencies.

- Rather than representing a more scientifically based guideline, at times they represent a political compromise between industry and labor regarding feasibility.

Industrial hygienists sometimes analyze sampling results and conclude that compliance with PELs is not sufficient to guarantee health in the workplace. Where NIOSH RELs and ACGIH TLVs differ from PELs, these guidelines provide additional benchmarks that represent conclusions from research designed to further control exposures.

RELs exist for nearly 200 chemicals; NIOSH tends to propose more conservative exposure limits and has criticized several OSHA PELs as being insufficiently protective (Robinson et al., 1991). Most RELs were developed in the 1970s, and some have been outdated by more recent findings. NIOSH recommends, for example, that exposure to occupational carcinogens be reduced to "the lowest feasible limit," acknowledging the absence of thresholds that will protect 100 percent of the population.

Roach and Rappaport wrote an article in 1990, titled "But they are not thresholds: A critical analysis of the documentation of Threshold Limit Values," in which they analyzed 127 TLVs. They found that the literature cited in the TLV documentation showed that one in seven workers experienced adverse health effects when exposure was limited to concentrations below the TLVs. They observed that "factors other than health appeared to have influence in assignment of particular TLVs...the TLVs represent levels of exposure that were perceived by the committee to be realistic and attainable at the time." Other critics have charged that some TLVs were based on insufficient data on chronically exposed animals, or on analogy to similar chemicals rather than chemical-specific data.

Such deficiencies do not necessarily mean that all TLVs are wrong. Nor do they deny the contribution to worker health that resulted from the TLVs. They do, however, limit the conclusions that can be drawn when comparing air-sampling results with them.

Another concern regards the concept that ACGIH limits are not expected to protect "hypersusceptible" workers, a group described by Mastromatteo (1981), who was a member of the TLV committee for many years, as including those with genetic disorders, nutritional deficiencies, parasitic diseases, or preexisting diseases such as asthma or chronic bronchitis; those consuming alcohol or drugs; and cigarette smokers. Others have stated that large groups may be especially susceptible to hemolytic chemicals or pesticides, and that female workers may react to certain chemical exposures differently than males (Sentes, 1992).

The TLV process has been criticized as unduly influenced by corporations by Castleman and Ziem (1988). They wrote that unpublished corporate communications that were largely unavailable for independent scientific review were influential in developing TLVs for 104 substances. With unverifiable data, it could be argued that TLVs might inadequately protect workers. The Board of Directors of the ACGIH wrote in response (1990),

> Many of the "personal communications" mentioned as a partial basis for some TLVs were from governmental health professionals and not corporations. Realizing the large gap in needed toxicity information for many substances, the ACGIH

made a concerted effort through the Notice of Intended Changes process to obtain information *from all possible sources.*

The ACGIH board also responded to the various criticisms with a defense of their process and intent. They reiterated "that TLVs are developed as guidelines and not for use as legal standards." They noted important improvements that had been made in the TLV procedures, documentation, and research. Some of these included the following:

- Assigning of experienced occupational health professionals to staff the TLV effort

- Development and implementation of conflict of interest procedures

- Development of policies concerning TLV committee meetings with outside groups

- Printing of the last revision date of each TLV in the booklet

- Publication of a booklet listing a number of different recommended exposure guidelines, in addition to the TLVs, such as RELs, PELs, and TLVs, etc.

- Initiating of the yearly publication of full text documentation for all intended changes for public comment

The task of providing guidance regarding risk in the absence of complete information often leads occupational health specialists to settle for existing data and guidelines without analyzing their adequacy. A minimum standard of care in many responsible industries is the maintenance of airborne chemical concentrations below all existing standards, guidelines, and internally generated standards. In addition, when employee health complaints persist when a standard is not exceeded, further investigation and action are often needed to ensure health. The Industrial Hygiene Code of Ethics requires placing employee health first in all considerations.

Some have proposed reliance on the EPA's Integrated Risk Information System (IRIS) data base for standard setting in the workplace. The IRIS data base was developed by the EPA in the late 1980s to early 1990s to systematically review human and animal toxicological data on chemicals of environmental concern. This may be an additional resource in considering worker exposure guidelines that are more inclusive.

### Comparison of Results with Other Data

Sampling may be conducted to investigate a problem or to measure the impact of changes in production processes or control measures. In these cases, current sampling results can be compared to previous results to determine the effectiveness of the new or modified control measure in reducing airborne concentrations. In other cases, when the intent of the sampling is to evaluate the effectiveness of in-place monitors used for regular surveillance, the current data collected should also be compared to previous results.

It can also be helpful to review sampling data from various types of workplaces that are available from NIOSH *Health Hazard Evaluation* reports, in order to evaluate the effectiveness of the current sampling or the health and safety oversight at the worksite in question.

Industrial hygienists are sometimes asked to evaluate the significance of chemical-sampling results for which there are no published standards or guidelines. In these situations, the manufacturer may have developed internal standards for the chemical's use within the company. Lacking any other guidelines, a review of anecdotal reports in the literature, health surveys among those exposed on the job, or a careful consideration of animal toxicology data can be helpful. The $LD_{50}$s (lethal dose for 50 percent of the exposed animal population in question during experiments) can be found in the Registry of Toxic Effects of Chemical Substances (RTECS) data base from NIOSH. (See Appendix A, Additional Resources, for more information.) Animal research that establishes a "no-effect level" (NOEL) is especially useful.

While debate continues over the adequacy of standards and guidelines, industrial hygienists must still conduct air sampling and analyze results, attempting to make the best use of all that is available to them as evaluation tools, including their skills in measurement, observation, interviewing, and communication with employers and employees. In addition, they often consult with other professionals in the fields of occupational medicine, infection control, ventilation, architecture, engineering, health physics, and others to ensure a broad and in-depth analysis of an industrial hygiene problem. This networking with other professionals is invaluable in the evaluation process.

## Summary

There are many factors to consider in evaluating the workplace. Evaluation is a process that must incorporate new research, advances in production technology, a changing work force, and alterations in air-sampling methodology, with the most fundamental concern being the lives and health of workers. This is at the core of the code of ethics for industrial hygienists and forms the basis for all interactions with other health professionals in the practice of prevention of occupational illness and injury.

## Bibliography

American Conference of Governmental Industrial Hygienists. *Threshold Limit Values and Biological Exposure Indices for 1994–1995.* Cincinnati: American Conference of Governmental Industrial Hygienists, 1995.

American Conference of Governmental Industrial Hygienists. *Documentation of Threshold Limit Values, Including Biological Exposure Indices and Issue of Supplements.* Cincinnati: American Conference of Governmental Industrial Hygienists, 1990.

American Conference of Governmental Industrial Hygienists Board of Directors. Threshold Limit Values: A more balanced appraisal. *Appl Occup Environ Hygiene* 5:340–344, 1990.

Borak J. ACGIH's Threshold Limit Values useful, but formulas are still controversial. *Occup Health Saf* 26–27, August 1994.

Brown MP, Froines JR, eds. *Technological Change in the Workplace, Health Impacts for Workers.* Los Angeles: Regents of the University of California, 1993.

Bureau of National Affairs. OSHA advance notice of proposed rulemaking on a generic standard for exposure monitoring. *Occup Saf Health Rptr* 956–959, September 1988a.

Bureau of National Affairs. OSHA advance notice of proposed rulemaking on medical surveillance programs. *Occup Saf Health Rptr* 960–963, September 1988b.

Burgess WA. *Recognition of Health Hazards in Industry, A Review of Materials and Processes,* 2nd ed. New York: John Wiley & Sons, 1995.

California Occupational Health Program. Model Contract for a Lead Medical Program. In *Prevent Lead Poisoning Before It Poisons Your Business.* Berkeley CA: Department of Health Services, 1990.

Castleman BI, Ziem GE. Corporate influence on Threshold Limit Values. *Am J Ind Med* 13:531–559, 1988.

Clayton GD, Clayton FE, eds. *Patty's Industrial Hygiene and Toxicology,* vols. I–IV. New York: John Wiley & Sons, 1994.

Cralley LJ, Cralley LV, eds. *Industrial Hygiene Aspects of Plant Operations,* vol. 1: *Process Flows,* vol. 2: *Unit Operations and Product Fabrication,* vol. 3: *Engineering Considerations in Equipment Selection, Layout, and Building Design.* Melbourne FL: Krieger, 1986.

Elkins HB. Letter to editor in response to corporate Threshold Limit Values. *Am J Ind Med* 14:737–740, 1988.

Gerhardsson L, Attewell R, Chettle DR, et al. In vivo measurements of lead in bone in long-term exposed lead smelter workers. *Arch Environ Health* 48:147–156, 1993.

Hu H, Milder FL, Bulger DE. X-ray fluorescence: Issues surrounding the application of a new tool for measuring burden of lead. *Environ Res* 49:295–317, 1989.

International Labour Organisation (ILO). *Encyclopaedia of Occupational Health and Safety,* 3rd ed. Geneva, Switzerland: The Organisation, 1983.

Iyaniwura TT. In vitro toxicology of organophosphorus pesticide combinations. *J Mol Cell Toxicol* 3:373–377, 1990.

Lauwerys RR, Hoet P. *Industrial Chemical Exposure, Guidelines for Biological Monitoring,* 2nd ed. Boca Raton FL: Lewis Publishers, 1993.

Levy BS, Wegman DH. *Occupational Health: Recognizing and Preventing Work-Related Disease,* 3rd ed. Boston: Little, Brown, 1994.

Lippmann M. Environmental toxicology and exposure limits for ambient air. *Appl Occup Environ Hygiene* 8:847–858, 1993.

Ma T-H, Sandhu SS, Peng Y, et al. Synergistic and antagonistic effects on genotoxicity of chemicals commonly found in hazardous waste sites. *Mutation Res* 270:71–77, 1992.

Mastromatteo, E. On the concept of threshold. *Am Ind Hyg Assoc J* 42:763–770, 1981.

Nicas M. Industrial hygiene sampling strategy. *State of the Workplace,* Internal Bulletin HESIS, California Hazard Evaluation System and Information Service 4:8–11, 1990.

Paul M, ed. *Occupational and Environmental Reproductive Hazards: A Guide for Clinicians.* Baltimore: Williams & Wilkins, 1993.

Proctor NH, Hughes JP, Fischman ML. *Chemical Hazards of the Workplace,* 3rd ed. Florence KY: Van Nostrand Reinhold, 1991.

Rabin R, Brooks DR, Davis LK. Elevated blood lead levels among construction workers in the Massachusetts Occupational Lead Registry. *Am J Public Health* 84:1483–1485, 1994.

Roach SA, Rappaport SM. But they are not thresholds: A critical analysis of the documentation of Threshold Limit Values. *Am J Ind Med* 17:727–753, 1990.

Robinson JC, Paxman DG, Rappaport SM. Implications of OSHA's reliance on TLVs in developing the air contaminants standard. *Am J Ind Med* 19:3–13, 1991.

Rose, VE. ACGIH meritorious achievement award honors Vernon E. Rose, DrPH, CIH. *Appl Occup Environ Hyg* 5:572–575, 1990.

Ryback LP. Hearing: The effects of chemicals. *Otolaryngol Head Neck Surg* 106:677–685, 1992.

Senn EP. "An evaluation of the effectiveness of OSHA's program to control chemical exposures in the workplace." Paper presented at 1992 American Industrial Hygiene Conference and Exposition, Boston, June 1992.

Sentes R. OSHA and standard-setting. *Am J Ind Med* 21:759–764, 1992.

Silverstein M. Analysis of medical screening and surveillance in twenty-one OSHA standards: Support for a generic medical surveillance standard. *Am J Ind Med* 26:283–295, 1994.

Tarlau ES. Playing industrial hygiene to win. *New Solutions* 4:72–81, 1991.

Tarlau ES. Industrial hygiene with no limits. *Am Ind Hyg Assoc J* 51:A9–A10, 1990.

U.S. Congress, Office of Technology Assessment. *Reproductive Health Hazards in the Workplace.* OTA-BA-266. Washington, DC: OTA, December 1985.

U.S. Congress, Office of Technology Assessment. *Preventing Illness and Injury in the Workplace.* OTA-H-256. Washington, DC: OTA, April 1985.

U.S. Department of Health and Human Services, NIOSH. *Pocket Guide to Chemical Hazards.* DHHS (NIOSH) Publication No. 94–116. Cincinnati, OH: NIOSH, June 1994.

U.S. Department of Health and Human Services, NIOSH. "Health Hazard Evaluation, 50 Standiford Street Office Building, Boston, Massachusetts," HETA 86–092–1870. Cincinnati, OH: NIOSH, February 1988.

U.S. Department of Health and Human Services, NIOSH. "Health Hazard Evaluation, Commercial Office Buildings, Boston, Massachusetts," HETA 86–472–1832. Cincinnati, OH: September 1987.

U.S. Department of Health and Human Services, NIOSH. *Occupational Safety and Health Guidance Manual for Hazardous Waste Site Activities,* DHHS (NIOSH) Publication No. 85–115. Washington, DC: GPO, October 1985.

*U.S. Department of Health and Human Services, NIOSH Manual of Analytical Methods,* 4th ed., DHHS (NIOSH) Publication No. 94–113. Washington, DC: GPO, August 1994.

U.S. Department of Health, Education, and Welfare, NIOSH. *Occupational Diseases, A Guide to Their Recognition.* DHEW

(NIOSH) Publication No. 77–181. Washington, DC: GPO, June 1977.

U.S. Department of Health, Education, and Welfare, NIOSH. *Occupational Exposure Sampling Strategy Manual.* Washington, DC: GPO, January 1977.

U.S. Department of Health, Education, and Welfare, NIOSH. *The Industrial Environment—Its Evaluation and Control.* Washington, DC: GPO, 1973.

USDOL-OSHA. 29 *Code of Federal Regulations (CFR),* Parts 1910 and 1926. Washington, DC: GPO, 1994.

USDOL-OSHA. "Framework for a Comprehensive Health and Safety Program in the Hospital Environment." Washington, DC: GPO, 1993.

USDOL-OSHA. *OSHA Technical Manual/*OSHA 3058. Washington, DC: GPO, March 1991.

Weeks JL, Levy BS, Wagner GR, eds. *Preventing Occupational Disease and Injury.* Washington, DC: American Public Health Association, 1991.

# Air Sampling

by Maureen A. Huey, MPH, CIH

INDUSTRIAL HYGIENISTS ARE RESPONSIBLE for the evaluation and control of employee exposure to occupational health hazards. For the evaluation and control of inhalation hazards, hygienists typically compare the measured concentration of an airborne chemical to a recognized exposure limit. Standardized methods for the collection of air samples have been developed to ensure that accurate and meaningful information is collected.

**Types of Air Sampling**
*Personal Versus Area Sampling* ▪ *Grab Versus Integrated Sampling*

**Air-Sampling Instruments**
*Sampling Train*

**Collection Devices for Gases and Vapors**
*Grab Sampling* ▪ *Integrated Air Sampling* ▪ *Passive Monitors*

**Collection Devices for Particulates**
*Filters* ▪ *Cyclones* ▪ *Electrostatic Precipitators* ▪ *Inertial Impactors* ▪ *Impinger* ▪ *Elutriators*

**Suction Pumps**

**Flow-Rate Meters**
*Pressure-Compensating Devices* ▪ *Critical-Flow Orifice*

**Sampling Method**

**Calibration**
*Primary Calibration: Mariotti Bottle* ▪ *Primary Calibration: Spirometer* ▪ *Primary Calibration: Soap-Bubble Meter* ▪ *Secondary Calibration: Wet-Test Meter* ▪ *Secondary Calibration: Dry-Gas Meter* ▪ *Secondary Calibration: Precision Rotameter* ▪ *Calibration Parameters*

**Sampling and Analytical Error**

**Record Keeping**

**Summary**

**Bibliography**

## Types of Air Sampling

Air sampling is used to evaluate employee exposure, assist in the design or evaluation of control measures, and document compliance with government regulations. These sampling objectives define the type of air sampling selected.

### Personal Versus Area Sampling

Personal air sampling is the preferred method of evaluating worker exposure to airborne chemicals. The worker wears a sampling device that collects an air sample. The sampling device is placed as close as possible to the breathing zone of the worker (defined as a hemisphere in front of the shoulders with a radius of 6–9 in.) so the data collected closely approximate the concentration inhaled. (Concentration is equal to the mass of the contaminant collected divided by the volume of air passed through the collection device.)

Area air samples can be used to evaluate background concentrations, locate sources of exposure, or evaluate the effectiveness of control measures. The sampling device is strategically placed in a fixed location in the area of interest.

For example, if a leak is suspected in a process, several area samples taken at key locations could be used to pinpoint the source. In general, this type of sampling is not used to provide an estimate of worker exposure because conditions at the fixed location may not be the same as those experienced by the worker.

## Grab Versus Integrated Sampling

Grab samples are taken to measure the airborne concentration of a substance over a short time period (usually less than 5 min). Personal or area grab samples are used to identify peak or ceiling concentrations.

Grab samples alone are rarely used to estimate an employee's 8-hour time-weighted average exposure. This is because they do not account for the time between samples. However, they can be used as a screening method to determine whether more extensive sampling is needed. For example, if grab samples and observations indicate that the concentration of a chemical is well below the 8-hour time-weighted average exposure limit, then sampling for the full shift *may* not be necessary.

Integrated air sampling is used to estimate a worker's 8-hr or 15-min exposure to a particular substance by collecting one or more personal air samples for the duration of a particular task or workshift. It is called integrated sampling because the result integrates all of the various concentrations to which the worker has been exposed during the sampling period. For example, if the worker was standing downwind from a welding fume plume during part of the day or if exposure during a few hours was particularly heavy due to production problems, these factors would be reflected in the final result.

## ■ Air-Sampling Instruments

There are two categories of air-sampling equipment: direct-reading instruments and sample collection devices. Direct-reading instruments provide an immediate measurement of concentration. These devices are covered in Chapter 17, Direct-Reading Instruments for Gases, Vapors, and Particulates. Sample collection devices collect a sample of air that is subsequently analyzed or weighed at a laboratory. These devices are the focus of this chapter.

### Sampling Train

Air-sample collection devices are made of five basic components: an air inlet orifice, a collection device, an airflow meter, a flow-rate control valve, and a suction pump (see Figure 16–1).

Air enters the sampling train through the orifice and the chemical is collected on a collection medium such as a filter. The airflow rate is set using the rate control valve. Airflows are usually in liters or cubic centimeters of air per

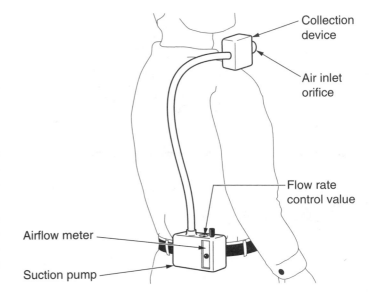

**Figure 16–1.** Components of a typical air-sampling train used to collect airborne particulates.

minute. Many air-sampling pumps have built-in flow-rate meters to visually gauge the flow rate. The suction pump moves air through all of the components of the sampling train.

## ■ Collection Devices for Gases and Vapors

Gases and vapors are formless fluids that completely occupy a space or enclosure. A substance is considered a gas if this is its normal physical state under standard temperature and barometric pressure conditions (70 F and 760 mm Hg). A vapor is the gaseous phase of a substance that under standard conditions exists as a liquid or a solid in equilibrium with its vapor.

In some cases, a chemical may exist as both a gas or vapor and a solid particle at the same time. Polyaromatic hydrocarbons are an example. For such chemicals, collection devices for both the gaseous and solid phases must be used.

Gases and vapors behave similarly. They follow the ideal gas laws in that their volume is affected by changes in temperature and pressure. They mix freely with the general atmosphere and quickly form homogeneous mixtures with other gases.

### Grab Sampling

Although direct-reading devices are usually used for grab sampling, the collection of a known volume of air for subsequent laboratory analysis is also used. Figures 16–2 and 16–3 are examples of the most commonly used collection devices, the evacuated container and the gas-sampling bag.

**Figure 16–2.** An evacuated container is used to collect air for analysis. (Courtesy MDA Scientific.)

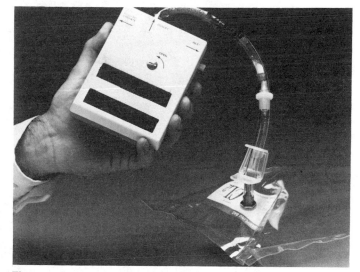

**Figure 16–3.** Portable, battery-operated pumps used to fill flexible plastic gas-sampling bags with air for analysis. (Courtesy MSA.)

The advantages of grab sampling are that it is inexpensive, it is simple to use, and it normally collects 100 percent of the chemical. The disadvantage is that usually it cannot be used to sample reactive gases such as hydrogen sulfide, nitrogen dioxide, and sulfur dioxide unless the samples are analyzed immediately. Reactive gases can react with atmospheric dust particles, other gases, moisture, container sealant compounds, or the container itself, producing erroneous results.

## Integrated Air Sampling

Integrated air sampling involves the extraction of a gas or vapor from a sample airstream followed by laboratory analysis. Two extraction techniques are normally used: absorption and adsorption

**Absorption.** In the absorption technique, a gas or vapor is removed from the airstream as it passes through an absorption liquid. The liquid can be highly soluble and nonreactive with the gas or vapor or it can contain a reactive reagent. Deionized water, for example, is a commonly used absorbing solution for acids because acids are highly soluble in water. A reactive absorbing reagent captures a gas or vapor by quickly reacting with it and creating a more stable compound, which can be analyzed by a laboratory. An example of these is Girard T reagent, which is used to sample for formaldehyde.

Absorption devices include gas wash bottles, spiral absorbers, and fritted bubblers (Figure 16–4). The simplest is the gas wash bottle, which forces air through a nozzle into the absorbing solution. Because absorbing solutions do not collect 100 percent of the gas or vapor passing through, sometimes two impingers are used in series. This increases the total amount of vapor or gas collected. The

spiral absorber forces the air to follow a spiral path through the liquid, which increases the amount of time the air and liquid are in contact with each other. The increased contact time increases the amount of material absorbed. The same concept is used in the fritted bubbler, where many tiny bubbles are formed as air is forced through the fritted surface. This increases the surface area of air in contact with the absorbing liquid. The spiral absorber and the fritted bubbler are used for gaseous substances that are only moderately soluble or that react slowly with the absorbing liquid.

The most commonly used gas wash bottle is the midget impinger. The impinger is placed inside a holster (Figure 16–5) and attached to the worker's shirt collar. During the course of work, the worker may accidentally invert the impinger, causing the absorbing liquid to be drawn into the sampling pump. To avoid this, an empty bottle or impinger is connected to the sampling train to collect spilled liquid. Although this precaution prevents damage to the pump, the recovered liquid is contaminated and cannot be used for an analysis. The use of spillproof midget impingers has minimized this problem. Table 16–A lists selected National Institute for Occupational Safety and Health (NIOSH) impinger sampling methods.

**Adsorption.** Air sampling for insoluble or nonreactive gaseous substances is commonly conducted using tubes filled with a granular sorbent such as activated charcoal or silica gel. The gas or vapor is retained or adsorbed, physically and chemically unchanged, onto the surface of the sorbent for subsequent laboratory extraction and analysis.

Activated charcoal is the most widely used solid sorbent for adsorbing organic vapors. The charcoal most commonly used is from coconut shells. Coconut shell

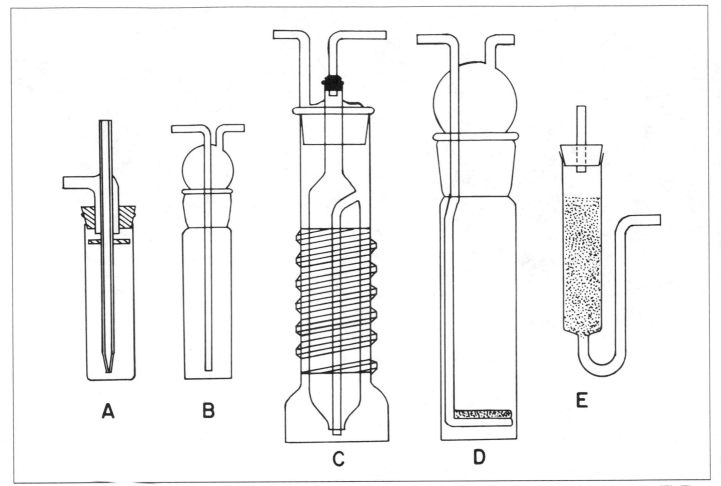

**Figure 16–4.** Basic absorbers are shown: gas washing (A and B), helical (C), fritted bubbler (D), and glass-bead column (E). They provide contact between sampled air and liquid surface for absorption of gaseous contaminants. (Reprinted from Powell CH, Hosey AD (eds.). *The Industrial Environment—Its Evaluation and Control,* PHS Publication No. 614.)

charcoal provides a large adsorptive surface area and is electrically nonpolar, meaning it preferentially adsorbs organic vapors rather than polar molecules such as water vapor. A standard charcoal tube is 7 cm long and 4 mm wide, and is divided into two sections. The first section contains 100 mg of charcoal and a fiberglass, glass wool, or urethane foam plug; the backup section contains 50 mg of charcoal (Figure 16–6). Other sizes are also available.

Although activated charcoal has a large adsorptive capacity, some contaminants invariably pass through the first section. The backup section increases collection efficiency by adsorbing some of the material that was initially missed.

*Breakthrough* describes a condition in which the mass of a collected gas or vapor in the backup section is greater than 10 percent of the mass in the front section. This means that a significant quantity of the contaminant may not have been collected. The calculated concentra-

tion, therefore, is of questionable validity. Sampling methods that specify the maximum sample volumes and recommended flow rates are designed to prevent breakthrough.

Charcoal tubes have a high adsorptive capacity for a large range of organic vapors. They can be used to sample several kinds of vapors at once. However, the analyzing laboratory should be consulted to determine whether there is a limit to the number of organic vapors that can be extracted or whether any of the sampled organic vapors must be collected separately.

Silica gel tubes are used to sample for gases and vapors that cannot be efficiently collected or extracted from activated charcoal. They are constructed in the same manner as charcoal tubes except that an amorphous form of silica is used as the adsorbent material. Silica gel is not as commonly used as activated charcoal because it is electrically polar and tends to attract interfering polar molecules such as water vapor.

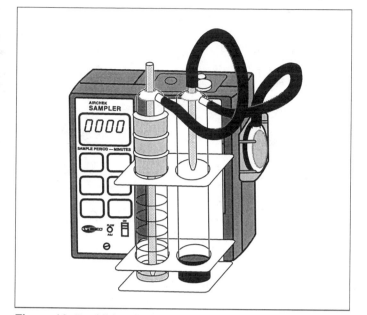

**Figure 16–5.** Midget impingers are sometimes used to collect personal air samples. They are placed in holsters so they can be worn in the worker's breathing zone. (Courtesy SKC, Inc.)

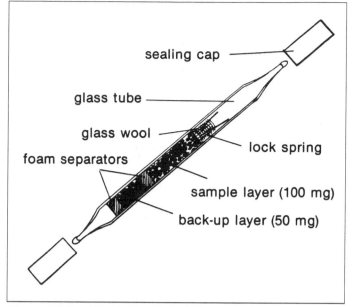

**Figure 16–6.** Standard activated charcoal tube used in organic vapor sampling. (Courtesy SKC, Inc.)

Many solid sorbent materials with chemical coatings have been developed to sample for reactive gases and vapors that are not efficiently collected by charcoal or silica gel. These include XAD-2, Tenac-GC, Ambersorb, and Chromosorb tubes (see Table 16–B).

## Passive Monitors

Passive monitors allow personal sampling without the use of sampling pumps. Whereas solid sorbent tubes rely on a sampling pump to draw air through the adsorbing material, passive monitors (Figure 16–7) rely on passive diffusion. Diffusion is the passage of molecules through a semipermeable barrier. It occurs because molecules tend to move from an area of high concentration to an area of low concentration. If the ambient concentration of a particular gas or vapor is greater than the concentration inside the monitor, then the gas or vapor molecules will diffuse across a barrier into the monitor and be collected by a sorbent material (Figure 16–8). As the sorbent adsorbs the gas or vapor, the concentration inside the monitor becomes less than the concentration outside. The rate of diffusion is determined by the manufacturer of the device.

Monitoring begins when the device's cover is removed; the time is recorded. The worker wears the monitor in his

or her breathing zone. When sampling is complete, the monitor is removed and resealed and the time is recorded. The badge is then sent to the laboratory for analysis.

Passive monitors are used because they are inexpensive and easy to use. Their accuracy has been studied extensively. Most commercially available monitors meet or exceed NIOSH accuracy requirements (±25 percent for 95 percent of samples tested between 0.5 and 2.0 times the exposure limit).

## ■ Collection Devices for Particulates

Airborne particulates can be either solid or liquid. Dusts, fumes, smoke, and fibers are dispersed solids; mists and fogs are dispersed liquids. They range in size from visible to microscopic.

### Filters

The filter is the most common collection device for particulates. There are several types, including glass fiber (GF), mixed cellulose ester fiber (MCE), and polyvinyl chloride (PVC) filters. They are selected based on their ability to collect material and their suitability for laboratory analysis. For

**Table 16–A.** Selected NIOSH Impinger Sampling Methods

| Chemical | NIOSH Sampling Method # | Impinging Solution | Analytical Method |
|---|---|---|---|
| Aminoethanol compounds II | 3509 | 15 ml of 2-mm hexanesulfonic acid | Ion chromatography |
| Phenol | 3502 | 01. N NaOH | Gas chromatography |
| Formaldehyde | 3501 | 15 ml Girard T reagent | Polarography |
| Isocyanates | 5521 | Solution of 1-(2-methoxyphenol)-piperazine in toluene | High-pressure liquid chromatography |

**Table 16–B.** NIOSH-Recommended Sorbent Tubes

| Chemical | NIOSH Method # | Tube | Lab Analytical Method |
|---|---|---|---|
| Methanol | 2000 | Silica gel | Gas chromatograph |
| Aromatic amines | 2002 | Silica gel | Gas chromatograph |
| Halogenated hydrocarbons | 1003 | Charcoal | Gas chromatograph |
| Naphthas | 1550 | Charcoal | Gas chromatograph |
| Phosphorus | 7905 | Tenac GC | Gas chromatograph |
| Nitroethane | 2526 | XAD-2 | Gas chromatograph |
| Methyl ethyl ketone | 2500 | Ambersorb XE-347 | Gas chromatograph |
| 1-butanethiol | 2525 | Chromosorb 104 | Gas chromatograph |

mineral and nuisance dusts, for example, the total weight of the collected particulate is of concern. In this case, PVC filters are used because they can be easily weighed. For metal dusts, the amount of a particular metal in the sample is of concern, so a chemical analysis must be done. MCE filters are generally used in this case. See Table 16–C for a list of NIOSH-recommended filters.

A typical collection device used for particulate sampling is a closed-face filter cassette, 37 mm in diameter, containing a filter supported with a cellulose backup pad (Figure 16–9). *Closed-face* means that the top of the cassette is not removed during the sampling; only the top and bottom caps are removed. The air inlet side of the cassette, opposite the filter, is usually marked so that the filter is not attached backwards.

There are some exceptions to the standard filter setup. Asbestos, for example, is collected using a 25-mm filter and cassette with an open-face 50-mm conductive extension cowl (Figure 16–10). The 25-mm filter improves the sensitivity of the test. The electrically conductive extension cowl reduces the number of asbestos fibers attracted to the sides of the cassette by static electricity.

## Cyclones

A cyclone is used to collect particles of respirable size. Respirable particles are those that are retained in the lung and are generally considered to be of an aerodynamic size below 10 μm. Cyclones have traditionally been used to sample for mineral dusts containing crystalline silica because of the strong association between the respirable dust fraction and lung disease silicosis.

Air is drawn into a cyclone tangentially through a small orifice. The centrifugal motion of the air inside the cyclone forces the larger particles to the periphery of the airstream, where they fall to the bottom of the cyclone. The respirable particles, in the center airstream, are drawn upward onto a preweighed filter. After sampling is completed, the filter is again weighed to determine how much material has been collected.

Use of a 10-cm nylon or stainless steel cyclone (Figure 16–11) is currently the most common method of collecting respirable dust samples in the United States. It meets the particle size selection efficiency guidelines (Table 16–D) of the American Conference of Governmental Industrial Hygienists (ACGIH) for respirable particulates. These

**Figure 16–7.** Passive diffusion monitors are an inexpensive and easy-to-use alternative to solid sorbent tubes. (Courtesy 3M.)

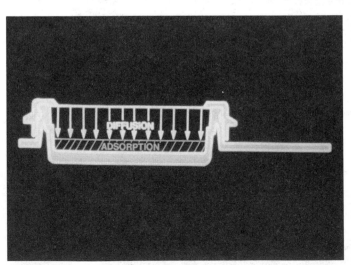

**Figure 16–8.** Gas or vapor molecules diffuse into a passive diffusion monitor across a permeable barrier and are collected by a sorbent material. (Courtesy 3M.)

**Table 16–C.** Selected NIOSH-Recommended Filters

| Chemical | NIOSH Method # | Filter | Analytical Method |
|---|---|---|---|
| Copper | 7029 | MCE | Atomic absorption (flame) |
| Carbon black | 5000 | PVC | Gravimetric |
| Mineral oil mist | 5026 | PVC, MCE or GF | Infrared spectrophotometry |
| Welding fumes | 7200 | MCE | X-ray fluorescence |
| Arsenic trioxide as As | 7901 | $Na_2CO_3$-impregnated MCE | Atomic absorption, graphite furnace |
| (2,4-dichlorophenoxy) acetic acid | 5001 | GF | High-pressure liquid chromatography |

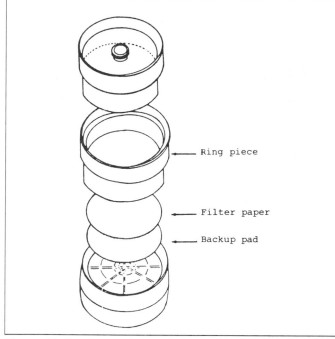

Ring piece

Filter paper

Backup pad

**Figure 16–9.** Standard filters are 37 mm in diameter and are placed in closed-face cassettes with a backup pad, which prevents contamination. (Reprinted from OSHA *Industrial Hygiene Technical Manual,* 1984.)

**Figure 16–10.** Air sampling for asbestos is conducted using a three-piece cassette, a 50-mm black conductive cowl, and a 25-mm filter. (Courtesy MSA.)

**Figure 16–11.** A cyclone attached to a filter cassette is used to sample for respirable dust. The filter cassette holder can be placed in the worker's breathing zone. (Courtesy SKC, Inc.)

**Table 16–D.** ACGIH Guidelines for Particle Size Collection

| Particle Aerodynamic Diameter (μm) | Respirable Particulate Mass (%) |
|---|---|
| 0 | 100 |
| 1 | 97 |
| 2 | 91 |
| 3 | 74 |
| 4 | 50 |
| 5 | 30 |
| 6 | 17 |
| 7 | 9 |
| 8 | 5 |
| 10 | 1 |

guidelines specify how much of each particle size range the cyclone must collect. These values were selected because they approximate the percentages of inhaled particles that are retained in the lung.

## Electrostatic Precipitators

Electrostatic precipitators use an electric charge to remove particles from the sampled air. As the particles pass through a high-voltage electric field, they acquire a charge and are attracted to an oppositely charged electrode. Collection efficiency increases with the length of passage through the collector, so precipitators are often used in series. Electrostatic precipitators are used when the required sample air volume is large, high-collection efficiency is required for very small particles (such as fumes), there is a possibility of filter clogging, or high-temperature airstreams must be sampled.

## Inertial Impactors

Inertial impactors collect particles by impacting them onto a surface. If an obstacle causes a moving airstream to deviate from a straight course, the particles in the airstream tend to leave the airstream and impact on the obstacle. Obstacles include filter paper, glass, stainless steel, and in the case of bioaerosol sampling, nutrient agar. The collection efficiency of this method is affected by the mass of the particles, the size and shape of the obstacle, and the velocity of the air.

Inertial impactors can be used to determine particle size distribution. The mini cascade impactor (Figure 16–12) is constructed with a series of stages, each of which is calibrated to collect particles of a certain aerodynamic size range.

## Impinger

The impinger is one of the oldest methods of particulate sampling, but it is little used today. It is used in situations where the number of particles must be expressed in millions of particles per cubic foot of air (mppcf).

Impingers use the same particle collection method as the inertial impactor, except that the particles are collected in a liquid (usually water). Air is drawn at high velocity into a liquid-filled flask through a glass nozzle or jet. The

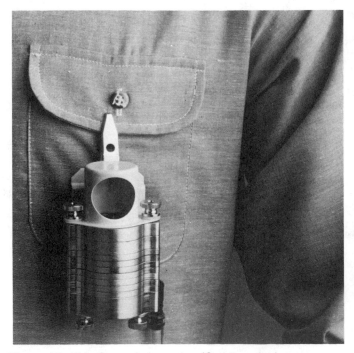

**Figure 16–12.** Cascade impactor. (Courtesy Anderson Sampler, Inc.)

particles impinge on a flat plate or the bottom of the flask, lose their velocity, and are trapped in the liquid. A small sample of the liquid is collected and then placed in a special cell that allows the particles to be counted and sized as they are viewed under a light microscope.

Impingers do not collect very small particles (less than 0.7 μm) well. For maximum collection efficiency, the air must be drawn at such a high velocity that larger particles are often shattered, thus producing erroneous results.

## Elutriators

Elutriators are used in front of a sampling train to remove coarse particles. The coarse particles are removed by gravity; the smaller particles remain suspended and are collected for subsequent analysis.

There are two types of elutriators: horizontal and vertical. The vertical elutriator is commonly used for cotton dust sampling. It consists of a large vertical tube through which the direction of airflow is opposite to the direction of gravity. Due to the airflow requirements and elutriator size, it must be operated in a stationary position. The flow rate is very important because if it is too high, the larger particles will not settle out but will be collected on the filter. If the flow rate is too low, some of the smaller particles may settle out and not be collected.

## ■ Suction Pumps

Suction pumps are responsible for the movement of air through the sampling train. To select the type of pump that

**Table 16–E.** Sampling Techniques for Collection of Airborne Particulates

| Sampling Technique | Force or Mechanism | Examples |
|---|---|---|
| Filters | Combination of inertial impaction, interception, diffusion, electrostatic attraction, and gravitational forces | Various types and sizes of fibrous, membrane, and nucleopore filters with holders |
| Impactors | Inertial—Impaction on a solid surface | Single- and multijet cascade impactors and single-stage impactors |
| Impingers | Inertial—Impingement and capture in liquid media | Greenburg–Smith and midget impingers |
| Elutriators | Gravitational separation | Horizontal and vertical elutriators |
| Electrostatic precipitation | Electrical charging with collection on an electrode of opposite polarity | Tube type, point-to-plane, and plate precipitators |
| Thermal precipitation | Thermophoresis—Particle movement under the influence of a temperature gradient in the direction of decreasing temperature | Various devices for particulate collection for microscopy analysis |
| Cyclones | Inertial—Centrifugal separation with collection on a secondary stage | Tangential and axial inlet cyclones in varying sizes |

(Reprinted from *Occupational Respiratory Diseases,* Pub. no. DHHS (NIOSH) 86–102, 1986.)

meets the needs of a particular sampling procedure, one must consider the airflow rate required, the pump's ease of use, and the pump's suitability for use in a potentially hazardous or flammable environment.

Most personal sampling pumps (Figure 16–13) are lightweight and quiet, use nickel/cadmium rechargeable batteries, and can be easily attached to the worker's belt. Each has a flow rate control valve and some are programmable. They must approved by the Mine Safety Administration, Underwriters Laboratory, or Factory Mutual Engineering Corp. for use in flammable or explosive atmospheres if they are to be used in such atmospheres.

Air-sampling pumps are generally available in the following airflow rate ranges: low-flow (0.5–500 ml/min), high-flow (0.5–5 l/min), and dual range (high- and low-flow).

Low-flow pumps are used for solid sorbent tube sampling. High-flow pumps are used for filter, cyclone, and impinger sampling.

However, there are situations where higher airflow rates are needed. The EPA, for example, requires a minimum air volume of 1,200 l for clearance area monitoring after an asbestos abatement project. Using a high-flow air-sampling pump with a flow rate of 5 l/min to sample 1,200 l of air would be very time-consuming. In this case, pumps that provide a flow rate of up to 10 l/min are used.

The low-power circuitry and sensors, amplifiers and microprocessors, and light plastic cases of the newer pumps have increased their susceptibility to radiofrequency (RF) interference. RF can be generated from facility or communications equipment and may alter the airflow rate or cause the pump to stop. Many manufacturers are now providing RF shielding.

## ■ Flow-Rate Meters

Maintaining a constant flow rate during sampling is critical. Devices that help maintain a constant flow rate are pressure-compensating devices and critical-flow orifices.

### Pressure-Compensating Devices

Pressure-compensating devices are designed to overcome the flow-rate variations inherent in many sampling situations. A sampling pump will slow down if the filter becomes loaded with dust or the hose is crimped. Pumps with pressure-compensating devices have sensors with feedback mechanisms that detect pressure changes and maintain the preset flow rate.

### Critical-Flow Orifice

Some pumps use critical or limiting orifices to regulate the airflow rate. A critical orifice is a precisely drilled hole in a metal plate through which the airstream being sampled is directed. When certain parameters are met, the flow rate through the orifice remains constant despite conditions at the inlet (such as a clogged filter). A critical orifice attached

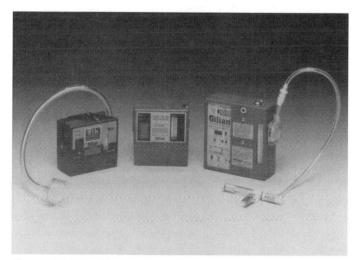

**Figure 16–13.** Personal sampling pumps must be lightweight and easy to use. (Courtesy Gillian Instrument Corp.)

to a sampling pump causes the pump to draw air at the desired flow rate.

The principle of the method is to draw air through the orifice under critical-flow conditions and constant upstream pressure. The volume flow rate of a gas through an orifice will increase with a decrease in the ratio of downstream pressure ($p_2$) to upstream absolute pressure ($p_1$) until the velocity through the opening reaches sonic velocity. The ratio, $p_2/p_1$, at which acoustic velocity is attained is called the critical pressure ratio. The velocity through the orifice will remain constant even if a much lower downstream pressure exists. Therefore, when the pressure ratio is less than critical, the rate of flow through the orifice is dependent only on upstream pressure.

For air flowing through an orifice where $p_2$ is less than $0.53p_1$ and the ratio of the upstream cross-sectional area ($s_1$) to the orifice area ($s_2$) is greater than 25, the flow rate is calculated as follows:

$$w = \frac{C_v S_2 p_1}{T_1^{1/2}} \tag{1}$$

where   $w$ = mass flow rate (lb/s)
$C_v$ = coefficient of velocity (normally 1 for well-rounded entrances and 0.6 for sharp-edged orifices)
$S_2$ = orifice area (in.$^2$)
$p_1$ = upstream pressure (lb/in.$^2$)
$T_1$ = upstream temperature (degrees K)

Orifices are calibrated under certain temperature and air pressure conditions. If the air sampling is conducted at a significantly different temperature and pressure, then a correction factor must be used to determine the actual airflow rate. (See the formula under Calibration Procedures later in this chapter.)

Some sampling pumps do not have a mechanism to maintain a constant airflow rate. In these cases, a calibration device such as a precision rotameter can be used to check the airflow rate during the sampling period. (Precision rotameters are discussed later in this chapter.) If the rate changes, it is manually adjusted to the desired rate using the flow-rate control valve.

## ■ Sampling Method

The selection of a sampling method depends on a number of factors:

- The sampling objective (documenting exposures, determining compliance, pinpointing sources of exposure)
- The physical and chemical characteristics of the chemical
- The presence of other chemicals that may interfere with the collection or analysis of the chemical
- The required accuracy and precision

- Regulatory requirements
- Portability and ease of operation
- Cost
- Reliability
- Type of sampling needed (area, personal, grab, integrated)
- Duration of sampling

Standardized sampling methods provide all the information needed to sample the air for a particular chemical. Analytical procedures are found in the NIOSH *Manual of Analytical Methods,* the American Public Health Association's *Methods of Air Sampling,* and the OSHA *Chemical Information Manual.* OSHA regulations do not specify a particular sampling method, but they do require that the method used have a specified and proven degree of accuracy. Some analytical laboratories have developed their own procedures that meet or exceed the OSHA criteria.

The sampling method used depends on the recommendations of the laboratory selected to analyze the samples. The laboratory must be experienced in industrial hygiene sampling methods and should be accredited by AIHA and involved in the NIOSH Proficiency Analytical Testing Program. These organizations monitor laboratory performance to ensure that the information and analytical results are accurate.

An example of a NIOSH sampling procedure is given in Figure 16–14. This sampling method for acetic acid provides information for both the industrial hygienist and the laboratory. The method requires a coconut shell charcoal tube (100 mg of charcoal with a 50-mg backup section), an airflow rate between 0.01 to 1.0 l/min and an air sample volume between 20 and 300 l. Precautions include analyzing the samples within 7 days and ensuring that the atmosphere being tested does not contain formic acid, which interferes with the analysis.

The recommended air sample volumes are important guidelines to follow. The minimum air sample volume is the minimum amount of air needed to ensure analytical accuracy. It also allows the laboratory to analyze the sample to a concentration well below the exposure limit for that chemical. This is called the sampling method's lower limit of detection and is the smallest amount of the chemical that the laboratory can detect.

Minimum sample volumes can be calculated if the lower limit of detection (LOD) of the analytical method is known. This can be useful if there is no listed minimum air sample volume or if the listed volume is quite large. Published values must assume worst-case conditions are present and have built-in safety factors to ensure that an adequate volume is collected. If the concentration of the contaminant can be estimated, then the following formula can be used:

$$SV = \frac{LOD}{EL \times F} \tag{2}$$

FORMULA: $CH_3COOH$; $C_2H_4O_2$

M.W.: 60.05

| | |
|---|---|
| OSHA: 10 ppm | PROPERTIES: liquid; d 1.049 g/mL @ 25 °C; |
| NIOSH: 10 ppm; STEL 15 ppm | BP 118 °C; MP 17 °C; |
| ACGIH: 10 ppm | VP 1.5 kPa (11 mm Hg; 1.4% v/v) @ 20 °C; |
| (1 ppm = 2.46 mg/m³ @ NTP) | explosive range 5.4 to 16% v/v in air |

SYNONYMS: glacial acetic acid; methane carboxylic acid; ethanoic acid; CAS #64-19-7.

| SAMPLING | MEASUREMENT |
|---|---|
| | !TECHNIQUE: GAS CHROMATOGRAPHY, FID |
| SAMPLER: SOLID SORBENT TUBE | ! |
|   (coconut shell charcoal, | !ANALYTE: acetic acid |
|   100 mg/50 mg) | ! |
| | !DESORPTION: 1 mL formic acid; stand 60 min |
| FLOW RATE: 0.01 to 1.0 L/min | ! |
| | !INJECTION VOLUME: 5 µL |
| VOL-MIN:  20 L @ 10 ppm | ! |
|   -MAX: 300 L | !TEMPERATURE-INJECTION: 230 °C |
| | !        -DETECTOR: 230 °C |
| SAMPLE STABILITY: at least 7 days | !          -COLUMN: 130 to 180 °C, 10°/min |
|   @ 25 °C [1,2] | !                 or 100 °C isothermal |
| | ! |
| BLANKS: 10% of samples | !CARRIER GASES: $N_2$ or He, 60 mL/min |
| | ! |
| | !COLUMN: 1 m x 4-mm ID glass; Carbopack B 60/80 |
| ACCURACY | !       mesh/3% Carbowax 20M/0.5% $H_3PO_4$ |
| | ! |
| RANGE STUDIED: 12.5 to 50 mg/m³ [2] | !CALIBRATION: standard solutions of acetic acid |
|   (173-L samples) | !         in 88 to 95% formic acid |
| | ! |
| BIAS: not significant [2] | !RANGE: 0.5 to 10 mg per sample |
| | ! |
| OVERALL PRECISION ($s_r$): 0.058 [2] | !ESTIMATED LOD: 0.01 mg per sample [3] |
| | ! |
| | !PRECISION ($s_r$): 0.007 @ 0.3 to 5 mg per |
| | !          sample [1,2] |
| | ! |

APPLICABILITY:  The working range is 2 to 40 ppm (5 to 100 mg/m³) for a 100-L air sample. High (90% RH) humidity during sampling did not cause breakthrough at 39 mg/m³ for 4.6 hrs [2].

INTERFERENCES:  Formic acid contains a small amount of acetic acid which gives a significant blank value.  High-purity formic acid must be used to achieve an acceptable detection limit. Alternate columns are 3 m glass, 2 mm ID, 0.3% SP-1000 + 0.3% $H_3PO_4$ on Carbopack A and 2.4 m x 2 mm ID glass, 0.3% Carbowax 20M/0.1% $H_3PO_4$ on Carbopack C.

OTHER METHODS:  This revises Method S169 [1], and Method 1603 (dated 2/15/84).

**Figure 16–14.**   NIOSH Sampling Method #1603 provides guidelines for acetic acid sampling. (Reprinted from *NIOSH Manual of Analytical Methods.*)

where  SV = Minimum sample volume (l)

LOD = Lower limit of detection (μg)

EL = Exposure limit (mg/m$^3$)

F = Anticipated fraction of threshold limit value (TLV) in atmosphere (decimal)

The LOD for acetic acid is 0.01 mg (10 μg). If the anticipated concentration is 25 percent of the TLV, then the minimum sample volume is calculated as follows:

$$SV = \frac{10 \text{ μg}}{25 \text{ mg/m}^3 \times 0.25}$$

$$= 1.6 \text{ l} \qquad (3)$$

Establishing a maximum air sample volume is necessary to prevent breakthrough when sampling for gases and vapors or overloading the filter when sampling for particles. Breakthrough occurs when a significant quantity of a gas or vapor passes uncollected through a collection device. It happens when the device is saturated with the chemical or interfering chemicals or the airflow rate is too fast. In particulate sampling, if the filter is overloaded it may cause the suction pump to slow down or quit, cause the loss of some of the sample as the filter is being handled in the laboratory, or make the analysis of the filter difficult. The maximum air sample volume is designed to minimize these problems.

Established maximum air sample volumes are designed to handle concentrations up to twice the exposure limit of a single contaminant. If the atmospheric concentration is well above twice the exposure limit or there are other interfering gases and vapors, saturation and breakthrough occur more quickly than anticipated. Maximum air sample volumes in these cases must be adjusted.

The flow rate specified in the air-sampling method provides the greatest collection efficiency for the chemical being sampled. For gases and vapors, it means that the analyte will be in contact with the absorbing or adsorbing material long enough to be captured. For particulates, it means that the particles will be effectively captured without damaging the collection device.

With the recommended flow rate and air sample collection volumes, the industrial hygienist can determine the time necessary to collect a sample. For example, if the recommended flow rate is 0.2 l/min and the minimum sample volume is 10 l, the sample time is at least 50 min. The formula is as follows:

$$\frac{\text{Required sample}}{\text{time (min)}} = \frac{\text{Minimum sample volume (l)}}{\text{Flow rate (l/min)}} \qquad (4)$$

Using the example above,

$$\text{Required sample time} = \frac{10 \text{ l}}{0.2 \text{ l/min}} = 50 \text{ min} \qquad (5)$$

It may be necessary, in this case, to use a series of samples to cover an 8-hour shift.

Laboratories require blanks for each set of samples submitted for analysis. The laboratory specifies the number and type of blanks needed. Two types of blanks may be used: a field blank or a media or lab blank. A field blank is a sample collection device that has been briefly opened and closed and is handled in the field identically to the other samples. The field blank is used to determine whether the air samples have been contaminated during handling. In contrast, a media blank is an *unopened* collection device used to determine whether the sampling collection device itself is contaminated.

# ■ Calibration

The suction pumps used for air sampling must be calibrated to the airflow recommended in the sampling method. Calibration is critical because the determination of air sample volume depends on the flow rate and the elapsed time. There are two categories of calibration devices: primary and secondary. Primary devices provide a direct measurement of airflow. They include soap-bubble meters, Mariotti bottles, and spirometers. Secondary calibration devices provide indirect measurements of airflow and must be periodically calibrated with a primary calibration device. They include rotameters, wet test meters, and dry test meters.

## Primary Calibration: Mariotti Bottle

The Mariotti bottle uses water displacement to measure air volume (Figure 16–15). To begin the calibration, the valve at the bottom of the water-filled bottle is opened. As the water drains out of the bottle, air is drawn through the sample collector and into the bottle. The volume of air drawn is equal to the change in water level multiplied by the cross-sectional area at the water surface.

## Primary Calibration: Spirometer

The spirometer is similar to a Mariotti bottle except that it measures displaced air rather than liquid (Figure 16–16). It is shaped like a cylindrical bell that contains a known volume of air with the open end sitting under a liquid seal. When the liquid is discharged, the air is displaced and forced out of the cylinder through the air-sampling instrument. The volume of air displaced is calculated based on the dimensions of the cylinder.

## Primary Calibration: Soap-Bubble Meter

The most commonly used primary calibration instrument is the soap-bubble meter. It consists of an inverted volumetric burette connected to the sampling train. The sampling train must contain the type of collection device that will be used to conduct air sampling because each device causes a unique pressure drop. The pressure drop will affect the

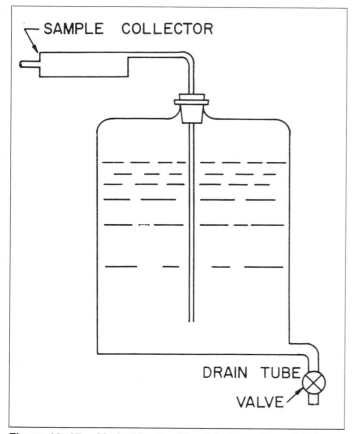

**Figure 16–15.** Mariotti bottle. (Reprinted from Powell CH, Hosey AD, eds. *The Industrial Environment—Its Evaluation and Control,* 2nd ed. PHS Pub. no. 614, 1965.)

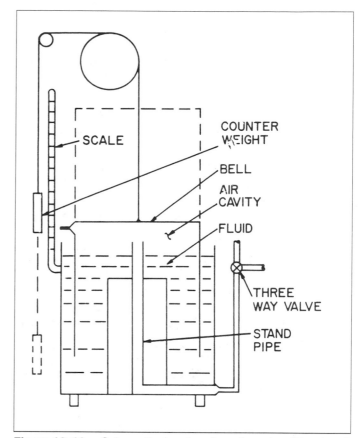

**Figure 16–16.** Schematic drawing of a spirometer. (Reprinted from Powell CH, Hosey AD, eds. *The Industrial Environment— Its Evaluation and Control,* 2nd ed. PHS Pub. no. 614, 1965.)

sampling pump's flow rate and must be accounted for during the calibration.

The general procedure for soap-bubble meter calibration is as follows:

1. Set up the apparatus as shown in Figure 16–17. Wet the inside of the burette with the soap solution or water before setup.

2. Allow the sampling pump to run for 5 min. Check the sampling pump's battery. If the battery is low, recharge the pump. Check the manufacturer's instructions for proper battery testing and recharging procedures.

3. Connect the sampling train to the burette.

4. To create a bubble, momentarily submerge the opening of the burette and then draw two or three bubbles up the length of the burette.

5. Adjust the pump to the nominal desired flow rate.

6. Create a soap bubble and, using a stopwatch, measure the time it takes to traverse a convenient calibration volume. For high-volume pumps, a 1,000-ml burette is used and the bubble is timed as it travels from 0 to the 1,000-ml mark. For low-flow pumps, a 100-ml burette is used.

7. Calculate the flow rate. The flow rate is determined by measuring the time required for the bubble to pass between two scale markings. For example, if 30 s were required for the bubble to go from the 0-ml to the 1,000-ml mark, then the flow rate is calculated as follows:

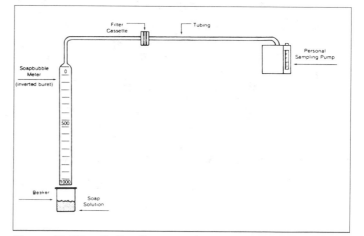

**Figure 16–17.** Calibration setup for personal sampling pump with filter cassette. (Reprinted from NIOSH *Manual of Analytical Methods,* 1984.)

$$\frac{1{,}000 \text{ ml}}{30 \text{ s}} \times \frac{60 \text{ s}}{1 \text{ min}} = 2{,}000 \text{ ml/min or } 2 \text{ l/min} \qquad (6)$$

8. If a different flow rate is desired, adjust the pump and repeat the procedure.

9. Repeat the determination at least twice. Calculate the average flow rate.

10. Record the following:
    a. Volumes measured
    b. Elapsed times
    c. Air temperature
    d. Atmospheric pressure
    e. Make, model, and serial number of the sampling pump
    f. Collection device used
    g. Name and date of person performing calibration

Electronic soap-bubble meters (Figure 16–18) calibrate in less time and with greater accuracy than the traditional burette method. A microprocessor is used to time a bubble as it traverses from the first to the second sensor and to calculate the volume per unit time. The margin for error is supplied by the manufacturer and is typically ±0.5 percent.

## Secondary Calibration: Wet-Test Meter

A typical wet-test meter is a partitioned drum, half submerged in a liquid (usually water), with openings in the center and periphery of each radial chamber (Figure 16–19). Air or gas enters at the center and flows into one compartment, causing the chamber to rise and rotate. The number of revolutions made by the chamber is recorded on a dial. Because the liquid is replaced by air, the measured volume depends on the height of the fluid, so a sight gauge is provided. Temperature and pressure gauges are also provided.

**Figure 16–18.** Electronic soap-bubble meter provides calibration results quickly and accurately. Calibration information can be stored on a computer database. (Courtesy Gillian Instrument Corp.)

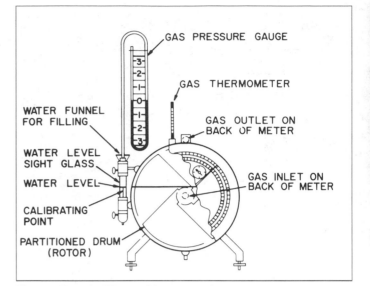

**Figure 16–19.** The working parts of a wet-test meter. (Reprinted from *The Industrial Environment—Its Evaluation and Control,* PHS Pub. no. 614.)

## Secondary Calibration: Dry-Gas Meter

A dry-gas meter, similar to a domestic gas meter, consists of two bellows connected by mechanical valves and a counting device (Figure 16–20). As air fills one bag, it mechanically empties another.

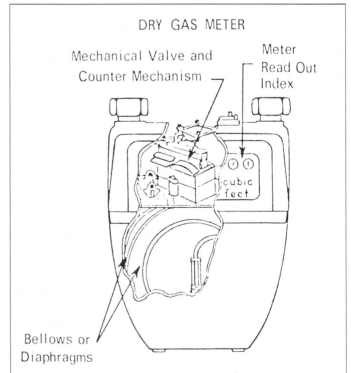

**Figure 16–20.** A dry-gas meter consisting of two bags connected by mechanical valves and a counting device. (Reprinted from *The Industrial Environment—Its Evaluation and Control,* PHS Pub. no. 614.)

## Secondary Calibration: Precision Rotameter

A rotameter consists of a float, or ball that is free to move, in a vertically tapered tube (Figure 16–21). Air is pulled through the tube so that the ball rises until there is an equilibrium between the force of gravity and the force of the air traveling upward. The flow rate is determined by reading the height of the float on an attached numerical scale.

The rotameter's numerical scale has no meaning until it has been calibrated against a primary calibration device. A soap-bubble meter is usually used as the primary calibration device. First, an air-sampling pump is calibrated with the soap-bubble meter. Then a rotameter is attached to the sampling train (see Figure 16–22) to determine what scale marking relates to this flow rate. This is done several times so that a graph (see Figure 16–23) or chart of measured flow rate versus rotameter scale reading can be made. For rotameters used on a regular basis, this process should be repeated monthly.

Rotameters that are part of an air-sampling pump are not precision rotameters and should not be used for calibration purposes. They provide only an approximate indication of the airflow rate.

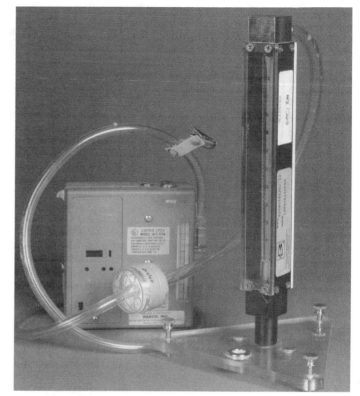

**Figure 16–22.** Precision rotameter connected to a sampling train. (Courtesy Fermilab Media Services Dept.)

## Calibration Parameters

**Temperature and Pressure.** Air volume is directly affected by temperature and pressure. If the conditions during air sampling are significantly different from those during calibration, then a correction factor must be used when calculating the sample air volume (the field volume). This can be done using the following expression:

$$V_{field} = V_{calibration} \times \frac{T_{field}}{T_{calibration}} \times \frac{P_{calibration}}{P_{field}} \qquad (7)$$

where  $V_{field}$ = air sample volume in liters obtained during sampling period

$V_{calibration}$ = air sample volume in liters obtained by multiplying the calibrated airflow rate by the elapsed sampling time

$T_{field}$ = Absolute temperature during sampling in degrees Kelvin or Rankin

$T_{calibration}$ = temperature during calibration in degrees Kelvin or Rankin

$P_{calibration}$ = atmospheric pressure during calibration in mm Hg or inches of water

$P_{field}$ = atmospheric pressure during sampling in mm Hg or inches of water

**Error.** The accuracy of calibration depends on the measuring limits of the equipment used. Each measuring instrument has a margin for error. This information can be obtained from the manufacturer.

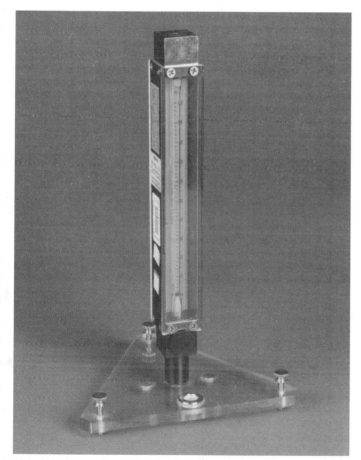

**Figure 16–21.** A single-column precision rotameter can be used as a secondary calibration device. (Courtesy Fermilab Visual Media Services Dept.)

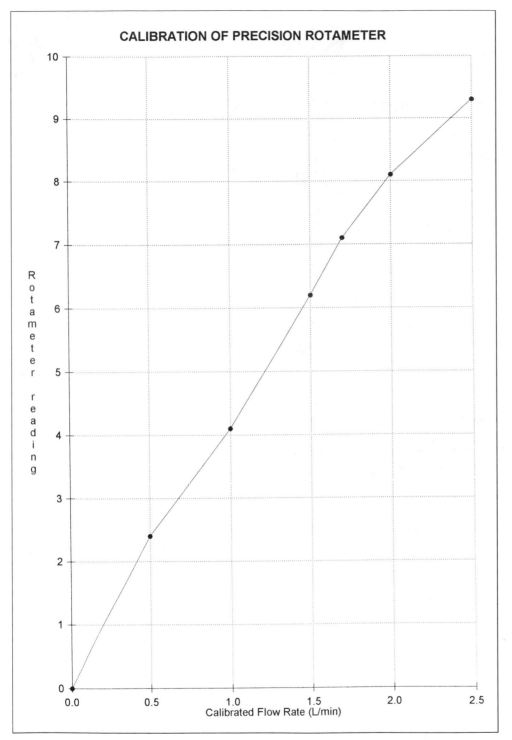

**CALIBRATION OF PRECISION ROTAMETER**

**Figure 16–23.** A calibration chart is needed when using a precision rotameter. The chart relates the rotameter's scale to a specific flow rate.

Soap-bubble meter calibration, for example, uses a volumetric burette and a stopwatch. The burette may be accurate to within ±5 ml (or 5 percent); when the 1,000-ml mark is read, the true volume could be anywhere between 1,005 and 995 ml. Stopwatches have a similar margin for error.

The margin for error associated with the calibration of airflow rate can be calculated using the error of measurement from each piece of equipment used in the calibration, using the following equation:

$$\text{Error} = [\,(\text{instrument error})^2 + (\text{instrument error})^2\,]^{1/2} \quad (8)$$

For the soap-bubble meter, for example, if the accuracy of the burette is 0.5 percent and the accuracy of the stopwatch is 0.5 percent, then the total margin for error is 0.7 percent. A calibrated flow rate of 2 l/min is more accurately reported as 2 l/min ± 0.7 percent.

**Sampling Technique.** The instructions for air-sampling techniques in Figures 16–24, 16–25, and 16–26 are from *Manual of Analytical Methods* (NIOSH, 1984).

Use these instructions for active personal sampling (i.e., pumped sample airflow) for substances which are retained on solid sorbents such as activated charcoal, silica gel, porous polymers, etc.

(1) Calibrate each personal sampling pump at the desired flow rate with a representative solid sorbent tube in line,* using a bubble meter or equivalent flow measuring device. (*Alternatively, use a flow restrictor to provide a pressure drop equal to that of the average solid sorbent tube.)

(2) Break the ends of the solid sorbent tube immediately before sampling to provide an opening at least one-half of the internal diameter at each end.

(3) Connect the solid sorbent tube to a calibrated personal sampling pump with flexible tubing with the smaller sorbent section (backup section) nearer to the pump. Do not pass the air being sampled through any hose or tubing before entering the solid sorbent tube. Position the solid sorbent tube vertically during sampling to avoid channeling and premature breakthrough.

(4) Prepare the field blanks at about the same time as sampling is begun. These field blanks should consist of unused solid sorbent tubes <u>from the</u> <u>same</u> <u>lot</u> used for sample collection. Handle and ship the field blanks exactly as the samples (e.g., break the ends and seal with plastic caps) but do not draw air through the field blanks. Two field blanks are required for each 10 samples with a maximum of 10 field blanks per sample set.

(5) Take the sample at an accurately known flow rate as specified in the method for the substance and for the specified air volume. Typical flow rates are in the range 0.01 to 0.2 L/min. Check the pump during sampling to determine that the flow rate has not changed. If sampling problems preclude the accurate measurement of air volume, discard the sample. Take two to four replicate samples for quality control for each set of field samples.

(6) Record pertinent sampling data including location of sample, times of beginning and end of sampling, initial and final air temperatures, relative humidity and atmospheric pressure or elevation above sea level.

(7) Seal the ends of the tube immediately after sampling with plastic caps. Label each sample and blank clearly with waterproof identification.

(8) Pack the tubes tightly with adequate padding to minimize breakage for shipment to the laboratory. In addition to the sample tubes and field blanks, ship at least six unopened tubes to be used as media blanks so that desorption efficiency studies can be performed on the same lot of sorbent used for sampling.

(9) Ship bulk samples in a separate package from the air samples to avoid contamination of the samples. Suitable containers for bulk samples are glass with a polytetrafluoroethylene (PTFE)-lined cap, e.g., 20-mL glass scintillation vials.

**Figure 16–24.** Sampling instructions for solid sorbent tube samplers. (Reprinted from NIOSH *Manual of Analytical Methods,* 1984.)

Use these instructions for personal sampling of total (respirable and non-respirable) aerosols. Methods requiring these instructions specify FILTER as the sampling method. These instructions are not intended for respirable aerosol sampling.

(1) Calibrate the personal sampling pump with a representative filter in line using a bubble meter or equivalent flow measuring device.

(2) Assemble the filter in the two-piece cassette filter holder. Support the filter by a stainless steel screen or cellulose backup pad. Close firmly to prevent sample leakage around the filter. Seal the filter holder with plastic tape or a shrinkable cellulose band. Connect the filter holder to the personal sampling pump with a piece of flexible tubing.

(3) Remove the filter holder plugs and attach the filter holder to the personal sampling pump tubing. Clip the filter holder to the worker's lapel. Air being sampled should not be passed through any hose or tubing before entering the filter holder.

(4) Prepare the field blanks at about the same time as sampling is begun. These field blanks should consist of unused filters and filter holders from the same lot used for sample collection. Handle and ship the field blanks exactly as the samples, but do not draw air through the field blanks. Two field blanks are required for each 10 samples with a maximum of 10 field blanks per sample set.

(5) Sample at a flow rate of 1 to 3 L/min until the recommended sample volume is reached. Set the flow rate as accurately as possible (e.g., within ± 5%) using the personal sampling pump manufacturer's directions. Take two to four replicate samples for quality control for each set of field samples.

(6) Observe the sampler frequently and terminate sampling at the first evidence of excessive filter loading or change in personal sampling pump flow rate. (It is possible for a filter to become plugged by heavy particulate loading or by the presence of oil mists or other liquids in the air.)

(7) Disconnect the filter after sampling. Cap the inlet and outlet of the filter holder with plugs. Label the sample. Record pertinent sampling data including times of beginning and end of sampling, initial and final air temperatures, relative humidity and atmospheric pressure or elevation above sea level. Record the type of personal sampling pump used and location of sampler.

(8) Ship the samples to the laboratory as soon as possible in a suitable container designed to prevent damage in transit. Ship bulk material to the laboratory in a glass container with a PTFE-lined cap. Never store, transport or mail the bulk sample in the same container as the samples or field blanks. In addition to the samples and field blanks, ship five unopened samplers from the same lot for use as media blanks.

**Figure 16–25.** Sampling instructions for filter samplers. (Reprinted from NIOSH *Manual of Analytical Methods,* 1984.)

Use these instructions for personal sampling of respirable aerosols (ACGIH definition [9]). Methods requiring these instructions specify CYCLONE + FILTER as the sampling method.

(1) Calibrate the pump to 1.7 L/min, with a representative cyclone sampler in line using a bubble meter or a secondary flow measuring device which has been calibrated against a bubble meter. The calibration of the personal sampling pump should be done close to the same altitude where the sample will be taken.

(2) Assemble the pre-weighed filter in the two-piece cassette filter holder. Support the filter with a stainless steel screen or cellulose backup pad. Close firmly to prevent sample leakage around the filter. Seal the filter holder with plastic tape or a shrinkable cellulose band.

(3) Remove the cyclone's grit cap and vortex finder before use and inspect the cyclone interior. If the inside is visibly scored, discard this cyclone since the dust separation characteristics of the cyclone might be altered. Clean the interior of the cyclone to prevent reentrainment of large particles.

(4) Assemble the two-piece filter holder, coupler, cyclone and sampling head. The sampling head rigidly holds together the cyclone and filter holder. Check and adjust the alignment of the filter holder and cyclone in the sampling head to prevent leakage. Connect the outlet of the sampling head to the personal sampling pump by a 1-m piece of 6-mm ID flexible tubing.

(5) Clip the cyclone assembly to the worker's lapel and the personal sampling pump to the belt. Ensure that the cyclone hangs vertically. Explain to the worker why the cyclone must not be inverted.

(6) Prepare the field blanks at about the same time as sampling is begun. These field blanks should consist of unused filters and filter holders from the same lot used for sample collection. Handle and ship the field blanks exactly as the samples, but do not draw air through the field blanks. Two field blanks are required for each 10 samples with a maximum of 10 field blanks per sample set.

(7) Turn on the pump and begin sample collection. If necessary, reset the flow rate to the pre-calibrated 1.7 L/min level, using the manufacturer's adjustment procedures. Since it is possible for a filter to become plugged by heavy particulate loading or by the presence of oil mists or other liquids in the air, observe the filter and personal sampling pump frequently to keep the flow rate within $\pm$ 5% of 1.7 L/min. The sampling should be terminated at the first evidence of a problem.

**Figure 16–26.** Sampling instructions for filter and cyclone sampler. (Reprinted from NIOSH *Manual of Analytical Methods*, 1984.) (*Continues*)

(8) Disconnect the filter after sampling. Cap the inlet and outlet of the filter holder with plugs. Label the sample. Record pertinent sampling data including times of beginning and end of sampling, initial and final air temperatures and atmospheric pressure or elevation above sea level. Record the type of personal sampling pump, filter, cyclone used and the location of the sampler.

(9) Ship the samples and field blanks to the laboratory in a suitable container designed to prevent damage in transit. Ship bulk samples in a separate package.

(10) Take two to four replicate samples for every set of field samples to assure quality of the sampling procedures. The set of replicate samples should be exposed to the same dust environment, either in a laboratory dust chamber or in the field. The quality control samples must be taken with the same equipment, procedures and personnel used in the routine field samples. The relative standard deviation, $s_r$, calculated from these replicates should be recorded on control charts and action taken when the precision is out of control.

**Figure 16–26.** (*Continued*)

## Sampling and Analytical Error

Once the air sample results are received from the analytical laboratory and the time-weighted averages are calculated, there should be a calculated margin of error associated with the results. This is called the sampling and analytical error (SAE), and can be calculated using the following formula:

$$SAE = [ \, (\text{airflow error})^2 + (\text{time error})^2 \\ + (\text{analytical error})^2 \, ]^{1/2} \qquad (9)$$

The airflow error is the error of measurement associated with the calibration of the air-sampling pump. The time error is associated with the instrument used to measure the time period over which the sample was collected. The analytical error is the error associated with the analytical methods used by the laboratory.

## Record Keeping

Complete and detailed records must be kept on sampling procedures, sampling conditions, and sample results. The hygienist must document that sampling was conducted according to accepted professional standards. Records should include the identity of the equipment and collection devices used, the calibration procedures and results, the identity of the analytical laboratory and related laboratory reports, and the air-sampling calculations.

The conditions under which the sampling was conducted should also be carefully documented to ensure the integrity and usefulness of the results. Anything that might help interpret or explain the final air sample result should be recorded. For example, in a production welding operation, the record should contain the name and location of the welder, the material being welded, the welding rods used, the number of pieces welded, the use of personal protective equipment, and the use and location of local exhaust ventilation.

Many industrial hygiene programs have developed air-sampling forms to ensure that all the necessary information is collected. OSHA has developed an air-sampling worksheet (Figure 16–27) for its Industrial Hygiene Compliance Officers.

OSHA Standard 29 *CFR* 1910.20, Access to Employee Exposure and Medical Records, requires that employee exposure records be preserved for at least 30 years. This information must be readily available to employees and their representative(s). Background data such as laboratory reports and field notes need only be retained for one year as long as information on the sampling method, the analytical and mathematical methods, and summary of other background information is retained for the required 30 years.

# Air Sampling Worksheet

| 1. Reporting ID | 2. Inspection Number | 3. Sampling Number ▶ 0741041 |
|---|---|---|

| 4. Establishment Name | 5. Sampling Date | 6. Shipping Date |
|---|---|---|

| 7. Person Performing Sampling (Signature) | 8. Print Last Name | 9. CSHO ID |
|---|---|---|

| 10. Employee (Name, Address, Telephone Number) | 14. Exposure Information | a. Number | b. Duration |
|---|---|---|---|
| | c. Frequency | | |
| | 15. Weather Conditions | 16. Photo(s) Y | |

| 11. Job Title | 12. Occupation Code |
|---|---|

| 13. PPE (Type and effectiveness) | 17. Pump Checks and Adjustments |
|---|---|

**18. Job Description, Operation, Work Location(s), Ventilation, and Controls**

Cont'd

**19. Pump Number:**                    **Sampling Data**

| | | | | | |
|---|---|---|---|---|---|
| 20. Lab Sample Number | | | | | |
| 21. Sample Submission Number | | | | | |
| 22. Sample Type | | | | | |
| 23. Sample Media | | | | | |
| 24. Filter/Tube Number | | | | | |
| 25. Time On/Off | | | | | |
| 26. Total Time (in minutes) | | | | | |
| 27. Flow Rate ☐ l/min ☐ cc/min | | | | | |
| 28. Volume (in liters) | | | | | |
| 29. Net Sample Weight (in mg) | | | | | |

| 30. Analyze Samples for: | 31. Indicate Which Samples To Include in TWA, Ceiling, etc. Calculations |
|---|---|

| 32. Interferences and IH Comments to Lab | 33. Supporting Samples | 34. Chain of Custody | Initials | Date |
|---|---|---|---|---|
| | a. Blanks: | a. Seals Intact? | Y    N | |
| | | b. Rec'd in Lab | | |
| | b. Bulks: | c. Rec'd by Anal. | | |
| | | d. Anal. Completed | | |
| | | e. Calc. Checked | | |
| | | f. Supr. OK'd | | |

Case File Page ___ / ___ of

OSHA-91A (Rev. 1/84)

**Figure 16–27.** Air-sampling worksheet used by OSHA to record all pertinent information related to the collection of an air sample. (*Continues*)

**Pre-Sampling Calibration Records**

| pre | | | | | |
|---|---|---|---|---|---|
| | 35. Pump Mfg. & SN | 38. Flow Rate Calculations | | | |
| | 36. Voltage Checked? ☐ Yes ☐ No | | | | |
| | 37. Location/T & Alt. | | | | |
| | | 39. Flow Rate | 40. Method ☐ Bubble ☐ PR | 41. Initials | 42. Date/Time |

**Post-Sampling Calibration Records**

| post | | | | |
|---|---|---|---|---|
| | 43. Location/T & Alt. | 44. Flow Rate Calculations | | |
| | | | | |
| | | | | |
| | 45. Flow Rate | 46. Initials | 47. Date/Time | |

**Sample Weight Calculations**

| | | | | | | |
|---|---|---|---|---|---|---|
| 48. Filter No. | | | | | | |
| 49. Final Weight *(mg)* | | | | | | |
| 50. Initial Weight *(mg)* | | | | | | |
| 51. Weight Gained *(mg)* | | | | | | |
| 52. Blank Adjustment | | | | | | |
| 53. Net Sample Weight *(mg)* | | | | | | |

54. Calculations and Notes:

☆ U.S. GOVERNMENT PRINTING OFFICE: 1984—448-944

**Figure 16–27.** (*Continued*)

## ■ Summary

Evaluating and controlling employee exposure to airborne occupational health hazards usually includes a comparison of the measured concentration of an airborne chemical to a recognized exposure limit. The various methods, instruments, and devices used for such air sampling have been described in this chapter.

## ■ Bibliography

Air Sampling Instruments Committee. *Air Sampling Instruments for Evaluation of Atmospheric Contamination,* 7th ed. Cincinnati, OH: American Conference of Governmental Industrial Hygienists, 1989.

American Conference of Governmental Industrial Hygienists. *1994–1995 Threshold Limit Values for Chemical Substances and Physical Agents.* Cincinnati, OH: American Conference of Governmental Industrial Hygienists, 1995.

Caravanos, J. *Quantitative Industrial Hygiene: A Formula Workbook.* Cincinnati, OH: American Conference of Governmental Industrial Hygienists, 1991.

Clayton GD, Clayton FE, eds. *General Principles.* In Clayton GD, Clayton FE, eds. *Patty's Industrial Hygiene and Toxicology,* 4th ed., Vol. I. New York: Interscience Publishers, 1991.

Glenn RE, Craft BF. Air sampling and analysis for particulates. In Merchant JA, ed. *Occupational Respiratory Diseases.* DHHS (NIOSH) Pub. No. 86–102. Washington, DC: Superintendent of Documents, U.S. Government Printing Office, 1986.

Leidel NA, Busch KA, Lynch JR. *Occupational Exposure Sampling Strategy Manual,* Pub. No. PB-274–792. Springfield, VA: National Technical Information Service (NTIS), 1977.

National Institute for Occupational Safety and Health. *Manual of Analytical Methods,* 3rd ed. (updated regularly). Springfield, VA: NTIS, 1984.

National Institute for Occupational Safety and Health. *The Industrial Environment—Its Evaluation and Control.* Washington, DC: U.S. Government Printing Office, 1973. Occupational Safety and Health Administration. OSHA CPL 2–2.20B, *OSHA Technical Manual* (updated regularly). Washington, DC, 1990.

Peach MJ, Carr WG. Air sampling and analysis for gases and vapors. In Merchant JA, ed. *Occupational Respiratory Diseases.* DHHS (NIOSH) Pub. No. 86–102. Washington, DC: Superintendent of Documents, U.S. Government Printing Office, 1986.

Powell CH, Hosey AD, eds. *The Industrial Environment—Its Evaluation and Control,* 2nd ed. Public Health Service Pub. no. 614, 1965.

*CHAPTER* **17**

# Direct-Reading Instruments for Gases, Vapors, and Particulates

by Rolf M.A. Hahne, PhD, CIH

$T$HE CONCENTRATION OF MANY GASES, vapors, and particulates in air can be measured readily by direct-reading instruments or other direct-reading devices, which perform both sampling and measurement. The user can read the concentration through a digital or analog readout or, for some colorimetric devices, by observing the length or intensity of a color stain.

A direct-reading instrument should be capable of sampling air in the breathing zone of the worker or in the work area of concern and should indicate the concentration of the substance (or class of substances, in the case of nonspecific instruments), either as an instantaneous concentration or as a time-weighted average, depending on the capabilities of the device.

This chapter is structured to first discuss direct-reading instruments that are intended for the measurement of a single compound or group of compounds, such as combustible gases. The second part of the chapter addresses direct-reading instruments that have wide applicability for the detection of many different compounds or substances, such as particulates.

**Monitors Intended for One Compound or Group of Compounds**
*Combustible Gas Monitors* ▪ *Oxygen Monitors* ▪ *Carbon Monoxide Monitors* ▪ *Indoor Air Quality Monitors* ▪ *Other Monitors Using Electrochemical or Metal Oxide Semiconductor Detectors* ▪ *Mercury Vapor Monitors* ▪ *Direct-Reading Colorimetric Tubes and Badges* ▪ *Other Colorimetric Direct-Reading Devices*

**Monitors Intended for a Broad Range of Compounds**
*Nonspecific Detectors* ▪ *Spectrophotometers and Spectrometers* ▪ *Gas Chromatographs* ▪ *Ion Mobility Spectrometer* ▪ *Particulate Monitors* ▪ *Calibration*

**Summary**

**Bibliography**

# Monitors Intended for One Compound or Group of Compounds

## Combustible Gas Monitors

Many portable direct-reading instruments for the measurement of combustible gases and vapors are commercially available. The operator must be thoroughly familiar with the calibration, use, and limitations of the instrument being used because an elevated level of combustible gases in the environment could be life-threatening. These instruments are based on one of two principles: the change in resistance of a conductor subjected to heat released by gas combustion, or the change in electrical conductivity of a metallic oxide semiconductor in the presence of a combustible gas. Both types require calibration using a reference gas (often pentane or hexane) and must be interpreted correctly.

**Explosive (flammable) limits (LEL/UEL).** When certain proportions of combustible vapor are mixed with air and a source of ignition is present, a fire or explosion can occur. The range of concentrations over which this occurs is called the explosive (or flammable) range. The low end of this range is called the lower explosive (or flammable) limit (LEL), and the high end is called the upper explosive (or flammable) limit (UEL). The explosive (flammable) range and the lower and upper explosive limits are expressed as volume percents. If the atmosphere is above the UEL, dilution with fresh air could bring the mixture into the flammable or explosive range, so any atmosphere with a flammable or explosive gas near or above the UEL should be considered a significant explosion hazard.

On the simplest type of combustible gas instrument, only one sensitivity is provided, usually with readings from 0 to 100 percent of the LEL. Different models of combustible gas meters are supplied with meters that range from 5 percent of the LEL to 100 percent by volume of the combustible gas, full scale.

**Instrument design.** There are three different detector types on which combustible gas monitors are based. One type is based on the release of heat when the combustible gas or vapor is burned (oxidized). The second is based on the change in electrical conductivity of a metal oxide semiconductor (MOS) sensor as a result of the adsorption of the combustible gas on its surface. The third, much less common, is based on the change in thermal conductivity of the atmosphere in the presence of a combustible gas. All three types are commercially available, usually in conjunction with other monitors built into the same instrument (multigas monitors).

In the first type of combustible gas monitor, the device has a heated, catalyst-coated filament, part of a Wheatstone bridge circuit, that causes the combustion of the gas, generating heat (see Figure 17–1). The heat released

**Figure 17–1.** A multigas monitor that uses catalytic decomposition to measure combustible gases. (Neotronics of North America, Inc.)

causes a change in the electrical resistance of the filament, producing an imbalance in the circuit that can be measured electrically and translated into a concentration. A schematic illustration of the basic flow system for a combustible gas meter and the wiring diagram of a Wheatstone bridge circuit are shown in Figure 17–2.

In the second type of combustible gas meter (Figure 17–3), an MOS sensor adsorbs the atmospheric contaminant and the electrical conductivity of the sensor is changed as a result. Again, this change in electrical conductivity is translated into a concentration of combustible gas. The sensitivity of the MOS sensor to a given com-

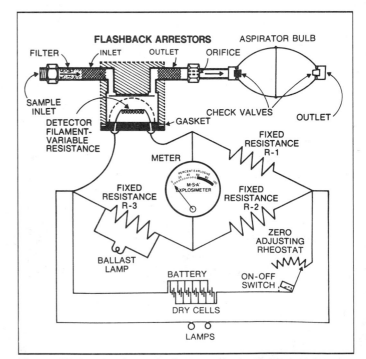

**Figure 17–2.** A schematic diagram of a typical hot-wire combustible gas monitor. (Courtesy MSA.)

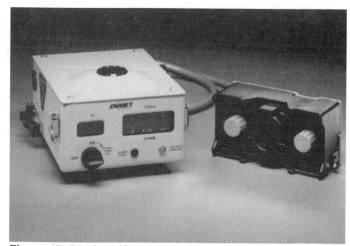

**Figure 17–3.** A multigas monitor that uses a metal oxide semiconductor detector to measure combustible gases. A detachable sensor head for remote sampling is one feature of this instrument. (Courtesy ENMET Corporation.)

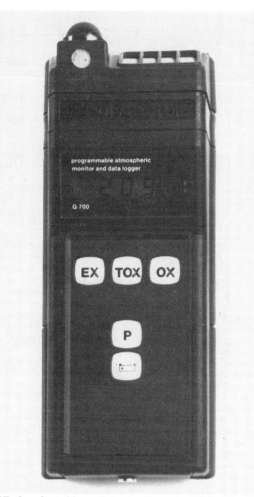

**Figure 17–4.** A multigas detector that can use a thermal conductivity probe for high concentrations of combustible gases (0–99.9 percent by volume). (Courtesy GfG Gas Electronics, Inc., 200 S. Hanley, St. Louis, MO 63105.)

pound can be altered by changing the surface temperature of the sensor through a resistance heater imbedded in the MOS bead.

In the third type of instrument (Figure 17–4), the atmosphere with the combustible gas passes over a heated filament, again part of a Wheatstone bridge circuit. The change in thermal conductivity of the atmosphere as a result of the presence of a combustible gas causes a change in the filament temperature, and thus the electrical resistance. This imbalance in the bridge circuit can be translated into a concentration.

Because the heat of combustion, the adsorptive properties of a combustible gas on the surface of an MOS, and thermal conductivity are all compound-dependent, instrument response for all these instruments is compound-dependent. Similarly, the lower explosive limits of combustible gases are also compound-dependent. Thus, a combustible gas meter is typically calibrated with a particular compound (often pentane or hexane) such that the concentration of other combustible gases present would be overestimated by an instrument calibrated with methane. Instrument manufacturers often provide calibration curves or tables for a variety of different combustibles, for use in correlating meter readings to the concentration of nonmethane gases and vapors. Figure 17–5 is an example of such a table.

All combustible gas meters that rely on a heated wire or filament—an ignition source—have a flashback arrestor that prevents the combustion in the detector from spreading to the atmosphere outside the instrument. This means that they are intrinsically safe (that is, they can be operated safely in flammable or explosive atmospheres). The manufacturer's instructions for operating a combustible gas meter should be carefully reviewed before the device is used. In general, combustible gas meters require a brief

initial warmup period so that the batteries can heat any components that operate at an elevated temperature.

Air is drawn through the sampling probe and into the detector by means of a small sampling pump or a hand-operated squeeze bulb. In some cases, air diffuses into the instrument without being actively drawn in. In most work areas, the concentration of combustible gas or vapor fluctuates constantly and it is necessary to observe the instrument carefully to determine average and peak readings. Some instruments have built-in data-logging features that can store and recall integrated average and peak measurements.

**Zero adjustment.** The zero adjustment must be made by taking the instrument to a location that does not contain combustible gases or by passing air into it through an activated carbon filter that removes all combustible vapors and gases (except methane). Because methane is not removed by activated charcoal filters, extra caution is required during zeroing if the presence of methane is suspected. In addi-

**CALIBRATION GAS**

| GAS BEING SAMPLED | ACETONE | ACETYLENE | BUTANE | HEXANE | HYDROGEN | METHANE | PENTANE | PROPANE |
|---|---|---|---|---|---|---|---|---|
| Acetone | 1.0 | 1.3 | 1.0 | 0.7 | 1.7 | 1.7 | 0.9 | 1.1 |
| Acetylene | 0.8 | 1.0 | 0.7 | 0.6 | 1.3 | 1.3 | 0.7 | 0.8 |
| Benzene | 1.1 | 1.5 | 1.1 | 0.8 | 1.9 | 1.9 | 1.0 | 1.2 |
| Butane | 1.0 | 1.4 | 1.0 | 0.8 | 1.8 | 1.7 | 0.9 | 1.1 |
| Ethane | 0.8 | 1.0 | 0.8 | 0.6 | 1.3 | 1.3 | 0.7 | 0.8 |
| Ethanol | 0.9 | 1.1 | 0.8 | 0.6 | 1.5 | 1.5 | 0.8 | 0.9 |
| Ethylene | 0.8 | 1.1 | 0.8 | 0.6 | 1.4 | 1.3 | 0.7 | 0.9 |
| Hexane | 1.4 | 1.8 | 1.3 | 1.0 | 2.4 | 2.3 | 1.2 | 1.4 |
| Hydrogen | 0.6 | 0.8 | 0.6 | 0.4 | 1.0 | 1.0 | 0.5 | 0.6 |
| Isopropanol | 1.2 | 1.5 | 1.1 | 0.9 | 2.0 | 1.9 | 1.0 | 1.2 |
| Methane | 0.6 | 0.8 | 0.6 | 0.4 | 1.0 | 1.0 | 0.5 | 0.6 |
| Methanol | 0.6 | 0.8 | 0.6 | 0.5 | 1.1 | 1.1 | 0.6 | 0.7 |
| Pentane | 1.2 | 1.5 | 1.1 | 0.9 | 2.0 | 1.9 | 1.0 | 1.2 |
| Propane | 1.0 | 1.2 | 0.9 | 0.7 | 1.6 | 1.6 | 0.8 | 1.0 |
| Styrene | 1.3 | 1.7 | 1.3 | 1.0 | 2.2 | 2.2 | 1.1 | 1.4 |
| Toluene | 1.3 | 1.6 | 1.2 | 0.9 | 2.1 | 2.1 | 1.1 | 1.3 |
| Xylene | 1.5 | 2.0 | 1.5 | 1.1 | 2.6 | 2.5 | 1.3 | 1.6 |

**Example:** The instrument has been calibrated on methane and is now reading 10% LEL in a pentane atmosphere. To find actual %LEL pentane, please multiply by the number found at the intersection of the methane column (calibration gas) and the pentane row (gas being sampled) ... in this case, 1.9. Therefore, the actual %LEL pentane is 19% (10 x 1.9).

**Multiplier accuracy is ±25%, subject to change without notice pending additional testing.**

**If the sensor is used in atmospheres containing unknown contaminants (silicone, sulfur, lead, or halogen compound vapors) methane is the recommended calibration gas. Periodic comparison of methane and pentane readings is recommended when using this chart. Contact Industrial Scientific for details.**

**Figure 17–5.** A table of correction factors for the catalytic combustion sensor, based on the gas used for calibration. The reading on the instrument is multiplied by the correction factor, using the appropriate column for the calibration gas used, to get the approximate reading for the combustible gas being measured. (Courtesy Industrial Scientific Corp.)

tion, the charcoal filter should be changed periodically because it becomes inactivated by moisture or hydrocarbon saturation.

**Interpretation of meter readings.** The user of any instrument should be thoroughly familiar with the necessary precautions. Users of combustible gas meters must be aware of interfering gases and vapors that could create discrepancies in instrument response. All instruments are subject, to some extent, to interferences from noncombustible and nonexplosive gases. For example, the presence of argon, which has a lower thermal conductivity than air, could create a false positive reading in combustion and thermal conductivity detectors. As a precaution, the least-sensitive LEL scale (generally 0–100 percent of the LEL) should be used first to determine whether an explosive atmosphere exists and to prevent overloading of a more sensitive (0–10 percent of LEL) scale. The typical meter responses to methane gas are shown in Figure 17–6 at the LEL, in the explosive range, and above the UEL.

If the indicator of the meter moves above the UEL and remains there, an explosive concentration of gas or vapor is present. However, if the meter climbs rapidly and then falls back to zero, there is either a concentration above the UEL or a gas mixture that lacks sufficient oxygen to support combustion. The instrument may read zero for several different reasons. Assuming that the instrument is functioning properly, the absence of an instrument response can mean either that there is little or no combustible gas in the space being tested or that the concentration is significantly above the UEL and combustion cannot occur because of insufficient oxygen.

Great care must be exercised to ensure that a reading above the UEL is not misinterpreted as a true zero reading. Figure 17–7 is a reminder of the importance of proper interpretation of the instrument readings, in this case in evaluation of the atmosphere in a confined space. A very high concentration of combustible gas can be identified by carefully watching the needle as the probe is moved into and withdrawn from the space being tested. At some point during entry and withdrawal, the instrument will exceed the LEL if a level above the LEL is actually present. These instruments should not be used to measure the concentration of combustible gases in steam or inert atmospheres because of the measurement uncertainties or interferences in nonair atmospheres.

**High-flash-point solvents.** Although it is relatively easy to operate a combustible gas indicator to detect a flammable gas or vapor, these instruments have some limitations. They respond only to combustible vapors drawn into the detector cell. If the vapor pressure of a combustible liquid is relatively low at room temperature, a relatively low concentration will be indicated. If a closed vessel holding a liquid contaminant is later heated (by welding or cutting, for example) the vapor concentrations will increase and the concentration of the substance in the atmosphere of the container, which originally was quite low, may increase and become explosive. Continuous monitoring may be recommended in this situation. When testing the atmosphere in drying ovens or other places where the temperature is unusually high, there may be some difficulty in measuring solvents (such as naphthas) that have a relatively high boiling point because the vapors may condense in the sampling

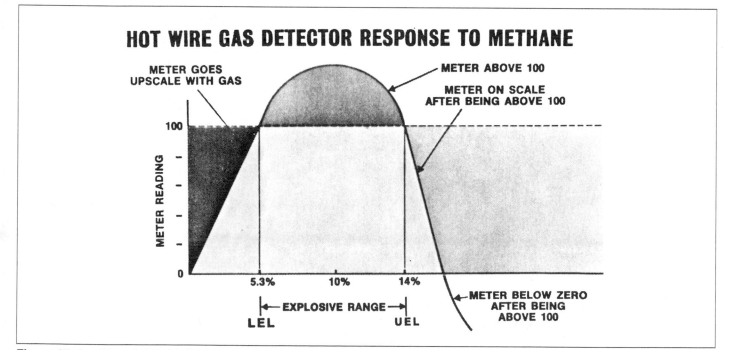

**Figure 17–6.** The relationship between meter reading and combustible gas concentration. (Courtesy MSA.)

**Figure 17–7.** A worker cautiously monitors a confined space in order to evaluate its atmosphere. (Courtesy National Draeger, Inc.)

line, thus giving a false indication of safety. In some instances, condensation can be prevented by heating the sampling line and the instrument to a temperature equal to or above that of the space to be tested.

Several types of combustible gas monitors have been designed to be calibrated so that specific combustibles can be measured. One variation of the instrument has adjustable calibration controls and can measure five different gases or vapors in the 0–100 percent LEL range for each. Another type has a dual-scale multiple-calibration curve in the 0–10 percent and 0–100 percent range of the LEL.

**Catalyst poisoning.** Because minute concentrations of silicone vapors—even 1 or 2 ppm—can rapidly poison the catalytic activity of the platinum filament, a hot-wire combustible gas indicator should not be used in areas where silicone vapors are present.

**Interferences.** Interfering gases and vapors can seriously affect instrument response; an experienced tester recog-

nizes the indications of their presence. Instrument manufacturers' instructions should be followed carefully because high concentrations of chlorinated hydrocarbons (such as trichloroethylene) or acid gases (such as sulfur dioxide) may cause depressed meter readings in combustion-type meters where high concentrations of combustibles are present. Trace amounts of these interferences may not affect the readings directly but can corrode the detector elements. High-molecular-weight alcohols in the atmosphere may burn out the filaments, rendering the instrument inoperative. When such limitations are understood, the tester can obtain reliable and accurate results.

**Other features.** Combustible gas monitors are available with audible and visual alarms. When the concentration reaches the preset limit, an alarm light and a loud alarm are activated, providing visible and audible warnings of a dangerous concentration. Figure 17–8 shows an instrument with a dual alarm. The audible alarm can be switched off; in this case, the pilot light will blink until the unit is reset and the combustible gas concentration falls below the set point. The manufacturer's instructions should be consulted for further details on the operation of such alarms.

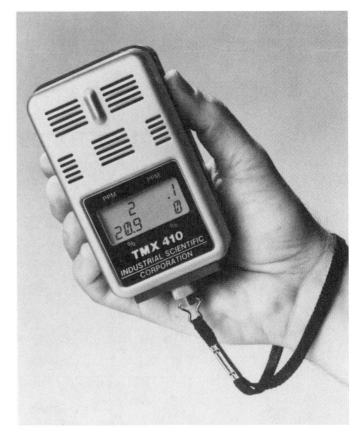

**Figure 17–8.** A multigas monitor that has an adjustable set point and audible and visible alarms. (Courtesy Industrial Scientific Corp., Oakdale, PA.)

## Oxygen Monitors

Although oxygen does not have a specific occupational exposure level, its level in workplace air must often be measured, particularly in enclosed areas where combustion or other processes may use up the available oxygen. Excess oxygen from oxyacetylene or oxyhydrogen flame operation should also be monitored to prevent a fire hazard. Air normally contains about 21 percent oxygen by volume. Sixteen percent oxygen is considered the minimum to support life. In some cases, however, air with less than 19.5 percent oxygen may be considered deficient, such as at high altitudes where atmospheric pressures are lower.

In many locations, such as mines, manholes, tunnels, or other confined spaces, the oxygen content can become low enough to be life-threatening. In such situations, it is necessary to determine the oxygen content of the air. In addition, it is necessary to take a sample to determine whether combustible gases are present in dangerous concentrations. Direct-reading oxygen monitors are small, lightweight, and easy to use. These instruments generally use either a coulometric or polarographic cell to detect oxygen. A few instruments that rely on the paramagnetic property of molecular oxygen are also available commercially.

**Coulometric detectors.** Coulometric detectors rely on the measurement of current flowing in an electrolyte between two electrodes, maintained at a controlled voltage difference, as a result of an oxidation-reduction reaction in the detector cell. The current flow is translated into an airborne concentration of the contaminant undergoing the oxidation or reduction. The most commonly used detector cell for oxygen is a coulometric cell, which has a semipermeable membrane that selectively allows oxygen to enter the cell. One of the electrodes is consumed during the flow of electrons, thus limiting the lifetime of the cell. The cells are temperature-compensated through the use of an external thermistor. Cells from different manufacturers have different response times, accuracies, and temperature and relative humidity performance ranges.

**Polarographic detectors.** Polarographic detectors rely on two parameters: the ability of the compound of interest to be chemically oxidized or reduced at an electrode at a given electrode potential, and the rate-determining step of the discharge of ions at a microelectrode that is determined by diffusion. Polarographic detectors are used to measure oxygen and carbon monoxide in ambient air.

## Carbon Monoxide Monitors

One of the most insidious toxic gas hazards in an industrial atmosphere is carbon monoxide. Odorless, tasteless, and colorless, carbon monoxide can be deadly even in small concentrations. Carbon monoxide can occur in many areas, including gas and utility properties, garages, bus terminals, sewers, vaults, blast furnaces, open-hearth furnaces, and mines. A number of instruments are available for measuring

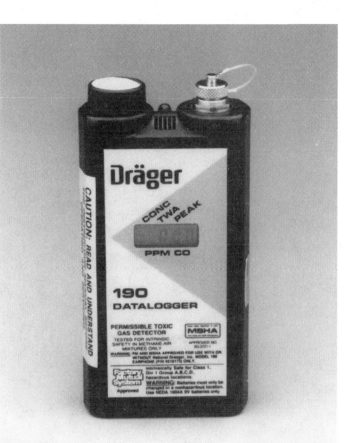

**Figure 17–9.** A direct-reading carbon monoxide monitor capable of data storage and readout. (Courtesy National Draeger, Inc.)

carbon monoxide (Figure 17–9). The most common instruments use a potentiometric or coulometric cell. The resulting voltage or current is translated into a concentration, with temperature compensation factored in. Samples are introduced to the detector cell through a diffusion barrier. Potentiometer cells rely on a change in voltage difference between two electrodes in the presence of a particular air contaminant. Typical portable carbon monoxide detectors feature both visible and audible alarms that alert the user when the danger level is reached. Battery-powered instruments can measure carbon monoxide in the atmosphere in the range of 1–2,000 ppm by volume.

## Indoor Air Quality Monitors

Industrial hygienists increasingly need to assess building air quality and require direct-reading recording instruments for measuring key parameters of indoor air quality. Commercially available devices measure temperature, relative humidity, carbon dioxide, and often several other parameters (chosen by the purchaser) simultaneously, as indicators of the quality of the indoor environment (Figure 17–10). The carbon dioxide monitor and other toxic gas monitors used in these devices are often of the types described in this section. The discussion of temperature and relative humidity

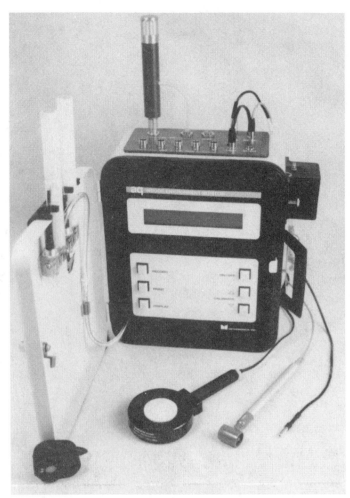

**Figure 17–10.** An indoor environmental monitor for measuring temperature, relative humidity, carbon dioxide, and one other parameter. (Courtesy Metrosonics, Inc.)

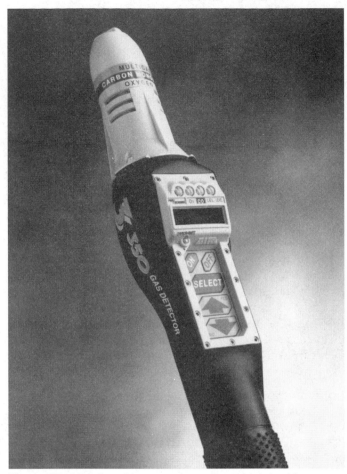

**Figure 17–11.** A multigas monitor using a metal oxide semiconductor sensor for combustible gases and electrochemical sensors for oxygen and toxic gases. (Courtesy AIM Safety USA, Inc.)

measurement is beyond the scope of this chapter, but both parameters can be measured with precision with probes that are readily available. (For more information, see Chapter 21, General Ventilation of Nonindustrial Occupancies.)

## Other Monitors Using Electrochemical or Metal Oxide Semiconductor Detectors

A number of instruments, usually containing multiple sensors, are available for detection of a large number of different compounds (Figures 17–1, 17–3, 17–4, 17–8, and 17–11). These instruments are typically based on electrochemical cells, either potentiometric (galvanic) or coulometric, described in the section on oxygen and carbon monoxide monitoring. They almost always have combustible gas and oxygen deficiency sensors, in addition to other toxic gas sensors. Sensors are commercially available for $H_2S$, $SO_2$, $Cl_2$, $NO$, $NO_2$, $H_2$, HCN, HCl, and $NH_3$.

All of these sensors are affected by other compounds, which interfere with the measurement of the compound of interest. However, sensor specificity for these compounds can be enhanced by adding filters that remove potential

interferences, controlling the voltage in a coulometric cell to minimize unwanted oxidation-reduction reactions, choosing an appropriate sensing electrode that catalyzes only the oxidation or reduction of the chemical species of interest, or introducing a semipermeable barrier into the cell to minimize the entry of interfering gases. For example, one manufacturer of $H_2S$ sensors provides a list of cross-sensitivities for its device, listing 11 compounds and 2 classes of compounds (saturated and unsaturated hydrocarbons) that may interfere. Among these, only HCN, HCl, $Cl_2$, and $COCl_2$ give positive interferences; $NO_2$ gives a negative interference. The sensitivity of the sensor to HCN is half that for $H_2S$, and for the others 0.2 times or less than that for $H_2S$.

At present, multiple-gas monitors are available that can accommodate up to five different detectors at a time. These instruments are typically configured to include combustible gases and vapors, oxygen, and carbon monoxide, all of which are of interest in confined spaces, but could include any compound. Two manufacturers currently manufacture detector cells for 11 different compounds or combustible gases and vapors. (See Carner et al., 1994.)

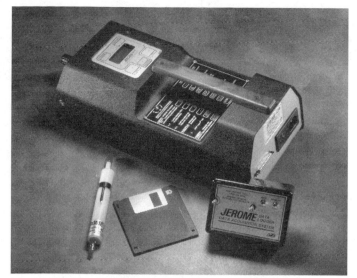

**Figure 17–12.** An instrument for monitoring mercury vapor based on the conductivity of a gold foil. (Courtesy Arizona Instrument Corporation.)

## Mercury Vapor Monitors

Portable, battery-operated ultraviolet analyzers are available for mercury vapor. The section on ultraviolet spectrophotometers has more information on the principles of operation of such devices. In this instrument, the wavelengths of ultraviolet light emitted from a mercury vapor lamp are absorbed by mercury vapor in the ambient air drawn into the instrument. In a dual-beam instrument, the ratio of the intensity of this absorption to that in a reference cell is translated electronically into a concentration of mercury vapor in the air. The specificity of absorption enables the instrument to detect well below 0.05 mg/m³.

Another type of mercury-specific direct-reading instrument (Figure 17–12) relies on the change in electrical conductivity of a gold foil when it comes into contact with mercury vapor to form an amalgam (a solid solution of another metal in mercury). Air containing mercury vapor is drawn into the cell and amalgamates the gold. The conductivity of the amalgam is different from that of the pure gold and the change in conductivity is related to the concentration of mercury in the air sampled during the fixed sampling periods of 1 or 10 seconds. Periodically, after the conductivity changes are significant, the gold foil is heated by an external power source and the amalgam is destroyed as the mercury vapor is driven off by the high temperature. The foil, thus renewed, is ready for a new series of measurements.

## Direct-Reading Colorimetric Tubes and Badges

Direct-reading colorimetric devices use the reaction of an airborne contaminant with a color-producing agent to yield a stain length or color intensity, which can be directly read to provide an instantaneous or time-weighted average value of the concentration of that contaminant. The colorimetric detector tube and badge are widely used by industrial hygienists and other health professionals. Their simplicity of operation, low initial cost, and the availability of multiple types for the detection of numerous contaminants make these popular devices for field use. Nevertheless, like nearly all direct-reading instruments, these devices are limited in applicability, specificity, and accuracy. The user must be familiar with these critical limitations if proper judgments are to be made about appropriate use and about the results.

**Detector tubes.** Colorimetric detector tubes provide a simple and economical method of measuring the exposure of workers to toxic vapors. The tubes are generally not specific for a single compound because nearly all have interferences. In atmospheres that are well-characterized for such interferences, they can be useful for estimating concentrations of certain airborne contaminants. The cost of chemical indicator tubes is considerably less than the cost of a chemical analysis of a sorbent tube in the laboratory. However, the sensitivity of the tubes, their lower accuracy, the possible presence of interferences, and the potential lack of appropriate tubes for determining anything more than instantaneous concentrations are all limitations that must be considered when using these devices.

**Principles of operation.** The hermetically sealed glass tubes contain an inert granular material impregnated with an agent that develops a color when it reacts with the contaminant (Figure 17–13). Sometimes there is a section in the tube or a separate tube that first causes a reaction to take place before the indicating section. For certain inert compounds, a pyrolyzer that thermally decomposes the compound into a form detectable by an indicator tube is available as an attachment to the hand pump. Chemical indicator tubes can be characterized by how the air reaches the active portion of the tube: by active sampling using a hand pump (for short-term measurements) or battery-operated pump (for longer-term measurements) or by passive sampling relying on diffusion. Tubes can be categorized as short-term measurement tubes or longer-term time-weighted average measurement tubes. Some brands of detector tubes are calibrated in milligrams per cubic meter. Conversion from milligrams per cubic meter to parts per million at 77 F (25 C) and 760 mm Hg (standard temperature and pressure) can be performed using the following equation:

$$\text{ppm} = \frac{\text{Milligrams per cubic meter} \times 24.45}{\text{Molecular weight (grams per mole)}} \quad (1)$$

**Active sampling.** In a test using an actively sampled tube, both ends of the indicator tube are broken off and a volume of air is drawn through the tube, using a hand pump or electrically-operated pump (Figure 17–14). Most

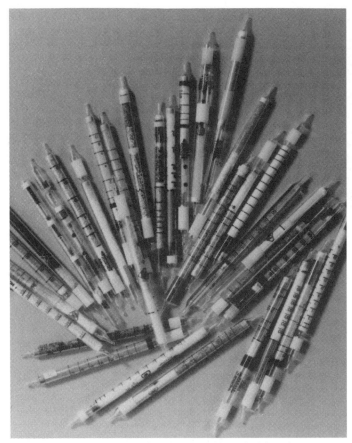

**Figure 17–13.** Length-of-stain tubes intended for short sampling periods. (Courtesy National Draeger, Inc.)

**Figure 17–14.** Manual and automatic pumps for short sampling period tubes. (Courtesy National Draeger, Inc.)

such tubes have an increased sensitivity if larger volumes of air are drawn through the tube. For tubes intended for multiple compound detection, a linear scale may be printed on the tube and the relationship of that scale to contaminant concentration is provided separately. The manufacturer prints a calibration curve on the tube and also provides instructions for interpretation of the stain length when multiples of the minimum volume of air (typically 50 or 100 ml) are drawn through the tube. Tubes used with hand pumps are usually designed to determine average concentrations of an airborne contaminant over periods of 0.5–10 min, depending on the total air volume drawn through the tube. See the American Society for Testing and Materials (ASTM) "Standard Practice for Measuring the Concentration of Toxic Gases or Vapors Using Length-of-Stain Dosimeters" (ASTM D 4599–86) for further information.

Tubes designed for use with a battery-operated pump have been developed for determination of longer-term (1–8 h) time-weighted average concentrations. The color development principle on which they are based is often identical to that for short-term tubes. However, the readings on the tube are often given as ppm-hours and the time-weighted average concentration of the contaminant is determined by dividing the reading by the sampling time (in hours).

**Flow rate.** Flow rates for length-of-stain devices must be maintained in accordance with the manufacturer's operating instructions and the flow rates of the pumps used with length-of-stain devices must be checked periodically. Proper flow rate ensures an appropriate residence time of the air sample in the device and provides sufficient time for the contaminant to react with the chemicals in the detector tube. To obtain meaningful test results, the residence time must be the same as that used to develop the color chart or length-of-stain chart supplied by the manufacturer.

**Passive monitors.** Sampling for contaminants using some colorimetric tubes can also be performed without using a pump to draw air into detector tube. These tubes are called passive or diffusional monitors. The driving force moving air into the tube is the difference in contaminant concentration between the ambient air and inside the tube at the point of reaction (where it is effectively zero). Some passive sampling tubes can be used to sample the atmosphere for several hours, whereas others are intended for only short sampling periods. Passive colorimetric monitors generally are calibrated in ppm-hours. The time-weighted average concentration during the period the tube is exposed to the air equals the ppm-hours indicated on the tube divided by the number of hours the tube was open.

Although most passive tubes function on the same chemical principle as actively pumped tubes, a few passive tubes are calibrated from the closed end of the tube: The air contaminant diffuses into the closed end of the tube and then creates the stain as it rediffuses back toward the open end, chemically modified.

One manufacturer of passively sampling chemical indicator tubes has shown that wind velocities below 0.011–0.022 mph (0.5–1.0 cm/s) result in undersampling of the contaminant, giving a lower reading than the true value. Therefore, these devices should not be used as area samplers where there is little or no air movement. Conversely, they recommend that the devices not be used in situations in which the air velocity past the tube opening exceeds 5.6 mph (250 cm/s), lest the contaminant be oversampled, giving a concentration higher than the true value.

**Interpreting the results.** It is important to recognize that some color stains fade or change with time. Thus, readings of stain length should be made as promptly as possible or in accordance with the manufacturer's recommendations. The ability to read color-change detector tubes and badges depends on the color perception of the observer and the lighting conditions. The exposed devices should be examined in an area with daylight or incandescent illumination rather than fluorescent lighting. Mercury vapor lamps should generally be avoided because the color change may not be visible and the end of the color stain may be difficult to perceive.

With most length-of-stain tubes, the stain front may not be sharp, so the exact length of stain cannot be readily determined. It could be helpful to obtain the results of a calibration test performed on known concentrations before using the tubes in the field. The National Institute for Occupational Safety and Health (NIOSH) has specified that such tubes must yield a concentration value within ±25 percent of the true value, as determined by a reference method, at the occupational exposure limit.

Obviously, performing reliable tests with indicating tubes requires careful use and thorough knowledge of their limitations. Experience has shown that the following measures help to minimize some errors:

- Test each batch of tubes with a known concentration of the air contaminant to be measured.
- Read the length of stain in a well-lighted area.
- Comply with the manufacturer's expiration date and discard outdated tubes.
- Store detector tubes in accordance with the manufacturer's recommendations.
- Refer to the manufacturer's data for a list of interfering materials.

**Specificity.** Most colorimetric detector tubes, both passive and active, are intended to measure a specific compound (or group of compounds) such as hydrogen sulfide, chlorine, mercury vapor, alcohols, or hydrocarbons. Because no device is completely specific for the substances of interest, care must be taken to ensure that interferences do not invalidate the sampling results. Specificity is one of the primary considerations in selecting a detector system. In most cases, the manufacturer has identified interfering substances and conditions and has included this information in the instructions enclosed with the tubes. Sometimes a preconditioning section is used in the detector tubes to remove potential contaminants, convert the gas or vapor of interest to a more suitable reacting compound, and react with the gas or vapor with the release of a new gas or vapor that can be measured by the second section.

Chemical reactions that occur in the detector tubes are temperature-dependent. The tube's instructions give an acceptable temperature range in which it is usable. Drastic differences in the temperature affect the volume of air going through the detector tube, but the uncertainty produced by the effect of temperature on volume is modest compared with other uncertainties in the measurement. Interchanging tubes obtained from various manufacturers will lead to erroneous results because the sampling rates of the various pumps are not the same, nor are the reaction rates of the chemical reagents in the indicator tubes. Each manufacturer produces, calibrates, and sells equipment as an integral system; tubes produced by one manufacturer and pumps produced by another should not be mixed.

**Shelf life.** The shelf life of detector tubes is a critical consideration because the tubes may not be used very often and, therefore, may not be used within the manufacturer's expiration date. Often, tube life can be extended by storage under refrigeration, but tubes should not be used beyond the expiration date unless the response of the expired tubes is compared with that of tubes that have not yet expired or tested with a calibration gas. Freezing temperatures should not adversely affect a tube's shelf life; however, the tubes must be warmed to room temperature before use. Detector tubes should be stored at temperatures below 86 F (30 C) and never in direct sunlight.

**Certification of chemical detector tubes.** Before September 1985, many chemical detector tubes were certified by NIOSH. The NIOSH program was designed to ensure that commercial detector tubes complied with established performance specifications. However, as a result of budget cuts, NIOSH eliminated the certification program for detector tubes. A private organization, the Safety Equipment Institute (SEI) of Arlington, Virginia, filled the void left by NIOSH's departure from the tube certification program. The SEI now certifies detector tube systems through a program similar to that established by the NIOSH, which involves product testing and quality assurance audits conducted by designated, third-party independent laboratories. In fact, the SEI program adheres to the same test standard established by the NIOSH program: Title 42, Part 84 of the *Code of*

*Federal Regulations* (42 *CFR* 84). The SEI offers certification of manufacturers' product models and grants the right to use the SEI certification mark if the testing laboratory has determined that the product models submitted have been tested according to the appropriate standard and the quality assurance auditor has determined that the manufacturer has complied with SEI quality-assurance requirements. The SEI certifies a manufacturer to produce a gas detector tube unit if it meets the minimum requirements set forth in the regulations (basically ±35 percent accuracy at one half the exposure limit and ±25 percent at one to five times the exposure limit). The quality of future production lots is secured by a quality assurance plan, which the SEI approves as part of the certification process. Adherence to the quality assurance plan is verified by periodic plant inspections and by testing samples obtained from actual inventory.

## Other Colorimetric Direct-Reading Devices

Both passive and active colorimetric indicator badges rely on the contaminant gas or vapor reacting with an indicating reagent to yield a uniform color change in the reactive portion of the badge. In some devices, the color changes as a function of time as well as concentration, so the user must note the duration of exposure and refer to a plot of color intensity versus time for a given concentration of the contaminant. The color may be compared against a color comparator, which provides a reference color intensity for a given number of ppm-hours. Both passively and actively pumped indicator badges are available for a number of contaminants, including toluene diisocyanate (TDI), carbon monoxide, phosgene, and ammonia. In an actively sampled badge, air is drawn with a sampling pump through a chemically treated porous paper for a fixed time period and a color appears if the contaminant is present. In a passively sampled badge, the ambient air simply diffuses up to the surface of the device and reacts to form the color. The color developed is a function of the concentration of the contaminant present and the exposure time of the device.

**Colorimetric tape samplers.** Another device that uses the color change resulting from a chemical reaction to measure air contaminants is a colorimetric tape sampler. In modern devices, a chemically treated paper tape is drawn at a constant rate over the sampling orifice; the air contaminant drawn through the tape reacts with the chemical to produce a stain. The intensity of color is directly related to the concentration of the contaminant and is read optically and then displayed on a digital readout. Chemically impregnated paper tape devices have been in use since the early 1950s. The first units were developed to detect hydrogen sulfide ($H_2S$). A filter paper was impregnated with a lead acetate solution that produced a dark stain (lead sulfide) when exposed to $H_2S$. The concentration of $H_2S$ was then determined by measuring the transmission of light through the stained paper.

Light of equal intensity (from a common source through matched fiber optics) is directed to both the top and bottom track and measured by a set of matched photoelectric detectors mounted at an angle of 45 degrees. The difference in reflected light is then measured. The system thereby compensates for slight tape variations. This is illustrated in Figure 17–15, which schematically illustrates the general principle of operation. This diagram also shows the capstan-driven cassette, which moves the impregnated paper tape past the exposure orifice and readout section of the optical block and gate assembly. The manufacturer of this particular instrument markets 20 different impregnated tapes, for compounds ranging from aliphatic amines/ammonia to sulfur dioxide.

Another type of direct-reading colorimetric monitor, used for monitoring TDI or methylene di-isocyanate (MDI), is shown in Figure 17–16. Field calibrations can be accomplished at any time by use of a test strip provided with each monitor. The strip has calibrated stains that are equivalent to the stains produced by known concentrations of the contaminant.

**Colorimetric analyzer.** One somewhat unusual direct-reading instrument is a colorimetric analyzer that takes advantage of the color-producing reactions of a number of airborne contaminants with appropriate reagents (Figure 17–17). The air is drawn into a trapping solution and mixed with the appropriate reagent, and then the colored species produced is analyzed photometrically and a digital readout provided. Each compound has a different reagent system associated with it, and the flow pattern may vary depending on the nature of the compound. Such instruments currently are capable of analyzing at least 16 different compounds.

## ■ Monitors Intended for a Broad Range of Compounds

### Nonspecific Detectors

A number of direct-reading instruments are not compound-specific or are specific for whole classes of compounds. These are of value as leak detectors or in atmospheres that are known to contain only a single contaminant. Nonspecific instruments used for detection of airborne contaminants include devices that contain flame ionization, photoionization, electron capture, and thermal conductivity detectors. The first three types rely on ionization of the contaminant molecules to produce a response. These detectors measure airborne contaminants directly and can be used alone, but are often used in conjunction with a gas chromatograph that separates multiple air contaminants.

**Flame ionization detectors.** Flame ionization detectors are highly sensitive to compounds that ionize in the presence of an oxyhydrogen flame. The ions are collected and

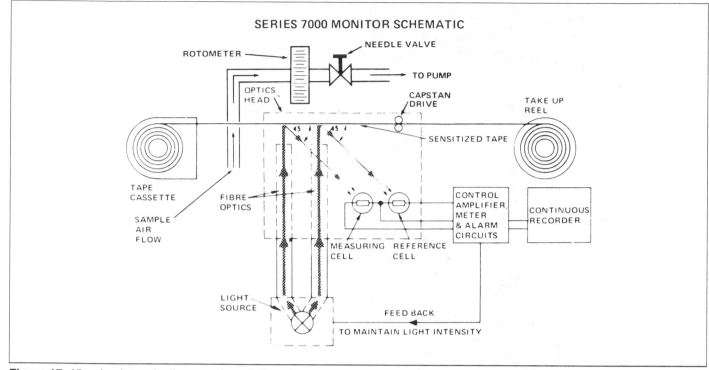

**Figure 17–15.** A schematic diagram of a colorimetric paper tape area monitor usable for a number of different compounds. (Courtesy MDA Scientific, Inc.)

the electric current generated for the compound of interest (whose response factor has been determined) can be translated into a concentration. Organic compounds that have a large number of carbon–hydrogen bonds are detected with great sensitivity with flame ionization detectors, but as the number of C–H bonds decreases (for example, with chloroform [$CHCl_3$]), the sensitivity decreases.

**Photoionization detectors.** Photoionization detectors are sensitive to compounds that are ionized by certain wavelengths of ultraviolet light. The ions produced by the ultraviolet lamp in a photoionization detector are collected and this current is translated electronically into a signal that can be read on the instrument (Figures 17–18 and 17–19). Aromatic hydrocarbons are particularly sensitively detected

**Figure 17–16.** A direct-reading, hand-held paper tape instrument for monitoring certain isocyanates. (Courtesy GMD Systems, Inc.)

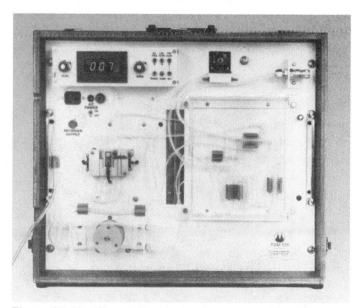

**Figure 17–17.** A portable colorimetric analyzer that can be configured to analyze 16 different compounds. (Courtesy CEA Instruments, Inc.)

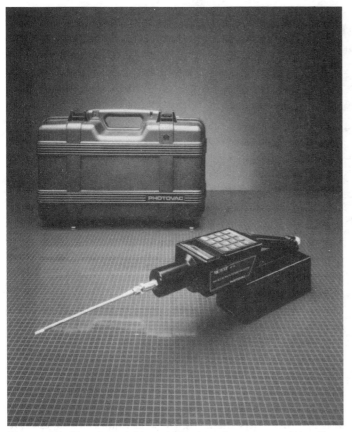

**Figure 17–18.** A portable volatile organic compound monitor using a photoionization detector. (Courtesy Photovac International, Inc.)

with a photoionization detector. There are several different wavelengths of ultraviolet lamp available for some direct-reading photoionization detectors, which can introduce some selectivity into the detection. Stable air constituents such as oxygen and nitrogen are not ionized by

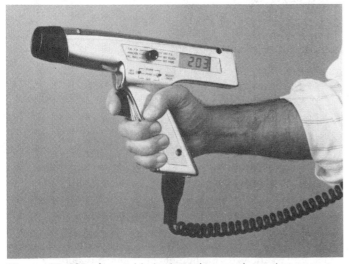

**Figure 17–19.** A portable hydrocarbon analyzer that uses a photoionization detector. (Courtesy Sentex Systems, Inc.)

photoionization detectors. Most hydrocarbons (except methane) cause a response on a photoionization detector. Photoionization detectors respond to water vapor, so changes in absolute humidity between where the instrument was calibrated and where it is used could introduce errors into the measurement.

**Electron capture detectors.** An electron capture detector relies on the ability of the compound of interest to capture primary and secondary electrons from a small radioactive source (typically tritium, $^3$H, or $^{63}$Ni) and thus attenuate a current flowing from the radioactive source to a collector electrode. The electronegativity of the elements (the most electronegative elements are in the upper-right portion of the periodic table) that make up the compound determines the sensitivity of the electron capture detector to the compound. Thus, halogen-containing compounds, as well as those containing nitrogen or oxygen, are detected with high sensitivity by an electron capture detector. Portable or transportable electron capture detectors are used to evaluate fume hood performance through the release of sulfur hexafluoride ($SF_6$) at the face of the hood. Sampling points outside the hood draw air samples into the electron capture detector, which indicates the hood's capture efficiency by comparing the external concentration of $SF_6$ with its internal concentration.

**Thermal conductivity detectors.** Thermal conductivity detectors rely on the change in the ability of contaminated air to transmit or conduct thermal energy. Air with the contaminant is passed over one leg of a Wheatstone bridge in which the filaments are heated by a current flowing through them. (See the section on combustible gas meters for more details.) The change in thermal conductivity of the measurement leg versus the reference leg causes a change in temperature in one leg, inducing an imbalance in the bridge circuit and a resultant measurable electrical voltage. This can be translated into a concentration for a known contaminant.

## Spectrophotometers and Spectrometers

A number of direct-reading instruments are based on the characteristic absorption of electromagnetic, ultraviolet, visible, or infrared radiation. Devices used to measure many organic compounds and mercury vapor are based on this principle.

**Infrared analyzers.** Many gases and vapors, both inorganic and organic, absorb certain characteristic frequencies of infrared radiation. This property and the resultant infrared spectrum can be used to identify and quantify compounds in the air that absorb in the infrared region. This includes most compounds except the diatomic molecules of hydrogen, nitrogen, and oxygen. An infrared source in the analyzer emits the full frequency range of infrared radiation. The window material in the cell may absorb certain frequencies, and thus limits the frequencies that can be used.

In a dispersive instrument, the radiation is separated into its component wavelengths with a prism or grating and the desired wavelengths are directed through the sample and onto a detector. In a nondispersive instrument, the infrared radiation is passed through a filter, then through the air sample and onto a detector.

In a double-beam instrument, the infrared energy passes through two cell paths simultaneously. At the opposite end is a detector that measures the energy transmitted through the two cells. One of the cells is the sample cell. The other is a sealed comparison cell with a special mixture inside. If the gas in the sample cell contains a gas that absorbs energy at the selected frequency, then the detector will detect less energy coming through the sample cell than through the comparison cell. The detector emits an electrical signal to alert the user to this imbalance. The same would be true with a single-beam instrument, but the infrared absorption background of uncontaminated air would not be subtracted out of the signal.

One battery-operated portable instrument can generate an entire infrared spectrum and has a preprogrammed library of compounds in its memory, so that the instrument automatically determines the correct wavelength to monitor for the compound of interest (Figure 17–20). When a single contaminant is present, identification and measurement are achieved easily. Lightweight instruments using filters are available for single specific compound detection. When a number of absorbing contaminants are present, separation of the contaminants may not be possible, depending on the differences in infrared absorption spectra among the compounds of interest. If the spectra do not overlap significantly, analysis of multiple compounds in the same sample, and thus compound specificity, is possible, especially with a dispersive instrument with a much narrower bandwidth.

**Figure 17–21.** A transportable, direct-reading photoacoustic spectrometer using 115 V AC power. (Courtesy Bruel & Kjaer Instruments, Inc.)

**Photoacoustic spectrometers.** Direct-reading instruments that are not yet portable or battery-operated, but are readily transportable, are photoacoustic spectrometers (Figure 17–21). The instruments rely on the absorption of a characteristic band of wavelengths of infrared radiation within the detector cell. The absorption of the infrared radiation causes slight heating, and thus expansion, of the gas contained in the cell. The measurement of the change of pressure in the cell is translated into a concentration of the contaminant present in the cell. The measurement is specific only if no other contaminant present absorbs infrared radiation significantly in the same band of wavelengths.

**Ultraviolet analyzer.** Just as many organic compounds absorb characteristic wavelengths of infrared radiation, there are also a number of compounds that absorb characteristic wavelengths of ultraviolet radiation. Whereas the absorption of infrared radiation by a molecule depends on the existence of a dipole moment in the molecule and causes excitation of the rotational and vibrational emissions of the molecule, ultraviolet radiation causes electronic excitation of atoms or molecules (sometimes enough to cause ionization, as in a photoionization detector) at ultraviolet wavelengths characteristic of the species. A portable ultraviolet spectrophotometer that operates in the wavelength range of 200–400 nm was developed in the early 1990s, but has not been commercialized. It uses a detector system (256-element diode array) capable of looking at the entire ultraviolet absorption spectrum simultaneously. A library of up to 30 different absorption spectra can be programmed into the analyzer. This library and the multielement detector enable the instrument to analyze 30 gases simultaneously in a single 3-minute analysis.

## Gas Chromatographs

Chromatography is a method of separating complex mixtures. In gas chromatography, the components of a volatile mixture migrate differentially through a separating column,

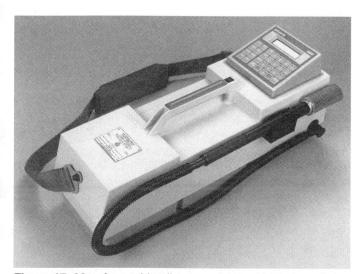

**Figure 17–20.** A portable, direct-reading ambient air analyzer with an infrared spectrophotometer. (Courtesy The Foxboro Company.)

transported by a carrier gas passing through the column. The extent of separation depends on the chemical and physical properties of the molecules being separated, the nature of the separating column, its temperature, and the flow rate of the carrier gas. Optimally, differential migration takes place and each component separates as a discrete substance. Detection of the separated components takes place as the carrier gas emerges from the column. Detectors used for portable gas chromatographs include flame ionization, photoionization, electron capture, argon ionization, flame-photometric, and thermal conductivity detectors. All but the argon ionization and flame photometric detectors are described in the section of this chapter on nonspecific detectors.

Argon ionization detectors rely on a somewhat more complicated energy transfer process to form ions. Argon gas, used as a carrier gas for the gas chromatograph, is electronically excited by decay of a radioactive material in the detector. This produces electrons that are accelerated to excite the Ar atoms to a particular higher-energy state. Collisions between the contaminant and the excited Ar atoms result in a transfer of energy from the Ar atom to the contaminant, resulting in ionization of the latter. The ions formed from the contaminant are collected and the resulting signal is translated into an instrument response.

Flame photometric detectors are based on the emission of certain wavelengths of light by an element introduced into a flame. Flame photometric detectors are generally used to detect compounds containing sulfur. Sulfur emits characteristic wavelengths of light in the range of 300–423 nm when heated to high temperatures; this light, passed through an optical filter or prism, generates a signal in a photosensitive detector. The intensity of the signal is related to the amount of the element in the flame.

Quantitative analysis for a specific component requires separation of the component from other compounds in the sample and identification and quantitation using a calibrated detector. The use of retention time (the length of time required for a compound to pass through the chromatographic column) is a common, though somewhat ambiguous, means of identification. If the possible components of the atmosphere being analyzed are understood, then misidentification possibilities are minimized. Calibration of the detector requires the introduction of known amounts of the compound into the detector and the determination of the relationship between the amount introduced and the instrument response.

Portable gas chromatographs may involve sample injection directly from the air, thermal desorption from a solid sorbent or preconcentration from an air sample followed by thermal desorption. Column backflush capability also exists, as do instruments that are preprogrammed for a specific group of compounds. Preprogrammed instruments recognize the retention time for a given compound

**Figure 17–22.** A hand-held, preprogrammed gas chromatograph with a photoionization detector. (Courtesy Photovac International, Inc.)

and apply a predetermined response factor to that compound, giving a direct reading of airborne concentration. Instrument reliability, ease of operation, ease of calibration, and instrument reliability are key considerations when defining the minimum technical skill required to operate the device. Two types of portable gas chromatographs are shown in Figures 17–22 and 17–23.

## Ion Mobility Spectrometer

A relatively new device available for air contaminant monitoring is the ion mobility spectrometer (Figure 17–24). In this device, ion–molecule reactions are initiated by the electrons emitted from a radioactive $^{63}$Ni source. Using an electric field to separate out ions of a certain charge before they recombine, the instrument injects these ions into a drift region, where they are separated by the time-of-flight through that region and then migrate to a collector electrode. The migration is related to the mass and size of the species and the temperature and pressure. This type of instrument was developed for rapid detection of toxic gases in combat situations, but is now used for the monitoring of other gases in the workplace. The instrument can be programmed for any number of compounds, but is extensively used for toluene diisocyanate monitoring because of the ceiling occupational exposure limit for this compound and the rapid response of the ion mobility spectrometer.

**Figure 17–23.** A portable gas chromatograph with a high-sensitivity photoionization detector. (Courtesy Photovac International, Inc.)

## Particulate Monitors

Several types of direct-reading monitors are used to measure airborne particulate concentrations. More precisely, these devices are generally aerosol monitors in which the aerosol is a solid (dust), liquid (mist), or condensed vapor from a high-

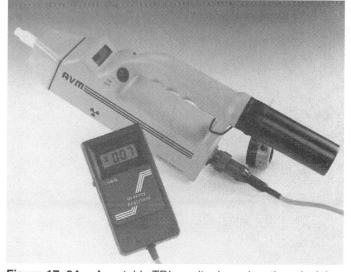

**Figure 17–24.** A portable TDI monitor based on the principles of ion mobility spectrometry. (Courtesy Graesby Environmental, Atlanta, GA.)

**Figure 17–25.** A direct-reading aerosol monitor using near-infrared light scattering to detect particles. (Courtesy MIE, Inc.)

temperature process such as combustion or welding (fume). Most of these devices are based on the light-scattering properties of particulate matter, and are sensitive to the size, shape, and refractive index of the particles (Figure 17–25). The light source could be monochromatic or polychromatic, and the scattered light detector a photomultiplier tube or photodiode. Most instruments have a pump that draws a sample into the sensing volume, but there are some in which convection is relied on to do that. The instrument must be calibrated with particulates of a size and refractive index similar to those to be measured in the ambient air. If this is not done, the results indicated on the instrument could easily be off by an order of magnitude or more.

Another type of particulate monitoring device relies on the behavior of a piezoelectric crystal. The frequency of the crystal's oscillations is changed by the amount of particulate matter deposited on it when it carries an electrostatic charge (Figure 17–26). After the sampling period is complete, the concentration of dust is displayed and the

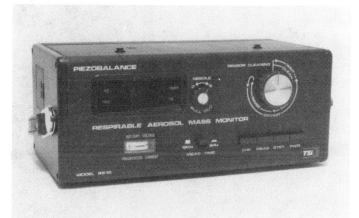

**Figure 17–26.** A direct-reading particulate monitor using a piezoelectric crystal for detection. (Courtesy TSI Incorporated, St. Paul, Minnesota.)

crystal is automatically cleaned and ready for the next cycle. The instrument measures particles ranging in size from 0.01 to 3.5 μm. (An inertial impactor in the device eliminates particles of larger size.)

## Calibration

All instruments used for sampling and analysis of gases, vapors, or particulates must be calibrated before use, and their limitations and possible sources of error must be fully understood. It is very important to establish that an instrument responds properly to the substance it is designed to sample. This is generally carried out by performing calibration procedures with standard concentrations of the substance of interest. In the case of particulate monitors, it may require a standard particle size, as well.

There are a number of commercially available static-type calibration kits for combustible and toxic gases. Typically, they contain gases that can be used for both types of instruments. These kits generally contain one or more cylinders filled with a known concentration of a specified gas–air mixture, a regulating valve, a pressure gauge for measuring the pressure in the container, and a hose adapter that connects the cylinder to the instrument to be checked (Figure 17–27). Once the container kit is attached to the instrument, a sample of the gas–air mixture from the container is permitted to flow into the device. The meter reading of the instrument is then compared with the known concentration of the sample to verify the proper response.

A somewhat more difficult way of calibrating gas or particulate detection instruments requires the generation of the desired mixture by adding the desired contaminants in a known quantity to a known quantity of air. The rate of airflow and the rate at which the contaminant is added to the sample stream must be carefully controlled in order to produce a known dilution ratio. Dynamic systems offer a continuous supply of contaminant, allow for rapid and predictable concentration changes, and minimize the effect of wall losses as the test substance comes into equilibrium with the interior surfaces of the system. A permanent record should be maintained of all calibration procedures, data, and results. The information to be kept for this record includes instrument identification, temperature, humidity, trial run results, and final results. It is important that the operator thoroughly understand how to operate the instrument and know the instrument's intended use and the calibration procedures recommended by the manufacturer.

## ■ Summary

The ultimate goal of the hazard evaluation process is to determine the exact amount of vapor, gaseous, or particulate contaminants present in the work environment. Proper operation of direct-reading instruments used in hazard evaluation is essential to ensure that the information obtained is accurate enough to provide a useful interpretation. Faulty operation of air-sampling instruments can result in either high or low readings. Low readings could falsely indicate that no hazard is present when dangerous conditions might exist; conversely, high instrument readings could lead to the implementation of unnecessary control measures.

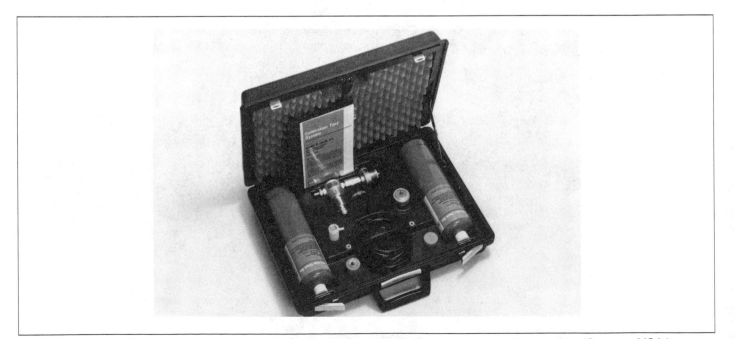

**Figure 17–27.** A calibration test kit for calibrating combustible gas and toxic gas sensors or instruments. (Courtesy MSA.)

# Bibliography

Carner KR, Mainga AM, Zhang X, et al. *Evaluation of Personal Multigas Monitors.* Springfield, VA: National Technical Information Service, 1994.

Clayton GD, Clayton FE, eds. *Patty's Industrial Hygiene and Toxicology,* 4th ed., vol. 1. New York: Wiley, 1994.

Hering SV, Cohen BS, eds. *Air Sampling Instruments for Evaluation of Atmospheric Contaminants,* 8th ed. Cincinnati, OH: American Conference of Governmental Industrial Hygienists, 1994.

Linch AL. *Evaluation of Ambient Air Quality by Personnel Monitoring,* 2nd ed. Boca Raton, FL: CRC Press, 1981.

National Institute for Occupational Safety and Health. *The Industrial Environment—Its Evaluation and Control.* Washington, DC: U.S. Government Printing Office, 1973.

Perper JB, Dawson BJ. *Direct Reading Colorimetric Indicator Tubes Manual,* 2nd ed. Fairfax, VA: American Industrial Hygiene Association, 1993.

Safety Equipment Institute. *Certified Product List, Safety Equipment List: Personal Protective Equipment.* Arlington, VA: Safety Equipment Institute, Oct. 1993.

Willard HH, et al. *Instrumental Methods of Analysis,* 7th ed. Belmont, CA: Wadsworth, 1988.

## PART 5

# Control of Hazards

# CHAPTER 18

# Methods of
# Control

Revised by Susan M. Raterman, CIH, REPA

**Methods of Control**

**Engineering Controls at Design Stage**
*Design* ▪ *Maintenance Considerations* ▪ *Design Specifications* ▪ *Hazardous Materials*

**Industrial Hygiene Control Methods**

**Principles of Engineering Controls**
*Substitution: Changing the Material* ▪ *Substitution: Changing the Process* ▪ *Isolation* ▪ *Ventilation*

**Administrative Controls**
*Reduction of Work Periods* ▪ *Wet Methods* ▪ *Personal Hygiene* ▪ *Housekeeping and Maintenance* ▪ *Maintenance Provisions*

**Special Control Methods**

**Waste Disposal**

**Personal Protective Equipment**
*Respiratory Protective Devices* ▪ *Protective Clothing* ▪ *Eye and Face Protection* ▪ *Hearing Protection*

**Education and Training**

**Health Surveillance**

**Summary**

**Bibliography**

THE GENERAL PRINCIPLES AND METHODS involved in controlling occupational health hazards will be discussed in this chapter. In the field of industrial hygiene, the objective of occupational health hazard control is to ensure that employees' exposure to harmful chemical stresses and physical agents does not result in occupational illness. The variables or quantities of interest that must be measured are the concentration or intensity of the particular hazard and the duration of exposure.

The types of industrial hygiene control measures to be instituted depend on the nature of the harmful substance or agent and its routes of entry into the body. An employee's exposure to airborne substances is related to the amount of contaminants in the breathing zone and the time interval during which the employee is exposed to this concentration. Reducing the amount of contaminant in the employee's breathing zone or the amount of time that an employee spends in the area will reduce the overall exposure.

With employment in the United States shifting from manufacturing to the service sector, many workplaces today present nontraditional occupational health hazards. Industrial hygienists need to possess the skills to implement control methodology in both traditional industrial settings and workplaces such as laboratories, offices, construction sites, and environmental hazard remediation projects. This requires an understanding of the toxicology of a broad range of potential hazards, including biological agents, chemicals, construction materials, and physical stressors, as well as an understanding of process technologies and work practices. A complete understanding of the

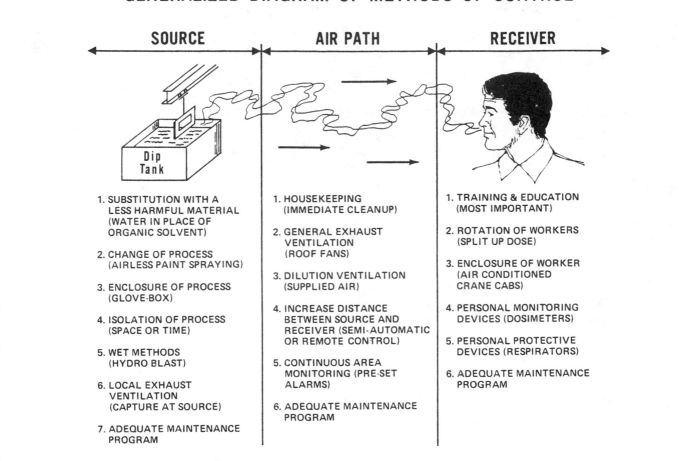

**Figure 18–1.** To determine the extent of exposure, locate the contaminant source, its path to the employee, and the employee's work pattern and use of protective equipment.

circumstances surrounding an exposure hazard is required in choosing methods that will provide adequate control. To lower exposures, the industrial hygienist must first determine the contaminant source, the path it travels to the worker, and the employee's work pattern and use of protective equipment (Figure 18–1). Hazards can change with time, so health hazard control systems require continuous review and updating.

## Methods of Control

The methods of control of health hazards in the work environment are divided into the following categories:

■ Engineering controls, which engineer out the hazard either by initial design specifications or by applying methods of substitution, isolation, or ventilation.

■ Administrative controls that control employees' exposures by scheduling reduced work times in contaminant areas, and/or good work practices and employee

training that includes hazard recognition and work practices specific to the employee's job that can help reduce exposures.

■ Personal protective equipment, which employees wear to protect them from their environment. Personal protective equipment can be used in conjunction with engineering controls and other methods.

Engineering controls are to be used as the first line of defense against workplace hazards wherever feasible. Such built-in protection, inherent in the design of a process, is preferable to a method that depends on continual human implementation or intervention. The federal regulations, and their interpretation by the Occupational Safety and Health Review Commission, mandate the use of engineering controls to the extent feasible, and if these are not sufficient to achieve acceptable limits of exposure, the use of personal protective equipment and other corrective measures may be considered.

Engineering controls include ventilation to minimize dispersion of airborne contaminants, isolation of a hazardous operation or substance by means of barriers, and substitution of a material, equipment, or process to provide control of a hazard. Although administrative control measures can limit the duration of individual exposures, they are not generally favored by employers because they are difficult to implement and maintain. For similar reasons, control of health hazards by using respirators and protective clothing is usually considered secondary to the use of engineering control methods.

# Engineering Controls at Design Stage

The best time to introduce engineering controls is when a facility is in the design phase. At that time, control measures can be integrated more readily into the design than after the facility has been built or the processes are on-line.

The systematic layout of the physical building, processes, and systems should comply with occupational safety and health standards. What is planned must be reconciled with what is permissible by law or advised by consensus standards. In any particular situation, jurisdiction and applicability of standards may become complex. When more than one agency or standard is involved, the more stringent standard can be assumed to be controlling. Consideration should be given to specifying design criteria that comply with proposed standards that may take effect after the facility goes on-line.

It is becoming increasingly common for facility and design engineers to consult with the industrial hygienist at the design phase of a new facility or process. Including industrial hygiene control measures at this point can be less costly than adding them later in the construction process. During the design phase, the proposed facility layout must be characterized with respect to construction type, proposed activities in all areas, and possible health hazards. The influence of one area on another and one work activity on another must be assessed. At this point, ergonomic concerns must be identified and corrected with proper workstation design.

It is also important to evaluate the finished materials within a facility for their propensity to generate hazardous air contaminants. For example, the installation of new carpets, flooring materials, adhesives, and paints can generate volatile organic compounds in concentrations that could result in respiratory and eye irritation in the building occupants.

When air contaminants are created, generated, or released in concentrations that can injure the health of workers, ventilation is the usual method of providing protection. However, other methods of protection should be investigated; one example is automatic operations.

Ideally, operations should be conducted in entirely closed systems, but not all processes lend themselves to this approach. When closed systems are used, raw materials can be brought to the processing site in sealed containers and their contents emptied into storage tanks or containers, minimizing employee contact with the material being processed.

Design all systems and components so that airborne contaminants are kept below their acceptable Threshold Limit Values (TLVs). Do not permit leaking of toxic chemicals from process equipment, such as pumps, piping, and containers, to the working environment to cause a condition in which the TLVs are routinely exceeded in any location where employees may be present. In industrial settings, isolate process equipment and vent to a scrubber, absorber, or incinerator. If feasible, remotely control the process from a protected control room.

Some work operations, if conducted separately, do not present a serious hazard, but when combined with other job operations they can become hazardous in certain situations. Two types of interrelationships can exist.

The first concerns accumulation, as can arise when additional welding stations are provided in a building of fixed general ventilation or when additional noise sources are added to an already noisy work area.

The second type of interrelationship concerns many activities going on in the same area. Activities that by themselves are safe can become hazardous in certain circumstances. For example, vapor degreasing with chlorinated solvents, even when the airborne concentration of the vapors is within permissible limits, may create major hazards when the activity is near work areas where ultraviolet (UV) radiation (from welding arcs, bright sunlight, or molten metal) exists. The decomposition of these solvents caused by the UV radiation can produce phosgene gas—a potent and toxic eye and lung irritant. Merely maintaining the concentration of solvent vapor below the TLV is not satisfactory. The most positive control is to prevent the chlorinated solvent vapors from entering the welding area in any detectable concentrations. If vapors cannot be reduced to a minimum, the UV field should be reduced to a minimum by shielding the welding arc.

The problem of considering safety and health with activity and workstation relationships becomes difficult when more than three or four activities must be considered, as in laying out workstations for new or relocated manufacturing operations where 20 or more activities might require consideration. Decisions should be made to arrive at either an optimum arrangement or a preferred compromise.

## Design

Occupational health hazards can best be minimized by workplace design that controls contaminants as much as possible. This requires close cooperation between the industrial hygienist and the design engineer and architect. The ideal situation would have the principles of health hazard

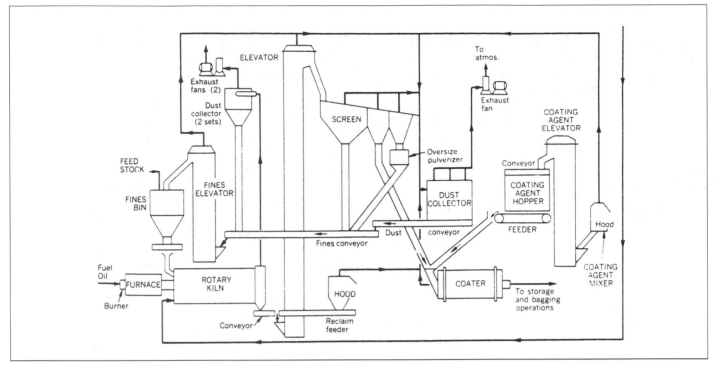

**Figure 18–2.** A simple process flow sheet showing the stepwise introduction of raw material and the product of each step. The extent of physical or chemical hazards that can occur at any step in the operation should be determined.

protection so thoroughly ingrained in the design professionals that the health and safety professional need only be a passive reviewer. However, the design team needs the help of the health and safety professional during the design process to make sure that a system can be set up so that it does not pose safety or health hazards to the operator or facility occupants.

Production processes in chemical plants should be designed so that hazardous materials are not released into the environment. It is important to keep the materials and the by-products and wastes within the closed system.

To maintain that integrity, a chemical process flow sheet should be reviewed from an overall material balance point of view (Figure 18–2).

The material that becomes airborne and gets into the work environment to cause problems can be an insignificant fraction of the total amount of material that is circulated through the system, so much so that in a material balance, the quantity of material that escapes and is released into the workplace that causes the hazard can be insignificant when compared with the total amount present.

Design factors that should be addressed include the following:

■ To what degree is it possible to remove hazardous residues from a piece of equipment before it is opened?

■ To what extent can a system be designed to be relatively maintenance free?

■ Can the system be designed so that the entire operation can be conducted as a closed system?

■ Can the process be conducted automatically without worker involvement?

A design engineer should have extensive knowledge not only of the main aspects of the process being created but also of the finer details, such as health hazard controls and safety devices. Design engineers are usually more familiar with the safety hazards because the effects of their being overlooked are much more obvious that those occurring when health hazards are overlooked.

The same importance should be assigned to minimize contaminant dispersion in other workplace settings such as mixed-use office buildings. The architect and engineer on the design team should address the following factors:

■ Are there any activities taking place in the building that use or generate hazardous materials?

■ Is the fresh air intake located away from any contaminant source or air pathway for these contaminant sources?

■ Has the HVAC system been designed to deliver an appropriate volume of air to each occupied space in accordance with the standards of the American Society of Heating, Refrigeration and Air Conditioning Engineers (ASHRAE)?

■ Is local exhaust ventilation required in any special-use areas, such as printing operations, photo developing, welding, or solvent degreasing?

■ Are any special filters required to clean incoming or recirculated air?

■ If tobacco smoking is permitted in the building, is there a separate room with a dedicated HVAC system planned in accordance with the ASHRAE standard? (See Chapter 21, General Ventilation of Nonindustrial Occupancies.)

When health professionals are involved early in the design process, it is possible to plan the development of sampling and analytical methods to yield exposure data concurrent with the development of the engineering design. Contaminant monitoring systems can be included as part of the engineering design. Elaborate automated leak-detection systems designed into the process can yield valuable information for evaluating health hazards in the operating unit (Figure 18–3).

Similarly, contaminant monitoring systems can be installed in the ventilation system to alert building engineers of high levels of carbon monoxide or carbon dioxide, which serve as a general indicator of degrading air quality.

Neglect of the health professional–engineer–architect interaction in facility design can lead to major management problems. What could have been an easy solution in the design phase can become an extremely difficult problem later. Changes that might have been readily accomplished during the design phase must now be done as a matter of equipment change and compromise. Worse yet, it may be necessary to shut down production or evacuate employees to correct a hazard that was overlooked. Consequently, management should consider that for certain processes and materials, the initial design of facilities to minimize the health hazards may be a significant and necessary part of the investment.

## Maintenance Considerations

It is important to look not only at planned operations but also at the fine details of what is not supposed to be happening. These untoward events may best be described in two general classes.

First, there may be releases of contaminants into the work environment that are relatively continuous, such as flange leaks, exhaust hoods that are not completely effective, pump seals that have weakened, diffusion that occurs along the valve stems, or noise emission from leaks in acoustic lagging on a machine that it does not fit. This general class of airborne contaminants or fugitive emissions may have begun as a low-level background that initially was not high enough to be of serious concern. Coupled with this is another kind of episodic exposure. As equipment becomes worn and starts to leak, the general level of background emissions may eventually result in major worker exposures. Much of this leakage can be dealt with by continuous, careful, intensive maintenance; however, much of it might have been avoided in the initial design. The degree to which any possibility of leakage is engineered or designed out of a system depends to a great extent on how much these potential leaks have been anticipated.

The second class of emissions of airborne contaminants arises when a closed system or control process becomes momentarily open or uncontrolled. For instance, the lagging has to be removed from the compressor in order to perform some adjustments, or perhaps samples have to be collected or filters replaced. These situations are common in chemical industries. A filter change operation may occur as infrequently as once every six months; when things are going badly, it may have to be done four times a shift. The system has to be designed so that it is possible to clean and purge the filter container so that an employee can perform needed maintenance without hazard.

From time to time, the system as a whole must be shut down for cleaning and purging and afterward opened for maintenance. Under these circumstances, most exposures tend to be brief, but exposure levels can be quite high and may be detected only by closely maintained industrial hygiene surveillance on a day-to-day basis.

Knowledge of the hazards that are present and the potential for the exposure that may exist in an operation gives an industrial hygienist an ideal starting point from which to develop the surveillance program of the operation. All too often this step is omitted and the industrial hygienist becomes aware of an engineering project only in the advanced stage of development. Waiting to make changes in the design when the system is about to go to construction can dramatically increase cost.

## Design Specifications

The design specifications are the drawings and documents that enable the engineers and architects to precisely define the building, systems, and processes. The industrial hygienist

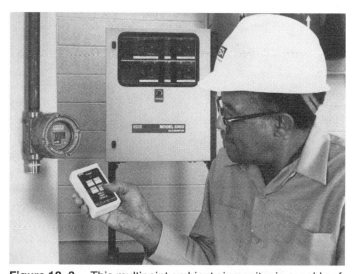

**Figure 18–3.** This multipoint ambient air monitor is capable of continually measuring up to five gases in as many as 24 remote locations. (Courtesy MSA.)

**Figure 18–4.** Loading or unloading of tank trucks can release airborne contaminants.

or safety professional should have a clear understanding of where in these specifications health hazards may occur as a result of the process, building materials, or system design.

**Review.** Before a new operation or process is begun, on-site engineering reviews that go over the whole process should be done to ensure that nothing was forgotten and everything will proceed as planned. Although these reviews are very detailed and time-consuming, it is worthwhile for an industrial hygienist and safety professional to be involved. Sometimes, last-minute changes in the process or equipment are made that can significantly increase or decrease the health hazard.

**Startup.** The industrial hygiene surveillance begins when a process is put into operation or a facility is brought on-line and should continue for as long as the operation continues. Problems in handling and operating procedures that were not anticipated during the design stage will now become apparent. Prompt correction of these problems is much easier during the early setup phase when procedures and people are still somewhat flexible.

**Sample taking.** In many industrial operations, such as steel mills and petrochemical facilities, taking product samples is a common procedure. The design engineer and the industrial hygienist can choose between a product sampling system that does not provide much control and a system that provides almost total control. Each of these choices probably has some cost increment associated with it. The choice should be based on assessment of the severity of the potential health hazard.

**Loading operations.** One of the most serious problems in the field of health hazard control is the loading and unloading of tank cars, tank trucks, and barges. Putting a liquid into a space previously occupied by air or vapor quickly saturates that air with vapor. It may become necessary to go to vented systems, enclosed systems, and automatic loading systems that include vapor recovery so that the vapor that is pushed out of the tank will be recovered (Figure 18–4).

Episodic exposures are difficult to control from an engineering point of view. Also, for these infrequent emergency or nonroutine events, personal protection can be the appropriate solution. However, design engineers should recognize that these exposure events will happen, that product samples must be taken, that equipment must be maintained, and that filters must be changed. The industrial hygienist working with the designer must consider how these operations can be conducted so that the worker need not be overexposed.

## Hazardous Materials

Some materials must be handled carefully because of their toxicity, flammability, reactivity, or corrosivity. The processes and practices to be used must be consistent with the standards applicable to materials with these characteristics.

Stringent controls regulating mutual proximities, ventilation, sources of ignition, and design are imposed on general industry by federal codes. When potentially photochemically reactive solvents are involved, process controls and discharges to the atmosphere are subject to regulation by air-quality regulatory authorities.

Compressed gas and equipment for its use in industry are extensively referenced in the Compressed Gas Association's standards. Methods of marking, hydrostatic testing of cylinders and vessels, labeling, metering, safety devices, and pipework and outlet and inlet valve-connecting are thoroughly described in pamphlets issued by the association.

Standards for the design and use of air receivers are promulgated based on the ASME Boiler and Pressure Vessel Codes. The provision and use of compressed gases in industrial settings must be carefully undertaken; otherwise, catastrophic situations may develop.

# Industrial Hygiene Control Methods

Industrial hygiene control methods for reducing or eliminating environmental hazards or stressors include the following:

- Substitution of a less hazardous material for one that is harmful to health

- Change or alteration of a process to minimize worker exposure

- Isolation or enclosure of a process or work operation to reduce the number of employees exposed, or isolation or enclosure of a worker in a control booth or area

- Wet work methods to reduce generation of dust and avoid dry sweeping of dust

- Local exhaust ventilation at the point of generation or dispersion of contaminants

- General or dilution ventilation to provide circulation of fresh air without drafts or to control temperature, humidity, or radiant heat load

- Personal protective devices, such as special clothing or eye and respiratory protection

- Good housekeeping and maintenance, including cleanliness of the workplace and adequate hygiene and eating facilities

- Administrative controls, including adjusting work schedules or rotating job assignments so that no employee receives an overexposure

- Special control methods for specific hazards, such as shielding, monitoring devices, and continuous sampling with preset alarms

- Employee training and education that is specific to the hazards and includes work methods that help reduce contaminant exposure

- Emergency response training and education

- Waste treatment and disposal

A generalized diagram of these methods is shown in Figure 18–1. Each of these industrial hygiene control methods will be discussed in turn.

# Principles of Engineering Controls

## Substitution: Changing the Material

An often effective industrial hygiene method of control is the substitution of nontoxic or less toxic materials for highly toxic ones. However, an industrial hygienist must exercise extreme caution when substituting one chemical for another, to ensure that some previously unforeseen hazard does not occur along with the substitution. Examples of this include fire hazards, synergistic interactions between chemical exposures, or previously unknown toxicity problems attributed to the "nontoxic" substitute chemical. The classic examples of substitution as an industrial hygiene control measure include replacement of white lead in paint pigments by zinc, barium, or titanium oxides; the use of phosphorus sesequisulfide instead of white phosphorus in match-making; shotblasting instead of sandblasting; and substitution of calcium silicates and mineral wool for asbestos as an insulating material.

As technology advanced and more toxicity information became available, the substitutions of degreasing solvents progressed from carbon tetrachloride to chlorinated hydrocarbons such as perchloroethylene and trichloroethylene. When studies revealed the possible carcinogenicity of chlorinated solvents, these solvents were replaced with fluorinated hydrocarbons. Because the fluorinated hydrocarbon solvents have been identified as ozone depleters, hydrochlorofluorocarbons are likely interim candidates for industrial degreasing. When substituting solvents, it is always advisable to experiment on a small scale before making the new solvent part of the operation or process. Detergent-and-water cleaning solutions or a steam-cleaning process should be considered for use in place of organic solvents.

Synthetic materials rather than sandstone can be used as grinding wheels and as nonsilica parting compounds in foundry molding operations. Removing beryllium phosphors from formulations for fluorescent lamps eliminated a serious pulmonary hazard to the workers making such lamps.

A change in the physical condition of raw materials received by a facility for further processing may eliminate health hazards. Pelletized or briquette forms of materials are less dusty and can drastically reduce atmospheric dust contamination in some processes.

However, there are instances when substitution of some toxic materials may be impossible or impractical, as in the manufacture of pesticides, drugs, or solvents, and processes producing ionizing radiation.

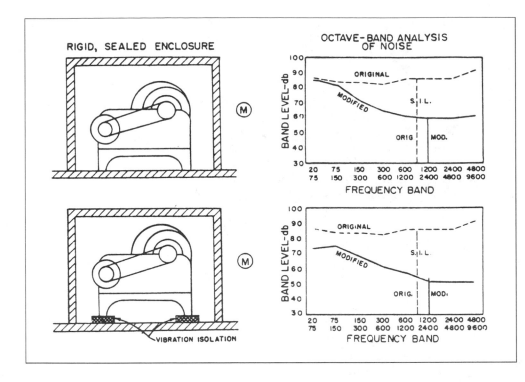

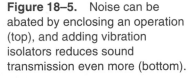

**Figure 18–5.** Noise can be abated by enclosing an operation (top), and adding vibration isolators reduces sound transmission even more (bottom).

Substituting less hazardous materials or process equipment may be the least expensive and most positive method of controlling many occupational health hazards and can often result in substantial savings. Exposure control by substitution is becoming more important from an environmental health and community air pollution perspective as well. Process materials should be selected only after review of their smog production and ozone depletion characteristics.

## Substitution: Changing the Process

A change in process offers an ideal chance to concomitantly improve working conditions. Most changes are made to improve quality or reduce the cost of production. However, in some cases, a process can be modified to reduce the dispersion of dust or fume and thus markedly reduce the hazard. For example, in the automotive industry, the amount of lead dust created by grinding solder seams with small, high-speed rotary sanding disks was greatly reduced by changing to low-speed, oscillating-type sanders. More recently, lead solder was replaced with tin solder and silicone materials.

Brush-painting or dipping instead of spray-painting can minimize the concentration of airborne contaminants. Other examples of process changes are employing arc welding to replace riveting, using vapor degreasing in tanks with adequate ventilation controls to replace hand washing of parts in open containers, using steam cleaning of parts instead of vapor degreasing, using airless paint-spraying techniques to minimize over-spray as replacements for compressed-air spraying, and employing machine application of lead oxide to battery grids,

which reduced lead exposure to operators making storage batteries.

Using automatic electrostatic paint-spraying instead of manual compressed-air paint-spraying and using mechanical continuous hopper-charging instead of manual batch-charging are additional examples of a change in process to control health hazards.

## Isolation

Potentially hazardous operations should be isolated to minimize exposure to employees. The isolation can be a physical barrier, such as acoustic panels used to minimize noise transmission from a whining blower or a screaming ripsaw (Figure 18–5).

The isolation can be in terms of time, such as providing remote control semiautomatic equipment so that an operator does not have to stay near the noisy machine constantly; or the worker may be isolated or enclosed in a soundproof control booth with a clean source of air supplied to the booth.

Isolation is particularly useful for jobs requiring relatively few workers and when control by other methods is difficult or not feasible. The hazardous job can be isolated from the rest of the work operations, thus eliminating exposures for the majority of workers. Additionally, the workers actually at workstations where contaminants are released should be protected by installing ventilation systems, which probably would not be satisfactory if the workstation were not isolated (Figure 18–6).

Exposure to employees may likewise be minimized by isolating hazardous materials in place. Exposure to asbestos-containing materials and lead-based paint can be abated

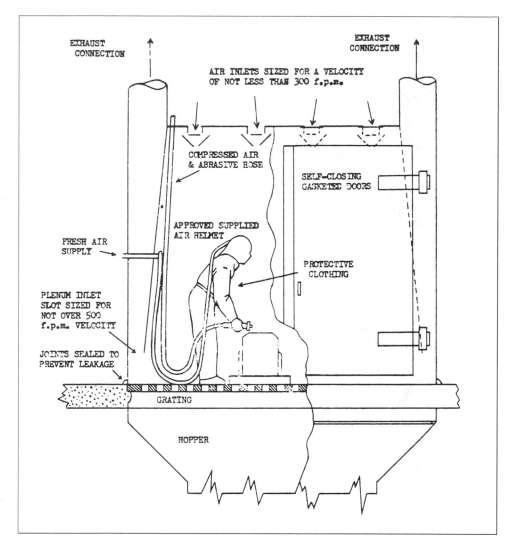

**Figure 18–6.** Air inlets and exhaust are arranged to sweep contaminated air away from the worker's breathing zone in this enclosed sandblast area. Downdraft averages 80 fpm over the entire floor area. Air should exhaust downward (as shown) or on two sides of the room at the floor line. (Courtesy Connecticut State Department of Health.)

in some instances by sealing these materials in airtight enclosures.

It may not be feasible to enclose and exhaust all operations. Abrasive blasting operations, such as those found in shipbuilding, are an example. The sandblasting should be done in a specified location, which is as far away as is practical from other employees. Another way to isolate the sandblasting is to do it when the least number of other employees would be exposed.

In some foundries, the shakeout operation may be performed during the swing shift after employees on the regular shift have gone for the day. The few shakeout workers can be provided with suitable respirators for the short time during which they are exposed to airborne dust.

Other work that can be scheduled to minimize the number of workers exposed to a hazard includes blasting in mines or quarries, which can be done at the end of or between shifts; and maintenance procedures, such as cleaning tanks and replacing filters on weekends when few workers are present. In offices, remodeling work should be performed during off hours when building occupants will not be exposed to construction dust and vapors from paints, adhesives, and finishes

In some operations, other methods of control cannot be relied on to maintain contaminants at desired levels, so these operations (such as asbestos and lead remediation projects) should be isolated. They may generate contaminants in large quantities that disperse throughout a work area or building to expose all workers to a hazard, although only a few of them are actually engaged in the operation.

Equipment isolation can be the easiest method of preventing hazardous physical contact. Insulating a hot water line may not be economical from a strictly heat conservation standpoint but may be necessary if that line is not sufficiently isolated from people.

When very toxic materials are to be processed, automation can be used to allow handling of equipment from a remote location. Robotic techniques can reproduce many industrial procedures, thus eliminating worker exposures. The work area can be viewed by remote-control television cameras or mirrors. The degree of isolation required depends on the toxicity of the contaminant, the

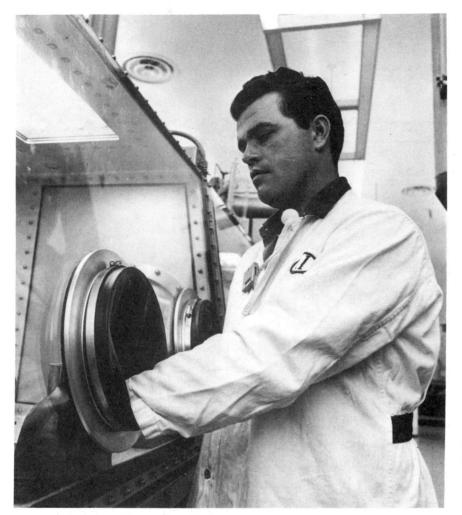

**Figure 18–7.** Some operations require complete enclosure. Here, a technician works with aluminum powder, used in atomic reactor fuel elements, at a glove box. He is wearing a film badge and air sampler on his lapel.

amount released, and work patterns around the process. Moving a process to another area is often sufficient. In other cases, a control room supplied with fresh air may be needed to isolate the process from employees monitoring the operation.

Many modern chemical facilities have centralized control rooms with automatic sampling and analysis, remote readout of various sensors, and on-line computer processing of the data and operation of the process. Some operations require complete enclosure and remote control so that nobody is exposed, as in many processes involving nuclear radiation (Figure 18–7).

Total enclosure can be accomplished by mechanization or automation to ensure that workers do not come into contact with toxic materials. The crane operators in a large foundry or in a bulk material storage building can be provided with a completely enclosed cab ventilated with filtered air under positive pressure to keep out contaminants. The same principle can be applied to heavy equipment operators in mines, coal yards, and soil remediation projects. In automatic stone-crushing, grinding, and conveying processes, only periodic or emergency attendance is required by an operator; therefore, small, well-ventilated

rooms, supplied with filtered air and strategically located within large workroom, can be occupied by the workers during the major part of the workshift.

Automated plating tanks, paint-dipping operations, and similar processes can be located in separate rooms. When continuous supervision of such operations by a worker is not necessary, general ventilation may be adequate to prevent buildup of air contamination in the workroom. If necessary, an exposed worker can be given a respirator for protection during the brief periods of exposure.

Segregating a hazardous operation or locating one or more such operations together in a separate enclosure or building not only sharply reduces the number of workers exposed but greatly simplifies the necessary control procedures.

Enclosing the process or equipment is a desirable method of control, because the enclosure prevents or minimizes the escape of contaminants into the workroom atmosphere. Enclosure should be one of the first control measures attempted, after substitution has been considered. Additional precautions must be taken when cleaning enclosed equipment or during start-up or shutdown to avoid exposure to high concentrations of the contaminant.

Enclosed equipment is usually tightly sealed and is opened only during cleaning or filling operations. Examples of such equipment include gloveboxes (Figure 18–7), airless-blast or shotblast machines for cleaning castings, and abrasive blasting cabinets.

In the chemical industry, the isolation of hazardous processes in closed systems is a widespread practice. This explains why the initial manufacture of toxic substances is often less hazardous than their subsequent use under less well-controlled conditions at other locations. In other industries, complete enclosure is often the best solution to severe dust or fume hazards, such as those from sandblasting or metal-spraying operations.

All equipment, whether enclosed or automated, requires maintenance and repair, during which control measures may have to be removed. In such circumstances, safety procedures must be specified, including lockout/tagout procedures (Grund, 1995), to work on such maintenance operations.

Isolation can also be provided by appropriate use of distance and time, for example, with respect to radiation and noise exposure. Both radiation and noise exposures decrease with an increase in the distance from the source and a decrease in the exposure time.

## Ventilation

Ventilation is a method of controlling the work environment by strategically supplying (adding) or exhausting (removing) air. Ventilation is used to dilute the concentration of contaminants to acceptable levels, to remove contaminants at their source, and to heat or cool the work environment. Ventilation can also serve to control humidity, odor, and other environmental conditions for worker comfort. (See Chapters 19, Local Exhaust Ventilation of Industrial Occupancies, Chapter 20, General Ventilation of Industrial Occupancies, and Chapter 21, General Ventilation of Nonindustrial Occupancies, for more information.)

**General ventilation.** General ventilation systems supply and exhaust large volumes of air from work spaces. They are used for temperature and humidity control or to dilute the concentration of an air contaminant below hazardous levels. This system uses natural convection through open doors or windows, roof ventilators, and chimneys, or air movement produced by mechanical fans or blowers. Exhaust fans mounted in roofs, walls, or windows constitute general ventilation.

With the exception of comfort control, general ventilation should be used only in situations meeting the following criteria:

- When small quantities of air contaminants are being released into the work environment at fairly uniform rates

- When there is sufficient distance between the worker and the contaminant source to allow sufficient air movement to dilute the contaminant to safe levels

- When only contaminants of low toxicity are being used

- When there is no need to collect or filter the contaminants before the exhaust air is discharged into the community environment

- When there is no possibility of corrosion or other damage to equipment from the diluted contaminants in the work environment air

The major disadvantage of general, or dilution, ventilation is that employee exposures can be very difficult to control near the source of the contaminant where sufficient dilution has not yet occurred. For this reason local exhaust ventilation is most often the proper method to control exposure to toxic contaminants.

When air is exhausted from a work area, consideration must be given to providing makeup, or replacement, air, especially during winter months. Makeup air volumes should be equivalent to the air being removed; it should be clean and humidified and the temperature regulated as required for comfort.

Care should be taken in selecting the makeup air intake locations so that toxic gases and vapors from discharge stacks, emergency vents, or operations outside of the building that generate hazardous contaminants are not brought back into work areas. When exhaust stacks and air supply inlets are not separated adequately, the exhaust air may be directed into the air inlet and recirculated to work areas. Inadvertent recirculation of exhaust air contaminants is a common problem, which ideally should be addressed in the design phase. It is not uncommon to find the air supply intake for a facility located adjacent to a loading dock or alley where gasoline and diesel engine vehicles idle. This can result in contamination of the "fresh" air supply and will almost certainly cause exposure or odor problems, or both.

Because equipment for moving, filtering, and tempering air is expensive, some engineers attempt to save money by recirculating some exhaust air into the supply system. Adequate monitoring of the recirculated air is necessary to prevent buildup of harmful contaminants. Recirculation of exhaust air may be forbidden in certain locations, such as smoking lounges. Check state and federal regulations.

Design of the general ventilation system in a nonindustrial or office environment must take into account conditions that affect worker comfort, such as temperature and humidity, odor level, the space provided per occupant, and concentrations of tobacco smoke. Construction practices, construction materials, and heightened public awareness have made indoor air quality an important ventilation design issue. American Society of Heating, Refrigerating, and Air

Conditioning (ASHRAE) Consensus Standard 62–1989 should be referred to for design parameters. (See Chapter 21, General Ventilation of Nonindustrial Occupancies.)

General ventilation should not be used where there are major localized sources of air contamination (especially highly toxic dusts and fumes); local exhaust ventilation is more effective and economical in such cases. More information on general ventilation is presented in Chapter 20, General Ventilation of Industrial Occupancies.

**Local exhaust ventilation.** Local exhaust ventilation is considered the classic method of control. Local exhaust systems capture or contain contaminants at their source before they escape into the work area environment. A typical system consists of one or more hoods, ducts, an air cleaner if needed, and a fan (Figure 18–8).

Local exhaust systems remove air contaminants rather than just dilute them, but removal of the contaminant is not always 100 percent effective. This method should be used when the contaminant cannot be controlled by substitution, changing the process, isolation, or enclosure. Although a process has been isolated, it still may require a local exhaust system.

A major advantage of local exhaust ventilation systems is that they require less airflow than dilution ventilation systems. The total airflow is important for plants that are heated or cooled, because heating and air-conditioning costs are an important operating expense. Also, local exhaust systems can be used to conserve or reclaim reusable materials.

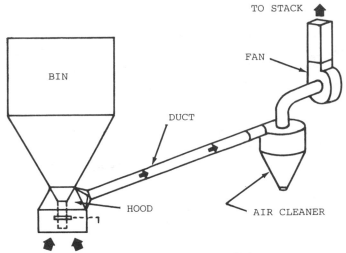

**Figure 18–8.** A typical local exhaust ventilation system consists of hoods, ducts, air cleaner, fan, and stack. (Courtesy American Conference of Governmental Industrial Hygienists.)

Two main principles govern the proper use of local exhaust ventilation to control airborne hazards. First, the process or equipment is enclosed as much as possible; and second, air is withdrawn at a rate sufficient to ensure that the direction of airflow is into the hood and that the airflow rate will entrain the contaminant into the airstream and thus draw it into the hood (Figure 18–9).

**Figure 18–9.** The fumes arising from lead-melting operations are controlled by local lateral-slot exhaust ventilation. (Courtesy Ford Motor Co.)

**Figure 18–10.** A typical local exhaust ventilation system—a dust collector—traps contaminants near their source, so the worker is not exposed to harmful concentrations.

The proper design of exhaust ventilation systems depends on many factors, such as the temperature of the process, the physical state of the contaminant (dust, fume, smoke, mist, gas, or vapor), the manner in which it is generated, the velocity and direction with which it is released to the atmosphere, and its toxicity (Figure 18–10).

Local exhaust systems can be difficult to design. The hoods or pickup points must be properly shaped and located to capture air contaminants, and the fan and ducts must be designed to draw the correct amount of air through each hood. Hood selection is based on the characteristics of the contaminants and how they are dispersed. The use and selection of an air cleaner are dependent on the contaminant, its concentration, and air pollution standards. (See Chapter 19, Local Exhaust Ventilation of Industrial Occupancies, for more details).

The low-volume, high-velocity exhaust system uses small volumes of air at relatively high velocities to control dust. Control is achieved by exhausting the air directly at the point of dust generation using close-fitting hoods. Capture velocities are relatively high, but the exhaust volume is low. For flexibility, small-diameter, lightweight plastic hoses are used with portable tools, resulting in very high duct velocities. This method allows the application of local exhaust ventilation to portable tools, which other-

wise require relatively large air volumes and large ductwork when controlled by conventional exhaust methods.

Portable local exhaust ventilation systems can be useful for facilities where dust- or fume-generating operations are not stationary. These machines capture contaminated air, filter particulate matter, and exhaust cleaned air into the work area. They can be a cost-effective solution for welding stations.

After the local exhaust ventilation system is installed and set in operation, its performance should be checked to see that it meets the engineering specifications—correct rates of airflow and duct velocities. Its performance should be rechecked periodically as a maintenance measure.

Full details on the design and operation of local exhaust ventilation systems are given in Chapter 19, Local Exhaust Ventilation of Industrial Occupancies.

# Administrative Controls

Engineering controls are to be used as the first line of defense against workplace hazards. Some circumstances require administrative controls, such as in cases when engineering controls are not technologically feasible, or during the installation of engineering controls. Administrative control of occupational hazards, such as work period reduction, job rotation, appropriate work practices, proper maintenance, and personal hygiene, depends on constant employee implementation or intervention, which makes them a less desirable form of control.

However, administrative controls are often useful in supplementing engineering controls to achieve acceptable exposure levels. The majority of the major OSHA health standards require administrative control measures including hygienic change rooms, regulated areas, and specific work practices.

## Reduction of Work Periods

Reduction of work periods is another method of control in limited areas where engineering control methods at the source are not practical. Heat stress can be managed by following a work-rest regimen that prevents excessive fatigue and reduces heart rate. For example, in the job forge industry, especially in hot weather, a shorter workday and frequent rest periods are used to minimize the effects of exposures to high temperatures, thereby lessening the danger of heat exhaustion or heatstroke.

For workers who must labor in a compressed-air environment, schedules of maximum length of workshift and length of decompression time have been prepared. The higher the pressure, the shorter is the workshift and the longer the decompression time period.

However, job rotation, when used as a way to reduce employee exposure to toxic chemicals or harmful physical agents, must be used with care. Rotation, although it may keep exposure below recommended limits, exposes more workers to the hazard.

## Wet Methods

Airborne dust hazards can often be minimized or greatly reduced by applying water or other suitable liquid. Wetting of floors before sweeping to keep down the dispersion of harmful dust is advisable when better methods, such as vacuum cleaning, cannot be used.

Wetting down is one of the simplest methods of dust control. Its effectiveness, however, depends on proper wetting of the dust. This may require the addition of a wetting agent (surfactant) to the water and proper disposal of the wetted dust before it dries out and is redispersed.

Significant reductions in airborne dust concentrations have been achieved by the use of water forced through the drill bits used in rock drilling operations. Many foundries successfully use water under high pressure for cleaning castings in place of sandblasting. Airborne dust concentrations can be kept down if molding sand is kept moist, molds with cooled castings can be moistened before shakeout, and the floors are wetted intermittently.

High-pressure water washing, used in a contained space or enclosure and with proper work practices, can effectively reduce airborne dusts and asbestos in the demolition and construction industry. In some instances it may be necessary to blanket the dust source completely. The particles must be thoroughly wetted by means of high-pressure sprays, wetting agents, deluge sprays, or other procedures while in the containment.

Batch charging of materials that are slightly moistened or that are packaged in paper bags rather than in a dry bulk state may eliminate or reduce the need for dust control in storage bins and batch mixers.

## Personal Hygiene

Personal hygiene is an important control measure. The worker should be able to wash exposed skin promptly to remove accidental splashes of toxic or irritant materials. If workers are to minimize contact with harmful chemical agents, they must have easy access to hand-washing facilities (Figure 18–11).

Inconveniently located washbasins invite such undesirable practices as washing at workstations with solvents, mineral oils, or industrial detergents, none of which is appropriate or intended for skin cleansing.

Many workplace hand cleansers are available as plain soap powders, abrasive soap powders, abrasive soap cakes, liquids, cream soaps, and waterless hand cleaners (Figure 18–12).

Powdered soaps provide a feeling of removing soils because of stimulation of the nerve endings in the skin by the abrasives. Waterless cleaners have become very popu-

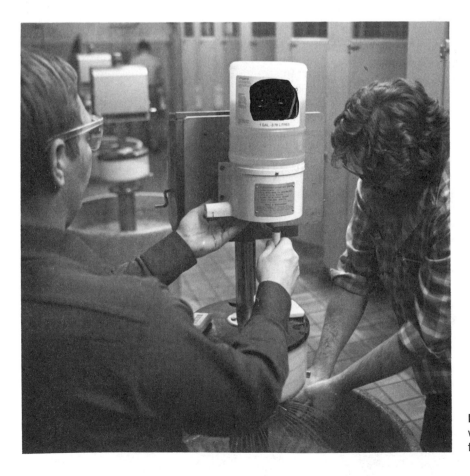

**Figure 18–11.** To minimize worker contact with harmful chemical agents, hand-washing facilities must be conveniently located.

**Figure 18–12.** Industrial hand cleansers are available as plain soap powders, abrasive soap cakes, liquids, cream soaps, and waterless hand cleaners.

lar because they remove most soils, such as greases, grimes, tars, and paint, with relative ease. Be aware, however, that some waterless hand cleaners have solvent bases. Soaps may also contribute to industrial dermatitis. Sensitive persons may require pH-neutral soaps or moisturizing agents. Antibacterial soaps are necessary in workplaces where infectious agents may be present.

The provision of washing facilities, emergency showers, and eyewash fountains is required in areas where hazardous or extremely toxic materials are handled. These should be located in an area convenient to employee workstations in case of accidental exposures.

When designated or suspected carcinogens are involved, stringent regulation of work areas and activities must be undertaken. The OSHA carcinogen regulations state that the employer must set aside a regulated area where only the particular carcinogen may be produced or handled. Only authorized and specially trained personnel with proper personal protection may be allowed to enter that area.

The eating, storage, or drinking of foods and liquids in areas where toxic materials are used should be forbidden.

All entrances to the regulated area where biohazards or suspected carcinogens are handled must be properly posted to inform employees of hazards and regular and emergency procedures required. Set aside special areas for employees to change clothing and protective equipment. Many of the major OSHA health standards, such as asbes-

tos and coke oven emissions, require hygienic change rooms and showers.

## Housekeeping and Maintenance

Good housekeeping plays a key role in the control of occupational health hazards. Good housekeeping is always important, but where there are toxic materials, it is of paramount importance, and often mandated by OSHA regulation. Remove dust on overhead ledges and on the floor before it can become airborne by traffic, vibration, and random air currents.

Immediate cleanup of any spills of toxic materials is a very important control measure. A regular cleanup schedule using vacuum cleaners is an effective method of removing dirt and dust from the work area. Never use compressed air to remove dust from rafters and ledges.

Good housekeeping is essential where solvents are stored, handled, and used. Immediately remedy leaking containers or spigots by transferring the solvent to sound containers or by repairing the spigots. Clean up spills promptly. Deposit all solvent-soaked rags or absorbents in airtight metal receptacles and remove daily to a safe location for proper disposal.

## Maintenance Provisions

If the thermostat on a vapor degreaser fails or is accidentally broken, excessive concentrations of trichloroethylene might quickly build up in the work area unless the equipment is shut down immediately and the necessary repairs made. Abnormal operating conditions can be detected by continuously monitoring airborne contaminants with instrumentation that triggers an alarm when concentrations exceed an established level. The workers or supervisors can then take steps to reduce airborne levels.

A key objective should be to provide for periodic shutdown of equipment for maintenance. Provisions should be made for cleaning the equipment and piping systems by flushing them with water, steam, or a neutralizing agent (depending on the conditions involved) to render them nonhazardous before dismantling.

Before any equipment is disassembled, it is essential that it be checked for the presence of toxic or hazardous materials. In cases in which this is not possible, employees involved in the disassembling operation should wear proper protective clothing and respirators, if needed. Contaminated equipment, tools, and protective clothing must be decontaminated before they are removed from the work area.

Operations and maintenance programs for hazardous materials or agents such as lead, asbestos, bioaerosols, and noise are important tools in the prevention of employee exposures. These programs are designed to identify and control hazardous conditions by means of periodic inspection, contaminant monitoring, and hazard abatement.

# Special Control Methods

Many of the general methods mentioned previously (either alone or in combination) can be used for the control of most occupational health hazards. A few special methods, however, deserve particular mention.

**Shielding.** This is one of the better control measures used to reduce or eliminate exposures to physical stresses such as heat and ionizing radiation. Lead and concrete are two materials commonly used to shield employees from high-energy ionizing radiation sources, such as particle generators and radioisotopes.

Shielding can also be used to protect employees against exposure to radiant heat sources. Furnaces can be shielded with shiny reflective aluminum panels. Nonreflective metal is not effective because it may act as a "black body," which absorbs and then reradiates the heat.

# Waste Disposal

Industrial hygiene controls include the proper disposal of wastes. Management and disposal of hazardous waste is regulated by the complex requirements of several governmental agencies (see Chapter 29, Government Regulations). To develop an appropriate waste management plan, employers must first make the following determinations:

- Are potentially regulated waste materials generated at this site?

- Are the wastes hazardous, special, infectious, or radioactive by regulatory definition?

- Can wastes be treated and rendered innocuous prior to disposal?

- Can wastes be recycled as part of the process?

- Given the quantities of waste generated, is the company a small-quantity or large-quantity generator?

- Is the company a waste generator, transporter, or treatment/storage/disposal facility?

These determinations will provide much of the information necessary to choose treatment and disposal alternatives. These decisions are particularly onerous, as generators of hazardous waste are perpetually responsible for on-site and off-site damages to the environment and worker and community health. This is often referred to as cradle-to-grave responsibility.

Disposal of hazardous materials must be done by highly trained individuals under strict supervision. Procedures should be established in accordance with the EPA's Resource Conservation and Recovery Act (RCRA) and other applicable regulations for the safe disposal of hazardous chemicals, toxic residues, and other contaminated waste, as well as containers of chemicals that are no longer needed and containers whose labels have been lost or obliterated.

A competent chemist can determine the best way to neutralize or detoxify small amounts of chemicals that are no longer needed. In some instances it may be appropriate to perform experimental investigations to determine a means of neutralizing and rendering waste products harmless before full-scale disposal operations are begun. There are a number of methods by which some dangerous chemicals can be rendered safe for disposal.

A number of facilities are available for off-site disposal of hazardous materials. All of them can be expensive and none of them provides a universal means of disposal for all hazardous materials. Landfills, incinerators, and chemical treatment facilities are the most commonly used disposal options for hazardous waste. Before a disposal facility is chosen, a determination should be made that the facility is competently managed, is in regulatory compliance, and has significant financial resources.

# Personal Protective Equipment

When it is not feasible to render the work environment free of occupational health hazards, it may be necessary to protect the worker from the environment with personal protective equipment. The use of personal protective equipment should be considered a last resort, when engineering or administrative controls are not possible or when they are not sufficient to achieve acceptable limits of exposure. Personal protective equipment may be appropriate during short exposures to hazardous contaminants, such as during nonroutine equipment maintenance or emergency responses to spills. The primary disadvantage of personal protective devices is that they do not eliminate the hazard from the workplace, and thus their failure results in immediate exposure to the hazard. A protective device may become ineffective without the wearer's knowledge, resulting in serious harm. The integrity and fit of a personal protective device is vital to its effectiveness.

Successful use of any personal protective equipment requires that a program be established and administered. The purpose of the program is to ensure that personal protective equipment is properly chosen, used, and maintained to protect workers. Record keeping as required by regulation must be part of this program.

## Respiratory Protective Devices

Respiratory protective devices are normally restricted for use in intermittent exposures or for operations that are not feasible to control by other methods. Respiratory protection should not be considered a substitute for engineering control methods.

Respiratory protection devices offer emergency or short-term protection. Respirators are a primary protective

**Figure 18–13.** This operator is provided with clean, respirable air.

device for normal operations only when no other method of control is possible (Figure 18–13).

Respirators should be used when it is necessary to enter a highly contaminated atmosphere for rescue or emergency repair work; as a means of escape from a suddenly highly contaminated atmosphere; for short-term maintenance or repair of equipment located in a contaminated atmosphere; and for normal operation in conjunction with other control measures when the containment is so toxic that other control measures, such as ventilation, cannot safely be relied on.

An approved respirator must be selected for the particular hazard and environment in which it is to be used (Figure 18–14).

The type of air contaminant, its expected maximum concentration, the possibility of oxygen deficiency, the useful life of the respirator, the escape routes available, and other factors must all be considered in selecting the proper type of respirator for emergency use or for standby purposes. When these factors are not known with certainty, the device providing the greatest factor of safety must be used.

There are two general types of respiratory protective devices: air-purified respirators, which remove the contaminant from the breathing air by filtering or chemical absorption, and air-supplied respirators, which provide clean air from an outside source or from a tank. Full

details of types of respirators certified by the National Institute of Occupational Safety and Health (NIOSH) and the Mine Safety and Health Administration (MSHA) should be obtained from the manufacturer. Only NIOSH/MSHA-certified respirators should be used. (See Chapter 22, Respiratory Protection, for more details.)

Half-mask cartridge respirators cover the mouth and nose. Full-facepiece respirators also protect the eyes. For dust protection, there are a large number of respirators that have met the requirements established by NIOSH/MSHA, which call for high filtering efficiency and low resistance to breathing. A smaller number of respirators have been certified for protection against metal fumes and mists.

Air-line respirators may be preferred by workers to chemical cartridge or mechanical filter respirators because they are cooler and offer no resistance to breathing; however, they require a proper source of Grade D breathing air (Compressed Gas Association CGA G-7–1988) and a suitable compressor located outside of the contaminated atmosphere.

Self-contained breathing apparatuses, which are mostly used for emergency and rescue work, have face masks attached by hoses to compressed air cylinders. Such apparatuses enable a worker to enter a contaminated or oxygen-deficient atmosphere, up to certain limits specified in the respirator certifications.

**Figure 18–14.** An approved respirator must be selected for the particular hazard and environment.

Selection of the proper type of respiratory protective equipment should be based on the following factors:

- Identification of the substance or substances for which respiratory protection is necessary and the activities of the workers
- Determination of the hazards of each substance and its significant physical and chemical properties
- Determination of the maximum levels of air contamination expected, probability of oxygen deficiency, and the condition of exposure
- Determination of the period of time for which respiratory protection must be worn
- Determination of the capabilities, physical characteristics, and limitations essential to the safe use of the respiratory protective device
- Identification of facilities needed for maintenance
- Determination of the location of the hazardous work area in relation to the nearest area with respirable-quality air
- Respirator assigned protection factors

Because wearing a respirator often becomes uncomfortable after extended periods, the worker must fully realize the need for protection or he or she will not wear it.

To obtain the worker's cooperation, the following factors are important:

- Prescribe respiratory protective equipment only after every effort has been made to eliminate the hazard.
- Explain the situation fully to the worker.
- Instruct the worker in the proper use and limitations of the respirator.
- Fit the respirator carefully according to OSHA guidelines.
- Provide for maintenance and cleanliness, including sterilization before reissue.

A respirator program is required by OSHA whenever respirators are used. The OSHA requirements for a respiratory protection program are contained in the *Code of Federal Regulations for General Industry* at 29 *CFR* 1910.134. Certain OSHA standards (such as asbestos and lead standards) have other specific regulations on respirator use. Check the *Code of Federal Regulations* for this information. (See Chapter 22, Respiratory Protection, for more details.)

## Protective Clothing

Chemical-protective clothing is worn as a barrier to a chemical, physical, or biological hazard that may cause injury if it contacts or is absorbed by the skin. Applications of chemical protective clothing include the following:

- Emergency response
- Hazardous waste site cleanup and disposal
- Asbestos removal
- Agricultural application of pesticides

A broad range of chemical-protective clothing is available to protect the body. Gloves, gauntlets, boots, aprons, and coveralls are available in a number of materials, each designed for protection against specific hazards. Choosing the most appropriate chemical-protective clothing depends on the hazards present and the tasks to be performed. Protective clothing is manufactured from different materials that protect against acids, alkalis, solvents, oils, and other chemical and physical agents. The selection should take into account the performance of the protective clothing in exposure reduction, the physical limitations created by using protective clothing, and site-specific factors. Physical and psychological stress, impaired mobility and vision, and heat stress influence or limit the selection of protective clothing.

The factors that should influence selection of protective clothing are as follows:

- Clothing design
- Material chemical resistance
- Physical properties
- Ease of decontamination
- Cost
- Chemical-protective clothing standards

Chemical-protective clothing is manufactured in a variety of styles and configurations to protect specific parts of the body or the entire body. Selection of the proper equipment should include design considerations such as clothing configuration and construction, sizes, ease of putting on and taking off, accommodation of other selected ensemble equipment, comfort, and restriction of mobility.

The effectiveness of protective clothing against chemical exposure depends on how well the material resists permeation, degradation, and penetration. Permeation is the process by which a chemical moves through a protective clothing material on a molecular level. Degradation occurs when chemical contact causes deterioration of the physical properties of the protective clothing material and causes, for example, discoloration, swelling, or loss of physical strength. Penetration is the direct flow of a chemical through closures, seams, pinholes, or other imperfections in the protective clothing material.

No material protects against all chemicals and combinations of chemicals, and no material currently available is an effective barrier to prolonged chemical contact. Protective clothing material recommendations for chemicals based on an evaluation of chemical resistance test data is available in *Quick Selection Guide to Chemical Protective Clothing* (Mansdorf, 1989) and from vendors. Many vendors and manufacturers supply charts with permeation and degradation test data and material recommendations. Protective garments constructed of rubber, neoprene, nitrile, polyvinyl chloride, and other synthetic fibers and coatings are available. It is important to select the material that protects most effectively against the specific hazard in question (acids, alkalis, oils, fibers, etc.). For mixtures of chemicals, materials having the broadest chemical resistance should be worn.

Chemical-protective materials offer wide ranges of physical qualities in terms of strength, resistance to physical hazards, and operation/effectiveness in extreme environmental conditions. The following parameters should be considered: physical strength; tear, puncture, cut, and abrasion resistance; flexibility to perform needed tasks; flame resistance; and integrity and flexibility under hot and cold extremes.

The difficulty involved in decontaminating protective clothing and the endurance of the material may dictate whether disposable or reusable clothing is selected. The relative cost of replacement and decontamination depend on the garment and the hazard. Limited-use/disposable chemical-protective clothing can be provided to minimize employee exposure to hazardous chemicals and at a reasonable cost. These types of garments are not designed to provide high levels of protection and should be used appropriately.

Body protection clothing, ranging from aprons to limited-use/disposable coveralls to totally encapsulating chemical-protective suits, are constructed of a flexible plastic or rubber film, sheet, coated plastic, or laminate. In contrast, totally encapsulating chemical-protective suits (TECP) are designed to prevent chemical exposure to the wearer.

Gloves are the most common form of chemical-protective clothing. Gloves should be selected for a specific job according to the guidelines above. Manufacturers provide a large selection of gloves made of butyl rubber, natural rubber, neoprene, nitrile rubber, polyvinyl alcohol, polyvinyl chloride, Teflon, Viton, and other construction materials. The material that has the highest level of protection should be used. A thicker glove will increase the level of protection but will result in loss of dexterity. Impregnated gloves protect against cuts and abrasions but are not liquid-proof, and they are therefore not chemical resistant. Cotton or leather gloves are useful for protecting the hands against friction and dust.

For intermittent protection against radiant heat, reflective aluminum clothing is available. These garments need special care to preserve their essential shiny surface. Air-cooled jackets and suits are available to minimize the risk of heat-related illnesses. For protection against ionizing radiation, garments constructed of lead-bearing materials are available.

More information on the subject of protection against skin hazards and the use of barrier creams is presented in Chapter 3, The Skin and Occupational Dermatoses. Also consult applicable OSHA regulations and ANSI standards.

## Eye and Face Protection

Eye and face protection includes safety glasses, chemical goggles, and face shields. The correct type of protector is chosen based on the hazard (such as corrosive liquids and vapors, foreign bodies, or ultraviolet radiation). Goggles fit snugly to the face, preventing chemical exposure in the event of a splash, and, depending on the style, may prohibit vapor exposure. Face shields are designed only to prevent direct splash exposures to the face and not to provide complete eye protection. Eye protection from exposure to ultraviolet radiation, such as that produced in welding operations, is accomplished with filter lenses of the correct shade mounted in the welding helmet.

Many chemicals in the workplace can cause significant eye damage and facial scarring from direct chemical contact. It is important that the protective device be worn at all times when the hazard is present. (Refer to Chapter 5, The Eyes, for further information. Also consult applicable OSHA regulations and ANSI standards.)

## Hearing Protection

Personal hearing protectors, such as earplugs or earmuffs, can provide adequate protection against noise-induced hearing impairment. The wearer is afforded effective protection only if the hearing protectors are properly selected, fitted, and worn. Like other types of personal protective equipment, these devices should be used as an exposure control alternative when noise exposures cannot feasibly be reduced below the OSHA permissible limit. When the noise level is 85 dBA or higher and the employee has suffered a significant threshold shift, hearing protection must be used.

There are primarily two forms of hearing protectors: insert types, which seal against the ear canal walls, and earmuffs, which seal against the head around the ear. Choice of the proper hearing protection should take into account the physiological and anatomical characteristics of the wearer, the noise exposure dose, the work activity, and environmental conditions (for example, dusty atmosphere). Refer to Chapter 9, Industrial Noise, for more information on hearing conservation.

# ■ Education and Training

Proper training and education are critical to supplement engineering controls and ensure the success of exposure controls in the workplace. It is important that all employees be provided the health and safety information and instruction needed to minimize their occupational health risk and that of their co-workers.

In a typical manufacturing plant, the primary responsibility for safe operation and control rests with the line organization of the operations department. This generally would include a first-line supervisor, a shift supervisor, and a facility area manager, all people familiar with every aspect of the day-to-day operation of the facility and the manufacturing process and readily available when critical decisions must be made.

The education of supervisors usually is process and equipment oriented. The aim of the safety and health professional should be to teach them about the safety and health hazards that may be found in their work areas. The supervisors should be told when and under what circumstances to request aid in solving the problems those hazards pose. Supervisors should be knowledgeable and well informed about hazardous processes, operations, and materials for which they are responsible.

Short courses on industrial hygiene can be an easy way to transmit a lot of valuable information with a small expenditure of time. Industrial hygiene short courses for managers should identify health hazards in broad areas. The courses should also consider the cost-benefit relationships of controlling health hazards in the work environment.

The worker must know the proper operating procedures that make engineering controls effective. If the worker performs an operation away from an exhaust hood, the purpose of the control measure will be defeated and the work area may become contaminated. Workers can be alerted to safe operating procedures through booklets, instruction signs, labels, safety meetings, and other educational devices.

The safety and health professional, by persuading a worker to position the exhaust hood properly or to change the manner of weighing a toxic material or of handling a scoop or shovel, can do much to minimize unnecessary exposure to air contaminants. For normal facility operations, a prescribed health hazard evaluation routine should be set up. This should include monitoring the exposures of the personnel involved. It can be accomplished by keeping a record of the exposures to chemical and physical agents in work areas.

In addition to the normal operating instructions that each employee is given when starting a new job, employees assigned to areas where exposures to toxic chemicals can occur must, by law, be given a special indoctrination program.

Also be sure to give employees training in how to respond to emergencies. Information on when *not* to respond is also critical. Many deaths have occurred when untrained workers rushed in to save fallen co-workers and were overcome themselves.

In order to minimize operator error, employees should be supplied with a detailed instruction manual outlining procedures for all foreseeable situations.

Health hazards affect the workers who are exposed and work directly with materials, process equipment, and processes. These employees should know about the effects of exposure to the materials and energies they work with so that controls can be installed before those problems become severe. A properly informed worker can often anticipate and take steps to control health hazards before they become serious. Once the hazard is known, the supervisor or facility engineers can issue work orders to eliminate the problem.

Workers should be given reasons for wearing respirators, protective clothing, and goggles. They also should be informed of the necessity of good housekeeping and maintenance. Because new materials are constantly being marketed and new processes being developed, reeducation and follow-up instruction must also be part of an effective industrial hygiene control program.

Over 100 specific OSHA standards contain training requirements. Some of these standards make it the employer's responsibility to limit certain job assignments to employees who have had special training that defines them as certified, competent, or qualified with respect to a particular hazard.

OSHA has developed Voluntary Training Guidelines to assist employers in determining training needs as well as developing and conducting the training. OSHA encourages employers to follow the model provided in the Voluntary Training Guidelines. The model can be used to develop training programs for a variety of hazards and to assist in compliance with training requirements in specific standards. The guidelines are as follows:

- Determine whether training is needed.
- Identify training needs.

- Identify goals and objectives.
- Develop learning activities.
- Conduct the training.
- Evaluate program effectiveness.
- Improve the program effectiveness.

Specific training requirements are set forth by OSHA in the Hazard Communication, Process Safety Management, Asbestos, Lead, and Bloodborne Pathogens Standards, among others. For example, the Hazard Communication Standard requires a training program that covers the following types of information:

- Requirements of the standard
- Identification of operations in the workplace where hazardous materials are present
- Methods and observations used to detect the presence of hazardous materials in the work area
- Physical and health hazards of those materials
- Hazards associated with chemicals in unlabeled pipes
- Hazards of nonroutine tasks
- Measures that employees can take to protect themselves from these hazards
- Explanation of the hazardous materials labeling system
- Explanation of Material Safety Data Sheets (MSDSs)
- Details on the availability and locations of Hazardous Material Inventory, MSDSs, and other printed Hazard Communication Program materials

The future of state and federal training requirements, led by California's Illness and Injury Prevention Act (Senate Bill 198), focuses on preventing rather than reacting to hazards. Under California's regulation, employers must identify the person responsible for implementing a Written Injury and Illness Prevention Program and provide training in health and safety matters to all employees. This approach is intended to improve efforts to prevent workplace hazards by identifying and evaluating hazards during periodic scheduled inspections.

## Health Surveillance

Health surveillance, although not an occupational exposure control, can be used to prevent health impairments by means of periodic evaluations. A health surveillance program includes preplacement, periodic, special purpose, and hazard-oriented examinations.

Medical surveillance is mandated by specific OSHA, MSHA, and Environmental Protection Agency (EPA) regulations. Over 30 OSHA standards and proposed standards contain medical surveillance requirements. Among these are the asbestos, lead, formaldehyde, and hazardous waste operations standards.

Hazard-oriented medical surveillance monitors biological indicators of absorption of chemical agents based on analysis of the agent or its metabolite in blood, urine, or expired air. Inorganic lead absorption is measured by blood lead levels, and carbon monoxide absorption is indicated by carboxyhemoglobin levels in blood or carbon monoxide in exhaled air. Refer to Chapter 25, The Occupational Physician, for a complete discussion of health surveillance.

## Summary

Control of occupational exposures to injurious materials or conditions may be accomplished by means of one or more of the following methods:

- Proper design engineering
- Substitution of less toxic materials or changes or process
- Isolation or enclosure of the source or the employee
- Local exhaust ventilation at the point of generation or dissemination of the air contaminant
- General ventilation or dilution with uncontaminated air
- Maintenance and housekeeping
- Personal protective equipment
- Employee information and training
- Proper waste disposal practices

One or a combination of these methods may be necessary to prevent excessive exposures to hazardous materials or physical agents.

Education of workers and periodic workplace inspections are paramount in the prevention of injury and illness. If engineering and administrative controls and the use of personal protective equipment are to be effective in minimizing occupational health risk, workers must be properly trained.

Management is responsible for furnishing the facilities and products required to keep the workplace healthful and safe. The worker also has responsibilities in a health hazard control program, including the following: to wear protective equipment if it is required, to use the local exhaust ventilation system properly, and to observe all company rules relating to cleanup and disposal of harmful materials.

## Bibliography

Allen RW, Ells MD, Hart AW. *Industrial Hygiene.* Englewood Cliffs, NJ: Prentice Hall, 1976.

American Conference of Governmental Industrial Hygienists, Committee on Industrial Ventilation. *Industrial Ventilation— A Manual of Recommended Practice,* 22nd ed. Lansing, MI: American Conference of Governmental Industrial Hygienists, 1995.

American National Standards Institute. *American National Standard for Respiratory Protection ANSI Z88.2-1992.* New York: American National Standards Institute, 1992.

ASHRAE 62–1989. *Ventilation for Acceptable Air Quality.*

Beddows NA. Safety and health criteria for plant layout. *National Safety News,* Nov. 1976.

Burgess WA. *Recognition of Health Hazards in Industry—A Review of Materials and Processes, 2nd ed.* New York: Wiley, 1995.

Cralley LE et al. *Industrial Environmental Health.* New York: Academic Press, 1972.

Cralley LV, Cralley LJ, et al. *Industrial Hygiene Aspects of Plant Operations,* vols. 1–3. New York: Macmillan, 1985.

Grund E. *Lockout/Tagout: The Process of Controlling Hazardous Energy.* Itasca, IL: National Safety Council, 1995.

Hazardous Waste Operations and Emergency Response. 51 *FR* Part 244, Dec. 19, 1986.

Key MM et al., eds. *Occupational Diseases—A Guide to Their Recognition,* rev. ed. DHHS (NIOSH) Publication No. 77-181. Washington, DC: GPO, June 1977.

Lindgren GF. *Managing Industrial Hazardous Waste.* Chelsea, MI: CRC/Lewis, 1989.

Lynch J. Industrial hygiene input into plant design. Paper presented at 65th National Safety Congress, Chicago, Oct. 18, 1977.

Mansdorf SZ. *Quick Selection Guide to Chemical Protective Clothing.* New York: Van Nostrand Reinhold, 1989.

McDermott HJ. *Handbook of Ventilation for Contaminant Control (Including OSHA Requirements).* Ann Arbor, MI: Ann Arbor Science Publisher, 1976.

McRae A et al., eds. *Toxic Substances Control Sourcebook.* Germantown, MD: Aspen Systems Corporation, 1978.

National Institute for Occupational Safety and Health. *A Guide to Industrial Respiratory Protection.* DHHS (NIOSH) Publication No. 76–189. Washington, DC: National Institute for Occupational Safety and Health, June 1976.

National Institute for Occupational Safety and Health. *Engineering Control Research Recommendations.* DHHS (NIOSH) Publication No. 76–180. Cincinnati, OH: National Institute for Occupational Safety and Health, Feb. 1976.

National Institute for Occupational Safety and Health. *The Industrial Environment—Its Evaluation & Control.* Washington, DC: GPO, 1973.

Occupational Safety and Health Reporter. *Hazard Communication Training Programs and Their Evaluations* 31:9701–9730. Washington, DC: Bureau of National Affairs.

Occupational Safety and Health Reporter. *Special Report: California Employers Face New Compliance Duties as State Moves to Strengthen Safety and Health Enforcement.* 1593–1596. Washington, DC: Bureau of National Affairs, 1991.

Occupational Safety and Health Administration. Hazard Communication Standard (50 *FR* 51852), Dec. 20, 1985.

Occupational Safety and Health Administration. *Training Requirements in OSHA Standards and Training Guidelines (OSHA 2254),* 1992.

Patty FA. *Industrial Hygiene and Toxicology, General Principles: Parts IA and IB.* New York: Wiley, 1991.

Patty FA. *Industrial Hygiene and Toxicology, Theory and Rationale of Industrial Hygiene Practice, The Work Environment, Part IIIA.* New York: Wiley, 1994.

Peterson JE. *Industrial Health.* Englewood Cliffs, NJ: Prentice Hall, 1977.

Schilling RSF, ed. *Occupational Health Practice.* Toronto, Canada: Butterworth, 1973.

# CHAPTER 19

# Local Exhaust Ventilation of Industrial Occupancies

Revised by D. Jeff Burton, PE, CIH, CSP

*Publication information on all references given in this chapter can be found in the Bibliography at the end of Chapter 21, General Ventilation of Nonindustrial Occupancies.*

**Local Exhaust Systems**

**Basic Terms and Formulas**

**Hoods**
*Capture Hoods ▪ Flanged Hood ▪ Face Velocity Versus Mass Air Movement ▪ Baffles and Enclosures ▪ Principles of Hood Design ▪ Hood Design for Specific Operations ▪ Receiving Canopy Hoods ▪ Employee Participation*

**Design Considerations**
*Ducts ▪ Multiple Ducts ▪ Common Piping Defects*

**Air Cleaners**
*Air-Cleaner Efficiency ▪ Centrifugal Collectors ▪ Air Cleaners for Fumes and Smokes ▪ Air Cleaners for Gases and Vapors*

**Fans**
*Centrifugal Fans ▪ Axial-Flow Fans ▪ Fire Prevention*

**Exhaust System Performance**
*Inspection ▪ Recirculation of "Cleaned" Air*

**Special Operations**
*Open-Surface Tanks ▪ Spray Booths ▪ Fume Control for Welding ▪ Grinding Operations ▪ Woodworking ▪ Oil Mist ▪ Materials Conveying ▪ Internal Combustion Engine Ventilation ▪ Fog Removal ▪ Kitchen Range Hoods*

THE PURPOSE OF THIS CHAPTER is to acquaint health and safety professionals with the basic principles of local exhaust ventilation systems.

Exhaust systems are usually designed by mechanical engineers under the supervision of management, such as the facility engineer or the industrial hygienist. All exhaust ventilation systems should be designed, built, and operated to meet current codes and standards. The health and safety professional can help attain this goal by becoming familiar with ventilation fundamentals, reviewing proposed plans and specifications, and conducting tests of existing systems. (The Bibliography for this chapter is found at the end of Chapter 21, General Ventilation for Nonindustrial Occupancies.)

## Local Exhaust Systems

A local exhaust system (LE) is used to control an air contaminant by collecting it at the source, as compared with a dilution ventilation system, which allows the contaminant to be diluted into the workroom air. (See Chapters 20 and 21 for more information on general ventilation.)

Local exhaust is often preferred to dilution because it removes the contaminant from the work area, handles lower volume flow rates, and is more energy efficient. (It removes less tempered air from the building and uses smaller fans.) An LE system, because of the greater concentration of contaminants, often has the ability to scrub or clean the contaminant from the exhaust air. As noted in Chapter 18, Methods of Control, other emission controls (such as substituting less harmful materials, changing the process, or isolating the process from employees) may be more effective, and these options should be explored before ventilation is installed.

A local exhaust system is usually appropriate for control when

- Other more cost-effective controls are not available.
- Air sampling or employee complaints suggest that the air contaminant is a health hazard or a fire hazard, impairs productivity, or creates unacceptable comfort problems.
- Federal, state, local, building, or consensus standards require (or suggest) local exhaust systems. For example, OSHA's 29 *CFR* 1910.94 regulation calls for local exhaust for open surface tank operations, spray finishing, grinding, and abrasive blasting; the ANSI Z9.5 standard on lab ventilation requires chemical operations in a lab to be conducted in an exhausted lab fume hood. (See the Bibliography at the end of Chapter 21 for more information.)
- Improvements will be seen in production rates, housekeeping, employee comfort or morale, and the operation and maintenance of equipment.
- Emission sources tend to be large, few, fixed, or widely dispersed.
- Emission sources are close to the breathing zones of employees.
- Emission sources tend to vary with time.

The local exhaust system consists of five important parts (Figure 19–1):

- *Hood.* Hoods capture, contain, or control the emission source.
- *Ductwork.* Ducts and piping carry the contaminant to the air cleaner or to the outside environment for dilution.
- *Air cleaner.* The air cleaner scrubs, separates, removes, or filters the air contaminant from the exhaust air, usually to meet permit requirements. It may also change the air contaminant to a less hazardous substance (for example, using an afterburner to convert hydrocarbon vapors to carbon dioxide and water vapor).
- *Fan and motor.* The fan generates static pressure and moves air.
- *Stack.* The stack assists in dispersing remaining air contaminants.

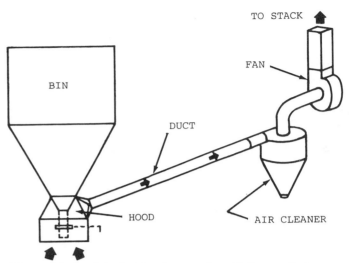

**Figure 19–1.** A typical local exhaust ventilation system consists of hoods, ducts, air cleaner, fan, and stack. (Courtesy American Conference of Industrial Hygienists.)

## Basic Terms and Formulas

**Standard conditions.** (ACGIH) 70 F, 29.92 in. Hg, dry air. (ASHRAE) 68 F, 29.92 in. Hg, 50 percent RH; air density = 0.075 lb/ft³; air density varies linearly with temperature and pressure (the density correction factor, *d*). (See Table 19–A).

**Pressure.** Air moves under the influence of differential pressures. A *fan* is commonly used to create a difference of pressure in *duct systems.*

**Static pressure.** Static pressure is the potential energy of the ventilation system. It is converted into kinetic energy (velocity pressure) or other energy types (*losses* in the form of heat, vibration, and noise). At sea level the standard *static pressure* is 14.7 psia, 29.92 in. Hg, or 407 in. of water column. If the fan generates one inch of water of negative static pressure (SP = –1 in. water gauge, wg), the absolute static pressure in the duct will be SP = 406 in. of water.

**Velocity pressure.** The velocity pressure is directly related to the velocity of air, as described by *Bernoulli's* equation:

$$V = 4{,}005 \cdot \left(\frac{\mathrm{VP}}{d}\right)^{0.5}$$

$$V = 4.043 \cdot \left(\frac{\mathrm{VP}}{d}\right)^{0.5} \quad \text{(metric)} \qquad (1)$$

where  $V$ = velocity of air (ft/min, fpm, m/s, mps)
 VP = velocity pressure (in. or mm wg)
  $d$ = density correction factor for nonstandard conditions (see Table 19–A)

**Table 19–A.**   Air Density Correction Factor, *d*

| Elevation *or* Barometric Pressure | | | 0 | 1,000 | 2,000 | 3,000 | 4,000 | 5,000 | 6,000 | 7,000 |
|---|---|---|---|---|---|---|---|---|---|---|
| *feet* | | | 0 | 1,000 | 2,000 | 3,000 | 4,000 | 5,000 | 6,000 | 7,000 |
| *meters* | | | 0 | 305 | 610 | 915 | 1,220 | 1,525 | 1,830 | 2,135 |
| *mm Hg* | | | 760 | 733 | 707 | 681 | 656 | 632 | 608 | 587 |
| *inch Hg* | | | 29.92 | 28.86 | 27.82 | 26.82 | 25.84 | 24.89 | 23.97 | 23.09 |
| *Air Temperature* | C | F | | | | | *d* | | | |
| | –18 | 0 | 1.15 | 1.11 | 1.07 | 1.03 | 0.998 | 0.959 | 0.921 | 0.882 |
| | 0 | 32 | 1.08 | 1.04 | 1.01 | 0.969 | 0.933 | 0.897 | 0.861 | 0.825 |
| | 21 | 70 | 1.00 | 0.966 | 0.933 | 0.900 | 0.866 | 0.833 | 0.799 | 0.766 |
| | 38 | 100 | 0.946 | 0.915 | 0.883 | 0.851 | 0.820 | 0.788 | 0.756 | 0.725 |
| | 66 | 150 | 0.869 | 0.840 | 0.811 | 0.782 | 0.753 | 0.723 | 0.694 | 0.665 |
| | 93 | 200 | 0.803 | 0.776 | 0.749 | 0.722 | 0.696 | 0.669 | 0.642 | 0.625 |

**Total pressure.** The *static pressure* (SP) and *velocity pressure* (VP) are related to total pressure through the following equation:

$$TP = SP + VP \qquad (2)$$

**Flow rate.** Volume flow rates are described by the conservation of mass formula (which can apply if we assume *incompressible flow* of air):

$$Q = VA \qquad (3)$$

where  $Q$ = volume flow rate (ft³/min, cfm, scfm, acfm; m³/s)
  $V$ = velocity (fpm, mps)
  $A$ = cross-sectional area of air flow (ft², m²)

**Hood entry loss.** *Hoods* capture, contain, or receive contaminants before they have a chance to dilute into the workroom air. The hood converts static pressure in the duct (SPh) into velocity pressure (VP) and hood entry losses (He) and is described mathematically as

$$He = |SPh| - VP \qquad (4)$$

where  He = hood entry loss (in. or mm wg)
  VP = average velocity pressure in duct serving the hood (in. or mm wg)
  |SPh| = positive value of static pressure in duct serving hood (in. or mm wg)

Note: He and SPh are usually measured four to six duct diameters downstream in the duct.

**Loss factor.** Losses in a ventilation system fittings (elbows, branch entries, friction loss) may be predicted using the loss factor equation:

$$SP \text{ loss} = K \cdot VP \cdot d \qquad (5)$$

where  SP loss = loss in system (in. or mm wg)
  $K$ = loss factor, determined experimentally, unit less (see the Bibliography at the end of Chapter 21)
  VP = average velocity pressure in duct (in. or mm wg)
  $d$ = density correction factor for nonstandard conditions

**Coefficient of entry.** A measure of the effectiveness of hood in converting static pressure to velocity pressure:

$$Ce = \left(\frac{VP}{SPh}\right)^{0.5} = \left[\frac{1}{1+K}\right]^{0.5} \qquad (6)$$

where  Ce = coefficient of entry (unitless)
  VP = average velocity pressure in duct serving hood (in. or mm wg)
  SPh = positive value of SPh (in. or mm wg)
  $K$ = loss factor for hood (see the Bibliography at the end of Chapter 21)

# Hoods

The local exhaust hood is the point of air entry into the duct system. The term *hood* is used in a broad sense to include all suction openings regardless of their shape or mounting arrangement.

No local exhaust system can succeed unless the contaminant is controlled by the hood. Clearly, no matter how well built the ducts and arrester are or how large the fan is, if the contaminant is not controlled by the hood, the overall value of the installation is nil. The hood is, therefore, a critical component (Figure 19–2).

There are three basic types of hoods: capture, enclosing, and receiving (canopy) (Figure 19–3).

## Capture Hoods

When a duct is connected to the inlet of an exhaust fan, *suction,* or an area of low pressure, is set up at the other end of the duct. Air from the room will move toward this low-pressure region. But the air moves in from all directions toward the hood face. Thus, as shown in Figure 19–4, air will move into a freely suspended duct opening from both front and back. The dashed lines going into the duct opening (stream lines) indicate the direction of airflow at that point. The solid curved lines represent spherical contours of equal velocity. What is needed for dust control is

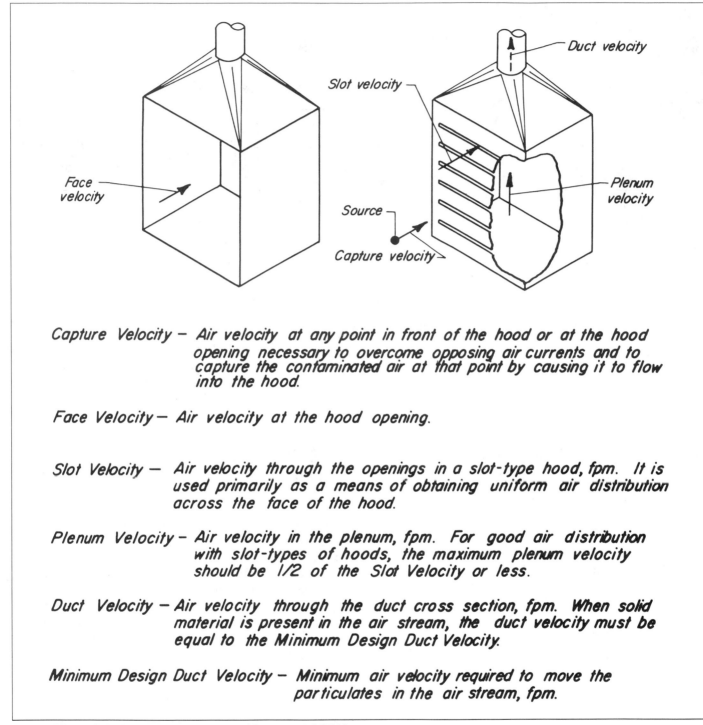

**Capture Velocity** – *Air velocity at any point in front of the hood or at the hood opening necessary to overcome opposing air currents and to capture the contaminated air at that point by causing it to flow into the hood.*

**Face Velocity** – *Air velocity at the hood opening.*

**Slot Velocity** – *Air velocity through the openings in a slot-type hood, fpm. It is used primarily as a means of obtaining uniform air distribution across the face of the hood.*

**Plenum Velocity** – *Air velocity in the plenum, fpm. For good air distribution with slot-types of hoods, the maximum plenum velocity should be 1/2 of the Slot Velocity or less.*

**Duct Velocity** – *Air velocity through the duct cross section, fpm. When solid material is present in the air stream, the duct velocity must be equal to the Minimum Design Duct Velocity.*

**Minimum Design Duct Velocity** – *Minimum air velocity required to move the particulates in the air stream, fpm.*

**Figure 19–2.** Principles of exhaust hoods. (Courtesy ACGIH.)

an air velocity $V$ at the point of dust release at a distance $x$ from the duct opening. The air velocity must be high enough to carry the particles into the hood (that is, into the duct). If the amount of air entering the pipe is $Q$, the velocity at the surface of the sphere (where the dust producer is located) is given by the equation $V = Q/A$, where $A$ is the surface of the sphere. The surface area of any sphere is calculated as follows:

Then

$$A = 4\pi x^2 \ (\text{ft}^2) \qquad (7)$$

$$V = \frac{Q}{4\pi x^2} \ (\text{fpm}) \qquad (8)$$

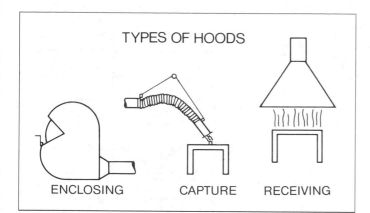

**Figure 19–3.**   Types of hoods include enclosing hoods, capture hoods, and receiving hoods.

This relation indicates that the velocity at a point where dust is being released is (1) proportional to volume of air, $Q$, flowing into the duct in cubic feet per minute (cfm) or cubic meters per second ($m^3$/s), and (2) inversely proportional to the square of the distance $x$ from the opening.

The basic equation to measure air velocity (Equation 1) has been modified empirically, and when $x$ is less than 1.5 hood diameters, it has the following form:

$$V = \frac{bQ}{x^2 + bA} \text{ (fpm)} \tag{9}$$

where   $V$ = centerline velocity at distance $x$ from hood
   $Q$ = airflow into duct (cfm)
   $x$ = distance outward along hood axis (ft)
   $A$ = area of hood opening (ft$^2$)
   $b$ = a constant that depends on the shape of the opening

For circular or square openings, $b$ is essentially 0.1; the equation then becomes

$$V = \frac{0.1\,Q}{x^2 + 0.1A} \tag{10}$$

Suppose the duct is 6 in. (15 cm) in diameter and the velocity of air passing through is 4,000 fpm (1,200 m/min), a common situation in dust exhaust systems. Because a circle 6 in. (15 cm) in diameter is 0.196 ft$^2$ (180 cm$^2$) in the area of cross section, $Q = 0.196 \cdot 4{,}000$, or 780 cfm (0.37 $m^3$/s); in ducts, $Q = AV_{\text{duct}}$.

Two inches (5 cm) out from the duct end, $V$ has fallen to 1,650 fpm (500 m/min).

At 4 in. (10 cm) from the duct, the velocity is 600 fpm (180 m/min); at 6 in. (15 cm) away, it is only 290 fpm (88 m/min), just enough to be felt by the hand.

Where x is very large compared with $A$, Equation (10) becomes

$$V = \frac{Q}{10x^2} \tag{11}$$

## Flanged Hood

If a flange is placed around the duct opening, as shown in Figure 19–4, it will reduce entrance or turbulence loss by preventing the hood from drawing air from behind the hood face. Although the same total amount of air is exhausted, a larger portion will come from the front of the duct. This is beneficial because air that moves from behind the hood does not help control the contaminant in front. A large flange will increase the useful airflow by 30 to 40 percent for the same total volume of air handled. Usually, the flange width can be estimated from the following:

$$W_f = x - .5D \tag{12}$$

where   $W_f$ = flange width (in. or cm)
   $x$ = distance in front of the capture hood along hood axis (in. or cm)
   $D$ = duct diameter, round ducts (in. or cm)

## Face Velocity Versus Mass Air Movement

The preceding equations show that hood capture velocity, $V$, depends on the total airflow entering the hood. This fact is often overlooked. A high face velocity or, in the case of a slot-shaped hood, a high slot velocity is not the important factor. The capture of air contaminants depends on mass air movement, $Q$, not on mere face velocity.

Regardless of face velocity, a source of suction has a woefully poor ability to reach out in a particular direction and induce an inflowing stream of air even a few inches from the usual hood face. Yet, often a person will attempt to improve the exhaust hood by decreasing its size, believing that this will raise the capture velocity. This does no good whatsoever, except in very close proximity to the hood face. It often, in fact, does harm because the total airflow is then reduced due to increased hood resistance. "Reaching out" can be achieved only from greater mass air movement.

Confusion on this point may occur because common experience shows that air can be blown from a pipe or hose for considerable distances. Air escaping from a compressed air line, for example, will be felt many feet away; air from a simple desk fan can be felt across the room. A person can blow out a match from several feet away, but try "sucking" or exhausting a match out. It is impossible at any distance beyond a couple of inches.

Figure 19–5 illustrates that air under pressure has a "throw" about 30 times farther than the "pull" on the suction side of the same fan or blower. Air from the pressure side is discharged in a specific restricted direction. Air on the suction side is drawn from all directions into the low-pressure area.

The purpose of capture hoods is to set up air movement at the point or area where the contaminant is released so that a sufficiently large percentage of the contaminant will be drawn to the hood and captured. In selecting the correct control velocity, Table 19–B may serve as a guide.

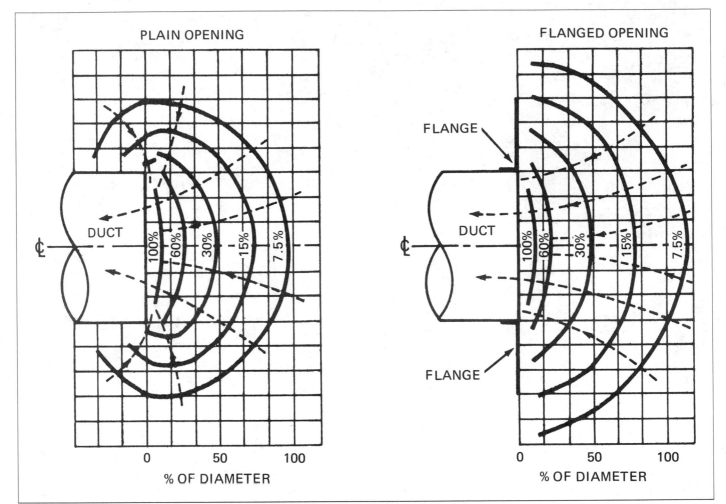

**Figure 19–4.** Velocity contours expressed as percentages of velocity at the opening (solid curved lines) and stream lines (dashed line) for both plain and flanged circular openings. (Courtesy ACGIH.)

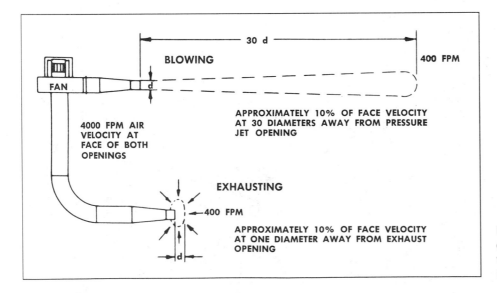

**Figure 19–5.** The "throw" of a blower can vary greatly, depending on whether it is exhausting or blowing. (Courtesy ACGIH.)

**Table 19–B.**  Capture, Control, Face, and Slot Velocities for Hood Design

| Velocity Range | | Type of Velocity | Operation or Hood Type | Typical Emission and Environmental Conditions |
|---|---|---|---|---|
| fpm | mps | | | |
| 25–50 | 0.12–0.25 | Random mixing | Any | Random motion of air in industrial operations |
| 50–100 | 0.25–0.50 | Capture | Degreasing | Release of emissions at no velocity into still air (no cross drafts) |
| 75–125 | 0.35–0.60 | Face | Lab fume hoods, drying ovens | Emission source is enclosed |
| | | Capture | Soldering | Release of emissions into quiet air (less than 75 fpm cross drafts) |
| | | Capture | Open-Surface tanks | For use with area approach to design; at edge of tank |
| 125–150 | 0.60–0.75 | Face | High-toxicity lab hood | |
| 100–200 | 0.5–1.0 | Capture | Spray booths, welding, container filling, pickling, low-speed conveyor transfer | Release of emission at low velocity into air with low cross drafts (both less than capture velocity) |
| | | Control | Open-surface tanks | For use with ANSI approach to open-surface tank hood design |
| 200–500 | 1.0–2.5 | Capture | Crushers, conveyor loading, shallow booths | Release of emissions at modest velocities into air with modest cross drafts (both less than capture velocity) |
| | | Control | Canopy hoods | For use with ANSI canopy hood, use lower end of range |
| 500–2,000 | 2.5–10 | Capture | Grinding, tumbling, | Release of emission at high velocities into air with high cross drafts (both less than capture velocity) |
| 1,200 | 6.0 | Slot | Lab fume hoods | Typical of commercial lab fume hoods |
| 1,500–2,000 | 7.5–10 | Slot | Slotted hoods, general | To ensure uniform flow through all slots, the plenum velocity should be designed for 1/4 to 1/3 of slot velocity |

(Reprinted with permission from IVE, Inc. *Industrial Ventilation Workbook,* 3rd edition, 1994.)

## Baffles and Enclosures

Most control velocities at the point of dust or fume emissions are relatively low and can be nullified by drafts from cross breezes or nearby moving equipment. For example, a person walking briskly (at a rate of 300 fpm, or 1.5 m/s) or a piece of moving equipment, such as a crane or lift truck, often stirs up a significant counter air movement. Unwanted cross drafts can easily blow the contaminant away before it ever comes under the influence of the hood. A cooling or pedestal fan, blowing across an operator working in front of a hood, can defeat the purpose of the hood. Cross drafts from windows, doors, or ventilators can have the same detrimental effect.

Eliminate cross drafts by using baffles, side shields, booths, and other semienclosures and full enclosures wherever possible. All too often such shields are left off or, if installed originally, are left open after maintenance jobs, or for the convenience of the operators. This often destroys the effectiveness of the capture hood.

A full enclosure acts differently and more efficiently than a capture hood because it completely envelops the operation. The only openings in the enclosure are the duct and unavoidable leaks (most enclosures cannot be completely airtight). With a capture hood, on the other hand, the air velocity at the point of emission must be high enough to transport the contaminant through free space to the hood and duct without being diverted by a cross draft.

## Principles of Hood Design

The following are important features of good exhaust hood design:

1. Enclose the operation as much as possible to reduce the rate of airflow needed to control the contaminant, and to prevent cross drafts from blowing the contaminant away from the field of influence of the hood (Figure 19–6).

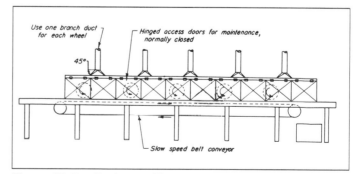

**Figure 19–6.**  Enclosure. The more completely the hood encloses the source, the less air is required for control in this straight-line automatic buffing operation. (Courtesy ACGIH.)

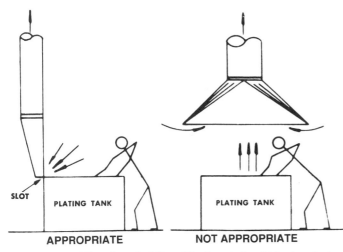

**Figure 19–7.** Direction of airflow. The hood should be located so that the contaminant is removed from the breathing zone of this worker. (Courtesy ACGIH.)

2. Locate the hood so that the contaminant is moved away from the breathing zone of the operator (Figure 19–7).

3. Locate and shape the hood so that the initial velocity of the contaminant will throw it into the hood opening (Figure 19–8).

4. Solvent vapors in health-hazard concentrations are not appreciably heavier than air. Capture them at their source rather than attempt to collect them at the floor level (Figure 19–9).

5. Locate the hood as close as possible to the source of the contaminant (Figure 19–10).

6. Design the hood so that it will not interfere with the worker.

## Hood Design for Specific Operations

A hood should be shaped and positioned to take advantage of any existing natural movement of the contaminant. For instance, if heat is released, the flow of convection currents should be toward the hood. If heavy particles are thrown off, as in rough grinding, their trajectory should be toward the hood intake.

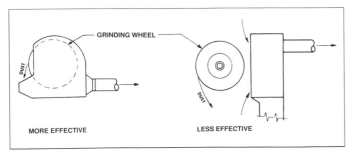

**Figure 19–8.** The hood should be located and shaped so that the initial velocity of the contaminant throws the contaminant into the hood opening.

For even distribution of the airflow through a large hood, flow can be controlled by passing the air through a narrow slot into a plenum chamber. The flow distribution along a hood is fixed by the ratio of the velocity through the slot to the velocity in the chamber. For most purposes, the distribution will be good enough if the velocity through the slot is about twice the average velocity through the plenum. Usually a slot velocity of 2,000 fpm (10 m/s) is suggested (Table 19–B).

Recommended designs for hoods for many special operations or applications are given in Section 10 of the ACGIH's *Industrial Ventilation—A Manual of Recommended Practice,* 22nd edition (see Bibliography). Designs define the size or proportions of hoods and are complete with recommended air quantities, duct sizes, air velocities, and hood entrance losses. This manual should be used for any exhaust hood design problem.

## Receiving Canopy Hoods

These seldom succeed as capture hoods. There are several reasons for this. Air velocity drops off tremendously (in proportion to the square of the distance) between the canopy and the source of contaminant evolution. A greater volume of airflow is required for contaminant control. Control may be lost due to room air cross currents. Finally, the canopy type of hood often forces the employee to work beneath it, that is, between the point of contaminant emission and the hood.

Receiving (canopy) hoods are used where rising hot air carries the contaminant into the exhaust system. However, this is not a safe solution where the operator works under the hood.

The amount of air required for a canopy hood can be approximated by the following equation:

$$Q = 1.4PHV \qquad (13)$$

where   $Q$ = rate of airflow (cfm)
   $P$ = tank perimeter (ft)
   $H$ = height of canopy above tank (ft)
   $V$ = desired control velocity (fpm)

$V$ ranges from 100 to 500 fpm (0.5 to 2.5 m/s), depending on cross drafts. (See Table 19–B.)

The effectiveness of receiving hoods can be improved by a vertical baffle or curtain wall on one or more sides. If the tank can be enclosed so that only one long side is left open and the hood can be kept at the same height from the tank surface, the air requirement for the same control velocity would be reduced by two-thirds. Even in this instance, the operator must not work under the hood.

This reduction in air requirement well illustrates the reason for designing and building hoods that partially or completely enclose the process. If such hoods must be constructed with channels to accommodate monorails or must be made to rise or swing to one side for access to the tank or other equipment, enclosing or baffling of the contaminant source is useful.

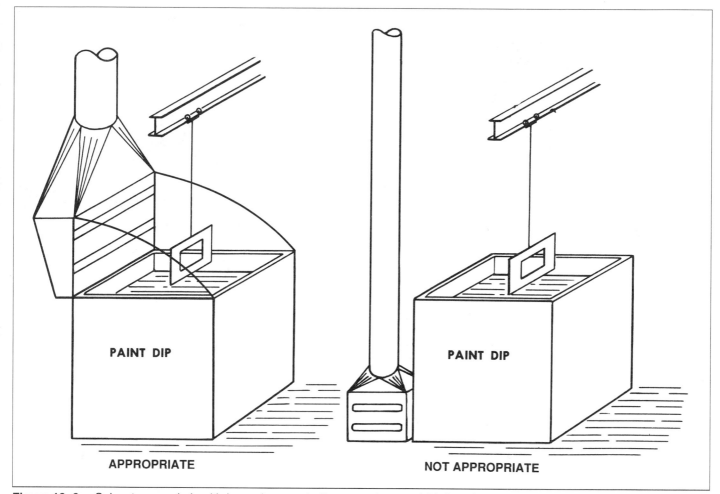

**Figure 19–9.** Solvent vapors in health-hazard concentration are not appreciably heavier than air. Exhaust from the floor usually gives fire protection only. (Courtesy ACGIH.)

If the canopy cannot be used because of the nature of the process, the most logical compromise is a side hood that overhangs the tank as much as possible.

A lateral exhaust hood, the type of hood shown in Figure 19–11, is used where the top of the tank must be fully accessible, as in electroplating (see the section on open-surface tanks later in this chapter).

With tanks up to 24 in. (0.7 m) wide, sufficient control may be obtained with a slot on only one long side. On wider tanks, two sides should have exhaust slots. Total airflow is given by this formula:

$$Q = \frac{\pi\,WL}{2} V_c \qquad (14)$$

where  $Q$ = airflow (cfm)
  $W$ = width of tank (ft) (single-sided slot), or ½ width of tank (slots on two sides)
  $L$ = length of tank (ft)
  $V_c$ = desired capture velocity at front edge of tank (fpm) (usually 75–150 fpm)

Because the velocity of air decreases rapidly at increasing distances away from the slot face, lateral exhaust

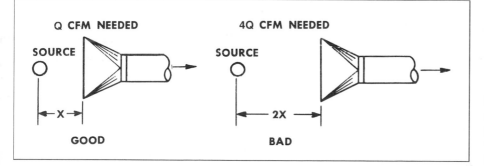

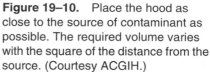

**Figure 19–10.** Place the hood as close to the source of contaminant as possible. The required volume varies with the square of the distance from the source. (Courtesy ACGIH.)

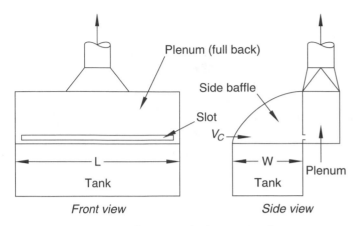

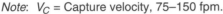

Note: $V_C$ = Capture velocity, 75–150 fpm.

**Figure 19–11.** Open-surface tank design.

should not be used when the tank is more than 4 ft (1.2 m) wide.

Control velocity over a wider tank can also be obtained by a push-pull system, that is, by blowing air from a nozzle at one side into a hood on the other side of the tank. A jet of air under pressure can be confined to a much narrower angle than can be obtained from a suction hood, so a wider tank can be served. The jets will entrain material rising from the tank surface and carry it into the hood.

Air jets should be used only with great caution, because they can spread the contamination into the working area. For instance, if material is lowered into and raised from the tank and passes through the air jet to dry, the air jet can be broken up and deflected so as to spread the contamination. The lateral exhaust hoods must be adequately sized and exhaust an adequate amount of air. The jet airstream must be pointed in the proper direction so that the contamination can be picked up by the hood. (See the Bibliography for design details for hoods.)

### Employee Participation

Finally, a hood should not interfere with the employee's job. The hood is there to help, not hinder, the employee. If an employee discovers that a hood inhibits the operation, the employee is likely to remove or alter it.

A job must be analyzed thoroughly in advance. The operator's motions must be studied, and process changes must be considered. Wherever possible, the person on the job should be consulted and opinions solicited. It may be wise to explain exactly how the proposed hood will work to help. Training of this sort is important to the optimum use of a hood.

After the hood is installed, the employee should use it to the fullest. A cooling or pedestal fan should not be placed where it interferes with the effectiveness of an exhaust hood, nor should any adjustments be made in the exhaust system by the employee. The employee should be instructed to tell the supervisor when there is a decline in exhaust control.

## ■ Design Considerations

### Ducts

After contaminated air has been drawn into a hood, ducts serve the purpose of carrying that air to an air cleaner or to the outdoors. When air passes through any duct or pipe, friction must be overcome; that is, energy must be expended. The amount of this friction loss must be calculated before the system is installed, so that the proper size fan and motor can be purchased.

Several useful references can remove the guesswork from duct design. (See the Bibliography at the end of Chapter 21, especially the *Industrial Ventilation Workbook,* latest edition, IVE, Inc., and the *Ventilation Manual—A Manual of Recommended Practice,* latest edition, ACGIH.)

The starting point in designing a local exhaust system is determining the volume of air per minute that must be handled by each hood to control the contaminant released in the workroom. Based on such data, careful duct design accomplishes the following objectives:

- Holds power consumption to a minimum
- Maintains proper transport velocity so the contaminant, if it is a dust or fume, does not settle out and plug the duct
- Keeps the system balanced at all times

### Multiple Ducts

Local exhaust systems with multiple hoods pose problems (Figure 19–12). After deciding the amount of airflow needed at each hood to control the contaminant, the duct designer is to select duct sizes and fittings (such as elbows, Ys, and reducers) so that air will flow from each hood as desired. When two branches coming from two hoods (hood A and hood B, for example) join at a Y to form a single main (or submain or header), the static pressure between this junction point and the face or inlet of hood A must be about the same as between the junction and the

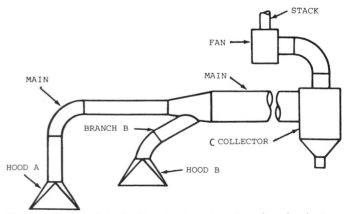

**Figure 19–12.** A typical local exhaust system: hoods, ducts, collector, fan, and stack.

face of hood B. In order to achieve the proper volume flow rate in each hood, balance must be achieved.

There are two approaches; the first is the preferred choice.

- Balance can be achieved by sizing the ductwork and selecting fittings that will give approximately equal pressure drops in the two branches when airflow through hood A and hood B is at the desired rates. For example, if one hood will exhaust more air than it should, smaller-diameter piping can be installed. This will increase the resistance, cause air velocity to drop, and allow the air flowing in that duct to be reduced. Once approximate balance has been achieved, the actual flow in the leg with the lower static pressure demand can be estimated. (See the Bibliography at the end of Chapter 21 for details.)

- Air balance may be achieved by use of dampers in each branch. These slide gates or dampers can be set to partially block the flow of air in order to lower the amount of air entering the hood. However, this is not the preferred method.

Advantages of the first method are as follows: If velocities are initially chosen properly, the ductwork will not become plugged; the system cannot be tampered with (tampering may deprive another workstation of the air needed for contaminant control there); and erosion will be reduced and there will be no buildup of dust or linty material caused by damper obstruction. This method is preferred where toxic materials are handled and is mandatory where explosives, magnesium, and radioactive dusts are exhausted.

The second method gives some leeway for correcting improperly estimated exhaust volumes and also provides some flexibility for future changes or additions. However, a good rule in local exhaust system design is that once a multihood layout is completed and balanced, additional hoods should not be added later. Additional hoods will alter the airflow and may make some hoods ineffective.

Table 19–C lists recommended minimum duct velocities for ducting different types of contaminants. Three terms that relate to duct velocity should be understood:

- *Static pressure.* Static pressure is created by the fan and is the energy source of the system. Static pressure is converted to velocity pressure and to heat in the form of friction loss and other losses.

- *Velocity pressure.* A definite pressure is created by moving air. Velocity pressure is created by converting static pressure to air movement at the duct serving the hood.

- *Duct friction loss.* For the same velocity of air, narrow-diameter ducts have higher friction loss than large-diameter ducts.

- *Other losses.* Elbows and tapered sections where a branch enters a main all add their loss. The sharper the bend and the more abrupt the taper in transition pieces, the higher the loss.

## Common Piping Defects

A quick inspection of an exhaust system will reveal a good deal about how well designed it is. For example, square and rectangular ducts, so common in the heating and ventilating field, are seldom used for local exhaust systems. Round piping is used because of its lower friction loss and resistance to collapsing.

Furnace-type elbows and T-fittings that come with hardware-store furnace pipe should be avoided. Because their sharp corners waste power, a fan must be sized larger than normal to handle the amount of air necessary to keep dust from dropping out of the air stream. This settling may plug ducts.

**Table 19–C.**   Range of Design Velocities

| Nature of Contaminant | Examples | Design Velocity |
|---|---|---|
| Vapors, gases, smoke | All vapors, gases, and smokes | Any desired velocity (economic optimum velocity usually 1,000–1,200 fpm) |
| Fumes | Zinc and aluminum oxide fumes | 1,400–2,000 |
| Very fine light dust | Cotton lint, wood flour, litho powder | 2,000–2,500 |
| Dry dusts and powders | Fine rubber dust, Bakelite molding powder dust, jute lint, cotton dust, shavings (light), soap dust, leather shavings | 2,500–3,500 |
| Average industrial dust | Sawdust (heavy and wet), grinding dust, buffing lint (dry), wool jute dust (shaker waste), coffee beans, shoe dust, granite dust, silica flour, general material handling, brick cutting, clay dust, foundry (general), limestone dust, packaging and weighing asbestos dust in textile industries | 3,500–4,000 |
| Heavy dusts | Metal turnings, foundry tumbling barrels and shakeout, sand blast dust, wood blocks, hog waste, brass turnings, cast iron boring dust, lead dust | 4,000–4,500 |
| Heavy or moist dusts | Lead dust with small chips, moist cement dust, asbestos chunks from transite pipe cutting machines, buffing lint (sticky), quick-lime dust | 4,500 and up |

(Adapted from ACGIH *Industrial Ventilation—A Manual of Recommended Practice.* 22nd edition.)

# Air Cleaners

Air cleaners fall into two broad classes according to their use.

- Industrial air cleaners remove airborne contaminants (dust, fumes, mist, vapor, gas, or odor) that would otherwise pollute the surroundings, either in-facility or outside the facility's neighborhood. These air cleaners are designed to function effectively under light to heavy loadings of contaminant. They can be cleaned of collected contaminant. They handle a moderate rate of airflow at high static pressure. This type of air cleaner is the principal concern of this section.

- Air cleaners that handle relatively high rates of airflow at low static pressure are associated with heating, ventilating, and air conditioning systems. This type of air cleaner removes particulates from incoming outdoor air and recirculated air to provide clean air for the building.

Industrial-type air cleaners include simple settling chambers, wet and dry centrifugals, wet and dry dynamic precipitators, electrostatic precipitators, wet-packed towers and venturi scrubbers, washers, and fabric filters.

A suitable air-cleaning device should be standard equipment in every exhaust system handling a contaminant that might result in a health hazard, be a nuisance, or cause air pollution. Increasing community awareness of air pollution and its cost to health and property in the neighborhood is making air cleaning necessary. All areas have pollution-control ordinances that require a permit. Be sure the collected contaminants are disposed of properly so as not to cause additional pollution.

Many types of air-cleaning devices are in use, and the selection of the proper one depends on several factors. The nature of the material to be removed, the quantity of the material, the degree of cleanliness of the outflow that must be achieved, and cost must all be considered. Most air cleaners are housed in unraveling enclosures, so the layperson generally has no idea what type they are or how to check their performance. General familiarity with types of cleaners, and familiarity with their gross shortcomings in particular, may be useful. Before discussing individual cleaners, we will look at how to measure and compare air-cleaner efficiency.

## Air-Cleaner Efficiency

Efficiency is the ratio of the amount of dust (or other contaminant) collected by the air cleaner to the amount of the dust that enters the device. Ordinarily, the units of measurement are in terms of weight of dust. The term *weight efficiency* is sometimes used because arresters are often used in the processing industries where the mass of the product

handled, shipped, or reclaimed is the important dollars-and-cents consideration.

Other factors enter the efficiency concept when air cleaners are used to arrest a contaminant that is a health hazard. Such contaminants may include silica, lead, beryllium, cadmium, or zinc fume, arsenic, or other toxic respirable dust.

Performance measures vary between manufacturers. The health and safety professional, if involved with the project, should make certain that the vendor agrees to a proper performance guarantee as part of the purchase contract. The health and safety professional must beware of that often misleading word *efficiency* as expressed as a percentage.

A comparison of important filter characteristics of ventilating system air filters is given in Table 19–D. A comparison of the size range of particles that can be collected by the various types of collectors and filters is shown in Figures 19–13 and 19–14.

## Centrifugal Collectors

The cyclone belongs to a class that relies on centrifugal force to throw the dust out of the airstream.

*Low-pressure cyclones* are sheet-metal cylinders set on top of a cone. Air enters at an angle on the side of the cylinder and swirls around inside, passing downward. It then rises through a center tube and passes out the top of the cylinder. Centrifugal force throws the particles out of the airstream to the bottom of the cone.

There are several advantages and disadvantages of the low-pressure cyclone. The advantages are low cost, low maintenance, and low pressure drop (0.75–1.5 in. [0.2–0.4 kPa] water gauge). The disadvantage is that the low-pressure cyclone is incapable of collecting fine particles. In fact, it is only about 75 percent efficient on particles that are as large as 40 μm. The low-pressure cyclone is worthless against dusts of hygienic significance (5 μm and less).

The primary use of this cyclone type is to collect woodworking dust (sawdust, shavings, and chips), paper scraps, and bulk materials after being pneumatically conveyed.

Collecting efficiency increases with an increasing pressure drop. Pressure drop results from making the diameter of the cyclone smaller. Small-diameter cyclones cannot carry large volumes of air. This disadvantage, however, is overcome if several are placed in parallel.

*High-efficiency cyclones* are made by combining many small-diameter, high-resistance units. Efficiencies of 75 percent against 10-μm particles and 90 percent against 15-μm particles can be obtained. However, efficiency may be inadequate for control of dusts of health importance (5 μm and less). Its pressure drop is high: 3–8 in. (0.75–2.0 kPa) water gauge. With abrasive dusts, castings or heavy sheet metal of special alloy must often be used to allow for the expected excessive wear.

**Table 19–D.** Comparison of Some Important Air Filter Characteristics

Air filters should be used only for supply air systems or other applications where dust loading does not exceed 1 grain per 1,000 cubic feet of air.

| Efficiency | Type | Pressure Drop in wg (Notes 1 & 2) | | ASHRAE Performance (Note 4) | | Face Velocity fpm | Maintenance (Note 5) | |
|---|---|---|---|---|---|---|---|---|
| | | Initial | Final | Arrestance | Efficiency | | Labor | Material |
| Low/ Medium | 1. Glass Throwaway (2″ Deep) | 0.1 | 0.5 | 77% | NA Note 6 | 300 | High | High |
| | 2. High Velocity (Permanent Units) (2″ Deep) | 0.1 | 0.5 | 73% | NA Note 6 | 500 | High | Low |
| | 3. Automatic (Viscous) | 0.4 | 0.4 | 80% | NA Note 6 | 500 | Low | Low |
| Medium/ High | 1. Extended Surface (Dry) | 0.15-0.60 | 0.5-1.25 | 90-99% | 25-95% | 300-625 | Medium | Medium |
| | 2. Electrostatic: a. Dry Agglomerator/ Roll Media | 0.35 | 0.35 | NA Note 7 | 90% | 500 | Medium | Low |
| | b. Dry Agglomerator/ Extended Surface Media | 0.55 | 1.25 | NA Note 7 | 95%+ | 530 | Medium | Medium |
| | c. Automatic Wash Type | 0.25 | 0.25 | NA Note 7 | 85-95% | 400-600 | Low | Low |
| Ultra High | 1. HEPA | 0.5-1.0 | 1.0-3.0 | Note 3 | Note 3 | 250-500 | High | High |

Note 1: Pressure drop values shown constitute a range or average, whichever is applicable.
Note 2: Final pressure drop indicates point at which filter or filter media is removed and the media is either cleaned or replaced. All others are cleaned in place, automatically, manually or media renewed automatically. Therefore, pressure drop remains approximately constant.
Note 3: 95–99.97% by particle count, DOP test.
Note 4: ASHRAE Standard 52-76 defines (a) arrestance as a measure of the ability to remove injected synthetic dust, calculated as a percentage on a weight basis, and (b) efficiency as a measure of the ability to remove atmospheric dust determined on a light-transmission (dust spot) basis.
Note 5: Compared with other types within efficiency category.
Note 6: Too low to be meaningful.
Note 7: Too high to be meaningful.

(Reprinted with permission from ACGIH. *Industrial Ventilation—A Manual of Recommended Practice*, 21st edition, ACGIH, 1992.)

*Dry dynamic precipitators* are a combined air cleaner and fan in one unit. They have a specially shaped impeller that precipitates the dust by centrifugal force while inducing airflow. The collection efficiency is about the same as that of the high-efficiency or high-pressure cyclone.

With abrasive dusts, the impeller blades are subjected to considerable wear. Normally the fan for inducing airflow is placed downstream from the collector to provide maximum protection.

**Wet collectors.** The principle employed in wet collectors is to get the dust into intimate contact with a liquid. Either the particles are trapped directly because they are in a bath or stream of water, or, because of the increased mass the liquid gives them, they can be better thrown out of the airstream by centrifugal force. Wet collectors can handle gases that are at a high temperature or are laden with moisture. With this type of collector, disposal of collected material presents little problem. However, wet collectors have the following disadvantages:

■ Unless cheap water is handy, settling tanks and accompanying equipment will be needed to reuse the water.

■ Freezing cannot be tolerated.

■ If corrosive chemicals are collected, expensive corrosion-resistant materials must be used.

Although the efficiency of wet collectors is usually only 75 percent or less, it can be as high as 90 percent against 1-$\mu$m particles. The efficiency averages less than 50 percent against 0.5-$\mu$m particles. The pressure drop of the wet collectors also varies over a wide range.

There are at least seven types of wet collectors.

*Spray chambers,* consisting of sprays and scrubber plates, are one of the oldest methods of air cleaning. Pressure drop varies, usually being in the range 2.5–6 in. (0.6–1.5 kPa) water gauge. Water consumption is about 3–5 gpm per 1,000 cfm (0.32–0.53 l/min per m³/min) of air handled.

*Packed towers,* extensively used by the chemical industry for gas absorption, are also used for collecting toxic dusts. In this type of collector, water usually flows downward over packing, which may be odd-shaped ceramic saddles, rings, coke, gravel, or similar material. Water flow is 5–10 gpm/1,000 cfm (0.53–1.1 l/min/m³/min) of air. Packing depth is usually about 4 ft (1.2 m). Pressure drop is 1.5–3.5 in. (0.4–0.9 kPa).

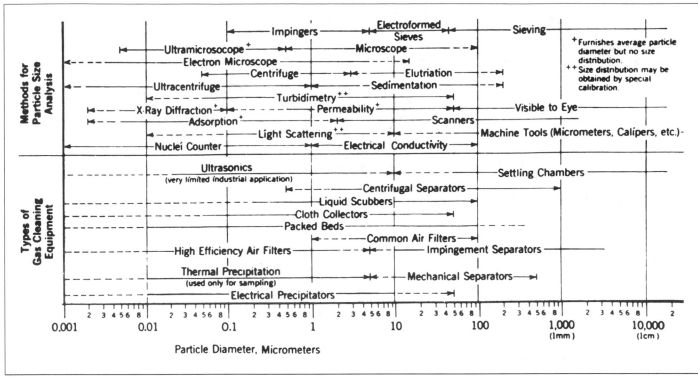

**Figure 19–13.** Particle size ranges that can be analyzed by various techniques and that can be collected by various types of gas-cleaning equipment. (Reprinted with permission from *Stanford Research Institute Journal* 5:95, 1961.)

*Wet centrifugals* use a combination of centrifugal force and water contact to effect collection. Air is introduced tangentially and frequently directed countercurrent to the flow of water by baffles or directional plates. Pressure losses range from 2.5–6 in. wg (6.0–0.15 kPa wg).

*Wet-dynamic collectors* are similar to the dry dynamic precipitator with the specially shaped impeller, except that water is sprayed on the blades. Water consumption is 0.5–1 gpm for 1,000 cfm (1.8–3.8 l/min for 0.47 m³/s) of air.

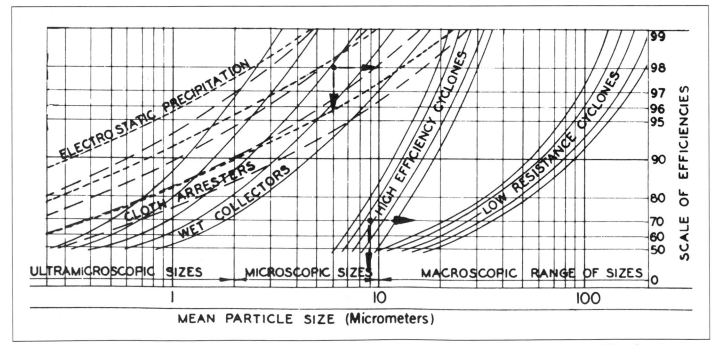

**Figure 19–14.** Relative performance and efficiencies of various dust collectors. (Copyright 1952, American Air Filter Co., Inc., from ACGIH, *Industrial Ventilation—A Manual of Recommended Practice.*)

*Orifice-type collectors* are arranged so that airflow through them is brought into contact with a sheet of water in a restricted passage.

*Venturi-type collectors* make use of the high velocity of air through a venturi throat to break up water fed into the throat. These fine water droplets collide with the dust in the airstream, and the dust is wetted. These are high-efficiency collectors, being 90–99 percent efficient in the size range 0.1–1 μm. However, their power consumption is high. The pressure drop is 10–15 in. wg (2.5 kPa–12.4 kPa).

*Fog filters* use many small high-pressure nozzles in centrifugal tower-type units to increase the probability of water droplets wetting the dust particles. High water pressure is used to form the droplets of high-energy fog. Good efficiency has been obtained.

**Electrostatic precipitators.** Air that contains solid or liquid particles is passed between a pair of electrodes—a discharge electrode at high negative potential and an electrically grounded collecting electrode. The potential difference must be high enough to cause a corona discharge to surround the discharge electrode. Gas ions formed there move rapidly under action of the electric field and toward the collecting electrode or plates. They transfer their charge to the particles by colliding with them. The electric field, interacting with the charge on the particles, causes them to move towards and be deposited on the collecting electrode (Figure 19–15).

These units are capable of achieving high collection efficiency. This high efficiency is not always profit producing, however, because of cost, worthlessness of collected material, and space limitations.

Electrostatic precipitators have the great advantage of negligible pressure drop. If the dust load is heavy, they are usually preceded by primary collectors that work on a different principle. Although unable to collect nonparticulate gases or vapors, they can collect mists, such as the oil mist given off during the cutting and machining of metals.

**Settling chambers.** Theoretically, it would be possible to settle out dust in a large chamber by dropping the conveying velocity to a point where the particles would no longer be conveyed. However, space requirements are excessive, and the presence of eddy currents and natural mixing nullifies the effective velocity. Settling chambers can be used only for bulky, coarse materials. Nevertheless, this solution to collection problems is offered perennially in manifold forms such as settling barrels, "dust boxes," and water drums. These appeal to the layperson because of their straightforwardness. A simple calculation will show that dust of health significance (5 μm and less) cannot possibly settle out in a reasonable time and space (see Chapter 8, Particulates).

Settling chambers have their place as chip traps and as precleaners for high-efficiency cleaners.

**Filters.** Two classes of filters are in common use: "throwaway" or furnace-type filters and cloth filters. The first type is not suitable for industrial exhaust air cleaning. Its principal use is in general ventilating systems.

The second filter type, cloth filters, can be made of a wide variety of fabrics such as cotton, wool, Dacron, Orlon, and, for high-temperature applications, glass. Cloth filters can be sewn together in the shape of tubes (hung vertically) or envelopes that fit over frames to keep the cloth sides from collapsing. In baghouses, the filters are equipped with manual or motor-driven shakers that dislodge the collected dust when the exhaust fan is shut off. This allows the dust to fall into hoppers below. From there it can be removed for disposal.

Cloth filters function by building up a layer of dust on the fabric, which then filters additional incoming dust. New cloth filters, therefore, may pass dust for a brief period—perhaps to the consternation of someone who has installed new bags for the first time—before the dust layer accumulates. As the filter operates, its resistance gradually rises. When the filter is cleaned, its resistance falls, but never to as low a point as it was when the filter was new. The residual dust mat necessary for efficient separation always remains in use.

Cloth filters operate at low velocities: 1–5 cfm/ft² (0.47–2.4 l/s/0.09 m²) of cloth. They have relatively high resistance: 2–15 in. (0.5–3.7 kPa) water gauge. They have high efficiencies, removing 99 percent or more of the incoming dust.

Cloth filters are relatively expensive, at least in the opinion of the heating and ventilating contractor who is accustomed to throwaway filters. Their cost may be $0.50 to $5.00 per cfm (per 28.2 l/min) filtered. Increased filter resistance means higher operating cost. For example, a system handling 10,000 cfm (4.72 m³/s) might expend 7.5 hp (5.6 kW) or more just overcoming the filter pressure drop.

Cloth filters can be used only against dry dust. Sticky or oily material clogs the bags, and when the temperature is below the dew point, condensation causes an exorbitant pressure drop and ineffective cleaning of the filter.

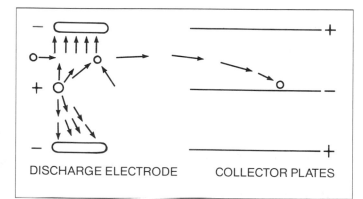

DISCHARGE ELECTRODE          COLLECTOR PLATES

**Figure 19–15.** In an electrostatic precipitator, particles passing through the electrostatic field (left) become ionized (electrically charged). They are then attracted to collector plates that have the opposite charge. Collector plates are cleaned periodically. (Courtesy ACGIH.)

Cloth filters are ordinarily used for intermittent operation (the airflow is shut off every few hours to permit cleaning of the cloth). They may also be equipped for automatic continuous use. Periodically, a section is shut off from the system and the bags in this section are cleaned. This requires a larger cloth area and additional air dampers and controls.

Regular cleaning is essential, and to ensure it, a U-tube water manometer should be permanently mounted on the outside of the casing. This will allow the pressure drop to be checked daily. A buildup of filter resistance, as shown by increased pressure drop, is the warning sign that the filter is becoming plugged and needs to be cleaned.

Access doors should be located conveniently so that the interior of the casing can be checked periodically for leaks in the bags, bridging of dust in the hopper, or any other problems. Too often filters are located where they are inaccessible, and are then neglected. Filters should be checked weekly.

Cloth filters of the reverse air jet type that can be periodically cleaned during filtration (no shutdown) are widely used. They consist of a felt-like tube with the dusty air entering on the inside. On the outside, a slotted ring made of pipe continuously travels up and down the bag surface. Compressed air is blown at high velocity out the slot, which, in passing through the bag, dislodges the dust so that it falls down in the dust hopper. The rate of airflow delivered by the exhaust system may be as high as 20 cfm/ft$^2$ (556 l/min/929.03 cm$^2$) of cloth, which is far higher than in the conventional cloth filter. The pressure drop and the efficiency of the two types are comparable.

High-efficiency particulate air (HEPA) filters, developed to handle low-inlet loadings in special applications, particularly where the air contaminant is radioactive, are available. They are made of various combinations of pads of glass fibers, compressed glass fibers, or resins carded into wool. They are generally disposable units, but they are quite different from the common disposable viscous filter used in heating and air conditioning.

## Air Cleaners for Fumes and Smokes

Fumes result from reactions such as burning, sublimation, distillation, and especially the condensing of a vapor given off by a molten metal. The composition of fumes may be different from that of the parent material. Lead oxide fume, cadmium fume, zinc oxide, and iron oxide (welding fume) are examples. Fume particles are generally less than 1 μm, so they show active Brownian movement (that is, they are far too small to settle), and they are remarkably uniform in size. Smoke is commonly organic, coming from incomplete combustion of coal, oil, wood, and other fuels. It is usually dark or jet black in color and obscures light. Its particles compare in size to the particles of metallurgical fumes.

The high-efficiency dust collectors already mentioned—the cloth filter, high-efficiency wet collector, and electrostatic precipitator—are best for arresting fumes. For smoke, the answer is improved combustion, or the high-voltage electrostatic precipitator, or both.

## Air Cleaners for Gases and Vapors

Vapors can be defined as the gaseous form of a material normally in the liquid or solid state at room temperature. Gases and vapors are not particles, but individual molecules dispersed among the molecules of the air. Because gases and vapors diffuse, arresters that rely on the straining action of filters or on centrifugal force are not applicable. Methods generally used are as follows:

**Absorption.** A liquid is used in a packed tower or scrubber that dissolves or reacts chemically with the gas or vapor and removes it from the air. The disposal of this liquid may present problems of waste control and water pollution.

**Adsorption.** Many solid particles have an adsorbing action for certain gases and vapors. The action takes place at the surface of the adsorbent where gas and solid come in contact with each other. The most widely used material for removing odors is activated carbon. The activated carbon is placed in granule form in trays, canisters, or perforated cans. Activated carbon can adsorb some vapors and gases up to 50 percent of its own weight. It can then be reactivated (for reuse) by heating. Activated carbon is frequently used for the removal of solvent vapors. This is a particularly appealing method because the solvent can later be reclaimed.

**Combustion.** If the gas or vapor can be oxidized into harmless or odorless products, combustion can be used for removal. All hydrocarbons can be so removed because on complete oxidation, water and carbon dioxide are the only end products.

**Condensation.** By reducing the temperature of the incoming air, vapor can be changed to the liquid state and removed. Several types of condensers and refrigeration units can be used to chill the incoming vapor.

# ■ Fans

We have discussed hoods, ducts, and air-cleaning devices for a local exhaust system. The fourth element in a local exhaust system is composed of the fan and motor. Two groups of fans are in use: centrifugal and axial flow. They are shown schematically in Figure 19–16. The radial-blade centrifugal fan can be used to move air that contains particulate material.

## Centrifugal Fans

Depending on how the blades are pitched, centrifugal fans are all modifications of the basic wheel type. They can be used where static pressure is medium to high (2 or more inches, or 7 kPa) water gauge and up.

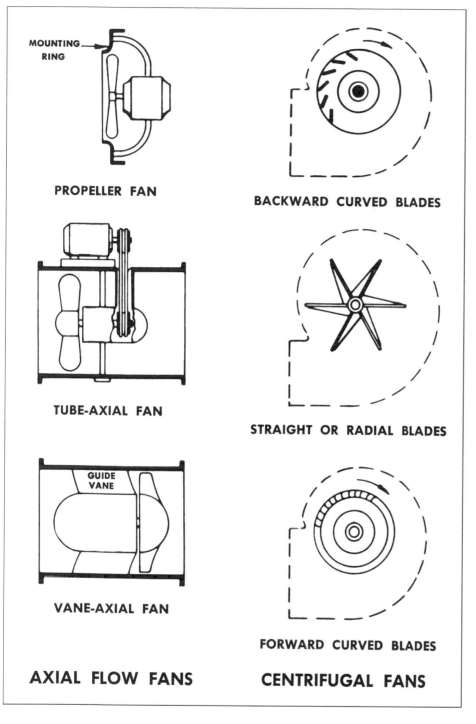

PROPELLER FAN

MOUNTING RING

TUBE-AXIAL FAN

GUIDE VANE

VANE-AXIAL FAN

**AXIAL FLOW FANS**

BACKWARD CURVED BLADES

STRAIGHT OR RADIAL BLADES

FORWARD CURVED BLADES

**CENTRIFUGAL FANS**

**Figure 19–16.** Types of exhaust fans. (Courtesy ACGIH, AMCA.)

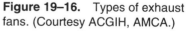

The straight-bladed radial wheel is the workhorse of fans in the industrial ventilation field. Its blades, made of steel or cast iron, do not readily clog with material passing through. They also withstand considerable abrasion. For decades, this fan has been used in buffing and woodworking shops where lint, chips, and shavings pass through the blades. It has medium tip speed, high noise factor, and low mechanical efficiency.

■ Backward-curved blades on a centrifugal-type fan permit higher tip speed and thus greater fan efficiency.

Material will build up on the blades, so it often has an air cleaner ahead of it. Although this type of blade has a higher noise factor, its high efficiency and nonoverloading feature makes it the choice for many large-volume exhaust and HVAC systems.

■ Forward blades, tipped in the direction of rotation, yield a fan that has low space requirements, low tip speed, and low noise factor. It is popular in small heating and air conditioning systems where the static pressures are low. Because airborne material will stick to the short, curved blades and throw the wheel out of balance, this fan may be preceded by an air cleaner.

- Fans with straight or forward-curved blades demand more power as airflow increases. If the actual duct system resistance is less than the estimate from which the fan was selected, the actual airflow will exceed the estimate, more power will be required, and the driving motor may be overloaded.
- Fans with backward-curved blades, on the other hand, have a power demand that reaches a maximum. If the driving motor is rated to carry this maximum, it cannot be overloaded at the given speed. For this reason, fans with backward-curved blades are called nonoverloading fans.

There are a number of intermediate designs between the extremes of forward and full-backward curved blades that are similar in varying degrees to the performance characteristics of each type.

## Axial-Flow Fans

Fans of this type are modifications of the familiar desk fan or propeller fan. Air leaves in the same straight line in which it enters, whereas in the centrifugal fan air leaves at right angles to the direction in which it enters.

- Propeller fans move large volumes of air against negligible resistance, and are generally mounted on pedestals for cooling of individuals and general circulation. They may also be set in window boards or in walls with no ducts connected. One limitation is that they cannot operate against the friction that a duct would entail; this is sometimes forgotten.
- A narrow-blade propeller fan is often provided for spray booth exhausts, but duct connections must be held to a minimum. These fans are sensitive to added resistance, and a small increase in resistance will drop the air volume handled markedly.
- Tube-axial fans are an improved version of the propeller fan and are mounted by the manufacturer in a short section of duct to which other piping can be attached. They are effective against only an inch or two of static pressure, water gauge.
- Vane-axial fans have guide vanes in the short pipe section that straighten the airflow. They will operate against low static pressures (up to 4 in., or 1.0 kPa, water gauge). However, they must be used only in clean air.

For roof or wall exhaust, there are available direct-discharge fans (no scroll) with backward-curved blades similar to the wheel used in centrifugal fans.

**Fan noise.** Except for low-speed fan units, fans usually are noisy. This can be distracting, irritating, or even damaging to the ear. Noise may interfere with speech. Fan noise can be a problem both in the facility area and to neighbors outside.

Fan manufacturers, through technical organizations such as the Air Moving and Control Association and the American Society of Heating, Refrigerating and Air-Conditioning Engineers, have developed noise ratings for fans based on considerations such as blade-tip speed, brake horsepower, and pressure. (See the discussion of these organizations in Appendix A, Additional Resources.)

When blower noise (inherent in the unit) is disturbing, one solution is to surround the fan by a sound-attenuating enclosure. Depending on the size, the enclosure may be made of masonry, heavy-gauge sheet metal, or even .75-in. (19-mm) plywood, if the unit is small. The enclosure should be lined fully with acoustically absorbent material.

## Fire Prevention

It is important to refer to available standards, such as NFPA Standard 91, *Blower and Exhaust Systems for Dust, Stock and Vapor Removal or Conveying* (see Bibliography).

Where fans handle flammable solid materials or vapors, the rotating element must be nonferrous or spark-resistant material, or the casing must consist of or be lined with such material. This requirement also applies to both the rotating element and the casing, where solid foreign material passing through the fan could produce a spark.

Fan motors located in rooms or areas in which flammable vapors or flammable dust is being generated and removed should be of the type approved for the particular conditions or hazard. When exhaust systems are used to handle flammable gases or vapors or combustible or flammable dust, stock, or refuse, static electricity must be removed from belts by grounded metal combs or other effective means.

## ■ Exhaust System Performance

No local exhaust system is foolproof. Its performance needs to be checked, and maintenance work is required. All too often, a system is well designed and properly installed but develops problems as time goes by. Perhaps the collectors have not been cleaned regularly, or the ducts have become partially plugged with settled material. Perhaps the hoods have become battered or abraded. It is possible that with facility expansion, additions have been made to the existing system that made the original hood ineffective. This fault is not uncommon and often goes unnoticed.

### Inspection

The system should be tested when it is new and clean, and all readings recorded so that they can be used as a level of acceptable performance when checks are made later during regular operation.

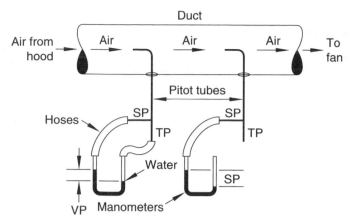

**Figure 19–17.** A water-filled manometer indicates pressure as inches of water column displaced in water gauge.

Local exhaust systems are primarily used to prevent a health or fire hazard. Their purpose is to keep people safe. Hence, the health and safety professional has a responsibility to see that exhaust systems perform as intended.

The health and safety professional can be equipped to make simple tests. First, with a simple U-tube or water manometer and permanently positioned test holes in branch ducts, at inlets and outlets of dust arresters, at fan inlets, and so on, the static pressure, or "suction," can be measured (see Figure 19–17). Comparison with earlier readings will tell whether or not anything has gone wrong. Second, a simple smoke tube can trace the flow of air into hoods and show visually whether the contaminated air is being removed from the work area.

Records of static pressure tests should be maintained. They may be supplemented by checks on fan speed and on weight of material collected in the dust arrester as needed (Figure 19–18).

## Recirculation of "Cleaned" Air

When the rate of air exhausted by a contaminant control system is high, either in absolute volume of air handled or in comparison with the volume of the room, cleaned air may need to be recirculated. It is often argued that the dollars-and-cents savings in heat will be such that recirculation is attractive; or the existing heating system may have so little reserve capacity that recirculation will be the only way to maintain a comfortable working temperature. There are some basic principles to follow in deciding whether to recirculate air or temper makeup air.

- If the contaminant is harmless, and if after passing through a dust, mist, fume, or vapor separator the processed air is no more contaminated than the outdoor air, recirculation would be safe.
- Savings in fuel or in heating facility usage are never a valid reason for recirculating air if the original air contaminant is a serious health hazard. For example, if the exhaust system is installed to control carcinogens, silica

dust, lead dust, and other toxic dusts, gases, or vapors, then the contaminated air from the air cleaner should be exhausted outdoors. Even if the dust filter or other arrester is operating at high efficiency so that the concentration of contaminant downstream is normally within safe limits, a breakdown can occur that would cause dangerous amounts of material to be blown back into the workroom.

- In every recirculation system, therefore, a Y-connection should be included in the discharge of the arrester. One leg should go to the outdoors, and the other should lead to the recirculation ductwork. Adjustable dampers should be arranged so that when the control is in one position, all of the air will be discharged outdoors, and when in the opposite position, all air will be sent back to the room. Various intermediate ratios can be adjusted depending on the outdoor air temperature.

A further advantage of this Y is that, should the arrester break down, the uncollected contaminant can immediately be blown outdoors temporarily instead of into the workroom.

Present federal and state laws, standards, and rules permit recirculation except in the specific instance of prohibiting recirculation from spray-finishing systems and with carcinogens.

Questions of application, operation, and maintenance of recirculation systems must be resolved on an individual basis; each set of circumstances should be evaluated and the expected or known contaminants considered. There cannot be a "cookbook" approach to recirculation except for the simplest systems involving innocuous or "nuisance" substances. The performance of most systems must be tested by a competent health and safety professional to evaluate worker exposure before and after the system is installed. The system must be carefully maintained and the system performance monitored to ensure optimum performance for adequate health protection. All recirculation systems should be equipped with automatic monitoring systems that will alert employees if the system fails.

## ■ Special Operations

### Open-Surface Tanks

A large variety of industrial operations, including surface treatment (such as anodizing and parkerizing), pickling, acid dripping, metal cleaning, plating, etching, and stripping, are conducted in open-surface tanks. Because some of these operations involve heat and gassing of the liquid, a simple solution may seem to be a canopy hood over the tank. This has several faults, however. With any canopy

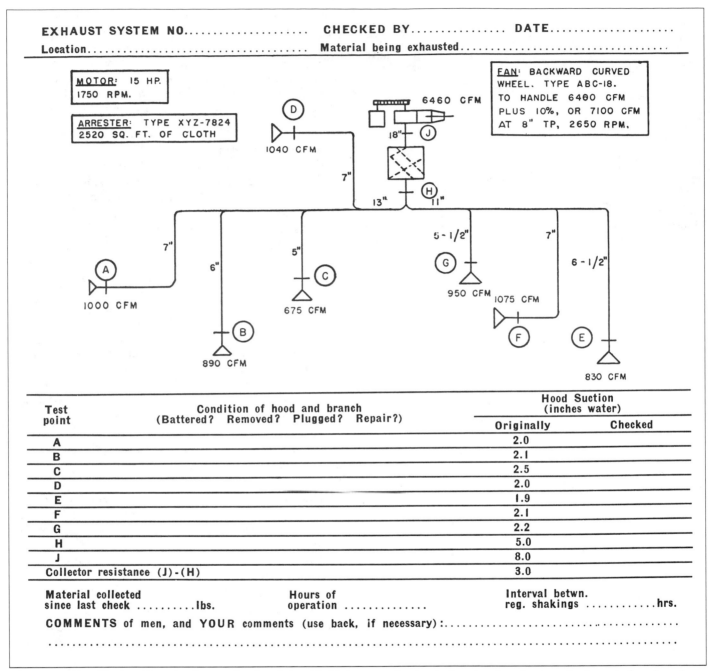

**Figure 19–18.** The health and safety professional should obtain or create simple line drawings to show the layout of each main duct and branch. General characteristics of the arrester, fan, and motor should be shown. Static pressure readings should be made at each test hole once the system has been balanced and dust counts, or air samples, made to show that the air contaminants are under control. These readings are indicated on the drawing by each test hole. A simple drawing should be made for each exhaust system.

hood the worker's head is likely to be between it and the tank surface, so he or she may have greater exposure with the hood than without it. Also, many of these operations require the work to be raised and lowered by a hoist on a monorail over the tank—the canopy hood would interfere with the operation. Furthermore, air currents caused by open doors and windows, as well as passing traffic, can seriously affect contaminant control (Figure 19–19). Can-

opy hoods should be used only as receiving hoods over hot processes.

The American National Standards Institute has formulated a code on ventilation of open-surface tanks, *Practices for Ventilation and Operation of Open-Surface Tanks*, Z9.1. This method of design is also included in *Industrial Ventilation—A Manual of Recommended Practice*. Figure 19–20 illustrates how air should flow away from the operator.

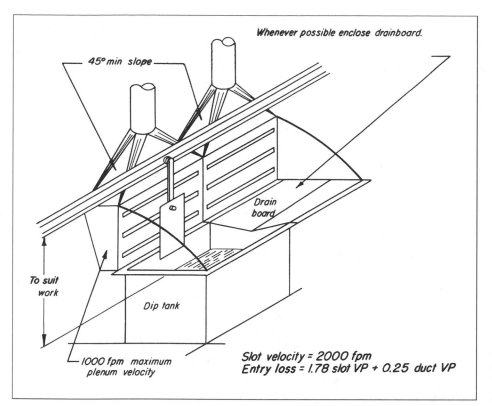

*Whenever possible enclose drainboard.*

45° min slope

Drain board

To suit work

Dip tank

1000 fpm maximum plenum velocity

*Slot velocity = 2000 fpm*
*Entry loss = 1.78 slot VP + 0.25 duct VP*

**Figure 19–19.** Exhaust for open-surface tank draws air and contaminants across the tank, away from any operator. (Courtesy ACGIH, ANSI.)

## Spray Booths

A spray booth is a partial enclosure through which a flow of air is drawn by an exhaust fan. The booths may vary in size from the small-bench type to huge installations large enough for a railroad car. A spray booth is simply an enclosure with sides, a rear, a top, and a bottom; the front, called the *face,* is open. The booth must, of course, be large enough to accommodate the work with space around it so that the operator has the necessary freedom of movement.

Most booths are used for spraying paint, enamel, or lacquer. Fewer booths are used for bleaching, cementing, glazing, metallizing, and cleaning. Another widespread use of booths is welding.

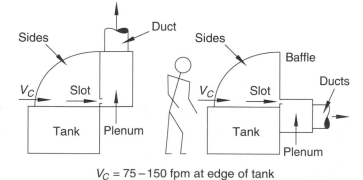

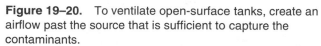

$V_C$ = 75 – 150 fpm at edge of tank

**Figure 19–20.** To ventilate open-surface tanks, create an airflow past the source that is sufficient to capture the contaminants.

A paint-spray booth has two main purposes: to protect the health of the painter and to reduce the fire and explosion hazard. These two interests are served by several regulatory bodies. The requirements of each must be met.

Some states have a code administered by its health or labor department designed to control occupational disease exposures, whereas another bureau may prescribe fire control aspects. The federal Occupational Safety and Health Administration (29 *CFR* 1910.94) has requirements. At the same time, the fire insurance carrier may have specific requirements that must be met. It may also have a secondary concern to reduce the total damage resulting from a fire, including water damage. The health and safety professional should be thoroughly aware of all requirements that apply.

When paint-spraying is in progress, the overspray must be carried away from the operator's breathing zone. This can usually be accomplished with an air velocity of 100–150 fpm (0.51–0.76 m/s) through the booth face. A small, shallow booth is usually designed for a face velocity of 200 fpm (1.0 m/s) or more. In metallizing, when the material is toxic, the velocity should also be at least 2,000 fpm (10 m/s).

Tempered makeup air to replace air exhausted by the booth should be supplied. This is needed for any exhaust system, but it deserves special attention here because some spray booths remove huge amounts of room air, which in turn should be supplied directly to the booth or to the area from which the air is drawn into the booth.

Routine and scheduled cleaning and maintenance of spray booths are a must. When allowed to accumulate, flammable materials increase the fire risk. In addition, paint in ducts also cuts down needed airflow. There are several products on the market that can be used to cover the inside of the booth that successfully prevent the sticking of the overspray to the booth surfaces.

In some high-production painting, much solvent is lost by evaporation from the wet, painted surfaces. If painted units are stacked out in the shop to dry, vapors may present a health and fire hazard. A separate ventilated area should be provided for preliminary drying of the finished materials before they are sent to the ovens for baking.

When the production level is not high, the material can be left in the booth for preliminary drying. In no instance should finished material be stacked in front of the booth so that the vapor-laden airflow drifts from the freshly finished material to the operator.

Dusts from some of the materials used in metallizing are explosive if finely divided and suspended in air in a critical concentration. When collecting dust from metallizing, it is sometimes necessary to provide wet-dust arrester units in the rear of the metallizing booth.

## Fume Control for Welding

Local exhaust hoods are useful in many indoor welding applications, and are a must when welding potentially toxic materials such as lead-containing or lead-painted metals and beryllium alloys. Except for production spot welding sometimes done at a fixed jig, a welder usually travels over a wide area. This necessitates some arrangement of flexible or movable ventilation.

The hood itself can have a circular or rectangular face with a flared section to which the duct is connected, thus giving it a streamlined airflow. Normally the hood is surrounded by a flange. The breadth of the face and the amount of the airflow exhausted should be such that fumes will be removed from the welding arc over the length of weld formed. The capacity of the exhaust, therefore, should be sufficient to reach laterally over the length of a weld of this distance.

Flexible metal duct branches or solid sections with swivel joints and counterbalances permit easy relocating of the hood. A duct branch should not be less than 6 in. (15 cm) in diameter and should exhaust sufficient air to control emissions. Average face velocities can exceed 1,500 fpm (7.6 m/s), and duct velocities should ensure a sufficiently high airflow into the hood to control fumes at the point of generation (usually several inches away from the hood face) as well as avoid settlement in the duct (Figure 19–21).

## Grinding Operations

Health, comfort, and good housekeeping suggest that grinding, buffing, and polishing equipment be equipped with local exhaust systems. Many state and provincial codes require them, as does OSHA, and where such codes exist, the health and safety professional should obtain copies so that these regulations can be met.

Hoods on grinding and cutting wheels serve a dual purpose: They protect the operator from hazards of bursting wheels and they provide for removal of dust and dirt generated. Most such wheels today are made of artificial abrasives that contain no free silica and so eliminate the silicosis hazard.

Uncleaned castings, on the other hand, may have mold sand adhering to them. This may be ground to dust, causing a silicosis hazard. In addition, grinding of some alloys, such as high manganese steel, cemented tungsten-tipped tools, and spark-resistant tools, presents other occupational disease hazards. (See Chapter 8, Particulates.)

Furthermore, grinding operations need local exhaust hoods to remove the dust and grindings that otherwise cause poor shop housekeeping.

The exhaust hoods must have sufficient structural strength and must enclose the wheel sufficiently. Details

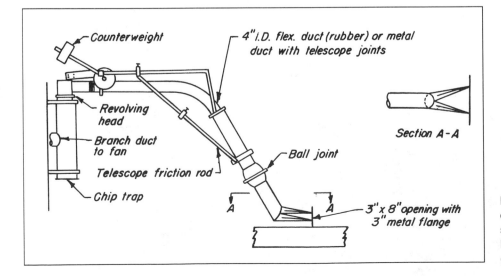

**Figure 19–21.** The flexible metal duct has swivel joints and telescoping sections to permit placement of the hood close to the point of contaminant generation. (Courtesy ACGIH.)

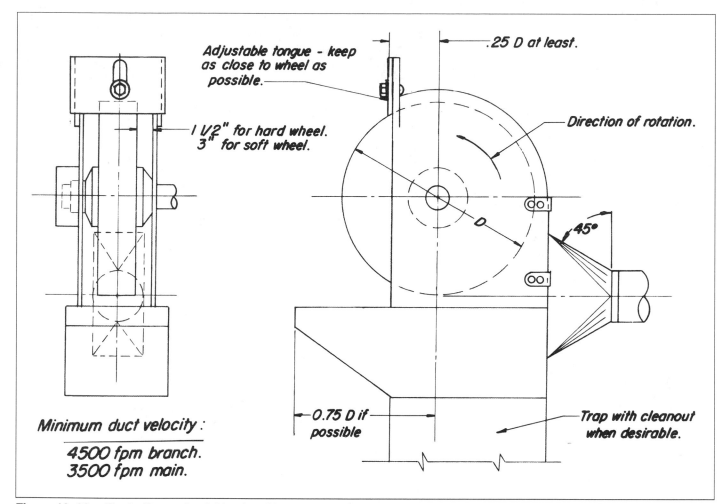

**Figure 19–22.** The buffing and polishing hood has an adjustable tongue that should be kept as close to the wheel as possible. (Courtesy ACGIH.)

are given in state codes, industry codes, and codes of the American National Standards Institute.

The hoods for a floor stand, pedestal, or bench grinder should have an adjustable tongue that can be placed so that it is within 0.125 in. (3 mm) of the periphery of the wheel at all times. This is an important adjunct because it tends to peel off the dust as it is carried around the wheel in the airstream set up by the wheel's rotation. No more than 25 percent of the wheel should be exposed (Figure 19–22).

The branch duct should attach to the hood at a tapered connecting piece. It should be inclined in the direction that material is thrown off the wheel. Chip trays may be provided if desired for heavy grinding.

Local codes should be consulted for hoods on horizontal and vertical spindle grinders, grinding and polishing straps and belts, and miscellaneous grinders.

Portable hand grinding can be done in a booth or on a table with downdraft ventilation (air drawn downward through a grating set in the tabletop). Shields should be attached to the back and sides of the table to reduce cross drafts and to ensure that the ventilation provided is more

effective (Figure 19–23). This can also be used for soldering and arc-welding.

## Woodworking

The woodworking industry was one of the first to recognize the value of local exhaust systems; they are applied almost universally in production shops. Their benefits include better housekeeping, improved working conditions, and decreased fire hazard. OSHA and many local jurisdictions have specific codes covering this type of ventilation, and these, of course, should be complied with.

Higher cutting speeds developed for woodworking machinery have led to higher production, which in turn means more shaving, chips, and dust. Exhaust ventilation must be increased correspondingly. In fact, with some of the newer machines, more air should be exhausted than is called for by the codes, which often lag behind technological progress.

The *ACGIH Industrial Venilation: A Manual of Recommended Practice* (see Bibliography) gives recommended exhaust volume and branch duct diameter for average-sized

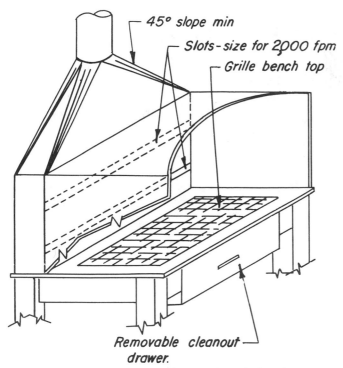

**Figure 19–23.** Exhaust booth for portable grinding operations, soldering, or arc welding. (Courtesy ACGIH.)

woodworking machines, with excellent drawings of typical hoods. Exhaust volume will vary with cutting speeds, length of cutter heads, length of sander drums, and diameter of saws. In general, saws require 4- to 6-in. (10- to 15-cm), and occasionally 8-in. (20-cm), branches; most cutting heads, a 5-in. (12.7-cm) branch.

## Oil Mist

Oil mist is a problem in shops where oil is used as a coolant; rooms can become filled with an oil haze, floors soaked with oil present a slipping hazard, and condensation on walls and ceilings becomes a fire hazard. The International Agency for Research on Cancer (IARC) has stated that there is sufficient evidence that mildly hydrotreated oils are potential carcinogens. Based on this, OSHA requires that such oils be considered as carcinogens under the Hazard Communication Standard (50 *FR* 51852).

Hood sides must act as splash guards, because airflow alone will not arrest large drops of oil thrown from the operation. The hood must incorporate such splash guards, but still permit the machine to be serviced and the operation observed when required. A few manufacturers have provided their machine tools with plastic enclosures that meet these requirements.

Hoods attach to machines in a variety of ways, and ingenuity is the best instruction manual. Usually, an exhaust volume of 400–600 cfm (0.2–0.3 m³/s) will provide the recommended 100–150 fpm (0.51–0.76 m/s) indraft. Sometimes, when considerable heat is generated

or where the hooding fails in being a complete enclosure, 1,000–1,500 cfm (0.5–0.7 m³/s) must be exhausted. Duct velocities of 2,500 fpm (12.7 m/s) are ample. Duct joints should be tight to prevent oil leakage and ducts should be pitched to allow condensed oil to flow back to the machine or to a special collector.

Electrostatic precipitators and wet (oil-filled) centrifugal collectors can be used as arresters for oil mist exhaust systems. To eliminate ductwork, units small enough to mount directly on the machine are available.

Collection and reuse of oil also may offer considerable savings.

## Materials Conveying

Dust that accompanies production processes when granular materials are conveyed, weighed, mixed, packaged, or otherwise handled falls in a different category from the dust generated during the operations considered thus far. Although dust that is generated can be detrimental both to health and comfort, its recovery carries no by-product value.

In materials conveying, the dust does not differ from the product being processed. The more dust removed, the greater the direct loss of usable product. The objective, therefore, is to trap the dust that would otherwise escape to the workroom air but not to remove useful material being processed.

Often the collected dust can be salvaged. In the handling of sand, lime, or flour, for example, the dust is chemically the same as the bulk material, and it can be reclaimed without downgrading.

Belt conveyors, bucket elevators, skip hoists, bin fillers and drum fillers, weighers, and mixers are all extremely common throughout industry, yet there is a scarcity of basic design data for related dust control. Many general guidelines are available. Pring et al. (1949) and Hemeon (1963) (see Bibliography) are among the few who have explored the fundamentals of needed control of these dusts.

No mixer, hopper, bin, silo, tank, or other container can be filled with a dry, dusty material without the air that is already inside the container being displaced and carrying with it dust to the outside. One solution to the dust problem is to make the container, chutes, and casing airtight (meaning completely enclosed) and run a hairpin-shaped air vent (breather) or duct from the container being filled to the container from which the dusty bulk material is draining, in order to let the dust-laden air escape back to the feed container. When the hairpin duct is mounted vertically, no dust will settle out in the piping, and so the breather cannot become plugged.

A basic fact, often overlooked, is that granular material falling through air (as when discharging from a belt into a storage bin) does not just displace its own volume of air, but induces additional airflow because air is entrained by the falling motion of the individual particles of the material. The aim of the engineer is to control this induced airflow by exhaust ducts.

Because the induced airflow is caused by the fall of particles, air enters with the solids at the top of a circuit, for example, the discharge of material through a chute at the head of a bucket elevator to a conveyor belt. The air, consequently, must be removed from the bottom (in this case, from the bottom of the elevator discharge chute) by properly located exhaust connections. To permit settling out of coarser particles before they are carried into the exhaust system, the volume of air exhausted must be kept at a minimum by the following methods:

- Restricting the opening through which incoming air can enter

- Reducing the height of fall as much as possible

- Effectively enclosing the impact zone (the region where the material lands on the belt)

Figure 19–24 shows how the exhausted air can be controlled. The enclosure around the head pulley has a restricting opening that the belt passes, following the first requirement above. The fall of material should be broken by a pocket in the chute, into which the material drops rather than splashing directly on the belt itself. This is in line with the second requirement and might be further improved if riffles or cleats could be built in, over which the material would cascade, resulting in practically no free fall. The third requirement can be met by providing a generous enclosure around the skirts on the belt at the foot of the chute. The single exhaust hood should be located well above the belt. This keeps the air velocity at the belt level so low that coarse, useful particles are not picked up. An exhaust connection is rarely required at the top of the enclosure. However, if the material falls directly to the conveyor, a second exhaust connection should be provided at the back of the chute, because dusty air dis-

lodged behind the impact zone cannot be captured by a single hood at the front without dragging coarse particles into the airstream.

Breaking the free fall of material when it is transferred (from belt to belt, elevator to belt, or chute to bin) results in a savings. "Choke feeding," the piling up of material in cone shapes under a discharge point, can often control the dust so that an exhaust hood is not required. For example, a hopper-bottom car might dump material into a hopper opening beneath the tracks, at the bottom of which is a screw conveyor that takes the product to some other point. If the material is discharged from the car at a faster rate than it is removed by the screw conveyor, it will "cone up" under the car hopper doors. The operation will then be essentially dustless, provided that stray cross breezes are screened off. This method of dust control is far superior to exhaust ventilation.

## Internal Combustion Engine Ventilation

Engines fueled by gasoline, diesel, and natural gas are used occasionally inside factory buildings. They are operated not only in garages, where maintenance checks and repairs are made, but also in warehouses, storage spaces, loading docks, aisleways, and other indoor areas. Engines are used as drives for lift trucks and for air conditioning and electricity-generating equipment. Their products of combustion contain such toxic materials as aldehydes, carbon monoxide, and nitrogen oxides.

Aldehydes in air are extremely irritating to the eyes. The Threshold Limit Value for formaldehyde, for example, is only 0.3 ppm. (Formaldehyde is designated as a suspect human carcinogen as well.) Some aldehydes may be generated if the engine is in poor operating condition and is burning oil or has a smoky exhaust. However, they usually are of less importance, from the industrial hygiene standpoint, than the carbon monoxide (CO) released.

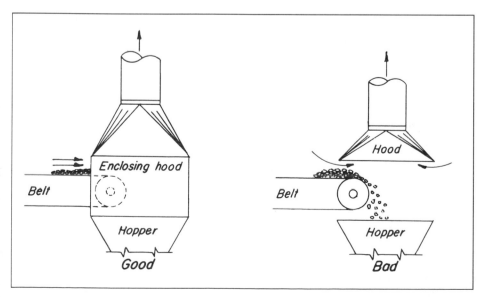

**Figure 19–24.** Enclose the operation as much as possible—the more completely enclosed the source, the less air is required for control. (Courtesy ACGIH.)

For an 8-hr exposure, the Threshold Limit Value of carbon monoxide gas in air is 25 ppm in the breathing zone area of employees. However, a ventilating system should be designed to keep a much lower level of CO. Exhaust gases released from a gasoline engine contain from 0.1 to more than 10 percent CO (1 percent = 10,000 ppm). If the engine is operating at full-rated horsepower, its exhaust gases will contain about 0.3 percent CO. They may contain more than 10 percent when idling.

Gasoline engines are built in many power levels, ranging from less than 1 hp (0.7 kW) to more than 200 hp (150 kW). Lift trucks commonly have a maximum delivery of 35–50 hp (26–37 kW). Because the engine normally discharges about 1 ft³/min (0.5 l/s) of exhaust gases per operating horsepower, a lift truck when operating near its maximum rated power would release 1.5 cfm (0.7 l/s or 0.03 × 50) CO. Fresh dilution air must be supplied to dilute emissions to below acceptable concentrations. (See Chapter 20, General Ventilation of Industrial Occupancies.)

The distribution of the incoming fresh air throughout the building is just as important as the ventilation rate. A warehouse, for example, may have a central runway with products or goods piled on each side in bays. Dead-end alleys or aisles may lead into these bays, and it is possible that these contaminant pockets are devoid of ventilation. The CO concentration may also build up to excessive levels in an aisle when a truck is stacking there, although the CO level might still be negligible along the central runway. Such a situation can best be solved by introducing fresh air by a duct having discharge grilles in the areas most likely to have CO concentrations, such as in aisles, rather than along the central runway. The latter generally receives adequate ventilation from the building doors.

Improvement is sometimes effected by extending the tailpipe of a warehouse truck upward so that it discharges vertically at a point above the operator's head. This solution is also effective in a situation where a dump truck backs up to a pit hopper, at the bottom of which there may be a conveyor. A worker may stand across the pit where the conveyor controls are located and get the full blast of a truck exhaust. A vertical discharge pipe would help prevent this.

There are areas where it may not be practical or economical to ventilate for the prevention of hazardous concentrations of CO, such as in frozen food storage rooms. Electric trucks are recommended for these areas.

Finally, there are garages with fixed repair stations. Here, a local exhaust system is preferred (Figure 19–25).

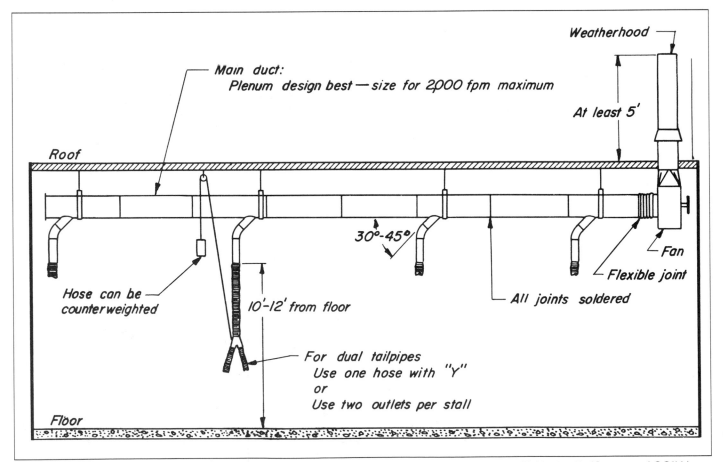

**Figure 19–25.** Local exhaust ventilation for service station garage. Specifications are given in Table 19–E. (Courtesy ACGIH.)

**Table 19–E.** Tailpipe Exhaust Requirements

| Type | Cfm per Tailpipe | Diam. (in.) Flexible Duct (minimum) |
|---|---|---|
| Auto and truck (up to 200 hp) | 100 | 3 |
| Auto and truck (over 200 hp) | 200 | 4 |
| Diesel | See reference below | |

(Reprinted with permission from American Conference of Governmental Industrial Hygienists, Committee on Industrial Ventilation, *Industrial Ventilation Manual*, 22nd edition.)

Three-inch (7.5-cm) flexible ducts are fitted over the tailpipes. The flexible ducts join 4-in. (10-cm) branches, which in turn are connected to a header located below floor level or overhead. (Exhaust requirements are shown in Table 19–E.)

For general garage ventilation where cars are in motion or are idling outside of the repair stall, the following ventilation guideline is provided for diluting carbon monoxide:

- Operating auto engine—5,000 cfm (2.4 $m^3$/s)
- Operating truck engine—10,000 cfm (4.7 $m^3$/s) or more
- Per horsepower (0.7 kW) for diesel—100 cfm

**Air sampling.** The ultimate effectiveness of any ventilation procedure for controlling CO should be verified by checking air samples for CO content. Several instruments are available that use detector tubes in which a color change is produced if CO is present in the air sample. Other instruments are available for continuous monitoring if necessary. (See Chapter 15, Evaluation.)

## Fog Removal

Hot water tanks are used for steaming purposes in various industries and operations. These may include foundries, veneer mills, plating, and metal cleaning. The steam that is released into a room may lead to excessive moisture condensation, lowered visibility, or discomfort. This is particularly true in winter weather, and at extensive vat operations such as those found in veneer mills. In vat operations, one way to attack fog problems is to install vertical stacks (not elbows) above the vat. If the vat is of the totally enclosed type, the stack inlet should be flush with the cover (centrally located, or proportionally located should there be two or more stacks). If the vat is used for dipping or quenching, it cannot be totally enclosed, but it is almost essential to have three sides baffled or enclosed (above the top edge of the vat); otherwise, cross drafts will nullify the purpose of the stacks. The stacks should extend above the high point of the roof, or that of the adjoining building, whichever is higher.

Exhaust fans should be installed in these stacks to ensure a positive outward flow so that airflow does not depend on outdoor weather conditions, thereby making "downdrafts" impossible.

## Kitchen Range Hoods

Some of the most elaborate exhaust hoods, and largest in terms of physical size and airflow, are kitchen range hoods, and yet as a class they receive a minimum of attention and research by health and safety professionals. With many company plants supporting their own well-equipped cafeterias, proper ventilation is as necessary in the kitchen as in any other production department. However, greater concern is often shown the aesthetic appearance of kitchen hoods than is given to their performance.

The investment in kitchen hoods may be wasted if installations are underdesigned. Because ceilings, walls, and fixtures become coated with solidified fat, maintenance of sanitary conditions becomes a problem and unappetizing odors pollute the dining area. If hoods are overventilated, original equipment cost is expensive, operating cost is high, and heating of the dining room may be a problem, because kitchen hoods sometimes exhaust a very large volume of air.

Range hoods, in large kitchens at least, are often extremely long for their width, because they must cover a row of ranges, ovens, grills, and deep-fat fryers.

The distance that a canopy hood should extend beyond the cooking equipment is regulated in many cities by ordinance. Where there is no code, it is customary to increase the dimension of a canopy hood 4–5 in. for every foot (0.33–0.41 m/m) of distance between the hood face and operating surface (range, fryer, or other heat source). Some designers use a minimum overhang of 12 in. (30.5 cm).

A rectangular exhaust hood placed along the wall at the back of deep-fat fryers and grills can be installed closer to the source of smoke and odor. In effect, it becomes a lateral exhaust for an open-surface tank, and may be designed as already outlined in the section under that heading. The smoke and oil vapor are trapped before they rise from the cooking unit, and they have less chance to spread through the room. End baffles further improve the operation.

Grease filters are desirable in all kitchen hoods, including the lateral exhaust type, for two reasons: They prevent fat from entering the duct, where it condenses and causes a fire hazard, and when placed in the hood itself, they act as excellent diffusers to ensure a uniform distribution of air over the hood opening, which is generally large in area. However, grease filters add resistance to the system, sometimes as much as 0.5 in. (0.12 kPa) water gauge. The exhaust fan should be able to handle the necessary volume of air over the range of filter resistance pressures. Grease filters should be cleaned regularly.

An approved fire damper with a fusible link should be installed in the main exhaust duct or branch adjacent to the range hood. This is required by code in many states.

Codes are often mandatory. These are not to be used as a crutch for poor or inadequate design, however. The designer should make certain that his or her plans and specifications meet, as a minimum, the requirements of any applicable codes. Additional requirements should be anticipated and met. The health and safety professional, the contractor, or the maintenance staff should not later need to salvage a poor engineering job.

A bibliography related to ventilation for this chapter and the following two chapters is at the end of Chapter 21, General Ventilation of Nonindustrial Occupancies). Particular emphasis is placed on the basic text on ventilation, *Industrial Ventilation—A Manual of Recommended Practice,* of the American Conference of Governmental Industrial Hygienists.

CHAPTER **20**

# General Ventilation of Industrial Occupancies

Revised by D. Jeff Burton, PE, CIH, CSP

*Publication information on all references given in this chapter can be found in the Bibliography at the end of Chapter 21, General Ventilation of Nonindustrial Occupancies.*

**Dilution Ventilation Fundamentals**
*Rate of Evaporation or Emission ▪ Acceptable Concentration for Exposure ▪ Properties of Air Mixing in the Space ▪ Dilution Volume Flow Rate*

**Comfort Ventilation**

**Degrading of Air in Industrial Environments**
*Odors ▪ High and Low Temperatures ▪ Chemical Air Contaminants ▪ Natural Ventilation ▪ General Exhaust Ventilation ▪ Freestanding Fans*

**Replacement, or "Makeup," Air**
*Replacement Air Volume ▪ Industrial Air-Supply Equipment*

**Heating of Replacement Air**
*Sources of Heat*

**Summary**

## Dilution Ventilation Fundamentals

LOCAL EXHAUST VENTILATION SYSTEMS (see Chapter 19, Local Exhaust Ventilation of Industrial Occupancies) are generally superior to general ventilation systems. They remove the contaminant from the work environment, reduce airflow rates, and are more economical. There are, however, many industrial operations where local exhaust ventilation is impractical, not cost-effective, or impossible to install. Such operations often use dilution ventilation for control because other controls are unavailable. (See Chapter 18, Methods of Control, for more information.)

Dilution ventilation in the industrial environment works best under the following conditions:

- Other more cost-effective controls are unavailable.
- Air contaminants are of relatively low toxicity.
- Air contaminants consist of gases, vapors, or small aerosols (D < 25 μm).
- Emissions occur uniformly in space and time.

- Emission sources are widely dispersed and not close to employee breathing zones.
- The facility is located in a moderate climate.
- The air used for dilution is not contaminated.
- The HVAC system used to condition dilution air is capable of maintaining appropriate temperatures and humidities.

The positioning of the supply and exhaust grills is important. Figure 20–1 shows different positioning scenarios.

Dilution ventilation is most often used for the control of gases and vapors. Hydrocarbon solvents and gases are used widely in synthetic varnishes, lacquers, adhesives, coatings, cleaning materials, and industrial processes. In many cases, the solvent or gas is evaporated from the material, leaving the desired end product; thus, the very heart of the process involves potentially contaminating the air. The goal of dilution ventilation is to maintain these concentrations in air to below acceptable levels.

In order to estimate the amount of dilution ventilation required, several important parameters must be determined or estimated: the rate of evaporation or emission, the acceptable concentration for exposure, the properties of air mixing in the space, and the time over which the emission occurs.

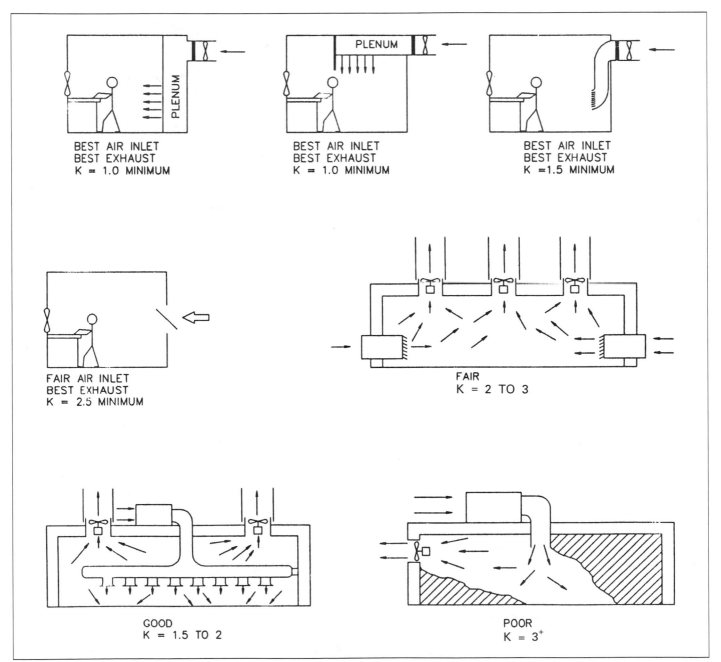

**Figure 20–1.** Fair, good, and best locations for fans and air inlets for dilution ventilation. (Adapted from American Conference of Governmental Industrial Hygienists [ACGIH], *Industrial Ventilation Manual,* 21st ed., 1992.)

## Rate of Evaporation or Emission

This rate is almost always exponential over time, but in many cases the emission rate can be approximated as linear. For example, we intuitively know that paint dries more quickly (with more solvent being evaporated) over the 1st hour as compared to the 24th hour or the 240th hour. It is possible, however, to approximate the average emission rate, say over the first eight hours. This makes calculations easier and more manageable.

It is known that a given volume of vapor forms from a given weight of solvent evaporated. Based on the volume of vapor formed, the dilution volume flow rate can be estimated. (See Equation 2.)

## Acceptable Concentration for Exposure

The acceptable concentration, or the concentration that the employer and employees are willing to be exposed to, is usually based on the American Conference of Governmental Industrial Hygienists' (ACGIH) Threshold Limit Value (TLV) or OSHA's permissible exposure limit (PEL). For example, many employers use OSHA's action level (one-half of the PEL) or less as the acceptable concentration.

## Properties of Air Mixing in the Space

Air never mixes completely at all times in every space. This requires choosing a mixing factor to account for poor or incomplete mixing. The mixing of air is sometimes called ventilation efficiency or ventilation effectiveness. Mathematically, it can be stated as

$$K_{mix} = \frac{\text{Actual } Q_{dilution} \text{ required to achieve the acceptable concentration}}{\text{Ideal } Q_{dilution} \text{ required to achieve the acceptable concentration}} \quad (1)$$

where $Q_{dilution}$ = volume flow rate of dilution air

The value of $K_{mix}$ ranges from 1.5 to 3.0 in most industrial occupancies. If $K_{mix} > 3$, mixing should be improved, or another form of control should be used—dilution is likely to be too expensive and the uncertainties too high. Typical values for $K_{mix}$ include the following:

| $K_{mix}$ | Condition |
|---|---|
| 1.5 | Wide open spaces with good supply and return locations; all ventilation equipment functions adequately. |
| 2.0 | Conditions not ideal, but wide use of freestanding fans to create mixing. |
| 2.5 | Poor placement of supply and return registers; partitioned spaces with generally adequate distribution of supply and return locations. |
| 3.0 | Crowded industrial spaces with many walls and partitions; poor supply and return locations. |

Dilution is more effective, and lower values of the mixing factor $K_{mix}$ are feasible, when the following conditions are met:

- Dilution air is routed first through the occupied zone and then to the emission sources.
- Supply air is distributed where it will be most effective.
- Exhausts are located close to contaminant sources.
- Auxiliary or freestanding fans are used to enhance mixing.

## Dilution Volume Flow Rate

Once the time period is determined, the dilution volume flow rate is determined as follows:

$$Q_{dilution} = \frac{387 \cdot lbs \cdot 10^6 \cdot K_{mix}}{MW \cdot t \cdot C_a \cdot d}$$
$$= \frac{0.0244 \cdot grams \cdot 10^6 \cdot K_{mix}}{MW \cdot t \cdot C_a \cdot d} \quad (2)$$

where $Q_{dilution}$ = volume flow rate of dilution air, in cubic feet per minute (cfm) (SI unit: m³/s)
$t$ = time, in minutes (SI unit: seconds)
MW = molecular weight of material emitted
lbs = amount evaporated, in lbs (SI unit: grams)
$C_a$ = the acceptable exposure concentration, in parts per million (ppm)
$K_{mix}$ = a mixing factor to account for incomplete or poor mixing in the space
$d$ = density correction factor to account for non standard conditions

Where the temperature is near normal room temperature and the elevation is near sea level, $d = 1$. (For more information on density correction, see Table 19–A in Chapter 19 and the references at the end of Chapter 21.)

### Example

What dilution volume flow rate should be used if 14 pounds of toluene are uniformly evaporated from a process during an 8-hour period? Assume that $C_a$ = 10 ppm, $K_{mix}$ = 1.5, $d$ = 1, and MW = 92.1.

### Solution

$$Q_{dilution} = \frac{387 \cdot lb \cdot 10^6 \cdot K_{mix}}{MW \cdot t \cdot C_a \cdot d} = \frac{387 \cdot 14 \cdot 10^6 \cdot 1.5}{92.1 \cdot 480 \cdot 10 \cdot 1}$$
$$= 18,000 \text{ cfm (rounded)} \quad (3)$$

# ■ Comfort Ventilation

Ventilation is defined as the process of supplying air to or removing air from a space by natural or mechanical means. As it is used, the word also implies adequate quantity and quality of the air supplied to the space. The term *air conditioning* implies control of physical and chemical qualities of the air. The American Society of Heating, Refrigerating, and Air Conditioning Engineers (ASHRAE) has defined air conditioning as "the process of treating air so as to control temperature, humidity, cleanliness, and distribution to meet the requirements of the space" (ASHRAE, 1991).

Comfort ventilation of the industrial work environment contributes to the health and comfort of employees, which in turn enhances the productivity and effectiveness of their work.

# ■ Degrading of Air in Industrial Environments

There are a number of reasons for air to become degraded in the industrial environment. (See Chapter 21, General Ventilation of Nonindustrial Occupancies.) The following paragraphs describe some of the most common problems.

## Odors

Odors are often harmless but can cause discomfort. They may also indicate the presence of other, undetectable air contaminants (for example, the odors detected in vehicle exhaust may be less hazardous than the carbon monoxide that accompanies them, which is odorless). Deodorants merely mask offensive odors and should not be used.

Industrial odors are most often caused by the processes used in the building, reentrainment of exhausted air, idling vehicles near air intakes, construction activities, and maintenance.

Some odors can be controlled by using filters of activated charcoal, which absorbs the odorous substance. However, the best approach is usually dilution ventilation.

## High and Low Temperatures

Environmental thermal factors profoundly influence everyday living, comfort, and often health itself. The body has remarkably complex and robust mechanisms that hold body temperature within extremely narrow limits despite a wide range of external air conditions. With body temperature regulated at 98.6 F (37 C) and skin temperature at perhaps 92 F (33 C) in winter and 95 F (35 C) in summer, the task of a heating and ventilating system is not to heat the body, but to permit the body heat to escape at a controlled rate. This is because air temperatures in the mid-70s F (mid-20s C) make up the comfortable range for humans. When-

ever air temperatures approach body temperature, an individual is no longer comfortable. (See Chapter 12, Thermal Stress, for more details.)

## Chemical Air Contaminants

Air contaminated with chemicals from industrial processes is often controlled with dilution ventilation. The rest of this chapter deals with dilution ventilation approaches and concerns.

## Natural Ventilation

Wind, aided by air currents from heating devices and higher indoor air temperatures, provides a certain amount of ventilation even when the doors and windows are closed, because of infiltration and exfiltration. When the doors and windows are open, a large volume of natural ventilation can be provided.

Hot process equipment such as ovens and furnaces cause the air to warm, expand, and rise. If openings are provided in the roof, the hot, rising air exits the building. This is known as the chimney effect.

The following formula can be used to estimate the flow rate through the building caused by wind blowing against the building:

$$Q \approx 25AV \qquad \text{Metric:} Q \approx 0.2AV \qquad (4)$$

where $Q$ = airflow rate through the building, in cfm (SI units: m³/s)

$A$ = upwind or downwind open area, whichever is smaller, in ft² (SI units: m²)

$V$ = average wind velocity, in miles per hour (mph) (SI units: kilometers/hour, or kph)

The upwind or downwind area includes any open doors; open windows; or cracks around doors, windows, and building materials.

■

### Example

A warehouse has opened its windows to create natural ventilation. For a wind of 8 mph and an upwind and downwind open window area of $A$ = 300 ft², find the estimated airflow rate through the building.

■

### Solution

$$Q \approx (25)(8)(300) \approx 60,000 \text{ cfm (rounded)} \qquad (5)$$

Because warm air is lighter than cooler air, warm air rises. This rising air is replaced by cooler air, which heats up and also rises, thus creating upward airflow. In buildings cooled by natural ventilation, the warm air rises and escapes from the openings near the top of the building. Cooler air enters the building at ground level.

The following formula can be used to estimate the air-flow rate through a building at steady state conditions:

$$Q \approx 10A(H\Delta T)^{0.5} \quad \text{Metric:} \quad Q \approx 0.12A(H\Delta T)^{0.5} \quad (6)$$

where   $A$ = area of building inlets or outlets, whichever is smaller, in ft² (SI units: m²)

$H$ = height of building between inlet and outlet, in ft (SI units: m)

$\Delta T$ = difference between average indoor and outdoor temperatures, in degrees F (SI units: degrees C)

## Example

Suppose that the average indoor air temperature is 10 F warmer than the outside air, the open area of both inlets and outlets $A$ = 400 ft², and $H$ = 30 ft.

## Solution

$$(H\Delta T)^{0.5} = (30 \times 10)^{0.5} = 17.3 \quad (7)$$

Therefore,

$$Q \approx 10(400)(17.3) \approx 70,000 \text{ cfm} \quad (8)$$

## General Exhaust Ventilation

General exhaust ventilation can be achieved by placing a fan in a window, exterior wall, or roof, thus removing the air from the room directly. (Windows that are near the fan should be kept closed to contribute to better mixing efficiency.) This type of ventilation involves low initial cost and comparatively low maintenance cost, but it is likely to cause drafts near doors and windows and temperature extremes in the occupied area.

## Freestanding Fans

Local, or freestanding, fans stir up the air and help prevent stagnant accumulations of heat and moisture in limited areas where workers are present. They do not, however, constitute a method of ventilation—the process of supplying or removing air, by natural or mechanical means, to or from any space. When fans are used for cooling, care must be taken to see that they do not interfere with hoods or other control equipment. (See Chapter 19, Local Exhaust Ventilation of Industrial Occupancies.) Also, too much airflow may be uncomfortable for workers (see Table 20–A).

Simple circulation alone does not remove impurities from the air, nor does it supply pure air, but power-driven fans placed in an open room can increase the comfort level for workers by increasing their rate of cooling and perspiration evaporation.

Cooling air should come from outdoors, because the outdoor temperature is usually 10–30 F (5–15 C) cooler

**Table 20–A.** Air Motion Acceptable to an Industrial Worker for Cooling

| | Air Velocity (fpm) |
|---|---|
| **Continuous Exposure** | |
| Air-conditioned space | 50–100 |
| Fixed workstation, general ventilation, or spot cooling: | |
| Sitting | 100–700 |
| Standing | 100–700 |
| **Intermittent Exposure, Spot Cooling, or Relief Stations** | |
| Light heat loads and activity | 1,000–2,000 |
| Moderate heat loads and activity | 2,000–3,000 |
| High heat loads and activity | 3,000–4,000 |

1 fpm = 0.005 m/sec.
(Adapted from ACGIH.)

than it is inside the facility. The duct supplying the air should not be placed near furnace walls or other hot areas, and it should be insulated to keep the air temperature down. Figure 20–2 shows proper placement of supply registers for winter and summer.

In arid areas, passing outside air through an evaporative cooler before it is delivered to the workers has been used with great success. When the air temperature inside a facility is high, say 85–95 F (30–35 C), if the air that is blowing on a person is 10–15 F (5–10 C) cooler, it gives extremely welcome relief. Some reference books suggest that the velocity of air striking workers should not exceed 700 fpm (3.5 m/s). However, if the heat is severe, experience indicates that in order to achieve any degree of relief the air must be moving at at least 1,800 fpm (9.0 m/s).

## ■ Replacement, or "Makeup," Air

Whenever air is exhausted, or removed, from a building, regardless of the method, outdoor air must enter to take its place. If the replacement system is not properly designed, the replacement, or "makeup," air may enter through random cracks in the walls or through open doors and windows. Mechanical makeup air systems should always be designed and installed at the same time exhaust systems are designed and installed. Well-designed and operated makeup air systems are necessary for the following reasons:

- They help ensure proper exhaust hood operation. A lack of replacement air creates a negative-pressure condition, which increases the static pressure the fans must overcome. This can create reduced exhaust flow rates, especially in low-pressure exhaust systems. (See Figure 20–3.) One caution: Replacement air should be introduced to the space in such a way that it does not disrupt capture or control conditions at the hood. For example, makeup air should not be directed toward the face of hood.

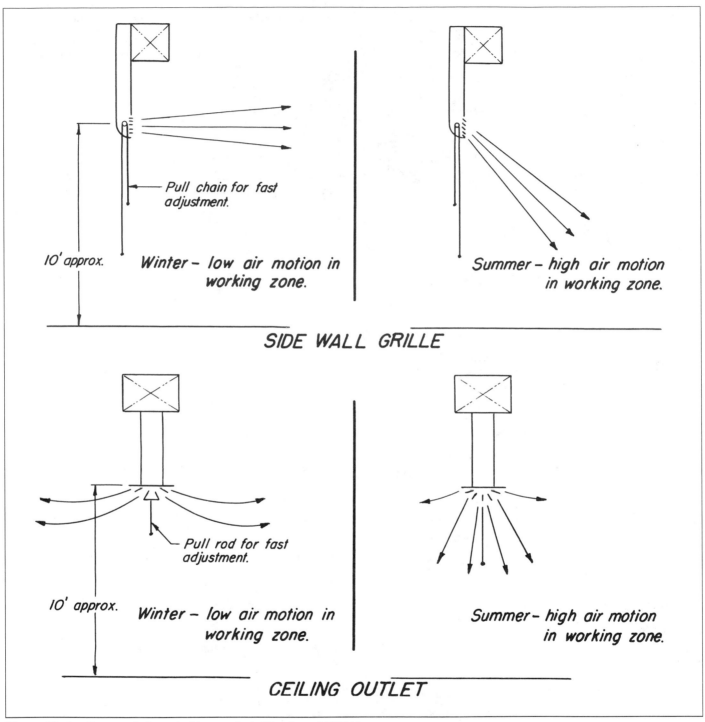

**Figure 20–2.** Seasonal air control for comfort. (Courtesy ACGIH.)

■ They eliminate high-velocity cross drafts. Cross drafts from windows and doors not only interfere with the proper operation of exhaust hoods but may also disperse contaminated air from one section of the building into another; this can interfere with the proper operation of process equipment such as solvent degreasers. In the case of dusty operations, settled material may be dislodged from beams and ledges, thereby recontaminating the work space.

■ They ensure natural-draft stack operation. Even moderate negative pressures can cause backdrafting in flues, which can result in a dangerous health hazard from the release of combustion products (principally carbon monoxide) into the work space. Backdrafting occurs in natural-draft stacks at negative pressures as low as 0.01 in. measured by a water gauge (wg) (Table 20–B). Secondary problems include difficulty in maintaining pilot lights in burners, poor operation of temperature

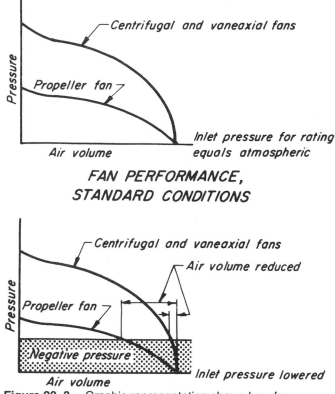

**Figure 20–3.** Graphic representation shows how fan performance falls off under negative pressure.

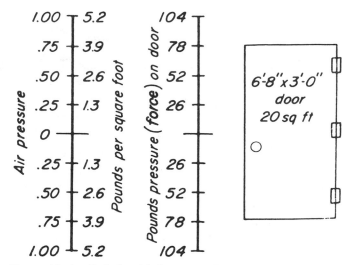

**Figure 20–4.** Relationship between air pressure and the amount of force needed to open or close an average-sized door (1 psf = 990 kPa; 1 lb force = 4.4 newtons). (Courtesy ACGIH.)

controls, and corrosion damage in stacks and heat exchangers resulting from condensation of water vapor.

- They eliminate cold drafts. Drafts not only cause discomfort and reduce working efficiency but can also result in lower overall ambient temperatures.

- They eliminate differential pressure. High differential pressures make doors difficult to open or shut and, in

some instances, can cause personnel safety hazards when the doors move in an uncontrolled fashion (see Table 20–B and Figure 20–4).

- They conserve fuel. Without adequate replacement air, uncomfortably cold conditions near the building perimeter often necessitate the installation of more heating equipment in those areas to correct the problem. Unfortunately, these heaters warm the air too much, and the overheated air that moves toward the building interior makes those areas uncomfortably warm (Figure 20–5). This in turn leads to the installation of more exhaust fans to remove the excess heat, which further aggravates the problem. Heat is wasted, and the problem still isn't solved. Fuel consumption with a well-designed replacement air system is usually lower than when attempts are made to achieve comfort without a proper replacement system.

## Replacement Air Volume

In most cases, replacement air volume should equal the total volume of air removed from the building by exhaust ventilation systems, process systems, and combustion processes. Determination of the actual volume of air removed usually involves a simple inventory of air exhaust locations and any necessary testing under atmospheric-pressure conditions.

When conducting the exhaust inventory, it is necessary to determine not only the quantity of air removed but also the need for a particular piece of exhaust equipment. At the same time, reasonable projections should be made of the total facility exhaust requirements for the next 1–2 years, particularly if process changes or facility expansions are being considered. In such a case, it may be practical to purchase a replacement air unit that is slightly larger than currently necessary, with the knowledge that the increased capacity will be needed within a

**Table 20–B.** Negative Pressures That May Cause Unsatisfactory Conditions in Buildings

| Negative Pressure Inches of Water | Adverse Conditions That May Result |
|---|---|
| 0.01–0.02 | **Worker draft complaints.** High-velocity drafts through doors and windows. |
| 0.01–0.05 | **Natural draft stacks ineffective.** Ventilation through roof exhaust ventilators, flow through stacks with natural draft greatly reduced. |
| 0.02–0.05 | **Carbon monoxide hazard.** Back drafting will take place in water heaters, unit heaters, furnaces, and other combustion equipment not provided with induced draft. |
| 0.03–0.10 | **General mechanical ventilation reduced.** Airflows reduced in propeller fans and low-pressure supply and exhaust systems. |
| 0.05–0.10 | **Doors difficult to open.** Serious injury may result from nonchecked, slamming doors. |
| 0.10–0.25 | **Local exhaust ventilation impaired.** Centrifugal fan fume exhaust flow reduced. |

1 in. wg = 0.25 kPa.
(Adapted from ACGIH.)

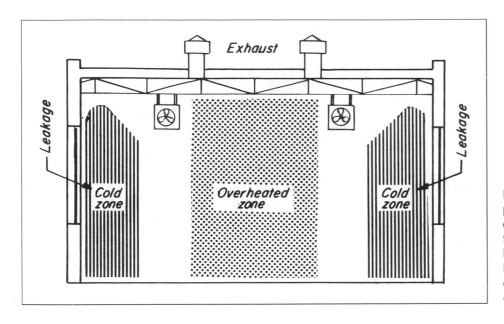

**Figure 20–5.** Under negative-pressure conditions, workers in the cold zones (left and right) turned up the thermostats in an attempt to get heat. Because this did nothing to stop leakage of cold air, they remained cold while center of facility was overheated. (Courtesy ACGIH.)

short time. The additional cost of a larger unit is relatively small, and in most cases the fan drive can be regulated to supply only the desired quantity of air.

Having established the minimum air supply quantity necessary for replacement air purposes, many engineers have found that it is wise to provide for additional supply air volume in order to overcome natural ventilation leakage and further minimize drafts at the perimeter of the building.

## Industrial Air-Supply Equipment

Industrial air heaters are usually designed to supply 100 percent outdoor air. The basic requirements for an air heater are that it be capable of continuous operation and of delivering constant air volume at a constant preselected discharge temperature. The heater must meet these requirements under varying conditions of service and be able to accommodate outside air temperatures that vary as much as 40 F (22 C) daily.

Standard-design heating and ventilating units are usually selected for mixed-air applications, that is, a mixture of outdoor air and recirculated air. Rarely, however, do the construction and operating capabilities of standard units meet the requirements of the manufacturing industry. Such units are better suited for use in commercial buildings and institutional facilities where the requirements are less severe and where mixed-air service is more common.

Air heaters are usually categorized according to their source of heat: steam and hot water units; indirect-fired gas and oil units; and direct-fired natural gas and LP-gas, or LPG, units. Each basic type is capable of meeting the first two requirements: constant operation and constant delivered air volume. As for the third requirement, that of constant, preselected discharge temperature, heat variations occur with most types of air heaters. One exception to the rule is the direct-fired air heater, which is designed

to provide a wide range of temperature control. Each type of air heater (direct-fired and indirect exchanger) has its own advantages and limitations, which must be understood by the designer when making the selection (Table 20–C). Direct-fired heaters are rarely used today in enclosed buildings because they present a fire hazard and may expose workers to combustion gases.

**Steam coil units.** Steam coil units were probably the earliest air heaters to be used in general industry as well as commercial and institutional buildings (Figure 20–6). When properly designed, selected, and installed, they are reliable and safe. They require a reliable source of clean steam at dependable pressure. For this reason, they are most widely used in large installations; smaller industrial facilities often do not have sufficient boiler or steam capacity for operating a steam air heater. The principal disadvantages of steam units are potential damage from freezing or water hammer in the coils, the complexity of controls when close temperature limits must be maintained, high cost, and excessive piping.

Multiple-coil steam units (Figure 20–7) and bypass steam units can help extend the temperature control range and help minimize freeze-up. In multiple-coil units, the first coil (preheat) is usually sized to raise the air temperature of outdoor air to at least 40 F (5 C). The coil is controlled with an on/off valve, which is fully open whenever outdoor temperature is below 40 F (5 C). The second (reheat) coil is designed to raise the air temperature from 40 F to the desired discharge condition. Temperature control is satisfactory for most outdoor conditions, but overheating can occur when the outside air temperature approaches 40 F; the rise through the preheat coil can result in temperatures of 79–89 F (26–32 C) of air entering the reheat stage. Refined temperature control can be accomplished by using a second preheat coil to split the preheat load.

**Table 20–C.** Advantages and Disadvantages of Direct-Fired Unvented Replacement Air Heaters and Indirect Exchanger-Type Replacement Air Heaters

| Direct-Fired Unvented Replacement Heater | Indirect Exchanger Replacement Air Heater |
|---|---|
| 1. Products of combustion in heater air stream (some $CO_2$, CO, oxides of nitrogen, and water vapor present). | 1. No products of combustion; outdoor air only is discharged into building. |
| 2. May be limited in application by state and municipal regulations. Consult local ordinances. | 2. Allowable in all types of applications and buildings, if provided with proper safety controls. |
| 3. Better heating capacity ratio 8:1 in small sizes, 25:1 in large sizes. Better control, lower operating cost. | 3. Heating capacity ratio limited: 3:1 usual, maximum 5:1. |
| 4. No vent stack, flue, or chimney necessary. Can be located in side walls of building. | 4. Flue or chimney required. Can be only located where flue or chimney is available. |
| 5. Higher efficiency (90 percent). Lower operating cost. (Efficiency based on available sensible heat.) | 5. Efficiency lower (80 percent). Higher operating cost. |
| 6. Can heat air over a wide temperature range. | 6. Can heat air over a limited range of temperature. |
| 7. Extreme care must be exercised to prevent minute quantities of chlorinated hydrocarbons from entering air intake, or toxic products may be produced in heated air. | 7. Small quantities of chlorinated hydrocarbon will normally not break down on exchanger to form toxic products in heated air. |
| 8. Can be used with gas only as a fuel. | 8. Can be used with both oil and gas as a fuel. |
| 9. Burner must be tested to ensure low CO and oxides of nitrogen content in airstream. | 9. No contaminants in airstream from combustion. |
| 10. First cost higher in small units and lower in large units. | 10. First cost lower in small units and higher in large units. |
| 11. No heat exchanger to corrode or leak. Burner plates are very durable. | 11. Heat exchanger subject to severe corrosion condition. Must be checked for leaks after a period of use. |
| 12. Can be easily adapted to take all combustion air from outdoors. This is important if corrosive or contaminated work room air is present. | 12. Difficult to adapt to take all combustion air from outdoors unless roof mounted or outdoor mounted. |
| | 13. Recirculation as well as replacement. |

(Adapted from ACGIH.)

Bypass units (Figure 20–8) incorporate dampers to direct the airflow. When maximum temperature rise is required, all air is directed through the coil. As the outdoor temperature rises, more and more air is diverted through the bypass section until finally all air is bypassed. The controls are relatively simple. The principal disadvantage is that the bypass is not always sized for full airflow at the same pressure drop as through the coil; thus the unit may deliver differing volumes of air depending on the damper position. Damper airflow characteristics are also a factor. An additional concern is that in some units, the air coming through the bypass and entering the fan compartment may have a nonuniform flow, which affects the fan's ability to deliver air.

A novel type of bypass design, the integral face and bypass (Figure 20–9), features alternating sections of coil and bypass. This design is said to promote more uniform mixing of the airstream, to minimize any nonuniform flow effect, and, through carefully engineered damper design, to permit minimum temperature pickup even at full steam flow and full bypass.

**Hot water.** This is an acceptable heating medium for air heaters. As with steam, there must be a dependable source of water at predetermined temperatures for accurate sizing of the coil. Hot water units are less susceptible to freezing than steam because they use forced convection, which

ensures that cooler water is fully removed from the coil. Practical difficulties and pumping requirements thus far have limited the application of hot water to relatively small systems—for a 100 F (55 C) air temperature rise and an allowable 100 F water temperature drop, 1 gallon per minute (gpm) (3.78 l/min) of water provides sufficient heat for only 460 cfm (0.227 m³/s) of air. This range can be extended with high-temperature hot water systems. Applications to date have been primarily in commercial and institutional buildings with mixed-air service.

Hybrid systems using an intermediate exchange fluid such as ethylene glycol have also been installed by industries with critical air-supply problems and a desire to eliminate all freeze-up dangers. A primary steam system provides the necessary heat to an exchanger, which supplies a secondary closed loop of the selected heat-exchange fluid. The added equipment cost is at least partially offset by the less complex control system.

**Indirect-fired gas and oil units.** These units (Figure 20–10) are widely used in small industrial and commercial applications. They appear to be economical up to approximately 10,000 cfm (4.7 m³/s)—above this size the capital cost of direct-fired air heaters is lower. Indirect-fired heaters incorporate a heat exchanger, commonly made of stainless steel, which effectively separates the incoming airstream from the products of combustion of the fuel being burned; thus,

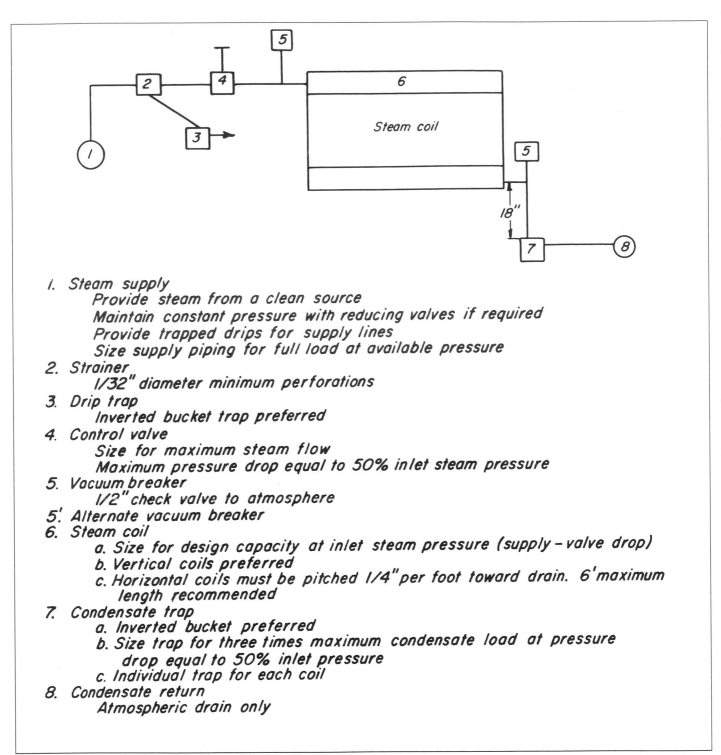

1. **Steam supply**
    Provide steam from a clean source
    Maintain constant pressure with reducing valves if required
    Provide trapped drips for supply lines
    Size supply piping for full load at available pressure
2. **Strainer**
    1/32" diameter minimum perforations
3. **Drip trap**
    Inverted bucket trap preferred
4. **Control valve**
    Size for maximum steam flow
    Maximum pressure drop equal to 50% inlet steam pressure
5. **Vacuum breaker**
    1/2" check valve to atmosphere
5.' **Alternate vacuum breaker**
6. **Steam coil**
    a. Size for design capacity at inlet steam pressure (supply – valve drop)
    b. Vertical coils preferred
    c. Horizontal coils must be pitched 1/4" per foot toward drain. 6' maximum
       length recommended
7. **Condensate trap**
    a. Inverted bucket preferred
    b. Size trap for three times maximum condensate load at pressure
       drop equal to 50% inlet pressure
    c. Individual trap for each coil
8. **Condensate return**
    Atmospheric drain only

**Figure 20–6.** Block diagram of a steam coil makeup air heating unit. (Courtesy ACGIH, *Industrial Ventilation Manual,* 21st ed., 1992.)

heated room air can be recirculated. Positive venting of combustion products is usually accomplished with induced-draft fans. These precautions are taken to minimize interior corrosion damage from condensation in the heat exchanger which is caused by the chilling effect of the incoming cold airstream. Another major advantage of the indirect-fired air heater is that it is economical in the smaller volume sizes

and can be widely applied as a package unit in small installations such as commercial kitchens and laundries.

Temperature control (turndown ratio) is limited to about 3:1 or 5:1 because of burner design limitations and the necessity to maintain minimum temperatures in the heat exchanger and flues. Temperature control can be made more versatile through the use of a bypass system

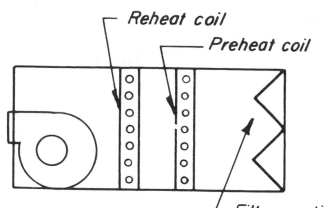

**Figure 20–7.** Multiple-coil steam unit. (Courtesy ACGIH.)

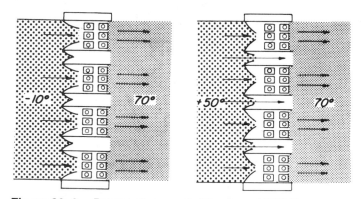

**Figure 20–8.** Bypass steam unit. (Courtesy ACGIH.)

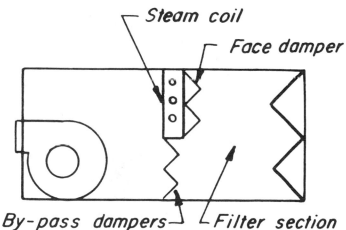

**Figure 20–9.** Integral face and bypass coil. (Courtesy ACGIH.)

similar to that described for single-coil steam air heaters. Bypass units of this design offer the same advantages and disadvantages as in the steam bypass units.

A newer design of indirect-fired unit incorporates a rotating heat exchanger. Temperature control is stated to be as high as 20:1. Presently, this unit is available using natural and liquefied petroleum gases.

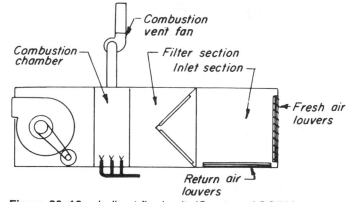

**Figure 20–10.** Indirect-fired unit. (Courtesy ACGIH.)

**Direct-fired air heaters.** Direct-fired air heaters (Figure 20–11), wherein the fuel (natural or LPG gas) is burned directly in the airstream and the products of combustion are released in the air supply, have been commercially available for the past 20–25 years. These units are economical to operate, because all of the net heating value of the fuel is available to raise the temperature of the air; this results in a net heating efficiency approaching 100 percent. Commercially available burner designs provide turndown ratios from approximately 25:1 to as high as 45:1, permitting excellent temperature control. In sizes above 10,000 cfm (4.7 m³/s), the units are relatively inexpensive on a cost per volume-of-air-handled basis; below this capacity, the costs of the additional combustion and safety controls weigh heavily against this design. A further disadvantage is that state and local codes prohibit the recirculation of room air across the burner. Controls on these units are designed to provide a positive proof of airflow before the burner can ignite, a timed preignition purge to make sure that any leakage gases are removed from the housing, and constantly supervised flame operation including both flame controls and high-temperature limit

Concerns are often expressed with respect to potentially toxic concentrations of carbon monoxide, oxides of nitrogen, aldehydes, and other contaminants produced by com-

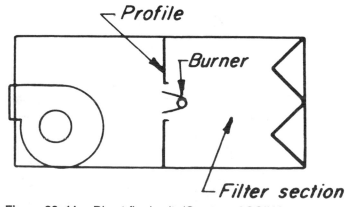

**Figure 20–11.** Direct-fired unit. (Courtesy ACGIH.)

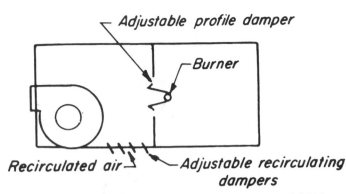

**Figure 20–12.** Direct-fired bypass unit. (Courtesy ACGIH.)

bustion and released into the supply airstream. Practical field evaluations and detailed studies show that with a properly operated, well-maintained unit, carbon monoxide concentrations do not normally exceed 5 ppm, and that oxides of nitrogen and aldehydes are well within acceptable limits.

A variation of this unit, known as a bypass design, has gained acceptance in larger buildings where there is a desire to circulate large volumes of air at all times (Figure 20–12). In this design, controls are arranged to reduce the flow of outside air across the burner and to permit the entry of room air into the fan compartment. In this way, the fan air volume remains constant, maintaining circulation in the work space. It is important to note that the bypass air does not cross the burner—100 percent outside air only is allowed to pass through the combustion zone. Controls are arranged to regulate outside air volume and also to ensure that burner profile velocity remains within the limits specified by the burner manufacturer, usually in the range of 2,000–3,000 fpm (10–15 m³/s). This is accomplished by providing a variable profile, which changes area as the damper positions change.

Inasmuch as there are advantages and disadvantages to both direct-fired and indirect-fired makeup air heaters, a careful consideration of characteristics of each heater should be made. A comparison of the heaters is given in Table 20–C.

# ■ Heating of Replacement Air

Replacement air heating should not be confused with space heating, which heats a building whether or not industrial operations are functioning within it. However, heated replacement air can be supplied to an entire building while directly compensating for air being exhausted at one point or being lost through such exits as loading-dock doors.

General factors that must be considered are the following:

- Proper location, size, and construction of air intakes.
- Filtering of replacement air.

- Air handling capability of equipment must be adequate to keep air in the building under positive pressure, even with exhaust systems in use. This reduces drafts and minimizes dispersal of contaminants into the work area.
- Automatic controls must be provided to ensure delivery of replacement air at temperatures acceptable to workers in the area. The controls should be located in a place as clean and free of vibration as possible.
- Proper start-up and shutdown procedures must be enforced.
- Controls and safety devices should be readily accessible to authorized personnel.
- Air distribution arrangement must provide adequate delivery of makeup air.
- Building codes should be checked to see what special precautions must be taken.

It is important that a qualified ventilation engineer be consulted when considering a new system.

## Sources of Heat

The three most common heating media for low-temperature industrial heating processes are oil, gas, and steam; all have some application in heating replacement air.

**Oil.** Factors that must be considered when using oil-fired heaters include the following:

- The possibility of formation of smoke and soot during start-up makes it necessary to transfer the heat through a tubular heat exchanger. Firing must be in closed combustion chambers.

**Gas.** Precautions that must be taken when using gas-fired heaters include the following:

- Be certain that gas is properly proportioned and mixed with excess air. Use burner systems designed for the operating conditions that will be present.
- The Threshold Limit Value for $CO_2$ (carbon dioxide) and CO (carbon monoxide) must not be exceeded. The TLVs recommended for a continuous 8-hour exposure by the ACGIH are 5,000 ppm for $CO_2$ and 25 ppm for CO (see Appendix B).

In choosing the type of gas combustion system, the following basic requirements should be kept in mind:

- Heat must be introduced into the airstream so as to facilitate uniform distribution and uniformity of temperature.
- The system must be capable of throttling smoothly over the total range of turndown (which can be as great as 25:1).

- The burner must be capable of producing complete combustion of the fuel throughout the turndown range.
- Flame retention and stability must be maintained in high-velocity airflow.

So-called line burners are almost exclusively used because they give uniform temperature distribution across the airstream. Line burners are sectional burners available in a variety of shapes; they are flanged so that they can be bolted together into assemblies to fit the size and shape of any unit. Air/gas-proportional and premixing equipment is required, as is a positive flame retention feature. A safeguard device for flame failure is usually satisfactory. The wide-range–type line burner was developed specifically to meet the needs of air heating that uses oxygen from the stream of air being heated to supplement that provided in the primary air/gas mixture in order to extend the range of heating capacity. The heater manufacturer's recommendations must be followed in installation, especially with regard to velocity of airflow across burners.

**Steam.** When excess steam capacity is available, use of steam coils or finned heaters should be considered because steam heaters are relatively simple, inexpensive, and easily maintained and require no combustion safety equipment.

Precautions that must be taken when using steam include the following:

- Avoid long heating coils. These may give uneven temperature distribution from side to side in the duct as the steam supply is throttled over the wide range of turndown required (20:1 to 25:1).
- Protect against freeze-up so that outside air at subfreezing temperatures can be drawn over the heating coils.
- Consider using oil- or gas-fired tempering air heaters ahead of steam coils.
- Watch for leaking hot water or steam lines—biological growths could be a problem.

The air exhausted from a room must be replaced. The current OSHA standards require that replacement air equal to that exhausted be supplied to each room that has exhaust hoods.

The air must enter the room in such a manner that it does not interfere with the operation of any exhaust hood. Air blowing out of a duct carries 30 times as far as the volume of air moving toward a similarly sized exhaust duct. This often necessitates the use of a diffuser at the end of a supply duct.

The airflow of the replacement system must be measured on installation and rechecked periodically. If replacement air falls below the required amount, corrective action must be taken.

Obviously, supply air can be measured inside the ducts in the same way as exhaust air. However, it is also possible to make rather good measurements at the discharge end of supply ducts, even when they are equipped with diffusers.

Major diffuser manufacturers have established "K factors"—effective area factors—for use with the diffuser probe and the various models and sizes of diffusers. Typical instructions illustrate the location at which the diffuser probe should be placed and explain how to calculate the K factor for various sizes. Four readings around the circumference should be taken and averaged (Volume per unit time = K × [average velocity of the air]).

## ■ Summary

There are some basic concerns that safety and health professionals should consider when evaluating industrial ventilation systems or when discussing them with engineering personnel:

- Replacement air must be supplied if the exhaust equipment is to function properly. Many industrial exhaust systems fail when mechanical supply for replacement air is lacking, because it is impossible for them to cope with the added force required to counter the resistance generated when drawing large volumes of air through cracks in the windows and doors.
- Whenever air pressure in a building is less than that outdoors, there is an increased inward flow of air through each and every crack. In fact, air can infiltrate through a brick wall. Whenever a door is opened, large quantities of outside air enter the building. In cold weather, where such indrafts occur the people located near the flow of cold air are chilled and cannot be made comfortable regardless of how much heat is added to the building itself.
- Whenever air enters a space and is exhausted, some dilution ventilation takes place. This alone may be sufficient to provide control in an area. The maximum dilution and benefits occur if the air is introduced under controlled conditions. Ideally, the best air available is always introduced to the cleanest part of the facility and is allowed to flow across the work space toward the contaminated areas, where it is exhausted.
- One of the most successful methods of improving the environment for the personnel in the summer is to direct air over their bodies at high velocities. Air-supply systems can, if properly equipped with suitable grills, provide such air motion. The air must be piped to the lower 10 ft (3 m) of the facility in order to function properly for this purpose. It is generally easier to control the environment within a space where air pressure is balanced or positive. When air is mechanically supplied to a space, it can, if desired, provide pressurization if sufficient volumes of air are introduced.

■ Most production facilities must be warmed in winter, and it is sometimes desirable to condition the air in summer. Air filtering is now common practice, either to provide a cleaner environment for the employees or to protect the equipment. Properly designed air supply systems can fulfill all of these responsibilities, thereby eliminating duplication of equipment.

References for this chapter and Chapter 19, Local Exhaust Ventilation of Industrial Occupancies, can be found at the end of Chapter 21, General Ventilation of Nonindustrial Occupancies.

# General Ventilation of Nonindustrial Occupancies

by D. Jeff Burton, PE, CIH, CSP

HEATING, VENTILATING, AND AIR CONDITIONING (HVAC) systems are built to provide adequate amounts of fresh, clean, and tempered air to employee occupants of a building. Fresh "outdoor" air (OA) is used to maintain concentrations of indoor airborne chemicals to below detectable or unhealthy levels.

The recommended quantity of fresh air (cubic feet of fresh outside air per minute per person, written commonly as cfm OA/person) has varied over the years. Figure 21–1 shows the history of ASHRAE OA recommendations.

**Energy Conservation Versus Indoor Air Quality (IAQ)**

**Terms**

**HVAC Systems**

**HVAC Sources and Causes of IAQ Problems**

**HVAC Standards for Maintaining Adequate IAQ**
*Consensus Standards* ▪ *Regulatory Standards*

**Testing, Troubleshooting, and Maintaining HVAC Systems**
*Simple Calculations for Characterizing the Airflow in a Space* ▪ *Troubleshooting HVAC Systems* ▪ *Operation and Maintenance*

**Summary**

**Bibliography**

## Energy Conservation Versus Indoor Air Quality (IAQ)

During the 1970s and 1980s, many buildings were built or remodeled to minimize air-handling energy costs. This often meant limiting the amount of outside air brought into the building, sometimes to as low as 0–5 cfm OA/person. During the same time, many new building methods and materials were introduced, some of which created new air contaminants. This has often resulted in under-ventilated buildings and occupants who complain about the air quality.

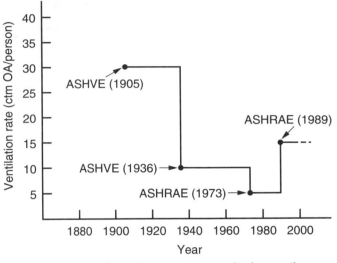

**Figure 21–1.** Outdoor air consensus standards over the years.

# Terms

*Air-handling unit (AHU)* is the ventilation equipment in HVAC systems.

*ASHRAE (American Society of Heating, Refrigeration, and Air Conditioning Engineers)* is the primary North American association dealing with IAQ issues; has developed a number of IAQ-related standards.

*Acceptable indoor air quality* is air in which there are no known contaminants at harmful levels as determined by appropriate authorities, *and* air with which 80 percent or more of the people do not express dissatisfaction based on several acceptability criteria (such as temperature, relative humidity, odors, and air movement).

*Commissioning* is the acceptance process in which an HVAC system's performance is determined, identified, verified, and documented to ensure proper operation in accordance with codes, standards, and design intentions.

*HVAC (heating, ventilating, and air conditioning, pronounced "H-Vac")* systems are air-handling systems designed primarily for temperature, humidity, odor, and air-quality control.

*Occupied zone* is usually the region within an occupied space between the floor and 72 in. above the floor and more than 2 ft from the walls.

*Outdoor air (OA)* is "fresh" air; the OA is mixed with return air (RA) to dilute contaminants in the supply air (SA); outdoor air is usually obtained from outside the building, but alternatives exist (such as from an acceptable hallway).

*Supply air (SA)* is air supplied to a space by the air-handling system.

*Constant air volume (CV)* is an HVAC system in which the supply air volume is constant; temperature and humidity are varied at the air-handling unit (AHU).

*Variable air volume (VAV)* refers to HVAC systems in which the air volume is varied by dampers or fan speed controls to maintain the temperature; primarily used for energy conservation.

# HVAC Systems

The term *HVAC* implies mechanically ventilating, heating, cooling, humidifying, dehumidifying, and cleaning air for comfort, safety, and health. HVAC systems also provide odor control and maintain oxygen and carbon dioxide levels at acceptable concentrations. Mechanical air-handling systems (as opposed to natural ventilation, which relies on wind and temperature differences to induce airflow through a building) range from a simple fan to complex, digitally controlled central air-handling units.

Individual units may be installed in the space they serve, or central units can be installed to serve multiple areas.

**Zones.** HVAC systems are built to serve specific *zones*, areas within a building that are served by an air handling system. The smaller the zone, the better chance there is of providing satisfactory conditions. But the costs increase as size decreases and the number of zones increases.

Most zones are defined by one thermostat. Some systems are designed to provide individual control of room air temperature in both single- and multiple-zone systems. (See Figure 21–2.) The following paragraphs describe basic systems.

**Single-zone constant-volume system.** The designer (or user) of an air-handling system must choose combinations of volume flow rate, temperature, humidity, and air quality that satisfy the needs of the space. Figure 21–3 shows a schematic of a simple, ideal commercial HVAC system. Systems vary in design and complexity, of course, but the single-zone constant-volume (SZCV) system is the simplest.

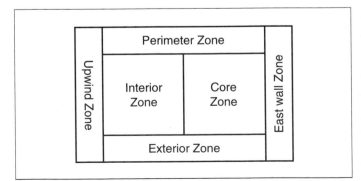

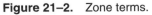

**Figure 21–2.** Zone terms.

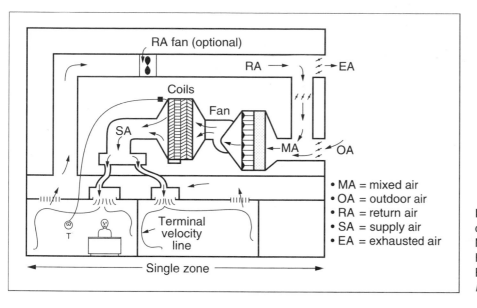

• MA = mixed air
• OA = outdoor air
• RA = return air
• SA = supply air
• EA = exhausted air

**Figure 21–3.** Typical single-zone constant-volume HVAC system. Note: MA = mixed air; OA = outside air; RA = return air; SA = supply air; EA = exhausted air. (Source: Burton, *IAQ and HVAC Workbook,* 1995.)

Every central HVAC system has an OA *intake,* usually a louvered opening on the top or side of the building. As air enters the intake, pushed by atmospheric pressure, a *damper* regulates the amount of OA taken into the system. The damper rarely closes completely (except in extreme weather conditions, such as subzero temperatures). (See Figure 21–3.)

Incoming outdoor air (OA) mixes with return air (RA) from the occupied space, forming recirculated air, or mixed air (MA). The MA usually passes through a coarse *filter* (arrestance filter), which removes bees, flies, bird feathers, leaves, and larger dust particles. Small particle collection requires a more efficient filter, the dust spot filter. The air then enters the *fan.*

Discharged at the fan outlet, the air, now under positive pressure, is pushed through coils that heat and cool the air depending on air temperature, the season, and the zone being served. A drain pan is fitted below the coils to collect water that condenses on the cooling coil. Leaving the coils, the air may be humidified (or dehumidified), which adjusts the relative humidity to 30–60 percent.

Typically, the conditioned air moves through metal ductwork (sometimes insulated with a sound-absorbing lining that must be kept dry and mold-free) at about 10–20 mph (1,000–2,000 fpm) to a distribution *box.* From there it travels through smaller ducts to the supply *terminals,* or *diffusers.*

Entering the room at 500–1,000 fpm, the air usually hugs the ceiling and walls, and its velocity slows to a terminal velocity of about 40–50 fpm. It should not flow past people until after it reaches terminal velocity.

In about 5–10 min, the air migrates through the space to the *return air register,* often a louvered opening into a return air duct or a return air plenum, the space above the ceiling tiles. From there it moves to the return air duct,

where it may be recirculated (RA) or be exhausted to the outdoors (EA). Many buildings are maintained under positive pressure, and air may exfiltrate anywhere along the building periphery. Sophisticated controls manage the system and determine how much OA will be introduced, how much RA will be exhausted, and what temperatures and humidities will be maintained.

Figure 21–4 shows how a single constant-value (CV) supply system can provide temperature control in multiple zones. Each zone has a reheat coil controlled by a thermostat. Air is first cooled at the central cooling coil and then reheated at the final distribution point to maintain proper temperature. These systems are less energy efficient because they both cool and reheat the air.

Figure 21–5 shows another approach, which serves many zones from one air handler. The air is split before the heating and cooling coils. Dual ducts carry heated and cooled air to final mixing boxes near the zones. Thermostats control dampers that mix the appropriate amounts of air to achieve the desired temperature. Energy use is high, but individual control is also high.

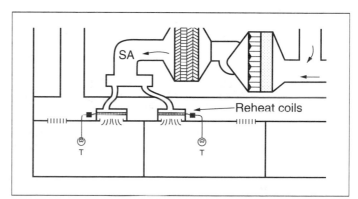

**Figure 21–4.** CV reheat system.

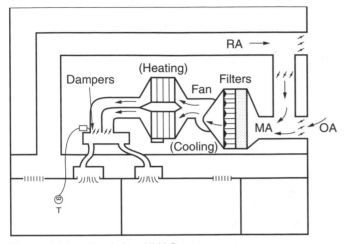

**Figure 21–5.** Dual-duct HVAC system.

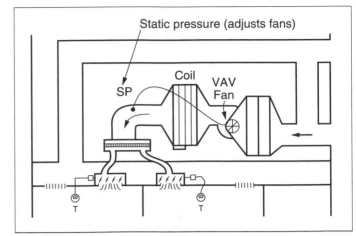

**Figure 21–7.** VAV with variable-air boxes.

**Variable air volume systems.** In contrast to the constant-volume system, which varies air temperature to maintain temperature, a *variable air volume* (VAV) system varies the amount of air delivered to the space to maintain temperature. (See Figure 21–6.) For example, in summer, if the temperature increases, additional cool air is provided. As the space temperature drops to desired levels, the airflow diminishes. These systems are energy efficient, but they may cause problems of insufficient outside air. For example, if the space does not call for cool air, the airflow will drop, perhaps to zero flow in poorly performing HVAC systems.

In cases where a minimum volume of air is required, reheat coils must be installed to provide temperature control.

The advantages of variable airflow are energy conservation and lower operating costs. Air volume and fan speed (in rpm) are linearly related. If the airflow is to be cut in half, then fan speed is cut in half. But horsepower and speed are related through a third-power relationship.

If the speed is reduced by half, the horsepower and the costs are reduced by eight times. The use of fan inlet dampers to reduce airflow also results in lower costs (but not by the same factors as reducing fan speed).

If variable-volume boxes are installed at the distribution point, more zones can be accommodated from a single fan. As the demand for air decreases, dampers at the boxes close, and the fan slows, saving energy. (See Figure 21–7.)

Reheat coils can be installed at each distribution box to regulate temperature where a minimum airflow is desired. This increases costs, however. Another way to maintain a minimum OA supply is to provide dual ducts with cool and warm air. Mixing boxes mix the air, always at the minimum flow required for temperature control and OA delivery, thus allowing the fan to run at the lowest rpm.

**Unit ventilators.** Unit systems can stand alone. (See Figure 21–8.) All of the air movement, OA delivery, and heating and cooling must be provided by the unit. These are often seen in offices with windows, hotel rooms, and other locations requiring self-contained operation.

Complex systems may have combinations of multiple zones, variable air volumes, and dual ducts. These are beyond the scope of this chapter. (See the Bibliography for more information.)

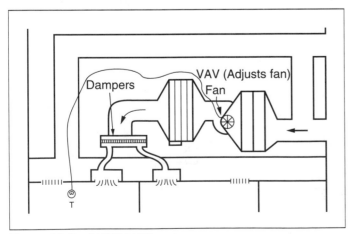

**Figure 21–6.** VAV HVAC system.

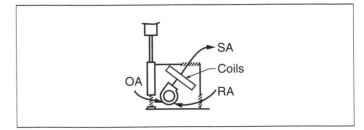

**Figure 21–8.** Unit ventilator.

**Air mixing in the occupied zone.** Delivery of air to a space does not guarantee that proper mixing will occur. Offices with partitions are often susceptible to this problem. The placement of bookshelves, furniture, windows, and walls changes the movement of air, often for the worse. This problem is often quantified by calculating the *mixing efficiency* or *ventilation effectiveness*. The *mixing factor* is defined as the ratio of the amount of air required to dilute a contaminant to the ideal amount of air that should reduce it.

$$K_m = \frac{Q_{actual}}{Q_{ideal}} \qquad (1)$$

*Example.* An office building houses 30 employees. ASHRAE Standard 62–1989 calls for 20 cfm OA per person to be delivered to the occupied zone.

$$Q_{ideal} = 20 \cdot 30 = 600 \text{ cfm} \qquad (2)$$

If because of poor mixing it actually takes 800 cfm to provide 20 cfm OA per person, then

$$K_m = \frac{Q_{actual}}{Q_{ideal}} = \frac{800 \text{ cfm}}{600 \text{ cfm}} = 1.3 \qquad (3)$$

In well-planned buildings served by good HVAC systems, $K_m$ is often close to 1.0 (perfect mixing). Good mixing can be attained by the following approaches:

- Provide and properly position an adequate number of supply and return registers.
- Place supply registers so as to circulate air to where people are located in the occupied zone.
- Provide free-standing fans for people located in areas of poor mixing. (This approach is often cost-effective and satisfying to building occupants.)

**Dampers.** Dampers are used in HVAC duct systems to control airflow. Damper positions can be set automatically or manually depending on the type and sophistication of the HVAC control system. Fire and smoke dampers are used to restrict the spread of heat and smoke during a fire. Dampers, actuators, and control systems should always be checked and maintained regularly to ensure proper flow of air through the system.

**Terminal devices.** Supply diffusers, return and exhaust grilles, and associated dampers and controls are used to produce proper distribution of the supply air. The number, location, and type of terminal devices determine the air distribution in the occupied space. Improper devices can lower the ventilation effectiveness and create stagnant areas, drafts, odor buildup, uneven temperatures, short-circuiting of the air, and air stratification.

**Return air systems.** Where the space above the ceiling (the return plenum) is used for returning air, certain practices should be followed:

- Electrical wiring should be in conduit or coated with a special fire-resistant covering.
- Plastic piping and plastic air registers, such as polyvinyl chloride, should not be used because of the smoke they produce in a fire.
- The space should be kept clean and dry to reduce bioaerosol amplification.
- Fire walls (which extend to the floor above) must be breached, sometimes with backdraft dampers, to allow the air to return to the AHU.
- Ceiling tiles and access doors must be kept in place to ensure proper airflow in the occupied space.

From the return plenum, air enters the return ducts and is ducted to the central AHU. Some systems use return fans to ensure proper pressure relationships between the supply and return ducts.

## ■ HVAC Sources and Causes of IAQ Problems

NIOSH, in studies of over 1,300 IAQ episodes since the late 1970s, categorized major causes or sources of IAQ problems (rounded to the nearest 10 percent):

- Fifty percent related to deficiencies in the ventilation of the building, such as lack of outside air, poor air distribution, uncomfortable temperatures and humidities, and sources of contaminants in the system.
- Thirty percent related to some indoor air contaminant, such as formaldehyde, solvent vapors, dusts, or microbiological agents.
- Ten percent could be attributed to an outdoor source, such as motor vehicle exhaust, pollen, fungi, smoke, or construction dusts.
- Ten percent had no observable cause.

Note that about half of all IAQ episodes had their origin in the HVAC system. According to NIOSH, common patterns emerge from HVAC-origin IAQ problems:

- Forced ventilation is common.
- Buildings are energy efficient.
- People perceive they have little control over their environment (for example, there is no thermostat in the room).
- There are more complaints when population densities are higher.

Table 21–A lists typical deficiencies and their potential causes and sources.

**Table 21–A.** Deficiencies in HVAC Systems and Their Causes

| Deficiency | Potential Causes/Typical Problems | Potential Corrections |
|---|---|---|
| Insufficient total air delivery to occupied space | Inadequate fan capacity<br>Worn fan blades<br>Faulty fan components<br>Imbalanced air-supply system<br>Increased number of occupants | Increase fan speed<br>Replace/repair wheel<br>Provide maintenance<br>Balance air-distribution system<br>Increase air capacity; redistribute occupants |
| Insufficient outdoor air (OA) delivered to occupied space | OA dampers set too low<br><br>Imbalanced supply and return systems<br>OA damper controls inoperative<br>Temperature-control capacity insufficient to meet space needs | Increase OA; provide fixed minimum OA delivery<br>Balance systems<br>Inspect, calibrate, reset controls<br>Increase system capacity |
| Air distribution within space not adequate; improper; insufficient | Improper supply system balancing<br><br>Poorly operating dampers, boxes<br><br>Maladjustment of thermostat/controls<br>Improper location of supply diffuser<br>Diffusers blocked<br>Office partitions resting on floor<br>Diffusers not attached to supply ducts | Rebalance<br><br>Repair, maintain, inspect boxes and control equipment<br>Calibrate thermostats<br>Relocate diffusers or occupants<br>Remove obstructions<br>Raise or remove partitions<br>Inspect, reattach connecting ductwork |
| AHU components not operating properly | System does not start before arrival of occupants; shuts down before departure<br>Filters inadequate<br><br><br>Room temperature and humidity controls inoperative | Reformat controls<br><br>Use appropriate filters; install filters in accordance with manufacturer's instructions; change filters on a regular basis<br>Monitor or calibrate; maintain |

Source: D. Jeff Burton, *IAQ and HVAC Workbook.*

# HVAC Standards for Maintaining Adequate IAQ

## Consensus Standards

The three most important consensus standards affecting IAQ are ASHRAE 62 on Ventilation for IAQ, ASHRAE 55 on Thermal Comfort, and ASHRAE 52 on Air Filtration. The latest standards (as of January 1995) are described in the following paragraphs.

**ASHRAE 62-1989: Ventilation for acceptable air quality.** The following paragraphs provide a list of the most important provisions of the standard. The "shoulds" and "shalls" are taken directly from the standard.

- When mechanical ventilation is used, provision for air-flow measurement should be provided. When natural ventilation is used, sufficient ventilation should be demonstrable. (p. 5.1)

- Ventilation air shall be supplied throughout the occupied zone. This implies delivery to where people actually are, as opposed to simply delivering air to the building. (p. 5.2)

- Where *variable air volume systems (VAV)* are used, and when the supply of air is reduced during times a space is occupied, indoor air quality shall be maintained throughout the occupied zone. (p. 5.4)

- Ventilation systems should be designed to prevent the growth and dissemination of microorganisms (e.g., limit use of fiber-lined ductwork where it may get dirty or wet). (pp. 5.5 and 5.6)

- Inlets and outlets shall be located to avoid contamination of intake air. (p. 5.5)

- Where practical, exhaust systems shall remove contaminants at the source. (p. 5.7)

- Where combustion sources, clothes dryers, or exhaust systems are used, adequate makeup air should be provided. (p. 5.8)

- Where necessary, particle filters and gas/vapor scrubbers should be sized and used to maintain air quality. (pp. 5.9 and 5.10)

- Relative humidities should be maintained between 30 and 60 percent. (p. 5.11)

- AHU condensate pans shall be designed for self-drainage. Periodic in-situ cleaning of cooling coils and condensate pans shall be provided. (p. 5.12)

- AHU shall be easily accessible for inspection and preventive maintenance. (p. 5.12)

- Steam is preferred for humidification. Standing water used in humidifiers and water sprays should be treated to avoid microbial buildup. (p. 5.12)

**Table 21–B.** Sample of OA Required by ASHRAE 62-1989 for Specific Occupancies

| Application | Estimated Maximum Occupancy per 1,000 ft² | cfm/person OA |
|---|---|---|
| Commercial dry cleaner | 30 | 30 |
| Office space | 7 | 20 |
| Smoking lounge | 70 | 60 |
| Conference room | 50 | 20 |
| Laboratories | 30 | 20 |

- Special care should be taken to avoid entrainment of moisture drift from cooling towers into makeup air and intake vents. (p. 5.12)

- Indoor air should not contain contaminants at concentrations known to impair health or cause discomfort. Outdoor air introduced to the building through the ventilation system should not exceed USEPA National Primary Ambient-Air Quality Standards. If the outdoor air contaminant levels exceed EPA values, the air should be treated. When confronted with air known to contain contaminants not on the EPA list, one should refer to other references for guidance. (p. 6)

The standard lists minimum OA requirements for about 100 occupancies, including offices and classrooms. (See the sample shown in Table 21–B.) The standard assumes good mixing and distribution of the outdoor air. It also assumes a certain occupant loading. Where these are not the case, additional OA may be required (p. 1.3.3).

### ASHRAE 55-1992: Thermal environmental conditions for human occupancy.

The standard specifies conditions in which 80–90 percent or more of the occupants should find the environment thermally acceptable. It does not address other environmental factors such as air quality and contaminants. This standard attempts to predict what conditions of temperature, humidity, activity, clothing, air movement, and radiant heat sources will satisfy 80–90 percent of the people. Satisfaction for any single parameter (such as temperature) should be 90 percent or more, and the satisfaction expressed for all parameters collectively should be 80 percent or more.

Figure 21–9 summarizes the standard's provisions for temperature and humidity. The figure is a simplified version of the original, which uses "operative temperatures" and concepts that can be understood only by reading the actual standard. (See the Bibliography.)

### ASHRAE 52: *1992 Methods of Testing Air Cleaning Devices Used in General Ventilation for Removing Particulate Matter.*

Most modern HVAC systems use two filters: a roughing filter to remove large particles and a dust spot filter to remove smaller particles. (Charcoal filters and HEPA filtration systems may be used in rare instances to

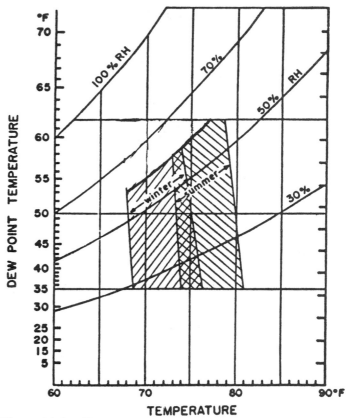

**Figure 21–9.** Thermal comfort standard. (Adapted from ASHRAE 55-1992.)

remove gases, vapors, and very small particles.) The three performance characteristics of greatest importance in selecting an air filter are as follows:

- The filter's efficiency in removing particles from the airstream
- The resistance to airflow through the filter
- The time interval between cleaning or replacement

ASHRAE 52 has established two testing or rating procedures:

- The ASHRAE Dust Spot Efficiency test, sometimes referred to as the 52 Atmospheric test, uses a fine dust as the test medium.
- The ASHRAE Arrestance test uses a coarse dust as the test medium.

Each has its place, but the dust spot efficiency test is used with filters that capture smaller particles. For example, a fine open foam filter with an arrestance rating of 80 percent may have a dust spot efficiency rating of 20–25 percent, and 0 percent for the very small DOP test aerosols used for testing HEPA filters.

Typical comparisons are shown in Table 21–C, by percent removal.

**Table 21–C.** Comparison of Filter Tests

| | Percent Removal for Specific Filter Media (%) | | |
|---|---|---|---|
| Filter Media | Arrestance Test | Dust Spot Test | DOP Test** |
| Fine open foams | 70–80 | 15–30 | 0 |
| Cellulose mats | 80–90 | 20–35 | 0 |
| Wool felt | 85–90 | 25–40 | 5–10 |
| Mats, 5–10 μm, 1/4 in. | 90–95 | 40–60 | 15–20 |
| Mats, 3–5 μm, 1/2 in. | >95 | 60–80 | 30–40 |
| Mats, 1–4 μm, fibers | >95 | 80–90 | 50–75 |
| Mats, 0.5–2 μm, glass | — | 90–98 | 75–90 |
| Wet-laid glass fibers, HEPA* | — | — | 95–99+ |

*HEPA: High-efficiency particulate air filters.
**A test used primarily with HEPA filters.

## Regulatory Standards

All existing buildings were built to comply with local building and fire codes in force at the time of construction (or rebuilding). Many building code authorities have adopted some version of ASHRAE 62 and 55. You should check your local codes.

As of January 1995, OSHA was in the process of developing an IAQ regulation for places of employment. The proposed OSHA rule on IAQ included the following major provisions related to ventilation:

- Keep carbon dioxide levels below 800 ppm; this translates into an OA requirement (in typical office spaces) of 25–30 cfm.

- Keep relative humidity below 60 percent (which is consistent with ASHRAE 55–1992).

- Obtain and maintain records on HVAC systems.

- Inspect, maintain, and operate the HVAC system so that it meets the criteria of the codes in force at the time the building was constructed.

- Exhaust designated smoking areas to the outdoors and keep the area under negative pressure.

- Locate intakes to prohibit uptake of contaminated air.

- Provide local exhaust of specific emitters if necessary.

OSHA also estimated that it would cost about $1.40 per square foot (for problem buildings) to upgrade existing HVAC systems to meet the proposed standard, for a total of $12 billion for the country. Call your local state or federal OSHA area office for the latest information on the status of this important regulation.

## ■ Testing, Troubleshooting, and Maintaining HVAC Systems

In order for any HVAC system to operate properly and consistently over its life span, *commissioning, testing, troubleshooting,* and *maintenance* are required.

*Commissioning* is a process in which a new HVAC system's performance is identified, verified, and documented to ensure proper operation and compliance with codes, standards, and design intentions. Commissioning often requires tests and demonstrations to verify that the system operates properly. Troubleshooting and maintenance activities also require system testing. A *commissioning agent* is often chosen by the building owner, the architect, or the contractor to oversee construction and commissioning activities.

*Testing and balancing* (TAB or T&B) is periodically required for all systems. This involves the testing and adjusting of system components (such as dampers) to ensure adequate air distribution to the occupied spaces. When hiring a TAB specialist, put requirements on paper in the form of a performance specification. Always check for references and certifications, such as those issued by the National Environmental Balancing Bureau (NEBB). You can obtain detailed information from your local Sheet Metal and Air Conditioning National Association (SMACNA).

Most health and safety professionals are not able to conduct in-depth testing of HVAC systems. Specialized knowledge of testing and balancing is required on the complex HVAC systems of today. This chapter provides guidelines for *simple* testing and troubleshooting that non-HVAC personnel might perform. For example, most people should be able to do the following:

- Become familiar with the HVAC system characteristics

- Determine the intended or desired operating parameters

- Perform cursory ventilation checks with smoke tubes, balometers, velometers, and pressure-measuring equipment

Early on, the investigator should contact the building engineer or a person who intimately knows the HVAC system. (Sometimes this person is the only one who really knows how the system operates "now that the system has been modified so much," a common problem.) Try to have that person available during the investigation.

Simple initial checks of the ventilation in a room that anyone can perform are shown in Table 21–D.

**Table 21–D.** Simple Checks of the HVAC System in a Room

- Does the room have a supply diffuser? A return?
- Is air moving through diffusers and return grills?
- Are air diffusers and grills open? Blocked? Attached to ductwork?
- Is supply air distributed throughout the occupied space?
- Do people actually feel air moving?
- Are there dead air spaces in the office or room?
- Do printers, copiers, and other equipment have adequate ventilation?
- Are mixing fans or portable heaters used by occupants?
- Does the HVAC system always operate when people are in the building?
- Is the air too hot? Too cold? Too humid? Too dry?
- Do people actually detect odors?
- Does the air make people uncomfortable? Sick?
- What contaminates the air?

Simple measurements of the ventilation in a room that anyone can perform are shown in Table 21–E.

One approach to becoming familiar with an HVAC system is to follow the system from start to finish. Go first to the air intakes and follow the air as it flows through the dampers, filters, fans, coils, ductwork, terminal boxes, and supply registers. Then identify return grills and follow the air back to the air handler. The following should be noted:

- Air intake and exhaust locations and damper settings
- What equipment is actually running and what is shut down
- Closed or jammed dampers
- Clogged or misplaced filters
- Settled water anywhere
- Thermostat locations and temperature levels
- Supply air quality (clean, odorless, properly humidified)
- Drafts or stuffiness in occupied space

**Table 21–E.** Simple Measurements of Indoor Air-Quality Parameters

| What to Measure | Typical Equipment |
|---|---|
| In occupied space<br>Dry-bulb temperature<br>Wet-bulb temperature<br>Relative humidity | Thermometer, psychrometer |
| System temperatures<br>Supply air (SA)<br>Outdoor air (OA)<br>Mixed air (MA)<br>Return air (RA) | Thermometer |
| $CO_2$ measurements<br>Supply air (SA)<br>Outdoor air (OA)<br>Return air (RA)<br>Occupied space air | Detector tubes, $CO_2$ monitors |
| Air movement<br>Any location | Smoke tubes, velometers |

- Supply and return register locations and damper settings
- Potential sources of contaminants (microbial or chemical) anywhere
- Provision for OA
- Positioning of OA and RA dampers
- Controls operation
- VAV system air delivery schedule, minimums

Next, review the as-built and as-modified drawings to become familiar with the HVAC system and the building. Read the original (or modified) specifications to become familiar with the intended operating parameters.

Forms 21–1 and 21–2 and Checklists 21–1, 21–2, and 21–3 (at the end of the chapter) can help you organize your information-gathering and testing efforts.

## Simple Calculations for Characterizing the Airflow in a Space

A number of simple calculations are available to the health and safety professional for determining air volume flow rates, the amount of OA being delivered to a space, and so forth. All of the equations presented in Chapter 20 can also be used in IAQ applications. Additionally, the following approximations are helpful.

**Percentage of outdoor air in the supply air.** It is possible to estimate the percentage of outdoor air (OA) in the supply air (SA) by measuring the temperatures of the air.

$$\%OA = \frac{T_{RA} - T_{MA}}{T_{RA} - T_{OA}} \cdot 100 \tag{4}$$

where  $T_{RA}$ = temperature of return air (dry-bulb)
 $T_{MA}$ = temperature of mixed return and outside air (dry-bulb)
 $T_{OA}$ = temperature of outdoor air (dry-bulb)

For example, assume the following temperatures were measured in an air-handling unit: $T_{RA}$ = 70 F, $T_{MA}$ = 66 F, and $T_{OA}$ = 50 F. The percentage of OA can be estimated by the following equation:

$$\%OA = \frac{T_{RA} - T_{MA}}{T_{RA} - T_{OA}} \cdot 100 = \frac{70 - 66}{70 - 50} \cdot 100 = 20\% \tag{5}$$

**Estimating the amount of OA in an office space.** Outdoor air (OA) reduces the indoor contaminants to concentrations that are acceptable to a majority (if not all) of the occupants in a space. Carbon dioxide is used as a surrogate for other air contaminants. ASHRAE and others have suggested that building owners and operators maintain $CO_2$ concentrations below 1,000 ppm. The concentrations of carbon dioxide in a building vary with the number of sources (people) and the volume of OA introduced and mixed in the space. (See Figure 21–10.) Note that concentrations tend to build and decay exponentially over time. This behavior

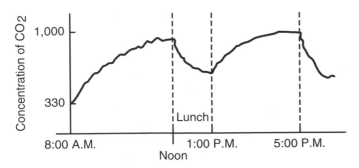

**Figure 21–10.** Typical carbon dioxide concentrations in today's buildings.

allows us to predict airflow rates from measurements of carbon dioxide.

**Tracer gas approaches to determining the outdoor airflow rate in a space.** The airflow through a building, space, or room can be estimated using tracer gas methods. Three methods are widely used: the *concentration-decay* method, the *constant-emission* method, and the *constant-concentration* method. All three methods are based on a simple relationship:

Change in tracer-gas concentration =

Amount introduced – Amount removed      (6)

The change in tracer gas is related to the nature of the airflow and the source of the tracer gas. Carbon dioxide, as well as other inert gases and vapors, can be used as tracer gas in occupied spaces. The sources listed in the Bibliography at the end of this chapter contain more information on tracer-gas calculations. The following simple estimating techniques are based on techniques using carbon dioxide as the tracer gas.

A useful approximation of the volume flow rate of outdoor air is based on the *constant-emission* technique. The number of people working in an office space and the measured carbon dioxide concentrations in the space (after the concentration has leveled off) are generally related to the volume flow rate of outdoor air as follows:

$$Q_{OA} \approx \frac{13,000\,n}{C_{indoors} - C_{OA}} \qquad (7)$$

where   $Q_{OA}$ = approximate volume flow rate of outdoor air (cfm)

   $n$ = number of people working in an office complex (with about seven people per 1,000 ft$^2$ of office space)

   $C_{indoors}$ = measured concentration of $CO_2$ in the office air after a long period of occupancy time, such as near lunch or near the end of the day (ppm)

   $C_{OA}$ = concentration of $CO_2$ in the outdoor air (ppm)

***Example.*** Suppose 12 employees work in an office complex. After several hours the measured $CO_2$ concentra-

tion has leveled off at about 1,000 ppm. The outside concentration of $CO_2$ is 340 ppm. The approximate volume of outdoor air being introduced to the space is calculated as follows:

$$Q_{OA} \approx \frac{13,000 \cdot n}{C_{indoors} - C_{OA}} \approx \frac{13,000 \cdot 12}{1,000 - 340} \approx 240 \text{ cfm} \qquad (8)$$

$$(20 \text{ cfm/person})$$

Perhaps the easiest and most useful method of measuring air-exchange rates over short periods of time is the *concentration-decay* tracer-gas approach. Human-origin carbon dioxide concentrations are used as the tracer gas. During the day $CO_2$ builds up naturally; after everyone leaves at 5:00 P.M., OA will dilute the carbon dioxide. (See Figure 21–10.) Knowing the initial and final concentrations and the time elapsed allows the use of purge formulas to predict the amount of OA delivered.

$$N = \frac{\ln(C_i - C_o) - \ln(C_a - C_o)}{h} \qquad (9)$$

where   $N$ = air exchange, air changes per hour, OA
   $C_i$ = Concentration of $CO_2$ at start of test
   $C_o$ = outdoor concentration, about 330 ppm
   $C_a$ = Concentration of $CO_2$ at end of test
   $h$ = time elapsed between start and end of test
   ln is the natural log.

When using this equation, be sure the HVAC system operates normally during the time of the test. (Many HVAC systems automatically reduce service at or before quitting time.) Be sure that almost all of the people have left the building or space being measured. Measure background carbon dioxide outside the building, if possible. A good location is at the air intake for the HVAC system. Be sure to subtract background concentrations from your measured values. If the decay line is flat, then no OA is being supplied. If the decay line is uneven (not smoothly exponential), then the air in the space is not being uniformly mixed in the space.

Volume flow rates can be calculated from the following formula:

$$Q_{OA} = \frac{N \cdot \text{Vol}}{60} \qquad (10)$$

***Example.*** Suppose the carbon dioxide concentration is $C_i$ = 1,200 ppm at 5:30 P.M., when all of the people have departed a building. By 7:30 P.M., the concentration has been reduced to $C_a$ = 400 ppm. The outside concentration is 330 ppm. How many air changes per hour of OA does this suggest? What is $Q_{OA}$ for a space volume of 50,000 ft$^3$?

$$N = \frac{\ln(C_i - C_o) - \ln(C_a - C_o)}{h} \qquad (11)$$

$$N = \frac{\ln(1{,}200 - 330) - \ln(400 - 330)}{2}$$

$$= 1.26 \text{ AC/Hr OA} \qquad (12)$$

$$Q_{OA} = \frac{N \cdot \text{Vol}}{60} = \frac{1.26 \times 50{,}000}{60} = 1{,}050 \text{ cfm} \qquad (13)$$

It is possible to estimate the percentage of outdoor air in the supply air by measuring the carbon dioxide concentrations at the AHU and using the following equation:

$$\% \text{ OA} = \frac{C_{RA} - C_{SA}}{C_{RA} - C_{OA}} \cdot 100 \qquad (14)$$

where  $C_{RA}$ = CO$_2$ concentration in return air
 $C_{SA}$ = CO$_2$ concentration in supply air (any point after the return air and outdoor air have mixed)
 $C_{OA}$ = CO$_2$ concentration in outdoor air

***Example.*** Assume the following concentrations of CO$_2$ are measured at the air-handling unit:

$$C_{RA} = 750 \text{ ppm}$$

$$C_{SA} = 650 \text{ ppm} \qquad (15)$$

$$C_{OA} = 330 \text{ ppm}$$

The percentage of OA is calculated as follows:

$$\% \text{ OA} = \frac{C_{RA} - C_{SA}}{C_{RA} - C_{OA}} \cdot 100 = \frac{750 - 650}{750 - 330} \cdot 100 = 24\% \qquad (16)$$

Additional information on estimating airflow rates can be found in the references listed in the Bibliography.

## Troubleshooting HVAC Systems

Invariably, something goes wrong with almost all ventilation systems. Simple troubleshooting usually involves three phases of study:

- Characterizing complaints and gathering background data
- Checking performance of ventilation systems and their controls
- Measuring carbon dioxide, temperature, and relative humidity

First, the troubleshooter should talk to those who are complaining to characterize the problem, gather background data, and try to establish causes or sources of the problem.

The most common causes and sources of trouble related to ventilation systems are as follows:

- Insufficient outdoor air (OA) introduced to the system
- Poor distribution/stratification of supply air in occupied space
- Draftiness—too much supply air or improper terminal settings

- Stuffiness—not enough air delivery or not delivered properly
- Improper pressure differences—doors hard to open
- Temperature extremes—too hot or too cold
- Humidity extremes—too dry or too humid
- Poor filtration—dirt, bugs, or pollen in the air-delivery system
- Poor maintenance
- Energy conservation the number-one priority
- Settled water in system
- Visual evidence of slime or mold
- Improper balance of distribution system
- Dampers at incorrect positions
- Terminal diffusers not at correct positions
- VAV systems in nondelivery or low-delivery mode

If after investigating the troubleshooter has identified one or more problems, he or she should not overreact; it is probably not a life-threatening situation. Furthermore, correction of the problems may or may not satisfy those who are registering complaints. Corrective measures don't hurt, of course, and any measures taken to solve the problem will probably be appreciated.

The following paragraphs list some common maladies or complaints and potential causes or sources of trouble.

*The temperature is too warm (or too cold).* Potential problems: thermostats misadjusted, supply air temperature setting too high or low, too much or too little supply air, supply diffuser blows air directly on occupants; temperature sensor malfunctioning or misplaced, cold air not mixing with occupied space air, HVAC system defective or undersized, and building under negative pressure, which causes infiltration of air at the building perimeter (building pressures should be +0.03 to +0.05 in. wg). Simple testing equipment: thermometer, velometer, smoke tubes.

*The air is too dry (or too humid).* Potential problems: Humidity controls not operating correctly or undersized. Simple testing equipment: sling psychrometer.

*The air is stuffy, stagnant* or *There is no air movement.* Potential problems: Nondelivery or low delivery of air to space, filters overloaded, VAV dampers malfunctioning, restrictions in ductwork, ductwork disconnected from supply diffusers, duct leaking, inadequate delivery of outside air, and blockage from furniture, partitions, or other barriers in the occupied space. Simple testing equipment: thermometer, velometer, smoke tube, CO$_2$ meter.

*There is no air movement when it gets cold.* Potential problems: VAV system set to deliver no air when system is not calling for cooling (common problem in older VAV systems). Simple testing equipment: thermometer, velometer, smoke tube, CO$_2$ meter.

*There are too many drafts.* Potential problems: Occupant outside of occupied zone, supply diffuser set to blow air directly on occupant, occupant near open door or window, free-standing fan blowing on occupant. Simple testing equipment: velometer, smoke tube.

*Air has a musty, "dirty sock" smell.* Potential problem: Microbiological contamination. Simple testing equipment: noses.

*Air smells like diesel exhaust.* Potential problems: Air intake near loading dock, other diesel engine exhaust source. Simple testing equipment: velometer, smoke tube, indicator tubes.

Checklists 21–4 and 21–5 (at the end of the chapter) are useful in checking and troubleshooting existing systems.

## Operation and Maintenance

Correct operating procedures and maintenance of the HVAC system will ensure its continued and consistent effectiveness. Maintenance is time-consuming and expensive but has been proven to be cost-effective. Labor-intensive maintenance (a general rule is one maintenance person per floor of building) requires trained workers, good materials, and good management. Preventive maintenance (PM) programs usually prevent problems before they arise.

Checklists 21–6, 21–7, and 21–8 (at the end of the chapter) can be helpful in establishing a good PM management program.

Lapses in maintenance activity require repair and renovation (for example, when ducts must be cleaned because filters have been left to deteriorate).

Dirt, debris, and microbiological growths in ductwork can be minimized by the following measures:

- Well-maintained filter systems (at least 40–60 percent efficiency, dust spot test)
- Regular HVAC maintenance
- Good housekeeping in the occupied space
- Locating air intakes in noncontaminated locations
- Keeping all HVAC system components dry (or drained)

Ducts can become both the source and the pathway for dirt, dust, and biological contaminants to spread through the building. ASHRAE 62–1989 and other standards suggest that efforts be made to keep dirt, moisture, and high humidity from ductwork. Filters must be used and kept in good working order to keep contaminants from collecting in the HVAC system.

Duct contaminants may include hair and dander, skin particles, insects and insect parts, organic and inorganic dust, carbon and oil particles, glass fibers, asbestos, pollen, mold, mildew, bacteria, leaves, dirt, and paper—all of which *may* contribute to IAQ problems. On the other hand, the mere presence of these contaminants has no effect on people if the contaminants do not leave the duct, if they do not generate other contaminants (for example, organic dust may support growth of mold or release adsorbed VOCs), or if they do not generate odors. Indeed, there have been cases where inert and inactive dusts were stirred up during duct cleaning, resulting in occupant complaints. Most ducts have small amounts of dust on their surfaces—a common occurrence that almost never necessitates duct cleaning.

Duct cleaning or replacement is generally warranted in the following conditions:

- There is slime growth in the duct.
- There is permanent water damage.
- There is debris that restricts airflow.
- Dust is actually seen issuing from supply registers.
- Offensive odors come from the ductwork.

Duct cleaning should be performed after one can answer "yes" to all of these questions:

- Are there contaminants in the ductwork?
- Has testing confirmed their type and quantity?
- Do they (or their odors or by-products) leave the duct and enter the occupied space?
- Is the source of these contaminants known? Can the source be controlled? (If not, cleaning is only a temporary measure.)
- Do the contaminants actually cause IAQ problems?
- Will duct cleaning effectively remove (neutralize, inactivate) the contaminant?
- Is duct cleaning the only (or the most cost-effective) solution?
- Has a qualified and reputable duct-cleaning firm been identified?
- Have the firm's references been checked?
- Does the duct-cleaning firm have a sensible, sound approach? Does it have the right kind of equipment? Will the cleaning process protect your HVAC equipment and the occupants of the space during cleaning?
- Will the firm give a guarantee?

## ■ Summary

The occupational health and safety professional has many tools with which to work when investigating, correcting, and controlling IAQ problems. The references that follow this chapter contain more detailed discussions of the particulars.

# ■ Bibliography

## Books

ACGIH, *Industrial Ventilation—A Manual of Recommended Practice* 1950–1995; Obtain latest copy, ACGIH, 1330 Kemper Meadow Dr., Cincinnati, OH 45240.

ASHRAE, *Handbook of Fundamentals,* 1993, about 600 pages; one of four basic handbooks on HVAC; 1791 Tullie Circle NE, Atlanta, GA 30329.

ASHRAE, *Handbook of HVAC Applications,* 1991.

ASHRAE, *Healthy Buildings,* Proceedings of the ASHRAE IAQ 91 Conference, 1991, 75 papers; 1791 Tullie Circle NE, Atlanta, GA 30329.

ASHRAE, *Indoor Air Quality,* 16 papers from the 1991 ASHRAE Annual Meeting, 1991; 1791 Tullie Circle NE, Atlanta, GA 30329.

Burgess, WA, Ellenbeck MJ, Treitman RD. *Ventilation for Control of the Work Environment.* New York: Wiley, 1989.

Burton DJ, *IAQ and HVAC Workbook,* 1995; *Industrial Ventilation Workbook,* 3rd ed., 1994; IVE, Inc.; each about 350 pages; available from ACGIH, ASSE, ASHRAE, AIHA, and NSC.

Cone JE, Hodgson MJ. Problem buildings: Building-associated illness and the sick building syndrome (State of the Art Reviews). *J Occup Med* 4 (4), Dec. 1989; 13 important papers summarize state of the art in 1989; about 220 pages.

EPA, *Building Air Quality: A Guide for Building Owners and Facility Managers,* EPA/400/1-91-033, Dec. 1991.

Godish T, *Indoor Air Pollution Control,* Lewis Publishers (121 S. Main St., Chelsea, MI 48118), 1989, 400 pages; good summary of recent literature, emphasis on residential indoor air quality; one of the better books on IAQ.

SMACNA, *HVAC Systems—Testing, Adjusting, and Balancing,* 1983, about 250 pages; *Indoor Air Quality,* 1989, about 200 pages; general information about IAQ and ductwork; 1385 Piccard Drive, Rockville, MD 20850; phone (301) 573-8330.

Weeks DM, Gammage RB, *The Practitioner's Approach to IAQ Investigations,* 1989 (Proceedings of the IAQ International Symposium); available from AIHA, 2700 Prosperity Ave., Suite 250, Fairfax, VA 22031-4319; phone (703) 849-8888.

## Standards and Codes

ASHRAE Standard 62-1989—*Ventilation for Acceptable Indoor Air Quality;* contact ASHRAE, 1791 Tullie Circle NE, Atlanta, GA 30329; about 30 pages.

ASHRAE 52-1992—*Methods of Testing Air Cleaning Devices Used in General Ventilation for Removing Particulate Matter;* contact ASHRAE, 1791 Tullie Circle NE, Atlanta, GA 30329; about 30 pages.

ASHRAE 55-1992—*Thermal Environmental Conditions for Human Occupancy;* contact ASHRAE, 1791 Tullie Circle NE, Atlanta, GA 30329; about 30 pages.

## Papers and Guidelines

American National Standards Institute. Standards for ventilation of open surface tanks, abrasive blasting, spray finishing, grinding, etc. American National Standards Institute, 11 W. 42nd St., New York, NY 10036.

ACGIH Committee on Bioaerosols, *Guidelines for the Assessment of Bioaerosols in the Indoor Environment.* Cincinnati, OH: ACGIH, 1989; about 100 pages.

ASHRAE Guideline 1-1989, *Guideline for the Commissioning of HVAC Systems,* 1989; 1791 Tullie Circle NE, Atlanta, GA 30329.

Fanger PO. The new comfort equation for indoor air quality. *ASHRAE J,* Oct. 1989, pp. 33–38; describes *olfs* and *decipols* and how they are used to estimate IAQ source strengths.

Hodgson MJ, Morey PR. Allergic and infectious agents in the indoor air. *Immunol Allerg Clin N Am* 9(2), Aug. 1989, pp. 399–412; overview of reactive chemicals and bioaerosols in the indoor environment.

Meckler M. The role of commissioning and building operations in maintaining acceptable indoor air quality. *Proceedings, 5th International Conference on IAQ,* Indoor Air '90, Toronto, Canada, 1990. (Copies available from ASHRAE.)

Morey PR. *Bioaerosols in the Indoor Environment: Current Practices and Approaches,* presented at the IAQ International Symposium, AIHA, St. Louis, May 1989.

Morey PR, Feeley JC. Microbiological aerosols indoors. *ASTM Standardization News,* Dec. 1988, pp. 54–58; overview of bioaerosols in the indoor environment.

## Agencies and Associations Involved in IAQ

ACGIH (publications, reports, committee activity), 1330 Kemper Meadow Dr., Cincinnati, OH 45240.

AIHA (list of industrial hygiene consultants, IAQ committee reports), 2700 Prosperity Ave, Suite 250, Fairfax, VA 22031-4319; phone (703) 849-8888.

ASHRAE (numerous books, articles, standards; journal), 1791 Tullie Circle NE, Atlanta, GA 30329.

Associated Air Balance Council (AABC; sets standards for TAB of HVAC systems), 1518 K St. NW, Suite 503, Washington, DC 20005; phone (202) 737-2926.

Columbus IAQ Resource Committee (study of energy impacts of new ASHRAE 62-1989), City of Columbus, Ohio, 90 W. Broad St., Columbus, OH 43215.

National Air Duct Cleaners Association (NADCA; publications on duct cleaning, recommended standards for duct cleaning), 1518 K St. NW, Suite 503, Washington, DC 20005; phone (202) 737-2926.

National Environmental Balancing Bureau (NEBB; list of certified HVAC balancing firms, publications, standards, and practice for TAB), 4201 LaFayette Center Drive, Chantilly, VA 22021; phone (703) 803-2980.

NIOSH (lists of publications, studies of IAQ, standards, research), 4646 Columbia Parkway, Cincinnati, OH 45226; phone (513) 841-4382.

SMACNA (publication *Indoor Air Quality,* sheet metal, ductwork), 1385 Piccard Drive, Rockville, MD 20850; phone (301) 573-8330.

U.S. Department of Energy (OSTI; energy conservation and IAQ), 1000 Independence Ave. SW, Washington, DC 20585; phone (202) 586-9455.

U.S. Department of Health and Human Services (information on smoking), Office on Smoking and Health, 1600 Clifton Road NE, Atlanta, GA 30333; phone (404) 488-5705.

USEPA (general publications on the subject, conducts research, training, information dissemination), 401 M St. SW, Washington, DC 20460; phone (202) 260-2080.

# Form 21–1

## Quick HVAC Survey Worksheet

Name _____ Date _____

Contact_____ Location _____

Phone_____ _____

| Potential Problem | Yes/No | Comments |
|---|---|---|
| Lack of outside air | _____ | _____ |
| Inadequate air distribution | _____ | _____ |
| Pressure difference between rooms | _____ | _____ |
| Air infiltration at perimeters | _____ | _____ |
| Detectable odors | _____ | _____ |
| Excessive tobacco smoke | _____ | _____ |
| Temperature too warm | _____ | _____ |
| Temperature too cold | _____ | _____ |
| Humidity too high | _____ | _____ |
| Humidity too low | _____ | _____ |
| Poorly vented heating equipment | _____ | _____ |
| Poorly located intakes | _____ | _____ |
| Visible mold, slime | _____ | _____ |
| Water visible | _____ | _____ |
| Water-damaged furnishings | _____ | _____ |
| Cleaning chemicals stored in mechanical room | _____ | _____ |
| Deteriorated insulation | _____ | _____ |
| Dirty, organic debris | _____ | _____ |
| Poor HVAC maintenance | _____ | _____ |
| Improper exhaust ventilation | _____ | _____ |
| Poorly located loading docks | _____ | _____ |
| Other | _____ | _____ |
| Other | _____ | _____ |

Source: D. Jeff Burton, *IAQ and HVAC Workbook*.

# Form 21–2

## Space Characterization Worksheet

Location_____ Date/time_____

Name _____ Address/phone _____

Contact_____ Number working in space_____

Sketch

Show doors, windows,
supply and return registers,
dimensions; show floor plan
or other necessary detail.
Use reverse side for more.

L_____ W _____ H_____ Room Volume_____

$T_{DB}$ _____ $T_{WB}$ _____ R.H._____

$C_{CO2}$ _____ $C_{other}$ _____ $C_{other}$ _____

$T_{MA}$_____ $T_{RA}$_____ $T_{OA}$ _____ % OA_____

$C_{SA}$ _____ ($CO_2$) $C_{RA}$ _____ ($CO_2$) $C_{OA}$ _____ ($CO_2$) % OA_____

Supply volume flow rates (SA)*: $Q_1$ _____ $Q_2$ _____ $Q_3$ _____

$Q_4$_____ $Q_5$ _____ $Q_6$ _____ $Q_{SAtotal}$_____

Return volume flow rates (RA)*: $Q_1$ _____ $Q_2$ _____ $Q_3$ _____

$Q_4$_____ $Q_5$ _____ $Q_6$ _____ $Q_{RAtotal}$_____

$Q_{OA}$_____ $Q_{OA/person}$_____ AC/hr_____

Terminal (draft) velocities_____ Location_____

Pressure relationship with hallways: (+) or (−)    Light_____ Noise _____

Source: D. Jeff Burton, *IAQ and HVAC Workbook.*

# Checklist 21–1

## Building Information Checklist

Building: _____  Contact Person: _____
Address: _____  Telephone: _____
Date: _____  Investigator: _____

*Building Description*

| | | |
|---|---|---|
| Year built _____ | HVAC type _____ | Interior layout _____ |
| Occupants _____ | Owner _____ | Date occupied _____ |
| Construction _____ | Number of floors _____ | Neighborhood type _____ |
| Emission sources _____ | Traffic pattern _____ | Location of garages _____ |
| Interior construction _____ | Tightness of doors, windows _____ | Insulation type _____ |

*Occupant Space Description*

| | | |
|---|---|---|
| Number of people _____ | $ft^2$ / person _____ | Type of activity _____ |
| Smoking policy _____ | Number of smokers _____ | Chemicals present _____ |
| Cleaning materials _____ | Furnishings _____ | Construction materials _____ |
| Recent construction _____ | Recent changes _____ | Free-standing fans _____ |
| Temperature, humidity _____ | Mold/dirt _____ | Wet surfaces _____ |
| Adjacent space _____ | Room pressure _____ | Carbon dioxide _____ |
| Drafts _____ | Stuffiness _____ | Interviews _____ |

*HVAC Systems*

| | | |
|---|---|---|
| Type of system _____ | Condition _____ | Windows _____ |
| Type of fuel _____ | Type of diffuser _____ | Location of intakes _____ |
| Location of exhaust _____ | OA provisions _____ | Distribution _____ |
| Terminal velocities _____ | Noise _____ | Dust/dirt _____ |
| Economizer cycle _____ | Controls _____ | Zones _____ |
| Total cfm _____ | Total OA _____ | Heat exchanger _____ |
| Local exhaust _____ | Makeup air _____ | Duct type _____ |
| Water in system _____ | Type humidifier _____ | Restroom exhaust _____ |
| Air-cleaner type _____ | Air-cleaner efficiency _____ | Person in charge _____ |

Source: D. Jeff Burton, *IAQ and HVAC Workbook.*

# Checklist 21–2

## Building Owner's HVAC Documentation and Programs

( ) References and calculations for required supply rate of OA
( ) Methods of measuring/monitoring OA supply
( ) Description of OA control systems
( ) Description of OA control systems for VAV systems
( ) Documentation of temperature/humidity control systems
( ) Filtration descriptions and SOP for use
( ) Filter efficiency documentation
( ) Written preventive maintenance program
( ) Maintenance record keeping
( ) Written Standard Operating Procedures
( ) Plans, drawings, specifications of building (as built/as is)
( ) Plans, drawings, specifications of building mechanical systems (as built/as is)
( ) Schematic drawings showing locations of all HVAC equipment for nonengineers
( ) Manufacturer literature for all operating equipment
( ) Testing, balancing, and monitoring records
( ) Building permits, stack permits, other applicable licences/permits
( ) History of changes to HVAC systems, occupancies
( ) Technical information about control systems
( ) Other _____
( ) Other _____

Note: In order to ensure IAQ, building operators should maintain these minimum programs, records, and documents.

Source: D. Jeff Burton, *IAQ and HVAC Workbook.*

# Checklist 21–3

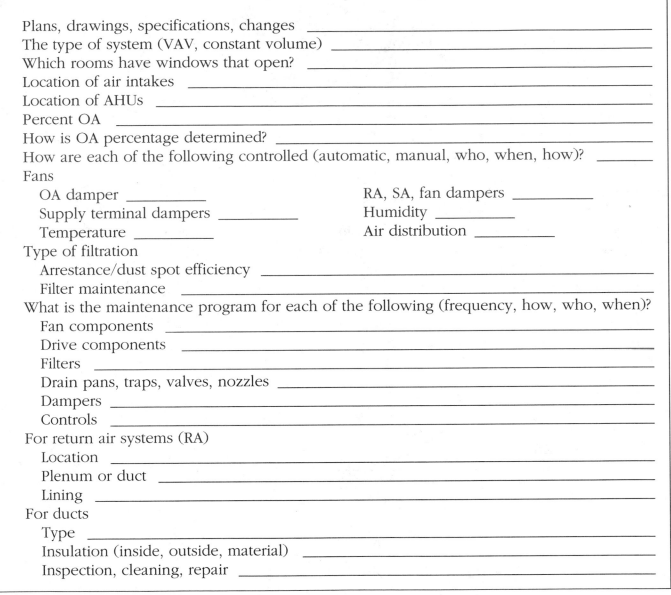

## Basic Information Checklist for HVAC Systems

Plans, drawings, specifications, changes _____

The type of system (VAV, constant volume) _____

Which rooms have windows that open? _____

Location of air intakes _____

Location of AHUs _____

Percent OA _____

How is OA percentage determined? _____

How are each of the following controlled (automatic, manual, who, when, how)? _____

Fans

   OA damper _____         RA, SA, fan dampers _____

   Supply terminal dampers _____    Humidity _____

   Temperature _____           Air distribution _____

Type of filtration

   Arrestance/dust spot efficiency _____

   Filter maintenance _____

What is the maintenance program for each of the following (frequency, how, who, when)?

   Fan components _____

   Drive components _____

   Filters _____

   Drain pans, traps, valves, nozzles _____

   Dampers _____

   Controls _____

For return air systems (RA)

   Location _____

   Plenum or duct _____

   Lining _____

For ducts

   Type _____

   Insulation (inside, outside, material) _____

   Inspection, cleaning, repair _____

Source: D. Jeff Burton, *IAQ and HVAC Workbook.*

# Checklist 21–4

## Inspection Checklist for HVAC Systems

Outdoor Air Intakes
  Location _____
  Open _____
  Controllable _____
  Outdoor contaminant sources nearby _____
  Type _____
  Location of exhausts _____
  Predominant wind direction and velocity _____

HVAC Equipment
  Intact _____
  Dry _____
  Clean _____
  Equipment running in accordance with specifications _____
  Filters in place, operating _____
  Slime, mold, dirt, soot removed _____

Ductwork
  Intact, connected _____
  Dry _____
  Balanced _____

Supply Air Diffusers
  Open _____          Set _____
  Airflow correct _____          Terminal velocities _____
  Air jet profile _____          Location _____
  Clean _____ _____            Quiet _____

Return Air Grills
  Location _____
  Air movement _____
  Open _____
  Attached to return system _____

Source: D. Jeff Burton, *IAQ and HVAC Workbook*.

# Checklist 21–5

## HVAC Troubleshooting Checklist Typical Problems

( )  Insufficient outdoor air (OA) introduced to the system
( )  Intake and exhaust dampers inoperative, malfunctioning
( )  Intake and exhaust at improper location

( )  Poor roughing filtration—dirt, bugs, pollen in air-delivery system
( )  Inadequate dust spot filtration
( )  Poor system maintenance

( )  Improper balance of distribution system
( )  Distribution dampers at incorrect positions
( )  Building under negative pressure

( )  Terminal diffusers not at correct positions
( )  VAV systems in nondelivery or low-delivery mode
( )  Terminal diffusers not attached to delivery system

( )  Poor distribution or stratification of supply air in occupant space
( )  Draftiness—too much supply air or improper terminal settings
( )  Placement of desks, personnel locations in high-velocity areas

( )  Stuffiness—not enough air delivery or not delivered properly
( )  Improper pressure differences between rooms—doors hard to open
( )  Temperature extremes—too hot or too cold

( )  Humidity extremes—too dry or too humid
( )  Energy conservation has become number 1 priority
( )  Settled water in system

( )  Visual evidence of slime or mold

Source: D. Jeff Burton, *IAQ and HVAC Workbook.*

# Checklist 21–6

## Checklist for Preventing and Minimizing IAQ Problems Through Good HVAC Practices

( )   Maintain the HVAC system in top working condition.
( )   Provide a written operating and maintenance plan for HVAC systems.
( )   Specify building materials with low VOC emissions.

( )   Provide appropriate volumes of outside air.
( )   Provide good distribution and mixing of supply air.
( )   Specify furnishings and materials with low VOC emissions.

( )   Restrict smoking to areas with dedicated ventilation systems.
( )   Use lowest temperatures consistent with energy and comfort.
( )   Provide relative humidities of 30–50 percent.

( )   Use high-efficiency filters (ASHRAE dust spot efficiency 50–70 percent).
( )   Involve and educate occupants.
( )   Lower occupant densities.

( )   Increase occupant control of environment (with, for example, personal fans, more thermostats, involvement in decisions).
( )   Involve professional assistance in IAQ problems.
( )   Provide monitoring of systems, air quality.

( )   Eliminate standing or stagnant water.
( )   Remove contaminated or emitting materials that cannot be controlled.
( )   Investigate bakeout for new buildings.

Source: D. Jeff Burton, *IAQ and HVAC Workbook.*

# Checklist 21–7

## Maintenance Checklist for Common HVAC Components

Unit Ventilators
( ) Brush and vacuum grills, coil, fan and unit interior.
( ) Remove any debris.
( ) Check belts and belt tightness.
( ) Repair any leaking water source.
( ) Inspect, clean, and/or replace filters.
( ) Adjust and lubricate dampers.
( ) Operation okay (controls, air delivery).

Induction Units
( ) Same as above, plus investigate air intake area for potential contaminants.

Reheat Coils, Mixing Boxes (Dual-Duct and VAV Systems)
( ) Clean coils.
( ) Inspect box and duct to ensure tight connection.
( ) Inspect, adjust, and lubricate dampers.
( ) Check operation and controls.
( ) Check minimum delivery, OA.

Humidifiers
( ) Determine type (water, water spray, wet stream, or dry stream).
( ) Check and clean water strainers, traps, and valves.
( ) Check and clean float equipment.
( ) Clean drains and drain pans.
( ) Inspect for slime, mold, odor.
( ) Clean spray nozzles.
( ) Check controls, instruments.

Cooling Towers
( ) Check for mold, slime.
( ) Wash down interior, plates.
( ) Clean head pans and nozzles.
( ) Drain and flush pipelines, pumps, tower pans.
( ) Clean strainers and screens.
( ) Operating in accordance with manufacturer's specs.

Source: D. Jeff Burton, *IAQ and HVAC Workbook.*

# Checklist 21–8

## Checklist for Reducing Microbial Problems in HVAC Systems

( ) Prevent buildup of moisture in occupied spaces.

( ) Prevent moisture collection in HVAC components.

( ) Remove stagnant water and slime from mechanical equipment.

( ) Use steam for humidifying.

( ) Avoid use of water sprays in HVAC systems.

( ) Maintain relative humidity less than 70 percent.

( ) Use filters with a 50–70 percent collection efficiency rating.

( ) Find and discard microbial-damaged furnishings and equipment.

( ) Remove or manage room humidifiers.

( ) Provide preventive maintenance of HVAC systems.

( ) Provide pigeon screens on intakes and exhausts (this will prohibit the contamination of the system by bird droppings, feathers, nesting materials, food, and so forth).

Source: D. Jeff Burton, *IAQ and HVAC Workbook.*

*CHAPTER* **22**

# Respiratory Protection

by Craig E. Colton, CIH

**Respiratory Protection Programs**
*Exposure Assessment ▪ Program Administration ▪ Standard Operating Procedures ▪ Medical Evaluations of Respirator Wearers ▪ Selection of Proper Respiratory-Protective Equipment ▪ Training ▪ Respirator Fit ▪ Cleaning, Maintenance, Inspection, and Storage*

**History of Respirator Regulations and Approvals**
*Voiding an Approval*

**Classes of Respirators**
*Air-Purifying Devices ▪ Atmosphere-Supplying Respirators ▪ Combination Air-Purifying and Atmosphere-Supplying Devices*

**Respirator Selection**
*Selection Requirements ▪ Hazard Determination ▪ Selection Steps*

**Respirator Fit Testing**
*Qualitative Fit Testing ▪ Qualitative Fit Test Protocols ▪ Quantitative Fit Testing ▪ Quantitative Fit Test Protocol ▪ Positive-Pressure Respirators*

**Summary**

**Bibliography**

**Addendum**
*29 CFR 1910.134: Respiratory Protection.*

A PRIMARY OBJECTIVE OF INDUSTRIAL HYGIENE programs in industry is the control of airborne contaminants by accepted engineering and work practice control measures. When effective engineering controls are not feasible, or while they are being implemented, appropriate respirators must be used. The Occupational Safety and Health Administration (OSHA) has established a standard, 29 *CFR* 1910.134 Respiratory Protection, for regulating the use of respiratory-protective equipment.

If the environment is still not completely safe after effective engineering and work practice controls have been fully used to reduce exposure to the lowest possible level, it will be necessary to use respirators to protect workers from contact with airborne contaminants or oxygen-deficient environments. Respiratory-protective equipment varies in design, specifications, application, and protective capability. Proper selection depends on the contaminant involved, conditions of exposure, human capabilities, and respirator fit.

Respirators are the least satisfactory means of exposure control because they provide good protection only if they are properly selected, fit tested, worn by the employees, and replaced when their service life is over. In addition, some employees may not be able to wear a respirator due to health or physical limitations. Respirators can also be cumbersome to use and hot to wear, and they may reduce vision and interfere with communication.

Despite these difficulties, respirators are the only form of protection available in the following situations: during the installation or implementation of feasible engineering and work practice controls; in work operations such as maintenance and repair activities for which engineering and work practice controls are not yet sufficient to reduce exposure to or below the PEL; and in emergencies.

# Respiratory Protection Programs

A respiratory protection program must be established when respiratory protection is needed. It should include the minimum requirements listed below; the order of importance may differ for each application:

- Exposure assessment
- Program administration
- Standard operating procedures
- Medical evaluation of respirator wearers
- Selection of proper respiratory-protective equipment
- Training
- Respirator fit
- Cleaning, maintenance, inspection, and storage

The OSHA standard requires that these points be addressed within a respiratory protection program. The *American National Standard for Respiratory Protection, Z88.2-1992,* is a voluntary consensus standard for the proper use of respiratory-protective equipment. This standard is published by the American National Standards Institute (ANSI). It specifies similar points for a program, but it enumerates them slightly differently. It is highly recommended that the American National Standard be consulted as well as the OSHA regulation.

## Exposure Assessment

Exposure assessments are basic to the proper use of respiratory-protective equipment. This information is used not only to identify the need for respirators but to identify the level of protection required, as well as when respirators must be used. The levels of worker exposure are determined by instruments and equipment designed to measure the concentrations of air contaminants and oxygen. Adequate air sampling and analysis should be carried out to determine time-weighted average concentrations and, when appropriate, compliance with ceiling and short-term exposure limits as well. Other chapters in this book should be consulted for more detail.

## Program Administration

Responsibility and authority for administration of a respiratory protection program must be assigned to one person who may and probably will have assistance from others. Centralizing authority and responsibility ensures that there is coordination and direction for the program. Respiratory protection programs will vary widely from company to company and depend upon many factors; a program may involve specialists such as safety personnel, industrial hygienists, health physicists, and physicians. In small facilities or companies having no formal industrial hygiene, health physics, or safety engineering department, the respiratory protection program should be administered by a qualified person responsible to the facility manager. The administrator must have sufficient knowledge to supervise the program properly. It is important that the administrator keep abreast of current issues, advances in technology, and regulations. In any case, overall responsibility must reside in a single individual if the program is to achieve optimum results.

The program administrator's responsibilities include the following:

- Conducting exposure assessments of the work area prior to respirator selection and periodically during respirator use to ensure that the proper respirator is being used
- Selecting the appropriate respirator that will provide adequate protection from all contaminants present or anticipated
- Maintaining records as well as the written procedures in a manner that documents the respirator program and allows for the evaluation of the program's effectiveness
- Evaluating the program's effectiveness through ongoing surveillance of the program

In addition to watching the program day to day, audits must be performed periodically (such as yearly) to ensure that the program reflects the written procedures and complies with current regulations and standards. The program must be periodically reevaluated to determine whether its goals are being met and changes are needed. It is recommended that the audit be conducted by a knowledgeable person not directly associated with the program instead of the respiratory protection program administrator. The "outside" individual brings a "new set of eyes" in an attempt to prevent overlooking deficiencies.

An audit checklist covering the entire program should be prepared and updated as required. There must be an effective means for correcting any defects found during the audit. A record should be kept of the findings along with plans and target dates for correction of deficiencies or problems and actual date completed.

## Standard Operating Procedures

Written standard operating procedures that cover the entire respiratory protection program need to be developed and implemented. The standard operating procedures need to cover all the elements of the program listed above as well as the issuance and purchasing of respirators and any company policies pertaining to respirator use.

In addition, written standard operating procedures must be developed for emergency and rescue operations. Although every situation cannot be anticipated, many of the needs for emergency and rescue use of respirators can be envisioned. This can be done by consideration of the following:

- An analysis of the emergency and rescue uses of respirators based on materials, equipment, work area, processes, and personnel involved
- A determination, based on the above analysis, of whether the available respirators can provide adequate protection to allow workers to enter the potentially hazardous environments
- Selection of the appropriate type and numbers of respirators
- Maintenance of these respirators so that they are readily accessible and operational when needed

Copies of these procedures must be available for the employees to read. The procedures need to be reviewed and revised as conditions and equipment change.

## Medical Evaluations of Respirator Wearers

Respirators can impose several physiological stresses ranging from very mild restriction of breathing to burdens of great weight and effort. The effects produced depend on the type of respirator in use. For this reason, a physician must determine whether or not an employee has any medical conditions that would preclude the use of a respirator.

To assist the physician, the program administrator must advise the physician of the types of respirators to be used for emergency or routine use; typical work activities; environmental conditions, such as high heat; frequency and duration of respirator use; and hazards for which the respiratory-protective equipment will be worn.

Although it is generally agreed that relatively few nondisabling medical conditions make respirator use dangerous, especially for employees who need respirators only briefly or occasionally to perform their tasks, it is important to identify those employees who may experience difficulties. For further information, consult the American National Standards for Respiratory Protection—*Physical Qualifications for Respirator Use* (See Bibliography).

## Selection of Proper Respiratory-Protective Equipment

Selection of the proper respirator is a very important task. The respirator must be National Institute for Occupational Safety and Health (NIOSH)/Mine Safety and Health Administration (MSHA) approved. Although it is obvious that the respirator selection must be based on the hazard to which the worker is exposed, there are many points that must be considered. These issues will be discussed later in this chapter.

## Training

For the safe use of any respirator it is essential that the user be properly instructed in its use. Supervisors as well as the person issuing respirators must be instructed by qualified persons. Emergency and rescue teams must be given adequate training to ensure proper respirator use. The OSHA standard requires that all employees be trained in the proper use of the device assigned to them. Many companies have their employees sign a document attesting to their having completed a training session with the respiratory-protective equipment. As a minimum, written records of the names of those trained and the dates when the training occurred must be kept. The workers need to be trained upon initial assignment of a respirator and followed up with annual training.

Each respirator wearer shall be given training that includes the following:

- An explanation of the need for the respirator, including an explanation of the respiratory hazard and what happens if the respirator is not used properly
- Instructions to inform their supervisor of any problems related to respirator use
- A discussion of what engineering and administrative controls are being used and why respirators are still needed for protection
- An explanation of why a particular type of respirator has been selected
- A discussion of the function, capabilities, and limitations of the selected respirator
- Instruction in how to put on the respirator and check its fit and operation
- Successful completion of either a qualitative or quantitative fit test
- Instruction in respirator maintenance
- Instruction in emergency procedures and the use of emergency escape devices and regulations concerning respirator use

The training in putting on the respirator must include an opportunity to handle the respirator with instructions for each wearer in the proper fitting of the respirator, including demonstrations and practice in how the respirator must be worn, how to adjust it, and how to determine whether it fits properly. Respirator manufacturers can provide training materials that tell and show how the respirator is to be adjusted, put on, and worn. The training session must also allow for time to practice. Hence, simply showing a videotape is not sufficient unless it is followed up with actual hands-on time. Close, frequent supervision can be useful to ensure that the workers continue to use the respirator in the manner they were trained. Supervisory personnel should periodically monitor the use of respirators to ensure that they are worn properly.

## Respirator Fit

Each wearer of a tight-fitting respirator must be provided with a respirator that fits. To find the respirator that fits, the worker must be fit tested. This will be discussed later in this chapter. In addition, each respirator wearer must be required to check the seal of the respirator by appropriate means before entering a harmful atmosphere. Each respirator manufacturer provides instructions on how to perform these fit checks. A fit check is a test conducted by the wearer to determine whether the respirator is properly adjusted to the face. The procedures may vary slightly from one respirator to another due to differ-ences in construction and design. In general the employee is checking either for pressure or flow of air around the sealing surface. Fit checks are not substitutes for qualitative or quantitative fit tests. Care must be taken in conducting fit checks. Thorough training in carrying out these tests must be given to respirator wearers.

A respirator equipped with a facepiece (tight or loose-fitting) (Figure 22–1) must not be worn if facial hair comes

**Figure 22–1.**  The three types of existing loose-fitting respiratory inlet coverings: a. Loose-fitting facepiece. b. Helmet. c. Hood. The loose-fitting facepiece is not suitable for workers with beards because it forms a partial seal with the face. The loose-fitting helmet and hood are acceptable for workers with beards. (Courtesy 3M.)

**Figure 22–2.** Full facepiece with provisions for prescription spectacles. (Courtesy 3M.)

between the sealing periphery of the facepiece and the face or if facial hair interferes with valve function. Only respirators equipped with loose-fitting hoods or helmets are acceptable with facial hair (Figure 22–1). If spectacles, goggles, or face shields must be worn with a half- or full-facepiece respirator, they must be worn so as not to adversely affect the seal of the facepiece to the face. Certain facepieces also enable the wearer to wear prescription spectacles without disturbing the facepiece seal (Figure 22–2).

## Cleaning, Maintenance, Inspection, and Storage

The respirator maintenance program includes cleaning and sanitizing of respirators where necessary, inspection of the equipment for defects, maintenance and repair of defects found, and proper storage of the respirator. When a program includes atmosphere-supplying respirators, then assurance of breathing air quality must be included. A maintenance schedule should be implemented that ensures that each worker is provided with a respirator that is clean, sanitary, and in good operating condition. The manufacturer's instructions should be followed. The precise nature of the program will vary because of such factors as size of the facility and the equipment involved.

**Cleaning and sanitizing.** Personally assigned respirators must be cleaned and sanitized regularly. Respirators that may be worn by different individuals must be cleaned and sanitized before being worn by a different individual. Cleaner-sanitizers that effectively clean the respirator and contain a bactericidal agent are commercially available. The bactericidal agent is frequently a quaternary ammonium compound. For personally assigned respirators, equipment wipes are available containing these compounds. Alternatively, respirators can be washed in a mild

detergent solution (such as a dishwashing liquid) and then immersed in a sanitizing solution. Commonly recommended sanitizing solutions are an aqueous hypochlorite (bleach) solution and aqueous iodine solution; 50 ppm of chlorine and iodine, respectively. The recommended immersion time is 2 min. Strong cleaning and sanitizing agents and many solvents can damage rubber and elastomeric respirator parts. These substances should be used with caution. Check the respirator manufacturer's instructions or contact them if there are questions.

**Inspection.** The respirator must be inspected by the wearer immediately prior to each use to ensure that it is in proper working order. In addition, emergency and rescue use respirators must be inspected at least monthly and a record of inspection dates and findings kept. All respirators that do not pass the inspection must be immediately removed from service and repaired or replaced. The respirators should also be inspected during cleaning to determine whether they are in good condition or if parts need to be replaced or repaired or whether they should be discarded. Respirator inspection must include a check for tightness of connections and for the condition of the respiratory inlet covering, head harness, valves, connecting tubes, harness assemblies, hoses, filters, cartridges, canisters, end-of-service-life indicator, electrical components, and shelf-life date(s). The inspection should also include a check for proper function of the regulators, alarms, and other warning systems. Compressed gas cylinders on self-contained breathing apparatus (SCBA) must be checked to ensure that they are fully charged according to the manufacturer's instructions.

**Repair.** Replacement of other than disposable parts and any repair should be done only by personnel with adequate training in the proper maintenance and assembly of the respirators. Replacement parts must be only those designated for the specific respirator being repaired. Failure to do so may result in malfunction of the respirator. In addition, it will void the NIOSH–MSHA approval.

**Storage.** The respirators must be properly stored in order to protect them from dust, sunlight, excessive heat, extreme cold, excessive moisture, damaging chemicals, and physical damage from things such as vibration and shock. Tool boxes, paint-spray booths, and lockers are not appropriate storage locations unless they are protected from contamination, distortion, and damage. In addition, emergency and rescue use respirators that are located in the work area must be readily accessible. Their location must be clearly marked.

**Air quality.** Compressed air, compressed oxygen, liquid air, and liquid oxygen used in atmosphere-supplying respirators must be of high purity. Oxygen must meet the requirements of the United States Pharmacopoeia for medical or breathing oxygen. Compressed gaseous air must meet the requirements for grade D as described in ANSI/CGA G-7.1-1989. The limiting characteristics are listed in Table 22–A.

**Table 22–A.** Grade D Breathing Air Requirements

| Limiting Characteristic | Allowable Maxima |
| --- | --- |
| Percent O$_2$ (balance predominantly N$_2$) | 19.5–23.5% |
| Water | Variable from very dry to saturated; no liquid water |
| Oil (condensed) | 5 mg/m³ |
| Carbon monoxide | 10 ppm |
| Odor | No pronounced odor |
| Carbon dioxide | 1,000 ppm |

(Reprinted with permission from Compressed Gas Association, CGA G-7.1-1989.)

# History of Respirator Regulations and Approvals

After enactment of the Occupational Safety and Health Act (OSHAct), the National Institute for Occupational Safety and Health (NIOSH) and the U.S. Bureau of Mines (USBM) jointly promulgated 30 *CFR* Part 11, which prescribes approval procedures, establishes test requirements, and sets fees for obtaining joint approval of respirators. Over the years, government reorganization has resulted in transfer of the USBM functions to the Mine Safety and Health Administration (MSHA). NIOSH has been named as the testing, approving, and certifying agency for respirators. The respirator approvals are issued jointly by NIOSH and MSHA (30 *CFR* 11).

The NIOSH Testing and Certification Laboratory has the following responsibilities:

- To publish certification requirements
- To test and certify products meeting those requirements
- To publish lists of certified products
- To audit respirator manufacturers' facilities to determine the acceptability of their quality-assurance programs
- To sample products from the open market and test them for continued conformance to certification requirements
- To perform research on the development of new test methods and requirements for product improvement where necessary to ensure worker protection

All NIOSH–MSHA–approved respiratory protection devices have an approval label similar to that shown in Figure 22–3.

## Voiding an Approval

Once approval has been granted to a device by NIOSH–MSHA, the user should become acquainted with the limitations of the device as set forth in the approval (Figure 22–3). The approval will be void if the device is used in conditions beyond the limitations set by NIOSH or those established by the manufacturer. The user should also guard against any alteration being made to the device. All parts, filters, canisters, cartridges, and anything else not specifically intended

to be used on the device by NIOSH–MSHA or the manufacturer will void the existing approval. If there is any question concerning parts, alteration, or limitation of the device, always check with the manufacturer. The employer should take care so as not to knowingly void the approval for a piece of equipment.

NIOSH has the authority to purchase and test respiratory-protective devices on the open market as a continuing check on manufacturers' quality-assurance standards and adherence to approvals. Manufacturers may not institute design changes of the device or its components without obtaining an extension of an existing approval or resubmitting a device for a new approval.

Passage of the OSHAct affected respiratory protection in another way besides leading to NIOSH approvals. Shortly after OSHA was established, OSHA promulgated a standard regulating the use of respiratory-protective devices. This standard, 29 *CFR* 1910.134, entitled "Respiratory Protection," established the requirements for a respiratory protection program (Table 22–B; see the Addendum to this chapter for the complete standard).

These program requirements are essentially identical to those discussed earlier in this chapter. NIOSH and OSHA requirements are interrelated in that OSHA requires approved respirators to be used and NIOSH certification establishes limitations on the use of the respirators; OSHA regulates the use, whereas NIOSH regulates the design and performance, of respiratory-protective equipment. NIOSH, however, sometimes makes recommendations regarding respiratory-protective equipment use that are different than OSHA requirements. These recommendations do not change or replace OSHA standards.

In addition to 29 *CFR* 1910.134, OSHA has promulgated other standards that address respiratory protection requirements that may be more specific to certain situations or more stringent. These include the ventilation standard for abrasive blasting respirator use requirements, the hazardous waste operations and emergency response, permit-required confined spaces, and the fire brigade standards, as well as the various substance-specific standards for substances such as asbestos, lead, benzene, and cadmium. This is not a comprehensive list; OSHA is continually promulgating new substance-specific standards. Consult the appropriate OSHA standards covering the industries or operations in question.

**Table 22–B.** OSHA Requirements for a Minimal Acceptable Respiratory Protection Program

| | |
| --- | --- |
| Standard operating procedures | Respirator inspection |
| Respirator selection | Air sampling |
| Training (including fit testing) | Program evaluation |
| Respirator cleaning | Medical evaluation of wearers |
| Respirator storage | Use of NIOSH–MSHA-approved respirators |

(Adapted from OSHA General Industry Standards, 29 *CFR* 1910.134.)

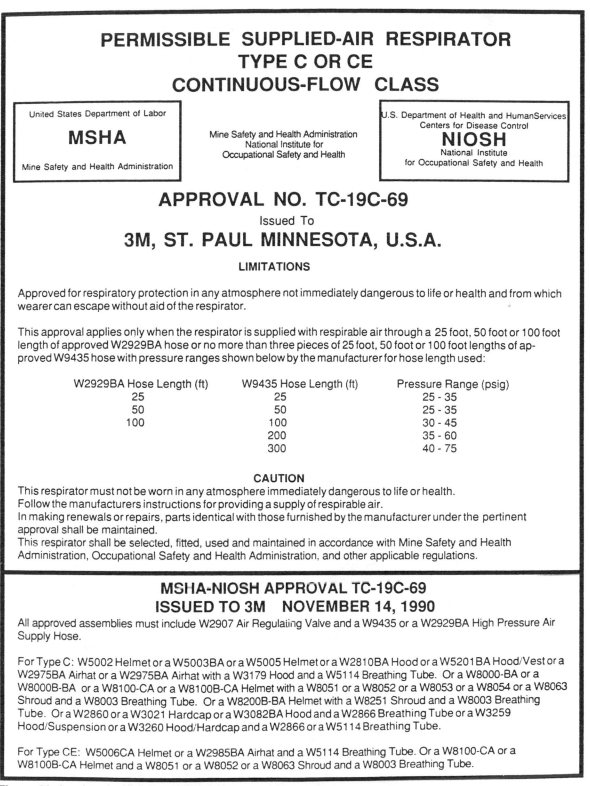

**PERMISSIBLE SUPPLIED-AIR RESPIRATOR**
**TYPE C OR CE**
**CONTINUOUS-FLOW CLASS**

United States Department of Labor
**MSHA**
Mine Safety and Health Administration

Mine Safety and Health Administration
National Institute for
Occupational Safety and Health

U.S. Department of Health and HumanServices
Centers for Disease Control
**NIOSH**
National Institute
for Occupational Safety and Health

## APPROVAL NO. TC-19C-69
Issued To
### 3M, ST. PAUL MINNESOTA, U.S.A.

**LIMITATIONS**

Approved for respiratory protection in any atmosphere not immediately dangerous to life or health and from which wearer can escape without aid of the respirator.

This approval applies only when the respirator is supplied with respirable air through a 25 foot, 50 foot or 100 foot length of approved W2929BA hose or no more than three pieces of 25 foot, 50 foot or 100 foot lengths of approved W9435 hose with pressure ranges shown below by the manufacturer for hose length used:

| W2929BA Hose Length (ft) | W9435 Hose Length (ft) | Pressure Range (psig) |
|---|---|---|
| 25 | 25 | 25 - 35 |
| 50 | 50 | 25 - 35 |
| 100 | 100 | 30 - 45 |
|  | 200 | 35 - 60 |
|  | 300 | 40 - 75 |

**CAUTION**

This respirator must not be worn in any atmosphere immediately dangerous to life or health.
Follow the manufacturers instructions for providing a supply of respirable air.
In making renewals or repairs, parts identical with those furnished by the manufacturer under the pertinent approval shall be maintained.
This respirator shall be selected, fitted, used and maintained in accordance with Mine Safety and Health Administration, Occupational Safety and Health Administration, and other applicable regulations.

**MSHA-NIOSH APPROVAL TC-19C-69**
**ISSUED TO 3M    NOVEMBER 14, 1990**

All approved assemblies must include W2907 Air Regulating Valve and a W9435 or a W2929BA High Pressure Air Supply Hose.

For Type C: W5002 Helmet or a W5003BA or a W5005 Helmet or a W2810BA Hood or a W5201BA Hood/Vest or a W2975BA Airhat or a W2975BA Airhat with a W3179 Hood and a W5114 Breathing Tube. Or a W8000-BA or a W8000B-BA or a W8100-CA or a W8100B-CA Helmet with a W8051 or a W8052 or a W8053 or a W8054 or a W8063 Shroud and a W8003 Breathing Tube. Or a W8200B-BA Helmet with a W8251 Shroud and a W8003 Breathing Tube. Or a W2860 or a W3021 Hardcap or a W3082BA Hood and a W2866 Breathing Tube or a W3259 Hood/Suspension or a W3260 Hood/Hardcap and a W2866 or a W5114 Breathing Tube.

For Type CE: W5006CA Helmet or a W2985BA Airhat and a W5114 Breathing Tube. Or a W8100-CA or a W8100B-CA Helmet and a W8051 or a W8052 or a W8063 Shroud and a W8003 Breathing Tube.

**Figure 22–3.** A typical approval label that accompanies each NIOSH–MSHA–approved respirator. The user should understand the limitations of the device. (Courtesy 3M.)

## ■ Classes of Respirators

Respiratory-protective devices can be described based on their capabilities and limitations and placed in three classes: air-purifying, atmosphere-supplying, and combination air-purifying and atmosphere-supplying devices.

### Air-Purifying Devices

The air-purifying device cleanses the contaminated atmosphere. Ambient air passes through an air-purifying element that can remove specific gases and vapors, aerosols, or a combination of these contaminants. This type of device is limited in its use to those environments where there is sufficient oxygen to support life and the air-contaminant level is within the maximum use concentration of the device. The useful life of an air-purifying device is limited by the concentration of the air contaminants, the breathing rate of the wearer, temperature and humidity levels in the workplace, and the removal capacity of the air-purifying medium.

**Aerosol-removing respirators (Figure 22–4).** Aerosol-removing respirators offer respiratory protection against airborne particulate matter, including dusts, mists, and fumes, but they do not protect against gases, vapors, or oxygen deficiency. These respirators are equipped with filters to remove aerosols (particles) from the air. The filter may be a replaceable part or a permanent part of the respirator. The respirators consist essentially of a facepiece, either quarter-face (above the chin), half-face (under the chin), or full-face design (Figure 22–5). Directly attached to the facepiece is one of several types of filters made up of a fibrous material that removes the particles by trapping them as air is inhaled through the filter.

There are many classes of filter respirators specifically designed for the various classes of airborne particulate matter. Although a single respirator can be made to provide effective protection against all aerosols, in most cases it would be too expensive and perhaps too cumbersome for the great majority of users. Therefore, proper filter selection depends on a knowledge of the material and the particle size.

In fibrous filters, various filtration mechanisms are at work. These filtration mechanisms include particle interception, sedimentation, impaction, and diffusion. In addition to these mechanical mechanisms some filters will also use electrostatic attraction. The filtration mechanisms work together in every filter to some degree as the filter manufacturer attempts to make an efficient filter with low

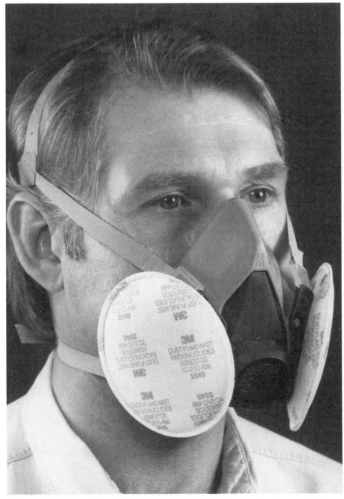

**Figure 22–4.** A half-facepiece HEPA filter respirator with two replaceable filters. (Courtesy 3M.)

**Figure 22–5.** A full-facepiece respirator with replaceable HEPA filters. The viewing lens has been adapted to accommodate a welding lens. Welding shrouds are available to protect the skin. (Courtesy 3M.)

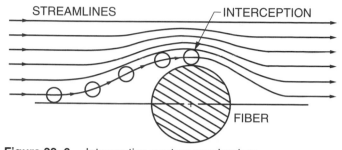

**Figure 22–6.** Interception capture mechanism.

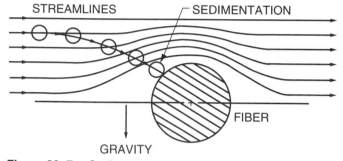

**Figure 22–7.** Sedimentation capture mechanism.

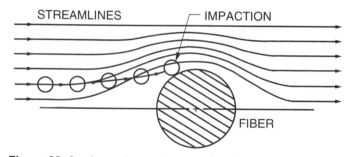

**Figure 22–8.** Impaction capture mechanism.

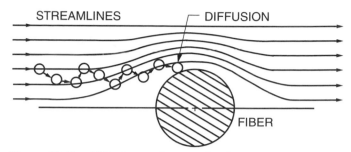

**Figure 22–9.** Diffusion capture mechanism.

Diffusion is particle movement due to air molecule bombardment and is important only for smaller particles (Figure 22–9). The particles can randomly cross the airstreams and encounter a filter fiber. This random motion is dependent on particle size and temperature. For example, as particle size decreases, diffusive activity of the particle increases. This increases the chance of capture. A lower flow rate through the filter also increases the chance of capture as the particle spends more time in the area of the fiber.

In electrostatic capture the charged particles are attracted to filter fibers or regions of the filter fiber having the opposite charge. Uncharged particles may also be attracted depending on the level of charge imparted on the filter fiber. This mechanism aids the other removal mechanisms, especially interception and diffusion. These filter types use electrical charges to enhance their mechanical filtering capabilities. Two types of electrostatic materials used in respirator filters are referred to as resin wool and electrets.

Resin wool was the earliest filter medium to incorporate electrostatic charges and is still used today by some manufacturers. Resin wool typically consists of wool and synthetic fibers combined with resin particles. Friction generated between the resin and wool during the filter carding process results in negatively charged resin particles and positively charged wool fibers. Due to this separation, local, nonuniform electrostatic fields develop throughout the filter. These electric fields are responsible for the enhanced aerosol collection ability of the filter.

*Electret fibers* are a recent development in filtration technology. Over the last several years, they have undergone many improvements. Electret fibers are plastic fibers that have a strong electrostatic charge permanently embedded into their surface during processing. They maintain a positive charge on one side of the fiber and a negative charge of equal magnitude on the opposite side of the fiber (Figure 22–10). Both charged and uncharged particles will be attracted to electret fibers. Charged particles are attracted to the parts of the fiber that have an opposite charge. Uncharged particles have equal internal positive and negative charges. The strong electrostatic forces of the electret fibers polarize these charges, inducing a dipole within the particle, and the particle is then attracted to the fiber by a polarization force. Long-term environmental testing of electrostatic filters using elevated temperatures and humidity has indicated that resin wool

breathing resistance. The exact contribution of each mechanism depends on flow rate and particle size.

In interception capture, the particles do not deviate from their original streamline of air in this mechanism (Figure 22–6). As the airstreams approach a fiber lying perpendicular to their path, they split and compress in order to flow around the fiber. The airstreams rejoin on the other side of the fiber. If the center of a particle in these airstreams comes within one particle radius of the fiber, it contacts the fiber surface and is captured. As particle size increases, the probability of interception increases. The particles do not deviate from their original streamline in this mechanism.

Sedimentation capture is due to the effect of gravity on the particle; therefore, the flow rate through the filter must be low (Figure 22–7). It is most significant for large particles, for example, larger than 3 μm.

Particles with sufficient inertia cannot change direction sufficiently to avoid the fiber. As the airstreams split and change direction suddenly to go around the fiber, these particles are captured due to impaction on the surface of the fiber (Figure 22–8). A particle's size, density, speed, and shape determine its inertia.

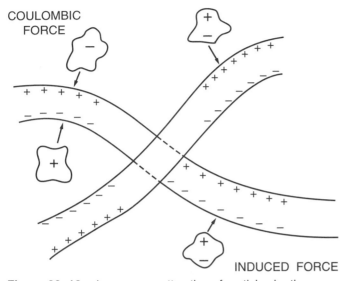

**Figure 22–10.** Long-range attraction of particles by the permanently charged electret fibers. The two charged particles on the left are attracted by a coulombic force, whereas the two uncharged particles on the right are attracted by a polarization force.

filters are susceptible to degradation at high humidities and elevated temperature conditions. Electret filters were not affected by exposure to these same conditions.

The exact combination of capture mechanisms depends on several factors. Generally, large, heavy particles are removed by impaction and interception; large, light particles are removed by diffusion and interception. Diffusion removes very small particles (Figure 22–11). When the fiber used in the explanation of the capture mechanisms is joined by other fibers to create a filter maze

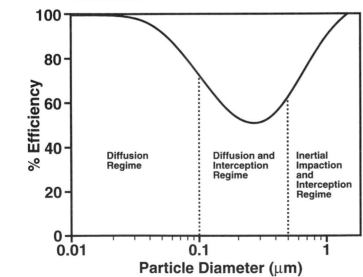

**Figure 22–11.** Filter efficiency versus particle size with mechanical filtration mechanisms identified.

of certain average porosity and thickness, the different filtration mechanisms will combine at different particle sizes to affect total filtration performance and efficiency. The capture mechanisms of sedimentation, interception, and inertial impaction combine effectively to remove nearly all particles sized above 0.6 μm. Additionally, the low flow rates through respirator filters of only a few centimeters per second let diffusion play its part very effectively for particles below 0.1 μm.

However, between these two particle size regions (0.1 to 0.6 μm), diffusion and impaction are not as effective, and a minimum filtration efficiency exists, as shown in Figure 22–11. The lowest point on this curve is called the most penetrating particle size and can be determined empirically in the laboratory. Figure 22–12 is an actual filter penetration curve determined in the laboratory. The most penetrating size range can vary with filter design and flow rate. The "dip" is inverted when compared with Figure 22–11 as percent penetration is plotted instead of percent efficiency. Percent penetration is the inverse of percent efficiency ($1/\% E$). The addition of an electrostatic charge to the fibers improves the filtering ability in this range by increasing the capture efficiency at the "most penetrating particle size." Most respirator filters have a most penetrating particle size between 0.2 and 0.4 μm. This is the basis for the widely used dioctyl phthalate (DOP) test for high-efficiency filters using a 0.3-μm particle. The filter efficiency for good filters will always be much better at any particle size other than the most penetrating particle size. Because a respirator filter has measurable penetration of particles in the range of 0.2–0.4 μm, it is easy to forget that anywhere else in the wide range of particle size in the workplace, filtration efficiency is essentially 100 percent. For the filter tested in Figure 22–12, the penetration at the most penetrating particle size is around 10 percent, whereas at 1 μm the penetration is only about 1.5 percent. It is the reduction of the entire actual work environment particulate challenge in particle number or mass that is important to protecting the worker.

The most desirable compromise must be worked out for each filter classification with respect to filter-surface area, resistance to breathing, efficiency in filtering particles of specific size ranges, and the time to clog the filter. Filters may be made of randomly laid nonwoven fiber materials, compressed natural wool or synthetic fiber felt, or fibrous glass that may be loosely packed in a filter container or made into a flat sheet of filter material that is pleated and placed in a filter container. Pleating is a way in which the filter surface area is increased, which can improve filter loading and lower breathing resistance.

NIOSH–MSHA certifies several types of filters. These filters may be either replaceable or an integral part of the respirator. At the end of service the filters are discarded or, in the case where they are a permanent part of the respirator, the entire respirator is disposed. The following are the most important and prevalent filter classifications on the market:

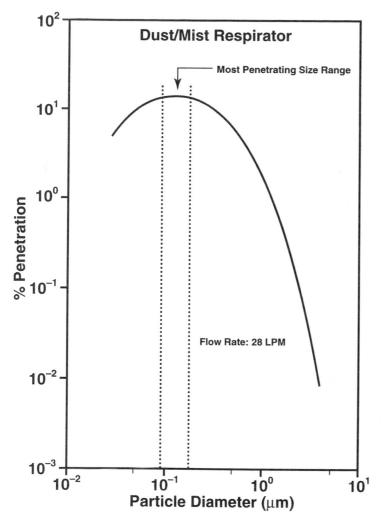

**Figure 22–12.** Aerosol penetration through a filter at 28 liters per minute for a dust/mist respirator. (Adapted from Fardi B, Liu BYH, Performance of disposable respirators. *Part Part Syst Charact* 8:308–314, 1991.)

- *Dust and mist.* These filter respirators (Figure 22–13) are designed for protection against dusts and mists having an occupational exposure limit not less than 0.05 mg/m³, or not less than 2 million particles per cubic foot of air (mppcf).

- *Dust, fume, and mist.* Respirators with filters designed for protection against dusts, fumes, and mists, of materials having an exposure limit not less than 0.05 mg/m³ or 2 mppcf.

- *Dust, fume, and mist or HEPA filters.* Respirators with filters designed for protection against dusts, fumes, and mists of materials having an occupational exposure limit *less* than 0.05 mg/m³ or 2 mppcf. These filters are often referred to as high-efficiency particulate air (HEPA) filters or simply high-efficiency filters. They are the most efficient respirator filters and are used on high-efficiency respirators. The respirators are available as disposable or with replaceable filters (Figure 22–14).

*High efficiency* refers to the filter test requirement they must meet. HEPA filters must be at least 99.97 percent efficient when tested against 0.3 μm dioctyl phthalate (DOP)

particles. DOP is the test material for the filter. The labels for these filters may also state that they can be used for particulate radionuclides. Radionuclides are materials that spontaneously emit ionizing radiation.

The efficiency of the dust/mist and dust/fume/mist filters is less than the HEPA filter. To get NIOSH approval the dust/mist filter minimum efficiency is about 99 percent when tested against a silica dust particle with a geometric mean diameter of 0.4 to 0.6 μm and a standard deviation not greater than 2. The dust/fume/mist filter minimum efficiency is about 99 percent when tested against lead fume. These efficiencies are difficult to compare, as the three filters are tested against different aerosols. It does indicate that when the filters are properly selected (for example, a dust/mist filter for a dust application), they are very efficient.

For a proper comparison of filters, the test conditions should be identical. Work reported in the industrial hygiene literature using DOP aerosol in the most penetrating size range showed minimum filter efficiencies of greater than 99.999 percent for a HEPA filter, 98–99 percent for a dust/fume/mist filter, and 88–89 percent for a dust/mist filter. All three filters were from the same respirator manufacturer. Although the test conditions are slightly

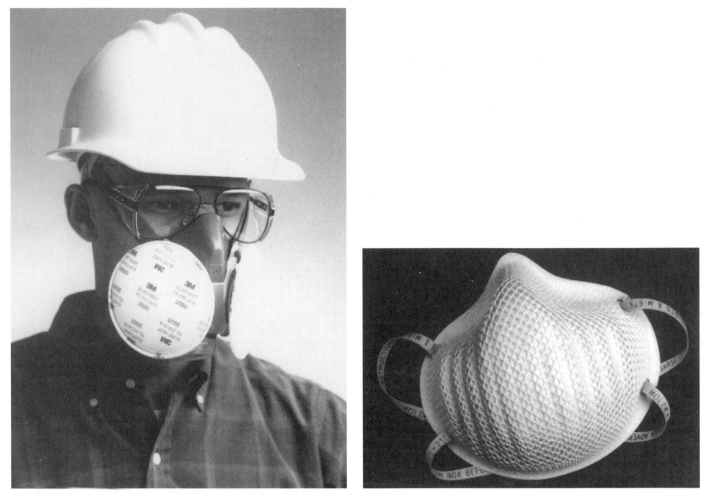

**Figure 22–13.** a. Dust and mist respirator with replaceable filters. (Courtesy 3M.) b. Disposable dust and mist respirator. (Courtesy Moldex Metric, Inc.)

different from the NIOSH–MSHA tests, these results do allow comparison of the classes of filters. Filters from other manufacturers may be better or worse than those reported here.

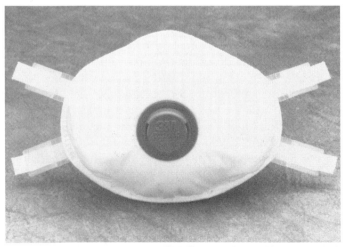

**Figure 22–14.** Disposable HEPA filter respirator. (Courtesy 3M.)

It is difficult to perceive the difference between aerosol filters visually. Therefore it is important to read the NIOSH–MSHA approval label to identify against which aerosols (dusts, mists, or fumes) the filter should be used. Then the occupational exposure limit of the material must be identified to determine whether a HEPA filter is required. At the time of this writing, the current classification scheme for particulate respirator filters certified by NIOSH–MSHA was in effect. However, on June 8, 1995, NIOSH published a final rule (42 *CFR* 84), which changes the particulate filter classes and the testing methods for filters. At this time, NIOSH has not certified any filters under this new rule. The filters in the classification scheme described in this chapter can be sold for 3 years after the effective date of this final rule. However, it is not known how long OSHA will allow these filters to be used beyond this 3-year period.

**Gas/vapor-removing respirators (Figure 22–15).** These air-purifying respirators protect against certain gases and vapors by using various chemical filters to purify the

**Figure 22–15.** Half-facepiece chemical cartridge respirator with interchangeable cartridges affords protection against light concentrations of organic vapors and certain gases. (Courtesy 3M.)

inhaled air. They differ from aerosol filters in that they use cartridges or canisters containing sorbents, generally carbon, to remove harmful gases and vapors. The cartridges may be replaceable or the entire respirator may be disposable. Sorbents are granular, porous materials that interact with the gas or vapor molecule to clean the air. In contrast to aerosol filters, which are effective to some degree no matter what the particle, cartridges and canisters are designed for protection against specific contaminants (such as ammonia gas or mercury vapor) or classes of contaminants (such as organic vapors). Activated carbon, which is commonly used for removal of organic vapors, is a carbon material that has its surface greatly enhanced using chemicals, heat, and steam. The most common starting materials for respirator cartridges are coconut and coal. Activated carbon has an extensive network of internal pores of near molecular dimensions and consequently large internal surface areas. The typical range of surface area is 1,000–2,000 m²/g of carbon. Generally, organic vapors of molecular weight (MW) greater than 50 or boiling points (BP) greater than 70 C are effectively adsorbed by activated charcoal. For gases and vapors that would otherwise be weakly adsorbed, sorbents can be impregnated with chemical reagents to make them more selective. Examples are activated charcoal impregnated with iodine to remove mercury vapor or with metal salts like nickel chloride to remove ammonia gas. These removal mechanisms are essentially 100 percent efficient until the sorbent's capacity is exhausted. At this point, "breakthrough" occurs as the contaminant passes through the cartridge or canister and into the respirator.

Cartridges and canisters should be changed when the worker detects chemical smell, taste, or irritation. Cartridges and canisters should not be used for gases and vapors with poor warning properties. Poor warning properties are an odor, taste, or irritation effect that is not detectable or persistent at concentrations at or below the occupational exposure limit.

When the chemical's taste, smell, or irritation is detected, the worker should exit to a clean area and replace the cartridges, canister, or respirator. Aerosol filters, on the other hand, become more efficient as particles are collected and plug the spaces between the filter fibers. Filters should be changed when they become difficult to breathe through.

Cartridges are similar to canisters. The basic difference is the volume of sorbent, not the function. Canisters have the larger sorbent volume (Figure 22–16). Service life of these respirators depends on the following factors: quality and amount of sorbent; packing uniformity and density; and exposure conditions, including breathing rate of the wearer, relative humidity, temperature, contaminant concentration, the affinity of the gas or vapor for the sorbent, and the presence of other gases and vapors. (Generally, high concentrations, a high breathing rate, and humid conditions adversely affect service life.) Because exposure conditions are subject to wide variation, it is difficult to estimate the service life of canisters and cartridges, even with other conditions being constant. Table 22–C shows various chemical cartridge breakthrough times for different organic gases and vapors.

Although the cartridge is approved for organic vapors by testing against carbon tetrachloride, the cartridge may last a longer (as with butanol) or a much shorter (as with methanol) time period than when compared with the test

**Table 22–C.** Organic Vapor Chemical Cartridge Breakthrough Times for Various Chemicals

| Chemical | Time to 1% (10 ppm) Breakthrough (min)* |
|---|---|
| Aromatics | |
| Benzene | 73 |
| Toluene | 94 |
| Xylene | 99 |
| Alcohols | |
| Methanol | 0.2 |
| Isopropanol | 54 |
| Butanol | 115 |
| 2-Methoxyethanol | 116 |
| Chlorinated hydrocarbons | |
| Methyl chloride | 0.05 |
| Vinyl chloride | 3.8 |
| Dichloromethane | 10 |
| Trichloroethylene | 55 |
| Carbon tetrachloride | 77 |
| Perchloroethylene | 107 |
| Ketones | |
| Acetone | 37 |
| 2-Butanone | 82 |

*Cartridges challenged with 1,000 ppm of the respective chemical. Tested at 50 percent relative humidity, 22 C, and 53.3 l/min.

(Adapted from Nelson GE, Harder CA. Respirator cartridge efficiency studies: V. Effect of solvent vapor. *AIHAJ* 35:391–410, 1974.)

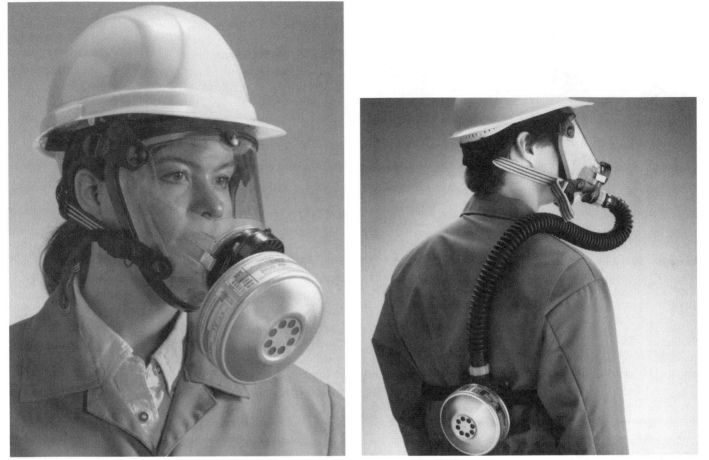

**Figure 22–16.** Gas masks provide longer service life than chemical cartridge respirators for many commonly encountered vapors and gases. a. Chin-style gas mask. b. Back-mounted gas mask. (Courtesy Scott Aviation Health and Safety Products.)

agent. Hence, an organic vapor cartridge may be recommended for use against butanol but not for methanol (MWT < 50; BP < 70 C) even though both compounds are classified as organic vapors.

Chemical cartridges and canisters are limited to use in concentrations that are no greater than the assigned protection factor of the respirator times the occupational exposure limit. This is called the maximum use concentration (MUC). At one time, maximum use limits were included on the cartridge or canister. These have been removed by NIOSH; thus the maximum use concentration limits are dependent upon the respirator's assigned protection factor. This topic will be discussed further later in this chapter. *Gas mask* is a term used often for a gas- or vapor-removing respirator that uses a canister. Although gas masks are limited by their assigned protection factor, they can be used for escape only from atmospheres immediately dangerous to life or health (IDLH) that contain adequate oxygen to support life (≥19.5 percent $O_2$). They must *never* be used for entry into an IDLH atmosphere.

*Cartridge and canister replacement.* Cartridges and canisters should be replaced under any one or more of the following conditions:

- If the end of service life indicator shows the specified color change
- If breakthrough is detected by smell, taste, or eye, nose, or throat irritation
- If the shelf life is exceeded
- If an OSHA regulation specifies a disposal frequency (as with formaldehyde and benzene)

A person wearing a cartridge or canister that needs replacement should return to fresh air as quickly as possible. In addition, if uncomfortable heat in the inhaled air is detected or the wearer has a feeling of nausea, dizziness, or ill-being, it is imperative that he or she return to fresh air. (A properly operating cartridge or canister may become warm on exposure to certain gases or vapors, but a device that becomes extremely hot indicates that concentrations greater than the device's limits have been reached.)

**Combination aerosol filter/gas or vapor-removing respirators (Figure 22–17).** These respirators use aerosol-removing filters with a chemical cartridge or canister for exposure to multiple contaminants or more than one physical form (for example, mist and vapor). The filter is generally a permanent part of the canister but can be

**Figure 22–17.** Combination chemical cartridge/aerosol filter respirators are designed to reduce exposure to multiple contaminants such as vapor and dust. (Courtesy 3M.)

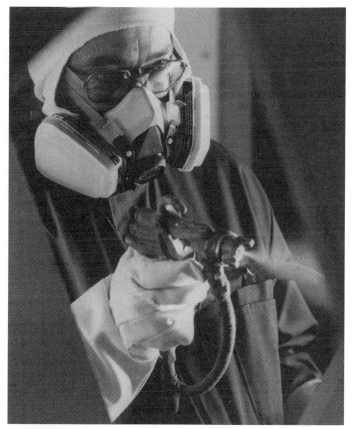

**Figure 22–19.** NIOSH–MSHA has approved specific air-purifying respirators for spray painting. (Courtesy 3M.)

either permanent or replaceable on the chemical cartridge. Replaceable filters are sometimes used because the filter and chemical cartridge are not exhausted at the same time (Figure 22–18). This allows for disposing only of the part that is in need of changing. Filters used in combination with cartridges must always be located on the inlet side of

**Figure 22–18.** Workers wearing combination chemical cartridge/dust and mist respirators for protection from solvent vapors and mists and glass fibers in a fiberglass operation. (Courtesy 3M.)

the cartridge. This way, any gas or vapor adsorbed onto a filtered particle is captured by the sorbent as it desorbs from the particle. The most often used combination aerosol filter/gas or vapor-removing respirators are the paint spray (Figure 22–19) and pesticide respirators (Figure 22–20). The paint spray and pesticide filter is a separate class of filters certified by NIOSH–MSHA. To be certified for paint spray and pesticides, these filters must pass specific tests. In other words, not all dust/mist and dust/fume/mist filters may be approved for pesticide and paint spray aerosols. The approval labels must be checked.

**Powered air-purifying respirators.** The air-purifying element of these respirators may be a filter, chemical cartridge, or canister. They protect against particles, gases and vapors, or particles and gases and vapors. The difference between these and the air-purifying respirators previously discussed is that the powered air-purifying respirator (PAPR) uses a power source (usually a battery) to operate a blower that passes air across the air-cleansing element to supply purified air to a respiratory inlet (mouth and nose) covering. To be certified as a powered air-purifying respirator by NIOSH–MSHA, the blower must provide at least 4 cubic feet per minute (cfm) of air to a tight-fitting facepiece (half face or full facepiece) and at least 6 cfm to a loose-fitting helmet, hood, or facepiece. Figure 22–21

**Figure 22–20.** The pesticide respirator consists of both a chemical cartridge and filter. (Courtesy 3M.)

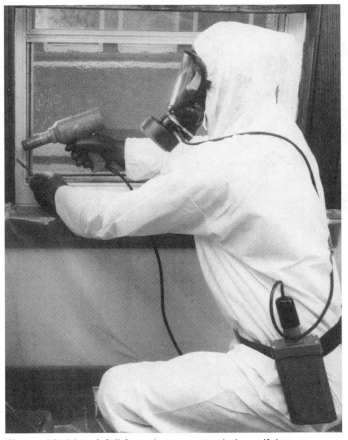

**Figure 22–21.** A full-facepiece powered air-purifying respirator with the motor and blower assembly mounted in the facepiece. (Courtesy 3M.)

shows a powered air-purifying respirator with a full face-piece, and Figure 22–22 shows one with a loose-fitting hood. The great advantage of the powered air-purifying respirator is that it usually supplies air at positive pressure, reducing inward leakage when compared with the negative-pressure respirators. This is why they have a higher assigned protection factor than their negative pressure counterpart. It is possible, however, at high work rates to create a negative pressure in the facepiece, thereby increasing facepiece leakage. This can be reduced by fit testing tight-fitting powered air-purifying respirators.

## Atmosphere-Supplying Respirators

Atmosphere-supplying devices are the class of respirators that provide a respirable atmosphere to the wearer independent of the ambient air. The breathing atmosphere is supplied from an uncontaminated source. The air source for an atmosphere-supplying respirator must conform to grade D requirements as specified in the Compressed Gas Association Standard, G-7.1-1989, *Commodity Specification for Air.* Table 22–A lists the air quality requirements for grade D breathing air. Atmosphere-supplying respirators fall into three groups: air-line respirators, self-contained breath-

ing apparatuses (SCBA), and combination air-line and SCBA.

**Air-line respirators (Figure 22–23).** Air-line respirators deliver breathing air through a supply hose connected to the wearer's facepiece or head enclosure. The breathing air is supplied through the hose from either a compressor or compressed air cylinders. If air is supplied by a compressor, it must be equipped with specific safety devices according to OSHA. All compressors must have an alarm to indicate overheating and compressor failure. If the compressor is oil lubricated, the air must be sampled frequently for carbon monoxide, or a carbon monoxide alarm must be installed, or both. A flow control valve, regulator, or orifice is provided to govern the rate of airflow to the worker. Depending on the certification, up to 300 feet of air supply hose is allowable. Hose supplied by the respirator manufacturer along with recommended hose lengths and operating pressures must be used. The maximum permissible inlet pressure is 125 pounds per square inch (psig). The approved pressure range and hose length are noted on the certification label or operating instructions provided with each approved device (Figure 22–3).

These devices should be used only in non-IDLH atmospheres or atmospheres in which the wearer can escape

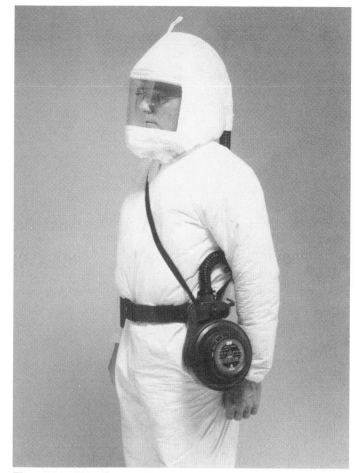

**Figure 22–22.** Powered air-purifying respirators with hoods or helmets can be worn by workers with beards or eyeglasses. (Courtesy 3M.)

**Figure 22–23.** Continuous-flow air-line respirators are used in conjunction with a compressor system. The manifold for connecting to the air source can be seen in the background. (Courtesy 3M.)

without the use of a respirator. This limitation is necessary because the air-line respirator is entirely dependent on an air supply that is not carried by the wearer of the respirator. If this air supply fails, the wearer may have to remove the respirator to escape from the area. Another limitation is that the air hose limits the wearer to a fixed distance from the air supply source.

Air-line respirators operate in three modes: demand, pressure demand, and continuous flow. The respirators are equipped with half facepieces, full facepieces, helmets, hoods, or loose-fitting facepieces. Some of these respiratory inlet coverings may provide eye protection.

**Demand.** Demand air-line respirators are equipped with either half or full facepieces. They deliver airflow only upon inhalation. Due to their design, a negative pressure with respect to the outside of the respirator is created in the facepiece upon inhalation. These respirators are negative-pressure devices. Although these respirators can still be found in worksites, they are not recommended if one is buying new respirators because the pressure-demand type is available. The pressure-demand air-line respirator is much more protective and the cost differential between the two is negligible.

**Pressure demand.** Pressure-demand respirators are very similar to the demand type except that because of their design, the pressure inside the respirator is generally positive with respect to the air pressure outside the respirator during both inhalation and exhalation. This positive pressure means that when a leak develops in the face seal due to head movement, for example, the leakage of air would be outward. Thus they provide a higher degree of protection to the user. They also are available only with half and full facepieces (Figure 22–24). Such respirators are normally used when the air supply is restricted to high-pressure compressed air cylinders. A suitable pressure regulator is required to ensure that the air pressure is reduced to the proper level for breathing.

**Continuous flow.** A continuous-flow unit has a regulated amount of air delivered to the facepiece or head enclosure and is normally used where there is an ample air supply such as that provided by an air compressor. These devices may be equipped with either tight-fitting or loose-fitting head enclosures. Those equipped with tight-fitting enclosures, a half or full facepiece, must provide at least 4

**Figure 22–24.** Pressure-demand air-line respirators are used with compressed air supplied by a compressor or a cascade of compressed air cylinders. (Courtesy 3M.)

cfm measured at the facepiece. When loose-fitting helmets, hoods, or facepieces are used, the minimum amount of air to be delivered is 6 cfm. In either case the maximum flow is not to exceed 15 cfm. Versions of these respirators may be designed for welding (Figure 22–25) or abrasive blasting (Figure 22–26). Respiratory-protective equipment designed

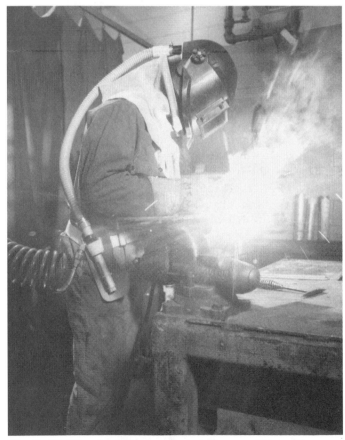

**Figure 22–25.** A continuous-flow air-line respirator with welding helmet. (Courtesy 3M.)

**Figure 22–26.** Abrasive blasting continuous-flow air-line respirator. (Courtesy 3M.)

for abrasive blasting is equipped to protect the wearer from impact of the rebounding abrasive material. A special hood may be used to protect the wearer's head and neck, and shielding material may be used to protect the viewing windows of the head enclosures.

**Self-contained breathing apparatus.** The self-contained breathing apparatus (SCBA) provides respiratory protection against gases, vapors, particles, and an oxygen-deficient atmosphere. The wearer is independent of the surrounding atmosphere because the breathing gas is carried by the wearer. SCBA may be used in IDLH and oxygen-deficient atmospheres either as escape-only devices or for entry into and escape from these atmospheres. A full facepiece is most commonly used with these devices. Half facepieces, hoods, and mouthpieces are available on some units. There are two major types of SCBAs: closed circuit and open circuit.

*Closed-circuit SCBA.* In closed-circuit SCBA (Figure 22–27), all or a percentage of the exhaled gas is scrubbed and rebreathed. Closed-circuit units have the advantage of lower weight for the same use duration as open-circuit devices. Units are available from 15 min to 4 hr. Disadvantages include increased complexity (for example, a carbon

**Figure 22–27.** Closed-circuit self-contained breathing apparatus designed for 60 min of service. (Courtesy Biomarine.)

**Figure 22–28.** Pressure-demand self-contained breathing apparatuses are designed for use in oxygen-deficient atmospheres that are immediately dangerous to life or health as well as for fire fighting. (Courtesy Scott Aviation Health and Safety Products.)

dioxide scrubber is required in many of the units) and cost. Due to the design of many of the devices, the air supply can become quite warm because of rebreathing of the air. Closed-circuit SCBAs are available as both negative- and positive-pressure devices. They may be designed as a stored oxygen system or an oxygen-generating system.

Stored oxygen systems supply oxygen compressed in cylinders or carried as a liquid. Oxygen is admitted to a breathing bag either as a continuous flow or controlled by a regulator governed by the pressure or degree of inflation of the bag. The wearer inhales from the bag and exhales into it. Exhaled breath is scrubbed of carbon dioxide by a chemical bed, usually a caustic such as sodium hydroxide.

Oxygen-generating systems rely on chemical reactions to provide the needed oxygen. Water vapor from the exhaled breath reacts with a solid chemical, usually potassium superoxide, in a canister-size container that releases oxygen. Carbon dioxide is removed from the exhaled breath in the canister also.

***Open-circuit SCBA.*** In an open-circuit SCBA the exhaled breath is released to the surrounding environment after use rather than being recirculated. The breathing gas is generally compressed air. Typically they are designed to provide 30–60 min of service (Figure 22–28). They are available in both demand (negative-pressure) and pressure-demand (positive-pressure) styles. Because of the greater protection provided by pressure-demand devices, they are recommended over negative-pressure systems.

***Escape SCBA.*** Some SCBAs are designed for escape only (Figure 22–29). They are similar in design to the types described above, except that the use duration tends to be shorter, typically 5, 7, or 10 min. Units approved for escape only must not be used to enter into a hazardous atmosphere. The fact that they are certified for escape only means that assigned protection factors have not been established for this category of respirator.

**The combination self-contained breathing apparatus (SCBA) and air-line respirator.** These units are air-line respirators with an auxiliary self-contained air supply combined into a single device (Figure 22–30). (An auxiliary SCBA is an air supply, independent of the one to the air-line respirator, that allows a person to evacuate a contaminated area.) Because they have escape provisions, these devices are usable in IDLH and oxygen-deficient atmospheres. The auxiliary air supply can be switched to in the event the primary air supply fails to operate. This allows the wearer to escape from the IDLH atmosphere.

**Figure 22–29.** Some self-contained breathing apparatuses are designed only for escape from hazardous atmospheres. (Courtesy Scott Aviation Health and Safety Products.)

**Figure 22–30.** Combination self-contained breathing apparatus and air-line respirators can be connected to an external air supply for working in a hazardous atmosphere and still allowing for escape without removing the respirator. (Courtesy Scott Aviation Health and Safety Products.)

An advantage of these devices is they can be used in situations requiring extended work periods where the self-contained air supply alone does not provide sufficient time. In this situation, the wearer may connect to an air line to provide longer service time. The longer service life and smaller SCBA cylinder make these devices particularly convenient for use in confined spaces.

The auxiliary self-contained air supply may be NIOSH–MSHA approved in one of two categories: 3-, 5-, or 10-min service time or for 15 min or longer. If the SCBA portion is rated for a service life of 3, 5, or 10 min, the wearer must use the air line during entry into a hazardous atmosphere; the SCBA portion is used for emergency egress only. When the SCBA is rated for service of 15 min or longer, the SCBA may be used for emergency entry into a hazardous atmosphere (to connect the air line) provided not more than 20 percent of the air supply's rated capacity is used during entry. This allows for enough air for egress when the warning device indicates a low air supply.

Combination SCBA/air-line respirators may operate in demand, pressure-demand or continuous-flow modes. These devices use the same principles as the respective air-line respirator. Demand mode is not recommended.

## Combination Air-Purifying and Atmosphere-Supplying Devices

Another type of respirator is gaining in popularity. It is a combination of an air-line respirator and an auxiliary air-purifying attachment, which provides protection in the event the air supply fails (Figure 22–31). NIOSH–MSHA has approved combination air-line and air-purifying respirators with the air line operating in either continuous flow or pressure-demand flow. These respirators can be used in either an air-purifying or atmosphere-supplying mode. The most popular versions are ones in which the air-purifying element is a HEPA filter, but devices are available with complete arrays of chemical cartridges as well.

These respirators have additional limitations:

- They are not for use in IDLH atmospheres.
- They are not for use in atmospheres containing less than 19.5 percent oxygen.

**Figure 22–31.** Combination air-purifying and air-line respirators provide protection in the event the air supply fails. (Courtesy 3M.)

- Use only the hose lengths and pressure ranges specified on the approval label.
- Use only in atmospheres for which the air-purifying element is approved.

The approval label must be consulted for proper use of the respirator in the air-purifying mode. The restrictions can vary from manufacturer to manufacturer depending on the respirator design.

# Respirator Selection

Proper selection of respirators must start with an assessment of the inhalation hazards present in the workplace. This assessment must include the following:

- The nature of the hazardous operation or process
- The type of respiratory hazard

- The location of the hazardous area in relation to the nearest respirable air source
- The time period that respirators must be worn
- The workers' activities

Respirators must then be selected for the situation after the physical characteristics and functional capabilities and limitations of the various types of respirators and their assigned protection factors (APF) have been considered.

## Selection Requirements

The OSHA Standard 29 *CFR* 1910.134 states that respirators shall be selected on the basis of the hazards to which workers are exposed and that ANSI Z88.2-1969 (revised in 1992) shall be used for guidance in their selection. For certain respiratory hazards, specific instructions about respirator use are given in other OSHA regulations (for example, Asbestos, 1910.1001 and 1926.1101; Vinyl Chloride, 1910.1017; and substances regulated after promulgation of vinyl chloride). The trend is toward regulations that specify the conditions of respirator use for each substance.

The new OSHA standards list respiratory protection equipment for various concentrations of a substance in the Respirator Selection Table (Table 22–D).

To provide additional protection, an employer may select a respirator prescribed for concentrations higher than those found in the workplace. However, the employer may not use respirators that are not listed.

These standards also call for a respiratory protection program as spelled out in 29 *CFR* 1910.134 of the OSHA regulations. Of course, the respirators must be approved by NIOSH–MSHA.

For the large number of chemicals for which OSHA does not have a substance-specific standard, the ANSI Z88.2 standard should be used for guidance. The 1969 edition is out of print, so even though the OSHA standard refers to it, the most recent edition (1992) should be consulted.

Approved or authorized respirators shall be selected. If there is no approved respirator commercially available

**Table 22–D.** Example of OSHA Respirator Selection Table: Respiratory Protection from Asbestos Fibers

| Airborne Concentration of Asbestos or Conditions of Use | Required Respirator |
|---|---|
| Not in excess of 1 f/cc (10 × PEL), or otherwise as required independent of exposure pursuant to (h)(2)(iv). | Half-mask air-purifying respirator other than a disposable respirator, equipped with high efficiency filters. |
| Not in excess of 5 f/cc (50 × PEL) | Full facepiece air-purifying respirator equipped with high efficiency filters. |
| Not in excess of 10 f/cc (100 × PEL) | Any powered air-purifying respirator equipped with high efficiency filters or any supplied air respirator operated in continuous flow mode. |
| Not in excess of 100 f/cc (1000 × PEL) | Full facepiece supplied air respirator operated in pressure demand mode. |
| Greater than 100 f/cc (1000 × PEL) or unknown concentration | Full facepiece supplied air respirator operated in pressure demand mode, equipped with an auxiliary positive pressure self-contained breathing apparatus. |

Note: Respirators assigned for high environmental concentrations may be used at lower concentrations, or when required respirator use is independent of concentration. A high-efficiency filter is one that is at least 99.97 percent efficient against mono-dispersed particles of 0.3 μm in diameter or larger.
(Reprinted from OSHA General Industry Standards, 29 *CFR* 1910.1001.)

that can do the required task, the user may seek authorization from the appropriate regulatory agency to use an unapproved device. Respirator selection involves determining the hazard and following a selection logic to choose the correct type or class of respirator that offers adequate protection.

## Hazard Determination

The steps for hazard determination are as follows:

1. Determine what contaminants may be present in the workplace.
2. Determine whether there are Threshold Limit Values (TLVs), permissible exposure limits (PELs), or any other available exposure limits.
3. Determine the IDLH level.
4. Measure the oxygen content if the potential for an oxygen-deficient atmosphere exists.
5. Measure or estimate the concentration of the contaminants.
6. Determine the physical state of the contaminant. If the contaminant is an aerosol, determine or estimate the particle size. Determine whether the vapor pressure of the aerosol is significant at the maximum expected temperature of the work environment. In these situations it may be possible to have a significant portion of the contaminant concentration in the vapor phase, requiring respiratory protection for both the particle and vapor phase of the contaminant.
7. Determine whether the contaminants can be absorbed through the skin, cause skin sensitization, or be irritating to or corrosive to the eyes or skin. Respirators that provide skin or eye protection or air-supplied suits may be required in addition to protection from the inhalation hazard.
8. For gases or vapors, determine whether the contaminants have adequate warning properties.

**Skin absorption.** Chemical absorption through skin can be a significant route of exposure. Depending on the chemical, this route of exposure may be more significant than absorption through the respiratory system. For example, assuming 100 percent skin absorption, two *drops* of aniline on the skin would be equivalent to an inhalation exposure at the threshold limit value (TLV) for eight hours. To avoid this possibility, selection of protective clothing may be required. Respirators may provide limited skin protection by a full facepiece or hood that protects the face area from absorption of gaseous contaminants and splashes. For total skin protection, either chemical clothing that encapsulates the respirator and worker or supplied-air suits must be selected (Figure 22–32). Encapsulation suits are available from several suppliers.

Supplied-air suits are usually custom-made for the intended purpose of the user. Generally, they consist of a

**Figure 22–32.** Totally encapsulating suits can be selected to provide skin protection and accommodate the respirator. (Courtesy ILC Dover.)

full body suit and an air line to supply air to the suit. For more information on supplied-air suits, consult *Respiratory Protection: A Manual and Guideline.*

**Warning properties.** Warning properties such as odor, eye irritation, and respiratory irritation that rely upon human senses are not foolproof. However, they do provide some indication to the wearer that the service life of the cartridge or canister is reaching the end. Warning properties may be assumed to be adequate when odor, taste, or irritation effects of the substance can be detected and are persistent at concentrations at or below the PEL or TLV.

If the odor or irritation threshold of the substance is many times greater than the PEL, the substance is considered to have poor warning properties, and atmosphere-supplying respirators would be specified (Table 22–E).

This same thinking is reflected in NIOSH–MSHA approvals for organic vapor, chemical cartridge respirators and gas masks. The approvals prohibit their use against organic vapors with poor warning properties.

There may be some situations where an atmosphere-supplying respirator may be impractical because of lack of feasible air supply or need for worker mobility. In these situations, air-purifying devices should be used only if one of the following conditions is met:

■ The air-purifying respirator has a reliable end of service life indicator that will warn the worker prior to contaminant breakthrough.

**Table 22–E.** Comparison of Odor Thresholds and 1993–94 TLVs for Selected Compounds

*Group 1 - Odor Threshold ≤ TLV (Adequate Warning Properties)*

| Chemical | Mean Odor Threshold (ppm)* | TLV-TWA (ppm)** |
|---|---|---|
| Acetone | 62 | 750 |
| Ammonia | 17 | 25 |
| Ethyl acetate | 18 | 400 |
| Methanol | 160 | 200 |
| Methyl ethyl ketone | 0.14 | 50 |
| Toluene | 1.6 | 50 |
| Xylene | 20 | 100 |

*Group 2 - Odor Threshold > TLV (Poor Warning Properties)*

| Benzene | 61 | 10 |
|---|---|---|
| Carbon tetrachloride | 250 | 5 |
| Methylene chloride | 160 | 50 |
| Methyl formate | 2,000 | 100 |
| 1-Nitropropane | 140 | 25 |
| Perchloroethylene | 47 | 25 |
| Trichloroethylene | 82 | 50 |

*Data are from *Odor Thresholds for Chemicals with Established Occupational Health Standards.* American Industrial Hygiene Association, 1989.

**Data are from *1993–1994 Threshold Limit Values for Chemical Substances and Physical Agents and Biological Exposure Indices.* Cincinnati, OH: ACGIH, 1993.

■ A cartridge change-out schedule is established based on cartridge service life data, desorption studies, expected concentrations, patterns of use, and duration of exposure. If the cartridges are changed out daily or more frequently, desorption studies may not be necessary.

Given the variability among people with respect to detection of odors and differences in measuring odor thresholds, a better practice is to establish cartridge change-out schedules even for chemicals with adequate warning properties. The warning properties in these cases should be used as a secondary indicator for cartridge change-out.

## Selection Steps

After information is collected in the hazard determination step, proper selection should be made as follows:

1. If the potential contaminants present were unable to be identified, consider the atmosphere IDLH (see below).

2. If no exposure limit or guideline is available and estimates of toxicity cannot be made, consider the atmosphere IDLH (see below).

3. If there is an oxygen-deficient atmosphere (<19.5 percent $O_2$), consider the atmosphere IDLH (see below).

4. If the measured or estimated concentration of the contaminants exceeds the IDLH levels, the atmosphere is IDLH (see below).

5. Divide the measured or estimated concentration by the exposure limit or guideline to obtain a hazard ratio. If more than one chemical is present, potential additive and synergistic effects of exposure should be consid-

ered. If the TLV for mixtures is used, a result greater than unity (that is, 1) is the hazard ratio. A respirator with an assigned protection factor greater than the value of the hazard ratio must be selected (Table 22–F). If an air-purifying respirator is selected, go on to step 6.

6. If the contaminant is a gas or vapor only, a respirator with a cartridge effective against the contaminant and with an assigned protection factor greater than the hazard ratio must be selected. If an aerosol (particle) contaminant is present, proceed to the next step.

7. If the contaminant is a paint, lacquer, or enamel, a respirator approved specifically for paint mists or an atmosphere-supplying respirator must be selected. It is important to note that limitations listed on the NIOSH–MSHA approval label or regulatory provisions may not allow the paint-spray respirator to be used.

8. If the contaminant is a pesticide, a respirator with a filter or filters specifically approved for pesticides or an atmosphere-supplying respirator must be selected. The NIOSH–MSHA approval label may preclude the pesticide respirator from being used for some pesticides (for example, fumigants).

9. If the contaminant is an aerosol with an unknown particle size or one with a mass median aerodynamic diameter (MMAD) less than 2 μm, a HEPA filter should be selected. OSHA requires a HEPA filter to be selected any time the PEL is less than 0.05 mg/m³; however, it makes more sense to select a HEPA filter whenever the probability of a large portion of the aerosol would exist as particles in the filters most penetrating particle size range regardless of the PEL. These guidelines use an aerosol distribution with a MMAD less than 2 μm as an indicator of this potential.

10. If the contaminant is an aerosol with a particle size distribution having a MMAD greater than 2 μm, any filter type (dust/mist, dust/mist/fume, or HEPA filter) may be used. Most aerosols produced as a result of industrial operations typically have a MMAD greater than 2 μm (Table 22–G).

11. If the contaminant is a fume, a filter respirator approved for fumes or a HEPA filter respirator should be selected without determining the MMAD.

**Immediately dangerous to life or health (IDLH).** Numerous definitions have been presented for IDLH atmospheres. ANSI defines any atmosphere that poses an immediate hazard to life or poses immediate, irreversible debilitating effects on health as being IDLH. The common theme in all the definitions is that IDLH atmospheres will affect the worker acutely as opposed to chronically. Thus, if the concentration is above the IDLH levels only highly reliable respiratory protective equipment is allowed. The only two devices that meet this requirement and provide escape provisions for the wearer are

- Pressure-demand or other positive-pressure self-contained breathing apparatus (SCBA)
- Combination-type, pressure-demand air-line respirators with auxiliary self-contained air supply

The IDLH limits are conservative, so any approved respirator may be used up to this limit as long as the maximum use concentration for the device has not been exceeded (Figure 22–33). IDLH limits have not been established by OSHA. NIOSH has recommended IDLH values for many chemicals in the *NIOSH Pocket Guide to Chemical Hazards* for the purpose of respirator selection. Two factors have been considered when establishing IDLH concentrations:

- The worker must be able to escape without losing his or her life or suffering permanent health damage within 30 min. Thirty minutes is considered by NIOSH as the maximum permissible exposure time for escape.
- The worker must be able to escape without severe eye or respiratory irritation or other reactions that could inhibit escape.

A location is considered IDLH when an atmosphere is known or suspected to have chemical concentrations above the IDLH level or if a confined space contains less than the normal 20.9 percent oxygen, unless the reason for the reduced oxygen level is known. Otherwise, according to OSHA, oxygen levels of less than 19.5 percent are IDLH. When there is doubt about the oxygen content, the contaminants present, or their airborne levels, the situation should be treated as IDLH. If an error in respirator selection is made, it should be on the side of safety. Thus in emergency situations, such as a spill, where the chemical or its airborne concentration are unknown, one of the above two respirators must be selected.

**Lower explosive limit (LEL) and fire fighting.** Concentrations in excess of the lower explosive limit (LEL) are considered to be IDLH. Generally, entry into atmospheres exceeding the LEL is not recommended except for lifesaving rescues. For concentrations at or above the LEL, respirators must provide maximum protection. Such devices include pressure-demand self-contained breathing apparatuses and combination positive-pressure air-line respirators with egress cylinders.

The ANSI standard Z88.5, *Practices for Respiratory Protection for the Fire Service,* 1981, defines fire fighting as immediately dangerous to life, so for fire fighting the only device providing adequate protection is the pressure-demand self-contained breathing apparatus. In addition to being NIOSH–MSHA approved, the SCBA used for fire fighting should comply with the most current edition of the National Fire Protection Association (NFPA) standard, NFPA 1981 (last published in 1992).

**Table 22–F.** Assigned Protection Factors

| Type of Respirator | Respiratory Inlet Covering | |
| --- | --- | --- |
| | Half Mask* | Full Facepiece |
| Air purifying | 10 | 100 |
| Atmosphere supplying | | |
| SCBA (demand)** | 10 | 100 |
| Air-line (demand) | 10 | 100 |

| Type of Respirator | Respiratory Inlet Covering | | | |
| --- | --- | --- | --- | --- |
| | Half Mask | Full Face | Helmet/Hood | Loose-Fitting Facepiece |
| Powered air purifying | 50 | 1,000[†] | 1,000[†] | 25 |
| Atmosphere-supplying air line | | | | |
| pressure demand | 50 | 1,000 | — | — |
| continuous flow | 50 | 1,000 | 1,000 | 25 |
| Self-contained breathing apparatus | | | | |
| Pressure demand | — | —[††] | — | — |
| open/closed circuit | | | | |

Note: Assigned protection factors are not applicable for escape respirators. For combination respirators, such as air-line respirators equipped with an air-purifying filter, the mode of operation in use dictates the assigned protection factor to be applied.

*Includes 1/4 mask, disposable half masks, and half masks with elastomeric facepieces.

**Demand SCBA shall not be used for emergency situations such as fire fighting.

[†]Protection factors listed are for high-efficiency filters and sorbents (cartridges and canisters). With dust filters, an assigned protection factor of 100 is to be used due to the limitations of the filter.

[††]Although positive-pressure respirators are currently regarded as providing the highest level of respiratory protection, a limited number of recent simulated workplace studies concluded that not all users may achieve protection factors of 10,000. Based on these limited data, a definitive assigned protection factor could not be listed for positive-pressure SCBAs. For emergency planning purposes where hazardous concentrations can be estimated, an assigned protection factor of no higher than 10,000 should be used.

(This material is reprinted with permission from American National Standard for Respiratory Protection, Z88.2, copyright 1992 by the American National Standards Institute. Copies of this standard may be purchased from the American National Standards Institute at 11 West 42nd Street, New York, NY 10036.)

**Table 22–G.** Aerosol Size Distributions for Various Industrial Operations

| Operation | MMAD, μm | GSD |
|---|---|---|
| **Mining** | | |
| Open pit, general environment | 2.5 | 4.7 |
| Open pit, in cab | 1.1 | 2.4 |
| Coal mine, continuous miner | 4.6 | 2.5 |
| Coal mine, continuous miner | 15.0 | 2.9 |
| Coal mine, continuous miner | 17.0 | 3.1 |
| Coal mine, other operations | 11.5 | 2.8 |
| Oilshale mine | 2.8 | 3.5 |
| **Smelting and foundry** | | |
| Lead smelter, sintering | 11.0 | 2.4 |
| Lead smelter, furnace | 3.3 | 15.7 |
| Brass foundry, pouring | 2.1 | 10.3 |
| Brass foundry, grinding | 7.2 | 12.9 |
| Iron foundry, general environment | 2.8 | 5.1 |
| Iron foundry, general environment | 16.8 | 4.4 |
| Be-Cu foundry, furnace | 5.0 | 2.4 |
| Nuclear fuel fabrication | 2.1 | 1.6 |
| **Nonmineral dust** | | |
| Bakery | 12.1 | 4.2 |
| Cotton gin | 47.1 | 2.7 |
| Cotton mill | 7.6 | 4.0 |
| Swine confinement building | 9.6 | 4.0 |
| Woodworking, machining, sanding | | |
|   fine mode | 1.3 | 2.7 |
|   coarse mode | 33.1 | 2.6 |
| Wood model shop | 7.2 | 1.4 |
| **Metal fume** | | |
| SMA (stick) welding | 0.38 | 1.8 |
| MIG welding | 0.48 | 2.3 |
| Lead fume (O₂–nat. gas) | 0.37 | 2.1 |
| **Mist and spray** | | |
| Pressroom, ink mist | 27.4 | 4.3 |
| Spray painting, lacquer | 6.4 | 3.4 |
| Spray painting, enamel | 5.7 | 2.0 |
| Aerosol spray products | 6.4 | 1.8 |
| **Other** | | |
| Forging | 5.5 | 2.0 |
| Refinery, fluid catalytic cracker | 6.2 | 2.4 |
| Cigarette smoke (diluted) | 0.4 | 1.4 |
| Pistol range | 2.6 | 3.8 |
| Diesel exhaust (age = 5–600 s) | 0.12 | 1.4 |

(Adapted from Hinds WC, Bellin P. Effect of facial-seal leaks on protection provided by half-mask respirators. *Appl Ind Hyg* 3(5):158–164, 1988.)

**Assigned protection factors.** Assigned protection factors (APFs) are a very important part of the selection process. The assigned protection factor is the expected workplace level of respiratory protection that would be provided by a properly functioning respirator or a class of respirators to properly fitted and trained users. Simply stated, APFs are a measure of the overall effectiveness of a respirator used in conjunction with a good respirator program. The APFs recommended by ANSI Z88.2-1992 (Table 22–F) are based on tests measuring the performance of respirators in the workplace or simulated workplace environments. These studies are sometimes referred to as workplace protection factor studies. In these studies measurements are taken simultaneously outside and inside the respirator as the worker does his or her normal job (Figure 22–34). The results of these studies indicate that an APF of 10 for both disposable and elastomeric half-facepiece respirators is appropriate. An APF of 10 means that the respirator will reduce the concentration actually breathed in by 10 times compared with the actual airborne concentration. Sometimes the APFs that OSHA uses will be different, as the OSHA standards do not keep up with the latest information. These recommended APFs should be used only when the employer has established an acceptable respiratory protection program meeting the requirements of 29 *CFR* 1910.134 and satisfactory fit testing of tight-fitting respirators has been performed. When evaluating recommended APFs, one should consider the basis for the APFs and the date of the recommendation. Research in this area is ongoing, and current data should be used. Quantitative fit test results should not be used for establishing APFs because studies have shown that these do not correlate to protection factors measured in the workplace.

**Health care settings.** One of the more recent worksites seeing increased respirator usage is health care settings. Respiratory-protective devices are being used to reduce exposure to aerosolized drugs (such as pentamidine, ribavirin, and antineoplastics) and bioaerosols (such as droplet nuclei containing *Mycobacterium tuberculosis* [TB]). This area presents many challenges including unknown safe levels of exposure for these agents and respirator efficacy for bioaerosols. Acceptable airborne levels have not been established for many pharmaceutical drugs or potentially infectious aerosols. NIOSH–MSHA-approved or -certified respirators have not been tested against bioaerosols such as TB. The particle size distribution of these contaminants may also be unknown.

This lack of information makes the respirator selection process difficult. Use of a properly selected respirator may reduce the risk due to exposure to these materials, but it cannot guarantee protection. Respirators with high assigned protection factors should be expected to reduce risk to a lower level than respirators with lower assigned protection factors when used within a respirator program and worn properly and diligently by the worker. On the other hand, respirators with higher assigned protection factors are more complex, burdensome to the worker, and costly. The proper balance needs to be achieved. For aerosols with unknown particle size, a respirator with a HEPA filter is recommended. For aerosolized drugs, particle size information may be provided by the manufacturer of the equipment used to administer the drug. Disposable particulate respirators mentioned earlier have been used in health care settings because of their simplicity, cost, and efficiency, and also because of the ease of disposal if they become contaminated. Reuse of a respirator or its disposal must also be consistent with the operating procedures of the infection-control program of the health care facility.

**Example of Maximum Use Concentration (MUC) Determination**

| **Problem:** |
| --- |
| What is the MUC for a half facepiece respirator with dust/mist filters for copper dust? |

**Solution:**

TLV for copper dust:                          $1 \text{ mg/m}^3$

APF for half facepiece respirator             10

$$
\begin{aligned}
\text{MUC} \quad &= \text{TLV X APF} \\
&= 1 \text{ mg/m}^3 \text{ X } 10 \\
&= 10 \text{ mg/m}^3
\end{aligned}
$$

**Explanation:**

If air sampling indicates an ambient concentration greater than $10 \text{ mg/m}^3$, this respirator does not provide sufficient protection!  Note that for half and full facepiece respirators, the filter or chemical cartridge type does not change the APF for the respirator.

**Figure 22–33.**   Assigned protection factors (APF) are used in the selection process to determine the maximum use concentration (MUC) for the respirator. It is determined by multiplying the TLV by the APF (Table 22–F). The APFs should be used only when the employer has established a respiratory protection program meeting the requirements stated in this chapter and satisfactory fit testing has been performed.

## ■ Respirator Fit Testing

After close consideration of all the details pertaining to respirator selection, proper protection will not be provided if the respirator facepiece does not fit the wearer properly. One make and model of respirator should not be expected to fit the entire workforce. Because of the great variety in face sizes and shapes encountered in male and female workers, most respirator manufacturers make their models of respirators available in more than one size. In addition, the size and shape of each facepiece varies among the different

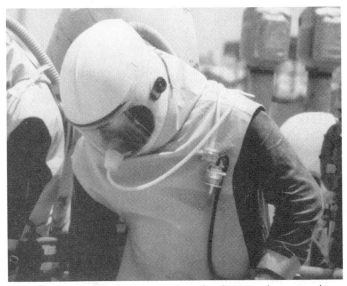

**Figure 22–34.**   Workplace studies simultaneously measuring air contaminants outside and inside the respirator are being used to establish assigned protection factors for the classes of respirators. (Courtesy 3M.)

manufacturers. In other words, the medium-size half facepiece of one manufacturer is not the same shape and size as the medium-size half facepiece from another manufacturer. For these reasons, it may be necessary to buy several commercially available respirators to conduct a good respirator fit testing program. The exact number of respirators to meet this requirement will vary. For a small number of respirator wearers (for example, four), one manufacturer's style and size may suffice. On the other hand, for a larger employer with hundreds of respirator wearers, several manufacturers' respirators in various sizes may be necessary.

All tight-fitting (half- and full-facepiece) respirators, whether negative or positive pressure, must be fit tested. This includes disposable respirators. This can be achieved by one of two fitting methods: qualitative or quantitative fit testing. In both cases, test agents or chemicals are used to detect leaks in the respirator facepiece-to-face seal. Fit testing should be conducted on all tight-fitting respirator wearers at least once every 12 months. Some OSHA substance-specific standards may require fit testing more often (for example, every 6 months). The fit test must be repeated when a worker has a new condition that may affect the fit, such as a significant change in weight (plus or minus 10 percent or more), significant scarring in the face seal area, dental changes, or reconstructive or cosmetic surgery.

The fit test conductor should be able to set up the test equipment, prepare any solutions, and maintain the test respirators. This individual should be familiar with the physical characteristics that may interfere with a face seal (such as beards) and should be able to recognize improper respirator donning and fit checking. The test conductor must also be able to perform the fit test and recognize a good test from an improper fit test. In addition, for quantitative fit testing, the fit tester should be able

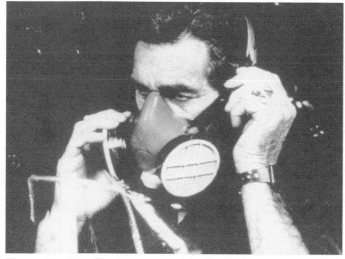

**Figure 22–35.** The isoamyl acetate qualitative fit test protocol requires respirators to be equipped with cartridges capable of removing organic vapors and uses a test enclosure. (Courtesy Survivair.)

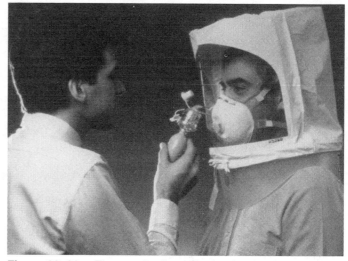

**Figure 22–36.** The saccharin qualitative fit test can be used with any particulate or gas/vapor respirator with a dust/mist filter to determine adequacy of fit. (Courtesy 3M.)

to perform preventive maintenance on the test equipment, check the system for leaks, and calibrate the equipment.

## Qualitative Fit Testing

A qualitative fit test relies on the wearer's subjective response. The test agent is a substance that typically can be detected by the wearer such as isoamyl acetate (banana oil), saccharin, or irritant smoke. The respirator must be equipped to remove the test agent (Figures 22–35 and 22–36). For example, if using isoamyl acetate, which is an organic chemical that gives off a vapor, an organic-vapor chemical cartridge must be used. With a respirator in good repair, if the wearer smells isoamyl acetate, the respirator does not fit well.

These tests are relatively fast and easily performed and use inexpensive equipment. Because these tests are based on the respirator wearer's subjective response to a test chemical, it is important that the purpose and importance of this test be thoroughly explained to the worker. Three qualitative tests are commonly used. Detailed protocols are available that should be followed when conducting fit testing. These tests have been shown to identify poor-fitting respirators by studies conducted in the laboratory. Two tests have had sufficient testing to be considered validated. An important point for validation is to be able to generate reliably low concentrations of the test agent to test the worker's ability to identify low levels of the test agent inside the respirator. The validated fit test methods have been designed to assess fit factors of 100 (≤1 percent face seal leakage). These test protocols that have been validated are the isoamyl acetate vapor test and the saccharin mist test. These tests have further been shown to be effective through workplace protection factor studies. Where these protocols have been used for fit testing, the results

show that the workers received adequate protection as indicated by the in-facepiece sampling results.

## Qualitative Fit Test Protocols

The qualitative fit test protocols consist of three steps: threshold screening, respirator selection, and fit testing. The threshold screening step is performed without wearing a respirator to determine whether the subject can detect low levels of the test agent. This level would be similar to the amount inside the respirator if the facepiece-to-face seal had a small leak. The test subject also learns what to expect if the respirator is leaking.

The purpose of the respirator selection step is to find one that provides the most comfortable fit. Every make and model of respirator has a different size and shape. If the respirator is correctly chosen, properly worn, and fit as indicated by a fit test, it should provide adequate protection. The respirators used in this step must be equipped with the filter or cartridge appropriate for the test agent. This is necessary to minimize the effects of filter or cartridge penetration so that only facepiece-to-face seal is evaluated. Respirators fit tested using isoamyl acetate must be equipped with organic vapor cartridges or canisters. Any respirator with a particulate filter can be fit tested with saccharin mist. The saccharin solution aerosol protocol is the only fit test protocol currently available that is validated and can be used with disposable particulate respirators not equipped with HEPA filters. Respirators must be equipped with HEPA filters to be tested with irritant smoke.

The fit test consists of the test subject's wearing the respirator while exposed to the test agent and performing facial movements (exercises) to test the facepiece-to-face seal. The following protocols have been taken from OSHA substance-specific standards. If the reader is conducting fit tests to comply with a specific standard, the standard should be consulted, as the protocol varies slightly from standard to standard.

**Isoamyl acetate protocol.** This protocol uses an organic vapor as the test agent, and therefore respirators equipped with organic vapor cartridges or canisters must be used. This minimizes the effects of cartridge penetration so that only the facepiece-to-face seal is evaluated. The protocol consists of three parts: odor threshold screening, respirator selection, and the fit test.

### Odor threshold screening.

1. Three 1-l glass jars with metal lids (such as Mason or Bell jars) are required.

2. Odor-free water (such as distilled or spring water) at approximately 25 C must be used for the solutions.

3. The isoamyl acetate (IAA) (also known as isopentyl acetate) stock solution is prepared by adding 1 cc of pure IAA to 800 cc of odor-free water in a 1-l jar and shaking for 30 s. This solution must be prepared fresh at least once a week.

4. The screening test must be conducted in a room separate from the room used for actual fit testing. The two rooms should be well ventilated, but may not be connected to the same recirculating ventilation system.

5. The odor test solution is prepared in a second jar by placing 0.4 cc of the stock solution into 500 cc of odor-free water using a clean dropper or pipette. Shake for 30 s and allow to stand for 2–3 min so that the IAA concentration above the liquid may reach equilibrium. This solution may be used for only one day.

6. A test blank is prepared in a third jar by adding 500 cc of odor-free water.

7. The odor test and test blank jars shall be labeled #1 and #2. If the labels are put on the lids, they can be periodically dried off and switched to prevent people from thinking that the same jar always has the IAA.

8. The following instructions should be typed on a card and placed on the table in front of the two test jars: "The purpose of this test is to determine if you can smell banana oil at a low concentration. The two bottles in front of you contain water. One of these bottles also contains a small amount of banana oil. Be sure the covers are on tight, then shake each bottle for two seconds. Unscrew the lid of each bottle, one at a time, and sniff at the mouth of the bottle. Indicate to the test conductor which bottle contains banana oil."

9. The mixtures used in the IAA odor detection test must be prepared in an area separate from where the test is performed in order to prevent olfactory fatigue in the subject.

10. If the test subject is unable to correctly identify the jar containing the odor test solution, the IAA qualitative fit test may not be used.

11. If the test subject correctly identifies the jar containing the odor test solution, the subject may proceed to respirator selection and fit testing.

### Respirator selection.

1. The test subject must be allowed to select the most comfortable respirator from a large array of various sizes and manufacturers.

2. The selection process must be conducted in a room separate from the fit test chamber to prevent olfactory fatigue. Prior to respirator selection, the test subject must be shown how to put on a respirator, how it should be positioned on the face, how to set strap tension, and how to assess if the respirator is comfortable. A mirror should be available to assist the subject in evaluating the fit and positioning of the respirator.

3. The test subject should understand that the purpose is to select the respirator that provides the most comfortable fit. Each respirator represents a different size and shape and, if fit properly, will provide adequate protection.

4. The test subject holds each facepiece up to his or her face and eliminates those that are obviously not giving a comfortable fit. Normally, selection will begin with a half facepiece, and if a satisfactory respirator cannot be found the subject will be asked to choose from full-facepiece respirators. (A small percentage of users will not be able to wear any half facepiece.)

5. The more comfortable facepieces are recorded; the most comfortable one is put on and worn at least 5 min to assess comfort. Assistance in assessing comfort can be given by discussing the points in step 6 below. If the test subject is not familiar with using a particular respirator, he or she should be directed to put on the facepiece several times and to adjust the straps each time to become adept at setting proper tension on the straps.

6. Assessment of comfort shall include reviewing the following points with the test subject:

    Chin properly placed

    Positioning of mask on nose

    Strap tension

    Fit across nose bridge

    Room for safety glasses

    Distance from nose to chin

    Room to talk

    Tendency to slip

    Cheeks filled out

    Self-observation in mirror

    Adequate time for assessment

7. The test subject must conduct the fit checks according to the manufacturer's instructions. Before conducting

the fit checks, the subject should be told to "seat" the mask by rapidly moving the head side to side and up and down, taking a few deep breaths.

8. The test subject is now ready for fit testing.

9. After successful completion of the fit test, the test subject should be questioned again regarding the comfort of the respirator. If it has become uncomfortable, another model of respirator should be tried.

10. The employee shall be given the opportunity to select a different facepiece and be retested if during the first two weeks of on-the-job wear the chosen facepiece becomes unacceptably uncomfortable.

### Fit test.

1. The fit test chamber is required to be constructed from a clear 55-gallon drum liner suspended inverted over a 2-ft diameter frame, so that the top of the chamber is about 6 in. above the test subject's head (Figure 22–35). The inside top center of the chamber needs to have a small hook attached.

2. Each respirator used for fit testing shall be equipped with organic vapor cartridges or offer protection against organic vapors. The cartridges or respirators (where the cartridges cannot be removed) must be changed at least weekly to prevent cartridge breakthrough from affecting the results.

3. After selecting, putting on, and properly adjusting a respirator without assistance, the test subject shall wear it to the fit testing room. This room must be separate from the room used for odor threshold screening and respirator selection, and be well ventilated to prevent general room contamination.

4. A copy of the following test exercises and rainbow (or equally effective) passage shall be taped to the inside of the test chamber.

### Test Exercises

   i. Normal breathing.
   ii. Deep breathing. Be certain breaths are deep and regular.
   iii. Turning head from side to side. Be certain movement is complete. Alert the test subject not to bump the respirator on the shoulders. Have the test subject inhale with the head at either side.
   iv. Nodding head up and down. Be certain motions are complete and made about every second. Alert the test subject not to bump the respirator on the chest. Have the test subject inhale with the head in the fully up position.
   v. Talking. Talk aloud and slowly. The following paragraph is called the Rainbow Passage. Reading it will result in a wide range of facial movements, and thus be useful to satisfy this requirement. Alternative passages that serve the same purpose may also be used.

**Rainbow Passage.** When the sunlight strikes raindrops in the air, they act like a prism and form a rainbow. The rainbow is a division of white light into many beautiful colors. These take the shape of a long round arch, with its path high above, and its two ends apparently beyond the horizon. There is, according to legend, a boiling pot of gold at one end. People look, but no one ever finds it. When a man looks for something beyond reach, his friends say he is looking for the pot of gold at the end of the rainbow.

   vi. Normal breathing.

5. Each test subject should wear the respirator for at least 10 min before starting the fit test.

6. Upon entering the test chamber, the test subject shall be given a 6-in. by 5-in. piece of paper towel or other porous absorbent single ply material, folded in half and wetted with three-quarters of one (0.75) cc of pure IAA. The test subject shall hang the wet towel on the hook at the top of the chamber.

7. Allow two minutes for the IAA test concentration to be reached before starting the fit test exercises. This would be an appropriate time to talk with the test subject, to explain the fit test, the importance of the test subject's cooperation, and the purpose for the head exercises, or to demonstrate some of the exercises.

8. Each exercise described above shall be performed for at least one minute.

9. Detection of the banana-like odor of IAA by the subject any time during the test is an indication to quickly exit from the test chamber and leave the test area to avoid olfactory fatigue.

10. Upon returning to the respirator selection area, the subject shall remove the respirator, repeat the odor sensitivity test, select and put on another respirator, return to the test chamber, and repeat the test. The process continues until a respirator that fits well, as indicated by not smelling the banana-like odor, has been found. Should the odor sensitivity test be failed, the subject must wait about 5 min before retesting. Odor sensitivity will usually have returned by this time.

11. If a person cannot be fitted with the selection of half-facepiece respirators, include full facepiece models in the selection process. When a respirator is found that provides an acceptable fit, its efficiency shall be demonstrated for the subject by having the subject break the face seal and take a breath before exiting the chamber.

12. When the test subject leaves the chamber, the saturated towel must be removed and returned to the test conductor. To keep the area from becoming contaminated, the used towels shall be kept in a self-sealing bag. There is no significant IAA concentration buildup in the test chamber from subsequent tests.

13. Persons successfully completing this IAA protocol are expected to achieve a protection factor of at least 10 from the tested respirator. For full-facepiece respirators where protection factors greater than 10 are required, it is recommended that quantitative fit testing be done to select respirators providing fit factors of 1,000 or greater.

**Saccharin solution aerosol protocol.** This protocol uses a test agent in the form of a fine mist. Therefore, the respirator must be equipped with particulate filters so that only face seal leakage is evaluated. The protocol consists of three parts; taste threshold screening, respirator selection (discussed above), and the fit test.

### Taste threshold screening.

1. The threshold screening as well as fit testing of employees uses an enclosure over the head and shoulders that is approximately 12 in. in diameter by 14 in. tall with at least the front portion clear. It must allow free movement of the head when a respirator is worn (Figure 22–36). An enclosure substantially similar to the 3M hood assembly of part # FT 14 and # FT 15 combined is adequate.

2. The test enclosure needs a three-quarter-inch hole in front of the test subject's nose and mouth area to accommodate the nebulizer nozzle.

3. The entire screening and testing procedure must be explained to the test subject prior to conducting the screening test.

4. For the threshold screening test, the test subject dons the test enclosure and breathes through his or her open mouth with tongue extended.

5. Using a DeVilbiss Model 40 Inhalation Medication Nebulizer or equivalent, the test conductor sprays the threshold sensitivity solution into the enclosure. This nebulizer must be clearly marked to distinguish it from the fit test solution nebulizer.

6. The threshold sensitivity solution consists of 0.83 g of sodium saccharin, USP, in water. It can be prepared by putting 1 cc of the test solution (from step 6 of the fit test) in 100 cc of water.

7. To produce the aerosol, the nebulizer bulb is firmly squeezed so that it collapses completely and then released and allowed to expand fully.

8. To generate the proper sensitivity test concentration, the bulb is squeezed rapidly 10 times and then the test subject is asked whether the saccharin can be tasted.

9. If the test subject cannot taste the saccharin, the bulb is squeezed rapidly 10 more times and the test subject is again asked whether the saccharin is tasted.

10. If the second response is still negative, the bulb is squeezed rapidly 10 more times and the test subject is again asked whether the saccharin is tasted.

11. The test conductor will take note of the number of squeezes (10, 20, or 30) required to elicit a taste response.

12. If the saccharin is not tasted after 30 squeezes of the nebulizer with sensitivity test solution, the test subject may not use the saccharin fit test.

13. If a taste response is elicited, the test subject must note the taste for reference in the fit test.

14. Correct use of the nebulizer requires approximately 1 cc of liquid placed in the nebulizer body at the start of testing.

15. The nebulizer must be thoroughly rinsed in water, shaken dry, and refilled at least each morning and afternoon or at least every four hours.

### Respirator selection.

Selection of respirators for fit testing is done in the same manner as described under respirator selection in the isoamyl acetate protocol, except that each respirator shall be equipped with a particulate filter, that is, dust/mist, dust/mist/fume, HEPA, paint-spray, or pesticide filter.

### Fit test.

1. The fit test uses the same enclosure described in steps 1 and 2 of the saccharin taste threshold screening protocol.

2. Each test subject should wear the selected respirator for at least 10 min before starting the fit test.

3. The test subject then puts on the enclosure while wearing the respirator. This respirator must be properly adjusted and equipped with a particulate filter.

4. The test subject may not eat, drink (except plain water), or chew gum for 15 min before the test.

5. A second DeVilbiss Model 40 Inhalation Medication Nebulizer or equivalent is used to spray the fit test solution into the enclosure. This nebulizer shall be clearly marked to distinguish it from the screening test solution nebulizer.

6. The fit test solution is prepared by adding 83 g of sodium saccharin to 100 cc of warm water.

7. As before, the test subject breathes through the open mouth with tongue extended.

8. The nebulizer is inserted into the hole in the front of the enclosure and the fit test solution is sprayed into the enclosure using the same technique as for the taste threshold screening and the same number of squeezes (10, 20, or 30) required to elicit a taste response in the screening.

9. After generation of the aerosol the test subject shall be instructed to perform the exercises listed in step 4 of the isoamyl acetate fit test protocol for 1 min each.

10. Every 30 s, the aerosol concentration shall be replenished using one-half the number of squeezes as initially (5, 10, or 15).

11. The test subject shall indicate to the test conductor if at any time during the fit test the taste of saccharin is detected.

12. If the saccharin taste as noted during the taste threshold screening protocol is detected, the fit is deemed unsatisfactory and a different respirator shall be tried.

13. Persons successfully completing this IAA protocol are expected to achieve a protection factor of at least 10 from the tested respirator. For full-facepiece respirators where protection factors greater than 10 are required, it is recommended that quantitative fit testing be done to select respirators providing fit factors of 1,000 or greater.

**Irritant fume protocol.** This protocol uses a particle produced by condensation. The process produces very small particles. Therefore, the respirator must be equipped with high-efficiency or HEPA filters so that only face seal leakage is evaluated. The protocol consists of two steps: respirator selection (discussed under the isoamyl acetate protocol) and the fit test. It is important to note that there is no threshold screening test as the threshold levels for irritant smoke have not been established. For this reason, respiratory protection experts do not consider the irritant fume protocol as a validated fit test method. It is acceptable for OSHA compliance, however.

***Respirator selection.*** Selection of respirators for fit testing is done in the same manner as described under respirator selection in the isoamyl acetate protocol, except that each respirator shall be equipped with high-efficiency (HEPA) filters.

### *Fit test.*

1. The test subject shall be allowed to breathe a weak concentration of the irritant smoke to become familiar with the characteristic cough or choke caused by the smoke.

2. The test subject shall properly don the selected respirator and wear it for at least 10 min before starting the fit test.

3. The test conductor must explain this testing protocol to the test subject before testing is started.

4. The test subject puts on the respirator and performs a fit check. Failure of the check is cause to select an alternative respirator.

5. The test conductor then breaks both ends of a ventilation smoke tube containing stannic oxychloride, such as the MSA part No. 5645 or equivalent. Next the test conductor attaches a short length of tubing to one end of the smoke tube and attaches the other end of the smoke tube to a low-pressure air pump set to deliver 200 ml/min.

6. The test subject should be advised that the smoke can be irritating to the eyes and instructed to keep eyes closed while the test is performed.

7. The test conductor shall direct the stream of irritant smoke from the tube toward the face seal area of the test subject, beginning at least 12 in. from the facepiece and gradually moving to within 1 in., moving around the whole perimeter of the respirator.

8. The same exercises described in step 4 of the isoamyl acetate fit test protocol shall be performed while the respirator seal is being challenged by the smoke. Each exercise shall be performed for 1 min. The rainbow passage in the talking exercise should be replaced with counting backwards, slowly and distinctly, from 100. The test subject will not be able to read a passage since his or her eyes will be closed.

9. If the irritant smoke produces an involuntary reaction (cough) by the test subject, the test conductor shall stop the test. In this case, the tested respirator is rejected and another respirator shall be selected.

10. Each test subject passing the smoke test without evidence of a response shall be given a sensitivity check of the smoke from the same tube to determine whether he or she reacts to the smoke. Failure to evoke a response shall void the fit test.

11. The generation of the smoke atmosphere should be performed in a location with exhaust ventilation sufficient to prevent general contamination of the testing area by the irritant smoke.

12. OSHA standards allow respirators successfully tested by this protocol to be used to assign protection factors not exceeding 10. For full-facepiece respirators where protection factors greater than 10 are required, it is recommended that quantitative fit testing be done to select respirators providing fit factors of 1,000 or greater.

## Quantitative Fit Testing

A quantitative fit test measures actual leakage of a test gas, vapor, or aerosol into the facepiece. Instrumentation is used to sample and measure the test atmosphere and the air inside the respirator facepiece. With this information, a quantitative fit factor (or fit factor) is calculated. The fit factor is the ratio of the outside concentration to the concentration inside the respirator facepiece. The advantage of this type of testing is that it does not rely on a subjective response. The disadvantages are cost of instrumentation, need for highly trained personnel to conduct the test, and use of probed respirators (Figure 22–37) to sample air from inside the respirator.

Commercially available quantitative fit testing equipment uses sodium chloride and corn oil mist generating systems (Figure 22–38). These devices use test enclosures to contain the test agent. A newer system is very portable and measures the penetration of ambient aerosols into the facepiece (Figure 22–39). No test enclosure is required in this system. Fit testing with either of these systems must use respi-

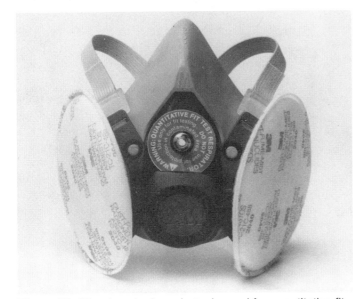

**Figure 22–37.** A probed respirator is used for quantitative fit testing to measure aerosol inside the facepiece. (Courtesy 3M.)

**Figure 22–39.** The ambient aerosol quantitative fit test system does not use a test enclosure and must be used with respirators equipped with high-efficiency filters. (Courtesy TSI.)

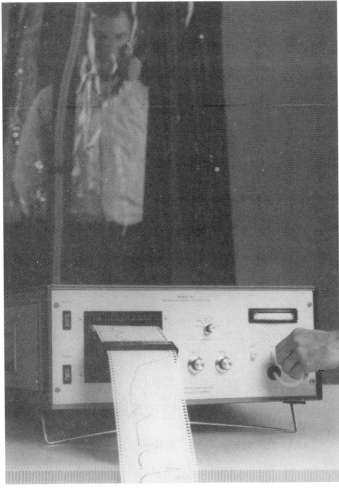

**Figure 22–38.** A portable corn oil quantitative fit test system uses a test enclosure. (Courtesy Dynatech Nevada, Inc.)

rators equipped with HEPA filters. HEPA filters are used to minimize particle penetration through the filters and allow facepiece fit evaluations. These instruments cannot be used to fit test non-HEPA-filter, disposable respirators.

A third method of quantitative fit testing does not involve aerosol measurement but rather determines leakage by creating a negative pressure inside the facepiece and measuring the leakage rate of air. This technique is sometimes referred to as the controlled negative-pressure method (Figure 22–40). The respirator does not need a probe, but test adapter manifolds are placed on the respirator in place of filters or cartridges. Only respirators that can be adapted with the manifolds can be tested by this method. These manifolds are specific for the brand of respirator. The manifolds seal the respirator inlets so that air cannot enter the respirator. One manifold contains a valve that allows air to enter the respirator so that the subject can breathe. The valve can be closed by squeezing a bulb hooked to this manifold. The second manifold contains ports so that air can be pumped from the respirator, creating a negative pressure inside the facepiece.

To perform a test, the test subject puts on the properly equipped respirator, takes a deep breath, and holds it. The person conducting the test squeezes the bulb to

**Figure 22–40.** This quantitative fit test equipment determines leakage by creating a negative pressure inside the facepiece similar to normal inspiratory pressures and measuring the leakage rate of air. (Courtesy Dynatech Nevada, Inc.)

"seal off" the respirator. The only way air can enter the respirator then is through face seal leaks. The instrument is then started, which contains a pump to pull air out of the facepiece. Air is drawn out of the respirator until a predetermined "challenge pressure" is reached. This negative pressure created inside the facepiece causes air to leak into the facepiece from around the seal. The pump speed is then controlled to maintain the "challenge pressure." The amount of air pumped out of the respirator is equal to the air that leaks into the facepiece; thus, the leak rate is measured. An equivalent fit factor is then calculated by the instrument comparing the airflow needed to maintain the negative pressure and the leak rate. A test takes about 12 s. The manufacturer has instructions on how to accommodate the desire to perform facial exercises while testing.

## Quantitative Fit Test Protocol

The following protocol is a generalized fit test procedure for quantitative fit test equipment designed to measure aerosol (for example, ambient particles) leakage into the respirator.

1. The equipment should be set up and tubing connected in accordance with the manufacturer's instructions.

2. Maintenance requirements (such as filling the unit with alcohol) must be performed as instructed.

3. The equipment should be started up and allowed to run the proper amount of time for warm-up. The zero check or calibration should be conducted as prescribed by the manufacturer's instructions.

4. All pertinent information about the test subject should be recorded, such as name, respirator model, date, and test conductor.

5. The test subject should select a respirator as described under the isoamyl acetate qualitative fit test protocol. The respirator must be equipped with HEPA filters.

6. The test subject should wear the respirator 10 min prior to the test.

7. The test subject should then be instructed to connect the respirator to the sample line in the test chamber or test area.

8. The test concentration should then be measured for 1 min.

9. Next a sample should be drawn from the respirator. While this is being done, the test subject should perform each of the exercises described under the IAA qualitative fit test protocol for 1 min.

10. In addition to the six exercises, the subject should grimace. This exercise is designed to break the seal. It should then be determined whether the respirator reseats on the subject's face. These results should not be used in calculating the fit factor.

11. The final challenge atmosphere should then be determined for 1 min.

12. The test subject should then disconnect the respirator from the sample line and leave the test chamber or area.

13. The fit factor should then be calculated by using the average outside concentration and average inside concentration.

OSHA regulations state that this procedure should be repeated two more times. The lowest of the three fit factors is the fit factor to be used. This fit factor should be at least 10 times the assigned protection factor (Table 22–F) of the negative-pressure respirator before that respirator is assigned to an individual. If that fit factor is achieved, the quantitative fit test is considered successful.

If the TSI Portacount® Respirator Fit Tester is used, the worker should not be allowed to smoke within 30 min of the fit test. This can result in low fit factors because this instrument counts particles in the air. It is also important that all connections to tubing be tight. The respirator probe must not leak around the outside. These situations can result in low fit factors that do not reflect face seal leakage.

## Positive-Pressure Respirators

Tight-fitting positive-pressure respirators must be either qualitatively or quantitatively fit tested in the negative-pressure mode. This can be done on tight-fitting powered air-purifying respirators by turning the blower off as long as the proper air-purifying element is on the respirator. Air-line respirators and SCBAs can be tested by obtaining the air-purifying model with the same facepiece used on the air-line respirator or SCBA. For some manufacturers, only filter or cartridge adapters that attach to the air-line respirator or SCBA facepiece need to be purchased. Combination air-line/air-purifying respirators can be fit tested in the negative-pressure mode by disconnecting from the air supply and placing the proper filter or cartridge in the holders. The purpose of this fit test is to ensure that an unacceptable fit of the respirator to the wearer does not occur. Only the fitting capability of the facepiece is being evaluated, not the performance of the respirator. A fit factor of at least 100 must be obtained. Remember that the validated qualitative fit test protocols were validated for fit factors of 100; therefore, either validated qualitative fit test methods or quantitative fit testing demonstrating a fit factor of 100 is acceptable. The assigned protection factors from Table 22–F can be issued after successful fit testing of the respirator.

# ■ Summary

The material presented in this chapter is intended for persons concerned with establishing and maintaining a respiratory protection program. It presents certain basic information for guidance purposes. However, it is not intended to be all-inclusive in content or scope.

Simplified interpretations of certain federal regulations pertaining to respiratory protection and monitoring were presented in this chapter. Although these interpretations convey background information about the regulations, under no circumstances should they be used as the sole basis of a respiratory protection program. In all cases, the current federal regulations, as published in the *Federal Register* and later collected in the *Code of Federal Regulations,* should be carefully studied and the rules and procedures in those regulations explicitly followed. Only they define the specific requirements that are in force. For additional information the reader should refer to the Bibliography.

# ■ Bibliography

Ackley MW. *Degradation of Electrostatic Filters at Elevated Temperature and Humidity.* Paper presented at World Filtration Congress III, September 1982.

American Conference of Governmental Industrial Hygienists. *1994–1995 Threshold Limit Values for Chemical Substances and Physical Agents and Biological Exposure Indices.* Cincinnati, OH: American Conference of Governmental Industrial Hygienists, 1994.

American Industrial Hygiene Association. *Odor Thresholds for Chemicals with Occupational Health Standards.* Fairfax, VA: American Industrial Hygiene Association, 1989.

American Industrial Hygiene Association. *Respiratory Protection: A Manual and Guideline,* 2nd ed. Fairfax, VA: American Industrial Hygiene Association, 1991.

American National Standards Institute. *Respiratory Protection,* Z88.2, 1992. New York: American National Standards Institute, 1992.

American National Standards Institute. *Physical Qualifications for Respirator Use,* Z88.6-1984. New York: American National Standards Institute, 1984.

American National Standards Institute. *Practices for Respiratory Protection,* Z88.2, 1969. New York: American National Standards Institute, 1969.

American National Standards Institute. *Practices for Respiratory Protection for the Fire Service,* Z88.5-1981. New York, American National Standards Institute, 1981.

*Code of Federal Regulations, Title 29, Labor, Parts* 1900–1926, and *Code of Federal Regulations, Title 30, Mineral Resources, Parts 11–14A.* Washington, DC: GPO, 1995.

Compressed Gas Association (CGA). *Commodity Specification for Air,* G-7.1-1989 (also ANSI/CGA G-7.1). Arlington, VA: Compressed Gas Association, 1989.

Compressed Gas Association. *Compressed Air for Human Respiration,* G-7, 1990. Arlington, VA: Compressed Gas Association, 1990.

Crutchfield CD, Eroh MP, Van Ert MD. A feasibility study of quantitative respirator fit testing by controlled negative pressure. *AIHAJ* 52(4):172–176, 1991.

Fardi B, Liu BYH. Performance of disposable respirators. *Particle and Particle Systems Characterization* 8:308–314, 1991.

Fiserova-Bergerova V. Relevance of occupational skin exposure. *Ann Occup Hyg* 37 (6):673–685, 1985.

Hinds WC, Bellin P. Effect of facial-seal leaks on protection provided by half-mask respirators. *Appl Ind Hyg* 3(5):158–164, 1988.

Japuntich DA. Respiratory particulate filtration. *Int Soc Respir Prot* 2:137–169, 1984.

National Fire Protection Association. *Self-Contained Breathing Apparatus for Fire Fighters.* Quincy, MA: National Fire Protection Association, 1992.

National Institute for Occupational Safety and Health. *A Guide to Industrial Respiratory Protection,* No. 87-116, Cincinnati, OH: National Institute for Occupational Safety and Health, 1987.

Nelson GE, Harder CA. Respirator cartridge efficiency studies: V. Effect of solvent vapor. *AIHAJ* 35:391–410, 1974.

Respiratory protective devices. *Federal Register* 60(110):30336–30398, June 8, 1995.

Stevens GA, Moyer ES. "Worst case" aerosol testing parameters: I. Sodium chloride and dioctylphthalate aerosol filter efficiency as a function of particle size and flow rate. *AIHAJ* 50:257–264, 1989.

# ■ Addendum

## 29 *CFR* 1910.134: Respiratory Protection.

(a) Permissible practice.
  (1) In the control of those occupational diseases caused by breathing air contaminated with harmful dusts, fogs, fumes, mists, gases, smokes, sprays, or vapors, the primary objective shall be to prevent atmospheric contamination. This shall be accomplished as far as feasible by accepted engineering

control measures (for example, enclosure or confinement of the operation, general and local ventilation and substitution of less toxic materials). When effective engineering controls are not feasible, or while they are being instituted, appropriate respirators shall be used pursuant to the following requirements.

(2) Respirators shall be provided by the employer when such equipment is necessary to protect the health of the employee. The employer shall provide the respirators which are applicable and suitable for the purpose intended. The employer shall be responsible for the establishment and maintenance of a respiratory protective program which shall include the requirements outlined in paragraph (b) of this section.

(3) The employee shall use the provided respiratory protection in accordance with instructions and training received.

(b) Requirements for a minimal acceptable program.

(1) Written standard operating procedures governing the selection and use of respirators shall be established.

(2) Respirators shall be selected on the basis of hazards to which the worker is exposed.

(3) The user shall be instructed and trained in the proper use of respirators and their limitations.

(4) [Reserved]

(5) Respirators shall be regularly cleaned and disinfected. Those used by more than one worker shall be thoroughly cleaned and disinfected after each use.

(6) Respirators shall be stored in a convenient, clean, and sanitary location.

(7) Respirators used routinely shall be inspected during cleaning. Worn or deteriorated parts shall be replaced. Respirators for emergency use such as self contained devices shall be thoroughly inspected at least once a month and after each use.

(8) Appropriate surveillance of work area conditions and degree of employee exposure or stress shall be maintained.

(9) There shall be regular inspection and evaluation to determine the continued effectiveness of the program.

(10) Persons should not be assigned to tasks requiring use of respirators unless it has been determined that they are physically able to perform the work and use the equipment. The local physician shall determine what health and physical conditions are pertinent. The respirator user's medical status should be reviewed periodically (for instance, annually).

(11) Respirators shall be selected from among those jointly approved by the Mine Safety and Health Administration and the National Institute for Occupational Safety and Health under the provisions of 30 *CFR* part 11.

(c) Selection of respirators. Proper selection of respirators shall be made according to the guidance of American National Standard *Practices for Respiratory Protection* Z88.2–1969.

(d) Air quality.

(1) Compressed air, compressed oxygen, liquid air, and liquid oxygen used for respiration shall be of high purity. Oxygen shall meet the requirements of the United States Pharmacopoeia for medical or breathing oxygen. Breathing air shall meet at least the requirements of the specification for Grade D breathing air as described in Compressed Gas Association Commodity Specification G-7.1–1966. Compressed oxygen shall not be used in supplied-air respirators or in open circuit self-contained breathing apparatus that have previously used compressed air. Oxygen must never be used with air-line respirators.

(2) Breathing air may be supplied to respirators from cylinders or air compressors.

(i) Cylinders shall be tested and maintained as prescribed in the Shipping Container Specification Regulations of the Department of Transportation (49 *CFR* Part 178).

(ii) The compressor for supplying air shall be equipped with necessary safety and standby devices. A breathing air-type compressor shall be used. Compressors shall be constructed and situated so as to avoid entry of contaminated air into the system and suitable in-line air purifying sorbent beds and filters installed to further assure breathing air quality. A receiver of sufficient capacity to enable the respirator wearer to escape from a contaminated atmosphere in event of compressor failure, and alarms to indicate compressor failure and overheating shall be installed in the system. If an oil-lubricated compressor is used, it shall have a high-temperature or carbon monoxide alarm, or both. If only a high-temperature alarm is used, the air from the compressor shall be frequently tested for carbon monoxide to insure that it meets the specifications in paragraph (d)(1) of this section.

(3) Air line couplings shall be incompatible with outlets for other gas systems to prevent inadvertent servicing of air line respirators with nonrespirable gases or oxygen.

(4) Breathing gas containers shall be marked in accordance with American National Standard Method of Marking Portable Compressed Gas Containers to Identify the Material Contained, Z48.1–1954; Federal Specification BB-A-1034a, June 21, 1968, Air, Compressed for Breathing Purposes; or Interim

Federal Specification GG-B-00675b, April 27, 1965, Breathing Apparatus, Self-Contained.

(e) Use of respirators.

(1) Standard procedures shall be developed for respirator use. These should include all information and guidance necessary for their proper selection, use, and care. Possible emergency and routine uses of respirators should be anticipated and planned for.

(2) The correct respirator shall be specified for each job. The respirator type is usually specified in the work procedures by a qualified individual supervising the respiratory protective program. The individual issuing them shall be adequately instructed to insure that the correct respirator is issued.

(3) Written procedures shall be prepared covering safe use of respirators in dangerous atmospheres that might be encountered in normal operations or in emergencies. Personnel shall be familiar with these procedures and the available respirators.

(i) In areas where the wearer, with failure of the respirator, could be overcome by a toxic or oxygen-deficient atmosphere, at least one additional man shall be present. Communications (visual, voice, or signal line) shall be maintained between both or all individuals present. Planning shall be such that one individual will be unaffected by any likely incident and have the proper rescue equipment to be able to assist the other(s) in case of emergency.

(ii) When self-contained breathing apparatus or hose masks with blowers are used in atmospheres immediately dangerous to life or health, standby men must be present with suitable rescue equipment.

(iii) Persons using air-line respirators in atmospheres immediately hazardous to life or health shall be equipped with safety harnesses and safety lines for lifting or removing persons from hazardous atmospheres or other and equivalent provisions for the rescue of persons from hazardous atmospheres shall be used. A standby man or men with suitable self-contained breathing apparatus shall be at the nearest fresh air base for emergency rescue.

(4) Respiratory protection is no better than the respirator in use, even though it is worn conscientiously. Frequent random inspections shall be conducted by a qualified individual to assure that respirators are properly selected, used, cleaned, and maintained.

(5) For safe use of any respirator, it is essential that the user be properly instructed in its selection, use, and maintenance. Both supervisors and workers shall be so instructed by competent persons. Training shall provide the men an opportunity to handle the respirator, have it fitted properly, test its face-piece-to-face seal, wear it in normal air for a long familiarity period, and, finally, to wear it in a test atmosphere.

(i) Every respirator wearer shall receive fitting instructions including demonstrations and practice in how the respirator should be worn, how to adjust it, and how to determine if it fits properly. Respirators shall not be worn when conditions prevent a good face seal. Such conditions may be a growth of beard, sideburns, a skull cap that projects under the facepiece, or temple pieces on glasses. Also, the absence of one or both dentures can seriously affect the fit of a facepiece. The worker's diligence in observing these factors shall be evaluated by periodic check. To assure proper protection, the facepiece fit shall be checked by the wearer each time he puts on the respirator. This may be done by following the manufacturer's facepiece fitting instructions.

(ii) Providing respiratory protection for individuals wearing corrective glasses is a serious problem. A proper seal cannot be established if the temple bars of eye glasses extend through the sealing edge of the full facepiece. As a temporary measure, glasses with short temple bars or without temple bars may be taped to the wearer's head. Systems have been developed for mounting corrective lenses inside full facepieces. When a workman must wear corrective lenses as part of the facepiece, the facepiece and lenses shall be fitted by qualified individuals to provide good vision, comfort, and a gas-tight seal.

(iii) If corrective spectacles or goggles are required, they shall be worn so as not to affect the fit of the facepiece. Proper selection of equipment will minimize or avoid this problem.

(f) Maintenance and care of respirators.

(1) A program for maintenance and care of respirators shall be adjusted to the type of plant, working conditions, and hazards involved, and shall include the following basic services:

(i) Inspection for defects (including a leak check),

(ii) Cleaning and disinfecting,

(iii) Repair,

(iv) Storage

Equipment shall be properly maintained to retain its original effectiveness.

(2) (i) All respirators shall be inspected routinely before and after each use. A respirator that is not routinely used but is kept ready for emergency use shall be inspected after each use and at least monthly to assure that it is in satisfactory working condition.

(ii) Self-contained breathing apparatus shall be inspected monthly. Air and oxygen cylinders shall be fully charged according to the manufacturer's instructions. It shall be determined that the regulator and warning devices function properly.

(iii) Respirator inspection shall include a check of the tightness of connections and the condition of the facepiece, headbands, valves, connecting tube, and canisters. Rubber or elastomer parts shall be inspected for pliability and signs of deterioration. Stretching and manipulating rubber or elastomer parts with a massaging action will keep them pliable and flexible and prevent them from taking a set during storage.

(iv) A record shall be kept of inspection dates and findings for respirators maintained for emergency use.

(3) Routinely used respirators shall be collected, cleaned, and disinfected as frequently as necessary to insure that proper protection is provided for the wearer. Respirators maintained for emergency use shall be cleaned and disinfected after each use.

(4) Replacement or repairs shall be done only by experienced persons with parts designed for the respirator. No attempt shall be made to replace components or to make adjustment or repairs beyond the manufacturer's recommendations. Reducing or admission valves or regulators shall be returned to the manufacturer or to a trained technician for adjustment or repair.

(5) (i) After inspection, cleaning, and necessary repair, respirators shall be stored to protect against dust, sunlight, heat, extreme cold, excessive moisture, or damaging chemicals. Respirators placed at stations and work areas for emergency use should be quickly accessible at all times and should be stored in compartments built for the purpose. The compartments should be clearly marked. Routinely used respirators, such as dust respirators, may be placed in plastic bags. Respirators should not be stored in such places as lockers or tool boxes unless they are in carrying cases or cartons.

(ii) Respirators should be packed or stored so that the facepiece and exhalation valve will rest in a normal position and function will not be impaired by the elastomer setting in an abnormal position.

(iii) Instructions for proper storage of emergency respirators, such as gas masks and self-contained breathing apparatus, are found in "use and care" instructions usually mounted inside the carrying case lid.

(g) Identification of gas mask canisters.

(1) The primary means of identifying a gas mask canister shall be by means of properly worded labels. The secondary means of identifying a gas mask canister shall be by a color code.

(2) All who issue or use gas masks falling within the scope of this section shall see that all gas mask canisters purchased or used by them are properly labeled and colored in accordance with these requirements before they are placed in service and that the labels and colors are properly maintained at all times thereafter until the canisters have completely served their purpose.

(3) On each canister shall appear in bold letters the following:

(i) Canister for _____
(Name for atmospheric contaminant)
or
Type N Gas Mask Canister

(ii) In addition, essentially the following wording shall appear beneath the appropriate phrase on the canister label: "For respiratory protection in atmospheres containing not more than _____ percent by volume of _____."
(Name of atmospheric contaminant)

(4) Canisters having a special high-efficiency filter for protection against radionuclides and other highly toxic particulates shall be labeled with a statement of the type and degree of protection afforded by the filter. The label shall be affixed to the neck end of, or to the gray stripe which is around and near the top of, the canister. The degree of protection shall be marked as the percent of penetration of the canister by a 0.3-micron-diameter dioctyl phthalate (DOP) smoke at a flow rate of 85 liters per minute.

(5) Each canister shall have a label warning that gas masks should be used only in atmospheres containing sufficient oxygen to support life (at least 16 percent by volume), since gas mask canisters are only designed to neutralize or remove contaminants from the air.

(6) Each gas mask canister shall be painted a distinctive color or combination of colors indicated in Table I–1. All colors used shall be such that they are clearly identifiable by the user and clearly distinguishable from one another. The color coating used shall offer a high degree of resistance to chipping, scaling, peeling, blistering, fading, and the effects of the ordinary atmospheres to which they may be exposed under normal conditions of storage and use. Appropriately colored pressure sensitive tape may be used for the stripes.

**Table I–1.** Gas Mask Canister Color Codes

| *Atmospheric contaminants to be protected against* | *Colors assigned(1)* |
| --- | --- |
| Acid gases | White. |
| Hydrocyanic acid gas | White with 1/2-inch green stripe completely around the canister near the bottom. |
| Chlorine gas | White with 1/2-inch yellow stripe completely around the canister near the bottom. |
| Organic vapors | Black. |
| Ammonia gas | Green. |
| Acid gases and ammonia gases | Green with 1/2-inch white stripe completely around the canister near the bottom. |
| Carbon Monoxide | Blue. |
| Acid gases and organic vapors | Yellow. |
| Hydrocyanic acid gas and chloropicrin vapor | Yellow with 1/2-inch blue stripe completely around the canister near the bottom. |
| Acid gases, organic vapors, and ammonia gases | Brown. |
| Radioactive materials, excepting tritium and noble gases | Purple (Magenta). |
| Particulates (dusts, fumes, mists, fogs, or smokes) in combination with any of the above gases or vapors | Canister color for contaminant, as designated above, with 1/2-inch gray stripe completely around the canister near the top. |
| All of the above atmospheric contaminants | Red with 1/2-inch gray stripe completely around the canister near the top. |

Gray shall not be assigned as a main color for a canister designed to remove acids or vapors. Orange shall be used as a complete body or stripe color to represent gases not included in this table. The user will need to refer to the canister label to determine the degree of protection the canister will afford.

(Approved by the Office of Management and Budget under control number 1218-0099. [39 *FR* 23502, June 27, 1974, as amended at 43 *FR* 49748, Oct. 24, 1978; 49 *FR* 5322, Feb. 10, 1984; 49 *FR* 18295, Apr. 30, 1984; 58 *FR* 35309, June 30, 1993].)

PART 6

# Occupational Health and Safety Programs

# The Industrial Hygienist

Revised by Jill Niland, MPH, CIH, CSP

THIS CHAPTER DISCUSSES THE BACKGROUND and definition of industrial hygiene, its interrelationships with other occupational health groups, the functions and characterization of industrial hygienists, personnel needs, and training programs.

**Background**
*Definition of Industrial Hygiene*

**Job Descriptions**
*Industrial Hygienist-in-Training (IHIT) ▪ Occupational Health and Safety Technologist (OHST) ▪ Industrial Hygienist ▪ Certified Industrial Hygienist (CIH)*

**Industrial Hygiene, Civil Service**
*Training Plan for Entry-Level OSHA Industrial Hygienists*

**Personnel Needs and Problems**
*Education and Training Programs ▪ Educational Resource Centers ▪ Professional Schooling ▪ Graduate Curricula ▪ Faculty ▪ Baccalaureate Programs ▪ Continuing Education*

**Summary**

**Bibliography**

**Addendum: Professional Societies and Courses of Interest to Industrial Hygienists**
*American Industrial Hygiene Association ▪ American Board of Industrial Hygiene ▪ American Academy of Industrial Hygiene ▪ American Conference of Governmental Industrial Hygienists ▪ American Public Health Association*

## Background

Industrial hygienists are scientists, engineers, and public health professionals committed to protecting the health of people in the workplace and the community. Industrial hygienists must be competent in a variety of scientific fields—principally chemistry, engineering, physics, toxicology and biology—as well as the fundamentals of occupational medicine. Trained initially in one of these fields, most industrial hygienists have acquired by experience and postgraduate study a knowledge of the other allied disciplines.

In traditional industrial organizations, industrial hygienists were required to relate to personnel in other functions including research and development, medical, management, safety, and production. Although the working relationships were close, it was understood that the industrial hygienist was not expected to have expertise in these areas. In today's downsized organization, the industrial hygienist may also act as the safety or environmental professional. Flattened management structures and the use of self-directed work teams have created the need for flexible industrial hygienists who understand not only technical and scientific issues, but also production and research concerns. Hygienists at all levels participate in management of cross-functional projects that draw on the expertise of all team members to develop and maintain a safe and healthful work environment. One of the challenges for this generation of industrial hygienists is maintaining a high level of technical expertise while broadening their roles in the activities just described.

## Definition of Industrial Hygiene

As stated in Chapter 1, the American Industrial Hygiene Association (AIHA) defines industrial hygiene as

> that science and art devoted to the anticipation, recognition, evaluation, and control of those environmental factors or stresses arising in or from the workplace which may cause sickness, impaired health and well-being, or significant discomfort among workers or among citizens of the community.
>
> An industrial hygienist is a person with a college or university degree or degrees in engineering, chemistry, physics, medicine, or related physical and biological sciences who, by virtue of special studies and training, has acquired competence in industrial hygiene. Such special studies and training must have been sufficient in all of the above cognate sciences to provide the abilities to anticipate and recognize environmental factors and to understand their effect on humans and their well-being, to evaluate (on the basis of experience and with the aid of quantitative measurement techniques) the magnitude of these stresses in terms of ability to impair human health and well-being, and to prescribe methods to eliminate, control, or reduce such stresses when necessary to alleviate their effects.

A person who excels in all of these activities would be ideal, but the capabilities to excel in every field of specialization cannot exist within a single individual. Specialists emerge in the fields of toxicology, epidemiology, chemistry, ergonomics, acoustics, ventilation engineering, and statistics, to name only a few.

Physicians, nurses, and safety professionals can move part or all of the way into industrial hygiene functions. Not to be ignored are overlaps into health physics, air pollution, water pollution, solid waste disposal, and disaster planning. The industrial hygienist also makes contributions in employee education and training, law and product liability, sales, labeling, and public information.

## ■ Job Descriptions

The job descriptions of industrial hygiene personnel are similar to those of safety personnel. The entry-level employee, normally called a safety or health technologist, regularly inspects operations using a few simple instruments, and investigates minor incidents.

The employee at the next higher level, normally called an industrial hygienist, is similar in function to a safety engineer. This person carries out more detailed studies of incidents, prepares recommendations and other reports, reviews new processes, machinery, and layouts from a health or safety engineering viewpoint, promotes health or safety education, and advises management in healthful practices, procedures, and equipment needs.

The industrial hygiene manager or supervisor is similar to the safety director, and manages the total hygiene program with responsibilities equivalent to those of the safety supervisor.

The American Board of Industrial Hygiene (ABIH) has established a level of competence in industrial hygiene, achieved through examination. Successful passage of this examination qualifies the person for membership in the American Academy of Industrial Hygiene; details are given in the Addendum.

Many certified industrial hygienists are also certified safety professionals and vice versa. Proficiency in industrial hygiene, by examination and experience, follows a route roughly comparable to that of occupational safety. Moreover, the type of organization that employs the industrial hygienist or safety professional often requires similar skills of each. The job descriptions for these two areas are grouped together because of their general similarity. Government agencies need capabilities in compliance, research, education and training, program development and management, and standards development. Consultants are engaged in a wide range of professional activities that are limited by the capabilities and specialties of personnel. Insurance and industry require a range of professional competency from the technologist level to the program manager. Universities require professional capabilities in research, teaching, and program administration; labor unions require skills in inspection, research, training, program development and administration, and standards development.

## Industrial Hygienist-in-Training (IHIT)

This designation is part of the ABIH's certification program. It is awarded to people with a college or university degree in industrial hygiene, chemistry, engineering, physics, medicine, or related biological sciences who, by virtue of special studies or training, have acquired competence in the basic principles of industrial hygiene and have successfully completed the core examination. The candidate for the examination must also have completed at least one full year of industrial hygiene practice acceptable to the board (see Addendum).

The ABIH instituted the IHIT category in 1972 because it recognized that recent graduates and people employed in industrial hygiene wanted to take the core examination before completing the 5 years of experience required for eligibility to take the comprehensive examination. At the time of this writing, the ABIH has certified about 800 IHITs in active status.

The IHIT performs field and laboratory work that gradually increases in variety as experience and proficiency develop. This person prepares technical reports that are reviewed by the next higher level of professional supervision.

During this period of apprenticeship, the IHIT is introduced to the elements of organization and management. The IHIT should also learn to understand a flowchart and a blueprint before progressing to anticipating, recognizing, evaluating, and controlling industrial hygiene hazards.

IHITs find that increasing emphasis is placed on communication skills, and should be encouraged to draft replies to letters, write reports, and prepare oral presentations that may be edited by the supervisor.

If the IHIT has not already done so, he or she should become involved in the local section or chapter of the professional association most appropriate. The IHIT should attend meetings and work on committees, and should begin to meet other professionals outside the immediate workplace.

## Occupational Health and Safety Technologist (OHST)

In 1976, the ABIH, in recognition of the growing group of technologists engaged in industrial hygiene activities, established an industrial hygiene technologist certification program.

An industrial hygiene technologist is a person who, by virtue of special studies and training, has acquired proficiency in an aspect or phase of industrial hygiene (such as air-sampling, monitoring, instrumentation, specialized investigations, or specialized laboratory procedures) and who performs such duties under the supervision of an industrial hygienist. The designation certified industrial hygiene technologist (CIHT) was awarded after the applicant passed the examination.

In 1985, the ABIH and the Board of Certified Safety Professionals (BCSP) began joint sponsorship of a new technologist certification: the occupational health and safety technologist (OHST), which replaced the industrial hygiene technologist designation.

This certification is not intended for Certified Safety Professionals (CSPs) or Certified Industrial Hygienists (CIHs), nor is it intended for those eligible to take the CSP or CIH examinations. It is a joint technologist certification. (See Chapter 24, The Safety Professional).

Candidates for OHST certification must show 5 years of occupational health or safety technologist experience or a combination of acceptable education credit and experience before being allowed to take the OHST examination. The OHST may be able to gain the education and experience to become a CSP or CIH. The OHST certification process is administered by the BCSP (see Chapter 24, The Safety Professional). At the time of this writing, there are about 1,100 OHSTs in the United States.

Industrial hygiene technologists and technicians have acquired knowledge, skills, and field experience, so they can function efficiently in their limited technical area. Industrial hygiene technicians can take samples and make measurements in the facility or community. Their data and observations can be used to provide information for an industrial hygiene plan or program.

The duties of a technician require thoroughness, dependability, and a concern for the accuracy of the data being collected. Industrial hygiene technicians should be given a detailed outline of their duties. Manuals should be available to technicians for reference. Technicians must see the relevance and value of their efforts; these should be reflected in the technician's salary and in workplace structure. The industrial hygiene technician is part of a team, a partner in an effort in which the technician and the industrial hygienist share responsibility. The technician should have access to the industrial hygienist.

Technology changes and adds new problems to the old ones. Rarely are old hazards totally replaced. New problems call for new approaches, new instruments, and new ways of recording, compiling, and integrating data. Thus, the technician must be willing to adjust to changing conditions. Technicians may become specialists in their own right. Some will be content to remain technicians, but many will move on and become industrial hygienists.

## Industrial Hygienist

Industrial hygienists are people who, because of their more generalized skills, should be able to make independent decisions. The industrial hygienist decides what information is available, what additional facts are needed, and how they will be used or acquired.

**Functions.** Over a generation ago, Radcliffe et al. (1959) described the sphere of responsibility of industrial hygienists, which remain current today. They stated that the industrial hygienist will:

- Direct the industrial hygiene program
- Examine the work environment

  Study work operations and processes and obtain full details of the nature of the work, materials and equipment used, products and by-products, number and sex of employees, and hours of work

  Make appropriate measurements to determine the magnitude of exposure or nuisance to workers and the public, devise methods and select instruments suitable for such measurements, personally (or through others under direct supervision) conduct such measurements, and study and test material associated with the work operations

  Using chemical and physical means, study the results of tests of biological materials, such as blood and urine, when such examination will aid in determining the extent of exposure

- Interpret results of the examination of the environment in terms of its ability to impair health, nature of health impairment, workers' efficiency, and community nuisance or damage, and present specific conclusions to appropriate parties such as management, health officials, and employee representatives

- Make specific decisions as to the need for or effectiveness of control measures and, when necessary, advise as to the procedures that are suitable and effective for both the work environment and the environment.

- Prepare rules, regulations, standards, and procedures for the healthful conduct of work and the prevention of nuisance in the community

- Present expert testimony before courts of law, hearing boards, workers' compensation commissions, regulatory agencies, and legally appointed investigative bodies

- Prepare appropriate text for labels and precautionary information for materials and products to be used by workers and the public

- Conduct programs for the education of workers and the public in the prevention of occupational disease and community nuisance

- Conduct epidemiological studies of workers and industries to discover the presence of occupational disease and establish or improve Threshold Limit Values or standards for the maintenance of health and efficiency

- Conduct research to advance knowledge concerning the effects of occupation on health and means of preventing occupational health impairment, community air pollution, noise, nuisance, and related problems

The industrial hygienist should be able to determine whether there are alternative solutions to a problem. Obviously, leadership and management skills are required.

Few problems are so unique that peer acceptance is not required. Thus, the industrial hygienist must be able to work with other industrial hygienists in the same functional area, whether it be industry, government, labor unions, insurance, consulting, or teaching.

The industrial hygienist should also have the experience, knowledge, and capability to specify corrective procedures to minimize or control environmental health hazards.

A difficult route for a company to take is that of "growing its own industrial hygienist"—that is, taking someone from inside the organization, with some scientific background and a knowledge of the firm's products, and exposing him or her to a crash program in industrial hygiene. A company initiating an industrial hygiene effort must recognize that company knowledge alone is not enough for the optimal solution of industrial hygiene

problems. The industrial hygienist must have the necessary professional expertise.

The capable industrial hygienist who has made the in-house adjustment to the company's problems should have the versatility and capability to deal with any industrial hygiene problem that may arise. In the development of a new product, for example, he or she should be able to meet with research and development personnel and find out what information is needed. This might include toxicological information, labeling requirements, assistance to customers, and any special engineering control requirements as the research effort progresses through pilot state to commercial production.

With the assistance of a qualified epidemiologist, the industrial hygienist can study an existing (or even a suspected) environmental health problem through epidemiological and biostatistical approaches, in addition to the usual sampling and measuring procedures. The industrial hygienist should know where to go (for example, personnel, purchasing, or process engineering) for the information he or she needs to investigate and solve a problem. If the industrial hygienist knows of another company engaged in making similar products, he or she can exchange information with its industrial hygienist and, under certain circumstances, exchange visits.

Industrial hygienists must work well with other professionals, such as physicians, nurses, safety engineers, toxicologists, health physicists, and others, in and out of the company. They must also communicate with and work very closely with employees. Employees have insights into potential health hazards in their work area that only those working with the processes every day can possess. They are a primary source of information and suggestions for the industrial hygienist.

**The industrial hygiene manager.** In an industry setting, the industrial hygiene manager supervises the technical and support staff in a health and safety department; prepares budgets and plans; is familiar with government agencies related to the operation; relates industrial hygiene operations to research and development, production, environmental, and other departments or functions; and prepares appropriate reports. He or she may be called on to assist the corporate legal department when regulatory and worker compensation issues arise. The industrial hygiene manager should be certified by ABIH (see the description of this organization in the Addendum to this chapter).

Many aspects of industrial hygiene expertise are unique. It makes sense for the industrial hygienist to extend his or her capabilities and sphere of activity by delegating responsibilities to others. This calls for supervisory and planning skills. The industrial hygienist must be able not only to plan, direct, and supervise technicians and assistants, but must also to plan, program, and budget the activities of the department and staff. As a manager, he or she must establish priorities and initiate appropriate corrective action. The industrial hygienist and the industrial hygiene manager must both be effective communicators. Many aspects of their work involve formal or impromptu training of employees, managers, and visitors to a facility. These professionals may also be called on to discuss an organization's health and safety goals and accomplishments with the media and other members of the public, and they must be articulate, knowledgeable, and able to convey technical information in nontechnical language.

## Certified Industrial Hygienist (CIH)

The designation of certified industrial hygienist by the ABIH indicates that a person has received special education, lengthy experience, and proven professional ability in the comprehensive practice or the chemical practice of industrial hygiene. Many industrial hygiene chemists who have experience in laboratory practice attain the certification in chemical practice.

The employer, employees, and the public have a right to be reasonably assured that the person to whom their lives are entrusted is professionally capable. The normal manner in which such protection is provided is through licensing; this licensing can be through a government agency, a peer review arrangement, or both. Certification by the ABIH provides this assurance.

For certification by the ABIH, an individual must meet rigorous standards of education and experience before proving, by written examination, competency in either the comprehensive practice of industrial hygiene or the chemical practice (see Addendum). Diplomates of the ABIH are eligible for membership in the American Academy of Industrial Hygiene.

Certification provides some assurance that this individual possesses a high level of professional competence. The certified industrial hygienist is the person most likely to direct an industrial hygiene program capably, to work with other professions and government agencies, and to provide the vision and leadership of an industrial hygiene program to keep occupational hazards at a minimum in a rapidly changing technology and society. At the time of this writing, there are about 5,600 active CIHs.

All CIHs must actively work to maintain their certification by earning a specified number of certification maintenance points during a 6-year cycle. These points are awarded for working as an industrial hygienist; participating in professional associations; attending approved meetings, seminars and short courses; participating on technical committees; publishing in peer-reviewed journals; teaching, when not part of their primary practice; and other ABIH-approved activities.

Of course, all of the previously described categories of ABIH certification—CIH, IHIT, and OHST—are open to all industrial hygiene personnel, whether they are employed in industry, government, labor unions, educational institu-

tions, or consulting, as long as they meet the qualifications. However, federally employed industrial hygienists also have their own unique training programs that reflect the structure and duties of their positions.

## ■ Industrial Hygiene, Civil Service

For industrial hygiene trainees, assignments are selected and designed to orient the new employee into the field of industrial hygiene, to determine areas of interest and potential, to relieve experienced industrial hygienists of detailed and simple work, and to develop the trainee's knowledge and competence. Specific assignments are carried out under direct supervision of a qualified industrial hygienist, including recognition of hazards, identification of controls, calibration of equipment, collection of samples, and initial preparation of reports. During inspections, the trainee observes specific safety items.

Under the general supervision of a senior industrial hygienist, the industrial hygienist conducts complete industrial hygiene inspections, including selection of sampling methods and locations, evaluation of controls and monitoring procedures, and preparation of reports. Completed work is reviewed for overall adequacy and conformance with policy and precedents. The industrial hygienist determines engineering feasibility, sets periods of abatement, interprets standards, and defends appeals under supervision of a senior industrial hygienist.

The senior industrial hygienist performs complete industrial hygiene inspections and prepares the final report. He or she determines engineering feasibility, sets periods of abatement, defends appeals, interprets standards, and provides offsite consultation. He or she receives general assignment of objectives and definition of policy from supervisors. The senior industrial hygienist differs from the industrial hygienist in that he or she receives more complex assignments and may act in place of the industrial hygiene supervisor when the supervisor is absent.

## Training Plan for Entry-Level OSHA Industrial Hygienists

On July 7, 1992, Assistant Secretary for Occupational Safety and Health Dorothy Y. Strunk issued an OSHA instruction specifying a revised training program for OSHA compliance personnel. The instruction provided policies and guidelines for the implementation of technical training programs and described a federal program change that also affects state OSHA programs. This revised training program applies to both newly hired and experienced compliance personnel.

The training program is designed to provide a series of training courses supported by on-the-job training and self-instructional activities to ensure that compliance personnel are able to apply technical information and skills to their work; however, the elements of the training program are not meant to be prerequisites for advancement.

**Objectives.** On completion of the developmental training program, the compliance safety and health officer (CSHO) will have the following skills:

■ A working knowledge of the fundamentals of hazard recognition, evaluation, and control

■ Adequate knowledge of the implementation of engineering controls, abatement strategies, and the interpretation of data

■ A reasonable comprehension of basic industrial processes and the ability to make quantitative observations and measurements

■ Field experience in the proper calibration and use of measuring instruments

■ The ability to perform solo or team inspections in most types of industries

■ Knowledge of regulations and laws that involve safety and health in the workplace

■ The ability to present inspection data in a legal proceeding efficiently

■ The ability to make a referral to other appropriate industrial hygienists or safety officers

**Organizational training responsibilities.** The mission of the OSHA Office of Training and Education is to provide a program to educate and train employers and employees in the recognition, avoidance, and prevention of unsafe and unhealthful working conditions and to improve the skill and knowledge levels of personnel engaged in work relating to the Occupational Safety and Health Act of 1970.

The Office of Training and Education consists of four components:

■ *Division of Training and Educational Programs.* This division is responsible for planning agency technical training programs and for managing the New Directions grants.

■ *Division of Training and Educational Development.* This division is responsible for developing and updating safety and health training programs and related materials.

■ *Division of Administration and Training Information.* This division is responsible for providing administrative and informational programs for the Office of Training and Education.

■ *OSHA Training Institute.* The Training Institute is responsible for the delivery of training to the populations served by the agency.

Specific responsibilities of the OSHA Training Institute include the following:

■ Conducting programs of instruction for federal and state compliance officers, state consultants, other federal agency personnel and private sector employers, employees, and their representatives

■ Participating in the development of course outlines, detailed lesson plans, and other educational aids necessary to carry out training programs

Each of OSHA's ten regions has a regional training officer, who assists the assistant regional administrator for training, education, and consultation in coordinating the management of all regionwide training programs. The regional training officer serves as the focal point in the regional office, ensuring the successful implementation of the training program for regional compliance personnel. Specifically, the regional training officer assists in providing resource material and current training information to area directors and supervisors concerning the implementation of the objectives of the training program and evaluates and monitors all records of training.

In OSHA area offices, the area director has overall responsibility for ensuring and implementing the development and training of newly hired and experienced CSHOs under his or her supervision. The supervisor, however, serves as the main focal point in the area office for ensuring training. The supervisor provides and coordinates instruction, assistance, and guidance to the CSHOs in order to meet the training program objectives. Reviewing and maintaining progress records for each CSHO and assigning senior CSHOs to assist in on-the-job training of new hires is also performed by the supervisor.

The program itself provides a well-articulated progression of training requirements for newly hired personnel. The elements include formal training at the OSHA Training Institute and informal training such as self-study and on-the-job training (OJT). Figure 23–1 illustrates the developmental training plan for new hires.

**Informational program.** The developmental training plan begins with the study of an informational package of materials developed jointly by the national office, the regional office, and the Office of Training and Education. Contents include information on the U.S. Department of Labor; an introduction, history, and purpose statement; the structures of regional and area offices, procedures, and libraries; common OSHA acronyms; individual training development programs; and such handout items as organizational charts, the *Field Operations Manual* (FOM), standards, directives, personal protective equipment, and instruments.

**Self-study program.** Before attending the initial compliance course at the OSHA Training Institute, each CSHO is required to complete three self-study programs on the OSHAct, Chapter III of the *Field Operations Manual,* and Integrated Management Information Systems (IMIS) forms 1, 1A, 1B, and 1B-IH. During these self-study assignments, the CSHO becomes familiar with the basic OSHAct requirements; studies basic inspection procedures in Chapter III of the FOM, and familiarizes him- or herself with the most commonly used forms.

**OSHA Training Institute.** After completing the basic self-study prerequisites, each CSHO is required to complete coursework in one of three tracks: safety, health, or construction.

*Initial Compliance Course.* This provides new CSHOs with an understanding of occupational safety and health programs, a working knowledge of the FOM, and OSHA policies.

*Standards Courses in Safety, Health, or Construction.* These courses provide new-hire CSHOs with a thorough introduction to the organization and content of the standards and to hazard recognition and documentation.

*Inspection Techniques and Legal Aspects.* This provides new CSHOs with an understanding of basic communication skills, formal requirements and processes of the legal system, and investigative techniques related to OSHA compliance activity.

*Technical Courses (at least two courses required).* These provide the CSHO with technical knowledge, skills, and information on hazard recognition as related to OSHA requirements. The specific courses are determined by the supervisor based on individual need. Figure 23–1 lists the technical courses in each track.

*Crossover Training.* Because CSHOs must be familiar with general concepts of safety and health, each CSHO is required to complete crossover training during the developmental period. CSHOs on the safety or construction track are encouraged to attend the introduction to health course; industrial hygienists are encouraged to attend the introduction to safety course.

**Area office.** The training plan incorporates alternative modes of instruction including self-instructional techniques and on-the-job training (OJT) assignments with supervi-

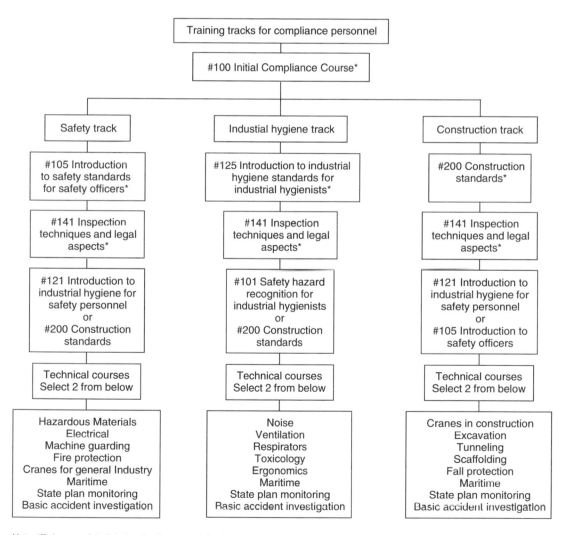

Note: *To be completed during the first year of the developmental period.

**Figure 23–1.** The OSHA training tracks for compliance personnel. (Source: OSHA Instruction TED 1.12A, Office of Training and Education.)

sion. The OJT and self-study programs are designed to reinforce formal training. Self-study is training that involves independently gained knowledge in the area office that will aid in preparation for formal training and coursework. OJT is training that relates principles and theories to work skills, which are then taught and applied in the field and office environment.

The OJT and self-study assignments are provided concurrently with formal training to emphasize and complement material covered in formal training courses. Time allowed to accomplish OJT and self-study assignments should be compatible with the new CSHO's current knowledge, skill, and experience level. The supervisor must document a CSHO's ability to successfully complete OJT and self-study assignments.

The expertise and judgment of the supervisor are required for assessing a CSHO's progress in the training program. The supervisor must make certain that the CSHO is ready to perform an assigned task independently. The program is flexible enough to afford the CSHO time for proper sequencing of training. Training assignments may be supplemented by other task assignments, as deemed necessary by the supervisor. Training in the following subject areas, at a minimum, is to be accomplished through both OJT and self-study assignments:

- Hazard recognition overview
- Inspection procedures
- Standards:

  General industry

  Construction

  Maritime (dependent on geographic location of area office)

  Agricultural

  American National Standards Institute (ANSI)

  National Electrical Code (NEC)

  National Fire Protection Association (NFPA)

  American Conference of Governmental Industrial Hygienists (ACGIH)

  International Agency for Research on Cancer (IARC)

  National Toxicology Program (NTP)

- *Field Operations Manual* (FOM)
- *IMIS Forms Manual*
- OSHA directives system
- *IH Technical Manual*
- Other regulations and procedures
- Common industrial processes
- Standards alleged violation elements (SAFEs) manuals:

  *Regulatory and General Industry*

  *Construction*

  *Maritime*

- Instrumentation
- Report writing
- Basic elements of safety and health programs

**Continuing maintenance of skills and knowledge.** Once the training period is completed, CSHOs typically require additional training to keep themselves current in the safety and health field. At a minimum, each CSHO is required to attend a technical course once every 3 years at the OSHA Training Institute. If an institute course has changed significantly during the years, the CSHO is permitted to repeat the course.

CSHOs are also encouraged to pursue other training opportunities available both within the Department of Labor and elsewhere.

# Personnel Needs and Problems

The number of practicing industrial hygienists in the United States in 1977 was estimated to be about 5,000 by John Short and associates under a National Institute for Occupational Safety and Health (NIOSH) contract.

Although no parallel survey data exist for 1994, membership figures in professional societies are helpful in assessing present trends.

The American Industrial Hygiene Association reports a national membership of 11,400. If local section AIHA members who are not also national members are included, the figure rises to approximately 16,000. Of these, there are currently about 5,600 active CIHs, 800 IHITs, and 1,100 OHSTs.

The American Conference of Governmental Industrial Hygienists (ACGIH) has a membership of over 5,000 industrial hygienists from 43 countries. Many hygienists belong to both organizations, limiting the data's usefulness as an estimate of the total number of professional industrial hygienists.

In 1975, OSHA, using 1973 NIOSH data, reported a national census of only 500 industrial hygienists, but 15,000 occupational safety and health specialists. The OSHA estimate indicated a then-current need for 5,500 industrial hygienists and 24,000 safety and health specialists. At that time, OSHA also predicted the need for 11,900 industrial hygienists and 62,300 occupational safety and health specialists by 1985.

Cycles of growth and contraction in industry and government will undoubtedly continue to affect the demand for industrial hygienists into the 21st century. In the 1980s, expansion of the need for hygienists came in nontraditional areas such as environmental remediation, indoor air quality, and a number of areas that many see as temporary trends; asbestos management and remediation projects are prime examples. In the 1990s, however, downsizing by many corporations has resulted in industrial hygienists often functioning as safety and environmental or even risk-management professionals, or delegating responsibilities such as safety training to less trained and credentialed personnel. Some industrial hygienists whose corporate jobs were eliminated now serve as private consultants to a variety of clients, including the corporations they left. Whereas 42 percent of AIHA's members are in private industry, consultants (in firms or self-employed) make up about 24–25 percent of the membership, up from about 10 percent in 1984.

Additionally, if a contraction of government agencies occurs because of a changing political climate, this may mute the demand for industrial hygienists in both industry and government.

The absolute need for individuals trained in the prevention of disease and preservation of health and safety will not change. Eventually, data such as worker compensation costs and illness and injury rates will reveal the need for prevention rather than repair of injury.

## Education and Training Programs

The education and training programs for industrial hygiene include professional school training, graduate curricula, and continuing education (short courses). Professional school curricula in industrial hygiene generally culminate in a Master of Science or a Master of Public Health degree.

## Educational Resource Centers

NIOSH's findings of shortages of trained occupational safety and health graduates were cited in successful efforts to expand training grants programs. One part of this expansion was the introduction of multidisciplinary educational resource centers (ERCs). The other part was growth of single-discipline training grants.

Congress authorized creation of up to 20 Educational Resource Centers for occupational safety and health in 1976. Funding increased from $2.9 million in 1977 to $12.9 million in 1980, and in 1995 the ERCs now number 14. These centers provide continuing education to occupational health and safety professionals; combine medical, industrial hygiene, safety, and nursing training so that graduates are better able to work effectively in complex and diverse conditions; conduct research; and conduct regional consultation services. All ERCs are located in universities. The centers are distributed as far as possible to give regional representation and to meet training needs for all areas of the nation. Research indicated that in 1990–1991, 438 master's degrees and 20 doctoral degrees were granted, some by non-ERC institutions.

## Professional Schooling

A program of study leading to a professional degree in industrial hygiene should start with two years of basic arts and sciences, two years of derivative sciences and advanced subjects,

and two years of professional courses. Such an advanced degree might appropriately be designated Doctor of Occupational Health, Doctor of Public Health, Doctor of Science, or Doctor of Engineering. Regardless of its name, however, it should be clearly understood that such a degree is a professional scholar's degree.

## Graduate Curricula

Graduate study programs have generally been developed to provide in-depth knowledge of a particular subject area and to develop scholarly research capabilities.

*Program Criteria for Industrial Hygiene and Similarly Named Engineering-Related Programs,* submitted by the American Academy of Industrial Hygiene and approved by the Accrediting Board of Engineering and Technology (ABET) in 1985, includes criteria for master's level programs in industrial hygiene. These criteria state that candidates for master's degree programs must have a baccalaureate degree that includes 63 or more semester hours of college-level mathematics, including technological courses, and a minimum of 21 semester hours in communications, humanities, and social sciences.

To be considered for accreditation, an industrial hygiene program must be designed to prepare students for the practice of industrial hygiene. Such a program must have an adequate foundation in mathematics and basic sciences, the humanities and social sciences, industrial hygiene science, and industrial hygiene practice; and a specialization in advanced industrial hygiene topics appropriate to the challenges presented by today's occupational health problems.

A minimum of 30 semester hours is required for a master's degree in industrial hygiene. The program must demonstrate an intensive and comprehensive level of inter-disciplinary instruction. Its content should include special projects, research, and a thesis or internship. Special emphasis also may be placed on the development of research capability, management skills, and interdisciplinary and governmental relationships.

The ABET calls engineering-related sciences industrial hygiene sciences. These expand topics of basic science for application in professional practice. A topic is identified as an industrial hygiene science if it amplifies basic science or mathematics, is taught by industrial hygiene faculty, solves closed-form problems, and contains quantitative expression.

To prepare students for practice in the industrial hygiene profession, the academic program must introduce them to the tools, methods, terminology, and professional services of the requisite interdisciplinary areas. Toward that goal, a master's level program in industrial hygiene must offer coursework in the following areas:

- Principles and practice of industrial hygiene
- Principles and practice of environmental sciences
- Epidemiology and biostatistics

The ABET also calls engineering-related specialties industrial hygiene practice where industrial hygiene sciences are applied to solve needs of society. A topic is properly placed in this category if it applies industrial hygiene sciences to these needs, employs open-form problems usually resulting in a written solution, involves cost and ethical considerations, and requires independent judgment to integrate specialty areas into a professional service. Typical topics include the following:

- Control of physical and chemical hazards
- Environmental health
- Occupational safety

A minimum of 18 semester hours must be in industrial hygiene sciences and industrial hygiene practice.

The unspecified 12-hour portion of a curriculum gives freedom to meet stated objectives without constraint by the accreditation process. Professional practice in industrial hygiene varies from state to state and country to country, depending on local law and custom. An industrial hygienist may be employed by a government agency; an industrial corporation; an insurance company; a community, labor union, or trade organization; or a consulting firm. The nature of the professional practice varies as much among these employers as it does with the area of the world with which the practice is concerned. The intent of "unspecified hours" is to allow specialization beyond that already embodied in the program. Typical topics include the following:

- Public health
- Environmental law
- Management techniques

## Faculty

To achieve sufficient breadth and depth, a minimum faculty equivalent to three full-time members is required.

Industrial hygiene faculty members not only lecture to students but must also generate new knowledge and demonstrate new ways to apply basic principles to real situations. The majority of faculty members should have advanced degrees appropriate to their area of expertise, extensive professional experience, and certification by the ABIH. All faculty members are expected to be active in research, consulting, and their professional societies. Currently, 12 master's-level programs (of approximately 50 programs in existence) are accredited.

## Baccalaureate Programs

Educational requirements for baccalaureate programs in occupational health are reviewed by Levine et al. (1977). NIOSH has offered training grants to academic institutions to develop educational programs in occupational health and safety. It was hoped that these programs might act as a bridge between present associate arts and graduate degree programs. Students completing the baccalaureate programs would be ideally qualified to function in the workplace or to pursue graduate study and specialization.

In his review, Levine grouped the activities of professional industrial hygienists into the following major areas of responsibility:

- Recognizing and identifying all chemical, physical, and biological agents that may adversely affect the physical, mental, and social well-being of the worker and the community
- Measuring and documenting levels of environmental exposure to specific hazardous agents
- Evaluating the significance of exposures and their relationships to the etiology of occupationally and environmentally induced diseases
- Establishing appropriate controls to prevent hazardous exposures and monitor their effectiveness
- Administering the occupational and environmental hygiene program
- Joining with medical, safety, and other members of the occupational health team in developing and presenting a comprehensive approach to prevention programs
- Developing procedures to ensure continuing professional development
- Participating in policy-making decisions

Specific tasks were then identified that must be performed to fulfill these responsibilities.

A recent proposal by the American Academy of Industrial Hygiene to ABET proposed an undergraduate curriculum that would solidify the requirements for undergraduate programs in industrial hygiene and occupational hygiene, as follows:

A minimum of 6 semester-credit hours are required in mathematics. The coursework in mathematics must include college algebra or courses more advanced, introductory level calculus through integrals, as well as statistics. Fundamental computer science courses must be completed to ensure facility with microcomputers and a range of software types, but any such courses must be in addition to the basic mathematics requirement. Access to computers must be sufficient to accommodate integration of computer work into coursework.

Industrial hygiene students must complete a minimum of 12 semester credit hours of chemistry courses that include laboratories, including organic chemistry; 6 semester credit hours of physics, with laboratories, including mechanics, sound, light, optics, and electricity; and 6 semester credit hours of biology.

Students must complete a minimum of 21 semester credit hours in communications, the humanities, and social sciences.

In order to practice in the industrial hygiene profession, the student must be introduced to the tools, methods, terminology, and professional services of each professional area. Toward that goal, a core program in industrial hygiene sciences of at least 15 semester credit hours is specified, consisting of the following:

- A minimum of 3 semester credit hours in each of the following four areas, however courses are titled or organized:

  Fundamentals (or principles) of industrial hygiene

  Industrial hygiene measurements, including laboratory work

  Industrial hygiene controls

  Toxicology

- A minimum of 3 additional semester credit hours in any other industrial hygiene science

Engineering-related specialties are considered industrial hygiene practice. A minimum of 15 semester credit hours must be in industrial hygiene practice. Typical topics include, but are not limited to, the following:

- Epidemiology
- Industrial hygiene field problems
- Industrial ventilation
- Noise control
- Ergonomics/human factors
- Air quality/indoor air quality
- Industrial hygiene problems
- Environmental health
- Radiation measurement and control/health physics
- Laboratory safety
- Structured industrial hygiene internship
- Industrial safety science/safety engineering
- Hazardous waste management
- Environmental engineering
- Environmental sciences

In addition to the 15 semester credit hours required in both industrial hygiene sciences and industrial hygiene practice, a minimum of 15 semester credit hours must be in technical subjects such as additional industrial hygiene sciences or practice, or mathematics, science, technology, computing, or engineering beyond the basic requirements. For the baccalaureate degree, a minimum of 120 semester credit hours is required.

**Unspecified hours.** The unspecified portion of the curriculum gives freedom to meet stated objectives without constraint by the accrediting process. Unique program objectives may be met elective courses, thus allowing for program specialization within the industrial hygiene profession. Also, unspecified hours allow for compliance with state statutory requirements for undergraduate education that are not completely met by the listings above.

**Faculty size.** The proper size of the undergraduate faculty is in general determined by enrollment in the program and the allocation of faculty time to such activities as teaching, laboratory supervision, research direction, direction of graduate work, extension or continuing education activities, and active participation in professional and technical societies. However, to achieve sufficient breadth and depth, a minimum full-time equivalent of two industrial hygiene instructors is required.

**Faculty qualifications.** Industrial hygiene faculty members not only lecture to students but also generate new knowledge and demonstrate new ways to apply basic principles to real situations. The majority of faculty members should have advanced degrees appropriate to their areas of expertise. The majority of primary industrial hygiene faculty members should be certified by the American Board of Industrial Hygiene. Other program faculty members in closely related areas should maintain relevant high-level certification or registration in their fields. Each faculty member should have had practical experience at a responsible level in professional practice or in research to ensure that current attitudes and methods are imparted to students. All faculty are expected to be active in their professional societies.

## Continuing Education

A wide variety of opportunities exist for industrial hygienists who want to remain technically current, receive training in previously unfamiliar aspects of industrial hygiene, pursue academic coursework leading to a more advanced degree, or earn certification maintenance points in order to maintain CIH certification.

A number of universities offer coursework leading to degrees. Also available at such universities are usually short courses (a few days or weeks long) on specific industrial hygiene topics. Summer institutes (1–4 weeks long) concentrating on a particular area of industrial hygiene are another continuing education opportunity.

NIOSH publishes an annual catalogue of all such training courses nationwide at universities that are funded as NIOSH ERCs. Most ERCs contain an industrial hygiene component that includes coursework leading to academic degrees and short courses. The catalogue can be obtained through the NIOSH publications dissemination office. A number of other not-for-profit and for-profit training organizations provide short courses in industrial hygiene and related topics. These include the National Safety Council and professional industrial hygiene and safety societies as well as consulting firms.

## ■ Summary

The need to control exposures to a rapidly rising number of chemicals and hazardous agents and to comply with and enforce OSHA regulations has brought about greater demand for industrial hygienists. This demand exists in private industry, labor unions, government, and academic organizations.

Individuals practicing industrial hygiene routinely work as a team; thus, the physician, the nurse, the safety professional, and the industrial hygienist are quite accustomed to working together. Other professions are included as needed; these include toxicologists, health physicists, epidemiologists, statisticians, professional trainers, and educators. A team approach, using the knowledge and skills of all these professionals, increases the effectiveness of programs to prevent occupational disease and injuries and helps to anticipate future requirements.

The need continues for industrial hygienists to interpret the findings of environmental investigations and to design and implement control measures. The industrial hygienist must, therefore, have the generalist's grasp of varied disciplines in order to interact with divergent groups to develop and maintain the most effective program.

Educational requirements for industrial hygienists will continue to expand with the increasing need to monitor and control hazardous agents and to comply with more stringent government regulations. Proposed curricula and course descriptions were included in this chapter. The training program for OSHA CSHOs was also discussed.

Personnel from three professional specialties—industrial hygiene, safety, and environmental health—will be working even more closely together in the future, their responsibilities overlapping in many instances. The separation between these professions has become increasingly blurred, and melding may eventually lead to the creation of a single profession whose scope is made up of what is currently recognized today as industrial hygiene and safety.

# ■ Bibliography

American Board of Industrial Hygiene. *Bulletin*. Lansing, MI: American Board of Industrial Hygiene, Aug. 1, 1994.

American Industrial Hygiene Association. *1994–1995 Membership Directory*. Fairfax, VA: American Industrial Hygiene Association, 1995.

Berry CM. Industrial hygiene manpower. *AIHAJ* 36:433–446, 1975.

Berry CM. What is an industrial hygienist? *National Safety News* 107:69–75, 1973.

Constantin MJ et al. Status of industrial hygiene graduate education in U.S. institutions. *AIHAJ* 55:537–545, 1994.

Corn M, Heath ED. OSHA response to occupational health personnel needs and resources. *AIHAJ* 38:11–17, 1977.

Hermann ER. Education and training of industrial hygienists. *National Safety Congress Trans* 12:64–66, 1975.

Levine MS, Watfa N, Hanna F, et al. A plan for baccalaureate education in occupational hygiene. *AIHAJ* 38:447–455, 1977.

Office of Technology Assessment, U.S. Congress. *Preventing Injury and Illness in the Workplace* (unpublished manuscript). New York: InfoSource, 1985.

Radcliffe JC, Clayton GD, Frederick WG, et al. Industrial hygiene definition, scope, function, and organization. *AIHAJ* 20:429–430, 1959.

Salzman BE. Adequacy of current industrial hygiene and occupational safety professional manpower. *AIHAJ* 43:254–260, 1982.

# ■ Addendum: Professional Societies and Courses of Interest to Industrial Hygienists

## American Industrial Hygiene Association

The AIHA is a nonprofit professional society for people practicing industrial hygiene in industry, government, labor, academic institutions, and independent organizations. Currently, close to 16,000 members are affiliated with AIHA. Of these, more than 11,000 are members of the National Association. Membership is drawn from the United States, Canada, and 43 other countries.

The AIHA was established in 1939 by a group of industrial hygienists to provide an association devoted exclusively to industrial hygiene. AIHA is a national society of professionals engaged in protecting the health and well-being of workers and the general public through the scientific application of knowledge concerning chemical, engineering, physical, biological, or medical principles to minimize environmental stress and to prevent occupational disease.

The AIHA promotes the recognition, evaluation, and control of environmental stresses arising in the workplace and encourages increased knowledge of occupational and environmental health by bringing together specialists in this field. The American Industrial Hygiene Conference and Exposition, cosponsored by AIHA, draws more than 10,000 industrial hygiene professionals each May.

**AIHA membership qualifications and types.** Membership in the AIHA is open to people engaged in industrial hygiene activities. The classes of membership are student, affiliate, associate, full, fellow, retired, honorary, and organizational. Application for membership may be obtained from the association office (see Bibliography).

A full-time student at the college undergraduate level may become a student member by submitting a yearly application and adequate matriculation documentation to the association. A student member may not serve on committees, vote, or hold office.

An affiliate member is a person who is employed full-time in an occupation requiring interaction with and cooperation of associate members and full members of the association. An affiliate member may not vote or serve on the Board of Directors, but may serve on committees.

An associate member is a graduate of an accredited school of college grade with a baccalaureate degree in industrial hygiene, chemistry, physics, engineering, biology, or a cognate discipline who is currently engaged a majority of time in industrial hygiene activities as defined by the Board of Directors, or is a full-time graduate student in industrial hygiene or a cognate discipline. Where an applicant does not have a baccalaureate degree, experience may be substituted on the basis of two years of qualifying experience for one year of undergraduate education. Membership as an associate member is limited to a maximum of 5 years. An associate member may serve on committees and vote but may not be elected to the Board of Directors.

A full member is a graduate of an accredited school of college grade with a baccalaureate degree in industrial hygiene, chemistry, physics, engineering, biology, or a cognate discipline who has been engaged a majority of time for at least 3 years in industrial hygiene-related activities as defined by the Board of Directors. The Board of Directors will consider, and may accept, any other degree proposed by a candidate who provides evidence of the scientific content of the curriculum. The social sciences are not considered qualifying sciences. Where an applicant does not have a baccalaureate degree, experience may be substituted on the basis of two years of qualifying experience for one year of undergraduate education. Full-time graduate study in industrial hygiene or a cognate discipline may be accepted on an equivalent-time basis for any portion of the required three years of experience. A full member may serve on committees, vote, and serve on the Board of Directors. The Board of Directors may grant fellow membership to full members who have at least 15 years of continuous membership and who significantly contributed to the field of industrial hygiene. Nominations to the fellow class may be made by a local section, committee, or the Board of Directors. Fellows have the same privileges as full members. A retired member is a full or fellow member who has retired from the practice of industrial hygiene. A retired member retains the privileges of a full or fellow member but does not pay dues and may not serve as an officer or a director.

Those who are particularly distinguished in the field of industrial hygiene or in a closely related scientific field may be granted honorary membership by the Board of Directors.

Any organization may apply for organizational membership, but organizational members cannot vote.

**Local section membership.** Any person with a professional interest in industrial hygiene may apply for membership in an AIHA local section. Application for membership in a local section should be made to the local section.

**Address:** American Industrial Hygiene Association, 2700 Prosperity Ave., Suite 250, Fairfax, VA 22031.

## American Board of Industrial Hygiene

The American Board of Industrial Hygiene (ABIH) was established to improve the practice and educational standards of the profession of industrial hygiene. To this end, the ABIH engages in the following activities:

- To receive and process applications for examinations and to evaluate the education and experience qualifications of the applicants for such examinations

- To grant and to issue (to qualified people who pass the board's examinations) certificates acknowledging their competence in industrial hygiene and to revoke certificates so granted or issued for cause

- To provide for maintenance of certification by requiring evidence of continued professional qualifications by certificate holders in the comprehensive or chemical practice of industrial hygiene

- To maintain a record of certificate holders

- To furnish to the public, and to interested people or organizations, a roster of certificate holders having special training, knowledge, and competence in industrial hygiene

The American Board of Industrial Hygiene issues three categories of certificates. The first certifies that the individual has the required education, experience, and professional ability in the comprehensive practice or chemical practice of industrial hygiene (CIH). The second category is the industrial hygienist in training (IHIT) certification. This refers to individuals who are permitted to take the core examination before completing the 5-year eligibility requirement. The third category, the occupational health and safety technologist (OHST) designation, is a joint certification with the Board of Certified Safety Professionals (BCSP). The OHST examination procedure is administered by the BCSP (see Bibliography).

Each applicant for a certificate must meet certain eligibility requirements and must pass an examination. The examination for certification consists of two parts.

- The first part, the core examination, covers general aspects of industrial hygiene to the degree that, in the opinion of the board, should be familiar to the candidate.

- The second part of the examination consists of different sets of questions for certification in the comprehensive practice or the chemical practice of industrial hygiene.

**Certification maintenance.** The ABIH also administers a certification maintenance program for CIHs. The purpose of this program is to ensure that CIHs continue to develop and enhance their professional industrial hygiene skills for the duration of their careers. The certificate is granted for a period of 6 years, after which time it expires unless renewed. Certificate holders must provide evidence to the board of their continued professional qualifications in order to renew the certificate. Activities that are accepted as evidence include continuing professional industrial hygiene practice; membership in an approved professional society (other than the American Academy of Industrial Hygiene); attendance at approved meetings, seminars, and short courses; participation in technical committees; publishing in peer-reviewed journals; teaching that is not part of the diplomate's primary practice; approved extracurricular professional activities; and reexamination or examination for an additional certification. Points for the approved activities are awarded and publicized by the board, as is a schedule for renewal of certificates.

Besides being entitled to use the CIH designation, people certified in either comprehensive practice or chemical practice become members of the American Academy of Industrial Hygiene and their names are published in the annual roster of the academy. The names of IHITs are also published in the academy roster.

**Address:** American Board of Industrial Hygiene, 4600 W. Saginaw, Suite 101, Lansing, MI 48917-2737.

## American Academy of Industrial Hygiene

The American Academy of Industrial Hygiene (AAIH) is a professional association of practicing industrial hygienists who have participated successfully in the certification program administered by the American Board of Industrial Hygiene (ABIH). Completion of this program demonstrates the highest degree of proficiency in the practice of industrial hygiene.

In 1957, the American Industrial Hygiene Association (AIHA) set out to establish a certification program for qualified industrial hygienists. The American Conference of Governmental Industrial Hygienists joined the effort in 1958. The ABIH was incorporated as an independent organization to develop and administer the certification program. Six members from each sponsoring organization made up the first board; its first annual meeting was held in 1960. In 1966, the diplomates activated the AAIH as a professional organization. Through 1994, 6,607 industrial hygienists have been certified.

The purpose of the AAIH is to establish high standards of professional conduct and professionalism among those practicing in the field of industrial hygiene. AAIH seeks to promote recognition of the need for high-quality industrial hygiene practice to ensure healthful work conditions in the occupations and industries its members serve.

Activities include establishment of a code of ethics to serve as a guide for professional conduct by industrial hygienists; promotion of the recognition of industrial hygiene as a profession by individuals, employers, and regulatory agencies; advancement of board certification as a basic qualification for employment as an industrial hygienist in both public and private organizations; accreditation of academic programs in industrial hygiene in cooperation with the Accreditation Board of Engineering and Technology; and recruitment of students into academic programs and training through initial education and continuing education for practicing industrial hygienists.

The AAIH sponsors the Professional Conference on Industrial Hygiene to provide a forum for exploring professional issues. Continuing education opportunities also are provided. The conference is aimed primarily at issues encountered by the more experienced industrial hygienist but is not restricted to members of AAIH.

## American Conference of Governmental Industrial Hygienists

The American Conference of Governmental Industrial Hygienists (ACGIH) was organized in 1938 by a group of government industrial hygienists who desired a medium for the free exchange of ideas and experiences and the promotion of standards and techniques in occupational and environmental hygiene.

As an organization devoted to the development of administrative and technical aspects of worker health protection, the ACGIH has contributed substantially to the development and improvement of official occupational health services to industry and labor. ACGIH endeavors to provide opportunities, information, and other resources needed by those who protect worker health and safety. Technical committees, publications, symposia, journals, and other programs work toward this goal. The committees on industrial ventilation and Threshold Limit Values are recognized throughout the world for their expertise and contributions to industrial hygiene. The ACGIH sets TLVs and annually updates these values.

**Objectives.** The objectives of the ACGIH are as follows:

- To promote and encourage the coordination of industrial hygiene, occupational and environmental health, and safety through federal, state, local, territorial, and international agencies
- To encourage the interchange of experience and knowledge among industrial hygiene, occupational and environmental health, and safety professionals and in the occupational and environmental health community at large
- To collect and make accessible to all those engaged in industrial hygiene, occupational and environmental health, and safety information and data that may assist them in the performance of their duties

- To collect and make available information, data, and reports to government agencies, international organizations, and the general public to assist in providing more adequate services

- To engage in activities and to hold annual and other meetings that may be necessary to carry out the objectives of the conference

**Membership.** ACGIH was originally formed as an organization of industrial hygienists who worked in government. It has recently expanded its scope to offer membership to a broader spectrum of practitioners. There are seven categories of membership in ACGIH: Today, anyone who is engaged in the practice of industrial hygiene or occupational and environmental health and safety is eligible for one of seven categories of membership.

A *full member* is an industrial hygienist or occupational health, environmental health, or safety professional whose full-time, primary employment is with a governmental agency or an educational institution and who is engaged in health or safety services, standard setting, enforcement, research, or education. Full members are accorded full voting privileges and can serve as officers or members-at-large of the Board of Directors as well as on any appointive committee.

An *associate member* is a person professionally employed in a full-time activity closely allied to industrial hygiene, occupational health, environmental health, or safety who is either an employee of a government agency or an educational institution, or who works more than 50 percent of his or her time on a government contract at a government facility. An associate member may also be a person with at least 10 years of membership in the full or technical category who has retired or is eligible for retirement benefits from a government agency or educational institution but who is employed in the field at least 25 percent of his or her time by a government agency or educational institution. Associate members may vote on all conference matters and may serve as a member-at-large on the Board of Directors or as a member of an appointive committee.

A *technical member* is a technician employed in a full-time activity by a government agency or an educational institution in industrial hygiene, occupational health, environmental health, or safety. A technical member may also be a person who works more than 50 percent of his or her time on a government contract at a government facility that is engaged in such services. Technical members have the same voting and service privileges as associate members.

*Student members* are people officially enrolled in a full-time course of study directly related to industrial hygiene, occupational health, environmental health, or safety. Evidence of academic enrollment must include one of the following: current transcript, current class schedule, or a letter of reference from an academic advisor (on uni-

versity or college letterhead) indicating that the applicant is a full-time student. Students may not vote or hold elected office but may serve on appointive committees.

An *emeritus member* is a full, associate, or technical member who has retired from the practice of industrial hygiene, occupational health, environmental health, or safety and who has been a member of the conference for a minimum of 10 consecutive years. Retirement is defined as employment less than 25 percent of full-time. These members retain the rights and privileges of the category from which they qualified for this status.

An *honorary member* is a person who is recognized by the conference as having advanced the science of industrial hygiene, occupational health, environmental health, or safety. Members designated as honorary may not vote or hold elective office but may serve as nonvoting consultants to appointive committees.

*Affiliate members* are people engaged in health or safety services who are not currently eligible for another category of membership. They may not vote on conference matters or hold elected office but they may serve as consultants on appointive committees.

**Address:** ACGIH, Kemper Meadow Center, 1330 Kemper Meadow Drive, Cincinnati, OH 45240.

## American Public Health Association

The American Public Health Association (APHA), established in 1872, is 50,000 strong in its collective membership, which represents all the disciplines and specialties in the public health spectrum. The APHA is devoted to the protection and promotion of public health. It achieves this goal in several ways:

- Sets standards for alleviating health problems

- Initiates projects designed for improving health, both nationally and internationally

- Researches health problems and offers possible solutions based on that research

- Launches public awareness campaigns about special health dangers

- Publishes materials reflecting the latest findings and developments in public health

The APHA has 23 special sections, including an occupational health and safety section that includes occupational health physicians and nurses, industrial hygienists, and other allied occupational health professionals. Each APHA section has its own professional meetings to provide a forum for the exchange of ideas.

**Membership.** Three types of membership are available:

- Regular membership is available to all health professionals.

- Contributing membership provides additional association benefits.

  Special memberships include the following:

- Student/trainee: people enrolled full-time in a college or university or occupied in a formal training program in preparation for entry into a health career

- Retired: APHA members who have retired from active public health practice and no longer derive significant income from professional health-related activities

- Consumer: people who do not derive income from health-related activities

- 15th Street NW, Washington, DC 20005.Special health workers: people employed in community health whose annual salary is less than $10,000 (or its equivalent for foreign nationals)

  **Address:** American Public Health Association, 1015

## An ABET-Accredited MS Curriculum in Industrial Hygiene*

| Required Course | Semester Credit |
|---|---|
| Environmental calculations | 1 |
| Fundamentals of industrial hygiene | 2 |
| Fundamentals of industrial hygiene: Methods | 1 |
| Industrial hygiene laboratory I | 2 |
| Industrial hygiene laboratory II (field studies) | 2 |
| Industrial hygiene: Engineering control | 2 |
| Principles of epidemiology | 2 |
| Biostatics I** | 3 |
| Biostatics II | 4 |
| Radiation protection | 3 |
| Air quality management I | 3 |
| Air quality laboratory | 1 |
| Industrial ventilation (1/2 semester) | 2 |
| Environmental acoustics (1/2 semester) | 2 |
| Operational safety science | 2 |
| Environmental and occupational health seminar | 1 |
| Water quality management I | 3 |
| Water quality laboratory | 1 |
| either Occupational and environmental diseases I | 2 |
| or Industrial toxicology | 2 |

*Research:*

| | |
|---|---|
| Research in public health sciences | 16 |
| Required + Research = 39 + 16 | = 55 semester credits |

\* Prerequisites for entering the IH program are a full year of general chemistry, at least one semester of organic chemistry, mathematics through differential and integral calculus, and a course in human physiology.

\*\* Many of our students have already taken a first course in statistics as undergraduates, and may substitute for these requirements.

## An ABET-Accredited MPH Curriculum in Industrial Hygiene*

| Required Course | Semester Credit |
|---|---|
| Environmental calculations | 1 |
| Environmental occupational health seminar | 1 |
| Behavioral sciences in public health | 2 |
| Principles of epidemiology | 2 |
| Biostatics I | 3 |
| Principles of management in public health | 3 |
| Fundamentals of industrial hygiene | 2 |
| Fundamentals of industrial hygiene: Methods | 1 |
| Industrial hygiene laboratory I | 1 |
| Air quality management I | 3 |
| Air quality laboratory | 1 |
| Water quality management I | 3 |
| Water quality laboratory | 1 |
| Management of solid/hazardous wastes | 3 |
| Radiological health | 3 |
| Public health concepts and practice | 2 |
| Field experience in public health** | 3–5 |
| Industrial hygiene laboratory II (field studies) | 2 |
| Industrial hygiene: engineering control | 2 |
| Industrial ventilation (1/2 semester) | 2 |
| Environmental acoustics (1/2 semester) | 2 |
| Occupational safety science | 2 |
| either Occupational Environmental Diseases I | 2 |
| or Industrial toxicology | 2 |

*Electives:*

| | |
|---|---|
| Quantitative methods in epidemiology I | 3 |
| Occupational/environmental epidemiology | 2 |
| Groundwater contamination | 3 |
| Hazardous water management II | 3 |
| Biostatics II | 4 |
| Required + Electives = 47–49 + 7–9 | = 56 semester credits |

\* Prerequisites for entering the IH program are a full year of general chemistry, at least one semester of organic chemistry, mathematics through differential and integral calculus, and a course in human physiology.

\*\* Often waived because of prior work experience.

# CHAPTER 24

# The Safety Professional

Revised by Peter B. Rice, CIH, CSP

**Definition of a Safety Professional**
*Accident Prevention Activities*

**Safety and Health Programs**

**Staff Versus Line Status**

**Codes and Standards**
*Policies and Procedures* ▪ *Engineering* ▪ *Machine and Equipment Design* ▪ *Purchasing*

**Safety and Health Inspections**
*Purpose of a Safety and Health Inspection Program* ▪ *Inspection of Work Areas* ▪ *Safety Inspectors or Technicians* ▪ *Safety Professionals* ▪ *Third-Party Inspections, or Audits*

**Accident and Occupational Illness Investigations**
*Purpose of Investigations* ▪ *Types of Investigations* ▪ *Who Conducts the Investigation?*

**Record Keeping and Reporting**
*Uses of Records* ▪ *Accident Reports and Illness Records*

**Education and Training**
*Employee Training* ▪ *Maintaining Interest in Safety* ▪ *Safety and Health Rule Enforcement* ▪ *Role of the Supervisor* ▪ *Supervisor Training* ▪ *Job Safety and Health Analysis*

**Risk Management**
*A Five-Step Program* ▪ *Damage Control*

**Systems Safety**
*Methods of Analysis*

**Safety Professional Certification**
*Current Requirements* ▪ *Academic and Experience Requisites*

**The Future of Safety as a Profession**

**Summary**

**Bibliography**

Tʜɪs ᴄʜᴀᴘᴛᴇʀ ᴅᴇᴀʟs ᴡɪᴛʜ ᴛʜᴇ ʀᴏʟᴇ of the safety professional in an effective occupational safety and health program. The duties and functions of safety professionals and how they work with other professionals are briefly discussed, as well as the many ways the safety professional can contribute to the success of an occupational safety and health program.

Safety is a multidisciplinary profession, drawing its professionals from many different fields such as education, engineering, psychology, medicine, biophysics, chemistry, and labor. Interestingly, many professionals in the field did not initially choose safety as a career, but instead became interested in accident prevention and loss control while working in other disciplines. Many have made the change to the safety profession after recognizing the fact that a well-defined safety program promotes management's objective of producing high-quality products at the lowest cost.

The safety profession today is a sophisticated discipline combining engineering, behavioral psychology, and a knowledge about such topics as systems safety analysis, human factors engineering, biomechanics, and product safety (Figure 24–1). In addition, the safety professional must possess a thorough knowledge of a facility's equipment, property, manufacturing processes, and employees. In many facilities the safety professional must be able to work with employees with varying linguistic and cultural backgrounds.

The safety professional must display tact, diplomacy, persuasiveness, and persistence. In short, today's safety professional must wear many hats and play many roles.

# The Scope of the Professional Safety Position

To perform their professional functions, safety professionals must have education, training and experience in a common body of knowledge. Safety professionals need to have a fundamental knowledge of physics, chemistry, biology, physiology, statistics, mathematics, computer science, engineering mechanics, industrial processes, business, communication and psychology. Professional safety studies include industrial hygiene and toxicology, design of engineering hazard controls, fire protection, ergonomics, system and process safety, safety and health program management, accident investigation and analysis, product safety, construction safety, education and training methods, measurement of safety performance, human behavior, environmental safety and health, and safety, health, and environmental laws, regulations and standards. Many safety professionals have backgrounds or advanced study in other disciplines, such as management and business administration, engineering, education, physical and social sciences and other fields. Others have advanced study in safety. This extends their expertise beyond the basics of the safety profession.

Because safety is an element in all human endeavors, safety professionals perform their functions in a variety of contexts in both public and private sectors, often employing specialized knowledge and skills. Typical settings are manufacturing, insurance, risk management, government, education, consulting, construction, health care, engineering and design, waste management, petroleum, facilities management, retail, transportation, and utilities. Within these contexts, safety professionals must adapt their functions to fit the mission, operations and climate of their employer.

Not only must safety professionals acquire the knowledge and skill to perform their functions effectively in their employment context, through continuing education and training they stay current with new technologies, changes in laws and regulations, and changes in the workforce, workplace and world business, political and social climate.

As part of their positions, safety professionals must plan for and manage resources and funds related to their functions. They may be responsible for supervising a diverse staff of professionals.

By acquiring the knowledge and skills of the profession, developing the mind set and wisdom to act responsibly in the employment context, and keeping up with changes that affect the safety profession, the safety professional is able to perform required safety professional functions with confidence, competence and respected authority.

## Functions of the Professional Safety Position

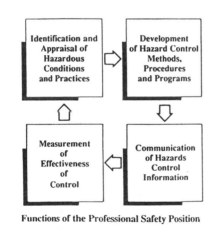

Functions of the Professional Safety Position

## The major areas relating to the protection of people, property and the environment are:

**A.** Anticipate, identify and evaluate hazardous conditions and practices.

**B.** Develop hazard control designs, methods, procedures and programs.

**C.** Implement, administer and advise others on hazard controls and hazard control programs.

**D.** Measure, audit and evaluate the effectiveness of hazard controls and hazard control programs.

**Figure 24–1a.** The scope and functions of the professional safety position are reprinted with permission from an American Society of Safety Engineers (ASSE) brochure. (*Continues*)

The safety professional should serve as a counselor to the company's chief executive. The safety professional must be able to enter the boardroom of a company as an equal; to do this, he or she must understand the basic technology of the industry.

An increasing number of institutions of higher education offer degrees in safety engineering and/or safety management. A listing of schools and of training courses is given in Appendix A, Additional Resources. Such courses are essential to the continuing development of safety and health professionals.

In its broadest sense, occupational health has come to mean not only freedom from disease but from injury as well. Because of this, the safety professional has become more closely aligned with the industrial hygienist and the field of occupational medicine. It is rare to find a safety professional who does not practice some traditional industrial hygiene or vice versa.

There is no question that accidents are painful and costly to the worker, the worker's family, and to society. (The term *accident* as used here is defined to mean any unexpected happening that interrupts the work sequence or process and that may result in injury, illness, or property damage to the extent that it causes loss.) Accidents produce economic and social loss, impair individual and group productivity, cause inefficiency, upset employee

## A. Anticipate, identify and evaluate hazardous conditions and practices.

**This function involves:**

1. Developing methods for
   a. anticipating and predicting hazards from experience, historical data and other information sources.
   b. identifying and recognizing hazards in existing or future systems, equipment, products, software, facilities, processes, operations and procedures during their expected life.
   c. evaluating and assessing the probability and severity of loss events and accidents which may result from actual or potential hazards.

2. Applying these methods and conducting hazard analyses and interpreting results.

3. Reviewing, with the assistance of specialists where needed, entire systems, processes, and operations for failure modes, causes and effects of the entire system, process or operation and any sub-systems or components due to
   a. system, sub-system, or component failures.
   b. human error.
   c. incomplete or faulty decision making, judgements or administrative actions.
   d. weaknesses in proposed or existing policies, directives, objectives or practices.

4. Reviewing, compiling, analyzing and interpreting data from accident and loss event reports and other sources regarding injuries, illnesses, property damage, environmental effects or public impacts to
   a. identify causes, trends and relationships.
   b. ensure completeness, accuracy and validity of required information.
   c. evaluate the effectiveness of classification schemes and data collection methods.
   d. initiate investigations.

5. Providing advice and counsel about compliance with safety, health and environmental laws, codes, regulations and standards.

6. Conducting research studies of existing or potential safety and health problems and issues.

7. Determining the need for surveys and appraisals that help identify conditions or practices affecting safety and health, including those which require the services of specialists, such as physicians, health physicists, industrial hygienists, fire protection engineers, design and process engineers, ergonomists, risk managers, environmental professionals, psychologists and others.

8. Assessing environments, tasks and other elements to ensure that physiological and psychological capabilities, capacities and limits of humans are not exceeded.

## B. Develop hazard control designs, methods, procedures and programs.

**This function involves:**

1. Formulating and prescribing engineering or administrative controls, preferably before exposures, accidents, and loss events occur, to
   a. eliminate hazards and causes of exposures, accidents and loss events.
   b. reduce the probability or severity of injuries, illnesses, losses or environmental damage from potential exposures, accidents, and loss events when hazards cannot be eliminated.

2. Developing methods which integrate safety performance into the goals, operations and productivity of organizations and their management and into systems, processes, and operations or their components.

3. Developing safety, health and environmental policies, procedures, codes and standards for integration into operational policies of organizations, unit operations, purchasing and contracting.

4. Consulting with and advising individuals and participating on teams
   a. engaged in planning, design, development and installation or implementation of systems or programs involving hazard controls.
   b. engaged in planning, design, development, fabrication, testing, packaging and distribution of products or services regarding safety requirements and application of safety principles which will maximize product safety.

5. Advising and assisting human resources specialists when applying hazard analysis results or dealing with the capabilities and limitations of personnel.

6. Staying current with technological developments, laws, regulations, standards, codes, products, methods and practices related to hazard controls.

**Figure 24–1b.** (*Continues*)

morale and public image, and generally retard progress. Also, in today's world, a company with a poor safety program often finds it difficult to compete.

Dedicated safety professionals continue to be accident prevention's most valuable asset. Their ranks have grown to the point where membership in the American Society of Safety Engineers (ASSE) is now approaching 29,000. This organization, dedicated to the interests and professional development of safety engineers, has approximately 135 chapters in the United States and Canada, and it has members worldwide. There are many other qualified safety professionals in addition to the ASSE members, who, together with thousands of specialists and technicians, carry out a limited scope of activities within the occupational safety and health field.

In 1968, the ASSE was instrumental in forming the Board of Certified Safety Professionals (BCSP). Its purpose is to provide the professional status of a Certified Safety Professional (CSP) to qualified safety professionals by certification after they have met strict education and experience requirements and passed an examination. As of March 1995, approximately 9,500 CSPs and approximately 3,500 Associate Safety Professionals (ASPs) had been certified by the BCSP. The ASP designation is awarded to those who pass the Safety Fundamentals Examination, the initial exam of the certification process, and indicates a recognition of a person's progress toward certification.

In 1985, the BCSP and the American Board of Industrial Hygiene (ABIH) began joint sponsorship of a certification program for occupational safety and health technologists (OSHTs). This designation is not intended for those who are certified industrial hygienists (CIHs) or CSPs, nor is it intended for those eligible to take either the CIH or CSP examination. Its purpose is to recognize technologists in the fields of safety and health.

**C.** Implement, administer and advise others on hazard controls and hazard control programs.

**This function involves:**

1. Preparing reports which communicate valid and comprehensive recommendations for hazard controls which are based on analysis and interpretation of accident, exposure, loss event and other data.

2. Using written and graphic materials, presentations and other communication media to recommend hazard controls and hazard control policies, procedures and programs to decision-making personnel.

3. Directing or assisting in planning and developing educational and training materials or courses. Conducting or assisting with courses related to designs, policies, procedures and programs involving hazard recognition and control.

4. Advising others about hazards, hazard controls, relative risk and related safety matters when they are communicating with the media, community and public.

5. Managing and implementing hazard controls and hazard control programs which are within the duties of the individual's professional safety position.

**D.** Measure, audit and evaluate the effectiveness of hazard controls and hazard control programs.

**This function involves:**

1. Establishing and implementing techniques, which involve risk analysis, cost, cost-benefit analysis, work sampling, loss rate and similar methodologies, for periodic and systematic evaluation of hazard control and hazard control program effectiveness.

2. Developing methods to evaluate the costs and effectiveness of hazard controls and programs and measure the contribution of components of systems, organizations, processes and operations toward the overall effectiveness.

3. Providing results of evaluation assessments, including recommended adjustments and changes to hazard controls or hazard control programs, to individuals or organizations responsible for their management and implementation.

4. Directing, developing, or helping to develop management accountability and audit programs which assess safety performance of entire systems, organizations, processes and operations or their components and involve both deterrents and incentives.

**Figure 24–1c.** (*Continued*)

There has been an orderly development of safety knowledge, which, when applied with sufficient skill and judgment, has produced significant reductions in occupational disease and in many types of accidents and accidental injuries. However, the tremendous increase in scientific knowledge and technological progress has added to the complexities of safety work.

The focus on the control of industrial disease and accident prevention has oscillated between environmental control or engineering and human factors. Some important trends in the pattern of the safety professional's development have emerged.

■ First, increasing emphasis on analyzing the loss potential of the activity with which the safety professional is concerned; such analysis requires greater ability to predict where and how loss- and injury-producing events will occur and to find the means of preventing such events.

■ Second, increased development of factual, unbiased, and objective information about loss-producing problems and accident causation, so that those who have ultimate decision-making responsibilities can make sound decisions.

■ Third, increasing use of the safety professional's help in developing safe products. The application of the principle of accident causation and control to the product being designed or produced has become more important because of product liability cases; legal aspects in the general field of safety and health, including negligent design; and the obvious impact that a safer product has on the overall safety and health of the environment.

## Definition of a Safety Professional

What, then, is a safety professional? In the broad sense, a safety professional is a person whose basic job function and responsibility is to prevent accidents and other harmful exposures and the personal injury, disease, or property damage that may ensue. In the narrower sense, the safety professional is also the person who has successfully met the requirements for certification developed by the Board of Certified Safety Professionals of the Americas, Inc., by other certifying groups, or by state certification boards. This chapter deals primarily with the broad definition.

Whether called a safety engineer, safety director, loss control manager, or some other title, the safety professional's job is to prevent—as nearly as possible—the human suffering brought about by exposure to health hazards and unsafe conditions. Secondary goals are to prevent damage to equipment or materials, to minimize interruptions in operations, and to reduce the costs associated with employee accidents or illness originating in operational processes. The scope of the safety professional's job is illustrated in Figure 24–1.

Safety professionals function as specialists whose authority is based on their specialized knowledge and the soundness of the information they provide. An ability to gather well-documented facts, based on valid and reliable methods of reducing accidents and losses, determines the validity of that authority. For the most part, the safety professional's efforts are directed at supplying management with accurate information to aid in decision making.

One of the more important skills necessary for achieving a reduction in the accident rate is the ability to see and identify safety and health hazards that others may overlook (Figure 24–2). Because of the safety professional's skill, management is provided with an evaluation of the safety and health problem and methods by which the problem can be resolved.

Based on the information collected and analyzed, together with recommendations based on specialized knowledge and experience, the safety professional can propose alternative solutions to those who have ultimate decision-making responsibilities.

The safety professional often has the staff responsibility for fire protection and prevention, may be involved with security, and, in small establishments, may have many additional nonsafety functions.

Knowledge of behavior, motivation, and communication is important, as is a command of management principles and the theory of business and government organization. This specialized knowledge must include a thorough understanding of accident causation as well as methods and procedures designed to control such events.

## Accident Prevention Activities

The basic accident prevention activities (in descending order of effectiveness and preference) are as follows:

- Eliminate the hazard from the machine, method, material, or facility structure.
- Control or contain the hazard by enclosing or guarding it at its source or exhausting an airborne hazard away from the operator.
- Train operating personnel to be aware of the hazard and to follow safe job procedures to avoid it.
- Prescribe personal protective equipment for personnel to shield them from the hazard (Figure 24–3).

It is beyond the scope of this section to describe completely all accident prevention activities of safety professionals at each operation. However, the primary responsibilities are outlined here:

- Provide advisory services on safety and health problems and other matters related to accident prevention.
- Develop a centralized program to control accident and fire hazards.

---

# HAZARD IDENTIFICATION CHECKLIST
## TYPE 1—SAFETY

**1A. INTERNAL ENERGY RELEASE**
- 1A1 ENERGY SOURCES   POTENTIAL   (PRESSURE VESSEL)
  KINETIC   (CENTRIFUGE)
  CHEMICAL   (FUEL, EXPLOSIVES)
- 1A2 UNSAFE CONDITIONS (INCLUDES HARDWARE UNRELIABILITIES)
- 1A3 HUMAN ERROR (CONSIDER PROGRAM OR TRAINING DEFICIENCY)

**1B. EXTERNAL ENERGY DAMAGE**
- 1B1 SYSTEM ENVIRONMENTS (SHOCK, TEMPERATURE, CONTAMINATION)
- 1B2 NATURAL ENVIRONMENTS (LIGHTNING, EARTHQUAKE, HURRICANE)

**1C. PHYSIOLOGICAL DAMAGE**
- 1C1 TOXIC SOURCES (POISON GASES, X-RAYS, NOISE)
- 1C2 DEPRIVATIONS (ANOXIA, STARVATION, DEHYDRATION)

**Figure 24–2.** Hazard evaluation and abatement checklists are very useful in pinpointing safety and health hazards. (Adapted from Workplace Injury and Illness Prevention, CS-1A. Revised January 1993. Cal. OSHA Consultation Service.)

**Figure 24–3.** ■ Personal protective equipment is commonly used to protect employees from potentially hazardous operations.

**Figure 24–4.** Educating employees about the importance of general health and safety principles and techniques is one of the mainstays of an effective health and safety program.

■ Keep informed of changes in federal, state, and local safety codes, and communicate such information to management.

■ Develop and apply safety standards both for production facilities (equipment, tools, work methods, and safeguarding) and for products, based on applicable legal and voluntary codes, rules, and standards.

■ Work closely with the engineering, industrial hygiene, medical, and purchasing departments during the development and construction of new equipment and facilities. See that a procedure is established to ensure that only safe tools, equipment, and supplies are purchased; advise the purchasing department on acceptable supplies and materials; and review and approve purchase requisitions for personal protective equipment and safety items.

■ Develop, plan, and implement the safety and health inspection program carried out by the operating super-

visors and field safety personnel to identify potential hazards, both in the workplace and in the use of the company's products. Inspect all new equipment in conjunction with engineering, operating, and personnel representatives for adequate safeguards and freedom from major safety and health hazards.

■ Guide operating supervision in accident investigation to determine the accident's cause and to prevent recurrence. Review nondisabling-injury accident reports on a sample basis to check the thoroughness of the accident investigation and corrective actions taken.

■ Collect and analyze data on illness and accidents for the purpose of instituting corrective action and to determine accident trends and provide targets for corrective action. Maintain such files as those of inspection records, employee training, OSHA injury and illness logs, a hazard log, and files of complaints and suggestions.

■ Ensure education and training of employees in general and specific safety and health principles and techniques. Maintain supervisory contacts for new instructions, follow-up, and general safety and health motivation (Figure 24–4).

■ Cooperate with medical personnel on matters of employee health and fitness to work, and with industrial hygiene or environmental quality control personnel on industrial hygiene problems.

## Safety and Health Programs

Management usually places administration of the accident prevention or safety and health program in the hands of a safety professional whose title is safety director, manager of safety, or loss control manager.

Full staff responsibility for the safety activities should be assigned to one person. The decision concerning proper placement of responsibility should be based on the size of the company and the nature of the hazards involved in its operation.

Employment of full-time safety professionals is increasing for the following reasons:

- The passage of the Occupational Safety and Health Act (OSHA) of 1970 requires that certain safety standards be met and maintained.

- A better understanding of the safety professional's services and functions is developing. To administer a safety program effectively, the individual in charge must be highly trained and/or have many years of experience in the safety field.

A safety and health program is not something that is imposed on company operations as an afterthought. Safety, an integral part of company operations, must be built into every process or product design and into every operation.

The prevention of illness, accidents, and injuries is basically achieved through control of the working environment and control of people's actions. The safety professional can assist management to implement such control.

A company with an effective health and safety program has a working environment in which operations can be conducted safely, economically, and efficiently, with a minimum of employee, customer, and public complaints.

## Staff Versus Line Status

In general, the safety and health program is administered by safety professionals or other persons holding line positions, in a small company, or staff positions, in a large company. In large corporations, the safety professionals and their organizations usually have staff status and authority. The exact organizational status of the safety staff is determined by each firm in terms of its own operating policies.

The safety and health program as a staff function should have the following objectives:

- To establish staff credibility to advise and counsel regarding safety or health matters

- To keep all affected personnel adequately informed regarding safety or health matters

- To ensure that responsibility and accountability for safety are properly assigned with every staff group and operating management

- To program activities that support harmonious supervisor/employee interaction on safety or health matters

- To establish and reinforce consistent attention to preventive practices and actions

Sometimes the safety professional is delegated authority that is usually reserved for line officials. On fast-moving and rapidly changing operations; operations on which delayed action would endanger the lives of workers or others, as in construction and demolition work, fumigation, and chemical processes or processes with other dangerous substances; or emergency work, it is common to find that the safety professional has authority to order immediate changes, including the shutting down of specific equipment or operations.

## Codes and Standards

The safety professional must be familiar with codes and standards applicable to equipment, material, environmental controls, and energy sources. Only by knowing which codes and standards apply can the safety professional give valid advice regarding company standards for purchasing specifications. The safety professional must know how to meet government agency regulatory requirements (such as those of OSHA, the Mine Safety and Health Administration [MSHA], and the EPA), but there are also many other guidelines and consensus standards that provide state-of-the-art models. Therefore, the safety professional should be familiar with the following:

- Codes and standards approved by the American National Standards Institute (ANSI) and other standards and specifications groups (see Bibliography).

- Codes and standards adopted or set by federal, state, and local government agencies. This is particularly important where local or state codes are more stringent than federal codes.

- Codes, standards, and lists of approved or tested devices published by such recognized authorities as Underwriters Laboratories, Inc., and the National Fire Protection Association.

- Safety practice recommendations of such organizations as the National Safety Council, insurance carriers or their associations, and trade and industrial organizations.

### Policies and Procedures

One of the main tasks confronting the safety professional is the development and implementation of company safety and health policies and procedures. Policies and procedures are necessary to ensure that OSHA and company requirements for safety and health are carried out uniformly within an organization. Examples of policies and procedures that many companies would have in a safety and health program might include visitor safety, accident investigation, safety meetings, new employee safety and health orientations, first aid/CPR, reporting injuries and illnesses, hazard communication, confined spaces, materials handling and lifting, personal protective equipment including respiratory protection, bloodborne pathogens, and fleet/vehicle safety.

Because safety and health policies and procedures often affect a number of departments and have far-ranging effects in terms of operations and costs, they must be reviewed by management as well as the safety professional.

Policies and procedures generally begin with a purpose statement. In other words, what is the policy and procedure intended to accomplish? The purpose statement is often followed by general requirements and a procedure including designation of individuals or positions along with their specific tasks or action steps.

An important duty of the safety professional should be that of checking plans for new or remodeled facilities and new, rebuilt, or rearranged equipment; changes in material used in product or processes and material-storage and -handling procedures; and plans for future products. This important function must be done early enough to afford an opportunity to discover health and safety hazards and to correct conditions that might otherwise be built into the facility and its equipment and that would later result in injuries or other casualty losses. There is also the opportunity at this planning stage to build in safety or fire protection features and to provide adequate space for exit aisles, janitor closets, waste-collection equipment, and other commonly overlooked functions.

Many companies do not permit a drawing or specification to be used until it has been approved by the safety professional. Hazards involved in making products should be eliminated insofar as possible. Instructions and warnings developed for employee use should be reviewed for safe manufacturing procedures. Plans that in any way affect current applicable hazard communication requirements must be reviewed carefully.

The safety professional should also make sure that company policies and applicable standards are followed in purchase specifications for new materials and equipment and for modification of existing equipment. Some companies have arranged for the purchasing department to notify the safety department when new materials or equipment are to be purchased, or when there is a new supplier of safety-related materials. For instance, when a new chemical is requested, the safety department should ensure that any applicable Material Safety Data Sheet (MSDS) is obtained from the manufacturer.

The engineering department, with the help of the safety professional, should check with the purchasing department to determine the necessary safety and health measures to be built on or into a machine before it is purchased. Purchasing agents in an industrial facility are necessarily cost-conscious. Consequently, the safety professional must know the occupational disease and accident losses to the company in terms of specific machines, materials, and processes. If the professional is to recommend the expenditure of several thousand dollars for protection of health or other safeguards to be used throughout the facility, for instance, there should be valid evidence that the investment is justified.

Because of highly competitive marketing, manufacturers of machine tools and processing equipment often list safety devices, such as guards and noise enclosures designed for the protection of operators, as separate auxiliary equipment. The supplier may not know the ultimate use of the product. The actual needs for guards and automatic controls depend on the proximity of the operator to the equipment and vary from one installation to another. The safety professional must evaluate each installation and be in a position to satisfy the purchasing agent of the need for health and safety equipment to be included in the original order, or to recommend the issuance of additional purchase orders to provide adequate protection to the operator.

In many organizations, particularly where certain personal protective equipment items such as goggles and safety shoes are reordered from time to time, standard lists have been prepared, and purchases are selected only from among the types and from the companies shown on these approved lists.

In many companies, safety and health functions are placed in three coordinate departments:

- The engineering department, where plans and specifications are prepared for all machinery and equipment purchased
- The safety and health department, where plans and specifications are carefully checked for safety and health
- The purchasing department, which has much latitude in making selections and determining standards of quality, efficiency, and price

Note that even in smaller organizations, someone must be responsible for these three functions for an effective safety and health program.

## Engineering

The ultimate objective of a company's engineering program is to design equipment and processes and to plan work procedures so that the company can produce the best product with the highest quality at the lowest cost. It is the safety professional's job to see that engineering personnel are acquainted with the particular safety and health hazards involved and to suggest methods of eliminating these hazards.

The goal is to design safe and healthful environments and equipment and to set up job procedures so that employee exposure to the hazards of illness and injury are either eliminated or controlled as completely as possible. This can be accomplished when safety and health are factors incorporated into the design of the equipment or the planning of the process, along with adequate training and supervision.

Company policy should specify that safety and health measures must be designed and built into the job or work instructions before the job is executed. To add safety and

**Figure 24–5.** As a production process is being planned, staff from engineering, production, and safety meet to review drawings and plans to incorporate safety and health features.

health features after work on a job has begun is usually less effective, less efficient, and more costly.

The most efficient time to engineer safety and health hazards out of the facility, product, process, or job is before building or remodeling, while a product is being designed, before a change in a process is put into effect, or before a job is started. Every effort, therefore, should be made to find and remove potential safety and health hazards at the blueprint or planning stage (Figure 24–5).

## Machine and Equipment Design

The machine manufacturer, like any other businessman, wants to have satisfied customers. If the machines cause accidents, customers are dissatisfied. If a customer's order for a machine specifies that the machine must meet specific regulations of OSHA (or another agency) and have safety built into it, the manufacturer's designers will regard such a specification as a design requirement that they must meet. If only a general statement such as "must meet OSHA standards" is used, the manufacturer does not know which standards apply, and the equipment may not be properly guarded.

In many instances, guards added to a machine after it has been installed in a facility are easily removed, and often are not replaced. If a guard or enclosure is an aid to production and efficiency rather than a hindrance, however, it is unlikely that the machine would be operated without having the guard in place. Machine safety must be improved without hindering the worker or reducing the efficiency of the equipment. (See Bibliography for more information.)

The best solution lies in a basic guard design that eliminates the safety and health hazard and, if possible, increases efficiency. There can be little prospect for safe operation of a machine unless the idea of building safety

and health measures into the machine's function is applied right on the drawing board for the establishment that is going to use the equipment.

## Purchasing

The safety professional is responsible for generating and documenting safety and health standards to guide the purchasing department. These standards should be set up so that the safety and health hazards associated with a particular kind of equipment or material being purchased are eliminated or, at the very least, substantially reduced.

The purchasing staff, although not directly involved with educational and enforcement activities, is vitally concerned with many phases of engineering activities. They select and purchase the various items of machinery, tools, equipment, and materials used in the organization, and it is to a considerable degree their responsibility to see that safety has received adequate attention in the design, manufacture, and particulars of shipment of these items.

The safety professional should be well prepared to advise the purchasing department when required to do so. The purchasing staff can reasonably expect the safety professional to offer the following:

- Specific information about safety and health hazards that can be eliminated by change in design or application of guarding by the manufacturer
- Information about equipment, tools, and materials that can cause injuries if misused
- Specific information about health and fire hazards at the facility's worksites
- Information on federal and state safety and health rules and regulations
- Information on accident experience with machines, equipment, or materials that are about to be reordered

**Safety and health considerations.** In purchasing items such as lifting devices and automatic packaging, chemical-processing, or storage equipment, safety and health concerns are extremely important. For example, extreme caution must be observed in the purchase of personal protective equipment, including eye protection, respirators, gloves, and the like; of equipment for the movement of suspended loads, such as ropes, chains, slings, and cables; of equipment for the movement and storage of materials; and of miscellaneous substances and fluids for cleaning and other purposes that might constitute or aggravate a fire or health hazard. Adequate labeling that identifies contents and calls attention to safety and health hazards should be specified. This labeling must comply with state or federal hazard communication (right-to-know) standards. Because the rules and regulations of federal and state agencies keep changing, the safety professional must keep up to date on both employee and community right-to-know regulations.

Many commonly unsuspected safety and health hazards must be considered when very ordinary items such as common hand tools, reflectors, tool racks, cleaning rags, and paint for shop walls and machinery are purchased. Among the factors to be considered are maximum load strength; long life without deterioration; sharp, rough, or pointed characteristics of articles; need for frequent adjustment; ease of maintenance; and ergonomic factors that result in excessive fatigue. Where toxic chemicals are involved, disposal of residue, scrap, and shipping containers must be considered. Safety professionals who are in day-to-day contact with the operating problems must give such information to the purchasing agent.

# Safety and Health Inspections

## Purpose of a Safety and Health Inspection Program

Safety inspections are one of the principal means of locating potential causes of accidents and illness and help determine what safeguarding is necessary to protect against safety and health hazards before accidents and personal injuries occur (Figure 24–6).

Just as inspections of a process are important functions in quality control, safety and health inspections are important in accident control.

Inspections should not be limited to a search for unsafe physical conditions but should also try to detect unsafe or unhealthful work practices. Finding unsafe conditions and work practices and promptly correcting them is one of the most effective methods of preventing accidents and safeguarding employees. Management can also show employees its interest and sincere effort in accident prevention by correcting unsafe conditions or work prac-

tices immediately. Inspections help to "sell" the safety and health program to employees. Each time a safety professional or an inspection committee passes through the work area, management's interest in safety and health is advertised. Regular facility inspections encourage individual employees to inspect their immediate work areas.

In addition, inspections facilitate the safety professional's contact with individual workers, thereby making it easier to obtain their help in eliminating accidents and illnesses. The workers can often point out unsafe conditions that might otherwise go unnoticed and uncorrected. When employee suggestions are acted on, all employees are made to feel that their cooperation is essential and appreciated.

Safety and health inspections should not be conducted primarily to find out how many things are wrong, but rather to determine whether everything is satisfactory. Their purpose should be to discover conditions that, if corrected, will bring the facility up to accepted and approved safety and health standards and result in making it a safer and more healthful place in which to work. When observed, inspectors should tactfully point out any unsafe work procedures to the employees involved. They should be certain to indicate the hazards. Inspectors may need to recommend new or continuing safety and health training for supervisors and employees.

## Inspection of Work Areas

Before the facility walk-through inspection, it is advisable to review reports of all accidents (including noninjury accidents and near misses, if possible) for the previous several years, so that special attention can be given to the conditions and locations known to be scenes of accidents.

Most facilities make use of irregularly scheduled inspections, which can include an unannounced inspection of a particular department, piece of equipment, or

**Figure 24–6.** Weekly safety inspections are conducted to spot potential health and safety hazards before an accident occurs.

small work area. Such inspections made by the safety department tend to keep the supervisory staff alert to find and correct unsafe conditions before they are found by the safety inspector.

The need for intermittent inspections is often indicated by accident report analysis. If the analysis shows an unusual number of accidents for a particular department or location or an increase in certain types of injuries, an inspection should be conducted to determine the reasons for the increase and to find out what corrections are necessary. All results of inspections must be discussed with operating supervision if any gain is to be made.

Supervisors should constantly ensure that tools, machines, and other department equipment are maintained properly and are safe to use. To do this effectively, they should use systematic inspection procedures and can delegate authority to others in a department.

Inspection programs should be set up for new equipment, materials, procedures, and processes. A process should not be put into regular operation until it has been checked for hazards; additional safeguards have been installed, if necessary; and safety instructions or procedures have been developed. This is also a good time to make a complete job safety analysis (JSA) of the operation. It takes less time and effort now than if done later.

## Safety Inspectors or Technicians

Inspectors should know how to locate safety and health hazards and should have the authority to act and make recommendations. A good safety inspector must know the company's accident experience, be familiar with accident potentials, have the ability to make intelligent recommendations for corrective action, and be diplomatic in handling situations and personnel.

Safety inspectors must be equipped with the proper personal protective equipment, protective clothing, and other required equipment to carry out duties. It would be difficult for a safety inspector to persuade an employee to wear eye protection or safety shoes if the inspector does not wear them, or to require workers to use respirators unless the inspector sets the example and uses one in a hazardous environment. It is essential that inspectors practice what they preach.

## Safety Professionals

The safety professional has a very productive role during safety inspections, coordinating the safety program and teaching by firsthand contact and on-the-spot examples.

The number of safety professionals and inspectors needed for adequate safety inspection activities depends a great deal on the size and complexity of the facility and the type of industry involved. Large companies with well-organized safety programs usually employ a staff of full-time safety professionals and inspectors who work directly

under a safety director or safety supervisor. Some large companies also have specially designated employees who spend part of their time on inspections, and some have employee inspection committees.

The safety professional should be fully in charge of developing safety inspection activities and should receive the reports of all inspectors. Special departmental inspectors should either make safety inspections personally or supervise the inspectors in their work. Although safety professionals often have a considerable amount of office work to do, they should get out into the production and maintenance areas as often as possible and make general as well as specific safety inspections. If there is more than one facility involved, there should be a plan to make at least an annual inspection survey of each facility.

## Third-Party Inspections, or Audits

The value of a third-party inspection of policies, procedures, and practices as well as an inspection of the physical facility and equipment is increasingly evident. The advantages of such audits are as follows:

- Objectivity of the inspecting party is less likely to lead to biased findings or their reporting.
- Results of external audits are usually directed to a higher level of decision-making authority and thus are more likely to be acted on promptly.
- Performance of the audit does not have to depend on the time or convenience of organization staff.
- Professionals contracted for such audits usually have much expertise in a given industry.

Many businesses currently find that an annual audit and inspection of their facilities to assess the state of their safety, health, and environmental affairs is as important as the traditional financial audit. Results of these third-party audits are often included in the company's annual report. More information on third-party audit services is often available from insurance carriers, independent safety/health consulting firms, OSHA Consultation Services, or the National Safety Council.

## ■ Accident and Occupational Illness Investigations

### Purpose of Investigations

Investigation and analysis are used by safety professionals to prevent accidents, both those that result in injury to personnel and those that don't. The investigation or analysis of an accident can produce information that leads to countermeasures to prevent accidents or reduce their number and their severity. The more complete the information, the easier it is for the safety professional to design effective control methods. For example, knowing that 40 percent of a facility's accidents involve ladders is useful, but it is not as useful as also knowing that 80 percent of the ladder accidents involve broken rungs.

An investigation of at least every disabling injury or illness (or every OSHA lost workday case) should be made. Incidents resulting in nondisabling injuries or no injuries and "near accidents" should also be investigated to evaluate their causes in relation to injury-producing accidents or breakdowns, especially if there is frequent recurrence of certain types of nondisabling injuries or if the frequency of accidents is high in certain areas of operations.

The consequences of certain types of accidents are so devastating that any hint of conditions that might lead to their occurrence warrants an investigation. In such cases, any change from standard safety specification that has been made warrants a thorough investigation.

For purposes of accident prevention, investigations must be fact finding, not fault finding; otherwise, they may do more harm than good. However, this is not to say that responsibility should not be fixed when personal failure or negligence has caused injury, or that such persons should be excused from the consequences of their actions.

## Types of Investigations

There are several accident investigation and analysis techniques available. Some of these techniques are more complicated than others. The choice of a particular method depends on the purpose and orientation of the investigation.

The accident investigation and analysis procedure focuses primarily on unsafe circumstances surrounding the occurrence of an accident, and it is the most often-used technique. Other similar techniques involve investigation within the framework of defects in man, machine, media, and management (the "four Ms") or education, enforcement, and engineering (the "three Es of safety").

These techniques involve classifying the data about a group of accidents into different categories for analysis. This is known as the statistical method of analysis. Control methods are designed on the basis of most frequent patterns of occurrence.

Other techniques are discussed later in this chapter under the systems approach to safety. Systems safety stresses an enlarged viewpoint that takes into account the interrelationships between the various events that could lead to an accident. Because accidents rarely have a single cause, the systems approach to safety can lead to the discovery of more than one place in a system where effective controls can be introduced. This allows the safety professional to choose the control methods that best meet criteria for such factors as effectiveness and speed of installment. Systems safety techniques also have the advantage of application before accidents or illnesses occur, and can be applied to new procedures and operations.

## Who Conducts the Investigation?

Depending on the nature of the accident and other conditions, the investigation can be conducted by the supervisor, the safety engineer or inspector, the workers' safety and health committee, the general safety committee, the safety professional, or a loss control specialist from the insurance company or other external source. Also, OSHA requires that fatalities and/or accidents resulting in serious injury be reported to them. Depending on the circumstances surrounding the fatality, serious injury, or illness, an OSHA inspection may result and should be anticipated by the employer. Regardless of who conducts the initial investigation, a representative of the company's safety department should verify the findings and direct a written report to the proper official or to the general safety committee.

The safety professional's value and ability are best shown in the investigation of an accident. Specialized training and analytical experience enable the professional to search for all the facts, both apparent and hidden, and to submit an unbiased report. The safety professional should have no interest in the investigation other than to get information that can be used to prevent a similar accident.

During an investigation, methods to prevent a recurrence can occur, but decisions about the specific course to take are best made after all the facts are well established. There are usually several alternatives; all must be fully understood in order for the most effective decision to be made. The safety professional should present every valid, feasible alternative to operating management for their consideration. At this stage, input from employees can be highly beneficial in determining the best corrective measure.

## Record Keeping and Reporting

The Williams-Steiger OSHAct of 1970 requires employers to maintain records of work-related employee injuries and illnesses, as well as many inspection reports of high–injury-potential equipment. In addition, many employers are also required to make reports to state compensation authorities.

Safety professionals are faced with two tasks: maintaining those records required by law and by their management, and maintaining records that are useful in an effective safety program. Unfortunately, the two are not always synonymous. A good record-keeping system necessitates more data than those called for in almost all OSHA-required forms.

There are many different safety records that must be maintained, and OSHA has established how long many of these records must be maintained by the employer. Records that must be generated and maintained include records of inspections; accident investigations; general and specific training; medical and exposure monitoring results; the OSHA log of injuries and illnesses; fatality and serious injury and illness reports to OSHA; insurance records such as the employer's and doctor's first reports of injury and illness; and respirator-fit test and other personal protective equipment records addressing the maintenance, use, selection, inspection, and storage of such equipment.

Records of accidents and injuries and the training experience of the people involved are essential to efficient and successful safety programs, just as records of production, costs, sales, and profits and losses are essential to efficient and successful operation of a business. Records supply the information necessary to transform haphazard, costly, ineffective safety and training efforts into a planned safety and health program that enables control of both conditions and acts that contribute to accidents. Good record keeping is the foundation of a scientific approach to occupational safety.

## Uses of Records

A good record-keeping system can help the safety professional in the following ways:

- It provides safety personnel with the means for an objective evaluation of the magnitude of occupational illness and accident problems and with a measurement of the overall progress and effectiveness of the safety and health program.

- It helps identify high-hazard units, facilities, or departments and problem areas so that extra effort can be made in those areas.

- It provides data necessary for an analysis of accidents and illnesses that can point to specific circumstances of occurrence, which can then be attacked by specific countermeasures.

- It can create interest in safety and health among supervisors by furnishing them with information about the accident and illness experience of their own departments.

- It provides supervisors and safety committees with hard facts about their safety and health problems so that their efforts can be concentrated.

- It helps in measuring the effectiveness of individual countermeasures and determining whether specific programs are doing the job that they were designed to do.

- It can help establish the need for, and the content of, employee and management training programs that can be tailored to fit the particular needs of that company or facility.

## Accident Reports and Illness Records

To be effective, preventive measures must be based on complete and unbiased knowledge of the causes of accidents and the knowledge of the supervisor and employee about the operation. The primary purpose of an accident report, like the inspection, is to obtain information, not to fix blame. Because the completeness and accuracy of the entire accident record system depend on information in the individual accident reports and the employee training history, it is important that the forms and their purpose are understood by those who must fill them out. Essential training or instruction by the safety professional should be given to those who are responsible for generating the information. (Illustrations of typical forms are given in the latest editions of the National Safety Council's *Accident Investigation*, *Accident Prevention Manual for Business & Industry*, vol. 1: *Administration & Programs*, and the *Supervisors Safety Manual*—see Bibliography.) Photographs, videotapes, and drawings of the accident or a depiction of the accident can be extremely useful.

**The first-aid report.** Collecting injury or illness data generally begins in the first-aid department. The first-aid attendant or the nurse fills out a first-aid report for each new case. Copies are sent to the safety department or safety committee, the worker's first-line supervisor, and other departments as management designates.

The first-aid attendant or the nurse should know enough about accident analysis and illness investigation to be able to record the principal facts about each case. Note that the questioning of the injured or sick person must be complete enough to establish whether the incident is or is not work related. Current emphasis on chemical air contaminants makes it necessary to include or exclude exposure to known health hazards. First-aid reports can be very helpful to the safety or industrial hygiene personnel. The company physician who treats injured employees should be informed of the basic rules for classifying cases because, at times, the physician's opinion of the severity of an injury is necessary to record the case accurately.

**The supervisor's accident report form.** This should be completed as soon as possible after an accident occurs, and copies sent to the safety department and to other designated persons. Information concerning unsafe or unhealthful work conditions and improper work practices is important in the prevention of accidents, but information that shows why the unsafe or unhealthful conditions existed can be even more important. This type of information is particularly difficult to get unless it is obtained promptly after the accident occurs. If the information is based on opinion, not on proven facts, it is still important, but should be so identified.

Generally, analyses of accidents are made only periodically, and often long after the accidents have occurred. Because it is often impossible to accurately recall the details of an accident, this information must be recorded accurately and completely at once or it may be lost forever.

**Injury and illness record of an employee.** The first-aid report and the supervisor's report contain information about the agency of injury (type of machine, tool, or material), the type of accident, and other factors that facilitate the use of the reports for accident prevention. Another form must be used to record the injury experience of individual employees.

The employee training record card should have space to record injury information such as the date, classification, days charged, and costs.

Much can be learned about accident causes from studying employee injury records. If certain employees or job classifications have frequent injuries or illnesses, a study of the work environment, job training, safety and health training, work practices, and the instructions and supervision given them may reveal more than a study of accident locations, agencies, or other factors.

# ■ Education and Training

Safety and health training begins at the time of hiring, before the employee actually starts work. An effective safety and health training program includes a carefully prepared and presented introduction to the company (Figure 24–4).

New employees immediately begin to learn and form attitudes about the company and their job, boss, and co-workers, whether or not the employer offers a training program. To encourage a new employee to form positive attitudes, it is important for the employer to provide a sound basis for them, and providing safety and health information is vital. Training about exposure to chemical hazards in the workplace is now mandated by state and federal hazard communication standards (right-to-know laws). (See Chapter 29, Government Regulations, for more information on these and other relevant regulations and standards.) In fact, most new OSHA regulations generally have a training requirement written into the standard.

An effective accident prevention and occupational health hazard control program is based on proper job performance. When people are properly trained to do their jobs, they do them safely. This means that supervisors must know what employee training needs to be given, which means knowing the requirements of the job; know how to train an employee in the safe way of doing a job; and know how to supervise. It also means the safety professional should be familiar with good training techniques. Although the professional is not always directly involved in the training effort, he or she should be able to recognize the elements of a practical training program.

A training program is needed for new employees, when new equipment or processes are introduced, when procedures are revised or updated, when new information must be made available, when employee performance must be improved, when new or unexpected hazards are uncovered, and on a periodic basis to refresh employees'

knowledge of the material. Employees with longer tenure also need training so that they have the same information about new equipment, products, or company policies that new employees are receiving.

Many supervisors acquired their present positions in organizations where some sort of safety and health program already existed, and their understanding of the program is firmly established. However, a safety professional undertaking the safety training of supervisors almost invariably finds that the first major job is to get supervisors at all levels to understand and accept their role in accident and illness prevention. This job cannot be done in a single meeting or through a single communication.

Simply getting supervisors to agree in theory that responsibility for safety and health is one of their duties is not enough. They must come to understand the many ways in which they can prevent illness and accidents, and they must become interested in improving their safety performance. For a safety and health program to be effective, all levels of management must be firmly committed to the program and express that commitment by action and example. Management is ultimately responsible for the safety and health of the employees. Much of the effort put into an industrial safety and health program by a safety professional is, therefore, directed toward educating and influencing management.

## Employee Training

The training of an employee begins the day the employee starts the job. As observed earlier in this chapter, whether or not the firm has a formal safety and health orientation program, the employee starts to learn about the job and to form attitudes about many things—including safety and health—on the first day.

The safety professional assists supervisors in instructing employees in the safe way of doing each job. Accidents can be prevented only when these recommended procedures are based on a thorough analysis of the job and when the procedures are followed. This is why a complete job safety analysis is so valuable (Figure 24–7). It provides a baseline for future comparison, and it details all necessary safety elements of the various job tasks.

The safety professional can provide supervisors with methods for observing all workers in the performance of their tasks to establish the job safety requirements, and should participate in follow-up observations to reinforce the supervisors' training. In this way, supervisors are informed of any weakness in the company's safety and health program and will have a common reference point for monitoring these problems.

### Training Tips

- Train small groups whenever possible. Employees seem to learn more and are more apt to ask questions in small training groups.

## JOB SAFETY ANALYSIS FORM

| | National Safety Council **JOB SAFETY ANALYSIS** | JOB TITLE (and number if applicable): Replacing a water bottle | PAGE 1 OF 2 JSA NO. 001 | DATE: Today | ☒ NEW ☐ REVISED |
|---|---|---|---|---|---|
| *INSTRUCTIONS ON REVERSE SIDE* | | TITLE OF PERSON WHO DOES JOB: Maintenance/Janitor | SUPERVISOR: Eric Utney | ANALYSIS BY: Mary Green | |
| COMPANY/ORGANIZATION: 123 Accounting Corp. | | PLANT/LOCATION: General Office | DEPARTMENT: Reception | REVIEWED BY: Bill Camp | |
| REQUIRED AND/OR RECOMMENDED PERSONAL PROTECTIVE EQUIPMENT: Protective footwear, non-slip gloves | | | | APPROVED BY: Greg Porter | |

| SEQUENCE OF BASIC JOB STEPS | POTENTIAL HAZARDS | RECOMMENDED ACTION OR PROCEDURE |
|---|---|---|
| 1. Lift and load the bottle | | |
| 2. Transport the bottle and place near the dispenser | | |
| 3. Remove empty bottle from dispenser | | |
| 4. Position full water bottle on stand | | |
| 5. Check system | | |

Printed in U.S.A.     156 15

## INSTRUCTIONS FOR COMPLETING THE JOB SAFETY ANALYSIS FORM

Job Safety Analysis (JSA) is an important accident prevention tool that works by finding hazards and eliminating or minimizing them *before* the job is performed, and *before* they have a chance to become accidents. Use JSA for job clarification and hazard awareness, as a guide in new employee training, for periodic contacts and for retraining of senior employees, as a refresher on jobs which run infrequently, as an accident investigation tool, and for informing employees of specific job hazards and protective measures.

Set priorities for doing JSA's: jobs that have a history of many accidents, jobs that have produced disabling injuries, jobs with high potential for disabling injury or death, and new jobs with no accident history.

Select a job to be analyzed. Before filling out this form, consider the following: The purpose of the job—What has to be done? Who has to do it? The activities involved—How is it done? When is it done? Where is it done?

In summary, to complete this form you should consider the purpose of the job, the activities it involves, and the hazards it presents. If you are not familiar with a particular job or operation, interview an employee who is. In addition, observing an employee performing the job, or "walking through" the operation step by step may give additional insight into potential hazards. You may also wish to videotape the job and analyze it.

Here's how to do each of the three parts of a Job Safety Analysis:

### SEQUENCE OF BASIC JOB STEPS

Examining a specific job by breaking it down into a series of steps or tasks, will enable you to discover potential hazards employees may encounter.

Each job or operation will consist of a set of steps or tasks. For example, the job might be to move a box from a conveyor in the receiving area to a shelf in the storage area. To determine where a step begins or ends, look for a change of activity, change in direction or movement.

Picking up the box from the conveyor and placing it on a handtruck is one step. The next step might be to push the loaded handtruck to the storage area (a change in activity). Moving the boxes from the truck and placing them on the shelf is another step. The final step might be returning the handtruck to the receiving area.

Be sure to list *all* the steps needed to perform the job. Some steps may not be performed each time; an example could be checking the casters on the handtruck. However, if that step is generally part of the job it should be listed.

### POTENTIAL HAZARDS

A hazard is a potential danger. The purpose of the Job Safety Analysis is to identify ALL hazards—both those produced by the environment or conditions and those connected with the job procedure.

To identify hazards, ask yourself these questions about each step:

Is there a danger of the employee striking against, being struck by, or otherwise making injurious contact with an object?

Can the employee be caught in, by, or between objects?

Is there potential for slipping, tripping, or falling?

Could the employee suffer strains from pushing, pulling, lifting, bending, or twisting?

Is the environment hazardous to safety and/or health (toxic gas, vapor, mist, fumes, dust, heat, or radiation)?

Close observation and knowledge of the job is important. Examine each step carefully to find and identify hazards—the actions, conditions, and possibilities that could lead to an accident. Compiling an accurate and complete list of potential hazards will allow you to develop the recommended safe job procedures needed to prevent accidents.

### RECOMMENDED ACTION OR PROCEDURE

Using the first two columns as a guide, decide what actions or procedures are necessary to eliminate or minimize the hazards that could lead to an accident, injury, or occupational illness.

Begin by trying to: 1) engineer the hazard out; 2) provide guards, safety devices, etc.; 3) provide personal protective equipment; 4) provide job instruction training; 5) maintain good housekeeping; 6) insure good ergonomics (positioning the person in relation to the machine or other elements in such a way as to improve safety).

List the recommended safe operating procedures. Begin with an action word. Say exactly what needs to be done to correct the hazard, such as, "lift using your leg muscles." Avoid general statements such as, "be careful."

List the required or recommended personal protective equipment necessary to perform each step of the job.

Give a recommended action or procedure for each hazard.

Serious hazards should be corrected immediately. The JSA should then be changed to reflect the new conditions.

Finally, review your input on all three columns for accuracy and completeness. Determine if the recommended actions or procedures have been put in place. Re-evaluate the job safety analysis as necessary.

**Figure 24–7.** A job safety analysis form is used to record information that will be used as a baseline for future comparison and includes information on all necessary safety elements of the various job tasks. (Reprinted with permission from the National Safety Council. *Job Safety Analysis: Participant Workbook.* Itasca: National Safety Council, 1994.)

■ Consider providing two levels of training, one for the supervisor and another for the workers. Generally, supervisory training needs to be more comprehensive than that given to workers.

■ Consider using outside trainers. Oftentimes employees perceive outside trainers and consultants as having a higher level of credibility.

■ Make use of commercially available audio/visual information (video tapes, slides, and films). It is important to screen these commercial products, because they are often very generic and must be supplemented with site-specific information and discussions. Many trade associations have produced audio/visual information for their member companies.

■ Keep records of employee attendance. Note the date, subjects covered, instructor, and training aids used (such as videos), and make a list of attendees (with their signatures if possible). A copy of any tests given and of the agenda should be kept on file as well.

■ Make the training as participatory as possible. Encourage discussions, use training aids, and practice the use of equipment and procedures discussed in the training.

## Maintaining Interest in Safety

A prime objective of a good safety and health program is to maintain interest in safety in order to prevent accidents. It is, however, as difficult to determine the degree of success achieved by an interest-maintaining effort as part of a safety program as it is to isolate the effectiveness of an advertising campaign separate from an entire marketing program. The reason is that companies with sound basic safety and health programs generally have working conditions that are safe, employees who are well trained and safety minded, and high-caliber supervision.

## Safety and Health Rule Enforcement

Obeying safety and health rules is actually a matter of education; employees must understand the rules and the importance of following them. In helping an employee understand, the possibility of language barriers should be considered. Language barriers are caused not only by national origins but also, and more often, by the jargon of a particular profession or industry. A considerable amount of confusion can occur when a new employee comes from a different industry or field of work.

## Role of the Supervisor

Supervisors are the key people in any program designed to create and maintain interest in safety and health, because they are responsible for translating management's policies into action and for promoting safe and healthful work practices directly among the employees. The supervisors' attitudes toward safety and health are a significant factor in the success of not only specific promotional activities but also the entire safety and health program, because their views will be reflected by the employees in their departments.

How well a supervisor meets this responsibility is determined to a large extent by how well the supervisor has been trained, and training and educating the supervisor in matters of safety and health is the responsibility of the safety professional.

## Supervisor Training

Supervisors are often responsible for providing safety and health training to employees. They may be the primary safety trainers and have the final responsibility for the effectiveness of training. If the employer chooses to put the responsibility of training on supervisors, the employer must clearly communicate that this is a discrete responsibility. Just as important, the employer must ensure that the supervisor has the time, interest, and training necessary to provide adequate employee training.

Generally speaking, the supervisor needs training at a level equal to or exceeding the training given to labor. Several recent OSHA regulations have been adopted that require employers to provide additional training to supervisors. To illustrate, supervisors who supervise hazardous waste cleanup workers are required (29 *CFR* 1910.120) to take an additional 8 hours of management/supervisor training covering such topics as the employer's safety and health program, employee training program, personal protective equipment, and health hazard–monitoring procedures and techniques.

There are many community colleges and independent training groups that offer supervisory safety and health courses to better prepare supervisors for their safety and health tasks. Also, there are many organizations that offer "train the trainer" courses.

The supervisor who is sincere and enthusiastic about accident prevention can do much to maintain interest because of a direct connection with the worker. Conversely, if the supervisor only pays lip service to the program or ridicules any part of it, this attitude offsets any good that might be accomplished by the safety professional.

Some supervisors are reluctant to change their mode of operation and slow to accept new ideas. It is the safety professional's task to sell these supervisors on the benefits of accident prevention, to convince them that promotional activities are not "frills" but rather projects that can help them do their job more easily and prevent illness and injuries, and to persuade them that their wholehearted cooperation is essential to the success of the entire safety and health program.

Setting a good example by wearing required personal protective equipment is an excellent way in which supervisors can promote the use of this equipment and demonstrate interest in safety. Teaching safety and health principles to supervisors is an important function of the safety professional; safety posters, a few warning signs, or merely general rules are not enough to do this job.

The safety professional should educate supervisors so that working conditions are kept as safe and healthful as possible and that the workers follow safe procedures consistently, as a routine part of good job performance. Supervisors are entitled to all of the help the safety professional can give through supplies of educational material for distribution and frequent visits to the jobsite, as circumstances permit. Supervisors should also receive adequate recognition for independent and original safety activity.

Supervisors can be very effective by giving facts and personal reminders on safety and health to employees as part of their daily work instructions. This procedure is particularly necessary in the transportation and utility industries, where the work crews are on their own.

In any case, supervisors should be encouraged to take every opportunity to exchange ideas on accident prevention with workers, to commend them for their efforts to do the job safely, and to invite them to submit suggestions for better ways to do the job that will prevent injuries or illness.

## Job Safety and Health Analysis

Job safety and health analysis (Figures 24–7 and 24–8) is a process used by safety professionals and supervisors to review job methods and uncover hazards. Once the safety and health hazards are known, the proper controls can be developed. Some controls are physical changes that control the hazard, such as enclosures to contain an air contaminant or a guard placed over exposed moving machine parts. Others are job procedures that eliminate or minimize the hazard; for example, safe stacking of materials. Procedure controls require training and supervision.

**Benefits of a job safety analysis.** The principal benefits that arise from job safety analysis are the establishment of the following practices:

- Individuals are given training in safe, efficient procedures.
- New employees are instructed on safety and health procedures.
- Preparations are made for planned safety and health observations.
- "Prejob" instructions are given on irregular jobs.
- Job procedures are reviewed after accidents occur.

New employees must be trained in the basic job steps. They must be taught to recognize the safety and health hazards associated with each job step and learn the necessary precautions. There is no better guide for this training than a well-prepared job safety analysis used with the job instruction training method.

**Job Title:** *Castings Grinding*

**Job Location:** *Machine Shop*

| Job Step | Potential Hazards | Recommended Action or Procedure |
|---|---|---|
| 1. Reach into metal box to right of machine, grasp casting and carry to wheel. | 1. Strike hand on edge of metal box or casting; cut hand on burr. Drop casting on toes. | 1. Provide gloves and safety shoes. |
| 2. Push casting against wheel to grind off burr. | 2. Strike hand against wheel. Flying sparks, dust or chips; wheel breakage. Not enough of wheel guarded. No dust removal system. Sleeves could get caught in machinery. | 2. Provide larger guard over wheel. Install local exhaust system. Provide safety goggles. Instruct worker to wear short or tight-fitting sleeves. |
| 3. Place finished casting in box to left of machine. | 3. Strike hand against metal box or castings. | 3. Provide for removal of completed stock. |

**Figure 24–8.** A job safety analysis, even a simple one, breaks the job into steps and identifies hazards leading to the recommended action or procedure. Here is an employee performing a castings grinding operation.

All supervisors are concerned with improving job methods to increase safety, reduce costs, and step up production. The job safety analysis is an excellent starting point for questioning the established way of doing a job.

## ■ Risk Management

### A Five-Step Program

Companies find they must control accidents if they are to continue to do business in a highly competitive market. One large company uses an approach consisting of five closely related, logically ordered steps for a coordinated program. These steps are hazard identification, hazard elimination, hazard protection, determining the maximum possible loss, and loss retention.

**Hazard identification.** To prevent accidents and control losses, first identify all safety and health hazards to determine those areas or activities in an operation where losses can occur. This requires studying processes at the research stage, reviewing design during engineering, checking pilot facility operations and start-up, and regularly monitoring normal production.

**Hazard elimination.** Toxic, flammable, or corrosive chemicals can sometimes be replaced by safer materials. Machines can be redesigned to eliminate danger points. Facility layouts can be improved by eliminating such hazards as blind corners and limited-visibility crossings.

**Hazard protection.** Hazards that cannot be removed must be protected against. Familiar examples include mechanical guards to keep fingers from pinch points, safety shoes to safeguard toes against dropped objects, and ventilation systems to control the buildup of air contaminants. Industry is concerned with all losses, injury to personnel, damage to products, and destruction of property.

**Maximum possible loss.** This step involves the determination of the maximum loss that could occur if everything went wrong. For example, entire buildings or areas can be lost as the result of a fire or explosion. The amount that a company could lose under the most adverse conditions can be estimated.

**Loss retention.** Having some idea of the amount that could be lost under a combination of unfavorable circumstances, one can then determine what portion of such a loss a company is willing to bear itself. Industrial companies can afford to retain a portion of each loss. The remaining loss potential is then insured through the company's insurance carrier. This proves a good incentive for management to institute strong safety and health programs. These activities can be consolidated in one department such as a risk management department, bringing together the safety professional, the fire protection manager, the security and facility protection manager, the industrial physician, the industrial hygienist, and the insurance manager. The administrator of a total loss control program does not need to know all the details of each function, but should be able to develop an atmosphere in which there is harmonious cooperation and mutual understanding. Primary concerns are the control of occupational disease and personnel safety.

## Damage Control

There should be a damage control program for investigating all accidents, not just those that produce injuries. This approach of studying accidents instead of injuries recognizes that a so-called no-injury accident, if repeated in the future, could result in personal injury, property damage, or both.

Ferreting out the causes of accidents reveals what unsafe conditions and/or work practices were responsible for the accident.

Three basic steps are successfully used to reduce property damage (and injuries): spot-checking, reporting by repair control centers, and auditing.

**Spot-checking.** Spot-checking consists of observing and taking notes to permit damage estimates by comparing total costs for a repair period with those found during sample observations.

**Reporting by repair control centers.** This step involves developing a system in which the repair or cost control center records property damage. The system should be designed to require the least possible amount of paperwork. No one system works in all companies, because repair cost–accounting methods vary greatly from company to company, and even from facility to facility within a single company.

**Auditing.** An effective reporting program necessitates complete auditing. Safety personnel should receive a copy of every original work order processed through the maintenance, planning, and cost control center. Safety professionals make on-the-spot checks to see if accidental damage was involved.

# Systems Safety

Recently, safety professionals have increasingly been exploring systems approaches to industrial accident prevention. Safety professionals are asked to find ways of implementing systems safety techniques. And although complete system safety analysis requires specially trained engineers and rather sophisticated mathematics, safety professionals find that some knowledge of these techniques can directly benefit them when it comes to codifying and directing their safety and health programs.

Through a system analysis, a safety professional can clarify a complex process by devising a chart or model that provides a comprehensive, overall view of the process, showing its principal elements and the ways in which they are interrelated (Figure 24–9).

Having established the concept of a system, the next step is the analysis of systems. Progress in the analysis of complex systems enables safety professionals to solve problems in accident prevention and the control of occupational illness.

## Methods of Analysis

There are four principal methods of analysis: failure mode and effect, fault tree, THERP (technique for human error prediction), and cost effectiveness. Each has variations, and two or more can be combined in a single analysis.

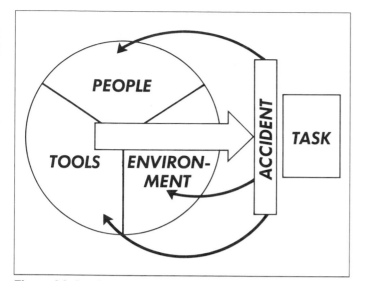

**Figure 24–9.** A system analysis can show how people, tools, and the environment can combine to produce an accident.

**Failure mode and effect.** In this method the failure or malfunction of each component is considered, including the mode of failure (such as a switch jammed in the on position). The effects of the failure are traced through the system, and the ultimate effect on task performance is evaluated.

**Fault tree.** In this method an undesired event is selected, and all the possible occurrences that can contribute to the event are diagrammed in the form of a tree. The branches of the tree are continued until independent events are reached. Probabilities are determined for the independent events, and, after simplifying the tree, both the probability of the undesired event and the most likely chain of events leading up to it can be computed.

**THERP.** This is a technique for human error prediction, developed by Scandia Corporation, that provides a means for quantitatively evaluating the contribution of human error to the degradation of product quality. It can be used for human components in systems and thus can be combined either with the failure mode and effect or the fault tree method.

**Cost effectiveness.** In the cost-effectiveness method, the cost of system changes made to increase safety and health measures is compared with either the decreased costs of fewer serious failures or with the increased effectiveness of the system to perform its task, in order to determine the relative value of these changes. Ultimately, all system changes have to be evaluated, and this method makes such cost comparisons explicit. Moreover, cost-effectiveness analysis is often used to help make decisions concerning the choice of one of several systems that can perform the same task.

In all of these analytical methods, the main point is to measure quantitatively the effects of various failures

within a system. In each case, probability theory is an important element.

The systems approach to safety can help to change the safety profession from an art to a science by codifying much of safety and health knowledge. It can help change the application of safety measures from piecemeal problem solving (such as putting a pan under a leak) to a safely designed operation (preventing the leak itself).

The safety professional determines what can happen if a component fails or the effects of malfunction in the various elements of the system, and provides solutions before the accident occurs instead of after the damage has been done.

## ■ Safety Professional Certification

Employers, employees, and the public deserve some assurance that the individuals practicing safety are professionals and are able to provide the safety expertise that, in turn, should provide adequate protection.

Usually, a candidate for professional status must complete a specified course of study, which is followed by practical experience in that field. The applicant must pass an examination to prove mastery of a specific body of knowledge. Finally, a board composed of members of the profession reviews that candidate's qualifications and grants professional certification.

Professional regulation usually results from the need to protect the public from potential harm at the hands of unqualified persons. Clearly, there was a need for professional regulation in the field of industrial safety, and the Board of Certified Safety Professionals was created to fill this need.

The Board of Certified Safety Professionals of the Americas was incorporated in Illinois in 1969 to establish criteria for professional certification, accept applications, evaluate the credentials of candidates, and issue certificates to those who met the requirements.

One method of determining professional abilities is to compare education and experience against a predetermined set of requirements. Once these criteria have been established, each application showing a candidate's education and experience is evaluated against that base. The applicant may be found to be eligible to go to the next step—to take the certification examinations; upon successful completion of the examinations, the candidate is granted certification as a certified safety professional, or CSP.

### Current Requirements

Currently, a candidate must take six specific steps before being designated a CSP (Figure 24–10):

1. Complete the application for certification by filling out the Board's application form.

## BOARD OF CERTIFIED SAFETY PROFESSIONALS

208 Burwash Avenue
Savoy, IL 61874-9571
(217) 359-9263
FAX: (217) 359-0055

# CERTIFICATION PROCEDURES and REQUIREMENTS

## CRITERIA FOR CERTIFICATION

### General Criteria

The General Criteria for certification are:

- graduation from college or university with an accredited baccalaureate degree in safety, *and*
- four or more years of professional safety experience acceptable to the Board, *and*
- achievement of passing scores on each of the two written examinations.
- Applicants are evaluated objectively without regard to age, sex, race, religion, national origin, handicap, or marital status.

### Substitutions or Modifications to the General Criteria

- Applicants without any college level academic work may substitute acceptable professional safety experience for the baccalaureate degree requirement at the rate of one year of experience for each academic year of a four-year degree.
- Applicants with baccalaureate degrees not meeting the Board requirements or who have completed college courses short of a degree will receive credit for their academic work based on the major area of study and other factors. Professional safety experience may be substituted for the remaining credit needed (See Methods of Determining Academic Credit).
- A master's degree may be substituted for a maximum of one year of professional safety experience, and a doctorate may be substituted for a maximum of two years of professional safety experience. The amount of experience credit allowed will depend upon the major area of study. Only one graduate degree will be credited. Applicants with more than one graduate degree will receive credit for the degree which results in the most credit (See methods for Determining Academic Credit).
- Certification as a Occupational Health and Safety Technologist by the ABIH/BCSP Joint Committee may be substituted for one year of the experience requirements.
- Applicants who hold one of the following current registrations or certifications may waive the Safety Fundamentals Examination:

Registered as a Professional Engineer in any U.S. State or Territory
Certified as an Industrial Hygienist by the American Board of Industrial Hygiene
Certified as a Health Physicist by the American Board of Health Physics
National Diploma in Occupational Safety and Health by the British National Examination Board
Registered as a Canadian Registered Safety Professional by the Association of Canadian Registered Safety Professionals.

## EXAMINATION ELIGIBILITY

### Safety Fundamentals Examination

- Applicants meeting the minimum academic requirements by education and/or experience will be eligible to sit for the Safety Fundamentals Examination.

### Specialty Examinations

- Applicants who meet all the academic and experience requirements for certification *and* who have passed the Safety Fundamentals Examination are eligible for the Specialty Examinations. Applicants who meet the Board's certification criteria and who hold one of the acceptable registrations or certifications listed above are eligible for the Specialty Examinations *without* taking the Safety Fundamentals Examination.

## METHODS OF DETERMINING ACADEMIC CREDIT

### Baccalaureate Degrees

Baccalaureate degrees are evaluated by the Board and credit is allowed varying from 12 units for unrelated degrees to 48 units for accredited degrees in safety. Professional safety experience may be substituted at the rate of one month of experience for each unit of credit needed to reach the maximum of 48 units. Experience used in meeting the baccalaureate degree requirement cannot be applied to the experience requirement. If an applicant has more than one undergraduate degree, credit will be allowed for only one of the degrees.

**Figure 24–10a.** Current certification procedures and requirements from the Board of Certified Safety Professionals. (Reproduced with permission from the Board of Certified Safety Professionals.) (*Continues*)

The units of credit allowed for baccalaureate degrees are as follows.

| Major | Units of Credit Allowed |
|---|---|
| Safety accredited by the Accreditation Board for Engineering and Technology (ABET) | 48 |
| Engineering (Accredited by ABET or equivalent foreign degree) | 42 |
| Non-ABET accredited safety | 36 |
| Engineering Technology and Industrial Technology accredited by ABET | 30 |
| Physical and Natural Sciences | 30 |
| Non-ABET accredited Engineering Technology and Industrial Technology | 24 |
| Business Administration, Industrial Education, and Psychology | 18 |
| Majors not listed above | 12 |

### Graduate Degrees

*Master's Degree.* All ABET accredited master's degrees in safety receive one year's experience credit. Other master's degrees receive one-fourth of the credit allowed for their baccalaureate counterparts.

*Doctorate.* All ABET accredited doctorate degrees in safety receive two years of experience credit. Other doctorates receive one-half of the credit allowed for their baccalaureate counterparts.

### Associate Degrees

Associate degrees in all majors receive one-half of the credit allowed for their baccalaureate counterparts.

### College Courses Short of a Degree

Applicants who have completed some college courses but who have not received a degree will receive some credit for the courses completed. The amount of credit is determined by multiplying the baccalaureate credit for the last major shown on the transcripts by the number of semester hours completed, divided by 130 and multiplied by 0.75.

## EXPERIENCE CRITERIA

For experience to be considered as acceptable for credit, it must be full-time professional level experience, safety must account for at least 50% of the position's functions and safety must be the primary function of the position. The first 6 months of the first safety position is considered non-acceptable trainee experience.

## EXAMINATION REQUIREMENTS

Applicants are required to achieve passing scores on two examinations. The first examination is the Safety Fundamentals Examination and covers the basic knowledge required for the broad spectrum of professional safety practice. The second examination is a Specialty Examination. This examination is taken in one of the following areas: Engineering Aspects, Management Aspects, System Safety Aspects, Construction Safety Aspects, and Comprehensive Practice.

## APPLICATION FOR CERTIFICATION

Applicants must first apply by completing the Application for Certification. After a review of the application, the Board will mail notification of eligibility for examination. A non-refundable application fee of eighty dollars ($80) is required at the time the application is submitted. Applicants who pass the Safety Fundamentals Examinations are not required to reapply for the Specialty Examinations.

## STUDENT APPLICANTS

Students in an ABET accredited baccalaureate or master's safety degree program may sit for the Safety Fundamentals examination in their final semester. Qualified students must submit the standard application and, in addition, must provide a faculty statement verifying that they are expected to graduate at the end of the semester.

## DESIGNATIONS AND CERTIFICATES

### Associate Safety Professional

- Those applicants who achieve passing scores on the Safety Fundamentals Examination will receive the designation of Associate Safety Professional (ASP) and will be issued a certificate confirming the designation. An individual may maintain this designation for a period of no more than three years beyond the projected eligibility for the Specialty Examination. At the end of this three-year period, the ASP designation will not be renewed and the individual will not be permitted to take the Specialty Examination except when authorized, after petition to the Board.

### Certified Safety Professional

- Individuals who have achieved passing scores on the Specialty Examination and meet the academic and experience qualifications will receive the designation of Certified Safety Professional. A certificate will be issued authorizing the use of the CSP® designation

**Figure 24–10b.** *(Continues)*

and indicating the area of specialization. The CSP designation is the registered certification mark of the Board of Certified Safety Professionals.

## FEE SCHEDULE

The fees for application, examinations, and renewals can be obtained from the Board of Certified Safety Professionals.

## EXAMINATION ADMINISTRATION

Applicants who are eligible for an examination will be notified of the examination and location approximately two months prior to the examination date. The Board schedules and administers examinations to candidates in locations determined by the geographical distributions of eligible applicants.

The Safety Fundamentals and Specialty Examinations are all full-day examinations and are given simultaneously on the scheduled examination days. Candidates will be able to take only one examination on each scheduled examination day. Thus, someone eligible for both examinations would be required to take and pass the Safety Fundamentals Examination before being scheduled for the Specialty Examination.

Examinations are normally held in the spring (May) and in the fall (October). Application deadlines for the examinations are January 10 and June 1, for the spring and fall examinations, respectively.

## RETAKING EXAMINATIONS FAILED

An applicant who does not pass an examination may retake the complete examination on regularly scheduled examination dates at the appropriate fee for that examination.

## CONTINUANCE OF CERTIFICATION

All Certified Safety Professionals are required to provide evidence of continued professional development in order to continue to renew their certificates. This requirement is called the Continuance of Certification Program. The program requires each CSP to accumulate a prescribed number of credits for professional development activities during five-year intervals.

Additional information and application forms can be obtained by writing to:

Board of Certified Safety Professionals
208 Burwash Avenue
Savoy, Illinois 61874-9571
(217) 359-9263
Fax: (217) 359-0055

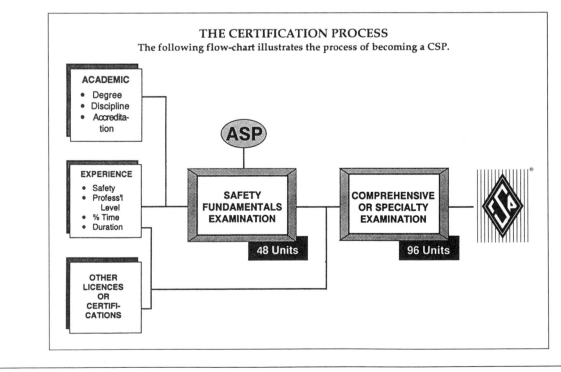

**THE CERTIFICATION PROCESS**
The following flow-chart illustrates the process of becoming a CSP.

**Figure 24–10c.** (*Continued*)

2. Pay the application fee.

3. Be of good character and reputation (in part, this is determined by the reference forms from individuals selected by the applicant).

4. Meet the educational requisite (or have acceptable experience in lieu of education).

5. Meet the experience requisite.

6. Successfully pass the professional safety examinations.

## Academic and Experience Requisites

The educational requirement states that the candidate must be a graduate of an accredited college or university with a bachelor's degree program in safety that meets the Board's requirements in order to receive full credit toward the academic requisites.

An additional paragraph on education states that applicants not meeting the academic requirements can substitute one year of professional safety experience of each academic year needed by the applicant to satisfy the requirements. The experience requirement calls for four years or more of professional safety experience. Graduate degrees may be recognized as professional experience. A doctoral degree may be recognized in lieu of up to two years, and a master's degree may be recognized in lieu of up to one year. These experience equivalents shall be acceptable at the highest level, and are not cumulative.

The Board recognized that there were many people already in safety work who were not college graduates, which is the reason for permitting one year of practical safety experience to be substituted for each year of required college work. Without a degree, to be eligible for certification, an applicant would need a total of four years of professional safety experience (to substitute for the educational requisite) plus four more years of professional safety experience and passing scores on the examinations. In contrast, a person with an acceptable bachelor's degree in safety plus a doctoral degree would need only two years of professional safety experience and passing scores on the examinations to be certified.

To be eligible to sit for the first, or "core," examination, the applicant need only meet the educational requirement. The examination can be taken at any time after meeting this requisite. The eligibility criterion for the second examination is that both the academic and the experience requisites have been met.

## The Future of Safety as a Profession

Problems, both predictable and unpredictable, can be expected to have an impact on the safety professional in the future. Some of these problems will call for reapplication of established safety techniques. Others will call for radical departures and the creation of new methods and new organizational forms. To be able to discriminate between the two solutions will, perhaps, be the safety professional's greatest test.

The field of occupational safety continues to progress and improve, largely through the continued application of techniques and knowledge that have been slowly and painfully acquired over the years. There appears to be no limit to the progress possible through the application of the universally accepted safety techniques of education, engineering, and enforcement.

Large and serious problems remain unsolved. A number of industries still have high accident rates. There are still far too many instances in which management and labor are not working together or have different goals for the safety program.

The resources of the safety movement are great and strong—an impressive body of knowledge, a corps of able professional safety people, a high level of prestige, and strong organizations for cooperation and exchange of information.

Well-trained workers are in high demand in practically all phases of safety. Growth in the trade and service industries and the expanding safety needs of educational institutions, construction, transportation, insurance, and governmental groups should further accentuate the demand for safety workers.

Obviously, there is a need in safety work for people with varying degrees of education and experience. The range of opportunities extends from what could be considered paraprofessional to the highly trained and skilled professional at the corporate management level, and includes safety educators and government safety inspectors and researchers.

The safety professional will also need diversified education and training to meet the challenges of the future. Growth in the population, the communication and information explosion, problems of urban areas and future transportation systems, and the increasing complexities of everyday life will create many problems and may extend the safety professional's creativity to the maximum to successfully provide the knowledge and leadership needed to conserve life, health, and property.

Training of the future safety professional can no longer be limited to the on-the-job experience but must include specialized undergraduate-level training leading to a bachelor's or higher degree.

The type of training needed will depend on the individual job requirements. This presents some difficulties for those preparing to enter the safety and health occupations. Some authorities view the safety and health specialist as a behaviorist and therefore would direct training toward the behavioral sciences; for example, psychology. Others see the specialist as a technician able to handle the technical problems of hazard control, and recommend a heavy background in engineering. Still others believe the safety worker's background should include both the engineering and behavioral aspects.

Future application of this knowledge in all aspects of our civilization—whether to industry or transportation, at home or in recreation—makes it imperative that those in this field be trained to use scientific principles and methods to achieve adequate results. The knowledge, skill, and ability to integrate machines, equipment, and environments with humans and their capabilities will be of prime importance.

## Summary

The work of the safety professional follows a pattern. Before taking any steps in the containment of illness or accidents, the safety professional first identifies and appraises all existing safety and health hazards, both immediate and potential. Once having identified the hazards, the necessary accident-prevention procedures are developed and put into operation. However, this is not enough; safety and health information must be communicated to both management and workers. Finally, the safety professional must evaluate the effectiveness of safety and health control measures after they have been put into practice. If conditions warrant, the safety professional can recommend changes in materials or operational procedures or, possibly, that additional enclosures or safety equipment be added to existing machinery.

Accurate records are essential in the search for the cause of an illness or an accident, and can aid in finding the means to prevent future similar incidents. When studying records to determine the cause or causes of accidents, the records of other companies with similar operations should not be overlooked. Upon determining the cause, the safety professional will have a firm basis on which to propose preventive measures.

Preventive measures are obviously better than corrective measures taken only after an accident has occurred. This means that one of the most valuable functions of the safety professional is to examine the specifications for materials, job procedures, new machinery and equipment, and new structures from the standpoint of safety and health well before installation or construction. In some cases, the safety professional can even help draft the necessary specifications.

As part of the overall safety and health program, the safety professional should recommend policies, codes, safety standards, and procedures that should become part of the operational policies of the organization.

The safety professional draws on specialized knowledge in both the physical and social sciences and applies the principles of measurement and analysis to evaluate safety performance. The safety professional should have a fundamental knowledge of statistics, mathematics, physics, chemistry, and engineering.

The safety professional should be a well-informed specialist who coordinates the safety and health program and supplies the ideas and inspiration while enlisting the wholehearted support of management, supervision, and workers.

## Bibliography

American National Standards Institute. *Injury Statistics—Recording and Measuring Employee Off-the-Job Injury Experience.* ANSI Z16.3–1989. 11 W. 42nd St., New York, NY 10036: American National Standards Institute, 1989.

American Society of Safety Engineers. *Scope and Function of the Professional Safety Position.* Des Plaines, IL: American Society of Safety Engineers, 1995.

Asfalh CR. *Industrial Safety and Health Management.* Englewood Cliffs, NJ: Prentice-Hall, 1990.

Bird FE, Germain GL. *Practical Loss Control Leadership.* Loganville, GA: International Loss Control Institute, 1990.

Boylston RP. *Managing Safety and Health Programs.* New York: Van Nostrand Reinhold, 1990.

Brauer RL. *Safety and Health for Engineers.* New York: Van Nostrand Reinhold, 1990.

Colvin RJ. *The Guidebook to Successful Safety Programming.* Chelsea, MI: CRC/Lewis Publishers, 1992.

DeReamer R. *Modern Safety and Health Technology.* New York: Wiley, 1980.

Hammer W. *Occupational Safety Management & Engineering,* 3rd ed. Englewood Cliffs, NJ: Prentice-Hall, 1985.

Heath ED, Ferry TS. *Training in the Work Place.* Goshen, NY: Aloray, 1990.

Hughes LM, LeBlanc J. The audit: a vital force in system safety. *Hazard Prevent* (Sept./Oct. 1980):13–15.

International Labour Office. *Encyclopedia of Occupational Health and Safety,* vols. 1 and 2. Geneva, Switzerland: International Labour Office, 1991.

Krause TR, Hidley JH, Hodson SJ. *The Behavior Based Safety Process.* New York: Van Nostrand Reinhold, 1990.

Manuele FA. *On the Practice of Safety.* New York: Van Nostrand Reinhold, 1993.

National Safety Council, 1121 Spring Lake Dr., Itasca, IL 60143–3201:
  *Accident Facts.* Published annually.
  *Accident Investigation,* 2nd ed. 1995.
  *Accident Prevention Manual for Business & Industry,* vol. 1: *Administration & Programs* (1992); vol. 2: *Engineering & Technology* (1992); vol. 3: *Environmental Management* (1995).
  *Management Safety Policies,* Data Sheet. No. 585, R(1995).
  *Safeguarding Concepts Illustrated,* 6th ed. 1993.
  *Supervisors Safety Manual,* 8th ed. 1993.

Peterson D. *Safe Behavior Reinforcement.* Goshen, NY: Aloray, 1989a.

Peterson D. *Techniques of Safety Management: A Systems Approach.* Goshen, NY: Aloray, 1989b.

Peterson D. *Safety Management: A Human Approach.* Goshen, NY: Aloray, 1988.

Simonds RH, Grimaldi JV. *Safety Management.* Homewood, IL: Irwin, 1984.

Slote L et al. *Handbook of Occupational Safety and Health*. New York: Wiley, 1987.

Surrey J. *Industrial Accident Research—A Human Engineering Appraisal*. Toronto: Toronto University Press, 1968.

Tarrants WE. *The Measurement of Safety Performance*. New York: Garland Press, 1980.

Tyler WW. Measuring unsafe behavior. *Prof Safety* 31:22–24.

U.S. Dept. of Labor, OSHA. *Recordkeeping Requirements Under the Williams-Steiger Occupational Safety and Health Act of 1970*. Washington, DC: OSHA.

Vincoli JW. *Basic Guide to Accident Investigation and Loss Control*. New York: Van Nostrand Reinhold, 1994.

Workman J. Safety sampling. *Natl Safety Cong Trans* 10, 1974.

CHAPTER **25**

# The
# Occupational
# Physician

by Carl Zenz, MD

**Provisions of the Americans with Disabilities Act**

**Government Programs**

**Objectives of an Occupational Health Program**

**Treating Occupational Injuries**
*Examples of Company Policies*

**Small Units of Large Organizations**

**Location of Employee Health Services**
*Health Service Office (Dispensary)* ▪ *First-Aid Room* ▪
*First-Aid Kits* ▪ *Stretchers*

**Planning for Emergency Care in the Occupational Setting**

**Medical Placement**
*The Preplacement or Post-Offer Medical Examination* ▪
*Evaluating Physical Capacity*

**Scope of the Examination**
*Fitness or Periodic Examination* ▪ *Low Back Problems* ▪
*Chest X Rays* ▪ *Pulmonary Function Testing* ▪ *History of
Personality Disorders*

**Health Education and Counseling**
*Substance Abuse* ▪ *Drug Screening* ▪ *Access to Medical Records*

**Medical Aspects of the Use of Respiratory Protective Equipment**

**A Program for Employee Respirator Use Assessment**
*Responsibilities of the Industrial Hygienist* ▪ *Responsibilities of the
Occupational Health Nurse* ▪ *Eyeglasses* ▪ *Speech Transmission*

**Potential Reproductive Hazards in the Workplace**
*Exposure of the Unborn Child* ▪ *Spontaneous Abortions* ▪
*Teratogenic Effects* ▪ *Guidelines for the Working Pregnant Woman*

**Biomechanical and Ergonomic Factors**

**Biological Monitoring**
*Explanation of the **Skin** Notation* ▪ *ACGIH Biological Exposure
Indices (BEIs)*

**Summary**

**Bibliography**
*American College of Environmental and Occupational Medicine
Publications* ▪ *NIOSH Publications*

OCCUPATIONAL MEDICINE WAS CLASSIFIED by the American Board of Preventive Medicine as a specialty in 1955. This board also includes the public health and aerospace medicine specialties.

As a subspecialty of preventive medicine, occupational medicine is concerned with the following:

▪ Appraisal, maintenance, restoration, and improvement of the workers' health through application of the principles of preventive medicine, emergency medical care, rehabilitation, and environmental medicine

▪ Promotion of a productive and fulfilling interaction of the worker and the job via application of principles of human behavior

▪ Active appreciation of the social, economic, and administrative needs and responsibilities of both the worker and work community

▪ Team approach to health and safety, involving cooperation of the physician with occupational or industrial hygienists, occupational health nurses, safety personnel, and other specialists

Requirements for qualification as a specialist in occupational medicine include completion of postgraduate courses in biostatistics and epidemiology, industrial toxicology, work physiology, radiation (ionizing and nonionizing), noise and hearing conservation, effects of certain environmental conditions such as high altitude and high pressures (hyperbaric and hypobaric factors), principles of occupational safety, fundamentals of industrial hygiene, occupational aspects of dermatology, psychiatric and psychological factors, occupational respiratory diseases, biological monitoring, ergonomics, basic personnel management functions, record and data collection, government regulations, and general environmental health (air, water, ground pollution, and waste management control).

Since 1911, workers' compensation laws have been enacted in all states. These laws require employers to provide medical care for occupationally injured employees and compensate employees or their heirs for occupational disability or death. In addition, most of these laws require employers to provide medical care for employees with occupational diseases.

These and other laws and regulations, especially those promulgated by the U.S. Department of Labor's Occupational Safety and Health Administration (OSHA), have obligated employers to maintain safe and healthful working environments. The problems associated with increasingly complex technology and with new, potentially hazardous physical and chemical agents act as an important stimulus to the development of occupational health programs. These developments caused a broadening of the earlier concept of curative occupational medicine. The field now includes and emphasizes preventive medicine and health maintenance. This new emphasis led to creation of the kind of occupational health program described in this chapter. Other social and administrative regulations have been promulgated, the most recent and complex of which is the Americans with Disabilities Act (ADA).

# Provisions of the Americans with Disabilities Act

The ADA defines a disabled person as one who:

■ Has a physical or mental impairment that limits one or more major life activities

■ Has a record of such an impairment

■ Is regarded as having such an impairment

A qualified disabled worker is defined as one who can perform the essential functions of a job, with or without reasonable accommodation.

Compliance with the ADA begins with documentation of the hiring and job placement procedures. This includes policies and procedures for every step in the hiring process and recommends written job descriptions.

Job descriptions should contain the following information:

■ Essential job functions: the fundamental purposes of the job (why does the job exist?)

■ Marginal job functions: the additional duties required that could be assigned to other workers or eliminated if the job structure were changed

■ Job structure: the specific tasks ordinarily needed to do the job (what functional capacities, skills, education, and experience are needed to perform these job tasks?)

Any tests must be clearly related to job performance and test methods may not discriminate against those with disabilities unrelated to job-required skills. Tests may involve a marginal job function unless the applicant has a disability that impairs performance of that marginal function.

The ADA prohibits an employer from inquiring into a job applicant's disability with questions concerning such areas as medical history, prior worker's compensation or health insurance claims, work absenteeism due to illness, and mental illness. Drug and alcohol use questions are limited and may not touch on addiction or rehabilitation.

Medical examinations are limited to examinations after the job offer has been made. The post-offer examination may ask about previous injuries and compensation claims and the scope can be expanded beyond usual if findings warrant. The post-offer examination can inquire about past and present drug and alcohol use. Periodic examinations must be job-related and necessary for business.

The ADA requires employers to make reasonable accommodations for disabled workers, which might include the following:

■ Acquiring or modifying work equipment

■ Providing qualified readers or interpreters

■ Making existing facilities, such as workstations, restrooms, telephones, and drinking fountains, accessible to a person with disabilities

■ Redesigning the job by eliminating problem tasks as long as the essential job functions are not changed

# Government Programs

The passage of the Occupational Safety and Health Act (OSHAct) of 1970 gave new and long-needed impetus to health and safety programs and to research in job-related hazards. The act seeks "to assure so far as possible every working man and woman in the nation safe and healthful working conditions." The law became effective on April 28, 1971.

Medical regulations first promulgated in 1972 under the OSHAct were applicable to all employers. Subpart K, §1910.151, Medical Services and First Aid (*Federal Register [FR]* 39:125. p. 23682, June 27, 1974) of these regulations states:

(a) The employer shall ensure the ready availability of medical personnel for advice and consultation on matters of plant health.

(b) In the absence of an infirmary, clinic, or hospital in near proximity to the workplace, which is used for the treatment of all injured employees, a person or persons shall be adequately trained to render first aid. First-aid supplies approved by the consulting physician shall be readily available.

(c) Where the eyes or body or any person may be exposed to injurious corrosive materials, suitable facilities for quick drenching or flushing of the eyes and body shall be provided within the work area for immediate emergency use.

Later, 1972 Regulations for Construction included everything in (a) above, and substituted the following portion of (b). (See Subpart D, Occupational Health and Environmental Controls; §1926.50, Medical Services and First Aid [*Federal Register* 37:243. p. 27510, Dec. 16, 1972]).

[A] person or persons who have a valid certificate in first-aid training from the U.S. Bureau of Mines, the American National Red Cross, or equivalent training that can be verified by documentary evidence, shall be available at the worksite to render first aid.

The regulations continue:

The first-aid kit shall consist of materials approved by the consulting physician. The materials shall be in a weatherproof container with individual sealed packages for each type of item. The contents of the first-aid kit shall be checked at least weekly to ensure that the expended items are replaced.

Proper equipment for prompt transportation of the injured person to a physician or hospital, and a communication system for contacting an ambulance service shall be provided. The telephone numbers of the physicians, hospitals, and ambulances shall be conspicuously posted.

Responsibilities for health standards are divided between two government agencies: the Department of Labor (DOL) and the Department of Health and Human Services (DHHS) (formerly the Department of Health, Education, and Welfare). The DOL has the responsibility for promulgating and enforcing new, mandatory occupational safety and health standards and has the authority to enter factories and workplaces to conduct inspections, to issue citations, and to impose penalties.

The DOL regulations require that employers maintain accurate records of work-related injuries, illnesses, and deaths. It initiated programs for educating employees and employers in the recognition, avoidance, and prevention of unsafe or unhealthy working conditions covered by the OSHAct.

NIOSH was created within the Department of Health and Human Services when OSHAct was enacted. NIOSH is required to develop criteria for new safety and health standards and to transmit them to the DOL for promulgation and enforcement. Related to the development of criteria for recommended standards are the requirements to conduct "research into the motivational and behavioral factors relating to the field of occupational safety and health" and to "conduct and publish industrywide studies

of the effect of chronic or low-level exposure to industrial materials, processes, and stresses on the potential for illness, disease, or loss of functional capacity in aging adults." NIOSH also "conduct[s] special research, experiments, and demonstrations relating to occupational safety and health as are necessary to explore new problems, including those created by new technology."

OSHA adopts standards and conducts inspections of workplaces to determine whether the standards are being met (see Chapters 29 and 30). OSHA requires each employer to provide a workplace free from safety and health hazards and to comply with the standards.

An OSHA standard is a legally enforceable regulation governing conditions, practices, or operations meant to ensure a safe and healthy workplace; these standards are published in the *Federal Register*. OSHA standards are divided into three major categories: general industry, maritime, and construction. An example of such a standard for general industry is §1910.1001, Asbestos, Paragraph (1) of the Asbestos Standard, Medical Examinations. The standard specifies requirements for preplacement, annual, and termination-of-employment examinations, and contains provisions for maintenance and access to medical records.

Using the asbestos standard as an example, the regulations call for annual comprehensive medical examinations of all employees exposed to airborne concentrations of asbestos fibers above the PEL of 0.1 fiber/cm$^2$ or excursion limit of 1 fiber/cm$^2$ for 30 min. Minimum requirements are a 14 × 17 in. (35 × 43 cm) chest roentgenogram, a "history to elicit symptomatology of respiratory disease, and pulmonary function tests including forced vital capacity (FVC) and forced expiratory volume at one second (FEV$_1$)." Examinations are paid for by the employer. Medical records must be retained by employers for at least 30 years and be accessible to OSHA, NIOSH, and certain physicians.

OSHA requires employers of 10 or more employees to maintain certain records of job-related fatalities, injuries, and illnesses. OSHAct requires that two simple forms be maintained: OSHA form no. 200 (the basic Log and Summary of Occupational Injuries and Illnesses) and a supplementary record (OSHA form no. 101) or a similar form that contains the same information.

On OSHA form no. 200 (Figure 25–1), employers can use a checkoff procedure to indicate the type of case. This provides the employer with a running summary of injuries and illnesses at any time during the year.

## ■ Objectives of an Occupational Health Program

The basic objectives of an occupational health program as set forth in the revision of the American Medical Association's (AMA's) *Scope, Objectives and Functions of Occupational Health Programs* (1972) are as follows:

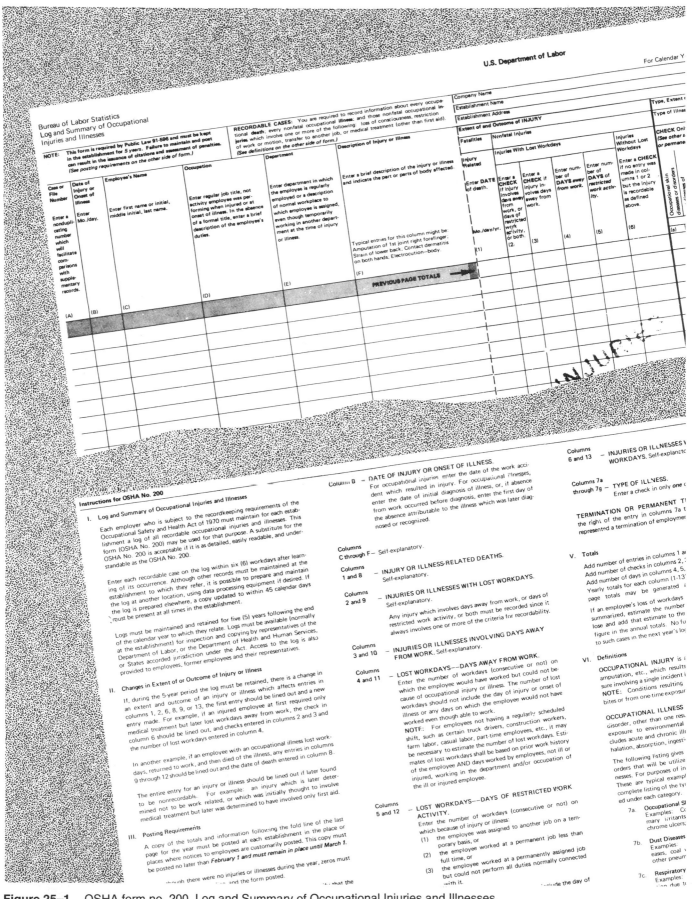

**Figure 25–1.** OSHA form no. 200, Log and Summary of Occupational Injuries and Illnesses.

1. To protect employees against health and safety hazards in their work situation

2. Insofar as practical and feasible, to protect the general environment of the community

3. To facilitate the placement of workers according to their physical, mental, and emotional capacities in work which they can perform with an acceptable degree of efficiency and without endangering their own health and safety or that of others

4. To ensure adequate medical care and rehabilitation of the occupationally ill and injured

5. To encourage and assist in measures for personal health maintenance, including the acquisition of a personal physician whenever possible

Achievement of these objectives benefits both employees and employers by improving employee health, morale, and productivity.

# Treating Occupational Injuries

This brief discussion assumes certain principles that depend on the available treatment facilities and medical staff. In all work settings, injured workers should be segregated from those who need preplacement examinations or other physical examinations, ensuring the immediate treatment of the injured workers and privacy. An onsite medical facility should be centrally located. Personnel trained in first aid should be available in all remote areas, and appropriate litters or other means of transportation available to transport ill or injured workers to the on-site medical facility or a nearby hospital.

A physician should act as medical director or medical advisor in order to properly conduct the employee health program and to provide medical direction for the occupational health nurse (a registered nurse working in the occupational health field; see Chapter 26, The Occupational Health Nurse). These principles were first stated in 1932, when the State Medical Society of Wisconsin published *Suggestions for the Guidance of the Nurse in Industry.* This was the first U.S. medical society to formally acknowledge the need for a well-defined physician–nurse relationship in occupational health programs (Figure 25–2).

All physicians concerned with an occupational health program should initiate written medical directives to guide the occupational health nurse in making a provisional diagnosis and determining the next step needed to aid an ill or injured employee. Although preventive health services are important, one of the nurse's major functions is the emergency care of ill and injured employees.

## Examples of Company Policies

Policies range from concise to elaborate.

**Concise policy.** The American Medical Association (AMA) has a concise statement:

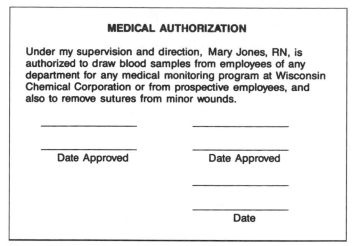

**MEDICAL AUTHORIZATION**

Under my supervision and direction, Mary Jones, RN, is authorized to draw blood samples from employees of any department for any medical monitoring program at Wisconsin Chemical Corporation or from prospective employees, and also to remove sutures from minor wounds.

_____        _____

_____        _____
Date Approved                Date Approved

                            _____

                            Date

**Figure 25–2.** This sample of a medical authorization is indicative of a well-defined relationship between the occupational physician and the occupational nurse.

The aim of the plant medical service is to provide a healthy, effective working force by encouraging personal health improvement, preventing illness and injury, and treating those industrial diseases and injuries that do occur.

**Elaborate policy.** An elaborate personnel statement is exemplified by the Volvo Company of Sweden, which has issued a policy statement to all employees concerning the physical environment.

In the design of products, workplaces, machines and equipment, attention must also be paid to the physical and mental conditions of the employees and the need for a good working environment. The working conditions to be aimed at are those in which the employees can carry out their jobs without any detrimental stress; systematic efforts are to be made to identify and eliminate health risks; and the working environment is to be designed in such a way that it provides opportunities for new forms of work and encourages cooperation and satisfaction at the workplace.

**Multiservice program.** Activities of a fully developed multiservice program in occupational medicine are outlined in the following seven sections quoted from the AMA's *Scope, Objectives and Functions of Occupational Health Programs.* This document is the most widely used guide for physicians concerned with planning and administering such programs. It also is used by many government agencies as a reference for recommendations.

1. Maintenance of a Healthful Work Environment—This requires that personnel skilled in industrial hygiene (environmental control) perform periodic inspections on the premises, including all facilities used by employees, and evaluate the work environment in order to detect and appraise health hazards, mental as well as physical. Such inspections and appraisals, together with the knowledge of processes and materials used, provide current information on health aspects of the work environment. This information will serve as the basis

for appropriate recommendations to management for preventive and corrective measures.

2. Preplacement (Post-Offer) Examinations and/or Screening—As an aid to suitable placement, some form of preplacement health assessment is desirable. The scope of such assessments will be influenced by such factors as size, nature, and location of the industry as well as by the availability of physician and nurse services. They may range from medical examination by a physician, through medical questionnaires and screening procedures by a nurse, to more simple screening by allied health personnel. The type of assessment to be performed as a routine, as well as the specific tests to be included in each category, should be determined by the physician in charge of the health program. In general, the assessment should include: (a) personal and family history, (b) occupational history, (c) such physical examinations and/or laboratory and appropriate screening tests as seem advisable for the particular industry and availability of professional personnel.

Unrealistic and needlessly stringent standards of physical fitness for employment defeat the purpose of health examination and of maximum utilization of the available work force and they are a major target of the Americans with Disabilities Act (ADA).

3. Periodic Health Appraisals—These health evaluations are performed at appropriate intervals to determine whether the employees' health remains compatible with his job assignment and to detect any evidence of ill health that might be attributable to his employment. Certain employees and groups may require more frequent examinations as well as additional procedures and tests, depending on their ages, their physical conditions, the nature of their work, and any special hazards involved. Health examinations and appraisals should be conducted or supervised by a physician with the assistance of such qualified allied health personnel as may be indicated and available. The examination may be made in any properly equipped medical facility, at the workplace.

The individual to be examined should be informed by appropriate means of the purpose and value of the examination. The physician should discuss the findings of the examination with the individual; he should explain the importance of further medical attention for any significant health defects found and make an appropriate medical referral as needed.

4. Diagnosis and Treatment:
   a. Occupational injury and disease—Diagnosis and treatment in occupational injury and disease cases should be prompt and should be directed toward rehabilitation. Workers' compensation laws, insurance coverage arrangements, and policies of medical societies usually govern the provision of medical services for such cases.

   b. Nonoccupational injury and illness—Every employee should be encouraged to use the services of a personal physician or medical service where these are available for care of off-the-job illness or injuries. Treatment of nonoccupational injury and illnesses never has been and is not ordinarily considered to be a routine responsibility of an occupational health program with these limited exceptions:

   In an emergency, of course, the employee should be given the attention required to prevent loss of life or limb or to relieve suffering until placed under the care or his personal physician. For minor disorders, first aid or palliative treatment may be given if the condition is one for which the employee would not reasonably be expected to seek the attention of his personal physician, or if it will enable the employee to complete his current work shift.

   Due to the shortage of physicians in some communities, it is recognized that some employees on occasion may find it impossible to locate or to obtain the services of a personal physician or health service. In such circumstances, limited to where treatment is otherwise unavailable, the occupational physician may undertake additional and continuing treatment of an employee's nonoccupational condition if requested to do so by the employee or his family. If such services might become ongoing within the occupational health program, approval of the employer should be obtained. In order to help assure high-quality medical care, discussion with the local medical society in developing such projects, including the methods of payment for services, is urged.

5. Immunization Programs—An employer may properly make immunization procedures available to his employees under the principles set forth in the General Recommendations on Immunization—Recommendations of the Advisory Committee on Immunization Practices (ACIP) (*MMWR*, 1994).

6. Medical Records—The maintenance of accurate and complete medical records of each employee from the time of his first examination or treatment is a basic requirement. The confidential character of these records, including the results of health examinations, should be rigidly observed by all members of the occupational health staff. Such records should remain in the exclusive custody and control of the medical personnel. Release of information from an employee's medical record must not be made without the employee's written authorization, except as required by law.

7. Health, Education, and Counseling—Occupational health personnel should educate employees in personal hygiene and health maintenance. The most favorable opportunity for reaching an employee with health education and counseling arises when he visits a health facility.

Health education appropriately goes hand in hand with safety education. The occupational health and safety personnel, therefore, should work cooperatively with supervisory personnel in imparting appropriate health and safety information to employees. Health and safety education should (a) encourage habits of cleanliness, orderliness, and safety, and (b) teach safe work practices, the use and maintenance of available protective clothing and equipment, and the use of available health services and facilities. Experience has shown that health education is most effective when the employer demonstrates his sincere and continuing interest in the health of his employees and when employees are encouraged to participate in the planning and conduct of health education activities.

## ■ Small Units of Large Organizations

See Figure 25–3 for example of functional organization.

Small operations belonging to large companies are the main users of partial in-house medical programs. At an AMA Congress on Occupational Health in 1971, Marcus B. Bond, MD, gave an excellent summary on this complex subject. The remainder of this section is taken from his report.

If small employee groups are part of a large company with headquarters elsewhere, advice and instructions can be supplied from the corporate offices where staffs of specialists exist. Physicians, nurses, industrial hygienists, safety experts, and laboratory relations representatives can combine their knowledge and provide specific instructions to supervisors and local physicians in any location covering any type work operation. Such instructions should, of course, be consistent with general company policies, insurance and benefit provisions, labor contracts, and the laws of the particular state. [See Figure 25–4.]

**Selection of a physician.** Selection of a physician obviously is a first step because a local physician, licensed in the state, is required for examinations, treatment, or any other professional activities. It is best if local management and the company medical director share the responsibility for obtaining a doctor's services. Usually local managers are acquainted with several physicians in the community through social, business, and civic activities and they can recommend one or

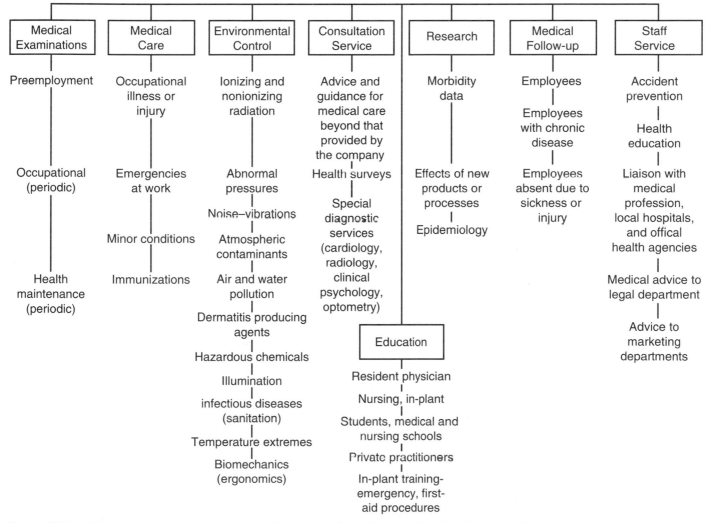

**Figure 25–3.** Functional activities of occupational health services. (Prepared by Carl Zenz, 1972; revised 1990.)

## Environmental Health & Hygiene Policy

As a responsible manufacturer of a wide variety of specialty chemicals and pesticides, this company's plants, laboratories and offices shall be maintained as clean and healthful places of employment.

At all times, this company's facilities shall be designed and operated in compliance with the spirit and letter of federal, state and local occupational health and hygiene regulations.

This company acknowledges and shall satisfy its repsonsibility to promptly provide current, comprehensive information on potential adverse health effects and appropriate handling procedures for chemicals handled by both our employees and our customers.

It is a basic responsibility of all this company's employees to make the health and safety of fellow human beings a part of their daily, hourly concern. This responsibility must be accepted by each one who conducts the affairs of the corporation, no matter in what capacity he may function.

A goal of this company is to achieve and maintain a corporate occupational health and hygiene program which is a model for the chemical industry.

Chairman of the Board &
Chief Executive Officer

President & Chief
Operating Officer

_____

_____

**Figure 25–4.** This statement is a good example of a health and hygiene policy. (Courtesy Velsicol Chemical Corporation.)

more physicians to the medical director. The medical director can check the physician's credentials by consulting with the *American Medical Directory,* published by the American Medical Association. Often the medical director will be personally acquainted with one or more doctors in the particular community or may know officials in the state or local medical society who can provide information about available physicians. When the choice of a physician has been agreed upon by the medical director and local management, the physician should be contacted to see if he or she is interested in serving the organization. If so, then the medical director should visit the physician personally. If this is not possible, the director could write a letter and provide the physician's manual or other written policies, standards for applicants, description of jobs, and the necessary periodic tests or examinations for employees.

**Work site visits.** The physician should visit the work sites and become familiar with all phases of the different jobs. The local manager should arrange for the physician to visit the plant or place of work and actually observe work conditions and operations and become acquainted with at least some of the supervisors and workers. The frequency of visits thereafter should be determined by the nature of the work and major changes in materials or processes. If significant hazards exist in the work environment, the physician should tour the work area every 1 or 2 months.

Often, raw materials, by-products, wastes (effluents), solvents, and related materials in the processing or manufacturing can be more hazardous than the final product. [See Chapter 6, Industrial Toxicology.]

Remuneration of the physician may be on a fee-for-service basis, including hourly charges for visits to the workplace. If the services provided are frequent, then a retainer or salary may be desirable for both parties—it increases the physician's attention to the employer's business and may provide certain fringe benefits associated with employment such as health and life insurance, pensions, Social Security, and others.

During the past decade, more physicians have become active in the private practice of occupational medicine, either as hospital-based practitioners, in partnerships in multispecialty clinics, in freestanding occupational medicine clinic groups, or internally as employees of a company. Often, several employers may receive professional guidance or consultations on a regular basis from one physician. In such circumstances, this externalization of services does not lend itself to accept employment status. Rather, these physicians find billing on a monthly basis more suitable and further enhancing neutrality, a vital aspect, often difficult to maintain in a salaried position, whether full- or part-time.

**Medical examinations.** After the physician has become familiar with the work environment and company policies, he or she should advise the employer how and when to send employees to the physician's office for nonemergency conditions such as applicant examinations, return-to-work checkups, and periodic tests and examinations such as blood counts, chest x-rays, etc. Instructions for these and for emergencies should be written and provided to the employer so the latter may post them on bulletin boards, distribute them to supervisors, place them in first-aid kits and in company trucks, etc. Emergency instructions should also include where to send patients at different times during the day and on special days such as weekends and days off; to be listed are the name address, and phone number of the hospital emergency rooms plus the name of an associate or colleague who should be contacted if the physician is not available.

The physician must then instruct his or her own nurses and office assistants as to what commitments have been made to the employer and provide them with the company forms or instructions. Reference books or articles pertaining to the hazards in the particular working environment should be obtained. The company medical director should furnish these or can advise on them.

At times, the local physician can have the advantage of working with the medical director on problem cases. This can be reassuring when an unusual illness or reaction occurs that could possibly be related to work exposure. The medical director should be familiar with products, materials, and processes in the company and be able to provide guidance, even if at long distance.

The physician should be given a list of all materials used throughout the workplace. The industrial hygienist and the responsible safety person will find the physician to be valuable professionally. They should ensure that the physician knows of all processes, intermediate products, effluents, and methods for handling or dealing with hazardous situations, and has all relevant data and documentation pertaining to this information (such as material safety data sheets [MSDSs]).

The physician who is new to occupational health practice should acquire certain reference literature, including this book, and *Occupational Health & Safety*. Both texts are published by the National Safety Council, and are valuable review and refresher sources for physicians.

# ■ Location of Employee Health Services

Although good occupational medicine can be practiced in any facility location that is clean and private, experience indicates that the medical unit commands respect only if careful attention is paid to suitable and efficient housing, appearance, and equipment. The entire unit should be painted in light colors and kept spotlessly clean. The dispensary should have hot and cold running water and be adequately heated, ventilated, and illuminated. Toilet facilities are necessary. Suitable provisions should be made for examining men and women if both are employed. Cleanliness and privacy are essential.

The medical department should be easily accessible and near the greatest number of employees so that distance does not impede the immediate reporting for treatment of even minor injuries. If possible, it should also be close to the human resources and safety departments; this facilitates prompt physical examinations of job applicants, the mutual use of clerical service, and the interchange of ideas and plans regarding employment, incident, and health problems.

Another location consideration is to have the facilities near the entrance so an ambulance can be brought to the door if necessary. Also, injured workers, who are off duty but under treatment, may be admitted through a separate entrance.

The medical department should be in a place of greatest safety in case of a major disaster that might otherwise destroy first-aid or dispensary supplies or facilities.

## Health Service Office (Dispensary)

A minimum of three rooms, consisting of a waiting room, a treatment room, and a room for consultation or for making physical examinations, is recommended. Rooms for special purposes can be added according to the needs and size of the company. As mentioned earlier, the layout of the dispensary should be such that applicants for employment waiting for physical examinations do not mingle with injured workers.

The surgical treatment room should be large enough to treat more than one person at a time; small dressing booths can be arranged to give some degree of privacy.

## First-Aid Room

It is always advisable to set aside a room at a convenient location for the sole purpose of administering first aid. It upsets morale to administer treatments in public; injured people prefer privacy. Furthermore, the person administering first aid should have a proper place to do his or her work.

The environment of a good first-aid room should be similar to that of the dispensary. At a minimum, the room should be equipped with the following items:

- Desk and chair
- Examining table
- Bed for emergency cases, enclosed by a movable curtain—hospital-type bed preferred (cot or sofa-style beds are too low and too narrow)
- Treatment table and containers for first-aid supplies
- Small table at bedside

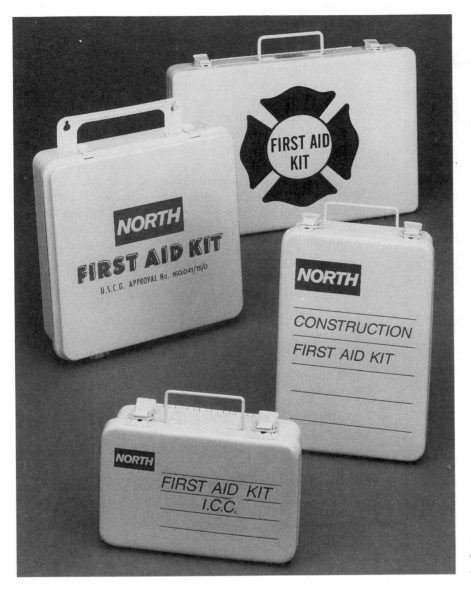

**Figure 25–5.** First-aid kit exteriors can be customized for specific users and applications.

- Chair with arms and one without arms
- Mayo stand
- Switchable lighting fixtures

Emergency oxygen is of great benefit in the treatment of some first-aid cases. It is inappropriate to allow smoking in a medical facility. Furthermore, when oxygen is in use, smoking is an immediate hazard because of the danger of fire or explosion. Any type of resuscitating device should only be used by trained personnel.

## First-Aid Kits

There are many types of emergency first-aid kits; they are designed to fill every need, depending on the special hazards that might occur. Commercial or cabinet-type first-aid kits, as well as unit-type kits, must meet OSHA requirements.

Kits vary in size from a pocket-size model to what amounts to a portable first-aid room. The size and content depend on the intended use and the types of injuries that could occur (Figure 25–5). For example, personal kits contain only essential articles for the immediate treatment of injuries. Departmental kits are larger—they are planned to cover a group of workers and the quantity of material depends on the size of the work force. Truck kits are the most complete—they can be carried easily to the accident site or stored near working areas that are distant from well-equipped emergency first-aid rooms. Although they include such bulky items as a wash basin, blankets, splints, and stretchers, they can be carried conveniently by two people.

First-aid kits should be distributed throughout the plant or job site. Each kit must be the responsibility of one trained individual who should understand that he or she is to care for the most trivial injuries only or to render only temporary treatment for more serious cases (Figure 25–6). It is believed that with this system many slight scratches and cuts are given attention that they would not otherwise receive.

**Figure 25–6.** First-aid kits should be distributed throughout the job site or facility. Each should be the responsibility of one trained person, who should understand that he or she is to care for the most trivial injuries only or to give only temporary treatment for more serious cases.

In industrial organizations such as mining companies and public utilities, where activities are widely scattered, the use of first-aid kits and some self-medication may be necessary. However, it is better to control first-aid service by having the attendant in charge properly instructed in first aid and by seeing that the service has proper medical supervision.

The maintenance and use of all first-aid kits should be supervised by medical personnel, and the kits should contain the materials approved by the consulting physician. A member of the medical services group should use a checklist to regularly inspect all first-aid materials and submit a report of content level and condition.

Maintenance of quantities of materials in the first-aid kit is easier if each kit lists the original contents and the quantities at which new materials should be ordered. All bottles or other containers should be clearly labeled.

Recommended materials for first-aid kits are listed in textbooks from the U.S. Mine Safety and Health Administration, the National Safety Council, and the American National Red Cross, and in the American National Standards Institute (ANSI) standard for first-aid kits, ANSI Z308.1–1978 (1984), and the National Safety Council's Data Sheet 12304–0202 Rev. 1991, *Unit First Aid Kits.* Suggestions also are available from the American Medical Association and manufacturers of first-aid materials (Figure 25–7). The physician or nurse should routinely inspect first-aid supplies and stretchers and check oxygen tanks for adequate pressure.

**Figure 25–7.** First-aid kits vary in size, depending on the specific requirements. This first-aid station is for use in mines, remote worksites, and places far from the dispensary in large facilities. It contains a clam-type (scoop) stretcher, blankets, splints, and a first-aid kit with emergency treatment supplies.

## Stretchers

It is essential to have adequate means of transporting a seriously injured person from the scene of the accident to a first-aid room or hospital. Unnecessary delays and inadequate transportation can add to the seriousness of the injury and may be the determining factor between life and death.

The stretcher provides the most acceptable method of hand transportation and it also can be used as a temporary cot at the scene of the accident, during transit in a vehicle, and in the first-aid room or dispensary.

There are several types of stretchers. The army-type is commonly used and is satisfactory. However, when it is necessary to hoist or lower the injured person out of awkward places, it is better to use specially shaped stretchers with straps to hold the patient immobile.

Stretchers should be conveniently located in all areas where employees are exposed to serious hazards. It is customary to keep stretchers, blankets, and splints in well-marked, conspicuous cabinets. Stretchers should be kept clean, not be exposed to destructive fumes, dust, or other substances, protected against mechanical damage, and ready for use at all times. If the stretcher is of a type that will deteriorate, it should be tested periodically for durability, strength, and cleanliness.

## ■ Planning for Emergency Care in the Occupational Setting

The following policies and procedures are helpful in planning for emergency care in the occupational setting. Some obviously are independent nursing functions and responsibilities; others are not and must be carried out in cooperation with the physician, management, safety personnel, industrial hygienists, first-aid teams, and other appropriate personnel (see Chapter 26, The Occupational Health Nurse). The names and phone numbers should be changed to fit local conditions.

Planning is necessary to maintain a safe working environment, proper health protection, prompt and definitive emergency care for injured and ill employees, and safe handling and transportation of injured and ill employees. Planning should include the following items:

- *Assessment of hazards in the work environment.* Hazards must be identified and assessed for corrosiveness, irritative capability, toxic or otherwise injurious qualities, flammability, and explosiveness; their location in the facility and their use; the symptoms that could be produced by exposure to these; and treatment to counteract effects. Treatment procedures should be posted and necessary medical supplies readily available in the employee health service areas.

  Since May 1986, the OSHA Hazard Communication Standard (HCS) has required employers to label in-plant containers of hazardous chemicals, to inform employees of workplace hazards, to make MSDSs available to employees, and to train workers in protective measures for specific chemical hazards. Employers must develop written hazard communication programs outlining their plans to accomplish these objectives. (See the discussion on hazard communication laws in Chapter 29, Government Regulations.)

- *Occupational and nonoccupational health emergencies.* A plan and procedure should be developed for meeting personal health emergencies of both minor and major magnitude.

- *Medical directives and nursing procedures.* These should be formulated to cover all emergency situations.

- *Emergency care equipment.* Plans should be drawn up for providing stationary and portable emergency equipment. Portable emergency equipment should be clearly labeled and stored in appropriate areas throughout the facility and in the employee health services.

- *Transportation of injured and ill employees.* There should be a written policy approved by management outlining the transportation of ill and injured employees.

- *Training of first aid teams and auxiliary personnel.* It is essential to have plans for providing well-trained first-aid teams with clearly identified responsibilities in emergency situations.

- *Cardiopulmonary resuscitation (CPR).* When a physician is available, he or she should take charge of the resuscitation of an employee undergoing cardiopulmonary arrest. In the absence of a physician, a professional nurse or another person specially trained and certified in the recognition of cardiopulmonary arrest and the technique of resuscitation may apply the appropriate emergency procedures (Figure 25–8). A policy must be clearly stated to cover this emergency and the requisite special training should be made available to the staff.

- *Critical illness and death.* A written policy should be prepared in collaboration with all concerned people and approved by management. This policy should cover the procedure to be followed in case of critical illness or death.

- *Employee health record.* Emergency medical identification information should be recorded on individual employee records. These records should be kept in the employee health service so the nurse giving emergency care has them available. (An example is shown in Figure 25–9.) It is especially important to have up-to-date information on each employee's immunization status against tetanus.

- *Anticipatory orders from employee's personal physician.* Special information and orders should be on file

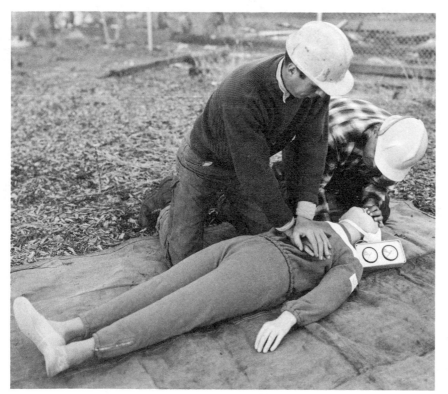

**Figure 25–8.** These linemen are practicing cardiopulmonary resuscitation (CPR) in the field, using a doll as the "victim." In the absence of a physician, a professional nurse or another person trained and certified in CPR can apply appropriate emergency procedures.

from an employee's personal physician covering emergency or routine care for special health problems, such as diabetes, asthma, and cardiovascular disease. These anticipatory orders may be obtained either by the employee or by the nurse with the employee's permission.

- *Personal emergency medical identification.* All employees with special health problems should be encouraged to wear or carry emergency medical identification information both on and off the job. Durable devices should be worn on wrist or around neck to identify employees with diabetes, epilepsy, and other problems for which emergency medical care may be needed. Further information on emergency medical identification is available from the American Medical Association, 515 North State St., Chicago, IL 60610. Also refer to Medic Alert Foundation, Turlock, CA. Such devices should include the following information:

Medical conditions (such as diabetes)

Allergies (such as horse serum, penicillin)

Medication regularly used (such as anticoagulants for heart disease)

Immunization status (such as date of tetanus booster)

Name, phone number, and address of personal physician and next of kin

- *Disaster planning and community resources.* It is recommended that all planning for disaster and emergency care be correlated with community health resources.

All factors listed above must be regularly reviewed by all appropriate personnel, never less than once yearly. For further details and sample directives, refer to the 5th edition of the *Occupational Health Guide for Medical and Nursing Personnel,* published by the State Medical Society of Wisconsin (see Bibliography). The objectives, principles, and directives found in this guide have worldwide acceptance and are adaptable to most jurisdictions.

# Medical Placement

## The Preplacement or Post-Offer Medical Examination

A preplacement or post-offer medical examination is performed to determine the individual's physical and emotional capacity to perform a particular job, to assess the individual's general health, and to establish a baseline record of physical condition for the personal needs of the employee as well as the employer. This is important not only in the event of job changes, but also for medical and legal purposes.

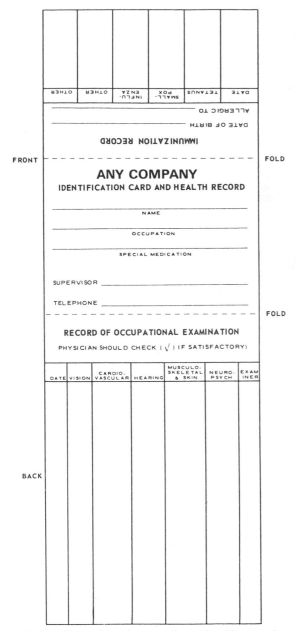

**Figure 25–9.** Illustration of an employee health record. (Courtesy M. B. Bond, MD.)

Determination of the physical and emotional capacity for job performance is the main purpose for a preplacement physical examination. Theoretically, this is how individuals can be matched to specific jobs according to their physical and emotional capabilities and aptitudes. With a good match, a happy, healthy employee can result; if not, the employee may not succeed.

Medical evaluation of physical capacity arose from the need to determine which individuals could do various levels of work without harming themselves or others. Although employers have a fear of an employee aggravating a preexisting condition with a resultant workers' compensation case and attendant costs, this fear has never been borne out as valid.

Unfortunately, many jobs involving considerable exposure to health hazards still remain. Industrial hygiene surveys and engineering controls have resulted in a gradual reduction of the physical demands of many jobs. Automation and high employee turnover rates have diminished the physical fitness aspect of the preplacement examination.

Nevertheless, jobs causing exposure to heat, noise, vibration, chemicals, dusts, and other adverse conditions still exist. Some jobs also require heavy labor. As described earlier in this chapter, with the advent of OSHA standards delineating physical requirements for certain jobs, especially those characterized by exposure to chemical and physical agents, it appears that the preplacement examination will assume a new and increasingly important role.

These examinations should not be undertaken with the intent to exclude a person from work, although that may have to be the case in some positions, such as piloting, some military duties, and commercial diving. (See previous discussion of the ADA in this chapter.)

## Evaluating Physical Capacity

In evaluating physical capacity for a particular job, the physician can categorize individuals with significant impairment as follows.

**Stable impairments.** These are conditions that cause some limitations but are unlikely to progress or be worsened by job activity if the employee is properly placed. Amputations, ankylosis (fixed joints), spasticity, blindness, deafness, cerebral palsy, and residual effects of polio are examples of these. Under these conditions, the major consideration is whether the employee can perform the job with reasonable efficiency and safety and without undue stress or harm to himself or herself and others. (The ADA permits productivity requirements.)

**Impairments, stable or unstable.** These are conditions that are likely to be made worse by work exposure. Examples of these are chronic skin diseases (such as psoriasis or certain eczemas) in an employee to be placed in a job that may involve solvent exposure and liver disease in an employee being placed where there might be excessive and uncontrolled exposure to solvents. Under these job conditions, a careful evaluation should be made of the prospective employee's medical condition and the job requirements to be sure the person's condition would not be made worse by the work.

**Progressive conditions.** Chronic obstructive lung disease (such as emphysema), congestive heart failure, and certain arthritic diseases of the spine are examples of progressive conditions. Although placement in other than sedentary work may be possible, progression of the condition may necessitate future placement in less demanding work.

**Conditions resulting in intermittent impairment.** These are conditions that may result in incapacity. Uncontrolled or inadequately treated epilepsy and poorly controlled diabetes subject to intermittent reactions such as dizziness or insulin shock are examples. An employee with these conditions could suddenly become unable to perform his or her job and become a grave risk to himself and co-workers.

Guidelines for the employment of individuals who are incapacitated have been outlined by several professional organizations (see Bibliography).

It is not the function of the physician, however, to inform the applicant whether he or she is to be employed. This is the prerogative and duty of management because there are many other factors in addition to physical qualification that bear on suitability for employment.

## Scope of the Examination

It is impossible to define what constitutes a complete examination or even a suitable examination because physicians have different opinions regarding the relative values of various test procedures. Therefore, the scope of physical examinations should be determined by the physician who is familiar with the job and working conditions. The nature of the industry, its inherent hazards, the variations in jobs, in physical demands and in health exposures are determinants. The values of different test procedures and their cost in time and dollars must be assayed. Perhaps examinations should be different in scope for different jobs. For example, the physical condition of an ironworker who will help build a multistory building is a far different problem than that of a sedentary garment worker. However, basic physical examination considerations are applicable to each.

It is estimated that 4–15 percent of diseases encountered by a practicing physician may be caused by the occupational environment or are naturally occurring diseases aggravated by occupational circumstances. Application of such statistics to the number of patients seen by physicians suggests that there are more occupationally induced medical problems than are reported by physicians each year. The missed conditions could better be diagnosed by increasing awareness of the large number of undiagnosed medical problems, developing a more adequate data base, especially in occupational medical history; and acquiring more information about the diseases of people at their workplaces.

The most common error made by the clinician who attempts to diagnose an occupational disease is to relate an ill-defined medical problem to a work environment about which no environmental information is available. The first step should be to diagnose the disease or observe the symptoms, signs, and abnormal laboratory parameters. As much as possible, this evaluation should be quantitative, both for symptoms and signs and for laboratory data. The second step is to obtain quantitative information about the work environment (such as opportunity for exposure, the materials used, and air and other types of sampling conducted by the company, a government

agency, or insurance carrier). This step often requires obtaining MSDSs from the employer, eliciting the employer's help in determining what exposures occurred, investigating trade names as possible toxic substances through the use of the poison index at the local poison control center, and writing or calling manufacturers. Finally, experience and published reports allow the investigation of any possible relationships between the medical problem and the occupational environmental circumstances.

Occupational history plays an important role in investigating the possibilities of relationship between disease or symptom complex and the work environment. At the same time, other environmental factors and possible exposures outside of the workplace, such as those experienced in hobbies, sports, and the home, must also be considered. The State Medical Society of Wisconsin's Committee on Environmental and Occupational Health suggests using the occupational history format as a guideline to help the practicing physician in acquiring the appropriate occupational medical data base. This suggested form can be adapted to meet clinical needs. It has been designed so that the worker-patient may complete the form; the physician should review the form with the patient to clarify and avoid incompleteness by helping the patient to understand parts of the form, if necessary (Figures 25–10 and 25–11).

The examinations are classified as preplacement or post-offer (described earlier), periodic, transfer, promotion, special, retirement, and termination. In general, medical examinations should be voluntary, but there are exceptions, such as the FAA physical for pilots and examinations for school bus drivers, interstate truck drivers, and overhead crane operators. It is important to note that the physical examination cannot be used in place of credible environmental monitoring.

### Fitness or Periodic Examination

A fitness examination is a health evaluation made any time during the career of an employee. These occupational evaluations of the fitness of an employee for his or her work may be done periodically, or only when there is (or appears to be) a health problem that might interfere with attendance or performance.

For certain jobs, periodic fitness examinations either ensure that the employee still has the special skills and abilities to safely continue the job or evaluate whether the employee has suffered potential harm from work exposures (noise, toxic substances in the environment, and the like).

It is recommended that all employees in potentially hazardous work be given a fitness examination at intervals varying from 1 or 2 to 4 years, depending on employee age and type of work. Requirements for examinations covering certain jobs are included in the OSHA regulations, and this list probably will be expanded. Federal regulations require operators of passenger or cargo vehicles to be periodically examined.

# SUGGESTED OCCUPATIONAL EXPOSURE HISTORY
## Part I

Please answer the following questions. Begin with your present job and list all jobs or military service you have held in order of date whether full or part time.

| TODAY'S DATE<br><br>NAME<br><br>SOCIAL SECURITY NUMBER<br><br>JOB CLASSIFICATION(S) | List potential hazards exposed to: (examples) | | | | Work related illnesses or injuries | |
|---|---|---|---|---|---|---|
| | **Physical** | **Chemical** | **Biological** | **Psychological** | | |
| | Noise<br>Radiation<br>Vibration<br>Electrical shock<br>Temperature extremes<br>Repetitive motion<br>Heavy lifting | Mercury<br>Lead<br>Dust<br>Gases<br>Fumes<br>Acids<br>Solvents<br>Caustics | Viruses<br>Bacteria<br>Parasites<br>Fungus<br>Animal bites<br>Etc | Boredom<br>Work shift fatigue<br>Risk of falling<br>Risk of being buried<br>Repetition | YES | NO |
| COMPANY NAME<br><br>CITY, STATE<br><br>JOB TITLE<br><br>FROM:    TO:    AVERAGE HR/WK | HAZARDS:<br><br>COMMENTS: | | | | | |
| COMPANY NAME<br><br>CITY, STATE<br><br>JOB TITLE<br><br>FROM:    TO:    AVERAGE HR/WK | HAZARDS:<br><br>COMMENTS: | | | | | |
| COMPANY NAME<br><br>CITY, STATE<br><br>JOB TITLE<br><br>FROM:    TO:    AVERAGE HR/WK | HAZARDS:<br><br>COMMENTS: | | | | | |
| COMPANY NAME<br><br>CITY, STATE<br><br>JOB TITLE<br><br>FROM:    TO:    AVERAGE HR/WK | HAZARDS:<br><br>COMMENTS: | | | | | |

**Figure 25–10.**   An exposure history should be kept in the employee's medical file. (Reprinted with permission from the *Wisconsin Medical Journal* 80, 1981.) (*Continues*)

# SUGGESTED OCCUPATIONAL EXPOSURE HISTORY
## Part II

| SECONDARY WORK (examples) Firefighting Civil defense Farming Gardening Civic activities Etc | List potential hazards exposed to: (examples) | | | | Work related illnesses or injuries | |
| | **Physical** Noise Radiation Vibration Electrical shock Temperature extremes Etc | **Chemical** Mercury Lead Dust Gases Fumes Acids Solvents Caustics | **Biological** Viruses Bacteria Parasites Fungus Animal bites Etc | **Psychological** Boredom Work shift fatigue Risk of falling Risk of being buried Repetition Etc | YES | NO |
|---|---|---|---|---|---|---|
| ORGANIZATION ___ CITY, STATE ___ JOB TITLE ___ FROM: TO: AVERAGE HR/WK | HAZARDS:  COMMENTS: | | | | | |
| ORGANIZATION ___ CITY, STATE ___ JOB TITLE ___ FROM: TO: AVERAGE HR/WK | HAZARDS:  COMMENTS: | | | | | |
| ORGANIZATION ___ CITY, STATE ___ JOB TITLE ___ FROM: TO: AVERAGE HR/WK | HAZARDS:  COMMENTS: | | | | | |
| **HOBBIES & ACTIVE SPORTS** | | | | | | |
| ACTIVITY ___ CITY, STATE ___ FROM: TO: AVERAGE HR/WK | HAZARDS:  COMMENTS: | | | | | |
| ACTIVITY ___ CITY, STATE ___ FROM: TO: AVERAGE HR/WK | COMMENTS: | | | | | |

Some chemicals have effects on the reproductive system. Have you or your present or former spouse had any problems with reproduction? If so, please indicate circumstances (e.g., stillborn, deformed, miscarriage, infertility).

EXPLAIN: _____

**Figure 25–10.** *(Continued)*

**WISCONSIN CHEMICAL CORPORATION**
**EMPLOYEE OCCUPATIONAL HEALTH RECORD**

| Name (Last | First | Middle) | Sex | SS# | Date of Birth | Location of Plant |
|---|---|---|---|---|---|---|

TO THE EXAMINING PHYSICIAN:

In order that Wisconsin Chemical Corporation may protect and maintain the occupational health of our employees, the attached examining physician's job information card is to appraise you of the demands of the job under consideration or now being performed.

Please complete the physical examination and test data evaluation and indicate the employee's fitness for work.

| DATE | JOB TITLE | FITNESS YES | NO | PHYSICIAN'S SIGNATURE | PHYSICIAN'S COMMENTS |
|---|---|---|---|---|---|
| | | | | | |
| | | | | | |
| | | | | | |
| | | | | | |
| | | | | | |
| | | | | | |
| | | | | | |
| | | | | | |

**Figure 25–11.** Employee occupational health record. (Courtesy Wisconsin Chemical Corporation.)

Fitness examinations requested by the employer are also important. These include those requested by the physician and nurse based on their knowledge of a particular health problem, those requested by the insurance or benefit department, and those requested by the supervisor.

Personality change, increased absence from work, tardiness, or other factors in the employee's behavior and appearance can help the supervisor identify a health problem early, when treatment may be most effective. When a supervisor refers an employee for a fitness examination, he or she should communicate factual information regarding deficiencies in job performance, attendance, or behavior so that the physician can properly evaluate the problem.

A physician who has the prior medical history of the particular employee is better prepared to conduct a fitness examination. Therefore, the examination need evaluate only the specific problem; this evaluation might be completed during an office visit or could require a very thorough examination, and even include referral to other specialists.

A report to the supervisor is necessary, and should state whether the employee is able to perform specific tasks and whether the employee has a health problem. If there is a problem, the physician should give the employee a general statement concerning the prognosis and advise him or her about proper medical care. No diagnosis or other detailed information should be revealed to the supervisor; this protects the confidentiality of information and is essential to obtaining the employee's respect and cooperation.

Fitness examinations should be given on return to work after a sickness or disability of one week or longer. At this time, it is determined whether the employee actually is able to return to work. Return to lighter work or shortened hours often is helpful in rehabilitation.

A fitness examination during an absence due to illness sometimes is indicated if return to work is delayed beyond what is expected. In this instance, the physician may discover that the diagnosis or treatment is improper and can make helpful recommendations, especially if present care is not working. It also may permit the physician to detect employees who would abuse the sickness disability payments plan by not returning to work when they have recovered adequately.

See Table 25–A for sample guidelines for required physical examinations and other special tests based on occupation.

It must be emphasized that periodic examinations and annual physicals may not necessarily reveal responses to

**Table 25–A.** Suggested Examples of Recommended Physical Examinations and Other Special Tests Based on Occupation

Employees in the following occupations or located in the listed work areas require periodic physical examination and/or special laboratory and x-ray studies.

| Occupation | Type of Examination | Frequency |
|---|---|---|
| Climbing—10 ft or more (includes hitchers, electricians, maintenance and repair, power house, plant engineering workers, and steam-fitters) | Examination by physician, orthorater, audiometer test, urinalysis, electrocardiogram (ECG) if over 40 years old or if otherwise medically indicated | Annually, if over 40 Every 3 yr, if 20-30 Every 2 yr, if 30-40 |
| Crane operators | As noted above | |
| Locomotive engineer and in train crew | As noted above | As noted above |
| Truck driver and vehicle test drivers | As noted above | As noted above |

Note: Not intended to be complete or to be regarded as official procedures; to be used as a guide only. Physical examinations and laboratory and x-ray studies are not to be used as substitutes for workplace hygienic studies such as air monitoring.

occupational exposures if the working conditions do not exceed the recommended ACGIH Threshold Limit Values for chemical or physical agents. Indeed, if exposures reach these levels, it is expecting too much to have a physician detect any chemical changes. Even with the use of multiple biochemistry blood tests, work exposures must be excessively high to cause overt physiological response (see Chapter 6, Industrial Toxicology, and other related chapters).

## Low Back Problems

In the past, employers hoped to detect individuals at risk for low back problems with preplacement lumbar x-rays. Many physicians claimed that such procedures were effective in reducing the frequency and severity of low back problems in industry, particularly during the 1950s and early 1960s. More controlled studies by experts have not justified the validity of low back x rays for employment screening and placement purposes when used as the sole criterion for employability. Back x rays should be reserved for diagnosis when history and physical examinations indicate that further workup is necessary. The American College of Occupational and Environmental Medicine states that preplacement screening using back x rays is not appropriate.

The alternative to x-ray-based screening seems to be to acquire both a better history of low back health and a better examination and functional evaluation of the person's back. However, a person could easily distort his or her own medical history, intentionally or unintentionally, to acquire a job.

Selection of people for materials-handling jobs based on their heights or weights is not justified according to the statistical evidence showing reduced low back pain incidence rates. The ADA puts the burden on the employer to show that jobs that require height and weight limitations cannot reasonably be modified.

## Chest X Rays

The routine use of preplacement chest x rays (or hospital admission x rays) has been proven to be redundant; many states have prohibited the taking of routine x rays unless clinically indicated and justified by the physician. Likewise, annual or other periodic chest x rays are not to be performed routinely.

## Pulmonary Function Testing

The routine use of pulmonary function tests (spirometry) is often overdone, especially in healthy workers under 40 years (see Chapter 2, The Lungs). The equipment used must be of an approved type (as recommended by the American Thoracic Society), proper calibration must be maintained, and the person performing the tests must have approved training, as well as be supervised by a physician, in order to produce credible test results (see Bibliography).

## History of Personality Disorders

Applicants with a history of treatment for mental disease should be carefully investigated. Before placement, written reports should be obtained from attending physicians and institutions. People with symptomatic phobias or certain psychophysiological reactions, especially cardiac reactions, may require special placement and may need psychiatric treatment either before or after employment. The employer must obtain written permission from the applicant before requesting medical information. The signed release form should be kept in the employer's files (Figure 25–12).

Applicants with manifestly serious personality disorders should be referred for evaluation by a psychiatrist or psychologist and special attention should be given to their histories. In many of these cases, it is necessary to recommend against their working in jobs that might involve too much freedom and responsibility; these people often do better in small groups with limited contact and responsibility with the public. Periodic follow-up visits are recommended and this may permit many with mental abnormalities to remain useful and productive citizens.

Applicants with a history of psychosis (mental derangement characterized by defective or lost contact with reality) often are unsatisfactory in the average work setting. Very careful investigation must be made of the health history. If all other factors appear favorable, such people may be able to work successfully if placed in positions with limited responsibility, at least initially, and with an understanding supervisor. Follow-up of these people on the job is indicated to make sure they pursue proper treatment and to detect recurrence of their disease.

**EMPLOYEE AUTHORIZATION FOR MEDICAL RECORD RELEASE TO HIS/HER DESIGNATED REPRESENTATIVE**

I _____ hereby authorize Wisconsin Chemical Corporation _____
(worker's full name)                                                                                    (location)

to release to _____ the following medical information from my personal medical records.
(authorized representative)

I give my permission for this medical information to be used for the following purpose.

I *do not* give my permission for any other use or re-disclosure of this information.

Full name of Employee/Legal Representative: _____

Signature of Employee/Legal Representative: _____

Date of Signature: _____

**Figure 25–12.** Employee release of personal medical information form. (Courtesy Wisconsin Chemical Corporation.)

## ■ Health Education and Counseling

Clinical experience has demonstrated that early detection, referral, and psychiatric care for emotionally troubled employees are essential in reducing prolonged disability. Psychiatrists and physicians associated with industry have long known that the work situation is unique in its potential for early detection of mental illness. There is a complex interplay involving the employee, family, supervisor, and co-workers. Some or all of these relationships can have a strong influence on job performance and successful rehabilitation.

Treatment delays, lack of follow-up, and outright disinterest are seen far too often. Medical departments, management personnel, and employees and their families are equally guilty. In an occupational injury, for example, it is routine practice to refer the injured employee to a physician or a specialist for prompt attention, especially when the employee elects to be directed by the facility's medical department. Frequent referrals are made to ophthalmologists, dermatologists, plastic surgeons, and other specialists, but rarely to psychiatrists. With emotional disorders,

on the other hand, busy practitioners usually are not eager to become involved in the care of people with depressive reactions, severe psychoneuroses, and alcoholism. The personal physician may appreciate the problem, but, because of the pace of his or her practice, often does not take the time to convince a sick person to seek the required social, psychological, and psychiatric assistance.

A resultant concern in this situation is the high cost of absences. Delays in obtaining treatment add a financial burden because a person's absence from the job may cost a company 4–6 times the actual wages. Employee assistance programs (EAPs) are valuable in such situations. Referral is triggered by performance, attendance, or behavior issues.

### Substance Abuse

Alcoholism and drug abuse problems are exceedingly troublesome, and it is essential that the physician and the employer have a good understanding of their potential for employees in the work setting. EAPs are increasingly available in many companies. They are designed to offer in-house services to troubled employees or to refer employees to the best source of treatment. Alcoholism and drug abuse continue to be among the nation's leading illnesses. Dependence on alcohol or other drugs is a major contributor to deterioration of family life, impaired job performance, morale and disciplinary problems, increased insurance rates, occupational accidents, and increased absenteeism. These illnesses know no boundaries: all ages and races are susceptible and socioeconomic status is no barrier.

It is estimated that one out of every ten U.S. workers may have a drinking problem, but fewer than 10 percent of such people actually receive treatment.

Every time an applicant or employee visits a physician or nurse, some health education or counseling should be given. Occupational medical departments provide an ideal opportunity for these services. Employers pay a large part of the costs for health care for employees and their dependents. It is economically important for the employer that employees have access to high-quality medical care.

Both physicians and nurses play an important role in guiding employees to the proper health care resources. An almost equally important role is played when the employee needs counseling, and physicians and nurses experienced in occupational health programs generally are skilled in the art of counseling. Employers often overlook the education and counseling functions when comparing the costs and benefits of occupational health programs because it is difficult to measure the units of service that can be defined. This is unfortunate because experience and common sense indicate these activities are among the most valuable to both employer and employees; these benefits are available to all employees, not just those exposed to potential hazards. This is an important point in justifying the costs of a health program in any sizable employee group.

## Drug Screening

The initiation of an employer-sponsored drug screening program is a delicate legal and social problem. If instituted, this is a legal decision and the direct involvement of medical and nursing personnel is inappropriate. In certain industries, however, drug screening and alcohol testing are required (DOT Regulations: 49 *CFR* Part 40, Procedures for Transportation Workplace Drug and Alcohol Testing Programs) and it is encouraged by regulations in several states (including Florida, Georgia, and Texas). If instituted, drug screening should be part of a comprehensive substance abuse program, but it may be necessary for a nurse or a physician to assist an employer in complying with the details of these regulations. A new type of specialist, the medical review officer (MRO), has been created to deal with the complexity of evaluating drug screening reports.

However, the reader is strongly advised to obtain and study the report "Health Promotion in the Workplace:

Alcohol and Drug Abuse," by a World Health Organization (WHO) expert committee.

[This report reviews] current approaches to health promotion in the workplace aimed at preventing and controlling alcohol- and drug-related problems and at mitigating their serious social and economic consequences in both industrialized and developing countries, taking account of the wide range of biological, psychological, social, and environmental factors that can contribute to such problems in different work settings, from large multinational enterprises to single homes. The report evaluates both older programmes and new initiatives applicable at national, community, and enterprise levels. While identifying specific concerns, such as the implications of drug screening, the regulatory context of drug and alcohol control, and the special requirements of developing countries, the report draws attention to the universal applicability of the health promotion concept to alcohol- and drug-related problems. It stresses the importance of collecting basic data on needs before any workplace programme is launched, training the personnel who will be responsible, and ensuring continuous evaluation of progress and outcome. Particular emphasis is placed on workers' participation in programme development and implementation, and on information dissemination and education that take into account the needs of specific occupational groups, such as migrant workers, seafarers and shift workers, and the diversity of cultural settings and attitudes to drinking and drug-taking around the world.

Additional WHO publications are listed in the Bibliography.

## Access to Medical Records

Part 1910.20 of the *CFR* is the standard covering an employee's right to have access to personal medical records. Its purpose is as follows:

§1910.20 Access to employee exposure and medical records.

(a) *Purpose.* The purpose of this section is to provide employees and their designated representatives a right of access to relevant exposure and medical records; and to provide representatives of the Assistant Secretary a right of access to these records in order to fulfill responsibilities under the Occupational Safety and Health Act. Access by employees, their representatives, and the Assistant Secretary is necessary to yield both direct and indirect improvements in the detection, treatment, and prevention of occupational disease. Each employer is responsible for assuring compliance with this section, but the activities involved in complying with the access to medical records provisions can be carried out, on behalf of the employer, by the physician or other health care personnel in charge of employee medical records. Except as expressly provided, nothing in this section is intended to affect existing legal and ethical obligations concerning the maintenance and confidentiality of employee medical information, the duty to disclose information to a patient/employee or any other aspect of the medical-care relationship, or affect existing legal obligations concerning the protection of trade secret information [Figure 25–12].

# Medical Aspects of the Use of Respiratory Protective Equipment

The problem of providing properly fitting respirators is complicated by the wide range of facial sizes and shapes that must be accommodated. Differences in facial sizes and shapes result from a wide variety of factors, the most significant of which are age, sex, and race.

Facial hair, such as beards or sideburns, makes it impossible to achieve an airtight seal between the facepiece and the face, particularly with the half-mask respirator. Even the stubble resulting from failure to shave daily can cause serious inward leakage of contaminated air. Therefore, workers who wear respirators with half or full facepieces should be cleanly shaven.

The wearer's comfort and acceptance of the distress caused by wearing a respirator are no less important than the devices's effectiveness. Some factors that can alter a respirator's acceptability are improperly fitted respirators, uncomfortable resistance to breathing (which may result from increased physical effort demanding respiratory air movement upwards of more than 50–60 l/min, and can become intolerable if continued for more than a few minutes), and limitations of vision and speech transmission.

A program to effectively assess the employee's ability to use respiratory protective equipment should be initiated. Assessment should be done on a case-by-case basis. Such a program consists of three parts:

- Employee evaluation
  Health history
  Health screening records
  Employee counseling interview
- Respirator-fitness testing performed by the company industrial hygienist
- Physician's review of data and employee fitness assessment

29 *CFR* 1910.134 (b)(10) states that "persons should not be assigned to tasks requiring the use of respirators unless it has been determined that they are physically able to perform the work and use the equipment" and notes that a physician must determine the health and physical conditions that are pertinent for an employee's ability to work while wearing a respirator. The following factors may be pertinent for this determination:

- Emphysema
- Chronic obstructive pulmonary disease
- Bronchial asthma
- X-ray evidence of pneumoconiosis
- Evidence of reduced pulmonary function
- Coronary artery disease or cerebral blood vessel disease

- Severe or progressive hypertension
- Epilepsy (grand mal or petit mal)
- Anemia
- Diabetes (insipidus or mellitus)
- Pneumomediastinum
- Communication of sinus through upper jaw to oral cavity
- Breathing difficulty when wearing a respirator
- Claustrophobia or anxiety when wearing a respirator

There are no guides or standards for the degree of reduced function that prohibits use of respiratory protective devices. For light work of short duration, 60 percent of predicted normal may be acceptable. Some examiners consider reduction to 70 percent, but the entire clinical aspect of the worker must be considered. In any event, nearly all workers can be instructed to use an air-supplied respirator, even with the previous health deficits (see Chapter 22, Respiratory Protection).

# A Program for Employee Respirator Use Assessment

The design and implementation of an effective respiratory protection program requires the expertise of preventive health disciplines, primarily those of occupational health nursing, medicine, and industrial hygiene.

## Responsibilities of the Industrial Hygienist

The industrial hygienist is responsible for the following:

- Identifying and listing the respirable occupational hazards present
- Respiratory fit testing and training each identified worker
- Documenting fit testing and training, noting the type of respirator recommended for hazard protection and indicating any difficulties encountered in fitting the employee with proper respiratory equipment

## Responsibilities of the Occupational Health Nurse

The occupational health nurse is responsible for the following (courtesy of Florence Ebert, RN, COHN, and *Occupational Health Guide for Medical and Nursing Personnel*, State Medical Society of Wisconsin):

- Identifying each employee alleging respiratory deficits
- Completing the individual employee worksheet
- Compiling the health screening data, chest x-rays, and valid pulmonary function test records
- Obtaining a completed health questionnaire

- Performing individual employee interviews and counseling
- Scheduling each employee for the physician's determination of respirator use approval or disapproval and medical counseling
- Performing periodic respiratory surveillance

## Eyeglasses

Respiratory protection for people wearing corrective glasses may present a serious problem. The ability to wear corrective glasses with a half-mask depends on the face fit. For a full-face mask, a proper face seal cannot be established if the temple bars of the eyeglasses extend through the sealing edge. Some full-facepiece designs provide for the mounting of special corrective lenses within the facepiece.

Any respirator affects the wearer's ability to see. The half-mask and the attached elements can restrict normal downward vision appreciably. Diminished vision in the full-face mask may be caused not only by the facepiece, but also by the design and displacement of the eyepieces (see Chapter 22, Respiratory Protection).

## Speech Transmission

Speech transmission through a respirator can be difficult, annoying, and even fatiguing. Moving the jaws to speak can cause leakage between the facepiece and face, especially with the half-mask respirator.

# Potential Reproductive Hazards in the Workplace

In the United States, women now make up nearly 50 percent of the labor force. There are important physiological elements of concern for the safety and health status of women of childbearing capability.

## Exposure of the Unborn Child

Of immediate concern is inadvertent exposure to a number of potentially toxic materials that could prove hazardous to a fetus during its development. Radiation exposure has long been recognized as an area of grave concern needing strict controls. Other potentially hazardous substances, suspected or known to adversely affect the embryo or fetus, include alcohol, aniline, arsenic, benzene, carbon disulfide, carbon monoxide, captan, formaldehyde, hexacholorobutadiene, hydrogen sulfide, lead, mercury, methyl mercury, nicotine, nitrates (and other chemicals capable of causing methemoglobinemia), nitrobenzene, polychlorinated biphenyls (PCBs), selenium compounds, vinyl chloride, and xylene.

In 1985, the AMA's Council on Scientific Affairs issued a detailed report, "Effects of Toxic Chemicals on the Reproductive System" (see Bibliography).

## Spontaneous Abortions

Concern about the possible chronic effects of anesthetic gases has been growing since 1967 reports of spontaneous abortions among Russian anesthetists working with ether in poorly ventilated operating rooms in the Soviet Union. In the United States, some operating rooms lack equipment for exhausting waste anesthetic gases. Other possible significant exposures in these occupational groups include ethyl chloride, ethyl ether, nitrous oxide, trichlorethylene, ionizing radiation, and infectious diseases.

## Teratogenic Effects

Götel and Stahl were the first to report teratogenic effects, liver damage, and addiction on exposure to halothane (2-bromo-2-chloro-1, 1, 1-trifluoroethane). Fatigue, headache, tiredness, and irritability are common symptoms of such exposure. Based on studies at the Regional Hospital in Örebro, Sweden, they stated that an anesthetist may be exposed to about two patient doses of halothane per year. It is easy to reduce exposure to halothane by connecting an evacuation unit to the anesthetic machine. With such equipment at work, the authors found that the gas in the anesthetist's breathing zone could be reduced 90 percent or more. Indeed, many women at work may be unaware of an underlying pregnancy, and the growing embryo or fetus is extremely susceptible to harm from agents external to the mother.

Because neither the mother nor anyone else is aware of the pregnancy until at least several days and usually a few weeks after conception, harm to the fertilized ovum or fetus can result from work exposures that are considered acceptable for nonpregnant workers. The courts have given priority to individual freedom of choice.

It is proper to attempt to define these risks, but information on which such definitions can be based is only partially available at this time. Whatever information is available to permit quantifying risks to at least the more common work exposures, such information seems appropriate for inclusion in this chapter as a guide to counseling women who may be faced with the choice. There will never be a complete definition of the range of risks because of the myriad substances and conditions that can potentially cause harm to the fertilized ovum or growing fetus. In addition, new substances are always being developed, which complicates the immense problem of determining whether an external agent will actually cause harm.

The following instructions should be accepted only as guidelines, and the variables of the individual workers, the pregnancy, and the job should be recognized. Jobs must be considered from the standpoint of their physical demands and potential hazards. (The efficiency of a woman in various stages of pregnancy as compared to a nonpregnant woman will not be considered, as this is a matter of employee–employer relations and not strictly a medical subject.)

An excellent, informative chapter, along with lucid diagrams, was published by Sorsa (see Bibliography).

## Guidelines for the Working Pregnant Woman

The following guidelines are recommendations set forth by joint efforts of the American College of Obstetricians and Gynecologist and NIOSH.

**Healthy woman, normal pregnancy.** A healthy woman who is experiencing a normal pregnancy should be given work that is neither strenuous nor potentially hazardous (the latter term refers both to the physical hazards and external agents in the work environment that constitute a significant risk). The most common jobs in this class are clerical and administrative, but include many craft, trade, and professional positions. The usual limiting factor on the work that is strenuous or potentially hazardous to a pregnant woman, fetus, or both is the pregnancy itself, as acceptable limits have been established in most positions for the nonpregnant worker. Work with video display terminals should be evaluated based on current epidemiological research in that area. (See Chapter 13, Ergonomics.)

**Healthy woman, abnormal pregnancy.** If an abnormal pregnancy is known to exist, a list of the abnormalities of the pregnancy should be developed and comments made on the ability to work in each instance, identifying the specific risk. Included in the evaluation should be appropriate intervals for the follow-up examinations while working.

**Woman with other medical conditions, normal pregnancy.** For pregnant women who have health conditions such as heart disease, diabetes, thyroid disease, and multiple sclerosis, guidelines should be developed for proper evaluation and monitoring during work.

**Woman with other medical conditions, abnormal pregnancy.** For this condition, the guidelines for the above last two conditions should be used.

**After-childbirth "disability."** A healthy woman who had a normal delivery should work only in nonstrenuous and nonhazardous jobs beginning about 4–6 weeks after delivery. If the woman has other medical conditions or the delivery is complicated, or both, a list of complications should be developed with an explanation of how each affects her ability to work.

**Hazardous exposure.** Without question, a pregnant woman should avoid unnecessary exposure to any of the following:

- Ionizing radiation
- Chemical substances that are mutagenic, teratogenic, or abortifacient
- Biological agents of potential harm

## Biomechanical and Ergonomic Factors

For detailed discussion of these important factors of concern, refer to Chapter 13, Ergonomics.

A number of devices are on the market that may be used to evaluate various task parameters such as strength, fatigue, and range of motion. The few controlled studies that have been done fail to show significant correlation between these parameters and future performance and injury (Newton and Waddell, 1993; Newton et al., 1993). Furthermore, the ADA places the burden on the employer to show that strength criteria are a measure of the essential functions of a job.

## Biological Monitoring

The use of biological monitoring in assessing workplace exposure to chemicals and other substances has gained in importance, particularly for observing workers in lead-using jobs, where, for example, blood lead level determinations predominate. Other examples include measurement of carbon monoxide in the blood (as carboxyhemoglobin), various solvents detectable in the air expired from lungs (including alcohol concentrations used in traffic law enforcement), phenol in urine, and mandelic acid in urine to detect styrene. Even the analysis for arsenic in hair, nails, and other tissues is a method of biological monitoring. Another example of biological monitoring is the study of seminal fluid in cases of exposure to the pesticide dibromochloropropane (DBCP), which proved a valuable indicator for site-specific toxicity in workers making this product.

These measurements can indicate that there was an exposure and absorption of a chemical or substance into the body, but do not indicate the quantity that the person was exposed to and absorbed or the source of the exposure. In very specific circumstances, fat biopsies can be taken to ascertain quantification and storage time of certain substances, such as DDT, PCBs, and solvents. Lung, liver, and other organ tissue samples can be used to augment or verify clinical findings. See Aitio et al. (1994), Guidotti (1994), Elinder et al. (1993) and Lauwerys & Hoet (1993) in the Bibliography.

Biological monitoring is based on the relationship between a measure of some absorbed dose and environmental exposure. Monitoring can offer information useful in studying a worker's individual response and in measuring overall exposure. Many factors can affect results, including the metabolism and toxicokinetics of the chemical or substance and individual factors such as the person's state of health, body size, workload, and lifestyle. Biological monitoring takes into account the uptake of a chemical or substance by all routes of exposure, including respiratory, cutaneous, oral, and nonoccupational sources (background levels). The main advantage of biological monitoring is that assessment of exposure by all routes can be made. A limitation is that these measurements are estimated and may not distinguish between occupational and nonoccupational exposures. Also, there is wide interpersonal variation due to factors outside the workplace such as diet, alcohol use, smoking, and other drug use; all potentially can affect the levels of a chemical or metabolite in a biological specimen, making interpretation difficult.

## Explanation of the *Skin* Notation

The *skin* notation used in the American Conference of Governmental Industrial Hygienists' (ACGIH) *Threshold Limit Values and Biological Exposure Indices* deals with the potential contribution of the cutaneous route of exposure to the overall bodily exposure (Appendix B). Although as yet there are not sufficient data from animal and human studies describing quantitative skin penetration, routing, and absorption into tissues or systemic uptake of materials such as dusts, vapors, and gases to be conclusive, the ACGIH substance listing describes the impact skin exposures have on overall exposure. The cutaneous route includes the mucous membranes and eye, by airborne or direct contact with substances that have a *skin* notation in the TLV booklet.

The rate of absorption is a function of the concentration of the chemical or substance and the duration to which the skin is exposed. Certain vehicles can alter this potential absorption. Substances with a *skin* notation and a low TLV can be problematic at high airborne concentrations, particularly if significant areas of skin are exposed for long periods. Respiratory tract protection, while the rest of the body surface is exposed to a high concentration, may not be adequate, but there are no standards for levels of skin exposure.

Biological monitoring is considered to include the contribution of skin exposure to the total dose. The *skin* designation is intended to suggest appropriate measures to prevent skin absorption, so that the respiratory threshold limit is not underestimated, misinterpreted, or otherwise misused.

## ACGIH Biological Exposure Indices (BEIs)

The ACGIH uses the term *Biological Exposure Index* (BEI) (Appendix B) for an *index chemical* that appears in a biological fluid or in expired air after exposure to a workplace chemical or substance. It acts as a warning of exposure, either by the appearance of the chemical or its metabolite, or by the appearance of a physiological response associated with exposure. The BEIs are intended as guidelines to assess total exposure, and compliance with BEIs is not a substitute for controlling the workplace environment. The proper role of biological monitoring is to complement environmental monitoring. OSHA regula-

tions require biological monitoring for only a few exposures (cadmium, lead, and benzene). Environmental monitoring should be used to assess workplace levels of chemicals and substance, whereas biological monitoring can be used to more accurately determine uptake of workplace chemicals. (See Guidotti [1994] and Aitio et al. [1994] in the Bibliography).

## Summary

Sound occupational health practice is based on two major fundamentals: control of hazards at the worksite by engineering and other industrial hygiene control measures (including environmental monitoring) and good medical surveillance (including biological monitoring). Neither of these should be considered a replacement for the other.

The elements of occupational environmental control are of utmost importance for protecting employee health in all work areas. Broad professional, occupational, or environmental control is most commonly the concern of industrial hygienists, who must recognize, evaluate, and use control measures to prevent illness or impaired health of workers.

The occupational health physician depends on the knowledge and techniques of the industrial hygienist, who provides insight on the type and magnitude of potentially hazardous or stressful occupational environment factors, be they chemical, physical, biological, ergonomic, or a combination of these. Without adequate industrial hygiene information, it may be difficult, if not impossible, for a physician to determine whether a relationship exists between symptoms and findings of occupational or nonoccupational disorders or diseases.

A history of exposure to a substance does not mean that occupational disease has occurred or that some form of poisoning to the worker is present. These especially complicated fine points constitute the heart of the practice of occupational medicine and industrial hygiene.

## Bibliography

*Accident Prevention.* Industrial Accident Prevention Association, 2 Bloor Street West, 31st Floor, Toronto, Ontario M4W 3NB, Canada.

Adams RM. *Occupational Dermatology,* 2nd ed. New York: Grune & Stratton, 1990.

Advisory Committee on Immunization Practices. General recommendations on immunization practices. *Morb Mortal Wkly Rep* 43(RR-1):1–38, 1994.

Aitio A, Riihimäki V, Liesivuori J, et al. Biologic monitoring. In Zenz C, Dickerson OB, Horvath EP, eds. *Occupational Medicine,* 3rd ed. St. Louis, MO: Mosby-Year Book, 1994.

Aitio A, Riihimäki V, Vainio H, eds. *Biological Monitoring and Surveillance of Workers Exposed to Chemicals.* New York: Hemisphere, 1984.

*Allergy at Work—On Hypersensitivity and Allergy.* The Swedish Work Environment Association, Birger Jarlsgatan 122, S-144 20 Stockholm, Sweden, 1989.

American Conference of Governmental Industrial Hygienists. *Threshold Limit Values and Biological Exposure Indices for 1994–1995.* Cincinnati, OH: American Conference of Governmental Industrial Hygienists, 1995.

American Conference of Governmental Industrial Hygienists. *Documentation of Threshold Limit Values,* 6th ed. Cincinnati, OH: American Conference of Governmental Industrial Hygienists, 1990.

American Industrial Hygiene Association. *Biohazards Reference Manual.* Fairfax, VA: American Industrial Hygiene Association, 1985.

American Medical Association. *Guides to the Evaluation of Permanent Impairment,* 4th ed. Chicago: American Medical Association, 1994.

American Medical Association. *Medical Conditions Affecting Drivers.* Chicago: American Medical Association, 1986.

American Medical Association. Effects of toxic chemicals on the reproductive system. *JAMA* 253:23, 1985.

American Medical Association. *Guide to Developing Small Plant Occupational Health Programs.* Chicago: American Medical Association, 1983.

*The Americans with Disabilities Act: Medical Director's Guidelines.* Boston: Liberty Mutual Insurance Group, 1992.

Beneson AS, ed. *Control of Communicable Diseases in Man,* 15th ed. Washington, DC: American Public Health Association, 1990.

Bond MB. Occupational health services for small businesses. In Zenz C, Dickerson OB, Horvath EP, eds. *Occupational Medicine,* 3rd ed. St. Louis, MO: Mosby-Year Book, 1994.

Bond MB, Messite J. Occupational health considerations for women at work. In Zenz C, ed. *Occupational Medicine: Principles and Practical Applications,* 2nd ed. Chicago: Year Book Medical Publishers, 1988a.

Bond MB, Messite J. Reproductive toxicology and occupational exposure. In Zenz C, ed. *Occupational Medicine: Principles and Practical Applications,* 2nd ed. Chicago: Year Book Medical Publishers, 1988b.

Briggs D. Trauma: Workplace principles. In Zenz C, Dickerson OB, Horvath EP, eds. *Occupational Medicine,* 3rd ed. St. Louis, MO: Mosby-Year Book, 1994.

Chaffin DB, Andersson GBJ. *Occupational Biomechanics,* 2nd ed. New York: Wiley, 1990.

*Chemical Toxicology/Carcinogenicity.* World Health Organization Publications, Distribution and Sales, 1211 Geneva 27, Switzerland, 1990.

Cooper CL, El-Batawi MA, Kalimo R, eds. *Psychosocial Factors at Work and Their Relation to Health.* Geneva, Switzerland: World Health Organization, 1987.

Cowell JWF, Guidotti TL, Jamieson CG, eds. *Occupational Health Services—A Practical Approach.* Chicago: American Medical Association, 1989.

Dodson VN, Lindesmith LA, Horvath EP, et al. Diagnosing occupationally induced diseases. *Wis Med J* 80:199, 1980.

*Early Detection of Occupational Diseases.* Geneva, Switzerland: World Health Organization, 1986.

Elinder C-G, Firberg L, Nordberg GF, et al. *Biological Monitoring of Metals in Man.* Geneva, Switzerland: WHO, 1994.

*Encyclopedia of Occupational Health and Safety,* 3rd ed. International Labor Office, 1828 L Street NW, Suite 801, Washington, DC 20036, 1989.

*Enforcement Policy and Procedures for Occupational Exposure to Tuberculosis.* U.S. Department of Labor, Occupational Safety and Health Administration, Washington, DC, 1993.

*Epidemiology of Work-Related Diseases and Accidents.* Tenth report of the Joint ILO/WHO Committee on Occupational Health. WHO Technical Report Series, No. 777, 1989.

Erdil M, Dickerson OB. Cumulative trauma disorders. In Zenz C, Dickerson OB, Horvath EP, eds. *Occupational Medicine,* 3rd ed. St. Louis, MO: Mosby-Year Book, 1994.

Felton JS. *Occupational Medical Management: A Guide to the Organization and Operation of In-Plant Occupational Health Services.* Boston: Little, Brown, 1989.

Gosselin RE, Smith RP, Hodge HC. *Clinical Toxicology of Commercial Products,* 5th ed. Baltimore: Williams & Wilkins, 1984.

Guidotti TL. Occupational toxicology. In Zenz C, Dickerson OB, Horvath EP, eds. *Occupational Medicine,* 3rd ed. St. Louis: Mosby-Year Book, 1994.

*Health Promotion for Working Populations.* Report of a WHO Expert Committee. WHO Technical Report Series no. 765. Geneva, Switzerland: WHO, 1988.

Hogstedt C, Reuterwall C, eds. *Progress in Occupational Epidemiology.* Proceedings of the Sixth International Symposium on Epidemiology in Occupational Health, Stockholm, Sweden, 1988.

Horvath EP. Occupational health programs in clinics and hospitals. In Zenz C, ed. *Occupational Medicine: Principles and Practical Applications,* 2nd ed. Chicago: Year Book Medical Publishers, 1988.

Horvath EP. *Manual of Spirometry in Occupational Medicine.* NIOSH, DHHS, PHS, CDC, Cincinnati, OH, Nov. 1981.

Horvath EP Jr., Andonian JJ, Rowe DM. Attending physician's return-to-work recommendations record. *Wis Med J* 83:202, June 1984.

*Industrial Toxicology News,* published quarterly by the Industrial Toxicology Laboratory of West Allis Memorial Hospital, 8901 West Lincoln Avenue, West Allis, WI 53227, phone 414–328–7940, fax 414–328–8560. Inquiries should be addressed to Leon A. Saryan, Technical Director.

Kalimo R, El-Batawi MA, Cooper CL, eds. *Psychosocial Factors at Work and Their Relation to Health.* Geneva, Switzerland: WHO, 1987.

LaDou J, ed. *Occupational Health and Safety,* 2nd ed. Itasca, IL: National Safety Council, 1993.

LaDou J, ed. *Occupational Medicine.* Norwald, CT/San Mateo, CA: Appleton & Lange, 1990. (Spanish edition, 1993.)

Lauwerys R, Hoet P. *Industrial Chemical Exposure—Guidelines for Biological Monitoring,* 2nd ed. Boca Raton, FL: Lewis Publishers, 1993.

Mathias CGT. Occupational dermatoses. In Zenz C, Dickerson OB, Horvath EP, eds. *Occupational Medicine,* 3rd ed. St; Louis, MO: Mosby-Year Book Publishers, 1994.

*Maximum Weights in Load Lifting and Carrying* (OS & H No. 59), ILO Publications Center, Suite OSH, Albany, NY 12210, 1988.

*Merck Index,* 11th ed. Rahway, NJ: Merck & Co., 1989.

*Merck Manual,* 17th ed. Rahway, NJ: Merck Sharp & Dohme Research Laboratories, 1994.

*Microelectronics and Change at Work,* ILO Publications Center, Suite OSH, Albany, NY 12210, 1989.

National Safety Council. Data Sheet 12304-0202 Rev. 1991, *Unit First Aid Kits*. Itasca, IL: National Safety Council, 1991.

Newton M, Thow M, Somerville D, et al. Trunk strength testing with iso-machines. Part 2: Experimental evaluation of the Cybex II back testing system in normal subjects and patients with chronic low back pain. *SPINE* 18:812–824, 1993.

Newton M, Waddell G. Trunk strength testing with iso-machines part 1: Review of a decade of scientific evidence. *SPINE* 18:801–811, 1993.

*NIOSH Recommended Guidelines for Personal Respiratory Protection of Workers in Health-Care Facilities Potentially Exposed to Tuberculosis*. Atlanta, GA: U.S. Department of Health and Human Services, Public Health Service, 1992.

Occupational Safety and Health Administration. Medical Services and First Aid (Subpart K, § 1910.151). *Federal Register* 39:125, June 27, 1974.

*The Organization of First Aid in the Workplace* (OS & H No. 63), ILO Publications Center, Suite OSH, Albany, NY 12210, 1989.

Paul M, Himmelstein J. A review—Reproductive hazards in the workplace: What the practitioner needs to know about chemical exposures. *Obstet Gynecol* 71:921, 1988.

*Physicians' Desk Reference*, 48th ed. Medical Economics Company, Inc., Oradell, NJ 07649, 1994 (published annually).

Plog B, ed. *Fundamentals of Industrial Hygiene*, 3rd ed. Itasca, IL: National Safety Council, 1988.

*Principles of Studies on Diseases of Suspected Chemical Etiology and Their Prevention*. Environmental Health Criteria, No. 72, World Health Organization Publications, Distribution and Sales, 1211 Geneva 27, Switzerland, 1987.

Proctor NH, Hughes JP, Fischman ML, eds. *Chemical Hazards in the Workplace*, 3rd ed. Philadelphia: Lippincott, 1991.

Rantanen J, ed. *Occupational Health Services: An Overview*. Geneva, Switzerland: WHO Regional Publications, European Series, No. 26, 1990.

*Reproductive Health Hazards in the Workplace*. Industrial Accident Prevention Association, 2 Bloor Street West, 31st Floor, Toronto, Ontario M4W 3N8, Canada, (Audio-cassette, 1991).

*Safety and Health Guide in the Use of Agrochemicals—A Guide*. ILO Publications Center, Suite OSH, Albany, NY 12210, 1991.

Salvaggio J. Special diagnostic and immunologic considerations (including "multiple chemical sensitivity"). In Zenz C, Dickerson OB, Horvath EP, eds. *Occupational Medicine*, 3rd ed. St. Louis, MO: Mosby-Year Book Publishers, 1994.

Sorsa M. Occupational genotoxicology. In Zenz C, ed. *Occupational Medicine: Principles and Practical Applications*, 2nd ed. Chicago: Year Book Medical Publishers, 1988.

U.S. Department of Transportation. *Procedures for Transportation in Workplace Drug and Alcohol Testing Programs*, 49 *CFR* Part 40, DOT, 1994.

Welter ES. The role of the primary care physician. In Zenz C, Dickerson OB, Horvath EP, eds. *Occupational Medicine*, 3rd ed. St. Louis, MO: Mosby-Year Book Medical Publishers, 1994.

*WHO Environmental Health and Chemical Safety Subscription Package*, World Health Organization Publications, Distribution and Sales, 1211 Geneva 27, Switzerland.

The WHO Environmental Health Criteria Series: Catalogue documenting volumes 1–100 in this series is available from WHO Distribution and Sales, 1211 Geneva 27, Switzerland. WHO Publications (books, not subscriptions) may be obtained from WHO Publications Center USA, 49 Sheridan Ave., Albany, NY 12210.

*WHO Technical Report Series 833: Health Promotion in the Workplace: Alcohol and Drug Abuse*. Report of a WHO Expert Committee World Health Organization, Geneva, 1993.

Whorton WD. Male occupational reproductive hazards. In Zenz C, Dodson V, eds. *Occupational Health Guide for Medical and Nursing Personnel*, 5th ed. Madison: State Medical Society of Wisconsin, Committee on Occupational Health, 1992.

*Working with Visual Display Units* (OS & H No. 61), ILO Publications Center, Suite OSH, Albany, NY 12210, 1989.

*Your Body at Work—Human Physiology and the Working Environment*. The Swedish Work Environment Association, Birger Jarlsgatan 122, S-144 20 Stockholm, Sweden, 1987.

Zenz C, ed. *Occupational Medicine*, 2nd ed. St. Louis, MO: Mosby-Year Book, 1994.

Zenz C. *Medical Evaluations and Determination for Fitness to Work in Hot Environments*. Cincinnati, OH: National Institute for Occupational Safety and Health, 1985.

Zenz C. Reproductive risks in the workplace. *National Safety News*, Sept. 1984.

## American College of Environmental and Occupational Medicine Publications

The American College of Environmental and Occupational Medicine (ACEOM) provides educational publications for occupational medical practitioners. ACEOM has developed a basic curriculum for practicing physicians who are interested in expanding their careers into occupational medicine. ACEOM has a list of over 200 publications on matters of occupational health, including articles on specific industrial exposures and diseases, cancer and mutagenicity, occupational medical practice, multiphasic screening, and ethics. Contact:

ACEOM
2340 South Arlington Heights Road
Arlington Heights, Illinois 60005

(Note: Any physician serving a workplace or with an interest in occupational medicine should apply for membership in the ACEOM; write for an application form.)

## NIOSH Publications

*Carpal Tunnel Syndrome—Selected References* (Order No. 013).
*Control Technology for Ethylene Oxide Sterilization in Hospitals*. DHHS (NIOSH) Publication No. 89–120.
*A Curriculum for Public Safety and Emergency Response Workers—Prevention of Transmission of Human Immunodeficiency Virus (HIV) and Hepatitis B Virus (HBV)*. DHHS (NIOSH) Publication No. 89–108.
*Guidelines for Prevention of Transmission of Human Immunodeficiency Virus (HIV) to Health Care and Public Safety Workers*. DHHS (NIOSH) Publication No. 89–107.
*Guidelines for Protecting the Safety and Health of Health Care Workers*. DHHS (NIOSH) Publication No. 88–119.

*NIOSH Basis for an Occupational Health Standard: Grain Dust-Health Hazards of Storing, Handling, and Shipping Grain.* DHHS (NIOSH) Publication No. 89–126.

*NIOSH Criteria for a Recommended Standard: Occupational Exposure to Hand–Arm Vibration.* DHHS (NIOSH) Publication No. 89–106.

*NIOSH Guide to Industrial Respiratory Protection.* DHHS (NIOSH) Publication No. 87–116; Publications Dissemination, DSDTT, National Institute for Occupational Safety and Health, 4676 Columbia Parkway, Cincinnati, OH 45226, 1987.

*NIOSH Pocket Guide to Chemical Hazards, 1990.* U.S. Department of Health and Human Services, Public Health Service Centers for Disease Control, National Institute for Occupational Safety and Health. (For sale by the Superintendent of Documents, U.S. Government Printing Office, Washington, DC 20402.)

*NIOSH Publications Catalog,* 9th ed., 1991. National Institute for Occupational Safety and Health Division of Standards Development and Technology Transfer, 4676 Columbia Pkwy, Cincinnati, OH 45226.

*Occupational Exposure to Hot Environments, Revised Criteria,* 1986.

*Proceedings of a NIOSH Workshop on Recommended Heat Stress Standards,* Dec. 1980. U.S. Dept. of Health and Human Services, Public Health Service, Center for Disease Control, NIOSH, Cincinnati, OH 45226.

*Surveillance in Occupational Health and Safety, 1989* (Order No. 014).

# Occupational Health Nursing

by Barbara J. Burgel, RN, MS, COHN

**OHN Definition**

**Standards of Practice**

**Scope of Practice**

**Professional Membership and Certification**

**The Practice of Occupational Health Nursing**
*Models of Occupational Health Services* ▪ *Current Trends in Health Service Staffing* ▪ *Occupational Health Nursing Role Functions*

**Summary**

**Bibliography**

**Addendum: Issues Affecting Health Care Service Staffing**
*Changing Demographics in the Workforce* ▪ *Research on Occupational Health Nursing Practice*

As WE ENTER THE 21ST CENTURY, it is critical to recognize the long-standing role of the occupational health nurse (OHN) in protecting the health of workers. Occupational health nurses have made over 100 years of solid contributions in this public health discipline and are committed to working with employees and their dependents, employers, and other health and safety team members to attain a safe and healthy workplace.

Not only is the OHN instrumental in managing health and safety programs in the worksite, but the OHN provides the critical link between employee health status, the work process, and the determination of the employee's ability to do the job. Knowledge of health and safety regulations, workplace hazards, direct care skills, counseling, teaching, and program management are but a few of the key knowledge areas for the OHN, with strong communication skills being of utmost importance. Knowledge of cost-control strategies in this era of rising direct and indirect health care costs is critical for the OHN in the 1990s.

The OHN often functions in multiple roles within one job position, including clinician, educator, manager, and consultant. Although the majority of OHNs practice on site in medium to large industries, there are many OHNs in other settings, such as free-standing or university-based occupational health clinics, workers' compensation insurance carriers, corporate occupational health and safety departments, hospital employee health units, and environmental/occupational health consulting groups.

This chapter provides an overview of the professional aspects of the OHN role and outlines key programmatic responsibilities for the OHN specific to the clinician, educator, manager, and consultant roles.

# OHN Definition

More than 23,000 registered nurses provide care to employees with the goal of preventing work-related injury and illness, preventing disability, and helping workers achieve and maintain the highest level of health throughout their lives. Occupational health nurses maintain a focus on the worksite when they deliver high-quality care, and philosophically support a primary prevention–based practice. If injuries do occur, a case management approach is used to return injured employees to appropriate work in a timely manner. The American Association of Occupational Health Nurses (AAOHN) maintains, as many industries have found, that "the occupational health nurse is the key to the delivery of comprehensive occupational health services" (American Association of Occupational Health Nurses, 1989).

Occupational health nursing, therefore, is the specialty practice that provides for and delivers health care services to workers and worker populations (American Association of Occupational Health Nurses, 1994). The practice of occupational health nursing is grounded in primary, secondary, and tertiary prevention public health principles and is focused on the promotion, protection, and restoration of workers' health in the context of a safe and healthy work environment (American Association of Occupational Health Nurses, 1994).

# Standards of Practice

AAOHN has published a document titled "Standards of Occupational Health Nursing Practice" (1994) that includes standards for the OHN in both the clinical and professional practice arenas.

The standards of clinical nursing practice include the following:

- *Standard I: Assessment.* The OHN systematically assesses the health status of the client.
- *Standard II: Diagnosis.* The OHN analyzes data collected to formulate a nursing diagnosis.
- *Standard III: Outcome identification.* The OHN identifies expected outcomes specific to the client.
- *Standard IV: Planning.* The OHN develops a comprehensive plan of care and formulates interventions for each level of prevention and for therapeutic modalities to achieve expected outcomes.
- *Standard V: Implementation.* The OHN implements interventions to promote health, prevent illness and injury, and facilitate rehabilitation, guided by the plan of care.

- *Standard VI: Evaluation.* The OHN systematically and continuously evaluates the client's responses to interventions and progress toward the achievement of expected outcomes.

Professional practice standards include the following:

- *Standard I: Professional development/evaluation.* The OHN assumes responsibility for professional development and continuing education and evaluates personal performance in relation to practice standards.
- *Standard II: Quality improvement/quality assurance.* The OHN monitors and evaluates the quality and effectiveness of occupational health practice.
- *Standard III: Collaboration.* The OHN collaborates with employees, management, other health care providers, professionals, and community representatives in assessing, planning, implementing, and evaluating care and occupational health services.
- *Standard IV: Research.* The OHN contributes to the scientific base in occupational health nursing through research, as appropriate, and uses research findings in practice.
- *Standard V: Ethics.* The OHN uses an ethical framework as a guide for decision making in practice.
- *Standard VI: Resource management.* The OHN collaborates with management to provide resources that support an occupational health program that meets the needs of the worker population (Figure 26–1).

**Figure 26–1.** An occupational health nurse presents information on lost time accidents to managers. (© Photo by Avery Photography per ABOHN, Inc.)

These standards guide the OHN in the evaluation of his or her own practice and in maintaining accountability to the public.

# Scope of Practice

The scope of the OHN role varies between industries. Size and type of industry often determine the scope of on-site health services, with over 87 percent of larger industries (greater than 500 employees) having an on-site health unit (U.S. Department of Health and Human Services, 1988). Manufacturing industries with more than 300 employees or service industries with over 750 employees could easily cost-justify one full-time OHN (American Association of Occupational Health Nurses, 1989).

If the industry is self-insured for health care benefit coverage or workers' compensation coverage, providing services for direct care and preventive health programs may also be a good business decision because of employee convenience and the increasing focus on managed care arrangements for health care delivery. For smaller industries at a greater distance from a hospital, an OHN may be a critical addition to a health and safety team to handle any emergencies. Many industries employ OHNs to implement regulatory requirements—such as the Americans with Disabilities Act, OSHA recordkeeping and hazard communication standards, and the bloodborne pathogens standard—and to keep abreast of changes in health care reform, workers' compensation reform, and managed care initiatives.

# Professional Membership and Certification

More than 12,000 OHNs are members of the professional specialty nursing organization, AAOHN. AAOHN has established a code of ethics and a position statement and guidelines on confidentiality of health information. Both documents, in addition to the *Standards of Occupational Health Nursing Practice,* are available from the AAOHN office, 50 Lenox Pointe, Atlanta, Georgia, 30324–3176.

More than 7,000 OHNs have been recognized for excellence in occupational health nursing practice through certification by the American Board for Occupational Health Nurses (ABOHN, 1994a). The Certified Occupational Health Nurse (COHN) credential is awarded based on over 5,000 hours of occupational health experience, 75 hours of occupational health and safety education, and successful completion of an examination.

Additional academic preparation as a manager, nurse practitioner, or clinical nurse specialist in occupational health nursing is available at the graduate level.

# The Practice of Occupational Health Nursing

## Models of Occupational Health Services

Most work-related injury and illness is preventable. Therefore, maintaining a focus on a safe and healthy workplace is critical to the design of an occupational health service. Additionally, the safety and environmental monitoring functions and health functions must be administratively linked in an organization for a successful and smoothly running program.

Other team members, depending on an assessment of the key industry variables listed below, may include an industrial hygienist, occupational and environmental medicine physician, safety professional, ergonomist, physical therapist, employee assistance program personnel, and rehabilitation counselor.

Depending on the education, expertise, and skills of the nurse, the OHN may take a more involved role in direct care (such as the nurse practitioner role), case management, employee assistance program activities, ergonomics, safety activities, and environmental monitoring.

Occupational health service models vary not only in the type and extent of health service personnel, but also in the degree to which they manage work-related and non–work-related health care conditions. Some programs manage only work-related injury and illness and give referrals for health problems not directly related to work. Others manage all work-related and a limited number of non–work-related health concerns. A growing number are offering full-service 24-hour managed care to employees and their dependents, for both work-related and non–work-related conditions.

There are several models for occupational health service delivery, ranging from on-site salaried personnel to off-site contractual arrangements. In 1993 AAOHN, with the American Nurses Association, published a monograph, *Innovation at the Worksite,* which explored various models to deliver more primary care services at the worksite (Burgel, 1993). The scope of occupational health services depends on the following key industry variables:

- Company size and demographics of the work force
- Geographic distance to a health care facility
- Type of industry (manufacturing versus service)
- Hazard profile (review of OSHA 200 log of recordable injuries and illnesses, emergency response needs, potential exposures/trends in claims)
- Risk management and health benefit philosophy of company
- Economic resources
- Self-insurance
- Organizational climate, specifically regarding health, hazard communication, and value of prevention activities

Models of occupational health service delivery include the following (Burgel, 1993):

- Model 1: One OHN, on-site
- Model 2: Multiple OHNs, on-site
- Model 3: Corporate OHNs, decentralized
- Model 4: OHN consultation, off-site

**Model 1: One OHN, on-site.** In more than 60 percent of companies employing an OHN, the OHN is the sole health care provider. This model is ideal for companies with limited resources, small work forces, or few workplace hazards.

The OHN in this setting acts as the in-house expert on health-related issues (Figure 26–2). Duties include assistance with policy development, such as establishing a nonsmoking workplace and helping to design health benefits, to ensure an emphasis on wellness, preventive services, and managed care approaches.

The nurse on site has the opportunity to develop powerful relationships with people including employees, supervisors, CEO, families, and community referrals. These relationships help to reinforce the importance of health and safety, with the emphasis on primary prevention for the reduction of both work-related and non–work-related injuries and illnesses. In addition, the OHN can help people become educated consumers in health care decision making. Provision of these services is at a convenient and familiar location—the worksite.

This direct access allows the OHN to design programs for the needs of the work force. For example, the evaluation of one case of wrist tendinitis may prompt a job analysis to assess for repetitious job tasks and for early identification of employees at risk. Ergonomic changes may then be tailored with an educational program designed to reduce repetitive motion disorders.

Due to limited resources, it is often not feasible in this setting to offer a comprehensive occupational health and safety program. Therefore, the OHN can develop a network of high-quality community-based occupational health and safety consultants and other referrals for employee assistance counselors, orthopedists, and physical and occupational therapists. Again, these contracts/referrals can be developed based on the demographic and health needs of the employees and their dependents and on community resources.

**Model 2: Multiple OHNs, on-site.** Based on the hazard profile and the demographics of the work force, OHNs with specialty backgrounds can be used in this model. For example, an OHN can manage the program, a nurse practitioner with occupational health expertise can direct the preplacement and health/medical surveillance programs, and an OHN with preparation in substance abuse and psychiatric nursing can manage the employee assistance function and case-manage job stress and psychiatric claims.

The strength of this model is that it offers the majority of occupational health and safety services on site (Figure 26–3), with the expected twin goals of improved quality control and ability to manage costs. This model is ideal for medium to large employers.

**Figure 26–3.** An occupational health nurse completes an audiogram. (© Photo by Avery Photography per ABOHN, Inc.)

**Figure 26–2.** An employee reviews one style of hearing protection with the laundry supervisor. (© Photo by Avery Photography per ABOHN, Inc.)

**Model 3: Corporate OHNs, decentralized.** The OHN as the corporate manager of occupational health and safety for national and international firms requires a more global view of health and safety needs. This may include knowledge of the various state, federal, and international health and safety regulations; policy and procedure development; and a "train the trainer" model for employee education.

**Model 4: OHN Consultation off-site.** Consulting arrangements with an occupational health clinic for treatment and surveillance activities are often the strategy chosen by smaller employers. This model is most successful when the off-site OHN visits the workplace to gain in-depth knowledge of the work process and the opportunities for modified work. Although more difficult in an off-site consultative arrangement, the establishment of a working relationship with key company personnel is critical to a successful occupational health and safety program.

## Current Trends in Health Service Staffing

Some industries are downsizing to zero or minimal on-site health personnel, or minimally staffing health units with emergency medical technicians. Cost is the driving reason behind this shift. What is lost in this situation is the on-site presence and knowledge of a professional who can help manage a changing risk environment.

Emerging health and safety regulations, stressful working environments, the increase in cumulative trauma disorders, the need to accommodate disabilities, the aging work force, the increased number of women at the worksite, the implementation of family leave, and the changing health care delivery structures are but a few current issues facing employers.

Knowledge of the key players—the employee, the supervisor, co-workers, and family members—in addition to knowledge of the work process, allows for worksite interventions to prevent, for example, a work-related stress claim.

Other industries, in contrast, are staffing their health units with advanced-practice occupational health nurses, such as occupational health nurse practitioners (Burgel, 1993; Dowrick & Rezents, 1993). Depending on state regulations, nurse practitioners can diagnose and treat many common health conditions in a consulting arrangement with a physician. Not only are cost and quality controlled with this salaried health care provider relationship, but this move seems timely as industries position themselves for health care reform and a managed care environment.

## Occupational Health Nursing Role Functions

The OHN often performs day-to-day activities that are organized into four major roles: clinician, educator, manager, and consultant (American Board for Occupational Health Nurses, 1994b). Table 26–A provides examples of common

OHN activities, organized into roles. The roles are not mutually exclusive and involve all levels of primary, secondary, and tertiary prevention. For example, establishment of an ergonomics program requires expertise in diagnosis and treatment options for repetitive strain injuries (clinician), program design with policy and procedures (manager), the ability to educate workers regarding neutral wrist position (educator), knowledge of the Americans with Disabilities Act so as to accommodate any permanent disabilities (manager), and a team approach that analyzes the workstations and institutes engineering controls (consultant). (See Figure 26–4.)

The following program components that are critical to the OHN role will be discussed in more detail:

- Assessment and management of health complaints
- Health/medical surveillance
- Preplacement (post-offer) evaluations
- Case management
- Modified duty programs
- Employee training
- Wellness programs
- Employee assistance programs
- Workers' compensation
- Americans with Disabilities Act
- Record keeping
- Bloodborne pathogens
- Ergonomics

**Assessment and management of health complaints.** Most commonly, an employee interacts with the OHN for a health complaint. He or she may have an acute problem (such as wrist pain or cough), a chronic health problem (such as high blood pressure), or a question regarding preventive measures (such as home glucose monitoring).

At all times, the OHN must evaluate the symptom in relation to the work tasks done by the employee:

- Is there a potential exposure that could contribute to, cause, or aggravate the complaint?
- Was there a change in the work process that could account for this symptom?
- Are other co-workers complaining of similar symptoms?

The OHN ideally should have access to the most recent environmental monitoring data, to fully evaluate the potential work relationships.

The OHN also knows the individual's prior health history, the family system, and the current department work group issues, and therefore can determine whether psychosocial issues (both at home and at work) may be influencing this complaint.

The OHN, depending on resources, skills, and expertise, may do the initial evaluation and treat within a first-aid/

**Table 26–A.** Occupational Health Nursing Practice Role Functions

*Clinician Role*

Assessment and management of work-related injury and illness
Assessment and management of non–work-related injury and illness
Emergency response
Health/medical surveillance
Preplacement (postoffer) evaluation
Case management of work- and non–work-related illness and injury
Modified duty programs
Rehabilitation

*Educator Role*

Employee training to reduce worksite hazards
Hazard communication
Wellness initiatives to control personal risk factors
Employee assistance program development
Fitness for duty program

*Manager Role*

Philosophy of occupational health unit, including mission statement, goals, and objectives
Analysis of health services trends for program design
Directing health unit with policy and procedure development
Managing resources (budget, personnel)
Negotiation of contracts, vendors, etc.
Management information systems with cost data
Workers' compensation
Compliance with regulations for, e.g., Americans with Disabilities Act and OSHA standards
Bloodborne pathogens
Record-keeping
Program evaluation
Confidentiality/ethics

*Consultant Role*

Safety committee
Hazard analysis/walk-through surveys
Job analysis
Biologic monitoring
Environmental sampling
Human resources/benefits
Ergonomics

self-care model and refer the person, if needed, for a more comprehensive medical evaluation. Or the OHN may do the initial evaluation and treat according to standardized procedures, without a physician referral, depending on state regulation of nursing practice.

The OHN is in a crucial position to monitor quality of care, recovery, and return to work by using a case management approach.

**Health/medical surveillance.** The goal of health/medical surveillance is early identification of biological markers or end points that may signify exposure. Health/medical surveillance, often a requirement of federal health and safety standards, is designed, coordinated, implemented, and evaluated by the OHN. Examples of surveillance activities are hearing conservation programs, respiratory protection programs (Figure 26–5), and asbestos programs.

The OHN reviews the environmental monitoring data, reviews the toxicology of the substance, and in consultation with an industrial hygienist or occupational physician,

outlines a health surveillance program that is exposure- and job-specific. Common to OHN practice is the communication of test results to the individual employee and to the primary care provider. The OHN develops policies and procedures in anticipation of the potential need for job rotation, job modification, confidentiality, and other ethical dilemmas that may arise when an abnormal finding is discovered during health/medical surveillance activities. Good communication skills to educate and counsel employees regarding the purpose and use of these test results are of paramount importance.

**Preplacement (post-offer) evaluations.** Preplacement evaluations are a primary prevention activity with the goal of placing workers in jobs based on physical capabilities and making reasonable accommodations, if needed, in compliance with the Americans with Disabilities Act (ADA). Critical to ADA compliance is the need to ensure that all evaluations are job-related and offered to all entering employees within the same job class.

**Figure 26–4.** An occupational health nurse uses an ergometer to determine strength needed for a job task. (© Photo by Avery Photography per ABOHN, Inc.)

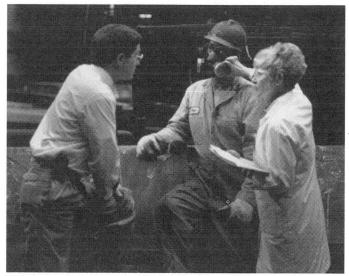

**Figure 26–5.** An occupational health nurse discusses respirator fit testing with employee and supervisor. (© Photo by Avery Photography per ABOHN, Inc.)

Another value of the preplacement evaluation is that the OHN can introduce the role of the occupational health service and establish a beginning relationship with the new employee. At this time, an expectation can be set for active participation of the employee in recognizing and reporting potentially hazardous working conditions.

Studies have documented the value of preplacement evaluations and the role of the OHN in determining appropriate placement based on job-specific criteria (Kemerer & Raniere, 1990). The prevention of just one back injury can often justify the cost of these preventive activities. One preplacement pilot study of just one job class projects the net annual savings to exceed $208,000 (Lukes & Bratcher, 1990).

**Case management.** Case management is the coordination of the efforts of the health care team with the goal of returning the patient to pre-illness or pre-injury function or easing the multiple burdens of chronic or terminal conditions (Mazoway, 1987). It is a system that aims to provide the right care, at the right time, in the right setting, by the right provider, at the right cost. The emphasis is on early intervention and coordination of care for targeted high-risk, high-cost cases (Burgel, 1992).

Communication with the consumer is an important component of case management. For example, suppose an employee or dependent sustains a spinal cord injury and the OHN recommends early transfer to a spinal cord rehabilitation facility; however, this facility is 100 miles from the family's home town. Clear OHN communication with the family is needed to explain the expected outcome in the rehabilitation facility and to gain their support of the transfer.

There are several tools OHNs use to establish a case management program:

■ The ability to flag catastrophic and chronic claims, such as premature births, spinal cord injuries, organ transplants, certain cancers, AIDS, or mental health disorders

■ Early identification of employees with high claim reserves and those at risk for delayed recovery

■ Establishment of a pool of modified-duty jobs

■ Establishment of a panel of qualified providers who support an aggressive medical rehabilitation plan with modified duty

■ Access to computerized information systems

■ Full knowledge of health benefit packages in case alternative benefits must be negotiated on behalf of the ill employee or dependent, such as home care with nursing assistance

**Modified duty programs.** Timely return to work in temporary modified duty assignments is a realistic outcome measure for a case management program targeting both non–work-related and work-related injuries and illnesses. This is a major role for the OHN. The OHN not only knows the work process, but also has been very involved in determining the level of care needed for an injured or ill employee. Therefore, the OHN is well-positioned to determine readiness to return to work and to support both the injured employee and his or her supervisor throughout the process. An aggressive sports medicine rehabilitative approach, in addition to modified duty, helps speed recovery of injured workers (Burgel & Gliniecki, 1986).

Establishment of a modified duty program requires a proactive approach with policies and procedures. Table 26–B outlines the data needed before development of a modified duty program.

When an injured employee returns, the OHN outlines the time-limited nature of the modified job. There is the expectation that as the employee progressively improves, the physical demands of the job will advance as well. Therefore, modified duty is, in reality, on-the-job work hardening. It is also imperative to have a panel of providers who philosophically support modified duty (Gliniecki, 1992).

**Employee training.** Worksite hazard recognition and control is a primary focus of OHNs in promoting a healthy work force. Most occupational illnesses and many worksite injuries are preventable by cooperative efforts of employers, employees, and the occupational health and safety team.

Employee training, either in one-on-one or group settings, encourages workers to engage in safe work practices and stimulates a level of understanding to help employees recognize and report potential hazards to employers. More than 100 OSHA standards require the employer to train employees in health and safety. Many of the occupational health and safety objectives in *Healthy People 2000: National Health Promotion and Disease Prevention Objectives* require training (U.S. Department of Health and Human Services, 1991).

A key OHN activity in all employee training is needs assessment. Analysis of the worksite is critical to determine efficacy of engineering, administrative, and personal protective controls (Figure 26–6). Baseline knowledge of the targeted group of workers is determined to plan the learning intervention. Risk communication principles require full participation of the target group in planning an intervention (Burgel, 1994).

OHNs are often involved in the employee training requirement in the OSHA Hazard Communication Standard.

---

**Table 26–B.** Modified-Duty Programs

To insure success of a modified duty program in your industry, the following areas need to be assessed:

*Workplace*

What areas have highest rate of lost time?  What type of injuries (i.e., what kind of jobs are needed?)? What percentage of cases are appropriate for modified duty?
What is the organization's philosophy and policy re: lost time? or job accommodation in general?
Have jobs been accommodated in the past?  If yes, what were the variables that contributed to success?  failure?
What fears/concerns have been expressed (formally or informally) by supervisors/managers?
How have supervisors used the health service in the past?  How successful has the employee assistance program been; for example, have supervisors appropriately referred employees to the EAP?
In areas of highest lost time, what are the group dynamics (values/norms, style of communication, etc.) among coworkers?
How are workers' compensation premiums expressed in dollars and cents to the individual department supervisor?  What is your experience modification rating?
How are budgets allocated within the organization?
What incentives could exist within your specific organization to support the modified duty program?
What is the extent of disability coverage?
Is there language in the union contract(s) specifying guidelines for modified duty?

*Employee Factors*

Helpful to review 3 to 5 recent cases of lost time to group those cases which could have benefited from modified duty programs;
Psychosocial indicators, including depression, loss, anger, expression of dependency needs;
Physical variables, including educational level, language competencies, technical expertise; nature of injury; culpability; type of work restriction;

*Workers' Compensation*

What type of education is given to injured workers re: explanation of benefits?
What have been the nature/extent of complaints re: the administration of the workers' compensation program in your organization?

*Medical Providers*

Based on a review of the most frequent or most severe lost time injuries, what top three job classifications have been involved?
Have job analyses been done on any of these job classifications?
How are return to work decisions currently made?
Is there a panel of medical providers that know the organization?

(This material is reproduced with permission of Continuing Professional Education Center, Inc., Skillman, New Jersey.)

**Figure 26–6.** An occupational health nurse prepares for a facility walk-through. (© Photo by Avery Photography per ABOHN, Inc.)

Using a team approach, the OHN educates employees on potential health effects from potential exposures with hazardous substances, including interpretation of the material safety data sheets

Other OHN employee training activities include the following:

- Education on how best to adjust workstations and work flow to decrease the number of forceful repetitions and thus prevent cumulative trauma disorders

- Education in hearing conservation relative to the long-term effects of noise on hearing and the need for hearing protection

- Education and demonstration on dividing lifting loads, in addition to teaching back strengthening and proper lifting techniques

**Wellness programs.** Health variables that employers have less control over are personal risk factors. These risk factors have received additional attention in the worksite because they contribute to rising health care costs. Opportunities to affect high-risk groups in an accessible location are a major strength of wellness programs at the worksite.

OHNs have long been involved in wellness initiatives in the workplace.

In 1985 nearly 66 percent of worksites with more than 50 employees reported offering at least one health promotion activity. These included programs related to smoking control (35.6 percent), health risk assessment (29.5 percent), back care (28.6 percent), stress management (26.6 percent), exercise and fitness (22.1 percent), and off-the-job accident prevention (19.8 percent) (U.S. Department of Health and Human Services, 1991).

Health promotion and disease prevention activities are prioritized by the OHN, based not only on the demographics of the work force but also on the *Healthy People 2000* objectives. Examples include smoking cessation programs with the establishment of nonsmoking work policies and breast health care for women. Elder care may be an issue for an older work force. A special emphasis is placed on self-care and individual responsibility.

Chenoweth (1989) noted that OHNs can facilitate risk reduction by effectively screening, educating, and monitoring high-risk employees. In his economic appraisal of a Fortune 500 manufacturing plant, he found a reduced incidence of smoking based on the health promotion efforts of the OHNs, with cost savings anticipated over time.

**Employee assistance programs.** Employee assistance programs (EAPs) use both primary and secondary prevention methods to recognize, assess, treat, and refer employees with personal and mental health problems that impact job performance. The OHN is often the first point of contact and is able to confidentially counsel and refer employees to an EAP resource. Although initially focused on substance abuse, EAPs are now assessing a full range of family and work issues that may affect an employee's ability to be a fully functioning member of a work team. These services may be crisis oriented and in response to changes in work performance, but they may also be preventive in nature, with emphasis on communication, conflict resolution, and stress management.

Emphasis on the appropriate use of psychiatric, chemical dependency, and other services by employees and their dependents is a current case management program in many industries. Growing mental health care costs are rapidly becoming the greatest concern for corporations. The OHN manages, implements, and cooperates with the EAP resource at the worksite and may serve as the gatekeeper to mental health care providers in the community. This strong link impacts mental health care in two ways: Employees are educated by the OHN in purchasing mental health care for themselves or their dependents, and the OHN becomes involved in supporting these individuals at the worksite (Burgel, 1993).

Stress at the workplace and at home, drug and alcohol abuse, and accommodation of the needs of employees with psychiatric diagnoses are just a few of the daily challenges facing OHNs. OHNs are often the first resource for the troubled employee and his or her supervisor. A common OHN role function is to complete the initial assessment and facilitate referral for long-term treatment (Thompson, 1989; Conry, 1991).

**Workers' compensation.** Job-related injuries in 1990 cost American employers an estimated $60 billion in direct workers' compensation expenses (Tillinghast, 1991), and these costs have doubled in the last 5 years. Not only are increasing medical costs fueling this growth in expenses, but increasing litigation, a widening definition of work-related injury and illness, and a need for insurance reform are also contributing to this crisis (Tillinghast, 1991).

Workers' compensation is a very complex system in many states. Injured workers often do not know how to access these benefits, and can become confused and angry if they attempt to negotiate this system alone. The OHN, on-site, is often the first contact for an injured worker and is able to explain the full scope of workers' compensation benefits. The OHN uses a case management approach to determine the appropriate care and work modifications, if required. Close communication and monitoring of workers' compensation cases is an important OHN role.

**Americans with Disabilities Act.** Some 43 million Americans have one or more physical or mental disabilities. This population is targeted by the Americans with Disabilities Act (ADA) (Public Law no. 101336). The ADA, signed into law in 1991 and put in effect in 1992, prohibits discrimination against people with disabilities in employment, transportation, public accommodation, activities of state and local government, and telecommunication relay services.

Employers must not only have nondiscriminatory selection criteria, but must make reasonable accommodation to the known limitations of the qualified applicant unless it causes undue hardship. OHNs advise employers on compliance with the ADA, ensuring that the preplacement (post-offer) program meets the requirements. In addition, OHNs often recommend reasonable accommodations and counsel employees with physical disabilities (Kaldor, 1992). Table 26–C outlines examples of reasonable accommodations suggested by OHNs.

In 1987, permanent impairments sustained on the job grew from 60,000 to 70,000 and total disabling injuries numbered 1.8 million (U.S. Department of Health and Human Services, 1991). This fact alone highlights the importance of prevention activities at the worksite. Those work-related permanent disabilities are also covered by the ADA, requiring reasonable accommodation efforts by the employer.

**Record keeping.** Record keeping is guided by OSHA standards, in addition to state workers' compensation record requirements. Maintenance of OSHA 200 logs, a key requirement of the OSHA standard, is often an OHN responsibility. The OHN determines the level of care and treatment. If the injury requires only first aid, it is not OSHA recordable. Recordable conditions include every death, every occupational illness and injury involving medical treatment beyond first aid, lost time, work modification, job transfer, and any loss of consciousness. The OSHA 200 log is posted every year from February 1 to March 1 (See Figure 26–7).

Reporting requirements for workers' compensation vary from state to state and are separate from OSHA record-keeping requirements. The employer's report of occupational injury and illness can be substituted, in many cases, for the supplemental OSHA form 101.

The OSHA 200 log is one data source for OHN analysis of trends in work-related disease. These data help prioritize walk-through surveys, periodic environmental sampling,

**Table 26–C.** Americans with Disabilities Act: Reasonable Accommodations

| Job | Essential Function | Disability | Possible Accommodations |
|---|---|---|---|
| Material handler | Routinely move pkgs. weighing 40–50 lb (18.14–22.68 kg) | No lifting over 25 lb (11.34 kg) | 1. Mechanical assist (roller racks, hoists)<br>2. Dual-person lifts |
| Computer operator | Rotation of shifts | Fatigue: Post chemotherapy | 1. Split-shift<br>2. Part-time job<br>3. Job-share |
| Mail carrier | Work outside 5 days/week | History of malignant melanoma | 1. Provide protective clothing (SPF shirts, hats, gloves) |
| Engineer | Communicate ideas to work-group | Speech impairment | 1. Voice-activated keyboard<br>2. Portable computer |
| Housekeeping | Follow instructions | Illiteracy due to retardation | 1. All directions in drawing form<br>2. Cassette tape of instructions with cassette player |

(This material is reprinted with permission of Continuing Professional Education Center, Inc., Skillman, New Jersey.)

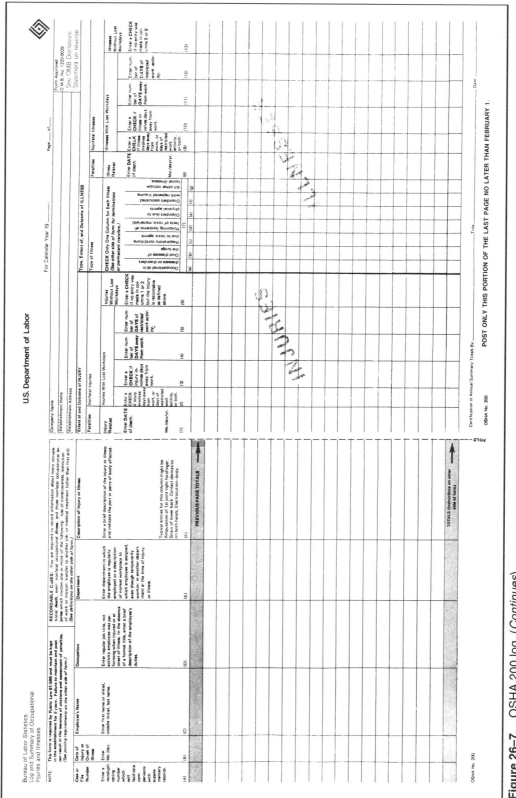

**Figure 26–7.** OSHA 200 log. (Continues)

Public reporting burden for this collection of information is estimated to vary from 8 to 30 minutes per line entry, including time for reviewing instructions, searching existing data sources, gathering and maintaining the data needed, and completing and reviewing the collection of information. Send comments regarding the burden estimate or any other aspect of this collection of information, including suggestions for reducing this burden, to the Office of Information Management, Department of Labor, Room N-1301, 200 Constitution Avenue, NW, Washington, DC 20210, and to the Office of Information and Regulatory Affairs, Office of Management and Budget, Washington, DC 20503.

**Instructions for OSHA No. 200**

I. **Log and Summary of Occupational Injuries and Illnesses**

Each employer who is subject to the recordkeeping requirements of the Occupational Safety and Health Act of 1970 must maintain for each establishment a log of all recordable occupational injuries and illnesses. This form (OSHA No. 200) may be used for that purpose. A substitute for the OSHA No. 200 is acceptable if it is as detailed, easily readable, and understandable as the OSHA No. 200.

Enter each recordable case on the log within six (6) workdays after learning of its occurrence. Although other records must be maintained at the establishment to which they refer, it is possible to prepare and maintain the log at another location, using data processing equipment if desired. If the log is prepared elsewhere, a copy updated to within 45 calendar days must be present at all times in the establishment.

Logs must be maintained and retained for five (5) years following the end of the calendar year to which they relate. Logs must be available (normally at the establishment) for inspection and copying by representatives of the Department of Labor, or the Department of Health and Human Services, or States accorded jurisdiction under the Act. Access to the log is also provided to employees, former employees, and their representatives.

II. **Changes in Extent of or Outcome of Injury or Illness**

If, during the 5-year period the log must be retained, there is a change in an extent and outcome of an injury or illness which affects entries in columns 1, 2, 6, 8, 9, or 13, the first entry should be lined out and a new entry made. For example, if an injured employee at first required only medical treatment but later lost workdays away from work, the check in column 6 should be lined out, and checks entered in columns 2 and 3 and the number of lost workdays entered in column 4.

In another example, if an employee with an occupational illness lost workdays, returned to work, and then died of the illness, any entries in columns 9 through 12 should be lined out and the date of death entered in column 8.

The entire entry for an injury or illness should be lined out if later found to be nonrecordable. For example, an injury which is later determined not to be work related, or which was initially thought to involve medical treatment but later was determined to have involved only first aid.

III. **Posting Requirements**

A copy of the totals and information following the fold line of the last page for the year must be posted at each establishment in the place or places where notices to employees are customarily posted. This copy must be posted no later than February 1 and must remain in place until March 1.

Even though there were no injuries or illnesses during the year, zeros must be entered on the totals line, and the form posted.

The person responsible for the annual summary totals shall certify that the totals are true and complete by signing at the bottom of the form.

IV. **Instructions for Completing Log and Summary of Occupational Injuries and Illnesses**

Column A – CASE OR FILE NUMBER. Self-explanatory.

Column B – DATE OF INJURY OR ONSET OF ILLNESS.

For occupational injuries, enter the date of the work accident which resulted in injury. For occupational illnesses, enter the date of initial diagnosis of illness, or, if absence from work occurred before diagnosis, enter the first day of the absence attributable to the illness which was later diagnosed or recognized.

Columns C through F – Self-explanatory.

Columns 1 and 8 – INJURY OR ILLNESS-RELATED DEATHS. Self-explanatory.

Columns 2 and 9 – INJURIES OR ILLNESSES WITH LOST WORKDAYS. Self-explanatory.

Any injury which involves days away from work, or days of restricted work activity, or both must be recorded since it always involves one or more of the criteria for recordability.

Columns 3 and 10 – INJURIES OR ILLNESSES INVOLVING DAYS AWAY FROM WORK. Self-explanatory.

Columns 4 and 11 – LOST WORKDAYS—DAYS AWAY FROM WORK.

Enter the number of workdays (consecutive or not) on which the employee would have worked but could not because of occupational injury or illness. The number of lost workdays should not include the day of injury or onset of illness or any days on which the employee would not have worked even though able to work.

NOTE: For employees not having a regularly scheduled shift, such as certain truck drivers, construction workers, farm labor, casual labor, part-time employees, etc., it may be necessary to estimate the number of lost workdays. Estimates of lost workdays shall be based on prior work history of the employee AND days worked by employees, not ill or injured, working in the department and/or occupation of the ill or injured employee.

Columns 5 and 12 – LOST WORKDAYS—DAYS OF RESTRICTED WORK ACTIVITY.

Enter the number of workdays (consecutive or not) on which because of injury or illness:

(1) the employee was assigned to another job on a temporary basis, or

(2) the employee worked at a permanent job less than full time, or

(3) the employee worked at a permanently assigned job but could not perform all duties normally connected with it.

The number of lost workdays should not include the day of injury or onset of illness or any days on which the employee would not have worked even though able to work.

Columns 6 and 13 – INJURIES OR ILLNESSES WITHOUT LOST WORKDAYS. Self-explanatory.

Columns 7a through 7g – TYPE OF ILLNESS.

Enter a check in only one column for each illness.

TERMINATION OR PERMANENT TRANSFER–Place an asterisk to the right of the entry in columns 7a through 7g (type of illness) which represented a termination of employment or permanent transfer.

V. **Totals**

Add number of entries in columns 1 and 8.

Add number of checks in columns 2, 3, 6, 7, 9, 10, and 13.

Add number of days in columns 4, 5, 11, and 12.

Yearly totals for each column (1-13) are required for posting. Running or page totals may be generated at the discretion of the employer.

If an employee's loss of workdays is continuing at the time the totals are summarized, estimate the number of future workdays the employee will lose and add that estimate to the workdays already lost and include this figure in the annual totals. No further entries are to be made with respect to such cases in the next year's log.

VI. **Definitions**

OCCUPATIONAL INJURY is any injury such as a cut, fracture, sprain, amputation, etc., which results from a work accident or from an exposure involving a single incident in the work environment.

NOTE: Conditions resulting from animal bites, such as insect or snake bites or from one-time exposure to chemicals, are considered to be injuries.

OCCUPATIONAL ILLNESS of an employee is any abnormal condition or disorder, other than one resulting from an occupational injury, caused by exposure to environmental factors associated with employment. It includes acute and chronic illnesses or diseases which may be caused by inhalation, absorption, ingestion, or direct contact.

The following listing gives the categories of occupational illnesses and disorders that will be utilized for the purpose of classifying recordable illnesses. For purposes of information, examples of each category are given. These are typical examples, however, and are not to be considered the complete listing of the types of illnesses and disorders that are to be counted under each category.

7a. **Occupational Skin Diseases or Disorders**
Examples: Contact dermatitis, eczema, or rash caused by primary irritants and sensitizers or poisonous plants, oil acne; chrome ulcers; chemical burns or inflammations; etc.

7b. **Dust Diseases of the Lungs (Pneumoconioses)**
Examples: Silicosis, asbestosis and other asbestos-related diseases, coal worker's pneumoconiosis, byssinosis, siderosis, and other pneumoconioses.

7c. **Respiratory Conditions Due to Toxic Agents**
Examples: Pneumonitis, pharyngitis, rhinitis or acute congestion due to chemicals, dusts, gases, or fumes; farmer's lung; etc.

7d. **Poisoning (Systemic Effect of Toxic Materials)**
Examples: Poisoning by lead, mercury, cadmium, arsenic, or other metals; poisoning by carbon monoxide, hydrogen sulfide, or other gases; poisoning by benzol, carbon tetrachloride, or other organic solvents; poisoning by insecticide sprays such as parathion, lead arsenate; poisoning by other chemicals such as formaldehyde, plastics, and resins; etc.

7e. **Disorders Due to Physical Agents (Other than Toxic Materials)**
Examples: Heatstroke, sunstroke, heat exhaustion, and other effects of environmental heat; freezing, frostbite, and effects of exposure to low temperatures; caisson disease; effects of ionizing radiation (isotopes, X-rays, radium); effects of nonionizing radiation (welding flash, ultraviolet rays, microwaves, sunburn); etc.

7f. **Disorders Associated With Repeated Trauma**
Examples: Noise-induced hearing loss, synovitis, tenosynovitis, and bursitis; Raynaud's phenomena; and other conditions due to repeated motion, vibration, or pressure.

7g. **All Other Occupational Illnesses**
Examples: Anthrax, brucellosis, infectious hepatitis, malignant and benign tumors, food poisoning, histoplasmosis, coccidioidomycosis, etc.

MEDICAL TREATMENT includes treatment (other than first aid) administered by a physician or by registered professional personnel under the standing orders of a physician. Medical treatment does NOT include first-aid treatment (one-time treatment and subsequent observation of minor scratches, cuts, burns, splinters, and so forth, which do not ordinarily require medical care) even though provided by a physician or registered professional personnel.

ESTABLISHMENT: A single physical location where business is conducted or where services or industrial operations are performed (for example: a factory, mill, store, hotel, restaurant, movie theater, farm, ranch, bank, sales office, warehouse, or central administrative office). Where distinctly separate activities are performed at a single physical location, such as construction activities operated from the same physical location as a lumber yard, each activity shall be treated as a separate establishment.

For firms engaged in activities which may be physically dispersed, such as agriculture, construction, transportation, communications, and electric, gas, and sanitary services, records may be maintained at a place to which employees report each day.

Records for personnel who do not primarily report or work at a single establishment, such as traveling salesmen, technicians, engineers, etc., shall be maintained at the location from which they are paid or the base from which personnel operate to carry out their activities.

WORK ENVIRONMENT is comprised of the physical location, equipment, materials processed or used, and the kinds of operations performed in the course of an employee's work, whether on or off the employer's premises.

**Figure 26–7.** (Continued)

health/medical surveillance programs, and employee training schedules.

Access to employee exposure and medical records, as required by OSHA standards, is an additional OHN responsibility. Establishment of a confidential record-keeping system is a high priority from a legal and ethical perspective. Employees and their designated representatives have access to aggregate exposure records of other employees with similar past or present job duties. Exposure records include environmental monitoring data and biological monitoring data. Access need not be provided to voluntary employee assistance records, which are maintained separately from the occupational health medical record.

**Bloodborne pathogens program.** The bloodborne pathogens OSHA standard, adopted in 1992, requires employers to establish an exposure control plan for all employees who have occupational exposure to blood or other potentially infectious materials. It mandates the use of universal precautions and the provision of personal protective equipment by the employer, in addition to safe needle disposal containers. The standard clarifies the employer's responsibility to provide, at no cost to the at-risk employee, the hepatitis B vaccine series. There are specific training requirements, often creating a role for the OHN. Postexposure policies and procedures are established by the OHN, as outlined in the OSHA bloodborne pathogen standard.

OSHA provides information on bloodborne pathogens to help employers comply with this standard. This standard applies not only to the hospital environment, but also to laboratories, dental offices, and on-site occupational health units. Questions arise as to whether volunteer first responders (employees who volunteer to respond to workplace emergencies when occupational health personnel are not available), need to be trained and offered the hepatitis B vaccine. If it is reasonably anticipated that there will be exposure to blood or body fluids, provision of the vaccine would be required for employees trained in first aid and designated by the employer as responsible for rendering medical assistance as part of their job duties.

Oversight of the bloodborne pathogens standard involves the clinician, educator, manager, and consultant OHN roles. Because of the confidential nature of an exposure, especially in the uncommon event of an HIV antibody conversion, the OHN must apply astute communication skills and strong professional ethics.

**Ergonomics.** The increase in cumulative trauma disorders (CTDs) has required the practice of occupational health nursing to expand to include ergonomics as a critical activity. Often an interdisciplinary activity involving the job design personnel of an industry, the practice of ergonomics involves workstation evaluation and job analysis. CTDs are common in an office, manufacturing, or hospital setting, and predominant in positions requiring repetitious and forceful extension, flexion, pinching, grasping, bending, or reaching.

Because of the waxing and waning of symptoms, employees need education and counseling about CTDs, with measures suggested to prevent and also treat the acute flare-ups. With ergonomic educational programs, there often is an increase in the number of symptomatic employees seeking a health evaluation. However, if engineering controls are introduced and subsequently reinforced on periodic walk-through surveys by the OHN, the severity of CTD cases should decrease over time. Program evaluation for ergonomic interventions, therefore, should detail not only the number of cases, but also indices of severity. The OHN manager would assume this program responsibility. Figure 26–8 (pp. 743–744) is an example of an ergonomic assessment tool for an office environment.

# Summary

The occupational health nurse, the predominant health care provider on site in industry, is key to a comprehensive occupational health and safety program. The OHN has a critical relationship not only with the employee, but also with the supervisor, the industrial hygienist, the safety professional, other co-workers, the union, family members, and the employee's primary care provider. The OHN has knowledge of the work process and potential hazards and provides a confidential and neutral analysis of the interaction between the worker's health status and his or her job.

Through the clinician, educator, manager, and consultant roles, the OHN uses a team approach to prevent work-related injury and illness and maintain the health of the work force. Case management, which includes coordination to ensure the best care at the right cost, is a key strategy used by the OHN in monitoring both work-related and non–work-related injury and illness care.

An on-site OHN, either full-time or part-time, is a valuable company asset in maintaining an emphasis on a safe and healthy workplace, managing occupational health and safety regulatory requirements, and monitoring the health care of employees.

# Bibliography

American Association of Occupational Health Nurses. "Standards of Occupational Health Nursing Practice—Booklet." Atlanta: American Association of Occupational Health Nurses, 1994.

American Association of Occupational Health Nurses. "Occupational Health Nursing: The Answer to Cost Containment—Booklet." Atlanta: American Association of Occupational Health Nurses, 1989.

American Board for Occupational Health Nurses, Inc. *COHN Rep* 15:1–11, 1994a.

American Board for Occupational Health Nurses, Inc. *Occupational Health Nursing Job Analysis.* Martinsville, NJ: Center for Nursing Education and Testing, June 1994b.

Bey JM, McGovern PM, Foley M. How management and nurses perceive occupational health nursing. *AAOHN J* 36:61–69, 1988.

Bey JM, Widtfeldt AK, Burns JM. The nurse as health manager. *Bus Health* 10:24–31, 1990.

Bureau of Labor Statistics. *Outlook 2000*. Washington, DC: U.S. Department of Labor, 1990.

Burgel BJ. Employee education. In LaDou J, ed. *Occupational Health and Safety*, 2nd ed. Chicago: National Safety Council, 1994.

Burgel BJ. *Innovation at the Worksite: Delivery of Nurse-Managed Primary Health Care Services*. Washington, DC: American Nurses Publishing, 1993.

Burgel BJ. Case management: A system of care delivery for the future. *AAOHN CE Update* 4:1–8, 1992.

Burgel BJ, Gliniecki CM. Disability behavior: Delayed recovery in employees with work compensable injuries. *AAOHN J* 34:26–30, 1986.

Chenoweth D. Nurses' intervention in specific risk factors in high risk employees: An economic appraisal. *AAOHN J* 37:367–373, 1989.

Conry PB. Drugs and alcohol in the workplace. *AAOHN J* 39:461–465, 1991.

Cordes DH, Rea DF. Farming: A hazardous occupation. *Occup Med: State of the Art Reviews* 6:327–333, 1991a.

Cordes DH, Rea DF. Preventive measures in agricultural settings. *Occup Med: State of the Art Reviews* 6:541–550, 1991b.

Dalton BA, Harris JS. A comprehensive approach to corporate health management. *J Occup Med:* 33:338–347, 1991.

Dowrick MA, Rezents K. Occupational health nurse practitioner has broader legal scope, accountability. *Occup Health Saf* 9:131–135, 1993.

Gliniecki C. Modified duty programs. *AAOHN CE Update* 4:1–8, 1992.

Ham FL. The future of dependent health benefits. *Bus Health* 9:19–26, 1989a.

Ham FL. How companies are making wellness a family affair. *Bus Health* 9:27–33, 1989b.

Johnston WB. Global workforce 2000: The new world labor market. *Harvard Bus Rev* March–April:115–127, 1991.

Kaldor C. The Americans with Disabilities Act: An invitation for OHN intervention. *AAOHN CE Update* 5:1–8, 1992.

Kemerer S, Raniere TM. Cost effective job placement physical examinations: A decision making framework. *AAOHN J* 38:236–242, 1990.

Lukes EN, Bratcher BP. Pre-employment physical examinations. *AAOHN J* 38:174–179, 1990.

Lusk SL, Disch JM, Barkauskas VH. Interest of major corporations in expanded practice of occupational health nurses. *Res Nurs Health* 11:141–151, 1988.

Mazoway JM. Early intervention in high cost care. *Bus Health* 1:12–16, 1987.

Meister JS. The health of migrant farm workers. *Occup Med: State of the Art Reviews* 6:503–518, 1991.

Meyer J, Sullivan S, Silow-Carroll S. *Private Sector Initiatives: Controlling Health Care Costs*. Washington, DC: Health Care Leadership Council, March 1991.

Miller MA. Social, economic, and political forces affecting the future of occupational health nursing. *AAOHN J* 37(9):361–366, 1989.

Miller RS. Reform of the health care system. Statement of Chrysler Corporation before the Senate Committee on Finance, Sub-

committee on Health for Families and the Uninsured. U.S. Senate. Washington, DC: U.S. Government Printing Office, 1991.

National Institute of Occupational Health and Safety. *Costs and Benefits of Occupational Health Nursing*. DHHS Pub. No. (NIOSH) 80–140. Cincinnati, OH: U.S. Department of Health and Human Services, 1980.

1990 National Executive Poll on Health Care Costs and Benefits. *Bus and Health* 4:24–38, 1990.

Pederson DH, Sieber WK. Some trends in worker access to health care in the United States (1974–1983). *Am J Ind Med* 15:151–165, 1989.

Thompson L. The occupational health nurse as an employee assistance program provider. *AAOHN J* 37:501–507, 1989.

Tillinghast Consulting Group. *Responding to the Workers' Compensation Crisis: Can Employers Manage and Control Costs?* Atlanta: Towers Perrin, 1991.

Touger GN, Butts J. The workplace: An innovative and cost-effective practice site. *Nurse Prac* 14:35–42, 1989.

U.S. Department of Health and Human Services. *Health United States 1987*. DHHS Pub. No. (PHS) 88–1232. Washington, DC: U.S. Government Printing Office, 1988.

U.S. Department of Health and Human Services. *Healthy People 2000: National Health Promotion and Disease Prevention Objectives*, DHHS Pub. No. (PHS) 91–50212. Washington, DC: U.S. Government Printing Office, 1991.

Wakefield DS. Rural hospitals and the provision of agricultural occupational health and safety services. *Am J Ind Med* 18:433–442, 1990.

Warshaw LJ, Barr JK, Schachter M. Care givers in the workplace: Employer support for employees with elderly and chronically disabled dependents. *J Occup Medicine* 29:520–525, 1987.

## ■ Addendum: Issues Affecting Health Care Service Staffing*

### Changing Demographics in the Workforce

**Women employees.** Women will continue to enter the workforce in large numbers. Currently 44% of the U.S. workforce are women and 66% of the female labor force in the United States are employed. Based on this trend, OHNs have developed new workplace initiatives, such as delivering pregnancy care closer to the worksite and developing health promotion and screening programs for breast and cervical cancer detection and menopause counseling. Further research into the area of reproductive hazards for both women and men will drive changes in the work environment. Parenting issues with policies exploring on-site child care, flex-time and work-at-home options for both women and men are current initiatives that promote health and reduce costs by decreasing absenteeism and improving morale and work performance. The OHN is centrally placed to design these health promoting programs.

| Name: | | Date: | Start Time: |
|---|---|---|---|
| Job Description: | | | End Time: |
| Location: | | Mail Stop: | Extension: |
| Dept.#: | Employee # | | Date of last move: |

Current Discomfort:



Seen M.D. ☐ or Intel Health Services ☐ for current symptoms?    ☐YES    ☐NO

Spends > 2 hrs./day at: Terminal ☐  Mouse ☐  Desk ☐   Phone ☐
Other:

Breaks:  Yes ☐  No ☐
Exercise: Yes ☐  No ☐
Training: Yes ☐  No ☐

☐ Glasses:                    (☐ Bifocals)   ☐ Monitor Glare:

☐Existing Ergonomic Accessories:
  ☐ Foot rest    ☐ Keybd tray    ☐ Document holder    ☐ Wrist rest
  ☐ Tele headset    ☐ Anti glare screen    ☐ Monitor stand
  ☐ Chair (type) _____    ☐ Other _____

☐Non-Neutral Posture Observed:



☐ Repetition:
☐ Forceful Movement:
☐ Mechanical Stress:

| Measure | Desk | Chair | Kybrd | Monitor | Mouse | Other | Anthro data |
|---|---|---|---|---|---|---|---|
| Too High | | | | | | | Overall height |
| Too Low | | | | | | | " |
| Too Far | | | | | | | |
| Too Close | | | | | | | Seated elbow |
| Wrong < | | | | | | | " |

Furniture type: ☐ WH components   ☐ GF   ☐ CF&A   ☐ Steelcase

Notes: _____
_____
_____

**Figure 26–8.**   Intel office ergonomics assessment tool. (Reprinted courtesy of Intel Corporate Environmental Health and Safety, Santa Clara, CA.) (*Continues*)

Name:_____ Date:_____

# Recommendations:

| Methods: | Service:  FWR or WIN# |
|---|---|
|  |  |
|  |  |
|  |  |
|  |  |
|  |  |
|  |  |
|  |  |
|  |  |

Suggested **Accessories** (Can order through FAR WEST Office Supplier)
☐  Need to Order

| Description | Item Number |
|---|---|
|  |  |
|  |  |
|  |  |
|  |  |

| ☐Need Chairs: | ☐ Sensor | ☐Criterion | ☐Neutral Posture |
|---|---|---|---|

Notes:

☐Educational Material Given:

☐Need Office Ergonomics Class:

☐Referred to Health Services:

Evaluated By:_____

Follow-Up Evaluation (by whom and date):_____

**Figure 26–8.** (*Continued*)

**Older employees.** U.S. workers over age 65 made up 10.3% of the labor force in 1985 with expectations of rising to 15% in 2000. Older employees are traditionally high users of expensive health care services, specifically in the areas of chronic illness. OHNs' use of the worksite for health education and monitoring of stable chronic illnesses is not only convenient, but cost effective. Establishing programs aimed at those employees with responsibilities to care for older or disabled dependents is a further example of health promoting initiatives by OHNs (Warshaw et al., 1987).

The size of the retired population, whose health care expenses are covered by many businesses, will grow. The cost burdens of this population are increased by new regulations requiring corporations to include the costs of funding retiree health as a liability on their balance sheets as of 1992. The workplace is an ideal setting for health screening, health promotion, education and case management of chronic diseases prevalent in elderly populations. OHNs screen for and provide case management for hypertension, diabetes, cancer and mental health problems (Miller, 1989). Southern California Edison, for example, has initiated a case management program for their retirees and their dependents, which includes prevention and social support. This program, called "Generation," is currently being evaluated for cost impact (Meyer et al., 1991).

**Cultural diversity.** By the year 2000, African-Americans will make up 12 percent of the U.S. labor force, up 1 percentage point from 1988, totalling 16.5 million. Asians will increase their share of the labor force by one percentage point to 4 percent by the year 2000, totalling 5.6 million. Hispanics will constitute 10 percent of the labor force, up 3 percentage points from 1988, totalling 14.3 million by the year 2000 (Bureau of Labor Statistics, 1990). It is also anticipated that people worldwide will be increasingly well educated and developing countries will produce a growing share of the world's high school and college graduates. The USA is expected to import this talent, and therefore have an increasingly culturally diverse workforce by the year 2000 (Johnston, 1991).

The OHN is prepared to address the needs of this changing workforce for culturally relevant and culturally sensitive health interventions. This includes awareness of health and disease patterns, health beliefs, and consideration of language and literacy issues. The OHN is well positioned to tailor health interventions based on cultural patterns of the workforce.

**Dependent care.** Dependent care costs are on the rise, particularly for mental health and substance abuse treatment. In 1989, 50 cents of every health care dollar went for medical services used by a spouse or child (Ham, 1989a). Fewer than one in five companies that offer on-site health promotion programs extend eligibility to family members of employees (Ham, 1989b). In the 1990 National Executive Poll on Health Care Costs and Benefits,

60% of corporate leaders identified rising dependent coverage costs as a health care issue of major concern. Rising dependent costs, especially in the area of mental health services, is another emergent area in need of a primary prevention and case management focus. Given sufficient resources, OHNs are well positioned to expand services to include dependent care.

Although the provision of a safe and healthy worksite must remain the primary value of the OHN, the comprehensive evaluation of an individual within a family system is critical when mounting disease prevention and health promotion programs at the worksite.

Kimberly-Clark Corporation has expanded its Health Management Program to spouses and retirees for use of an on-site exercise facility and participation in wellness classes. "K-C Kids Care" teaches fitness, nutrition, health habits and safety to children ages 5–8 and to their parents (Ham, 1989b).

**Small employers.** For small employers the burden to provide a comprehensive occupational health and safety program is particularly great, and it is complicated by the lack of universal health insurance. Rising costs make even the provision of minimal employee health insurance coverage extremely difficult. Twenty million of the nation's 34 million uninsured are employees or their dependents, and most of these work for small businesses. Up to 76 percent of employees in service industries, such as retail, eating and drinking establishments, are uninsured (Miller, 1991).

Small employers have specific needs. Not only do small employers need assistance in reducing insurance rates to offer health coverage to employees and their dependents, they often do not have the resources for an occupational health and safety program. *Healthy People 2000: National Health Promotion and Disease Prevention Objectives* describes the need for small employers to implement more programs on worker health and safety and back injury prevention and rehabilitation. Additionally, states are encouraged to improve occupational health and safety consultative services to small employers (USDHHS, 1991).

It is also more difficult for a small employer to participate fully in a modified duty program. Opportunities for job modification and reasonable accommodation may be limited because of the smaller number of job positions. The OHN can therefore work with the small employer to prioritize primary prevention activities, and to creatively problem solve modified duty when an injury does occur.

**Agricultural workers.** Of special interest are the estimated 6.5 million agricultural workers, recognizing that agriculture/farming is still one of the most dangerous occupations today (Cordes & Rea, 1991a). Difficulty in access to health care is well documented for rural communities. Small family farms and a migrant farm workforce often involve children and adolescents being exposed to potentially hazardous working conditions, such as exposure to pesticides, gases, agricultural grain and fiber dust, traumatic injuries and work related stress (Wakefield, 1990).

Challenges to meeting the health care needs of migrant farm workers include developing strategies to address transience, physical and social isolation, language and cultural differences (Meister, 1991). The OHN, in coordination with a rural hospital structure, can meet some of these needs. In fact, recent funding from the National Institute for Occupational Safety and Health (NIOSH) has targeted the high mortality rate of agricultural workers and the use of OHNs to develop prevention programs to decrease injury and illness (Cordes & Rea, 1991b).

## Research on Occupational Health Nursing Practice

With the changing economy, companies have increased their awareness of the dollars spent on health care, both for personal health care and for workers' compensation benefits. As companies look at ways to control health care costs, occupational health nurses are assuming increased responsibility for benefit design, primary care and health services management (Bey et al., 1990). Many of these efforts are focused at the secondary and tertiary prevention levels, for example, contracting for a capitated rate with a physical therapy provider. It is critical for the OHN, while designing quality and cost-effective health care treatment options for employees and their dependents, to maintain an emphasis on primary prevention of work related injury and illness, with an aim towards a safe work environment.

In more than 60% of companies, the occupational health nurse is the sole health care provider at the worksite. In 1978, recognizing that the OHN was the predominant provider of health care in small industry, NIOSH conducted a study on the costs and benefits of occupational health nursing. Four pairs of manufacturing facilities with less than 1000 employees were studied, with documented direct and indirect benefits for both employers and employees. NIOSH concluded that the occupational health nurse was cost-effective, especially in those industries with hazardous work processes and in those companies who had not already established cost-effective alternatives to the delivery of occupational medical care (NIOSH, 1980).

Additional surveys conducted by NIOSH, the National Occupational Hazard Survey in 1972 and the National Occupational Exposure Survey in 1981–83, described the health and safety conditions of the American workforce, and documented trends of worker access to on-site health services. In 1981–83, 3.8 percent of employers with less than 100 employees reported the presence of a health unit at the worksite, 32 percent of employers with 100–499 employees reported an on-site health unit, and 87 percent of industries with 500 or more employees reported an on-site health unit (USDHHS, 1988). Additionally, there was an increase in the access to nursing services on-site, a slight decrease in on-site physician services with an associated

increase of contractual agreements with off-site physicians noted over the study period (Pederson & Sieber, 1989). This data further documents the established and growing role of the occupational health nurse in meeting the health care needs of small and large employers.

The Tillinghast Report (1991), in discussing how best to respond to the workers' compensation crisis, notes employer ranking of the most effective cost control initiatives as:

- Safety programs (82 percent)
- Pre-employment screening (67 percent)
- Light duty programs (78 percent)
- Use of case managers (72 percent)

In addition, claims administration audits, litigation management, fee schedule compliance and other cost-control measures were reported. Of note, 87 percent of these employers report coordination with the group health program as one of the most effective cost-control initiatives. The occupational health nurse is well prepared to participate in these cost-control programs, especially with the national trend towards 24-hour care for both work- and non–work-related health conditions.

Large employers are very aware of what they need from on-site occupational health nurses. Lusk et al. (1988) found in their survey of Fortune 500 companies ($N = 173$) that 90% employed registered nurses, 61% employed safety engineers, 61% employed physicians, and 45% employed industrial hygienists, documenting the predominance of nursing expertise in large industries. This survey notes the four most frequent occupational health nursing activities as:

- Supervising the provision of nursing care for job related emergency and minor illness episodes (90 percent)
- Counseling employees regarding health risks (88 percent)
- Providing case management for employees with workers' compensation claims (67 percent) and
- Performing periodic health assessments (63 percent)

Lusk also evaluated desires of employers relative to future role functions of the OHN. The four most frequently reported desired activities were for the OHN to:

- Generate analyses on trends in health promotion, risk reduction, and health care expenditures (36 percent)
- Develop special health programs particular to the needs of the corporation (29 percent)
- Make recommendations for more efficient and cost effective operation of the health care department (29 percent) and
- Conduct research to determine cost effective alternatives to health care programs and services (29 percent)

In another study of a Fortune 500 company and its national sites (Bey et al., 1988), an analysis of the OHN role was studied from both the OHN ($N = 26$) and the manager ($N = 15$) perspective. Both nurses and managers ranked direct care of employees as the highest priority, followed by health education and counseling (#2 for occupational health nurses, #3 ranking for managers), medical management (#3 for occupational health nurses, #2 for managers), recordkeeping (#4), health promotion (#5) and environmental hazard recognition and control (#6). Of these functions, managers ranked direct injury/illness care, health education and counseling and medical management as nursing benefits to the company.

These studies document the current complex role of the OHN and the desires of large employers to have OHNs be part of the solution to the health care cost crisis.

Several recent studies have documented the growing trend for more on-site salaried health care providers to provide 24 hour care to employees, their dependents and retirees.

In 1984, the SAS Institute, a software research and development company in North Carolina, established a nurse practitioner managed health care center on-site for employees and covered dependents. The program offers quality primary care, identification of risk factors, early intervention and health education. A fitness center and child care services are located on site. Under the program, all employees and covered dependents can be seen for diagnosis and treatment of acute problems, referrals and counseling and a comprehensive physical with health appraisal. Emphasis is placed on early intervention, comprehensive care and wellness programs. Through the combination of a self-funded insurance fund and the on-site nurse managed primary care center, the company saved more than $791,000 in 1987 (Touger & Butts, 1989).

Northern Telecom, a telecommunications company in Tennessee, has taken a comprehensive, integrated approach to managing health care costs. This includes all levels of prevention, including a focus on occupational health and safety, and on-site primary care by nurse clinicians and nurse practitioners. Education of employees and dependents about self care and the appropriate use of health care is a priority. A case management/managed care approach monitors health care services, including workers' compensation claims. The work environment promotes healthy lifestyles, for example, a no smoking work environment. Cost data showed substantial savings particularly from the health promotion and primary care projects. In 1988, 42% of visits to the health center were for primary care reasons; 37% for occupational injury. Comparisons of costs for on-site occupational and primary care, when contrasted with comparable community rates, showed estimates of an annual net benefit of over $2.4 million (Dalton & Harris, 1991). Evaluation data also showed annual decreases in accident frequency and severity, with a 40% decline in lost work days. Additionally, there was a 50% improvement in internal safety audits.

# CHAPTER 27

# The Industrial Hygiene Program

by Maureen A. Huey, MPH, CIH

**Benefits of an Industrial Hygiene Program**

**Establishing an Industrial Hygiene Program**
*Written Program and Policy Statement* ▪ *Hazard Recognition and Evaluation* ▪ *Hazard Control* ▪ *Employee Training* ▪ *Record Keeping* ▪ *Employee Involvement* ▪ *Program Evaluation and Program Audit*

**Organizational Responsibilities**
*Medical Program* ▪ *Engineering* ▪ *Safety* ▪ *Purchasing* ▪ *Facility Manager* ▪ *Supervisor* ▪ *Employee* ▪ *Safety and Health Committee*

**Summary**

**Bibliography**

THE CHIEF GOAL OF AN INDUSTRIAL HYGIENE program is the prevention of occupational disease and injury through the anticipation, recognition, evaluation, and control of occupational health hazards. Providing a safe and healthful workplace benefits both employees and employers by improving health, morale, and productivity.

The industrial hygiene program comprises several key components: a written program and policy statement, hazard recognition, hazard evaluation and exposure assessment, hazard control, employee training, employee involvement, program evaluation, and record keeping. There is no established format, however, for the development and implementation of a program. Its form depends on a variety of factors such as the type and size of the organization, its management philosophy, the range of workplace hazards at the facility, and the resources devoted to the program. Small companies, for example, may not have a formal program but instead rely on insurance carriers or consulting agencies for their industrial hygiene needs. On the other hand, large corporations or government organizations often have comprehensive industrial hygiene programs.

The implementation of an effective industrial hygiene program depends on the commitment of senior management. Serious commitment is demonstrated when management is visibly involved and provides the resources and, most importantly, the authority necessary to carry out the program. The health and safety function should be given the same level of importance and accountability as the production function.

## Benefits of an Industrial Hygiene Program

Senior management's commitment to the industrial hygiene program is imperative for its success. To gain their support, both the social and economic benefits of the program must be made apparent. The following are some of the benefits that have been cited in well-established programs:

- They provide a place of employment in which employees are protected from all known occupational health hazards at the workplace.
- Compensable injuries or illnesses are reduced, thus lowering insurance premiums and associated medical and record-keeping costs.
- Productivity is usually increased by the improving working conditions. Improved working conditions reduce lost time from accidents and illnesses, reduce absenteeism, and improve morale and labor relations.
- Operating costs are reduced by anticipating and controlling potential occupational health hazards during the design phase of new projects.
- The Occupational Safety and Health Administration (OSHA) and other government regulations concerning industrial hygiene are quickly assessed and implemented.
- They provide a liaison with government regulatory agencies dealing with occupational health issues.
- They provide an important source of health and safety information and training for employees.
- They provide a liaison between the production areas and the medical department.
- They assist the medical department in determining whether a worker's medical condition is associated with a job-related health hazard.

## Establishing an Industrial Hygiene Program

### Written Program and Policy Statement

The purposes of a written industrial hygiene program are to provide a general statement of industrial hygiene policy, clearly delineate program authority and responsibilities, and establish standard operating procedures for the program components.

**Policy statement.** The written program should begin with a brief policy statement from and signed by the chief administrator of the organization. It should state the purpose of the program and require the active participation of all employees. The statement should reflect the following:

- The importance that management places on the health and safety of its employees.
- Management's commitment to occupational safety and health, which is demonstrated by its placing health and safety at the same level of authority and accountability as production.
- The company's pledge to comply with all federal, state, and local occupational safety and health regulations.
- The necessity for active leadership, direct participation, and the enthusiastic support of the entire organization.

**Goals and objectives.** The establishment of clear long- and short-range goals and objectives is vital to the development of an effective industrial hygiene program. These goals and objectives should also be part of the written program. They are often established by a committee, ideally a joint labor–management health and safety committee.

A goal is a desired outcome, whereas an objective is a specific activity or means of achieving a goal. Goals should be realistic and, when possible, measurable. For example, if ergonomics-related injuries are a problem, the goal may be to reduce the number of accidents by 25 percent within a 3-year period. The objectives/activities to achieve this goal could include establishing an ergonomics committee, providing ergonomic training for the committee and affected personnel, and selecting an ergonomics consulting firm to provide initial workplace surveys.

Goals and objectives should not be static—they should be evaluated and updated on a regularly scheduled basis (Table 27–A). The evaluation process may determine that the objectives are inadequate or that the goals are not well enough defined. In addition, as conditions change, there may be new problems to address, in which case new goals and objectives should be developed. The written program thus becomes a continually updated document.

**Standard operating procedures for program components.** Each program component should have a standard operating procedure that identifies what should be done, how it should be done, and who should do it. Besides effectively communicating the program to the rest of the organization, it also documents how the organization has identified and dealt with industrial hygiene problems. This can be very important information during an audit by an outside agency such as OSHA. Written procedures also provide measurable performance guidelines for those responsible for the implementation of the program.

**Table 27–A.** Summary of Criteria and Activities

| Program Element Activity | Activity | Measurement Criteria | Goal |
|---|---|---|---|
| Policy | Write, prepare/present for management acceptance | Is policy complete? Is policy understood and supported by management/employees? Does policy carry authority needed for implementation? | An accepted and working policy that clearly states the scope, responsibilities and authority of the program. |
| Education | New employee orientation Periodic information and education sessions Written safety and health guidelines Posting of dangerous areas Labeling of materials handled by employees | No. educational materials produced and distributed Increase in employee knowledge of safety and health issues Employee avoidance of hazards | Increased employee awareness of health and safety in the workplace. |
| Health hazard recognition | Plant survey Chemical inventory Process and equipment review Health hazard review procedures Process change review procedures | No. surveys Completion and procedure update Procedures and staff in place for review, etc. | Identify all present and potential hazards in the workplace. |
| Health hazard evaluation | Environmental monitoring (area, personal) Sample analysis Statistical analysis of data Biological monitoring Records of data Establishment of criteria | No. samples collected No. analyses performed Statistical significance of sample data Well-documented record-keeping system Established criteria for each stress | Measure and quantitatively evaluate stresses and hazards, determine their impact upon the work environment. |
| Health hazard control | Design and/or recommend administrative and engineering controls Procedural mechanism for implementing controls Procedural mechanism for including controls as a part of planning for new processes and changes in existing processes Administrative review of rejected procedures | Controls implemented and working Administrative procedures in place | Control or reduce to the lowest level all potential workplace hazards. |
| OSHA compliance | Review all present and future regulations, standards Determine level of compliance obtained via compliance inspections | No violations present Program positioned to comply with regulations | Complete compliance with all laws, regulations, standards, etc. |

(Printed with permission from Toca FM. Program evaluation: Industrial hygiene. *AIHAJ* 42.213–210, 1981.)

## Hazard Recognition and Evaluation

Hazard recognition is the identification of workplace occupational health hazards. These include chemical, physical, and biological hazards. The identification depends on the professional judgment of the industrial hygienist, which is based on information gathered during walk-through surveys, interviews with employees and management, and reviews of applicable documentation. Hazard evaluation is the determination of whether worker exposure to these environmental hazards is acceptable, or if engineering, administrative, or work practice control measures are necessary.

Various industrial hygiene systems have been developed to systematically and comprehensively identify and evaluate the occupational health hazards at a facility. Most recently, systems called exposure assessment strategies have been used.

The American Industrial Hygiene Association (AIHA), in their publication *A Strategy for Occupational Exposure Assessment,* has listed the following goals for an exposure assessment strategy:

- To assess potential health risks faced by all workers, to differentiate between acceptable and unacceptable exposures, and to control unacceptable exposures

- To establish and document a historical record of exposure levels for all workers and to communicate exposure-monitoring results to each worker

- To ensure and demonstrate compliance with governmental and other exposure guidelines

- To accomplish the above goal with efficient and effective allocation of time and resources

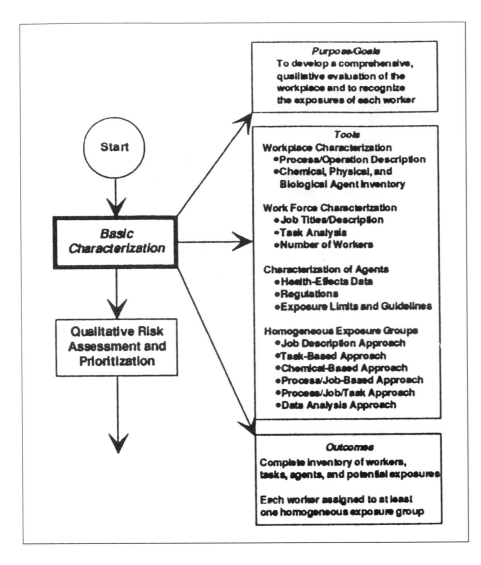

**Figure 27–1.** The occupational exposure assessment strategy—basic characterization component. (Reprinted with permission from the American Industrial Hygiene Association [from *A Strategy for Occupational Exposure Assessment.* Fairfax, VA: AIHA, 1991].)

The AIHA's publication contains a suggested approach for exposure assessments. It begins with a hazard recognition stage they call basic characterization (Figure 27–1), which has four major components:

■ *Workplace characterization* is a description of the processes and operations in the workplace, with particular attention paid to those areas with potential exposure to an environmental hazard (Figure 27–2).

■ *Work force characterization* groups and describes employees with similar work duties or job classifications (Figure 27–3).

■ *Agent characterization* is the construction of an inventory of environmental agents and includes a description of their potential adverse health effects, how they are used, how much is used, and their physical properties.

■ *Determination of homogeneous exposure groups* is a grouping of workers who are expected to have the same or similar exposure profiles to an environmental agent, based on the information gathered during the workplace, work force, and agent characterizations.

The evaluation of each hazard listed for a homogeneous group involves two stages. First, a subjective determination is made as to whether the exposure to each environmental agent listed for a homogeneous exposure group is low, moderate, high, or very high relative to an exposure limit. This determination is based on such factors as the frequency and duration of the exposure, estimated exposure level, and the severity of the health effects resulting from the exposure. Second, the exposures are monitored and the results are compared to the exposure limits. Exposures rated very high are monitored first, and those rated low are monitored last or not at all. Sampling plans should specify the number and duration of samples to be taken, in order to ensure that a true representation of employee exposure is obtained. Table 27–B illustrates a sampling strategy developed for a fungicide used in the lumber industry.

A written procedure for sampling methods is necessary to ensure that samples are collected in a proper and uniform manner. Procedures should include information on calibration, field use and maintenance of the equipment, and quality

---

### Sodium Chloride Production Plant

### Process Description

Chlorine gas is received directly by pipeline from an adjacent vendor plant; the gas arrives at a line pressure of 200 psig, passes through a pressure letdown valve, and enters T-1 column at 20 psig. 25% hydrogen peroxide is received in tank cars and is diluted with process water to 19% before being pumped to T-2 column. In Columns T-1 and T-2, chlorine and hydrogen peroxide go through a counterflow reaction to produce a 29% aqueous hydrochloric acid (HCl) solution; trace amounts of chlorine are vented to the atmosphere at T-2 column.

A 50% sodium hydroxide solution is received by pipeline from an adjacent vendor plant and is stored in S-2 tank before being pumped to R-1 reactor. From T-2 column, the HCl solution is pumped to R-1 reactor to react with the caustic and produce a 35% sodium chloride solution (brine). The brine is contacted with heated air in F-1 fluid bed dryer to make sodium chloride granules with a size range of 100 to 200 microns. The finished product is transferred to storage hopper H-1 via an airvey system.

The industrial grade sodium chloride is packaged in 50-pound bags and palletized in 1,000-pound loads. Each pallet is film-wrapped in an automatic shrink-film apparatus. A tow motor is used to move pallets from the bagging station to the shrink-film station and then to the warehouse. At the warehouse, the tow motor is also used to load the wrapped pallets into truck trailers.

The plant is currently producing about 50 million pounds per year of industrial grade sodium chloride; the plant operates about 300 days per year with the remaining time used for scheduled maintenance work. On a daily basis, the plant receives about 53 tons of chlorine gas, 120 tons of caustic solution, and 306 tons of hydrogen peroxide solution.

---

**Figure 27–2.** Process description developed for an example of an exposure assessment strategy for a hypothetical sodium chloride production facility. Development of such a workplace description is fundamental to completing the workplace characterization portion of the basic characterization step of the strategy. (Reprinted with permission from the American Industrial Hygiene Association [from *A Strategy for Occupational Exposure Assessment*. Fairfax, VA: AIHA, 1991].)

---

control. In addition, procedures should be developed to ensure that employees receive copies of the monitoring results and are afforded the opportunity to observe monitoring.

A periodic reevaluation of exposure assessments should be done to determine whether conditions have significantly changed. The frequency and scope of the reevaluation depend on the severity of the hazards. A new assessment should be done if a new process or potentially hazardous agent is introduced into the workplace.

## Hazard Control

If exposure levels are judged to be unacceptable, measures must be taken to reduce or eliminate the exposure. The bulk of the written industrial hygiene program concerns control measures including such programs as hearing conservation, respiratory protection, hazard communication, and ergonomics. These should be written in a concise manner, with responsibilities clearly assigned. They should be reviewed and approved by the safety and health committee.

## Employee Training

The industrial hygienist has become increasingly involved in employee training programs. Employees need information and training to become actively involved in protecting their health. Training has become a standard part of most OSHA regulations.

There are many helpful courses and publications available on the development of effective training programs. The following summary was taken from *Training Requirements in OSHA Standards and Training Guidelines* (OSHA, 1992).

*Determine if Training is Needed.* Determine whether a problem can be solved by training. Training can address lack of knowledge or incorrect knowledge, but cannot effectively address lack of motivation or attention to the job.

*Identify Training Needs.* Determine what is expected of the employee and what kind of training is needed to accomplish this. Consult with the safety committee and employees in the area to get their ideas to help make the training more effective and tailored to their needs.

## Sodium Chloride Production Plant

### Job Descriptions
### Operations Personnel

**Superintendent:** Spends about 10% of time in general process areas observing operations, checking equipment conditions, and supervising maintenance work; remaining time spent in office environments on administrative, supervisory, and planning activities.

**Engineer:** Spends about 35% of time in general process areas troubleshooting process problems, supervising maintenance work, and collecting industrial hygiene samples; remaining time spent on training, computer program development, and other office activities.

**Shift Supervisor:** Spends about 5% of time in general process areas checking on operations and investigating possible process problems; remaining time spent in control room areas overseeing operation of the acid, reactor, and dryer systems.

**Relief Operator:** Spends about 20% of time covering each of the shift supervisor, acid system operator, reactor system operator, and assistant operator job classifications; remaining time spent on various maintenance activities.

**Operator, Acid System:** Spends about 40% of time in the HCl production areas checking equipment, adjusting flows, and preparing equipment for maintenance; about 10% of time is spent collecting process samples and another 25% is spent in lab running analyses on all process samples; remaining time is spent in control room areas.

**Operator, Reactor System:** Spends about 60% of time in the reactor and fluid bed dryer areas checking equipment, adjusting flows, and preparing equipment for maintenance. About 10% of time is spent collecting process samples; remaining time is spent in control room areas.

**Assistant Operator:** Spends about 50% of time at bagging station loading product into 50-pound bags; another 25% of time is spent using tow motor to move pallets to and from the warehouse and to load product into truck trailers; about 5% of time is spent loading product directly from storage into hopper cars; remaining time is spent in control room areas.

**Electrical/Instrument Technician:** Spends about 5% of time in the HCl production areas maintaining and calibrating in-line chlorine analyzer; approximately 25% of time is spent in general process areas and switchgear room maintaining electrical equipment; about 50% of time is spent in control room or maintenance shop repairing or modifying process control instruments; remaining time is spent on office activities.

**Figure 27–3.** Sample job description for operations personnel in a hypothetical sodium chloride production facility. The descriptions were developed to fulfill the work force characterization requirements of the basic characterization step of the strategy and to be used as an approach for determining homogeneous exposure groups. (Reprinted with permission from the American Industrial Hygiene Association [from *A Strategy for Occupational Exposure Assessment.* Fairfax, VA: AIHA, 1991].) (*Continues*)

*Identify Goals and Objectives.* Instructional objectives tell the employee what is expected of them. Clear and measurable objectives must be established before training begins. Using specific, action-oriented language, the instructional objectives should describe the preferred practice or skill and its observable behavior.

*Developing Learning Activities.* Learning activities enable employees to demonstrate that they have acquired the desired skills and knowledge. The learning situation should simulate the actual job as closely as possible using participatory training techniques such as hands on work or opportunities to engage in case studies.

*Conducting Training.* The training should be presented so that its organization and meaning are clear to the employees. An effective program allows employees to participate in the training process and to practice their skills or knowledge.

*Evaluation of Program.* To make sure that the program is accomplishing its goals, an evaluation of the training is necessary. Methods of evaluation include student opinions surveys, supervisor's observations, and student tests.

*Improving the Program.* If the training did not give the employees the necessary level of knowledge and skills, then the training program must be revised. It may be necessary to repeat the steps in the training process.

**Pipe Fitter/Welder:** Spends about 30% of time in maintenance shop on welding work associated with process piping and equipment repairs; about 40% of time is spent in general process areas removing or replacing piping and process equipment; remaining time is spent in shop office or control room.

**Utility Mechanic:** Spends approximately 20% of time helping process operators with preparation of equipment for maintenance; about 50% of time is spent repairing equipment in shop; another 10% of time is spent in general process areas making minor repairs to process equipment; remaining time is spent in shop office or in control room.

### Personnel Roster

| Name | Company ID No. | Job Classification | Job Code |
|------|----------------|--------------------|----------|
| J. R. Smith | 20066 | Superintendent | 0062 |
| M. C. Jones | 10081 | Engineer | 0085 |
| | 10018 | Shift Supervisor | 0105 |
| | 10025 | Shift Supervisor | 0105 |
| | 10038 | Shift Supervisor | 0105 |
| | 10040 | Shift Supervisor | 0105 |
| | 10019 | Relief Operator | 0121 |
| | 10050 | Relief Operator | 0121 |
| | 10029 | Operator, Acid System | 0123A |
| | 10042 | Operator, Acid System | 0123A |
| | 10066 | Operator, Acid System | 0123A |
| | 10071 | Operator, Acid System | 0123A |
| | 10021 | Operator, Reactor | 0123R |
| | 10048 | Operator, Reactor | 0123R |
| | 10052 | Operator, Reactor | 0123R |
| | 10065 | Operator, Reactor | 0123R |
| | 10088 | Assistant Operator | 0129 |
| | 10096 | Assistant Operator | 0129 |
| | 10099 | Assistant Operator | 0129 |
| | 10100 | Assistant Operator | 0129 |
| | 10044 | Elec/Instr Technician | 0210 |
| | 10030 | Pipe Fitter/Welder | 0230 |
| | 10061 | Utility Mechanic | 0255 |
| | 10078 | Utility Mechanic | 0255 |

**Figure 27–3.** (*Continued*)

*Document the Training.* A written record is needed to document how identified training needs have been met. It should include attendance records, course outlines or lesson plans, student exams, and handout materials.

## Record Keeping

Industrial hygiene–related documentation must be maintained. The decisions made by industrial hygienists can have legal as well as regulatory consequences. OSHA 29 *CFR* 1910.20 mandates that exposure records must be maintained for at least 30 years. The documentation is needed to demonstrate that the work has been conducted in accordance with professional standards, and it may be useful for future industrial hygiene or medical evaluations. Documentation of other programs such as training, respiratory protection programs, and hearing conservation programs must be maintained as well. Many organizations have developed their own forms, record-keeping procedures, and data bases to efficiently handle the large amount of documentation that is generated.

## Employee Involvement

An effective industrial hygiene program includes a commitment by the employer to encourage employee involvement

**Table 27–B.** Proposed Strategy for Assessing Exposures

- Identify all work sites using antisapstain agent of interest.
- Ask managers at each work site to tally number of workers in each of five strata:
    —Graders or lumber pullers who handle wet wood
    —Elevator or forklift dip-tank operators
    —Others who handle wet wood
    —Maintenance workers who operate the fungicide supply system or maintain machinery downstream of treatment
    —Employees who handle dry treated lumber
- Randomly select 30 workers from complete population of each stratum.
- Contact each worksite whose workers have been selected for exposure measurement.
- Randomly select two measurement days within a 1-year period for each selected worker.

(Reprinted with permission from American Industrial Hygiene Association. Strategies for determining occupational exposure assessment. *AIHAJ* 55:447, 1994.)

in decisions that affect worker safety and health. In addition to employee participation on the safety and health committee, an employee complaint or suggestion program should be established that allows employees the opportunity to notify management of safety and health issues without fear of reprisal and also provides timely feedback on their concerns. Prompt reporting of any signs or symptoms of exposure to a hazardous agent must be encouraged so that such medical and workplace conditions can be quickly evaluated and, if necessary, remedied.

## Program Evaluation and Program Audit

Methods must be developed to periodically evaluate the effectiveness of the industrial hygiene program. Audits are commonly used to determine whether the elements of the program have been implemented in accordance with established procedures, and whether these procedures have been effective in achieving their goal. Auxiliary benefits include a reassessment of priorities and resources allocation, and an increased awareness of and commitment to the program by management.

An audit is usually requested by senior management and is done by a health and safety specialist or a team of specialists from outside the facility being audited. This ensures the objectivity of the auditor and also provides fresh insight into the program. The audit team is usually from the corporate or headquarters staff, but in some cases, an independent third party such as an insurance loss control representative or independent consultant is used. Self-audits, though not independent, can nonetheless be useful evaluation tools.

The scope of an audit depends on the time and resources devoted to it. For a small facility, a comprehensive audit of all industrial hygiene program components can be easily accomplished, whereas at a large facility it would be very difficult to complete. For this reason, larger organizations often concentrate their audits on high-priority items.

Auditors prepare for an audit by researching the requirements of the program components and developing a plan to evaluate compliance. Audit checklists are often developed to guide and focus the collection of information. Extensive lists are usually developed for each program component. In the interest of time, these are often sent in advance to the facility. This allows management time to collect the necessary written documentation, schedule interviews with key personnel, and, if necessary, ensure that certain processes or tasks of concern will be operational during the audit.

There are generally five phases to an audit. It usually begins with an opening conference with the management of the facility, during which the purpose, scope, and schedule for the audit is discussed. Then there is the information-gathering stage. Next, the information is analyzed, key facts confirmed, and contradictions resolved.

During this phase the auditor can usually generalize from specific situations to underlying program deficiencies. Then the auditors present their findings to management during a closing conference, at which time any remaining concerns can be discussed. Finally, a report of findings is issued.

Much of the value of the audit is lost if there is no established mechanism for follow-up, which can be accomplished with follow-up audits and/or by requiring the facility to develop written action plans and submit periodic progress reports.

# ■ Organizational Responsibilities

Organizational responsibilities for the program should be clearly defined. Industrial hygiene may be part of the safety department or another department, or it may be a department by itself. There must be a statement that clearly indicates where the industrial hygiene program gets its authority and to whom it reports.

The success of safety and industrial hygiene programs requires the cooperation of many organizations and groups. Figure 27–4 illustrates how each step of the industrial hygiene process of recognition, evaluation, and control requires the expertise of many other functional areas. The role of all areas need not be defined in the written program, but the roles of the main players must be in writing in order to avoid confusion and ensure efficient implementation of the program.

## Medical Program

Modern occupational health programs ideally are composed of elements and services designed to maintain the overall health of the work force and to prevent and control occupational and nonoccupational diseases and injuries. A large corporation may have a full-time staff of occupational health physicians and nurses, equipped with a model clinic; a small manufacturer, on the other hand, may need to rely on a nearby occupational health clinic.

Medical programs usually offer the following services:

- Health examinations
- Diagnosis and treatment
- Medical record keeping
- Medical or biological monitoring
- Health education and counseling
- Wellness activities

The industrial hygiene program should provide information on the working conditions in the facility to the medical department. To do an effective job, the health professional must have a good understanding of what is made, how it is made, the potential safety and health hazards associated with these manufacturing processes, and

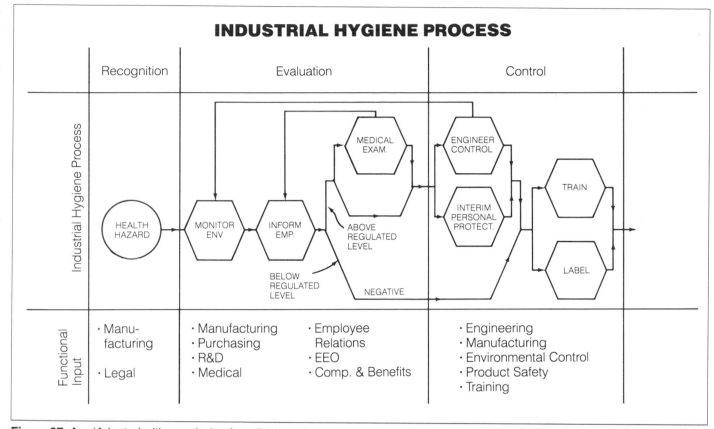

**Figure 27–4.**    (Adapted with permission from Bridge DP. Developing and implementing an industrial hygiene and safety program in industry. *AIHAJ* 40:255–263, 1979.)

the physical requirements of the various jobs. This information is necessary to adequately perform preplacement and periodic medical examinations, to detect conditions that might be work related, and to conduct health education programs.

The medical department works with the industrial hygiene department in developing adequate, effective measures to prevent exposure to harmful agents. They periodically examine employees who are working with or exposed to hazardous agents or materials, and, if warranted, they restrict employees from further exposure and notify the industrial hygiene staff of their findings. Maintenance of medical records associated with all medical examinations and findings is the responsibility of medical personnel.

## Engineering

Engineering professionals are involved in both the design and modification of manufacturing processes. Because these processes may introduce health and safety hazards into the workplace, engineers must coordinate their plans with the safety professional and the industrial hygienist. It costs much less to anticipate and eliminate a hazard in the planning stage than it does to manage it afterward. In cases where there is an existing staff hazard, the industrial

hygienist must work with the engineer to develop control methods to reduce or eliminate the hazard.

## Safety

The safety professional and industrial hygienist are concerned with the same goal: maintaining a safe and healthful workplace. Because safety programs tend to be older and more established than industrial hygiene programs, industrial hygiene is often part of the safety department.

The safety professional's main responsibility is to run an effective safety program. An effective safety program lends credibility and builds support for all health- and safety-related work at the facility. It also enhances the safety program's recognition of industrial hygiene issues and will work them into such safety activities as workplace inspections, accident investigations, and accident trend analysis and make appropriate referrals to the industrial hygienist. If industrial hygiene staffing is limited, safety professionals may have to accept responsibility for the implementation of the industrial hygiene at their facility.

## Purchasing

The purchasing department has the responsibility to ensure that only equipment and material approved by the industrial hygiene, safety, environmental, or other responsible reviewing organization are purchased. Purchasing should obtain material safety data sheets for all chemicals purchased.

## Facility Manager

Facility managers (also known by such titles as location managers, operations managers, and general managers) have the ultimate responsibility for the industrial hygiene program at their facilities; they must ensure that their facilities comply with applicable corporate policies and government regulations by providing the necessary resources and support that they need to be successful.

## Supervisor

The supervisor is a key person in the implementation and maintenance of safety and health requirements on a day-to-day basis. His or her responsibilities include setting a good example, ensuring that safety and health rules are followed, ensuring that employees are provided training concerning potential safety and health hazards and control measures associated with their jobs, ensuring that all necessary personal protective equipment is provided and used, ensuring that employees receive all required medical examinations, and promptly reporting any operation or condition that might present a hazard to employees.

## Employee

Employees have the responsibility to perform their work in a manner that ensures their own personal safety as well as the safety of fellow employees. Employees must notify their supervisor immediately of hazardous work conditions or work practices, observe all safety and health rules, properly use and maintain personal protective equipment and other safety devices, maintain their work area in a neat and clean manner, and immediately report all accidents and near-miss incidents.

## Safety and Health Committee

The safety and health committee provides a forum for securing the cooperation, coordination, and exchange of ideas among those groups involved in the safety and health program. It typically has three major functions: It examines company safety and health issues and recommends policies to management, conducts periodic workplace inspections, and evaluates and promotes interest in the program. It also provides a means of involving employees in the program.

Joint management–labor health and safety committees are often used if the employees are represented by a union. At the committee meeting, key industrial hygiene program matters should be discussed and policies formulated.

## ▌ Summary

To accomplish the goal of the prevention of occupational illness and injury, there must be an effective industrial hygiene program. It requires the cooperation of employees and all levels of management. The program consists of a written program and policy statement, hazard identification, hazard evaluation and exposure assessment, hazard control, employee training and involvement, program evaluation and audit, and record keeping. Each component must be periodically evaluated to determine its effectiveness.

## ▌ Bibliography

American Industrial Hygiene Association. *Industrial Hygiene Auditing; A Manual for Practice.* Fairfax, VA: American Industrial Hygiene Association, 1994.

Bridge DP. Developing and implementing an industrial hygiene and safety program in industry. *AIHAJ* 40:255–263, 1979.

Grimaldi JV, Simonds RH. *Safety Management.* Homewood, IL: Irwin, 1975.

Hawkins NC, Noorwood SK, Rock JC, eds. *A Strategy for Occupational Exposure Assessment.* Fairfax, VA: American Industrial Hygiene Association, 1991.

Lynch JR, Sanderson JT. Program evaluation and audit. *Industrial Hygiene Management.* Cralley LJ, Cralley LJ, Garrett JT, eds. New York: Wiley, 1988.

Nelson TJ, Holmes RS, Gordon T, et al. Workplace exposure assessment workshop: An integrated approach for the '90s. *AIHAJ* 54:633–637, 1993.

Occupational Safety and Health Administration. *Training Requirements in OSHA Standards and Training Guidelines.* OSHA 2254 (revised). Washington, DC: GPO, 1992.

Occupational Safety and Health Administration. *Ergonomics Program Management Guidelines for Meatpacking Plants.* OSHA 3123. Washington, DC: GPO, 1990.

Teschke K, Marion SA, Jin A, et al. Strategies for determining occupational exposures in risk assessments: A review and a proposal for assessing fungicide exposures in the lumber industry. *AIHAJ* 55:443–449, 1994.

Toca FM. Program evaluation: Industrial hygiene. *AIHAJ* 42:213–216, 1981.

CHAPTER **28**

# Computerizing an Industrial Hygiene Program

by Adrienne A. Whyte, PhD

**Information Systems**

**Advantages of Automated Information Systems**
*Source Data Collection ▪ Elimination of Duplication ▪ Improved Communications ▪ Data Standardization and Accuracy ▪ Improved Analytical Capabilities ▪ Cost Savings*

**Health and Safety Functions**
*Incident Management ▪ Workplace Conditions ▪ Environmental Agents ▪ Protective Measures ▪ Employee Health ▪ Employee Demographics and Job Histories ▪ Regulatory, Administrative, and Action Items ▪ Scheduling ▪ Standard and Ad Hoc Reporting ▪ Statistical Analysis*

**Development and Implementation of a Computer System**
*Understanding Needs ▪ Identify and Evaluate Systems ▪ Purchase and Customization ▪ System Development ▪ System Implementation ▪ Evaluating the System*

**System Software**

**System Hardware**

**System Communications**

**Summary**

**Bibliography**

INFORMATION TECHNOLOGY HAS CHANGED the way health and safety professionals do business. The widespread use of computers and the networks that link them and the availability of inexpensive, easy-to-use software have provided solutions to yesterday's analytical, record-keeping, and reporting challenges.

The passage of the Occupational Safety and Health Act (OSHAct) in 1970 was followed by the passage of environmental laws such as the Toxic Substances Control Act (TOSCA) and the Resource Conservation and Recovery Act (RCRA). In the late 1970s and early 1980s, hazard communication or right-to-know laws and regulations added to regulatory requirements with complex procedural, reporting, and record-keeping requirements for industry and, by executive order, government agencies. As safety, industrial hygiene, and medical surveillance requirements increased, so did the requirements for documentation, reporting, and analysis. However, the information technology infrastructure was not in place to support these new requirements and the resources of some occupational health and safety programs were strained.

## ▮ Information Systems

Today, computer systems are an integral part of the business environment, and health and safety professionals use them to help prevent work-related deaths, injuries, and illnesses, manage risks, and control losses.

The following are examples of how safety and health professionals use computers to support their activities.

- Collect and record environmental monitoring data in the workplace.
- Perform calculations, such as time-weighted averages, and model exposure scenarios.
- Compile, analyze, and produce reports, such as summary reports of workplace sampling results for chemical, physical, and biological agents.
- Produce follow-up reminders and manage work schedules.
- Correlate data on workplace exposures with employee medical records.
- Maintain the records needed to document regulatory compliance, such as training required by the OSHA hazard communication standard.
- Access data bases that contain information useful to health and safety professionals. Within minutes, a user of a data base can obtain lists of needed references, abstracts of articles, or properties of a substance used in a manufacturing process, to name only some of the information available on-line. A detailed listing of currently available data bases is included in Appendix A, Additional Resources.

Computers are tools that help health and safety professionals obtain the information necessary to make decisions, make their jobs easier and their efforts more productive, and help protect the organization and the employees (Figure 28–1).

**Figure 28–1.** Computers today facilitate management of industrial hygiene programs. Safety, health, and environmental professionals can use them to perform several tasks at once. (Courtesy IBM Corp.)

## ■ Advantages of Automated Information Systems

### Source Data Collection

Personal computers, especially hand-held and laptop models, are being used to enter data collected in the field, replacing the log books and worksheets that were once used to document workplace conditions and the details associated with air sampling. Time is saved and the data are usually more accurate.

### Elimination of Duplication

The time spent maintaining duplicate records can be reduced with an automated system. For example, health and safety, medical, human resources, and area supervision maintain records generated and maintained by the other departments. With a shared or distributed information system, each department can get access to the information it needs directly from the source files.

### Improved Communications

Local and wide-area networks have revolutionized the way businesspeople communicate. Electronic mail and file sharing capabilities facilitate communication within and between facilities. Health and safety personnel can send notifications to employees and their supervisors quickly and easily.

### Data Standardization and Accuracy

Standardizing the data input and the way in which data are organized is particularly important in organizations with many system users. When the same data are being input in the same way by each user, the analysis and decision-making capabilities of the data base are enlarged. For instance, accident and incident rates can be analyzed for all locations at the same time, rather than analyzing each location separately and then standardizing and entering the collective data for final analysis.

Accuracy of data can be vastly improved by customized entry screens, editing routines, and checks for the completeness of records. For instance, a computer system can alert the medical department that part of an OSHA-required physical examination was not conducted.

### Improved Analytical Capabilities

Compilation and analysis of data from various records consume professional time that could be better spent on program management. A good automated system performs analytical tasks quickly and provides flexibility in formatting data for reports. For example, time-weighted averages can be calculated and assigned to employee exposure records.

Trends in employee health can become apparent in time to prevent injuries and illnesses.

## Cost Savings

This advantage is probably the hardest to document, but automated systems can save organizations money above and beyond the costs of system development and operation. Increased employee productivity, decreased workers' compensation costs, and more effective program management can be achieved.

## ■ Health and Safety Functions

Occupational health and safety systems can support single functions, such as management of training records, or a comprehensive set of functions. This section is an overview of commonly used system functions and tools.

### Incident Management

Many organizations use automated systems to store, analyze, and report incident data. Data for all types of incidents are entered, such as occupational injury, property damage, near miss, and transportation. These data usually parallel the organization's first report and incident investigation forms. The system can produce individual incident reports; summary reports that categorize incidents by location, type, rate, severity, loss, and other factors; OSHA 200 logs; and analytical reports that pinpoint major causes and types of accidents within organizational subsets. For example, an organization might determine through use of its automated system that a disproportionate number of accidents are occurring in a particular operation, and then arrange for a health and safety inspection and evaluation of the operation.

### Workplace Conditions

Many systems manage safety, industrial hygiene, health physics, environmental and other sampling, audit, and inspection data to describe and quantify workplace conditions. Industrial hygiene sampling results can be entered and used to produce exposure profiles for each workplace. Also, if health and safety inspection findings, recommendations, target dates for completion, and completion dates are automated, a system can produce follow-up reports highlighting identified deficiencies that have not been corrected.

In comprehensive systems, workplace data can be correlated with employee health data to support health surveillance programs.

### Environmental Agents

Automated systems greatly facilitate environmental agent monitoring, including the preparation and maintenance of facility and corporate inventories of environmental agents, toxicology data, and material safety information on toxic agents in the workplace. These agents include chemical, biological, radiological, and physical hazards. Inventories of agents can be maintained by department, area, process, job, or any combination of these.

Right-to-know and hazard communication requirements make it mandatory that organizations identify agents in each workplace, train employees in their safe use, maintain material safety data sheets, and promptly respond to requests for information. Automated systems can facilitate compliance by maintaining agent inventories, training records, material safety data sheets, and information for right-to-know requests.

### Protective Measures

Computer systems can maintain information on measures taken to protect employee health and safety through engineering controls such as hoods for ventilation, administrative controls for health hazards, personal protective equipment assignment and fit-testing, and individual employee training and group training programs. These functions allow an organization to maintain the information needed to facilitate regulatory compliance and document the measures taken to protect employees.

### Employee Health

Monitoring employee health is a key function of comprehensive occupational health and safety systems. The protection of employees is the major goal of health and safety programs. This function is used primarily by occupational medicine specialists to manage and store records related to employee health surveillance. These records, shown in Table 28–A, result from both scheduled and unscheduled events. Scheduled health events include all physical examinations (preplacement, periodic, certification, and termination examinations and biological monitoring). Unscheduled

**Table 28–A.**  Occupational Medicine Records

Physical Examination Records
    Personal and Family Medical History
    Occupational History
    Physical Measurements (such as vision
      tests and blood pressure levels)
    Blood and Urine Laboratory Tests
    Clinical Evaluations and Diagnoses
    Audiometry Results
    Spirometry Results
    Electrocardiogram Results
    X-Ray Results
    Other or Special Test Results
Immunization Records
Clinic Visit Records
Occupational Injury and Illness Records
Disability Data
Workers' Compensation Records
Biological Monitoring Results
Sickness and Absence Records
Mortality Records

health events are those that cannot be anticipated. It is necessary to maintain a record of both scheduled and unscheduled health events to compile a historical profile of an employee's health.

As with other system functions, individual components of employee health programs can be automated. Hearing conservation program records are a common candidate. Noise monitoring, audiogram, and employee history data can be entered and stored. These data can be used to produce monitoring schedules, threshold shift evaluations, notices to employees, and summary reports for program management.

## Employee Demographics and Job Histories

In addition to maintaining health records, this function (or an independent but related function) must also maintain data that identify and describe individual employees and maintain a history of job assignments and locations. Employees are typically identified by name, social security number, employee number, or badge number. These normally are found in personnel information systems and can be downloaded to a health and safety system.

Job assignments and location tell where an employee works and what he or she does. These data represent a necessary bridge between employee health records and workplace and environmental records. They allow health professionals to correlate medical records with workplace records and are an integral part of comprehensive occupational health and safety surveillance systems.

## Regulatory, Administrative, and Action Items

Another function can provide information on regulatory requirements, including OSHA permissible exposure limits (PELs) or action levels, internally adopted standards, or such advisory guidelines as the American Conference of Governmental Industrial Hygienists (ACGIH) Threshold Limit Values (TLVs). This function also can be used to track ongoing events or problems requiring follow-up, such as hazard abatement plans and schedules.

## Scheduling

The computerized system can be used to schedule physical examinations, workplace monitoring or inspections, fit testing of personal protective equipment, training, or any other event documented in an occupational health and safety system. The schedules can be generated by system logic, or they can be generated by users and maintained by the system.

## Standard and Ad Hoc Reporting

Reporting is the most important function of an automated system. Standard reports are predefined and routinely used, and they are often preformatted. They can be prepared with one or few instructions (often through menu choices). Examples of typical standard reports are audit reports and lists of occupational injuries by workplace. All applications software packages should have a standard reporting function.

Ad hoc reports cannot be predefined or preformatted, but are designed by users for special analyses and reporting. There are several methodologies for producing ad hoc reports. Two of them, the question-and-answer and fill-in-the-blanks approaches, prompt the system user to name the variables for study, indicate how they are to be manipulated, and specify the format for the report. Another method requires the use of a simple computer language for queries.

Some applications have built-in capabilities for ad hoc reporting, often the same tools associated with the underlying data base management system. Multiple methods are often available, such as an easy query-by-example tool and a language such as industry-standard SQL. Most systems today offer export tools, allowing the user to specify a data set, create a file, and move that file into another software package, such as SAS, for analysis.

Whatever tools are used, query capabilities that are easy to learn and use are among the most important features of occupational health and safety systems.

## Statistical Analysis

Statistical analyses are necessary for many health and safety studies. Commercial statistical software packages are available for this function, but they often are separate from the occupational health and safety software. These are available on personal computers and mini and mainframe computers. It is usually an easy task to pass data from newer data base management systems to these packages for analysis. In addition, many packages are available for preparation of graphics from health and safety data systems.

# Development and Implementation of a Computer System

There are five major steps to the successful development and implementation of an automated system:

1. Understand the operations and needs
2. Identify and evaluate software and hardware meeting those needs
3. Purchase and customize a system or develop a system
4. Implement the system
5. Evaluate the system

## Understanding Needs

When considering a new information management system, conduct a *requirements study,* a careful assessment of the current and future needs that a system should fulfill. The requirements study should say why a system is needed, what system features will satisfy the needs, and how the system should be constructed.

To do this, the requirements analysis must address three subjects:

- *Needs analysis.* This is a high-level description of the reasons why the information system should be created—how the system will solve problems and why certain technical, operational, and economic constraints are the criteria for the system.

- *Functional specification.* This is a description of what functions the system must accomplish, the processes it will support, and the data it will manage. Functional specifications describe, from the user's perspective, what the system will do. Teams of potential users should work together in joint application design sessions to identify current and future business processes and business rules, develop a data model, determine input and output requirements, and identify other functional requirements, such as security and record access and performance.

- *Physical specifications and design constraints.* This is a detailed description of how the system is to be constructed and implemented, usually developed by systems analysts. What hardware, software, and network tools will be used to develop the system infrastructure? How will the programs work? How will interfaces with other data bases and instruments be accomplished?

Each of these subjects should be fully explored and documented during the requirements analysis. Collectively, these analysis components should provide all of the information needed to design or evaluate a system.

## Identify and Evaluate Systems

After the requirements and specifications for a system have been defined, there should be enough information to identify the alternatives capable of meeting the organization's needs. There are three common alternatives: purchase and customization of commercially available applications software, in-house development of software, and contracted development of software.

Many good occupational health and safety packages are commercially available. They come in every variety, from single-function packages to cross-functional, comprehensive systems. They operate in many hardware and software environments, such as *FoxPro* on personal computers and *Oracle* on a range of computers.

The advantage of purchasing software is the time and expense that can be saved. The disadvantage is that modifications and work practice adaptations are often necessary when commercial applications are used. Carefully study the software features before purchasing. See the Bibliography and Appendix A, Additional Resources, for additional information.

Software programs available from the National Safety Council that have many applications for safety and health professionals include *ACCUSAFE for Windows* OSHA injury and illness reporting software, *Training Tracker* safety and health training manager, revised *1994 NIOSH Lift Guide*, *ErgoKnowledge* office ergonomics CD-ROM, *REASON* root cause analysis software, *EPA/OSHA Right-to-Know* software, and *Confined Space Entry Tracking*.

The major advantage of developing customized software, either in-house or through a contractor, is the exact fit afforded between the requirements and the system. The users can determine their most efficient work practices and model the system to their own needs. Often, users and developers work hand in hand to develop prototypes of system functions, and users can tweak the system during development. Many new development tools and methods are available, making custom software development a viable alternative.

When comparing the available alternatives, remember two cardinal rules. First, the alternatives should be judged on the basis of criteria established in the requirements analysis. A commercially available application cannot be evaluated until the requirements for the system have been thoroughly defined. Second, life-cycle costs should be used to compare the costs of the considered alternatives. Life-cycle costs are all costs associated with the system's design, development, implementation, and operation over the expected useful life of the system. The life of a system is closely tied to the amount of time the software, hardware, and network configuration can be expected to perform without major modification.

## Purchase and Customization

A software system may have to be modified before it meets all requirements. When evaluating systems, look at the ease with which they can be changed. Once it is purchased, treat the package as a prototype. Let system users work with it and define its deficiencies, then make any required changes. There may have to be several rounds of modifications before all users are completely happy with the system.

## System Development

If developing a system, participate in the design and development process. At this stage, obtain assistance from a systems analyst. The intricacies of system development depend on the size and complexity of the system and the hardware and software environment in which it will be built. Naturally, a small PC-based system for management of one or more unrelated functions requires less effort and systems expertise than a comprehensive health and safety system.

The first step is to establish the system structure. This activity, often called general design, consists of defining the subparts of the system and the interfaces between them. In the detailed design step that follows, the precise algorithms or system processes and data structures are defined. Detailed design may involve several iterations. Then system production begins; this involves programming or using system tools to define file structures and report formats, testing individual pieces, integrating various pieces, testing the system, writing documentation, and, finally, evaluating system performance.

Many organizations build and implement large systems in phases. This can work well if an overall plan for total system design is prepared and followed during the phased development. Too often, groups have built system modules without considering total system requirements and necessary connections between modules, thus leading to project failure.

## System Implementation

Implementing the new information system should be a joint effort between system developers and users. Although there are many activities associated with implementation, the most important one is user training. This training should establish realistic expectations among users about the system's capabilities and requirements. Any changes required in record-keeping procedures and potential transitional difficulties should be thoroughly explained.

User training should not be left to programmers or systems analysts. Ideally, it should be carried out by user representatives. If a system has been purchased, some training will, by necessity, have to be conducted by the supplier. However, a core of system users should be trained by the supplier; then, if possible, these users should train others.

Training should be conducted as part of the everyday work routine. Naturally, some general training sessions must be held, but the most effective type of training is on-the-job training, with health and safety professionals learning to use terminals for data input and output in the context of their actual work.

Training should be timed so there is no gap between training and actual use of the system. When something entirely new is learned, most people need to practice it to remember it. The longer the gap between training and system use, the less learning retention there will be.

Follow-up to training also is important. Periodically, the system manager should return to the users to determine whether their expectations of the system are realistic and whether they are using the system correctly. Error rates should be studied. If they are too high, the reasons should be identified and the users retrained. System success depends just as much on follow-up training as it does on initial training.

An effective user's manual is important to user training. During the development process for any information system, system documentation is written. Much of this documentation is for the use of systems analysts and programmers. However, a user's manual is also prepared, and great care must be taken in its preparation. Both the system developers and user representatives must participate in writing the user's manual. It must be simple to understand and direct in its instructions. If it is properly written, it will be a permanent, valued reference for all users. If it is not well-prepared, it either will not be used or will promote errors.

Every system should have a manager or data base administrator. This person has ultimate responsibility for the day-to-day operation of the system, including management of system security, supervision of the data base content and data quality, problem solving, coordination of changes to the system, archiving of data, and planning for future needs and applications. The system manager should know the application well and work closely with users to make sure the system meets user expectations.

## Evaluating the System

After an information system has been implemented, it should be evaluated annually. Estimating an information system's value to an occupational health and safety program can be difficult for the same reasons that conducting a cost-benefit analysis is difficult. However, evaluation is an important follow-up, and the following factors should be considered:

- Completeness
- Reliability
- User acceptance
- Costs
- Improved availability of information
- New capabilities

If problems are identified in any but the last area, corrective steps should be taken. Problems can result from human, hardware, or software elements of the system.

# System Software

Personal computers are ideal tools for information management because they are easy to use and relatively inexpensive. Most companies make personal computers available to managers for information and data management, word processing, electronic mail, and other office automation activities.

*Word processing,* an integral part of most office environments, is the ability to create, edit, format, save, and print documents. The most popular packages, such as Microsoft *Word* and *WordPerfect,* are easy to use and loaded with features, such as spreadsheet tools, drawing, charting, and integration with other applications.

Word processors are perfect tools for creating letters, memos, notifications, reports, newsletters, presentations, and other written communication. Users can incorporate charts and other graphics, giving their documents and presentations a professional appearance. They can create, store, and use mailing lists for correspondence. Users can also manage their documents with indexing and text retrieval capabilities.

The most popular word processors can be purchased as single applications or within integrated office automation packages that include electronic mail, calendar, work group scheduling, and other management tools. Many word processors have multiple versions and operate in either DOS or Windows on IBM PCs and compatibles and on Macintosh computers.

*Electronic mail* allows users to compose, send, save, and forward information. Sending a memo to one or every person in a department can be done easily. Many health and safety professionals use electronic mail today to send notices of appointments, work restrictions, and occupational injuries and illnesses.

*Spreadsheets,* such as Microsoft *Excel* and *Lotus 1–2–3,* are popular applications with professionals who manipulate numbers and produce mathematical models. They perform quick calculations and are ideal for industrial hygiene use.

*Data base management systems* are an integral component of most contemporary occupational health and safety systems. It is important to understand the distinction between applications and data base management software. Today, most applications, including health and safety systems, are written with data base management systems. All data base management systems allow the storage, sorting, and retrieval of information in useful ways. They have tools for building files and reports, and they offer programming languages or commands for building complex programs. A data base management system looks like a generic computer facility until its tools are used to build a specific application.

*Applications* are programs or sets of programs that provide tailored menus, data entry screens, and reports to support specific functions such as occupational health and safety record keeping and analysis. The commercially available packages for health and safety are applications software, and most of them were built with data base management systems.

Many different data base management systems are available. Some health and safety professionals use them to develop automated functions on personal computers. Data base management systems are available for all types of computers; some are easier to use than others. If an organization plans to develop its own applications software, the resources in the organization's information center or data processing department should be investigated to determine what is available and whether there is any development support.

Popular data base management systems for personal computers include Microsoft's *FoxPro* and *Access* and Borland's *Paradox* and *dBase*. These products provide easy-to-get-started capabilities for novice users and sophisticated features for programmers. Other data base management systems operate in a variety of computing environments, such as Oracle, which runs on personal, mini, and mainframe computers. Today many systems support client–server environments, meaning that different computers work together in a network to support an application.

## System Hardware

The types and placement of computers in an occupational health and safety information system must be determined during the requirements study. Today, there are many alternatives, ranging from large mainframe computers to personal computers, and information processing can be accomplished on one or many of these.

Many health and safety departments use personal computers for information management. Personal computer technologies are evolving at a rapid pace. Every year, vendors offer greater speed, more memory, and better connectivity. New operating systems permit users to run multiple tasks simultaneously, and new networking capabilities let several PCs share disks, programs, and printers.

Because rapid advances in personal computer technologies have occurred and prices have decreased, many companies have developed occupational health and safety systems relying solely on personal computers, and some use local area networks to link the PCs.

Personal computers also can be used in system configurations that dedicate the PCs to local information management and tie them into central mainframe or minicomputers for long-term storage, cross-functional analyses, and corporate data base management. This type of system configuration distributes system tools and mimics the decentralized organizational structure of many corporations. It is ideally suited to the seemingly opposing needs for local records management and control and corporate data base management.

Many factors influence an organization's choice and location of computer hardware for its information system, including currently available hardware; corporate standards for computer hardware, telecommunications, and software; and the information requirements of system users. All system options and constraints should be studied during the requirements study so an economical, efficient, and integrated system can be developed.

## System Communications

Many telecommunications options are available to professionals who want to send, retrieve, or access information stored on other computers. Most large companies have already established local and wide-area networks that link computers within and across sites.

Local area networks (LANs) tie computers together so that they can easily share information. The network is local in the sense that the computers are generally in the same building or within a few thousand feet of each other. Wide-area networks (WANs) also tie computers or LANs together, but usually the elements of the WAN are geographically distant. LANs and WANs are often bridged to create enterprise networks, appearing to the user as one big corporate network.

A great deal of communication today occurs over regular telephone lines using modems and fax technologies. Modems allow their users to connect to other computers with modems and share files and data. Software packages usually manage the communications. Fax cards in computers and stand-alone fax machines have revolutionized the way we send documents.

Many health and safety professionals use applications that operate on networks. Because a network can connect two or more computers, each computer has the power to work on its own and the ability to use information and programs from other computers connected to the network. For example, in a client–server environment, each user has a client PC that can store and operate the programs for entering and retrieving data. The data are stored on the server computer, which can be a PC, mini, or mainframe computer. The data base on the server is accessible to all authorized users, but the processing occurs on each user's PC—an efficient use of each computer. Networks make this transparent to the user.

In addition to the benefits of information sharing, networking computers permits the sharing of devices. For example, several different computer users can share a laser printer, color plotter, or modem. This sharing of equipment saves money.

Similarly, a network enables a number of users to access the same software. This is useful when network versions of popular software are available. The LAN permits the establishment of an information network, integrating departments in useful ways, while effortlessly increasing communications.

Network development, management, and administration requires specialized expertise in network protocols, topologies, software, and equipment. The health and safety professional wishing to set up a network should bring in experts to plan and deploy the network and ensure that those responsible for maintenance are properly trained. Large companies have departments dedicated to this purpose.

It is a good idea to assign someone in health and safety to administer the network. This person assigns passwords and access rights, writes log-in scripts, trains people to use the network, and makes sure the network is ready to serve the users.

## Summary

The passage of occupational safety, health, and environmental regulations mandated meticulous documentation, reporting, and analysis for industry and government agencies. The widespread use of computers, linking networks, and inexpensive, easy-to-use software has provided solutions to these record-keeping and reporting challenges.

## Bibliography

*Best's Loss Control Engineering Manual.* Oldwick, NJ: A.M. Best, 1993.

Dobin D, ed. *Microcomputer Applications in Occupational Health and Safety.* Chelsea, MI: CRC-Lewis, 1987.

Helander MG, ed. *Handbook of Human/Computer Interaction.* New York: Elsevier, 1989.

Klonicke DW. Loss control in the computer room. *Prof Safety* (April):17–20, 1983.

Miller E, O'Hern C. Microcomputers can make it work for you. *Safety & Health* 35:28–32, 1987.

National Fire Protection Association, Batterymarch Park, Quincy, MA 02269.
*Fire Protection Handbook,* 17th ed., 1986.
Standards:
*Electronic/Data Processing Equipment,* NFPA 75.
*Halogenated Fire Extinguishing Agent Systems—HALON 1301, Fire Extinguishing System,* NFPA 12A.
*Protection of Records,* NFPA 232.

National Safety Council. Computers and information management. In *Accident Prevention Manual for Business & Industry, vol. 1: Administration & Programs.* Itasca, IL: National Safety Council, 1992.

## PART 7

# Government Regulations and Their Impact

# CHAPTER 29

# Government Regulations

Revised by Gabriel J. Gillotti, PE

**The Occupational Safety and Health Administration**
*Major Authorities, Functional Areas, and Responsibilities* ▪ *Major Duties Delegated by the Secretary of Labor* ▪ *General Duties and Obligations* ▪ *Key Provisions*

**State Plans**

**OSHA Standards**
*Categories* ▪ *Standards Development*

**Hazard Communication**

**Enforcement of the OSHAct**
*Highlights*

**OSHA Field Operations**
*Field Inspection Reference Manual (FIRM)* ▪ *OSHA's Reorganization*

**OSHA Industrial Hygiene Inspections**
*Preinspection Planning* ▪ *Opening Conference* ▪ *Walk-Through Inspection* ▪ *Collecting Samples* ▪ *Closing Conference* ▪ *Employer's Occupational Health Programs* ▪ *Compliance Programs* ▪ *Issuance of Citations* ▪ *Engineering Controls* ▪ *Contest of OSHA Citations*

**National Institute for Occupational Safety and Health**
*Responsibilities* ▪ *Testing and Certification* ▪ *Research* ▪ *Training* ▪ *Recommendations for Standards* ▪ *Field Investigations*

**Other U.S. Government Regulatory Agencies**
*Mining Enforcement and Safety Administration* ▪ *Federal Mine Safety and Health Amendments Act of 1977* ▪ *Environmental Protection Agency*

**Summary**

**Bibliography**

[*Editor's note:* **This information is current as of June 1995. However, as we go to press, congressional legislation has been introduced that may substantially change both the mandate and function of the federal safety and health agencies.**]

Before 1970, GOVERNMENT REGULATIONS of safety and health matters were largely the concern of state agencies. There was little uniformity of application of codes and standards from one state to another and almost no enforcement proceedings were undertaken against violators of those standards. Some states adopted as guidelines the Threshold Limit Values (TLVs) for exposure to toxic materials as recommended by the American Conference of Governmental Industrial Hygienists (ACGIH). However, enforcement of those guidelines was minimal.

The federal government had some safety and health standards for its contractors and for the stevedoring industry. Enforcement of those standards rested with the Bureau of Labor Standards in the U.S. Department of Labor. Although there were thousands of federal contractors, inspection and enforcement activities were restricted by the U.S. Department of Labor's limited budget and staff.

In 1970 and 1977, Congress enacted two new safety and health laws. These legislative efforts continue to have a significant impact on industrial hygiene activities in the United States. These laws are as follows:

- Public Law 91-596, December 29, 1970: the Occupational Safety and Health Act of 1970, popularly known as OSHAct
- Public Law 91-173, November 9, 1977: the Federal Mine Safety and Health Act of 1977

# ■ The Occupational Safety and Health Administration

The Occupational Safety and Health Administration (OSHA) came into official existence on April 28, 1971, the date the OSHAct became effective. This organization was created by the Department of Labor to discharge the responsibilities assigned to it by the Act. (See Chapter 30, History of the Federal Occupational Safety and Health Administration.)

## Major Authorities, Functional Areas, and Responsibilities

The OSHAct grants the secretary of labor the authority to promulgate, modify, and revoke safety and health standards; to conduct inspections and investigations and to issue citations, including proposed penalties; to require employers to keep records of safety and health data; to petition the courts to restrain imminent danger situations; and to approve or reject state plans for programs under the act.

The secretary of labor's authority regarding federal agencies includes the right to inspect agency worksites based on compensation data and to issue notices of violation when appropriate. Annual reports are filed with the respective agency heads, citing the deficiencies and positive elements of the agency's program. Under Executive Order 12196, the secretary must ensure that all federal agencies comply with OSHA's standards.

The act authorizes the secretary to have the Department of Labor train personnel in the duties related to their responsibilities under the act and, in consultation with the U.S. Department of Health and Human Services (DHHS) (formerly the Department of Health, Education, and Welfare), to provide training and education to employers and employees. The secretary and his or her designees are authorized to consult with employers, employees, and organizations regarding prevention of injuries and illnesses. The secretary of labor, after consulting with the secretary of the DHHS, may grant funds to the states to identify program needs and plan development, experiments, demonstrations, administration, and operation of programs. In conjunction with the secretary of the DHHS, the secretary of labor is charged with developing and maintaining a statistics program for occupational safety and health.

## Major Duties Delegated by the Secretary of Labor

In establishing the Occupational Safety and Health Administration, the secretary of labor delegated to the assistant secretary for occupational safety and health the authority and responsibility for the safety and health programs and activities of the Department of Labor, including responsibilities derived from the following legislation:

- Occupational Safety and Health Act of 1970
- Walsh–Healey Public Contracts Act of 1936, as amended
- Service Contract Act of 1965
- Public Law 91-54 of 1969 (Construction Safety Amendments)
- Public Law 85-742 of 1958 (Maritime Safety Act)
- National Foundation on the Arts and Humanities Act of 1965

The delegated authority includes responsibility for organizational changes, for coordination with other officials and agencies with responsibilities in the occupational safety and health area, and for contracting.

At the same time, the commissioner of the Bureau of Labor Statistics was delegated the authority and given the responsibility for developing and maintaining an effective program for collection, compilation, and analysis of occupational safety and health statistics, providing grants to the states to assist in developing and administering the statistics programs, and coordinating functions with the assistant secretary for occupational safety and health.

The solicitor of labor is assigned responsibility for providing legal advice and assistance to the secretary and all officers of the Department of Labor in the administration of statutes and executive orders relating to occupational safety and health. In enforcing the act's requirements, the solicitor of labor also has the responsibility for representing the secretary in litigation before the Occupational Safety and Health Review Commission and, subject to the control and direction of the attorney general, before the federal courts.

The Department of Labor regulations dealing with OSHA are published in Title 29 of the *Code of Federal Regulations (CFR)* as follows:

- 29 *CFR* Part 1910—General Industry Standards
- 29 *CFR* Part 1915—Shipyard Standards
- 29 *CFR* Part 1917—Marine Terminal Standards
- 29 *CFR* Part 1918—Longshoring Standards
- 29 *CFR* Part 1926—Construction Standards

## General Duties and Obligations

OSHAct sets out two duties for employers and one for employees. The general duty provisions are as follows:

- Each employer shall furnish to each employee a place of employment free from recognized hazards that cause or are likely to cause death or serious physical harm to the employee.

- Each employer shall comply with occupational safety and health standards under the act.

- Each employee shall comply with occupational safety and health standards and all rules, regulations, and orders issued pursuant to the act that are applicable to his or her own actions and conduct.

The significance of the General Duty Provision (Section 5[a]) is that it authorizes the enforcement of a recognized industry safety or health standard when identified hazards are not covered by an existing OSHA standard. Only violations viewed as serious may be cited under the general duty clause. This interpretation of the general duty clause for providing a safe and healthful working environment adds new dimensions to the protection of employee health.

## Key Provisions

Some of the key provisions of the act are as follows:

Assure, insofar as possible, that every employee has safe and healthful working conditions.

Require employers to maintain accurate records of exposures to potentially toxic materials or harmful physical agents that are required, under the various safety and health standards, to be monitored or measured, and inform employees of the monitoring results.

Provide for employee walkaround or interview of employees during the inspection process.

Provide procedures for investigating alleged violations at the request of any employee or employee representative, issuing citations, and assessing monetary penalties against employers.

Empower the secretary of labor (through the Occupational Safety and Health Administration) to issue safety and health regulations and standards that have the force and effect of law.

Provide for establishment of new rules and regulations for new or anticipated hazards to health and safety (Section 6[b] of OSHAct).

Establish a National Institute for Occupational Safety and Health (NIOSH), with the same right of entry as OSHA representatives, to undertake health studies of alleged hazardous conditions and to develop criteria to support revisions of health standards or recommendations to OSHA for new health standards.

Provide up to 50/50 funding with states that wish to establish state programs that are at least as effective as the federal program in providing safe and healthful employment (Section 18 of OSHAct).

## State Plans

As provided for under Section 18 of OSHAct, a state agency can assert jurisdiction under state law over safety and health

if a state plan that meets the criteria set forth in Section 18(c) is submitted for approval by federal OSHA.

As of this date, there are 23 approved state plans as well as plans from Puerto Rico and the Virgin Islands. Federal OSHA officers submit periodic written reports to the assistant secretary of labor addressing the quality of state performance as measured by established criteria. The data are gathered and analyzed on an ongoing basis, dialogue is carried on, and meetings are held routinely with state representatives to gather data and resolve problems identified by the measures.

After the plan is approved, the administering agency must promulgate standards that are at least as effective as (ALAEA) the federal standards, not only at the time of approval but on a continuing basis. Although most states adopt the federal standards verbatim, some modify and expand the federal standards to make them more applicable to the state's industries. State compliance policies must be ALAEA federal penalty-determining formulas and inspection procedures.

To ensure that the level of protection across states is consistent with the intent of OSHAct, the regulations also provide for the filing of complaints against a state program's administration (CASPA). Any person can call on federal OSHA to investigate any state program deficiency that may render the state program less effective than the federal program. The filing of a CASPA is a rare occurrence but it acts as a monitoring tool along with the performance measures.

There is a provision in the statute for withdrawing a deficient state plan approval, but there has not been adequate cause to do so to date.

Recently when a disaster occurred in one state, the program deficiencies that allegedly contributed to the incident in the form of inadequate enforcement were identified and discussed by the responsible federal and state OSHA officials and a satisfactory resolution was implemented. The federal staff assumed jurisdiction in the state for a limited time, but the state officials succeeded in increasing their staff and modifying their enforcement policies until both parties agreed that the probability of such an incident recurring had been reduced substantially.

## OSHA Standards

Health standards are promulgated under the OSHAct by the Department of Labor with technical advice from NIOSH. A review of OSHA's standard setting process will be helpful in understanding how regulations are derived.

Most of the safety and health standards now in force under OSHAct for general industry were promulgated 30 days after the law went into effect on April 28, 1971, as 29 *CFR* Part 1910 of Department of Labor regulations (Title 29). They represented a compilation of material authorized by the act from existing federal, state, and consensus standards (ANSI and NFPA). These, with some amendments, deletions, and additions, remain the body of standards under the OSHAct.

The act prescribes procedures for use by the secretary of labor in promulgating regulations. It is of special interest that the 1968 ACGIH Threshold Limit Values for exposures to toxic materials and harmful agents have been adopted in the regulations and have the effect of law. Although procedures are given for measuring exposure levels to specific materials and agents in the standards promulgated by the Department of Labor, professional skills and judgments are still required in applying the intent of the many aspects of the act.

## Categories

The OSHA standards consist of the following categories:

**Design standards.** Examples of these detailed design criteria are the ventilation design details contained in Section 1910.94 of the initial standards.

**Performance standards.** Such standards are the Threshold Limit Values (TLVs) of the ACGIH, which are contained in Section 1910.1000. A performance standard states the objective that must be obtained and leaves the method for achieving it up to the employer.

**Vertical standards.** A vertical standard applies to a particular industry, with specifications that relate to individual operations. Section 1910.261 (Subpart R) of the initial standards is in this category—it applies only to pulp, paper, and paperboard mills.

**Horizontal standards.** A horizontal standard is one that applies to all workplaces and relates to broad areas, such as Sanitation (1910.141) or Walking and Working Surfaces (Part 1910 Subpart D).

## Standards Development

The development of standards is a continuing process. NIOSH provides information and data about health and safety hazards, but the final authority for promulgation of the standards remains with the secretary of labor.

Section 6 of the OSHAct defines how safety and health standards are to be set. The secretary of labor promulgates standards "based upon research, demonstrations, experiments, and such other information as may be appropriate. In addition to the attainment of the highest degree of health and safety protection for the employee, other considerations shall be the latest scientific data in the field, the feasibility of the standards, and experience gained in this and other health and safety laws. Whenever practicable, the standard promulgated shall be expressed in terms of objective criteria and of the performance desired" (Section 6[b][5]).

There is a mechanism in the act (Section 6[c]) by which the labor secretary can promulgate emergency standards if he or she believes the evidence supports it. Emergency health standards have been promulgated for asbestos, carcinogens, and acrylonitrile. Following the time period required by the act, these have then been followed by the public rule-making process with promulgation of final standards.

As mentioned earlier, the first health standards were the 8-hour time-weighted average (TWA) values of air contaminants from the 1968 ACGIH TLV list; they now have the force of legal requirements. Guidance for specific sampling strategies, medical surveillance, and protective measures were lacking. However, Section 8(c)(3) of the OSHAct requires employers to measure contaminants, maintain records, and notify employees of overexposures and corrective action to be taken for all future health standards.

Because there has been a significant effort in the Department of Labor to increase the rate at which health standards are promulgated, NIOSH has prepared a number of criteria documents for use as a basis for various health standards. In turn, OSHA has proposed some health standards that include the areas of action described in Section 8(c)(3).

For reference purposes, the indices of OSHA Health Standards 1910, Subpart G—Occupational Health and Environmental Control, and Subpart Z—Toxic and Hazardous Substances, are given in Table 29–A. Other standards of interest to health professionals are 1910.20 *Access to Employee Exposure and Medical Records,* 1910.120 *Hazardous Waste Operations and Emergency Response,* and 1910.132 *Personal Protective Equipment for General Industry.*

**Action level.** Of interest to the industrial hygienist is the action level concept. In 1976, OSHA defined the action level as typically one-half the permissible exposure limit (PEL). Where exposures reach or exceed the action level, additional requirements apply, including medical surveillance and a full air-monitoring program. Exposures to an airborne concentration above the PEL trigger still further requirements, including reduction of exposures to (or below) the PEL by means of engineering controls supplemented by work practice controls, use of specified respirators, and use of other appropriate protective clothing and equipment.

**Employee protection.** OSHA decided as a policy matter that the action level, which triggers the measurement requirements, be set below the PEL to better protect employees from overexposure. OSHA reasoned that this method was the most reasonable approach to a recurring problem, that is, how to provide the maximum employee protection necessary with the minimum burden to the employer. Thus, where the results of employee exposure measurements demonstrate that no employee is exposed

**Table 29–A.** Part 1910—Occupational Safety and Health Standards

*Subpart G—Occupational Health and Environmental Control*

| | |
|---|---|
| 1910.94 | Ventilation |
| 1910.95 | Occupational noise exposure |
| 1910.96 | Ionizing radiation |
| 1910.97 | Nonionizing radiation |
| 1910.98 | Additional delay in effective date |
| 1910.99 | Sources of standards |
| 1910.100 | Standards organizations |

*Subpart Z—Toxic and Hazardous Substances*

| | |
|---|---|
| 1910.1000 | Air contaminants |
| 1910.1001 | Asbestos |
| 1910.1002 | Coal tar pitch volatiles; interpretation of term |
| 1910.1003 | 4-nitrobiphenyl |
| 1910.1004 | alpha-naphthylamine |
| 1910.1005 | Reserved |
| 1910.1006 | Methyl chloromethyl ether |
| 1910.1007 | 3,3'-dichlorobenzidine (and its salts) |
| 1910.1008 | bis-chloromethyl ether |
| 1910.1009 | beta-naphthylamine |
| 1910.1010 | Benzidine |
| 1910.1011 | 4-aminodiphenyl |
| 1910.1012 | Ethyleneimine |
| 1910.1013 | beta-propiolactone |
| 1910.1014 | 2-acetylaminofluorene |
| 1910.1015 | 4-dimethylaminoazobenzene |
| 1910.1016 | N-nitrosodimethylamine |
| 1910.1017 | Vinyl chloride |
| 1910.1018 | Inorganic arsenic |
| 1910.1025 | Lead |
| 1910.1027 | Cadmium |
| 1910.1028 | Benzene |
| 1910.1029 | Coke oven emissions |
| 1910.1030 | Bloodborne pathogens |
| 1910.1043 | Cotton dust |
| 1910.1044 | 1,2-dibromo-3-chloropropane |
| 1910.1045 | Acrylonitrile |
| 1910.1047 | Ethylene oxide |
| 1910.1048 | Formaldehyde |
| 1910.1050 | Methylenedianiline |
| 1910.1200 | Hazard communication |
| 1910.1450 | Occupational exposure to hazardous chemicals in laboratories |
| 1910.1499 | Sources of standards |

to airborne concentrations of a substance in excess of the action level, employers are exempted from major provisions of the particular standard.

A duty to measure employee exposure and provide medical surveillance only when the employee exposure was equal to or greater than the PEL was rejected by OSHA as not providing sufficient protection for the exposed employees. Among other things, such a scheme would not protect employees from overexposure because the employer would have no way of knowing when airborne concentrations of a regulated substance approached the TWA. It is not possible to ensure that all exposures are within the permissible limits simply because sampling was done when an employee's exposure was at the PEL.

**Other requirements.** It has been determined, therefore, that three key duties should be triggered when an action

level is reached—exposure measurement, medical surveillance, and employee training. All three actions are considered necessary by OSHA before employee exposure reaches the PEL. It is important to initiate measurement procedures periodically to monitor whether levels are approaching the PEL. One must do so to ensure that employee exposure does not exceed it. Similarly, employees should be screened for preexisting medical conditions and trained in suitable precautions against dangerous properties of the substance when there is some chance that their exposure will become significant.

OSHA has claimed that an alternative to establishing action levels would be to require medical and measurement procedures at any level of exposure, no matter how low. This alternative would burden employers unnecessarily because they would be required to implement medical and measurement provisions even where concentrations were so low that they presented no health problem.

The action level concept will, no doubt, continue to be a subject of discussion between the Department of Labor and practicing industrial hygienists. Some hygienists argue that a new permissible exposure level is created by arbitrarily setting 50 percent of the PEL as the action level. Some maintain that the blanket concept should not be imposed unless there is clear evidence of toxicity and there should be a differentiation between toxic substances and irritating substances. Others have stated that the action level approach used by OSHA is overly conservative because it assumes a single lognormal distribution for all cases and uses only one measured 8-hour TWA to estimate the mean of the distribution of other 8-hour TWAs.

Exposures at or just above the action level are not citable by OSHA, but failure to take the specified actions at the action level is citable.

## ■ Hazard Communication

The Hazard Communication Standard, 29 *CFR* 1910.1200, is one of OSHA's most significant standards. This standard was promulgated as a final rule on November 25, 1983, and contains three important compliance dates. The first, November 25, 1985, was the date on which all chemical manufacturers, importers, and distributors were required to label chemical shipping containers, assess chemical hazards, and provide material safety data sheets (MSDSs) to recipients of their chemicals.

The second compliance date, May 26, 1986, affected all employers covered by the standard under manufacturing division Standard Industrial Classification (SIC) codes 20 through 39. Beginning on this date, all covered employers were to have had a written and operating hazard communication program that provides the necessary information and training to affected employees, including chemical container warnings, MSDSs, and other warnings.

On August 24, 1987, OSHA expanded the scope of the hazard communication standard to include the nonmanufacturing sector. As of the third compliance date, May 24, 1988, all employers were to comply with the standard.

The intent of the hazard communication standard is to provide employees with information about the potential health hazards from exposure to workplace chemicals. The objective of the training is to provide employees with enough information to allow them to make more knowledgeable decisions with respect to the risks of their work and to impress on them the need for safe work practices. The standard requires the following employee training measures:

- Explanations of the requirements of the standard
- Identification of workplace operations where hazardous chemicals are present
- Knowledge of the methods and observations used to detect the presence of hazardous workplace chemicals
- Assessment of the physical and health hazards of those chemicals
- Warnings about hazards associated with chemicals in unlabeled pipes
- Descriptions of hazards associated with nonroutine tasks
- Details about the measures employees can take to protect themselves against these hazards, including specific procedures
- Explanation of the labeling system
- Instructions on location and use of material safety data sheets (MSDSs)
- Details on the availability and location of the hazardous material inventory, MSDSs, and other written hazard communication material

The standard applies to any chemical known to be present in the workplace that employees may be exposed to under normal conditions of use or may be exposed to in a foreseeable emergency. Pesticides, foods, food additives, cosmetics, distilled spirits, certain consumer products, and hazardous wastes are all covered under other federal legislation and, therefore, are exempt from this OSHA standard.

Numerous state and local governments have promulgated similar legislation, also known as right-to-know laws. These laws differ in some ways from the hazard communication standard.

In states with a federal OSHA enforcement program, the federal hazard communication standard preempts the state right-to-know law. This preemption was challenged once the standard went into effect, but the standard has not been modified as a result.

To date, the most frequently cited sections of the hazard communication standard are those dealing with the lack of a written hazard communication program, lack of a training program, and lack of labels on hazardous chemical containers.

The important role of the industrial hygienist in ensuring compliance with hazard communication legislation is obvious from initial risk assessment through employee training and MSDS interpretation.

# ■ Enforcement of the OSHAct

The secretary of labor is the principal administering officer of the OSHAct. OSHA is authorized to conduct inspections and, when alleged violations of safety and health standards are found, to issue citations and, when necessary, to assess penalties.

## Highlights

- OSHA schedules inspections on a priority system: First, in response to fatalities and multiple (three or more) hospitalization incidents and imminent danger situations; second, in response to employee complaints; third, random inspections of high hazard industries; fourth, follow-up inspections.
- OSHA compliance officers may enter the employer's premises without delay to conduct inspections and usually without advance notice (Section 8[a]); however, if the employer refuses entry, a search warrant may be requested.
- OSHA's right to inspect includes records of injuries and illnesses, including certain medical records.
- OSHA compliance officers who find conditions of imminent danger can only request, not demand, shutdown of an operation. If shutdown is refused, the compliance officer notifies employees of the hazard and the Department of Labor may seek court authority to shut down the operation.
- Criminal penalties can be invoked only by court action and in extreme cases (usually willful violations leading to death).
- An appeal system has been set up under which employers and employees can appeal certain OSHA actions to the independent Occupational Safety and Health Review Commission (OSHRC) (Section 12).

# ■ OSHA Field Operations

## *Field Inspection Reference Manual (FIRM)*

The *Field Inspection Reference Manual (FIRM)* was issued in September 1994 by OSHA as a major revision to what was previously referred to as the *Field Operations Manual*

(FOM). This manual contains general instructions and policies on field compliance operations. The OSHA *Technical Manual* sets forth the technical industrial hygiene practices and procedures used by OSHA personnel. This section summarizes these procedures and gives background about OSHA functions at the regional and local levels.

Health compliance operations involve several technical and professional disciplines possessed by industrial hygienists, safety engineers, and safety specialists. Therefore, the processing of all health inspections and citations requires close coordination between industrial hygienists, engineers, and local, regional, and national staff. In addition, nonagency assistance, either from outside consultants or from other agencies such as NIOSH, may be required. (NIOSH is discussed later in this chapter.)

**Responsibilities.** The national OSHA office, through its technical and analytical units, coordinates the technical aspects of health programming among the regions. An industrial hygienist in each regional office is responsible for coordinating the technical aspects of the health program within the region. This responsibility includes, but is not limited to, providing guidelines for inspections, evaluating and assisting in contested cases, and guidance in using technical equipment in accordance with criteria provided by the national office.

*The Area Director.* This official administers the field compliance program in the designated geographic area. Each area director designates an industrial hygiene supervisor, who is responsible for the technical aspects of the health compliance program and for recommending health inspection priorities in his or her area.

In state plan states, area directors and staffs are involved in monitoring the activities of the state staffs using performance measures for significant state plan activities. Simultaneously, the staffs in those offices enforce federal standards in workplaces not covered by the state plan, such as the maritime industry and military bases. *All* workers are covered in state plan states, including public sector, state, and local employees not covered in federal jurisdiction states.

Health inspections are conducted in industries in accordance with priorities outlined earlier. A health inspection can be either a complete survey of a particular workplace for all health hazards, or a special survey such as an accident investigation. Some inspections require a team effort because they involve more than one specialty. In such cases, the area director designates a team leader to coordinate the efforts.

*Compliance Officers.* The officers doing inspections are provided cross-training in either safety or health, depending on their job classification, so that they are better able to recognize potential safety or health hazards.

Safety officers trained to recognize and evaluate health hazards are expected to collect information for possible referral to an industrial hygienist. Health referrals are incorporated into the regular inspection schedule.

Health complaints are investigated by an industrial hygiene compliance officer trained to recognize health hazards and evaluate conditions. Complaints involving the appropriateness of unusual medical testing or questionable results of medical findings are discussed with the regional industrial hygienist and the technical and analytical assistance unit.

## OSHA's Reorganization

As of this writing, OSHA's headquarters, regional offices, and area offices are reorganizing and reducing staff. The goals are to serve all of OSHA's constituents more efficiently, to improve agency productivity, to reduce the number of middle managers, and to reduce staff size as part of the governmentwide 250,000-employees cutback.

Region IX of OSHA (Arizona, California, Hawaii, and Nevada) was the first region to implement the changes in the organization. The new units, collectively, are continuing to fulfill the OSHA mission but the emphasis and the balance of effort and staff are different.

One unit, the Analysis and Evaluation Group, is responsible for ensuring that each of the four states of Region IX performs according to federal benchmark fiscal and performance criteria. That group also assists the states in their effort to achieve their mission.

The Enforcement and Investigation Unit is responsible for conducting inspections and for investigating accidents in the private and federal sectors.

The Program Planning and Support Unit is responsible for providing assistance in planning, providing equipment, allocating and tracking budgets, and handling personnel matters for the entire regional organization.

The Voluntary Programs and Outreach Unit has as its mission the responsibility to provide technical information over a toll-free phone line or in writing, working with federal agencies and employees with their efforts to comply with federal standards, encouraging voluntary compliance by seeking out Voluntary Protection Program (VPP) candidates, and providing on-site consultation to any organization that cannot be served by the State 7(c) Consultation Services.

The remaining nine regions, with their field offices and the Washington headquarters office, are proceeding to design and implement their reorganization. Each has a uniqueness that warrants different approaches in their reorganization effort, so not all offices will mirror Region IX. However, the new Congress has endorsed all of these efforts.

# ■ OSHA Industrial Hygiene Inspections

An industrial hygiene inspection is conducted by an industrial hygienist and often is complex and time-consuming. The essential elements of a visit are preinspection planning, opening conference, walk-through inspection, sample collection, and closing conference. These guidelines are given to all OSHA industrial hygienists.

## Preinspection Planning

The OSHA industrial hygienist should become familiar with the particular industry and general process information and size of the facility. The industrial hygienist should also review appropriate standards and select sampling methods. If the inspection is a referral visit to a workplace previously visited by a compliance officer, the industrial hygienist should review all information contained in the previous inspection reports. Based on both experience and specific study, he or she should select the necessary field instrumentation. The hygienist can then prepare instruments and equipment according to standard methods of sampling and calibration.

## Opening Conference

The instructions state that on entering the establishment, the OSHA industrial hygienist presents identification credentials. An opening conference with facility management is arranged to discuss the purpose and scope of the inspection. If the employees are represented by a labor union, they must be notified of OSHA's presence and invited to participate in the inspection. At the beginning of inspection, the industrial hygienist usually requests a complete process flow diagram or facility layout or, if no layout chart is available, he or she can ask that a sketch be made to identify the operations, distribution of equipment including engineering controls, and approximate layout of the facility.

A brief examination is made of all required records kept at the establishment, such as the nature of any injuries or illnesses shown on OSHA form no. 200, Record-keeping Log. Record-keeping requirements can be discussed at the closing conference after the inspection has been made or after sampling results have been analyzed. At this time, preliminary information is gathered about the occupational health program.

## Walk-Through Inspection

A walk-through inspection is required for all health inspections regardless of whether the establishment was previously inspected. The main purpose of the walk-through inspection is to identify potential workplace health hazards; during the walk-through, the industrial hygienist becomes familiar with work processes, collects information on chemical and physical agents, and observes workers' activities. The industrial hygienist obtains information concerning raw materials used, intermediates (if any), and final products. Estimated amounts of substances present in the facility and a complete inventory are also obtained. In addition, the hygienist requests a list of raw materials received at the loading dock. He or she also checks for hazardous physical agents present in the facility, such as noise and excessive heat. He or she observes work activity throughout the facility, but concentrates particularly on potential health hazard areas.

The approximate number of workers in each area is written on the sketch of the facility. The industrial hygienist observes and records the general mobility of the workers and indicates whether they are engaged in stationary or transient activities. Existing engineering controls are marked on the facility layout or sketch. Ventilation measurements are made at strategic locations in the duct system and recorded on the facility sketch.

Information usually is requested from the facility manager concerning a preventive maintenance program for engineering controls. The industrial hygienist keeps alert for any imminent dangers during the walk-through inspections and takes appropriate action if necessary. Photographs can be taken to document the survey. Employee interviews are encouraged at this phase of the inspection.

In a small facility, health hazards sampling can be initiated after the opening conference. A final sampling schedule can be prepared using the information collected.

## Collecting Samples

Representative jobs should be selected and personal sampling devices prepared. Operations with the highest expected exposure should be monitored first. The sampling program should be planned to follow industrial processes as closely as possible in order to keep information obtained in a logical sequence.

The OSHA industrial hygienist determines compliance with air quality and noise standards based on one or more days of full-shift, TWA concentration measurements. These findings are tempered by the hygienist's professional judgment of the data and conditions at the facility.

The TWA concentration must be determined for 8-hour exposures. Sampling devices monitoring full-shift exposures must operate for a minimum of 7 hours. This implies that the concentration is calculated using the air volume sampled during the shift for a time greater than 7 hours for some chemicals. Spot samples must be taken throughout the work shift to represent at least 7 hours of exposure and to represent periods of exposure for TWA calculations. When the actual work shift exceeds 8 hours, the TWA can be calculated using the results of 7- or 8-hour sampling period and separate samples taken to determine any additional exposure. This value is compared with the standard for compliance determination.

Materials in Table Z–1 of OSHA Standard 1910.1000 preceded by a *C* have maximum peak ceiling limits to which an employee can be exposed; these values are never to be exceeded. Generally, a 15-minute sampling period should be applied to ceiling measurements (except for imminent danger situations where immediate escape from the atmosphere is necessary).

A minimum number of samples must be taken. These can be a single 8-hour sample for full-shift assessment, several spot samples that represent on a TWA a full-shift assessment, or one 15-minute sample for a ceiling assessment. The measurements must exceed the sum of the allowable limit and a calculated margin of error before a citation can be issued.

The criteria for determining full-shift TWA concentrations have been developed because legal considerations require a degree of certainty that the standard has been violated when a citation is issued. A citation shall not be issued unless the measured level exceeds the calculated upper confidence level of the permissible level based on a single day's sampling results. Measurements below the confidence limit indicate that overexposure may have occurred.

All sampling equipment must be checked and calibrated (in accordance with standard procedures described in the *OSHA Technical Manual*) before sampling. The sampling for both physical and chemical contaminants in the environment must relate to the worker's exposure, unless otherwise specified in a standard. In order that measurements represent, as far as possible, actual exposure of the employee, the measurements for air pollutants should be taken from the employee's breathing zone; measurements for noise exposures should be taken near the ear.

## Closing Conference

Because the industrial hygienist may not have the results of the environmental measurements at the end of the inspection or during the closing conference, a second closing conference may be held; for this, a telephone call or a letter can be used in place of a personal visit. Employees or their representatives are also informed of the inspection results at that time. If the results indicate noncompliance, the alleged violations can be discussed at the second conference, along with abatement procedures and methods of control.

## Employer's Occupational Health Programs

Information on the following aspects of the employer's occupational health program is gathered during the inspection for evaluation. These aspects may be discussed during the closing conference and later considered in relation to the standard requirements and also as evidence of good faith when penalties are being proposed.

**Monitoring program.** Detailed information concerning the industrial hygiene program is obtained. Especially valuable data that are asked for include number and qualifications of personnel and availability of necessary sampling and calibration equipment, ventilation-measuring equipment, and laboratory services.

**Medical program.** The industrial hygienist should determine whether the employer provides employees with pre-employment and regular medical examinations as required by certain OSHA standards. The medical examination protocol is reviewed to determine the extent of the medical examination.

**Education and training program.** A special check is made to evaluate the company's efforts to comply with the Hazard Communication Standard (1910.1200).

**Record-keeping program.** The employer's record-keeping program is checked, including types of records, how long they are maintained, and the accessibility of these records to employees.

## Compliance Programs

The industrial hygienist determines the employer's implementation of engineering and administrative controls and checks appropriate equipment and preventive maintenance. A specific engineering control should be evaluated for effectiveness in its present application.

**Work-practice and administrative controls.** Control techniques include the redesign of or modification of equipment, isolation of hazardous operations, rotation of employee job assignments, use of personal protective equipment, and sanitation and housekeeping practices. A detailed description of such controls should be obtained. It is essential that work practice controls and the education program be implemented simultaneously because the overall effectiveness of such practices is enhanced by employees' knowledge of their exposures.

**Protective devices.** The industrial hygienist determines whether protective devices are effectively used in the facility. A detailed investigation of the personal protection program includes a determination of compliance with 29 *CFR* 1910.134, Respiratory Protection. Special emphasis is given to the sanitary condition and maintenance of the equipment and to the use of training programs on selection and use of protective devices. The program must be available as a written procedure. See Chapter 22, Respiratory Protection, for more information.

**Regulated areas.** Some standards require the establishment of regulated areas, where concentrations exceed the PELs. Sampling is used to determine the limits of areas that must be regulated because employee exposure is expected to be greater than prescribed levels. The industrial hygienist makes sure that regulated areas meet the following standards:

- They must be clearly identified and known to affected employees.
- Regulated areas shall be demarcated and segregated from the rest of the workplace in a manner that minimizes the number of people who will be exposed to a chemical. The regulated area designations must be maintained according to the criteria of the standard.
- Daily rosters of authorized personnel entering and leaving the area must be maintained. Summaries of such rosters are acceptable.

**Emergency procedures.** Some standards provide for specific emergency procedures to be followed when handling certain hazardous substances. Company procedures should be checked for the inclusion of potential emergency conditions in the written plan, explanation of such emergency conditions to the employees, training schemes for the protection of affected employees, and delegation of authority for the implementation of the operational plan in emergency situations.

**Evaluation of sampling data.** OSHA industrial hygienists use professional judgment when evaluating the data and conditions to confirm the sampling results. Sampling results should reasonably correlate with each other. For example, an occupation that is expected to be dusty should have higher dust concentrations measured than those expected to be less dusty. Sampling results must reasonably correlate with previous measurements made by OSHA, the company, and others (taking into account variations due to weather, production, breakdowns, and the like). The sampling results must take into account accuracy of the instrument. The industrial hygienist's record must state that all sampling was performed according to OSHA standard methods.

## Issuance of Citations

The industrial hygienist uses OSHA guidelines when classifying violations. Here are some examples.

The workplace may be found to be in compliance with OSHA standards. In this case, no citations are issued or penalties proposed.

Violations may be found in the establishment. In that case, citations may be issued and civil penalties may be proposed. In order of significance, these are the types of violations or conditions normally considered on a first inspection.

**Imminent danger.** A condition where there is reasonable certainty a hazard exists that can be expected to cause death or serious physical harm immediately or before the hazard can be eliminated through regular procedures. If the employer fails to abate such conditions immediately, the compliance officer, through his or her area director, can go directly to the nearest federal district court for legal action as necessary.

**Serious violations.** A serious violation has a substantial probability that death or serious physical harm could result and that the employer knew, or should have known, of the hazard. An example is the absence of point-of-operation guards on punch presses or saws. A serious penalty may be adjusted downward based on the employer's good faith, history of previous violations, and size of the business.

**Willful violations.** A willful violation exists where the evidence shows either an intentional violation of the act or indifference to its requirements.

The employer commits an intentional and knowing violation under either of the following conditions:

- An employer representative is aware of the requirements of the act or the existence of an applicable standard or regulation and is also aware of a condition or practice in violation of those requirements.
- An employer representative is not aware of the requirements of the act or standards but is aware of a comparable legal requirement (such as state or local law) and is also aware of conditions in violation of that requirement.

The employer commits a violation with plain indifference to the law under any of the following conditions:

- Higher management officials are aware of an OSHA requirement applicable to the company's business but make little or no effort to communicate the requirement to supervisors and employees.
- Company officials are aware of continuing compliance problems but make little or no effort to avoid violations (an example is repeated issuance of citations addressing the same or similar conditions).
- An employer representative is not aware of any legal requirement but is aware that a condition or practice is hazardous to the safety or health of employees and makes little or no effort to determine the extent of the problem or to take the corrective action. Knowledge of a hazard may be gained from insurance company reports, safety committee or other internal reports, the occurrence of illnesses or injuries, media coverage, or, in some cases, complaints of employees or their representatives.
- In particularly flagrant situations, willfulness can be found despite lack of knowledge of a legal requirement or a hazard if the circumstances show that the employer would not place importance on such knowledge even if he or she possessed it.

It is not necessary that the violation be committed with a bad purpose or an evil intent to be deemed willful. It is sufficient that the violation was deliberate, voluntary, or intentional as distinguished from inadvertent, accidental, or ordinarily negligent.

The determination of whether to issue a citation for a willful or repeated violation often raises difficult issues of law and policy and requires the evaluation of complex factual situations.

**Criminal/willful violations.** An employer who willfully violates any standard, rule, or order, if that violation caused death to any employee, will, on conviction, be punished by a fine or by imprisonment for not more that six months, or both. If the conviction is for a violation committed after a first conviction, the punishment will be a fine or imprisonment for not more than one year, or both.

In order to establish a criminal/willful violation OSHA must prove that the employer violated an OSHA standard. A criminal/willful violation cannot be based on violation of Section 5(a)(1). OSHA must also prove that the violation was willful in nature. See discussion of *willful* above.

**Repeated violations.** An employer may be cited for a repeated violation if that employer has been cited previously for a substantially similar condition and the citation has become a final order. Generally, similar conditions can be demonstrated by showing that in both situations the identical standard was violated. In some circumstances, similar conditions can be demonstrated when different standards are violated.

**Repeated versus willful.** Repeated violations differ from willful violations in that they may result from an inadvertent, accidental, or ordinarily negligent act. If a repeated violation also meets the criteria for willfulness, but not clearly so, a citation for a repeated violation is normally issued.

**Repeated versus failure to abate.** A failure to abate situation exists when an item of equipment or condition previously cited has never been brought into compliance and is noted at a later inspection. However, if the violation was not continuous (i.e., if it had been corrected and then recurred), the subsequent occurrence is a repeated violation.

**Egregious citations.** Cases under consideration for treatment as egregious must be classified as willful (as described above) and meet one of the following criteria:

- The violations resulted in a worker fatality, a worksite catastrophe, or a large number of injuries or illnesses.
- The violations resulted in persistently high rates of worker injuries of illnesses.
- The employer has an extensive history of prior violations of the act.
- The employer has intentionally disregarded its safety and health responsibilities.
- The employer's conduct taken as a whole amounts to clear bad faith in the performance of his or her duties under the act.
- The employer has committed a large number of violations so as to undermine significantly the effectiveness of any safety and health program in place.

**Other-than-serious violations.** Other-than-serious violations are those that have a direct relationship to job safety and health but probably would not cause death or serious physical harm, such as a tripping hazard. A nonserious penalty may be adjusted downward depending on the severity of the hazard, employer's good faith, his or her history of previous violations, and the size of the business. Congress has told OSHA that, exclusive of serious violations, it must find more than 10 other violations before any penalty can be imposed.

**De minimis.** A de minimis violation is a condition that has no direct or immediate relationship to job safety and health.

If respirators and other personal protective equipment are not properly fitted or are not worn and the affected employee is exposed to a toxic agent above the PEL, a citation will be issued, classified as serious.

Employers may be cited for an other-than-serious violation if, for example, they have not established written operating procedures governing the use of respirators, have not trained and instructed employees in their proper use, or have not regularly cleaned and disinfected the respirators even though such respirators are properly fitted and worn.

Sometimes, when appropriate personal protective equipment is being used properly, citations are issued for failure to use administrative or engineering controls if the industrial hygienist believes that such controls are feasible. Generally, such citations will be other-than-serious unless available data indicate that the personal protective equipment, although properly fitted and worn, is not effective in fully reducing exposure to acceptable limits as required by the standard.

OSHA has technical support units and compliance organizational units in its regional and national offices, with staffs that provide guidance to the field compliance officers. The goal is to establish and maintain uniform compliance activities across the nation so that hazardous conditions are not cited differently by different inspectors. These support groups help determine which engineering controls are feasible and give this information to OSHA industrial hygienists to help them judge whether an item is serious or nonserious.

During an inspection, the industrial hygienist carefully investigates the source or cause of observed hazards to determine whether some type of engineering or administrative control (or combination) may be applied that would significantly reduce employee exposure. In order to issue a citation, an OSHA representative need not show that the controls would reduce exposure to the limits prescribed by the applicable standard.

## Engineering Controls

Engineering controls include any procedure, other than administrative controls or use of personal protective equipment, that reduces exposure at its source or close to the employee's breathing or hearing zone. Proper work practices and personal hygiene facilities are defined as engineering controls. For a particular engineering control to be feasible, a general technical knowledge about the materials or methods available or adaptable to the specific conditions must exist. There must be a reasonable possibility that these materials and methods will reduce employee exposure to violative conditions of noise, dust, or other substances.

## Contest of OSHA Citations

If an employer disagrees with a citation or proposed penalty, he or she can request an informal meeting with the area director to discuss the case.

If the employer decides to contest the citation, the act contains a specific appeal procedure, guaranteeing full review of the case by an agency separate from the Department of Labor. That agency is the independent Occupational Safety and Health Review Commission (OSHRC), which has no connection with the U.S. Department of Labor.

The OSHRC was created to adjudicate enforcement actions initiated under the act when they are contested by employers, employees, or representatives of employees.

The commission's functions are strictly adjudicatory; however, it is more of a court system than a simple tribunal, for within the OSHRC there are two levels of adjudication. All cases that require a hearing are assigned to an OSHRC judge, who decides the case. Each such decision is subject to discretionary review by the three members of the OSHRC on the motion of any one of the three.

Employers may contest a citation, a proposed penalty, a notice of failure to correct a violation, the time allotted for abatement of a violation, or any combination of these. Employees or employee representatives can contest the time allotted for abatement.

A notice of contest must be filed by certified mail. The OSHA area office initiates the action being contested within 15 federal working days of receipt of the notice of the enforcement action, so if the notice of contest is not filed within 15 working days, the proposed penalties and citations are final and not subject to review by any court or agency.

There is no prescribed form for the notice of contest. However, the notice must clearly indicate what is being contested, whether it is the citation, the proposed penalty, or the time for abatement.

Employees working on the site of the alleged violation must be notified of a contest filed by their employer. If they are members of a union, the union must be given a copy of the notice of contest. The employer must notify nonunion employees either by posting a copy of the notice of contest where they will see it or by serving each a copy.

The notice of contest must contain the names of those to whom it has been served or the address of the place it was posted. The posted or individually served copy of the notice of contest must be accompanied by a warning that the employees or employee representatives may not be allowed to participate in the case unless they identify themselves before the hearings begin to the commission or hearing examiner. This is called filing for party status. To file for party status, an employee or employee representative must send a notice of intent to the OSHRC. If an employer contests an alleged violation in good faith and not solely for delay or variance of penalties, the abatement period does not begin until the entry of the final order by the OSHRC.

When a notice of contest reaches the area director, the executive secretary of the OSHRC is notified of the contest and its details. The latter gives the case a docket number.

In time, a hearing will be held by an OSHRC judge. OSHA first presents its case, subject to cross-examination by the other parties. The defendant then presents the case, subject to cross-examination by the other parties. Employees or their representatives may participate in this hearing if they have filed for party status. The judge is allowed to consider only what is on the record. Any statements that go unchallenged are therefore considered to be fact.

The hearing is ordinarily held in or near the community where the alleged violation occurred. At the hearing, the secretary of labor has the burden of proving the case.

After the hearing, the judge must issue a report based on findings of fact that affirms, modifies, or vacates the secretary's citation or proposed penalty or that directs other appropriate relief. The report becomes a final order of the commission 30 days thereafter unless, within such period, a commission member directs that such report shall be reviewed by the commission itself. When this occurs, the members of the commission issue their own decisions on the case.

Once a case is decided, any person adversely affected or aggrieved thereby may obtain a review of such decision in the United States Court of Appeals.

# ■ National Institute for Occupational Safety and Health

The National Institute for Occupational Safety and Health (NIOSH) is the principal federal agency engaged in research to eliminate on-the-job hazards to the health and safety of American workers. It was established within the Department of Health, Education, and Welfare (now the DHHS) under the provision of Public Law 91-596, the Occupational Safety and Health Act of 1970. Administratively, NIOSH is located within DHHS's Centers for Disease Control of the Public Health Service.

## Responsibilities

NIOSH is responsible for identifying occupational safety and health hazards and for recommending changes in the regulations limiting them. It also is responsible for training occupational health personnel.

The institute's main research laboratories are in Cincinnati, Ohio, where studies include not only the effects of exposure to hazardous substances used in the workplace, but also the psychological, motivational, and behavioral factors involved in occupational safety and health. Much of the institute's research centers on specific hazards, such as asbestos and other fibers, beryllium, coal tar pitch volatiles, silica, noise, and stress.

At the NIOSH Appalachian Laboratory for Occupational Safety and Health (ALOSH) in Morgantown, West Virginia, research has primarily focused on coal workers' pneumoconiosis (black lung disease), but the program has been expanded to include other occupational respiratory diseases. Also located in Morgantown is the NIOSH Testing and Certification Branch, which evaluates and certifies the performance of respirators.

## Testing and Certification

Certification tests are performed under the authority of the Federal Coal Mine Health and Safety Act of 1969, the Occupational Safety and Health Act of 1970, and the numerous regulations and standards issued by the Mine Safety and Health Administration (MSHA) and the Occupational Safety and Health Administration (OSHA).

These tests have three purposes. First, they determine whether currently available protective devices conform or fail to conform with existing performance standards for such devices. Second, they encourage manufacturers of such devices and instruments to improve their performance and quality when NIOSH finds them to be out of conformance or of marginal quality. Third, the results of the tests are used by NIOSH in determining the need for future NIOSH certification projects and in establishing priorities for research directed at product improvements through increased quality and scope of performance testing.

The NIOSH personnel, in addition to doing respirator research, serve as consultants to OSHA and industry on respirator selection, use, and maintenance. NIOSH personnel regularly participate in OSHA hearings and submit new respirator information to OSHA and industry on a routine basis.

## Research

In addition to conducting its own research, NIOSH funds supportive research activities at a number of colleges, universities, and private facilities.

Not all of NIOSH's work is done in the laboratories. A legislatively mandated activity involves Health Hazard Evaluations (HHEs); these are on-the-job investigations of reported worker exposures to toxic or potentially toxic substances. Performed as a direct response to requests by management or authorized representatives of employees, HHEs are usually initiated through NIOSH's representatives, although scientists from other institute facilities often are involved.

Under the authority of the OSHAct, NIOSH conducts research for new occupational safety and health standards. Its recommended standards are transmitted to the Department of Labor, which has the responsibility for development, promulgation, and enforcement of the standards.

## Training

NIOSH has training grant programs in colleges and universities across the nation. These are located at designated NIOSH-funded Educational Resource Centers (ERCs) for the purpose of training occupational physicians and nurses, industrial hygienists, engineers, and others in the safety and health field. Typical ERCs establish cooperative arrangements with medical schools and hospitals with established programs in occupational medicine and incorporate and expand on such existing programs.

NIOSH also maintains a limited number of Regional Offices throughout the United States. The Regional Offices are focal points for special surveys and evaluations of existing occupational safety and health problems, consultative services to the states, and other activities. NIOSH headquarters are in Atlanta.

## Recommendations for Standards

One of NIOSH's most important responsibilities under this act is to transmit recommended standards to OSHA. NIOSH recommendations are intended to assist OSHA in developing new standards and in revising the approximately 400 consensus health standards (most of them consist only of a numerical exposure limit) that were promulgated when the act was passed.

NIOSH recommendations, called Criteria Documents, include an environmental limit for workplace exposure as well as recommendations on the use of labels and other forms of warning, the type and frequency of medical examinations to be provided by the employer, sampling and analytical methods, procedures for technological controls of hazards, and suitable personal protective equipment. In addition to the criteria documents, NIOSH developed technical standards for most of the consensus health standards. These standards supplement the existing environmental limits with procedures for informing employees of hazards, monitoring techniques, engineering and control mechanisms, and medical surveillance programs. These recommendations should protect workers from many of the more serious occupational exposures. These recommendations are based on laboratory and epidemiologic research conducted by NIOSH and other organizations.

## Field Investigations

NIOSH has promulgated regulations governing field investigations (42 *CFR* Part 85). Under these regulations, it is the practice to meet with company management and employee representatives before initiating a study in order to explain its purpose and scope. Before conducting medical examinations, investigators must receive specific approval from the NIOSH Human Subjects Review Board and obtain the informed consent of each employee examined. All participating employees and their designated physicians are given the results of these medical examinations. Before final reports on group data (with individual identifiers removed) are released, draft copies are provided to employers and employee representatives for their comments on technical accuracy. The results of the epidemiologic studies are then presented in NIOSH criteria documents and technical reports, in scientific journals, at scientific meetings, and at OSHA hearings on workplace standards.

Workplace investigations also are conducted in a health hazard evaluation program. Under this program, NIOSH responds to requests from employers and employee representatives to investigate a workplace, collect environmental samples, make toxicity determinations, and provide medical examinations for workers. The results of these investigations, including recommendations for work practices, personal protective equipment, and engineering controls, are reported back to company or facility management, employee representatives, and OSHA.

# Other U.S. Government Regulatory Agencies

In addition to OSHA and NIOSH, there are other federal government regulatory agencies and commissions.

## Mining Enforcement and Safety Administration

The Mining Enforcement and Safety Administration (MESA) was established on May 7, 1973, by the secretary of the interior; the administration became operative on July 16, 1973. The secretary's order assigned to MESA the responsibility for administering the enforcement provisions of the Federal Coal Mine Health and Safety Act of 1969 and the Federal Metal and Nonmetallic Mine Safety Act.

MESA administered the enforcement provisions of the public laws and related standards and training programs in a manner that guards the health and safety of American miners.

The secretary's order of May 7, 1973, designated the Bureau of Mines of the Department of the Interior to continue its research functions for mine health and safety.

## Federal Mine Safety and Health Amendments Act of 1977

On November 9, 1977, President Carter signed the Federal Mine Safety and Health Amendments Act of 1977. The act transfers authority for enforcement of mining safety and health from the Department of Interior to the Department of Labor. Most of the provisions of the act became effective March 9, 1978.

The Federal Mine Safety and Health Act of 1977 repeals the Metal and Nonmetallic Mine Safety Act and establishes a single mine safety and health law for all mining operations under an amended Coal Mine Health and Safety Act of 1969.

The Mine Safety and Health Administration (MSHA) was created to replace the Department of the Interior's Mining Enforcement and Safety Administration. This agency is headed by the assistant secretary of labor for mine safety and health. The new agency is separate from OSHA and the secretary of labor and is authorized to settle jurisdictional disputes between the two agencies.

The 1977 act defines *mine* broadly to include all underground or surface areas from which a mineral is extracted, all surface facilities used in preparing or processing minerals, and all roads, structures, dams, impoundments, tailing ponds, and similar facilities related to mining activity. Included under the act is the protection of miners from radiation hazards connected with the milling of certain radioactive materials. Mine construction activity on the surface is included in the scope of the act.

The act does not amend the section of the Coal Act dealing with benefits for victims of black lung disease.

**Provisions.** The Federal Mine Safety and Health Act of 1977 joins all mines (coal and noncoal) under one statute, transfers administration from the Department of the Interior to the Department of Labor, provides that existing and new standards applicable to metal and nonmetal mining remain separate from existing and new standards for coal mining, and establishes statutory timetables for each step of the standard-setting process.

This act vests in the secretary of labor all authority for developing safety and health standards; it authorizes NIOSH to prepare criteria documents for the development of health standards, with the secretary of labor required to act within 60 days of receipt of criteria.

**Federal Mine Safety and Health Review Commission.** This act established an independent Federal Mine Safety and Health Review Commission. It requires each mine operator to have a safety and health training program that is approved by the secretary of labor (with such training to be provided at the operator's expense and during normal working hours) and authorizes miners' representatives to participate in inspections not only during the actual inspection, but also in pre- and postinspection conferences held at the mine. More than one miner representative may participate, but only one representative, who is also an employee of the operator, is to be paid by the operator for participation in the inspection and conferences.

The act authorizes a withdrawal order (an order to stop operations) based on a pattern of standards violations that could "significantly or substantially contribute to the cause and effect of a mine hazard and health risk," authorizes miners or their representatives to make written requests for inspection based on suspected violations of standards or conditions of imminent danger, and mandates a minimum of two inspections per year of all surface mines.

The secretary of DHHS is required to appoint an advisory committee on mine health research composed of representatives of the director of the Bureau of Mines, the director of the National Science Foundation, and the director of the National Institutes of Health (NIH), as well as other people knowledgeable in the field of mine health research. The secretary of DHHS also designates the committee chairman.

The purpose of the advisory committee is to consult with and make recommendations to the secretary of

DHHS on matters relating to mine health research. The secretary is required to consider the recommendations of the committee in the conduct of such research and the awarding of research grants and contracts. The chair and the majority of committee members are not permitted to have economic interests in mining or be miners, mine operators, or government employees. In effect, the new mining law expands the existing Coal Mine Health Research Advisory Committee to cover all mine health research.

Under the existing Federal Coal Mine Health and Safety Act of 1969, NIOSH conducted research on occupational diseases of coal miners, established coal mine health standards, and ensured the availability of medical examinations for underground coal miners. MESA established coal mine safety standards and enforced both health and safety standards. In addition, NIOSH has established joint regulations with MESA to test and certify respirators and coal mine dust personal samplers.

In the past, NIOSH has not had specific legislative authority to conduct research in noncoal mines but has done so under the general authority of the Public Health Service Act and as designees of the Department of the Interior. Under the OSHAct, NIOSH has developed recommended standards that could apply to some of the most important health hazards facing workers in metal and nonmetal mines, such as noise, silica, asbestos, beryllium, and arsenic.

The new legislation changed NIOSH research authority into occupational diseases of coal miners significantly. The law does provide specific authority to conduct health hazard evaluations and establish a list of toxic substances and hazardous physical agents found in mines. Significantly, however, instead of actually setting coal mine health standards, as provided by the Coal Act, NIOSH would submit recommended standards to the Department of Labor, as is done under the OSHAct.

Research on health hazards in metal and nonmetallic mines has been expanded considerably because NIOSH had not previously had specific authority to investigate conditions in such workplaces and develop standards.

## Environmental Protection Agency

**Toxic Substances Control Act.** In 1976, Congress enacted the Toxic Substances Control Act (TSCA), PL 94-469. The act provides the Environmental Protection Agency (EPA) with the authority to require testing of chemical substances entering the environment and to regulate them when necessary. The regulatory actions include toxicity testing and environmental monitoring. This authority supplements and closes the loop of already existing hazardous substance laws in the EPA and other federal agencies. Title I of TSCA also included a provision requiring the EPA to take specific steps to control the risk for polychlorinated biphenyls (PCBs). Three titles have been added to TSCA to address

concerns about other specific toxic substances, including asbestos (Title II), radon (Title III), and lead (Title IV).

Title II of TSCA, the Asbestos Hazard Emergency Response Act (PL 99-519), was enacted in 1986 and amended in July 1988. It required the EPA to set standards for responding to the presence of asbestos in schools. The standard set responses based on the physical condition of asbestos and schools were required to inspect for asbestos-containing material and develop a management plan for such material. The title also requires asbestos contractors and analytical laboratories to be certified and requires schools to use certified people for asbestos work. The title was later amended to extend training and accreditation requirements to include inspectors, contractors, and workers performing asbestos abatement work in all public and commercial buildings. However, the mandate for inspecting buildings for asbestos was not extended to nonschool buildings.

In 1988, Title III, Indoor Radon Abatement (PL 100-551), was added to TSCA. The basic purpose of the amendment is to provide financial and technical assistance to states that support radon monitoring and control; however, neither monitoring nor abatement is required by the act. The title required the EPA to update its pamphlet on radon, develop model constructions standards and techniques for controlling radon levels in new buildings, and provide technical assistance to the states.

Title IV of TSCA, the Residential Lead-Based Paint Hazard Reduction Act (PS 102-550), was enacted in 1992. The purpose of this title is to reduce the risks to young children who are exposed to lead-based paint in their homes. The law aims to stimulate development of lead inspection and hazard abatement services in the private sector. The EPA is directed to develop definitions of lead-contaminated dust, lead-contaminated soil, and lead-based paints hazards; requirements for accreditation of training programs for lead abatement work; criteria to evaluate the effectiveness of commercial products used to detect or reduce risks associated with lead-based paint; protocols for laboratory analysis of lead in paint soils, films, and dust; and certification requirements for laboratories performing such analyses. Also, the EPA is directed to conduct a study of lead hazards due to renovation and remodeling activities that may disturb lead-based paint. It must also promulgate guidelines for the renovation and remodeling of buildings or other structures when these activities might create a hazard.

**Resource Conservation and Recovery Act.** The Resource Conservation and Recovery Act (RCRA) of 1976 established the federal program regulating solid and hazardous waste management. RCRA actually amends earlier legislation (the Solid Waste Disposal Act of 1965), but the amendments were so comprehensive that the act is commonly called RCRA rather than its official title. The act greatly expanded the federal government's role in solid waste disposal management, with emphasis on hazardous waste disposal. RCRA continued the federal facilities guidelines under the program established by the 1970 Solid Waste Disposal Act, created a major hazardous waste regulatory program, and prohibited the practice of open dumping. The act provides for extensive federal aid through grants to state and regional agencies for solid waste planning and information programs.

In the cradle-to-grave program of control established by RCRA, custody and responsibility moves with a waste material from the generator and transporter to its final disposal site. However, the generator never loses liability for the waste created. Although the original RCRA regulations (1980–1984) exempted small-quantity generators if they produced less than 1,000 kg of hazardous waste per calendar month, the 1984 amendments to RCRA require generators producing between 100 and 1,000 kg per month to meet certain procedural standards.

Since 1980, general waste management requirements of RCRA have included proper notification and recording of hazardous waste activities, along with adequate packaging, labeling, and manifesting of wastes for shipment off site. A RCRA permit is required for treatment, storage, and disposal (TSD) of hazardous waste on-site or off-site. Standards for TSD facilities include rigorous facility management plans, preparedness and prevention of emergencies and releases, contingency plans, operating records and reports, groundwater protection for land disposal facilities, and closure and postclosure plans with financial responsibility assurance.

The new RCRA requirements, following the amendments of 1984, make TSD facilities responsible for assessing human exposure to current and past waste management operations and for corrective action needed to remedy releases of hazardous constituents to the environment. The RCRA regulations also require worker training, development of safe handling procedures, and emergency response measures. Documentation of training is required and inspectors may require a review of the documentation.

The 1984 amendments required that land disposal of specified highly hazardous wastes be phased out over the period 1986–1990. The EPA was directed to review all wastes that it has defined as hazardous and to determine the appropriateness of land disposal for them. Minimal technological standards were set for new landfills, generally requiring double liners, a leachate collection system, and groundwater monitoring. Under the new amendments, states were encouraged to assume the EPA's hazardous waste program. As of September 1992, 46 states were authorized to run the pre-1984 elements of the programs and 6 had received authorization to run most post-1984 components as well.

The third major amendment to RCRA, the Federal Facility Compliance Act, was passed in 1992. This act allows the states, the EPA, and the Department of Justice to enforce the provisions of RCRA against federal facilities; federal departments and agencies can be subjected to injunction, administrative orders, or penalties for noncompliance. The act also contains special provisions applicable to mixtures of radioactive and hazardous waste at Department of Energy facilities and to munitions, military ships, and military sewage treatment facilities handling hazardous wastes.

**Comprehensive Environmental Response, Compensation and Liability Act.** In 1980, Congress enacted the Comprehensive Environmental Response, Compensation and Liability Act (CERCLA), PL 96-510. This act established the Superfund Program to handle emergencies at uncontrolled waste sites, to clean up the sites, and to deal with related problems. In 1986, CERCLA was reauthorized by Congress as the Superfund Amendment and Reauthorization Act (SARA) of 1986, PL 99-499. The purpose of the amendments was to provide additional funding and additional provisions. The new authorities and programs included in this reauthorization include underground storage tanks, emergency planning, risk assessment, community right-to-know, research, development, demonstrations, and training. In 1990, the Superfund Extension Act authorized appropriations for SARA through 1995.

These regulations include requirements that owner-operators of leaking underground storage tanks undertake corrective action to protect human health and the environment. SARA also established a comprehensive federal program to promote various research, development, demonstration, and training activities, including the following:

- Techniques to detect, assess, and evaluate health effects of hazardous substances
- Methods to assess human health risks
- Methods and technologies to detect hazardous substances and to reduce volume and toxicity

CERCLA created the Agency for Toxic Substances and Disease Registry (ATSDR) in the Public Health Service to carry out the health-related authorities in the act. In 1986, SARA created new duties for ATSDR. The agency and EPA are to prepare a list of at least 275 of the hazardous substances most commonly found at National Priority List (NPL) sites. The agency is to prepare toxicological profiles of these substances. Where there is insufficient information on a substance, ATSDR is to conduct research. In addition, ATSDR is required to perform a health assessment at each facility on the NPL list. Finally, ATSDR is to provide consultations to the EPA and state and local officials on health issues related to hazardous substances.

With the RCRA amendments and the requirements for cleanup under the CERCLA legislation, the time of total accountability for current and past waste management practices has arrived. Quantitative liabilities for waste management practices are directly proportionate to how much a facility is affected by these requirements and remedial actions. The 1986 law added a provision limiting the amount of coverage specified in the policy. SARA also authorized companies to form risk retention groups as a means of insuring themselves.

Companies covered under the OSHA Hazard Communication standard are also subject to the EPA Hazardous Chemical Reporting Rules under Title III of Superfund. Covered facilities are required to submit either copies of the MSDSs they prepare for OSHA compliance or a list of all chemicals for which MSDSs are required to the state emergency response commission, the local emergency planning committee, and the local fire department.

Covered facilities must also submit emergency and hazardous chemical inventory forms to the same state and local authorities. Information on the maximum daily amounts and chemical locations (designated Tier I information) submittal date was March 1, 1988, and annually thereafter. The more detailed Tier II information would be submitted on request.

## Summary

This chapter provides an overview of the national, regional, state, and local government agencies and regulations concerned with occupational health, safety, and environmental issues. The U. S. Congress is reevaluating the roles of the Occupational Safety and Health Administration and NIOSH. The material herein reflects the situation as of March 1995. The history of the Occupational Safety and Health Administration and relevant legislation are detailed in Chapter 30.

## Bibliography

Occupational Safety and Health Act of 1970. U.S. Congress (91st) S2193. Public Law 91-596. Washington, DC: U.S. Government Printing Office, 1977.

Occupational Safety and Health Administration. *OSHA Technical Manual* (OSHA Instruction #TED 1.14). Washington, DC: U.S. Government Printing Office, 1995.

Occupational Safety and Health Administration. *OSHA Field Inspection Reference Manual,* OSHA Instruction CPL 2.103. Washington, DC: U.S. Government Printing Office, September 1994.

*Safety and Health Program Management Guidelines: Issuance of Voluntary Guidelines.* US DOL–OSHA *CFR* vol. 54, no. 16 published January 26, 1989.

U.S. Department of Labor, OSHA. *All About OSHA,* OSHA Pub. no. 2056. Washington, DC: U.S. Government Printing Office, 1994.

Congressional Research Service. *Summaries of Environmental Laws Administered by the Environmental Protection Agency,* 93-53 ENR. Washington, DC: Library of Congress, January 14, 1993.

# History of the Federal Occupational Safety and Health Administration

by Benjamin W. Mintz, LLB

**1968–1970: The Agency Is Created**
*The Legislative Battle* ▪ *OSHA Program Structure*

**1971–1973: Early Agency Activities**
*The New Agency* ▪ *Compliance Activity Commences* ▪ *Start-Up Standards* ▪ *Regulating Asbestos* ▪ *Involving the States* ▪ *Involving Workers* ▪ *Injury and Illness Rates*

**1973–1976: The Agency Continues to Grow**
*OSHA's First Transition* ▪ *The Kepone Incident* ▪ *OSHA in Court* ▪ *Disputes over Jurisdiction* ▪ *Amendments, Riders, and Oversight* ▪ *The State Plan Framework and the Benchmarks Controversy* ▪ *Standards for Construction, Agriculture, and Other Industries* ▪ *Health Standards: An Overview* ▪ *Health Standards Proceedings* ▪ *Federal Safety: Trends in Injury Rates*

**1977–1981: Giving Teeth to the Tiger**
*A New Assistant Secretary* ▪ *Health Standards: The New Priority* ▪ *Deletion of De Minimis Standards* ▪ *"Common-Sense" Enforcement Policy* ▪ *Increasing Worker Participation* ▪ *Continuing Congressional Oversight* ▪ *Interagency Cooperation and Controversy* ▪ *A New Executive Order* ▪ *The Bingham Administration Ends*

**1981 to 1987: OSHA's Balanced Approach**
*A New Era in Compliance* ▪ *The States: A Partnership Once Again* ▪ *Health Standards, Continued* ▪ *Safety Standards* ▪ *The Federal Program* ▪ *BLS Statistical Survey*

**1987 to the Present: Reevaluating OSHA**
*Supreme Court Decisions* ▪ *Standards Activity: Health Standards* ▪ *Standards Activity: Safety Standards* ▪ *Standards Activity: Overview* ▪ *Compliance Activity* ▪ *State Programs* ▪ *BLS Statistical Survey* ▪ *Legislative Activity* ▪ *Occupational Safety and Health Review Commission* ▪ *Other OSHA Activities* ▪ *Studies and Evaluations of OSHA*

**Summary**

**Bibliography**

Testifying before a congressional subcommittee in 1968, Secretary of Labor Willard Wirtz spoke dramatically of the industrial casualty rate. "Each day," he stated, "there will be 55 dead, 8,500 disabled and 27,200 hurt in America's workplaces." Secretary Wirtz was pressing the subcommittee to approve the comprehensive federal occupational safety and health bill, which was being sponsored by the administration of President Lyndon B. Johnson. The issue, Secretary Wirtz asserted, is "simply whether the Congress will act to stop a carnage which continues only because people don't realize its magnitude, and can't see the blood on the things they buy, on the food they eat, and the services they get." Dr. Irving Selikoff, a pioneer in the field of occupational medicine, testified that it was an "unhappy reflection" on the country that in the 1960s in the United States, 7 percent of all insulation workers would die of asbestosis, a "completely preventable disease" (*OSHAct: Hearings on S. 2864 Before the Subcomm. on Labor of the Senate Comm. on Labor and Public Welfare*, 90th Cong., 2nd Sess. 69 [1968]).

Despite this testimony, Congress did not pass occupational safety and health legislation that year. Not until 1970, after a difficult legislative battle, was an OSHA law enacted. It was signed into law by Richard Nixon on December 29, 1970, and became effective 120 days later.

Fifteen years later, in April 1985, a major report on OSHA entitled *Preventing Illness and Injury in the Workplace* was issued by the Office of Technology Assessment (OTA) of the Congress. The OTA estimated that there were between 2.5 and 11.3 million nonfatal occupational injuries each year and 6,000 deaths annually resulting from workplace injuries. Noting the great disagreement about the number of workplace illnesses, OTA refused even to estimate the correct number (OTA, 1985).

This chapter covers the history of OSHA from 1970 to 1995, in a framework that details OSHA's achievements and failures during its 25-year history.

## ■ 1968–1970: The Agency Is Created

### The Legislative Battle

The 91st Congress is remembered as the occupational safety and health Congress. In 1969, it passed two landmark statutes: the Coal Mine Safety Act and the Construction Safety Act. In December 1970, the most comprehensive statute of all, the OSHAct, was adopted by overwhelming votes in the Senate and the House of Representatives (OSHAct, 29 U.S.C. §651 and following). Earlier in 1968, a federal OSHA bill was introduced but did not reach a vote, largely because of strong opposition from the business community. A representative of the American Iron and Steel Institute told a subcommittee that voluntary employer efforts, along with state activity and federally sponsored research, would accomplish far more in reducing workplace injuries "than would a program of federal penalties and other attributes of overwhelming federal authority reaching into hundreds of thousands of large and small business operations" (*OSHAct: Hearings on S. 2864 Before the Subcomm. on Labor of the Senate Comm. on Labor and Public Welfare*, 90th Cong., 2nd Sess. 347, 349–352 [1968]).

In 1969, with support from newly elected President Nixon, a broad consensus had emerged on the need for federal legislation to prevent illnesses and injuries in the workplace. Several important factors contributed to this consensus:

■ In the past, the workers' compensation system had not provided sufficient financial incentives for employers to undertake efforts to improve workplace safety and health.

■ The accident rate was rising.

■ Illness in the workplace was a serious and rapidly increasing problem.

■ State efforts had proven inadequate (MacLaury, 1981, p. 18).

The main issue was what the substance of the legislation would be. There was basic agreement on several points; for example, broad coverage by the new law, the need for occupational safety and health standards prescribing the employer conduct necessary to achieve safety and health, government enforcement of these standards, and a role for the states in the program. The disagreement centered on the extent of the powers to be assigned to the secretary of labor. The unions, generally supported by Democratic members of Congress, favored a strong role for the secretary of labor. They argued in favor of that office's authority to issue standards, to enforce the standards, and to adjudicate violations of the standards, as well as to administer closing down employer operations in the event of imminent dangers. With considerable support from the Republican party, the business community vigorously objected. In the words of Senator Dominick of Colorado, who favored the legislation but supported a "division of responsibilities," the "concentration of authority" advocated by the Democrats was not "balanced" and was "objectionable because concentration of power gives rise to a great potential for abuse" (116 *Cong. Rec.* 37,336 [1970]).

The Democratic view was embodied in the Williams bill, which was adopted by the Senate early in 1970. The Republican approach was taken in the bill sponsored by Congressman William A. Steiger of Wisconsin, also an advocate of OSHA legislation, that passed the House of Representatives. The two bills went to a congressional conference committee late in 1970, and its version became the OSHAct. The conference OSHA bill was a compromise, but the thrust of the legislation was much closer to the stringent Senate bill than to the House version. The Democratic chairman of the House Labor Committee selected representatives to the conference committee who favored the Senate bill, and who no doubt often joined their Senate counterparts in voting on critical issues (Page & O'Brien, 1973). The OSHA law as passed has generally been viewed as a stringent regulatory statute; its subsequent history describes the process by which the courts, the Congress, and the Department of Labor joined in an often uncomfortable alliance to make this new and historic legislation—borrowing the words of Senator Dominick—"workable and effective."

### OSHA Program Structure

The statutory structure of the OSHAct is often articulated; therefore, only the basic policy principles underlying the new law will be stated.

**Universal coverage.** The OSHAct applies to all private employers, without exception. Federal employees and state and local employees in states with approved plans are covered by separate occupational safety and health programs. A limited number of particular working conditions subject to the enforcement of occupational safety and health standards of other federal agencies such as the Coast Guard are also not covered by the OSHAct.

**Employer obligations.** The employer's obligation to provide safe and healthful working conditions is defined primarily by means of standards promulgated by the secretary. The standards are generally promulgated after public rulemaking proceedings designed to elicit data and views on which these standards are based. The standards define employer obligations prospectively and in considerable detail. To the extent that a hazard is not covered by a standard, the employer must comply with the general duty obligation to provide a workplace "free from recognized hazards likely to cause death or serious physical harm."

**Enforcement.** Enforcement of the OSHAct through workplace inspections, citations, and assessment of civil penalties is designed to achieve safe and healthful workplaces in two significant ways:

- With respect to workplaces actually inspected, OSHA imposes legally enforceable abatement requirements and conducts follow-up inspections. These are expected to bring about neutralization of hazards at that workplace. The OSHA system for determining priorities for inspection, usually called "worst-first," should thus result in substantial elimination of hazards in the most dangerous workplaces.

- As for other workplaces, the enforcement program is intended to constitute an incentive to employers to abate hazards without regard to whether an inspection takes place. Two elements in the statutory structure are particularly designed to achieve this result: Sanctions are imposed when violations are disclosed at the first inspection, and no advance notice is given of workplace inspections.

**Employee participation.** The OSHAct is primarily designed to protect workers, and the statute gives them a crucial role in virtually every aspect of the program. For example, an employee representative has the right to request a workplace inspection, to participate in the walkaround inspection, to participate in adjudicatory and court review proceedings and rulemaking hearings, and, significantly, to be free from employer reprisal for the exercise of rights protected under the act. Regulatory requirements have also been imposed on employers to ensure that employees are informed of workplace hazards; this has been called worker right to know.

**Checks and balances.** Reflecting the concern of Congress that the OSHA program be balanced, the act contains both broad agency authority and constraints on that authority. Discretion is given to the secretary to issue

occupational safety and health standards; however, these standards can generally be issued only if the agency follows detailed rulemaking proceedings. Specific criteria are set forth in the statute, notably in Sections 6(b)(5) and (7), for the content of OSHA health standards, which must be based on the best available evidence. All standards are subject to court of appeals review based on the record in the rulemaking proceeding. Similarly, OSHA enforcement actions can be contested and are adjudicated before a neutral and independent commission; employees have the right to intervene and protect their interest in these adjudicatory proceedings. Other constraints on the agency, though not explicitly stated in the act, are implicit in our governmental system. Noteworthy among them is the right of an employer to refuse a workplace inspection unless preceded by a warrant based on probable cause. Congressional and White House scrutiny of OSHA implementation of the act, particularly in the area of standards development, has been pervasive.

**Role of the states.** The OSHA program is primarily a federal program. After a brief transition period, state occupational safety and health enforcement was preempted. However, the act assigns a role to the states to develop their individual OSHA programs, which are supported by 50 percent federal funding; the state programs must be "at least as effective as" the federal program and are subject to continuing federal monitoring and evaluation. The history of OSHA has been replete with sharp controversies over the proper role of the states in the enforcement of the OSHAct.

**Information and data.** Congress was acutely aware of the inadequacy of the then available fund of information on occupational injuries and illnesses. In order to more accurately determine injury and illness rates, the act mandates that employers maintain accurate records on all but minor work-related injuries and illnesses, and that they periodically report that information. Furthermore, OSHA was required to develop "an effective program of collection, compilation, and analysis" of occupational safety and health statistics. This responsibility was assigned to the Bureau of Labor Statistics (BLS) in the Department of Labor, which publishes annually the results of its statistical survey of occupation-related injuries and illnesses.

**Training and education.** In addition to its power to impose sanctions, OSHA has a wide range of authority that includes education, training, and consultation, which is designed to elicit voluntary employer and employee cooperation in bringing about safe and healthful workplaces. These voluntary programs have been undertaken by OSHA throughout its history and have received special emphasis during the administration of President Reagan. The program of enforcement-free workplace consultation visits, which has been financed mostly by OSHA and operated by the states, has been particularly important.

# ■ 1971–1973: Early Agency Activities

## The New Agency

The secretary of labor first created an agency within the Department—the Occupational Safety and Health Administration—to implement the new program. Its head, OSHA's first assistant secretary, was George C. Guenther, who served until January 1973. Guenther had formerly been head of the Department of Labor's Bureau of Labor Standards, which had responsibility for administering the pre-OSHA occupational safety and health programs under the Walsh–Healey Act, the Longshoremen and Harbor Workers Act, and related legislation. An administrative structure was established for enforcing the act; this included 10 regional offices and 49 area offices, staffed with compliance officers assigned to the area offices who were responsible for conducting workplace inspections. At that time, most of the OSHA inspectors had expertise in the area of safety. In June 1973, OSHA had 456 safety inspectors and 68 industrial hygienists. Not unexpectedly, the bulk of OSHA inspection activity, 93.4 percent in fiscal year 1973, was safety related (OTA, 1985). This reflected the primary emphasis that had been placed in the past, both in and out of government, on industrial accidents, about which a great deal more was known than work-related illnesses. It would be another five years, when Assistant Secretaries Corn and Bingham shifted OSHA's emphasis to health hazards, before industrial hygiene inspectors constituted a significant portion of OSHA's compliance staff.

## Compliance Activity Commences

A regulatory framework for OSHA activity was established in the first two years of OSHA's existence. Two major regulations were issued in 1972 after public comment. The first, known as Part 1903, dealt with agency inspection procedures (29 *CFR* 1903). The OSHA statute sets forth the basic outline for OSHA enforcement activity. The inspection regulations fleshed out inspection procedures in numerous ways and provided interpretations of key provisions. The regulations in Part 1903 have remained substantially intact throughout the history of OSHA.

Another major element in OSHA's regulatory framework was the first *Compliance Manual,* originally issued in April 1971 and periodically revised thereafter (OSHA, 1972). It is currently titled the *Field Inspection Reference Manual* (FIRM) (see Chapter 29). Although intended as instructions to field staff for implementing the OSHA program, the *Compliance Manual* from the start was a public document and, in practice, has constituted the primary means by which the agency advised interested parties of its policies in administering the statute. Many interpretations of statutory terms—what constitutes a serious, willful, and repeated violation—are included in the manual; a general definition of imminent danger is also included.

A particularly crucial component of OSHA's *Compliance Manual* was the establishment of inspection priorities. Under the act, OSHA has general inspection authority; however, inspections are mandated only in response to certain written employee complaints. It has therefore been essential for OSHA to define the criteria for its selecting for inspection by its limited compliance resources among the approximately 5 million workplaces.

In its first *Compliance Manual,* OSHA gave instructions on what it called compliance programming. The first priority (after dealing with imminent dangers) was to investigate workplaces where fatalities or catastrophes had occurred. The second priority was response to employee complaints. The third priority was the special hazard elimination programs, which included target industries and health hazards. These priorities brought OSHA inspectors to the most hazardous workplaces—those hazardous as determined in part in the annual statistical surveys of the BLS.

Fourth, to the extent resources were available, OSHA said it would conduct inspections of all other workplaces in order to make clear that the act's obligations are applicable to all private employers. (The latter two categories are usually called programmed inspections [OSHA, 1972, Chapter 4]). Although these broad priority categories have remained substantially the same since 1972, they have been modified in detail in many respects, particularly in regard to agency response to complaints. Considerable controversy later developed, in Congress and elsewhere, as to whether employers with above-average safety records should be exempt from inspection activity.

Field staff were required to report regularly on enforcement activity to OSHA's national office in Washington. One of the required statistical reports is the number and type of OSHA inspections. In fiscal year 1973, for example, OSHA conducted about 48,000 inspections; 5.1 percent of these were fatality/catastrophe inspections, 13.7 percent were complaint inspections, 66.5 percent were programmed inspections (target-industry and general-industry inspections), and follow-up inspections in workplaces where violations had been found constituted 14.7 percent. Since then, the total number of inspections grew to a high of about 90,000 in fiscal year 1974, leveling to about 60,000 or fewer after 1977, which may reflect the new emphasis on more time-consuming health inspections. Although the number of fatality/catastrophe inspections has remained fairly constant over the years, there have been significant variations in the number of complaint inspections, resulting from important agency policy changes either emphasizing or deemphasizing complaint inspections. A significant deemphasis in follow-up inspections occurred after 1981 (OTA, 1985).

The enforcement program began at the end of August 1971, when OSHA's newly adopted standards came into effect. The first OSHA citation was issued at the beginning of May; it was a general duty violation based on excessive employee exposure to mercury. During the remainder of the calendar year, 9,507 citations were issued, with proposed penalties amounting to $737,486. In 1972, the number of issued citations increased to 23,900, and the amount of the proposed penalties was more than $3 million (*The President's Report on Occupational Safety and Health,* May 1972, Dec. 1973).

## Start-Up Standards

One of OSHA's most important tasks after the act became effective was to issue standards that would provide the agency with a basis for promptly commencing its enforcement program. Congress gave the agency authority under Section 6(a) to promulgate certain standards without rulemaking—that is, without the delays inherent in public comment proceedings. These standards have been called start-up standards and included national consensus standards and established federal standards, both of which had gone through at least some public comment process before they were issued. The OSHA quickly—many have said too quickly—promulgated these standards, and on May 29, 1971, barely a month after the act's effective date, OSHA issued a large body of these start-up standards. Most were in the safety area and were taken from standards issued by the American National Standards Institute (ANSI) and the National Fire Protection Association (NFPA), both determined by Congress to be national consensus organizations; others were pre-OSHA standards issued under the Walsh–Healey Act, the Longshoremen's and Harbor Workers' Compensation Act, and the Construction Safety Act, all of which had already gone through rulemaking proceedings. Some health standards were issued in 1971. Among them were the Threshold Limit Value (TLV) levels of the American Conference of Governmental Industrial Hygienists (ACGIH) for about 400 substances, which had become established federal standards under the Walsh–Healey Act. These contained no more than the Permissible Exposure Limit (PEL) and a requirement for implementation of engineering and protective-equipment controls to reach the limit. But these limits were often based on inadequate information; in 1970, Congress recognized that the standards "may not be as effective and up-to-date as desirable," and that they would provide only a "minimum level of health and safety" (S. Rep. No. 1282, 91st Cong., 2nd Sess. 6 [1970]).

There is no question that OSHA's adoption of start-up standards in 1971 got enforcement off to a rapid start. Even today, the bulk of OSHA safety standards appearing in the *Code of Federal Regulations* is the original material that was issued at the start of the program. But OSHA's hasty action led to no end of controversy and criticism of the young regulatory agency. Litigation ensued when OSHA began to enforce its national consensus safety standards, and many citations and penalties were reversed by the Review Commission and the courts. Indeed, one of the most outspoken critics of OSHA's start-up standards

was Robert D. Moran, chairman of the Review Commission, who argued that these standards "violate the spirit and purpose of the Act" (Moran, 1976, pp. 19–20). Even the president of the AFL–CIO, testifying many years later, said, "Because of the hue and cry over these 'nit-picking' aspects of the program, the more serious and important goals of the OSHA program became lost in a morass of largely unintelligible debate and political animosity" (*Oversight on the Administration of the Occupational Safety and Health Act, 1980: Hearings Before the Senate Comm. on Labor and Human Resources,* 96th Cong., 2nd Sess. 730–731 [1980] [testimony of Lane Kirkland, president of the AFL–CIO]). Largely in response to the criticism, a presidential task force was organized in 1976 to develop a "model approach to safety standards." The task force reported a year later that OSHA should avoid design or specification requirements, and recommended a performance approach, by means of which employers would have had considerable flexibility to achieve the safety goals of the standard "by any appropriate means" (MacAvoy, 1977, pp. 17–21). The OSHA Appropriations Act in 1976 directed the agency to undertake a "review and simplification" of existing standards and to eliminate "nuisance" standards (*Department of Labor and Health, Education and Welfare Appropriations Act of 1976,* Pub. L. No. 94-206, 90 Stat. 3 [1976]).

The pressure continued to mount, and in 1978, after receiving public comment, OSHA eliminated about 600 safety standards, using seven criteria for determining which should be revoked; among those deleted were "obsolete" or "inconsequential" standards and those "encumbered by unnecessary detail" (43 *FR* 49, 726 [1978]). OSHA also undertook a broad revision of all its national consensus standards, a project that has moved slowly, with only the standards on fire protection and electrical hazards revised by 1985. Other issues arising from the initial promulgation continued to receive OSHA's attention. One of these was whether so-called "should" standards—those stating that employers "should" take a particular action rather than "shall" do so—were mandatory in effect. This issue was litigated extensively and was not resolved until February 1984, when OSHA revoked a group of its "should" standards (49 *FR* 5318 [1984]).

## Regulating Asbestos

In enacting OSHA, Congress was well aware of the dangers of employee exposure to asbestos fibers. The Senate Labor Committee in its 1970 report described asbestos as "another material which continues to destroy lives of workers," noting that as many as 3.5 million workers are at risk of asbestosis, pulmonary cancer, and mesothelioma as a result of asbestos exposure (S. Rep. No. 1282, 91st Cong., 2nd Sess. 2–4 [1970]). In 1971 OSHA adopted, as an established federal standard, the Walsh–Healey Asbestos Standard, requiring the implementation of engineering controls to achieve a

PEL of 12 fibers per cubic centimeter of air. However, it was immediately apparent that this standard provided inadequate protection. Several months after the effective date of the asbestos start-up standard, on December 7, 1971, OSHA invoked its emergency authority to issue an emergency temporary standard (ETS) for asbestos. The ETS mandated, effective immediately, a 5-fiber per cubic centimeter of air 8-hour time-weighted average (TWA), with a 10-fiber per cubic centimeter of air ceiling for any 15-minute period (36 *FR* 23,207 [1971]).

The standard was based on OSHA's finding that there was a "grave danger" to employees from asbestos exposure. Unlike many later OSHA emergency standards that were challenged, and the challenges sustained, the asbestos ETS was not challenged and remained in effect for the entire six-month period. In June 1972, after receiving written comment and testimony at a public hearing, OSHA issued a permanent asbestos standard. In its preamble to the standard—which explains the basis for the standard—OSHA asserted that there was no dispute that exposure to asbestos is "causally related" to cancers and asbestosis; the only issue, it said, was the specific level below which exposure is safe. This was an issue because the agency did not have "accurate measures" of the levels of exposure occurring 20–30 years ago that caused the disease. However, OSHA concluded that in view of the undisputed "grave consequences" of exposure to asbestos fibers, it is essential that the exposure be regulated "on the basis of the best evidence available now, even though it may not be as good as scientifically desirable" (37 *FR* 11,318 [1972]). The agency therefore determined that the PEL should be set at 2 fibers per cubic centimeter of air. However, concluding that "many work operations" would meet "varying degrees of difficulty," such as the necessity for extensive redesign and relocation of equipment, in complying with the 2-fiber per cubic centimeter of air standard, OSHA delayed the effective date for four years, applying a less strict 5-fiber standard in the interim.

The AFL–CIO's Industrial Union Department challenged the standard in the Court of Appeals for the District of Columbia Circuit. Among other provisions, the union attacked the delay in applying the stricter PEL. The court's decision in the case was a landmark in the development of OSHA law. It affirmed OSHA's decision to adopt a 2-fiber per cubic centimeter standard. In much-quoted language, the court, in an opinion by Judge Carl McGowan, recognized that some of the questions involved in OSHA standards development are "on the frontiers of scientific knowledge," and therefore not susceptible to precise factual determination. Court review, therefore, cannot be the factual type of review that is typically undertaken by the courts, but rather must be deferential to the agency, calculated only "to negate the dangers of arbitrariness and irrationality." In the matter of asbestos, the court said, the choice of a lower level was "doubtless sound," inasmuch "as the protection of the health of employees is the over-riding concern of OSHA" (*Industrial Union Department v. Hodgson,* 499 F.2d 467, 478 [D.C. Cir. 1974]).

However, although it affirmed the agency's 2-fiber per cubic centimeter of air level, the court rejected OSHA's delay of the effective date. The court granted that "feasibility" considerations, both economic and technical, can play a role in OSHA decision making, but OSHA had not shown in the record why it was necessary to impose an across-the-board delay in needed protection for workers in all industries. The case was remanded to OSHA for further development of a record on the feasibility of the lower PELs in specific industries. The language of the court on the feasibility issue has also served as a significant precedent for later decisions on OSHA standards. "Practical considerations can temper protective requirements," Judge McGowan said. At the same time, standards may be economically feasible "even though, from the standpoint of employers, they are financially burdensome and affect profit margins adversely." However, the court determined that a standard was not feasible if it required "protective devices unavailable under existing technology or by making financial viability generally impossible" (*Industrial Union Department v. Hodgson,* 499 F.2d 467, 478 [D.C. Cir. 1974]).

## Involving the States

The dissatisfaction of organized labor with OSHA's asbestos standard was exceeded in intensity by the controversy engendered by the agency's initial policy in the area of state plans. The act contains a detailed scheme defining the OSHA relationship between the federal government and the states. Primary authority was assigned to the federal agency; the states were preempted except if they submitted "at least as effective" plans that are approved by OSHA and found in practice to be equally effective. Finally, in order to prevent preemption of ongoing state programs while the states were still developing plans for submission, Section 18(h) authorized two-year agreements permitting continued state enforcement.

OSHA acted quickly to encourage state participation in the OSHA program. The secretary of labor quickly wrote to the governors, urging the states to enter into Section 18(h) agreements and to submit "at least as effective" plans. All but three states accepted the secretary's invitation and entered into Section 18(h) agreements. By the end of 1972, 44 states, the District of Columbia, and four territories had submitted plans. However, it was apparent that these plans as submitted did not meet the statutory criteria for equal effectiveness. Thus, many states lacked occupational safety and health–enabling statutes that could provide authority for their state OSHA programs. To meet this problem, apparently unanticipated by Congress, and "to allow the transition that states must undergo to upgrade an existing program," OSHA developed a new concept—the developmental plan—in its State Plan Regulations (Part 1902). This authorized approval of state plans would trigger 50 percent federal funding granted on the basis of commitments by the states to effectively meet goals in the future, generally within three years, even though, on submission, the plans were concededly not equally effective (29 *CFR* 1902.2).

Organized labor was skeptical from the start about state efforts, arguing before the committees considering OSHA bills that state OSHA programs were largely understaffed and ineffective. When the developmental concept appeared by means of OSHA interpretation in 1971, it was vigorously opposed by the unions. The main forum for debate was the National Advisory Committee on Occupational Safety and Health (NACOSH), created by the act, where Jack J. Sheehan, legislative representative of the United Steelworkers of America, said, "On the developmental plan, it's a non-plan as far as I can understand it" (Proc. NACOSH, Sept. 24, 1971, pp. 119–126).

The debate over developmental plans led to further debate on OSHA's attempt to extend the effective period of Section 18(h) agreements by means of so-called temporary orders. These orders were even more "temporary" than OSHA planned, because in January 1973 a federal district court found them beyond OSHA authority. By the end of 1972, only three plans—those of South Carolina, Oregon, and Montana—had been approved. In explaining the need for temporary orders, Assistant Secretary George Guenther argued that according to an "absolutely strict interpretation," after 1972 there would be a "protection gap" for employees in states with plans pending approval. The temporary orders would close the gap by allowing state enforcement for an additional six months (Proc. NACOSH, Nov. 16, 1972, pp. 44–46). Judge Barrington Parker did not find this policy imperative sufficiently compelling and refused to approve OSHA's expansive interpretation of its authority. Noting congressional and union objections to temporary orders, Judge Parker relied particularly on the "express language of the Act, which mandated exclusive federal jurisdiction after December 28, 1972"; he also took the position that even without temporary orders, state OSHA jurisdictions would not be "seriously" disrupted, which turned out to be the case (*AFL–CIO v. Hodgson*, 1971–1973 OSHD [CCH] ¶ 15,353 [D.D.C., 1973]). OSHA did not pursue an appeal of Judge Parker's decision, and temporary orders were abandoned.

## Involving Workers

The act pervasively provides for the participation of employees in the OSHA program. The right of employees or their representatives to request an inspection and their right to participate in the physical inspection of the workplace are foremost. Under Section 8(f)(1), OSHA must conduct an inspection if an employee or representative files a written complaint alleging that a hazard is threatening physical harm or that there is an imminent danger, and OSHA deter-

mines that there are "reasonable grounds" to believe there is merit to the complaint. Early on, an issue arose concerning OSHA's response to complaints that did not meet the formality requirements of the act. These, usually known as informal complaints, include oral communications by employees—such as telephone calls to an area office—alleging hazardous workplace conditions. OSHA, emphasizing its limited inspection resources, determined in its first *Compliance Manual* that, as a general rule, informal complaints would not trigger workplace inspections except in situations apparently involving imminent dangers (*OSHA Compliance Manual,* Jan. 1972, Chapter 6). This inspection policy continued essentially unchanged for the next four years, until the traumatic "Kepone incident," to be discussed later in this chapter, forced OSHA to a major reversal of its policy on informal complaints.

The right of employees to participate in the physical inspection is known as the walkaround right. In reporting favorably on the OSHA bill in 1970, the Senate Labor Committee observed that under pre-OSHA laws, "workers tend[ed] to be cynical regarding the thoroughness and efficacy" of inspections, because they were usually not advised that an inspection was taking place. The walkaround right was therefore added to "provide an appropriate degree of involvement of employees themselves in the physical inspections of their own places of employment" (S. Rep. No. 1281, 91st Cong., 2nd Sess. 11–12 [1970]). The question that arose almost immediately was, Are employees selected as representatives in the walkaround entitled to the wages they would otherwise have received if they had continued actual work? In a proceeding involving an inspection of the Mobil Oil Company, the Oil, Chemical and Atomic Workers Union complained that pay was withheld from its members, thus interfering with a basic statutory right. The solicitor of labor rejected the claim, deciding that time spent in accompanying an inspector did not constitute hours worked, and therefore no discrimination had taken place. The Court of Appeals for the District of Columbia upheld the solicitor's view in a related case involving the Fair Labor Standards Act (*Leone v. Mobil Oil Corp.*, 523 F.2d 1153 [D.C. Cir. 1975]). The OSHA policy remained the same until 1977, when it was reexamined and reversed by Assistant Secretary Bingham.

Many statutes, both federal and state, contain whistle-blower provisions protecting the exercise by individuals of their protected rights against reprisal and retaliation. Section 11(c) of the OSHAct is a whistle-blower provision: It prohibits discrimination in conditions against employees "because of the exercise of any rights afforded by the Act." Specifically mentioned are the protected rights to file complaints, to institute OSHA proceedings, and to testify at those proceedings. In its initial interpretation of the provision, OSHA said that the provision also protected "other rights [that] exist by necessary implication." That interpretation confronted a particularly difficult issue: whether an employer could discharge or discipline an employee for refusing to work in particularly dangerous circumstances. OSHA expansively interpreted the act and determined that an employee was protected against discipline if "with no reasonable alternative" he or she "in good faith" refuses to expose himself to imminently dangerous conditions (29 *CFR* 1977.12).

## Injury and Illness Rates

One of Congress's important goals in enacting OSHA legislation in 1970 was to improve the reliability of the work-injury and work-illness data. Although the BLS had been collecting and publishing these data for more than 30 years, there were serious limitations on the usefulness of the statistics that were based on the Z.16.1 ANSI standard, the *American Standard Method of Measuring and Recording Work Injury Experience.* Only disabling injuries had been counted in the injury rates; occupational illnesses were seldom, if ever, recorded, except in the most obvious and extreme cases, and the information was limited by its dependence on voluntary reports from employers.

The OSHA statute required the secretary of labor to issue regulations requiring employers "to maintain accurate records of, and to make periodic reports on, work-related deaths, injuries and illnesses other than minor injuries requiring only first aid treatment and which do not involve medical treatment, loss of consciousness, restriction of work or motion or transfer to another job." The secretary was also required "to develop and maintain an effective program of collection, compilation, and analysis of occupational safety and health statistics." Soon after the act was passed, the secretary directed the BLS to continue to collect and publish work-injury statistics under the new law. Mandatory record-keeping regulations (Part 1906) were adopted by OSHA and published on July 2, 1971, following public written comment, advice and consultation by an interagency group, and a public hearing held by the Office of Management and Budget (OMB). Under the regulations, employers were required to maintain a log of occupational injuries and illnesses, a supplemental record containing more detailed information on each injury and illness, and a yearly statistical summary of injuries and illnesses. These documents were maintained by all employers at individual establishments, but only a statistical sample of employers was required to report to BLS on injury and illness experience.

The BLS's first full-year survey in 1972 included data for all employments outside of farms and government (which were separately surveyed) and showed almost 5.7 million recordable work-related injuries and illnesses; that is, one out of every 10 workers experienced a job-related injury or illness. Then, as it has in each survey since 1972, BLS emphasized that "underreporting of occupational illnesses is prevalent due to problems of identification and measurement." These problems include "lack of facilities

and trained medical personnel for proper diagnosis; long latency periods thwarting timely detection; questions of occupational illness coverage under workers' compensation; and factors outside the work environment that cloud the work relationship concept." In addition to reporting on numbers of injuries and illnesses, BLS also reports on incidence rates—that is, the number of injuries and illnesses per 100 full-time employees. In 1972, the incidence rate for all recordable cases was 10.9. The incidence rate for lost-workday cases was 3.3, and for nonfatal cases without lost workdays, it was 7.6 (*The President's Report on Occupational Safety and Health,* Dec. 1973).

Additional OSHA activities during this period included establishing a training institute in Rosemont, IL (now located in Des Plaines, IL). The institute opened in January 1972 to provide training primarily for federal and state inspection staff and employees in the private sector. On July 28, 1971, President Nixon signed Executive Order 11,612, implementing Section 19 of the act and setting up the framework for the safety and health program for federal employees. OSHA's role in the federal safety program was essentially advisory; primary responsibility for carrying out the requirements of the law and the executive order was assigned to the individual federal agency heads. Finally, OSHA embarked on a program of cooperation with 28 other federal agencies that had authority relating to occupational safety and health in the private sector. Despite an agreement with the Department of Transportation in May 1972 that was designed to avoid duplication in the handling of complaints in the railroad industry, it became clear very quickly that major jurisdictional disagreements among the various federal agencies were emerging that would be difficult to resolve.

## 1973–1976: The Agency Continues to Grow

### OSHA's First Transition

George C. Guenther left office in January 1973, at the beginning of President Nixon's second term. After a brief hiatus, John H. Stender, a former union official and a Republican legislator from the state of Washington, became assistant secretary of OSHA. He served until July 1975 in an often stormy term of office. There was no assistant secretary from June until December 2, 1975, when Dr. Morton Corn, who had been a professor of industrial hygiene and engineering at the University of Pittsburgh, took office. As the first health professional to head OSHA, Dr. Corn brought about major changes, emphasizing professionalism and effecting a major reorientation toward health regulation and enforcement. Dr. Corn left office in January 1977, when the newly elected President Jimmy Carter chose Dr. Eula Bingham, also a health professional, as assistant secretary.

During the period between 1973 and 1976, OSHA continued to grow, both in size and in the magnitude of the controversy it engendered. The OSHA budget in fiscal year 1972 in current dollars was 33.9 million; by fiscal 1977 it was $130.2 million. (As adjusted for inflation, the growth rate, of course, was less marked: $33.9 million in 1972 and $90.0 million in 1977.) The OSHA inspection staff also grew. At the end of 1972, the field enforcement staff included 456 compliance officers and 68 industrial hygienists. Almost all compliance officer positions authorized for fiscal year 1976 were industrial hygienists, so that in December 1976 there were 358 hygienists in OSHA, or 27 percent of the total staff (*The President's Report on Occupational Safety and Health*, Dec. 1976).

There were parallel increases in the number of OSHA inspections; in particular, Assistant Secretary Stender emphasized the importance of numbers of inspections. In fiscal year 1973, there were 48,409 establishment inspections; by fiscal 1976 the number had grown to 90,482.

The total number of OSHA inspections was reduced in fiscal 1977, largely as a result of the new emphasis on health enforcement that was introduced by Dr. Morton Corn. In fiscal 1976, 8.4 percent of OSHA inspections were listed as health inspections; in fiscal 1977, 15.2 percent, and in fiscal 1978 (the beginning of the administration of Dr. Bingham) the number of health inspections grew further, to 18.6 percent. Health inspections are typically far more time-consuming, requiring, among other things, calibration of equipment, monitoring of workplace atmospheres, and collection and analysis of monitoring results. In 1977, OSHA separated from its *Compliance Manual,* which had come to be known as the *Field Operations Manual (FOM),* a detailed instructions manual for health inspections—then named the *Industrial Hygiene Field Operations Manual (IHFOM).* The *IHFOM* specifies the technical procedures for making an industrial hygiene inspection to ensure uniformity. OSHA also embarked on a three-year apprenticeship program for entry-level health compliance officers.

## The Kepone Incident

The basic OSHA priorities for selecting workplaces for inspection remained basically the same during the Stender and Corn Administrations. Special emphasis programs, focusing on inspections of specific hazardous industries, were adopted. In March 1973, the emphasis was placed on trenching and excavation inspections; a new program was instituted in 1975, called the National Emphasis Program (NEP), targeting foundries for inspection. The purpose of these programs was to combine a variety of OSHA resources including training and education, consultation, and enforcement in concentrating on a single high-hazard area. The most important change in OSHA's inspection priority system was its major modification in response to employee complaints, which took place in 1976 as a result of the

"Kepone incident." In September 1974, OSHA received a complaint from a former employee of Life Science Products Company of Hopewell, VA, a chemical-manufacturing company, alleging his exposure to pesticide fumes and dust. The OSHA area office did not conduct an inspection of the facility because of the complaint's informality, treating it instead as a discrimination complaint under Section 11(c). The matter rested there for about 10 months, when it became known that Life Science employees had been massively exposed to a pesticide, Kepone, which caused serious illness in seven of these employees. The facility was quickly closed down by the state of Virginia, and it became known that pervasive ecological damage had been caused by the company's irresponsible and unlawful disposal of Kepone into the James River.

The great human and environmental damage, caused—on the surface, at least—by OSHA's failure to inspect, led to great public outrage. In March 1976 a subcommittee of the House Labor Committee held a hearing in Hopewell, VA, on the Kepone incident. Dr. Corn had just assumed office, and he was thoroughly and angrily questioned on OSHA's handling of the Kepone complaint, which had preceded his association with OSHA. Although OSHA sometimes was blamed for workplace accidents with little or no justification, the agency's performance in the Kepone matter required much explanation. Dr. Corn wrote to Chairman Gaydos, saying that the "episode has pointed up a distinct need for improvements in OSHA's response to employee complaints of hazardous working conditions" (Corn letter, 1976). Soon thereafter, OSHA revised its field instructions, instructing staff to conduct a workplace inspection "whenever information comes to the attention of the Area Director without regard to its source and without regard to whether it meets the formality requirements of Section 8(f), indicating that safety or health hazards exist at a workplace." The directive also shortened the time frame for complaint responses: no more than 24 hours for imminent dangers, 3 days for serious violations, and 7 working days for nonserious violations (OSHA Field Information Memo No. 76-9, 1976). The Kepone incident also solidified Dr. Corn's intention to focus a greater portion of OSHA's resources on health regulation.

It was expected that the new policy would substantially increase the number of complaint inspections. In fiscal year 1976, complaints were 10.2 percent of total inspections; one year later, complaint inspections constituted 32.4 percent (19,415 out of 60,004 inspections). There were similar high percentages of complaint inspections for the next two years. The impact that this new emphasis had on OSHA enforcement activity was not entirely anticipated, however. In December 1977 OSHA said that the number of complaints received and investigated "overtaxed the resources available—introduced complaint backlogs, reduced inspection activity at some field offices in several important safety and health programs, and severely decimated planned regional inspection programs" (OSHA Prog. Dir. No. 200-69 [1977]). The complaint policy was criticized in a 1979 report of the comptroller general, as will be discussed later, and was significantly modified by Dr. Bingham in 1979.

With the increase in OSHA inspections, there were substantial increases in the number of OSHA citations issued, the numbers of violations of all kinds, and the rate of employer contest of citations and penalties. For example, in fiscal year 1973, 3.2 percent of inspections resulted in serious violations (1,535 violations); by fiscal year 1977 this increased to 18.5 percent of inspections (11,092 violations). The increase in the number of serious citations resulted in part from a series of issue papers by the General Accounting Office (GAO) at the request of the Senate Committee of Labor and Human Resources, which criticized OSHA for its citation policy (*OSHA Review, 1974: Hearings Before the Subcomm. on Labor of the Senate Comm. on Labor and Public Welfare*, 93rd Cong., 2nd Sess., apps. 941–1238 [1974]). The percentages of OSHA inspections finding willful and repeated violations, which are subject to penalties of up to $10,000, were substantially smaller (in fiscal 1977, 0.3 percent willful violations and 3.9 percent repeated violations), but they too increased over the four-year period. The contest rate almost tripled between 1973 and 1977, reflecting the growing adversarial quality of the OSHA program. During 1975, OSHA implemented a comprehensive field performance evaluation system, establishing qualitative and quantitative criteria for federal field activities (*The President's Report on Occupational Safety and Health*, 1975, pp. 49–53).

## OSHA in Court

With increased litigation over citations and penalties, the OSHA enforcement case law evolved at a rapid rate. Some of the proceedings challenged the constitutionality of the enforcement scheme. The issue was decided favorably to OSHA in 1977 in *Atlas Roofing Company Inc. v. OSHRC*, 430 U.S. 442 (1977). In this major constitutional decision, the U.S. Supreme Court held that Congress could constitutionally establish a flexible administrative procedure for the imposition of civil penalties, even though no jury trial was provided.

The general duty clause also resulted in numerous court decisions. In a major case decided in 1974, the Court of Appeals for the Eighth Circuit, *American Smelting and Refining Company v. OSHRC*, 501 F.2d 504 (8th Cir. 1974), held that high levels of airborne concentrations of lead were a "recognized" hazard under Section 5(a)(1), even though they could not be detected except through the use of monitoring equipment. Although a lead standard was soon to become effective, and the general duty clause no longer applied to lead hazards, the basic principle had a major impact. In *National Realty and Construction Co. v. OSHRC*, 489 F.2d 1257 (D.C. Cir. 1973), decided a year before by the Court of Appeals for the District of Columbia Circuit, several principles of general duty law were set forth:

- In determining whether a hazard was "recognized," the standard "would be the common knowledge of safety experts who were familiar with the circumstances of the industry or activity in question."
- A hazard was "likely to cause" death or serious physical harm if the result "could eventuate" "upon other than a freakish or utterly implausible concurrence of circumstances."
- The general duty clause imposes sanctions only on "preventable hazards"; hazardous conduct by employees is not "preventable" "if it is so idiosyncratic in motive or means that conscientious employers familiar with the industry could not take it into account in prescribing a safety program."

This last issue, known as the employee misconduct defense, followed OSHA in its commission and court litigation throughout the agency's history.

The increased court litigation, involving both citations and penalties and, as we shall see, litigation on the review of OSHA standards, raised a critical question as to which group of attorneys would be responsible for this litigation. The act stated that OSHA litigation (except for litigation in the U.S. Supreme Court) would be handled by Department of Labor attorneys in the Office of the Solicitor, "subject to the direction and control of the Attorney General." Almost immediately after the effective date of the act, a dispute arose as to the extent of this Department of Justice "direction and control." The Labor Department claimed that its attorneys, who were experienced and expert in OSHA affairs, were more qualified to handle its litigation with minimum supervision by the Department of Justice. Justice argued, on the other hand, that they were more experienced litigators and that they should handle OSHA litigation, just as they handle most other Executive Department litigation. The Department of Justice's view prevailed at first, causing considerable bad feeling and morale deterioration on the part of OSHA and its attorneys. However, in 1975, an agreement was reached between the Departments of Labor and Justice relating to OSHA and two other enforcement programs, assigning primary litigating authority to the Office of the Solicitor and establishing procedures to determine the rare circumstance when the Department of Justice would become involved. This arrangement has continued to the present time on courts of appeals litigation; Supreme Court litigation continues to be handled by the Solicitor General, and procedures for handling federal district court litigation are normally worked out on a local level between the U.S. attorneys and the regional solicitors' offices.

## Disputes over Jurisdiction

OSHA is not the only federal agency with authority on occupational safety and health. Other agencies, notably the Department of Transportation (DOT), also enforce statutes which, although less comprehensive in scope than OSHA, concern worker protection. To avoid duplication of effort of agencies, as well as to ensure that there is no hiatus in protection, Section 4(b)(1) of the act provides that OSHA does not apply to "working conditions" with respect to which another federal agency "exercise[s] statutory authority to prescribe or enforce standards or regulations affecting occupational safety or health." The meaning of the provision and the delineation of jurisdiction among agencies have been yet another source of continuing controversy and litigation from the start of the program.

OSHA has construed the section to mean that the exemption from OSHA applies only to specific working conditions as to which another agency exercised its authority; that is, only when the agency issues standards affecting occupational safety and health. The railroad industry, among others, often supported by federal regulatory agencies eager to protect their jurisdictions, argued that Section 4(b)(1) provides an "industry exemption" for any industry subject to any occupational safety or health standards. The Review Commission, in 1974, and several courts of appeals, in 1976, rejected the "industry exemption" argument (see, for example, *Southern Railway v. OSHRC,* 539 F.2d 335 [4th Cir.], *cert. denied,* 429 U.S. 999 [1976]), but litigation continued on a variety of related issues under Section 4(b)(1). One of those issues was the meaning of *working condition.* The courts split on whether it means physical surroundings or hazards. Another issue was the definition of *exercise;* how much "exercise" does the other federal agency have to undertake to preempt OSHA (see, for example, *Northwest Airlines,* 8 OSHC 1982 [Review Commission, 1980]).

Litigation, it was generally thought, was a poor and expensive way to decide jurisdictional issues, and considerable effort was then invested by the affected agencies to negotiate jurisdictional agreements to clarify to employers and employees which agency was responsible for occupational safety and health enforcement activity. These agreements were useful in principle, but seldom succeeded in clarifying any jurisdictional issues. In 1972, OSHA's memorandum with the Federal Railroad Administration provided an expeditious method of handling employee complaints in the railroad industry, without attempting to decide jurisdictional questions. However, the agreement was cancelled in 1974 by Assistant Secretary Stender, with no reasons given, and was never renewed. Various agreements were negotiated between OSHA and the U.S. Coast Guard in the DOT in respect to jurisdiction during activities and in the nation's waterways and on the outer continental shelf. In 1983, OSHA conceded authority to the Coast Guard respecting "working conditions of seamen aboard inspected vessels" (48 *FR* 11,365 [1983]). The most successful interagency agreements have been between OSHA and the Department of Interior regarding mine safety, with any remaining jurisdictional issues regarding mines largely resolved when Congress passed the Mine Safety and Health Act of 1977, which broadened the pro-

tection of mine employees and transferred enforcement authority to the newly created Mine Safety and Health Administration in the Department of Labor (30 U.S.C. §801 and following).

Anticipating the possibility of disputes in this sensitive area, Congress in the OSHAct directed the agency to submit a report to the Congress within three years of the act's effective date with recommendations to avoid "unnecessary duplication" and to achieve coordination between agencies. The report was filed six years late, and only after OSHA was sued by members of Congress in a federal court for not submitting a report. The report stated that it was satisfied with the status quo, concluding that court opinions will give "an even clearer picture" of jurisdiction, that cooperative efforts among agencies "will continue to expand," and, therefore, that no new legislation was needed (Mintz, 1984, p. 485).

## Amendments, Riders, and Oversight

The involvement of Congress with the OSHAct did not end with its enactment in 1970. One of the main techniques available to Congress for monitoring agency performance is the oversight function; this activity includes public hearings and, less formally, contacts, letters, speeches, and telephone calls between members of Congress and staff and the agency, all designed to monitor and influence agency action. The source of Congress's influence is its ultimate authority to amend the act, and, on a continuing basis, its power to control the agency's annual appropriations.

The OSHAct was not amended substantively until 1990, but, as will be discussed, appropriations bills have often been accompanied by legislative riders. Oversight hearings have frequently taken place, often with notable impact on agency activity. Although the Senate and House Labor Committees have primary jurisdiction over OSHA issues, and have frequently conducted oversight hearings, other committees have also held oversight hearings on OSHA. Among these are the Small Business Committees, concerned with the impact of OSHA enforcement on small business, and the Government Operations Committees, which have dealt particularly with regulation of toxic chemicals, hazard communication, and the safety of federal employees.

One of the earliest oversight hearings took place in 1972. A House subcommittee on agricultural labor critically questioned Assistant Secretary Guenther on the absence of OSHA standards to protect farm workers and the lack of enforcement activity in agriculture. At the close of that hearing, Chairman O'Hara said to Guenther in an exercise in understatement, "I want you to know that it is not my intention that you leave this hearing with the impression that we want you to slow down on your enforcement of OSHA in agriculture" (*Farm Workers Occupational Safety and Health: Oversight Hearings Before the Subcomm. on Agricultural Labor of the House*

*Comm. on Education and Labor,* 92nd Cong., 2nd Sess. 15–20 [1972]). At an oversight hearing before the Senate Labor Committee in 1974, Assistant Secretary Stender was pressed on OSHA's delay in setting up a standards advisory committee for coke oven emissions. Stender emphasized his frustration with "bureaucratic delays," an explanation that has often been heard at oversight hearings (*OSHA Review 1974: Hearings Before the Subcomm. on Labor of the Senate Comm. on Labor and Public Welfare,* 93rd Cong., 2nd Sess. 221–250 [1974]). Oversight hearings sometimes lead to committee reports, but often they do not. However, the absence of a report does not mean that the hearing was without impact. After the Kepone hearing, at which Assistant Secretary Corn appeared, OSHA changed its complaint policy, even though no committee report was issued.

Sometimes Congress makes its views known to the agency without a hearing. In 1974, the Senate Labor Committee, with the assistance of the GAO, sent 17 issue papers to OSHA that were critical of various aspects of its enforcement activity. One of these, on classification of violations, suggested that the "number of total serious violations should be far greater than that which have been reported." Shortly thereafter, OSHA issued a "major clarification" of policy on serious violations to regional and area offices in order to "focus attention" of field staff on detecting during inspections hazards involving a significant probability of serious harm (*OSHA Review, 1974: Hearings Before the Subcomm. on Labor of the Senate Comm. on Labor and Public Welfare,* 93rd Cong., 2nd Sess., apps. 941–1238 [1974]). In the next three years, the percentage of serious violations cited by OSHA increased from 2.1 percent of all violations in fiscal year 1976 to 29 percent in fiscal year 1979. Other factors, including new and stricter enforcement policies already discussed, played a part; but congressional pressure, it seems clear, was a crucial factor in making sure that field staff would not overlook the need for serious citations.

The appropriation process is another way that Congress influences the policies of administrative agencies. Appropriations committee reports often give "instructions" to an agency on how the money should or should not be spent. An example discussed previously was Congress's direction to OSHA in the 1976 appropriations act to undertake a "review and simplification of existing [national consensus] standards" and to eliminate "nuisance" standards.

Congressional interest in expanding OSHA's on-site consultation program was also transmitted to OSHA through the appropriations process. Section 21(c) provides specific authorization for OSHA to provide for the education and training of employers and employees in the recognition, avoidance, and prevention of unsafe and unhealthful working conditions. From the beginning, OSHA undertook programs to educate and train individuals in the private sector in OSHA matters. Some of these programs were conducted by OSHA at its Des Plaines, IL, training institute or at regional area offices; others were implemented by OSHA contractors such as the National Safety Council and schools for workers at universities. Numerous educational books, pamphlets, and audiovisual materials were distributed as part of this activity. However, an insistent demand continued for an additional component in employer education: on-site consultation. In these consultations, OSHA would provide information on hazards and controls at the worksite without threat of citation or penalty. Early in its history, OSHA determined that when OSHA personnel enter a workplace and observe hazards, the statute requires that a citation must be issued. This determination elicited criticism, but attempts to amend the OSHAct to authorize on-site consultation were opposed by unions and were not enacted.

States with approved plans generally included on-site consultation programs in their plans (the legal interpretation didn't apply to the states), and in 1974 Congressman William A. Steiger, one of the main sponsors of the act in 1970, sponsored an amendment to the OSHA appropriations bill that would authorize additional funds for the specific purpose of financing agreements between OSHA and states without approved plans for on-site consultation. The funds were appropriated and OSHA entered agreements with 12 states, providing 50 percent federal financing for the on-site consultation. The idea was politically attractive, and the Senate Appropriations Committee later directed OSHA to increase the level of funding; OSHA responded, amending its regulations to provide 90 percent financing of these agreements. By 1980, on-site consultation had been made available to employers in all states, either through state plans, through agreements between OSHA and states without plans, or, in limited instances, by private consultants (*Oversight on the Administration of the OSHAct: Hearings Before the Senate Comm. on Labor and Human Resources,* 96th Cong., 2nd Sess., Pt. 1 at 24 [1980] [Testimony of Basil Whiting, deputy assistant secretary of OSHA]).

More commonly, the appropriations process is used by Congress as a check on the agency, through the enactment of limitations on the expenditure of funds—so-called limitations riders. Although the rules of the Congress restrict use of appropriations riders to some extent, in practice they are often introduced and enacted and provide a potent weapon for members of Congress to control agency action without resorting to the formal amendment process. This means that limitations riders are not preceded by committee hearings held by the standing committee with jurisdiction over the statute (in the case of OSHA, the labor committees), which are typically opposed to limitation of agency authority. Often, even the appropriations committee does not consider the rider, which is introduced on the floor. Finally, the pressure to enact appropriations laws is often a strong impetus for Congress to accept the riders. As early as 1972, riders limiting OSHA enforcement activity were passed by one or both houses of Congress, but for a variety of reasons never became law. In 1974, a rider was enacted exempting employers with 10 or fewer employees from OSHA's record-keeping requirements. The provision was incorporated into OSHA regulations and dropped from the appropriation law (29 *CFR* 1904.15).

The breakthrough in OSHA riders took place in 1976, when two major riders were passed. The first exempted small farms with 10 or fewer employees from OSHA enforcement. This rider was Congress' reaction to OSHA's proposal to regulate field sanitation (the history of OSHA's regulation of field sanitation will be discussed later in this chapter) and Congress' irritation at what it viewed to be a patronizing educational pamphlet published by OSHA on farm safety. The debate in the House was remarkable in its extreme hostility to the agency. For example, one congressman from a farm state, a sponsor of the rider, responded to an accusation that he wanted to "castrate OSHA" by saying, "Believe me, my colleagues, I do not want to castrate OSHA because if I do it might grow more rapidly. But if castration is the only solution I would sooner castrate the zealots who are drawing up regulations at OSHA than let them destroy the smaller farmers of America" (122 *Cong. Rec.* 20,366–20,372 [1976]). The other rider that was enacted that year eliminated OSHA authority to impose penalties on companies that are cited for fewer than 10 nonserious violations during an inspection.

Since 1976, five more riders have been added to the OSHA appropriations acts and, with minor changes, have remained in effect to the present. These riders relate for the most part to enforcement against small businesses; one deals with state plan monitoring; one with OSHA–Coast Guard jurisdiction; and another, demonstrating the legislative power of special small interest groups, limits OSHA enforcement in recreational hunting and fishing (Pub. L. no. 98-619, 98 Statutes at Large 3305).

## The State Plan Framework and the Benchmarks Controversy

With the approval of the Virginia Plan in 1976, 23 states and one territory, the Virgin Islands, had received initial approval of their plans. Eleven states that had originally submitted plans had withdrawn them by 1976; six of these (New Jersey, New York, Illinois, Wisconsin, Montana, and North Dakota) had already received OSHA approval when their plans were withdrawn. Five states had never submitted plans. Under the statutory scheme, a state first receives "initial" approval of its plan. After initial approval, federal OSHA has concurrent enforcement authority with the state. This concurrent jurisdiction was designed to ensure that workers in the state would continue to receive occupational safety and health protection during this transition period. This concurrent federal authority is discretionary, so that federal OSHA legally could discontinue its enforcement either in whole or in part at any time between initial and final approval. The transition period ends when the state's plan receives final approval, which is granted only after the state demonstrates in fact that implementation of its occupational safety and health program has been at least as effective as the federal program. After "final" approval, federal authority must end, unless formal proceedings are undertaken. However, because OSHA had decided to give initial approval to plans on the basis of promises to later meet their effectiveness goal (developmental plans), in 1972 OSHA committed itself not to exercise its discretionary authority to terminate federal enforcement during the developmental stage.

This commitment was reexamined and overridden by Assistant Secretary Stender in 1973, who expressed his special concern to avoid redundant state and federal enforcement. He decided to enter into operational agreements with states, thus ending federal enforcement. In order to achieve this sought-after operational status, a state was required to have in place legislation, standards that are at least as effective, a procedure to review enforcement actions, and sufficient number of enforcement personnel. This new policy drew strong criticism, not only from organized labor, which from the start had distrusted OSHA for what the unions believed was its abdication of federal enforcement responsibility to the states, but also from public interest groups and NACOSH. One of NACOSH's subcommittees issued a report objecting to the new policy and urging that federal enforcement continue until there had been evaluation by OSHA of the "in fact" effectiveness of the state plan (*Report of the NACOSH Subcomm. on State Programs*). The NACOSH criticism elicited a strong negative response from Stender, who recommended that NACOSH abolish its subcommittees and concentrate on training and education issues. Congressman Steiger, whose relations with Assistant Secretary Stender were strained, said this step in restricting NACOSH's role would be a "tragic mistake." Stender's operational policy was carried out despite the opposition, and by the end of 1975 there were 13 operational agreements in effect (*The President's Report on Occupational Safety and Health*, 1975, p. 57).

Once a state has met all of its developmental requirements, it is certified by OSHA. After a period of further concentrated monitoring, the state plan is eligible for final approval, which can take place no sooner than one year after certification and three years after initial approval. The determination of whether a state in practice has been at least as effective, and therefore entitled to final approval, or deficient, and therefore subject to withdrawal proceedings, is based on OSHA's monitoring of the plan. OSHA's continuing monitoring of state plan effectiveness is required by Section 18 of the act and has included collection of statistical data, the examination of state case files, and investigation of complaints about state plan performance, known as CASPAs. On the basis of this monitoring, OSHA prepares reports with recommendations to the states for changes to improve effectiveness. Major problems with state plans in the past have been the lack of sufficient health enforcement personnel and lack of thoroughness in inspection activity.

By the end of 1976, five states had received certification: South Carolina, Iowa, Minnesota, North Carolina, and Utah. Fourteen states had operational status agreements (*The President's Report on Occupational Safety and Health*, 1976). However, it would be eight more years before OSHA would give final approval to a state plan. The primary reason for this was the benchmarks litigation, which began in 1974 and did not end until 1983.

The benchmarks litigation involved the validity of OSHA's numerical requirements for state compliance personnel under state plans. OSHA's interpretations of the "at least as effective" requirements were contained in the state plan regulations Part 1902, issued in 1972. Under OSHA's interpretation of the act, states were required to achieve staffs and budgets equal to those federal OSHA would have provided in that state in the absence of a state plan. Because of budget restrictions and resource limitations, it was conceded that the federal OSHA compliance staff was not as large as optimally needed. This meant that OSHA set state staffing requirements—or benchmarks, as they were called—at a relatively low level. Thus, a state could receive initial approval, certification, and final approval without providing—in the words of Section 18(c)(4) of the act, as understood by organized labor—"satisfactory assurances" that the state ultimately (at the end of the developmental period) would have "qualified personnel necessary for the enforcement" of the state program. The AFL–CIO challenged the OSHA interpretation, urging that state programs be required to provide adequate funds and staff "to ensure that normative standards are in fact enforced"; in other words, a judgment would be made as to whether the necessary staff had been hired by the state.

In 1978, the Court of Appeals for the District of Columbia decided *AFL–CIO v. Marshall*, 570 F.2d 1030. It rejected the notion that "at least as effective" means "at least as ineffective" and agreed with the AFL–CIO that personnel and funding benchmarks must be "part of a coherent program to realize a fully effective enforcement effort at some point in the foreseeable future." The Court remanded the case to OSHA for the establishment of criteria in accordance with the decision. In 1980, Assistant Secretary Bingham submitted benchmarks substantially increasing the personnel requirements for the states. Particularly in the health area, these became the basis for continuing disagreement among OSHA, the states, and the AFL–CIO (U.S. Department of Labor news release, April 25, 1980).

## Standards for Construction, Agriculture, and Other Industries

The Stender and Corn Administrations devoted much attention to dealing with criticism of the so-called nuisance safety standards adopted in 1971. Many of these were revoked in 1978; in addition, revisions of all other national consensus standards were undertaken, and a complete revision of subparts on fire protection and electrical hazards was issued in 1980 and 1981.

A number of safety standards projects were completed by OSHA between 1973 and 1976; some of these involved the construction industry. Whereas generally OSHA standards were vertical, applied across the board to all industries, in the construction and maritime industries OSHA's standards were horizontal, applying only to those industries. This special treatment was largely the result of historical reasons; OSHA's start-up standards in these areas were adopted as established federal standards from pre-OSHA statutes applying to these specific industries. Another unique feature of the construction industry's standards was OSHA's obligation to consult with the Construction Safety and Health Advisory Committee, which had been established under the Construction Safety Act to advise OSHA on standards for the construction industry. The construction committee was a standing committee, whereas most other standards advisory committees were ad hoc, meaning that OSHA had discretion as to whether to seek their advice on specific standards.

In 1972 OSHA issued a major addition to the construction standards that required rollover protection structures on construction equipment. Also in 1972, a subpart on power transmission and distribution lines was added to the construction standards. In 1976, OSHA required ground fault circuit interrupters (GFCIs) on electrical circuits in the construction industry.

Another industry that OSHA treated vertically was agriculture. The OSHA start-up standards contained only four limited standards applicable to agriculture. In 1972, the Migrant Legal Action Program and other groups petitioned OSHA to promulgate additional standards for agriculture. Shortly after the petition, in 1973, OSHA issued an ETS protecting farm workers from pesticide exposure. There was protest from Congress and from agricultural employers against OSHA's finding that pesticides create a "grave danger" to workers. The chairman of the Pesticide Subcommittee of OSHA's Agriculture Advisory Committee wrote to the assistant secretary that he was "shocked" by the action because there was "no disagreement" in the subcommittee "regarding the absence of any need" for emergency action (Arant–Stender letter, May 9, 1973). In 1974, the Court of Appeals for the Fifth Circuit vacated the standard, finding that the "easily curable and fleeting effects" of pesticide exposure on health did not meet the statutory requirements for a finding of "grave danger" (*Florida Peach Growers Association v. Department of Labor*, 489 F.2d 120 [5th Cir. 1974]). Although OSHA started rulemaking for a permanent pesticide standard to impose field reentry times for pesticide exposure, sharp disagreement arose with the EPA, which claimed jurisdiction over pesticide regulation under the Federal Insecticide, Fungicide and Rodenticide Act (FIFRA). OSHA conceded to EPA authority, and ultimately, EPA jurisdiction was upheld by the Court of Appeals for the District of Columbia in a suit brought by a migrant worker public interest group against OSHA for abdicating its regulatory responsibility (*Organized Migrants in Community Action v. Brennan*, 520 F.2d 1161 [D.C. Cir. 1975]). OSHA issued two major agriculture standards, which were not challenged in court. The first, requiring rollover protection for agricultural tractors, was issued in 1975; the other, requiring guarding for farm equipment, was issued in early 1976.

Two other OSHA safety standards led to court decisions that were important in the evolution of the law governing OSHA standards. In partially vacating OSHA's action in reducing the number of lavatories required for office employees, the Court of Appeals for the Second Circuit faulted OSHA for the lack of substantial evidence and because of the inadequacy of the statement of reasons. The court said that when the public opposes a provision on "substantial" grounds, OSHA "has the burden of offering some reasoned explanation" (*Associated Industries v. Department of Labor* 487 F.2d 342 [2nd Cir. 1973]).

In *AFL-CIO v. Brennan* (530 F.2d 109 [3rd Cir. 1975]), the Court of Appeals for the Third Circuit followed the precedent of the asbestos case in upholding OSHA's elimination of the so-called no-hands-in-dies machines-guarding requirement. The court agreed with OSHA that the standard was not feasible as originally written and said that "an economically impossible standard would in all likelihood prove unenforceable, inducing employers faced with going out of business to evade rather than to comply with the regulation."

## Health Standards: An Overview

A number of important actions on occupational health standards were taken by OSHA during this period, with a mixed record of success in the courts. Four emergency standards were promulgated—on vinyl chloride, pesticides, 14 carcinogens, and diving. Several were challenged, and all that were challenged were vacated or stayed by a court of appeals. OSHA issued four "permanent," or final, standards—on vinyl chloride, 14 carcinogens, diving, and coke oven emissions. These standards, issued after rulemaking, were affirmed in court of appeals decisions, except the standard for one of the carcinogenic substances, known as MOCA (4,4-methylene(bis)-2-chloroaniline), and one provision—medical surveillance in the diving standard. Several rulemakings were begun but not completed until later; these included proposed revisions of OSHA standards on lead, hearing conservation, arsenic, and asbestos. A number of rulemakings were begun and abandoned; among them were those for trichloroethylene, beryllium, and sulfur dioxide (*The President's Report on Occupational Safety and Health,* 1975, pp. 22–23).

A number of general observations should be made regarding developments in health standards rulemaking and in court review of these proceedings.

**Issues.** All standards rulemaking, particularly on health standards, had become much more lengthy, complex, and controversial. (In the 1974 Annual Report for OSHA, 22 steps were listed in the standards development process [*The President's Report on Occupational Safety and Health,* 1974, pp. 9–11]). Two overriding issues were argued and resolved in standards proceedings: the PEL necessary to protect employees and the economic and technological feasibility of reaching that level through engineering controls. The question as to whether OSHA should require engineering controls as the primary method of achieving the permissible limit recurred in each proceeding. Although the details of the argument varied, the thrust of the business community's contention was that adequate protection from toxic substances could be achieved through the much less costly means of protective equipment. OSHA argued, and continues to maintain, that engineering controls are the preferred method of compliance, and that protective equipment should be used only when the preferred engineering controls were inadequate or not feasible. The basis for their argument is that protective equipment is unreliable because of the uncertainty as to whether it will be worn and whether it will afford complete protection.

Other issues relating to the nature of the protective equipment, medical surveillance, and monitoring and other requirements were often raised in standards proceedings. Affected parties uniformly presented witnesses who gave public testimony on the issues in the proceedings. Examination of witnesses, under the direction of the administrative law judge, routinely took place. At first OSHA allowed public comment but did not itself present witnesses or ask questions, but the agency soon found it necessary to offer expert witnesses and to question other witnesses to better defend the standard in court. Beginning with the Vinyl Chloride proceeding, OSHA contracted for feasibility studies, which became part of the record; often, as in the coke oven emissions and cotton dust proceedings, additional feasibility studies were introduced by employer associations. Presidential orders issued beginning with the administration of President Gerald Ford, who insisted on inflationary impact statements, required economic or regulatory analysis; more recently, President Reagan's Executive Order 12,291 required an agency to prepare a proposed and final regulatory impact analysis (RIA) for all "major" actions.

**Preambles.** As the issues in the health standards proceedings became more complex, and the records more lengthy, the preambles to OSHA's proposed and final standards became considerably more detailed, each including a detailed analysis of the record, a summary of the contentions of the parties, and OSHA's resolution of each issue, with a section-by-section analysis discussing the basis of each provision and presenting preliminary interpretations of these provisions. Some recent standard preambles have been longer than 100 three-columned, printed pages in the *Federal Register.*

The significant trend toward very detailed statements of reasons was also made necessary by the increasing scrutiny that the courts of appeals were giving to OSHA's statements of reasons for the standards. The Court of Appeals for the Third Circuit, for example, vacated the challenged portion of OSHA's ETS on 14 carcinogens because of the inadequacy of the statement of reasons (*Dry Color Manufacturers Association v. Department of Labor,* 486 F.2d 98 [3rd Cir. 1973]).

**Requirements.** The basic content of a health standard remained similar to the content of OSHA's first health standard, which was for asbestos. The key provision is the appropriate PEL, the level above which the employer is not permitted to expose employees. The PEL is usually expressed as an 8-hour TWA. Sometimes a ceiling, or short-term exposure level (STEL), is added; the issue of whether to include a more protective STEL has been critical in the recent ethylene oxide proceeding. An action level, usually one-half the PEL, is also often included in the standard, and is defined as the point where certain provisions of the standard, such as medical surveillance, are mandated. Other requirements (such as medical surveillance, monitoring, training, and record keeping) continued to be included, with refinements and elaborations reflecting agency experience and knowledge. In the area of medical surveillance, for example, OSHA evolved toward a statement of policy that employers were required to make medical examinations available to employees but that OSHA would not require that employees take the examination. (An employer could, of course, make the examination a condition of employment.) Questions on the type and frequency of monitoring and the extent and availability of employer records were often significant in the rulemaking proceedings, both during this period and continuing into the administration of Dr. Bingham. As will be discussed, the issues of mandatory transfer of employees who are at increased risk from exposure and wage retention became major issues after 1976.

**Pace of promulgation.** Because of resource limitations, and the increasing length of the proceedings, OSHA was falling farther and farther behind in its regulation of toxic substances. A variety of strategies were discussed and tried, with little noticeable impact. The so-called standards completion project to fill out the bare-boned health standards adopted in 1971 was initiated by OSHA and NIOSH as a cooperative venture in 1975 but never completed. Later, during the administration of Dr. Bingham, the Carcinogens Policy was issued for the purpose of facilitating the issuance of standards on carcinogens; for a variety of reasons, the policy was never implemented. The OSHA

policy for determining standards' priorities was therefore extremely critical in its overall regulatory effort. Regulation of carcinogens was almost always OSHA's first priority, although other noncarcinogenic toxic substances that have serious health effects and are pervasively present in workplaces were also regulated. Examples of the latter efforts were the cotton dust and lead standards. Court suits requiring OSHA to rearrange its priorities and to initiate rulemaking were undertaken, usually by unions or public interest groups. The field sanitation suit in 1973 is an example; later, suits were brought to require OSHA to start rulemaking on ethylene oxide and formaldehyde. By 1987, court involvement in decisions on OSHA standards priorities were a regular feature of standards activity.

**Court precedents.** The courts of appeals, with only one exception, followed the lead of the Court of Appeals for the District of Columbia in the asbestos case, generally deferring to OSHA's policy judgments so long as they were within the bounds of rationality, particularly when the agency was acting to afford greater protection for workers. Thus, in the vinyl chloride case, the Court of Appeals for the Second Circuit upheld OSHA's one-part-per-million PEL. Even though the Court said "the factual finger points, [but] it does not conclude," it decided that "under the command of OSHA, it remains the duty of the Secretary to act to protect the working man, and to act even in circumstances where existing methodology or research is deficient." In that case, and in other standards review cases, the court upheld OSHA's policy judgment that evidence of carcinogenicity from animal studies should be extrapolated "from mouse to man." The court also applied the doctrine of "technology forcing" in *Vinyl Chloride,* in deciding that despite lack of substantial evidence establishing the technological feasibility of the one-part-per-million PEL, the standard was feasible because employers "simply need more faith in their own technological potentialities" (*Society of Plastics Industry, Inc. v. OSHA,* 509 F.2d 1301 [2nd Cir.], *cert. denied sub nom. Firestone Plastics Co. v. Department of Labor,* 421 U.S. 992 [1975]). On the other hand, in the Court of Appeals for the Fifth Circuit, the view was evolving toward closer scrutiny of OSHA's actions; this appeared first in that court's decisions on the pesticide and diving emergency standards and was proclaimed fully in 1978 when the court set aside OSHA's benzene standard, disagreeing with the deferential view of other courts of appeals (*American Petroleum Inst. v. OSHA,* 581 F.2d 493 [5th Cir. 1978]). This decision was later upheld by the Supreme Court, but on somewhat more narrow grounds.

**Procedural scrutiny.** Although they gave the agency substantial deference on policy judgments, the courts nevertheless insisted on rigorous adherence to the procedural requirements on rulemaking as stated in the OSHAct and the Administrative Procedure Act. As the court of appeals later said in the cotton dust case, the courts' role in the partnership is to ensure that the regulations resulted from a process of reasoned decision making, including "notice to the interested parties of issues presented in the proposed rule,…opportunities for these parties to offer contrary evidence and arguments," and assurance that the agency has "explicated" the basis for its decision (*AFL–CIO v. Marshall* 617 F.2d 636 [D.C. Cir. 1979]). A number of OSHA standards, particularly during the early years, were vacated because of procedural defects.

**Emergency standards.** Throughout OSHA's history, the courts of appeals applied particularly rigorous scrutiny to ETSs. The view was expressed by the Court of Appeals for the Fifth Circuit in the *Pesticides* case as follows: "Extraordinary power is delivered to the Secretary under the emergency provisions of the Occupational Safety and Health Act. That power should be delicately exercised, and only in those emergency situations which require it" (*Florida Peach Growers Association v. Department of Labor,* 489 F.2d 120 [5th Cir. 1974]). Other courts agreed with the Fifth Circuit Court of Appeals, at least on principle; for example, the Third Circuit, in *Dry Color Manufacturers Association v. Department of Labor,* 486 F.2d 98 (3rd Cir. 1973). Although it is clear that some emergency standards, such as the carcinogens standards, were vacated because OSHA had failed to follow proper procedures, increasingly, there were questions on whether any challenged emergency standard could be upheld, particularly after the Fifth Circuit Court of Appeals vacated OSHA's second asbestos emergency standard (*Asbestos Information Association v. OSHA,* 727 F.2d 415 [5th Cir. 1984]). Indeed, throughout the history of OSHA, only in one case, that of *Acrylonitrile,* when the judicial challenge was withdrawn after the court of appeals refused a stay of the standard (*Visitron v. OSHA,* 6 OSHC 1483 [6th Cir. 1978]), did OSHA prevail in court in a proceeding on an emergency standard.

## Health Standards Proceedings

In March 1974, OSHA issued an ETS on vinyl chloride based on recently discovered evidence, both in animal studies and in humans, that the substance causes an unusual form of liver cancer. The ETS was preceded by a brief fact-finding hearing held by the agency, though this was not required by law. The emergency standard was not challenged and, after rulemaking, a permanent standard was issued in October 1974, six months after the ETS, which is within the statutory period. After a stay that lasted for a brief period, the court of appeals affirmed the standard and dissolved the stay. The Supreme Court refused to hear the case. This proceeding was generally recognized as successful both in terms of result and of speed in achieving protection for employees from a life-threatening hazard. In the words of Dr. Irving Selikoff, the regulation was "a success for science in having defined the problem; success for labor in rapid mobilization of concern; success for govern-

ment in urgently collecting data, evaluating it, and translating it into necessary regulations; and success for industry in preparing the necessary engineering controls to minimize or eliminate the hazard" (OTA, 1985, pp. 230–231).

Some other OSHA health standard proceedings were not quite as successful. The regulation of field reentry time for pesticides, which has already been discussed, did not result in any OSHA regulation. The OSHA emergency standard on 14 carcinogens was set aside in part; after rulemaking, OSHA issued 14 permanent standards. The Court of Appeals for the Third Circuit vacated the standard for one of the carcinogens, MOCA, for procedural reasons: The court held that OSHA had failed to use the proper sequence of procedures in obtaining the recommendation of an advisory committee. The court upheld the standard as it applied to another carcinogen, ethyleneimine, deferring to OSHA's policy judgments on the interpretation of scientific evidence, and no challenge was filed as to the standards for other substances. OSHA has not issued another standard on MOCA (*Synthetic Organic Chemical Manufacturers Association v. Brennan,* 503 F.2d 1155; 506 F.2d 385 [3rd Cir. 1974]).

In 1977, OSHA completed rulemaking on the permanent diving standard, and on challenge from the Diving Contractors Association the Court of Appeals for the Fifth Circuit set aside the medical surveillance provisions. The court held that a portion of the medical requirements was beyond the agency's authority because the purpose was to protect the jobs of workers rather than to protect their occupational safety and health (*Taylor Diving & Salvage Co. v. Department of Labor,* 599 F.2d 622 [5th Cir. 1979]).

The widely praised coke oven emission proceeding was completed during Dr. Corn's term. The United Steelworkers of America petitioned OSHA for rulemaking based on strong epidemiological evidence demonstrating that coke oven emissions were carcinogenic. A standards advisory committee was formed in November 1974, under the chairmanship of Dr. Eula Bingham, who later became assistant secretary. (Dr. Bingham's statements favoring medical removal protection at a coke oven advisory committee meeting became an issue in the lead standard court proceeding in 1980, which will be discussed later in this chapter). Based on the advisory committee's recommendations, but differing substantially from them, OSHA issued a proposed coke oven emissions standard in July 1975. After a comment period and a public hearing, OSHA issued a final standard in October 1976. The standard regulated the benzene-soluble fraction of total particulate matter present during the coking process, establishing a PEL of 150 $\mu g/m^3$. It also specified the engineering controls required; most health standards are different in this respect, mandating only a level and permitting the employer to select the specific controls. The standard was upheld by the Court of Appeals of the Third Circuit against an industry challenge; the court found the PEL necessary for the protection of employees and economically and technologically feasible. The court also upheld OSHA's authority to prescribe specific engineering controls (*American Iron and Steel Institute v. OSHA,* 577 F.2d 825 [3rd Cir. 1978]). Industry appealed the case to the Supreme Court, and although the Court agreed to hear the appeal, the petition was withdrawn before decision, and the court of appeals ruling thus became final.

## Federal Safety: Trends in Injury Rates

In an effort to strengthen the federal agency safety and health program, Executive Order 11,807 was issued, effective September 28, 1974, superseding the order that had been issued in 1971. At the end of 1974, OSHA published guideline regulations implementing the new executive order and specifying the responsibilities of the federal agencies and OSHA's Office of Federal Agency Safety Programs (29 *CFR* 1960). The order was based on section 19 of the act, which requires the head of each federal agency, after consulting with employee representatives, to establish and maintain an effective and comprehensive occupational safety and health program consistent with the standards promulgated by OSHA for the private sector.

The underlying premise of the Federal Agency Program was that employees of the federal government need and are also entitled to protection. In addition, federal efforts to require private sector compliance would be severely hampered if the government were shown to have failed to provide a model by keeping its own house in order. However, despite the good intentions, criticism of the federal government's effort continued. In March 1973, the GAO published a report showing the need for improvements in the Federal Agency Program (GAO Report, March 15, 1973). In 1975, hearings were held in Charleston, SC, dealing with the U.S. Navy Shipyard and in Washington DC, with testimony received from various government executive departments, unions, and the GAO (*Safety in the Federal Workplace: Hearings Before a Subcomm. of the House Comm. on Government Operations,* 94th Cong., 1st Sess. [1975]). The Committee on Government Operations' report, entitled "Safety in the Federal Workplace," was issued in 1976 (H.R. rep. No. 94–784 [1976]). The report was sharply critical of the federal effort, pointing to the fact that agency policy directives were "vague and ambiguous," that "deficiencies" existed in consultation with employee representatives, and that the Department of Labor was "not staffed to perform the evaluations" of agency programs mandated by the law. The report noted that agencies had made commitments to improve their programs, but that because of "bureaucratic infighting," little progress had been made in meeting these commitments. The committee made numerous recommendations, including a suggestion that OSHA accelerate its efforts to improve the accident and illness reporting system for federal employees. Congressional hearings and committee reports criticizing the federal government's

internal safety and health effort have been a recurring component of the federal government program.

Meanwhile, the BLS continued to make its annual surveys of injuries and illnesses in the private sector. Based on reports submitted by private employers in 1976, the 1975 statistical survey showed that the overall incidence rate (the number of injuries and illnesses per 100 full-time workers) dropped from 10.4 in 1974 to 9.1 in 1975; however, OSHA said that the reduction could be explained due to the disproportionate decline in manufacturing and contract construction employment from 1974–1975. Despite the improvements, on the average, 1 out of every 11 workers experienced a job-related injury or illness, and the report itself again recognized the deficiency of the illness statistics. Four out of every 10 recorded illnesses were for skin diseases or disorders (*The President's Report on Occupational Safety and Health,* Dec. 1976, pp. 86–105). The decline in incidence rates did not continue, which lead to renewed and bitter controversy over the effectiveness of the OSHA program.

# ■ 1977–1981: Giving Teeth to the Tiger

## A New Assistant Secretary

At an oversight hearing on OSHA held early in the Bingham Administration, Senator Harrison A. Williams, chairman of the Senate Labor Committee, observed that although OSHA originally enjoyed "broad support among legislators and the public," it was now perceived either as a "meddling, mischievous intrusion by the Government into the affairs of our Nation's businesses" or by others as a "paper tiger" (*Oversight on the Administration of the OSHAct: 1978 Hearings Before the Subcomm. on Labor of the Senate Comm. on Human Resources,* 95th Cong., 2nd Sess. [1978]). Jimmy Carter, elected president in 1976, gave high priority to improving the deteriorated image of OSHA. He selected Ray Marshall, a labor economist and professor, as secretary of labor and Dr. Eula Bingham, professor of toxicology at the University of Cincinnati, as assistant secretary. Dr. Bingham was no stranger either to occupational safety and health or to OSHA; she had served in 1973 as a member of the Carcinogens Advisory Committee and in 1974 as chairperson of the Coke Oven Emissions Advisory Committee. The new assistant secretary sought to give teeth to the OSHA tiger and, at the same time, to eliminate the agency as an irritant to business; her basic policy was a "shift to common-sense priorities." The reorganization of the national office staff, begun in 1976 and designed to improve coordination of field activity and to improve the agency's technical support activities, was fully implemented. The executive staff of the agency was almost completely changed. Dr. Bingham remained assistant secretary during President

Carter's entire term of office. This was the longest administration in the history of OSHA. During this four-year period, there were numerous major shifts in OSHA policy and new initiatives. However, it is by no means clear that in 1981 OSHA enjoyed wider support than at the start of the administration.

## Health Standards: The New Priority

From the start, Dr. Bingham emphasized the importance of standards activity, particularly standards regulating health hazards. In April 1977, soon after she came to office, Dr. Bingham told a subcommittee of the House Committee on Government Operations, "Quite honestly, I plan to stretch the resources of the agency in putting out health standards, and I intend to use the ETS authority whenever employees are exposed to grave danger." She added, "All I can say is watch the *Federal Register*" (*Performance of the OSH Administration: Hearings Before the Subcomm. on Manpower and Housing of the House Comm. on Government Operations*, 95th Cong., 1st Sess., 1977, pp. 77–78, 92).

Dr. Bingham's first standards effort was to lower the PEL for benzene, and although it was ultimately unsuccessful in terms of practical protective results, to many, the benzene proceeding was a turning point in the history of OSHA standards rulemaking.

One of OSHA's start-up standards regulated benzene exposure; the standard established a permissible level of 10 parts per million (ppm). More epidemiological evidence demonstrating that benzene causes leukemia had become available, and soon after reaching office, Dr. Bingham's attention was directed to the pressing need for increased protection to workers from benzene hazards. On April 29, 1977, Dr. Bingham signed an ETS lowering the PEL of benzene to 1 ppm, with a ceiling level of 5 ppm.

Both the Industrial Union Department of the AFL–CIO (IUD) and the American Petroleum Institute (API) filed petitions to review the emergency standard. Courts of appeals are divided into separate circuits, and both the IUD and the API sought review in a court that each thought would be favorable to its view, the IUD in the District of Columbia Circuit and API in the Fifth Circuit. A federal statute provides that when petitions are filed in different circuits, the court of appeals that decides the case is the one in which the first petition for review is filed. As a result, interested parties rush to file the first petition, in order to win what has been called the "race to the courthouse." The "race," at best, is unseemly and wasteful; in the case of benzene, the litigation was particularly complex because factual and legal disputes made it difficult to determine which was the first-filed petition. The litigation over the venue of the benzene appeal lasted for five months, during which time the standard was stayed, and therefore not in effect. In September 1977, the Court of Appeals for the District of Columbia issued a decision

transferring the proceeding to the Court of Appeals for the Fifth Circuit. Although the judges agreed that the case should be transferred, they disagreed on the theory, and three separate opinions were written, each presenting a different legal approach. In any event, because an emergency standard remains in effect no longer than six months, OSHA decided that it was not worthwhile to pursue the litigation any farther, and the ETS expired without ever being effective. The agency decided instead to concentrate its efforts on a permanent standard (*Industrial Union Department v. Bingham*, 570 F.2d 965 [D.C. Cir. 1977]).

After full rulemaking OSHA issued a permanent benzene standard, modifying the PEL to 1 ppm as a TWA, and a 5-ppm ceiling. There were no animal data on the carcinogenic effect of benzene, and the human data were at exposure levels considerably higher than the prior PEL of 10 ppm. In its preamble to the final benzene standard, OSHA asserted that the conclusions to be derived from the available data were that higher exposures to a toxic substance carry a greater risk, and because a determination of the precise level of benzene that presents no hazard cannot be made, the question of whether there is a safe level "cannot be answered" on the basis of present knowledge. Prudent health policy, the agency said, requires that the limit be set at the "lowest feasible level," which was found to be 1 ppm (43 *FR* 5918 [1978]). Thus, the lowest-feasible-level policy for carcinogens, as it was called, was the culmination of the regulatory theory, first expressed by OSHA in regulating asbestos, to resolve all doubts in favor of workers' protection.

The permanent standard was challenged primarily by API in the Court of Appeals for the Fifth Circuit, this time without any race to the courthouse. The standard was immediately stayed, and in October 1978 the Court of Appeals vacated the standard on two grounds:

- The agency had failed to provide an estimate, supported by substantial evidence, of the expected benefits from reducing the PEL.

- OSHA did not assess the reasonableness of the relationship between expected costs and benefits.

In more familiar terminology, the court of appeals decided that OSHA must do a cost–benefit analysis. The court's decision was noteworthy because it departed markedly from the decisions of other courts of appeals, which gave great deference to OSHA decisions on health standards. The Fifth Circuit, however, defined its partnership with OSHA in a completely different way, insisting that OSHA "regulate on the basis of more knowledge and fewer assumptions" (*American Petroleum Institute v. OSHA*, 581 F.2d [5th Cir. 1978]).

Both OSHA and the IUD sought review, and the Supreme Court agreed to hear the case. On July 2, 1980, in one of the major regulatory decisions of the decade, the

Supreme Court, in a sharply divided vote, invalidated the benzene standard. The lengthy written opinions of the justices of the Supreme Court, including those concurring on the results but disagreeing on rationale, are almost as significant as the decision itself. There was no majority opinion. The plurality—four justices—ruled that the act addresses only "significant risks" and does not seek to provide a "risk free" workplace; and that in developing standards, OSHA has the burden of showing by substantial evidence that it is addressing a "significant risk" of harm in the workplace, and that the proposed standard would eliminate or reduce that significant risk. Because OSHA had not met that burden (indeed, not knowing of the requirement, it had not even tried to do so), the benzene standard was vacated. The practical impact of the decision was that in the future, OSHA would have to establish the extent of risk from the toxic substance, even of carcinogens, quantitatively—usually by means of quantitative risk assessments—and then find that the risk was "significant." The OSHA "lowest-feasible-level" policy for carcinogens was rejected outright by the Supreme Court, and quantitative risk assessments are now a routine part of OSHA health standards development.

Because the Supreme Court had decided the case on other grounds, it did not have to consider the cost–benefit issue, leaving it for the next case. Both the plurality and the dissenting opinions were sharply worded and partly ideological in their thrust. For example, Justice John Paul Stevens, who wrote the plurality opinion, asserted that it would be "unreasonable to assume that Congress intended to give the Secretary the unprecedented power over industry that would result from the Government's view"; he refused to agree with OSHA that the "mere possibility that some employee somewhere in the country may confront some risk" as a basis for the secretary's requiring "the expenditure of hundreds of millions of dollars to minimize that risk." The dissent written by Justice Marshall was equally strident, accusing the majority of deciding the case not on the basis of congressional intent but rather "in line with the plurality's own view of proper regulatory policy" (*Industrial Union Department v. American Petroleum Institute,* 448 U.S. 607 [1980]). OSHA did not issue a new proposal on benzene until December 1985.

The cost–benefit issue was not resolved by the Supreme Court until the cotton dust proceeding. A proposed cotton dust standard had been published at the end of December 1976, shortly before Dr. Corn left. The rulemaking was completed, and after a major controversy with economists in the White House, who exerted pressure for a less costly final standard, Dr. Bingham issued the standard in June 1978 with only minor changes to accommodate the economists' view. The standard established a PEL of 200 µg/m$^3$ for yarn manufacturing and cotton washing, 750 µg/m$^3$ for slashing and weaving in the textile industry, and 500 µg/m$^3$ for textile mill waste house operations or

for exposure to dust from "lower grade washed cotton" in yarn manufacturing. Rejecting industry arguments during the rulemaking, OSHA refused to perform a cost–benefit analysis, saying that prior attempts to quantify benefits as an aid to decision making had not proven "fruitful"; it based the PEL in the textile industry on an excellent epidemiological study performed by Dr. James Merchant and found that the standard was technologically and economically feasible. The Court of Appeals for the District of Columbia Circuit affirmed the standard for the most part, rejecting the argument that the OSHAct requires cost–benefit analysis. The court said, "Especially where a policy aims to protect the health and lives of thousands of people, the difficulties in comparing widely dispersed benefits with more concentrated and calculable costs may overwhelm the advantages of such analysis" (*AFL v. Marshall,* 617 F.2d 636 [D.C. Cir. 1979]).

Industry appealed to the Supreme Court, and it withdrew its appeal to the Supreme Court in the coke oven emissions case, apparently so that the Supreme Court could direct its full attention to the cotton dust case. By this time, cost–benefit had emerged as the major OSHA regulatory issue. On the one hand, industry argued that cost–benefit would provide a "potential legislative check on what might otherwise amount to the exercise of virtually untrammeled authority" and allow the "correction, through the political process, of actions that are deemed by the Congress to be extreme, unwarranted, and inconsistent with congressional intent" (Brief, *American Textile Manufacturers Institute in American Textile Manufacturers' Institute v. Donovan,* 452 U.S. 490 [1981]). To other sectors of the public, unions and public interest groups, in particular, cost–benefit analysis was an anathema because it "places a monetary value on human life, thereby obliterating the moral purpose which led Congress to pass the OSHAct." (Comments of the United Steelworkers of America on the Advance Notice of Proposed Rulemaking for Cotton Dust, Docket No. 052B, May 29, 1981). The Supreme Court in June 1981 agreed with OSHA and the unions, and upheld the cotton dust standard as it applied to the textile industry. Relying in particular on the fact that neither the act nor the legislative history mentions cost–benefit analysis, the Court held that the act did not require that it be used. In his majority opinion, Justice William Brennan looked to the legislative history and concluded that Congress viewed the costs of safety as a cost of doing business. He quoted Senator Yarborough's statement, which was still relevant to the OSHA program: "We are talking about people's lives, not the indifference of some cost accountants" (*American Textile Manufacturers Institute v. Donovan,* 452 U.S. 490 [1981]). The Supreme Court opinion addressed only the issue in the case: whether the act requires cost–benefit analysis. The decision was generally understood to mean that cost–benefit was prohibited. The Supreme Court, however, opened the regulatory door somewhat, saying that "cost–effectiveness" analysis could be used; that is, once the agency decided what level of protection was necessary, it could adopt the least expensive means to achieve that level. As we shall see, cost–effectiveness analysis in the 1980s became one of the keystones in OSHA standards development.

Dr. Bingham issued two other emergency standards during the first years of her administration:

- The first, in September 1977, was for 1.2-dibromo-3-chloropropane (DBCP); this was based on findings that the substance was a carcinogen and caused sterility, and therefore created a "grave danger." The standard was not challenged and was replaced by a permanent standard in March 1978.

- In January 1978, OSHA issued another ETS, this time for acrylonitrile, also a carcinogen. The standard was challenged in the Court of Appeals for the Sixth Circuit, but the court refused a stay, and the court challenge was withdrawn. The ETS was superseded by a permanent standard in November 1978, which was not challenged.

Although Dr. Bingham had urged Congress to watch the *Federal Register* for emergency standards, none were published by her after January 1978. Indeed, in 1980, OSHA asserted a different policy approach. It said that there might be other occasions when the "level of the Agency's resources including compliance, legal and technical personnel, at a given time, may suggest that employee health may be more effectively protected by concentrating those resources in work on permanent standards" (Carcinogens Policy, 45 *FR* 5002, 5215–5216 [1980]). Undoubtedly, this judgment was also predicated on the legal vulnerability of emergency standards.

That statement was made in OSHA's carcinogens policy, issued in 1980. Its purpose was to provide a framework for the regulation of carcinogens in a "timely and efficient manner." The carcinogens policy contained policy determinations on the issues in the regulation of carcinogens that could be questioned only in specified circumstances in later substance-specific rulemaking proceedings. This, the agency said, was to avoid reargument of the same policy issues in each rulemaking proceeding. Although the lengthy preamble to the carcinogens policy was an important contribution to the principles of identification, classification, and regulation of occupational carcinogens, mainly because of the change of administration, the policy was never used and never resulted in the promulgation of a carcinogen standard; crucial portions of the policy were stayed by the agency in 1981. Thus, the considerable resources expended on the promulgation of the policy have failed at least for the present to achieve the saving of resources, for which it was designed.

Several other health standards were issued during Dr. Bingham's administration. An arsenic standard was issued in May 1978, lowering the PEL for the carcinogenic substance to 10 μg/m³. In 1985, after reconsideration by the agency in light of the intervening Supreme Court decisions in the benzene case and the reaffirmation of the original PEL, the standard was affirmed by the Court of Appeals for the Ninth Circuit (*ASARCO v. OSHA,* 746 F.2d 483, [9th Cir. 1984]).

Another major proceeding was OSHA's regulation of lead hazards. Although not an occupational carcinogen, lead exposure affects numerous employees and industries and had long been known to result in serious illness. A proposed revised lead standard was issued in November 1975, during the Corn Administration; it proposed reducing the PEL for lead from 200 μg/m³ to 100 μg/m³. A final lead standard was issued in November 1978 by Dr. Bingham, further reducing the PEL to 50 μg/m³, requiring the use of engineering controls but affording affected industries periods of up to 10 years to comply with these requirements. The standard also contained a novel provision requiring employers to transfer employees who were at excess risk from lead exposure to lower-exposure jobs, and to maintain their wage levels and seniority generally for a period of up to 18 months while the employees were on the other jobs or laid off. This program came to be known as medical removal protection (MRP). It was designed to protect employees by encouraging participation in the medical surveillance program; the MRP provision had not been a subject of the original proposal, and a reopened hearing on MRP was held before the issuance of the final standard.

The standard was challenged in the court of appeals by the Steelworkers Union and the Lead Industries Association (LIA); two petitions for review were filed simultaneously in two different circuit courts of appeal. After a preliminary round of litigation on venue, the case was transferred to the Court of Appeals for the District of Columbia Circuit. In a lengthy opinion in August 1980, the court, with a vigorous dissent by Judge McKinnon on some issues, rejected numerous procedural challenges raised by LIA, based on lack of adequate notice by OSHA on the permissible level, *ex parte* communications, and bias of Dr. Bingham in prejudging the issue of MRP. The court found that substantial evidence supported the new PEL and that it was feasible for the major industries affected. The court also interpreted OSHA's authority expansively and ruled that it could require wage guarantees under the MRP provision. However, the court directed the agency to determine the feasibility of engineering controls for 38 other industries. The Supreme Court refused *certiorari* (*United Steelworkers v. Marshall,* 647 F.2d 1189, [D.C. Cir. 1980] *cert. denied* 453, U.S. 913 [1981]).

On the last day of her term, Dr. Bingham issued another final health standard regulating occupational noise. The original proposal had been issued in 1974 by Mr. Stender; after numerous studies and reopenings of the record, OSHA decided on January 16, 1980, to retain the

90-decibel level, to be achieved by engineering controls, but to require a hearing conservation program for employees exposed above 85 decibels. Mr. Auchter later suspended the standard and modified it, and extended litigation ensued. The Court of Appeals for the Fourth Circuit, *en banc,* upheld the standard in 1985. This litigation will be discussed later in this chapter.

## Deletion of De Minimis Standards

One of the major accomplishments of the Bingham Administration was the deletion of approximately 600 safety standards in November 1978 (Revocation of Selected General Industry Safety and Health Standards, 43 *FR* 49, 726–49, 727, 1978). These standards were determined by OSHA to be unsuitable for regulatory purposes for a variety of reasons such as that they were obsolete, inconsequential, or directed to public safety. The broad review of OSHA national consensus standards promulgated in 1971 began in 1977, and two complete subparts were revised and simplified: the fire protection standards in 1980 and the electrical standards in 1981. In addition to eliminating provisions unrelated to worker safety and health, lengthy provisions of the reference materials were removed and placed in nonmandatory appendices. OSHA emphasized the "performance" approach, that is, giving employers the flexibility of selecting from among a variety of methods to provide the required protection. A standard regulating commercial diving operations was issued in 1977. A provision on medical surveillance in the standard was vacated by the Court of Appeals for the Fifth Circuit (*Taylor Diving and Salvage Co. v. Department of Labor,* 599 F.2d 622 [5th Cir. 1979]), and OSHA's enforcement of the diving standard was largely superseded when the Coast Guard issued a parallel standard in 1978. Two other safety standards were issued during the Bingham Administration:

- A standard on servicing of multipiece wheel rims.
- A standard on the guarding of low-pitched roof perimeters. This was issued in response to a series of unfavorable court decisions on the issue of whether OSHA's perimeter guarding standard covered roofs (see, for example, *Diamond Roofing Co., Inc. v. OSHRC,* 528 F.2d 645 [5th Cir. 1976]).

## "Common-Sense" Enforcement Policy

The OSHA program of "common-sense priorities" was intended to focus workplace inspections on health hazards, larger workplaces, and the more serious health hazards. From fiscal year 1976 to 1977, the percentage of total inspections that were health inspections rose from 8.4 percent to 15.2 percent, and then to 18.6 percent and 19.2 percent in 1978 and 1979. The focus on health hazards was demonstrated by the continuing growth of the industrial hygiene staff. In 1976, OSHA had 967 safety inspectors and 314 industrial hygienists, but by the end of 1980, there were 972 safety inspectors and 548 hygienists. A critical aspect of Dr. Bingham's common-sense priorities was her emphasis on serious hazards. Under the program, 95 percent of OSHA's programmed inspections targeted the industries with the most serious health and safety hazards; the determination of hazards was based on the annual BLS survey. Similar targeting was initiated for health hazards, but the lack of adequate illness data limited the effectiveness of this targeting. Finally, OSHA directed that the field compliance officers spend the bulk of their time on actual inspection activity, with only 30 percent of their time permitted for support work. The National Emphasis Program continued to be implemented at least through 1977, but new crises brought forth new emphases. A series of grain elevator explosions at the end of 1977, leading to more than 50 employee deaths, resulted in OSHA's targeting that industry for inspection activity and initiating a review of standards in the industry (*The President's Report on Occupational Safety and Health,* 1977). The new OSHA targeting resulted in an increase in inspections in the more hazardous manufacturing sector (43.7 percent in fiscal year 1976, 52.1 percent in fiscal 1977, and 52.3 percent in fiscal 1978). The percentage of inspections in the hazardous construction sector also grew, but more slowly, rising from 25.9 percent in fiscal 1977 to 45.5 percent in fiscal 1981. By 1983, the construction percentage had grown even more, to 58.1 percent. The percentage of inspections in other industries continued to drop as the number in the targeted industries increased.

During this period OSHA continued to give priority to inspections in response to employee complaints; however, a major change was made in 1979 on the issue of informal complaints. Because a great proportion of OSHA inspection resources was being expended on inspecting in response to nonwritten, sometimes nonemployee, complaints, many of which did not result in the discovery of workplace hazards, OSHA revised its Kepone-inspired policy and decided that inspections would be conducted in response to informal complaints only when they appeared to involve imminent dangers or extremely serious hazards. Otherwise, OSHA said, it would send a letter to the employer, advising him or her of the complaint "and of the action required." If the employer response was satisfactory, no inspection would be conducted; if unsatisfactory, or no response was received, an inspection would be conducted. In addition, as a check, random inspections would be conducted in the case of every 10th informal complaint. As a result of this change, the percentage of complaint inspections dropped significantly, from a high of 37.6 percent in 1978 to 23.4 percent in fiscal 1981. This made possible substantial increases in the numbers of targeted inspections, which, of course, was the major goal of the new policy. The issue of response to employee complaints again became a major issue—this time legislative—in 1980 with the introduction of the Schweiker bill, as will be discussed in the section relating to OSHA and Congress.

During Dr. Bingham's administration, OSHA also embarked on a serious effort to give added credibility to the sanctions imposed for violations disclosed during workplace inspections. In the past, no sanctions were imposed for nonserious violations (largely because of congressional action in appropriations riders), and even when fines were imposed for more serious hazards, they were rarely high enough to constitute a deterrence to future violation. The charge that OSHA was a "paper tiger" was based in major part on the lack of meaningful sanctions. Under Dr. Bingham, the amounts of proposed penalties for serious, willful, repeated, and failure-to-abate violations all rose sharply. For example, in fiscal 1977, the proposed penalties for serious violations were approximately $6 million; in fiscal 1980, they were just over $11.3 million. In February 1980, OSHA proposed a record penalty (up until then, the record had been $786,190) against Newport News Shipbuilding and Dry Dock Company for 551 alleged safety and 66 alleged health violations. A major factor in the increase in total penalties was the parallel increases in the percentages of serious and willful violations being cited. Increased emphasis was also placed by OSHA on sending criminal cases to the U.S. Justice Department. Under the OSHAct, criminal penalties can be imposed for willful violations of a standard causing the death of an employee. Instructions emphasizing the importance of the criminal sanctions were sent to field staff, and in 1980, for example, nine cases were referred to the Justice Department for possible criminal action, bringing the total since OSHA began to 27 referrals (*The President's Report on Occupational Safety and Health*, 1980). However, there were few convictions in these cases, and the effectiveness of the criminal provisions have been seriously questioned.

Not surprisingly, this more vigorous enforcement brought with it a sharp rise in both the rate and the number of contests. In fiscal 1973, the rate of contest was 2.7 percent, with 1,315 cases being contested; in fiscal 1980 the rate was 11.7 percent, and there were 7,391 contested cases. This meant that the backlog of cases for decision by the Review Commission, which had always been a problem, increased at an alarming rate. In addition, regional solicitors (the attorneys for OSHA) found it increasingly difficult to handle the greater litigation load, particularly because many of the administrative proceedings were complex and time-consuming, involving expert testimony on the feasibility of engineering controls. Attempts by regional solicitors to settle cases with reduced penalties and sometimes by reducing the nature of the violation were often criticized by unions that were parties to the proceedings and even by OSHA officials. Increasing tension between lawyers and clients resulted until 1980, when Dr. Bingham issued a directive to field staff authorizing area directors themselves to adjust citations and penalties before the contest period. This informal settlement policy had the desired effect, and the contest rate

dropped sharply in fiscal 1981 to 6.3 percent. The policy was well received and particularly attractive to the next OSHA administration of Auchter, which gave added emphasis to the policy, leading to an even greater reduction in the contest rate, to 1.9 percent in fiscal 1983.

A major legal development affecting enforcement activity during the Bingham Administration was the Supreme Court decision that Fourth Amendment protection against unreasonable search and seizure applied to OSHA inspections. OSHA argued that Congress had intended prompt and unannounced inspections and that a warrant requirement would delay entry to the workplace, defeat the legislative purpose, and "significantly impede the implementation of OSHA." The Supreme Court, in *Marshall v. Barlow's Inc.*, 436 U.S. 307 (1978) (decided in May 1978) rejected OSHA's arguments, saying that it was unconvinced that the warrant requirement would impose "serious burdens on the inspection system or the courts." In the first place, the Supreme Court said, the "great majority" of businessmen could be expected to consent to inspection. Furthermore, "probable cause in the criminal law sense is not required." According to the Court, OSHA could obtain the warrant by showing that the establishment was selected on an "administrative plan for the enforcement of the Act derived from neutral sources"; the Court referred specifically to OSHA's targeting plan, which was based on accident experience and dispersal of employees. Finally, the Court said, there was no reason why OSHA could not change its regulations to authorize its obtaining *ex parte* warrants; that is, without first notifying the employer or holding a hearing before the district court. Since *Barlow's*, about 3 percent of OSHA inspections have resulted in warrant proceedings.

OSHA had consistently emphasized that the "compliance assistance" aspects of its program (education, consultation, and informational assistance) were an integral part of the total OSHA program. In April 1978, OSHA launched its "New Directions" grants program. Its purpose was to use labor unions, trade associations, educational institutions, and nonprofit organizations to provide job safety and health education and training to employers and employees, including assistance in hazard recognition and control and training in employer and worker rights. The amounts funded by OSHA in the New Directions program increased during the Bingham Administration, and in August 1980, OSHA awarded $3.5 million to 66 organizations for training and education in hazard abatement. This was in addition to 82 continuing grants re-funded at $13.4 million (*The President's Report on Occupational Safety and Health*, 1980). In 1980 oversight hearings before the Senate Labor Committee, Deputy Assistant Secretary Basil Whiting testified on three "success stories" of the New Directions program: health and safety seminars for high-level foundry management sponsored by the Pennsylvania Foundrymen's Association; formal instruction in safety and health to employees of a large electronics manufacturing company by Indiana University; resolution of a problem causing job illnesses by the Machinists Union, in cooperation with management. At the same oversight hearing, Mr. Whiting testified on the continued expansion of the on-site consultation program. He noted that employer demand for those services had grown and that in 1979 one out of every six OSHA-related visits to worksites was a consultative visit and not an inspection (*Oversight on the Administration of the Occupational Safety and Health Act: Hearings Before the Senate Comm. on Labor and Human Resources,* 96th Cong., 2nd Sess. 24 [1980] [referred to hereafter as *Oversight Hearings, 1980*]).

## Increasing Worker Participation

Secretary Ray Marshall said in 1980, "During its first 7 years, the Agency dealt almost exclusively" with the rights of the employer "with little attention to the role of the workers in recognition and abatement of hazards." This, he said, had been remedied by the Bingham Administration, which concentrated equally on the contributions "all parties can make" (*Oversight Hearings,* 1980, pp. 1034-1044). One of OSHA's first steps in this direction was the revision of instructions in the FOM in 1978 to assure employees and to encourage employee participation in the opening, closing, and informal conferences, "when practical." If it was not practical to hold a joint conference with employers and employees, the instructions required separate conferences to be held (Program Directive No. 200-82, Aug. 15, 1978). The issue of the payment of employee wages for walkaround time reemerged at the start of Dr. Bingham's term of office. Early in the history of OSHA, a legal determination was made that employees were not entitled to pay for walkaround time. One of Dr. Bingham's first actions in 1977 was to reverse that decision; the solicitor of labor issued a new interpretation that the employer's refusal to pay for walkaround was discrimination (because the exercise was a protected right) and therefore illegal. The U.S. Chamber of Commerce challenged this interpretation, and in 1980 the Court of Appeals for the District of Columbia upset the walkaround pay rule on the grounds that it had been issued by OSHA without OSHA first having gone through public notice and comment proceedings. Sharply criticizing OSHA for its "high handed" action and for treating the procedural obligations as "meaningless ritual," Judge Tamm noted for the court that public comment serves the practical purposes of reducing the risk of factual errors, arbitrary actions, and unforeseen detrimental actions (*Chamber of Commerce of the United States v. OSHA,* 636 F.2d 464 [D.C. Cir. 1980]). OSHA immediately started notice-and-comment rulemaking in January 1981, shortly before the end of Dr. Bingham's administration. The 1981 regulation was short-lived, however. On assuming office, Assistant Secretary Auchter delayed the effective date of the rule, and, following a period of public comment, revoked it. At present, there is no legal requirement on walkaround pay,

although such pay is required by some collective bargaining agreements.

A particularly troublesome issue was employee participation in commission and court litigation. Although the OSHAct expressly states that workers have a right "to participate as parties to hearings," major disagreements arose on the role of employee parties with respect to the withdrawal of a citation and penalty by OSHA, and employees' right to participate in the settlement of contested citations and penalties. Employee groups argued that they have the right to object to withdrawals and prejudicial settlements, which derives from their statutory right to participate as parties in commission proceedings. They argued further that it is essential that they—for whose protection the law was passed—have the right to prevent OSHA's lawyers from reducing or eliminating citations and penalties for reasons often unrelated to worker participation, such as resource limitations.

The solicitor of labor argued, on the other hand, that it has "prosecutorial discretion" to decide not only whether to issue citations but also whether to prosecute them or settle them. The controversy became particularly sensitive because the OSHA view was being advanced by its lawyers, who were committed to the view that they should control litigation, even though OSHA as agency did not uniformly side with its attorneys on the issue. Attempts to settle the controversy by providing for informal consultation between unions and the solicitor's office were not successful, and the issue was litigated through the commission and the courts. In 1985 the Supreme Court sided with the solicitor, giving the agency exclusive control over withdrawal and settlement of citations (*Cuyahoga Valley Ry. Co. v. United Transportation Union,* 106 S. Ct. 286 [1985]). Nonetheless, the courts have held that employees have the right to appeal adverse commission decisions to the court of appeals, even if OSHA does not wish to pursue the case (*Oil, Chemical & Atomic Workers International Union v. OSHRC [American Cyanamid],* 671 F.2d 643 [D.C. Cir. 1982], *cert. denied* sub nom. *American Cyanamid Co. v. OCAW,* 456 U.S. 969 [1982]).

Section 11(c) of the act, like analogous provisions in other regulatory statutes, protects employees against reprisal in the exercise of their statutory rights. One Section-11(c) case, *Whirlpool Corp. v. Marshall,* reached the Supreme Court. It involved the question of whether employees have the right to refuse to work under conditions that are reasonably believed to be imminently dangerous. The Supreme Court unanimously held that the secretary's regulation affirming this right was valid, saying that the regulation "on its face appears to further the overriding purpose of the Act, and rationally to complement its remedial scheme" (445 U.S. 1 [1980]). The *Whirlpool* case involved a safety hazard—the fall from potentially dangerous heights. The application of the refusal-to-work principle to health hazards, with long latency periods, is less clear and has not been definitively resolved. The issue is analogous to the question of whether health hazards would constitute imminent dangers. This, too, has not been resolved; indeed, throughout the history of OSHA, the number of imminent danger proceedings, either safety or health, has been extremely limited. It is not entirely clear that, as has sometimes been claimed, voluntary action by employers in apparent imminent danger situations has prevented the need for employees to resort to court proceedings.

Criticism of OSHA's implementation of the Section 11(c) program has continued. One of the major issues has been the delays in the processing of employee Section-11(c) complaints by OSHA. Another is the delays in cases reaching trial, even if OSHA decides to prosecute the case. In 1980, a representative of the Oil, Chemical and Atomic Workers International Union told a Senate committee "without exaggeration" that Section 11(c) "no longer works" (*Oversight Hearings*, 1980, pp. 873–879). To help remedy the situation, an attempt was made to imply the right of individual employees to sue an employer under Section 11(c), at least when OSHA refuses to act, but the Court of Appeals for the Sixth Circuit in 1980 decided that the act means what it says: that only the secretary of labor can sue in court to vindicate Section-11(c) rights (*Taylor v. Brighton*, 616 F.2d 256 [6th Cir. 1980]).

The right of employees to know of conditions in the workplace was addressed in the act and amplified considerably during the administration of Dr. Bingham. Under the statute, an employer who is cited must post the citation prominently "at or near" the place where the violation occurred. Also, under the mandates of Sections 6(b)(7) and 8(b)(3), all OSHA substance-specific health standards include provisions requiring labels or other forms of warning on toxic substances, access of employees to their medical records, and provisions requiring that employees have an opportunity to observe workplace monitoring and have access to monitoring records. In July 1978, OSHA amended its record-keeping regulations, originally issued in 1971, to give employees, former employees, and their representatives access to the employer's injury and illness log and to the summary of recorded occupational injuries and illnesses (29 *CFR* 1904.7[b]).

In 1980, another major step in employees' right to know occurred, with the issuance of a regulation providing for employee access to the employer's existing monitoring and medical records. The new rule did not mandate the creation of new records; it applied only to those that employers had already developed under the employer's own ongoing programs. However, the new rule applied to records relating to a broad range of toxic substances—not just to the limited number of substances that OSHA had regulated with specific health standards. In 1982, a federal district court upheld the access rule in all respects, rejecting, among others, arguments based on employer trade secret rights and employee rights of privacy. In May 1984, the Court of Appeals for the Fifth Circuit summarily affirmed the rule without opinion (*Louisiana Chemical Association v. Bingham*, II OSHC 1922 [5th Cir. 1984]).

Probably the most important area of employees' right to know addressed by OSHA was hazard communication. The agency's involvement with requirements for employer identification and communication of hazards in the workplace began almost at the beginning of the program. An advisory committee recommended a standard in 1975, as did the National Institute for Occupational Safety and Health, and the House Government Operations Committee held a number of hearings aimed, at first unsuccessfully, at pressing OSHA to issue a hazard communication standard. Dr. Bingham published a proposed standard on hazards identification, as it was then called, just prior to the end of her administration. The standard would have required employers to assess the hazards in their workplace; and labels containing extensive information about hazards would have been required on all containers, including pipes. The proposal was withdrawn just after the beginning of Auchter's administration, as part of the regulatory reevaluation that was undertaken under Executive Order 12,291. As will be discussed, a significantly revised hazards communication standard was issued in 1983, and in 1985 it was largely upheld by the Court of Appeals for the Third Circuit.

## Continuing Congressional Oversight

Many amendments have been proposed throughout the history of OSHA, almost all curtailing the agency's authority, but none has been passed. OSHA has invariably opposed these weakening amendments, a position that has been generally supported by the labor committees of the two houses, which, as strong proponents of the OSHA program, have consistently refused to report out "anti-OSHA" bills. The pressure for amendment of the OSHAct has been significantly relieved through the device of appropriations riders limiting OSHA authority in various respects, which have functioned as a catharsis for congressional frustration with OSHA and opposition to certain of its policies.

The greatest legislative threat to OSHA was the bill introduced by Senator Richard Schweiker of Pennsylvania in 1979. The bill primarily would have completely revamped OSHA inspection priorities; it sought to reduce OSHA safety inspection activity in "safe" workplaces, "safe" being determined by establishment injury data for past years in that workplace. OSHA's targeting programs had been based on industry-wide rates rather than individual establishments' injury inspection rates. In introducing the bill, which would also have limited OSHA response to employee complaints—even those meeting formality requirements—in "safe" workplaces, Senator Schweiker said, "The bottom line is this: After 9 years under the Act's present safety regulatory scheme, we are left with no demonstrable record that it works and with a bad taste all around from the experience" (125 *Cong. Rec.* 37,135–37,137 [1979]).

Senator Schweiker's assertion that OSHA does not "work" seemed to be borne out by the statistical record of injuries. According to BLS surveys, beginning with 1976 and continuing to 1980, there were increases in both the lost workday case incidence rate and the lost workday incidence rate. In the 1975 survey, the case incidence rate per 100 full-time workers for lost workday cases was 3.3; four years later, it was 4.3. The incidence rate for lost workdays in 1975 was 56.1 and in 1979 it was 67.7. Although the incidence rate for cases without lost workdays went down somewhat during that period, the statistics did not establish the marked improvement in workplace safety that had been anticipated by the OSHAct's sponsors. Dr. Bingham and others countered by saying that statistics "provide only a partial picture of the true state of safety and health in the workplace" (U.S. Department of Labor news release, Nov. 20, 1980). The testimony of Lloyd McBride, president of the United Steelworkers of America, at the Senate oversight hearing also argued that the wider benefits of OSHA were not measurable through statistics (*Oversight Hearings,* 1980).

Assistant Secretary Bingham testified vigorously in opposition to the Schweiker bill on both practical and philosophical grounds. The targeting system would decrease protection, she said. "Is an air carrier not inspected for safety because it had no accidents the prior year?" she asked. Dr. Bingham also sharply criticized the bill for its change in complaint policy, which overturned OSHA's "fundamental policy judgment" that (*Hearings on X. 2153 Before the Senate Comm. on Labor and Human Resources,* 96th Cong., 2nd Sess. 33 36 [1980])

> if an employee actually working in the facility and exposed on a daily basis to hazards cares about his safety and health sufficiently to write the Secretary of Labor about it, and is courageous enough to ask for direct help from OSHA by signing his name to the complaint, that employee deserves and should receive an on-site inspection if the complaint, in OSHA's judgment, appears to have merit.

Organized labor joined OSHA's major effort to defeat the bill, which was never reported by the Senate Labor Committee. In 1979, Senator Schweiker introduced an appropriations rider that was enacted, which served a similar purpose, limiting OSHA inspections of "safe" employers with 10 or fewer workers. An appropriations rider does not require the approval of the agency's standing committee (for OSHA, the labor committees), thus avoiding a major stumbling block to passage. The Schweiker rider has remained in effect.

Simultaneous with its legislative hearings on the Schweiker bill, the Senate Labor Committee also held general oversight hearings on the administration of the OSHAct. As is usually the case in oversight hearings, testimony covering many topics related to the program was presented by representatives of a variety of interests: unions, employers associations, academicians, state officials, professional organizations, public interest groups, and many others. Representatives of OSHA, the agency, typically also appear at oversight hearings; their testimony constitutes a report on the state of the agency, often anticipating issues of concern to the committee. Questioning by members of the committee, particularly of OSHA witnesses, is thorough and often sharp.

In 1979, for example, OSHA testified at six congressional hearings, including one held by the subcommittee on investigations of the House Post Office and Civil Service Committee held a hearing on the Federal Agency Program; a hearing was held by a subcommittee of the Home Labor and Education Committee on the effectiveness of the OSHA enforcement program in the Philadelphia area. In July, OSHA testified on the concerns of small businesses before the Judiciary Committee's Subcommittee on Administrative Practice and Procedure. The OSHA representatives also testified before three different subcommittees on cost–benefit analysis on the issue of extending OSHA coverage to legislative and executive employees and on OSHA enforcement of standards affecting migrant workers (*The President's Report on Occupational Safety and Health,* 1979). Tragedies causing multiple deaths of employees often lead to congressional oversight hearings to determine the cause of the accident and whether action is needed to avoid recurrences. Examples include the hearings after a series of grain elevator explosions that caused many employee deaths in December 1977 and January 1978, and the hearing held by a subcommittee of the House Labor and Education Committee in June 1978 in St. Mary's, West Virginia, following the Willow Island cooling tower collapse, which caused the death of 78 employees (*OSHA Oversight—Willow Island, West Virginia Cooling Tower Collapse: Hearings Before the Subcomm. on Compensation, Health and Safety of the House Comm. on Education and Labor,* 95th Cong., 2nd Sess. [1978]).

The 1980 oversight hearing elicited wide interest, particularly because of the parallel legislative hearing. Many of the witnesses were prominent: Secretary of Labor Marshall appeared in addition to Dr. Bingham and Basil Whiting. Lane Kirkland, president of the AFL–CIO; Howard Samuel, president of the IUD; and Lloyd McBride, president of the United Steelworkers of America, testified for organized labor. (The witnesses representing business interests had appeared for the most part in the prior legislative hearings.) McBride's testimony was a broad-ranging evaluation of OSHA activity over the period of 10 years. He commented, "Probably the major impact of the first decade of OSHA has been the development of an occupational safety and health infrastructure, which, in turn, generates its own ameliorating influence upon the hard conditions of work and the workplace." He cited, among other things, safety and health clauses in bargaining agreements and the increase in the number of professionals in the area (*Oversight Hearings,* 1980, 698–699, 745–751). Of course, not all testimony was favorable to the agency;

many groups and interests including organized labor criticized various facets of OSHA activity. Mike McKevitt, representing the National Federation of Independent Business, claimed that OSHA would continue to have little positive impact unless it gave up its "steadfast adherence" to specification standards. In sum, oversight and legislative activity, including consideration of proposed amendments and the appropriations process, have served as the major means for Congress's continuous monitoring of the administration of regulatory programs.

## Interagency Cooperation and Controversy

OSHA continued to work closely with other federal agencies that had occupational safety and health responsibilities, but disputes continued and it was often necessary for parties to resort to litigation in order to settle jurisdictional issues. An example is *Northwest Airlines,* 8 OSHC 1982 (Review Comm. 1980), involving safety responsibility for maintenance work on airplanes, in which the Review Commission held that the Federal Aviation Administration had exercised authority and preempted OSHA. Some jurisdictional agreements were reached, notably with the Coast Guard in the Department of Transportation (45 *FR* 9142 [1980]) on jurisdiction over employees working on the outer continental shelf and the Mine Safety and Health Administration in the Department of Labor (44 *FR* 22, 827 [1979]). Dr. Bingham was particularly committed to the cooperative governmental efforts of the Interagency Regulatory Liaison Group (IRLG). Established in 1977, the IRLG consisted of five major social regulatory agencies: OSHA, the EPA, the Consumer Product Safety Commission, the Food and Drug Administration, and the Food Safety and Quality Service of the Department of Agriculture. Because these agencies have common regulatory responsibilities, their directors determined that it would be beneficial for them to share their research facilities' knowledge and personnel. The IRLG established seven work groups with the responsibility of developing consistent approaches among the agencies in such areas as compliance and enforcement, epidemiological activity, risk assessments, and testing standards (*The President's Report on Occupational Safety and Health,* 1978). One of the major products of IRLG activity was its publication of a major policy report on procedures for the determination of whether chemicals were carcinogenic and the extent of the risk. The report was prepared by the Risk Assessment Group of IRLG, with the assistance of scientists from the National Cancer Institute and the National Institute of Environmental Health Sciences, and was published in February 1979. (*The President's Report on Occupational Safety and Health,* 1979).

## A New Executive Order

In February 1980, President Jimmy Carter signed Executive Order 12,196, adding many new features to the Federal Employee Safety and Health Program (45 *FR* 45,235 [1980]).

Among the provisions in the order, which greatly strengthened the program, federal agency heads are required to comply with OSHA standards applicable to the private sector unless the secretary of labor approves compliance with alternative standards. For the first time, OSHA was given authority to conduct on-site inspections of federal agency facilities in specified circumstances and to recommend abatement measures. Agency heads were authorized to establish safety and health committees comprising an equal number of management and employee representatives. Such a committee, by majority vote, can request an OSHA inspection if it is not satisfied with the agency response to a report of hazardous working conditions. The order duplicated OSHA's policy on private employers in prohibiting discrimination against employees for the exercise of protected rights and providing for employee walkaround rights. The order also authorizes "official time" for employees participating in activities under the order. A substantial revision of the Department of Labor's regulations, newly entitled "Basic Program Elements for Federal Employee Occupational Safety and Health Programs," that reflected the new executive order was also issued by OSHA in 1980 (29 *CFR* 1960).

## The Bingham Administration Ends

Dr. Bingham's administration ended as it began—with a flurry of activity. During the last several weeks, a number of standards and regulations were issued, on walkaround pay, hearing conservation, hazard communication, and several other items of somewhat less major consequence involving supplemental decisions in lead and the carcinogens policy. Not all were destined to be long-lived, but the assistant secretary's commitment to vigorous standards and enforcement action continued to the last day of her term of office.

# 1981 to 1987: OSHA's Balanced Approach

The election of Ronald Reagan in 1980 inaugurated a new era in the history of OSHA. President Reagan appointed Ray Donovan as secretary of labor and Thorne Auchter as assistant secretary for OSHA. Auchter, who had been a construction executive, served as assistant secretary from March 1981 to March 1984. After Auchter's resignation, Roberts Rowland, a Texas attorney who was previously chairman of the Occupational Safety and Health Review Commission, served as assistant secretary under a recess appointment for about nine months. Rowland's term was controversial, and he resigned in mid-1985, soon after William Brock became secretary of labor. Patrick Tyson, a former attorney in the Office of the Solicitor, headed OSHA in the absence of an assistant secretary until early 1986. John Pendergrass, a certified industrial hygienist, took office on May 22, 1986, and served as assistant secretary until March 31, 1989.

Assistant Secretary Auchter summarized his approach to OSHA to a subcommittee of the House Labor Committee in 1982:

> Only by working together can business, labor and government achieve their common goal of safe and healthful workplaces. The varied authorities granted under the Act allow a wide range of agency activities and programs that involve the government and the private sector in cooperative efforts. We have accordingly developed a balanced program mix that focuses not only on standards-setting and enforcement, but also on ways of helping employers and employees solve safety and health problems in the workplace.

He listed as examples of "self-help approaches" on-site consultation, education and training, and other voluntary protection methods (*OSHA Oversight—Agency Report by Assistant Secretary of Labor for OSHA: Hearings Before the Subcomm. on Health Safety of the House Comm. on Education and Labor,* 97th Cong., 2nd Sess. 2–4 [1982]).

## A New Era in Compliance

Consistent with his philosophy of "working together," Auchter instituted a number of changes in enforcement priorities and procedures. In the past, follow-up inspections (that is, OSHA reinspections of establishments under a commission order to abate hazards to determine if abatement in fact had taken place) were more or less routine. This policy was revised by Assistant Secretary Auchter; he stated that because experience showed that almost all firms visited in follow-up inspections were in compliance, the agency was deemphasizing follow-ups, keeping them to an "essential minimum." This would permit OSHA inspection resources to be focused in higher-priority areas. Sharp reductions in the number of follow-up inspections took place: In fiscal year 1980, follow-ups were 18.4 percent of all inspections; in 1981, 9.5 percent; and in 1982, 2.5 percent. In fiscal year 1986, 3 percent of OSHA inspections were follow-ups (16 *OSHR* 911, 1987). Also noting the high contest rate, Auchter underscored the existing policy of informal conferences between OSHA regional staff and employers, so that, "whenever possible," settlement agreements under which the employer agrees to comply would be reached. As anticipated, the result was, in Auchter's words, a "dramatic drop" in contested cases: from 25 percent of all OSHA inspections in fiscal 1980 to approximately 8 percent in fiscal 1981 and 2.8 percent in 1982. In fiscal 1986, the contest rate was 3.6 percent (16 *OSHR* 912, 1987). OSHA also issued guidelines on the use of the general duty clause to avoid its unwarranted use. In early 1982, one of Auchter's earliest actions was to withdraw the walkaround pay regulation after notice and comment.

Another innovation by OSHA was to target high-hazard establishments in the manufacturing sector for inspections. Under the new procedures, applicable to programmed inspections for general industry in the safety area, compliance officers continued to visit establishments based on the high-hazard industry list. At the beginning of the inspection, however, the compliance officer would inspect the firm's lost workday injury rate (the number of lost workdays per 100 workers). If the particular firm's rate was below the most recently published national lost workday rate for manufacturing, the inspector would not walk through the workplace and conduct a full-scale safety inspection. The purpose of this system, according to Auchter, was to identify and inspect "only those workplaces where there is a high likelihood of finding serious problems." The program did not apply to health inspections, nor did it affect OSHA's response to complaints. In fiscal 1983 there were 71,303 inspections and 8,444 records inspections—a new category, inspections of records and not of workplaces. In fiscal 1986, there were 64,071 inspections, of which 4,619 were records inspections (16 *OSHR* 911, 1987).

Responding to criticism that excluding companies with low lost workday injury rates from the threat of inspection eliminates the incentive for compliance, OSHA announced in 1986 that it was allocating 5 percent, or about 700, of its programmed safety inspections to manufacturing establishments with below-average lost workday rates. In addition, OSHA announced its intention to undertake a comprehensive inspection of every 10th high-hazard industry manufacturer that had undergone only a records review by the agency (15 *OSHR* 867, 1986).

The new, "balanced" approach of OSHA was viewed differently by all groups. Shortly after the program was instituted, the AFL–CIO stated that the system was "ill-conceived" and "unsound," removing numerous manufacturing employers from "one of OSHA's most effective compliance tools . . . the threat of general scheduled safety inspections" (*Oversight on the Administration of the Occupational Safety and Health Act: Joint Hearings Before the Subcomm. on Investigations and General Oversight and Subcomm. on Labor of the Subcomm. on Labor and Human Resources,* 97th Cong., 1st Sess. 171 [1981]).

OSHA has also continued to implement special-emphasis targeting programs in response to new information showing particular hazards to employees. The grain elevator inspection, in effect since 1977, was renewed. In August 1985, OSHA instituted a targeting program for the fireworks industry, after 30 employee deaths occurred in fireworks-manufacturing facilities (16 *OSHR* 133–134, 1986). Following the Bhopal, India, catastrophe, OSHA instituted a pilot Special Emphasis Inspection Effort in the chemical industry in 1985. Congress called for a report on the program, and on July 15, 1985, OSHA preliminarily noted instances disclosed in its inspections of chemical facilities where hazardous conditions were not addressed by OSHA standards (16 *OSHR* 147–148, 1986).

Consultation, education, and training were major components of the Auchter compliance effort. On-site consultation activity was emphasized and expanded. By 1985, 52 states and jurisdictions provided consultation services to employers. Under revised consultation regulations issued in 1984, an employer is given a one-year exemption from OSHA general-schedule inspection (but not complaint inspections or accident investigations) if it undergoes a comprehensive consultation visit, corrects all identified hazards, and demonstrates that it has an effective safety and health program in operation (49 *FR* 25, 082, 1984). The regulations also broadened the scope of consultation to include advice on the "effectiveness of the employer's total management system" to ensure safety and health at the workplace. In 1985, the program's first full year of operation, OSHA granted a total of 382 exemptions to participating employers (*The President's Report on Occupational Safety and Health,* 1985). In light of the diminished chance that an employer would receive an OSHA programmed inspection, however, some have questioned whether there is any great incentive for employers to participate in the consultation program (OTA, 1985).

Three new Voluntary Protection Programs—Star, Praise, and Try—were instituted in 1982. (The Praise Program has since been discontinued.) Each of the programs was directed at a different category of employers, but all were "based on the premise that [an employer] having a comprehensive safety and health program which operates effectively can provide a greater worker protection than the chance of an OSHA enforcement inspection." For employers accepted under one of these programs, OSHA programmed inspections are discontinued but complaint inspections and accident investigations are handled in accordance with regular procedures (*The President's Report on Occupational Safety and Health,* 1985, pp. 60–61). In early 1986, there were 26 general-industry and three construction employers in Star and six general-industry employers in Try. A construction company with injury rates nearly four times the injury average was removed from the Try Program in November 1985—the first time, according to OSHA, that an employer was removed (15 *OSHR* 845, 1985).

The New Directions grant program, which was designed to develop competence in nonprofit organizations for safety and health training and education for employers and employees, continued. Significant changes were made, however. The program funding was cut down substantially, from $13.9 million in fiscal year 1981 to $6.8 million in fiscal years 1982 and 1983, and $5.6 million in 1985 (OTA, 1985; *The President's Report on Occupational Safety and Health,* 1985). In 1981 OSHA stopped using a peer review process under which persons affiliated with grant recipients evaluated applicants for new grants. The reason given was that peer review was "too costly and resulted in a possible conflict of interest" because those who had received grants evaluated applicants (*How OSHA Controls and Monitors Its New Directions Program,* Draft report, GAO No. HRD, 85–29, p. 7 [1984]).

In 1983, OSHA made a major change in New Directions program requirements, making educational and other nonprofit organizations ineligible to receive grants unless they had previously been approved for a planning grant. Educational and other nonprofit organizations can be members of a consortium eligible for a grant, but there must be a labor or employer organization in the consortium that assumes responsibility for submitting the proposal and administering the grant (*The President's Report on Occupational Safety and Health,* 1985). The COSH groups (local safety and health coalitions) were no longer funded, and a 1986 report issued by the public interest group Public Citizen sharply criticized OSHA for eliminating its funding of these occupational safety and health coalitions.

In order to better focus the activity of New Direction grantees in high-priority areas, in November 1985 OSHA announced that in the future it would award grants only to organizations that proposed to develop education and training programs in one of four specific areas: chemical industry, chemical and toxic substances, hazardous waste sites, and new OSHA standards (Notice of Grant Program, 50 *FR* 47, 294 [1985]; *The President's Report on Occupational Safety and Health,* 1985).

The debate over the wisdom and success of OSHA's new orientation continued. In 1983 the Center for Responsive Law, a public interest group, published a report critical of the Auchter Administration, asserting that the terms *voluntary, cooperative,* and *nonadversarial* are "clear code words for regulatory abdication." Much the same view was expressed in another public interest report, *Retreat from Safety,* issued in 1984. Auchter responded to the earlier report of the Center for Responsive Law, saying it was "flawed and biased," that it relied on "inaccurate and misleading statements, undocumented opinions, and unrepresentative anecdotes to support its preconceived conclusions" (13 *OSHR* 408, 1983). When the BLS injury statistics showed improvement during the early years of Auchter's administration, Auchter referred to this fact as demonstrating that the administration's new approach to enforcement had succeeded whereas the prior "tough" enforcement approaches had failed, as evidenced by the growing injury rate before 1981.

## The States: A Partnership Once Again

Assistant Secretary Auchter also took a radically different approach to state programs. He said in 1981 that it was his "firm intention" to "resolve differences" that have existed in the past between the federal and state OSHAs and "to develop a management and policy framework that in the future will integrate the states into the overall OSHA program." In the last analysis, he said, "local problems are best addressed by those closest to them" (*State Implementation of Federal Standards: Hearings Before the Subcomm. on Intergovernmental Regulations of the Senate Comm. on Governmental Affairs,* 97th Cong., 1st Sess. 14–25 [1981]). Among the steps taken by Mr. Auchter to implement this partnership was to terminate the withdrawal proceedings against Indiana, deny the petition for withdrawal of the Virginia plan, and enter into operational agreements with the remaining states with approved plans, thus ending discretionary federal enforcement in those states. One of the major initiatives of the Auchter administration was to radically reverse Dr. Bingham's benchmarks, which Mr. Auchter viewed as the major impediment to final approval because of their "stringent requirement." OSHA supported a rider to its appropriations bill that would preclude the expenditure of agency funds to implement the court of appeals benchmarks decision; this was an unusual response, because the agency had traditionally opposed riders as a means of legislating. The rider was passed, but elicited opposition and resentment, and was deleted by Congress in December 1982, never to be renewed.

At this point, OSHA embarked a broad reconsideration of Bingham benchmarks. At the request of OSHA, a State Plan Task Group was established in August 1983 to work with OSHA in reviewing and revising the benchmarks. The task group decided that the original benchmark formula used in 1980 was "conceptually sound" but that modifications in input were necessary to incorporate, where available, state-specific data and "to build flexibility into the formula to accommodate differences among states" (Kentucky, Final Approval Notice, 50 *FR* 24,884, 24,886 [1985]). Twelve states completed revision of their benchmarks. During 1985, eight states (Arizona, Iowa, Kentucky, Maryland, Minnesota, Tennessee, Utah, and Wyoming) obtained approval of their revised benchmarks and final approval of their state plans, making 11 final approvals in all (*The President's Report on Occupational Safety and Health,* 1985). In January 1986, OSHA approved revised benchmarks for four more states (Indiana, North Carolina, South Carolina, and Virginia), and final approval was granted to the Indiana state plan in September 1986 (51 *FR* 34,215). The AFL–CIO voiced strong opposition to the new benchmarks because of the substantial reductions in the number of inspectors required for final approval. In the 12 states, for example, 171 health inspectors would make the states eligible, instead of 746, as in 1980. The unions argued that the decrease conflicted with the court of appeals decision and that the number of inspectors required failed to provide for, among other things, enforcement of new standards (15 *OSHR* 903–904, 1986). In commenting on the revised benchmarks, the United Steelworkers of America complained that the revision enabled each state to "manipulate" the number of workplaces to be covered by the inspection program in order to justify current staff levels (O'Brien, 1985, p. 1).

Assistant Secretary Auchter also changed OSHA's state monitoring system. Section 19(f) of the OSHAct requires OSHA to make a continuing evaluation of how the state is carrying out its approved plan, which would be the basis for determining whether the state should receive final approval. The main components of OSHA's monitoring system had been case file reviews, spot-check visits by federal inspectors of establishments previously inspected by state enforcement personnel, and accompanied visits by federal inspectors of actual state inspection and consultation visits. An appropriations rider in 1980 radically limited OSHA's spot-check monitoring authority; during that year, state representatives testified at House oversight hearings, strongly criticizing OSHA's monitoring activity.

Under Auchter's new monitoring system, first implemented in August 1983, there was a shift in emphasis in state plan evaluation from "intrusive on-site monitoring to analysis of the state-submitted statistical data" (*The President's Report on Occupational Safety and Health,* 1983, p. 41). The monitoring system compared state statistical information in 11 major program areas such as standards, variances, consultation, and enforcement to federal performance, also as statistically measured, although states were "not necessarily" expected to equal federal performance in respect to each measure. A primary component of the system was the identification and analysis of outliers, that is, "state performance on a particular performance measure that falls outside the established level or range of performance" required in comparison to the federal program (OSHA Instruction STP 2.22A "State Plan Policies and Procedures," 1986, Chapter 3). An outlier is not necessarily a deficiency but requires "further explanation." Most data used to evaluate state performance are obtained through state participation in the Integrated Management Information System (IMIS). As of 1986, 22 states were participating in IMIS. These states provide data to the system either by forms for OSHA entry or by direct entry using OSHA-provided equipment.

On July 1, 1987, the California state program covering private employers ceased operations when funding ended because of a budget dispute (17 *OSHR* 199, 1987).

## Health Standards, Continued

Auchter early on directed his attention to OSHA's health standards activity. The administration's attempt to modify OSHA's historic position against cost–benefit analysis was rejected by the U.S. Supreme Court in the cotton dust case, which held that the act prohibited cost–benefit analysis. This was only the beginning; the health standards area continued to be characterized by controversy, including some major litigation, throughout the Reagan Administration.

Among the major trends in standard development during this period were the following:

■ In 1986, OSHA stated that its approach to standards encompassed the following elements:

  Adopting "less rigid, performance-oriented standards."

  Addressing "existing, significant risks" and adopting requirements that will "significantly" reduce such risks.

Adopting requirements that are technologically and economically feasible.

Using the most "cost-effective" approach (*The President's Report on Occupational Safety and Health*, 1985, pp. 7–8).

■ In determining "significant risk" for health standards, OSHA typically would prepare one or more quantitative risk assessments, by means of which OSHA would quantify the excess risk from the disease for various PELs. Based on these numerical calculations, OSHA would determine whether the risk was significant under the benzene case guidelines.

■ On February 17, 1981, President Reagan issued Executive Order 12,291 "to reduce the burdens of existing and future regulations, increase agency accountability for regulatory actions, provide for presidential oversight of the regulatory process, minimize duplication and conflict of regulations, and ensure well-reasoned regulations" (E.O. 12,291). Significant requirements of the order were the preparation by agencies of a preliminary and final Regulatory Impact Analysis (RIA) for each "major" regulatory action and the review of these analyses by officials in the OMB. As a result of Executive Order 12,219, the OMB was assigned a major role in reviewing OSHA standards activity. In several standards proceedings, OSHA standards actions were questioned and often delayed by OMB. In the ethylene oxide proceeding, the legality of OMB involvement was sharply challenged by the Public Citizen Group (a public interest group). As will be discussed, the court of appeals vacated OSHA's standard partly, but without considering the OMB question.

■ In this period, OSHA undertook only limited new health standard actions, and in some cases it was pressured to do so by litigation. Two new final standards were issued, for asbestos and ethylene oxide, and several rulemaking proceedings are under way at this writing. At the same time, OSHA's attempts to significantly cut back the stringency of previously issued standards were mostly unsuccessful. A notable example is the revision of the cotton dust standard, which, although revised, was retained substantially as originally promulgated in 1979.

■ Somewhat reluctantly, the courts of appeals have been thrust into deciding cases raising the question of whether OSHA should be ordered to commence rulemaking or complete previously delayed rulemaking. Two major decisions were issued by the Court of Appeals for the District of Columbia in this area. The first directed OSHA to promulgate an ethylene oxide proposal within 30 days. Ultimately, the agency issued a final standard, which was further challenged in the court of appeals. Early in 1987, the court of appeals sharply criticized OSHA for its 14-year delay in issuing a field sanitation standard, and directed that a final standard be promulgated within 30 days. The standard was issued in April 1987 (52 *FR* 16,050).

■ OSHA's overall past success in defending its standards in court continued throughout this period. However, this success, as before, did not extend to the Court of Appeals for the Fifth Circuit, which in 1983 set aside OSHA's emergency standard on asbestos. OSHA issued no other emergency standard during the Reagan Administration.

■ A major area of new litigation was the preemption of state standards activity; this arose particularly in connection with OSHA's hazard communication standard.

■ OSHA sought to use "negotiated" rulemaking procedures in developing a revised benzene standard. This cooperative approach had earlier been attempted briefly and unsuccessfully with the coke ovens standard. Although negotiations ultimately failed in the benzene proceeding (OSHA later proposed its own revised standard), the negotiated approach later succeeded in developing a proposed standard for MDA (4,4'-methylene dianiline) (16 *OSHR* 1451, 1987). However, skepticism continued to be expressed about the usefulness, or even legality, of these efforts (15 *OSHR* 942, 1986).

■ The assistant secretary and many others continued to express concern over OSHA's slow pace in standards development. Early in 1987, the Administrative Conference published a report noting "severe management problems" in OSHA's standards-setting process (16 *OSHR* 995, 1987). Congress made clear its concern about OSHA's promptness in issuing standards when it required in the Superfund Amendments and Reauthorization Act of 1986 (SARA) that OSHA issue a protective standard for hazardous waste site operations within 60 days of the date of enactment. The law was enacted on October 17, 1986; the OSHA interim standard, effective immediately, was published on December 19, 1986 (51 *FR* 45,654 [1986]).

The major individual OSHA rulemakings during the Reagan Administration were related to cotton dust, hearing conservation, hazard communication, asbestos, field sanitation, and ethylene oxide.

**Cotton dust.** One of the requirements of Executive Order 12,291 pertained to regulations pending when the Order was issued in 1981. These had to be reviewed by the agency to determine their consistency with the policies of the new executive order. The cotton dust case, involving the issue of cost–benefit analysis, was then pending before the U.S. Supreme Court, briefs having been submitted and oral arguments held. In March 1981, OSHA, through the solicitor general, filed a supplemental memorandum with the Supreme Court, asking the Court to refrain from deciding the case so that the agency would be able to "reconsider the cotton dust standard and the role of cost–benefit analysis under the Act" in light of the new order. The Supreme Court, in deciding the case and prohibiting cost–benefit analysis, rejected OSHA's request for a second chance in a footnote, without giving reasons. Although the Supreme Court thus affirmed the cotton dust standard as it applied to the textile industry, the agency proceeded with its reevaluation of the standard, publishing a notice of proposed revisions in June 1983. Prior to the issuance of the proposal, there was an extended dispute between OSHA and OMB over the scope of the reconsideration—particularly as to whether engineering controls should continue to be required—but in its proposal, OSHA ultimately adhered to its established policy of requiring engineering controls. Revisions in the cotton dust standard were eventually issued in December 1985, following, according to reports, another dispute between OSHA and OMB over the scope of the medical surveillance requirements in the standard. The revised standard, according to OSHA, contained substantial cost savings while maintaining health protection for workers (50 *FR* 51,120 [1985]). Two challenges to the cotton dust standard were rejected by the Court of Appeals for the District of Columbia in 1987.

### Hearing conservation.

The OSHA hearing conservation amendment, issued by Dr. Bingham in January 1981, was also significantly affected by the change of administration. Auchter quickly stayed the effective date of the amendment, and then stayed it a second time; there followed a suit by the AFL–CIO challenging the stays, because they were issued without notice and comment. In the meantime, Auchter issued an interim revised hearing conservation amendment in August 1981 and, after rulemaking, a final revised amendment in March 1983. A variety of changes were made in the Bingham standard, mostly emphasizing a more flexible "performance" approach. In other words, employers were given flexibility in complying, and compliance would therefore be encouraged "in the manner that is easiest under the circumstances present in the particular work environment" (48 *FR* 9738 [1983]).

The standard was challenged by an employer association, and in November 1984, in an unexpected and remarkable decision, the court of appeals, with one dissent, vacated the hearing conservation amendment on the ground that the agency failed to distinguish between hearing losses caused by workplace noise and those caused by nonworkplace noise, and therefore "clearly imposes responsibilities on employers based on nonwork-related hazards." This, the court said, was "not a problem that Congress delegated to OSHA to remedy." OSHA and the AFL–CIO moved for a hearing by the full court; the union emphasized the impact of the court's decision not only on the noise standard but on all other OSHA health standards in which it is often impossible to separate between workplace-related illness and illness caused outside the work-

place. The full court reconsidered the case and in October 1985 reversed the original decision, completely rejecting its reasoning and finding the standard feasible and supported by substantial evidence (*Forging Industry v. Secretary*, 773 F.2d 1436 [4th Cir. 1985]).

### Hazard communication.

The proposed hazard communication standard was also published at the end of Dr. Bingham's administration. The standard was quickly withdrawn by the new administration in February 1981 "for further consideration of regulatory alternatives," and in March 1982 a revised hazard communications proposal was published, limited to the manufacturing sector. The standard also "accommodated" the health interest and the economic interest in trade secret protection by "narrowly defining the circumstances under which specific chemical identity must be disclosed." It is significant that the standard was challenged only by a union (the United Steelworkers of America), a public interest group (Public Citizen), and by several states. Employers did not challenge the standard. Indeed, they generally supported a federal OSHA hazard communication standard; employers hoped its effect would be to preempt the "multiplicity of differing and conflicting State and local hazard communication laws" which "impose an undue burden on products moving in interstate commerce and on multistate employers" (Final Hazard Communication Standard, 48 *FR* 53,280 [1983]).

In May 1985, the Court of Appeals for the Third Circuit issued an important decision, ruling in the favor of the challenging parties on several critical issues. First, the court ruled that OSHA had failed to adequately justify its limitation of the hazard communication requirements to the manufacturing sector, thus ignoring the recorded evidence that workers in sectors outside of manufacturing are also exposed to toxic materials hazards. Secondly, the court found that OSHA afforded unduly broad trade secret protection. Suggesting that a rule that protected only "formula and process information but [required] disclosure of hazardous ingredients" would adequately protect trade secrets, the court of appeals directed OSHA to reconsider this issue as well as the scope of the standard. Finally, the Court rejected OSHA's limitation of access to certain confidential information to health professionals (*United Steelworkers of America v. Auchter*, 763 F.2d 728 [3rd Cir. 1985]).

On September 30, 1986, OSHA issued a final rule amending the hazard communication standard to provide wider access to trade secrets in nonemergency situations, and to eliminate trade secret protection for chemical information that can be discovered readily through reverse engineering processes (51 *FR* 34,590 [1986]). However, OSHA delayed issuing even a proposal on extending the scope of the hazard communication standard for more than 18 months and the court of appeals issued a second decision ordering OSHA to broaden the standard within 60 days, or to explain why it was not feasible to do so (*United Steelworkers v. Pendergrass,* 13 *OSGC* 1305 [3rd Cir. 1987]).

In the meantime, in October 1985, the Court of Appeals for the Third Circuit decided another case involving New Jersey's Right-to-Know Law (New Jersey does not have an OSHA state plan). A variety of business groups claimed that the New Jersey law was preempted by the OSHA hazard communication rule. Under the OSHAct, the laws and regulations relating to occupational safety and health in states without approved OSHA plans are preempted as to issues covered by the federal OSHA program. The court made two distinctions: between the manufacturing and nonmanufacturing sectors and between workplace and environmental hazards. It decided that the New Jersey law was preempted by the OSHA standard only in respect to the identification and disclosure of workplace hazardous substances in the manufacturing area—that is, on the "issue" that was covered by OSHA's standard. There was no preemption on nonmanufacturing activities, the court held, because the OSHA standard did not apply to those activities. Finally, the court ruled that there were unresolved factual questions about whether there was preemption as to state labeling requirements for environmental hazards (*New Jersey State Chamber of Commerce v. Hughey,* 774 F.2d 587 [3rd Cir. 1985]).

In September 1986, courts of appeals handed down two additional decisions further outlining the complex rules on the effect of the hazard communication standard in preempting state and municipal requirements. The first related to the Pennsylvania Right-to-Know Ordinance (*Manufacturers Association of Tri-County v. Krepper,* 801 F.2d 130 [3rd Cir. 1986]; *Ohio Manufacturers' Association v. City of Akron,* 801 F.2d 824 [6th Cir. 1986]). (Note: Effective May 4, 1988, all employees were required to follow the rules of the hazard communication standard.)

**Asbestos.** Assistant Secretary Auchter took a major regulatory action in November 1983, issuing an ETS lowering the PEL for asbestos to 0.5 fibers per cubic centimeter of air. The existing PEL was 2 fibers per cubic centimeter, which was established in 1976 at the expiration of the four-year delayed effective date of OSHA's first asbestos standard, issued in 1972. In 1975, meanwhile, OSHA proposed a 0.5 PEL for asbestos, but no further action had been taken on the proposal. In 1983, OSHA performed a risk assessment and concluded that for all workers exposed to 0.5 fibers per cubic centimeter, there would be an estimated 196 excess cancer deaths per 1,000 workers with 45 years of exposure, 139 deaths for those with 20 years, 10 for one year, and 6 extra deaths per 1,000 workers with six months of exposure. On the basis of these statistics, OSHA said, "The overall extraordinary degree of risk, the extent that very high risk is found in many asbestos using industries, and the unusually high quality of the data utilized to make these assessments present a very strong evidentiary basis for a 'grave danger' finding" (ETS

on Asbestos, 48 *FR* 51,086 [1983]). The ETS was quickly challenged by a trade association representing 47 employers and by a number of individual employers. On November 23, the Court of Appeals for the Fifth Circuit granted the stay, and in March 1984 held that the standard was invalid.

The decision of the court of appeals was significant not only because it vacated a particular ETS, but because of its negative implications for any OSHA emergency standards, at least in the Fifth Circuit. This standard, unlike some earlier ones that were vacated, would have regulated a substance that is known to cause death and serious physical harm; OSHA had performed extensive and careful risk assessments, statistically demonstrating the excess risk at various levels of asbestos exposure, and the preamble set forth the agency data, its reasoning, and findings in detail. Yet the court of appeals, once again warning that the emergency authority must be "delicately exercised," faulted OSHA for its mistake in calculating the number of deaths that could be avoided over a six-month period— "substantially less than 80," the court said—and held that "evidence based on risk assessment analysis is precisely the type of data that may be more uncritically accepted after public scrutiny, through notice-and-comment rulemaking especially when the conclusions it suggests are controversial or subject to different interpretations" (*Asbestos Information Association v. OSHA,* 727 F.2d 415 [5th Cir. 1984]).

After rulemaking, including a public hearing that lasted from June 19 to July 10, 1984, and a printed record of 55,000 pages on June 20, 1986, OSHA issued two final standards regulating exposure to asbestos and related materials; one standard applied to general industry (including maritime) and the second to the construction industry. The standards lowered the PEL for asbestos to 0.2 fibers per cubic centimeter (OSHA's 1972 standard was 12 fibers per cubic centimeter). The requirements of the construction standard were tailored to the unique characteristics of the industry—notably, the fact that construction industry worksites are nonfixed and are temporary in nature (Final Asbestos Standard, 51 *FR* 22,612 [1986] [to be codified at 29 *CFR* 1910.1001]). Challenges to the new asbestos standard were filed quickly. Two departments of the AFL–CIO sought review, claiming that OSHA's regulation was not stringent enough. The Asbestos Information Association, an employer organization, sought review because it had "questions about the feasibility, particularly of monitoring, at the new Permissible Exposure Limit." Petitions for review were also filed by R. T. Vanderbilt and the National Stone Association, challenging OSHA's regulation of nonasbestiform tremolite, actinolite, and anthophyllite. In July 1986, OSHA stayed the standard insofar as it applied to these substances, pending reconsideration. The review proceedings were all transferred to the Court of Appeals for the District of Columbia (16 *OSHR* 885 1986).

**Field sanitation.** Throughout the history of OSHA, courts have been petitioned to review standards already issued by OSHA. However, more recently the courts have been pressed into another role, that of deciding suits brought to compel OSHA (or other regulatory agencies) to initiate rulemaking actions for the issuance of a standard, or to issue standards when the completion of rulemaking is unduly delayed. In a period when OSHA standards development, in the view of many groups, has slowed down, it is not surprising that this type of proceeding has become more common.

The courts' authority to review standards that have been issued is explicitly granted in the OSHAct. It is less clear whether a court can or should decide whether a standards proceeding should be initiated. A decision to undertake rulemaking typically involves the weighing of competing demands for agency resources; the importance of a variety of projects; and considerations of general policy, priorities, and politics. In such situations, the court obviously has preferred to give the agency great discretion in determining what actions to take. Also, in a suit to compel agency action, there is typically no record before the court on the basis of which a decision could reasonably be made. Despite these strong considerations against review, courts on occasion have found the agency actions so arbitrary as to justify a finding of abuse of agency discretion, and have directed that action be taken. As an alternative, the court may require the agency to make a decision—whether to publish a proposal or not—without determining the particular action that should be undertaken.

The field sanitation proceeding raised the issue of OSHA's failure to promulgate standards requiring toilet and drinking facilities for agricultural workers parallel to the sanitation requirements for nonagricultural workers that it imposed in 1971. A suit was brought in December 1974 by the Migrant Legal Action Program to require OSHA to initiate rulemaking. First OSHA published a proposed standard, but angry congressional pressure forced it to drop the rulemaking. The case reached the Court of Appeals for the District of Columbia for the first time in 1977. The court was willing to give broad deference to OSHA's discretion on rulemaking priorities, saying, "With its broader perspective, and access to a broad range of undertakings, and not merely the program before the Court, the agency has a better capacity than the Court to make the comparative judgments involved in determining priorities and allocating resources." However, the court of appeals insisted that OSHA, which had never said that a field sanitation standard should not be issued, develop a timetable indicating when it would be in a position to complete rulemaking (*National Congress of Hispanic American Citizens [El congreso] v. Usery,* 554 F.2d 1196 [D.C. Cir. 1977]). The agency developed timetables and on several occasions in 1981 said that it could not work on field sanitation for the next two years because of higher priorities, and that it would be between 58 and 63 months before the standard would be issued. The district judge, who had been sympathetic to the suit of the migrant workers from the outset, rejected the timetable, saying that the existence of other OSHA work doesn't justify "relegating a simple Standard of Field Sanitation to the dust bin" (*National Congress of Hispanic American Citizens v. Donovan,* 2142–73 [D.D.C., 1981]). The case was heading for the court of appeals again, but was settled, with OSHA promising that it would make a good faith effort to issue a final standard in 31 months.

Consistent with its commitments, OSHA commenced rulemaking, and public hearings were held in Washington DC, Florida, Texas, Ohio, and California. Over 200 witnesses were heard, and 4,000 pages of transcript were received. On April 16, 1985, with Robert Rowland under a recess appointment, OSHA published a *Federal Register* notice saying that no final field sanitation standard would be issued. OSHA gave three reasons: Other enforcement priorities would make it difficult for the agency to enforce a field sanitation standard, and therefore future expenditure of resources on the standard would serve no useful purpose; it would be more appropriate under principles of federalism for the field sanitation issue to be regulated by the states; and, finally, OSHA was reluctant to preempt field sanitation standards already issued by states with approved plans (Decision, 50 *FR* 15,086 [1985]). (Because OSHA up to that time had no standard on the issue, there was no preemption.) There was a storm of angry criticism of OSHA, including congressional oversight hearings, and litigation was renewed in the court of appeals in a new proceeding brought by the Farmworker Justice Fund. The public interest brief to the court begins by saying "It is difficult to imagine an agency action less supported by substantial evidence in the record than OSHA's refusal to adopt the Field Sanitation Standard" (Brief of Migrant Legal Action Program in *Farmworker Justice Fund Inc. v. Brock,* D.C. Cir. No. 85–1349, p. 27 [1985]).

Meanwhile, William E. Brock became secretary of labor, and Robert Rowland left the agency. In October 1985, in a notice signed by Deputy Assistant Secretary Patrick R. Tyson, OSHA set aside its original determination not to issue a standard. The new decision again concludes that state regulation of field sanitation "would be preferable to, and more effective than, federal actions," but this time OSHA gave the states 18 months (until April 1987) to develop and implement field sanitation standards of their own; if the state response was inadequate, as measured by certain criteria established by OSHA, OSHA would within six months issue a federal field sanitation standard (Comment Period Reopened, 50 *FR* 42,660 [1985]). OSHA stated that the state field sanitation standards would have to provide "protection equivalent to the Federal Field Sanitation Proposal of 1984. . . . At the same time," according to OSHA, "specific requirements may vary from the Federal Proposal." Finally, OSHA insisted that the states must have "adequate enforcement programs."

Testifying on November 6, 1985, before a subcommittee of the House Government Operations Committee, Department of Labor representatives stated confidently that the secretary of labor "has ended, not prolonged, the 13-year debate over field sanitation because he has made an unequivocal commitment to provide additional needed protections to American agricultural field workers" (*OSHA's Failure to Establish a Farmworker Field Sanitation Standard: Hearing Before a Subcomm. of the House Comm. on Government Operations,* 99th Cong., 1st Sess. 79 [1985]).

OSHA's statement was premature, for on February 6, 1987, the Court of Appeals for the District of Columbia decided that OSHA must issue a final field sanitation standard within 30 days. The court angrily stated that the agency had utilized an "arsenal of administrative law doctrines" as a justification for "ricocheting" the case between OSHA and the courts for over a decade. It expressed the hope that its decision would "bring to an end this disgraceful chapter of legal neglect." In particular, the court concluded that OSHA was legally wrong in relying on state action, and had acted unreasonably in delaying the standard for yet another two years in the "unsupported and unrealistic" hope that the states would move *en masse* to issue field sanitation standards (*Farmworker Justice Fund v. Brock,* 811 F.2d 613 [D.C. Cir. 1987]). OSHA issued the field sanitation standard on April 28, and the court then vacated its decision as "moot."

**Ethylene oxide.** The ethylene oxide (ETO) proceeding raised several important issues concerning OSHA rulemaking. At the proceedings, on petition from the public interest group Public Citizen, the court of appeals in 1983 directed OSHA to initiate rulemaking on the carcinogenic substance within 30 days of the decision. The court said, "Three years from announced intent to regulate to final rule is simply too long given the significant risk of grave danger that ETO poses." (*Public Citizen Health Research Group v. Auchter,* 702 F.2d 1150 [D.C. Cir. 1983]). OSHA conducted a rulemaking and at the end of the proceeding was prepared to issue a standard containing a TWA of 1 part per million (ppm) and an STEL of 5 ppm. As required by Executive Order 12,291, OSHA submitted the draft standard for OMB review. Following OMB review, OSHA deleted the STEL. Public Citizen challenged the standard in the court of appeals and argued, among other things, that OMB interference with OSHA rulemaking violated the OSHAct, which gives authority to issue standards to the agency and not to OMB, and that the off-the-record communications between OSHA and OMB violate the procedural requirements for rulemaking by denying parties their right of rebuttal. Significantly, a number of chairpersons of House committees filed a brief as a friend of the court, also arguing against OMB control of health and safety regulation. The Justice Department filed a brief for OSHA, defending the right—indeed, the obligation—of the presi-

dent to oversee the Executive Department's regulatory actions, to ensure its consistency with national policy. The proper role of OMB as representative of the president in monitoring the Executive Department Agencies has been widely debated and, in the OSHA context, has also been the subject of congressional hearings. For example, the House Government Operations Committee, following hearings in 1983, issued a report entitled "OMB Interference with OSHA Rulemaking." The majority report said that Executive Order 12,291 had been "used as a backdoor, unpublicized channel of access to the highest levels of political authority in the Administration for industry alone." Twelve members of the Committee filed a dissent arguing in defense of the OMB role (*OMB Interference with OSHA Rulemaking, House Government Operations Comm.,* H.R. REP. NO. 98, 98th Cong., 1st Sess. [1983]).

On July 25, 1986, the Court of Appeals for the District of Columbia issued a major decision, *Public Citizen Health Research Group v. Tyson,* 796 F.2d 1479 (D.C. Cir. 1986), affirming OSHA's long-term exposure limit for ethylene oxide but concluding that OSHA's decision not to set an STEL was not supported by the record and remanding the issue to OSHA for further proceedings. In view of this disposition, the court found it unnecessary to decide the "difficult Constitutional questions concerning the Executive's proper role in administrative proceedings" and the proper scope of power delegated by Congress to certain executive agencies. Meanwhile, OMB's activity in the rulemaking led to consideration in Congress in the form of legislation proposed by Senators Levin, Durenburger, and Rudman. The bill, entitled "The Rulemaking Information Act of 1986" (S. 2033), would require the disclosure to the public of a number of documents exchanged between OMB and executive agencies in the process of review by OMB of agency rules under Executive Order 12,291. At the same time, the House Energy and Commerce Subcommittee on Oversight and Investigation was investigating allegations that OMB interfered with cancer risk assessment guidelines being prepared by OSHA and other regulatory agencies. In response to these actions, OMB agreed in June 1986 to disclose to the public certain information pertaining to its review process.

On July 21, 1987, the court of appeals directed OSHA to issue a final rule on the short-term exposure issue by March 1988 or face a contempt citation (17 *OSHR* 237–238, 1987).

## Safety Standards

A number of actions have been taken on safety standards since 1981. Some were the promulgation of new standards: on marine terminals in July 1983, servicing single and multipiece wheel rims in February 1984, and electrical standards for the construction industry in 1986 (the electrical standard has been challenged in court). Other actions were amendments limiting earlier standards—in particular, exemptions to the diving standards for educational diving issued in November

1982 and revocation of "should" standards in February 1984. In September 1986, OSHA amended its accident tag requirements to provide employers with more flexibility in meeting the standard (51 *FR* 33,251 [1986]). Finally, rulemaking continued in other safety areas, but disputes with OMB led to extensive delays, and no final standards were promulgated; these included oil and gas drilling and grain elevators, the latter an issue that had been high on OSHA's agenda since the grain elevator explosions in 1975 (*Oversight of the OMB Regulatory Review and Planning Process: Hearings Before the Subcomm. on Inter-Governmental Relations of the Senate Governmental Affairs Comm.*, 99th Cong., 2nd Sess. 1–58 [1986]: 16 *OSHR* 260, 15 *OSHR* 557 1985).

## The Federal Program

OSHA recently has undertaken a number of new initiatives in the area of federal employee safety and health, primarily on the basis of the expansion of its authority in a 1980 executive order. In 1983, OSHA revised its staff manual, outlining three types of federal agency inspections that would be conducted by OSHA: unannounced inspections in response to employee reports of hazardous conditions, fatality and catastrophe inspections, and inspections of targeted high-hazard workplaces. The number of OSHA inspections has continued to rise; in fiscal 1985, OSHA conducted 1,883 inspections (*OSHA Oversight, Status of Federal Agency, Health and Safety Programs: Hearings Before the Subcomm. on Health and Safety of the House Comm. on Education and Labor*, 99th Cong., 1st Sess. 2–3 [1985] [referred to hereafter as *OSHA Oversight, Status of Federal Agency*, 1985]; *The President's Report on Occupational Safety and Health*, 1985). OSHA not only inspects federal agency workplaces but evaluates the "complete safety and health programs" of federal agencies. For example, in 1985 OSHA conducted a full evaluation of the Agriculture and Interior Departments and conducted a follow-up evaluation of the programs of the Tennessee Valley Authority, the Postal Service, and the Navy Department (*OSHA Oversight, Status of Federal Agency*, 1985, pp. 8–9).

In 1983, important changes were made to OSHA's practice in reporting on federal injury and illness rates. Previously, the annual statistical report was based on data submitted to OSHA by the agencies. Beginning in 1984, however, OSHA no longer required the submission of data and used as a basis for its surveys compensation claims data that had already been submitted to the Department of Labor by the agencies. Using this new method, OSHA concluded that the incidence rate for all federal civilian employee injuries and illnesses stayed essentially the same from fiscal year 1984 to 1985 (5.77 in 1984; 5.65 in 1985). However, the lost workday case incidence rate dropped from 2.91 to 2.6 percent, continuing a downward trend that began in 1980. OSHA also reported on related statistics on employee compensation charge-back costs. In 1985, the rate of growth increased substantially to 10 percent, although OSHA in its 1986 statistical report asserted that this was not caused by an increase in the number of claims filed.

Criticism of OSHA's efforts continued, with Michael Urquhart, president, Local 12, American Federation of Government Employees, telling a House subcommittee in 1985 that OSHA "has abandoned the goal of reducing worker injuries and illnesses and instead is pursuing the goal of reducing workers' compensation costs." He asserted that the reduction of workplace risks involves commitment, resources, and leadership, "three things we find sorely lacking in OSHA's federal agency program today" (*OSHA Oversight, Status of Federal Agency*, 1985, p. 17).

## BLS Statistical Survey

In November 1986, BLS reported that its 1985 injury and illness survey showed virtually no change in the injury and illness rates between 1984 and 1985; in 1985, there were 7.9 injuries and illnesses for every 100 full-time workers as compared to 8 in 1984. The actual number of injuries and illnesses increased by 1.6 percent, but the rate did not increase because of the increased number of workers and hours of work. Assistant Secretary Pendergrass saw the data as affording encouragement to OSHA on the "course we have mapped out" for the agency. The AFL–CIO, on the other hand, termed the results disappointing, focusing on increases in the service industries, where employment growth had occurred (16 *OSHR* 628, 1986).

In the meantime, serious questions have arisen over the accuracy of the data of injuries and illnesses that are being submitted by employers to BLS. An expert panel under the National Academy of Sciences was asked to report at the end of 1987 on a number of issues related to the record-keeping and reporting system. Labor unions have charged that employers seriously underreport injuries and illnesses, which not only distorts BLS statistics but also results in inappropriate exemptions from OSHA inspection activity. The unions argue that their assertions are confirmed by a number of major OSHA citations and penalties for willful violations of the record-keeping regulations; in January 1987, for example, a large automobile company agreed to pay a $295,000 penalty for record-keeping violations, reduced from $910,000, and in July, OSHA proposed a record $2.59 million fine against a meatpacking company for alleged intentional failure to record 1,000 worker injuries (17 *OSHR* 235, 1987).

In an effort to study the accuracy of BLS statistics, OSHA conducted a pilot program to examine the injury and illness records of 200 employers and compare them with a "reconstructed" picture of the actual number of injuries and illnesses, based on interviews and other records (16 *OSHR* 880–82, 1987).

# 1987 to the Present: Reevaluating OSHA

On December 26, 1990, OSHA observed its twentieth anniversary. (The OSHAct was approved by President Nixon on December 26, 1970, and took effect April 28, 1971.) On this occasion, the agency was saluted for its many accomplishments, having raised the consciousness of the public on the issue of occupational safety and health. Concurrently, however, these commendations were tempered by broad recognition that much remained for OSHA to do, and that in a number of areas, the agency's performance was disappointing.

John Pendergrass, a health professional, became assistant secretary of OSHA on May 22, 1986. During his term of office, the agency moved away from the deregulatory policies that had been emphasized during much of the Reagan Administration, particularly during the tenure of Assistant Secretary Thorne Auchter. The major project undertaken during Pendergrass's administration was the rulemaking to update the more than 400 OSHA air contaminant standards originally issued in 1971. Although this precedent-setting standards project was completed by the agency in an expedited fashion, it ended in disappointment when the entire rule was invalidated in 1992 by the United States Court of Appeals for the Eleventh Circuit. As a result, OSHA was forced to reinstitute its largely outdated 1971 permissible exposure limits. This legal setback cast serious doubt on all of OSHA's efforts to use generic rulemaking to expedite the promulgation of standards and, it has been argued, greatly added to the urgency of the need to amend the OSHAct (which is discussed later in the chapter).

Pendergrass resigned on March 31, 1989, and was replaced by Gerard Scannell, who served as assistant secretary from November 1989 until January 1992. Scannell had served as a high official in early OSHA administrations and, later, as a safety director in private industry. During his administration as assistant secretary, OSHA gave special emphasis to its compliance efforts, and in 1990 the OSHAct was substantively amended for the first time, with the maximum amounts of penalties increased by a factor of seven. Scannell also expanded the use of OSHA's "egregious" penalty policy, which, together with the amended penalty provisions, led to greatly increased penalties in certain cases, thus significantly increasing the credibility of the agency's enforcement program. During the Scannell Administration, the agency responded to a series of tragic petrochemical facilities explosions with a special-emphasis compliance program for that industry and, simultaneously, by developing and issuing a new chemical industry process safety management standard. In 1991, OSHA issued a bloodborne pathogens standard, directed at the hazards of the HIV and hepatitis-B viruses. The agency has also initi-

ated a number of activities directed at ergonomics hazards, which increasingly have been recognized as a serious occupational safety and health issue.

When Scannell left office in 1992, his administration of the OSHA program was widely praised by representatives of labor, management, and a wide variety of other interested groups. There was no appointment of a new assistant secretary during the later part of the Bush Administration, and it was not until June 1993 that President Clinton announced his intention to appoint Joseph Dear, a former official of the Oregon state OSHA program, as assistant secretary.

Dear was not sworn in as assistant secretary of OSHA until November 1993, 22 months after Scannell left office. During the interim period, OSHA was run by a number of acting directors.

On assuming office, Dear described his vision for the agency, which included streamlining the standards process, more effective targeting, and promotion of worker–management cooperation in the workplace. He also said that he would emphasize criminal enforcement and implementation of "egregious" penalties. In the early months of the Dear Administration, there was considerable discussion of the possibility that OSHA reform legislation would finally be passed. However, with the bitter partisan controversy in Congress over a variety of matters, particularly health care reform, the OSHA legislation never reached the floor, and barely one year after Dear assumed office, after the Republican victory in Congress, he spoke of the need to reassess the OSHA program in light of "political realities."

As we approach the beginning of 1996, the future of OSHA is, to say the least, uncertain.

## Supreme Court Decisions

Throughout the history of OSHA, decisions of the U.S. Supreme Court have played a key role in the direction of OSHA policy. During the 1987–1993 period, the Court decided four cases under the OSHAct or involving issues closely related to occupational safety and health.

The first of these, decided on February 21, 1990, involved the authority of the Office of Management and Budget (OMB) under the Paperwork Reduction Act (*Dole v. United Steelworkers of America*). OMB, in reviewing OSHA's expanded hazard communication standard under the Paperwork Reduction Act, disapproved three of its provisions relating to material safety data sheets (MSDSs). This disapproval was challenged by the Steelworkers Union, and the Supreme Court in a split decision agreed with the union and ruled that OMB had acted beyond its statutory authority.

Under OSHA's hazard communication standard, MSDSs must be prepared by employers and made available to employees to inform them of the chemical hazards to which they are exposed. OMB disapproved of OSHA's MSDS requirements applicable to multiemployer worksites, concluding that they increased the paperwork burdens of employers and were not necessary to protect employees. The Supreme Court rejected OMB's interpretation of the Paperwork Reduction Act and ruled that OMB authority under that law was limited to the review of information-collection requirements imposed by a federal agency where the information was for the agency's own use. In OSHA's hazard communication standard, however, the information in the required MSDSs was for the benefit of employees, and the MSDSs were not submitted to OSHA. These requirements were therefore not subject to OMB review, the Court held. Justice White and Chief Justice Rehnquist dissented.

The second case was that of *United Auto Workers v. Johnson Controls Inc.* Handed down by the Supreme Court on March 20, 1991, it was decided under Title VII of the Civil Rights Act of 1964. In that case, the employer had instituted a policy, called the fetal protection policy, that excluded women of childbearing age from areas of the facility involving exposure to lead. The employer argued that the policy was necessary to protect fetuses from harmful exposures to lead, relying on studies showing that lead in a pregnant woman's bloodstream could be transmitted to the fetus through the placenta. The Court rejected the employer's justification for the policy and invalidated the fetal protection policy under Title VII.

The court concluded that the fetal protection policy was discriminatory because it singled women out for special treatment in the face of scientific findings that lead exposure also had adverse reproductive effects on males. The fetal protection policy was not a bona fide occupational qualification under Title VII, the Court further held, because both fertile and infertile women were equally capable of efficient work at a lead battery facility. In thus holding that employment discrimination based on reproductive potential is equivalent to explicit sex discrimination and unlawful, the Supreme Court resolved disagreements among several courts of appeals, and its conclusion was consistent with statements in OSHA's 1978 lead standard. In that standard, OSHA recognized the dangers of exposure to lead for both men and women, and explicitly stated that the standard was not intended to justify the exclusion of women of childbearing age from the workplace.

Under the OSHA statutory scheme, agency citations and penalties are initially reviewed by the Occupational Safety and Health Review Commission (OSHRC), whose decisions are further reviewed by U.S. courts of appeal. Such review often involves the interpretation by both OSHRC and the courts of standards promulgated by OSHA. A well-established principle in administrative law requires that courts defer to the interpretations by an administrative agency of its own regulations and standards. In the case of OSHA, however, there was a difference of opinion between OSHA and the Review Commission and among the courts as to whether the courts should defer to the interpretations of OSHA or to those of OSHRC, because both are administrative agencies under the OSHAct.

The issue was resolved in *Martin v. Occupational Safety and Health Review Commission,* decided on March 20, 1991, which involved an interpretation of OSHA's coke oven standard. OSHRC had rejected OSHA's interpretation of the standard and dismissed OSHA's citation, and the Court of Appeals for the Tenth Circuit deferred to the Review Commission's interpretation. The Supreme Court reversed the decision. The Court held that a "necessary adjunct" to OSHA's statutory authority to issue standards was the power to render authoritative interpretations of those standards. This conclusion was based partly on the statutory structure of the OSHAct and, further, on the practical consideration that because OSHA develops and promulgates standards and enforces these standards, it is in a far better position than the Review Commission, which merely adjudicates cases, to interpret the standards. Accordingly, the Supreme Court held that a court must in the first instance defer to "reasonable" interpretations of standards by OSHA; only if OSHA's interpretation is deemed unreasonable should the court rely on OSHRC's interpretation.

The fourth case, *Gade v. National Solid Wastes Management Association,* was decided on June 19, 1992. It involved the issue of OSHA's preemption of state OSHA enforcement. OSHA had promulgated a final standard protecting employees engaged in hazardous waste operations. The state of Illinois, which does not have an approved state plan, enacted two statutes requiring the licensing of certain hazardous waste equipment operators and laborers. The stated purpose of the legislation was to protect both the employees involved and the general public. (This became known as a dual-purpose statute.) The issue in the case was whether the federal OSHAct preempted the Illinois statutes.

The Supreme Court held that the state statutes were preempted. It first rejected the argument offered by Illinois and unions that OSHA preempts only conflicting state occupational safety and health regulations. Following lower court precedent, the Court held that federal OSHA preempts all state occupational safety and health regulation on an issue covered by an OSHA standard. The Court went on to reject the state argument that dual-purpose state regulations are not preempted because they also have a nonoccupational purpose. The Court said that a state law that "directly, substantially, and specifically" regulates occupational safety and health is an occupational safety and health law, and it is not any less so for preemption analysis by virtue of the fact that it may also have a nonoccupational impact. Four Supreme Court justices dissented.

## Standards Activity: Health Standards

**Air contaminants.** OSHA's major effort to update its 1971 air contaminants standards ended in disappointment when the Court of Appeals for the Eleventh Circuit in July 1992 decided that the agency's action was invalid under the OSHAct.

In 1971, OSHA adopted as start-up health standards approximately 400 PELs for various toxic substances. These were based for the most part on the Threshold Limit Values (TLVs) developed by the American Conference for Government Hygienists (ACGIH) in 1969 and subsequently adopted by the Department of Labor as established federal standards under the Walsh–Healey Act, a predecessor to the OSHAct. Although these PELs were important in providing the agency with a basis for prompt enforcement after enactment of the then-new OSHA statute, over the years significant problems developed as OSHA relied on these standards as a primary basis for OSHA enforcement. By the 1980s, these PELs were in many cases outmoded, providing completely inadequate protection to employees. Although permissible levels for some substances had been updated in substance-specific proceedings, most had not, and it was generally agreed that it would be highly impractical for the agency to embark on individual rulemaking proceedings to update PELs for hundreds of toxic substances. Furthermore, the 1971 health standards contained none of the ancillary provisions such as requirements for medical surveillance and exposure monitoring that have since been routinely contained in separately issued OSHA health standards.

In 1988, OSHA, responding to this unsatisfactory situation, began one of the most extensive rulemakings in its history, proposing to update the 1971 PELs. In January 1989, after expedited rulemaking proceedings, OSHA published a final revised air contaminants standard, in which it required lowered permissible limits for 212 substances included in the 1971 standard, and established new limits for 164 substances not previously regulated in the earlier standard. The new air contaminant standard became effective in September 1989, but the requirement for implementation of engineering controls to meet the new limits was not scheduled to be effective until January 1993. (The new standard did not adopt ancillary requirements for the regulated substances, reserving those for further generic or individual rulemakings.)

Both industry and union groups sought review of the rule updating the PELs, and the review proceeding was transferred to the Court of Appeals for the Eleventh Circuit.

On July 7, 1992, in a far-reaching decision, the court of appeals unanimously vacated the entire air contaminants rule. Although concluding that generic rulemaking was not an inappropriate way for OSHA to issue standards, the court held that in the proceeding under review the agency had failed to meet the specific statutory requirements for rulemaking. The court ruled that, in effect, OSHA had issued 428 individual standards but that it had failed to establish that each of the standards addressed a significant risk in the workplace, as required by the Supreme Court benzene decision, or that the levels were feasible for the industrial operations covered. The court later denied a motion for reconsideration of its decision, and further efforts by OSHA to delay the effective date of the decision were not successful.

OSHA then held further discussions with the Office of the Solicitor General, urging appeal to the U.S. Supreme Court, but the solicitor general ultimately decided that the case was not appropriate for Supreme Court review. So, in March 1993, OSHA advised the public and OSHA field staff that the 1989 revised PELs were not in effect, and that the agency would return to the 1971 levels for enforcement purposes.

Meanwhile, shortly before the decision of the Court of Appeals for the Eleventh Circuit, OSHA had published in the *Federal Register* a lengthy proposal to update the permissible levels for air contaminants in the construction and maritime industries and agriculture. The air contaminant standard issued in 1989 did not apply to these sectors, and many permissible exposure limits in construction, maritime, and agriculture either did not exist or were out of date. When the court of appeals vacated the general-industry air contaminant standard, OSHA suspended further action on the proposal to update the levels in the other three industries.

**Bloodborne pathogens.** An important new OSHA priority in the health standards area has been on workplace hazards resulting from infectious diseases. In 1991, OSHA successfully completed rulemaking and issued a final standard addressing occupational exposure to bloodborne pathogens, particularly the hepatitis-B virus and the human immunodeficiency virus (HIV). The standard was upheld for the most part by the Court of Appeals for the Seventh Circuit in February 1993.

OSHA was petitioned in September 1986 to issue an emergency temporary standard for bloodborne pathogens. OSHA rejected the petition but commenced rulemaking on the issues, publishing an advanced notice requesting information in 1987 and a proposed bloodborne pathogen standard in May 1989. The agency announced that during the rulemaking it would enforce existing regulations and the general duty clause to help prevent workplace transmission of bloodborne diseases. Great public interest was shown in the rulemaking; hearings were held in five cities and more than 400 persons testified at the hearings. The record was extensive.

OSHA promulgated the final Bloodborne Pathogens Standard on December 6, 1991 (56 *FR* 64004). The standard contains detailed provisions for the protection of employees, mainly health care workers, from hazards resulting from exposure to the hepatitis-B and HIV viruses. Among these provisions are the availability, without charge, of hepatitis-B vaccines to all employees within 10 days of assignment to jobs involving exposure to blood; engineering controls such as puncture-resistant containers for used needles; work practice controls such as hand-washing for exposed employees; requirements for use of personal protective equipment including gloves, masks, and gowns; postexposure evaluations; hazard communication requirements, including signs and warning labels; and record-keeping requirements.

In a decision dated January 28, 1993, the Court of Appeals for the Seventh Circuit upheld OSHA's bloodborne pathogens standard except in one respect. Judge Coffey dissented. In his opinion for the court, Judge Posner first noted that most health care employers had already voluntarily accepted the provisions of the OSHA standard, which was based largely on the recommendations of the Centers for Disease Control (CDC) of the National Institutes of Health, even before they had become OSHA requirements. The court analyzed in detail and rejected all of the arguments of the American Dental Association, which challenged the standard, concluding that the OSHA standard, "accepted as it has been by most health care industries and based as it is on the recommendations of the nation's, perhaps the world's, leading repository of knowledge of infectious diseases," does not cross the "boundary of reasonableness." The majority, however, remanded to OSHA the question raised on appeal by those employers who provide to employees worksites not under their control, as, for example, home health care workers. The court directed the agency to address more specifically the extent of the responsibility of these employers for compliance with the standard.

**Asbestos.** Asbestos has been the subject of OSHA regulation since the earliest days of the agency. In June 1986, OSHA issued a revised asbestos standard, sharply lowering the PEL from 2 fibers per cubic centimeter (cc) to 0.2 fibers per cc. The new standard was quickly challenged by both employer groups and unions, but, in an important decision, the Court of Appeals for the District of Columbia upheld the standard in most respects. However, on several issues, the court returned the proceeding to OSHA for further consideration to determine, generally, whether more protective provisions should be added.

OSHA's response to the court remand has continued over a period of several years and is not yet complete. The first issue to be reconsidered by OSHA was the short-term exposure limit (STEL). In September 1988, the agency promulgated a 1-fiber/cc short-term limit averaged over a 30-minute sampling period. In 1989 and in 1990, OSHA responded to several additional remand issues by amending the asbestos standard. Among other things, they expanded the ban on workplace smoking and required

employer training programs on smoking and warning signs and labels.

The most controversial aspect in the court remand was the issue of the appropriate 8-hour PEL. The court of appeals had directed the agency to consider whether a permissible limit more protective than the 0.2 fiber/cc PEL established in the standard was feasible, either generally or for specific industrial operations. On remand, OSHA initially decided that it could not determine this issue without further rulemaking, and in July 1990 it published a proposal lowering the PEL from 0.2 to 0.1 fibers/cc for all industries. Hearings on the proposals were held and public comment received. Final action was taken on the proposed revision on August 10, 1994. At that time, OSHA issued amended standards for asbestos in general industry and in the construction industry and a new asbestos standard for the shipyard industry. The principal revision in the new standard was a reduced PEL of 0.1 fiber/cc for asbestos work in all industries, which was half of the earlier 0.2 fiber/cc limit. Also significant was the new requirement that the building and facility owners communicate their knowledge of the location and presence of asbestos-containing materials or of certain high-risk materials presumed to contain asbestos, to employers of workers who may be exposed to those materials.

A number of industry groups and two unions challenged the new asbestos standard, and the appeal is pending in the Court of Appeals for the Fifth Circuit.

Meanwhile, OSHA has also reconsidered the application of the asbestos standard to the nonasbestiform substances tremolite, anthophyllite, and actinolite. They were covered by the original 1986 asbestos standard, but in response to legal challenge, the application of the standard to these substances was stayed. In June 1990, OSHA, following additional rulemaking, amended the asbestos standard by removing the nonasbestiform substances from its scope. The agency concluded that substantial evidence was lacking that these substances present the "same type or magnitude of health effects as asbestos." In the future, the agency said, these nonasbestiform substances will be regulated under other applicable OSHA dust standards.

**Ethylene oxide.** OSHA's ethylene oxide standard was originally issued in June 1984. In 1986 the Court of Appeals for the District of Columbia upheld the standard but determined that OSHA's decision not to issue a short-term exposure limit was not supported by the record, and remanded the proceeding to OSHA for reconsideration of the STEL issue. When after a year OSHA had still failed to take action on the court remand, the court of appeals, showing considerable impatience, ordered OSHA to act on the STEL by March 1988 or face contempt action. On April 6, 1988, OSHA promulgated a 5-ppm STEL, measured over a 15-minute sampling period, for ethylene oxide (53 *FR* 11414).

**Lead.** OSHA's original revised lead standard was promulgated in 1978. The standard established a limit of 50 $\mu g/m^3$ of air averaged over an 8-hour period for all industries except construction. Although the Court of Appeals for the District of Columbia generally upheld the standard, in a wide-ranging 1980 opinion it remanded the proceeding to OSHA for consideration of the feasibility of the standard in nine specified industries. After a number of delays and further rulemaking in 1989 and 1990, the agency determined that the lead standard's 50-$\mu g/m^3$ PEL was feasible in all remanded industries except for the small foundry segment of the nonferrous industry. OSHA established a 75-$\mu g/m^3$ limit in that industrial segment. This supplemental determination by OSHA was challenged, but in another decision on the lead standard, the Court of Appeals for the District of Columbia upheld OSHA's feasibility determinations, except as to the brass and ingot industry. For that industry, the court vacated the requirements for engineering controls and held that, in the meantime, the 50-$\mu g/m^3$ limit could be met by a combination of engineering, work practice, and respirator controls.

OSHA has also been concerned about lead hazards in the construction industry—the revised 1978 standard did not cover construction operations, and the applicable construction PEL was four times higher than the general industry standard. OSHA had begun work on a proposed standard on lead for construction when, in October 1992, Congress concluded that the matter was urgent, and passed the Housing and Community Development Act, which required OSHA to issue a construction industry interim lead standard within six months of enactment. Acting under this legislation, OSHA, on May 4, 1993, without rulemaking, published an interim lead standard that lowered the PEL for lead in construction to a 50-$\mu g/m^3$ level—the same level applicable to general industry (58 *FR* 26590). The standard also includes the same monitoring and medical removal provisions as for general industry. Under the 1992 statute, this interim standard would be effective immediately.

**Formaldehyde.** In December 1987, after extensive rulemaking, OSHA published a comprehensive standard regulating formaldehyde. Among other requirements, the standard established a 1-ppm 8-hour PEL and a 2-ppm short-term limit. The standard was challenged both by industry and by a number of unions, and the Court of Appeals for the District of Columbia remanded the proceeding to OSHA for reevaluation of two issues:

- The 1-ppm PEL—the court said OSHA must establish a lower PEL if it determined that a significant risk remained at 1 ppm and a lower PEL was feasible.
- Provision for medical removal protection, which OSHA had failed to include in the standard.

Following the remand, the opposing parties to the proceeding made recommendations to the agency for resolution of the issues pending under the remand. OSHA then proposed changes in the standard that were based on the parties' recommendations, and on May 27, 1992, issued a final rule amending the Formaldehyde Standard (587 *FR* 22290). The rule incorporated the substance of the parties' recommendations and lowered the PEL from 1 ppm to 0.75 ppm. The final standard also provided medical removal protection to employees suffering from certain illnesses resulting from exposure to formaldehyde.

**Hazard communication.** OSHA's 1983 hazard communication standard has been subject to extensive litigation and has been amended by the agency several times. In 1987, following a decision of the Court of Appeals for the District of Columbia, OSHA expanded the coverage of the hazard communication standard to include all industries and all employees exposed to hazardous chemicals. In November 1988, the Court of Appeals for the Third Circuit upheld the expanded hazard communication standard against challenges filed by the construction and grain industries. Earlier, OMB had disapproved three provisions of the expanded standard, notably the requirement that material safety data sheets be provided to all employers on multi-employer worksites. OMB's authority to disapprove a standard's provision in these circumstances was challenged, and ultimately the U.S. Supreme Court rejected OMB's argument that the Paperwork Reduction Act gave OMB adequate authority to review these requirements (*United Steelworkers v. Dole,* discussed previously).

In August 1988, OSHA published a proposal asking for comment on the issues raised by OMB in its disapproval of provisions in the hazard communication standard and for general comment on the expanded coverage of the standard to nonmanufacturing industries. Public hearings were held and comments received on this proposal, and on February 9, 1994, OSHA promulgated a final rule modifying in several respects the 1987 hazard communication standard.

Meanwhile, concerns had been expressed to OSHA about the usefulness of the material safety data sheets (MSDSs) required under the hazard communication standard. In May 1990, OSHA responded to these concerns by publishing a Request for Information that asked for comment on improving the presentation and quality of chemical hazard information under the hazard communication standard. Many comments were received, and OSHA also commissioned consultant studies on the accuracy of the MSDSs.

**Laboratory safety and health.** In July 1990, OSHA published a final standard (29 *CFR* 1910.1450) regulating employee exposure to hazardous chemicals in laboratories (55 *FR* 3300). The standard requires continued compliance with PELs for hazardous substances, and although it does not itself establish any new PELs, it requires the development and implementation of a chemical hygiene plan, tailored to the individual workplace, that includes work practices and procedures to protect laboratory employees.

**Cotton Dust.** Although the U.S. Supreme Court had affirmed OSHA's cotton dust standard in major respects in a 1980 decision, the agency proceeded to reevaluate the standard, publishing proposed revisions in 1983. In 1985, OSHA published final amendments to the cotton dust standard, but with respect to the most controversial issue in the rulemaking, the agency adhered to its existing policy that engineering controls were the preferred means of reaching the PEL. In two decisions in 1987, the Court of Appeals for the District of Columbia upheld provisions of the amended cotton dust standard. Rejecting a challenge of the Minnesota Mining and Manufacturing Company, the court upheld the standard's respiratory selection provisions. The court of appeals also affirmed OSHA's decision to retain medical surveillance requirements for the cottonseed industry, even though OSHA did not find a significant risk in that industry and, as a consequence, did not establish a PEL for the industry.

**Cadmium.** In June 1986, OSHA was petitioned to issue an emergency standard to reduce employee exposure to cadmium. OSHA rejected the request, and this action was upheld by the Court of Appeals for the District of Columbia, after which OSHA undertook a section 6(b) rulemaking on cadmium. In February 1990, OSHA published a proposal suggesting two alternative PEL levels for cadmium, $5 \mu g/m^3$ or $1 \mu g/m^3$, both based on the carcinogenicity of the substance. When, following public comment and hearings, OSHA delayed the issuance of a final cadmium standard, the agency was ordered by the Court of Appeals of the District of Columbia, on the petition of a union and a public interest group, to issue a final cadmium standard by August 31, 1992. The final standard was published in the *Federal Register* on September 14, 1992 (57 *FR* 42102).

OSHA's final cadmium standard includes one tier for general industry, maritime, and agriculture and a separate tier for construction. Both tiers of the standard provided a PEL of $5 \mu g/m^3$ and an action level of $2.5 \mu g/m^3$, with the action level triggering certain protective requirements such as exposure monitoring and medical surveillance. In light of the different working conditions in the construction industry, the construction portion of the standard contained specific provisions uniquely tailored to conditions in that industry.

Seven challenges to the cadmium standard were filed in court. Six were settled without litigation, and, on March 22, 1994, the Court of Appeals for the Eleventh Circuit upheld the standard except in one respect; the court ruled that although OSHA was justified in including cadmium pigments in the standard, the record did not establish the technological and economic feasibility of the standard in the dry color formulator industry. The proceeding was remanded to OSHA for further consideration of the feasibility issue.

**Other health standards.** In July 1980, the U.S. Supreme Court vacated OSHA's benzene standard in one of its most significant regulatory decisions of the decade. More than seven years later, in an action described as "landmark" by Assistant Secretary Pendergrass, the agency, after rulemaking, issued a new benzene standard reducing the PEL from 10 ppm to 1 ppm, the same PEL that had been vacated by the Supreme Court almost a decade earlier. The new standard also established a short-term exposure limit for benzene of 5 ppm (52 *FR* 34460). In its preamble to the standard, OSHA stated that it had found a significant risk to employees who were exposed to benzene at existing levels, and that promulgation of the new PEL would substantially reduce that risk. (The Supreme Court in its 1980 decision had for the first time imposed the "significant risk" requirement, and vacated the original benzene standard because of OSHA's failure to make the significant risk finding.) Several court challenges were filed to the 1987 benzene standard but were later withdrawn, and the 1-ppm PEL for benzene finally went into effect on December 10, 1987.

On May 12, 1989, OSHA published a proposed standard regulating methylenedianiline (MDA). The proposal was based substantially on the recommendations of a mediated rulemaking committee established by the agency and representing interested parties. This was the first successful attempt by OSHA to use mediated rulemaking in the development of a proposed standard. Mediated rulemaking is intended to help OSHA reach consensus in rulemaking and speed up the issuance of standards. However, largely because of OMB review, there was considerable delay before the proposed MDA standard was published, and the advantages of mediated rulemaking were partly vitiated (57 *FR* 35630). The final standard was for the most part based on the committee recommendations. Separate standards were issued for general industry and construction, but they were similar in important respects, both establishing a PEL for MDA of 10 million parts per billion. This PEL is based on animal and human studies demonstrating that MDA poses carcinogenic risks.

Several health standard proceedings are currently pending before the agency. On August 10, 1990, after rejecting a petition for an emergency standard, OSHA published a proposed standard on butadiene. The proposed standard would lower the PEL for butadiene to 2 ppm, with a short-term limit of 10 ppm.

In November 1991, OSHA published a proposed standard regulating methylene chloride. Based on animal studies showing the carcinogenicity of the chemical, OSHA's proposal would reduce the PEL to 25 ppm, delete the existing ceiling, and provide a short-term limit of 125 ppm for a 15-minute period. OSHA has received public comment and held hearings on both of these proposals, but no final standards have been issued.

The issue of indoor air pollutants is a significant one, particularly in light of studies showing the harmful effects of passive smoking. After the Court of Appeals for the District of Columbia upheld OSHA's decision not to issue an emergency standard on smoking in the workplace, the agency, in September 1991, published a Request for Information on indoor air pollutants. The request asked for data on four air contaminants: chemical agents, bioaerosols, passive tobacco smoke, and radon. During the last several sessions of Congress, legislation has been introduced to regulate indoor air quality.

On April 5, 1994, in the *Federal Register,* OSHA promulgated a Notice of Proposed Rulemaking (NPRM) to address air quality in indoor work environments. The provisions of the proposed indoor air quality (IAQ) standard apply to all indoor "nonindustrial work environments." In addition, all worksites, both industrial and nonindustrial, within OSHA's jurisdiction are covered with respect to the proposed provisions addressing control of environmental tobacco smoke (ETS).

OSHA's IAQ proposal would require covered employers to develop a written indoor air quality compliance plan and implement that plan through actions such as inspection and maintenance of heating, ventilation, and air-conditioning (HVAC) systems. Employers would be required to implement controls for specific contaminants and their sources, such as outdoor air contaminants, microbial contamination, maintenance and cleaning materials, pesticides, and other hazardous chemicals within indoor work environments.

Designated smoking areas, which are to be separate, enclosed rooms exhausted directly to the outside, are required in buildings in which the smoking of tobacco products is not prohibited. Specific provisions are also proposed to limit the degradation of IAQ during the performance of renovation, remodeling, and similar activities. Provisions for information and training of HVAC system maintenance and operations workers and other employees within the facility are also included in OSHA's IAQ proposal.

Finally, proposed provisions in this notice address the establishment, retention, availability, and transfer of records; for example, inspection and maintenance records of written compliance programs and employee complaints of building-related illnesses.

A record-breaking number of comments were received on the proposed standard, and after a postponement, the public hearing commenced on September 20, 1994. Meanwhile, on July 12, 1994, the Court of Appeals for the District of Columbia rejected as unripe for judicial review a claim by Action for Smoking and Health, a public interest group, that OSHA acted arbitrarily in proposing to regulate environmental tobacco smoke together with other air contaminants. The group's argument, which the court found to be premature, was that the regulation of tobacco smoke would be unreasonably delayed by the omnibus regulation.

In other actions, OSHA has denied petitions for emergency standards for chromium and tuberculosis hazards, but has asserted that both have a high priority and that rulemaking would be initiated to regulate each of these hazards.

## Standards Activity: Safety Standards

**Process safety management for hazardous chemicals.** In response to a series of tragic workplace explosions at a number of chemical facilities in 1989 and 1990, which resulted in more than 40 employee deaths and numerous injuries, OSHA took a number of major steps in both the enforcement area and in the development of standards to protect employees from such occurrences. In July 1990, the agency issued a proposed chemical process safety management standard based on data it had received from a variety of sources including industry and labor. Meanwhile, Congress held oversight hearings to investigate the causes of the facility explosions. In the Clean Air Act Amendments, enacted into law in November 1990, Congress directed OSHA to complete its rulemaking on a process safety management (PSM) standard within one year of enactment. The statute also prescribed 14 provisions that the agency was required to include in the PSM standard.

Hearings on OSHA's proposed standard were held in 1990 and 1991 in Washington DC and in Houston, Texas, and an extensive rulemaking record was compiled. One of the major issues in the rulemaking was the adequate protection of employees of contractors working at chemical facilities. Controversy on this issue forced the agency to delay promulgation of the final standard beyond the congressional deadline in order to allow full public comment on a study commissioned by OSHA on contractor employees. The final PSM standard was issued on February 24, 1992 (57 *FR* 6356).

Under the standard, employers are required to conduct systematic analyses of potential hazards in every step of the chemical process and, based on these analyses, to take appropriate steps to prevent chemical explosions or releases. In addition, the standard requires companies to consider safety records in choosing contractors and to institute policies and procedures to ensure the training of contract workers in safety procedures. The new PSM standard also requires that contractors themselves be responsible for training their workers in various chemical processes, in order to protect both the contractor's own employees and the employees of the chemical manufacturer. The standard requires each employer to prepare a hazard analysis covering each chemical process and written operating procedures, including steps for each operating phase and safety systems, and that these be readily available to employees.

The PSM standard was originally scheduled to go into effect on May 26, 1992. OSHA delayed the effective date of seven provisions in the standard until August 26 but, on August 25, announced that it was rejecting a request by various industry groups for a further stay of four provisions until 1994. The PSM standard has been challenged by both industry and unions, but this litigation was settled by the parties and the standard is currently in effect.

**Lockout/tagout.** On September 1, 1989, OSHA issued a final standard on hazardous energy sources, known as the lockout/tagout (LOTO) standard. The standard requires employers to implement practices and procedures to disable machinery or equipment and to prevent the release of potentially hazardous energy while maintenance and servicing activities are being performed. The standard applies the lockout provisions to equipment that is designed with a lockout capability, and tagout provisions to other types of equipment.

The standard was challenged by both industry and labor, and on July 12, 1991, the Court of Appeals for the District of Columbia remanded the LOTO standard to OSHA. In a lengthy and, in many respects, difficult decision, the court agreed with OSHA that the specific statutory requirements for standards contained in section 6(b)(5) of the act were applicable only to health standards and not to safety standards such as LOTO. At the same time, the court indicated its concern with OSHA's interpretation of its authority to issue safety standards, because, the court indicated, the broad discretionary authority claimed by OSHA would raise serious constitutional questions with respect to delegation of legislative authority. The court suggested that the agency might wish to consider using cost–benefit analysis as a criterion in safety standards development, in order to limit the scope of its discretion and avoid questions of constitutionality. The court remand directed OSHA to articulate its view of the scope of its authority to issue safety standards, and to reconsider specified portions of the standard in light of its new interpretation of the standard. In September 1991, the court of appeals decided that the standard would remain in effect during the remand.

In suggesting that cost–benefit analysis would be permissible in the development of safety standards, the Court of Appeals for the District of Columbia agreed with the view previously expressed by the Court of Appeals for the Fifth Circuit in its amended decision in the grain dust case. (The Supreme Court cotton dust decision barred cost–benefit analysis for health standards). On March 30, 1990, OSHA published a response to the court's remand in the LOTO case, rejecting the use of cost–benefit analysis in safety standards. The agency stated in its *Federal Register* notice that there were adequate limits on the scope of its safety standards authority even if it did not resort to cost–benefit analysis. These included feasibility as a lower limit and significant risk as an upper limit in establishing the PEL.

On October 21, 1994, the Court of Appeals for the District of Columbia accepted OSHA's supplemental statement as responsive to the remand and dismissed the petition for review of the LOTO standard. The court held that OSHA, without having to adopt cost–benefit analysis, has recognized adequate constraints on its discretion to issue safety standards, which could meet with objections on constitutional grounds. The court did not rule directly on whether the cost–benefit issue was lawful in standards proceedings.

**Grain handling.** OSHA's original grain-handling standard was issued on December 31, 1987. The standard was designed to prevent or minimize employee deaths and injuries from fires, explosions, and other hazards in grain facilities. The standard was challenged by both industry and labor, and in two decisions in October 1988 and January 1989 the Court of Appeals for the Fifth Circuit upheld the standard in part but remanded the standard to OSHA for reconsideration of two issues. In its standard, OSHA had imposed a ⅛-inch action level of dust in certain particularly hazardous areas in grain facilities. The court directed OSHA to consider whether the ⅛-inch level was feasible and whether, as the unions had claimed, the stricter requirement should be applied on a facility-wide basis.

One of the major issues in the litigation was whether cost–benefit was lawful under the OSHA statute in safety standard development. In its initial decision, the court held that it was not, but in 1990 the court reversed itself, deferring to OSHA's interpretation that safety standards such as grain handling are not subject to the same restrictions on the use of cost–benefit as health standards.

In response to the remand, OSHA subsequently found the ⅛-inch action level feasible for priority areas, and the court of appeals upheld this determination. However, OSHA decided that it did not have sufficient information in the record on the issues of the scope of the ⅛-inch action level. Accordingly, in December 1990, the agency published a Request For Information on whether the action level should be extended to all portions of grain dust facilities. On April 1, 1994, OSHA published a final decision, concluding that the existing record was adequate for decision, and determining on the basis of this record that the action-level trigger for the housekeeping provisions of the grain dust standard should not be expanded beyond the priority areas.

**Hazardous waste operations.** In the 1986 Superfund Amendments and Reauthorization Act (SARA), Congress required OSHA to issue, by a specified deadline, a final standard to protect employees engaged in hazardous waste operations. In a related action, Congress in 1987, in the Omnibus Budget Reconciliation Act (OBRA), required OSHA to provide for certification of training programs for employees involved in emergency response. On March 6, 1989, OSHA issued a final standard on hazardous waste operations and emergency response (known as HAZWOPER). The standard protects employees involved in cleanup operations at uncontrolled hazardous waste sites under government mandate; in certain hazardous waste treatment, storage, and disposal operations; and in any emergency response to incidents involving hazardous substances.

In January 1990, acting under the mandate of the 1987 OBRA, OSHA published proposed rules to establish procedures for accrediting hazardous waste operation training programs. Public hearings were held on the proposal in 1991.

**Standards-related testing.** Under a number of OSHA safety standards, testing for safety is required to determine whether specified equipment and materials are acceptable for workplace use. The original OSHA standards listed two specific national laboratories as qualified to perform these standards-related tests. Over a number of years, several other testing laboratories sued OSHA, claiming that the safety standards were discriminatory in favoring the two specifically listed laboratories. On December 4, 1987, a U.S. district judge gave OSHA 180 days to promulgate procedures that would allow for OSHA's recognition of other laboratories as qualified to certify equipment. OSHA's final standard on the testing and certification of workplace equipment was promulgated on April 12, 1988 (53 FR 121020). The standard deletes the two named laboratories from all safety standards and establishes a procedure by which other testing organizations can be accredited as "nationally recognized testing laboratories" for purposes of OSHA standards. A number of laboratories have subsequently gained accreditation by OSHA under the new procedures.

**Confined spaces.** OSHA's long-delayed final Standard on Confined Spaces was published in the *Federal Register* on January 14, 1993 (58 FR 4462). The proceeding was begun by an advance notice of proposed rulemaking in 1975, and it continued through several OSHA administrations. (The standard's preamble describes in detail the lengthy procedural steps that preceded its issuance.) The standard imposes various safety requirements, including a permit system regulating employee entry into certain confined spaces that are designated by the standard as "permit-required confined spaces" and defined as posing special dangers because their "configurations hamper efforts to protect entrants from serious hazards, such as toxic, explosive or asphyxiating atmospheres." The standard also requires that employers implement a comprehensive confined-space entry program, which must include ongoing monitoring, testing, and communication with employees.

Court challenges to the confined spaces standard were filed by both employer and union groups, but were dismissed by the court on the basis of understandings between OSHA and the parties on the meaning and implementation of the standard.

**Construction standards.** Because construction is one of this country's most hazardous industries, OSHA has traditionally accorded special attention to standards to protect construction workers.

In June 1988, OSHA issued a final standard on concrete and masonry construction, removing ambiguities and redundancies in the existing standard and filling gaps in coverage. The new standard incorporates new technology, and specifies all requirements within the text of the standard rather than incorporating national consensus standards by reference (53 FR 22612). While the rulemaking on this standard was underway, a tragic accident took place at L'Ambiance Plaza in Bridgeport, Connecticut, involving lift-slab construction. In light of the special hazards involved in that type of construction, OSHA decided to deal with lift-slab construction in a separate rulemaking, and a final standard on that subject was later issued in October 1990 (55 FR 42306). The record in the later proceeding included a special report completed by the National Institute of Standards and Technology (formerly the National Bureau of Standards) at the request of OSHA on the causes of the L'Ambiance collapse.

In 1989, OSHA issued three other construction safety standards: one on underground construction, published on June 1, 1989 (54 FR 23824); one on powered platforms for building maintenance, published on July 28, 1989 (54 FR 31408); and one on excavations, published on October 31, 1989 (54 FR 45894). On November 14, 1990, OSHA issued a construction standard covering stairways and ladders in the construction industry (55 FR 47660). This standard consolidates into one subpart scattered provisions in other standards and adds a number of new safety provisions. On August 9, 1994, OSHA promulgated a final standard consolidating all of its requirements for protection against falls in the construction industry into a single subpart of the construction safety standards and revising these provisions in a number of respects. In particular, the final standard allowed alternative methods of fall-protection compliance in cases where traditional approaches are inappropriate or unreasonable.

**Other safety standards.** In March 1988, OSHA promulgated a rule permitting presence-sensing devices on mechanical power presses (53 FR 8322). Under the original mechanical power press standard, the operator of the press was required to initiate the stroke of the press by hand or pedal. Under the new standard, a presence-sensing device would be permitted to initiate the stroke when the operator's body was no longer exposed to harm.

In August 1990, OSHA issued a final rule on electrical safety-related work practices (55 FR 31984). The standard establishes required work practices for employees whose jobs require them to work with, on, or near electrical equipment, in order to protect them from electric shock, burns, and other electrical accidents. On April 6, 1994, OSHA promulgated a final standard covering personal protective equipment for eyes, face, head, and feet in general industry.

OSHA has also initiated rulemaking in two major areas involving priority safety issues. These are ergonomic safety and health management, and motor vehicle safety. On August 3, 1992, OSHA published an advance notice of proposed rulemaking requesting data and comments on ergonomic hazards in the workplace related to the absence of ergonomic safety and health programs (57 FR 34192). In addition, OSHA has undertaken a priority compliance effort in this area. Earlier, in April 1992, OSHA rejected a petition filed by a number of labor organizations for an emergency temporary standard on ergonomics, concluding that there was no basis for emergency action. Although OSHA has stated that the publication of a proposed ergonomics standard is a high standards-setting priority, the future of the rulemaking is unclear in view of strong employer opposition and controversy in Congress over the standard.

Motor vehicle accidents constitute the single greatest cause of occupational fatalities in the country, and in July 1990, OSHA proposed a standard for occupant protection in motor vehicles (55 *FR* 28728). The standard would require employers to mandate the use of safety belts for employee occupants of motor vehicles when on official business, and would require employer-sponsored safety awareness programs for employees. The proceeding is pending.

## Standards Activity: Overview

In viewing OSHA standards activity for the years from 1987 to 1994 in retrospect, it is difficult to resist the clichéd conclusion, the more things have changed, the more they have stayed the same. The following observations, highlighting trends for the seven-year period, should be noted:

- OSHA has been confronted with increasingly sophisticated safety and health problems in the workplace, raising issues that were scarcely contemplated when the OSHAct was enacted. Among the more important of the new types of hazards are infectious diseases, specifically hepatitis-B, HIV, and tuberculosis; ergonomics hazards; and violence in the workplace. OSHA has already issued a standard on bloodborne diseases, and has promised that it would propose an ergonomics standard in the near future; however, no standards action seems near on tuberculosis or violence.

- Emergency temporary standards (ETSs) are no longer a factor in the OSHA program. Largely because of a number of court decisions vacating OSHA emergency standards on a variety of grounds, the agency now almost routinely denies petitions for ETSs. Members of the public continue to request ETSs, in order, it would seem, to add impetus to their requests and to help ensure that, at least, section 6(b) rulemaking be initiated.

- Congress has continued its involvement in the standards process. In recent years, Congress has directed the agency, by statute, to issue several specific standards, and it has even legislated timetables for their issuance. Although OSHA complies with these statutory directives, as it must, it does so at the expense of other standards priorities. As a result, the congressional action should be seen not as a measure to increase, but rather to redirect, OSHA standards activity.

- As in earlier years, all major OSHA standards, with limited exceptions, have been challenged in the courts of appeals. The courts have continued to look closely at

the substance of the standards, and although many have passed muster, in some significant instances the standards have been remanded to OSHA for further action. Significantly, in a number of important proceedings such as asbestos and formaldehyde, the courts decided that the OSHA permissible exposure limit was insufficiently stringent, and that a lower limit should be set, if feasible. OSHA appears to have mastered the rulemaking process, and procedural issues are not often raised in standards challenges. The most significant standards decision of the period was the ruling of the Court of Appeals for the Eleventh Circuit setting aside OSHA's air contaminants standard.

- Although OSHA has recently completed a number of standards proceedings that had been pending for considerable periods of time, notably the asbestos and hazard communication standards, delays in the issuance of standards continue to be an agency problem. After 1987, the rate of standards promulgation was much improved over that during the earlier deregulatory era; however, there were many disappointments, with some standards proceedings dragging on for years, and numerous important standards priorities awaiting agency action. In light of these realities, interested parties looked to Congress and to the courts for help in triggering agency standards action, and the courts of appeals several times lost patience with OSHA and ordered the agency to take action within a specified deadline. Congress has also dealt with the matter legislatively. No solution to the standards problem has appeared on the horizon, particularly because generic rulemaking has been called into question by the failure of the major air contaminants standard effort.

- In August 1994, Assistant Secretary Joseph Dear initiated a new standards-planning process. Under the new procedures, OSHA has solicited the views of top officials in labor, industry, and academia, known as "stakeholders," in an effort to help the agency determine what workplace hazards it should first address in its standards program.

- On April 23, 1994, Linda Rosenstock, Director of the Occupational and Environmental Medicine Program at the University of Washington, was appointed to head the National Institute for Occupational Safety and Health (NIOSH), the research agency of the federal occupational safety and health program. At the same time, it was announced that NIOSH headquarters would be moved from Atlanta, Georgia, back to Washington DC.

## Compliance Activity

Enforcement programs played an increasingly important role in the OSHA program in the period beginning with the administration of Assistant Secretary Pendergrass.

Civil penalties are a major component of OSHA enforcement. These penalties, imposed by the agency when an employer is found to have violated the act or OSHA standards, serve as an incentive to employers to abate hazards disclosed by OSHA inspections and, in the future, to abate hazards prior to inspection. However, it has been apparent for some time that the original statutory penalty amounts were scarcely sufficient to provide the necessary incentives to employers to undertake costly abatement measures. The need for statutory change was widely recognized and, finally, as part of the Omnibus Budget Reconciliation Act of 1990, Congress adopted a seven-fold increase in the penalties OSHA was authorized to propose for violations of the act. The statute, as amended, authorized penalties of up to $70,000 for willful and repeated violations and $7,000 for other violations, and established a $5,000 minimum penalty for willful violations.

In addition, in 1986 OSHA implemented what has come to be known as the "egregious" penalty policy. In the earliest days of the act, OSHA adopted a compliance policy under which separate instances of the same violation would be grouped for penalty purposes; thus, if 45 machines lacked the same machine guarding, all these instances would be only one violation for purposes of imposing a penalty. However, in 1986, the agency directed its field staff to impose separate penalties for different instances of violations if the inspection disclosed flagrant and widespread violations at the workplace. The stated purpose of the new policy was to make OSHA penalties more credible rather than merely being a slap on the wrist. Originally implemented for record-keeping violations, the "egregious" policy was later applied to violations in the petrochemical and construction industries, as well as to ergonomics and other program areas in other industries. The lawfulness of the "egregious" penalty policy under the OSHAct was challenged, and the Review Commission has upheld its validity on substantive and procedural grounds.

As might be expected, as a result of the statutory changes and the new administrative policies, OSHA civil penalties have markedly increased, with OSHA, for the first time in its history, proposing penalties amounting to millions of dollars. According to agency statistics for the fiscal year ending September 30, 1992, the first full year in which OSHA used the increased maximum penalties mandated by Congress and the "egregious" penalty policy, the total amount of penalties imposed reached $116.1 million, well over the previous high of $91.7 million that was recorded for the preceding fiscal year. There were sharp increases in penalties between 1990 and 1992, both in aggregate penalties and in the average amount of penalties. From 1990 to 1992, the average penalty for serious violations tripled and appears to have risen sharply for willful violations. OSHA statistics on willful violations are less clear.

On June 14, 1994, OSHA issued a directive to field staff to increase the minimum penalty for willful serious violations to $25,000 from the prior $5,000 minimum. Assistant Secretary Dear in announcing the change said that its pur-

pose was to make it "more difficult for those few bad actors to regard penalties as simply a cost of doing business." In November 1994, OSHA published statistics on its compliance activity for fiscal year 1994. The statistics showed that federal OSHA penalties imposed were nearly $120 million, a record level, and that the number of federal inspections increased to 42,377, more than half of which were in the construction industry. In the prior fiscal year OSHA had conducted 39,536 inspections. The rate of contested federal cases was stable, continuing at approximately 11 percent.

Another major OSHA enforcement initiative during this period was corporate-wide settlement agreements. Whereas the agency and the Office of the Solicitor in the Department of Labor always encouraged informal settlement of citations and penalties, OSHA after 1987 began to seek, in the case of national companies, settlements covering not only the facility or facilities that were the subject of the citations but all or many of the corporation's facilities on a nationwide basis. Nation-wide settlements have numerous advantages. Like single-facility settlements, they avoid the burdens of administrative and court litigation and the delays in employer abatement that necessarily result from litigation, during which abatement requirements are suspended. Corporate-wide agreements also apply to facilities that were not inspected and often include requirements for specific abatement measures that are typically not included in citations themselves, or even in Review Commission orders. In fact, specific abatement requirements included in nation-wide settlements have formed the nucleus of OSHA's voluntary abatement guidelines in the field of ergonomics, and the basis of OSHA's new process safety management standard for the petrochemical industry.

The OSHAct contains criminal penalties, but these are applicable only when a willful violation leads to the death of an employee, and they generally do not apply to corporate employees. In addition, there has been little use of these criminal sanctions at the federal level. As a result, throughout the history of OSHA there has been ongoing criticism of both OSHA and the U.S. Department of Justice, which prosecutes the criminal violations, over the lack of significant criminal enforcement in the field of occupational safety and health.

The decade of the 1980s witnessed a considerable increase at the state level in the prosecution of criminal cases involving death, injury, or illness in the workplace. In these proceedings, states would typically base their prosecutions on state general criminal laws such as the manslaughter or assault statutes. In states without approved state plans, a major legal issue arose as to whether these state prosecutions were preempted by the federal OSHA statute. The highest courts in several states have ruled that there was no preemption, and the U.S. Supreme Court has refused to hear appeals on this issue.

In the meantime, federal OSHA has taken steps to improve its criminal enforcement program. These steps include OSHA staff training in criminal investigation technique and improved criminal enforcement coordination both within the Department of Labor and with the Department of Justice. Since 1990, a significantly greater number of OSHA enforcement cases have been referred to the Department of Justice for criminal enforcement, and some of these cases have resulted in convictions and jail sentences.

In establishing inspection priorities, OSHA has continued to direct its limited inspection staff resources to businesses in which it believes the greatest hazards exist; these inspection priorities are implemented in part through special-emphasis programs. A series of chemical explosions in 1989 and 1990 that resulted in the deaths and injuries of many employees led OSHA to institute a petrochemical industry special-emphasis program (PetroSEP). In 1990 this program involved in-depth inspection activity in the industry, special training for compliance officers in chemical industry–inspection techniques, and cooperation with the EPA, which has parallel responsibility in the chemical industry under environmental statutes.

Over the last several years, there has been a significant increase in injuries and illnesses caused by repetitive motion and other ergonomic-related factors, cutting across a wide spectrum of industries and workplaces. In 1990, OSHA, responding to this new and serious problem, established a special-emphasis enforcement program for ergonomics. This program was first directed to the red meat industry, where injuries were common and related specifically to the absence of ergonomic programs. Special program management guidelines for this industry were developed by OSHA, and in August 1990, copies of the guidelines were sent to every red meat–processing facility in the country. Simultaneously, a comprehensive nationwide inspection program was initiated, targeting large meatpacking facilities. These inspections were conducted by teams consisting of safety, health, and ergonomics experts, joined in some cases by medical personnel. To coordinate and support these and other agency activities (such as standards development) in the ergonomics area, OSHA established a new Office of Ergonomic Support in its Technical Support Directorate.

The construction industry is one of the most hazardous, and OSHA has continued to conduct approximately one half of its federal inspections in this industry. The agency has taken several other steps to enhance its compliance efforts in construction. A new Office of Construction and Engineering was created in February 1990. One of its major responsibilities is to use its expertise to investigate construction accidents and determine their causes. The office prepared a special report analyzing construction fatalities investigated by OSHA between 1985 and 1989. The reports were used both in developing standards and in fashioning enforcement strategies. OSHA has also conducted several pilot programs to improve its targeting of construction inspections, so that inspections will be made when the greatest hazards are present. OSHA reform legislation, as well as other pending bills, would greatly expand OSHA activity in the construction area.

Workplace accidents attract public attention and have always been a matter of great concern in the OSHA program. The recent accidents in the petrochemical industry, the L'Ambiance Plaza collapse, and the fire in the poultry-processing facility in North Carolina, each resulting in many worker deaths, have underscored the importance of efforts by both federal and state OSHAs to prevent such occurrences in the future. One effort is toward better coordination of standards and enforcement activity. For example, in the development of its lift-slab construction standard, OSHA used data obtained in its investigation of the L'Ambiance Plaza accident.

OSHA has also attempted to improve the quality of its investigation of the causes of accidents, and in 1991 promulgated a directive containing detailed instructions to field staff on accident investigations. Previously, both as a matter of compassion and as an investigation technique, OSHA in April 1990 had issued instructions to field staff requiring greater involvement with families of victims of workplace accidents in agency investigations of fatal accidents. Under the directive, OSHA field staff are required to contact victims' families, to explain OSHA's investigation and enforcement procedures, to invite family members to meet with OSHA officials, and to share with them information regarding the accident. Moreover, members of OSHA staff must keep family members informed "at appropriate times" of the steps OSHA is taking in connection with the investigation and enforcement proceeding. Procedures for involvement of victims and their families in OSHA enforcement are also the subject of pending legislation. On April 1, 1994, OSHA published a final rule requiring employers to report to the agency within eight hours all employee fatalities or hospitalizations of three or more workers. Previously, the time frame for reporting was 48 hours, and reports were mandatory only for five or more hospitalizations. Because OSHA inspections are triggered by these employer reports, OSHA stated, the amended rule would result in more timely inspection of serious accidents.

In September 1994, OSHA released its new *Field Inspection Reference Manual,* previously called the *Field Operations Manual,* which contains guidance to field staff in conducting workplace inspections and issuing citations and penalties. The new manual is much shorter, down from 369 pages to 102 pages, and it gives field staff greater discretion in making enforcement decisions themselves.

## State Programs

Section 18 of the OSHAct encourages states to develop their own occupational safety and health programs. If a state plan is approved by federal OSHA, the state is authorized to implement its program and to receive 50 percent of the cost of the program from federal OSHA. At the present time, 21 states and 2 territories have approved plans covering approximately 40 percent of the nation's work force. Two states, New York and Connecticut, have approved plans that cover only state and municipal employees.

In July 1987, the governor of California suspended operation of the California state plan, one of the largest in the country, because of budget constraints. While the California state plan was not being implemented, federal OSHA resumed its private-sector enforcement in California. The governor's action was challenged by a number of California groups, including environmentalists and labor unions, and state court litigation on the lawfulness of the governor's action ensued. Eventually, the issue of the continuation of the state program was placed on the ballot, and in November 1988, the voters in California passed Proposition 97, reinstituting California's state occupational safety and health enforcement. In May 1989, the California program resumed operation and federal OSHA suspended its enforcement activity.

The Michigan state plan experienced a crisis, also for budget reasons, when the governor of Michigan in 1991 proposed elimination of funding for the plan. However, after a legislative compromise, adequate funds for the state plan were appropriated, and the plan has continued operating.

On September 3, 1991, there was a major fire at a poultry-processing facility in Hamlet, North Carolina, as a result of which 25 employees died. (North Carolina's state plan was approved by OSHA in January 1983, and an operational agreement giving North Carolina sole enforcement authority in the state was in effect.) Following the incident, Secretary of Labor Lynn Martin directed OSHA to conduct a complete reevaluation of the North Carolina program, as well as an evaluation of all other state programs. On September 12, 1991, a subcommittee of the House Labor Committee conducted an oversight hearing on OSHA reform legislation, at which testimony on the North Carolina fire was presented. On October 24, OSHA terminated North Carolina's operational agreement and resumed partial federal concurrent enforcement in the state.

In January 1992, OSHA completed its evaluation of the North Carolina program and concluded that the state plan was experiencing "serious operational difficulties." The state was given 90 days to correct its deficiencies, and on April 24, 1992, OSHA issued a show-cause order giving North Carolina 45 days to demonstrate why proceedings to withdraw approval of the state plan should not be initiated. In explaining its action, OSHA said that although the state had dealt with some of the difficulties in its program, others were "addressed only by assurances," without firm commitments to timetables. On June 18, OSHA suspended further action on the withdrawal of the North Carolina state plan, concluding that additional state commitments had been received, by which the state promised to take action to correct seven major deficiencies. Representatives

of the labor movement criticized this OSHA action as being "outrageous" and politically motivated.

Meanwhile, in December 1991, North Carolina proposed fines of more than $800,000 against Imperial Food Products, the employer at the poultry-processing facility. Imperial did not contest the citations and penalties, and on September 14, 1992, the owner of the then-defunct company pleaded guilty to involuntary manslaughter charges and was sentenced to prison for 20 years. This was believed to be the most stringent sentence ever imposed for a state criminal violation involving occupational safety and health.

Under section 18 of the OSHAct, states with approved plans, in order to remain as effective as the federal OSHA program, are required to impose criminal sanctions for violation of their OSHA laws. However, in states without approved plans, there is a serious legal question as to whether the federal OSHAct preempts those states from enforcing state general criminal statutes for murder, manslaughter, and assault for occupational safety and health offenses. Early state court decisions suggested that such state criminal enforcement activity was preempted by the federal statute. However, the U.S. Department of Justice ultimately expressed the view that the federal act had no preemptive effect, and three later decisions of the highest courts of Illinois, Michigan, and New York ruled that state criminal enforcement was not preempted. In the first of these, *People v. Chicago Magnet Wire Corp.*, the state supreme court of Illinois emphasized the policy considerations underlying these decisions, reasoning that if the preemption argument were accepted it would in effect "convert the [OSHA] statute, which was enacted to create a safe work environment for the nation's workers, into a grant of immunity for employers responsible for serious injuries or deaths of employees." Pending OSHA reform bills would explicitly confirm the nonpreemption of state criminal enforcement.

The U.S. Supreme Court decided an important case, *Gade v. National Solid Wastes Management Association,* in which it upheld OSHA preemption of so-called dual-purpose state statutes.

## BLS Statistical Survey

The most recently published results of the survey of the Bureau of Labor Statistics (BLS) of occupational injuries and illnesses in the United States was for calendar year 1992. According to the survey, the rate of work-related injuries and illnesses in the United States increased to 8.9 cases per 100 full-time workers in 1990 from 8.4 cases per 100 full-time workers, the largest one-year increase in injuries and illnesses since the BLS survey began. Virtually all of the increase consisted of less serious cases that did not involve any lost workdays. The rate of lost workday injuries and illnesses for 1992 was the same as in 1991, 3.9 per 100 workers.

The first BLS survey was completed in 1973. In that year, the injury/illness incidence rate was 11.0 per 100 workers. Since then, the rate has fluctuated, and was 8.9 in 1992. On the other hand, the lost workday case incidence rate (the number of incidents involving lost workdays) has gone up slightly since 1973 (from 3.4 to 3.9 per 100 workers).

Although BLS also reports on occupational illnesses, it has repeatedly emphasized that because of the difficulty of collecting accurate information on illnesses resulting from long-term health hazards, as in the case of carcinogens, its statistics markedly understate the number of work-related illnesses.

The results of the 1992 BLS survey did not include the number of work-related fatalities in the totals. These are now separately calculated by BLS, and on August 10, 1994, BLS published the results of its National Census of Fatal Occupational Injuries for 1993. According to the survey, a total of 6,271 workers were killed in job-related incidents in 1993, with vehicle crashes and homicide leading all other causes. Also significant was the fact that violence in the workplace was the single largest cause of death for women on the job in 1993, accounting for 39 percent of all female job-related fatalities. 1993 was the second year for which data were produced from the BLS Census of Fatal Occupational Injuries program.

BLS includes in its statistical survey all employers who are required to maintain injury and illness records under the OSHAct. In 1990, the survey covered a sample of 250,000 establishments in private industry. Parallel data for mines and railroads are compiled by the Department of Labor's Mine Safety and Health Administration and by the Federal Railroad Administration in the U.S. Department of Transportation. BLS combines all the private-sector data into its final published statistical report. Data on injuries and illnesses in the public sector, both federal and state, are published separately.

In January 1991, a major change was made in the relationship between OSHA and the Bureau of Labor Statistics. Under a new policy, OSHA assumed responsibility for determining employer record-keeping obligations under the OSHAct. BLS will continue to conduct the annual national statistical survey of occupational injuries and illnesses as it has done in the past, which is based on the employer-maintained records. Both agencies are constituent parts of the Department of Labor and consult on occupational injury and illness issues.

Also in 1991, OSHA created a new Office of Statistics. One of its responsibilities is the revision of the regulations on employer responsibilities for keeping records on occupational injuries and illnesses. The new office also works with the agency's compliance staff in Washington DC and field staff to develop better statistics for the purpose of more effectively targeting inspections to priority workplaces.

Meanwhile, BLS, in consultation with OSHA, redesigned the national data collection system. One of the main purposes of the new system is to provide more detailed demographic and case-characteristic information on workplace injuries and illnesses. In 1987, the National Academy of Sciences of the National Research Council issued a report recommending an overhaul of the BLS statistical system. After considerable study and consultation, BLS developed a new system, known as ROSH (Redesigned Occupational Safety and Health), which was first implemented for the 1992 calendar year survey. The new survey forms were mailed out by BLS in January 1993 to 280,000 employers. BLS indicated that the revision of the survey, the first in 20 years, would provide more specific and understandable data on occupational injuries and illnesses, such as the nature of the disability and pertinent worker characteristics. In addition, the revision would save employers time in completing the survey by avoiding duplication of workers' compensation data.

## Legislative Activity

Between 1970, when the OSHAct was passed, and December 1990, when legislation increased seven-fold the maximum penalties that OSHA was authorized to propose, no amendments to the OSHAct were enacted. The agency, supported mainly by labor organizations, opposed any amendments to statute, fearing that even measures purporting to strengthen the law would open the statute to a rash of changes. The labor committees of the House and Senate generally agreed with this view, and as a result the statute remained unchanged for a twenty-year period.

This does not mean that there was total satisfaction with the law or its implementation. On the contrary, during this period, particularly in the early years of OSHA, there were many complaints about agency performance. In Congress, these translated into appropriations riders prohibiting the expenditure of appropriated funds by OSHA for particular purposes. In many years, there were as many as seven appropriations riders attached to OSHA appropriations legislation. However, commencing in 1989, as its hostility to OSHA decreased, Congress began a process of deleting various riders from the appropriations bills. Currently only three appropriations riders remain:

- OSHA funds may not be spent for enforcement in a farming operation with fewer than eleven employees that does not maintain a temporary labor camp.

- Funds may not be expended to develop or enforce standards affecting recreational hunting, shooting, or fishing.

- OSHA may not conduct programmed safety inspections in firms employing ten or fewer employees that are included in an industrial category that has an occupational injury lost workday case rate below the national average.

In the closing years of the 1980s, there was increasing pressure to amend the OSHAct. The traditional opposition to any amendment activity by labor organizations came to an end, and instead, labor organizations urged that the act be strengthened. Several areas of change were assigned priority for amendment; among these were criminal enforcement, the dollar amounts of penalties, the agency program in the construction industry, the procedures for issuing standards, and criteria for standards. In November 1990, the Omnibus Budget Reconciliation Act increased maximum OSHA penalty levels seven-fold and established a minimum penalty of $5,000 for willful violations. This was the first substantive amendment in the history of OSHA.

However, even after the 1990 amendment was enacted, bills continued to be introduced in Congress to further strengthen the OSHAct. Among the most important were those relating to criminal enforcement and construction standards and enforcement. Eventually, efforts to amend the OSHAct came to center around comprehensive legislation that would reform OSHA in virtually all of its areas of activity. The Senate OSHA reform bill was introduced in August 1991 by Senators Metzenbaum and Kennedy, and the House of Representatives bill was simultaneously introduced by Chairman Ford of the House Labor Committee. Extensive hearings were held in the 102nd Congress on these bills; generally, the bills were supported by labor and opposed by industry, whereas OSHA, for the most part, took the position that any needed changes in the implementation of the act could be achieved administratively. The House bill was reported favorably by the Labor Committee in July 1992, and the Senate bill by the Senate Committee in October 1992. Nonetheless, Congress adjourned before either bill reached the floor.

Similar but not identical legislation was introduced in both the House of Representatives and the Senate in 1993. Hearings in the 103rd Congress were held in both the House of Representatives and the Senate, but the bills again did not reach the floor of either house. With the results of the 1994 election giving the Republican party a majority in Congress, any action in the foreseeable future on OSHA reform seems highly unlikely.

Congress in recent years has become actively involved in the OSHA standards program. In a number of instances, Congress has mandated that OSHA issue an occupational safety or health standard. The congressionally required standards were for employee exposure to hazardous wastes, certification for training programs given to individuals providing emergency response, process safety management, bloodborne pathogens, hazardous labeling, and lead in construction work. Particularly noteworthy was the action by Congress in October 1992 in passing the Housing and Community Development Act, which required within 180 days of enactment that OSHA issue an "interim final lead standard" covering the construction industry. Under the law, the standard would take effect upon issuance and would be "as protective" as the working protection guidelines for identification and abatement of lead-based paint

issued by the U.S. Department of Housing and Urban Development. OSHA published the interim lead standard on May 4, 1993.

During the 1987–1993 period, Congress considered other legislation that affected occupational safety and health but did not amend the OSHAct itself. In 1986, Congress passed the Superfund Amendments and Reauthorization Act (SARA), requiring OSHA to issue a hazardous waste operations standard, with a specified timetable imposed. In 1987, Congress enacted as part of the Omnibus Budget Reconciliation Act a requirement that OSHA, as part of the hazardous waste standard, also provide for the accreditation of training programs for employees who work with hazardous waste. These provisions are particularly significant because the OSHAct itself does not provide for deadlines for the issuance of final standards. Since 1987, Congress on other occasions has stipulated (or tried to stipulate), through legislation or riders, time frames for the issuance of standards; a notable example was the statute requiring OSHA to issue a process safety management standard within one year.

In October 1987, the House of Representatives approved the High Risk Occupational Disease Notification and Prevention Act. The main responsibilities under this legislation were assigned to the Department of Health and Human Services, which would identify and notify employees at increased risk of disease because of their occupational exposures to toxic chemicals. OSHA would be required to submit annual reports to Congress on its enforcement of the hazard communication standard and to investigate alleged discrimination against employees at increased risk. A parallel bill reached the Senate floor in March 1988, but attempts to limit debate through cloture failed, and the bill was ultimately withdrawn.

Enacted in 1990 by Congress, the Negotiated Rulemaking Act encourages various federal agencies, including OSHA, to utilize negotiated rulemaking procedures in developing occupational safety and health standards. Although it does not require that negotiated rulemaking be used, the law facilitates its use and establishes certain procedures that must be followed if negotiated rulemaking is used. OSHA has used negotiated rulemaking successfully once, in developing an MDA standard, and unsuccessfully on other occasions. On May 11, 1994, OSHA announced that it had established a negotiated rulemaking advisory committee to make recommendations for revision of OSHA's construction standard regulating steel erection. The committee is currently continuing its deliberations. On several other occasions, notably in the grain dust and formaldehyde proceedings, OSHA made use of the recommendations of interested parties in deciding its official action.

On November 15, 1991, after considerable debate, Congress passed the Clean Air Act Amendments of 1990. Although the law related primarily to the responsibilities of the Environmental Protection Agency, two provisions directly involved OSHA. One required OSHA, within 12 months of enactment, to issue a chemical process safety standard with certain provisions as specified by Congress. The second provision established a Chemical Safety and Hazard Investigation Board with responsibility to investigate chemical releases and required OSHA to enter into a memorandum of understanding with the board to limit duplication of activity between the two.

In the last several years, Congress has also considered so-called whistle-blower legislation. These bills would provide protection for employees who exercise their rights under federal statutes, and establish prompt procedures for the vindication of those rights. These bills have not been passed. Several specific regulatory statutes such as the OSHAct and the Fair Labor Standards Act have long-standing antireprisal provisions. Because of its experience in enforcing the OSHAct antireprisal provisions, OSHA was given responsibility for enforcing the whistle-blower provisions in three non-OSHA statutes. The most important are the Surface Transportation Assistance Act, protecting trucking employees, and the Asbestos Hazard Emergency Response Act (AHERA), protecting individuals complaining about asbestos hazards in primary and secondary schools.

A recent congressional action affecting OSHA was the Workers' Family Protection Act (§209 of Pub. L. No. 102–522), passed October 26, 1992. It deals with potential hazards to families resulting from chemicals and other substances being transported on workers' clothing and persons. The law mandates a study of the problem and regulatory action by OSHA within four years based on the study.

## Occupational Safety and Health Review Commission

The OSHAct established the independent three-member Occupational Safety and Health Review Commission (OSHRC) to adjudicate citations and penalties issued by OSHA. OSHRC has had a troubled history and has often found it difficult to carry out its assigned statutory role. For a period of almost two years, in 1988 and 1989, the Review Commission lacked a quorum and was unable to decide cases. This situation resulted from delays in the presidential appointment of OSHRC members and brought about a large backlog of undecided cases. In the spring of 1990, President Bush appointed three OSHRC members: Edwin G. Foulke, Jr., as chairman of the commission, and Donald G. Wiseman and Velma Montoya as members. By September 1990, after Senate action, the Commission had a full roster of members and began to decide cases on a regular basis. The commission first directed itself to the backlog of undecided cases, and in fiscal year 1991 made 86 dispositions, more than triple the number during the prior fiscal year. At the close of fiscal year 1991, 86 cases were awaiting decision at the review level, as compared with 108 at the end of the prior fiscal year.

Review Commission decisions cover a broad range of issues under the act. During the 1987–1993 period, the commission decided cases on such issues as the application of the OSHAct's general duty clause (section 5[a][1]), the right of employees to participate in the settlement of cases, the scope of medical removal protection under OSHA's lead standard, citation of employers at multiemployer worksites, and interpretation of the complex requirements of the hazard communication standard.

Because an important aspect of the Review Commission's adjudicator role is to interpret OSHA standards, the issue has arisen in court review of Review Commission decisions as to whether the court should defer to the commission's interpretation when it differs from OSHA's interpretation of the same standard. In *Martin v. Occupational Safety and Health Review Commission,* the Supreme Court ruled in 1990 that reasonable interpretations of standards by OSHA were entitled to court deference.

In September 1992, the Review Commission, after notice and comment, adopted revised procedural rules for commission administrative proceedings. These new procedures, the commission said, were designed to speed up litigation before administrative law judges and the commission. Among other changes, the rules allow for mandatory scheduling of prehearing conferences at the administrative law judge's discretion, and for the imposition of simplified procedures that cannot be unilaterally vetoed by one of the parties.

On February 23, 1994, Stuart E. Weisberg was sworn in as chair of the commission, replacing Edwin G. Foulke, Jr. Mr. Weisberg, a former attorney with the National Labor Relations Board and staff member in Congress, was appointed to the commission by President Clinton, only the second commission member to be appointed by a Democratic president. Weisberg stated that he intends to emphasize prompt and even-handed decision making by the commission.

## Other OSHA Activities

As in prior years, OSHA continued its efforts to coordinate its activities with those of other agencies with responsibilities related to occupational safety and health. In November 1990, OSHA and the EPA entered into a memorandum of understanding under which the two agencies established a framework for coordinating their compliance activities, for exchanging data and training, and for providing one another with technical and professional assistance. Under the agreement, OSHA and the EPA have conducted a number of joint inspections in the petrochemical industry. In July 1992, OSHA and the EPA entered into a work plan implementing the 1990 memorandum, which provided for the continuation of the joint petrochemical inspections and for joint activity to remove toxic hazards from lead-smelting facilities. As already noted, under the Clean Air Act Amendments of

1990, OSHA is required to enter into an agreement with the newly established Chemical Safety and Hazard Investigation Board to eliminate duplication in the investigation of accidental chemical releases. OSHA continues to be active on the Interagency National Response Team, which coordinates federal policies and procedures in responding to oil spills and other hazardous material releases.

In April 1990, OSHA entered into an agreement with the Employment Standards Administration in the U.S. Department of Labor, which implements various labor regulatory statutes including the Fair Labor Standards Act, for the purpose of establishing procedures for the exchange of information among the compliance personnel of the two agencies. In another significant action, following the fire in the North Carolina poultry-processing facility, OSHA and the U.S. Department of Agriculture's Food Safety and Inspection Service entered into an agreement that provided for, among other things, the training of Department of Agriculture (DOA) meat inspectors in the recognition of safety hazards and for the establishment of procedures to permit USDA inspectors to report safety hazards directly to OSHA. At an oversight hearing following the fire, members of Congress expressed concern upon learning that DOA inspectors visited the facility almost daily and yet took no action to bring observed hazards to the attention of appropriate authorities.

Under an agreement between the secretaries of labor and energy, OSHA in 1990 conducted an on-site evaluation of occupational safety and health conditions at the Department of Energy's government-owned, contractor-operated (GOCO), nuclear defense facilities. A report on the investigation was submitted by OSHA to the secretary of energy, who subsequently announced that the department would make changes in operations to implement OSHA's recommendations.

Most recently, in January 1993, the U.S. Department of Labor published a proposed rule to coordinate the inspection activity of several Department of Labor agencies with regulatory responsibility affecting farm workers. The agencies enforcing various health, safety, housing, and wage laws applicable to farm workers are OSHA, the Employment and Training Administration, the Employment Standards Administration, and the Office of the Solicitor. Whereas OSHA standards are generally not applicable to agriculture, several specific standards do apply, including the temporary labor standard and field sanitation standards.

OSHA has also increased its participation in international occupational safety and health programs. The agency lent its expertise to help emerging democracies in Europe establish occupational safety and health programs, has participated in a number of international conferences, and has been involved in deliberations relating to the North American Free Trade Agreement (NAFTA).

Although OSHA gave major emphasis to compliance activity during this period, it has continued a number of programs encouraging voluntary compliance by employers. The most important of these are state-run, on-site, sanction-free consultation services, and OSHA's three Voluntary Protection Programs, called the Star Program, the Merit Program, and the Demonstration Program. Training and education have always been an agency priority, and OSHA devoted significant resources to the New Directions grant program and to the new Targeted Training Grant Program. The OSHA Training Institute in Des Plaines, Illinois, operates a variety of training programs for OSHA compliance personnel, state personnel, and safety and health personnel from the private sector and public agencies. In the fall of 1989, as part of its outreach program OSHA began again to publish its magazine, renamed *Job Safety and Health Quarterly*. The magazine's articles provide in-depth information on an array of agency projects. OSHA also began to publish four series of one-page information sheets, distributed to the public, on the following topics: construction accidents, ergonomic hazards, consultation, and chemical industry manufacturing hazards.

## Studies and Evaluations of OSHA

Since the earliest days of the OSHA program, the agency's implementation of the act has been a subject of study and evaluation, both within the government and by outside scholars and observers. At the request of congressional committees and members of Congress, the General Accounting Office (GAO) completed numerous reports on the OSHA program. The normal procedure used by the General Accounting Office is to conduct an investigation, and then prepare a draft report, which is submitted to the affected agency for comment. The final published report includes the GAO's findings, its recommendations for action, and the agency's comments. Among a number of others, the GAO published reports on the following topics relating to OSHA: "OSHA Contracting for Rulemaking Activity" (June 1989), "Options for Improving Safety and Health in the Workplace" (August 1990), "Inspectors' Opinions On Improving OSHA Effectiveness" (November 1990), "OSHA Action Needed to Improve Compliance With Hazard Communication Standard" (November 1991), "Worksite Safety and Health Programs Show Promise" (May 1992), and "Uneven Protections Provided to Congressional Employees" (October 1992). Thus, the GAO's activity covers a broad range of topics, both evaluating what OSHA does and making proposals for legislative change.

In a 1990 major report on options for improving safety and health in the workplace, the GAO presented a number of options for administrative and legislative change. In the area of standards, for example, they suggested an expedited process of revising OSHA's 1970 start-up standards; an amendment giving the agency separate authority to require substance testing by manufacturers; and a statutory requirement that OSHA act on new information (as, for example, from NIOSH) and give public explanations for its decisions to take particular courses of action. In order to enhance the deterrent effect of enforcement, the GAO proposed better administrative targeting of inspections, more inspections of hazardous worksites, an increase in the size of civil penalties, expanding criminal sanctions, and barring violators from federal contracts. GAO reports are often the basis for congressional oversight and legislative activity, and have been an important part of the development of legislative proposals for OSHA reform.

Evaluations of OSHA activity are also periodically prepared by the Department of Labor's in-house inspector general (IG). The inspector general, although part of the Labor Department, is independent of OSHA and has often criticized OSHA performance. For example, in a final report dated September 11, 1987, the inspector general found "a pattern of systemic weaknesses" in OSHA management control, construction targeting, and penalty assessment policy. The report was based on an IG investigation of OSHA's New York and Philadelphia offices. In response to the report, OSHA's Assistant Secretary Pendergrass disagreed with the IG's "sweeping" conclusions, noting that the scope of the investigation was limited to offices already identified by OSHA as having "severe problems." In April 1992, the IG commented more favorably, concluding that OSHA's "egregious" penalty was a bold step and that OSHA's settlement of "egregious" penalty cases brings about broader and more timely abatement action by employers and makes possible the collection of more penalties than would take place if the citations and penalties were litigated.

The Administrative Conference of the United States is a federal agency that studies and makes recommendations regarding the operation of federal administrative agencies. At the request of OSHA, the conference in 1986 undertook a study of the agency's operations. In 1987, law professors Sidney Shapiro and Thomas McGarity submitted a comprehensive two-phase report on OSHA. The report, largely critical of OSHA's standards activity, dealt with internal agency management and priority setting, and in the second part presented proposals for alternative approaches to development and promulgation of standards. Based on this report, the Administrative Conference formed a set of recommendations on OSHA, first recommending administrative changes in agency operations, and then, if they do not succeed, recommending legislative change. The proposed changes include the agency use of generic standards, expedited rulemaking to update start-up standards, and renewed use of advisory committees.

Private organizations have also periodically evaluated OSHA's performance, often critically. One of the more active of these groups is the National Safe Workplace Institute, based in Chicago. In a report released on Labor Day 1988, the Institute criticized the agency's enforcement policy, including the "megafine" policy, claiming that it was a cloak for a generally weak enforcement program

and that any benefits of the high penalties were offset by "sweetheart" settlement deals with large corporations.

In response, OSHA emphasized that settlements avoid the burdens of administrative and court litigation and the delays in abatement that necessarily result from litigation, during which abatement requirements are suspended. In subsequent reports issued during the administration of Assistant Secretary Scannell, the institute found less to criticize and generally offered more favorable evaluations of agency efforts.

The Bureau of National Affairs' (BNA) weekly publication, the *Occupational Safety and Health Reporter (OSHR),* celebrated its twentieth anniversary in 1990. The publication offers a full report on OSHA news each week as well as the text of OSHA decisions, standards, and other official documents. In September 1990, it published a Special Supplement entitled, "OSHA After 20 Years: a BNA Survey of Safety and Health Professionals." The BNA, in conjunction with the American Bar Association's subcommittee on OSHA, published a major treatise entitled *Occupational Safety and Health Law* in March 1988 and a First Supplement to the treatise in March 1990. A second supplement will be published in 1995. These volumes represent a comprehensive summary of OSHA law.

Important scholarly literature on OSHA continues to be produced. Professor John E. Mendeloff, whose earlier work on OSHA standards, *Regulating Safety,* was widely recognized as a thoughtful work, in 1988 published another text on OSHA standards activity, *The Dilemma of Toxic Substance Regulation—How Overregulation Causes Underregulation.* As the title suggests, Professor Mendeloff proposed that OSHA regulate less stringently and more extensively.

In 1993, professors McGarity and Shapiro published a critical evaluation of the OSHA program entitled *Workers at Risk—The Failed Promise of the Occupational Safety and Health Administration.* The volume, which focuses on the Reagan–Bush Administrations, concludes that even when "under the leadership of professionals committed to occupational safety and health, OSHA has not risen to its potential." The book proposes some possible solutions to OSHA's problems, emphasizing programs that "give workers more power to protect themselves." In 1993, professors Wayne B. Gray and John T. Scholz published a study in *Law and Society Review* on the impact of OSHA enforcement on the workplace injury rate. Disagreeing with some of the conventional wisdom on the subject, the authors conclude that inspections resulting in penalties induced a 22 percent reduction in injury rate in the affected workplaces studied during the few years following inspection. They asserted that more pessimistic assumptions about enforcement effectiveness have resulted from "narrow deterrence perspectives."

In two other challenging studies, professor David Weil of Boston University's School of Management concluded that the presence of a union in a facility dramatically increases enforcement activity by OSHA. One study was conducted in the manufacturing sector. He found that union impact is greatest in larger facilities, where the employees are "best organized" and that, as a general matter, "if workers do not become partners in this regulatory process, the chances for OSHA success seem dim indeed." In the second study, Weil concluded that unions in the construction industry play a critical role in directing the agency's attention to union job sites "in a world of limited OSHA resources." The studies were published, respectively, in two journals, *Industrial Relations* and the *Journal of Labor Research*.

# Summary

The year 1995 marks the 25th anniversary of the OSHA program. Looking ahead, OSHA confronts an uncertain future, as the new Republican majority in Congress proclaims its intention to establish an entirely new direction for government regulation. Looking back, however, OSHA can quietly boast of many important accomplishments. These accomplishments, in such areas as enforcement, standards setting, training and education, and statistics and research, have been described in detail in this chapter. But beyond any specific accomplishments, OSHA's central significance is that it has changed the way Americans think about occupational safety and health.

Before 1970, the concept of occupational safety and health was on the fringe of the national consciousness. There were hardly any individuals or organizations that paid any particular attention to the increasingly serious safety and health hazards in the workplace and the moral and economic imperatives that demanded that workers be protected from those hazards. Because of OSHA, all this has changed. Although some admire OSHA and others harbor animosity toward the program, almost no one with any connection to America's workplaces would think of ignoring the responsibility of the community "to assure so far as possible every working man and woman in the Nation safe and healthful working conditions" (OSHAct Section 2(b), 1970).

The recognition of this responsibility is the lasting contribution of OSHA to the forging of the American moral conscience. It will not be quickly forgotten, whatever changes in regulatory fashion may take place in the years to come.

# Bibliography

American Bar Association. *Occupational Safety and Health Law.* Chicago: American Bar Association, 1988.

Berman DM. *Death on the Job.* New York: Monthly Rev, 1978.

Cherrington DR. Comment: The race to the courthouse: Conflicting views toward the judicial review of OSHA standards. 1994 *B.Y.U. L. Rev.* 95.

Claybrook, J. *Retreat from Safety.* New York: Pantheon, 1984.

Gombar RC, Collins JA, eds. *Occupational Safety and Health Law, First Supplement: 1987–1988.* Washington, DC: BNA, 1990.

HHS DOL. *The President's Report on Occupational Safety and Health.* Washington, DC: U.S. Government Printing Office, May 1972, December 1973, December 1976, 1974, 1979, 1980, 1983, 1985.

Johnson GR Jr. The split enforcement model: some conclusions from the OSHA and MSHA experiences. 39 *Admin. L. Rev.* 315, 1987.

Lofgren DJ. *Dangerous Premises: An Insider's View of OSHA Enforcement.* Ithaca NY: ILR Pr, 1989.

MacAvoy P, ed. *Report of the Presidential Task Force, OSHA Safety Regulation.* Washington, DC: GPO, 1977.

McGarity TO. Reforming OSHA: Some thoughts for the current legislative agenda. Presented at symposium: *New Challenges in Occupational Health.* 31 *Houston Law Review* 99, 1994.

McGarity TO, Shapiro SA. *Workers at Risk: The Failed Promise of the Occupational Safety and Health Administration.* Westport, CT: Praeger, 1993.

MacLaury J. The Job Safety Law of 1970: Its passage was perilous. *Monthly Labor Rev* 18 (March), 1981.

Mendeloff JE. *The Dilemma of Toxic Substance Regulation—How Overregulation Causes Underregulation.* Cambridge, MA: MIT Press, 1988.

Mendeloff JE. *Regulating Safety.* Cambridge, MA: MIT Press, 1979.

Mintz B. *OSHA: History, Law, and Policy.* Washington, DC: Bureau of National Affairs, 1984.

Moran R. Cite OSHA for violations. *OSHR* (March–April 1976):19–20.

Note: Getting away with murder: Federal OSHA preemption of state prosecution for industrial accidents. 101 *Harv. L. Rev.* 535, 1987.

O'Brien MW, Assistant General Counsel, United Steelworkers of America. Letter to OSHA on state programs. March 15, 1985.

Occupational Health and Safety Administration (OSHA). *Field Inspection Reference Manual.* OSHA Instruction CPL 2. 103. Washington DC: U.S. Government Printing Office, September 1994.

Page J, O'Brien MW. *Bitter Wages.* New York: Viking-Penguin, 1973.

Rothstein MA. *Occupational Safety and Health Law,* 3rd ed. St. Paul, MN: West Publishing, 1983.

Shapiro SA, McGarity TO. Reorienting OSHA: regulatory alternatives and legislative reform. *Yale Journal on Regulation* 6, 1989.

Stader KD. Casenote: OSHA's air contaminants standard revision succumbs to the substantial evidence test: *AFL–CIO v. OSHA* 965 F.2d 962 (11 Cir. 1993). 62 *U. Cin. L. Rev.* 351, 1993.

U.S. Congress, Office of Technology Assessment. *Preventing Illness and Injury in the Workplace.* OTA-H-256. Washington, DC: GPO, 1985.

U.S. Dept. of Labor. *Protecting People at Work—A Reader in Occupational Safety and Health.* Washington, DC: GPO, 1983.

**Professional Organizations**

**Scientific and Service Organizations**

**U.S. Government Agencies**
*NIOSH ▪ Key NIOSH Services ▪ OSHA ▪ Regional Offices ▪ Other Federal Agencies*

**Association of Occupational and Environmental Clinics (AOEC)**
*Alabama ▪ California ▪ Colorado ▪ Connecticut ▪ District of Columbia ▪ Georgia ▪ Illinois ▪ Iowa ▪ Kentucky ▪ Louisiana ▪ Maine ▪ Maryland ▪ Massachusetts ▪ Michigan ▪ Minnesota ▪ New Jersey ▪ New York ▪ North Carolina ▪ Ohio ▪ Oklahoma ▪ Pennsylvania ▪ Rhode Island ▪ Texas ▪ Utah ▪ Washington ▪ West Virginia*

**Canadian Clinics**
*Alberta ▪ Manitoba*

**Community Organizations**
*Alaska ▪ California ▪ Connecticut ▪ Illinois ▪ Maine ▪ Massachusetts ▪ Michigan ▪ New Hampshire ▪ New Mexico ▪ New York ▪ North Carolina ▪ Ohio ▪ Pennsylvania ▪ Rhode Island ▪ Tennessee ▪ Texas ▪ Washington, DC ▪ Wisconsin ▪ Canada*

**Industry Organizations**
*Labor Unions ▪ University-Based Research, Training, and Continuing Professional Education Programs*

**Computer-Based Sources of Information**
*CD-ROM ▪ Environmental and Safety Regulations ▪ MSDS ▪ Online Resources*

**Newsletters and Reports**

**Journals and Magazines**

**Bibliography**
*AIDS ▪ Biological Hazards ▪ Biological Monitoring and Medical Surveillance ▪ Engineering Controls ▪ Ergonomics ▪ Exposure and Risk Assessment ▪ History and Critiques of Industrial Hygiene Practice ▪ Noise and Hearing Conservation ▪ Occupational Epidemiology ▪ Occupational Health Policy ▪ Occupational Medicine ▪ People of Color ▪ Radiation Hazards ▪ Reproductive Hazards ▪ Resource Materials ▪ Right to Know and Hazard Communication ▪ Safety and Health Programs ▪ Sampling and Laboratory Methods ▪ Toxicology and Chemical Hazards ▪ Ventilation ▪ Video Display Terminals ▪ Women*

# Additional Resources

Revised by Deborah Gold, MPH,
and Donna Iverson

The twenty years since the passage of the Occupational Safety and Health Act have seen an almost explosive broadening in the practice of industrial hygiene. Industrial hygiene practice now includes such diverse issues as the control of airborne chemical contaminants, noise, ionizing and nonionizing radiations, ergonomics, environmental pollution, infectious and communicable diseases, safety, and indoor air quality.

It is almost impossible for any one professional to be an expert on every aspect of industrial hygiene practice. Fortunately, there are many organizations to which the industrial hygienist or safety professional can turn for help, including professional organizations, scientific and service organizations, government agencies, occupational health clinics, community groups, industry organizations, labor unions and other employee organizations, and university-based research and training programs.

The first part of this appendix includes information about those types of organizations. The second part of this appendix includes information about on-line sources of information and a bibliography for selected topics.

## Professional Organizations

American Association of Occupational Health Nurses (AAOHN)
50 Lenox Pointe
Atlanta, GA 30324
609-848-1000 or 800-257-8290

American Board of Industrial Hygiene
4600 W. Saginaw, Suite 101
Lansing, MI 48917-2737
517-321-2638

The ABIH administers a national certification program for industrial hygienists in general practice and in several specialty areas.

American Conference of Governmental Industrial Hygienists (ACGIH)
Kemper Woods Center, 1330 Kemper Meadow Dr.
Cincinnati, OH 45240
Phone: 513-742-2020
Fax: 513-742-3355

The ACGIH is an association whose seven categories of membership are open to anyone engaged in the practice of industrial hygiene or occupational and environmental health and safety. The ACGIH annually publishes Threshold Limit Values for air contaminants, and other consensus standards, and publishes and distributes a wide range of manuals and other materials regarding industrial hygiene practice.

American Industrial Hygiene Association (AIHA)
2700 Prosperity Ave., Suite 250
Fairfax, VA 22031
Phone: 703-849-8888
Fax: 703-207-3561

The AIHA provides continuing education programs and opportunities for industrial hygienists to meet and exchange ideas. It also maintains a laboratory certification program.

American Public Health Association (APHA)
1015 Fifteenth St. NW
Washington, DC 20005
202-789-5600

The Occupational Health Section of the APHA makes policy recommendations regarding occupational safety and health issues, and views industrial hygiene practice in a public health context.

American Society of Heating, Refrigerating, and Air Conditioning Engineers (ASHRAE)
1791 Tullie Circle NE
Atlanta, GA 30329
Phone: 404-636-8400
Fax: 404-321-5478

ASHRAE publishes consensus standards regarding the performance of building ventilation systems and other indoor air quality issues.

American Society of Safety Engineers
1800 E. Oakton St.
Des Plaines, IL 60018
Phone: 847-699-2929
Fax: 847-296-3769

Board of Certified Safety Professionals
208 Burwash Ave.
Savoy, IL 61874
217-359-9263

The Board of Certified Safety Professionals administers a national certification program that is open to industrial hygienists.

Health Physics Society
8000 Westpark Dr., Suite 130
McLean, VA 22102
703-790-1745

Human Factors and Ergonomics Society
P.O. Box 1369
Santa Monica, CA 90406-1369
Phone: 310-394-1811
Fax: 310-394-2410

Illuminating Engineering Society of North America
345 E. 47th St.
New York, NY 10017
212-705-7913

Society of Manufacturing Engineers
P.O. Box 930, One SME Drive
Dearborn, MI 48121
800-733-4763

Society of Toxicology
1101 14th St. NW, Suite 1100
Washington, DC 20005
202-371-1393

## Scientific and Service Organizations

American National Standards Institute (ANSI)
11 W. 42nd St.
New York, NY 10036
Phone: 212-642-4900
Fax: 212-302-1286

National Council on Radiation Protection and Measurements
7910 Woodmont Ave., Suite 800
Bethesda, MD 20814
301-657-2652

National Fire Protection Association
Batterymarch Park, P.O. Box 9101
Quincy, MA 02269-9101
800-344-3555

National Safety Council
1121 Spring Lake Dr.
Itasca, IL 60143-3201
800-621-7619

Underwriters Laboratories Inc.
333 Pfingsten Rd.
Northbrook, IL 60062
708-272-8800

# ■ U.S. Government Agencies

## NIOSH

The National Institute for Occupational Safety and Health (NIOSH) was established by the Occupational Safety and Health Act of 1970. OSHAct made NIOSH responsible for conducting research to make the nation's workplaces healthier and safer.

To identify hazards, NIOSH conducts inspections, publishes its findings, and makes recommendations for improved working conditions to regulatory agencies such as the Occupational Safety and Health Administration and the Mine Safety and Health Administration. It also conducts epidemiological research and develops sampling and analytical techniques.

NIOSH works with groups and individuals who share its concern for protecting the health of all workers. It plays a vital role in training occupational health and safety experts and communicating the latest information to those most concerned.

## Key NIOSH Services

**Health hazard evaluations.** Employers, employees, or their representatives who suspect a health problem in the workplace can request a NIOSH Health Hazard Evaluation (HHE) to assess the problem; call 1-800-35-NIOSH.

**Miners' x-rays.** NIOSH administers periodic chest x-rays to coal miners to facilitate early detection of coal workers' pneumoconiosis; call 304-291-4301.

**Fatal accident investigations.** NIOSH identifies risk factors for work-related fatalities and injuries through its Fatal Accident Circumstances and Epidemiology project (FACE); call 304-291-4575.

**Extramural grants.** NIOSH sponsors extramural research in priority areas and coordinates this with its intramural and contract research and that of other HHS and U.S. departments; call 404-639-3343.

**Databases.** NIOSH has access to extensive databases of occupational safety and health information from around the world; call 513-533-8326.

**Respirators.** NIOSH tests and certifies respirators to ensure their compliance with federal requirements; call 304-291-4331.

**Educational resource centers.** NIOSH supports educational resource centers (ERCs) at 14 U.S. universities to help ensure an adequate supply of trained occupational safety and health professionals; call 513-533-8241.

**Publications.** NIOSH publishes and distributes a variety of materials related to occupational safety and health; call 513-533-8287.

**Toll-free number.** NIOSH operates a toll-free number (8:00 A.M.–4:30 P.M. E.S.T. weekdays) in the United States to receive requests for technical information and health hazard evaluations; call 1-800-35-NIOSH (1-800-356-4674).

In addition to NIOSH, many state health departments maintain their own occupational health and safety divisions. You should contact them for more information.

## OSHA

The Occupational Safety and Health Administration (OSHA) was created under the Occupational Safety and Health Act (OSHAct) of 1970, within the Department of Labor "to assure so far as possible every working man and woman in the Nation safe and healthful working conditions and to preserve our human resources."

U.S. Department of Labor
Occupational Safety and Health Administration (OSHA)
200 Constitution Ave. NW
Washington, DC 20210
Phone: 800-321-OSHA (to report an emergency)
202-219-8148 (for information)
Fax: 202-219-5986

## Regional Offices

(Please Note: In addition to federal OSHA, many states maintain their own federally approved occupational safety and health plans, indicated with *).

Region I (CT*, MA, ME, NH, RI, VT*)
133 Portland St., 1st Floor
Boston, MA 02114
617-565-7164

Region II (NJ, NY*, PR*, VI*)
201 Varick St., Room 670
New York, NY 10014
212-337-2378

Region III (DC, DE, MD*, PA, VA*, WV)
Gateway Building, Suite 2100
3535 Market St.
Philadelphia, PA 19104
215-596-1201

Region IV (AL, FL, GA, KY*, MS, NC*, SC*, TN*)
1375 Peachtree St. NE, Suite 587
Atlanta, GA 30367
404-347-3573

Region V (IL, IN*, MI*, MN*, OH, WI)
230 S. Dearborn St., Room 3244
Chicago, IL 60604
312-353-2220

Region VI (AR, LA, NM*, OK, TX)
525 Griffin St., Room 602
Dallas, TX 75202
214-767-4731

Region VII (IA*, KS, MO, NE)
911 Walnut St., Room 406
Kansas City, MO 64106
816-426-5861

Region VIII (CO, MT, ND, SD, UT*, WY*)
Federal Building, Room 1576
1961 Stout St.
Denver, CO 80294
303-844-3061

Region IX (American Samoa, AZ*, CA*, Guam, HI*,
NV*, Trust Territories of the Pacific)
71 Stevenson St., Room 415
San Francisco, CA 94105
415-744-6670

Region X (AK*, ID, OR*, WA*)
1111 Third Ave., Suite 715
Seattle, WA 98101-3212
206-553-5930

* These states and territories operate their own OSHA-approved job safety and health programs (the Connecticut and New York plans cover public employees only). States with approved programs must have a standard that is identical to, or at least as effective as, the federal standard. Address of state OSHA offices are available through the corresponding regional office.

## Other Federal Agencies

Agency for Toxic Substances and Disease Registry
(ATSDR)
Department of Health and Human Services
Public Health Service
1600 Clifton Rd. NE (E-33)
Atlanta, GA 30333
404-639-6204

Center for Devices and Radiological Health
Food and Drug Administration
5600 Fishers Ln.
Rockville, MD 20857
301-443-2444

Centers for Disease Control
Department of Health and Human Services
Public Health Service
1600 Clifton Road NE
Atlanta, GA 30333
404-639-8063

Environmental Protection Agency (EPA)
EPA Publications and Information Center
P.O. Box 42419
Cincinnati, OH 45242-0419
Emergency Planning and Community Right-to-Know
Hotline: 800-535-0202
SARA and CERCLA Hotline (Nonemergency): 800-424-9346

Health Resources and Service Administration
Food and Drug Administration
5600 Fishers Lane
Rockville, MD 20857
301-443-4690

Library of Congress
Science and Technology Division
Washington, DC 20540
202-707-5000

Mine Safety and Health Administration
Department of Labor
4015 Wilson Blvd., Room 601
Arlington, VA 22203
703-235-1452

National Aeronautical and Space Administration
300 E Street SW
Washington, DC 20546
202-358-1000

National Cancer Institute
National Institutes of Health
9000 Rockville Pike
Bethesda, MD 20892
301-496-5583

National Center for Health Statistics
Public Health Service
3700 East–West Hwy
Hyattsville, MD 20782
301-436-8500

National Center for Toxicological Research
Hwy 365 N, County Rd. 3
Jefferson, AR 72079
501-541-4000

National Council on Radiation Protection
and Measurement (NCRP)
7910 Woodmont Ave., #800
Bethesda, MD 20814
301-657-2652

National Institute for Occupational Safety
and Health (NIOSH)
Centers for Disease Control
1600 Clifton Road NE
Atlanta, GA 30333
800-356-4647

National Institute of Standards and Technology
Rte I-270 and Quince Orchard Rd.
Gaithersburg, MD 20899
301-975-3058

National Technical Information Service
Department of Commerce
5285 Port Royal Rd.
Springfield, VA 22161
Phone: 703-487-4650
Fax: 703-321-8547

Nuclear Regulatory Commission
Washington, DC 20555-0001
301-492-7000

Office of Energy Research
Department of Energy
1000 Independence Ave. SW
Washington, DC 20585
301-903-4944

Office of Hazardous Materials Transportation
Department of Transportation
400 Seventh St. NW
Washington, DC 20590
202-366-0656

U.S. Government Printing Office
Washington, DC 20402-9325
Phone: 202-783-3238
Fax: 202-275-7810

# ■ Association of Occupational and Environmental Clinics (AOEC)

1010 Vermont Ave. NW, #513
Washington, DC 20005
202-347-4976

## Alabama

University of Alabama at Birmingham
930 20th Street South
Birmingham, AL 35205
Phone: 205-934-7303
Fax: 205-975-4377
Timothy Key, MD, MPH
Alt. contact: Brian G. Forrester, MD, MPH

## California

Irvine Occupational Health Center
UC Irvine
19722 MacArthur Blvd.
Irvine, CA 92717
Phone: 714-824-8641
Fax: 714-824-2345
Dean Baker, MD, MPH

Occupational and Environmental Health Clinic
University of California at Davis
ITEH
Davis, CA 95616
Phone: 916-752-3317
Fax: 916-752-5300
Stephen McCurdy, MD, MPH

Occupational and Environmental Health Clinic
University of California at San Francisco
1515 Scott Street
San Francisco, CA 94115
Phone: 415-885-7700
Fax: 415-206-8949
Patricia Quinlan, MPH
Alt. contacts: Diane Liu, MD, MPH
Robert Harrison, MD, MPH

## Colorado

Occupational and Environmental Medicine Division
National Jewish Center for Immunology
and Respiratory Medicine
1400 Jackson Street
Denver, CO 80206
Phone: 303-398-1520
Fax: 303-398-1452
Kathleen Kreiss, MD
Alt. contact: Cecile Rose, MD, MPH

## Connecticut

University of Connecticut Occupational Medicine
Program
263 Farmington Ave.
Farmington, CT 06030
Phone: 203-679-2893
Fax: 203-679-4587
Eileen Storey, MD, MPH
Alt. contact: Michael Hodgson, MD, MPH

Waterbury Occupational Health
140 Grandview Ave. Suite 101
Waterbury, CT 06708
Phone: 203-573-8114
Fax: 203-755-3823
Gregory McCarthy, MD, MPH

Yale Occupational/Environmental Medicine Program
School of Medicine
135 College Street, 3rd Floor
New Haven, CT 06510
Phone: 203-785-5885
Fax: 203-785-7391
Mark Cullen, MD, MPH

## District of Columbia

Division of Occupational and Environmental Medicine
School of Medicine, George Washington University
2300 K St. NW
Washington, DC 20037
Phone: 202-994-1734
Fax: 202-994-0011
Laura Welch, MD
Alt. contact: Rosemary Sokas, MD, MOH

## Georgia

Environmental and Occupational Program—
The Emory Clinic
Rollins School of Public Health, Emory University
1518 Clifton Road
Atlanta, GA 30329
Phone: 404-248-5978
Fax: 404-727-8744
Howard Frumkin, MD, Dr. PH

## Illinois

Occupational Medicine Clinic
Cook County Hospital
720 South Wolcott
Chicago, IL 60612
Phone: 312-633-5310
Fax: 312-633-6442
Stephen Hessl, MD, MPH
Alt. contact: Ann Naughton, RN, MPH

Sinai Occupational Health Program
Mt. Sinai Hospital Medical Center
California Ave. at 15th Street, Room N727
Chicago, IL 60608
Phone: 312-257-6480
Fax: 312-257-6213
Edward Mogabgab, MD

University of Illinois
Occupational Medicine Program
840 S. Wood
P.O. Box 6998 M/C 678
Chicago, IL 60612
Phone: 312-996-1063
Fax: 312-996-1286
Linda Forst, MD, MS, MPH
Alt. contact: Stephen Hessl, MD, MPH

## Iowa

University of Iowa Occupational Medicine Clinic
Department of Internal Medicine, College of Medicine
T304, GH 200 Hawkins Drive
Iowa City, IA 52242
Phone: 319-356-8269
Fax: 319-356-6406
David Schwartz, MD, Dr. PH

## Kentucky

University of Kentucky Occupational Medicine Program
Warren Wright Medical Plaza
800 Rose Street
Lexington, KY 40536-0084
Terence Collins, MD, MPH
Phone: 606-257-5166
Fax: 606-258-1038

## Louisiana

Ochsner Center for Occupational Health
1514 Jefferson Hwy.
New Orleans, LA 70121
Phone: 504-842-3955
Fax: 504-842-3977
Peter G. Casten, MD, MPH
Alt. contact: Douglas A. Swift, MD, MSPH

## Maine

Center for Health Promotion
1600 Congress Street
Portland, ME 04102
Phone: 207-774-7751
Fax: 207-828-5140
Stephen Shannon, DO, MPH
Alt. contact: Betsy Buehrer, DO

## Maryland

Johns Hopkins University
Center for Occupational and Environmental Health
5501 Hopkins Bayview Circle
Baltimore, MD 21224
Phone: 410-550-2322
Fax: 410-550-3355
Edward J. Bernacki, MD, MPH
Alt. contact: Theresa Pluth, MSN, MPH

Occupational Health Project/School of Medicine
Division of General Internal Medicine,
University of Maryland
405 Redwood Street
Baltimore, MD 21202
Phone: 410-706-7464
Fax: 410-706-4078
James Keogh, MD

## Massachusetts

Center for Occupational and Environmental Medicine
Massachusetts Respiratory Hospital
2001 Washington Street
South Braintree, MA 02184
Phone: 617-848-2600
Fax: 617-849-3290
Dianne Plantamura, Coordinator
Alt. contact: David Christiani, MD, MPH, Director

Occupational and Environmental Health Center
Cambridge Hospital
1493 Cambridge Street
Cambridge, MA 02139
Phone: 617-498-1580
Fax: 617-498-1671
Rose Goldman, MD, MPH
Alt. contact: Susan Rosenwasser, MEd

Occupational Health Service
Department of Family and Community Medicine,
University of Massachusetts
55 Lake Avenue North
Worcester, MA 01655
Phone: 508-856-2734
Fax: 508-856-1680
Glenn Pransky, MD, MOH
Alt. contact: Jay Himmelstein, MD, MPH

Pulmonary Associates (Occupational Medicine)
Ambulatory Care Center
Parkman Street
Boston, MA 02114
Phone: 617-726-3741
Fax: 617-726-6878
L. Christine Oliver, MD, MPH, MS
Alt. contact: Elisha Atkins, MD

## Michigan

Center for Occupational and Environmental Medicine
22255 Greenfield Rd. Suite 440
Southfield, MI 48075
Phone: 810-559-6663
Fax: 810-559-8254
Margaret Green, MD, MPH
Alt. contact: Michael Harbut, MD, MPH

Division of Occupational Health
Wayne State/Department of Family Medicine
4201 St. Antoine, Suite 4-J
Detroit, MI 48201
Phone: 313-577-1420
Fax: 313-577-3070
Ray Demers, MD, MPH
Alt. contact: Mark Upfal, MD, MPH

Michigan State University
Department of Medicine
117 West Fee
East Lansing, MI 48824-1316
Phone: 517-353-1846
Fax: 517-432-3606
Kenneth Rosenman, MD, MPH

Occupational Health Program
School of Public Health, University of Michigan
1420 Washington Heights
Ann Arbor, MI 48109-2029
Phone: 313-764-2594
Fax: 313-763-8095
David Garabrant, MD, MPH
Alt. contact: Alfred Franzblau, MD, MPH

Occupational Health Service
Work and Health Institute
St. Lawrence Hospital
1210 W. Saginaw
Lansing, MI 48915

Phone: 517-377-0309
Fax: 517-377-0310
R. Michael Kelly, MD, MPH
Alt. contact: John McPhail

## Minnesota

Columbia Park Medical Group
6401 University Ave. NE #200
Fridley, MN 55432
Phone: 612-572-5710
Fax: 612-571-3008
Donald Johnson, MD, MPH
Alt. contact: Dorothy Quick

Ramsey Clinic
Occupational and Environmental Health
and Occupational Medicine Residency Training
640 Jackson St.
St. Paul, MN 55101-2595
Phone: 612-221-3771
Fax: 612-221-3874
Paula Gelger
Alt. contact: William H. Lohman, MD

## New Jersey

Environmental and Occupational Health
Clinical Center
Environmental and Occupational Health
Sciences Institute
P.O. Box 1179
Piscataway, NJ 08855-1179
Phone: 908-932-0123
Fax: 908-932-0127
Howard Kipen, MD, MPH
Alt. contact: Gail Buckler, RN, MPH

## New York

Center for Occupational and Environmental
Medicine
Health Sciences Center, Level 3–086
University at Stony Brook
Stony Brook, NY 11794
Phone: 516-444-2167
Fax: 516-444-7525
Wajdy Hailoo, MD, MPH

Central New York Occupational Health
Clinical Center
6712 Brooklawn Parkway Suite 204
Syracuse, NY 13211-2195
Phone: 315-432-8899
Fax: 315-431-9528
Michael B. Lax, MD, MPH

Eastern New York Occupational Health Program
1201 Troy Schenectady Road
Latham, NY 12110
Phone: 518-783-1518
Fax: 518-783-1827
Anne Tencza, RN, COHN
Alt. contact: Eckardt Johanning, MD, MPH

Finger Lakes Occupational Health Services
601 Elmwood Avenue
Box EHSC
Rochester, NY 14642
Phone: 716-275-1335
Fax: 716-256-2591
Julie Cataldo
Alt. contact: Marc Utell, MD

HHC Bellevue Occupational/Environmental
Health Clinic
Bellevue Hospital Room CD 352
1st Ave. and 27th Street
New York, NY 10016
Phone: 212-561-4572
Fax: 212-561-4574
George Friedman-Jimenez, MD

Mount Sinai
Irving J. Selikoff Occupational Health
Clinic Center
P.O.B. 1058 Gustave Levy Pl.
New York, NY 10029
Phone: 212-241-6173
Fax: 212-996-0407
Stephen Mooser, MPH
Alt. contact: Stephen Levin, MD

## North Carolina

Division of Occupational and Environmental
Medicine
Duke University Medical Center
Box 2914
Durham, NC 27710
Phone: 919-286-3232
Fax: 919-286-1021
Dennis J. Darcey, MD, MPSH
Alt. contact: Gary Greenberg, MD, MPH

## Ohio

Center for Occupational Health
Holmes Hospital—Tate Wing
University of Cincinnati College of Medicine
Eden and Bethesda Ave.
Cincinnati, OH 45267-0182
Phone: 513-558-1234
Fax: 513-558-1010
James Donovan, MD, MPH
Alt. contact: Douglas Linz, MD, MS

Greater Cincinnati Occupational Health Center
Jewish Hospital Evandale
10475 Reading Road, Suite 405
Cincinnati, OH 45241
Phone: 513-769-0561
Fax: 513-769-0766
Harriet Applegate
Alt. contact: Margaret Atterbury, MD, MPH

WorkLink
Occupational and Environmental Health Clinic
2500 MetroHealth Drive
Cleveland, OH 44109-1998
Phone: 216-778-8087
Fax: 216-778-8225
Kathleen Fagan, MD, MPH
Alt. contact: Seth Foldy, MD

## Oklahoma

University Occupational Health Services
Oklahoma Memorial Hospital
800 NE 15th St., Room 520
Oklahoma City, OK 73104
Phone: 405-271-3100
Fax: 405-271-4125
Roy DeHart, MD, MPH
Alt. contact: Lynn Mitchell, MD, MPH

WorkMed
9330 E. 41st Street, #102
Tulsa, OK 74145-3718
Phone: 918-627-4646
Fax: 918-669-4425
James W. Small, MD, MPH
Alt. contact: Andrew Floren, MD, MPH

## Pennsylvania

Center for Occupational and Environmental Health
3901 Commerce Ave., Suite 101
Willow Grove, PA 19090-1109
Phone: 215-881-5904
Fax: 881-5920
Jessica Herztein, MD, MPH

Occupational and Environmental Medicine Program
University of Pittsburgh
130 DeSoto Street, Room A729
Pittsburgh, PA 15261
Phone: 412-624-3155
Fax: 412-624-3040
David Tollerud, MD, MPH

Occupational Health Service
Department of Community and Preventive Medicine/MCP
3300 Henry Ave.
Philadelphia, PA 19129

Phone: 215-842-6540
Fax: 215-843-2448
Eddy Bresnitz, MD, MS
Alt. contact: Harriet Rubenstein, JD, MPH

## Rhode Island

Memorial Hospital of Rhode Island Occupational
Health Service
Brown University Program in Occupational Medicine
111 Brewster St.
Pawtuckett, RI 02860
Phone: 401-729-2859
Fax: 401-729-2950
David G. Kern, MD, MOH

## Texas

Texas Institute of Occupational Safety and Health
Highway 271 and 155
Tyler, TX 75710
Phone: 903-877-7262
Fax: 903-877-7982
Jeffrey Levin, MD, MSPH

## Utah

Rocky Mountain Center for Occupational and Environmental Health
Bldg. 512
University of UT
Salt Lake City, Utah 84112
Phone: 801-581-5056
Fax: 801-581-3841
Anthony Suruda, MD, MPH
Alt. contact: Royce Moser, MD, MPH

## Washington

Occupational Medicine Program
University of Washington Harborview Medical Center
325 9th Ave. ZA-66
Seattle, WA 98104
Phone: 206-223-3005
Fax: 206-223-8247
Drew Brodkin, MD, MPH
Alt. contact: Scott Barnhart, MD, MPH

## West Virginia

Division of Occupational and Environmental Health
Department of Family and Community Medicine
Marshall University School of Medicine
1801 6th Ave.
Huntington, WV 25755
Phone: 304-696-7045
Fax: 304-696-7048
Chris McGuffin, MS
Alt. contact: James Becker, MD

# ■ Canadian Clinics

## Alberta

University of Alberta Faculty of Medicine
13–103 Clinical Science Bldg.
Edmonton, Alberta, CD T6G 2G3
Phone: 403-492-7849
Fax: 403-492-0364
Linda Cocchiarella, MD, MSc
Tee Guidotti, MD, MPH

## Manitoba

MFL Occupational Health Centre, Inc.
102–275 Broadway
Winnipeg, Manitoba, CD R3C 4M6
Phone: 204-949-0811
Fax: 204-956-0848
Judy Cook, Executive Director
Alt. contact: Ahmod Randeree, MD

# ■ Community Organizations

Committees for Occupational Safety and Health (COSH)
and similar organizations are made up of rank-and-file
workers, labor leaders, occupational safety and health professionals, medical professionals, and community activists.
They are an excellent source of current information. Many
of them maintain libraries and have professionals on staff
who can provide technical assistance.

## Alaska

Alaska Health Project
1818 W. Northern Lights Boulevard, #103
Anchorage, AK 99517
907-276-2864

## California

BACOSH
c/o Robin Baker, LOHP
School of Public Health
University of California at Berkeley
2515 Channing Way
Berkeley, CA 94720
Phone: 510-642-5507
Fax: 510-643-5698

LACOSH
600 S. New Hampshire Ave.
Los Angeles, CA 90005
213-383-4416

SACOSH (Sacramento)
3101 Stockton Blvd.
Sacramento, CA 95820
916-924-8060

SCCOSH (Santa Clara)
760 N. First St., Second Floor
San Jose, CA 95112
408-998-4050

## Connecticut

ConnectiCOSH
Box 231107
Hartford, CT 06123
203-549-1877

## Illinois

CACOSH (Chicago)
37 S. Ashland
Chicago, IL 60607
312-666-1611 or −1721

Midwest Center for Labor Research
3411 W. Diversey Ave., Suite 10
Chicago, IL 60647
312-278-5418

The National Safe Workplace Institute
122 S. Michigan Ave., Suite 1450
Chicago, IL 60603
312-661-0690

## Maine

Maine Labor Group on Health
Box V
Augusta, ME 04332
207-622-7823

## Massachusetts

MassCOSH
555 Amory St.
Boston, MA 02130
617-524-6686

Western MassCOSH
P.O. Box 55
Hatfield, MA 01038
413-247-9413

## Michigan

Labor Education and Research Project
P.O. Box 20001
Detroit, MI 48220
313-842-6262

SEMCOSH (SE Michigan)
2727 Second Ave.
Detroit, MI 48201
313-961-3345

## New Hampshire

NHCOSH
c/o NH AFL-CIO
110 Sheep Davis Rd.
Pembroke, NH 03275
603-224-4789

## New Mexico

Southwest Research and Information Center
P.O. Box 4524
Albuquerque, NM 87106

## New York

ALCOSH (Allegheny)
100 E. 2nd St., Suite 3
Jamestown, NY 14701
716-488-0720

CNYCOSH (Central NY)
615 W. Genessee St.
Syracuse, NY 13204
315-471-6187

ENYCOSH (Eastern NY)
c/o Larry Rafferty
121 Erie Blvd.
Schenectady, NY 12305
518-372-4308

NYCOSH
275 Seventh Ave., 8th Floor
New York, NY 10001
212-627-3900
914-939-5612 (Westchester office)

ROCOSH (Rochester)
797 Elmwood Ave. #4
Rochester, NY 14620
716-244-0420

WNYCOSH (Western NY)
450 Grider St.
Buffalo, NY 14215
716-897-2110

## North Carolina

North Carolina Occupational Safety and Health Project
(NCOSH)
P.O. Box 2514
Durham, NC 27715
919-286-9249

## Ohio

9 to 5—National Association of Working Women
614 Superior Ave. NW
Cleveland, OH 44113
216-566-9308

## Pennsylvania

PhilaPOSH
3001 Walnut St., 5th Floor
Philadelphia, PA 19104
215-386-7000

## Rhode Island

RICOSH
741 Westminster St.
Providence, RI 02903
401-751-2015

## Tennessee

Highlander Research and Education Center
1959 Highlander Way
New Market, TN 37820
615-933-3443

TNCOSH
1515 E. Magnolia, Suite 406
Knoxville, TN 37917
615-525-3147

## Texas

TexCOSH
c/o Karyl Dunson
5735 Regina Lane
Beaumont, TX 77706
409-898-1427

## Washington, DC

Alice Hamilton Occupational Health Center
410 Seventh St., SE
Washington, DC 20003
202-543-0005

Public Citizen's Health Research Group
2000 P Street, NW
Washington, DC 20036
202-833-3000

## Wisconsin

WisCOSH
734 W. 26th St.
Milwaukee, WI 53233
414-933-2338

## Canada

WOSH (Windsor, Canada)
1731 Wyandotte St. E.
Windsor, Ontario, N8Y 1C9
519-254-5157

## ■ Industry Organizations

Industry and trade associations allow their members to share information and work to promote the interests of the industry as a whole. Many of these associations actively participate in the development of regulations, and some sponsor research into the health and safety issues in their industry. These associations can also be a source of information regarding current practices in the industry. A few of the organizations that are active in health and safety issues are listed below. Most public libraries carry directories of these organizations, such as the annual *National Trade and Professional Associations of the U.S.,* which cross-indexes organizations by subject.

American Hospital Association
840 N. Lake Shore Dr.
Chicago, IL 60611
312-280-6000

American Iron and Steel Institute
1101 7th St. NW
Washington, DC 20036-4700
202-452-7100

American Petroleum Institute
1220 L St. NW
Washington, DC 20005-8029
202-682-8000

Chemical Manufacturers Association
2501 M St. NW
Washington, DC 20037-1303
202-887-1100

Compressed Gas Association
1725 Jefferson Davis Hwy., Ste 1004
Arlington, VA 22202-4100
703-412-0900

Electric Power Research Institute
Box 10412
Palo Alto, CA 94303-0813
415-855-2000

Industrial Safety Equipment Association
1901 N. Moore St., Ste 808
Arlington, VA 22209
703-525-1695

Semiconductor Industry Association
4300 Stevens Creek Blvd., Ste 271
San Jose, CA 95129-1249
408-246-2711

Society of Plastics Engineers
14 Fairfield Dr.
Brookfield, CT 06804-0403
203-775-0471

## Labor Unions

Many labor unions maintain health and safety departments and are active in promoting and developing health and safety regulations. They can often provide information about the hazards in specific industries. Some organizations are listed below. Others can be found in the *Directory of U.S. Labor Organizations* or other directories of associations.

American Federation of Labor–Congress of Industrial Organizations (AFL–CIO)
815 16th St., NW
Washington, DC 20006
202-637-5000

American Federation of State, County and Municipal Employees (AFSCME)
1625 L Street, NW
Washington, DC 20036-5687
202-429-1000

Chemical Workers Union
1655 W. Market St.
Akron, OH 44313
216-867-2444

International Association of Machinists and Aerospace Workers (IAM)
9000 Machinists Place
Upper Marlboro, MD 20722
301-967-4500

International Brotherhood of Teamsters
25 Louisiana Ave., NW
Washington, DC 20001
202-624-6800

International Ladies' Garment Workers' Union (ULGWU)
1710 Broadway
New York, NY 10019
212-265-7000

International Longshoremen's and Warehousemen's Union (ILWU)
1188 Franklin St.
San Francisco, CA 94109
415-775-0533

Laborers' International Union of North America
905 16th St. NW
Washington, DC 20006
202-737-8320

Oil, Chemical and Atomic Workers International Union (OCAW)
255 Union Blvd.
Lakewood, CO 80228
303-987-2229

Service Employees' International Union (SEIU)
1313 L St. NW
Washington, DC 20005
202-898-3200

United Auto Workers (UAW)
8000 E. Jefferson Ave.
Detroit, MI 48214
313-926-5000

United Farmworkers of America (UFW)
P.O. Box 62-LaPaz
Keene, CA 93570
805-822-5771

United Food and Commercial Workers International Union (UFCW)
1775 K Street NW
Washington, DC 20006
202-223-3111

United Mine Workers of America (UMW)
900 15th St. NW
Washington, DC 20005
202-842-7200

## University-Based Research, Training, and Continuing Professional Education Programs

**NIOSH educational resource centers.** NIOSH has developed a program to establish centers of learning for occupational safety and health throughout the United States. Educational resource centers (ERCs) are located in 27 universities, serving all 10 Department of Health and Human Services (DHHS) regions.

Deep South Center for Occupational Health and Safety
Cherie Hunt, University of Alabama at Birmingham
Phone: 205-934-7178
Fax: 205-975-6341

Harvard ERC
Daryl Bichel, Boston
Phone: 617-432-3314
Fax: 617-432-0219

Illinois ERC
Dick Lyons, School of Public Health, Chicago
Phone: 312-996-7473
Fax: 312-413-7369

Johns Hopkins ERC
Linda A. Lamb, Baltimore
Phone: 410-955-2609
Fax: 410-955-9334

Michigan ERC
Randy Rabourne, University of Michigan
Phone: 313-936-0148
Fax: 313-764-3451

Minnesota ERC
Jeanne F. Ayers, Midwest Center for Occupational Health and Safety
Phone: 612-221-3992
Fax: 612-292-4773

New York/New Jersey ERC
Barbara Young, EOHSI Centers for Education
and Training
Phone: 908-235-5062
Fax: 908-235-5133

NIOSH
Marsha Striley, Cincinnati
Phone: 513-533-8225
Fax: 513-533-8560

North Carolina ERC
Larry D. Hyde, Chapel Hill
Phone: 919-962-2101
Fax: 919-966-7579

Northern California ERC
Linda Elwood, University of California at Berkeley
Phone: 510-231-5645
Fax: 510-231-5648

Northwest Center for Occupational Health and Safety
Sharon Morris, University of Washington
Phone: 206-543-1069
Fax: 206-685-3872

Southern California ERC
Ramona Cayuela, University of Southern California
Phone: 213-740-3995
Fax: 213-740-8789

Southwest Center for Occupational and Environmental
Health
Pam Parker, Houston
Phone: 713-792-4648
Fax: 713-792-4407

University of Cincinnati ERC
Judy L. Jarrell
Phone: 513-558-1730
Fax: 513-558-1756

Utah ERC
Kim Gianelo, University of Utah
Phone: 801-581-5710
Fax: 801-585-5275

## Computer-Based Sources of Information

A wide range of health and safety information is now available on computer. Many of these sources can be accessed through university or medical school libraries at no charge, or you can contact the publisher for subscription information.

### CD-ROM

*CCInfo Disc:* NIOSHTIC, MSDS/FTSS, and more, in French and English.
Canadian Centre for Occupational Health and Safety (CCOHS)

250 Main Street East
Hamilton, Ontario, Canada L8N 1H6
Phone: 800-668-4284 or 416-570-8094
Fax: 416-572-2206
Occupational Safety and Health Exchange:
416-572-2307

*Chem Sources:* Includes U.S. and international data.
Information: 803-646-7840
Orders: 800-222-4531

*CiDATA:* Lists chemical substances cross-referenced by regulating body.
Phone: Susan Sapia at 310-429-9055
Fax: 310-429-4233

OSHA regulations, documents, and technical information, published by the U.S. Department of Labor Occupational Safety and Health Administration.
Superintendent of Documents
U.S. Government Printing Office
Washington, DC 20402-9325
Phone: 202-783-3238
Fax: 202-275-7810

## Environmental and Safety Regulations

*Computer-Aided Management of Emergency Operations (CAMEO):* Computer software for emergency responders and emergency planners.
National Safety Council
1019 19th St. NW, Suite 401
Washington, DC 20036-5105
202-293-2270

*Environmental Software Report:* Lists several companies that offer environmental regulations on diskettes and compact disc.
Donley Technology
Box 335
Garrisonville, VA 22463
703-659-1954

*Information Handling Services* (CD-ROM)
800-241-7824 or 303-790-0600 x59

Cross-references of RCRA, SARA, OSHA, etc.
Technical Services Associates, Inc.
800-388-1415

## MSDS

*MSDS Collection* and *MSDS Engine:* Enables user to store, retrieve, and track an MSDS from the supplier.
Genium Publishing Corporation
Phone: 518-377-8854
Fax: 518-377-1891

*MSDS Reference for Crop Protection Chemicals* and *MSDSs for Pesticides*
John Wiley & Sons

Occupational Health Services, Inc. (OHS): MSDS, CFRs, Summary sheets, Labeling information on CD-ROM
Phone: 212-789-3655
Fax: 212-789-3646

## Online Resources

CCINFOline: Online information service in French and English.
Canadian Centre for Occupational Health and Safety
250 Main Street East
Hamilton, Ontario, Canada, L8N 1H6
Phone: 416-572-4400 or 800-263-8466
Fax: 416-572-4500

Lexis/Nexis
Bureau of National Affairs, Inc. (BNA)
Phone: 800-372-1033
Fax: 800-253-0332

MEDLARS and TOXNET
National Library of Medicine
Bldg. 38, Rm. 4N421 8600 Rockville Pike
Bethesda, MD 20894
301-496-6193 or 800-638-8480

### Medlars Databases

AIDSLINE: Acquired immunodeficiency syndrome and related topics

AVLINE: Audiovisuals

BIOETHICSLINE: Ethics and related public policy issues in health care and biomedical research

CANCERLIT: Major cancer topics

CATLINE: Biomedical sciences

CHEMLINE: Dictionary of chemicals

CLINPROT: Summaries of clinical investigations of new anticancer agents and treatment modalities

DIRLINE: Directory of information resources

DOCUSER: Document delivery

HEALTH: Health planning and administration

MEDLINE: Biomedicine

MeSH VOCABULARY FILE: Thesaurus of biomedical terms

TOXLINE: Toxicology information

TOXLIT: Toxicology literature from special sources

### TOXNET Data Network

CCRIS: Chemical Carcinogenesis Research Information System

CDF: Chemical Dictionary File

DBIR: Directory of Biotechnology Information Resources

EMICBACK: Environmental Mutagen Information Center Backfile

ETICBACK: Environmental Teratology Information Center Backfile

GENE-TOX: Short-term mutagenicity test activity profiles

HSDB: Hazardous Substances Data Bank

IRIS: Integrated Risk Information System

RTECS: Registry of Toxic Effects of Chemical Substances

TRI: Toxic Chemical Release Inventory

U.S. Department of Labor *Labor News* (202-219-4784). An electronic information bulletin board offering text of daily news releases and other downloadable text files; free except for toll call.

## Newsletters and Reports

*ACOEM Report.* American College of Occupational and Environmental Medicine, 55 W Seegers Rd., Arlington Heights, IL 60005.

*Environmental Health Letter.* Business Publishers, 951 Pershing Dr., Silver Springs, MD 20910-4464.

*Environmental Health and Safety News.* Department of Environmental Health, School Public Health and Community Medicine, University of Washington, F-461 Health Sciences Bldg., Seattle, WA 98195.

*Industrial Hygiene News Bimonthly.* Rimbach Publishing, 8650 Babcock Blvd., Pittsburgh, PA 15237.

*Industrial Section Newsletter* (bimonthly). National Safety Council, 1121 Spring Lake Dr., Itasca, IL 60143-3201.

*Occupational Health and Safety Letter* (semimonthly). Business Publishers, 951 Pershing Dr., Silver Springs, MD 20910-4464.

*Occupational Safety and Health Reporter.* Bureau of National Affairs, 1231 25th St. NW, Washington, DC 20037.

*OSHA Up-to-Date* (monthly). National Safety Council, 1121 Spring Lake Dr., Itasca, IL 60143-3201.

*World Health Organization Technical Report Series.* WHO, Distribution and Sales, CH-1211 Geneva 27, Switzerland.

## Journals and Magazines

Many articles on industrial hygiene can be found in the following journals and magazines.

*American Industrial Hygiene Association Journal.* American Industrial Hygiene Association. 2700 Prosperity Dr., Fairfax, VA 22031.

*American Journal of Epidemiology.* Johns Hopkins University School of Hygiene and Public Health, 2007 E. Monument St., Baltimore, MD 21205.

*American Journal of Industrial Medicine.* John Wiley & Sons, Inc., Journals, 60 Third Ave., New York, NY 10158.

*American Journal of Public Health* (monthly). American Public Health Association, 1015 Fifteenth St., Washington, DC 20005.

*Annals of Occupational Hygiene.* Pergamon Press Journals Division, 660 White Plains Rd., Tarrytown, NY 10591-5153.

*Applied Occupational and Environmental Hygiene* (monthly; formerly *Applied Industrial Hygiene*), 6500 Glenway Ave., D-7, Cincinnati, OH 45211.

*Archives of Environmental Health* (bimonthly). Heldref Publications, 1319 18th St. NW, Washington, DC 20036-1802.

*ASHRAE Journal.* American Society of Heating, Refrigerating and Air Conditioning Engineers, 1791 Tullie Circle NE, Atlanta, GA 30329.

*Cal-OSHA Reporter.* Sten-O-Press, P.O. Box 36, San Pablo, CA 94806.

*Chemical and Engineering News.* American Chemical Society, 1155 16th St. NW, Washington, DC 20036.

*Environmental Health Perspectives.* National Institute of Environmental Sciences, P.O. Box 12233, Research Triangle Park, NC 27709.

*Environmental Health and Safety News.* University of Washington, Department of Environmental Health, School of Public Health and Community Medicine, F-461 Health Sciences Bldg., Seattle, WA 98195.

*Environmental Research.* Academic Press, Journal Division, 1250 Sixth Ave., San Diego, CA 92101.

*Environmental Science and Technology* (monthly). American Chemical Society, 1155 16th St. NW, Washington, DC 20036.

*Ergonomics.* Taylor & Francis, 242 Cherry St., Philadelphia, PA 19106.

*Excerpta Medica* (monthly). Section 35: Occupational Health and Industrial Medicine, Elsevier, P.O. Box 882, Madison Sq. Sta., New York, NY 10159.

*Health Physics* (monthly). Williams & Wilkins, 428 E. Preston St., Baltimore, MD 21202. Sponsored by the Health Physics Society.

*Industrial Hygiene Digest* (monthly). Industrial Health Foundation, 34 Penn Circle W, Pittsburgh, PA 15206.

*International Archives of Occupational and Environmental Health* (quarterly). Springer-Verlag, 175 Fifth Ave., New York, NY 10010.

*Job Safety and Health* (biweekly). The Bureau of National Affairs, 1231 25th Street NW, Washington, DC 20037.

*Job Safety & Health Quarterly.* Occupational Safety and Health Administration, U.S. Department of Labor. Available from the Superintendent of Documents, U.S. Government Printing Office, Washington, DC 20402.

*Journal of the American Medical Association.* American Medical Association, 515 N. State St., Chicago, IL 60610.

*Journal of Aviation, Space and Environmental Medicine* (monthly; formerly *Aerospace Medicine*). Aerospace Medical Association, 320 S. Henry St., Alexandria, VA 22314-3579.

*Journal of Occupational Medicine* (monthly). American Occupational Medicine Association, Williams & Wilkins, 428 E. Preston St., Baltimore, MD 21202.

*Journal of Toxicology: Clinical Toxicology* (bimonthly). Marcel Dekker, 270 Madison Ave., New York, NY 10016.

*Journal of Toxicology & Environmental Health.* Hemisphere, 1900 Frost Rd., Ste. 101, Bristol, PA 19007-1598.

*Lancet.* Williams & Wilkins, 428 E. Preston St., Baltimore, MD 21202.

*Monthly Labor Review.* Bureau of Labor Statistics, U.S. Department of Labor. Available from the Superintendent of Documents, U.S. Government Printing Office, Washington, DC 20402.

*New England Journal of Medicine.* New England Journal of Medicine, 1440 Main St., Waltham, MA 02254.

*New Solutions.* P.O. Box 2812, Denver, CO 80201.

*Noise Control Engineering Journal.* Institute of Noise Control Engineering, Department of Mechanical Engineering, Auburn University, Auburn, AL 36849.

*Noise Regulation Report* (formerly *Noise Control Report*). Business Publishers, 951 Pershing Dr., Silver Springs, MD 20910-4464.

*Occupational and Environmental Medicine* (monthly; formerly *British Journal of Industrial Medicine*). British Medical Association, Tavistock Sq., London WC1 H9JP, UK.

*Occupational Hazards.* Penton, 1100 Superior Ave., Cleveland, OH 44114.

*Occupational Health.* Available from U.S. National Technical Information Service, 5825 Port Royal Rd., Springfield, VA 22161.

*Occupational Health Nursing Newsletter* (bimonthly). National Safety Council, 1121 Spring Lake Dr., Itasca, IL 60143-3201.

*Occupational Health and Safety.* P.O. Box 2573, Waco, TX 76702-2573.

*Occupational Health and Safety Letter.* Business Publishers, 951 Pershing Dr., Silver Springs, MD 20910-4464.

*Occupational Medicine* (formerly *Journal of the Society of Occupational Medicine*). Butterworth-Heinemann, Linacre House, Jordan Hill, Oxford OX2 8DP, UK.

*Occupational Medicine: State of the Art Reviews.* Hanley & Belfus, Inc., 210 S. 13th St., Philadelphia, PA 19107.

*Occupational Safety & Health Reporter* (weekly). Bureau of National Affairs, 1231 25th St. NW, Washington, DC 20037.

*OSHA Up-to-Date* (monthly). National Safety Council, 1121 Spring Lake Dr., Itasca, IL 60143-3201.

*Professional Safety* (formerly *American Society of Safety Engineers Journal*). ASSE, 1800 E. Oakton St., Des Plaines, IL 60018.

*Public Health Reports.* U.S. Public Health Services, Department of Health and Human Services, Parklawn Bldg., Rm. 13C-26, 5600 Fishers Ln., Rockville, MD 20857.

*Safety & Health.* National Safety Council, 1121 Spring Lake Dr., Itasca, IL 60143-3201.

*Scandinavian Journal of Work, Environment & Health.* Finnish Institute of Occupational Health, Topeliuksenkatu 41aA, SF-00250 Helsinki, Finland.

*Sound and Vibration.* Acoustical Publications, P.O. Box 40416, Bay Village, OH 44140.

*VDT News.* P.O. Box 1799, Grand Central Station, New York, NY 10163.

# ■ Bibliography

There is a wealth of printed material available on various issues in industrial hygiene. The following bibliography is not meant to be exhaustive, but to point readers in the direction of a few interesting sources. Other references can be found in the individual chapters in this book.

## AIDS

*AIDS in the Workplace: Resource Material,* 3rd ed. Washington, DC: Bureau of National Affairs, 1989.

Banta WF. *AIDS in the Workplace: Legal Questions and Practical Answers.* New York: Lexington Books, 1993.

Becher CE, ed. *Occupational HIV Infection: Risks and Risk Reduction*. Philadelphia: Hanley & Belfus, 1989. (Published as *Occup Med 4*, special issue.)

Brown KC, Turner JG. *AIDS: Policies and Programs for the Workplace*. New York: Van Nostrand Reinhold, 1989.

Miike LH, Ostrowsky J. *HIV in the Health Care Workplace*. Washington, DC: Office of Technology Assessment, 1991.

Puckett SB, Emery AR. *Managing AIDS in the Workplace*. Reading: Addison-Wesley, 1988.

## Biological Hazards

Benenson AS, ed. *Control of Communicable Diseases in Man,* 15th ed. Washington DC: American Public Health Association, 1990.

Collins CH, Grange JM. *The Microbiological Hazards of Occupations*. Leeds, UK: Science Reviews in association with H & H Scientific Consultants, 1990.

Moran RO, Katzeff K. *The Healthcare Industry Guide to the OSHA Bloodborne Pathogens Standard*. Washington, DC: Moran Associates, 1992.

Putnam LD, Langerman N. *OSHA Bloodborne Pathogens Exposure Control Plan*. Chelsea, MI: CRC/Lewis, 1992.

U.S. Dept. of Health and Human Services. *Biosafety in Microbiological and Biomedical Laboratories,* 2nd ed. Washington, DC: U.S. Government Printing Office, 1988.

## Biological Monitoring and Medical Surveillance

American Conference of Governmental Industrial Hygienists. *Documentations for the Biological Exposure Indices (BEIs)*. Cincinnati, OH: American Conference of Governmental Industrial Hygienists.

Ashford NA, et al. *Monitoring the Worker for Exposure and Disease: Scientific, Legal, and Ethical Considerations in the Use of Biomarkers*. Baltimore: Johns Hopkins University Press, 1990.

Rempel D, ed. *Medical Surveillance in the Workplace*. Philadelphia: Hanley & Belfus, 1990. (Published as *Occup Med 5[3]*.)

## Engineering Controls

BOHS Technology Committee, Working Group on Ventilation Design. *Controlling Airborne Contaminants in the Workplace*. Northwood, Middx., UK: Science Reviews; Leeds, UK: H & H Scientific Consultants, 1987.

Cralley LV, Cralley LJ, eds. *In-Plant Practices for Job Related Health Hazards*. New York: Wiley, 1989.

Cralley LV, Cralley LJ, eds. *Industrial Hygiene Aspects of Plant Operations*. New York: Macmillan, 1985.

National Institute for Occupational Safety and Health. *Advanced Industrial Hygiene Engineering* (552). Cincinnati, OH: National Institute for Occupational Safety and Health, 1986.

National Institute for Occupational Safety and Health. *Applied Industrial Hygiene* (549). Cincinnati, OH: National Institute for Occupational Safety and Health, 1986.

Perkins JL, Rose VE. *Case Studies in Industrial Hygiene*. New York: Wiley, 1988.

Talty JT, ed. *Industrial Hygiene Engineering: Recognition, Measurement, Evaluation, and Control,* 2nd ed. Park Ridge, IL: Noyes Data Corp., 1988.

Wadden RA, Scheff PA. *Engineering Design for the Control of Workplace Hazards*. New York: McGraw Hill, 1987.

## Ergonomics

American Conference of Governmental Industrial Hygienists. *Ergonomic Interventions to Prevent Musculoskeletal Injuries in Industry*. Cincinnati, OH: Lewis Publishers, 1987.

Bullock MI, ed. *Ergonomics: The Physiotherapist in the Workplace*. New York: Churchill Livingstone, 1990.

Bureau of National Affairs. *Cumulative Trauma Disorders in the Workplace: Costs, Prevention, and Progress*. Washington, DC: Bureau of National Affairs, 1991.

Chaffin DB, Andersson BJ. *Occupational Biomechanics*. New York: Wiley, 1984.

Deyo RA, ed. *Back Pain in Workers*. Philadelphia: Hanley & Belfus, 1988. (Published as *Occup Med 3[1]*.)

Grandjean E. *Fitting the Task to the Man: A Textbook of Occupational Ergonomics,* 4th ed. New York: Taylor & Francis, 1988.

Hadler NM. *Occupational Musculoskeletal Disorders*. New York: Raven Press, 1993.

Millender LH, Louis DS, Simmons BP. *Occupational Disorders of the Upper Extremity*. New York: Churchill Livingstone, 1992.

Moore JS, Garg A, eds. *Ergonomics: Low-Back Pain, Carpal Tunnel Syndrome, and Upper Extremity Disorders in the Workplace*. Philadelphia: Hanley & Belfus, 1992.

National Institute for Occupational Safety and Health. *Criteria for a Recommended Standard: Occupational Exposure to Hand–Arm Vibration*. Cincinnati, OH: National Institute for Occupational Safety and Health, 1989.

Sanders MS, McCormick EJ. *Human Factors in Engineering and Design,* 7th ed. New York: McGraw-Hill, 1993.

Scott AJ, ed. *Shiftwork*. Philadelphia: Hanley & Belfus, 1990. (Published as *Occup Med 5[2]*.)

Stevenson M, ed. *Readings in RSI: The Ergonomics Approach to Repetition Strain Injuries*. Kensington: New South Wales University Press, 1987.

## Exposure and Risk Assessment

Alavanja MCR, Brown C, Spirtas R, et al. Risk assessment for carcinogens: A comparison of approaches of the ACGIH and the EPA. *Appl Occup Environ Hygiene* 5(8):510–517, 1990.

Hallenbeck WH. *Quantitative Risk Assessment for Environmental and Occupational Health,* 2nd ed. Boca Raton: Lewis Publishers, 1993.

Nicas M, Spear RC. A task-based statistical model of a worker's exposure distribution. Part I—Description of the model. *AIHAJ* 53:411–418, 1992.

Nicas M, Spear RC. A task-based statistical model of a worker's exposure distribution. Part II—Application to sampling strategy. *AIHAJ* 53:419–426, 1992.

Paustenbach DJ. Health risk assessment and the practice of industrial hygiene. *AIHAJ* 51(7):339–351, 1990.

## Practice

Bayer R, ed. *The Health and Safety of Workers: Case Studies in the Politics of Professional Responsibility.* New York: Oxford University Press, 1988.

Board of Directors, American Conference of Governmental Industrial Hygienists. Threshold Limit Values: A more balanced appraisal. *Appl Occup Environ Hygiene* 5(6):340–344, 1990.

Castleman BI, Ziem GE. Corporate influence on Threshold Limit Values. *Am J Ind Med* 13:531–559, 1988.

Corn JK. *Response to Occupational Health Hazards: A Historical Perspective.* New York: Van Nostrand Reinhold, 1992.

Corn JK. *Protecting the Health of Workers: The American Conference of Governmental Industrial Hygienists, 1938–1988.* Cincinnati, OH: American Conference of Governmental Industrial Hygienists, 1989.

Cralley LJ, Cralley LV, eds. *Theory and Rationale of Industrial Hygiene Practice,* 2nd ed. New York: Wiley, 1985.

Kumashiro M, Megaw ED, eds. *Towards Human Work: Solutions to Problems in Occupational Health and Safety.* Bristol, UK: Taylor & Francis, 1991.

Rappaport SM. Threshold Limit Values, permissible exposure limits, and feasibility: the bases for exposure limits in the United States. *Am J Ind Med* 23(5):683–694, 1993.

Roach SA, Rappaport SM. But they are not thresholds: A critical analysis of the documentation of Threshold Limit Values. *Am J Ind Med* 17(6):727–753, 1990.

Robinson JC, Rappaport SM. Implications of OSHA's reliance on TLVs in developing the Air Contaminants Standard. *Am J Ind Med* 19:3–13, 1991.

Tarlau ES. Industrial hygiene with no limits. *AIHAJ* 51(11):A9–A10, 1990.

Yaffe CD, ed. *Some Pioneers of Industrial Hygiene.* Cincinnati, OH: American Conference of Governmental Industrial Hygienists, 1984.

## History of Worker Health and Safety

Atack J, Bateman F. *Louis Brandeis, Work and Fatigue at the Start of the Twentieth Century: Prelude to Oregon's Hours Limitation Law.* Cambridge, MA: National Bureau of Economic Research, 1991.

Brodeur P. *Outrageous Misconduct: The Asbestos Industry on Trial.* New York: Pantheon, 1985.

Cherniak M. *The Hawk's Nest Incident: America's Worst Industrial Disaster.* New Haven, CT: Yale University Press, 1986.

Hamilton A. *Exploring the Dangerous Trades: The Autobiography of Alice Hamilton, M.D.* Boston: Northeastern University Press, 1985.

Light K, DiPerna P. *With These Hands.* New York: Pilgrim, 1986.

Markowitz G, Rosner D, eds. *Slaves of the Depression: Workers' Letters About Life on the Job.* Ithaca, NY: Cornell University Press, 1987.

McFeely MD. *Lady Inspectors: The Campaign for a Better Workplace, 1893–1921.* Athens: University of Georgia Press, 1991.

Page JA, O'Brien M. *Bitter Wages: Ralph Nader's Study Group Report on Disease and Injury on the Job.* New York: Grossman, 1973.

Palladino G. *Dreams of Dignity, Workers of Vision: A History of the International Brotherhood of Electrical Workers.* Washington, DC: International Brotherhood of Electrical Workers, 1991.

Rosner D, Markowitz G, eds. *Dying for Work: Workers' Safety and Health in Twentieth-Century America.* Bloomington: Indiana University Press, 1987.

Scott R. *Muscle and Blood.* New York: E.P. Dutton, 1974.

Weindling P, ed. *The Social History of Occupational Health.* London: Croom Helm, 1985.

## Indoor Air Quality

*Building Air Quality: A Guide for Building Owners and Facility Managers.* Washington, DC: Environmental Protection Agency/National Institute for Occupational Safety and Health, 1991.

Burton, DJ. *IAQ and HVAC Workbook.* Bountiful, VT: IAQ, 1995.

Cone JE, Hodgson MJ, eds. *Problem Buildings: Building-Associated Illness and the Sick Building Syndrome.* Philadelphia: Hanley & Belfus, 1989. (Published as *Occup Med: State of the Art Reviews* 4[4].)

Fisk WJ, Spencer RK, Grimsud DT, et al. *Indoor Air Quality Control Techniques.* Park Ridge, IL: Noyes Data Corporation, 1987.

*Introduction to Indoor Air Quality: A Reference Manual.* Washington, DC: Environmental Protection Agency, Available from the National Environmental Health Association, 303-756-9090.

*Introduction to Indoor Air Quality: A Self-Paced Learning Module.* Washington, DC: Environmental Protection Agency, Available from the National Environmental Health Association, 303-756-9090.

*National Institute for Occupational Safety and Health: Guidance for Indoor Air Quality Investigations.* Cincinnati, OH, Hazard Evaluation and Technical Assistance Branch, Division of Surveillance, Hazard Evaluations and Field Studies, NIOSH, 1987.

## International Aspects

*Crisis at Our Doorstep: Occupational and Environmental Health Implications for Mexico–U.S.–Canada Trade Negotiations.* Chicago: National Safe Workplace Institute, 1991.

Elling RH. *The Struggle for Workers' Health: A Study of Six Industrialized Countries.* Amityville, NY: Baywood Publishing, 1986.

International Labour Office. *Safety, Health, and Working Conditions in the Transfer of Technology to Developing Countries*. Geneva, Switzerland: International Labour Office, 1988.

International Labour Office. *Safety and Health Practices of Multinational Enterprises*. Geneva, Switzerland: International Labour Office, 1984.

Jeyaratnam J. *Occupational Health in Developing Countries*. New York: Oxford University Press, 1992.

Karpilow C. *Occupational Medicine in the International Workplace*. New York: Van Nostrand Reinhold, 1991.

Kogi K. *Improving Working Conditions in Small Enterprises in Developing Asia*. Geneva, Switzerland: International Labour Office, 1985.

Neal AC, Wright FB. *The European Communities' Health and Safety Legislation*. New York: Chapman & Hall, 1992.

Occupational Safety and Health Administration. *A Comparison of Occupational Safety and Health Programs in the United States and Mexico: An Overview*. Washington, DC: Occupational Safety and Health Administration, 1992.

Phoon WO, Ong CN, eds. *Occupational Health in Developing Countries in Asia: Service, Training, and Research in Eight Countries*. Tokyo: Southeast Asian Medical Information Center, 1985.

Reich MR, Okubo T, eds. *Protecting Workers' Health in the Third World: National and International Strategies*. New York: Auburn House, 1992.

## Noise and Hearing Conservation

American Industrial Hygiene Association. *Noise and Hearing Conservation Manual*, 4th ed. Fairfax, VA: American Industrial Hygiene Association, 1986.

Burns W. *Noise and Man*, 2nd ed. London: John Murray, 1973.

Occupational Safety and Health Administration. *Hearing Conservation*. Washington, DC: Occupational Safety and Health Administration, 1992.

Sataloff RT, Sataloff J. *Hearing Loss*, 3rd ed. New York: Dekker, 1993.

Suter AH, Franks JR. *A Practical Guide to Effective Hearing Conservation Programs in the Workplace*. Washington, DC: National Institute for Occupational Safety and Health, 1990.

## Occupational Epidemiology

Band P, ed. *Occupational Cancer Epidemiology*. New York: Springer-Verlag, 1990.

Hernberg S. *Introduction to Occupational Epidemiology*. Chelsea, MI: Lewis Publishers, 1992.

Karvonen M, Mikheev MI, eds. *Epidemiology of Occupational Health*. Copenhagen, Denmark: World Health Organization, 1986.

Lalich N. *A Guide for the Management, Analysis, and Interpretation of Occupational Mortality Data*. Cincinnati, OH: National Institute for Occupational Safety and Health, 1990.

Monson RR. *Occupational Epidemiology*, 2nd ed. Boca Raton, FL: CRC Press, 1990.

Olsen J. *Searching for Causes of Work-Related Diseases: An Introduction to Epidemiology at the Work Site*. New York: Oxford University Press, 1991.

Pollack ES, Keimig DG, eds. *Counting Injuries and Illnesses in the Workplace: Proposals for a Better System*. Washington, DC: National Academy Press, 1987.

Steenland K. *Case Studies in Occupational Epidemiology*. New York: Oxford University Press, 1993.

## Occupational Health Policy

Bezold C, Carlson RJ, Peck JC. *The Future of Work and Health*. Westport: Auburn House, 1986.

General Accounting Office. *Occupational Safety and Health: OSHA Policy Changes Needed to Confirm that Employers Abate Serious Hazards*. Washington, DC: General Accounting Office, 1991.

Hemenway D. *Monitoring and Compliance: The Political Economy of Inspection*. Greenwich: JAI Press, 1985.

Judkins BM. *We Offer Ourselves as Evidence: Toward Workers' Control of Occupational Health*. New York: Greenwood Press, 1986.

Lofgren DJ. *Dangerous Premises: An Insider's View of OSHA Enforcement*. Ithaca: ILR Press, 1989.

McGarity TO, Shapiro SA. *Workers at Risk: The Failed Promise of the Occupational Safety and Health Administration*. Westport, CT: Praeger, 1993.

Mintz BW. *OSHA: History, Law, and Policy*. Washington, DC: BNA Books, 1984.

Nelkin D, Brown MS. *Workers at Risk: Voices from the Workplace*. Chicago: University of Chicago Press, 1984.

Noble C. *Liberalism at Work: The Rise And Fall of OSHA*. Philadelphia: Temple University Press, 1986.

Polakoff PL. *Work and Health: It's Your Life: An Action Guide to Job Hazards*. Washington, DC: Press Associates, 1984.

Rosner D, Markowitz G. *Deadly Dust: Silicosis and the Politics of Occupational Disease in Twentieth-Century America*. Princeton, NJ: Princeton University Press, 1991.

Ruttenberg, R. *The Role of Labor–Management Committees in Safeguarding Worker Safety and Health*. Washington, DC: U.S. Department of Labor, Bureau of Labor–Management Relations and Cooperative Programs, 1989.

Siskind FB. *Twenty Years of OSHA Federal Enforcement Data: A Review and Explanation of the Major Trends*. Washington, DC: U.S. Department of Labor, Office of the Assistant Secretary for Policy, 1993.

Wallick F. *Don't Let Your Job Kill You*. Washington, DC: Progressive Press, 1984.

Watts TJ. *OSHA Enforcement of Workplace Safety in the 1980s: A Bibliography*. Monticello, IL: Vance Bibliographies, 1990.

Wilson GK. *The Politics of Safety and Health: Occupational Safety and Health in the United States and Britain*. New York: Oxford University Press, 1985.

## Occupational Medicine

Becker CE, Coye MJ. *Cancer Prevention: Strategies in the Workplace.* Washington, DC: Hemisphere Publishing, 1986.

Brandt-Rauf PW, ed. *Occupational Cancer and Carcinogenesis.* Philadelphia: Hanley & Belfus, 1987. (Published as *Occup Med* 2[1].)

Harber P, Balmes JR, eds. *Prevention of Pulmonary Disease in the Workplace.* Philadelphia: Hanley & Belfus, 1991. (Published as *Occup Med* 6[1].)

Harber P, Schenker M, Balmes JR. *Occupational and Environmental Respiratory Diseases.* St. Louis, MO: Mosby-Year Book, 1995.

Hunter D. *Hunter's Diseases of Occupations,* 7th ed. Boston: Little, Brown, 1987.

International Labour Office. *Occupational Cancer: Prevention and Control,* 2nd ed. Geneva, Switzerland: International Labour Office, 1988.

Johnson BL, ed. *Prevention of Neurotoxic Illness in Working Populations.* New York: Wiley, 1987.

LaDou J, ed. *Occupational Medicine.* Norwalk, CT: Appleton & Lange, 1990.

Levy BS, Wegman DH, eds. *Occupational Health: Recognizing and Preventing Work-Related Disease,* 2nd ed. Boston: Little, Brown, 1988.

McCunney RJ, ed. *Handbook of Occupational Medicine.* Boston: Little, Brown, 1988.

National Safe Workplace Institute. *Beyond Neglect: The Problem of Occupational Disease in the U.S.* Chicago: National Safe Workplace Institute, 1990.

Parkes WR. *Occupational Lung Disorders,* 2nd ed. Stoneham, MA: Butterworth, 1981.

Rom WN, ed. *Environmental and Occupational Medicine,* 2nd ed. Boston: Little, Brown, 1992.

Shusterman DJ, Blanc PD, eds. *Unusual Occupational Diseases.* Philadelphia: Hanley & Belfus, 1992. (Published as *Occup Med* 7[3].)

Tver DF, Anderson KA. *Industrial Medicine Desk Reference.* New York: Chapman and Hall, 1986.

Walsh DC. *Corporate Physicians: Between Medicine and Management.* New Haven: Yale University Press, 1987.

Zenz C, ed. *Occupational Medicine,* 3rd ed. St. Louis, MO: Mosby-Year Book, 1994.

## People of Color

Allen RL. *The Port Chicago Mutiny: The Story of the Largest Mass Mutiny Trial in U.S. Naval History.* New York: Amistad, 1993.

Davis M, Roland A. *Occupational Disease Among Black Workers: An Annotated Bibliography.* Berkeley, CA: Labor Occupational Health Program, 1980.

Robinson JC. Exposure to occupational hazards among hispanics, blacks, and non-hispanic whites in California. *Am J Public Health* 79:629–630, 1989.

*Work and Health in the Latino Community* (a slide/tape). Berkeley, CA: Labor Occupational Health Program, 1990.

*Workers Kaleidoscope: 2001.* Washington, DC: Food and Allied Service Trades Department, AFL-CIO, 1991.

## Radiation Hazards

Cember H. *Introduction to Health Physics,* 2nd ed. New York: Pergamon, 1983.

Charron D. *Radiofrequency Radiation in the Workplace.* Hamilton, Ontario: Canadian Centre for Occupational Health and Safety, 1989.

Daw HT. *Guidelines for the Radiation Protection of Workers in Industry (Ionising Radiations): Requirements for Control of Exposure to Radiation of Workers Engaged in Radiation Work in Specific Installations and Practices.* Geneva, Switzerland: International Labour Office, 1989.

Eichholz GG, Poston JW. *Principles of Nuclear Radiation Detection.* Chelsea, MI: Lewis Publishers, 1985.

International Atomic Energy Agency. *Radiation Protection in Occupational Health: Manual for Occupational Physicians.* Vienna: International Atomic Energy Agency, 1987.

International Commission on Radiological Protection. *Individual Monitoring for Intakes of Radionuclides by Workers: Design and Interpretation.* New York: Pergamon, 1988.

National Council on Radiation Protection and Measurements. *Implementation of the Principle of As Low As Reasonably Achievable (ALARA) for Medical and Dental Personnel.* Bethesda, MD: The Council, 1990.

Scherer E, Streffer C, Trott K, eds. *Radiation Exposure and Occupational Risks.* New York: Springer-Verlag, 1990.

Waxler M, Hitchens VM. *Optical Radiation and Visual Health.* Boca Raton, FL: CRC Press, 1986.

## Reproductive Hazards

Hazard Evaluation System and Information Service. *Workplace Chemical Hazards to Reproductive Health: A Resource for Worker Health and Safety Training and Patient Education.* Berkeley, CA: Hazard Evaluation System and Information Service, California Occupational Health Program, 1990.

Paul M. *Occupational and Environmental Reproductive Hazards: A Guide for Clinicians.* Baltimore: Williams & Wilkins, 1993.

U.S. Office of Technology Assessment. *Reproductive Health Hazards in the Workplace.* Washington, DC: U.S. Government Printing Office, 1985.

Steen ZA, Hatch MC, eds. *Reproductive Problems in the Workplace.* Philadelphia: Hanley & Belfus, 1986. (Published as *Occup Med* 1[3].)

Women's Bureau, Labour Canada. *Annotated Bibliography on Reproductive Health Hazards in the Workplace in Canada.* Ottawa: Women's Bureau, Labour Canada, 1988.

Workplace Reproductive Hazards Policy Group. *Reproductive Health Hazards in the Workplace: Policy Options for California.* Berkeley: California Policy Seminar, University of California, 1992.

World Health Organization. *Effects of Occupational Factors on Reproduction: Report on a WHO Meeting.* Copenhagen, Denmark: World Health Organization, Regional Office for Europe, 1985.

## Resource Materials

International Commission on Occupational Health. *International Directory of Research Institutions in Occupational Health.* Geneva, Switzerland: International Commission on Occupational Health, 1988.

International Labour Office. *International Directory of Occupational Safety and Health Institutions.* Geneva, Switzerland: International Labour Office, 1990.

National Institute for Occupational Safety and Health. *Basic NIOSH Bookshelf.* Washington, DC: National Institute for Occupational Safety and Health, 1984.

Rest KM, Manzone T, Rommel V. *Directory of Educational Resources and Training Opportunities in Occupational Health.* Washington, DC: Association of Teachers of Preventive Medicine, 1987.

Pease ES. *Occupational Safety and Health: A Sourcebook.* New York: Garland, 1985.

Sax NI, Lewis RJ. *Hawley's Condensed Chemical Dictionary,* 11th ed. New York: Van Nostrand Reinhold, 1987.

*Stedman's Medical Dictionary,* 25th ed. Baltimore: Williams & Wilkins, 1990.

Tucker ME. *Industrial Hygiene: A Guide to Technical Information Sources.* Fairfax, VA: American Industrial Hygiene Association, 1984.

## Right to Know and Hazard Communication

Brown MP. *A Worker's Guide to Right to Know About Hazards in the Workplace.* Los Angeles: Institute of Industrial Relations, University of California, 1987.

Chess C. *Winning the Right to Know: A Handbook for Toxics Activists.* Philadelphia: Delaware Valley Toxics Coalition, 1984.

Lowry GG, Lowry RC. *Lowrys' Handbook of Right-to-Know and Emergency Planning: Handbook of Compliance for Worker and Community, OSHA, EPA, and the States.* Boca Raton, FL: Lewis Publishers, 1988.

Robinson JC. *Toil and Toxics: Workplace Struggles and Political Strategies for Occupational Health.* Berkeley: University of California Press, 1991.

Shepard J. *Working in the Dark: Reagan and the "Right to Know" About Occupational Hazards.* Washington, DC: Public Citizen's Open Government Project, 1986.

Wallerstein N, Rubenstein HL. *Teaching About Job Hazards: A Guide for Workers and their Health Providers.* Washington, DC: American Public Health Association, 1993.

## Safety and Health Programs

Blosser F. *Primer on Occupational Safety and Health.* Washington, DC: Bureau of National Affairs, 1992.

Bryant M. *Success with Occupational Safety Programmes.* Geneva, Switzerland: International Labour Office, 1984.

Grund E. *Lockout/Tagout: The Process of Controlling Hazardous Energy.* Itasca, IL: National Safety Council, 1995.

Irwin MHK. *Risks to Health and Safety on the Job.* New York: Public Affairs Committee, 1986.

LaDou J, ed. *Occupational Health & Safety,* 2nd ed. Itasca, IL: National Safety Council, 1993.

Manual, F. *On the Practice of Safety.* New York: Van Nostrand Reinhold, 1993.

National Institute for Occupational Safety and Health. *Guidelines for Protecting the Safety and Health of Health Care Workers.* Washington, DC: National Institute for Occupational Safety and Health, 1988.

National Safety Council. *Accident Prevention Manual for Business & Industry,* Vol. 1: *Administration & Programs,* Vol. 2: *Engineering & Technology,* and Vol. 3: *Environmental Management (1995).* Chicago: National Safety Council, 1992.

Office of Technology Assessment. *Preventing Illness and Injury in the Workplace.* Washington, DC: U.S. Government Printing Office, 1985.

Rekus, J. *Complete Confined Spaces Handbook.* Boca Raton, FL: CRC/Lewis, 1994.

U.S. Department of Labor. *Evaluating your Firm's Injury and Illness Record: Construction Industries.* Washington, DC: U.S. Department of Labor, Bureau of Labor Statistics, 1992.

U.S. Department of Labor. *Evaluating your Firm's Injury and Illness Record: Manufacturing Industries.* Washington, DC: U.S. Department of Labor, Bureau of Labor Statistics, 1992.

U.S. Department of Labor. *Evaluating your Firm's Injury and Illness Record: Service Industries.* Washington, DC: U.S. Department of Labor, Bureau of Labor Statistics, 1992.

## Sampling and Laboratory Methods

American Conference of Governmental Industrial Hygienists. *Air Sampling Instruments,* 8th ed. Cincinnati, OH: American Conference of Governmental Industrial Hygienists, 1995.

Berlin A, Yodaiken RE, Henmman BA, eds. *Assessment of Toxic Agents at the Workplace: Roles of Ambient and Biological Monitoring.* Boston: M. Nijhoff, for the Commission of the European Communities, 1984.

Coffman MA, Singh J. *Sampling and Analysis of Gases and Vapors.* Washington, DC: National Institute for Occupational Safety and Health, 1991.

National Institute for Occupational Safety and Health. *Industrial Hygiene Laboratory Quality Control* (587). Cincinnati, OH: National Institute for Occupational Safety and Health, 1987.

National Institute for Occupational Safety and Health. *Manual of Analytical Methods.* Springfield, VA: National Technical Information Service, 1984.

National Institute for Occupational Safety and Health. *Industrial Hygiene Sampling Decision-Making, Monitoring and Recordkeeping: Sampling Strategies* (553). Cincinnati, OH: National Institute for Occupational Safety and Health, 1983.

Ness SA. *Air Monitoring for Toxic Exposures.* Cincinnati, OH: American Conference of Governmental Industrial Hygienists, 1991.

Occupational Safety and Health Administration. *OSHA Technical Manual.* Washington, DC: Occupational Safety and Health Administration, 1990.

Royal Society of Chemistry. *Measurement Techniques for Carcinogenic Agents in Workplace Air.* Letchworth, UK: Royal Society of Chemistry, 1989.

World Health Organization. *Evaluation of Exposure to Airborne Particles in the Work Environment.* Geneva, Switzerland: World Health Organization, 1984.

## Toxicology and Chemical Hazards

Amdur MO, Doull J, Klaassen CD, eds. *Casarett and Doull's Toxicology,* 4th ed. New York: Pergamon, 1991.

American Conference of Governmental Industrial Hygienists. *1993–1994 Threshold Limit Values for Chemical Substances and Physical Agents and Biological Exposure Indices.* Cincinnati, OH: American Conference of Governmental Industrial Hygienists, 1993.

American Conference of Governmental Industrial Hygienists. *Documentation of the Threshold Limit Values and Biological Exposure Indices,* 6th ed. Cincinnati, OH: American Conference of Governmental Industrial Hygienists, 1991.

Clayton GD, Clayton FE. *Patty's Industrial Hygiene and Toxicology,* 4th ed. New York: Wiley, 1991. (Volume 1A, 1B: General Principles, 4th ed., 1991; Volume 2A, 2B, 2C, 2D, 2E, 2F: Toxicology, 4th rev. ed., 1993, 1994; Volume 3A: The Work Environment, 2nd ed. 1985; Volume 3B: Biological Responses, 2nd ed. 1985).

Danse IR. *Common Sense Toxics in the Workplace: A Manual for Doctors, Nurses, Emergency Responders, Employers, Industrial Hygienists, Risk Managers, Claims Adjusters, and Lawyers.* New York: Van Nostrand Reinhold, 1991.

Finkel AJ. *Hamilton and Hardy's Industrial Toxicology,* 4th ed. Boston: J. Wright PSG, 1983.

Grandjean P. *Skin Penetration: Hazardous Chemicals at Work.* New York: Taylor & Francis, 1990.

Lewis P, ed. *Health Protection from Chemicals in the Workplace.* New York: E. Horwood, 1993.

Lu FC. *Basic Toxicology: Fundamentals, Target Organs and Risk Assessment.* Bristol, PA: Hemisphere, 1991.

National Institute for Occupational Safety and Health. *NIOSH Pocket Guide to Chemical Hazards.* Cincinnati, OH: National Institute for Occupational Safety and Health, 1990.

National Institute for Occupational Safety and Health. *Occupational Safety and Health Guidelines for Chemical Hazards.* Cincinnati, OH: National Institute for Occupational Safety and Health, 1988.

O'Donoghue JL. *Neurotoxicity of Industrial and Commercial Chemicals.* Boca Raton, FL: CRC Press, 1985.

Ottoboni, A. *The Dose Makes the Poison: A Plain Language Guide to Toxicology,* 2nd ed. New York: Van Nostrand Reinhold, 1991.

Plunkett ER. *Handbook of Industrial Toxicology,* 3rd ed. New York: Chemical Publishing, 1987.

Proctor NH. *Proctor and Hughes' Chemical Hazards of the Workplace,* 3rd ed. New York: Van Nostrand Reinhold, 1991.

Sax NI, Lewis RJ. *Dangerous Properties of Industrial Materials,* 7th ed. New York: Van Nostrand Reinhold, 1989.

Scott RM. *Chemical Hazards in the Workplace.* Chelsea, MI: Lewis Publishers, 1989.

Stellman JM, Daum SM. *Work is Dangerous to Your Health.* New York: Vintage Books, 1974.

## Ventilation

Alden JL, Kane JM. *Design of Industrial Ventilation Systems,* 5th ed. New York: Industrial Press, 1982.

American Conference of Governmental Industrial Hygienists. *Industrial Ventilation,* 22nd ed. Cincinnati, OH: American Conference of Governmental Industrial Hygienists, 1995.

Bergess WA, Ellenbacker MJ, Treitman RD. *Ventilation for the Control of the Work Environment.* New York: Wiley, 1989.

Heinsohn RJ. *Industrial Ventilation: Engineering Principles.* New York: Wiley, 1991.

McDermott HJ. *Handbook of Ventilation for Contaminant Control.* Ann Arbor, MI: Ann Arbor Science, 1981.

McQuisten FC, Parker JD. *Heating, Ventilating and Air Conditioning,* 3rd ed. New York: Wiley, 1988.

## Video Display Terminals

DeMatteo B. *Terminal Shock: The Health Hazards of Video Display Terminals.* Toronto: NC Press; Port Washington, NY: Distributed in the U.S. by Independent Publishers Group, 1985.

Human Factors Society. *American National Standard for Human Factors Engineering of Visual Display Terminal Workstations.* Santa Monica: Human Factors Society, 1988.

9 to 5, National Association of Working Women. *VDT Syndrome: the Physical and Mental Trauma of Computer Work.* Cleveland, OH: 9 to 5, National Association of Working Women; Washington, DC: Service Employees International Union, 1988.

Sauter SL, Chapman LJ, Knutson SJ. *Improving VDT Work: Causes and Control of Health Concerns in VDT Use.* Lawrence, KS: Report Store, 1985.

Scalet EA. *VDT Health and Safety.* Lawrence, KS: Ergosyst Associates, 1987.

## Women

Chavkin W, ed. *Double Exposure: Women's Health Hazards on the Job and at Home.* New York: Monthly Review Press, 1984.

Denning JV. *Women's Work and Health Hazards: A Selected Bibliography.* London: Department of Occupational Health, London School of Hygiene & Tropical Medicine, 1984.

Howe LK. *Pink Collar Workers: Inside the World of Women's Work.* New York: Avon, 1977.

Kessler-Harris A. *Out to Work: A History of Wage-Earning Women in the United States.* Oxford: Oxford University Press, 1982.

Nelson L, Kenen R, Klitzman S. *Turning Things Around: A Women's Occupational and Environmental Health Resource Guide.* Washington, DC: National Women's Health Network, 1990.

Zones JS. *Women's Occupational Health and Safety in California: Safe at Work?* Sacramento: California Elected Women's Association for Education and Research, 1993.

# Threshold Limit Values and Biological Exposure Indices (ACGIH)

THIS APPENDIX GIVES THE THRESHOLD LIMIT VALUES (TLVs) for Chemical Substances and Physical Agents and the Biological Exposure Indices (BEIs) that were adopted by the American Conference of Governmental Industrial Hygienists (ACGIH). It is reprinted with permission from the American Conference of Governmental Industrial Hygienists, Inc., *1994–1995 Threshold Limit Values for Chemical Substances and Physical Agents and Biological Exposure Indices,* ©1994.

## INTRODUCTION TO THE CHEMICAL SUBSTANCES

Threshold Limit Values (TLVs) refer to airborne concentrations of substances and represent conditions under which it is believed that nearly all workers may be repeatedly exposed day after day without adverse health effects. Because of wide variation in individual susceptibility, however, a small percentage of workers may experience discomfort from some substances at concentrations at or below the threshold limit; a smaller percentage may be affected more seriously by aggravation of a pre-existing condition or by development of an occupational illness. Smoking of tobacco is harmful for several reasons. Smoking may act to enhance the biological effects of chemicals encountered in the workplace and may reduce the body's defense mechanisms against toxic substances.

Individuals may also be hypersusceptible or otherwise unusually responsive to some industrial chemicals because of genetic factors, age, personal habits (smoking, alcohol, or other drugs), medication, or previous exposures. Such workers may not be adequately protected from adverse health effects from certain chemicals at concentrations at or below the threshold limits. An occupational physician should evaluate the extent to which such workers require additional protection.

TLVs are based on available information from industrial experience; from experimental human and animal studies; and, when possible, from a combination of the three. The basis on which the values are established may differ from substance to substance; protection against impairment of health may be a guiding factor for some, whereas reasonable freedom from irritation, narcosis, nuisance, or other forms of stress may form the basis for others. Health impairments considered include those that shorten life expectancy, compromise physiological function, impair the capability for resisting other toxic substances or disease processes, or adversely affect reproductive function or developmental processes.

The amount and nature of the information available for establishing a TLV varies from substance to substance; consequently, the precision of the estimated TLV is also subject to variation and the latest TLV *Documentation* should be consulted in order to assess the extent of the data available for a given substance.

**These limits are intended for use in the practice of industrial hygiene as guidelines or recommendations in the control of potential health hazards and for no other use, e.g., in the evaluation or control of community air pollution nuisances; in estimating the toxic potential of continuous, uninterrupted exposures or other extended work periods; as proof or disproof of an existing disease or physical condition; or adoption or use by countries whose working conditions or cultures differ from those in the United States of America and where substances and processes differ. These limits *are not* fine lines between safe and dangerous concentration nor are they a relative index of toxicity. They *should not* be used by anyone untrained in the discipline of industrial hygiene.**

The TLVs, as issued by the American Conference of Governmental Industrial Hygienists, are recommendations and should be used as guidelines for good practices. In spite of the fact that serious injury is not believed likely as a result of exposure to the threshold limit concentrations, the best practice is to maintain concentrations of all atmospheric contaminants as low as is practical.

**The American Conference of Governmental Industrial Hygienists disclaims liability with respect to the use of TLVs.**

**Notice of Intended Changes.** Each year, proposed actions of the Chemical Substances TLV Committee for the forthcoming year are issued in the form of a "Notice of Intended Changes." This Notice provides an opportunity for comment and *solicits suggestions of substances to be added to the list. The suggestions should be accompanied by substantiating evidence.* The "Notice of Intended Changes" is presented after the Adopted Values in this section. Values listed in parentheses in the "Adopted" list are to be used during the period in which a proposed change for that Value is listed in the Notice of Intended Changes.

**Definitions.** Three categories of Threshold Limit Values (TLVs) are specified herein, as follows:

*a) Threshold Limit Value–Time-Weighted Average (TLV–TWA)*—the time-weighted average concentration for a normal 8-hour workday and a 40-hour workweek, to which nearly all workers may be repeatedly exposed, day after day, without adverse effect.

*b) Threshold Limit Value–Short-Term Exposure Limit (TLV–STEL)*—the concentration to which workers can be exposed continuously for a short period of time without suffering from 1) irritation, 2) chronic or irreversible tissue damage, or 3) narcosis of sufficient degree to increase the likelihood of accidental injury, impair self-rescue or materially reduce work efficiency, and provided that the daily TLV–TWA is not exceeded. It is not a separate independent exposure limit; rather, it supplements the time-weighted average (TWA) limit where there are recognized acute effects from a substance whose toxic effects are primarily of a chronic nature. STELs are recommended only where toxic effects have been reported from high short-term exposures in either humans or animals.

A STEL is defined as a 15-minute TWA exposure which should not be exceeded at any time during a workday even if the 8-hour TWA is within the TLV–TWA. Exposures above the TLV–TWA up to the STEL should not be longer than 15 minutes and should not occur more than four times per day. There should be at least 60 minutes between successive exposures in this range. An averaging period other than 15 minutes may be recommended when this is warranted by observed biological effects.

*c) Threshold Limit Value–Ceiling (TLV–C)*—the concentration that should not be exceeded during any part of the working exposure.

In conventional industrial hygiene practice if instantaneous monitoring is not feasible, then the TLV–C can be assessed by sampling over a 15-minute period except for those substances that may cause immediate irritation when exposures are short.

For some substances, e.g., irritant gases, only one category, the TLV–Ceiling, may be relevant. For other substances, one or two categories may be relevant, depending upon their physiologic action. It is important to observe that if any one of these types of TLVs is exceeded, a potential hazard from that substance is presumed to exist.

The Chemical Substances TLV Committee holds to the opinion that TLVs based on physical irritation should be considered no less binding than those based on physical impairment. There is increasing evidence that physical irritation may initiate, promote, or accelerate physical impairment through interaction with other chemical or biologic agents.

***Time-Weighted Average (TWA) vs Ceiling (C) Limits.*** TWAs permit excursions above the TLV provided they are compensated by equivalent excursions below the TLV–TWA during the workday. In some instances, it may be permissible to calculate the average concentration for a workweek rather than for a workday. The relationship between the TLV and permissible excursion is a rule of thumb and in certain cases may not apply. The amount by which the TLVs may be exceeded for short periods without injury to health depends upon a number of factors such as the nature of the contaminant, whether very high concentrations—even for short periods—produce acute poisoning, whether the effects are cumulative, the frequency with which high concentrations occur, and the duration of such periods. All factors must be taken into consideration in arriving at a decision as to whether a hazardous condition exists.

Although the TWA concentration provides the most satisfactory, practical way of monitoring airborne agents for compliance with the TLVs, there are certain substances for which it is inappropriate. In the latter group are substances which are predominantly fast acting and whose TLV is more appropriately based on this particular response. Substances with this type of response are best controlled by a ceiling limit that should not be exceeded. It is implicit in these definitions that the manner of sampling to determine noncompliance with the limits for each group must differ; a single, brief sample, that is applicable to a ceiling limit, is not appropriate to the TWA; here, a sufficient number of samples are needed to permit a TWA concentration throughout a complete cycle of operations or throughout the workshift.

Whereas the ceiling limit places a definite boundary that concentrations should not be permitted to exceed, the TWA requires an explicit limit to the excursions that are permissible above the listed TLVs. It should be noted that the same factors are used by the Chemical Substances TLV Committee in determining the magnitude of the value of the STEL or whether to include or exclude a substance for a ceiling listing.

*Excursion Limits.* For the vast majority of substances with a TLV–TWA, there is not enough toxicological data available to warrant a STEL. Nevertheless, excursions above the TLV–TWA should be controlled even where the 8-hour TLV–TWA is within recommended limits. Earlier editions of the TLV list included such limits whose values depended on the TLV–TWAs of the substance in question.

While no rigorous rationale was provided for these particular values, the basic concept was intuitive: in a well-controlled process exposure, excursions should be held within some reasonable limits. Unfortunately, neither toxicology nor collective industrial hygiene experience provide a solid basis for quantifying what those limits should be. The approach here is that the maximum recommended excursion should be related to variability generally observed in actual industrial processes. In reviewing large numbers of industrial hygiene surveys conducted by the National Institute for Occupational Safety and Health, Leidel, Busch, and Crouse[1] found that short-term exposure measurements were generally lognormally distributed with geometric standard deviations mostly in the range of 1.5 to 2.0.

While a complete discussion of the theory and properties of the lognormal distribution is beyond the scope of this section, a brief description of some important terms is presented. The measure of central tendency in a lognormal description is the antilog of the mean logarithm of the sample values. The distribution is skewed, and the geometric mean is always smaller than the arithmetic mean by an amount which depends on the geometric standard deviation. In the lognormal distribution, the geometric standard deviation ($sd_g$) is the antilog of the standard deviation of the sample value logarithms and 68.26% of all values lie between $m_g/sd_g$ and $m_g \times sd_g$.

If the short-term exposure values in a given situation have a geometric standard deviation of 2.0, 5% of all values will exceed 3.13 times the geometric mean. If a process displays a variability greater than this, it is not under good control and efforts should be made to restore control. This concept is the basis for the following excursion limit recommendations which apply to those TLV–TWAs that do not have STELs:

> Excursions in worker exposure levels may exceed 3 times the TLV–TWA for no more than a total of 30 minutes during a workday, and under no circumstances should they exceed 5 times the TLV–TWA, provided that the TLV–TWA is not exceeded.

The approach is a considerable simplification of the idea of the lognormal concentration distribution but is considered more convenient to use by the practicing industrial hygienist. If exposure excursions are maintained within the recommended limits, the geometric standard deviation of the concentration measurements will be near 2.0 and the goal of the recommendations will be accomplished.

When the toxicological data for a specific substance are available to establish a STEL, this value takes precedence over the excursion limit regardless of whether it is more or less stringent.

*"Skin" Notation.* Listed substances followed by the designation "Skin" refer to the potential significant contribution to the overall exposure by the cutaneous route, including mucous membranes and the eyes, either by contact with vapors or, of probable greater significance, by direct skin contact with the substance. Vehicles present in solutions or mixtures can also significantly enhance potential skin absorption. It should be noted that while some materials are capable of causing irritation, dermatitis, and sensitization in workers, these properties are *not considered relevant* when assigning a skin notation. It should be noted, however, that the development of a dermatological condition can significantly affect the potential for dermal absorption.

While limited quantitative data currently exist with regard to skin absorption of gases, vapors, and liquids by workers, the Chemical

Substances TLV Committee recommends that the integration of data from acute dermal studies and repeated dose dermal studies in animals and/or humans, along with the ability of the chemical to be absorbed, be used in deciding on the appropriateness of the skin notation. In general, available data which suggest that the potential for absorption via the hands/forearms during the workday could be significant, especially for chemicals with lower TLVs, could justify a skin notation. From acute animal toxicity data, materials having a relatively low dermal $LD_{50}$ (1000 mg/kg of body weight or less) would be given a skin notation. Where repeated dermal application studies have shown significant systemic effects following treatment, a skin notation would be considered. When chemicals penetrate the skin easily (higher octanol–water partition coefficients) and where extrapolations of systemic effects from other routes of exposure suggest dermal absorption may be important in the expressed toxicity, a skin notation should be considered.

Substances having a skin notation and a low TLV may present special problems for operations involving high airborne concentrations of the material, particularly under conditions where significant areas of the skin are exposed for a long period of time. Under these conditions, special precautions to significantly reduce or preclude skin contact may be required.

Biological monitoring should be considered to determine the relative contribution of exposure via the dermal route to the total dose. The TLV/BEI Booklet contains a number of adopted biological exposure indices, which provide an additional tool when assessing the worker's total exposure to selected materials. For additional information, refer to "Dermal Absorption" in the "Introduction to the Biological Exposure Indices," 6th edition of the *Documentation of Threshold Limit Values and Biological Exposure Indices*, and to Leung and Paustenbach.[2]

Use of the skin designation is intended to alert the reader that air sampling alone is insufficient to accurately quantitate exposure and that measures to prevent significant cutaneous absorption may be required.

*Mixtures.* Special consideration should be given also to the application of the TLVs in assessing the health hazards that may be associated with exposure to mixtures of two or more substances. A brief discussion of basic considerations involved in developing TLVs for mixtures and methods for their development, amplified by specific examples, are given in Appendix C.

*Respirable and Inhalable Dust.* For solid substances and liquified mists, TLVs are expressed in terms of inhalable dust, except where the term "respirable dust" is used. See Appendix D, Particle Size-Selective Sampling Criteria for Airborne Particulate Matter, for the definition of respirable dust (respirable particulate mass).

*Particulates Not Otherwise Classified (PNOC).* There are many substances on the TLV list, and many more that are not on the list, for which there is no evidence of specific toxic effects. Those that are particulates have frequently been called "nuisance dusts." Although these materials may not cause fibrosis or systemic effects, they are not biologically inert. At high concentrations, otherwise nontoxic dusts have been associated with the occasionally fatal condition known as alveolar proteinosis. At lower concentrations, they can inhibit the clearance of toxic particulates from the lung by decreasing the mobility of the alveolar macrophages. Accordingly, the Chemical Substances TLV Committee recommends the use of the term "Particulates Not Otherwise Classified (PNOC)" to emphasize that all materials are potentially toxic and to avoid the implication that these materials are harmless at all exposure concentrations. Particulates identified under the PNOC heading are those containing no asbestos and <1% crystalline silica. To recognize the adverse effects of exposure to otherwise nontoxic dusts, a TLV–TWA of 10 $mg/m^3$ for inhalable particulate and a TLV–TWA of 3 $mg/m^3$ for respirable particulate have been established and are included in the main TLV list. Refer to the Documentation for Particulates Not Otherwise Classified (PNOC) for a complete discussion of this subject.

*Simple Asphyxiants—"Inert" Gases or Vapors.* A number of gases and vapors, when present in high concentrations in air, act primarily as simple asphyxiants without other significant physiologic effects. A TLV may not be recommended for each simple asphyxiant because the limiting factor is the available oxygen. The minimal oxygen content should be 18% by volume under normal atmospheric pressure (equivalent to a partial pressure, $pO_2$ of 135 torr). Atmospheres deficient in $O_2$ do not provide adequate warning and most simple asphyxiants are odorless. Several simple asphyxiants present an explosion hazard. Account should be taken of this factor in limiting the concentration of the asphyxiant.

*Biological Exposure Indices (BEI).* A cross reference is indicated for those substances for which there are also Biological Exposure Indices. For such substances, biological monitoring should be instituted to evaluate the total exposure, e.g., dermal, ingestion, or nonoccupational. See the BEI section in this Booklet.

*Physical Factors.* It is recognized that such physical factors as heat, ultraviolet and ionizing radiation, humidity, abnormal pressure (altitude), and the like may place added stress on the body so that the effects from exposure at a TLV may be altered. Most of these stresses act adversely to increase the toxic response of a substance. *Although most TLVs have built-in safety factors to guard against adverse effects to moderate deviations* from normal environments, the safety factors of most substances are not of such a magnitude as to take care of gross deviations. For example, continuous work at temperatures above 32°C (90°F), or overtime extending the workweek more than 25%, might be considered gross deviations. In such instances, judgment should be exercised in the proper adjustments of the TLVs.

*Unlisted Substances.* The list of TLVs is by no means a complete list of all hazardous substances or of all hazardous substances used in industry. For a large number of materials of recognized toxicity, little or no data are available that could be used to establish a TLV. Substances that do not appear on the TLV list should not be considered to be harmless or nontoxic. When unlisted substances are introduced into a workplace, the medical and scientific literature should be reviewed to identify potentially dangerous toxic effects. It may also be advisable to conduct preliminary toxicity studies. In any case, it is necessary to remain alert to adverse health effects in workers which may be associated with the use of new materials. The TLV Committee strongly encourages industrial hygienists and other occupational health professionals to bring to the Committee's attention any information which would suggest that a TLV should be established. Such information should include exposure concentrations and correlated health effects data (dose–response) that would support a recommended TLV.

*Unusual Work Schedules.* Application of TLVs to workers on work schedules markedly different from the conventional 8-hour day, 40-hour week requires particular judgement in order to provide, for such workers, protection equal to that provided to workers on conventional workshifts.

As tentative guidance, field hygienists are referred to the "Brief and Scala model" which is described and explained at length in Patty.[3]

The Brief and Scala model reduces the TLV proportionately for both increased exposure time and reduced recovery (nonexposure) time. The model is generally intended to apply to work schedules longer than 8 hours/day or 40 hours/week. The model should not be used to justify very high exposures as "allowable" where the exposure periods are short (e.g., exposure to 8 times the TLV–TWA for one hour and zero exposure during the remainder of the shift). In this respect, the general limitations on TLV excursions and STELs should be applied to avoid inappropriate use of the model with very short exposure periods or shifts.

Since adjusted TLVs do not have the benefit of historical use and long-time observation, medical supervision during initial use of adjusted TLVs is advised. In addition, the hygienist should avoid unnecessary exposure of workers even if a model shows such exposures to be "allowable" and should not use models to justify higher-than-necessary exposures.

The Brief and Scala model is easier to use than some of the more complex models based on pharmacokinetic actions. However, hygienists thoroughly familiar with such models may find them more appropriate in specific instances. Use of such models usually requires knowledge of the biological half-life of each substance, and some models require additional data.

Short workweeks can allow workers to have two full-time jobs, perhaps with similar exposures, and may result in overexposure even if neither job by itself entails overexposure. Hygienists should be alert to such situations.

*Conversion of TLVs in ppm to mg/m³.* TLVs for gases and vapors are usually established in terms of parts per million of substance in air by volume (ppm). For convenience to the user, these TLVs are also listed here in terms of milligrams of substance per cubic meter of air (mg/m³). The conversion is based on 760 torr barometric pressure at 25°C (77°F), and where 24.45 = molar volume in liters, giving a conversion equation of:

$$\text{TLV in mg/m}^3 = \frac{(\text{TLV in ppm}) \ (\text{gram molecular weight of substance})}{24.45}$$

Conversely, the equation for converting TLVs in mg/m³ to ppm is:

$$\text{TLV in ppm} = \frac{(\text{TLV in mg/m}^3) \ (24.45)}{(\text{gram molecular weight of substance})}$$

Resulting values are rounded to two significant figures below 100 and to three significant figures above 100. This is not done to give any converted value a greater precision than that of the original TLV, but to avoid increasing or decreasing the TLV significantly merely by the conversion of units.

The above equation may be used to convert TLVs to any degree of precision desired. When converting TLVs to mg/m³ units for other temperatures and pressures, the reference TLVs should be used as a starting point. When converting values expressed as an element (e.g., as Fe, as Ni), the molecular value of the element should be used, not that of the entire compound.

In making conversions for substances with variable molecular weights, appropriate molecular weights have been estimated or assumed (see the TLV *Documentation*).

*Biologically-derived Airborne Contaminants.* The ACGIH Bioaerosols Committee has developed Guidelines for evaluating biological-source air contaminants in indoor environments (*Guidelines for the Assessment of Bioaerosols in the Indoor Environment*, ACGIH, 1989). The Guidelines rely on medical assessment of symptoms, evaluation of building performance, and professional judgement. For the reasons identified in the following, there are no numerical guidelines or TLVs that allow ready interpretation of bioaerosol data and routine sampling for bioaerosols is not recommended. If sampling is necessary (e.g., to document the contribution of identified sources), standard protocols are recommended in the Guidelines.

Biologically derived airborne contaminants include bioaerosols (airborne particulates composed of or derived from living organisms) and volatile organic compounds released from living organisms. Bioaerosols include microorganisms (culturable, nonculturable, and dead microorganisms) and fragments, toxins, and particulate waste products from all varieties of living things. Biologically derived airborne contaminant mixtures are ubiquitous in nature and may be modified by human activity. All persons are repeatedly exposed, day after day, to a wide variety of such contaminants. At present, gravimetric Threshold Limit Values (TLVs) exist for some wood dusts, which are primarily of biological origin, and for cotton dust, which is at least in part biological. There are no TLVs for concentrations of total culturable or countable organisms and particles (e.g., "bacteria" or "fungi"); specific culturable or countable organisms and particles (e.g., *Aspergillus fumigatus*); infectious agents (e.g., *Legionella pneumophila*; or assayable biological-source contaminants (e.g., endotoxin or volatile organic compounds).

**A.** A general TLV for a concentration of culturable (e.g., total bacteria and/or fungi) or countable bioaerosols (e.g., total pollen, fungal spores, and bacteria) is not scientifically supportable because:

1. Culturable organisms or countable spores do not comprise a single entity, i.e., bioaerosols are complex mixtures of different kinds of particles.

2. Human responses to bioaerosols range from innocuous effects to serious disease and depend on the specific agent and susceptibility factors within the person.

3. Measured concentrations of culturable and countable bioaerosols are dependent on the method of sample collection and analysis. It is not possible to collect and evaluate all of these bioaerosol components using a single sampling method.

**B.** Specific TLVs for individual culturable or countable bioaerosols, established to prevent irritant, toxic, or allergic responses have not been established. At present, information relating culturable or countable bioaerosol concentrations to irritant, toxic, or allergic responses consists largely of case reports containing only qualitative exposure data. The epidemiologic data that exist are insufficient to describe exposure–response relationships. Reasons for the absence of good epidemiologic data on exposure–response relationships include:

1. Most data on concentrations of specific bioaerosols are derived from indicator measurements rather than from measurement of actual effector agents. For example, culturable fungi are used to represent exposure to allergens. In addition, most measurements are either from reservoir or from ambient air samples. These approaches are unlikely to accurately represent human exposure to actual effector agents.

2. The components and concentrations of bioaerosols vary widely. The most commonly used air sampling devices collect only "grab" samples over short periods of time and these single samples may not represent human exposure. Short-term grab samples may contain an amount of a particular bioaerosol that is orders of magnitude higher or lower than the average environmental concentration. Some organisms release aerosols as "concentration bursts" and can be detected only rarely using grab samples. Yet, such episodic bioaerosols may produce significant health effects.

**C.** Dose–response data are available for some infectious bioaerosols. At present, air sampling protocols for infectious agents are limited and suitable only for research endeavors. Traditional public health methods, including immunization, active case finding, and medical treatment, remain the primary defenses against infectious bioaerosols. Certain public and medical facilities with high risk for transmission of infection (e.g., tuberculosis) should employ exposure controls to reduce possible airborne concentrations of virulent and opportunistic pathogens.

**D.** Assayable, biologically derived contaminants are substances produced by living things that can be detected using either chemical, immunological, or biological assay and include endotoxin, mycotoxins, allergens, and volatile organic compounds. Evidence does not yet support TLVs for any of the assayable substances. Assay methods for certain common aeroallergens and endotoxin are steadily improving. Also, innovative molecular techniques are rendering assayable the concentration of specific organisms currently detected only by culture or counting. Dose–response relationships for some assayable bioaerosols have been observed in experimental studies and occasionally in epidemiologic studies. Validation of these assays in the field is also progressing.

The ACGIH Bioaerosols Committee actively solicits information, comments, and especially data that will assist it in evaluating the role of bioaerosols in the environment.

### References

1. Leidel, N.A.; Busch, K.A.; Crouse, W.E.: Exposure Measurement Action Level and Occupational Environmental Variability. DHEW (NIOSH) Pub. No. 76-131; NTIS Pub. No. PB-267 509. National Technical Information Service, Springfield, VA (December 1975).

2. Leung, H.; Paustenbach, D.J.: Techniques for Estimating the Percutaneous Absorption of Chemicals Due to Occupational and Environmental Exposure. Appl. Occup. Environ. Hyg. 9(3):187–197 (March 1994).

3. Paustenbach, D.J.: Occupational Exposure Limits, Pharmacokinetics, and Unusual Work Schedules. In: Patty's Industrial Hygiene and Toxicology, 3rd ed., Vol. 3A, The Work Environment, Chap. 7, pp. 222–348. R.L. Harris, L.J. Cralley and L.V. Cralley, Eds. John Wiley and Sons, Inc., New York (1994).

| Substance | [CAS #] | ADOPTED VALUES TWA ppm[a] | TWA mg/m[3b] | STEL/CEILING (C) ppm[a] | STEL/CEILING (C) mg/m[3b] |
|---|---|---|---|---|---|
| •■ Acetaldehyde [75-07-0] (1993) .. | | — | — | C 25,A3 | C 45,A3 |
| Acetic acid [64-19-7] (1976)....... | | 10 | 25 | 15 | 37 |
| • Acetic anhydride [108-24-7] (1993) | | 5 | 21 | — | — |
| ◄■ Acetone [67-64-1] (1982)........... | | 750 | 1780 | 1000 | 2380 |
| *• Acetone cyanohydrin [75-86-5], as CN—Skin (1994).............. | | — | — | C 4.7 | C 5 |
| • Acetonitrile [75-05-8] (1976)...... | | 40 | 67 | 60 | 101 |
| Acetophenone [98-86-2] (1993). | | 10 | 49 | — | — |
| Acetylene [74-86-2] (1981) ........ | | —[c] | — | — | — |
| Acetylene dichloride, see 1,2-Dichloroethylene | | | | | |
| Acetylene tetrabromide [79-27-6] (1986)..................... | | 1 | 14 | — | — |
| Acetylsalicylic acid (Aspirin) [50-78-2] (1980)................ | | — | 5 | — | — |
| Acrolein [107-02-8] (1976)......... | | 0.1 | 0.23 | 0.3 | 0.69 |
| ■ Acrylamide [79-06-1]—Skin (1987) | | — | 0.03,A2 | — | — |
| Acrylic acid [79-10-7]—Skin (1990) | | 2 | 5.9 | — | — |
| •■ Acrylonitrile [107-13-1]— Skin (1984) ........................... | | 2,A2 | 4.3,A2 | — | — |
| Adipic acid [124-04-9] (1993)..... | | — | 5 | — | — |
| * Adiponitrile [111-69-3] — Skin (1994)........................... | | 2 | 8.8 | — | — |
| ■ Aldrin [309-00-2]—Skin (1986) . | | — | 0.25 | — | — |
| Allyl alcohol [107-18-6]— Skin (1976).......................... | | 2 | 4.8 | 4 | 9.5 |
| ■ Allyl chloride [107-05-1] (1976)... | | 1 | 3 | 2 | 6 |
| Allyl glycidyl ether (AGE) [106-92-3] (1976) ......... | | 5 | 23 | 10 | 47 |
| Allyl propyl disulfide [2179-59-1] (1976) ................ | | 2 | 12 | 3 | 18 |
| α-Alumina, see Aluminum oxide | | | | | |
| Aluminum [7429-90-5] | | | | | |
| Metal dust (1986) ................ | | — | 10 | — | — |
| Pyro powders, as Al (1979) .. | | — | 5 | — | — |
| Welding fumes, as Al (1979) .. | | — | 5 | — | — |
| Soluble salts, as Al (1979) .... | | — | 2 | — | — |
| Alkyls (NOC[d]), as Al (1979) | | — | 2 | — | — |
| Aluminum oxide [1344-28-1] (1986) ............................. | | — | 10[e] | — | — |
| ■ 4-Aminodiphenyl [92-67-1]— Skin (1972) | | — | A1 | — | — |
| 2-Aminoethanol, see Ethanolamine | | | | | |
| 2-Aminopyridine [504-29-0] (1986) | | 0.5 | 1.9 | — | — |
| 3-Amino-1,2,4-triazole, see Amitrole | | | | | |
| ■ Amitrole [61-82-5] (1986) ......... | | — | 0.2 | — | — |
| Ammonia [7664-41-7] (1976)..... | | 25 | 17 | 35 | 24 |
| Ammonium chloride fume [12125-02-9] (1976) ............. | | — | 10 | — | 20 |
| * Ammonium perfluorooctanoate [3825-26-1]—Skin (1994)....... | | — | 0.01,A3 | — | — |
| Ammonium sulfamate [7773-06-0] (1986) ................. | | — | 10 | — | — |
| Amosite, see Asbestos | | | | | |
| n-Amyl acetate [628-63-7] (1987) | | 100 | 532 | — | — |
| sec-Amyl acetate [626-38-0] (1987) | | 125 | 665 | — | — |
| ◄■ Aniline [62-53-3] and homologues—Skin (1986)... | | 2 | 7.6 | — | — |
| ■ Anisidine [29191-52-4] (o-, p- isomers)—Skin (1977)......... | | 0.1 | 0.5 | — | — |
| Antimony [7440-36-0] and compounds, as Sb (1980).... | | — | 0.5 | — | — |

**ADOPTED VALUES**

| Substance [CAS #] | TWA ppm[a] | TWA mg/m[3b] | STEL/CEILING (C) ppm[a] | STEL/CEILING (C) mg/m[3b] |
|---|---|---|---|---|
| ■ Antimony trioxide [1309-64-4] handling and use, as Sb (1978) | — | 0.5 | — | — |
| Production (1980) | — | A2 | — | — |
| ANTU [86-88-4] (1986) | — | 0.3 | — | — |
| Argon [7440-37-1] (1981) | —(c) | — | — | — |
| ◄■■ Arsenic, elemental [7440-38-2] and inorganic compounds (except Arsine), as As (1993) | — | 0.01,A1 | — | — |
| ■■ Arsine [7784-42-1] (1977) | 0.05 | 0.16 | — | — |
| ‡■■ Asbestos(f) | | | | |
| ‡●Amosite [12172-73-5] (1980) | (0.5 fiber/cc, A1) | | | |
| ‡●Chrysotile [12001-29-5] (1980) | (2 fibers/cc, A1) | | | |
| ‡●Crocidolite [12001-28-4] (1980) | (0.2 fiber /cc, A1) | | | |
| ‡●Other forms (1980) | (2 fibers/cc, A1) | | | |
| ■ Asphalt (petroleum) fumes [8052-42-4] (1987) | — | 5 | — | — |
| Atrazine [1912-24-9] (1983) | — | 5 | — | — |
| ◄ Azinphos-methyl [86-50-0]— Skin (1986) | — | 0.2 | — | — |
| Barium [7440-39-3], soluble compounds, as Ba (1977) | — | 0.5 | — | — |
| Barium sulfate [7727-43-7] (1986) | — | 10(e) | — | — |
| Benomyl [17804-35-2] (1986) | 0.84 | 10 | — | — |
| Benz[a]anthracene [56-55-3] (1993) | A2 | A2 | — | — |
| ‡◄■■ Benzene [71-43-2] (1987) | (10,A2) | (32,A2) | | |
| ■ Benzidine [92-87-5]—Skin (1982) | — | A1 | | |
| Benzo[b]fluoranthene [205-99-2] (1992) | — | A2 | | |
| p -Benzoquinone, see Quinone | | | | |
| Benzoyl peroxide [94-36-0] (1977) | — | 5 | — | — |
| ■ Benzo[a]pyrene [50-32-8] (1976) | — | A2 | — | — |
| ■ Benzyl chloride [100-44-7] (1977) | 1 | 5.2 | — | — |
| ■■ Beryllium [7440-41-7] and compounds, as Be (1979) | — | 0.002,A2 | — | — |
| Biphenyl [92-52-4] (1987) | 0.2 | 1.3 | — | — |
| Bismuth telluride, as $Bi_2Te_3$ | | | | |
| Undoped [1304-82-1] (1986) | — | 10 | — | — |
| Se-doped (1986) | — | 5 | — | — |
| Borates, tetra, sodium salts [1303-96-4] | | | | |
| Anhydrous (1977) | — | 1 | — | — |
| Decahydrate (1977) | — | 5 | — | — |
| Pentahydrate (1977) | — | 1 | — | — |
| Boron oxide [1303-86-2] (1986) | — | 10 | — | — |
| Boron tribromide [10294-33-4] (1986) | — | — | C 1 | C 10 |
| Boron trifluoride [7637-07-2] (1977) | — | — | C 1 | C 2.8 |
| Bromacil [314-40-9] (1986) | — | 10 | — | — |
| * Bromine [7726-95-6] (1994) | 0.1 | 0.66 | 0.2 | 1.3 |
| Bromine pentafluoride [7789-30-2] (1986) | 0.1 | 0.72 | — | — |
| Bromochloromethane, see Chlorobromomethane | | | | |
| Bromoform [75-25-2]— Skin (1977) | 0.5 | 5.2 | — | — |
| *■■ 1,3-Butadiene [106-99-0] (1994) | 2,A2 | 4.4,A2 | — | — |
| Butane [106-97-8] (1981) | 800 | 1900 | — | — |
| Butanethiol, see Butyl mercaptan | | | | |
| n-Butanol [71-36-3]— Skin (1977) | — | — | C 50 | C 152 |
| sec-Butanol [78-92-2] (1990) | 100 | 303 | — | — |
| ‡ tert-Butanol [75-65-0] (1993) | (100) | (303) | — | — |
| 2-Butanone, see Methyl ethyl ketone (MEK) | | | | |
| 2-Butoxyethanol (EGBE) [111-76-2]—Skin (1987) | 25 | 121 | — | — |
| ‡ n-Butyl acetate [123-86-4] (1976) | (150) | (713) | (200) | (950) |
| sec-Butyl acetate [105-46-4] (1987) | 200 | 950 | — | — |
| tert-Butyl acetate [540-88-5] (1987) | 200 | 950 | — | — |
| n-Butyl acrylate [141-32-2] (1978) | 10 | 52 | — | — |
| n-Butylamine [109-73-9]— Skin (1976) | — | — | C 5 | C 15 |
| ■■ tert-Butyl chromate, as $CrO_3$ [1189-85-1]—Skin (1977) | — | — | — | C 0.1 |

**ADOPTED VALUES**

| Substance [CAS #] | TWA ppm[a] | TWA mg/m[3b] | STEL/CEILING (C) ppm[a] | STEL/CEILING (C) mg/m[3b] |
|---|---|---|---|---|
| • n-Butyl glycidyl ether (BGE) [2426-08-6] (1981) | 25 | 133 | — | — |
| n-Butyl lactate [138-22-7] (1977) | 5 | 30 | — | — |
| Butyl mercaptan [109-79-5] (1977) | 0.5 | 1.8 | — | — |
| o-sec-Butylphenol [89-72-5]—Skin (1980) | 5 | 31 | — | — |
| p-tert-Butyl toluene [98-51-1] (1993) | 1 | 6.1 | — | — |
| ◄■● Cadmium, elemental [7440-43-9] and compounds, as Cd (1993) | — | 0.01,(i)A2 | — | — |
| | — | 0.002,(i)A2 | — | — |
| Calcium carbonate [1317-65-3] (1986) | — | 10(e) | | |
| Calcium chromate [13765-19-0], as Cr (1991) | — | 0.001,A2 | | |
| Calcium cyanamide [156-62-7] (1986) | — | 0.5 | | |
| Calcium hydroxide [1305-62-0] (1978) | — | 5 | | |
| Calcium oxide [1305-78-8] (1978) | — | 2 | | |
| Calcium silicate (synthetic) [1344-95-2] (1991) | — | 10 (e) | | |
| Calcium sulfate [7778-18-9] (1986) | — | 10 (e) | | |
| Camphor, synthetic [76-22-2] (1976) | 2 | 12 | 3 | 19 |
| Caprolactam [105-60-2] | | | | |
| Dust (1974) | — | 1 | — | 3 |
| Vapor (1992) | 5 | 23 | 10 | 46 |
| ■ Captafol [2425-06-1]— Skin (1977) | — | 0.1 | — | — |
| ■ Captan [133-06-2] (1986) | — | 5 | — | — |
| Carbaryl [63-25-22] (1986) | — | 5 | — | — |
| Carbofuran [1563-66-2] (1986) | — | 0.1 | — | — |
| ■ Carbon black [1333-86-4] (1986) | — | 3.5 | — | — |
| Carbon dioxide [124-38-9] (1986) | 5000 | 9000 | 30,000 | 54,000 |
| ◄● Carbon disulfide [75-15-0]— Skin (1980) | 10 | 31 | — | — |
| ◄ Carbon monoxide [630-08-0] (1992) | 25 | 29 | — | — |
| Carbon tetrabromide [558-13-4] (1976) | 0.1 | 1.4 | 0.3 | 4.1 |
| ■■ Carbon tetrachloride (Tetrachloromethane) [56-23-5]—Skin (1993) | 5,A3 | 31,A3 | 10,A3 | 63,A3 |
| Carbonyl chloride, see Phosgene | | | | |
| Carbonyl fluoride [353-50-4] (1986) | 2 | 5.4 | 5 | 13 |
| Catechol [120-80-9]— Skin (1981) | 5 | 23 | — | — |
| Cellulose [9004-34-6] (1986) | — | 10 | — | — |
| Cesium hydroxide [21351-79-1] (1977) | — | 2 | — | — |
| ■ Chlordane [57-74-9]— Skin (1990) | — | 0.5 | — | — |
| ■ Chlorinated camphene (Toxaphene) [8001-35-2]—Skin (1977) | — | 0.5 | — | 1 |
| Chlorinated diphenyl oxide [55720-99-5] (1990) | — | 0.5 | — | — |
| Chlorine [7782-50-5] (1989) | 0.5 | 1.5 | 1 | 2.9 |
| Chlorine dioxide [10049-04-4] (1976) | 0.1 | 0.28 | 0.3 | 0.83 |
| Chlorine trifluoride [7790-91-2] (1977) | — | — | C 0.1 | C 0.38 |
| Chloroacetaldehyde [107-20-0] (1977) | — | — | C 1 | C 3.2 |
| Chloroacetone [78-95-5]— Skin (1989) | — | — | C 1 | C 3.8 |
| α-Chloroacetophenone [532-27-4] (1977) | 0.05 | 0.32 | — | — |
| Chloroacetyl chloride [79-04-9] — Skin (1991) | 0.05 | 0.23 | 0.15 | 0.69 |

| Substance | [CAS #] | TWA ppm[a] | TWA mg/m³[b] | STEL/CEILING (C) ppm[a] | STEL/CEILING (C) mg/m³[b] |
|---|---|---|---|---|---|
| **ADOPTED VALUES** | | | | | |
| ◄ Chlorobenzene [108-90-7] (1991) | | 10 | 46 | — | — |
| o-Chlorobenzylidene malononitrile [2698-41-1]—Skin (1983) | | — | — | C 0.05 | C 0.39 |
| Chlorobromomethane [74-97-5] (1990) | | 200 | 1060 | — | — |
| 2-Chloro-1,3-butadiene, see β-Chloroprene | | | | | |
| Chlorodifluoromethane [75-45-6] (1990) | | 1000 | 3540 | — | — |
| •■ Chlorodiphenyl (42% chlorine) [53469-21-9]—Skin (1990) | | — | 1 | — | — |
| •■ Chlorodiphenyl (54% chlorine) [11097-69-1]—Skin (1990) | | — | 0.5 | — | — |
| 1-Chloro-2,3-epoxy propane, see Epichlorohydrin | | | | | |
| 2-Chloroethanol, see Ethylene chlorohydrin | | | | | |
| Chloroethylene, see Vinyl chloride | | | | | |
| •■ Chloroform [67-66-3] (1986) | | 10,A2 | 49,A2 | — | — |
| ■ bis(Chloromethyl) ether [542-88-1] (1981) | | 0.001,A1 | 0.0047,A1 | | |
| ■ Chloromethyl methyl ether [107-30-2] (1983) | | A2 | A2 | | |
| 1-Chloro-1-nitropropane [600-25-9] (1981) | | 2 | 10 | | |
| Chloropentafluoroethane [76-15-3] (1981) | | 1000 | 6320 | — | — |
| Chloropicrin [76-06-2] (1990) | | 0.1 | 0.67 | — | — |
| •■ β-Chloroprene [126-99-8]—Skin (1980) | | 10 | 36 | — | — |
| 2-Chloropropionic acid [598-78-7]—Skin (1991) | | 0.1 | 0.44 | — | — |
| o-Chlorostyrene [2039-87-4] (1976) | | 50 | 283 | 75 | 425 |
| o-Chlorotoluene [95-49-8] (1990) | | 50 | 259 | — | — |
| 2-Chloro-6-( trichloromethyl) pyridine, see Nitrapyrin | | | | | |
| Chlorpyrifos [2921-88-2]—Skin (1990) | | — | 0.2 | — | — |
| • Chromite ore processing (Chromate), as Cr (1978) | | — | 0.05,A1 | | |
| * Chromium, metal [7440-47-3], and inorganic compounds, as Cr | | | | | |
| *■Metal and Cr III compounds (1994) | | — | 0.5,A4 | — | — |
| *◄■Water-soluble Cr VI compounds, NOC(d) (1994) | | — | 0.05,A1 | | |
| *•■ Insoluble Cr VI compounds, NOC(d) (1994) | | — | 0.01,A1 | | |
| •■ Chromyl chloride [14977-61-8] (1982) | | 0.025 | 0.16 | — | — |
| ■ Chrysene [218-01-9] (1981) | | A2 | A2 | — | — |
| Chrysotile, see Asbestos | | | | | |
| Clopidol [2971-90-6] (1990) | | — | 10 | | |
| Coal dust (1987) | | — | 2,(g,j) | | |
| •■ Coal tar pitch volatiles [65996-93-2], as benzene solubles (1981) | | — | 0.2,A1 | | |
| *■ Cobalt, elemental [7440-48-4], and inorganic compounds, as Co (1994) | | — | 0.02,A3 | | |
| Cobalt carbonyl [10210-68-1], as Co (1983) | | — | 0.1 | | |
| Cobalt hydrocarbonyl [16842-03-8], as Co (1983) | | — | 0.1 | | |
| Copper [7440-50-8] | | | | | |
| Fume (1977) | | — | 0.2 | — | — |
| Dusts & mists, as Cu (1986) | | — | 1 | — | — |
| Cotton dust, raw (1986) | | — | 0.2(h) | — | |
| • Cresol [1319-77-3], all isomers—Skin (1977) | | 5 | 22 | | |
| Cristobalite, see Silica—Crystalline | | | | | |
| Crocidolite, see Asbestos | | | | | |
| ■ Crotonaldehyde [4170-30-3] (1987) | | 2 | 5.7 | — | — |

| Substance | [CAS #] | TWA ppm[a] | TWA mg/m³[b] | STEL/CEILING (C) ppm[a] | STEL/CEILING (C) mg/m³[b] |
|---|---|---|---|---|---|
| **ADOPTED VALUES** | | | | | |
| Crufomate [299-86-5] (1990) | | — | 5 | — | — |
| Cumene [98-82-8]—Skin (1987) | | 50 | 246 | — | — |
| Cyanamide [420-04-2] (1977) | | — | 2 | — | — |
| Cyanogen [460-19-5] (1977) | | 10 | 21 | — | — |
| Cyanogen chloride [506-77-4] (1980) | | — | — | C 0.3 | C 0.75 |
| Cyclohexane [110-82-7] (1987) | | 300 | 1030 | — | — |
| Cyclohexanol [108-93-0]—Skin (1977) | | 50 | 206 | — | — |
| Cyclohexanone [108-94-1]—Skin (1987) | | 25 | 100 | — | — |
| Cyclohexene [110-83-8] (1977) | | 300 | 1010 | — | — |
| Cyclohexylamine [108-91-8] (1977) | | 10 | 41 | — | — |
| Cyclonite [121-82-4]—Skin (1990) | | — | 1.5 | — | — |
| Cyclopentadiene [542-92-7] (1987) | | 75 | 203 | — | — |
| Cyclopentane [287-92-3] (1987) | | 600 | 1720 | — | — |
| Cyhexatin [13121-70-5] (1986) | | — | 5 | — | — |
| 2,4-D [94-75-7] (1986) | | — | 10 | — | — |
| •■ DDT (Dichlorodiphenyltrichloroethane) [50-29-3] (1986) | | — | 1 | — | — |
| Decaborane [17702-41-9]—Skin (1976) | | 0.05 | 0.25 | 0.15 | 0.75 |
| ◄ Demeton [8065-48-3]—Skin (1986) | | 0.01 | 0.11 | — | — |
| Diacetone alcohol [123-42-2] (1987) | | 50 | 238 | — | — |
| 1,2-Diaminoethane, see Ethylenediamine | | | | | |
| Diatomaceous earth, see Silica—Amorphous | | | | | |
| ◄ Diazinon [333-41-5]—Skin (1986) | | — | 0.1 | — | — |
| ■ Diazomethane [334-88-3] (1977) | | 0.2 | 0.34 | — | — |
| Diborane [19287-45-7] (1977) | | 0.1 | 0.11 | — | — |
| 1,2-Dibromoethane, see Ethylene dibromide | | | | | |
| * 2-N-Dibutylaminoethanol [102-81-8]—Skin (1994) | | 0.5 | 3.5 | — | — |
| Dibutyl phenyl phosphate [2528-36-1]—Skin (1990) | | 0.3 | 3.5 | — | — |
| Dibutyl phosphate [107-66-4] (1976) | | 1 | 8.6 | 2 | 17 |
| Dibutyl phthalate [84-74-2] (1987) | | — | 5 | — | — |
| ‡■ Dichloroacetylene [7572-29-4] (1970) | | — | — | (C 0.1) | (C 0.39) |
| o-Dichlorobenzene [95-50-1] (1992) | | 25 | 150 | 50 | 301 |
| p-Dichlorobenzene [106-46-7] (1993) | | 10,A3 | 60,A3 | — | — |
| ■ 3,3'-Dichlorobenzidine [91-94-1]—Skin (1976) | | — | A2 | — | — |
| 1,4-Dichloro-2-butene [764-41-0]—Skin | | 0.005,A2 | 0.025,A2 | — | — |
| Dichlorodifluoromethane [75-71-8] (1986) | | 1000 | 4950 | — | — |
| 1,3-Dichloro-5,5-dimethyl hydantoin [118-52-5] (1976) | | — | 0.2 | — | 0.4 |
| 1,1-Dichloroethane [75-34-3] (1992) | | 100 | 405 | — | — |
| 1,2-Dichloroethane, see Ethylene dichloride | | | | | |
| 1,1-Dichloroethylene, see Vinylidene chloride | | | | | |
| 1,2-Dichloroethylene [540-59-0] (1987) | | 200 | 793 | — | — |
| ■ Dichloroethyl ether [111-44-4]—Skin (1976) | | 5 | 29 | 10 | 58 |
| Dichlorofluoromethane [75-43-4] (1986) | | 10 | 42 | — | — |
| Dichloromethane, see Methylene chloride | | | | | |
| 1,1-Dichloro-1-nitroethane [594-72-9] (1986) | | 2 | 12 | — | — |
| 1,2-Dichloropropane, see Propylene dichloride | | | | | |

| Substance [CAS #] | ADOPTED VALUES TWA ppm[a] | mg/m[3b] | STEL/CEILING (C) ppm[a] | mg/m[3b] |
|---|---|---|---|---|
| ■ 1,3-Dichloropropene [542-75-6]—Skin (1986) | 1 | 4.5 | — | — |
| 2,2-Dichloropropionic acid [75-99-0] (1980) | 1 | 5.8 | — | — |
| Dichlorotetrafluoroethane [76-14-2] (1986) | 1000 | 6990 | — | — |
| ◄ Dichlorvos [62-73-7]— Skin (1986) | 0.1 | 0.90 | — | — |
| Dicrotophos [141-66-2]— Skin (1977) | — | 0.25 | — | — |
| Dicyclopentadiene [77-73-6] (1977) | 5 | 27 | — | — |
| Dicyclopentadienyl iron [102-54-5] (1986) | — | 10 | — | — |
| ■ Dieldrin [60-57-1]—Skin (1986) | — | 0.25 | — | — |
| * Diethanolamine [111-42-2]—Skin (1994) | 0.46 | 2 | — | — |
| * Diethylamine [109-89-7]— Skin (1994) | 5,A4 | 15,A4 | 15,A4 | 45,A4 |
| * 2-Diethylaminoethanol [100-37-8]—Skin (1994) | 2 | 9.6 | — | — |
| Diethylene triamine [111-40-0]—Skin (1977) | 1 | 4.2 | — | — |
| Diethyl ether, see Ethyl ether | | | | |
| Di(2-ethylhexyl)phthalate, see Di-sec-octyl phthalate | | | | |
| Diethyl ketone [96-22-0] (1981) | 200 | 705 | — | — |
| Diethyl phthalate [84-66-2] (1987) | — | 5 | — | — |
| Difluorodibromomethane [75-61-6] (1986) | 100 | 858 | — | — |
| ■ Diglycidyl ether (DGE) [2238-07-5] (1981) | 0.1 | 0.53 | — | — |
| Dihydroxybenzene, see Hydroquinone | | | | |
| Diisobutyl ketone [108-83-8] (1977) | 25 | 145 | — | — |
| Diisopropylamine [108-18-9]— Skin (1977) | 5 | 21 | — | — |
| Dimethoxymethane, see Methylal | | | | |
| N,N-Dimethyl acetamide [127-19-5]—Skin (1986) | 10 | 36 | — | — |
| Dimethylamine [124-40-3] (1992) | 5 | 9.2 | 15 | 27.6 |
| Dimethylaminobenzene, see Xylidene | | | | |
| ◄ Dimethylaniline [121-69-7] (N,N-Dimethylaniline)— Skin (1976) | 5 | 25 | 10 | 50 |
| Dimethylbenzene, see Xylene | | | | |
| ■ Dimethyl carbamoyl chloride [79-44-7] (1978) | A2 | A2 | — | — |
| Dimethyl-1,2-dibromo-2,2-dichloroethyl phosphate, see Naled | | | | |
| ◄ Dimethylformamide [68-12-2]— Skin (1986) | 10 | 30 | — | — |
| 2,6-Dimethyl-4-heptanone, see Diisobutyl ketone | | | | |
| ‡•■ 1,1-Dimethylhydrazine [57-14-7]—Skin (1976) | (0.5,A2) | (1.2,A2) | — | — |
| Dimethylnitrosoamine, see N-Nitrosodimethylamine | | | | |
| Dimethylphthalate [131-11-3] (1986) | — | 5 | — | — |
| ■ Dimethyl sulfate [77-78-1]—Skin (1977) | 0.1,A2 | 0.52,A2 | — | — |
| Dinitolmide [148-01-6] (1976) | — | 5 | — | — |
| ◄ Dinitrobenzene [528-29-0; 99-65-0; 100-25-4] (all isomers)—Skin (1986) | 0.15 | 1.0 | — | — |
| Dinitro-o-cresol [534-52-1]—Skin (1986) | — | 0.2 | — | — |
| 3,5-Dinitro-o-toluamide. see Dinitolmide | | | | |
| ◄■ Dinitrotoluene [25321-14-6]—Skin (1992) | — | 0.15,A2 | — | — |
| •■ Dioxane [123-91-1]— Skin (1986) | 25 | 90 | — | — |

| Substance [CAS #] | ADOPTED VALUES TWA ppm[a] | mg/m[3b] | STEL/CEILING (C) ppm[a] | mg/m[3b] |
|---|---|---|---|---|
| ◄ Dioxathion [78-34-2]— Skin (1977) | — | 0.2 | — | — |
| Diphenyl, see Biphenyl | | | | |
| Diphenylamine [122-39-4] (1986) | — | 10 | — | — |
| Diphenylmethane diisocyanate, see Methylene bisphenyl isocyanate | | | | |
| Dipropylene glycol methyl ether [34590-94-8]—Skin (1976) | 100 | 606 | 150 | 909 |
| Dipropyl ketone [123-19-3] (1981) | 50 | 233 | — | — |
| Diquat [2764-72-9]—Skin (1993) | — | 0.5(i) | — | — |
| | — | 0.1(j) | — | — |
| ■ Di-sec-octyl phthalate [117-81-7] (1976) | — | 5 | — | 10 |
| Disulfiram [97-77-8] (1986) | — | 2 | — | — |
| Disulfoton [298-04-4]— Skin (1986) | — | 0.1 | — | — |
| 2,6-Di-tert-butyl-p-cresol [128-37-0] (1987) | — | 10 | — | — |
| Diuron [330-54-1] (1977) | — | 10 | — | — |
| Divinyl benzene [1321-74-0] (1980) | 10 | 53 | — | — |
| Emery [1302-74-5] (1986) | — | 10 (e) | — | — |
| Endosulfan [115-29-7]— Skin (1986) | — | 0.1 | — | — |
| Endrin [72-20-8]—Skin (1988) | — | 0.1 | — | — |
| Enflurane [13838-16-9] (1988) | 75 | 566 | — | — |
| Enzymes, see Subtilisins | | | | |
| ‡•■ Epichlorohydrin [106-89-8]—Skin (1986) | (2) | (7.6) | — | — |
| *◄ EPN [2104-64-5]—Skin (1994) | — | 0.1 | — | — |
| 1,2-Epoxypropane, see Propylene oxide | | | | |
| 2,3-Epoxy-1-propanol. see Glycidol | | | | |
| Ethane [74-84-0] (1981) | —(c) | — | — | — |
| Ethanethiol. see Ethyl mercaptan | | | | |
| Ethanol [64-17-5] (1977) | 1000 | 1880 | — | — |
| Ethanolamine [141-43-5] (1978) | 3 | 7.5 | 6 | 15 |
| ◄ Ethion [563-12-2]—Skin (1977) | — | 0.4 | — | — |
| ◄• 2-Ethoxyethanol (EGEE) [110-80-5]—Skin (1984) | 5 | 18 | — | — |
| ◄ 2-Ethoxyethyl acetate (EGEEA) [111-15-9]—Skin (1984) | 5 | 27 | — | — |
| Ethyl acetate [141-78-6] (1977) | 400 | 1440 | — | — |
| ■ Ethyl acrylate [140-88-5] (1990) | 5,A2 | 20,A2 | 15,A2 | 61,A2 |
| Ethyl alcohol, see Ethanol | | | | |
| * Ethylamine [75-04-7]— Skin (1994) | 5 | 9.2 | 15 | 27.6 |
| Ethyl amyl ketone [541-85-5] (1977) | 25 | 131 | — | — |
| ◄ Ethyl benzene [100-41-4] (1976) | 100 | 434 | 125 | 543 |
| Ethyl bromide [74-96-4]— Skin (1992) | 5,A2 | 22,A2 | — | — |
| Ethyl butyl ketone [106-35-4] (1987) | 50 | 234 | — | — |
| ‡ Ethyl chloride [75-00-3] (1986) | (1000) | (2640) | — | — |
| Ethylene [74-85-1] (1981) | —(c) | — | — | — |
| Ethylene chlorohydrin [107-07-3]—Skin (1977) | — | — | C 1 | C 3.3 |
| Ethylenediamine [107-15-3]—Skin (1977) | 10 | 25 | — | — |
| ■ Ethylene dibromide [106-93-4]—Skin (1982) | A2 | A2 | — | — |
| •■ Ethylene dichloride [107-06-2] (1986) | 10 | 40 | — | — |
| ‡ Ethylene glycol [107-21-1] Vapor and mist (1981) | — | — | (C 50) | (C 127) |
| Ethylene glycol dinitrate [628-96-6]—Skin (1985) | 0.05 | 0.31 | — | — |
| Ethylene glycol methyl ether acetate, see 2-Methoxyethyl acetate | | | | |

## ADOPTED VALUES

| Substance [CAS #] | TWA ppm[a] | TWA mg/m[3b] | STEL/CEILING (C) ppm[a] | STEL/CEILING (C) mg/m[3b] |
|---|---|---|---|---|
| ■ Ethylene oxide [75-21-8] (1984). | 1,A2 | 1.8,A2 | — | — |
| •■ Ethylenimine [151-56-4]— | | | | |
| Skin (1977) | 0.5 | 0.88 | — | — |
| Ethyl ether [60-29-7] (1976) | 400 | 1210 | 500 | 1520 |
| Ethyl formate [109-94-4] (1987). | 100 | 303 | — | — |
| Ethylidene chloride, *see* 1,1-Dichloroethane | | | | |
| Ethylidene norbornene | | | | |
| [16219-75-3] (1977) | — | — | C 5 | C 25 |
| Ethyl mercaptan [75-08-1] (1986) | 0.5 | 1.3 | — | — |
| N-Ethylmorpholine | | | | |
| [100-74-3]—Skin (1986) | 5 | 24 | — | — |
| Ethyl silicate [78-10-4] (1986) | 10 | 85 | — | — |
| ◄ Fenamiphos [22224-92-6]— | | | | |
| Skin (1984) | — | 0.1 | — | — |
| ◄ Fensulfothion [115-90-2] (1977) | — | 0.1 | — | — |
| ◄ Fenthion [55-38-9]— | | | | |
| Skin (1983) | — | 0.2 | — | — |
| Ferbam [14484-64-1] (1986) | — | 10 | — | — |
| Ferrovanadium dust | | | | |
| [12604-58-9] (1983) | — | 1 | — | 3 |
| ■ Fibrous glass dust (1978) | — | 10 | — | — |
| ◄ Fluorides, as F (1977) | — | 2.5 | — | — |
| • Fluorine [7782-41-4] (1976) | 1 | 1.6 | 2 | 3.1 |
| Fluorotrichloromethane, *see* Trichlorofluoromethane | | | | |
| ◄ Fonofos [944-22-9]—Skin (1977) | — | 0.1 | — | — |
| ■ Formaldehyde [50-00-0] (1992) | — | — | C 0.3,A2 | C 0.37,A2 |
| Formamide [75-12-7]— | | | | |
| Skin (1988) | 10 | 18 | — | — |
| Formic acid [64-18-6] (1991) | 5 | 9.4 | 10 | 19 |
| ◄ Furfural [98-01-1]—Skin (1987) | 2 | 7.9 | — | — |
| Furfuryl alcohol [98-00-0]— | | | | |
| Skin (1982) | 10 | 40 | 15 | 60 |
| •■ Gasoline [8006-61-9] (1982) | 300 | 890 | 500 | 1480 |
| Germanium tetrahydride | | | | |
| [7782-65-2] (1986) | 0.2 | 0.63 | — | — |
| Glass, fibrous or dust, *see* Fibrous glass dust | | | | |
| Glutaraldehyde [111-30-8] (1979) | — | — | C 0.2 | C 0.82 |
| Glycerin mist [56-81-5] (1981) | — | 10[(i)] | — | — |
| Glycidol [556-52-5] (1987) | 25 | 76 | — | — |
| Glycol monoethyl ether, *see* 2-Ethoxyethanol | | | | |
| Grain dust (oat, wheat, barley) (1986) | — | 4[(i)] | — | — |
| Graphite (all forms except graphite fibers) | | | | |
| [7782-42-5] (1991) | — | 2[(j)] | — | — |
| Gypsum, *see* Calcium sulfate | | | | |
| Hafnium [7440-58-6] (1986) | — | 0.5 | — | — |
| Halothane [151-67-7] (1988) | 50 | 404 | — | — |
| Helium [7440-59-7] (1981) | —[(c)] | — | — | — |
| *■ Heptachlor [76-44-8] and Heptachlor epoxide | | | | |
| [1024-57-3]—Skin (1994) | — | 0.05,A3 | — | — |
| Heptane [142-82-5] (n-Heptane) (1976) | 400 | 1640 | 500 | 2050 |
| 2-Heptanone, *see* Methyl n-amyl ketone | | | | |
| 3-Heptanone, *see* Ethyl butyl ketone | | | | |
| *■ Hexachlorobenzene [118-74-1]— | | | | |
| Skin | — | 0.025,A3 | — | — |
| ■ Hexachlorobutadiene | | | | |
| [87-68-3]—Skin (1982) | 0.02,A2 | 0.21,A2 | — | — |
| Hexachlorocyclopentadiene | | | | |
| [77-47-4] (1986) | 0.01 | 0.11 | — | — |
| ■ Hexachloroethane | | | | |
| [67-72-1]—Skin (1992) | 1,A2 | 9.7,A2 | — | — |
| Hexachloronaphthalene | | | | |
| [1335-87-1]—Skin (1986) | — | 0.2 | — | — |

## ADOPTED VALUES

| Substance [CAS #] | TWA ppm[a] | TWA mg/m[3b] | STEL/CEILING (C) ppm[a] | STEL/CEILING (C) mg/m[3b] |
|---|---|---|---|---|
| Hexafluoroacetone | | | | |
| [684-16-2]—Skin (1986) | 0.1 | 0.68 | — | — |
| Hexamethylene diisocyanate | | | | |
| [822-06-0] (1988) | 0.005 | 0.034 | — | — |
| ■ Hexamethyl phosphoramide | | | | |
| [680-31-9]—Skin (1978) | A2 | A2 | — | — |
| 1,6-Hexanediamine | | | | |
| [124-09-4] (1992) | 0.5 | 2.3 | — | — |
| ◄ Hexane (n-Hexane) | | | | |
| [110-54-3] (1982) | 50 | 176 | — | — |
| • Other isomers (1982) | 500 | 1760 | 1000 | 3500 |
| 2-Hexanone, *see* Methyl n-butyl ketone | | | | |
| Hexone, *see* Methyl isobutyl ketone | | | | |
| sec-Hexyl acetate | | | | |
| [108-84-9] (1977) | 50 | 295 | — | — |
| Hexylene glycol | | | | |
| [107-41-5] (1977) | — | — | C 25 | C 121 |
| ‡•■ Hydrazine [302-01-2]— | | | | |
| Skin (1977) | (0.1,A2) | (0.13,A2) | — | — |
| Hydrogen [1333-74-0] (1981) | —[(c)] | — | — | — |
| Hydrogenated terphenyls | | | | |
| [61788-32-7] (1977) | 0.5 | 4.9 | — | — |
| Hydrogen bromide | | | | |
| [10035-10-6] (1986) | — | — | C 3 | C 9.9 |
| Hydrogen chloride | | | | |
| [7647-01-0] (1977) | — | — | C 5 | C 7.5 |
| * Hydrogen cyanide and Cyanide salts as CN | | | | |
| *Hydrogen cyanide [74-90-8]— | | | | |
| Skin (1994) | — | — | C 4.7 | C 5 |
| *Calcium cyanide [592-01-8]— | | | | |
| Skin (1994) | — | — | | C 5 |
| *Potassium cyanide [151-50-8]— | | | | |
| Skin (1994) | — | — | | C 5 |
| *Sodium cyanide [143-33-9]— | | | | |
| Skin (1994) | — | — | | C 5 |
| Hydrogen fluoride | | | | |
| [7664-39-3], as F (1986) | — | — | C 3 | C 2.6 |
| Hydrogen peroxide | | | | |
| [7722-84-1] (1986) | 1 | 1.4 | — | — |
| Hydrogen selenide | | | | |
| [7783-07-5], as Se (1977) | 0.05 | 0.16 | — | — |
| • Hydrogen sulfide | | | | |
| [7783-06-4] (1976) | 10 | 14 | 15 | 21 |
| Hydroquinone [123-31-9] (1987) | — | 2 | — | — |
| 4-Hydroxy-4-methyl-2-pentanone, *see* Diacetone alcohol | | | | |
| 2-Hydroxypropyl acrylate | | | | |
| [999-61-1] — Skin (1980) | 0.5 | 2.8 | — | — |
| Indene [95-13-6] (1987) | 10 | 48 | — | — |
| Indium [7440-74-6] & compounds, as In (1986) | — | 0.1 | — | — |
| Iodine [7553-56-2] (1977) | — | — | C 0.1 | C 1.0 |
| Iodoform [75-47-8] (1986) | 0.6 | 10 | — | — |
| Iron oxide dust & fume (Fe$_2$O$_3$) | | | | |
| [1309-37-1], as Fe (1986) | B2 | 5 | — | — |
| Iron pentacarbonyl | | | | |
| [13463-40-6], as Fe (1982) | 0.1 | 0.23 | 0.2 | 0.45 |
| Iron salts, soluble, as Fe (1986) | — | 1 | — | — |
| Isoamyl acetate [123-92-2] (1987) | 100 | 532 | — | — |
| Isoamyl alcohol [123-51-3] (1976) | 100 | 361 | 125 | 452 |
| Isobutyl acetate [110-19-0] (1990) | 150 | 713 | — | — |
| Isobutyl alcohol [78-83-1] (1987) | 50 | 152 | — | — |
| Isooctyl alcohol [26952-21-6]— | | | | |
| Skin (1982) | 50 | 266 | — | — |
| ‡ Isophorone [78-59-1] (1977) | — | — | (C 5) | (C 28) |
| Isophorone diisocyanate | | | | |
| [4098-71-9] (1988) | 0.005 | 0.045 | — | — |
| Isopropoxyethanol | | | | |
| [109-59-1]—Skin (1987) | 25 | 106 | — | — |

| | | ADOPTED VALUES | | | |
|---|---|---|---|---|---|
| | | TWA | | STEL/CEILING (C) | |
| Substance | [CAS #] | ppm[a] | mg/m³[b] | ppm[a] | mg/m³[b] |
| Isopropyl acetate [108-21-4] (1976) | | 250 | 1040 | 310 | 1290 |
| Isopropyl alcohol [67-63-0] (1976) | | 400 | 983 | 500 | 1230 |
| Isopropylamine [75-31-0] (1976) | | 5 | 12 | 10 | 24 |
| N-Isopropylaniline [768-52-5]—Skin (1986) | | 2 | 11 | — | — |
| Isopropyl ether [108-20-3] (1976) | | 250 | 1040 | 310 | 1300 |
| • Isopropyl glycidyl ether (IGE) [4016-14-2] (1976) | | 50 | 238 | 75 | 356 |
| Kaolin [1332-58-7] (1992) | | — | 2 (j) | — | — |
| Ketene [463-51-4] (1976) | | 0.5 | 0.86 | 1.5 | 2.6 |
| ‡◄•■Lead [7439-92-1], inorg. dusts & fumes, as Pb (1986) | | — | (0.15) | — | — |
| Lead arsenate [7784-40-9], as Pb₃(AsO₄)₂ (1985) | | | 0.15 | | |
| •■ Lead chromate [7758-97-6], as Pb (1991) | | — | 0.05,A2 | — | — |
| as Cr (1991) | | — | 0.012,A2 | — | — |
| Limestone, see Calcium carbonate | | | | | |
| ■ Lindane [58-89-9]—Skin (1986) | | — | 0.5 | — | — |
| Lithium hydride [7580-67-8] (1977) | | — | 0.025 | — | — |
| L.P.G. (Liquified petroleum gas) [68476-85-7] (1987) | | 1000 | 1800 | — | — |
| Magnesite [546-93-0] (1986) | | — | 10(e) | — | — |
| Magnesium oxide fume [1309-48-4] (1977) | | — | 10 | — | — |
| ◄ Malathion [121-75-5]— Skin (1977) | | — | 10 | — | — |
| Maleic anhydride [108-31-6] (1977) | | 0.25 | 1.0 | — | — |
| ‡ Manganese [7439-96-5], as Mn | | | | | |
| ‡•Dust & compounds (1988)— | | — | (5) | — | — |
| † Fume (1979) | | — | (1) | — | (3) |
| Manganese cyclopentadienyl tricarbonyl [12079-65-1], as Mn—Skin (1986) | | — | 0.1 | — | — |
| Marble, see Calcium carbonate | | | | | |
| *◄ Mercury [7439-97-6], as Hg—Skin | | | | | |
| Alkyl compounds (1980) | | — | 0.01 | — | 0.03 |
| *Aryl compounds (1982) | | — | 0.1 | | |
| *Inorganic forms including metallic mercury (1994) | | — | 0.025,A4 | — | — |
| • Mesityl oxide [141-79-7] (1981) | | 15 | 60 | 25 | 100 |
| Methacrylic acid [79-41-4] (1981) | | 20 | 70 | — | — |
| Methane [74-82-8] (1981) | | —(c) | — | — | — |
| Methanethiol, see Methyl mercaptan | | | | | |
| ◄ Methanol [67-56-1]— Skin (1976) | | 200 | 262 | 250 | 328 |
| ◄ Methomyl [16752-77-5] (1977) | | — | 2.5 | — | — |
| •■ Methoxychlor [72-43-5] (1977) | | — | 10 | — | — |
| • 2-Methoxyethanol (EGME) [109-86-4]—Skin (1984) | | 5 | 16 | — | — |
| • 2-Methoxyethyl acetate (EGMEA) [110-49-6]—Skin (1984) | | 5 | 24 | — | — |
| 4-Methoxyphenol [150-76-5] (1982) | | — | 5 | — | — |
| Methyl acetate [79-20-9] (1976) | | 200 | 606 | 250 | 757 |
| Methyl acetylene [74-99-7] (1990) | | 1000 | 1640 | — | — |
| Methyl acetylene-propadiene mixture (MAPP) (1976) | | 1000 | 1640 | 1250 | 2050 |
| ‡ Methyl acrylate [96-33-3]— Skin (1977) | | (10) | (35) | — | — |
| Methylacrylonitrile [126-98-7]—Skin (1986) | | 1 | 2.7 | — | — |
| Methylal [109-87-5] (1987) | | 1000 | 3110 | — | — |

| | | ADOPTED VALUES | | | |
|---|---|---|---|---|---|
| | | TWA | | STEL/CEILING (C) | |
| Substance | [CAS #] | ppm[a] | mg/m³[b] | ppm[a] | mg/m³[b] |
| Methyl alcohol, see Methanol | | | | | |
| Methylamine [74-89-5] (1992) | | 5 | 6.4 | 15 | 19 |
| Methyl amyl alcohol, see Methyl isobutyl carbinol | | | | | |
| Methyl n-amyl ketone [110-43-0] (1987) | | 50 | 233 | — | — |
| ◄ N -Methyl aniline [100-61-8]— Skin (1986) | | 0.5 | 2.2 | — | — |
| •■ Methyl bromide [74-83-9]— Skin (1986) | | 5 | 19 | — | — |
| ‡*■ Methyl-tert butyl ether [1634-04-4] (1994) | | (40) | (144) | — | — |
| • Methyl n-butyl ketone [591-78-6]—Skin (1981) | | 5 | 20 | — | — |
| •■ Methyl chloride [74-87-3]— Skin (1981) | | 50 | 103 | 100 | 207 |
| •◄ Methyl chloroform [71-55-6] (1976) | | 350 | 1910 | 450 | 2460 |
| Methyl 2-cyanoacrylate [137-05-3] (1976) | | 2 | 9.1 | 4 | 18 |
| Methylcyclohexane [108-87-2] (1987) | | 400 | 1610 | — | — |
| Methylcyclohexanol [25639-42-3] (1987) | | 50 | 234 | — | — |
| o-Methylcyclohexanone [583-60-8]—Skin (1976) | | 50 | 229 | 75 | 344 |
| 2-Methylcyclopentadienyl manganese tricarbonyl [12108-13-3], as Mn —Skin (1986) | | — | 0.2 | — | — |
| ◄ Methyl demeton [8022-00-2]— Skin (1986) | | — | 0.5 | — | — |
| Methylene bisphenyl isocyanate (MDI) [101-68-8] (1988) | | 0.005 | 0.051 | — | — |
| •■ Methylene chloride (Dichloromethane) [75-09-2] (1988) | | 50,A2 | 174,A2 | — | — |
| ◄•■4,4'-Methylene bis (2-chloroaniline) [MOCA] [101-14-4]—Skin (1993) | | 0.01,A2 | 0.11,A2 | — | — |
| Methylene bis(4-cyclo-hexylisocyanate) [5124-30-1] (1988) | | 0.005 | 0.054 | — | — |
| ■ 4,4'-Methylene dianiline [101-77-9]—Skin (1986) | | 0.1,A2 | 0.81,A2 | — | — |
| ◄ Methyl ethyl ketone (MEK) [78-93-3] (1976) | | 200 | 590 | 300 | 885 |
| Methyl ethyl ketone peroxide [1338-23-4] (1977) | | — | — | C 0.2 | C 1.5 |
| Methyl formate [107-31-3] (1976) | | 100 | 246 | 150 | 368 |
| 5-Methyl-3-heptanone, see Ethyl amyl ketone | | | | | |
| ‡•■ Methyl hydrazine [60-34-4]—Skin (1976) | | — | — | (C 0.2,A2) | (C 0.38,A2) |
| ■ Methyl iodide [74-88-4]— Skin (1986) | | 2,A2 | 12,A2 | — | — |
| Methyl isoamyl ketone [110-12-3] (1982) | | 50 | 234 | — | — |
| Methyl isobutyl carbinol [108-11-2]—Skin (1976) | | 25 | 104 | 40 | 167 |
| ◄ Methyl isobutyl ketone [108-10-1] (1981) | | 50 | 205 | 75 | 307 |
| Methyl isocyanate [624-83-9]—Skin (1977) | | 0.02 | 0.047 | — | — |
| Methyl isopropyl ketone [563-80-4] (1981) | | 200 | 705 | — | — |
| • Methyl mercaptan [74-93-1] (1977) | | 0.5 | 0.98 | — | — |
| Methyl methacrylate [80-62-6] (1987) | | 100 | 410 | — | — |
| ◄ Methyl parathion [298-00-0]—Skin (1986) | | — | 0.2 | — | — |
| • Methyl propyl ketone [107-87-9] (1976) | | 200 | 705 | 250 | 881 |

| Substance | [CAS #] | TWA ppm^a | TWA mg/m^3b | STEL/CEILING (C) ppm^a | STEL/CEILING (C) mg/m^3b |
|---|---|---|---|---|---|
| Methyl silicate [681-84-5] (1986) | | 1 | 6 | — | — |
| α-Methyl styrene [98-83-9] (1981) | | 50 | 242 | 100 | 483 |
| Metribuzin [21087-64-9] (1984) | | — | 5 | — | — |
| ◄ Mevinphos [7786-34-7]— Skin (1976) | | 0.01 | 0.092 | 0.03 | 0.27 |
| Mica [12001-26-2] (1986) | | — | 3^(j) | — | — |
| Mineral wool fiber (1974) | | — | 10^(e) | — | — |
| Molybdenum [7439-98-7], as Mo | | | | | |
|   Soluble compounds (1986) | | — | 5 | — | — |
|   Insoluble compounds (1986) | | — | 10 | — | — |
| Monochlorobenzene, see Chlorobenzene | | | | | |
| Monocrotophos [6923-22-4]—Skin (1977) | | — | 0.25 | — | — |
| Morpholine [110-91-8]— Skin (1991) | | 20 | 71 | — | — |
| ◄ Naled [300-76-5]—Skin (1986) | | — | 3 | — | — |
| Naphthalene [91-20-3] (1976) | | 10 | 52 | 15 | 79 |
| ■ β-Naphthylamine [91-59-8] (1972) | | — | A1 | — | — |
| Neon [7440-01-9] (1981) | | — (c) | — | — | — |
| ‡■ Nickel [7440-02-0] | | | | | |
|   ‡Metal (1966) | | — | (1) | — | — |
|   ‡Insoluble compounds, as Ni (1974) | | — | (1) | — | — |
|   ‡■Soluble compounds, as Ni (1976) | | — | (0.1) | — | — |
| ‡•■ Nickel carbonyl [13463-39-3], as Ni (1977) | | (0.05) | (0.12) | — | — |
| ‡■ Nickel sulfide roasting, fume & dust, as Ni (1978) | | — | (1,A1) | — | — |
| Nicotine [54-11-5]— Skin (1986) | | — | 0.5 | — | — |
| Nitrapyrin [1929-82-4] (1982) | | — | 10 | — | 20 |
| Nitric acid [7697-37-2] (1976) | | 2 | 5.2 | 4 | 10 |
| ◄ Nitric oxide [10102-43-9] (1986) | | 25 | 31 | — | — |
| ◄ p-Nitroaniline [100-01-6]— Skin (1982) | | — | 3 | — | — |
| ◄ Nitrobenzene [98-95-3]— Skin (1986) | | 1 | 5 | — | — |
| ◄■ p-Nitrochlorobenzene [100-00-5]—Skin (1988) | | 0.1 | 0.64 | — | — |
| ■ 4-Nitrodiphenyl [92-93-3]— Skin (1976) | | — | A1 | — | — |
| Nitroethane [79-24-3] (1986) | | 100 | 307 | — | — |
| Nitrogen [7727-37-9] (1989) | | — (c) | — | — | — |
| • Nitrogen dioxide [10102-44-0] (1981) | | 3 | 5.6 | 5 | 9.4 |
| ◄ Nitrogen trifluoride [7783-54-2] (1986) | | 10 | 29 | — | — |
| • Nitroglycerin (NG) [55-63-00]—Skin (1985) | | 0.05 | 0.46 | — | — |
| * Nitromethane [75-52-5] (1994) | | 20 | 50 | — | — |
| 1-Nitropropane [108-03-2] (1986) | | 25 | 91 | — | — |
| ■ 2-Nitropropane [79-46-9] (1987) | | 10,A2 | 36,A2 | — | — |
| ■ N-Nitrosodimethylamine [62-75-9]—Skin (1972) | | — | A2 | — | — |
| ◄ Nitrotoluene [88-72-2; 99-08-1; 99-99-0] —Skin (1982) | | 2 | 11 | — | — |
| Nitrotrichloromethane, see Chloropicrin | | | | | |
| Nitrous oxide [10024-97-2] (1989) | | 50 | 90 | — | — |
| Nonane [111-84-2], all isomers (1976) | | 200 | 1050 | — | — |
| Nuisance particulates, see Particulates Not Otherwise Classified (PNOC) | | | | | |
| Octachloronaphthalene [2234-13-1]—Skin (1976) | | — | 0.1 | — | 0.3 |
| Octane [111-65-9] (1976) | | 300 | 1400 | 375 | 1750 |
| ‡ Oil Mist, mineral (1976) | | — | 5^(k) | — | (10) |
| Osmium tetroxide [20816-12-0], as Os (1976) | | 0.0002 | 0.0016 | 0.0006 | 0.0047 |
| Oxalic acid [144-62-7] (1976) | | — | 1 | — | 2 |
| Oxygen difluoride [7783-41-7] (1986) | | — | — | C 0.05 | C 0.11 |
| ‡ Ozone [10028-15-6] (1989) | | (—) | (—) | (C 0.1) | (C 0.20) |
| Paraffin wax fume [8002-74-2] (1987) | | — | 2 | — | — |
| Paraquat [4685-14-7], total dust (1978) | | — | 0.5 | — | — |
|   respirable fraction (1978) | | — | 0.1 | — | — |
| ◄• Parathion [56-38-2]— Skin (1986) | | — | 0.1 | — | — |
| Particulate polycyclic aromatic hydrocarbons (PPAH), see Coal tar pitch volatiles | | | | | |
| ‡ Particulates Not Otherwise Classified (PNOC) (1989) | | — | 10^(e) | — | — |
| Pentaborane [19624-22-7] (1976) | | 0.005 | 0.013 | 0.015 | 0.039 |
| Pentachloronaphthalene [1321-64-8]—Skin (1986) | | — | 0.5 | — | — |
| Pentachloronitrobenzene [82-68-8] (1991) | | — | 0.5 | — | — |
| ◄■ Pentachlorophenol [87-86-5]—Skin (1986) | | — | 0.5 | — | — |
| Pentaerythritol [115-77-5] (1986) | | — | 10 | — | — |
| • Pentane [109-66-0] (1976) | | 600 | 1770 | 750 | 2210 |
| 2-Pentanone, see Methyl propyl ketone | | | | | |
| ◄•■ Perchloroethylene (Tetrachloroethylene) [127-18-4] (1993) | | 25,A3 | 170,A3 | 100,A3 | 685,A3 |
| Perchloromethyl mercaptan [594-42-3] (1977) | | 0.1 | 0.76 | — | — |
| Perchloryl fluoride [7616-94-6] (1976) | | 3 | 13 | 6 | 25 |
| Perfluoroisobutylene [382-21-8] (1992) | | — | — | C 0.01 | C 0.082 |
| Precipitated silica, see Silica—Amorphous | | | | | |
| Perlite [93763-70-3] (1986) | | — | 10^(e) | — | — |
| Petroleum distillates, see Gasoline; Stoddard solvent; VM&P naphtha | | | | | |
| Phenacyl chloride, see α-Chloroacetophenone | | | | | |
| ◄ Phenol [108-95-2]—Skin (1987) | | 5 | 19 | — | — |
| Phenothiazine [92-84-2]— Skin (1986) | | — | 5 | — | — |
| ■ N-Phenyl-beta-naphthylamine [135-88-6] (1979) | | A2 | A2 | — | — |
| o-Phenylenediamine [95-54-5] (1991) | | — | 0.1,A2 | — | — |
| m-Phenylenediamine [108-45-2] (1991) | | — | 0.1 | — | — |
| p-Phenylenediamine [106-50-3] (1991) | | — | 0.1 | — | — |
| Phenyl ether [101-84-8], vapor (1976) | | 1 | 7 | 2 | 14 |
| Phenylethylene, see Styrene, monomer | | | | | |
| *•■ Phenyl glycidyl ether (PGE) [122-60-1]—Skin (1994) | | 0.1,A3 | 0.6,A3 | — | — |
| •■ Phenylhydrazine [100-63-0]—Skin (1991) | | 0.1,A2 | 0.44,A 2 | — | — |
| • Phenyl mercaptan [108-98-5] (1978) | | 0.5 | 2.3 | — | — |
| Phenylphosphine [638-21-1] (1977) | | — | — | C 0.05 | C 0.23 |
| Phorate [298-02-2]—Skin (1976) | | — | 0.05 | — | 0.2 |

| Substance | [CAS #] | ADOPTED VALUES TWA ppm[a] | TWA mg/m³[b] | STEL/CEILING (C) ppm[a] | STEL/CEILING (C) mg/m³[b] |
|---|---|---|---|---|---|
| Phosdrin, *see* Mevinphos | | | | | |
| Phosgene [75-44-5] (1978)........ | | 0.1 | 0.40 | — | — |
| Phosphine [7803-51-2] (1976) ... | | 0.3 | 0.42 | 1 | 1.4 |
| Phosphoric acid [7664-38-2] (1976) | | — | 1 | — | 3 |
| Phosphorus (yellow) [7723-14-0] (1986) | | 0.02 | 0.1 | — | — |
| Phosphorus oxychloride [10025-87-3] (1990) | | 0.1 | 0.63 | — | — |
| Phosphorus pentachloride [10026-13-8] (1980) | | 0.1 | 0.85 | — | — |
| Phosphorus pentasulfide [1314-80-3] (1976) | | — | 1 | — | 3 |
| Phosphorus trichloride [7719-12-2] (1982) | | 0.2 | 1.1 | 0.5 | 2.8 |
| Phthalic anhydride [85-44-9] (1987) | | 1 | 6.1 | — | — |
| m-Phthalodinitrile [626-17-5] (1977) | | — | 5 | — | — |
| Picloram [1918-02-1] (1990) | | — | 10 | — | — |
| Picric acid [88-89-1] (1990) | | — | 0.1 | — | — |
| Pindone [83-26-1] (1987) | | — | 0.1 | — | — |
| Piperazine dihydrochloride [142-64-3] (1982) | | — | 5 | — | — |
| 2-Pivalyl-1,3-indandione, *see* Pindone | | | | | |
| Plaster of Paris, *see* Calcium sulfate | | | | | |
| Platinum [7440-06-4] | | | | | |
|   Metal (1981) | | — | 1 | — | — |
|   Soluble salts, as Pt (1970).... | | — | 0.002 | — | — |
| Polychlorobiphenyls, *see* Chlorodiphenyls | | | | | |
| Polytetrafluoroethylene decomposition products (1972) | | — | B1 | — | — |
| Portland cement [65997-15-1] (1986) | | — | 10(e) | — | — |
| Potassium hydroxide [1310-58-3] (1977) | | — | — | — | C 2 |
| • Propane [74-98-0] (1981) | | —(c) | | | |
| ■ Propane sultone [1120-71-4] (1977) | | A2 | A2 | — | — |
| Propargyl alcohol [107-19-7]—Skin (1987) | | 1 | 2.3 | — | — |
| ■ β-Propiolactone [57-57-8] (1987) | | 0.5,A2 | 1.5,A2 | — | — |
| Propionic acid [79-09-4] (1990) | | 10 | 30 | — | — |
| Propoxur [114-26-1] (1987) | | — | 0.5 | — | — |
| n-Propyl acetate [109-60-4] (1976) | | 200 | 835 | 250 | 1040 |
| n-Propyl alcohol [71-23-8]—Skin (1976) | | 200 | 492 | 250 | 614 |
| Propylene [115-07-1] (1976) | | —(c) | — | — | — |
| ■ Propylene dichloride [78-87-5] (1976) | | 75 | 347 | 110 | 508 |
| ◄ Propylene glycol dinitrate [6423-43-4]—Skin (1985) | | 0.05 | 0.34 | — | — |
| Propylene glycol monomethyl ether [107-98-2] (1976) | | 100 | 369 | 150 | 553 |
| ■ Propylene imine [75-55-8]—Skin (1983) | | 2,A2 | 4.7,A2 | — | — |
| ■ Propylene oxide [75-56-9] (1981) | | 20 | 48 | — | — |
| ◄ n-Propyl nitrate [627-13-4] (1978) | | 25 | 107 | 40 | 172 |
| Propyne, *see* Methyl acetylene | | | | | |
| Pyrethrum [8003-34-7] (1981) | | — | 5 | — | — |
| Pyridine [110-86-1] (1987) | | 5 | 16 | — | — |
| Pyrocatechol, *see* Catechol | | | | | |
| Quartz, *see* Silica—Crystalline | | | | | |
| Quinone [106-51-4] (1987) | | 0.1 | 0.44 | — | — |
| RDX, *see* Cyclonite | | | | | |
| Resorcinol [108-46-3] (1976) | | 10 | 45 | 20 | 90 |
| • Rhodium [7440-16-6] | | | | | |
|   •Metal (1982) | | — | 1 | — | — |
|   •Insoluble compounds, as Rh (1984) | | — | 1 | — | — |

| Substance | [CAS #] | ADOPTED VALUES TWA ppm[a] | TWA mg/m³[b] | STEL/CEILING (C) ppm[a] | STEL/CEILING (C) mg/m³[b] |
|---|---|---|---|---|---|
|   • Soluble compounds, as Rh (1984) | | — | 0.01 | — | — |
| Ronnel [299-84-3] (1977) | | — | 10 | — | — |
| Rosin core solder thermal decomposition products, as resin acids—colophony [8050-09-7] (1993) | | Sensitizer; reduce exposure to as low as possible | | | |
| Rotenone (commercial) [83-79-4] (1987) | | — | 5 | — | — |
| Rouge (1986) | | — | 10(e) | — | — |
| • Rubber solvent (Naphtha) [8030-30-6] (1977) | | 400 | 1590 | — | — |
| Selenium [7782-49-2] and compounds, as Se (1977) | | — | 0.2 | — | — |
| Selenium hexafluoride [7783-79-1], as Se (1979) | | 0.05 | 0.16 | — | — |
| Sesone [136-78-7] (1986) | | — | 10 | — | — |
| Silane, *see* Silicon tetrahydride | | | | | |
| Silica—Amorphous | | | | | |
| ‡• Diatomaceous earth (uncalcined) [61790-53-2] (1986) | | — | 10(e) | — | — |
| • Precipitated silica [112926-00-8] (1987) | | — | 10 | — | — |
| Silica, fume [69012-64-2] (1992) | | — | 2(j) | — | — |
| ■■Silica, fused [60676-86-0] (1992) | | — | 0.1(j) | — | — |
| • Silica gel [112926-00-8] (1987) | | — | 10 | — | — |
| ■ Silica—Crystalline | | | | | |
|   Cristobalite [14464-46-1] (1986) | | — | 0.05(j) | — | — |
| •  Quartz [14808-60-7] (1986) | | — | 0.1(j) | — | — |
|   Tridymite [15468-32-3](1986).. | | — | 0.05(j) | — | — |
| •  Tripoli [1317-95-9] (1985) | | — | 0.1,(j) of contained respirable quartz | — | — |
| Silicon [7440-21-3] (1986) | | — | 10(e) | — | — |
| Silicon carbide [409-21-2] (1986) | | — | 10(e) | — | — |
| Silicon tetrahydride [7803-62-5] (1983) | | 5 | 6.6 | — | — |
| Silver [7440-22-4] | | | | | |
| •  Metal (1981) | | — | 0.1 | — | — |
|   Soluble compounds, as Ag (1981) | | — | 0.01 | — | — |
| Soapstone | | | | | |
|   Respirable dust (1985) | | — | 3(j) | — | — |
|   Inhalable dust (1985) | | — | 6(e) | — | — |
| Sodium azide [26628-22-8] (1977) | | | | | |
|   as Sodium azide | | — | — | — | C 0.29 |
|   as Hydrazoic acid vapor | | — | — | C 0.11 | — |
| Sodium bisulfite [7631-90-5] (1980) | | — | 5 | — | — |
| Sodium 2,4-dichloro-phenoxyethyl sulfate, *see* Sesone | | | | | |
| * Sodium fluoroacetate [62-74-8] — Skin (1994) | | — | 0.05 | — | — |
| Sodium hydroxide [1310-73-2] (1977) | | — | — | — | C 2 |
| Sodium metabisulfite [7681-57-4] (1980) | | — | 5 | — | — |
| Sodium perfluoroacetate, *see* Sodium fluoroacetate | | | | | |
| Starch [9005-25-8] (1986) | | — | 10 | — | — |
| Stearates[l] (1988) | | — | 10 | — | — |
| Stibine [7803-52-3] (1986) | | 0.1 | 0.51 | — | — |
| • Stoddard solvent [8052-41-3] (1987) | | 100 | 525 | — | — |
| Strontium chromate [7789-06-2], as Cr (1992) | | — | 0.0005,A2 | — | — |
| Strychnine [57-24-9] (1986) | | — | 0.15 | — | — |
| ■◄ Styrene, monomer [100-42-5] — Skin (1981) | | 50 | 213 | 100 | 426 |
| Subtilisins [1395-21-7; 9014-01-1] (Proteolytic enzymes as 100% pure crystalline enzyme) (1977) | | — | — | — | C 0.00006(m) |

| Substance [CAS #] | TWA ppm[a] | TWA mg/m³[b] | STEL/CEILING (C) ppm[a] | STEL/CEILING (C) mg/m³[b] |
|---|---|---|---|---|
| Sucrose [57-50-1] (1986) | — | 10 | — | — |
| * Sulfometuron methyl [74222-97-2] (1994) | — | 5,A4 | — | — |
| ◄ Sulfotep [3689-24-5]— Skin (1986) | — | 0.2 | — | — |
| Sulfur dioxide [7446-09-5] (1986) | 2 | 5.2 | 5 | 13 |
| Sulfur hexafluoride [2551-62-4] (1986) | 1000 | 5970 | — | — |
| Sulfuric acid [7664-93-9] (1989) | — | 1 | — | 3 |
| Sulfur monochloride [10025-67-9] (1986) | — | — | C 1 | C 5.5 |
| Sulfur pentafluoride [5714-22-7] (1986) | — | — | C 0.01 | C 0.10 |
| Sulfur tetrafluoride [7783-60-0] (1986) | — | — | C 0.1 | C 0.44 |
| Sulfuryl fluoride [2699-79-8] (1976) | 5 | 21 | 10 | 42 |
| Sulprofos [35400-43-2] (1984) | — | 1 | — | — |
| Systox, see Demeton | | | | |
| 2,4,5-T [93-76-5] (1986) | — | 10 | — | — |
| Talc (containing no asbestos fibers) [14807-96-6] (1983) | — | 2[j] | | |
| ■ Talc (containing asbestos fibers) (1985) | Use asbestos TLV–TWA[n] | | | |
| Tantalum [7440-25-7], metal and oxide [1314-61-0] dusts, as Ta (1988) | — | 5 | — | — |
| TEDP, see Sulfotep | | | | |
| Tellurium [13494-80-9] and compounds, as Te (1977) | — | 0.1 | — | — |
| Tellurium hexafluoride [7783-80-4], as Te (1977) | 0.02 | 0.10 | — | — |
| ◄ Temephos [3383-96-8] (1986) | — | 10 | — | — |
| Terephthalic acid [100-21-0] (1993) | — | 10 | — | — |
| ◄ TEPP [107-49-3]—Skin (1986) | 0.004 | 0.047 | — | — |
| Terphenyls [26140-60-3] (1980) | — | — | C 0.53 | C 5 |
| 1,1,1,2-Tetrachloro-2,2-difluoroethane [76-11-9] (1986) | 500 | 4170 | — | — |
| 1,1,2,2-Tetrachloro-1,2-difluoroethane [76-12-0] (1986) | 500 | 4170 | — | — |
| ■ 1,1,2,2-Tetrachloroethane [79-34-5]—Skin (1986) | 1 | 6.9 | — | — |
| Tetrachloroethylene, see Perchloroethylene | | | | |
| Tetrachloromethane, see Carbon tetrachloride | | | | |
| Tetrachloronaphthalene [1335-88-2] (1986) | — | 2 | — | — |
| ● Tetraethyl lead [78-00-2], as Pb—Skin (1986) | — | 0.1[o] | — | — |
| Tetrahydrofuran [109-99-9] (1976) | 200 | 590 | 250 | 737 |
| ● Tetramethyl lead [75-74-1], as Pb — Skin (1986) | — | 0.15[o] | — | — |
| Tetramethyl succinonitrile [3333-52-6]—Skin (1986) | 0.5 | 2.8 | — | — |
| Tetranitromethane [509-14-8] (1993) | 0.005,A2 | 0.04,A2 | — | — |
| Tetrasodium pyrophosphate [7722-88-5] (1980) | — | 5 | — | — |
| ● Tetryl [479-45-8] (1986) | — | 1.5 | — | — |
| Thallium, elemental [7440-28-0], and soluble compounds, as Tl—Skin (1977) | — | 0.1 | — | — |
| 4,4'-Thiobis(6-tert-butyl-m-cresol) [96-69-5] (1986) | — | 10 | — | — |
| Thioglycolic acid [68-11-1]—Skin (1978) | 1 | 3.8 | — | — |
| Thionyl chloride [7719-09-7] (1986) | — | — | C 1 | C 4.9 |

| Substance [CAS #] | TWA ppm[a] | TWA mg/m³[b] | STEL/CEILING (C) ppm[a] | STEL/CEILING (C) mg/m³[b] |
|---|---|---|---|---|
| Thiram [137-26-8] (1990) | — | 1 | — | — |
| Tin [7440-31-5] | | | | |
| Metal (1982) | — | 2 | — | — |
| Oxide & inorganic compounds, except SnH₄, as Sn (1982) | — | 2 | — | — |
| Organic compounds, as Sn—Skin (1992) | — | 0.1 | — | 0.2 |
| ■ Titanium dioxide [13463-67-7] (1986) | — | 10 | — | — |
| ■ o-Tolidine [119-93-7]— Skin (1982) | A2 | A2 | — | — |
| Toluene [108-88-3]— Skin (1992) | 50 | 188 | — | — |
| ■ Toluene-2,4-diisocyanate (TDI) [584-84-9] (1983) | 0.005 | 0.036 | 0.02 | 0.14 |
| ◄■ o-Toluidine [95-53-4]— Skin (1984) | 2,A2 | 8.8,A2 | — | — |
| ◄ m-Toluidine [108-44-1]— Skin (1986) | 2 | 8.8 | — | — |
| ◄■ p-Toluidine [106-49-0]— Skin (1986) | 2,A2 | 8.8,A2 | — | — |
| Toluol, see Toluene | | | | |
| Toxaphene, see Chlorinated camphene | | | | |
| Tributyl phosphate [126-73-8] (1986) | 0.2 | 2.2 | — | — |
| Trichloroacetic acid [76-03-9] (1980) | 1 | 6.7 | — | — |
| 1,2,4-Trichlorobenzene [120-82-1] (1978) | — | — | C 5 | C 37 |
| 1,1,1-Trichloroethane, see Methyl chloroform | | | | |
| ■ 1,1,2-Trichloroethane [79-00-5]—Skin (1986) | 10 | 55 | — | — |
| ◄■ Trichloroethylene [79-01-6] (1993) | 50,A5 | 269,A5 | 100,A5 | 537,A5 |
| Trichlorofluoromethane [75-69-4] (1982) | — | — | C 1000 | C 5620 |
| Trichloromethane, see Chloroform | | | | |
| Trichloronaphthalene [1321-65-9]—Skin (1986) | — | 5 | — | — |
| Trichloronitromethane, see Chloropicrin | | | | |
| ■ 1,2,3-Trichloropropane [96-18-4]—Skin (1987) | 10 | 60 | — | — |
| 1,1,2-Trichloro-1,2,2-trifluoroethane [76-13-1] (1976) | 1000 | 7670 | 1250 | 9590 |
| Tricyclohexyltin hydroxide, see Cyhexatin | | | | |
| Tridymite, see Silica—Crystalline | | | | |
| Triethanolamine [102-71-6] (1993) | — | 5 | — | — |
| ‡* Triethylamine [121-44-8]— Skin (1994) | (1) | (4.1) | (5) | (20.7) |
| Trifluorobromomethane [75-63-8] (1986) | 1000 | 6090 | — | — |
| Trimellitic anhydride [552-30-7] (1993) | — | — | — | C 0.04 |
| Trimethylamine [75-50-3] (1992) | 5 | 12 | 15 | 36 |
| Trimethyl benzene [25551-13-7] (1987) | 25 | 123 | — | — |
| Trimethyl phosphite [121-45-9] (1986) | 2 | 10 | — | — |
| 2,4,6-Trinitrophenol, see Picric acid | | | | |
| 2,4,6-Trinitrophenylmethylnitramine, see Tetryl | | | | |
| ■ 2,4,6-Trinitrotoluene (TNT) [118-96-7]—Skin (1986) | — | 0.5 | — | — |
| Triorthocresyl phosphate [78-30-8]—Skin (1986) | — | 0.1 | — | — |
| Triphenyl amine [603-34-9] (1980) | — | 5 | — | — |
| Triphenyl phosphate [115-86-6] (1986) | — | 3 | — | — |

| Substance | [CAS #] | TWA ppm[a] | TWA mg/m³[b] | STEL/CEILING (C) ppm[a] | STEL/CEILING (C) mg/m³[b] |
|---|---|---|---|---|---|
| ADOPTED VALUES | | | | | |
| Tripoli, see Silica—Crystalline | | | | | |
| Tungsten [7440-33-7], as W | | | | | |
|   Insoluble compounds (1976) . | | — | 5 | — | 10 |
|   Soluble compounds (1976) | | — | 1 | — | 3 |
| Turpentine [8006-64-2] (1987) | | 100 | 556 | — | — |
| •▪ Uranium (natural) [7440-61-1] | | | | | |
|   Soluble & insoluble | | | | | |
|     compounds, as U (1976) | | — | 0.2 | — | 0.6 |
| n-Valeraldehyde | | | | | |
|   [110-62-3] (1978) | | 50 | 176 | — | — |
| Vanadium pentoxide [1314-62-1], | | | | | |
|   as V₂O₅; respirable | | | | | |
|   dust or fume (1982) | | — | 0.05 | — | — |
| Vegetable oil mists[p] (1972) | | — | 10 | — | — |
| • Vinyl acetate | | | | | |
|   [108-05-4] (1993) | | 10,A3 | 35,A3 | 15,A3 | 53,A3 |
| Vinyl benzene, see Styrene | | | | | |
| •▪ Vinyl bromide [593-60-2] (1980) | | 5,A2 | 22,A2 | — | — |
| •▪ Vinyl chloride [75-01-4] (1980) | | 5,A1 | 13,A1 | — | — |
| Vinyl cyanide, see Acrylonitrile | | | | | |
| 4-Vinyl cyclohexene | | | | | |
|   [100-40-3] (1992) | | 0.1,A2 | 0.4,A2 | — | — |
| ▪ Vinyl cyclohexene dioxide | | | | | |
|   [106-87-6]—Skin (1977) | | 10, A2 | 57,A2 | — | — |
| •▪ Vinylidene chloride | | | | | |
|   [75-35-4] (1984) | | 5 | 20 | 20 | 79 |
| Vinyl toluene [25013-15-4] (1981). | | 50 | 242 | 100 | 483 |
| VM & P Naphtha | | | | | |
|   [8032-32-4] (1987) | | 300 | 1370 | — | — |
| Warfarin [81-81-2] (1987) | | — | 0.1 | — | — |
| ▪ Welding fumes | | | | | |
|   (NOC[d]) (1977) | | — | 5,B2 | — | — |
| ▪ Wood dust (certain hard woods as | | | | | |
|   beech & oak) (1981) | | — | 1 | — | — |
|   ▪ Soft wood (1981) | | — | 5 | — | 10 |
| ◄ Xylene [1330-20-7; 95-47-6; 108-38-3; | | | | | |
|   106-42-3] (o-, m-, p-isomers) | | | | | |
|   (1976) | | 100 | 434 | 150 | 651 |
| m-Xylene α,α'-diamine | | | | | |
|   [1477-55-0]—Skin (1977) | | — | — | — | C 0.1 |
| ◄ Xylidine (mixed isomers) | | | | | |
|   [1300-73-8]—Skin (1990) | | 0.5,A2 | 2.5,A2 | — | — |
| Yttrium [7440-65-5] metal & | | | | | |
|   compounds, as Y (1988) | | — | 1 | — | — |
| Zinc chloride fume | | | | | |
|   [7646-85-7] (1976) | | — | 1 | — | 2 |
| •▪ Zinc chromates [13530-65-9; 11103-86-9; | | | | | |
|   37300-23-5], as Cr (1988) | | — | 0.01,A1 | — | — |
| Zinc oxide [1314-13-2] | | | | | |
|   Fume (1976) | | — | 5 | — | 10 |
|   Dust (1976) | | — | 10[e] | — | — |
| Zirconium [7440-67-7] and compounds, | | | | | |
|   as Zr (1976) | | — | 5 | — | 10 |

## NOTICE OF INTENDED CHANGES
## (for 1994–1995)

These substances, with their corresponding values, comprise those for which either a limit has been proposed for the first time, for which a change in the "Adopted" listing has been proposed, or for which retention on the Notice of Intended Changes has been proposed. In all cases, the proposed limits should be considered trial limits that will remain in the listing for a period of at least one year. If, after one year no evidence comes to light that questions the appropriateness of the values herein, the values will be reconsidered for the "Adopted" list. Documentation is available for each of these substances and their proposed values.

| Substance | [CAS #] | TWA ppm[a] | TWA mg/m³[b] | STEL/CEILING (C) ppm[a] | STEL/CEILING (C) mg/m³[b] |
|---|---|---|---|---|---|
| Asbestos, all forms [1332-21-4] | | 0.2 f/cc,[f]A1 | | — | — |
| † Benzene [71-43-2]—Skin | | 0.3,A1 | 0.96,A1 | — | — |
| † Benzoyl chloride [98-88-4] | | — | — | C 0.5 | C 2.8 |
| † Benzyl acetate [140-11-4] | | 10,A4 | 61,A4 | — | — |
| † tert-Butanol [75-65-0] | | 100,A4 | 303,A4 | — | — |
| n-Butyl acetate [123-86-4] | | 20 | 95 | — | — |
| † Dichloroacetylene [7572-29-4] | | — | — | C 0.1,A3 | C 0.39,A3 |
| Dimethylethoxysilane [14857-34-2] | | 0.5 | 2.1 | 1.5 | 6.4 |
| † 1,1-Dimethylhydrazine | | | | | |
|   [57-14-7]—Skin | | 0.01,A3 | 0.025,A3 | — | — |
| Epichlorohydrin | | | | | |
|   [106-89-8]—Skin | | 0.1,A2 | 0.38,A2 | — | — |
| † Ethyl chloride [75-00-3]—Skin . | | 100,A3 | 264,A3 | — | — |
| † Ethylene glycol [107-21-1], | | | | | |
|   as an aerosol | | — | — | C 39.4,A4 | C 100,A4 |
| † Hydrazine [302-01-2]—Skin | | 0.01,A3 | 0.013,A3 | — | — |
| † Isophorone [78-59-1] | | — | — | C 5,A3 | C 28,A3 |
| Lead, elemental [7439-92-1], | | | | | |
|   and inorganic compounds, | | | | | |
|   as Pb☆ | | — | 0.05,A3 | — | — |
| Manganese, elemental | | | | | |
|   [7439-96-5],and inorganic | | | | | |
|   compounds, as Mn | | — | 0.2 | — | — |
| † Methyl acrylate [96-33-3]— | | | | | |
|   Skin | | 2,A4 | 7,A4 | — | — |
| † Methyl-tert butyl ether | | | | | |
|   [1634-04-4] | | 40,A3 | 144,A3 | — | — |
| † Methyl hydrazine[60-34-4]— | | | | | |
|   Skin | | 0.01,A3 | 0.019,A3 | — | — |
| Nickel, elemental [7440-02-0], insoluble and | | | | | |
|   soluble compounds, as Ni . | | — | 0.05,A1 | — | — |
| Nickel carbonyl [13463-39-3], | | | | | |
|   as Ni | | Delete listing; included in listing for Nickel, elemental, insoluble and soluble compounds | | | |
| Nickel sulfide roasting, | | | | | |
|   fume & dust, as Ni | | Delete listing; included in listing for Nickel, elemental, insoluble and soluble compounds | | | |
| Oil Mist, mineral | | | | | |
|   Severely refined | | — | 5[k] | — | — |
|   Mildly refined, as cyclohexane soluble particulate | | | | | |
|     containing polynuclear aromatic | | | | | |
|     hydrocarbons (PNAs) | | — | 0.2[k],A1 | — | — |
| Ozone [10028-15-6] | | 0.05 | 0.1 | 0.2 | 0.4 |
| Particulates Not Otherwise Classified (PNOC) | | | | | |
|   Inhalable particulate | | — | 10[e] | — | — |
|   †Respirable particulate | | — | 3[e] | — | — |
| Silica—Amorphous | | | | | |
|   Diatomaceous earth (uncalcined) | | | | | |
|   [61790-53-2] | | | | | |
|     Inhalable particulate | | — | 10[e] | — | — |
|     † Respirable particulate | | — | 3[e] | — | — |
| † Triethylamine [121-44-8]— | | | | | |
|   Skin | | 1,A4 | 4.1,A4 | 3,A4 | 12.4,A4 |

☆A value for blood Pb is under review.

## ADOPTED APPENDICES

## APPENDIX A: Carcinogenicity

The Chemical Substances TLV Committee has been aware of the increasing public concern over chemicals or industrial processes that cause or contribute to increased risk of cancer in workers. More sophisticated methods of bioassay, as well as the use of sophisticated mathematical models that extrapolate the levels of risk among workers, have led to differing interpretations as to which chemicals or processes should be categorized as human carcinogens and what the maximum exposure levels should be. The goal of the Committee has been to synthesize the available information in a manner that will be useful to practicing industrial hygienists, without overburdening them with needless details. The categories for carcinogenicity are:

A1 — *Confirmed Human Carcinogen:* The agent is carcinogenic to humans based on the weight of evidence from epidemiologic studies of, or convincing clinical evidence in, exposed humans.

A2 — *Suspected Human Carcinogen:* The agent is carcinogenic in experimental animals at dose levels, by route(s) of administration, at site(s), of histologic type(s), or by mechanism(s) that are considered relevant to worker exposure. Available epidemiologic studies are conflicting or insufficient to confirm an increased risk of cancer in exposed humans.

A3 — *Animal Carcinogen:* The agent is carcinogenic in experimental animals at a relatively high dose, by route(s) of administration, at site(s), of histologic type(s), or by mechanism(s) that are not considered relevant to worker exposure. Available epidemiologic studies do not confirm an increased risk of cancer in exposed humans. Available evidence suggests that the agent is not likely to cause cancer in humans except under uncommon or unlikely routes or levels of exposure.

A4 — *Not Classifiable as a Human Carcinogen:* There are inadequate data on which to classify the agent in terms of its carcinogenicity in humans and/or animals.

A5 — *Not Suspected as a Human Carcinogen:* The agent is not suspected to be a human carcinogen on the basis of properly conducted epidemiologic studies in humans. These studies have sufficiently long follow-up, reliable exposure histories, sufficiently high dose, and adequate statistical power to conclude that exposure to the agent does not convey a significant risk of cancer to humans. Evidence suggesting a lack of carcinogenicity in experimental animals will be considered if it is supported by other relevant data.

Substances for which no human or experimental animal carcinogenic data have been reported are assigned no carcinogen designation.

Exposures to carcinogens must be kept to a minimum. Workers exposed to A1 carcinogens without a TLV should be properly equipped to eliminate to the fullest extent possible all exposure to the carcinogen. For A1 carcinogens with a TLV and for A2 and A3 carcinogens, worker exposure by all routes should be carefully controlled to levels as low as possible below the TLV. Refer to the "Guidelines for the Classification of Occupational Carcinogens" in the Introduction to the 6th Edition of the *Documentation of the Threshold Limit Values* for a more complete description and derivation of these designations.

## APPENDIX B: Substances of Variable Composition

**B1. *Polytetrafluoroethylene\* decomposition products.*** Thermal decomposition of the fluorocarbon chain in air leads to the formation of oxidized products containing carbon, fluorine, and oxygen. Because these products decompose in part by hydrolysis in alkaline solution, they can be quantitatively determined in air as fluoride to provide an index of exposure. No TLVs are recommended at this time, but air concentration should be controlled as low as possible.
(\*Trade names include: Algoflon, Fluon, Teflon, Tetran)

**B2. *Welding Fumes—Total Particulate (NOC[(d)]); TLV–TWA,*** 5 mg/m[3]
Welding fumes cannot be classified simply. The composition and quantity of both are dependent on the alloy being welded and the process and electrodes used. Reliable analysis of fumes cannot be made without considering the nature of the welding process and system being examined; reactive metals and alloys such as aluminum and titanium are arc-welded in a protective, inert atmosphere such as argon. These arcs create relatively little fume, but they do create an intense radiation which can produce ozone. Similar processes are used to arc-weld steels, also creating a relatively low level of fumes. Ferrous alloys also are arc-welded in oxidizing environments that generate considerable fume and can produce carbon monoxide instead of ozone. Such fumes generally are composed of discreet particles of amorphous slags containing iron, manganese, silicon, and other metallic constituents depending on the alloy system involved. Chromium and nickel compounds are found in fumes when stainless steels are arc-welded. Some coated and flux-cored electrodes are formulated with fluorides and the fumes associated with them can contain significantly more fluorides than oxides. Because of the above factors, arc-welding fumes frequently must be tested for individual constituents that are likely to be present to determine whether specific TLVs are exceeded. Conclusions based on total fume concentration are generally adequate if no toxic elements are present in welding rod, metal, or metal coating and conditions are not conducive to the formation of toxic gases.

## APPENDIX C: Threshold Limit Values for Mixtures

When two or more hazardous substances which act upon the same organ system are present, their combined effect, rather than that of either individually, should be given primary consideration. In the absence of information to the contrary, the effects of the different hazards should be considered as additive. That is, if the sum of

$$\frac{C_1}{T_1} + \frac{C_2}{T_2} + \cdots \frac{C_n}{T_n}$$

exceeds unity, then the threshold limit of the mixture should be considered as being exceeded. $C_1$ indicates the observed atmospheric concentration and $T_1$ the corresponding threshold limit (see Example A.1 and B.1).

Exceptions to the above rule may be made when there is a good reason to believe that the chief effects of the different harmful substances are not in fact additive, but are independent as when purely local effects on different organs of the body are produced by the various components of the mixture. In such cases, the threshold limit ordinarily is exceeded only when at least one member of the series ($C_1/T_1 +$ or $+ C_2/T_2$, etc.) itself has a value exceeding unity (see Example B.1).

Synergistic action or potentiation may occur with some combinations of atmospheric contaminants. Such cases at present must be determined individually. Potentiating or synergistic agents are not necessarily harmful by themselves. Potentiating effects of exposure to such agents by routes other than that of inhalation are also possible, e.g., imbibed alcohol and inhaled narcotic (trichloroethylene). Potentiation is characteristically exhibited at high concentrations, less probably at low.

When a given operation or process characteristically emits a number of harmful dusts, fumes, vapors or gases, it will frequently be only feasible to attempt to evaluate the hazard by measurement of a single substance. In such cases, the threshold limit used for this substance should be reduced by a suitable factor, the magnitude of which will depend on the number, toxicity, and relative quantity of the other contaminants ordinarily present.

Examples of processes that are typically associated with two or more harmful atmospheric contaminants are welding, automobile repair, blasting, painting, lacquering, certain foundry operations, diesel exhausts, etc.

### Examples of TLVs for Mixtures

**A. *Additive effects.*** The following formulae apply only when the components in a mixture have similar toxicologic effects; they should not be used for mixtures with widely differing reactivities, e.g., hydrogen cyanide and sulfur dioxide. In such case, the formula for **Independent Effects** should be used.

1. General case, where air is analyzed for each component, the TLV of mixture =

$$\frac{C_1}{T_1} + \frac{C_2}{T_2} + \frac{C_3}{T_3} + \cdots = 1$$

*Note:* It is essential that the atmosphere be analyzed both qualitatively and quantitatively for each component present in order to evaluate compliance or noncompliance with this calculated TLV.

*Example A.1:* Air contains 400 ppm of acetone (TLV, 750 ppm), 150 ppm of sec-butyl acetate (TLV, 200 ppm) and 100 ppm of methyl ethel ketone (TLV, 200 ppm).

Atmospheric concentration of mixture = 400 + 150 + 100 = 650 ppm of mixture.

$$\frac{400}{750} + \frac{150}{200} + \frac{100}{200} = 0.53 + 0.75 + 0.5 = 1.78$$

Threshold Limit is exceeded.

2. Special case when the source of contaminant is a liquid mixture and the atmospheric composition is assumed to be similar to that of the original material, e.g., on a time-weighted average exposure basis, all of the liquid (solvent) mixture eventually evaporates. When the percent composition (by weight) of the liquid mixture is known, the TLVs of the constituents must be listed in mg/m³. TLV of mixture =

$$\frac{1}{\dfrac{f_a}{TLV_a} + \dfrac{f_b}{TLV_b} + \dfrac{f_c}{TLV_c} + \cdots \dfrac{f_n}{TLV_n}}$$

*Note:* In order to evaluate compliance with this TLV, field sampling instruments should be calibrated, in the laboratory, for response to this specific quantitative and qualitative air-vapor mixture, and also to fractional concentrations of this mixture, e.g., 1/2 the TLV; 1/10 the TLV; 2 × the TLV; 10 × the TLV; etc.)

*Example A.2:* Liquid contains (by weight):
50% heptane: TLV = 400 ppm or 1640 mg/m³
$\quad$ 1 mg/m³ ≡ 0.24 ppm
30% methyl chloroform: TLV = 350 ppm or 1910 mg/m³
$\quad$ 1 mg/m³ ≡ 0.18 ppm
20% perchloroethylene: TLV = 25 ppm or 170 mg/m³
$\quad$ 1 mg/m³ ≡ 0.15 ppm

$$\begin{aligned}
\text{TLV of Mixture} &= \frac{1}{\dfrac{0.5}{1640} + \dfrac{0.3}{1910} + \dfrac{0.2}{170}} \\[2mm]
&= \frac{1}{0.00030 + 0.00016 + 0.00118} \\[2mm]
&= \frac{1}{0.00164} = 610 \text{ mg/m}^3
\end{aligned}$$

of this mixture
$\quad$ 50% or (610)(0.5) = 305 mg/m³ is heptane
$\quad$ 30% or (610)(0.3) = 183 mg/m³ is methyl chloroform
$\quad$ 20% or (610)(0.2) = 122 mg/m³ is perchloroethylene

These values can be converted to ppm as follows:

$\quad$ heptane: 305 mg/m³ × 0.24 = 73 ppm
$\quad$ methyl chloroform: 183 mg/m³ × 0.18 = 33 ppm
$\quad$ perchloroethylene: 122 mg/m³ × 0.15 = 18 ppm

TLV of mixture = 73 + 33 + 18 = 124 ppm, or 610 mg/m³

**B.** *Independent effects.* TLV for mixture =

$$\frac{C_1}{T_1} = 1; \quad \frac{C_2}{T_2} = 1; \quad \frac{C_3}{T_3} = 1; \text{ etc.}$$

*Example B.1:* Air contains 0.15 mg/m³ of lead (TLV, 0.15) and 0.7 mg/m³ of sulfuric acid (TLV, 1).

$$\frac{0.15}{0.15} = 1; \qquad\qquad \frac{0.7}{1} = 0.7$$

Threshold limit is not exceeded.

**C.** *TLV for mixtures of mineral dusts.* For mixtures of biologically active mineral dusts, the general formula for mixtures given in A.2 may be used.

## APPENDIX D: Particle Size-Selective Sampling Criteria for Airborne Particulate Matter

For chemical substances present in inhaled air as suspensions of solid particles or droplets, the potential hazard depends on particle size as well as mass concentration because of: 1) effects of particle size on the deposition site within the respiratory tract, and 2) the tendency for many occupational diseases to be associated with material deposited in particular regions of the respiratory tract.

The Chemical Substances TLV Committee has recommended particle size-selective TLVs for crystalline silica for many years in recognition of the well established association between silicosis and respirable mass concentrations. The Committee is now re-examining other chemical substances encountered in particulate form in occupational environments with the objective of defining: 1) the size-fraction most closely associated for each substance with the health effect of concern, and 2) the mass concentration within that size fraction which should represent the TLV.

The Particle Size-Selective TLVs (PSS–TLVs) are expressed in three forms:

1. *Inhalable Particulate Mass TLVs* (IPM–TLVs) for those materials that are hazardous when deposited anywhere in the respiratory tract.

2. *Thoracic Particulate Mass TLVs* (TPM–TLVs) for those materials that are hazardous when deposited anywhere within the lung airways and the gas-exchange region.

3. *Respirable Particulate Mass TLVs* (RPM–TLVs) for those materials that are hazardous when deposited in the gas-exchange region.

The three particulate mass fractions described above are defined in quantitative terms in accordance with the following equations:[1,2]

A. *Inhalable Particulate Mass* consists of those particles that are captured according to the following collection efficiency regardless of sampler orientation with respect to wind direction:

$$SI(d) = 50\% \times (1 + e^{-0.06d})$$
$$\text{for } 0 < d \le 100 \ \mu m$$

where: SI(d) = the collection efficiency for particles with aerodynamic diameter d in μm

B. *Thoracic Particulate Mass* consists of those particles that are captured according to the following collection efficiency:

$$ST(d) = SI(d) \, [1 - F(x)]$$

where: $x = \dfrac{\ln (d/\Gamma)}{\ln (\Sigma)}$
$\quad \Gamma = 11.64 \ \mu m$
$\quad \Sigma = 1.5$
$\quad F(x) =$ the cumulative probability function of a standardized normal variable, x

C. *Respirable Particulate Mass* consists of those particles that are captured according to the following collection efficiency:

$$SR(d) = SI(d) [1 - F(x)]$$

where F(x) has the same meaning as above with $\Gamma = 4.25$ μm and $\Sigma = 1.5$

The most significant difference from previous definitions is the increase in the median cut point for a respirable dust sampler from 3.5 μm to 4.0 μm; this is in accord with the International Standards Organization/European Standardization Committee (ISO/CEN) protocol[3,4]. At this time, no change is recommended for the measurement of respirable dust using a 10-mm nylon cyclone at a flow rate of 1.7 liters per minute. Two analyses of available data indicate that the flow rate of 1.7 liters per minute allows the 10-mm nylon cyclone to approximate the dust concentration which would be measured by an ideal respirable dust sampler as defined herein.[5,6]

Collection efficiencies representative of several sizes of particles in each of the respective mass fractions are shown in Tables I, II, and III. References 2 and 3 provide documentation for the respective algorithms representative of the three mass fractions.

### TABLE I. Inhalable

| Particle Aerodynamic Diameter (μm) | Inhalable Particulate Mass (IPM) (%) |
|---|---|
| 0 | 100 |
| 1 | 97 |
| 2 | 94 |
| 5 | 87 |
| 10 | 77 |
| 20 | 65 |
| 30 | 58 |
| 40 | 54.5 |
| 50 | 52.5 |
| 100 | 50 |

### TABLE II. Thoracic

| Particle Aerodynamic Diameter (μm) | Thoracic Particulate Mass (TPM) (%) |
|---|---|
| 0 | 100 |
| 2 | 94 |
| 4 | 89 |
| 6 | 80.5 |
| 8 | 67 |
| 10 | 50 |
| 12 | 35 |
| 14 | 23 |
| 16 | 15 |
| 18 | 9.5 |
| 20 | 6 |
| 25 | 2 |

### TABLE III. Respirable

| Particle Aerodynamic Diameter (μm) | Respirable Particulate Mass (RPM) (%) |
|---|---|
| 0 | 100 |
| 1 | 97 |
| 2 | 91 |
| 3 | 74 |
| 4 | 50 |
| 5 | 30 |
| 6 | 17 |
| 7 | 9 |
| 8 | 5 |
| 10 | 1 |

### References

1. American Conference of Governmental Industrial Hygienists: Particle Size-Selective Sampling in the Workplace. ACGIH, Cincinnati, OH (1985).
2. Soderholm, S.C.: Proposed International Conventions for Particle Size-Selective Sampling. Ann. Occup. Hyg. 33:301–320 (1989).
3. International Organization for Standardization (ISO): Air Quality—Particle Size Fraction Definitions for Health-Related Sampling. Approved for publication as CD 7708. ISO, Geneva (1991).
4. European Standardization Committee (CEN): Size Fraction Definitions for Measurement of Airborne Particles in the Workplace. Approved for publication as prEN 481. CEN, Brussels (1992).
5. Bartley, D.L.: Letter to J. Doull, TLV Committee, July 9, 1991.
6. Lidén, G.; Kenny, L.C.: Optimization of the Performance of Existing Respirable Dust Samplers. Appl. Occup. Environ. Hyg. 8(4):386–391 (1993).

## CHEMICAL SUBSTANCES AND OTHER ISSUES UNDER STUDY

Information, data especially, and comments are solicited to assist the Committee in its deliberations and in the possible development of draft documents. Draft documentations are used by the Committee to decide what action, if any, to recommend on a given question.

### Chemical Substances

Acetomethylchloride
Aluminum alkyls
tert-Amyl methyl ether (TAME)
Antimony
Attapulgite/Palygorskite/ Sepiolite
Aviation fuel
Bentonite
Borax and boron compounds
Bromochloromethane
Bromodichloromethane
Bromoform
1,2,3,4-Butanetetracarboxylic acid
sec-Butanol
sec-Butyl acetate
2-t-Butylazo-2-hydroxy-5-methylhexane
Carbon disulfide
Chlorine
Chlorodiphenyls (42% & 54% chlorine)
Copper fume
Cristobalite
Crystalline silicas
Cyanamide
Dichlorocyclopentadiene
Dichlorodiphenyl sulfone
1,2-Dichloroethane
2,4-D (2,4-Dichlorophenoxy-acetic acid)
1,3-Dichloropropene
Diesel fuel
1,4-Diethyl benzene
N,N-Dimethyl acetamide
Dimethylformamide
Dimethyl disulfide
Dimethylterephthalate
2-Ethoxyethanol (EGEE)
2-Ethoxyethyl acetate (EGEEA)
Ethyl bromide
Ethyl tert-butyl ether

2-Ethyl hexaoic acid
Fibrous glass dust (synthetic inorganic fibers)
Fluorine
Furfural
Gallium arsenide
Gasoline (unleaded)
Glycidol
Glycol ethers
Graphite fibers
1-Hexene
Hexachlorocyclopentadiene
Isobutene
Isopropyl glycidyl ether (IGE)
Lead, organic compounds
2-Methoxyethanol (EGME)
2-Methoxyethyl acetate (EGMEA)
Methyl n-butyl ketone
Methyl chloride
Methylene chloride
Methylene diamine
4,4'-Methylene dianiline
Methyl propyl ketone
α-Methyl styrene
Methyl vinyl ketone
Mineral spirits
Naled
Nitrogen dioxide
Pentachlorophenol
Pentane
2,4-Pentanedione
Perlite
Petroleum solvents
Phosphates (including mining)
Picoline
Propylene dichloride
Quartz
Styrene
Synthetic inorganic fibers (man-made mineral fibers)
Tantalum

1,1,2,2-Tetrachloroethane
Tetrahydrofuran
Tetrakis (hydroxymethyl)
   phosphonium chloride
Tetrakis (hydroxymethyl)
   phosphonium sulfate
Tetrasodium pyrophosphate
   (Phosphates)

Tridymite
Trona
Uranium
Vanadium
Vinylidene chloride
Vinyl cyclohexene dioxide
Xylene

### Other Issues

1. Ceiling Limit, Excursion Limit, and Short-Term Exposure Limit (STEL).
2. Reproductive Effects Notation. Under further review, but for the present, a notation will not be included in the TLV/BEI Booklet listing.
3. Risk Assessment.
4. Neurotoxicity.

### 1993–94 CHEMICAL SUBSTANCES TLV COMMITTEE

John Doull, Ph.D., M.D., University of Kansas Medical Center—Chair
Michael J. Blotzer, CIH, CSP, National Aeronautics and Space Administration
Dennis M. Casserly, Ph.D., CIH, University of Houston, Clear Lake
D. Dwight Culver, M.D., University of California–Irvine
Richard E. Fairfax, CIH, Occupational Safety and Health Administration
Lora E. Fleming, M.D., University of Miami School of Medicine
S. Katharine Hammond, Ph.D., CIH, University of Massachusetts
Jesse Lieberman, PE, CIH, Retired–U.S. Navy
Ernest Mastromatteo, M.D., Retired–University of Toronto
Ronald S. Ratney, Ph.D., CIH, Occupational Safety and Health Administration
Karl Rozman, Ph.D., The University of Kansas Medical Center
Meier Schneider, PE, CIH, Retired–Metro Water Dist. of So. California
Raghubir Sharma, Ph.D., CIH, Utah State University
Robert Spirtas, Dr. P.H., National Institutes of Health
Thomas F. Tomb, Mine Safety and Health Administration
Elizabeth K. Weisburger, Ph.D., Retired–National Cancer Institute
Calvin Willhite, Ph.D., State of California
Margie E. Zalesak, CIH, Mine Safety and Health Administration

### NONVOTING MEMBERS

James S. Bus, Ph.D., Dow Chemical Company
Bernard L. Fontaine, Jr., CIH, CSP, Atlantic Mutual Insurance
Stanley Haimes, M.D., CIH, Martin-Marietta
Gregory L. Kedderis, Ph.D., Chemical Industry Institute of Toxicology
Gerald L. Kennedy, Jr., E.I. duPont de Nemours and Company, Inc.
Michael S. Morgan, Sc.D., CIH, University of Washington,
   BEI Committee Liaison
Robert A. Scala, Ph.D., Retired–EXXON

## INTRODUCTION TO THE BIOLOGICAL EXPOSURE INDICES

Biological monitoring provides occupational health personnel with a tool for assessing a worker's exposure to chemicals. **Workplace air monitoring** consists of assessment of inhalation exposure to chemicals in the workplace through measurement of the chemical concentration in the ambient air. **TLVs** serve as a reference value. **Biological monitoring** consists of an assessment of overall exposure to chemicals that are present in the workplace through measurement of the appropriate determinant(s) in biological specimens collected from the worker at the specified time. **BEIs** (Biological Exposure Indices) serve as a reference value.

The **determinant** can be the chemical itself or its metabolite(s), or a characteristic reversible biochemical change induced by the chemical. The measurement can be made in exhaled air, urine, blood, or other biological specimens collected from the exposed worker. Based on the determinant, the specimen chosen, and the time of sampling, the measurement indicates either the intensity of a recent exposure, an average daily exposure, or a chronic cumulative exposure.

**Biological Exposure Indices (BEIs)** are reference values intended as guidelines for the evaluation of potential health hazards in the practice of industrial hygiene. BEIs represent the levels of determinants which are most likely to be observed in specimens collected from a healthy worker who has been exposed to chemicals to the same extent as a worker with inhalation exposure to the TLV. The exceptions are the BEIs for some chemicals, namely those for which TLVs are based on protection against nonsystemic effect (e.g., irritation or respiratory impairment) and biological monitoring is desirable because of their potential for significant absorption via an additional route of entry (usually the skin). The BEIs for these chemicals may be based on protection against systemic effect, thus allowing the internal dose to exceed the pulmonary intake resulting from exposure at the TLV. BEIs do not indicate a sharp distinction between hazardous and nonhazardous exposures. Due to biological variability, it is possible for an individual's measurements to exceed the BEI without incurring an increased health risk. If, however, measurements in specimens obtained from a worker on different occasions persistently exceed the BEI, or if the majority of measurements in specimens obtained from a group of workers at the same workplace exceed the BEI, the cause of the excessive values must be investigated and proper action taken to reduce the exposure.

BEIs apply to eight-hour exposures, five days a week. However, BEIs for altered work schedules can be extrapolated on pharmacokinetic and pharmacodynamic bases. BEIs should not be applied either directly or through a conversion factor, in the determination of safe levels for nonoccupational exposure to air and water pollutants or food contaminants. The BEIs are not intended for use as a measure of adverse effects or for diagnosis of occupational illness.

**The data base for each BEI recommendation** consists of available information on absorption, elimination, and metabolism of chemicals, and on the correlation between exposure intensity and biological effect in workers. The BEI is based either on the relationship between intensity of exposure and biological levels of the determinant or on the relationship between biological levels and health effects. Human data from controlled and field studies are used to find such relationships. Animal studies usually do not provide data suitable for establishing a BEI.

In the alphabetical listing of the TLVs for Chemical Substances, a special symbol (◄) is used to note that a particular substance also has a BEI. Included are those substances identified in the BEI documentations for methemoglobin and organophosphorous cholinesterase inhibitors.

**Implementation:** Biological monitoring should be considered complementary to air monitoring. It should be conducted when it offers an advantage over the use of air monitoring alone. Biological monitoring should be used to substantiate air monitoring, to test the efficacy of personal protective equipment, to determine the potential for absorption via the skin and the gastrointestinal system, or to detect nonoccupational exposure. The existence of a BEI does not indicate the need for conducting biological monitoring. Occupational health personnel must exercise professional judgment in designing monitoring protocols. The documentations are intended to provide helpful background information.

**Interpretation of data:** When biological monitoring data are interpreted, intraindividual and interindividual differences in tissue levels of determinants occurring at the same exposure conditions must be considered. Such differences arise on account of variation in pulmonary ventilation, hemodynamics, body composition, efficacy of excretory organs, and activity of enzyme systems that mediate metabolism of the chemical. Multiple sampling is necessary to reduce the effects of variable factors. Biological monitoring may confirm the results of air monitoring, but where there is a discrepancy between the results, the entire exposure situation should be carefully reviewed and an explanation found.

**The main source of inconsistency** in information on exposure intensity by air monitoring and biological monitoring is the variability of the following factors: ***physiological and health status of the worker,*** such as body build, diet (water and fat intake), enzymatic activity, body fluid composition, age, sex, pregnancy, medication, and disease state; occupational ***exposure sources,*** such as the intensity of the physical work load and fluctuation of exposure intensity, skin exposure, temperature and humidity, coexposure to other chemicals; ***environmental sources,*** such as community and home air pollutants, and water and food contaminants; ***individual life style sources,*** such as after work activities, personal hygiene, working and eating habits, smoking, alcohol and drug intake, exposure to household products, or exposure to chemicals from hobbies or from another workplace; and ***methodological sources,*** which include specimen contamination and deterioration during collection, storage and analysis, and bias of the selected analytical methods. The significance of these effects must be assessed individually for each situation. Drugs, pollutants or coexposure to another chemical can alter the relationship between the intensity of occupational exposure and the level of the determinant in the specimen either by adding to the level of the determinant or by altering the metabolism or elimination of the studied chemical. The BEI documentations provide specific information on the effects of these factors.

***The timing*** indicates when the sample should be collected with respect to the exposure. It must be carefully observed because distribution and elimination of a chemical or its metabolic products, as well as biochemical changes induced by exposure to the chemical, are kinetic events. The listed BEIs are applicable only if collection is conducted at the specified time.

***The quality control program*** of the laboratory performance is essential to reduce analytical errors and bias in the results.

***Comments on the Table of BEIs:*** The Table specifies, for indicated chemicals, the determinant, the specimen to be collected, the time at which the specimen must be collected, and the BEI. Additional important information is indicated by notation(s).

***Determinants:*** The Table comprises BEIs for all determinants where a sufficient data base is available and on which the BEI Committee took action. Occupational health personnel must use professional judgment to decide which determinant should be used in specific instances in order to meet the objectives of the monitoring.

***Biological specimens:*** Urine, exhaled air, and blood specimens are recommended. Each type of specimen has distinctive causes of variability which affect the level of the determinant in the specimen. Other specimens, such as hair or nails, are not recommended at this time.

For data based on ***urine analysis,*** variation in urine volume is the most significant. Measurements of elimination rate usually provide more precise information. However, quantitative collection of urine during a precise time period is rarely possible. A simple measurement of concentration can provide information on exposure, but the quantitative measure of exposure is weakened by the variability of rate of urine output. The urinary concentration related to excretion of the solute provides some correction for fluctuation of urine output. BEIs for determinants whose excretion is dependent on urine output are given in relation to creatinine excretion. Some determinants, however, are excreted by diffusion and adjustment for solute is inappropriate. BEIs for these determinants are given as concentration. Highly diluted and highly concentrated urine specimens are usually not suitable for monitoring, and a new specimen should be taken. The excretion mechanism of determinants can be altered when the urine specimen is very concentrated (specific gravity >1.030, creatinine >3 g/L) or dilute (specific gravity <1.010, creatinine <0.5 g/L). In this case, measurements in urine samples are not reliable and should be repeated in a specimen collected on some other occasion.

For data based on ***exhaled air analysis,*** the rapid changes of the concentration with time are critical; moreover, the concentration changes during the expiration phase. Therefore, the sampling of end-exhaled air (which usually represents alveolar air) or mixed-exhaled air is specified. In general, during exposure the concentrations in end-exhaled air are smaller than in mixed-exhaled air, and during postexposure the concentration in mixed-exhaled air accounts for about two-thirds of the concentration in end-exhaled air. Exhaled air specimens collected from workers with altered pulmonary function may not be suitable for exposure monitoring.

For data based on ***blood analysis,*** the plasma-erythrocyte ratio and the distribution of some determinants among blood constituents can affect the outcome of some measurements. Therefore, the analysis of whole blood, plasma, serum, or erythrocytes is specified. The protein binding of some determinants should be considered when the analytical method is selected. The concentration difference between arterial and venous blood induced by pulmonary uptake or pulmonary clearance has to be considered when blood is collected for measurements of volatile chemicals. Unless otherwise indicated, the BEIs for volatile chemicals relate to venous blood and cannot be applied to capillary blood, which mainly represents arterial blood.

***Sampling time:*** In many instances, when the level of the determinant changes rapidly or when accumulation occurs, the sampling time is very critical and must be carefully observed. The sampling time is specified in the table according to the differences in the uptake and elimination rates of chemicals and their metabolites, and according to the persistence of induced biochemical changes, as follows:

1. Determinants with timing "prior to shift" (meaning after 16 hours without exposure), "during shift," or "end of shift" (meaning the last two hours of exposure) are eliminated rapidly with a half-time less than five hours. Such determinants do not accumulate in the body and, therefore, their timing is critical only in relation to the exposure and postexposure periods.

2. Determinants with timing "beginning of workweek," or "end of workweek" (meaning after two days without exposure or after four or five consecutive working days with exposure, respectively) are eliminated with half-times longer than five hours. Such determinants accumulate in the body during the workweek; therefore, their timing is critical in relation to previous exposures. For chemicals with multiphase elimination, the timing is given in relation to the workday exposure (shift) as well as to the workweek exposure.

3. Determinants with timing "not critical," or "discretionary," have very long elimination half-times and accumulate in the body over years, some for a lifetime. After a couple of weeks of exposure, specimens for measurements of such determinants can be collected any time.

**It is essential to consult the specific BEI documentation published in the *Documentation of Threshold Limit Values and Biological Exposure Indices,* 6th edition, 1991, before designing biological monitoring and interpreting BEIs. Action on unexpected values should not be based on a single isolated measurement but on measurements of multiple sampling.**

***Notations*** provide the following information:

***"Sc" Notation.*** This notation indicates that an identifiable population group might have an increased ***susceptibility*** to the effect of the chemical, thus leaving it unprotected by the recommended BEI. The specific BEI documentation should be consulted for information.

***"B" Notation.*** This notation indicates that the determinant is usually present in a significant amount in biological specimens collected from subjects who have not been occupationally exposed. Such background levels are included in the BEI value. For information on ***background levels,*** consult the specific documentation.

***"Ns" Notation.*** This notation indicates that the determinant is ***nonspecific,*** since it is observed after exposure to some other chemicals. These nonspecific tests are preferred because they are easy to use and usually offer a better correlation with exposure than specific tests. In such instances, a BEI for a specific, less quantitative biological determinant is recommended as a confirmatory test. The documentation should be consulted for information on factors affecting interpretation of these BEIs.

***"Sq" Notation.*** This notation indicates that the biological determinant is an indicator of exposure to the chemical, but the quantitative interpretation of the measurement is ambiguous (semiquantitative). These biological determinants should be used as a screening test if a quantitative test is not practical or as a confirmatory test if the quantitative test is not specific and the origin of the determinant is in question.

In some instances, BEIs for confirmatory and screening tests are not listed, but the pertinent documentation provides information for estimation of the reference value. Examples are measurements of inhaled chemicals (which are extensively metabolized) in exhaled air. BEIs for some screening tests are derived as an upper (or lower) limit of the levels observed in unexposed populations. Examples are measurements of cholinesterase or methemoglobin.

## ADOPTED BIOLOGICAL EXPOSURE DETERMINANTS

| Chemical [CAS #]<br>Determinant | Sampling Time | BEI | Notation |
|---|---|---|---|
| *ACETONE [67-43-1] (1994)<br>Acetone in urine | End of shift | 100 mg/L | B, Ns |
| ANILINE [62-53-3] (1991)<br>Total p-aminophenol in urine<br>Methemoglobin in blood | End of shift<br>During or end of shift | 50 mg/g creatinine<br>1.5% of hemoglobin | Ns<br>B, Ns, Sq |
| ARSENIC AND SOLUBLE COMPOUNDS<br>INCLUDING ARSINE [7784-42-1] (1993)<br>Inorganic arsenic metabolites in urine | End of workweek | 50 μg/g creatinine | B |
| BENZENE [71-43-2] (1987) [See note below]<br>Total phenol in urine<br>Benzene in exhaled air:<br>  mixed-exhaled<br>  end-exhaled | End of shift<br>Prior to next shift | 50 mg/g creatinine<br><br>0.08 ppm<br>0.12 ppm | B, Ns<br><br>Sq<br>Sq |
| CADMIUM AND INORGANIC COMPOUNDS (1993)<br>Cadmium in urine<br>Cadmium in blood | Not critical<br>Not critical | 5 μg/g creatinine<br>5 μg/L | B<br>B |
| CARBON DISULFIDE [75-15-0] (1988)<br>2-Thiothiazolidine-4-carboxylic acid<br>(TTCA) in urine | End of shift | 5 mg/g creatinine | |
| CARBON MONOXIDE [630-08-0] (1993)<br>Carboxyhemoglobin in blood<br>Carbon monoxide in end-exhaled air | End of shift<br>End of shift | 3.5% of hemoglobin<br>20 ppm | B, Ns<br>B, Ns |
| CHLOROBENZENE [108-90-7] (1992)<br>Total 4-chlorocatechol in urine<br>Total p-chlorophenol in urine | End of shift<br>End of shift | 150 mg/g creatinine<br>25 mg/g creatinine | Ns<br>Ns |
| CHROMIUM (VI), Water-Soluble Fume (1990)<br>Total chromium in urine | Increase during shift<br>End of shift at end of workweek | 10 μg/g creatinine<br>30 μg/g creatinine | B<br>B |
| ‡N,N-DIMETHYLFORMAMIDE (DMF) [68-12-2] (1988)<br>N-Methylformamide in urine | End of shift | (40 mg/g creatinine) | |
| *2-ETHOXYETHANOL (EGEE) [110-80-5] and<br>2-ETHOXYETHYL ACETATE (EGEEA) [111-15-9] (1994)<br>2-Ethoxyacetic acid in urine | End of shift at end of workweek | 100 mg/g creatinine | |
| ETHYL BENZENE [100-41-4] (1986)<br>Mandelic acid in urine<br>Ethyl benzene in end-exhaled air | End of shift at end of workweek | 1.5 g/g creatinine | Ns<br>Sq |
| FLUORIDES (1990)<br>Fluorides in urine | Prior to shift<br>End of shift | 3 mg/g creatinine<br>10 mg/g creatinine | B, Ns<br>B, Ns |
| FURFURAL [98-01-1] (1991)<br>Total furoic acid in urine | End of shift | 200 mg/g creatinine | B, Ns |
| n-HEXANE [110-54-3] (1987)<br>2,5-Hexanedione in urine<br>n-Hexane in end-exhaled air | End of shift | 5 mg/g creatinine | Ns<br>Sq |

Note: The Chemical Substances TLV Committee has proposed a revision of the TLV for benzene. See Chemical Substances Notice of Intended Changes and TLV Documentation.

## Chemical [CAS #]

| Determinant | Sampling Time | BEI | Notation |
|---|---|---|---|
| ‡LEAD (1987) [See note below] | | | |
| Lead in blood | Not critical | (50 μg/100 ml) | B |
| Lead in urine | Not critical | 150 μg/g creatinine | B |
| Zinc protoporphyrin in blood | After 1 month exposure | 250 μg/100 ml erythrocytes or 100 μg/100 ml blood | B |
| MERCURY (1993) | | | |
| Total inorganic mercury in urine | Preshift | 35 μg/g creatinine | B |
| Total inorganic mercury in blood | End of shift at end of workweek | 15 μg/L | B |
| ‡METHANOL [67-56-1] (1991) | | | |
| Methanol in urine | End of shift | 15 mg/L | B, Ns |
| (Formic acid in urine) | (Prior to the last shift of workweek) | (80 mg/g creatinine) | (B, Ns) |
| METHEMOGLOBIN INDUCERS (1990) | | | |
| Methemoglobin in blood | During or end of shift | 1.5% of hemoglobin | B, Ns, Sq |
| METHYL CHLOROFORM [71-55-6] (1989) | | | |
| Methyl chloroform in end-exhaled air | Prior to the last shift of workweek | 40 ppm | |
| Trichloroacetic acid in urine | End of workweek | 10 mg/L | Ns, Sq |
| Total trichloroethanol in urine | End of shift at end of workweek | 30 mg /L | Ns, Sq |
| Total trichloroethanol in blood | End of shift at end of workweek | 1 mg/L | Ns |
| METHYL ETHYL KETONE (MEK) [78-93-3] (1988) | | | |
| MEK in urine | End of shift | 2 mg/L | |
| METHYL ISOBUTYL KETONE (MIBK) [108-10-1] (1993) | | | |
| MIBK in urine | End of shift | 2 mg/L | |
| NITROBENZENE [98-95-3] (1991) | | | |
| Total p-nitrophenol in urine | End of shift at end of workweek | 5 mg/g creatinine | Ns |
| Methemoglobin in blood | End of shift | 1.5% of hemoglobin | B, Ns, Sq |
| ORGANOPHOSPHORUS CHOLINESTERASE INHIBITORS (1989) | | | |
| Cholinesterase activity in red cells | Discretionary | 70% of individual's baseline | B, Ns, Sq |
| PARATHION [56-38-2] (1989) | | | |
| Total p-nitrophenol in urine | End of shift | 0.5 mg/g creatinine | Ns |
| Cholinesterase activity in red cells | Discretionary | 70% of individual's baseline | B, Ns, Sq |
| PENTACHLOROPHENOL (PCP) [87-86-5] (1988) | | | |
| Total PCP in urine | Prior to the last shift of workweek | 2 mg/g creatinine | B |
| Free PCP in plasma | End of shift | 5 mg/L | B |
| ‡PERCHLOROETHYLENE [127-18-4] (1989) | | | |
| Perchloroethylene in end-exhaled air | Prior to the last shift of workweek | (10 ppm) | |
| Perchloroethylene in blood | Prior to the last shift of workweek | (1 mg/L) | |
| Trichloroacetic acid in urine | End of workweek | (7 mg/L) | Ns, Sq |
| PHENOL [108-95-2] (1987) | | | |
| Total phenol in urine | End of shift | 250 mg/g creatinine | B, Ns |
| STYRENE [100-42-5] (1986) | | | |
| Mandelic acid in urine | End of shift | 800 mg/g creatinine | Ns |
| | Prior to next shift | 300 mg/g creatinine | Ns |
| Phenylglyoxylic acid in urine | End of shift | 240 mg/g creatinine | Ns |
| | Prior to next shift | 100 mg/g creatinine | |
| Styrene in venous blood | End of shift | 0.55 mg/L | Sq |
| | Prior to next shift | 0.02 mg/L | Sq |
| ‡(TOLUENE [108-88-3]) (1986) | | | |
| (Hippuric acid in urine) | (End of shift) (Last 4 hrs of shift) | (2.5 g/g creatinine) | (B, Ns) |
| (Toluene in venous blood) | (End of shift) | (1 mg/L) | (Sq) |
| (Toluene in end-exhaled air) | | | (Sq) |
| TRICHLOROETHYLENE [79-01-6] (1986) | | | |
| Trichloroacetic acid in urine | End of workweek | 100 mg/g creatinine | Ns |
| Trichloroacetic acid and trichloroethanol in urine | End of shift at end of workweek | 300 mg/g creatinine | Ns |

Note: The Chemical Substances TLV Committee has proposed a revision of the TLV for elemental lead and inorganic compounds. See Chemical Substances Notice of Intended Changes and TLV Documentation.

| Chemical [CAS #]<br>Determinant | Sampling Time | BEI | Notation |
|---|---|---|---|
| Free trichloroethanol in blood<br>Trichloroethylene in blood (1993)<br>Trichloroethylene in end-exhaled air | End of shift at end of workweek | 4 mg/L | Ns<br>Sq<br>Sq |
| XYLENES [13307] (Technical Grade) (1986)<br>Methylhippuric acids in urine | End of shift | 1.5 g/g creatinine | |

## NOTICE OF INTENT TO ESTABLISH OR CHANGE

| Chemical [CAS #]<br>Determinant | Sampling Time | BEI | Notation |
|---|---|---|---|
| COBALT [7440-48-4] (1993)<br>Cobalt in urine<br>Cobalt in blood | End of shift at end of workweek<br>End of shift at end of workweek | 15 μg/L<br>1 μg/L | B<br>B |
| N,N-DIMETHYLACETAMIDE [127-19-5] (1993)<br>N-Methylacetamide in urine | End of shift at end of workweek | 30 mg/g creatinine | |
| N,N-DIMETHYLFORMAMIDE (DMF) [68-12-2] (1993)<br>N-Methylformamide in urine | End of shift | 20 mg/g creatinine | |
| †LEAD (1994)<br>Lead in blood | Not critical | 30 μg/100 ml | B |
| METHANOL [67-56-1] (1993)<br>Methanol in urine | End of shift | 15 mg/L | B, Ns |
| PERCHLOROETHYLENE [127-18-4] (1993)<br>Perchloroethylene in end-exhaled air<br>Perchloroethylene in blood<br>Trichloroacetic acid in urine | Prior to last shift of workweek<br>Prior to last shift of workweek<br>End of workweek | 5 ppm<br>0.5 mg/L<br>3.5 mg/L | <br><br>Ns, Sq |
| TOLUENE [108-88-3] (1993) | BEIs withdrawn based on reduction of Chemical Substances TLV<br>from 100 ppm to 50 ppm. BEI revision under review. | | |
| †VANADIUM PENTOXIDE [1314-62-1] (1994)<br>Vanadium in urine | End of shift | 50 μg/g creatinine | |

## CHEMICAL SUBSTANCES AND OTHER ISSUES UNDER STUDY TO ESTABLISH OR CHANGE BIOLOGICAL EXPOSURE INDICES

### Chemical Substances

Acrylonitrile
Benzene
2-Butoxyethanol
Dichlorobenzene
2,4-Dichlorophenoxyacetic acid
(2,4-D)
Hydrazine
Lead, inorganic
Lead, organic
2-Methoxyethanol (EGME)
2-Methoxyethyl acetate
(EGMEA)

4,4'-Methylene bis
(2-chloroaniline) [MOCA]
Methylene chloride
Nickel
Styrene
Synthetic pyrethrins
Tetrahydrofuran
Toluene: under study due to
adoption of revised TLV
(50 ppm; STEL deleted)
Vanadium

### Other Issues

1. Quality control.
2. Biological monitoring strategies.
3. Effect of smoking on biological levels of exposure indicators.
4. Ethnic differences in biological levels of exposure indicators.
5. Creatinine correction of urine samples.
6. Development of BEI documents tabled because of insufficient databases on:

| | |
|---|---|
| Alkylating agents | Manganese |
| Beryllium | Polychlorinated biphenyls |
| Ethanol* | (PCBs) |
| Malathion | Propylene glycol dinitrate |

*A BEI is not recommended because of interference from consumption of alcoholic beverages.

### 1993–94 BIOLOGICAL EXPOSURE INDICES COMMITTEE

Robert J. Sherwood, CIH, Harvard School of Public Health—Interim Chair
Pierre O. Droz, Ph.D., Lausanne University, Switzerland
Philip A. Edelman, M.D., University of California—Irvine
Michael S. Morgan, Sc.D., CIH, University of Washington
Masana Ogata, M.D., Okayama University Medical School, Japan

## INTRODUCTION TO THE PHYSICAL AGENTS

These Threshold Limit Values (TLVs) refer to levels of physical agents and represent conditions under which it is believed that nearly all workers may be repeatedly exposed day after day without adverse health effects. Because of wide variations in individual susceptibility, exposure of an occasional individual at, or even below, the TLV may not prevent annoyance, aggravation of a pre-existing condition, or physiological damage. Individuals may also be hypersusceptible or otherwise unusually responsive to some physical agents at the workplace because of genetic factors, age, personal habits (smoking, alcohol, or other drugs), medication, or previous exposures. Such workers may not be adequately protected from adverse health effects from exposures to certain physical agents at or below the TLVs. An occupational physician should evaluate the extent to which such workers require additional protection.

These TLVs are based on the best available information from industrial experience, from experimental human and animal studies, and when possible, from a combination of the three.

**These limits are intended for use in the practice of industrial hygiene and should be interpreted and applied only by a person trained in this discipline. They are not intended for use, or for modification for use, 1) in the evaluation or control of the levels of physical agents in the community, 2) as proof or disproof of an existing physical disability, or 3) for adoption by countries whose working conditions differ from those in the United States of America.**

These values are reviewed annually by the Committee on Threshold Limits for Physical Agents for revision or additions, as further information becomes available.

**The ACGIH disclaims any liability with respect to the use of TLVs.**

*Notice of Intended Changes*—At the beginning of each year, proposed actions of the Committee for the forthcoming year are issued in the form of a "Notice of Intended Changes." This notice provides not only an opportunity for comment but also solicits suggestions of physical agents to be added to the list. The suggestions should be accompanied by substantiating evidence.

Materials on the Notice of Intended Changes have been incorporated into the text and are indicated by a † preceding the revision/addition and by a vertical rule in the margin.

*Definitions*—Categories of Threshold Limit Values (TLVs) are specified herein, as follows:

*a)* Threshold Limit Value–Time-Weighted Average (TLV–TWA)—the time-weighted average exposure for a normal 8-hour workday and a 40-hour workweek, to which nearly all workers may be repeatedly exposed day after day, without adverse health effects.

*b)* Threshold Limit Value–Ceiling (TLV–C)—the exposure that should not be exceeded even instantaneously.

***Physical and Chemical Factors.*** It is recognized that combinations of such physical factors as heat, ultraviolet and ionizing radiation, humidity, abnormal pressure (altitude), and the like, as well as the interaction of physical factors with chemical substances in the workplace, may place added stress on the body so that the effects from exposure at a TLV may be altered. Also, most of these stresses may act adversely to increase the toxic response to a foreign substance. Although most threshold limits have built-in safety factors to guard against adverse health effects to moderate deviations from normal environments, the safety factors of most exposures are not of such a

magnitude as to take care of gross deviations. For example, continuous work at WBGT temperatures above 30°C (86°F), or overtime extending the workweek more than 25%, might be considered gross deviations. In such instances, judgment must be exercised in the proper adjustments of the TLVs.

***Mutagenicity, Carcinogenicity, and Adverse Reproductive Effects.*** The Physical Agents TLV Committee is examining the data related to mutagenicity, cancer, and adverse reproductive effects of physical agents. For example, it has been clearly demonstrated that ionizing radiation is a cause of cancer in humans. Likewise, ultraviolet (UV) radiation shows mutagenic and carcinogenic effects in the skin. Heat stress, infrared energy, radiofrequency (RF) hyperthermia, and whole-body vibration have been associated with adverse reproductive effects.

More speculative is the scientific literature suggesting a connection between exposure to RF, sub-RF energy, and/or magnetic fields with cancer. Neither an etiologic role nor a risk analysis is established for these agents.

The Physical Agents TLV Committee will continue to examine the bioeffects literature relating to cancer and adverse reproductive effects.

## ADOPTED THRESHOLD LIMIT VALUES

### AIRBORNE UPPER SONIC AND ULTRASONIC ACOUSTIC RADIATION

These TLVs refer to sound pressure levels that represent conditions under which it is believed that nearly all workers may be repeatedly exposed without adverse health effects. The values listed in Table 1 should be used as guides in the control of noise exposure and, due to individual susceptibility, should not be regarded as fine lines between safe and dangerous levels. The levels for the third-octave bands centered below 20 kHz are below those which cause subjective effects. Those levels for 1/3 octaves above 20 kHz are for prevention of possible hearing losses from subharmonics of these frequencies.

**TABLE 1. Permissible Airborne Upper Sonic and Ultrasound Acoustic Radiation Exposure Levels**

| Mid-Frequency of Third-Octave Band kHz | One-Third Octave—Band Level in dB re 20 μPa |
|---|---|
| 10 | 80 |
| 12.5 | 80 |
| 16 | 80 |
| 20 | 105 |
| 25 | 110 |
| 31.5 | 115 |
| 40 | 115 |
| 50 | 115 |

Subjective annoyance may occur in some sensitive individuals at levels between 75 and 105 dB at 20 kHz 1/3 octave band and hearing protection or engineering controls may be needed to minimize or prevent the annoyance.

### COLD STRESS

The cold stress TLVs are intended to protect workers from the severest effects of cold stress (hypothermia) and cold injury and to describe exposures to cold working conditions under which it is believed that nearly all workers can be repeatedly exposed without adverse health effects. The TLV objective is to prevent the deep body temperature from falling below 36°C (96.8°F) and to prevent cold injury to body extremities (deep body temperature is the core temperature of the body determined by conventional methods for rectal temperature measurements). For a single, occasional exposure to a cold environment, a drop in core temperature to no lower that 35°C (95°F) should be permitted. In addition to provisions for total body protection, the TLV objective is to protect all parts of the body with emphasis on hands, feet, and head from cold injury.

## Introduction

Fatal exposures to cold among workers have almost always resulted from accidental exposures involving failure to escape from low environmental air temperatures or from immersion in low temperature water. The single most important aspect of life-threatening hypothermia is the fall in the deep core temperature of the body. The clinical presentations of victims of hypothermia are shown in Table 1. Workers should be protected from exposure to cold so that the deep core temperature does not fall below 36°C (96.8°F); lower body temperatures will very likely result in reduced mental alertness, reduction in rational decision making, or loss of consciousness with the threat of fatal consequences.

Pain in the extremities may be the first early warning of danger to cold stress. During exposure to cold, maximum severe shivering develops when the body temperature has fallen to 35°C (95°F). This must be taken as a sign of danger to the workers and exposure to cold should be immediately terminated for any workers when severe shivering becomes evident. Useful physical or mental work is limited when severe shivering occurs.

Since prolonged exposure to cold air, or to immersion in cold water, at temperatures well above freezing can lead to dangerous hypothermia, whole body protection must be provided.

1. Adequate insulating dry clothing to maintain core temperatures above 36°C (96.8°F) must be provided to workers if work is performed in air temperatures below 4°C (40°F). Wind chill cooling rate and the cooling power of air are critical factors. [Wind chill cooling rate is defined as heat loss from a body expressed in watts per meter squared which is a function of the air temperature and wind velocity upon the exposed body.] The higher the wind speed and the lower the temperature in the work area, the greater the insulation value of the protective clothing required. An equivalent chill temperature chart relating the actual dry bulb air temperature and the wind velocity is presented in Table 2. The equivalent chill

**TABLE 1. Progressive Clinical Presentations of Hypothermia***

| Core Temperature | | Clinical Signs |
|---|---|---|
| °C | °F | |
| 37.6 | 99.6 | "Normal" rectal temperature |
| 37 | 98.6 | "Normal" oral temperature |
| 36 | 96.8 | Metabolic rate increases in an attempt to compensate for heat loss |
| 35 | 95.0 | Maximum shivering |
| 34 | 93.2 | Victim conscious and responsive, with normal blood pressure |
| 33 | 91.4 | Severe hypothermia below this temperature |
| 32 } 31 } | 89.6 } 87.8 } | Consciousness clouded; blood pressure becomes difficult to obtain; pupils dilated but react to light; shivering ceases |
| 30 } 29 } | 86.0 } 84.2 } | Progressive loss of consciousness; muscular rigidity increases; pulse and blood pressure difficult to obtain; respiratory rate decreases |
| 28 | 82.4 | Ventricular fibrillation possible with myocardial irritability |
| 27 | 80.6 | Voluntary motion ceases; pupils nonreactive to light; deep tendon and superficial reflexes absent |
| 26 | 78.8 | Victim seldom conscious |
| 25 | 77.0 | Ventricular fibrillation may occur spontaneously |
| 24 | 75.2 | Pulmonary edema |
| 22 } 21 } | 71.6 } 69.8 } | Maximum risk of ventricular fibrillation |
| 20 | 68.0 | Cardiac standstill |
| 18 | 64.4 | Lowest accidental hypothermia victim to recover |
| 17 | 62.6 | Isoelectric electroencephalogram |
| 9 | 48.2 | Lowest artificially cooled hypothermia patient to recover |

*Presentations approximately related to core temperature. Reprinted from the January 1982 issue of *American Family Physician*, published by the American Academy of Family Physicians.

temperature should be used when estimating the combined cooling effect of wind and low air temperatures on exposed skin or when determining clothing insulation requirements to maintain the deep body core temperature .

2. Unless there are unusual or extenuating circumstances, cold injury to other than hands, feet, and head is not likely to occur without the development of the initial signs of hypothermia. Older workers or workers with circulatory problems require special precautionary protection against cold injury. The use of extra insulating clothing and/or a reduction in the duration of the exposure period are among the special precautions which should be considered. The precautionary actions to be taken will depend upon the physical condition of the worker and should be determined with the advice of a physician with knowledge of the cold stress factors and the medical condition of the worker.

**TABLE 2. Cooling Power of Wind on Exposed Flesh Expressed as Equivalent Temperature (under calm conditions)★**

| Estimated Wind Speed (in mph) | Actual Temperature Reading (°F) | | | | | | | | | | | |
|---|---|---|---|---|---|---|---|---|---|---|---|---|
| | 50 | 40 | 30 | 20 | 10 | 0 | −10 | −20 | −30 | −40 | −50 | −60 |
| | Equivalent Chill Temperature (°F) | | | | | | | | | | | |
| calm | 50 | 40 | 30 | 20 | 10 | 0 | −10 | −20 | −30 | −40 | −50 | −60 |
| 5 | 48 | 37 | 27 | 16 | 6 | −5 | −15 | −26 | −36 | −47 | −57 | −68 |
| 10 | 40 | 28 | 16 | 4 | −9 | −24 | −33 | −46 | −58 | −70 | −83 | −95 |
| 15 | 36 | 22 | 9 | −5 | −18 | −32 | −45 | −58 | −72 | −85 | −99 | −112 |
| 20 | 32 | 18 | 4 | −10 | −25 | −39 | −53 | −67 | −82 | −96 | −110 | −121 |
| 25 | 30 | 16 | 0 | −15 | −29 | −44 | −59 | −74 | −88 | −104 | −118 | −133 |
| 30 | 28 | 13 | −2 | −18 | −33 | −48 | −63 | −79 | −94 | −109 | −125 | −140 |
| 35 | 27 | 11 | −4 | −20 | −35 | −51 | −67 | −82 | −98 | −113 | −129 | −145 |
| 40 | 26 | 10 | −6 | −21 | −37 | −53 | −69 | −85 | −100 | −116 | −132 | −148 |

| (Wind speeds greater than 40 mph have little additional effect.) | *LITTLE DANGER* In < hr with dry skin. Maximum danger of false sense of security | *INCREASING DANGER* Danger from freezing of exposed flesh within one minute. | *GREAT DANGER* Flesh may freeze within 30 seconds. |
|---|---|---|---|
| | Trenchfoot and immersion foot may occur at any point on this chart. | | |

* Developed by U.S. Army Research Institute of Environmental Medicine, Natick, MA.

☐ Equivalent chill temperature requiring dry clothing to maintain core body temperature above 36°C (96.8°F) per cold stress TLV.

## Evaluation and Control

For exposed skin, continuous exposure should not be permitted when the air speed and temperature results in an equivalent chill temperature of –32°C (–25.6°F). Superficial or deep local tissue freezing will occur only at temperatures below –1°C (30.2°F) regardless of wind speed.

At air temperatures of 2°C (35.6°F) or less, it is imperative that workers who become immersed in water or whose clothing becomes wet be immediately provided a change of clothing and be treated for hypothermia.

TLVs recommended for properly clothed workers for periods of work at temperatures below freezing are shown in Table 3.

Special protection of the hands is required to maintain manual dexterity for the prevention of accidents:

1. If fine work is to be performed with bare hands for more than 10–20 minutes in an environment below 16°C (60.8°F), special provisions should be established for keeping the workers' hands warm. For this purpose, warm air jets, radiant heaters (fuel burner or electric radiator), or contact warm plates may be utilized. Metal handles of tools and control bars should be covered by thermal insulating material at temperatures below –1°C (30.2°F).

2. If the air temperature falls below 16°C (60.8°F) for sedentary, 4°C (39.2°F) for light, –7°C (19.4°F) for moderate work, and fine manual dexterity is not required, then gloves should be used by the workers.

To prevent contact frostbite, the workers should wear anticontact gloves.

1. When cold surfaces below –7°C (19.4°F) are within reach, a warning should be given to each worker to prevent inadvertent contact by bare skin.

2. If the air temperature is –17.5°C (0°F) or less, the hands should be protected by mittens. Machine controls and tools for use in cold conditions should be designed so that they can be handled without removing the mittens.

Provisions for additional total body protection are required if work is performed in an environment at or below 4°C (39.2°F). The workers should wear cold protective clothing appropriate for the level of cold and physical activity:

1. If the air velocity at the job site is increased by wind, draft, or artificial ventilating equipment, the cooling effect of the wind should be reduced by shielding the work area or by wearing an easily removable windbreak garment.

2. If only light work is involved and if the clothing on the worker may become wet on the job site, the outer layer of the clothing in use may be of a type impermeable to water. With more severe work under such conditions, the outer layer should be water repellent, and the outerwear should be changed as it becomes wetted. The outer garments should include provisions for easy ventilation in order to prevent wetting of inner layers by sweat. If work is done at normal temperatures or in a hot environment before entering the cold area, the employee should make sure that clothing is not wet as a consequence of sweating. If clothing is wet, the employee should change into dry clothes before entering the cold area. The workers should change socks and any removable felt insoles at regular daily intervals or use vapor barrier boots. The optimal frequency of change should be determined empirically and will vary individually and according to the type of shoe worn and how much the individual's feet sweat.

3. If exposed areas of the body cannot be protected sufficiently to prevent sensation of excessive cold or frostbite, protective items should be supplied in auxiliary heated versions.

4. If the available clothing does not give adequate protection to prevent hypothermia or frostbite, work should be modified or sus-

pended until adequate clothing is made available or until weather conditions improve.

5. Workers handling evaporative liquid (gasoline, alcohol or cleaning fluids) at air temperatures below 4°C (39.2°F) should take special precautions to avoid soaking of clothing or gloves with the liquids because of the added danger of cold injury due to evaporative cooling. Special note should be taken of the particularly acute effects of splashes of "cryogenic fluids" or those liquids with a boiling point that is just above ambient temperature.

### Work–Warming Regimen

If work is performed continuously in the cold at an equivalent chill temperature (ECT) or below –7°C (19.4°F), heated warming shelters (tents, cabins, rest rooms, etc.) should be made available nearby. The workers should be encouraged to use these shelters at regular intervals, the frequency depending on the severity of the environmental exposure. The onset of heavy shivering, frostnip, the feeling of excessive fatigue, drowsiness, irritability, or euphoria are indications for immediate return to the shelter. When entering the heated shelter, the outer layer of clothing should be removed and the remainder of the clothing loosened to permit sweat evaporation or a change of dry work clothing provided. A change of dry work clothing should be provided as necessary to prevent workers from returning to work with wet clothing. Dehydration, or the loss of body fluids, occurs insidiously in the cold environment and may increase the susceptibility of the worker to cold injury due to a significant change in blood flow to the extremities. Warm sweet drinks and soups should be provided at the work site to provide caloric intake and fluid volume. The intake of coffee should be limited because of the diuretic and circulatory effects.

For work practices at or below –12°C (10.4°F) ECT, the following should apply:

1. The worker should be under constant protective observation (buddy system or supervision).

2. The work rate should not be so high as to cause heavy sweating that will result in wet clothing; if heavy work must be done, rest periods should be taken in heated shelters and opportunity for changing into dry clothing should be provided.

3. New employees should not be required to work fulltime in the cold during the first days of employment until they become accustomed to the working conditions and required protective clothing.

4. The weight and bulkiness of clothing should be included in estimating the required work performance and weights to be lifted by the worker.

5. The work should be arranged in such a way that sitting still or standing still for long periods is minimized. Unprotected metal chair seats should not be used. The worker should be protected from drafts to the greatest extent possible.

6. The workers should be instructed in safety and health procedures. The training program should include as a minimum instruction in:

   a. Proper rewarming procedures and appropriate first aid treatment.

   b. Proper clothing practices.

   c. Proper eating and drinking habits.

   d. Recognition of impending frostbite.

   e. Recognition of signs and symptoms of impending hypothermia or excessive cooling of the body even when shivering does not occur.

   f. Safe work practices.

### Special Workplace Recommendations

Special design requirements for refrigerator rooms include the following:

**TABLE 3. Threshold Limit Values Work/Warm-up Schedule for Four-Hour Shift**☆

| Air Temperature—Sunny Sky | | No Noticeable Wind | | 5 mph Wind | | 10 mph Wind | | 15 mph Wind | | 20 mph Wind | |
|---|---|---|---|---|---|---|---|---|---|---|---|
| °C (approx.) | °F (approx.) | Max. Work Period | No. of Breaks | Max. Work Period | No. of Breaks | Max. Work Period | No. of Breaks | Max. Work Period | No. of Breaks | Max. Work Period | No. of Breaks |
| −26° to −28° | −15° to −19° | (Norm. Breaks) | 1 | (Norm. Breaks) | 1 | 75 min | 2 | 55 min | 3 | 40 min | 4 |
| −29° to −31° | −20° to −24° | (Norm. Breaks) | 1 | 75 min | 2 | 55 min | 3 | 40 min | 4 | 30 min | 5 |
| −32° to −34° | −25° to −29° | 75 min | 2 | 55 min | 3 | 40 min | 4 | 30 min | 5 | Non-emergency work should cease | |
| −35° to −37° | −30° to −34° | 55 min | 3 | 40 min | 4 | 30 min | 5 | Non-emergency work should cease | | | |
| −38° to −39° | −35° to −39° | 40 min | 4 | 30 min | 5 | Non-emergency work should cease | | | | | |
| −40° to −42° | −40° to −44° | 30 min | 5 | Non-emergency work should cease | | | | | | | |
| −43° & below | −45° & below | Non-emergency work should cease | | | | | | | | | |

*Notes for Table 3 :*

1. Schedule applies to any 4-hour work period with moderate to heavy work activity, with warm-up periods of ten (10) minutes in a warm location and with an extended break (e.g., lunch) at the end of the 4-hour work period in a warm location. For Light-to-Moderate Work (limited physical movement): apply the schedule one step lower. For example, at −35°C (−30°F) with no noticeable wind (Step 4), a worker at a job with little physical movement should have a maximum work period of 40 minutes with 4 breaks in a 4-hour period (Step 5).

2. The following is suggested as a guide for estimating wind velocity if accurate information is not available:
   5 mph: light flag moves; 10 mph: light flag fully extended; 15 mph: raises newspaper sheet; 20 mph: blowing and drifting snow.

3. If only the wind chill cooling rate is available, a rough rule of thumb for applying it rather than the temperature and wind velocity factors given above would be: 1) special warm-up breaks should be initiated at a wind chill cooling rate of about 1750 W/m²; 2) all non-emergency work should have ceased at or before a wind chill of 2250 W /m². In general, the warmup schedule provided above slightly under-compensates for the wind at the warmer temperatures, assuming acclimatization and clothing appropriate for winter work. On the other hand, the chart slightly over-compensates for the actual temperatures in the colder ranges because windy conditions rarely prevail at extremely low temperatures.

4. TLVs apply only for workers in dry clothing.

☆Adapted from Occupational Health & Safety Division, Saskatchewan Department of Labour.

1. In refrigerator rooms, the air velocity should be minimized as much as possible and should not exceed 1 meter/sec (200 fpm) at the job site. This can be achieved by properly designed air distribution systems.

2. Special wind protective clothing should be provided based upon existing air velocities to which workers are exposed.

   Special caution should be exercised when working with toxic substances and when workers are exposed to vibration. Cold exposure may require reduced exposure limits.

   Eye protection for workers employed out-of-doors in a snow and/or ice-covered terrain should be supplied. Special safety goggles to protect against ultraviolet light and glare (which can produce temporary conjunctivitis and/or temporary loss of vision) and blowing ice crystals should be required when there is an expanse of snow coverage causing a potential eye exposure hazard.

   Workplace monitoring is required as follows:

1. Suitable thermometry should be arranged at any workplace where the environmental temperature is below 16°C (60.8°F) so that overall compliance with the requirements of the TLV can be maintained.

2. Whenever the air temperature at a workplace falls below −1°C (30.2°F), the dry bulb temperature should be measured and recorded at least every 4 hours.

3. In indoor workplaces, the wind speed should also be recorded at least every 4 hours whenever the rate of air movement exceeds 2 meters per second (5 mph).

4. In outdoor work situations, the wind speed should be measured and recorded together with the air temperature whenever the air temperature is below −1°C (30.2°F).

5. The equivalent chill temperature should be obtained from Table 2 in all cases where air movement measurements are required; it should be recorded with the other data whenever the equivalent chill temperature is below −7°C (19.4°F).

   Employees should be excluded from work in cold at −1°C (30.2°F) or below if they are suffering from diseases or taking medication which interferes with normal body temperature regulation or reduces tolerance to work in cold environments. Workers who are routinely exposed to temperatures below −24°C (−11.2°F) with wind speeds less than five miles per hour, or air temperatures below −18°C (0°F) with wind speeds above five miles per hour, should be medically certified as suitable for such exposures.

   Trauma sustained in freezing or subzero conditions requires special attention because an injured worker is predisposed to cold injury. Special provisions should be made to prevent hypothermia and freezing of damaged tissues in addition to providing for first aid treatment.

## HAND–ARM (SEGMENTAL) VIBRATION SYNDROME (HAVS)

The TLVs in Table 1 refer to component acceleration levels and durations of exposure that represent conditions under which it is believed that nearly all workers may be exposed repeatedly without progressing beyond Stage 1 of the Stockholm Workshop Classification System for Vibration-induced White Finger (VWF), also known as Raynaud's Phenomenon of Occupational Origin (Table 2). Since there is a paucity of dose–response relationships for VWF, these recommendations have been derived from epidemiological data from forestry, mining, and metal working. These values should be used as guides in the control of hand–arm vibration exposure; because of individual susceptibility, they should not be regarded as defining a boundary between safe and dangerous levels.

It should be recognized that control of hand–arm vibration syndrome (HAVS) from the workplace cannot occur simply by specifying and adhering to a given TLV. The use of: 1) antivibration tools, 2) antivibration gloves, 3) proper work practices which keep the worker's hands and remaining body warm and also minimize the vibration coupling between the worker and the vibration tool are necessary to minimize vibration exposure, and 4) a conscientiously applied medical surveillance program are ALL necessary to rid HAVS from the workplace.

### Continuous, Intermittent, Impulsive, or Impact Hand–Arm Vibration

The measurement of vibration should be performed in accordance with the procedures and instrumentation specified by the International Standard ISO 5349 (1986), Guide for the Measurement and the Assessment of Human Exposure to Hand Transmitted Vibration, or ANSI S3.34-1986, Guide for the Measurement and Evaluation of Human Exposure to Vibration Transmitted to the Hand, and summarized below:

The acceleration of a vibration handle or work piece should be determined in three mutually orthogonal directions at a point close to where vibration enters the hand. The directions should preferably be those forming the biodynamic coordinate system but may be a closely related basicentric system with its origin at the interface between the hand and the vibrating surface (Figure 1) to accommodate different handle or work piece configurations. A small and lightweight transducer should be mounted so as to record accurately one or more orthogonal components of the source vibration in the frequency range from 5 to 1500 Hz. Each component should be frequency-weighted by a filter network with gain characteristics specified for human-response vibration measuring instrumentation, to account for the change in vibration hazard with frequency (Figure 2).

Assessment of vibration exposure should be made for EACH applicable direction ($X_h$, $Y_h$, $Z_h$) since vibration is a vector quantity (magnitude and direction). In each direction, the magnitude of the vibration during normal operation of the power tool, machine or work piece should be expressed by the root-mean-square (rms) value of the

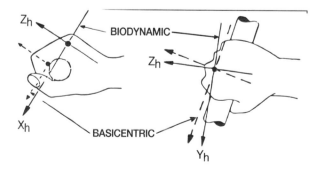

**Figure 1**—Biodynamic and basicentric coordinate systems for the hand, showing the directions of the acceleration components (ISO 5349 and ANSI S3.34–1986).

frequency-weighted component accelerations, in units of meters per second squared (m/s²), or gravitational units (g), the largest of which, $a_K$, forms the basis for exposure assessment.

For each direction being measured, linear integration should be employed for vibrations that are of extremely short duration or vary substantially in time. If the total daily vibration exposure in a given direction is composed of several exposures at different rms accelerations, then the equivalent, frequency-weighted component acceleration in that direction should be determined in accordance with the following equation:

$$\left(a_{K_{eq}}\right) = \left[\frac{1}{T} \sum_{i=1}^{n} \left(a_{K_i}\right)^2 T_i\right]^{1/2}$$

$$= \sqrt{\left(a_{K_1}\right)^2 \frac{T_1}{T} + \left(a_{K_2}\right)^2 \frac{T_2}{T} + \dots \left(a_{K_n}\right)^2 \frac{T_n}{T}}$$

where: $T = \sum_{i=1}^{n} T_i$

T = total daily exposure duration

$a_{K_i}$ = ith frequency-weighted, rms acceleration component with duration $T_i$

These computations may be performed by commercially available human-response vibration measuring instruments.

**TABLE 1. Threshold Limit Values for Exposure of the Hand to Vibration in Either $X_h$, $Y_h$, or $Z_h$ Directions**

| Total Daily Exposure Duration☆ | Values of the Dominant,★ Frequency-Weighted, rms, Component Acceleration Which Shall not be Exceeded $a_K$,($a_{K_{eq}}$) | |
|---|---|---|
| | m/s² | g△ |
| 4 hours and less than 8 | 4 | 0.40 |
| 2 hours and less than 4 | 6 | 0.61 |
| 1 hour and less than 2 | 8 | 0.81 |
| less than 1 hour | 12 | 1.22 |

☆ The total time vibration enters the hand per day, whether continuously or intermittently.

★ Usually one axis of vibration is dominant over the remaining two axes. If one or more vibration axes exceeds the Total Daily Exposure then the TLV has been exceeded.

△ g = 9.81 m/s².

### Notes for Table 1:

1. The weighting network provided in Figure 2 is considered the best available to frequency weight acceleration components. However, recent studies suggest that the frequency weighting at higher frequencies (above 16 Hz) may not incorporate a sufficient safety factor, and CAUTION must be applied when tools with high-frequency components are used.

2. Acute exposures to frequency-weighted, root-mean-square (rms), component accelerations in excess of the TLVs for infrequent periods of time (e.g., 1 day per week or several days over a 2-week period) are not necessarily more harmful.

3. Acute exposures to frequency-weighted, rms, component accelerations of three times the magnitude of the TLVs are expected to result in the same health effects after 5 to 6 years of exposure.

4. Preventive measures, including specialized preemployment and annual medical examinations to identify persons susceptible to vibration, should be implemented in situations in which workers are or will be exposed to hand–arm vibration.

5. To moderate the adverse effects of vibration exposure, workers should be advised to avoid continuous vibration exposure by cessation of vibration exposure for approximately 10 minutes per continuous vibration hour.

6. Good work practices should be used and should include instructing workers to employ a minimum hand grip force consistent with safe operation of the power tool or process, to keep their body and hands warm and dry, to avoid smoking, and to use antivibration tools and gloves when possible. As a general rule, gloves are more effective for damping vibration at high frequencies.

7. A vibration measurement transducer, together with its device for attachment to the vibration source, should weigh less than 15 grams and should possess a cross-axis sensitivity of less than 10%.

8. The measurement by many (mechanically underdamped) piezoelectric accelerometers of repetitive, large displacement, impulsive vibrations, such as those produced by percussive pneumatic tools,

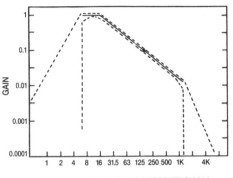

Figure 2—Gain characteristics on the filter network used to frequency-weight acceleration components (continuous line). The filter tolerances (dashed lines) are those contained in ISO 5349 and ANSI S3.34-1986.

is subject to error. The insertion of a suitable, low-pass, mechanical filter between the accelerometer and the source of vibration with a cut-off frequency of 1500 Hz or greater (and cross-axis sensitivity of less than 10%) can help eliminate incorrect readings.

9. The manufacturer and type number of all apparatus used to measure vibration should be reported, as well as the value of the dominant direction and frequency-weighted, rms, component acceleration.

**TABLE 2. Stockholm Workshop HAVS Classification System for Cold-induced Peripheral Vascular and Sensorineural Symptoms**

| | Vascular Assessment | |
|---|---|---|
| Stage | Grade | Description |
| 0 | —— | No attacks |
| 1 | Mild | Occasional attacks affecting only the tips of one or more fingers |
| 2 | Moderate | Occasional attacks affecting distal and middle (rarely also proximal) phalanges of one or more fingers |
| 3 | Severe | Frequent attacks affecting ALL phalanges of most fingers |
| 4 | Very Severe | As in Stage 3, with trophic skin changes in the finger tips |

Note: Separate staging is made for each hand, e.g., 2L(2)/1R(1) = stage 2 on left hand in 2 fingers; stage 1 on right hand in 1 finger.

| Sensorineural Assessment | |
|---|---|
| Stage | Symptoms |
| 0SN | Exposed to vibration but no symptoms |
| 1SN | Intermittent numbness, with or without tingling |
| 2SN | Intermittent or persistent numbness, reducing sensory perception |
| 3SN | Intermittent or persistent numbness, reducing tactile discrimination and/or manipulative dexterity |

Note: Separate staging is made for each hand.

# HEAT STRESS

The heat stress TLVs specified in Table 1 and Figure 1 refer to heat stress conditions under which it is believed that nearly all workers may be repeatedly exposed without adverse health effects. These TLVs are based on the assumption that nearly all acclimatized, fully clothed (e.g., lightweight pants and shirt) workers with adequate water and salt intake should be able to function effectively under the given working conditions without exceeding a deep body temperature of 38°C (100.4°F).

Where there is a requirement for protection against other harmful substances in the work environment and additional personal protective clothing and equipment must be worn, a correction to the Wet Bulb Globe Temperature (WBGT) TLV values, as presented in Table 2, must be applied.

Since measurement of deep body temperature is impractical for monitoring the workers' heat load, the measurement of environmental factors is required which most nearly correlate with deep body temperature and other physiological responses to heat. At the present time, the WBGT Index is the simplest and most suitable technique to measure the environmental factors. WBGT values are calculated by the following equations:

1. Outdoors with solar load:
   $$WBGT = 0.7\ NWB + 0.2\ GT + 0.1\ DB$$

2. Indoors or Outdoors with no solar load:
   $$WBGT = 0.7\ NWB + 0.3\ GT$$

where: WBGT = Wet Bulb Globe Temperature Index
NWB = Natural Wet-Bulb Temperature
DB = Dry-Bulb Temperature
GT = Globe Temperature

The determination of WBGT requires the use of a black globe thermometer, a natural (static) wet-bulb thermometer, and a dry-bulb thermometer.

Higher heat exposures than those shown in Table 1 and Figure 1 are permissible if the workers have been undergoing medical surveillance and it has been established that they are more tolerant to work in heat than the average worker. Workers should not be permitted to continue their work when their deep body temperature exceeds 38°C (100.4°F).

**Evaluation and Control**

*I. Measurement of the Environment*

The instruments required are a dry-bulb, a natural wet-bulb, a globe thermometer, and a stand. The measurement of the environmental factors should be performed as follows:

**TABLE 1. Examples of Permissible Heat Exposure Threshold Limit Values [Values are given in °C and (°F) WBGT]***

| | Work Load | | |
|---|---|---|---|
| Work–Rest Regimen | Light | Moderate | Heavy |
| Continuous work | 30.0 (86) | 26.7 (80) | 25.0 (77) |
| 75% Work — 25% Rest, each hour | 30.6 (87) | 28.0 (82) | 25.9 (78) |
| 50% Work — 50% Rest, each hour | 31.4 (89) | 29.4 (85) | 27.9 (82) |
| 25% Work — 75% Rest, each hour | 32.2 (90) | 31.1 (88) | 30.0 (86) |

*As workload increases, the heat stress impact on an unacclimatized worker is exacerbated (see Figure 1). For unacclimatized workers performing a moderate level of work, the permissible heat exposure TLV should be reduced by approximately 2.5°C.

**A.** The range of the dry and the natural wet-bulb thermometer should be –5°C to +50°C (23°F to 122°F) with an accuracy of ± 0.5°C. The dry bulb thermometer must be shielded from the sun and the other radiant surfaces of the environment without restricting the airflow around the bulb. The wick of the natural wet-bulb thermometer should be kept wet with distilled water for at least 1/2 hour before the temperature reading is made. It is not enough to immerse the other end of the wick into a reservoir of distilled water and wait until the whole wick becomes wet by capillarity. The wick should be wetted by direct application of water from a syringe 1/2 hour before each reading. The wick should extend over the bulb of the thermometer, covering the stem about one additional bulb length. The wick should always be clean and new wicks should be washed before using.

**B.** A globe thermometer, consisting of a 15-cm (6-inch) diameter hollow copper sphere painted on the outside with a matte black finish or equivalent, should be used. The bulb or sensor of a thermometer (range –5°C to +100°C [23°F to 212°F] with an accuracy of ± 0.5°C)

**TABLE 2. TLV WBGT Correction Factors in °C for Clothing**

| Clothing Type | Clo Value* | WBGT Correction |
|---|---|---|
| Summer work uniform | 0.6 | 0 |
| Cotton coveralls | 1.0 | –2 |
| Winter work uniform | 1.4 | –4 |
| Water barrier, permeable | 1.2 | –6 |

*Clo: Insulation value of clothing. One clo unit = 5.55 kcal/m²/hr of heat exchange by radiation and convection for each °C of temperature difference between the skin and adjusted dry-bulb temperature.

must be fixed in the center of the sphere. The globe thermometer should be exposed at least 25 minutes before it is read.

**C.** A stand should be used to suspend the three thermometers so that they do not restrict free air flow around the bulbs, and the wet-bulb and globe thermometers are not shaded.

**D.** It is permissible to use any other type of temperature sensor that gives a reading identical to that of a mercury thermometer under the same conditions.

**E.** The thermometers must be placed so that the readings are representative of the conditions under which the employees work or rest, respectively.

**II.** *Work Load Categories*

Heat produced by the body and the environmental heat together determine the total heat load. Therefore, if work is to be performed under hot environmental conditions, the workload category of each job should be established and the heat exposure limit pertinent to the workload evaluated against the applicable standard in order to protect the worker exposure beyond the permissible limit.

**A.** The work load category may be established by ranking each job into light, medium, or heavy categories on the basis of type of operation:

(1) light work (up to 200 kcal/hr or 800 Btu/hr): e.g., sitting or standing to control machines, performing light hand or arm work,

(2) moderate work (200–350 kcal/hr or 800–1400 Btu/hr): e.g., walking about with moderate lifting and pushing, or

(3) heavy work (350–500 kcal/hr or 1400–2000 Btu/hr): e.g., pick and shovel work.

Where the work load is ranked into one of said three categories, the permissible heat exposure TLV for each workload can be estimated from Table 1 or calculated using Tables 3 and 4.

**B.** The ranking of the job may be performed either by measuring the worker's metabolic rate while performing a job or by estimating the worker's metabolic rate with the use of Tables 3 and 4. Additional

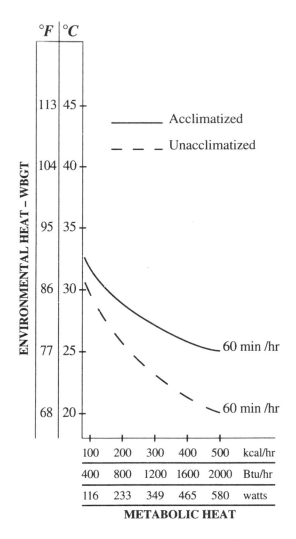

**Figure 1**—Permissible heat exposure Threshold Limit Values for heat acclimatized and unacclimatized workers.

tables available in the literature[1–4] may be utilized also. When this method is used, the permissible heat exposure TLV can be determined by Figure 1.

**III.** *Work–Rest Regimen*

The TLVs specified in Table 1 and Figure 1 are based on the assumption that the WBGT value of the resting place is the same or very close to that of the workplace. Where the WBGT of the work area is different from that of the rest area, a time-weighted average value should be used for both environmental and metabolic heat.

The time-weighted average metabolic rate (M) should be determined by the equation:

$$\text{Av. } M = \frac{M_1 \times t_1 + M_2 \times t_2 + ... + M_n \times t_n}{t_1 + t_2 + ... + t_n}$$

**TABLE 3. Assessment of Work Load**

Average values of metabolic rate during different activities.

| A. Body position and movement | kcal/min |
|---|---|
| Sitting | 0.3 |
| Standing | 0.6 |
| Walking | 2.0–3.0 |
| Walking up hill | add 0.8 per meter (yard) rise |

| B. Type of Work | | Average kcal/min | Range kcal/min |
|---|---|---|---|
| Hand work | *light* | 0.4 | 0.2–1.2 |
| | *heavy* | 0.9 | |
| Work with one arm | *light* | 1.0 | 0.7–2.5 |
| | *heavy* | 1.7 | |
| Work with both arms | *light* | 1.5 | 1.0–3.5 |
| | *heavy* | 2.5 | |
| Work with body | *light* | 3.5 | 2.5–15.0 |
| | *moderate* | 5.0 | |
| | *heavy* | 7.0 | |
| | *very heavy* | 9.0 | |

where $M_1$, $M_2$...and $M_n$ are estimated or measured metabolic rates for the various activities and rest periods of the worker during the time periods $t_1$, $t_2$...and $t_n$ (in minutes) as determined by a time study.

The time-weighted average WBGT should be determined by the equation:

$$\text{Av. WGBT} = \frac{\text{WBGT}_1 \times t_1 + \text{WBGT}_2 \times t_2 + ... + \text{WBGT}_n \times t_n}{t_1 + t_2 + ... + t_n}$$

where $\text{WBGT}_1$, $\text{WBGT}_2$ ... and $\text{WBGT}_n$ are calculated values of WBGT for the various work and rest areas occupied during total time periods and $t_1$, $t_2$ ... and $t_n$ are the elapsed times in minutes spent in the corresponding areas which are determined by a time study. Where exposure to hot environmental conditions is continuous for several hours or the entire work day, the time-weighted averages should be calculated as an hourly time-weighted average, i.e., $t_1 + t_2 + ... + t_n = 60$ minutes. Where the exposure is intermittent, the time-weighted averages should be calculated as two-hour time-weighted averages, i.e., $t_1 + t_2 + ... + t_n = 120$ minutes.

The TLVs for continuous work are applicable where there is a work–rest regimen of a 5-day work week and an 8-hour work day with a short morning and afternoon break (approximately 15 minutes) and a longer lunch break (approximately 30 minutes). Higher exposure values are permitted if additional resting time is allowed. All breaks, including unscheduled pauses and administrative or operational waiting periods during work, may be counted as rest time when additional rest allowance must be given because of high environmental temperatures.

### IV. *Water and Salt Supplementation*

During the hot season or when the worker is exposed to artificially generated heat, drinking water should be made available to the workers in such a way that they are stimulated to frequently drink small amounts, i.e., one cup every 15–20 minutes (about 150 ml or 1/4 pint).

The water should be kept reasonably cool, 10°C to 15°C (50°F to 60°F) and should be placed close to the workplace so that the worker can reach it without abandoning the work area.

The workers should be encouraged to salt their food well during the hot season and particularly during hot spells. If the workers are unacclimatized, salted drinking water should be made available in a concentration of 0.1% (1 g salt to 1.0 liter or 1 level tablespoon of salt to 15 quarts of water). The added salt should be completely dissolved before the water is distributed, and the water should be kept reasonably cool.

### TABLE 4. Activity Examples

- Light hand work: writing, hand knitting
- Heavy hand work: typewriting

- Heavy work with one arm: hammering in nails (shoemaker, upholsterer)
- Light work with two arms: filing metal, planing wood, raking of a garden
- Moderate work with the body: cleaning a floor, beating a carpet
- Heavy work with the body: railroad track laying, digging, barking trees

### Sample Calculation

Assembly line work using a heavy hand tool.

| | | |
|---|---|---|
| A. Walking along | | 2.0 kcal/min |
| B. Intermediate value between heavy work with two arms and light work with the body | | 3.0 kcal/min |
| | Subtotal: | 5.0 kcal/min |
| C. Add for basal metabolism | | 1.0 kcal/min |
| | Total: | 6.0 kcal/min |

### V. *Other Considerations*

**A.** *Clothing:* The permissible heat exposure TLVs are valid for light summer clothing as customarily worn by workers when working under hot environmental conditions. If special clothing is required for performing a particular job and this clothing is heavier or it impedes sweat evaporation or has higher insulation value, the worker's heat tolerance is reduced, and the permissible heat exposure TLVs indicated in Table 1 and Figure 1 are not applicable. For each job category where special clothing is required, the permissible heat exposure TLV should be established by an expert.

Table 2 identifies TLV WBGT correction factors for representative types of clothing.

**B.** *Acclimatization and Fitness:* Acclimatization to heat involves a series of physiological and psychological adjustments that occur in an individual during the first week of exposure to hot environmental conditions. The recommended heat stress TLVs are valid for acclimated workers who are physically fit. Extra caution must be employed when unacclimated or physically unfit workers must be exposed to heat stress conditions.

**C.** *Adverse Health Effects:* The most serious of heat-induced illnesses is heat stroke because of its potential to be life threatening or result in irreversible damage. Other heat-induced illnesses include heat exhaustion which in its most serious form leads to prostration and can cause serious injuries as well. Heat cramps, while debilitating, are easily reversible if properly and promptly treated. Heat disorders due to excessive heat exposure include electrolyte imbalance, dehydration, skin rashes, heat edema, and loss of physical and mental work capacity.

If during the first trimester of pregnancy, a female worker's core temperature exceeds 39° C (102.2°F) for extended periods, there is an increased risk of malformation to the unborn fetus. Additionally, core temperatures above 38°C (100.4°F) may be associated with temporary infertility in both females and males.

### References

1. Astrand, P-O.; Rodahl, K.: Textbook of Work Physiology. McGraw–Hill Book Co., New York, San Francisco (1970).
2. Ergonomics Guide to Assessment of Metabolic and Cardiac Costs of Physical Work. Am. Ind. Hyg. Assoc. J. 32:560 (1971).
3. Energy Requirements for Physical Work. Research Progress Report No. 30. Purdue Farm Cardiac Project, Agricultural Experiment Station, West Lafayette, IN (1961).
4. Durnin, J.V.G.A.; Passmore, R.: Energy, Work and Leisure. Heinemann Educational Books, Ltd., London (1967 ).

## IONIZING RADIATION STATEMENT

The Physical Agents TLV Committee accepts the occupational exposure guidance of the National Council on Radiation Protection and Measurements (NCRP) for ionizing radiation.

For the purpose of this TLV, ionizing radiation includes particulate and electromagnetic radiations having an energy exceeding 12.4 electron volts (eV) [1 eV = 1.6 × 10⁻¹⁹ joule (J)]. Animal and human studies have shown that exposure to ionizing radiation can result in carcinogenic, teratogenic, or mutagenic effects, as well as other sequelae. The scientific literature is summarized in critical reviews by the United Nations Scientific Committee on the Effects of Atomic Radiation (UNSCEAR) and the Biological Effects of Ionizing Radiation (BEIR) Committee of the U.S. National Academy of Science. The NCRP has formulated occupational exposure guidance that limits the risk of carcinogenesis and mutagenesis; further guidance is available regarding exposure of the pregnant woman to limit the *in utero* dose to the fetus. The TLV Committee also strongly recommends that all occupational exposure be minimized.

The NCRP and the International Commission on Radiological Protection (ICRP) disseminate information and recommendations about radiation protection and related issues, including the development of basic concepts. Their occupational exposure guidance is provided in NCRP Report No. 116 ("Recommendations on Limits for Exposure to Ionizing Radiation") and ICRP Report No. 60 ("The 1990 Recommendations of the ICRP"). Other ICRP and NCRP reports address specific areas of radiation protection and, collectively, provide an excellent basis for establishing a radiation control program.

## LASERS

The laser TLVs are for exposure to laser radiation under conditions to which nearly all workers may be exposed without adverse health effects. The values should be used as guides in the control of exposures and should not be regarded as fine lines between safe and dangerous levels. They are based on the best available information from experimental studies.

### Limiting Apertures

The TLVs expressed as radiant exposure or irradiance in this section may be averaged over an aperture of 1 mm, 3.5 mm, or 7 mm, depending upon the exposure condition as given in Table 1.

### Extended Sources

The TLVs for "extended sources" are obtained through the use of a correction factor $C_E$ provided at the bottom of Table 2 and are only applied at wavelengths in the retinal hazard region (400–1400 nm) for sources that subtend a linear angle greater than $\alpha_{min}$ and less than $\alpha_{max} = 100$ mrad. The angle $\alpha_{min}$ is defined as:

$\alpha_{min} = 1.5$ mrad for t < 0.7 s,
$\alpha_{min} = 2 \cdot t^{3/4}$ mrad for 0.7 s < t < 10 s, and
$\alpha_{min} = 11$ mrad for t > 10 s.

Normally, a laser source is not an "extended source," and the TLVs in Table 2 can be applied directly without $C_E$. Any sources whose centers are separated by an angle greater than $\alpha_{min}$ are treated as extended sources; examples would be some large laser diode arrays.

### Correction Factors A, B, C ($C_A$, $C_B$, $C_C$)

The TLVs for ocular exposure in Table 2 are to be used as given for all wavelength ranges. The TLVs for wavelengths between 700 nm and 1400 nm are to be increased by the factor $C_A$ (to account for reduced absorption of melanin) as given in Figure 1. For certain exposure times at wavelengths between 550 nm and 700 nm, a correction factor $C_B$ (to account for reduced photochemical sensitivity for retinal injury) must be applied. The correction factor $C_C$ is applied from 1150 to 1400 nm to account for pre-retinal absorption of the ocular media.

**TABLE 1. Limiting Apertures Applicable to Laser TLVs**

| Spectral Region | Duration | Eye | Skin |
|---|---|---|---|
| 180 nm–400 nm | 1 ns to 0.25 s | 1 mm | 3.5 mm |
| 180 nm–400 nm | 0.25 s to 30 ks | 3.5 mm | 3.5 mm |
| 400 nm–1400 nm | 1 ns to 0.25 s | 7 mm | 3.5 mm |
| 400 nm–1400 nm | 0.25 s to 30 ks | 7 mm | 3.5 mm |
| 1400 nm–0.1 mm | 1 ns to 0.25 s | 1 mm | 3.5 mm |
| 1400 nm –0.1 mm | 0.25 s to 30 ks | 3.5 mm | 3.5 mm |
| 0.1 mm–1.0 mm | 1 ns to 30 ks | 11 mm | 11 mm |

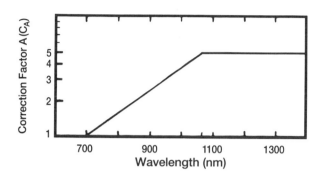

**Figure 1**—TLV correction factor for λ = 700–1400 nm* (*For λ = 700–1049 nm, $C_A = 10^{[(0.002(\lambda-700)]}$; for λ = 1050–1400 nm, $C_A = 5$ )

The TLVs for skin exposure are given in Table 3. The TLVs are to be increased by a factor $C_A$ as shown in Figure 1 for wavelengths between 700 nm and 1400 nm. To aid in the determination for exposure durations requiring calculations of fractional powers, Figures 2 and 3 may be used.

### Repetitively Pulsed Exposures

Scanned CW lasers or repetitively pulsed lasers can both produce repetitively pulsed exposure conditions. The TLV for intrabeam viewing which is applicable to wavelengths between 400 and 1400 nm and a single-pulse exposure (of pulse duration t) is modified in this instance by a correction factor determined by the number of pulses in the exposure. First, calculate the number of pulses (n) in an expected exposure situation; this is the pulse repetition frequency (PRF in Hz) multiplied by the duration of exposure. Normally, realistic exposures may range from 0.25 second (s) for a bright visible source to 10 s for an infrared source. The corrected TLV on a per-pulse basis is:

$$TLV = (n^{-1/4}) \text{ (TLV for single-pulse)} \quad (1)$$

This approach applies only to thermal-injury conditions, i.e., all exposures at wavelengths greater than 700 nm and for many exposures at shorter wavelengths. For wavelengths less than or equal to 700 nm, the corrected TLV from equation 1 above applies if the average irradiance does not exceed the TLV for continuous exposure. The average irradiance (i.e., the total accumulated exposure for nt s) shall not exceed the radiant exposure given in Table 2 for exposure durations of 10 s to $T_1$.

It is recommended that the user of the TLVs for laser radiation consult *A Guide for Control of Laser Hazards*, 4th Edition, 1990, published by ACGIH for additional relevant information.

**TABLE 2. Threshold Limit Values for Direct Ocular Exposures (Intrabeam Viewing) from a Laser Beam**

| Spectral Region | Wave Length | Exposure, (t) Seconds | TLV |
|---|---|---|---|
| UVC | 180 nm to 280 nm* | $10^{-9}$ to $3 \times 10^4$ | 3  mJ/cm$^2$ |
| UVB | 280 nm to 302 nm | " | 3 " |
| | 303 nm | " | 4 " |
| | 304 nm | " | 6 " |
| | 305 nm | " | 10 " |
| | 306 nm | " | 16 " |
| | 307 nm | " | 25 " |
| | 308 nm | " | 40 " |
| | 309 nm | " | 63 " |
| | 310 nm | " | 100 " |
| | 311 nm | " | 160 " |
| | 312 nm | " | 250 " |
| | 313 nm | " | 400 " |
| | 314 nm | " | 630 " |
| UVA | 315 nm to 400 nm | $10^{-9}$ to 10 | $0.56t^{1/4}$ J/cm$^2$ |
| | " " | 10 to $10^3$ | 1.0 J/cm$^2$ |
| | " " | $10^3$ to $3 \times 10^4$ | 1.0 mW/cm$^2$ |
| Light | 400 nm to 700 nm | $10^{-9}$ to $1.8 \times 10^{-5}$ | $5 \times 10^{-7}$ J/cm$^2$ |
| | 400 nm to 700 nm | $1.8 \times 10^{-5}$ to 10 | $1.8 \, (t/\sqrt[4]{t})$ mJ/cm$^2$ |
| | 400 nm to 549 nm | 10 to $10^4$ | 10 mJ/cm$^2$ |
| | 550 nm to 700 nm | 10 to $T_1$ | $1.8 \, (t/\sqrt[4]{t})$ mJ/cm$^2$ |
| | 550 nm to 700 nm | $T_1$ to $10^4$ | $10 \, C_B$ mJ/cm$^2$ |
| | 400 nm to 700 nm | $10^4$ to $3 \times 10^4$ | $C_B$ µW/cm$^2$ |
| IRA | 700 nm to 1049 nm | $10^{-9}$ to $1.8 \times 10^{-5}$ | $5 \, C_A \times 10^{-7}$ J/cm$^2$ |
| | 700 nm to 1049 nm | $1.8 \times 10^{-5}$ to $10^3$ | $1.8 \, C_A \, (t/\sqrt[4]{t})$ mJ/cm$^2$ |
| | 1050 nm to 1400 nm | $10^{-9}$ to $5 \times 10^{-5}$ | $5 \, C_C \times 10^{-6}$ J/cm$^2$ |
| | 1050 nm to 1400 nm | $5 \times 10^{-5}$ to $10^3$ | $9 \, C_C \, (t/\sqrt[4]{t})$ mJ/cm$^2$ |
| | 700 nm to 1400 nm | $10^3$ to $3 \times 10^4$ | $320 \, C_A \cdot C_C$ µW/cm$^2$ |
| IRB & C | 1.401 µm to 1.5 µm | $10^{-9}$ to $10^{-3}$ | 0.1 J/cm$^2$ |
| | 1.401 µm to 1.5 µm | $10^{-3}$ to 10 | $0.56 \, t^{1/4}$ J/cm$^2$ |
| | 1.501 µm to 1.8 µm | $10^{-9}$ to 10 | 1.0 J/cm$^2$ |
| | 1.801 µm to 2.6 µm | $10^{-9}$ to $10^{-3}$ | 0.1 J/cm$^2$ |
| | 1.801 µm to 2.6 µm | $10^{-3}$ to 10 | $0.56 \, t^{1/4}$ J/cm$^2$ |
| | 2.601 µm to $10^3$ µm | $10^{-9}$ to $10^{-7}$ | 10 mJ/cm$^2$ J/cm$^2$ |
| | 2.601 µm to $10^3$ µm | $10^{-7}$ to 10 | $0.56 \, t^{1/4}$ J/cm$^2$ |
| | 1.400 µm to $10^3$ µm | 10 to $3 \times 10^4$ | 100 mW/cm$^2$ |

The UVB and UVC rows are bracketed with the note: **not to exceed $0.56 \, t^{1/4}$ J/cm$^2$ for $t \le 10$ s.**

*Ozone (O$_3$) is produced in air by sources emitting ultraviolet (UV) radiation at wavelengths below 250 nm. Refer to Chemical Substances TLV for ozone.

### Notes for Table 2

$C_A$ = Fig. 1; $C_B$ = 1 for $\lambda$ = 400 to 549 nm; $C_B = 10^{[0.015 \, (\lambda - 550)]}$ for $\lambda$ = 550 to 700 nm; $C_C$ = 1.0 from 700 to 1150 nm;

$C_C = 10^{[0.0181 \, (\lambda - 1150)]}$ for wavelengths greater than 1150 nm and less than 1200 nm; $C_C$ = 8.0 from 1200 to 1400 nm;

$T_1$ = 10 s for $\lambda$ = 400 to 549 nm; $T_1 = 10 \times 10^{[0.02 \, (\lambda - 550)]}$ for $\lambda$ = 550 to 700 nm.

* For extended-source laser radiation (e.g., diffuse reflection viewing) at wavelengths between 400 nm and 1400 nm, the intrabeam viewing TLVs can be increased by the following correction factor ($C_E$) provided that the angular subtense of the source (measured at the viewer's eye) is greater than $\alpha_{min}$ (e.g., greater than 1.5 mrad for $t < 0.7$ s; $\alpha_{min} = 2 \cdot t^{3/4}$ mrad for $0.7 < t < 10$ s, and $\alpha_{min}$ = 11 mrad for $t > 10$ s).

The value of $C_E$ is equal to 1.0 for all angles $\alpha$ less than $\alpha_{min}$ and is never less than 1.0.

$C_E = (\alpha/\alpha_{min})$ for $\alpha_{min} < \alpha < \alpha_{max}$ = 100 mrad.

$C_E = \alpha^2 /(\alpha_{min} \cdot \alpha_{max})$ for $\alpha > \alpha_{max}$ = 100 mrad for $\alpha$ expressed in mrad.

The angle of 100 mrad may also be referred to as $\alpha_{max}$ at which point the TLVs may be expressed as a constant radiance and the last equation can be rewritten in terms of radiance L.

$L_{TLV} = (8.5 \times 10^3) \times (TLV_{pt \, source})$ J(cm$^2$ • sr) for 0.7 s

$L_{TLV} = (6.4 \times 10^3 \, t^{-3/4}) \times (TLV_{pt \, source})$ J(cm$^2$ • sr) for 0.7 s < t < 10 s

$L_{TLV} = (1.2 \times 10^3) \times (TLV_{pt \, source})$ J(cm$^2$ • sr) [or expressed in W(cm$^2$ • sr) as applicable] for t > 10 s

The measurement aperture should be placed at a distance of 100 mm or greater from the source.

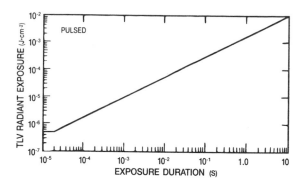

Figure 2a—TLV for intrabeam (direct) viewing of laser beam (400–700 nm)

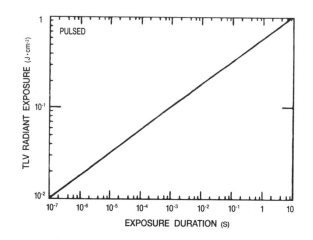

Figure 3a—TLV for laser exposure of skin and eyes for far-infrared radiation (wave-lengths greater than 1.4 μm).

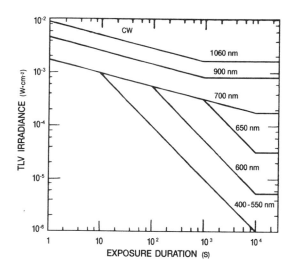

Figure 2b—TLV for intrabeam (direct) viewing of CW laser beam (400–1400 nm)

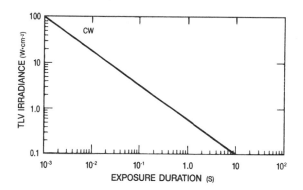

Figure 3b—TLV for CW laser exposure of skin and eyes for far-infrared radiation (wave-lengths greater than 1.4 μm).

### TABLE 3. Threshold Limit Value for Skin Exposure from a Laser Beam

| Spectral Region | Wave Length | Exposure, (t) Seconds | TLV |
|---|---|---|---|
| UV* | 180 nm to 400 nm | $10^{-9}$ to $3 \times 10^4$ | Same as Table 2 |
| Light & | 400 nm to 1400 nm | $10^{-9}$ to $10^{-7}$ | $2\,C_A \times 10^{-2}$ J/cm² |
| IRA | " " | $10^{-7}$ to 10 | $1.1\,C_A \sqrt[4]{t}$ J/cm² |
| | " " | 10 to $3 \times 10^4$ | $0.2\,C_A$ W/cm² |
| IRB & C | 1.401 μm to $10^3$ μm | $10^{-9}$ to $3 \times 10^4$ | Same as Table 2 |

*Ozone ($O_3$) is produced in air by sources emitting ultraviolet (UV) radiation at wavelengths below 250 nm. Refer to Chemical Substances TLV for ozone.

$C_A = 1.0$ for λ = 400 – 700 nm; see Figure 1 for λ = 700 to 1400 nm .

At wavelengths greater than 1400 nm, for beam cross-sectional areas exceeding 100 cm² the TLV for exposure durations exceeding 10 seconds is: TLV = $(10,000/A_s)$ mW/cm²; where $A_s$ is the irradiated skin area for 100 to 1000 cm². and the TLV for irradiated skin areas exceeding 1000 cm² is 10 mW/cm² and for irradiated skin areas less than 100 cm² is 100 mW/cm².

# *LIGHT AND NEAR-INFRARED RADIATION

These TLVs refer to values for visible and near-infrared radiation in the wavelength region of 400 nm to 3000 nm and represent conditions under which it is believed that nearly all workers may be exposed without adverse health effects. The values are based on the best available information from experimental studies and should be used only as guides in the control of exposures to light and should not be regarded as fine lines between safe and dangerous levels. For purposes of specifying this TLV, the optical radiation spectrum has been divided into regions shown in Table 1.

**TABLE 1. Regions of the Optical Radiation Spectrum**

| Region | Wavelength Range |
|---|---|
| Ultraviolet (UV) | 100 to 380–400 nm |
| UV-C | 100 to 280 nm |
| UV-B | 280 to 315–320 nm |
| UV-A | 315–320 to 380–400 nm |
| Visible (Light) | 380–400 to 760–780 nm |
| Infrared (IR) | 760–780 to 1 mm |
| IR-A | 760–780 nm to 1.4 µm |
| IR-B | 1.4–3.0 µm |
| IR-C | 3.0 µm to 1 mm |

## Recommended Values

The TLV for occupational exposure to broad-band light and near-infrared radiation for the eye apply to exposure in any 8-hour workday and require knowledge of the spectral radiance ($L_\lambda$) and total irradiance (E) of the source as measured at the position(s) of the eyes of the worker. Such detailed spectral data of a white-light source are generally required only if the luminance of the source exceeds 1 cd/cm$^2$. At luminances less than this value, the TLV would not be exceeded.

The TLVs are:

1. *To protect against retinal thermal injury from a visible light source,* the spectral radiance of the lamp weighted against the function $R_\lambda$ (given in Table 2) should not exceed:

$$\sum_{400}^{1400} L_\lambda \cdot R_\lambda \cdot \Delta\lambda \le \frac{5}{\alpha t^{1/4}} \qquad (1)^\Delta$$

where $L_\lambda$ is in W/(cm$^2$ · sr · nm), t is the viewing duration (or pulse duration if the lamp is pulsed) expressed in seconds, but restricted to durations of 1 µs to 10 s, and $\alpha$ is the angular subtense of the source in radians. If the lamp is oblong, $\alpha$ refers to the longest dimension that can be viewed. For instance, at a viewing distance r = 100 cm from a tubular lamp of length l = 50 cm, the viewing angle $\alpha$ is:

$$\alpha = l/r = 50/100 = 0.5 \text{ rad} \qquad (2)$$

2. *To protect against retinal photochemical injury from chronic blue-light (light [400 < λ < 700 nm]) exposure,* the integrated spectral radiance of a light source weighted against the blue-light hazard function $B_\lambda$ (as given in Table 2) should not exceed:

$$\sum_{400}^{700} L_\lambda \cdot t \cdot B_\lambda \cdot \Delta\lambda \le 100 \text{ J/(cm}^2 \cdot \text{sr) } (t \le 10^4 \text{s)} \qquad (3a)$$

$$\sum_{400}^{700} L_\lambda \cdot B_\lambda \cdot \Delta\lambda \le 10^{-2} \text{ W/(cm}^2 \cdot \text{sr) } (t > 10^4 \text{s)} \qquad (3b)$$

The weighted product of $L_\lambda$ and $B_\lambda$ is termed $L_{blue}$. For a source radiance L weighted against the blue-light hazard function ($L_{blue}$) which exceeds 10 mW/cm$^2$/sr in the blue spectral region, the permissible exposure duration $t_{max}$ in seconds is simply:

$$t_{max} \le \frac{100 \text{ J/(cm}^2 \cdot \text{sr)}}{L_{blue}} \text{ (for } t \le 10^4 \text{s)} \qquad (4)$$

The latter limits are greater than the maximum permissible exposure limits for 440 nm laser radiation (see Laser TLV) because of the need for caution related to narrow-band spectral effects in the case of the Laser TLV. For a light source subtending an angle less than 11 mrad (0.011 radian), the above limits are relaxed such that the spectral irradiance weighted against the blue-light hazard function $B_\lambda$ should not exceed $E_{blue}$.

$$\sum_{400}^{700} E_\lambda \cdot t \cdot B_\lambda \cdot \Delta\lambda \le 10 \text{ mJ/cm}^2 \ (t \le 10^4 \text{s}) \qquad (5a)$$

$$\sum_{400}^{700} E_\lambda \cdot B_\lambda \cdot \Delta\lambda \le 1.0 \text{ µW/cm}^2 \ (t > 10^4 \text{s}) \qquad (5b)$$

For a source where the blue-light-weighted irradiance $E_{blue}$ exceeds 1 µW/cm$^2$, the maximum permissible exposure duration $t_{max}$ in seconds is:

$$t_{max} \le \frac{10 \text{ mJ/cm}^2}{E_{blue}} \text{ (for } t \le 10^4 \text{s)} \qquad (6)$$

3. *To protect the worker having a lens removed (cataract surgery), against retinal photochemical injury from chronic exposure,* $B_\lambda$ may not adequately provide an indication of an increased blue-light hazard. Unless an ultraviolet (UV)-absorbing intraocular lens has been surgically inserted into the eye, an adjusted $B_\lambda$ function should be used in equations 3a, 3b, 5a, and 5b and the function summed from 300 to 700 nm. This alternative $B_\lambda$ function is termed the Aphakic Hazard Function, $A_\lambda$.

4. Infrared (IR) radiation:

   a. *To Protect the Cornea and Lens:* To avoid thermal injury of the cornea and possible delayed effects upon the lens of the eye (cataractogenesis), the infrared radiation (770 nm < λ < 3 µm) exposure should be limited for lengthy periods (≥ 1000 s) to 10 mW/cm$^2$, and to:

$$\sum_{770}^{3000} E_\lambda \cdot \Delta\lambda \le 1.8 \, t^{-3/4} \text{ W/cm}^2 \text{ (for } t < 1000 \text{ s)} \qquad (7)$$

   b. *To Protect the Retina:* For an infrared heat lamp or any near-IR source where a strong visual stimulus is absent, the IR-A, or near-IR (770 nm < λ < 1400 nm) radiance as viewed by the eye should be limited to:

$$\sum_{770}^{1400} L_\lambda \cdot \Delta\lambda \le \frac{0.6}{\alpha} \qquad (8)$$

for extended duration viewing conditions. This limit is based upon a 7-mm pupil diameter (since the aversion response may not exist due to an absence of light) and a detector field-of-view of 11 mrad.

---

$^\Delta$Equations (1) and (8) are empirical and are not, strictly speaking, dimensionally correct. To make the equations dimensionally correct, one would have to insert a dimensional correction factor k in the right-hand numerator in each equation. For equation (1), this would be $k_1 = 1$ W · rad · s$^{1/2}$/(cm$^2$ · sr), and for equation (8), $k_2 = 1$ W · rad/(cm$^2$ · sr).

**TABLE 2. Retinal and UVR Hazard Spectral Weighting Functions**

| Wavelength (nm) | Aphakic Hazard Function $A_\lambda$ | Blue-Light Hazard Function $B_\lambda$ | Retinal Thermal Hazard Function $R_\lambda$ |
|---|---|---|---|
| 305 | 6.00 | — | |
| 310 | 6.00 | — | |
| 315 | 6.00 | — | |
| 320 | 6.00 | — | |
| 325 | 6.00 | — | |
| 330 | 6.00 | — | |
| 335 | 6.00 | — | |
| 340 | 5.88 | — | |
| 345 | 5.71 | — | |
| 350 | 5.46 | — | |
| 355 | 5.22 | — | |
| 360 | 4.62 | — | |
| 365 | 4.29 | — | |
| 370 | 3.75 | — | |
| 375 | 3.56 | — | |
| 380 | 3.19 | — | |
| 385 | 2.31 | — | |
| 390 | 1.88 | — | |
| 395 | 1.58 | — | |
| 400 | 1.43 | 0.100 | 1.0 |
| 405 | 1.30 | 0.200 | 2.0 |
| 410 | 1.25 | 0.400 | 4.0 |
| 415 | 1.20 | 0.800 | 8.0 |
| 420 | 1.15 | 0.900 | 9.0 |
| 425 | 1.11 | 0.950 | 9.5 |
| 430 | 1.07 | 0.980 | 9.8 |
| 435 | 1.03 | 1.000 | 10.0 |
| 440 | 1.000 | 1.000 | 10.0 |
| 445 | 0.970 | 0.970 | 9.7 |
| 450 | 0.940 | 0.940 | 9.4 |
| 455 | 0.900 | 0.900 | 9.0 |
| 460 | 0.800 | 0.800 | 8.0 |
| 465 | 0.700 | 0.700 | 7.0 |
| 470 | 0.620 | 0.620 | 6.2 |
| 475 | 0.550 | 0.550 | 5.5 |
| 480 | 0.450 | 0.450 | 4.5 |
| 485 | 0.400 | 0.400 | 4.0 |
| 490 | 0.220 | 0.220 | 2.2 |
| 495 | 0.160 | 0.160 | 1.6 |
| 500 | 0.100 | 0.100 | 1.0 |
| 505 | 0.079 | 0.079 | 1.0 |
| 510 | 0.063 | 0.063 | 1.0 |
| 515 | 0.050 | 0.050 | 1.0 |
| 520 | 0.040 | 0.040 | 1.0 |
| 525 | 0.032 | 0.032 | 1.0 |
| 530 | 0.025 | 0.025 | 1.0 |
| 535 | 0.020 | 0.020 | 1.0 |
| 540 | 0.016 | 0.016 | 1.0 |
| 545 | 0.013 | 0.013 | 1.0 |
| 550 | 0.010 | 0.010 | 1.0 |
| 555 | 0.008 | 0.008 | 1.0 |
| 560 | 0.006 | 0.006 | 1.0 |
| 565 | 0.005 | 0.005 | 1.0 |
| 570 | 0.004 | 0.004 | 1.0 |
| 575 | 0.003 | 0.003 | 1.0 |
| 580 | 0.002 | 0.002 | 1.0 |
| 585 | 0.002 | 0.002 | 1.0 |
| 590 | 0.001 | 0.001 | 1.0 |
| 595 | 0.001 | 0.001 | 1.0 |
| 600–700 | 0.001 | 0.001 | 1.0 |
| 700–1050 | — | — | $10^{[(700-\lambda)/500]}$ |
| 1050–1400 | — | — | 0.2 |

## *NOISE

These TLVs refer to sound pressure levels and durations of exposure that represent conditions under which it is believed that nearly all workers may be repeatedly exposed without adverse effect on their ability to hear and understand normal speech. Prior to 1979, the medical profession had defined hearing impairment as an average hearing threshold level in excess of 25 decibels (ANSI S3.6-1989)[1] at 500, 1000, and 2000 hertz (Hz). The limits that are given here have been established to prevent a hearing loss at 3000 Hz and 4000 Hz also. The values should be used as guides in the control of noise exposure and, due to individual susceptibility, should not be regarded as fine lines between safe and dangerous levels.

It should be recognized that the application of the TLV for noise will not protect all workers from the adverse effects of noise exposure. The TLV should protect the median of the population against a noise-induced hearing loss exceeding 2 dB after 40 years of occupational exposure for the average of 0.5, 1, 2, and 3 kHz. A hearing conservation program with all its elements including audiometric testing is necessary when workers are exposed to noise at or above the TLV levels.

**TABLE 1. Threshold Limit Values for Noise**☆

| | Duration per Day | Sound Level dBA★ |
|---|---|---|
| Hours | 24 | 80 |
| | 16 | 82 |
| | 8 | 85 |
| | 4 | 88 |
| | 2 | 91 |
| | 1 | 94 |
| Minutes | 30 | 97 |
| | 15 | 100 |
| | 7.50△ | 103 |
| | 3.75△ | 106 |
| | 1.88△ | 109 |
| | 0.94△ | 112 |
| Seconds△ | 28.12 | 115 |
| | 14.06 | 118 |
| | 7.03 | 121 |
| | 3.52 | 124 |
| | 1.76 | 127 |
| | 0.88 | 130 |
| | 0.44 | 133 |
| | 0.22 | 136 |
| | 0.11 | 139 |

No exposure to continuous, intermittent, or impact noise in excess of a peak C-weighted level of 140 dB.

★ Sound level in decibels are measured on a sound level meter, conforming as a minimum to the requirements of the American National Standards Institute Specification for Sound Level Meters, S1.4 (1983)[2] Type S2A, and set to use the A-weighted network with slow meter response.

△ Limited by the noise source—not by administrative control. It is also recommended that a dosimeter or integrating sound level meter be used for sounds above 120 decibels.

### Continuous or Intermittent Noise

The sound pressure level should be determined by a sound level meter or dosimeter conforming, as a minimum, to the requirements of the American National Standards Institute (ANSI) Specification for Sound Level Meters, S1.4-1983, Type S2A,[2] or ANSI S1.25-1991 Specification for Personal Noise Dosimeters.[3] The measurement device should be set to use the A-weighted network with slow meter response. The duration of exposure should not exceed that shown in Table 1. These values apply to total duration of exposure per working day regardless of whether this is one continuous exposure or a number of short-term exposures.

When the daily noise exposure is composed of two or more periods of noise exposure of different levels, their combined effect should be considered rather than the individual effect of each. If the sum of the following fractions:

$$\frac{C_1}{T_1} + \frac{C_2}{T_2} + \dots \frac{C_n}{T_n}$$

exceeds unity, then the mixed exposure should be considered to exceed the TLV. $C_1$ indicates the total duration of exposure at a specific noise level, and $T_1$ indicates the total duration of exposure permitted at that level. All on-the-job noise exposures of 80 dBA or greater should be used in the above calculations. With sound level meters, this formula should be used for sounds with steady levels of at least 3 seconds. For sounds in which this condition is not met, a dosimeter or an integrating sound level meter must be used. The limit is exceeded when the dose is more than 100% as indicated on a dosimeter set with a 3 dB exchange rate and an 8-hour criteria level of 85 dBA.

The TLV is exceeded on an integrating sound level meter when the average sound level exceeds the values of Table 1.

### Impulsive or Impact Noise

By using the instrumentation specified by the ANSI S1.4,[2] S1.25,[3] or IEC 804,[4] impulsive or impact noise is automatically included in the noise measurement. The only requirement is a measurement range between 80 and 140 dBA and the pulse range must be at least 63 dB. No exposures of an unprotected ear in excess of a C-weighted peak sound pressure level of 140 dB should be permitted. If hearing protection that reduces the C-weighted peak level to levels below 140 dB and if instrumentation is not available to measure a C-weighted peak, an unweighted peak measurement below 140 dB may be used to imply that the C-weighted peak is below 140 dB.

### References

1. American National Standards Institute: ANSI S3.6-1989, Specification for Audiometers. ANSI, New York (1989).
2. American National Standards Institute: ANSI S1.4-1983, Specification for Sound Level Meters. ANSI, New York (1983).
3. American National Standards Institute: ANSI S1.25-1991, Specification for Personal Noise Dosimeters. ANSI, New York (1991).
4. International Electrotechnical Commission: Integrating-Averaging Sound Level Meters. IEC 804. IEC, New York (1985).

## *RADIOFREQUENCY/MICROWAVE RADIATION

These TLVs refer to radiofrequency (RF) and microwave radiation in the frequency range 30 kHz to 300 GHz and represent conditions under which it is believed that nearly all workers may be repeatedly exposed without adverse health effects. The TLVs, in terms of RMS electric (E) and magnetic (H) field strengths, the equivalent plane-wave free-space power densities (PD), and induced currents in the body which can be associated with exposure to such fields or contact with objects exposed to such fields, are given in Table 1 and Figure 1 as a function of frequency.

(a) Access should be restricted to limit the RMS RF body current and potential for RF shock or burn as follows:

(i) For freestanding individuals (no contact with metallic objects), RF current induced in the human body, as measured through each foot, should not exceed the following values:

$$I = 1000f \text{ mA for } (0.03 < f < 0.1 \text{ MHz})$$

$$I = 100 \text{ mA for } (0.1 < f < 100 \text{ MHz})$$

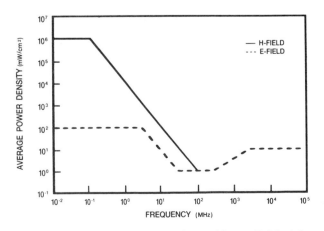

**Figure 1**—Threshold Limit Values (TLV) for Radiofrequency/Microwave Radiation in the workplace (whole body SAR < 0.4 W/kg).

### TABLE 1. Radiofrequency/Microwave Threshold Limit Values

#### Part A—Electromagnetic Fields☆
f = frequency in MHz

| Frequency | Power Density, S (mW/cm²) | Electric Field Strength (V/m) | Magnetic Field Strength (A/m) | Averaging Time E², H² or S (minutes) |
|---|---|---|---|---|
| 30 kHz–100 kHz | | 614 | 163 | 6 |
| 100 kHz–3 MHz | | 614 | 16.3/f | 6 |
| 3 MHz–30 MHz | | 1842/f | 16.3/f | 6 |
| 30 MHz–100 MHz | | 61.4 | 16.3/f | 6 |
| 100 MHz–300 MHz | 1 | 61.4 | 0.163 | 6 |
| 300 MHz–3 GHz | f/300 | | | 6 |
| 3 GHz–15 GHz | 10 | | | 6 |
| 15 GHz–300 GHz | 10 | | | 616,000/f^{1.2} |

☆The exposure values in terms of electric and magnetic field strengths are the values obtained by spatially averaging values over an area equivalent to the vertical cross-section of the human body (projected area).

#### Part B—Induced and Contact Radio Frequency Currents★
Maximum Current (mA)

| Frequency | Through Both Feet | Through Each Foot | Contact |
|---|---|---|---|
| 30 kHz–100 kHz | 2000f | 1000f | 1000f |
| 100 kHz–100 MHz | 200 | 100 | 100 |

★It should be noted that the current limits given above may not adequately protect against startle reactions and burns caused by transient discharges when contacting an energized object. See text for additional comment.

(ii) For conditions of possible contact with metallic bodies, maximum RF current through an impedance equivalent to that of the human body for conditions of grasping contact as measured with a contact current meter should not exceed the following values:

$$I = 1000f \text{ mA for } (0.03 < f < 0.1 \text{ MHz})$$

$$I = 100 \text{ mA for } (0.1 < f < 100 \text{ MHz})$$

The means of compliance with this current limit can be determined by the user of the TLV as appropriate. The use of protective gloves, the prohibition of metallic objects, or training of personnel may be sufficient to assure compliance with this aspect of the TLVs. Evaluation of the magnitude of the induced currents will normally require a direct measurement.

(b) The TLVs refer to exposure values obtained by spatially averaging over an area equivalent to the vertical cross-section of the human body (projected area). In the case of partial body exposure, the TLVs can be relaxed. In nonuniform fields, spatial peak values of field strength may exceed the TLVs if the spatially averaged value remains within the specified limits. The TLVs may also be relaxed by reference to SAR limits by appropriate calculations or measurements.

(c) For near-field exposures at frequencies less than 300 MHz, the applicable TLV is in terms of RMS electric and magnetic field strength, as given in Table 1, columns 3 and 4. Equivalent plane-wave PD can be calculated from the field strength measurement data as follows:

$$PD \text{ (mW/cm}^2) = \frac{E^2}{3770}$$

where: $E^2$ is in volts squared ($V^2$) per meter squared ($m^2$); and

$$PD \text{ (mW/cm}^2) = 37.7H^2$$

where: $H^2$ is in amperes squared ($A^2$) per meter squared ($m^2$).

(d) For exposures to pulsed RF fields of pulse duration less than 100 msec and frequencies in the range of 0.1 to 300,000 MHz, the TLV in terms of peak power density for a single pulse is given by the TLV multiplied by the averaging time in seconds and divided by 5 times the pulse width in seconds, that is,

$$\text{Peak TLV} = \frac{\text{TLV} \times \text{Avg. Time (sec)}}{5 \times \text{Pulsewidth (sec)}}$$

A maximum of five such pulses is permitted during any period equal to the averaging time. If there are more than 5 pulses during any period equal to the averaging time, then the peak TLV is limited by the normal time-averaging process. For pulse durations greater than 100 msec, normal time-averaging calculations apply.

These values should be used as guides in the evaluation and control of exposure to radiofrequency/microwave radiation and should not be regarded as a fine line between safe and dangerous levels.

*Notes:*

1. It is believed that workers may be exposed repeatedly to fields up to these TLVs without adverse health effects. Nevertheless, personnel should not be needlessly exposed to higher levels of RFR, approaching the TLVs, when simple measures will prevent it.
2. For mixed or broadband fields at a number of frequencies for which there are different values of the TLV, the fraction of the TLV (in terms of $D^2$, $H^2$, or power density) incurred within each frequency interval should be determined and the sum of all such fractions should not exceed unity.

   In a similar manner, for mixed or broadband induced currents at a number of frequencies for which there are different values of the TLV, the fraction of the induced current limits (in terms of $I^2$) incurred within each frequency interval should be determined and the sum of all such fractions should not exceed unity.
3. The TLV refers to values averaged over any 6-minute (0.1-hour) period for frequencies less than 15 GHz and over shorter periods for higher frequencies down to 10 seconds at 300 GHz as indicated in Table 1.
4. At frequencies between 100 kHz and 1.5 GHz, the TLV may be exceeded if the radiofrequency input power of the radiating device is 7 watts or less. This exclusion does not apply to devices that are attached to the body on a continual basis.
5. At frequencies between 100 kHz and 6 GHz, the TLV for electromagnetic field strengths may be exceeded if (a) the exposure conditions can be shown by appropriate techniques to produce SARs below 0.4 W/kg as averaged over the whole body and spatial peak SAR values not exceeding 8 W/kg as averaged over any one gram of tissue (defined as a tissue volume in the shape of a cube), except for the hands, wrists, feet, and ankles where the spatial peak SAR shall not exceed 20 W/kg as averaged over any ten grams of tissue (defined as a tissue volume in the shape of a cube), and (b) the induced currents in the body conform with the guide in Table 1. The SARs are averaged over any 6 minutes. Above 6 GHz, the relaxation of the TLV under partial body exposure conditions is permitted.

   At frequencies between 0.03 and 0.1 MHz, the SAR exclusion rule, stated above, does not apply. However, the TLV can still be exceeded if it can be shown that the peak RMS current density as averaged over any 1 cm² area of tissue and 1 second does not exceed 35f mA/cm² where f is the frequency in MHz.
6. No measurement of emitted fields should be made within 5 cm of any object.
7. All exposures should be limited to a maximum (peak) electric field intensity of 100 kV/m.

## *STATIC MAGNETIC FIELDS

These TLVs refer to static magnetic flux densities to which it is believed that nearly all workers may be repeatedly exposed day after day without adverse health effects. These values should be used as guides in the control of exposure to static magnetic fields and should not be regarded as a fine line between safe and dangerous levels.

Routine occupational exposures should not exceed 60 milliteslas (mT)—equivalent to 600 gauss (G)—whole body or 600 mT (6000 G) to the extremities on a daily, time-weighted average basis [1 tesla (T) = 10⁴ gauss (G)]. A flux density of 2 T is recommended as a ceiling value. Safety hazards may exist from the mechanical forces exerted by the magnetic field upon ferromagnetic tools and medical implants. Cardiac pacemaker and similar medical electronic device wearers should not be exposed to field levels exceeding 0.5 mT (5 G). Perceptible adverse effects may also be produced at higher flux densities resulting from forces upon other implanted ferromagnetic devices, e.g., suture staples, aneurism clips, prostheses, etc.

## *SUB-RADIOFREQUENCY (30 kHz and below) MAGNETIC FIELDS

These TLVs refer to the amplitude of the magnetic flux density (B) of sub-radiofrequency magnetic fields in the frequency range of 30 kHz and below to which it is believed that nearly all workers may be exposed repeatedly without adverse health effects. The magnetic field strengths in these TLVs are root-mean-square (rms) values. These values should be used as guides in the control of exposure to sub-radiofrequency magnetic fields and should not be regarded as a fine line between safe and dangerous levels.

Occupational exposures in the extremely-low-frequency (ELF) range from 1 Hz to 300 Hz should not exceed the ceiling value given by the equation:

$$B_{TLV} \text{ in mT} = \frac{60}{f}$$

where f is the frequency in Hz.

For frequencies in the range of 300 Hz to 30 kHz (which includes the voice frequency [VF] band from 300 Hz to 3 kHz and the very-low-frequency [VLF] band from 3 kHz to 30 kHz), occupational exposures should not exceed the ceiling value of 0.2 mT.

These ceiling values for frequencies of 300 Hz to 30 kHz are intended for both partial-body and whole-body exposures. For frequencies below 300 Hz, the TLV for exposure of the extremities can be increased by a factor of 5.

The magnetic flux density of 60 mT/f at 60 Hz corresponds to a maximum permissible flux density of 1 mT. At 30 kHz, the TLV is 0.2 mT which corresponds to a magnetic field intensity of 160 A/m.

*Notes:*

1. This TLV is based on an assessment of available data from laboratory research and human exposure studies. Modifications of the TLV will be made if warranted by new information. At this time, there is insufficient information on human responses and possible health effects of magnetic fields in the frequency range of 1 Hz to 30 kHz to permit the establishment of a TLV for time-weighted average exposures.
2. For workers wearing cardiac pacemakers, the TLV may not protect against electromagnetic interference with pacemaker function. Some models of cardiac pacemakers have been shown to be susceptible to interference by power-frequency (50/60 Hz) magnetic flux densities as low as 0.1 mT. It is recommended that, lacking specific information on electromagnetic interference from the manufacturer, the exposure of persons wearing cardiac pacemakers or similar medical electronic devices be maintained at or below 0.1 mT at power frequencies.

## *SUB-RADIOFREQUENCY (30 kHz and below) AND STATIC ELECTRIC FIELDS

These TLVs refer to the maximum unprotected workplace field strengths of sub-radiofrequency electric fields (30 kHz and below) and static electric fields that represent conditions under which it is believed that nearly all workers may be exposed repeatedly without adverse health effects. The electric field intensities in these TLVs are root-mean-square (rms) values. The values should be used as guides in the control of exposure and, due to individual susceptibility, should not be regarded as a fine line between safe and dangerous levels. The electric field strengths stated in this TLV refer to the field levels present in air, away from the surfaces of conductors (where spark discharges and contact currents may pose significant hazards).

Occupational exposures should not exceed a field strength of 25 kV/m from 0 Hz (DC) to 100 Hz. For frequencies in the range of 100 Hz to 4 kHz, the ceiling value is given by:

$$E_{TLV} \text{ in V/m} = \frac{2.5 \times 10^6}{f}$$

where f is the frequency in Hz.

A value of 625 V/m is the ceiling value for frequencies from 4 kHz to 30 kHz. These ceiling values for frequencies of 0 to 30 kHz are intended for both partial-body and whole-body exposures.

*Notes:*

1. This TLV is based on limiting currents on the body surface and induced internal currents to levels below those that are believed to produce adverse health effects. Certain biological effects have been demonstrated in laboratory studies at electric field strengths below those permitted in the TLV; however, there is no convincing evidence at the present time that occupational exposure to these field levels leads to adverse health effects.

   Modifications of the TLV will be made if warranted by new information. At this time, there is insufficient information on human responses and possible health effects of electric fields in the frequency range of 0 to 30 kHz to permit the establishment of a TLV for time-weighted average exposures.

2. Field strengths greater than approximately 5–7 kV/m can produce a wide range of safety hazards such as startle reactions associated with spark discharges and contact currents from ungrounded conductors within the field. In addition, safety hazards associated with combustion, ignition of flammable materials and electro-explosive devices may exist when a high-intensity electric field is present. Care should be taken to eliminate ungrounded objects, to ground such objects, or to use insulated gloves when ungrounded objects must be handled. Prudence dictates the use of protective devices (e.g., suits, gloves, and insulation) in all fields exceeding 15 kV/m.

3. For workers with cardiac pacemakers, the TLV may not protect against electromagnetic interference with pacemaker function. Some models of cardiac pacemakers have been shown to be susceptible to interference by power-frequency (50/60 Hz) electric fields as low as 2 kV/m. It is recommended that, lacking specific information on electromagnetic interference from the manufacturer, the exposure of pacemaker and medical electronic device wearers should be maintained at or below 1 kV/m.

## ULTRAVIOLET RADIATION

These TLVs refer to ultraviolet (UV) radiation in the spectral region between 180 and 400 nm and represent conditions under which it is believed that nearly all workers may be repeatedly exposed without adverse health effects. These values for exposure of the eye or the skin apply to UV radiation from arcs, gas and vapor discharges, fluorescent and incandescent sources, and solar radiation, but they do not apply to UV lasers (see the TLVs for Lasers). These values do not apply to UV radiation exposure of photosensitive individuals or of individuals concomitantly exposed to photosensitizing agents. These exposures to the eye do not apply to aphakics. (See Light and Near-Infrared Radiation TLVs.) These values should be used as guides in the control of exposure to continuous sources where the exposure durations shall not be less than 0.1 sec.

These values should be used as guides in the control of exposure to UV sources and should not be regarded as a fine line between safe and dangerous levels.

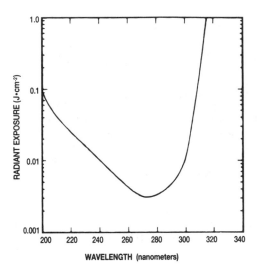

**Figure 1**—Threshold Limit Values (TLV) for Ultraviolet Radiation.

*Recommended Values*

The TLVs for occupational exposure to UV radiation incident upon skin or eye where irradiance values are known and exposure time is controlled are as follows:

1. For the near UV spectral region (320 to 400 nm), total irradiance incident upon the unprotected eye should not exceed 1.0 mW/cm$^2$ for periods greater than 10$^3$ seconds (approximately 16 minutes) and for exposure times less than 10$^3$ seconds should not exceed 1.0 J/cm$^2$.

2. The UV radiant exposure incident upon the unprotected skin or eye should not exceed the values given in Table 1 or Figure 1 within an 8-hour period.

3. To determine the effective irradiance of a broad band source weighted against the peak of the spectral effectiveness curve (270 nm), the following weighting formula should be used:

$$E_{eff} = \Sigma \, E_\lambda \, S_\lambda \, \Delta\lambda$$

   where: $E_{eff}$ = effective irradiance relative to a monochromatic source at 270 nm in W/cm$^2$ [J/(s • cm$^2$)]
   $E_\lambda$ = spectral irradiance in W/(cm$^2$ • nm)
   $S_\lambda$ = relative spectral effectiveness (unitless)
   $\Delta\lambda$ = band width in nm

4. For most white-light sources and all open arcs, the weighting of spectral irradiance between 200 and 315 nm should suffice to determine the effective irradiance. Only specialized UV sources designed to emit UV-A radiation would normally require spectral weighting from 315 to 400 nm.

5. Permissible exposure time in seconds for exposure to actinic UV radiation incident upon the unprotected skin or eye may be computed by dividing 0.003 J/cm$^2$ by $E_{eff}$ in W/cm$^2$. The exposure time may also be determined using Table 2 which provides exposure times corresponding to effective irradiances in $\mu$W/cm$^2$.

6. All of the preceding TLVs for UV energy apply to sources which subtend an angle less than 80°. Sources which subtend a greater angle need to be measured only over an angle of 80°.

   Conditioned (tanned) individuals can tolerate skin exposure in excess of the TLV without erythemal effects. However, such conditioning may not protect persons against skin cancer.

   Ozone ($O_3$) is produced in air by sources emitting UV radiation at wavelengths below 250 nm. Refer to the Chemical Substances TLV for ozone.

## TABLE 1. Ultraviolet Radiation Exposure TLV and Spectral Weighting Function

| Wavelength☆ (nm) | TLV ($J/m^2$) | TLV ($mJ/cm^2$) | Relative Spectral Effectiveness, $S_\lambda$ |
|---|---|---|---|
| 180 | 2500 | 250 | 0.012 |
| 190 | 1600 | 160 | 0.019 |
| 200 | 1000 | 100 | 0.030 |
| 205 | 590 | 59 | 0.051 |
| 210 | 400 | 40 | 0.075 |
| 215 | 320 | 32 | 0.095 |
| 220 | 250 | 25 | 0.120 |
| 225 | 200 | 20 | 0.150 |
| 230 | 160 | 16 | 0.190 |
| 235 | 130 | 13 | 0.240 |
| 240 | 100 | 10 | 0.300 |
| 245 | 83 | 8.3 | 0.360 |
| 250 | 70 | 7.0 | 0.430 |
| 254* | 60 | 6.0 | 0.500 |
| 255 | 58 | 5.8 | 0.520 |
| 260 | 46 | 4.6 | 0.650 |
| 265 | 37 | 3.7 | 0.810 |
| 270 | 30 | 3.0 | 1.000 |
| 275 | 31 | 3.1 | 0.960 |
| 280* | 34 | 3.4 | 0.880 |
| 285 | 39 | 3.9 | 0.770 |
| 290 | 47 | 4.7 | 0.640 |
| 295 | 56 | 5.6 | 0.540 |
| 297* | 65 | 6.5 | 0.460 |
| 300 | 100 | 10 | 0.300 |
| 303* | 250 | 25 | 0.120 |
| 305 | 500 | 50 | 0.060 |
| 308 | 1200 | 120 | 0.026 |
| 310 | 2000 | 200 | 0.015 |
| 313* | 5000 | 500 | 0.006 |
| 315 | $1.0\times10^4$ | $1.0\times10^3$ | 0.003 |
| 316 | $1.3\times10^4$ | $1.3\times10^3$ | 0.0024 |
| 317 | $1.5\times10^4$ | $1.5\times10^3$ | 0.0020 |
| 318 | $1.9\times10^4$ | $1.9\times10^3$ | 0.0016 |
| 319 | $2.5\times10^4$ | $2.5\times10^3$ | 0.0012 |
| 320 | $2.9\times10^4$ | $2.9\times10^3$ | 0.0010 |
| 322 | $4.5\times10^4$ | $4.5\times10^3$ | 0.00067 |
| 323 | $5.6\times10^4$ | $5.6\times10^3$ | 0.00054 |
| 325 | $6.0\times10^4$ | $6.0\times10^3$ | 0.00050 |
| 328 | $6.8\times10^4$ | $6.8\times10^3$ | 0.00044 |
| 330 | $7.3\times10^4$ | $7.3\times10^3$ | 0.00041 |
| 333 | $8.1\times10^4$ | $8.1\times10^3$ | 0.00037 |
| 335 | $8.8\times10^4$ | $8.8\times10^3$ | 0.00034 |
| 340 | $1.1\times10^5$ | $1.1\times10^4$ | 0.00028 |
| 345 | $1.3\times10^5$ | $1.3\times10^4$ | 0.00024 |
| 350 | $1.5\times10^5$ | $1.5\times10^4$ | 0.00020 |
| 355 | $1.9\times10^5$ | $1.9\times10^4$ | 0.00016 |
| 360 | $2.3\times10^5$ | $2.3\times10^4$ | 0.00013 |
| 365* | $2.7\times10^5$ | $2.7\times10^4$ | 0.00011 |
| 370 | $3.2\times10^5$ | $3.2\times10^4$ | 0.000093 |
| 375 | $3.9\times10^5$ | $3.9\times10^4$ | 0.000077 |
| 380 | $4.7\times10^5$ | $4.7\times10^4$ | 0.000064 |
| 385 | $5.7\times10^5$ | $5.7\times10^4$ | 0.000053 |
| 390 | $6.8\times10^5$ | $6.8\times10^4$ | 0.000044 |
| 395 | $8.3\times10^5$ | $8.3\times10^4$ | 0.000036 |
| 400 | $1.0\times10^6$ | $1.0\times10^5$ | 0.000030 |

☆ Wavelengths chosen are representative; other values should be interpolated at intermediate wavelengths.

*Emission lines of a mercury discharge spectrum.

## TABLE 2. Permissible Ultraviolet Exposures

| Duration of Exposure Per Day | Effective Irradiance, $E_{eff}$ ($\mu W/cm^2$) |
|---|---|
| 8 hrs | 0.1 |
| 4 hr | 0.2 |
| 2 hrs | 0.4 |
| 1 hr | 0.8 |
| 30 min | 1.7 |
| 15 min | 3.3 |
| 10 min | 5 |
| 5 min | 10 |
| 1 min | 50 |
| 30 sec | 100 |
| 10 sec | 300 |
| 1 sec | 3000 |
| 0.5 sec | 6000 |
| 0.1 sec | 30000 |

### PHYSICAL AGENTS UNDER STUDY

The Physical Agents TLV Committee has examined the current literature on the following agents and has not found sufficient data upon which to propose a new or revised TLV. However, these agents will remain under study during the coming year to examine new evidence indicating the need and feasibility for establishing a proposed TLV. Comments and suggestions, accompanied by substantive documentation, are solicited and should be forwarded to the Technical Affairs Office, ACGIH.

1. *Contact Currents.*

2. *Electromagnetic Pulses and Radiofrequency Radiation (RFR). Peak power limits.*

3. *Infrasound.*

4. *Laser radiation with pulse duration less than 1 ns.*

5. *Lifting.*

6. *Physical Agents as Carcinogens — Known and Suspect.*

7. *Pressure Variations.*

8. *Repetitive Motion Stresses.*

9. *Whole-body Vibration.*

### 1993–94 PHYSICAL AGENTS TLV COMMITTEE

David H. Sliney, Ph.D., U.S. Army Environ. Hygiene Agency–Chair
Gerald V. Coles, Deakin University, Australia
Irving H. Davis, Retired–Michigan Dept. of Public Health
David N. Erwin, Ph.D., USAF Armstrong Laboratory
Zory R. Glaser, Ph.D., M.P.H., U.S. Pharmacopia
Carla L. Hattel, Navy Bureau of Medicine and Surgery
Allan P. Heins, Ph.D., CIH, OSHA
Daniel L. Johnson, Ph.D., EG&G
John C. Mitchell, U.S. Air Force
Anthony M. Muc, Ph.D., Retired–Ontario Hydro
Robert M. Patterson, Sc. D., CIH, Temple University–Secretary
Thomas S. Tenforde, Ph.D., Battelle Pacific Northwest Laboratories

### NONVOTING MEMBERS

Richard A. Ilka, M.D., Occupational Health Consultant
Donald E. Wasserman, Vibration Consultant

## POLICY STATEMENT ON THE USES OF TLVs AND BEIs

The Threshold Limit Values (TLVs™) and Biological Exposure Indices (BEIs™) are developed as guidelines to assist in the control of health hazards. These recommendations or guidelines are intended for use in the practice of industrial hygiene, to be interpreted and applied only by a person trained in this discipline. They are not developed for use as legal standards, and the American Conference of Governmental Industrial Hygienists (ACGIH℠) does not advocate their use as such. However, it is recognized that in certain circumstances individuals or organizations may wish to make use of these recommendations or guidelines as a supplement to their occupational safety and health program. The ACGIH will not oppose their use in this manner, if the use of TLVs and BEIs in these instances will contribute to the overall improvement in worker protection. However, the user must recognize the constraints and limitations subject to their proper use and bear the responsibility for such use.

The Introductions to the TLV/BEI Booklet and the TLV/BEI Documentation provide the philosophical and practical bases for the uses and limitations of the TLVs and BEIs. To extend those uses of the TLVs and BEIs to include other applications, such as use without the judgment of an industrial hygienist, application to a different population, development of new exposure/recovery time models, or new effect endpoints, stretches the reliability and even viability of the database for the TLV or BEI as evidenced by the individual documentations.

It is not appropriate for individuals or organizations to impose on the TLVs or the BEIs their concepts of what the TLVs or BEIs should be or how they should be applied or to transfer regulatory standards requirements to the TLVs or REIs.

---

The Policy Statement on the Uses of TLVs/BEIs was approved by the Board of Directors of ACGIH on March 1, 1988.

## FOOTNOTES

\*   1995–1996 Adoption.

‡   See Notice of Intended Changes.

( )   Adopted values enclosed are those for which changes are proposed. Consult the Notice of Intended Changes for current proposal.

†   1995–1996 Revision or Addition to the Notice of Intended Changes.

◄   Identifies substances for which there are also BEIs (see BEI section). Substances identified in the BEI documentations for methemoglobin inducers (for which methemoglobin is the principle toxicity) and organophosphorus cholinesterase inhibitors are part of this notation.

•   Substance for which the TLV is higher than the OSHA Permissible Exposure Limit (PEL) and/or the NIOSH Recommended Exposure Limit (REL). See Fed. Reg. 58(124):35338–35351, June 30, 1993, for revised OSHA PELs.

■   Substance identified by other sources as a suspected or confirmed human carcinogen.

A   Refers to Appendix A — Carcinogens.

B   Refers to Appendix B — Substances of Variable Composition.

C   Denotes Ceiling limit.

(a) Parts of vapor or gas per million parts of contaminated air by volume at 25°C and 760 torr.

(b) Milligrams of substance per cubic meter of air.

(c) Simple asphyxiant; see definition in the "Introduction to the Chemical Substances."

(d) NOC = not otherwise classified.

(e) The value is for inhalable (total) particulate matter containing no asbestos and < 1% crystalline silica.

(f) Fibers longer than 5 $\mu$m and with an aspect ratio equal to or greater than 3:1 as determined by the membrane filter method at 400–450X magnification (4-mm objective) phase contrast illumination.

(g) The value is for particulate matter containing < 5% crystalline silica. For particulates containing more than this percentage of crystalline silica, the environment should be evaluated against the TLV–TWA of 0.1 mg/m$^3$ for respirable quartz. The concentration of respirable particulates for the application of this limit is to be determined from the fraction passing a size-selector with the characteristics defined in the "C" paragraph of Appendix D.

(h) Lint-free particulate matter as measured by the vertical elutriator cotton-dust sampler described in the Transactions of the National Conference on Cotton Dust, p. 33, by J.R. Lynch (May 2, 1970).

(i) Inhalable (total) dust/particulate.

(j) These TLVs are for the respirable fraction of particulate matter for the substance listed. The concentration of respirable dust for the application of this limit is to be determined from the fraction passing a size-selector with the characteristics defined in the "C" paragraph of Appendix D.

(k) As sampled by method that does not collect vapor.

(l) Does not include stearates of toxic metals.

(m) Based on "high-volume" sampling.

(n) However, should not exceed 2 mg/m$^3$ respirable particulates.

(o) For greater assurance of worker protection, biological monitoring is recommended.

(p) Except castor, cashew nut, or similar irritant oils.

*APPENDIX* **C**

# Conversion
# of Units

ALL PHYSICAL UNITS OF MEASUREMENT can be reduced to one or more of three dimensions: mass, length, and time. Reducing units to basic dimensions simplifies problem solving and makes comparison between operations, or operations and standards, easier and more accurate.

For example, three airflows could be measured: the first in liters per second, the second in cubic meters per second, and the third in cubic feet per minute. Then the total volume of air in each of the three samplings could be converted to cubic meters or cubic feet, and the airflows could be compared. In another situation, the results of atmospheric pollution studies and stack sampling surveys are often reported as grains per cubic foot, grams per cubic foot, or pounds per cubic foot. The degree of contamination is usually reported in the standard unit of parts of contaminant per million parts of air.

If physical measurements are made or reported in different units, they must be converted to the standard units if any comparisons are to be meaningful.

In order to achieve a uniform system of measurement, governments representing 98 percent of the world's population have committed to using the *Système International d'Unités* (SI) version of the metric system (McQueen MJ. Conversion to SI units; The Canadian experience. *JAMA* 256:3001–3002, 1986.) In 1975, Congress passed the Metric Conversion Act, which endorsed a voluntary conversion to SI, but the English system is still in popular use in the United States. The SI system, however, is the standard for the international scientific community.

## Fundamental Units

Because of the need to conserve time and space when reporting data, universally accepted abbreviations are often used in place of unit names. This appendix shows the abbreviations used throughout this book and those generally agreed on by industrial hygiene practitioners. Conversion factors are provided when data are reported in nonstandard units.

Each measurement unit, such as length, area, and flow, has a table of conversion factors. To use the table to find the numerical value of the quantity desired, locate the unit to be converted in the first column. Then multiply this value by the number appearing at the intersection of the row and the column containing the desired unit. The answer will be the numerical value in the desired unit. Various English system and metric system units are given for your convenience.

An explanation of the SI system and official conversion factors are given to a 6- or 7-place accuracy in ASTM standard E380–76 (ANSI Z210.1–1976). (This standard is available, although not listed in the ANSI Catalog.)

**Table C–A.** Base Système International (SI) Units

| Physical Quantity | Base Units | SI Symbol |
|---|---|---|
| Length | Meter | m |
| Mass | Kilogram | kg |
| Time | Second | s |
| Amount of substance | Mole | mol |
| Thermodynamic temperature | Kelvin | K |
| Electric current | Ampere | A |
| Luminous intensity | Candela | cd |

**Table C–B.** Units Derived from Combinations of Base Units

| Derived Unit | Name and Symbol | Expressed as SI Base Derived Unit |
|---|---|---|
| Area | Square meter | $m^2$ |
| Volume | Cubic meter | $m^3$ |
| Force | Newton (N) | $kg \cdot m \cdot s^{-2}$ ($kg \cdot m/s^2$) |
| Frequency | Hertz (Hz) | $s^{-1}$ |
| Work, energy, heat | Joule (J) | $N \cdot m$ |
| Power | Watt (W) | $J \cdot s^{-1}$ ($J/s$) |
| Pressure | Pascal (Pa) | $kg \cdot m^{-1} \cdot s^{-2}$ ($N/m^2$) |
| Electric potential | Volt (V) | $W \cdot A^{-1}$ ($W/A$) |
| Electric charge | Coulomb (C) | $A \cdot s$ |
| Electric capacitance | Farad (F) | $A \cdot sV^{-1}$ ($A \cdot s/V$ or $C/V$) |
| Inductance | Henry (H) | $V \cdot s \cdot A^{-1}$ ($V \cdot s/A$) |

**Table C–C.** Multiples and Submultiples of SI Units

| Factor | Prefix | Symbol |
|---|---|---|
| $10^{12}$ | tetra | T |
| $10^{9}$ | giga | G |
| $10^{6}$ | mega | M |
| $10^{3}$ | kilo | k |
| $10^{-3}$ | milli | m |
| $10^{-6}$ | micro | $\mu$ |
| $10^{-9}$ | nano | n |
| $10^{-12}$ | pico | p |
| $10^{-15}$ | femto | f |
| $10^{-18}$ | atto | a |

Tables C–A through C–C are reprinted with permission from *JAMA* 256(21), Dec. 5, 1986, pp. 3001–3002, ©1986, American Medical Association.)

**Table C–D.** Area

| To Obtain → Multiply Number of ↓ by ↘ | Square Meter $(m^2)$ | Square Inch $(in.^2)$ | Square Foot $(ft^2)$ | Square Centimeter $(cm^2)$ | Square Millimeter $(mm^2)$ |
|---|---|---|---|---|---|
| Square meter | 1 | 1,550 | 10.76 | 10,000 | $10^6$ |
| Square inch | $6.452 \times 10^{-3}$ | 1 | $6.94 \times 10^{-3}$ | 6.452 | 645.2 |
| Square foot | 0.0929 | 144 | 1 | 929.0 | 92,903 |
| Square centimeter | 0.0001 | 0.115 | 0.001 | 1 | 100 |
| Square millimeter | $10^{-6}$ | 0.00155 | 0.00001 | 0.01 | 1 |

**Table C–E.** Length

| To Obtain → Multiply Number of ↓ by ↘ | Meter (m) | Centimeter (cm) | Millimeter (mm) | Micron ($\mu$) or Micrometer | Angstrom Unit (Å) | Inch (in.) | Foot (ft) |
|---|---|---|---|---|---|---|---|
| Meter | 1 | 100 | 1,000 | $10^6$ | $10^{10}$ | 39.37 | 3.28 |
| Centimeter | 0.01 | 1 | 10 | $10^4$ | $10^8$ | 0.394 | 0.0328 |
| Millimeter | 0.001 | 0.1 | 1 | $10^3$ | $10^7$ | 0.0394 | 0.00328 |
| Micron | $10^{-6}$ | $10^{-4}$ | $10^{-3}$ | 1 | $10^4$ | $3.94 \times 10^{-5}$ | $3.28 \times 10^{-6}$ |
| Angstrom | $10^{-10}$ | $10^{-8}$ | $10^{-7}$ | $10^{-4}$ | 1 | $3.94 \times 10^{-9}$ | $3.28 \times 10^{-10}$ |
| Inch | 0.0254 | 2.540 | 25.40 | $2.54 \times 10^4$ | $2.54 \times 10^8$ | 1 | 0.0833 |
| Foot | 0.305 | 30.48 | 304.8 | 304,800 | $3.048 \times 10^9$ | 12 | 1 |

**Table C–F.**  Density

| To Obtain → <br> Multiply Number of ↓ | by ↘ | gm/cm³ | lb/ft³ | lb/gal |
|---|---|---|---|---|
| Gram/cubic centimeter | | 1 | 62.43 | 8.345 |
| Pound/cubic foot | | 0.01602 | 1 | 0.1337 |
| Pound/gallon (U.S.) | | 0.1198 | 7.481 | 1 |

1 grain/ft³ = 2.28 mg/m³

**Table C–G.**  Force

| To Obtain → <br> Multiply Number of ↓ | by ↘ | Dyne | Newton (N) | Kilogram-Force | Pound-Force (lbf) |
|---|---|---|---|---|---|
| Dyne | | 1 | $1.0 \times 10^{-5}$ | $1.02 \times 10^4$ | $2.248 \times 10^4$ |
| Newton | | $1.0 \times 10^5$ | 1 | 0.1020 | 0.2248 |
| Kilogram-force | | $9.807 \times 10^{-5}$ | 9.807 | 1 | 2.205 |
| Pound-force | | $4.448 \times 10^{-5}$ | 4.448 | 0.4536 | 1 |

**Table C–H.**  Mass

| To Obtain → <br> Multiply Number of ↓ | by ↘ | Gram (gm) | Kilogram (kg) | Grains (gr) | Ounce (avoir) (oz) | Pound (avoir) (lb) |
|---|---|---|---|---|---|---|
| Gram | | 1 | 0.001 | 15.432 | 0.03527 | 0.00220 |
| Kilogram | | 1,000 | 1 | 15,432 | 35.27 | 2.205 |
| Grain | | 0.0648 | $6.480 \times 10^{-5}$ | 1 | $2.286 \times 10^{-3}$ | $1.429 \times 10^{-4}$ |
| Ounce | | 28.35 | 0.02835 | 437.5 | 1 | 0.0625 |
| Pound | | 453.59 | 0.4536 | 7,000 | 16 | 1 |

**Table C–I.**  Volume

| To Obtain → <br> Multiply Number of ↓ | by ↘ | ft³ | Gallon <br> (U.S. Liquid) | Liters | cm³ | m³ |
|---|---|---|---|---|---|---|
| Cubic foot | | 1 | 7.481 | 28.32 | 28,320 | 0.0283 |
| Gallon (U.S. liquid) | | 0.1337 | 1 | 3.785 | 3,785 | $3.79 \times 10^{-3}$ |
| Liter | | 0.03531 | 0.2642 | 1 | 1,000 | $1 \times 10^{-3}$ |
| Cubic centimeters | | $3.531 \times 10^{-5}$ | $2.64 \times 10^{-4}$ | 0.001 | 1 | $10^{-6}$ |
| Cubic meters | | 35.31 | 264.2 | 1,000 | $10^6$ | 1 |

**Table C–J.** Velocity

| To Obtain → <br> Multiply Number of ↓ by ↘ | cm/s | m/s | km/hr | ft/s | ft/min | mph |
|---|---|---|---|---|---|---|
| Centimeter/second | 1 | 0.01 | 0.036 | 0.0328 | 1.968 | 0.02237 |
| Meter/second | 100 | 1 | 3.6 | 3.281 | 196.85 | 2.237 |
| Kilometer/hour | 27.78 | 0.2778 | 1 | 0.9113 | 54.68 | 0.6214 |
| Foot/second | 30.48 | 0.3048 | 18.29 | 1 | 60 | 0.6818 |
| Foot/minute | 0.5080 | 0.00508 | 0.0183 | 0.0166 | 1 | 0.01136 |
| Mile per hour | 44.70 | 0.4470 | 1.609 | 1.467 | 88 | 1 |

**Table C–K.** Flow Rates

| To Obtain → <br> Multiply Number of ↓ by ↘ | l/min | $m^3/s$ | $m^3/hr$ | gal/min | $ft^3/min$ | $ft^3/s$ |
|---|---|---|---|---|---|---|
| Liter/minute | 1 | $1.67 \times 10^{-5}$ | 0.06 | 0.2640 | 0.0353 | $5.89 \times 10^{-4}$ |
| Cubic meters/second | $4.63 \times 10^{-3}$ | 1 | $2.77 \times 10^{-4}$ | $1.22 \times 10^{-3}$ | $1.63 \times 10^{-4}$ | $2.7 \times 10^{-6}$ |
| Cubic meter/hour | 16.67 | $2.78 \times 10^{-4}$ | 1 | 4.4 | 0.588 | $9.89 \times 10^{-3}$ |
| Gallon (U.S.)/minute | 3.78 | $6.3 \times 10^{-5}$ | 0.227 | 1 | 0.1338 | $2.23 \times 10^{-3}$ |
| Cubic foot/minute | 28.32 | $4.71 \times 10^{-4}$ | 1.699 | 7.50 | 1 | 0.01667 |
| Cubic foot/second | $1.69 \times 10^3$ | $2.83 \times 10^{-3}$ | $1.02 \times 10^2$ | 448.8 | 60 | 1 |

**Table C–L.** Heat, Energy, or Work

| To Obtain → <br> Multiply Number of ↓ by ↘ | Joule | ft-lb | kwh | hp-hour | kcal | cal | Btu |
|---|---|---|---|---|---|---|---|
| Joules | 1 | 0.737 | $2.773 \times 10^{-7}$ | $3.725 \times 10^{-7}$ | $2.39 \times 10^{-4}$ | 0.2390 | $9.478 \times 10^{-4}$ |
| Foot-pound | 1,356 | 1 | $3.766 \times 10^{-7}$ | $5.05 \times 10^{-7}$ | $3.24 \times 10^{-4}$ | 0.3241 | $1.285 \times 10^{-3}$ |
| Kilowatt-hour | $3.6 \times 10^6$ | $2.66 \times 10^6$ | 1 | 1.341 | 860.57 | 860,565 | 3,412 |
| Hp-hour | $2.68 \times 10^6$ | $1.98 \times 10^6$ | 0.7455 | 1 | 641.62 | 641,615 | 2,545 |
| Kilocalorie | 4,184 | 3,086 | $1.162 \times 10^{-3}$ | $1.558 \times 10^{-3}$ | 1 | 1,000 | 3.9657 |
| Calorie | 4.184 | 3.086 | $1.162 \times 10^{-6}$ | $1.558 \times 10^{-6}$ | 0.001 | 1 | 0.00397 |
| British thermal unit | 1,055 | 778.16 | $2.930 \times 10^{-4}$ | $3.93 \times 10^{-4}$ | 0.252 | 252 | 1 |

**Table C–M.** Emission Rates

| To Obtain → <br> Multiply Number of ↓ by ↘ | gm/s | gm/min | kg/hr | kg/day | lb/min | lb/hr | lb/day |
|---|---|---|---|---|---|---|---|
| Gram/second | 1.0 | 60.0 | 3.6 | 86.40 | 0.13228 | 7.9367 | 190.48 |
| Gram/minute | 0.016667 | 1.0 | 0.06 | 1.4400 | $2.2046 \times 10^{-3}$ | 0.13228 | 3.1747 |
| Kilogram/hour | 0.27778 | 16.667 | 1.0 | 24.000 | 0.036744 | 2.2046 | 52.911 |
| Kilogram/day | 0.011574 | 0.69444 | 0.041667 | 1.0 | $1.5310 \times 10^{-3}$ | $9.1860 \times 10^{-2}$ | 2.2046 |
| Pound/minute | 7.5598 | 453.59 | 27.215 | 653.17 | 1.0 | 60.0 | 1440 |
| Pound/hour | 0.12600 | 7.5598 | 0.45359 | 10.886 | $1.6667 \times 10^{-2}$ | 1.0 | 24.0 |
| Pound/day | $5.2499 \times 10^{-3}$ | 0.31499 | $1.8900 \times 10^{-2}$ | 0.45359 | $6.9444 \times 10^{-4}$ | $4.1667 \times 1^{-2}$ | 1.0 |

**Table C–N.** Pressure

| To Obtain → | lb/in.² (psi) | atm | in. (Hg) 32 F 0 C | mm (Hg) 32 F 0 C | k Pa (k N/m²) | ft (H₂0) 60 F 15 C | in. (H₂0) | lb/ft² |
|---|---|---|---|---|---|---|---|---|
| Multiply Number of by ↓ ↘ | | | | | | | | |
| Pound/square inch | 1 | 0.068 | 2.036 | 51.71 | 6.895 | 2.309 | 27.71 | 144 |
| Atmospheres | 14.696 | 1 | 29.92 | 760.0 | 101.32 | 33.93 | 407.2 | 2,116 |
| Inch (Hg) | 0.4912 | 0.033 | 1 | 25.40 | 3.386 | 1.134 | 13.61 | 70.73 |
| Millimeter (Hg) | 0.01934 | 0.0013 | 0.039 | 1 | 0.1333 | 0.04464 | 0.5357 | 2.785 |
| Kilopascals | 0.1450 | $9.87 \times 10^{-3}$ | 0.2953 | 7.502 | 1 | 0.3460* | 4.019 | 20.89 |
| Foot (H₂0)(15C) | 0.4332 | 0.0294 | 0.8819 | 22.40 | 2.989* | 1 | 12.00 | 62.37 |
| Inch (H₂0) | 0.03609 | 0.0024 | 0.073 | 1.867 | 0.2488 | 0.0833 | 1 | 5.197 |
| Pound/square foot | 0.0069 | $4.72 \times 10^{-4}$ | 0.014 | 0.359 | 0.04788 | 0.016 | 0.193 | 1 |

* at 4 C

**Table C–O.** Radiant Energy Units

| To Obtain → | Erg | Joule (J) | W-s | µW-s | g-cal |
|---|---|---|---|---|---|
| Multiply by ↓ ↘ | | | | | |
| Erg | 1 | $10^{-7}$ | $10^{-7}$ | 0.1 | $2.39 \times 10^{-8}$ |
| Joule | $10^7$ | 1 | 1 | $10^6$ | 0.239 |
| Watt-second | $10^7$ | 1 | 1 | $10^6$ | 0.239 |
| Micro-watt second | 10 | $10^{-6}$ | $10^{-6}$ | 1 | $2.39 \times 10^{-7}$ |
| Gram-calorie | $4.19 \times 10^7$ | 4.19 | 4.19 | $4.19 \times 10^6$ | 1 |

**Table C–P.** Energy/Unit Area (Dose Units)

| To Obtain → | erg/cm² | J/cm² | W-s/cm² | µW-s/cm² | g-cal/m² |
|---|---|---|---|---|---|
| Multiply by ↓ ↘ | | | | | |
| Erg/square centimeter | 1 | $10^{-7}$ | $10^{-7}$ | 0.1 | $2.39 \times 10^{-8}$ |
| Joule/square centimeter | $10^7$ | 1 | 1 | $10^6$ | 0.239 |
| Watt-second/square centimeter | $10^7$ | 1 | 1 | $10^6$ | 0.239 |
| Microwatt-second/square centimeter | 10 | $10^{-6}$ | $10^{-6}$ | 1 | $2.39 \times 10^{-7}$ |
| Gram-calorie/square centimeter | $4.19 \times 10^7$ | 4.19 | 4.19 | $4.19 \times 10^6$ | 1 |

**Table C–Q.** Temperature Equivalents

| Scale | Symbol | Freezing Point of Water (1 atm) | Boiling Point of Water (1 atm) |
|---|---|---|---|
| Celsius | C | 0 | 100 deg |
| Fahrenheit | F | 32 | 212 |
| Thermodynamic } Kelvin Absolute Celsius | K, A | $273.16 \pm 0.01$* | $373.16 \pm 0.01$* |
| Approximate absolute | AA | 273 | 373 |
| Rankine } Absolute Fahrenheit | R | 491.69 | 671.69 |

Conversion formulae

$$C = (5/9)(F - 32) = K - 273.16 = AA - 273$$
$$F = (9/5)C + 32 = (9/5)(K - 273.16) + 32$$
$$K = C + 273.16 = AA + 0.16 = (5/9)(F - 32) + 273.16$$
$$AA = C + 273 = K - 0.16 = (5/9)(F - 32) + 273$$
$$Rankine = F + 459.69$$

(From Birge RT, *Rev Mod Phys* 13:233, 1941.)

# Review of Mathematics

## ▮ Significant Figures

Measurements often result in what are called *approximate numbers,* in contrast to *discrete counts.* For example, the dimensions of a table can be reported as 29.6 in. (75.2 cm) by 50.2 in. (127.5 cm). This implies that the measurement is to the nearest tenth of an inch (or centimeter) and that the table is less than 50.25 in. (127.6 cm) and more than 50.15 in. (127.4 cm) in length. One can show the same thing for the width, using the following symbolic notations:

75.0 cm (29.55 in.) < width < 75.3 cm (29.65 in.)

If, on the other hand, one knows the degree of precision of the measurement (say 0.03 cm or ± 0.08 cm), one may write·

50.2 ± 0.3 or 50.2 ± 0.8

to indicate the degree of accuracy of the measurement of the length.

In reporting results, the number of significant digits that can be recorded is determined by the precision of the instruments used.

## Rules

- In any approximate number, the significant digits include the digit that determines the degree of precision of the number and all digits to the left of it, except for zeros used to place the decimal.

- All digits from 1 to 9 are significant.

- All zeros between significant digits are significant.

- Final zeros of decimal numbers are significant. For example:

**Significant Figures**
*Rules* ▪ *Scientific Notation* ▪ *Addition and Subtraction* ▪
*Multiplication and Division*

**Logarithms**
*Common Logarithms* ▪ *How to Use Logarithms* ▪ *How to Use
Logarithm Tables* ▪ *Decibel Notation*

**Normal and Lognormal Frequency Distributions**
*Variability* ▪ *Coefficient of Variation*

**Exposure Concentration**

**Geometric Standard Deviation**

| Number | Number of Significant Digits |
|---|---|
| 0.0702 | 3 |
| 0.07020 | 4 |
| 70.20 | 4 |
| 7,002. | 4 |
| 7,020. | 3 |

## Scientific Notation

One case where it is difficult to determine the number of significant digits is the figure 7,000. In general, it is considered to have only one significant digit. It is better to use scientific notation.

In standard scientific notation, the number is written as a number between 1 and 10, in which only the significant digits are shown, multiplied by an exponential number to the base 10. For example:

| Number | Number of Significant Digits |
|---|---|
| $5{,}320{,}000 = 5.32 \times 10^6$ | 3 |
| $= 5.320 \times 10^6$ | 4 |
| $= 5.3200 \times 10^6$ | 5 |
| $0.00000532 = 5.32 \times 10^{-6}$ | 3 |

## Addition and Subtraction

The result must not have more decimal places than the number with the fewest decimal places. For example:

| | | |
|---|---|---|
| 21.262 | should be | 21.3 |
| 23.74 | should be | 23.7 |
| 139.6 | should be | 139.6 |
| 184.602 | should be | 184.6 |

## Multiplication and Division

The result must not have more significant places than are possessed by the number with the fewest significant digits. For example:

$$(50.20)(29.6) = 1485.92$$
$$= 1490$$
$$= 1.49 \times 10^3$$

# Logarithms

Logarithms are exponents. The logarithm of any number is the power to which a selected base must be raised to produce the number. The laws of exponents apply to logarithms.

The following two equations:

$$a^x = y$$

and

$$x = \log_a y$$

are two ways of expressing the same thing, that is, the exponent applied to $a$ to give $y$ is equal to $x$. The value $a$ is called the base of the system of logarithms.

Although any positive number greater than 1 can be used as the base of some system of logarithms, there are two systems in general use. These are the *common* (or *Briggs'*) system and the *natural* (or *Napierian*) system. In the common system, the base is 10; in the natural system, the base is the irrational number e = 2.71828 . . .

## Common Logarithms

Common logarithms use the base 10 and are identified by the notation *log*. The common logarithm of a number consists of a characteristic, which locates the decimal point in the number, and a mantissa, which defines the numerical arrangement of the number.

A bar over a characteristic indicates a negative characteristic and a positive mantissa. The log may be written $\bar{4}.7$ or 6.7 – 10 or –3.3. The form –3.3 does not contain a characteristic and mantissa.

The integral part of a logarithm is called the *characteristic* and the decimal part is called the *mantissa*. In log 824, the characteristic is 2 and the mantissa is 0.9162. For convenience in constructing tables, it is desirable to select the mantissa as positive even if the logarithm is a negative number. For example, log 1/2 = –0.3010; but because –0.3010 = 9.6990 – 10, this may be written log 1/2 = 9.6990 – 10 with a positive mantissa. This is also the log of 0.5, which we could have looked up in the first place. The following illustration shows the method of writing the characteristic and mantissa:

| | | |
|---|---|---|
| log | 8245 | = 3.9162 |
| log | 824.5 | = 2.9162 |
| log | 82.45 | = 1.9162 |
| log | 8.245 | = 0.9162 |
| log | 0.8245 | = 9.9162 – 10 |
| log | 0.08245 | = 8.9162 – 10 |

By using scientific notation, we can easily find logarithm characteristics, as shown in the table in the next section.

## How to Use Logarithms

If the laws of exponents are rewritten in terms of logarithms, they become the *laws of logarithms*:

$$\log_a(x^n) = n \log_a x$$
$$\log_a\left(\frac{x}{y}\right) = \log_a x - \log_a y$$
$$\log_a(x^a) = n \log_a x$$

Logarithms derive their main usefulness in computation from these laws because they allow multiplication, division, and exponentiation to be replaced by the simpler operations of addition, subtraction, and multiplication, respectively.

| Number | Exponential Form | Common Logarithmic Form | | |
|---|---|---|---|---|
| | | Characteristic | Mantissa | Complete Log |
| 0.0005 | $5 \times 10^{-4}$ | −4 | 0.7 | $\overline{4}.7$ |
| 0.05 | $5 \times 10^{-2}$ | −2 | 0.7 | $\overline{2}.7$ |
| 5.0 | $5 \times 10^{0}$ | 0 | 0.7 | 0.7 |
| 500.0 | $5 \times 10^{2}$ | 2 | 0.7 | 2.7 |
| 50,000.0 | $5 \times 10^{4}$ | 4 | 0.7 | 4.7 |

## How to Use Logarithm Tables

In this appendix is a four-place table of logarithms. In this table, the mantissas of the logarithms of all integers from 1 to 999 are recorded correct to four decimal places, which is all one needs to work with decibels, which have three significant digits at most.

To find the logarithm of a given number, use the table as follows: To find the logarithm of 63.5, glance down the column headed $N$ for the first two significant digits (63), and then along the top of the table for the third figure (5). In the row across from 63 and in the column under 5 is found 8028. This is the mantissa. Adding the proper characteristic 1, the logarithm (or log) of 63.5 is 1.8028.

Conversely, one can find the number that corresponds to a given logarithm (the antilogarithm). For example, find the number whose logarithm is 1.6355. The mantissa 6355 corresponds to the number in the table that is in the column below 2 and in the row across from 43. Thus, the mantissa corresponds to the number 432. Because the characteristic is 1, the number whose logarithm is 1.6355 is 43.2.

Because in measuring sound we are concerned only with three significant digits, the number whose logarithm is 1.6360 would also be 43.2. The number whose logarithm is 1.6361 would be 43.3.

## Decibel Notation

Again, using the measurement of sound as an example, if two sound intensities $P_1$ and $P_2$ are to be compared according to the ability of the ear to detect intensity differences, we may determine the number of decibels that expresses the relative value of the two intensities by

$$N_{dB} = 10 \log_{10} \frac{P_1}{P_2}$$

where $P_1$ is greater than $P_2$.

The factor 10 comes into this picture because the original unit devised was the *bel*, which is the logarithm of 10 to the base 10 and represents 10 times as many decibels in any expression involving the relation between two sound intensities as there are bels.

The decibel is a logarithmic unit. Each time the amount of power is increased by a factor of 10, we have added 10 decibels (abbreviated dBA).

To determine the number of decibels by which two powers differ, we must *first determine the ratio of the two powers;* we look up this ratio in a table of logarithms to the base 10 and then we multiply the figure obtained by a factor of 10.

If we want to find the relative loudness of 10,000 people who can shout louder than 100 people can, we use the following reasoning.

The logarithm (to the base 10) of any number is merely the number of times 10 must be multiplied by itself to be equal to the number. In the example here, 100 represents 10 multiplied by itself, and the logarithm of 100 to the base 10, therefore, is 2. For example, the number of decibels expressing the relative loudness of 10,000 people shouting compared with 100 is

$$\begin{aligned} N_{dB} &= 10 \log_{10}(10,000 \div 100) \\ &= 10 \log_{10} 100 \\ &= 10 \times 2.0 \\ &= 20 \end{aligned}$$

Now let us see what happens if we double the number of people to 20,000.

$$\begin{aligned} N_{dB} &= 10 \log_{10}(20,000 \div 100) \\ &= 10 \log_{10} 200 \\ &= 10 \times 2.3010 \\ &= 23 \text{ (rounded to significant digits)} \end{aligned}$$

It can be seen, therefore, that decibels are logarithm ratios. In their use in sound measurement, P (the usual reference level) is 20 micropascals or 0.0002 dynes/square centimeter, which approximates the threshold of hearing, the sound that can just be heard by a young person with excellent hearing.

## ■ Normal and Lognormal Frequency Distributions

The statistical methods discussed here assume that measured concentrations of random occupational environmental samples are lognormally and independently distributed within one 8-hour period and over many daily exposure averages.

Before sample data can be statistically analyzed, we must have knowledge of the frequency distribution of the measurements or some assumptions must be made. Most community air pollution environmental data can be described by a lognormal distribution. That is, the logarithms (either base e or base 10) of the data are approximately normally distributed.

What are the differences between normally and lognormally distributed data? A normal distribution is completely determined by the parameters: the arithmetic mean ($\mu$); the standard deviation ($\sigma$) of the distribution. A lognormal distribution is completely determined by the median or geometric mean (GM) and the geometric standard deviation (GSD). For lognormally distributed data, a logarithmic transformation of the original data is normally distributed. The GM and GSD of the lognormal distribution are the antilogs of the mean and standard deviation of the logarithmic transformation. Normally distributed data have a symmetrical distribution curve whereas lognormally distributed environmental data are generally positively skewed (long "tail" to the right indicating a larger probability of very large concentrations than for normally distributed data.)

## Variability

The variability of occupational environmental data (differences between repeated measurements at the same site) can usually be broken into three major components: random errors of the sampling method, random errors of the analytical method, and variability of the environment with time. The first two components of the variability are known in advance and are approximately normally distributed. However, the environmental fluctuations of a contaminant in a facility usually greatly exceed the variability of known instruments (often by factors of 10 or 20).

When several samples are taken in a facility to determine the average concentration of the contaminant to estimate the average exposure of an employee, then the lognormal distribution should be assumed. However, the normal distribution may be used in the special cases of taking a sample to check compliance with a ceiling standard, and when a sample (or samples) is taken for the entire time period for which the standard is defined (be it 15 minutes or 8 hours). In these cases, the entire time interval of interest is represented in the sample, and only sampling and analytical errors are present.

## Coefficient of Variation

The relative variability of a normal distribution (such as the random errors of the sampling and analytical procedures) is commonly measured by the coefficient of variation (CV). The CV is also known as the *relative standard deviation*. The CV is a useful index of dispersion in that limits consisting of the true mean of a set of data plus or minus twice the CV will contain about 95 percent of the data measurements.

Thus, if an analytical procedure with a CV of 10 percent is used to repeatedly measure some nonvarying physical property (as the concentration of a chemical in a beaker of solution), then about 95 percent of the measurements will fall within plus or minus 20 percent (two times the CV) of the true concentration.

## Exposure Concentration

Unfortunately, the property we are trying to measure, the employee's exposure concentration, is not a fixed, nonvarying physical property. The exposure concentrations are fluctuating in a lognormal manner. First, the exposure concentrations are fluctuating over the 8-hour period of the time-weighted average (TWA) exposure measurement. Breathing zone grab samples (samples of less than about 30 minutes' duration, typically only a few minutes) tend to reflect this intraday environmental variability so that grab sample results have relatively high variability.

Intraday variability in the sample results can be eliminated from measurement variability by going to a full-period sampling strategy. The day-to-day (interday) variability of the true 8-hour TWA exposures is also lognormally distributed. This interday variability creates a need for an action level where only one day's exposure measurement is used to draw conclusions regarding compliance on unmeasured days.

## Geometric Standard Deviation

The parameter often used to express either the intraday or interday environmental variability is the *geometric standard deviation* (GSD). A GSD of 1.0 represents absolutely no variability in the environment. GSDs of 2.0 and above represent relatively high variability.

The shape of lognormal distributions with low variabilities, such as those with GSDs less than about 1.4, roughly approximate normal distribution shapes. For this range of GSDs, there is a rough equivalence between the GSD and CV as follows:

| GSD | Approximate CV |
|-----|-----|
| 1.40 | 35 percent |
| 1.30 | 27 percent |
| 1.20 | 18 percent |
| 1.10 | 9.6 percent |
| 1.05 | 4.9 percent |

## COMMON LOGARITHMS

| 9 | 8 | 7 | 6 | 5 | 4 | 3 | 2 | 1 | 0 | N |
|---|---|---|---|---|---|---|---|---|---|---|
| 7067 | 7059 | 7050 | 7042 | 7033 | 7024 | 7016 | 7007 | 6998 | 6990 | 50 |
| 7152 | 7143 | 7135 | 7126 | 7118 | 7110 | 7101 | 7093 | 7084 | 7076 | 51 |
| 7235 | 7226 | 7218 | 7210 | 7202 | 7193 | 7185 | 7177 | 7168 | 7160 | 52 |
| 7316 | 7308 | 7300 | 7292 | 7284 | 7275 | 7267 | 7259 | 7251 | 7243 | 53 |
| 7396 | 7388 | 7380 | 7372 | 7364 | 7356 | 7348 | 7340 | 7332 | 7324 | 54 |
| 7474 | 7466 | 7459 | 7451 | 7443 | 7435 | 7427 | 7419 | 7412 | 7404 | 55 |
| 7551 | 7543 | 7536 | 7528 | 7520 | 7513 | 7505 | 7497 | 7490 | 7482 | 56 |
| 7627 | 7619 | 7612 | 7604 | 7597 | 7589 | 7582 | 7574 | 7566 | 7559 | 57 |
| 7701 | 7694 | 7686 | 7679 | 7672 | 7664 | 7657 | 7649 | 7642 | 7634 | 58 |
| 7774 | 7767 | 7760 | 7752 | 7745 | 7738 | 7731 | 7723 | 7716 | 7709 | 59 |
| 7846 | 7839 | 7832 | 7825 | 7818 | 7810 | 7803 | 7796 | 7789 | 7782 | 60 |
| 7917 | 7910 | 7903 | 7896 | 7889 | 7882 | 7875 | 7868 | 7860 | 7853 | 61 |
| 7987 | 7980 | 7973 | 7966 | 7959 | 7952 | 7945 | 7938 | 7931 | 7924 | 62 |
| 8055 | 8048 | 8041 | 8035 | 8028 | 8021 | 8014 | 8007 | 8000 | 7993 | 63 |
| 8122 | 8116 | 8109 | 8102 | 8096 | 8089 | 8082 | 8075 | 8069 | 8062 | 64 |
| 8189 | 8182 | 8176 | 8169 | 8162 | 8156 | 8149 | 8142 | 8136 | 8129 | 65 |
| 8254 | 8248 | 8241 | 8235 | 8228 | 8222 | 8215 | 8209 | 8202 | 8195 | 66 |
| 8319 | 8312 | 8306 | 8299 | 8293 | 8287 | 8280 | 8274 | 8267 | 8261 | 67 |
| 8382 | 8376 | 8370 | 8363 | 8357 | 8351 | 8344 | 8338 | 8331 | 8325 | 68 |
| 8445 | 8439 | 8432 | 8426 | 8420 | 8414 | 8407 | 8401 | 8395 | 8388 | 69 |
| 8506 | 8500 | 8494 | 8488 | 8482 | 8476 | 8470 | 8463 | 8457 | 8451 | 70 |
| 8567 | 8561 | 8555 | 8549 | 8543 | 8537 | 8531 | 8525 | 8519 | 8513 | 71 |
| 8627 | 8621 | 8615 | 8609 | 8603 | 8597 | 8591 | 8585 | 8579 | 8573 | 72 |
| 8686 | 8681 | 8675 | 8669 | 8663 | 8657 | 8651 | 8645 | 8639 | 8633 | 73 |
| 8745 | 8739 | 8733 | 8727 | 8722 | 8716 | 8710 | 8704 | 8698 | 8692 | 74 |
| 8802 | 8797 | 8791 | 8785 | 8779 | 8774 | 8768 | 8762 | 8756 | 8751 | 75 |
| 8859 | 8854 | 8848 | 8842 | 8837 | 8831 | 8825 | 8820 | 8814 | 8808 | 76 |
| 8915 | 8910 | 8904 | 8899 | 8893 | 8887 | 8882 | 8876 | 8871 | 8865 | 77 |
| 8971 | 8965 | 8960 | 8954 | 8949 | 8943 | 8938 | 8932 | 8927 | 8921 | 78 |
| 9025 | 9020 | 9015 | 9009 | 9004 | 8998 | 8993 | 8987 | 8982 | 8976 | 79 |
| 9079 | 9074 | 9069 | 9063 | 9058 | 9053 | 9047 | 9042 | 9036 | 9031 | 80 |
| 9133 | 9128 | 9122 | 9117 | 9112 | 9106 | 9101 | 9096 | 9090 | 9085 | 81 |
| 9186 | 9180 | 9175 | 9170 | 9165 | 9159 | 9154 | 9149 | 9143 | 9138 | 82 |
| 9238 | 9232 | 9227 | 9222 | 9217 | 9212 | 9206 | 9201 | 9196 | 9191 | 83 |
| 9289 | 9284 | 9279 | 9274 | 9269 | 9263 | 9258 | 9253 | 9248 | 9243 | 84 |
| 9340 | 9335 | 9330 | 9325 | 9320 | 9315 | 9309 | 9304 | 9299 | 9294 | 85 |
| 9390 | 9385 | 9380 | 9375 | 9370 | 9365 | 9360 | 9355 | 9350 | 9345 | 86 |
| 9440 | 9435 | 9430 | 9425 | 9420 | 9415 | 9410 | 9405 | 9400 | 9395 | 87 |
| 9489 | 9484 | 9479 | 9474 | 9469 | 9465 | 9460 | 9455 | 9450 | 9445 | 88 |
| 9538 | 9533 | 9528 | 9523 | 9518 | 9513 | 9509 | 9504 | 9499 | 9494 | 89 |
| 9586 | 9581 | 9576 | 9571 | 9566 | 9562 | 9557 | 9552 | 9547 | 9542 | 90 |
| 9633 | 9628 | 9624 | 9619 | 9614 | 9609 | 9605 | 9600 | 9595 | 9590 | 91 |
| 9680 | 9675 | 9671 | 9666 | 9661 | 9657 | 9652 | 9647 | 9643 | 9638 | 92 |
| 9727 | 9722 | 9717 | 9713 | 9708 | 9703 | 9699 | 9694 | 9689 | 9685 | 93 |
| 9773 | 9768 | 9763 | 9759 | 9754 | 9750 | 9745 | 9741 | 9736 | 9731 | 94 |
| 9818 | 9814 | 9809 | 9805 | 9800 | 9795 | 9791 | 9786 | 9782 | 9777 | 95 |
| 9863 | 9859 | 9854 | 9850 | 9845 | 9841 | 9836 | 9832 | 9827 | 9823 | 96 |
| 9908 | 9903 | 9899 | 9894 | 9890 | 9886 | 9881 | 9877 | 9872 | 9868 | 97 |
| 9952 | 9948 | 9943 | 9939 | 9934 | 9930 | 9926 | 9921 | 9917 | 9912 | 98 |
| 9996 | 9991 | 9987 | 9983 | 9978 | 9974 | 9969 | 9965 | 9961 | 9956 | 99 |
| 0039 | 0035 | 0030 | 0026 | 0022 | 0017 | 0013 | 0009 | 0004 | 0000 | 100 |
| 9 | 8 | 7 | 6 | 5 | 4 | 3 | 2 | 1 | 0 | N |

| 9 | 8 | 7 | 6 | 5 | 4 | 3 | 2 | 1 | 0 | N |
|---|---|---|---|---|---|---|---|---|---|---|
| 9542 | 9031 | 8451 | 7782 | 6990 | 6021 | 4771 | 3010 | 0000 | .... | 0 |
| 2788 | 2553 | 2304 | 2041 | 1761 | 1461 | 1139 | 0792 | 0414 | 0000 | 1 |
| 4624 | 4472 | 4314 | 4150 | 3979 | 3802 | 3617 | 3424 | 3222 | 3010 | 2 |
| 5911 | 5798 | 5682 | 5563 | 5441 | 5315 | 5185 | 5051 | 4914 | 4771 | 3 |
| 6902 | 6812 | 6721 | 6628 | 6532 | 6435 | 6335 | 6232 | 6128 | 6021 | 4 |
| 7709 | 7634 | 7559 | 7482 | 7404 | 7324 | 7243 | 7160 | 7076 | 6990 | 5 |
| 8388 | 8325 | 8261 | 8195 | 8129 | 8062 | 7993 | 7924 | 7853 | 7782 | 6 |
| 8976 | 8921 | 8865 | 8808 | 8751 | 8692 | 8633 | 8573 | 8513 | 8451 | 7 |
| 9494 | 9445 | 9395 | 9345 | 9294 | 9243 | 9191 | 9138 | 9085 | 9031 | 8 |
| 9956 | 9912 | 9868 | 9823 | 9777 | 9731 | 9685 | 9638 | 9590 | 9542 | 9 |
| 0374 | 0334 | 0294 | 0253 | 0212 | 0170 | 0128 | 0086 | 0043 | 0000 | 10 |
| 0755 | 0719 | 0682 | 0645 | 0607 | 0569 | 0531 | 0492 | 0453 | 0414 | 11 |
| 1106 | 1072 | 1038 | 1004 | 0969 | 0934 | 0899 | 0864 | 0828 | 0792 | 12 |
| 1430 | 1399 | 1367 | 1335 | 1303 | 1271 | 1239 | 1206 | 1173 | 1139 | 13 |
| 1732 | 1703 | 1673 | 1644 | 1614 | 1584 | 1553 | 1523 | 1492 | 1461 | 14 |
| 2014 | 1987 | 1959 | 1931 | 1903 | 1875 | 1847 | 1818 | 1790 | 1761 | 15 |
| 2279 | 2253 | 2227 | 2201 | 2175 | 2148 | 2122 | 2095 | 2068 | 2041 | 16 |
| 2529 | 2504 | 2480 | 2455 | 2430 | 2405 | 2380 | 2355 | 2330 | 2304 | 17 |
| 2765 | 2742 | 2718 | 2695 | 2672 | 2648 | 2625 | 2601 | 2577 | 2553 | 18 |
| 2989 | 2967 | 2945 | 2923 | 2900 | 2878 | 2856 | 2833 | 2810 | 2788 | 19 |
| 3201 | 3181 | 3160 | 3139 | 3118 | 3096 | 3075 | 3054 | 3032 | 3010 | 20 |
| 3404 | 3385 | 3365 | 3345 | 3324 | 3304 | 3284 | 3263 | 3243 | 3222 | 21 |
| 3598 | 3579 | 3560 | 3541 | 3522 | 3502 | 3483 | 3464 | 3444 | 3424 | 22 |
| 3784 | 3766 | 3747 | 3729 | 3711 | 3692 | 3674 | 3655 | 3636 | 3617 | 23 |
| 3962 | 3945 | 3927 | 3909 | 3892 | 3874 | 3856 | 3838 | 3820 | 3802 | 24 |
| 4133 | 4116 | 4099 | 4082 | 4065 | 4048 | 4031 | 4014 | 3997 | 3979 | 25 |
| 4298 | 4281 | 4265 | 4249 | 4232 | 4216 | 4200 | 4183 | 4166 | 4150 | 26 |
| 4456 | 4440 | 4425 | 4409 | 4393 | 4378 | 4362 | 4346 | 4330 | 4314 | 27 |
| 4609 | 4594 | 4579 | 4564 | 4548 | 4533 | 4518 | 4502 | 4487 | 4472 | 28 |
| 4757 | 4742 | 4728 | 4713 | 4698 | 4683 | 4669 | 4654 | 4639 | 4624 | 29 |
| 4900 | 4886 | 4871 | 4857 | 4843 | 4829 | 4814 | 4800 | 4786 | 4771 | 30 |
| 5038 | 5024 | 5011 | 4997 | 4983 | 4969 | 4955 | 4942 | 4928 | 4914 | 31 |
| 5172 | 5159 | 5145 | 5132 | 5119 | 5105 | 5092 | 5079 | 5065 | 5051 | 32 |
| 5302 | 5289 | 5276 | 5263 | 5250 | 5237 | 5224 | 5211 | 5198 | 5185 | 33 |
| 5428 | 5416 | 5403 | 5391 | 5378 | 5366 | 5353 | 5340 | 5328 | 5315 | 34 |
| 5551 | 5539 | 5527 | 5514 | 5502 | 5490 | 5478 | 5465 | 5453 | 5441 | 35 |
| 5670 | 5658 | 5647 | 5635 | 5623 | 5611 | 5599 | 5587 | 5575 | 5563 | 36 |
| 5786 | 5775 | 5763 | 5752 | 5740 | 5729 | 5717 | 5705 | 5694 | 5682 | 37 |
| 5899 | 5888 | 5877 | 5866 | 5855 | 5843 | 5832 | 5821 | 5809 | 5798 | 38 |
| 6010 | 5999 | 5988 | 5977 | 5966 | 5955 | 5944 | 5933 | 5922 | 5911 | 39 |
| 6117 | 6107 | 6096 | 6085 | 6075 | 6064 | 6053 | 6042 | 6031 | 6021 | 40 |
| 6222 | 6212 | 6201 | 6191 | 6180 | 6170 | 6160 | 6149 | 6138 | 6128 | 41 |
| 6325 | 6314 | 6304 | 6294 | 6284 | 6274 | 6263 | 6253 | 6243 | 6232 | 42 |
| 6425 | 6415 | 6405 | 6395 | 6385 | 6375 | 6365 | 6355 | 6345 | 6335 | 43 |
| 6522 | 6513 | 6503 | 6493 | 6484 | 6474 | 6464 | 6454 | 6444 | 6435 | 44 |
| 6618 | 6609 | 6599 | 6590 | 6580 | 6571 | 6561 | 6551 | 6542 | 6532 | 45 |
| 6712 | 6702 | 6693 | 6684 | 6675 | 6665 | 6656 | 6646 | 6637 | 6628 | 46 |
| 6803 | 6794 | 6785 | 6776 | 6767 | 6758 | 6749 | 6739 | 6730 | 6721 | 47 |
| 6893 | 6884 | 6875 | 6866 | 6857 | 6848 | 6839 | 6830 | 6821 | 6812 | 48 |
| 6981 | 6972 | 6964 | 6955 | 6946 | 6937 | 6928 | 6920 | 6911 | 6902 | 49 |
| 7067 | 7059 | 7050 | 7042 | 7033 | 7024 | 7016 | 7007 | 6998 | 6990 | 50 |
| 9 | 8 | 7 | 6 | 5 | 4 | 3 | 2 | 1 | 0 | N |

# European Union Initiatives in Occupational Health and Safety

**Occupational Health and Safety Regulation in the European Union Countries**
*Belgium ▪ Denmark ▪ France ▪ Germany ▪ Greece ▪ Ireland ▪ Italy ▪ Luxembourg ▪ The Netherlands ▪ Portugal ▪ Spain ▪ United Kingdom*

**The European Union**
*Executive Branch ▪ Legislative Branch ▪ Judicial Branch*

**Legal Instruments**
*Regulation ▪ Directive*

**European Union Directives**
*Directive 77/576 ▪ Directive 78/610 ▪ Directive 80/1107 ▪ Directive 82/605 ▪ Directive 86/188 ▪ Directive 89/391 ▪ Directive 89/654 ▪ Directive 89/655 ▪ Directive 89/656 ▪ Directive 89/686 ▪ Directive 90/269 ▪ Directive 90/270 ▪ Directive 90/394 ▪ Directive 90/679 ▪ Directive 91/322 ▪ Directive 91/382 ▪ Directive 91/383 ▪ Directive 92/29 ▪ Directive 92/57 ▪ Directive 92/58 ▪ Directive 92/85 ▪ Directive 92/91/EEC ▪ Directive 92/104/EEC*

**New Chemical Notification—The Seventh Amendment**

**ISO 9000**

**Information Sources**

**Bibliography**

**Addendum**

by Robin S. Coyne, CIH, ROH, LIH

THE EUROPEAN UNION COMPRISES Belgium, Denmark, Germany, France, Greece, Ireland, Italy, Luxembourg, Netherlands, Portugal, Spain, and the United Kingdom of Great Britain and Northern Ireland (United Kingdom). These twelve countries have had occupational health and safety regulation since the 1800s. Early statutes such as the United Kingdom's Health and Morals of Apprentices Act of 1802 dealt primarily with protecting child labor. Child labor laws were then followed by more comprehensive occupational health and safety regulations. These regulations provided for factory and mine inspection, accident investigation, medical treatment for work-related accidents, and accident insurance. One of the more progressive statutes was the German Industrial Code, introduced in 1900. This law required employers to maintain workplaces that were not hazardous to the health and life of workers. Today, workers throughout the European Union are protected from workplace hazards through an extensive set of occupational health and safety regulations. Because each government has a slightly different approach to regulating occupational health and safety, it has been recognized that harmonization of standards within the Union is necessary to ensure that all employees within the Union are adequately protected from workplace hazards and to minimize unfair trade advantage.

# Occupational Health and Safety Regulation in the European Union Countries

## Belgium

The principal law is the *Law on Health and Safety of Workers and the Cleanliness of Work and Workplaces 1952.* Subsidiary laws and regulations are collected together in *General Regulations for the Protection of Labor (RGPT),* which is updated annually. ACGIH TLVs are published as guidelines for compliance by the Administration for Hygiene and Medicine at Work. The regulatory bodies are the Ministry of Employment and Labor and the Ministry of Economic Affairs.

## Denmark

The principal law is the *Working Environment Act 1975.* This act is implemented through framework orders and specific orders. The Danish occupational exposure limits (OELs) are a combination of ACGIH TLVs and exposure limits published in German Research Society and World Health Organization (WHO) technical reports. The regulatory body is the Ministry of Labor, Directorate of Labor Inspection.

## France

The primary laws can be found in the Labor Code, in *Law No. 77–771 on Control of Chemicals,* and in the Public Health Code. In addition, through the Social Security Code, the Sickness Insurance Funds can draft general rules, which are enforceable by financial sanctions relating to contributions to the fund. Occupational exposure limits, VMEs and VLEs, are determined by the Superior Council for the Prevention of Occupational Hazards. The regulatory body is the Ministère du Travail, de l'Emploi et de la Formation Professionnelle.

## Germany

The original framework law was the Industrial Code. Other primary statutes today include the *Order on Dangerous Substances in the Workplace,* the *Occupational Safety Act,* and the *Workplaces Order.* Accident prevention regulations are issued by the Employers' Associations for Accident Insurance and are enforced by fines or increased contributions. Occupational exposure limits, or maximum acceptable concentrations (MACs), are established by a commission under the German Research Society. MAC values are published in the official bulletin of the Ministry of Labor. The regulatory bodies are the Federal Ministry of Labor and Social Affairs and the National Federation of Industrial Employment Accident Insurance Funds. The individual states within the republic also have enforcement authority.

## Greece

The general principle of protecting the health and safety of individuals in the workplace is established in the Greek Constitution. The primary laws are codified in the *Royal Decree of 25 August 1920,* the *Presidential Decree of 13 March 1934* on health and safety of manual and clerical workers, and the *Royal Decrees of 29 September/4 October* and *15/21 October 1922* on establishment of industrial plants and mechanical installations. The regulatory body is the Ministry of Labor.

## Ireland

The principal laws are the *Factories Act 1955* and the *Safety in Industry Act 1980,* which are generally called the *Safety in Industry Acts.* Provisions of these acts are legislated through subordinate regulations and orders. Occupational exposure limits are regulated by the Safety in Industry Act of 1980 and are primarily the most recent ACGIH TLVs. The regulatory body is the Department of Labour.

## Italy

General precepts for protecting the health and safety of individuals in the workplace are established in the Italian Constitution and in the Civil and Penal Codes. Detailed provisions are found under the *Law Giving Powers to Introduce Health and Safety Regulation, 1955* and the framework law on the *National Health Service 1978.* Implementation of the framework law has been done on a regional level through the promulgation of regional laws. Occupational exposure limits are established and periodically revised through a Prime Minister decree on proposal of the Minister of Health. The regulatory bodies are the Ministry of Labor and Social Security; the Ministry of Industry, Commerce and Crafts; the Directorate General of Mines; and the Central Medical Labor Inspectorate.

## Luxembourg

The right of employees to have a safe and healthy workplace is granted in the constitution. Principal legislation is the *Law on Health and Safety of Workers 1924* and the *Law on Dangerous, Dirty and Noxious Installations, 1979.* Provisions are enacted through the Gran Ducal Decrees and a series of substance-specific laws. Under the Social Insurance Code, the Accident Insurance Association issues rules that it enforces through financial sanctions. The association recommends use of the German MAC values as exposure limits. The regulatory bodies are the Ministry of Labor and Social Security, the Ministry of Public Health, and the Accident Insurance Association.

## The Netherlands

General principles for protecting the health and safety of individuals in the workplace are established in the civil code. The principal framework law is the *Working Environment Act 1980*. This act embodies the *Labor Act 1919* and the *Safety Act 1934*. The provisions of these acts are enacted by decrees and orders. Occupational exposure limits are published as the *National MAC-list Arbeidsinspectie P no 145*. They are based on ACGIH TLVs, NIOSH recommendations, and German MACs. The regulatory body is the Ministry of Social Affairs and Employment.

## Portugal

Occupational health and safety are regulated primarily by the general decrees of 1918, 1922, and 1959; the *General Health and Safety Regulations of 1971;* and the *General Regulations on Safety and Health in Mines and Quarries of 15 January 1985*. The regulatory bodies are the General Inspectorate of Labor and the Directorate General of Occupational Safety and Health.

## Spain

Occupational health and safety were originally regulated under the *General Industrial Safety and Health Regulation of 31 January 1940*. These regulations were replaced by the *General Occupational Safety and Health Ordinance through Order of 9 March 1971,* which established employers' duties and workers' rights and obligations. The Ministry of Labor and Social Security is the regulatory body, having responsibility for preparing and issuing legislation and inspecting workplaces. ACGIH TLVs are used as practical guides for occupational exposure limits.

## United Kingdom

Under *Common Law*, which rules the United Kingdom, employers are obligated to make reasonable provisions for the safety of employees. The principal legislation is the *Health and Safety at Work Act 1974,* provisions of which are enacted in subsidiary regulations. Occupational exposure limits, known in the UK as maximum exposure limits (MELs) and occupational exposure standards (OESs), are regulated under the *Control of Substances Hazardous to Health (COSHH)* legislation. OESs and MELs are set on the recommendation of the Advisory Committee on Toxic Substances following evaluation of the substances by the Working Group on the Assessment of Toxic Chemicals. The regulatory body is the Health and Safety Executive.

## The European Union

The European Union, as we know it today, evolved out of the European Community for Coal and Steel (1951, Treaty of Paris); the European Economic Community and European Atomic Energy Community (1957, Treaties of Rome); the Single European Act (1987); and the Maastricht Treaty, which was implemented November 1, 1993. The objectives of forming the union were best expressed in the preambles to the treaties that established the European Union: to create an "organized and vital Europe," "to lay the foundations of an ever-closer union among the peoples of Europe," and to combine their efforts for "the constant improvement of the living and working conditions of their peoples" (Borchardt, 1986).

To promulgate legislation that applies to all 12 member states, the European Union has a constitution, an executive branch, a legislative branch, and a judicial branch.

### Executive Branch

The executive branch comprises the European Council, subject-specific councils, and the Committee of Permanent Representatives (COREPER). Prime ministers or heads of state of the member states make up the European Council. The European Council's role is to discuss broad issues and policy and to grant final approval to proposals—it is not involved in the initiation of legislative action. The council must approve virtually all significant legislation. Vote is by qualified majority.

Specific issues are delegated for discussion and recommendations to subject-specific councils. These councils are formed by the responsible ministers from each member-state government. The councils are then supported by working groups of experts from the 12 governments.

Another essential body is COREPER, which is made up of representative members from each of the member states. COREPER reviews all proposals submitted to the council and resolves any problems related to the proposals.

In 1993 the European Council chartered the European Agency for Health and Safety at the Workplace. This agency is charged with monitoring, reinforcing, and simplifying existing legislation within the European Community; providing information, training, and education for both labor and management; promoting occupational health and safety in countries involved in commerce with the member states; and conducting research and developing health and safety guidelines. The agency is located in Bilbao, Spain.

### Legislative Branch

Legislation can be proposed only by the commission. Under the direction of the European Parliament, 17 individuals are appointed to the commission on the grounds of their general competence and ability to act independently of their government or other authority. There must be at least one commissioner from each member state, but not more than two from each. The Commission is supported by 23 directorates general and a number of services. In the case of occupational health and safety, the Advisory Committee on Safety, Hygiene and Health Protection at Work was set up by a council decision in 1974 to assist the commission in matters relating to health and safety. This committee is composed of 72 members comprising two government, two trade union, and two employers' representatives from each member state. They are appointed by the European Council, based on recommendations from the individual governments.

All matters raised by the commission that deal with issues such as occupational health and safety, protection of the environment, human health, or use of natural resources must be sent to the Economic and Social Committee for their opinion. This committee is a tripartite body representing employers, workers, and general interests. Members are nominated by the member states and appointed by the council.

The European Parliament, consisting of 518 members elected by universal suffrage in the member states, does not have true legislative power but acts in consultation with the council. Its major responsibility is overseeing the commission.

## Judicial Branch

The judicial branch is represented by the Court of Justice, which is responsible for interpreting union or community law and for issuing judgments in disputes between the union institutions, member states, and/or third parties.

# Legal Instruments

Two different legal instruments are available within the European Union for issuing legislation: regulation and directive.

## Regulation

Regulations have general application, are binding in their entirety, and are directly applicable in all member states.

They are used extensively in the areas of customs affairs, market regulation, and agriculture and when data must be collected in a standardized format.

## Directive

A directive is a performance-oriented instrument. It is binding in all member states, but each government can achieve the desired result within its own constitutional and legislative framework.

The procedure used by the commission to prepare a proposal is summarized in Figure E–1 and the legislative process is illustrated in Figure E–2. Once a regulation or directive is adopted by the council, each member state must implement it within a prescribed amount of time.

# European Union Directives

Legislation pertaining to occupational health and safety has been in the form of two types of directives. One type pertains to the safety and standardization of equipment, and the other applies to employee health and safety. A brief summary of the directives follows.

## Directive 77/576

**On the provision of safety signs at work.** This directive provides for the following six types of safety signs: prohibition (of such things as smoking), warning (fire, for example), mandatory (such as use of ear protection), emergency (such as emergency exits), information (for example, health hazard), and additional signs such as those used to give supplemental information to another sign. The direc-

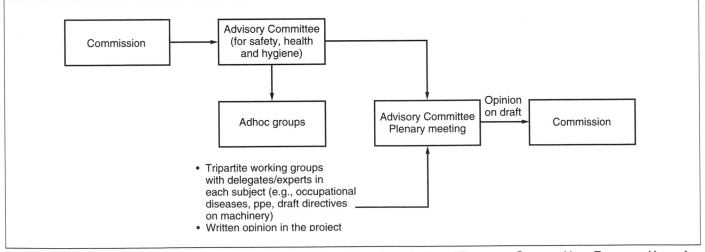

**Figure E–1.** Development of a commission proposal. (Adapted from Commission of European Communities. *European Year of Safety, Hygiene and Health Protection at Work.* Luxembourg: Office for Official Publications of the European Communities, Catalog no. CE–BT–91–001–EN–CO, 1992.)

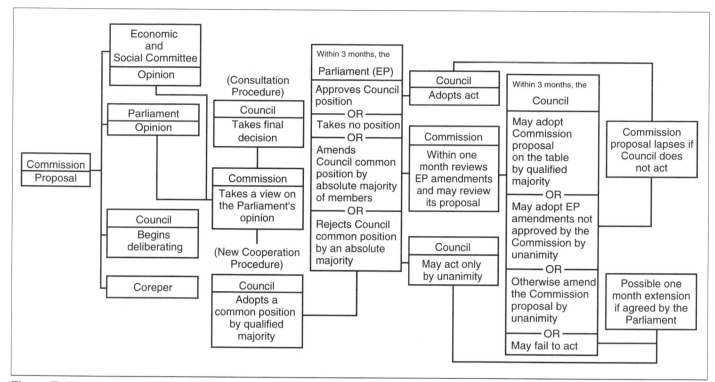

**Figure E–2.** Flowchart—Legislative process in the EC. (Adapted from the Oil Companies' European Organization for Environmental and Health Protection [OCEOEHP]. *European Community Legislative Process.* Brussels, Belgium: OCEOEHP. Rept. no. 91–65, April 1991.)

tive specifies sign color, geometric form, pictograms, and design. Directive 79/640 specifies sign size.

## Directive 78/610

**On the approximation of the laws, regulations, and administrative provisions of the member states on the protection of the health of workers exposed to vinyl chloride monomer.** This directive sets down a specific set of provisions for health protection for workers exposed to vinyl chloride monomer, including exposure monitoring, an exposure limit of 3 parts per million (ppm), personal protective equipment, and medical monitoring.

## Directive 80/1107

**On the protection of workers from the effects of biological, chemical, and physical agents.** This directive is a framework for the control of worker exposure to chemical, physical, and biological agents. Elements of the directive include establishment of limit values, sampling procedures, procedures for evaluating monitoring data, methodology for minimizing employee exposures, hygiene and decontamination measures, medical surveillance, training and information, emergency measures, and warning and safety signs. Additional measures are specified for acrylonitrile, asbestos, arsenic, benzene, cadmium, mercury, nickel, lead, and chlorinated hydrocarbons.

## Directive 82/605

**On the protection of workers from risks related to exposure to metallic lead and its ionic compounds at work.** This directive establishes exposure limit values for lead in both air (40 $\mu g/m^3$) and biological fluids (40 $\mu g$ Pb/ 100 ml blood) and provides for exposure monitoring, health surveillance, employee information, and other preventive measures.

## Directive 86/188

**On the protection of workers from risks related to exposure to noise at work.** This directive establishes exposure limits for noise: 85 dBA for hearing-conservation requirements and 90 dBA for engineering/administrative controls and warning signs. Hearing-conservation requirements include periodic noise measurements, hearing protection, audiometric testing, and employee training and information programs. Procedures to be used to measure both noise levels and hearing loss are specified in the directive.

## Directive 89/391

**Introduction of measures to encourage improvements in the safety and health of workers at work.** This directive, adopted in 1989, is known as the Framework Directive. It was the first major new policy to emerge under the new legislative process initiated by the

Single European Act. The directive extends its scope to both public and private sector employment. EC directives passed prior to the Framework Directive will eventually have to be modified in accordance with its principles. The aim of this directive is to introduce measures to encourage improvements in the health and safety of workers at work. It obligates employers to ensure the safety and health of workers through training; exchanging information with employees; performing risk assessments; developing measures to minimize the risks and informing workers and their representatives of these risks and actions; designating workers to carry out preventive health and safety activities; providing appropriate health surveillance; adapting the work to the individual through workplace design; protecting particularly sensitive groups; protecting workers who refuse work in situations of imminent, serious, and unavoidable danger; and logging and reporting workplace accidents. Also, guidelines are provided for the commission to develop future directives. Most importantly, the Framework Directive requires member states to put mechanisms in place to ensure legal enforcement of its provisions.

## Directive 89/654

**Minimum safety and health requirements for the workplace.** This directive prescribes the minimum safety and health requirements for workplaces or modifications to older facilities put into operation after December 31, 1992. It covers such topics as safety, structural integrity, electrical, means of egress, fire protection, ventilation, room temperature, lighting, sanitary facilities, and provisions for handicapped workers. The effective date for compliance was December 31, 1992. Workplaces in existence on or prior to December 31, 1992 are covered under a second annex, which has an effective date of January 1, 1996.

## Directive 89/655

**Directive on minimum safety and health requirements for the use of work equipment by workers at work.** This directive applies to any machine, apparatus, tool, or installation used at work, and the area within or around the aforementioned equipment in which a worker may be subject to harm. Under this directive employers are obligated to ensure the suitability of equipment for the intended use, maintain the equipment in safe working order, and provide workers with adequate instructions and training for using the equipment. Nineteen specific safety requirements for equipment are also included in the directive. The directive applies to equipment being used for the first time after December 31, 1992. Equipment in use prior to December 31, 1992 must comply with the directive by December 31, 1996.

## Directive 89/656

**Directive on the minimum health and safety requirements for the use of personal protective equipment at the workplace.** This directive classifies personal protective equipment used in the workplace and makes recommendations as to the types of work and conditions for which the equipment is appropriate. Under this directive employers must evaluate the appropriateness of the personal protective equipment with regard to both the work being performed and the worker; determine the compatibility of all the pieces of personal protective equipment if multiple items are recommended; and provide training and information to workers on the equipment to be used, its purpose, and detailed instructions for proper use. There are also requirements that workers and/or their representatives be consulted on the use of personal protective equipment and provisions for providing equipment free of charge and for maintaining it. Specifications for personal protective equipment are given in Directive 89/686.

## Directive 89/686

**Approximation of laws relating to personal protective equipment.** This directive harmonizes the European standards for the design, manufacture, specifications, and test methods applicable to personal protective equipment. Personal protective equipment that meets the requirements and is used within the European Union must display the symbol specified in the directive. The effective date for this directive is June 30, 1995.

## Directive 90/269

**Minimum safety and health requirements for the manual handling of loads.** The purpose of this directive is to reduce the risks of back injury resulting from manual handling of loads. Under the directive employers are obligated to minimize manual handling of loads through the use of mechanical equipment, evaluate the characteristics of loads to be handled, determine whether the workers involved are capable of performing the necessary operations, ensure that workers are made aware of the characteristics of the load, and train workers on how to accomplish the operation without injury.

## Directive 90/270

**Minimum safety and health requirements for work with display screen equipment.** This directive specifies measures that should be taken to improve the safety and comfort of workers using video display equipment. Under the directive employers are obligated to analyze workstations for conditions that could affect eyesight, produce mental stress, or produce repetitive trauma, and to provide workstation equipment designed to minimize the aforementioned risks (specifications for equipment, lighting, temperature, and humidity are listed in the annex to the directive). In addition, workers must be informed of the possible risks of working on video display equipment, allowed periodic breaks from video display work, provided with ophthalmological examinations and corrective lenses (if necessary), and trained in the proper use of their workstations. All equipment put into service after December 31, 1992 must comply with the requirements specified in the directive. Equipment put into service prior to December 31, 1992 must be upgraded to comply by December 31, 1996.

## Directive 90/394

**Protection of workers from risks related to exposure to carcinogens at work.** This directive expands Directive 88/364, which regulated the use and handling of carcinogenic substances and banned the production or use of four specific carcinogenic materials: 4-aminodiphenyl and its salts; 2-naphthylamine and its salts; benzidine and its salts; and 4-nitrodiphenyl. Directive 90/384 expanded the definition of *carcinogen* from specific substances or processes that may cause cancer to include by-products of carcinogen-producing processes, and it put in place provisions to prevent exposure to carcinogens. Under this directive employers are obligated to substitute less hazardous materials or processes for those determined to pose a carcinogenic risk; measure and record workplace concentrations of carcinogens annually; minimize employee exposure to carcinogens; institute medical surveillance for exposed workers; post warning signs; inform workers of risks associated with exposure; and provide training in proper handling, decontamination, and emergency procedures.

## Directive 90/679

**Protection of workers from risks related to exposure to biological agents at work.** This directive creates four categories of biological agents into which all agents must be classified, and measures that must be taken to minimize employee risk. Endeavors covered include activities that involve direct contact with specific agents as well as accidental exposure, such as health care and farming. The four categories are agents that are unlikely to cause human disease; agents that are hazardous and cause disease, but are unlikely to spread into the community; agents that cause severe disease and may be spread into the community, but effective prophylaxis or treatment is available; agents that present the most serious hazard to workers, with a high risk of spreading into the community, and for which there is no effective treatment available. Under this directive employers are obligated to perform risk assessments on work areas and activities; take measures to minimize risk, provide medical surveillance; furnish personal protection and hygiene

facilities; maintain lists of exposed workers; develop emergency procedures; post areas with biohazard warnings; and provide workers with risk information, proper instruction, and training.

## Directive 91/322

**Establishing indicative limit values for implementing council directive 80/1107/EEC on the protection of workers from risks related to exposure to chemical, physical, and biological agents at work.** This directive furthers Directives 80/1107 and 88/642 by establishing the first harmonized list of exposure limits for chemical, physical, and biological agents. The list is currently being developed.

## Directive 91/382

**Amending directive 83/477/EEC on the protection of workers from risks related to exposure to asbestos at work.** This directive applies recent scientific knowledge, technology, and experience to the action levels and limit values for occupational exposure to asbestos. Directive 83/477 established exposure limits (an action level of 0.2 fiber/cc for chrysotile and 0.1 fibers/cc for other forms; an exposure limit of 0.6 fibers/cc chrysotile and 0.3 fibers/cc for other forms), banned the spraying of asbestos, prescribed employee exposure assessment and medical surveillance, obligated employers to control employee exposures, specified measures that must be taken to minimize risks, and mandated that employees be made aware of the risks involved.

## Directive 91/383

**Supplementing the measures to encourage improvements in the safety and health at work of workers with a fixed-duration or temporary employment relationship.** This directive provides for training, medical surveillance, and the dissemination of hazard information to workers engaged in temporary employment. Provisions are included for workers to be informed, prior to employment, as to the nature of the activities to be performed and the type of establishment in which the work will occur. Also, the qualifications necessary to perform the activities must be clearly established prior to employment.

## Directive 92/29

**Minimum health and safety requirements for improved medical treatment onboard vessels.** The object of this directive is to provide emergency medical assistance for workers onboard ship. Provisions of this directive include training for all employees in medical and emergency procedures and inclusion of a medical kit aboard all vessels.

## Directive 92/57

**Implementation of minimum health and safety requirements at temporary and mobile worksites.** This directive provides for the inclusion of health and safety requirements in the planning of mobile worksites and specifies safety requirements for certain jobs. It also requires that the health and safety responsibilities of all persons assigned to the worksite be defined.

## Directive 92/58

**Minimum requirements for the provision of safety and/or health signs at work.** This directive is a revision of Directives 77/575 and 79/640. It provides for harmonization of safety and/or health warning signs in workplaces. The objective of the directive is to minimize the potential for injury or illness that result from miscommunication. Minimum requirement for printed or illuminated signs, hand signals, verbal or acoustic communication, and combinations thereof are specified. Provisions for hearing- or sight-impaired workers and for technical progress are also included.

## Directive 92/85

**Concerning measures to encourage improvements in the safety and health of pregnant workers, women workers who have recently given birth, and women who are breast-feeding.** The purpose of this directive is to minimize workplace risks unique to pregnant women, women that have recently given birth, and women who are breast-feeding and to provide protection from discrimination. Specific provisions include workplace assessment, actions to eliminate possible risks (including removal from the workplace, if necessary), restriction of night work, paid maternity leave, and prohibition of dismissal (except under specific conditions).

## Directive 92/91/EEC

**Concerning the minimum requirements for improving the safety and health protection of workers in the industries of mineral extraction through drilling.** See next directive.

## Directive 92/104/EEC

**The minimum requirements for improving the safety and health protection of workers in surface and underground mineral-extracting industries.** These two directives were issued in response to the North Sea Piper Alpha rig explosion, which resulted in 167 fatalities, and the Stolzenbach mine explosion in Germany, which resulted in 51 fatalities. Directive 92/91/EEC specifically covers exploration and extraction operations by means of boreholes. Directive 92/104/EEC pertains to mines and quarries. Mini-

mum requirements for electrical and mechanical equipment, fire and explosion protection, monitoring for hazardous atmospheres, sanitation, and emergency response are specified in these two directives.

## ■ New Chemical Notification— The Seventh Amendment

One of the key directives for protecting worker health in the workplace is Directive 67/548/EEC. This directive has been amended seven times, each time becoming increasingly more protective of the workplace and the environment. The amendments to this directive that have had the greatest impact on the workplace have been the sixth (Directive 79/831/EEC) and seventh (Directive 92/32/EEC) amendments. New chemical notification is commonly called the seventh amendment.

The seventh amendment obliges chemical manufacturers to submit dossiers indicating specific chemical, physical, toxicological, and ecotoxicological properties of the new chemical. This information is then used to complete a material safety data sheet, a proposed classification, and a label with the proper phrases and pictograms. Classifications, risk phrases, and pictograms are specified in the directive. Material safety data sheets and labels are important components of the dossier. For chemicals manufactured outside of the European Union, either the manufacturer or the importer submits the notification dossier. The important point is that every chemical placed into commerce in the European Union, except where exempted by the directive, must be notified. Data requirements depend on the quantity of material that is being manufactured and the results of the initial "base" tests. Once a new chemical notification has been accepted by the competent authorities, it is placed on the European List of Notified Chemical Substances (ELINCS). Chemicals manufactured prior to notification requirements (1971–1981) are listed in the European Inventory of Existing Commercial Substances (EINECS).

## ■ ISO 9000

An international initiative that has been gaining in momentum over the last few years is implementation of the ISO 9000 standards and subsequent ISO registration. The standards are generated by the International Organization for Standardization, the body that inspects and grants the registration. The purpose of ISO 9000 is not occupational health and safety, but it is a manufacturing or operating system that, if used properly, greatly improves worker health and safety along with product quality and productivity. Basically, ISO 9000 standards and guidelines mandate that a registered workplace must have detailed standard operating procedures (SOPs) for all of their operations, including those pertaining to health and safety. Once these SOPs are in place, they must be strictly complied with, and any changes or deviations from them must be clearly and completely documented. Workers must be trained in how to perform their work according to the procedures. The advantages of this system are that the manufacturing process undergoes a complete evaluation and is documented, deviations to this process are not tolerated (unless evaluated and documented), product quality is standardized, and occupational health and safety are built into the process. Efforts are under way to produce ISO standards for environmental control in the chemical industry.

## Information Sources

The information presented here is merely an overview to provide an idea of the occupational health and safety requirements within the European Union and how these requirements are promulgated. Additional information can be obtained from the Delegation of the Commission of the European Communities in Washington DC (telephone: [202]862–9500) or any of the European Union Depository Libraries (see Addendum).

## Bibliography

Borchardt K. *The ABC of Community Law,* 2nd ed. Luxembourg: Office for Official Publications of the European Communities, February 1986.

Confederation of British Industry (CBI). *Euro-Laws—Understanding and Influencing Them.* London: CBI, June 1983.

Borchardt K. *Eurofacts—Getting Around Brussels.* London: CBI, March 1983.

Commission of the European Communities. *Health and Safety at Work: A Challenge to Europe.* Luxembourg: Office for Official Publications of the European Communities, February 1992.

Commission of the European Communities Directorate-General for Employment, Industrial Relations and Social Affairs. *Safety and Health Protection at Work.* EEC Directives. Luxembourg: Commission of the European Communities, 1992.

Cook WA. *Occupational Exposure Limits—Worldwide.* Fairfax, VA: American Industrial Hygiene Association, 1987.

Council of the European Communities. *Official Journal of the European Communities.*

Health and Safety Executive. *Memorandum on the Harmonisation of European Health and Safety Law.* Printed in UK for HMSO (Her Majesty's Stationary Office), UK: Select Committee on Employment, October 1992.

Hecker S. Occupational health and safety policy in the European Community: a case study of economic integration and social policy, Part 1—Early initiatives through the Single European Act. *New Solutions* 3(4):56–69, 1993a.

Hecker S. Occupational health and safety policy in the European Community: a case study of economic integration and social policy, Part 2—The Framework Directive, whither harmonization? *New Solutions* 4(1):57–67, 1993b.

Hopper WC. *The European Community Legislative Process,* Report no. 91/55. Brussels: Concawe, April 1991.

International Labor Organisation. *International Directory of Occupational Safety and Health Institutions.* Occupational safety and health series. Geneva: International Labour Office, 1990.

Murray M. *A Guide to the European Community Legislation on Health and Safety at Work.* London: The Association of the British Pharmaceutical Industry, September 1990.

Sanderson JT. *EC Health & Safety Legislation—Update.* Luxembourg: Office for Official Publications of the European Community, August 1993a.

Sanderson JT. *Health and Safety Regulatory Issues.* Luxembourg: Office for Official Publications of the European Community, August 1993b.

## Addendum

**European Community Depository Libraries and Documentation Centers**

University of Arkansas
Documents Department
UALR Library
2801 S. University Avenue
Little Rock, AR 72204

University of Arizona
International Documents
University Library
Tucson, AZ 85721

Stanford University
Central Western European College
The Hoover Institution
Stanford, CA 94305

University of California
Documents Department
Central Library
La Jolla, CA 92093

University of California
Documents Department
General Library
Berkeley, CA 94720

University of California
International Documents
Public Affairs Service
Research Library
Los Angeles, CA 90024

University of Southern California
International Documents
Von Kleinschmidt Library
Los Angeles, CA 90089

University of Colorado
Government Publications
University Library
Box 184
Boulder, CO 80309–0814

Yale University
Government Documents Center
Seeley G. Mudd Library
38 Mansfield
New Haven, CT 06520

American University
Law Library
4400 Massachusetts Avenue NW
Washington, DC 20016

Library of Congress
Serial Division
Madison Bldg.
10 First Street, SE
Washington, DC 20540

University of Florida
Documents Department
Libraries West
Gainesville, FL 32611

Emory University
Law Library
School of Law
Athens, GA 30322

University of Georgia
Law Library
Law School
Athens, GA 30602

University of Hawaii
Government Documents
University Library
2550 The Mall
Honolulu, HI 96822

Illinois Institute of Technology
Law Library
77 South Wacker Drive
Chicago, IL 60606

Northwestern University
Government Publications
University Library
Evanston, IL 60201

University of Chicago
Government Documents
Regenstein Library
1100 E. 57th Street
Chicago, IL 60637

University of Illinois
Law Library
School of Law
504 E. Pennsylvania Avenue
Champaign, IL 61820

Indiana University
Government Documents
University Library
Bloomington, IN 47405

University of Notre Dame
Document Center
Memorial Library
Notre Dame, IN 46556

University of Iowa
Government Publications Library
Iowa City, IA 52242

University of Kansas
Government Documents & Maps
University Library
6001 Malott Hall
Lawrence, KS 66045

University of Kentucky
Government Publications
Margaret I. King Library
Lexington, KY 40506

University of New Orleans
Business Reference
Earl K. Long Library
New Orleans, LA 70148

University of Maine
Law Library
246 Deering Avenue
Portland, ME 04102

Harvard University
Law School Library
Langdell Hall—Law 431
Cambridge, MA 02138

Michigan State University
Documents Department
University Library
East Lansing, MI 48824–1048

University of Michigan
Serials Department
Law Library
Ann Arbor, MI 48109–1210

University of Minnesota
Government Publications
Wilson Library—409
Minneapolis, MN 55455

Washington University
John M. Olin Library
Campus Box 1061
1 Brookings Drive
St. Louis, MO 63130

University of Nebraska
Acquisitions Division
University Libraries
Lincoln, NE 68588–0410

Princeton University
Documents Division Library
Princeton, NJ 08544

University of New Mexico
Social Science Coll. Dev.
Zimmerman Library
Albuquerque, NM 87131

Council on Foreign Relations Library
58 E. 68th Street
New York, NY 10021

New York Public Library
Research Library, Economic & Public Affairs
Grand Central Station
PO Box 2221
New York, NY 10017

New York University
Law Library
School of Law
40 Washington Square South
New York, NY 10012

State University of New York
Government Documents
Lockwood Library Bldg.
Buffalo, NY 14260

State University of New York
Government Publications Library
1400 Washington Avenue
Albany, NY 12222

Duke University
Public Documents Department
University Library
Durham, NC 27706

Ohio State University
Information Services Dept.
University Library
1858 Neil Avenue Mall
Columbus, OH 43210

University of Oklahoma
Government Documents
Bizzell Memorial Library
Room 440
401 West Brooks
Norman, OK 73019

University of Oregon
Documents Section
University Library
Eugene, OR 97403

Pennsylvania State University
Documents Section
University Library
University Park, PA 16802

University of Pennsylvania
Serials Department
Van Pelt Library
Philadelphia, PA 19104

University of Pittsburgh
Hillman Library
Pittsburgh, PA 15260

University of South Carolina
Documents/Microforms
Thomas Cooper Library
Columbia, SC 29208

University of Texas
Law Library
School of Law
727 E. 26th Street
Austin, TX 78705

University of Utah
International Documents
Marriott Library
Salt Lake City, UT 84112

Georgia Mason University
Center for European Studies
4001 N. Fairfax Drive
Suite 450
Arlington, VA 22203

University of Virginia
Government Documents
Alderman Library
Charlottesville, VA 22903

University of Washington
Government Publications
University Library FM-25
Seattle, WA 98195

University of Wisconsin
Documents Department
Memorial Library
728 State Street
Madison, WI 53706

University of Puerto Rico
Law Library
PO Box 23349
Rio Piedras, PR 00931

APPENDIX **F**

# Glossary

Every industry has its own terminology. The health and safety professional must be aware of the precise meanings of certain words commonly used in industrial hygiene, occupational health, and chemistry to communicate effectively with other professionals in these areas.

A fume respirator, for instance, is worthless as protection against gases or vapors. Too often, these terms are used interchangeably; each term has a definite meaning and describes a certain state of matter that can be achieved only by certain physical changes to the given substance.

This glossary defines words and terms, some of which are peculiar to a single industry and others common to many industries. Terms were taken or adapted from the latest editions of *The Chemical Industry Facts Book*, published by the Manufacturing Chemists Association, Washington, DC; *Occupational Diseases and Industrial Medicine*, by RT Johnstone and SE Miller, published by WB Saunders, Philadelphia; *Guide for Industrial Audiometric Technicians*, published by the Safety and Health Services, Employers Insurance of Wausau, WI; American National Standards S1.1: *Acoustical Terminology* and Z88.2: *Respiratory Protection; 101 Atomic Terms and What They Mean*, by the Esso Research and Engineering Company, Linden, NJ; *Paramedical Dictionary*, by JE Schmidt, published by Charles C. Thomas, Springfield, IL; *The Condensed Chemical Dictionary*, published by Van Nostrand Reinhold Publishing, New York; *Stedman's Medical Dictionary*, 26th ed., published by W. B. Saunders Company, Philadelphia; and *Dictionary of Scientific and Technical Terms*, 4th ed., edited by S Parker, published by McGraw-Hill.

# A

**A-, an-** (prefix). Absent, lacking, deficient, without. Anemia: deficient in blood.

**AAOHN.** American Association of Occupational Health Nurses.

**ABIH.** American Board of Industrial Hygiene.

**Abrasive blasting.** A process for cleaning surfaces by means of such materials as sand, alumina, or steel grit in a stream of high-pressure air.

**Absorption.** In air sampling, the capture of a gas or vapor accomplished by passing an airstream containing the gas or vapor through a liquid.

**Absorption coefficient.** See Sound absorption coefficient.

**AC.** See Alternating current.

**Accelerator.** A device for imparting very high velocity to charged particles such as electrons or protons. Also, a chemical additive that increases the speed of a chemical reaction.

**Acclimation.** The process of becoming accustomed to new conditions (such as heat).

**Accommodation.** The ability of the eye to adjust focus for various distances.

**Accuracy (instrument).** Often used incorrectly as precision (see Precision). Accuracy is the agreement of a reading or observation obtained from an instrument or a technique with the true value.

**ACGIH.** American Conference of Governmental Industrial Hygienists. An association whose membership is open to *anyone* who is engaged in the practice of industrial hygiene or occupational and environmental health and safety.

**Acid.** A proton donor.

**Acid pickling.** A bath treatment to remove scale and other impurities from metal surfaces before plating or other surface treatment. Sulfuric acid is commonly used.

**Acne.** See Oil dermatitis.

**Acoustic, acoustical.** Containing, producing, arising from, actuated by, related to, or associated with sound.

**Acoustic trauma.** Hearing loss caused by sudden loud noise in one ear or by a sudden blow to the head. In most cases, hearing loss is temporary, although there may be some permanent loss.

**Acro-** (prefix). Topmost; outer end. An extremity of the body. Acro-osteolysis is degeneration of the terminal or distal end of bone tissue.

**Acrylic.** A family of synthetic resins made by polymerizing esters of acrylic acids.

**Action level.** A term used by OSHA and NIOSH (see entries) to express the level of toxicant that requires medical surveillance, usually one half of the permissible exposure limit.

**Activated charcoal.** Charcoal is an amorphous form of carbon formed by burning wood, nutshells, animal bones, and other carbonaceous materials. Charcoal becomes activated by heating it with steam to 800–900 C. During this treatment, an aporous, submicroscopic internal structure is formed that gives it an extensive internal surface area. Activated charcoal is commonly used as a gas or vapor adsorbent in air-purifying respirators and as a solid sorbent in air sampling.

**Activation.** Making a substance artificially radioactive in an accelerator or by bombarding it with protons or neutrons in a reactor.

**Activity.** Often used as a shortened form of radioactivity; refers to the radiating power of a radioactive substance. Activity may be given in terms of atoms disintegrating per second.

**Acuity.** This sense pertains to the sensitivity of receptors used in hearing or vision.

**Acute.** Health effects that show up a short length of time after exposure. An acute exposure runs a comparatively short course.

**ADA.** Americans with Disabilities Act: a 1991 federal law prohibiting discrimination against people with disabilities in most public activities, including the workplace.

**Additives.** An inclusive name for a wide range of chemical substances that are added in low percentage to stabilize certain end products, such as antioxidants in rubber.

**Aden-** (prefix). Pertaining to a gland. Adenoma is a tumor of gland-like tissue.

**Adenoma.** An epithelial tumor, usually benign, with a gland-like structure (the cells lining gland-like depressions or cavities in the stroma).

**Adhesion.** The ability of one substance to stick to another. There are two types of adhesion: mechanical, which depends on the penetration of the surface, and molecular or polar adhesion, in which adhesion to a smooth surface is obtained because of polar groups such as carboxyl groups.

**Administrative controls.** Methods of controlling employee exposures by job rotation, work assignment, time periods away from the hazard, or training in specific work practices designed to reduce the exposure.

**Adsorption.** The condensation of gases, liquids, or dissolved substances on the surfaces of solids.

**AEC.** Atomic Energy Commission. Now called Nuclear Regulatory Commission in the U.S. Department of Energy.

**Aerobe.** Microorganisms that require the presence of oxygen.

**Aerodynamic diameter.** The diameter of a unit density sphere having the same settling velocity as the particle in question of a given shape and density.

**Aerodynamic forces.** The forces exerted on a particle in suspension by either the movement of air or gases around the particle or the resistance of the gas or air to movement of the particle through the medium.

**Aerosols.** Liquid droplets or solid particles dispersed in air that are of fine enough particle size (0.01–100 μm) to remain so dispersed for a period of time.

**Agglomeration.** Implies consolidation of solid particles into larger shapes by means of agitation alone, that is, without application of mechanical pressure in molds, between rolls, or through dies. Industrial agglomeration is usually implemented in balling devices such as rotating discs, drums, or cones, but it can occur in a simple mixer. Agglomeration has also been used to describe the entire field of particulate consolidation.

**AIDS.** Acquired Immunodeficiency Syndrome.

**AIHA.** American Industrial Hygiene Association.

**Air.** The mixture of gases that surrounds the earth; its major components are as follows: 78.08 percent nitrogen, 20.95 percent oxygen, 0.03 percent carbon dioxide, and 0.93 percent argon. Water vapor (humidity) varies. See Standard air.

**Air bone gap.** The difference in decibels between the hearing levels for a particular frequency as determined by air conduction and bone conduction.

**Airborne microorganisms.** Biologically active contaminants suspended in air either as free-floating particles surrounded by a film or organic or inorganic material, or attached to the surface of other suspended particulates.

**Air cleaner.** A device designed to remove atmospheric airborne impurities, such as dusts, gases, vapors, fumes, and smokes.

**Air conditioning.** The process of treating air to control its temperature, humidity, cleanliness, and distribution to meet requirements of the conditioned space.

**Air conduction.** The process by which sound is conducted to the inner ear through air in the outer ear canal.

**Air filter.** An air-cleaning device to remove light particulate matter from normal atmospheric air.

**Air hammer.** A percussion-type pneumatic tool fitted with a handle at one end of the shank and a tool chuck at the other, into which a variety of tools may be inserted.

**Air horsepower.** The theoretical horsepower required to drive a fan if there are no losses in the fan, that is, if it is 100 percent efficient.

**Air monitoring.** The sampling for and measuring of pollutants in the atmosphere.

**Air mover.** Any device that is capable of causing air to be moved from one space to another. Such devices are generally used to exhaust, force, or draw gases through specific assemblies.

**Air quality criteria.** The amounts of pollution and lengths of exposure at which specific adverse effects to health and welfare take place.

**Air-regulating valve.** An adjustable valve used to regulate airflow to the facepiece, helmet, or hood of an air-line respirator.

**Air, standard.** See Standard air.

**Air-purifying respirator.** Respirators that use filters or sorbents to remove harmful substances from the air.

**Air-supplied respirator.** Respirator that provides a supply of breathable air from a clean source outside of the contaminated work area.

**Albumin.** A protein material found in animal and vegetable fluids, characterized by being soluble in water.

**Albuminuria.** The presence of albumin or other protein substance, such as serum globulin, in the urine.

**-algia** (suffix). Pain. A prefix such as *neur-* tells where the pain is (*neuralgia,* for example).

**Algorithm.** A precisely stated procedure or set of instructions that can be applied stepwise to solve a problem.

**Aliphatic.** (Derived from the Greek word for *oil.*) Pertaining to an open-chain carbon compound. Usually applied to petroleum products derived from a paraffin base and having a straight or branched chain, saturated or unsaturated molecular structure. Substances such as methane and ethane, are typical aliphatic hydrocarbons. See Aromatic.

**Alkali.** A compound that has the ability to neutralize an acid and form a salt. Sodium hydroxide, known as caustic soda or lye, is an example. Used in soap manufacture and many other applications. Turns litmus paper blue. See Base.

**Alkaline earths.** Usually considered to be the oxides of alkaline earth metals: barium, calcium, strontium, beryllium, and radium. Some authorities also include magnesium oxide.

**Alkyd.** A synthetic resin that is the condensed product of a polybasic acid such as phthalic, a polyhydric alcohol such as glycerin, and an oil fatty acid.

**Alkylation.** The process of introducing one or more alkyl radicals by addition or substitution into an organic compound.

**Allergy.** An abnormal response of a hypersensitive person to chemical or physical stimuli. Allergic manifestations of major importance occur in about 10 percent of the population.

**Alloy.** A mixture of metals (and sometimes a nonmetal), as in brass.

**Alpha-emitter.** A radioactive substance that gives off alpha particles.

**Alpha-particle (alpha-ray, alpha-radiation).** A small, positively charged particle made up of two neutrons and two protons and of very high velocity, thrown off by many radioactive materials, including uranium and radium.

**Alternating current (AC).** Electric current that reverses direction. Ordinary house current in the United States reverses direction 60 times per second.

**Aluminosis.** A form of pneumoconiosis due to the presence of aluminum-bearing dust in the lungs, especially that of alum, bauxite, or clay.

**Alveoli.** Tiny air sacs of the lungs, formed at the ends of bronchioles; through the thin walls of the alveoli, the blood takes in oxygen and gives up carbon dioxide in respiration.

**Alveolus.** A general term used in anatomical nomenclature to designate a small sac-like dilation.

**Amalgamation.** The process of alloying metals with mercury. This is one process used in extracting gold and silver from their ores.

**Ambient noise.** The all-encompassing noise associated with a given environment; usually a composite of sounds from many sources.

**Amorphous.** Noncrystalline.

**Anaerobe.** A microorganism that grows without oxygen. Facultative anaerobes are able to grow with or without oxygen; obligate anaerobes grow only in the absence of oxygen.

**Anaerobic bacteria.** Any bacteria that can survive in a partial or complete absence of air.

**Anaphylaxis.** Hypersensitivity resulting from sensitization following prior contact with a chemical or protein.

**Andro-** (prefix). Man, male. An androgen is an agent that produces masculinizing effects.

**Anechoic room (free-field room).** One whose boundaries effectively absorb all the sound incident therein, thereby affording essentially free-field conditions.

**Anemia.** Deficiency in the hemoglobin and erythrocyte content of the blood. Term refers to a number of pathological states that may be attributed to a large variety of causes and appear in many different forms.

**Anemometer.** A device to measure air speed.

**Anesthesia.** Loss of sensation; in particular, the temporary loss of feeling induced by certain chemical agents.

**Angi-, angio-** (prefix). Blood or lymph vessel. Angiitis is the inflammation of a blood vessel.

**Angle of abduction.** Angle between the longitudinal axis of a limb and a sagittal plane.

**Angstrom (Å).** Unit of measure of wavelength equal to 1010 m or 0.1 nm.

**Anneal.** To treat by heat with subsequent cooling for drawing the temper of metals, that is, to soften and render them less brittle. See Temper.

**Anode.** The positive electrode.

**Anorexia.** Lack or loss of the appetite for food.

**ANSI.** American National Standards Institute: a voluntary membership organization (run with private funding) that develops consensus standards nationally for a wide variety of devices and procedures.

**Antagonist.** A muscle opposing the action of another muscle. An active antagonist is essential for control and stability of action by a prime mover.

**Antagonistic interaction.** Interaction of two chemicals in which the resultant toxic effect is lower than the chemicals' individual actions.

**Anthracosilicosis.** A complex form of pneumoconiosis; a chronic disease caused by breathing air containing dust that has free silica as one of its components and that is generated in the various processes in mining and preparing anthracite (hard) coal, and, to a lesser degree, bituminous coal.

**Anthracosis.** A disease of the lungs caused by prolonged inhalation of dust that contains particles of carbon and coal.

**Anthrax.** A highly virulent bacterial infection picked up from infected animals and animal products.

**Anthropometry.** The part of anthropology having to do with measurement of the human body to determine differences in individuals or groups of individuals.

**Anti-** (prefix). Against. An antibiotic is "against life" in the case of a drug—against the life of disease-causing germs.

**Antibiotic.** A substance produced by a microorganism that in dilute solutions kills other organisms, or retards or completely represses their growth, normally in doses that do not harm higher orders of life.

**Antibody.** Any of the body globulins that combine specifically with antigens to neutralize toxins, agglutinate bacteria or cells, and precipitate soluble antigens. It is found naturally in the body or produced by the body in response to the introduction into its tissues of a foreign substance.

**Antigen.** A substance that when introduced into the body stimulates antibody production.

**Antioxidant.** A compound that retards deterioration by oxidation. Antioxidants for human food and animal feeds, sometimes referred to as freshness preservers, retard rancidity of fats and lessen loss of fat-soluble vitamins (A, D, E, K). Antioxidants also are added to rubber, motor lubricants, and other materials to inhibit deterioration.

**Antiparticle.** A particle that interacts with its counterpart of the same mass but opposite electric charge and magnetic properties (e.g., proton and antiproton), with complete annihilation of both and production of an equivalent amount of radiation energy. The positron and its antiparticle, the electron, annihilate each other upon interaction and produce gamma-rays.

**Antiseptic.** A substance that prevents or inhibits the growth of microorganisms; a substance used to kill microorganisms on animate surfaces, such as skin.

**Aplastic anemia.** A condition in which the bone marrow fails to produce an adequate number of red blood corpuscles.

**Approved.** Tested and listed as satisfactory by an authority having jurisdiction, such as U.S. Department of HHS, NIOSH-MSHA; or U.S. Department of Agriculture.

**Aqueous humor.** Fluid in the anterior chamber of the eye.

**Arc welding.** A form of electrical welding using either uncoated or coated rods.

**Arc-welding electrode.** A component of the welding circuit through which current is conducted between the electrode holder and the arc.

**Argyria.** A slate-gray or bluish discoloration of the skin and deep tissues caused by the deposit of insoluble albuminate of silver, occurring after the medicinal administration for a long period of a soluble silver salt; formerly fairly common after the use of insufflations of silver-containing materials into the nose and sinuses. Also seen with occupational exposure to silver-containing chemicals.

**Aromatic.** Applied to a group of hydrocarbons and their derivatives characterized by the presence of the benzene nucleus (molecular ring structure). See Aliphatic.

**Arthr-** (prefix). Joint. Arthropathy is a disease affecting a joint.

**Artificial abrasive.** Materials such as carborundum or emery substituted for natural abrasive such as sandstone.

**Artificial radioactivity.** That produced by bombardment of a target element with nuclear particles. Iodine-131 is an artificially produced radioactive substance.

**Asbestos.** A hydrated magnesium silicate in fibrous form.

**Asbestosis.** A disease of the lungs caused by inhalation of fine airborne asbestos fibers.

**Asepsis.** Clean and free of microorganisms.

**Aseptic technique.** A procedure or operation that prevents the introduction of septic material.

**ASHRAE.** American Society of Heating, Refrigeration, and Air Conditioning Engineers.

**Aspect ratio.** Length to width ratio.

**Asphyxia.** Suffocation from lack of oxygen. Chemical asphyxia is produced by a substance such as carbon monoxide that combines with hemoglobin to reduce the blood's capacity to transport oxygen. Simple asphyxia is the result of exposure to a substance, such as methane, that displaces oxygen.

**Asphyxiant.** A gas whose primary or most acute health effect is asphyxiation. There are two classes of asphyxiant: Simple asphyxiants, such as nitrogen or methane, which act by replacing oxygen; and chemical asphyxiants, such as carbon monoxide, which cause asphyxiation by preventing oxygen uptake at the cellular level.

**Assigned Protection Factor (APF).** The level of respiratory protection expected from a respirator that is properly functioning, has been properly fitted, and is worn by a worker trained in its use. APFs can be used to help provide an estimate of the maximum concentrations of a contaminant in which a particular respirator can be used.

**Asthma.** Constriction of the bronchial tubes in response to irritation, allergy, or other stimulus.

**Ataxia.** Lack of muscular coordination caused by any of several nervous system diseases.

**Atmospheric pressure.** The pressure exerted in all directions by the atmosphere. At sea level, mean atmospheric pressure is 29.92 in. Hg, 14.7 psi, or 407 in. wg.

**Atmospheric tank.** A storage tank designed to operate at pressures from atmospheric through 0.5 psig (3.5 kPa).

**Atom.** All materials are made of atoms. The elements, such as iron, lead, and sulfur, differ from each other because their atomic structures are different. The word *atom* comes from the Greek word meaning indivisible. We now know it can be split and consists of an inner core (nucleus) surrounded by electrons that rotate around the nucleus. As a chemical unit, it remains unchanged during any chemical reaction, yet may undergo nuclear transmutations to other atoms, as in atomic fission.

**Atom smasher.** Accelerator that speeds up atomic and subatomic particles so that they can be used as projectiles to literally blast apart the nuclei of other atoms.

**Atomic energy.** Energy released in nuclear reactions. Of particular interest is the energy released when a neutron splits an atom's nucleus into smaller pieces (fission) or when two nuclei are joined together under millions of degrees of heat (fusion). *Atomic energy* is a popular misnomer; it is more correctly called nuclear energy.

**Atomic hydrogen welding.** A shielded gas-electric welding process using hydrogen as the reducing atmosphere.

**Atomic number.** The number of protons found in the nucleus of an atom. All elements have different atomic numbers. The atomic number of hydrogen is 1, that of oxygen 8, iron 26, lead 82, uranium 92. The atomic number is also called charge number and is usually denoted by Z.

**Atomic power.** The name given to the production of thermal power in a nuclear reactor or power facility.

**Atomic waste.** The radioactive ash produced by the splitting of uranium fuel, as in a nuclear reactor. It may include products made radioactive in such a device.

**Atomic weight.** The atomic weight is approximately the sum of the number of protons and neutrons found in the nucleus of an atom. This sum is also called mass number. The atomic weight of oxygen is approximately 16, with most oxygen atoms containing 8 neutrons and 8 protons. Aluminum is 27; it contains 14 neutrons and 13 protons.

**Atrophy.** Arrested development or wasting away of cells and tissue.

**Attenuate.** To reduce in amount. Usually refers to noise or ionizing radiation.

**Attenuation.** The reduction of intensity at a designated first location as compared with intensity at a second location, which is farther from the source.

**Attenuation block.** A block or stack, having dimensions 20 cm by 20 cm by 3.8 cm, of Type 1100 aluminum alloy or aluminum alloy having equivalent attenuation.

**Audible range.** The frequency range across which normal ears hear: approximately 20 Hz to 20,000 Hz. Above the range of 20,000 Hz, the term ultrasonic is used. Below 20 Hz, the term subsonic is used.

**Audible sound.** Sound containing frequency components lying between 20 and 20,000 Hz.

**Audiogram.** A record of hearing loss or hearing level measured at several different frequencies—usually 500 to 6,000 Hz. The audiogram may be presented graphically or numerically. Hearing level is shown as a function of frequency.

**Audiologist.** A person with graduate training in the specialized problems of hearing and deafness.

**Audiometer.** A signal generator or instrument for measuring objectively the sensitivity of hearing. Pure-tone audiometers are standard instruments for industrial use for audiometric testing.

**Audiometric technician.** A person who is trained and qualified to administer audiometric examinations.

**Audiometric zero.** The threshold of hearing: 0.0002 microbars of sound pressure. See Decibel.

**Auditory.** Pertaining to or involving the sense or organs of hearing.

**Auricle.** Part of the ear that projects from the head; medically, the pinna. Also, one of the two upper chambers of the heart.

**Autoclave.** An apparatus using pressurized steam for sterilization.

**Autoignition temperature.** The lowest temperature at which a flammable gas-air or vapor-air mixture ignites from its own heat source or a contacted heated surface without necessity of spark or flame. Vapors and gases spontaneously ignite at a lower temperature in oxygen than in air, and their autoignition temperature may be influenced by the presence of catalytic substances.

**Avogadro's number.** The number of molecules in a mole of any substance; it equals $6.02217 \times 10^3$. At 0 C and

29.92 in. Hg, 1 mole of any gas occupies 22.414 liters of volume.

**Axial-flow fan.** A propeller-type fan useful for moving large volumes of air against little resistance.

**Axis of rotation.** The true line about which angular motion takes place at any instant. Not necessarily identical with anatomical axis of symmetry of a limb, nor necessarily fixed. Thus, the forearm rotates about an axis that extends obliquely from the lateral side of the elbow to a point between the little finger and ring finger. The elbow joint has a fixed axis maintained by circular joint surfaces, but the knee has a moving axis as its cam-shaped surfaces articulate. Axis of rotation of tools should be aligned with true limb axis of rotation. System of rotation of tools should be aligned with true limb axis of rotation. Systems of predetermined motion times often specify such an axis incorrectly.

**Axis of thrust.** The line along which thrust can be transmitted safely. In the forearm, it coincides with the longitudinal axis of the radius. Tools should be designed to align with this axis.

## B

**Babbitt.** An alloy of tin, antimony, copper, and lead used as a bearing metal.

**Babbitting.** The process of applying babbitt to a bearing.

**Bacillus.** A rod-shaped bacterium.

**Background radiation.** The radiation coming from sources other than the radioactive material to be measured. This background is primarily because of cosmic rays that constantly bombard the earth from outer space.

**Background noise.** Noise coming from sources other than the particular noise source being monitored.

**Bacteria.** Microscopic organisms living in soil, water, organic matter, or the bodies of plants and animals characterized by lack of a distinct nucleus and lack of ability to photosynthesize. Singular: Bacterium.

**Bactericide.** Any agent that destroys bacteria.

**Bacteriophage.** Viruses that infect bacteria and lyse the bacterial cell.

**Bacteriostat.** An agent that stops the growth and multiplication of bacteria but does not necessarily kill them. Usually growth resumes when the bacteriostat is removed.

**Bag house.** Many different trade meanings. Commonly connotes the housing containing bag filters for recovery of fumes of arsenic, lead, sulfur, and others from the flues of smelters.

**Bagasse.** Sugar cane pulp residues.

**Bagassosis.** Respiratory disorder believed to be caused by breathing fungi found in bagasse.

**Balancing by dampers.** Method for designing local exhaust system ducts using adjustable dampers to distribute airflow after installation.

**Balancing by static pressure.** Method for designing local exhaust system ducts by selecting the duct diameters that generate static pressure to distribute airflow without dampers.

**Ball mill.** A grinding device using balls usually made of steel or stone in a revolving container.

**Banbury mixer.** A mixing machine that permits control over the temperature of the batch; commonly used in the rubber industry.

**Band-pass filter.** A wave filter that has a single transmission band extending from a lower cutoff frequency greater than zero to a finite upper cutoff frequency.

**Band-pressure level.** Band-pressure level of a sound for a specified frequency band is the sound-pressure level for the sound contained within the restricted band. The reference pressure must be specified.

**Bandwidth.** When applied to a band-pass filter, bandwidth is determined by the interval of transmitted waves between the low and high cutoff frequencies.

**Baritosis.** An inert pneumoconiosis produced by the inhalation of insoluble barium compounds.

**Barotrauma.** An injury to the ear caused by a sudden alteration in barometric (atmospheric) pressure; aerotitis.

**Basal metabolism.** A measure of the amount of energy required by the body at rest.

**Base.** A compound that reacts with an acid to form a salt; another term for alkali. It turns litmus paper blue.

**Basilar.** Of, relating to, or situated at the base.

**Bauxite.** Impure mixture of aluminum oxides and hydroxides; the principal source of aluminum.

**Bauxite pneumoconiosis.** Shaver's disease. Found in workers exposed to fumes containing aluminum oxide and minute silica particles arising from smelting bauxite in the manufacture of corundum.

**Beam axis.** A line from the source through the centers of the x-ray fields.

**Beam divergence.** Angle of beam spread measured in mrad (1 mrad = 3.4 min of arc).

**Beam-limiting device.** A device that provides a means to restrict the dimensions of an x-ray field.

**Beat elbow.** Bursitis of the elbow; occurs from use of heavy vibrating tools.

**Beat knee.** Bursitis of the knee joints caused by friction or vibration; common in mining.

**Becquerel (Bq).** One disintegration per second; a measure of the rate of radioactive disintegration. There are 37 billion Bqs per curie.

**Beehive kiln.** A kiln shaped like a large beehive usually used for calcining ceramics.

**BEI.** See Biological exposure indices.

**Bel.** A unit of sound level based on a logarithmic scale.

**Belding-Hatch index.** (See also Heat stress index.) Estimate of the body heat stress of a standard man for various degrees of activity; also relates to sweating capacity.

**Benign.** Not malignant. A benign tumor is one that does not metastasize or invade tissue. Benign tumors may still be lethal because of pressure on vital organs.

**Benzene, CH.** A major organic intermediate and solvent derived from coal or petroleum. The simplest member of the aromatic series of hydrocarbons.

**Beryl.** A silicate of beryllium and aluminum.

**Berylliosis.** Chronic beryllium intoxication.

**Beta decay.** The process whereby some radioactive emitters give off a beta particle. Also called beta disintegration.

**Beta particle (beta-radiation).** A small electrically charged particle thrown off by many radioactive materials; identical to the electron. Beta particles emerge from radioactive material at high speeds.

**Betatron.** A large doughnut-shaped accelerator in which electrons (beta particles) are whirled through a changing magnetic field gaining speed with each trip and emerging with high energies. Energies of the order of 100 million electron volts have been achieved. The betatron produces artificial beta radiation.

**Biceps brachii muscle.** The large muscle in the front of the upper arm. Supinates the forearm.

**Bicipital tuberosity.** A protuberance on the medial surface of the radius to which the biceps brachii attaches.

**Billet.** A piece of semifinished iron or steel, nearly square in section, made by rolling and cutting an ingot.

**Binder.** The nonvolatile portion of a coating vehicle that is the film-forming ingredient used to bind the paint pigment particles together.

**Binding energy.** The energy that holds the neutrons and protons of an atomic nucleus together. Represents the difference between the mass of an atom and the sum of the masses of protons and neutrons that make up its nucleus.

**Biohazard.** An abbreviation of *biological hazard*. Organisms or products of organisms that present a risk to humans.

**Biohazard area.** Any area (a complete operating complex, a single facility, a room within a facility, and so on) in which work has been or is being performed with biohazardous agents or materials.

**Biohazard control.** Any set of equipment and procedures used to prevent or minimize the exposure of humans and their environment to biohazardous agents or materials.

**Biological Exposure Indices (BEI).** Advisory biological limit values adopted by the ACGIH for some substances. Indices are based on urine, blood, or expired air samples. A BEI may be a value for the substance itself or it may refer to a level of a metabolite. BEIs represent the value of the biological determinant that is most likely to be the value of that determinant obtained from a worker exposed at the 8-hour TLV-TWA for the substance in question.

**Biological half-life.** The time required to reduce the amount of an exogenous substance in the body by half.

**Biological oxygen demand (BOD).** Quantity of oxygen required for the biological and chemical oxidation of waterborne substances under test conditions.

**Biomechanics.** The study of the human body as a system operating under two sets of laws: the laws of Newtonian mechanics and the biological laws of life.

**Biopsy.** Careful removal of small bits of living tissue from the body for further study and examination, usually under the microscope.

**Black liquor.** A liquor composed of alkaline and organic matter resulting from digestion of wood pulp and cooking acid during the manufacture of paper.

**Bleaching bath.** Chemical solution used to bleach colors from a garment preparatory to dyeing it; a solution of chlorine or sodium hypochlorite is commonly used.

**Bleph-** (prefix). Pertaining to the eyelid.

**Blind spot.** Normal defect in the visual field due to the position at which the optic nerve enters the eye.

**Bloodborne pathogen program.** A 1992 OSHA standard mandates exposure control plans and the use of universal precautions for places of employment where there is risk of employee exposure to blood or other potentially infectious material. Hepatitis B and HIV are the most often-discussed pathogens, but the program is not limited to these two areas.

**Blood count.** A count of the number of corpuscles per cubic millimeter of blood. Separate counts may be made for red and white corpuscles.

**BLS.** Bureau of Labor Statistics.

**Body burden.** The amount of noxious material in the body at a given time.

**Body burden, maximum permissible.** The body burden of a radionuclide that if maintained at a constant level would produce the maximum permissible dose equivalent in the critical organ.

**Boiling point.** The temperature at which the vapor pressure of a liquid equals atmospheric pressure.

**Bombardment.** Shooting neutrons, alpha particles, and other high-energy particles at atomic nuclei, usually in an attempt to split the nucleus or to form a new element.

**Bone conduction test.** A special test conducted by placing an oscillator on the mastoid process to determine the nerve-carrying capacity of the cochlea and the eighth cranial (auditory) nerve.

**Bone marrow.** A soft tissue that constitutes the central filling of many bones and that produces blood corpuscles.

**Bone-seeker.** Any element or radioactive species that lodges in the bone when introduced into the body.

**Brachialis muscle.** Short, strong muscles originating at the lower end of the humerus and inserting into the ulna. Powerful flexor of forearm; employed when lifting.

**Brady-** (prefix). Slow. Bradycardia is slow heartbeat.

**Bradycardia.** Abnormal slowness of the heartbeat, as evidenced by slowing of the pulse rate to 50 or less.

**Brake horsepower.** The horsepower required to drive a unit; it includes the energy losses in the unit and can be determined only by actual test. It does not include drive losses between the motor and unit.

**Branch (or path) of greatest resistance.** The path from a hood to the fan and exhaust stack in a ventilation system that causes the most pressure loss.

**Brass.** An alloy of copper and zinc that may contain a small amount of lead.

**Brattice.** A partition constructed in underground passageways to control ventilation in mines.

**Braze.** To solder with any relatively infusible alloy.

**Brazing furnace.** Used for heating metals to be joined by brazing. Requires a high temperature.

**Breathing tube.** A tube through which air or oxygen flows to the facepiece, helmet, or hood.

**Breathing zone.** Imaginary globe of two foot radius surrounding the head.

**Breathing zone sample.** An air-sample collected in the breathing zone of workers to assess their exposure to airborne contaminants.

**Bremsstrahlung.** Secondary x-radiation produced when a beta particle is slowed down or stopped by a high-density surface.

**Briquette.** Coal or ore dust pressed into oval or brick-shaped blocks.

**Broach.** A cutting tool for cutting non-round holes.

**Bronch-, broncho-** (prefix). Pertaining to the air tubes of the lung.

**Bronchial tubes.** Branches or subdivisions of the trachea (windpipe). A bronchiole is a branch of a bronchus, which is a branch of the windpipe.

**Bronchiectasis.** A chronic dilation of the bronchi or bronchioles marked by fetid breath and paroxysmal coughing, with the expectoration of mucopurulent matter. It may affect the tube uniformly, or may occur in irregular pockets, or the dilated tubes may have terminal bulbous enlargements.

**Bronchiole.** The slenderest of the many tubes that carry air into and out of the lungs.

**Bronchiolitis.** See Bronchopneumonia.

**Bronchitis.** Inflammation of the bronchi or bronchial tubes.

**Bronchoalveolitis.** Bronchopneumonia.

**Bronchopneumonia.** A name given to an inflammation of the lungs that usually begins in the terminal bronchioles. These become clogged with a mucopurulent exudate forming consolidated patches in adjacent lobules. The disease is essentially secondary in character, following infections of the upper respiratory tract, specific infectious fevers, and debilitating diseases.

**Bronzing.** Act or art of imparting a bronze appearance with powders, painting, or chemical processes.

**Brownian motion.** The irregular movement of particles suspended in a fluid as a result of bombardment by atoms and molecules.

**Brucella.** A genus of short, rod-shaped to coccoid, encapsulated, gram-negative, parasitic, pathogenic bacteria.

**Brucellosis.** A group of diseases caused by an organism of the Brucella genus. Undulant fever. One source is unpasteurized milk from cows suffering from Bang's disease (infectious abortion).

**Bubble chamber.** A chamber containing a liquefied gas such as liquid hydrogen, under conditions such that a charged particle passing through the liquid forms bubbles that make its path visible.

**Bubble tube.** A device used to calibrate air-sampling pumps.

**Buffer.** Any substance in a fluid that tends to resist the change in pH when acid or alkali is added.

**Bulk facility.** That portion of a property where flammable or combustible liquids are received by tank vessel, pipeline, tank car, or tank vehicle, and are sorted or blended in bulk for the purpose of distributing such liquids.

**Burn-up.** The extent to which the nuclear fuel in a fuel element has been consumed by fission, as in a nuclear reactor.

**Burns.** Result of the application of too much heat to the skin. First degree burns show redness of the unbroken skin; second degree, skin blisters and some breaking of the skin; third degree, skin blisters and destruction of the skin and underlying tissues, which can include charring and blackening.

**Burr.** The thin rough edges of a machined piece of metal.

**Bursa.** A synovial lined sac that facilitates the motion of tendons; usually near a joint.

**Bursitis.** Inflammation of a bursa.

**Byssinosis.** Disease occurring to those who experience prolonged exposure to heavy air concentrations of cotton or flax dust.

## C

**Calcination.** The heat treatment of solid material to bring about thermal decomposition, to lose moisture or other volatile material, or to oxidize or reduce.

**Calender.** An assembly of rollers for producing a desired finish on paper, rubber, artificial leather, plastic, or other sheet material.

**Caulking.** The process or material used to fill seams of boats, cracks in tile, etc.

**Calorimeter.** A device for measuring the total amount of energy absorbed from a source of electromagnetic radiation.

**Cancer.** A cellular tumor the natural course of which is fatal and usually associated with formation of secondary tumors.

**Capitulum of humerus.** A smooth hemispherical protuberance at the distal end of the humerus articulating with the head of the radius. Irritation caused by pressure between the capitulum and head of the radius may be a cause of tennis elbow.

**Capture velocity.** Air velocity at any point in front of the hood necessary to overcome opposing air currents and to capture the contaminated air by causing it to flow into the exhaust hood.

**Carbohydrate.** An abundant class of organic compounds, serving as food reserves or structural elements for plants and animals. Compounded primarily of carbon, hydrogen, and oxygen, they constitute about two thirds of the average daily adult caloric intake. Sugar, starches, and plant components (cellulose) are all carbohydrates.

**Carbon black.** Essentially a pure carbon, best known as common soot. Commercial carbon black is produced by making soot under controlled conditions. It is sometimes called furnace black, acetylene black, or thermal black.

**Carbon monoxide.** A colorless, odorless, toxic gas produced by any process that involves the incomplete combustion of carbon-containing substances. It is emitted through the exhaust of gasoline-powered vehicles.

**Carbonizing.** The immersion in sulfuric acid of semiprocessed felt to remove any vegetable matter present.

**Carborundum.** A trade name for silicon carbide, widely used as an abrasive.

**Carboy.** A large glass bottle, usually protected by a crate.

**Carboxyhemoglobin.** The reversible combination of carbon monoxide with hemoglobin.

**Carcinogenic.** Cancer-producing.

**Carcinoma.** Malignant tumors derived from epithelial tissues, that is, the outer skin, the membranes lining the body cavities, and certain glands.

**Cardi-, cardio-** (prefix). Denoting the heart.

**Cardiac.** (1) Pertaining to the heart; (2) a cordial or restorative medicine; (3) a person with heart disorder.

**Carding.** The process of combing or untangling wool, cotton, and so on.

**Carding machine.** A textile industry machine that prepares wool, cotton, or other fibers for spinning.

**Cardiovascular.** Relating to the heart and to the blood vessels or circulation.

**Carp-** (prefix). The wrist.

**Carpal tunnel.** A passage in the wrist through which the median nerve and many tendons pass to the hand from the forearm.

**Carpal tunnel syndrome.** A common affliction caused by compression of the median nerve in the carpal tunnel. Often associated with tingling, pain, or numbness in the thumb and first three fingers—may be job-related.

**Carrier.** A person in apparent good health who harbors a pathogenic microorganism.

**Carrier gas.** A mixture of gases that contains and moves a contaminant material. Components of the carrier gas are not considered to cause air pollution or react with the contaminant material.

**CAS number.** Identifies a particular chemical by the Chemical Abstract Service, a service of the American Chemical Society that indexes and compiles abstracts of worldwide chemical literature called Chemical Abstracts.

**Case-hardening.** A process of surface-hardening metals by raising the carbon or nitrogen content of the outer surface.

**Cask (or coffin).** A thick-walled container (usually lead) used for transporting radioactive materials.

**Casting.** Pouring a molten material into a mold and permitting it to solidify to a desired shape.

**Catalyst.** A substance that changes the speed of a chemical reaction but that undergoes no permanent change itself. In respirator use, a substance that converts a toxic gas (or vapor) into a less toxic gas (or vapor). Usually catalysts greatly increase the reaction rate, as in conversion of petroleum to gasoline by cracking. In paint manufacture, catalysts, which hasten film-forming, sometimes become part of the final product. In most uses, however, they do not, and can often be used again.

**Cataract.** Opacity in the lens of the eye that may obscure vision.

**Cathode.** The negative electrode.

**Catwalk.** A narrow suspended footway usually used for inspection or maintenance purposes.

**Caustic.** Something that strongly irritates, burns, corrodes, or destroys living tissue. See Alkali.

**Ceiling Limit (C).** An airborne concentration of a toxic substance in the work environment that should never be exceeded.

**Cell.** A structural unit of which tissues are made. There are many types: nerve cells, muscle cells, blood cells, connective tissues cells, fat cells, and others. Each has a special form to serve a particular function.

**Cellulose.** A carbohydrate that makes up the structural material of vegetable tissues and fibers. Its purest forms are chemical cotton and chemical pulp; it is the basis of rayon, acetate, and cellophane.

**Celsius.** The Celsius temperature scale is a designation of the scale previously known as the centigrade scale.

**-cele** (suffix). Swelling or herniation of a part, as in *rectocele* (prolapse of the rectum).

**Cement, portland.** Portland cement commonly consists of hydraulic calcium silicates to which the addition of certain materials in limited amounts is permitted. Ordinarily, the mixture consists of calcareous materials such as limestone, chalk, shells, marl, clay, shale, blast furnace slag, and so on. In some specifications, iron ore and limestone are added. The mixture is fused by calcining at temperatures usually up to 1,500 C.

**Centrifugal fan.** Wheel-type fan useful where static pressure is medium to high.

**Centrifuge.** An apparatus that uses centrifugal force to separate or remove particulate matter suspended in a liquid.

**Cephal-** (prefix). Pertaining to the head. *Encephal-*, "within the head," pertains to the brain.

**Ceramic.** A term applied to pottery, brick, and tile products molded from clay and subsequently calcined.

**Cerumen.** Earwax.

**Cervi-** (prefix). Neck.

**CFR.** See *Code of Federal Regulations*.

**Chain reaction.** When a fissionable nucleus is split by a neutron it releases energy and one or more neutrons. These neutrons split other fissionable nuclei releasing more energy and more neutrons, making the reaction self-sustaining for as long as there are enough fissionable nuclei present.

**Charged particles.** A particle that possesses at least a unit electrical charge and that does not disintegrate upon a loss of charge. Charged particles are characterized by particle size, number, and sign of unit charges and mobility. See also Ion.

**Chelating agent or chelate.** (Derived from Greek word *kelos* for claw.) Any compound that inactivates a metallic ion with the formation of an inner ring structure in the molecule, the metal ion becoming a member of the ring. The original ion, thus chelated, is effectively out of action.

**Chemical cartridge.** The type of absorption unit used with a respirator for removal of low concentrations of specific vapors and gases.

**Chemical engineering.** That branch of engineering concerned with the development and application of manufacturing processes in which chemical or certain physical changes of materials are involved. These processes usually may be resolved into a coordinated series of unit physical operations and unit chemical processes. The work of the chemical engineer is concerned primarily with the design, construction, and operation of equipment and facilities in which these unit operations and processes are applied.

**Chemical burns.** Generally similar to those caused by heat. After emergency first aid, their treatment is the same as that for thermal burns. In certain instances, such as with hydrofluoric acid, special treatment is required.

**Chemical hygiene plan.** Required by OSHA to protect laboratory employees from hazardous chemicals.

**Chemical reaction.** A change in the arrangement of atoms or molecules to yield substances of different composition and properties. Common types of reactions are combination, decomposition, double decomposition, replacement, and double replacement.

**Chemotherapy.** Use of chemicals of particular molecular structure in the treatment of specific disorders on the assumption that known structures exhibit an affinity for certain parts of malignant cells or infectious organisms, and thereby tend to destroy or inactivate them.

**Chert.** A microcrystalline form of silica. An impure form of flint used in abrasives.

**Cheyne-Stokes respiration.** The peculiar kind of breathing usually observed with unconscious or sleeping individuals who seem to stop breathing altogether for 540 seconds, then start up again with gradually increasing intensity, stop breathing once more, and then repeat the performance. Common in healthy infants.

**Chloracne.** Caused by chlorinated naphthalenes and polyphenyls acting on sebaceous glands.

**Chol-, chole-** (prefix). Relating to bile. Cholesterol is a substance found in bile.

**Chon-, chondro-** (prefix). Cartilage.

**Chromatograph.** An instrument that separates and analyzes mixtures of chemical substances.

**Chromosome.** Important rod-shaped constituent of all cells. Chromosomes contain the genes and are made up of deoxyribonucleic acids (DNA).

**Chronic.** Persistent, prolonged, repeated.

**Cilia.** Tiny hairlike whips in the bronchi and other respiratory passages that aid in the removal of dust trapped on these moist surfaces.

**Ciliary.** Pertaining to the cilium (pl. cilia), a minute vibratile hairlike process attached to the free surface of a cell.

**Clays.** A great variety of aluminum-silicate–bearing rocks that are plastic when wet and hard when dry. Used in pottery, stoneware, tile, bricks, cements, fillers, and abrasives. Kaolin is one type of clay. Some clay deposits may include appreciable quartz. Commercial grades of clays may contain up to 20 percent quartz.

**Clostridium botulinum.** Human pathogenic bacteria that produce an exotoxin, botulinin, which causes botulism.

**Cloud chamber.** A glass-domed chamber filled with moist vapor. When certain types of atomic particles pass through the chamber they leave a cloud-like track much like the vapor trail of a jet plane. This permits scientists to see these particles and study their motion. The cloud chamber and bubble chamber serve the same purpose.

**CNS.** Central nervous system.

**Coagulase.** An enzyme produced by pathogenic staphylococci; causes coagulation of blood plasma.

**Coagulation.** Formation of a clot or gelatinous mass.

**Coalesce.** To unite into a whole; to fuse; to grow together.

**Coated welding rods.** The coatings of welding rods vary. For the welding of iron and most steel, the rods contain manganese, titanium, and a silicate.

**Coccidiomycosis.** A fungal disease (also known as valley fever or San Joaquin Valley fever) that can affect agricultural, horticultural, construction, and any workers who disturb soil containing spores. Although most often a respiratory disease, in rare cases it can be systemic and fatal. It is transmitted by inhalation of dust containing spores of *Coccidioides immitis.*

**Coccus.** A spherical bacterium. Plural: cocci.

**Cochlea.** The auditory part of the internal ear, shaped like a snail shell. It contains the basilar membrane on which the end organs of the auditory nerve are distributed.

***Code of Federal Regulations.*** The rules promulgated under U.S. law, published in the *Federal Register,* and actually enforced at the end of a calendar year are incorporated in this code (*CFR*).

**Coefficient of discharge.** A factor used in figuring flow through an orifice. The coefficient takes into account the facts that a fluid flowing through an orifice contracts to a cross-sectional area that is smaller than that of the orifice, and there is some dissipation of energy caused by turbulence.

**Coefficient of entry.** The actual rate of flow caused by a given hood static pressure compared to the theoretical flow that would result if the static pressure could be converted to velocity pressure with 100 percent efficiency; it is the ratio of actual to theoretical flow.

**Coefficient of variation.** The ratio of the standard deviation to the mean value of a population of observations.

**Coffin.** A thick-walled container (usually lead) used for transporting radioactive materials.

**Cohesion.** Molecular forces of attraction between particles of like compositions.

**Colic.** A severe cramping, gripping pain in or around the abdomen.

**Collagen.** An albuminoid, the main supportive protein of skin: tendon, bone, cartilage, and connective tissue.

**Collection efficiency.** The percentage of a specific substance removed and retained from air by an air cleaning or sampling device. A measure of the cleaner or sampler performance.

**Collimated beam.** A beam of light with parallel waves.

**Colloid.** Generally a liquid mixture or suspension in which the particles of suspended liquid or solid are very finely divided. Colloids do not appreciably settle out of suspension.

**Colloid mill.** A machine that grinds materials into a very fine state of suspension, often simultaneously placing this suspension in a liquid

**Colorimetry (colorimetric).** The term applied to all chemical analysis techniques involving reactions in which a color is developed when a particular contaminant is present in the sample and reacts with the collection medium. The resultant color intensity is measured to determine the contaminant concentration.

**Coma.** A level of unconsciousness from which a patient cannot be aroused.

**Combustible liquids.** Combustible liquids are those having a flash point at or above 100 F (37.8 C).

**Comedones.** Blackheads. Blackened, oily masses of dead epithelial matter clogging the openings of oil glands and hair follicles.

**Comfort ventilation.** Airflow intended to maintain comfort of room occupants (heat, humidity, and odor).

**Comfort zone.** The range of effective temperatures over which the majority of adults feels comfortable.

**Communicable.** A disease whose causative agent is readily transferred from one person to another.

**Compaction.** The consolidation of solid particles between rolls or by tamp, piston, screw, or other means of applying mechanical pressure.

**Compound.** A substance composed of two or more elements joined according to the laws of chemical combination. Each compound has its own characteristic properties different from those of its constituent elements.

**Compressible flow.** Flow of high-pressure gas or air that undergoes a pressure drop resulting in a significant reduction of its density.

**Compton effect.** The glancing collision of a gamma-ray with an electron. The gamma-ray gives up part of its energy to the electron.

**Concentration.** The amount of a given substance in a stated unit of measure. Common methods of stating concentration are percent by weight or by volume, weight per unit volume, normality, and so on.

**Conchae.** See Turbinates.

**Condensate.** The liquid resulting from the process of condensation. In sampling, the term is generally applied to the material that is removed from a gas sample by means of cooling.

**Condensation.** Act or process of reducing from one form to another denser form such as steam to water.

**Condensoid.** A dispersoid consisting of liquid or solid particles formed by the process of condensation. The dispersoid is commonly referred to as a condensation aerosol.

**Conductive hearing loss.** Type of hearing loss; not caused by noise exposure, but by any disorder in the middle or external ear that prevents sound from reaching the inner ear.

**Confined space.** Any enclosed area not designed for human occupancy that has a limited means of entry and egress and in which existing ventilation is not sufficient to ensure that the space is free of a hazardous atmosphere, oxygen deficiency, or other known or potential hazards. Examples are storage tanks, boilers, sewers, and tank cars. A permit-required confined space, as defined by the OSHA standard, is one that requires a permit process and implementation of a comprehensive confined space entry program prior to entry.

**Congenital.** Some problem that originates before birth.

**Conjunctiva.** The delicate mucous membrane that lines the eyelids and covers the exposed surface of the eyeball.

**Conjunctivitis.** Inflammation of the conjunctiva.

**Contact dermatitis.** Dermatitis caused by contact with a substance—gaseous, liquid, or solid. May be caused by primary irritation or an allergy.

**Control rod.** A rod (containing an element such as boron) used to control the power of a nuclear reactor. The control rod absorbs neutrons that would normally split the fuel nuclei. Pushing the rod in reduces the release of atomic power; pulling out the rod increases it.

**Controlled areas.** A specified area in which exposure of personnel to radiation or radioactive material is controlled and that is under the supervision of a person who knows appropriate radiation protection practices, including pertinent regulations, and who is responsible for applying them.

**Convection.** The motions in fluids resulting from differences in density and the action of gravity.

**Converter.** A nuclear reactor that uses one kind of fuel and produces another. For example, a converter charged with uranium isotopes might consume uranium-235 and produce plutonium from uranium-238. A breeder reactor produces more atomic fuel than it consumes; a converter does not.

**Coolants.** Transfer agents used in a flow system to convey heat from its source.

**Copolymers.** Mixed polymers or heteropolymers. Products of the polymerization of two or more substances at the same time.

**Core.** (1) The heart of a nuclear reactor where the nuclei of the fuel fission (split) and release energy. The core is usually surrounded by a reflecting material that bounces stray neutrons back to the fuel. It is usually made up of fuel elements and a moderator. (2) A shaped, hard-baked cake of sand with suitable compounds that is placed within a mold, forming a cavity in the casting when it solidifies. (3) The vital centers of the body—heart, viscera, brain—as opposed to the shell—the limbs and integument.

**Corium.** The deeper skin layer containing the fine endings of the nerves and the finest divisions of the blood vessels, the capillaries. Also called the derma.

**Cornea.** Transparent membrane covering the anterior portion of the eye.

**Corpuscle.** A red or white blood cell.

**Corrected effective temperature (CET).** An index of thermal stress similar to the effective temperature index except that globe temperature is used instead of dry-bulb temperature.

**Corrective lens.** A lens ground to the wearer's individual prescription.

**Corrosion.** Physical change, usually deterioration or destruction, brought about through chemical or electrochemical action, as contrasted with erosion, caused by mechanical action.

**Corrosive.** A substance that causes visible destruction or permanent changes in human skin tissue at the site of contact.

**Corundum.** An impure form of aluminum oxide.

**Cosmic rays.** High-energy rays that bombard the earth from outer space. Some penetrate to the earth's surface and others may go deep into the ground. Although each ray is energetic, the number bombarding the planet is so small that the total energy reaching the earth is about the same as that from starlight.

**Costo-** (prefix), **costal.** Pertaining to the ribs.

**Cottrell precipitator.** A device for dust collection using high-voltage electrodes.

**Coulometry.** Measurement of the number of electrons that are transferred across an electrode solution interface when a reaction in the solution is created and carried to completion. The reaction is usually caused by a contaminant in a sample gas that is drawn through or onto the surface of the solution. The number of electrons transferred in terms of coulombs is an indication of the contaminant concentrations.

**Count.** A click in a Geiger counter or the numerical value for the activity of a radioactive specimen.

**Counter.** A device for counting. See Geiger counter and Scintillation counter.

**Count median size.** The size of the particle in a sample of particulate matter containing equal numbers of particles larger and smaller than the stated size.

**Covered electrode.** A composite filler metal electrode consisting of a core of bare electrode or metal-cored electrode to which a covering (sufficient to provide a slag layer on the weld metal) has been applied. The covering may contain materials providing such functions as shielding from the atmosphere, deoxidation, and arc stabilization and can serve as a source of metallic additions to the weld.

**Cps.** Cycles per second, now called hertz.

**Cracking.** Used almost exclusively in the petroleum industry, cracking is thermal or catalytic decomposition of organic compounds, usually for the manufacture of gasoline. Petroleum constituents are also cracked for the purpose of manufacturing chemicals.

**Cramps.** Painful muscular contractions that may affect almost any voluntary or involuntary muscle.

**Cranio-** (prefix). Skull. As in *craniotomy,* incision through a skull bone.

**Cristobalite.** A crystalline form of free silica, extremely hard and inert chemically, and very resistant to heat. Quartz in refractory bricks and amorphous silica in diatomaceous earth are altered to cristobalite when exposed to high temperatures (calcined).

**Critical mass.** The amount of nuclear fuel necessary to sustain a chain reaction. If too little fuel is present, too many neutrons will stray, and the reaction will die out.

**Critical pressure.** The pressure under which a substance may exist as a gas in equilibrium with the liquid at the critical temperature.

**Critical temperature.** The temperature above which a gas cannot be liquefied by pressure alone.

**Crucible.** A heat-resistant barrel-shaped pot used to hold metal during melting in a furnace or in other applications.

**Crude petroleum.** Hydrocarbon mixtures that have a flash point below 150 F (65.6 C) and that have not been processed in a refinery.

**Cry-, cryo-** (prefix). Very cold.

**Cryogenics.** The field of science dealing with the behavior of matter at very low temperatures.

**CTD.** See Cumulative trauma disorder.

**Cubic centimeter (cm³).** A volumetric measurement that is equal to one milliliter (ml).

**Cubic meter (m³).** A measure of volume in the metric system.

**Culture** (biology). A population of microorganisms or tissue cells cultivated in a medium.

**Culture medium.** Any substance or preparation suitable for the growth of cultures and cultivation of microorganisms. Selective medium, a medium composed of nutrients designed to allow growth of a particular type of microorganism; broth medium, a liquid medium; agar medium, solid culture medium.

**Cumulative trauma disorder (CTD).** A disorder of a musculoskeletal or nervous system component caused or aggravated by repeated and/or forceful movements of the same musculoskeletal systems.

**Curie.** A measure of the rate at which a radioactive material decays. The radioactivity of one gram of radium is a curie. It is named for Pierre and Marie Curie, pioneers in radioactivity and discoverers of the elements radium, radon, and polonium. One curie corresponds to 37 billion disintegrations per second.

**Cutaneous.** Pertaining to or affecting the skin.

**Cuticle.** The superficial scarfskin or upper strata of skin.

**Cutie-pie.** A portable instrument equipped with a direct-reading meter used to determine the level of ionizing radiation in an area.

**Cutting fluids (oils).** The cutting fluids used in industry today are usually an oil or an oil–water emulsion used to cool and lubricate a cutting tool. Cutting oils are usually light or heavy petroleum fractions.

**CW laser.** Continuous wave laser.

**Cyan-** (prefix). Blue.

**Cyanide (as CN).** Cyanides inhibit tissue oxidation upon inhalation or ingestion and cause death.

**Cyanosis.** Blue appearance of the skin, especially on the face and extremities, indicating a lack of sufficient oxygen in the arterial blood.

**Cyclone separator.** A dust-collecting device that has the ability to separate particles by size. Typically used to collect respirable dust samples.

**Cyclotron.** A particle accelerator. In this atomic "merry-go-round," atomic particles are whirled around in a spiral between the ends of a huge magnet, gaining speed with each rotation in preparation for their assault on the target material.

**Cyst** (prefix). Pertaining to a bladder or sac, normal or abnormal, filled with gas, liquid, or semisolid material. The term appears in many words concerning the urinary bladder (*cystocele, cystitis*).

**Cyto-** (prefix). Cell.

**Cytoplasm.** Cell plasma (protoplasm) that does not include the cell's nucleus.

**Cytotoxin.** A substance, developed in the blood serum, having a toxic effect upon cells.

## D

**Damage risk criterion.** The suggested baseline of noise tolerance, which, if not exceeded, should result in no hearing loss due to noise. A damage risk criterion may include in its statement a specification of such factors as time of exposure, noise level, frequency, amount of hearing loss considered significant, percentage of the population to be protected, and method of measuring the noise.

**Damp.**  A harmful gas or mixture of gases occurring in coal mining.

**Dampers.**  Adjustable sources of airflow resistance used to distribute airflow in a ventilation system.

**Dangerous to life or health, immediately (IDLH).**  Used to describe very hazardous atmospheres where employee exposure can cause serious injury or death within a short time or serious delayed effects.

**Daughter.**  As used in radioactivity, this refers to the product nucleus or atom resulting from decay of the precursor or parent.

**dBA.**  Sound level in decibels read on the A scale of a sound-level meter. The A scale discriminates against very low frequencies (as does the human ear) and is therefore better for measuring general sound levels. See also Decibel.

**dBC.**  Sound level in decibels read on the C scale of a sound-level meter. The C scale discriminates very little against very low frequencies. See also Decibel.

**DC.**  See Direct current.

**Decay.**  When a radioactive atom disintegrates, it is said to decay. What remains is a different element. An atom of polonium decays to form lead, ejecting an alpha particle in the process.

**Decibel (dB).**  A unit used to express sound-power level $(L_W)$ and sound-pressure level $(L_P)$. Sound power is the total acoustic output of a sound source in watts (W). By definition, sound-power level, in decibels, is: $L_W = 10 \log W/W_O$, where $W$ is the sound power of the source and $W_O$ is the reference sound power of $10^{-12}$. Because the decibel is also used to describe other physical quantities, such as electrical current and electrical voltage, the correct reference quantity must be specified.

**Decomposition.**  The breakdown of a chemical or substance into different parts or simpler compounds. Decomposition can occur because of heat, chemical reaction, decay, etc.

**Decontaminate.**  To make safe by eliminating poisonous or otherwise harmful substances, such as noxious chemicals or radioactive material.

**Deltoid muscle.**  The muscle of the shoulder responsible for abducting the arm sideways and for swinging the arm at the shoulder. Overuse of the deltoid muscle may cause fatigue and pain in the shoulder.

**Density.**  The ratio of mass to volume.

**Dent-, dento-** (prefix). Pertaining to a tooth or teeth, from Latin.

**Derma.**  The dermis. The corium or true skin.

**Dermatitis.**  Inflammation of the skin from any cause.

**Dermatology.**  Branch of medicine concerned with the diagnosis and treatment, including surgery and prevention, of diseases of the skin, hair, and nails.

**Dermatophytosis.**  Athlete's foot.

**Dermatosis.**  A broader term than dermatitis, it includes any cutaneous abnormality. Thus it encompasses folliculitis, acne, pigmentary changes, and nodules and tumors.

**Desiccant.**  Material that absorbs moisture.

**Deuterium.**  Heavy hydrogen. The nucleus of heavy hydrogen is a deuteron. It is called heavy hydrogen because it weighs twice as much as ordinary hydrogen.

**Deuteron.**  The nucleus of an atom of heavy hydrogen containing one proton and one neutron. Deuterons are often used for the bombardment of other nuclei.

**Diagnostic x-ray system.**  An x-ray system designed for irradiation of any part of the human body for the purpose of diagnosis or visualization.

**Diaphragm.**  (1) The musculomembranous partition separating the abdominal and thoracic cavities. (2) Any separating membrane or structure. (3) A disk with one or more openings, or with an adjustable opening, mounted in relation to a lens, by which part of the light may be excluded from the area.

**Diatomaceous earth.**  A soft, gritty amorphous silica composed of minute siliceous skeletons of small aquatic plants. Used in filtration and decolorization of liquids, insulation, filler in dynamite, wax, textiles, plastics, paint, and rubber. Calcined and flux-calcined diatomaceous earth contains appreciable amounts of cristobalite, and dust levels should be controlled the same as for cristobalite.

**Die.**  A hard metal or plastic form used to shape material to a particular contour or section.

**Differential pressure.**  The difference in static pressure between two locations.

**Diffuse sound field.**  One in which the time average of the mean-square sound pressure is everywhere the same and the flow of energy in all directions is equally probable.

**Diffusion, molecular.**  A process of spontaneous intermixing of different substances attributable to molecular motion and tending to produce uniformity of concentration.

**Diffusion rate.** A measure of the tendency of one gas or vapor to disperse into or mix with another gas or vapor. This rate depends on the density of the vapor or gas as compared with that of air, which is given a value of 1.

**Diluent.** A liquid blended with a mixture to reduce concentration of the active agents.

**Dilution.** The process of increasing the proportion of solvent or diluent (liquid) to solute or particulate matter (solid).

**Dilution ventilation.** See General ventilation.

**Diopters.** A measure of the power of a lens or prism, equal to the reciprocal of its focal length in meters.

**Direct current (DC).** Electric current flowing in one direction only.

**Direct-reading instrumentation.** Instruments that give an immediate indication of the concentration of aerosols, gases, or vapors or magnitude of physical hazard by some means such as a dial or meter.

**Disease.** A departure from a state of health, usually recognized by a sequence of signs and symptoms.

**Disinfectant.** An agent that frees from infection by killing the vegetative cells of microorganisms.

**Disintegration.** A nuclear transformation or decay process that results in the release of energy in the form of radiation.

**Dispersion.** The general term describing systems consisting of particulate matter suspended in air or other fluid; also, the mixing and dilution of contaminant in the ambient environment.

**Distal.** Away from the central axis of the body.

**Distal phalanx.** The last bony segment of a toe or finger.

**Distillery.** A facility or that portion of a facility where flammable or combustible liquids produced by fermentation are concentrated and where the concentrated products may also be mixed, stored, or packaged.

**Diuretic.** Anything that promotes excretion of urine.

**DNA.** Deoxyribonucleic acid. The genetic material within the cell.

**DOP.** Dioctyl phthalate, a powdered chemical that can be aerosolized to an extremely uniform size, i.e., 0.3 $\mu$m for a major portion of any sample.

**Dose.** (1) Used to express the amount of a chemical or of ionizing radiation energy absorbed in a unit volume or an organ or individual. Dose rate is the dose delivered per unit of time. (See also Roentgen, Rad, Rem.) (2) Used to express amount of exposure to a chemical substance.

**Dose, absorbed.** The energy imparted to matter in a volume element by ionizing radiation divided by the mass of irradiated material in that volume element.

**Dose equivalent.** The product of absorbed dose, quality factor, and other modifying factors necessary to express on a common scale, for all ionizing radiations, the irradiation incurred by exposed persons.

**Dose equivalent, maximum permissible (MPD).** The largest dose equivalent received within a specified period that is permitted by a regulatory agency or other authoritative group on the assumption that receipt of such a dose equivalent creates no appreciable somatic or genetic injury. Different levels of MPD may be set for different groups within a population. (In popular usage, "dose, maximum permissible" is an accepted synonym.)

**Dose-response relationship.** Correlation between the amount of exposure to an agent or toxic chemical and the resulting effect on the body.

**Dosimeter (dose meter).** An instrument used to determine the full-shift exposure a person has received to a physical hazard.

**DOT.** Department of Transportation.

**Drier.** Any catalytic material that, when added to a drying oil, accelerates drying or hardening of the film.

**Drop forge.** To forge between dies using a drop hammer or drop press.

**Droplet.** A liquid particle suspended in a gas. The liquid particle is generally of such size and density that it settles rapidly and remains airborne for an appreciable length of time only in a turbulent atmosphere.

**Dross.** The scum that forms on the surface of molten metals, consisting largely of oxides and impurities.

**Dry-bulb thermometer.** An ordinary thermometer, especially one with an unmoistened bulb, not dependent on atmospheric humidity. The reading is the dry-bulb temperature.

**Duct.** A conduit used for conveying air at low pressures.

**Duct velocity.** Air velocity through the duct cross section. When solid particulate material is present in the airstream, the duct velocity must exceed the minimum transport velocity.

**Ductile.** Capable of being molded or worked, as metals.

**Dust collector.** An air-cleaning device to remove heavy particulate loadings from exhaust systems before discharge to outdoors; usual range is loadings of 0.003 g/ft³ (0.007 mg/m³) and higher.

**Dusts.** Solid particles generated by handling, crushing, grinding, rapid impact, detonation, and decrepitation of organic or inorganic materials, such as rock, ore, metal, coal, wood, and grain. Dusts do not tend to flocculate, except under electrostatic forces; they do not diffuse in air but settle under the influence of gravity.

**Dynometer.** Apparatus for measuring force or work output external to a subject. Often used to compare external output with associated physiological phenomena to assess physiological work efficiency.

**Dys-** (prefix). Difficult, bad. This prefix occurs in a large number of medical words because it is attachable to a term for any organ or process that is not functioning as well as it should.

**Dysfunction.** Disturbance, impairment, or abnormality of the functioning of an organ.

**Dyspnea.** Shortness of breath, difficult or labored breathing. More strictly, the sensation of shortness of breath.

**Dysuria.** Difficulty or pain in urination.

## E

**EAP.** Employee Assistance Program.

**Ear.** The entire hearing apparatus, consisting of three parts: external ear, the middle ear or tympanic cavity, and the inner ear or labyrinth. Sometimes the pinna is called the ear.

**Ecology.** The science of the relationships between living organisms and their environments.

**-ectomy** (suffix). A cutting out; surgical removal. Denotes any operation in which all or part of a named organ is cut out of the body.

**Eczema.** A skin disease or disorder. Dermatitis.

**Edema.** A swelling of body tissues as a result of being waterlogged with fluid.

**Effective temperature (ET).** An arbitrary index that combines into a single value the effects of temperature, humidity, and air movement on the sensation of warmth and cold on the human body.

**Effective temperature index.** An empirically determined index of the degree of warmth perceived on exposure to different combinations of temperature, humidity, and air movement. The determination of effective temperature requires simultaneous determinations of dry-bulb and wet-bulb temperatures.

**Efficiency, fractional.** The percentage of particles of a specified size that are removed and retained by a particular type of collector or sampler. A plot of fractional efficiency values versus the respective sized particles yields a fractional efficiency curve that may be related to the total collecting efficiency of air-cleaning or air-sampling equipment.

**Efflorescence.** A phenomenon whereby a whitish crust of fine crystals forms on a surface. These are usually sodium salts that diffuse from the substrate.

**Effluent.** Generally something that flows out or forth, like a stream flowing out into a lake. In terms of pollution, an outflow of a sewer, storage tank, canal, or other channel.

**Ejector.** An air mover consisting of a two-flow system wherein a primary source of compressed gas is passed through a Venturi and the vacuum developed at the throat of the Venturi is used to create a secondary flow of fluid. In the case of air movers for sampling applications, the secondary flow is the sample gas.

**Elastomer.** In a chemical industry sense, a synthetic polymer with rubber-like characteristics; a synthetic or natural rubber or a soft, rubbery plastic with some degree of elasticity at room temperature.

**Electrical precipitators.** A device that removes particles from an airstream by charging the particles and collecting the charged particles on a suitable surface.

**Electrolysis.** The process of conduction of an electric current by means of a chemical solution.

**Electromagnetic radiation.** The propagation of varying electric and magnetic fields through space at the speed of light, exhibiting the characteristics of wave motion.

**Electron.** A minute atomic particle possessing a negative electric charge. In an atom the electrons rotate around a small nucleus. The weight of an electron is so infinitesimal that it would take 500 octillion (500 followed by 27 zeros) of them to make a pound. It is only about a two-thousandth of the mass of a proton or neutron.

**Electron volt (eV).** A small unit of energy. An electron gains this much energy when it is acted upon by one volt. Energies of radioactive materials may be millions of electron volts (MeV), whereas particle accelerators generate energies of billions of electron volts (BeV).

**Electroplate.** To cover with a metal coating (plate) by means of electrolysis.

**Element.** Solid, liquid, or gaseous matter that cannot be further decomposed into simpler substances by chemical means. The atoms of an element may differ physically but do not differ chemically. All atoms of an element contain a definite number of protons and thus have the same atomic number.

**ELF.** Extremely low frequency electromagnetic field.

**Elutriator.** A device used to separate particles according to mass and aerodynamic size by maintaining a laminar flow system at a rate that permits the particles of greatest mass to settle rapidly while the smaller particles are kept airborne by the resistance force of the flowing air for longer times and distances. The various times and distances of deposit may be used to determine representative fractions of particle mass and size.

**Embryo.** The name for the early stage of development of an organism. In humans, the period from conception to the end of the second month.

**Emergent beam diameter.** Diameter of the laser beam at the exit aperture of the system.

**Emery.** Aluminum oxide, natural and synthetic abrasive.

**Emission factor.** Statistical average of the amount of a specific pollutant emitted from each type of polluting source in relation to a unit quality of material handled, processed, or burned.

**Emission inventory.** A list of primary air pollutants emitted into a given community's atmosphere, in amounts per day, by type of source.

**Emission standards.** The maximum amount of pollutant permitted to be discharged from a single polluting source.

**Emmetropia.** A state of perfect vision.

**Emphysema.** A lung disease in which the walls of the air sacs (alveoli) have been stretched too thin and have broken down.

**Emulsifier or Emulsifying agent.** A chemical that holds one insoluble liquid in suspension in another. Casein, for example, is a natural emulsifier in milk, keeping butterfat droplets dispersed.

**Emulsion.** A suspension, each in the other, of two or more unlike liquids that usually do not dissolve in each other.

**Enamel.** A paint-like oily substance that produces a glossy finish to a surface to which it is applied, often containing various synthetic resins. It is lead free, in contrast to the ceramic enamel, that is, porcelain enamel, which contains lead.

**Endemic.** (1) Present in a community or among a group of people; usually refers to a disease prevailing continually in a region. (2) The continuing prevalence of a disease, as distinguished from an epidemic.

**Endo-** (prefix). Within, inside of, internal. The endometrium is the lining membrane of the uterus.

**Endocrine.** Secreting without the means of a duct or tube. The term is applied to certain glands that produce secretions that enter the bloodstream or the lymph directly and are then carried to the particular gland or tissue whose function they regulate.

**Endothermic.** Characterized by or formed with absorption of heat.

**Endotoxin.** A toxin that is part of the wall of a microorganism and is released when that organism dies.

**Energy density.** The intensity of electromagnetic radiation per unit area per pulse expressed in joules per square centimeter.

**Engineering controls.** Methods of controlling employee exposures by modifying the source or reducing the quantity of contaminants released into the work environment.

**Enteric.** Intestinal.

**Entero-** (prefix). Pertaining to the intestines.

**Enterotoxin.** A toxin specific for cells of the intestine; gives rise to symptoms of food poisoning.

**Entrainment velocity.** The gas flow velocity, which tends to keep particles suspended and cause deposited particles to become airborne.

**Entrance loss.** The loss in static pressure of a fluid that flows from an area into and through a hood or duct opening. The loss in static pressure is caused by friction and turbulence resulting from the increased gas velocity and configuration of the entrance area.

**Entry loss.** Loss in pressure caused by air flowing into a duct or hood.

**Enzymes.** Delicate chemical substances, mostly proteins, that enter into and bring about chemical reactions in living organisms.

**EPA.** Environmental Protection Agency.

**EPA number.** The number assigned to chemicals regulated by the Environmental Protection Agency.

**Epicondylitis.** Inflammation of certain bony prominences in the area of the elbow, for example, tennis elbow.

**Epidemiology.** The study of disease in human populations.

**Epidermis.** The superficial scarfskin or upper (outer) layer of skin.

**Epilation.** Temporary or permanent loss of body hair.

**Epithelioma.** Carcinoma of the epithelial cells of the skin and other epithelial surfaces.

**Epithelium.** The purely cellular, avascular layer covering all the free surfaces—cutaneous, mucous, and serous—

including the glands and other structures derived therefrom; for example, the epidermis.

**Equivalent chill temperature (ECT).** Also known as wind chill index. A temperature index used to account for heat loss from skin exposed to the combined effects of cold temperatures and air speed.

**Erg.** The force of one dyne acting through a distance of one centimeter. It would be equivalent to the work done by a June bug climbing over a stone 0.5 in. (1 cm) high, or the energy required to ionize about 20 billion molecules of air.

**Ergonomics.** A multidisciplinary activity dealing with interactions between humans and their total working environment plus stresses related to such environmental elements as atmosphere, heat, light, and sound as well as all tools and equipment of the workplace.

**Erysipeloid.** A bacterial infection affecting slaughterhouse workers and fish handlers.

**Eryth-, erythro-** (prefix). Redness. Erythema is indicated by redness of the skin (including a deep blush). An erythrocyte is a red blood cell.

**Erythema.** Reddening of the skin.

**Erythemal region.** Ultraviolet light radiation between 2,800 and 3,200 angstroms (280–320 millimicrons); it is absorbed by the cornea of the eye.

**Erythrocyte.** A type of red blood corpuscle.

**Eschar.** The crust formed after injury by a caustic chemical or heat.

**Essential oil.** Any of a class of volatile, odoriferous oils found in plants and imparting to the plants odor and often other characteristic properties. Used in essence, perfumery, etc.

**Esters.** Organic compounds that may be formed by interaction between an alcohol and an acid, or by other means. Esters are nonionic compounds, including solvents and natural fats.

**Etch.** To cut or eat away material with acid or another corrosive substance.

**Ethylene oxide.** A carcinogenic hospital sterilant regulated by OSHA. Ethylene oxide is also a reproductive hazard.

**Etiologic agent.** Refers to organisms, substances, or objects associated with the cause of disease or injury.

**Etiology.** The study or knowledge of the causes of disease.

**Eu-** (prefix). Well and good. A euthyroid person has a thyroid gland that couldn't be working better. A euphoric person has a tremendous sense of well-being.

**Eukaryote.** An organism whose cells contain mitochondria and a nuclear membrane. Describes organisms from yeasts to humans.

**Eustachian tube.** A structure about 2.5 in. (6 cm) long leading from the back of the nasal cavity to the middle ear. It equalizes the pressure of air in the middle ear with that outside the eardrum.

**Evaporation.** The process by which a liquid is changed to the vapor state.

**Evaporation rate.** The ratio of the time required to evaporate a measured volume of a liquid to the time required to evaporate the same volume of a reference liquid (ethyl ether) under ideal test conditions. The higher the ratio, the slower the evaporation rate.

**Exhalation valve.** A device that allows exhaled air to leave a respirator and prevents outside air from entering through the valve.

**Exhaust ventilation.** The removal of air, usually by mechanical means, from any space. The flow of air between two points is because of a pressure difference between the two points. This pressure difference causes air to flow from the high-pressure to the low-pressure zone.

**Exothermic, Exothermal.** Characterized by or formed with evolution of heat.

**Exotoxin.** A toxin excreted by a microorganism into the surrounding medium.

**Explosive limit.** See Flammable limit.

**Exposure.** Contact with a chemical, biological, or physical hazard.

**Extension.** Movement whereby the angle between the bones connected by a joint is increased. Motions of this type are produced by contraction of extensor muscles.

**Extensor muscles.** A muscle that, when active, increases the angle between limb segments, for example, the muscles that straighten the knee or elbow, open the hand, or straighten the back.

**Extensor tendon.** Connecting structure between an extensor muscle and the bone into which it inserts. Examples are the hard, longitudinal tendons found on the back of the hand when the fingers are fully extended.

**External mechanical environment.** The synthetic physical environment, for example, equipment, tools, machine controls, clothing. Antonym: internal (bio)mechanical environment.

**Extravasate.** To exude a substance from the body's vessels into tissues.

**Extrusion.** The forcing of raw material through a die or a form in either a hot or cold state, in a solid state, or in partial solution. Long used with metals and clays, it is now extensively used in the plastic industry.

**Eyepiece.** Gas-tight, transparent window in a full facepiece through which the wearer may see.

## F

**Face velocity.** Average air velocity into the exhaust system measured at the opening into the hood or booth.

**Facepiece.** That portion of a respirator that covers the wearer's nose and mouth in a half-mask facepiece, or the nose, mouth, and eyes in a full facepiece. It is designed to make a gas-tight or dust-tight fit with the face and includes the headbands, exhalation valves, and connections for air-purifying device, or respirable gas source, or both.

**Facing.** In foundry work, the final touch-up work of the mold surface to come in contact with metal is called the facing operation, and the fine powdered material used is called the facing.

**Fainting.** Technically called syncope, a temporary loss of consciousness as a result of a diminished supply of blood to the brain.

**Fallout.** Dust particles that contain radioactive fission products resulting from a nuclear explosion. The wind can carry fallout particles many miles.

**Fan laws.** Statements and equations that describe the relationship between fan volume, pressure, brake horsepower, size, and rotating speed.

**Fan rating curve or table.** Data that describe the volumetric output of a fan at different static pressures.

**Fan static pressure.** The pressure added to the system by the fan. It equals the sum of pressure losses in the system minus the velocity pressure in the air at the fan inlet.

**Far field (free field).** In noise measurement, this refers to the distance from the noise source where the sound-pressure level decreases 6 dBA for each doubling of distance (inverse square law).

**Farmer's lung.** Fungus infection and ensuing hypersensitivity from grain dust.

**Federal Register.** Publication of U.S. government documents officially promulgated under the law, documents whose validity depends upon such publication. It is published on each day following a government working day. It is, in effect, the daily supplement to the *Code of Federal Regulations (CFR)*.

**Feral animal.** A wild animal, or a domestic animal that has reverted to the wild state.

**Fertilizer.** Plant food usually sold in a mixed formula containing basic plant nutrients: compounds of nitrogen, potassium, phosphorus, sulfur, and sometimes other minerals.

**Fetus.** The term used to describe the developing organism (human) from the third month after conception to birth.

**FEV.** Forced expiratory volume.

**Fever.** A condition in which the body temperature is above its regular or normal level.

**Fibrillation.** Very rapid irregular contractions of the muscle fibers of the heart resulting in a lack of synchronism of the heartbeat.

**Fibrosis.** A condition marked by an increase of interstitial fibrous tissue. Exposures to contaminants via inhalation can lead to fibrosis or scarring of the lung, a particular concern in industrial hygiene.

**Film badge.** A piece of masked photographic film worn by nuclear workers. It is darkened by nuclear radiation, and radiation exposure can be checked by inspecting the film.

**Filter.** (1) A device for separating components of a signal on the basis of its frequency. It allows components in one or more frequency bands to pass relatively unattenuated, and it greatly attenuates components in other frequency bands. (2) A fibrous medium used in respirators to remove solid or liquid particles from the airstream entering the respirator. (3) A sheet of material that is interposed between patient and the source of x rays to absorb a selective part of the x rays. (4) A fibrous or membrane medium used to collect dust, fume, or mist air samples.

**Filter efficiency.** The efficiency of various filters can be established on the basis of entrapped particles (that is, collection efficiency), or on the basis of particles passed through the filter (that is, penetration efficiency).

**Filter, HEPA.** High-efficiency particulate air filter, one that is at least 99.97 percent efficient in removing thermally generated monodisperse dioctyl phthalate smoke particles with a diameter of 0.0003 mm.

**Firebrick.** A special clay that is capable of resisting high temperatures without melting or crumbling.

**Fire damp.** In mining, the accumulation of an explosive gas, chiefly methane gas. Miners call all dangerous underground gases "damps."

**Fire point.** The lowest temperature at which a material can evolve vapors to support continuous combustion.

**Fission.** The splitting of an atomic nucleus into two parts accompanied by the release of a large amount of radioactivity and heat. Fission reactions occur only with heavy isotopes, such as uranium-233, uranium-235, and plutonium-239.

**Fissionable.**   A nucleus that undergoes fission under the influence of neutrons, even very slow neutrons.

**Fission product.**   The highly radioactive nuclei into which a fissionable nucleus splits (fissions) under the influence of neutron bombardment.

**Flagellum.**   A flexible, whip-like appendage on cells used as an organ of locomotion.

**Flame ionization detector (FID).**   A direct-reading monitoring device that ionizes gases and vapors with an oxyhydrogen flame and measures the differing electrical currents thus generated.

**Flameproofing material.**   Chemicals that catalytically control the decomposition of cellulose material at flaming temperature. Substances used as fire retardants are borax–boric acid, borax–boric acid diammonium phosphate, ammonium bromide, stannic acid, antimony oxide, and combinations containing formaldehyde.

**Flame propagation.**   See Propagation of flame.

**Flammable aerosol.**   An aerosol that is required to be labeled *Flammable* under the Federal Hazardous Substances Labeling Act (15 *USC* 1261).

**Flammable limits.**   Flammables have a minimum concentration below which propagation of flame does not occur on contact with a source of ignition. This is known as the lower flammable explosive limit (LEL). There is also a maximum concentration of vapor or gas in air above which propagation of flame does not occur. This is known as the upper flammable explosive limit (UEL). These units are expressed in percent of gas or vapor in air by volume.

**Flammable liquid.**   Any liquid having a flash point below 100 F (37.8 C).

**Flammable range.**   The difference between the lower and upper flammable limits, expressed in terms of percentage of vapor or gas in air by volume, also often called the explosive range.

**Flange.**   A rim or edge added to a hood to reduce the quantity of air entering from behind the hood.

**Flash blindness.**   Temporary visual disturbance resulting from viewing an intense light source.

**Flash point.**   The lowest temperature at which a liquid gives off enough vapor to form an ignitable mixture with air and produce a flame when a source of ignition is present. Two tests are used: open cup and closed cup.

**Flask.**   In foundry work, the assembly of the cope and the drag constitutes the flask. It is the wooden or iron frame containing sand into which molten metal is poured. Some flasks may have three or four parts.

**Flexion.**   Movement whereby the angle between two bones connected by a joint is reduced. Motions of this type are produced by contraction of flexor muscles.

**Flexor muscles.**   A muscle that, when contracting, decreases the angle between limb segments. The principal flexor of the elbow is the brachialis muscle. Flexors of the fingers and the wrist are the large muscles of the forearm originating at the elbow. See Extensor muscles.

**Flocculation.**   The process of forming a very fluffy mass of material held together by weak forces of adhesion.

**Flocculator.**   A device for aggregating fine particles.

**Flora, microflora.**   Microorganisms present in a given situation (such as intestinal flora, soil flora).

**Flotation.**   A method of ore concentration in which the mineral is caused to float due to chemical frothing agents while the impurities sink.

**Flotation reagent.**   Chemical used in flotation separation of minerals. Added to a pulverized mixture of solids and water and oil, it causes preferential nonwetting by water of certain solid particles, making possible the flotation and separation of nonwet particles.

**Flow coefficient.**   A correction factor used for figuring the volume flow rate of a fluid through an orifice. This factor includes the effects of contraction and turbulence loss (covered by the coefficient of discharge), plus the compressibility effect and the effect of an upstream velocity other than zero. Because the latter two effects are negligible in many instances, the flow coefficient is often equal to the coefficient of discharge (see Coefficient of discharge).

**Flow meter.**   An instrument for measuring the rate of flow of a fluid or gas.

**Flow, turbulent.**   Fluid flow in which the fluid moves transversely as well as in the direction of the tube or pipe axis, as opposed to streamline or viscous flow.

**Fluid.**   A substance tending to flow or conform to the outline of its container. It may be liquid, vapor, gas, or solid (such as raw rubber).

**Fluorescence.**   Emission of light from a crystal, after the absorption of energy.

**Fluorescent screen.**   A screen coated with a fluorescent substance so that it emits light when irradiated with x rays.

**Fluoroscope.**   A fluorescent screen mounted in front of an x-ray tube so that internal organs may be examined through their shadow cast by x rays. It may also be used for inspection of inanimate objects.

**Fluoroscopy.**   The practice of examining through the use of an x-ray fluoroscope.

**Flux.**   Usually refers to a substance used to clean surfaces and promote fusion in soldering. However, fluxes of varying chemical nature are used in the smelting of ores, in the ceramic industry, in assaying silver and gold ores, and in other endeavors. The most common fluxes are silica, various silicates, lime, sodium and potassium carbonate, and litharge and red lead in the ceramic industry. See also Soldering, Galvanizing, and Luminous flux.

**Fly ash.**   Finely divided particles of ash entrained in flue gases arising from the combustion of fuel.

**Focus** (pl. foci).   A center or site of a disease process.

**Follicle.**   A small anatomical cavity or deep, narrow-mouthed depression; a small lymph node.

**Folliculitis.**   Infection of a hair follicle, often caused by obstruction by natural or industrial oils.

**Fomites.**   Clothing or other substances that can absorb and transmit contaminants, as in the case of poison ivy.

**Footcandle.**   A unit of illumination. The illumination at a point on a surface that is one foot from, and perpendicular to, a uniform point source of one candle.

**Foot-pounds of torque.**   A measurement of the physiological stress exerted upon any joint during the performance of a task. The product of the force exerted and the distance from the point of application to the point of stress. Physiologically, torque that does not produce motion nonetheless causes work stress, the severity of which depends on the duration and magnitude of the torque. In lifting an object or holding it elevated, torque is exerted and applied to the lumbar vertebrae.

**Force.**   That which changes the state of rest or motion in matter. The SI (International System) unit of measurement is the newton (N).

**Fovea.**   A depression or pit in the center of the macula of the eye; it is the area of clearest vision.

**Fractionation.**   Separation of a mixture into different portions or fractions, usually by distillation.

**Free sound field (free field).**   A field in a homogeneous, isotropic medium free from boundaries. In practice, it is a field in which the effects of the boundaries are negligible over the region of interest. See Far field.

**Frequency (in Hz).**   Rate at which pressure oscillations are produced. One hertz is equivalent to one cycle per second. A subjective characteristic of sound related to frequency is pitch.

**Friction factor.**   A factor used in calculating loss of pressure due to friction of a fluid flowing through a pipe or duct.

**Friction loss.**   The pressure loss caused by friction.

**Fuller's earth.**   A hydrated silica–alumina compound associated with ferric oxide. Used as a filter medium and as a catalyst and catalyst carrier and in cosmetics and insecticides.

**Fume.**   Airborne particulate formed by the condensation of solid particles from the gaseous state. Usually, fumes are generated after initial volatilization from a combustion process, or from a melting process (such as metal fume emitted during welding). Usually less than 1 μm in diameter.

**Fume fever.**   Metal fume fever is an acute condition caused by a brief high exposure to the freshly generated fumes of metals, such as zinc or magnesium, or their oxides.

**Functional anatomy.**   Study of the body and its component parts, taking into account structural features directly related to physiological function.

**Fundamental frequency.**   The lowest component frequency of a periodic quantity.

**Fundus.**   The interior surface of a hollow organ, such as the retina of the eye.

**Fungus (pl. fungi).**   Any of a major group of lower plants that lack chlorophyll and live on dead or other living organisms. Fungi include molds, rusts, mildews, smuts, and mushrooms.

**Fusion.**   (1) The joining of atomic nuclei to form a heavier nucleus, accomplished under conditions of extreme heat (millions of degrees). If two nuclei of light atoms fuse, the fusion is accompanied by the release of a great deal of energy. The energy of the sun is believed to be derived from the fusion of hydrogen atoms to form helium. (2) In welding, the melting together of filler metal and base metal (substrate), or of base metal only, which results in coalescence.

**FVC.**   Forced vital capacity.

## G

**Gage pressure.**   Pressure measured with respect to atmospheric pressure.

**Galvanizing.**   An old but still used method of providing a protective coating for metals by dipping them in a bath of molten zinc.

**Gamete.**   A mature germ cell. An unfertilized ovum or spermatozoon.

**Gamma-rays (gamma radiation).** The most penetrating of all radiation. Gamma-rays are very high-energy x rays.

**Ganglion (pl. ganglia).** A knot or knot-like mass; used as a general term to designate a group of nerve cell bodies located outside of the central nervous system. The term is also applied to certain nuclear groups within the brain or spinal cord.

**Gangue.** In mining or quarrying, useless chipped rock.

**Gas.** A state of matter in which the material has very low density and viscosity, can expand and contract greatly in response to changes in temperature and pressure, easily diffuses into other gases, and readily and uniformly distributes itself throughout any container. A gas can be changed to the liquid or solid state only by the combined effect of increased pressure and decreased temperature (below the critical temperature).

**Gas chromatography.** A gaseous detection technique that involves the separation of mixtures by passing them through a column that enables the components to be held up for varying periods of time before they are detected and recorded.

**Gas metal arc-welding (GMAW).** An arc-welding process that produces coalescence of metals by heating them with an arc between a continuous filler metal (consumable) electrode and the work; shielding is obtained entirely from an external supplied gas or gas mixture. Some methods of this process are called MIG or $CO_2$ welding.

**Gas tungsten arc-welding (GTAW).** An arc-welding process that produces coalescence of metals by heating them with an arc between a tungsten (nonconsumable) electrode and the work; shielding is obtained from a gas or gas mixture. Pressure may or may not be used, and filler metal may or may not be used. This process has sometimes been called TIG welding.

**Gastr-, gastro-** (prefix). Pertaining to the stomach.

**Gastritis.** Inflammation of the stomach.

**Gate.** A groove in a mold to act as a passage for molten metal.

**Geiger counter.** A gas-filled electrical device that counts the presence of an atomic particle or ray by detecting the ions produced. Sometimes called a Geiger–Müller counter.

**General ventilation.** System of ventilation consisting of either natural or mechanically induced fresh air movements to mix with and dilute contaminants in the workroom air. This is not the recommended type of ventilation to control contaminants that are toxic.

**Genes.** The ultimate biological units of heredity.

**Genetic effects.** Mutations or other changes produced by irradiation of the germ plasm.

**Genetically significant dose (GSD).** The dose that, if received by every member of the population, would be expected to produce the same total genetic injury to the population as the actual doses received by the various individuals.

**Germ.** A microorganism; a microbe usually thought of as a pathogenic organism.

**Germicide.** An agent capable of killing germs.

**GI.** Gastrointestinal.

**Gingival.** Pertaining to the gingivae (gums), the mucous membrane, with the supporting fibrous tissue, that overlies the crowns of unerupted teeth and encircles the necks of those that have erupted.

**Gingivitis.** Inflammation of the gums.

**Gland.** Any body organ that manufactures some liquid product and secretes it from its cells.

**Globe thermometer.** A thermometer set in the center of a metal sphere that has been painted black in order to measure radiant heat.

**Globulin.** General name for a group of proteins that are soluble in saline solutions but not in pure water.

**Glossa-** (prefix). Pertaining to the tongue.

**Glove box.** A sealed enclosure in which all handling of items inside the box is carried out through long, impervious gloves sealed to ports in the walls of the enclosure.

**Gob.** Gob pile is waste mineral material, such as from coal mines, that contains sufficient coal that gob fires may arise from spontaneous combustion.

**Gonads.** The male (testes) and female (ovaries) sex glands.

**Grab sample.** A sample taken within a very short time period to determine the constituents at a specific time.

**Gram (g).** A metric unit of weight. One ounce equals 28.4 grams.

**Grams per kilogram (g/kg).** This indicates the dose of a substance given to test animals in toxicity studies.

**Granuloma.** A mass or nodule of chronically inflamed tissue with granulations; usually associated with an infective process.

**Graticule.** See Reticle.

**Gravimetric.** Pertaining to measurement by weight.

**Gravimetric method.**    A procedure dependent upon the formation or use of a precipitate or residue, which is weighed to determine the concentration of a specific contaminant in a previously collected sample.

**Gravitation.**    The universal attraction existing between all material bodies. The gravitational attraction of the earth's mass for bodies at or near its surface is called gravity.

**Gravity, specific.**    The ratio of the mass of a unit volume of a substance to the mass of the same volume of a standard substance at a standard temperature. Water at 39.2 F (4 C) is the standard substance usually referred to. For gases, dry air, at the same temperature and pressure as the gas, is often taken as the standard substance.

**Gravity, standard.**    A gravitational force that produces an acceleration equal to 32.17 ft (9.8 m) per second. The actual force of gravity varies slightly with altitude and latitude. The standard was arbitrarily established as that at sea level and 45-degree latitude.

**Gray (Gy).**    Unit of absorbed radiation dose equal to one joule of absorbed energy per kilogram of matter; also equal to 100 rad.

**Gray iron.**    The same as cast iron; in general, any iron with high carbon content.

**Grooving.**    Designing a tool with grooves on the handle to accommodate the fingers of the user—a bad practice because of the great variation in the size of workers' hands. Grooving interferes with sensory feedback. Intense pain may be caused by the grooves to the arthritic hand.

**Gyn-, gyne-** (prefix). Woman, female.

**Gynecology.**    The medical specialty concerned with diseases of women.

**Gyratory crusher.**    A device for crushing rock by means of a heavy steel pestle rotating in a steel cone, with the rock being fed in at the top and passing out of the bottom.

## H

**Half-life, radioactive.**    For a single radioactive decay process, the time required for the activity to decrease to half its value by that process.

**Half-thickness.**    The thickness of a specified absorbing material that reduces the dose rate to one half its original value.

**Half-value layer (HVL).**    The thickness of a substance necessary to reduce the intensity of a beam of gamma or x rays to half its original value. Also known as half-thickness.

**Halogenated hydrocarbon.**    A chemical material that has carbon plus one or more of these elements: chlorine, fluorine, bromine, and iodine.

**Hammer mill.**    A machine for reducing the size of stone or other bulk material by means of hammers usually placed on a rotating axle inside a steel cylinder.

**Hardness.**    A relative term to describe the penetrating quality of radiation. The higher the energy of the radiation, the more penetrating (harder) is the radiation.

**Hardness of water.**    A degree of hardness is the equivalent of one grain of calcium carbonate, $CaCO_3$, in one gallon of water.

**Hazardous material.**    Any substance or compound that has the capability of producing adverse effects on the health and safety of humans.

**Hazwoper.**    Hazardous waste operations and emergency response—an OSHA standard intended to protect workers engaged in hazardous waste operations.

**Heading.**    In mining, a horizontal passage or drift of a tunnel, also the end of a drift or gallery. In tanning, a layer of ground bark over the tanning liquor.

**Health physicist.**    A professional person specially trained in radiation physics and concerned with problems of radiation damage and protection.

**Hearing conservation.**    The prevention or minimizing of noise-induced deafness through the use of hearing protection devices, the control of noise through engineering methods, annual audiometric tests, and employee training.

**Hearing level.**    The deviation in decibels of an individual's threshold from the zero reference of the audiometer.

**Heat cramps.**    Painful muscle spasms as a result of exposure to excess heat.

**Heat exhaustion.**    A condition usually caused by loss of body water because of exposure to excess heat. Symptoms include headache, tiredness, nausea, and sometimes fainting.

**Heat, latent.**    The quantity of heat absorbed or given off per unit weight of material during a change of state, such as ice to water or water to steam.

**Heat of fusion.**    The heat given off by a liquid freezing to a solid or gained by a solid melting to a liquid, without a change in temperature.

**Heat of vaporization.**    The heat given off by a vapor condensing to a liquid or gained by a liquid evaporating to a vapor, without a change in temperature.

**Heat rash.**    Itchy rash caused by sweating and inadequate hygiene practices.

**Heat, sensible.** Heat associated with a change in temperature; specific heat exchange with environment, in contrast to a heat interchange in which only a change of state (phase) occurs.

**Heat, specific.** The ratio of the quantity of heat required to raise the temperature of a given mass of any substance one degree to the quantity required to raise the temperature of an equal mass of a standard substance (usually water at 59 F [15 C]) one degree.

**Heat stress.** Relative amount of thermal strain from the environment.

**Heat stress index (HSI).** Also known as the Belding–Hatch heat stress index, this index combines the environmental heat and metabolic heat into an expression of stress in terms of requirement for evaporation of sweat.

**Heatstroke.** A serious disorder resulting from exposure to excess heat. It results from sweat suppression and increased storage of body heat. Symptoms include hot dry skin, high temperature, mental confusion, convulsions, and coma. Heatstroke is fatal if not treated promptly.

**Heat syncope.** A heat-related disorder characterized by symptoms of blurred vision and brief fainting spells, heat syncope is caused by pooling of blood in the legs or skin during prolonged static postures in a hot environment.

**Heat treatment.** Any of several processes of metal modification, such as annealing.

**Heavy hydrogen.** Same as deuterium.

**Heavy metals.** Metallic elements with high molecular weights.

**Heavy water.** Water containing heavy hydrogen (deuterium) instead of ordinary hydrogen. It is widely used in reactors to slow down neutrons.

**Helmet.** A device that shields the eyes, face, neck, and other parts of the head.

**Hem-, Hemato-, -em-** (prefix). Pertaining to blood. *Hematuria* means blood in the urine. When the roots occur internally in a word, the *h* is often dropped for the sake of pronunciation, leaving *em* to denote blood, as in anoxemia (deficiency of oxygen in the blood).

**Hematology.** Study of the blood and the blood-forming organs.

**Hematuria.** Blood in the urine.

**Hemi-** (prefix). Half. The prefix is straightforward enough in *hemiplegia,* "half paralysis," affecting one side of the body. It is not so plain in *migraine* (one-sided headache), a word that shows how language changes through the centuries. The original word was *hemicrania,* "half-head."

**Hemoglobin.** The red coloring matter of the blood that carries the oxygen.

**Hemolysis.** Breakdown of red blood cells with liberation of hemoglobin.

**Hemoptysis.** Bleeding from the lungs, spitting blood, or blood-stained sputum.

**Hemorrhage.** Bleeding; especially profuse bleeding, as from a ruptured or cut blood vessel (artery or vein).

**Hemorrhagic.** Pertaining to or characterized by hemorrhage.

**HEPA filter.** High efficiency particulate air filter. A disposable, extended-medium, dry-type filter with a particle removal efficiency of no less than 99.97 percent for 0.3 μm particles.

**Hepatitis.** Inflammation of the liver.

**Hepatitis B.** A virus causing hepatitis. The virus may also cause liver cancer in some of those infected by it. The virus is bloodborne and as such is one of the agents targeted by OSHA's bloodborne pathogen standard.

**Hepatotoxin.** Chemicals that produce liver damage.

**Herpes.** An acute inflammation of the skin or mucous membranes, characterized by the development of groups of vesicles on an inflammatory base.

**Hertz.** The frequency measured in cycles per second. 1 cps = 1 Hz.

**High frequency loss.** Refers to a hearing deficit starting with 2000 Hz and beyond.

**HIV.** Human immunodeficiency virus. Held to be the initiating cause of acquired immunodeficiency syndrome (AIDS).

**Homeotherm.** Uniform body temperature, or a warm-blooded creature remaining so regardless of environment.

**Homogenizer.** A machine that forces liquids under high pressure through a perforated shield against a hard surface to blend or emulsify the mixture.

**Homoiotherm.** See Homeotherm.

**Hood.** (1) Enclosure, part of a local exhaust system. (2) A device that completely covers the head, neck, and portions of the shoulders.

**Hood entry loss.** The pressure loss from turbulence and friction as air enters the ventilation system.

**Hood, slot.** A hood consisting of a narrow slot leading into a plenum chamber under suction to distribute air velocity along the length of the slot.

**Hood static pressure.** The suction or static pressure in a duct near a hood. It represents the suction that is available to draw air into the hood.

**Hormones.** Chemical substances secreted by the endocrine glands, exerting influence over practically all body activities.

**Horsepower.** A unit of power, equivalent to 33,000 foot-pounds per minute (746 W). See Brake horsepower.

**Host.** A plant or animal harboring another as a parasite or as an infectious agent.

**Hot.** In addition to meaning having a relatively high temperature, this is a colloquial term meaning highly radioactive.

**HSI.** See Heat stress index.

**Human–equipment interface.** Areas of physical or perceptual contact between person and equipment. The design characteristics of the human–equipment interface determine the quality of information. Poorly designed interfaces may lead to excessive fatigue or localized trauma, e.g., calluses.

**Humerus.** The bone of the upper arm that starts at the shoulder joint and ends at the elbow. Muscles that move the upper arm, forearm, and hand are attached to this bone.

**Humidify.** To add water vapor to the atmosphere; to add water vapor or moisture to any material.

**Humidity.** (1) Absolute humidity is the weight of water vapor per unit volume: pounds per cubic foot or grams per cubic centimeter. (2) Relative humidity is the ratio of the actual partial vapor pressure of the water vapor in a space to the saturation pressure of pure water at the same temperature.

**Humidity, specific.** The weight of water vapor per unit weight of dry air.

**HVAC system.** Heating, ventilating, and air conditioning system.

**Hyalinization.** Conversion into a substance resembling glass.

**Hydration.** The process of converting raw material into pulp by prolonged beating in water; to combine with water or the elements of water.

**Hydrocarbons.** Organic compounds composed solely of carbon and hydrogen. Several hundred thousand molecular combinations of C and H are known to exist. Basic building blocks of all organic chemicals. Main chemical industry sources of hydrocarbons are petroleum, natural gas, and coal.

**Hydrogenation.** A reaction of molecular hydrogen with numerous organic compounds. An example is the hydrogenation of olefins to paraffins or of the aromatics to the naphthenes or the reduction of aldehydes and ketones to alcohols.

**Hydrolysis.** The interaction of water with a material resulting in decomposition.

**Hydrometallurgy.** Science of metal recovery by a process involving treatment of ores in an aqueous medium, such as acid or cyanide solution.

**Hydrophobic.** Repelled by water, or water-hating.

**Hygroscopic.** Readily absorbing or retaining moisture.

**Hyper-** (prefix). Over, above, increased. The usual implication is overactivity or excessive production, as in *hyperthyroidism.*

**Hyperkeratosis.** Hypertrophy of the horny layer of the skin.

**Hypertension.** Abnormally high tension; especially high blood pressure.

**Hypertrophy.** Increase in cell size causing an increase in the size of the organ or tissue.

**Hypnotic.** Anything that induces sleep or that produces the effects ascribed to hypnotism.

**Hypo-** (prefix). Under, below; less, decreased. The two different meanings of this common prefix can be tricky. *Hypodermic* might reasonably be interpreted to mean that an unfortunate patient has too little skin. The actual meaning is "under or beneath the skin," a proper site for an injection. The majority of *hypo-* words, however, denote an insufficiency, lessening, or reduction from the norm, as in *hypoglycemia,* meaning too little glucose in the blood.

**Hypothermia.** A systemic effect of cold stress; condition of reduced body temperature.

**Hysteresis.** A retardation of the effect when the forces acting upon a body are changed (as if from viscosity or internal friction). Specifically, the magnetization of a sample of iron or steel actually lags behind the magnetic field that induced it, when the field varies.

## I

**IAQ.** Indoor air quality.

**IARC.** International Agency for Research on Cancer.

**Iatro-** (prefix). Pertaining to a doctor. A related root, *-iatrist,* denotes a specialist, as in *psychiatrist.*

**Iatrogenic.** Caused by the doctor.

**ICC.** Interstate Commerce Commission.

**ICRP.** International Commission on Radiological Protection and Measurements.

**Idio-** (prefix). Peculiar to, private, or distinctive, as in *idiosyncrasy*.

**Idiopathic.** Disease that originates in itself.

**Idiosyncrasy.** A special susceptibility to a particular substance introduced into the body.

**IDLH.** Immediately dangerous to life or health.

**IES.** Illumination Engineering Society.

**Iliac crest.** The upper rounded border of the hip bone. No muscles cross the iliac crest, which lies immediately below the skin. It is an important anatomical reference point because it can be felt through the skin. Seat backrests should clear the iliac crest.

**Image.** The fluorescent picture produced by x rays hitting a fluoroscopic screen.

**Image receptor.** Any device, such as a fluorescent screen or radiographic film, that transforms incident x-ray photons either into a visible image or into another form that can be made into a visible image by further transformations.

**Immiscible.** Not miscible. Any liquid that does not mix with another liquid, in which case the result is two separate layers or cloudiness or turbidity.

**Immune.** Resistant to disease.

**Immunity.** The power of the body to successfully resist infection and the effects of toxins. This resistance results from the possession by the body of certain "fighting substances," or antibodies. To immunize is to confer immunity. Immunization is the process of acquiring or conferring immunity.

**Impaction.** The forcible contact of particles of matter; a term often used synonymously with impingement, but generally reserved for the case where particles are contacting a dry surface.

**Impingement.** As used in air sampling, impingement refers to a process for the collection of particulate matter in which a particle-containing gas is directed against a wetted glass plate and the particles are retained by the liquid.

**Impinger.** A device containing an absorbing liquid used in air sampling for the collection of gaseous or particulate constituents of an airstream directed by the device through the liquid. The impinger draws air at high velocity through a glass nozzle or jet. A commonly used type is called the midget impinger.

**Inches of mercury column.** A unit used in measuring pressures. One inch of mercury column equals a pressure of 0.491 lb/in.$^2$ (1.66 kPa).

**Inches of water column.** A unit used in measuring pressures. One inch of water column equals a pressure of 0.036 lb/in.$^2$ (0.25 kPa).

**Incompatible.** A term applied to liquid and solid systems to indicate that one material cannot be mixed with another specified material without the possibility of a dangerous reaction.

**Incubation.** Holding cultures of microorganisms under conditions favorable to their growth.

**Incubation time.** The elapsed time between exposure to infection and the appearance of disease symptoms, or the time period during which microorganisms inoculated into a medium are allowed to grow.

**Induration.** Heat hardening that may involve little more than thermal dehydration.

**Inert (chemical).** Not having active properties.

**Inert gas.** A gas that does not normally combine chemically with the base metal or filler metal.

**Inert gas welding.** An electric welding operation using an inert gas such as helium to flush away the air to prevent oxidation of the metal being welded.

**Inertial moment.** As related to biomechanics, that moment of force-time caused by sudden accelerations or decelerations. Whiplash of the neck is caused by an inertial moment. In an industrial setting, sidestepping causes application of a lateral inertial moment on the lumbosacral joint, which may cause trauma, pain, and in any case lowers performance efficiency. The inertial moment is one of the seven elements of a lifting task.

**Infection.** Entrance into the body or its tissues of disease-causing organisms with the effect of damage to the body as a whole or to tissues or organs. It also refers to the entrance into the body of parasites, like certain worms. On the other hand, parasites such as mites and ticks that attack the surface of the body are said to infest, not infect.

**Infectious.** Capable of invading a susceptible host, replicating and causing an altered host reaction, commonly referred to as a disease.

**Infestation.** Invasion of the body surface by parasites. See Infection.

**Inflammation.** The reaction of body tissue to injury, whether by infection or trauma. The inflamed area is red, swollen, hot, and usually painful.

**Infrared.** Wavelengths of the electromagnetic spectrum longer than those of visible light and shorter than radio waves, $10^{-4}$–$10^{-1}$ cm wavelength.

**Infrared radiation.** Electromagnetic energy with wavelengths from 770 nm to 12,000 nm.

**Ingestion.** (1) The process of taking substances into the stomach, such as food, drink, or medicine. (2) With regard to certain cells, the act of engulfing or taking up bacteria and other foreign matter.

**Ingot.** A block of iron or steel cast in a mold for ease in handling before processing.

**Inguinal region.** The abdominal area on each side of the body occurring as a depression between the abdomen and the thigh; the groin.

**Inhalation valve.** A device that allows respirable air to enter the facepiece and prevents exhaled air from leaving the facepiece through the intake opening.

**Inhibition.** Prevention of growth or multiplication of microorganisms.

**Inhibitor.** An agent that arrests or slows chemical action or a material used to prevent or retard rust or corrosion.

**Injury.** Damage or harm to the body, as the result of violence, infection, or anything else that produces a lesion.

**Innocuous.** Harmless.

**Inoculation.** The artificial introduction of microorganisms into a system.

**Inorganic.** Used to designate compounds that generally do not contain carbon, whose source is matter other than vegetable or animal. Examples are sulfuric acid and salt. Exceptions are carbon monoxide and carbon dioxide.

**Insomnia.** Inability to sleep; abnormal wakefulness.

**Instantaneous radiation.** The radiation emitted during the fission process. These instantaneous radiations are often called prompt gamma-rays or prompt neutrons. Most fission products continue to emit radiation after the fission process.

**Inter-** (prefix). Between.

**Intermediate.** A chemical formed as a middle step in a series of chemical reactions, especially in the manufacture of organic dyes and pigments. In many cases, it may be isolated and used to form a variety of desired products. In other cases, the intermediate may be unstable or used up at once.

**Internal biomechanical environment.** The muscles, bones and tissues of the body, all of which are subject to the same Newtonian force as external objects in their interaction with other bodies and natural forces. When designing for the body, one must consider the forces that the internal biomechanical environment must withstand.

**Interphalangeal joints.** The finger or toe joints. The thumb has one interphalangeal joint; the fingers have two interphalangeal joints each.

**Interstitial.** (1) Pertaining to the small spaces between cells or structures. (2) Occupying the interstices of a tissue or organ. (3) Designating connective tissue occupying spaces between the functional units of an organ or a structure.

**Intoxication.** Either drunkenness or poisoning.

**Intra-** (prefix). Within.

**Intraperitoneal.** Inside the space formed by the membrane that lines the interior wall of the abdomen and covers the abdominal organs.

**Intravenous.** Into or inside the vein.

**Intrinsically safe.** Said of an instrument that is designed and certified to be operated safely in flammable or explosive atmospheres.

**Inverse square law.** The propagation of energy through space is inversely proportional to the square of the distance it must travel. An object 3 m away from an energy source receives one-ninth as much energy as an object 1 m away.

**Inversion.** Phenomenon of a layer of cool air trapped by a layer of warmer air above it so that the bottom layer cannot rise. This is a special problem in polluted areas because the contaminating substances cannot be dispersed.

**Investment casting.** There are numerous types of investment casting, and the materials include fire clay, silicon dioxide, silica flour, stillimanite, cristobalite, aluminum oxide, zirconium oxide, and others. The Mercast process uses mercury poured into a steel die. A ceramic shell mold is built around the pattern, and then the pattern is frozen. The mercury is subsequently recovered at room temperature. The potential harm from exposure to mercury often is unrecognized.

**Ion.** An electrically charged atom. An atom that has lost one or more of its electrons is left with a positive electrical charge. Those that have gained one or more extra electrons are left with a negative charge.

**Ion-exchange resin.** Synthetic resins containing active groups that give the resin the property of combining with or exchanging ions between the resin and a solution.

**Ionization.** The process whereby one or more electrons is removed from a neutral atom by the action of radiation. Specific ionization is the number of ion pairs per unit distance in matter, usually air.

**Ionization chamber.** A device roughly similar to a Geiger counter and used to measure radioactivity.

**Ionizing radiation.** (1) Electrically charged or neutral particles. (2) Electromagnetic radiation that interacts with gases, liquids, or solids to produce ions. There are five major types: alpha, beta, x- (or x-ray), gamma, and neutrons.

**Ion pair.** A positively charged atom (ion) and an electron formed by the action of radiation on a neutral atom.

**Irradiation.** The exposure of something to radiation.

**Irritant.** A substance that produces an irritating effect when it contacts skin, eyes, nose, or respiratory system.

**Ischemia.** Loss of blood supply to a particular part of the body.

**Ischial tuberosity.** A rounded projection on the ischium. It is a point of attachment for several muscles involved in moving the femur and the knee. It can be affected by improper chair design and by situations involving trauma to the pelvic region. When seated, pressure is borne at the site of the ischial tuberosities. Chair design should provide support to the pressure projection of the ischial tuberosity through the skin of the buttocks.

**Isometric work.** Refers to a state of muscular contraction without movement. Although no work in the "physics" sense is done, physiologic work (energy use and heat production) occurs. In isometric exercise, muscles are tightened against immovable objects. In work measurements, isometric muscular contractions must be considered as a major factor of task severity.

**Isotope.** One of two or more atomic species of an element differing in atomic weight but having the same atomic number. Each contains the same number of protons but a different number of neutrons. Uranium-238 contains 92 protons and 146 neutrons; the isotope U-235 contains 92 protons and 143 neutrons. Thus the atomic weight (atomic mass) of U-238 is 3 higher than that of U-235. See also Radioisotope.

**Isotropic.** Exhibiting properties with the same values when measured along axes in all directions.

**-itis** (suffix). Inflammation.

## J

**Jaundice.** Icterus. A serious symptom of disease that causes the skin, the whites of the eyes, and even the mucous membranes to turn yellow.

**Jigs and fixtures.** Often used interchangeably; precisely, a jig holds work in position and guides the tools acting on the work, whereas a fixture holds but does not guide.

**Joint.** Articulation between two bones that may permit motion in one or more planes. They may become the sites for work-induced trauma (such as tennis elbow or arthritis) or other disorders.

**Joule.** Unit of energy used in describing a single pulsed output of a laser. It is equal to one watt-second or 0.239 calories. It equals $1 \times 10^7$ ergs.

**Joule/cm² (J/cm²).** Unit of energy density used in measuring the amount of energy per area of absorbing surface or per area of a laser beam. It is a unit for predicting the damage potential of a laser beam.

## K

**Kaolin.** A type of clay composed of mixed silicates and used for refractories, ceramics, tile, and stoneware. In some deposits, free silica may be present as an impurity.

**Kaolinosis.** A condition induced by inhalation of the dust released in the grinding and handling of kaolin (china clay).

**Kelvin scale.** The fundamental temperature scale, also called the absolute or thermodynamic scale, in which the temperature measure is based on the average kinetic energy per molecule of a perfect gas. The zero of the Kelvin scale is –273.18 degrees Celsius.

**Keratin.** Sulfur-containing proteins that form the chemical basis for epidermis tissues; found in nails, hair, and feathers.

**Keratinocyte.** An epidermal cell that produces keratin.

**Keratitis.** Inflammation of the cornea.

**Kev.** A unit of energy equal to 1,000 electron volts.

**Kilocurie.** 1,000 curies. A unit of radioactivity.

**Kilogram (kg).** A unit of weight in the metric system equal to 2.2 lb.

**Kinesiology.** The study of human movement in terms of functional anatomy.

**Kinetic energy.** Energy due to motion. See Work.

**Kyphosis.** Abnormal curvature of the spine of the upper back in the anteroposterior plane.

## L

**Laboratory-acquired infection.** Any infection resulting from exposure to biohazardous materials in a laboratory environment. Exposure may be the result of a specific accident or inadequate biohazard control procedure or equipment.

**Lacquer.** A colloidal dispersion or solution of nitrocellulose or similar film-forming compounds, resins, and plasticizers in solvents and diluents used as a protective and decorative coating for various surfaces.

**Laminar airflow.** Streamlined airflow in which the entire body of air within a designated space moves with uniform velocity in one direction along parallel flow lines.

**LAN.** Local area network. A network of computers linked electronically and by software. Located geographically locally, usually in one office or office building.

**Lapping.** The operation of polishing or sanding surfaces such as metal or glass to a precise dimension.

**Laryngitis.** Inflammation of the larynx.

**Larynx.** The organ by which the voice is produced. It is situated at the upper part of the trachea.

**Laser.** Light amplification by stimulated emission of radiation. Lasers may operate in either pulsed or continuous mode.

**Laser light region.** A portion of the electromagnetic spectrum including ultraviolet, visible, and infrared light.

**Laser system.** An assembly of electrical, mechanical, and optical components that includes a laser.

**Latent period.** The time that elapses between exposure and the first manifestation of damage.

**Latex.** Originally, a milky extract from the rubber tree, containing about 35 percent rubber hydrocarbon, with the remainder being water, proteins, and sugars. Also applied to water emulsions of synthetic rubbers or resins. In emulsion paints, the film-forming resin is in the form of latex.

**Lathe.** A machine tool used to perform cutting operations on wood or metal by the rotation of the workpiece.

**Latissimus dorsi.** A large, flat muscle of the back that originates in the lower back and inserts into the humerus near the armpit. It adducts the upper arm, and when the elbow is abducted, it rotates the arm medially. It is actively used in operating equipment such as the drill press, where a downward pull by the arm is required.

**LC$_{50}$.** Lethal concentration that kills 50 percent of the test animals within a specified time. See LD$_{50}$.

**LD$_{50}$.** The dose required to produce the death in 50 percent of the exposed population within a specified time.

**Leakage radiation.** Radiation emanating from the diagnostic source assembly, except for the useful beam and radiation, produced when the exposure switch or timer is not activated.

**Lens, crystalline.** Lens of the eye—a transparent biconvex body situated between the anterior chamber (aqueous) and the posterior chamber (vitreous) through which the light rays are further focused on the retina. The cornea provides most of the refractive power of the eye.

**Lesion.** Injury, damage, or abnormal change in a tissue or organ.

**Lethal.** Capable of causing death.

**Leuk-, leuko-** (prefix). White.

**Leukemia.** A group of malignant blood diseases distinguished by overproduction of white blood cells.

**Leukemogenic.** Having the ability to cause leukemia.

**Leukocyte.** White blood cell.

**Leukocytosis.** An abnormal increase in the number of white blood cells.

**Leukopenia.** A serious reduction in the number of white blood cells.

**Lig-** (prefix). Binding. A ligament ties two or more bones together.

**Linear accelerator.** A machine for speeding up charged particles such as protons. It differs from other accelerators in that the particles move in a straight line at all times instead of in circles or spirals.

**Line-voltage regulation.** The difference between the no-load and the load-line potentials expressed as a percent of the load-line potential.

**Lipo-** (prefix). Fat, fatty.

**Liquefied petroleum gas.** A compressed or liquefied gas usually composed of propane, some butane, and lesser quantities of other light hydrocarbons and impurities; obtained as a by-product in petroleum refining. Used chiefly as a fuel and in chemical synthesis.

**Liquid.** A state of matter in which the substance is a formless fluid that flows in accord with the law of gravity.

**Liter (l).** A measure of capacity; one quart equals 0.9 l.

**Liver.** The largest gland or organ in the body, situated on the right side of the upper part of the abdomen. It has many important functions, including regulating the amino acids in the blood; storing iron and copper for the body; forming and secreting bile, which aids in absorption and digestion of fats; transforming glucose into glycogen; and detoxifying exogenous substances.

**Live room.** A reverberant room that is characterized by an unusually small amount of sound absorption.

**Local exhaust ventilation.** A ventilation system that captures and removes the contaminants at the point at which they are being produced before they escape into the workroom air.

**Localized.** Restricted to one spot or area in the body, and not spread all through it; contrasted with systemic.

**Lockout/tagout.** A basic safety concept and OSHA standard requiring implementation of practices and procedures to prevent the release of potentially hazardous energy from machines or parts of machines and equipment while maintenance, servicing, or alteration activity is performed. The energy in question may be electrical, mechanical, chemical, or any other form. Also called lockout/tagout/blockout.

**Lordosis.** The curvature of the lower back in the anteroposterior plane.

**Loudness.** The intensive attribute of an auditory sensation, in terms of which sounds may be ordered on a scale extending from soft to loud. Loudness depends primarily upon the sound pressure of the stimulus, but it also depends upon the frequency and wave form of the stimulus.

**Louver.** A slanted panel.

**Low-pressure tank.** A storage tank designed to operate at pressures between 0.5 and 15 psig (3.5 to 103 kPa).

**Lower confidence limit (LCL).** In analyzing sampling data, a statistical procedure used to estimate the likelihood that the true value of the sampled quantity is lower than that obtained.

**Lower explosive limit (LEL).** The lower limit of flammability of a gas or vapor at ordinary ambient temperatures expressed by a percentage of the gas or vapor in air by volume. This limit is assumed constant for temperatures up to 250 F (120 C); above this, it should be decreased by a factor of 0.7, because explosibility increases with higher temperatures.

**LP gas.** See Liquefied petroleum gas.

**Lumbar spine.** The section of the lower spinal column or vertebral column immediately above the sacrum. Located in the small of the back and consisting of five large lumbar vertebrae, it is a highly stressed area in work situations and in supporting the body structure.

**Lumbosacral joint.** The joint between the fifth lumbar vertebrae and the sacrum. Often the site of spinal trauma from lifting tasks.

**Lumen.** The flux on one square foot of a sphere—one foot in radius—with a light source of one candle at the center that radiates uniformly in all directions.

**Luminous flux.** The rate of light flow measured in lumens.

**Lux.** A unit of illumination—1 footcandle = 10.8 lux.

**Lyme disease.** A disease transmitted to humans by the deer tick.

**Lymph.** A pale, coagulable fluid consisting of a liquid portion resembling blood plasma and containing white blood cells (lymphocytes).

**Lymph node.** Small oval bodies with a gland-like structure scattered throughout the body in the course of the lymph vessels. Also known as lymphatic nodes, lymph glands, and lymphatic glands.

**Lymphoid.** Resembling lymph.

**Lyophilized.** Freeze-dried, as in freeze-dried bacterial cultures.

**Lysis.** The distribution or breaking up of cells by internal or external means.

# M

**MAC.** Maximum allowable concentration.

**Maceration.** Softening of the skin by action of a liquid.

**Macrophage.** Immune system cell whose normal function is to engulf and remove foreign matter from the body's tissues.

**Macroscopic.** Visible without the aid of a microscope.

**Macula.** An oval area in the center of the retina devoid of blood vessels; the area most responsible for color vision.

**Magnification.** The number of times the apparent size of an object has been increased by the lens system of a microscope.

**Makeup air.** Clean, tempered outdoor air supplied to a work space to replace air removed by exhaust ventilation or by some industrial process.

**Malaise.** A vague feeling of bodily discomfort.

**Malignant.** As applied to a tumor, cancerous and capable of undergoing metastasis (invasion of surrounding tissue).

**Manometer.** Instrument for measuring pressure; essentially a U-tube partially filled with a liquid (usually water, mercury, or a light oil) and constructed in such a way that the amount of displacement of the liquid indicates the pressure being exerted on the instrument.

**Maser.** Microwave amplification by stimulated emission of radiation. When used in the term *optical maser,* it is often interpreted as molecular amplification by stimulated emission of radiation.

**Masking.**   The stimulation of a person's ear with controlled noise to prevent that person from hearing with one ear the tone or signal given to the other ear. This procedure is used when there is at least a 15 to 20 dBA difference in the hearing level between ears.

**Mass.**   Quantity of matter; measured in grams or pounds.

**Material safety data sheet (MSDS).**   As part of hazard communication standards (right-to-know laws), federal and state OSHA programs require manufacturers and importers of chemicals to prepare compendia of information on their products. Categories of information that must be provided on MSDSs include physical properties, recommended exposure limits, personal protective equipment, spill-handling procedures, first aid, health effects, and toxicological data.

**Matter.**   Anything that has mass or occupies space.

**Maximum evaporative capacity.**   The maximum amount of evaporating sweat from a person that an environment can accept.

**Maximum line current.**   The rms current in the supply line of an x-ray machine operating at its maximum rating.

**Maximum permissible concentration (MPC).**   Concentrations set by the National Committee on Radiation Protection (NCRP); recommended maximum average concentrations of radionuclides to which a worker may be exposed assuming that he works 8 hours a day, 5 days a week, and 50 weeks a year.

**Maximum permissible dose (MPD).**   A dose of ionizing radiation not expected to cause appreciable bodily injury to a person at any time during his or her life.

**Maximum permissible power or energy density.**   The intensity of laser radiation not expected to cause detectable bodily injury to a person at any time during his or her life.

**Maximum use concentration (MUC).**   The product of the protection factor of the respiratory protection equipment and the permissible exposure limit (PEL).

**Mechanical efficiency curve.**   A graphical representation of a fan's relative efficiency in moving air at different airflow rates and static pressures.

**Mechanotactic stress.**   Stress caused by contact with a mechanical environment.

**Mechanotaxis.**   Contact with a mechanical environment consisting of forces (pressure, moment), vibration, and so on; one of the ecological stress vectors. Improper design of the mechanotactic interface may lead to instantaneous trauma, cumulative pathogenesis, or death.

**Median nerve.**   A major nerve controlling the flexor muscles of the wrist and hand. Tool handles and other grasped objects should make solid contact with the sensory feedback area of this nerve, located in the palmar surface of the thumb, index finger, middle finger, and part of the ring finger.

**Medium.**   See Culture medium.

**Medulla.**   The part of the brain that controls breathing.

**Mega.**   One million. For example, a megacurie = one million curies.

**Mega-, megalo-** (prefix).   Large, huge. The prefix *macro-* has the same meaning.

**Meiosis.**   The process whereby chromosome pairs undergo nuclear division as the germ cell matures.

**Melanoderma.**   Abnormal darkening of the skin.

**Melanocyte.**   An epidermal cell containing dark pigments.

**Melt.**   In the glass industry, the total batch of ingredients that may be introduced into pots or furnaces.

**Melting point.**   The transition point between the solid and liquid states. Expressed as the temperature at which this change occurs.

**Membrane.**   A thin, pliable layer of animal tissue that covers a surface, lines the interior of a cavity or organ, or divides a space.

**Membrane filter.**   A filter medium made from various polymeric materials such as cellulose, polyethylene, and tetrapolyethylene. Usually exhibit narrow ranges of effective pore diameters and are therefore useful in collecting and sizing microscopic and submicroscopic particles and in sterilizing liquids.

**Men-, meno-** (prefix).   Pertaining to menstruation; from a Greek word for *month*.

**Ménière's disease.**   A combination of deafness, tinnitus, and vertigo.

**Meson.**   A particle that weighs more than an electron but generally less than a proton. Mesons can be produced artificially or by cosmic radiation (natural radiation from outer space). Mesons are not stable and disintegrate in a fraction of a second.

**Mesothelioma.**   Cancer of the membranes that line the chest and abdomen.

**Metabolism.**   The flow of energy and the associated physical and chemical changes constantly taking place in the billions of cells that make up the body.

**Metal fume fever.** A flu-like condition caused by inhaling heated metal fumes.

**Metallizing.** Melting wire in a special device that sprays the atomized metal onto a surface. The metal can be steel, lead, or another metal or alloy.

**Metastasis.** Transfer of the causal agent (cell or microorganism) of a disease from a primary focus to a distant one through the blood or lymphatic vessels. Also, spread of malignancy from a site of primary cancer to secondary sites.

**Methemoglobinemia.** The presence of methemoglobin in the blood. (Methemoglobin is a compound formed when the iron moiety of hemoglobin is oxidized from the ferrous to the ferric state.) This protein inactivates the hemoglobin as an oxygen carrier.

**Mev.** Million electron volts.

**Mica.** A large group of silicates of varying composition that are similar in physical properties. All have excellent cleavage and can be split into very thin sheets. Used in electrical insulation.

**Microbar.** A unit of pressure commonly used in acoustics; equals one dyne/cm$^2$. A reference point for the decibel, which is accepted as 0.0002 dyne/cm$^2$.

**Microbe.** A microscopic organism.

**Microcurie (μc).** One-millionth of a curie. A still smaller unit is the micromicrocurie (μμc).

**Micron (micrometer).** A unit of length equal to $10^{-4}$ cm, approximately 1/25,000 in.

**Microorganism.** A minute organism—microbes, bacteria, cocci, viruses, and molds, among others.

**Microphone.** An electroacoustic transducer that responds to sound waves and delivers essentially equivalent electric waves.

**Midsagittal plane.** A reference plane formed by bisecting the human anatomy into a right and left aspect. Human motor function can be described in terms of movement relative to the midsagittal plane.

**Miliary.** Characterized or accompanied by seedlike blisters or inflamed raised portions of tissue.

**Milligram (mg).** A unit of weight in the metric system. One thousand milligrams equal one gram.

**Milligrams per cubic meter (mg/m$^3$).** Unit used to measure air concentrations of dusts, gases, mists, and fumes.

**Milliliter (ml).** A metric unit used to measure volume. One milliliter equals one cubic centimeter.

**Millimeter of mercury (mmHg).** The unit of pressure equal to the pressure exerted by liquid mercury one-millimeter-high column at a standard temperature.

**Milliroentgen.** One one-thousandth of a roentgen.

**Millwright.** A mechanic engaged in the erection and maintenance of machinery.

**Mineral pitch.** Tar from petroleum or coal as opposed to wood tar.

**Mineral spirits.** A petroleum fraction with a boiling range between 300 and 400 F (149 and 240 C).

**Miosis.** Excessive smallness or contraction of the pupil of the eye.

**Mists.** Suspended liquid droplets generated by condensation from the gaseous to the liquid state or by breaking up a liquid into a dispersed state, such as by splashing, foaming, or atomizing. Formed when a finely divided liquid is suspended in air.

**Mitosis.** Nuclear cell division in which resulting nuclei have the same number and kinds of chromosomes as the original cell.

**Mixture.** A combination of two or more substances that may be separated by mechanical means. The components may not be uniformly dispersed. See also Solution.

**Moderator.** A material used to slow neutrons in a reactor. These slow neutrons are particularly effective in causing fission. Neutrons are slowed when they collide with atoms of light elements such as hydrogen, deuterium, and carbon—three common moderators.

**Mold.** (1) A growth of fungi forming a furry patch, as on stale bread or cheese. See Spore. (2) A hollow form or matrix into which molten material is poured to produce a cast.

**Molecule.** A chemical unit composed of one or more atoms.

**Moment.** Magnitude of force times distance of application.

**Moment concept.** A concept based on theoretical and experimental bases that lifting stress depends on the bending moment exerted at susceptible points of the vertebral column rather than depending on weight alone.

**Monaural hearing.** Hearing with one ear only.

**Monochromatic.** Single fixed wavelength.

**Monomer.** A compound of relatively low molecular weight that, under certain conditions, either alone or with another monomer, forms various types and lengths of molecular chains called polymers or copolymers of high molecular weight. Styrene, for example, is a monomer that polymerizes readily to form polystyrene. See Polymer.

**Morphology.** The branch of biological science that deals with the study of the structure and form of living organisms.

**Motile.** Capable of spontaneous movement.

**MPE.** Maximum permissible exposure.

**MPL.** May be either maximum permissible level, limit, or dose; refers to the tolerable dose rate for humans exposed to nuclear radiation.

**Mppcf.** Million particles per cubic foot.

**mr.** Millirem.

**mR.** Milliroentgen.

**MSHA.** The Mine Safety and Health Administration; a federal agency that regulates safety and health in the mining industry.

**MSDS.** Material safety data sheet.

**Mucous membranes.** Lining of the hollow organs of the body, notably the nose, mouth, stomach, intestines, bronchial tubes, and urinary tract.

**Musculoskeletal system.** The combined system of muscles and bones that comprise the internal biomechanical environment.

**Mutagen.** Anything that can cause a change (mutation) in the genetic material of a living cell.

**Mutation.** A transformation of the gene that may result in the alteration of characteristics of offspring.

**MWD.** Megawatt days, usually per ton. The amount of energy obtained from one megawatt power in one day, normally used to measure the extent of nuclear fuel burnup. 10,000 MWD per ton is about 1 percent burnup.

**My-, myo-** (prefix). Pertaining to muscle. Myocardium is the heart muscle.

**Myelo-** (prefix). Pertaining to marrow.

## N

**Nanometer.** A unit of length equal to $10^{-7}$ cm.

**Naphthas.** Hydrocarbons of the petroleum type that contain substantial portions of paraffins and naphthalenes.

**Narcosis.** Stupor or unconsciousness produced by chemical substances.

**Narcotics.** Chemical agents that completely or partially induce sleep.

**Narrow band.** Applies to a narrow band of transmitted waves, with neither the critical or cutoff frequencies of the filter being zero or infinite.

**Nasal septum.** Narrow partition that divides the nose into right and left nasal cavities.

**Nascent.** Just forming, as from a chemical or biological reaction.

**Nasopharynx.** Upper extension of the throat.

**Natural gas.** A combustible gas composed largely of methane and other hydrocarbons with variable amounts of nitrogen and noncombustible gases; obtained from natural earth fissures or from driven wells. Used as a fuel in the manufacture of carbon black and in chemical synthesis of many products. Major source of hydrogen for the manufacture of ammonia.

**Natural radioactivity.** The radioactive background or, more properly, the radioactivity that is associated with the heavy naturally occurring elements.

**Natural uranium.** Purified from the naturally occurring ore, as opposed to uranium enriched in fissionable content by processing at separation facilities.

**Nausea.** An unpleasant sensation, vaguely referred to the epigastrium and abdomen. Often precedes vomiting.

**NCRP.** National Committee on Radiation Protection; an advisory group of scientists and professionals that makes recommendations for radiation protection in the United States.

**Near field.** In noise measurement, refers to a field in the immediate vicinity of the noise source where the sound-pressure level does not follow the inverse square law.

**Necro-** (prefix). Dead.

**Necrosis.** Death of body tissue.

**Neoplasm.** A cellular outgrowth characterized by rapid cell multiplication; may be benign (semicontrolled and restricted) or malignant.

**Nephr-, nephro-** (prefix). From the Greek for *kidney.* See also Ren-.

**Nephrotoxins.** Chemicals that produce kidney damage.

**Nephritis.** Inflammation of the kidneys.

**Neur-, neuro-** (prefix). Pertaining to the nerves.

**Neural loss.** Hearing loss. See also Sensorineural.

**Neuritis.** Inflammation of a nerve.

**Neurological (neurology).** The branch of medical science dealing with the nervous system.

**Neurotoxin.** Chemicals that produce their primary effect on the nervous system.

**Neutrino.** A particle, resulting from nuclear reactions, that carries energy away from the system but has no mass or charge and is absorbed only with extreme difficulty.

**Neutron.** A constituent of the atomic nucleus. A neutron weighs about as much as a proton, and has no electric charge. Neutrons make effective atomic projectiles for the bombardment of nuclei.

**NFPA.** The National Fire Protection Association; a voluntary membership organization whose aim is to promote and improve fire protection and prevention. The NFPA publishes the *National Fire Codes.*

**NIOSH.** The National Institute for Occupational Safety and Health; a federal agency that conducts research on health and safety concerns, tests and certifies respirators, and trains occupational health and safety professionals.

**Nitrogen fixation.** Chemical combination or fixation of atmospheric nitrogen with hydrogen, as in the synthesis of ammonia. Bacteria fixates nitrogen in soil. Provides an industrial and agricultural source of nitrogen.

**Node.** (1) A point, line, or surface in a standing wave where some characteristic of the wave field has essentially zero amplitude. (2) A small, round, or oval mass of tissue; a collection of cells. (3) One of several constrictions occurring at regular intervals in a structure.

**Nodule.** A small mass of rounded or irregularly shaped cells or tissue; a small node.

**Nodulizing.** Simultaneous sintering and drum balling, usually in a rotary kiln.

**NOEL.** See No observable effect level.

**Noise.** Any unwanted sound.

**Noise-induced hearing loss.** Slowly progressive inner-ear hearing loss resulting from exposure to continuous noise over a long period of time, as contrasted to acoustic trauma or physical injury to the ear.

**Nonauditory effects of noise.** Refers to stress, fatigue, health, work efficiency, and performance effects of loud, continuous noise.

**Nonferrous metal.** Metal such as nickel, brass, or bronze that does not include any appreciable amount of iron.

**Nonionizing radiation.** Electromagnetic radiation that does not cause ionization. Includes ultraviolet, laser, infrared, microwave, and radiofrequency radiation.

**Nonpolar solvents.** The aromatic and petroleum hydrocarbon groups characterized by low dielectric constants.

**Nonvolatile matter.** The portion of a material that does not evaporate at ordinary temperatures.

**No Observable Effect Level (NOEL).** In toxicology, the concentration of a substance at (and below) which exposure produces no evidence of injury or impairment.

**Normal pulse (conventional pulse).** Heartbeat; also, a single output event whose pulse duration is between 200 microseconds and one millisecond.

**Nosocomial.** (1) Pertaining to a hospital. (2) Disease caused or aggravated by hospital life.

**NRC.** Nuclear Regulatory Commission of the U.S. Department of Energy.

**NTP.** National Toxicology Program.

**Nuclear battery.** A device in which the energy emitted by decay of a radioisotope is first converted to heat and then directly to electricity.

**Nuclear bombardment.** The shooting of atomic projectiles at nuclei, usually in an attempt to split the atom or to form a new element.

**Nuclear energy.** The energy released in a nuclear reaction such as fission or fusion. Nuclear energy is popularly, though mistakenly, called atomic energy.

**Nuclear explosion.** The rapid fissioning of a large amount of fissionable material; creates intense heat, a light flash, a heavy blast, and a large amount of radioactive fission products. These may be attached to dust and debris forming fallout. Nuclear explosions also result from nuclear fusion, which does not produce radioactive fission products.

**Nuclear reaction.** Result of the bombardment of a nucleus with atomic or subatomic particles or very high energy radiation. Possible reactions are emission of other particles, fission, fusion, and the decay of radioactive material.

**Nuclear reactor.** A machine for producing a controlled chain reaction in fissionable material. It is the heart of nuclear power facilities, where it serves as a heat source. See Reactor.

**Nucleonics.** The application of nuclear science and techniques in physics, chemistry, astronomy, biology, industry, and other fields.

**Nucleus.** The inner core of the atom; consists of neutrons and protons tightly locked together.

**Nuclide.** A type of atom characterized by its mass number, atomic number, and energy state of the nucleus, provided that the mean life in that state is long enough to be observable.

**Nuisance dust.** Dust with a long history of little adverse effect on the lungs; does not produce significant organic disease or toxic effect when exposures are kept at reasonable levels.

**Null point.** The distance from a contaminant source at which the initial energy or velocity of the contaminants is dissipated, allowing the material to be captured by a hood.

**N-unit (or n-unit).** A measure of radiation dose caused by fast neutrons.

**Nutrient.** A substance that can be used for food.

# O

**Occupational health nursing (OHN).** Specialized nursing practice providing health care service to workers and worker populations.

**Occupational Safety and Health Review Commission (OHSHRC).** An independent body established to review actions of federal OSHA that are contested by employers, employees, or their representatives.

**Octave.** The interval between two sounds having a basic frequency ratio of two.

**Octave band.** An arbitrary spread of frequencies. The top frequency in an octave band is always twice the bottom one. The octave band may be referred to by a center frequency.

**Ocul-, oculo-, ophthalmo-** (prefixes). Refer to the eye; *ophth-* words refer more often to cyc diseases.

**Odor.** That property of a substance that affects the sense of smell.

**Odor threshold.** The minimum concentration of a substance at which a majority of test subjects can detect and identify the characteristic odor of a substance.

**Ohm.** The unit of electrical resistance.

**Ohm's Law.** Voltage in a circuit is equal to the current times the resistance.

**Oil dermatitis.** Blackheads and acne caused by oils and waxes that plug the hair follicles and sweat ducts.

**Olecranon fossa.** A depression in the back of the lower end of the humerus in which the ulna bone rests when the arm is straight.

**Olefins.** A class of unsaturated hydrocarbons characterized by relatively great chemical activity. Obtained from petroleum and natural gas. Examples are butene, ethylene, and propylene. Generalized formula: $C_nH_{2n}$.

**Olfactory.** Pertaining to the sense of smell.

**Olig-, oligo-** (prefix). Scanty, few, little. *Oliguria* means scanty urination.

**Oncogenic.** Tumor-generating.

**Oncology.** Study of causes, development, characteristics, and treatment of tumors.

**Opacity.** The condition of being nontransparent; a cataract.

**Ophthalmologist.** A physician who specializes in the structure, function, and diseases of the eye.

**Optical density (OD).** A logarithmic expression of the attenuation afforded by a filter.

**Optically pumped laser.** A type of laser that derives its energy from a noncoherent light source, such as a xenon flash lamp; usually pulsed and commonly called a solid-state laser.

**Organ.** An organized collection of tissues that have a special and recognized function.

**Organ of Corti.** The heart of the hearing mechanism; an aggregation of nerve cells in the ear lying on the basilar membrane that picks up vibrations and converts them to electrical energy, which is sent to the brain and interpreted as sound.

**Organic.** Chemicals that contain carbon. To date, nearly one million organic compounds have been synthesized or isolated. See also Inorganic.

**Organic disease.** Disease in which some change in the structure of body tissue could either be visualized or positively inferred from indirect evidence.

**Organic matter.** Compounds containing carbon.

**Organism.** A living thing, such as a human being, animal, germ, plant, and so on, especially one consisting of several parts, each specializing in a particular function.

**Orifice.** (1) The opening that serves as an entrance and/or outlet of a body cavity or organ, especially the opening of a canal or a passage. (2) A small hole in a tube or duct. A critical, or limiting, orifice is used to control rate of flow of a gas in rotometers and other air-sampling equipment.

**Orifice meter.** A flow meter, employing as the measure of flow rate the difference between pressures measured on the upstream and downstream sides of a restriction within a pipe or duct.

**Ortho-** (prefix). Straight, correct, normal. Orthopsychiatry is the specialty concerned with "straightening out" behavioral disorders.

**Orthoaxis.** The true anatomical axis about which a limb rotates, as opposed to the assumed axis. The assumed axis is usually the most obvious or geometric one; the orthoaxis is less evident and can only be found by the use of anatomical landmarks.

**Os-, oste-, osteo-** (prefix). Pertaining to bone. The Latin *os-* is most often associated with anatomical structures, whereas the Greek *osteo-* usually refers to conditions involving bone. *Osteogenesis* means formation of bone.

**Oscillation.** The variation, usually with time, of the magnitude of a quantity with respect to a specified reference when the magnitude is alternately greater and smaller than the reference.

**OSHA.** U.S. Occupational Safety and Health Administration.

**OSHA 200 Log.** Record keeping of employee injuries and illnesses is required by OSHA standard; OSHA 200 Log is a format that contains the necessary required details. It may be used by employers and is available from OSHA.

**Osmosis.** The passage of fluid through a semipermeable membrane as a result of osmotic pressure.

**Osseous.** Pertaining to bone.

**Ossicle.** Any member of a chain of three small bones from the outer membrane of the tympanum (eardrum) to the membrane covering the oval window of the inner ear.

**ot-, oto-** (prefix). Pertaining to the ear. *Otorrhea* means ear discharge.

**Otitis media.** An inflammation and infection of the middle ear.

**Otologist.** A physician specializing in surgery and diseases of the ear.

**Otosclerosis.** A condition of the ear caused by a growth of body tissue about the foot plate of the stapes and oval window of the inner ear; results in a gradual loss of hearing.

**Output power and output energy.** Power is used primarily to rate CW lasers, because the energy delivered per unit time remains relatively constant (output measured in watts). In contrast, pulsed lasers deliver their energy output in pulses and their effects may be best categorized by energy output per pulse. The output power of CW lasers is usually expressed in milliwatts or watts, pulsed lasers in kilowatts, and q-switch pulsed lasers in megawatts or gigawatts. Pulsed energy output is usually expressed as joules per pulse.

**Overexposure.** Exposure beyond the specified limits.

**Oxidation.** Process of combining oxygen with some other substance; technically, a chemical change in which an atom loses one or more electrons. Opposite of reduction.

# P

**PAH.** Polynuclear aromatic hydrocarbons.

**Pair production.** The conversion of a gamma ray into a pair of particles: an electron and a positron. This is an example of direct conversion of energy into matter according to Einstein's famous formula, $E = mc^2$: energy = mass·velocity of light squared.

**Palmar arch.** Blood vessels in the palm of the hand from which the arteries supplying blood to the fingers are branched. Pressure against the palmar arch by poorly designed tool handles may cause ischemia of the fingers and loss of tactile sensation and precision of movement.

**Palpitation.** Rapid heartbeat of which a person is acutely aware.

**Papilloma.** A small growth or tumor of the skin or mucous membrane; warts and polyps, for example.

**Papule.** A small, solid, usually conical elevation of the skin.

**Papulovesicular.** Characterized by the presence of papules and vesicles.

**Para-** (prefix). Alongside, near, abnormal; as in *paraproctitis,* inflammation of tissues near the rectum. A Latin suffix with the same spelling, *-para,* denotes bearing or giving birth, as in *multipara,* a woman who has given birth to two or more children.

**Paraffins, Paraffin series.** (From *parum affinis*—small affinity.) Straight- or branched-chain hydrocarbon components of crude oil and natural gas whose molecules are saturated (that is, carbon atoms attached to each other by single bonds) and therefore very stable. Examples are methane and ethane. Generalized formula: $C_nH_{2n+2}$.

**Parasite.** An organism that derives its nourishment from a living plant or animal host. Does not necessarily cause disease.

**Parenchyma.** The distinguishing or specific (working) tissue of a bodily gland or organ, contained in and supported by the connective tissue framework, or stroma.

**Parent.** Precursor; the name given to a radioactive nucleus that disintegrates to form a radioactive product or daughter.

**Partial barrier.** An enclosure constructed so that sound transmission between its interior and its surroundings is minimized.

**Particle.** A small discrete mass of solid or liquid matter.

**Particle concentration.** Concentration expressed in terms of number of particles per unit volume of air or other gas. When expressing particle concentrations, the method of determining the concentration should be stated.

**Particle size.** The measured dimension of liquid or solid particles, usually in microns.

**Particle size distribution.** The statistical distribution of the sizes or ranges of size of a population of particles.

**Particulate.** A particle of solid or liquid matter.

**Particulate matter.** A suspension of fine solid or liquid particles in air, such as dust, fog, fume, mist, smoke, or sprays. Particulate matter suspended in air is commonly known as an aerosol.

**Path-, patho-** (prefix), **-pathy** (suffix). Feeling, suffering, disease. *Pathogenic* means producing disease; *enteropathy* means disease of the intestines; pathology is the medical specialty concerned with all aspects of disease. The root appears in the everyday word *sympathy* (feeling with).

**Pathogen.** Any microorganism capable of causing disease.

**Pathogenesis.** Describes how a disease takes hold on the body and spreads.

**Pathogenic.** Producing or capable of producing disease.

**Pathognomonic.** Distinctive or characteristic of a specific disease or pathological condition; a sign or symptom from which a diagnosis can be made.

**Pathological.** Abnormal or diseased.

**Pathology.** The study of disease processes.

**Pelleting.** In various industries, powdered material may be made into pellets or briquettes for convenience. The pellet is a distinctly small briquette. See Pelletizing.

**Pelletizing.** Refers primarily to extrusion by pellet mills; also refers to other small extrusions and to some balled products. Generally regarded as being larger than grains and smaller than briquettes.

**Percent impairment of hearing** (percent hearing loss). An estimate of a person's ability to hear correctly; usually determined by the pure tone audiogram. The specific rule for calculating this quantity varies from state to state according to law.

**Percutaneous.** Performed through the unbroken skin, as by absorption of an ointment through the skin.

**Peri-** (prefix). Around, about, surrounding. Periodontium is tissue that surrounds and supports the teeth.

**Periodic table.** Systematic classification of the elements according to atomic numbers (nearly the same order as by atomic weights) and by physical and chemical properties.

**Peripheral neuropathy.** Deterioration of peripheral nerve function; affects the hands, arms, feet, and legs. Certain hydrocarbon solvents are known to cause peripheral neuropathies in overexposed individuals.

**Permeation.** Process by which a chemical moves through a protective clothing material on a molecular level.

**Permissible dose.** See MPC, MPL.

**Permissible exposure limit (PEL).** An exposure limit published and enforced by OSHA as a legal standard.

**Personal protective equipment.** Devices worn by the worker to protect against hazards in the environment. Respirators, gloves, and hearing protectors are examples.

**Pesticides.** General term for chemicals used to kill such pests as rats, insects, fungi, bacteria, weeds, and so on, that prey on humans or agricultural products. Among these are insecticides, herbicides, fungicides, rodenticides, miticides, fumigants, and repellents.

**Petrochemical.** A term applied to chemical substances produced from petroleum products and natural gas.

**Pink noise.** Noise that has been weighted, especially at the low end of the spectrum, so that the energy per band (usually octave band) is approximately constant over the spectrum.

**pH.** The degree of acidity or alkalinity of a solution, with neutrality indicated as 7.

**Phagocyte.** A cell in the body that engulfs foreign material and consumes debris and foreign bodies.

**Phalanx (pl. phalanges).** Any of the bones of the fingers or toes. Often used as anatomical reference points in ergonomic work analysis.

**Pharmaceuticals.** Drugs and related chemicals reaching the public primarily through drug suppliers. In government reports, this category includes not only such medicinals as aspirin and antibiotics but also such nutriments as vitamins and amino acids for both human and animal use.

**Pharyngeal.** Pertaining to the pharynx (the musculo-membranous sac between the mouth, nares, and esophagus).

**Phenol.** $C_6H_5OH$. Popularly known as carbolic acid. Important chemical intermediate and base for plastics, pharmaceuticals, explosives, antiseptics, and many other end products.

**Phenolic resins.** A class of resins produced as the condensation product of phenol or substituted phenol and formaldehyde or other aldehydes.

**Phosphors.** Fluorescent or luminescent materials.

**Photochemical process.** Chemical changes brought about by radiant energy acting upon various chemical substances. See Photosynthesis.

**Photoelectric effect.** Occurs when an electron is thrown out of an atom by a light ray or gamma-ray. This effect is used in an "electric eye;" light falls on a sensitive surface throwing out electrons that can then be detected.

**Photoionization detector (PID).** A direct-reading monitoring instrument that operates by detecting and distinguishing between ions of vapors and gases following ionization by the instrument's ultraviolet light source.

**Photomultiplier tube.** A vacuum tube that multiplies electron input.

**Photon.** A bundle (quantum) of radiation. Constitutes, for example, x rays, gamma-rays, and light.

**Photophobia.** Abnormal sensitivity to light.

**Photosynthesis.** The process by which plants produce carbohydrates and oxygen from carbon dioxide and water.

**Physiology.** The study of the functions or actions of living organisms.

**Physiopathology.** The science of functions in disease or modified by a disease.

**Pig.** (1) A container (usually lead) used to ship or store radioactive materials. The thick walls protect workers from radiation. (2) In metal refining, a small ingot from the casting of blast furnace metal.

**Pigment.** A finely divided, insoluble substance that imparts color to a material.

**Pilot facility.** Small scale operation preliminary to major enterprises. Common in the chemical industry.

**Pinna.** Ear flap; the part of the ear that projects from the head. Also known as the auricle.

**Pitch.** The attribute of auditory sensation in terms of which sounds may be ordered on a scale extending from low to high. Pitch depends primarily on the frequency of the sound stimulus, but also on the sound pressure and wave form of the stimulus.

**Pitot tube.** A device consisting of two concentric tubes, one serving to measure the total or impact pressure existing in the airstream, the other to measure the static pressure only. When the annular space between the tubes and the interior of the center tube are connected across a pressure-measuring device, the pressure difference automatically nullifies the static pressure, and the velocity pressure alone is registered.

**Plasma.** (1) The fluid part of the blood in which the blood cells are suspended. Also called protoplasm. (2) A gas that has been heated to a partially or completely ionized condition, enabling it to conduct an electric current.

**Plasma arc welding (PAW).** A process that produces coalescence of metals by heating them with a constricted arc between an electrode and the workpiece (transferred arc) or between the electrode and the constricting nozzle (nontransferred arc). Shielding is obtained by the hot, ionized gas issuing from the orifice, which may be supplemented by an auxiliary source of shielding gas. Shielding gas can be an inert gas or a mixture of gases. Pressure may or may not be used, and filler metal may or may not be supplied.

**Plastics.** Any one of a large group of materials that contains as an essential ingredient an organic substance of large molecular weight. Two basic types are thermosetting (irreversibly rigid) and thermoplastic (reversibly rigid). Before compounding and processing, plastics often are referred to as (synthetic) resins. Final form may be a film, sheet, solid, or foam-flexible or rigid.

**Plasticizers.** Organic chemicals used in modifying plastics, synthetic rubber, and similar materials to facilitate compounding and processing, and to impart flexibility to the end product.

**Plenum.** Pressure-equalizing chamber.

**Plenum chamber.** An air compartment connected to one or more ducts or connected to a slot in a hood; used for air distribution.

**Pleura.** The thin membrane investing the lungs and lining the thoracic cavity, completely enclosing a potential space known as the pleural cavity. There are two pleurae, right and left, entirely distinct from each other. The pleura is moistened with a secretion that facilitates the movements of the lungs in the chest.

**Plumbism.** One name for lead intoxication.

**Pleurisy.** Caused when the outer lung lining (visceral pleura) and the chest cavity's inner lining (parietal pleura) lose their lubricating properties; the resultant friction causes irritation and pain.

**Plume trap.** An exhaust ventilation hood designed to capture and remove the plume given off the target on impact of a laser beam.

**Plutonium.** A heavy element that undergoes fission under the impact of neutrons. It is a useful fuel in nuclear reactors. Plutonium cannot be found in nature, but can be produced and "burned" in reactors.

**Pneumo-** (Greek), **pulmo-** (Latin) (prefix). Pertaining to the lungs.

**Pneumoconiosis.** Dusty lungs; a result of the continued inhalation of various kinds of dust or other particles.

**Pneumoconiosis-producing dust.** Dust, which when inhaled, deposited, and retained in the lungs, may produce signs, symptoms, and findings of pulmonary disease.

**Pneumonitis.** Inflammation of the lungs.

**Poison.** (1) A material introduced into the reactor core to absorb neutrons. (2) Any substance that, when taken into the body, is injurious to health.

**Polarography.** A physical analysis method for determining certain atmospheric pollutants that are electroreducible or electro-oxidizable and are in true solution and stable for the duration of the measurement.

**Polar solvents.** Solvents (such as alcohols and ketones) that contain oxygen and that have high dielectric constants.

**Pollution.** Synthetic contamination of soil, water, or atmosphere beyond that which is natural.

**Poly-** (prefix). Many.

**Polycythemia.** A condition marked by an excess in the number of red corpuscles in the blood.

**Polymer.** A high molecular-weight material formed by the joining together of many simple molecules (monomers). There may be hundreds or even thousands of the original molecules linked end to end and often crosslinked. Rubber and cellulose are naturally occurring polymers. Most resins are chemically produced polymers.

**Polymerization.** A chemical reaction in which two or more small molecules combine to form larger molecules (polymers) that contain repeating structural units of the original molecules. A hazardous polymerization is one with an uncontrolled release of energy.

**Polystyrene resins.** Synthetic resins formed by polymerization of styrene.

**Popliteal clearance.** Distance between the front of the seating surface and the popliteal crease. This should be about 5 in. in good seat design to prevent pressure on the popliteal artery.

**Popliteal crease (or line).** The crease in the hollow of the knee when the lower leg is flexed. Important anatomical reference point.

**Popliteal height of chair.** The height of the highest part of the seating surface above the floor.

**Popliteal height of individual.** The distance between the crease in the hollow of the knee and the floor.

**Porphyrin.** One of a group of complex chemical substances that forms the basis of the respiratory pigments of animals and plants; hemoglobin and chlorophyll are other examples.

**Portal.** Place of entrance.

**Portland cement.** See Cement, portland.

**Positive displacement pump.** Any type of air mover pump in which leakage is negligible, so that the pump delivers a constant volume of fluid, building up to any pressure necessary to deliver that volume.

**Positron.** A particle that has the same weight and charge as an electron but is electrically positive rather than negative. The positron's existence was predicted in theory years before it was actually detected. It is not stable in matter because it reacts readily with an electron to give two gamma-rays.

**Potential energy.** Energy due to position of one body with respect to another or to the relative parts of the same body.

**Power.** Rate at which work is done; measured in watts (one joule per second) and horsepower (33,000 foot-pounds per minute). One horsepower equals 746 watts.

**Power density.** The intensity of electromagnetic radiation per unit area, expressed as watts/cm.

**Power level.** 10 times the logarithm to the base 10 of the ratio of a given power to a reference power; measured in decibels.

**ppb.** Parts per billion.

**ppm.** Parts per million parts of air by volume of vapor or gas or other contaminant.

**PPE.** See Personal protective equipment.

**Precision.** The degree of agreement (expressed in terms of distribution of test results about the mean result) of repeated measurements of the same property, obtained by repetitive testing of a homogeneous sample under specified conditions. The precision of a method is expressed quantitatively as the standard deviation, computed from the results of a series of controlled determinations.

**Presby-** (prefix). Old. As in *presbyopia*—eye changes associated with aging.

**Presbycusis.** Hearing loss caused by age.

**Pressure.** Force applied to or distributed over a surface; measured as force per unit area. See Absolute pressure, Atmospheric pressure, Gage pressure, Standard temperature and pressure, Static pressure, Total pressure, Vapor pressure, and Velocity pressure.

**Pressure drop.** The difference in static pressure measured at two locations in a ventilation system; caused by friction or turbulence.

**Pressure loss.** Energy lost from a pipe or duct system through friction or turbulence.

**Pressure, static.** The normal force per unit area that would be exerted by a moving fluid on a small body immersed in it if the body were carried along with the fluid. Practically, it is the normal force per unit area at a small hole in a wall of the duct through which the fluid flows or on the surface of a stationary tube at a point where the disturbances, created by inserting the tube, cancel. The potential pressure exerted in all directions by a fluid at rest. It is the tendency to either burst or collapse the pipe, usually expressed in inches of water gauge when dealing with air.

**Pressure, total.** In the theory of the flow of fluids, the sum of the static pressure and the velocity pressure at the point of measurement. Also called dynamic pressure.

**Pressure, vapor.** The pressure exerted by a vapor. If a vapor is kept in confinement at a constant temperature over its liquid so that it can accumulate above the liquid, the vapor pressure approaches a fixed limit called the maximum, or saturated, vapor pressure, dependent only on the temperature and the liquid.

**Pressure vessel.** A storage tank or vessel designed to operate at pressures greater than 15 psig (103 kPa).

**PRF laser.** A pulsed recurrence frequency laser, which is a pulsed-typed laser with properties similar to a CW laser when the frequency is very high.

**Probe.** A tube used for sampling or for measuring pressures at a distance from the actual collection or measuring apparatus; commonly used for reaching inside stacks or ducts.

**Process safety management (PSM).** Encompassing safety concept for the chemical processing industry that is mandated and regulated in OSHA's process safety management standard. In PSM, potential hazards are systematically analyzed for each step of a chemical process.

**Prokaryote.** Single-celled organism lacking mitochondria and a defined nucleus. Usually has a cell wall. Describes primarily bacterial organisms.

**Proliferation.** The reproduction or multiplication of similar forms, especially of cells and morbid cysts.

**Pronation.** Rotation of the forearm in a direction to face the palm downward when the forearm is horizontal, and backward when the forearm is in a vertical position.

**Propagation of flame.** The spread of flame through the entire volume of a flammable vapor-air mixture from a single source of ignition. A vapor-air mixture below the lower flammable limit may burn at the point of ignition without propagating from the ignition source.

**Prophylactic.** Preventive treatment for protection against disease.

**Protection factor (PF).** In respiratory protective equipment, the ratio of the ambient airborne concentration of the contaminant to the concentration inside the facepiece.

**Protective atmosphere.** A gas envelope surrounding an element to be brazed, welded, or thermal-sprayed, with the gas composition controlled with respect to chemical composition, dew point, pressure, flow rate, and so on.

**Protective coating.** A thin layer of metal or organic material, applied as paint to a surface to protect it from oxidation, weathering, and corrosion.

**Proteins.** Large molecules found in the cells of all animal and vegetable matter containing carbon, hydrogen, nitrogen, and oxygen, and sometimes sulfur and phosphorus. The fundamental structural units of proteins are amino acids.

**Proteolytic.** Capable of splitting or digesting proteins into simpler compounds.

**Proton.** A fundamental unit of matter having a positive charge and a mass number of one.

**Protoplasm.** The basic material from which all living tissue is made. Physically it is a viscous, translucent, semifluid colloid, composed mainly of proteins, carbohydrates, fats, salts, and water.

**Protozoa.** Single-celled microorganisms belonging to the animal kingdom.

**Proximal.** The part of a limb that is closest to the point of attachment. The elbow is proximal to the wrist, which is proximal to the fingers.

**Psittacosis.** Parrot fever. An infectious disease of birds to which poultry handlers and other workers exposed to dried bird feces are at risk. Caused by *Chlamydia psittaci*. The most noted symptom of the disease among humans is fever.

**Psych-, psycho-** (prefix). Pertaining to the mind, from the Greek word for *soul*.

**Psychogenic deafness.** Loss originating in or produced by the mental reaction of an individual to their physical or social environment. It is sometimes called functional deafness or feigned deafness.

**Psychrometer.** An instrument consisting of wet- and dry-bulb thermometers for measuring relative humidity.

**Psychrometric chart.** A graphical representation of the thermodynamic properties of moist air.

**Pterygium.** A growth of the conjunctiva caused by a degenerative process brought on by long, continued irritation (as from exposure to wind, dust, and possibly to ultraviolet radiation).

**Pulmonary.** Pertaining to the lungs.

**Pulse length.** Duration of a pulsed laser flash; may be measured in milliseconds, microseconds, or nanoseconds.

**Pulsed laser.** A class of laser characterized by operation in a pulsed mode; that is, emission occurs in one or more flashes of short duration (pulse length).

**Pumice.** A natural silicate from volcanic ash or lava. Used as an abrasive.

**Pupil.** The variable aperture in the iris through which light travels toward the interior regions of the eye. The pupil size varies from 2 mm to 8 mm.

**Pur-, pus-** (Latin), **pyo-** (Greek) (prefixes). Indicates pus, as in *purulent, suppurative, pustulant,* and *pyoderma.*

**Pure tone.** A sound wave characterized by its singleness of frequency.

**Purpura.** Extensive hemorrhage into the skin or mucous membrane.

**Push-pull hood.** A hood consisting of an air supply system on one side of the contaminant source blowing across the source and into an exhaust hood on the other side.

**Putrefaction.** Decomposition of proteins by microorganisms, producing disagreeable odors.

**Pyloric stenosis.** Obstruction of the pyloric opening of the stomach caused by hypertrophy of the pyloric sphincter.

**Pylorus.** The orifice of the stomach leading to the small intestine.

**Pyel-, pyelo-** (prefix). Pertaining to the urine-collecting chamber of the kidney.

**Pyr-, pyret-** (prefix). Fever.

**Pyrethrum.** A pesticide obtained from the dried, powdered flowers of the plant of the same name; mixed with petroleum distillates, it is used as an insecticide.

**Pyrolysis.** The breaking apart of complex molecules into simpler units by the use of heat, as in the pyrolysis of heavy oil into gasoline.

## Q

**QF.** See Quality factor.

**Q fever.** Disease caused by a rickettsial organism that infects meat and livestock handlers; similar but not identical to tick fever.

**Q-switched laser.** (Also known as Q-spoiled). A pulsed laser capable of extremely high peak powers for very short durations (pulse length of several nanoseconds).

**Qualitative fit testing.** A method of assessing the effectiveness of a particular size and brand of respirator based on an individual's subjective response to a test atmosphere. The most common test agents are isoamyl acetate (banana oil), irritant smoke, and sodium saccharin. Proper respirator fit is indicated by the individual reporting no indication of the test agent inside the facepiece during the performance of a full range of facial movements.

**Quality.** A term used to describe the penetrating power of x rays or gamma-rays.

**Quality factor.** A linear energy transfer-dependent factor by which absorbed radiation doses are to be multiplied to obtain the dose equivalent.

**Quantitative fit testing.** A method of assessing the effectiveness of a particular size and brand of respirator on an individual. Instrumentation is used to measure both the test atmosphere (a gas, vapor or aerosol, such as DOP) and the concentration of the test contaminant inside the facepiece of the respirator. The quantitative fit factor thus obtained is used to determine if a suitable fit has been obtained by referring to a table or to the software of the instrumentation. Quantitative fit factors obtained in this way do not correlate well with Assigned Protection Factors, which are based on actual measurements of levels of contaminant inside the facepiece during actual work.

**Quantum.** "Bundle of energy"; discrete particle of radiation. Pl. quanta.

**Quartz.** Vitreous, hard, chemically resistant, free silica, the most common form in nature. The main constituent in sandstone, igneous rocks, and common sands.

**Quenching.** A heat-treating operation in which metal raised to the desired temperature is quickly cooled by immersion in an oil bath.

## R

**Rabbit.** A capsule that carries samples in and out of an atomic reactor through a pneumatic tube in order to permit study of the effect of intense radiation on various materials.

**Rad.** Roentgen absorbed dose or radiation absorbed dose; a standard unit of absorbed ionizing radiation dose equal to 100 ergs absorbed per gram.

**Radial deviation.** Flexion of the hand that decreases the angle between its longitudinal axis and radius. Tool design should minimize radial deviation. Strength of grasp is diminished in radial deviation.

**Radian.** An arc of a circle equal in length to the radius.

**Radiant temperature.** The temperature resulting from a body absorbing radiant energy.

**Radiation (nuclear).** The emission of atomic particles or electromagnetic radiation from the nucleus of an atom.

**Radiation protection guide (RPG).** The radiation dose that should not be exceeded without careful consideration of the reasons for doing so; every effort should be made to encourage the maintenance of radiation doses as far below this guide as practicable.

**Radiation (radioactivity).** See Ionizing radiation.

**Radiation source.** An apparatus or material emitting or capable of emitting ionizing radiation.

**Radiation (thermal).** The transmission of energy by means of electromagnetic waves longer than visible light. Radiant energy of any wavelength may, when absorbed, become thermal energy and result in an increase in the temperature of the absorbing body.

**Radiator.** That which is capable of emitting energy in wave form.

**Radioactive.** The property of an isotope or element that is characterized by spontaneous decay to emit radiation.

**Radioactivity.** Emission of energy in the form of alpha-, beta-, or gamma-radiation from the nucleus of an atom. Always involves change of one kind of atom into a different kind. A few elements, such as radium, are naturally radioactive. Other radioactive forms are induced. See Radioisotope.

**Radioactivity concentration guide (RCG).** The concentration of radioactivity in the environment that is determined to result in organ doses equal to the radiation protection guide (RPG).

**Radiochemical.** Any compound or mixture containing a sufficient portion of radioactive elements to be detected by a Geiger counter.

**Radiochemistry.** The branch of chemistry concerned with the properties and behavior of radioactive materials.

**Radiodiagnosis.** A method of diagnosis that involves x-ray examination.

**Radiohumeral joint.** Part of the elbow. Not truly a joint, but a thrust bearing.

**Radioisotope.** A radioactive isotope of an element. A radioisotope can be produced by placing material in a nuclear reactor and bombarding it with neutrons. Many of the fission products are radioisotopes. Sometimes used as tracers, as energy sources for chemical processing or food pasteurization, or as heat sources for nuclear batteries. Radioisotopes are at present the most widely used outgrowth of atomic research and are one of the most important peacetime contributions of nuclear energy.

**Radionuclide.** A radioactive nuclide; one that has the capability of spontaneously emitting radiation.

**Radioresistant.** Relatively invulnerable to the effects of radiation.

**Radiosensitive.** Tissues that are more easily damaged by radiation.

**Radiotherapy.** Treatment of human ailments with the application of relatively high roentgen dosages.

**Radium.** One of the earliest-known naturally radioactive elements. It is far more radioactive than uranium and is found in the same ores.

**Radius.** The long bone of the forearm in line with the thumb; the active element in the forearm during pronation (inward rotation) and supination (outward rotation). Also provides the forearm connection to the wrist joint.

**Rale.** Any abnormal sound or noise in the lungs.

**Random noise.** A sound or electrical wave whose instantaneous amplitudes occur as a function of time, according to a normal (Gaussian) distribution curve. Random noise is an oscillation whose instantaneous magnitude is not specified for any given instant of time. The instantaneous magnitudes of a random noise are specified only by probability functions giving the fraction of the total time that the magnitude, or some sequence of the magnitudes, lies within a specific range.

**Rare earths.** Originally, the elements in the periodic table with atomic numbers 57 through 71. Often included are numbers 39 and, less often, 21 and 90. Emerging uses include the manufacture of special steels and glasses.

**Rash.** Abnormal reddish coloring or blotch on some part of the skin.

**Rated-line voltage.** The range of potentials, in volts, of the supply line specified by the manufacturer at which an x-ray machine is designed to operate.

**Rated output current.** The maximum allowable lead current of an x-ray high-voltage generator.

**Rated output voltage.** The allowable peak potential, in volts, at the output terminals of an x-ray high-voltage generator.

**Raynaud's syndrome or phenomenon.** Abnormal constriction of the blood vessels of the fingers on exposure to cold temperature.

**RBE.** Relative biological effectiveness; the relative effectiveness of the same absorbed dose of two ionizing radiations in producing a measurable biological response.

**Reactivity (chemical).** A substance's susceptibility to undergo a chemical reaction or change that may result in dangerous side effects, such as an explosion, burning, and corrosive or toxic emissions.

**Reactor.** An atomic "furnace" or nuclear reactor. In a reactor, nuclei of the fuel undergo controlled fission under the influence of neutrons. The fission produces new neutrons in a chain reaction that releases large amounts of energy. This energy is removed as heat that can be used to make steam. The moderator for the first reactor was piled-up blocks of graphite. Thus, a nuclear reactor was formerly referred to as a pile. Reactors are usually classified now as research, test, process heat, and power, depending on their principal function. No workable design for a controlled fusion reactor has yet been devised.

**Reagent.** Any substance used in a chemical reaction to produce, measure, examine, or detect another substance.

**REL.** Recommended exposure limit. An exposure limit, generally a time-weighted average, to a substance; developed by NIOSH based on toxicological and industrial hygiene data.

**Recoil energy.** The energy emitted and shared by the reaction products when a nucleus undergoes a nuclear reaction such as fission or radioactive decay.

**Reduction.** Addition of one or more electrons to an atom through chemical change.

**Refractories.** A material exceptionally resistant to the action of heat and hence used for lining furnaces; examples are fire clay, magnesite, graphite, and silica.

**Regenerative process.** Replacement of damaged cells by new cells.

**Regimen.** A regulation of the mode of living, diet, sleep, exercise, and so on for a hygienic or therapeutic purpose; sometimes mistakenly called regime.

**Reid method.** Method of determining the vapor pressure of a volatile hydrocarbon by the *Standard Method of Test for Vapor Pressure of Petroleum Products,* ASTM D323.

**Relative humidity.** The ratio of the quantity of water vapor present in the air to the quantity that would saturate it at any specific temperature.

**Reliability.** The degree to which an instrument, component, or system retains its performance characteristics over a period of time.

**Rem.** Roentgen equivalent man; a radiation dose unit that equals the dose in rads multiplied by the appropriate value of relative biological effect or Quality Factor for the particular radiation.

**Renal.** Having to do with the kidneys.

**Replication.** A fold or folding back; the act or process of duplicating or reproducing something.

**Resin.** A solid or semisolid amorphous (noncrystalline) organic compound or mixture of such compounds with no definite melting point and no tendency to crystallize. May be of vegetable (gum arabic), animal (shellac), or synthetic (celluloid) origin. Some resins may be molded, cast, or extruded. Others are used as adhesives, in the treatment of textiles and paper, or as protective coatings.

**Resistance.** (1) Opposition to the flow of air, as through a canister, cartridge, particulate filter, or orifice. (2) A property of conductors, depending on their dimensions, material, and temperature, that determines the current produced by a given difference in electrical potential.

**Resonance.** Each object or volume of air resonates or strengthens a sound at one or more particular frequencies. The frequency depends on the size and construction of the object or air volume.

**Respirable-size particulates.** Particulates in a size range that permits them to penetrate deep into the lungs upon inhalation.

**Respirator.** A device to protect the wearer from inhalation of harmful contaminants.

**Respiratory system.** Consists of the nose, mouth, nasal passages, nasal pharynx, pharynx, larynx, trachea, bronchi, bronchioles, air sacs (alveoli) of the lungs, and muscles of respiration.

**Reticle.** A scale or grid or other pattern located in the focus of the eyepiece of a microscope.

**Retina.** The light-sensitive inner surface of the eye that receives and transmits images formed by the lens.

**Retro-** (prefix). Backward or behind.

**Reverberatory furnace.** A furnace in which heat is supplied by burning fuel in a space between the charge and the low roof.

**Rheumatoid.** Resembling rheumatism, a disease marked by inflammation of the connective tissue structures of the body, especially the membranous linings of the joints, and by pain in these parts; eventually the joints become stiff and deformed.

**Rhin-, rhino-** (prefix). Pertaining to the nose.

**Rhinitis.** Inflammation of the mucous membrane lining in the nasal passages.

**Rickettsia.** Rod-shaped microorganisms characterized by growing within the cells of animals. These human pathogens are often carried by arthropods.

**Riser.** In metal casting, a channel in a mold to permit escape of gases.

**Roasting of ores.** A refining operation in which ore is heated to a high temperature, sometimes with catalytic agents, to drive off certain impurities; an example is the roasting of copper ore to remove sulfur.

**Roentgen (R).** A unit of radioactive dose or exposure. See Rad.

**Roentgenogram.** A film produced by exposing x-ray film to x rays.

**Roentgenography.** Photography by means of roentgen rays. Special techniques for roentgenography of different areas of the body have been given specific names.

**Route of entry.** A path by which chemicals can enter the body. There are three main routes of entry: inhalation, ingestion, and skin absorption.

**Rosin.** Specifically applies to the resin of the pine tree and chiefly derives from the manufacture of turpentine. Widely used in the manufacture of soap and flux.

**Rotameter.** A flow meter consisting of a precision-bored, tapered, transparent tube with a solid float inside.

**Rotary kiln.** Any of several types of kilns used to heat material, as in the portland cement industry.

**Rouge.** A finely powdered form of iron oxide used as a polishing agent.

**RTECS.** Registry of Toxic Effects of Chemical Substances.

## S

**SAE.** Sampling and analytical error. The reason a particular sampling result may vary from the true value. Quantitative estimates of SAE are often used to develop a clear picture of the potential range of a given exposure.

**Safety can.** An approved container of not more than 5 gal (19 l) capacity having a spring-closing lid and spout cover and designed to safely relieve internal pressure when subjected to fire exposure.

**Sagittal plane.** A plane from back to front vertically dividing the body into the right and left portions. Important in anthropometric definitions. Midsagittal plane is a sagittal plane symmetrically dividing the body.

**Salamander.** A small furnace, usually cylindrical in shape, without grates.

**Salivation.** An excessive discharge of saliva; ptyalism.

**Salmonella.** A genus of gram-negative, rod-shaped pathogenic bacteria.

**Salt.** A product of the reaction between an acid and a base. Table salt, for example, is a compound of sodium and chlorine. It can be made by reacting sodium hydroxide with hydrochloric acid.

**Sampling.** The withdrawal or isolation of a fractional part of a whole. In air analysis, the separation of a portion of an ambient atmosphere with subsequent analysis to determine concentration.

**Sandblasting.** A process for cleaning metal castings and other surfaces with sand by a high-pressure airstream.

**Sandhog.** Any worker doing tunneling work requiring atmospheric pressure control.

**Sanitize.** To reduce the microbial flora in or on articles such as eating utensils to levels judged safe by public health authorities.

**Saprophyte.** An organism living on dead organic matter.

**Sarcoma.** Malignant tumors that arise in connective tissue.

**Scattered radiation.** Radiation that is scattered by interaction with objects or within tissue.

**Scintillation counter.** A device for counting atomic particles by means of the tiny flashes of light (scintillations) that particles produce when they strike certain crystals or liquids.

**Scler-** (prefix). Hard, tough.

**Sclera.** The tough white outer coat of the eyeball.

**Scleroderma.** Hardening of the skin.

**Sebum.** Oily lubricating secretion of the sebaceous glands.

**Scotoma.** A blind or partially blind area in the visual field.

**Sealed source.** A radioactive source sealed in a container or having a bonded cover, in which the container or cover has sufficient mechanical strength to prevent contact with and dispersion of the radioactive material.

**Sebaceous.** Of, related to, or being fatty material.

**Seborrhea.** An oily skin condition caused by an excess output of sebum from the sebaceous glands of the skin.

**SCBA.** Self-contained breathing apparatus.

**Semicircular canals.** The special organs of balance closely associated with the hearing mechanism and the eighth cranial nerve.

**Semiconductor or junction laser.** A class of laser that normally produces relatively low CW power outputs; can be tuned in wavelength and has the greatest efficiency.

**Sensation.** The translation into consciousness of the effects of a stimulus exciting a sense organ.

**Sensible.** Capable of being perceived by the sense organs.

**Sensitivity.** The minimum amount of contaminant that can repeatedly be detected by an instrument.

**Sensitization.** The process of rendering an individual sensitive to the action of a chemical.

**Sensitizer.** A material that can cause an allergic reaction of the skin or respiratory system.

**Sensorineural.** Type of hearing loss that affects millions of people. If the inner ear is damaged, the hearing loss is sensory; if the fibers of the eighth nerve are affected, it is a neural hearing loss. Because the pattern of hearing loss is the same in either case, the term *sensorineural* is used.

**Sensory end organs.** Receptor organs of the sensory nerves located in the skin. Each end organ can sense only a specific type of stimulus. Primary stimuli are heat, cold, or pressure, each requiring different end organs.

**Sensory feedback.** Use of external signals perceived by sense organs to indicate quality or level of performance of an event triggered by voluntary action. On the basis of sensory feedback information, decisions may be made; for instance, permitting or not permitting an event to run its course or enhancing or decreasing activity levels.

**Septum.** A dividing wall or partition; used as a general term in anatomical nomenclature.

**Septicemia.** Blood poisoning; growth of infectious organisms in the blood.

**Sequestrants.** Chelates used to deactivate undesirable properties of metal ions without removing these ions from solution. Sequestrants have many uses, including application as antigumming agents in gasoline, antioxidants in rubber, and rancidity retardants in edible fats and oils.

**Serum.** (1) The clear fluid that separates from the blood during clotting. (2) Blood serum–containing antibodies.

**Shakeout.** In the foundry industry, the separation of the solid—but still not cold—casting from its molding sand.

**Shale.** Many meanings in industry, but in geology, a common fossil rock formed from clay, mud, or silt; somewhat stratified but without characteristic cleavage.

**Shale oil.** Tarry oil distilled from bituminous shale.

**Shaver's disease.** Bauxite pneumoconiosis.

**Shell.** The electrons around the nucleus of an atom are arranged in shells—spheres centered on the nucleus. The innermost shell is called K-shell, the next is called the L-shell, and so on to the Q-shell. The nucleus itself may also have a shell-type structure.

**Shield, shielding.** Interposed material (such as a wall) that protects workers from harmful radiations released by radioactive materials.

**Shielded-metal arc welding (SMAW).** An arc-welding process that produces coalescence of metals by heating them with an arc between a covered metal electrode and the work. Shielding is obtained from decomposition of the electrode covering. Pressure is not used and filler metal is obtained from the electrode.

**Shock.** Primarily, the rapid fall in blood pressure following injury, operation, or the administration of anesthesia.

**Shotblasting.** A process for cleaning metal castings or other surfaces by small steel shot in a high-pressure airstream; a substitute for sandblasting to avoid silicosis.

**SI.** The *Système International d'Unités* (International System of Units), the metric system that is being adopted throughout the world. It is a modern version of the MKSA (meter, kilogram, second, ampere) system, whose details are published and controlled by an international treaty organization financed by member states of the Metre Convention, including the United States.

**Siderosis.** The deposition of iron pigments in the lung—can be associated with disease.

**Sievert.** Unit of absorbed radiation dose in Gray times the Quality Factor of the radiation in comparison to gamma-radiation. A Sievert equals 100 rem.

**Silica gel.** A regenerative absorbent consisting of amorphous silica manufactured by the action of HCl on sodium silicate. Hard, glossy, quartz-like in appearance. Used in dehydrating and drying and as a catalyst carrier.

**Silicates.** Compounds of silicon, oxygen, and one or more metals with or without hydrogen. These dusts cause nonspecific dust reactions, but generally do not interfere with pulmonary function or result in disability.

**Silicon.** A nonmetallic element being, next to oxygen, the chief elementary constituent of the earth's crust.

**Silicones.** Unique group of compounds made by molecular combination of silicon (or certain silicon compounds) with organic chemicals. Produced in a variety of forms, including silicone fluids, resins, and rubber. Silicones have special properties, such as water repellency, wide temperature resistance, high durability, and great dielectric strength.

**Silicosis.** A disease of the lungs caused by the inhalation of silica dust.

**Silver solder.** A solder of varying components but usually containing an appreciable amount of cadmium.

**Simple tone (pure tone).** (1) A sound wave whose instantaneous sound pressure is a simple sinusoidal function of time. (2) A sound sensation characterized by its singularity of pitch.

**Sintering.** Process of making coherent powder of earthy substances by heating without melting.

**Skin dose.** A special instance of tissue dose referring to the dose immediately on the surface of the skin.

**Slag.** The dross of flux and impurities that rise to the surface of molten metal during melting and refining.

**Slot velocity.** Linear flow rate through the opening in a slot-type hood (plating, degreasing operations, and so on).

**Short-term exposure limit (STEL).** ACGIH-recommended exposure limit. Maximum concentration to which workers can be exposed for a short period of time (15 min) only four times throughout the day with at least 1 h between exposures.

**Sludge.** Any muddy or slushy mass. Specifically, mud from a drill hole in boring, muddy sediment in the steam boiler, or precipitated solid matter arising from sewage treatment processes.

**Slug.** A fuel element for a nuclear reactor; a piece of fissionable material. Slugs in large reactors consist of uranium coated with aluminum to prevent corrosion.

**Slurry.** A thick, creamy liquid resulting from the mixing and grinding of limestone, clay, water, and other raw materials.

**SMACNA.** Sheet Metal and Air Conditioning National Association.

**Smog.** Irritating haze resulting from the sun's effect on certain pollutants in the air, notably automobile and industrial exhaust.

**Smoke.** An air suspension (aerosol) of particles originating from combustion or sublimation; generally contains droplets as well as dry particles. Tobacco, for instance, produces a wet smoke composed of minute tarry droplets.

**Soap.** Ordinarily a metal salt of a fatty acid, usually sodium stearate, sodium oleate, sodium palmitate, or some combination of these.

**Soapstone.** Complex silicate of varied composition, similar to some talcs, with wide industrial application, including rubber manufacture.

**Solder.** A material used for joining metal surfaces together by filling a joint or covering a junction. The most commonly used solder contains lead and tin; silver solder may contain cadmium. Zinc chloride and fluorides are commonly used as fluxes to clean the soldered surfaces.

**Solid-state laser.** A type of laser that uses a solid crystal such as ruby or glass; commonly used in pulsed lasers.

**Solution.** Mixture in which the components lose their individual properties and are uniformly dispersed. All solutions are composed of a solvent (water or other fluid) and a solute (the dissolved substance). A true solution is homogeneous, as salt in water.

**Solvent.** A substance that dissolves another substance. Usually refers to organic solvents.

**Soma.** Body, as distinct from psyche (mind).

**Somatic.** Pertaining to all tissue other than reproductive cells.

**Somnolence.** Sleepiness; also unnatural drowsiness.

**Soot.** Agglomerations of carbon particles impregnated with tar; formed in the incomplete combustion of carbonaceous material.

**Sorbent.** (1) A material that removes toxic gases and vapors from air inhaled through a canister or cartridge. (2) Material used to collect gases and vapors during air-sampling.

**Sound.** An oscillation in pressure, stress, particle displacement, particle velocity, and so on, propagated in an elastic material, in a medium with internal forces (elastic or viscous, for example); or, the superposition of such propagated oscillations. Also the sensation produced through the organs of hearing usually by vibrations transmitted in a material medium, commonly air.

**Sound absorption.** The change of sound energy into some other form, usually heat, on passing through a medium or striking a surface. Also, the property possessed by materials and objects, including air, of absorbing sound energy.

**Sound absorption coefficient.** The ratio of the sound energy absorbed by the surface of a medium (or material) exposed to a sound field (or to sound radiation) to the sound energy incident on that surface.

**Sound analyzer.** A device for measuring the band-pressure level or pressure-spectrum level of a sound as a function of frequency.

**Sound level.** A weighted sound-pressure level obtained by the use of metering characteristics and the weighting A, B, or C specified in ANSI S1.4.

**Sound-level meter and octave-band analyzer.** Instruments for measuring sound-pressure levels in decibels referenced to 0.0002 microbars. Readings can also be made in specific octave bands, usually beginning at 75 Hz and continuing through 10,000 Hz.

**Sound-pressure level, SPL.** The level, in decibels, of a sound is 20 times the logarithm to the base 10 of the ratio of the pressure of this sound to the reference pressure, which must be explicitly stated.

**Sound transmission.** The word *sound* usually means sound waves traveling in air. However, sound waves also travel in solids and liquids. These sound waves may be transmitted to air to make sound we can hear.

**Sound transmission loss.** A barrier's ability to block transmission; measured in decibels.

**Sour gas.** Slang for either natural gas or a gasoline contaminated with odor-causing sulfur compounds. In natural gas, the contaminant is usually hydrogen sulfide; in gasoline, usually mercaptans.

**Source.** Any substance that emits radiation. Usually refers to a piece of radioactive material conveniently packaged for scientific or industrial use.

**Spasm.** Tightening or contraction of any set of muscles.

**Specific Absorption (SA).** Quantity of radiofrequency energy in joules per kilogram.

**Specific Absorption Rate (SAR).** Radiofrequency dosage term expressed as watts of power per kilogram of tissue.

**Specific gravity.** The ratio of the mass of a unit volume of a substance to the mass of the same volume of a standard substance at a standard temperature. Water at 39.2 F (4 C) is usually the standard for liquids; for gases, dry air (at the same temperature and pressure as the gas) is often taken as the standard substance. See Density.

**Specific ionization.** See Ionization.

**Specific volume.** The volume occupied by a unit mass of a substance under specified conditions of temperature and pressure.

**Specific weight.** The weight per unit volume of a substance; same as density.

**Specificity.** The degree to which an instrument or detection method is capable of accurately detecting or measuring the concentration of a single contaminant in the presence of other contaminants.

**Spectrography—spectral emission.** An instrumental method for detecting trace contaminants using a spectrum formed by exciting the subject contaminants by various means, causing characteristic radiation to be formed, which is dispersed by a grating or prism and photographed.

**Spectrophotometer.** A direct-reading instrument used for comparing the relative intensities of corresponding electromagnetic wavelengths produced by absorption of ultraviolet, visible, or infrared radiation from a vapor or gas.

**Spectroscopy.** Observation of the wavelength and intensity of light or other electromagnetic waves absorbed or emitted by various materials. When excited by an arc or spark, each element emits light of certain well-defined wavelengths.

**Spectrum.** The frequency distribution of the magnitudes (and sometimes phases) of the components of the wave. Also used to signify a continuous range of frequencies, usually wide in extent, within which waves have some specified common characteristics. Also, the pattern of red-to-blue light observed when a beam of sunlight passes through a prism and then projects upon a surface.

**Speech interference level (SIL).** The average, in decibels, of the sound-pressure levels of a noise in the three octave bands of frequency: 600–1,200, 1,200–2,400, and 2,400–4,800 Hz.

**Speech perception test.** A measurement of hearing acuity by the administration of a carefully controlled list of words. The identification of correct responses is evaluated in terms of norms established by the average performance of normal listeners.

**Speech reading.** Lip reading or visual hearing.

**Sphincter.** A muscle that surrounds an orifice and functions to close it.

**Sphygmomanometer.** Apparatus for measuring blood pressure (and a good word for testing spelling ability).

**Spore.** A resistant body formed by certain microorganisms; resistant resting cells. Mold spores: unicellular reproductive bodies.

**Spot size.** Cross-sectional area of laser beam at the target.

**Spot welding.** One form of electrical-resistance welding in which the current and pressure are restricted to the spots of metal surfaces directly in contact.

**Spray coating painting.** The result of the application of a spray in painting as a substitute for brush painting or dipping.

**Squamous.** Covered with or consisting of scales.

**Stain.** A dye used to color microorganisms as an aid to visual inspection.

**Stamping.** A term with many different usages in industry; a common one is the crushing of ores by pulverizing.

**Standard air.** Air at standard temperature and pressure. The most common values are 70 F (21.1 C) and 29.92 in. Hg (101.3 kPa). Also, air with a density of 0.075 lb/ft$^3$ (1.2 kg/m$^3$) is substantially equivalent to dry air at 70 F and 29.92 in. Hg.

**Standard air density.** The density of air—0.075 lb/ft$^3$ (1.2 kg/m$^3$), at standard conditions.

**Standard conditions.** In industrial ventilation, 70 F (21.1 C), 50 percent relative humidity, and 29.92 in. Hg (101.3 kPa) atmospheric pressure.

**Standard gravity.** Standard accepted value for the force of gravity. It is equal to the force that produces an acceleration of 32.17 ft/s (9.8 m/s).

**Standard Industrial Classification (SIC) Code.** Classification system for places of employment according to major type of activity.

**Standard temperature and pressure.** See Standard air.

**Standing wave.** A periodic wave having a fixed distribution in space that is the result of interference of progressive waves of the same frequency and kind. Such waves are characterized by the existence of nodes or partial nodes and antinodes that are fixed in space.

**Stannosis.** A form of pneumoconiosis caused by the inhalation of tin-bearing dusts.

**Static pressure.** The potential pressure exerted in all directions by a fluid at rest. For a fluid in motion, it is measured in a direction normal (at right angles) to the direction of flow; thus it shows the tendency to burst or collapse the pipe. When added to velocity pressure, it gives total pressure.

**Static pressure curve.** A graphical representation of the volumetric output and fan static pressure relationship for a fan operating at a specific rotating speed.

**Static pressure regain.** The increase in static pressure in a system as air velocity decreases and velocity pressure is converted into static pressure according to Bernoulli's theorem.

**Sterile.** Free of living microorganisms.

**Sterility.** Inability to reproduce.

**Sterilization.** The process of making sterile; the killing of all forms of life.

**Sterilize.** To perform any act that results in the absence of all life on or in an object.

**Sternomastoid muscles.** A pair of muscles connecting the breastbone and lower skull behind the ears, which flex or rotate the head.

**Stink damp.** In mining, hydrogen sulfide.

**Stp flow rate.** The rate of flow of fluid, by volume, corrected to standard temperature and pressure.

**Stp volume.** The volume that a quantity of gas or air would occupy at standard temperature and pressure.

**Stress.** A physical, chemical, or emotional factor that causes bodily or mental tension and may be a factor in disease causation or fatigue.

**Stressor.** Any agent or thing causing a condition of stress.

**Strip mine.** A mine in which coal or ore is extracted from the earth's surface after removal of overlayers of soil, clay, and rock.

**Stupor.** Partial unconsciousness or nearly complete unconsciousness.

**Sublimation.** A process in which a material passes directly from a solid to a gaseous state and condenses to form solid crystals, without liquefying.

**Sulcus (pl. sulci).** A groove, trench, or furrow; used in anatomical nomenclature as a general term to designate such a depression, especially on the surface of the brain, separating the gyri; also, a linear depression in the surface of a tooth, the sloping sides of which meet at an angle.

**Supination.** Rotation of the forearm about its own longitudinal axis. Supination turns the palm upward when the forearm is horizontal, and forward when the body is in anatomical position. Supination is an important element of available motions inventory for industrial application, particularly where tools such as screwdrivers are used. Efficiency in supination depends on arm position. Workplace design should provide for elbow flexion at 90 degrees.

**Supra-** (prefix). Above, on.

**Surface-active agent; surfactant.** Any of a group of compounds added to a liquid to modify surface or interfacial tension. In synthetic detergents, which is the best known use of surface-active agents, reduction of interfacial tension provides cleansing action.

**Surface coating.** Paint, lacquer, varnish, or other chemical composition used for protecting and/or decorating surfaces. See Protective coating.

**Suspect carcinogen.** A material believed to be capable of causing cancer, based on limited scientific evidence.

**Sweating.** (1) Visible perspiration. (2) The process of uniting metal parts by heating solder so that it runs between the parts.

**Swing grinder.** A large power-driven grinding wheel mounted on a counterbalanced swivel-supported arm guided by two handles.

**Symptom.** Any bit of evidence from a patient indicating illness; the subjective feelings of the patient.

**Syncope.** Fainting spell.

**Syndrome.** A collection, constellation, or concurrence of signs and symptoms, usually of disease.

**Synergism.** Cooperative action of substances whose total effect is greater than the sum of their separate effects.

**Synergistic.** Pertaining to an action of two or more substances, organs, or organisms to achieve an effect greater than the additive effects of the separate elements.

**Synonym.** Another name by which a chemical may be known.

**Synthesis.** The reaction or series of reactions by which a complex compound is obtained from simpler compounds or elements.

**Synthetic.** (From the Greek word *synthetikos,* "that which is put together.") "Man-made 'synthetic' should not be thought of as a substitute for the natural," according to *Encyclopedia of the Chemical Process Industries,* which adds: "Synthetic chemicals are frequently more pure and uniform than those obtained naturally." A classic example is synthetic indigo.

**Synthetic detergents.** Chemically tailored cleaning agents soluble in water or other solvents. Originally developed as soap substitutes; because they do not form insoluble precipitates, they are especially valuable in hard water. They may be composed of surface-active agents alone, but generally are combinations of surface-active agents and other substances, such as complex phosphates, to enhance detergency.

**Synthetic rubber.** Artificial polymer with rubber-like properties. Types have varying composition and properties. Major types are designated as S-type, butyl,

neoprene (chloroprene polymers), and N-type. Several synthetics duplicate the chemical structure of natural rubber.

**Systemic.** Spread throughout the body; affecting all body systems and organs, not localized in one spot or area.

## T

**Tachy-** (prefix). Indicates fast or speedy, as in *tachycardia,* abnormally rapid heartbeat.

**Tailings.** In mining or metal recovery processes, the gangue rock residue after all or most of the metal has been extracted.

**Talc.** A hydrous magnesium silicate used in ceramics, cosmetics, paint, and pharmaceuticals, and as a filler in soap, putty, and plaster.

**Tall oil.** (Derived from the Swedish word *tallolja;* a material first investigated in Sweden—not synonymous with U.S. pine oil.) Natural mixture of rosin acids, fatty acids, sterols, high-molecular weight alcohols, and other materials, derived primarily from waste liquors of sulfate wood pulp manufacture. Dark brown, viscous, oily liquid often called liquid rosin.

**Tar.** A loose term embracing wood, coal, or petroleum exudations. In general represents complex mixture of chemicals of top fractional distillation systems.

**Tar crude.** Organic raw material derived from distillation of coal tar and used for chemicals.

**Tare.** A deduction of weight, made in allowance for the weight of a container or medium. The initial weight of a filter, for example.

**Target.** The material into which the laser beam is fired or at which electrons are fired in an x-ray tube.

**Temper.** To relieve the internal stresses in metal or glass and to increase ductility by heating the material to a point below its critical temperature and cooling slowly. See Anneal.

**Temperature.** The condition of a body that determines the transfer of heat to or from other bodies. Specifically, it is a manifestation of the average translational kinetic energy of the molecules of a substance caused by heat agitation. See Celsius and Kelvin scale.

**Temperature, dry-bulb.** The temperature of a gas or mixture of gases indicated by an accurate thermometer after correction for radiation.

**Temperature, effective.** An arbitrary index that combines into a single value the effect of temperature, humidity, and air movement on the sensation of warmth or cold felt by the human body. The numerical value is the temperature of still, saturated air that would induce an identical sensation.

**Temperature, mean radiant (MRT).** The temperature of a uniform black enclosure in which a solid body or occupant would exchange the same amount of radiant heat as in the existing nonuniform environment.

**Temperature, wet-bulb.** Thermodynamic wet-bulb temperature is the temperature at which liquid or solid water, by evaporating into air, can bring the air to saturation adiabatically at the same temperature. Wet-bulb temperature (without qualification) is the temperature indicated by a wet-bulb psychrometer.

**Tempering.** The process of heating or cooling makeup air to the proper temperature.

**Temporary threshold shift (TTS).** The hearing loss suffered as the result of noise exposure, all or part of which is recovered during an arbitrary period of time when one is removed from the noise. It accounts for the necessity of checking hearing acuity at least 16 hours after a noise exposure.

**Tendon.** Fibrous component of a muscle. It often attaches to bone at the area of application of tensile force. When its cross section is small, stresses in the tendon are high, particularly because the total force of many muscle fibers is applied at the single terminal tendon. See Tenosynovitis.

**Tennis elbow.** Sometimes called lateral epicondylitis, an inflammatory reaction of tissues in the lateral elbow region.

**Tenosynovitis.** Inflammation of the connective tissue sheath of a tendon.

**Teratogen.** An agent or substance that may cause physical defects in the developing embryo or fetus when a pregnant female is exposed to that substance.

**Terminal velocity.** The terminal rate of fall of a particle through a fluid as induced by gravity or other external force; the rate at which frictional drag balances the accelerating force (or the external force).

**Tetanus.** A disease of sudden onset caused by the toxin of the bacterium called *Clostridium tetani*. It is characterized by muscle spasms. Also called lockjaw.

**Therm.** A quantity of heat equivalent to 100,000 Btu.

**Thermal pollution.** Discharge of heat into bodies of water to the point that the increased warmth activates all sewage, depletes the oxygen the water must have to cleanse itself, and eventually destroys some of the fish and other organisms in the water. Eventually, thermal pollution makes the water smell and taste bad.

**Thermonuclear reaction.** A fusion reaction, that is, a reaction in which two light nuclei combine to form a heavier atom, releasing a large amount of energy. This is believed to be the sun's source of energy. It is called thermonuclear because it occurs only at a very high temperature.

**Thermoplastic.** Capable of being repeatedly softened by heat.

**Thermoplastic plastics.** Plastics that can repeatedly melt or that soften with heat and harden on cooling. Examples: vinyls, acrylics, and polyethylene.

**Thermosetting.** Capable of undergoing a chemical change from a soft to a hardened substance when heated.

**Thermosetting plastics.** Plastics that are heat-set in their final processing to a permanently hard state. Examples are phenolics, ureas, and melamines.

**Thermostable.** Resistant to changes by heat.

**Thinner.** A liquid used to increase the fluidity of paints, varnishes, and shellac.

**Threshold.** The level where the first effects occur; also, the point at which a person begins to notice a tone becoming audible.

**Thromb-** (prefix). Pertaining to a blood clot.

**Timbre.** The quality given to a sound by its overtones; the tone distinctive of a singing voice or a musical instrument. Pronounced "TAMbra" or "TIMber."

**Time-weighted average concentration (TWA).** Refers to concentrations of airborne toxic materials weighted for a certain time duration, usually 8 hours.

**Tinning.** Any work with tin such as tin roofing; in particular, in soldering, the primary coating with solder of the two surfaces to be united.

**Tinnitus.** A perception of sound arising in the head. Most often perceived as a ringing or hissing sound in the ears. Can be the result of high frequency hearing loss.

**Tissue.** A large group of similar cells bound together to form a structural component. An organ is composed of several kinds of tissue, and in this respect it differs from a tissue as a machine differs from its parts.

**TLV.** Threshold Limit Value. A time-weighted average concentration under which most people can work consistently for 8 hours a day, day after day, with no harmful effects. A table of these values and accompanying precautions is published annually by the American Conference of Governmental Industrial Hygienists. See Appendix B.

**Tolerance.** (1) The ability of the living organism to resist the usually anticipated stress. (2) The limits of permissible inaccuracy in the fabrication of an article above and below its design specifications.

**Tolerance dose.** See Maximum permissible concentration and MPL.

**Toluene, $C_6H_5CH_3$.** Hydrocarbon derived mainly from petroleum but also from coal. Source of TNT, lacquers, saccharin, and many other chemicals.

**Tone deafness.** The inability to discriminate between fundamental tones close together in pitch.

**Topography.** Configuration of a surface, including its relief and the position of its natural and man-made features.

**Total pressure.** The algebraic sum of the velocity pressure and the static pressure (with due regard to sign).

**Toxemia.** Poisoning by the way of the bloodstream.

**Toxicant.** A poison or poisonous agent.

**Toxin.** A poisonous substance derived from an organism.

**Tracer.** A radioisotope mixed with a stable material. The radioisotope enables scientists to trace the material as it undergoes chemical and physical changes. Tracers are used widely in science, industry, and agriculture today. When radioactive phosphorus, for example, is mixed with a chemical fertilizer, the radioactive substance can be traced through the plant as it grows.

**Trachea.** The windpipe, or tube that conducts air to and from the lungs. It extends between the larynx above and the point where it divides into two bronchi below.

**Trade name.** The commercial name or trademark by which a chemical is known. One chemical may have a variety of trade names depending on the manufacturing or distributors involved.

**Transducer.** Any device or element that converts an input signal into an output signal of a different form; examples include the microphone, phonograph pickup, loudspeaker, barometer, photoelectric cell, automobile horn, doorbell, and underwater sound transducer.

**Transmission loss.** The ratio, expressed in decibels, of the sound energy incident on a structure to the sound energy that is transmitted. The term is applied both to building structures (walls, floors, etc.) and to air passages (muffler, ducts, etc.).

**Transmutation.** Any nuclear process that involves a change in energy or identity of the nucleus.

**Transport (conveying) velocity.** Minimum air velocity required to move the suspended particulates in the airstream.

**Trauma.** An injury or wound brought about by an outside force.

**Tremor.** Involuntary shaking, trembling, or quivering.

**Triceps.** The large muscle at the back of the upper arm that extends the forearm when contracted.

**Tridymite.** Vitreous, colorless form of free silica formed when quartz is heated to 1,598 F (870 C).

**Trigger finger.** Also known as snapping finger, a condition of partial obstruction in flexion or extension of a finger. Once past the point of obstruction, movement is eased. Caused by constriction of the tendon sheath.

**Tripoli.** Rottenstone. A porous, siliceous rock resulting from the decomposition of chert or siliceous limestone. Used as a base in soap and scouring powders, in metal polishing, as a filtering agent, and in wood and paint fillers. A cryptocrystalline form of free silica.

**Tritium.** Often called hydrogen-3, extra-heavy hydrogen whose nucleus contains two neutrons and one proton. It is three times as heavy as ordinary hydrogen and is radioactive.

**Tuberculosis.** A contagious disease caused by infection with the bacterium *Mycobacterium tuberculosis*. It usually affects the lung, but bone, lymph glands, and other tissues may be affected.

**Tularemia.** A bacterial infection of wild rodents, such as rabbits. It may be generalized or localized in the eyes, skin, lymph nodes, or respiratory tract. It can be transmitted to humans.

**Tumbling.** An industrial process, as in founding, in which small castings are cleaned by friction in a revolving drum (tumbling mill, tumbling barrel), which may contain sand, sawdust, stone, etc.

**Turbid.** Cloudy.

**Turbidity.** Cloudiness; disturbances of solids (sediment) in a solution, so that it is not clear.

**Turbinates.** A series of scroll-like bones in the nasal cavity that serves to increase the amount of tissue surface exposed in the nose, permitting incoming air to be moistened and warmed prior to reaching the lungs. Also called conchae.

**Turbulence loss.** The pressure or energy lost from a ventilation system through air turbulence.

**Turning vanes.** Curved pieces added to elbows or fan inlet boxes to direct air and so reduce turbulence losses.

**TWA.** Time-weighted average.

**Tympanic cavity.** Another name for the chamber of the middle ear.

## U

**UCL.** Upper confidence limit.

**Ulcer.** The destruction of an area of skin or mucous membrane.

**Ulceration.** The formation or development of an ulcer.

**Ulna.** One of the two bones of the forearm. It forms the hinge joint at the elbow and does not rotate about its longitudinal axis. It terminates at the wrist on the same side as the little finger. Task design should not impose thrust loads through the ulna.

**Ulnar deviation.** A position of the hand in which the angle on the little finger side of the hand with the corresponding side of the forearm is decreased. Ulnar deviation is a poor working position for the hand and may cause nerve and tendon damage.

**Ultrasonics.** The technology of sound at frequencies above the audio range.

**Ultraviolet.** Wavelengths of the electromagnetic spectrum that are shorter than those of visible light and longer than x rays, $10^{-5}$ cm to $10^{-6}$ cm wavelength.

**Unstable.** Refers to all radioactive elements, because they emit particles and decay to form other elements.

**Unstable (reactive) liquid.** A liquid that in the pure state or as commercially produced or transported, vigorously polymerizes, decomposes, condenses, or becomes self-reactive under conditions of shocks, pressure, or temperature.

**Upper confidence limit (UCL).** In sampling analysis, a statistical procedure used to estimate the likelihood that a particular value is above the obtained value.

**Upper explosive limit (UEL).** The highest concentration (expressed as the percentage of vapor or gas in the air by volume) of a substance that will burn or explode when an ignition source is present.

**Uranium.** A heavy metal. The two principal isotopes of natural uranium are U-235 and U-238. U-235 has the only readily fissionable nucleus, which occurs in appreciable quantities in nature—hence its importance as nuclear fuel. Only one part in 140 of natural uranium is U-235. Highly toxic and a radiation hazard that requires special consideration.

**Urethr-, urethro-** (prefix). Relating to the urethra, the canal leading from the bladder for discharge of urine.

**Urticaria.** Hives.

**USC.** United States Code. The official compilation of federal statutes. New editions are issued approximately every 6 years. Cumulative supplements are issued annually.

## V

**Vaccine.** A suspension of disease-producing microorganisms modified by killing or attenuation so that it does not cause disease and can facilitate the formation of antibodies upon inoculation into humans or animals.

**Valence.** A number indicating the capacity of an atom and certain groups of atoms to hold others in combination. The term also is used in more complex senses.

**Valve (air oxygen).** A device that controls the direction of air or fluid flow or the rate and pressure at which air or fluid is delivered, or both.

**Vapor pressure.** Pressure (measured in pounds per square inch absolute-psia) exerted by a vapor. If a vapor is kept in confinement over its liquid so that the vapor can accumulate above the liquid (the temperature being held constant), the vapor pressure approaches a fixed limit called the maximum (or saturated) vapor pressure, dependent only on the temperature and the liquid.

**Vapors.** The gaseous form of substances that are normally in the solid or liquid state (at room temperature and pressure). The vapor can be changed back to the solid or liquid state either by increasing the pressure or decreasing the temperature alone. Vapors also diffuse. Evaporation is the process by which a liquid is changed to the vapor state and mixed with the surrounding air. Solvents with low boiling points volatilize readily.

**Vapor volume.** The number of cubic feet of pure solvent vapor formed by the evaporation of one gallon of liquid at 75 F (24 C).

**Vasoconstriction.** Decrease in the cross-sectional area of blood vessels. This may result from contraction of a muscle layer within the walls of the vessels or may be the result of mechanical pressure. Reduction in blood flow results.

**Vat dyes.** Water-insoluble, complex coal tar dyes that can be chemically reduced in a heated solution to a soluble form that can impregnate fibers. Subsequent oxidation then produces insoluble color dyestuffs that are remarkably fast to washing, light, and chemicals.

**Vector.** (1) Term applied to an insect or any living carrier that transports a pathogenic microorganism from the sick to the well, inoculating the latter; the organism may or may not pass through any developmental cycle. (2) Any quantity (for example, velocity, mechanical force, electromotive force) having magnitude, direction, and sense that can be represented by a straight line of appropriate length and direction.

**Velocity.** A vector that specifies the time rate of change of displacement with respect to a reference.

**Velocity, capture.** The air velocity required to draw contaminants into the hood.

**Velocity, face.** The inward air velocity in the plane of openings into an enclosure.

**Velocity pressure.** The kinetic pressure in the direction of flow necessary to cause a fluid at rest to flow at a given velocity. When added to static pressure, it gives total pressure.

**Velometer.** A device for measuring air velocity.

**Vena contracta.** The reduction in the diameter of a flowing airstream at hood entries and other locations.

**Veni-, veno-** (prefix). Relating to the veins.

**Ventilation.** One of the principal methods to control health hazards, may be defined as causing fresh air to circulate to replace foul air simultaneously removed.

**Ventilation, dilution.** Airflow designed to dilute contaminants to acceptable levels. Also called general ventilation.

**Ventilation, local exhaust.** Ventilation near the point of generation of a contaminant.

**Ventilation, mechanical.** Air movement caused by a fan or other air-moving device.

**Ventilation, natural.** Air movement caused by wind, temperature difference, or other nonmechanical factors.

**Vermiculite.** An expanded mica (hydrated magnesium-aluminum-iron silicate) used in lightweight aggregates, insulation, fertilizer, and soil conditioners; as a filler in rubber and paints; and as a catalyst carrier.

**Vertigo.** Dizziness; more exactly, the sensation that the environment is revolving around you.

**Vesicant.** Anything that produces blisters on the skin.

**Vesicle.** A small blister on the skin.

**Vestibular.** Relating to the cavity at the entrance to the semicircular canals of the inner ears.

**Viable.** Living.

**Vibration.** An oscillation motion about an equilibrium position produced by a disturbing force.

**Vinyl.** A general term applied to a class of resins such as polyvinyl chloride, acetate, butyryl, etc.

**Virulence.** The capacity of a microorganism to produce disease.

**Virulent.** Extremely poisonous or venomous; capable of overcoming bodily defensive mechanisms.

**Viruses.** A group of pathogens consisting mostly of nucleic acids and lacking cellular structure.

**Viscera.** Internal organs of the abdomen.

**Viscose.** Term applied to viscous liquid composed of cellulose xanthate.

**Viscose rayon.** The type of rayon produced from the reaction of carbon disulfide with cellulose and the hardening of the resulting viscous fluid by passing it through dilute sulfuric acid, this final operation causing the evolution of hydrogen sulfide gas.

**Viscosity.** The property of a fluid that resists internal flow by releasing counteracting forces.

**Viscosity, absolute.** A measure of a fluid's tendency to resist flow, without regard to density. The product of a fluid's kinematic viscosity times its density, expressed in dyne-seconds per centimeter or poises (or pascal-seconds).

**Viscosity, kinematic.** The relative tendency of a fluid to resist flow. The value of the kinematic viscosity is equal to the absolute viscosity of the fluid divided by the fluid density and is expressed in units of stoke (or square meters per second).

**Visible radiation.** The wavelengths of the electromagnetic spectrum between $10^{-4}$ cm and $10^{-5}$ cm.

**Vision, photopic.** Vision attributed to cone function characterized by the ability to discriminate colors and small details; daylight vision.

**Vision, scotopic.** Vision attributed to rod function characterized by the lack of ability to discriminate colors and small details and effective primarily in the detection of movement and low luminous intensities; night vision.

**Visual acuity.** Ability of the eye to sharply perceive the shapes of objects in the direct line of vision.

**Volatility.** The tendency or ability of a liquid to vaporize. Such liquids as alcohol and gasoline, because of their well-known tendency to evaporate rapidly, are called volatile liquids.

**Volume flow rate.** The quantity (measured in units of volume) of a fluid flowing per unit of time, such as cubic feet per minute, gallons per hour, or cubic meters per second.

**Volume, specific.** The volume occupied by one pound of a substance under specified conditions of temperature and pressure.

**Volumetric analysis.** A statement of the various components of a substance (usually applied to gases only), expressed in percentages by volume.

**Vulcanization.** The process of combining rubber (natural, synthetic, or latex) with sulfur and accelerators in the presence of zinc oxide under heat and usually pressure in order to change the material permanently, from a thermoplastic to a thermosetting composition, or from a plastic to an elastic condition. Strength, elasticity, and abrasion resistance also are improved.

**Vulcanizer.** A machine in which raw rubber that has been mixed with chemicals is cured by heat and pressure to render it less plastic and more durable.

# W

**WAN.** Wide-area network of linked computers or LANs, whose elements are usually geographically distant.

**Wart.** A characteristic growth on the skin, appearing most often on the fingers; generally regarded as a result of a virus infection. Synonym: verruca.

**Water column.** A unit used in measuring pressure. See also Inches of water column.

**Water curtain or waterfall booth.** A term with many different meanings in industry; but in spray painting, a stream of water running down a wall into which the excess paint spray is drawn or blown by fans, and which carries the paint downward to a collecting point.

**Waterproofing agents.** Usually formulations of three distinct materials: a coating material, a solvent, and a plasticizer. Among the materials used in waterproofing are cellulose esters and ether, polyvinyl chloride resins or acetates, and variations of vinyl chloride–vinylidine chloride polymers.

**Watt (W).** A unit of power equal to one joule per second. See Erg.

**Watts/cm².** A unit of power density used in measuring the amount of power per area of absorbing surface, or per area of a CW laser beam.

**Wavelength.** The distance in the line of advance of a wave from any point to a like point on the next wave. It is usually measured in angstroms, microns, or nanometers.

**Weight.** The force with which a body is attracted toward the earth. Although the weight of a body varies with its location, the weights of various standards of mass are often used as units of force. See Force.

**Weighting network (sound).** Electrical networks (A, B, C) associated with sound level meters. The C network provides a flat response over the frequency range 20–10,000 Hz; the B and A networks selectively discriminate against low (less than 1 kHz) frequencies.

**Weld.** A localized coalescence of metals or nonmetals produced either by heating the materials to suitable temperatures, with or without the application of pressure, or by the application of pressure alone, and with or without the use of filler material.

**Welding.** The several types of welding are electric arc-welding, oxyacetylene welding, spot welding, and inert or shielded gas welding using helium or argon. The hazards involved in welding stem from the fumes from the weld metal such as lead or cadmium metal, the gases created by the process, and the fumes or gases arising from the flux.

**Welding rod.** A rod or heavy wire that is melted and fused to metals in arc-welding.

**Wet-bulb globe temperature index.** An index of the heat stress in humans when work is being performed in a hot environment.

**Wet-bulb temperature.** Temperature as determined by the wet-bulb thermometer or a standard sling psychrometer or its equivalent. This temperature is influenced by the evaporation rate of the water, which in turn depends on the humidity (amount of water vapor) in the air.

**Wet-bulb thermometer.** A thermometer having the bulb covered with a cloth saturated with water.

**Wheatstone bridge.** A type of electrical circuit used in one type of combustible gas monitor. Combustion of small quantities of the ambient gas are detected as changes in electrical resistivity by this circuitry.

**White damp.** In mining, carbon monoxide.

**White noise.** A noise whose spectrum density (or spectrum level) is substantially independent of frequency over a specified range.

**Wide band.** Applied to a wide band of transmitted waves, with neither of the critical or cutoff frequencies of the filter being zero or infinite.

**Work.** When a force acts against resistance to produce motion in a body, the force is said to do work. Work is measured by the product of the force acting and the distance moved against the resistance. The units of measurement are the erg (a joule is $1 \times 10^7$ ergs) and the foot-pound.

**Work hardening.** The property of metal to become harder and more brittle on being worked (bent repeatedly or drawn).

**Work strain.** The natural physiological response of the body to the application of work stress. The locus of the reaction may be remote from the point of application of work stress. Work strain is not necessarily traumatic but may appear as trauma when excessive, either directly or cumulatively, and must be considered by the industrial engineer in equipment and task design.

**Work stress.** Biomechanically, any external force acting on the body during the performance of a task. It always produces work strain. Application of work stress to the human body is the inevitable consequence of performance of any task, and is therefore synonymous with stressful work conditions only when excessive. Work stress analysis is an integral part of task design.

**Working level (WL).** Any combination of radon daughters in one liter of air that result in the ultimate emission of $1.3 \times 10^5$ MeV of alpha energy.

## X

**Xanth-** (prefix). Yellow.

**Xero-** (prefix). Indicated dryness, as in *xerostomia,* dryness of the mouth.

**Xeroderma.** Dry skin; may be rough as well as dry.

**X rays.** Highly penetrating radiation similar to gamma-rays. Unlike gamma-rays, x rays do not come from the nucleus of the atom but from the surrounding electrons. They are produced by electron bombardment. When these rays pass through an object, they give a shadow picture of the denser portions.

**X-ray diffraction.** Because all crystals act as three-dimensional gratings for x rays, the pattern of diffracted rays is characteristic for each crystalline material. This method is of particular value in determining the presence or absence of crystalline silica in an industrial dust.

**X-ray tube.** Any electron tube designed for the conversion of electrical energy into x-ray energy.

## Z

**Z.** Symbol for atomic number. An element's atomic number is the same as the number of protons found in one of its nuclei. All isotopes of a given element have the same Z number.

**Zinc protoporphyrin (ZPP).** Hematopoietic enzyme used as a measure of recent lead exposure.

**Zoonoses.** Diseases biologically adapted to and normally found in lower animals, but that under some conditions also infect humans.

**Zygote.** Cell produced by the joining of two gametes (sex or germ cells).

# Index

## A

Abortion, spontaneous, 723

Absorbed dose, 249

Absorption, 21, 57–58, 125, 725; through eye, 112, 725; through hair follicles, 56; of ionizing radiation, 269; through mucous membrane, 725; personal protective equipment for, 640; of radiofrequency and microwave radiation, 290; Threshold Limit Value for, 26, 70; of solvents, 170; through sweat glands, 56

Acanthamoeba, 113

Acanthoma, 69

Accessible emission limit (AEL), for lasers, 302–3, 307, 311

Accident investigation, 685–86

Accident report, 687–88

Acclimation, to heat stress, 335, 338

Accrediting Board of Engineering and Technology (ABET), 667–68, 672

Accumulation (in tissues), 127

Accumulation (in workplace), 533

*ACCUSAFE for Windows,* 763

Acetic acid, 60

Acetone, 61, 167

Acetylcholine, 132–33

Acid: chemical composition of, 156; for disinfection, 423; hazard from, 28, 60, 63, 70–71, 112, 165, 179–80; interference with direct-reading monitor from, 514; irreversibility of effect of, 131

Acne, 56, 64, 68

Acoustic nerve, 90

Acoustic trauma, 98, 198

Acrolein, 130

Acro-osteolysis, 70

Acrylonitrile, 803, 807

Actinic degeneration, 57

Actinic keratosis, 69

Actinolite, 182, 819, 826

*Actinomyces,* 405

Action level (AL), 466, 583, 762, 772–73, 802

Action limit (AL), for lifting, 368

Action for Smoking and Health, proposed smoking standard opposed by, 829

Activities of daily living (ADL), and impairment guidelines, 73

Activity of ionization, 248

Adenoid, 38

Adjustability, and workplace design, 377–79, 388, 389

Administrative control, 29, 30, 31, 532–33, 537, 543–45; of asbestos, 185; of cold stress, 343–44; of cumulative trauma disorders, 397; of heat stress, 338; housekeeping, 31, 74, 75, 189; of lasers, 306; of noise, 219–20; OSHA requirements for, 777; of particulates, 194; of radioactive particulates, 189; of skin hazards, 74, 75. *See also* Cleaning and maintenance

Aerosol, 22; biological, 409, 434, 438, 545, 643; direct-reading monitor for, 525–26; infectious, 409, 411, 412–13, 415, 434; nuisance, 133; respirator for, 626–30; respiratory hazards from, 24, 47, 125, 420–21. *See also* Particulate

Affordance, 349

AFL–CIO, challenges to OSHA by, 791, 792, 800, 805, 814, 816, 818, 819, 822

*AFL–CIO v. Brennan,* 801

*AFL–CIO v. Hodgson,* 793

*AFL–CIO v. Marshall,* 800, 803, 806

Age: and dermatosis, 64; and hearing loss, 92, 93, 198, 204, 207; and vision, 106, 108, 110

Agency for Toxic Substances and Disease Registry (ATSDR), 784

Agriculture: and biological hazards, 409–10; and field sanitation, 817, 820–21; and occupational health program, 745–46; history of OSHA regulation of, 800–1, 839

AIDS. *See* Human immunodeficiency virus

Airborne contaminant. *See* Aerosol; Dust; Fume; Gas; Particulate; Vapor

Air cleaner, 30, 554, 564–68

Air conditioning, 337, 584; vs. air cleaning, 30. *See also* HVAC system

Air filtration, 438

Airflow rate, 555, 583, 603–5

Air-handling unit, 596

Air heater, industrial, 588–93. *See also* HVAC system

Air mixing, 583, 599

Air movement: and air sampling, 468–69; and cold stress, 14, 321, 341; and exhaust ventilation hood, 556–57; and freestanding fan, 585; for heat stress control, 321, 337; HVAC system problems with, 605; measurement of, 13, 326–27, 554–55. *See also* Ventilation

Air Moving and Control Association, noise ratings for fans by, 570

Air pollution, 167

Air pressure (ventilation): calculation of, 554; and duct velocity, 563; static, 554–55, 563; total, 555; velocity, 554–55, 563

Air quality. *See* Air pollution; Indoor air quality

Air sampling, 27–28, 467–74; absorption, 487; active, 517–18; adsorption, 487–89; area, 485–86; for biological hazards, 436–37; vs. biological sampling, 144; for carbon monoxide, 579; direct-reading instrument for, 510–26; and exposure guidelines, 479–81; for gases and vapors, 486–89; grab, 486–87; integrated, 486, 487–89; for ionizing radiation, 269; for lead, 145; method, 494–96; for particulates, 190–92, 489–92, 525–26; passive, 489, 518–19; personal, 485–86; record keeping for, 504–6; sample collection device for, 486–94; for skin exposure monitoring, 75

Air-sampling suction pump, 496–503

Akerblom pad, 382

Albinism, 55

Alcohol: and air pollution, 167; aliphatic, 132; chemical composition of, 155, 156; for disinfection, 423, 424; and sampling monitor, 514, 519; toxicological effects of, 8, 29, 61, 132, 163–64, 723

Alcoholism, 720–21

Aldehyde: and air pollution, 167, 577; chemical composition of, 156; for disinfection, 423, 424; toxicological effects of, 8, 164, 577

Algae. *See* Microorganism

Aliphatic hydrocarbon, 8, 155, 156; biological effects of, 29, 162; breath analysis for, 146; and solvent hazard control, 169

Alkali: for disinfection, 423; hazard from, 28, 60, 62, 70, 112, 179–80; irreversibility of effect of, 131; natural defense against, 58; skin permeability to, 54

Alkane, 155, 156, 162, 167

Alkene, 156, 162, 167

Alkyne, 156, 162

Allergen, biological, 20, 29, 64, 403, 404, 408, 410, 412; dermatitis from, 28, 55, 59, 62, 63, 65, 66–67, 80, 190; exposure guidelines for, 434; particulate, 179, 190. *See also* Sensitizer

Allergic alveolitis, 190

Allergic contact dermatitis. *See* Dermatitis, allergic contact

Alopecia, 69–70

Alpha-radiation, 14, 64, 248, 251, 252

*Alternaria,* 438

Alternating current, 275. *See also* Electric field, time-varying

Alveoli, 40–41, 47

Amar, contributions to ergonomics by, 348

Ambient water vapor pressure, 326

American Academy of Industrial Hygiene (AAIH), 672; and ethics, 4–5; membership qualifications in, 660, 663; *Program Criteria for Industrial Hygiene and Similarly Named Engineering-Related Programs,* 667–68

American Academy of Ophthalmology and Otolaryngology (AAOO), 95, 207

American Association of Occupational Health Nurses (AAOHN), 730, 731

American Biological Safety Association, 416, 440

American Board for Occupational Health Nurses (ABOHN), 731

American Board of Industrial Hygiene (ABIH), 671–72; certification of industrial hygiene professionals by, 472, 660–61, 662, 663, 669, 671–72, 677; and ethics, 4–5

American Board of Ophthalmology, 105

American Chemical Society, 151

American College of Obstetricians and Gynecologists, 724

American College of Occupational and Environmental Medicine, 719

American Conference of Governmental Industrial Hygienists (ACGIH), 151, 672–73; Bioaerosol Committee, 434–35, 440; Biological Exposure Index (BEI), 140, 146, 466, 725; carcinogen, categories of, 142; committees, 440; *Documentation of Threshold Limit Values,* 143, 146; and ethics, 4–5; *Industrial Ventilation—A Manual of Recommended Practice,* 170, 572, 575; membership in, 666, 673; and noise exposure measurement, 206; pacemaker recommendations by, 284; particle size selection efficiency guidelines by, 490–92; standard air conditions of, 554; *Threshold Limit Values and Biological Exposure Indices,* 13, 25, 70, 146, 725. *See also* Threshold Limit Value

*American Cyanamid Co. v. OCAW,* 810

American Dental Association, 826

American Federation of Government Employees, 822

American Industrial Hygiene Association (AIHA), 151, 670–71; Biosafety Committee, 440; *An Ergonomics Guide to Carpal Tunnel Syndrome,* 395; and ethics, 4–5; *Hygienic Guide Series,* 161; industrial hygiene defined by, 660; membership in, 666,

670–71; sampling laboratory accreditation by, 468, 494; *A Strategy for Occupational Exposure Assessment,* 751–52

American Iron and Steel Institute, 788

*American Iron and Steel Institute v. OSHA,* 804

*American Journal of Infection Control,* 408

American Medical Association (AMA): "Effects of Toxic Chemicals on the Reproductive System" (report), 723; and emergency medical identification, 713; impairment guidelines of, 49–50, 72–74, 100, 118–20, 207; *Scope, Objectives and Functions of Occupational Health Programs,* 703–7

American National Red Cross, 711

*American National Standard for Respiratory Protection,* 620, 621

*American National Standard for Safe Use of Lasers,* 301–8

*American National Standard Method of Measuring and Recording Work Injury Experience,* 794

American National Standards Institute (ANSI), 150; audiometric standards of, 209, 214, 224, 226, 227; cumulative trauma disorder standards of, 398; eye and face protection standards of, 113–14, 172, 299; first-aid kit standards of, 711; laser safety standard of, 296, 299, 301–8, 309, 311; lighting standards of, 110, 311–14; OSHA standards adopted from, 26, 791; radiofrequency and microwave exposure standards of, 290, 292, 293; respiratory protection standards of, 139, 620, 621, 642; safety professionals' familiarity with, 681; ventilation standards of, 170, 572, 575; video display terminal workstation standards, 380; work injury reporting standard of, 794

American Nurses Association, 731

American Petroleum Institute (API), 805

*American Petroleum Institute v. OSHA,* 802, 805

American Public Health Association, 407, 494, 673–74

*American Smelting and Refining Company v. OSHRC,* 796

American Society for Testing and Materials (ASTM), 160, 170, 518

American Society of Heating, Refrigerating and Air Conditioning Engineers (ASHRAE), 440; air conditioning defined by, 584; *Methods of Testing Air Cleaning Devices Used in General Ventilation for Removing Particulate Matter,* 601; noise ratings for fans by, 570; thermal comfort standard of, 344, 601; ventilation standards of, 534, 541–42, 554, 595, 596, 600–1

American Society of Microbiology (ASM), 404, 440

American Society of Safety Engineers (ASSE), 151, 677

Americans with Disabilities Act (ADA), 367, 702, 734, 738

*American Textile Manufacturers' Institute v. Donovan,* 806

American Thoracic Society, 431–32

Amide, 156

Amine, 67, 156, 423

Ammonia: direct-reading monitor for, 516, 520; hazards from, 24, 48, 60, 130; reversibility of effect of, 131

Amosite, 182

Amplification, 88–89

Anaphylactic protein. *See* Allergen, biological

Anaphylaxis. *See* Sensitizer

Anatomical/anthropometric model for material-handling personnel selection, 367

Anemia, 134, 186, 722

Anemometer, 327

Anesthetic, 8, 132, 161, 162, 164, 723

Angioedema, 67

Anhydride, 28

Aniline, 723

Animal experimentation, 137–38, 143

Animals, hazards from, 408–10, 411, 425

Annihilation (electromagnetic), 248–49

Annual limit of intake (ALI) of radiation, 249

*Annual Report on Carcinogens,* 163

Anoxia, 23, 131

Antagonistic action, 142, 466–67, 480, 533

Anthophyllite, 182, 819, 826

Anthrax, 409, 410

*Anthropometric Sourcebook,* 357

Anthropometry, 19, 357–59

Antibiotics, 97

Antimony, 60

Antioxidant, 61

Aphakia, 119–20, 296, 299

Apocrine sweat, 56

Aqueous humor, 104–5, 296

Aqueous solution, 155, 161, 169

Arachnid, 20, 403

Arene, 162

Argon ionization detector, 524

Armrests, for static work, 19

Armstrong, T. J., 395

Aromatic hydrocarbon, 8, 131, 155, 156; and air pollution, 167; biological effects of, 29, 63, 132, 162; breath analysis for, 146; photoionization detector for, 521; and solvent hazard control, 169

Arrectores pilorum, 56

Arsenic: biological sampling for, 724; hazards from, 60, 61, 62, 134, 723; history of OSHA regulation of, 801, 807

Arthropod, 20, 403

*ASARCO v. OSHA,* 807

Asbestos: control methods for, 184–85, 545, 639; exposure monitoring of, 184, 490; government regulation of, 183–84, 783, 791–92, 801, 817, 819, 826; hazards from, 47, 133, 182–83, 466; Threshold Limit Value for, 141

Asbestos-Containing Materials in Schools Rule, 183

Asbestos Hazard Emergency Response Act (AHERA), 183, 783, 838

*Asbestos Information Association v. OSHA,* 803, 819

Asbestosis, 46, 47, 133, 182

Asbestos Schools Hazard Abatement Reauthorization Act (ASHARA), 183

As low as reasonably achievable (ALARA) concept, 257, 284

ASME Boiler and Pressure Vessel Codes, 537

*Aspergillus,* 405, 410, 438

Asphalt. *See* Petroleum products

Asphyxia, 23–24, 131–32, 159, 165; chemical, 8, 24, 48, 131–32, 160; simple, 131, 159–60

*Associated Industries v. Department of Labor,* 801

Association of Professionals in Infection Control (APIC), 408, 440

Asthma, 47, 67, 179, 190, 403; atopic, 65

Asthmatic bronchitis, 130

Astigmatism, 108

Atelectasis, 47

*Atlas Roofing Company Inc. v. OSHRC,* 796

At least as effective as (ALAEA) state plan provision, 771, 792, 799, 800

Atmospheric pressure: and air sampling, 474, 499, biological effects of, 16, 17, 44, 92, 159–60; and sound, 199

Atom, 251, 273

Atomic absorption spectroscopy, 191

Atomic number, 249, 254

Atomic weight, 249

Atopy, 65

Auchter, Thorne: deregulatory policies of, 808, 810, 814, 818, 823; emergency temporary standard for asbestos issued by, 819; OSHA assistant secretary term of, 813; and OSHA state plans, 815–16; response to OSHA criticism by, 815

Audiogram. *See* Audiometry

Audiometry, 94–95, 199, 206, 224–27, 229–30, 233–35, 240–42; bone-conduction, 95; specifications for, 224, 226, 227

Audit: of industrial hygiene program, 756; safety inspection, 685

Auditory nerve, 90, 97

Aural insert protector. *See* Earplug

*Aureobasidium,* 438

Autoimmunodeficiency syndrome. *See* Human immunodeficiency virus

Automated information system. *See* Computerized industrial hygiene system

Autonomic nervous system, 349

Autophonia, 92

*Auto Workers v. Johnson Controls,* 136

## B

*Bacillus,* 405, 438; *anthracis,* 411

Background radiation, 249, 255

Bacteria: concentrations of, 438; hazards from, 20, 64, 403; identification of, 404–5; natural defense against, 58; from particulates, 179, 189–90. *See also* Microorganism

Baffle, 559

Bagassosis, 24, 47, 190

Balance. *See* Equilibrium

Barium, 60

Barotrauma, 16

Barrier cream, 76–77, 172

Basal ganglia, 350

Bauxite dust, 188

Becquerel, Antoine-Henri, 256

Becquerel (Bq), 249

BEIR V report, 257

Bends, 17, 159–60

Benzene, 8; and air pollution, 167; chemical composition of, 156; hazards from, 61, 134, 162, 723; history of OSHA regulation of, 802, 805–6, 817, 828. *See also* Aromatic hydrocarbon

Benzine, 8

Benzoyl peroxide, 62

Bernoulli's equation, 554

Berylliosis, 187

Beryllium, 187, 801

*Beta-Particle Sealed Sources,* 266

Beta-radiation, 14, 64, 249, 251, 252

Bingham, Eula: OSHA policy of, 790, 793, 805, 808, 809, 811, 813; OSHA standards advisory committee created by, 803; OSHA standards issued by, 806, 807, 811; and state plan benchmarks litigation, 800, 816; term as OSHA assistant secretary, 794, 804–5, 813; testimony at OSHA oversight hearing of, 812

Binocular vision, 105, 109

Bioaerosol. *See* Aerosol, biological

Bioassay. *See* Biological sampling

Bioelectromagnetics Society, 283

Biological Defense Research Program (BDRP), 404

Biological exposure index (BEI), 25, 140, 146, 466, 725

Biological hazard, 7, 20–21; assessment of, 410–13, 427–28, 433–38; classification of, 403–4, 413–16; control of, 417–25, 427–28, 445–49; exposure guidelines for, 414–17, 439–40; route of entry of, 406, 411; sources of, 404–7; and workplace, 407–10

Biological safety. *See* Biosafety

Biological safety officer (BSO), 449

Biological sampling, 143–46, 464–65, 724–25; limitations of, 467

Biological toxin, 20, 403, 410; dermatosis from, 29, 62, 63, 64

Biological warfare agents, 404, 416

Biomechanical model, for material-handling personnel selection, 367

Biomechanics, 18, 360–64, 724

Biosafety, 404, 421–22, 439

Biosafety cabinet, 419–20, 431–32, 438

Biosafety equipment, 418–21

*Biosafety in Microbiological and Biomedical Laboratories (BMBL),* 414, 415, 418, 426, 429, 431, 439, 445

*Biosafety in the Laboratory,* 416, 426, 439

Biosafety level (BSL), 414–18, 445–49

Biosafety manual, 425–26

Biosafety program, 425–28

Biosafety specialist, 416, 425

Biotechnology. *See* Recombinant DNA

Birds. *See* Animals

Birth defect. *See* Teratogenesis

Bis (chloromethyl) ether (BCME), 134

Black lung, 187

Blastomogen. *See* Carcinogen

Blastomycosis, 409

Bleach. *See* Chlorine; Sodium hypochlorite

Blindness. *See* Vision, impairment of

Blinking, 111, 296, 297

Blister, 63

Blood analysis, 49, 50, 145

Bloodborne pathogen, 126, 428–29; controls for, 429, 741; history of OSHA regulation of, 823, 825–26; standards for, 404, 407, 428, 429, 439, 741

Blood circulation, 322, 340

Blood disease, 134, 164, 256. *See also* Anemia; Leukemia

Blood vessel, 56, 58, 393, 394

*Blower and Exhaust Systems for Dust, Stock and Vapor Removal or Conveying,* 570

Board of Certified Safety Professionals (BCSP), 661, 677, 693–97

Body core temperature, 324, 328, 333

Body dimensions, 358–59

Bond, Marcus B., 707

Bone, 256, 392

Bone-conduction hearing, 95, 221

Bone marrow, 256

Bony labyrinth, 86, 89

Borelli, Giovanni Alfonso, 348, 360

Borg's rating of perceived exertion, 356

*Borrelia,* 405

Brachial plexus neuritis, 394

Brain, 349–50; damage, 164, 186

Braune, biomechanical research of, 360

Breakthrough, and sampling, 488

Breath analysis, 145–46

Breathing, 41–46

Bremsstrahlung, 249, 253
Brennan, William, opinion in *American Textile Manufacturers'
Institute v. Donovan,* 806
Brock, William E., 813, 820
Bromine, 29, 60
Bronchial asthma, 722
Bronchial contraction, 48
Bronchiole, 39, 40, 46, 47
Bronchitis, 47, 179, 187
Bronchus, 39, 40, 46, 47
Browning, E., 161
*Brucella,* 405
Brucellosis, 406, 413
Bruising, 63
Building-related illness (BRI), 20; control methods for, 438–39;
investigation of, 433–38
Bureau of Labor Standards, 228, 769
Bureau of Labor Statistics (BLS): occupational disorder statistics
by, 58, 59, 789, 794, 804, 822, 835–36; responsibilities under
OSHAct, 770
Bureau of Mines, 624, 782
Burn, 63, 70, 71; chemical, 70–71, 112, 165; from electric shock,
63; from electromagnetic radiation, 252, 289, 290, 297, 301; to
eye, 111, 112, 296; from hot environments, 339; incidence of, 58
Bursa, 393
Bursitis, 18
Bush, George, 838
Butadiene, 828

# C

Cadmium, 828
Cadmium oxide, 48
Calcium cyanide, 60
Calcium oxide, 60
Calibration, 496–503, 526; parameters for, 499–503; primary, 496–98;
secondary, 498–99
California Illness and Injury Prevention Act, 551
California Occupational Health Program, 467
Callus, 55, 63
Campus Safety Association, 440
Canal cap, 223
Cancer: from asbestos, 133, 182, 183; from beryllium, 187; of
bone, 134; bronchogenic, 47, 182; from coal tar, 68; environ-
mental factors in, 135; from ionizing radiation, 64, 256; of liver,
134; of lung, 179, 180, 181, 183, 187, 256; mesothelioma, 182,
183; occupational causes of, 134; from particulates, 47, 179,
180; of skin, 57, 64, 68, 69, 256, 297–98; from sunlight and
ultraviolet radiation, 57, 64, 69, 297–98. *See also* Carcinogen;
Leukemia
Canons of Ethical Conduct, 4–5
Caplan's syndrome, 181
Captan, 723
Capture hood, 555–57
Carbolic acid. *See* Phenol
Carbon dioxide, 17, 41–43, 45, 167, 515, 592, 604–5
Carbon disulfide, 8, 61, 146, 723
Carbon monoxide: diffusion capacity ($D_{CO}$), 50; hazards from,
24, 48, 131, 132, 160, 723; from internal combustion engines,
577–79; sampling for, 515, 520, 724; Threshold Limit Value for,
578

Carbon tetrachloride, 163, 169
Carboxyhemoglobin, 131, 160, 724
Carcinogen, 134–35, 142; administrative control methods for, 545;
biological, 20; chemical, 8, 162, 163, 186, 187, 576; history of
OSHA regulation of, 801, 802, 803, 805–6; safety guidelines for,
439, 805–6. *See also* Cancer
Carcinoma. *See* Cancer
Cardiac sensitization, 132
Cardiopulmonary resuscitation (CPR), 712
Cardiovascular system, 353
Carlsson, B., 79, 80
Carpal tunnel syndrome, 18, 392, 394–95, 396
Carter, Jimmy, 794, 804, 813
Cartilage, 393
Cascade impactor, 191
Cashew nut oil, 62
Cataract, 110, 256, 287, 289, 296
Cathode ray tube (CRT). *See* Video display terminal
Caustic. *See* Alkali
Ceiling limit, 25, 141; and chemical inventory, 455; for heat
stress, 328; OSHA determination of, 776, 802
Cement, burns from, 71
Center for Devices and Radiological Health (CDRH), 301–2, 311
Center for Responsive Law, 815
Centers for Disease Control (CDC), 7; Bacterial and Mycotic Dis-
eases Branch, 414; and biological hazard assessment and classi-
fication, 412, 413, 416; biosafety guidelines, 407–8, 423, 430–31,
439, 826; Biosafety in Microbiological and Biomedical Laborato-
ries (BMBL), 414, 415, 418, 426, 429, 431, 439, 445; Hospital
Infections Branch, 432, 440; NIOSH organized under, 780
Central nervous system, 349, 350, 351; damage to, from lead, 186
Central nervous system depressant (CNSD), 132, 161, 162, 163
Centrifugal collector, 564–65; wet, 566
Centrifugal separator, 192
Centrifugation, and biosafety, 420–21
Cerebellum, 350
Cerebral blood vessel disease, 722
Cerumen. *See* Ear wax
Ceruminal gland, 85
Cervicobrachial disorder, 394
*Chamber of Commerce of the United States v. OSHA,* 810
Checklist for hazard evaluation, 458–59, 462
Chemical analogy for exposure guidelines, 137
Chemical hazard, 7–11; burns from, 70–71, 112, 165; cataracts
from, 110; control methods for, 81, 455, 459–60, 462; dermato-
sis from, 28, 59–63; evaluating, 124; and hearing loss, 467; and
indoor air quality, 584. *See also* Toxicity
*Chemical Information Manual,* 494
Chemical Safety and Hazard Investigation Board, 838, 839
Chemical safety committee (CSC) and recombinant DNA, 425
Chemical Warfare Service, 404
Chilblain, 341
*Chlamydia,* 405; *psittaci,* 409, 410
Chloracne, 68
Chlorinated ethylene, 167
Chlorinated hydrocarbon. *See* Halogenated hydrocarbon
Chlorine: for disinfection of biological hazard, 424; hazards from,
29, 130; sampling for, 516, 519
Chlorofluorocarbon (CFC), 163
Choroid, 104, 105

Chromic acid, 29, 60

Chromium, 61, 829

Chrysolite, 182

CIE, 296

Cigarette smoking: biological effects of, 46, 131; OSHA standard proposed for, 828–29; synergistic effects of, 48, 133, 187, 466

Cilia, 37, 39, 48, 179

Ciliary body, 104, 105

Circadian rhythm, 282, 283

Circulating air system, 338

Circulating water system, 339

Circumaural protector. *See* Earmuff

Ciriello, V. M., 370–72

Civil Rights Act, 824

*Cladosporium,* 438

*Classification of Etiologic Agents on the Basis of Hazard,* 416

Claustrophobia, 722

Clayton, F. E., 161

Clayton, G. D., 161

Clean Air Act Amendments, 829, 837–39

Clean bench vs. biosafety cabinet, 420

Cleaning and maintenance, 31, 74, 75, 189, 218, 423, 458, 537, 545; and hazard evaluation, 454, 459; and workplace design, 535

Cleveland Open-Cup Tester, 160

Climatic conditions, 320, 321, 328; measurement of, 13, 326–27

Clinical Laboratory Improvement Act (CLIA), 439

*Clinical Toxicology of Commercial Products,* 155

Clinton, Bill, 823, 838

Closed-cup flash point, 160

Clothing and thermal balance, 320, 321, 328, 337, 343

Coal dust, 24, 187

Coal Mine Health Research Advisory Committee, 783

Coal tar, 61, 62, 134, 162

Coast Guard, 797, 808, 813

*Coccidioides,* 405

Coccidioidomycosis, 407, 409

Cochlea, 87, 89–90, 97

Code of Ethics for the Practice of Industrial Hygiene, 4–5, 481

Coefficient of entry, 555

Coefficient of variation (CV), 473

Coke oven emissions, 801, 803–4, 824

Cold-related disorders, 341

Cold stress, 14, 320, 340–42; control methods for, 343–44; dermatitis from, 29; exposure guidelines, 342–43

Collection device. *See* Sample collection device

Colorimetric sampling device, 517–20

Combustible gas, monitoring device for, 510–14

Combustible liquid, 160–61

Commissioning (of HVAC system), 596, 602

Commission Internationale d'Eclairage (CIE), 296

Complaints against state program's administration (CASPA), 771, 799

Compliance officer, 664, 774, 775

Comprehensive Environmental Response, Compensation and Liability Act (CERCLA), 784–85

Compressed gas, 537

Compressed Gas Association, 537, 547

Compton effect, 249

Computerized industrial hygiene system, 759–60; advantages of, 760–61; communications, 765–66; development and implementation of, 762–64; functions of, 761–62; hardware for, 765; software for, 763, 764–65

Computer workstation. *See* Office workstation; Video display terminal

Concentration, 27, 126, 154, 165, 168, 177–78, 438; calculation of, 474–77; and dilution ventilation, 583; lethal, 127–28; no-effect level, 126–27, 138. *See also* Time-weighted average

Concentration–decay tracer gas calculation, 604

Conchae, 37

*Condensed Chemical Dictionary,* 155

Conductive heat exchange rate, 320–21

Cones (of retina), 105, 109, 296

Confidence limit, 473–74

Confined space, 831

*Confined Space Entry Tracking* software, 763

Congenital malformation. *See* Teratogenesis

Conjunctiva, 104, 109

Conjunctivitis, 109

Connective tissue, 393

Constant air volume (CV) system, 596–97

Constant-concentration tracer gas calculation, 604

Constant-emission tracer gas calculation, 604

Construction Industry Noise Standard, 228

Construction industry regulation, 228, 770, 788, 800, 819, 831, 834

Construction Safety Act, 770, 788, 800; OSHA standards adopted from, 791

Construction Safety and Health Advisory Committee, 800

Consultation service, 31, 243–44, 685, 733; by OSHA, 775, 814–15, 839

Consumer Product Safety Commission, 813

Contact dermatitis. *See* Dermatitis, allergic contact; Dermatitis, irritant contact

Contact lenses, 116, 172

Containment: of biological hazards, 417–21, 445–49; for tuberculosis control, 431–32. *See also* Enclosure; Isolation; Shielding

Control (of hazard), 29–31, 532–33; appraisal during field survey, 461; of asbestos, 184–85; of biological hazards, 417–25, 438–39; of building-related biological hazards, 438–39; of cold stress, 343; of cumulative trauma disorders, 397–98; for eye safety, 117; of heat stress, 334–40; and industrial hygiene program, 753; of ionizing radiation, 264–69; of lasers, 304–8; of noise, 216–24; of nonionizing radiation, 285–86, 295, 299; of skin hazards, 75–77; of solvents, 168–72. *See also* Administrative control; Engineering control; Personal hygiene; Personal protective equipment

Controlled area, 249, 777

*Control of Communicable Diseases in Man,* 407

Controls (equipment), 19–20, 364, 389–91

Convective heat exchange rate, 320; by respiration, 320, 321

Coolant, 59, 62

Coordinated Framework for Biotechnology, 416

Copper, 111

Corn, Morton: OSHA policies of, 790, 795, 798; term as OSHA assistant secretary, 794

Cornea, 104, 105, 109, 296

Coronary artery disease, 722

Cor pulmonale, 180, 187

Corrected effective temperature, 327

Corrective eyewear, 108, 116, 172, 389, 723

Corrosive, 11

Cortex, 349

Cost–benefit analysis for OSHA standards, 805–7, 816, 817–18, 830

Cost-effectiveness safety analysis, 693; for OSHA standards, 807, 817

Costoclavicular syndrome, 394

Cotton dust, 802, 806, 817–18, 828

Cough, 38, 48

Coulometric detector, 515, 516

Council for Accreditation in Occupational Hearing Conservation, 226

Coupling and hand tools, 372–75

*Coxiella,* 405; *burnetii,* 409

*CRC Handbook of Chemistry and Physics,* 155

Creosote. *See* Petroleum products

Cresol, 16

Cresylic acid, 60

Creutzfeldt–Jakob dementia, 405

Cristobalite, 182

Criteria document, 140, 161, 772, 781

*Criteria Document on Hot Environments,* 13, 325–26, 335, 337

Criticality, 267–68

Crocidolite, 182

Cross-sensitivity, 67

Crustacean, 20, 403

Cryogenic liquid, 159

*Cryptococcus,* 405

Cryptosporidiosis, 407

Cubital tunnel syndrome, 396

Cumulative trauma disorder, 17–18, 391–98, 741

Curie (Ci), 249

Current Intelligence Bulletin (CIB), 140

Cutting oil, 59, 62, 64, 80–81, 134

*Cuyahoga Valley Ry. Co. v. United Transportation Union,* 810

Cyanide gas, 24

Cyclic hydrocarbon, 155, 156; and solvent hazard control, 169; toxicological effects of, 162

Cycloalkane, 162

Cyclone: for particulate sampling, 191, 192, 490–92; for air cleaning, 564

Cytomegalovirus (CMV), 412

**D**

Damage control program, 692

Damper, 597, 599

Data base of toxic substances, 150

DDT, 724

Dead finger. *See* Raynaud's phenomenon

Deafness. *See* Hearing loss

Dear, John, 823, 832, 833

Decay (radioactive), 249, 255

Decibel, 202–4

Decompression sickness, 17, 159–60

Decontamination, 422–23

Dehydration, 58, 324–25, 334, 343

Department of Agriculture, 416, 439; Food Safety and Inspection Service, 839; Food Safety and Quality Service, 813

Department of Commerce, 416, 439

Department of Health and Human Services, 703, 770; *Healthy People 2000,* 736, 737, 745; mine health advisory committee appointed by, 782–83; NIOSH established by, 228, 780; Registry of Toxic Effects of Chemical Substances, 139

Department of Labor, 140, 228, 703, 770, 771, 772; mine safety enforcement by, 782, 783; NIOSH recommended standards transmitted to, 781, 783; OSHA under jurisdiction of, 770, 840; OSHA training information about, 664. *See also* Bureau of Labor Statistics; Mine Safety and Health Administration

Department of the Interior, 797; Bureau of Mines, 624, 782

Department of Transportation, 794, 796–797, 813; and infectious agents, 413, 423, 439

Dependent care, 745

Depth perception, 105, 109

DeQuervain's syndrome, 396

Derived air concentration (DAC), 249

Dermatitis: allergic contact, 28, 55, 62, 63, 66–67, 80, 190; atopic, 65; causes of, 28–29, 162, 163, 164, 256, 403, 545; vs. dermatosis, 58; incidence of, 59; irritant contact, 28, 58, 62, 64, 65, 66, 92, 161; nonoccupational, 65; prevention of, 29. *See also* Dermatosis

Dermatoglyphics, 55

Dermatology, 54

Dermatosis, 58; causes of, 59–64; classification of, 66–70; vs. dermatitis, 58; diagnosis of, 71–72; evaluation of, 72–74; incidence of, 58–59; predisposing factors for, 64–65; and susceptibility to infection, 412. *See also* Dermatitis

Dermis, 54, 55

Descriptive study, for exposure guidelines, 138

Dessicator, 28

Detergent, 28

Detoxification, 126

Diabetes, 714–15, 722

*Diamond Roofing Co., Inc. v. OSHRC,* 808

Diaphragm, 44–45

Dibromochloropropane (DBCP), 724, 807

Digestive tract, 256

*The Dilemma of Toxic Substance Regulation—How Overregulation Causes Underregulation,* 840

Dilution ventilation, 30, 581–83; as control method, 165, 170, 192–94, 337, 537, 541–42; and replacement air, 585–88, 592–93; efficiency of, 583, 599

Dinitrobenzene, 61

Diopters, 108

Diplopia, 118, 120

Dipole/diode field survey instrument, 292

Direct-fired air heater, 591–92

Direct-reading instrument, 486; calibration of, 511–13, 526; catalyst poisoning of, 514; colorimetric, 517–20; design of, 510–11; gas chromatographic, 520, 523–24; for indoor air quality, 515–16; interference with, 514; metal oxide semiconductor, 510–11, 516; nonspecific, 520–22; oxided gas type, 510; for particulates, 525–26; reading, 513; for specific compounds, 510–20; spectrometric, 522–23, 524; thermal conductivity, 511

Disability, 73

Disfigurement, 73–74

Disinfection, 423, 424, 432

Disintegration. *See* Decay (radioactive)

Dispensary. *See* Employee health care facility

Display, 19, 364, 377, 389–91. *See also* Video display terminal

Disposal. *See* Hazardous waste management

Distal stimulus, 350

Diving, 16–17, 92; federal regulation of, 801, 802, 803, 808, 821

Diving Contractors Association, 803

Dizziness, 90

*Documentation of Threshold Limit Values,* 143, 146

*Dole v. United Steelworkers of America,* 823, 827

Donovan, Ray, 813

Dose: and hazard evaluation, 453; infectious, 411; of radiation, 249, 255, 256, 257, 264, 288

Dose–response relationship, 126–28, 256, 257, 434

Dosimeter: for hazard evaluation, 15, 75, 189, 249, 269, 288–89, 461–62; noise, 211–12, 216; types of, 258–59

Double vision, 118, 120

Drug abuse, 720–21

Drug screening, 721

Dry bulb temperature, 13, 326–27, 342

*Dry Color Manufacturers Association v. Department of Labor,* 802, 803

Dry dynamic precipitator, 565

Dry-gas meter, 498

Duct, 554, 562–63

Dukes-Dobos, heat stress research by, 328

Duration of exposure, 27, 124, 126, 168, 177, 207; and lethal concentration, 127–28. *See also* Time-weighted average

Dust, 21–22; bauxite, 188; coal, 24, 187; control methods for, 30, 80, 576–77, 626–30; epoxy, 80; fused silica, 182; hazards from, 41, 47, 66, 133, 175; inert, 133, 188; kaolin, 181, 188; lead, 186–87; mica, 188; monitoring for, 191, 476–77, 525–26; radioactive, 189; silica, 24, 133, 180–82; toxic, 188. *See also* Particulate

Dusty lung. *See* Pneumoconiosis

Dye, 61; laser, 308–9, 310

Dynamic technique, for material-handling personnel selection, 368

Dyspnea, 182

**E**

Ear: anatomy of, 84–87; atmospheric pressure effects on, 17, 92; pathology of, 92–93; physiology of, 87–91

Ear canal, 85, 92; and earplug fitting, 222; cap for hearing protection, 223

Eardrum: anatomy of, 85, 86; pathology of, 92; physiology of, 87–89, 93

Earmuff, 220, 223, 549–50

Earplug, 220, 221–22, 549–50

Ear wax, 92; and earplug fitting, 222

Ebola, 406, 409

Eccrine sweat, 56

*E. coli,* 405

Eczema, atopic, 65

Educational Resource Center (ERC), 7, 667, 669, 781

Effective temperature, 13, 327; corrected, 327

Effector, 350, 352

"Effects of Toxic Chemicals on the Reproductive System" (AMA report), 723

Ekman's rating of perceived exertion, 355–56

Electrical hazard, 29, 63, 289, 308; federal regulation of, 821, 831

Electric field, 273–74; biological effects of, 280–84, 287–88; control of, 285–86; direct current (DC), 282; exposure guidelines for, 282–84, 287–92; measurement of, 284, 292; microwave, 286–92; pulsed, 292; vs. radiation, 279; radiofrequency, 286–92; strength, 279–80; subradiofrequency, 279–86; time-varying, 283–84; types of, 279; in video display terminal, 294. *See also* Nonionizing radiation

Electric Power Research Institute, 329

Electrocardiogram (ECG), 282

Electromagnetic device, 279

Electromagnetic radiation, 251, 275–79; and interaction with matter, 287; spectrum, 277–78. *See also* Ionizing radiation; Nonionizing radiation

Electromyography (EMG), 362

Electron, 249, 251, 273–74

Electron capture detector, 522, 524

Electron volt (eV), 249

Electrostatic precipitator, 492, 567

Element, 249, 254

ELF radiation. *See* Extremely low frequency radiation

Elutriator, 191, 492

Emergency care, 712–13; for eyes, 113; OSHA evaluation of, 778; for temperature-related disorders, 324, 341

Emergency medical identification, 713

Emphysema, 47, 48, 722

Employee: diversity of, 742, 745; and heat stress behaviors, 325; medical removal protection (MRP), 807; participation under OSHAct, 770–71, 789, 793–94, 810–11; role in occupational health, 6, 80, 755–56, 758; walkaround right, 793, 810; whistle-blower protection, 793–94, 838

Employee assistance program (EAP), 720–21, 737–38

Employee health care facility, 705, 709

Employee health care program. *See* Occupational health program

Employer, responsibilities under OSHAct, 770–71, 789

Employment Standards Administration, 839

Encephalopathy, 186

Enclosure, 30, 532, 537, 540–41; for asbestos control, 185; for exhaust ventilation hood, 559; hearing protective, 221; for laser control, 304–6; for noise control, 217–18; for particulate control, 192; for plutonium control, 268. *See also* Containment; Isolation; Shielding

Endolymph, 89–90

Endotoxin, 410, 434

Engineering control, 29–30, 532–43; of asbestos, 185; of cold stress, 343; of cumulative trauma disorders, 398; at design stage, 533–37; of heat stress, 336–38; of lasers, 304–6; of noise, 218–19; OSHA requirements for, 779, 801; of particulates, 192; of radioactive particulates, 189; of skin hazards, 74–75; of solvents, 169–70

Entropion, 118

Environmental control. *See* Engineering control

Environmental Protection Agency (EPA): asbestos regulation by, 183, 184, 783; and biological hazards, 416, 439; and Comprehensive Environmental Response, Compensation and Liability Act, 784–85; emission requirements, 458; Hazardous Chemical Reporting Rules, 785; Integrated Risk Information System, 481; lead regulation by, 783; medical surveillance requirements of, 551; National Primary Ambient-Air Quality Standards, 601; noise regulation by, 206, 224, 228; and OSHA jurisdiction, 800–1, 813, 838–39; radon regulation by, 783; and Resource Conservation and Recovery Act, 546, 783–84; *Right-to-Know* software, 763; and Toxic Substances Control Act, 139, 783

Environmental sampling, 436–38

Environmental stresses, 7–21

*Epicoccum,* 438

Epicondylitis, 396

Epidemiology, 138, 143, 454

Epidermis, 54–55, 57, 58, 297

Epiglottis, 38–39

Epilepsy, 714–15, 722

Epiphora, 118

Epoxide, 164

Equal-loudness contour, 205

Equilibrium, 90; impairment of, 93, 100

Equipment: design of, 364, 683; evaluation of hazard from, 458–59

Equivalent chill temperature (ECT), 14, 341

*ErgoKnowledge* CD-ROM, 763

Ergonomic hazard, 7, 17–20; OSHA regulation of, 823, 831–32, 834

Ergonomics, 17, 347; and anthropometry, 357–59; and biomechanics, 360–64; and cumulative trauma disorders, 397–98; and equipment design, 372–76, 389–91; evaluation of, 741, 743–44; and material handling, 364–72; and physical capacity, 724; and workplace design, 377–79; and workstation design, 376–77, 379–89

*An Ergonomics Guide to Carpal Tunnel Syndrome,* 395

Erysipelas, 409

*Erysipelothrix rhusiopathiae,* 410

Erythema. *See* Sunburn

Esophagus, 37, 38, 39, 40

Ester: biological effects of, 29, 164; breath analysis for, 146; chemical composition of, 155, 156

Ethanol, 163–64

Ether: biological effects of, 29, 132, 164, 723; breath analysis for, 146; chemical composition of, 155, 156

Ethics, 4–5, 481

Ethyl chloride, 723

Ethyleneimine, 803

Ethylene oxide: for disinfection, 424; federal regulation of, 802, 817, 821, 826

Ethyl ether, 723

Eukaryote, 404–5

European Federation of Biotechnology, 414

Eustachian tube, 17, 37, 85–86, 92

Evaluation (of hazard), 7, 26–28, 453–55; field survey for, 460–61; and industrial hygiene program, 751–53; interview for, 461; by process analysis, 455–60; and Toxic Substances List, 139–40. *See also* Monitoring; Sampling

Evaporative heat loss, 320, 321; by respiration, 320, 321

Exhaust ventilation, 30, 537, 542–43, 553–55; air cleaner, 564–68; for asphyxiant control, 165; baffles and enclosures for, 559; design of, 562–63; fan, 568–70; for fog removal, 579; general, 585; for grinding operations, 574–75; hood, 555–62; for internal combustion engines, 577–79; for materials conveying, 576–77; for oil mist, 576; for open-surface tanks, 571–72; for particulate control, 192; for solvent hazard control, 170; for spray booths, 573–74; system components, 554; system performance, 570–71; for welding, 574; for woodworking, 575–76; and workplace design, 534

Expiratory reserve volume (ERV), 46

Explosive, 11, 62

Explosive range, 161, 510, 513

Exposure: acute, 128–29; chronic, 128–29; and mode of use, 154; and vapor pressure, 154

Exposure guidelines, 31; for airborne contaminants, 24–26, 140–43, 154, 177–78; basis for, 137–38; for biological hazards, 414–17, 430–31, 439–40; for cold stress, 342–43; for cumulative trauma, 398; for heat stress, 13, 325–34; for ionizing radiation, 257–58; for noise, 12, 206–7, 215–16, 228–29, 232–45; for nonionizing radiation, 282–84, 287–92, 298–99, 301–4; and sampling results, 479–81. *See also* Permissible Exposure Limit; Recommended Exposure Limit; Threshold Limit Value

External ear, 84–85, 87, 92

External work rate, 320

Exteroceptor, 350

Extremely low frequency (ELF) radiation, 278; biological effects of, 280–84; field strength, 279–80; magnetic field measurement, 285

Eye: absorption through, 725; anatomy of, 103–5; burn of, 111, 112, 296; chemical hazards to, 61, 112; defects, 106–8; disorders, 109–10; emergency care for, 113; infection, 111; inflammation, 61, 256; ionizing radiation effects on, 256; natural defenses of, 296–97; nonionizing radiation effects on, 16, 296–97; physical hazards to, 110–12; and radiofrequency and microwave exposure guidelines, 289, 291. *See also* Corrective eyewear; Protective eyewear; Vision

Eyeglasses. *See* Corrective eyewear

Eye-hazard area concept vs. job approach, 113

Eyestrain, 110

Eyewash fountain, 113

## F

Face velocity, 557–59

Failure mode and effects analysis (FMEA), 460, 693

Fair Labor Standards Act, 793, 839

Fan, 554, 568–70; axial-flow, 570; centrifugal, 568–70; freestanding, 585; for HVAC system, 597; noise from, 570. *See also* Ventilation

Faraday cage, 295

*Farmworker Justice Fund Inc. v. Brock,* 820, 821

*Farm Workers Occupational Safety and Health Oversight Hearings,* 797

Farsightedness, 106–8

Fascia, 393

Fatigue, 357

Fault tree analysis, 460, 693

Federal Agency Safety Programs, 804, 812

Federal Aviation Administration, 813

Federal Coal Mine Health and Safety Act, 187, 781, 782, 783, 788

Federal Employee Health and Safety Program, 794, 804, 813, 822

Federal Facility Compliance Act, 784

Federal Insecticide, Fungicide and Rodenticide Act (FIFRA), 139, 800–1

Federal Metal and Nonmetallic Mine Safety Act, 782

Federal Mine Safety and Health Act, 770, 782–83, 797

Federal Mine Safety and Health Review Commission, 782

Federal Radiation Council, 257

Federal Railroad Administration, 797

Federal regulations, 6–7; for asbestos exposure, 183–84; for biological hazards, 439–40; for coal dust exposure, 187; and engineering control methods, 532; and exposure guidelines, 26, 138–40; for eye and face protection, 113–14; for insecticides, 139; for laser exposure, 309–11; for lead, 186–87; for medical examinations, 715; for noise exposure, 215–16, 228–31, 232–45; for occupational medicine, 702–3; before OSHAct, 769; for radiofrequency and microwave exposure, 292; for respirators, 194–95; safety professional's familiarity with, 681; and sampling results, 479. *See also specific regulations*

Ferric chloride, 29

Fertility. *See* Reproductive hazard

Fetal protection policy, 824

Fiber, 175

Fibrosis, 24, 47, 48, 133, 175, 179, 180

Field sanitation, 817, 820–21

Field survey, 435, 460–61

Film badge, 15, 249, 258

Filovirus, 406, 409

Filter: for aerosol-removing respirator, 626–30, 632–33; for air cleaner, 567–68; for air disinfection, 432; for asbestos, 183, 185; for biosafety cabinets, 419; diffusion, 627; electret fiber, 627–28; electrostatic capture, 627; high-efficiency particulate air (HEPA), 568, 629; for HVAC system, 597; interception capture, 627; for measuring particulate exposure, 191; NIOSH–MSHA certification of, 628–30; for particulates, 194, 489–90; for plutonium, 268; resin wool, 627; for respirator, 431; sedimentation capture, 627; for tuberculosis control, 432; for vacuum, 183, 185, 194, 459

Fire fighter, respirator for, 642

*Fire Hazard Properties of Flammable Liquids, Gases and Volatile Solids,* 159, 160

Fire point, 161

Fire safety, 308, 458, 570, 573

*Firestone Plastics Co. v. Department of Labor,* 802

First aid. *See* Emergency care

First-aid kit, 710–11

First-aid room, 709–10

Fischer, biomechanical research of, 360

Fishing, 410

Flame ionization detector, 520–21, 524

Flame photometric detector, 524

*Flammable and Combustible Liquids Code,* 160

Flammable liquid, 11, 160–61

Flammable materials and lasers, 308

Flammable range. *See* Explosive range

Flanged hood, 557

Flash lamp, 309

Flash point, 160–61

Fletcher–Munson contour, 205

*Florida Peach Growers Association v. Department of Labor,* 800, 803

Flow-rate meter, 493–94

Fluorescent tube, 314

Fluoride, 61

Fluorinated hydrocarbon. *See* Halogenated hydrocarbon

Fluorocarbon, 167

Fog removal, 567, 579

Folliculitis, 64

Food and Drug Administration (FDA), 295, 416, 813. *See also* Center for Devices and Radiological Health

Food Safety and Inspection Service, 839

Food Safety and Quality Service, 813

Force, 360–62

Forced expiratory flow (FEF), 46

Forced expiratory volume (FEV), 46, 49, 50

Forced vital capacity (FVC), 46, 49, 50, 180

Ford, Gerald, 801

Forebrain, 349

Foreign bodies, in eye, 111

Forestry, 410

Formaldehyde: biological effects of, 29, 62, 130, 723; for disinfection, 424; from internal combustion engines, 577; OSHA regulation of, 802, 827; Threshold Limit Value for, 577

Formic acid, 60

Forsberg, K., 79

Fort Detrick Biological Defense Research Program (BDRP), 404

Foulke, Edwin G., Jr., 838

Fovea, 109, 296

Freon, 163

Freon TF, 156, 163

Friction, 29, 63

Fritted bubbler, 487

Frostbite, 14, 63–64, 92, 341

Frostnip, 341

Fuel, for air heating, 592–93

Fume, 48, 175; air cleaner for, 568; concentration calculation, 477; lead, 186–87; respirators for, 626–30; toxic, 188; from welding, 188–89, 574

Functional residual capacity (FRC), 46

Fungus: allergenic, 410; biological effects of, 20, 24, 29, 47, 64, 403; concentrations of, 438; endotoxic, 410; identification of, 404–5; from particulates, 179, 189–90. *See also* Microorganism

*Fusarium,* 410, 438

## G

*Gade v. National Solid Wastes Management Association,* 824, 835

Gamma-radiation, 14, 249, 251, 254–55; dermatosis from, 64; properties of, 278; shielding for, 263–64

Ganglion, 396

Gas, 22; absorption of, 55; asphyxiant, 131–32; biological effects of, 8, 24, 47, 48, 112, 129–31, 165; concentration calculation, 474–76; control methods for, 568, 582, 630–34; inert, 131; inhalation of, 125; irritant, 8, 48, 129–31; in lasers, 309; mode of use and exposure hazard, 154; properties of, 159; sampling devices for, 486–89, 509; toxic, 24, 47, 48, 159, 309

Gas chromatograph, 520, 523–24

Gas diffusion study, 49, 50

Gas exchange, in respiration, 42–43

Gas wash bottle, 487

Geiger–Mueller counter, 15, 260

General Industry Noise Standard, 228

General ventilation. *See* Dilution ventilation

Gerarde, H. W., 161

Germicidal lamps, 16

*Giardia,* 405, 410

Glare, 311–12, 313, 380

Glasses. *See* Corrective eyewear; Protective eyewear

Glaucoma, 109–10, 112

Gleason, M. N., 155

Global warming, 167

Globe temperature, 13, 327

Glove box, 268, 541

Gloves, 78–80; vs. barrier cream, 77; latex, 67; permeability of, 170–72; for solvents, 170–72

Glutaraldehyde, 424

Glycol, 156, 164–65

Goggles, 114, 172, 308

Golgi organ, 350

Gonioscope, 106

Good large-scale practices (GLSP), 414–15, 445

Goose bumps, 56

Government regulations. *See* Federal regulations; State and local regulations

Grab sampling, 468, 486–87

Grain handling, 822, 830

Grandjean, E., 382

Granuloma, 69
Gravimetric analysis, 191
Gravitational separator, 192
Gray, Wayne B., 840
Gray (Gy), 249
Greenburg–Smith impinger, 191
Greenhouse effect, 167
Grinding operations, 574–75
Guenther, George C., 790, 793, 794, 797
*Guides to the Evaluation of Permanent Impairment:* of hearing, 100, 207; of respiration, 49–50; of skin, 72–74; of vision, 118
Guyon tunnel syndrome, 397

# H

Hair, 56; loss from ionizing radiation, 256
Hair follicle, 55, 56, 58
Half-life, 249, 254, 255
Half-value layer, 249, 254
Hall effect, 285
Halogen, 423
Halogenated hydrocarbon, 8, 155, 156; and air pollution, 167; biological effects of, 29, 132, 162–63; breath analysis for, 146; interference with direct-reading monitor from, 514; and solvent hazard control, 169
Halogen-containing compound, 522
Halothane, 723
Hamann, C. P., 80
Hand, 18
Handle, 370, 372–76
Hand tool, 372–76
Hantavirus, 409, 414
Hantavirus pulmonary syndrome, 407
Hardware. *See* Computerized industrial hygiene system
Harless, biomechanical research of, 360
Hay fever, 179, 190; atopic, 65
Hazard, 7–21, 27; and reactivity, 154; recognition of, 454, 751–53; vs. toxicity, 11, 124
Hazard and operability study (HAZOP), 460
Hazard communication program, 773–74
Hazard Communication Standard, 8, 138–39, 146, 773–74; history of, 811, 818–19; and Paperwork Reduction Act, 823–24, 827; training program required by, 551, 777
Hazardous waste management, 458, 537, 546; federal regulation of, 772, 783–84, 824, 830
Health care facility: biological hazards in, 407–8, 432; regulation of, 440; respirators for use in, 643. *See also* Employee health care facility
Health education and counseling, 720–21
Health Hazard Evaluation (HHE), 7, 481, 781
*Health Physics and Radiological Handbooks,* 263
"Health Promotion in the Workplace: Alcohol and Drug Abuse" (WHO report), 721
Health service office, 705, 709
Health technologist, 660
*Healthy People 2000: National Health Promotion and Disease Prevention Objectives,* 736, 737, 745
Hearing, 83, 93–94; measurement of, 94–95, 97, 199, 206–7, 224–27, 229–30, 233–35, 240–42; speech, 206; standard threshold shift (STS), 97–98, 207, 229–30; threshold of, 94–95, 202, 224–25; and threshold of pain, 202
Hearing aid, 99–100

Hearing Conservation Amendment (HCA), 12, 100–1, 228–31, 232–45; history of, 801, 818
Hearing conservation program, 12, 100–1, 226–27, 228–31; record keeping for, 227, 230–31, 236–37
Hearing loss, 95–97, 198; causes of, 92, 97; and chemical hazard synergism, 467; communication problems from, 98–100; conductive, 93, 94, 95, 96–97; measurement of, 94–95, 97, 100, 199, 206–7, 224–27, 229–30, 233–35, 240–42; mixed, 96–97; noise-induced, 98, 99, 198; risk factors for, 207–8; sensorineural, 94, 95–97, 99; and tinnitus, 93; treatment for, 99–100
Hearing-protective device, 220–24, 230, 235–36, 239–40, 549–50
Heart rate, 322–25, 333–34, 354–55
Heat-alert program, 336
Heat balance analysis, 329–33
Heat cramps, 13, 63, 324
Heat exhaustion, 13, 63, 323, 324
Heating. *See* HVAC system
Heat loss, 13, 14, 340–43
Heat rash, 63, 69, 324
Heat-related disorders, 323, 324
Heat storage rate, 320
Heat stress, 12, 320, 321–23, 333–34; and chemical exposure, 142; control methods for, 334–40; dermatitis from, 29; environmental measurements of, 13; exposure guidelines, 325–34; recognition of, 322–25
Heat Stress Index (HSI), 13–14, 329
Heat stroke, 12–13, 63, 323, 324
Heat syncope, 324
Helicotrema, 87
Helmet, hearing-protective, 221
Hemoglobin, 41, 42–43
Henry, N., 79
HEPA filter. *See* High-efficiency particulate air (HEPA) filter
Hepatitis, 406, 407; A virus, 410; B virus (HBV), 404, 412, 428; B virus, history of OSHA regulation of, 823, 825–26; B virus, vaccination program, 741; C virus (HCV), 428
Herpes simplex, 109
*Herpes virus: simiae* (B), 406, 409; viability of, 411
Hertz, Heinrich, 275
Hertz (Hz), 200, 275
Hexachlorobutadiene, 723
Hiccup, 38–39
High-efficiency particulate air (HEPA) filter, 568, 629; for air disinfection, 432; for asbestos, 183, 185; for biosafety cabinets, 419; for handling plutonium, 268; for particulates, 194; for respirator, 431; for tuberculosis control, 432; for vacuum, 183, 185, 194, 459
High Risk Occupational Disease Notification and Prevention Act, 837
Hindbrain, 349
*Histoplasma,* 405
Histoplasmosis, 409
HIV. *See* Human immunodeficiency virus
Hive. *See* Urticaria
Holding. *See* Coupling; Handle
Hood, 554, 555–62; kitchen range, 579–80; and replacement air, 585–86
Horny layer. *See* Stratum corneum
Hospital. *See* Health care facility
Hospital waste, 423
Hot water air heater, 589

Housekeeping. *See* Cleaning and maintenance

Housing and Community Development Act, 827, 837

Human Factors and Ergonomics Society, 347

Human immunodeficiency virus (HIV), 406, 407, 428–29; exposure guidelines for, 414; history of, 404, 823, 825–26

Human subject review board (IRB), 425

Humidity, 65, 321; HVAC system problems with, 605; measurement of, 13, 326–27, 515

HVAC system, 534, 596–99; for building-related illness control, 438; and indoor air quality problems, 599–600; and legionnaires' disease, 433; maintenance, 602–6; standards, 600–2, 829

Hybridoma technology. *See* Recombinant DNA

Hydrocarbon: and air pollution, 167; colorimetric monitor for, 519; oxygen-containing, 156, 163–65

Hydrochloric acid, 24, 60

Hydrochlorofluorocarbon (HCFC), 163

Hydrocyanic acid, 516

Hydrofluoric acid, 60, 71

Hydrogen, 516

Hydrogen chloride, 516

Hydrogen cyanide, 131–32

Hydrogen fluoride, 24, 48, 165

Hydrogen peroxide, 29, 62, 424

Hydrogen sulfide, 131, 132, 723; direct-reading monitor for, 516, 519, 520

*Hygienic Guide Series,* 161

Hygroscopic agent, 28

Hyperabduction syndrome, 394, 397

Hyperbaric environment. *See* Atmospheric pressure

Hyperhidrosis, 64–65

Hyperopia, 106–8

Hyperpigmentation, 68, 69

Hypertension, 722

Hypochlorite, 29

Hypopigmentation, 68–69

Hypothermia, 14, 340–42

Hypoxia, 131

## I

Ice garment, 339

Ideal gas law, 159

Illumination. *See* Lighting

Illumination Engineering Society (IES), 15, 311–14

Immediately dangerous to life or health (IDLH), 641–42

Immunization, 412, 426, 438

Impactor, 191, 192, 492

Impact-resistant eyewear. *See* Protective eyewear

Imperial Food Products poultry-processing facility fire, 835

Impinger, 191, 487, 492

Incus, 85, 86, 87

Index chemical, and biological sampling, 725

Indirect-fired air heater, 589–91

Indoor air quality, 584, 595, 596; calculation of, 603; monitor for, 515–16; problems, 599–600; standards, 600–2, 828–29

Indoor Radon Abatement Act, 783

Industrial hygiene, 3–4, 660

Industrial hygiene manager, 660, 662–63

Industrial hygiene program, 680–81, 750; and employee involvement, 755–56, 758; and employee training, 753–55; evaluation of, 756; and hazard evaluation and control, 751–53; implementation of, 749, 750; organizational responsibilities, 756–58; and record keeping, 755

Industrial hygiene technologist, certified (CIHT), 661

Industrial hygienist, 3–4, 660, 661–62; and biosafety, 440; certified (CIH), 472, 660, 663; civil service, 663–66; education and training of, 667–69; ethics of, 4–5; and hospital infection control, 408; number of, 666; in respiratory protection program, 722; and safety professional, 757; and skin protection, 80

Industrial hygienist-in-training (IHIT), 660–61

*Industrial Union Department v. American Petroleum Institute,* 806

*Industrial Union Department v. Bingham,* 805

*Industrial Union Department v. Hodgson,* 792

*Industrial Ventilation—A Manual of Recommended Practice,* 170, 572, 575

Infection, 20, 29, 403, 404, 405–7, 410; of ear, 92, 93; epidemiology of, 406–7; of eye, 111; hearing loss from, 97; modes of transmission, 410–11; and particulate inhalation, 179; and reproductive hazards, 412, 723; secondary, 63, 64–65

*Infection Control and Hospital Epidemiology,* 408

Infectious dose, 411

Infectious waste, 423

Inflammation: of eye, 61, 109, 256; of skin, 403; of soft tissue, 393

Influenza vaccine, 414

Information processing, human, 349–52

Information system. *See* Computerized industrial hygiene system

Infrared analyzer, 190, 522–23

Infrared radiation, 278; biological effects of, 15, 110, 112, 296–98; controls for, 114–15, 299; exposure guidelines for, 298–99; thermal effects of, 296, 299

Ingestion, 21, 125–26; of biological agent, 406, 411, 415; of particulates, 176–77

Inhalation, 21, 125; of biological aerosols, 189–90, 406, 411, 415, 434; and hazard control, 169, 415; of ionizing radiation, 180, 269; of irritant, 179–80; of particulates, 176

Injection: of biological hazard, 411, 415; of ionizing radiation, 269; of toxin, 126

Inner ear: anatomy of, 86–87; pathology of, 93, 287; physiology of, 89–90

*Innovation at the Worksite,* 731

Inoculation. *See* Injection; Immunization

Insect, 20, 403

Insecticide. *See* Pesticide

Inspiratory capacity (IC), 46

Inspiratory reserve volume (IRV), 46

Institute of Electrical and Electronic Engineers (IEEE), 290, 292, 293

Institutional animal care and use committee (IACUC), 425

Institutional biosafety committee (IBC), 425

Insulation, 337, 340

Intake (for HVAC system), 597

Integrated Risk Information System (IRIS), 481

Integrated sampling vs. grab sampling, 486

Interagency National Response Team, 839

Interagency Regulatory Liaison Group (IRLG), 813

Internal combustion engine, 577–79

International Agency for Research on Cancer (IARC), 163, 576

International Atomic Energy Agency (IAEA), 266

International Commission for Non-Ionizing Radiation Protection (ICNIRP), 282, 283–84, 314

International Commission on Radiological Protection and Measurements (ICRP), 257, 258
International Committee of Contamination Control Societies, 169
International Lighting Commission (CIE), 296
International Organization for Standardization (ISO), 328, 330–32, 344
International Union of Pure and Applied Chemistry, 155
Interoceptor, 350
Intersociety Committee on Guidelines for Noise Exposure Control, 207
Intertrigo, 69
Interview, for hazard evaluation, 454, 461
Intrapulmonic pressure, 44
*Introduction of Recombinant DNA-Engineered Organisms into the Environment*, 416
Intrusive task, 353
Inventory, of chemicals, 455
Iodine, 29
Iodophor, 423, 424
Ion, 250, 274
Ionization chamber, 15, 250, 259
Ionizing radiation, 250, 278; biological effects from, 14–15, 29, 64, 68, 69, 110, 134, 136, 255–57, 723; and chemical exposure, 142; control methods for, 264–69; exposure factors, 261–62; monitoring exposure to, 15, 189, 249, 269; exposure standards for, 257–58; external hazard from, 14, 15, 253, 254–55, 269; internal hazard from, 14, 15, 252, 253, 269; latent period, 256; measurement of, 15, 258–61; nuclear, 251; operational safety factors, 268–69; from particulates, 180, 189; record keeping for, 269; shielding from, 262–64, 546; sources of, 265–68; types of, 251–55
Ion mobility spectrometer, 524
Ion pair, 250, 251
Iridoplegia, 118
Iris, 105, 296
Iritis, 112
Iron oxide, 188
Irritant, 8, 161, 162, 164; of respiratory system, 129–31, 179–80; vs. sensitizer, 66; strong vs. weak, 66. *See also* Dermatitis, irritant contact
Irritant contact dermatitis. *See* Dermatitis, irritant contact
ISO. *See* International Organization for Standardization
Isocyanate, 62
Isoinertial technique, for material-handling personnel selection, 368
Isokinematic technique, for material-handling personnel selection, 368
Isolation, 30, 532, 537, 538–41. *See also* Containment; Enclosure; Shielding
Isomer, 155
Isometric work, 19, 326
Isotope, 250, 254
Itching, 73

**J**

Jaeger notation, 118, 119
Jet dust sampler, 191
Job design, 364
Job safety and health analysis, 691
*Job Safety and Health Quarterly*, 839

Johannson, contributions to ergonomics by, 348
Joint Commission on the Accreditation of Healthcare Organizations (JCAHO), 407–8, 440

**K**

Kaolin dust, 181, 188
Kepone incident, 793, 795
Keratin, 56
Keratin layer. *See* Stratum corneum
Keratinocyte, 55, 297
Keratin solvent, 62
Keratin stimulant, 62–63
Keratitis, 61, 112, 287
Keratoacanthoma, 69
Keratoconjunctivitis, 68, 112
Keratosis, 69
Ketone, 155, 156; and air pollution, 167; aliphatic, 132; breath analysis for, 146; for disinfection, 423; toxicological effects of, 8, 29, 132, 164
Keyboard user, cumulative trauma disorders of, 392
Kick, S. A., 80
Kidney damage, 164, 186
Kirkland, Lane, 791, 812
Konimeter, 191
Korean hemorrhagic fever, 406, 409
Kuru, 405
Kyphosis, 382

**L**

Label (for controls and displays), 390–91
Laboratory: biological hazards in, 406–7; history of OSHA regulation of, 827; tuberculosis controls for, 431–32
Laboratory animal allergy (LAA), 408
Labyrinthitis, 93
Laceration, 58, 63; to eye, 110–11
Lacrimal gland, 104
Lacrimation, 104, 112, 118
Lactic acid, 60, 355
Lagophthalmia, 118
L'Ambiance Plaza construction accident, 831, 834
Landolt's broken-ring chart, 118, 119
Langerhans' cells, 55
Laryngitis, 39, 47, 93
Larynx, 37, 39
Laser, 299–301; biological effects of, 16, 301; and blinking, 297; continuous wave (CW), 301; controls for, 115, 304–8; dermatosis from, 64; exposure guidelines for, 301–4; nonbeam hazards of, 308–9; pointers, 308; pulsed, 301, 304
Laser Institute of America (LIA), 299, 301–8, 311
Latex allergy, 65, 67, 80
Lavoisier, contributions to ergonomics by, 348
Lawrence Livermore National Laboratory, 309, 310
Lead, 186–87; biological effects of, 133, 136, 723; biological sampling for, 465, 724; blood analysis for, 145; cleaning and maintenance program for, 545; evaluation of hazard of, 186; federal regulation of, 783, 801, 802, 807, 824, 827
Lead Industries Association (LIA), 807
Leather. *See* Animals
*Legionella*, 405, 433; *pneumophila*, 410
Legionnaires' disease, 433

Length-of-stain dosimeter, 518, 519
Lens (of eye), 104, 105, 296; injury to, 109, 112. *See also* Aphakia
*Leone v. Mobil Oil Corp.,* 793
*Leptospira,* 405; *interrogans,* 410
Lethal concentration (LC), 127–28
Lethal dose (LD), 127, 138, 481
Leukemia, 134, 162
Leukoderma, 69
Lichen planus, 65
*Lift Guide* software, 763
Lifting. *See* Material handling
Lifting index (LI), 370
Ligament, 393
Lighting, 15–16, 117, 311–14; colored, 313–14, 390; and contrast, 312–13; excessive, 110; and glare, 311–12, 313, 380; for video display terminal use, 314
Light signal, colors for, 390
*Limitation of Exposure to Ionizing Radiation,* 269
Limit of detection (LOD), 494
Lind, Alexander, 328
Lipid-soluble material, 55, 62
Literature review, and hazard evaluation, 454–55
Liver damage, 163
Load constant (LC), 368–69
Load handling. *See* Material handling
Loading operation, 458, 536
Local exhaust ventilation. *See* Exhaust ventilation
Lockout/tagout (LOTO), 829–30
Longshoremen and Harbor Workers Act, 790, 791
Lordosis, lumbar, 382
Loss factor (ventilation), 555
*Louisiana Chemical Association v. Bingham,* 811
Low back problem, 719
Lower confidence limit (LCL), 473–74
Lower explosive limit (LEL), 161, 510, 513, 642
Lowest-feasible-level policy, for carcinogens, 805–6
Lung, 40–41, 44–45; volumes and capacities of, 45–46, 49–50
Lupus erythematosus, 65
Lyme disease, 405, 410
Lymphatic vessel, 55
Lymphocytic choriomeningitis, 406

**M**

Macula, 296, 297
Magnetic field, 274–75; biological effects of, 281, 282–84, 287–88; control of, 285–86; exposure guidelines for, 282–84, 287–92; measurement of, 284–85, 292–93; microwave, 286–93; pulsed, 292; vs. radiation, 279; radiofrequency, 286–93; static, 282–83; strength of, 280; subradiofrequency, 280, 281, 284–86; time-varying, 283–84; types of, 279; in video display terminal, 294
Magnetohydrodynamic (MHD) voltage, 282
Maintenance. *See* Cleaning and maintenance
Malaria, 411
Malignancy. *See* Cancer
Malleus, 85, 86, 87
Management, role in occupational health, 4, 80, 758
Manganese poisoning, 133–34
Mansdorf, S. Z., 79
*Manual of Analytical Methods,* 494–95, 500–3
*Manufacturers Association of Tri-County v. Krepper,* 819
Marburg virus, 409

Marine terminal, 821
Mariotti bottle, 496
Maritime Safety Act, 770
Marshall, Ray, 804, 812
Marshall, Thurgood, 806
*Marshall v. Barlow's Inc.,* 809
Martin, Lynn, 835
*Martin v. Occupational Safety and Health Review Commission,* 824, 838
Maser, 16, 299–300
Mastoid air cells, 85, 86, 93
Mastoiditis, 93
Material handling, 18–19, 364–72
Material Safety Data Sheet (MSDS), 8, 9–10, 11, 146–49, 773–74; for dermatosis diagnosis, 72; and EPA Hazardous Chemical Reporting Rules, 785; in hazard evaluation, 27; infectious agent exclusion, 413; and the Paperwork Reduction Act, 823–24, 827; for organic solvents, 155, 157–58; training for, 551
Materials conveying, 576–77
Maximum permissible dose (MPD), 264
Maximum permissible exposure (MPE), for lasers, 303, 305, 306
Maximum permissible load, 368
McBride, Lloyd, 811, 812
McGarity, Thomas, 840
McGowan, Carl, 792
McKevitt, Mike, 813
McNamara–O'Hara Service Contracts Act, 228, 770
Mechanical hazard, 63
Mediastinum, 40
Medic Alert Foundation, 713
Medical examination, 715–20; for hazard monitoring, 184, 186, 335, 435, 464; periodic, 715–19; preplacement, 29, 74, 367, 706, 713–15, 734–35; required by OSHA, 777
Medical facility. *See* Employee health care facility; Health care facility
Medical records: computerized, 761–62; employee access to, 504, 721, 772
Medical review officer (MRO), 721
Medical surveillance: for hazard monitoring, 335–36, 343, 410, 426, 465–66, 551, 621; occupational health nurse's role in, 734; OSHA requirement for, 773, 801, 802
MEDLARS (Medical Literature Analysis and Retrieval System), 150
Meissner's corpuscle, 350
Melanocyte, 55, 297
Mellstrom, G. A., 79, 80
Membranous labyrinth, 86, 89–90
Mendeloff, John E., 840
Ménière's disease, 93
Mental workload, 353
Mercaptobenzothiazole, 61
Merchant, James, 806
*The Merck Index,* 155
Mercuric chloride, 29
Mercury, 61, 133, 136, 723
Mercury vapor monitor, 517, 519
Merkel cell, 55
Mesothelioma, 133, 182, 183
Metabolic cost (of work), 354–55, 356
Metabolic rate, 320, 321; assessment of, 326; time-weighted average (TWA) for, 328–29
Metabolite, 144, 145

Metal, 48, 133; radioactive, 267

Metal fume fever, 179, 189

Metallic salt: biological effects of, 28–29, 60–61, 62, 63, 134; for disinfection, 423

Metal oxide semiconductor monitor, 510–11, 516

Metamorphopsia, 118

Methanol, 163–64

*Methods of Air Sampling,* 494

*Methods of Testing Air Cleaning Devices Used in General Ventilation for Removing Particulate Matter,* 601

4,4-methylene(bis)-2-chloroaniline (MOCA), 801, 803

Methylene chloride, 828

Methylenedianiline (MDA), 817, 828

Methylene diisocyanate (MDI), 520

Methyl mercury, 723

Mica dust, 181, 188

*Micrococcus,* 438

Microorganism: concentrations of, 438; hazards from, 20; identification of, 404–5; routes of entry, 411; safety guidelines for, 439–40. *See also* Bacteria; Fungus; Virus; *specific microorganism*

Microscopic counting, 191

Microtrauma, 391

Microwave oven, 292, 295

Microwave radiation, 278, 286–87; biological effects of, 15, 16, 287–88; control of, 295; dosimetry, 288–89; exposure guidelines for, 287–92; field vs. radiation, 279; measurement of, 292–93

Midbrain, 349

Middle ear, 85–86, 87–89, 93

Migrant Legal Action Program, 820

Miliaria, 63, 69, 324

Mine Safety and Health Administration (MSHA): certification standards of, 547, 624–25, 628–30, 781; and coal dust exposure, 187; establishment of, 782, 797; first-aid kit recommendations of, 711; medical surveillance requirements of, 551; and OSHA jurisdiction, 813

Mining, 24, 187, 410

Mining Enforcement and Safety Administration (MESA), 782, 783

Mist, 22, 175; biological effects of, 66, 112; monitoring, 477, 525–26; respiratory protection for, 576, 626–30

Model Accreditation Plan (for asbestos), 183, 184

Moderator, 250

Modified duty program, 735–36

Moisture control (of particulates), 194

Molecule, 250

Moment, 360–62

Monitoring, 461–67; asbestos exposure, 184, 490; ionizing radiation exposure, 15, 189, 249, 269; noise exposure, 207–8, 212–15, 229, 232–33, 242–43; OSHA evaluation of, 777; OSHA requirement for, 773, 801, 802; particulate exposure, 190–92. *See also* Evaluation; Sampling

Monocular vision, 105, 109

Montoya, Velma, 838

Montreal Protocol on Substances that Deplete the Ozone Layer, 167

Moran, Robert D., 791

*Morbidity and Mortality Weekly Report,* 414

"De Morbis Artificium Diatriba," 197

Motion-related injury. *See* Cumulative trauma disorder

Motion sickness, 90

Motion time, 352

Motor vehicle safety, 832

*Mucor,* 438

Mucous membrane, 37, 38, 39; absorption through, 725; hazards to, 60, 69, 131; of middle ear, 85, 86; as natural defense, 48, 179; penetration by infectious agent, 411, 415

Mucus, 37, 48

Muscle, 362, 393

Muscle strength, 360–64

Mutagen, 135–36

*Mycobacterium,* 405, 407, 410; *bovis,* 413, 430; *marinum,* 410; *tuberculosis,* 407, 411, 429

*Mycoplasma,* 405

Mycosis, 410

Mycotoxin, 434

Myopia, 108

# N

*Naegleria,* 410

Nail (finger and toe), 56–57, 60, 70

Narcotic, 8, 162, 164

Nasal cavities, 37, 39

Nasal mucosa. *See* Mucous membrane

Nasal septum, 37, 61

Nasopharynx, 37, 38

National Academy of Science, 404, 822. *See also* National Research Council

National Advisory Committee on Occupational Safety and Health (NACOSH), 793, 799

National Cancer Institute, 439

National Committee for Clinical Laboratory Standards, 440

*National Congress of Hispanic American Citizens (El congreso) v. Donovan,* 820

*National Congress of Hispanic American Citizens (El congreso) v. Usery,* 820

National Council on Radiation Protection and Measurements (NCRP), 257–58, 262, 266, 269

National Environmental Balancing Bureau (NEBB), 602

National Executive Poll on Health Care Costs and Benefits, 745

National Federation of Independent Business, 813

National Fire Protection Association (NFPA): fire hazard standards of, 159, 160, 161, 308; OSHA standards adopted from, 791; respirator standards of, 642; safety professionals' familiarity with, 681; ventilation standards of, 570

National Foundation on the Arts and Humanities Act, 770

National Institute for Occupational Safety and Health (NIOSH), 7, 780–82, 783; Appalachian Laboratory for Occupational Safety and Health (ALOSH), 780; carcinogen, definition of, 134; catalogue of industrial hygiene continuing education programs, 669; criteria documents, 140, 772, 781; Current Intelligence Bulletin (CIB), 140; Educational Resource Center (ERC), 7, 667, 669, 781; establishment of, 138, 228, 771, 780; and Federal Coal Mine Health and Safety Act, 782; filter certification, 491, 628–30; Health Hazard Evaluation (HHE) by, 7, 481, 781–82; heat exposure guidelines, 13, 325–26, 328–29, 335, 336; Human Subjects Review Board, 781; and indoor air quality problem sources, 599; *Lift Guide* software, 763; *Manual of Analytical Methods,* 468, 470, 472–73, 494–95, 500–3; material handling guidelines, 366, 368–72; *Occupational Respiratory Disease,* 180; occupational health nursing study by, 746; *Pocket Guide to Chemical Hazards,* 455, 642; pregnancy guidelines, 724; Proficiency Analytical

Testing Program, 494; Recommended Exposure Limit (REL), 7, 26, 140, 455, 467, 480; Registry of Toxic Effects of Chemical Substances (RTECS), 161, 481; respirator certification by, 194, 431, 547, 624–25; sampling device certification by, 519; sampling guidelines of, 466, 468, 470, 472–73, 487, 494–95, 500–3; skin injury statistics by, 58–59; solvents, guidelines of, 161; sorbent tube recommendations of, 490; standards development by, 703, 781, 783; technical assistance to OSHA by, 771, 772, 775; Testing and Certification Branch, 781; training by, 781

National Institute of Standards and Technology, 831

National Institutes of Health (NIH): biosafety guidelines, 407, 412, 415–16, 425, 439, 445; *Biosafety in Microbiological and Biomedical Laboratories (BMBL),* 414, 415, 426, 429, 439, 445; on mine health advisory committee, 782

National Priority List (NPL) hazardous sites, 784

*National Realty and Construction Co. v. OSHRC,* 796

National Research Council: biosafety publications of, 416, 426, 439; and Bureau of Labor Statistics reporting, 836; Committee on Hazardous Biological Substances in the Laboratory, 416; Committee on Human Factors, 347

National Safety Council, 440; exposure guidelines of, 151, 266; first-aid kit recommendations of, 711; resources available from, 685, 763; safety professional's familiarity with, 681

National Safe Workplace Institute, 840

National Sanitation Foundation, 420, 440

National Science Foundation, 782

National Society to Prevent Blindness, 106

National Stone Associations, 819

National Toxicology Program, 163

Natural wet bulb temperature, 326

Nearsightedness, 108

Neck tension syndrome, 385, 396

Negative pressure enclosure (NPE), 185

Negotiated Rulemaking Act, 837

*Neisseria,* 405

Neoplasm. *See* Cancer

Nerve, 350, 394

Neurasthenia, 133

Neurotoxic effect, 132–33

Neutron, 250, 251

Neutron radiation, 14, 253

*New Jersey State Chamber of Commerce v. Hughey,* 819

Newton's laws of motion, 360, 362

Nickel allergy, 65

Nickel salt, 61

Nicotine, 723

Night blindness, 110

Nitrate, 29, 723

Nitric acid, 60

Nitric oxide, 167, 516

Nitrobenzene, 723

Nitro compound, 156; biological effects of, 29, 61, 163; sampling device for, 522; and solvent hazard control, 169

Nitrogen, liquid, 159

Nitrogen dioxide: and air pollution, 167; respiratory hazards from, 24, 48, 130; sampling device for, 516

Nitrogen narcosis, 159–60

Nitrogen oxide, 577

Nitrous oxide, 723

Nixon, Richard, 788, 794

*Nocardia,* 405

Noise, 207, 208, 210–11; cleaning and maintenance program for, 545; continuous, 208, 215–16; control methods for, 216–24; effects of, 11, 92–93, 97–98, 99, 198, 467; exposure guidelines, 11–12, 206–8, 228–29, 232–45, 807–8; exposure levels, in manufacturing, 197, 198; exposure monitoring, 207–8, 213–15, 225–26, 229, 232–33, 242–43; from fan, 570; impact, 208, 216; intermittent, 216; measurement of, 206, 208–13; nonoccupational, 92; permissible levels of, 12; protective equipment for, 220–24, 230, 235–36, 239–40; vs. sound, 199. *See also* Hearing; Sound

Noise dosimeter, 211–12, 216

Noise Reduction Rating, 224

Nominal hazard zone (NHZ), for lasers, 307

Nomogram, 327

Nonionizing radiation, 278, 279–80; biological effects of, 15–16, 112, 280–84, 287–88, 296–98; controls for, 285–86, 295, 299; exposure guidelines for, 283, 287–92, 298–99; field vs. radiation, 279; industrial, scientific, and medical (ISM) bands, 286–87; infrared, 296–99; measurement of, 284–85, 292–93, 311; microwave, 287–93, 295; optical, 296–99, 311; protective eyewear for, 114–15; pulsed, 292; radiofrequency, 287–93, 295; subradiofrequency, 279–86; ultraviolet, 296–99

No Observable Effect Level (NOEL), 127, 481

*Northwest Airlines OSHRC review,* 797, 813

Nose, 37

Nuclear Regulatory Commission, 257

Nucleus (of atom), 251, 273–74

Nystagmus, 90, 110

**O**

Obstructive bronchopulmonary disease, 46, 722

*Occupational Exposure Sampling Strategy Manual,* 470

Occupational health and safety team, 4–6, 168

Occupational health and safety technologist (OHST), 661

*Occupational Health Guide for Medical and Nursing Personnel,* 713

Occupational health hazards, definition of, 4

Occupational health nurse (OHN), 730; certified (COHN), 731; and demographics, 742, 745–46; functions of, 729, 733–41, 746–47; role in occupational health program, 6, 80, 705, 722–23, 731–33; scope of practice of, 731; standards of practice of, 730

Occupational health program, 426, 703–9; and industrial hygiene program, 756–57; models of, 731–33; organization of, 707–9; OSHA evaluation of, 777

Occupational hearing conservationist (OHC), 226, 229

Occupational medicine, 701

Occupational physician, 6, 80, 707–9

*Occupational Respiratory Disease,* 180

Occupational Safety and Health Act. *See* OSHAct

Occupational Safety and Health Administration. *See* OSHA

*Occupational Safety and Health Law,* 840

*Occupational Safety and Health Reporter,* 840

Occupational Safety and Health Review Commission (OSHRC), 292, 532, 770, 774, 780, 809, 824, 833, 838

Occupational safety and health specialist, 666

Occupational safety and health technologist (OSHT), 677

Occupational skin disease. *See* Dermatosis

Occupied zone, 596

Octave-band analyzer, 210–11

Ocular motility, 120

Oculist, 105

Odor, 584

Office chair. *See* Seat

Office of Management and Budget (OMB), 817, 821, 823–24, 827

Office of Technology Assessment (OTA), 788

Office workstation: design of, 379–89; health problems from, 379–80; and posture, 380–86

O'Hara (House subcommittee chair) and OSHA farm workers oversight hearing, 797

*Ohio Manufacturers' Association v. City of Akron,* 819

Ohm's law, 275, 277

Oil and gas drilling, 822

Oil, Chemical and Atomic Workers International Union, 793, 811

*Oil, Chemical & Atomic Workers International Union v. OSHRC,* 810

Oil gland, 55, 56

Oil mist, 576

Olefinic hydrocarbon, 146

Olfactory nerve, 37

OMB. *See* Office of Management and Budget

Omnibus Budget Reconciliation Act (OBRA), 830, 833, 837

Oncogen. *See* Carcinogen

Open-surface tank, 571–72

Ophthalmologist, 105

Ophthalmoscope, 106

Optical density (OD), 299

Optical radiation, 278; biological effects of, 112, 296–98; controls for, 299; exposure guidelines for, 298–99; measurement of, 311; photochemical effects of, 296, 299; protective eyewear for, 110, 114–15, 116; thermal effects of, 296, 299. *See also* Infrared radiation; Laser; Ultraviolet radiation

Optician, 105

Optic nerve, 104, 110, 164

Optometrist, 105

Oral temperature. *See* Body core temperature

Ora serrata, 104

Organic chemistry, 155

Organic compound, 155, 161–65, 434, 533

Organic solvent, 155–59; biological effects of, 63, 132, 161–65; breath analysis for, 146

*Organized Migrants in Community Action v. Brennan,* 801

Organ of Corti, 87, 90

Orifice-type collector, 567

Orthopedic model, for material handling personnel selection, 367

OSHA, 6–7, 138–39, 228, 770–71; air-sampling worksheet, 504–6; Analysis and Evaluation Group, 775; and biological hazard assessment, 416; *Chemical Information Manual,* 494; citation appeals, 779–80; citation by, 778–79, 790, 796, 809; "common-sense" enforcement policy of, 808–10; compliance activity, 774, 775, 832–34; *Compliance Manual,* 790, 795; compliance programs for, 777–78; consultation service, 243–44, 685, 775, 814–15, 839; cumulative trauma disorder program, 391; Division of Administration and Training Information, 664; Division of Training and Educational Development, 664; Division of Training and Educational Programs, 664; Enforcement and Investigation Unit, 775; evaluation of, 839–41; Federal Agency Safety Programs, 804, 812; *Field Inspection Reference Manual* (FIRM), 664, 774, 790, 834; field operations, 774–75; form 200 record-keeping log, 738–41, 776; *Industrial Hygiene Field Operations Manual* (IHFOM), 795; injury and illness reporting software, 763; inspection by, 774, 775–78, 790, 795–96, 809, 814; Integrated Management Information System (IMIS), 816; *Job Safety and Health Quarterly,* 839; jurisdictional issues, 794, 796–97, 813, 838–39; medical surveillance of PEL effectiveness, 465; membership in Interagency Regulatory Liaison Group (IRLG), 813; National Emphasis Program (NEP), 795, 808; New Directions grant program, 809–10, 815, 839; occupational health program evaluation by, 777; Office of Training and Education, 664; petrochemical industry special-emphasis program (PetroSEP), 834; Program Planning and Outreach Unit, 775; program structure, 788–89; reorganization of, 775; respiratory protection program, 439, 548, 620; *Right-to-Know* software, 763; sampling by, 776–77, 778; State Plan Task Group, 816; state plans, 771, 789, 792–93, 799–800, 815–16, 819, 834–35; *Technical Manual,* 775; testing laboratory qualifications, 831; Toxic Substances List, 139–40; Training Institute, 664, 665; training program for industrial hygienists, 663–66; *Training Requirements in OSHA Standards and Training Guidelines,* 753; tuberculosis infection control plan, 432; Voluntary Protection Program (VPP), 775, 815, 839; Voluntary Training Guidelines, 550–51

OSHA Appropriations Act, 791

OSHAct, 6–7, 138–39, 228, 770, 787–88; amendments, riders, and oversight, 797–99, 811–13, 837; checks and balances for, 789; employee participation under, 789, 793–94, 810–11; employer responsibilities under, 789; enforcement of, 774, 789, 832–34; general responsibilities, 770–71; NIOSH certification authorized by, 781; and occupational medicine, 702–3; provisions of, 771; record-keeping burden from, 759; standard-setting process defined by, 772; universal coverage, 788–89; whistle-blower provision, 793–94, 838

*OSHA's Failure to Establish a Farmworker Field Sanitation Standard Hearing,* 821

OSHA standard: Access to Employee Exposure and Medical Records, 504, 721, 772; for acrylonitrile, history of, 803, 807; for agriculture, 800–1, 839; for arsenic, history of, 801, 807; for asbestos, 183–84, 791–92, 801, 817, 819, 826; for benzene, history of, 802, 805–6, 817, 828; for beryllium, 187, 801; for blood-borne pathogens, 404, 407, 428, 429, 439, 741, 823, 825–26; for carcinogens, 545, 801, 802, 803; for cleaning, 459, 545; for coke oven emissions, history of, 801, 803–4, 824; for compressed air for cleaning, 459; computerized information on, 762; for confined spaces, history of, 831; for construction industry, 770, 800, 819, 831, 834; cost–benefit analysis for, 805–7, 816, 817–18, 830; cost-effectiveness approach for, 817; for cotton dust, history of, 802, 806, 817–18, 828; criminal violations of, 833–34; for 1,2-dibromo-3-chloropropane (DBCP), history of, 807; for diving, 801, 802, 803, 808, 821; for electricity, history of, 821, 831; emergency temporary standard (ETS), 800–1, 802, 803, 807, 832; for ergonomic hazards, history of, 823, 831–32, 834; for ethylene oxide, history of, 802, 817, 821, 826; for ethyleneimine, history of, 803; for exhaust ventilation, 554; for eye and face protection, 113–14, 172, 299; for field sanitation, history of, 817, 820–21; for flammable and combustible liquid handling, 161; for general industry, 770; for grain elevators, history of, 822, 830; Hazard Communication, 8, 138–39, 146, 413, 455, 551, 773–74, 777, 811, 818–19, 823–24, 827; for hazardous waste management, history of, 772, 824, 830; Hearing Conservation Amendment, 12, 100–1, 228–31, 232–45, 801, 818; for heat stress, 335; history of, 796, 801–4, 809, 816–22,

825–32; horizontal, 772, 800; for hydrotreated oil, 576; for indoor air quality, 602; for laser hazard, 16, 309–11; for lead, 186–87, 801, 802, 807, 827, 837; for lighting, 312; limitations of, 413, 480–81; for lockout/tagout (LOTO), history of, 829–30; maritime, 770, 800, 819, 821; for medical surveillance, 466, 551, 773; for methylenedianiline (MDA), history of, 817, 828; for monitoring exposure, 773; for motor vehicle safety, history of, 832; NIOSH recommendations for, 781, 783; for noise, 12, 206, 215–16, 228–29, 232–45, 807–8; for oil and gas drilling, history of, 822; Occupational and Environmental Control, 772; performance, 772; for personal protective equipment, 439, 772; for pesticides, history of, 800–1, 802, 803; for power presses, history of, 831; for process safety management, 459–60, 829; for protective clothing, 78; for radiofrequency and microwave exposure, 292; for record keeping, 686–87, 777, 789, 794; Regulatory Impact Analysis (RIA) required for, 801, 817; requested for chromium, 829; requested for tuberculosis, 829; for respirators, 194–95, 624; for respiratory protection, 439, 548, 619, 620, 639, 652–56, 777; for sanitation facilities regulation, 20–21, 75; setting process, 771–74, 801, 802, 832, 837; for shipyards, 770; for sound level meters, 210; for spray booth ventilation, 573; start-up, 791, 825; for sulfur dioxide, history of, 801; Toxic and Hazardous Substances, 772; for training, 550–51, 773, 777, 789; for trichloroethylene, history of, 801; vertical, 772, 800; for vinyl chloride, history of, 801, 802, 803; violation of, 778–79; for wheel rim servicing, history of, 821; for woodworking ventilation, 575. *See also* Action limit; Permissible exposure limit; Short-term exposure limit; Time-weighted average
Osseous labyrinth, 86, 89
Ossicles, 85, 86, 87–89, 93–94
Osteoarthrosis. *See* Cumulative trauma disorder
Otalgia, 100
Otitis media: nonsuppurative, 92; suppurative, 93
Otomycosis, 92
Otorrhea, 100
Otosclerosis, 93
Outdoor air, 595, 596, 603–5
Oval window (of ear), 85, 86, 87–89
Overexertion, 365, 367
Overloading, 348
Overuse disorder. *See* Cumulative trauma disorder
Oxalic acid, 60
Oxidizing material, 11, 29, 62
Oxygen: direct-reading monitor for, 515; in respiration, 35–36, 41–43, 354
Oxygen-containing compound, 522
Oxygen tension, 42–43
Ozone: atmospheric, 167; hazards from, 24, 29, 48, 130

**P**

Pacemaker, 282, 284
Pacinian corpuscle, 350
Packed tower, 565
Padosimeter, 75
Pair production, 250, 251
Paperwork Reduction Act, 823–24, 827
Papilla, 55, 56
Paraformaldehyde, 424
Parasite, 20, 29, 64, 404, 410
Paresis of accommodation, ocular, 118
Parker, Barrington, 793

Paronychia, 70
Partial pressure, 474–75
Particulate: allergenic, 179; control methods for, 192–95; definition of, 175; direct-reading monitor for, 509, 525–26; effects of, 23, 24, 41, 47–48, 133, 179–90; exposure limits for, 129, 140–43; infectious, 409, 411, 412–13, 415, 434; ingestion of, 176–77; inhalation of, 125; inorganic, 177; lethal concentration of, 127; monitoring exposure to, 190–92; natural defense against, 48; not otherwise classified (PNOC), 188; nuisance, 133, 188; organic, 177; radioactive, 189; respirators for, 626–30, 633–34; sample collection device for, 489–92; size, 177, 178–79; types of, 21–23, 47
Passive monitor, 489
Patch test, 67, 72, 74
*Patty's Industrial Hygiene and Toxicology,* 161
PCB. *See* Polychlorinated biphenyl
Peak expiratory flow (PEF), 46
Peak expiratory flow rate (PEFR), 46, 49
Peak flow meter, 46
Pedal, 377
Pendergrass, John, 813, 822, 823, 828
*Penicillium,* 410, 438
Pensky–Martens Closed Tester, 160
Pentachlorophenol, 61
*People v. Chicago Magnet Wire Corp.,* 835
Perchloroethylene, 163, 167
Peres, N. J. V., 391–92
Perilymph, 89
Perimeter, 106, 118, 120
Peripheral nervous system, 349, 350, 351, 393
Peripheral neuropathy, 133, 162, 164
Permanent threshold shift (PTS), 98
Permanganate, 29
Permissible exposure limit (PEL), 26, 31, 138–39, 140, 215, 762, 791; and action level, 772–73; for airborne contaminants, 467, 583, 825; for arsenic, history of, 807; for asbestos, 184, 819, 826; for benzene, 162, 805, 828; for cadmium, history of, 828; for carbon tetrachloride, 163; and chemical inventory, 455; for coal dust, 187; for coke oven emissions, history of, 803–4; and confidence limits, 473; for cotton dust, history of, 806, 828; for formaldehyde, history of, 827; for glycol, 165; for lead, 186, 807, 827; and medical surveillance, 465, 466; for methylenedianiline, history of, 828; for particulates, 177; proposed for butadiene, 828; proposed for methylene chloride, 828; and sampling, 470, 479–81; setting procedure for, 801, 802; for silica dust, 181–82; for solvents and gases, 154;
Permissible heat exposure limit (PHEL), 329
Personal hygiene, 544–45; and biological hazard control, 422; and dermatosis, 65, 75–77; for radioactive particulate control, 189; for solvent hazard control, 172; for thermal stress control, 334, 343
Personality disorder, 719
Personal protective equipment, 29, 30, 31, 532–33, 537, 546–50; for asbestos, 184–85; for biological hazards, 422; for building-related illness control, 439; clothing, 170, 548–49; documentation of use of, 470; OSHA requirement for, 772, 777, 801; for particulates, 194–95; for radioactive particulates, 189; for skin, 75–76, 77–80; for solvents, 170–72; supplied-air suit, 640; for thermal stress, 338–39, 344; for tuberculosis, 431; and weather, 65; for zoonotic diseases, 410. *See also* Gloves; Hearing-protective device; Protective eyewear; Respirator
Perspiration. *See* Sweat

Pesticide, 61, 139, 466; history of OSHA regulation of, 800–1, 802, 803

Petroleum products, 61, 62, 134, 162

Phagocyte, 48

Phalen's test for carpal tunnel syndrome, 394

Pharynx, 37–39, 86

Phenol, 29, 60, 156, 724

Phenolic, 423, 424

Phenyl hydrazine, 61

Phenylmercury compound, 61

Phosgene, 48, 130, 520

Phosphorus pentoxide, 28

Photoacoustic spectrometer, 523

Photoallergy, 57, 67

Photoelectric effect, 250

Photoionization detector, 521–22

Photon, 250, 277

Photopatch test, 72

Photosensitizer, 63, 64, 67–68

Phototoxicity, 67

Phototropic lenses, 116

Physical capacity, 714–15

Physical examination. *See* Medical examination

Physical hazard, 7, 11–17, 63–64, 110–12

Physiological model for material-handling personnel selection, 367

Picric acid, 29, 60

Pigmentary abnormalities, 68–69

Pinna, 84

Pitch. *See* Petroleum products

Plant Pest Act, 439

Plant toxin. *See* Biological toxin

*Plasmodium,* 405

Plethysmography, 354–55

Pleura, 40, 47

Pleurisy, 40, 47

Plutonium, 250, 268

Pneumoconiosis, 8, 46, 47, 133, 175; coal workers', 187, 780; from dust, 188; and respirator use, 722

Pneumomediastinum, 722

Pneumonia, 24, 47

Pneumonitis, 47, 131, 187

Pneumothorax, 40

*Pocket Guide to Chemical Hazards,* 455, 642

Polarity (of electric field), 274

Polarization, 288

Polarographic detector, 515

Poliovirus vaccine, 414

Polychlorinated biphenyl (PCB), 136, 723, 724, 783

Polycyclic aromatic hydrocarbon (PAH), 308

Polymer, 8

Pontiac fever, 433

Positron, 252

Posture, 382–86, 395

Potash, 28

Potassium cyanide, 60

Potassium hydroxide, 60

Potentiometer, 515, 516

Power press, 831

*Practice for Occupational and Educational Eye and Face Protection,* 172

*Practices for Respiratory Protection for the Fire Service,* 642

*Practices for Ventilation and Operation of Open-Surface Tanks,* 572

Precision rotameter, 499

Pregnancy. *See* Reproductive hazard; Teratogenesis

Presbycusis, 92, 93, 198, 204

Presbyopia, 106, 108

Pressure, 29

*Preventing Illness and Injury in the Workplace,* 788

Prickly heat, 63, 69, 324

Primary irritant: dermatitis from, 28, 58, 62, 64, 65, 66, 92, 161; respiratory effects of, 131

Prion, 405

*Proceedings of the 4th International Symposium on Contamination Control,* 169

Process analysis, 453, 455–60

Process flow sheet, 456–58, 534

Process modification, 30, 533, 537, 538; for noise control, 218; for skin hazard control, 74–75; for solvent hazard control, 168–69

Process safety management, 459–60, 829

*Program Criteria for Industrial Hygiene and Similarly Named Engineering-Related Programs,* 667–68

Progressive massive fibrosis (PMF), 181

Prokaryote, 404–5

Pronator (teres) syndrome, 396

Proprioceptor, 350

Prospective study, 138

Protective clothing. *See* Personal protective equipment

Protective eyewear, 113–16, 549; fitting, 116, 117; goggles, 114; for lasers, 306–7, 308; for optical radiation hazards, 114–15, 299, 300; plastic vs. glass, 115–16; for solvent hazard control, 172; spectacles, 114; sunglasses, 110, 116

Protein allergen, 20, 29, 64, 403, 408, 410, 412

Protein participant, 28, 63

Proton, 250, 251, 273–74

Protozoa. *See* Microorganism

Proximal stimulus, 350, 352

Pruritus, 73

*Pseudomonas,* 405

Psittacosis, 40, 413

Psoriasis, palmar, 65

Psychophysical model, for material-handling personnel selection, 367

Psychrometric wet bulb temperature, 326

Public Citizen, 815, 817, 818

*Public Citizen Health Research Group v. Auchter,* 821

*Public Citizen Health Research Group v. Tyson,* 821

Public Health Service, 439, 780, 784

Public Law 91–173. *See* Federal Mine Safety and Health Act

Public Law 91–596. *See* OSHAct

Pulmonary edema, 24, 47, 48, 130

Pulmonary mycosis, 413

Pulmonary ventilation, 49–50, 125, 719

Puncture. *See* Laceration

Pupil, 105, 296

Pyrethrum, 61

Pyroelectric detector, 311

**Q**

Q fever, 406, 409, 410, 413

Q-switching, 301, 309

Quality factor (Q), 250, 253

Quantum detector, 311
Quartz. *See* Silica dust
Quaternary ammonium compound, 423, 424
*Quick Selection Guide to Chemical Protective Clothing,* 549

# R

Rabies, 410
Race, 480
Rad, 250
Radian, 298
Radiant heat, 13, 320, 546
Radiation. *See* Ionizing radiation; Nonionizing radiation; *specific radiation*
Radiation counter, 15, 249, 250, 259, 260–1
Radiation-producing machine, 266
*Radiation Protection Guides,* 262
Radiation safety committee (RSC), 425
Radioactive decay, 249, 255
Radioactivity. *See* Ionizing radiation; *specific radiation*
Radiodermatitis, 64
Radiofrequency radiation, 278, 286–87; biological effects of, 15, 287–88; control of, 295; dosimetry, 288–89; exposure guidelines for, 287–92; field vs. radiation, 279; measurement of, 292–93; uses of, 280
Radioisotope, 250, 266
Radium, 250
Radon, 783
Ramazzini, Bernadino, 197, 391
Rate of evaporation, 583
Rating of perceived exertion, 355–56
Raynaud's phenomenon, 18, 65, 69, 341, 394, 397
Reaction time, 352
Reactivity, 154
Reagan, Ronald, 801, 813, 817
*REASON* software, 763
Receiving canopy hood, 560–62
Recirculation of air, 571
Recombinant DNA, 404, 408, 425; exposure guidelines for, 414, 415, 416
*Recommendations for the Safe Use and Regulation of Radiation Sources in Industry, Medicine, Research and Teaching,* 258
Recommended alert limit (RAL) for heat stress, 328–29
Recommended exposure limit (REL), 7, 26, 140, 467; and chemical inventory, 455; and difference from PEL, 480; for heat stress, 328–29
Recommended weight limit (RWL), 368
Record keeping, 755; for hearing conservation program, 227, 230–31, 236–37; for ionizing radiation exposure, 269; OSHA form 200 log, 738–41; OSHA requirement for, 504–6, 777, 789, 794, 802; for sampling, 470, 504–6; of work-related illnesses and injuries, 686–88
Red Cross, 711
Reducer, 63
Reflective clothing, 339
Refraction equipment, 106, 118
Refrigerant, 163
Regional musculoskeletal disorder. *See* Cumulative trauma disorder
Registry of Toxic Effects of Chemical Substances (RTECS), 139, 161, 481

Regulated area, 249, 777
*Regulating Safety,* 840
Regulatory Impact Analysis (RIA), 801, 817
Reissner's membrane, 87
Renal injury, 48
Renovation workers, 410
Repetitive motion injury. *See* Cumulative trauma disorder
Repetitiveness, 395, 398
Replacement air, 585–88, 592–93
Reproductive hazard, 136–37, 723–24; from glycol, 164; from ionizing radiation, 258; from lead, 186; from radiofrequency and microwave radiation, 287; and *United Auto Workers v. Johnson Controls Inc.* decision, 824. *See also* Teratogenesis
Residential Lead-Based Paint Hazard Reduction Act, 783
Residual volume (RV), 46
Resin, 8, 62
Resource Conservation and Recovery Act (RCRA), 546, 759, 783–84
Respirable mass sampling, 191–92
Respiration, 41–46
Respirator, 340, 546–48, 619–20; aerosol-removing, 626–30, 632–33; air-line, 634–36, 637–38; air-purifying, 547, 626–34, 638–39; air-supplying, 547, 634–39; for asbestos, 184–85; assigned protection factor (APF), 643, 644; combination aerosol filter/gas or vapor removing, 632–33; combination air-purifying and air-supplying, 638–39; combination self-contained breathing apparatus and air-line, 637–38; for fire fighting, 642; fit, 621–23, 644–52; gas/vapor-removing, 630–34; for health care facilities, 643; maintenance and storage, 623; medical aspects of use of, 722–23; for particulates, 194–95; powered air-purifying, 633–34; for radioactive particulates, 189; regulations for, 624–25; selection of, 548, 621, 639–43; self-contained breathing apparatus, 547, 636–38; for solvents, 170; for tuberculosis control, 431
Respiratory center, 45
Respiratory hazard, 23–24, 47–48, 133; causes of, 176, 179, 180, 187; evaluation of, 640
Respiratory protection program, 620–23, 722–23
Respiratory system, 36–37, 48–50, 353
Reticular dermis, 55
Retina, 104, 105; inflammation of, 109; radiation effects on, 112, 296–97
Retinal pigmented epithelium (RPE), 296
*Retreat from Safety,* 815
Retrospective study, 138
Return air register, 597, 599
Rheumatic disease. *See* Cumulative trauma disorder
Rhinitis, 47, 403
*Rhizopus,* 438
*Rhototorula,* 438
*Rickettsia,* 405
Rickettsial agent, 409
Ridder, seat designs of, 382
*Right-to-Know* software, 763
Risk assessment. *See* Evaluation
Risk management, 691–92
Rocky Mountain spotted fever, 410, 411
Rods (of retina), 105, 109, 296
Roentgen, Wilhelm, 256
Roentgen (R), 250
Roentgen absorbed dose (rad), 250

Roentgen equivalent man (rem), 250
Roentgenogram. *See* X ray, diagnostic
Root-mean-square (rms) sound pressure, 202
Rosenstock, Linda, 832
Rosin, 62
Rotenone, 61
Round window (of ear), 85, 86, 87
Route of entry (of hazardous material), 11, 21, 124, 125–26, 406. *See also* Absorption; Ingestion; Inhalation; Injection
Rowland, Robert, 813, 820
Rubber manufacturing, 61, 81
Rubella, 136, 412
Rubner, contributions to ergonomics by, 348
Ruffini organ, 350
Russian spring summer fever, 410

## S

*Safe Handling of Radioactive Materials,* 266
*Safe Handling of Radionuclides,* 266
Safety and health committee, 6, 80, 750, 753, 758
Safety Equipment Institute (SEI), 519–20
Safety glasses. *See* Protective eyewear
Safety inspection, 684–85
*Safety in the Federal Workplace Hearings,* 804
*Safety Levels with Respect to Human Exposure to Radio Frequency Electromagnetic Fields,* 290
Safety professional, 4–6, 678–80; associate (ASP), 677; certified (CSP), 660, 677, 693–97; duties of, 80, 679–85; future demands for, 697–98; and industrial hygienist, 757
Safety program, 757
Safety technologist, 660
*Salmonella,* 405
Salt, 334. *See also* Metallic salt
Sample collection device, 486; for gases and vapors, 486–89; for particulates, 489–92
Sampling, 467–74; absorption, 487; accuracy and precision of, 472–74; active, 517–18; adsorption, 487–89; area, 463–64, 468, 485–86; collection device for, 486–94; direct-reading instrument for, 510–26; and exposure guidelines, 479–81; grab, 468, 486–87; integrated, 486, 487–89; method, 494–96; by OSHA, 776–77, 778; passive, 489, 518–19; personal, 461–63, 468, 485–86; record keeping for, 470, 504–6; and workplace design, 535, 536. *See also* Air sampling; Biological sampling; Evaluation; Monitoring
Sampling and analytical error (SAE), 473–74, 504
Sampling pump, 462–63
Sampling train, 486
Samuel, Howard, 812
Sanitation facilities, 20–21, 75, 76; field, 817, 820–21
Scala media, 87
Scala tympani, 87
Scala vestibuli, 87
Scannel, Gerard, 823
*Schistosoma,* 410
Schneider wedge, 382
Scholz, John T., 840
Schweiker, Richard, 811–12
Schwope, A. D., 79
Scintillation counter, 250, 260–61
Sclera, 104, 105

*Scope, Objectives and Functions of Occupational Health Programs,* 703–7
Scotoma, 297
Scrapie, 405
Scuba diving. *See* Diving
Seat, 379, 382–83, 385–86, 387–89; stand-seat, 378
Sebaceous gland, 55, 56
Secondary irritant, 131
Secretary of Labor. *See* Department of Labor
Selenium compound, 723
Selikoff, Irving, 787, 803
Sensitization dermatitis. *See* Dermatitis, allergic contact
Sensitizer, 65, 67, 80, 164; and chronic exposure, 129; dermatitis from, 59, 63, 66–67; vs. irritant, 66; particulate, 179, 190. *See also* Allergen, biological
Service Contracts Act, 228, 770
Setaflash Closed Tester, 160
Settling chamber, 567
Settling rate, 179
Sewage, 410
Sex, 65, 480
Shapiro, Sidney, 840
Shaver's disease. *See* Pneumoconiosis
Sheehan, Jack J., 793
Sheet Metal and Air Conditioning National Association (SMACNA), 602
Shellac, 62
Shielding, 250, 262–64; for heat stress control, 337; for radiation and radiant heat, 285–86, 546. *See also* Containment; Enclosure; Isolation
Shivering, 340
Short-term exposure limit (STEL), 25, 141, 455; for asbestos, 184, 826; for benzene, 828; for carbon tetrachloride, 163; for ethylene oxide, 821, 826; for formaldehyde, 827; OSHA setting procedure for, 802; and sampling time, 470
Sick-building syndrome (SBS), 20, 434
Sievert (Sv), 250
Sight. *See* Vision
Silica dust, 24, 133, 180–82, 190
Silicates, 181
Silicon dioxide. *See* Silica dust
Silicosis, 47, 180–81, 187
Sin nombre hantavirus, 409, 414
Sinus, 17, 39, 722
Sitting, 377, 382–86, 387–89
Skin, 53–58; cold hazard to, 341; penetration by infectious agent, 411, 415; radiation effects on, 256, 297–98; and thermal balance, 322. *See also* Dermatitis; Dermatosis
Skin irritation. *See* Dermatitis; Dermatosis
Skin notation, 26, 70, 725
Small employers, 745
Smoke, 22, 568
Smoking. *See* Cigarette smoking
Sneeze, 48
Snellen chart, 106, 108, 118, 119
Snook, S. H., 370–72
Soap, 61, 545
Soap-bubble meter, 496–98
Society for Healthcare Epidemiologists of America, 408, 440
*Society of Plastics Industry, Inc. v. OSHA,* 802

Sodium, 61
Sodium cyanide, 60
Sodium hydroxide, 60
Sodium hypochlorite, 423
Software. *See* Computerized industrial hygiene system
Solid Waste Disposal Act, 783, 784
Solvent, 153–54; biological effects of, 8–11, 28, 29, 61, 63, 132, 161–66; biological sampling for, 146, 724; classification of, 155–59; control methods for, 168–72, 582; evaluating hazard of, 167–68; organic, 63, 132, 146, 155–59, 161–65; sampling monitor for, 513–14
Solvent-repellent cream, 77
Somatic nervous system, 349
Sorbent cartridges and canisters, 631–33
Sound, 199; generation, 199–200; vs. noise, 199; pressure, 201–4; pressure level, 202–4, 205, 207; properties of, 94–95, 199–206; weighting, 205–6, 209. *See also* Hearing; Noise
Sound level contour, 214
Sound level meter, 208–10, 214
Sound power level, 202–4
*Southern Railway v. OSHRC,* 797
Specific absorption (SA), 288
Specific absorption rate (SAR), 288, 289, 290
Spectacles, protective, 114
Spectrometer, 522–23, 524
Spectrophotometer, 522–23
Speech, 39, 723; hearing, 99, 206
Spill. *See* Cleaning and maintenance
Spinal cord, 350, 351
Spiral absorber, 487
Spirometry, 46, 49, 50, 496, 719
Spray booth, 573–74
Spray chamber, 565
Stack, exhaust, 554
Standard operating procedure (SOP): and hazard evaluation, 456; for industrial hygiene program, 620–21, 750
*Standard Practice for Measuring the Concentration of Toxic Gases or Vapors Using Length-of-Stain Dosimeters,* 518
*Standards of Occupational Health Nursing Practice,* 731
*Standard Test Method for Resistance of Protective Clothing Materials to Permeation by Liquids or Gases Under Conditions of Continuous Contact,* 170
*Standard Test Methods for Flash Point,* 160
Standing, 377, 389
Stapediius, 89
Stapes, 85, 86, 87–89, 90
*Staphylococcus,* 405, 438; *aureus,* 411
State and local regulations: ANSI and NFPA standards, 772; California Illness and Injury Prevention Act, 551; before OSHAct, 769; under OSHAct, 771, 789, 792–93, 799–800, 815–16, 819, 824, 834–35; right-to-know law, 774, 819
*State Implementation of Federal Standards Hearings,* 815
State Medical Society of Wisconsin, 705, 713, 715
Static strength, 363
Static technique, for material handling personnel selection, 368
Static work, 19, 326
Statistical analysis, computerized, 762
Statistical model, for material handling personnel selection, 367
Steam coil air heater, 588–89
Steiger, William A., 788, 798, 799
Stender, John H., 794, 795, 797, 798, 799, 807
Stenosis, 92
Steradian, 298
Stereoscopic vision, 105, 109
Sterilization, 423
Stevens, John Paul, 806
Stevens's rating of perceived exertion, 355–56
Storage, 11, 169, 458
Strain, 393
*A Strategy for Occupational Exposure Assessment,* 751–52
Stratum corneum, 54–55, 57–58, 63, 66, 297
Stratum malphigii, 297
Strength, 360–64
*Streptomyces,* 405
Stretcher, 712
Strontium-90, 250
Strunk, Dorothy Y., 663
Styrene, 724
Subcutaneous layer, 54, 55
Subradiofrequency radiation: biological effects of, 280–84; control methods for, 285–86; field strength, 279–80; measurement of, 284–85; properties of, 278; static magnetic field, 282–83; time-varying, 283–84; uses of, 280
Substance abuse, 720–21
Substitution of equipment, 30, 218, 533, 537–38
Substitution of material, 30, 75, 533, 537
Substitution of process. *See* Process modification
Suction pump (for air-sampling device), 492–93; calibration parameters for, 499–503; primary calibration, 496–98; secondary calibration, 498–99
*Suggestions for the Guidance of the Nurse in Industry,* 705
Sulfur dioxide, 28, 48, 130, 516, 801
Sulfuric acid, 24, 28, 60
Sunburn, 16, 57, 92, 297–98
Sunglasses, 110, 116
Sunlight, 16, 57; effects on skin, 16, 29, 57, 64, 69, 297–98; and eye protection, 110; natural defense against, 58, 297
Superaural protector, 223
Superfund Amendments and Reauthorization Act (SARA), 784–85, 817, 830, 837
Superfund Extension Act, 784
Supplied-air suit, 640
Supply air, 596
Supply diffuser, 597, 599
Supreme Court, decisions affecting OSHA, 805–7, 809, 810, 817–18, 823–24, 827, 828, 835, 838
Surface lipid film, 57, 63
Surface Transportation Assistance Act, 838
Susceptibility, 412
Swallowing, 38
Sweat, 57; and dermatosis, 64–65; and thermal balance, 321, 322, 334
Sweat gland, 55–58
Swedish Board for Technical Accreditation (SWEDAC), 294–95
Swedish Confederation of Professional Employees (TCO), 295
Synergism. *See* Antagonistic action
Synovial fluid, 393
*Synthetic Organic Chemical Manufacturers Association v. Brennan,* 803
Systemic toxin, 8, 186, 403; particulate, 131, 179; and skin, 68, 70
Systems safety, 686, 692–93

# T

Tag (Tagliabue) Closed Tester, 160
Tag (Tagliabue) Open-Cup Apparatus, 160
Talc, 181
Tannic acid, 29
Target organ, 289, 291
Tarsal gland, 104
*Taylor Diving and Salvage Co. v. Department of Labor,* 803, 808
*Taylor v. Brighton,* 811
Tears, 104, 112, 118
Technique for human error prediction, 693
Teeth, 17
Telangiectasia, 69
Temperature: and concentration calculation, 474; extremes, 12–14, 29, 58, 63, 341; and hazard monitoring equipment, 499, 515, 519; and indoor air quality, 584, 605; measurement of, 13, 326–27; outdoor, 319. *See also* Body core temperature; Cold stress; Heat stress
Temperature regulation. *See* Thermal balance
Temporary threshold shift (TTS), 97–98
Tendinitis, 396
Tendon, 393
Tendosynovitis, 18, 397
Tennis elbow, 396
Tensor tympani, 89
Teratogenesis, 136, 723; and fetal infection, 412; from lead, 186; from radiation, 256, 258, 287, 289; and *United Auto Workers v. Johnson Controls Inc.* decision, 824
Testes, 289, 291
Tetrachloroethylene, 163, 167
Textile workers, 410
Thermal balance, 53–54, 58, 320–21, 340–41; and radiofrequency and microwave radiation, 287, 291
Thermal comfort zone, 319, 344
Thermal conductivity detector, 511, 522, 524
Thermal detector, 311
Thermal precipitation, 191
Thermal stress. *See* Cold stress; Heat stress; Temperature, extremes
Thermoluminescence detector (TLD), 258
Thermoregulation. *See* Thermal balance
THERP, 693
Thoracic outlet syndrome, 394, 397
Threshold Limit Value (TLV), 26, 139, 140–43, 470, 479; adopted as OSHA standard, 772, 791, 825; adopted by states before OSHAct, 769; for aliphatic hydrocarbons, 162; by animal experimentation, 138; for benzene, 162; for beryllium, 187; for biological aerosols, 434; for cadmium oxide, 48; for carbon dioxide, 592; for carbon monoxide, 578; for carbon tetrachloride, 163; for chemical asphyxiants, 132; and chemical inventory, 455; for cold stress, 342–43; computerized information on, 762; and concentration of airborne contaminant, 583; for direct current electric fields, 282; for formaldehyde, 577; for heat stress, 13, 325–26, 328–29; for impact noise, 216; for mixtures, 142; for noise, 11; for nuisance dusts and aerosols, 133; for optical and infrared radiation, 298–99; for particulates, 24–26, 177–78; and permissible exposure limit (PEL), 139, 480; for refrigerants, 163; for respiratory irritants, 130; for silica dust, 181–82; for solvents and gases, 154; for static magnetic fields, 282; for subradiofrequency fields, 283; and synergism, 467; and unlisted substances, 142–43

*Threshold Limit Values and Biological Exposure Indices,* 13, 25, 70, 146; skin notation, 725
Threshold of effect, 126–27
Tidal volume (TV), 45, 46
Tillinghast Report, 746
Time-weighted average (TWA), 25, 26, 140–41; for asbestos, 184; for beryllium, 187; calculation of, 477–79; and chemical inventory, 455; for ethylene oxide, history of, 821; as legal requirement, 772; for metabolic rate, 326, 328–29; for noise, 12, 213, 228, 229, 237–39; OSHA inspector's determination of, 776–77; OSHA setting procedure for, 802; for radiofrequency and microwave radiation, 291–92; and recommended exposure limit, 140; and sampling, 468, 470, 479; for silica dust, 181; for static magnetic fields, 282; for wet bulb globe temperature, 328–29
Tinnitus, 93, 100
Toluene, 61, 156. *See also* Aromatic hydrocarbon
Toluene diisocyanate (TDI), 520, 524
Tonometer, 106
Tonsil, 38
Torque, 360–62
Total lung capacity (TLC), 46
"Total" mass concentration, 191
Toxic chemical, 11, 123. *See also* Chemical hazard; Toxicity
Toxicity, 123, 124, 481; acute, 128–29; chronic, 128–29; vs. hazard, 11, 124; and hazard evaluation, 168, 453; local, 124; systemic, 8, 68, 70, 131, 179, 186, 403
*Toxicity and Metabolism of Industrial Solvents,* 161
Toxicological screening, 138
Toxicology, 123
*Toxicology and Biochemistry of Aromatic Hydrocarbons,* 161
Toxic Substances Control Act (TOSCA), 139, 439, 759, 783
Toxic Substances List, 139–40
Toxic use reduction (TUR), 458
*Toxoplasma,* 405
Toxoplasmosis, 412
Tracer (radioactive), 250
Tracer gas airflow calculation, 604–5
Trachea, 37, 38, 39, 40–41, 44–45
Tracheitis, 39
Training: for asbestos handling, 184; and biosafety, 426–27; for cold stress control, 343; for computerized industrial hygiene system, 764; for hearing conservation, 230, 236; for heat stress control, 334; as industrial hygienist, 663–66, 667–69; for lifting, 365–66; for OSHA industrial hygienists, 663–66; OSHA requirement for, 773, 774, 777, 789, 802; by NIOSH, 781; for respirator use, 621; safety and health, 537, 550–51, 688–91, 736–37, 753–55
*Training Requirements in OSHA Standards and Training Guidelines,* 753
*Training Tracker* software, 763
Transducer, 350
Transport, 458, 536
Trauma, 29, 58, 63
Treatment, storage, and disposal (TSD) facility, 784
Tremolite, 182, 819, 826
Trench foot, 341
1,1,1-trichloroethane, 169
Trichloroethylene, 61, 723, 801
Tridymite, 182

Trigger finger, 18, 397

Trioxide, 28

Tripoli, 182

Trisodium phosphate, 60

Tritium, 251

*Trypanosoma,* 405

Tuberculosis, 47, 406, 407, 413, 429–32; control methods for, 431–32; exposure guidelines for, 430–31; OSHA standard requested for, 829; and silicosis, 181

Tularemia, 406, 409, 410

Tumor, cutaneous, 69

Tumorigen. *See* Carcinogen

Turbinates, 37

Turpentine, 61

T wave, 282

Tympanic membrane. *See* Eardrum

Typhoid fever, 406

Tyson, Patrick R., 813, 820

## U

*UAW, Brock v. General Dynamics Land Systems Division,* 292, 309

Ulceration: of mucous membrane, 69, 180; of skin, 69, 187

Ulnar artery aneurysm, 397

Ulnar nerve entrapment, 397

Ultraviolet analyzer, 523

Ultraviolet light-sensitive disease, 65

Ultraviolet radiation, 57, 112–13, 278; biological effects of, 16, 29, 64, 68, 110, 112, 296–98; and chemical exposure, 142; control methods for, 114–15, 299; exposure guidelines for, 298–99; germicidal, 432, 438; photochemical effects of, 296, 299; protective eyewear for, 114–15

Ultraviolet spectrophotometer, 517

Underloading, 348

Underwriters Laboratories, 681

*United Auto Workers v. Johnson Controls Inc.,* 824

*United Steelworkers of America v. Auchter,* 818

*United Steelworkers of America v. Pendergrass,* 819

*United Steelworkers v. Marshall,* 807

Unit size principle, 365

Unit ventilator, 598

Unreasonable risk, 139

Upper confidence limit (UCL), 473–74

Upper explosive limit (UEL), 161, 510, 513

Uranium, 251, 267

Urinary system, 186, 405

Urine testing, 145, 724

Urquhart, Michael, 822

Urticaria, 65, 67, 80, 190

## V

Vaccination, 412, 426, 438

Vanderbilt, R. T., 819

Vanishing cream, 77

Vapor, 23, 125, 159; biological effects of, 24, 66, 112; concentration calculation, 474–76; control methods for, 568, 582, 630–34; sampling monitors for, 486–89, 509, 525–26

Vapor/hazard ratio, 165–66

Vapor pressure, 124, 154, 474–75

Variability of response, 454

Variable air volume (VAV) system, 596, 600

Vasodilator, 162

Ventilation, 30, 532, 537, 541–43; for asphyxiant control, 165; for building-related illness control, 438–39; comfort, 584; for heat stress control, 337, 338; natural, 584–85; for particulate control, 192–94; for solvent hazard control, 169–70; and tuberculosis control, 432. *See also* Dilution ventilation; Exhaust ventilation; HVAC system

Venturi-type collector, 567

Vertigo, 90

Vestibular sensor, 350

Vestibule (of inner ear), 87, 89–90

Veterinary practice. *See* Animals

Viability, 411

Vibration, 199; hazards from, 18, 70, 392, 394; and sound generation, 200

Vibration syndrome. *See* Raynaud's phenomenon

Vibrissa, 85, 94, 99

Video display terminal (VDT), 115, 286, 294–95, 314. *See also* Office workstation, design of

Vinyl chloride, 62, 723, 801, 802, 803

Virulence, 411–12

Virus, 20, 64, 403. *See also* Microorganism

Visceroceptor, 350

Visible light. *See* Lighting; Optical radiation

Vision: acuity of, 108–9, 117, 118–20; and dark adaptation, 109; and depth perception 105, 109; evaluation of, 106, 118–20; impairment of, 106–8, 164. *See also* Eye

Vision conservation program, 117–18

Vision screening device, 106

*Visitron v. OSHA,* 803

Visual field, 120

Vital capacity (VC), 46, 49

Vitiligo, 55, 69

Vitreous humor, 105, 296

Vocal cord, 39, 93

Voluntary Training Guidelines, 550–51

von Meyer, biomechanical research of, 360

## W

Walkthrough, 435, 460–61

Walsh–Healey Asbestos Standard, 791

Walsh–Healey Public Contracts Act, 228, 770, 790, 791, 825

Wart, 69

Water, potable, 21

Water loss. *See* Dehydration

Water-repellent cream, 77

Water system, 410, 433

Wedum, Arnold G., 404

Weil, David, 840–41

Weisberg, Stuart E., 838

Welding arc: eye protection for, 114–15, 299, 300; hazards from, 16, 112, 296

Welding fumes, 188–89, 574

Wellness program, 737

Wet bulb globe temperature (WBGT), 13, 327, 328–29

Wet bulb temperature, 13; natural, 327; psychrometric, 327

Wet collector, 565–67

Wet process, 30, 194, 537, 544

Wet-test meter, 498

What-if scenario, for process safety management, 460

Wheatstone bridge circuit, 510, 511, 522

*Whirlpool Corp. v. Marshall,* 810

White finger. *See* Raynaud's phenomenon
Whiting, Basil, 798, 809–10, 812
Williams, Harrison A., 804
Williams–Steiger OSHAct of 1970. *See* OSHAct
*Willow Island, West Virginia Cooling Tower Collapse Hearings,* 812
Windchill index, 14, 342
Windholz, M., 155
Wipe samples, 75
Wirtz, William, 787
Wiseman, Donald G., 838
Woodprocessing, 410
Woodworking, 575–76
Work, classification of, 355–56
Work capacity, 353–57
Work demands, and thermal balance, 320, 321, 328
*Workers at Risk—The Failed Promise of the Occupational Safety and Health Administration,* 840
Workers' compensation: hearing conservation program, 228; for injury or impairment, 72–74, 92, 118, 198–99; and occupational medicine, 702, 738
Workers' Family Protection Act, 838
Workload, mental, 353
Workplace: and associated hazards, 407–10, 457; design of, 19, 364, 377–79; design of, for hazard control, 417–18, 533–37. *See also* Building-related illness

Work-practice control, 421–22, 438, 777
*Work Practices Guide for Manual Lifting,* 368–72
Work/rest cycles, 356–57
Work schedule as hazard control method, 31, 338, 343, 543–44
Work seat. *See* Seat
Workstation, 376–77. *See also* Office workstation
World Health Organization (WHO), 150, 283, 325–26, 333, 418, 721
Written Injury and Illness Prevention Program, 551, 750

**X**

X-radiation, 14, 251, 253–54, 278; dermatosis from, 29, 64
X ray, diagnostic, 465, 719
X-ray diffraction, 190, 191
X-ray fluorescence (XRF), 465
X-ray machine, 266
Xylene, 61, 723

**Y**

Yellow fever, 411

**Z**

Zinc chloride, 61
Zoonotic disease, 408–10